ENCYCLOPAEDIA OF OCCUPATIONAL HEALTH AND SAFETY

INTERNATIONAL LABOUR OFFICE GENEVA

ENCYCLOPAEDIA OF OCCUPATIONAL HEALTH AND SAFETY

THIRD (REVISED) EDITION

Technical editor:
Dr. Luigi Parmeggiani

VOLUME 1 A-K

First published 1930
Second edition 1971
Third edition 1983
Fourth impression 1991

ILO
Encyclopaedia of occupational health and safety
Technical editor: L. Parmeggiani
Third edition
Geneva, International Labour Office, 1983. 2 vols.
/Encyclopaedia/ of /Occupational health/ and /Occupational safety/. 13.04
ISBN for complete set of two volumes: 92-2-103289-2
ISBN for Volume 1: 92-2-103290-6

ILO Cataloguing in Publication Data

ILO publications can be obtained through major booksellers or ILO local offices in many countries, or direct from ILO Publications, International Labour Office, CH-1211 Geneva 22, Switzerland. A catalogue or list of new publications will be sent free of charge from the above address.

Printed in Italy

SAG

PREFACE TO THE SECOND EDITION

Occupational accidents and diseases remain the most appalling human tragedy of modern industry and one of its most serious forms of economic waste. The best estimates currently available on a world basis reckon the number of fatal injuries at the workplace at close to 100 000 annually. In some highly industrialised countries industrial accidents are responsible for the loss of four or five times as many working days as industrial disputes. In certain cases their cost is comparable to that of national defence. Industrialisation and the mechanisation of agriculture have made the problem acute in a much wider range of countries and occupations.

The economic burden on the community cannot be expressed in compensation costs alone. It also includes loss of production, disruption of production schedules, damage to productive equipment and—in the case of large-scale accidents—major social dislocations. But the economic burden is by no means the full measure of the human cost.

Protection "against sickness, disease and injury arising out of ... employment" is one of the essential aims of the ILO as defined in its Constitution. It calls for a far more vigorous effort than has yet been devoted to it.

Originally, the main thrust of preventive action was to improve the unhealthiest working conditions and remedy the appalling lack of physical protection against the most dangerous occupational hazards. The first international standards were designed either to do away with the more flagrant abuses impairing health, such as the employment of very young children, over-long hours of work, the absence of any form of maternity protection, and night work by women and children, or to combat the risks most commonly encountered by industrial workers—anthrax, and lead or chronic phosphorus poisoning.

When the ILO passed beyond formulating these basic standards to grapple with the problem of social security, the first question it considered was compensation for occupational accidents and diseases. Workmen's compensation legislation already existed in many countries; it was developed on the basis of ILO standards and its financial implications gave a powerful impetus to preventive measures. The ILO did much to bring about the standardisation of industrial injury and occupational disease statistics and the systematic collection of data on accident frequency.

Special attention was given at an early stage to the industries with the highest incidence of accident and disease. The prevention of accidents and elimination of dust in the mining industry had a high priority. In nearly every country, mines have the highest incidence of serious accidents, not to mention disasters; they are also one of the main sources of pneumoconiosis and silicosis, the occupational diseases which cripple more workers than any other. Docks and building were also given special consideration by reason of their high accident rates.

Gradually this concentration of attention upon the most flagrant abuses and the highest accident and disease rates broadened into a more comprehensive approach designed to promote the highest standards of safety and health in all industries and occupations. The monumental Model Code of Safety Regulations for Industrial Establishments for the Guidance of Governments and Industry, first issued in 1949 on the basis of work initiated during the Second World War and periodically revised since, was an important step in this direction. It furnished an impetus which has now found expression in a wide range of codes of practice and guides to practice which are complementary to it. In the 1950s this broader approach was reflected in new comprehensive international standards for the protection of workers' health, welfare facilities and occupational health services.

In the 1960s these were supplemented by a new series of specific provisions dealing with particular risks which had assumed increased importance. In factories, one accident in six is caused by machinery; hence the importance of international standards on the guarding of moving parts which regulate not only the use, sale and hire of machinery having dangerous parts but also its manufacture. Ionising radiations require stringent safety precautions and preventive measures, and the ILO, in collaboration with the International Atomic Energy Agency and the World Health Organisation, plays its part in ensuring that such measures are taken in member countries, especially in general industry, which is much less well equipped in this respect than the nuclear industry.

In more recent years, ILO action in respect of industrial safety and health has also taken new forms. Technical co-operation with the developing countries is increasingly covering the field of occupational safety and health. The Industrial Committees, which bring together government representatives and delegates from employers' and workers' organisations, play an important role in the exchange of information on safety and health and in the solution of some of the problems arising at the plant level. The introduction of modern machinery and new chemicals, particularly pesticides, in agriculture has given a new urgency to the problem in the greatest of the world's industries.

As the human factor has become ever more prominent, it has become apparent that safety and health must be treated as an indivisible whole. This has made closer co-operation between the ILO and WHO ever more necessary.

The organisation of occupational health services within the undertaking lies at the heart of a positive policy to promote workers' health that was formulated by the ILO-WHO Joint Committee on Occupational Health in 1950. Ergonomics is perhaps the most recent expression of this approach, bringing results both socially in the form of improvements in workers' physical and mental well-being, and economically in the form of higher productivity.

Modern industrial medicine has outgrown the stage where it merely involved first aid in the event of an accident and the diagnosis of occupational diseases; nowadays it is concerned with all the effects of work upon physical and mental health, and even with the impact of man's physical or psychological disabilities upon his work. Such a broadening of the field has inevitably led to a rapid growth in the volume of research and publication by national institutes; more than 10 000 publications on occupational safety and health are issued every year by some 1 000 institutes all over the world.

The need to supplement the reference facilities provided by certain national institutions prompted the ILO to set up an international documentation service, catering in a practical way for industry's needs and covering every area of economic activity in which workers' health problems arise. Since 1959, the International Occupational Safety and Health Information Centre (CIS) has been collating, sifting, analysing and disseminating the material assembled by its network of national centres. The usefulness of this written material is enhanced by special symposia and technical meetings at which individual topics are explored thoroughly and the ground prepared for future action at the international level.

The impact of this periodical information is necessarily limited unless it can be used in the context of the whole body of existing knowledge and experience. A general encyclopaedia is therefore a necessary foundation for the effective use of a current information service.

To provide a comprehensive survey of the whole question of workers' health and safety the ILO published, between 1930 and 1934, its first encyclopaedia, *Occupation and Health*, the value of which was widely recognised. Since then, the whole concept of industrial medicine has changed; scientific knowledge itself has, of course, made enormous strides, particularly in such fields as biochemistry, toxicology, radiobiology and physiology and therapeutics. Yet this new *Encyclopaedia of Occupational Health and Safety* is not a scientific treatise designed to add to the sum of knowledge; its aim is to make the knowledge now available more readily accessible to those directly responsible for promoting improved conditions for the worker. Its approach is eminently practical, stress being laid throughout on risks and their prevention. It includes the basic essentials of occupational safety for the benefit of all persons concerned with workers' protection, and deals with the specific problems facing the industrialising countries. For this reason, too, the information, presented in summary form, is purposely free of theoretical digressions and excludes points about which too little is known. The encyclopaedia sets out, first, to provide industrial physicians and nurses with a compendium of the available practical knowledge in their field and to familiarise them with the technical features of prevention, and secondly, to make technicians aware of the more important biological and social aspects of their work. It is not intended exclusively for specialists, but for all who are responsible in one way or another for workers' health—employers, workers or their organisations, public authorities or inspectors. Matters which are of special interest to the developing countries, but about which adequate knowledge is often lacking, have been dealt with in detail. The reader will thus have at his disposal material only rarely found in standard reference books.

This is the first work on occupational safety and health with contributions from as many as 60 countries all over the world. The eminent specialists without whose help this new edition could not have been prepared made their services available virtually free of charge. In expressing to them the gratitude of the International Labour Organisation I am confident that I also speak for all whose suffering will be averted or relieved by the knowledge which they have made available. The ILO is also grateful to the many international and regional organisations which have collaborated with us in this important task, and in particular to the World Health Organisation, which has revised a number of the articles dealing with general medical questions.

In the preface to the first encyclopaedia, Albert Thomas wrote:

"It was a difficult task, and one which was bound to be open to the reproach of being neither complete nor final. But how could it be otherwise? No one can hope to fix once for all something which is living, evolving, progressive. Although ... the evolution of technical practice in industry may create new dangers for the worker every day, yet the progress of this same technique and of industrial hygiene may, on the following day, do away with certain existing dangers, which must, notwithstanding, be recorded and analysed in this work. One of the virtues of this work is just the fact that it is not final. It seizes one moment in social life and in the progress of industrial hygiene, but it requires to be kept constantly up to date precisely because it is a scientific as well as a practical work."

Technological progress now moves far more swiftly than it did 40 years ago. There is every reason to believe that the pace will quicken still further. This new encyclopaedia will therefore be merely the next stage in our work. But each stage is the indispensable foundation for its successor. During the coming years the *Encyclopaedia of Occupational Health and Safety* will be an essential tool for humanising the working environment and improving the lot of workers the world over. In human and economic terms alike higher health and safety standards are a primary responsibility of enlightened social policy and efficient management. Neither can be effective without the comprehensive body of knowledge necessary to appraise the relevance of current information to policy and action. The present encyclopaedia, which was prepared under the technical responsibility of Dr. Luigi Parmeggiani, Chief of the Occupational Safety and Health Branch, is designed to make readily accessible to all the comprehensive knowledge of these matters which is now available. In editing the encyclopaedia, Dr. Parmeggiani has worthily maintained the traditions established by Dr. Luigi Carozzi, who laid the foundations of the industrial health work of the ILO.

Geneva, 1971

WILFRED JENKS,
Director-General.

PREFACE TO THE THIRD EDITION

The decision to publish the second edition of the *Encyclopaedia of Occupational Health and Safety* was taken some 15 years ago, and its preparation lasted throughout the years 1966 to 1971. Since then a great deal of progress has been made in the knowledge and activities covered by this publication. Side by side with technological progress there have been great advances in methods of identifying, evaluating and controlling occupational hazards and providing health protection in the workplace. Toxic substances, dust in industry, mineral fibres, non-ionising radiation, allergy and occupationally induced cancer have been the subject of intensive experimental research and important epidemiological studies. Nevertheless, the changes that took place in working environments in the 1970s were not due merely to wider technical knowledge and awareness; a new trend began to take shape: the workers' claim for a better quality of life at work and the increasing involvement of trade unions in health and safety protection in the workplace; the fuller support by employers of comprehensive occupational health and safety programmes; and increasing efforts by governments to apply far-reaching measures in this field. This trend has been reflected in national and international legislation concerning the working environment and working conditions, which has advanced to an unprecedented extent. Thus the panorama of occupational health and safety, industrial hygiene and ergonomics has undergone profound changes in many member countries of the ILO, not only as regards the state of the art, but also as regards the practical application of these disciplines in the workplace.

The International Labour Office has been encouraged to publish this new edition of the encyclopaedia on account of the success of the first edition, which exceeded all expectations—it was reprinted five times, and orders for copies are still flowing in.

The basic criteria for this work are still the same, particularly its over-all practical nature giving priority to preventive action in the field of occupational health and safety; most of the authors are the same as those who contributed to the previous edition. The encyclopaedia is still intended for all those who have administrative or moral responsibilities for safeguarding workers' health and safety, especially in the developing countries, and for all those who have technical responsibilities in this field but do not have at their disposal adequate library facilities or international selections of recent documents on the many and vast fields covered by, or related to, occupational health and safety. Each article from the first edition has been revised, updated, supplemented or in some cases entirely rewritten. Approximately 200 new articles have been added covering new knowledge or new subjects, particularly in the fields of toxicology and occupational hygiene, occupational cancer, occupational diseases of agricultural workers, occupational safety, psycho-social problems, and institutions and organisations active in the field of occupational health and safety. There are a number of annexes of a practical nature, and over 6 000 bibliographical references to the world-wide literature published in this field over the last five years. More than 1 000 authors from the member States of the ILO and over 15 specialised international organisations have generously and gratuitously participated in this joint effort to promote occupational health and safety throughout the world. The ILO is much indebted to them. Once again, the WHO has given its support, and several of its specialists have undertaken to revise articles on medical questions of a general nature. Many members of the Permanent Commission and International Association on Occupational Health are to be found among the authors who have contributed to this work. The present edition was planned and revised by Dr. Luigi Parmeggiani, who was also the Technical Editor of the second edition.

It is 63 years since the ILO first established as one of its basic objectives "the protection of the worker against sickness, disease and injury arising out of his employment". The objective is still the same, but the form and methods of this protection have evolved along with technical progress and economic development. In particular a broader approach is henceforth given to this protection through the International Programme for the Improvement of Working Conditions and Environment (PIACT) launched by the ILO in 1976. International dissemination of the most recent scientific and practical knowledge in this field is an integral part of ILO activity—together with the traditional modes of action: standard-setting and technical co-operation—to promote the increased effectiveness of health and safety protection at work throughout the world. The new edition of the encyclopaedia will make an important contribution to that great endeavour.

Geneva, 1983

FRANCIS BLANCHARD,
Director-General.

INTRODUCTION

The ILO's first encyclopaedia, entitled *Occupation and Health*, was published between 1930 and 1934 in accordance with a resolution adopted by the First Session of the International Labour Conference in 1919—the year in which the International Labour Organisation itself was founded. It was prepared under the supervision of Dr. L. Carozzi, in close collaboration with the Correspondence Committee on Industrial Hygiene, and contained 416 articles by 95 contributors in 16 different countries. Its appearance was widely welcomed, especially by industrial physicians, who found it an invaluable reference work and the only international publication in their field. Before long the Office was receiving requests to bring it up to date and by 1944 it had issued six supplements containing 52 articles. With the Second World War, however, there were so many changes in techniques as well as in living and working conditions that the need for a complete recasting became apparent.

Accordingly the ILO Governing Body decided in 1966 that a new encyclopaedia should be prepared. The work was carried out under the editorship of Dr. Luigi Parmeggiani and the *Encyclopaedia of Occupational Health and Safety* was issued in 1972-73 in the English version and in 1973-74 in the (partially revised and expanded) French version. In 1974-75 a Spanish version was published by the Instituto Nacional de Previsión of Spain. The new encyclopaedia differed from the first by its wider scope, which embraced principles and examples of occupational safety, ergonomics, physiology and psychology at work, as well as various medico-social topics related to occupational health and safety. It included 851 articles in the English version (872 in the French) and nine appendices. It was prepared with the collaboration of 714 authors from 60 countries and a dozen international organisations and with the assistance of the World Health Organisation. The publication was mainly designed for all those concerned with the protection of workers at the workplace, whether with or without a medical or technical background, and gave special consideration to the conditions prevailing in developing countries. It was well received and widely distributed, the English version being reprinted five times.

As a result of continuing developments, technical and social progress and the emergence of new knowledge in the field of occupational health and safety, much of the information contained in the encyclopaedia has become outdated. The present, second edition accordingly contains approximately 50% more matter than the previous edition; in addition, all former articles have been reviewed, many entirely recast and the remainder updated; and new bibliographies have been compiled. It has been prepared with the collaboration of 913 authors. All this work has again been carried out under the editorship of Dr. L. Parmeggiani.

How to use the encyclopaedia

Articles are classified in strict alphabetical order. When a title consists of several words, the most significant of these is used as a heading. Because of their complexity the different aspects of some subjects, such as "Accidents", are dealt with in several articles, which are, as far as possible, usually grouped together.

Wherever necessary for the understanding of the subject, the articles themselves contain cross references to other articles, in particular to those whose direct relation to the entry might escape the reader.

However, the analytical index remains the main tool for guiding the reader to a given item. It lists, again in strict alphabetical order, the subjects dealt with in the articles, including those which, though treated more briefly or sometimes even only mentioned in the bibliography attached to an article, are nevertheless of some importance for occupational health and safety. The page indicated against each entry in the index is the first page of the article concerned.

Chemical and physical data

Normally, the nomenclature of chemical substances follows that of the International Union of Pure and Applied Chemistry, with the exception of a few products for which the old terminology is still very widely used in occupational safety and health practice. For organic compounds the main synonyms have also been provided.

Certain physical and chemical data of interest from the standpoint of occupational safety and health are given at the beginning of articles dealing with dangerous substances. For the sake of uniformity the main sources of these data have been the following publications: *CRC handbook of chemistry and physics.* Weast, R. C., and Astle, M. J. (eds.) (West Palm Beach, CRC Press Inc., 59th ed., 1978-79); *Dangerous properties of industrial materials.* Sax, I. N. (New York, London, Toronto, Melbourne, Van Nostrand Reinhold Company, 5th ed., 1979); *Handling chemicals safely* (Amsterdam, Het Veiligheids-instituut, 2nd ed., 1980).

The information given is usually as follows:

m.w. molecular weight
or alternatively
a.w. atomic weight
sp.gr. specific gravity (water = 1) or density (kg/m^3)
m.p. melting point
b.p. boiling point
fr.p. freezing point
v.d. vapour density (air = 1)
v.p. vapour pressure
f.p. flash point (closed cup, unless indicated oc = open cup)
e.l. explosive limits in % by volume, lower and upper
i.t. auto-ignition temperature
Solubility: slightly soluble = less than 10 g/100 cm^3;
soluble = 10-100 g/100 cm^3;
very soluble = more than 100 g/100 cm^3
Description
Exposure limits for concentrations of toxic substances in the air (see below).

Exposure limits

These include the United States Occupational Safety and Health Administration (OSHA) values and when they differ or are the only ones in existence, those of the National Institute of Occupational Safety and Health (NIOSH) and the American Conference of Governmental Industrial Hygienists (ACGIH), as well as the MAC values (maximum allowable concentrations) fixed by the Ministry of Public Health in the USSR. The values quoted are those valid in 1980. They are subject to periodical adjustment and should not be taken over-rigidly.

The following further details may be given:

TWA time-weighted average for a normal 8-h work-day and 40-h work-week, unless otherwise indicated;
TLV time-weighted average adopted by the ACGIH for a normal 8-h work-day and 40-h work-week;
ceil ceiling, i.e. the concentration that should not be exceeded even instantaneously;
skin possibility of absorption in significant amounts through the skin, mucous membranes and eye;
STEL short-term exposure limit of the ACGIH, i.e. the maximum concentration to which workers can be exposed for a period of up to 15 min continuously, provided that no more than four excursions per day are permitted, with at least 60 min between exposure periods and provided that the daily TLV/TWA is not exceeded;
IDLH concentration immediately dangerous to life or health from which a worker could escape without any escape-impairing symptoms or any irreversible health effects (NIOSH/OSHA Standards Completion Programme);
MAC USSR maximum allowable concentration not to be exceeded;
TSRAL USSR temporary safe reference action level.

In view of their general usefulness for preventive purposes, the above-mentioned exposure limits have been added by the editor; their inclusion in an article does not imply that the author of the article accepts their accuracy.

Units of measurement and abbreviations

The units of measurement used in the encyclopaedia are those of the International System (SI), with the exception of a few that are recognised by the International Organisation for Standardisation and still widely used. A list of the symbols for units and of the their abbreviations, together with conversion tables, is to be found in appendices to the second volume.

Abbreviations other than those of units or measurement and the physical and chemical data mentioned above are given alongside the full expression the first time they are used in an article.

Some abbreviations commonly used in occupational health and safety are, however, employed without further amplification. These include:

mmHg	millimetres of mercury
ppm	parts per million
ppb	parts per billion
$ppcm^3$	particles per cubic centimetre
ppcf	particles per cubic foot
LC_{50}	lethal concentration 50
LD_{50}	lethal dose 50
%w/w	percentage, weight in weight, i.e. number of grammes of active substance in 100 g of product

Text of articles

Almost 70% of the entries consist of revised articles from the 1972 edition. Any significant addition to or alteration of the author's original text by the editor is included in square brackets. In several articles use has been made of the bibliography in order to enlarge the scope of the entry beyond the treatment given by the author; thus, on controversial matters that have not yet been settled, suggested readings may express views different from those of the author.

Throughout the encyclopaedia, whenever workers in general are referred to, for the sake of brevity only pronouns of the masculine gender have been used. Unless the context requires a restrictive interpretation any such reference should be read as applying equally to both women and men.

Transliteration

The recommendations of the ISO's International System for the Transliteration of Slavic Cyrillic Characters (2nd ed., R 9-1968) have been followed throughout the encyclopaedia.

Bibliographical references

Each article is usually accompanied by a short bibliography of suggested readings, the purpose of which is much more to supplement the information in the article and to develop individual points or different approaches than to support statements or figures quoted by the author.

Where possible and appropriate, care has been taken to include documents published in different countries so as to cater for as wide a circle of readers as possible. No mention is usually made of sources of general information, such as textbooks and standard works of reference; the information concerned is generally easily accessible. References preceded by a CIS number have been abstracted by the International Occupational Safety and Health Information Centre (CIS), and in those areas the CIS is available for further bibliographical research and services.

In addition to the language of publication, the choice of the reader may be further guided by the details given of the length of the suggested reading, the number of references it contains, and whether it is illustrated or not. Furthermore, in a number of cases the references have been grouped under subject headings according to their main content.

LIST OF ARTICLES AND APPENDICES

VOLUME 1

Abattoirs
Abrasive cleaners
Abrasives
Absenteeism, causes and control of
Absenteeism, definition and statistics of
Accelerators, particle
Accident analysis
Accident cost
Accident investigation
Accident proneness
Accidents
Accidents and diseases, notification of
Accidents, commuting
Accidents (human factors)
Accidents, occupational domestic
Accidents, off-the-job
Accident statistics
Acetaldehyde
Acetic acid
Acetone and derivatives
Acetylaminofluorene
Acetylene
Acid-base balance
Acids and anhydrides, inorganic
Acids and anhydrides, organic
Acridine and derivatives
Acrolein
Acro-osteolysis, occupational
Acrylic acid and derivatives
Acrylic resins
Acrylonitrile
Actinide elements
Actors and players
Adaptation
Adhesives
Aerosols
Aerospace (ground operations)
Aerospace (space operations)
Agricultural chemicals
Agricultural implements
Agricultural machinery
Agricultural work
Agricultural workers
Agriculture, occupational health and safety measures in
Air
Airborne micro-organisms in the workplace
Air conditioning
Aircraft and aerospace industry
Air ionisation
Air pollution
Air pollution control
Airports
Alcoholism
Alcohols
Aldehydes and ketals
Algae
Alkaline materials
Alkaloids
Alkyd resins
Alkylating agents
Allergy
Allergy, screening and treatment of
Allyl compounds
Altitude
Aluminium alloys and compounds
Alveolitis, allergic extrinsic
Amides
Amines, aliphatic
Amines, aromatic
Aminothiazole
Aminotriazole
Ammonia
Anaesthetists
Ancylostomiasis
Aniline
Animal experimentation
Animals, aquatic
Animals, venomous
Anthracene and derivatives
Anthrax
Anthropometry
Antibiotics
Antidotes
Antimony, alloys and compounds
Apprentices
Arsenic and compounds
Arsines
Artists
Asbestos
Asbestosis
Asbestos (mesothelioma and lung cancer)

Ashes and cinders
Asphalt
Asphyxia
Asthma, occupational
Audiometry
Audiovisual aids
Automation
Automobile industry
Automobiles, safety and health design of
Avalanches and glaciers
Aviation–flying personnel
Aviation–ground personnel
Aziridines
Azo and diazo dyes

Back pain
Bagasse
Bakeries
Bamboo and cane
Bananas
Band saws
Barium and compounds
Bark
Barrier creams and lotions
Batteries, dry
Batteries, dry (mercury)
Batteries, secondary or rechargeable, or accumulators
BCG
Beat diseases
Benzanthrone
Benzene
Benzoyl peroxide
Benzyl chloride
Beryllium, alloys and compounds
Beverage or soft drink industry
Beverages at work
Biological monitoring
Biological rhythms
Biomechanics
Biometrics
Biotransformation of toxic substances
Biscuit making
Bismuth and compounds
Bladder cancer
Blasting and shotfiring
Blastomycosis
Bleaching and bleaching agents
Blind workers
Blood diseases, occupational
Body reference man
Boilers and pressure vessels
Boilermaking
Boilers and furnaces, cleaning and maintenance of
Bolt guns
Bone and bone meal
Boranes
Boron, alloys and compounds
Brewing industry
Brick and tile manufacture
Briquette manufacture
Bromine and compounds
Bromomethane
Bronchitis, chronic, and emphysema
Brucellosis
Building construction
Burns and scalds
Burns, chemical
Bursitis and tenosynovitis
Business machine operation
Butadiene
Butchery trade
Button manufacture
Byssinosis

Cable transport
Cadmium and compounds
Calcium and compounds
Calcium carbide
Calcium cyanamide
Calibration
Camphor
Campylobacter infections
Cancer, environmental
Cancer, occupational
Cancer, occupational (legislation)
Cancer, occupational (statistics and registration)
Candle manufacture
Canning and food preserving
Caplan's syndrome
Carbamates and thiocarbamates
Carbon black
Carbon dioxide
Carbon disulphide
Carbon monoxide
Carbon tetrachloride
Carcinogenic substances
Carcinogenic substances, epigenetic
Cardiacs at work
Cardiovascular diseases
Cardiovascular exercise tests
Cardiovascular system
Carpets, handwoven
Carpets, machine-made
Castor oil and bean
Catalysts
Cataract, occupational
Catering, industrial (operation)
Catering, industrial (organisation)
Cellulose and derivatives
Cement
Centrifuges
Cervicobrachial disorders, occupational
Charcoal burning
Chelating agents
Chemical industry
Chemical reactions, dangerous
Chemicals, new

Chlorinated nitroparaffins
Chlorine and inorganic compounds
Chlorobenzene and derivatives
Chloroethylamines
Chloroform
Chloromethane
Chloronaphthalenes
Chloropicrin
Chloroprene
Chromium, alloys and compounds
Chromosome aberrations
Cinema industry
Civil engineering
Clay
Climate and meteorology
Clothing industry
Coal and derivatives
Coalworkers' pneumoconiosis
Cobalt, alloys and compounds
Cocoa cultivation
Cocoa industry and chocolate production
Coconut cultivation
Codes of practice
Coffee cultivation
Coffee industry
Coir
Coke industry
Cold and work in the cold
Collective agreements (occupational safety and health)
Colour in industry
Colour vision
Commerce
Commission of the European Communities (CEC)
Compressed-air work
Computers
Concrete and reinforced concrete work
Conditioned reflexes
Confectionery industry
Confined spaces
Contact dermatitis or eczema, occupational
Contagious ecthyma
Control devices, isolating and switching
Control technology for occupational safety and health
Copper, alloys and compounds
Copra
Copy paper, carbonless
Coral and shell
Cork
Corrosive substances
Cosmetics
Cotton cultivation
Cotton industry
Coumarins and derivatives of indandione
Council for Mutual Economic Assistance (CMEA)
Cowpox and pseudocowpox
Cramps
Cranes and lifting appliances
Cresols, creosote and derivatives
Cryogenic fluids
Cumene
Cuts and abrasions
Cyanogen, hydrocyanic acid and cyanides
Cybernetics
Cycloparaffins

Dairy products industry
Dangerous substances
Dangerous substances, labelling and marking of
Dangerous substances, storage of
Dangerous substances, transportation of
Date palms
Day nurseries and nursery schools
DDT
Deafness, occupational
Decompression sickness
Degreasing
Dentists
Detection and analysis of airborne contaminants (chemical laboratory methods)
Detection and analysis of airborne contaminants (field methods)
Detection and analysis of airborne contaminants (instrumental methods)
Detergents
Diabetics at work
Diatomaceous earth
Diazomethane
Dibromochloropropane
1,2-Dibromoethane
Dichloromethane
Diesel engines, underground use of
Digestive system
Dimethylaminoazobenzene
Dimethyl carbamoyl chloride
Dimethyl sulphate
Dinitro-*o*-cresol
Dinitrophenols
Dioxane
Dioxin, tetrachlorodibenzopara
Diphenyls and terphenyls
Dirty occupations
Disability evaluation
Disability prevention and rehabilitation
Disasters
Diving
Dock work
Domestic workers
Doping
Dose-response relationship
Double-jobbing
Drilling, oil and water
Drilling, rock
Drinking water
Drug dependence
Dry cleaning
Dupuytren's contracture

Dust, biological effects of
Dust control in industry
Dust explosions
Dust sampling
Dusts, vegetable
Dyeing industry
Dyes and dyestuffs

Earth-moving equipment
Ecotoxicity
Effects, combined
Electrical accidents
Electrical equipment industry
Electrical installations, fixed
Electrical installations, temporary
Electric cable manufacture
Electric current, physiology and pathology of
Electric fields
Electricity distribution
Electricity, static
Electric lamp and tube manufacture
Electric power tools, portable
Electronics industry
Electroplating
Embryotoxic, fetotoxic and teratogenic effects
Emergency exits
Employers' and workers' co-operation
Enamels and glazes
Energy expenditure
Energy sources, comparative risks of
Engine testing
Entertainment industry
Enzymatic changes
Enzymes in industry
Epidemiology
Epilepsy
Epoxy compounds
Equilibrium
Ergonomics
Erysipeloid
Esters
Ethers
Ethics
Ethyl alcohol
Ethylene
Ethylene dichloride
Ethylene glycol dinitrate
Ethylene oxide
Executives
Exhaust systems
Explosives industry
Explosive substances
Exposure limits
Exposure limits, biological
Eye
Eye and face protection

Factory premises and workplaces
Falls
Falls from heights, personal protection against
Farmer's lung
Fatigue
Feathers
Felt hat manufacture
Felt industry
Fermentation, industrial
Ferroalloys
Fertilisers
Fettling
Fibres, man-made
Fibres, man-made glass and mineral
Fibres, man-made synthetic
Fibres, natural
Fibres, natural mineral
Fire
Fire fighting
Firemen
Fire prevention and protection
Fire protection equipment, in-plant
First-aid organisation
Fishing
Fitness for employment
Flammable substances
Flax and linen industry
Floors and stairways
Flour milling
Fluorine and compounds
Fluoroacetic acids and compounds
Fluorocarbons
Foam resins
Food-borne infections and intoxications
Food industries
Foot and leg protection
Foot and mouth disease
Footwear industry
Forestry industry
Forges
Formaldehyde and derivatives
Foundries
Frostbite
Frozen food industry
Fruit ripening
Fuel and oil additives
Fungicides
Furfural and derivatives
Fur industry
Furnaces, kilns and ovens

Gallium and compounds
Galvanising
Garages
Gardening and market gardening
Gas cylinders
Gases and air, compressed
Gases and vapours, biological effects of
Gases and vapours, irritant
Gas manufacture
Genetic manipulation

Genital system and sex-linked occupational characteristics of women
Genital system, male
Geology and safety
Germanium, alloys and compounds
Glanders
Glass industry
Glove manufacture
Glycerol and derivatives
Glycols and derivatives
Gold, alloys and compounds
Graphite
Grinding and cutting fluids
Grinding and polishing
Group medical services
Gum arabic
Gypsum

Hair and bristle
Hair-cutting and shearing
Hairdressers
Halogens and compounds
Hand and arm protection
Handicapped and disabled persons
Handicrafts and craftsmen
Hand injuries
Hand tools, ergonomic design of
Hand tools, safety of
Hardeners
Hazardous areas, electrical apparatus and wiring in
Head protection
Health physics
Hearing protection
Heat acclimatisation
Heat and hot work
Heat disorders
Heating of workplaces
Heat protective clothing
Helium-group gases
Helminthiasis
Hemp
Hepatitis, infectious
Herbicides
Hernia
Hexachlorobutadiene
n-Hexane
High risk groups
Histoplasmosis
Home work
Hormones, sex
Horn
Hospitals
Hotels and restaurants
Hours of work
Housekeeping and maintenance
Housing of workers
Human engineering
Human relations in industry
Hunting and trapping
Hydrazine and derivatives
Hydrazoic acid and azides
Hydrocarbons, aliphatic
Hydrocarbons, aliphatic: olefins
Hydrocarbons, aromatic
Hydrocarbons, halogenated aliphatic
Hydrochloric acid
Hydrofluoric acid
Hydrogen
Hydrogen peroxide
Hydrogen sulphide
Hydroxylamine
Hygiene, personal
Hypoxia and anoxia

Immunisation and vaccination
Incentives and productivity
Incentives, safety
Indicators and control panels
Indium, alloys and compounds
Industrialisation, impact on health and safety of
Influenza
Information and documentation
Injection, accidental
Inks
Inland navigation
Inspection, safety and health
International Agency for Research on Cancer
International and regional organisations
International Association of Labour Inspection (IALI)
International Atomic Energy Agency (IAEA)
International Ergonomics Association (IEA)
International Labour Organisation (ILO)
International Labour Organisation (application of Conventions on occupational safety and health)
International Labour Organisation (occupational safety and health activities)
International Occupational Safety and Health Hazard Alert System
International Occupational Safety and Health Information Centre (CIS)
International Organisation for Standardisation (ISO)
International Programme on Chemical Safety (IPCS)
International Social Security Association (ISSA)
Iodine
Iridium and compounds
Iron and compounds
Iron and steel industry
Isocyanates
Isolated work
Ivory

Jewelry manufacture
Jute

Kapok
Kepone
Ketones
Kidney
Kienböck's disease

VOLUME 2

Laboratory work
Laboratory work, chemical
Laboratory work, microbiological
Labour legislation
Lactones
Ladders
Larynx cancer
Lasers
Lathes
Laundries
Lead alkyl compounds
Lead, alloys and inorganic compounds
Lead arsenate
Lead control in the working environment
Leather goods industry
Leisure time
Leptospirosis
Lifting and carrying
Lifts, escalators and hoists
Light, artificial
Lighting
Lightning
Limestone and lime
Liver
Livestock confinement
Livestock feed preparation
Live work (low voltage)
LNG and LPG (liquefied natural gas and liquefied petroleum gas)
Lubricants
Lung function tests

Machinery guarding
Machine tools
Magnesium, alloys and compounds
Maintenance of machinery and equipment
Major hazards control
Malaria
Manganese, alloys and compounds
Manganese: chronic poisoning
Manpower redeployment
Marble
Match industry
Maternity protection
Maximum weights
Mechanical engineering, work organisation in
Mechanical handling
Mechanisation
Median-nerve compression (carpal tunnel syndrome)
Medical care at sea
Medical care of workers
Medical examination of workers
Medical history
Medical inspection
Medical practitioners
Medical records
Melamine
Mental health
Mentally handicapped, rehabilitation of
Mental work
Merchant marine
Mercury
Mercury: chronic poisoning
Mercury, organic compounds
Metal carbonyls
Metal fume fever
Metallising
Metals, alkali, and compounds
Metals, hard
Metals, heat treatment of
Metals, surface treatment of
Metals, toxicity of
Metal-working industry
Methaemoglobinaemia
Methyl alcohol
Mica
Migrant workers
Miners' nystagmus
Mines, coal
Mines, dust control in
Mines, metal
Mines, occupational health in
Mines, opencast
Mines, safety in
Mines, small-scale
Mines, ventilation of
Minimum age
Mirror manufacture
Mobile units
Molybdenum, alloys and compounds
Monotonous work
Morpholine
Motion sickness
Mouth and teeth
Multiphasic screening
Muscular work
Musicians
Mutagenic effects
Mycoses

Nails
Naphthalene
Nasal cancer, occupational
National Safety Council
Nervous system, central and autonomous
Nervous system, peripheral

Neuro-endocrine system
Neurosis, post-traumatic
Newcastle disease
Nickel and compounds
Nicotine
Night work
Niobium, alloys and compounds
Nitric acid and nitrates
Nitriles and cyanates
Nitrobenzene
Nitro-compounds, aliphatic
Nitro-compounds, aromatic
Nitrofuran
Nitrogen
Nitrogen chloride
Nitrogen oxides
Nitroglycerin
Nitropropane
N-Nitroso compounds
Noise
Noise measurement and control
Nose
Nuclear reactors
Nurses
Nutrition and food

Obesity
Occupational diseases: international list
Occupational health
Occupational health in developing countries
Occupational health institutes
Occupational health legislation
Occupational health, medico-legal aspects of
Occupational health organisation
Occupational health, teaching and training of
Occupational hygiene
Occupational hygiene laboratory
Occupational hygiene practice
Occupational hygiene, systematic approach and strategy of
Occupational hygiene, teaching and training of
Occupational nurse
Occupational physician
Occupational safety and health, teaching and training of
Occupational safety, development and implementation of
Occupational safety in developing countries
Occupational safety in undertakings
Occupational safety, national policies of
Offices
Offshore oil operations
Oil palms
Oils and fats, animal and vegetable
Older workers
Opium
Organisation for Economic Co-operation and Development (OECD)
Ornithosis and psittacosis
Osmium, alloys and compounds
Oxalic acid and derivatives
Oxidising substances
Oximes
Oxygen
Ozone

Packaging
Painting and varnishing
Paints, lacquers and varnishes
Palladium, alloys and compounds
Paper and paper pulp industry
Parathion
Partial Agreement of the Council of Europe
Patch tests
Peanuts
Pearls
Peat and lignite
Pedals and levers
Pencils and ball-point and felt pens
Pentyl alcohols
Peptic ulcer
Perfumes and essences
Permanent Commission and International Association on Occupational Health
Permit-to-work systems
Peroxides, organic
Pest control at the plant
Pesticides
Pesticides, halogenated
Pesticides, organophosphorus
Petrochemicals
Petroleum and petroleum products
Petroleum, extraction and transport by sea of
Petroleum products, storage and transport of
Petroleum refineries
Pharmaceutical industry
Pharmacist
Phenolic and amino resins
Phenols and phenolic compounds
Phenothiazine and derivatives
Phenylhydrazine
Phosgene
Phosphates and superphosphates
Phosphine
Phosphorus and compounds
Photographic laboratories, industrial
Photographic processing
Photographic sensitive materials
Phthalates
Phthalic anhydride and some derivatives
Physical training
Physiology of work
PIACT (International Programme for the Improvement of Working Conditions and Environment)
Pickling
Picric acid and derivatives
Pineapples

Pipelines
Planing machines, wood
Plantations
Plastics industry
Plastics processing industry
Platinum, alloys and compounds
Plutonium
Pneumatic tools
Pneumoconioses
Pneumoconioses, international classification of
Pneumoconiosis, infective
Poison centres
Poisoning, acute, treatment of
Polyacrylonitrile fibres
Polyamides
Polychlorinated biphenyls
Polycyclic aromatic hydrocarbons
Polyester fibres
Polyester resins
Polyfluorines
Polyolefins
Polypropylene fibres
Polystyrene
Porphyrins
Postal services
Postures and movements, occupational
Pottery industry
Power stations
Power transmissions
Precious stones
Precision instrument manufacture
Presses
Preventive medicine
Printing
Productivity and safety and health
Project design in industry: engineering interaction
Protective clothing
Psychogenic (mass) illness or hysteria, epidemic
Psychology, industrial
Psychopathology, occupational
Psychotechnics
Pumice
Pyridine, homologues and derivatives
Pyrolysis
Pyrotechnics industry
Pyrrole and pyrrolidine

Q fever
Quarrying
Questionnaire design

Rabies
Radar
Radiation, ionising: accidents and emergencies
Radiation, ionising: agricultural uses
Radiation, ionising: biological effects
Radiation, ionising: decontamination
Radiation, ionising: detectors and portable survey instruments
Radiation, ionising: equivalence of chemicals
Radiation, ionising: industrial uses
Radiation, ionising: medical surveillance
Radiation, ionising: medical uses
Radiation protection
Radiation protection philosophy
Radiation, radiofrequency
Radiation, ultraviolet, visible and infrared
Radioactive luminous compounds
Radioactive materials
Radioactive materials: transport and storage of
Radioactive ore, mining and milling of
Radioactive waste management
Radio and television broadcasting
Radium
Radon and thoron
Raffia
Railways
Rare earths
Raynaud's phenomenon
Recovery industry
Refractories
Refrigerating plants
Refuse collectors
Resins, natural
Respiratory cancer, occupational
Respiratory protective equipment
Respiratory system
Rest periods
Resuscitation
Retirement from work
Rhenium, alloys and compounds
Rheumatic diseases
Rhodium, alloys and compounds
Rice
Ringworm
Risk acceptable in industrial society
Road accidents: human factor
Road maintenance
Robots and automatic production machinery
Rocket propellants
Rodenticides
Rodents
Rolling mills
Rolls
Roofing
Rope (fibre) industry
Routing machines
Rubber cultivation
Rubber industry, natural
Rubber, synthetic
Rubella
Ruthenium, alloys and compounds

Safety and health education
Safety approval and certification
Safety belts
Safety colours
Safety factor: exposure limits

Safety nets
Salt industry
Sampling, personal
Sandblasting and shotblasting
Sanitary facilities
Sawmills
Saws
Scaffolding
Schistosomiasis
Seasonal workers
Seats, tables and desks
Selenium and compounds
Sewers
Shale oil industry
Shift work
Shipbuilding
Siderosis
Silica and silicates
Silicon and organosilicon compounds
Silicosis
Silicotuberculosis
Silk industry
Silos
Silver and compounds
Sisal
Skeleton
Skin cancer, occupational
Skin diseases, non-occupational
Skin diseases, occupational
Slag, basic
Slate
Small undertakings
Smell, sense of
Smelting and refining
Smoking
Soap industry
Social security
Social security and occupational risks
Sociology, industrial
Solvents, industrial
Somatotypes
Soot
Spice industry
Spindle moulders
Sport, professional
Starch industry
Statistics
Stigmata, occupational
Stockfarming and breeding
Stone industry
Straw
Street cleaning
Stress
Strontium and compounds
Students
Styrene and ethylbenzene
Sugar-beet industry
Sugar cane cultivation
Sugar cane industry
Sulphur
Sulphur compounds
Sulphuric acid
Supervisors and foremen
Susceptibility and hypersensitivity
Switchboard operators
Syphilis
System analysis (safety): methods and analysis
Systems reliability

Talc
Tank trucks
Tanning and leather finishing
Tantalum and compounds
Tar and pitch
Taxidermy
Teaching
Tea cultivation
Tea industry
Telecommunications operation
Tellurium and compounds
Temporary work
Tetanus
Tetrachloroethane
Tetrafluoroethylene
Tetrahydrofuran
Tetramethylthiuram disulphide
Tetryl
Textile industry
Thallium and compounds
Thioglycolic acid and derivatives
Thiols
Thorium and compounds
Timber floating
Tin, alloys and compounds
Titanium, alloys and compounds
Tobacco cultivation
Tobacco industry
Toluene and derivatives
Toxicity and structure
Toxicokinetics
Toxicometry
Toxoplasmosis
Toys and fancy goods manufacture
Tractors
Trade unions
Transport by road
Transport, in-plant
Transport of workers
Transport, urban
Traumatic injuries
Trenching
Trichloroethanes
Trichloroethylene
Tricresyl phosphates
Trinitrotoluene
Tripe and gut
Tuberculosis and employment
Tuberculosis, occupational

Tularaemia
Tungsten, alloys and compounds
Tunnelling
Turpentine

Ultrasound
Underwater work
Unemployment
United Nations Environment Programme (UNEP)
Uranium, alloys and compounds

Vanadium, alloys and compounds
Vanilla
Varicose veins
Ventilation, industrial
Veterinarians
Vibration
Vigilance
Vinyl and polyvinyl chloride
Vinyl compounds
Vinylcyclohexene dioxide
Viscose
Vision
Visual acuity
Visual strain in modern technology
Voice, occupational diseases of the

Waste water treatment and disposal
Watch and clock making
Water and electrolyte balance
Water pollution
Water supply and treatment
Wax
Welding and thermal cutting
Welfare in industry
Whaling
White oil
White spirits
Wines and spirits industry
Wireworks
Women in employment
Wood
Woodworking industries
Wool industry
Workers' education
Working clothes
Works safety and health committees
Work study
World Health Organisation

Xylene

Young persons

Zinc, alloys and compounds
Zirconium and hafnium
Zoonoses

APPENDICES

Basic data

I. The elements
   1. Symbols and valencies
   2. Periodic table
II. Conversion table for gases and vapours
III. Units of measurement and definitions
IV. Conversion factors

International documentation

V. ILO Conventions and Recommendations
VI. Select bibliography of material published by international organisations

Some operational guidelines

VII. Evaluation of heat stress
VIII. Air pollution monitoring equipment
IX. Threshold limit values

LIST OF AUTHORS

INDEX

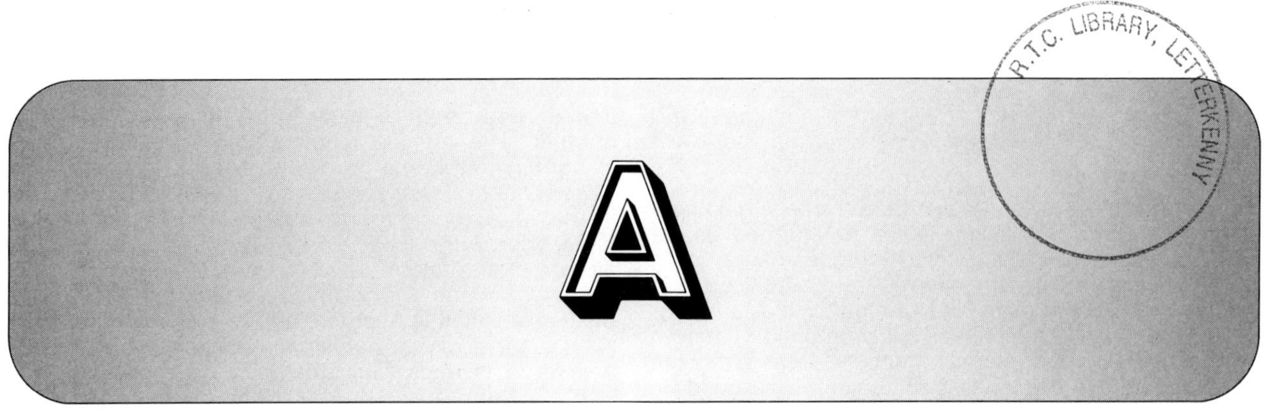

Abattoirs

An abattoir is a slaughter-house for animals intended for human consumption, such as horned cattle, pigs, sheep and, in some countries, horses and camels. The abattoir may be a self-contained unit existing solely for the supply of meat for sale to butchers' shops or it may be part of a larger factory in which the meat is processed, such as in meat-packing or canning factories or in bacon-curing factories. A special type of abattoir is known as a knacker's yard. This is the term applied to premises of a factory type, in which cattle are killed and processed but the meat is not for human consumption but for dog food, etc. The chief animals killed are old horses, mules and asses and occasionally diseased cows and sheep not fit for human consumption. In some countries stray dogs collected by the police authorities or animal protection societies are killed by gassing under police supervision and then sent to a knacker's yard for processing. The dead dogs are boiled to extract the fat and the residue is sold for use as a fertiliser.

Types of abattoir. The size and the production capacity of abattoirs vary considerably. The smallest ones are found in country and rural areas. They are often housed in poor types of buildings and with primitive equipment. Each of the small number of workers employed will usually work on all the operational processes. The sanitary conditions are usually poor and liquid and solid wastes are often left untreated on the premises, resulting in offensive smells and attracting rats and flies which breed in abundance. The larger type of abattoir is found in towns and cities and is usually subject to strict controls by the local health authority and in some cases by one or more government departments. These controls were designed originally to ensure that the foodstuffs produced should be fit for human consumption, that the disposal of waste materials should not constitute a danger to public health and that, as far as practicable, the abattoirs should not be a nuisance to nearby residents. But more recent legislation in many countries has also been aimed at giving protection to the workers against the occupational safety and health hazards of work in abattoirs. In these larger abattoirs the work is more specialised, often being done on a production-line basis with one worker doing only one operation. The trend is towards more rationalisation and automation.

Types of building. Some countries have detailed regulations regarding the type of building that may be used. Adequate space must be provided for keeping the various kinds of animals in separate pens or compounds on arrival from the farms. Drinking water for livestock must be available. In a hot climate, where no shelter is provided, the cattle should be sprayed with water from time to time to avoid overheating. There must be sufficient room in the cattle pens for the government or municipal inspector to examine the animals to see that they are not diseased. Animals unfit for human consumption are segregated into a separate pen and then taken directly to the inedible-rendering department and slaughtered under the supervision of the inspector. The pens should be cleaned daily. In some countries the abattoir buildings are single storey but in a number of countries they are of three or more storeys with the killing of the cattle taking place on the top floor and the various

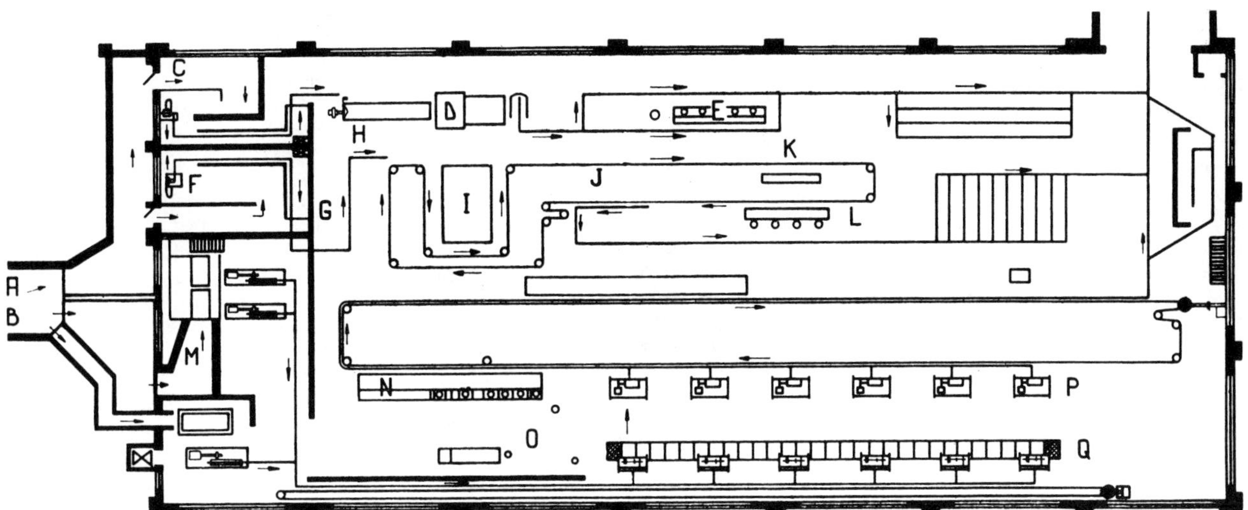

Figure 1. Layout of slaughter floor for cattle, sheep and pigs. A. Pigs and sheep; B. Cattle; C. Pig-stunning pen; D. Dehairing machine; E. Viscera table; F. Sheep-stunning pen; G. First legging and foot removal; H. Transfer to moving conveyor; I. Sheep-dressing ring; J. Dressing conveyor; K. Sheep-pelt pulling; L. Viscera table; M. Stunning pens; N. Viscera and inspection table; O. Horn saw bench; P. Dressing hoists; Q. Dropper.

processes of separation of edible and inedible parts and cutting up of the carcasses being done on the lower floors.

Processes. The cattle are driven from the pens or lairage to the slaughter. The animal must be stunned before being bled unless slaughtered in accordance with Jewish or Muslim rites. In some countries the use of the cruder type of instrument for stunning such as a bolt or spike, a mallet or hammer or club is prohibited. Usually a special type of slaughtering hatchet or an automatic pistol is used. In some cases the animal is stunned by means of an electric shock applied to the ears from an electrolethaler.

After stunning, the rear legs of the animal are secured by a chain hooked to an overhead travelling runway which transfers the animal to the next room in which it is killed by severance of the throat blood vessels. The bleeding-out process follows and the blood is drained through pipes to collecting tanks on the floor below. The animal is then scalded but, in the case of pigs, the hair on the skin is sometimes removed either by passing the carcass through a large flame of the Bunsen burner type or by passing through a dehairing machine. The skinning of the animal follows, then the extraction of the intestines. The carcass is cut into two parts vertically and given a veterinary inspection. The edible parts are removed to the cold-storage room and the other parts removed to the inedible-rendering department which must be completely separate from other departments to prevent any contamination of the edible products (see BUTCHERY TRADE).

HAZARDS

A worker may be injured by the animal when leading it into the abattoir or by a falling animal after stunning. The most frequent injuries are cuts and abrasions to hands and upper parts of the body caused by knives and cutting tools, including mechanical saws. Eye injuries can occur from bone splinters thrown off during sawing of the bones. Where mechanical handling is not used there may be strains caused by carrying heavy carcasses. Falls may be caused by slippery floors covered with blood, fat and water, and burns and scalds from the use of steam and steam-heated fat-rendering vessels. Electric shock from the electrolethaler may result from improper use or if the insulation is defective.

Health hazards. In spite of the handling of animals, the number of cases of zoonosis is low. The most common is brucellosis in countries where this disease exists among cattle. Pulmonary tuberculosis has proved a considerable hazard in some countries; however, in Argentina the mortality rate does not exceed that of the general population. Acute or chronic respiratory syndrome cases are very common, with chronic catarrh, sinusitis and bronchitis. As far as the large abattoirs are concerned, cases of anthrax or pemphigus are rare. Simple skin diseases are common, especially athletes' foot which is sometimes contracted by new workers. Workers in high temperatures may have folliculitis. There have been some cases of leptospirosis but it is not clear whether these were contracted in abattoir work. Erysipeloid, glanders, tularaemia, Q-fever and Pseudomonas infection are also health hazards in abattoirs and knacker's yards; in some countries, Q-fever is recognised as an occupational disease. The incidence of toxoplasma antibodies amongst abattoir workers is high in certain countries.

Atmospheric conditions. The humidity in abattoirs is high owing to steam used in various operations and to quantities of hot or cold water on the floors, and room temperatures vary considerably from one place to another. The highest temperatures are found in the slaughter-house and in workplaces where there are water-heating vessels, while the lowest temperatures are in the cold-storage rooms. As the workplaces are very large, often without partition walls, there is considerable air movement creating numerous different temperature zones in which the work is carried on. There is a distinctive smell in abattoirs due to a mixture of odours such as those of wet leather, blood, vomit, urine and the dejecta of animals. This smell spreads throughout the whole building and its surroundings. If there are residues or remains of blood, flesh, bones or fat, they rapidly become putrid and give off additional smells. Offal, such as horns and hooves, should be stored in well ventilated places and covered with quicklime. Exhaust ventilation is necessary to remove the odours and the fumes and steam from digestors and the fat-rendering processes.

SAFETY AND HEALTH MEASURES

Accident prevention. Floors should be kept free from water and waste materials. Sand is recommended for the floor in killing rooms to prevent slipping. Mechanical handling should be provided for heavy loads. All electrical installations and equipment should be in accordance with the standards of the national electrical code. Employees engaged in meat cutting should wear wire mesh hand protection on the hand not holding the cutting tool as a protection against accidental cuts. All knives should be provided with guards on the handles to prevent the person's hands from slipping on to the blade. Sheaths or pouches should be used for carrying knives. Workers should be properly trained to use the appropriate cutting instruments in the correct manner, and taught how to sharpen and keep them in a good state of repair. There are a number of power-driven saws such as brisket saws, beef saws, breaking-down saws and portable circular saws. The correct type of machinery guarding should be used on each of these saws and the workers should be fully instructed as to the method of safe use. Pig dehairing machines and scalding tanks should be self-loading and discharging or so arranged that the rotating scrapers cannot be contacted whilst in motion. They should also be guarded to prevent scalding of persons by either splashing or contact. Hide-strippers should be guarded so that if any clamp or other restraining attachment slips or releases, the carcass cannot strike any worker. Meat-grinding machines of the worm type should be so constructed that it is impossible for the operator to reach the moving parts through the feed opening.

Protective clothing. Butchers who use knives on cutting operations which require the knife to be drawn backward should be provided with a special type of abdominal apron made of heavy leather or leather reinforced with steel. All workers should be provided with suitable overalls and head covers. Eye and face protection is required for all employees who are subject to the risk of eye injury as in the use of bone-cutting saws and the treatment of skins and hides with acids. Safety hats are recommended for employees exposed to falling objects or likely to come into contact with articles suspended from overhead trolleys and conveyors.

Hygiene. Very high standards of cleanliness, maintenance and good housekeeping are required as a protection for the health of the worker and also to ensure that the meat products are fit for human consumption. The interior walls should be ceramic surfaced to facilitate cleaning and the floors should be of impervious material resistant to blood, water and mechanical damage and also have a non-skid finish. Washing down of the floors and inside walls by water under pressure should be carried out at the end of work, and also of the floors from time to time during the day to remove accumulations of waste. Hot water is essential for washing down the floor of the slaughter department as cold water does not remove deposits of fat. Disinfected water should be used periodically. There should be adequate provision for the drainage of waste liquids. Catch basins are provided to prevent fat entering the drain pipes to the sewerage. The fat rises to the surface and is skimmed off regularly and taken to the inedible-rendering department. The catch basins should be cleaned out daily.

Waste water charged with organic matter should not be discharged into rivers but should be decanted into a septic tank in which it is subjected to the action of anaerobic microbes before being passed successively over bacteria beds by means of a pipe or channel distribution which runs the water on to the fields. Waste offal is usually burnt but in some tropical countries it is thrown on to an open space either inside or near the outside of the abattoir yard and is immediately devoured by scavenger birds, usually vultures.

Sanitary and welfare facilities. In view of the exceptionally dirty nature of the processes, very good sanitary and washing facilities including bathing accommodation should be provided. Eating, drinking and smoking in the workplaces should be prohibited and a separate messroom be provided outside the working area.

Medical services. Pre-employment and periodic medical examinations of all workers should be compulsory. There should be an appointed doctor who will visit the abattoir to maintain a strict survey over the health of the workers. In very large abattoirs it may be necessary for him to visit daily and for there also to be a full-time nurse. In the smaller abattoirs where there is no nurse in charge of first aid it is essential that at least one well trained first aider is available. In some countries immunisation of abattoir workers against brucellosis is undertaken.

Legislation

France was the first country to enact legislation for the control of abattoirs, following a report of a government commission in 1807. Practically all countries now have very detailed legislation usually under public health codes or pure food codes and, in addition, in many countries, the occupational safety and health provisions of the factory legislation also apply. There is a very large international trade in frozen meat and canned meat. Such imported meat products if not prepared under the best hygienic conditions may result in food poisoning or in the spread of foot-and-mouth disease in the country of consumption. In order to prevent such occurrences and to maintain the health standards laid down by the legislation of the importing country, an abattoir in an exporting country is sometimes required to have permanently on its premises an expert from the public health department of the importing country who checks that no diseased meat is processed and that the strictest hygienic controls are observed in the abattoir.

In most countries the legislation for abattoirs covers all aspects of the processes and is very strictly enforced; in addition, the abattoir must be registered and licensed and there may be restrictions on the location of the premises because of the possible nuisance of unpleasant odours.

A number of countries prohibit the employment of young persons under 18 years of age in abattoir premises and, in some countries, women are not allowed to be employed in certain very dirty operations.

KAPLAN, J.

Health and safety guides:

CIS 77-1741 *Health and safety guide for meat packing, poultry dressing, and sausage manufacturing plants.* DHEW (NIOSH) publication No. 77-127 (National Institute for Occupational Safety and Health, 4676 Columbia Parkway, Cincinnati, Ohio, 1977), 67 p. Illus.

CIS 76-1443 *Safety in industrial abattoirs* (La seguridad en los mataderos industriales). Gimeno, F. G. (Servicio Social de Higiene y Seguridad del Trabajo, Instituto Territorial, Barcelona, 1974), 120 p. Illus. (In Spanish)

Brucellosis:

"Brucellosis in the United States, 1960-1972: An abattoir associated disease. Part III. Epidemiology and evidence for acquired immunity". Buchanan, T. M.; Hendricks, S. L.; Patton, C. M.; Feldman, R. A. *Medicine* (Baltimore), 1974, 53/6 (427-439).

Erysipeloid:

"On erysipeloid in slaughterhouse workmen" (Erysipeloid bei Schlachthofarbeitern). Marinescu-Dinizvor, G. *Medizinische Klinik* (Munich), 9 Nov. 1979, 74/45 (1686-1688). 21 ref. (In German)

Poultry:

CIS 78-856 *Workplace studies in industrial poultry slaughtering* (Etude de quelques postes de travail en abattage industriel de volailles). Seguela, J. L. (Paris, Université de Paris VI, Faculté de médecine Broussais, 1978), 64 p. Illus. 22 ref. (In French)

Turkeys:

CIS 76-1135 "Human psittacosis associated with commercial processing of turkeys". Durfee, P. T.; Pullen, M. M.; Currier, R. W.; Parker, R. L. *Journal of the American Veterinary Medical Association* (Chicago), Nov. 1975, 167/9 (804-808). Illus. 11 ref.

Abrasive cleaners

This article deals with the abrasive cleaners (sometimes known as scouring powders) which are widely used for the household cleaning of hard metal, porcelain or enamel surfaces (sinks, baths, stoves, cooking utensils, floors, etc.). Abrasive soaps are sometimes used for hand cleaning in dirty occupations.

Raw materials. An abrasive cleaner is obtained by mixing detergent powder with an insoluble abrasive. The abrasive is usually powdered silica in one of its forms although insoluble silicates, like pumice, calcium carbonate, or combinations of these substances may be used. The abrasive content is often high (70-90%). The detergent is usually either alkyl benzene sulphonate or soap powder. Other constitutents may be sodium carbonate, sodium silicate, and bisodium phosphate.

Preparation. In some plants, processing starts with the raw materials; silica may be calcined, ground and pulverised, often in ball mills. The soap or detergent powder may also be prepared on the same premises. In other plants, the prepared constituents are merely mixed and packaged.

HAZARDS AND THEIR PREVENTION

Accidents. Moving parts of mills, disintegrators, mixers, etc., can cause serious injury unless secure machinery guarding is maintained. Outlets as well as intakes should always be made safe.

Handling accidents may be caused by lifting and carrying of heavy loads or by bad methods of lifting. Sharp edges of packages may cause serious cuts. Mechanisation and training in safe methods of lifting and handling will reduce accident incidence. Hazards arising from soap and detergent manufacture are dealt with more specifically elsewhere but the following may be mentioned. Chemical burns of the skin and, possibly, the subcutaneous tissue may result from contact with strong alkalis or acid. Chemical eye injuries and possibly corneal opacities may be caused by corrosive gases, mists or dusts. Personal protective equipment including suitable eye and face protection and hand and arm protection should be provided for those at risk.

Diseases. Where the abrasive element is silica, the risk of silicosis is high: the association of silica with an alkali appears to accelerate its action on the respiratory system. Especially where calcining, crushing, grinding or pulverising are carried out, there may be exposure to massive quantities of small-particle dust. Filling of cartons and packets produces "puffs" of dust; breakage or spillage of containers may contaminate floors and work benches.

The risk is intensified when manufacture or packing is carried out in old, unsuitable factories, using inferior equipment and with little care for industrial hygiene of any kind. Cases of rapid, disabling or even fatal silicosis have been reported.

The classic recipe for prevention is substitution of a less dangerous material, such as the use of amorphous instead of a crystalline silica or substitution of a silicate, such as pumice.

Enclosure of processes, provision of exhaust ventilation at all dust sources, rigorous dust control in the ambient atmosphere by strict cleanliness of walls, floors, work benches and fittings are essential. A high standard of maintenance of all equipment will prevent dust leakage and spillage. Mechanical handling and automation of processes also reduces the risk.

Other health hazards are similar to those occurring in soap and detergent manufacture; irritation of the eyes and upper respiratory tract or chronic bronchitis from repeated exposure to irritants, and contact dermatitis should be mentioned here.

Personal protective equipment, overalls and head coverings should be provided and kept clean. Synthetic materials are to be preferred but may not be suitable in hot climates. Cloakrooms and good washing and sanitary facilities including shower baths, where practicable, should be provided together with a messroom or canteen where meals can be taken; these facilities should be located away from sources of contamination by dust or irritants. Personal hygiene is most important, as is also changing from work clothes to street clothes at the end of each shift.

Pre-placement and periodic medical examinations are desirable: workers with chronic generalised obstructive pulmonary disease or a history of pulmonary tuberculosis should not be exposed to irritants or silica dust. Periodical X-ray examinations may be necessary to detect early signs of silicosis. Conditions such as atopic or seborrhoeic skin or a history of eczema make exposure to irritants undesirable.

There is no risk from silica in the use of abrasive cleaners since these are normally used wet. [On the other hand, the abrasion of the horny layer and the alkali neutralisation of the acid protective mantle of the skin may, in the long run, result in dermatitis with swelling and roughness of the skin, and may contribute to the causation of housewife's eczema (see CONTACT DERMATITIS OR ECZEMA, OCCUPATIONAL). In industry the use of abrasive cleansing agents should be limited to those cases where it is absolutely necessary (such as in radioactive contamination).]

Thorough rinsing and drying of the hands and application of an emollient after the use of an abrasive cleaner are to be recommended.

EL-SAMRA, G. H.

Raw materials:

CIS 78-1249 "Silica flour exposures in Ontario". Nelson, H. M.; Rajhans, G. S.; Morton, S.; Brown, J. R. *American Industrial Hygiene Association Journal* (Akron, Ohio), Apr. 1978, 39/4 (261-269). Illus. 28 ref.

Use by workers:

CIS 2122-1971 "The use of abrasive products as skin-cleansing agents in industry" (Berufliche Hautreinigung mit abrasiven Präparaten). Tronnier, H.; Martin, U. *Arbeitsmedizin–Sozialmedizin–Arbeitshygiene* (Stuttgart), May 1971, 6/5 (108-110). 6 ref. (In German)

Abrasives

Abrasives are materials of great hardness used to shape other materials such as metals, wood, glass, ceramics, etc., by a process of grinding or polishing.

Materials. Abrasives may be classified as either natural or synthetic; natural abrasives are widely used for certain applications although the importance of the synthetic materials is growing rapidly. The principal natural abrasives are diamond, the natural aluminium oxides (corundum, emery), garnet, feldspar and various forms of silica including sandstone, sand, flint and diatomite. A wide range of "softer" materials, such as chalk, chromium oxide, magnesium oxide, etc., are used for polishing, and materials such as steel shot and steel grit are used for the blast-cleaning of castings. The more important of the synthetic abrasives are silicon carbide (SiC), fused aluminium oxide (Al_2O_3), boron carbide (B_4C), and boron nitride (BN), probably one of the hardest substances known, which is used on high-speed wheels. Synthetic diamond is now also an important abrasive material.

Abrasives manufacture. The two principal synthetic abrasives–aluminium oxide and silicon carbide–are made as follows. Aluminium oxide is made from calcined bauxite or purified alumina and silicon carbide by the firing of a mixture of sand, coke, sawdust and salt in an electric resistance furnace.

The abrasive materials may be used in the form of loose grains, compacted and bonded, as in the case of grinding wheels, or bonded to a backing material such as paper or cloth.

Synthetic abrasive wheels and whetstones are produced by bonding together individual grains of, e.g., aluminium oxide or silicon carbide; these binding materials may be mineral (a silicate or magnesite), ceramic (clay, kaolin or feldspar) or organic (rubber, resinoid, shellac, etc.).

Abrasives are classified by their hardness and their grain size.

HAZARDS

The hazard of abrasives themselves (rather than those of grinding wheel bursting, flying particles, sparks, etc., which are dealt with in the article GRINDING AND POLISHING) is that of the inhalation of fine dust produced by the gradual wear of the abrasive material during use.

The hazard was first fully appreciated during the Industrial Revolution when it was found that cutlery grinders in Sheffield, Solingen and other metal manufacturing centres, working with dry grindstones, suffered from grinders' asthma or grinders' rot and had a low expectation of life; on the other hand, persons doing wet grinding were more long-lived. Grinding on a dry siliceous stone gave rise to high concentrations of dust, and in workshops, often underground and devoid of ventilation, the grinders rapidly contracted silicosis.

The recognition of the hazard of the siliceous grindstone led, during the early years of the 20th century, to the replacement of hazardous natural abrasives, especially sandstone, by artificial abrasives such as silicon carbide and aluminium oxide. This was an outstanding example of replacement of hazardous by less hazardous materials. The use of sand for sandblasting has diminished considerably and, in many instances, sand has been replaced by non-siliceous materials such as steel grit and in some countries it is prohibited.

Silicon carbide. A limited pneumoconiosis hazard exists in the production of silicon carbide during the charging of the furnace with sand or crushed quartz or during the handling of that part of a processed charge which has not been fully converted and may consequently contain substantial quantities of hazardous dust. During processing the furnace itself may give off appreciable amounts of carbon monoxide. Although a considerable quantity is converted to carbon dioxide before release into the workplace atmosphere, furnacemen and charger-crane drivers may be exposed to a risk of carbon monoxide poisoning. The pig of silicon carbide extracted from the furnace is crushed and the grains are sorted and classified.

Studies on the effects of silicon carbide dust on the lungs have given divergent results. Early experimental studies indicated that, when compared with quartz, silicon carbide dust had virtually no effect on the lungs. However, a later examination of 53 silicon carbide crushers showed 15 cases of pulmonary fibrosis and 17 cases of nodular opacities, and a report has appeared of 10 cases of pneumoconiosis in workers employed exclusively on the crushing, sieving and packing of silicon carbide. There are still no conclusive data on the fibrogenic effect of silicon carbide although prolonged exposure to high concentrations of commercial silicon carbide dust have produced pulmonary X-ray changes in certain individuals.

No cases of pneumoconiosis due to silicon carbide dust produced during the use of silicon carbide grinding wheels have been reported. This may be due, to some extent, to the fact that, being extremely hard, silicon carbide grinding wheels produce only very small quantities of dust.

Emery. Natural emery varies in its composition within the following limits: 50-70% aluminium oxide (Al_2O_3), 15-30% haematite or magnetite, up to 10% quartz and a number of complex aluminium compounds. It is used mainly for metal polishing.

A number of cases of pulmonary X-ray changes have been described in emery polishers, crushers and emery-cloth manufacturers. However, it is possible that, in these cases, the changes were the result of combined effects of components other than the aluminium oxide. In addition, emery polishers are also exposed to the dust of the adhesive and the textile backing used in emery cloth and the dust of the material being polished.

Corundum. Natural corundum and artificial corundum (alundum or artificial emery) are usually relatively pure. The artificial material is produced from bauxite by smelting in an electric furnace. Shaver and Riddell reported severe pulmonary disability in workers employed on smelting bauxite in combination with coke, iron and very small amounts of silica. The fumes contained aluminium oxide and 16-54% silica; later reports implicated the silicate, mullite ($3Al_2O_3.2SiO_2$). Other cases of "Shaver's disease" have since been reported.

As with silicon carbide, the evidence of fibrogenic activity is inconclusive. In one report pneumoconiosis in 13 of 20 workers grinding iron and steel was attributed to emery dust and in experimental studies corundum in fine dusts was found to be acutely fibrogenic in rats; nevertheless, pulmonary X-rays of workers employed for over 20 years on aluminium oxide bagging revealed no changes.

SAFETY AND HEALTH MEASURES

Natural abrasives, such as sand and sandstone, which produce fibrogenic dusts should be replaced by the harder and less hazardous abrasives such as silicon carbide and aluminium oxide. In addition, although the use of these materials has considerably reduced lung disease in abrasives workers, the dust concentrations produced during their application should be kept to a minimum by the implementation of strict dust-control measures such as local exhaust ventilation, wet grinding where possible, or approved respiratory protective equipment where ventilation is impracticable.

Pre-employment and periodic medical examinations for workers manufacturing and using abrasives are advisable; these examinations should include an X-ray examination to detect pulmonary changes at the earliest possible moment.

BRUUSGAARD, A.

"Emery pneumoconiosis". Bech, A. O.; Kipling, M. D.; Zundel, W. E. *Transactions of the Association of Industrial Medical Officers* (London), July 1965, 15/3 (110-116). Illus. 27 ref.

CIS 76-1858 "Hygiene properties of chromium-containing electrolytic corundum dust" (Gigieničeskaja harakteristika pyli hromistogo ȇlektrokorunda). Latuškina, V. B.; Zelenkin, S. N.; Lihačev, Ju. P. *Gigiena truda i professional'nye zabolevanija* (Moscow), Oct. 1975, 10 (28-31). 5 ref. (In Russian)

CIS 77-1015 "The analysis of aluminium in serum and urine for the monitoring of exposed persons" (Die Analyse von Aluminium im Serum und Urin zur Überwachung exponierter Personen). Valentin, H.; Preusser, P.; Schaller, K. H. *International Archives of Occupational and Environmental Health* (West Berlin), 21 Oct. 1976, 38/1 (1-17). Illus. 39 ref. (In German)

CIS 81-375 "Fibrogenic potential of slags used as substitutes for sand in abrasive blasting operations". Mackay, G. R.; Stettler, L. E.; Kommineni, C.; Donaldson, H. M. *American Industrial Hygiene Association Journal* (Akron, Ohio), Nov. 1980, 41/11 (836-842). Illus. 13 ref.

"Causes of death among employees of a synthetic abrasive product manufacturing company". Wegman, D. H.; Eisen, E. A. *Journal of Occupational Medicine* (Chicago), Nov. 1981, 23/11 (748-754). 20 ref.

Absenteeism, causes and control of

It has been said that workers are absent for only two reasons: either they cannot attend because of incapacity, or they choose not to attend for personal reasons. This simplistic black and white view is widely held but the reality is rarely so clear cut; it is made up of a wide variety of different shades of grey. Even the fact that the great majority of days lost due to absenteeism are covered by medical certificates does not mean that they were solely caused by medical conditions. All of us know some people who insist upon going to work despite quite serious disease and others who seem to stay off sick with the most trivial of ailments. Sick absence rates in virtually every industrialised country in the world have risen substantially (perhaps by 30% or more) over the past 25 years despite the considerable improvements in the scale and quality of health care and in the socioeconomic circumstances of life. What factors have caused this and what should be the role of the occupational health and safety professional in this problem?

Apart from the relatively easy computation of sick pay or social insurance benefit, few organisations or governments have attempted seriously to cost absenteeism. One estimate in the United Kingdom suggested that the cost was similar to that of running the National Health Service, and for employers in industry the cost could be more than 10% of total wages, salaries and related overheads. Although a few days' sick absence of a senior manager may cost little since his work will await his return, absence of a shop floor employee may have to be covered immediately by another or by the expensive use of overtime. If absence also affects sales it can then be extremely costly.

Concern with, and thus investigation into, the problems of absenteeism has tended to be greater when labour is scarce (as in war) or when it is expensive. There is no reason to believe that the recent international rise in absence levels is likely to stop, and thus absenteeism and its associated costs will continue to be a major problem in the future.

This short article will indicate some of the problems of sick absence, its causes and solutions, since it is a matter that concerns not merely doctors but also managers, workers and trade unionists as well as administrators of social insurance schemes and thus taxpayers.

Causes of absenteeism

The declared cause of the great majority of absenteeism in industrial employment is incapacity, and definitions of sickness and other absenteeism are given in ABSENTEEISM: DEFINITIONS AND STATISTICS. Simple classifications are attractive, but in the case of sick absence all the studies undertaken since the early days of the Industrial Revolution point toward a multi-factorial aetiology. The proportion of spells of sick absence due solely to unequivocal and total incapacity of the worker is small. Even in such cases the duration of the absence, in other words the timing of the worker's return to his job, is influenced by non-medical considerations. A selection of factors, all of which have been shown to influence sick absence spells or duration, is set out in table 1 and the list is certainly not complete. Although not every factor influences every case of absence, there can be few that are attributable only to one. It should also be noted that of all the factors listed, only one in each of the three groups is strictly related to incapacity.

Table 1. Some factors known to influence sick absence

Geographical	Organisational	Personal
Climate	Nature	Age
Region	Size	Sex
Ethnic	Industrial relations	Occupation
Social insurance	Personnel policy	Job satisfaction
Health services	Sick pay	Personality
Epidemics	Supervisory quality	Life crises
Unemployment	Working conditions	Medical conditions
Social attitudes	Environmental hazards	Alcohol
Pension age	Occupational health service	Family responsibility
	Labour turnover	Journey to work
		Social activities

It would be impracticable in an article of this length to consider each point in detail but further references can be found in the selected bibliography.

Geographical factors. These are mostly concerned with the political, social, economic and regional factors of the country or region in which the organisation is placed. Multinational companies are well aware of such differences and even within countries sick absence rates often vary considerably between different regions. In the United Kingdom for example, Wales, Northern Ireland and the North have had consistently higher rates than the Midlands or South East (see table 2). These differences are only partly due to the type of industry and it is cause for concern that such regional differences are becoming more marked. Similar regional differences have been recorded in several other countries and they probably apply universally.

Table 2. Indices of days of social security sickness and invalidity benefit per man, and percentage rise in 20 years, by area of the United Kingdom, 1974

Area	Index	Rise in 20 years (%)
All areas	100	30
Wales	193	59
North	152	54
Yorks and Humber	129	52
North West	126	49
Scotland	120	36
South West	96	28
East Midlands	95	34
West Midlands	89	25
South East and Anglia	64	6

The influence of social insurance benefits is well recognised; the nature and provision of health care services and the rules for certification of incapacity have also been shown to be of great relevance in a number of countries. The economic situation and levels of unemployment do not seem to have such a straightforward influence today as used to apply. Probably because of wider and more adequate health and social insurance benefit, a rise in unemployment is no longer associated with a fall in sick absence.

The one medical factor in this category–epidemics–is included because of the substantial effect of large scale outbreaks of influenza such as were experienced in 1957 and 1970-71.

Organisational factors. These include the nature and function of the particular enterprise and the personnel policies it adopts. Mining and heavy industry have traditionally had higher levels of absenteeism than service industries such as transport or communications. The public health requirements which affect processed food factories may increase the amount of sick absence for preventive reasons, while research laboratories tend to have low rates.

The relationship between organisational size and absenteeism was recognised many years ago, but it is now clear that it is the size of the working group or work unit rather than the size of the whole organisation that matters. Large groups tend to make the worker feel anonymous and, coupled with poor quality of supervision, they may develop the attitude known by sociologists as "alienation", which often goes with high rates of absenteeism. Labour turnover rates often relate directly to absence rates and there is wide acceptance of the theory that both are stages in a worker's means of withdrawal from what he may see as an unsatisfactory working situation.

The organisation of work–production line or batch process, round-the-clock shift working or discontinuous shift systems—also influences absenteeism. The balance of evidence shows that shift work is associated with lower rates of absence. Improvements in the general work environment sometimes seem to have paradoxical effects due to other factors. It is, for example, not uncommon for employees who have been moved from old, small and inadequate offices into new spacious, open-plan accommodation to show a temporary rise in short-term sick absence.

The specific health-related factor in the organisational group is listed as "environmental hazards", but largely due to the influence of international and national health and safety legislation, such hazards are seldom major causes of sick absence. Industrial injuries, for example, are numerically much less important in this context than are injuries sustained away from work. The main type of industrial disease affecting absence is dermatitis, but this too is usually far less prevalent than other non-industrial causes of skin disease.

The existence of an occupational health service may influence sick absence rates. In countries where the employer is required to pay for the general health care of workers the presence of an occupational health service can show an economic saving. In countries where this does not apply this cost-benefit aspect of a service is difficult or impossible to demonstrate.

Personal factors. These are the most numerous in table 1 and are also the most important. The over-riding influence of sex, age and occupational status is described in the article ABSENTEEISM: DEFINITIONS AND STATISTICS. Although females at every age are less likely to die than males, the view that their higher rates of sick absence were only due to their lower wage rates and less congenial work has not been borne out since equal pay was introduced. Age exerts opposite effects upon rates of spells and of days, the first fall with increasing age but the second rise, particularly for those over the age of 50 years. Occupational status exerts a clear effect on both spells and days once age is allowed for and, as might be expected, unskilled workers have about three times as much absence as managerial grades (see table 3). It is important to recognise that much of this difference is due to the physical demands of the work and that a man with a sprained ankle could attend if his work is at an office desk, but he could certainly not if he was required to dig up the road. Other, more objective, measures of health also favour senior managers since despite the hypochondriacal fears of some, their expectation of life is also better than that of workers on the shop floor.

Table 3. Absence from work due to illness or injury: men in employment in the United Kingdom, 1971[1]

Socio-economic group	Percentage absent in a 2-week period	Average work-days lost per man per year
All groups	5.2	9.1
Professional	3.7	3.9
Employers/managers	3.7	7.2
Junior non-manual	4.4	6.7
Skilled manual	5.7	9.3
Semi-skilled manual	5.6	11.5
Unskilled manual	8.8	18.4

[1] From General Household Survey 1973.

Job satisfaction, which is made up of a number of different factors, concerns the extent to which a worker is motivated to attend for work. The employee who dislikes his job and who feels that his supervisor does not mind whether or not he attends is quite likely to stay off sick if he has a relatively minor respiratory infection. Personality, whether measured by an extroversion-introversion-neuroticism scale or by much more complicated criteria can also exert a strong influence on an individual's decision whether or not to attend for work, as can the extent to which he adheres to what Max Weber described as the "Protestant work ethic".

Last but not least is the presence or absence of a medical condition. It might be supposed that this should be the only important factor, but this is certainly not true because it is primarily *absence* not sickness that is under consideration. It may not be widely appreciated just how prevalent some form of ill health is in all communities. Complete health as defined by the World Health Organisation is relatively rare, most of us suffer from minor ailments of one sort or another, for much of the time. Field studies in various countries have suggested that only 5-10% are "WHO healthy" for periods of up to two weeks, even though the proportion with objectively serious and incapacitating conditions may be equally small. Doctors are rarely trained to assess fitness, their skills lie in the detection of signs and symptoms of disease and, hopefully, in its treatment. Furthermore, the presence of an objective injury or disease does not necessarily imply that the individual is incapable of work. In the great majority of situations the relationship between ill health and incapacity for work involves subjective judgements which are largely influenced by factors totally unrelated to medicine. The requirement by employers and by social insurance agencies that a medical certificate shall state whether a person is fit or unfit for work makes matters more difficult for all the parties concerned. In practical terms it is usually the worker who decides whether or not to consult a doctor in the first place, what he tells the doctor about his symptoms and the type of job he does, and whether or not he feels able to attend for work. It is probably true to say that a doctor is better placed to interpret these matters than anyone else, but the decision is seldom easy or incontrovertible.

Control of absenteeism

Given these complex and wide-ranging factors which influence absenteeism, there can clearly be no single method which can be expected to control this phenomenon. One point is, however, quite clear: occupational physicians and nurses should never be expected to undertake this executive responsibility. Their role should be to advise both management and the sick or injured employees on how they may minimise or tolerate their disabilities and what type of work they may safely be able to undertake. Despite the explicit statement included in the ILO's Occupational Health Services Recommendation, 1959 (No. 112), that occupational physicians should not be called upon to verify the justification of another doctor's certificate of incapacity, there are still some employers who try to recruit doctors for this purpose. There is also a mistaken belief on the part of some that pre-employment medical examinations can reliably predict sickness absence. This is not so; for although they may be used to identify physical diseases in applicants, they cannot predict whether or not the people will take time off work, the only reliable predictor is previous absence.

An effective control programme must be designed to meet the particular problems of the organisation, but it should have three main features: firstly adequate information, that is to say a reliable and sufficiently comprehensive system of individual absence records together with the ability to measure both frequency and severity rates in groups of employees. The establishment of acceptable standards of attendance must depend upon local evidence, not wishful thinking or the pronouncements of others.

The second principle of effective absence control is the establishment throughout the organisation of the right attitude. Is senior management really concerned and do the employees know this? Who is responsible for providing the information and are supervisors aware of the rates for their own groups and how they compare? Do they realise their own role and are they given both guidance and responsibility to undertake it?

Finally, but of equal importance, there must be appropriate company personnel policies and procedures designed to meet local needs and to cover all aspects of absenteeism. This is where the role of an occupational health service can be of considerable assistance both in terms of preventive health for the whole workforce and by providing support and advice to individuals. One most effective means of providing such help is for the occupational physician or nurse to see all employees on the first day back at work following a period of sick absence lasting perhaps three or four weeks, or after several shorter spells. The purpose of such an interview is to assess the employee's fitness to return to his or her normal work or whether modified duties would be advisable. Health advice and recommendations to the employee and, as necessary, to his supervisor, can reduce the risks of repeated absences and improve efficiency and productivity.

TAYLOR, P. J.

"Aspects of sickness absence". Taylor, P. J. *Current approaches to occupational medicine.* Ward Gardner, A. (ed.) (Bristol, John Wright and Son, 1979), 368 p. Illus. 65 ref.

"Short term absence from industry". Froggatt, P. *British Journal of Industrial Medicine* (London), July 1970, 27/3 (199-224), Oct. 1970, 27/4 (297-312). Illus. 144 ref.

"The physician's role in sickness absence certificate". Coe, J. *Journal of Occupational Medicine* (Chicago), Nov. 1975, 17/11 (722-724). 11 ref.

"Social determination of sick-absenteeism. Part 1". Indulski, J. *Santé publique* (Bucarest), 1966, 8/4 (403-416). Illus. 32 ref.

"Some characteristics of repeated sickness absence". Ferguson, D. *British Journal of Industrial Medicine* (London), Oct. 1972, 29/4 (420-431). Illus. 43 ref.

"Sickness absence of alcoholics". Pell, S.; D'Alonzo, C. A. *Journal of Occupational Medicine* (Chicago), June 1970, 12/6 (198-210). Illus. 6 ref.

Absenteeism and social security. Studies and research, No. 16 (Geneva, International Social Security Association, 1981), 157 p.

Absenteeism, definitions and statistics of

People may be absent from the workplace for a variety of reasons ranging from holidays or illness to attending football matches and so on. The word "absenteeism" is used to describe absence when an employee is normally expected to attend for work and therefore excludes holidays and strikes. The main type of absenteeism is that attributed to incapacity (illness or injury), and this usually accounts for not less than three-quarters and often almost all industrial absenteeism. The attribution to incapacity may be supported by a medical certificate depending upon local rules or social insurance regulations, but the definition also requires that this attribution to incapacity must be accepted by the employer or the social insurance system.

The term "sick absence" is widely used to describe absence attributed to incapacity, but its brevity seems to encourage over-simplification and misunderstanding of what is a very complex phenomenon. Sometimes misunderstandings arise because authors fail to define what is or is not being measured. Organisations that include only sick absences lasting for one week or more will probably report absence rates appreciably lower than others that include all absences of one day or more. It is essential to be precise in defining the data used, and the Absenteeism Subcommittee of the Permanent Commission and International Association on Occupational Health has published revised recommendations to encourage comparability.

Occupational physicians and nurses are often required to advise on problems of sick absence, not only for individuals but also for groups of workers, and for this there must be properly defined and reliable systems of recording the data. There are two aspects to the preparation of any index or rate of absenteeism, the numerator which concerns the absence, and the denominator concerning the population at risk. The latter often provides the greater problem. Sick absence analyses are usually done for 12-month periods since, in most countries, there are strong relationships of illness with season.

Data concerning the absence

Sick absence. This is all absence from work accepted as attributable to incapacity, except that due to normal pregnancy or confinement. It is often useful to distinguish between absence covered by a doctor's certificate and absence that is not. The latter is usually of relatively brief duration, but in most organisations it is more frequent than medically certificated absence.

Spell of absence. This is an uninterrupted period of absence from its commencement, irrespective of its duration. In annual analyses a spell commencing in one year and continuing into the next contributes its spell only to the first year.

Duration of absence. Duration of absence should preferably be counted in calendar days although some analyses have used working days (or shifts). Management in industry is interested in the number of work-days lost, social insurance administrations may count the number of days for which benefit is paid, but in the context of health care all calendar days of incapacity are important. Confusion may easily arise unless an author clearly defines the method used. The first day or substantial part of a day a person is away is counted as the initial day. The final day is either the day preceding the day on which he returns to work, is pensioned or dismissed, or it is the day on which the sick person dies. There is some difference in practice between authors on the maximum duration of a spell. Most count up to 365 days in a year but some stop at 182. The Subcommittee on Absenteeism recommended the use of 365 calendar days. Similar problems may arise when working days are used and here too the maximum number (e.g. 250) must be specified in the report. Short spells of sick absence of one day or more should also be included, and if they are not, this *must* be mentioned. Finally, spells of absence extending beyond the end of the 12-month period will incur days in the second year even though no new spell is counted.

Diagnosis of absence. Whenever it is possible to do so, the main cause of the incapacity, determined at the end of the spell if the diagnosis has changed, should also be recorded. Although it may seem easier to restrict this to main diagnostic groups such as the 17 main sections of the International Classification of Diseases (ICD), the Subcommittee on Absenteeism recommended that classification should be to the 3- or 4-digit rubric of the ICD. The reason is that subsequent research into specific conditions, such as influenza, back injury or contact dermatitis, would never be possible if broad diagnostic groups only had been recorded. The increasing use of computers to handle sick absence data makes this recommendation more practicable. Nevertheless, broad diagnostic groups are usually sufficient for routine annual analyses, and respiratory tract conditions are almost invariably found to be the most prevalent. In the case of injuries it is advisable to distinguish those sustained at work from those sustained elsewhere.

Data concerning the absentee and the population at risk

It is much easier to collect information about persons who have been absent (sex, age, occupation and so on) than it is to obtain the same information about all the population at risk. Unfortunately it is seldom of value to know one without the other. Only if the data are known for the whole population can useful rates be calculated. Some studies of sick absence have proved to be almost worthless because of this problem. The basic data required for routine annual analyses of absence are set out below.

Person-years. Although with a stable population a mid-year census may be sufficient, any appreciable expansion or contraction can make this unreliable as a denominator for the calculation of rates. The mean of four quarterly population figures can improve reliability, but when a computerised payroll is available, the method of choice is to obtain a mean annual population derived from person-months for each individual employed for all or some of the 12-month period. Mean person-year computations will also be required for each of the other population subgroups for which rates are to be calculated.

Full-year persons. For some analyses, notably frequency distribution of spells and of days, a person-year population is not appropriate. Those who join or leave during the year cannot have been at risk throughout the year and sick rates and median values for such people must be separately calculated. It is usually found that sick rates of those employed throughout the year are lower than those for mean staff.

Sex. Even in this era of sex equality, sick rates for women are often found to be higher than for men. It follows that absence analyses must take this into account and separate rates must be calculated. Some organisations distinguish between married and single women (the latter are better termed "Other women") but for most purposes this additional classification will not be necessary.

Age. There is usually a markedly inverse relationship between spells of absence and age, particularly for brief or uncertificated absences. On the other hand, severity or duration rises with age. Where populations at risk are of a size sufficient to justify it, five-yearly groups are recommended (e.g. < 20, 20-24, 25-29, 30-34 and so on), for smaller populations ten-year groups would suffice. When fewer than 100 people are involved the investigator should at least distinguish between those up to 39 years and those aged 40 or more.

Occupational status. This is the third variable that must always be considered in sick absence analyses. Occupational status in an organisation can itself account for threefold differences in sick rate even when age and sex have been taken into account. The degree to which different groups should be distinguished will, like age groups, depend upon numbers at risk. For medium-sized organisations groups such as managers, skilled manual, skilled non-manual, semi-skilled and unskilled groups might be appropriate. At the very least distinction must be made between white-collar or "staff" employees on the one hand and blue-collar or "labour" on the other hand.

Other factors. The very wide range of factors already known to influence sick absence rates (see ABSENTEEISM, CAUSES AND CONTROL OF) provides many optional alternatives for further study. These may be used by the investigator to study particular problems for ad hoc rather than routine annual analyses. It must be emphasised that these can only be studied after allowance has been made for the three primary variables of sex, age and occupational status. In most circumstances it will be necessary to have the information for the population at risk, but where this is not possible a sample or a matched control group may be used.

Sick absence indices

Although the maintenance of or access to individual sick absence records is certainly necessary for the occupational physician or nurse to assist them in their important role as advisers to individual workers with problems related to their health and their work, this alone is insufficient to permit useful analysis of sick absence in the working population. Thus, for example, the information that 12 mechanical fitters have had dermatitis (or back pain) in the previous year needs to be extended to a rate to enable comparison with other occupational groups; it might then be found that a higher rate occurred in a different occupation.

Sickness absence, unlike births or deaths, is a repetitive phenomenon with the additional complication of variable duration. A single rate of index can thus never adequately describe absenteeism. As has long been appreciated with industrial safety statistics, a minimum of two indices (frequency and severity) must be used and others may be valuable to clarify certain problems. There are four main types of index which are most frequently used for sick absence analyses: prevalence, severity, spell frequency, and frequency distributions.

Prevalence rates are the simplest form of index and, because the data are simple to obtain, such rates are often the first to be used when commencing an investigation. The *point prevalence rate* is defined as the number of people absent on a day expressed as a percentage of the total population who should have attended on that day. This rate, however, gives no indication at all about the duration of the absences or their frequency.

Period prevalence rates, the proportion away at any time during a defined period of, for example, one month, are of less practical value in routine sick absence analysis but they may be helpful in studies of certain diseases.

Severity rates are most widely used in industry and are of two main forms, which are different methods of expressing the same thing. The one most often used by personnel management and preferred by accountants is the *lost time percentage*. This expresses the hours of working time lost due to sick and other types of absence as a percentage of the potential normal working hours, excluding overtime, that would have been worked had all employees supposed to attend actually done so. This is usually calculated for consecutive weekly or monthly periods and can be used to provide a mean annual rate. The second type of severity rate, most often used in occupational health literature, is the *average annual duration per person.* This expresses total calendar (or sometimes working) days lost due to sick absence in a year as a rate per person at risk and thus the denominator must be "person-years" at risk (see definition above). Since both these indices are measurements of severity, knowing the normal expected working days each year and the normal hours worked per day, the two rates can be more or less interchangeable.

Spell frequency rates. The one most often used is that known as the "inception rate for spells", usually over one year. This rate is calculated as the total number of new spells (irrespective of their duration) commencing in a year divided by person-years at risk, and is thus complementary to the average annual duration per person. It can be seen that by dividing the total number of days by the number of spells the *mean length of a spell* can easily be obtained.

Frequency distributions of sick absence. The prevalence, severity and frequency rates described above are convenient to use but since they are mean rates they give no indication of the very wide range of values that can always be observed when individual records are considered. In most organisations it will be found that half the total time lost is caused by no more than 5-10% of the workforce. Many employees will take no absence at all in any one year, but a few will have taken many spells or days. It is this highly skewed distribution that accounts for the reluctance of statisticians to permit the use of the usual parametric tests of statistical significance with mean sick absence rates. The distribution of absence in any large group of workers will never follow the Poisson or normal distribution, rather it will be found similar to the negative binomial distribution of uneven risk described 70 years ago for industrial injuries. A similar skewed distribution has been noted for non-sickness absenteeism and also lateness in attendance at work. The median value for spells or for days of absence will always be a good deal lower than the mean value. If sick absence indices are to be of any help to the occupational physician or nurse, it is as important to know the median as the mean. Is it more useful to recognise that half the men in a factory took less than four days of sick absence or that the mean value for all men was 14 days? The preparation of a frequency distribution can permit measurement of quartiles, deciles and so on in order to identify individuals who have the highest 5 or 10% of absence experience and who might be helped by occupational health physicians and nurses. Note that the population base for such frequency distributions must always be those employed throughout the year since only they were at risk for the full period.

Comparisons between groups

Most of the advances in our understanding of sick absence patterns in industry have been derived from comparisons between groups of workers. The overwhelmingly powerful relationship between absence and the three crucial factors of sex, age and occupational status must always be allowed for before the investigator can attempt to determine the role of other factors. There are three main methods by which this may be achieved.

Standardisation of sick rates. Most usually effected by the indirect method, this is a relatively simple technique that can be found in standard textbooks of medical statistics in relation to age standardisation for mortality rates. The method can then be applied in sequence to occupational status and sex.

Matched pairs. This is a useful method since the cases and controls can be matched at least for sex, age and occupation, and possibly also for other factors. To obtain an adequate person-year population, absence data may be collected for two or more years. As a rule of thumb it is seldom worth looking at data involving less than 50 person-years and larger group sizes are usually desirable, given the skewed nature of the sick absence phenomenon.

Analysis of variance. This more complicated technique has recently been used in some research papers, but many occupational physicians and nurses might prefer to seek professional statistical assistance.

Finally a cautionary note concerning *statistical tests of significance.* The skewed nature of sick absence spells, and even more so of days, means that many of the ordinary tests which require that the variable be normally distributed must *not* be used for mean rates of spells or of days. It is, however, permissible to use χ^2 tests to compare the proportions of two groups of people having, for example, more than one, more than two and so on spells or days. For more sophisticated statistical analysis it would be prudent to seek the help of a statistician with experience of sick absence analysis, since non-parametric tests or logarithmic transformation of the data may well be required.

TAYLOR, P. J.

"Permanent Commission and International Association on Occupational Health, Subcommittee on Absenteeism, Draft recommendations". *British Journal of Industrial Medicine* (London), Oct. 1973, 30/4 (402-403). 2 ref.

A short textbook of medical statistics. Bradford Hill, Sir. A. (London, Sydney, Auckland and Toronto, Hodder and Stoughton, 1977), 325 p. Illus.

"Individual variations in sickness absence". Taylor, P. J. *British Journal of Industrial Medicine* (London), July 1967, 24/3 (169-177). Illus. 22 ref.

How to monitor absence from work, from head count to computer. Behrend, H. (London, Institute of Personnel Management, 1978), 62 p. Illus. 20 ref.

Accelerators, particle

The discovery made in 1919 by Sir Ernest Rutherford that atomic nuclei disintegrate under the impact of α-particles from natural radioactive substances led to a search for sources of particle radiation. Instruments designed to produce such radiation are called particle accelerators.

The principle of particle accelerators is basically that a particle which carries an electrical charge is pulled along an electrical field like a stone falling in the field of gravity. The particle then acquires an energy equal to its charge multiplied by the difference in the electrical potential through which it falls. The unit is the electron volt = eV, based on the unit charge of an electron (e) and the volt (V) (10^3 eV = l keV; 10^6 eV = l MeV; 10^9 eV = 1 GeV).

A particle accelerator consists of an ion source, in which the charged particles are produced, normally from hydrogen gas for protons, helium for α-particles, etc., and electrons from a heated filament; a vacuum tube, in which the particle can travel under the influence of the electrical field, and a target, which contains the specimen to be irradiated.

Classification and characteristics. The acceleration of particles can be made by providing the electrical field as one single potential, as in the case of Van de Graaff generators and Cockroft-Walton accelerators, or by repeated application of a relatively low voltage, as in the case of cyclotrons or synchro-cyclotrons and synchrotrons. The acceleration of electrons in betatrons occurs by electromagnetic induction much in the same

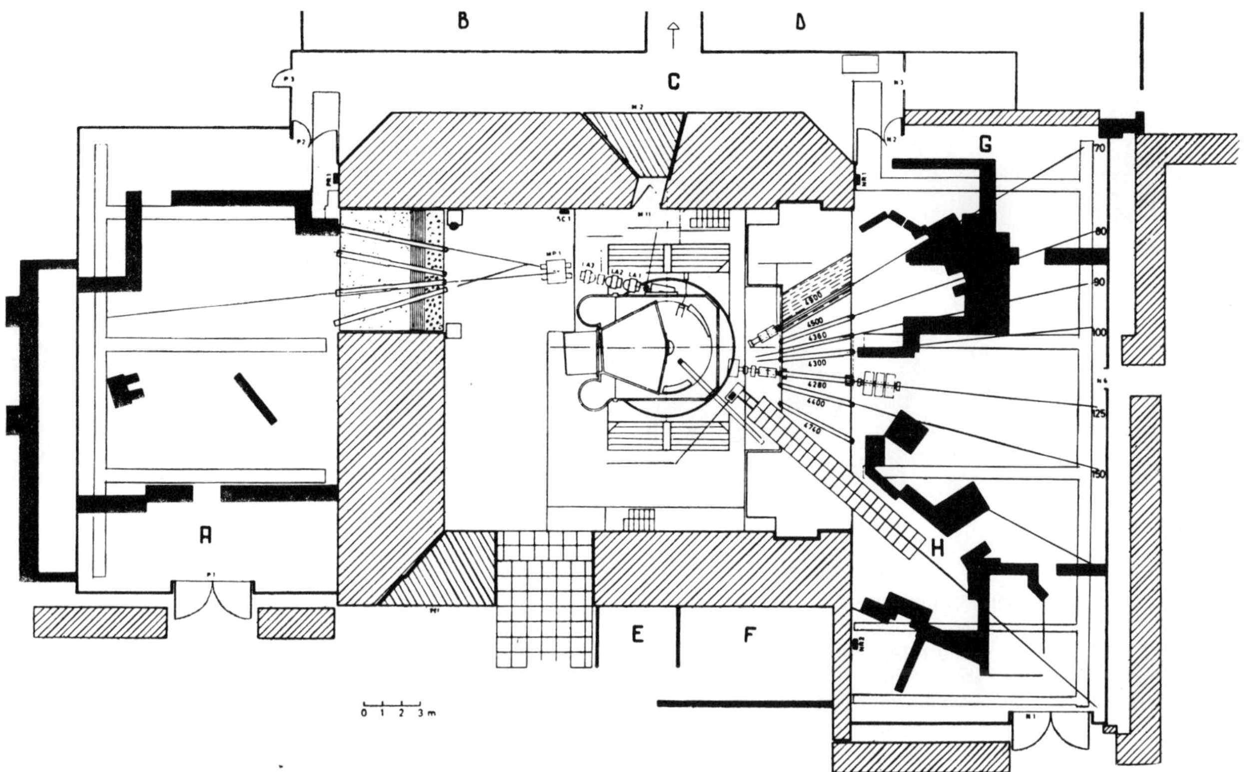

Figure 1. The CERN 600 MeV synchro-cyclotron with its shielding enclosure, beam pipes and shielding arrangements of the neutron experimental hall. A. Proton room. B. Electrical equipment. C. Control room. D. Cooling. E. Gas depot. F. Liquefied gas depot. G. Neutron room. H. Channel.

way as the current is produced in the windings of a transformer. The acceleration of charged particles in linear accelerators occurs by the fact that the cylindrical vacuum tube is constructed as a wave guide with its electrical component along its axis. The charged particles injected into the vacuum tube therefore travel under the influence of this component and with the same speed as the radio wave.

Operation. Accelerators are normally operated in one of two ways. Either the beam is used to irradiate a target located inside the vacuum chamber or the beam is ejected and utilised outside the accelerator. In the first case, secondary particles produced in the target (neutrons, pions, muons, kaons, γ-rays, etc.) are used, while in the second case the whole beam is applied for irradiation or experiments. High-energy accelerators may make use of several targets simultaneously (beam sharing). See RADIATION, IONISING–INDUSTRIAL USES.

Accelerators are also used in combination with each other as particle injectors for larger accelerators. Very high-energy accelerators consist usually of a complex of several accelerators arranged in a chain. The CERN 400 GeV SPS (Superproton Synchrotron) for example consists of five accelerators in a chain to produce the final energy of protons of 400 GeV.

An additional family of accelerator complexes are the high-energy particle storage rings, which are used to accumulate and store high-energy particle beams from accelerators in circular vacuum tubes. The stored particles in storage rings are also accelerated, but only to compensate for the energy lost by the curved pathway of

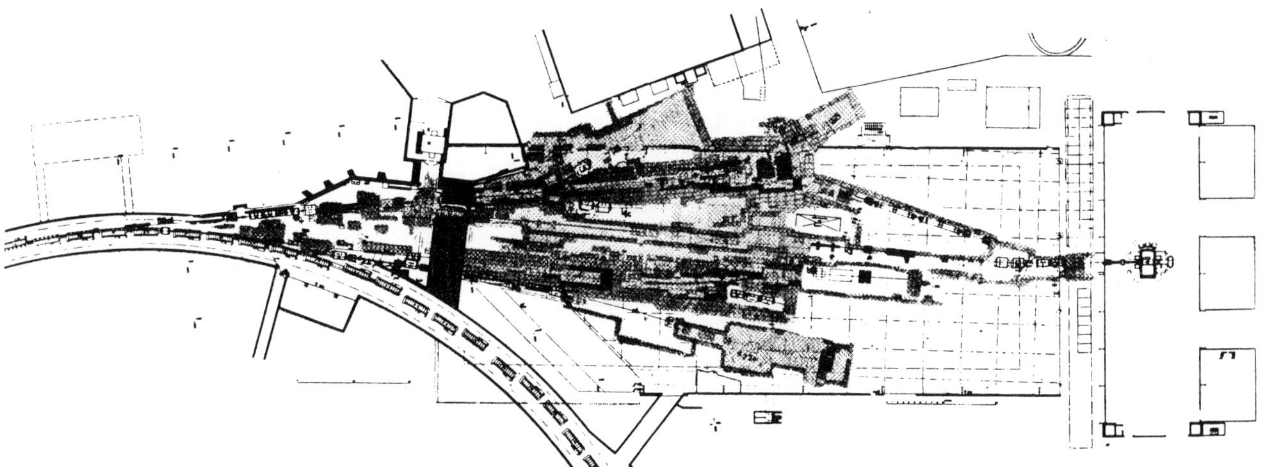

Figure 2. Shielding arrangements of ejected proton beams from the CERN 28 GeV proton synchrotron.

Figure 3. The target region of the CERN 28 GeV proton synchrotron.

the beam. Storage rings have been built and are in operation for electrons, positrons, protons and antiprotons. They may accumulate large intensities (several tens of amperes) and are normally arranged in two circular orbits with opposite beam directions. Such installations, known as Intersecting Storage Rings (ISR), are used for colliding beam experiments, with which the highest colliding particle energies may be achieved.

Existing particle accelerators for research cover energies up to 400 GeV of protons (the CERN SPS in Geneva) with a vacuum chamber of a diameter of 2200 m. Electrons are accelerated up to 40 GeV (SLAC-Stanford, the two-mile linear accelerator). Colliding particle energies with proton intersecting storage rings have operated since 1971 with beam energies at 31 x 31 GeV (the CERN ISR in Geneva), and a 400 x 400 GeV proton ISR has recently been constructed at Brookhaven, United States (ISABELLE).

Storage rings for electron-electron or electron-positron beams have been constructed in Hamburg (PETRA) and one is under construction in Stanford (PEP), both with 19 x 19 GeV maximum colliding beam energies. Plans are being studied (1979) for a much larger electron-electron storage ring for Europe (LEP) with colliding beam energies of 70 x 70 GeV.

Heavy ions are usually accelerated with linear accelerators or synchrocyclotrons. An accelerator complex of a linear accelerator (HILAC) and a synchrocyclotron (BEVATRON) at Berkeley, USA (the BEVALAC), produces beams of a variety of elements to energies of 0.25 to 2.6 GeV per nucleon.

Application. Particle accelerators have found wide application in research. The purpose of producing beams of protons and electrons that collide under the highest possible energies is to study the ultimate structure and composition of matter and the forces existing when nuclear and other aggregates exist as stable systems. At lower energies particle accelerators are used to study the more general property of atomic nuclei or in medicine to treat cancer or to produce isotopes for medical diagnosis. Other applications are in industry for sterilisation of containers or equipment. Accelerators are also used for geological purposes in petroleum prospecting.

Synchrotron radiation from the large electron storage rings has found a wide application in molecular biology, in particular for biomolecular structural studies. Such use of accelerators is becoming increasingly important in fundamental research in biology and medicine.

HAZARDS AND THEIR PREVENTION

The radiation protection of accelerators during their operation is ensured by enclosing the accelerator itself and its primary ejected beams within a shielding enclosure. The shielding enclosure may be made of earth, concrete or steel, depending on the size of the accelerator. In figure 1 is shown the shielded enclosure (concrete) of the CERN 600 MeV synchro-cyclotron and the shielding arrangement of secondary beams, and figure 2 illustrates the beam layouts and shielding of ejected proton beams and secondary beams from the 28 GeV CERN proton synchrotron. Access to regions inside the shielded areas is by interlocking doors which cannot be opened when the accelerators are in operation.

The radiological protection problems near high-energy accelerators are found in the unpredictable composition of the radiation escaping from the shield. Normally it is assumed that neutrons and γ-rays contribute the major part of the danger near present-day accelerators, but muons and other elementary particles might also contribute.

In addition to the radiation danger during operation of high-energy accelerators, a considerable quantity of radioactivity is produced in the targets and in places where the accelerated particles have been lost. This fact presents severe radiation protection problems during maintenance work on the accelerator after it has been turned off (figure 3). The main hazard is found near targets, vacuum chambers, magnets, beam stoppers and shielding materials. The severity of this problem increases with particle energy and intensity. Since very high energy particles normally are able to produce a variety of radioisotopes within the same material, a waiting period ranging from hours to days is imposed in order to let the short-lived radioisotopes decay before maintenance work is undertaken.

Dose rates found near targets, etc., in a 28 GeV accelerator with beam intensities of 10^{12} particles per second range up to 50 rad/h. At lower energies these dose rates are substantially reduced. The radioactivity of the air and dust (loose radioactivity) does not normally present a hazard that exceeds the danger from the induced β- and γ-radioactivity of the machine components.

Electron accelerators present a less severe radiation protection problem than do proton accelerators.

Medical supervision of workers. Personnel protection near high-energy accelerators is ensured by personnel radiation monitoring, together with medical supervision. Personnel radiation monitoring is done by individual γ- and neutron-sensitive detectors. The particular hazard caused by nuclear processes is evaluated from observing interactions (nuclear stars) in the neutron-sensitive film emulsions. The medical supervision includes, in particular, examination of the lens of the eye with respect to cataract formation.

BAARLI, J.

Particle accelerators. Livingstone, M. S.; Blewell, J. P. (New York, McGraw-Hill, 1962).

Principles of particle accelerators. Perisco, E.; Ferrari, E.; Segie, S. E. (New York and Amsterdam, W. A. Benjamin Inc., 1968).

Accelerator health physics. Patterson, H. W.; Thomas, R. H. (New York and London, Academic Press, 1973).

Radiation protection design guidelines for 0.1-100 MeV particle accelerator facilities (National Council on Radiation Protection and Measurements, 7910 Woodmont Avenue, Washington DC, 1977).

"Individual radioprotection at Saclay accelerators" (Radioprotection individuelle auprès des accélérateurs de Saclay). Brochen; Delsaut; Drouet; Vialettes; Zerbib. *Rayonnements ionisants. Techniques de mesures et de protection* (Verfeuil, Goudargues, France), Mar. 1982, 1 (13-52). Illus. (In French)

Accident analysis

Every notified accident results not only in the official notification, whose usual purpose is the compensation of the victim, but also in a report or inquiry in the undertaking, whose purpose is prevention.

These reports long remained very succinct and empirical, generally taking the form of a short account of the accident, concluding at best with the recommendation of a measure to prevent the recurrence of the same accident.

The causes of accidents brought to light by such analyses, moreover, were few and they generally reflected the analyst's conception of the "accident phenomenon". Thus, the accident was at first conceived as a simple phenomenon, due to a single, or a main, cause and then as something due to a small number of causes. For example, the conception of the accident as resulting from a "dangerous act" and "dangerous conditions" was prevalent for a long time. This led the analyst to give prominence, among the causes of the accident, to a "human" cause (disregard of instructions, failure to use personal protection, etc.), and a "technical" cause (a machine without a guard or not running properly, etc.). That is to say, only the direct causes of the injury were the subject of preventive measures, measures that, although they were doubtless necessary, did not suffice to eliminate working conditions conducive to accidents.

Aims and stages of analysis

At present an occupational accident is generally regarded as an index (or a symptom) of dysfunctions in a system formed by a production unit, such as a factory, a workshop, a shift or a workplace. The notion of the system leads the analyst to study not only the elements making it up but also the relations between them.

From this point of view the first purpose of accident analysis is to trace back to its origin the chain of elementary dysfunctions that has led to the injury and, more generally, the whole pattern of antecedents to the unwelcome event, whether it is an accident, a near accident or an incident.

Accident analysis is thus an "inverse *a posteriori*" process, since it is carried out after the unwelcome event has occurred and consists in working back step by step from this event to its origin (see SYSTEM ANALYSIS (SAFETY); SYSTEM RELIABILITY).

The reconstitution of the network of antecedents to the unwelcome event is represented by a diagram generally known as the causal tree. This stage is followed by the drawing up of as complete as possible a list of preventive measures. Later the information recorded during successive analyses may also provide the basis for a quantitative treatment designed, in particular, to bring out factors common to several accidents.

Information of use in analysis

The analysis starts with the collection of information, which must make it possible to describe the successive stages of the accident in concrete, precise and objective terms.

The analyst therefore sets about recording tangible facts without allowing himself to interpret them or express an opinion on them. These are the antecedents to the accident. They may be of two types:

- those of an occasional character (change, variation) in relation to the "normal" course of work;
- those of a permanent character, which play an active part in causing the accident through or in combination with the occasional antecedents.

For example the inadequate guarding of a machine (a permanent antecedent) can be a factor in an accident by enabling the operator to accede to a dangerous area to deal with an incident (an occasional antecedent).

Information is collected on the spot as soon as possible after the accident has happened. It is best collected by a person familiar with the undertaking, who should endeavour to obtain an accurate description of the work without confining himself to the immediate causes of the injury. He depends mainly on interviews, of the victim if possible, of eye-witnesses, of workmates and of seniors at various levels. The information obtained is supplemented where necessary by reports of technical experts.

The analyst then attempts first to extract the occasional antecedents and establish the logical connections between them. In doing this he also detects the permanent antecedents that have made it possible for the accident to happen. He is thus taken far back before the immediate antecedents of the injury.

Antecedents may concern persons (who they are), their duties (what they do), the material they use and the environment in which they carry on their activities.

In this way it is generally possible to draw up a list showing numerous antecedents from which it is difficult to draw immediate conclusions. Interpretation is based on a graphic representation of all the antecedents giving rise to the accident: this is the causal tree.

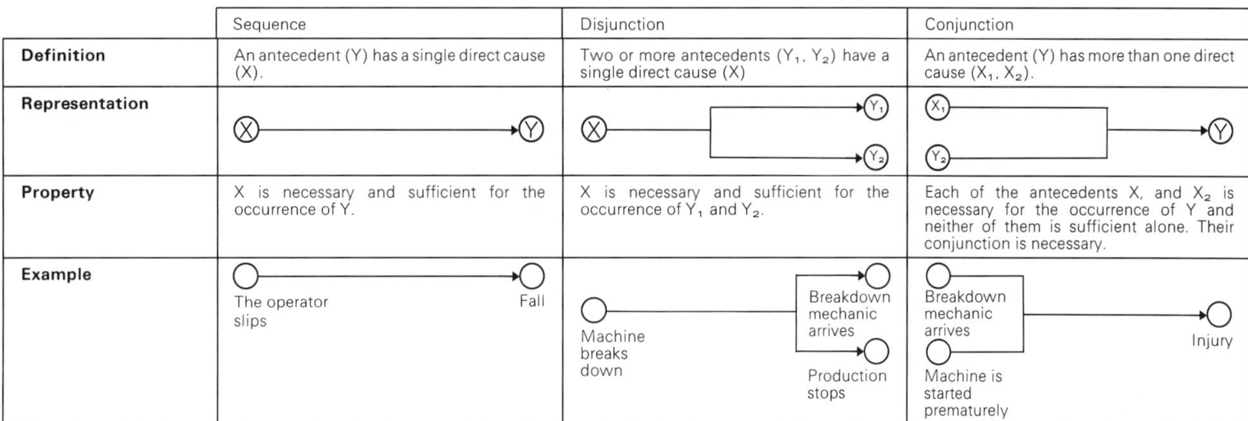

	Sequence	Disjunction	Conjunction
Definition	An antecedent (Y) has a single direct cause (X).	Two or more antecedents (Y_1, Y_2) have a single direct cause (X)	An antecedent (Y) has more than one direct cause (X_1, X_2).
Representation	X → Y	X → Y_1; X → Y_2	X_1, X_2 → Y
Property	X is necessary and sufficient for the occurrence of Y.	X is necessary and sufficient for the occurrence of Y_1 and Y_2.	Each of the antecedents X_1 and X_2 is necessary for the occurrence of Y and neither of them is sufficient alone. Their conjunction is necessary.
Example	The operator slips → Fall	Machine breaks down → Breakdown mechanic arrives; Machine breaks down → Production stops	Breakdown mechanic arrives, Machine is started prematurely → Injury

Figure 1. Types of logical connection.

Drawing up the causal tree

The causal tree shows all the antecedents detected that have led to the accident and indicates the logical and chronological connections between them: it represents the network of antecedents that have directly or indirectly caused the injury.

The causal tree starts with the last stage in the affair, namely the injury, and for each antecedent detected the following questions are asked systematically:

- By what antecedent (*X*) has antecendent *Y* been directly caused?
- Has antecedent *X* alone sufficed to cause antecedent *Y*?
- If not, what other antecedents (X_1, X_2 .... X_n) are also necessary to cause *Y* directly?

This set of questions may bring to light the three types of logical connection between the antecedents shown in figure 1.

The logical consistence of the causal tree is checked by asking the following questions for each antecedent:

- If *X* had not occurred would *Y* have occurred nevertheless?
- For *Y* to occur was *X* necessary and was it sufficient in itself?

The very fact of drawing up the causal tree, moreover, leads the analyst to pursue, and if necessary supplement, the collection of information and so the analysis, often very far back from the injury.

When the causal tree is completed, it shows the network of the antecedents that have led to the injury, which are, indeed, so many factors in the accident.

The following specimen analysis illustrates the above remarks:

Report of an accident

In the yard of an industrial establishment a lorry driver was preparing to couple a waiting trailer to a tractor. The operation was difficult because of the difference in height between the tractor and the trailer and the lorry driver got down to find out why he was unable to take up the right position, but forgot to apply the hand brake. The tractor, moreover, was not the one that was usually attached to the trailer in question. When the driver was between the trailer and the tractor, the tractor, with its engine running, backed down a gentle slope and crushed him against the framework of the trailer (see figure 2).

Application of the causal tree to accident prevention

The application of the causal tree to the prevention of accidents fulfils two purposes:

- to make the recurrence of the same accident impossible;
- to avoid the occurrence of more or less comparable accidents, that is to say accidents that analysis would show to have factors in common with accidents that have already occurred.

The logical structure of the causal tree is such that in the absence of a single one of the antecedents, the accident would not have occurred. A suitable preventive measure would thus be enough to prevent the recurrence of the same accident. The second purpose, however, can be fulfilled only where all the factors brought to light are eliminated. In practice the antecedents are not all equally important to prevention, and so it is advisable to draw up a list of the antecedents that call for possible and realistic preventive measures. If the list is long, a selection must be made, and it is more likely to be relevant if it is reached through discussion between the partners concerned in the accident. The discussion gains in precision, moreover, where it is possible to estimate the cost/efficiency ratio of each measure proposed. The efficiency of a preventive measure can be assessed by means of several criteria, as set out below:

The durability of the measure. The effects of a preventive measure must not disappear with time in the manner of measures such as the training of operators or the repetition of instructions, which are not durable since their effects are often fleeting. This is also true of certain material forms of protection when they are easy to move.

The integration of safety. Many preventive measures are extraneous to the production process and in these cases it is said that safety is not integrated with this process. The measure in question is then invariably abandoned more or less quickly. Generally speaking, any preventive measure that involves extra cost to the operator must be avoided, whether the cost is physiological (increased physical or nervous strain), financial (in the case of incentive wages) or merely the time lost.

The absence of transferred risk. Certain preventive measures may have secondary effects that are harmful to safety, and so it is always necessary to consider the consequences that a preventive measure taken at one place in a system may have at other places in it (workplaces, shifts, workshops).

General applicability (the notion of the potential accident factor). This criterion relates to the preoccupation that a single preventive action should concern the greatest number of workplaces. Whenever possible, an effort must be made to go beyond the particular case that has led to the analysis, though this often calls for a restatement of the problems that have been brought to light. For example, it is not enough to place a guard on the part of the machine that has proved dangerous; consideration must be given to the broader problem of all the accessible moving parts and their guarding.

It can be seen, then, that the lesson drawn from an accident may result in preventive action concerning

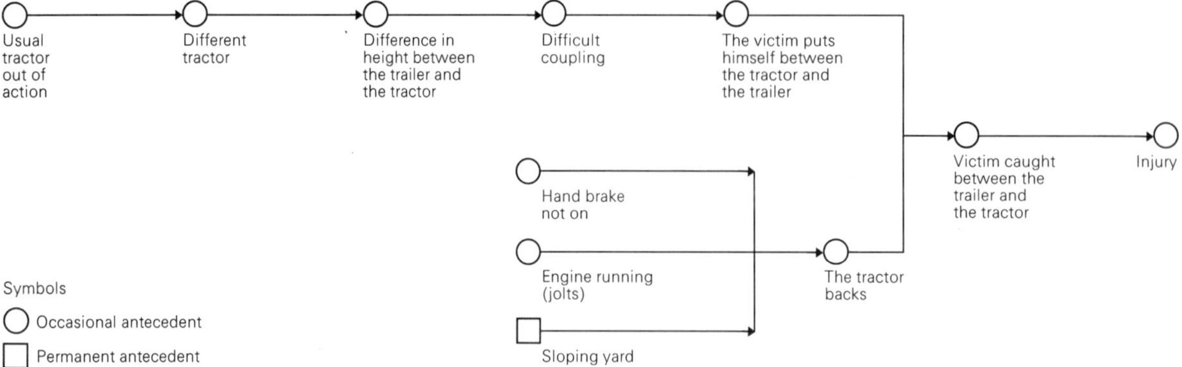

Figure 2. Causal tree.

unknown factors in other working situations, where they have not yet contributed to the occurrence of accidents. These factors are known as potential accident factors. This conception opens the way to the early detection of risks, which is dealt with further on.

Effect on underlying causes. Where there are many accidents, analysis shows that this is generally due to a work situation that deteriorates, very gradually at first and then more rapidly, until an injury suddenly occurs. For this reason the prevention of accident factors closely concerning the injury eliminates certain effects of the dangerous situations, and if the process is pushed back far enough the dangerous situations may not even arise.

The time required. Action must obviously be taken as quickly as possible after an accident has occurred in order to prevent its recurrence. This necessity often leads to the adoption of a simple preventive measure (an instruction, for example). But the emergency measure cannot be substituted for fuller action. In other words, every accident must give rise to a whole set of proposals and the action to be taken on them must be kept under examination.

The purpose of the criteria that have just been mentioned is a better assessment of the value of the preventive measures proposed after the analysis of each accident. The final choice, however, does not depend on them alone; economic and social considerations, for example, may also have to be taken into account. Lastly, the measures adopted must not conflict with the regulations in force.

Towards the organisation of safety

The lessons derived from the analysis of each accident are worth recording systematically with a view to facilitating the transition from knowledge to action. For example, a set of tables might be drawn up to record the lessons derived from the analysis of each accident (table 1) and the follow-up of the measures adopted (table 2). These tables are proposed as general patterns to be modified by those using them.

Table 1 has three columns. At the left the analyst records the accident factors to which preventive measures can be applied. The possible preventive measures are entered in the central column opposite the accident factors. After the discussion referred to above, the measures to be adopted are ticked on the table. The right-hand column gives the potential accident factors deduced from the factors recorded in the left-hand column. In other words, each accident factor brought to light is regarded as being simply a particular case of a more general factor known as a potential accident factor. The transition from the particular case to the more general case is often spontaneous. Whenever an accident factor is expressed, however, in such a way that it can be found only in the situation it has appeared in, an effort must be made to express it in a different way. When this is done with a view to making effective use of the conception of the potential accident factor in the early detection of risks, care must be taken to avoid two opposite errors. If an accident factor is expressed too narrowly, systematic detection is impossible; and if it is too broadly expressed, the notion becomes unworkable and void of all practical interest. The detection of potential accident factors thus depends on their being properly expressed. Detection can be carried out in two ways, which are in fact complementary:

- either by seeking the possible presence of known potential accident factors at, for example, a given workplace;
- or by seeking the workplaces where a given factor may be observed.

Preventive measures taken after or before an accident must be carefully followed up. Table 2 provides a means of recording the measures under consideration and also indicates the practical methods of putting them into effect (Who? When? Where? How much?). One part is used during the verification of the effect given to the measures. This verification makes it possible to detect and explain any differences between the expected and actual results of a preventive measure. Lastly, the success of the whole procedure depends also on the information imparted to the staff of the undertaking. This information is essential if it is desired to develop a positive attitude among the staff concerned.

Under a perfect system of prevention, risks would always be identified and prevented at the planning stage, as in the production of material and equipment presenting the risk of explosion or fire. In fact, many risks arise during the organisation or in the course of work and are revealed by incidents, by material damage or by accidents.

Table 1. Lessons derived from an accident (hypothetical example)

Accident factors established	Possible preventive measures	Potential accident factors
Sloping yard	Reconstruction of the yard	Unsuitable parking area
Motor running Hand brake off	Training driver and acquainting him with the facts	Inadequate training of operators
Difference in height between trailer and tractor	Standardisation of coupling devices	Technical incompatibility of materials
Usual tractor out of action	Preventive overhauling of vehicles	Breakdown of equipment

Table 2. Follow-up of preventive measures under consideration

1st stage: recording						2nd stage: verification		
Preventive measures under consideration	Already proposed? Yes or no	Respecting (workplace, shift, workshop)	Expected time required	In charge of execution	Expected cost	Date	Put into effect? Yes or no	Results observed or Reasons for not putting into effect

The analysis of accidents or incidents thus loses nothing of its necessity: it provides not only the opportunity of eliminating accident factors that have led to an unwelcome event but also a means of identifying potential accident factors that can be found and neutralised in work situations where they might contribute to the occurrence of other accidents.

MONTEAU, M.

A practical method of investigating accident factors. Principles and experimental application. Monteau, M. (Luxembourg, Commission of the European Communities, 1977), 59 p. Illus.

Techniques of safety administration. An analysis of industrial accidents and the use of the concept of the potential accident factor in the prevention of professional hazards (Techniques de gestion de la sécurité – L'analyse des accidents du travail et l'emploi de la notion de facteurs potentiels d'accidents pour la prévention des risques professionnels). Meric, M.; Monteau, M; Szekely, J. (Institut national de recherche et de sécurité, avenue de Bourgogne, 54500 Vandœuvre-lès-Nancy, France, 1976), 85 p. Illus. (In French)

Industrial accidents (Les accidents du travail). Leplat, J.; Cuny, X. (Paris, Presses Universitaires de France, 1974), 124 p. Illus. 15 ref. (In French)

AIM: Accident investigation methodology (Safety Sciences Division of MIOSH, WSA Inc., 11772 Sorrento Valley Road, San Diego, California, 1978), 106 p. Illus. 17 ref.

"Accident analysis and work analysis". Leplat, J. *Journal of Occupational Accidents* (Amsterdam), Apr. 1978, 1/4 (331-340). Illus. 12 ref.

Accident cost

Workers who are the victims of accidents at work suffer from material consequences, which include expenses and loss of earnings and which may be of short or long duration. They also suffer intangible consequences including pain and suffering, which likewise may last over a long period. These consequences include:

(a) doctor's fees, cost of ambulance or other transport, hospital charges or fees for home nursing, payments made to persons who gave assistance, cost of artificial limbs, etc.;

(b) the immediate loss of earnings during absence from work;

(c) loss of future earnings if the injury precludes the victim's normal advancement in his career or occupation;

(d) permanent afflictions resulting from the accident such as mutilation, lameness, loss of vision, ugly scars or disfigurement, mental changes, etc., which may reduce life expectation and give rise to physical or psychological suffering and to further expenses arising from the victim's need to find new interests;

(e) subsequent economic difficulties with the family budget if other members of the family have to give up their employment in order to look after the victim;

(f) anxiety for the rest of the family and detriment to their future, especially in the case of children.

There may be additional loss of income in certain cases where the victim was engaged in private work outside normal working hours and he is no longer able to perform it.

Workers who become victims of accidents frequently receive allowances both in cash and in kind. These do not affect the intangible consequences of the accident (except in exceptional circumstances) and they constitute a more or less important part of the material consequences, inasmuch as they affect the sole income which will take the place of the salary. There is no doubt that part of the over-all costs of an accident must, except in very favourable circumstances, be borne directly by the victims.

Considering the national economy as a whole, it must be admitted that the interdependence of all its members is such that the consequences of an accident affecting one individual will have repercussions upon the others.

These repercussions have an adverse effect on the general standard of living and among them may be included:

(a) an increase in the price of manufactured products, since the expenses and losses resulting from an accident will be added to the costs of the producer;

(b) a decrease in the gross national product as a result of the adverse effects of accidents on people and on materials; these effects will vary according to the availability in a country of manpower and material resources;

(c) additional expenses incurred to cover the cost of compensating accident victims and the amount necessary to provide safety measures.

Covering these latter expenses is one of the functions of society, in that it must protect the health and income of its members. It meets these obligations through the creation of social security institutions and of a safety system (including legislation, inspection, assistance, research, etc.), the administrative costs of which are a charge on society.

As a result of all this a considerable amount of capital is no longer available for productive investment; nevertheless, the money devoted to preventive action does result in certain economic benefits to the extent that there is a reduction in the total number of accidents and their cost. In addition, much of the effort devoted to the prevention of accidents takes place outside the undertaking itself; this includes the incorporation of higher safety standards into machinery and equipment and the general education of the population before working age. Activities of this nature are equally useful outside the workplace, and this is of increasing importance because, as is well known, the number and cost of accidents occurring at home, on the road and in other activities of modern life continues to grow (see ACCIDENTS, OCCUPATIONAL DOMESTIC; ACCIDENTS, OFF-THE-JOB). This growth may be attributed to the widespread penetration of the products of modern technology into the daily life of the individual and results in the imposition of a further heavy burden on the national finances.

The level of compensation benefits and the amount of resources devoted to accident prevention by governments are limited for two reasons. They depend firstly on the value placed on human life and suffering which varies from one country to another and from one era to another, and secondly on the funds available and the priorities allotted to the protection of the public against famine, sickness, a hostile environment and general poverty as well as to the creation of employment and the provision of housing, educational facilities and the like.

The total cost of accidents may be said to be the sum of the cost of prevention and of the subsequent charges; it is not the same for every branch of economic activity. The value of the subsequent charges has been estimated in a number of industrialised countries as varying between 1% and 3% of the gross national product

(GNP). The cost of accident prevention is more difficult to estimate; it would not seem unreasonable to assume that it amounts to at least twice the value of the subsequent charges.

The necessary financial resources are drawn from the economically active section of the population, which is made up of workers (including the victims of accidents), employers, undertakings and other members of society who, though not part of the labour force, are economically active in that they are contributors. The system works either on the basis of contributions to the institutions that provide the benefits, or through taxes collected by the State and other public authorities. In many cases both systems are in force.

At the level of the undertaking the cost of accidents includes expenses and losses. These are made up of:

(1) expenses incurred while setting up the system of work and the related equipment and machinery with a view to ensuring safety in the production process. Estimation of these expenses is difficult because it is not possible to draw a line between the safety of the process itself and that of the workers. Major sums are involved which are entirely expended before production commences and are included in general or special costs to be amortised over a period of years;

(2) expenses incurred during production, which in turn include:

(a) fixed charges related to accident prevention, notably for medical, safety and educational services and for arrangements for the workers' participation in the safety programme;

(b) fixed charges for accident insurance, plus variable charges in schemes where premiums are based on the number of accidents;

(c) varying charges for activities related to accident prevention. These depend largely on accident frequency and severity, and include the cost of training and information activities, safety campaigns, safety propaganda and research, as well as of the workers' participation in these activities;

(d) costs arising from personal injuries. These include the cost of medical care, transport, grants to accident victims and their families, administrative and legal consequences of accidents, salaries paid to injured persons during their absence from work and to other workers during interruptions to work after an accident and during subsequent inquiries and investigations, etc.;

(e) costs arising from material damage and loss. These need not be accompanied by personal injury. In fact the most typical and expensive material damage in certain branches of industry arises in circumstances other than those which give rise to personal injury. The techniques of material damage control and those required for the prevention of personal injury have only a few points in common and attention should be concentrated upon these;

(3) losses arising out of a fall in production or from the cost of introducing special counter-measures, both of which are heavy. In addition to affecting the place where the accident occurred, successive losses may occur at other points in the plant or in associated plants. Apart from economic losses which result from work stoppages due to accidents or injuries, account must be taken of the losses resulting when the workers stop work or come out on strike during industrial disputes concerning serious, collective or repeated accidents.

The total value of these costs and losses are by no means the same for every undertaking. The most obvious differences depend on the particular hazards that go with each branch of industry or type of occupation and on the extent to which appropriate safety precautions are applied. Rather than trying to place a value on the initial costs incurred while incorporating accident prevention measures into the system at the earliest stages, many authors have tried to work out the consequential costs. Among these may be cited:

- Heinrich, who proposed that they be divided into "direct costs" (particularly insurance) and "indirect costs" (expenses incurred directly by the manufacturer);
- Simonds, who proposed dividing them into insured costs and non-insured costs;
- Wallach, who proposed a division under the different headings used for analysing production costs, viz. labour, machinery, maintenance and time;
- Compes, who divided them into general costs and individual costs.

In all these examples (with the exception of Wallach), two groups of costs are described which, although differently defined, have many points in common.

In view of the difficulty of estimating over-all costs, attempts have been made to arrive at a suitable value for this figure by expressing the indirect cost (uninsured or individual costs) as a multiple of the direct cost (insured or general costs). Heinrich was the first to attempt to obtain a value for this figure and proposed that the indirect costs amounted to four times the direct costs, in other words that the total cost amounts to five times the direct cost.

This estimation is valid for the group of undertakings studied by Heinrich, but is not valid for other groups and is even less so in the case of individual factories. In a number of industries in various industrialised countries this value has been found to be of the order of 1-7 ($4 \pm 75\%$) but individual studies have shown that this figure can be considerably higher or lower and may even vary over a period of time for the same undertaking.

As far as the cost of incorporating accident prevention measures into the system during the initial stages of a manufacturing project is concerned, there is no doubt that money spent in this way will be offset by the reduction of losses and expenses that would otherwise have been incurred. This saving is not, however, subject to any particular law or fixed proportion and will vary from case to case. It may in fact be found that a small expenditure may result in very substantial savings, whereas in another case a much greater expenditure would result in very little apparent gain. In making calculations of this kind, however, allowance should always be made for the time factor, which works in two ways: on the one hand, current expenses may be reduced by amortising the initial cost over several years and, on the other hand, the probability of an accident occurring, however rare it may be, will increase with the passage of time.

Taking industry as a whole, it is clear that there is no financial incentive to reduce accidents in view of the fact that their cost is added to the production cost and is thus passed on to the consumer.

Considered from the point of view of an individual undertaking, however, it is a different matter. There may be a great incentive for an undertaking to take steps to avoid the serious economic effects of accidents involving key personnel or essential equipment. This is particularly so in the case of small plants which do not have a reserve of qualified staff or those engaged in certain specialised activities. There may also be cases where a larger undertaking can be more competitive and thus increase its profits by taking steps to reduce accidents.

The following data may be of interest in this connection. Hundred (100) monetary units spent on accidents nullify a profit on:

10 000 monetary units of sales at a profit rate of 1%
5 000 monetary units of sales at a profit rate of 2%
2 000 monetary units of sales at a profit rate of 5%

Furthermore, no undertaking can afford to overlook the financial advantages that stem from maintaining good relations with workers and their trade unions.

As a final point, passing from the abstract concept of an undertaking to the concrete reality of those who occupy senior positions in the business (i.e. the employer or the senior management), there is a personal incentive which is not only financial and which stems from the desire or the need to further their own career and to avoid the penalties, legal and otherwise, which may befall them in the case of certain types of accident.

The cost of occupational accidents, therefore, has repercussions on both the national economy and that of each individual member of the population: there is thus an over-all and an individual incentive for everybody to play his part in reducing this cost.

ANDREONI, D.

CIS 74-891 *A cost-effectiveness approach to industrial safety*. Sinclair, T. C. (London, HM Stationery Office, 1972), 59 p. Illus. 48 ref.

CIS 76-2087 *The economic costs of occupational accidents* (Die volkswirtschaftlichen Kosten der Arbeitsunfälle). Franke, A.; Joki, S. Forschungsbericht Nr. 148 (Dortmund-Marten, Bundesanstalt für Arbeitsschutz und Unfallforschung, 1975), 336 p. Illus. 69 ref. (In German)

CIS 77-1499 "Cost of accidents–Cost of safety measures–Cost-effectiveness" (Unfallkosten–Sicherungskosten–Wirtschaftlichkeit). Rieger. *Die Tiefblau-Berufsgenossenschaft* (Munich), Sep. 1976, 88/9 (518-522); Oct. 1976, 88/10 (599-601); Jan. 1977, 89/1 (30-32). (In German)

CIS 79-876 *The cost of a human life* (Le prix de la vie humaine). Le Net, M. Notes et études documentaires, No. 4455 (La Documentation française, 29-31 quai Voltaire, Paris, 1978), 151 p. 59 ref. (In French)

"Accident costing in industry". Bamber, L. *Health and safety at work* (Croydon), Dec. 1979, 2/4 (32-34).

"Safety–the cost of accidents and their prevention". Collinson, I. L. *The Mining Engineer* (London), Jan. 1980, 139/220 (561-571). 3 ref.

Accident investigation

The immediate aim of an accident investigation is to obtain the most accurate and full information about the circumstances and causes of the accident, whereas the ultimate objective is to prevent the occurrence of similar accidents in the future, to uncover new hazards where such exist and to devise adequate measures to control them. A good knowledge of the pattern of accidents is obviously essential to the formulation and application of a sound accident prevention policy at all levels.

The value of an accident investigation is enhanced if the data recorded in the course of the investigation are subsequently collated with other relevant data to form the basis of a statistical evaluation, or if the report of an unusual accident is given wide circulation to attract the attention of persons confronted with similar situations. This requires of course that the person conducting the investigation always bears in mind the importance of an objective examination of all the facts and circumstances leading to the accident and does not let himself be influenced by considerations of guilt or blame.

Accidents to be investigated

Management should arrange for an investigation to be conducted without delay into the circumstances of any accident which has resulted in an injury or in serious material damage. Dangerous occurrences (the overturning of a crane or the failure of a safety device, for instance) should also be investigated promptly and thoroughly. The thoroughness of the investigation will depend upon the nature and severity of the accident. If a carpenter hits his thumb with a hammer, a telephone interview should suffice; if a scaffold collapses, it will usually be necessary to supplement the investigation on-site with, among other things, analysis and tests of the materials used and structural computations, and to hand over some of these supplementary investigations to specialised institutions. All investigations can be placed somewhere between the above-mentioned extremes.

The investigator

According to the nature and circumstances of the accident or dangerous occurrence, the in-plant investigation should be carried out by the safety department or safety officer (if there is one), by the safety committee or by a qualified member of the plant management. It should not normally be entrusted to the person in charge of the work or operation at the time of the accident, although this person will obviously be asked to place all relevant information at the disposal of the investigator.

The employer should in addition give prompt notice of the accident or dangerous occurrence to the local labour inspectorate or other competent authority, whenever he is statutorily required to do so. The inspectorate is rarely, if ever, staffed to investigate every notifiable accident or dangerous occurrence, and the inspector has to accept the fact that his investigations will cover a small percentage of the total.

Contrary to in-plant investigations, which are largely centred on the practical measures to prevent recurrence, those undertaken by the labour inspectorate or other competent authority may have to consider legal implications. Another purpose of these investigations is to examine whether the relevant statutory provisions are adequate and, if not, how they should be amended or supplemented. In cases where there has been a breach of the law, it will be for the labour inspectorate to decide whether it is necessary to institute legal proceedings. As a rule, accidents involving death or serious injuries should all be investigated. Minor injuries which keep recurring should also be investigated, for they reveal dangerous conditions and/or practices which should be corrected before they result in a major accident. Finally, the labour inspectorate may wish to investigate accidents involving unusual features as well as selected dangerous occurrences brought to its attention.

Conduct of investigation

Once it has been decided that an accident should be investigated, the investigation should be carried out as promptly as possible. It is essential that full and accurate information should be obtained while the circumstances are fresh in the memory of the witnesses. Also, important evidence may be removed or destroyed soon after the accident, deliberately or not. Labour inspectors or other official investigators, as the case may be, should concentrate on the objective collection, evaluation and recording of all pertinent facts and of the evidence submitted by the witnesses interviewed by them. The injured person must always be interviewed as soon as possible at the place of employment, at the hospital or elsewhere, particularly where it appears to the investigator that this person played an active part in the events which led to the accident.

Witnesses should never be questioned as a group, the questioning procedure itself is an important factor obtaining a true account of the facts. The witness should not be asked to submit his own views as to how the accident was caused, but merely to state the conditions prevailing before and up to the time of its occurrence. This does not preclude the possibility of getting sound judgement from experienced persons. The investigator must always be aware of the risk of getting erroneous evidence from persons who may experience a certain feeling of guilt. The investigation may require the making of dimensioned sketches or the taking of photographs. Photographs are preferable as a rule, particularly where legal proceedings are contemplated. It is essential in this case that some witnesses should be able to prove that the sketches or photographs reproduce the material conditions prevailing at the time of the investigation. It is desirable for accident investigations (but obviously not for the personal interviews) to be carried out with the employer or some other person in authority being in attendance. The question of who is to be regarded as a person in authority must be left to the discretion of the inspector; in general, this is a person who commands the confidence of the employer, has direct access to him and has the authority to carry out his instructions. It is nevertheless permissible to proceed with an accident investigation in the absence of the employer or a person in authority, if every effort to contact them has proved unsuccessful. Whenever legal proceedings are contemplated, signed statements should be obtained from the persons who saw the accident or had first-hand knowledge of the conditions prevailing at the time. Investigations following fatal accidents should be carried out with particular care because of their possible legal implications, and inspectors' reports on them should be written in such a way that they can be understood by laymen. Technical matters should be expanded and simplified in a way which is not normally required for non-fatal accident reports which are seen at headquarters only.

Facts to be collected

The facts to be collected can be divided into two main groups: first, all the facts bearing on the identification, registration, filing and recording of the accident or dangerous occurrence; second, the facts of significance in the occurrence itself.

The first group should include all useful information about time, injured person and place: the date, day of the week and exact time of the accident and when the injured person started work on the day concerned; the person's name, age, sex, social security number (if any) and occupation (including how long in the present employment); whether the injured person was a skilled worker, an unskilled labourer or an apprentice; the method of payment (piecework or time-rate, for instance); the name and type of the undertaking, department, workshop and workplace. The nature and site of the injury should also be given, together with the probable duration of disability (as stated in the medical certificate if available).

The second group of facts concerns the accident and the circumstances under which it occurred. It is reasonable to commence by stating all the facts that can be read and measured. As far as the machine or equipment involved are concerned, note should be made of make, type, number and year. With regard to the materials being used, it is necessary to determine their manufacturer or supplier, nature, dimensions, quantity and quality; it may also be necessary to take samples for further examination, perhaps by a laboratory or testing institution. Adequate and relevant information should be gathered about the work environment (access to workplace, cubic space, temperature, ventilation, lighting, noise, floor surface, standard of housekeeping, etc.). Where personal protective equipment or protective clothing is involved, notes should be made about its availability, quality and actual use. Further information about the injured person's physical and mental condition may be necessary (illness, fatigue or depression, for instance).

The final and most difficult facts to obtain are those concerned with what actually happened, facts which will rely mainly upon pieces of evidence from the injured person and others.

Analysis

The collected facts will, by and large, give the answers to the two questions: what? and how? The answer to the next question–why?–will depend upon the analysis of the facts (see ACCIDENT ANALYSIS).

Every accident has as its preconditions a number of causes, causes which can be roughly divided into those that have to do with working conditions and environment and those connected with acts of work.

Neither of the two sorts of causes must be overlooked. The superficial analysis is to call the incident fortuitous–"just one of those accidents". It is almost as bad to limit the analysis to pointing out the injured person's or a third party's negligence; this is not profitable and often unjust. It is safer to use some form of a check-list, e.g. on the following lines–was the process arranged and designed to operate reasonably safely?; were the environmental conditions beyond reproach?; had the man been instructed and trained properly?; had there been supervision to ensure that safety precautions were observed?; had the man followed instructions?; had he notified any faults?; had he acted carefully? Often these questions cannot be answered exactly, but a judgement can be based upon whether there were divergencies from legal requirements or–perhaps better–from current practice.

The analysis of causes can be a very simple matter, e.g. if a man is injured by an inadequately guarded machine; however, it can also be very difficult. The greatest difficulties which meet the investigator are when he is faced with an accident in connection with an unusual situation–breakdown of machinery, reorganisation, repairs or injuries to workers employed by outside contractors.

Preparation of report

Each accident report should be given a serial number and should carry the date of notification, investigation and of the accident report. It should be signed by the person in charge of the investigation.

The report should contain all the relevant facts that have been collected. A report form where as many of the facts as possible can be placed in numbered spaces is advantageous for future recording and statistical treatment. The description of the conditions and circumstances leading to the accident should be in a narrative which can be read and understood independently of the rest of the report. There should be a clear distinction between the observed facts and the information based upon oral evidence and assumptions.

In reports from official services, comments should be made on possible violations of the law; this may be

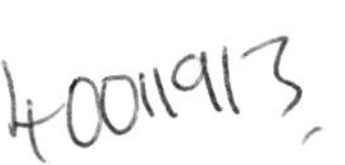

necessary for a possible prosecution, but has less value from an accident prevention point of view.

MOLLERHOJ, B.
ROBERT, M.

Methods of investigation:

CIS 77-574 "A new method of accident analysis" (Une nouvelle méthode d'étude d'accident). *Vigilance* (Paris), Sep. 1976, 54 (3-25) Illus. (In French)

CIS 78-561 *Inquiries into occupational accidents* (Untersuchung von Arbeitsunfällen). Rudloff, M.; Schladebach, H. (Verlag Tribüne, Am Treptower Park 28-30, DDR-1193 Berlin, 1976), 69 p. (In German)

CIS 79-294 *A practical method of investigating accident factors–Principles and experimental application.* Monteau, M. (Luxembourg, Commission of the European Communities, 1977), 84 p. Illus.

Accident proneness

The theory

The frequency with which occupational accidents occur is not the same for every worker: some of them suffer injuries relatively frequently (they are sometimes called "accident repeaters") and others only rarely. This has led to the coining of the term "accident proneness" and to the formulation of a theory according to which certain individuals are more accident prone than others. This theory can be summarised as follows: under conditions of equal risk, there exists a statistically significant difference in the number of accidents that occur to those persons falling in the accident-prone group as compared with the rest. The difference stems from the fact that the members of the first group present certain individual physical and/or psychological features which predispose them to accidents. These features being inborn or acquired in infancy, a process of careful selection at the time of recruitment would result in a substantial reduction of the frequency of occupational accidents.

This theory, which was widely acclaimed in the years between the two world wars, at a time when importance was being attached to the part played by the human factor in accident causation and when progress was being made in differential psychology, has today become most controversial. It will be critically examined below.

It should be observed, in the first place, that the fact that a worker has sustained several accidents does not necessarily mean that he is accident prone; these accidents may be due to the fact that he has been assigned to a particularly dangerous machine or process in respect of which adequate safety measures have not been taken; any other worker sent to work under these conditions would also have met with a number of accidents. It has further been shown that some accident repeaters are persons who have been involved in two or more accidents without any demonstrated responsibility on their part. For these reasons, the study of personal accident statistics should not lead to hard and fast conclusions in respect of accident proneness without a very careful investigation of all factors involved.

The question whether accident proneness can be demonstrated statistically has been investigated along three lines:

(1) By using the percentage method: this involves the observation, over a given period, of the way in which accidents are distributed among a group of workers. It is found that this distribution is unequal; for example, 10% of the subjects incur 32% of the accidents, or 50% of the total, 83%. However, such a distribution can be the result of one single hazard.

(2) By comparing the real distribution with the theoretical distribution: in order to take into account the preceding objection, it would have to be shown that a statistically significant difference exists between the real distribution of accidents in a given group and the distribution that would be obtained if the accidents were distributed haphazardly. Greenwood and Woods (1919), followed by several other authors, found that such a difference exists. Schulzinger (1954), however, showed that the accident-prone group is a variable one from which persons are continually leaving to be replaced by others. If this were to remain a stable group, a hypothesis which could not be excluded is that if the first accident had taken place haphazardly, the workers thus affected would as a result, and because of the very fact that this first accident would put them more at risk, become more susceptible to accidents; this would not correspond in fact with their own constitutional susceptibility.

(3) By a comparison of two periods of observation: the accidents experienced by the same group of workers are recorded over two periods (consecutive or not), and note is taken whether those who were subject to several accidents during the first period are among those who met with frequent accidents during the second period. Marbe (1926), followed by others, verified this agreement. The correlation, however, was weak in most cases; if, for example, from among a group of 104 workers the ten having had the greatest number of accidents are withdrawn, the subsequent accident rate of the members of the group is not changed by this withdrawal.

To sum up, if an individual predisposition towards accidents exists–and this has not been completely demonstrated–the phenomenon does not carry much statistical weight.

Individual factors of predisposition

Contrary to what has been affirmed for a long time, constitutional psycho-physiological characteristics very seldom play any role. Psychometric tests in particular, which should in theory be able to show up any accident proneness, have given contradictory results (visual and auditory acuity, reaction time, vigilance tests) or have not been sufficiently positive to be conclusive (psychomotor tests and intelligence tests); an exception could perhaps be made in the case of certain highly responsible jobs (public transport drivers) or where practical intelligence tests and emotional stability tests have enabled valid conclusions to be drawn.

The exploration of personality traits has given results that are no more convincing: following Freud, some psychoanalysts have attributed certain accidents to a subconscious self-punishment mechanism, but this has been applied more to road accidents than to occupational accidents. Other authors have made use of questionnaires, based on attitudes, or even experimental simulated situations, to try and show up in workers a tendency to caution or to lack of caution, but this approach has not led to any valid grounds on which selection can be based.

The only valid individual factors which can be accepted with any certitude are, finally, those which are not constitutional, but acquired: they include age (adults have many fewer accidents than the young, and somewhat less frequently than the older workers) and training and experience (which reduce the number of accidents). Above all is the onset of illness, of which accidents represent only one of the symptoms; there is a statistical correlation between the individual frequency of illnesses and that of accidents; there is no doubt that the presence of certain symptoms such as hypertension or of a diseased condition such as alcoholism contributes to accidents.

What is certain is that the concept of "predisposition" to accidents, being some kind of constitutional defect which affects some individuals and not others, should be replaced by that of "disposition" in the sense of "an acquired susceptibility" to accidents, which becomes apparent during working life and symbolises a temporary weakness in the organic defences of the human body, or the lasting effects of age, or the intercurrent effects of illness.

Practical consequences

In view of the fact that the principal individual factors which increase the susceptibility of an individual to accidents usually appear after the commencement of his working life, a process of selection at the time of recruitment, however rigorous it may be, will have no notable effect on the number of accidents he will meet with, exception being made in the case of certain high-risk jobs or jobs where the safety of other people is at risk. In such cases special selection procedures should be applied. Nevertheless, the value of periodical medico-psychological supervision of persons at work is to be stressed, as this enables any alterations in the state of health to be detected at an early stage. Such alterations can be the root cause of an accident and in this way a change of job can be prescribed in good time.

It should be recognised at any rate that there are many more work situations which are conducive to accidents as a result of poor design, layout or maintenance than accident-prone workers. The bulk of the accident prevention effort should therefore be concentrated on the elimination or improvement of such situations; selection and health supervision of workers, useful as they may be, can only play an ancillary role in a comprehensive safety programme.

CAZAMIAN, P.

"Individual predisposition to work accidents" (La prédisposition individuelle aux accidents du travail). Cazamian, P.; Chich, Y.; Deveze, G.; Faure, G. *La revue du praticien* (Paris), 1968, 18/4 (513-523). (In French)

"Accident proneness–does it exist?" Lindsay, F. *Occupational Safety and Health* (Birmingham), Feb. 1980, 10/2 (8-9). 10 ref.

CIS 74-883 "The theory of accident proneness and the role of the Poisson distribution". Schugsta, P. M. *ASSE Journal* (Park Ridge, Illinois), Nov. 1973, 18/11 (24-28). Illus. 33 ref.

CIS 77-1782 "Use of psychotechnical tests in occupational safety" (Los exámenes psicotécnicos aplicados a la prevención). Arenal, F. A.; Fernández Pereira, P.; Martín Val, A. *Salud y trabajo* (Madrid), June 1977, 7 (38-44). Illus. 26 ref. (In Spanish)

CIS 78-1485 "Influence of the emotive structure of the personality on occupational accidents" (Uticaj emocionalne strukture ličnosti na pojavu nesreća u toku rada). Pavlović, S. *Ergonomija* (Belgrade), 1978, 5/4 (51-56). Illus. (In Serbocroatian)

CIS 74-885 "Science sheds new light on accident proneness". *Occupational Hazards* (Cleveland), Sep. 1973 (61-64). Illus.

Accidents

An accident may be defined as an unexpected, unplanned occurrence which may involve injury. There is a possibility of accident in every sphere of human life, at home, whilst travelling, at play and at work (see ACCIDENTS, COMMUTING; ACCIDENTS, OCCUPATIONAL DOMESTIC; ACCIDENTS, OFF-THE-JOB). The victims of an accident may not be directly involved in the activity which gives rise to it–they may be nearby workers, bystanders or those living in the vicinity. This article is mainly concerned with accidents in industrial occupations but many similar risks and injuries occur in other spheres. Specific accident risks in the different occupations, or associated with particular substances, machinery, etc., are dealt with in the appropriate articles and are too many and varied to be listed here.

Many countries collect data on accidents at work, during travel or in the home. It is often obligatory to notify certain categories of accidents (e.g. boiler explosions or crane failures) that cause no injury, or accidents that cause injury of particular severity (e.g. prevent a person from doing his normal job for more than three days). The number of notified accidents causing injury can often be compared with the number of workers engaged in a particular activity to produce a frequency rate (e.g. number of notified accidents per 100 000 workers or working shifts) and so enable more hazardous occupations and activities to be identified and priorities for preventive action assessed. Because of the differences in compiling accident statistics in various countries, international comparisons are generally not valid (see ACCIDENTS AND DISEASES, NOTIFICATION OF; ACCIDENT STATISTICS).

Causes of accidents

Very rarely does an accident arise from a single cause: more frequently there is a combination of factors which must all be simultaneously present; a potentially unsafe situation does not give rise to an accident until someone is exposed to it, and a cloud of flammable vapour cannot explode until there is some means of ignition. In particular, accidents result from the combined effects of physical circumstances, which can often be recognised and hazards engineered out of the working system, or human factors, which can be influenced by training, instruction or supervision. However, it must be accepted that workers may be unthinking, make mistakes, be clumsy, lack concentration or even deliberately take risks, albeit for the best of motives, and that these human factors may be aggravated by stresses in the physical environment, especially temperature, ventilation and noise. Both physical and psychological factors must be considered together in recognising that accidents result from unsafe systems of work either by error in design or by default (see ACCIDENTS (HUMAN FACTORS); ACCIDENT PRONENESS).

The hazardous situation, even if compounded by human factors, is not in itself the cause of an accident but the indicator of some other deficiency and the end product of an underlying malaise. The crane that fails through overload is not simply the result of physical stress but the consequences of a system of work which failed to determine the weight to be lifted and plan the operation, failed to check the crane and its safety devices to ensure that it was suitable for the purpose and failed to see that the crane driver knew what was expected of him and was trained to do it (see ACCIDENT ANALYSIS; SYSTEM ANALYSIS (SAFETY); SYSTEM RELIABILITY).

Above all, failure to maintain good housekeeping standards is responsible for a high proportion of falling, tripping and impact injuries. It may also contribute to many machinery accidents, as when a slippery floor causes a fall onto dangerous moving machinery. Inadequate or unsuitable lighting is another contributory cause in many accidents of every kind. These environmental hazards are also indications of some other failing, which should be recognised as the real cause of any accident associated with them.

Consequences of accidents

Damage, disorganisation, distress, disablement, death: any or all of these may result from a limited accident or major incident. Even where the consequences of an accident are restricted to damage of the plant or equipment, a machine or tool, or work in progress, without causing personal injury, considerable disorganisation and loss will arise. In industrialised countries time lost as a result of accidents at work often exceeds time lost by industrial disputes, although these receive much more publicity (see ACCIDENT COST).

Every injury brings a measure of distress to the victim; death or serious injury affect all other members of his family. Where the accident results in permanent disability the consequences may well be disastrous for the family and the victim, who loses his earning capacity and ability to enjoy a normal active life, and for society, which will be deprived of his skill and contribution to production while continuing to support him and his dependants.

It is necessary to distinguish between the accidental occurrence and the injury sustained. Similar occurrences may produce most divergent injuries and the extent of a personal injury is often fortuitous, e.g. a man may fall 2 m in such a way as to incur fatal injuries while another may fall 6 m and suffer no more than shock or bruising; a foreign body in the eye may cause only temporary discomfort for one person while causing blindness in another; failure of a guard on a power press may cause anything from loss of a finger tip to loss of hand or forearm. Sometimes a dangerous incident may cause no personal injury at all, e.g. the fall of a crane jib or the bursting of a centrifugal vessel, if there is no one in the danger area. In developing accident prevention measures it is necessary to consider both the possibility of the accident and the severity of its consequences (see ACCIDENT INVESTIGATION).

Types of accident

The time a person is away from work following an accident may not be directly related to the severity of the injury: the duration of absence may be influenced by social and other factors, which could then determine whether an accident is reportable and included in accident statistics. Attempts are being made to categorise accidents more objectively according to severity, but considerable dependence must still be placed upon reported accidents as an indication of success in accident prevention. Use may also be made of accident data collected at the workplace, such as the number of lost-time accidents which do not have to be reported, or the number of accidents requiring first-aid treatment.

Whatever accident data are used it is clear that the incidence and type of accident varies with the occupation: the frequency rate in the heavy industries, such as shipbuilding, will naturally be higher than in light trades such as the clothing industry; falls from a height are naturally more prevalent in the building industry than in most other occupations; a trade with a high ratio of machinery to men (e.g. some branches of woodworking or the engineering industry) will have a higher proportion of machinery accidents than a trade where there are many non-mechanical operations. It may be said that each industry has its own characteristic type of accident, but at the same time the general over-all pattern remains the same.

As an example of this pattern, figure 1 shows the main causes of industrial accidents in the United Kingdom for the year 1978: comparison with other years shows that percentages vary little and there is no reason to suppose that there is any essential difference in the relative percentages in other industrialised countries. A noteworthy feature everywhere is the relatively low proportion that can be ascribed to machinery in motion and the large numbers that arise from ordinary, everyday causes. A brief comment on the more important of these causes will also indicate possible methods of prevention of occupational risks.

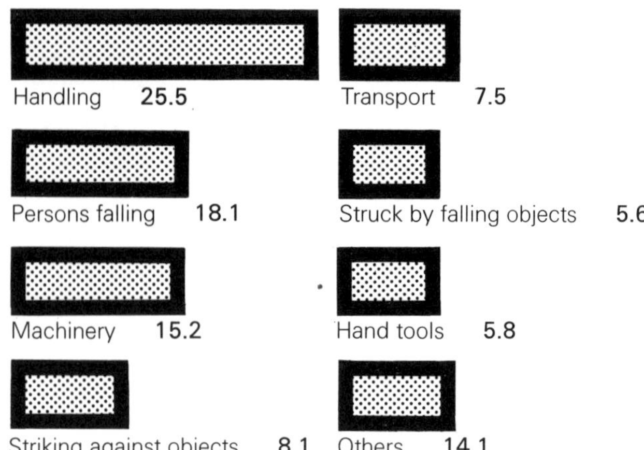

Figure 1. Causes of reported industrial accidents in the United Kingdom for the year 1978.

N.B. These figures are provisional and are not directly comparable with earlier years due to a change in the method of accident classification.

Handling accidents. The manual lifting and carrying of objects gives rise to more accidents than any other activity. Most of these accidents can be traced to one or more of the following factors: (1) faulty lifting techniques; (2) load too heavy or too awkard; (3) failure to wear personal protective equipment, especially for hands and feet.

Falls. Common causes of falls on the level include badly maintained, uneven or slippery floors and unsuitable footwear; defective or insecure ladders are responsible for many falls from a height as are also badly maintained steps and stairs, inadequately constructed or inadequately protected working platforms, unprotected openings in floors, fragile roofs. Insufficient lighting often plays a part.

Striking against objects. Impact or collision accidents are often caused by overcrowding of factory premises and workplaces and by obstructions left in gangways. Maintenance and good housekeeping, proper storage and good lighting could prevent many of these.

Falling objects. Heavy objects may fall from inadequately protected elevated working places; badly stacked materials or goods may overturn; sometimes injury could have been prevented by head protection.

Hand tools. Injuries are usually caused by defective, unsuitable, inadequately maintained or misused hand tools.

Machinery in motion. The range of machinery in common use is so vast that only general comment can be made here. It can be said that most of the accidents could be prevented by the provision and maintenance of secure guarding and reasonable care in use. Machinery guarding is dealt with at length in another article but it may be convenient to indicate here some of the main categories of risk:

(a) entanglement of clothing, hair, jewelry, etc., on smooth shafting, spindles, etc., or on projections on such shafts;

(b) trapping of any part of the body between in-running nips, between belts and pulleys, gear wheels, process rollers, etc.;

(c) crushing, again usually of the hand or arm, at machines where one part closes on another or on material being processed, e.g. power presses, garment presses, pie-making machines;

(d) trapping by a moving part of the machine, e.g. worms, beaters, paddles; or contact with a cutting edge, e.g. of a paper or metal-cutting guillotine, metal and wood-cutting saws, etc.

Electrical accidents. In general industry (as distinct from the electrical industry) a high proportion of electrical accidents are caused by failure to provide or maintain efficient earthing for portable apparatus and to ensure that circuits are safe before maintenance work is started.

Accidents by burning or explosion. Although catastrophic fires with much material damage may also cause loss of life and injury, most burning accidents do not occur in this way. They are caused rather by smaller incidents, such as the ignition of spilled liquid flammable substances, waste rags or saturated clothing by a match or often a spark. However, the injuries caused by major explosions of flammable vapours or combustible dusts are often caused by small incidents such as the welding or cutting of vessels containing flammable vapour or liquid or by misuse of gas-burning equipment and gas cylinders.

Works transport accidents. The risks involved with delivery lorries and vans, lines and sidings are much the same as those in general road transport and railways. Special risks arise from in-plant transport, hand and powered trucks—overloaded or misused, driven by untrained drivers—and also from obstructed gangways; on construction sites bulldozers, tractors, etc., may overturn, especially if the ground is uneven and the driver unskilled.

Plant and machinery failures. Failure of hoists, cranes and lifting machinery, explosion of pressure vessels, bursting of abrasive wheels, flywheels, centrifuges may cause great material damage but the extent of human injury is often fortuitous: they should not be regarded less seriously for that. Regular examination, inspection, testing and a high standard of maintenance, together with strict adherence to maximum limits on loads, speeds, pressures can prevent risks and occurrences of this kind.

Major hazards such as the risk of fire or explosion in chemical works or large-scale liquid installations which could endanger a substantial area around or the possible release of clouds of toxic gas or vapour, the escape of dangerous pathogens or the release of radioactive substances are now widely recognised. There have been some dramatic incidents with considerable damage, loss of life or disease but the over-all number of injuries is still low as compared with other accidents. However, it must be remembered that the effects of some of these accidents may be long delayed (see MAJOR HAZARDS CONTROL).

Accident prevention

An accident is a physical occurrence and prevention must include the physical measures such as maintaining plant, equipment and buildings, providing secure machinery guarding to the best known standards, ensuring good housekeeping, cleanliness, controlling environmental conditions and the application of ergonomic principles. However, accident prevention must be more than this: it requires a deliberate policy of promoting safety by ensuring safe systems of work not only for normal operations but particularly for irregular activities such as cleaning and maintenance (see OCCUPATIONAL SAFETY DEVELOPMENT AND IMPLEMENTATION).

This positive approach to safety and health at work demands a decision by management that safety and health will be accorded due importance and the co-operative design, introduction and operation of arrangements and organisation to give effect to it. It is essential that senior management should accept responsibility for accident prevention and have a definite safety programme implemented at all levels. Training and supervision of workers, especially new entrants, in safe methods of work and provision of personal protective equipment can minimise the risk of injury. Safety engineers, safety officers, joint accident prevention committees to associate the workers or their representatives in co-operation all have a part to play. Although accident investigation will indicate where attention is required, special campaigns, competitions and safety incentive schemes may from time to time be stimulating. There are undoubted advantages in pre-employment medical examinations to guide on placement especially for those occupations such as driving vehicles, or operating cranes which demand physical fitness, good vision and unimpaired hearing. The co-operation and advice of industrial medical officers may be important in dealing with some of the environmental and personal factors involved in accident causation.

EVANS, D. J.

Accident prevention manual for industrial operations (Chicago, National Safety Council, 7th ed., 1964), 1 523 p. Illus.

Accident prevention. A workers' education manual (Geneva, International Labour Office, 1976), 174 p.

Accidents and diseases, notification of

Occupational accidents and cases of occupational disease must be notified under specified conditions—

(a) to the management of the undertaking;

(b) to the national or local authorities concerned with compensation and prevention, respectively.

Notification to the management is intended to alert the medical department, the safety department or the personnel department, as the case may be, with the object if necessary of conducting a thorough accident investigation or a study of the circumstances producing the disease, in order to prevent recurrence of similar accidents or cases of disease.

Notification to the competent occupational safety and health authority has, in addition to the same basic purpose, the object of deciding whether there has been any breach of statutory requirements and, if this is the case, whether any sanction should be applied.

Notification to the workmen's compensation institution or the insurance company is obviously required in all cases where the victim or the undertaking is entitled to compensation for the injury or the damage sustained.

Finally, notification is essential if adequate statistics are to be compiled at the level of the undertaking, the branch of economic activity or the country.

Notifiable accidents and diseases

The legal requirements regarding the notification of occupational accidents and diseases vary from country to country.

Usually, an accident is required by legislation to be notified to the competent authority if it is fatal or if the injured worker is incapacitated from carrying on his normal work for a period of time. There is no universal rule as to how many days of work incapacity make the accident notifiable. In the USSR and New York State it is 1 day; in France and India it is 2 days; in the Federal Republic of Germany and the United Kingdom it is 3 days; and in Malaysia it is 4 days.

Many countries have in their legislation a schedule of notifiable occupational diseases. [This schedule may not necessarily be identical with that of compensatable occupational diseases.] These usually include lead, phosphorus, arsenic, mercury, manganese, aniline, carbon disulphide and chronic benzene poisoning, anthrax, caisson disease, epitheliomatous ulcerations due to pitch, tar, bitumen, petroleum oil or paraffin, and chrome ulceration due to chromic acid or potassium, sodium or ammonium bichromate. Article 15 of ILO Recommendation No. 97 (1953) contains the following:

National laws or regulations should–

- *(a)* specify the persons responsible for notifying cases and suspected cases of occupational disease, and
- *(b)* prescribe the manner in which cases of occupational disease should be notified and the particulars to be notified and, in particular, specify–
  - (i) in which cases immediate notification is required and in which cases notification at specified intervals is sufficient;
  - (ii) in respect of cases in which immediate notification is required, the time limit after the detection of a case or suspected case of occupational disease within which notification is required;
  - (iii) in respect of cases in which notification at specified intervals is sufficient, the intervals at which notification is required.

Accidents notification form

In many countries there is a prescribed standard form which must be completed by the management of the undertaking and sent to the government authority or authorities dealing with occupational safety and industrial injury benefits. The details required on the form usually include the name and address of the undertaking, name and address of the injured person, age, sex, occupation, nature of work being carried out at the time of the accident, how the accident happened, the agency causing the accident and the nature and site of the injury (e.g. skull fracture, laceration of arm, amputation of finger, foot burn, eye injury, etc.).

Diseases notification form

In some countries a special prescribed form is used very similar to that used for the notification of accidents but with additional information as to the length of employment or exposure and any toxic materials involved. There is, in some countries, an additional legal requirement that every medical practitioner attending a patient whom he believes to be suffering from certain scheduled diseases which may have been contracted in an industrial undertaking must send a notification of the case to the competent authority.

Notification of dangerous occurrences

Some dangerous occurrences do not result in any worker sustaining a physical injury, usually because no worker happens to be in the vicinity at the time. These occurrences are, however, of a very dangerous nature and may recur in different circumstances, resulting in loss of life or serious injury to workers. In many countries certain dangerous occurrences must, by law, be notified to the competent authority in the same way as occupational accidents. Examples of dangerous occurrences are:

- *(a)* collapse of a crane, hoist or other appliance used for raising or lowering persons or goods;
- *(b)* explosion or serious fire resulting in complete stoppage of ordinary work;
- *(c)* electrical short-circuit or failure of electrical machinery or plant attended by fire or explosion or causing structural damage thereto;
- *(d)* explosion of a pressure vessel used for storage at a pressure greater than atmospheric pressure of any gas or gases (including air) or of any solid or liquid resulting from the compression of gas;
- *(e)* structural collapse of a building.

GENOT, R.
HUBLET, P.

CIS 78-2092 "Accident reporting–An exercise in futility?" MacCollum, D. V. *National Safety News* (Chicago), Aug. 1978, 118/2 (80-82). 5 ref.

The notification of accidents and dangerous occurrences. Health and Safety Executive publication HS(R)5 (London, HM Stationery Office, 1980), 43 p.

CIS 77-261 *Notification concerning diseases which may be work-related* (Anmälan on sjukdom som kan ha samband med arbete). Meddelanden 1975: 15, Kungliga Arbetarskyddsstyrelsen (Fack, Stockholm), 30 June 1975, 6 p. (In Swedish)

Accidents, commuting

In this article a commuting accident means an off-the-job accident sustained by a worker during his travel from his home to his place of work or during his return journey home. The journey may be made on foot or on any type of vehicle either owned by himself or his employer or by a transport undertaking. The commuting is in the large majority of cases daily, but for some employments there is an interval of several weeks or even months between the time the worker leaves home and the time of his return, e.g. lighthouse keepers taken by boat for duty spells of 2-3 weeks, workers flown by helicopters to work for at least a week on off-shore oil drilling rigs, merchant navy seamen flown distances of several thousand kilometres to join ships at the end of their annual leave. Workers on building and civil engineering sites such as dams in remote areas, lumber camp workers, and workers in oil fields in hot desert regions usually are housed on the site of work but are given transport facilities at intervals of one or several weeks for them to return to their homes or to the nearest large town for one or more days. Originally workers lived within a short walking distance of their place of employment but, with the growth of large cities and rapid transport, the number of workers living at some distance increases yearly, with a corresponding increase in the number of commuting accidents.

Commuting accident statistics

It is difficult to obtain valid accident statistics on the proportion of all occupational accidents that are due to commuting, because conditions vary so much from one industrial undertaking to another. In some cases all the workers may have to travel some distance to work while in other cases most of them live near or even on the premises, and are not exposed to any travel risks. The length of time of exposure to an accident at work is generally the same for all employees but the length of time spent in travelling to and from work may vary from 1-2 min to over 2 h, and in varying circumstances according to whether the journey is on foot, by private or by public urban transport, in the hours of darkness or in daylight. Only if all the workers in one undertaking travelled the same distance by the same means, could one ascertain reliable data. Statistics obtained in one country show that about two-thirds of commuting accidents occurred to workers travelling in private vehicles. It has been suggested that the inference to be drawn from these figures is that more workers should use public transport; but this is not always possible, as many workers live in places not served by public transport.

In France, accidents while travelling to and from work add up to an extremely heavy financial burden for the Social Security Scheme. The cost of these accidents is greater than that of accidents at work, since they are more frequent. (For every fatal accident at work, there are two fatal accidents while commuting or moving in traffic.) These accidents are two to three times as serious and are responsible for more deaths than are the most dangerous occupations (building and civil engineering and the iron and steel industry).

Over 50% of the commuting accidents in France occur on the way to work in the morning, at a time when traffic is generally scarce, and more than half of these accidents happen to two-wheeled vehicles without involving a collision.

It has been ascertained in numerous countries that while the frequency (the number of accidents divided by the number of hours of exposure to risk) of reported commuting accidents is less than that of working accidents, the situation is totally reversed as regards serious accidents. A difficult task which should be undertaken by the competent authorities is the preparation of statistics on traffic accidents showing figures of those due to commuting accidents. These should show without doubt, if they exist, any specific blackspots on the roads which are causing commuting accidents. In certain countries, investigations have been made into the danger points of the road transport route to and from work. With the help of the workers, a systematic inventory of these points has been established.

Another characteristic of these accidents that can be concluded from the statistics is that their highest frequency occurs on Monday mornings and that they decrease in frequency with the days of the week.

The high average severity of this type of accident, most often involving multiple injuries, significantly influences the cost of compensation. In Belgium, for example, in 1977 the average cost of compensation for an accident at work was US$1 525, while for a commuting accident it was US$4 040.

SAFETY MEASURES

There are three parties involved in commuting accidents: the workers, the employers, and the public authorities concerned with traffic safety. All these can make some contribution towards reducing the number of commuting accidents.

Workers. They should be conversant with the rules of the national traffic safety code even if they are only pedestrians. They should know that at night it is very difficult for the driver of a vehicle to distinguish bicycles without rear reflectors or pedestrians wearing dark clothing in badly lit areas. Unsuitable footwear, especially that of women, with high narrow heels presents dangers. Where the worker provides his own transport, it is especially important that it is maintained in good road condition. Two factors tending to cause commuting accidents are:

(a) a journey begun immediately without a pause after an abrupt break in the biological rhythm (sleep or the cessation of engrossing or tiring work);

(b) attenuation of safety consciousness due to familiarity with the route.

Employers. Unfortunately in most cases the employer does little, since he believes wrongly that he is powerless because the commuting accident occurs at a time when the worker is out of his care. Experience has, however, shown that despite this handicap the industrial undertaking can make a number of improvements in several ways such as:

(a) removing dangerous conditions in the immediate vicinity of the factory with the co-operation of the competent public authorities;

(b) checking, during the hours of parking at the factory, the individual means of transport used by the workers, i.e. checking the condition of tyres and testing of brakes;

(c) providing for transport of workers to and from work;

(d) a constant inculcation of safety-mindedness;

(e) training and propaganda with repeated reminders concerning road safety.

Public authorities. Commuting accidents are part of traffic accidents in general. The measures imposed by public authorities regarding the safety of the traveller can only be partially effective as far as they concern commuting accidents. The ordinary risks of general traffic are actually augmented by the special circumstances relating to journeys to and from work, these being chiefly:

(a) a concentration of traffic during a limited period and on routes ill-suited to deal with abnormally heavy traffic of all kinds;

(b) an overloaded transport timetable for both the outward and homeward journeys due to various causes.

As commuting in many countries takes place in the hours of darkness for a considerable part of the year, the provision of good street and road lighting is very important. Better road traffic control, improved roads and road signs will all help towards reducing the accident rate, but the increased speed of traffic, the ever-growing number of road vehicles and the tendency towards heavier motor trucks carrying up to 40 t will result in the problem of commuting accidents continuing to be a serious one.

Legislation for compensation

While in some countries a workman meeting with a commuting accident is considered to have been injured at work and is entitled to the same compensation benefits, there are still many countries in which such benefits are not given or in which there are alternative compensation arrangements. First, there are those countries in which no compensation or benefits of any kind are given to the victim of a commuting accident under workmen's compensation law or any other labour law. In some countries the workmen's compensation law only gives benefits for commuting accidents if the employer provides the transport. Where the worker injured in a commuting accident has no legal redress entitling him to compensation from his employer, he may in some countries have a right to compensation under a national social security scheme, but the benefits or compensation are usually less than in the case of an accident sustained at work. A worker injured in a commuting accident while travelling on public transport may in many countries be entitled to compensation from a public transport organisation but there may be statutory or company limitations on the amount of such compensation. If the worker travels on cheap tickets such as workmen's fare tickets, he may find that the amount of compensation he can claim in respect of a commuting accident is very small. If the commuting accident is caused by the driver of a vehicle, the injured worker may claim against the driver or his insurance company and under such circumstances it is possible to get very large damages but there is usually a delay of up to several years in obtaining court judgements, and the actions are risky as they may fail on a legal technicality or for want of reliable witnesses. Usually the ordinary worker cannot afford to initiate such legal action unless there is a legal aid scheme or his trade union is willing to pay the legal costs in case of the action being lost.

Large numbers of workers are now recruited to work abroad either permanently or as seasonal workers for a fixed period such as harvesting, or actors or other entertainment industry workers on a tour of several countries. Some governments have strict regulations governing the terms of contract for such labour, including the right to compensation for commuting accidents.

International instruments

The ILO Employment Injury Benefits Recommendation, 1964 (No. 121), stipulates that all Members should, under prescribed conditions, treat as industrial accidents those accidents sustained while on the direct way between the place of work and:

(i) the employee's principal or secondary residence; or

(ii) the place where the employee usually takes his meals; or

(iii) the place where he usually receives his remuneration.

Also in the ILO Plantations Convention, 1958 (No. 110), there are a number of safety and health provisions relating to the transport of recruited plantation workers and their rights to compensation for industrial injuries.

ASSOCIATION NATIONALE
POUR LA PRÉVENTION
DES ACCIDENTS DU TRAVAIL (ANPAT)

CIS 81-543 "The journey to work–Some advice for accident-free commuting" (Les chemins du travail–Quelques conseils pour un trajet sans embûches). *Travail et sécurité* (Paris), Aug. 1980, 8 (436-440, 465). Illus. 4 ref. (In French)

CIS 76-1172 "Commuting accidents: hall-marks and means of prevention" (Arbeidswegongevallen: hun kenmerken en preventiemogelijkheden). Pote, R. J. *Technical aspects of road safety* (Brussels), 1975, 16/62 (4.1-4.10). Illus. 11 ref. (In Dutch)

"The commuting accident" (Der Wegeunfall). Pistulka, G. *Sicherheitsingenieur* (Heidelberg), 1974, 4 (162-168). Illus. (In German)

"Cost and prevention of commuting accidents" (Coût et prévention des accidents de trajet). Freycenet, A. *Cahiers des comités de prévention du bâtiment et des travaux publics* (Issy-les-Moulineaux, France), 1972, 6 (208-211). Illus. (In French)

Accidents (human factors)

In attempting to describe an accident and explain what happened, it is necessary to bring to light firstly the sequence of events which culminated in a worker being injured (he was knocked down by a wagon, an object fell on him, etc.), and secondly, those circumstances surrounding the situation which contributed to the probability of an accident or the "risk factors" (the worker lacked experience, the workshop was encumbered with material, etc.). These factors concern every accident and must be the targets for deliberate action designed to prevent its recurrence. Among them are a certain number which are related more or less directly to the man at work and these are known as human factors.

Classification of risk factors. The work situation is made up of one (or more) *individuals I*, a *job* or *jobs J*, the *material and equipment used M* (machines, tools, raw materials, etc.), and a *socio-technical environment E*. This is sometimes described as a man-machine system *IJME*. An attempt will be made to classify these factors by grouping them under one or other of the above terms or under an interaction of more than one within the system *IJME*. Taking into consideration only the human factors and not going beyond interactions of the second order, we find the factors *I*, *II*, *IJ*, *IM* and *IE*. Eight examples of these factors will follow together with what would appear to be the most appropriate classification in each case.

Individual susceptibility to accidents (I). By susceptibility is meant a certain individual tendency or disposition to have or to cause accidents. This important factor is discussed elsewhere (see ACCIDENT PRONENESS).

Lack of experience (I). Generally speaking, a worker either enters the factory after having received some training or he is given this training upon his arrival; this consists of the standard apprenticeship in the trade or the work that he will be called upon to perform; he will have been taught at the workbench or on a machine and he will become part of a production system whose parts are interrelated. It is now up to him to gain the necessary experience, i.e. a knowledge of these interrelationships and how to live with them. To the extent that he lacks this experience, he is exposed to serious risks: he is unable to interpret correctly the various signals which reach him from all sides and in many different ways, particularly those which indicate danger. Consequently it is found that new workers in a factory meet with more accidents than older ones; this finding is frequently mentioned in the safety literature. It is also to be observed that persons who are moved from one workplace to another or who act as temporary replacements often meet with accidents.

Lack of experience may influence the four facets of the work system *IJME* in the following ways:

(1) *the individual:* not being acquainted with other members of the team, with the ways in which they work, with their way of speaking and with the informal signals they make to each other. Workers may also be in surroundings where they do not understand or only poorly understand the language; they may not grasp the significance of gestures or signs which the others are in the habit of using, or of abbreviated phrases that have grown out of continual repetition of the same actions;

(2) *the job:* lack of familiarity with various occurrences, malfunction or irregularity and the signals which give warning of such incidents, how to prevent them or set them right, not knowing what is happening earlier or later on in the production process. The new worker has not had the time to adapt his behaviour to these

occurrences, to avoid snags or sharp objects, to face contingencies, etc.;

(3) *the equipment:* lack of knowledge of the machines in use and of their weaknesses or defects, of tools and how they should be used, of the materials being worked with;

(4) *the environment:* lack of knowledge about the surrounding hazards, of general safety instructions, of the significance of events which may take place nearby, etc.

This lack of experience should be countered by:

- describing and explaining all these matters which contribute to such experience, providing a list of the informal signals and codes used by the operators to communicate, making notes on the erratic behaviour of machines which gives rise to repeated incidents;
- teaching all this to the new worker; this is an essential part of the task of supervision and should even involve his fellow workers.

Interaction between the worker and the job (IJ). The job places various demands on the worker that tend to inhibit his natural feelings and impulses; as a result of these constraints, he in turn is placed under a certain degree of stress. Different kinds of constraint exist which include: heavy manual work, jobs where there may be numerous minor incidents which can lead to more serious consequences, those which call for sustained attention, the need for a high rate of production, etc. It has always been stressed that constraints which lead the worker to neglect safety instructions represent a risk factor. Similarly, the safe way of working may be more tiring or may simply require a little more trouble, and the worker neglects the precautions that he should be taking as, for example, in the case where he does not wear protective clothing because it is not comfortable. The demand for production is frequently one of the principal sources of constraint, and the worker is inclined to neglect safety precautions in order to satisfy production requirements. The conflict between productivity and safety is often referred to in cases where the worker is said to have disobeyed safety instructions or to have adopted a more dangerous way of working in order to save time. In extreme cases workers claim that it is impossible to observe safety rules and maintain the required output; such a situation places a worker under a double constraint, as though he were expected to observe all the rules and to break them at the same time. For example, he has to take a risk that he has been forbidden to take, as in the case of a travelling crane where an unqualified driver has been put in charge in order not to lose time, although the rules require only qualified drivers to operate the crane; or where the driver is required to operate the crane while a maintenance worker is on the platform, which is also forbidden by the rules; cases even occur where the crane has been used to pull or push a load obliquely on the request of a foreman.

It has been shown that the system of remuneration can affect safety. In a Swedish iron mine serious accidents were reduced when the men were put on to monthly payments (although minor accidents increased, which was no doubt due a redefinition of accidents in terms of their seriousness). Changing to monthly payment led to a reduction in the constraint resulting from the demand for productivity. Many studies have shown that unsafe conduct is encouraged by the system of payment by results.

In the interests of accident prevention situations should be sought where this safety-productivity conflict exists and efforts made to eliminate it by improving the safety arrangements in such a way that they do not hinder the work or by changing the method of payment.

Restoration of a situation (IJ). When it is necessary to restore the situation after any departure from the normal state of affairs, a worker may have to perform his task, or another task, in a way that is unusual and unforeseen in order to return to the normal course of events.

Two kinds of accident are to be noted in connection with this situation: there are the ordinary accidents where injuries occur as an immediate result of the initial incident, and there are those which occur in the course of the work necessary to restore the situation after departure from normal as the result of a technical problem. This distinction is made because there is an increased danger of accidents during the period of restoration.

As both these types of accident stem from a technical problem within the production system, it can be seen that, from the point of view of safety, it is most important that the system be reliable. Since improved reliability also improves productivity by eliminating lost time, there is an additional argument in favour of the adoption of appropriate remedial measures. The combination of safety, reliability and productivity is sometimes described as "integral safety".

In this kind of situation, the initial incident may trigger off secondary incidents or at least be followed by other events because it reveals the existence of a weakness, or because its occurrence has upset the normal stability of the system or weakened some related component of it, or again because the man concerned faces an unusual task which he does not know properly how to perform and which places an excessive demand upon him requiring additional effort, all of which tends to increase his nervousness and anxiety. A sequence of events has thus been set in motion which may end with an accident.

It is also to be noted that when work is recommenced after such an event, there is often an attempt to make up lost time. Thus there is a productivity constraint which increases the risk of accident.

Accidents of the type which occur when unusual tasks are being performed after some departure from the normal course of events are best prevented by improving the reliability of the system and by providing special training for the workers in the operations that are likely to have to be performed in the course of the most typical and the most frequent incidents of this kind.

Misapplication (IM). The situation described as misapplication occurs when a tool is used for a purpose other than that for which it was intended, or when a machine is worked beyond its normal or designed capacity. The incorrect use of tools has been the subject of much study, but the misuse of machines is also a common occurrence as, for example, in the case of a tool being sharpened on a grinding wheel which is not designed to be used for such a tool, or overloading of industrial trucks and lifting gear. As far as tools are concerned, particular attention has been drawn to the danger of not using the correct tool, and especially that of using some article such as a bar or pipe that may by lying conveniently nearby to shift a piece of equipment.

Dilapidation (IM). Equipment that has fallen into a state of disrepair, through wear and tear or damage, or has a particular weakness, can give rise to this kind of situation. The repeated occurrence of some small fault is one indication of more extensive trouble to come. Such incidents may end up in a complete breakdown of a whole installation. The danger lies in the fact that an additional workload is imposed and unusual measures have to be taken in order to ensure that work is not held up. Such conditions, whether they concern machines and their surroundings or premises, are considered by workers to contribute largely to fatigue and to accidents;

numerous examples can be quoted. The best way of preventing such accidents is to ensure the prompt replacement or repair of all damaged equipment or premises.

Interference (IE). This situation may arise where one more or less independent process is likely to hinder or obstruct a second process. The following examples illustrate the sort of danger spot that should be identified:

- conflicting activities: such situations exist where there are two or more categories of workman performing work within different systems, such as when a team working on production finds another team of, say, construction or repair and maintenance workers working in the same area. Undesirable incidents and accidents are more frequent in these situations;
- interface situations: these relate to those areas where successive activities in a production process involve some type of alteration or change of circumstances. It may be represented by a change of department or branch, a change of surroundings, a change in the rules applicable, or of the operation and number of persons involved. Examples are: the point at which a construction site opens on to the road; the areas on either side of any hold-up in a production process;
- intersecting situations, as for example where two lanes cross each other.

The prevention of accidents falling under these headings usually requires the improvement of signalling systems and communications. This requirement is also important in the case of the next factor to be mentioned: lack of information.

Lack of information (IE). This concerns situations where the operator does not possess the requisite information about the surrounding conditions. It may be that no signalling system exists, but it can also be a matter of a signal that is difficult to see because it is badly placed, or because it does not appear at the right time, or it may be a signal that is ambiguous.

A typical situation falling under this heading would be one where a given team is being replaced by another to continue the same work. If the first team leaves behind an abnormal situation, particularly if there is something wrong which is not clearly apparent, it can be particularly dangerous for the oncoming team if the appropriate information is not handed on.

In general, the need for satisfactory communication is most important from the point of view of safety, be it between the members of a team, a workshop or a branch of activity, or between supervisory staff and the workers, or between the workers themselves. It is in this way that all those concerned become informed of any incidents which have occurred, any temporary or long-term dangerous conditions, the action to be taken or the precautions to be adopted. In some cases there may be psychological reasons for lack of communication, such as when a worker fails to draw the attention of his supervisor to a dangerous condition because he thinks he will not be listened to, or that nothing will be done anyway. This is also the case when the message is distorted by the person receiving it: an example of this would be provided when workers fail to believe in the good intentions of management in its efforts to improve safety conditions and find an ulterior motive therein.

Workers' participation

The basis of accident prevention is the identification and the elimination of hazards. All accidents should be carefully investigated in order to bring to light all those factors and circumstances which contributed to their occurrence. Work situations should be studied and all danger points noted so that action can be taken to remove them and generally improve the situation.

Corrective action consists of training the workers, establishing appropriate instructions, the rational organisation of the work and its surroundings, the application of ergonomic methods, the installation of safety devices and the provision of protective clothing, which should not be uncomfortable and should not hinder the worker.

The workers should be invited to co-operate in this programme of accident prevention, and this is particularly important because no prevention programme can be effective without their participation.

FAVERGE, J. M.

CIS 77-1178 *The human side of accident prevention– Psychological concepts and principles which bear on industrial safety.* Margolis, B. L.; Kroes, W. H. (Springfield, Illinois, Charles C. Thomas, 1975), 144 p. 398 ref.

CIS 75-289 *Accident research: the human element in industrial accidents.* Dieterly, D. L. (Wright-Patterson Air Force Base, Ohio, Air Force Institute of Technology, 1973), 55 p. Illus. 61 ref. Available from National Technical Information Service, 5285 Port Royal Road, Springfield, Virginia (Accession No. AD 768 342).

"Safety at work as a problem of acceptance from the viewpoint of motivation theory" (Arbeitssicherheit als Akzeptanzproblem aus motivationstheoretischer Sicht). Zink, J. *Zentralblatt für Arbeitsmedizin, Arbeitsschutz und Prophylaxe* (Heidelberg), Feb. 1980, 30/2 (39-48). Illus. 44 ref. (In German)

CIS 79-1785 "The unsafe act. Exploring the dark side of accident control". Denton, D. K. *Professional Safety* (Park Ridge, Illinois), July 1979, 24/7 (34-37). 12 ref.

Studies in industrial physiology and psychology (Collection d'études de psychologie et de physiologie du travail). Commission of the European Communities (Luxembourg, 1967-1972), Nos. 1 to 7. (In French and German)

Accidents, occupational domestic

Occupational domestic accidents (ODA), result like all other types of accident, from an interaction of three main factors: the person (host), the agent, and the (home) environment. Considering all classes of accidents, most non-fatal injuries in North America occur in the home and in the domestic surroundings. In the United Kingdom more than a million persons are affected annually by home accidents (HA). In France about 5 000 annual deaths have been reported recently in this accident class. In Canada there are about 2 000 and in the United States around 24 000 fatalities a year due to HA. A 1978 study from the United Kingdom showed that the active population (aged 15-64) accounted for 45%, children (aged 0-14) for 42.7% and older people (aged 65 and over) for 12.3% of all HA; about 47% of those affected by such injuries were men and 53% were women.

Epidemiology

There is a marked lack of reliable data for proper epidemiological information about ODA injuries and fatalities on world-wide and national levels. Socio-economic as well as cultural and regional differences have made it virtually impossible to draw a reliable picture of the topic. Studies from various countries cannot be compared in practice since they differ widely in the quality of sampling, methods and the accuracy of the facts reported. The same applies to our present knowledge about the characteristics of the accident victims, the working environment, and the medical, social and economic consequences of ODA. The situation in North America seems quite different in this respect from that in Europe and cannot be compared in its extent to the problem in developing countries. We may assume that in Europe, Asia, Africa and both South and Central America, the injury and death rates due to ODA are higher than in the United States and Canada, merely because of the relatively less frequent involvement of the North American population in professional domestic work activities. Differences in the ratio of domestic to non-domestic workers in the world regions, in the design of households and their equipment, in the hazards involved, in the risk-taking attitude of the people, and in many other pertinent factors render it difficult to find a sound world-wide scheme of the actual status of ODA.

Nevertheless, data from hospital records and detoxication centres, mortality statistics and reports from safety councils, public health departments, police and fire departments, insurance companies, general practitioners and other sources make it possible to sketch an approximate picture of the situation.

Working and living conditions depend largely upon the socioeconomic make-up of the country involved and may vary also from one place to another. Occupations of domestic workers are regulated by codes and legislation in several countries; however, not all of them make occupational domestic injuries compensable.

In recent years HA fatalities have decreased due to modernisation of homes and improved facilities for cooking, heating, lighting and laundering.

The nature of ODA most frequently encountered have been cuts (33%), dislocations and fractures (12%), contusions (12%), sprain and strain (9%), burns (4%), etc. Non-fatal injuries occur most frequently in the kitchen.

Epidemiological studies have revealed that single, widowed, separated and divorced persons have an excess of mortality from all accidents; ODA may not be an exception. In contrast, many handicapped workers have learned good risk-taking attitudes in their professional activities and may thus have a relatively favourable ODA record. According to Dalton, women seem to be more vulnerable to accidents during the second half of their menstrual cycle, and this may be related to reduced attention and perhaps to a decline in good risk-taking attitudes in the pre-menstrual phase.

Certain pathologic conditions may increase the risk of occupational domestic accidents, mainly in unsafe homes: vertigo, epilepsy, hypertension, hypotension, loss of balance, arthritis and osteoporosis in the lower limbs (proneness to tripping and falling), drugs and alcoholism. Alcohol is a major contributing factor to domestic accident death and injury, mainly from burns and falls.

Considerable difficulties stem from the overlap of accidents and masked suicides, particularly in terms of forensic and administrative considerations (life insurance, compensation, etc.). In some countries where suicide is considered a crime, suicides are recorded as accidents.

The types of work giving rise to occupational domestic accidents include–

(a) various manual and mechanical tasks related to the daily running of a residence and all its annexes (see DOMESTIC WORKERS);

(b) domestic duties and related outdoor and indoor tasks, including baby-sitting;

(c) care of persons, goods, household linen, family clothes, furniture and carpets;

(d) cleaning of the residence and annexes, including floors and windows;

(e) handling of combustibles and potentially toxic products, electrical equipment, gas equipment for cooking and heating and mechanical tools and machines;

(f) kitchen work related to the preparation of meals, handling tableware, washing and cleaning;

(g) shopping and delivery of food and various products; and

(h) transportation to and from work: "commuting accidents", in many countries not compensable (see ACCIDENTS, COMMUTING).

The hazards, agents and external courses of ODA are classified in table 1.

Table 1. Hazards, agents and external causes of ODA

Air conditioning:	Draughts, respiratory infections, allergies.
Allergens:	By contact, inhalation and ingestion.
Animals:	Domestic pets (dogs, cats) or wild animals kept at home. Animals found in human habitations (flies, mosquitoes, spiders, scorpions, wasps, etc.).
Burns:	Hot or burning substances, acids, lyes, electricity.
Cans:	Causing cuts while being opened.
Carbon monoxide (the silent killer):	Faulty installation, leakage of gas, inadequate ventilation (e.g. in garages).
Carelessness:	In the disposal of toxic substances of all kinds, caustic chemicals, incandescent and burning materials, broken glass, sharp tools, etc.
Distractions during work:	Telephone, television, radio, door bells, persons, animals.
Drowning:	Bathtubs, swimming pools, wells, cisterns, tanks, lakes, rivers, ponds, ditches, open drains, cesspools, pit latrines.
Electricity:	Dangerous, overloaded, misused, incorrectly wired, ungrounded or inappropriate (wrong wattage or voltage) tools and appliances. Faulty insulation, improper or inadequate wiring (poorly connected, worn out, ungrounded, with exposed strands) causing electrocution by high voltage (220-250 volts is dangerously high). Carelessness (wet hands, wet floors in contact with electric current).
Ergonomics:	Poor design of housing (low ceilings, dark aisles and stairways), of tools, machines and appliances; dangerous and flammable toys.
Explosion:	Combustible materials, gas tanks, compressed air or gas containers, cooking and heating installation and appliances.
Falling objects:	See Objects.
Falls:	On the same level while walking, standing or sitting; from a different level (from windows, ladders, roofs); caused by darkness, unprotected stairs, wet slippery floors or floor coverings.
Fire:	Smoking; combustible materials, paints, solvents; flammable clothes (particularly sleeves, nylon, artificial silk) and celluloid toys ignited on a stove or at an open fireplace; open or unguarded domestic fires; old and faulty gas, oil or electric heating installations; rags soaked with oil-based materials (self-ignition in poorly ventilated places).
Firearms:	Ignorance in handling, cleaning and servicing.

Fireplaces and stoves:	Unattended or poorly designed; badly ventilated; misused (e.g. for burning trash).
Food poisoning:	Infected or rotten food and beverages; expired preserves.
Furniture:	Poor ergonomic design; sharp, pointed, angular and unstable.
Gardening hazards:	Ladders, electrical equipment (lawnmowers, saws) and outlets, tools, sprays.
Gas:	For cooking and heating or as by-product of fermentation (causing irritation of eyes, skin and mucosae).
Glass:	Windows, screens, utensils, doors.
Housing:	Inadequately designed (unsafe floors, ceilings, walls, doors, windows and stairways; unguarded stairways and balconies); insufficiently ventilated; crowded, poor and unhealthy; faulty elevators; lack of or improper hydrants, fire extinguishers and other fire fighting equipment and emergency exits.
Lighting:	Insufficient or inappropriate (causing accidents; damaging vision).
Meteorological factors:	Cold; snow- and ice-covered aisles and stairways; lightning.
Noise:	Noisy appliances (causing hearing loss).
Objects:	Falling or moving; heavy, stored in high places; sharp, pointed, angular, containing glass.
Overexertion:	Improper lifting (causing hernias, ruptures, back injuries).
Poisoning and gassing:	Utility gas; sprays, caustics, cleaning substances, solvents, pesticides, bleaching agents, paints; food, drugs (aspirin, barbiturates), detergents, other household liquid and solid substances; insufficient ventilation of stoves and fireplaces.
Smoking:	Carelessness resulting in fires and burns.
Stoves:	Gas, oil and electric (causing burns; poisoning by carbon monoxide and utility gas; asphyxia by lack of ventilation).
Tableware and kitchenware:	Glass, pottery, porcelain, metal cans.
Suffocation:	By gas or mechanical means; food inhalation.
Tools and machines:	Sharp or faulty objects (nails, needles, scissors, knives, forks, electrical appliances and machines); faulty or misused handtools or portable power tools (drills, chainsaws).
Ventilation:	Insufficient or faulty (causing asphyxia by lack of oxygen; gas poisoning from fires, stoves, solvents).

Deaths related to falls, fire and poisoning account for 75% of all fatal ODA; more than 50% of all fatal ODA are related to fires, poisoning and electric current and machinery (see figure 1); 60-90% of all deaths due to burns and 50-80% of those due to falls are HA.

PREVENTION

The frequent occurrence of home accidents as well as the nature and extent of multiple hazards encountered by domestic workers elicited interest in the ergonomic approach in the design of housing and of all household equipment. Prevention should be applied at the community level and at the family level. It includes health and safety measures, ergonomic approaches in housing, furniture, dishes, tools, machines, installations, appliances, equipment and work conditions, as well as safety education.

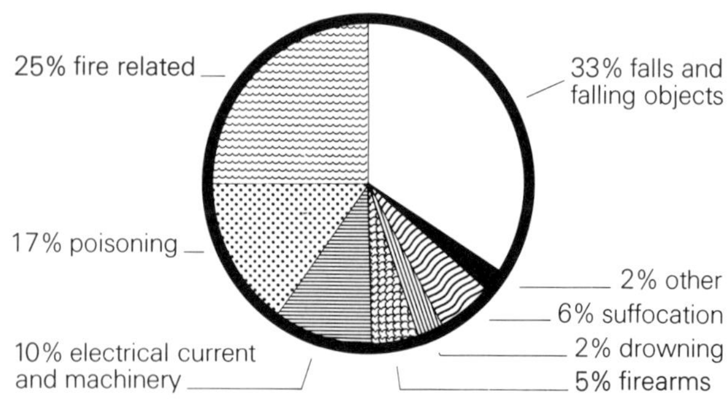

Figure 1. Estimated world-wide distribution of occupational domestic fatalities by type of accident.

Some useful safety precautions are as follows. Safe clothing, shoes and even gloves should be used and protective goggles should be worn when handling hot fats, boiling liquids, acids, caustics and flammable materials. Safety belts should be worn for window-cleaning. Gas, gas tanks, combustible materials and toxic substances should be carefully stored and handled. Heavy objects should be stored in low places. Proper ventilation should be provided for garages, heating and cooking units and storage spaces. Smoke detectors and fire extinguishers should be fitted. Emergency exits, adequate illumination and safe ceilings, floors, windows, doors, stairways and elevators and properly guarded ladders and balconies should be ensured. Electrical and other household, garden and garage equipment should be regularly checked. Strict safety measures should be enforced for swimming pools, and safety check lists, warnings and proper instructions given to domestic workers. Consumer legislation should provide for non-toxic, non-combustible products.

As regards electrical hazards, separate wiring should be required for electric irons, hair dryers, vacuum cleaners and other high wattage appliances. All electrical installations, except double-insulated ones, should be grounded. An emergency fire plan and list of emergency telephone numbers should be foreseen. Manifestations of a faulty or overloaded dangerous electric connection needing immediate attention are: a hot cord, plug or wall outlet; dimming or flickering of the lights; a shrinking picture on the television; appliances that do not run at full power (e.g. an iron or toaster that heats slowly); deteriorated insulation; fuses that blow often; an outlet, light or entire circuit that fails to work; strange odours, especially resembling burning plastic; smoke or sparks coming from a switch or outlet; a fixture that causes an electric shock. Electrical appliances should never be used near water. Switches in bathrooms and kitchens should be of non-conductive material. Cords and plugs should be kept in excellent condition.

Cigarettes and matches should never be dropped into a waste basket, onto a carpet or in an elevator. Spray cans and other aerosol products should not be thrown into unprotected garbage bags or into the fire, or explosion, fire, poisoning and burns will result.

The prevention of occupational domestic accidents is a difficult task, as it often requires interference with people and their homes. The role of safety education needs clarification and promotion; more studies of home accidents in general, and particularly of occupational domestic accidents, are needed to prevent their devastating consequences.

ROHAN, P.

Accident facts (National Safety Council, 425 N. Michigan Avenue, Chicago, Illinois, 1979), 96 p.

"Family safety, domestic hazards" (La sécurité familiale, les risques domestiques). Rambeau, D. *Revue de la sécurité* (Paris), Mar. 1979, 15/155 (5-14). Illus. (In French)

Domestic accidents (Unfälle im Hausbereich). Hadjimanolis, E.; Seiler, G. (Dortmund, Bundesanstalt für Arbeitsschutz und Unfallforschung, 1973). Forschungsbericht No. 104, 184 p. Illus. 45 ref. (In German)

Accidents, off-the-job

Off-the-job safety is a term used by employers to identify that part of their safety programme directed to the employee when he is not at work. It is an extension of the on-the-job safety programme and not a separate programme. By definition, off-the-job accidents occur only to employees, and operating costs and production schedules are affected as much when employees are injured away from work as when they are injured on the job: all injuries are a waste of valuable manpower. With conditions that take some time to develop, e.g. dermatitis, strained backs, it is not always easy to determine whether the first cause was off or on the job but the loss of working time is the same.

In the United States three out of four accidental deaths and more than half of the injuries suffered by workers in 1978 were the result of accidents off the job. These totals, and the corresponding rates per million hours' exposure are shown in table 1.

Table 1. Fatal and disabling accidents in the United States (1978)

Place	Deaths (in '000s)		Disabling injuries (in '000s)	1978 rates[1]	
	1978	1977	1978	Deaths	Injuries
All accidents	54.8	53.0	5 400	0.10	9.8
At work	13.0	12.9	2 200	0.07	11.6
Away from work	41.8	40.2	3 200	0.11	8.8
Motor vehicles	25.0	23.2	1 000	0.72	28.9
Public non-motor vehicles	9.2	9.3	1 100	0.07	8.4
Home	7.6	7.7	1 100	0.04	5.5

[1] Per 1 million hours' exposure, by place.

Production time lost due to off-the-job accidents totalled nearly 70 million days in 1978 compared with 45 million days lost by workers injured on the job. However, indirect losses of time, which seldom arise out of off-the-job accidents, totalled an estimated 200 million days from accidents on the job.

Types of accident

Off-the-job accidents can be divided into three major classes–transportation, home, and public–and it may be desirable to use sub-divisions to assist in fitting activities to needs, e.g. seasons of the year, accident types and accident areas.

When people think about off-the-job accidents, they generally think of traffic injuries. While traffic accidents are serious and require all the attention possible, they do not form the entire problem; experience in the United States, based on records collected over the years, shows that, while the vast majority of fatalities are in the transportation category, most injuries occur in domestic surroundings. In fact, the "public" sector category, which principally covers recreation and leisure, has more injuries than transportation.

Accident experience will differ, according to location and the spare-time activities of employees, between a metropolitan area and a rural area: where there is great interest in water sports, shooting or hunting, etc., this will affect the type of accident.

In the United States it has been found possible to collect accident and injury information from employees about off-the-job accidents: special forms are used for this purpose. Some sort of detailed record is needed to provide material for evaluation of programme areas and to serve as a yardstick to measure progress.

Since the Second World War accidental deaths of workers on the job have decreased by 21% in the United States. In 1978 deaths caused by off-the-job accidents were 39% higher than in 1945, but with an increase in the number of workers the *rate* was 27% lower. The ratio of off-the-job to on-the-job accidents resulting in death in 1978 was 3.22 : 1 (see table 2 overleaf).

PREVENTIVE MEASURES

The principal aim in preventing off-the-job accidents is to get the employee to use the same safe practices in his outside activities as those he uses on the job. Experience indicates that the average employee tends to leave his safety training at the workplace every time he goes home.

The same principles and techniques used to put across safety on the job are employed in preventing off-the-job accidents: climbing a ladder at home is the same as climbing one at work: driving the family car is the same as driving a company road transport vehicle.

Through education and persuasion, the employee can be convinced that accidents do not always happen to other people and persuaded to accept the responsibility of being the safety supervisor in the home, to set a good example and to control the safety practices of family members. A check list is an effective method of obtaining co-operation: this helps the employee to put into practice off the job the safety message received from the company, by becoming aware of hazards and taking corrective action.

An important element for a programme to prevent off-the-job accidents is to keep the safety activities seasonal: people are more receptive to subjects when they coincide with their normal routine, e.g. water sports in summer, ice-skating and skiing in winter. Some subjects–seat belts, falls, poison prevention, to name a few–can be used at any time.

Help in preventive work can often be obtained from community associations such as sports clubs, the fire department, voluntary first-aid associations. Most of these organisations have literature that can be distributed and some have speakers and films available for meetings.

The same methods used to sell safety on the job can also be used to sell safety off the job–meetings, articles in a house magazine, bulletin board notices, and audiovisual aids such as films, displays and posters. While some companies have held elaborate open days or open houses, safety fairs and other large programmes, others obtain equally good results with simple activities. Off-the-job accident prevention can be worked into regular plant safety programmes; as an example, one company selected one month to cover lifting and carrying: in-plant problems were covered in departmental meetings and notices, and it was pointed out that the same principles and practices for safe lifting could be used at home, when moving furniture and other objects round the house.

Since the aim of safety education is to change the employee's attitude, companies have discovered that the

Table 2. Trends in on-the-job and off-the-job deaths and injuries in the United States (1945-78)

Year	Deaths ('000s)					Disabling injuries ('000s)		
	On the job		Off the job		Ratio Off:on	On the job	Off the job	Ratio Off:on
	No.	Rate[1]	No.	Rate[1]				
1945	16.5	33	30.0	60	1.82	2 000	2 750	1.38
1950	15.5	27	31.5	56	2.03	1 950	2 500	1.28
1955	14.2	24	31.3	53	2.20	1 950	2 400	1.23
1957	14.2	23	31.7	52	2.23	1 900	2 450	1.29
1958	13.3	22	29.0	48	2.18	1 800	2 250	1.25
1959	13.8	23	29.0	47	2.10	1 950	2 200	1.13
1960	13.8	22	29.2	47	2.12	1 950	2 250	1.15
1961	13.5	21	28.9	45	2.14	1 950	2 200	1.16
1962	13.7	21	30.0	46	2.19	2 000	2 250	1.13
1963	14.2	21	31.7	48	2.23	2 000	2 350	1.18
1964	14.2	21	34.6	51	2.44	2 050	2 500	1.22
1965	14.1	20	36.5	52	2.59	2 100	2 700	1.29
1966	14.5	20	39.6	55	2.73	2 200	2 900	1.32
1967	14.2	19	40.0	54	2.82	2 200	3 000	1.36
1968	14.3	19	41.9	54	2.93	2 200	3 100	1.41
1969	14.3	18	43.3	55	3.03	2 200	3 200	1.45
1970	13.8	17	43.7	55	3.17	2 200	3 250	1.48
1971	13.7	17	41.5	52	3.03	2 300	3 200	1.39
1972	14.0	17	42.5	52	3.04	2 400	3 200	1.33
1973	14.3	17	43.7	51	3.06	2 500	3 300	1.32
1974	13.5	16	39.4	45	2.92	2 300	3 200	1.39
1975	13.0	15	37.8	44	2.91	2 200	3 200	1.45
1976	12.5	14	38.1	44	3.05	2 200	3 100	1.41
1977	12.9	14	40.2	44	3.12	2 300	3 200	1.39
1978	13.0	14	41.8	44	3.22	2 200	3 200	1.45
Change 1945-1978 (%)C	−21	−58	+39	−27	+77	+10	+16	+5

[1] Deaths per 100 000 workers.

Source: *Accident facts 1979*.

more the safety message gets to the home the better the results. Some measures that have proved effective are mailing booklets to the home, holding social gatherings for wives and families, sponsoring posters and essay contests for children. In general, a company can realise many benefits by preventing off-the-job accidents: a reduction in lost production time and operating costs, an increased interest in the job safety programme and a general improvement of relationships.

BELKNAP, R. G.

CIS 1175-1968 *Off-the-job safety*. Data Sheet No. 601 (National Safety Council, 425 N. Michigan Avenue, Chicago, Illinois, Nov. 1967), 16 p. Illus. 5 ref.

Accident facts 1979 (National Safety Council, 425 N. Michigan Avenue, Chicago, Illinois, 1979).

Accident statistics

Employment injuries include both industrial and commuting accidents resulting in death or personal injury, and all occupational diseases. Industrial accidents, though definitions vary widely, are those occurring at a place of work, whereas commuting accidents are those occurring on the way to and from work.

It is more difficult to devise a definition of occupational disease that is valid for all countries. In some countries the principle in force is that of the "closed schedule", with the result that only diseases clearly laid down and specified in the law are considered to be occupational diseases, whereas in others there are more flexible systems based on an "open schedule"; these take various forms but have in common the fact that they allow the sick worker in certain cases the possibility of proving that the disease he is suffering from, although it is not listed, is of occupational origin. There are even more flexible systems under which the concept of occupational disease is becoming less and less rigid.

The unit of enumeration for statistics of employment injuries should be the victim. If one person is the victim of two or more accidents during the period covered by the statistics, each accident should be counted separately.

Purpose of employment injury statistics

Industrial accidents, commuting accidents and occupational diseases cause much suffering and loss of life and are a heavy financial burden on industry and on social security schemes. From the economic point of view the loss is in fact much greater if the so-called indirect costs are taken into account, that is to say all the items of expense that accompany the direct costs of the employment injury. These include the fall in output, the economic consequences of the shock to workmates, the costs of training the replacement, medical and legal expenses and sometimes penalties, and their aggregate amount is estimated at about twice the cost of insurance in Europe and even more in the United States (see ACCIDENT COST).

In order to take practical steps towards reducing their number by the establishment and application of safety and health measures, it is very useful to have detailed statistics showing the causes and types of accidents, the industries or occupations in which they occur and the severity of the resultant injuries or occupational diseases. The compilation of employment injury statistics is the first step towards prevention of occupational risks; this factor was recognised in the ILO's Prevention of Industrial Accidents Recommendation, 1929 (No. 31). Employment injury statistics provide essential information for government labour inspectorates, industrial organisations and other agencies dealing with occupational safety and health and also for managements. Similarly, the establishment and operation of an occupational accidents and diseases compensation programme will have to be based on numerical data on the characteristics and incidence of the risks covered.

Very often, in fact, the statistics in this field are not based on the necessary preventionist approach but are rather derived from the activities of insurance, their characteristics being determined by its requirements. There are thus very often large gaps in the quantitative information concerning those exposed to risk and the way in which the employment injury occurs. It is only in recent years and in certain countries that special methods have been introduced to obtain statistics based on a more purely preventionist approach.

Reliability of statistics

It is not always possible to obtain 100% accuracy in statistical returns. Some managements fail to notify accidents to the competent authorities either in ignorance of their legal obligation, through forgetfulness or, deliberately, because they have failed to comply with some legal occupational safety or health requirements and wish to conceal the fact from the authority responsible for inspection.

The rules in the matter are extremely varied and so, accordingly, are the subtleties of the expedients adopted to find some way of circumventing the law or concealing failures to observe it or somehow or other deriving illicit profits from them. From time to time an investigation brings to light the difference existing between the official data and the real situation, a difference that in some countries assumes far from negligible proportions.

Statistics, however, must be understood in relation to what can be expressed by them. Thus, with respect to the coverage of insurance, there is striking variety in the corresponding rules. In some cases even highly important sectors of activity are excluded from compulsory insurance, whereas in others the criterion is the size of the undertaking and below a certain number of employees insurance is not compulsory. Yet again, certain classes of more highly paid workers may be excluded from the protection of insurance, for example, office workers or workers whose remuneration exceeds determined limits. With regard to the meaning of the data, it must be remembered that accidents resulting in temporary incapacity of very brief duration, regarded as falling within the waiting period, are usually not covered by insurance and therefore not even compulsorily notifiable. The duration in question varies and may be one to three or more days, not counting that of the accident, of absence from work.

It is thus extremely difficult to reach an exact assessment of the total number of cases of employment injury and this uncertainty leaves the field wide open to deductions and manipulations of every kind.

Finally, some countries do not yet have a fully developed and trained governmental labour statistics department, and reliable employment statistics are, therefore, not available.

Among the most difficult problems in this connection are those of the international comparability of accident statistics. The differences between the laws of different countries, their rules, their systems of classifying activities and labour and collecting and preparing statistics, and even the effect of custom and local practice make this almost impossible. The search for a way of overcoming the difficulty has in some cases produced data–at a high cost, indeed–that may be regular or sporadic, or drawn up on an ad hoc basis (for example, the annual statistical data of the EEC in the steel industry) and in others led to doubtful attempts at obtaining through certain limited aspects (such as fatal accidents) factors that, rather than ensure genuine comparability, offer a set of indications on the ways in which the differences between the data can be explained and interpreted.

Compilation of data

The methods used for the preparation of statistics on industrial injuries vary widely from country to country and even from one type of agency to another, due to differences in national legislation. The agency compiling such statistics will usually be one of the following:

(a) National statistical offices: in some countries a copy of the official accident notification must be forwarded to the national statistical office. In a certain number of countries this office obtains its information by means of sample surveys.

(b) Compensation agencies: some countries make compensation compulsory but do not set up a relevant national agency. In such cases a government supervising authority may collect the necessary data from private insurance companies, mutual benefit societies, occupational insurance associations, etc. These data are often related primarily to financial aspects such as premiums, compensation and total number of accidents; there may be little information on accident causes and types and on the number of workers at risk.

(c) National insurance and social security agencies: these compile statistics from accident compensation claims sent in by the employer or, in some countries, by the injured worker; the data obtained are usually more concerned with the financial aspects than with accident causation.

(d) Labour inspectorates: these prepare statistics from the accident notification forms received from employers. In some countries labour inspectors codify accidents and transmit data to the central inspectorate at regular intervals. This information may contain no details concerning duration of incapacity or degree of disablement.

(e) Accident prevention agencies: in some countries these agencies use similar methods to those followed by the labour inspectorates; in others their statistics are obtained from the same sources as those of the insurance and social security agencies. Some large establishments such as nationalised industries, transport undertakings and industrial groups have safety organisations which compile their own accident statistics.

Notifiable data. The information to be given on the notification form varies from country to country; in some countries one notification form may be required by the labour inspectorate and another with different particulars by the workmen's compensation authority. In view of the many different ways of collecting accident data and the different conditions in the same industry in countries in various stages of development, the international organisations are not yet in a position to recommend a model international employment accident notification form (see ACCIDENTS AND DISEASES, NOTIFICATION OF).

Sampling methods. Many agencies, particularly those in developing countries, do not have the staff and resources to tabulate and analyse the employment accident data collected. These agencies could profitably employ sampling methods, which provide statistical information that is sufficiently accurate for the purposes of accident prevention and social security.

Standardisation of accident statistics

There is, as yet, no international definition of employment injuries. However, a resolution adopted in 1962 by the Tenth International Conference of Labour Statisticians and still applicable establishes new basic standards in the field of employment injury statistics and provides international definitions of fatalities, permanent disablement and temporary disablement for statistical purposes. These definitions are as follows:

(a) fatalities: accidents resulting in death;

(b) permanent disablement: accidents resulting in permanent physical or mental limitation or impairment;

(c) temporary disablement: accidents resulting in incapacity for work for at least one full day beyond the day on which the accident occurred, irrespective of whether the days of incapacity were days on which the victim would otherwise have been at work;

(d) other cases: accidents resulting in incapacity for work lasting less than the period defined under *(c)*, and not involving permanent disablement.

As far as statistical methods are concerned, complete uniformity has not been attained in the international field; however, the ILO has, as a result of its International Conferences of Labour Statisticians, been instrumental in the adoption by many countries of the resolutions of these Conferences recommending uniform statistical methods. The resolution concerning statistics of employment injuries, for instance, adopted in 1962 by the Tenth International Conference of Labour Statisticians, has been used as a basis for the criteria then adopted by very many countries.

The Eleventh (Geneva, 1966) and Twelfth (Geneva, 1973) International Conferences of Labour Statisticians did not deal directly with the matter but provided useful secondary information concerning the standardising of data through the revision of the International Standard Classification of Occupations and through special instructions on statistics of manpower and relevant data.

Comparative measures

Sound comparisons between periods, industries and countries can be made (in the last case, with the very severe limitations already mentioned above) only if the statistics of industrial accidents are considered in conjunction with data on employment, hours of work, production, etc. For such purposes it may be useful to resort to relative measures, such as frequency, incidence and severity rates.

Accident frequency rate. The accident frequency rate is an expression relating the number of specific accidents to a number of man-hours worked. The resolution adopted in 1962 by the Tenth International Conference of Labour Statisticians recommended that the frequency rate of industrial accidents should be calculated by dividing the number of accidents (multiplied by 1 000 000) which occurred during the period covered by the statistics by the number of man-hours worked by all persons exposed to risk during the same period.

$$\text{Frequency rate} = \frac{\text{total number of accidents} \times 1\,000\,000}{\text{total number of man-hours worked}}$$

Example 1. An industrial undertaking with 850 employees had 100 disabling accidents in one year. Assuming that there were 300 working days in a year, each of 8 working hours, and a total of 40 000 working days lost through holidays, absenteeism, sickness and accidents, etc., what would be the annual accident frequency rate?

The total number of man-hours of exposure is: (850 employees x 300 working days x 8 h) − (40 000 lost days x 8 h) = 1 720 000.

$$\text{Frequency rate} = \frac{100 \times 1\,000\,000}{1\,720\,000} = 58.1$$

Accident incidence rate. For countries which do not know the number of man-hours worked, the resolution suggested that incidence rates of industrial accidents should be calculated by dividing the number of accidents (multiplied by 1 000) which occurred during the period covered by the statistics by the average number of workers exposed to risk during the same period.

$$\text{Incident rate} = \frac{\text{total number of accidents} \times 1\,000}{\text{average number of persons exposed}}$$

Accident severity rate. The object of a severity rate is to give some indication of the loss in terms of incapacity resulting from industrial accidents. The severity rate should be calculated by dividing the number of working days lost (multiplied by 1 000) by the number of hours of working time of all persons covered and, where practicable, rates should be calculated for principal industries, for each sex and for different age groups.

$$\text{Severity rate} = \frac{\text{total number of days lost} \times 1\,000}{\text{total number of man-hours worked}}$$

At present, few countries publish severity rates for industrial injuries. Comparison between those national rates that are available is difficult because of the diversity of methods used to calculate the number of days lost in the event of death and permanent total or permanent partial disability. The resolution of the Sixth International Conference of Labour Statisticians in 1947 stated that each accident resulting in death or permanent total disability should be considered for this purpose as 7 500 working days lost. A large number of countries which compute severity rates have not adopted this figure: the majority of them prefer to assume a loss of 6 000 days, as proposed by the American National Standards Institute. The Tenth International Conference of Labour Statisticians, considering the lack of a uniform scale of time charges for fatal or permanent disablement cases, concluded that further research was needed before international recommendations on the method of compiling severity rates could be developed.

Example 2. Taking the same figures as in example 1, i.e. 850 employees, 100 disabling accidents a year, 300 working days of 8 hours each and 40 000 working days lost through holiday, absenteeism, sickness and accidents, etc., and assuming that the 100 accidents caused a loss of 3 000 days, then the severity rate would be calculated as follows:

$$\text{Severity rate} = \frac{3\,000 \times 1\,000}{1\,720\,000} = 1.74$$

The severity rate is 1.74 days loss per 1 000 man-hours. The need for both frequency and severity rates becomes more apparent when fatal accidents or accidents causing permanent disability occur.

Example 3. If one fatal accident is added to the 100 disabling accidents given in example 2, the calculation of the severity rate, with an allowance of 7 500 working days for a fatal accident or for injuries resulting in permanent total disability as adopted by the Sixth International Conference of Labour Statisticians, would be:

$$\text{Severity rate} \times \frac{(7\,500 + 3000) \times 1\,000}{1\,720\,000} = 6.1$$

This means that the addition of one fatal or permanent total disability has increased the severity rate nearly 350%, whereas the frequency rate would rise only from 58.4 to 58.7; an increase of only 0.3%. To obtain a true picture, it is preferable to calculate frequency and severity rates separately for fatal and non-fatal injuries.

Application of statistics

In one large factory in the Republic of Korea, it was found from an analysis of the accident data that most accidents occurred during the hot summer months, so steps were taken to improve the ventilation and reduce the temperature of the workrooms during these months. Some of the large oil companies evaluate their accident data by plotting on graphs the number of accidents and the costs of compensation against oil production. A cotton industry employers' federation in one country ascertained from its accident statistics and the amounts paid in compensation that a high proportion of the compensation paid was in respect of workers sustaining small cuts and abrasions or bruises. These would not have entailed absence from work if prompt first-aid treatment had been given; however, no treatment had been provided and each resultant case of sepsis entailed an average of 15 days' work absence. The federation consequently appointed two safety officers to check all machinery and plant likely to cause such cuts and bruises, and first-aid facilities were improved.

These fragmentary examples show the great and varied use that can be made of accident statistics when they are properly prepared and, even more, when they are properly understood and interpreted. [In particular, statistical distribution and statistical significance should be understood. An example of normal distribution of accidents among a group of workers is given under ACCIDENT PRONENESS. An example of normal distribution of accidents according to time periods is the following: 52 accidents a year may statistically occur as follows:

0 accidents a week for 19 weeks
1 accidents a week for 19 weeks
2 accidents a week for 10 weeks
3 accidents a week for 3 weeks
4 accidents a week for 1 week.]

One thing that is essential in a modern preventionist analysis is the quantitative knowledge of the way in which accidents happen, for this knowledge, in conjunction with other parameters, makes it possible to draw up what are known as risk maps based on the frequency shown by calculating the probability, at the territorial level, of an employment injury with given characteristics and given causes.

Presentation of statistics at plant level

Well designed statistical tables and graphs are very useful in stimulating safety consciousness among both workers and managements. Such tables and graphs should be well designed and self explanatory; complicated presentations should be avoided. Tables should not contain too many columns, the scales of graphs should be properly selected and the symbols used should be readily understood. In some cases, it is necessary to symbolise statistics in figurative pictures so that they can be understood by illiterate workers.

BRANCOLI, M.

Year Book of Labour Statistics (Geneva, ILO, annual).

General report on progress of labour statistics. 8th International Conference of Labour Statisticians. Report I. Part II: *Methods of statistics of occupational diseases* (Geneva, ILO, 1954), 80 p.

Final report. 10th International Conference of Labour Statisticians (Geneva, ILO, 1962), 81 p.

Statistics of industrial injuries. D.17.1970/X CIST/II/SAT (Geneva, ILO, 1970), 56 p.

Manual of the international statistical classification of diseases, injuries, and causes of death. Based on the Recommendations of the Ninth Revision Conference, 1975 and adopted by the 29th World Health Assembly. Volume 1 (Geneva, WHO, 1977), 773 p. Volume 2: Alphabetical Index (Geneva, WHO, 1978), 659 p.

"Occupational accident frequency rates: the possibility and the reliability of international comparison" (Indici di frequenza degli infortuni sul lavoro: possibilità e validità di confronti a livello internazionale). Brancoli, M.; Cassanelli, A.; Ortolani, G. *Rivista degli infortuni e delle malattie professionali* (Rome), May-June 1979, 66/3 (187-205). 4 ref. (In Italian)

A system of basic periodical statistics of occupational injuries. Meeting of Experts on Statistics of Occupational Injuries, Geneva, 21-25 January 1980. Paper No. 2 (Geneva, ILO, 1980), 45 p.

Acetaldehyde (CH_3CHO)

ETHANAL; ETHYL ALDEHYDE

m.w.	44
sp.gr.	0.78
m.p.	−124.6 °C
b.p.	20.8 °C
v.d.	1.5
f.p.	−38 °C
e.l.	4.1-60%
i.t.	175 °C

soluble in all proportions in water, ethanol, ether and benzene

a colourless, flammable liquid with a pungent fruity odour.

TWA OSHA	200 ppm 360 mg/m³
TLV ACGIH	100 ppm 180 mg/m³
STEL ACGIH	150 ppm 270 mg/m³
IDLH	10 000 ppm
MAC USSR	5 mg/m³

Production. Acetaldehyde is manufactured by the following processes:

(a) acetylene hydration using mercury catalysts;

(b) ethanol oxidation or dehydrogenation using copper catalysts;

(c) paraffin oxidation: non-catalytic process using butane-rich saturated hydrocarbon feedstocks;

(d) ethylene oxidation using palladium and copper chlorides as catalysts.

This latter process is industrially the most important. In 1974 a single-stage process for converting synthesis gas to acetaldehyde was developed, viz.:

$$CO+H_2 \rightarrow CH_3CHO+CH_3COOH+CH_3CH_2OH$$
$$\qquad 24\% \qquad 20\% \qquad 16\%$$

This process could become more important in the future if coal gasification programmes continue to develop.

Polymers. Acetaldehyde may be converted to its polymer, paraldehyde ($(C_2H_4O)_3$), in the presence of traces of acid acting as a catalyst and reaches an equilibrium point at 15 °C with a mixture containing 94.3% paraldehyde and 5.7% acetaldehyde. Paraldehyde is a colourless liquid (b.p. 124 °C) of characteristic odour. The second polymer of acetaldehyde, metaldehyde ($(C_2H_4O)_4$), can be obtained by the action of an acid catalyst on acetaldehyde, providing the temperature of the reaction is carefully maintained at or below 0 °C. It is a white powder, melting point 246 °C.

Uses. Acetaldehyde is a most important chemical intermediate and takes part in a wide range of reactions. Since the war the expansion in the manufacture and use of acetaldehyde has been increasing very rapidly. Total world capacity is about 3 million tonnes per year and US capacity 1 million tonnes per year. At the present time some 60% of all acetaldehyde is used for acetic acid manufacture. Other important uses of acetaldehyde include the manufacture of–

(a) ethyl acetate $2CH_3CHO \rightarrow CH_3CH_2COOCH_3$ (the reaction is carried out in the presence of sodium, aluminium or magnesium alkoxide);

(b) peracetic acid (catalytic oxidation of acetaldehyde);

(c) pyridine derivatives (condensation of acetaldehyde with ammonia);

(d) pentaerythritol (reacting acetaldehyde with formaldehyde in an alkaline media. Pentaerythritol is used mainly in surface-coating compositions);

(e) butanol (acetaldehyde is converted to crotonaldehyde via aldol, and the crotonaldehyde hydrogenated to butanol).

Paraldehyde, a trimer of acetaldehyde, is used in medicine as a hypnotic. Industrially it has been used as a solvent and also as a rubber activator and antioxidant. Metaldehyde is used as a fuel in portable cooking stoves and also in gardening and market gardening for slug control.

HAZARDS

Fire and explosion. This aldehyde is a highly reactive material, its chief hazard in practice being that of fire and explosion. Both liquid and vapour phases are extremely flammable. The vapour forms flammable and explosive mixtures with air over a very wide range of concentrations. The low boiling point of the liquid makes it difficult to handle without considerable escape of vapour. In the presence of traces of acid, polymerisation may occur, a reaction which is exothermic, and may lead to serious pressure rise in containing vessels, and indeed to fire and explosion. When acetaldehyde is diluted with water, heat of solution will be generated. Paraldehyde in the presence of heat or sunlight will decompose to acetaldehyde.

Health hazards. Acetaldehyde is an irritant of mucous membranes and also has a general narcotic action on the central nervous system. Low concentrations cause irritation of the eyes, nose and upper respiratory passages, as well as bronchial catarrh. High concentrations cause headache, stupor, bronchitis and pulmonary oedema. Ingestion causes nausea, vomiting, diarrhoea, narcosis and respiratory failure; death may result from damage to kidneys and fatty degeneration of the liver and heart muscle. Acetaldehyde is produced in the blood as a metabolite of ethyl alcohol, and will give rise to facial flushing, palpitations and other disagreeable symptoms. This effect is enhanced by the drug disulphiram (Antabuse) and by exposure to the industrial chemicals disulphiram (an antioxidant), cyanamide and dimethylformamide. Ingestion of paraldehyde ordinarily induces sleep without depression of respiration, although deaths occasionally occur from respiratory and circulatory failure after doses of 10 cm^3 or more. Metaldehyde, if ingested, may cause nausea, severe vomiting, abdominal pain, muscular rigidity, convulsions, coma, and death from respiratory failure. Repeated exposure to the vapours of acetaldehyde causes dermatitis and conjunctivitis. In chronic intoxication, the symptoms resemble those of chronic alcoholism, such as loss of weight, anaemia, delirium, hallucinations of sight and hearing, loss of intelligence and psychic disturbances. Amounts of metaldehyde less than those necessary to produce acute poisoning are without effect.

SAFETY AND HEALTH MEASURES

Fire prevention. It is recommended that steel storage tanks of a suitable standard be used; they should be equipped with cooling coils to maintain the temperature below 20 °C, or they should be pressure tanks, the pressure being maintained by a blanket of nitrogen or other inert gas. Acetaldehyde stored under pressure should be kept from contact with air to prevent the formation of highly explosive peroxides. Storage vessels should be fitted with temperature gauges and automatic water sprays. All precautions must be taken against ignition by flames, sparks, electrical discharges, etc. All tanks and equipment must be earthed. Transfer of the material by pipeline must be by pressure of nitrogen. There is a danger of spontaneous ignition if liquid acetaldehyde spills on to hot pipework or vessels. Drums containing acetaldehyde should never be stored in direct sunlight or other warm areas. Spillages should be washed away to a drain by using copious amounts of water. Care must be taken to prevent the accumulation of acetaldehyde vapour in sewers, where it could form an explosive mixture with air. If possible, the drainage system should have an automatic explosimeter. Plant operators should have specific instruction in the hazards associated with the handling of acetaldehyde and in the precautions to be taken for its safe handling.

Health protection. Contact with acetaldehyde should be minimised by attention to plant design and handling procedure. Spillages should be avoided where possible and, where they occur, adequate water and drainage facilities should be available. An open-plan type of plant is probably the most satisfactory; however, if this is not possible, ventilation must be adequate and the prescribed atmospheric concentration should not be exceeded. Protective clothing should be freely available and operators instructed in its use. Chemical eye and face protection of an approved design should be mandatory in the plant area and, for maintenance work, plastic face shields should also be worn.

Suitable protective clothing, aprons, hand protection and impervious foot protection should be provided. Water showers and eye irrigation systems should be available on the plant area and operators informed of their location. A pre-placement medical examination to exclude those with chronic skin and lung conditions is desirable.

Treatment. Acute poisoning from fume exposure is treated by early removal of the victim to fresh air and administration of oxygen if available. There is no specific antidote. Pulmonary oedema is treated by accepted medical methods in hospital. Acute poisoning by ingestion requires the recognised emergency measures to remove the poison by gastric lavage or emesis. General measures for treatment of coma and shock may be necessary. Workers with evidence of chronic ill effects due to acetaldehyde should be removed to alternative work. Chronic poisoning from paraldehyde ingestion requires psychiatric care and the avoidance of further exposure.

For the derivatives of acetaldehyde: chloroacetaldehyde and cloral, see ALDEHYDES AND KETALS.

COOKE, W. G.

General:

CIS 76-1375 "Acetaldehyde" (Aldéhyde acétique). Morel, C.; Cavigneaux, A.; Protois, J. C. Fiche toxicologique nº 120, *Cahier de notes documentaires–Sécurité et hygiène du travail* (Paris), 2nd quarter 1976, 83, Note No. 1014-83-76 (319-322). 19 ref. (In French)

"Dealing with acetaldehyde". Cooper, P. *Food and Cosmetics Toxicology* (Oxford), Dec. 1975, 13/6 (668-670). 11 ref.

Inhalation toxicity:

"Repeated exposure to acetaldehyde vapour". Kruysse, A.; Feron, V. J.; Til, H. P. *Archives of Environmental Health* (Chicago), Sep. 1975, 30/9 (449-452). Illus. 21 ref.

Protection:

"Protective action of ascorbic acid and sulphur compounds against acetaldehyde toxicity: implications in alcoholism and smoking". Sprince, H.; Parker, C. M.; Smith, G. G.; Gonzales, L. J. *Agents and Actions* (Basel), 1975, 5/2 (164-173).

Acetic acid (CH_3COOH)

ETHANOIC ACID; METHANE CARBOXYLIC ACID

m.w.	60
sp.gr.	1.04
m.p.	16.6 °C
b.p.	117.9 °C
v.d.	2.07
v.p.	11 mmHg (1.46·10³ Pa) at 20 °C
f.p.	42.8 °C
e.l.	5.4-16%
i.t.	516 °C

soluble in all proportions in water, ethanol and ether

a colourless liquid with a pungent vinegar-like odour (threshold 1 ppm); in the presence of air it attacks a wide range of metals evolving hydrogen.

TWA OSHA	10 ppm 25 mg/m³
STEL ACGIH	15 ppm 37 mg/m³
IDLH	1 000 ppm
MAC USSR	5 mg/m³

Production. Weak concentrations of acetic acid (vinegar contains about 4-6%) are produced by aerobic fermentation *(Acetobacter)* of alcohol solutions. Acetic acid is mainly obtained by:

(a) catalytic oxidation (usually in the presence of a manganese salt) of acetaldehyde in liquid phase with pure oxygen, according to:
$2CH_3CHO + O_2 = 2CH_3COOH$.
The product of the reaction is distilled to obtain pure acetic acid;

(b) hydration of acetic anhydride;

(c) reacting methyl alcohol with carbon monoxide.

Uses. Acetic acid is one of the most widely used organic acids. It is employed in the production of cellulose acetate, vinyl acetate (by means of a catalytic reaction with gaseous acetylene), inorganic acetates (aluminium, lead, copper, etc.), organic acetates (esters), and acetic anhydride; acetic acid itself is used in the dyeing industry, pharmaceutical industry, the canning and food preserving industry, pigment production, etc. The complexity of acetic acid production and end uses can be seen from figure 1.

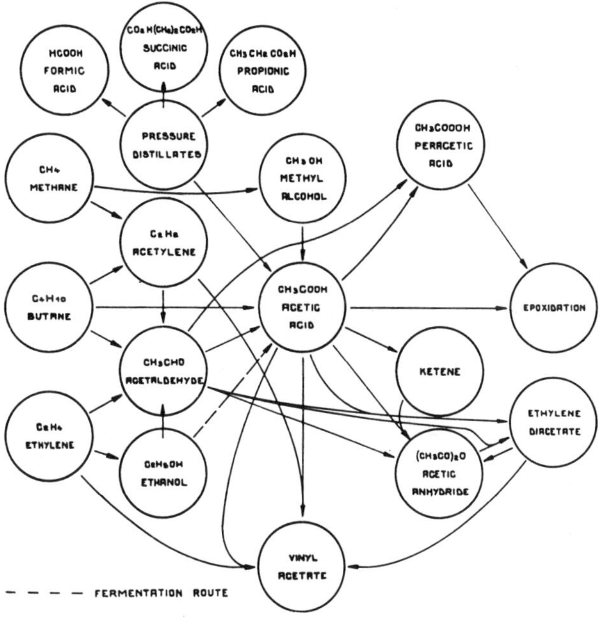

Figure 1. The acetic acid complex (reprinted from *Chemical and Process Engineering*).

HAZARDS

Acetic acid vapour may form explosive mixtures with air and constitute a fire hazard either directly or by the release of hydrogen.

Glacial acetic acid or acetic acid in concentrated form are primary skin irritants and will produce erythema, chemical burns and blisters. In cases of accidental ingestion, severe ulceronecrotic lesions of the upper digestive tract have been observed with bloody vomiting, diarrhoea, shock, haemoglobinuria followed by anuria and uraemia.

The vapours have an irritant action on exposed mucous membranes, particularly the conjunctivae, rhinopharynx and upper respiratory tract. Acute bronchopneumonia developed in a woman who was made to inhale acetic acid vapours following a fainting attack.

Workers exposed for a number of years to concentrations of up to 200 ppm have been found to suffer from palpebral oedema with hypertrophy of the lymph nodes, conjunctival hyperaemia, chronic pharyngitis, chronic catarrhal bronchitis and, in some cases, asthmatic bronchitis and traces of erosion on the vestibular surface of the teeth (incisors and canines).

The extent of acclimatisation is remarkable; however, following repeated exposure, workers may complain of digestive disorders with pyrosis and constipation. The skin on the palms of the hands is subject to the greatest exposure and becomes dry, cracked and hyperkeratotic, and any small cuts and abrasions are slow to heal.

SAFETY AND HEALTH MEASURES

Acetic acid should be stored away from all sources of ignition and oxidising substances. Storage areas should be well ventilated to prevent the accumulation of dangerous concentrations. Containers should be of stainless steel or glass. In the event of leakage or spillage, acetic acid should be neutralised by application of alkaline solutions. Eye-wash fountains and emergency showers should be installed for dealing with cases of skin or eye contact. Marking and labelling of containers is essential and, for all forms of transport, acetic acid is classified as a dangerous substance.

To prevent damage to the respiratory system and mucous membranes, the atmospheric concentration of acetic acid should be kept below maximum permissible levels using standard industrial hygiene practices such as local exhaust ventilation and general ventilation, backed up by periodical determination of atmospheric acetic acid concentrations. Detection and analysis, in the absence of other acid vapours, is by means of bubbling in an alkaline solution and determination of residual alkali; in the presence of other acids, fractional distillation was necessary; however, a gas chromatographic method is now available for determination in air or water.

Persons working with the pure acid or concentrated solutions should wear protective clothing, eye and face protection, hand and arm protection and respiratory protective equipment. Adequate sanitary facilities should be provided and good personal hygiene encouraged. Pre-employment and periodic medical examinations should be carried out to ensure that workers with respiratory ailments, skin disorders or keratoconjunctivitis are not exposed to acetic acid.

Treatment. Ingestion should be treated with an emetic such as a weak solution of sodium bicarbonate. Skin and eye contact should be followed by copious irrigation with water.

PARMEGGIANI, L.

Toxicology:

Acetic acid (Essigsäure). Gesundheitsschädliche Arbeitsstoffe. Toxikologisch-arbeitsmedizinische Begründung von MAK-Werten (Verlag Chemie, Weinheim, Postfach 1260/1280, Federal Republic of Germany, 5 Dec. 1973), 5 p. 27 ref. (In German)

Safety:

CIS 74-1693 *Acetic acid.* Chemical Safety Data Sheet SD-41 (Manufacturing Chemists Association, 1825 Connecticut Avenue NW, Washington, DC, 1973), 15 p.

Analysis:

"Gas chromatographic determination of acetic acid in industrial atmosphere and waste water". Esposito, G. G.; Schaeffer, K. A. *American Industrial Hygiene Association Journal* (Akron), May 1976, 37/5 (268-273). Illus. 22 ref.

Acetone and derivatives

Acetone (CH_3COCH_3)

2-PROPANONE; DIMETHYLKETONE; KETONE PROPANE

m.w.	58
sp.gr.	0.79
m.p.	−95.3 °C
b.p.	56.2 °C
v.d.	2.0
v.p.	226.3 mmHg (30.09·10³ Pa) at 25 °C
f.p.	−17.8 °C
e.l.	2.6-12.8%
i.t.	465 °C

soluble in all proportions in water, ethanol and ether

a volatile, colourless, flammable liquid with a pungent odour.

TWA OSHA	1 000 ppm 2 400 mg/m³
TLV ACGIH	750 ppm 1 780 mg/m³
STEL ACGIH	1 000 ppm 2 375 mg/m³
IDLH	20 000 ppm
MAC USSR	200 mg/m³

level recommended by the Japanese Association of Occupational Health 200 ppm 480 mg/m³.

Production. Acetone is produced commercially by the dry distillation of calcium acetate, the distillation of crude wood spirit, the fermentation of carbohydrates using bacteria such as *B. macerans* and *B. acetoethylicum* and by the catalytic oxidation of isopropyl alcohol. About six times as much acetone was produced by the oxidation process as by the fermentation process during the Second World War.

Uses. Acetone is used as an industrial solvent in resins, lacquers, oils, fats, collodion, cotton, cellulose acetate and acetylene. It is found in the paint, lacquer and varnish industry, rubber industry, plastics industry, dyeing industry, celluloid industry, photographic industry and explosives industry, and in the manufacture of artificial silk and synthetic leather. It is also used in the production of lubricating oils and as a solvent in the chemical industry for the production of ketene, acetic anhydride, methyl methacrylate, 4,4'-isopropylidenediphenol, diacetone alcohol, mesityl oxide, methyl isobutyl ketone, hexylene glycol, isophorone, chloroform, iodoform, vitamin C, etc.

HAZARDS

Acetone is one of the least hazardous of the industrial solvents as far as health is concerned. However, it is a highly flammable and explosive substance.

Health hazards. Acetone is highly volatile and may be inhaled in large quantities when it is present in high concentrations. It may be absorbed into the blood through the lungs and diffused throughout the body. Small quantities may be absorbed through the skin.

Typical symptoms following acetone exposure are: narcosis, slight skin irritation and more pronounced mucous membrane irritation. Exposure to high concentrations produces a feeling of unrest, followed by progressive collapse accompanied by stupor and periodic breathing and, finally, coma. Nausea and vomiting may also occur and are sometimes followed by haematemesis. In some cases, albumin and red and white blood cells in the urine indicate the possibility of kidney damage and, in others, liver damage can be presumed from the high levels of urobilin and the early appearance of bilirubin reported. The longer the exposure, the lower the respiratory rate and pulse; these changes are roughly proportionate to the acetone concentration. Cases of chronic poisoning resulting from prolonged exposure to low concentrations of acetone are rare; however, in cases of repeated exposure to low concentrations, complaints were received of headache, drowsiness, vertigo, irritation of the throat and coughing.

Exposure at 9 300 ppm could not be tolerated by a man more than 5 minutes on account of throat irritation. At 2 000 ppm a slight narcotic effect was observed. Workers having been exposed to 1 000 ppm, 3 hours per day for 7-15 years, complained of chronic inflammation of airways, stomach and duodenum; some of them complained also of dizziness and asthenia. Similar complaints were reported after exposure down to 700 ppm.

Exposure at 500 ppm, 6 hours a day for 6 days, caused irritation of mucous membranes, an unpleasant smell, heavy eyes, overnight headache, and general weakness accompanied by blood changes such as increase of leucocyte and eosinophil counts and decrease of the phagocytic activity of neutrophils. The recovery took several days. At 250 ppm subjective symptoms and blood changes were very slight. Exposure at over 0.5 mg/l (211 ppm) for 8 hours running has been reported to make acetone detectable in blood all through the weekdays.

Fire and explosion. Acetone is an extremely flammable liquid with a potentially severe fire hazard. Aqueous solutions of acetone are also highly flammable and the acetone must be considerably diluted if the flash point is to be brought to a relatively safe level.

SAFETY AND HEALTH MEASURES

Acetone should be stored in tightly closed steel containers in a dark, cool, well ventilated area, and all sources of ignition should be excluded from such an area. Large stocks should be situated at a distance from inhabited buildings. Acetone should be transported in steel tank cars, tank trucks or drums and the containers should not be subjected to impacts. Non-sparking tools should be used to open containers which should then be effectively earthed before decanting. Fire extinguishers of the carbon dioxide, dry chemical, etc., type should be at hand.

Inhalation of acetone can be prevented by the careful handling and inspection of all containers to eliminate leaks and spillage, and the replacement of all covers, bungs, etc., to reduce vaporisation. In the presence of atmospheric concentrations of acetone, the use of respiratory protective equipment with an activated charcoal filter is recommended. Enclosure and ventilation should be installed in the case of processes employing acetone.

Atmospheric concentrations may be measured by either colorimetry of acetone in a sodium nitrite solution with the addition of salicylaldehyde and potassium hydroxide, or by detector tube.

Treatment. In the event of prolonged exposure to very high concentrations followed by narcotic symptoms, remove the patient to an uncontaminated atmosphere at once and apply resuscitation if breathing has stopped. If there has been extensive liquid contact with the skin or clothing, remove clothing and flush underlying areas with water. In the case of eye contact with the liquid, flush the eyes with copious amounts of water.

Chloroacetone ($CH_2ClCOCH_3$)

MONOCHLOROACETONE; 1-CHLORO-2-PROPANE

m.w. 92.5
sp.gr. 1.15
m.p. -44.5 °C
b.p. 119 °C

slightly soluble in water, ethyl alcohol and ethyl ether

a colourless liquid with a pungent odour.

Chloroacetone is obtained by the chlorination of acetone and is used in couplers for colour photography, as a pesticide and as a tear gas.

The vapour of this liquid is a strong lacrimator and is irritating to the skin and respiratory tract. A concentration of 0.018 mg/l is sufficient to produce lacrimation and a concentration of 0.11 mg/l will normally not be supported for more than 1 min. The same precautions should be respected in handling and storing as those applicable to chlorine.

Bromoacetone ($CH_2BrCOCH_3$)

m.w. 137
sp.gr. 1.63
m.p. −54 °C
h.p. 136 ° C (partial decomposition)
v.d. 4.75
v.pr. 9 mmHg ($1.1 \cdot 10^3$ Pa) at 20 °C

slightly soluble in water; readily soluble in ethyl alcohol and acetone

a liquid which is colourless when pure but which rapidly turns violet even in the absence of air; it has a pungent odour.

It is produced by treating aqueous acetone with bromine and sodium chlorate at 30-40 °C and is used in organic synthesis and as a tear gas.

Bromoacetone is poisonous and is intensely irritating to the skin and mucous membranes. It should be stored in a ventilated area, and personnel exposed to its vapours should wear gastight chemical safety goggles and respiratory protective equipment. Wherever possible, this material should be used in enclosed systems. Containers should be kept closed and plainly labelled.

Hexafluoroacetone (CF_3COCF_3)

PERFLUOROACETONE; HEXAFLUORO-2-PROPANONE

m.w. 166
sp.gr. 1.65
m.p. −122 °C
b.p. −27.5 °C

a colourless, hygroscopic, non-flammable gas; musty, becomes acrid in presence of acidic impurities.

TLV ACGIH 0.1 ppm 0.7 mg/m³
STEL ACGIH 0.3 ppm 2 mg/m³

This is a very irritant gas, particularly to the eyes. Exposure to relatively high concentrations causes respiratory impairment and conjunctival haemorrhages. Repeated exposure at 12 ppm can impair spermatogenesis and erythropoiesis and cause changes in liver, kidneys and lymphopoietic system. At 1 ppm such changes may be reversible; at 0.1 ppm they can be hardly detected after 90 days' exposure.

INOUE, T.

"Exposure to acetone. Uptake and elimination in man". Wigaens, E.; Holm, S.; Astrand, I. *Scandinavian Journal of Work, Environment and Health* (Helsinki), June 1981, 7/2 (84-94). Illus. 14 ref.

CIS 79-431 *Criteria for a recommended standard–Occupational exposure to ketones.* DHEW (NIOSH) publication No. 78-173 (National Institute for Occupational Safety and Health, 4676 Columbia Parkway, Cincinnati, Ohio) (June 1978), 244 p. 186 ref.

Acetylaminofluorene

2-Acetylaminofluorene ($C_{15}H_{13}O$)

N-2-FLUORENYLACETAMIDE; 2-ACETAMIDOFLUORENE; AAF

CH_3CONH–

m.w. 223.3
m.p. 194 °C

crystals insoluble in water; soluble in several organic solvents.

Uses. AAF was produced as a pesticide but it has no use as such. Small amounts are used in cancer research.

HAZARDS AND THEIR PREVENTION

AAF has a different teratogenic and carcinogenic power on animals, being a procarcinogen which requires metabolic activation before expressing its carcinogenic potential. Human liver microsomal and cytosol fractions can convert N-hydroxy-2-acetylaminofluorene into a mutagen, and although to date no human cancers are known that can be attributed to AAF, it has been included in the US Federal Standard for carcinogens. In experimental animals it produces bladder, kidney, liver and pancreas cancers. Taking into account the above data, work with AAF should be carried out in a closed system and, when this is impossible, under exhaust ventilation in areas to which only authorised employees may have access. Exposed workers should wear clean full-body protective clothing including gloves; shoe covers and a filter-type respirator for dusts, mists and fumes. After use, clothing and equipment should be placed in an impervious container for decontamination or disposal. On leaving the restricted area, workers should take a shower.

Pre-employment and periodical medical examinations should focus on the kidney, bladder, liver, skin and respiratory system. Periodical monitoring by means of sputum and urine cytological examinations is recommended. Special precautions should be taken as regards pregnant women.

PARMEGGIANI, L.

"On the correlation between hepatocarcinogenicity of the carcinogen N-2-fluorenylacetamide and its metabolic activation by the rat". Gutmann, H. R.; Malejka-Giganti, D.; Barry, E. J.; Rydell, R. E. *Cancer Research* (Chicago), July 1972, 32/7 (1 554-1 561). Illus. 35 ref.

"Mammary carcinogenesis in the rat by topical application of fluorenylhydroxamic acids". Malejka-Giganti, D.; Gutmann, H. R.; Rydell, R. E. *Cancer Research* (Chicago), Oct. 1973, 33/10 (2 489-2 497). Illus. 44 ref.

"In vitro metabolism and activation of carcinogenic aromatic amines by subcellular fractions of human liver". Dybing, E.; von Bahr, C.; Aune, T.; Glaumann, H.; Levitt, D. S.; Thorgeirsson, S. S. *Cancer Research* (Chicago), Oct. 1979, 39/10 (4 206-4 211). 38 ref.

"2-Acetylaminofluorene" (570-575) 1910-1014. *General industry.* OSHA Safety and Health Standards (29 CFR 1910) OSHA 2206 (revised Jan. 1976) (Washington, DC, US Department of Labor, 1976), 649 p.

Acetylene (HC⁝CH)

ETHINE; ETHYNE

m.w.	26
sp.gr.	0.62(−82 °C)
m.p.	−81.8 °C
b.p.	−75 °C
v.d.	0.9
v.p.	40 atm ($4052 \cdot 10^3$ Pa) at 16.8 °C
f.p.	−32 °C
e.l.	2.5-81%
i.t.	305 °C

soluble in water and many organic materials

a colourless gas with a faint odour of ether but most commercial grades contain such impurities as phosphine, hydrogen sulphide and ammonia, and have a garlic-like odour.

TLV ACGIH a simple asphyxiant

Production. Acetylene is produced industrially by two methods:

(a) by reaction of calcium carbide with water;

(b) by an endothermic process.

In the first method calcium carbide (CaC_2) reacts in the presence of water to give gaseous acetylene and a residue of slaked lime. The reaction, which is effected in specially designed acetylene generators, is highly exothermic. The generators are designed with a heat-dissipation coefficient which is matched to the speed and heat output of the acetylene generation, to ensure that the reaction does not get out of hand. Specific safety precautions exist for the use and maintenance of generators; in some countries, apparatus must be officially approved before use.

In the second method acetylene is obtained from gaseous hydrocarbons (e.g. methane) or liquid hydrocarbons (e.g. petroleum products) by partial combustion, electric arc or high temperature regenerative processes. The acetylene must then be separated from the resultant gas mixture.

Uses. Acetylene burns in air or oxygen with an intense flame. The flame from an acetylene/air blowpipe may have a temperature of up to 2 100-2 400 °C and it is used for brazing, local heating and in the glass industry. The oxygen/acetylene flame may have a temperature up to 3 100 °C; it is used for welding, cutting, brazing, metallising, building up worn surfaces, surface hardening, localised hardening, flame scarfing, localised heating (shrink-fitting, etc.). In all these cases, the flame is produced with a blowpipe which is manually or automatically regulated to ensure correct combustion.

In the chemical industry acetylene is used as a feedstock in the manufacture of vinyl chloride, acrylonitrile, synthetic rubber, vinyl acetate, trichloroethylene, acrylate, butyrolactone, 1,4-butanediol, vinyl alkyl ethers, pyrrolidone, etc.

HAZARDS

Acetylene is non-poisonous and a mild anaesthetic. The phosphine contained in crude acetylene is highly toxic but is generally present in too low a concentration to be considered dangerous and, in most cases, acetylene is, moreover, purified before use. The gas is non-irritant to skin and mucous membranes, and ingestion is not possible.

Acetylene forms explosive mixtures with air or oxygen and under certain conditions will react with copper, silver or mercury to produce acetylides which, when subject to impact, friction or a rise in temperature in the dry state may decompose violently; it also reacts explosively with chlorine and fluorine.

Gaseous acetylene at high temperature and pressure may decompose explosively into its constituent elements. The danger grows with increasing pressure although container size and shape are contributory factors; the presence of a diluent (e.g. water vapour) reduces the degree of risk. The heating of acetylene may lead to polymerisation reactions which are highly exothermic.

SAFETY AND HEALTH MEASURES

Any leakage of acetylene will constitute a fire or explosion hazard and must be controlled immediately. Due to its characteristic odour, leaking gas is relatively easy to detect; the precise location of the leak should be determined by wiping soapy water over the suspected area—naked lights should never be used.

Storage. Acetylene can be stored as a gas or in solution. Gaseous acetylene becomes highly explosive when compressed or heated and consequently it is stored under low pressure in large tanks or gasholders. In most countries the maximum storage pressure for gaseous acetylene is fixed at 1.5 bar gauge, in others it is 1 bar (10^5 Pa) gauge.

Dissolved acetylene can be stored at much higher pressures (15-20 bar gauge at room temperature). The most common solvents are acetone and dimethyl formamide. The gas cylinders are filled with porous material to prevent decomposition. Cylinder capacity is usually 4-8 m³ gas at normal pressure and temperature (760 mmHg or 1.013.10⁵ Pa at 15 °C).

Production and use. Areas should be kept dry, well ventilated and shielded from direct sunlight; ventilation openings should never be blocked in cold weather. Smoking, naked lights and other sources of ignition must be strictly forbidden, electrical equipment and lighting and heating facilities should be explosion proof and any hand tools employed should be of the non-sparking type. Containers, pipes, valves or fittings made of copper or copper alloys (brass, bronze) should not be used.

Cylinders of dissolved acetylene should be handled carefully to prevent shocks. If a gas cylinder shows signs of internal heating the valve must be closed if possible and the cylinder liberally sprinkled using a fire extinguisher. Before any repairs or adjustments are made to containers or piping which have held acetylene, they must be well purged (with nitrogen for example) and, if necessary, completely filled with water. Dissolved acetylene should not be withdrawn from the cylinder at an hourly rate greater than a value depending on the type of the cylinder and in particular its diameter. A too fast withdrawal may carry over solvent, leaving gaseous acetylene at high pressure or it may cause static electricity sparks.

In installations for the manufacture or use of acetylene, safety devices must be provided to reduce pressure and flashbacks and to prevent the formation of explosive mixtures. Personnel should be informed about the safety rules to be followed and not depart in any way from instructions regarding the use of the material.

Health precautions. No formal official permissible limits have been established for acetylene; however, the American Industrial Hygiene Association recommends that, since acetylene is relatively non-toxic, any such value should be based on its explosive characteristics and, consequently, the atmospheric concentration should be maintained below 0.5% or 5 000 ppm to minimise the explosion hazard. It is also recommended that to protect against the effects of phosphine, the most dangerous impurity in carbide-generated acetylene, the acceptable concentration may range from 500 ppm for acetylene containing 0.05% phosphine to 5 000 ppm for acetylene containing smaller amounts of phosphine.

Although acetylene is non-toxic, where conditions of use are such that the presence of impurities may constitute a health hazard, the composition of the acetylene should be determined. Respiratory protective equipment is not normally required; however, where plant failure may cause an immediate high acetylene concentration, respiratory protective equipment of the self-contained type should be available for emergency and rescue use.

Treatment. If symptoms develop after exposure to acetylene, the patient should be removed to uncontaminated air. If breathing has stopped, resuscitation should be applied. Particularly where acetylene from a generator is involved, acute phosphine poisoning should be suspected.

EVRARD, M.

General:

CIS 77-1628 "Acetylene". Data Sheet 494, Revision A (Extensive) (National Safety Council, 425 North Michigan Avenue, Chicago, 60611, 1977). *National Safety News* (Chicago), May 1977, 115/5 (79-83). Illus. 8 ref.

Occupational health:

CIS 77-452 *Criteria for a recommended standard–Occupational exposure to acetylene.* DHEW (NIOSH) publication No. 76-195 (National Institute for Occupational Safety and Health, 4676 Columbia Parkway, Cincinnati, Ohio) (July 1976), 84 p. 90 ref.

Safety:

CIS 77-1536 *Dissolved acetylene cylinders–Basic requirements.* International Standard ISO 3807-1977 (E) (International Organisation for Standardisation, Case postale 56, 1211 Genève 20, 15 May 1977), 7 p. Illus. 1 ref.

CIS 76-241 *Safety manual for personnel of plants producing acetylene and gases extracted from the atmosphere–Manuel de sécurité pour le personnel ouvrier travaillant dans les usines de production d'acétylène et des gaz extraits de l'air* (Permanent International Commission for Acetylene, Gas Welding and Allied Industries, 32 boulevard de la Chapelle, Paris, 2nd ed., 1973), 22 p. Illus.

Acid-base balance

Constancy of the "milieu intérieur", i.e. of the fluid surrounding the cells of an organism, is a basic requirement for normal function. This fundamental principle, first enunciated by the great French physiologist Claude Bernard, is the keystone of current concepts of physiology and pathophysiology; its importance is particularly evident in acid-base regulation.

Physiology of acid-base balance

The plasma concentration of protons or hydrogen ions, $[H^+]$, is exceedingly low and must be kept around 40 nanoequivalents/litre, which corresponds to a pH of 7.40. This set-point value is constantly threatened by the arrival in the extracellular fluid of acids (proton-donor substances) and of bases (proton-acceptor substances), which derive from constituents of the diet and from cellular metabolism. Two categories of end-products of cell metabolism represent a continuous challenge to acid-base equilibrium: CO_2 and non-volatile, "fixed" acids, mainly sulphuric and phosphoric acids. Regardless of the exact nature of the acid-base perturbation, the defence of normal blood pH is assured by three main elements: (i) the body buffers; (ii) the lungs; (iii) the kidneys.

Body buffers. Changes in $[H^+]$ of a given solution can be minimised by the presence of buffers, which are substances capable of converting strong, highly dissociated acids or bases into weaker, low-dissociated, counterparts. There is a variety of buffers in different compartments of the body fluids, among which are proteins, phosphates and bicarbonate. Of all body buffers, the system formed by bicarbonate (HCO_3^-) and carbonic acid (H_2CO_3) is by far the most important physiologically. Such a prevalence is explained by the fact that two organs control independently the components of this buffer pair: the kidneys maintain plasma $[HCO_3^-]$ at 24 millimoles/l, while the lungs keep plasma $[H_2CO_3]$ at 1.2 millimoles/l by controlling the partial pressure of CO_2 (PCO_2) at 5.3·10³ Pa or 40 mmHg. The ratio $[HCO_3^-] : [H_2CO_3]$ is the major determinant of blood pH and, in normal conditions, has a value of 20 : 1. In pathophysiological conditions $[HCO_3^-]$ and/or $[H_2CO_3]$ can be altered in different ways, but the kidney and the lungs adjust the parameter under their specific control, so that the ratio $[HCO_3^-] : [H_2CO_3]$ remains as close as possible to 20 : 1.

Lungs. The metabolic end-product, CO_2, is not an acid itself. However, by partially dissolving in body water, a fraction of CO_2 is hydrated and converted into H_2CO_3, a relatively strong acid. The amount of H_2CO_3 + dissolved CO_2 in body fluids depends on PCO_2, a parameter regulated by lung ventilation. The daily load of CO_2 resulting from metabolism is huge (15 to 20 moles) and would cause a lethal acidosis if it were not completely excreted by the lungs.

Kidneys. The non-volatile moiety of endogenous acid production is much smaller (60 to 100 mEq/day of H^+) but requires, to be excreted, the specialised transport systems of the kidney. In short, the contribution of the kidney to acid-base regulation can be summarised as follows: (1) reabsorption of filtered bicarbonate; (2) excretion of "fixed" acid; (3) replenishment of extracellular bicarbonate stores, consumed in the buffering of endogenous acid.

According to current concepts, a single mechanism accounts for this major regulatory function of the kidney, the main steps of which are: *(a)* the liberation of H^+ in the cytosol of the tubular cells, a process activated by the enzyme carbonic anhydrase; *(b)* the transport of H^+ across the apical cell membrane by means of a H^+ pump (primarily active transport) or of an antiport exchanging Na^+ for H^+ (secondarily active transport of H^+); *(c)* the buffering of H^+ transported into the tubular fluid by three main proton acceptors: bicarbonate, phosphate and ammonia (NH_3). Acceptance of H^+ by bicarbonate present in the tubular fluid allows for the virtually complete reabsorption of the 5 000 mEq of bicarbonate filtered each day. Acceptance of H^+ by monohydrogen phosphate and NH_3 leads to the formation of dihydrogen phosphate (titratable acidity) and NH_4^+, respectively. These two urinary constituents account for the bulk of the renal acid excretion and match the 60 to 100 mEq/day of endogenous acid production. By the same token, for every mEq of H^+ liberated inside the tubular cells and subsequently excreted as titrable acidity and NH_4^+, an equivalent amount of bicarbonate is formed; that amount crosses the baso-lateral membrane and returns to the extracellular fluid. In this way the kidney regenerates bicarbonate and replenishes the extracellular stores that are continuously titrated by the cellular release of metabolic "fixed" acid. To emphasise the importance of the renal mechanisms just described, it should be remembered at this point that the amount of free H^+ excreted by the kidney is negligible (less than 0.02 mEq/day). By virtue of the urinary buffers, 100 mEq of H^+ can be readily concealed in the normal daily output of 1-2 litres of urine at a pH between 5 and 6.

Pathophysiology of acid-base balance

The disturbances of acid-base balance can be conveniently analysed and systematised with reference to the fundamental equation of acid-base regulation

$$pH = constant + \log\frac{[HCO_3^-]}{[H_2CO_3]}$$

A state of acidosis is characterised by a positive balance of H^+, usually accompanied by a decrease in blood pH. The latter results from either a decrease in $[HCO_3^-]$ – *metabolic acidosis,* or an increase in $[H_2CO_3]$ – *respiratory acidosis.* Conversely, a state of alkalosis is characterised by a negative H^+ balance, usually accompanied by an increase in blood pH. The latter results from either an increase in $[HCO_3^-]$ – *metabolic alkalosis,* or a decrease in $[H_2CO_3]$ – *respiratory alkalosis.*

Metabolic acidosis. Several pathophysiological mechanisms can lead to the decrease in plasma bicarbonate: *(a)* loss of bicarbonate, via the kidney or the gastrointestinal tract (e.g. diarrhoea); *(b)* consumption of bicarbonate either by an abnormal acid load exceeding the maximal capacity of the kidney to excrete H^+, or by a normal acid load partially retained because of a defective renal acid excretion. Metabolic diseases such as diabetic ketoacidosis and lactic acidosis are frequent causes of severe acidosis. Less common, although extremely severe, acidotic states can occur in association with poisoning by methyl alcohol, paraldehyde, salicylate, ethylene glycol and ammonium chloride. Both in these metabolic and toxic situations, large amounts of strong acids (some of which remain chemically unidentified) are released into the extracellular fluid and are the direct cause of the acidosis. In contrast, in chronic renal failure of different aetiologies, there is no excessive generation of protons but the diseased kidney is unable to cope with the normal production of endogenous acid, thus creating a positive H^+ balance. In many instances the defect in acid excretion simply reflects an important loss of renal mass. In other situations, known under the general designation of renal tubular acidosis (RTA), more subtle functional lesions may hamper the H^+ transport machinery and cause a urinary leak of bicarbonate or a decrease in the excretion of titratable acidity and NH_4^+. Workers dealing with heavy metals (cadmium, mercury, lead) are exposed to kidney injuries which include the RTA syndrome. In metabolic acidosis, a partial compensation in the deviation of blood pH is achieved by hyperventilation, with the consequent decrease in PCO_2 and improvement of the bicarbonate : carbonic acid ratio.

Respiratory acidosis. In this type of acidosis the primary disturbance is a retention of CO_2, leading to increased values of PCO_2 and $[H_2CO_3]$ and a drop in blood pH. In acute situations, $[HCO_3^-]$ is only slightly elevated. In chronic situations the kidney minimises the deviation of blood pH by generating a surplus of bicarbonate that will tend to restore the normal $[HCO_3^-]$: $[H_2CO_3]$ ratio of 20 : 1. Acute respiratory acidosis is found in neuromuscular disorders (brain stem lesions, botulism, overdose of sedatives), airway obstruction, severe pneumonia and cardiac arrest. Chronic respiratory acidosis is present in several types of progressive lung disease. Acute and chronic forms of respiratory acidosis result sometimes from occupational hazards; such is the case in pneumoconioses due to inhalation of mineral dusts (asbestosis, berylliosis, coal worker's pneumoconiosis, silicosis) or organic dusts (byssinosis, bagassosis, suberosis and "farmer's lung").

Respiratory alkalosis. The cardinal feature of this disturbance is a primary hyperventilation causing a decrease in PCO_2 and an elevation of blood pH. Among the most common causes we shall mention the stimulation of the medullary respiratory centre (emotional stress, head injuries, fever, salicylate), hepatic failure and certain lung diseases.

Metabolic alkalosis. In this clinical setting blood chemistry typically shows an increase in both $[HCO_3^-]$ and pH. The genesis of the alkalotic state is usually the loss of H^+ by the gastro-intestinal tract (vomiting, gastric drainage) or by the kidney (diuretics, excess of mineralocorticoids, liquorice). The maintenance of the metabolic alkalosis is due to a complex interplay among chloride depletion, volume regulation and potassium homeostasis, leading to persistent renal acid excretion despite the negative H^+ balance. Provision of chloride (as NaCl or KCl) dramatically and quickly corrects most situations of metabolic alkalosis.

One of the first successful attempts in computer-aided diagnosis concerned acid-base disturbances. In view of the wide clinical spectra of diseases associated with each major acid-base syndrome, which include many environmental and toxic diseases, such computer programmes already available in the literature are of particular value in occupational medicine.

de SOUSA, R. C.

Body fluids and the acid-base balance. Christensen, H. N. (Philadelphia and London, W. B. Saunders Co., 1964), 506 p.

Renal and electrolyte disorders. Schrier, R. W. (ed.). (Boston, Little, Brown and Co., 1976), 500 p.

Acid-base and electrolyte balance. Schwartz, A. B.; Lyons, H. (eds.). (New York, San Francisco and London, Grune and Stratton, 1977), 320 p.

Clinical physiology of acid-base and electrolyte disorders. Rose, B. D. (New York, McGraw-Hill Book Co., 1977), 549 p.

Renal pathophysiology. Leaf, A.; Cotran, R. (New York, Oxford University Press, 1976), 387 p.

The kidney. Brenner, B. M.; Rector Jr., F. C. (eds.). (Philadelphia, London and Toronto, W. B. Saunders Co., 1975), 2 volumes, 1 948 p.

Acids and anhydrides, inorganic

An inorganic acid is a compound of hydrogen and one or more other element (with the exception of carbon) that dissociates or breaks down to produce hydrogen ions when dissolved in water or other solvents. The resultant solution has certain characteristics such as the ability to neutralise bases, turn litmus paper red and produce specific colour changes with certain other indicators. Inorganic acids are often termed mineral acids. The anhydrous form may be gaseous or solid.

An inorganic anhydride is an oxide of metalloid which can combine with water to form an inorganic acid. It can be produced by synthesis such as: $S + O_2 \rightarrow SO_2$, or by eliminating water from the corresponding acid, such as:

$$2HMnO_4 \xrightarrow[-H_2O]{} Mn_2O_7.$$

Inorganic anhydrides share in general the biological properties of their acids.

The inorganic acids of greatest industrial significance are each dealt with in a separate article, e.g. CHROMIUM AND COMPOUNDS (chromic acid); HYDROCHLORIC ACID; HYDROFLUORIC ACID; HYDROGEN SULPHIDE; NITRIC ACID; PHOSPHORUS AND COMPOUNDS (phosphoric acid); SULPHURIC ACID.

Some inorganic anhydrides are dealt with in separate articles, e.g. CHLORINE AND COMPOUNDS; NITROGEN OXIDES; SULPHUR COMPOUNDS.

HAZARDS

The specific hazards of the industrially important inorganic acids will be found in the separate articles indicated above; however, all these acids have certain dangerous properties in common.

Fire and explosion. Solutions of inorganic acids are not flammable in themselves; however, when they come into contact with certain other chemical substances or combustible materials, a fire or explosion may result. These acids react with certain metals with the liberation of hydrogen, which is a highly flammable and explosive substance when mixed with air or oxygen. They may also act as oxidising agents and, when in contact with organic or other oxidisable materials, may react destructively and violently.

Health hazards. The inorganic acids are corrosive, especially in high concentrations; they will destroy body tissue and cause chemical burns when in contact with the skin and mucous membranes. In particular, the danger of eye accidents is pronounced. Inorganic acid vapours or mists are respiratory tract and mucous membrane irritants although the degree of irritation depends to a large degree on the concentration; discoloration or erosion of the teeth may also occur in exposed workers. See MOUTH AND TEETH. Repeated skin contact may lead to dermatitis. Accidental ingestion of concentrated inorganic acids will result in severe irritation of the throat and stomach and destruction of the tissue of internal organs, perhaps with fatal outcome, when immediate remedial action is not taken. Certain inorganic acids may also act as systemic poisons.

SAFETY AND HEALTH MEASURES

Wherever possible, highly corrosive acids should be replaced by acids which present less hazard; it is essential to use only the minimum concentration necessary for the process. Wherever inorganic acids are used, appropriate measures should be instituted concerning storage, handling, waste disposal, ventilation, personal protection and first aid.

Storage. Inorganic acids should be stored in fire-resistant buildings having acid-resistant floors fitted with retaining sills and adequate drainage. They should be kept in a cool, well ventilated area away from contact with other acids and combustible or oxidisable materials. Electrical installations should also be of the acid-resistant type.

Glass or plastics containers should be adequately protected against impact; they should be kept off the floor to facilitate flushing in the event of leakage. Drums should be stored on cradles or racks and chocked in position. Gas cylinders of gaseous anhydrous acid should be stored upright with the cap in place. All containers should be clearly labelled, and empty and full containers should preferably be stored apart. Maintenance and good housekeeping is essential.

Handling. Wherever possible acids should be pumped through sealed systems to prevent all danger of contact. Wherever individual containers have to be transported or decanted, the appropriate equipment should be employed and only experienced persons allowed to undertake the work. Large containers should be transported on hand trucks or power trucks; small containers, such as bottles, should be carried in safety bottle carriers which provide protection against impact and retain acid in the event of damage. Decanting should be done by means of special syphons, transfer pumps, or drum or carboy tilting cradles, etc. Cylinders of anhydrous acid gas require special discharge valves and connections.

Where acids are mixed with other chemicals or water, workers must be fully aware of any violent or dangerous reaction that may take place, e.g. when water is poured into nitric acid, rather than vice versa; the necessary precautions should be taken and the workers should wear suitable personal protective clothing.

Ventilation. Where processes produce acid mists or vapours, such as in electroplating, exhaust ventilation should be installed.

Personal protection. Persons exposed to dangerous splashes or inorganic acids should be required to wear acid-resistant personal protective equipment including hand and arm protection, eye and face protection and aprons, overalls or coats. Provided safe working procedures are adopted, the use of respiratory protective equipment should not be necessary; however, it should be available for emergency use in the event of leakage or spillage.

When tanks that have contained inorganic acids require to be entered for maintenance or repair purposes, they should first be purged; persons entering should wear full protective equipment and a safety belt and lifeline, and a person should always be stationed outside the tank for rescue purposes.

Personal hygiene is of utmost importance where there is contact with inorganic acids. Adequate washing and sanitary facilities should be provided and workers encouraged to wash thoroughly before meals and at end of shifts.

First aid. Essential treatment for inorganic acid contamination of skin or eyes is immediate and copious flushing with running water. Emergency showers and eye-wash fountains, baths or bottles should be strategically located. For more detailed consideration of treatment, see under the individual acids.

KARPOV, B. D.

CIS 76-1916 *Caution–Inorganic metal cleaners can be dangerous.* DHEW (NIOSH) publication No. 76-110 (National Institute for Occupational Safety and Health, 4676 Columbia Parkway, Cincinnati, Ohio) (1975), 21 p. Illus.

CIS 75-1700 "Occupational dental disease due to acids" (Berufsbedingte Säureschäden der Zähne). Reinhardt J.; Kittner, E. *Zentralblatt für Arbeitsmedizin und Arbeitsschutz* (Heidelberg), Mar. 1975, 25/3 (72-75). Illus. (In German)

CIS 78-15 "A case history report: An investigation into the safe handling of acid carboys and drums". Hodnick, H. V. *Protection* (London), Mar. 1977, 14/3 (6-10). Illus.

CIS 79-476 "Determination of acid vapours and mists at the workplace with the use of miniature absorbers" (Die Bestimmung von Säuredämpfen und Säurenebeln am Arbeitsplatz unter Verwendung von Kleinabsorbern). Schaffernicht, H. *Zeitschrift für die gesamte Hygiene und ihre Grenzgebiete* (Berlin), June 1978, 24/6 (425-428). Illus. (In German)

Acids and anhydrides, organic

Acids

Organic acids and their derivatives cover a wide range of substances. They are used in nearly every type of chemical manufacture. Because of the variety in the chemical structure of the members of the organic acid group, several types of toxic effects may occur. These compounds have a primary irritant effect, the degree determined in part by acid dissociation and water solubility. Some may cause severe tissue damage similar to that seen with strong mineral acids. Sensitisation may also occur but is more common with the anhydrides than the acids.

For the purpose of this article, organic acids may be divided into saturated monocarboxylic and unsaturated monocarboxylic acids, aliphatic dicarboxylic acids, halogenated acetic acids, miscellaneous aliphatic monocarboxylic acids and aromatic carboxylic acids.

Saturated monocarboxylic acids

Formic acid (see below) ($HCOOH$)
Acetic acid (CH_3COOH)
Propionic acid (see below) (CH_3CH_2COOH)
Butyric acid ($CH_3CH_2CH_2COOH$)
Valeric acid ($CH_3(CH_2)_3COOH$)
Caproic acid ($CH_3(CH_2)_4COOH$)
Heptanoic acid ($CH_3(CH_2)_5COOH$)

These low molecular weight acids are primary irritants and produce severe damage to tissues. Strict precautions are necessary in handling, suitable protective equipment should be available and any skin or eye splashes irrigated with copious amounts of water. The most important acids of this group are acetic acid (dealt with in a separate article) and formic acid.

Caprylic acid ($CH_3(CH_2)_6COOH$)
Pelargonic acid ($CH_3(CH_2)_7COOH$)
Undecylic acid ($CH_3(CH_2)_9COOH$)

These acids are used as fungicides, flavouring agents and in the preparation of plasticisers. They produce relatively mild irritation to the skin and mucous membranes. No hazard is likely in industrial use.

Lauric acid ($CH_3(CH_2)_{10}COOH$)
Myristic acid ($CH_3(CH_2)_{12}COOH$)
Palmitic acid ($CH_3(CH_2)_{14}COOH$)
Stearic acid ($CH_3(CH_2)_{16}COOH$)
Arachic acid ($CH_3(CH_2)_{18}COOH$)
Behenic acid ($CH_3(CH_2)_{20}COOH$)

The long chain saturated monocarboxylic acids are the fatty acids and are in the main derived from natural sources. Synthetic fatty acids may also be manufactured by air oxidation of paraffins (aliphatic hydrocarbons) using metal catalysts. They are also produced by the oxidation of alcohols with caustic soda. They are non-irritant and have a wide application in soaps, detergents, lubricants, protective coatings and intermediate chemicals. They are of a very low order of toxicity and no problems are likely in industrial use.

Formic acid ($HCOOH$)

METHANOIC ACID

m.w.	45
sp.gr.	1.22
m.p.	8.4 °C
b.p.	100.7 °C
v.d.	1.59
v.p.	40 mmHg ($5.32 \cdot 10^3$ Pa) at 24 °C
f.p.	68.9 °C (90% solution)
e.l.	18-57% (90% solution)
i.t.	434 °C (90% solution)

soluble in all proportions in water, ethanol and ether

a colourless fuming liquid with a pungent penetrating odour.

TWA OSHA	5 ppm 9 mg/m³
IDLH	100 ppm
MAC USSR	1 mg/m³ skin

The sodium formate route is the manufacturing process currently in use.

$$CO + NaOH \xrightarrow[\text{pressure}]{125\text{-}150\,°C} HCOONa \text{ (sodium formate)}$$

sodium formate → (sulphuric acid in presence of 85-95% formic acid) → HCOOH (formic acid 85-90%)

Mainly used in the textile and leather industry. It acts as a dye-exhausting agent for a number of natural and synthetic fibres, and in chrome-dyeing it serves as a reducing agent.

Formic acid is used as a deliming agent and neutraliser in the leather industry, and as a coagulant for rubber latex. It is also used as an intermediate for production, and as a component of nickel plating baths.

The principal hazard is that of severe primary damage to the skin, eye or mucosal surface. Sensitisation is rare, but may occur in a person previously sensitised to formaldehyde. Accidental injury in man is the same as for other relatively strong acids. No delayed or chronic effects have been noted. Formic acid is a flammable liquid and its vapour forms flammable and explosive mixtures with air.

Adequate ventilation should be provided and suitable personal protective equipment should be worn to give protection against splashes and acid burns; respiratory protective equipment should be available. Open flames and other sources of ignition should not be allowed in the immediate vicinity of the acid, particularly when it is at a temperature above 69 °C.

Propionic acid (CH_3CH_2COOH)

METHYLACETIC ACID; PROPENOIC ACID

m.w. 74.1
sp.gr. 0.99
m.p. −22 °C
b.p. 141 °C
v.d. 2.56
v.p. 10 mmHg (1.33·10³ Pa) at 37 °C
f.p. 54 °C
e.l. 2.1-12%
i.t. 485 °C

soluble in water in all proportions

an oily liquid with a pungent rancid odour.

TLV ACGIH 10 ppm 30 mg/m³
STEL ACGIH 15 ppm 45 mg/m³
MAC USSR 2 mg/m³

Propionic acid is produced by different processes including the catalytic synthesis of ethanol and carbon monoxide or the wood pyrolysis. It is used in organic synthesis, as mould inhibitor and a food preservative.

In solution it has corrosive properties towards several metals. It is irritant to eye, respiratory system and skin. The same precautions recommended for exposure to formic acid are applicable, taking into account the lower flash point of propionic acid.

Unsaturated monocarboxylic acids

Acrylic acid (CH_2:CHCOOH)
Methacrylic acid (CH_2:C(CH_3)COOH)
Crotonic acid (CH_3CH:CHCOOH)
(See ACRYLIC ACID AND DERIVATES)

These acids have a wide application in the manufacture of resins, plasticisers and drugs. They are highly reactive substances and are recognised as severe irritants of the skin, eye and respiratory tract in concentrated solution. No cumulative toxic reactions are known.

Aliphatic dicarboxylic acids

Oxalic acid (HOOCCOOH.2H_2O)
Malonic acid (HOOCCH_2COOH)
Succinic acid (HOOCCH_2CH_2COOH)
Malic acid (HOOCCH(OH)CH_2COOH)
Thiomalic acid (HOOCCH(SH)CH_2COOH)
Tartaric acid (HOOC(CHOH)$_2$COOH.H_2O)
Adipic acid (HOOC(CH_2)$_4$COOH) (see below)
Pimelic acid (HOOC(CH_2)$_5$COOH)
Azelaic acid (HOOC(CH_2)$_7$COOH)
Sebacic acid (HOOC(CH_2)$_8$COOH)
Citric acid (HOOCCH_2COHCOOHCH_2COOH)
Maleic acid (HOOCCH:CHCOOH) (see below)
Fumaric acid (HOOCCH:CHCOOH) (see below)
Itaconic acid (HOOCC(:CH_2)CH_2COOH)

The dicarboxylic acids are of importance because of their use in food, beverages, drugs and a range of manufacturing processes. The majority present no hazard from low level chronic exposure and are normally present in human metabolic processes. Primary irritant effects are present with a number of these acids, particularly in concentrated solutions or as dusts. Sensitisation is rare. As the materials are all solids at room temperature, contact is usually in the form of dust or crystals. Citric acid is a tricarboxylic acid, but has similar properties to the others in this group. OXALIC ACID is dealt with in a separate article.

Maleic acid (HOOCCH:CHCOOH) *(cis)*

cis-BUTENEDIOIC

m.w. 115
sp.gr. 1.59
m.p. 137-138 °C
b.p. decomposes
v.d. 4

colourless crystals possessing a characteristic astringent taste and a faint odour.

Maleic acid is a by-product of phthalic anhydride manufacture and may also be made by the oxidation of benzene. It is used in the manufacture of synthetic resins; its salts are used in the dyeing of cotton, wool and silk; it is employed in the organic syntheses of malic, succinic, aspartic, tartaric, propionic, lactic, malonic, acrylic and hydroacrylic acids; it is a preservative for oils and fats.

Maleic acid is a strong acid and produces marked irritation of the skin and mucous membranes. Severe effects, particularly in the eye, can result from concentrations as low as 5%. There are no reports of cumulative toxic effects in man. The hazard in industry is of primary irritation of exposed surfaces, and this should be averted where necessary by the provision of appropriate personal protective equipment, generally in the form of impermeable gloves or gauntlets.

Fumaric acid (HOOCCH:CHCOOH) *(trans)*

Trans-BUTENEDIOIC

m.w. 116
sp.gr. 1.63

sublimes at 200 °C

colourless, odourless crystals with a fruit-acid taste.

Fumaric acid may be produced from the fermentation of molasses or by the oxidation of benzene. It may also be obtained through the isomerisation of maleic acid. The United States capacity in 1976 was 50 000 t. It is used in polyesters and alkyd resins, plastics surface coatings, food acidulants, inks and organic syntheses.

Fumaric acid is a relatively weak acid and has a low solubility in water. It is a normal metabolite and is less toxic orally than tartaric acid. Humans tolerate 500 mg per day for a year without ill effect. It is a mild irritant of skin and mucous membranes and no problems of industrial handling are known.

Adipic acid (HOOC(CH_2)$_4$COOH)

HEXANEDIOIC ACID; 1,4-BUTANEDICARBOXYLIC ACID

m.w. 146
sp.gr. 1.36
m.p. 153 °C
b.p. 337 °C
v.d. 5.04
v.p. 1 mmHg (0.13·10³ Pa) at 159.5 °C
f.p. 210 °C (oc)
i.t. 420 °C

white crystalline solid soluble in water particularly at higher temperatures.

Manufacture. Adipic acid can be produced by either nitric acid or air oxidation of cyclohexane, cyclohexanol or cyclohexanone. Most production uses a two-stage process from cyclohexane, viz.:

(the mixture below is usually known as KA)

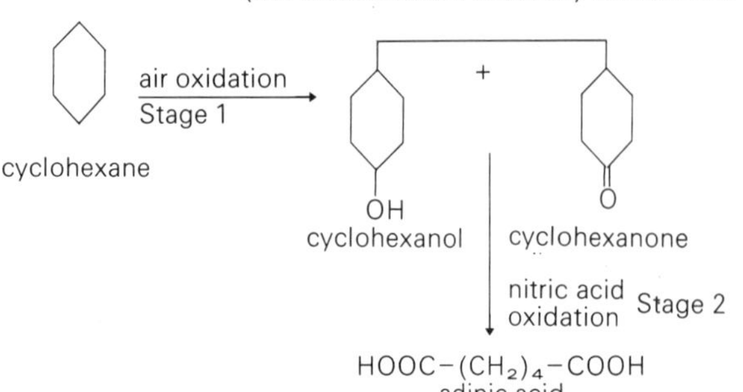

Stage 1 uses either a cobalt catalyst at moderate temperatures and pressures or a boron catalyst. Alternatively the cyclohexane oxidation can be carried out in two steps. The initial reaction is non-catalytic and converts the cyclohexane into cyclohexanehydroperoxide. This is followed by the decomposition of the peroxide in the presence of a metal such as cobalt, chromium, vanadium, etc., into cyclohexanol. In Stage 2 the KA mixture is oxidised by nitric acid, in the presence of vanadate-copper catalyst, to adipic acid. The yield is reported to be of the order of 90-95%.

Alternatively cyclohexanol is oxidised to adipic acid by nitric acid under the same conditions as previously described.

Production and uses. Total world adipic acid capacity is about 1 775 500 tonnes a year. In the United States 98% of adipic acid is produced from cyclohexane feedstock.

Some 90% of manufacture is used for nylon production and smaller quantities in plasticisers and synthetic lubricants, polyurethanes and food acidulants.

Adipic acid is non-irritant and of very low toxicity when ingested.

Halogenated acetic acids

Chloroacetic acid (see below) ($ClCH_2COOH$)
Dichloroacetic acid ($Cl_2CHCOOH$)
Trichloroacetic acid (Cl_3CCOOH)
Bromoacetic acid ($BrCH_2COOH$)
Iodoacetic acid (ICH_2COOH)
Fluoroacetic acid (FCH_2COOH)
Trifluoroacetic acid (F_3CCOOH)

These acids are highly reactive chemically and are widely used as chemical intermediates and in the manufacture of pharmaceuticals and herbicides.

They cause severe damage to the skin and mucous membranes and, when ingested, may interfere with essential enzyme systems in the body. Strict precautions are necessary for their handling. They should be prepared and used in enclosed plant, the openings in which should be limited to the necessities of manipulation. Exhaust ventilation should be applied to the enclosure to ensure that fume or dust does not escape through the limited openings. Personal protective equipment should be worn by persons engaged in the operations and eye protective equipment and respiratory protective equipment should be available for use when necessary. FLUOROACETIC ACID is dealt with in a separate article.

Chloroacetic acid ($ClCH_2COOH$)

CHLOROETHANOIC ACID; MONOCHLOROACETIC ACID

m.w.	94.5
sp.gr.	1.40
m.p.	63 °C (α) 56.2 °C (β) 52.5 °C (γ)
b.p.	187.8 °C
v.d.	3.26
v.p.	0.065 mmHg (8.66 Pa) at 25 °C
f.p.	126.2 °C

colourless crystalline solid, readily soluble in water, acetone, benzene and carbon tetrachloride.

MAC USSR 1 mg/m³ skin

Manufacture. The two main processes used *(a)* mainly in the United States and Canada and *(b)* mainly in Europe, are as follows:

(a) $2CH_3COOH + 3Cl_2 \xrightarrow[\text{or red phosphorus}]{\text{sulphur}} ClCH_2COOH + Cl_2CHCOOH + 3HCl$

The mixture can be selectively dechlorinated with hydrogen and a heavy metal catalyst, e.g. Pb, to give 99+% of the mono acid.

(b) The hydrolysis of trichloroethylene with sulphuric acid; this process yields high purity chloroacetic acid.

Production and uses. United States capacity in 1977 was 50 000 tonnes a year.

Chloroacetic acid is a highly reactive chemical and is used industrially as a chemical intermediate. It is reacted with ammonia to form glycine, and with aniline it forms a precursor for indigo dyes.

The acid and its derivatives are also used in the production of chemicals such as barbiturates, herbicides (2,4-D and 2,4,5-T), carboxymethyl cellulose, thioglycolic acid and vitamin A.

This material is a highly reactive chemical and should be handled with care. Gloves, goggles, rubber boots and impervious overalls are mandatory when workers are in contact with concentrated solutions.

Miscellaneous aliphatic monocarboxylic acids

Glycolic acid ($HOCH_2COOH$)

This acid is used in the leather, textile, electroplating, adhesives and metal-cleaning industries. It is stronger than acetic acid and produces very severe chemical burns of the skin and eyes. No cumulative effects are known, and it is believed to be metabolised by glycine. Strict precautions are necessary for its handling. These are similar to those required for acetic acid.

Aminoacetic acid (NH_2CH_2COOH)

Aminoacetic acid is used as a buffering agent and in syntheses. It is a normal constituent of protein. No precautions are necessary in handling it.

Sulphoacetic acid ($HOSO_2CH_2COOH$)

It is a primary irritant producing severe skin and eye injury in concentrated form. Prevention of skin and eye contact with the solid or with concentrated solutions is necessary in industrial handling. Personal protective equipment should be worn by persons handling the material and eye protective equipment should be available.

Peracetic acid (CH_3COOOH)

This substance is used as a bleach, catalyst and oxidant. It is a strong skin and eye irritant, and personal protective equipment, including, when necessary, eye protection equipment, should be worn by persons handling the material. See PEROXIDES, ORGANIC.

In high concentrations it may decompose with explosive violence. It is an oxidising material and will support the combustion of organic materials, thus giving rise to a severe fire hazard. Whenever possible, it should be used in dilute solutions. It should not be stored in the vicinity of combustible materials.

Lactic acid ($CH_3CHOHCOOH$)

This substance is used as a food acidulant and also in adhesives, plastics and textiles. Concentrated solutions can cause burns of the skin and eye. No cumulative effects are known. Personal protective equipment should be worn by persons handling concentrated solutions of this acid.

Sorbic acid ($CH_3(CH)_4COOH$)

Sorbic acid is used as a fungicide in foods. It is a primary irritant of the skin, and individuals may develop sensitivities to it. For these reasons contact with the skin should be avoided.

Thioglycolic acid

See separate article.

Mercaptopropionic acid ($HSCH_2CH_2COOH$)

This substance is used in cold-wave preparations and syntheses. It is toxic orally and a skin irritant. For persons who handle it regularly, adequate ventilation should be provided and, in certain circumstances, gloves should be worn to prevent contact with the skin.

Thioacetic acid (CH_3COSH)

Its vapours are irritating to the eyes, nose and throat. Skin contact and inhalation of vapour should be avoided in the manner indicated for mercaptopropionic acid.

Aromatic acids

The aromatic carboxy and sulphonic acids comprise one of the largest and most important groups of industrial chemicals. They are extensively used in the synthesis of dyes, elastomers, medicinals, pesticides and plastics.

The majority of these substances are of a low order of toxicity and present little hazard in industrial processes. They are rapidly excreted in the urine either unchanged or conjugated with glycine or glycuronic acid. Their primary irritant effects vary and, although they are often crystalline solids of low water solubility, they may be used industrially in such a manner that the operatives are exposed to heat and vapour and there is, consequently, irritation of the respiratory tract, eyes and skin. Adequate ventilation should therefore be provided and appropriate personal protective equipment should be worn.

Benzenesulphonic acid ($C_6H_5SO_3H.\ 1\frac{1}{2}\ H_2O$)

m.w. 158.2

m.p. (anhydrous acid) 65.6 °C

colourless deliquescent solid very soluble in water and ethanol, sparingly soluble in benzene.

Benzenesulphonic acid is commercially produced by sulphonating benzene with sulphuric acid, oleum or sulphur trioxide.

It is mainly used as an intermediate for the synthesis of phenol, and in the dyestuffs industry, in the production of various resins and in the manufacture of resorcinol.

In solutions it forms an extremely corrosive liquid. Dangerous when heated to decomposition or on contact with acid or acid fumes, when the highly toxic fumes of SO_2 would be emitted.

Gloves, goggles, rubber boots and impervious overalls are required when handling. Self-contained breathing sets should be available where fume may be evolved.

Naphthenic acids (C_5H_9COOH, $C_6H_{10}COOH$, etc.)

Commercial naphthenic acid is usually a dark-coloured malodorous mixture of naphthenic acids.

Naphthenic acids are derived from cycloparaffins in petroleum, probably by oxidation. Commercial acids are usually viscous liquid mixtures and may be separated as low and high boiling fractions. The molecular weights vary from 180 to 350. They are used principally in the preparation of paint dryers, where the metallic salts, such as lead, cobalt and manganese, act as oxidising agents. Metallic naphthenic acids are used as catalysts in chemical processes. An industrial advantage is their solubility in oil. The United States capacity in 1976 was 30 000 tonnes.

Salicylic acid ($C_6H_4(OH)\ (COOH)$)

O-HYDROXYBENZOIC ACID

sp.gr.	1.44
m.p.	159 °C
b.p.	211 °C
v.d.	4.8
v.p.	1 mmHg ($0.13{\cdot}10^3$ Pa) at 113.7 °C
f.p.	156.1 °C
i.t.	545 °C

very soluble in ethanol and ether

white needle crystals or powder.

Salicylic acid is produced by the reaction of carbon dioxide on phenol, or by the treatment of a hot solution of sodium phenolate with carbon dioxide and acidification of the sodium salt thus formed. Up to 60% of all salicylic acid produced in the United States is used to produce acetylsalicylic acid (aspirin). Other medicinal products (e.g. salicylamide) consume a further 25%, and the remainder is used by various industries such as rubber and dyestuffs. It has a limited use in medicine in dermatological ointments and corn applications.

Salicylic acid is a strong irritant when in contact with skin or mucous membranes. Strict precautions are necessary for plant operatives.

Anhydrides

Anhydride is defined as an oxide which, when combined with water, gives an acid or a base. Acid anhydrides are derived from the removal of water from two molecules of the corresponding acid:

$$\begin{array}{lcl} R-COOH & & R-C=O \\ + & = & \quad\;\; >O + H_2O \\ R-COOH & & R-C=O \end{array}$$

Acid anhydrides have higher boiling points than the corresponding acids. Their physiological effects generally resemble those of the corresponding acids, but they are more potent eye irritants in the vapour phase, and may produce chronic conjunctivitis. They are slowly hydrolysed on contact with body tissues and may occasionally cause sensitisation. Adequate ventilation should be provided and suitable personal protective equipment should be worn. In certain circumstances, particularly those associated with maintenance work, suitable eye protection equipment and respiratory protective equipment are necessary. Industrially, the most important anhydrides are acetic and phthalic. In recent years the production of maleic anhydride has increased considerably. PHTHALIC ANHYDRIDE is dealt with in a separate article.

Acetic anhydride ($(CH_3CO)_2O$)

ACETIC ACID ANHYDRIDE; ETHANOIC ANHYDRIDE; ACETYLE OXIDE

m.w.	102
sp.gr.	1.08
m.p.	−73.1 °C
b.p.	139.6 °C
v.d.	3.52
v.p.	5 mmHg ($0.65{\cdot}10^3$ Pa) at 25 °C
f.p.	49 °C

i.t. 390 °C
e.l. 2.9-10.3%

soluble in all proportions in ether; soluble in water forming acetic acid; soluble in ethyl alcohol

a colourless, flammable, strongly refractive liquid, with a pungent acetic odour.

TWA OSHA 5 ppm 20 mg/m³
TLV ACGIH 5 ppm 20 mg/m³ ceil
IDLH 1 000 ppm

Acetic anhydride is mainly produced from acetylene and acetic acid in the presence of mercuric oxide or by oxidation of acetaldehyde. It is the aliphatic anhydride most largely used in industry, in the manufacture of cellulose esters, in the production of aspirin, in the production of dyes and perfumes, as a dehydrating agent, and as an analytical reagent.

When exposed to heat acetic anhydride can emit toxic fumes and its vapours can explode in the presence of flame. It can react violently with strong acids and oxidisers such as sulphuric acid, nitric acid, hydrochloric acid, permanganates, chromium trioxide, hydrogen peroxide, as well as with soda.

Acetic anhydride is a strong irritant and has corrosive properties on contact with eyes, usually with delayed action; contact is followed by lacrimation, photophobia, conjunctivitis and corneal oedema. Inhalation can cause nasopharyngeal and upper respiratory tract irritation, with burning sensations, cough and dyspnoea; prolonged exposure may lead to pulmonary oedema. Ingestion causes pain, nausea and vomiting. Dermatitis can result from prolonged skin exposure.

When contacts are possible, protective clothing and goggles are recommended and eyewash and shower facilities should be available. Chemical cartridge respirators are appropriate for protection against concentrations up to 250 ppm, supplied air respirators with a full eyepiece are recommended for concentrations of 1 000 ppm; self-contained breathing apparatus is necessary in case of fire.

Propionic anhydride $(CH_3CH_2CO)_2O$

PROPANOIC ANHYDRIDE; METHYLACETIC ANHYDRIDE

m.w. 130.1
sp.gr. 1.01
m.p. −45 °C
b.p. 168.1 °C
v.d. 4.49
v.p. 1 mmHg (0.13·10³ Pa) at 20 °C
f.p. 74 °C (oc)

soluble in all proportions in ether; decomposes in water

a colourless liquid with a rancid odour.

Propionic anhydride is produced either by dehydration of propionic acid or by oxidation of propionaldehyde in the presence of cobalt and copper catalysts. It is used in the manufacture of perfumes, alkyd resins, drugs and dyes.

Butyric anhydride $(CH_3CH_2CH_2CO)_2O$

BUTANOIC ACID ANHYDRIDE

m.w. 158.2
sp.gr. 0.97
m.p. −75
b.p. 199.4 °C
v.d. 5.4
f.p. 87.8 °C

soluble in ether, it decomposes in water

a liquid with a pungent, rancid odour.

MAC USSR 1 mg/m³

Butyric anhydride is manufactured by catalytic hydrogenation of crotonic acid.

Butyric anhydride and propionic anhydride present hazards similar to those of the acetic anhydride.

Maleic anhydride $(C_4H_2O_3)$

cis-BUTENEDIOIC ANHYDRIDE; 2,5-FURANEDIONE

m.w. 98
sp.gr. 1.31 (60 °C)
m.p. 56 °C
b.p. 197-199 °C
v.d. 3.4
v.p. 1 mmHg (0.13·10³ Pa) at 44 °C
f.p. 101.7 °C
e.l. 3.4-7.1%
i.t. 421 °C

soluble in water, ether and acetone

white needles or powder with a penetrating odour.

TWA OSHA 0.25 ppm 1 mg/m³
MAC USSR 1 mg/m³

Maleic anhydride is manufactured by vapour-phase oxidation of benzene over vanadium pentoxide catalyst, and by catalytic vapour-phase oxidation of butylenes. Minor quantities of maleic anhydride are also produced as a by-product of phthalic anhydride production from naphthalene. United States capacity in 1978 was 180 000 tonnes. It is used in the manufacture of polyester resins, alkyd resins, agricultural chemicals and fumaric acid. It is a copolymer in a wide range of chemical syntheses (see ALKYD RESINS).

Maleic anhydride can produce severe eye and skin burns. These may be produced either by solution of maleic anhydride or by flakes of the material in the manufacturing process coming into contact with a moist skin. Skin sensitisation has occurred. Strict precautions should be taken to prevent contact of the solution with skin or eyes. Suitable goggles and other protective clothing must be worn by plant operatives; ready access to eye irrigation solution bottles is essential.

When suspended in air in a finely divided condition, maleic anhydride is capable of forming explosive mixtures with the air. Condensers in which the sublimed material settles in the form of fine crystals should be situated in a safe position outside an occupied room.

Trimellitic anhydride $(C_9H_4O_5)$

1,2,4-BENZENETRICARBOXYLIC ACID ANHYDRIDE; 4-CARBOXYPHTHALIC ANHYDRIDE; 1,3-DIHYDRO-1,3-DIOXO-5-ISOBENZOFURANCARBOXYLIC ACID; TMAN

m.w. 192
m.p. 168 °C
b.p. 240 °C

white crystalline solid which reacts with water and ethanol and is soluble in acetone and in dimethylformamide.

TLV ACGIH 0.005 ppm 0.04 mg/m³

It is produced by liquid phase air oxidation of pseudocumene to trimellitic acid and subsequent dehydration of the acid to the anhydride.

The major use of the material is for trimellitate plasticisers, the United States production of which was 11 000 tonnes in 1977 when about 4000 to 4 500 tonnes of trimellitic anhydride were used for the manufacture of water-based alkyd surface coatings.

The anhydride is also reacted with hydroquinone and aromatic diamines to produce poly(ester-imide) resins and poly(amide-imide) resins respectively.

Trimellitic anhydride has been reported to have caused pulmonary oedema in workers after severe acute exposure, and airways sensitisation after exposure periods of weeks to years with rhinitis and/or asthma.

COOKE, W. G.

General:

CIS 75-1052 *Dangerous properties of industrial materials.* Sax, N. I. (New York, London, Toronto and Melbourne, Van Nostrand Reinhold Company, 5th ed., 1979), 1 118 p. Illus. 392 ref.

Handling chemicals safely 1980 (Amsterdam, Het Veiligheidsinstituut, 2nd ed., 1980), 1 013 p.

Further information can be found in data sheets currently published by a number of institutions such as:

American Industrial Hygiene Association, 475 Wolf Ledges Parkway, Akron, OH 44311;

National Safety Council, 444 North Michigan Avenue, Chicago, IL 60611;

National Fire Protection Association, 470 Atlantic Avenue, Boston, MASS 02210;

Manufacturing Chemists Association, 1825 Connecticut Avenue NW, Washington, DC 20009;

Institut national de recherche et de sécurité, 30 rue Olivier Noyer, 75680 Paris Cedex 14;

to which should be added the Guidance Notes of the Health and Safety Executive, Baynards House, 1 Chepstow Place, London W2 4TF, and the NIOSH/OSHA Current Intelligence Bulletin, National Institute for Occupational Safety and Health, Rockville, MD 20857.

Acridine and derivatives

Acridine ($C_{13}H_9N$)

DIBENZOPYRIDINE; 10-AZAANTHRACENE

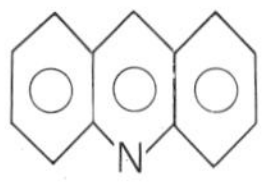

m.w.	179.2
sp.gr.	1.00
m.p.	111 °C
b.p.	345 °C (sublimes)
v.p.	1 mmHg (0.13·10³ Pa) at 129 °C

small, colourless or faintly yellow crystals.

Production. Acridine is present in coal tar and can be extracted from high boiling tar oils. It can also be obtained from benzylaniline.

Uses. Acridine and aminoacridines are used as raw materials or as intermediates, in a substitute form, in the synthesis of various dyestuffs and drugs. The orange-red dye is a mixture of two derivatives (acriflavine and proflavine); it is used as an antiseptic. Quinacrine is an antimalarial commercially known as atabrine, mepacrine, palusan, etc. The 6,9-diamino-2-ethoxyacridine is used as an antiseptic.

A wide range of colours is provided by the acridine vat dyes. The benzoflavine and chrysaniline vat dyes are derived from phenyl acridine.

HAZARDS

Acridine is a powerful irritant which, in contact with the skin or mucous membrane, causes itching, burning, sneezing, lacrimation and irritation of the conjunctiva.

Workers exposed to acridine crystal dust in concentrations of 0.02-0.6 mg/m³ complained of headache, disturbed sleep, irritability and photosensitisation, and presented oedema of the eyelids, conjunctivitis, skin rashes, leucocytosis and increased red cells sedimentation rates. These symptoms did not appear at an acridine airborne concentration of 1.01 mg/m³. When heated, acridine emits toxic fumes. Acridine, and a large number of its derivatives have been shown to possess mutagenic properties and to inhibit DNA repair and cell growth in several species.

SAFETY AND HEALTH MEASURES

The airborne concentration of acridine (dust or vapour) should be kept as low as possible by appropriate technical means and should not exceed 0.01 mg/m³. In addition precautions should be directed to preventing this substance from coming in contact with the skin and eyes. When necessary, they should include the wearing of personal protective equipment, safety glasses, and the application of a lanoline ointment on the skin. Workers should wash thoroughly at the end of the working period and working clothes should be changed and cleaned regularly. Fire-fighting equipment and respiratory protective equipment should be readily available to prevent exposure to toxic fumes in cases of fire. The facilities should include emergency showers and eye-washing bottles for decontamination and first aid.

Benz(c)acridine ($C_{17}H_{11}N$),

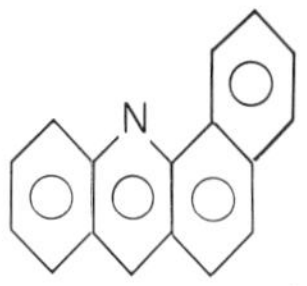

Dibenz(a,h)acridine ($C_{21}H_{13}N$), and

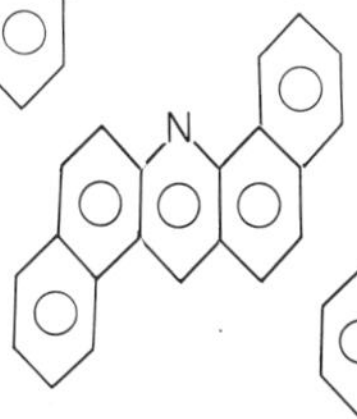

Dibenz(a,j)acridine ($C_{21}H_{13}N$)

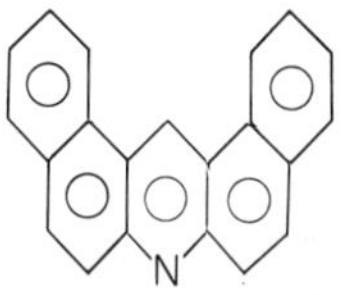

occur in effluents of various air pollution sources (petroleum refinery incinerators, domestic coal combustion, coal tar pitch volatiles from coke plants), motor exhaust gases, urban airborne particle samples. Dibenzacridines are also found in cigarette smoke and in products of pyrolysis of pyridine and nicotine.

They all induce skin tumors in mice following topical application and some of them can induce sarcomas and bladder papillomas in rats following implantation of paraffin wax pellets. No case reports or epidemiological studies on humans are available; however, it is possible that acridines contribute to the over-all environmental carcinogenic risk.

PARMEGGIANI, L.

Acridine:

CIS 1045-1971 "Data for the toxicological evaluation of acridine" (Materialy k toksikologiceskoj ocenke akridina). Kapitul'skij, V. B.; Kogan, F. M.; Dorinovskaja, A. P.; Pačašev, E. N. *Gigiena truda i professional'nye zabolevanija* (Moscow), Sep. 1970, 14/9 (56-57). 2 ref.

Acridine derivatives:

IARC monographs on the evaluation of carcinogenic risk of chemicals to man. Volumes 3, 13, 16 (Lyons, International Agency for Research on Cancer, 1973, 1977, 1978).

"Genetic effects of acridine compounds". Nasim, A.; Brychcy, T. *Mutation Research* (Amsterdam), 1979, 65/4 (261-268).

Acrolein (CH_2:CHCHO)

ACRYLIC ALDEHYDE; ACRALDEHYDE; 2-PROPENAL

m.w.	56
sp.gr.	0.86
m.p.	−87.7 °C
b.p.	52.5 °C
v.d.	1.94
v.p.	100 mmHg (13.3·10³ Pa) at 22.8 °C
f.p.	< 17.8 °C
e.l.	2.8-31%
i.t.	234 °C

soluble in water; highly soluble in organic solvents

a clear, yellowish liquid with a pungent odour.

TWA OSHA	0.1 ppm 0.25 mg/m³
STEL ACGIH	0.3 ppm 0.8 mg/m³
IDLH	5 ppm
MAC USSR	0.2 mg/m³

Production. Formerly by glycerol hydration or the catalytic condensation of formaldehyde and acetaldehyde. Since 1960, it has been manufactured by catalytic oxidation of propylene. Being a petrochemicals by-product, it is available in large quantities and at a low price.

Uses. It is used as a starting material for the manufacture of many organic compounds including plastics, plasticisers, acrylates, textile finishes, synthetic fibres, the pharmaceutical methionine, and in animal foodstuffs. Acrolein vapours are also given off when oils and fats containing glycerol are heated to high temperatures such as in the reduction of animal fat and bone, and in the manufacture of soap, fatty acids, stearine, linseed oil, linoleum, oil cloth, etc.

HAZARDS

Atmospheric pollution. The exhaust fumes of internal combustion engines contain many and varied aldehydes, including acrolein, particularly when diesel oil or fuel oil is used. In addition acrolein is found in tobacco smoke in considerable quantities, not only in the particulate phase of the smoke, but also, and even more, in the gaseous phase. Accompanied by other aldehydes (acetaldehyde, propionaldehyde, formaldehyde, etc.) it reaches such a concentration (50 to 150 ppm) that it seems to be the most dangerous aldehyde in tobacco smoke.

Fire and explosion. Acrolein is highly flammable and its vapours form explosive mixtures with air. It should be treated as a serious fire hazard, and ignition sources should be avoided whenever acrolein is handled in contact with an air atmosphere. It is also a very reactive compound and polymerises spontaneously at around 200 °C, the speed of the reaction increasing as the temperature rises. Normal inhibitors are ineffective at these temperatures. In the presence of alkaline or strong acid contamination, which acts as a catalyst, acrolein undergoes a condensation reaction liberating around 300 kJ/kg acrolein reacted; this reaction may be very rapid and violent.

Health hazards. Acrolein is toxic and very irritating, and its high vapour pressure may result in the rapid formation of hazardous atmospheric concentrations. Vapours are capable of causing injury to the respiratory tract, and the eyes can be injured by both liquid and vapours. Skin contact may produce severe burns. However, acrolein has excellent warning properties and severe irritation occurs at concentrations less than those expected to be hazardous (its powerful lacrimatory effect in very low concentrations in the atmosphere (1 mg/m³) compels people to run away from the polluted place in search of protective devices). Consequently, exposure is most likely to result from leakage or spillage from pipes or vessels.

Inhalation presents the most serious hazard. It causes irritation of nose and throat, tightness of the chest and shortness of breath, nausea and vomiting. The bronchopulmonary effect is very severe; even if the victim recovers from acute exposure, there will be permanent radiological and functional damage. Animal experiments indicate that acrolein has a vesicant action, destroying respiratory tract mucous membranes to such an extent that respiratory function is fully inhibited within 2-8 days.

No cases of chronic toxicity are known but repeated skin contact may cause dermatitis, and skin sensitisation has been observed.

The discovery of the mutagenic properties of acrolein is not recent. Rapaport pointed it out as long ago as 1948 in *Drosophila*. Research has been carried out to establish whether cancer of the lung, whose connection with the abuse of tobacco is unquestionable, can be traced to the presence of acrolein in the smoke, and whether certain forms of cancer of the digestive system that are found to have a link with the absorption of burnt cooking oil are due to the acrolein contained in the burnt oil. Recent studies have shown that acrolein is mutagenic for certain cells (*Drosophila*, *Salmonella*, algae such as *Dunaliella bioculata*) but not for others (yeasts such as *Saccharomices cerevisiae*). Where acrolein is mutagenic for a cell, ultrastructural changes can be identified in the nucleus which are reminiscent of those caused by X-rays in algae. It also produces various effects on the synthesis of DNA by acting on certain enzymes. Experimental studies, both published and unpublished, that have been carried out on rats, mice and hamsters give no proof of a carcinogenic effect.

It is none the less true that, of all the aldehydes tested, acrolein remains the most effective in inhibiting the activity of the cilia of the bronchial cells, that help to keep the bronchial tree clear. This, added to its action favouring inflammation, indicates it as one of the more important factors causing chronic bronchial lesions.

SAFETY AND HEALTH MEASURES

Wherever possible, acrolein should be stored, handled and processed in the open or in roofed areas with open sides. When stored indoors, it must be in a well ventilated area.

Processing equipment should be of the totally enclosed type with an oxygen-free atmosphere. Where acrolein is handled, exhaust and/or general ventilation should be fitted. All sources of ignition should be avoided by use of flameproof electrical equipment, earthing to prevent large charges of static electricity, and prohibition of smoking. Furthermore, the acrolein should be inhibited by addition of 0.1% hydroquinone.

Provisions should be made for retaining massive spillage by means of kerbs or sills, and a warning system should ensure the evacuation of all workers in the event of leakage. Where there is risk of exposure (during e.g. decanting), workers should wear protective clothing, hand and arm protection, eye and face protection and respiratory protective equipment. Before a tank having contained acrolein is entered, it should be purged with nitrogen and the precautions relevant to entering confined spaces should be observed.

Treatment. Skin contact should be treated by emergency showering, and eye contact by irrigation. All contaminated clothing should be removed. In the event of inhalation, remove the worker from the contaminated area and administer oxygen. Treatment for initial and secondary pulmonary inflammation should be given (corticosteroids with oxolamine) even when respiratory distress is not evident. This should be followed by oxygen therapy and symptomatic treatment (analeptics, analgesics, bleeding).

CATILINA, P.

"Physiopathological effects of atmospheric pollution by aldehydes" (Action physiopathologique de la pollution atmosphérique par les aldéhydes). Catilina, P.; Champeix, J. *Revue d'épidémiologie, médecine sociale et santé publique* (Paris), 1974, 22/6 (461-475). (In French)

CIS 78-1020 "Experimental irritation effect of acrolein in man" (Experimentelle Reizwirkungen von Akrolein auf den Menschen). Weber-Tschopp, A.; Fischer, T.; Gierer, R.; Grandjean, E. *International Archives of Occupational and Environmental Health* (West Berlin), 9 Nov. 1977, 40/2 (117-130). Illus. 18 ref. (In German)

"Effects of acrolein on the synthesis of nucleic acids in vivo" (Action de l'acroléine sur les synthèses d'acides nucléiques). *Biochimie* (Paris), 1971, 53 (243-248). (In French)

CIS 79-1032 *Some monomers, plastics and synthetic elastomers, and acrolein. IARC monographs on the evaluation of the carcinogenic risk of chemicals to humans.* Vol. 19 (Lyons, International Agency for Research on Cancer, 1979), 513 p. 1 623 ref.

CIS 79-1328 *Acrolein.* Data Sheet 1-436-78, Revised 1978 (National Safety Council, 444 North Michigan Avenue, Chicago) (1978), 5 p. Illus. 4 ref.

CIS 78-1981 "A solid sorbent personal sampling method for the determination of acrolein in air". Hurley, G. F.; Ketcham, N. H. *American Industrial Hygiene Association Journal* (Akron), Aug. 1978, 39/8 (615-619). 5 ref.

Acro-osteolysis, occupational

When the industrial production of polyvinyl chloride started in the 1930s, the monomer vinyl chloride (VC) was considered capable of health impairment only as a narcotic at high concentrations. After Raynaud's phenomenon symptoms were found in 1954 among Japanese workers engaged in the production of PVC, the discovery, in October 1963, of osteolytic lesions of the ungueal process in the distal phalanges of two Belgian plastics industry workers suffering from Raynaud's phenomenon led to the individualisation of a new occupational syndrome termed "occupational acro-osteolysis". This was the first well recognised toxic effect of vinyl chloride; and today it rings like an alarm bell as regards the most severe effects which were at that time still unknown.

Epidemiology

This hand syndrome appeared to be found only in workers exposed to vinyl chloride in the polyvinyl chloride (PVC) resin manufacturing and polymerisation processes; cases have not been found amongst workers handling or processing the finished resin. In addition, the affected workers were found almost exclusively amongst those employed on removing deposits on the inside of polymerisation vessels by hand scraping ("polyscrapers"). [An investigation carried out in 1966-67 in 32 plants of the American and Canadian VC industry showed 1 302 cases of X-ray hand abnormalities out of 5 011 exposed employees. However, most of them were unrelated to acro-osteolysis. The diagnosis was based on Raynaud's phenomenon symptoms and characteristic X-ray changes, and the ratio of patients with the disease to those exposed to VCM was 1 : 31. In various studies, 3-5% of polyscrapers presented the syndrome. The average exposure time ranged from 5 to 42 months and the ages of the affected people from 20 to 47 years. The average time weighted VC concentration before 1950 must have been 300 ppm. Acro-osteolysis was reproduced in experimental animals. In 1977 acro-osteolysis did not already show an excessive prevalence in workers exposed to VCM as compared with industrial control workers.]

Three factors were thought to combine in the causation: a chemical insult; a physical insult; and personal reactivity.

Description

Most cases had two common features: symptoms akin to those of Raynaud's phenomenon and acro-osteolysis of the distal phalanges. There have been rare instances of urticaria-like small nodular eruptions; these are accompanied by tegumental thickening but not by eczematous lesions. In some cases there have been persistent nodules localised symmetrically on the dorsal surfaces of hands and forearms reminiscent of scleroderma.

In the United Kingdom and the United States rheumatoid pain, sometimes with arthrotic lesions, decalcification of the patellar and cystic lesions of the sacro-iliac joint, have been reported.

Thrombocytopenia occurred much earlier than any other symptom.

Raynaud's phenomenon. This may occur in varying degrees, and in both hands. It is accompanied by rheumatoid pain and acroparaesthesia, and is aggravated by contact with cold tools or cold weather. The disorder persists even after osseous consolidation (see RAYNAUD'S PHENOMENON).

Skeletal changes. The X-ray findings characteristic of this syndrome are those of acro-osteolysis, a rare clinical entity, usually of a familial or hereditary nature involving non-regressive deprivation or removal of bone calcium from the extremities. Occupational acro-osteolysis may be differentiated from the familial form by its intermittent progression and the fact that recalcification occurs after cessation of exposure. X-ray findings in healed fingers include clubbing of the distal phalanges, which may be shortened with, sometimes, a dorsal incurvation of the ungueal process. Texture is denser than in the unaffected distal phalanges. The ring (fourth) finger is often unaffected. The affected fingers have a podgy appearance, the nail is shortened, ovalised transversely and slightly dished.

PREVENTIVE MEASURES

It had been suggested that before being employed as polycleaners, workers should be examined for evidence of collagen disease, osteolysis of the hands, or abnormal response of hands to cold. Evidence of the existence of such factors contraindicated employment on this work.

[Yearly physical examination, periodic thrombocyte count and, when necessary, X-ray of the hands were the essence of medical measures aiming at a diagnosis early enough to remove the affected worker from exposure when he could still fully recover. Frequent washing with water, wearing of gloves and an essential reduction of reactor cleaning hours and reduction of PVM concentration in the air inhaled by the workers were first recommended as the basic preventive measures for polyscrapers, and 50 ppm was suggested as the allowable concentration to prevent acro-osteolysis among workers entering the reactor. Then a much more drastic reduction was required (see VINYL CHLORIDE) for the prevention of carcinogenic effects. Present recommendations are: develop techniques to minimise employee exposure to VCM when opening any closed vessel; provide employees entering the reactor with a full-face supplied-air respirator of the continuous flow or pressure demand type, and with clean full-body protective clothing, gloves, footwear, and head covering; have the employees engaged in reactor cleaning take a shower at the end of their work shift. The present preventive measures have been accompanied by the disappearance of acro-osteolysis.]

Treatment. In the absence of general phenomena, cessation of polycleaning work produced spontaneous healing. Workers were often unaware of the acro-osteolytic lesions and noticed only the pains from the

Raynaud's phenomenon, which themselves disappear following cervical or perihumeral sympathectomy.

BASTENIER, H.
CORDIER, J. M.
LEFEVRE, M. J.

"Occupational acro-osteolysis. III: A clinical study". Dodson, V.; Dinman, B. D.; Whitehouse, W. M.; Nasr, A. N. M.; Magnuson, H. J. *Archives of Environmental Health* (Chicago), Jan. 1971, 22/1 (83-91). Illus. 11 ref.

"Prevalence of disease among vinyl chloride and polyvinyl chloride workers". Lilis, R.; Anderson, H.; Nicholson, W. J.; Daum, S.; Fischbein, A. S.; Selikoff, I. J. *Annals of the New York Academy of Sciences* (New York), 31 Jan. 1975, 246 (22-41). Illus. 26 ref.

"Bone lesions among polyvinyl chloride production workers in Japan". Sakabe, H. *Annals of the New York Academy of Sciences* (New York), 31 Jan. 1975, 246 (78-79). 6 ref.

CIS 75-493 "Sclerodermal aspects of occupational acro-osteolysis (polymerisation of vinyl chloride)" (Aspects sclérodermiques de l'acroostéolyse professionnelle (polymérisation du chlorure de vinyle)). Moulin, G.; Régy, J.; Paliard, P.; Vouillon, G.; Guttin, G. *Annales de dermatologie et de syphiligraphie* (Paris), 1974, 101/1 (33-44). Illus. 19 ref. (In French)

"Dermatological aspects of so-called vinyl chloride monomer disease". Czernielewski, A.; Kiec-Swierczynska, M.; Gluszcz, M.; Wozniak, L. *Dermatosen in Beruf und Umwelt* (Aulendorf, Federal Republic of Germany), 1979, 27/4 (108-112). Illus. 32 ref.

Acrylic acid and derivatives

Acrylic acid (CH_2:CHCOOH)

PROPENOIC ACID; ACROLEIC ACID; ETHYLENECARBOXYLIC ACID

m.w. 72
sp.gr. 1.05
m.p. 12.3 °C
b.p. 141.6 °C
v.d. 2.5
v.p. 10 mmHg (1.33·10³ Pa) at 39.9 °C
f.p. 54.4 °C (oc)

soluble in water, ethyl alcohol and ethyl ether

an unstable, colourless liquid with a sharp odour.

MAC USSR 5 mg/m³

Methacrylic acid (CH_2:C(CH_3)COOH)

2-METHYLPROPENOIC ACID; α-METHYLACRYLIC ACID

m.w. 86.1
sp.gr. 1.01
m.p. 16 °C
b.p. 162-163 °C
v.p. < 0.1 mmHg (0.01·10³ Pa) at 20 °C
f.p. 77.2 °C (oc)

soluble in water; soluble, in all proportions, in ethyl alcohol and ethyl ether

a colourless corrosive liquid with a sharp odour.

MAC USSR 10 mg/m³

Acrylic acid and methacrylic acid readily polymerise in the presence of light, heat and oxygen, and also under the action of oxidising agents such as peroxides. The polymer may be a white powder or a translucent mass according to the conditions under which it is prepared. The polymers generally have higher viscosities than the monomers from which they are derived. High-temperature polymers dissolve readily in water but when polymerisation occurs at low temperatures, the resultant polymer may merely swell. Such polymers are used as thickeners for paints and varnishes and as binders for printing and other pastes.

Acrylic esters

Acrylic esters may be formed from the combination of an acrylic acid with an alcohol, the resultant ester corresponding with the alcohol used in the reaction (e.g. methyl alcohol is used to produce the methyl ester), and the physical characteristics vary somewhat according to the chain length of the alcohol.

Methyl acrylate (CH_2:$CHCOOCH_3$)

2-PROPENOIC ACID METHYL ESTER; METHYL PROPENOATE; METHOXYCARBONYL ETHYLENE

m.w. 86.1
sp.gr. 0.95
m.p. −75 °C
b.p. 80.5 °C
v.d. 3
v.p. 100 mmHg (13.3·10³ Pa) at 28 °C
f.p. −2.8 °C (oc)
e.l. 2.8-25%

soluble in water and organic solvents

a colourless liquid with a sharp fruity odour.

TWA OSHA 10 ppm 35 mg/m³ skin
IDLH 1 000 ppm
MAC USSR 20 mg/m³

Ethyl acrylate (CH_2:$CHCOOC_2H_5$)

2-PROPENOIC ACID ETHYL ESTER; ETHYL PROPENOATE; ETHOXYCARBONYL ETHYLENE

m.w. 100.1
sp.gr. 0.94
m.p. −71.2 °C
b.p. 99.8 °C
v.d. 3.44
v.p. 29.3 mmHg (3.89·10³ Pa) at 20 °C
f.p. 15.5 °C (oc)
e.l. 1.8%

soluble in organic solvents and slightly soluble in water

a colourless liquid with a sharp acrid odour.

TWA OSHA 25 ppm 100 mg/m³ skin
TLV ACGIH 5 ppm 20 mg/m³ skin
STEL ACGIH 25 ppm 100 mg/m³
IDLH 2 000 ppm

The characteristics of some of the esters of methacrylic acid are tabulated below:

Ester	Softening temperature °C	Characteristics
Methyl	125	clear, hard, strong
Ethyl	65	clear, hard, strong
n-Propyl	38	clear, hard
Isopropyl	95	clear, strong, flexible
n-Butyl	33	clear, brittle
Isobutyl	70	clear, brittle
tert-Amyl	76	brittle
n-Octyl	below room temperature	gel
n-Dodecyl	below room temperature	thick liquid

It can be seen that the lower esters have the more desirable qualities of toughness, strength and transparency, whereas the higher esters, although still transparent, tend to show brittleness. For methylmethacrylate see ACRYLIC RESINS.

Production. Acrylic acid may be prepared by the oxidation of acrolein produced by oxidation of propylene and, in fact, was first made in this way. Other methods, more suitable for commercial production, have been devised, some of which are shown in the equations below, in which the conventional notation of ROH is used for the common alcohols and COOR for the corresponding ester:

(a)
$$\underset{\text{Ketene}}{CH_2{:}CO} + \underset{\text{Formaldehyde}}{HCHO} \longrightarrow \underset{\text{Propiolactone}}{OCH_2CH_2C{:}O} \xrightarrow[H_2O]{ROH} \underset{\text{Acrylate}}{CH_2{:}CHCOOR}$$

(b)
$$\underset{\text{Acetylene}}{CH{\vdots}CH} + \underset{\text{Carbon monoxide}}{CO} \xrightarrow[H_2O]{\text{Catalyst}} \underset{\text{Acrylic acid}}{CH_2{:}CHCOOH} \xrightarrow[H_2O]{ROH} \underset{\text{Acrylate}}{CH_2{:}CHCOOR}$$

(c)
$$\underset{\text{Acetylene}}{CH{\vdots}CH} + \underset{\text{Carbon monoxide}}{CO} \xrightarrow[ROH]{\text{Catalyst}} \underset{\text{Acrylate}}{CH_2{:}CHCOOR}$$

The production of methacrylic acid and methacrylates is as follows:

(a)
$$\underset{\text{Isobutylene}}{CH_2{:}C(CH_3)_2} \xrightarrow[O_2]{\text{Catalyst}} \underset{\text{Methacrylic acid}}{CH_2{:}C(CH_3)COOH} + \xrightarrow[H_2O]{ROH} \underset{\text{Methacrylate}}{CH_2{:}C(CH_3)COOR}$$

(b)
$$\underset{\text{Acetone}}{CH_3COCH_3} \xrightarrow{HCN} \underset{\text{Acetone-cyanohydrin}}{(CH_3)_2C(OH)CN} \xrightarrow[H_2SO_4]{CH_3OH} \underset{\text{Methyl methacrylate}}{CH_2{:}C(CH)_3COOCH_3}$$

Other methods of importance are:

(a) β-chloroethyl alcohol may be reacted with sodium cyanide to produce ethylene cyanohydrin and a further reaction with methyl alcohol and sulphuric acid gives methacrylate;

(b) Acrylonitrile may be hydrolysed to acrylamide which is then esterified to methyl or ethyl acrylate by reaction with the appropriate alcohol. However, no purpose would be served by further enumerating the many methods which have been proposed.

Methyl and ethyl acrylates are produced by the same processes as the acrylic acid, followed by reaction of it with methanol or ethanol.

Uses. [The main use of acrylic acid is as a precursor of acrylates; it is also used in water-soluble resins and in acrylic emulsion and solution polymers. The copolymers of acrylamide and acrylic acid are used as flocculants in the manufacture of paper and sugar, and in water treatment and oil drilling. Methyl acrylate is used in the manufacture of acrylic fibres as a comonomer of acrylonitrile, because its presence facilitates the spinning of fibres. Ethyl acrylate is a component of emulsion and solution polymers for surface coating textiles, paper and leather; it is also used in the production of acrylic fibres, adhesives and binders. Polyacrylic acid is used in rubber manufacture, latex paints, and as a flocculant.]

HAZARDS

The polymerised acrylic acids and their esters are not toxic, but some of the monomers, and particularly some of the chemicals used in their preparation, are extremely so and therefore must be considered as attendant hazards. Three of these chemicals are dealt with in the articles ACROLEIN; ACRYLONITRILE, CYANOGEN, HYDROCYANIC ACID AND CYANIDES.

Acrylic and methacrylic acids have a pungent smell and are rather more irritating than acetic acid. [Skin and eye contact should be avoided (a 1% solution of acrylic acid caused significant injuries to the rabbit eye and severe corneal burns have been reported after acute exposure to methacrylic acid). Acrylic acid is more irritant than methacrylic acid. Acrylic acid has also proved to be embryotoxic and teratogenic in rats. No data are available thus far on the carcinogenicity to humans of acrylic and methacrylic acids. Methyl and ethyl acrylate are mild to moderate irritants to eye and skin, and may cause contact dermatitis.]

Intermediates. β-Chloroethyl alcohol (ethylene chlorohydrin) is a highly toxic product (see ALCOHOLS).

Ethylene cyanohydrin (β-hydroxypropionitrile, hydracrylonitrile) differs from some of the other nitriles in being of relatively low toxicity because it is not hydrolysed with the liberation of cyanide (see NITRILES AND CYANATES).

[**Ketene** ($CH_2{:}CO$)

CARBOMETHANE; ETHENONE; KETEN

m.w.	42
m.p.	−151 °C
b.p.	−56 °C
v.d.	1.45

decomposes in water and alcohol

a colourless gas with a disagreeable taste.

TWA OSHA	0.5 ppm 0.9 mg/m^3
STEL ACGIH	1.5 ppm 3 mg/m^3
IDLH	25 ppm

Ketene is an extremely irritant gas to the respiratory system. Its toxicity and delayed action on the lungs resemble those of phosgene. Pulmonary oedema of a severe degree is the most common sequel to the inhalation of ketene. In repeated exposure acquired tolerance to acute effects has been reported; however emphysema and lung fibrosis may ensue.]

SAFETY AND HEALTH MEASURES

Apart from the toxicity of the materials mentioned and of some of their intermediates, an additional hazard is the generation of considerable exothermic heat in some of the reactions, so that high pressures and temperatures may develop. This danger, as well as that of toxicity, should be borne in mind when designing plant.

Awareness of the dangers and good engineering design are essential to safety. Employees should be instructed about the necessity of cleansing the skin if it is contaminated by materials which are irritant or skin-absorbed. With careful design, however, and complete enclosure of those processes where toxic chemicals or intermediates occur, dangerous exposures can be avoided. Suitable protective clothing and self-contained respiratory protective apparatus should be available for the use of those who may have to go to the rescue of persons overcome by fumes.

Treatment. The medical arrangements for protection of employees will vary according to the process used. Acrylonitrile and cyanide produce similar symptoms though there is some difference of opinion as to their metabolism. Treatment, however, is the same in either case, and where these substances are used, the customary first aid kit containing amyl nitrite capsules and intravenous apparatus for administering sodium nitrite and sodium thiosulphate, should be immediately available (see CYANOGEN, HYDROCYANIC ACID AND CYANIDES). A ready means of administering oxygen should also be at hand not only for cyanide poisoning, but for the relief of persons who may be affected by the other respiratory irritants mentioned.

TRAINOR, D. C.

Experimental data:

CIS 78-1041 "Toxicity of acrylic compounds used in industry" (Toxicita akrylových sloučenin užívaných v průmyslu).

Sokal, J.; Knobloch, K.; Majka, J.; Sapota, A.; Szendzikowski, S. *Pracovní lékarství* (Prague), June 1977, 29/4-5 (157-161). Illus. 14 ref. (In Czech)

Clinical data:

"Acrylates in industry". Calnan, C. D. *Contact Dermatitis* (Copenhagen), Jan. 1980, 6/1 (53-54).

"Contact dermatitis from polyfunctional acrylic monomers". Emmett, E. A. *Contact Dermatitis* (Copenhagen), May 1977, 3/5 (245-248). 8 ref.

"Delayed irritation: hexanediol diacrylate and butanediol diacrylate". Malten, K. E.; Den Arend, J. A. C. J.; Wiggers, R. E. *Contact Dermatitis* (Copenhagen), Mar. 1979, 5/3 (178-184). Illus. 11 ref.

"Acrylic acid, methyl acrylate, ethyl acrylate and polyacrylic acid". *IARC monographs on the evaluation of the carcinogenic risk of chemicals to humans.* Volume 19: *Some monomers, plastics, and synthetic elastomers, and acrolein* (Lyons, International Agency for Research on Cancer, 1979) (47-71). 65 ref.

Industrial hygiene:

"An air sampling and analysis method for monitoring personnel exposure to vapors of acrylate monomers". Bosserman, M. W.; Ketcham, N. H. *American Industrial Hygiene Association Journal* (Akron), 1980, 41/1 (20-26). Illus. 7 ref.

Acrylic resins

The most important example of this group of resins is polymethyl methacrylate–the homopolymer of methyl methacrylate. This is produced as sheet ("Perspex", "Plexiglas", etc.) and also in the form of moulding powder ("Diakon", "Lucite", etc.).

Other monomers in the group can also be polymerised on their own (e.g. acrylonitrile to polyacrylonitrile or methacrylic acid to polymethacrylic acid) or with other monomers to form copolymers (e.g. methyl and ethyl acrylate can be copolymerised with methyl methacrylate; or acrylonitrile with styrene, butadiene, vinyl chloride, etc.).

[**Methyl methacrylate** (CH_2:CCH_3COOCH_3)

2-METHYL-2-PROPENOIC ACID METHYL ESTER; METHACRYLIC ACID METHYL ESTER; METHYL-2-PROPENOATE; MME

m.w.	100.1
sp.gr.	0.94
m.p.	−48 °C
b.p.	100-101 °C
v.p.	40 mmHg ($5.32 \cdot 10^3$ Pa) at 25.5 °C
f.p.	10 °C (oc)
e.l.	1.7-8.2%

slightly soluble in water; very soluble in organic solvents

a colourless liquid with a fruity smell.

TWA OSHA	100 ppm 410 mg/m³
STEL ACGIH	125 ppm 510 mg/m³
IDLH	4 000 ppm
MAC USSR	10 mg/m³

Production. Methyl methacrylate is made from acetone by reaction with hydrocyanic acid to form acetone cyanohydrin which, when hydrolysed with sulphuric acid and in the presence of methyl alcohol, yields methyl methacrylate.

Uses. More than 50% of the produced amount of MME is used for the production of acrylic polymers as polymethylmethacrylate and other resins, mainly as sheets, moulding and extrusion powders, surface coating resins, emulsion polymers, fibres, inks and films.]

Polymethyl methacrylate preparation. In the preparation of polymethyl methacrylate sheet, a catalyst (together with any dye or pigment required) is added to the monomer and the mixture is then thickened by stirring in a heated vessel. When a certain viscosity is obtained the liquid ("syrup") is cooled and poured into glass "cells"–two sheets of heat-resistant plate glass separated by a gasket. The cells are then placed in ovens where polymerisation takes place–the temperature and duration depending on the thickness of the sheet.

For extrusion and injection moulding purposes, material of a lower molecular weight is required. This is achieved by granular polymerisation in which a methyl methacrylate/water mixture is heated and stirred in an autoclave in the presence of a catalyst. The polymerisation of the individual droplets yields a slurry of polymethyl methacrylate in water. When the slurry is dried, polymer powder is left. In order to control the properties of the melt to ensure ease of working in injection and extrusion machines, it is customary to add a small proportion of comonomer–for example ethyl acrylate–to the reaction system and thus to produce a copolymer in which methyl methacrylate is the major constituent.

Uses. Acrylic resins are of considerable commercial importance. They are widely used as substitutes for glass and in certain situations are preferred to it since they are almost as transparent as glass but weigh only half as much. They are used for inspection windows and face shields. This "organic glass" has the property of total internal reflection and can "pipe" light, which makes it useful for surgical retractors and throat lights. Because acrylics are not irritating to tissues, they are also widely used for dentures and surgical protheses. The polymers may be cast into sheets or moulded, depending on the manner of their preparation. In emulsion form they are used in paints or may be precipitated as coatings for fabrics. Dissolved in volatile solvents, they make effective adhesives by forming firm joints when the solvent has evaporated. On account of its properties–excellent light transmission (92%), toughness, good chemical resistance and outstanding weather resistance, and relatively high softening point (100 °C)–polymethyl methacrylate is used in a wide variety of applications. Among the more important of these are its use for aircraft canopies, traffic signs and advertising displays, light fittings (domestic, street and automobile), television implosion guards and guards for industrial machinery, sanitary ware, disposable hypodermic syringes and, in special grades, for artificial dentures and replacement lenses for eyes.

HAZARDS

Methyl methacrylate is a mild irritant to the skin and mucous membranes and, on inhalation, the vapour is a narcotic. Both the monomer and the finely divided polymer may cause an allergic skin reaction. In the jointing of acrylic materials to each other, adhesives containing organic solvents (1,2-dichloroethane, chloroform, etc.) with toxic properties may be used. When processing polymethyl methacrylate on machine tools, a certain amount of swarf is produced which may be irritant to the skin or which may enter the eyes. No data are thus far available for the evaluation of carcinogenic power to humans of acrylic resins.

SAFETY AND HEALTH MEASURES

Adequate exhaust ventilation should be provided to protect the worker from the inhalation of fumes. Workers should be provided with hand protection and eye and face protection. Maintenance and good housekeeping should be ensured.

WINTER, D.

Methyl methacrylate:

A study of methyl methacrylate exposures and employee health. Cromer, J.; Kronoveter, K. (Cincinnati, DHEW (NIOSH) Division of Surveillance, Hazard Evaluations, and Field Studies, Nov. 1976), 54 p. 32 ref.

"Methyl methacrylate and polymethyl methacrylate". *IARC monographs on the evaluation of the carcinogenic risk of chemicals to humans.* Volume 19: *Some monomers, plastics and synthetic elastomers, and acrolein* (Lyons, International Agency for Research on Cancer, Feb. 1979), (187-211). 101 ref.

CIS 77-1375 "Internal changes in workers exposed to methyl methacrylate" (Nekotorye izmenenija vnutrennyh organov u rabotajuščih v kontakte s metilovym ėfirom metakrilovoj kisloty). Dorofeeva, E. D. *Gigiena truda i professional'nye zabolevanija* (Moscow), Aug. 1976, 8 (31-35). 17 ref. (In Russian)

"On the adipogenic effect of some industrial poisons" (K voprosu ob adipozogennom dejstvii nekotoryh promyšlennyh jadov). Makarov, I. A.; Makarenko, K. I.; Desyatnikova, N. V. *Gigiena truda i professional'nye zabolevanija* (Moscow), Dec. 1981, 12 (29-32). 21 ref. (In Russian)

CIS 77-1613 "Establishment of the maximum allowable concentration of methyl methacrylate in workplace air" (Obosnovanie predel'no dopustimoj koncentracii metilovogo ėfira metakrilovoj kisloty v vozduhe rabočej zony). Blagodatin, V. M.; Smirnova, E. S.; Dorofeeva, E. D.; Golova, I. A.; Arzjaeva, E. Ja. *Gigiena truda i professional'nye zabolevanija* (Moscow), June 1976, 6 (5-8). 10 ref. (In Russian)

Polymethyl methacrylate:

CIS 77-1947 "Polymethyl methacrylate" (Le polyméthacrylate de méthyle). Fillassier, G. *Cahiers de médecine interprofessionnelle* (Paris), 1st quarter 1977, 65 (39-42). 7 ref. (In French)

Polyethyl acrylate:

CIS 74-1331 "Occupational health case report No. 3–Ethyl acrylate". Cohen, S. R.; Maier, A. A.; Flesch, J. P. *Journal of Occupational Medicine* (Downers Grove, Illinois), Mar. 1974, 16/3 (199-200).

Other polyacrylates:

CIS 78-1041 "Toxicity of acrylic compounds used in industry" (Toxicita akrýlovych sloučenin užívaných v prumyslu). Sokal, J.; Knobloch, K.; Majka, J.; Sapota, A.; Szendzjkowski, S. *Pracovní lékařství* (Prague), June 1977, 29/4-5 (157-161). Illus. 14 ref. (In Czech)

Acrylonitrile (CH_2:CHCN)

VINYL CYANIDE; PROPENE NITRILE; AN

m.w.	53
sp.gr.	0.80
m.p.	−83.5 °C
b.p.	77.5 °C
v.d.	1.83
v.p.	100 mmHg (13.3·10³ Pa) at 22.8 °C
f.p.	−5 °C
e.l.	3-17%
i.t.	481 °C

soluble in water; soluble in all proportions in ethanol and ether

a colourless mobile liquid with a faint odour.

TWA NIOSH	4 ppm 9 mg/m³ 10 h
ACGIH	human carcinogen 2 ppm 4.5 mg/m³
IDLH	4 ppm
MAC USSR	0.5 mg/m³ skin irritant

Production. The principal route for the production of acrylonitrile is the SOHIO process in which propylene and ammonia are oxidised in the presence of a catalyst. Hydrogen cyanide, acetonitrile and oxides of carbon are produced as by-products. The early process of reacting acetylene with hydrogen cyanide is now of little commercial importance. For the future, interest is being shown in propane rather than propylene as a starting point.

When assessing the hazards of an acrylonitrile installation, it should be remembered that hydrogen cyanide is likely to be present either as feedstock or by-product.

Uses. In 1975 world capacity was 2.7 million tonnes per year and demand is expected to increase at approximately 10% annually. The major uses are in the manufacture of polymers and copolymers for the synthetic fibre industry, for resins and plastics and for nitrile rubbers.

HAZARDS

Fire and explosion. The dominant hazard is fire. The low flash point indicates that sufficient vapour is evolved at normal temperatures to form a flammable mixture with air. Acrylonitrile has the ability to polymerise spontaneously under the action of light or heat leading to explosion of closed containers. It must therefore never be stored uninhibited. These dangers associated with fire and explosion are intensified by the lethal nature of the fumes and vapours evolved, e.g. ammonia and hydrogen cyanide.

Health hazards: acute. Acrylonitrile is a chemical asphyxiant like hydrogen cyanide. It is irritant to mucous membranes and a severe eye irritant capable of causing corneal damage if not rapidly irrigated.

Inhalation is clearly the most important route of absorption in the industrial context. Skin absorption can also take place but is slower to cause effects. Leather gloves and footwear readily absorb acrylonitrile and can lead to skin absorption and blistering over a period.

Symptoms of poisoning may be slower to develop than in exposure to hydrogen cyanide but when they appear the victims can have significant levels of cyanide in their blood. Technical difficulties make blood acrylonitrile levels impractical as a measure of absorption but the author has measured 1 to 3 mg/l of cyanide in accidentally exposed people using a spectrophotometric method for free cyanide. Clinical symptoms in these cases were surprisingly few despite the cyanide content of the blood. When poisoning cases do occur symptoms derive from tissue anoxia and are in order of onset: limb weakness, dyspnoea, burning sensation in the throat, dizziness and impaired judgement, cyanosis, nausea, collapse, irregular breathing, convulsions and death. In the latter stages collapse, irregular breathing or convulsions and cardiac arrest may occur without warning. Some patients appear hysterical or may even be violent. Any deviation of this sort from normal is deeply suspect.

Health hazards: chronic effects. The metabolic pathway of acrylonitrile detoxification in the body is still unclear but multiple routes are suggested, as both N-acetyl-s-cyanoethyl cysteine and thiocyanates have been traced in the urine from labelled ^{14}C acrylonitrile experiments.

The detection of cyanide in poisoning cases may also give some guidance on metabolism.

Health hazards: mutagenesis. Although the evidence available on the mutagenic potential is still inconclusive, recent studies on rats showed an increase in the number of tumours developed by the animals under the test conditions. European workers did not produce a significant increase in the incidence of tumours in parallel experiments.

Further animal studies are taking place in a number of centres. Although the interpretation of the results of recent animal experiments is equivocal they have added importance to an epidemiological report from Dupont suggesting a higher incidence of lung and gastrointestinal tumour in workers exposed at a particular plant between 1950 and 1952. Similar studies are taking place in the Federal Republic of Germany, the United Kingdom and the United States, which will lead to a greater understanding of human risk.

As regards the carcinogenic risk and other chronic effects see POLYACRYLONITRILE FIBRES.

SAFETY AND HEALTH MEASURES

The usual precautions for highly flammable liquids are necessary. Steps must be taken to eliminate the risk of ignition from sources such as electrical equipment, static electricity and friction. Because of the toxic, as well as the flammable, nature of the vapour, precautions must be taken by enclosure of the plant and by means of exhaust ventilation to prevent vapour escaping into the workplace air. A programme of monitoring of the workplace air is necessary to ensure that these engineering controls remain effective. Personal respiratory protection preferably of the positive pressure type and impermeable protective clothing is necessary when there is the possibility of exposure resulting from a normal but non-routine operation such as a pump replacement. Respiratory protective equipment, protective clothing and fire fighting equipment must be available for emergency use.

Medical supervision

Acrylonitrile plants require effective medical cover. Skilled medical attention is required in emergencies. The doctor must assess occupational health needs in close association with management. The principal requirements are an alarm system and plant personnel trained in rescue, decontamination and life support procedures to support the activities of the occupational nurse and doctor. Supplies of specific antidotes should be available on site and at adjacent hospital centres.

The practical value of routine biological monitoring of acrylonitrile workers as a means of preventing illness must be questioned. A number of companies carry out routine monitoring but no effective or predictive procedures have been reported. Similarly, routine analysis of blood and urine appear to have little specific merit in assessing total body burden of acrylonitrile under normal conditions. A carefully documented record of the atmospheric levels on plant, and readings taken on personal samplers, should be maintained plus details on the morbidity and mortality experience of the group for epidemiological assessment.

Treatment. Treatment of the acrylonitrile victim is basically that for cyanide. Although it is possible that other metabolic toxins may be in the patient's body, free cyanide is present.

Since specific antidotes exist for cyanide ions, and since cyanide is a known toxic agent, these antidotes should be used plus supportive measures.

Since the antidotes are in themselves toxic, cases of poisoning, although requiring rapid treatment, need a judicious approach. The decision to treat is always clinical, biochemistry being too tardy, and the antidote must only be given where the patient is ill and is becoming comatose.

Treatment will therefore be:

(a) Specific—either sodium nitrite intravenously followed by sodium thiosulphate, or cobalt edetate intravenously again followed by sodium thiosulphate if required. Both regimens may be preceded by amyl nitrite which can easily be administered on the plant.

(b) Supportive—oxygen, rest and reassurance will suffice for minor cases. Where poisoning is severe, mechanical respiratory support, cardiac stimulant, and cardiac care techniques including defibrillation may be required.

BRYSON, D. D.

"Toxicology and mode of action of acrylonitrile" (Toxikologie und Wirkungsweise von Acrylnitril). Grahl, R. *Zentralblatt für Arbeitsmedizin und Arbeitsschutz* (Heidelberg), Dec. 1970, 20/12 (369-378). 112 ref. (In German)

"Mutagenic activity of acrylonitrile". Meester, de C.; Poncelet, F.; Roberfroid, M.; Mercier, M. *Toxicology* (Amsterdam), Sep. 1978, 11/1 (19-27). 18 ref.

"Carcinogenic bioassays on rats of acrylonitrile administered by inhalation and by ingestion". Maltoni, C.; Ciliberti, A.; Di Maio, V. *La Medicina del Lavoro* (Milan), Nov.-Dec. 1977, 68/6 (401-411). Illus. 1 ref.

CIS 1161-1972 *Acrylonitrile experiments—Report of the Committee for the Prevention of Dangerous Substance Disasters* (Experimenten met acrylnitril—Rapport van de Commissie preventie van rampen door gevaarlijke stoffen). (Voorburg (Netherlands), Directoraat-Generaal van de Arbeid, 1971), 45 p. Illus. (In Dutch)

"Epidemiologic study of workers exposed to acrylonitrile". O'Berg, M. T. *Journal of Occupational Medicine* (Chicago), Apr. 1980, 22/4 (245-252). 11 ref.

Occupational exposure to acrylonitrile—final standard (Department of Labor, Occupational Safety and Health Administration, US Federal Register, 3 Oct. 1978).

Properties and essential information for safe handling and use of acrylonitrile. Chemical Safety Data-Sheet SD-31, Revised 1974 (Manufacturing Chemists Association, 1825 Connecticut Avenue NW, Washington, DC, 1974), 19 p.

Actinide elements

The actinide elements consist of the group of "rare-earth-like" elements that start with actinium, atomic number 89, and end with lawrentium, atomic number 103. They include:

Atomic No.	Name	Symbol
89	Actinium	Ac
90	Thorium	Th
91	Protactinium	Pa
92	Uranium	U
93	Neptunium	Np
94	Plutonium	Pu
95	Americium	Am
96	Curium	Cm
97	Berkelium	Bk
98	Californium	Cf
99	Einsteinium	Es
100	Fermium	Fm
101	Mendelevium	Md
102	Nobelium	No
103	Lawrentium	Lw

Chemically, the actinides are similar because the configurations of the outermost electronic shells are the same. The elements progress by sequentially adding electrons to the 5f energy level; thorium displays a slight irregularity in this sequence. Plutonium, thorium and uranium are dealt with in greater detail in separate articles. All the actinide elements are radioactive.

Sources. Only the first four actinides, actinium, thorium, protactinium and uranium, are found in nature, and until

1940 they were the only known actinides. All the others are produced synthetically through nuclear reactions. Thorium and uranium are both plentiful and widespread; it is estimated that the average Th and U contents of the top 30 cm of soil are respectively 7 t/km^2 and 2.3 t/km^2 (see URANIUM, ALLOYS AND COMPOUNDS; THORIUM AND COMPOUNDS). Actinium and protactinium are found in nature in minute quantities as short-lived decay products in the uranium series, in equilibrium with their long-lived progenitors. Actinium is found mainly as ^{227}Ac and protactinium as ^{231}Pa, both being members of the ^{235}U series. Because of their short half-lives, only 0.2 mg actinium and 300 mg protactinium are found in 1 ton of natural uranium.

Production. The first trans-uranium element, neptunium, was discovered in 1940. Neptunium was created when a ^{238}U atom captured a neutron to become ^{239}U, and then decayed by β-emission to $^{239}_{93}$Np. Neptunium itself is radioactive, and decays by β-emission with a half-life of 2.35 days to form an isotope of the second man-made trans-uranium element, plutonium. The reactions leading to $^{239}_{94}$Pu are:

$$^{238}_{92}\text{U} + ^{1}_{0}\text{n} \rightarrow ^{239}_{92}\text{U} \xrightarrow[T\frac{1}{2} = 23.5\text{ min}]{\beta\text{-decay}} ^{239}_{93}\text{Np} \xrightarrow[T\frac{1}{2} = 2.35\text{ days}]{\beta\text{-decay}} ^{239}_{94}\text{Pu}$$

Shortly after the discovery of ^{239}Pu, it was learned that this plutonium isotope is fissile; that is, ^{239}Pu can undergo nuclear fission and can sustain a nuclear chain reaction.

Element 95, americium, was prepared in 1944 by neutron irradiation of plutonium. The Pu absorbed two successive neutrons, then decayed by β-emission to $^{241}_{95}$Am:

$$^{239}_{94}\text{Pu} + 2^{1}_{0}\text{n} \rightarrow ^{241}_{94}\text{Pu} \xrightarrow[T\frac{1}{2} = 13.2\text{ years}]{\beta\text{-decay}} ^{241}_{95}\text{Am}(T\tfrac{1}{2} = 458\text{ years})$$

Americium 241 is an α-γ-emitting isotope that is used commercially as a source of radiation. It is extracted in large quantities from the ^{241}Pu in spent reactor fuel. Neutron irradiation of ^{241}Am and subsequent β-decay of ^{242}Am led to the discovery, in 1947, of an isotope of curium, element 96, $^{242}_{96}$Cm. Later, another curium isotope, $^{244}_{96}$Cm, was produced from ^{239}Pu. Curium 244 can be produced in large quantities and it, too, has a good potential for use as an isotopic heat source for direct conversion of heat to electricity in systems designed to operate for a long period of time. The next element, called berkelium, was produced in 1949 by irradiating ^{241}Am with high energy helium ions (α-rays), according to the reaction:

$$^{241}_{95}\text{Am} + ^{4}_{2}\text{He} \rightarrow ^{245}_{98}\text{Bk}\ (T\tfrac{1}{2} = 44\text{ min}) + ^{1}_{0}\text{n}$$

Shortly thereafter, in 1950, element number 98, californium (Cf) was produced by bombarding ^{242}Cm with high-energy α-rays:

$$^{242}_{96}\text{Cm} + ^{4}_{2}\text{He} \rightarrow ^{245}_{98}\text{Cf}\ (T\tfrac{1}{2} = 44\text{ min}) + ^{1}_{0}\text{n}$$

Later, other isotopes of Bk and Cf were produced by neutron irradiation, in a nuclear reactor, of plutonium and its transmutation products. Among the Cf isotopes that are potentially useful is ^{252}Cf. This isotope, whose half-life is 2.6 years and which disintegrates mainly by α-decay, also undergoes spontaneous fission and one curie emits 4.4 x 10^9 neutrons per second.

Isotopes of the succeeding actinides have been discovered and some of their properties measured. However, the yields have been so low that, at this time, they are only of scientific interest.

Uses. Prior to the advent of nuclear reactor technology, the actinides were of relatively little commercial value. Uranium salts were used in the ceramic industry to produce various colours in the glaze, while the main industrial use for thorium was in the production of incandescent gas mantles. Colloidal thorium oxide (Thorotrast) was used by physicians as a radiographic contrast medium. This use was discontinued when it was discovered that the injected thorium had carcinogenic properties. Although thorium is still used in the manufacture of incandescent gas mantles, its chief use today is in metallurgy as an alloying element. Thorium is important as a potential source of nuclear fuel. Although it is not capable of sustaining a nuclear chain reaction, it can readily be converted into ^{233}U, which is fissile, by neutron irradiation in a reactor.

The chief use of uranium is as the fuel in thermal reactors. Uranium 238, too, can be made to undergo nuclear fission under certain conditions. Its main contribution to the nuclear fuel cycle, however, is as the fertile material from which fissile ^{239}Pu can be produced. In a breeder reactor, the ^{238}U, under neutron irradiation, undergoes a series of nuclear reactions that lead to the production of fissile ^{239}Pu.

Americium 241, an α-γ-emitter (5.4 MeV) whose half-life is 458 years, is widely used as an ionisation source; its main application is in static electricity eliminators and in smoke detectors.

Californium 252, because of its prolific neutron emission, has widespread industrial and medical applications. Industrial uses include monitoring of moisture content in numerous materials, and monitoring sulphur content in oil and in coal. It is also used by geologists in petroleum exploration and by hydrologists in the location of water supplies. In medicine, ^{252}Cf is used as an implantable neutron source for cancer therapy, in neutron activation analysis of body fluids and tissues, and for the production of short-lived clinically useful radioisotopes, such as 25-minute ^{128}I.

One of the isotopes of uranium, ^{235}U, which comprises 0.71% of uranium (most of the remaining 99.29% is ^{238}U) is the only naturally occurring substance that is directly suitable for nuclear fuel. Accordingly, a major part of nuclear technology deals with enrichment of uranium, that is the concentration of the ^{235}U isotope in excess of its natural abundance. Most commercial reactors that are used for electrical power generation use fuels that are enriched to about 3.5-5%. For other applications, such as in research or in propulsion systems, enrichment may exceed 90%.

HAZARDS AND THEIR PREVENTION

All the actinides are radioactive; most are intensely radioactive, and hence are highly toxic and must be handled, transported and stored with extreme care. Generally, the intensity of radioactivity varies inversely with the half-life. Thus, the long-lived naturally occurring isotopes of uranium and thorium are less hazardous radiologically than the transuranium elements. One of the transuranics, ^{252}Cf, poses a unique potential hazard. First, because it is a neutron emitter, its shielding requirements are qualitatively different from those of the usual β-γ-emitting radioisotopes. Furthermore, induced radioactivity in materials that are exposed to the neutrons must be considered in the safety analysis of all proposed uses of ^{252}Cf.

Health hazards. Most of the actinides are α-emitters, and hence are especially toxic when deposited in the body. All the actinides accumulate mainly in the bone, and may lead to bone cancer in excessive amounts; uranium and americium also accumulate in the kidneys and may do damage there. Once deposited, most of the actinides are cleared from the body very slowly. Maximum environmental concentrations that are considered safe are very small. For example the annual intake limit (AIL) for relatively soluble inhaled ^{239}Pu is 200 Bq (5.4×10^{-3} μCi), which corresponds to 8.5×10^{-8} g. The mean annual atmospheric concentration (DAC) that leads to this maximum allowable intake is 0.08 Bq/m^3 (2.2×10^{-6} μCi/m^3), which corresponds to 3.5×10^{-11} g/m^3. For ^{241}Am, the AIL and DAC are the same as those for ^{239}Pu in terms of activity, 200 Bq and 0.08 Bq/m^3. In terms of mass, these limits correspond to 1.6×10^{-9} and 6.4×10^{-13} g/m^3 respectively.

Radiation protection. Because of their extreme toxicity, the trans-uranium elements must be completely sealed off from the environment to ensure against transmission to man. Extensive radiation protection, safety planning and adequate monitoring techniques must be employed when manipulating even microgramme quantities of the trans-uranium elements. When α-radiation is the main mode of radioactive decay from the isotopes that are being worked on, then a glove box maintained under negative pressure, so that a leak will cause air to flow into the glove box rather than out into the laboratory, may suffice to isolate the hazardous material. If the isotope emits penetrating radiation, such as gammas or neutrons, then adequate shielding and remote handling equipment must be used. The caves in which these dangerous materials are handled are constructed of shielding materials such as concrete, iron, lead and high density glass. They are often highly sophisticated in design to permit the performance of complicated operations like machining, grinding, and dust-collecting as well as accurate operations like weighing. Manipulator tongs are used to carry out these operations. These "master-slave" manipulators are fitted in such a manner that they make no breach in the shielding.

Criticality. Another potential hazard that is unique to fissile materials is criticality. Criticality may be defined as the attainment of physical conditions such that the fissile material will sustain a chain reaction. Under controlled conditions, as in an operating nuclear reactor, the tremendous amount of energy liberated during the chain reaction is safely harnessed. Under uncontrolled conditions, as in the case of an accidental criticality, the consequences to life and property are grave. Accordingly, the utmost degree of care and control, at both the technical and the administrative levels, must be exercised when handling or transporting fissile materials if a criticality accident, which may lead to loss of life and extensive property damage, is to be avoided.

Fire hazard. The actinide metals constitute a fire hazard in that they oxidise rapidly, and finely divided powders may be pyrogenic. When handling the metals in a powdered form, it is therefore necessary to maintain an atmosphere of a non-reactive gas. Argon is most often used for this purpose, since nitrogen reacts with metals to form the actinide nitrides.

CEMBER, H.

Transuranium nuclides in the environment (Vienna, International Atomic Energy Agency, 1976), 724 p. Illus. Ref.

Biological implications of radionuclides released from nuclear industries (Vienna, International Atomic Energy Agency, 1979), 2 vols. 479 and 442 p. Illus. Ref.

Treatment of incorporated transuranium elements. Technical Report Series No. 184 (Vienna, International Atomic Energy Agency, 1978), 168 p. Illus. 353 ref.

Some physical dosimetry and biomedical aspects of Californium-252 (Vienna, International Atomic Energy Agency, 1976), 278 p. Illus. Ref.

Actors and players

Actors and players cover a wide section of the entertainment industry and include dramatic artists, dancers, singers, acrobats and jugglers, animal trainers, etc. Musicians and cinema industry actors are dealt with in separate articles.

In addition to a wide variety of physical hazards presented by the unusual working environment or the risks inherent in the performance itself, actors and players are exposed to intense physical and psychological stress due entirely to the nature of their work.

In all kinds of entertainment, every effort is concentrated on a single short period of intense physical and mental stress–the performance–which takes place in the late afternoon or evening and at a time when the human body is fatigued and, under normal circumstances, is preparing for sleep. The actor has to concentrate intensely to be at his best during the performance and, when he goes to bed well after midnight, he may often have difficulty in sleeping, especially following the tension of a "first-night" or the emotion of a highly dramatic play or ballet. These unusual hours of work may also affect eating habits and aggravate the need for strict dietary control resulting from aesthetic factors.

These time factors may increase the psychological stress occasioned by the performance itself. At each performance the actor or player is exposed to a stress situation, the peak of which probably occurs at the moment of entry on to the stage. Telemetric measurements of pulse rate indicate that appearance before an audience may produce rates of up to 160 pulses per minute. In addition, during the performance itself the actor is confronted with a wide variety of stress situations such as those presented by the possibility of faulty or missing properties, for example, ill-fitting wigs, jammed doors, etc.

Actors and players are also exposed to a wide range of physical hazards. Firstly there are the hazards involved in the performance itself, and the dangers of tight-rope walking, lion-taming, aerial acrobatics or juggling are obvious. However, since the artist is trained to meet these dangers, they may present less of a hazard that high dust concentrations, poor ventilation, glare from stage lights, defective stage fittings or decorations that may fall or collapse on the unsuspecting artist. The actor may be required to perform on multi-level or revolving stages, fight sword battles on staircases, sometimes in semi-darkness or in glare, and serious falls are not uncommon. Cosmetics used by actors and players may cause skin reactions as may wigs and the adhesives used for fixing facial hair. Finally, the physical effort entailed by a presentation may produce great bodily fatigue; extremely high energy expenditures have been measured in conductors and ballet dancers.

[Injuries of bones and joints are much more frequent among ballet dancers than in many other occupations: they include meniscus injuries, ruptures of the Achilles tendon, patellar bursitis, multiple stress fractures of the femoral necks and tibiae, stress hypertrophy of leg bones and metatarsal deformation.]

A high standard of physical fitness should be maintained by all actors and players and warming-up exercises should be carried out especially by dancers to prevent sprains and strains and minimise the danger of a dancer losing control of his partner during a lifting movement, etc. [Singers and dramatic artists should keep ear, nose and throat disorders under medical control because of their importance for the performance.]

To reduce the problems of stress in these professions, considerable attention should be paid to personal well-being and welfare, particularly when the show is travelling from town to town. Regular meals with readily digestible food rich in vitamins are essential and the excessive consumption of stimulating beverages such as black coffee and alcohol should be discouraged. Adequate accommodation should be available to ensure good rest but the use of sedatives should be avoided.

Efficient stage management and supervision behind the scenes with effective control of properties will reduce the hazards of accidents and relieve the actors and players of much nervous tension. Scenery and wardrobes should be kept clean and, where necessary, disinfected, and cases of sensitivity to cosmetics, etc., should be immediately investigated.

Various countries have passed special legislation aimed at protecting young people employed as actors or players, in particular as far as general and occupational health are concerned; certain international organisations have issued recommendations to the same effect.

GLUCKSMANN, J.

Health impairment:

"Skin injuries of artists" (Les accidents cutanés chez les artistes). Bolgert, M. *Archives des maladies professionnelles, de médecine du travail et de sécurité sociale* (Paris), Sep. 1976, 37/9 (635-640). (In French)

"Considerations on the pelvis of classical dancers" (Considérations sur le bassin de la danseuse classique). Rougier, G.; Mohand Cherif, A. *Archives des maladies professionnelles, de médecine du travail et de sécurité sociale* (Paris), Nov. 1979, 40/11 (1 063-1 065). Illus. (In French)

Protection of children:

"Children in entertainment" (Les enfants du spectacle). Loriot, J.; Leclercq, A.; Philbert, A.; Proteau, J. *Archives des maladies professionnelles, de médecine du travail et de sécurité sociale* (Paris), Sep. 1976, 37/9 (649-655). (In French)

Neuroendocrine impairment in dancers:

"Fatness, puberty and ovulation". Fishman, J. *New England Journal of Medicine* (Boston), 3 July 1980, 303/1 (42-43). 9 ref.

Adaptation

Adaptation is the adjustment of living organisms to a changing environment.

True physiological adaptation in preventive medicine is to be understood as adjustment to a minimal effect which is not attended by a strain of defence mechanisms exceeding the limits of homoeostatic variations. A compensation of the pathological process or "tolerance" (pseudo-adaptation) is observed when the defence mechanisms are under a considerable strain, exceeding the limits of physiological variations. The mechanisms and limits of adaptation are determined by hereditary transmission of morphological and physiological properties which organisms have acquired under known conditions of life on earth and by adjustment to the environment (I. P. Pavlov). An active part in biological adjustment to chemical factors of the environment is played by the neuro-endocrine regulatory systems which accelerate the elimination and fixation of exogenous chemical compounds (immunoglobulins are also involved), and by biochemical mechanisms of detoxication (oxidation by specific oxidases, reduction, hydrolytic decomposition, conjugation, etc.).

Since the changes of the environment may go on more rapidly than the processes of biological adjustment consolidated by heredity, *social adaptation* is needed first and foremost, i.e. the conscious action of society to limit environmental pollution by chemical substances, to control the levels of harmful factors whether physical (noise, vibration, electromagnetic radiation), biological (micro-organisms) or social factors (mental stress, lack of physical exercise), and also to increase the resistance of man and human populations. An important aspect of social adaptation is the establishment of sanitary standards, i.e. tolerable levels of harmful environmental factors, by differentiating physiological adaptation on the one hand and compensation of pathological processes on the other. The levels of action of environmental factors causing physiological adaptation or compensation of pathological processes can be defined by applying the laws of the unity of organism and environment (functional strains), numerical constance of population (balance of reproductive processes and population-reducing processes), and unity of the organism as a biological system (complex and synchronous research into the effects of harmful factors at different levels of the biological organisation of the organism).

Comparative study of the state of different organs and systems often brings to light signs of harmful action of substances. "Tolerance" of the effects of irritative poisons is widely known. However, it is attended by morphological changes of the upper airways. For instance, the Lim_{ir} (the threshold level of the irritative effect) of bromoacetopropylacetate, a typical irritant ($Z_{ir} = 3$), is determined by observing the change in respiration rate in rats. (Z_{ir} is the zone of the specific irritative effect; it is evaluated by the ratio of the threshold limit of the acute irritative effect ($Lim_{ac\,ir}$) to the threshold limit of the acute integrated effect ($Lim_{ac\,int}$).) After prolonged exposure of the animals to the same concentration of this irritant no further changes in rate and type of respiration are observed, but an impairment of the peripheral neurons of the olfactory analyser has taken place, i.e. a loss in sensitivity of the animals to a number of odoriferous substances such as camphor, tar, thymol, rosemary. After an exposure of 4 months there were no changes in external respiration of the animals, but their sensitivity to odoriferous substances decreased on the average by a factor of 8. This decrease was due to significant changes of the mucous membranes of the nose. Apart from neuron impairment, the goblet cells were filled with mucus, the boundaries between the cells vanished, and there were haemorrhage foci in the mucous membrane.

The stage of true adaptation can be discriminated from the compensation of pathological processes by determining metabolites in biosubstrates and comparing them with functional changes.

Valuable results are obtained by drawing biochemical-physiological and morphological parallels. If functional changes are attended by structural changes of tissues and organs, the modifications may be considered deleterious. To evaluate the modifications it is important to watch their evolution. Progressive changes and changes which do not disappear during recovery should be considered deleterious.

It should be emphasised that adjustment to substances with a preponderantly specific action is achieved above all by perfecting the adaptation to the specific action, and therefore the most intimate mechanisms of poison-cell interaction may be involved. For instance the adaptive reactions to the effects of CCl_4 at its Lim_{ac} level start from the cells and subcellular structures of the liver (according to results from changes of organ- and organelle-specific enzymes), and subsequently cause modifications to organ and organ-change indices. These and other examples illustrate that molecular biology is a most promising field for elucidating prognostically important changes.

IZMEROV, N. F.
SANOCKIJ, I. V.

Methodological problems in establishing hygiene standards for industrial factors (Metodologičeskie voprosy gigieničeskogo normirovanija proizvodstvennyh faktorov). Izmerov, N. F.; Kasparov, A. A. (eds.) (Moscow, 1976). (In Russian)

Hygienic and toxicological health hazard criteria for the evaluation of new harmful chemicals (Kriterii vrednosti v gigiene i toksikologii pri ocenke opasnosti novyh himičeskih soedinenij). Sanockij, I. V.; Ulanova, I. P. (Moscow, 1975). (In Russian)

Methods used in the USSR for establishing biologically safe levels of toxic substances. Papers presented at a WHO meeting held in Moscow from 12 to 19 December 1972. (Geneva, World Health Organisation, 1975). 171 p. Ref.

"Development of workplace environment standards in foreign countries. Pt.2–Concepts of higher nervous functions in the USSR". Dinman, B. D. *Journal of Occupational Medicine* (Chicago), July 1976, 18/7 (477-484). Illus. 24 ref.

Adhesives

These are substances used for bonding two or more similar or dissimilar materials.

The advent of synthetic resins has brought about great changes in the adhesives industry in recent years. Natural adhesives, such as glues from animal products, are less and less used in the more highly industrialised countries, and are being increasingly supplanted by synthetic adhesives, which offer a wider range of technical characteristics and are generally cheaper. There are, however, still many adhesives that are made from natural materials such as animal and vegetable tissues. The composition and hazards of this group of adhesives differ considerably from those of the group based on synthetic resins, and natural and synthetic adhesives are therefore regarded as two distinct branches of the same industry.

Natural adhesives

Many natural products have adhesive properties under certain conditions, but not all of them are commonly manufactured or used by industry.

Glue. Glue designates a crude, impure, amber-coloured form of commercial gelatine of unknown composition and with an indefinite melting point. It is produced by hydrolysis of animal collagen. Hides, bones and other collagenous materials are extracted and hydrolysed with hot water; the resulting liquor is concentrated to a predetermined moisture content and, after fat has been skimmed off, is run off at the bottom of the extraction vessel to containers in which it is allowed to set. A liquid glue of a similar nature is commonly made from fish collagen by aqueous extraction of fish skins. Animal glue is used as an adhesive for porous materials, chiefly wood, cloth and paper.

In addition to the obvious hazards associated with the manufacture and use of hot liquids requiring care in handling, these glues may have slightly allergenic properties, which makes it advisable for susceptible persons to wear protective gloves.

Casein adhesive. This material possesses a greater water resistance than fish glue. It is made by treating casein with an alkaline agent such as lime. Because of its water resistance, it is used as an adhesive for wood joints exposed to the weather.

Casein adhesives often contain a flammable petroleum solvent. In such cases the adhesive should not be used in the vicinity of a flame or other possible sources of ignition.

Soybean adhesive. This adhesive is made by dispersing soybean flour in an alkaline solution. It is widely used in the woodworking industry and is free from any hazard.

Dextrines. This group of mixtures belongs to the widely used vegetable gums. The yellow or white powders or granules are soluble in water and insoluble in alcohol. Dextrines are intermediate products formed by hydrolysis of starch. They are manufactured by processing various starches with dilute acids or by heating dry starch. Dextrines are used as adhesives where pleasant, innocuous properties are particularly desirable, e.g. for stamps and envelopes.

While these adhesives are harmless in use, the hydrolysis of starch is an exothermic reaction, and disastrous dust explosions have been associated with their production. Safety measures include scrupulous cleanliness in dextrine factories to ensure that dust does not accumulate in the process rooms. Segregation of processes and explosion reliefs help to prevent the propagation of dust explosions (see EXPLOSIVE SUBSTANCES; DUST EXPLOSIONS).

Gum arabic. See separate article.

Latex. This name was at one time applied exclusively to the milky sap of the rubber tree, but it is now used to describe uncured natural and synthetic rubbers.

Natural latex as collected from the tree is mixed with ammonia to prevent its decomposition or coagulation until it is processed. One of its numerous uses after preliminary processing is the manufacture of an adhesive especially designed for rubber articles where the bond is subsequently cured by vulcanisation.

Little danger is associated with the handling of natural latex even when it is in ammoniacal suspension. However, certain health and fire hazards may arise from vulcanisation, particularly if a cold curing process involving carbon disulphide or sulphur chloride is employed.

Rubber cement. It is obtained by dissolving natural rubber in organic solvents (generally solvent naphtha) with some additives (see below).

Mucilages. Many harmless adhesive formulations containing starch and dextrine mucilages are available for industrial and domestic use.

Synthetic adhesives

These are compositions containing synthetic polymers and other ingredients. Synthetic adhesives are nowadays extremely widely used in all branches of industry, because bonded assemblies present great advantages over riveted, bolted, welded, sewn and other joints. They offer the possibility of bonding together the most heterogenous materials: metals and their alloys, various plastics, synthetic fibres, polymeric and silicate glass, natural and artificial leather, natural and synthetic rubber, porcelain, earthenware, concrete, various kinds of wood, paper, cotton and wool fabrics. The biological effects of these adhesives on workers depend on their chemical composition and conditions of use.

Synthetic rubber cements. These adhesives are generally manufactured by dissolving synthetic rubber (polychloropropene, butadiene-styrene, polyurethane, etc.) in organic solvents (solvent naphtha, ethyl or butyl acetate, chlorinated hydrocarbons, etc.) and by adding vulcanising agents, vulcanisation accelerators and catalysts.

The biological effects of these adhesives depend on the properties of the organic solvents used, the most harmful ones of which are benzene, carbon disulphide and chlorinated hydrocarbons. Inhalation of solvent vapours may cause acute or chronic poisoning, and skin contact may give rise to dermatitis. The dermatological effects of the solvents contained in these adhesives are potentiated by certain additives (resins, catalysts) which have sensitising properties.

Synthetic rubber latexes. They are intermediate products in the manufacture of synthetic rubbers, i.e. colloids which precede the precipitation of the solid elastomer. Their toxicity depends on the amount of free, unpolymerised monomer (chloroprene, styrene, acrylonitrile) contained in the latex. Skin disease may be caused by direct contact.

[A number of cases of polyneuropathy due to natural and synthetic rubber adhesives and their solvents used in the footwear industry have been observed in the last 20 years. The solvents most frequently involved were paraffin hydrocarbons with low boiling point such as n-hexane, cyclohexane, 2-methylpentane, 3-methylpentane and methylcyclopentane. Triorthocresyl phosphate has also been suspected of being responsible for these injuries; however, no conclusive evidence has thus far been provided. The aetiology of the polyneuropathy seems to be rather complex and may differ in different observations (see FOOTWEAR INDUSTRY; PERIPHERAL NERVE EFFECTS).]

Urea- and melamine-formaldehyde resins. These resins belong to the most widely used synthetic adhesives. They are obtained by condensation polymerisation of urea (or melamine) with formaldehyde and serve as adhesives for wood, paper and fabrics. They release formaldehyde and, at high temperatures, carbon oxides and ammonia.

Cases of irritation of the upper airways and dermatitis due to exposure to both formaldehyde and resins have been observed among workers.

Phenol-formaldehyde resins. They are obtained by reaction of phenol (sometimes also cresol and resorcin) with formaldehyde in the presence of a catalyst. They give off phenol and formaldehyde, and the solution of these resins releases solvent vapours (e.g. acetone). These substances may cause both general and local effects (dermatitis and eczema of the hands and face and respiratory impairment).

Epoxy resins. Liquid epoxy resins are obtained by condensation of epichlorhydrin with bivalent alcohols or phenols, and commonly also with diphenylpropane or resorcin. Amines (hexamethylenediamine, polyethylenepolyamine) and anhydrides (maleic or phthalic acid anhydride) serve generally as hardeners for these adhesives which are widely used in various industries for bonding glass, ceramics and metals. A peculiarity of the industrial use of epoxy resins is that they harden immediately upon application and give off volatile ingredients such as epichlorhydrin, toluene, diphenylolpropane and others. They frequently cause skin disease in exposed workers, resulting from both direct contact and sensitisation by the resin. The dermatological effects are potentiated by the hardeners. Dermatitis is sometimes attended by irritation of the eyes and upper airways. Cases of asthmatic bronchitis and bronchial asthma have been observed.

Polyurethanes. They are obtained by condensation and polymerisation from diisocyanates, dicarboxylic acids and polyvalent alcohols or polyesters containing two or more hydroxyl groups. The diisocyanates most frequently used in the manufacture of polyurethane adhesives are tolylene diisocyanate and 4,4'-diphenylmethane diisocyanate. Chlorinated hydrocarbons serve as hardeners, but it is recommended to replace them by less toxic products, e.g. by ketones. Large amounts of volatile diisocyanates and solvent vapours are released when the adhesive is mixed with the hardener. Mixing at the workplace therefore constitutes a health hazard.

Diisocyanate vapours may cause severe lung impairment (asthmatic bronchitis) and also give rise to conjunctivitis and inflammatory skin disease.

Polyvinyl acetate. It is obtained by polymerisation of vinyl acetate. This adhesive possesses a slight biological activity which is due to the irritative effect of the residual vinyl acetate monomer. The toxicity of the adhesive is potentiated when it contains plasticisers (dioctyl or dibutyl phthalate, tricresylphosphate, etc.).

Hot melts. They are made from high-molecular compounds such as polyamides and polyesters. They acquire their adhesive properties when they are heated to the melting point prior to application to the surface to be sealed. These adhesives are advantageous from the occupational hygiene viewpoint because they contain no solvents, need no drying after application and do not contaminate the hands. However, they set free thermolysis products when they are heated, such as esters of unsaturated fatty acids, carbon monoxide, ammonia, etc.

PREVENTIVE MEASURES

Many types of adhesive contain organic solvents, which present the main occupational hazard and necessitate special preventive measures.

Benzene is no longer used in many countries for the manufacture of adhesives because of its highly toxic properties. Most organic solvents are flammable and more or less harmful to health. Flammable liquids and solvents should be transported in special containers. Jars or pots with adhesives used at the workplace must be of special design and have an opening which should be as small as possible to limit evaporation. It should be forbidden to keep open jars or pots with adhesives containing organic solvents at the workplace. Such adhesives must not be used where there is a fire hazard and smoking should be prohibited in premises where they are employed.

The use of harmful solvents such as benzene, carbon disulphide and the most dangerous among the chlorinated hydrocarbons should be discontinued. These solvents should be replaced by less harmful ones (acetone and other ketones, acetic acid esters, perchloroethylene, solvent naphthas containing a minimum of aromatic hydrocarbons). Polyurethane adhesives should preferably contain non-volatile isocyanates. Since many processes may give rise to the release of large amounts of solvent vapours, monomers or polymer decomposition products, the process equipment should be tightly enclosed and provided with an exhaust system to prevent contamination of the workplace air by harmful substances. Many adhesive applications require heating and are associated with the spread of harmful vapours and gases. Preference should therefore be given to techniques which dispense with heating of adhesive compounds.

Most of the solvents used may degrease the skin and cause dermatitis on the hands of workers wearing no protective gloves and using no barrier creams. Skin protection must also be worn when working with many other adhesives containing no organic solvents. Protective creams should be chosen according to the properties of the adhesive handled; they can be either hydrophilic or water-repellent.

VOLKOVA, Z. A.

Harmful substances in industry. Vol. II: *Organic substances. Manual for chemists, engineers and physicians* (Vrednye veščestva v promyšlennosti, t.II. Organičeskie veščestva. Spravočnik dlja himikov, iňenerov i vračej). Lazarev, N. V.; Levina, E. N. (eds.) (Moscow, Izdatel'stvo "Himija", 1976). (In Russian)

Occupational skin diseases in the principal sectors of the national economy (Professional'nye zabolevanija kŏi v veduščih otrasljah narodnogo hozjajstva). Somov, B. A.; Dolgov, A. P. (Moscow, 1976). (In Russian)

CIS 80-317 "Determination of flashpoint of adhesives" (Opredelenie temperatury vspyški kleevyh kompozicij). Putilov, A. V.; Kuz'min, V. I. *Koženno-obuvnaja promyšlennost,* Apr. 1979, 4 (63). (In Russian)

CIS 80-1017 *Petroleum based adhesives in building operations.* Health and Safety Executive, Guidance Note EH7 (London, HM Stationery Office, Sep. 1977), 3 p.

CIS 79-1971 "Polyneuropathy due to glues and solvents" (La polineuropatia da collanti e solventi). Mazzella di Bosco, M. *Rivista degli infortuni e delle malattie professionali* (Rome), Nov.-Dec. 1978, 117/6 (1 163-1 169). 33 ref. (In Italian)

CIS 79-1646 "Allergic properties of p-tert-butylphenol-formaldehyde resins" (Zur Allergennatur der para-tert. Butylphenolformaldehydharze). Schubert, H.; Agatha, G. *Dermatosen in Beruf und Umwelt* (Aulendorf (Federal Republic of Germany)), 1979, 27/2 (49-52). 22 ref. (In German)

Aerosols

Aerosols are defined as suspensions of solid or liquid particles in gases. The particle diameters range from about 0.001 to about 100 µm. Typical mass concentrations may range from 10^{-9} to 10 g per cubic metre of gas. Aerosol systems are of great importance in environmental control. Accurate methods are required to assess the air quality.

Analysis of an aerosol system may require the simultaneous consideration of the interactions of the particles with the gas phase, the interactions between particles, the kinetics of particle behaviour in the flow field, and the motion of particles with respect to constraining flow boundaries or obstacles in the flow stream.

In industrial hygiene surveys of air contaminants in the working environment a distinction is made between dusts, fumes, smokes, and mists and fogs.

Dusts are generally formed by disintegration processes, such as in mining and ore-reduction operations. Examples are silica and asbestos dusts.

Fumes usually result from chemical reactions, such as oxidation, or from sublimation or distillation processes followed by condensation. Examples are oxides of iron and copper.

Smokes result from the combustion of fossil fuels, asphaltic materials, and wood. Smokes consist of soot, liquid droplets and, as in the case of wood and coal, a significant material-ash fraction.

Mists and *fogs* consist of liquid droplets produced by atomisation or condensation processes. Examples are oil mists from cutting and grinding operations, mists from spraying operations.

Emissions

Emissions include in occupational health terms the formation of dust, fumes, gases, radiation, micro-organisms, etc. They arise from both primary and secondary sources and can be either mobile or stationary. The source strength or emission rate is expressed in quantities per unit time.

Emission control in the working environment is used to locate and quantify all relevant sources, whether point sources or surfaces and whether mobile or stationary. Variations in source strength are determined. Analysis may deal with workshop layout, process and flow charts of workplaces and production areas, and transport. All work operations and processes releasing toxic materials into the working environment should be investigated. The first step is to map all materials that may be used or produced in the work operations or processes. Many raw materials are first identified by trade names. Additional information on the composition of these materials must be obtained from the supplier so that the constituents can be identified and evaluated properly. Checklists are very useful (table 1).

Table 1. Potentially hazardous operations and air contaminants

Process types	Contaminant type	Contaminant examples
Hot operations:		
Welding, chemical reactions, soldering, burning	Gases (g), particulates (p) (dust, fumes, mists)	Chromates (p), zinc and compounds (p), manganese and compounds (p), carbon monoxide (g), ozone (g), vinyl chloride (g)
Liquid operations:		
Painting, degreasing, spraying, coating, cleaning, galvanising	Vapours (v), gases (g), mists (m)	Benzene (v), trichlorethylene (v), sulphuric acid (m), cyanide salts (m), hydrogen cyanide (g), sulphur dioxide (g)
Solid operations:		
Pouring, extraction, crushing, conveying, loading, bagging	Dusts	Cement, quartz (free silica), fibrous glass
Pressurised spraying:		
Cleaning parts, applying pesticides, degreasing, sand blasting, painting	Vapours (v), dusts (d), mists (m)	Organic solvents (v), parathion (m), quartz (free silica) (d)
Shaping operations:		
Cutting, grinding, filing, milling, moulding, sawing, drilling	Dusts	Asbestos, beryllium, uranium, zinc, lead

Emission monitoring systems must be integrated into the technical process as a whole. Emission control systems (which are important in all workplaces, not only in large ones) can be designed, installed and operated so as to initiate safety measures automatically or manually. For example they may shut down a process, raise an evacuation alarm, isolate faulty filtering systems and commission back-up systems, etc. Emission profiles can be drawn continuously for each process and used to determine the most appropriate emission control measures.

In the case of continuous emissions, the simplest approach is to instal filters and air-cleaning devices as well as monitoring and measuring systems. Extreme fluctuations in the emission level increase the difficulties of designing air cleaners and specifying methods of measurement. Automatic monitoring systems offer the advantage of reducing the risk of human error. However, such systems can be rendered inoperative by mechanical or electrical faults. Manual systems are therefore needed, but must be supplemented with suitable control functions.

A knowledge of the emission levels of machines and installations is basic to designing technical preventive measures. Standard measurement criteria are necessary for calculating and evaluating these levels.

Immissions

Immissions are levels or doses expressing exposure. The exposure is often related to exposure limits. Efficient management of air contaminants must be based on an understanding of how pollutants flow, disperse and are transformed into other physical and chemical forms as they move from source to receptor.

The behaviour of aerosol particles is complex and multifaceted. The diffusion of aerosol particles is due to bombardment by molecules of the gas, i.e. Brownian motion. Aerosol particles coagulate when they collide with one another and larger particles are formed. The forces causing such collisions are thermal, electrical, molecular, gravitational, inertial, acoustic, etc. Many factors affect the rate of coagulation, such as degree of heterogeneity, mixtures of particles of various substances, turbulence, particle shape, electrical effects, temperature, pressure and viscosity.

In order to evaluate exposure it is necessary to monitor the environment. This requires an adequate strategy and a variety of instruments. The choice of measuring method is dependent on the type of airborne pollutant to be monitored. Both direct-indicating and non-direct-indicating methods are available. The characteristic feature of non-direct-indicating methods is that the pollutants are first sampled and then analysed. Direct-indicating methods provide results at the actual site. The following are some examples of the various types of pollutants and some suitable methods of measurement.

TECHNICAL PREVENTIVE MEASURES

Knowledge of the contaminants and concentrations present at the workplace are important for adequate preventive measures. In addition, it is necessary to know the characteristics of workplaces, processes and ventilation systems.

Airborne pollutants can be classified into hazard classes in terms of permissible limits and the emission rate of the substances concerned. Some examples are given in table 3.

Depending upon the type of operation and the nature of other activities in the work area, exhaust volumes may contain only particulate matter, or particulate matter together with gases or vapours.

Table 2. Examples of pollutants associated with conventional material working methods[1]

Operation	Source of pollution	Type of pollution	Sampling method	Method of analysis
Grinding	Grinding wheel, workpiece	Dust	Filtration	Weighing, etc.
	Grinding fluid (if used)	Oil mist / Oil vapour	GD tube bubblers	IR spectro-photometry
Cutting operations	Cutting fluid	Oil mist / Oil vapour	GD tube bubblers	IR spectro-photometry
	Workpiece	Dust	Filtration	Weighing, etc.
Hot forging	Lubricant	Lubricant mist / Oil vapour	GD tube bubblers	IR spectro-photography
	Scale	Dust	Filtration	Weighing, etc.
	Fuel	Combustion gases	Detector tubes, etc.	

[1] Report summaries from the Swedish Work Environment Fund (ASF No. 76).

Table 3. Typical classification into hazard classes[1]

Intensity class (emission rate)	Hazard potential (threshold limit values–TLV)	Hazard class	Minimum recommended capture-air velocity (m/s)
1. Strong	A. 0-10 ppm	A1, A2	0.76
2. Average	B. 11-100 ppm	A3, B1, B2, C1	0.50
3. Low	C. 101-500 ppm	B3, C2, D1	0.40
4. Negligible	D. above 500 ppm	A4, C3, D2	0.25
		B4, C4, D3, D4	Good general ventilation sufficient

Surface treatment process	Process chemicals which may produce airborne pollution	Temperature (°C)	Type of pollution	Hazard class
Degreasing (dipping) using trichloroethylene	Trichoroethylene	20-30	Trichloroethylene vapour	B1
Electrolytic degreasing	Sodium hydroxide (69-120 g/1)	50-90	Sodium hydroxide-bearing mist	C1
Pickling (hydrochloric acid)	Hydrochloric acid (5-20% by weight)	20-40	Hydrochloric acid gas	A1
Cadmium plating	Cyanides, cadmium	20-30	Bath mist containing cadmium and cyanides	A3

[1] Swedish Work Environment Fund report ASF No. 31.

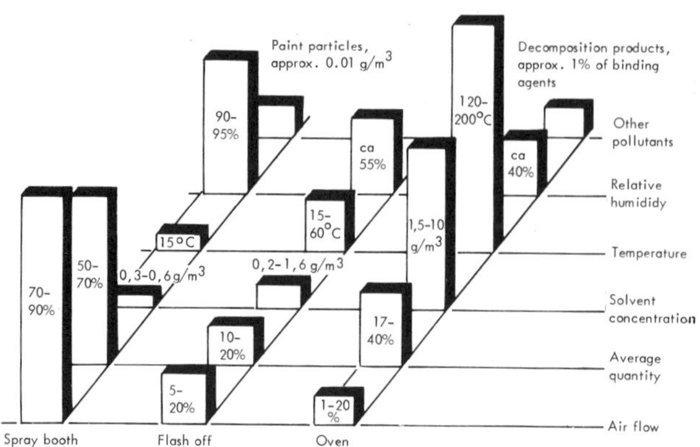

Figure 1. Variations in exhaust air flow–calculated example. (From *Färg och Fernissa* (Paint and Varnishes), 2/1979.)

Exhaust air cannot be defined in unique terms. The exhaust air from a painting installation, for example, may be described in terms of a variety of parameters depending on the particular unit–spray booth, flash-off zone or oven–from which it is extracted. The same considerations govern the choice of purification method. Figure 1 illustrates how the quality of the exhaust air differs depending on the three sources, expressed in terms of the various parameters.

Consequently the design procedure must include a number of steps based on determining the concentration of each contaminant in each specified location. Proposals for modification may concern workplaces, process, equipment, instructions and forms of work and are based on hazard descriptions and exposure levels. Possible means of reducing the emission to the lowest possible level are evaluated and one or more air cleaning devices which, in series, and/or parallel, will significantly reduce the concentration of each emitted contaminant are selected.

Special importance attaches to single contaminants the concentration of which it is difficult to reduce to an acceptable level because of their high toxicity or the lack of sufficiently efficient air-cleaning devices. In such cases a special system should be designed to achieve the desired breathing zone concentration for the contaminant concerned.

New methods of sampling and analysis

Remarkable advances have been made in recent years in the development of new and improved methods of sampling and analysis of aerosols. Sensitivity, precision and accuracy have improved. Advanced techniques have been developed such as chromatography coupled with mass spectrometry and aided by computer processing, chemiluminiscence, X-ray fluorescence and photoelectron spectroscopy, proton-induced X-ray emission. Light-scattering methods are very suitable for routine measurements. The laser offers new possibilities for small-particle measurements. The hologram technique allows the recording of the images of many individual particles in a cloud simultaneously. In regular photography, diffraction from the small particles makes size determination impossible. This effect, however, is being used in holography to determine the particle size. Continuous sampling and monitoring instruments facilitate long-term measurements.

Sampling methods are related to aerosol category, sampling rate and particle rate (table 4).

Size analysis has several purposes and the methods used depend on size range (table 5), physical possibilities and chemical composition.

Table 4. Examples of sampling category

Sampling category	Volumetric rate, l/min	Diameter of applicability, µm
Horizontal elutriator	0.5-2.5	1-20
Filter		
Fibrous	1000	> 0.001
Moving tape	26	> 0.5
Membrane	0.5-100	> 0.001
Impactor		
Fixed stage	20-1 000	0.5-100
Moving stage	75	0.5-100
Impinger		
Liquid	3-28	0.1-50
Bacterial	28-500	0.1-50
Centrifuge	1-8	0.1-10
Cyclone	100-2 000	> 5
Precipitation		
Electric	4-15 000	0.05-5
Thermal	0.5	0.005-2
Condensation	l-3	0.001-0.1

Table 5. Size measurements

Method	Diameter of applicability, µm
Optical	
Light imaging	0.5+
Electron imaging	0.001-15
Light scanning	1+
Electron scanning	0.1+
Direct photography	5+
Laser holography	3+
Sieving	2+
Light scattering	
Right angle	0.5+
Forward	0.3-10
Polarisation	0.3-3
With condensation	0.01-0.1
Laser scan	5+
Electric	
Current alteration	0.5+
Ion counting, unit charge	0.01-0.1
Ion counting, corona charging	0.015-1.2
Impaction	0.5+
Centrifugation	0.1+
Diffusion battery	0.001-0.5
Acoustical	
Orifice passage	15+
Sinusoidal vibration	1+
Thermal	0.1-1
Spectrothermal emission	0.1+

Characterisation of particulate composition is important in several fields: research, assessment of health effects, identification of sources. The composition and effects of particles often vary with their size. Numerous organic compounds occur in particulate matter or are adsorbed on the surface of inorganic particles. Methods to meet these different requirements have been developed but should be used with skill.

Relevant methods should be carefully selected, more sophisticated and accurate methods than are necessary for the purpose being avoided. Analytical instrumentation has increased the importance of calibration and collaborative testing (see for details DUST SAMPLING; FIBRES, NATURAL (MINERAL)).

Personal respiratory protection

In situations where local or general ventilation systems are not convenient or effective in reducing aerosol levels, personal respiratory protection is required, the type depending on the aerosols concerned. The most common hazards requiring respirators are divided into the following categories:

(a) Particulate contaminants: immediately harmful; not immediately harmful.

(b) Gas or vapour contaminants: immediately harmful; not immediately harmful.

(c) Combination of gas, vapour and particulate contaminants: immediately harmful; not immediately harmful.

Respirators may be divided into two general categories, air supply systems and air cleaning devices. Type and degree of exposure determine the choice between one of the following types:

Hose mask respirators. These comprise a full face-piece fitted with a length of relatively large-bore air hose, through which air from a clean source is drawn by the normal breathing action of the wearer.

Air line respirators. These may be of full face-piece, half face-piece, hood or helmet type. Their common feature is the supply of clean breathable air at suitable pressure from a remote source.

Self-contained breathing apparatus. This equipment, being entirely self-contained, with cylinder-fed air, the user is not dependent on an air compressor subject to failure. It is recommended for use in confined spaces.

Dust respirators. These may be of full-face type or the half-face type and are fitted with dust cartridges.

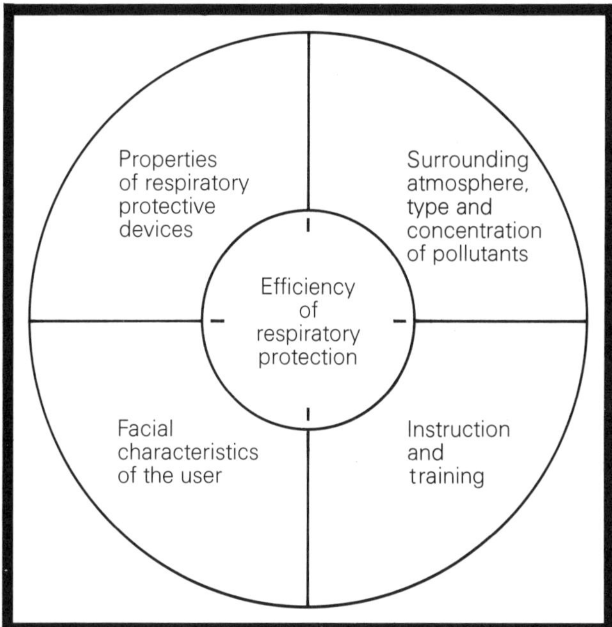

Figure 2. Important factors to observe when using respirators.

In order to obtain a satisfactory respiratory protection all factors outlined in figure 2 should be considered. See also RESPIRATORY PROTECTIVE EQUIPMENT.

GERHARDSSON, G.

Handbook on Aerosols. Dennis, R. (ed.) (Oak Ridge (Tennessee), TIC, 1977).

CIS-79-970 *Fundamentals of aerosol science.* Ward, D. T.; Fuchs, N. A.; Mercer, T. T.; Davies, C. N.; Kirsch, A. A.; Stechkina, I. B.; Spurny, K. R.; Gentry, J. W.; Stöber, W.; Pich, J. (Bognor Regis (West Sussex), John Wiley and Sons, 1978), 372 p. Illus. 520 ref.

Particulate air pollution. Problems and solutions. Theordore, L.; Sosa, F. Y.; Fajardo, P. L.; Buonicore, A. J. (Boca Raton (Florida), CRC Press, 1980), 112 p.

CIS 80-960 *Aerosol measurement.* Lundgren, D. A.; Harris, F. S.; Marlow, W. H.; Lippmann, M.; Clark, W. E.; Durham, M. D. (Gainesville, University Presses of Florida, 1979), 716 p. Illus. 690 ref.

Aerospace (ground operations)

Space vehicles, both manned and unmanned, are utilised and will be utilised in the future for the scientific exploration of space and the Earth itself. With respect to study of the Earth, they provide technologically superior methods of world-wide weather observation and forecasting, relay radio and television broadcasting at great distances between countries, provide useful global navigation beacons independent of weather and time of day, and enhance our knowledge of the physical properties of the Earth through photographic, televised or visual observations. Direct contributions to scientific research are found in the fields of biology, geophysics, astronomy and astrophysics and cosmology.

Launching and recovery. In order to place a space vehicle in orbit around the Earth or out of orbit towards distant planets, powerful rocket engines are used. The rocket engines required for launch are components of the "launch vehicle" which does not include the spacecraft or satellite (payload). Launch vehicles have four elements in common:

(a) Rocket engine. A chemical engine which burns a fuel and oxidiser to generate large volumes of hot gases that produce a thrust when expanded through a nozzle.

(b) Fuel and oxidisers. These are, if liquid in form, stored in tanks and fed to the engine by various means. If solid in form, the solid propellant grains are stored in the body of the launch vehicle where they burn when ignited.

(c) A superstructure which supports the above components.

(d) Guidance and control system. This stabilises the launch vehicle and keeps it on the predetermined path. The path is computer controlled, either from the ground or from the space vehicle itself.

Launch and recovery operations vary with the type of satellite or space vehicle used. Many satellites are not intended for recovery, and disintegrate from heat when they re-enter the Earth's atmosphere from declining orbit, or continue in space indefinitely. These vehicles are instrumented to transmit signals back to Earth or perform relaying operations for limited periods, and are eventually lost. Manned vehicles and vehicles with biological and other payloads, for which recovery is desirable, must be equipped to withstand the stress of re-entry and are recovered after landing by specially designated ships or during parachute descent by aircraft designed to snatch the spacecraft from the sky.

Launch facilities. A typical launch facility consists of industrial areas for assembly of the missiles and their payloads, areas for production and/or storage of propellants, a computerised launch control area which may serve several launch pads, and the launch pads themselves. Large spacecraft centres, such as Cape Kennedy, employ thousands of technical and support personnel. Small facilities designed to handle unmanned spacecraft or satellites are less extensive and require fewer personnel. Ground equipment for support of the missiles and their payloads perform all functions to support the airborne equipment and to simplify airborne equipment, thus increasing reliability and improving launch efficiency. This equipment is also used to prepare, transfer and control the raw materials that go into propellants, as well as to display the status of launch preparation and execution (countdown) and range safety.

A wide variety of occupational skills is found in the ground personnel who perform all the tasks essential to preparation, launch and recovery of space vehicles. The personnel and equipment comprise a large industrial complex and the industrial hygiene and occupational medical problems for the greatest portion of the operation differ only in magnitude from those of other industries of comparable size.

HAZARDS AND THEIR PREVENTION

Maintenance and launch crews are subjected to a variety of environmental influences which may affect them in many ways. Figure 1 depicts the major personal, environmental and occupational factors, and the discussion that follows outlines control features.

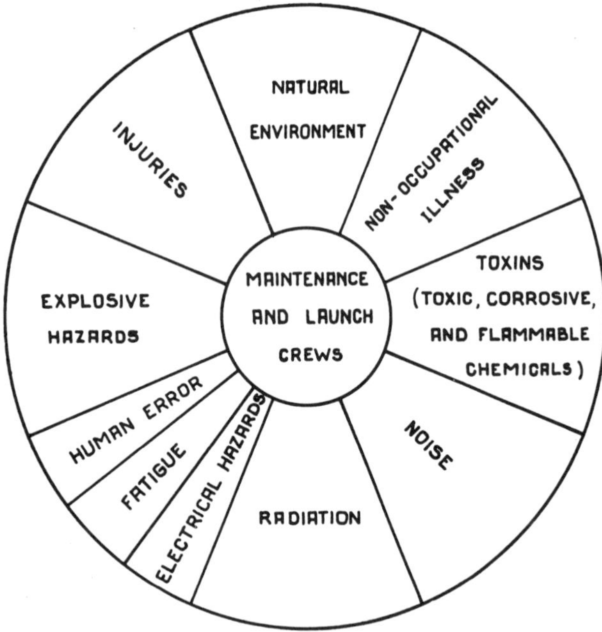

Figure 1. Personal and environmental factors affecting aerospace ground personnel. The various factors act alone or in concert at varying intervals. The magnitude of these factors varies with time and circumstances of exposure.

All occupational hazards peculiar to aerospace ground personnel may be viewed as adverse environmental factors impinging on the worker and thereby affecting the efficient performance of his duties. The chart exemplifies the various environmental factors which can have an adverse effect on aerospace ground personnel.

Natural environment. Starting at the top of the circle, the "natural environment" refers to the micrometeorologic conditions extant at the base site. The temperature variations, humidity, and transient weather conditions, such as storms, heavy rainfall, snow, etc., can all affect the workers. These factors must be taken into account in site planning, but control of these environmental factors is not a reality at present. Prevention of health hazards to persons living on or in the vicinity of missile launch sites from environmental pollution (air, soil, water) is vital and must be a major aim in planning the site and site location, as well as control techniques.

Non-occupational illnesses. Diseases of particular interest to operational planning are the infectious diseases, especially those that may cause epidemics. Recommended immunisation procedures will vary from one geographical locale to another. Cardio-pulmonary disorders which may be permanently disabling require careful medical surveillance.

Toxic substances. These include toxic and corrosive chemicals used in the operation of the launch site as well as the fuels and oxidisers used in the rockets themselves.

Propellants are grouped into two major types, liquid and solid. These types are used separately or in combination to power launch vehicles. Higher specific impulse from liquid propellants may dictate their use in certain cases. Newer developments in production of solid propellants may make their use more universal. This would be most desirable from the point of view of hazards reduction since solids are easier to handle and present only minor toxicity problems for the most part.

The high vapour pressure of liquid propellants is attended by greater health hazards in handling. Some typical high energy storable liquid propellants are: hydrazine, 1,1-dimethylhydrazine (UDMH), Aerozine 50, RP-1, nitrogen tetroxide, the fluorine derivatives, and other less frequently used compounds such as cryogenic propellants. Use of liquid propellants requires more complex engines and difficult ground handling operations in the storage and transfer of fuels to the missile. Further developments in highly dangerous chemicals will be accompanied by new hazards, so research and development on new propellants must be accompanied by a parallel development of biological data on the health hazards of these propellants.

Control of propellant hazards depends on the development of carefully planned techniques for transferring the fuel from trucks or railroad cars to storage facilities, and the eventual transfer of the liquids to the rocket. Maximal use of automated systems for transfer will reduce exposure possibilities to a minimum. However, there are practical limits to the degree of automation possible, and human handling is possible, hence the liability to exposure. Protective clothing in the form of rubberised materials which contain compressed air supplies for breathing in contaminated atmospheres are available and in common use. In the event that accident exposure does occur, medical facilities on hand must be adequate to deal with the emergency or even to render definitive treatment if the base is isolated.

Hydrazines may produce haemolytic anaemia, renal damage, and hepatotoxicity. Individuals selected for duties as missile fuel handlers should undergo pre-employment or pre-placement medical evaluation. Since the effects of exposures to organic solvents and excess ethanol intake may produce liver changes indistinguishable from hydrazine hepatotoxicity, and since pre-existing liver disease is believed to increase susceptibility to hepatotoxins, individuals with previous history of liver disease and heavy alcohol intake should be excluded. Prior exposure to hepatotoxic chemicals will also confuse results of periodic examinations, and persons so exposed may also be excluded. Pre-employment and periodic evaluations should include, but not necessarily be limited to, a careful occupational history, general physical examination, haematocrit determination, white blood cell count, urinalysis, chest X-ray, pulmonary function studies and indices of liver function (such as serum glutamic pyruvic transaminase determination) (see ROCKET PROPELLANTS; HYDRAZINE AND DERIVATIVES).

Noise. This is an unavoidable partner in rocket operations. Noise level profiles of rockets to be used and estimates of noise levels of rockets yet to be built must be considered in the planning and construction of a rocket base. Distance attenuation is by far the most important control device now used to protect the public from blast-off noise. Workers at the site must be protected inside structures which have noise attenuation characteristics built in by virtue of their explosion-protective construction. The noise associated with actual launch of space vehicles is dramatic and massive in proportions. Also of serious import, however, is the noise resulting from daily operations of the equipment and machinery at the base. Noise levels at operating bases currently in existence reach above the threshold of damage risk criteria for workers in the environment. Protective measures are necessary in these environments and must be prescribed on the basis of noise level surveys conducted by industrial hygiene engineers.

Radiation hazards. These may be of several types. Radio frequency radiation may well be present in the vicinity of missile bases in close proximity to radar instrumentation. Maintenance personnel should be warned against dangers of working on equipment with power on. Radioactivity from other sources, such as power cells for space operation (krypton-85 batteries are γ-ray and Bremsstrahlung X-ray emitters) must be under constant surveillance and protection should be afforded to all workers. α-Emitters become a problem only when conditions permitting inhalation exist, such as fires. The thorium-magnesium alloys used in fabricating missile airframes can yield inhalable α-sources when burned.

Industrial radiography, using X-rays or radioisotope power sources (Cobalt-60, Iridium-192), requires personnel protection and monitoring.

The health physicist must play a key role here. He should maintain a radiological locator log, mapping all areas where hazards exist. He must pursue a vigorous protection and monitoring programme, ensuring that proper exclusion distances are observed as well as proper use of shielding and limitation of exposure times. Use of personal dosimeters, survey instruments, and medical evaluation through bioassay are important elements of the health physicist's work.

Electrical hazards. These include base current, static electricity and missile current exposure possibilities. Good industrial safety practices should provide adequate protection against these electrical hazards.

Fatigue and human error. These are mentioned together since fatigue is often a contributing factor to human error. These factors are of paramount importance and the industrial physician and his team will be taxed to the utmost to detect and control them and the elements contributing to their exacerbation. The launch of missiles requires the accomplishment of a number of detailed interconnected, technical tasks. Proper accomplishment of the tasks requires a great deal of high technical skill, efficiency and physical and mental well-being. The tension produced by a missile launch offers significant deterring effects to human precision performance. Hours of work and manning requirements must be monitored; excessive overtime should be avoided and pre-launch preparation should be so planned as not to exceed the limitations of launch personnel. Coping with problems as they arise will ensure that the health team is always cognisant of developments and may assist in reducing the basic elements in otherwise complex problems.

Explosive hazards. These may result from fuels and oxidisers, explosive mechanisms used for rocket and payload components separation during or after launch, from an aborted launch where the missile drops on the pad or in the vicinity, and from those explosive elements used in the missile for intentional destruction if the flight path is not correct. Rupture or disconnection of high-pressure lines and tanks provide a real and constant threat to safety. Here again, the safety engineer must be constantly alert to all possibilities. Medical planning for disasters is an absolute requirement. Distance requirements and explosion-protective construction will minimise injuries. Explosive damage is the combined result of air blast, shock waves and debris which is hurled by the blast. Planning for explosion protection requires studies of terrain, shielding, distance separation, estimation of explosive and combustible materials which could be involved and human reactions and tolerance levels to blast effects. Blast effects as they apply to humans are conveniently categorised as primary (from blast induced variations in environmental pressure), secondary (injuries due to impact of missiles from the detonating device, from debris energised by blast pressures, winds, ground shock or gravity), tertiary (whole body displacement) and miscellaneous effects (dust, thermal damage, toxic gases such as carbon monoxide).

Miscellaneous injuries. At missile bases, other injuries follow the same pattern as for comparable industrial operations. There is clear indication from analysis of records of operational missile launch sites that there is no essential difference between the accident experience at launch sites and that in industrial operations. Only the magnitude of the problems varies.

Hazard control

Advance planning is the key to control of environmental and occupational hazards. This planning must commence in the earliest phases of designing an operational facility. The occupational medicine specialist and industrial hygiene engineer (biomedical engineer) must work together in a co-operative effort with engineers responsible for over-all design and construction of the facility. A team must be formed which consists of, in addition to the above, chemical and rocket engineers, safety engineers, toxicologists, micrometeorologists and consultants in various highly specialised disciplines. The team personnel must decide the most favourable site for construction of the facility, toxic exclusion radii, safe distance for explosive protection, detection and monitoring of propellants, and sources and prevention of acute and long-term environmental pollution. The team must continue to function after the facility becomes operational to detect and monitor problem areas and to take rapid, effective action to correct difficulties which arise (both unexpected and expected). A sound programme of preventive medical practice is an asset of inestimable significance. Pre-employment and pre-placement medical evaluations will contribute to the general health and welfare of workers and improve the efficiency of the operation. No amount of pre-planning will obviate all problems, hence the continued operation of the team must be assured.

MASTERS, R. L.

"Biologic risks resulting from the electromagnetic environment in aerospace activities" (Risques biologiques résultant de l'environnement electromagnétique dans les activités aérospatiales). Martin, A. V. J. *Aéronautique et astronautique* (Paris), 1976, 61 (53-72). (In French)

"Rocket-propellant inhalation in the Apollo-Soyuz astronauts". Dejournette, R. L. *Radiology* (Syracuse (New York)), 1977, 125/1 (21-24).

CIS 77-616 "Heat radiation from fires of organic peroxides as compared with propellant fires". Groothuizen, T. M.; Romijn, J. *Journal of Hazardous Materials* (Amsterdam), Nov. 1976, 1/3 (191-198). Illus. 3 ref.

"Hearing levels of aerospace workers as affected by duration of employment". Martin, O. E.; Crowder, W. F. *American Industrial Hygiene Association Journal* (Akron), Nov. 1978, 39/11 (860-865). Illus. 3 ref.

Aerospace (space operations)

The propulsion of man into space has presented a wide range of complex safety and health problems, some of which had already been encountered, although in a less acute form, in other fields, and of others of which man had had no previous direct experience.

These problems that have faced the scientists preparing the entry of man into space include:

(a) compensation of atmospheric pressure and the provision of respirable gas mixtures;

(b) the effect of the massive acceleration to which the astronaut is exposed on leaving and returning to the Earth;

(c) the absence of gravity experienced in freefall in space or in a satellite orbiting the earth—in the latter case, the force of gravity is equal and opposite to the centrifugal force;

(d) exposure to radiation normally absorbed by the Earth's ionosphere, troposphere and stratosphere;

(e) exposure to meteorites and other interplanetary bodies which normally burn out on entry in the Earth's atmosphere.

In dealing with these problems, the astronautical engineer must contend with the environment of space, which is vastly different from that found in the Earth's atmosphere. Here it is necessary for the engineer to learn and apply new information and, in many cases, to rid himself of old concepts and rules of thumb.

Atmospheric pressure

Atmospheric pressure decreases with increasing altitude. At sea level, it is 760 mmHg ($1.013{\cdot}10^5$ Pa); at an altitude of 5 000 m, it has fallen to about one-half of its sea-level value, at 10 000 m—one-quarter, at 15 000 m—one-eighth, etc. Since the composition of the atmosphere remains constant up to an altitude of 20 000 m, the partial pressure of oxygen, which is 160 mmHg ($0.21{\cdot}10^5$ Pa) at sea level, decreases in line with barometric pressure.

The minimum partial pressure of oxygen compatible with life in air saturated with water vapour at 37 °C is about 50 mmHg ($0.066{\cdot}10^5$ Pa) for a man who is not acclimatised. This condition is encountered at an altitude of about 7 000 m in a normal air atmosphere and about 15 000 m in a pure oxygen atmosphere. Consequently, if these altitudes are to be exceeded, a pressurisable cabin must be employed so that the pressure inside the cabin can be maintained at a higher level than that outside.

At an altitude of 19 500 m, the environmental pressure is 47 mmHg ($0.062{\cdot}10^5$ Pa), i.e. the pressure of saturated water-vapour at 37 °C and, consequently, at or below this pressure, the water in organic liquids boils, and the pressure of the other gases dissolved in these liquids falls to zero. At this altitude the hazard is not only a lack of oxygen and consequent hypoxia, which would not be fatal until after a few minutes exposure, but also this other and much more serious factor: as the distance from the Earth's surface increases, the pressure continues to fall until it approaches zero.

The problem of maintaining in the cabin a total pressure and an oxygen partial pressure compatible with life is mainly an engineering one and is easy to solve. The air in the cabin should have an oxygen partial pressure similar to that at sea level and should be continuously cleaned of the carbon dioxide produced by the occupants. It should also contain a certain percentage of nitrogen, since breathing pure oxygen may produce atelectatic zones in the lungs due to the total absorption of alveolar oxygen by the blood and to the subsequent collapse of the alveoli.

Acceleration

To achieve an orbit round the Earth, a space vehicle must reach a velocity of at least 8 km/s, and a velocity of over 10 km/s is required to send a vehicle to another planet. To attain an orbit-escape velocity, the vehicle and its occupants must be subjected to an acceleration in the order of 800 g ($g = 9.8\ m/s^2$) for a period of 1 s or of 4 g for 200 s.

The human body cannot withstand high accelerations, and a level of 100 g applied for only a fraction of a second will be sufficient to break bones or rupture internal organs. The accelerations employed for space flights are in the region of 7-8 g, which a human body in a reclining position can tolerate for several tens of seconds. However, experiments with animals and man have shown that, when immersed in water, the human body can withstand over ten times this acceleration.

Gravity

Once the space vehicle is in orbit or free-fall, the problem of acceleration is replaced by that of weightlessness due to the absence of gravitational pull. This phenomenon has caused functional disorders, in particular disturbances of labyrinthine function.

In space flight a state of weightlessness ensues when the gravitational pull of the Earth or other planets is at zero; in such a state the human body and other unrestrained objects tend to float. Under these conditions the otolithic apparatus and the mechanical receptors of the skin and muscles, which provide for vertical orientation and posture regulation, become inoperative and individuals, because of the loss of position sense, tend to become disorientated, confused, and suffer from malaise, nausea and giddiness. However, it has been found that, as long as tactile and visual references remain intact, adjustment is possible.

Under conditions of zero gravity it is impossible to eat and drink unless water and food are placed sufficiently far back in the mouth so that the normal reflex of deglutition can propel the substances down the oesophagus. In addition, liquids in particular are difficult to convey to the mouth since, once released from physical restraint, they will tend to float and disperse; the technique adopted is to suck liquids from a compressible plastics container.

Gravity also performs an air-cleaning function by causing the deposition of airborne particulate material. In the gravity-less space vehicle, dust remains suspended indefinitely in air and is inspired by the vehicle occupants, and any bacteria attached to the dust may attain a pathogenic concentration in the respiratory system. In addition, the gravity-dependent phenomenon of convection is also non-existent in a space craft; consequently, warm exhaled air remains near the nose and mouth and is repeatedly re-inhaled, and air warmed by the body remains on the body surface making it difficult to ensure adequate thermo-regulation and presenting a danger of heat stroke.

On Earth, the majority of physical work consists in moving the body against the pull of gravity. Consequently, the weightless astronaut will have an energy expenditure lower than that of a bed-bound subject on Earth. As a result, on return to Earth, the astronaut will feel as though he has been bedridden for the duration of his flight unless he regularly performs physical fitness exercises during the period of weightlessness.

Weightlessness affects, in particular, the normal working of the central nervous system and the psychological functions. The visual, auditory, tactile and other sensations to which the body is exposed, and which stimulate the sensory organs on Earth, are vastly different during space flight. [Space sickness has been described, similar to motion sickness (see separate article).] If the astronaut is unable to adapt to this new situation, and respond appropriately to new stimuli, by creating new conditioned reflexes and inhibiting many others that he has formed in the course of his life, he may find himself in a state of psychological stress which will easily lead to anxiety or even neurosis and seriously impair all the higher psychological functions indispensable to the success of the undertaking.

Of particular importance for the purpose of maintaining the astronaut in the best possible mental health are his intelligence and his cultural preparation in all matters affecting the space operation. Only by a thorough knowledge of his new condition and confidence in his ability to deal successfully with any emergency will he be able to preserve the psychological stability that is indispensable in the performance of the difficult tasks facing him. This cultural preparation is perhaps even more necessary than physical and athletic training.

[In space flight, in addition to loss of muscular tissue, loss of water, loss of body fat and loss of red blood cells have been observed, the last effect being attributed to both haemolysis and inhibition of erythropoiesis caused by pure oxygen. Prolonged space flight produces metabolic changes whose patterns and levels depend on the duration of the flight and the type of crew activities. The most persistent effect appears to be the loss of bone mineral–up to 1%, with local losses up to 8% and consequent osteoporotic changes with an increased tendency to pathological fracture. Deficiency in exercise due to weightlessness and possible deficiency in diet are the main but not perhaps the only aetiological factors involved.]

Ionising radiations

Ionising radiation is probably the greatest danger in space travel. [The surface of the Earth is protected by the upper layers of the atmosphere against the primary cosmic rays. The secondary radiation, both electromagnetic and particulate, to which humans are exposed on the Earth, is produced by the interaction of cosmic rays and gaseous nuclei. Exposure to it differs according to latitude, longitude and more significantly to altitude. At sea level the dose-rate in intermediate latitudes amounts approximately to 0.5 mGy/year. Such exposure nearly doubles for each 1 000 metres increase in altitude.] In the Van Allen belts found at 3 000 km and 25 000 km above the Earth's equator, radiation rates may be as high as 0.2-0.25 Gy/h and 5 Gy/h respectively. Since a total dose of 5 Gy is lethal and 0.2-0.25 Gy may cause damage, passage through these belts is extremely dangerous even when a considerable portion of the radiation is absorbed by the vehicle shell. During periods of solar activity, radiation increases and aggregate doses of 0.1 Gy may reach the upper layers of the Earth's atmosphere about 24 h after the solar eruption has been detected by visual means.

Conquest of space

Moon, the planet nearest the Earth, has no atmosphere and, consequently, the lunar explorer has to wear a form of pressurised diver's suit and carry his own breathing-air supply. The lunar day/night cycle lasts 28 days and temperatures are as high as 100 °C during the day and as low as −150 °C during the night; consequently, contact with the lunar soil can be made only at dawn or dusk when the temperature is within tolerable limits. The gravitational force of the Moon is one-sixth of that of the Earth, so that a person who weights 70 kg on Earth would weigh little more than 11 kg on the Moon. He could amuse himself by jumping considerable heights because, whilst his muscular strength would be unchanged, resistance to the body's weight would be much reduced. The mechanics of locomotion are necessarily greatly altered, and the most rational method of locomotion would be jumping, somewhat like that adopted on Earth by kangaroos and grasshoppers.

On other planets, conditions are even less amenable to human life. On Jupiter, for example, the temperature is about −250 °C and the force of gravity is 2.6 times that on Earth. A person weighing 70 kg would weigh about 200 kg on Jupiter, and he would not even have the strength to raise himself from the ground. Mars has the environmental conditions that most closely resemble those found on Earth; however, this planet has a very rarefied atmosphere and the Mars explorer would have to live permanently in a space suit.

Voyages to planets more distant than the Moon present considerable technical problems. For example, the trip to Mars would last about 3 years, whereas a journey to a planet in another solar system would last several decades even at a speed approaching that of light.

There is considerable divergence of opinion concerning the presence of life on other planets. However, the author considers that life on other planets is so highly improbable as to be practically impossible because the chemical sequences requisite for the creation of the first living being, that is an organism capable of transforming energy and reproducing itself, would entail a succession of reactions such that the chances of their producing successful results would be exceedingly remote. Accordingly, the possibility that somewhere in the universe there are living beings similar to human beings with whom relations would be established cannot be taken into serious consideration.

MARGARIA, R.

Physiopathology of altitude and aerospace flying. (Physiopathologie liée à l'altitude et aux vols dans l'atmosphère et l'espace). Colin, J. *Encyclopédie médico-chirurgicale. Instantanés médicaux* (Paris), 1981, 52/76, 16506 A 10, 16 p. Illus. 30 ref.

"Prevention and treatment of space sickness in shuttle orbiter missions". Graybiel, A. *Aviation, Space and Environmental Medicine* (Elmhurst (Illinois)), Feb. 1979, 50/2 (171-176). 5 ref.

CIS 80-1478 *Psychological problems of space flight* (Psihologičeskie problemy kosmičeskih poletov). Lomov, B. F. (Moscow, Izdatel'stvo Nauka, 1979), 239 p. Illus. 289 ref. (In Russian)

"Mineral and nitrogen metabolic studies, experiment M071". Whedon, G. D.; Lutwak, L.; Rambaut, P. C.; Wittle, M. W.; Smoth, M. C.; Reid, J.; Leach, C.; Stadler, C. R.; Stanford, D. D. *Biomedical results from Skylab.* Johnston, R. J.; Dietlein, L. F. (eds.) (1977), NASA-377, 164 p.

"Peculiar characteristics of crewmembers' metabolism on the second expedition aboard orbital station Salyut-4". Tigranian, R. A.; Ushakov, A. S. *Aviation, Space and Environmental Medicine* (Elmhurst (Illinois)), Sep. 1978, 49/9 (1074-1079). 6 ref.

Agricultural chemicals

One of the most significant developments of the present century has been the extensive application of chemical techniques to the solution of agricultural problems. To counter the action of plant pests that would otherwise reduce the yield from agricultural effort, chemical manufacturers specialising in pesticide production have been synthesising new substances of great chemical complexity and biological potency. So great has been the effect of this activity that the economic well-being of many agricultural countries now depends on the higher yields obtained by the use of these pesticides. Equally effective in many respects have been the synthesis and use of chemical fertilisers, the most advanced applications of which originate from the results of soil research by chemical means. There are many other aspects of agricultural activity in which chemicals play a part (nutritional, pathological, biological, etc.), but those associated with the use of pesticides and fertilisers are the most significant in terms of the tonnage quantities involved and the hazards associated with these substances in the farms or plantations where they are used. Many of the pesticides are toxic and, in addition, disastrous explosions have occurred in the manufacture of ammonium nitrate to serve as a fertiliser. For this reason a number of articles in the Encyclopaedia have been devoted to the consideration of chemical fertilisers and the various groups of pesticides (see PESTICIDES; PESTICIDES, HALOGENATED; PESTICIDES, ORGANOPHOSPHOROUS; FUNGICIDES; HERBICIDES; RODENTICIDES; FERTILISERS).

Many of the above substances are produced exclusively for agricultural use, but other chemical substances employed in numerous industries and occupations are also used by the agricultural worker.

Cleaning and disinfecting agents

Detergents are necessarily used for many purposes in agriculture. Caustic soda is used for cleaning milking equipment and for milk containers on a dairy farm, and calcium or sodium hypochlorite is used for disinfecting them. Sodium carbonate is used for cleaning and degreasing the trays and fittings of hatcheries and incubators, while, to prevent the spread of disease among the chicks, thoroughly hygienic conditions must be preserved by treating these washed fittings with a solution of sodium hypochlorite. Formaldehyde is used for disinfecting purposes and ammonia solution is subsequently applied to neutralise this chemical when it has served its purpose.

Concentrated solutions of sodium hydroxide and ammonium hydroxide are corrosive and can cause severe injury in the course of almost instantaneous contact with the skin, while the more dilute solutions of these substances and solutions of calcium hypochlorite, sodium hypochlorite and of formaldehyde can cause dermatitis on prolonged contact.

In the modern agricultural industry, there are certain buildings inside which the highest standards of hygiene should be observed. The interiors of these buildings are periodically painted with white paint containing 5% DDT. Paint is also used on the farm for the purposes of preservation and maintenance. The hazards associated with these substances and the appropriate preventive measures are described in the relevant separate articles.

Antibiotic and hormonal products

Side by side with the now well established practice of promoting healthy plant life by the use of pesticides and of trace elements in fertilisers, a somewhat parallel development has been taking place in animal husbandry in the use of antibiotic and hormone preparations—a development which is largely, but not entirely, associated with intensive stockfarming and breeding systems.

Thus antibiotic drugs such aureomycin and procaine penicillin are being used to check the incidence and control the development of disease. It has been found that the incorporation of small amounts of these drugs in animal fodder increases growth rate and the efficiency of food conversion.

Hormones are used to direct growth in a desired manner. For example a pellet of the synthetic female hormone known as oestrogen, when inserted under the skin of a young cock, acts temporarily as a substitute for caponisation, the castration operation being one that is sometimes dangerous to the bird.

SAFETY AND HEALTH MEASURES

Generally, care must be exercised to ensure against accidental ingestion of these materials. They should, therefore, be stored in a safe place where there is no likelihood that they could be mistaken as foodstuffs for human consumption. A label on the container should state the hazards and the precautions. Considerable attention has necessarily been given to this form of labelling in the case of pesticides where the possibilities of serious and even fatal injury are great, but concern has been expressed that not all antibiotics (such as tylosin) and hormonal products (such as stilboestrol) intended for veterinary use on farms are receiving adequate control. A common form which such control may take is the issue of recommendations for safe use agreed jointly by official organisations and the manufacturers. This arrangement is necessary only when the product is on unrestricted sale, and not when it is obtained on a prescription by a veterinary surgeon.

Washing facilities. In general, wherever chemical products are used on the farm, sanitary and washing facilities should be readily available for workers who may come into injurious contact with the substances.

MATHESON, D.

Chemical compounds used in agriculture and food storage in Great Britain. A permanent loose-leaf publication. Ministry of Agriculture, Fisheries and Food, Safety, Health and Welfare Branch, Great Westminster House, Horseferry Road, London, United Kingdom.

Agricultural implements

The agricultural implements dealt with in this article are those tools and instruments which are operated manually or are worked by draft animals such as oxen, water buffaloes or horses as distinct from accessory implements used on power-driven machines, which are dealt with in the article AGRICULTURAL MACHINERY.

Although more machinery is being used every year, the majority of the workers employed in agriculture, especially in developing countries, make extensive use of implements of which there is a very large variety. The production of these implements is enormous; for example in India one factory produces over 2 million chopping hoes a year, as well as another 2 million pieces a year of shovels, hammers, crowbars, picks and beaters.

In many cases the design of these implements has been the same for hundreds of years. In the poorer countries the poverty of the small farmers contributes to the continued use of the tools and implements of their ancestors: the hoe and the sickle, primitive ploughs, drags and threshing sledges, pulled by draft animals such as the ox, the buffalo, the camel and the mule.

The number of agricultural implements is so large that space permits of a description of only a few of the more common ones.

Land clearance

Hand tools and implements used in land clearance should comply with the requirements mentioned below in the paragraph on safety measures. Pulley blocks and shear legs used for extracting tree roots should be substantially constructed and of sound material. The following tackle should not be used: wire ropes having broken wires, or showing signs of corrosion or wear, fibre ropes not in good condition, and chains, rings, hooks, shackles and swivels showing signs of stretch, excessive wear, cracks or open welds.

Seed-bed preparation

The most important implement is the plough. In some developing countries ploughs are of crude design, of heavy construction and made chiefly of wood. They are usually pulled by bullocks or water buffaloes. Some of the *desi* or country ploughs used in India and South East Asia weigh up to 100 kg as compared with the hill-land ploughs of about 12 kg. These heavy ploughs are gradually being replaced by the steel mouldboard plough which, owing to its smooth shape and better design, can plough through the soil more easily, requiring less effort on the part of the beast or beasts pulling it and on the man steering it. Disc ploughs consisting of disc blades attached to an axle are used for rapid shallow ploughing particularly where stones are present.

In the preparation of paddy fields, either a plough or puddler is used to produce a churning action to mix the soil with the water, the field being flooded to a depth of 5-15 cm. The plough is often of a crude wooden type. The puddler consists of two or three rotors each having six blades. Water buffaloes are used to draw such ploughs or puddlers. The work of steering these implements requires considerable exertion, especially as the operator works barefooted on wet ground. Cultivators and harrows are secondary implements used for breaking the clods of ploughed land, levelling the surface and destroying the weeds. These implements are equipped with spikes, pegs, blades, discs or projecting sharp points which cut through the clods and bring the roots of weeds to the surface for destruction (figure 1). For land levelling, a flat log of wood is sometimes used. Improved methods of levelling are by means of a heavy stone roller or a scraper.

Figure 1. A disc harrow of Indian manufacture.

Workers should not stand on levellers, cultivators or harrows when it is necessary to weight them down. Weights should be used for this purpose.

Planting

Attachments to ploughs and cultivators are now in use which can sow seeds in several furrows at the same time, permitting better control of seed distribution and depth of sowing, and even simultaneous placement of fertilisers. If a platform is fixed to a seed drill on which a worker has to stand, it should be provided with a hand grip and with a toe-board on the front edge. If an implement is trailed behind the drill, the platform should have a rail at least 1 m high at the back (figure 2).

Figure 2. Seed drill with operator's platform protected by handrail in front, rail at the back, toe-board and wheel-guard.

In paddy transplanting in developing countries the paddy shoots are laboriously planted in the flooded paddy fields, the workers, of whom a large proportion are women, standing barefooted in water in a stooping position for long periods.

Intercultivation

This is carried out between rows of plants to kill weeds, loosen the soil and prepare soil mulch. Intercultivation is usually performed manually with simple tools such as hand hoes, spades, rakes, forks and pick-axes. Mechanical hoes are also used in some countries. Rotary cultivators should be provided with top protection hoods to prevent flying stones and other objectives being thrown against the operator, and also to safeguard the operator against risk of injury through accidental contact with the rotating parts (figure 3).

Plant protection

Various types of sprayers for liquid pesticides and of dusters for applying similar chemical compounds in powder form are in use. Other chemical agricultural equipment now widely employed are fog or smoke generators, flame throwers, soil injectors, fumigators and bird scarers. In view of the highly toxic nature of many of the chemical substances used with this type of equipment, special precautions in their handling are necessary. These are discussed in more detail in AGRICULTURAL CHEMICALS and PESTICIDES.

Figure 3. Rotary cultivator properly designed and protected.

Harvesting and threshing

Although mechanised methods of harvesting (using mowers, reapers and binders) and threshing are employed in many countries, there is still widespread use of hand tools and equipment, especially on small-scale farms. For the harvesting of crops by manual methods, cutting implements such as sickles, scythes, cutlasses and machetes are used. Rakes and pitchforks are employed for handling cut forage crops. The sickle, which consists of a curved steel blade attached to a wooden handle, is made in various patterns. One type has serrated edges, another has both edges sharpened. The blade of the Japanese sickle has a middle section of hard steel with outer layers of mild steel and is self-sharpening.

For harvesting sugar-cane and afterwards cutting it into lengths, either a cutlass or machete or a sickle-knife with a long wooden handle may be used. The tangled growth of the sugar-cane makes it difficult to cut, and accidents occur if a worker holding a bunch of canes in one hand accidentally strikes it with the cutting implement. In some countries a special protective glove with a thick covering of rubber is worn as a precaution against such accidents.

Where machinery is not available, root and similar crops such as groundnuts are harvested by means of a blade-harrow; potatoes are lifted by a special plough or are simply dug out with a digging fork. A special two-pronged fork known as a beet-lifter is used to extract sugar-beet from the ground.

On small farms in poor countries threshing is often carried out by means of a flail consisting of two pieces of wood, one a light rod about 1.5 m long to which is attached by a thong a cylinder about 0.8 m long and 30 mm in diameter. Other methods of threshing are to pass heavy loaded planks or drags or crude rollers over the grain. These may be driven by draft animals, and the driver sits on the equipment to weight it down. Another method is to have animals trample on the harvested crop.

Winnowing to separate the grain from the chaff may be done by throwing the produce into the air so that the wind blows the chaff to one side. To assist in winnowing, a multi-bladed fan driven by hand or pedal may be used to create a breeze for removing the chaff and dust from the grain.

For chopping fodder a fodder-cutter or fodder-knife may be used. This consists of a cutting tool about 25 cm long attached to a wooden handle about 20 cm long. A fodder-cutting roller similar to a lawn mower can be run over the fodder to cut it into smaller pieces.

Other processing machines are groundnut decorticators, corn (maize) shellers, paddy rice shellers, turmeric polishers and sugar-cane crushers. The simpler forms of these implements can be driven by hand or foot or by draft animals. The in-running nip or intake of sugar-cane crushers should be guarded so that the operator feeding the material cannot have his fingers trapped.

Miscellaneous implements

The wooden yoke of a pair of draft animals used for hitching various kinds of equipment is very heavy and of crude construction. A recent development is an animal-drawn "universal toolbar carrier". The toolbar carried on a pair of pneumatic wheels can be used with a variety of ground tools such as ploughs, cultivators and harrows. The carrier can also be converted into a goods or passenger vehicle or used for carrying water or manure. Many carts in use in agriculture still have wooden wheels with steel rims and are of crude design. The new type of cart in use has pneumatic tyres, roller bearings, good brakes and reflectors. A high ground clearance is desirable for negotiating uneven ground.

One common practice in developing countries is the carrying of loads on the head. The use of a wheel-barrow where the terrain permits would enable the transport of at least ten times the weight carried on the head and reduce correspondingly the large number of journeys necessary to move the same amount of material.

Sheep shearing is done by hand clipping where flocks are small but in Australia, where the total number of sheep is about 100 million, mechanical aids are used. The mechanical shears or clippers are usually driven from an internal combustion engine. The men engaged on this work are highly skilled and can shear a sheep in a few minutes. They are employed full time all the year travelling from one sheep station to another. Their daily close contact with sheep exposes them to the risk of contracting brucellosis, tuberculosis or various zoonoses.

In some hot countries the watering of growing vegetable crops, where no irrigation system or automatic water-spraying devices are available, is done manually. A worker has a bamboo rod about 1 m long which he places across the back of his neck and shoulders. At each end of the rod there is a watering can or a bucket of water. As he walks between the rows of growing vegetables he tilts the containers forward to discharge the water. This kind of work is very heavy due to the long distances to be travelled from the source of water supply.

HAZARDS

The main hazards from agricultural implements are as follows:

(a) Physical injuries from the cutting edges of implements such as sickles and machetes. As first-aid facilities are often non-existent or very poor and medical attention may not be available, there is the risk of small injuries becoming serious and there is also the danger of tetanus. Working barefooted and without hand and arm protection with implements such as machetes increases the accident risk.

(b) The cumulative ill effects to the health of the worker over a period of time owing to the handling of heavy and poorly designed implements especially in climates with extreme climatic conditions.

SAFETY MEASURES

All implements should be of good quality material, well designed for the work to be performed and requiring the least possible effort on the part of the operator. They should be used only for the purpose for which they are designed. Instruction in the proper safe operation of equipment should be given to the workers. Wooden handles of implements should be of hard straight-grained wood, free from cracks and knots, and securely fitted. The handles of cutting implements such as machetes, billhooks and large knives should have a projection which prevents the hand slipping on to the blade. All implements should be examined at suitable intervals and not used if found defective. The edges of cutting tools should be kept sharp and, where necessary, be dressed, tempered and repaired by competent persons. While being transported, the edges or points of implements such as scythes, pitchforks, billhooks, rakes and cutting tools should be so placed or so sheathed as to prevent danger (figure 4). This is especially so if implements are carried on a bicycle. Implements should not be left in places where they are likely to be of danger to anyone and should be kept out of the reach of children.

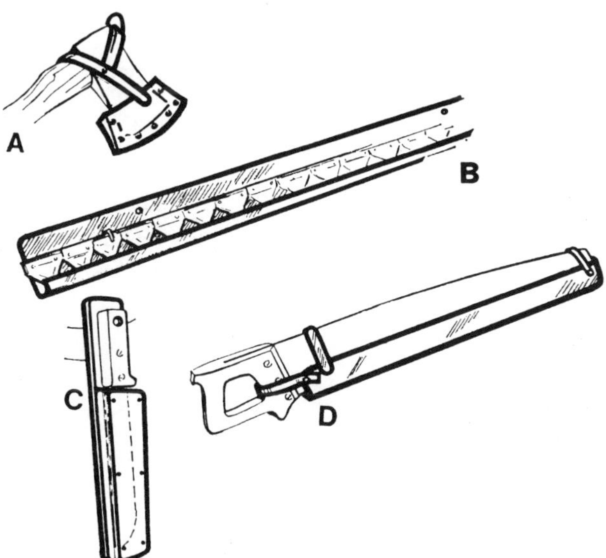

Figure 4. Guards for the transport of sharp-edged or pointed tools. A. Axes. B. Cutter blades. C. Knives. D. Hand saws.

Ergonomics of agricultural work

Some of the implements used are of a crude design and made of wood. Much of this work could be made easier and performed more efficiently if ergonomically designed implements were used. A great deal of research in the ergonomics of tool and implement design has been made during the last few years but it will be some considerable time before its practical applications are seen in the fields. Peasants and agricultural workers in communities where the same type of implements have been used for hundreds of years are, to begin with, slow to change their methods. Educational programmes and demonstrations by agricultural experts will assist in such changes, which will benefit both the health of the worker and the economy of the country.

JAIN, B. K. S.

Tools for agriculture, a buyer's guide to low cost agricultural implements (London, Intermediate Technology Publications Ltd., 1976), 173 p. Illus.

Agricultural safety handbook (Arbeitsschutztechnisches Auskunftsbuch Landtechnik) (Verlag Tribüne, Am Treptower Park 28-30, DDR-1193 Berlin) (1972), 541 p. Illus. 50 ref. (In German)

Agricultural machinery industry in developing countries, a report of the expert group, Document ID/47 (Vienna, UNIDO, Aug. 1969), 159 p.

Agricultural machinery

Agricultural machinery is designed to till the soil and render it more suitable for crop growth, to sow seeds, to apply agricultural chemicals for improved plant growth and control of pests and diseases, and to harvest and store the mature crops. There is an extremely wide variety of agricultural machines, but all are essentially a combination of gears, shafts, chains, belts, knives, shakers, etc., assembled to perform a certain task. These parts are usually suspended in a frame which may be either stationary or, as if more often the case, mobile, and performs the desired operation while moving across a field.

The major groups of agricultural machines are: soil tillage machines; planting machines; cultivating machines; forage harvesting machines; grain, fibre, vegetable, and fruit and nut harvesting machines; agricultural chemical applicators; transport and elevating machines; and sorting and packaging machines.

Figure 1. Tractor pulling a soil cultimulcher machine.

Soil tillage machines. These include ploughs, tillers, subsoilers, harrows, rollers, levellers, graders, etc. They are designed to turn, agitate, level and compact the soil to prepare it for planting. They may be small in size and require only a small power source (as in the case of a one-man roto-tiller for tilling a rice paddy), or they may be large and require a considerable power source (as in the case of a combined subsoiler, drill and harrow).

Planting machines. These include planters, drills, broadcast seeders, etc., and are designed to take seeds from a hopper or bin and insert them in the soil at a predetermined depth and spacing or spread them uniformly over the ground. Planters may be of simple design and comprise a single-row seeding mechanism or they may be highly complex (as is the case with the multi-row planter with attachments that simultaneously add fertiliser, pesticides and herbicides).

Cultivating machines. These include rotary hoes, cultivators, weeders (mechanical and flame), etc. They are used to eradicate undesirable weeds or grass which competes with the plant for soil moisture and makes the

Figure 2. Appropriate shields, hand holds and steps facilitate the safe movement of the operator on and about this combine.

harvest of the crop more difficult. They also improve the soil tilth so as to make it more absorptive of rain.

Forage harvesting machines. These include mowers, choppers, balers, etc., and are designed to sever the stems of roughage crops from their roots and prepare them for storage or immediate use. The machines also vary in their complexity: the simple mower merely cuts the crop, whereas the chopper will not only separate the stalk from the root but will also chop the entire plant into small pieces and load it into a vehicle, which may be a towed wagon. Crimpers, which crush or break the stems of plants, are often used to expedite the field drying process of fodder crops to prevent spoilage, especially of legumes that will be placed in dry storage or baled. Pelleting machines are used to compress fodder crops into compact cubes for mechanical feeding of livestock. Balers are used to compress fodder into square or round bales to facilitate storage and handling. Some bales are small enough (20 to 40 kg) to handle manually, while others may be so large (400 to 500 kg) as to require mechanical handling systems.

Figure 3. A tomato-harvester picking tomatoes and loading them into wagons as it moves across the field. A number of workers under the sun shade are assisting in the sorting operation.

Grain and fibre harvesting machines. These include reapers, binders, corn pickers, combines, threshers, etc. They are used to remove the ripe grain or fibre from the plant and place it in a bin or bag for transport to the storage area. Grain harvesting may involve the use of a number of machines such as a reaper or binder to cut the standing grain, a wagon or truck to transport the crop to the threshing or separating machines, and vehicles to transport the grain to a storage area. In other cases many of these functions may be performed by a single machine, the combine harvester, which cuts the standing grain, separates it from the stalk, cleans it and collects it in a bin, all while moving through the field. Such machines will also load the grain into transport vehicles. Some machines such as cotton pickers and corn pickers may operate selectively and only remove the grain or fibre boll from the stem or stalk.

Vegetable harvesting machines. These include diggers and lifters and are designed either to dig the crops from the earth and separate them from the soil or to lift or pull the plant free. The potato digger, for example, may form part of a potato-combine comprising a sorting, grading device, polisher, bagger and elevator. At the other extreme is the simple two-wheeled, bladed sugar-beet lifter which is followed by hand labourers.

Fruit and nut harvesting machines. These machines are used to harvest berries, fruit and nuts. They may be as simple as a tractor-mounted, vibrating tree shaker which separates the ripe fruit from the tree. Or they may be as complex as the ones which harvest the fruit, catch the falling fruit, place it in a storage container and later transfer it to transport vehicles.

Transport and elevating machines. These also vary considerably in size and complexity ranging, for example, from a simple wagon comprising merely a platform on wheels to a self-loading and stacking transport unit.

Figure 4. Pineapple harvester–workers with face shields and protective clothing including gloves.

Inclined chain, flight or belt conveyors or other mechanical handling devices are used to move bulky material (hay, straw, ear corn, etc.) from wagon to storage or from one location in a building to another. Screw conveyors are used to move granular material and grain from one level to another, and blowers or pneumatic conveyors are used to move light materials horizontally or vertically.

Agricultural chemical applicators. These are used to apply fertilisers to stimulate plant growth, or herbicides, pesticides, etc., to control weeds and pests. The chemicals may be liquid, powdered or granular and the applicator distributes them either by pressure through a nozzle or by centrifugal force. Applicators may be portable or vehicle-mounted; the use of aircraft for chemical application is growing rapidly.

Sorting and packaging machines. These machines are usually stationary. They may be as simple as a fanning-mill which grades and cleans grains merely by passing it over a series of screens, or as complex as a seed mill which will not only grade and clean but also, for example, separate different types of seeds. Packaging machines usually form part of a sophisticated grading system. They are used primarily for fruit and vegetables and may wrap the produce in paper, bag it or insert it into a plastic container.

Power plants. Electric motors may be used to drive stationary equipment permanently located near a mains supply; however, since many agricultural machines are mobile and required to operate in remote areas, they are usually powered by an integral petrol engine or by a separate engine such as that of a tractor. Power from a tractor may be transmitted to the machine via belt, chain, gear or shaft drives; most tractors are fitted with a power take-off coupling specially designed for this purpose (see TRACTORS).

Figure 5. A farm worker wearing protective equipment while filling one of the chemical dispensers on his multirow corn planter.

HAZARDS AND THEIR PREVENTION

The agricultural machine operator or the farmhand working with a machine is exposed to three main types of hazard:

(a) traumatic injuries such as cuts, burns, electrocution, fractures and amputations caused by contact with moving machine parts, falls from or collision with the machine and flying particles projected by the machine;

(b) organic injuries caused by noise and vibration from the machine;

(c) health impairment caused by toxic substances such as agricultural chemicals and engine gases.

Traumatic injuries

In some countries more than 10% of all accidents occurring on the farm involve agricultural machines of one description or another; the proportion varies widely according to the degree of mechanisation and the methods of working. More often than not these accidents cause serious or fatal injuries. The results of some farm accident surveys indicate that machinery accounts for nearly half of all fatal accidents. The most dangerous types of machine include threshing machines, chaff cutters and root choppers. The main risk is coming into contact with moving parts (shafts, belts, pulleys, gears, tools, etc.). All dangerous parts should therefore be fully protected by adequate guards or so positioned behind bars or rods of such strength, spacing and distance from the moving part or danger point as to prevent access to them. Projecting pins, bolts, etc., on rotating parts should be so designed, sunk or protected as to prevent entanglement of clothing; workers should avoid wearing loose or torn clothing. Feed openings should be so designed or guarded as to prevent accidental contact with the working parts. Mobile machines working in the fields have risks of their own: people may fall off them or be run over by them. Covers and shields should be strong enough to retain or deflect stones, broken tools or other flying objects. Wherever possible, guards should be built-in by the manufacturer. Warning decals may be useful in reminding operators of the presence of danger spots; acoustic signals should be provided on large machines to warn the operator of continuing machine motion after the power has been disconnected. Before any maintenance or repair work is carried out, the machine should be halted; simply disengaging the engine should not be depended upon. (On small machines motors which can be turned by hand may under certain circumstances easily start if the machine is moved while the motor is engaged.)

To prevent falls from and collisions with moving machines, operators' stands should be safely and easily accessible and, if at a height of more than 1.5 m, they should be fenced and fitted with toe-boards. Wherever access to the operating station necessitates steps or gangways, these should be equipped with non-slip floor surfaces, hand and guard rails and, where the access ways pass close to moving parts, adequate fencing should be provided. Drivers of moving machines should have good visibility and be able to emit a warning signal. They should be responsible and mature persons who have received adequate training and, under normal

Figure 6. An integrally mounted power take-off shield for operator protection connects the farm machinery to the power source (tractor).

circumstances, they alone should be allowed on the machine; other persons should be allowed to travel on the machine only where a specific and properly equipped station is provided for them. All electrically operated farm machines should be adequately grounded and extreme care must be exercised when operating high-lift trucks or moving agricultural elevators near power lines to avoid inadvertent contact and electrocution of the operator or worker.

Forage crops and small grains and shelled maize are sometimes dried by passing them on conveyors through drying machines heated by oil-burning furnaces or liquefied petroleum gas burners. After drying, the forage crop may be passed through a grinding machine or hammermill in which it is pulverised into a very fine dust which is compressed into pellets for use as cattle food. In the event of any small piece of metal or stone being fed with the forage to the hammermill, there is the danger of a spark being formed which may cause the dry dust to explode. An explosion relief panel of easily breakable material should form part of the housing of the hammermill so that in case of a dust explosion, the blast escapes through this panel. It should be so situated that the blast travels away from any place where anyone is working.

Health impairment

Organic injuries due to noise and vibration are difficult to diagnose and may be irreversible. Where noise levels on machines exceed acceptable limits, operators and those working in the close vicinity should wear hearing protection devices such as ear plugs or muffs to prevent permanent hearing loss. The high levels of vibration encountered on certain agricultural machines may cause fatigue, nausea, pain, temporary loss of sensation in the hands, etc. In the case of new machines, efforts should be made at the design stage to reduce the levels of vibration encountered and, in the case of existing machines, the effects of vibration on operators should be minimised by the installation of vibration damping seats, etc., or by limiting the worker's duration of exposure.

Inhalation or ingestion of, or skin contact with, certain agricultural chemicals may produce serious intoxication or skin disorders. Chemical dispensers and broadcasters should consequently be so designed as to minimise operator exposure during filling, application, adjustment and maintenance. Such a measure does not eliminate the need for the use of suitable personal protective equipment during such operations.

First-aid organisation

In all cases, measures should be taken to ensure that first-aid facilities are immediately available in the event of an accident. Firstly, each machine or operational area should be provided with a first-aid kit containing antiseptic, tourniquet, bandages, stimulant and, in the case of chemical application, the appropriate antidotes. Secondly, adequate means of transport and trained personnel should be readily available during working hours at places where agricultural machinery is used.

KNAPP, L. W. Jr.

CIS 79-1126 *Occupational safety and health in mechanised crop production* (Arbeitsschutz beim Einsatz der Technik in der Pflanzenproduction). Arfert, G.; Beutel, K. D. (Verlag Tribüne, Am Treptower Park 28-30, DDR-1193 Berlin) (1978), 64 p. Illus. 6 ref. (In German)

CIS 77-506 *Rotary mowers* (Slaghøstere). Verneregler Nr. 23 (Direktoratet for arbeidstilsynet, Postboks 8103, Oslo-Dep., Norway) (Nov. 1974), 5 p. Illus. (In Norwegian)

CIS 78-1408 "Hazards of motor mowers and 2-wheeled tractors when pointed directly uphill or downhill" (Dangers présentés par les motofaucheuses et les tracteurs à un essieu quand on travaille selon le sens de la pente et en montant). *Technique agricole* (Brugg, Switzerland), May 1978, 40/7 (266-268). Illus. (In French)

Agricultural work

The development of mechanisation and automation and of the use of chemical products and biological preparations during recent years has brought essential changes to agricultural work. It has made physical work lighter, but increased the risk of accidents, poisoning and allergies. This situation is particularly acute in the highly developed countries where the proportion of new physical factors (ionising radiation, electromagnetic fields, laser radiation) has grown. The situation observed in industry, where workers may be exposed simultaneously to both traditional and new biological, physical and chemical factors, now also exists in agriculture. This trend is liable to strengthen in future.

In most countries, however, and in particular in the developing ones, the character of agricultural work has not undergone drastic changes and is essentially concerned with the production of foodstuffs and certain types of raw materials for industry. A large portion of agricultural work consists in soil tillage, harvesting, crop protection, livestock rearing, and the production and conversion of products of animal origin.

Safety and health conditions in agricultural work are determined by certain distinctive features of agricultural production, viz.—

(a) the seasonal nature of the work and the consequent urgency of certain tasks such as grape-harvesting, which must be carried out rapidly and necessitates not only long work-days, but also the employment of a considerable seasonal workforce;

(b) the fact that work is carried out for the most part in the open air, exposing the workers to different climatic and meteorological conditions depending on the season and the climatic zone;

(c) frequent changes in the type of task carried out by the same person, particularly in small-scale enterprises where the duties cannot be strictly defined and it is difficult to provide regular periods of work and rest;

(d) multifarious contacts with animals and plants which may give rise to infectious or parasitic diseases, bites or other accidents due to animals, exposure to dust containing spores or other fungal allergenic matter or to toxic or irritant vegetable saps;

(e) the use of a large variety of agricultural chemicals (pesticides, fertilisers, fungicides, herbicides, seed dressings) presenting a serious hazard of poisoning by skin absorption or inhalation;

(f) the considerable distances between the living quarters and workplace, e.g. where there are extensive pastures, which entail a great expenditure of energy and time, and present the risk of commuting accidents, disturb eating habits and render medical surveillance difficult;

(g) the often primitive conditions of life, particularly on small farms where the workplaces and living quarters are under the same roof, and where sanitary conditions are sometimes unsatisfactory;

(h) the large variety of working methods, the same task being performed either manually or mechanically depending on the level of economic development of the country or region, local habits or size of the farm, with consequent differences in the degree of risk;

(i) the difficulty of imposing and complying with occupational safety and health standards and regulations in small farms;

(j) the frequent employment of casual seasonal labour without any real occupational qualifications and ill informed of risks and preventive measures, and also the frequent use of children to work in small farms (see AGRICULTURAL WORKERS).

HAZARDS AND THEIR PREVENTION

This article can only give a general idea of the main hazards and brief comments on their prevention. For more detailed information on preventive measures see the other articles on agriculture and particularly AGRICULTURE, OCCUPATIONAL HEALTH AND SAFETY MEASURES IN.

Crop farming

Field-crop cultivation represents the largest sector of agriculture in most countries and covers cereal farming (the most widespread branch of agriculture), vegetable growing, farming of industrial crops, fodder-plant growing, horticulture, etc. Basic types of work in this sector are seedbed preparation (ploughing, harrowing, fertiliser spreading); planting and crop protection (potting, pest control, disease and weed control, watering, etc.); harvesting and some types of preservation and preparatory processing; storage of fodder and grain; storage and treatment of seeds.

Mechanised agricultural work is carried out by tractor drivers or machine operators and, where trailed equipment is used, by auxiliary personnel (sowers, planters). The type of machine used largely determines the working conditions and safety measures (see AGRICULTURAL MACHINERY).

Mechanised seedbed preparation (ploughing, grubbing, harrowing) for cereal and fodder-plant growing does not demand great physical effort, whereas sowing and care of maize and sugar beet and planting of potatoes involve considerable nervous and mental strain because of the vigilance required for keeping several machines in accurate alignment. These types of work frequently encountered in farming of industrial crops give rise to particularly severe nervous stress.

The negative factors of agricultural work increase as a function of the operating speeds of agricultural machines; furthermore, the need to perform each operation in a minimum of time involves additional strain. As machines are often used before sunrise and after sunset, artificial lighting is required which adds to the hazards for the operators and the workers near the machines.

Apart from tractors and machines, both small and large farms use a variety of mechanical equipment needed for maintenance and current work, such as mechanical saws, drilling machines, crushers, mills, etc. This equipment is often old, inadequately guarded, badly installed and operated by insufficiently trained people. It presents more sources of risk than are generally found in industry. Badly fitted and poorly maintained electrical equipment and installations may also cause accidents.

When employing unskilled workers for mechanised tasks, it is very important to familiarise them with safe working methods and to give them the necessary vocational training.

Where chemicals are applied with mechanical equipment, the risk of poisoning is generally negligible. However, there have been cases where pilots of aeroplanes or helicopters have been poisoned when spraying pesticides or dusting other toxic substances.

Non-mechanised work is carried out with the aid of draught animals or by hand. Draught animals are used for soil tillage, transport and starting machines and engines. Tillage is often done with implements that may be very primitive and therefore entails considerable physical effort; merely walking on ploughed, uneven soil requires great effort. The energy expenditure during ploughing and sowing with horse draught may exceed 5 000 kcal per day.

Manual planting and sowing involve uncomfortable postures; the working methods and implements are often a legacy of the past or products of local traditions or conditions. In general, neither the tools nor the methods are rational from the ergonomic viewpoint: forks and hoes with short handles entail working in a bent or crouching posture, whereas longer handles would render work much easier. Scything by hand carried out in a bent position requires a great deal of energy (5 500-6 400 kcal per day), though it can be made easier by increasing the angle between the blade and the handle.

Mention should also be made of the injuries that frequently result from the use of hand tools, particularly sharp ones, by inexperienced people (young persons, town dwellers). It is noteworthy that most stings due to plants, insect and reptile bites, and cases of irritation or dermatitis from vegetable saps, occur during manual agricultural work.

The use of agricultural chemicals, especially pesticides, is particularly dangerous where the work is done by hand. In countries where the sale and use of these products are strictly regulated, the risks are minimal and are related mainly to the carelessness of ill informed persons or to the re-use of empty and contaminated containers. Far too often these products are handled, prepared and used by people completely unaware of their toxicity, taking none of the necessary precautions, wearing no protective clothing and equipment, and neglecting their personal hygiene (see AGRICULTURE, OCCUPATIONAL HEALTH AND SAFETY MEASURES IN).

Handling of fodder, straw, cereals and industrial crops generally results in the release of large quantities of vegetable dust which can cause pneumoconiosis, mycosis, irritation of the respiratory system, or respiratory and cutaneous allergies. This danger exists practically everywhere, but is probably greatest during mechanised work where the dust concentrations are particularly high if no technical control measures are taken or they are insufficient (see FARMER'S LUNG).

Finally, in areas where crops require irrigation and watering, working in water can bring about excessive chilling, resulting in respiratory tract infections and skin irritation; in certain regions (e.g. the Middle East), where stagnant or slowly flowing water is used, helminthiasis may be contracted. Similarly, schistosomiasis is most often contracted after contact with water infested by snails in swamps or grass where they are prevalent, and ankylostomiasis is common where human faeces are used as a fertiliser. Stagnant waters can also be breeding places for insect vectors of infectious diseases. For the special hazards of work in plantations see PLANTATIONS.

Stock farming

Here too, work is either mechanised or carried out by hand. In large enterprises a considerable portion of the work required by fodder preparation and distribution, water supply, milking, housekeeping of cattle sheds, shearing, egg collecting, etc., is mechanised. Performing such tasks by hand demands considerable physical effort, as well as prolonged close contact with animals. Manual milking, shearing or plucking a large number of animals causes excessive strain of certain small neuromuscular regions of the hands and arms, which may result in neuromyositis, peripheral angioneurosis, bursitis, tenosynovitis and neuritis.

Contact with sick animals can be the cause of infection or parasitic disease in man. It is therefore very important to look out for early signs of disease in domestic animals so that the animals affected may be isolated and given special treatment, or, if necessary, killed and buried or incinerated in strict compliance with the relevant legislation. The prevention of diseases transmissible from animals to man is a field where collaboration between veterinarians and doctors is of great importance (see LIVESTOCK CONFINEMENT; ZOONOSES).

Some workers display, from the very beginning or after prolonged exposure, an intolerance to contact with certain animals, giving rise to respiratory or cutaneous allergies.

The risk of contracting infectious diseases is generally greater in the countryside than in urban areas. For example the fact that water and soil can be contaminated by domestic animals and that there are insects capable of spreading many viral, rickettsial, bacterial, fungal and parasitic diseases constitutes a hazard that is rarely encountered in towns.

These problems occur constantly, especially in poor or isolated areas where social welfare and health services are inadequate or located a long way from the agricultural community.

The use of animals in agricultural work may also be the cause of serious injuries. Those in charge of draught animals should therefore be well aware of their habits and observe their behaviour. A worker who ill-treats animals is much more liable to be attacked.

Work organisation

Any rational organisation of agricultural work must be based on scientific research. This applies in particular to the establishment of work and rest schedules and permissible physical workloads, and to the design of machines and tools. Finding solutions to these problems helps to achieve optimal results with a minimum of effort.

During the hot summer period, the work schedules should allow for essential duties to be performed in the morning and evening, leaving the hottest part of the day for eating and a siesta. In addition, short rest periods are indispensable where work is particularly exhausting. Special attention should be paid to the nutritional requirements of workers, a matter which is often neglected because some everyday food products are produced in abundance on the spot. It is not so much that the caloric content is insufficient as that the diet is unbalanced owing to an understandable tendency to eat what happens to be produced near the workplace. This monotony is aggravated where one crop only is grown.

Moreover, the provision of beverages is important. Agricultural workers consume not only water but also various traditional drinks (tea, coffee, fruit juices, infusions, broths, etc.). These should be sufficient in quantity and wholesome. It should be ensured that alcoholic beverages are not consumed in excess.

Lastly, great importance should be attached to providing clothing suitable for the protection of workers exposed to unfavourable outdoor conditions. Protective clothing and adequate personal protective equipment must be worn when dangerous work is performed, in particular when handling toxic chemicals and applying pesticides.

Housing is a problem when seasonal workers need accommodation. It must be hygienic, of easy access and well ventilated; adequate sanitary facilities including washing arrangements must be provided.

Medical supervision

The medical supervision of agricultural workers should be concerned mainly with the prevention of general and occupational diseases, accidents, poisoning and infections. In rural areas this supervision may take different forms, but the essential thing is that the village inhabitants, as well as the seasonal workers, should have qualified and easily accessible medical care.

Education and training

The maintenance of health in the countryside depends to a large extent on the spread of knowledge of sanitation and health. Agricultural workers should receive instruction and training on good safety and health practices, e.g. in the use of machines and hand tools, in the management of animals and in the handling of chemicals. They should be assisted in planning an appropriate diet. In small farms it is only through worker training and education that hygienic conditions and safety and health can be improved. For the purpose use can be made not only of health education posters but also of special radio and television transmissions for the rural population.

KUNDIEV, Ju. I.

Agricultural workers

Agricultural workers are all those persons who are engaged in cultivating, harvesting and treating or processing agricultural produce, and in stock farming and breeding. Even nowadays, agricultural workers still account for more than half of the working population of the world, though there are enormous differences from one country to another; in Botswana, Zaire, the Ivory Coast, Nepal, Niger and the Sudan, for example, more than 85% of the workforce is active in the agricultural sector, whereas in the Netherlands Antilles, the United States, Hong Kong, Kuwait, the United Kingdom and Zambia, the proportion does not exceed 5%.

In the majority of countries, nevertheless, the agricultural worker has no well defined status, nor does he enjoy anything approaching the same facilities and advantages as the industrial worker. Moreover, the differences between the various kinds of agricultural work are far more marked than are those between the operations characterising other branches of productive activity. This is explained first and foremost by the environmental factor, which plays an overriding role in agriculture. For land–be it fertile or arid–is tilled in all climates, tropical or polar, rainy or dry, at all altitudes, in mountains or in the steppe, and according to intensive or extensive methods. The plethora of smallholdings dotted throughout vast regions, the inaccessibility of workplaces, the dearth of financial resources, the lack of general education and vocational training among agricultural workers, the high percentage of women and children compared with the workforce as a whole, the use of primitive tools, which is still to be encountered, the poor housing conditions and inadequate diet, the attraction exercised by industry and the exodus towards the towns that it provokes–all these factors tend to keep agricultural productivity at a low level, greatly inferior to industrial productivity, and account for the serious economic difficulties standing in the way of the social protection of rural workers.

Yet in most countries there is still great scope for development, sometimes by bringing new land under cultivation, but more often by increasing productivity. Today's demographic pressures, under which some 120 000 persons are daily being added to the world's population, and the state of undernourishment in which some 3 000 million human beings now exist, call for an urgent increase in agricultural production, which will bring a substantial improvement in the conditions of work and life of agricultural workers in its train.

If productivity is to be increased in the agricultural sector, there are a number of socio-economic factors to be taken into consideration: land reform, improved technology, education, organisation of rural workers. Each of these embraces a vast programme, which should be developed in an orderly and harmonious fashion.

It has been shown that land reform alone, without education of the workers and without improved technology, will not provide a solution to the problem of increasing productivity. The introduction of scientific farming methods in the developing countries is primarily a question of economics. The present low standard of productivity creates an obstacle to the acquisition of new technology, machinery, equipment, materials for improving the land, pesticides, irrigation, etc., which could be overcome through the provision of adequate credit facilities, but even so, all this requires to be matched by a vast programme of education and training.

In this connection the ILO Rural Workers' Organisations Convention, (No. 141) 1975, would, if applied in the developing countries, form the basis for an ambitious programme for the solution of these problems.

This Convention applies to all types of organisations of rural workers, including organisations not restricted to but representative of rural workers; it does not, however, apply to those proprietors who permanently employ workers, or who employ a substantial number of seasonal workers, or have any land cultivated by sharecroppers or tenants.

Categories of agricultural workers

In a good many countries it is possible to find all the different categories of agricultural workers, from the smallholder who cultivates his few acres with the help of his family to the powerful man of property who owns vast expanses of land and usually hires wage earners to work it. A rather typical example is provided by the distribution of land in Latin America, where a few years ago, 62% of the land was in the hands of 98 000 big landowners, and 13% was parcelled out amongst 9 million people, 7.5 million of whom owned plots of less than 20 hectares apiece. Twenty million workers did not own the land that they worked.

The principal categories of agricultural workers are:

(a) the sharecroppers, who work land owned by a person to whom they owe a part of their harvest; they live and work in a condition of semi-independence;

(b) the tenants, who pay a cash rent for a lease usually concluded for from 3 to 5 years;

(c) the small landowners who cultivate their land with the help of the family and do not employ permanent workers;

(d) the daily paid labourers, some of whom are permanent, while others take temporary contracts during harvest time (sugar-cane, maté, cotton and rice, as well as plantations and vineyards, etc.). There are also those who take daily contracts.

Although precise information on the number of tenants, share-croppers and other similar categories of agricultural workers is lacking, it is estimated that these workers and their families account for at least two-fifths of the world's agricultural population.

Amongst these workers the active participation of women and children is worthy of special note. In the production of such crops as cotton, rice, tea, grapes and fruit, they are sometimes more numerous than men. In many countries children work in the fields virtually from infancy, to the certain detriment of their physical and intellectual development; this practice gives rise to a serious problem of absenteeism from school.

As regards the conditions of life (health, diet, housing) of workers in the lowest categories, we find little variation; the diet is generally inadequate, and housing very often lacks the most elementary sanitary facilities. The situation is usually better, however, on large industrial-type holdings, which operate medical care and housing assistance schemes.

HAZARDS

The principal risks to the health of workers stem from the following causes: environment, kind of work, tools, machines and human factors.

The particular conditions in which land is worked do not admit of any sharp distinction between the working and living environment, since in practice the two are merged. This makes it more difficult to differentiate between occupational pathology and the morbidity common to all the inhabitants of a region, and the organisational and administrative problems created by this situation have led the countries concerned to adopt various solutions, most of which are as yet still relatively unsatisfactory.

Occupational diseases appear to occur less frequently than in industry, either because they are more difficult to identify or because work is performed in the open air and is of a varied nature. Similarly, occupational accidents are less serious than in industry, except in areas where farming is highly mechanised.

Work with animals exposes the agricultural worker to the risk of accidents and zoonoses (see AGRICULTURAL WORK; AGRICULTURE, OCCUPATIONAL HEALTH AND SAFETY MEASURES IN). Work in the open air, with the concomitant exposure to all kinds of climatic conditions, permanent contact with the soil, plants and animals, unsatisfactory conditions of life, a frequent scarcity of fresh water, and inadequate medical care–all these factors favour the continuing existence of a vast endemic and epidemic pathology, made worse by the lack of elementary notions of hygiene. In Latin America, for example, virtually the entire range of known pathology is to be encountered: infectious and parasitic diseases are one of the five prime causes of mortality; in some regions, the death rate is primarily ascribable to diseases due to impure water. Respiratory ailments, aggravated by the inhalation of dust heavily laden with all kinds of microbes and fungi, are frequent, as are dermatitis and eye diseases. Contact with the soil exposes the agricultural worker to the bites of venomous animals, and to ankylostomiasis and tetanus.

Very frequently the land is worked in hot weather, under the blazing sun, in very dry or very damp regions where the heat sometimes becomes unbearable and gives rise to serious disorders. The worker counters it by drinking large quantities of beverages, sometimes alcoholic, sometimes even in the form of polluted water which exposes him to the full range of gastro-enteritis.

Still other factors tend to aggravate the hazards common to agricultural workers in many lands; of these, one may mention illiteracy (in some areas, up to 70% of the population is illiterate), poverty–sometimes without hope of any improvement–and resignation, which leads workers to pay no attention to their health, let alone to safety considerations.

SAFETY AND HEALTH MEASURES

The use of personal protective equipment is effective in the case of particular operations, such as crop treatment with pesticides; with the increasing mechanisation of farming techniques, machinery guarding represents an important safety measure. But the mainstay of prevention will always remain the inculcation of safety-mindedness at all levels (employers, supervisors and workers). Similarly, the attack on disease cannot rely solely on immunisation against endemic and epidemic diseases, but demands a thorough grounding in public hygiene, farm hygiene and personal hygiene. Even nowadays, health education is still hampered by numerous obstacles in many countries, including illiteracy, primitive beliefs and customs, scattered populations, poor housing, the lack of communications and of medical services close at hand, and the precarious conditions of existence. All these problems are very difficult to overcome; hence the most effective training media are schools and radio programmes.

Although they look to the future rather than to the present, school curricula are of inestimable value because of their direct influence on the child–tomorrow's worker–and their indirect influence on rural homes. A schoolteacher who is thoroughly conversant with the rural environment can act to great effect.

Backing up this action, radio broadcasting is a medium which has not yet been used to the full. It provides contact with every rural unit, and if the programmes are entertaining, simple and suitable for the regions covered, it can stimulate the listener's interest in occupational safety and health questions. In certain circumstances the programmes can be supplemented by illustrated pamphlets, talks, films and all the other possible approaches made available to educators by modern technology.

In a more general context there is a need to raise the level of living of the rural worker to the greatest extent possible, and to upgrade his skills and ability and supplement them by providing him with adequate technical and mechanical aids. The industrialisation of agriculture is at present fraught with many complex technical and social problems, but there is no doubt that in the next half-century it will bring about a radical change in the conditions of work and life of agricultural workers in many countries.

BAZTARRICA, J.

Incomes of agricultural workers with particular reference to developing countries. Report III, Advisory Committee on Rural Development, 8th Session (Geneva, International Labour Office, 1974), 99 p.

Organisations of rural workers and their role in economic and social development. Report VI (1), 59th Session of the International Labour Conference (Geneva, International Labour Office, 1973), 67 p.

Manpower and training needs for rural development. Report II, Advisory Committee on Rural Development, 8th Session (Geneva, International Labour Office, 1974), 34 p.

Special services of rural workers' organisations. Workers' education manual (Geneva, International Labour Office, 1978), 89 p.

Improvement of conditions of life and work of tenants, sharecroppers and similar categories of agricultural workers. Reports VII (1) and (2), 51st Session of the International Labour Conference (Geneva, International Labour Office, 1967), 92 p. and 119 p.

Agriculture, occupational health and safety measures in

Rapid scientific and technical progress has brought about changes in traditional agricultural work, made it possible to cultivate soils hitherto considered unsuitable for crop growing, and facilitated the application of industrial methods to agriculture. However, traditional agricultural methods still predominate in many countries.

HAZARDS

Environmental hazards

The geographical location of the farm can be the source of various types of accident. This factor must be analysed so as to avoid locating farms in "dangerous" areas. There is a danger of inundation in depressions or near water courses in the event of heavy rainfall or floods. In arid regions, on the other hand, it will be necessary to create installations to ensure an adequate water supply for the farm inhabitants, cattle and crops. Where there is a danger of landslides, it is important to sound the soil thoroughly before erecting new buildings. Forests present the hazard of fires, and appropriate fire protection measures must be taken for farms located in forest areas.

The location of farms near highways or railway lines may have a bearing on both production (noise can negatively affect domestic animals) and safety (cattle crossing the road, self-propelled machines driving on to the road or leaving it). Gases, vapours and dust from transport equipment or industrial plant may also constitute a hazard. Plans for locating factories or similar premises in agricultural areas should therefore be as severely controlled as they are in urban areas.

Farms should not be located near airports where people and animals would be exposed not only to noise, but also to high-frequency electromagnetic radiation from radar installations. The same hazard exists in the close neighbourhood of powerful television and radio broadcasting stations. Adverse biological effects for humans and animals also result from exposure to low-frequency electromagnetic waves. Agricultural premises and pastures should therefore not be located below high-voltage lines, which also present the hazards of fire and electrical accidents in the event of a line breakage.

Finally, climatic factors such as predominant winds, rain, relative humidity, the regional prevalence of hurricanes and storms should be evaluated from the viewpoint of their effect on safety and health.

Technological hazards

Statistics show that falls from heights or on the level are the most frequent accidents in farms. Multi-storey buildings for agricultural premises should therefore be avoided despite the advantages they may present. Barns, haylofts and other storage buildings must have safe means of access and be equipped with implements for lifting and lowering heavy loads. Floors should have non-slip surfaces.

Effective fire prevention and protection measures are essential. Straw and hay must be stored in places with fire-resistant partitions, which must also be so designed as to prevent draughts promoting the spread of the fire. Fire fighting equipment must be available at strategic points, maintained in good condition and regularly checked. Electrical equipment must be of high quality so as to withstand the wear and tear inherent to rough farm work. Fuel stores and dumps must be planned and located in accordance with fire prevention rules.

Since there are more and more machines for all sorts of agricultural work, it is important to provide them all with adequate guards and to design them for safe operation (see AGRICULTURAL IMPLEMENTS; AGRICULTURAL MACHINERY). Precautions are also required in the use of tractors (see TRACTORS).

A high level of work organisation is indispensable for the prevention of accidents, particularly when introducing new techniques in agriculture and stockfarming. No hazardous work should be performed when vigilance, dexterity and physical force are diminished. For example no heavy physical work should be undertaken immediately after having a meal with alcoholic drinks or on an empty stomach when the blood sugar level is low. The work must be so organised as not to increase the risks inherent in subsequent operations, e.g. water should not be poured on the floors of sheds and cattle yards before work involving the transport of loads on these surfaces.

First-aid kits, which must be available in all farms, should contain everything necessary for the emergency treatment of bleeding wounds and fractures, antidotes against toxic substances, sera against snake bites, etc. Farms should be connected with each other and with the nearest hospital by telephone or radio communication, and should also be provided with facilities for the immediate evacuation of the injured.

Measures must be taken to prevent the transmission of disease from animal to man and the spread of disease among animals. Workers looking after sick animals must wear appropriate protective clothing, thoroughly observe the rules of personal hygiene and undergo disinfection before starting other work such as picking fruit or harvesting vegetables. Cattle sheds should be designed so as to provide for isolation of diseased animals; the excrement and litter from sick animals must not become sources of infection, and the udders of milch cows should be thoroughly washed before milking (see separate articles on specific occupational diseases in agriculture and also LIVESTOCK CONFINEMENT; ZOONOSES).

Chemicals used in agriculture

There are many aspects of farm work involving the use of chemical products. Pesticide dusting and fertiliser spreading are by far the most important as regards quantity and hazard, because most pesticides are toxic. A number of articles in this Encyclopaedia deal with this question (see PESTICIDES; PESTICIDES, HALOGENATED; PESTICIDES, ORGANOPHOSPHORUS; FUNGICIDES; HERBICIDES; RODENTICIDES; FERTILISERS).

Apart from the above-mentioned chemicals, other products are also used in agriculture, such as detergents for many purposes: caustic soda for cleaning milking equipment and milk containers, and sodium hypochlorite for disinfecting them; sodium carbonate for cleaning and degreasing chicken hatching equipment and incubators (see AGRICULTURAL CHEMICALS).

PREVENTIVE MEASURES

To avoid accidental ingestion of these drugs, they must be kept so as to exclude any possibility of confusion with drugs intended for people. The type of danger involved and precautions to be taken must be printed on the packages. Great attention should be paid to the presentation of the labels on pesticide containers, especially in cases where there is a great probability of serious or even fatal poisoning. Unfortunately, not all antibiotics and hormone preparations intended for veterinary purposes in farms are under corresponding control. In general, printed recommendations are handed out which have been worked out by public authorities in conjunction with the manufacturing firms and inform about the hazards and safe use of the preparations. Such recommendations are also necessary when a preparation is on sale without restrictions, and not only when a veterinarian's prescription is required to obtain it.

In all cases where chemicals are used in farms the people handling them must work under hygienic conditions and be given adequate detergents.

A common chemical hazard in agriculture is carbon monoxide contained in the exhaust gases of internal combustion engines. The premises in which engine-powered machines are operated or kept must be well ventilated.

It is advisable to provide for an adequate separation between living quarters and workplaces in order to diminish the hazards of poisoning or infection. The availability of clean drinking water is essential.

The scientific and technical revolution does not do away with the traditional character of farm work and sometimes even intensifies it. The spread of agriculture to zones with a more extreme climate not only involves direct exposure to more rigorous weather but also longer work-days on account of the shorter periods during which agricultural work can be performed.

On the other hand, the mechanisation of auxiliary tasks in large farms demands greater vocational skills. Particular progress has been made in trades such as chicken farming. There is a trend towards increasing specialisation in agriculture, and the number of specialised farms will consequently grow. Thanks to the development of means of transport, the distance between the workplace and living quarters no longer plays an important part. Chemical, physical and biological factors affecting man will grow in importance, and alongside with them the problems of their immediate and delayed effects.

KUNDIEV, Ju. I.

CIS 79-525 "Typical health risks in socialist agriculture" (Typische Gesundheitsrisiken in der sozialistischen Landwirtschaft). Mönnich, H. T.; Jürgens, W. W. *Zeitschrift für ärztliche Fortbildung* (Jena), 1978, 72/5 (229-233). Illus. 11 ref. (In German)

"Occupational pathology in agricultural activities. An attempt at a statistical approach. I: Accidental pathology. II: Diseases of occupational origin" (La pathologie professionnelle dans les activités agricoles; essai d'approche statistique. I: Pathologie accidentelle. II: Les maladies d'origine professionnelle). Dubrisay, J.; Fages, J. *Archives des maladies professionnelles, de médecine du travail et de sécurité sociale* (Paris), June 1978, 39/6 (339-355) and July 1978, 39/7-8 (459-468). (In French)

CIS 79-518 *Occupational safety and health in agriculture—Check list* (Sikkerhet i landbruket–Sjekkliste), Bestillningsnr. 356 (Statens arbeidstilsyn, Postboks 8103 Dep., Oslo 1) (no date), 15 p. Illus. (In Norwegian)

CIS 80-811 *Guide to health and hygiene in agricultural work* (Geneva, International Labour Office, 1979), 309 p. Illus.

Hygiene of village dwellers (Gigiena žitelej sela). Zarubin, G. P.; Kundiev, Ju. I.; Nikitin, D. P. (Moscow, Znanie, 1975), 96 p. (In Russian)

CIS 78-201 *Occupational health and safety for agricultural workers–Agricultural health and safety considerations for a rural primary health care system.* Bondy, M. K.; Lebow, R. H.; O'Malley, M.; Reilly, T. DHEW (NIOSH) publication No. 77-150 (National Institute for Occupational Safety and Health, 4676 Columbia Parkway, Cincinnati, Ohio) (1976), 128 p. Illus.

CIS 78-203 "Health care of people at work–Agricultural workers". Smith, D. M. *Journal of the Society of Occupational Medicine* (Bristol), July 1977, 27/3 (87-92). 8 ref.

Air

Air is a predominantly physical mixture of a variety of individual gases enveloping the terrestrial globe to form the Earth's atmosphere. Pure dry air is a mixture of gases in relatively stable proportions: nitrogen–78.09%; oxygen–20.95%; argon–0.93%; and carbon dioxide –0.03% by volume, plus trace amounts of other gases such as hydrogen, ozone and nitrogen oxides. Its specific gravity at 0 °C and 760 mmHg is 1.293. The purity of air, however, is relative. Its moisture content varies up to 4% by volume with changing climatic and meteorological conditions, and the atmospheric concentration of other substances is dependent upon the environment and the nature of man's activities. Thus ambient air may contain up to 0.5% carbon dioxide, variable amounts of aerosols (suspended particulate matter such as water droplets, ice crystals, dust, fumes or mists), and up to 1% of organic and inorganic contaminants.

The oxygen in air supports the most common type of combustion which occurs when the kindling temperature of a combustible substance has been reached and the reaction then proceeds without the application of additional outside heat. This process, essential to the support of animal life, is the means by which the body acquires its heat and energy from foodstuffs and eliminates dying cells from the body tissues. When the concentration of oxygen drops below 17%, symptoms of distress appear; at 12% or lower, danger to life is imminent with unconsciousness occurring at levels below 11% and cessation of breathing below 6% (see HYPOXIA AND ANOXIA).

The thermal conductivity of air is 60.34×10^{-6} cal/s.cm. °C at 15.6 °C and increases uniformly to a value of 126.4×10^{-6} cal/s.cm. °C at 537.8 °C. Except for heat added by compressing equipment, the temperature of air tends to equal that of the environment. While the thermal conductivity of gases is not affected by pressure, certain changes in heat loss from the body may result from pressure changes in the atmosphere, because the changed heat capacity of the air tends to alter the temperature differential required to maintain a constant heat transfer from the body to the environment.

The dielectric constant of air at 0 °C and 760 mmHg is 1.000590 (see AIR IONISATION).

Contamination

Laboratory research with animals has demonstrated repeatedly the importance of pure air for respiration. Certain pollutants, dependent upon their nature and mechanism of action, may reduce or even stop ciliary activity–a lung clearance mechanism–in the air passages, thus increasing susceptibility to respiratory infection, induce bronchogenic cancer, and increase the mortality rate of the exposed animals (see BRONCHITIS, CHRONIC, AND EMPHYSEMA).

Air is the gaseous sea in which man lives and works, and its purity is of utmost importance to his health. The respiratory route provides the most common mode of entry of harmful chemical substances into the body. Discharge of dusts, fumes and vapours of hazardous substances into the atmosphere surrounding the myriad activities of civilised man requires effective control measures to reduce the concentration of these contaminants to levels which will not produce any deleterious effects (see AIR POLLUTION; DUST, BIOLOGICAL EFFECTS OF).

Compressed air

Compressed air and liquefied air are used on a wide scale industrially. For example compressed air in caissons and diving bells makes underwater construction and repair of tunnels and bridge piers possible, although the workers are subject to an occupational disease known as the "bends"; this condition is produced in compressed-air workers and divers following too rapid decompression as a result of which nitrogen bubbles are formed in the bloodstream and body tissues. Other uses of compressed air include such applications as pneumatic tools, pneumatic hoists, rock cutting and sandblasting, spray painting, etc. (see DECOMPRESSION SICKNESS; PNEUMATIC TOOLS).

With the introduction of modern cryogenic processes, liquefied air has become the chief source of commercial oxygen. The economic importance of the latter is illustrated by its large-scale use in the new basic oxygen process in the iron and steel industry. The majority of industrial processes are now considering the use of oxygen in place of combustion or process air, at least in part (CRYOGENIC FLUIDS).

CLAYTON, G. D.

Air quality–General aspects–Vocabulary. International Standard ISO 4225 (Geneva, International Organization for Standardization, 1980), 7 p.

Airborne micro-organisms in the workplace

HAZARDS

Farmers handling mouldy hay or straw can be exposed to an aerosol of thermophilic actinomycete spores (*Micropolyspora faeni* or *Thermoactinomyces vulgaris* in particular) in gross concentrations. Several hours after exposure the farmer may fall ill with extrinsic allergic alveolitis.

Cases of extrinsic allergic alveolitis usually occur in industries where animal or vegetable matter is processed. Thermophilic actinomycetes, fungi or other micro-organisms may grow on these raw materials especially when the environmental conditions are warm and humid. Handling such mouldy material liberates spores in massive quantities and large numbers of them are inhaled by the handlers and deposited in the alveolar region of their lungs, where they elicit the above allergic reaction. Forms of extrinsic allergic alveolitis due to micro-organisms are (1) bagassosis caused by *T. vulgaris* growing on mouldy, overheated sugar cane pulp (bagasse), (2) mushroom worker's lung by *M. faeni* and *T. vulgaris* occurring in mushroom compost, (3) maple bark stripper's lung by the fungus *Cryptostroma (Coniosporum corticale)* found on the mouldy bark, (4) malt worker's lung by *Aspergillus clavatus* infecting mouldy barley or malt dust, (5) sequoiosis by *Graphium aureobasidium pollulans (Pollularia)*, the mould in redwood dust, (6) cheese worker's lung by *Pencillium* of cheese mould, (7) suberosis by mouldy oak bark or cork dust, (8) paprika splitter's lung possibly by *Mucor stolifer* growing on the paprika fruit when picked late in the wet season and (9) wood-pulp worker's lung by *Alternaria* infecting mouldy wood pulp. The causative agent is yet to be identified in several other forms of allergic alveolitis such as animal food handler's lung caused by fish meal, smallpox handler's lung presumably by the crusts from smallpox lesions, furrier's lung by fox fur, blackfat tobacco smoker's lung by blackfat tobacco, etc.

Outbreaks of extrinsic allergic alveolitis were also described in workplaces that were ventilated through air ducts contaminated with moulds such as thermophilic actinomycetes. On Mondays, or the first working day after the weekend, workers complained of chills, fever and breathlessness, which sometimes subsided on subsequent days of the working week. The workers' sera were found to contain precipitating antibodies against thermophilic actinomycetes, and radiography, lung function tests and medical examination indicated a condition closely resembling extrinsic allergic alveolitis, which was resolved completely once the air conditioning system was altered and the ducts were cleaned. In other outbreaks of this disease, referred to as "humidifier fever", humidifiers, humidification plants of air conditioning systems, cooling water or oil, airborne dust deriving from sewage sludge, etc., were found to be heavily infected with a variety of micro-organisms such as bacteria, fungi, protozoa but not necessarily with thermophilic actinomycetes, which grow at an elevated temperature only. Sera of the affected persons contained precipitating antibodies against many organisms present in the infected plant or machine, but people did not necessarily develop any symptoms when they inhaled an aerosol of the suspected organism in inhalation challenge tests. However, when the sources of airborne micro-organisms were eliminated by altering the relevant industrial process or method of humidification, all respiratory symptoms vanished overnight. In one particular outbreak the antigen was traced through precipitin tests to protozoa such as *Neaglaria gruberi*, which were found growing on wet man-made fibre (rayon) deposited on a suspended ceiling in an office.

Larger, non-respirable particles which on inhalation tend to deposit in the upper respiratory tract may sensitise this region (sometimes after many years of exposure) and cause asthma.

People who work for a number of years in spinning mills processing cotton, flax or hemp may develop byssinosis, a condition which is considered by many authorities to be a pathological reaction of the bronchial airways (mediated through the release of histamine) to a still unidentified agent occurring in the factory atmosphere. The condition prevails mainly in blowrooms and cardrooms and becomes increasingly rarer as the fibres are further processed. The prevalence of byssinotic symptoms in cardrooms was found to correlate with the concentration of airborne cotton dust, especially with particles of respirable size and with airborne bacteria such as Gram-negative rods, the spores of which can constitute a significant portion of the respirable dust. Several agents have been proposed as the causative factor including the endotoxins of Gram-negative bacteria.

Micro-organisms occur in low concentration in the atmosphere of several industries where vegetable or animal material is processed. Gram-positive bacteria have been found in wool mills, Gram-negative bacteria in sewage treatment plants and fungal spores in tobacco factories, flour mills, furniture factories, saw mills, tea packing plants, etc. Although workers in these industries do not usually complain of chest-tightness or breathlessness that could be attributed to the working conditions, their ventilatory capacity (FEV_1, FVC) has been found to decrease during workshifts (especially on Mondays) by a small but statistically highly significant amount.

All workrooms in a cotton spinning mill are usually dusty. Micro-organisms occur, however, mainly in cardrooms, which is mostly where workers complain of byssinotic symptoms. The chief source of dust and micro-organisms in these workrooms is undoubtedly the carding machine, especially when heavily contaminated, coarse cotton is processed. The method of dust control used commonly at present consists usually of a number of suction points incorporated into the carding machines. This is sometimes referred to as the "Shirley pressure point" system and has proved to be efficient enough in suppressing "fly" (or fluff) and aerodynamically large particles but failed altogether to control respirable dust associated with the byssinotic symptoms. A more recent approach to dust control in cardrooms utilises local exhaust booths which almost fully enclose the carding machines. Organic dusts, especially those containing microbes or microbial derivatives, should be carefully controlled.

People in their workplace, like anyone else in contact with the public, may be infected by the common cold, influenza, etc., through inhaling the airborne micro-organisms. Certain highly infectious diseases are, however, restricted mainly to the working population. For example workers handling skins, hide, fleece or hair of animals infected with anthrax may contract pulmonary anthrax (also known as wool-sorter's disease) through inhaling anthrax spores. Stringent control of the disease in animal stocks is the best form of prevention.

Health workers such as doctors, pathologists and microbiologists in particular can be exposed to some highly pathogenic organisms unless conditions in a microbiology laboratory are stringently controlled.

PREVENTIVE MEASURES

Occupational diseases due to the inhalation of airborne micro-organisms are best prevented by the eradication of the source of the micro-organisms, which usually grow on organic matter under warm and wet, or highly humid conditions. Consequently materials of animal or vegetable matter such as straw, corn, wood, hide, textiles, coffee, tea, tobacco, bagasse, etc., should be stored under comparatively dry conditions (below 70% relative humidity) to prevent moulding. If a material becomes mouldy, it is often best discarded because the microbial spores or toxic agents that such a material carries cannot easily be removed or destroyed without damaging the material. Only pure, fresh water should be used for humidification by sprinkling. Water must not be allowed to condense into the air supply ducts of an air conditioning system which together with the humidification plant must be kept perfectly clean. Air from water coolers, or any machine containing stale water, oils or a fermenting suspension in a state of agitation, must not be allowed to vent into the workplace. The addition of bacteriocides or fungicides to an aqueous or oily medium does not necessarily prevent organic growth in that medium.

In industries where dusty organic matter is processed, as in textile spinning, significant quantities of microbial spores can be released into the work atmosphere together with the dust. This is best prevented by the application of local exhaust ventilation at the points of dust release. Local exhaust "booths" are to be preferred to "hoods" or "slots" from the point of view of capture efficiency. If microbes are released at too many points in the workplace their airborne concentration can only be controlled by general ventilation, which is considerably less effective than local exhaust ventilation and more expensive. Usually the air from the workplace is collected through return grills and ducts, carefully filtered through banks of filter bags, conditioned and recirculated into the workplace through supply ducts and grills, to save on thermal energy. In an industrial situation ultraviolet irradiation cannot normally replace filtration to "sterilise" air, because (1) ultraviolet light does not affect either microbial spores or their toxins and (2) it generates unacceptable amounts of ozone, a highly toxic gas which is difficult to get rid of.

The use of personal protection (i.e. dust masks or respirators) against airborne micro-organisms is not normally recommended because it cannot altogether prevent exposure but does cause considerable discomfort to the wearer. If a person develops pulmonary hypersensitivity to some airborne micro-organism in the workplace or to its product, he or she should change employment immediately to prevent further damage to his or her health.

CINKOTAI, F. F.

Endotoxin levels and bacteria contamination of bale cotton. Rylander, R.; Lundholm, M. Beltwide Cotton Producers Research Conference, Dallas, 1978 (Memphis, National Cotton Council).

"Investigation of a respiratory disease associated with an air-conditioning system". Pickering, C. A. C.; Moore, W. K. S.; Lacey, J.; Holford-Stevens, C. V.; Pepys, J. *Clinical Allergy* (Oxford), Mar. 1976, 6/2 (109-118). Illus. 6 ref.

"Humidifier fever". Medical Research Council Symposium. *Thorax* (London), Dec. 1977, 32/6 (653-663). 20 ref.

CIS 79-506 "An outbreak of humidifier fever in a foundry" (Befeuchterfieber in einer Giesserei). Scherrer, M.; Imhof, K.; Weickhardt, U.; Lebek, G. *Schweizerische Rundschau für Medizin– Revue suisse de médecine* (Berne), 12 Dec. 1978, 67/50 (1 855-1 861). Illus. 27 ref. (In German)

"Humidifier-associated extrinsic allergic alveolitis". Van Assendelft, A.; Forsen, K. O.; Keskinen, H.; Alanko, K. *Scandinavian Journal of Work Environment and Health* (Helsinki), Mar. 1979, 5/1 (35-41). Illus. 24 ref.

"Microbial and immunological investigations and remedial action after an outbreak of humidifier fever". Edwards, J. H. *British Journal of Industrial Medicine* (London), Feb. 1980, 37/1 (55-62). 21 ref.

Air conditioning

Air conditioning involves establishing an indoor microclimate conducive to the health and comfort of the human or animal occupants or a microclimate which will provide an optimum atmospheric environment for carrying out particular chemical or industrial processes. [In other words, air conditioning is defined by the American Society of Heating, Refrigerating and Air Conditioning Engineers (ASHRAE), as the process of treating air so as to control simultaneously its temperature, humidity, cleanliness, and distribution to meet the requirements of the conditioned space.]

Thermal environment

One major requirement of an air-conditioning system is to control, within specified limits, the thermal environment; this requires the establishment of a given relationship between the four variables that define a given thermal environment:

(*a*) temperature of air;

(*b*) average inside temperature of the enclosing walls, floor and ceiling;

(*c*) moisture content (humidity) of air;

(*d*) air movement within the enclosure with respect both to average velocity and to uniformity of properties at all points within the enclosure.

In addition to establishing a controlled thermal environment, a complete air-conditioning system must also control the concentration of gases, vapours, and solid particles (dust) within the enclosure and, when the system is one for human comfort, the level of body odour must also be controlled.

Air conditioning for process and industrial purposes (as textile factories, instrument shops, computer rooms, and "clean" rooms) is a specialised subject with a wide range of specified control conditions. This is comprehensively covered in the periodical trade literature, and current standards are available in guides produced by organisations such as the American Society of Heating, Refrigerating and Air Conditioning Engineers. The coverage in this article will therefore be limited to air-conditioning systems designed specifically to provide healthy and/or comfortable conditions within spaces occupied by humans.

Body heat

The first requirement in designing an air-conditioning system is a knowledge of how the human body produces and loses heat, since this knowledge leads to a definition of thermal environments which will provide a neutral, or equilibrium, atmosphere for the occupants.

The human being, whether working or resting, is thermally similar to a heat engine, whether working or idling. If the heat engine is unable to lose waste heat as rapidly as it is produced the excess will go into storage within the parts and structure, leading to a rising temperature with ultimate structural failure. If the human body is unable to lose heat as rapidly as it is produced, the excess heat will go into storage in the tissues with resultant rise in body temperature and eventual physical collapse (see HEAT AND HOT WORK; HEAT DISORDERS).

When ambient air temperature is equal to or higher than body surface temperature and the vapour pressure in the air equals or exceeds the vapour pressure of saturated air at body surface temperature, the only way to prevent a continuous rise in body temperature would be to provide some type of cooled surface "visible" to the body and to which it could lose energy by radiation at a rate sufficient to maintain equilibrium. Cooling panels of this type have found industrial applications in situations where operators of specialised equipment (such as crane operators working over molten metal in steel mills) cannot be protected against very hot environmental conditions except by use of such a refrigerated umbrella.

Equivalent air temperature

Up to that thermal level at which evaporative regulation (sweating) starts, body heat loss by evaporation from the lungs and by insensible perspiration remains practically constant, regardless of air temperature, at about 25 kcal/h. Under these circumstances, equilibrium can be maintained by adjustment in opposite directions of the conductive-convective body heat loss and the radiative loss. It has been found that a very simple relationship exists between the air temperature and the average surface temperature of any room as compared with the equivalent temperature, for equal sensation of warmth, in a room in which air and average surface temperatures are equal. Thus,

$$t_{eat} = (t_{dbt} + t_{ast})/2$$

where t_{eat} = equivalent air temperature;
t_{dbt} = actual air temperature as measured by a dry thermometer, shielded from radiant transfer
t_{ast} = average surface temperature (weighted by

areas of surfaces at different temperatures) of the interior of the enclosure.

Consequently, a room with air and all inside surfaces at 20 °C would have an equivalent air temperature also of 20 °C, whereas rooms with air at 25 °C and surfaces at 15 °C [t_{eat} = (25 + 15)/2 = 20 °C] or air at 15 °C with surfaces at 25 °C [t_{eat} = (15 + 25)/2 = 20 °C] would have thermal environments in which the average occupant would experience essentially the same feeling of warmth as in the 20 °C air, 20 °C surfaces room. This means that an occupant of any room who experiences thermal comfort at any particular air temperature in that room would be equally thermally comfortable in that room if, for a specified increase or decrease in air temperature, there were an equal and opposite decrease or increase in average surface temperature.

Properties of moist air

The engineering design of an air-conditioning system consists in determining the design state or states of the outside air, the design state or states of the air in the conditioned space and the required state or states of the conditioned air to be supplied to the room. With these three states known, the procedure is to select a thermodynamic process or series of processes which will permit conditioning of the outside air to the supply state; equipment is then purchased, designed, or combined which will permit the selected processes to be carried out.

A first requirement is therefore to fix air states, and this requires a knowledge of the properties of moist air and the means of measuring or calculating those properties. Among the many thermodynamic properties of moist air the following are the ones of greatest interest to the air-conditioning engineer:

t_{dbt} = dry bulb temperature, as measured with a dry thermometer shielded from radiation;

t_{dpt} = dew point temperature, or the dry bulb temperature to which unsaturated moist air must be reduced before it reaches saturation (if cooled to any lower temperature, "dew" will form on the thermometer);

W = humidity ratio, varying from zero for dry air to W_s for air saturated with water vapour, expressed in kilogrammes of water vapour, per kilogramme of dry air;

rh = relative humidity, or ratio of vapour pressure in air of a given unsaturated state to vapour pressure in saturated air at the same dry bulb temperature (often symbolised as ϑ);

t^* = thermodynamic wet bulb temperature, more correctly known as temperature of adiabatic saturation (often shown as t_{ast}). When water at t^* evaporates without external heat transfer into unsaturated air at t_{dbt}, the resultant temperature of the saturated air will be t^*;

v = specific volume, m³/kg, of dry air with associated water vapour; the reciprocal of the density;

H = enthalpy, kcal/kg, of dry air with associated water vapour (sometimes incorrectly referred to as total heat).

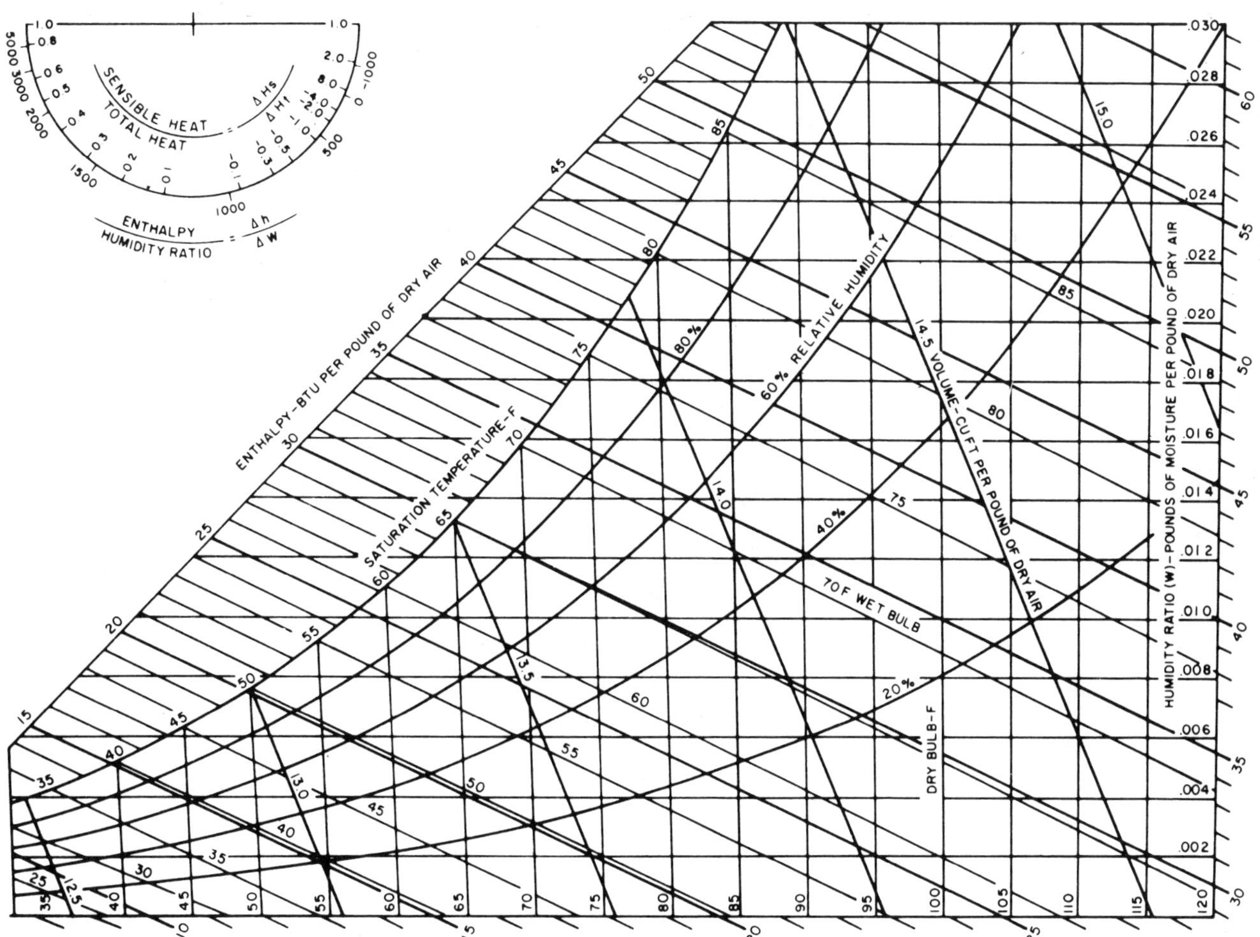

Figure 1. Abridgement of Psychrometric Chart 1 of the ASHRAE (from *ASHRAE Handbook 1981. Fundamentals*, by courtesy of the ASHRAE).

Of the above seven thermodynamic properties only three, t_{dbt}, t_{dpt}, and rh are subject to direct measurement under field conditions. For this reason, an additional accurately measurable psychrometric property is defined: t_{wbt} = wet bulb temperature, or the equilibrium temperature reached by a thermometer with wetted bulb when moved through unsaturated moist air at moderate velocity. The values of t_{wbt} and t^* do not agree, but for the purposes of this article it can be considered that the measured value of t_{wbt} will differ inappreciably (less than

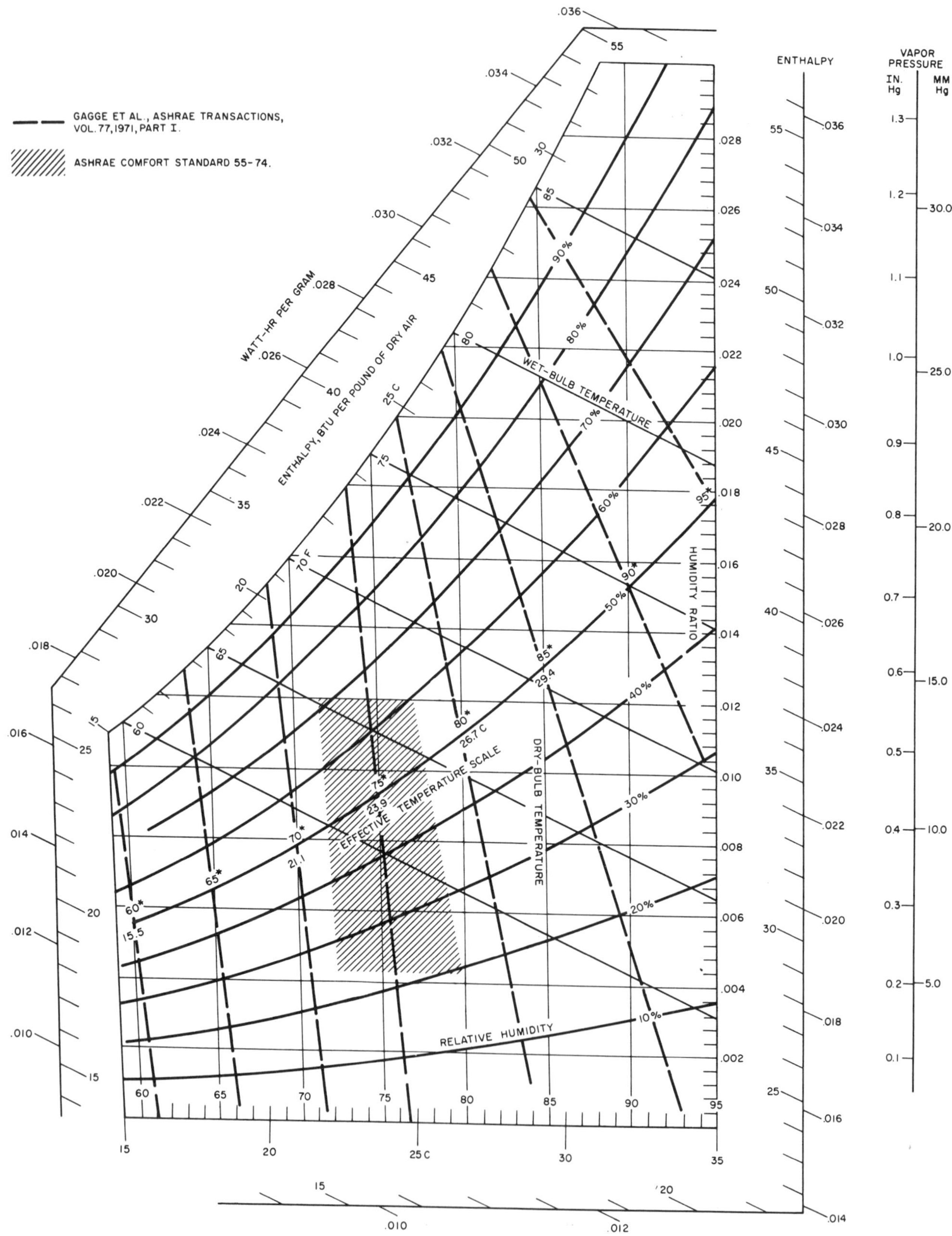

Figure 2. New effective temperature scale (ET*) of the ASHRAE (from *ASHRAE Handbook 1981. Fundamentals*, by courtesy of the ASHRAE).

3%) from the calculated value of t^* for moist air at the same state.

Psychrometric chart

[Since a functional relationship exists among the eight properties defined above, the thermodynamic properties of moist air can be represented in graphical form in a psychrometric chart. Figure 1 is an abridged version of Psychrometric Chart 1 (for normal temperatures) of the ASHRAE. It includes co-ordinates of enthalpy and humidity ratio and lines of constant dry-bulb and thermodynamic wet-bulb temperature, relative humidity and volume. When any two of the four factors: t_{dbt}, t_{wbt}, t_{dpt} and *rh* are known, the other two may be predicted by the chart.]

Thermal comfort

[By means of research carried out over half a century, the region on the psychrometric chart in which the average adult will experience thermal comfort has been deeply investigated. It has been recognised that various physical and physiological factors are involved in such a sensation, which was defined by the ASHRAE Standard 55-66 in 1966 as "that condition of mind which expresses satisfaction with the thermal environment". Several environmental indices have been proposed (see HEAT AND HOT WORK; HEATING OF WORKPLACES; APPENDICES) intended to predict the conditions associated with comfort.

Comfort zone. The comfort zone proposed by the ASHRAE is based on the new effective temperature which is defined as the dry-bulb temperature of a uniform enclosure at 50% *rh* in which humans would have the same net heat exchange by radiation, convection and evaporation as they would in the varying humidities of the test environment. For the ASHRAE ET* Scale, clothing is standardised at 0.6 clo (the unit of thermal insulation of clothing: 1 clo (standard clothing) ensures a thermal insulation of 0.155 $K.m^2.W^{-1}$, where K is the heat exchange by conduction ($W.m^{-2}$)), air movement (still) at 0.2 m/s, time of exposure 1 h and the chosen activities as sedentary ($\simeq$ 1 met, i.e. the unit of metabolism: 1 met = 50 $Kcal/m^2.h$). The comfort zone recommended in ASHRAE Comfort Standard 55-74 is shown in figure 2, which generally applies to altitudes from sea level to 2 134 m and to a thermal environment (indoor) in which mean radiant temperature is nearly equal to dry-bulb air temperature and air velocity is less than 0.2 m/s. For this case the thermal environment is well specified by the two variables shown–dry-bulb air temperature and *rh*. The data are useful for individuals with light clothing and seated or during sedentary activity. Comfort and heat tolerance during light or medium activity work are graphically presented in figures 3 and 4.

Comfort equation. The comfort equation by P. O. Fanger takes into account the following variables: ambient temperature, mean radiant temperature, relative air velocity, vapour pressure in ambient air, activity level, thermal resistance of clothing (tables 1 and 2). The main physiological factors of thermal comfort have been identified in the mean skin temperature and the evaporative weight loss. The comfort equation is rather complex because it is based on the analysis of heat transfer processes; however, for direct practical application several diagrams are available that are applicable to nude persons or to persons with high medium and heavy clothing at different activity levels. The comparison between comfort equations from different geographical locations shows that if a difference exists between

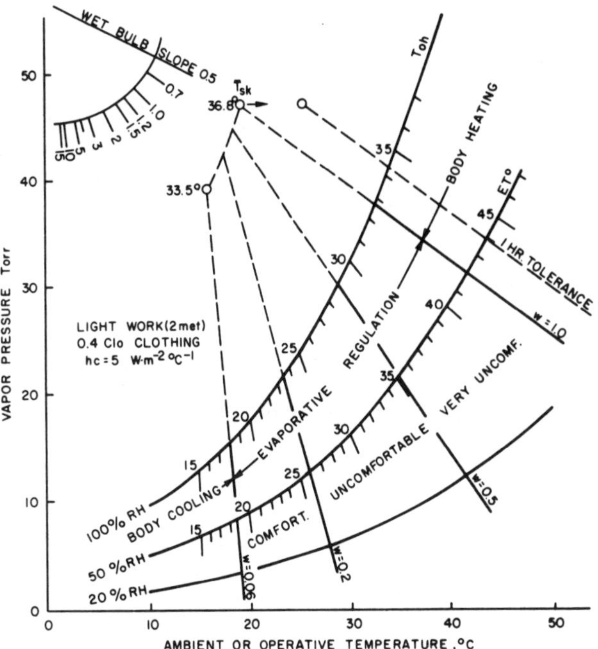

Figure 3. Comfort and heat tolerance during light work (2 met) (from *ASHRAE Handbook 1981. Fundamentals*, by courtesy of the ASHRAE).

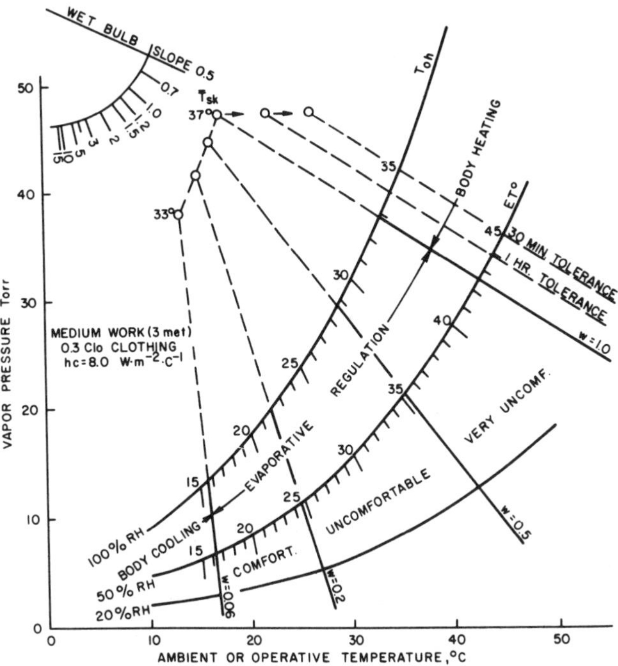

Figure 4. Comfort and heat tolerance during medium activity (3 met) (from *ASHRAE Handbook 1981. Fundamentals*, by courtesy of the ASHRAE).

temperate and tropical areas, then it is slight and probably not of engineering significance.]

Air-conditioning systems

For the usual summer conditioning problems requiring cooling and dehumidification, there are three types of system:

(a) hermetically sealed units are refrigerant-charged at the factory and must be cut into to recharge refrigerant or to repair or replace parts; these "unit coolers" are widely used for residential and single-office use;

(b) semi-hermetic units are factory assembled, but can be recharged or repaired through access openings;

(c) split systems for stores and large offices have separate high sides (compressor and condenser) and low sides (evaporator and expansion valve); large split systems are often referred to as central air conditioning and are used in office buildings, hotels, hospitals, and factory premises or other industrial buildings; in these systems extensive duct work with air distribution outlets must be provided.

The mechanical elements of all air-conditioning systems are of limited number and include: some means of filtering out solid matter, usually consisting of straining-type or electrostatic filters; a means of heating or cooling air, requiring mechanical, absorption, or steam-jet refrigeration in summer, and a combustion process or electrical source in winter; and a means of humidifying in winter as by direct evaporation of water, or steam injection, and of dehumidifying in summer, as by use of a refrigerated coil on which water from the moist input air condenses, or a refrigerated air washer in which the moist air is "dried" in passing over droplets that are at a temperature below the dew point.

For residential, hospital, commerce and office use, air conditioning is designed to provide comfort at times when the outdoor conditions are too hot. For industrial and factory applications comfort air conditioning is desirable, but when this would be too costly a compromise is to design for indoor thermal conditions which at least do not represent a hazard to either health or working effectiveness.

Industry by industry statistical data are not yet available, but at present one can say qualitatively that maintenance of industrial environments in or near the comfort zone for summer work increases productivity, tends to reduce accidents, decreases absenteeism, and in general leads to an improvement in morale and human relations.

[The improper maintenance and use of air conditioning equipment and in particular of humidifiers has been recently found to be a rather common source of disturbances and health impairments (see AIRBORNE MICRO-ORGANISMS IN THE WORKPLACE).]

HUTCHINSON, F. W.

Table 1. Thermal comfort temperature (°C) in an environment where air temperature is assumed to equal wall temperature (Based on Fanger's equation)

		Air temperature = wall temperature					
Metabolism (W)		105		157		210	
Relative humidity (%)		20	80	20	80	20	80
Clothing (clo) 0.5 V_a (m.5^{-1})	0.2	27.3	26.0	21.5	20.3	15.0	13.0
	0.5	28.3	27.0	22.8	21.7	17.0	16.0
	1.5	29.0	28.0	24.3	23.3	19.0	18.3
Clothing (clo) 0.5 V_a (m.5^{-1})	0.2	24.7	23.3	16.7	15.5	8.5	7.8
	0.5	25.7	24.0	18.0	17.0	10.6	9.6
	1.5	26.7	25.0	19.5	18.3	12.7	11.7

Source: Reproduced by permission from Vogt, J. J.: "Thermal working environment" (L'ambiance thermique de travail, 1979). *Archives des Maladies professionnelles* (Paris), Jan.-Feb. 1979, 40/1-2 (131-173).

Basic readings:

ASHRAE handbook 1981 fundamentals (American Society of Heating, Refrigerating and Air-Conditioning Engineers, Inc., 1791 Tullie Circle NE, Atlanta) (1981). Illus. Ref.

CIS 74-705 "Assessment of man's thermal comfort in practice". Fanger, P. O. *British Journal of Industrial Medicine* (London), Oct. 1973, 30/4 (313-324). Illus. 48 ref.

Technique:

CIS 75-1897 *Heating and cooling for man in industry.* Allan, R. E.; Anania, T. L.; Brief, R. S.; Clarke, J. H.; Cheever, C. L.; Confer, R. G.; Dieringer, L. F.; Humphreys, C. M.; Soule, R. D.; Starkey, R. H.; Thompson, H. K.; Thompson, R. M.; Waid, D. E. (American Industrial Hygiene Association, 66 S. Miller Road, Akron, Ohio) (2nd ed., 1975), 147 p. Illus. 113 ref.

Central cooling air conditioners. Standard for safety (Underwriters Laboratories Inc., 333 Pfingsten Road, Northbrook, Illinois) (1978), 72 p. Illus.

Hazards:

CIS 78-997 "Medical significance of air conditioning installations" (Die medizinische Bedeutung von Klimaanlagen). Müller, H. E. *Deutsche medizinische Wochenschrift* (Stuttgart), 29 Apr. 1977, 102/17 (639-643). 71 ref. (In German)

Energy saving:

CIS 77-1309 *Workplace ventilation and heating seen from the angle of energy saving and recovery* (Ventilation et chauffage des locaux de travail associés à l'économie et à la récupération d'énergie). Brunet, R. Edition INRS No. 532 (Institut national de recherche et de sécurité, 30 rue Olivier-Noyer, 75680 Paris) (Sep. 1976), 62 p. Illus. 3 ref. (In French)

The recirculation of industrial exhaust air. Symposium proceedings. DHEW(NIOSH) publication No. 78-141 (National Institute for Occupational Safety and Health, 4676 Columbia Parkway, Cincinnati, Ohio) (Apr. 1978), 152 p. Illus. Ref.

Table 2. Thermal comfort temperature (°C) in an environment where relative humidity is assumed to equal 50% (Based on Fanger's equation)

		Relative humidity = 50%								
Metabolism (W)		105			157			210		
Radiation temperature (°C)		20	25	30	20	25	30	20	25	30
Clothing (clo) 0.5 V_a (m.5^{-1})	0.2	30.7	27.5	24.3	21.0	17.7	14.0	11.0	8.0	4.0
	0.5	30.5	29.0	27.0	23.0	20.7	18.3	15.0	13.0	7.4
	1.5	30.6	29.5	28.3	23.5	23.3	22.0	18.3	17.0	16.0
Clothing (clo) 0.5 V_a (m.5^{-1})	0.2	26.0	23.0	20.0	13.3	10.0	6.5	-7.0	/	/
	0.5	26.7	24.3	22.7	16.0	14.0	11.5	-1.5	-3.0	/
	1.5	27.0	25.7	24.5	18.3	17.0	15.7	5.0	2.0	1.0

Source: Reproduced by permission from Vogt, J. J.: "Thermal working environment" (L'ambiance thermique de travail, 1979). *Archives des Maladies professionnelles* (Paris), Jan.-Feb. 1979, 40/1-2 (131-173).

Aircraft and aerospace industry

After only 70 years of existence the aircraft and aerospace industry rivals any in the world in terms of growth and sales. Despite cyclic business trends, production and sales have grown geometrically in the past half-century. With continuing tendencies for heavy governmental subsidisation of military systems and space programmes (and of commercial aviation in most countries) the industry remains a major contributor to world trade. At the end of 1979 general aviation manufacturers in non-socialist countries employed more than 1.5 million workers.

Much of this growth can be attributed to increased market demand resulting from the expansion of air traffic and the replacement of aircraft. Sales have climbed steadily since before 1975 and continue to do so at the start of the 1980s. Industry sales estimates for 1980-81 range between US$70 and 80 billion. This represents a significant rise above the world sales figure for 1979 which reached US$60 billion (excluding socialist countries).

Diversification in conjunction with high technological achievements gave the industry the flexibility to offer the market a more advanced and wider range of products and services in the 1970s. With emphasis on fuel efficiency and over-all performance, it is expected the industry will be able to maintain this performance through the 1980s.

Main technological developments

Since the beginnings of recorded time man has yearned to fly and to explore space. Because of the phenomenal growth of technology in the 20th century these dreams have now been realised. Several unique technological achievements stimulated this remarkable development.

The internal combustion engine. Despite resourceful (albeit sometimes ludicrous) attempts to utilise rubber bands, clockwork and steam engines, lack of a suitable motor continued to frustrate all efforts to build a flying machine until the invention of the gasoline fuelled internal combustion engine. The Wright brothers' first flights were made using a 4-cylinder, 200 lb, 24-horsepower gasoline engine, minute in comparison with modern piston-driven aircraft engines. Nevertheless, a major technologic barrier had been broken; by the late 1920s propulsion capability had far outstripped airframe technology.

Lightweight materials. Canvas, wire and wood were the mainstays of airframe fabrication until the 1930s when sufficient amounts of electricity and advances in metallurgy provided strong lightweight alloys of aluminium, magnesium and other metals. Tough and durable, these alloys could be easily formed and machined precisely for rapid assembly-line manufacturing using only rivets and other simple metal fasteners. For similar reasons synthetic fibres, plastics and composite materials are now used extensively to fabricate aircraft interior fittings.

Turbojet engine. The Second World War provided the impetus for the development of jet propulsion. The turbojet's relatively lightweight, small aerodynamic profile and very powerful thrust permitted high-speed flight at great altitudes. Modern jet airplane engines commonly can produce up to 44 000 lb of thrust (a pound of thrust is roughly equivalent to 1 horsepower) while consuming only about 0.5 lb of fuel per hour per lb of thrust. The powerful and efficient turbojet shortened travel time and revolutionised air transportation.

Space technology. A new era began in 1957 with the first successful launching of an Earth-orbiting satellite, Sputnik. Russian and American aerospace industries were plunged into intercontinental ballistic missile and space exploration races. A scant decade later thousands of ICBMs were operational, and man was soon to walk on the surface of the moon. Emphasis had shifted from production to research; the result was that the precision of aerospace technology increased by several orders of magnitude. Much of industry benefited from technological advances resulting from aerospace research in electronics, materials, computers and surveillance systems during the "space race".

Current technological advances. Fuel-efficient, high-bypass, turbofan engines are being fitted with inlet and fan exhaust acoustic liners and combined with wings having a lower lift co-efficient and simplified flap systems to permit very substantial improvements in jet aircraft noise reduction. A continuing revolution in electronics and avionics is contributing to improved systems reliability. Improvements are continuing to be made in corrosion resistance and durability. Carbon fibres imbedded in polymer matrices are making substantial contributions to improvements in structural efficiency.

Technology of the future. During the 1980s the aerospace industry's greatest challenge will be to find new economical sources of propulsion power that will have no adverse effect upon man's global environment.

Raw materials

Major categories within this extensive list include metals, plastics, paints, natural and synthetic rubbers, petroleum products, fuels and a large group of miscellaneous chemical mixtures.

Predominant structural metals used in aircraft and space hardware include alloys of aluminium, magnesium, steel, nickel and titanium. A large number of other metals including chromium, copper, zinc, manganese, beryllium, selenium, thallium and mercury are used in lesser amounts. Lead, kirksite and tungsten carbide are examples of commonly used manufacturing metals which do not normally become part of manufactured products. There are few plastics which are not in some way used by the industry. The list includes ethoxylin resins (epoxy compounds), phenoplasts (resins which are polycondensates of phenols and aldehydes), aminoplasts (urea and other resins), saturated and unsaturated polyesters, polyurethanes (isocyanates) and vinyl, acrylic and fluorine-containing polymers. The fillers which give the polymers strength are amorphous silicon dioxide, and fibres made of synthetics (polyamides and polyesters), and glass. Stabilisers, catalysts, accelerators, antioxidants and plasticisers act as accessories to produce a desired consistency. Primer, lacquer and enamel paints protect vulnerable surfaces from extreme temperatures and corrosive conditions. The most common primer paint is composed of synthetic resins pigmented with zinc chromate and extender pigment. It dries very rapidly, improves adhesion of top coats, and prevents corrosion of aluminium, steel, and their alloys. Enamels and lacquers are applied to primed surfaces as exterior protective coatings and finishes, and for colour purposes. Aircraft enamels are made of drying oils, natural and synthetic resins, pigments and appropriate solvents. Depending on their application, lacquers may contain resins, plasticisers, cellulose esters, zinc chromate, pigments, extenders and appropriate solvents.

Some natural and many synthetic rubbers are used in manufacturing processes and often appear in manufactured aircraft and aerospace products. Rubber mixtures find common use in paints, fuel cell lining materials, lubricants and preservatives, engine mountings, protective clothing, hoses, gaskets and seals. Natural and synthetic oils are used to cool, lubricate and reduce friction in engines, hydraulic systems and machine tools. Aviation gasoline and jet fuel are derived from petroleum-based hydrocarbons. High energy liquid and solid fuels have space flight applications and contain materials with inherently hazardous physical and chemical properties; such materials include liquid oxygen, hydrazines, peroxides and fluorine.

More than 5 000 chemicals and mixtures of chemical compounds are commonly used in the aircraft and aerospace industry. On the average each compound contains between five and ten ingredients. The exact composition of any particular mixture purchased from a chemical supplier is usually a proprietary or trade secret, which lends additional complexity to this enormously heterogeneous group.

Manufacturing processes

The rather diverse, somewhat complex and often changing manufacturing processes encountered in the aircraft and aerospace industry reflect continuing efforts to improve product performance and production efficiency by applying new technologies to existing manufacturing methods.

Some heavy non-ferrous metals and numerous lightweight alloys are poured in sand or phenolic resin shell moulds, or dye cast. Thin pieces of metal work are forged and hammered into shape, while thicker ones are precision machined or chemically milled and then normalised, tempered, annealed or hardened by heat treatment. Ferrous metals are carbonised and carbonitridised. Formed metal parts are cleaned by mechanical abrasion, emulsions, acid or alkaline cleaning agents, chemical etching solutions, solvent degreasing, steam cleaning, ultrasonic cleaning and vapour degreasing. Surfaces of metal parts are rendered more corrosion resistant by barrel finishing, chemical and electrochemical surface treatments, and electroplating. Chromate containing, corrosion resisting synthetic resin paint primers are applied by spraying or dipping the parts, which are then fitted on precision tools or laser-aligned assembly jigs for joining by means of riveting application of bonding agents, or welding.

Riveting is accomplished after drilling, reaming and coldworking holes to close tolerances using manually operated single or multiple impact pneumatic riveting guns, manual or numerically controlled, hydraulically operated squeeze riveting devices. For smooth contouring on aerodynamic surfaces rivet heads are smoothed by countersinking, dimpling and microshaving. A wide spectrum of welding techniques are in common use: gas, arc, plasma arc, inert gas shielded arc, electron beam and various forms of resistance welding. Structural adhesive bonding using synthetic resins (phenolic or epoxy) or rubber (neoprene or polyacrylic nitrile) has found many applications and has the important advantages of greater strength, superior fatigue characteristics, smooth aerodynamic surfaces, joining of dissimilar metals, leakproof joints, less corrosion and lower production costs.

Fibreglass or other synthetic fibre batting is glued and sewn between plain or aluminised Mylar sheets to make heat-reflective, fire-resistant and noise-absorbing thermal insulating blankets for aircraft. Fuel cells, weather surfaces, firewalls and pressurised compartments are sealed with silicone, polyurethane and polysulphide sealants. Because of their resilience, durability and ease of application, polyurethane paint systems have largely replaced epoxy and other synthetic resin systems for final exterior (decorative) painting. Corrosion resistant ("stainless") steel tubing is formed, shaped, fitted, colour coded, meticulously cleaned by solvent and vapour degreasing, blown dry with clean compressed air, sealed, packaged and shipped to assembly areas for final installation. Colour coded electrical wire is machine stamped with numerical codes, measured, cut and wound in intricate patterns around pegs laid out on form-boards. Groups of wires which will ultimately traverse common channels are tied together in bundles and fitted with manually and machine soldered electrical fittings prior to shipment to assembly areas.

Graphite, fibreglass and other synthetic fibrous materials are combined with various synthetic resin systems to form two-phase, strong, composite, lightweight linings, mouldings, bulkheads and passenger service unit exteriors. A variety of chemical materials and other additives are used to alter strength, rigidity, machinability, chemical resistance and curing time. Fibresheet material pre-impregnated with resin is maintained at low temperature, until it is cut and placed on forms and mandrels, sealed in polyvinyl alcohol film, atmospheric or other pressure applied, and curing completed by heating the mould, use of infrared light, or placement in ovens or temperature, pressure and atmosphere controlled autoclaves. Complex shapes are formed using two-stage, heated, matched, dye-moulding processes. Finished laminates may be flame sprayed with metal to make them more heat and abrasion resistant; or they may be embossed or embellished with colourful decorative designs using silk screen printing processes.

SAFETY AND HEALTH HAZARDS

A quite substantial number of potentially serious hazards can be encountered in the aircraft and aerospace industry, largely because of the sheer physical size and complexity of the products produced and the diverse and changing array of manufacturing and assembly processes utilised. Inadvertent or inadequately controlled exposure to these hazards can produce immediate obvious traumatic injuries or sudden acute forms of occupational illness. Non-occupational illnesses can be aggravated. Subtle exposures over longer periods of time can produce insidious forms of subacute and chronic occupational disease. Tables 1, 2 and 3 provide an overview of this industry's recognised safety and health hazards.

Safety hazards. Immediate direct trauma can result from dropped rivet bucking bars or other falling objects; tripping on irregular, slippery or littered work surfaces; falling from overhead crane catwalks, ladders, aerostands and major assembly jigs; touching ungrounded electrical equipment, heated metal objects and concentrated chemical solutions; contact with knives, drill bits and router blades; hair, hand or clothes entanglement or entrapment in milling machines, lathes and punch presses; flying chips, particles and slag from drilling, grinding and welding; misapplication of body mechanics in bending, lifting and carrying heavy or cumbersome objects.

Health hazards. Heat stress is not a common source of occupational disease in the aerospace industry. However, in warm climates it may be encountered in connection with hot metal processes such as foundry work. Cold stress is a rare source of problems, but may occasionally occur when employees enter sub-zero storage rooms or similar rooms maintained to feed extremely cold rivets into automatic riveting machines.

Table 1. Aircraft and aerospace industry safety hazards

Type of hazard	Common examples	Possible effects
Physical		
Falling objects	Rivet guns, bucking bars, fasteners, hand tools	Contusions, head injuries
Moving equipment	Trucks, tractors, bicycles, fork-lift vehicles, cranes	Contusions, fractures, lacerations
Hazardous heights	Ladders, scaffolding, aerostands, assembly jigs	Multiple serious injuries, death
Sharp objects	Knives, drill bits, router and saw blades	Lacerations, puncture wounds
Moving machinery	Lathes, punch presses, milling machines, metal shears	Amputations, avulsions, crush injuries
Airborne fragments	Drilling, sanding, sawing, reaming, grinding	Occular foreign bodies, corneal abrasions
Heated materials	Heat treated metals, welded surfaces, boiling rinses	Burns, keloid formation, pigmentation changes
Hot metal, dross, slag	Welding, flame cutting, foundry operations	Serious skin, eye and ear burns
Electrical equipment	Hand tools, cords, portable lights, junction boxes	Contusions, strains, burns, death
Pressurised fluids	Hydraulic systems, airless grease and spray guns	Eye injuries, serious subcutaneous wounds
Altered air pressure	Aircraft pressure testing, autoclaves, test chambers	Ear sinus and lung injuries; bends
Temperature extremes	Hot metal working, foundries, cold metal fabrication work	Heat exhaustion, frost bite
Loud noises	Riveting, engine testing, high speed drilling, drop hammers	Temporary or permanent loss of hearing
Ionising radiation	Industrial radiography, accelerators, radiation research	Sterility, cancer, radiation sickness, death
Non-ionising radiation	Welding, lasers, radar, micro-wave ovens, research work	Cornea burns, cataracts, retinal burns, cancer
Chemical		
Chemical irritants	Acid baths, hydraulic fluids, ammonia, plating solutions	Eye, nose and throat irritation, dermatitis
Simple asphyxiants	Inert gas welding, fire extinguishers, isolated spaces	Confusion, disorientation, unconsciousness, death
Chemical asphyxiants	Cyanide solutions, engine exhaust research laboratories	Confusion, unconsciousness, convulsions, death
Organic solvents	Solvent and vapour degreasing, painting, silk screening	Narcosis, hepatitis, sudden death
Ergonomic		
Lifting and carrying	Moving raw materials, parts, tools and equipment	Back injuries, other musculoskeletal injuries
Work in confined spaces	Aircraft fuel cells	Oxygen deprivation, entrapment, narcosis, anxiety
Poor man-machine fit	Assembly jigs, wire crimping tools, rivet guns	Excess fatigue, musculoskeletal injuries
Work surfaces	Spilled lubricants, disarranged tools, hoses and cords	Contusions, lacerations, strains, fractures

Sources of both ionising and non-ionising radiation are very prevalent in the industry and require stringent controls to prevent serious injuries and illness. A limited number of workers will be exposed to both hyperbaric and hypobaric conditions in the course of doing research, testing aircraft and spacecraft hardware, and occasionally even of undersea operations.

High levels of riveting, jet engine and metal-forming and machining noise are encountered by many employees in the industry. Dust generated while machining beryllium alloys and the fumes generated by heating cadmium and lead must be carefully controlled if intoxication is to be prevented. Spillage of metallic mercury in confined spaces occasionally creates a significant hazard. The relatively insoluble salts of hexavalent chromium, but not chromic acid itself, require handling as potential carcinogens. Although sulphur dioxide, hydrogen cyanide and carbon monoxide are potentially very serious hazards, they are relatively infrequent sources of problems in well controlled aircraft and aerospace industry establishments. Scrupulous attention to the layout of work processes and the provision of adequate ventilation is required in the industry to prevent serious problems among welders resulting from exposure to the oxides of nitrogen, phosgene and ozone. Aliphatic hydrocarbon cutting oils still produce outbreaks of dermatitis among machinists, unless careful attention is given to the make-up and cleanliness of machine lubricants and cutting oils. Nitrites added as preservatives for cutting fluids can react to form carcinogenic nitrosamines. Aromatic and halogenated hydrocarbons used in the aerospace industry

Table 2. Aircraft and aerospace industry physical agent health hazards

Agent of stress	Frequency	Sources	Potential disease
Temperature			
Heat	Uncommon	Melting and heat treating metals	Heat stress with or without cramps, exhaustion, stroke
Cold	Rare	Cold rivet feed rooms	Frostbite, hypothermia
Radiation			
Ionising	Common	Industrial radiography, accelerators, tests	Adverse genetic and reproductive effects, leukaemia, cancer, cataracts, accelerated ageing
Ultraviolet	Common	Welding, quality control tests	Burns of skin and cornea, skin keratoses and cancer
Infrared	Rare	Welding (some), furnaces	Cataracts
Microwave and radar	Common	Antennas, ovens	Interference with some cardiac pacemakers, cataracts, radiomimetic effects(?)
Pulsed electromagnetic	Rare	Research and special tests	Cataracts(?), radiomimetic effects(?)
Laser/maser	Common	Tool and jib alignment, metal cutting	Skin burns, retinal burns, cataracts(?)
Air pressure			
Increased	Common	Aircraft testing, test chambers, diving	Nitrogen narcosis, air embolism, barotitis, decompression sickness, aseptic necrosis of bone
Decreased	Common	Aircraft testing, test chambers, altitude chambers	Aerotitis, air embolism, decompression sickness
Noise			
Continuous	Common	Manufacturing machinery	Noise induced hearing loss (occupational deafness), other functional or organic effects(?)
Impulse	Uncommon	Pneumatic and hydraulic equipment (some)	Same as continuous noise(?)
Impact	Uncommon	Drop hammers, riveting (some)	Same as continuous noise(?)

must be very carefully controlled, if the serious health problems outlined in table 3 are to be prevented. Most synthetic resins probably present some degree of carcinogenic hazard and must be used with appropriate precautions. In addition, epoxy and other resins utilising polyamine hardeners are extremely strong skin sensitisers in the uncured state. Diisocyanates used in polyurethane systems are sensitisers associated with a high incidence of occupational asthma, unless very stringent ventilation controls are implemented and maintained in good working condition. Obvious forms of asbestos have been rapidly eliminated from this industry, but it still presents a significant hazard as a "minor ingredient" or "inert filler" in the materials provided by vendors of other industries. Use of silica as a filler in resin systems has been largely eliminated in the aerospace industry, but abrasive blasting using crystalline forms of silicon dioxide still carries a potential for causing silicosis among inadequately protected sandblasters.

Fibreglass is a common source of mechanical skin and respiratory tract irritation. Scientific concern still exists that particles of fibreglass approaching the dimensions of asbestos may be fibrogenic and/or carcinogenic. Because of their apparent carcinogenic potential the dust of western red cedar and certain other wood dusts require careful control measures.

SAFETY AND HEALTH MEASURES

Precision manufacturing of aerospace products requires clean, organised and well controlled work environments. This requirement itself has contributed significantly to the health and safety of employees. In addition, the following techniques are commonly used to minimise health and safety hazards:

- Proposed new materials and processes are reviewed by health and safety professionals prior to their actual introduction into manufacturing environments. Potential hazards are identified, safe usage standards developed and necessary controls implemented as the material or process is introduced.
- Chemical mixtures are carefully labelled.
- Hazardous work areas are prominently placarded.
- To the extent feasible local exhaust ventilation and other engineering controls are used to control workplace exposure.
- At times administrative controls (rotation of employees) can be used to control exposure, but this approach has found only very limited application in the aircraft and aerospace industry because of skill factors and collective bargaining agreements.
- Safe work procedures are clearly established and compliance with them is mandatory.
- When necessary, employees are provided with protective gloves and clothing, eye and face protection, ear defenders, respirators and other protective equipment.
- Industrial hygiene and radiation health protection surveys are performed at frequent intervals to ensure that controls are adequate.

Table 3. Aircraft and aerospace industry chemical agent health hazards

Chemical agent	Sources	Potential disease
Metals		
Beryllium dust	Machining beryllium alloys	Skin lesions, acute or chronic lung disease
Cadmium fume	Welding and burning	Delayed acute pulmonary oedema, chronic kidney disease
Chromium dust/mist	Paint primer sanding, spraying paint primer	Cancer of the respiratory tract(?)
Lead fumes	Melting lead, burning lead primers	Lead poisoning
Mercury vapours	Laboratories, engineering tests	Mercury poisoning
Gases		
Sulphur dioxide	Furnace operations, foundry work	Pulmonary oedema (rare), chronic lung disease(?)
Hydrogen cyanide	Electroplating solutions	Chemical asphyxiation, chronic effects(?)
Carbon monoxide	Heat treating, engine work, furnace operation	Chemical asphyxiation, chronic effects(?)
Oxide of nitrogen	Pickling metals, welding, electroplating	Delayed acute pulmonary oedema, permanent lung damage (possible)
Phosgene	Welding decomposition of solvent vapours	Delayed acute pulmonary oedema, permanent lung damage (possible)
Ozone	Arc welding, high altitude flights (some)	Acute and chronic lung damage, cancer of respiratory tract(?)
Organic compounds		
Aliphatic	Machine lubricants, cutting oils	Follicular dermatitis
Aromatic, nitro and amino	Constituents: rubber, plastic, paint, dyes	Anaemia, cancer, skin sensitisation, nervous irritability
Aromatic, other	Solvents, chemical contaminants	Narcosis, anaemias(?), leukaemia
Halogenated	Solvents, propellants, refrigerants	Narcosis, cardiac arrhythmias, cancer(?)
Plastics		
Epoxy (amine hardeners)	Aircraft interior fabrication	Dermatitis, allergic sensitisation, cancer(?)
Polyurethane (diisocyanates)	Paint systems and plastic foams	Allergic pulmonary sensitisation
Other	Same as epoxies and polyurethanes	Allergic dermatitis, cancer(?)
Fibrogenic dusts		
Asbestos	Insulating materials, "inert fillers", plastics	Asbestosis, mesothelioma, lung cancer, other cancer
Silica	Abrasive blasting, foundries, fillers, plastics	Silicosis
Tungsten carbide	Precision tool grinding	Pneumoconiosis ("hard metal disease")
Benign dusts(?)		
Fibreglass	Blanket insulation, interior fabrication	Skin and respiratory irritation, chronic disease(?)
Wood	Aircraft mock-up and model making	Allergic sensitisation, respiratory cancer(?)

- Physical examinations, bio-assays, tests of pulmonary function, audiometric examinations and other test procedures are used by medical personnel to detect, evaluate and control early forms of occupational illness.
- Industrial hygiene, radiation health protection, safety and medical records are maintained and periodically analysed to identify areas and processes requiring more effective hazard control measures.

Work organisation

Health and safety programmes tend to be highly structured in the aircraft and aerospace industry. Company executives are "safety conscious" and directly involved in systematic health and safety programmes. Employees and their union representatives tend to be interested and active participants in this sphere of activity. Supporting staffs of health and safety professionals (industrial hygienists, health physicists, safety engineers, nurses, physicians and technicians) are common in the larger aircraft and aerospace companies. The activities of these staffs tend to be integrated with companies' research, design, engineering and manufacturing activities. Such integrated programmes are usually quite effective in detecting, evaluating and controlling all forms of occupational injury and illness.

DUNPHY, B. E.
GEORGE, W. S.

CIS 79-1138 "Occupational hazards in the aeronautics industry" (Risques professionnels rencontrés dans l'industrie aéronautique). Pepersack, J. P. *Cahiers de médecine du travail–Cahiers voor arbeidsgeneeskunde* (Gerpinnes, Belgium), Dec. 1978, 15/4 (265-270). 11 ref. (In French and Dutch)

CIS 78-1069 "Long-term exposure to jet fuel. II. A cross-sectional epidemiologic investigation on occupationally exposed industrial workers with special reference to the nervous sytem". Knave, B.; Olson, B. A.; Elofsson, S.; Gamberale, F.; Isaksson, A.; Mindus, P.; Persson, H. E.; Struwe, G.; Wennberg, A.; Westerholm, P. *Scandinavian Journal of Work Environment and Health* (Helsinki), Mar. 1978, 4/1 (19-45). Illus. 49 ref.

CIS 78-691 *Riveting noise and riveters in aircraft construction* (Mémoire sur le rivetage et les riveurs dans la construction aéronautique). Reilhac, G. (Société d'hygiène industrielle et de médecine du travail, 15 rue de l'Ecole-de-Médecine, 75006, Paris) (1977). 54 p. Illus. (In French)

CIS 76-102 "Lighting for the aircraft/airline industries–Manufacturing and maintenance". Aircraft Industry Subcommittee, Illuminating Engineering Society, New York. *Journal of the Illuminating Engineering Society* (New York), Apr. 1975, 4/3 (207-219). Illus. 3 ref.

Air ionisation

Any free atmosphere, like the air of workplaces, living space or public rooms, always contains varying amounts of particles carrying electrical charges. These particles–called air ions–move in the electrical field conditioning conductivity. By air ionisation we understand both the process of production and evolution of the electrically charged particles, and the aeroionic spectrum. The present article refers to air ionisation of the low atmosphere.

Air ions differ from one another by charge sign, chemical substratum, aggregation state, dimension, number of charges, mobility in the electrical field (measured in cm/s for 1 V/cm), and length of life. The *natural ionising agents* are cosmic (cosmic rays, ultraviolet rays, etc.) and/or telluric (radioactive substances; processes involving liquid-spraying, bubbling or evaporation; electrical phenomena such as discharges, or storms; combustion; etc.). Burning and various industrial or nuclear processes constitute *artificial ionising factors*.

Under the influence of the ionising agents, gaseous molecules lose or gain an electron. The *negative ions* take the form particularly of oxygen ions. Molecules that have lost an electron become *positive ions*. These ions of molecular size rapidly polarise around the uncharged molecules and in this way *small ions* are formed, having a relatively high mobility, of over 1 cm/s for 1 V/cm.

The small ions disappear by recombination, or discharge on conductors, or adsorption on aerosols; in this way *heavy ions or large ions* are formed (with a greater mass, and a lower mobility, than small ions).

Determination of air ions is made by means of ion-counters. Air ion concentrations are expressed per 1 cc of air.

Table 1. Concentrations and ratios of air ions in the free atmosphere of unpolluted areas over land (mean values)

Symbol	Denomination of air ion factor	Mean value
$(n^+);(n^-)$	Concentration of positive or negative small ions/cc	400-700 pairs
$q = (n^+)/(n^-)$	Unipolarity coefficient of small ions	1.10-1.20
$(N^+);(N^-)$	Concentration of positive or negative large ions/cc	1 500-3 000 pairs
$K = \frac{(N^+)+(N^-)}{(n^+)+(n^-)}$	Ratio of large ions of both signs to small ions of both signs	2.0-5.0

Outdoor and indoor air ionisation

In a given area the natural air ionisation (small ion concentration) undergoes a seasonal cycle: it is greater during summer and smaller during winter. At high altitudes small ion concentration increases. In very unpolluted areas (mountains, woods, health resorts) or near radioactive sources the small ion concentration may rise to nearly 3 000 pairs/cc. Near waterfalls or rapid mountain springs tens of thousands of small ions/cc can be formed. A high concentration of small ions and a prevalence of negative ions are considered favourable for the organism.

Near the seaside, especially in the absence of vegetation, a significant deficit of negative small ions often exists. In polluted areas (work, inhabited, or outdoor spaces) the number of small ions can decrease to 100 pairs/cc, or even below this level. Also, in polluted or very wet air the concentration of large ions can increase to 80 000-100 000, or even more. As a consequence of these changes in polluted or very wet air the K ratio also increases. It is generally considered that when K is lower than 50 the atmosphere is relatively pure, and when K is over 50 the air is polluted. Thus, air ionisation can constitute an *unspecific indicator* of air pollution or purity of air.

Of course, such modifications are significant when the polluting process does not generate air ions. In the case of a process generating air ions, the detection of modifications may constitute a *specific indicator* of pollution, for example by radioactive substances.

Air ionisation in workplaces

Generally in workplaces the concentration of large ions increases, and that of small ions decreases. In hermetically closed spaces or in mines these changes are very marked. It is often the negative small ions that are most substantially reduced. Exceptions in this respect are workplaces in which ionising factors have an influence, e.g. radioactive substances, nuclear reactors, X-rays, ultraviolet rays, electric discharges or electrical devices under high voltage, voltaic arcs, incandescent metals, pulverisation, friction of materials (weaving mills, printing industry), etc. In such cases the number of small ions may amount to tens of thousands, or as many as 100 000-130 000 (industrial radiography), or 200 000-400 000 (electrical welding).

Mechanical ventilation or air conditioning devices usually de-ionise the air, the small ions being retained by pipes or by dust filters.

Techniques for artificial ionisation of the air

By the use of air ion generators the air ion content can be normalised, or increased to very high densities (millions of ions/cc). The generators are based upon one of the following physical phenomena: radioactivity (α), electrical discharges, water spraying, thermal-electronic effect, ultraviolet rays. The generators produce small ions of both signs in close concentrations. In order to study or apply the action of unipolar ions, those of one sign must be retained by using electrical separators.

Distilled water spraying practically gives negative ions only.

Air ion generators should not produce secondary factors (radioactive or ultraviolet radiations, ozone, high electric field, air currents, alterations of air temperature, humidity, chemical substances) that could have a separate and sometimes unfavourable effect.

Besides aeroionotherapy, electro-aerosolotherapy (aerosols with air ion loading) is also used.

Air ion influence on the organism

Generally speaking, small negative air ions exert a favourable influence on many biological functions; examples are cell development and division; circulatory, secretory and trophic processes; changes in the tracheal mucosa; vegetative and central nervous system tonicity; allergic and anaphylactic phenomena; external behaviour of animals (motility, etc.); physical and intellectual work capacity; and fatigue.

The application of negative air ionisation gave favourable therapeutic results in certain respiratory diseases (chronic bronchitis, asthma), cardiovascular diseases (hypertension), digestive diseases (ulcer), neuropsychic diseases (neurosis, insomnia, anxiety), endocrine and allergic diseases, burns, etc. Prophylactic applications were made in the case of workers exposed to silicosis or other respiratory risks, sportsmen with a view to improving their accommodation to effort, and pupils with the aim of increasing work capacity and decreasing the incidence of respiratory diseases.

Improving the air ion content in workplaces

No health standards exist regarding air ionisation, but it is considered that the high levels found in nature are favourable for the human organism. The most appropriate level is thus considered to be that prevalent in certain climatic resorts, particularly in the mountains, of 2 000-2 500 pairs of small ions/cc, with a slight negative predominance.

In workplaces or rooms in which air ionisation is reduced or excessively increased, natural ventilation can bring the ionic condition to the outdoor level.

In the absence of toxic substances, the air conditioning system should ensure a natural level of air ionisation in workplaces similar to that of mountain resorts. If work is carried out in a toxic environment, the air ion deficit could be compensated by means of preventive and curative treatment with air ions in special rooms (inhalatoria) before and/or after the work shift.

For health purposes air ionisation can also be used to reduce air pollution produced by dust, smoke, gases and micro-organisms; to measure certain particles in the air; and to reduce electrostatic charges.

DELEANU, M.

General:

Problems concerning ionisation and air ionisation (Problèmes d'ionisation et d'aéro-ionisation). Rager, G. R. (ed.) (Paris, Maloine SA, 1975), 244 p. Illus. 687 ref.

Biological effects:

Ionisation of the air and electrical field effects in biology. Bibliography of published references, 1960-1978. 7th edition. King, G. W. K. (compiled by) (Philadelphia, American Institute of Medical Climatology, 1979), 114 p. 367 ref.

"The effects of negative air ions on human performance". Inbar, O.; Rotstein, A.; Dotan, R.; Dlin, R. *Proceedings of the Seventh Annual IRA Symposium on Human Engineering and Quality of Work Life* (Ramat Efal (Israel), IRA Memorial Foundation, 1980), (314-343). Illus. 30 ref.

"Air ions and human performance". Hawkins, L. H.; Barker, T. *Ergonomics* (London), Apr. 1978, 21/4 (273-278). 24 ref.

"Effect of air ionization on heart rate and perceived exertion during a bicycle exercise test. A double-blind cross-over study". Sovijärvi, A. R. A.; Rosset, S.; Hyvärinen, J.; Fraussila, A.; Graeffe, G.; Letimäki, M. *European Journal of Applied Physiology* (Berlin, Heidelberg), Aug. 1979, 41/4 (285-291). Illus. 19 ref.

Air pollution

The moisture content of air varies up to 4% by volume with changing climatic and meteorological conditions, and the atmospheric concentration of other substances is dependent upon the environment and the nature of man's activities. Thus ambient air may contain up to 0.5% carbon dioxide, variable amounts of aerosols (suspended particulate matter such as water droplets, ice crystals, dust, fumes or mists), and up to 1% of organic and inorganic contaminants. The term "air pollution" is commonly used today to mean pollution of the ambient air by products of combustion and industrial wastes rather than the contamination of the atmospheres in factories and other places of work.

Sources

Air pollution may be caused by a wide variety of substances and processes. However, under normal circumstances pollution from natural phenomena (such as volcanoes, forest fires), by radioactive material from weapon testing and by bacteria, is usually excluded; moulds and spores are sometimes included because of their allergenic properties.

Properties of combustion. The "traditional" problem is pollution by the products of combustion, complete and incomplete, of coal and, latterly, of petroleum products. Ideally, fossil fuels containing carbon and hydrogen should be burned to produce only carbon dioxide and water. This desirable state of affairs is found in some places where natural gas is burned for heating and power production. More commonly, coal and oil, containing compounds of sulphur as impurities, are burned with the production of undesirable pollutants. Probably the most wasteful use of fuel seen in modern times is the burning of raw coal in open domestic grates. This inefficient process results in offensive pollution being discharged from houses at low level, where it may be inhaled. In the early stages of burning in domestic grates, coal is subject to a process of destructive distillation during which an aerosol of tarry droplets mixed with carbonaceous particles is discharged. As combustion proceeds the smoke comes to contain more carbon and sulphur dioxide with small amounts of sulphuric acid. The tar is rich in polycyclic aromatic hydrocarbons, phenols and many other organic compounds. Where coal is burned in industrial plants, such inefficiency would be wholly intolerable and, indeed, tarry smoke is difficult to produce. Black carbonaceous smoke is, however, not uncommonly emitted for short periods when firing is uneven or inefficient.

The use of smokeless solid fuels such as cokes in domestic grates is an advantage, though sulphurous effluents are still emitted at low level. Industrial steam-raising plants discharge their effluents from higher stacks and this fact, together with the necessarily higher efficiency of combustion, should lead to less pollution at ground level. In many countries there is now legislation to control emissions of smoke from industries and from domestic sources.

Industrial waste. Industrial wastes discharged to atmosphere may be as varied as the processes from which they originate. Some, such as cement dust, may constitute a great nuisance but a negligible hazard, whereas others, such as lead and beryllium compounds, may be dangerous.

Internal combustion engine. This causes pollution of the air at low level. Gasoline engines (spark ignition engines) burn a "rich" mixture in which the fuel is frequently present in excess over the air available for complete combustion. Carbon monoxide, unburnt or partially burned fuel and "cracked" hydrocarbons are emitted. In addition there are usually present in the exhausts inorganic compounds of lead and its halides derived from antiknock additives. The diesel engine (compression ignition) burns its fuel in a large excess of air and usually produces virtually no carbon monoxide or unburnt fuel. When badly adjusted, worn, inappropriately operated or overloaded, the diesel engine emits black smoke and carbon monoxide together with aldehydes and other malodorous compounds. Both types of engines emit nitrogen oxides derived from fuel and from the fixation of atmospheric nitrogen in the high temperature and pressure obtaining in the combustion chamber. Again, both types of engine may, if worn, emit lubricating oil.

Photochemical pollution

The "traditional" forms of air pollution discussed above, derived largely from the combustion of coal or heavy oil for the purposes of power generation and heating, impart reducing properties to the air by virtue of their sulphur dioxide content. In contrast, the "photochemical" pollution which first became prominent in Los Angeles, but which may now be found in many parts of the world, has oxidising properties. It is the result of complex interactions, in the presence of strong sunlight, between volatile hydrocarbons and oxides of nitrogen leading to the formation of ozone and many complex organic compounds among which is peroxyacetylnitrate (PAN), which is irritant to the eyes. While gasoline-engine powered motor vehicles are by far the most common sources of the hydrocarbon and nitrogen oxide precursors, contributions may also come from oil refineries and other industrial plants. Since the reactions take place comparatively slowly, maximum concentrations of ozone and other secondary products may occur at some distance from the sources in rural as well as in urban areas. While many results of a vast amount of research on the possible effects of "photochemical" pollution on health remain equivocal, there is no doubt that it contains powerful lachrimators and is most unpleasant. It can cause serious damage to many species of plants. Photochemical reactions take place in air which is polluted mainly by sulphur dioxide. Again the reactions are complex and lead to the formation of an aerosol of very small particles of ammonium sulphate, which can impair visibility. Ozone is a conspicuous feature of this type of pollution.

Dispersion and dilution

There usually exists in the atmosphere a temperature gradient which, together with the efflux velocity of pollutants or their buoyancy, ensures that emissions are dispersed and diluted; the air near the ground is usually warmer than that aloft: it therefore rises carrying pollutants with it and acts as a scavenging mechanism.

Smog

This normal scavenging mechanism fails during anticyclonic weather in which there are temperature inversions. Then cold air comes to lie under warmer air with resultant cessation of the upward currents. In such weather a lid is in effect put over towns and the pollutants emitted therein accumulate in the still air. Fog not infrequently accompanies such weather and the mixture of fog, coal smoke and irritant gases found in many temperate cities during winter anticyclones has been termed "smog" (an abbreviation of "smoke polluted fog"). This term is now used to describe the photochemically produced Los Angeles-type pollution and, because of its imprecision and the consequent confusion, its use should be discontinued.

In such episodes of calm winter weather the concentrations of pollutants may rise to levels of over 20 times the average winter values. In such concentrations new atmospheric reactions may occur: for example the oxidation of sulphur dioxide to sulphuric acid may be facilitated. It is in such conditions of high pollution that the effect on health of contamination of the air may most easily be seen.

HAZARDS

In cities increases in mortality are seen to follow quickly on rises in pollution. The excess mortality occurs mainly among the elderly and infirm and weak. Particularly vulnerable are respiratory and cardiac cripples. The larger the population the more easily are rises in mortality seen in relation to relatively small increases in pollution; in small towns pollution must reach extraordinarily high levels to produce unequivocal peaks in mortality.

Respiratory and cardiac diseases. Commonly one finds during episodes of high pollution symptoms and signs of respiratory irritation with embarrassment of lungs and heart in those who by virtue of age or disease are already weakened. It is likely that high pollution provides a final and intolerable stress to some people, this being reflected in a rise in mortality in a community if it contains enough "susceptible" people. The pollutant responsible for the mortal effect has not yet been identified with certainty. Sulphur dioxide has received much attention from investigators but experimental work has hitherto failed to demonstrate significant ôr consistent impairment of lung function when the gas is administered in concentrations such as are found in conditions of high pollution. Similar results have followed like experiments with sulphuric acid. The results of current epidemiological studies exploiting the alteration of the ratio of smoke to sulphur dioxide and supplementary experimental work tend to throw the blame on particulate pollutants, though the effects may yet be proved to be due to a synergistic action of gases and particles. But the effects are not likely to be due to a simple or single substance. The frequent association of cold weather with high pollution adds a further complication since, obviously, both are stressful factors.

There is no doubt that degrees of pollution less dramatic than those considered above can produce deterioration in the health of patients with established respiratory or cardiac disease which, in populations of adequate size and constitution, can be recognised as peaks in some chosen index of morbidity (such as sickness absence or applications for admission to hospital).

Chronic bronchitis. This is a disease especially common in the United Kingdom; it is more prevalent in towns than in country districts and much effort has gone into research to determine the part played by air pollution in the genesis of the disease. There would appear to be two main stages in the development of progression of the disease: "simple" bronchitis is due to hypersecretion of mucus in the bronchial tree; the mucus and the hypertrophic mucosa impedes air flow in the respiratory tract. Later, this stage, which is characterised clinically by chronic cough and the expectoration of mucoid sputum, is followed by the establishment of infection of the respiratory tract with subsequent destruction of lung substance and the development of emphysema and, frequently, ultimate heart failure. There is little evidence to suggest that air pollution contains irritants powerful enough to produce "simple" bronchitis which is seen (by the results of several studies) to be closely related to cigarette smoking. Modern work suggests that an urban factor is involved in the second "complicated" stage and this factor may well be air pollution. It is important to remember that many factors are involved in the production of the disease, which is especially related to social conditions.

Lung cancer. It has been found that lung cancer occurs more commonly in towns than in the country and many workers have claimed that the disease is caused by air pollution. Town air in coal-burning communities is relatively rich in polycyclic aromatic hydrocarbons, some of which are carcinogenic and some studies have shown lung cancer mortality in parts of Britain to be closely related to smoke content of the air. This relationship is, however, not consistently seen; the urban/rural gradient is steep in countries such as Finland and Norway in which the air contains little pollution. A further objection to the suggestion that air pollution is responsible for the lung cancer now seen in such profusion is that lung cancer has increased at the same time as there has been a steady decline in pollution. The results of a large number of excellent studies leave no doubt that the rise in lung cancer is due, overwhelmingly, to the smoking of cigarettes.

The motor car and diesel vehicles were at one time thought to be responsible for the rise in lung cancer since their increasing use was, in part at least, contemporaneous with the rise in lung cancer mortality. Fortunately there is no evidence to indict motor traffic. Workers who have been exposed to asbestos fibres have an enhanced risk of developing carcinoma of the bronchus especially if they are smokers; exposure to very small amounts of blue asbestos (crocidolite) has been shown to be related to the development of a rare malignant tumour (mesothelioma) of the pleura or of the peritoneum. Cases of mesothelioma have been seen in persons who have lived near to crocidolite mines or to factories in which asbestos has been used or processed. Though there is no reason to believe that the amount of asbestos found in the general air in towns is hazardous it must be borne in mind that there are few data which enable a dose/response curve to be constructed for low concentrations. Recent research tends to throw blame on the actual physical dimensions and properties of the crocidolite fibre (in contrast to those of chrysotile and other asbestos minerals) and attention is being paid to the physical properties of man-made fibres being used as "safe substitutes" for asbestos in case these may have similar long-term effects.

The Los Angeles-type photochemically produced pollution has not been shown unequivocally to have any serious effect on health, though in view of the vast amount of current work being done, any judgement would be premature. There are some claims that lung function might be impaired or infection of the respiratory tract encouraged when oxidant concentrations are high; experimental work has shown that exposure to ozone at concentrations not much in excess of those liable to occur in the photochemical complex may impair lung function especially when combined with exercise.

Monitoring

Many national and international networks have been established to monitor concentrations of the principal air pollutants. Some of these use simple equipment sampling for 24-hour periods with manual changes of sampler; others involve the use of more complex recording instruments which give continuous records. Within each network procedures are carefully standardised and arrangements are usually made for the results to be collated centrally. The range of pollutants monitored depends on local circumstances; sulphur dioxide and smoke (or total suspended particulates) being of prime interest in areas where there is "traditional" pollution from burning of coal or heavy fuel oil. Ozone, oxides of nitrogen and volatile hydrocarbons are more relevant pollutants where there are problems related to the photochemical reactions associated with traffic or refineries (see AIR POLLUTION CONTROL).

LAWTHER, P. J.

Air quality criteria and guides for urban air pollutants. Technical Report Series No. 506 (Geneva, World Health Organisation, 1972), 35 p. Illus. 82 ref.

Environmental health criteria (a series of reports on individual pollutants) (Geneva, World Health Organisation, various dates), including:

Oxides of nitrogen, No. 4 (1977), 79 p. Illus. 193 ref.

Photochemical oxidants, No. 7 (1978), 110 p. Illus. 303 ref.

Sulfur dioxide and suspended particulate matter, No. 8 (1979), 107 p. Illus. 249 ref.

Selected methods of measuring air pollutants. Offset publication No. 24 (Geneva, World Health Organisation, 1976), 112 p. Illus. 88 ref.

Air monitoring programme design for urban industrial areas. Offset publication No. 33 (Geneva, World Health Organisation, 1977), 46 p. 33 ref.

Air quality in selected urban areas, 1975-76. Offset publication No. 41 (Geneva, World Health Organisation, 1978), 42 p. 2 ref.

Glossary on air pollution. WHO Regional Publications European series No. 9 (Copenhagen, World Health Organisation, Regional Office for Europe, 1980), 114 p.

Further information and current research is available in the following periodicals:

Atmospheric Environment (Oxford, Pergamon Press);

Journal of the Air Pollution Control Association (Pittsburgh);

Environmental Research (New York, Academic Press);

Archives of Environmental Health (Washington, DC, Heldref Publications).

Air pollution control

Air pollution control aims at the elimination, or reduction to acceptable levels, of agents (e.g. gaseous materials, particulate matter, physical agents and, up to a certain extent, biological agents), whose presence in the atmosphere can cause adverse effects on human health or welfare (e.g. irritation, cancer, odours, interference with visibility, etc.), deleterious effects on animal or plant life, damage to materials of economic value to society and damage to the environment (e.g. climatic modifications). The serious hazards associated with radioactive pollutants, as well as the special procedures required for their control and disposal, deserve careful attention (see RADIOACTIVE WASTE MANAGEMENT).

The importance of efficient air pollution control cannot be overemphasised. Unless there is adequate control, the multiplication of pollution sources in the modern world may lead to irreparable damage to the environment and mankind. International agencies such as the World Health Organisation, the World Meteorological Organisation and the United Nations Environment Programme have instituted monitoring and research projects in order to clarify the issues involved in air pollution and to promote measures to prevent further deterioration of environmental and climatic conditions.

The objective of this article is to give a general overview of the possible approaches to the control of air pollution, particularly from industrial sources. Industrial air pollution starts at the workplace, therefore it is there that its control should start. There are many analogies and there should be close co-ordination between in-plant and community air pollution control; many preventive measures can solve both problems at the same time. However, air pollution control involves consideration of additional factors, such as topography and meteorology, mobile pollution sources, community and government participation, among many others, all of which must be integrated into a comprehensive programme. For example, meteorological conditions can greatly affect the ground-level concentrations resulting from the same pollutant emission. Besides, air pollution sources may be scattered over a community or a region and their effects may be felt by, or their control may involve, more than one administration. Air pollution control requires a multidisciplinary approach as well as a joint effort by different entities, private and governmental.

Sources of air pollution

The sources of man-made air pollution (or emission sources) are of basically two types.

(1) Stationary, which can be subdivided into–

(a) industrial, e.g. factories, mills, power plants, mines and quarries, refineries, cement plants, industrial incinerators;

(b) community, e.g. heating of homes and buildings, incinerators, fireplaces, cooking facilities.

(2) Mobile, comprising any form of combustion-engine vehicles, e.g. automobiles, planes, trains.

There are also natural sources of pollution, e.g. certain plants which release great amounts of pollen, sources of bacteria, spores and viruses, etc. Physical, biological and vegetable agents are not discussed in this article.

Types of air pollutants

Air pollutants are usually classified into particulate matter (dusts, fumes, mists, smokes), gaseous pollutants (gases and vapours) and odours. Although types of air pollution are discussed under AIR POLLUTION, some examples of usual pollutants are presented in what follows.

Particulate matter: coal fly-ash, mineral dusts (e.g. coal, asbestos, limestone, cement), metal dusts and fumes (e.g. zinc, copper, iron, lead), acid mists (e.g. sulphuric acid), fluorides, paint pigments, pesticide mists, carbon black, tobacco smoke, oil smoke, etc.

Particulate pollutants, besides their effects of corrosion, toxicity, irritation, carcinogenicity, destruction to plant life, etc., can also act as a nuisance (e.g. accumulation of dirt), interfere with sunlight (e.g. formation of smog and haze due to light scattering) and, also, act as catalytic surfaces for reaction of adsorbed chemicals.

Gaseous pollutants: sulphur compounds (e.g. SO_2 and SO_3), carbon monoxide, nitrogen compounds (e.g. nitric oxide, nitrogen dioxide, ammonia), organic compounds (e.g. hydrocarbons including polycyclic aromatic hydrocarbons and halogen derivatives, aldehydes, etc.), halogen compounds (HF and HCl),

Table 1. Common atmospheric pollutants and their sources[1]

Industry	Source	Emissions	
		Particulates	Others
Iron and steel mills	Blast and open hearth furnaces	Iron oxide and carbon fumes	CO and combustion gases
Iron foundries	Cupolas, shake-out systems and core machines	Iron oxide fumes, sand, smoke, oil	Hydrocarbons and combustion gases
Power plants	Combustion furnaces	Coal and oil fly-ash	SO_x and NO_x
Petrol refineries	Catalyst furnace and sludge incinerator	Catalyst fines and fly-ash	H_2SO_4 mist, oil fumes and combustion gases
Cement manufacturing	Kilns, coolers, dryers and transfer equipment	Limestone and cement dusts	Combustion gases
Kraft pulp mills	Recovery furnaces, kilns and digestors	Salt cake and lime fumes	SO_x, H_2S, and mercaptans
Asphalt plants	Dryers, coolers, batch operations and transfer equipment	Rock dusts and oil fumes	Combustion gases
Phosphoric acid plants	Grinders, reactors and filtration equipment	Phosphate rock dust and fumes	H_3PO_4 mist, HF and SiF_4 fumes
Coke processing	Storage, transfer, oven charging and quenching	Coal and coke dusts, tar fumes	Phenols, H_2S and combustion gases
Glass manufacturing	Melt furnaces and materials and handling	Raw material dusts, sulphate fumes	NO_x and combustion gases
Aluminium processing	Melt furnaces and machining operations	Al_2O_3 and $AlCl_3$ fumes, Al dusts	HCl and Cl_2 gases
Coffee processing	Roasters, spray dryers and coolers	Coffee, chaff and ash dusts	Oil fumes
Coal cleaning	Dryers, coolers and transfer equipment	Coal dusts	Combustion gases
Automotive vehicles	Gasoline storage and exhaust system	Carbon fumes	NO_x, CO, and HC

[1] From *Control techniques for particulate air pollutants* (Washington, DC, US Department of Health, Education and Welfare, 1969).

hydrogen sulphide, carbon disulphide and mercaptans (odours).

Secondary pollutants may be formed by thermal, chemical or photochemical reactions. For example, by thermal action sulphur dioxide can oxidise to sulphur trioxide which, dissolved in water, gives rise to the formation of sulphuric acid mist (catalysed by manganese and iron oxides). Photochemical reactions between nitrogen oxides and reactive hydrocarbons can produce ozone, formaldehyde and peroxyacetyl nitrate; reactions between HCl and formaldehyde can form *bis*-chloromethyl ether.

Odours: While some odours are known to be caused by specific chemical agents such as hydrogen sulphide, carbon disulphide and mercaptans, others are difficult to define chemically.

Examples of the main pollutants associated with some industrial air pollution sources are presented in table 1.

AIR POLLUTION CONTROL APPROACHES AND MEASURES

The aim of a health-oriented environmental pollution control programme is to promote a better quality of life by reducing pollution to the lowest level possible. Environmental pollution control programmes and policies, whose implications and priorities vary from country to country, cover all aspects of pollution (air, water, etc.) and involve co-ordination among areas such as industrial development, city planning, water resources development and transportation policies. Such aspects are beyond the scope of this article.

As well presented by de Koning *(Air pollution and human health)*, in air pollution basically two approaches can be used for control and prevention of harmful effects: (1) air quality management and (2) best practicable means.

Air quality management aims at the preservation of environmental quality by prescribing the tolerated degree of pollution, leaving it to the local authorities and polluters to devise and implement actions which will ensure that this degree of pollution will not be exceeded. An example of legislation within this approach is the adoption of ambient air quality standards for different pollutants; these are accepted maximum levels of pollutants (or indicators) in the target area (e.g. at ground level at a specified point in a community) and can be either primary or secondary standards. Primary standards, to use the World Health Organisation definition *(Glossary on air pollution)*, are the maximum levels consistent with an adequate safety margin and with the preservation of public health, and must be complied with within a specific time limit; secondary standards are those judged to be necessary for protection against known or anticipated adverse effects other than health hazards (mainly on vegetation) and must be complied with "within a reasonable time". Air quality standards are for 24 hours per day, 7 days per week exposure of all living subjects (including children, the elderly and the sick) as well as non-living subjects; this is in contrast to maximum permissible levels for occupational exposure, which are for a partial weekly exposure (e.g. 8 hours per day, 5 days per week) of adult and supposedly healthy workers. Typical measures in air quality management are land-use planning and "shut-down" of factories during unfavourable weather conditions.

The best practicable means approach stresses that the air pollutant emissions should be kept to a minimum; this is basically defined through emission standards for single sources of air pollution and could be achieved, for example, through closed systems and high-efficiency collectors. An emission standard is a limit on the amount or concentration of a pollutant emitted from a source. This type of legislation requires a decision, for each industry, on the best means of controlling its emissions. In practice, although the two approaches mentioned differ, it is possible to combine elements from both in one air pollution control programme. The option for one or the other approach, or a combined approach, depends on the country or region concerned.

Among the many factors that must be considered in order to select the most adequate air pollution control strategy for a given situation, the following can be mentioned:

- geographical situation and meteorology;
- number of sources and their relative location to one another and to communities;
- type of source(s) and effluents;
- characteristics of the pollutants involved (physio-chemical effects, etc.);
- degree of control required;
- socio-economic aspects and priorities.

Air quality and emission standards can be achieved through one air pollution control measure or, more often, a combination of measures.

Specific air pollution control measures are well described in the specialised literature; however, some of the most widely used are briefly presented below.

Land-use planning

Land-use planning aims basically at reducing the impact of pollution generated; it requires that future situations and problems be foreseen and it involves many aspects such as zoning of industries and other stationary air pollution sources, planning for waste disposal and transportation, and the reservation of parks and open spaces.

Adequate zoning of industries is of fundamental importance in order to reduce future problems. Factors to be considered include topography of the region, water stream characteristics, meteorological factors (particularly prevailing winds and probability of temperature inversion), location of populated areas, already existing pollution sources, etc.

Situations that may favour the accumulation of pollutants and prevent their dispersion should be avoided as much as possible, for example, concentration of industries in a valley where there is the possibility of temperature inversion. Also, the way prevailing winds promote the transportation of pollutants to inhabited areas, farms or plantations should be carefully studied.

Control at the source

The control of industrial air pollution at the source—that is, at the workplace level—can be achieved by one or a combination of the following approaches:

Prevention or reduction of pollutants generation

Whenever facing an industrial air pollution problem, the first question to be asked is the following: "Is it possible to eliminate or reduce the generation of the air pollutant(s) in question?".

Before considering the utilisation of air cleaning devices or high stacks, the possibility of eliminating or reducing air pollutants at their source of generation, by controlling operations at the workplace level, should be investigated. This approach, which at the same time protects the health of workers, is discussed under CONTROL TECHNOLOGY FOR OCCUPATIONAL SAFETY AND

HEALTH. The fundamental measures involved are the following:

(a) Substitution of materials
Examples: substitution of less toxic solvents for highly toxic ones used in certain industrial processes; use of fuels with lower sulphur content (e.g. washed coal) therefore giving rise to less sulphur compounds, etc.

(b) Modification or change of the industrial process or equipment
Examples: in the steel industry, a change from raw ore to pelleted sintered ore (to reduce the dust released during ore handling), use of closed systems instead of open ones; change of fuel heating systems to steam, hot water or electrical systems; use of catalysers at the exhaust air outlets (combustion processes), etc.
Modifications in processes, as well as in plant layout, may also facilitate and/or improve the conditions for dispersion and collection of pollutants. For example a different plant layout may facilitate the installation of a local exhaust system; the performance of a process at a lower rate may allow the use of a certain collector (with volume limitations but otherwise adequate). Process controls that concentrate pollutants in smaller air volumes offer an advantage since the cost of control equipment is closely related to the volume of effluent handled and the efficiency of some air cleaning equipment increases with the concentration of pollutants in the effluent.
Both the substitution of materials and the modification of processes may have technical and/or economic limitations, and these should be considered.

(c) Adequate housekeeping and storage
Examples: strict sanitation in food and animal product processing; avoidance of open storage of chemicals (e.g. sulphur piles) or dusty materials (e.g. sand), or, failing this, spraying of the piles of loose particulates with water (if possible) or application of surface coatings (e.g. wetting agents, plastic) to piles of materials likely to give off pollutants.

(d) Adequate disposal of wastes
Examples: avoidance of simply piling up chemical wastes (such as scraps from polymerisation reactors), as well as of dumping pollutant materials (solid or liquid) in water streams. This latter practice not only causes water pollution but can also create a secondary source of air pollution as in the case of liquid wastes from sulphite process pulp mills, which release offensive odorous gaseous pollutants.

(e) Maintenance
Example: well maintained and well tuned internal combustion engines produce less carbon monoxide and hydrocarbons.

(f) Work practices
Example: taking into account meteorological conditions, particularly winds, when spraying pesticides.
By analogy with adequate practices at the workplace, good practices at the community level can contribute to air pollution control, e.g. changes in the use of motor vehicles (more collective transportation, small cars, etc.), control of heating facilities (better insulation of buildings in order to require less heating, better fuels, etc.).

Control of emissions

Once the pollutants have been generated and are likely to disperse throughout the workplace and, beyond its boundaries, throughout the surrounding community, their control falls basically into two categories:

- atmospheric dispersion;
- containment in a closed system or capture by means of a local exhaust ventilation system and treatment of the effluent.

Polluting operations and processes can be enclosed, or the pollutants can be captured as they are generated by means of suction hoods. In both cases the pollutants would be removed by means of local exhaust ventilation. However, the pollutants removed from the working environment should not be simply discharged into the general environment. This would mean transferring the problem from the workplace to the community without solving it, and the health hazards involved would continue to exist.

The exhaust air (effluent) can be controlled either by the utilisation of air cleaning equipment (collectors), by being discharged through sufficiently high stacks (dispersion), or by a combination of both methods. Closed systems (as in chemical plants and petroleum refineries) have vent stacks which should be fitted with adequate air cleaning devices (e.g. vent scrubbers) or whose height and location should be such as to promote adequate dispersion of the chemicals vented out.

The selection of the most suitable method to control emissions will depend on several factors which include the nature and concentration of pollutants in the effluent, the discharge rate, location of the industrial source in relation to other sources and urban areas, the degree of control required, economical feasibility, etc.

Air cleaning devices (collectors). The most efficient way to control emissions is by the use of adequate, well designed, well installed, efficiently operated and maintained air cleaning devices, also called separators or collectors.

According to the WHO *Glossary on air pollution*, a separator or collector can be defined as an "apparatus for separating any one or more of the following from a gaseous medium in which they are suspended or mixed: solid particles (filter and dust separators), liquid particles (filter and droplet separator), and gases (gas purifier)". The basic types of air pollution control equipment are the following:

(a) For particulate matter:
inertial separators (e.g. cyclones);
fabric filters (baghouses);
electrostatic precipitators;
wet collectors (scrubbers).

(b) For gaseous pollutants:
wet collectors (scrubbers);
adsorption units (e.g. adsorption beds);
afterburners, which can be direct-fired (thermal incineration) or catalytic (catalytic combustion).

Wet collectors (scrubbers) can be used to collect, at the same time, gaseous pollutants and particulate matter. Also, certain types of combustion devices can burn combustible gases and vapours as well as certain combustible aerosols. Depending on the type of effluent, one or a combination of more than one collector can be used (see AIR POLLUTION CONTROL EQUIPMENT in the Annex).

The control of odours that are chemically identifiable relies on the control of the chemical agent(s) from which they emanate (e.g. by absorption, by incineration).

However, when an odour is not defined chemically or the producing agent is found at extremely low levels, other techniques may be used such as masking (by a stronger, more agreeable and harmless agent) or counteraction (by an additive which counteracts or partially neutralises the offensive odour).

It should be kept in mind that adequate operation and maintenance are indispensable to ensure the expected efficiency from a collector. This should be ensured both from the know-how and financial points of view, at the planning stage. Energy requirements must not be overlooked. Whenever selecting an air cleaning device, not only the initial cost but also operational and maintenance costs should be considered.

Whenever dealing with radioactive or high-toxicity pollutants, high efficiency should be ensured, as well as special procedures for maintenance and disposal of waste materials. Workers involved in these activities (e.g. change of filters) should be adequately trained and protected.

One aspect to consider is that, while some emissions contain only undesirable wastes, others contain materials of economical value (fuels, metals, etc.) whose recovery can be financially interesting. The possibility of product recovery may be a consideration when selecting one type of air pollution control equipment rather than another. As an example could be mentioned the collection of fluorine and alumina pollutants, discharged from reduction cells in the aluminium industry, and their chemical conversion to cryolite (which can be recharged to the cells), which is accomplished in wet scrubbers.

Atmospheric dispersion

Dispersion involves the utilisation of meteorological factors such as winds, horizontal and vertical diffusion parameters, temperatures, etc., as well as topographical factors and characteristics of the pollutant carrier gases (temperature, velocity, etc.).

Dispersion and dilution of air pollutants in the atmosphere reduce their concentration and may be used as a control measure or as a complement to other measures.

There are, in the atmosphere, mechanisms which operate in the sense of removing pollutants, such as chemical reactions, physical mechanisms (e.g. rain) and biological mechanisms.

However, the environmental capacity of dealing with pollutants is limited and this becomes more critical as industrialisation takes place and pollution from all sources increases. Therefore, dispersion and dilution, as means of achieving acceptable ground level concentrations of pollutants, have limitations, particularly in heavily industrialised and/or heavily populated areas.

Tall stacks. Since air quality standards refer to ground pollution levels, it may be helpful if pollutants are emitted at increased heights and are dispersed over very large areas. A way of improving the degree of dispersion is the use of tall discharge stacks which may be installed at the outlet of a ventilation system, a closed system or a particular operation (e.g. a furnace).

Adequate stack heights for specific situations can be determined taking into account characteristics of the effluent (temperature, discharge rate, type and concentration of pollutants, etc.), topography and prevailing meteorological conditions, as well as air quality standards.

The effective stack height (equal to stack height plus the height that the effluent plume initially rises above the stack) increases with the effluent temperature and velocity of discharge. The most convenient are tall wide stacks.

If the total amount of pollutants to be dispersed is very large and there are many sources in the area, effects such as reduction of sunlight may occur. Besides, in such a situation the ground pollution levels obtained through dispersion and dilution might still be unacceptable. Atmospheric dispersion, particularly tall stacks, may be used as a complement to air cleaning devices should these not be efficient enough to achieve the required degree of air pollution control.

It should be realised that tall stacks may reduce the pollution problem temporarily; in the long run, as the pollution sources increase, they will no more provide a solution since the pollutants, although dispersed, are not removed and remain in the atmosphere where they will gradually accumulate. The best and most effective method of controlling air pollutants is either to eliminate or reduce their generation at the source or to contain them and use collectors before emission.

Prevention of air pollution episodes

The combination of fluctuations in emissions and meteorological conditions may lead to dangerous build-ups in the concentration of air pollutants. With adequate air monitoring, changes in pollutant levels can be detected and upward trends observed; these data, combined with adequate weather forecast, can predict such episodes, which can then be prevented by reduction in pollutant generation or, in extreme cases, partial or complete shut-down of the sources.

In order to prevent air pollution episodes, alert levels can be used. Alert levels are concentrations of pollutants indicative of imminent or actual danger to health. Several different alert levels may be defined, ranging from a concentration at which a preliminary warning is issued to one that necessitates emergency action. As an example can be mentioned a three alert level scheme as follows: a first level, which constitutes an initial warning indicating that conditions have started to deteriorate; a second level, which triggers instructions for the curtailment of certain significant air pollution sources; and, a third level, which requires that emergency action be taken, e.g. shut-down of factories.

What in occupational health are usually called "administrative controls" also apply to air pollution. "Limitation of exposure" would be equivalent to "period of air pollution sources shut-down".

Monitoring data, emission records, pollution transport and meteorological data must be collected, statistically analysed, stored and, when needed, retrieved and disseminated. Surveillance networks with adequate and efficient lines of communication are needed to collect the data and to transmit warnings and shut-down orders whenever necessary.

Monitoring in air pollution

Monitoring programmes in air pollution are concerned with pollutant sources, pollutant levels in environmental health media and effects of pollutants. Such programmes are of fundamental importance and are related to air pollution control programmes.

The basic components of a complete monitoring and surveillance programme are: continuous and systematic collection of data on pollutant sources and levels, as well as on resulting health and other effects; analysis and evaluation of data, use and dissemination of data with a view to taking the necessary control measures.

Evaluation of an industrial emission source requires, for example, stack sampling; air quality monitoring requires the measurement of ground-level concentrations at representative sites around the source(s) and in the community. Both types of sampling are important for air pollution monitoring and control programmes;

reliance on one or another depends on the legislative approach adopted.

Whenever emissions are to be controlled, the nature of pollutants present and their concentration should be determined, through adequate air sampling, in order to establish the need for control, the degree of control required and the most suitable techniques to achieve it. Once the control measures have been introduced, monitoring is necessary to check on their efficiency, which should not only be initially adequate but should be maintained in the long run.

Air quality monitoring is closely related to air pollution management actions and is performed with objectives which include:

(a) to collect environmental data which, combined with monitoring of heatlh and other effects, provides useful information for the establishment of air quality standards;

(b) to determine the nature and ground-level concentrations of pollutants (at certain selected points) in order to assess compliance with adopted air quality standards and, in this way, to establish the need for control measures and the degree of control required;

(c) to evaluate the efficiency of the control measures and strategies adopted;

(d) to observe long-term air pollution trends, particularly with a view to detecting any deterioration in air quality (resulting, for example, from increased number of pollution sources, relaxing of controls, etc.);

(e) to make initial assessment surveys;

(f) to provide the data base for land-use planning;

(g) to predict, in combination with weather forecast, unusual and dangerous air pollution episodes and activate emergency control procedures;

(h) to evaluate health hazards to populations exposed;

(i) to investigate specific complaints, among others.

The strategy for environmental sampling and monitoring, that is, *when* monitoring programmes should be started, *which* pollutants should be monitored, *where* and *how* monitoring should be carried out, as well as the objectives of monitoring, should be carefully considered before launching a programme, since these are determinant factors for its design and implementation. These aspects, as well as methods of measuring specific pollutants in the atmosphere are described in the specialised literature, including the WHO Offset Publications, mentioned in the select bibliography at the end of this article.

Whenever considering factors such as pollutants to be measured, number of stations, sampling instruments and procedures, etc., the specific conditions of the place or region concerned must be kept in mind.

The importance of data storage, analysis and retrieval facilities, as well as efficient communication lines among all involved in a monitoring and surveillance network, should not be overlooked.

To conclude, it is obvious that in order to apply and co-ordinate air pollution control measures appropriate legislation (involving air quality standards and/or emission standards), as well as adequate management, are essential. Appropriate legislation provides a basis for effective pollution control; however, it should be realistic in terms of achievements and time. In addition to legislation, administrative provisions to ensure its enforcement are also needed.

Not only is it necessary to determine the approach to be followed, the degree of pollution control to be achieved and which are the most suitable techniques to achieve it, but it is also important to establish strategies for the implementation of control measures, and to decide at which stage this should be done, keeping in mind the priorities to be considered and how the responsibility for their design and implementation should be shared among all concerned.

Only a well co-ordinated, continuous and efficient multidisciplinary effort by industry, community and government can avoid the great damage which air pollution, if uncontrolled, can cause to the environment and mankind.

GOELZER FERRARI, B. I.

Control techniques for particulate air pollutants (Washington, DC, US Department of Health, Education and Welfare, Public Health Service, 1969).

"Air pollution and human health". de Koning, H. W. *Changing disease patterns and human behaviour.* Stanley, N. F.; Joske, R. A. (eds.). (London, Academic Press, 1980) (436-445). 9 ref.

Health aspects of environmental pollution control: planning and implementation of national programmes. Report of a WHO expert committee, Technical Report Series No. 554 (Geneva, World Health Organisation, 1974), 57 p.

Glossary on air pollution. WHO regional publications, European series No. 9 (Copenhagen, World Health Organisation, 1980), 114 p.

Air pollution engineering manual. Danielson, J. A. (ed.). Public Health Service Publication No. 999-AP-40 (Washington, DC, US Department of Health, Education and Welfare, Public Health Service, National Center for Air Pollution Control, 1967), 892 p. Illus. 334 ref.

Industrial pollution control handbook. Lund, H. F. (ed.). (New York, McGraw-Hill, 1971), 792 p. Illus. Ref.

Selected methods of measuring air pollutants. WHO offset publication No. 24 (Geneva, World Health Organisation, 1976), 112 p.

Air monitoring programme design for urban and industrial areas. WHO offset publication No. 33 (Geneva, World Health Organisation, 1977), 46 p.

Airports

This article deals with the personnel who are employed at the airport for the maintenance and loading of aircraft and the handling of air freight. The airport employees who are directly concerned with the movement and control of the aircraft from the moment it leaves the loading quay until the moment it leaves the air traffic control corridors are dealt with in the article AVIATION, GROUND PERSONNEL. The importance of civil airports all around the world has been rapidly increasing with the development of air transportation.

[According to presently available data of the International Civil Aviation Organisation (ICAO), the number of kilometres flown and the volume of passengers, freight and mail carried all over the world (excluding China and USSR) by domestic and international scheduled services operated by airlines registered in each country have grown as follows (in millions):

Year	km flown	Passengers carried	Freight and mail carried (tonne-km)
1970	7 010	311	47 800
1975	7 520	436	71 190
1977	8 090	517	86 170]

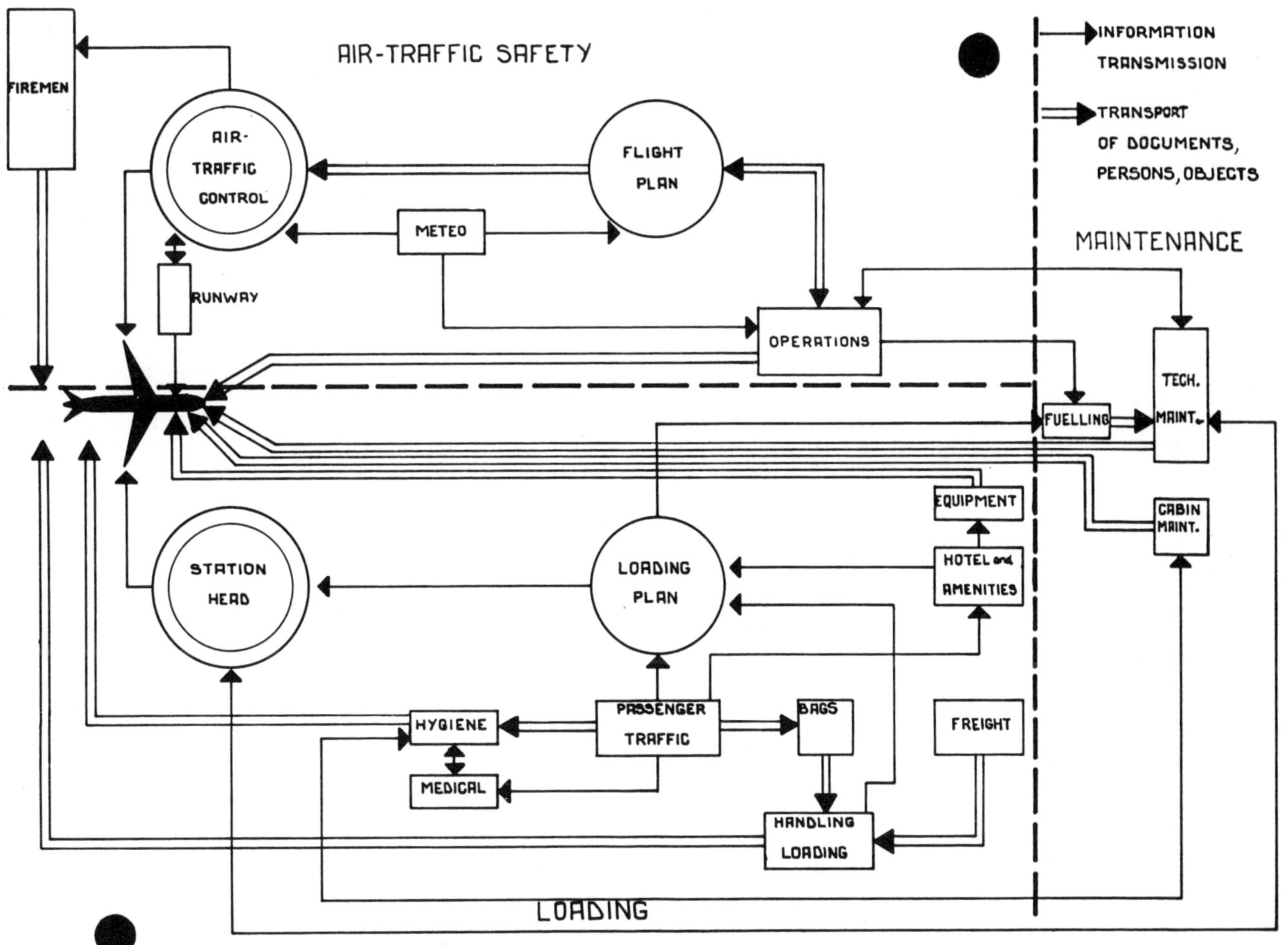

Figure 1. The complex of operations involved in the maintenance and loading of aircraft and in the handling of passengers and freight.

Aircraft maintenance. The basic maintenance and overhaul of aircraft and aircraft engines is, in its general outline, identical to the normal maintenance of machinery and plant and is carried out under true engineering works conditions. However, the routine turnround maintenance has characteristics which are specific to the aviation industry. This turnround maintenance includes: refuelling, equipment checks, small adjustments, internal and external cleaning, restocking, etc.

As soon as the aircraft comes to a halt in the unloading bays, the team of mechanics starts a series of operations which, although they follow a set routine, vary with each different type of aircraft. These mechanics refuel the aircraft, check a number of safety systems which must be inspected after each landing, investigate any anomalies or defects that the crew may have noticed during the flight and, where necessary, make repairs. It is not until this work has been completed that the mechanics can declare the aircraft flightworthy once again. Climatic and meteorological conditions play a considerable role: in cold climates, de-icing of wings, landing gear, flaps, etc., may be necessary whereas in hot climates special attention is paid to tyre condition.

Refuelling is carried out by petroleum company employees under the supervision of the mechanics. The amount of fuel to be loaded is determined by a special service on the basis of factors such as flight duration, take-off weight, flight path and possible diversions.

A team of cleaners deals with the cabins, and cleans, replaces dirty or damaged material (cushions, blankets, etc.), empties the toilets and refills the water tanks; under the supervision of official public health authorities, the cleaners may also disinfect or disinfest the aircraft.

Loading. The personnel of the stores department equip the aircraft with the material needed for passenger comfort, e.g. cushions, rugs, newspapers, magazines, films, the crockery needed for the in-flight meals and the emergency equipment such as inflatable dinghies and first-aid kits. Where invalids or sick persons are being carried, the personnel may also be required to load on board stretchers, oxygen cylinders, medicaments, etc.

Great care is taken in loading food and drink into the galley; food hygiene is an important problem in air transport and all risk of food poisoning which may affect passengers and, in particular, endanger the flight crew, must be eliminated. Meals are normally prepared under high standards of hygiene and by staff subject to strict supervision. Foodstuffs are carefully examined for quality and hygiene; certain meals are deep frozen to −40 °C, stored at −20 °C and reheated in flight.

Freight, ranging from radioisotopes to fresh fruit and vegetables, machinery and live animals, is received by specialist staff and passed through the necessary formalities with customs officials, etc. The freight is then stored in warehouses from which freight handlers, using up-to-date mechanical handling equipment and in-plant transport, load it into the aircraft hold under the supervision of freight loaders who then fasten the cargo down.

Flight passengers are received and registered by traffic agents and pass through official administrative formalities (passport, public health control, etc.). Passenger baggage is stored in the hold along with freight.

All loading operations (passengers, freight, baggage, fuel, stores, etc.) are controlled and integrated by a supervisor who draws up the "loading plan". This plan is given to the pilot prior to take-off. When all the operations have been completed and any checks he considers necessary have been made, the airport controller gives authorisation for take-off.

HAZARDS AND THEIR PREVENTION

Most of the airport workers described above are subject to the same hazards. They all work out of doors and are subject to considerable variations in climatic conditions. They are exposed to the very high noise levels produced by the aircraft power plants, especially jet engines, and by the servicing vehicles. Vehicle movement on the apron is particularly heavy and the risk of traffic accidents is high. During refuelling, enormous quantities of kerosene and gasoline are pumped into the aircraft and spillage or leakage may lead to catastrophic fires. The large amount of mechanical handling equipment needs particular attention. Finally, shift work is the rule and all operations are carried out under pressure of time.

Technical prevention. Strict regulations should be drawn up and enforced for vehicle movements on the tarmac and all equipment should be regularly maintained and inspected. Protection from high noise levels and blast can be obtained by blast deflectors or by the wearing of appropriate hearing protection equipment. Suitable work clothing for work in different climatic conditions (heat, cold, rain, etc.) should be provided together with personal protective equipment such as hand and arm protection, non-slip reinforced-toecap foot protection, eye and face protection, etc. During refuelling, rigorous fire prevention and protection measures should be observed and fire-fighting equipment maintained at the ready.

Reports have been made of aircraft fuselage and wing surfaces becoming contaminated after having flown through clouds of radioactive fallout from nuclear explosions. When such a possibility has occurred, the aircraft should be monitored and if necessary decontaminated immediately.

Medical prevention. All employees should be vaccinated against yellow fever or cholera, depending on the region, to protect them against possible infection from passengers or freight. Antitetanus vaccination should be compulsory for all workers handling freight or organic waste matter. Strict medical supervision is essential, with pre-employment and periodical medical examinations, which should include audiometry to detect hearing loss.

Airport management staff includes executives with considerable responsibility and subject to prolonged high levels of stress. Other executive, management and supervisory workers are required to work shifts and also come into direct contact with passengers from all parts of the globe. They should therefore be subject to medical supervision and receive the same vaccinations.

ALLARD, A.
VLEMINCKX, L. M.
BROUNS, H.

CIS 75-1455 *An occupational health survey of selected airports.* Larsen, L. B. DHEW (NIOSH) publication No. 74-123 (National Institute for Occupational Safety and Health, Post Office Building, Cincinnati, Ohio) (1974), 73 p. 47 ref.

CIS 74-839 *Safety in apron loading* (British Airways, Ground Safety Branch, PO Box 10, Hounslow, Middlesex) (1973), 20 p. Illus.

CIS 79-249 "Aircraft refuellers–Hazards and occupational diseases" (Les ravitailleurs d'avion–Risques et maladies professionnelles). Noland, R. *Médecine aéronautique et spatiale, médecine subaquatique et hyperbare* (Paris), 1977, 16/64 (330-334). (In French)

CIS 75-237 "Danger zones during refuelling of aircraft" (Gefahrbereiche bei der Flugzeugbetankung). Degener, C. H. *Arbeitsschutz* (Cologne), Feb. 1974, 2 (48-50). Illus. 6 ref. (In German)

CIS 79-429 "Radiation exposure of air cargo workers at the St. Louis international airport". Luszczynski, K.; Borgwald, J.; Grimmer, D. P.; Ringermacher, H.; Sutton, S.; Glasgow, G. P.; Oliver, G. D. *Health Physics* (Oxford), Oct. 1978, 35/4 (523-527). 6 ref.

CIS 78-221 *Airport vehicular traffic.* Data Sheet 539, Revision A (National Safety Council, 444 North Michigan Avenue, Chicago) (1977), 6 p. Illus.

Alcoholism

It is difficult to deal with the subject of alcoholism without running the risk of arousing emotional reactions since everyone, in most parts of the world, is more or less affected, either personally or for family or professional reasons. The alcohol content of many drinks, mainly fermented or distilled, is considered when used in moderation to have beneficial effects on one's mood, one's health and in developing the art of living. It takes its place in religious and cultural rites and among the popular traditions of many groups of people. There is a risk, nevertheless, of its giving rise to toxic phenomena and of leading to a physical and psychological dependence, i.e. the loss of freedom to control one's intake of alcohol. Thus in this sense it is truly a drug (see DRUG DEPENDENCE). According to a definition given by an Expert Subcommittee of the World Health Organisation: "Alcoholics are those excessive drinkers whose dependence upon alcohol has attained such a degree that it shows a noticeable mental disturbance or an interference with their bodily and mental health, their inter-personal relations, and their smooth social and economic functioning; or who show the prodromal signs of such developments. They therefore require treatment."

Alcoholism presents itself in different ways according to the customs and the most commonly used drinks in different countries; in the Anglo-Saxon and Scandinavian countries, where the drinking of spirits is prevalent, the term "alcoholism" tends to be restricted to cases of alcohol addiction, while in the Mediterranean countries where wine is the main alcoholic beverage, the word "alcoholism" generally covers all forms of excessive consumption of alcoholic drinks whether or not they reach the stage of addiction.

Because of the different unfortunate consequences resulting from an excessive consumption of alcohol, and in order to avoid any confusion in the terminology, the tendence today is to speak of "alcohol-related problems" or, "alcohol-related disabilities" rather than "alcoholism".

A characteristic feature of our present time is the considerable increase in the production and consumption of the alcohol drug in the form of many varied kinds of drinks. The general consumption of alcohol, which is traditionally a characteristic of Western civilisation, is at present spreading throughout the world, mainly as beer, and is to a large extent taking the place of those drugs traditionally used by the other groups of civilisation.

The harmful effects, both for the individual and for the community, of excessive addiction to alcohol is considered in the United States to be the fourth most important public health problem, surpassed in importance only by heart disease, cancer, and mental disease. These effects are in many countries among the essential causes of excessive mortality among middle-aged males. In those countries where studies have been made, it has been estimated that 3 to 20% of adult workers are affected, and in some cases many more. All of which points to alcohol as being one of the most widespread causes of disease in the world and one of the biggest threats to the future of mankind.

Effects of alcohol on the body

The effect of alcoholic beverages on the body is complex. A distinction should be drawn between acute alcoholic intoxication or drunkenness and chronic intoxication or alcoholism, although the two conditions overlap to some extent.

Acute intoxication. Acute intoxication passes through various phases according to the degree of alcoholaemia, i.e. the concentration of ethyl alcohol in the blood. Nevertheless, the relationship is not constant and a given quantity of alcohol may produce different effects in different subjects or different effects in the same subject at different times.

The extent of intoxication is determined by two factors–alcohol absorption and alcohol metabolism.

Alcohol absorption, i.e. the passage of alcohol from the digestive system into the bloodstream, involves the following factors:

(a) alcohol concentration–the higher the concentration of alcohol in the drink, the higher will be the blood alcohol;

(b) chemical composition of beverages–the more complex the chemical composition of the beverage ingested the less rapidly will the alcohol be absorbed;

(c) the presence of food in the stomach–alcohol taken on an empty stomach will produce a higher blood alcohol concentration, a more rapid rise to the maximum level and more readily obvious effects;

(d) the subject's body weight–for a given quantity of absorbed alcohol, the lighter the subject the higher the blood alcohol;

(e) the sex of the subject–women are constitutionally more sensitive to alcohol than men;

(f) habituation–the symptoms of alcoholaemia are less apparent in the regular drinker in a good state of health than in the person unaccustomed to the effects of alcohol; however–and this is one of the signs of damage to the body of alcohol–advanced stages of alcoholism are frequently characterised by a reduction of alcohol tolerance.

Once the alcohol has been absorbed into the bloodstream it is metabolised or oxidised in a cycle of complex reactions which take place mainly in the liver. The end products of these reactions are carbon dioxide and water, with the release of 7 cal/g alcohol; intermediate products include acetaldehyde and acetic acid, which block sugar and fat utilisation and reduce the ingestion of certain nutrients necessary for a balanced diet. This is the origin of the deficiency disorders and acidosis often observed in heavy drinkers. A very small proportion of the alcohol (5-15%) is eliminated directly in the urine, the sweat and expired air. The rate of oxidation varies but averages 0.1 g/h.kg body weight in man, independent of total energy expenditure. Claims that physical effort or exposure to cold accelerates the

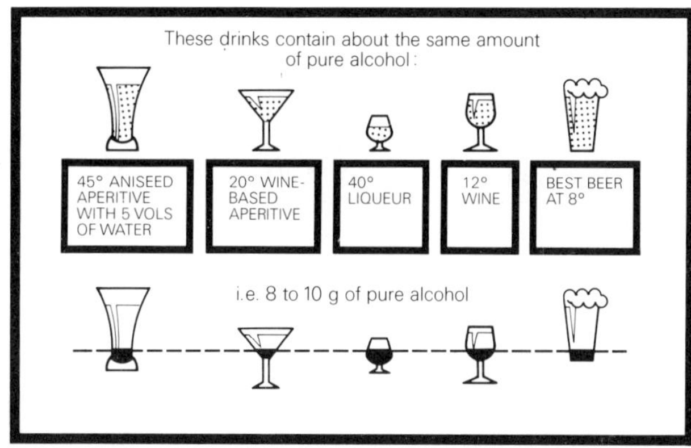

Figure 1. Alcohol content per glass of common alcoholic drinks.

Note: 10 g of alcohol will result in a peak blood alcohol level of 0.20 for a man of 70 kg if taken on an empty stomach and one-third less when taken during a meal. Thus two glasses taken by such a person on an empty stomach would be sufficient to bring him to the "safety limit" of 0.30 to 0.50 g, and four glasses taken during a meal would bring him within a breach of regulations at a level of 0.80 g/l of blood.

oxidation process are, consequently, incorrect and the obvious conclusion is that alcohol is as toxic for the manual worker as it is for the office worker. The safety limit is reached when the calorific content of the alcohol ingested reaches a level equal to 10% of energy expenditure, i.e. in the case of a manual worker, a consumption of 0.5-0.75 l of wine with a 10% alcohol content during a 24-h period.

The symptoms of acute alcohol poisoning can generally be broken down into four phases.

(a) Prodromal phase–from a blood alcohol level as low as 0.25 g/l. Even before the subject has noticed the slightest change in his physical or mental state, certain psychomotor and aptitude tests can reveal some disturbances. These include changes in visual acuity with diminution of stereoscopic sense and contraction of the peripheral field of vision; slowing and irregularity of reflexes; an effect on peripheral nervous functions including reduction in the sense of smell, diminution of the sense of touch and heat (and also of pain), disturbed kinaesthetic sense and balance control; lowering of alertness and impaired performance in psychotechnical testing. At a blood alcohol level of only 0.3 g/l, the majority of subjects present electroencephalographic changes.

(b) Excitation phase–from 0.5-1 g/l. This can be defined as a loss of inhibition of the primitive mental processes with progressive loss of self control and self criticism and a progressive paralysis of the higher mental processes. This state of euphoria (or aggressiveness) can induce a belief in enhanced intelligence; however, this is in fact the first stage in a breakdown of the personality.

(c) Incoordination phase–from 1.5-3 g/l, depending on the subject. This phase is characterised by balance disorders, motor incoordination, tremor, digestive disorders, mental confusion and reduction of sensitivity to stimuli. This is classical drunkenness. The intoxication will usually terminate in long and profound sleep of some duration; wakening is accompanied by characteristic symptoms.

(d) Coma phase–from 3-4 g/l. At this stage the subject is "dead drunk" and insensitive to external stimuli.

The reflexes are absent, the skin is cold and the body temperature low. Levels of over 4 g/l may prove fatal.

Chronic intoxication or alcoholism. Repeated acute intoxication or excessive alcoholic consumption over a number of years, even if the subject is seldom really "drunk", may result in chronic intoxication or alcoholism. The disease may present itself in widely differing ways depending on the subject's drinking habits, preferred beverages, temperament and predispositions. However, it is the digestive system and nervous system that are principally affected.

The digestive form is characterised by mucosal changes in the upper digestive tract. Liver cirrhosis is the most serious digestive complication and the most frequent of the nutritional cirrhoses. In spite of recent therapeutic advances, almost all cases have a fatal outcome after a few years unless the subject does not refrain completely from alcohol consumption.

Also to be noted is the frequency of acute and chronic relaxing pancreatitis, the latter often with irregular calcification of the pancreas, and of overweight which is itself related to hypercholesterolemia, diabetes and arterial hypertension, cardio-vascular complications, and atherosclerosis. [In addition, data collected under the aegis of IARC indicate an increased risk of oesophageal cancer among alcohol drinkers which is closely related to alcohol daily consumption (see table 1). Tobacco smoking involves a multiplicative effect on the risk.]

This is one of the aspects which is little known or realised by the public concerning the dangers of alcohol. It is no less serious for the many drinkers who consider that their consumption of alcohol does not go beyond the danger limit.

There is a constant *effect on the nervous system.* The mechanism of this destructive action upon the nerve cells is now better understood. A personality change occurs which is more or less rapid according to whether or not it has been preceded by a pathological consumption of alcohol, most constant features being perception and memory disorders, euphoria, verbosity, severe affective disorders, abnormal suggestibility with feebleness of will, intellectual fatiguability and egocentricity. Psychoaffective reactions complete the picture: conflicts with associates, jealousy, lack of conscientiousness, social deterioration. Polyneuritis is seen with sensory and motor disorders, especially of the upper limbs, or even delirium tremens (an acute crisis of alcoholic insanity) and other forms of meta-alcoholic psychosis. In the final stage, alcoholic dementia results in confinement in a psychiatric hospital or a home for incurables. Other organs, especially the cardiovascular system and the endocrine glands, may also be affected.

Finally, stress should be laid on the particularly bad effects of alcohol, and even worse, of alcohol allied with tobacco in relation to cancer, especially, as previously mentioned, of the upper digestive tract (mouth, larynx, oesophagus), and in the development of tuberculosis.

Table 1. Relative risks of oesophageal cancer in relation to alcohol consumption in Calvados (France)[1]

Total daily average consumption of ethanol (g/day)	Cases of oesophageal cancer	Hospital controls	Relative risks	
			Crude	Adjusted for smoking
0-20	20	228	1.00	1.00
21-40	20	208	1.10	1.11
41-60	33	157	2.40	2.54
61-80	41	123	3.80	3.59
81-100	52	55	10.78	9.83
101-120	38	39	11.11	10.90
121-140	27	25	12.31	11.28
141+	81	34	27.16	23.26

[1] From A. J. Tuyns, G. Péquignot and J. S. Abbatucci: "Oesophageal cancer and alcohol consumption: importance of types of beverage", in *International Journal of Cancer* (Geneva), 15 Apr. 1979, 23/4, pp. 443-447.

Detection of intoxication

The presence of alcohol in the bloodstream can be detected and the level determined by various measurement techniques. The onset of chronic poisoning is not, however, detectable in such a simple manner.

Table 2. Approximate blood alcohol levels (in grammes per litre) produced in men and women by specific quantities of various common alcoholic beverages

Type of alcoholic beverage	Quantity in cm³	Body weight in kg							
		Men				Women			
		60	75	90	105	45	60	75	90
Beers:									
Normal-strength	300	0.25	0.20	0.15	0.15	0.40	0.30	0.25	0.20
Extra-strong	300	0.40	0.30	0.25	0.15	0.70	0.50	0.40	0.30
Wines:									
Light	100	0.20	0.15	0.10	0.10	0.30	0.20	0.15	0.15
Medium	100	0.15	0.15	0.15	0.10	0.35	0.25	0.20	0.20
Dry	100	0.25	0.20	0.15	0.15	0.40	0.30	0.25	0.20
Fortified wines or aperitives:									
Vermouth, port, malaga	50	0.15	0.15	0.10	0.10	0.30	0.20	0.15	0.15
Bitters	50	0.30	0.25	0.20	0.15	0.50	0.35	0.30	0.25
Sherry, aniseed aperitives	50	0.40	0.30	0.25	0.20	0.65	0.50	0.40	0.30
Spirits:									
Brandy, rum, kirsch	25	0.20	0.15	0.15	0.10	0.35	0.25	0.20	0.15
Whisky, gin	40	0.30	0.25	0.20	0.20	0.55	0.40	0.35	0.25
Liqueurs:									
Cherry brandy, Apricot brandy	25	0.15	0.10	0.10	0.10	0.25	0.20	0.15	0.10

Measurement of blood alcohol. Ingestion of only relatively small quantities of alcohol are sufficient to produce the blood alcohol levels at present considered the upper safety limits for men carrying out difficult or dangerous work or driving a mechanical device. Table 2 gives some indicative average figures in this respect.

The measurement of blood alcohol has become common laboratory practice and various methods are used which are all sufficiently valid if the sampling and analysis are correctly done. The measurement of alcohol in the expired air is obviously more convenient but it gives less accurate results when the classical methods are used. However, some of the modern instruments give reliable results. In connection with road safety the legislation in certain countries specifies a threshold blood alcohol level and a subject whose blood alcohol exceeds this threshold level is considered to be under the influence of alcohol; this level varies from 0.5-1.5 g/l according to country. There is a general tendency to lower the limit, the average at present being 0.8 g. In the Scandinavian countries the level is fixed at 0.50, while in the Eastern European countries it has been brought down to zero. The use of such tests in industry, apart from introducing precise prohibitive legislation, would present personnel problems; they should only be used within the occupational medical service.

Detection of chronic intoxication. The detection of alcoholism at an early stage, i.e. whilst there is still time for effective remedial measures, is very difficult. Early alcoholism has no characteristics or standard symptoms but it is a disease of behaviour and the subject's wife or foreman will be aware of the situation before the doctor.

The following "signs of probability" often precede the overt signs of fully developed alcoholism: frequent absence or late arrival on Mondays or the day after holidays, and slowness and irregularity of work on these days; small repeated absenteeism for slight disorders or for slight domestic or commuting accidents; progressive change in the attitude of a man hitherto regarded as a good worker. Such alarm signals should lead to the industrial physician being informed, and he can then discover true physical and mental symptoms blending usefully with the probability diagnosis arrived at in the workplace. A social investigation, discreet and careful, will complete the picture and define the type and degree of effect. Considerable progress has been made in recent times in relation to the early diagnosis of alcohol-related problems–

- in the clinical field: by systematic examinations carried out on the basis of precise techniques and plans as, for example, the Le Gô grid in France, or Feuerlein's malt test in the Federal Republic of Germany;
- by laboratory tests such as the determination of the amount of hepatic enzyme or gamma glutamyl transpeptidase in the blood or by measuring the mean corpuscular volume.

These techniques should be introduced with the help of management and of the staff representatives, who should be given adequate information about them beforehand.

Such action does, of course, raise delicate problems concerning, for example, personal liberty and professional secrecy. Nevertheless, such obstacles can be overcome if it is made clear that this action at the plant level is not repressive in nature but is aimed solely at providing assistance in order to reintegrate the subject into society as a citizen, a worker, a head of a family and a healthy individual.

ALCOHOL AS A HAZARD

The study of the man-alcohol relationship has been the subject of considerable research in the fields of psychology, sociology and psychiatry, as well as from the point of view of its somatic and biological effects.

There is general agreement that many danger factors must be taken into account in one way or another when trying to determine the behaviour of a person in regard to alcohol. Four main types of attitude may be distinguished: abstinence, sobriety or moderation, excessive consumption, and alcohol dependence or toxicomania.

The principal danger factors to be taken into account are briefly described below:

(a) Access to alcohol is encouraged when it is plentiful and cheap, by the laxity or absence of restrictive legislation, or by persuasive and misleading advertising.

(b) Factors linked with the personality traits of the individual, which may be mental as well as physical or biological. While it is difficult to define the alcoholic type, or even several types of alcoholic, certain behavioural patterns are frequently exhibited by these subjects, frequently from childhood.

A number of works seem to point to the strong probability of there being a genetically based predisposition, particularly in regard to the enzymatic endowment of the individual; however, if the expressions "weakness", "craving for alcohol", "sensitivity to poison" are to be used, hereditary alcoholism as such, together with a fatalistic view of the problem are to be rejected.

(c) Psycho-social factors which relate to the family and the associates of the individual, with their traditions, customs and prejudices.

At this point consideration should be given to the relation between the consumption of alcohol and the working environment.

Alcoholism and occupation

Alcoholism may be regarded as a social disease by reason of its frequency and the severity of its effects and of what it costs the individual and the community.

But it can be called asocial because of the personal, family and social disturbances it causes; numerous statistical investigations have shown that alcohol is at the origin of offences and accidents of all kinds, especially on the highway and that it is the cause of family quarrels, divorce and bankruptcy and that alcohol is a common cause of suicide. Its occupational consequences are numerous.

Alcoholism and work capacity. Only over the last 30 years has industry recognised the disastrous effects of alcohol on its workers and found that excessive drinkers have a higher rate of sickness absenteeism and a higher accident proneness (the risk increases by 50% for a blood alcohol over 0.25 g/l), that their professional ability is impaired and they age and die prematurely.

The absence of a parallel between mental and physical disorders is frequent. There are cases where although the physical damage has reached an advanced stage, mental and especially artistic faculties are apparently unaffected; conversely there are heavy drinkers whose physical strength seems undiminished and who continue to do very heavy muscular work.

Nevertheless, modern industry demands such high psychosensory and mental capacities and places such heavy responsibility on the individual that the alcoholic's reduced physical and mental ability makes him less and less employable in industry. Sooner or later he is left behind and becomes a burden upon society.

While the cost of alcohol-related disabilities is very difficult to estimate, it is nevertheless clear that the costs which arise from its consumption by a fairly large proportion of the active population greatly exceed the returns which may result from alcohol in the form of employment, commercial profits and state taxes.

Working conditions and alcoholism. It may be asked to what extent the physical and the psychological conditions of work encourage alcoholism. The initial effects of alcohol are those of an anaesthetic and a mood improver. Although his muscular and sensory capacities are weakened, the subject under the influence of alcohol feels and believes himself to be stronger, more capable, and braver; however, fatigue is not long in reappearing and with it dietary disorders and an increased risk of accidents at work or on the highway.

In many cases persons doing heavy work, especially those exposed to heat, have the habit of consuming alcoholic beverages; consequently it is among the more thirsty workers that the risk is greatest. This thirst caused by physical effort and heat is exacerbated by industrial dusts and fumes, which dry the mucosa, and certain occupations have therefore been particularly affected, e.g. iron and steel workers, miners, dockers, etc.

Some industrial substances (e.g. certain solvents) can intensify the action of alcohol on the nervous system. Other substances such as calcium cyanamide may affect the alcohol metabolism and cause the appearance of toxic metabolites. Finally, since alcohol tends to stimulate intestinal resorption, it may potentiate the action of pharmaceutical products, such as antibiotics, hypnotics, neuroleptics, anthelminthics, etc., and industrial substances like lead, mercury, arsenic, aniline, etc.

Among the psychological factors that contribute to alcoholism is the custom of offering alcoholic beverages to visitors and suppliers of all kinds, such as delivery men, postmen, gas and electricity meter and repair men, commercial travellers and canvassers, as well as senior management with their business luncheons, not to mention politicians, who are exposed to frequent temptation. Furthermore, the nature of the work itself trmay be a contributory factor. Many modern jobs are repetitive and monotonous, and boredom is one of the factors conducive to excess drinking. Other jobs demand intense vigilance and mental effort; these are both likely to cause nervous fatigue, which the worker tries to diminish by resorting to the euphoric and anaesthetic effect of alcohol.

PREVENTION

General opinion is coming to accept that action must be taken within industry itself to reduce the problem of alcoholism. This action is justified by the facts that: alcoholism, detected and treated at a sufficiently early stage, is usually curable; the rehabilitation of skilled workers who have become alcoholics is an economy in terms of better productivity; the emotional stresses generated in the plant by the presence of alcoholics may lead to a deterioration in human relations; the solution which consists in penalising and dismissing the alcoholic worker merely transfers the problem to the community, the expenses of which are ultimately paid by industry.

Prevention should be at three levels: education, improvement of working conditions, and catering for workers' food and drink requirements.

Worker education must be done with skill and persistence and the assistance of both management and unions is invoked. It should take place on an individual basis or be directed at small groups of workers who run a high risk of alcoholism, particularly the younger ones, migrant workers or those who have been transferred from their usual surroundings, and heavy manual workers. The information put across must be simple, consistent with scientific data, and adapted to the intelligence and motivations of the listener; it must avoid all taint of moralising and should stress that all social classes are involved. Educational action should be undertaken by the industrial medical officer, the industrial nurse, the safety officer, etc., and should have two main aims: to counter the pro-alcohol prejudice still widespread in many communities and to propose practical remedial measures. These measures are primarily: not to drink alcohol on an empty stomach, especially at work and when driving; to drink less alcohol with meals or to ensure that the quantities do not raise the blood alcohol to dangerous levels; and never to be insistent when offering alcoholic drinks to others. Groups of former drinkers, such as Alcoholics Anonymous in the United States, the Blue Cross or the Gold Cross in Europe, have a very important role to play. Certain institutions like the Comrades Tribunal in Eastern European countries can make a useful contribution towards the rehabilitation of relapsed drinkers.

The improvement of working conditions can also contribute to the reduction of alcoholism but, on the other hand, pre-employment medical examination should ensure that the worker's strength, psychosensory aptitude and state of health, etc., are matched to his job.

Attention to the workers' nutritional requirements and dietary intake is an important measure in preventing alcoholism in the factory. Hunger and thirst at work present complex problems, which must be dealt with in the context of the workers' dietary habits. The provision of a catering service such as a canteen with well balanced menus, the introduction of rest periods sufficient for refreshment, the provision of drinking water or soft drinks at work and the establishment of workshop rules limiting or prohibiting the consumption of alcoholic drinks are indispensable elements in the policy of food hygiene.

Treatment

Treatment is most often concerned with action taken outside the factory bearing in mind the available anti-alcoholic resources. Disintoxication, whether it is done on an ambulatory basis or in a specialised establishment, is the most effective measure. Numerous checks or relapses often arise on return to a less favourable environment.

Obviously, close collaboration is needed between the medical and social services of the undertaking towards making the working environment as protective as possible: by ensuring the regularity of the course of treatment. In certain cases the medical service, in agreement with the patient's own doctor, may keep a watchful eye on the regularity of the cure, but the main task is to provide medical supervision and follow-up, to obtain the discreet collaboration of the foreman and work colleagues and to ensure that the job is a suitable one.

Alcoholism must be tackled at the workplace as elsewhere. For the management of an undertaking of any form, size or technical characteristics to overlook it will prove a costly error from both the economic and human points of view.

GODARD, J.

Seminar on the medico-social risks of alcohol consumption, Luxembourg 16, 17 and 18 November 1977 (Brussels-Luxembourg, Commission of the European Communities, 1979), 225 p. Illus. Ref.

"Alcoholism wins the 'Black Oscar'". Emery, M. *Occupational Safety and Health* (Birmingham), Sep. 1979, 9/9 (8-13).

"Sessions of 5th, 6th, 7th May at Belle-Ile". Société de médecine du travail, d'hygiène industrielle et d'ergonomie de l'Ouest. *Archives des maladies professionnelles, de médecine du travail et de sécurité sociale* (Paris), June 1978, 39/6 (384-399). (In French)

CIS 76-1772 "The problem and management of alcoholism in industry". Madden, J. S. *Journal of the Society of Occupational Medicine* (Bristol), 2 Apr. 1976, 26/2 (61-64). 15 ref.

"Alcoholism and the GGT in occupational medicine" (Alcoolisme et GGT en médecine du travail). Cirodde, M. *Revue de médecine du travail* (Paris), 1978, 6/2 (95-98). (In French)

"The Unions look at alcohol and drug dependency". Morris, J. *International Labour Review* (Geneva), Oct. 1972, 106/4 (335-346).

"Oesophageal cancer and alcohol consumption: importance of types of beverage". Tuyns, A. J.; Péquignot, G.; Abbatucci, J. S. *International Journal of Cancer* (Geneva), 15 Apr. 1979, 23/4 (443-447). Illus. 14 ref.

Alcohols

Alcohols are a class of organic compounds formed from hydrocarbons by the substitution of one or more hydroxyl groups for an equal number of hydrogen atoms; the term is extended to various substitution products which are neutral in reaction and which contain one or more of the alcohol groups.

The alcohols of greatest interest for occupational health are dealt with in separate articles, e.g. ALLYL COMPOUNDS; ETHYL ALCOHOL; FURFURAL AND DERIVATIVES; METHYL ALCOHOL; PENTYL ALCOHOLS.

n-Propyl alcohol ($CH_3CH_2CH_2OH$)

1-PROPANOL

m.w.	60.1
sp.gr.	0.78
m.p.	−127 °C
b.p.	97.1 °C
v.d.	2.1
v.p.	21 mmHg ($2.79 \cdot 10^3$ Pa) at 25 °C
f.p.	25 °C
e.l.	2.1-13.5%

soluble in water, ethyl alcohol, ethyl ether, benzene

a colourless, volatile liquid.

TWA OSHA	200 ppm 500 mg/m³
STEL ACGIH	250 ppm 625 mg/m³ skin
IDLH	4 000 ppm
MAC USSR	10 mg/m³

It is a by-product in the synthesis of methyl alcohol by high pressure and in the propane-butane oxidation process. *n*-Propyl alcohol finds use in organic synthesis and in the manufacture of cosmetics; it is used as a solvent for natural and synthetic resins, vegetable oils, waxes, ethyl cellulose and pesticides.

Ill-effects from the industrial usage of *n*-propyl alcohol have not been reported. In animals it is moderately toxic via inhalation, oral and dermal routes. It is an irritant of the mucous membranes and a depressant of the central nervous system. After inhalation slight irritation of the respiratory tract and ataxia may occur. It is slightly more toxic than isopropyl alcohol, but it appears to produce the same biological effects. There is evidence of one fatal case after ingestion of 400 ml of *n*-propyl alcohol. The pathomorphological changes were mainly brain oedema and lung oedema, which have also been often observed in ethyl alcohol poisoning. It is also flammable and a moderate fire hazard.

Isopropyl alcohol ($CH_3CHOHCH_3$)

2-PROPANOL; SECONDARY PROPYL ALCOHOL; DIMETHYL CARBINOL

m.w.	60.1
sp.gr.	0.78
m.p.	−89.5 °C
b.p.	82.4 °C
v.d.	2.1
v.p.	44 mmHg ($5.87 \cdot 10^3$ Pa) at 25 °C
f.p.	11.7 °C
e.l.	2.0-12%

soluble in water, ethyl alcohol, ethyl ether

a colourless, volatile liquid.

TWA OSHA	400 ppm 980 mg/m³
NIOSH	400 ppm/10 h
	800 ppm/15 min ceil
STEL ACGIH	500 ppm 1 225 mg/m³ skin
IDLH	20 000 ppm

It is synthesised from propylene and a small percentage is produced by the reaction of natural gas hydrocarbons. Isopropyl alcohol is an important industrial solvent and disinfectant, and is used as a substitute for ethyl alcohol in cosmetics, i.e. skin lotions, hair tonics and rubbing alcohol. It cannot be used for oral pharmaceuticals.

Isopropyl alcohol in animals is slightly toxic via dermal and moderately toxic via oral and intraperitoneal routes. No case of industrial poisoning has been reported. An excess of sinus cancers and laryngeal cancers has been found among workers producing isopropyl alcohol. This could be due to the by-product isopropyl oil. Clinical experience shows that isopropyl alcohol is more toxic than ethyl alcohol but less toxic than methyl alcohol. Isopropyl alcohol is metabolised to acetone, which can reach high concentrations in the body and is in turn metabolised and excreted by the kidneys and lungs. In man, concentrations of 400 ppm produce mild irritation of the eyes, nose and throat.

The clinical course of isopropyl alcohol poisoning is similar to that of ethanol intoxication. The ingestion of up to 20 ml diluted with water has caused only a sensation of heat and slight lowering of the blood pressure. However, in two fatal cases of acute exposure, within a few hours after ingestion respiratory arrest and deep coma were observed and also hypotension, which is regarded as a bad prognostic sign. Isopropyl alcohol is a flammable liquid and a dangerous fire hazard.

n-Butyl alcohol ($C_2H_5CH_2CH_2OH$)

1-BUTANOL; *n*-BUTANOL; *n*-PROPYLCARBINOL

m.w.	74.1
sp.gr.	0.81
m.p.	−89.8 °C
b.p.	117.5 °C
v.d.	2.6
v.p.	5.5 mmHg ($0.73 \cdot 10^3$ Pa) at 20 °C
f.p.	28.9 °C
e.l.	1.4-11.2%
i.t.	365 °C

slightly soluble in water, soluble in ethyl alcohol and ethyl ether

a colourless, volatile liquid.

TWA OSHA	100 ppm 300 mg/m³
TLV ACGIH	50 ppm 150 mg/m³ ceil skin
IDLH	8 000 ppm
MAC USSR	10 mg/m³

n-Butyl alcohol is synthesised from acetaldehyde or by the fermentation of carbohydrates. It is employed as a solvent for paints, lacquers and varnishes, natural and synthetic resins, gums, vegetable oils, dyes, and alkaloids. It is used as an intermediate in the manufacture of pharmaceuticals and chemicals. Exposure to butyl alcohol may also occur in the production of artificial leather, safety glass, rubber cement, shellac, raincoats, photographic films and perfumes.

n-Butyl alcohol is potentially more toxic than any of its lower homologues but the practical hazards associated with its industrial production and use at ordinary temperature are substantially reduced by its lower volatility. High vapour concentrations produce narcosis and death in animals. Exposure of human beings to the vapour may induce irritation of the mucous membranes. The reported levels at which irritation occurs are conflicting and vary between 50 and 200 ppm. Transient mild oedema of the conjunctiva of the eye and a slightly reduced erythrocyte count may occur above 200 ppm. Contact of the liquid with skin may result in irritation, dermatitis and absorption. It is slightly toxic when ingested. It is also a dangerous fire hazard.

***sec*-Butyl alcohol** ($C_2H_5CHOHCH_3$)

2-BUTANOL; METHYL ETHYL CARBINOL

m.w. 74.1
sp.gr. 0.81
b.p. 99.5 °C
v.d. 2.6
v.p. 23.9 mmHg ($3.18 \cdot 10^3$ Pa) at 30 °C
f.p. 23.9 °C
e.l. 1.7-9.8%
i.t. 406 °C

soluble in water

a colourless, volatile liquid.

TWA OSHA 150 ppm 450 mg/m³
TLV ACGIH 100 ppm 305 mg/m³
STEL ACGIH 150 ppm 455 mg/m³
IDLH 10 000 ppm

It is synthesised by the hydration of butylene and used as a solvent, as well as in hydraulic brake fluids, industrial cleaning compounds, polishes, paint removers, ore-flotation agents, fruit essences, perfumes, dyestuffs, and as a chemical intermediate.

The response of animals to its vapours is similar to that to *n*-butyl alcohol, but it is more narcotic and lethal. It is a flammable liquid and a dangerous fire hazard.

Isobutyl alcohol ($(CH_3)_2{:}CHCH_2OH$)

2-METHYL-1-PROPANOL; ISOPROPYL CARBINOL; ISOBUTANOL

m.w. 74.1
sp.gr. 0.80
b.p. 108.3 °C
v.d. 2.6
v.p. 10 mmHg ($1.33 \cdot 10^3$ Pa) at 21.7 °C
f.p. 27.8 °C
e.l. 1.2-10.9%
i.t. 427 °C

soluble in water

a colourless, volatile liquid with a sweet odour.

TWA OSHA 100 ppm 300 mg/m³
TLV ACGIH 50 ppm 150 mg/m³
STEL ACGIH 75 ppm 225 mg/m³
IDLH 8 000 ppm

It is obtained from propylene by the oxo process. It is employed in lacquers, paint strippers, cleaners, hydraulic fluid and as a chemical intermediate.

At high concentrations the action of its vapour, like the other alcohols, is primarily narcotic. It is irritating to the human eye above 100 ppm. Contact of the liquid with the skin of man may result in erythema. It is slightly toxic when ingested. This liquid is flammable and a dangerous fire hazard.

***tert*-Butyl alcohol** ($(CH_3)_3COH$)

2-METHYL-2-PROPANOL; TRIMETHYLCARBINOL

m.w. 74.1
sp.gr. 0.79
m.p. 25.5 °C
b.p. 82.2-82.3 °C
v.d. 2.6
v.p. 42 mmHg ($4.34 \cdot 10^3$ Pa) at 25 °C
f.p. 11.1 °C
e.l. 2.4-8.0%
i.t. 480 °C

soluble in water, ethyl alcohol, ethyl ether

a colourless, volatile liquid.

TWA OSHA 100 ppm 300 mg/m³
STEL ACGIH 150 ppm 450 mg/m³
IDLH 8 000 ppm

It is obtained by the hydration of isobutylene and is used for removal of water from products, as a solvent in the manufacture of drugs, perfumes and chemicals, and as a chemical intermediate.

Although the vapour is more narcotic to mice than that of *n*- or isobutyl alcohol, no industrial ill-effects have been reported, other than occasional slight irritation of the skin. It is slightly toxic when ingested. In addition, it is flammable and a dangerous fire hazard.

Methylamyl alcohol ($(CH_3)_2CHCH_2CHOHCH_3$)

METHYLISOBUTYL CARBINOL; 4-METHYL-2-PENTANOL

m.w. 102.2
sp.gr. 0.80
b.p. 131 °C
v.d. 3.5
v.p. 3.5 mmHg ($0.45 \cdot 10^3$ Pa) at 20 °C
f.p. 41 °C
e.l. 1.0-5.5%

slightly soluble in water, soluble in ethyl alcohol and ethyl ether

a colourless liquid.

TLV ACGIH 25 ppm 100 mg/m³ skin
STEL ACGIH 40 ppm 160 mg/m³

It is prepared by the reduction of mesityl oxide and finds use in brake fluids, as a frothing agent in ore flotation, and in the manufacture of ore-floating agents, lubricant additives, solvents, plasticisers, lacquers, and in organic syntheses.

Although prolonged exposure to saturated vapour has proved lethal to experimental animals, the industrial hazard of the vapour is low, due to its low pressure. The liquid may be very slightly irritating to the skin and is slightly toxic when ingested. This material is also flammable and a moderate fire hazard.

2-Ethylbutyl alcohol ($C_2H_5CH(C_2H_5)CH_2OH$)

2-ETHYL-1-BUTANOL

m.w. 102.2
sp.gr. 0.83
m.p. < -15 °C
b.p. 149 °C
v.d. 3.5
v.p. 0.9 mmHg ($0.12 \cdot 10^3$ Pa) at 20 °C
f.p. 57 °C (oc)

slightly soluble in water, soluble in ethyl alcohol and ethyl ether

a colourless liquid.

It is prepared by condensation of acetaldehyde and butyraldehyde, and employed in the manufacture of printing inks, lacquers, surfactants and synthetic lubricants.

The vapour hazard is extremely low. The liquid may irritate the skin and, when ingested, it is slightly toxic. It is flammable and a moderate fire hazard.

2-Ethylhexanol ($CH_3(CH_2)_3CH(C_2H_5)CH_2OH$)

m.w. 130.2
sp.gr. 0.83
m.p. <−76 °C
b.p. 185 °C
v.d. 4.5
f.p. 85 °C (oc)

insoluble in water, soluble in ethyl alcohol and ethyl ether

a colourless liquid.

It is made by the condensation of butyraldehyde followed by hydrogenation. It is an intermediate in the manufacture of plasticisers, a solvent, wetting agent and defoaming agent. It is used in the production of lacquers, enamels, cellulose nitrate, varnishes, ceramics, paper coatings, rubber, latex and textiles.

No apparent injury has been reported from its use in industry. It is moderately irritating to the skin and slightly toxic when ingested, in addition to being a moderate fire hazard.

Nonyl alcohol ($CH_3(CH_2)_7CH_2OH$)

m.w. 144.3
sp.gr. 0.83
b.p. 215 °C

insoluble in water, miscible with ethanol and ether

a colourless to yellowish liquid with an odour of citronella oil.

Nonyl alcohol has numerous isomers. A product rich in primary alcohols, primarily 3,5,5-trimethyl hexanol, with some dimethyl heptanols is manufactured by the catalytic addition of carbon monoxide and hydrogen to an olefin (oxo process). Another commercial product is rich in a secondary nonyl alcohol, namely, diisobutylcarbinol (or 2,6-dimethyl-4-heptanol).

Because of the low vapour pressure, nonyl alcohol vapour does not constitute an industrial hazard. The liquid may irritate the skin and is slightly toxic when ingested.

Cyclohexanol ($C_6H_{11}OH$)

HEXAHYDROPHENOL; CYCLOHEXYL ALCOHOL

m.w. 100.16
sp.gr. 0.95
m.p. 25.1 °C
b.p. 161.1 °C
v.d. 3.5
v.p. 1 mmHg (0.13·10³ Pa) at 21 °C
f.p. 68 °C
i.t. 300 °C

soluble in water, ethyl alcohol and ethyl ether

a colourless liquid.

TWA OSHA	50 ppm 200 mg/m³
IDLH	3 500 ppm

It is manufactured by either oxidation of cyclohexane or reduction of phenol. Cyclohexanol is a solvent and is used in leather processing, lacquers, paint and varnish removers, polishes, and as an intermediate in the preparation of plasticisers and other chemicals.

Although headache and conjunctival irritation may result from prolonged exposure to its vapour, no serious industrial hazard exists. Irritation to the eyes, nose and throat of human subjects results at 100 ppm. Prolonged contact of the liquid with the skin results in irritation, and the liquid is slowly absorbed through the skin. It is slightly toxic when ingested. Cyclohexanol is excreted in the urine, conjugated with glucoronic acid. The liquid is flammable and a moderate fire hazard.

Methylcyclohexanol ($CH_3C_6H_{10}OH$)

HEXAHYDROCRESOL; HEXAHYDROMETHYLPHENOL; METHYLHEXALIN

m.w. 114.1
sp.gr. 0.92
b.p. 173-175 °C
v.d. 3.94
v.p. 1.5 mmHg (0.2·10³ Pa) at 30 °C
f.p. 67.78 °C
i.t. 296 °C

slightly soluble in water

a colourless liquid with a faint odour of coconut oil.

TWA OSHA	100 ppm 470 mg/m³
TLV ACGIH	50 ppm 235 mg/m³
STEL ACGIH	75 ppm 350 mg/m³
IDLH	10 000 ppm

It is manufactured by the hydrogenation of *m*- and *p*-cresols and consists essentially of two (*m*- and *p*-) of the three (ortho, meta and para) structural isomers. This product finds use as a solvent for gums, resins, oils and waxes, lacquers and textile soaps.

Headaches and irritation of the eye and upper respiratory tract may result from prolonged exposure to the vapour. Prolonged contact of the liquid with the skin results in irritation and the liquid is slowly absorbed through the skin. It is slightly toxic when ingested. Methylcyclohexanol, conjugated with glucuronic acid, is excreted in urine. It is a moderate fire hazard.

Benzyl alcohol ($C_6H_5CH_2OH$)

PHENYL METHANOL; PHENYL CARBINOL; α-HYDROXYTOLUENE

m.w. 108.1
sp.gr. 1.04
m.p. −15.3 °C
b.p. 205.3 °C
v.d. 3.7
v.p. 1 mmHg (0.13·10³ Pa) at 58 °C
f.p. 100.6 °C
i.t. 436 °C

soluble in water, ethyl alcohol, ethyl ether, acetone

a colourless liquid with a weak aromatic odour.

It is produced by the hydrolysis of benzyl chloride and used in the preparation of perfumes, pharmaceuticals, cosmetics, dyestuffs, inks and benzyl esters.

Other than temporary headache, vertigo, nausea, diarrhoea and loss of weight during exposure to a high vapour concentration resulting from a mixture containing benzyl alcohol, benzene, and ester solvents, no industrial illness is known from benzyl alcohol. It is slightly irritating to the skin and produces a mild lacrimating effect. The liquid is flammable and a moderate fire hazard.

β-Chloroethyl alcohol (CH_2ClCH_2OH)

ETHYLENE CHLOROHYDRIN; 2-CHLOROETHANOL

m.w. 80.5
sp.gr. 1.20
m.p. −67.5 °C
b.p. 128 °C
v.d. 2.8
f.p. 60 °C (oc)
e.l. 4.9-15.9%

soluble in water, ethyl alcohol, slightly soluble in ethyl ether

a colourless liquid with a weak ether-like odour.

TWA OSHA	5 ppm 16 mg/m³ skin
NIOSH	15 ppm/15 min ceil
TLV ACGIH	1 ppm 3 mg/m³ ceil skin
IDLH	10 ppm
MAC USSR	0.5 mg/m³ skin

It may be produced by simultaneously passing chlorine and ethylene into water. It is not stable to heat and decomposes into 1,2-dichloroethane and acetaldehyde at 184 °C and into glycol and aldehyde with water at 100 °C.

β-Chloroethyl alcohol finds use as a solvent for acetylcellulose, resins, waxes, dyes, and lacquers and also in cleaning.

It is highly toxic via all routes. It can induce irritation of the eyes and the skin, dizziness, visual disturbances, weakness, nausea and vomiting, dyspnoea, shock, cyanosis and coma. Several fatal cases with brain oedema and lung oedema have been reported from industrial exposure to this compound by inhalation or by the dermal route.

In a submammalian system (*S. thyphimurium* TA 1530) β-chloroethyl alcohol induces a mutagenic response.

It is flammable and a moderate fire hazard.

TREON, J. F.
STASIK, M. J.

"Alcohols". Treon, J. F. *Patty's industrial hygiene and toxicology.* Second revised edition. Vol. II (New York, London, Sydney, Inter-science Publishers, 1963) (1409-1496). 437 ref.

Isopropyl alcohol and isopropyl oil (Isopropylalkohol und Isopropylöl) Gesundheitsschädliche Arbeitsstoffe. Toxikologisch-arbeitsmedizinische Begründung von MAK-Werten (Weinheim, Federal Republic of Germany, Verlag Chemie, 1977), 5 p. 13 ref. (In German)

CIS 77-735 "Carbon tetrachloride toxicity potentiated by isopropyl alcohol". Folland, D. S.; Schaffner, W.; Ginn, H. E.; Crofford, O. B.; McMurray, D. R. *Journal of the American Medical Association* (Chicago), Oct. 1976, 236/16 (1 853-1 856). Illus. 20 ref.

CIS 78-19 "Isopropyl alcohol". H45, Information sheets on hazardous materials, Fire Prevention Association, London. *Fire Prevention* (London), Mar. 1976, 113 (29-30).

CIS 75-169 "Determination of the maximum allowable concentration, time-weighted for a day, of propyl and isopropyl alcohols in the air" (Gigieničeskoe normirovanie srednesutočnyh predel'no dupostimyh koncentracij propilovogo i izopropilovogo spirtov v atmosfernom vozduhe). Bajkov, B. K.; Gorlova, O. E.; Gusev, M. I.; Novikov, Ju., V.; Judina, T. V.; Sergeev, A. N. *Gigiena i sanitarija* (Moscow), Apr. 1974, 4 (6-13). Illus. 4 ref. (In Russian)

CIS 78-26 "n-Butanol". H64, Information sheets on hazardous materials, Fire Protection Association, London. *Fire Prevention* (London), Dec. 1977, 122 (47-48).

CIS 76-168 "Isobutyl alcohol" (Alcool isobutylique). Morel, C.; Cavigneaux, A.; Protois, J. C. *Cahiers de notes documentaires–Sécurité et hygiène du travail* (Paris), 3rd quarter 1975, 80, Note No. 979-80-75 (411-414). 15 ref. (In French)

Cyclohexanol (Cyclohexanol). Gesundheitsschädliche Arbeitsstoffe. Toxikologisch-arbeitsmedizinische Begründung von MAK-Werten (Weinheim, Federal Republic of Germany, Verlag Chemie, 1971), 3 p. 10 ref. (In German)

Aldehydes and ketals

Aldehydes

These are members of a class of organic chemical compounds represented by the general formula RCHO. In this formula, R may be hydrogen or a hydrocarbon radical–substituted or unsubstituted. The important reactions of aldehydes include oxidation (whereby carboxylic acids are formed), reduction (with the formation of alcohol), aldol condensation (when two molecules of an aldehyde react in the presence of a catalyst to produce a hydroxy aldehyde), and the Cannizzaro reaction (with the formation of an alcohol and the sodium salt of an acid).

Production. Aldehydes can be produced readily from freely available raw materials such as saturated or unsaturated hydrocarbons (e.g. by means of the oxo reaction) or by the oxidation of corresponding alcohols. A number of aromatic aldehydes occur naturally in such products as essential oils.

Uses. Because of their high chemical reactivity, they are important intermediates for the manufacture of resins, plasticisers, solvents and dyes; the aromatic aldehydes and the higher aliphatic aldehydes are associated with the manufacture of perfumes and essences.

HAZARDS

Many aldehydes are volatile, flammable liquids which evolve, at normal room temperatures, vapours in explosive concentrations in the air. They have a tendency to cause primary irritations of the skin, eyes and respiratory system–a tendency which is most pronounced in the lower members of a series, in members that are unsaturated in the aliphatic chain, and in halogen-substituted members. With very few exceptions, the effect of any anaesthetic property is overtaken by the irritant properties; in other words, the small quantities that can be tolerated are metabolised before they can accumulate. This rapid metabolism of aromatic aldehydes and certain aliphatic aldehydes accounts for their safety as foods and flavourings. In certain instances (paraldehyde and chloral) their physiological action is to produce a hypnotic effect.

Because of their widespread industrial use and their significance in the context of safety and health a number of aldehydes have been described in the following separate articles: ACETALDEHYDE; ACROLEIN; FORMALDEHYDE; FURFURAL AND DERIVATIVES; VANILLA.

Propyl aldehyde (CH_3CH_2CHO)

PROPIONALDEHYDE; PROPANAL

m.w.	58.1
sp.gr.	0.81
m.p.	−81 °C
b.p.	48.8 °C
v.d.	2.0
v.pr.	300 mmHg (39.9·10³ Pa) at 25 °C
f.p.	−7.2 °C
e.l.	3.7-16.1%
i.t.	207 °C

soluble in water; soluble in all proportions in ethanol and ether

a colourless liquid with a suffocating odour.

MAC USSR 5 mg/m³

This aldehyde is produced mainly by synthesis from ethylene, carbon monoxide and hydrogen, but smaller quantities are made by the oxidation of propyl alcohol. Propyl aldehyde is an intermediate in the manufacture of trimethylolethane for use in alkyd resin systems and, in certain cases, it is oxidised to propionic acid and reduced to propyl alcohol. It has minor uses in medicinal and agricultural chemical preparations. It is a fire and respiratory irritation hazard.

[**Butyl aldehyde** ($CH_3(CH_2)_2CHO$)

BUTYRALDEHYDE; BUTANAL

m.w.	72.1
sp.gr.	0.80
m.p.	−96 °C
b.p.	76 °C
v.d.	2.48
v.p.	88 mmHg (11.7·10³ Pa) at 20 °C
f.p.	−7 °C
e.l.	2.5-12.5%
i.t.	230 °C

a colourless liquid with a pungent odour, slightly soluble in water, soluble in organic solvents; is readily oxidised in air to form butyric acid.

MAC USSR 5 mg/m³

It is usually produced by catalytic dehydrogenation of metanol or by catalytic hydrogenation of crotonaldehyde. It is used in organic synthesis, mainly in the manufacture of rubber accelerators.

It is a dangerous fire hazard. Its vapour may travel a considerable distance in low air layers.]

Isobutyl aldehyde ($(CH_3)_2CHCHO$)

ISOBUTYRALDEHYDE; 2-METHYLPROPANOL

m.w.	72.1
sp.gr.	0.79
m.p.	−66 °C
b.p.	62 °C
v.d.	2.5
v.pr.	170 mmHg ($22.6 \cdot 10^3$ Pa) at 20 °C
f.p.	−40 °C
e.l.	1.6-10.6%
i.t.	254 °C

a colourless liquid with a pungent odour.

This aldehyde is prepared by the oxo process from propylene, carbon monoxide and hydrogen. It is an intermediate in the manufacture of amino acids and vitamins. Its derivatives are used in medicines, mould inhibitors and repellants. It is also used in the synthesis of perfumes, flavourings, plasticisers and gasoline additives.

Crotonaldehyde ($CH_3CH{:}CHCHO$)

2-BUTENAL; BETAMETHYLACROLEIN; PROPYLENE ALDEHYDE

m.w.	70
sp.gr.	0.86
b.p.	104 °C
v.d.	2.04
v.p.	30 mmHg ($3.99 \cdot 10^3$ Pa) at 20 °C
f.p.	12.8 °C (oc)
e.l.	2.1-15.5%
i.t.	232.2 °C

soluble in water; very soluble in ethanol and ether

a water-white liquid with a pungent, suffocating odour.

TWA OSHA	2 ppm 6 mg/m³
STEL ACGIH	6 ppm 18 mg/m³
IDLH	400 ppm
MAC USSR	0.5 mg/m³

The most widely used method for the synthesis of this aldehyde is the aldol condensation of acetaldehyde accompanied or followed by dehydration. The largest use of crotonaldehyde is in the manufacture of *n*-butyl alcohol. It is also used in the manufacture of crotonic acid and in the preparation of surface active agents, pesticidal compounds and chemotherapeutic agents. It is a solvent for polyvinyl chloride and acts as a short-stopper in vinyl chloride polymerisation.

[A strongly irritant substance and a definite corneal burn hazard, resembling acrolein in toxicity.]

[**Chloroacetaldehyde** ($ClCH_3CHO$)

MONOCHLOROACETALDEHYDE; 2-CHLOROETHANAL

m.w.	78.5
sp.gr.	1.19
m.p.	−16.3 °C
b.p.	90 °C
v.p.	100 mmHg ($13.3 \cdot 10^3$ Pa) at 45 °C (40% sol)
f.p.	87.8 °C

soluble in ether

a colourless liquid with a pungent odour.

TWA OSHA	1 ppm 3 mg/m³ ceil
IDLH	250 ppm

Chloroacetaldehyde is manufactured by chlorination of acetaldehyde. It is used in the manufacture of 2-aminothiazole.

Chloroacetaldehyde has very irritant properties not only with regard to mucous membranes (it is dangerous to the eyes even in the vapour phase), but also to the skin. It can cause burnlike injuries on contact at 40% solution, and an appreciable irritation at 0.1% solution on prolonged or repeated contact. Prevention should be based on the avoidance of any contact and the control of atmospheric concentration.]

Chloral (CCl_3CHO)

TRICHLOROACETALDEHYDE

m.w.	147.4
sp.gr.	1.51
m.p.	−57.5 °C
b.p.	97.7 °C
v.d.	5.1
v.p.	35 mmHg ($4.65 \cdot 10^3$ Pa) at 20 °C
f.p.	75 °C

an oily liquid with an irritating odour.

Chloral is prepared by the chlorination of absolute alcohol and the subsequent treatment of the product with sulphuric acid. It is used in the preparation of chloral hydrate.

Chloral hydrate ($CCl_3CH(OH)_2$)

TRICHLOROACETALDEHYDE MONOHYDRATE

m.w.	165.4
sp.gr.	1.91
m.p.	51.6-7 °C
b.p.	96.3 °C (decomposes)

soluble in water

a colourless, transparent, crystalline substance with an acrid, aromatic odour and a bitter taste.

This substance is prepared from chloral by the addition of a small quantity of water. It is used as a hypnotic and a sedative.

[In man it is mainly excreted as trichloroethanol with the early urines, then as trichloroacetic acid, which may reach up to half the dose in repeated exposure.

On severe acute exposure chloral hydrate acts like a narcotic and impairs the respiratory centre.]

[**Glutaric aldehyde** ($OCH(CH_2)_3CHO$)

GLUTARALDEHYDE; 1-5,PENTANEDIAL; SUCCIN-DIALDEHYDE

m.w.	100.1
b.p.	187 °C

miscible in all proportions in water and ethanol.

TLV ACGIH	0.2 ppm 0.7 mg/m³ ceil

It is used as a tanning agent for leather, as a chemical disinfectant for cold sterilisation in dentistry. It causes allergic contact dermatitis.]

Ketals (acetals)

These diesters of aldehyde or ketone hydrates are produced by reactions of aldehydes with alcohols. They are now coming into industrial use and are likely to serve as solvents, plasticisers and intermediates. They are capable of hardening natural adhesives like glue or casein. The following two typical examples are representative of the general picture at the present time.

Methylal ($CH_2(OCH_3)_2$)

DIMETHOXYMETHANE

m.w.	76
sp.gr.	0.86
m.p.	−104.8 °C
b.p.	43 °C
v.d.	2.6
v.p.	400 mmHg ($53.2 \cdot 10^3$ Pa) at 25 °C
f.p.	−18 °C (oc)

e.l. 1.6-17.6%
i.t. 237 °C
a colourless, flammable liquid with a pungent odour.

TLV ACGIH	1 000 ppm 3 100 mg/m³
STEL ACGIH	1 250 ppm 3 875 mg/m³
IDLH	10 000 ppm

It is prepared by the reaction of formaldehyde and methyl alcohol with an acidic catalyst. Methylal is used medicinally, externally as an ointment and internally as an anaesthetic and hypnotic. [It can produce liver and kidney impairment and acts as a lung irritant on acute exposure.]

Dichloroethyl formal ($CH_2(OCH_2CH_2CH_2Cl)_2$)
m.w. 1.73
sp.gr. 1.23
b.p. 218 °C
v.d. 5.9
f.p. 110 °C (oc)
a colourless liquid.

It is used as a solvent and an intermediate for polysulphide synthetic rubber.

SAFETY AND HEALTH MEASURES

Precautions in respect of fire and explosion must be most rigorous in the case of the lower members of the aldehyde family, and safeguards in respect of irritant properties must be most extensive for the lower members and for those with an unsaturated or substituted chain.

Fire prevention. When the flash point of an aldehyde is at, or below, the working temperature there is a danger that an explosive mixture of vapour and air will form in the vicinity. For this reason, care must be taken to exclude open lights and other agencies capable of igniting the vapour. The presence of explosive mixtures of vapour and air in plant and in the workrooms should be prevented by exhaust ventilation systems which ensure that sufficient air is drawn through plant and ducting to reduce the concentration of flammable vapour to below the lower flammable limit (see FLAMMABLE SUBSTANCES).

Health precautions. To prevent injury from occurring as a result of contact with an aldehyde in the form of liquid or vapour, the process in which the substance is present should, if possible, be conducted in enclosed plant. If openings are necessary for the purposes of manipulation, vapour should be prevented from escaping into the workroom atmosphere by the application of exhaust ventilation to the plant or enclosure. When the process cannot be enclosed, exhaust ventilation hoods should be fitted to trap and extract escaping vapour before it can diffuse into the workroom atmosphere.

Workers who have to handle aldehydes or ketals, or who may be liable to accidental exposure to them, should wear personal protective equipment appropriate to the extent of the possible exposure and the irritant characteristics of the particular aldehyde. When there is a risk that eyes may be splashed, eye protection should be provided.

MATHESON, D.

Chloroacetaldehyde (p. 82). Documentation of the Threshold Limit Values. Fourth edition (Cincinnati, American Conference of Governmental Industrial Hygienists Inc., 1980), 486 p.

CIS 74-780 *Toxicological evaluation of chloral liberated during spraying and moulding of polyurethane foam* (Evaluation toxicologique du chloral libéré lors du pistoletage et du coulage de mousses de polyuréthane). Bojcov, A. N.; Rotenberg, Ju., S.; Mulenkova, V. G. Translation INRS 123 B-72, Institut national de recherche et de sécurité, 30 rue Olivier-Noyer, 75680 Paris Cedex 14, France, 1972, 9 p. 2 ref. (In French)

Glutaraldehyde (204-205). Documentation of the Threshold Limit Values. Fourth edition (Cincinnati, American Conference of Governmental Industrial Hygienists Inc., 1980), 486 p.

"Contact dermatitis from glutaraldehyde". Jordan, W. P.; Dahl, M. V.; Albert, H. L. *Archives of Dermatology* (Chicago), Jan. 1972, Vol. 105 (94-95). 19 ref.

"The subacute inhalation toxicity of 109 industrial chemicals". Gage, G. C. *British Journal of Industrial Medicine* (London), Jan. 1970, 27/1 (1-18). 28 ref.

CIS 74-1395 *Total aldehydes–Sampling: cumulative–Analysis: titrimetric (bisulfite method).* Skare, I. Method Report T 105/73, National Board of Occupational Safety and Health, Fack, 100 26 Stockholm 34, Sweden, 1973, 12 p. Illus. 7 ref.

Algae

Algae include organisms of diverse nature, interrelationships, structure and reproduction, though they have some features in common. Most algae are plants but some can be called bacteria, others are animals and a minority have features common both to animals and plants. They are all of relatively simple structure; for example, *Chlorella*, much used in the study of photosynthesis and plant biochemistry, consists of a single, minute, globular green cell whose only form of multiplication and reproduction is the subdivision of its cell contents into replicas of itself. Whereas *Chlorella* is about 0.00001 m or less in diameter, the seaweed *Macrocystis* may be over 50 m in length and has a plant body with structures superficially resembling the stems and leaves of flowering plants. *Macrocystis* and its allies are called kelps and can form large underwater "forests". Yet this giant kelp has a separate sexually reproducing generation of microscopic size. All algae capable of photosynthesis can liberate oxygen in the process, unlike the photosynthetic bacteria.

There is no general agreement about their classification but they are commonly divided into 13 major groups whose members may differ markedly from one group to another in colour. The blue-green algae *(Cyanophyta)* are also considered by many microbiologists to be bacteria *(Cyanobacteria)* because they are procaryotes, that is they lack the membrane-bounded nuclei and other organelles of eucaryotic organisms. They are probably descendants of the earliest photosynthetic organisms and their fossils have been found in rocks some 2 000 million years old. Green algae *(Chlorophyta)*, to which *Chlorella* belongs, have many of the characteristics of other green plants. Some are seaweeds, as are most of the red *(Rhodophyta)* and brown *(Phaeophyta)* algae. *Chrysophyta*, usually yellow or brownish in colour, include the diatoms, algae with walls made of polymerised silicon dioxide. Their fossil remains form industrially valuable deposits (Kieselguhr, diatomite, diatomaceous earth). Diatoms are the main basis of life in the oceans and contribute about 20-25% of the world's plant production. Dinoflagellates *(Dinophyceae)* are free-swimming algae especially common in the sea; some are toxic.

Algae are subject to diseases caused by viruses, bacteria and fungi.

Algae are found wherever there is moisture, for a time at least. Plant plankton consists almost exlusively of algae. In lakes, rivers and on the seashore algae abound. The slipperiness of stones and rocks, the slimes, tangles and discolourations of water usually are formed by aggregations of microscopic algae. They are found in hot springs, snowfields and antarctic ice. On mountains they can form dark slippery streaks (Tintenstriche) dangerous to climbers. On drier rocks and on trees they are commonest as partners of fungi in the dual organisms called lichens.

Uses. Some seaweeds are valuable commercially as sources of alginates, carragenin and agar which are used in industry and medicine (textiles, food additives, cosmetics, pharmaceuticals, emulsifiers, etc.). Agar is the standard solid medium on which bacteria and other micro-organisms are cultivated. In the Far East, especially in Japan, a variety of seaweeds are used as human food. Seaweeds are good fertilisers but their use is decreasing because of the labour costs and the availability of relatively cheap artificial fertilisers. Algae play an important part in tropical fish farms and in rice fields. The latter are commonly rich in *Cyanophyta*, some species of which can utilise nitrogen gas as their sole source of nitrogenous nutrient. As rice is the staple diet of the majority of the human race, the growth of algae in rice fields is under intensive study in countries such as India and Japan. Certain algae have been employed as a source of iodine and bromine.

The use of microscopic algae, cultivated industrially, for human food has often been advocated and is theoretically possible with very high yields per unit area. However, it has not been possible, so far, to make such algae as cheap as traditional agricultural crops because of the cost of dewatering. Research has been carried out to see if such algae can be used in spacecraft, since they can be grown in a confined space, need only light and simple salts and can utilise human wastes.

If the climate is good enough and land is cheap, algae can be used as part of the process of purification of sewage and harvested for animal food. While a useful part of the living world of reservoirs, too many can seriously impede, or increase the cost of, water supply. In swimming pools algal poisons (algicides) can be used to control algal growth but, apart from copper in low concentrations, such substances cannot be added to water for domestic supply. Overenrichment of water with nutrients, notably phosphorus, with consequent excessive growth of algae is a major trouble in some regions and has led to bans on the use of phosphorus-rich detergents, though the best solution is to remove the excess phosphorus chemically in a sewage plant.

HAZARDS

Abundant growths of fresh-water algae often contain potentially toxic blue-green algae. Such "waterblooms" are unlikely to harm man because the water is so unpleasant to drink that swallowing a large and hence dangerous amount of algae is unlikely. On the other hand, cattle may be killed, especially in hot, dry areas where no other source of water may be available to them. Paralytic shellfish poisoning is caused by algae (dinoflagellates) on which the shellfish feed and whose powerful toxin they concentrate in their bodies with no apparent harm to themselves. Man, as well as marine animals, can be harmed or killed by the toxin.

Prymnesium (Chrysophyta) is very toxic to fish and flourishes in weakly or moderately saline water. It presented a major threat to fish farming in Israel till research provided a practical method of detecting the presence of the toxin before it reached lethal proportions. A colourless member of the green algae *(Prototheca)* infects man and other mammals from time to time. There have been a few records of algae causing skin irritations but little that is certain is known about this.

LUND, J. W. G.

General:

Introduction to the algae. Bold, H. C.; Wynne, M. J. (Englewood Cliffs, New Jersey, Prentice-Hall, 1978), 706 p. Illus. Over 5 000 ref.

Algal cultures and phytoplankton ecology, Second edition. Fogg, G. E. (Madison, Wisconsin, University of Wisconsin Press, 1975), 175 p. Illus. 383 ref.

The blue-green algae. Fogg, G. E.; Stewart, D. W. P.; Fay, P.; Walsley, A. E. (London and New York, Academic Press, 1973), 259 p. Illus. Over 2 000 ref.

"The technical production of microalgae and its prospects in marine agriculture". Soeder, C. J. *Harvesting polluted waters.* Devik, O. (ed.) (New York and London, Plenum Press Publishing Corp., 1976) (11-38). 51 ref.

Marine algae in pharmaceutical science. Hoppe, H. A.; Levring, T.; Tanaka, Y. (eds.). (Berlin and New York, Walter de Gruyter, 1979).

Toxic effects:

Paralytic shellfish poisoning in Eastern Canada. Trakash, A.; Medcof, J. C.; Tennant, A. D. Bulletin 177 of the Fisheries Research Board of Canada (Ottawa, 1971), 88 p. Illus. 109 ref.

Alkaline materials

Alkalis are caustic substances which dissolve in water to form a solution with a pH substantially higher than 7. These include ammonia, ammonium hydroxide, calcium hydroxide and oxide, potassium, potassium hydroxide and carbonate, sodium, sodium carbonate, hydroxide, peroxide and silicates and trisodium phosphate.

The alkalis, whether in solid form or concentrated liquid solution, are more destructive to tissues than most acids. The free caustic dusts, mists and sprays may cause irritation of the eyes and respiratory tract and lesions of the nasal septum. Strong alkalis combine with tissue to form albuminates, and with natural fats to form soaps. They gelatinise tissue to form soluble compounds which may result in deep and painful destruction. Potassium and sodium hydroxide are the most active materials in this group. Even dilute solutions of the stronger alkalis tend to soften the epidermis and emulsify or dissolve the skin fats. First exposures to atmospheres slightly contaminated with alkalis may be irritating, but this irritation soon becomes less noticeable. Workmen often work in such atmospheres without showing any effect, while this exposure will cause coughing and painful throat and nasal irritation in unaccustomed persons. The greatest hazard associated with these materials is the splashing or splattering of particles or solutions of the stronger alkalis into the eyes.

Since some of the alkaline materials are included in other articles, as for example ammonia and limestone and lime, descriptions will be limited here to the hydroxides and carbonates of potassium and sodium.

Potassium hydroxide (KOH)

CAUSTIC POTASH; POTASSIUM HYDRATE

m.w. 56.1
sp.gr. 2.04
m.p. 360.4 ± 0.7 °C
b.p. 1 320-1 324 °C
pH 13.5 (0.1 molar aqueous solution)
soluble in water and ethyl alcohol, insoluble in ethyl ether
white or slightly yellow lumps, rods, or pellets.
TLV ACGIH 2 mg/m³ ceil

Production. By the electrolysis of potassium chloride solution or by boiling potassium carbonate with milk of lime followed by filtration and vaporisation.

Uses. For the manufacture of liquid soap, as a mordant for wood, for absorbing carbon dioxide, mercerising cotton, in paint and varnish removers, for electroplating, photo-engraving and lithography, in printing inks, for the production of other potassium compounds, and in analytical chemistry and organic syntheses.

Sodium hydroxide (NaOH)

CAUSTIC SODA; LYE; SODIUM HYDRATE; LIQUID CAUSTIC

m.w.	40.1
sp.gr.	2.13
m.p.	318.4 °C
b.p.	1 390 °C

soluble in water, ethyl alcohol; insoluble in ethyl ether

white deliquescent flakes, pellets, sticks or cake.

TWA OSHA	2 mg/m^3
NIOSH	2 mg/m^3/15 min ceil
TLV ACGIH	2 mg/m^3 ceil
IDLH	200 mg/m^3
MAC USSR	0.5 mg/m^3

Liquid caustic is a solution of 45-75% sodium hydroxide in water.

Production. By treating sodium carbonate solution with calcium hydroxide or by electrolysis of solutions of sodium salts (chiefly the chloride) in mercury or diaphragm-type cells.

Uses. In the manufacture of rayon, mercerised cotton, soap, paper, explosives and dyestuffs. It is also used in the chemical industries, in metal cleaning, electrolytic extraction of zinc, tin plating, oxide coating, laundering and bleaching.

HAZARDS

These compounds are very dangerous to the eyes, both in liquid and solid form. As strong alkalis, they destroy tissues and cause severe chemical burns. Inhalation of dusts or mists of these materials can cause serious injury to the entire respiratory tract and ingestion can severely injure the digestive system. Even though they are not flammable and will not support combustion, much heat is evolved when the solid material is dissolved in water. Therefore, cold water must be used for this purpose, otherwise the solution may boil and splatter corrosive liquid over a wide area.

SAFETY AND HEALTH MEASURES

Safety in handling caustic alkalis depends to a great extent upon the effectiveness of employee education, proper safety instructions, intelligent supervision and the use of safe equipment. Workers should be thoroughly informed of the hazards that may result from improper handling. They should be cautioned to prevent spills and well instructed about the proper action to take if spills do occur. Each employee should know what to do in an emergency and should be fully informed about proper first-aid measures. Hazards from spills and leaks should be minimised by an adequate supply of water for washing-down. Drainage of hard-surfaced or diked areas should be directed so as to minimise the exposure of personnel and equipment. Adequate ventilation should be provided in areas where caustic potash or soda mist or dust is present. There should be no evidence of skin, eye, nose or throat irritation.

For the protection of the eyes, chemical safety goggles should be worn, as well as face shields, if complete face protection is necessary. Eyewash fountains and safety showers must be available at any location where eye and/or skin contact can occur. Protection against mist or dust of this compound can be provided by filter or dust-type respiratory protective equipment. The wearing of protective clothing is also advisable to avoid skin contact. This may consist of rubber gloves, aprons, shoes or boots, and cotton coveralls which fit snugly. Safety shoes or boots made of rubber, chlorobutadiene, or other caustic-resistant materials with built-in steel toecaps are recommended for workers handling alkalis in drums or in process areas where leakage may occur. Containers should be stored in rooms with trapped floor drains towards which the floors should be slanted. Where floor drains are not provided, curbs or a drained gutter, covered with an appropriate grill, should be constructed at door openings. Tanks should be entered for cleaning or repairing only after these have been drained and flushed thoroughly with water and ventilated. Men entering tanks should be observed by someone on the outside and a supplied-air respirator or self-contained breathing apparatus, together with rescue harness and lifeline, should be kept outside for rescue purposes.

Treatment. Speed in removing caustic alkali is of primary importance. In case of contact with skin or mucous membranes, the safety shower should be used immediately. Clothing can be removed later under the shower. Contaminated skin areas should be washed with very large quantities of water for as long as 1-2 h or until medical help arrives. Contaminated clothing and shoes should be thoroughly washed and decontaminated before re-use. If the caustic hydroxide enters the eyes, they should be copiously irrigated with water for at least 15 min and a physician consulted. If ingested, the chemical may be diluted by drinking large quantities of water or milk, followed by diluted vinegar or fruit juice to accomplish neutralisation. If discomfort is experienced from inhalation of caustic mist or dust, the employees should leave the contaminated atmosphere until proper ventilation is restored.

Carbonates and bicarbonates

The salts of carbonic acid (H_2CO_3), or carbonates, are widespread in nature as minerals. They are used in the building trade, in the glass, ceramics and chemical industries, and in agriculture. The principal carbonates are: calcium carbonate ($CaCO_3$), magnesite ($MgCO_3$), soda ash ($NaCO_3$), sodium bicarbonate ($NaHCO_3$), potash (K_2CO_3). The normal carbonates (with the anion CO_3) and the acid or bicarbonates (with the anion HCO_3) are the most important compounds. All bicarbonates are water-soluble; of the normal carbonates only the salts of alkaline metals are soluble. The solutions give rise to alkaline reactions because of the considerable hydrolysis involved. The bicarbonates are converted to normal carbonates by heating: $2\ NaHCO_3 = Na_2CO_3 + H_2O + CO_2$. Anhydrous carbonates decompose when being heated before reaching the melting point. Formation of crystal hydrates is little characteristic of carbonates; they are known to form with the derivatives of a few elements only (Na, Se). The normal carbonates are decomposed by strong acids (H_2SO_4, HCl) and set free CO_2.

The sodium carbonates occur in the following forms: soda ash–anhydrous sodium carbonate (Na_2CO_3); crystallised soda–sodium bicarbonate ($NaHCO_3$); sodium carbonate decahydrate ($Na_2CO_3.10\ H_2O$).

Sodium carbonate (Na_2CO_3)

SODA ASH

m.w.	106
sp.gr.	2.53 (aqueous solution 1.10 at 10 °C)

m.p. 853 °C
b.p. decomposes
soluble in water, insoluble in most organic solvents
pH of the 1% solution 11
saturated solution (31.13%) b.p. 105 °C
an odourless, white hygroscopic powder.
MAC USSR 2 mg/m³

In solution, $Na_2CO_3.10\,H_2O$ starts crystallising at temperatures below 32 °C; the crystals are monoclinic with a density of 1.45. $Na_2CO_3.7\,H_2O$ forms rhombic crystals between 32 and 34.8 °C. Anhydrous Na_2CO_3 crystallises above 112.5 °C.

Production. Sodium carbonate exists in nature as natural brines, saline water in lakes and solid stratified deposits. It is produced industrially by the ammonium-chloride process (NaCl and NH_3 being raw materials), the Le Blanc process (from NaCl and H_2SO_4), or by leaching out rock-salt deposits. The industrial product contains at least 98% Na_2CO_3, 1% NaCl and 0.1% Na_2SO_4. It may also contain traces of cyanogen, arsenic and sulphur salts as impurities.

Uses. Sodium carbonate is widely used in the manufacture of glass, caustic soda, sodium bicarbonate, aluminium, detergents, salts and paints, as well as for the desulphurisation of pig iron and purification of petroleum.

Sodium bicarbonate ($NaHCO_3$)

BAKING SODA; SODIUM ACID CARBONATE

m.w. 84
sp.gr. 2.16-2.22
m.p. loses CO_2 at 50 °C; at 100-150 °C it disintegrates and turns into carbonate
soluble in water
monoclinic colourless crystals.

Production. Sodium bicarbonate is obtained industrially by carbonating dissolved soda under pressure at 75 °C. It is produced on a very large scale in inorganic chemistry.

Uses. Sodium bicarbonate is used in the confectionery, pharmaceutical, leather and rubber industries, for the manufacture of fire extinguishers, mineral waters, non-alcoholic drinks, and also in everyday life.

Potassium carbonate (K_2CO_3)

SALT OF TARTAR; PEARL ASH; POTASH

m.w. 138
sp.gr. 2.42
m.p. 891 °C
b.p. decomposes
soluble in water, insoluble in ethyl alcohol
density of the aqueous solutions: 1.0904 (10%), 1.899 (20%), 1.4141 (40%)
forms a hydrate ($K_2CO_3 \cdot 1.5\,H_2O$) consisting of shiny, glass-like crystals which disappear completely on a heated surface at 130-160 °C
hygroscopic, odourless granules or a granular powder.

Production. Potassium carbonate is produced by carbonating KOH solutions obtained by electrolysis; a $NaCO_3$ suspension in dissolved KCl is also used. In earlier times potassium carbonate was obtained from plant ashes.

Uses. Potassium carbonate is widely used in the glass industry, as potash fertilisers, in the soap industry, for wool dyeing, and in the pharmaceutical industry.

HAZARDS AND THEIR PREVENTION

Alkaline carbonates may cause harmful irritation of the skin, the conjunctivae and the upper airways during various industrial operations (handling and storage, processing). Workers who load and unload bagged carbonates may present cherry-sized necrotic skin portions on their arms and shoulders. Rather deep ulcerated pitting is sometimes observed after the black-brown scabs have fallen off. Prolonged contact with soda solutions may cause eczema, dermatitis and ulceration.

Health engineering measures are mechanisation and hermetic enclosure of the processes, as well as efficient exhaust ventilation. Wash basins and showers must be provided near the workplaces, and the workers should wear personal protective equipment (respirators, gloves, arm protectors, aprons). Skin care after work should be encouraged (2% boric acid vaseline or cold cream are recommended).

Treatment. First aid in the case of chemical eye burns consists in immediate abundant flushing with water to remove all soda from the conjunctival folds. It is recommended to apply an antiseptic medication after flushing (20% sodium aldehyde, potassium permanganate 1 : 5 000).

In case of accidental ingestion of a saturated soda solution gastric lavage is the most effective treatment.

Cessation of exposure to soda helps restore health. In case of disease a physician should be consulted.

KARPOV, B. D.

General:

"Clinical aspects and prophylaxy of ocular burns of occupational origin" (Aspects cliniques et prophylaxie des brûlures oculaires survenant dans le processus de travail). Schwartzenberg, T. *Annales d'oculistique* (Paris), 1975, 208/7-8 (507-520). (In French)

CIS 75-1702 "Sensitivity of the skin to alkali in occupational dermatosis due to abrasion" (Über die Alkaliempfindlichkeit der Haut bei der Abnützungsdermatose). Tronnier, H. *Berufs-dermatosen* (Aulendorf in Württemberg), Apr. 1975, 23/2 (48-54). Illus. 4 ref. (In German)

CIS 76-1806 *Recommended practice–Loading and unloading liquid caustic–Tank cars (caustic soda and caustic potash).* Technical Bulletin TC-28 (Manufacturing Chemists' Association, 1825 Connecticut Avenue, NW, Washington, DC) (1975), 11 p. Illus.

Sodium hydroxide:

CIS 76-757 *Criteria for a recommended standard: Occupational exposure to sodium hydroxide.* DHEW (NIOSH) publication No. 76-105 (Washington, DC, Government Printing Office, 1975), 93 p. 91 ref.

Potash mining:

CIS 74-1636 "Potash ore and perforation of the nasal septum". Williams, N. *Journal of Occupational Medicine* (Chicago), June 1974, 16/6 (383-387). Illus. 4 ref.

CIS 74-505 "Mortality of potash workers". Waxweiler, R. J.; Wagoner, J. K.; Archer, V. E. *Journal of Occupational Medicine* (Chicago), June 1973, 15/6 (486-489). 18 ref.

Alkaloids

Alkaloids are a group of organic compounds with a certain number of characteristics in common:

- origin: generally vegetable; very rarely animal;
- chemical structure: all alkaloid molecules contain at least one atom of nitrogen with a fairly pronounced basic character;

- pharmacodynamic activity: often intense, even in small doses.

Chemical constitution and properties

Alkaloids are simple or complex-structure molecules whose nitrogen atom (or atoms) is:

- usually heterocyclic, sometimes extracyclic;
- generally a tertiary base, rarely a secondary base or quaternary ammonium compound;
- carrying an amine or imine function.

Alkaloids can be divided into two groups, depending on their general physical characteristics:

- non-oxygenated alkaloids, usually found in liquid form at normal temperature, volatile, strong-smelling, may be separated by steam distillation;
- oxygenated alkaloids, almost always in solid form at normal temperature and capable of crystallisation.

Alkaloids are insoluble or slightly soluble in water but can be easily dissolved in organic solvents. Most alkaloids have an optical rotatory activity.

As regards their chemical properties, alkaloids form a pure homogeneous group:

- basic properties: the very term alkaloid, or vegetable alkali as they used to be called, indicates that they all possess fairly pronounced alkaline properties;
- precipitation may be provoked by various reagents known as general alkaloid reagents.

Occurrence

Alkaloids are found in a very large number of plant families (about 40). They are principally encountered in dicotyledons, sometimes in gymnosperms, and occasionally in pteridophytes.

In nature they are rarely found in a free-base form but rather in association with cytologic constituents, generally in the form of salts as a result of the action of organic or mineral acids, or combined with tannin.

Production

Most alkaloids are produced by extraction from the plant material, though some (usually the simple-structure alkaloids) are manufactured synthetically.

There are two principal methods for obtaining alkaloids:

- extraction by means of a polar solvent (watery acid solution or alcohol);
- extraction by means of a non-polar solvent (organic solvent) after separation of the alkaloids by alkalinisation.

Specific methods of extraction exist for alkaloids with special physicochemical properties. Non-oxygenated, volatile alkaloids, for example, are extracted by means of hydrodistillation.

Pharmacodynamic activities and therapeutic and toxicological uses

The behaviour and characteristics of alkaloids are highly varied, as exemplified in the following account of some of the more important and typical substances in this group.

Aconitine ($C_{34}H_{47}NO_{11}$)

ACETYL BENZOYL ACONINE

m.w. 645.7
m.p. 204 °C

soluble in chloroform and benzene; less soluble in ethyl alcohol and ethyl ether; only very slightly soluble in water

colourless crystals or white crystalline powder, with bitter taste.

It is obtained from *Aconitum napellus* (monkshood), *Ranunculus* and other aconites.

It is used as an antineuralgic or as a sedative in cough mixtures.

It is extremely toxic and is one of the most potent and quick-acting poisons; toxic doses are very close to therapeutic doses. First symptoms of poisoning are nervous disorders, formication, followed by anaesthesia of the extremities, loss of co-ordination, vertigo, followed in turn by hyper-salivation, nausea, vomiting, polyuria and diarrhoea. As the heart becomes affected, bradycardia—at first regular then irregular—occurs, accompanied by signs of myocardic hyperexcitability and possible ventricular fibrillation. As little as 2 mg may rapidly prove fatal as a result of cardio-respiratory failure, unless intensive reanimation is practised.

Atropine ($C_{17}H_{23}NO_3$)

dl-HYOSCYAMINE; DATURINE; TROPINE; TROPATE

m.w. 289.4
m.p. 114-116 °C

soluble in ethyl alcohol, benzene and chloroform; only very slightly soluble in water

odourless, colourless crystals or white crystalline powder.

It is obtained by extraction from various solanaceous plants and sometimes by racemisation of 1-hyoscyamine. Atropine is mainly found in belladonna *(Atropa belladonna L.)* and stramonium *(Datura stramonium L.)*.

Atropine has a parasympatholytic action, with antisecretory and spasmolytic effect on the digestive tube, a depressant effect on Parkinsonian tremors, hypersialorrhea and a cycloplegic mydriatic effect on the eye. Toxic doses cause dilation of the pupils, paralysis of the adjustment function and dryness of the mucous membranes, especially of the mouth. These initial symptoms are followed by a state of excitement with general agitation, confusion, hallucinations and delirium, accompanied by hyperventilation, tachycardia and hyperthermia with flushing of the face and neck. Poisoning terminates with general depression of the central nervous system and coma. Its toxic effects are particularly acute in children, less frequent and milder in adults.

Caffeine ($C_8H_{10}N_4O_2$)

TRIMETHYLXANTHINE; METHYLTHEOBROMINE; THEINE; GUARANINE

m.w. 194.2
m.p. 237 °C

soluble in chloroform, slightly soluble in water and ethyl alcohol

white, prismatic crystals which are odourless and bitter-tasting.

Caffeine is an alkaloid deriving from various species of plant: coffee, tea, cocoa, cola nuts, maté. It can be obtained either by synthesis or by semisynthesis from theobromine, or else by extraction from tea or coffee in the course of the decaffeination process.

In medicine it is used as a cardiovascular and psychostimulant analeptic.

Acute poisoning is characterised by nausea, vomiting, headaches, vertigo, tremors, manic excitement and, occasionally, even convulsive coma. Additional symptoms are tachycardia, polyuria, sometimes followed by oliguria. The prognosis is particularly serious in the case of children.

Caffeine is well known as a cause of chronic caffeinism, inducing acute nervousness, tremors, arrhythmia and insomnia.

Cocaine ($C_{17}H_{21}NO_4$)
METHYL BENZOYL ECGONINE
m.w. 303.3
m.p. 98 °C
soluble in ethyl alcohol, chloroform, ethyl ether; slightly soluble in water
colourless crystals or white crystalline powder.

Cocaine is one of the alkaloids of the coca plant *(Erythroxylon coca Lam.)*. It is extracted from its leaves for use as a local anaesthetic.

Acute poisoning is characterised by nausea, vomiting, abdominal pains, dilation of the pupil of the eye, excitement and formication in the extremities; serious cases also lead to confusion, hallucinations, delirium, convulsions, hypothermia, respiratory failure and circulatory insufficiency.

Because of its euphoric effects, the use of cocaine may lead to addiction resulting in weakness, visual disturbances, manic attacks, alterations of libido and cachexia.

Codeine ($C_{18}H_{21}NO_3$)
METHYLMORPHINE
m.w. 299.4
m.p. 154-156 °C
very soluble in chloroform; soluble in ethyl alcohol, benzene and ethyl ether;slightly soluble in water
colourless crystals.

Codeine is extracted from opium or obtained by the methylation of another opium alkaloid, morphine.

Codeine and its salts are used as a sedative in cough mixtures.

An overdose of codeine may be responsible for respiratory failure, accompanied by gastric pains, constipation, flushing of the face, tremors and excitement. Addiction to codeine has been observed.

Emetine ($C_{29}H_{40}N_2O_4$)
CEPHALINE-O-METHYL ETHER
m.w. 480.6
m.p. 74 °C
soluble in ethyl alcohol and ethyl ether; slightly soluble in water
white amorphous powder of very bitter taste.

Emetine is the alkaloid of the ipecac *(Uragoga ipecacuanha (Brot.) Baill.)*. It is obtained solely by extraction from the root of the ipecac.

Emetine is used in medicine as a tissular amoebicide (effective against trophozoites; slightly effective against cysts).

The toxic effects of emetine are cumulative and cause digestive disorders (nausea, diarrhoea, vomiting), neuro-muscular disorders (peripheral neuropathy with pains and oedemas), cardiovascular disturbances such as lowered blood pressure with subacute cardiac insufficiency (degenerative myocarditis).

Morphine ($C_{17}H_{19}NO_3$)
m.w. 285.3
sp.gr. 1.31
m.p. 254 °C (anhydrous)
slightly soluble in ethyl alcohol, chloroform and water; insoluble in benzene
white crystalline or amorphous powder.

Morphine is an alkaloid obtained by extraction from opium, which contains about 10%, or from the capsule of the poppy *(Papaver somniferum L.)*.

Morphine is used in medicine as an analgesic.

The use of morphine can lead to addiction and to acute poisoning. Respiratory failure is the principal danger of poisoning and its most serious manifestation; it is accompanied by fainting fits, or by coma and acute myosis. Other signs of poisoning are lowered blood pressure (even shock), oliguria, hypothermia and muscular hypotonia.

Nicotine ($C_{10}H_{14}N_2$)
See separate article.

Opium
See separate article.

Papaverine ($C_{20}H_{21}NO_4$)
m.w. 339.4
sp.gr. 1.34
m.p. 147 °C
soluble in ethyl alcohol, benzene and acetone; slightly soluble in chloroform and carbon tetrachloride
white crystalline powder.

Papaverine is essentially obtained by synthesis or sometimes by extraction from opium as a by-product of morphine and codeine production (approximately 1% concentration in opium).

Papaverine is a musculo-tropic spasmolytic particularly effective against digestive or biliary spasms and bronchial spasms in people suffering from asthma. It is also used in medicine for its vasodilator effect on the blood vessels of the brain.

This alkaloid has very low toxicity. Excessive doses may however lead to such disorders as constipation, headache, increased breathing rate and hepatitis.

Pilocarpine ($C_{11}H_{16}N_2O_2$)
m.w. 208.3
m.p. 34 °C
b.p. 200 °C
soluble in ethyl alcohol, chloroform and water; slightly soluble in ethyl ether and benzene
colourless crystals (sometimes a colourless oil).

Pilocarpine is obtained by extraction from the leaves of various species of pilocarpus, especially *Pilocarpus microphyllus Stapf* and *Pilocarpus jaborendi Holmes*.

Pilocarpine is a direct parasympathomimetic, mainly used in opthalmology to produce contraction of the pupil of the eye and to reduce intra-ocular tension. It is also used as an antidote for atropine poisoning.

The symptoms of pilocarpine poisoning are characterised by the muscarinic effects of the alkaloid: hypersecretion of saliva, sweat and tears, abdominal pains, diarrhoea, nausea, vomiting, contraction of the pupils. Poisoning is also accompanied by neurological symptoms such as excitement and muscular symptoms such as twitching, tachycardia and lowered blood pressure. When death occurs, it is usually due to cardiac arrest and respiratory paralysis with pulmonary oedema.

Quinine ($C_{20}H_{24}N_2O_2$)
m.w. 324.4
m.p. 177 °C
very soluble in ethyl alcohol and chloroform; slightly soluble in ethyl ether and benzene; very slightly soluble in water
intensely bitter colourless crystals.

Quinine is one of numerous quinquina alkaloids and is extracted from the bark of *Cinchona Ledgeriana Molns*.

Quinine is used in medicine against malaria and for its febrifugal and tonic effects.

Severe poisoning by quinine is characterised by vomiting, followed by neurosensory disorders such as vertigo, buzzing of the ears, clouded vision, double vision and scotoma. Additional signs are dilation of the pupils and neurological disorders such as headaches, excitement and sometimes coma. In prognosis, the main danger is its cardiovascular effects (sharp drop in blood pressure). Death is rare and occurs only from massive doses.

Strychnine ($C_{21}H_{22}N_2O_2$)

m.w.	334.4
sp.gr.	1.36
m.p.	268 °C
b.p.	270 °C

soluble in chloroform; slightly soluble in ethyl alcohol and benzene; insoluble in cold water

odourless, colourless, translucent crystals or white crystalline powder, with a very bitter taste.

TWA OSHA	0.15 mg/m³
STEL ACGIH	0.45 mg/m³
IDLH	3 mg/m³

Strychnine is obtained from *Strychnos nux-vomica* and the seeds of other species of *Strychnos*.

It is very occasionally used medicinally as a psychotonic in cases of general weakness and as a bitter eupeptic; it is mostly used as a pesticide to poison small mammals and birds.

The toxic effects of strychnine are generally characterised by medullar hyper-excitability; convulsions with attacks of tetaniform contractures which may be spontaneous or provoked by the slightest stimulus; particularly severe attacks may be accompanied by respiratory arrest. Sensory disorders (green-coloured vision) may also occur, but consciousness is not affected. After a few seizures, death usually follows by medullary paralysis.

SAFETY AND HEALTH MEASURES

The risk of alkaloid poisoning arises from accidental ingestion of the dangerous material. Workers who have to handle these dangerous substances in the course of manufacture, packing or use should be thoroughly informed concerning the nature of the risk and the means of avoiding it. Containers should carry warning labels.

Thorough cleanliness should be enforced in the workrooms and every care should be exercised to prevent even small accumulations of dangerous materials where they could contaminate the hands of unwary workers. Persons exposed to these materials should be educated in personal hygiene and in the use of showers at the end of the working day. Washing and sanitary facilities should be readily available. Eating and smoking should be forbidden in the workrooms.

Personal protective equipment should be provided to ensure that wearing apparel does not become contaminated with dangerous alkaloids.

Treatment. Where alkaloids are made and handled there should be a prearranged procedure to deal with accidental poisoning. The treatment will include the use, by trained personnel, of gastric lavage, and the application of such symptomatic measures as those needed to overcome central depression or to maintain renal function. The administration of large amounts of fluids is recommended in acute poisoning by caffeine, lobeline, pilocarpine, quinine. True antidotes do not exist; however, antagonistic effects may be used for treatment as in the case of atropine poisoning (pilocarpine, eserine) or of pilocarpine (atropine).

FERRY, S.
VIGNEAU, C.

Natural alkaloids:

CIS 1704-1965 "Skin lesions incurred in the manufacture of Devincan" (Die bei der Erzeugung von Devincan auftretenden Hautläsionen). Valer, M. *Berufsdermatosen* (Aulendorf in Württemberg), Apr. 1965, 13/2 (96-110). 10 ref. (In German)

Synthetic alkaloids:

CIS 690-1973 "Description of a case of allergy to lysergic acid amide derivatives" (Descrizione di un caso di allergia a lisergidi (derivati dell'amide dell'acido lisergico). Nava, C. *Medicina del lavoro* (Milan), Jan.-Feb. 1972, 63/1-2 (57-61). Illus. 6 ref. (In Italian)

Alkyd resins

An alkyd resin is a synthetic macromolecular material formed by the reaction of a polybasic organic acid and a polyhydric alcohol. Most alkyd resins are modified, usually with a vegetable oil or fatty acids; if one speaks about alkyd resins, generally the oil-modified alkyds are meant. A typical alkyd might contain phthalic acid, glycerol, pentaerythritol, linseed or soyabean oil.

Production

There are two ways to produce oil-modified alkyd resins depending on the type of raw material used.

(a) The *fatty acid process*, in which polybasic acids, polyols and fatty acids are heated together to temperatures of 200-280 °C, at which esterification and polymerisation takes place.

(b) The *alcoholysis process*, in which a vegetable (and sometimes animal) oil is heated with the polyol and a catalyst to temperatures of 230-260 °C, at which temperature re-esterification to partial esters takes place. After cooling down, a polybasic acid is added and esterified at a temperature of 200-280 °C. The reaction water has to be removed continually. This can be achieved by using vacuum or by employing the so-called solvent process in which a solvent is used which is capable of forming a pseudo-azeotropic mixture with water. The vapours are cooled in a condensor and after separation from the water the solvent is refluxed back in the process. In both processes inert gas, CO_2 or N_2, is used.

Alkyd resins are made in stainless-steel kettles equipped with stirrer, connections for the raw material supply, inert gas, safety valve and a condensor. Heating may be by direct fire, thermal oil, steam or induction. Most alkyds are thinned down with a solvent after production.

Common raw materials are—

polyols: glycerol; pentaerythritol; ethylene glycol.

polybasic acids: phthalic anhydride; isophthalic acid; azelaic acid; maleic anhydride.

oils: soyabean oil; linseed oil; tung oil; castor oil; coconut oil; dehydrated castor oil.

monobasic acids: benzoic acid; mixtures of fatty acids; linolic acid; isononanic acid.

catalysts: litharge; lithium hydroxide or soaps; sodium hydroxide or soaps.

Uses. In paints and varnishes.

HAZARDS

As all the materials used are combustible, fire control during production is important. In the kettle the fire hazards are small, because with the use of an inert gas there is little or no oxygen present. However, condensed reaction water falling back into the reaction mixture can cause such an excessive foam, and build up so high a pressure, that a considerable amount of the contents can be pressed out of the kettle. The temperature of this material can be as high as 280 °C.

Health hazards

Polyols. Glycerol and pentaerythritol present no toxic problems at exposure during production; ethylene glycol is toxic by ingestion and inhalation. At ambient temperature, however, the vapour pressure is so low that the latter is only serious when vapours escape from heated kettles.

Polybasic acids. Phthalic and isophthalic acid vapours are irritant. Prolonged inhalation should be avoided. In modern plants these products are used in closed kettles so there is little danger of vapours. More care should be taken to avoid dust when handling the bagged materials. Long and intensive contact can cause skin and/or lung diseases. Maleic anhydride vapours can be lethal or cause permanent injury even after relatively short exposure. Because of the low boiling point care should be taken when adding this material to a hot reaction mixture (see ACIDS AND ANHYDRIDES, ORGANIC).

Oils. Soyabean oil and linseed oil do not present health hazards. Ingestion of castor oil should be avoided.

Monobasic acids. Benzoic acid vapour or dust is pungent and irritating, but not a major health danger.

Catalysts. Litharge can cause lead poisoning, but because of the small amounts used there is not much danger if the normal hygienic precautions are used.

SAFETY AND HEALTH MEASURES

As in modern plants the production takes place in closed equipment, the main preventive measures to be taken are avoiding the escape of vapour and the handling of material when filling and emptying the reactors. Liquids are normally pumped directly into the kettles by pipeline. Adding solid material via the manhole will cause an excessive vapour escape into the working room when the kettle is warm, so this should be done via a closed system or when the kettle has cooled to room temperature. Vapour displaced from the kettle by raw material should escape via a vent connected with a stack or a catalytic or thermal combustion unit. By no means should it escape into the working room. The same precautions are to be taken with the letdown tank, because at the higher temperature at which the dissolving of the resins usually takes place vapours from boiling solvents can escape. The letdown tanks should therefore be equipped with a reflux cooler and a stirrer. Good ventilation of the whole working place is necessary. Fire and explosion protection is important. Heating of the reactors should preferably be indirect, by heating oil or induction. The reactor should be equipped with a good stirrer, thermometers, pressure meter, safety valve and breaking disc. It is advisable that the vent of the latter should be wide and connected to an area or tank where, in case of frothing over, material can escape without causing any danger to persons or equipment. An internal cooling coil is useful to cool the reaction mixture in case of an exothermic reaction. Inert gas has to be used for discharging and during the process in order to avoid explosive vapour mixtures. Escaping solvent vapours can easily be ignited, so all electrical equipment should be explosion-proof.

Pumping solvents can generate static electricity which can discharge via a spark and so ignite a solvent/air mixture. The speed of pumping combustible liquids should therefore not surpass the limit of 1 m/s.

When a person has to enter a vessel for repairs, this should be clean, free of inert gas and organic vapours, the connections to inert gas blocked, the power supply to the stirrer disconnected and the person himself tied to a lifeline (see CONFINED SPACES).

KLEIN, W. A.

CIS 75-1672 *Properties and essential information for safe handling and use of maleic anhydride* (Chemical Safety Data Sheet SD-88, Manufacturing Chemists Association, 1825 Connecticut Avenue, NW, Washington, DC) (revised 1974), 13 p.

Alkylating agents

Of many organic compounds used as alkylating agents, a small number of those that are implicated as carcinogens are discussed. These are aliphatic chlorinated ethers, a five-membered sultone ring, an organic sulphate and iodide, and a four-membered lactone ring.

The specific substances discussed are as follows:

Chloromethyl methyl ether (CMME)
Bis(chloromethyl) ether (BCME)
Bis(2-chloroethyl) ether (BCEE)
Bis(1-chloroethyl) ether
1,3-Propane sultone (PS)
Diethyl sulphate (DS)
Methyl iodide (MI)
Beta-propiolactone (BPL)

Other alkylating agents covered in separate articles are: CHLOROMETHANE; DIAZOMETHANE; DIMETHYL SULPHATE; 1,2-DIBROMOETHANE, DIBROMO-CHLOROPROPANE; and epichlorohydrin under EPOXY COMPOUNDS.

Production. CMME is produced by the action of hydrogen chloride on a solution of formaldehyde in methanol and also by direct chlorination of methyl ether.

BCME results from the action of hydrogen chloride on a sulphuric acid solution of paraformaldehyde. An extensive research has been published (1979) on the formation of BCME from chlorides and formaldehyde. In the gas phase this was found experimentally to be extremely slow, is unlikely to be spontaneous in ambient conditions but can be formed in an aqueous medium. A survey of eight plants in five industries having sizeable quantities of both formaldehyde and chloride-containing substances showed no BCME in workroom air within the detectable limit of 0.1 ppb.

BCEE is produced by the reaction of 1,2-dichloroethane with chloromethyl alcohol and as a by-product in the production of ethylene glycol from beta-chloroethyl alcohol. PS results from the dehydrogenation of gamma-hydroxypropane sulphonic acid and BPL from the action of ketene on formaldehyde. DS is produced by the action of concentrated sulphuric acid on ethylene and of fuming sulphuric acid on ethyl ether or ethyl alcohol.

Uses. The chloromethyl and chloroethyl ethers have been and, except for the BCME, continue to be used largely in the chloroalkylation of organic compounds in the production of anionic exchange resins. Although BCME is no longer used for this purpose in industry, it is used as the monitoring indicator for CMME. The 1,3-propane sultone introduces the sulphopropyl group into a wide variety of organic compounds to provide water solubility and anionic properties. DS, as well as other ethylating agents, is used to produce ethyl derivatives of such compounds as phenols, amines and thiols, but DS is the only ethylating agent effective for the production of quarternary ammonium ethosulphate salts. MI is used for the introduction of the methyl radical in organic synthesis. Large quantities of beta-propiolactone are employed as an intermediate in acrylic acid and ester production; much smaller quantities are used in sterilisation operations for such materials as vaccines and blood plasma.

GENERAL HAZARDS AND THEIR PREVENTION

All of these alkylating agents are suspected of having carcinogenic potential for man on the basis of experiments with mice or rats or as human carcinogens on the basis of epidemiological studies together with animal experimentation. In view of the carcinogenic potential, exposure by all routes of entry should be kept at an irreducible minimum where complete elimination of exposure is not feasible. Physical examination of workers for placement in operations involving possible exposure to these alkylating agents should include consideration of increased personal risk due to cigarette smoking, pregnancy, or treatment with steroids or cytotoxic agents.

Rodents with vitamin A deficiency have been found especially susceptible to effects of alkylating agents.

Chloromethyl methyl ether ($ClCH_2OCH_3$)

CMME; DIMETHYLCHLOROETHER; CHLOROMETHOXYMETHANE

m.w.	80.5
sp.gr.	1.06
m.p.	−103.5 °C
b.p.	59.1 °C
v.d.	2.9
v.p.	163 mmHg (21.7·10³ Pa) at 120 °C
f.p.	−8 °C

a colourless liquid with an ethereal odour, reacts with water.

HAZARDS AND THEIR PREVENTION

In experimental animal exposure and in human work exposure, CMME is contaminated with 1 to 8% of the highly carcinogenic BCME. Such contaminated CMME caused carcinomas on subcutaneous injection of the mouse but equivocal evidence of this action in the rat. Epidemiological investigations indicated increased incidence of pulmonary carcinoma of exposed workers.

No TLV has been suggested for CMME but exposure that causes irritation of the skin and eyes is certainly highly excessive and should be brought under immediate control.

Determination of concentration of CMME in air is conducted by electron capture gas chromatography of a hexane extraction of air samples collected by glass impingers containing 2,4,6-trichlorophenol. The sensitivity of the method is 0.5 ppb (v/v) when a 10 l air sample is used (1977).

Bis(chloromethyl) ether ($ClCH_2OCH_2Cl$)

BCME; DIMETHYL-1,1-DICHLOROETHER; CHLOROCHLOROMETHOXY-METHANE

m.w.	115
sp.gr.	1.31
m.p.	−42 °C
b.p.	104 °C
v.d.	4
f.p.	42 °C

a colourless liquid with pungent odour, reacts with water.

TLV ACGIH	0.001 ppm, a human carcinogen
MAC USSR	0.5 mg/m³

HAZARDS AND THEIR PREVENTION

On experimental exposure of mice, BCME has been found carcinogenic on inhalation, subcutaneous administration and skin application. An industry-wide retrospective cohort study of six United States producers of chloromethyl ethers (CME) found an increased risk of respiratory cancer deaths in only the one firm where high exposures had occurred. Thirty-four cases of lung cancer associated with exposure to CME have been reported in the United States, eight in the Federal Republic of Germany, and five in Japan. Although no increased risk of carcinoma was found in five of the six US plants having lesser exposure, and in the positive plant a dose-response relationship with risk ratios exceeding ten for the longest duration and greatest exposure subgroups was demonstrated, the data were not considered adequate to support a threshold effect.

Air analysis should be conducted periodically to ascertain that exposure is maintained at less than 0.001 ppm. The method referred to under CMME is applicable to BCME, for which it also has a sensitivity of 0.5 ppb.

Bis(2-chloroethyl) ether ($ClCH_2CH_2OCH_2CH_2Cl$)

BCEE; SYM-DICHLOROETHYLETHER; 1-CHLORO-2(BETACHLOROETHOXY) ETHANE; CHLOREX; 1,1-OXYBIS(2-CHLOROETHANE)

m.w.	143
sp.gr.	1.22
m.p.	−51.7 °C
b.p.	178.5 °C
v.d.	4.93
v.p.	0.77 mmHg (0.10·10³ Pa) at 20 °C
f.p.	55 °C
i.t.	370 °C

a colourless liquid with pungent odour, reacts with water.

TWA OSHA	15 ppm 90 mg/m³ ceil
TLV ACGIH	5 ppm 30 mg/m³ skin
IDLH	250 ppm
MAC USSR	2 mg/m³

Uses. In addition to its action as an alkylating agent, BCEE has extensive industrial use in paint, varnish, and lacquer production, in petroleum refineries as a selective solvent for naphthenes and for dewaxing lubricating oils, in the textile industry, and in agriculture as a pesticide for soil fumigation.

HAZARDS AND THEIR PREVENTION

The flash point of dichloroethyl ether is well above room temperature and it introduces no appreciable fire or vapour-air explosion hazard. However, if scoured textiles wet with this solvent are placed in an enclosure heated above the flash point, static electricity or other ignition sources could cause an explosion. Where dichloroethyl ether is evaporated above its flash point, as may occur in a textile drying oven, means of ignition including sources of static electricity should be eliminated.

BCEE is considered "animal positive" for carcinoma on the basis of positive results of oral administration to the mouse. No epidemiological study has been reported.

Dichloroethyl ether has little, if any, irritative effect on the skin but it can penetrate the skin sufficiently to cause serious and even fatal poisoning. Contact with the eye can cause corneal lesions. In the event of excessive exposure, as in a spill, irritative action on the respiratory tract may be sufficiently severe to cause delayed pulmonary oedema.

The liquid should be kept off the skin, but if contact occurs it should be immediately washed off. Concentrations of dichloroethyl ether should be kept not only below the level where nasal or eye irritation occurs but below the threshold limit of 5 ppm as shown by air analysis. Analytical methods available include collection by silica gel adsorption or absorption by alcoholic potassium hydroxide in a fritted glass bubbler. In the former method the sample may be removed from the silica gel by heated air, then pyrolysed and the chloride determined by titration. In the latter method the chlorinated ether is hydrolysed by refluxing for 2 h and the chloride determined. Infrared absorption and gas chromatography are faster methods and are recommended where such instrumentation is available. No cases of injury from industrial exposure have been reported.

Bis(1-chloroethyl) ether ($CH_3CH_2ClOCH_2ClCH_3$)

BIS(ALPHA-CHLOROETHYL) ETHER; 1,1-OXYBIS(1-CHLOROETHANE)

HAZARDS AND THEIR PREVENTION

Bis(1-chloroethyl) ether has been found to be carcinogenic to the mouse on subcutaneous injection of 2 400 mg/kg over 60 weeks. Accordingly, it is to be considered suspect of carcinogenic potential for man and exposure should be kept as low as feasible. No epidemiological study has been reported. No mouse subcutaneous administration of bis(2-chloroethyl) ether has been reported to serve as a comparison of the carcinogenicity of the two isomers.

1,3-Propane sultone ($C_3H_6O_3S$)

PS; 1,2-OXATHIOLANE-2,2-DIOXIDE; 3-HYDROXY-1-PROPANE SULPHONIC ACID SULTONE

m.p. 31 °C
b.p. 112 °C

H₂C——CH₂
H₂C O O
S
O

HAZARDS AND THEIR PREVENTION

PS has produced carcinoma in the mouse and rat and is considered suspect of carcinogenic potential to man. No epidemiological study has been reported. Injury to the respiratory passages can occur but is unlikely owing to the low vapour pressure of PS at normal temperatures. At higher temperatures control measures should be taken to prevent dissemination of PS vapour into workplace air. Although no TLV has been assigned, exposures should be limited as completely as feasible.

Diethyl sulphate ($(C_2H_5)_2SO_4$)

DS; ETHYL SULPHATE

m.w.	154.2
sp.gr.	1.2
m.p.	−24.5 °C
b.p.	208 °C with slight decomposition
v.d.	5.3
v.p.	$0.13 \cdot 10^3$ Pa at 47 °C
f.p.	104 °C
i.t.	436

a colourless oily liquid with ethereal odour, slightly soluble in water.

HAZARDS AND THEIR PREVENTION

In experimentation with the rabbit, application of 500 mg on the uncovered skin caused mild irritation and 2 mg in the eye produced severe injury; an LD_{50} of 600 mg/kg resulted from skin application. The lowest lethal concentration on 4-h exposure of the rat is 250 ppm.

Carcinoma has been produced in the rat on subcutaneous administration and prenatal exposure. No epidemiological studies nor cases of human injury have been reported. Although no TLV has been proposed, exposures should be kept as low as feasible since animal experimentation indicates DS as suspect of carcinogenic potential for man.

Methyl iodide (CH_3I)

MI; IODOMETHANE

m.w.	142
sp.gr.	2.3
m.p.	−66.1 °C
b.p.	42.5 °C
v.d.	4.9
v.p.	400 mmHg ($53.2 \cdot 10^3$ Pa) at 25 °C

a colourless liquid with a characteristic odour, reacts with water

industrial substance suspected of carcinogenic potential to man.

TWA OSHA	5 ppm 28 mg/m³ skin
TLV ACGIH	2 ppm 10 mg/m³ skin
STEL ACGIH	5 ppm 30 mg/m³

HAZARDS AND THEIR PREVENTION

MI acts primarily as a central nervous system depressant with indication of lung irritation on acute exposure. It causes mild skin irritation on animal experimentation. Exposure of 3 800 ppm for 15 min was fatal to rats. Subcutaneous administration to the rat and intraperitoneal administration to the mouse resulted in carcinoma. On the basis of carcinogenic action on these animals, MI is considered suspect of carcinogenic potential for man. The TLV is considered low enough to prevent neurotoxic effects but with the carcinogenic potential, exposures should be kept as low as feasible. A fatal case of poisoning has been reported but there is no record of human carcinoma of exposed workers. No epidemiological study has been reported.

No fire or explosion hazard exists; in fact, MI has been suggested as a fire extinguishing agent. See also HYDROCARBONS, HALOGENATED ALIPHATIC.

Beta-propiolactone ($C_3H_4O_2$)

BPL; 2-OXETANONE; HYDRACRYLIC ACID; 3-HYDROPROPIONIC ACID

H₂C——C=O
H₂C——O

BPL is carcinogenic to the mouse and rat but results with the hamster and guinea-pig are equivocal. On the basis of human experimentation, BPL is considered suspect of carcinogenic potential for man. No epidemiological study has been reported. Added information is included in the article on LACTONES.

COOK, W. A.

"Potential carcinogenic chemicals. I: Alkylating agents". Fishbein, L. *Occupational cancer and carcinogens.* Vainio, H.; Sousa, M.; Hemminki, K. (eds.). (Washington, DC, Hemisphere Publishing Corporation, 1979) (213-257). Illus. 299 ref.

IARC monographs on the evaluation of the carcinogenic risk of chemicals to man. Vol. 4: *Some aromatic amines, hydrazine and related substances, N-nitroso compounds and miscellaneous alkylating agents* (Lyons, International Agency for Research on Cancer, 1974), 286 p.

Methods of air sampling and analysis. Katz, M. (ed.) (American Public Health Association, 1015 Eighteenth St., NW, Washington, DC) (2nd ed., 1977), 984 p.

Research study on bis(chloromethyl) ether formation and detection in selected work environments. Yao, C. C.; Miller, G. C. DHEW (NIOSH) publication No. 79-118 (Cincinnati, National Institute for Occupational Safety and Health, 1979), 151 p.

"Occupational exposure to chloromethyl ethers. A retrospective cohort study (1948-1972)". Pasternack, B. S.; Shore, R. E.;

Albert, R. E. *Journal of Occupational Medicine* (Chicago), Nov. 1977, 19/11 (741-746). 25 ref.

CIS 80-471 "Lung cancer in chloromethyl ether workers". Weiss, W.; Moser, R. L.; Auerbach, O. *American Review of Respiratory Disease* (New York), Nov. 1979, 120/5 (1031-1037). 29 ref.

CIS 77-152 "Acute impairment due to methyl iodide" (Akutní poškození metyljodidem). Skutilová, J. *Pracovní lékařství* (Prague), Nov. 1975, 27/10 (341-342). 8 ref. (In Czech)

"The cigarette factor in lung cancer due to chloromethyl ethers". Weiss, W. *Journal of Occupational Medicine* (Chicago), Aug. 1980, 22/8 (527-529). 7 ref.

Allergy

The considerable reduction taking place in the incidence and/or severity of cases of intoxication by substances encountered in the work process, which is due to the limitation of exposure to harmful substances, is accompanied by an increase in cases of adverse immune response at work, even in environments hitherto regarded as presenting little risk. This is particularly true, for example, of the pharmaceutical industry.

The number of occupational allergenic substances is extremely high: it is possible, indeed, that almost all the substances recorded in industrial use are allergenic.

The term "allergy" is taken to mean an altered capacity to react to a substance foreign to the organism. This altered capacity results from the exposure of the immunological system to foreign substances (allergens or antigens). An exception to the rule is auto-immunity, through which it is possible to become allergic even to substances forming part of the subject's own body.

During the sensitisation stage the T and B lymphocytes memorise the chemical structure of the foreign substance and start to produce the antibodies. Exposure at a subsequent stage sets off the immunoreaction, in which antigens and antibodies interact to give rise, through the liberation of chemical intermediates, to the allergic disease. The recognition of the antigen by the antibody is highly specific in the physicochemical sense. The antibody identifies certain parts of the matter of the antigen, and so reacts only with them. It is possible, however, for other molecules to have the same determinants sterically arranged in the same positions. In that case they too interact with the antibody. What is known as a cross-reaction is usual between haptens. These are substances with a low molecular weight that are immunogenic only in combination with a proteic substance. They can start an immunoreaction in a sensitised subject.

The groups of substances that can take part in cross-reactions (the term is derived from a chemical structure containing antigen determinants) are the following: the betalactamic, desoxyaminic, phenazonic, para-phenolic, phenothiazinic and para-amino group, the morphine derivatives, the hydrazides, the oligosaccharides, the rifamycin, the piperazinic group, and others.

Inadequate knowledge of cross-reacting substances can lead to manifestations in a sensitised person exposed to them.

On the basis of the first four types of immunoreaction under the Gell and Coombs classification, a brief list follows of occupational allergies and some of the substances causing them.

Type 1: immediate or anaphylactic. Short-term anaphylactic allergies are produced by the coming together of antigens and IgE or IgG_4 antibodies. Atopic (predisposed) subjects are particularly liable to them (see ALLERGY, SCREENING AND TREATMENT OF):

- *(a)* anaphylactic shock: this may be produced by antibiotics or enzymes;
- *(b)* asthma, rhinitis: these are caused by natural resins (gum arabic, colophony, tragacanth), synthetic resins (epoxides), isocyanates, wood, flour from cereals or the castor bean, various drugs, platinum salt and phenylenediamine (see ASTHMA, OCCUPATIONAL);
- *(c)* anaphylactic purpura (see also type 2): this is caused by enzymes, antibiotics and pyrazoles;
- *(d)* dermatitis bullosa: this is caused by acrylates;
- *(e)* photo-dermatitis (see also type 4): this is caused by eosin, carbanilide, promethazine, salicylanilide;
- *(f)* urticaria, erythematous dermatitis (see also type 3): these are caused by cyanides, gold salts, natural and synthetic resins, drugs, organic pigments (azo dyes) and inorganic pigments (for example, titanium oxide).

Type 2: cytolytic or cytotoxic. Allergies of this type involve IgM and IgG antibodies. The immunoreaction takes place when both the antigen and the complement are present: agranulocytosis, thrombocytopenia, leucopenia, haemoloytic anaemia due to pyrazoles, chloramphenicol.

A toxic mechanism may be set in motion at the same time as the allergy.

Type 3: from immune complexes. Precipitating IgE or IgM antibodies are present in a polymeric chain with the antigens. The complement takes part in the reaction:

- *(a)* fever: this may be caused by polymers or ampicillin;
- *(b)* serum diseases: these may be caused by hormones or serums;
- *(c)* nephropathy: this may be caused by rifampicine;
- *(d)* alveolitis: this may be caused by mycetes, germs, diisocyanates, skins and furs, animal proteins, epoxide resins;
- *(e)* contact dermatitis (see also type 4): this may be caused by phenolphthalein, phenylenediamine, salicylates, quarternary ammonium salts.

Type 4: retarded. The T lymphocytes are involved together with sessile antibodies and lymphokines:

- *(a)* contact dermatitis, eczema: even an abridged list would contain a huge number of items (at least 2 000) (see CONTACT DERMATITIS OR ECZEMA, OCCUPATIONAL);
- *(b)* granulomatosis: this may be caused by beryllium or mycetes.

Epidemiology

Tables 1 and 2 reproduce the data given by Pepys in the previous edition of this encyclopaedia.

Thus, the presence of asthmogens was already reported 30 years ago in various industries. The substances to be found in the work process give rise to asthma through different pathogenic mechanisms (immunologic, toxicologic, pharmacologic). Even today there is no certain method of diagnosis to distinguish the presence of the various mechanisms. Studies have been carried out more recently in different sectors of work. Some examples are given in tables 3, 4 and 5.

The study of occupational allergy

A correct evaluation of these affections calls for: (1) acquaintance with the substances used and the

Table 1. Incidence of allergic diseases in the United Kingdom for the period June 1953-June 1954 (population: 19 785 000)[1]

Ilnesses	Days lost	
	in millions	%
All causes	280.64	100
Allergic diseases	7.79	2.8
Asthma	4.60	1.6
Urticaria	0.20	
Other allergic disorders	0.04	
Eczema and dermatitis	2.95	1.0
Diabetes	1.57	0.6
Pneumonia	1.89	0.7
Appendicitis	2.56	0.9
Duodenal ulcer	3.11	1.1 } 2.4
Gastric ulcer	3.78	1.3 }

[1] Abstracted and calculated from *Reports of the Ministry of Pensions and National Insurance for 1953-54* by Williams, D. A.

Table 2. Incidence of asthma in relation to occupation[1]

Occupations	Population at risk (in thousands)	Spells of incapacity (in thousands)	Spells of incapacity (%)
1. Administrators, directors, managers	373	–	–
2. Professional and technical	695	1	0.144
3. Commerce, finance and insurance	1 328	2	0.151
4. Agriculture, horticulture and forestry	1 052	2	0.190
5. Engineering, metal manufacture	2 521	5	0.198
6. Workers in wood, cane, cork	484	1	0.207
7. Persons engaged in personal service: hotels, clubs, institutions	491	1	0.204
8. Workers in building and contracting	902	2	0.222
9. Fitters, machine erectors	813	2	0.246
10. Clerks, typists	793	2	0.252
11. Road transport workers	788	2	0.254
12. Warehousemen, storekeepers, packers	363	1	0.275
13. Electricians, electrical apparatus makers and fitters	358	1	0.279
14. Painters and decorators	332	1	0.201
15. Railway transport workers	316	1	0.316
16. Water, air and other workers in transport and communications	293	1	0.341
17. Workers in unskilled occupations	1 233	5	0.406
18. Coal miners	630	3	0.476
All occupations	14 400	34.9	0.242

[1] Information obtained from *Ministry of Pensions and National Insurance Digest of Statistics Analysing Certificates of Incapacity, 1951/52* by Williams, D. A.

manner of exposure to them; (2) an analysis of the immunological behaviour of the exposed worker.

If it is to be of anamnestic value, any substance encountered in work should be identified by the following characteristics:

Table 3. Allergy in 68 workers employed on galvanising processes

Eczematous dermatitis	2
Desquamative erythematous dermatitis	2
Urticaria	1
Psychogenic pruritus	2
Rhinitis, oculorhinitis	4
Bronchial asthma	2
Laryngeal oedema	1
Total	14/68 (21%)

Table 4. Allergy in 380 affected workers exposed to the epoxide system (epoxide prepolymer plus additives)

Eczematous dermatitis	229
Urticaria	57
Bronchial asthma	85
Rhinitis, oculorhinitis	9

Table 5. The asthmogenic capacity of certain substances in industrial use

Substance	Industry investigated	Asthmatics/cases investigated
Coffee	Roasting	4/70 (5.71%)
Hair, dandruff	Pharmaceutical research	13/97 (13.40%)
Spiramycin	Pharmaceutical	17/305 (5.57%)
Streptomycin	Pharmaceutical	11/400 (2.75%)
Phenylglycine HCl	Pharmaceutical	8/100 (8.00%)
Diisocyanates	Wood painting or varnishing	95/182 (52.19%)
Epoxide system	Various[1]	85/490 (17.34%)
Protease	Detergents	33/207 (15.94%)
Amylase	Detergents	26/207 (12.56%)
Lysozyme	Pharmaceutical	10/40 (25.00%)
Papain	Pharmaceutical, veterinary	11/120 (9.16%)
Bromelin	Pharmaceutical	8/76 (10.52%)
Trypsin	Pharmaceutical	11/536 (2.05%)

[1] Electromechanics, painting and varnishing and the manufacture of skis, abrasives and varnishes.

(a) active components;

(b) vehicle (solvent, emulsion, etc.);

(c) degree of purity of the components and of the vehicle;

(d) stability of the preparation during storage and in ordinary use;

(e) other physicochemical characteristics (pH, evaporation rate, etc.);

(f) conditions of exposure (quantity, volume, concentration, frequency, duration and nature of exposure, possible absorption by the mucus or the skin);

(g) biological studies of the irritant, immunological, teratogenic, carcinogenic and toxic properties;

(h) derivatives (intermediate and final).

Lastly, it is necessary to know in which sectors of activity the substance is encountered and which other substances cross-react with it. All this is essential not only to accurate diagnosis but also to adequate prevention (see ALLERGY, SCREENING AND TREATMENT OF).

The study of predisposition and of the immunological behaviour of the subject exposed to immunogens is essential in secondary prevention. The workers must be selected, and this entails methodological and socio-economic limitations that can be overcome only partially. On the basis of present knowledge, it is possible to evaluate genetic predisposition only in respect of the first type of immunopathy (atopy). For the other types of immunoreaction, such evaluation is doubtful or, in the view of certain writers, impossible. The production of the IgEs is under monofactorial control.

In an allergic subject the level of IgE in the serum may be high and the condition thus assumes a recessive autosomic character. But an allergy may also be found where the IgE level is low. In that case, it is believed that there may be one or more genes of the Ir locus (linked to the HL-A system) that can determine the optimal recognition of the allergen. It is possible that the process calls for modulation through the T system, with the helper and/or suppressor conditioning of the T lymphocytes.

It is thus necessary, in order to evaluate the atopy, not only to measure the total IgEs but also to determine the specific IgEs. The study of the other immunoglobulins may be desirable for the evaluation of quantitative changes. Although a correlation is not always evident between the IgEs in the serum and those of the surface, there is believed to be a lack of the former in subjects exposed to the risk of asthma. If there is any sign of atopy, the subject may contract an allergic affection of the first type more easily from exposure to occupational immunogenic substances. In any case, it is also possible for non-atopic subjects to contract these affections.

Although this is not a predisposing immunological condition, it should be mentioned here that subjects with increased broncho-motor tonus are more liable to occupational asthma. For the methods to employ in evaluating predisposing states, see ALLERGY, SCREENING AND TREATMENT OF.

Predisposing factors in contact dermatitis include ichthyosis, dyshidrosis in general and hyperhidrosis in particular, xerotic skin and asteatosic skin. There are other less definite factors such as the odour of the skin. This applies to the White race, for dark-skinned people are less subject to these affections.

NAVA, C.

"Immunology today" (L'immunologie aujourd'hui). Katz, D. *Semaine des Hôpitaux* (Paris), 8-15 May 1980, 56/17-18 (882-890). Illus. (In French)

CIS 80-448 "Allergy in occupational environment" (L'allergie en milieu de travail). Stevens, E.; Gervais, P.; Diamant-Berger, O.; Pariente, E.; Dean, G.; Groentenbriel, G.; Marcelle, R. *Cahiers de médicine du travail–Cahiers voor arbeidsgeneeskunde* (Brussels), June 1979, 16/2 (155-198). Illus. 16 ref.

CIS 80-1086 "Occupational allergopathies–Prevention, diagnosis, therapy" (Le allergopatie professionali–La prevenzione, la diagnosi, la terapia). Nava, C.; Briatico-Vangosa, G. (eds.). *Medicina del lavoro* (Milan), Jan.-Feb. 1980, 71/1 (2-105). Illus. 279 ref. (In Italian)

"Allergic occupational diseases" (Berufsbedingte allergische Erkrankungen). Stroehmann, I. *Arbeitsmedizin, Sozialmedizin, Präventivmedizin* (Stuttgart), Aug. 1980, 15/8 (173-177). Illus. 19 ref. (In German)

"Treatment of the allergic occupational diseases" (Behandlung berufsbedingter allergischer Erkrankungen). Fuchs, E. *Arbeitsmedizin, Sozialmedizin, Präventivmedizin* (Stuttgart), Aug. 1980, 15/8 (185-188). (In German)

CIS 79-1948 "Experimental evaluation of sensitising properties of new chemicals" (Expermentální hodnocení senzibilizacních vlastnosti nových chemických látek). Znojemská, S.; Janecková, V.; Pekárek, J. *Pracovní lékařství* (Prague), July, 1979, 31/6-7 (218-222). 17 ref. (In Czech)

Allergy, screening and treatment of

Aetiological diagnosis of a suspected case of allergy must be based on a full understanding of the substances used in the technological process and those present in the working environment (the case history principle). In particular, information is needed on the chemical composition of the following groups of substances:

(a) substances encountered in work (raw materials, solvents and diluents; various additives; intermediates; end products and degradation products (whether pyrolytic, chemical or biological);

(b) substances used for environmental and personal cleaning and disinfecting;

(c) substances used for environmental and personal protection.

There are at present obstacles in the way of an accurate diagnosis, which may be summarised as follows:

(a) Information on the chemical composition of the substances in use is lacking for economic reasons (trade secrets), owing to gaps in legislation and because there are similar or identical trade names for various substances used in the same technological process. Data sheets are also often defective since they indicate the presence in a product for commercial use of a toxic or carcinogenic substance but nearly always fail to mention an immunogenic risk.

(b) There are few biological studies (whether on laboratory animals or on man) concerning the allergenic properties of a chemical.

(c) There are also few epidemiological studies on allergies in industry and agriculture.

(d) There is a lack of standardisation as regards titration of the antigenic capacity of the biological extracts used in diagnosis and treatment (Bethesda Workshop, 1978).

Tables 1, 2, 3 and 4 show the models used in the study and prevention of immunopathy due to substances used in industry and agriculture.

Table 1. Methods of evaluating the immunogenic properties of a substance

1. Biological studies of animals with a similar immunity system to that of the human being (Draize test, maximisation test, optimisation test) (Marignac et al., 1978).
2. Biological studies to show other properties (non-immunological) of the substance (irritative, pharmacological, carcinogenic, etc.).
3. Epidemiological studies based on exposure to similar substances.
4. Search for a threshold value below which allergic sensitisation is difficult (allergological exposure limit) (Nava, 1980).

The concept of allergological exposure limit is applicable only in the introductory stage of the immunity process, while the allergic sensitising is still taking place through the joint effect of the various properties of a substance (immunogenic, irritative, pharmacological, etc.). It may be possible by restricting dissemination in the environment and contact with the worker to reach a level of exposure low enough to eliminate or reduce the risk of sensitisation. This has, in fact, been demonstrated in work involving exposure to isocyanates, papain and other proteases (from *Bacillus subtilis*) used in the production of detergents containing proteolytic enzymes.

Table 2. Proposals for the reduction of environmental allergens

1. Replacement of dangerous technologies and substances by more suitable ones.
2. Introduction of automation.
3. Adoption of suitable cleaning methods.
4. Introduction of the notion of allergenic risk in the productive process and the administrative budget (Nava, 1980).

The genetic predisposition of the worker to allergies must be evaluated, in relation to immunopathies of the first type only, by means of the tests listed in table 3.

Table 3. Tests for the evaluation of genetic predisposition (non-specific immunological tests)

Recommended test	Technique to be employed
Total IgE dosage	RIST or PRIST
IgA, IgG, IgM immunoglobulin dosage	Radial immunodiffusion
Search for a specific IgE against widespread antigens	Intradermal reaction

Medical checking before and during the performance of the work should also take place in accordance with the scheme set out in table 4.

Table 4. Predisposition and diagnostic checks to be carried out on workers exposed to the risk of allergies

1. Pre-employment medical examination accompanied by:
   (a) non-specific immunological tests (see table 3)
   (b) specific immunological tests (to be carried out with the substances encountered in the work)
   (c) bronchial irritability tests (through non-specific bronchial provocation)[1]
2. Annual or other periodical medical examination accompanied by:
   (a) specific immunological tests
   (b) bronchial irritability tests[1]

[1] Where there is a risk of asthma.

Checks departing from the annual frequency (see table 4) must be carried out:

(a) on the appearance of symptoms;
(b) before absence on maternity leave;
(c) on return to work after a long absence;
(d) at the end of the work in question;
(e) during a change of department or duties;
(f) before treatment (only for activities involving contact with pharmaceutical products).

The non-specific bronchial provocation test is carried out before and after exposure to an asthmogen in order to assess increased bronchial reactivity. The inhalation of acetylcholine, methacholine or twice distilled water is a suitable method.

The techniques recommended for immunological diagnostic screening consist in direct tests (skin tests, exposure tests) and indirect tests (serological tests, cellular tests).

Skin tests can be carried out for conditions of the types 1, 3 and 4 (see ALLERGY; ANTIBIOTICS). They can also be carried out in works clinics with modest equipment.

Exposure tests with the substances encountered in work call for a sheltered environment (hospital) and means of detecting variations in the respiratory function. They are to be avoided if alveolitis is suspected (Pepys and Hutchcroft, 1975).

Serological and cellular tests are seldom used in occupational medicine for mass screening. They are to be used in doubtful cases (when the medical history or the stop-start test indicates a suspected allergy and the direct tests are negative), for checking the validity of the skin and exposure tests or for forming a diagnosis where other tests are not suitable (during immunoreactions of type 2).

Among the tests used in occupational medicine may be mentioned those for the study of the complement, of cytolysis, of lymphocytic transformations and of quantification of lymphokines. The radio-immunological methods adopted are the RIA and the RAST, which are at present confined to a few substances (mycotic antigens, penicillin, phenylglycine, diisocyanates, bromelin and papain).

THE PREVENTION AND TREATMENT OF OCCUPATIONAL IMMUNOPATHY

Prevention and treatment nearly always consist in symptomatic treatment and the advice to remove the worker from the dangerous environment and contact with the harmful substance. The use of chromoglycate before exposure in cases of allergy may be dangerous because protection is incomplete and there may be other immunological and extra-immunological troubles that are not inhibited by this drug.

Causal treatment (specific immunotherapy) has proved useful in eliminating or reducing symptoms and permitting the continuation of the previous work only in respect of allergies to wood and natural resins. It is of value, of course, in agriculture and where there are airborne particles such as pollen, dust from the skin of livestock and mycophytes.

In some cases permanent or temporary hyposensitisation has been obtained by using salts of nickel and of hexavalent chromium.

NAVA, C.

"Methods of evaluating predisposition and diagnosis in occupational allergy" (Metodi di valutazione di predisposizione e di diagnosi delle allergopatie professionali). Nava, C. "Le allergopatie professionali", Nava, C.; Briatico Vangosa, G. (eds.). *Medicina del lavoro* (Milan), Jan. 1980, 71/1 (4-16). 34 ref. (In Italian)

"In-vivo diagnosis of occupational allergies" (In-vivo-Diagnostik bei berufsbedingten Allergien). Düngemann, H. *Arbeitsmedizin, Sozialmedizin, Präventivmedizin* (Stuttgart), Aug. 1980, 15/8 (177-182). Illus. (In German)

"In-vitro diagnosis of allergic occupational diseases" (In-vitro-Diagnostik allergischer Berufskrankheiten). Baenkler, H. W. *Arbeitsmedizin, Sozialmedizin, Präventivmedizin* (Stuttgart), Aug. 1980, 15/8 (182-185). 12 ref. (In German)

"Bronchial provocation tests in etiologic diagnosis and analysis of asthma". Pepys, J.; Hutchcroft, B. J. *American Review of Respiratory Disease* (New York), Dec. 1975, 112/6 (829-859). Illus. 68 ref.

"Inhalation tests" (Les tests par inhalation). Chrétien, J. *Revue française des maladies respiratoires* (Paris), 1976, 4/suppl. 1, 127 p. Illus. Ref. (In French)

"Use of the maximisation test on guinea-pigs to show the power of industrial products to cause skin allergies"

(Utilisation du test de maximisation chez le cobaye pour la mise en évidence du pouvoir allergisant cutané de produits industriels). Marignac, B.; Poitou, P.; Gradinski, D. *Revue française d'allergologie* (Paris), 1978, 61/4. (In French)

"Workshop on antigens in hypersensitivity pneumonitis (Bethesda, Maryland)". *Journal of Allergy and Clinical Immunology* (St. Louis), Apr. 1978, 61/4 (199-239). Illus. 60 ref.

Allyl compounds

The allyl compounds are unsaturated analogues of corresponding propyl compounds and are represented by the general formula, CH_2:$CHCH_2X$, where X in the present context is usually a halogen, hydroxyl or organic acid radical. As in the case of the closely allied vinyl compounds, the reactive properties associated with the double bond have proved useful for the purposes of chemical synthesis and polymerisation.

Certain physiological effects of significance in industrial hygiene are also associated with the presence of the double bond in the allyl compounds. It has been observed that unsaturated aliphatic esters exhibit irritant and lacrimatory properties which are not present (at least to the same extent) in the corresponding saturated esters; and the acute LD_{50} by various routes tends to be lower for the unsaturated ester than for the saturated compound. Striking differences in these respects are found between allyl acetate and propyl acetate. These irritant properties, however, are not confined to the allyl esters; they are found in different classes of allyl compounds.

Allyl alcohol (CH_2:$CHCH_2OH$)

2-PROPEN-1-OL

m.w.	58
sp.gr.	0.85
m.p.	−129 °C
b.p.	97 °C
v.d.	2.0
v.p.	23.8 mmHg ($3.17 \cdot 10^3$ Pa) at 25 °C
f.p.	21.1 °C
e.l.	2.5-18%
i.t.	378 °C

miscible in all proportions in water, ethanol and ether

a colourless liquid with a pungent odour detectable at 0.8 ppm.

TWA OSHA	2 ppm 5 mg/m³ skin
STEL ACGIH	4 ppm 10 mg/m³
IDLH	150 ppm
MAC USSR	2 mg/m³

Production. Propylene is chlorinated at high temperature to yield allyl chloride which is then hydrolysed to produce the alcohol. Allyl alcohol is also obtained by the dehydration and reduction of glycerol.

Uses. Next to allyl chloride, allyl alcohol is the most important of the allyl compounds in industry. It is useful in the manufacture of pharmaceuticals and in general chemical syntheses, but the largest single use of allyl alcohol is in the production of various allyl esters of which the most important are diallyl phthalate and diallyl isophthalate which serve as monomers and prepolymers.

HAZARDS AND THEIR PREVENTION

Allyl alcohol is a flammable and irritant liquid. It causes irritation in contact with the skin and absorption through the skin gives rise to deep pain in the region where absorption has occurred in addition to systemic injury. Severe burns may be caused by the liquid if it enters the eye. The vapour does not possess serious narcotic properties but it has an irritant effect on the mucous membranes and the respiratory system when it is inhaled as an atmospheric contaminant. Its presence in a factory atmosphere has given rise to lacrimation, pain in the eye and blurred vision [necrosis of the cornea, haematuria and nephritis].

In places where this substance is produced or used the precautions should take account of the risks of fire and explosion and of the risks associated with injury to health (see FLAMMABLE SUBSTANCES; GASES AND VAPOURS, IRRITANT). Storage should take place in steel containers.

Allyl chloride (CH_2:$CHCH_2Cl$)

3-CHLOROPROPENE

sp.gr.	0.94
m.p.	−134.5 °C
b.p.	45 °C
v.d.	2.64
v.p.	368 mmHg ($48.94 \cdot 10^3$ Pa) at 25 °C
f.p.	−32 °C
e.l.	3.3-11.1%
i.t.	392 °C

insoluble in water; miscible in all proportions in ethanol and ether

a colourless liquid with an unpleasant pungent odour dectable at 3 ppm.

TWA OSHA	1 ppm 3mg/m³
NIOSH	15 min ceil 3 ppm 9 mg/m³
STEL ACGIH	2 ppm 6 mg/m³
IDLH	300 ppm
MAC USSR	0.3 mg/m³

Allyl chloride is produced by the chlorination of propylene at high temperature. It is used in the production of epichlorohydrin, a step in the manufacture of epoxy resins and in the production of commercial glycerol. It is also used in the synthesis of intermediates for the manufacture of polymers, resins, and plastics used by themselves or incorporated in surface coatings and adhesives.

Allyl chloride has flammable and toxic properties. It is only weakly narcotic but is otherwise highly toxic. It is very irritating to the eyes and upper respiratory tract. Both acute and chronic exposure can give rise to lung, liver and kidney injury, but liver injury is more significant in the case of chronic exposure. [Chronic exposure has also been associated with decrease in the systolic pressure and in the tonicity of the brain blood vessels.] In contact with the skin it causes mild irritation, but absorption through the skin causes deep-seated pain in the contact area. Systemic injury may be associated with skin absorption.

The precautions are similar to those recommended for allyl alcohol.

Allyl trichloride

See HYDROCARBONS, HALOGENATED ALIPHATIC.

Allyl bromide (CH_2:$CHCH_2Br$)

m.w.	121
sp.gr.	1.40
m.p.	−119.4 °C
b.p.	71.3 °C
v.d.	4.17
f.p.	−1 °C
e.l.	4.4-7.3%
i.t.	295 °C

insoluble in water; miscible in all proportions in ethanol and ethyl ether

colourless to light-yellow liquid with a pungent odour.

Allyl bromide is prepared by treating allyl alcohol with a bromide and sulphuric acid, or by the partial dehydrobromination of dibromopropane in a high-temperature cracking reaction. It is used as a fumigant and as an intermediate in the synthesis of other compounds.

Its local irritant effect on contact with the skin is slightly less pronounced than that of allyl chloride, but it has a similar systemic effect.

The precautions against the risk of fire and explosion and those to prevent skin and eye irritation and systemic injury are similar to those recommended for allyl alcohol.

Chloroallyl bromide

See HYDROCARBONS, HALOGENATED ALIPHATIC.

[**Allyl iodide** (CH_2:$CHCH_2I$)

3-IODOPROPENE; 3-IODOPROPYLENE

m.w. 168
sp.gr. 1.85
m.p. −99 °C
b.p. 103 °C
v.d. 5.8

insoluble in water

a yellowish liquid which darkens on exposure to light and air, with a pungent odour.

Health effects and safety and health precautions are similar to those mentioned for allyl bromide.

Allyl cyanide (CH_2:$CHCH_2CN$)

VINYL ACETONITRILE

m.w. 67
sp.gr. 0.83
m.p. −87 °C
b.p. 116 °C

a colourless liquid with an onion-like odour.

MAC USSR 0.3 mg/m³ skin

It is a dangerous disaster hazard because it can emit highly toxic fumes when decomposed on heating or acid contact.]

Other allyl compounds

Reference has already been made to allyl esters. Allyl formate and allyl acetate are flammable esters which evolve flammable and irritant vapours. Allyl ether (diallyl ether) has similar properties. The important ester, diallyl phthalate, has only a slight fire hazard at room temperature (f.p. 165 °C); its irritant effect on the skin is more pronounced than that of the other phthalates.

Allyl ether, allyl ethyl ether and allyl phenylether are flammable liquids that irritate eyes and airways and possess narcotic properties. When flowing, they can generate electrostatic charges due to low electric conductivity. All of them react violently with oxidants.

For allyl glycidyl ether, see ETHERS.

Allyl amine is an extremely irritant substance capable of producing lung oedema on high exposure (see AMINES, ALIPHATIC). Allyl hydrazine hydrochloride has shown carcinogenic properties on mice. Allyl sulphide or "oil garlic" is used in the manufacture of flavours and shares the toxic properties of allyl chloride. Allyl propyl disulphide is a component of the onion oil; it is an irritant and lacrimatory liquid for which ACGIH recommends a TLV of 2 ppm, 12 mg/m³ and a STEL value of 3 ppm, 18 mg/m³.

Where the necessity is indicated by the properties described above, precautions similar to those for allyl alcohol should be adopted.

MATHESON, D.

CIS 74-1661 *Properties and essential information for safe handling and use of allyl chloride.* Chemical Safety Data Sheet SD-99 (Manufacturing Chemists Association, 1825 Connecticut Avenue, NW, Washington, DC) (1973), 17 p.

CIS 77-703 *Criteria for a recommended standard–Occupational exposure to allyl chloride.* DHEW (NIOSH) publication No. 76-204 (National Institute for Occupational Safety and Health, 4676 Columbia Parkway, Cincinnati, Ohio) (Sep. 1976), 87 p. 54 ref.

CIS 80-795 "A fertility study of male employees engaged in the manufacture of glycerine". Venable, J. R.; McClimans, C. D.; Flake, R. E.; Dimick, D. B. *Journal of Occupational Medicine* (Chicago), Feb. 1980, 22/2 (87-91). 5 ref.

Altitude

Atmospheric pressure decreases with height. By way of example, the atmospheric pressure in Mexico City at an altitude of around 2 230-2 240 m is approximately 578.6 mmHg ($0.7714 \cdot 10^5$ Pa) in comparison with a pressure of 760 mmHg ($1.0133 \cdot 10^5$ Pa) at sea level. Rising air thus expands and the conversion of internal energy into the work of expansion leaves the air cooler. Rising dry air cools adiabatically (literally "with no heat flow") 9.8 °C/km height. Ultraviolet radiation originating from the sun is to a large extent absorbed by the earth's atmosphere; at high altitudes the intensity of ultraviolet radiation is significantly higher than at sea level. Around 25 million people live and work at altitudes over 3 000 m and are therefore subject to the above factors occupationally.

Atmospheric pressure

The physiological problems of reduced atmospheric pressure at altitude are twofold but interrelated:

(a) the smaller number of oxygen molecules per unit volume;

(b) lower oxygen partial pressure.

Work capacity depends on the oxidation of foodstuffs to provide energy for the muscles. A given amount of work done at altitude requires the same quantity of oxygen as when done at sea level; in addition, the oxygen molecules must reach the muscle cell in a continous flow if work is to be maintained. However, oxygen flow from atmosphere to tissues is possible only if there is sufficient pressure gradient to overcome resistance in its path.

At 5 500 m the pressure of the inspired oxygen is only half that at sea level. In a person suddenly exposed to this altitude the diminished pressure can be only partially compensated for by respiratory and cardiovascular responses to acclimatisation and to high demands of exercise. The tissue capillaries receive less oxygen at lower pressure, maximum oxygen uptake is reduced and capacity for aerobic work falls.

Processes of altitude adaptation lead to: a smaller pressure drop along the oxygen pathway (figure 1); smaller diffusion distance between capillary and cell; and enzymatic and other changes within the cell.

Working capacity at altitude

Three factors play a part in determining quantitative and qualitative working capacity at altitude:

(a) actual height (low, moderate, high altitude);

(b) duration of exposure (acute, chronic or for generations);

(c) individual factors including state of health and, an as yet undefined, physiological adaptive capacity which results in wide variations in ease and degree of adaptation.

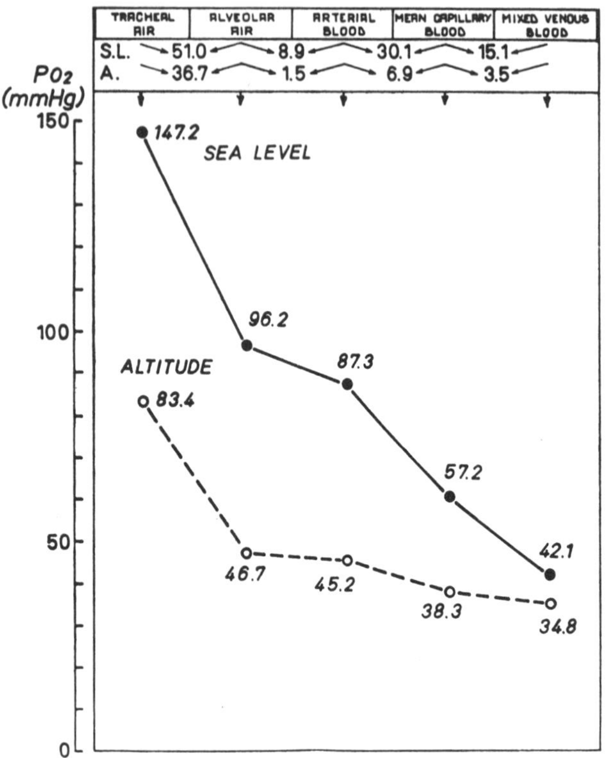

Figure 1. Mean PO_2 gradients from tracheal air to mixed venous blood, at sea level and at altitude. Differences in the air are large but differences in the capillaries are almost nil.

Height factor. Low altitudes (1 000-2 000 m) seem to affect physical performance, as was clearly seen at the 1968 Olympic Games in Mexico City, where prolonged endurance effort was affected in particular. However, at this level, acclimatisation is expected to be rapid.

Studies on exercise at moderate altitudes up to 3 000 m are scarce. Studies of high altitude work are numerous but often with contradictory findings, possibly resulting from a combination of the three above factors. Nevertheless, men have been found living at 5 800 m and in the mines of Aucanquilcha (Chile) at 6 000 m men are employed on heavy physical work.

Acute exposure. Acute exposure to altitude reduces working capacity in proportion to degree of altitude, although the individual factor and time of climb may play important roles.

Above 3 500 m hypoxia may affect the nervous system on initial exposure and altitude sickness may occur; and, although individual response as regards severity and duration may vary, working capacity is always affected.

Impaired working capacity is demonstrated by a definite reduction in maximal oxygen consumption and in physical endurance. Oxygen uptake in light and moderate work is constant at different altitudes and at different stages of acclimatisation; however, submaximal work at sea level could be maximal at altitude because the compensating mechanisms are more heavily taxed. Physiological response to exercise in newcomers to altitude follows the same lines as that at sea level but, in some functions, maximum limits are lowered while, in others, the response is exaggerated and closer to the limits. A special form of pulmonary oedema, not related to heart failure, may develop a few hours after arrival at altitude, and usually occurs after exercise. It may attack a man ascending for the first time but is most common in acclimatised persons returning from a sojourn at sea

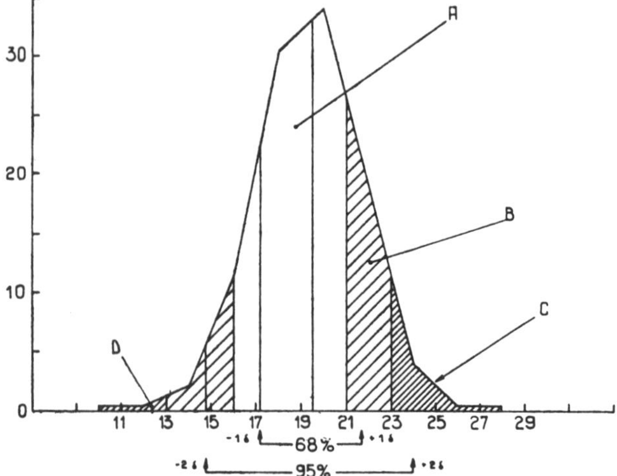

Figure 2. Distribution curve for haemoglobin values in 3 138 miners working in mines located at over 4 400 m. The curve can be divided into five zones in accordance with the haemoglobin counts in relation to altitude adaptation. *(From Cosio, G., 1967.)* Ordinate: Percentage of subjects. Abscissa: Haemoglobin in grammes. A. Haematic compensation zone. B. Intermediate zone (borderline cases). C. Mong's disease. D. Anaemia.

level. The genesis is still obscure but oxygen administration and sometimes only rest produce dramatic recovery.

Long-term acclimatisation. After a few weeks sojourners at altitude feel well although working capacity remains impaired and improves only with acclimatisation; recovery of physical capacity could be used as an index of adaptation. Studies of mountaineers, climbers, etc., at high altitudes show reductions in working capacities even after prolonged exposure; factors limiting capacity are dyspnoea due to extremely high ventilation, heavily taxed cardiovascular mechanisms and limited diffusion across alveolar and tissue barriers. One series of experiments on lowland inhabitants who were moved to a high altitude region (4 540 m), showed that after 12 months only half the subjects had recovered their initial sea-level capacity and endurance. The other half presented clear improvement in oxygen consumption but at the cost of higher and higher ventilation.

Maximum adaptation is found in persons descending from numerous generations of high-altitude natives, and these persons are able to perform very heavy work at high

Figure 3. Bolivian Indian women employed as labourers in a mine located at an altitude of over 4 500 m in the Andes Mountains.

altitudes with the same ease that lowlanders perform similar tasks at sea level.

A number of special physiological characteristics have been observed amongst high-level natives. Maximum working capacity (maximum oxygen uptake) is the same at 4 540 m as at sea level; heart rate may rise to 200 beats per minute; maximum ventilatory capacity is high (115 l/m^2/min); maximum diffusing capacity of lungs is 30% higher than in lowlanders, as is also the oxygen-carrying capacity of the blood; haemoglobin affinity for oxygen is diminished and the oxygen is released more readily at tissue level; activity of respiratory enzymes is higher. These and other factors involved in the physiology of exercise at altitude give natives the ability to perform very heavy tasks in agriculture, mining, etc., with greater endurance than at sea level. However, studies on energy expenditure in occupational activities at altitude have not been made and consequently it is necessary to extrapolate results of running and walking studies.

Chronic mountain sickness. This condition, also known as Monge's disease occurs when natives or people with long residence at altitude lose their adaptation and develop intolerance (see plate 1). The reasons are unknown and, fortunately, the incidence is low. Three parameters dominate the picture and are needed as diagnosis criteria: symptomatology, excessive blood volume and hypoventilation.

The clinical picture depends mainly upon deep hypoxia of the nervous system which produces a very wide symptomatology. Physiological findings in cardiocirculatory and respiratory systems are deceptively small and scarce, although increased cardiac output and pulmonary pressure should be mentioned. In other body functions the changes are mild or non-existent. Complaints of diminished physical capacity are contradicted very often by surprisingly good performances; however, much lower PO_2 at the capillaries does result in somewhat impaired working capacity.

Larger than normal blood volume and hypoventilation are interrelated as the cause and effect of each other. Hypoxia will depress the respiratory centre causing hypoventilation and deeper hypoxia. Where this vicious circle starts is unknown. Haematological response to the deeper hypoxia gives a larger than normal blood volume; haematocrits of 85% and Hb of 27 g have been seen.

Symptoms disappear when subject is moved to a lower level but they usually will develop again upon return to altitude.

Individual factors. Certain pathological conditions impair ability to acclimatise to high altitudes. Anaemia and respiratory and cardiovascular diseases are contraindications to heavy work at moderate altitudes or for any exercise at high altitudes. As hypoxia and exercise both increase blood pressure, the combination of these two factors makes work at altitudes very hazardous for hypertensive newcomers.

Occupational diseases and altitude

Silicosis is the most common occupational disease in mountain regions; it is also the most incapacitating since it affects the lungs. Silicosis at altitude appears earlier and develops more rapidly than at sea level, often commencing after 2-5 years' exposure. Two factors seem to be responsible for this:

(a) higher pulmonary ventilation and larger functional residual capacity of the lungs, which result in a greater amount of dust in contact with alveolar membrane;

(b) a very definite tendency at altitude to develop fibrotic tissue.

Diagnosis and grading for compensation are greater problems at altitude. The increased lung blood volume gives, on the X-ray film, shadows that could be misinterpreted as silicosis or modify the grading (see PNEUMOCONIOSES, INTERNATIONAL CLASSIFICATION OF). Furthermore, diffused fibrosis, not seen with X-rays, seems common and is more incapacitating.

A combination of silicosis and chronic mountain sickness has been described; however, the condition is very difficult to assess because both components produce the same physiological picture.

Working capacity in silicotic subjects is of course impaired; the few studies made show no close relationship between physical performance and the grading of disease; it is not rare to see persons doing heavy work with 60% arterial saturation.

PREVENTIVE MEASURES

Pre-employment medical examinations are recommended for workers who are to be employed at high altitudes; those members of the worker's family who are to accompany him should also be examined. Cardiovascular or pulmonary conditions and hypertension should be considered contraindications. On arrival at altitude, time should be allowed for acclimatisation, and medical supervision should be ensured so that pathological reactions may be dealt with swiftly.

The lower temperatures, especially at night, and higher levels of ultraviolet radiation also require adequate precautionary measures. Warm clothing for cold periods and the use of ultraviolet-filter barrier creams during exposure to sunshine should be considered.

VELASQUEZ, T.

General physiology:

"Factors affecting performance". Astrand, P. O.; Rodahl, K. *Textbook of work physiology* (New York, MacGraw-Hill, 1970), Ch. 17 (559-596). 84 ref.

Acclimatisation:

"Altitude". *Environmental stress. Individual human adaptations.* Proc. Symposium University of California, Santa Barbara, Aug. 31-Sept. 3 1977. Folinsbee, L. J.; Wagner, J. A.; Borgia, J. F.; Drinkwater, B. L.; Gliner, J. A.; Bedi, J. F. (eds.). (New York, San Francisco, London, Academic Press, 1978), 393 p. Illus. Ref.

"Oxygen transport during early altitude acclimatization: a perspective study". Hannon, J. P.; Vogel, J. A. *European Journal of Applied Physiology and Occupational Physiology* (Munich), 10 May 1977, 36/4 (285-297). Illus. 25 ref.

"Pulmonary gas exchange, diffusing capacity in natives and newcomers at high altitude". Vincent, J.; Hellot, M. F.; Vargas, E.; Gautier, H.; Pasquis, P.; Lefrançois, R. *Respiration Physiology* (Amsterdam), Aug. 1978, 34 (219-231). 39 ref.

Mining:

"Mining work in high altitude". Cosio, G. *Archives of Environmental Health* (Chicago), Oct. 1969, 19/4 (540-547). Illus. 21 ref.

Health effects:

"Some aspects of high-altitude respiratory disorders" (Aspectos y problemas respiratorios en la altura). Gumiel, A. *Medicina y seguridad del trabajo* (Madrid), 1969, 17/67 (67-72). (In Spanish)

High altitude diseases. Mechanisms and management. Monge, M. C.; Monge, C. C.; (Springfield, Baltimore, C. C. Thomas, 1966).

Aluminium, alloys and compounds

Aluminium (Al)

a.w. 26.98
sp.gr. 2.7

m.p. 660.4 °C
b.p. 2 467 °C

a silvery, ductile, non-magnetic metal.

Aluminium metal and oxide
TLV ACGIH 10 mg/m³
STEL ACGIH 20 mg/m³
Aluminium pyropowders
TLV ACGIH 5 mg/m³
Aluminium welding fumes
TLV ACGIH 5 mg/m³
Aluminium soluble salts
TLV ACGIH 2 mg/m³
Aluminium and alloys
MAC USSR 2 mg/m³

Occurrence. Aluminium is the most abundant metal in the earth's crust, where it is found in combination with oxygen, fluorine, silica, etc., but never in the metallic state. Bauxite is the principal source of aluminium. It consists of a mixture of minerals formed by the weathering of aluminium-bearing rocks. Bauxites are the richest form of these weathered ores containing up to 55% alumina. Some lateritic ores (containing higher percentages of iron) contain up to 35% Al_2O_3. The commercial deposits of bauxite are mainly gibbsite ($Al_2O_33H_2O$) and boehmite ($Al_2O_3H_2O$) and are found in Australia, Guyana, France, Brazil, Ghana, Guinea, Hungary, Jamaica and Surinam. World production of bauxite in 1979 was more than 85 million t. Gibbsite is more readily soluble in sodium hydroxide solutions than boehmite and is therefore preferred for alumina production.

Extraction. Bauxite is extracted by open-cast mining. The richer ores are used as mined. The lower grade ores may be beneficiated by crushing and washing to remove clay and silica waste.

Production. The production of the metal comprises two basic steps:

(a) Refining: Production of alumina from bauxite by the Bayer process in which bauxite is digested at high temperature and pressure in a strong solution of caustic soda. The resulting hydrate is crystallised and calcined to the oxide in a kiln (figure 1);

(b) Reduction: Reduction of alumina to virgin aluminium metal by the Hall-Heroult electrolytic process using carbon electrodes and cryolite flux.

Experimental development suggests that aluminium may be in the future reduced to the metal by either direct reduction from the ore or by chemical treatment of alumina.

There are presently two major types of Hall-Heroult electrolytic cells in use. The so-called "pre-bake" process (figure 2) utilises electrodes manufactured as noted below. In such smelters exposure to polycyclic hydrocarbons normally occurs in the electrode manufacturing facilities, especially about mixing mills and presses. The smelter utilising the Soderberg-type cell (figure 3) does not require facilities for the manufacture of baked carbon anodes. Rather, the mixture of coke and pitch binder is put into hoppers whose lower end is inserted into the molten cryolite-alumina bath mixture. As the pitch-coke mixture is heated by this mixture, it bakes into a hard graphitic mass *in situ*. Metal rods are inserted into the anodic mass as conductors of the direct flux. These rods must be replaced periodically; in extracting these, considerable amounts of coal tar pitch volatiles are evolved into the cell room environment. To this exposure is added those pitch volatiles generated as the baking of the pitch-coke mass proceeds.

Carbon electrode manufacture. The electrodes required for the electrolytic reduction to pure metal are normally made by a facility associated with the aluminium smelting plant. The anodes and cathodes are made from a mixture of ground petroleum or coal-derived coke and pitch. Coke first is ground in ball mills, then conveyed and mixed mechanically with the pitch and is cast into blocks in a pressure mould. Such anode or cathode blocks are then heated in a gas-fired furnace for several days until they form hard graphitic masses with essentially all volatiles having been driven off. Finally they are attached to anode rods or saw-grooved to receive the cathode bars.

It should be noted that the pitch used to form such electrodes represents a distillate derived from coal and/

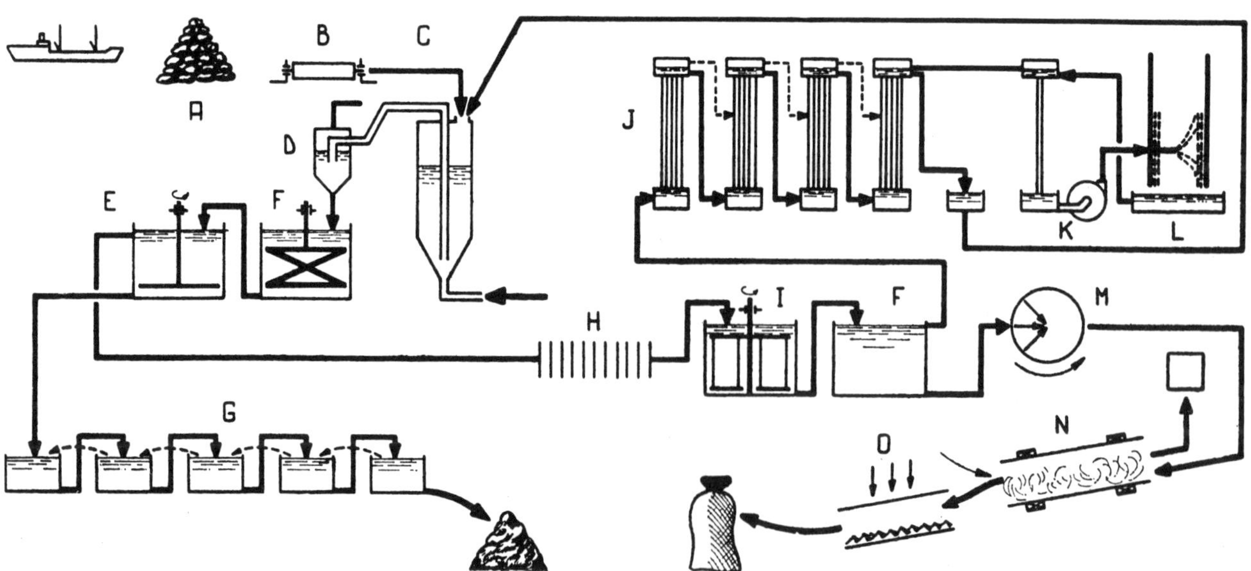

Figure 1. Alumina processing. A. Storage (bauxite). B. Crusher. C. Autoclave. D. Separator. E. Digester. F. Decanter. G. Washers. H. Filter press. I. Precipitator. J. Vaporisers. K. Pump. L. Refrigerant. M. Rotary filter. N. Rotary calcining kiln. O. Cooler.

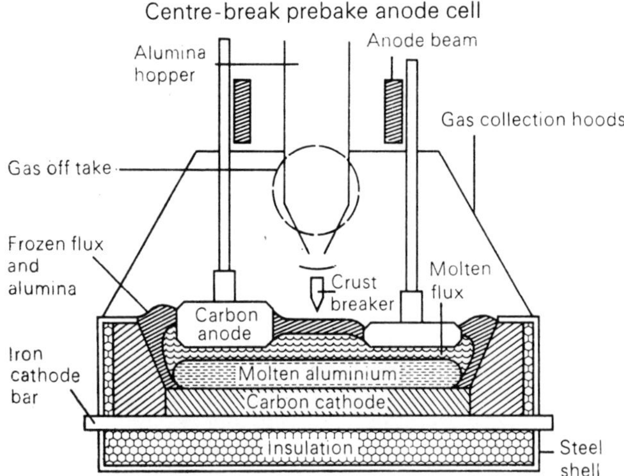

Figure 2. Diagram of prebake anode reduction cell, centre-break type.

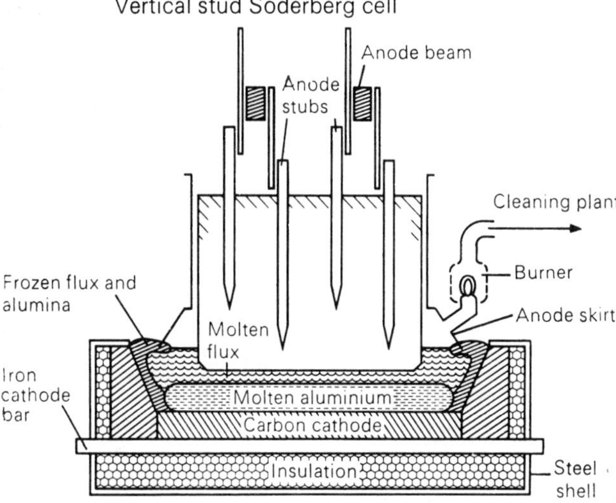

Figure 3. Diagram of Soderberg reduction cell, vertical stud type.

or petroleum tar. In the conversion of this tar to pitch by heating, the final pitch product has had boiled off essentially all of its low-boiling point inorganics, e.g. SO_2 as well as aliphatic compounds and one- and two-ring aromatic compounds. Thus, such pitch should not present the same hazards in its use as coal or petroleum tars since these classes of compounds ought not be present. There are some indications that the carcinogen potential of such pitch products may not be as great as the more complex tars associated with the incomplete combustion of coal.

Uses. Aluminium is used widely throughout industry and in larger quantities than any other non-ferrous metal. It is alloyed with a variety of other material including copper, zinc, silicon, magnesium, manganese and nickel and may contain small amounts of chromium, lead, bismuth, titanium, zirconium and vanadium for special purposes. Aluminium and aluminium alloy ingots can be extruded or processed in rolling mills, wire-works, forges or foundries. The finished products are used in shipbuilding for internal fittings and superstructures; the electrical industry for wires and cables; building industry for house and window frames, roofs and cladding; aircraft industry for airframes and aircraft skin, etc.; automobile industry for bodywork and pistons; light engineering for domestic appliances and office equipment; and in the jewelry industry. A major application of sheet is in beverage or food containers, while aluminium foil is used for packaging; a fine particulate form of aluminium is employed in paints and in the pyrotechnics industry. Articles manufactured from aluminium are frequently given a protective and decorative surface finish by anodisation.

It was thought that inhalation of powdered metallic aluminium tended to protect miners from silicosis; claims have been made that the inhalation of powdered aluminium by silicotic patients retarded further development of silicosis or even led to regression. However, comprehensive studies have found no foundation for these therapeutic claims. Besides, untoward consequences of the breathing of rather massive doses of the large flake aluminium employed have not been reported by the advocates of this treatment.

HAZARDS

These are basically those of smelting and refining in general; however, the individual processes present certain specific hazards.

Mining. Although sporadic references to "bauxite lung" occur in the literature, there is little evidence that such an entity has been convincingly demonstrated. However, the possibility of the presence of free crystalline silica in bauxite ores should be considered.

Bayer process. The use of caustic soda in this process may result in chemical burns to the skin and eyes. Descaling of tanks by pneumatic hammers is responsible for severe noise exposure among such operatives.

Electrolytic reduction. This exposes workers to the potential for skin burns and eye accidents due to molten metal and cryolite and hydrofluoride acid fumes. The electrolytic cells may emit large quantities of fluoride dust and alumina. The hazards to workers, the general population and the environment resulting from the emission of fluoride-containing gases, smokes and dusts due to the use of cryolite flux have been widely reported. In children living in the vicinity of poorly controlled aluminium smelters, variable degrees of mottling of permanent teeth have been reported if exposure occurred during the developmental phase of permanent teeth growth. In smelters prior to 1950, or where inadequate control of fluoride effluents was the case, variable degrees of bony fluorosis have been seen. The first stage of this condition consists simply of an increase in bone density, particularly marked in the vertebral bodies and pelvis. As fluoride is further absorbed into bone, calcification of the ligaments of the pelvis may be seen. Finally, in the event of extreme and protracted exposure to fluoride, calcification of the paraspinal and other ligamentous structures as well as about joints are noted. While this last stage has been seen in its severe form in cryolite processing plants, such advanced stages are rarely if ever seen in aluminium smelter workers. Furthermore, the less severe X-ray changes in bony and ligamentous structures are not associated with alterations of the architectural or metabolic function of bone. By proper work practices and adequate ventilatory control, workers in such reduction operations can be readily prevented from developing any of the foregoing X-ray changes, despite 25-40 years of such work.

Because of the occasional need to expend in excess of 300 kcal/h in the course of changing anodes or performing other strenuous work in the presence of molten cryolite and aluminium, heat disorders may be seen during periods of hot weather. Those workers who are poorly heat acclimatised, whose salt intake is inadequate, or who have intercurrent or recent illness are particularly prone in such arduous tasks to develop heat exhaustion and/or heat cramps; heat stroke occurs but rarely under unusual conditions among workers with predisposing aberrations, e.g. alcoholism.

Exposure to the polycyclic aromatics associated with breathing of pitch fume and particulates appears to place Soderberg-type reduction cell operators in particular at excess risk of developing lung cancer. Workers in carbon electrode plants where mixtures of heated coke and tar are heated might also be supposed to be at such risk. However, after electrodes have been baked for several days at about 1 200 °C, polycyclic aromatic compounds are practically totally combusted or volatilised and no longer associated with such anodes or cathodes. Hence, the reduction cells utilising prebaked electrodes should not present an undue risk of development of lung neoplasia.

In the vicinity of the electrolytic furnaces, the use of pneumatic crust breakers in the furnace rooms produce noise levels of the order of 100 dB. The electric furnaces are run in series from a low-voltage high-amperage current supply and, consequently, cases of electric shock are not usually severe. However, in the power house and at the point where the high-voltage supply joins the series-connection network, severe electrical shock accidents may occur, especially as the supply is direct current. The hazards of the low-temperature process are limited mainly to burns due to molten metal and flux.

Electrode manufacture. Workers in contact with pitch fume may develop erythema, and exposure to sunlight induces photosensitisation with increased irritation. Cases of epitheliomata have occurred among carbon electrode workers where inadequate personal hygiene was practised, but after excision and change of job no further involvement has been noted. During electrode manufacture, considerable quantities of carbon and pitch dust can be generated. There have been occasional reports that carbon electrode makers may develop simple pneumoconiosis with focal emphysema, complicated by the development of massive fibrotic lesions where exposure is severe and poorly controlled. Both the simple and complicated pneumoconioses are indistinguishable from the corresponding condition of coalworkers' pneumoconiosis. The grinding of coke in ball mills produces noise levels of up to 100 dB. The risk of lung neoplasm development has been previously discussed.

Powdered aluminium. Powdered aluminium, used in pyrotechnic devices and paints, will ignite readily; hence stringent fire precautions should be taken in its use. Pyrotechnic aluminium powders function more effectively in that use if they consist of unusually fine material; hence, such powders are doubly ground and considerably finer than aluminium flakes used as a paint pigment. It is only among workers breathing this fine pyrotechnic aluminium flake powder that cases of pulmonary fibrosis have been reported, for which the term aluminosis has been adopted.

SAFETY AND HEALTH MEASURES

All workers involved in the Bayer process should be well informed of the hazards associated with handling caustic soda. In all sites at risk, eyewash bottles and basins with running water should be provided, with notices explaining their use. Personal protective equipment, e.g. goggles, gloves, aprons and boots, should be supplied. Showers and double locker accommodations (one locker for work clothing, the other for personal clothing) should be provided and all employees encouraged to wash thoroughly at the end of the shift. All furnacemen and carbon electrode workers should be supplied with visors, respirators, gauntlets, aprons, armlets and spats to protect them against burns, dust and fumes. Workers employed on the Gadeau low-temperature process should be supplied with special gloves and suits to protect them from hydrochloric acid fumes given off when the cells start up; wool has proved to have a good resistance to these fumes. Respirators with charcoal filters or alumina-impregnated masks give adequate protection against pitch and fluorine fumes; efficient dust masks are necessary for protection against carbon dust. In carbon electrode manufacturing shops, exhaust ventilation equipment with bag filters should be installed and regular checks on atmospheric dust concentrations should be made with a suitable sampling device. Periodic X-ray examinations should be carried out on workers exposed to dust, and these should be followed up by clinical examinations when necessary.

In order to reduce the risk of handling pitch, transport of this material should be mechanised as far as possible (e.g. heated road tankers can be used to transport liquid pitch to the works where it is pumped automatically into heated pitch tanks). Regular skin examinations to detect erythema, epitheliomata or dermatitis are also prudent and extra protection can be provided by an alginate-base barrier cream.

Workers doing hot work should be instructed in hot weather to increase fluid intake and heavily salt their food. They should also be trained to recognise incipient heat-induced disorders and the proper measures to prevent these problems. Workers exposed to high noise levels should be supplied with hearing protection equipment such as earplugs which allow the passage of low-frequency noise (to allow perception of orders) but reduce the transmission of high-frequency sound. Moreover, workers should undergo regular audiometric examination to detect hearing loss. Finally, personnel should also be trained to give cardiopulmonary resuscitation to victims of electric shock accidents.

Aluminium alloys

For the production of aluminium alloys, refined aluminium is melted in an oil-fired furnace. A regulated amount of hardener containing aluminium blocks with a percentage of manganese, silicon, zinc, magnesium, etc., is added; the melt is then mixed and half is passed into a holding furnace for degassing either by passing chlorine through the metal or adding solid hexachlorethane (CCl_3CCl_3); the resultant gas emission (hydrochloric acid, hydrogen and chlorine) should be exhausted and captured before release into the atmosphere. A further method is to blow nitrogen through the melt and exhaust the resulting fumes. Dross is skimmed off the surface of the melt and placed in containers which should be moved into the open air as soon as possible. A flux containing fluoride is added which increases melt temperature and allows the pure aluminium content to be drained off. Dense aluminium oxide and fluoride fumes are given off and production workers should wear respirators and goggles. In the casting shops, sulphur dioxide concentrations of 1-2 ppm may occur.

Aluminium oxide (Al_2O_3)

m.w. 101.9
sp.gr. 3.96
m.p. 2 045 °C
b.p. 2 980 °C
TLV ACGIH 30 million particles/ft^3 or 10 mg/m^3, whichever is smaller, of total dust < 1% quartz, or 5 mg/m^3 of respirable dust
MAC USSR 2 mg/m^3

Various forms of aluminium oxide are used as abrasives, refractories and catalysts. Reports have been made of the appearance of progressive, non-nodular interstitial fibrosis in the aluminium abrasives industry in which aluminium oxide and silicon are processed. This condition, known as Shaver's disease, is rapidly progressive and often fatal; however, since the exposure in the recorded cases was to aluminium oxide, silicon dioxide and iron, each as a fume, and since the great mass of evidence indicates the comparative harmlessness of aluminium oxide, the role of aluminium in the aetiology of Shaver's disease remains unclear. Animal studies indicate that even under extreme exposure conditions, aluminium oxide does not cause chronic lung pathology or pneumoconiosis. Only an especially fine aluminium oxide (0.02 to 0.04 μm), which is rarely used commercially, caused lung changes in such animals. Likewise, studies on the use of alumina in the pottery industry have produced no evidence that the inhalation of alumina dust produces chemical or radiographic signs of pulmonary dysfunction.

Aluminium chloride ($AlCl_3$)

m.w.	133.3
sp.gr.	2.44
m.p.	190 °C (2.5 atm)
b.p.	182.7 °C (sublimes 177.8 °C)

Aluminium chloride is used in petroleum cracking and in the rubber industry. It fumes in air to form HCl and combines explosively with water; consequently containers should be kept tightly closed and protected from moisture.

Alkyl aluminium compounds

TLV ACGIH 2 mg/m³

These are growing in importance as catalysts for the production of low-pressure polyethylene. They present a toxic, burn and fire hazard. They are extremely reactive with air, moisture, and compounds containing active hydrogen and must be kept under a blanket of inert gas.

DINMAN, B. D.

CIS 76-783 "Some aspects of industrial fluorosis in Switzerland–I. The industrial hygienist's viewpoint; II. Radiology and bone fluoride (Preliminary study)" (Quelques aspects de la fluorose industrielle en Suisse–I. Le point de vue du médecin du travail; II. Radiologie et fluor osseux (Etude préliminaire)). Maillard, J. M.; May, P.; Boillat, M. A.; Dettwiler, W.; Rouget, A.; Curati, W.; Demeurisse, C. *Archives des maladies professionnelles, de médecine du travail et de sécurité sociale* (Paris), July-Aug. 1975, 36/7-8 (409-420). Illus. 14 ref. (In French)

Health protection in primary aluminium production. Proceedings of a Seminar, Copenhagen, 28-30 June 1977. Hughes, J. P. (ed.). (London, IPAI, 1978), 158 p. Illus. Ref.

"Lung cancer mortality in aluminium reduction plant workers". Gibbs, G. W.; Horowitz, I. *Journal of Occupational Medicine* (Chicago), May 1979, 21/5 (347-353). 7 ref.

CIS 78-320 *Prevention of dust explosions during grinding and polishing of aluminium and aluminium alloy workpieces* (Schutzmassnahmen beim Schleifen und Polieren von Aluminium und seinen Legierungen zur Vermeidung von Staubexplosionen). Beck, H. A. J. STF report 3-77 (Staubforschungsinstitut des Hauptverbandes der gewerblichen Berufsgenossenschaften, Postfach 5040, Bonn), 14 p. 10 ref. (In German)

"Control technology for health in alumina processing". Haag, W. A.; Sheey, J. W. *Industry and Environment* (Paris), July-Aug.-Sep. 1981, 4/3 (2-4). Illus. 7 ref.

Alveolitis, allergic extrinsic

Extrinsic allergic alveolitis (EAA) is a diffuse, interstitial lung disease characterised by an allergic reaction of lung tissue several hours after the inhalation of organic dusts.

The EAA can be induced by the inhalation of a large number of organic dusts containing particles with a diameter of 2 to 5 μm. These dust particles are able to penetrate deep into the respiratory airways and lung tissue. They often have a strict causal relationship to a specific disease form such as farmer's lung, hen breeder's lung or cheese washer's lung. The total number of known antigens able to induce EAA in adequate concentrations and time of exposure is very large. The spectrum includes proteins as well as enzymes from various birds (pigeons, hens, budgerigars, canaries and wild birds), mammals (rats), fish, insects and bacteria; it also comprises antigens from fungi and thermophibic actinomycetes, possibly also plant-derived antigens and haptens (see AIRBORNE MICRO-ORGANISMS IN THE WORKPLACE).

EAA is one of those diseases in which extreme diagnostic vigilance is the basis for further relatively simple diagnostic measures. The final diagnosis is based on anamnesis, clinical symptoms, immunological examinations, chest X-ray, lung function test and histological findings.

Anamnesis

Once suspicion of EAA arises, then a precise case history becomes extremely important. An exact record of working conditions and questions concerning employment, as well as activities after work which may involve inhaled dusts, most often lead to identification of etiologically most important antigens.

Of great significance is the time dependence of the symptoms, which generally occur approximately 4 to 10 h after exposure. Simple immediate reactions, as with atopic extrinsic asthma, are of little value. A case history questionnaire is very useful and can be recommended.

Clinical symptoms

The clinical features are various and depend on the massiveness of the exposure and the stage of the disease. In the case of budgerigar breeder's disease the symptoms are frequently only minimal, because budgerigar fanciers are usually exposed continuously to birds kept indoors. The amount of inhaled antigen is therefore low, whereas pigeon breeders are normally exposed intermittently to high doses of antigen when cleaning out their pigeon lofts. As a result, pigeon fanciers often present acute symptoms after exposure, whereas budgerigar fanciers often show no symptoms until the disease is far advanced.

The prominent symptoms in most patients with acute disease are breathlessness and dry cough, sometimes accompanied by influenza-like symptoms such as general malaise, fever, aches and pains in the muscles, occurring after an interval of about 4 to 10 h following more or less heavy exposures. A peripheral leucocytosis is also commonly observed. Fine crepitations are heard on auscultation in this stage. Symptoms tend to persist for 12 to 24 h.

At first, they occur intermittently and only some hours after exposure, but after repeated exposures breathlessness becomes more continuous. Usually there is no wheezing, as there is in the case of asthma. Not uncommonly, the symptoms are more insidious so that no clear time relationship with exposure is observed. In this chronic stage auscultatory signs may be very slight, neither râles nor clubbing being prominent–as in cryptogenic fibrosing alveolitis–but sometimes a rapid and marked loss of weight is observed. Patients may ignore their symptoms for months or years before reporting to a doctor and even then the offending antigen is easily overlooked.

Immunological examinations

The diagnosis of EAA also requires immunological examinations to determine the sensitisation of the individual. This examination includes detection of antibodies, skin tests, inhalation provocation tests or cell mediated reactions *in vitro*.

Precipitating antibodies can be found in the serum of 90 to 100% of all cases by means of simple Ouchterlony double diffusion in agar or by the use of the much more sensitive and rapid counter-immunoelectrophoresis. In non-exposed groups antibodies are not detectable, but they occur in asymptomatic exposed individuals too. Therefore, on the one hand the absence of detectable antibodies cannot exclude EAA and on the other their presence in the serum does not necessarily indicate the disease.

The frequency of antibodies in exposed persons seems to be dependent on the duration of exposure, but there is no sure relation between antibody titre and the occurrence of EAA. Antibodies remain detectable for one year after last exposure.

With EAA, intracutaneous or prick tests with avian antigens (sterile, inactivated pigeon, chicken or budgerigar serum, diluted 1 : 10) or some mouldy extracts yield immediate Arthus as well as delayed reactions; but sometimes these tests are also positive in asymptomatic exposed and sensitised persons. In contrast, non-exposed persons show non-specific irritative immediate reactions but no Arthus or delayed reactions with these antigens.

Of special diagnostic relevance is the inhalation provocation test by which Arthus reactions 6 to 10 h after antigen challenge cause a bronchial response, including obstruction and restriction or corresponding reduction in diffusion capacity. In addition respiratory and general symptoms occur during the test (figure 1). As a rule the provocation test must be performed under hospital conditions. Its performance is especially indicated in all cases with a typical anamnesis and proven sensitisation but without X-ray changes or actual lung function disturbances. The test is also recommended in cases where an expert opinion is required.

The detection of cell-mediated immunity (CMI) by use of lymphocyte blast transformation or leucocyte migration inhibition tests with specific antigens indicates sensitisation, even when antibodies are not demonstrable, but CMI alone does not seem to be a reliable diagnostic tool, since elevated specific stimulation has also been observed in some asymptomatic exposed individuals.

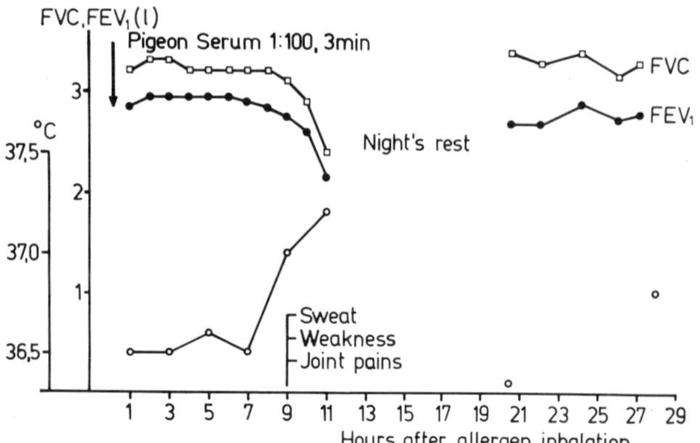

Figure 1. Results of an inhalation test carried out on a sensitised person.

Chest X-ray

The radiographic changes in EAA are not specific and cover a wide range of appearances.

In the acute stages the radiograph shows fine nodular or reticulonodular shadows in most cases, either widespread or predominantly in the mid zones. These clear spontaneously on withdrawal from exposure (see plates 2 and 3).

In the later stages, fibrosis, especially of the upper lobes, is characteristic. There are sometimes honeycomb shadows (small ring shadows) and lung shrinkage appears, also affecting the upper lobes more severely. In fibrotic stages no regression was observed.

Lung function tests

The lung function test is an essential component in the diagnosis of EAA. All phases of respiration can be involved, including ventilation, diffusion, perfusion and distribution.

In acute stages obstructive phenomena with lowered FEV_1 and increased bronchial resistance are observed, as well as inhomogeneity of ventilation and perfusion on the basis of disturbed lung mechanisms. In some cases the typical changes take the form of restricted ventilation, with a reduction of static lung volume. Disturbances of diffusion manifest themselves as membrane diffusion disturbances following alveolar-capillary block already in this stage. The inequality of ventilation and perfusion is reflected by a reduced pO_2 with a normal or even low pCO_2. At first the changes revert to normal after removal from exposure but, as irreversible fibrosis develops, so the changes with lowered diffusion cross-section becomes untreatable. Pulmonary hypertension and *cor pulmonale* are late signs.

Histological examinations

Generally, biopsies performed on patients during or after an acute episode show two types of inflammatory process—mononuclear inflammatory infiltrate of the alveolar walls, and granulomas found in the walls of the alveoli and in the bronchioles.

The mononuclear infiltrate is largely composed of lymphocytes and plasma cells, but lymphoid collections containing antibody-secreting germinal follicles are prominent. The non-caseating sarcoid-like granulomas have certain special characteristics which distinguish them from sarcoid lesions, for instance multinucleate giant cells with characteristic clefts. Surrounding the central epitheloid cells is a mononuclear infiltrate largely made up of lymphocytes and plasma cells, together with fibroblasts and associated with variable proliferation of collagen. The granulomas can disappear some months after an acute episode, this contrasts with the persisting granulomata of sarcoidosis. Also in contrast, granulomata are seen only occasionally in hilar lymph nodes.

In the chronic stages there is a fine fibrosis distributed irregularly through the alveolar walls and larger focal areas of fibrosis with destruction of the lung architecture. Irregular destruction of alveoli may be manifest as areas of emphysema. The fibrotic changes are usually more prominent in the upper than in the lower zones, and this contrasts also with the later stages of cryptogenetic fibrosing alveolitis where the lower zones tend to bear the brunt of damage. Chronic inflammatory changes seen around the pulmonary arteries presumably relate to the later appearance of pulmonary hypertension and *cor pulmonale*.

Diagnosis

As we have seen, no single parameter appears to be a reliable diagnostic tool: the diagnosis of EAA must be composed like a mosaic of different parameters.

Therefore, it is recommended that diagnosis of EAA should be based on the following simultaneously appearing criteria:

1. Proof of special exposure by detection of an antigen or its source.
2. Respiratory and/or general symptoms about 4 to 10 h after exposure.
3. Proof of sensitisation by antibodies or sensitised lymphocytes *in vivo* or *in vitro*.
4. X-ray changes for which other causes cannot be found.
5. Disturbances of lung function or positive inhalation provocation test.

The diagnosis of EAA seems to be sure if criteria 1 to 4, or 1 to 3 and 5 are fulfilled; it can be supported by histological findings on biopsy material. In particular, notification as an occupational disease requires careful consideration of these criteria.

There are exposed individuals with antibodies or CMI showing functional disorders without respiratory or general symptoms. Such asymptomatic sensitised persons are at risk of developing disease and therefore require medical monitoring.

BERGMANN, K. C.

Hypersensitivity diseases of the lungs due to fungi and other organic dusts. Monographs on Allergy 4. Pepys, J. (Basel, Karger, 1969), 148 p. Illus.

"Recommendation for a definition of extrinsic allergic alveolitis" (Empfehlung zur Definition des Krankheitbildes Allergische Alveolitis). Bergmann, K. C.; Branski, H.; Götz, M.; Kurzawa, R.; Luther, P.; Molina, C. *Zeitschrift für Erkrankungen der Atmungsorgane* (Leipzig), 1978, 151/2 (167-171). Illus. 6 ref. (In German)

"Pigeon breeder's lung lacking detectable antibodies". Sennekamp, J.; Niese, D.; Strochmann, J.; Rittner, C. *Clinical Allergy* (London), May 1978, 8/3 (305-310). Illus. 14 ref.

"The radiological appearances of allergic alveolitis due to bird sensitivity (bird fancier's lung)". Hargreave, F.; Hinson, K. F.; Reid, L.; Simon, G.; McCarthy, D. S. *Clinical Radiology* (Edinburgh), 1972, 23/1 (1-10). Illus. 39 ref.

"Long-term occupational inhalation of organic dust–effect on pulmonary function". Petro, W.; Bergmann, K.-Ch.; Heinze, R.; Müller, E.; Wuthe, H.; Vogel, J. *International Archives of Occupational and Environmental Health* (West Berlin), 1978, 42/2 (119-127). Illus. 32 ref.

Amides

The amides are a class of organic compounds which can be regarded as having been derived from either acids or amines. For example the simple aliphatic amide, acetamide (CH_3CONH_2), is related to acetic acid in the sense that the $-OH$ group of acetic acid is replaced by an $-NH_2$ group. Conversely, acetamide can be regarded as being derived from ammonia, by replacement of one ammonia hydrogen by an acyl group. Amides can be derived not only from aliphatic or aromatic carboxylic acids but also from other types of acids, for example sulphur- and phosphorus-containing acids.

$$R\overset{\overset{O}{\|}}{C}-$$

The term substituted amides may be used to describe those amides having one or both hydrogens on the nitrogen replaced by other groups, for example, N,N-dimethylacetamide. This compound could also be regarded as an amine, acetyl dimethyl amine.

$$CH_3\overset{\overset{O}{\|}}{C}-\underset{\underset{CH_3}{|}}{\overset{\overset{CH_3}{|}}{N}}$$

Amides are generally quite neutral in reaction compared with the acid or amine from which they are derived and they are occasionally somewhat resistant to hydrolysis. The simple amides of aliphatic carboxylic acids (except formamide) are solids at room temperature, while the substituted aliphatic carboxylic acid amides may be liquids with relatively high boiling points. The amides of aromatic carboxylic or sulphonic acids are usually solids. A wide variety of methods are available for the synthesis of amides.

Uses. The unsubstituted aliphatic carboxylic acid amides have wide use as intermediates, stabilisers, release agents for plastics, films, surfactants and soldering fluxes. The substituted amides such as dimethylformamide and dimethylacetamide have powerful solvent properties. Some unsaturated aliphatic amides, e.g. acrylamide, are reactive monomers used in polymer synthesis. Aromatic amide compounds form important dye and medicinal intermediates. Some have insect repellent properties.

HAZARDS

The wide variety of possible chemical structures of amides is reflected in the diversity of their biologic effects. Some appear entirely innocuous, for example the longer chain simple fatty acid amides such as stearic or oleic acid amides. On the other hand, acetamide and thioacetamide have been shown to be liver toxins and liver carcinogens in high doses experimentally in the rat. Neurologic effects have been noted in humans and experimental animals with acrylamide. Dimethylformamide and dimethylacetamide have produced liver injury in animals and formamide and monomethylformamide have been shown experimentally to be teratogens.

Although a considerable amount of information is available on the metabolism of various amides, the nature of their toxic effects has not yet been explained on a molecular or cellular basis. Many simple amides are probably hydrolysed by non-specific amidases in the liver and the acid produced, excreted or metabolised by normal mechanisms.

Some aromatic amides, for example N-phenylacetamide (acetanilide) are hydroxylated on the aromatic ring and then conjugated and excreted. The ability of a number of amides to penetrate the intact skin is especially important in considering safety precautions.

Neurotoxic effects

Acrylamide (CH_2:$CHCONH_2$)

PROPENOIC ACID AMIDE; PROPENAMIDE

m.w.	71
sp.gr.	1.12
m.p.	84.5 °C
b.p.	125 °C
v.d.	2.45
v.p.	0.007 mmHg (0.9 Pa) at 25 °C
f.p.	138 °C

very soluble in water; soluble in ethanol, and ethyl ether, methanol, acetone

colourless odourless leaf crystals from benzene

thermally stable it polymerises rapidly in the molten state with evolution of heat.

TWA OSHA	0.3 mg/m³ skin
STEL ACGIH	0.6 mg/m³ skin

Acrylamide is produced from acrylonitrile by treatment with acid. With improved methods of manufacture, the increased availability of the monomer in recent years has led to a markedly increased use of acrylamide polymers. It is estimated that at the present time annual use is greater than 31.5 million kg. The polyacrylamides find extensive use as flocculants in water and sewage treatment, as strengthening agents during paper manufacture in the paper and pulp industry, and for other purposes.

Acrylamide was initially made in Germany in 1893. Practical use of this compound had to wait until the early 1950s before commercial manufacturing processes became available. This development occurred primarily in the United States. By the mid-1950s it was recognised that workers exposed to acrylamide developed characteristic neurologic changes primarily characterised by both postural and motor difficulties. Reported findings included tingling of the fingers, tenderness to touch, coldness of the extremities, excessive sweating of the hands and feet, a characteristic bluish-red discolouration of the skin of the extremities, and a tendency toward peeling of the skin of the fingers and hands. These symptoms were accompanied by weakness of the hands and feet which led to difficulty in walking, climbing stairs, etc. Recovery generally occurs with cessation of exposure. The time for recovery varies from a few weeks to as long as one year.

Neurologic examination of individuals suffering from acrylamide intoxication shows a rather typical peripheral neuropathy with weakness or absence of tendon reflexes, a positive Romberg test, a loss of position sense, a diminution or loss of vibration sense, ataxia, and atrophy of the muscles of the extremities.

Following recognition of the symptom complex associated with acrylamide exposure, animal studies were carried out in an attempt to document these changes. It was found that a variety of animal species including rat, cat and baboon were capable of developing peripheral neuropathy with disturbance of gait, disturbance of balance, and a loss of position sense. Histopathologic examination revealed a degeneration of the axons and myelin sheaths. The nerves with the largest and longest axons were most commonly involved. There did not appear to be involvement of the nerve cell bodies.

Several theories have been advanced as to why these changes occur. One of these has to do with possible interference with the metabolism of the nerve cell body itself. Another theory postulates interference with the intracellular transport system of the nerve cell. The most recent explanation is that there is a local toxic effect on the entire axon which is felt to be more vulnerable to the action of acrylamide than is the cell body. The most recent studies of the changes taking place within the axons and myelin sheaths has resulted in a description of the process as a "drying back" phenomenon. This term is used to describe more accurately the progression of changes observed in the peripheral nerves.

While the described symptoms and signs of the characteristic peripheral neuropathy associated with acrylamide exposure are widely recognised from exposure in industry and from animal studies, it appears that in man with the ingestion of acrylamide, as has occurred from drinking water contaminated with this chemical, the symptoms and signs are of involvement of the central nervous system. In these instances drowsiness, disturbance of balance and mental changes characterised by confusion, memory loss, and hallucinations, were paramount. Peripheral neurological changes did not appear until later.

Skin penetration has been demonstrated in rabbits and this may have been a principal route of absorption in those cases reported from industrial exposures to acrylamide monomer. It is felt that the hazard from inhalation would be primarily from exposure to aerosolised material.

Hepatotoxic effects

N,N-dimethylformamide ($HCON(CH_3)_2$)

N,N-DIMETHYL FORMIC ACID; DMF; DMFA

m.w.	73
sp.gr.	0.94
m.p.	61 °C
b.p.	153 °C
v.d.	2.51
v.p.	3.7 mmHg ($0.48 \cdot 10^3$ Pa) at 25 °C
f.p.	57.7 °C
e.l.	2.2-15.2%
i.t.	445 °C

miscible with water and most solvents

a colourless liquid with a faint fishy odour detectable between 10 and 15 ppm.

TWA OSHA	10 ppm 30 mg/m³ skin
STEL ACGIH	20 ppm 60 mg/m³ skin
IDLH	3 500 ppm
MAC USSR	10 mg/m³

It is produced from dimethylamine and hydrogen cyanide. Dimethylformamide, while not currently as extensively employed as acrylamide is, nevertheless, becoming a commonly used industrial solvent. It is used primarily as a solvent in organic synthesis and has also found use in the preparation of synthetic fibres.

The good solvent action of N,N-dimethylformamide results in drying and defatting of the skin on contact, with resultant itching and scaling. Some complaints of eye irritation have resulted from vapour exposure in industry. Subjective complaints of exposed workmen have included nausea, vomiting, and anorexia. Intolerance to alcoholic beverages after exposure to dimethylformamide has been reported.

Animal studies with N,N-dimethylformamide have shown experimental evidence of liver and kidney damage in rats, rabbits, and cats. These effects have been seen from both intraperitoneal administration and inhalation studies. Dogs exposed to high concentrations of the vapour exhibited polycythemia, decrease of the pulse rate, and a decline in systolic pressure, and showed histologic evidence of degenerative changes in the myocardium.

In man this compound is capable of being readily absorbed through the skin, and repeated exposures can lead to cumulative effects. In addition, like dimethylacetamide, it may facilitate the percutaneous absorption of substances dissolved in it.

It should be mentioned that dimethylformamide will readily penetrate rubber (both natural and neoprene) gloves so that prolonged use of such gloves or prolonged emersion with gloves is inadvisable. Polyethylene provides better protection; however, any gloves used with this solvent should be washed after each contact and discarded frequently.

See also POLYACRYLONITRILE FIBRES.

N,N-dimethylacetamide ($CH_3CON(CH_3)_2$)

ACETIC ACID DIMETHYLAMIDE; DMAC; ACETYL DIMETHYL AMINE

m.w.	87.1
sp.gr.	0.94
m.p.	−20 °C
b.p.	166 °C
v.d.	3.01
v.p.	1.3 mmHg ($0.17 \cdot 10^3$ Pa) at 25 °C
f.p.	66 °C
e.l.	1.8-11.5%

a liquid miscible with water and most organic solvents.

TWA OSHA 10 ppm 35 mg/m³ skin
STEL ACGIH 15 ppm 50 mg/m³ skin

It is produced from acetic anhydride and dimethylformamide and is used as a solvent for many organic reactions.

Dimethylacetamide has been studied in animals and has been shown to exhibit its principal toxic action in the liver on repeated or continued excessive exposure. Skin contact may cause the absorption of dangerous quantities of the compound.

Acetamide (CH_3CONH_2)

ACETIC ACID AMIDE; ETHANAMIDE

m.w. 59
sp.gr. 1
m.p. 82.3 °C
b.p. 221.2
v.p. 1 mmHg (0.13·10³ Pa) at 65 °C

soluble in water; very soluble in ethanol

deliquescent trigonal monoclinic crystals, often with a mousy odour.

It is produced by fractional distillation of ammonium acetate and is used as a solvent for many organic compounds, as a plasticiser and for denaturing ethanol.

Thioacetamide (CH_3CSNH_2)

m.w. 75.2
m.p. 115.5 °C

plates or crystals with a faint odour of thiols.

It is prepared by heating ammonium acetate and aluminium sulphide, or from acetamide, and is used in the laboratory as an analytical reagent.

Both compounds have been shown to produce hepatomas in rats on prolonged dietary feeding. Thioacetamide is more potent in this respect, is carcinogenic also to mice and can also induce bile ducts tumours in rats. Neither of these substances has been reported to be carcinogenic in humans. Their ability to penetrate the skin has not been reported.

Teratogenic effects. These have been reported in mice from exposure to formamide or monomethylformamide administered either by mouth or percutaneously. Dimethylformamide did not cause this effect.

SAFETY AND HEALTH MEASURES

The potential toxic properties of any amide should be carefully considered before use or exposure commences. Owing to the general tendency of amides (especially those of lower molecular weight) to be absorbed percutaneously, skin contact should be prevented. Inhalation of dusts or vapours should be controlled. It is desirable that persons with exposure to amides be under regular medical observation with particular reference to the functioning of the nervous system and liver.

FASSETT, D.
JONES, W. H.

Acrylamide:

CIS 77-440 *Criteria for a recommended standard—Occupational exposure to acrylamide.* DHEW (NIOSH) publication No. 77-112 (National Institute for Occupational Safety and Health, 4676 Columbia Parkway, Cincinnati, Ohio) (Oct. 1976), 127 p. 78 ref.

Assessment of testing needs: Acrylamide. Support document for decision not to require testing for health effects. Toxic Substances Control Act, Section 4. TSCA Chemical Assessment Series. EPA (Washington, DC, US Environmental Protection Agency, Office of Pesticides and Toxic Substances, July 1980), 34 p. 65 ref.

CIS 76-423 "Urinary excretion of metabolite following experimental human exposures to DMF or to DMAC". Maxfield, M. E.; Barnes, J. R.; Azar, A.; Trochimowicz, H. T. *Journal of Occupational Medicine* (Chicago), Aug. 1975, 17/8 (506-511). Illus. 20 ref.

N,N-*dimethylformamide:*

"Relation of exposure to dimethylformamide vapour and the metabolite, methylformamide, in urine of workers". Yonemoto, J.; Suzuki, S. *International Archives of Occupational and Environmental Health* (West Berlin), June 1980, 46/2 (159-165). Illus. 9 ref.

"Dimethylformamide and alcohol intolerance". Lyle, W. H.; Spence, T. W. M.; McKinneley, W. M.; Duckers, K. *British Journal of Industrial Medicine* (London), Feb. 1979, 36/1 (63-66). 12 ref.

Thioacetamide:

"Thioacetamide". *IARC monographs on the evaluation of carcinogenic risk of chemicals to man.* Vol. 7: *Some antithyroid and related substances, nitrofurans and industrial chemicals* (Lyons, International Agency for Research on Cancer, 1974) (77-83). 16 ref.

Other amides:

*2-Chloro-*N,N*'-diallylacetamide*

"Herbicide dermatitis". Spencer, M. C. *Journal of the American Medical Association* (Chicago), 19 Dec. 1966, 198/12 (1307-1308). Illus. 4 ref.

Dimethylcyanamide

CIS 76-1942 "Toxicity of dimethylcyanamide" (O toksičnosti dimetilcianamida). Gurova, A. I.; Alekseeva, N. P.; Gorlova, O. E.; Černyšova, R. A. *Gigiena truda i professional'nye zabolevanija* (Moscow), Nov. 1975, 11 (23-27). 14 ref. (In Russian)

Fluoroacetamide

CIS 2130-1965 *Use of fluoroacetamide and sodium fluoroacetate as rodenticides. Precautionary measures.* Ministry of Agriculture, Fisheries and Food (London, HM Stationery Office, 1965), 4 p.

Amines, aliphatic

Alkyl amines

These compounds are formed when one or more hydrogen atoms in ammonia (NH_3) are replaced by one, two or three alkyl or alkanol radicals. The lower aliphatic amines are gases like ammonia and freely soluble in water but the higher homologues are insoluble in water. All the aliphatic amines are basic in solution and form salts. The salts are odourless, non-volatile solids freely soluble in water.

According to the number of hydrogens substituted, the amines may be primary (NH_2R), secondary (NHR_2) or tertiary (NR_3).

Production. General methods of manufacture include the alkylation of ammonia by the alkyl halide or hydrogenation of the corresponding nitrite.

Uses. The variety of uses for aliphatic amines has greatly increased in recent years—in the chemical industry, in such fields as pharmaceutical and dyestuff manufacture, rubber products, curing agents or catalysts in plastics, ion exchange resins, synthetic cutting fluids, corrosion inhibitors and flotation agents. They are used also in the cold box process in foundries.

GENERAL HAZARDS

Since the amines are bases and may form strongly alkaline solutions, they can be damaging if splashed in the eye or if allowed to contaminate the skin. Otherwise they have no specific toxic properties and the lower aliphatic amines are normal constituents of body tissues, so that they occur in a large number of foods, particularly fish to which they impart a characteristic odour. One area of concern at present is the possibility that some aliphatic amines may react with nitrate or nitrite *in vivo* to form nitroso compounds, many of which are known to be potent carcinogens in animals. The more important members are considered individually.

Methylamine (CH_3NH_2)

AMINOMETHANE

m.w. 31.1
sp.gr. 0.70
m.p. −93.5 °C
b.p. −6.3 °C
v.d. 1.1
v.p. 1 520 mmHg (202.2·10³ Pa) at 25 °C
f.p. 0 °C
e.l. 4.9-20.7%
i.t. 430 °C

very soluble in water

a colourless gas with a strong ammoniacal odour.

TWA OSHA 10 ppm 12 mg/m³
MAC USSR 1 mg/m³

Made from methyl alcohol and ammonium chloride, this is a flammable gas at normal temperature and pressure. It may be available as a liquid under pressure or as a strong solution in water, and is used in tanning and organic synthesis. It is a stronger base than ammonia and the vapour is irritating to the eyes and respiratory tract. Otherwise it is not poisonous.

Dimethylamine ($(CH_3)_2NH$)

m.w. 45
sp.gr. 0.68 (liquid)
m.p. −96 °C
b.p. 7.4 °C
v.d. 1.6
v.p. 1 520 mmHg (202.2·10³ Pa) at 25 °C
f.p. 12.2 °C
e.l. 2.8-14.4%

very soluble in water

a colourless liquid or a gas with a pungent odour.

TWA OSHA 10 ppm 18 mg/m³
IDLH 2 000 ppm
MAC USSR 1 mg/m³

Prepared by the interaction of methyl alcohol and ammonia over a catalyst at high temperature or by the addition of soda to nitrosodimethyl aniline. It is a gas at ordinary temperatures and pressures but is available as a liquid under pressure or as a strong aqueous solution. It is used in the rubber industry as an accelerator and in the manufacture of soaps. The vapour is both flammable and irritant, and solutions strongly alkaline.

Ethylamine ($C_2H_5NH_2$)

AMINOETHANE

m.w. 45
sp.gr. 0.69
m.p. −84 °C
b.p. 16.6 °C
v.d. 1.6
v.p. 760 mmHg (101.1·10³ Pa) at 16.6 °C
f.p. 0 °C
e.l. 3.5-14%

a colourless liquid with a strong ammoniacal odour.

TWA OSHA 10 ppm 18 mg/m³
IDLH 4 000 ppm

Made from chloroethane heated with alcoholic ammonia, it forms strongly alkaline solutions in water and is used in the rubber industry as a stabiliser for rubber latex and as a dye intermediate. Eye irritation and corneal damage may occur in those exposed to the vapour. Other toxic effects have not been recorded and the compound is excreted unchanged by man.

Propylamine ($CH_3CH_2CH_2NH_2$)

1-AMINO PROPANE

m.w. 59.1
sp.gr. 0.72
m.p. −83 °C
b.p. 49 °C
v.d. 2.0
v.p. 400 mmHg (53.2·10³ Pa) at 31.5 °C
f.p. < −37 °C (oc)
e.l. 2.0-10.4%
i.t. 318 °C

soluble in all proportions in water, ethanol and ether

a colourless liquid with a strong ammoniacal odour.

MAC USSR 5 mg/m³

Made by reducing ethyl cyanide with sodium in ethyl alcohol, propylamine is a strongly alkaline liquid smelling like ammonia and is used as an intermediate in the chemical industry. The vapour may injure the eyes and respiratory tract but small quantities are readily metabolised and are not toxic.

Butylamine

n-Butylamine is the most important isomer commercially.

***n*-Butylamine** ($CH_3CH_2CH_2CH_2NH_2$)

1-AMINO BUTANE

m.w. 73.1
sp.gr. 0.76
m.p. −50.5 °C
b.p. 77.8 °C
v.d. 2.5
v.p. 100 mmHg (13.3·10³ Pa) at 18.8 °C
f.p. 7 °C (oc)
e.l. 1.7-9.8%
i.t. 312 °C

a colourless liquid with an ammoniacal odour.

TWA OSHA 5 ppm 15 mg/m³ ceil skin
IDLH 2 000 ppm
MAC USSR 10 mg/m³

A strongly alkaline liquid used in the pharmaceutical, dyestuff and rubber industries and as a pesticide. Although the vapour is said to have severe effects on the central nervous system of animals exposed to it, the outstanding toxic effect on man is irritation of the eyes and respiratory tract. Any *n*-butylamine which is absorbed is readily metabolised.

Allylamine ($CH_2{:}CHCH_2NH_2$)

2-PROPENYLAMINE

m.w. 57.1
sp.gr. 0.76
b.p. 58 °C
v.d. 2.0
f.p. −29 °C
e.l. 2.2-22%
i.t. 374 °C

a colourless liquid with an ammoniacal odour.

The vapour is intensely irritating and this appears to be its main toxic effect. Definite risk of explosion over a wide range of concentrations in air (see also ALLYL COMPOUNDS).

Cyclohexylamine ($C_6H_{11}NH_2$)

HEXAHYDROANILINE; AMINOCYCLOHEXANE

sp.gr. 0.87
m.p. −17.7 °C

b.p. 134.5 °C
v.d. 3.4
f.p. 32.2 °C (oc)
i.t. 293 °C
miscible in all proportions with water, forming strongly alkaline solutions
a liquid with a fishy odour.

TLV ACGIH	10 ppm 40 mg/m^3 skin
MAC USSR	1 mg/m^3

It is prepared by hydrogenation of aniline. It is used as an intermediate in chemical manufacture and its salts are used by themselves as corrosion inhibitors under a variety of different circumstances including the packaging of metal equipment. Its main toxic effect is as an irritant and it may damage and sensitise the skin. This amine is also a principal metabolite of cyclamate.

Ethylenediamine ($H_2NCH_2CH_2NH_2$)
1,2-DIAMINO ETHANE
m.w. 60.1
sp.gr. 0.90
m.p. 8.5 °C
b.p. 116.5 °C
v.d. 2.1
v.p. 10.7 mmHg (1.42·10^3 Pa) at 20 °C
f.p. 43.3 °C (oc) (anhydrous)
i.t. 385 °C
a volatile hygroscopic liquid with an ammoniacal odour, very soluble in water, insoluble in ether, miscible in all proportions with ethanol.

TWA OSHA	10 ppm 25 mg/m^3
IDLH	2 000 ppm
MAC USSR	2 mg/m^3

Prepared from dichloroethane and ammonia, ethylenediamine is a strongly alkaline liquid and is used during the preparation of dyes, rubber accelerators, fungicides, synthetic waxes, resins, insecticides and asphalt wetting agents. It is also used in drugs such as aminophylline. Its main toxic effect is as an eye irritant and it damages the eyes, skin and respiratory tract. Sensitisation may follow vapour exposure.

Ethanolamine ($HOCH_2CH_2NH_2$)
COLAMINE; MONOETHANOLAMINE; HYDROXYETHYLAMINE; BETAAMINOETHYL ALCOHOL
m.w. 61
sp.gr. 1.02
m.p. 10.3 °C
b.p. 170 °C
v.d. 2.1
v.p. < 1 mmHg (0.13·10^3 Pa) at 20 °C
f.p. 85 °C (oc)
e.l. 5.5-17%
a viscous hygroscopic liquid with an ammoniacal odour.

TWA OSHA	3 ppm 6 mg/m^3
STEL ACGIH	6 ppm 15 mg/m^3
IDLH	1 000 ppm

Prepared by the ammonolysis of ethylene oxide; despite its widespread use in industry to remove carbon dioxide and hydrogen from natural gas, in the synthesis of surface active agents, etc., toxic effects on man have not been recorded.

Diethanolamine ($(HOCH_2CH_2)_2NH$)
2,2'-IMINODIETHANOL; DIETHYLOLAMINE; 2,2'-DIHYDROXYDIETHYLAMINE
m.w. 105.1
sp.gr. 1.09
m.p. 28 °C
b.p. 271 °C (decomposes)
v.d. 3.65
v.p. 5 mmHg (0.67·10^3 Pa) at 138 °C
f.p. 148.5 °C
i.t. 629 °C
miscible with water and acetone
a viscous liquid with a faint ammoniacal odour.

Produced along with mono- and tri-ethanolamine by ammonolysis of ethylene oxide, it is a strong base. It is used to scrub gases and in the manufacture of rubber chemicals, surface active agents, herbicides and demulsifiers. It is also employed as an emulsifier and dispersing agent in agricultural chemicals, cosmetics and pharmaceuticals. It is irritating to the skin and mucous membranes.

Triethanolamine ($(HOCH_2CH_2)_3N$)
2,2'2''-NITROLOTRIETHANOL
m.w. 149.2
sp.gr. 1.12
m.p. 20-21.2 °C
b.p. 277 °C (150 mmHg)
v.d. 5.1
v.p. 10 mmHg (1.33·10^3 Pa) at 205 °C
f.p. 179 °C
a pale yellow, viscous, very hygroscopic liquid with a faint ammoniacal odour, soluble in all proportions in water and acetone.

Prepared like ethanolamine, it is a more viscous alkaline liquid. It is used extensively in industry for the manufacture of surface active agents, waxes, polishes, herbicides, cutting oils, etc., but any toxic effects are confined to skin irritation (see also CATALYSTS).

MAGOS, L.
MANSON, M. M.

CIS 79-766 "Hazardous chemical reactions–60. Amines, imines" (Réactions chimiques dangereuses–60. Amines, imines). Leleu, J. Cahiers de notes documentaires–*Sécurité et hygiène du travail* (Paris), 1st quarter 1979, 94, Note No. 1167-94-79 (127-132). (In French)

"Toxicity of aliphatic amines" (Zur Toxizität von aliphatischen Aminen). Bittersohl, von C.; Heberer, H. *Zeitschrift für die gesamte Hygiene und ihre Grenzgebiete* (Berlin), 7 July 1978, 24/7 (529-534). 208 ref. (In German)

CIS 79-1066 "Nitrosamine impurities in amines" (Verunreinigung von Aminen mit N-Nitrosaminen). Spiegelhalder, B.; Eisenbrand, G.; Preussmann, R. *Angewandte Chemie* (Weinheim/Bergstr., Federal Republic of Germany, 1978), 90/5 (379-380). 5 ref. (In German)

"Allergy to ethylenediamine" (Äthylendiamin-Allergie). Pevny, I.; Schäfer, U. *Dermatosen in Beruf und Umwelt* (Aulendorf, Federal Republic of Germany, 1980), 28/2 (35-40). 43 ref. (In German)

"Monitoring personal exposure to ethylenediamine in the occupational environment". Vincent, W. J.; Hahn, K. J.; Ketcham, N. H. *American Industrial Hygiene Association Journal* (Akron, Ohio), June 1979, 40/6 (512-516). Illus. 2 ref.

CIS 79-482 *Development of air-monitoring techniques using solid sorbents.* Wood, G. O.; Nickols, J. W. LA-7295-PR, NIOSH-IA-77-12, Progress Report (Los Alamos Scientific Laboratory, PO Box 1663, Los Alamos, New Mexico) (1978), 66 p. Illus. 28 ref.

Amines, aromatic

The aromatic amines are a class of chemicals derived from aromatic hydrocarbons, e.g. benzene, toluene, naphthalene, anthracene, diphenyl, etc., by the replacement of at least one hydrogen atom by an amino ($-NH_2$)

group. A compound with a free amino group is described as a primary amine. When one of the hydrogen atoms of the $-NH_2$ group is replaced by an alkyl or aryl group, the resultant compound is a secondary amine; when both hydrogen atoms are replaced, a tertiary amine results. The hydrocarbon may have one amino group or two, more rarely three. It is thus possible to produce a considerable range of compounds and, in effect, the aromatic amines constitute a large class of chemicals of great technical and commercial value.

Aniline is the simplest aromatic amine, consisting of one $-NH_2$ group attached to a benzene ring; it is the most widely used in industry. Other common single-ring compounds include dimethylaniline and diethylaniline, the chloroanilines, nitroanilines, toluidines, the chlorotoluidines, the phenylenediamines, and acetanilide. Benzidine, *o*-tolidine, *o*-dianisidine, 3,3'-dichlorobenzidine and 4-aminodiphenyl are the most important conjoined ring compounds from the point of view of occupational health. Of compounds with ring structures, the naphthylamines and aminoanthracenes have attracted much attention because of problems of carcinogenicity.

Production. The manufacturing processes commonly involve two steps; first nitration of the parent hydrocarbon to form the nitro-compounds, and then catalytic reduction to the amine. Sometimes direct amination may be effected by reacting a chloro- or hydroxy-derivative of the parent hydrocarbon with ammonia in suitable conditions.

Uses. Aromatic amines rarely occur naturally; they are synthetic compounds used mainly in the synthesis of other chemicals. Today the most important aromatic amines on the market are aniline and 2,4-toluenediamine which are used as intermediates in the manufacture of isocyanates, basic raw materials for the production of polyurethanes. Aromatic amines, especially aniline, are also used as intermediates in the manufacture of a wide range of dyes and pigments. The largest class of dyestuffs is that of the azo colours, which are made by diazotisation, a process by which a primary aromatic amine reacts with nitrous acid in the presence of excess mineral acid to produce a diazo ($-N{=}N-$) compound; this compound is subsequently coupled with a phenol or an amine. Another important class of dyestuffs, the triphenylmethane colours, is also manufactured from aromatic amines. Another use of aromatic amines is in the manufacture of certain rubber antioxidants and antiozone agents, which are used to improve resistance to deterioration and ageing of rubber. Mercaptobenzothiazole and its derivatives are among the compounds used as accelerators in the vulcanisation of rubber, while among the antioxidants and antiozone agents the principal aniline derivatives used are the para-phenylenediamines, phenyl-β-naphthylamine, aniline-acetone condensates and aniline diphenylamine.

p-Phenylenediamine and *p*-aminophenol are also used as hair dyes and for dyeing fur. Diaminodiphenylmethane is commonly used as a hardener or curing agent in epoxy resin systems. N-acetyl-*p*-aminophenol is used as an analgaesic and antipyretic agent, and N-methyl-*p*-aminophenol has had widespread use as a photographic developer. Magenta (fuchsin, rosaniline) and auramine are used as colours in printing inks. Some of the more dangerous aromatic amines have been and, in some countries, still are used for routine laboratory tests. Examples are benzidine and *o*-tolidine in medical laboratories for detecting occult blood in biological specimens, and 1-naphthylamine in the routine testing of water supplies for nitrite content.

HAZARDS

The manufacture and use in industry of certain aromatic amines may constitute a grave and sometimes unexpected hazard. However, since these hazards have become better known, there has, over recent years, been a tendency to substitute other substances or to take precautions which have reduced the hazard.

Discussion has also taken place concerning the possibility of aromatic amines having health effects either when they exist as impurities in a finished product, or when they may be restored as the result of a chemical reaction taking place during the use of a derivative, or–and this is a totally different case–as the result of metabolic degradation within the organism of persons who may be absorbing more complex derivatives.

Absorption pathways

Generally speaking, the principal risk of absorption lies in skin contact: the aromatic amines are nearly all lipid-soluble. This particular hazard is all the more important because in industrial practice it is one that is not easily appreciated.

There is, nevertheless, also a considerable risk of absorption by inhalation. This may be the result of inhaling the vapours, even though most of these amines are of low volatility at normal temperatures; or it may result from breathing in dust from the solid products. This applies particularly in the case of the amine salts such as sulphates and chlorhydrates, which have a very low volatility and lipid solubility: the occupational hazard from the practical point of view is less but their over-all toxicity is about the same as the corresponding amine, and thus the inhalation of their dust and even skin contact must be considered dangerous.

Ingestion by way of the digestive tract is not considered to be an occupational hazard in the true sense of the word. However, the absorption of aromatic amines in this way does represent a serious danger.

Metabolism

The amines undergo a process of metabolisation within the organism and the real active agents are the metabolites, some of which induce methaemoglobinaemia while others are carcinogenic. These metabolites generally take the form of hydroxylamines ($R-NHOH$), changing to aminophenols ($H_2N-R-OH$) as a form of detoxification; their excretion provides a means of estimating the degree of contamination when the level of exposure has been such that they are detectable.

Health effects

Aromatic amines have various pathological effects and these are not common to all of them. Each one must be considered on its own account. They may be resumed as follows:

- the principal hazard, in the long term, is cancer of the urinary tract, particularly the bladder;
- the danger of acute poisoning is represented by methaemoglobinaemia, and this can lead to adverse effects on the red cells;
- a number of the amines may act as skin sensitisers;
- respiratory sensitivity is less frequent; and, lastly
- hepatic effects have been seen.

Cancer of the urinary tract

Such cases are concerned almost entirely with bladder cancer. It is of interest to note that the aromatic amines make up one of the most important chapters in the study of occupational cancer, as may be illustrated by the following points.

In the first place, their discovery was entirely empirical in that the abnormally high incidence of this disease among employees in a dye factory led to its being originally described as "dye cancer", but further analysis very soon pointed to their origin being in the raw materials, of which the most important was aniline, and they became known as "aniline cancers". It was only later that the true cause was shown to be β-naphthylamine and benzidine, which are polycyclic aromatic amines. However, the experimental confirmation of this was long in coming, difficult, and in fact, was indirect because the experimental work was based on the knowledge of cancer in man, which was not entirely conclusive. Finally, an opposite case was noted when 4-aminodiphenyl was shown to be carcinogenic for animals (in the liver), following which a number of cases of bladder cancer in man were brought to light.

Some of the aromatic amines are certainly carcinogenic, others are probably not so, and some uncertainty remains regarding the occupational hazard that is represented by the many products derived from this large family of chemicals. In the first place, all these derivatives must be suspect. However, the significance of short-term tests is very much a matter for discussion and while more or less typical results have been gained from animal experimentation, their extrapolation to man and to the hazard is uncertain. Epidemiological studies do not reflect the anomalies and are open to doubt, and very frequently it is not possible to assemble a numerically adequate collection of facts. Cancers of the bladder of non-occupational origin are identical, as also is their evolution, and the treatment is the same for both. They evolve quietly over a long period and frequently the first sign takes the form of a macroscopic, painless haematuria. The latent period between exposure and the first clinical signs is generally of the order of ten years, but in some cases it may be only a few years and in others as much as 40 years or more. Specific systematic examinations, including cytological examination of the urine and, if necessary, cytoscopy, have made early diagnosis possible and improved the prognosis, which must, however, remain subject to reservations (see BLADDER CANCER).

Methaemoglobinaemia. Some degree of methaemoglobinaemia may be induced experimentally by many aromatic amines, but for practical purposes it can be regarded as a hazard of industrial exposure only in respect of single-ring compounds (see ANILINE; METHAEMOGLOBIN).

Dermatitis

Because of their alkaline nature, certain amines, particularly the primary ones, constitute a direct risk of dermatitis. Many aromatic amines can cause allergic dermatitis, such as that due to sensitivity to the "para-amines" (*p*-aminophenol and particularly *p*-phenylenediamine). Cross-sensitivity is also possible.

Respiratory allergy

A number of cases of asthma due to sensitisation to *p*-phenylenediamine have been reported.

Haemorrhagic cystitis

Haemorrhagic cystitis can result from heavy exposure to *o*- and *p*-toluidine, particularly the chlorine derivatives, of which chloro-5 *o*-toluidine is the best example. These haematuria are short-lived and no abnormal incidence of bladder tumours have been noted afterwards.

Liver injuries

Certain diamines (e.g. toluenediamine and diaminodiphenylmethane) have potent hepatotoxic effects in experimental animals, but despite widespread use in industry for many years scarcely a single case of liver damage resulting from industrial exposure has been attributed to either compound. In 1966, however, 84 cases of toxic jaundice were reported from eating bread baked from flour contaminated with 4,4'-diaminodiphenylmethane, and cases of toxic hepatitis have also been reported after occupational exposure.

Some of the aromatic amines are set out schematically below. Because the members of this chemical family are very numerous, it is not possible to include them all, and there may be others, not included below, which also have toxic properties.

Aniline

See separate article. (Note that certain simple derivatives of aniline are carcinogenic in the case of some laboratory animals but have not been recognised as such in the case of man. Examples are: anisidine, trimethylaniline, diaminoanisole, etc.)

Aminophenols ($NH_2C_6H_4OH$)

HYDROXYANILINE; HYDROXYAMINOBENZENE

Aminophenol exists in three isomeric forms. However, the *m*-isomer is little used, and the *o*-isomer only on a moderate scale. *p*-Aminophenol is the most important and has the following properties:

m.w.	109.1
m.p.	186 °C
b.p.	284 °C (decomposes)

slightly soluble in water, ethanol and ether
white plates.

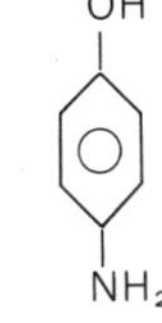

All three isomers are crystalline solids of low volatility and are not absorbed through the skin. Both *o*- and *p*-aminophenol may act as skin sensitisers and cause contact dermatitis, and this appears to be the greatest hazard arising from their use in industry. Although both isomers can cause methaemoglobinaemia, this seldom arises from industrial exposure, since their physical properties are such that neither compound is readily absorbed into the body. *p*-Aminophenol is the major metabolite of aniline in man and is excreted in the urine in conjugated form.

4-Aminodiphenyl ($C_6H_5C_6H_4NH_2$)

4-AMINOBIPHENYL; *p*-BIPHENYLAMINE; XENYLAMINE

m.w.	169.2
sp.gr.	1.16
m.p.	53 °C
b.p.	302 °C
i.t.	450 °C
ACGIH	human carcinogen without an assigned TLV, skin

It was the first compound in which the demonstration of carcinogenic activity in experimental animals preceded the first reports of bladder tumours in exposed workmen, when it was used as an antioxidant in rubber manufacture.

Benzidine ($NH_2C_6H_4C_6H_4NH_2$)

4,4'-DIAMINODIPHENYL

m.w.	184.2
sp.gr.	1.25
m.p.	128 °C
b.p.	400 °C

white or slightly reddish crystals, powder or leaflets.

ACGIH the substance associated with benzidine production is a human carcinogen without an assigned TLV, skin

It is manufactured by reduction of nitrobenzene to hydrazobenzene, with subsequent conversion to benzidine. The major use of benzidine is in the manufacture of azo dyestuffs. It is tetrazotised and coupled with other intermediates to form colours. Its use in the rubber industry has been abandoned.

Benzidine is generally accepted as a dangerous carcinogen, the manufacture and industrial use of which has caused cases of papilloma and carcinoma of the urinary tract.

Benzidine base is a crystalline solid with a significant vapour pressure. Penetration through the skin seems to be the most important pathway for the absorption of benzidine. There is also a hazard from the inhalation of vapour or fine particles. The carcinogenic activity of benzidine has been established by the many reported cases of bladder tumour in exposed workmen and by experimental induction in animals.

Dichloro-3,3′-benzidine ($C_{12}H_{10}Cl_2N_2$)

m.w. 253
m.p. 133 °C

crystalline powder, insoluble in water, soluble in alcohol and benzene.

ACGIH industrial substance suspect of carcinogenic potential for man, skin

***o*-Tolidine** ($C_{14}N_{16}H_2$)

3,3′-DIMETHYLBENZIDINE

m.w. 213.3

white to red crystals, slightly soluble in water, soluble in alcohol and ether.

***o,o*′-Dianisidine** ($(NH_2(OCH_3)C_6H_3)_2$)

DIMETHOXYBENZIDINE; 3,3′-DIMETHOXY-4,4′-DIAMINOBIPHENYL

m.w. 244.3
m.p. 137 °C
v.d. 8.5
f.p. 206 °C

white to violet crystals, insoluble in water, soluble in alcohol and benzene.

The question of the toxicity of the benzidine derivatives has not been fully clarified at the present time. For example several experiments have shown that dichloro-3,3′-benzidine, *o,o*′-dianisidine and *o*-tolidine are potential carcinogens for laboratory animals. Inquiries in the industrial field, however, have failed to reveal any abnormal incidence of cancer among workers who have not been exposed to some other dangerous substance. Nevertheless, this practical information relating to man, which has up to the present been judged as being adequate, is not sufficient to eliminate the suspicion that has arisen as a result of the experiments, and the potential hazard demands that very special preventive measures be observed.

Chloroanilines ($NH_2C_6H_4Cl$)

Chloroaniline exists in three isomeric forms: ortho, meta and para; of these only the first and the last have some importance for manufacturing dyes, drugs and pesticides. Both can be absorbed through the skin.

***o*-Chloroaniline**

2-CHLOROPHENYLAMINE; ORTHOAMINOCHLOROBENZENE

m.w. 127.6
sp.gr. 1.21
m.p. α-14 °C, β-3.5 °C
b.p. 208.8 °C
v.d. 1 mmHg (0.13·10³Pa) at 46.3 °C

a liquid.

NH_2 Cl

***m*-Chloroaniline**

2-CHLOROPHENYLAMINE; METAAMINOCHLOROBENZENE

m.w. 127.6
sp.gr. 1.22
m.p. −10 °C
b.p. 230.5 °C
v.p. 1 mmHg (0.13·10³ Pa) at 63.5 °C

a colourless liquid, insoluble in water.

MAC USSR 0.05 mg/m³ skin

NH_2 Cl

***p*-Chloroaniline**

2-CHLOROPHENYLAMINE; PARAAMINOCHLOROBENZENE

m.w. 127.6
sp.gr. 1.43
m.p. −10.4 °C
b.p. 230 °C
v.p. 1 mmHg (0.13·10³ Pa) at 60 °C

a yellow crystalline solid.

MAC USSR 0.3 mg/m³ skin

NH_2 Cl

p-Chloroaniline is a potent methaemoglobin-former and is irritating to the eyes. Animal experiments have provided no evidence of carcinogenicity.

Diamino-4,4′-diphenylmethane ($CH_2(C_6H_4-NH_2)_2$)

p-p′-METHYLENE DIANILINE; MDA

m.w. 198.3
m.p. 92-93 °C
f.p. 226 °C

yellowish crystals with a faint amine-like odour.

This is used particularly in the production of isocyanates and polyisocyanates and in the manufacture of polyurethane. The most striking example of the toxicity of this compound was when 84 persons contracted toxic hepatitis as a result of eating bread baked from flour that was contaminated with the substance. Other cases of hepatitis were noted after occupational exposure where the substance had been absorbed through the skin. Apart from this, it may also give rise to allergic dermatitis.

Animal experiments have led to its being a suspected potential carcinogen, but conclusive results have not been obtained.

Diaminodiphenylmethane derivatives and particularly dichloro-3,3′diamino-4,4′diphenylmethane (known also as methylene *o*-chloroaniline or MOCA) have been shown to be carcinogens for laboratory animals. No cases of cancer in man have been attributed to these compounds, but the experimental carcinogenic potentiality of MOCA justifies the implementation of strict preventive measures (see CATALYSTS).

N,N-diethylaniline ($C_6H_5N(C_2H_5)_2$)

N-PHENYLDIETHYLAMINE

m.w. 149.2
sp.gr. 0.93
m.p. −38.8 °C
b.p. 216.3 °C
v.d. 5.15
v.p. 1 mmHg (0.13·10³ Pa) at 49.7 °C
f.p. 85 °C
i.t. 630 °C

an oily liquid.

$N(C_2H_5)_2$

N,N-dimethylaniline ($C_6H_5N(CH_3)_2$)

N-PHENYLDIMETHYLAMINE

sp.gr. 0.96
m.p. 2.4 °C
b.p. 194.1 °C

a brownish liquid with a characteristic amine-like odour.

TWA OSHA 5 ppm 25 mg/m³ skin
STEL ACGIH 10 ppm 50 mg/m³
IDLH 100 ppm
MAC USSR 0.2 mg/m³ skin

$N(CH_3)_2$

These two compounds are used in the synthesis of dyestuffs and other intermediates (particularly the dimethylaniline). They are readily absorbed through the skin, but poisoning may also occur through inhalation of vapours. Their hazards may be considered as similar to those of aniline; they are, in particular, potent methaemoglobin-formers.

Diphenylamine ($(C_6H_5)_2NH$)

PHENYLANILINE; ANILINOBENZENE

m.w. 169.2
sp.gr. 1.16
m.p. 52.8 °C
b.p. 302 °C
v.d. 5.82
v.p. 1 mmHg (0.13·10³ Pa) at 108.3 °C
f.p. 153 °C
i.t. 634 °C

a pale crystalline solid with a floral odour.

TLV ACGIH 10 mg/m³
STEL ACGIH 20 mg/m³

It is used mainly in the manufacture of dyestuffs, antioxidants, pharmaceuticals and explosives and as a pesticide. There is no experimental evidence of methaemoglobin formation and despite its widespread use in industry no definite cases of poisoning from industrial exposure have been described. In ordinary industrial conditions it offers little hazard, but the carcinogen 4-aminodiphenyl may be present as an impurity during the manufacturing process. This may be concentrated to significant proportions in the tars produced at the distillation stage and will constitute a hazard of bladder cancer.

Modern manufacturing procedures have enabled the amount of impurities in this compound to be considerably reduced in the commercial product.

Naphthylamines ($C_{10}H_7NH_2$)

Naphthylamine occurs in two isomeric forms, 1-naphthylamine and 2-naphthylamine.

1-Naphthylamine

α-NAPHTHYLAMINE

m.w. 143.2
sp.gr. 1.12
m.p. 50 °C
b.p. 300.8 °C (sublimes)
v.d. 4.93
v.p. 1 mmHg (0.13·10³ Pa) at 104.3 °C
f.p. 157.2 °C

a pale crystalline solid with a characteristic monkey-like odour.

1-Naphthylamine is made by the nitration of naphthalene to form nitro-naphthalene, followed by reduction to the amine. It is used in the manufacture of dyestuffs, naphthionic acid, α-naphthol and α-naphthylthiourea (a rodenticide). 1-Naphthylamine is absorbed through the skin and by inhalation. Acute poisoning does not arise from its industrial use, but exposure to commercial grades of this compound in the past has resulted in many cases of papilloma and carcinoma of the bladder. It is still a matter of dispute whether these tumours were caused by 1-naphthylamine *per se*, or whether they were brought about by the substantial 2-naphthylamine impurity. This matter is not merely of academic interest, as 1-naphthylamine with greatly reduced levels of 2-naphthylamine impurity is now available. Up to the present time, numerous experiments have been conducted and it has not been possible to show more than one case of cancer in an animal.

2-Naphthylamine

β-NAPHTHYLAMINE

m.w. 143.2
sp.gr. 1.06
m.p. 113 °C
b.p. 306.1 °C
v.p. 1 mmHg (0.13·10³ Pa) at 108 °C

a pinkish crystalline solid.

ACGIH human carcinogen without an assigned TLV

Although at one time extensively used as an intermediate in the manufacture of dyestuffs and antioxidants, its manufacture and use has been almost entirely abandoned throughout the world, and it has been condemned as too dangerous to make and handle without prohibitive precautions. It is readily absorbed through the skin and by inhalation. The question of its acute toxic effects does not arise because of its high carcinogenic potency. A high incidence of bladder tumours resulting from its manufacture and use has been reported from many countries.

Nitroanilines ($NO_2C_6H_4NH_2$)

Of the three mono-nitroanilines, the most important is *p*-nitroaniline. All are used as dye intermediates, but the *o*- and *m*- isomers only on a small scale.

***p*-Nitroaniline**

1-AMINO-4-NITROBENZENE

m.w. 138.1
sp.gr. 1.43
m.p. 147.8 °C
b.p. 331.7 °C
v.p. 1 mmHg (0.13·10³ Pa) at 142 °C
f.p. 199 °C

a bright yellow solid.

TWA OSHA 1 ppm 6 mg/m³ skin
STEL ACGIH 2 ppm 12 mg/m³ skin
IDLH 300 mg/m³
MAC USSR 0.1 mg/m³ skin

p-Nitroaniline is readily absorbed through the skin and also by inhalation of dust or vapour. It is a powerful methaemoglobin-former, and is alleged, in serious cases, also to bring about haemolysis, or even liver damage.

The chloronitranilines are also potent methaemoglobin-formers, leading to haemolysis, and are hepatotoxic. They may give rise to dermatitis by sensitisation.

***p*-Nitroso-N,N-dimethylaniline** ($NOC_6H_4N(CH_3)_2$)

m.w. 150.1
sp.gr. 1.14
m.p. 92.5-93.5 °C

a greenish-yellow solid.

It is used as a dyestuffs intermediate. It possesses both primary irritant and skin sensitising properties, and it is a common cause of contact dermatitis. Although, occasionally, workmen who develop dermatitis may subsequently work with this compound without further trouble, most will suffer a severe recurrence of the skin lesions on re-exposure, and, in general, it is wise to transfer them to other work to avoid further contact.

Phenylenediamines ($(C_6H_4NH_2)_2$)

The phenyldiamines exist in various isomeric forms but only the *m*- and *p*- isomers are of industrial importance.

***m*-Phenylenediamine**

1,3-DIAMINOBENZENE

m.w. 108.2
sp.gr. 1.14

m.p. 63-64 °C
b.p. 282-284 °C
v.p. 1 mmHg (0.13·10³ Pa) at 99.8 °C
a pale crystalline solid.

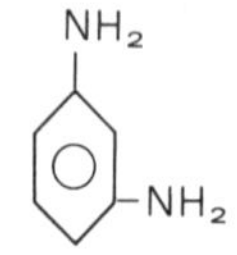

p-Phenylenediamine

1,4-DIAMINOBENZENE
m.w. 108.2
m.p. 146 °C
b.p. 267 °C
v.d. 3.72
f.p. 155.5 °C
a pale crystalline solid.
TWA OSHA 0.1 mg/m³ skin
IDLH 25 mg/m³

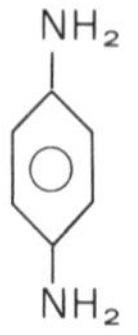

m-Phenylenediamine is used mainly as a dyestuff intermediate and as a curing agent for synthetic resins. *p*-Phenylenediamine is used as a dyestuff intermediate, photographic chemical, and for the dyeing of fur and hair. While *p*-phenylenediamine can act as a methaemoglobin-former *in vitro*, methaemoglobinaemia arising from industrial exposure is unknown. It is notorious for its sensitising properties of the skin and respiratory tract. Regular skin contact readily causes contact dermatitis. The former problem of "fur dermatitis" is much less frequent now owing to improvements in the dyeing process having the effect of removing all traces of *p*-phenylenediamine. Similarly, asthma, at one time common among fur dyers using this substance, is now relatively rare after improvements in the control of airborne dust. *p*-Phenylenediamine is still widely used as an oxidation dye for hair, but is banned in some countries. A preliminary skin test should always precede its use. *m*-Phenylenediamine has been known to cause skin sensitisation, but it is much less potent in this respect than the *p*- isomer.

Experiments conducted on the compounds in this series and their derivatives (e.g. N-phenyl or 4- or 2-nitro) relating to their carcinogenic potential are, up to the present time, either insufficient, inconclusive or negative. Chlorine derivatives that have been tested seem to have a carcinogenic potential in animal tests.

N-phenylnaphthylamines ($C_{10}H_7NHC_6H_5$)

Phenylnaphthylamines exist in two isomers: alpha and beta.

Phenyl-α-naphthylamine

N-PHENYL-1-NAPHTHYLAMINE
m.w. 219.3
m.p. 62 °C
b.p. 335 °C
colourless crystals.

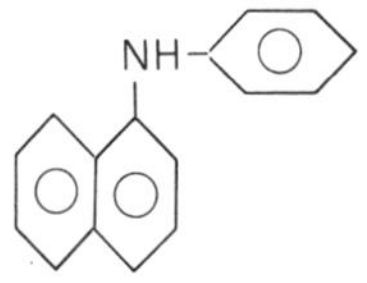

Phenyl-β-naphthylamine

PBNA; N-PHENYL-2-NAPHTHYLAMINE
m.w. 219.3
sp.gr. 1.2
m.p. 108 °C
b.p. 395 °C
a grey crystalline powder.

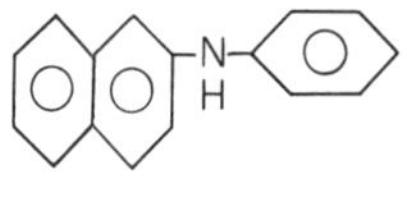

These are used mainly as antioxidants in the rubber industry. They give rise to dermatitis by sensibilisation, while acne and leukoderma have also been described. Questions as to their carcinogenic potential have arisen because of the presence of β-naphthylamine, which had been found to exist as an impurity in considerable quantities (running into tens or even hundreds of ppm) in some of the older preparations, and by the discovery, in the case of PBNA, of β-naphthylamine as a metabolic excretion, though in infinitesimal quantities. The experiments point to a carcinogenic potential for the animals tested but do not permit a conclusive judgement to be made, and the degree of significance of the metabolic findings is not yet known. Epidemiological investigations on a large number of persons working under different conditions have not shown any significant increase in the incidence of cancer among workers exposed to these compounds. The amount of β-naphthylamine that is present in the marketed products today is very low—less than 1 ppm and frequently 0.5 ppm.

At the present time it is not possible to draw any conclusions as to the true cancer hazard, and for this reason every precaution should be taken, including the elimination of impurities that may be suspect, and technical protective measures in the manufacture and use of these compounds. At the same time, experimental and epidemiological investigations should continue.

Toluidines ($CH_3C_6H_4NH_2$)

Toluidine exists in three isomeric forms but only the *o*- and *p*- isomers are of industrial importance.

o-Toluidine

o-METHYLANILINE
m.w. 107.2
sp.gr. 0.99
m.p. −23.7 °C
b.p. 199.7 °C
v.d. 3.69
v.p. 1 mmHg (0.13·10³ Pa) at 44 °C
f.p. 85 °C
i.t. 482 °C
a pale liquid.
TWA OSHA 5 ppm 22 mg/m³ skin
TLV ACGIH 2 ppm 9 mg/m³
IDLH 100 ppm
MAC USSR 3 mg/m³ skin (toluidine)

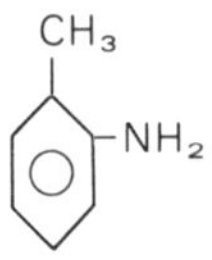

p-Toluidine

p-METHYLANILINE
m.w. 107.2
sp.gr. 0.96
m.p. 43.7 °C
b.p. 200.5 °C
v.d. 3.90
v.p. 1 mmHg (0.13·10³ Pa) at 42 °C
f.p. 86.6 °C
i.t. 482 °C
a white solid.
MAC USSR 3 mg/m³ skin (toluidine)

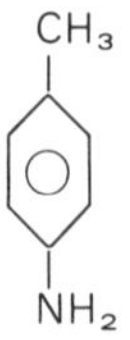

o-Toluidine and *p*-toluidine are extensively used in the manufacture of dyestuffs and other organic chemicals. They are readily absorbed through the skin, or inhaled as dust, fume or vapour. They are powerful methaemoglobin-formers, and acute poisoning may be accompanied by microscopic or macroscopic haematuria, but they are much less potent as bladder irritants than 5-chlor-*o*-toluidine. Animal experiments and epidemiological studies at workplaces have caused ortho-toluidine to come under suspicion as a potential carcinogen. Present results have not enabled a conclusion to be drawn as to its potential either for laboratory animals or for humans.

Chlorotoluidines ($CH_3C_6H_3(NH_2)Cl$)

The most important isomer is 5-chloro-*o*-toluidine.
m.w. 141.6
m.p. 29-30 °C
b.p. 241 °C
a greyish-white solid.

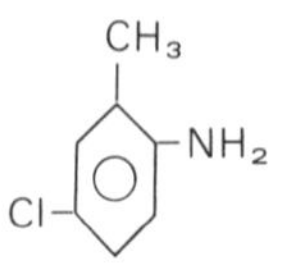

This is used in the synthesis of organic dyestuffs. This compound is readily absorbed through the skin or by inhalation. Although it (and some of its isomers) may cause methaemoglobin formation, the most striking feature is its irritant effect on the urinary tract, resulting in haemorrhagic cystitis characterised by painful haematuria and frequency of micturition. Microscopic haematuria may be present in men exposed to this compound before the cystitis is manifest, but there is no carcinogenic hazard to man (as for other isomers, certain experiments have cast doubts on its carcinogenicity for certain species of laboratory animals).

Toluenediamines ($CH_3C_6H_3(NH_2)_2$)
TOLYLENEDIAMINES

Among the six isomers of toluenediamine the one most frequently encountered is the -2,4 which accounts for 80% of the intermediate product in the manufacture of toluene diisocyanate, a further 20% being the -2,6 isomer, which is one of the basic substances for the polyurethanes. Attention was drawn to this compound following the experimental discovery of a carcinogenic potential in laboratory animals. There has not been a single pathology nor any statistics reported in the case of man.

Xylidines ($(CH_3)_2C_6H_3NH_2$)
DIMETHYLANILINE

TWA OSHA	5 ppm 25 mg/m³ skin
TLV ACGIH	2 ppm 10 mg/m³ skin
IDLH	150 ppm
MAC USSR	3 mg/m³ skin

These are not extensively used in industry. Results of animal experiments indicate that they are primarily liver poisons and act only secondarily on the blood. However, other experiments demonstrated that methaemoglobinaemia and Heinz body formation were readily induced in cats, though not in rabbits.

SCOTT, T. S.
MUNN, A.
SMAGGHE, G.

IARC monographs on the evaluation of the carcinogenic risk of chemicals to man. Vol. 16: *Some aromatic amines and related nitro compounds. Hair dyes, colouring agents and miscellaneous industrial chemicals* (Lyons, International Agency for Research on Cancer, 1978), 400 p. Ref.

Benzidine, its congeners, and their derivative dyes and pigments. Preliminary risk assessment phase I. Jones, T. (Washington, DC, Office of Testing and Evaluation, Office of Pesticides and Toxic Substances, US Environmental Protection Agency, June 1980), 55 p. 83 ref.

Benzidine-o-Tolidine-, and o-Dianisidine-based dyes. Health Hazard Alert (Washington, DC, US Department of Labor Occupational Safety and Health Administration; US DHEW, National Institute for Occupational Safety and Health, 1980), 20 p. 35 ref.

"Paraphenylenediamine in allergic contact dermatitis" (La paraphénylènediamine dans les dermites allergiques de contact). Meneghini, C. L.; Angelini, G. *Médecine et Hygiène* (Geneva), 30 Apr. 1980, 38/1976 (1 577-1 581). 18 ref. (In French)

"Permeation of methanolic aromatic amines solutions through commercially available glove materials". Weeks, R. W.; Dean, B. J. *American Industrial Hygiene Association Journal* (Akron), Dec. 1977, 38/12 (721-725). Illus. 9 ref.

Aminothiazole

2-AMINOTHIAZOLE; 2-THIAZYLAMINE; ABADOL

Aminothiazole (SCH:NCH:CNH₂)

m.w.	100.1
m.p.	92 °C
b.p.	sublimes

slightly soluble in cold water, ethyl alcohol, ethyl ether; readily soluble in hot water and dilute mineral acids
white-to-yellow crystals.

Production. Aminothiazole is made by chlorination of vinyl acetate and condensation with thiourea.

Uses. Aminothiazole has long been used as an intermediate in the synthesis of the sulpha-drug, sulphathiazole. It had relatively brief use as an antithyroid drug but is not generally used for that purpose any longer.

HAZARDS

In the manufacture of aminothiazole, precautions should be taken in the handling of the precursors, vinyl acetate and thiourea (see VINYL COMPOUNDS).

Aminothiazole itself causes reduced thyroid activity and adenomatous hyperplasia in laboratory animals and man, and such changes have been observed in men manufacturing aminothiazole. Animal experiments indicate that there is the possibility of damage to the liver from excessive absorption of this compound but the effect has not been observed in man.

There is one report of workmen exposed to atmospheric aminothiazole concentrations of 3.6-110 mg/m³, who developed brown staining of the skin, brown discolouration of the urine during periods of exposure, loss of appetite and occasionally nausea and vomiting; some primary skin irritation was also observed. A small number of workers developed illness of sudden onset, which was characterised by cutaneous itching and joint or muscle pain. Urticaria or a maculopapular rash was present in several cases associated with the itching.

SAFETY AND HEALTH MEASURES

Engineering controls including enclosure and local exhaust and general ventilation should be applied where there is danger of exposure to atmospheric concentrations.

In the case of the precursor, thiourea, particular attention should be given to the hazard of skin absorption by the provision of protective clothing and adequate sanitary facilities including showers, and by encouraging workers to cleanse the skin thoroughly after work.

Treatment. In the event of symptoms developing, the worker involved should be removed from further exposure and treated symptomatically. If thyroid hyperplasia has occurred, removal from exposure should result in reversion to the normal state.

ZAVON, M. R.

Aminotriazole

Aminotriazole ($C_2H_4N_4$)
3-AMINO-1,2,4-TRAZOLE; AMITROLE; AMIZOL; 3AT; ATA

m.w.	84.1
m.p.	156-159 °C

soluble in water and ethyl alcohol; insoluble in acetone
odourless white crystalline powder.

Production and uses. It is obtained by condensation of formic acid with aminoguanidine and is used as a non-selective herbicide.

Toxicity. The acute oral toxicity (LD_{50}) of aminotriazole is 1 100 mg/kg for rats and 14 700 mg/kg for mice. The substance reduces the catalase enzyme in the liver and kidneys of rats treated intraperitoneally with 1 g/kg. The addition of 50 250 and 1 250 ppm aminotriazole to the diet fed to rats during 2 years produced a considerable enlargement of the thyroid and thyroid tumours. The carcinogenic effects encountered after the occurrence of thyroid hyperplasia appear not to be due to a genetic mechanism. In another experiment aminotriazole caused thyroid tumours and hepatomas in two strains of mice after administration of 2 192 ppm with the food during 50-60 weeks, and thyroid adenomas in rats fed for 104 weeks on diets to which 10, 50 and 100 ppm aminotriazole had been added.

^{14}C-labelled aminotriazole administered orally to rats in a single dose of 50 mg/kg was rapidly absorbed in the gastrointestinal tract: during the first 24 h most of the radioactivity had been eliminated with the urine as a non-metabolised substance, while approximately 6% was excreted in the form of the two metabolites 3-amino-5-mercapto-1,2,4-triazole and 3-amino-1,2,4-triazolyl-5-mercapturic acid. In mice there was no evidence of teratogenic effects; however, after injection of 20-40 mg aminotriazole into chicken embryos and 0-96 h of incubation anomalies of the beak and, at times, curvatures of the tibia were observed. No mutagenic effects have so far been detected.

Effects on humans. A study of 348 Swedish railway workers (1957-78) exposed to various herbicides revealed 17 cases of tumours compared with 11.8 expected. The excess mortality from tumours occurred particularly among persons with combined exposure to aminotriazole and phenoxy acids. The study was, however, not conclusive, and a carcinogenic effect in man of aminotriazole could not be corroborated, especially since the most seriously affected were the workers exposed to a combination of aminotriazole and phenoxy acids. A woman who had ingested 20 mg/kg of the substance displayed no symptoms of poisoning; the substance was found unchanged in the urine with which 50% was eliminated within a few hours.

The substance appears in the ACGIH list as a suspected carcinogen. On heating it can develop dangerous toxic gases.

SAFETY AND HEALTH MEASURES

The production and use of aminotriazole demand—like all potentially carcinogenic substances—work premises and plant offering the maximum of safety for the workers involved. The technical process must be designed so as to prevent pollution of the workroom air and contamination of surfaces with which the workers may come into contact. Maintaining a slight negative pressure in the premises liable to be polluted by the toxic substance will avoid the involvement of contiguous or adjacent workplaces and limit the hazard to the personnel assigned to the process. Another important measure is the installation of exhaust ventilation where the substance may leak into the atmosphere. The workers should be familiar with the main properties and effects of aminotriazole in order to be in a better position to protect their health. Good working practices should include the daily changing of work clothing, taking a shower at the end of the shift, separation of work clothing from the ordinary clothes, prohibition of eating and smoking at the workplace, compulsory wearing protective equipment for the eyes, hands and respiratory tract, cleaning of contaminated premises and equipment, etc.

Periodic medical examinations associated with the necessary laboratory tests of workers occupationally exposed should be carried out.

ARMELI, G.
De RUGGIERO, D.

"Herbicide exposure and tumor mortality: an updated epidemiologic investigation on Swedish railroad workers". Axelson, O.; Sundell, L.; Andersson, K.; Edling, C.; Hogstedt, C.; Kling, H. *Scandinavian Journal of Work, Environment and Health* (Helsinki), Mar. 1980, 6/1 (73-79). 14 ref.

"Amitrole" (31-43). 42 ref. Vol. 7: *Some antithyroid and related substances, nitrofurans and industrial chemicals. IARC Monographs on the evaluation of carcinogenic risk of chemicals to man* (Lyons, International Agency for Research on Cancer, 1974), 326 p.

Pesticide residues in food. Report of the joint FAO/WHO meeting. Technical report series 574 (Geneva, World Health Organisation, 1975), 37 p. 30 ref.

Ammonia

Ammonia (NH_3)

m.w.	17
sp.gr.	0.77
m.p.	−77.7 °C
b.p.	−33.3 °C
v.d.	0.59 at 25 °C
v.p.	10 atm ($1013 \cdot 10^3$ Pa) at 25.7 °C
e.l.	16-25%
i.t.	651 °C

soluble in water, ethyl alcohol, ethyl ether and organic solvents

a colourless, easily liquefied gas with a very sharp characteristic odour.

TWA OSHA	50 ppm 35 mg/m³
NIOSH	50 ppm/5 min ceil
TLV ACGIH	25 ppm 18 mg/m³
STEL ACGIH	35 ppm 27 mg/m³
IDLH	500 ppm
MAC USSR	20 mg/m³

Ammonia is lightly reactive, easily undergoing oxidation, substitution (of hydrogen atoms) and additional reactions. For example it burns in air or in hydrogen to form nitrogen; an example of substitution would be the formation of amides of alkaline and alkaline-earth metals; as a result of addition it forms ammoniates (e.g. $CaCl_28NH_3$, $AgCl3NH_3$) and other compounds. When ammonia dissolves in water, it forms ammonium hydroxide (NH_4OH) which is a weak base and dissociates as follows: $NH_4OH \rightleftarrows NH_4^+ + OH-$. The radical NH_4^+ does not exist in free form since it decomposes into ammonia and hydrogen when an attempt is made to isolate it.

Occurrence. Ammonia is present in small amounts in the air, water, earth, and particularly in decomposing organic matter. It is the product of normal human, animal and plant metabolism. Muscular effort and excitement of the nervous system result in the formation of an increased amount of ammonia, an accumulation of which in the tissues would result in poisoning. Endogenous formation of ammonia increases also in the course of many diseases. Through vital processes it is combined and excreted from the organism, mainly via urine and sweat, in the form of ammonium sulphate and urea. Ammonia is also of primary importance in the nitrogen metabolism of plants.

Production. The industrial production of ammonia is based on the use of nitrogen, air and the organic compounds of nitrogen in coal and peat. Coal contains about 20% of nitrogen, which is partly (up to 25%) converted into ammonia by dry distillation and absorbed in baths of water or sulphuric acid. Nowadays, all coke ovens are fitted with appliances designed to extract the ammonia in exhaust gases. It is very difficult to fix nitrogen from atmospheric air, owing to its chemically inert nature. This was not done until early in the 20th century with the use of calcium cyanamide, but in the meantime the importance of the process had declined. The principal method is now the Haber-Bosch process, or its modifications, which basically consists of synthesis from nitrogen and hydrogen: $3H_2 + N_2 \rightleftarrows 2NH_3$. The mixture of hydrogen and nitrogen is passed through a complex catalyst, under a pressure of 200-1 000 atm and a temperature of about 600 °C. The activity of the catalysts is reduced by impurities such as sulphur and oxygen compounds, which are present in the hydrogen that is produced for reaction with nitrogen. The removal of these is an essential but complicated procedure. Liquid ammonia is stored or transported under pressure in steel cylinders or tank cars.

Uses. Ammonia is an important source of various nitrogen-containing compounds. An enormous quantity of ammonia is used in the production of ammonium sulphate and nitrate which are used as fertilisers. It is further used for oxidation into nitric acid, for the production of soda by the ammonia process, and of synthetic urea, and for the preparation of water solutions used in chemical and pharmaceutical industries, as well as in medicine and in agriculture for fertilising. In refrigeration, ammonia is used to lower temperatures below the freezing point and in the manufacture of synthetic ice (see REFRIGERATING PLANTS).

HAZARDS

Ammonia poisoning may occur in the production of ammonia and in the manufacture of nitric acid, ammonium nitrate and sulphate, liquid fertilisers (ammoniates), urea and soda, in refrigeration, synthetic ice factories, cotton printing mills, fibre dyeing, electroplating processes, organic synthesis, heat treatment of metals (nitriding), chemical laboratories, and in a number of other processes. It is formed and emitted into the air during the processing of guano, in the purification of refuse, in sugar refineries and tanneries, and it is present in unpurified acetylene.

Industrial poisoning is usually acute, while chronic poisoning, although possible, is less common. The irritant effect of ammonia is felt especially in the upper respiratory tract and in large concentrations it affects the central nervous system, causing spasms. Irritation of the upper respiratory tract occurs at concentrations of above 100 mg/m³, while the maximum tolerable concentration in 1 hour is between 210 and 350 mg/m³. Splashes of ammonia water into the eyes are particularly dangerous. The rapid penetration of ammonia into the ocular tissue may result in perforation of the cornea and even in death of the eye-ball. Particular health hazards exist in each section of an ammonia plant. In the sections where the gas is generated, converted (oxidation of CO to CO_2), compressed and purified, the main problem is the emission of carbon monoxide and hydrogen sulphide. Considerable quantities of ammonia may escape during its synthesis. Escaping ammonia in the atmosphere may reach explosive limits.

SAFETY AND HEALTH MEASURES

Ammonia production is characterised by great fire and explosion hazards. Process plant should therefore be installed in the open air at a safe distance from other equipment. Only the compressors should be housed in buildings. In order to prevent the escape of ammonia, containers and plant should be made from materials which are not affected by high pressure and are not subject to corrosion. Corrosion of cooling systems can be reduced by using air instead of water for cooling. Rigid rules must be established about the regular maintenance of the plant, which must be done with utmost care. All parts of such a plant should be equipped with general and local exhaust ventilation and warning devices. Adequate measures should be taken for explosion prevention and fire fighting. The workers involved in the generation, compression and conversion of the gas must be supplied with proper filter-type respiratory protective equipment with devices to absorb carbon monoxide, hydrogen sulphide and ammonia. Work performed inside converters, such as emptying and cleaning, must be done with the use of supplied-air respiratory equipment and a safety belt and lifeline. A second worker must remain on the outside in order to help in case of emergency.

Treatment. If ammonia water is splashed into the eyes, first-aid consists of immediate washing with a large amount of water or a solution of 0.5-1% alum. An ophthalmologist should immediately be consulted, even if the injured worker complains of no pain. Affected parts of the skin should be washed with clean water, and a lotion applied consisting of a 5% solution of acetic, citric, tartaric or salicylic acid. In the event of ammonia poisoning through the respiratory tract, the person should breathe fresh air and inhale warm water vapour (if possible with the addition of vinegar or citric acid) and a 10% solution of menthol in chloroform. He should drink warm milk. In the event of asphyxia, oxygen should be inhaled, preferably under low pressure, until the dyspnoea or cyanosis lessens, followed by a subcutaneous injection of 1 cm³ of a 1% solution of atropine. Resuscitation must be applied if breathing is interrupted or stops. Cardiac preparations or tranquillisers may be given, if advised by a physician. If pulmonary oedema develops, the person must be kept as quiet as possible and kept warm, and oxygen must be administered as soon as possible.

GADASKINA, I. D.

Health effects:

CIS 77-1968 "Human physiological response and adaption to ammonia". Ferguson, W. S.; Koch, W. C.; Webster, L. B.; Gould, J. R. *Journal of Occupational Medicine* (Chicago), May 1977, 19/5 (319-326). Illus. 5 ref.

Ammonia. Subcommittee on Ammonia, Committee on Medical and Biological Effects of Environmental Pollutants, National Research Council (Baltimore, University Park Press, 1979), 384 p.

CIS 78-1332 "Ammonia" (Ammoniak). Lazarev, N. V.; Gadaskina, I. D. (88-92). *Harmful substances in industry–III. Inorganic and hetero-organic compounds* (Vrednye veščestva v promyšlennosti–III: Neorganičeskie i elementorganičeskie soedinenija) (Leningrad, Izdatel'stvo "Himija", 7th ed., 1977), 608 p. 58 ref. (In Russian)

CIS 77-482 "Detection of chronic ammonia poisoning in joint plant medical services" (L'intoxication chronique par l'ammoniaque est-elle observable en service médical interentreprises?). Andanson, J.; Berenger, J.; Castela, R.; Catoir, J.; Delmon, P.; Graille, M.; Guidoni, B.; D'Ortoli, J. P.; Raulot-Lapointe, H.; Vigneau, J. J. *Revue de médecine du travail* (Paris), 1976, 4/4 (293-300). Illus. 22 ref. (In French)

Safety:

Ammonia plant safety (and related facilities). Vol. 22 (New York, American Institute of Chemical Engineers, 1980), 230 p. Illus. Ref.

CIS 77-1937 *Code of practice for the safe handling and transport of anhydrous ammonia in bulk by road in the UK* (Chemical Industries Association Ltd., Alembic House, 93 Albert Embankment, London) (1976), 30 p.

Workers' education:

CIS 79-1368 *A guide for developing a training program for anhydrous ammonia workers—Working safely with anhydrous ammonia.* DHEW (NIOSH) publications Nos. 79-119 and 79-120 (National Institute for Occupational Safety and Health, 4676 Columbia Parkway, Cincinnati, Ohio) (2 booklets, Dec. 1978, and Jan. 1979), 85 and 24 p. Illus.

Anaesthetists

Anaesthesiology is a relatively new clinical speciality, which may be one reason why the occupational health problems connected with it were not seriously discussed until the beginning of the 1970s. The speciality has experienced very rapid development during the past 30 years with the introduction of new techniques, apparatus and anaesthetic agents to meet the needs of increasingly demanding surgery. Modern clinical anaesthesia started with the use of inhalants and today inhalation anaesthesia, in combination with intravenous drugs, is still the most common anaesthetic technique throughout the world. Some of the inhalants, at high concentrations, have been shown to have toxic properties. This is apparently why chronic inhalation of waste anaesthetics has been suspected of being responsible for all kinds of health troubles. The epidemiological health studies have been critically reviewed and it seems that conclusive evidence for the proposed risks is lacking (Ferstanding, 1978; Vessey, 1978).

Table 1. Distribution of deaths among anaesthetists by cause[1]

Cause of death	Number of deaths	Percentage of all deaths
Total cardiovascular	347	56.9
Arteriosclerotic heart disease	241	39.5
Other heart disease	37	6.1
Cerebrovascular disease	45	7.4
Other circulatory	24	3.9
Total cancer	114	18.7
Cancer of the–		
Lung	20	3.3
Colon and rectum	15	2.5
Stomach	5	0.8
Liver	7	1.1
Pancreas	10	1.6
Kidney	3	0.5
Bladder	4	0.7
Prostate	12	2.0
Lymphomas and Hodgkin's disease	5	0.8
Leukaemia	6	1.0
Lymphosarcoma	10	1.6
Other cancer	17	2 8
Hepatic disease	6	1.0
Renal disease	10	1.6
Suicide	38	6.2
Accidents	42	6.9
All other causes	53	8.7
	610	100.0
Cause not ascertained	27	
	637	

[1] Reprinted by permission from Lew, *Anaesthesiology* (Philadelphia), 1979, 51/3 (195-199).

Mortality and morbidity

In the United Kingdom the mortality of anaesthetists in the period from 1951 to 1971 has been found to be 92% of expected death-rate among doctors, and in the United States the mortality among male anaesthetists during 1954-76 was 93% of the expected rate. The distribution of causes of death from the US study is shown in table 1. There was no suggestion of rates of cancer or hepatic or renal disease above those expected for all physicians. There was, however, a considerable excess in the suicide rate, 6.2%, as opposed to the expected 1.5%.

There have been reports of a variety of illnesses and diseases among operating theatre personnel. Excluding minor subjective complaints like headaches, fatigue, etc., results from a combination of the largest retrospective survey in the United States and two large retrospective surveys in the United Kingdom showed that male anaesthetists have an increased incidence of hepatic disease compared with non-anaesthetist physicians (Spence et al., 1977). However, in the United States male nurse anaesthetists and technicians in operating theatres did not show increased rates of hepatic disease (Cohen et al., 1974). Statistically significant increases in rates of hypertension, arrhythmias, disk disease and peptic ulcer were also found in the United Kingdom (table 2).

Reproduction and teratogenicity

A large number of retrospective questionnaire studies have found increased rates of diagnosed spontaneous abortions among women working in operating theatres (Vessey, 1978). In the majority of studies the incidence has ranged from 16 to 38% (percentage of pregnancies

Table 2. Disease rates among anaesthetists and non-anaesthetist physicians in the United Kingdom and the United States[1]

	United Kingdom		United States	
	Anaesthetists	Non-anaesthetist physicians[2]	Anaesthetists	Paediatricians
Sample size	1 407	3 502	5 828[3]	2 337[3]
Cancer (excl. skin)	1.07	0.79	0.70	0.70
Gall bladder	1.34	0.48[4]	0.93	0.99
Liver disease	3.09	1.79[5]	4.90	2.60[4]
Hepatitis	2.68	1.60[5]	3.20	1.70[4]
Serum hepatitis			0.51	0.11[4]
Other liver disease	0.41	0.19	1.20	0.80
Myocardial infarction	1.72	1.78	1.75	1.61
Hypertension	1.80	0.80[4]	2.31	2.47
Arrhythmias	0.70	0.30[4]	0.75	0.70
Disk disease	1.44	0.53[4]	1.27	1.47
Peptic ulcer	2.30	1.20[4]	1.95	1.67
Ulcerative colitis	0.32	0.24	0.24	0.08[5]
Migraine	0.15	0.00	0.23	0.06[5]
Kidney disease (excl. pyelonephritis and infections)	1.70	2.10	4.20	4.60
Renal lithiasis	1.27	1.14	3.07	3.27

[1] Rates per 100 respondents for exposed male anaesthetists and non-anaesthetist physicians; UK data adjusted to US population. Reprinted by permission from A. A. Spence et al., *Journal of the American Medical Association* (Chicago), 1977, 238/9 (955-959). [2] Excluding surgeons and radiologists; 74% of the physicians were working outside the hospital. [3] Sample size varies slightly with type of disease. [4] Significant, $p < 0.01$. [5] Significant, $p < 0.05$.

ending in abortion) compared to 9 to 15% in control subjects. Nurses working in intensive care units also had an incidence of spontaneous abortion similar to anaesthesia nurses.

Based on retrospective information it has been suggested that the incidence of congenital malformation is higher in children of occupationally exposed nurse anaesthetists than in children of those not exposed during or just before pregnancy (Cohen et al., 1974). In the same study female anaesthetists who were occupationally exposed during pregnancy were reported to have given birth to children with a similar incidence of congenital malformations than those who did not work during pregnancy. In the above-mentioned combined United States-United Kingdom study it was further calculated that the incidence of malformations in children of wives of anaesthetists (5.0%) was higher than in those of wives of control subjects (3.7%) (Spence et al., 1977). To exclude the influence of bias an analysis of registration data was made in Sweden of the outcome of deliveries during 1973 and 1975 of women working during pregnancy in operating theatres. When these were compared with all other women employed in medical work it was found that there was no difference in the incidence of congenital malformations, perinatal death rate, birth weight or incidence of threatened abortions.

Possible causes of health hazards

Gas pollution. Gaseous atmospheric pollution in an operating theatre not only consists of waste anaesthetic gases; the propellants of different sprays, scrubbing agents, cleansing agents, methylmethacrylate (released from surgical cement) and possible decomposition products of the volatile or gaseous agents may also contribute to the atmospheric contamination.

During anaesthetic work in unscavenged operating theatres the average exposure of anaesthetists to N_2O over a 3-h period (anaesthetists not continuously present in operating theatre) has been found to be 269 ppm (range 108-430 ppm), while during a 10-min sampling period (Magill circuit: N_2O 6 l/min, O_2 3 l/min; halothane 2%; anaesthetists present) the average N_2O exposure level has been as high as 3 038 ppm (600-5 380 ppm). Corresponding halothane levels were 3.6 ppm (1-8 ppm) and 52 ppm (12.5-115 ppm) respectively. Valid comparisons between levels obtained in different studies are difficult to make as there are marked variations in gas flow, air conditioning and sampling techniques, but in unscavenged operating theatres the mean levels during anaesthetic work are approximately 600 ppm N_2O and 10 ppm halothane.

Magnitude of gas flow, type of flow circuit and scavenging of waste gases significantly influence the levels of waste gases in room air. In a controlled study, the mean halothane level in the room air during the use of a semi-closed circuit system was 57% of that when a non-rebreathing system was used (table 3). Scavenging reduced the halothane levels by 85% and 91% respectively. In the above-mentioned 3-h study the halothane level decreased to 0.57 ppm when scavenging was used.

The published reports on anaesthetic exposure in recovery rooms show great variation. Calculated "actual" exposure of N_2O has been found to be 10-34 ppm, while individual very high levels (up to 1 660 ppm) have been observed. Halothane levels in recovery rooms have ranged from 0 to about 8 ppm.

During the process of mixing plastic surgical cement, the operating theatre personnel are temporarily exposed to toxic methylmethacrylate vapours. During a hip join operation a peak concentration of 227 ppm has been

Table 3. Mean combined concentrations of halothane determined at 1-, 2-, and 3-ft perpendiculars to the exhalation port and at 0- through 6-ft levels. Samples (n = 423) collected at 30-min intervals

System	Concentration of halothane in ppm	Percentage reduction with scavenging
Non-rebreathing system:		
No scavenging	8.69 ± 0.91	–
With scavenging	0.79 ± 0.15	91
Semiclosed circuit system:		
No scavenging	4.93 ± 0.96	–
With scavenging	0.73 ± 0.10	85

[1] Reprinted by permission from C. E. Whitcher et al., *Anaesthesiology* (Philadelphia), 1971, 35/4 (348-353).

recorded immediately upon mixing, but the level decreased rapidly to below 10 ppm in 11 min (the US Occupational Safety and Health Act sets a standard of 100 ppm in 8 h). Its possible contribution to health problems in operating theatre staff is uncertain, but its inhalation may cause symptoms such as headache, nausea and gastrointestinal upset.

The freon gases, dichlorodifluoromethane and trichlorofluoromethane, used in the aerosol sprays of wound dressing agents and local anaesthetics, have a low degree of toxicity. The scrubbing and cleansing agents vary from hospital to hospital but commonly comprise a high proportion of ethanol. In a Swedish study operating theatre nurses were shown to be exposed to mean levels of 12-15 ppm of ethanol.

During surgical procedures in which nitrous oxide is administered, increased concentrations of nitric oxide (NO) and nitrogen dioxide (NO_2) may occur in the operating theatre air. The levels of these toxic oxides, which are apparently oxidisation products of N_2O during interaction with energy-releasing devices, are far below accepted American industrial hygiene limit levels.

Other causes of health hazards. These other causes are largely speculative and include infections, stress, radiation (X-ray, UV, microwaves), smoking habits and selected population.

It has not been possible to quantify these other causes, which have been suggested by the fact that they experimentally affect reproduction, with respect to health hazards. The widely discussed problem of stress in anaesthetic work is possibly associated with the above-mentioned high rate of suicide among American anaesthetists and the high rates of certain types of disease (hypertension, peptic ulcer, migraine, ulcerative colitis) (table 2). Mean exposure of anaesthetists to X-rays has been shown to average a low level of 13 milliroentgens/week during normal activities.

The effect of smoking is also difficult to evaluate, but smoking mothers who work in operating theatres appear to have a high risk of spontaneously aborting.

Psychomotor performance

Results from psychomotor tests are controversial. Slight impairment of ability to perform certain psychomotor tests during exposure to 500 ppm N_2O with or without 10-15 ppm halothane, or 50 ppm N_2O with or without 1 ppm halothane in volunteers has been reported. This has not been substantiated by more recent, partially duplicating studies. As a matter of fact, the latter investigations showed no adverse effects on the psychomotor study tasks. Whether such laboratory test conditions are comparable to operating theatre work and environment has been questioned.

Animal studies

Although much has been published about anaesthetic effects on experimental animals, very little can be considered relevant for the evaluation of occupational health hazards. In one of the most significant series of animal experiments in this respect rats were exposed to trace concentrations of N_2O and halothane (50 or 500 ppm N_2O with or without 1 or 10 ppm halothane) 7 h per day, 5 days per week, for up to 104 weeks. No teratological effects and abortifacient effects were observed following exposure of pregnant female rats during organogenesis, nor were these complications seen when males were pre-exposed to contaminated environment. On the other hand, in long-term exposed males some damage to chromosomes in bone marrow cells and in spermatogonial cells was noted.

In rats exposed for 104 weeks to the above-mentioned concentrations of the anaesthetics no evidence was found for exposure-related effects on body weight, appearance, behaviour or survival rates. No increases in rate of cancer were observed.

The Ames test for mutagenicity has been negative for halothane and modern ether inhalation anaesthetics (methoxyflurane, enflurane, isoflurane). Only with help of microsomal homogenates from induced livers did the discontinued anaesthetic fluroxene, at high concentrations, give a positive test response.

Control of waste gas pollution

Although the association between chronic occupational exposure to trace concentrations of inhalation anaesthetics and health hazards remains unproven much effort has already been spent on reducing and controlling exposure. In some countries governmental agencies have recommended "safe" levels of the commonest anaesthetics. The Hospital Engineering Co-operative Group in Denmark (1974) has recommended permissible average concentrations in the breathing zone of anaesthesia personnel of 10 ppm N_2O and 1 ppm halothane. The National Institute of Occupational Safety and Health in the United States (1977) has recommended that occupational exposure to nitrous oxide, when used as the sole anaesthetic agent, shall be controlled so that no worker is exposed to time-weighted average concentrations greater than 25 ppm. For the halogenated anaesthetic agents, the recommendation is 2 ppm. In Sweden the Labour Protection Board (1978) has issued a regulation allowing a highest average halothane level of 5 ppm.

The technical details of how the waste gases should be reduced are beyond the scope of this article. Useful recommendations and methods are presented in a recent review (Whitcher, 1980). It appears that maximal reduction of waste gases is achieved by having a proper non-recirculating air conditioning system (minimum of 20 exchanges per hour), a safe scavenging system, which includes collection of waste gases at the anaesthetic breathing system and their removal to the outside, and by using low flows of anaesthetic gases. Filling of vapourisers should preferably be done through closed attachments and spilling of volatile agents must be avoided. Leak tests should be performed on the anaesthetic circuits and a regular programme for testing the quality of the air in the operating theatres has been considered mandatory.

HEALTH MEASURES

No particular group of disease can be regarded as specific or typical to anaesthetists. Thus it may be recommended that the medical check-up should be the same as for other medical hospital employees, performed at regular intervals. Severe infectious diseases, which are transmitted through contact with blood or blood products, offer a potential health hazard to anaesthetists and therefore tests for liver function and serum hepatitis have been included in many anaesthesia departments. Anaesthetists working in orthopaedic theatres or elsewhere frequently exposed to X-rays should wear a dosimeter.

Female operating theatre personnel known to have had habitual abortions or with threatening obstetrical complications must be allowed a lighter work load and sufficient rest when pregnant (see EMBRYOTOXIC, FETOTOXIC AND TERATOGENIC EFFECTS).

ROSENBERG, P. H.

"Trace concentrations of anaesthetic gases: a critical review of their disease potential". Ferstanding, L. L. *Anesthesia and Analgesia: Current Researches* (Cleveland), 1978, 57 (328-345). 126 ref.

"Epidemiological studies of the occupational hazards of anaesthesia–a review". Vessey, M. P. *Anaesthesia* (London), May 1978, 33/5 (430-438). 14 ref.

"Occupational disease among operating room personnel: a national study". Cohen, E. N.; Brown Jr., B. W.; Bruce, D. L.; Cascorbi, H. F.; Corbett, T. H.; Jones, T. W.; Whitcher, C. E.; *Anaesthesiology* (Philadelphia), Oct. 1974, 41/4 (321-340). 44 ref.

"Occupational hazards for operating-room-based physicians. Analysis of data from the United States and the United Kingdom". Spence, A. A.; Cohen, E. N.; Brown Jr., B. W.; Knill-Jones, R. P.; Himmelberger, D. U. *Journal of the American Medical Association* (Chicago), 29 Aug. 1977, 238/9 (955-959). 10 ref.

"Occupational hazards to reproduction and health in anesthetists and paediatricians". Rosenberg, P. H.; Vänttinen, H. *Acta Aanesthesiologica Scandinavica* (Aarhus, Denmark), 1978, 22/3 (202-207).

"Methods of control". Whitcher, C. E. (117-148). *Anesthetic exposure in the workplace.* Cohen, E. N. (ed.) (Littleton, Massachussetts, PSG Publishing Company Inc., 1980).

Ancylostomiasis

Of all the helminthic infections, ancylostomiasis or hookworm infection is considered to be one of the worst because of its pathogenic effects and widespread dissemination. There are at least two species of hookworms–*Ancylostoma duodenale* and *Necator americanus*–these two main types can be easily differentiated by examining the adult worm (figure 1). This is of interest to parasitologists and epidemiologists. Clinically their infection is similar. They produce a debilitating disease characterised by malnutrition and anaemia, which are more serious in previously undernourished individuals. The blood sucking activity of the parasite leads to a hypochromic microcytic anaemia. If infection is heavy this can impede mental and physical development. Although this is rarely a direct cause of death, infection with hookworms makes one vulnerable to a wide variety of diseases.

Life cycle. Adult hookworms (male and female) are present in the small intestine of infected men and women. The mature female lays eggs that are discharged in the faeces. The eggs are not infectious; they need warm moist soil with suitable nutrients to develop into the larvae that are infectious. This takes about 7 to 10 days. The larvae can penetrate the intact skin of persons coming into contact with them. This can produce a severe dermatitis commonly called ground itch. The larvae migrate along the lymphatics and bloodstream and reach the lungs. From the alveoli, they

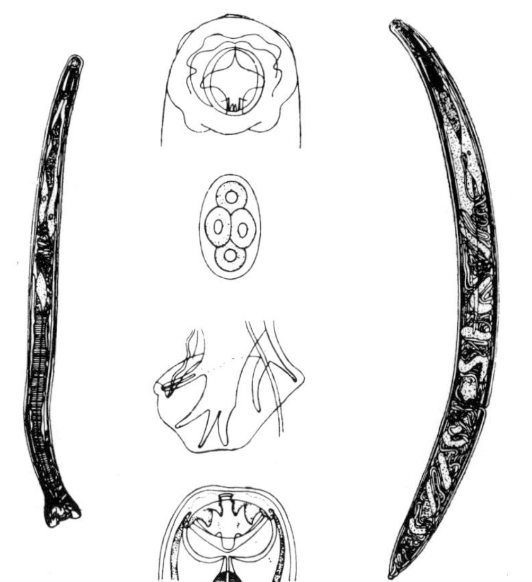

Figure 1. *Ancylostoma duodenale:* left, male; right, female; centre, from bottom to top, mouth, anus, egg. *Necator americanus:* top centre, mouth.

reach the air passages, go up the trachea to the throat and are swallowed. They reach the small intestine and it takes about five weeks for them to mature and lay eggs.

Occupational risk. Workers in any occupation in tropical and temperate regions that brings them in contact with hookworm larvae are likely to develop the infection. This has been a classical occupational infection among miners and underground workers because the poor standards of sanitation as well as temperature and the humidity of the soil are well suited for the development of the larvae. Historically, outbreaks among miners were reported in Hungary and France towards the end of the 18th century. During the construction of the St. Gotthard tunnel in Switzerland in 1880 there was a severe outbreak of anaemia with high fatality affecting about 10 000 workmen.

Hookworm infection among agricultural workers has been reported in China, Egypt, Southeast Asia, Africa, Latin America and the southern part of the United States. Agricultural workers and plantation workers develop the disease whenever they work under insanitary conditions and come into contact with soil that promotes the growth of the larvae. People who work in rice fields or in plantations that grow coffee, sugar, bananas, cocoa, cotton and tobacco have often developed hookworm disease. Brick workers, pottery workers, tile workers, kitchen workers, gardeners, florists and a variety of farm workers may get hookworm disease as an occupational illness. Many of these workers are undernourished and for them hookworm infection represents a serious disease. Large numbers of women work as agricultural workers in many parts of the world and suffer from severe anaemia as a result of this infection. Their haemoglobin level is reduced to critical levels and they are susceptible to a wide variety of infections. Due to poor sanitary conditions they are often infected by more than one parasite, and the heavy parasitic load, combined with malnutrition, lowers their life expectancy, especially at early reproductive ages. Child labour is also common in many of these communities and children pay a severe price in health; their physical and mental development is arrested and it is not uncommon to find whole communities severely debilitated by hookworm infection.

Hookworm infection is diagnosed by examining the faeces for hookworm eggs or larvae. Special techniques are available for detecting mild infections. The counting of eggs in the faeces can give an indication of the severity of the worm burden. A count of 1 000 eggs per gramme of faeces is the minimum permitting a diagnosis of disease. Healthy carriers may have up to 500 eggs per gramme of faeces.

PREVENTIVE MEASURES

The community prevention of hookworm disease requires study of the local conditions that favour infection or reinfection. An adequate number of infected people in a community are the reservoir of infection. The environmental conditions are important; sandy humus with abundant nutrients, shade, moisture and warmth provide an ideal environment for the larvae to develop. Once the soil is loaded with these larvae, every contact with human skin is an opportunity for infection. Being barefooted or handling of the soil, the habit of indiscriminate defecation or using human excreta as fertiliser are all conditions favouring development of the infection. The provision of adequate sanitary toilets, the education of the people and the use of footwear will reduce the risk of infection. A diet rich in iron and other nutrients will minimise its effects.

Treatment. The treatment for individual cases is by Mebendasole, Pyrantel or Bephenium. Iron supplements are of value to combat the anaemia. The frequency of reinfection makes individual treatment of cases questionable in endemic areas.

Larva migrans. In addition to the two parasites mentioned there are other species of hookworms that occur in dogs and cats. Creeping eruption is a skin condition resulting from exposure to the infective larvae of *Ancylostoma braziliense* or *Ancylostoma caninum.* These are intradermal lesions produced by the larvae that are unable to penetrate the complete skin. They produce a serpiginous tunnel between the layers of the skin. The lesion becomes erythematous, elevated and filled with fluid. Since the larvae move a few millimetres a day this is also called *Larva migrans.* Cases have been reported among tea plantation workers in India, and in many other parts of the world. In the United States people who repair the plumbing under summer cottages frequently get this condition. Treatment is by local application of ethyl chloride spray or symptomatic drugs.

MANOHARAN, A.

Bibliography of hookworm disease, 1920-1962 (Geneva, World Health Organisation, 1965), 168 p.

CCTA/WHO African Conference on Ancylostomiasis. WHO Technical Report Series No. 255 (Geneva, World Health Organisation, 1963), 30 p.

Soil-transmitted helminths. Report of a WHO Expert Committee on Helminthiases. Technical Report Series No. 277 (Geneva, World Health Organisation, 1964), 70 p.

Drug treatment in intestinal helminthiasis. Davis, A. (Geneva, World Health Organisation, 1973), 125 p.

Aniline

Aniline ($C_6H_5NH_2$)

AMINOBENZENE; PHENYLAMINE; ANILINE OIL

m.w.	93.1
d.	1.02
m.p.	−6.2 °C
b.p.	184.3 °C
v.d.	3.2
v.p.	7 mmHg (0.91·10³ Pa) at 20 °C

NH_2

f.p. 70 °C
e.l. 1.3%
i.t. 619 °C

moderately soluble in water and miscible with most organic solvents

pure aniline is a clear, almost colourless, oily liquid with a characteristic odour and which darkens with age to a brown colour.

TWA OSHA	5 ppm 19 mg/m³ skin
TLV ACGIH	2 ppm 10 mg/m³ skin
STEL ACGIH	5 ppm 20 mg/m³
IDLH	100 ppm
MAC USSR	0.1 mg/m³ skin

Production. There are two principal commercial processes at present in use for the manufacture of aniline.

The first and most commonly used of these is the catalytic hydrogenation of nitrobenzene at a temperature of 250-300 °C under a pressure slightly above atmospheric (0.5 to 1 bar). As catalyst, a neutral support of silica or kaolin impregnated with a metal is employed. The metal used may be nickel, iron or copper or any of the other hydrogenation catalysts such as the oxides of the heavy metals (chrome, vanadium, molybdenum, etc.). The second common commercial process is the amination of phenol at a temperature of around 400-500 °C over activated aluminium.

Uses. Aniline is widely used in the manufacture of synthetic dyestuffs. It is also used in the manufacture of rubber as a vulcanising agent in the form of mercaptobenzothiazole and its derivatives, as an antioxidant in the form of aniline-acetone condensate or phenyl β-naphthylamine, and as an antiozone agent such as the *p*-phenylenediamines and the diphenylamines, etc. A further important use of aniline is in the manufacture of *p,p'*-methylenebisphenyldiisocyanate (MDI), and this accounts for at least half of its world consumption.

HAZARDS

Aniline is a flammable liquid and a moderate fire hazard. The danger of industrial exposure to aniline arises from the ease with which it can be absorbed either by inhalation or from skin absorption. Absorption through ingestion, which can have serious consequences, does not fall within the field of occupational pathology. Because it is moderately volatile, hazardous concentrations of vapour can easily arise in industrial conditions. It is lipid soluble, and can be readily absorbed through the intact skin, particularly in the form of a liquid, but also to a lesser extent as the vapour. The risk is influenced by temperature, sweating, lipid and water solubility, skin damage and the use of solvents. It was demonstrated experimentally that aniline vapour can be absorbed via the skin and respiratory tract in approximately equal amounts; however, the rate of absorption of the liquid through the skin is about 1 000 times greater than that of the vapour. The most frequent cause of industrial poisoning was found to be accidental skin contamination, either directly through accidental contact, or indirectly through handling soiled clothing or footwear. The use of clean and suitable protective clothing and rapid washing in case of accidental contact constitute the best protection.

Acute poisoning by aniline and its homologues and by most of its derivatives results from the inhibition of the haemoglobin function through the formation of methaemoglobin. Methaemoglobin is normally present in the blood at a level of about 1-2% of the total haemoglobin. Cyanosis at the oral mucosae begins to become apparent at levels of 10-15%, though subjective symptoms are normally not experienced until methaemoglobin levels of the order of 30% are reached. With increases above this level, the patient's skin colour deepens; later, headache, weakness, malaise and anoxia occur, to be succeeded, if absorption continues, by coma, cardiac failure and death. Most cases of acute poisoning react favourably to treatment and the methaemoglobin disappears completely after 2 to 3 days. The consumption of alcohol is conducive to and aggravates acute methaemoglobin poisoning. Haemolysis of the red blood cells can be detected after severe poisoning, and is followed by a process of regeneration which is demonstrated by the presence of reticulocytes. The presence of Heinz bodies in the red blood corpuscles may sometimes also be detected.

In cases of relatively mild poisoning the administration of oxygen can lead to an improvement in the condition of the patient. In more serious cases intravenous injection of 10 ml of methylene blue leads to a recession of the condition that is frequently spectacular, but this treatment should not be repeated more than once or twice as it has been known to lead to haemolysis in cases where there is a genetic deficiency of glucose-6-phosphate dehydrogenase. Vitamin C injections have been suggested as a means of helping the release of the haemoglobin and likewise the injection of glucose solution.

The biological effects of many aniline derivatives are similar to aniline itself. Some may vary in degree and others may differ in their action or in the organs they attack; these compounds are dealt with under AMINES, AROMATIC. Although aniline salt has a very similar toxicity to aniline itself, it is water soluble and not lipid soluble and is not readily absorbed through the skin, or by inhalation. Poisoning from industrial exposure is rare.

The degree of exposure may be determined by means of the quantitative analysis of the methaemoglobin in the bloodstream. Although this measurement may sometimes be useful in cases of acute poisoning, it is less useful for routine checking. It has nevertheless been suggested that cases where the methaemoglobin level exceeds 5% should receive close attention, and where it exceeds 10% action should be taken.

A relative assessment of the degree of exposure can be based on the metabolite level. The content of *p*-aminophenol in the urine should not exceed 50 mg/l (for reasons of safety the biological MAC level is fixed at 10 mg/l). Other authors have, however, described levels of 88 or 115 mg/l in cases of regular exposure to concentrations of aniline not exceeding 10 mg/l³ of air, where there have been no harmful consequences.

Where the conditions of work are unfavourable, haematological examinations with the estimation of haemoglobin levels and investigation for reticulocytes should be performed.

SAFETY AND HEALTH MEASURES

The prevention of aniline poisoning requires high standards of industrial hygiene. Both workers and supervisors should be educated to be aware of the nature and extent of the hazard and to carry out the work in a clean, safe manner. The most important specific measure for the prevention of spillage or contamination of the working atmosphere with aniline vapour is proper plant design. Work clothing should be changed daily and a bath or shower should be obligatory at the end of the working period. Any contamination of skin or clothing should be washed off immediately and the individual kept under medical supervision. The plant should be made safe before maintenance work is carried out. Fire-fighting materials include foam, carbon dioxide, carbon tetrachloride, and dry chemical, but not water.

Treatment. Since most cases of aniline poisoning result from contamination of the skin or clothing, leading to absorption through the skin, clothing should be removed (and laundered). Even when the intoxication results from inhalation, the clothing is likely to be contaminated and should be removed. The entire body surface, including hair and finger nails, should be carefully washed with soap and tepid water.

SCOTT, T. S.
MUNN, A.
SMAGGHE, G.

Occupational health:

"Aniline". *IARC monographs on the evaluation of carcinogenic risk of chemicals to man.* Vol. 4: *Some aromatic amines, hydrazine and related substances, N-nitroso compounds and miscellaneous alkylating agents* (Lyons, International Agency for Research on Cancer, 1974) (27-39). 56 ref.

Monitoring:

CIS 80-1318 "The development of a passive dosimeter for airborne aniline vapors". Campbell, J. E.; Konzen, R. B. *American Industrial Hygiene Association Journal* (Akron), Mar. 1980, 41/3 (180-184). Illus. 16 ref.

"Biological monitoring for industrial exposure to cyanogenic aromatic nitro and amino compounds". Linch, A. L. *American Industrial Hygiene Association Journal* (Akron), July 1974, 35/7 (426-432). Illus. 8 ref.

"Continuous investigation of paraaminophenol in urine of workers dealing with aniline during two shifts including breaks" (Průběžné sledováni para-aminofenolu v moči u pracujicich s anilinem v obdobi dvou pracovnich směn včetně pracovniho volna). Kuzelova, M.; Kunor, V.; Merhaut, J. *Pracovni Lékařství* (Prague), 1970, 22/4 (126-129). Illus. 21 ref. (In Czech)

Safety and health:

CIS 80-1304 *Aniline.* Data Sheet 1-409-79, Revised 1979 (National Safety Council, 444 North Michigan Avenue, Chicago) (1979), 6 p. 12 ref.

Animal experimentation

The principles applicable to animal experimentation in the fields of occupational medicine and industrial hygiene are similar to those in the field of drugs and food additives. It is common practice to attempt the prediction of human health hazards from experimental studies in several species of laboratory animals and to base on them rational guidelines for the safe handling of toxic materials.

Awareness of the use of a potentially dangerous material is the first essential. Secondly, when making predictions relevant to man, emphasis is put on observations in those species most likely to react to and to metabolise the foreign material in the same way as man. Thirdly, the effect of any exposure is a function of duration of exposure and concentration of the substance in the environment examined. Fourthly, a full knowledge of the biological effects exerted in animals permits an estimate of the dose-response relationships likely to be encountered during human exposure and may give a clue to possible interactions with other important environmental agents. Fifthly, however well designed and expertly executed the initial animal laboratory work and despite apparent knowledge of the biological properties and expected side-reactions, there is always a need to supervise closely every person exposed to the material tested in order to detect unexpected human reactions.

Choice of animals

The selection of a suitable species of laboratory animal for toxicological investigations depends on the type of human health hazard being studied and the way in which human exposure could occur. It will also depend on whether a biological, pharmacological, metabolic or biochemical experiment is planned. No single animal species can be regarded as the ideal model for elucidating the human response to toxic agents. A species which handles the active agent biologically in a similar manner to man is to be preferred. There is no clear scientific evidence for according the results of tests in non-human primates, e.g. monkeys, preferential consideration to those obtained in other species. Nevertheless monkeys might be advantageous when assessing neurophysiological and behavioural hazards. The serious disadvantages of primates are their vastly more troublesome husbandry problems, the real hazard of infection with organisms pathogenic to man, and their high cost.

Because species differ widely in their reactions to toxic agents, it is essential in animal testing to use several species and at least two from different phyla. A variety of strains of small laboratory animals is now commercially available. Highly inbred lines are genetically fairly homogeneous and are useful for comparative investigations, while randomly bred animals from a closed colony or F/1 (first-generation) hybrids are preferable for general toxicity testing. Caesarean-section-derived, specific-pathogen-free animals, e.g. rats, mice, guinea-pigs, rabbits, ensure good survival and are well documented as regards their liability to spontaneous disease. Usually, young weanling rats or mice, young adult beagles (1 year) or young adult rhesus monkeys (3.5-4 years) are employed but other species such as guinea-pigs, golden or Syrian hamsters, ferrets, rabbits, cats and pigs have been used. Most of these suffer from uncertainty regarding the purity of the genetic make-up of the colony, and their liability to natural disease; they also require the strict maintenance of healthy, well managed colonies and careful stocking up.

For acute single-dose and cumulative-dose studies it is usual to investigate several species, e.g. rat, mouse, guinea-pig, rabbit, dog, by the oral and parenteral route of administration. For testing dermal toxicity the rat and rabbit are preferred, though the cat is sometimes also chosen. Cutaneous sensitisation and antigenicity are best examined in the guinea-pig because of the genetic predisposition found in some strains to develop sensitisation easily.

Irritancy is best observed in the intact or abraded skin, the conjunctival sac and the cornea of the rabbit. The fowl is useful for detecting some neurotoxic effects while behavioural studies make use of rats, mice, cats, dogs and monkeys, the final choice depending on the particular nervous function to be investigated. Inhalational toxicity may be studied in any common laboratory species.

Subchronic studies, which by definition extend over 10% of the life-span of the species, are normally carried out in a rodent and non-rodent species, e.g. rats, mice and beagle dogs, but other species may be used. For long-term (chronic) studies the rat, mouse and, sometimes, the Syrian hamster are preferred. Reproduction and teratogenicity studies are usually performed in the rat and rabbit, though the mouse is sometimes also used. For investigation of the carcinogenic potential of a substance, studies in rats, mice and Syrian hamsters are acceptable, although exceptionally dogs and monkeys have been used despite the grave disadvantages of the long duration necessary (7 years in dogs, 10-15 years in monkeys) and the difficult husbandry problems involved. Mice have traditionally been used for studying carcinogenic action on the skin.

Recent developments in molecular biology and genetic toxicology have afforded simpler, cheaper and more rapid procedures for assessing the carcinogenic and/or mutagenic potential of a substance. Transmissible body cell mutations are now considered to be one of the important mechanisms underlying the process of carcinogenesis, while heritable germ cell mutations, whether or not phenotypically expressed, induce permanent changes in the genome of the species. For any environmental substance it is therefore important to be as aware of its mutagenic as of its carcinogenic potential, because most mutations adversely affect the genetic pool of a species. Evidence for mutagenic potential rests on demonstrating either point mutations in genetic loci of the DNA, the genetic material of germ or body cells, or on demonstrating numerical or structural changes in the chromosomes of dividing cell nuclei. Existing *in vitro* procedures use specially constructed bacteria, certain yeasts or cell cultures of plant, mammalian or human origin as indicator systems. These serve to determine the capability of a substance to react with the DNA of the nuclear genome so as to induce recognisable DNA alterations or frank mutations in the test organisms employed. The advantages of this *in vitro* approach are the comparative simplicity of the test design, the short duration until results become apparent, and the saving in laboratory animals. The scientific basis for the use of these tests is the accumulating evidence of a close correlation between mutagenic and carcinogenic potential. The *in vivo* procedures presently in use employ a variety of species ranging from insects such as *Drosophila* to laboratory rodents to demonstrate directly or indirectly the occurrence of somatic or gene mutations in the host or its progeny.

Toxicity tests used

The full toxic potential of any substance in the human environment to which exposure can occur may be gleaned from the results of a sequence of experimental animal studies supplemented by relevant epidemiological and clinical observations in man. Such animal tests should always precede human exposure as they are likely to reveal the parameters for subsequent human studies, e.g. target organs, function tests on affected organs, early visible reversible changes, and biochemical abnormalities.

Toxicological studies in animals may be divided into the following five groups:

(1) *(a)* Acute toxicity tests represent a rapid assessment of the immediate hazard to be expected from a substance on acute exposure. Results are frequently expressed as LD_{50} (mean dose killing 50% of the test animals) or in analogous terms. These tests offer some clue of the likely target organs and the nature of the expected toxic effects.

(1) *(b)* Subchronic toxicity studies extend over 10% of the life-span, while chronic toxicity studies extend over the greater part of the life-span of the experimental animal species (e.g. 24 months in mice, 30 months in rats). They aim at determining a no-adverse-effect dose level by comparison with contemporary controls kept under similar conditions but without being administered any test substance. Life-span studies, designed with emphasis on the detection of tumorigenic potentialities, are essentially bioassays for carcinogenicity.

(2) Metabolic and pharmacokinetic studies are designed to elucidate the pathways and the dynamics by which the animal body deals with the foreign substance administered. They reveal the mechanisms of action which underlie the observed abnormal biological effects.

(3) Genetic toxicity studies comprise a selection (battery) from among *in vitro* tests involving bacteria (e.g. Ames bacterial reversion assay, *E. coli* rec assay), yeasts (e.g. mitotic recombination assay, sister chromatid exchange test, cytogenetic analysis) and *in vivo* procedures in laboratory mammals (e.g. micronucleus assay, host-mediated assay, cytogenetic analyses of bone marrow and germ cells, dominant lethality assay, heritable translocation test, specific locus test). They are designed to detect the potential of a substance either to induce mutations in germ cells or body cells by covering different genetic endpoints or to produce numerical or structural abnormalities in the chromosomes of the indicator cells.

(4) Reproduction and teratogenicity studies aim specifically at detecting effects on reproductive capacity and on the developing embryo, foetus and offspring.

(5) Special studies, e.g. irritancy of skin and mucosal surfaces, skin sensitisation, immunotoxicity, inhalational toxicity, neurotoxicity, behavioural disturbances, alterations in organ function, etc., are designed to elucidate effects on special organ systems or those resulting from specific types of exposure.

Technical details on toxicological studies in experimental animals may be found in the relevant literature.

Predictive value and extrapolation

Although laboratory models are useful for studying some human diseases and many of the toxic effects seen in man, animal studies suffer from the following drawbacks:

(a) the variability of reactions to toxic agents due to the wide variations in the susceptibility of individual species and strains tested;

(b) the impossibility of detecting and measuring in experimental animals the equivalents of human subjective experience and certain functional changes, visual interpretation, and reaction-time changes;

(c) the existence of toxic effects peculiar to some human individuals such as hypersensitivity, agranulocytosis, and peripheral neuritis.

In addition, animal studies reveal considerable variability due to other factors such as sex, age, nutritional status, hormonal status, and the effects of circadian rhythms.

Species and strain differences are most often due to differences in the metabolic handling of a given substance, hence the necessity for assuming that man will react like the most sensitive animal species tested unless the available metabolic evidence suggests otherwise. Despite phylogenetic differences laboratory rodents have been found to be the most suitable test species for predicting human reactions to foreign chemicals. Larger mammals may occasionally be more appropriate as experimental species, but where larger numbers of animals are required, they are obviously unsuitable. If several unrelated species of vertebrates show a similar pattern of response to a test substance, man is likely to react in the same way. Parallel reasoning applies to biotransformations and organ function tests.

If the metabolism, including pharmacokinetics, of a toxic agent is similar in man and a particular animal species, the latter should be selected for subchronic and chronic studies because the results would then be of higher predictive value and would allow an intelligent forecast to be made about the health hazard of repeated human exposure to doses which are known not to be acutely poisonous. Animal studies are also useful for devising suitable tests in man to detect early clinical disturbances and may suggest ways of treatment of toxic effects.

Particular importance nowadays attaches to the outcome of genetic toxicity studies. Single *in vitro* tests, using one or other indicator system, e.g. Ames bacterial reversion test, cultured cell cytogenetics, are insufficient evidence for establishing clearly the carcinogenic or mutagenic potential of a substance. However, the existing considerable evidence for a close correlation between a positive outcome in these tests and the finding of carcinogenic activity in mammalian bioassays demands that the substance be further investigated to determine the nature of the observed genetic toxicity. If further positive evidence is found, carcinogenicity bioassays must be performed. It is preferable to choose an appropriate selection (battery) of genetic toxicity tests which covers a sufficient number of different genetic endpoints in order to determine whether the substance under test has a mutagenic potential at least for laboratory mammals. If the test battery reveals evidence of chemical reactivity with DNA, of induction of point mutations or chromosomal abnormalities, and of ability to transform normal cultured cells into cells exhibiting malignant appearance and behaviour, the substance is very likely to be a carcinogen for mammals and possibly also for man. In these circumstances the substance should be subjected to carcinogenicity bioassays. If these assays are positive the substance should be controlled like any other carcinogen. If these assays are negative, the substance should be regarded at least as a likely animal mutagen.

The predictive value for human hazard varies for different animal experiments. The discovery of any toxic effect points to the need for close medical supervision of exposed persons in order to detect similar changes in man at an early stage. Although accurate quantitative extrapolation to man of the results of animal experiments is not feasible, chronic studies usually yield information on the no-adverse-effect dose level at which no obvious toxicity can be detected in test animals. From such a no-effect level, a safe dose may be estimated by the use of arbitrary safety factors, which may vary from 10 to 5 000 or more. These safety factors allow for individual variations in human age, weight, genetic make-up, nutritional and health status, and other parameters. Alternatively, the dose-response relationship obtained in chronic tests may allow an estimate of a virtually safe dose of exposure by extrapolation from the experimental high dose situation to the actual lower dose exposure. A variety of mathematical models have been developed for these low dose extrapolations which relate the probability of an effect occurring to the exposure levels encountered in practice. Such a calculated safe dose may then allow the establishment of basic conditions of exposure which are unlikely to lead to any hazard to the health of the individual. There is at present no method available to make quantitative extrapolations of human risk for substances showing mutagenic potential.

Acute percutaneous, irritancy and inhalation tests indicate likely local toxicity effects and are means for comparing the reactivity of individual substances. Organ function tests are of limited value because these studies will only detect gross malfunction. The most reliable information comes from those animal experiments which demonstrate the biochemical basis for any of the observed biological phenomena. Such information is not useful, however, for making any predictions about mutagenic or carcinogenic potential. Specific studies are needed to detect these biological activities. Biochemical mechanisms are now recognised by which certain compounds may inhibit or potentiate the biological effects observed in animal studies, and similar processes are known to occur also in man. Experimental evidence for such activity is important for interpretation (see also CHEMICALS, NEW; CARCINOGENIC SUBSTANCES; MUTAGENIC EFFECTS).

ELIAS, P. S.

Environmental health criteria 6. Principles and methods for evaluating the toxicity of chemicals, Part I (Geneva, World Health Organisation, 1978), 272 p. Illus.

Handbook of mutagenicity test procedures. Kilbey, B. J.; Legator, M.; Nichols, W.; Ramel, C. (eds.). (Amsterdam, New York, Oxford, Elsevier Scientific Publishing Company, 1977), 485 p. Illus.

"Basic principles in selecting animal species for research projects". Hughes Jr., H. C.; Lang, C. M. *Clinical Toxicology* (San Francisco), 1978, 13/5 (611-621). 20 ref.

Pathology of tumours in laboratory animals. Vol. I: *Tumours of the rat.* Part 1. Scientific Publications Series No. 5 (Lyons, International Agency for Research on Cancer, 1973), 216 p.

Pathology of tumours in laboratory animals. Vol. I: *Tumours of the rat.* Part 2. Scientific Publications Series No. 6 (Lyons, International Agency for Research on Cancer, 1976), 315 p.

Pathology of tumours in laboratory animals. Vol. II: *Tumours of the mouse.* IARC Scientific Publications No. 23 (Lyons, International Agency for Research on Cancer, 1979), 671 p.

Animals, aquatic

Among practically all of the divisions (phyla) of aquatic animals are to be found those that are dangerous to man. Men may come into contact with these animals in the course of various activities including surface and subaqua fishing, the installation and handling of equipment in connection with the exploitation of petroleum under the sea, underwater construction, and scientific research, and thus be exposed to the risk of accidents. Bathers and subaqua enthusiasts are also exposed to the same risk. Most of the dangerous species inhabit warm or temperate waters.

Characteristics and behaviour

Porifera. The common sponge belongs to this phylum. Fishermen including helmet divers, scuba and other subaqua swimmers who handle sponges may contract contact dermatitis with skin irritation, vesicles or blisters (colour plate 1 under DANGEROUS SUBSTANCES). The "sponge diver's sickness" of the Mediterranean region is caused by the tentacles of a small coelenterate *(Sagartia rosea)* that is a parasite of the sponge. A form of dermatitis known as "red moss" is found among North American oyster fishers resulting from contact with a scarlet sponge found on the shell of the oysters. Cases of type 4 allergy have been reported. The poison secreted by the sponge *Suberitus ficus* (colour plate 1 under DANGEROUS SUBSTANCES) contains histamine and antibiotic substances.

Coelenterata. These are represented by many families of the class known as *Hydrozoa* which includes the *Millepora* or coral (stinging coral, fire coral), the *Physalia* (*Physalia physalis,* sea wasp, Portuguese man-of-war), the *Scyphozoa* (jellyfish) and the *Actiniaria* (stinging anemone), all of which are found in all parts of the ocean. Common to all these animals is their ability to produce an urticaria by the injection of a strong poison that is retained in a special cell (the cnidoblast) containing a hollow thread which explodes outwards on contact being made with the tentacle and which penetrates the skin. The various substances contained in this structure are responsible for such symptoms as severe itching, congestion of the liver, pain and depression of the central nervous system, and have been identified as thalassium, congestine, equinotoxin (which contains 5-hydroxy-tryptamine and tetramine) and hypnotoxin respectively.

Effects on the individual depend upon the extent of the contact made with the tentacles and hence on the number of microscopic punctures, which may amount to many thousands, up to the point where they may cause the death of the victim within a few minutes. In view of the fact that these animals are dispersed so widely throughout the world, many incidents of this nature occur but the number of fatalities is relatively small. Effects on the skin are characterised by intense itching and the formation of papules having a bright red, mottled appearance (colour plate 2), developing into pustules and torpid ulceration. Intense pain similar to electric shock may be felt. Other symptoms include difficulty in breathing, generalised anxiety and cardiac upset, collapse, nausea and vomiting, loss of conscience and primary shock.

Echinoderma. This group includes the starfishes and sea urchins both of which possess poisonous organs (pedicellariae), but are not dangerous to man. The spine of the sea urchin can penetrate the skin leaving a fragment deeply imbedded; this can give rise to a secondary infection followed by pustules and persistent granuloma (colour plate 3) which can be very troublesome if the wounds are close to tendons or ligaments. Among the sea urchins, only the *Acanthaster planci* seems to have a poisonous spine which can give rise to general disturbances such as vomiting, paralysis and numbness.

Mollusca. Among the animals belonging to this phylum are the cone shells and these can be dangerous. They live on a sandy sea-bottom and appear to have a poisonous structure consisting of a radula with needle-like teeth, which can strike at the victim if the shell is handled incautiously with the bare hand. The poison acts on the neuromuscular and central nervous systems. Penetration of the skin by the point of a tooth is followed by temporary ischaemia, cyanosis, numbness, pain, and paraesthesia as the poison spreads gradually through the body. Subsequent effects include paralysis of the voluntary muscles, lack of co-ordination, double vision and general confusion. Death can follow as a result of respiratory paralysis and circulatory collapse. Some 30 cases have been reported of which 8 were fatal.

Platyhelminthes. These include the *Eirythoe complanata* and the *Hermodice caruncolata*, known as "bristle worms". They are covered with numerous bristle-like appendages or setae containing a poison (nereistotoxin) with a neurotoxic and local irritant effect.

Polyzoa (Bryozoa). These are made up of a group of animals which form plant-like colonies resembling gelatinous moss and which frequently encrust rocks or shells. One variety, known as *Alcyonidium*, can cause an urticarious dermatitis on the arms and face of fishermen who have to clean this moss off their nets. It can also give rise to an allergic eczema.

Selachiis (Chondrichthyes). Animals belonging to this phylum include the sharks and sting-rays. The sharks live in fairly shallow water where they search for prey and may attack man if they encounter him swimming on the surface. Many varieties have one or two large poisonous spines in front of the dorsal fin which contain a weak poison that has not been identified, and these can cause a wound giving rise to immediate and intense pain with reddening of the flesh, swelling and oedema. A far greater danger from these animals is their bite which, because of several rows of sharp pointed teeth, causes severe laceration and tearing of the flesh leading to immediate shock, acute anaemia and drowning of the victim. The danger that sharks represent is a much-discussed subject, each variety seeming to be particularly aggressive. There seems no doubt that their behaviour is unpredictable, although it is said that they are attracted by movement and by the light colour of a swimmer, as well as by blood and by vibrations resulting from a fish or other prey that has just been caught.

Figure 1. Tail fin of a sting-ray.

Sting-rays have large flat bodies with a long tail having one or more strong spines or saws (figure 1), which can be poisonous. The poison contains serotonine, 5-nucleotidase, and phosphodiesterase, and can cause generalised vasoconstriction and cardio-respiratory arrest. Sting-rays live in the sandy regions of coastal waters where they are well hidden, making it easy for bathers to step on one without seeing it. The ray reacts by bringing over its tail with the projecting spine impaling the spike keep into the flesh of the victim. This may cause piercing wounds in a limb or even penetration of an internal organ such as the peritoneum, lung, heart or liver, particularly in the case of children. The wound can also give rise to great pain, swelling, lymphatic oedema and various general symptoms such as primary shock and cardio-circulatory collapse. Injury to an internal organ may lead to death in a few hours. Sting-ray incidents are among the most frequent, there being some 750 every year in the United States alone. They can also be dangerous for fishermen, who should immediately cut off the tail as soon as the fish is brought aboard. Various species of rays such as the torpedo and the narcine possess electric organs on their back, which, when stimulated by touch alone, can produce electric shocks ranging from 8 up to 220 volts; this may be enough to stun and temporarily disable the victim, but recovery is usually without complications.

Osteichthyes. Many fishes of this phylum have dorsal, pectoral, caudal and anal spines which are connected with a poison system and whose primary purpose is defence. If the fish is disturbed or stepped upon or handled by a fisherman, it will erect the spines, which can pierce the skin and inject the poison. Not infrequently they will attack a diver seeking fish or if they are disturbed by accidental contact. Numerous incidents of this kind are reported because of the widespread distribution of fish of this phylum, which includes the catfish, which are also found in fresh water (South America, West Africa and the Great Lakes), the scorpion fish *(Scorpaenidae)* (figure 2), the weever fish *(Trachinus)* (figure 3), the toadfish, the surgeon fish and others.

Figure 2. Dorsal fin of the scorpion fish.

Figure 3. Dorsal fin of the weever fish.

Wounds from these fishes are generally painful, particularly in the case of the catfish and the weever fish, causing reddening or pallor, swelling, cyanosis, numbness, lymphatic oedema and haemorrhagic suffusion in the surrounding flesh (colour plate 4). There is a possibility of gangrene or phlegmonous infection and peripheral neuritis on the same side as the wound. Other symptoms include faintness, nausea, collapse, primary shock, asthma, and loss of consciousness. They all represent a serious danger for underwater swimmers.

A neurotoxic and haemotoxic poison has been identified in the catfish, and in the case of the weever fish a number of substances have been isolated such as 5-hydroxytryptamine, histamine and catecholamine. Some catfishes and stargazers that live in fresh water as well as the electric eel *(Electrophorus)* have electric organs (see under *Selachii* above).

Hydrophiidae. This group (sea snakes) is to be found mostly in the seas around Indonesia and Malaysia and some 50 species have been reported including *Pelaniis platurus*, *Enhydrina schistosa* and *Hydrus platurus*. The venom of these snakes is very similar to that of the cobra, but is 20 to 50 times as poisonous; it is made up of a basic protein of low melocular weight (erubotoxin) which affects the neuromuscular junction blocking the acetylcholine and provoking myolysis. Fortunately sea snakes are generally docile and bite only when stepped on, squeezed or dealt a hard blow; furthermore, they inject little or no venom from their teeth. Fishermen are among those most exposed to this hazard and account for 90% of all reported incidents, which results either from stepping on the snake on the sea bottom or from encountering them among their catch. Snakes are probably responsible for thousands of the industrial accidents attributed to aquatic animals, but few of these are serious, while only a small percentage of the serious accidents turn out to be fatal. Symptoms are mostly slight and not painful. Effects are usually felt within two hours, starting with muscular pain, difficulty with neck movement, lack of dexterity, and sometimes including nausea and vomiting. Within a few hours myoglobinuria will be seen. Death can ensue from paralysis of the respiratory muscles, from renal insufficiency due to tubular necrosis, or from cardiac arrest due to hyperkalaemia.

PREVENTION

Every effort should be made to avoid all contact with the spines of these animals when they are being handled, unless strong gloves are worn, and the greatest care should be taken when wading or walking on a sandy sea bottom. The wet suit worn by skin divers offers protection against the jellyfish and the various *Coelenterata* as well as against snake bite. The more dangerous and aggressive animals should not be molested and zones where there are jellyfish should be avoided, as they are difficult to see. If a sea snake is caught on a line, the line should be cut and the snake allowed to go.

If sharks are encountered, there are a number of principles that should be observed. The feet and legs should be kept out of the water, and the boat gently brought to shore and kept still; a swimmer should not stay in the water with a dying fish or with one that is bleeding; a shark's attention should not be attracted by the use of bright colours or by making a noise or explosion, by showing a bright light, or by waving the hands towards it. A diver should never dive alone.

Treatment

In general treatment should be aimed at alleviating pain, treating any local or general effects of poison, and endeavouring to obviate any complications, particularly primary shock.

In the case of injury by *Coelenterata* or *Annelida*, any tentacles should be detached, using gloves, cotton wool or rags, and then the cnidoblasts or explosive cells, for which an adhesive bandage should be used. The wound should then be washed out with alcohol or oil, a dilute solution of acetic acid, ammonia or bicarbonate of soda. To alleviate the pain a local injection should be given of 0.5-2% procaine or, if necessary, morphine. In the case of a sting from a jellyfish, the administration of cortisone, antihistamine or wasp antivenom would be indicated.

Large wounds from fishes should be washed abundantly with cold salt water or sterilised saline solution and a tourniquet applied above the wound (the tourniquet being eased off every 10-15 min). The wounded limb should be immersed in water at the highest temperature that can be withstood for 30-60 min, failing which a hot compress should be applied; magnesium sulphate can be usefully added.

It is generally advisable to administer antibiotics to prevent infection. The general symptoms such as shock and collapse usually disappear after the administration of cortisone.

Pricks or punctures from cone-shells should be treated by administering neostigmina against the respiratory paralysis; it may be necessary to resort to artificial respiration.

Sea snake bites should be treated by applying a tourniquet immediately–it is not recommended to incise the wound and apply suction–after which the victim should be kept as still as possible, since any movement will tend to accelerate the destruction of the muscles, and removed with all possible haste to a treatment centre. If the symptoms do not amount to more than fright, it is sufficient to administer a placebo.

Bites from sharks and fishes in the *Squalidae* group will result in wounds that have suffered buffeting or stretching with tearing of the flesh and will require treatment for profuse bleeding. Transfusion and surgical intervention will be necessary with the least possible delay.

ZANNINI, D.

Poisonous and venomous marine animals of the world (revised edition). Halstead, B. W. (Princeton, Darwin Press, revised ed., 1978), 1 043 p. Illus. Ref.

Dangerous marine animals. Halstead, B. W. (Cambridge, Maryland, Cornell Maritime Press, 1959), 146 p. 23 ref. Illus.

The biology of sea snakes. Dunson, W. A. (ed.). (Baltimore, London, Tokyo, University Park Press, 1975), 530 p. Illus. Ref.

Man and the under-water world (L'uomo e il mondo sommerso). Molfino, F.; Zannini, D. (Turin, Minerva Medica, 1964), 454 p. Illus. (In Italian)

"Pulmonary hypersensitivity in prawn workers". Gaddie, J.; Legge, J. S.; Friend, J. A. R. *Lancet* (London), 20-27 Dec. 1980, 2/8280 (1 350-1 353). 7 ref.

"An experimental study on the sensitising power of *Alcyonidium gelatinosum (L.)* (a marine bryozoan). I–Skin hypersensitivity reactions" (Etude expérimentale du pouvoir sensibilisant d'Alcyonidium gelatinosum (L.) (Bryozoaire marin). Dubos, M.; Susperregui, A.; Drouet, J.; Niaussat, P. M. *Archives des maladies professionnelles, de médecine du travail et de sécurité sociale* (Paris), 1980, 41/1 (9-13). 5 ref.

Animals, venomous

The animals that can inflict injury on man by the action of their venom include: invertebrates such as *Arachnida* (spiders, scorpions and sun spiders), *Acarina* (ticks and mites), *Chilopoda* (centipedes), and *Hexapoda* (bees, wasps, butterflies, and midges); and vertebrates such as snakes and lizards. *Coelenterata* (medusas and polyps) and certain fish are dealt with in the article ANIMALS, AQUATIC.

Arachnida

Spiders *(Aranea)*

All species are venomous but, in practice, only a few types produce injury in man. Spider poisoning may be of two types:

(a) cutaneous poisoning in which the bite is followed after a few hours by oedema centred round a cyanotic mark, and then by a blister–extensive local necrosis may ensue and healing may be slow and difficult in cases of bites from spiders of the *Lycosa* genus (e.g. the tarantula);

(b) nerve poisoning due to the exclusively neurotoxic venom of the mygales *(Latrodectus ctenus)*, which produces serious injury, with early onset, tetany, tremors, paralysis of the extremities and, possibly, fatal shock–this type of poisoning is relatively common amongst forestry and agricultural workers and is particularly severe in children: in the Amazonas, the venom of the "black-widow" spider *(Latrodectus mactans)* is used for poison arrows.

Prevention. In areas where there is a danger of venomous spiders, sleeping accommodation should be provided with mosquito nets and workers should be equipped with footwear and working clothes that give adequate protection.

Treatment. In Brazil the high incidence of spider poisoning has led to the production of an antitoxic serum which seems to be effective against all species of *Latrodectus.* Topical treatment comprises removal of the venom by excision, suction and irrigation with Dakin's antiseptic (a solution of sodium hypochlorite and boric acid). Additional treatment may include the administration of cardiotonics, analgaesics and antihistamines.

Scorpions *(Scorpionida)*

These arachnids have a sharp poison claw on the end of the abdomen with which they can inflict a painful sting, the seriousness of which varies according to the species, the amount of venom injected and the season (the most dangerous season being at the end of the scorpions' hibernation period). In the Mediterranean region, South America and Mexico, the scorpion is responsible for more deaths than poisonous snakes. Many species are nocturnal and are less aggressive during the day. The most dangerous species *(Buthidae)* are found in arid and tropical regions; their venom is neurotropic and highly toxic. In all cases, the scorpion sting immediately produces intense local signs (acute pain, inflammation) followed by general manifestations such as tendency to lipothymia, salivation, sneezing, lacrimation, diarrhoea. The course in young children is often fatal. The most dangerous species are found amongst the genera *Androctonus* (sub-Saharan Africa), *Centrurus* (Mexico), *Tituus* (Brazil).

The scorpion will not spontaneously attack man and stings only when he considers himself endangered, as when trapped in a dark corner or when boots or clothes in which he has taken refuge are shaken or put on. Scorpions are highly sensitive to halogenated pesticides (e.g. DDT).

Treatment. Persons who have been bitten by a scorpion should receive an immediate subcutaneous antivenom serum in the vicinity of the wound; if the latent period is greater than 20 min, the injection should be at a distance from the wound. Provided the serum is specific, it will prevent the occurrence of severe symptoms and death.

Symptomatic treatment includes the application of a tourniquet above a wound which has occurred not more than 20 min beforehand, removal of venom by suction or incision, and the intravenous injection of 1 mg atropine. Resuscitation may sometimes be necessary (adrenalin, injectable corticosteroids, transfusions).

Sun spiders *(Solpugida)*

This order of arachnid is found chiefly in steppe and sub-desert zones such as the Sahara, Andes, Asia Minor, Mexico and Texas, and is non-venomous; nevertheless, sun spiders are extremely aggressive, may be as large as 10 cm across and have a fearsome appearance. In exceptional cases, the wounds they inflict may prove serious due to their multiplicity. Solpugids are nocturnal predators and may attack a sleeping man.

Ticks and mites *(Acarina)*

Ticks are blood-sucking arachnids at all stages of their life cycle and the "saliva" they inject through their feeding organs may have a toxic effect. Poisoning may be severe, although mainly in children (tick paralysis), and may be accompanied by reflex suppression. In exceptional cases death may ensue due to bulbar paralysis (in particular where a tick has attached itself to the scalp).

Mites are haematophagic only at the larval stage and their bite produces pruriginous, afebrile erythema; the incidence of mite bites is high in tropical regions.

Treatment. Ablation should be carried out following anaesthesia with a drop of benzene, ethyl ether or xylene. Prevention is based on the use of organophosphorous pesticides, pest repellents and local anaesthetics.

Centipedes *(Chilopoda)*

Centipedes differ from millipedes *(Diplopoda)* in that they have only one pair of legs per body segment and that the appendages of the first body segment are poison fangs. The most dangerous species are encountered in the Philippines. Centipede venom has only a localised effect (painful oedema).

Treatment. Bites should be treated with topical applications of dilute ammonia, permanganate or hypochlorite lotions. Antihistamines may also be administered.

Insects *(Hexapoda)*

Insects may inject venom via the mouthparts (*Simuliidae*–blackflies, *Culicidae*–mosquitoes, *Phlebotomus*–sandflies) or via the sting (bees, wasps, hornets, carnivorous ants); they may cause urtication by their hairs (caterpillars, butterflies); or they may produce vesication by their haemolympth (*Cantharidae*–blister flies and *Staphylinidae*–rove beetles).

Black fly bites produce necrotic lesions, sometimes with general disorders; mosquito bites produce diffuse pruriginous lesions.

The stings of *Hymenoptera* (bees, etc.) produce intense local pain with erythema, oedema and, sometimes, necrosis. General accidents may result from sensitisation or multiplicity of stings (shivering, nausea, dyspnoea, chilling of the extremities). Stings on the face or the tongue are particularly serious and may cause death by asphyxiation due to glottal oedema.

Caterpillars and butterflies may cause generalised pruriginous skin lesions of an urticarial or oedematous type (Quincke's oedema), sometimes accompanied by conjunctivitis. Superimposed infection is not infrequent.

The venom from blister flies produces vesicular or bullous skin lesions *(Poederus)*. There is also the danger of visceral complications (toxic nephritis).

Certain insects such as *Hymenoptera* and *Lepidoptera* (butterflies) are found in all parts of the world; other suborders are more localised however. Dangerous butterflies are found mainly in Guyana and the Central African Republic; blister flies are found in Japan, South America, and Kenya; black flies live in the intertropical regions and in central Europe; sandflies are found in the Middle East.

Prevention. Workers who are very exposed to insect bites can be desensitised in cases of allergy by the administration of increasingly large doses of insect body extract.

Treatment. In the case of bee stings, the sting should be removed with tweezers. Various types of supportive treatment can be applied, including: topical or general antihistamine administration, dilute hypochlorite or bicarbonate solutions for conjunctivitis, intravenous calcium, cardiotonics, adrenalin, corticosteroids, and antibiotics for cases of superimposed infection. In the case of multiple stings with severe shock, ACTH or cortisone may be necessary.

Squamata

Serpentes (snakes)

In hot and temperate zones, snake bites may constitute a definite hazard for certain categories of workers: agricultural workers, woodcutters, building and civil engineering workers, fishermen, mushroom gatherers, snake charmers, zoo attendants, and laboratory workers employed in the preparation of antivenom serums.

The vast majority of snakes are harmless to man although a number are capable of inflicting serious injury with their venomous bites; dangerous species are found among both the terrestrial snakes (*Colubridae* and *Viperidae*) and aquatic snakes *(Hydrophiidae)*.

The dangerous *Colubridae* are the proteroglyphodonts (i.e. with fixed anterior fangs with a groove for the injection of venom). These snakes are slender, with an oblong head covered with large flat scales; the pupil is round and the tail relatively long in comparison with the body. The main venomous types are the *Naja* (or cobra) and the *Bungaris*. The venom is usually neurotoxic. The immediate local signs after venom injection are relatively discreet, but general signs appear rapidly and are dominated by paralytic manifestation, with drowsiness, respiratory difficulty, weakened pulse, vomiting, fainting, and relaxation of the sphincters. Following a stage of numbness punctuated by fainting fits, coma sets in and death occurs 5-6 h after the bite.

The *Viperidae* are solenoglyphodonts (i.e. have grooved fangs inserted in the moveable maxilla so that they may be rotated forward when the mouth is open and folded back against the roof of the mouth when not in use), and are always very dangerous. They are squat with a head which is spear-shaped and covered with small scales like the rest of the body (figure 1). The pupil is elongated vertically, the tail is extremely short and the over-all length is 30-75 cm. The main species are the *Vipera* and *Crotalus*. Their venom is extremely haemotoxic and the bite is accompanied by very marked local signs with intense pain and hard oedema. About 6 h after the bite, the oedema becomes haemorrhagic and after 12 h there is extensive, livid blotching with blisters, lymphangitic streaks and adenopathy. The lesions reach their maximum towards the second day with diffuse purpura and necrosis. General signs develop concomitantly and include anxiety, vertigo, respiratory difficulty, increased pulse rate, cooling of the extremities, vomiting, diarrhoea, tachycardia, hypotension and fall in body temperature. Death may occur due to bulbar involvement in 1-3 days. Bites from aquatic snakes are discussed in the article ANIMALS, AQUATIC.

Prevention. Snakes do not usually attack man unless they feel menaced, are disturbed or are trodden on. In regions infested with venomous snakes workers should wear foot and leg protection and be provided with

Figure 1. Above: head of *Vipera aspis*; below: the same, showing the venom gland.

monovalent or polyvalent antivenom serum. It is recommended that persons working in a danger area at a distance of over half-an-hour's travel from the nearest first-aid post should carry an antivenom kit containing a sterilised syringe. However, it should be explained to workers that bites even from the most venomous snakes are seldom fatal since the amount of venom injected is usually small.

Certain snake charmers achieve immunisation by repeated injections of venom, but no scientific method of human immunisation has yet been developed.

Treatment. There are two prime factors that influence prognosis: speed of specific treatment and the proximity of a centre equipped to provide treatment of shock.

Immediately a bite accident has occurred, a tourniquet should be placed above the wound; however, a tourniquet is of value only if applied within 20 min of the accident–afterwards it may produce serious consequences. The patient should make no unnecessary effort, the part of the body affected should be immobilised and the bite washed in cold water. An incision should be made between the two points of the bite, which are always clearly visible on the oedema; the venom should then be sucked out and spat out by a person with no lip or mouth lesions. The wound should be bathed in a hypochlorite or potassium permanganate solution. However, the prime measure is the subcutaneous injection of antivenom serum (in the region of the bite if within 20 min of the accident). If possible, antitetanus serum and antibiotics should also be administered.

Additional treatment may often be essential and will include soluble corticosteroids, analeptics, analgaesics and, in severe cases, resuscitation and oxygen therapy. Finally, isogroup transfusions against collapse, extrarenal dialysis and exsanguinotransfusion may be necessary.

Sauria (lizards)

There are only two species of venomous lizards, both members of the genus *Heloderma, H. suspectum* (Gila monster) and *H. horridum* (beaded lizard). Venom similar to that of the *Viperidae* penetrates wounds inflicted by the anterior curved teeth but bites in man are uncommon and recovery is generally rapid.

RIOUX, J. A.
JUMINER, B.

General:

"Poisoning by animals". George, C. *Medicine* (London), 1972, 4 (317-321). 31 ref.

"Poisonings by terrestrial animals" (Envenimations par animaux terrestres). D'Imeux, A. *Encyclopédie Médico-chirurgicale*, 2-1979, 16078 A-10, 10 p. Illus. (In French)

Snakes:

"Symposium: Snake venoms and envenomation". *Clinical Toxicology* (New York), Sep. 1970, 3/3 (343-511). Illus. Ref.

"Medical toxicologist's notebook: snakebite treatment and International Antivenin Index". Rappolt, R. T.; Quinn, H.; Curtis, L.; Minton, S. A.; Murphy, J. B. *Clinical Toxicology* (New York), Oct. 1978, 13/3 (409-438). 11 ref.

"Snake bites (study concerning the haematotoxic syndrome due to *Viperidae*)" (Morsures des serpents (étude orientée sur le syndrome hématotoxique dû aux vipéridés)). Robert, M. *Médecine et Hygiène* (Geneva), 4 Feb. 1981, 39/1410 (378-384). Illus. 8 ref. (In French)

Bees:

"Acute inflammatory polyradiculoneuropathy following Hymenoptera stings". Bachman, D. S.; Paulson, G. W.; Mendell, J. R. *Journal of the American Medical Association* (Chicago), 12 Mar. 1982, 247/10 (1 443-1 445). 19 ref.

"Occupational accidents in bee-keeping" (Accidentes de trabalho na apicultura). *Saude Ocupacional e Segurança* (São Paulo), 1977, 12/4 (251-261). Illus. 4 ref. (In Portuguese)

Insects:

Insects as a cause of inhalant allergies. A bibliography. Bellas, T. E. Division of Entomology Report No. 25 (Canberra City, Commonwealth Scientific and Industrial Research Organisation, 1981), 64 p. 328 ref.

Anthracene and derivatives

Anthracene ($C_6H_4(CH)_2C_6H_4$; $C_{14}H_{10}$)

m.w.	178.2
sp.gr.	1.28
m.p.	216.4 °C
b.p.	340 °C
v.d.	6.15
e.l.	0.6%
i.t.	590 °C

slightly soluble in most organic solvents and in water

colourless crystals with blue fluorescence when pure, yellow with green fluorescence when impure.

Anthracene is a polynuclear aromatic hydrocarbon with condensed rings which forms anthraquinone by oxidation and 9,10-dihydroanthracene by reduction.

Production. Anthracene oil containing up to 20% pure anthracene is extracted from the heavy fractions of coal tar with boiling points between 270 and 360 °C. Anthracene is obtained by purifying the anthracene oil from admixtures (chiefly carbazole and phenanthrene) with the aid of selective solvents and by subsequent crystallisation.

Uses. Anthracene is used for the production of anthraquinone, an important raw material for the manufacture of fast dyes, and also in the production of synthetic fibres, plastics and monocrystals.

HAZARDS

The toxic effects of anthracene are similar to those of coal tar and its distillation products, and depend on the proportion of heavy fractions contained in it. Anthracene is photosensitising. It can cause acute and chronic dermatitis with symptoms of burning, itching and oedema which are more pronounced in the exposed bare skin regions. Skin damage is associated with irritation of the conjunctiva and upper airways. Other symptoms are lacrimation, photophobia, oedema of the eyelids and conjunctival hyperaemia. The acute symptoms disappear within several days after cessation of contact. Prolonged exposure gives rise to pigmentation of the bare skin regions, cornification of its surface layers and telangioectasis. The photodynamic effect of industrial anthracene is more pronounced than that of pure anthracene, which is evidently due to admixtures of acridine, carbazole, phenanthrene and other heavy hydrocarbons. Systemic effects manifest themselves by headache, nausea, loss of appetite, slow reactions and adynamia. Prolonged effects may lead to inflammation of the gastrointestinal tract.

It has not been established that pure anthracene is carcinogenic, but some of its derivatives and industrial anthracene (containing impurities) have carcinogenic effects. 1,2-Benzanthracene and certain monomethyl and dimethyl derivatives of it are carcinogens. The dimethyl and trimethyl derivatives of 1,2-benzanthracene are more powerful carcinogens than the monomethyl ones, especially 9,10-dimethyl-1,2-benzanthracene which causes skin cancer in mice within 43 days. The 5,9- and 5,10-dimethyl derivatives are also very carcinogenic. The carcinogenicity of 5,9,10- and 6,9,10-trimethyl derivatives is less pronounced. 20-Methylcholanthrene, which has a structure similar to that of 5,6,10-trimethyl-1,2-benzanthracene, is an exceptionally powerful carcinogen. All dimethyl derivatives which have methyl groups substituted on the additional benzene ring (in the 1, 2, 3, 4 positions) are non-carcinogenic. It has been established that the carcinogenicity of certain groups of alkyl derivatives of 1,2-benzanthracene diminishes as their carbon chains lengthen.

Anthracene is also moderately flammable.

Toxicity

Intragastric administration of industrial anthracene (containing 20% pure anthracene) has revealed that it is slightly toxic because of its impurities, whereas pure anthracene does not cause animals to die after a single administration of the maximum possible dose (17 g/kg). The LD_{50} of industrial anthracene is 4.88 g/kg (the animals cease moving after 10-15 min and die 6-24 h later). Repeated poisoning of albino rats gives rise to a decrease in haemoglobin, reticulocytosis, leukopenia, and increase in residual blood nitrogen. Chronic inhalation of anthracene aerosol in concentrations of 0.05 and 0.01 mg/l is associated with a reduced gain in body weight and the same blood changes observed after intragastric administration. After subcutaneous administration of 20 mg anthracene per day for 33 weeks, 5 of 9 animals surviving more than 17 months developed fibromas and sarcomas in the region of injection. Daily skin applications of 40% anthracene with vaseline caused only reddening in guinea-pigs, whereas industrial anthracene gave rise to swelling and soreness. Both pure and industrial anthracene are skin sensitisers. Skin resorption is associated with a slowdown of body weight gain, dystrophic changes in the liver, kidneys and myocardium, and with neutrophilia. Application during 3 weeks caused death in some animals.

Rats eliminate 70-80% of the anthracene given by mouth unchanged with the faeces. Numerous metabolites can be observed in the urine: N-acetyl-S-(1,2-dihydro-2-hydroxy-1-anthryl) cysteine and conjugates of 1,2-dihydroanthracene and trans-1,2-dihydroanthracene-1,2-diol. The latter is a precursor of anthrone and of a number of hydroxylated metabolites.

HEALTH PROTECTION AND MEDICAL SUPERVISION

It is extremely important to avoid skin contact with anthracene and its derivatives, and to prevent the release of anthracene vapours and dust into the workplace atmosphere. Protective measures include tight enclosure of equipment and mechanisation of processes, and replacement or dilution of the most carcinogenic products by less harmful ones. The recommended maximum allowable concentration of anthracene at the workplace is 0.1 mg/m^3. It is advisable to wear work clothing with a protective layer (e.g. polyvinyl chloride), protective footwear and tight goggles; to use skin cleansers after work (55% kaolin, 25% neutral soap, 20% bran); and to take a warm (not hot) shower. Underwear should be supplied and changed daily.

To prevent photosensitisation, it is recommended to limit work in sunlight as far as possible. Photoprotective creams or pastes must be applied to the bare skin regions.

Workers handling anthracene and its derivatives must undergo annual medical examinations. Highly sensitive workers and persons suffering from skin diseases or neoplasms must not be admitted to work with anthracene.

VOLKOVA, N. I.

"Sanitary and hygienic conditions of work in a pilot plant producing pure anthracene and toxicological evaluation of anthracene by experiments" (Sanitarno-gigieničeskie uslovija truda opytnopromyšlennogo ceha čistogo antracena i toksikologičeskaja ocenka antracena v èksperimente). Gudz', Z. A.; Volodčenko, V. A.; Timčenko, A. M. *Proceedings of a Republican meeting of industrial physicians and scientific sessions of the Harkov Institute of Occupational Hygiene and Diseases* (Materialy respublikanskogo soveščanija promsanvračej i naučnye sessii Har'kovskogo NII gigieny truda i profzabolevanij) (Kiev, 1968) (53-55). (In Russian)

"Comparative study of the toxicity of pure and industrial anthracene" (Sravnitel'noe izučenie toksičnosti čistogo i tehničeskogo antracena). Nagornyj, P. A. *Gigiena truda i profzabolevanija* (Moscow), 1969, 5 (59-62). (In Russian)

"Methods of air monitoring and air pollution when introducing new techniques in coal by-product plants" (Metody kontrolja i sostojanie vizdušnoj sredy pri vvedenii novoj tehnologii no koksohimzavodah). Jarym-Agaeva, N. T.; Gorskaja, R. V.; Čubar', L. V. *Aktual'nye voprosy kraevoj gigieny truda Donbassa i fiziologija truda v glubokih ugol'nyh šahtah* (Moscow, 1972). (In Russian)

Criteria for a recommended standard–Occupational exposure to coal tar products. DHEW (NIOSH) publication No. 78-107 (National Institute for Occupational Safety and Health, 4676 Columbia Parkway, Cincinnati) (1977), 189 p. 115 ref.

Anthrax

Anthrax is an infectious disease, primarily of animals from which man may be secondarily infected.

Aetiology

The causal micro-organism is *Bacillus anthracis*, a spore-forming bacterium. It is Gram-positive, rod-shaped, 3-8 μm in length by 1-1.2 μm in diameter, and belongs to the *Bacillaceae* family. In infected animals, the micro-organism occurs as chains of 2-8 bacilli surrounded by a large capsule. When carcasses are dissected and skinned the anthrax bacilli form spores. The spores are resistant to temperature extremes and dehydration, and when they get into the soil, they remain capable of growth for many decades.

Epidemiology

B. anthracis is pathogenic to herbivora, especially sheep, cattle, horses, pigs and goats. Infection in animals is caused by feeding on pasture or fodder contaminated by the excreta or carcasses of sick animals. With animals anthrax always takes the form of acute sepsis and death follows. The sudden death of herbivora, with enlarged spleen and dark, hardly clotting blood are typical manifestations of anthrax infection.

Infection in man occurs most frequently from contact with sick animals or infected animal products, e.g. in agricultural workers, stock farming and breeding, abattoirs, butchery trade, bone and bone meal processing, tanneries, wool industry, hair and bristle processing, ivory and horn processing, etc. The route of infection may be by skin contact, inhalation of dust containing spores or ingestion of infected meat. Anthrax may also be transmitted by blood-sucking insects.

Anthrax is found throughout the world. In the developing countries occupational anthrax morbidity is related to the incidence of the disease in agricultural animals. In industrialised countries occupational infection is almost exclusively the result of contact with imported contaminated animal products. It is reported that conditions for the development of stationary anthrax foci include a mean daily temperature of over 21 °C in the summer months. In Europe the following countries have a high anthrax morbidity: Albania, Bulgaria, Greece, Italy, Portugal, Rumania and Spain. In Asia, anthrax is widely distributed in Iran, Iraq and Turkey, whilst on the American Continent the largest number of registered cases has been in Bolivia, Chile and Venezuela. It is also widely encountered in Africa.

Clinical symptoms

Anthrax may occur in one of three clinical forms: cutaneous, pulmonary and gastrointestinal; all forms may develop into anthrax sepsis. The cutaneous form is by far the most common, accounting for about 95% of all cases; the pulmonary and gastrointestinal forms are relatively rare.

Cutaneous anthrax is the result of the virulent organism being deposited below the skin surface; the incubation period is 1-8 days. At the site of infection a red spot appears, which soon develops into a papule, with a necrotic centre forming after 3-4 days (see colour plates 5 and 6). Large collateral oedema develops around the pustule, often with blisters containing serous fluid. Multiple pustules occur in 4-10% of cases. The cutaneous form usually runs its course without affecting the patient's general state; one report indicated that 12% of the cases studied were non-febrile and 34% sub-febrile.

Pulmonary anthrax is the result of inhalation of dust containing anthrax spores. It is known as "wool sorters disease". It begins suddenly after a short incubation period and assumes the form of severe haemorrhagic pneumonia and within 24-48 h death follows.

Gastrointestinal anthrax is met more frequently in Africa and Asia as a result of consuming infected meat (not boiled or baked enough). With cases not subject to medical treatment, general sepsis occurs quickly and death follows. Our experience proves that early diagnosis of intestinal anthrax helps to control the toxi-infectious phase of the disease by energetic treatment with antibiotics–penicillin and tetracycline. Necrotic changes of the intestinal wall, oedema, mesenteric lymphatic knots and effusions in the body cavities are typical for cases of gastroinestinal anthrax. In cases where the acute phase is under control, perforation of the intestinal wall might endanger the patient's life. Figure 1 shows temperature curves of three patients with intestinal anthrax. The rate of complications does not always correspond to the severeness of the initial toxi-infectious stage. The general condition of the third patient was satisfactory during the initial stage of the illness, but two ulcerous perforations of the intestinal wall occurred and peritonitis developed. All three patients were cured completely. For that reason patients with gastrointestinal anthrax must be observed for at least 14 days with readiness for surgical intervention if perforation occurs.

Diagnosis

Cutaneous anthrax is diagnosed on the basis of epidemiological data (i.e. contact with infected animals or their products) and the characteristic clinical symptoms (indolence of anthrax carbuncle, haemorrhagic necrosis and absence of suppuration). Diagnosis of the intestinal and pulmonary forms of anthrax is difficult and is based mainly on epidemiologic data–consumption of contaminated or suspected meat or data for the patient's occupation–textile worker, etc.

In the case of cutaneous infection examination should be made of cultures from the pustule; sputum should be used in the pulmonary form, faeces in the gastrointestinal form and blood in the septic form. The detection of Gram-positive bacteria situated in double or short chains with cut ends and surrounded by capsules provides sufficient grounds for preliminary anthrax diagnosis. Immunofluorescence has recently been used as a rapid diagnostic technique. The most rapid and sensitive method of anthrax diagnosis is the skin allergic test with "anthraxin". The test is positive on the third to fifth day of the illness. Positive skin reaction occurs with all patients who have suffered, even a number of years previously.

PREVENTIVE MEASURES

The first step in anthrax control is the elimination of anthrax among farm animals. In known anthrax districts susceptible animals should be treated with anti-anthrax vaccine. Animals that die as a result of anthrax must be destroyed by incineration. Disinfection of premises and material can be carried out using a 5-10% activated hypochloride solution. When cases of anthrax occur among farm animals, preventive measures should be taken to eliminate soil pollution. In the Soviet Union and Bulgaria systematic control of anthrax morbidity in endemic regions is carried out, and the influence of geographic and economic factors favourable for the elimination of infection in such regions is considered.

As far as animal products are concerned, veterinary certificates should be demanded for all imported hides or wool testifying that they originate from healthy animals. Hides which have been found by the Ascoli reaction to be infected with anthrax must be destroyed. Wool imported from countries with stationary anthrax should be subject to decontamination. In the United Kingdom the disinfection of wool is carried out by washing at 40 °C, immersion in 2% formaldehyde and then drying at 100 °C. Bacteriological examination of the materials before and after disinfection is made to ensure that the process is efficient.

The International Labour Organisation recommends that arrangements should be made for the disinfection of wool infected with anthrax spores, either in the country exporting such wool or, if that is not practicable, at the port of entry in the country importing such wool. Other control measures include education of employees on the hazard of anthrax and in particular on the early appearance of the lesions. In some countries a cautionary placard has to be posted up in certain factories where this risk may exist. In factories where there is an anthrax risk the dust from opened bales of fibre should be removed by downwards exhaust ventilation. However, in certain circumstances, personal respiratory protective equipment may also be necessary.

Shepherds, shearers, butchers and others at risk in the endemic areas and persons handling animal products, particularly goat hair, in industrial areas have been vaccinated against anthrax and reduction in morbidity has been claimed. In the USSR a live attenuated vaccine is employed. In some countries chemical anti-anthrax vaccine is applied, manufactured on the basis of protective anthrax antigen. In Bulgaria prophylactics with antibiotics is obligatory for all persons who have had contact with sick animals or infected meat. The treatment consists of penicillin (2 x 500 000 daily) or tetracycline (1.5 g daily), for 3 to 5 days.

Treatment

Modern therapy is based on antibiotics. Anti-anthrax serum removes the intoxication and is recommended in severe forms of anthrax (200-400 ml native hyperimmune serum, or 20-60 ml concentrated and purified serum). Antibiotics do not shorten the duration of the cutaneous forms.

KEBEDJIEV, G. N.

General:

Anthrax: health hazards— Guidance Note EH 23 from the Health and Safety Executive (London, HM Stationery Office, 1979), 3 p.

Clinical observations:

CIS 75-1712 "Cutaneous anthrax: observations in 18 cases". Meneghini, C. L.; Lospalluti, M.; Angelini, G. *Berufs-Dermatosen* (Aulendorf in Württemberg), Nov.-Dec. 1974, 6 (233-237). Illus. 7 ref.

Diagnosis:

"Results of the use of the anthraxin test in the epidemiological studies on anthrax in the USSR and Bulgaria" (Rezultati ot izpolzuraneto na antraksinovata proba pri epidemiologičnite provčranija na antraksa v SSSR i NRB). Slyahov, E. N.; Kebedjiev, G. *Epidemiologija, Mikrobiologija i Infekciozni Bolesti* (Sofia), 1978, 15/2 (140-144). 12 ref. (In Bulgarian)

Prevention:

The Anthrax Prevention Recommendation, 1919 (No. 3) (Geneva, International Labour Office).

Abstracts with temperature curves of patients with intestinal anthrax

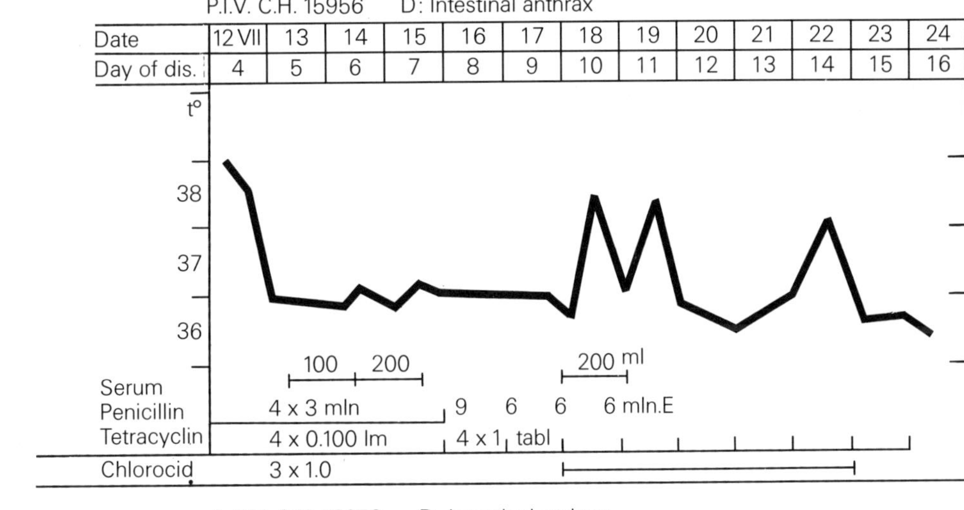

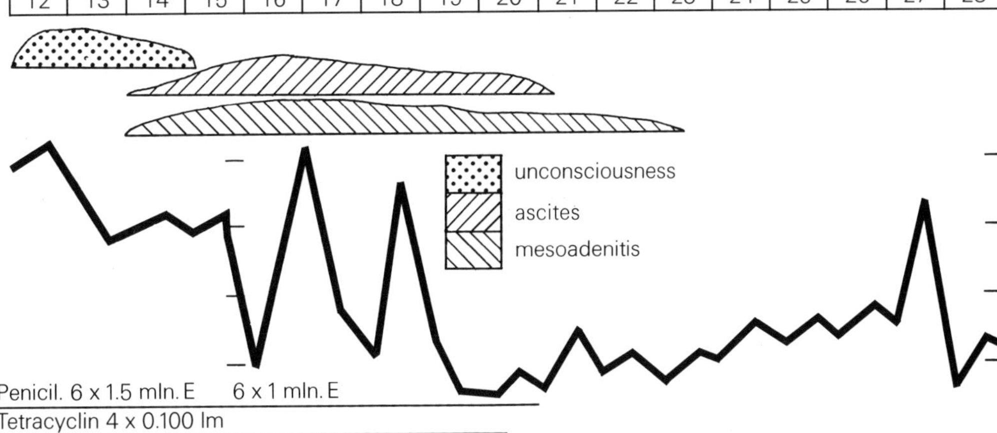

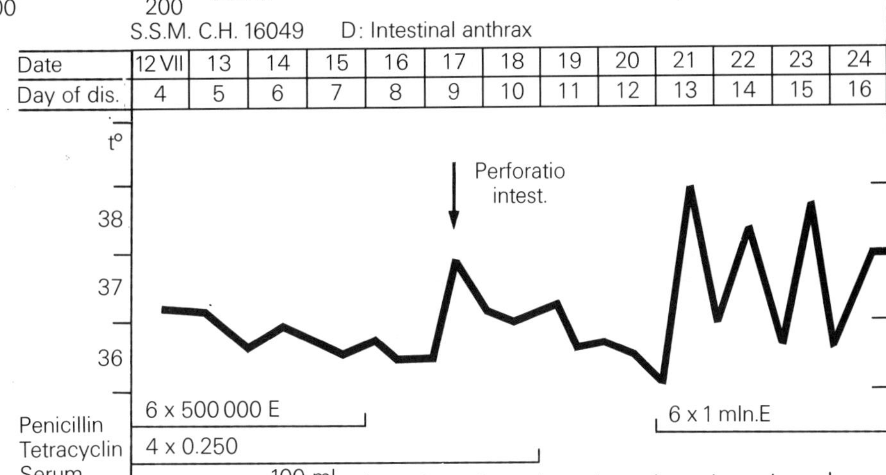

Figure 1. Clinical course of 3 cases of intestinal anthrax.

Anthropometry

Anthropometry is a fundamental branch of physical anthropology of which it represents the quantitative aspect. A wide system of theories and practice is devoted to defining methods and variables to relate the aims in the different fields of application. In the field of occupational health, safety and ergonomics anthropometric systems are mainly concerned with body build, composition and constitution, and to the dimensions of the human body in relation to machines, the industrial environment and clothing.

Anthropometric variables

An anthropometric variable is a measurable characteristic of the body that can be defined, standardised and referred to a unit of measurement. Linear variables are generally defined by landmarks that can be precisely traced on the body. Landmarks are generally of two types: skeletal-anatomical, that may be found and traced by feeling bony prominences through the skin, and virtual landmarks that are simply found as maximum or minimum distances with the caliper branches.

Anthropometric variables have both genetic and environmental components and may be used to define individual and populational variability. The choice of variables must be related to the specific research purpose and standardised with other research in the same field, as the number of variables described in the literature is extremely large, up to 2 200 having been described for the human body by Garret and Kennedy (1971).

The anthropometric variables are mainly: linear straight measures, namely heights, as distances from landmarks to soil or seat with subject in standardised posture, diameters, as distances between bilateral landmarks, and lengths, as distances between two different landmarks; and linear curved measures, namely arcs, as distances on the body surface between two landmarks, and girths, as closed all-around measures on body surfaces generally positioned on at least one landmark or at a defined height.

Other variables may require special methods and instruments. For instance skinfold thickness is measured by means of special constant-pressure calipers and volumes by calculation or immersion in water, while to obtain full information on body surface characteristics a computer matrix of surface points may be plotted using a biostereometric technique.

Instruments

Although sophisticated anthropometric instruments have been described and used with a view to automated data collection, basic anthropometric instruments are quite simple and easy to use. Conversely a lot of care must be taken to avoid common errors resulting from misinterpretation of landmarks and incorrect postures of subjects.

The standard anthropometric instrument is the anthropometer—a rigid rod 2 m long, with two counter-reading scales, with which vertical body dimensions, such as heights of landmarks from floor or seat, and transverse dimensions, such as diameters, can be taken.

Commonly the rod can be split into three or four sections which fit into one another. A sliding branch with a straight or curved claw makes it possible to measure distances from the floor for heights, or from a fixed branch for diameters. More elaborated anthropometers have a single scale for heights and diameters to avoid scale errors, or are fitted with digital mechanical or electronic reading devices (figure 1).

A stadiometer is a fixed anthropometer, generally used only for stature and frequently associated with a weight beam scale.

For transverse diameters a series of calipers may be used: the pelvimeter for measures up to 600 mm and the cephalometer up to 300 mm. The latter is particularly suitable for head measurements when used together with a sliding compass (figure 2).

The foot-board is used for measuring the feet and the head-board provides cartesian co-ordinates of the head when oriented in the "Frankfort plane" (a horizontal plane passing through porion and orbitale landmarks of

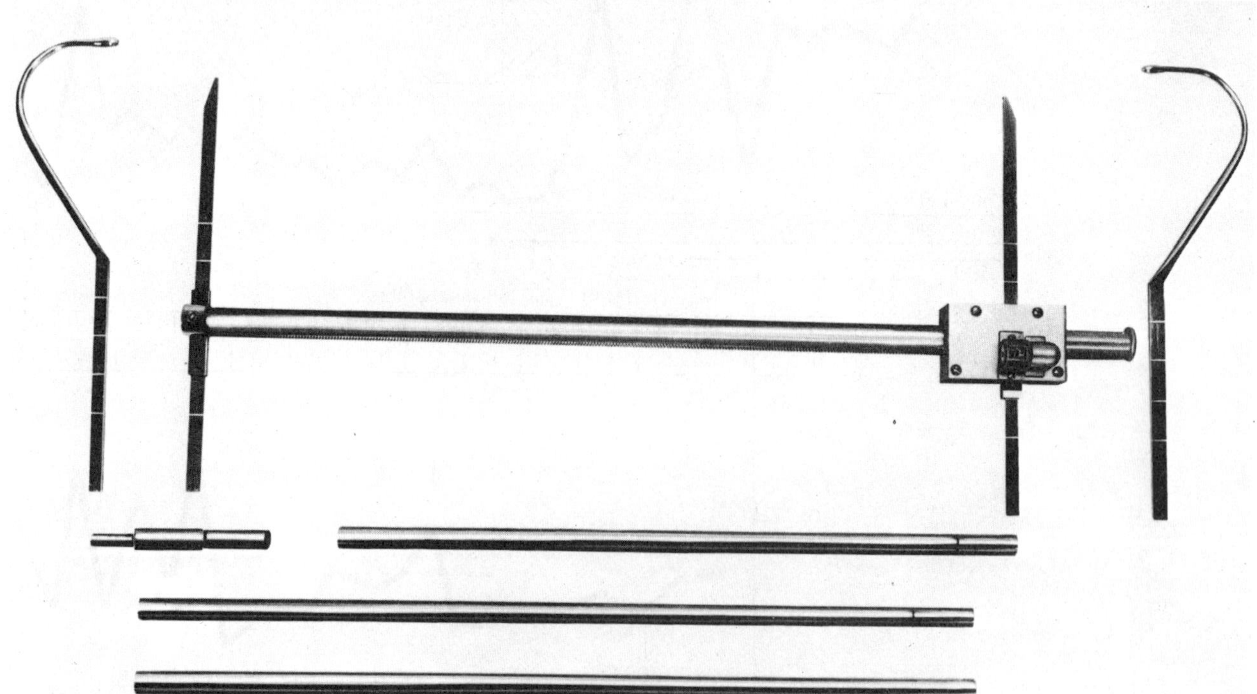

Figure 1. An anthropometer.

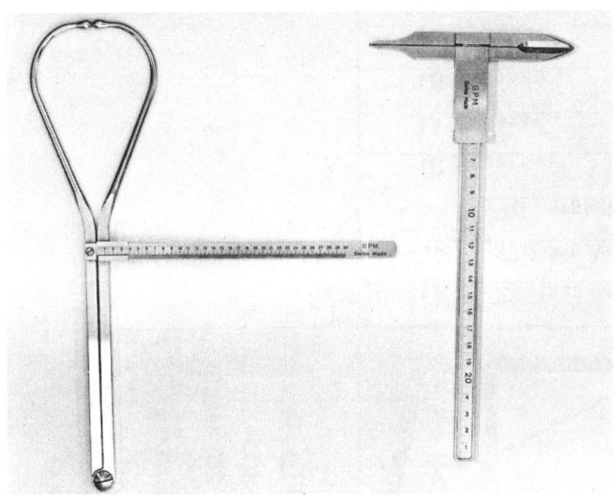

Figure 2. A cephalometer together with a sliding compass.

the head). The hand may be measured with a caliper, or with a special device composed of five sliding rulers.

Skinfold thickness is measured with a constant-pressure skinfold caliper generally with a pressure of 10 g/mm² ($9.81 \cdot 10^4$ Pa).

For arcs and girths a narrow, flexible steel tape with flat section is used. Self-straightening steel tapes must be avoided.

Systems of variables

A system of anthropometric variables is a coherent set of body measurements to solve some specific problem.

In the field of ergonomics and safety the main problem is fitting equipment and work space to humans and tailoring clothes to the right size.

Equipment and work space require mainly linear measures of limbs and body segments that can easily be calculated from landmark heights and diameters, whereas tailoring sizes are based mainly on arcs, girths and flexible tape lengths. Both systems may be combined according to need.

In any case it is absolutely necessary to have a precise space reference for each measurement. The landmarks must therefore be linked by heights and diameters and every arc or girth must have a defined landmark reference. Heights and slopes must be indicated.

In a particular survey the number of variables has to be limited to the minimum so as to avoid undue stress on the subject and operator.

A basic set of variables for work space has been reduced to 33 measured variables (figure 3) plus 20 derived by simple calculation. For a general purpose military survey, Hertzberg and collaborators use 146 variables. For clothes and general biological purposes the Italian Fashion Board (Ente Italiano della Moda) use a set of 32 general purpose variables and 28 technical ones (figure 4). The German norm (DIN 61 516) of control body dimensions for clothes includes 12 variables. The recommendation of the International Organisation for Standardisation (ISO) for anthropometry includes a core list of 36 variables (see table 1).

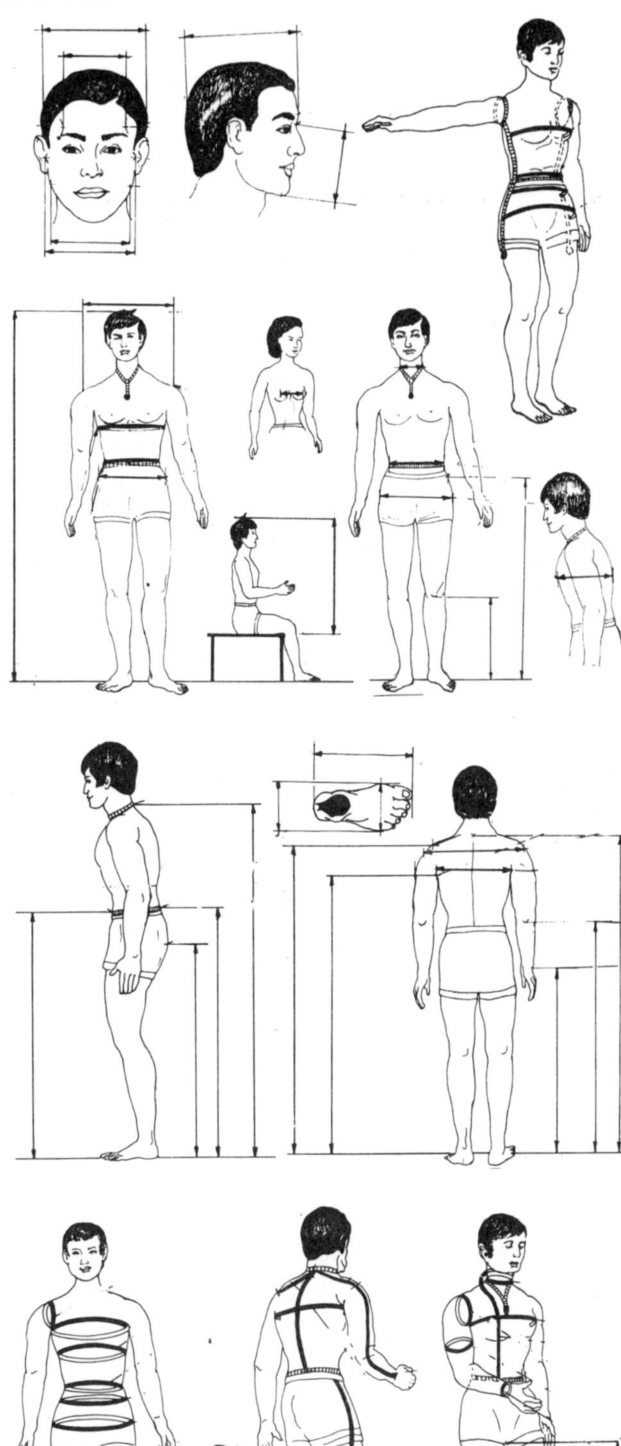

Figure 3. Basic set of anthropometric variables.

Precision and errors

Precision of living body dimensions must be considered in a stochastic manner, because the human body is highly unpredictable both as a static and as a dynamic structure.

A single individual may grow or change in muscularity and fatness; undergo skeletal changes as a consequence of ageing, disease or accidents; or modify his behaviour or his posture. Different subjects differ by proportions, not only by general dimensions. Tall stature subjects are not mere enlargements of short ones; constitutional types and somatotypes probably vary more than general dimensions.

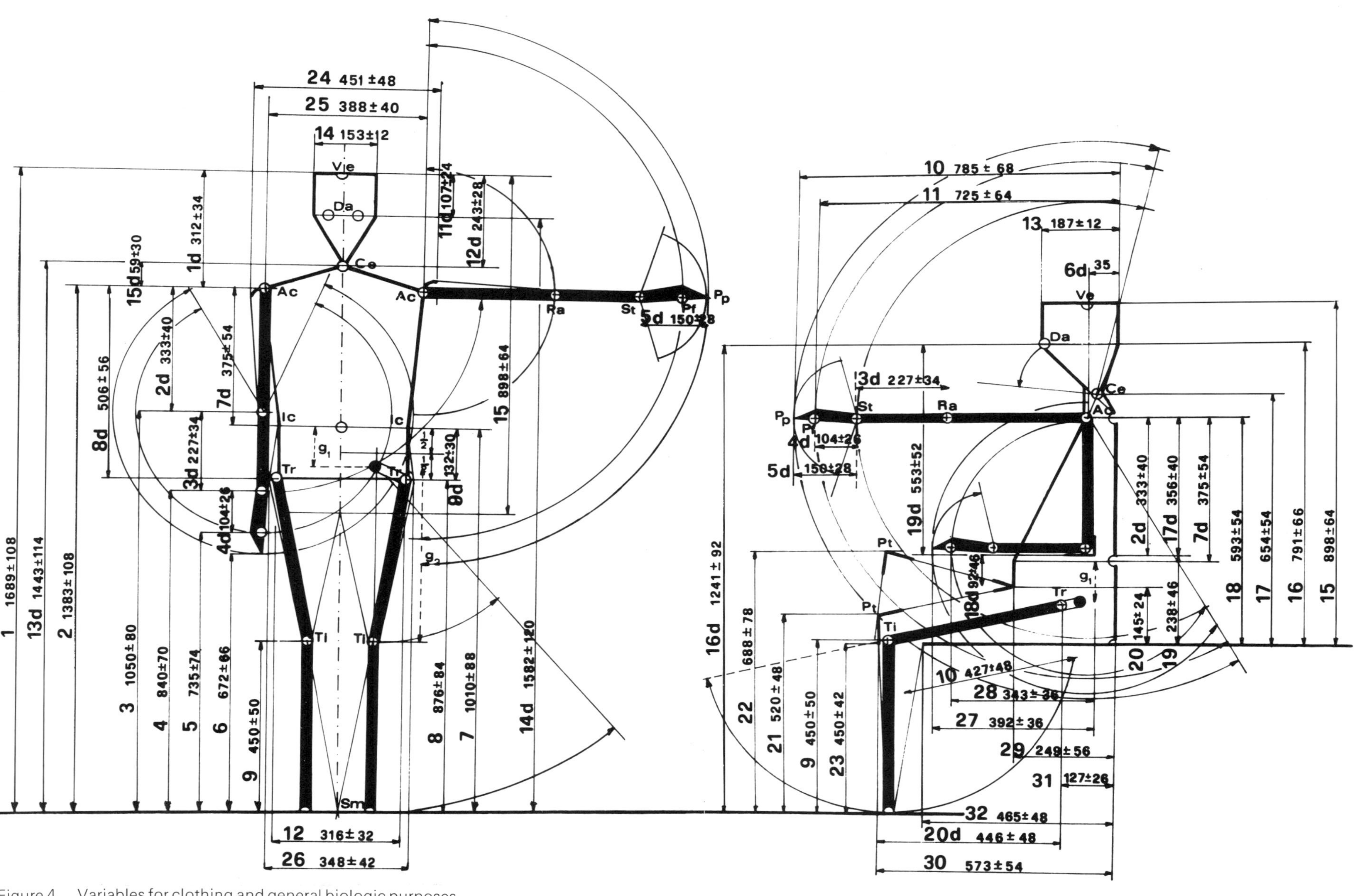

Figure 4.   Variables for clothing and general biologic purposes.

Table 1. Basic anthropometric core list[1]

1.1	Forward reach (to hand grip with subject standing upright against a wall)
1.2	Stature (vertical distance from floor to head vertex)
1.3	Eye height (from floor to inner eye corner)
1.4	Shoulder height (from floor to acromion)
1.5	Elbow height (from floor to radial depression of elbow)
1.6	Crotch height (from floor to pubic bone)
1.7	Finger tip height (from floor to grip axis of fist)
1.8	Shoulder breadth (biacromial diameter)
1.9	Hip breadth, standing (the maximum distance across hips)
2.1	Sitting height (from seat to head vertex)
2.2	Eye height, sitting (from seat to inner corner of the eye)
2.3	Shoulder height, sitting (from seat to acromion)
2.4	Elbow height, sitting (from seat to lowest point of bent elbow)
2.5	Knee height (from foot-rest to the upper surface of thigh)
2.6	Lower leg length (height of sitting surface)
2.7	Forearm-hand length (from back of bent elbow to grip axis)
2.8	Body depth, sitting (seat depth)
2.9	Buttock-knee length (from knee-cap to rearmost point of buttock)
2.10	Elbow to elbow breadth (distance between lateral surface of the elbows)
2.11	Hip breadth, sitting (seat breadth)
3.1	Index finger breadth, proximal (at the joint between medial and proximal phalanges)
3.2	Index finger breadth, distal (at the joint between distal and medial phalanges)
3.3	Index finger length
3.4	Hand length (from tip of middle finger to styloid)
3.5	Hand breadth (at metacarpals)
3.6	Wrist circumference
4.1	Foot breadth
4.2	Foot length
5.1	Head circumference (at glabella)
5.2	Sagittal arc (from glabella to inion)
5.3	Head length (from glabella to opisthocranion)
5.4	Head breadth (maximum above the ear)
5.5	Bitragion arc (over the head between the ears)
6.1	Waist circumference (at the umbilicus)
6.2	Tibial height (from the floor to the highest point on the antero-medial margin of the glenoid of the tibia)
6.3	Cervical height sitting (to the tip of the spinous process of the 7th cervical vertebra).

[1] Condensed from *Draft proposal for core list of anthropometric measurements* (Geneva, ISO, 1980).

The use of manikins, particularly those representing the standard 5th, 50th and 95th percentiles for fitting trials may be highly misleading, if body variations in body proportions are not taken into consideration.

Errors result from misinterpretation of landmarks and incorrect use of instruments (personal error), imprecise or inexact instruments (instrumental error), or changes in subject posture (subject error); this latter may be due to difficulties of communication if the cultural or linguistic background of the subject differs from that of the operator.

Statistical treatment

Being mainly a matter of stochastics, anthropometric data must be treated by statistical procedures mainly in the field of inference methods applying univariate (mean, mode, percentiles, histograms, variance analysis, etc.), bivariate (correlation, regression) and multivariate (multiple correlation and regression, factor analysis, etc.) methods. Computer facilities may be of great aid in solving applied anthropometric problems.

Various methods based on statistical applications have been devised to classify human types (anthropometrograms, morphosomatograms).

Sampling and survey

As anthropometric data cannot be collected for the whole population (except in the rare case of particularly small populations), sampling is generally necessary.

A basically random sample should be the starting point of any anthropometric survey. To keep the number of measured subjects to a reasonable level it is generally necessary to have recourse to multiple-stage stratified sampling. This allows the most homogeneous subdivision of the population into a number of classes or strata.

The population may be subdivided by sex, age group, geographical area (post office zip-codes may be used as a geographical variable), social variables, physical activity and so on.

Survey forms have to be designed keeping in mind both measuring procedure and data treatment. An accurate ergonomic study of the measuring procedure should be made in order to reduce the operator's fatigue and possible errors. For this reason variables must be grouped according to the instrument used and ordered in sequence so as to reduce the number of body flexions the operator has to make.

To reduce effect of personal error, the survey should be carried out by one operator. If more than one operator has to be used, co-ordinate training is necessary.

Population anthropometrics

Disregarding the highly criticised concept of "race", human populations are nevertheless highly variable in size of individuals and in size distribution. Generally human populations are not strictly Mendelian; they are commonly the result of admixture. Sometimes two or more populations, with different origins and adaptation, live together in the same area without interbreeding. This complicates the theoretical distribution of traits. From the anthropometric viewpoint, sexes are different populations. Industrial populations may not correspond exactly to the biological population of the same area as a consequence of possible aptitudinal selection or autoselection due to job choice.

Populations of different areas may differ as a consequence of different adaptation conditions or biological and genetic structures.

When close fitting is important a survey on a random sample is necessary.

Fitting trials and regulation

The adaptation of work space or equipment to the user may depend not only on the bodily dimensions, but also on such variables as tolerance of discomfort and nature of activities, clothing, tools and environmental conditions. For this purpose J. C. Jones suggested a combination of a *check list of relevant factors*, a *simulator* and a series of *fitting trials* using a sample of subjects chosen to represent the range of body sizes of the expected user population.

The aim is to find tolerance ranges for all subjects. If the ranges overlap it is possible to select a narrower final range which is not outside the tolerance limits of any subject. If there is no overlap it will be necessary to make the structure adjustable or to provide it in different sizes.

If more than two dimensions are adjustable a subject may not be able to decide which of the possible adjustments will fit him best.

Adjustability can be a complicated matter, especially when uncomfortable postures result in fatigue. Precise indications must therefore be given to the user who frequently knows little or nothing about his own anthropometric characteristics. In general an accurate design should reduce the need for adjustment to the minimum. In any case it should constantly be kept in mind what is involved is anthropometrics and not merely engineering.

Automated anthropometry

The gathering of anthropometric data and their subsequent handling is a rather laborious job. The automation of such data collection has been suggested by Garn and Helmrich (1968) and illustrated by means of a special caliper.

A complete system of automated anthropometry has been proposed by Prahl-Andersen and his collaborators (1972), a number of special devices being used to collect data on distances, circumferences, stature, weight, skinfold and hand-grip muscular strength to an accuracy of ±1 mm per thousand. The data are punched automatically onto paper tape.

Dynamic anthropometrics

Static anthropometrics may give wide information about movement if an adequate set of variables has been chosen. Nevertheless when movements are complicated and a close fit with the industrial environment is desirable, as in most man-machine and man-vehicle interfaces, an exact survey of postures and movements is necessary. This may be done with suitable mock-ups that allow tracing of reach lines or by photography. In this case a camera fitted with telephoto lens and an anthropometric rod, placed in the sagittal plane of the subject, allows standardised photographs with little distortion of image to be taken. Small labels on subjects' articulations make exact tracing of movements possible.

Another way of studying movements is to formalise postural changes according to a series of horizontal and vertical planes passing through the articulations.

MASALI, M.

"Automated anthropometry". Prahl-Andersen, B.; Pollmann, A. J.; Raaben, D. J.; Peters, K. A. *American Journal of Physical Anthropometry*, 1972, 37, 151 p.

"Setting up a method for the measurement of anthropometric parameters with a view to the ergonomic design of workplaces". (Messa a punto di una metodologia per la misura dei parametri antropometrici ai fini della progettazione ergonomica dei posti di lavoro). Grieco, A.; Masali, M. *Medicina del Lavoro* (Milan), Nov. 1971, 62/11 (505-531). Illus. 37 ref.

CIS 74-802 *Anthropometry for respirator sizing*. Webb Associates, Yellow Springs, Ohio (Cincinnati, National Institute for Occupational Safety and Health, 1972), 112 p. Illus. 32 ref.

CIS 77-1774 "An anthropometer for use in developing countries". Davies, B. T.; Shahnawaz, H. *Ergonomics* (London), May 1977, 20/3 (317-320). Illus. 1 ref.

CIS 78-2068 "Determination of the spatial reach area of the arms for workplace design purposes". Nowak, E. *Ergonomics* (London), July 1978, 21/7 (493-507). Illus. 16 ref.

"Anthropometric elements of ergonomics" (Les aspects anthropométriques de l'ergonomie). Rohmert, W. *Bulletin de l'AISS* (Geneva), 1975 (25-35). Illus.

Draft proposal for core list of anthropometric measurements. ISO/TC 159/SC 3 N 28. DP 7250 (Geneva, International Organisation for Standardisation, 1980), 15 p.

Antibiotics

Antibiotics are substances produced by micro-organisms (usually bacteria or fungi) which, in diluted solution, are capable of inhibiting the multiplication of, or of destroying, other micro-organisms or of causing some chemical rearrangement in their structure. These effects are achieved by inhibition of cell-wall or protein synthesis. In addition to effects on micro-organisms, some antibiotics have auxinic effect; others (anthranilics) are used as anti-tumoral drugs.

To date, antibiotics are among the most effective weapons available in medicine. They can be divided into two groups according to their most important effects:

(a) bactericidal antibiotics (e.g. penicillins, cephalosporins, streptomycins, gentamycins, bacitracins, neomycins);

(b) bacteriostatic antibiotics (e.g. chloramphenicol, tetracyclines, erythromycins, lincomycins, spiramycins).

The micro-organism may block the effect of the antibiotic by creating a resistance against one (one-step) or more (multi-step) antibiotics. Antibiotics now number several hundreds, but only certain of them are currently produced by the pharmaceutical industry. Table 1 summarises the range of action of some of these and their possible additive or synergistic effects. The data given are subject to variation according to strain and strain mutations.

Production. Following laboratory studies and pilot production, the substance goes into production on an industrial scale. As an example of extraction by fermentation, we can take penicillin G. This antibiotic is obtained from fungi (specially selected so as to optimise production qualities), immersed in fermentation liquids in large tanks of up to 100 000 l capacity, at constant temperature and ventilation. After fermentation the liquid is filtered and the antibiotic is extracted by solvent (e.g. with amyl acetate, or acetone), and purified by centrifugation, repeated filtration, and grinding. The substance obtained is again dissolved in acetone and the previous operations repeated. The final product is then weighed and, after biological, sterility and functional controls, goes to the packaging department. If semi-synthetic antibiotics are to be produced, the process is modified after fermentation by the introduction of chemical reagents capable of modifying the molecule of the original substance.

HAZARDS

The accident and fire and explosion hazards in antibiotics production are, in principle, the same as those discussed in the articles FERMENTATION, INDUSTRIAL and PHARMACEUTICAL INDUSTRY. The specific health hazards are directly related to the nature of the product.

Accidents. Burns and scalds may occur due to contact with hot piping and the steam or chemicals used for cleaning and sterilisation. Pipes and other exposed hot surfaces should, therefore, be lagged and personnel should be provided with personal protective equipment when working with steam or chemicals. Falls may occur on wet, slippery floors and these may be prevented by installation of non-slip floor and stairway surfaces and correct maintenance and housekeeping. Amyl acetate and acetone, which are used in solvent extraction, are highly flammable substances and the appropriate fire protection and prevention measures should be taken for their storage and use. All centrifuges should be fitted

Table 1. Effects of some antibiotics and their possible additive or synergistic effects[1]

Antibiotic	Spectrum of selective action against	Additive or synergistic effects
Amoxycillin	*Haemophilus influenziae, Escherichia coli, Proteus mirabilis, Shigella* species	Other beta-lactamase-resistant pencillins Cephalosporins Aminoglucosides
Ampicillin	(see Amoxycillin)	(see Amoxycillin)
Carbenicillin	*Proteus* species *Pseudomonas aeruginosa*	Penicillins Beta-lactamase-resistant penicillins Cephalosporins Aminoglucosides
Cephaloridine	*Klebsiella pneumoniae*	Aminoglucosides Beta-lactamase-resistant penicillins
Chloramphenicol	*Salmonella typhy* *Shigella* species	Tetracyclines
Dicloxacillin	*Staphylococci* producing beta-lactamase	Aminoglucosides Cephalosporins Penicillins Beta-lactamase-resistant penicillins
Erythromycin	Not elective for any species. Useful against *Staphylococci, Streptococci, Pneumococci, Haemophilus, Neisseriae*	Tetracyclines
Gentamycin	*Escherichia coli* *Proteus mirabilis* *Klebsiella pneumoniae* *Pseudomonas aeruginosa*	Cephalosporins Penicillins Beta-lactamase-resistant penicillins
Lincomycin	Not elective against any strain. Useful against *Staphylococci, Streptococci, Pneumococci*	Tetracyclines
Methicillin	*Staphylococci*	Aminoglucosides Cephalosporins Penicillins Beta-lactamase-resistant penicillins
Penicillin G.	Beta-haemolytic *Streptococci* *Pneumococci* *Neisseria* species	Aminoglucosides Cephalosporins Beta-lactamase-resistant penicillins
Spiramycin	Not elective against any species. Useful against *Staphylococci, Pneumococci, Neisseriae*	Tetracyclines
Tetracycline	*Neisseria* species *Shigella* species	Chloramphenicol Macrolides

[1] From: *Martindale Extra Pharmacopoeia* (London, Pharmaceutical Press, 27th ed., 1977).

with suitable protective equipment and bottling, pelleting and packaging machinery should be fitted with machinery guarding.

Allergic diseases. Allergic symptoms of the anaphylactic shock type due to exposure to antibiotics during work are very rare. However, they may occur when there is exposure to airborne dust or the spray from aerosols. The pattern of response includes pronounced perspiration, pallor, collapse perhaps with loss of consciousness, possibly dyspnoea and cough, sneezing and urticaria. In such cases the victim should be removed from exposure and given supportive and symptomatic therapy (central and peripheral analeptics, post-pituitary extracts, adrenaline, cardiotonics, corticosteroids, oxygen resuscitation). Hospital observation is advisable.

Eye reactions include, most frequently, conjunctivitis, blepharo-conjunctivitis and dacryocystitis. Hyperaemia with conjunctival inflammation often accompanied with peri-orbital oedema and intense lacrimation are the most obvious symptoms and the patient complains of intense itching of the eyelids. Lacrimation is often intensified as a result of naso-lacrimal canal dysfunction (obstruction due to concomitant rhinitis). Treatment entails lavage with antiseptic and astringent solutions and cortisone eye drops.

Upper respiratory tract involvement is found mainly in workers exposed to airborne particulate. Hyperaemia and capillary paralysis may produce nasal obstruction and profuse rhinorrhea. Tetracycline exposure may produce haemorrhagic rhinitis, although this is rare. Frequent sneezing and endonasal pruritis are common and the *velum pendulum palati* and oropharynx are often affected by the hyperaemia.

The most frequent lesions of the skin and adnexa due to contact with antibiotics are of the exfoliative dermatitis type and dry or only slightly moist eczema (possibly with mycotic superinfection). The most frequent sites are the forearms, face, neck and hands, and hair often falls out at the sites of this dermatitis. Treatment should include application of mild disinfectants such as boric acid solution, vitamin therapy and topical treatment with antimycotic, anti-inflammatory, emollient and barrier creams.

As far as the larynx, trachea and bronchi are concerned, oedema of the epiglottis accompanied by laryngospasm has been encountered in medical practitioners administering antibiotic aerosol inhalations. In most cases the oedema is of moderate severity and involves all the mucosae (larynx, trachea, large bronchi). Haemorrhagic symptoms are rare and generally of the petechial or haemorrhagic suffusion type. Asthmatic attacks may occur with oedema and bronchospasm: the patient being dyspnoeic and upset, complains of insistent dry cough with production of slight mucous excretion, rarely haemorrhagic in nature. The clinical picture is typical and the symptoms may strike the patient even 2-6 h after exposure to the antibiotic. Treatment requires removal from the contaminated premises; use of ephedrine-like aerosols, aminophenyl derivatives, corticosteroids by the general route and, sometimes, sedatives are useful. Hospitalisation may be necessary for the more severe cases.

Cases have been reported of cardiovascular disorders of suspected toxico-allergic pathogenesis in workers exposed to penicillin and streptomycin. Special attention should be paid to the possibility of onset of autoimmune phenomena due to exposure to adriamycin (Doxorubicin). Myocarditis during antineoplastic therapy may be attributed to such pathogenesis; there are no reports of similar disease in workers.

Effects on saprophytic flora and vitamin metabolism. Changes of the vitamin deficiency type may affect the skin and mucosae, and pellagroid-type acrodermatitis lesions may also occur together with loss of hair.

Glossitis and gingivitis have been observed. These lesions may be treated with multivitamin or, specifically, vitamin A preparation.

Affections of the colitic or gastroenteric type are frequent amongst workers involved in the production of antibiotics; however, their aetiology is uncertain but may be due to avitaminosis or action on the saprophytic flora.

Mycotic infections are found in animal experimentation departments and result from exposure to infected animals (strains of the genera *Trichophyton, Monilia, Mucor, Penicillium*) or to prolonged exposure in production departments due to the action of specific antibiotics on sensitive micro-organisms (pathological states caused by antibiotic-resistant micro-organisms should also be borne in mind). The commonest clinical pictures are cutaneous (dermatomycosis of the interdigital folds, at the elbows, armpits and groin) or of the cutaneous adnexa (trichophytosis, alopecia, onychomycosis). Treatment is topical, with application of iodised alcohol, salicylate ointment, undecylenic acid, tolnaphthates, corticosteroids, and/or general (nystatin, griseofulvin), in association with multivitamin preparations. Ultraviolet ray applications are beneficial.

Toxic effects. Workers exposed to antibiotics have been found to suffer from damage to the vestibule of the ear (streptomycin, dihydrostreptomycin and kanamycin), to the bone-marrow (chloramphenicol), kidneys and liver (all antibiotics) and digestive system (neomycin, tetracyclines, colimycin). Mention should also be made of the following effects in connection with therapeutic use of adriamycin and actinomycin D (Dactinomycin): acne, pigmentation of the skin, alopecia, ulceration of the mucosa, effects on haemopoiesis due to reproductive abnormalities of bone marrow cells, resulting in anaemia, leukopenia. In workers, alterations in the chromosome karyogram may occur. At the therapeutic level, immunodepression is possible.

The lesions are generally reversible when at the initial stage and treatment, which is symptomatic, requires removal from contact. Chronic poisoning due to exposure to solvents such as amyl acetate, acetones and chloroform is uncommon although acute exposure may occur as the result of spillage or leakage, etc.

Psychological effects. Psychological problems have been encountered in workers required to work in hermetically sealed sterile chambers or in airline hoods or suits. Symptoms of claustrophobia may be transformed into psychosomatic equivalents (dyspnoea, persistant cough, feeling of suffocation, faintness, etc.), psychomotor agitation and fits of depression. Cases of hysteria have also been recorded, especially among women workers and in departments where an accident has occurred or toxic or allergic symptoms observed. Workers who have a history of psychological disturbances should not be employed in these conditions. However, where such cases arise, psychotherapy may prove useful.

SAFETY AND HEALTH MEASURES

Direct contact with antibiotics should be avoided and atmospheric contamination due to particulate antibiotics or solvent vapours should be kept to a minimum. Wherever possible the dustiest processes (loading of hoppers with antibiotic for grinding or tableting) should be automated and enclosed, and less dusty processes carried out in sealed chambers or boxes under negative pressure or on benches, etc., fitted with exhaust ventilation. Dust deposits should be collected by vacuum cleaning. Workers in sterile areas should wear polyamide-fibre (nylon) overalls, head protection and trousers and water-resistant (silicon) barrier creams, and cotton and rubber latex hand protection. Hours of work in sterile premises should be limited, e.g. to 4 h with the remainder of the shift devoted to other work.

To prevent contamination of the work premises by antibiotic-resistant microbes or mycetes, spraying with formaldehyde should be performed at the end of each working day. The walls should be washed with sodium hypochlorite solutions or with other bactericidal detergents. Premises used for sterile processes should be fitted with ultraviolet lamps and with electrostatic filters supplying sterile air. However, ultraviolet radiations may produce considerable ozone concentrations and may also constitute an eye hazard unless suitable eye protection is worn. In spite of these precautions, job rotation is advisable both from the physiological and psychological point of view.

The measures given above for the prevention of atmospheric contamination dangerous to health will, under normal circumstances, prevent the formation of flammable and explosive solvent concentrations.

Medical prevention. Pre-employment medical examinations are essential to ensure that persons likely to be particularly susceptible to antibiotics are not exposed to these substances. This examination should include:

(a) A medical history: a personal or family history of allergies, gout, diabetes, etc., and the presence of bronchitis (particularly with an asthmatic component), eczematous or urticaroid skin diseases or oculorhinitis all contraindicate exposure to antibiotics. Persons with hepatic and digestive disorders should not work in areas in which high concentrations of ozone have been produced by ultraviolet radiation and upper respiratory tract disorders contraindicate prolonged use of respiratory protective equipment. Finally a check should be made on neuropsychic conditions such as neurovegative dystonia or psychoses which may indicate susceptibility to claustrophobia.

(b) Special examinations should be carried out according to exposure. An audiogram should be made when the job is expected to involve contact with streptomycin or other ototoxic substances. A blood count–including a count of erythrocytes, leukocytes and platelets–should be performed in the case of jobs involving contact with myelotoxics. A chromosome karyogram is useful if there is a risk of chromosome damage. An electrocardiogram is indicated for exposure to cardiotoxics. A check of the state of the immune system would be advisable in the case of exposure to immunodepressants and immunogenic substances. Similarly, the atopy of workers exposed to the risk of sensitisation may usefully be determined by means of intradermal tests with pools of ubiquitous (common) allergens, possibly supplemented by determination of IgA, IgG, IgM and IgE immunoglobulins. Where risk of asthma exists, assessment of bronchomotor tone with acetylcholine or methacholine tests will be of use. Workers with atopy and/or increased bronchomotor tone should be considered as high risk subjects and exposure to sensitising antibiotics should be avoided. Exposure is absolutely contraindicated in cases of hypersensitivity to one or more of the substances used in the course of work.

Immunological investigations can be carried out by direct skin tests with suitable preparations (see table 2). Patch and intradermal tests are used to study skin reactivity; readings are made 20-30 min and 24-48 h after application. Aspecific skin reactivity tests should be made with phosphate-buffered physiological solution (pH 7.4), with histamine dihydrochloride 1 : 1 000 and with the solvents used, other than water. In normal response conditions, 20 min after inoculation of 0.02 ml of the preparations described, there will be a weal of about 15 mm diameter with histamine, and no reaction with physiological solution.

(c) Periodic check-ups should be made every year or more frequently if necessary. Such check-ups should comprise a medical examination and all the appropriate tests (see *(b)* above). All employees, including maintenance staff, should be covered.

Immunological tests should be made with the same substances to which the worker has been exposed. Further tests are advisable in the case of transfer, changes in production, long absence from work and resignation. It would also be advisable for women to have a check-up before maternity leave.

Workers identified as allergic should be advised to avoid therapeutic use of the antibiotic that caused the sensitisation and to discontinue occupational exposure.

The written report given to the worker should include the above recommendations and also details of possible

Table 2. Preparations for direct skin tests[1]

Antibiotic	Preparations for epicutaneous tests	Preparations for intradermal tests[2]
Amoxycillin	25% l.par. & vas. 1 : 1	0.5% phys.
Ampicillin	50% l.par. & vas. 1 : 1	0.5% phys.
Bacitracin	25% l.par. & vas. 1 : 1	-
Benzylpenicillin	50% l.par. & vas. 1 : 1	5 000 U.Ox./ml/ phys.
Carbenicillin	50% l.par. & vas. 1 : 1	0.5% phys.
Cephaloridine	25% l.par. & vas. 1 : 1	0.5% phys.
Chloramphenicol	60% l.par. & vas. 1 : 1	0.5% phys.
Dicloxacillin	30% l.par. & vas. 1 : 1	0.5% phys.
Erythromycin	25% l.par. & vas. 1 : 1	0.5% w.[3]
Flucloxacillin	30% l.par. & vas. 1 : 1	0.5% w.
Gentamycin	4% w.	0.2% phys.
Lincomycin	30% w.	0.2% phys.
Methicillin	30% l.par. & vas. 1 : 1	0.5% phys.
Neomycin	30% l.par. & vas. 1 : 1	0.5% w.
Novobiocin	25% l.par. & vas. 1 : 1	0.5% w.
Oleandomycin	25% l.par. & vas. 1 : 1	0.5% w.
Pencillin G.	(see Benzylpenicillin)	(see Benzyl-penicillin)
Polymyxin B.	25% l.par. & vas. 1 : 1	-
Rifamycin	20% l.par. & vas. 1 : 1	0.25% w.[3]
Rifampicin	20% l.par. & vas. 1 : 1	0.25% in polyethylene glycol 200 and w. 1 : 1[3]
Spiramycin	60% l.par. & vas. 1 : 1	0.5% w.[3]
Streptomycin	30% l.par. & vas. 1 : 1	0.5% phys.

Abbreviations:
l.par.: in liquid paraffin
vas.: in Vaseline (petroleum jelly)
phys.: in physiological phosphate buffer (pH 7.4)
w.: in bidistilled water

[1] The antibiotics used for the preparations must be pharmaceutical quality. See also the *Specifications for quality control* published from time to time by the World Health Organisation. [2] The quantity for injection is 0.02 ml of the preparation. [3] The solutions may have primary irritant action at these concentrations, due to histamine-releasing substances and to pH. Reading of tests at 20 min from inoculation should be compared with results from histamine. The test is positive only if the weal with the antibiotic is equal to or greater than the weal obtained with histamine.
NB: For all beta-lactamase antibiotics, an intradermic test should also be performed with penicilloyl-polylysin at a concentration of $0.6.10^{-6}$ M.

cross-reactions between the antibiotic in question and others with the same chemical structure that may, in an allergic subject, trigger off situations similar to those that could be caused by a sensitising drug. Thus, for an individual allergic to penicillin, all the other betalactamics (cephalosporins, penicillins) will be contraindicated; for an individual allergic to streptomycin, all other oligosaccharides (neomycins, streptomycins, tobramycins) will be contraindicated.

Prophylaxis. Treatment with polyvitamins containing C- and B-group vitamins may be useful in limiting or preventing the effects of hypovitaminosis. Physiological thirst-quenchers (drinks) may be of help in avoiding the excessive loss of liquid and salts that may occur in individuals working in hot environments with low humidity.

BARTALINI, E.
NAVA, C.

Health effects:

CIS 2747-1970 "Modification of the bacteria of the intestine and other organs following occupational exposure to antibiotics (streptomycin, tetracyclin, penicillin)" (Formirovanie disbakterioza kišečnika i drugih organov u lic, kontaktirujuščih s antibiotikami (streptomicinom, tetraciklinom, penicillinom) v proizvodstvennyh uslovijah). Vil'šanskaja, F. L.; Stejnberg, G. B. *Gigiena truda i professional'nye zabolevanija* (Moscow), May 1970, 14/5 (25-28). (In Russian)

CIS 709-1972 "Ear, nose and throat affections among workers employed in the production of antibiotics" (Sostojanie Lorganov rabočih proizvodstva antibiotikov). Alieva, N. K. *Gigiena truda i professional'nye zabolevanija* (Moscow), Sep. 1971, 15/9 (20-23). 16 ref. (In Russian)

"Occupational exposure to drugs—Antibiotics" (L'esposizione professionale a farmaci: Antibiotici). Farina, G.; Alessio, L.; Bulgheroni, C. *Medicina del Lavoro* (Milan), May-June 1980, 71/3 (228-234). 23 ref. (In Italian)

CIS 78-158 "Occupational hazards of antibiotics: contribution to the study of occupational disease due to spiramycin" (Rischi lavorativi da antibiotici: contributo allo studio della patologia professionale da spiramicina). Nava, C. *Securitas* (Rome), May-June 1976, 61/5-6 (275-280). 12 ref. (In Italian)

CIS 78-1391 "Allergic myocarditis due to occupational contact with antibiotics" (Toksiko-allergičeškie miokarditi, voznikajuščie pri professional'nom kontakte s antibiotikami). Bogoslovskaja, I. A.; Gerasimova, E. A.; Parfenova, E. S.; Sokolova, V. G.; Filjušina, Z. G.; Hil', R. G. *Sovetskaja medicina* (Moscow), Jan. 1978, 1 (11-15). 10 ref. (In Russian)

Bacterial resistance:

CIS 2375-1972 "Resistance of microbes isolated from individuals occupationally exposed to chlortetracycline" (Rezistence mikrobů izolovaných u osob professionálně exponovanych chlortetracyklinu). Schön, E.; Wagner, V.; Wagnerova, W. *Časopis lékařů českých* (Prague), 1971, 110/34 (796-801). Illus. 28 ref. (In Czech)

Exposure limits:

CIS 78-1971 "Toxicological features of oleandomycin and establishment of hygiene standards for antibiotics" (Sur les caractéristiques toxicologiques de l'oléandomycine et l'établissement de normes hygiéniques pour les antibiotiques). Spasovski, M.; Hinkova, L.; Djejev, A.; Stamova, N.; Burkova, T. *Archives des maladies professionnelles, de médecine du travail et de sécurité sociale* (Paris), Apr.-May 1978, 39/4-5 (259-264). 22 ref. (In French)

International standards:

Biological substances. International standards, reference preparations, and reference reagents (Geneva, World Health Organisation, 1979), 81 p.

Antidotes

Antidotes are drugs or special prescriptions administered in order to prevent or treat both acute and chronic cases of occupational poisoning thanks to their specific antitoxic effects (see POISONING, ACUTE, TREATMENT OF). The use of antidotes is a fundamental part of prophylactic or therapeutic measures aimed at neutralising the toxic effect of a chemical substance. Because many chemicals have multiple mechanisms of toxic action, there are cases where different antidotes have to be administered simultaneously, and where also therapeutic measures have to be taken which are not directed against the causes, but only against certain symptoms of poisoning. Indeed, since the intimate mechanisms of action of most chemical compounds are insufficiently known, treatment of poisoning is frequently limited to symptomatic therapy. Experience gained in clinical toxicology shows that certain preparations, such as vitamins and hormones, can be classified as universal antidotes because of the positive prophylactic and therapeutic effects they produce on various types of poisoning. This is due to the fact that there are general pathogenic mechanisms at the basis of intoxications (and of many other pathological processes) which have been investigated by G. Selye and his disciples.

There is so far no universally recognised classification of antidotes. Analysis of the literature suggests that the most rational system is one that subdivides the antidotes into basic groups according to the mechanism of their antitoxic effects, which may be physical, chemical, biochemical or physiological. As regards the conditions under which antidotes react with poisons, a distinction is made between locally acting antidotes, which react with the poison before it is absorbed by the tissues, and absorptive antidotes, which react with the poison even after it has been absorbed by tissues and physiological liquors. Antidotes having a physical action are used exclusively for the prevention of poisoning, whereas absorptive antidotes serve the purpose of both prevention and treatment of poisoning.

Antidotes with a physical action

These antidotes protect mainly by adsorption of the poison. Thanks to their great surface activity, adsorbents bind the molecules of the toxic substance and impede its absorption by the surrounding tissue. However, molecules of the adsorbed poison may later part from the adsorbent and again reach the stomach tissues. This parting phenomenon is called desorption. When administering physical antidotes, it is therefore most important to combine them with measures for the subsequent elimination of the adsorbent from the body. This can be achieved by gastric lavage, or by administering laxatives if the adsorbent has entered the intestine. Preference should be given to saline laxatives (e.g. sodium sulphate), which are hypertonic solutions and stimulate the admission of the liquid into the intestine, so that absorption of the toxic substance is practically excluded. Fatty laxatives (e.g. castor oil) may contribute to the absorption of fat-soluble chemicals, which may result in additional poison uptake by the body. It is advisable to use saline laxatives when the nature of the chemical compound is not exactly known.

The most typical antidotes of this group are activated carbon and kaolin. They are very efficient in the event of acute poisoning by alkaloids (organic substances of vegetable origin, e.g. atropine) or by salts of heavy metals.

Antidotes with a chemical action

The underlying principle of their mechanism is the direct reaction between poison and antidote. Chemical antidotes can be used for both local and absorptive action.

Local action. While physical antidotes have little specific antidotal effect, chemical antidotes have a rather high specificity owing to the very nature of the chemical reaction. The local effects of chemical antidotes include reactions such as neutralisation, formation of insoluble compounds, oxidation, reduction, concurrent antagonism, and formation of complexes, the first three mechanisms being the more important and better known.

A good example of *neutralisation* is the use of alkalis to counter strong acids swallowed accidentally or contaminating the skin. Neutralising antidotes are also used when the reaction results in compounds with little biological activity. When, for instance, strong acids have been swallowed, it is advisable to carry out a gastric lavage with warm water to which calcined magnesia (20 g/l) has been added. In the event of poisoning with hydrofluoric acid or citric acid, a slurry of calcium chloride with calcined magnesia is given by mouth. When caustic lyes (caustic soda, caustic potash) have been swallowed, a gastric lavage with a 1% solution of citric or acetic acid is carried out. In all cases where strong acids or lyes have been swallowed it must be borne in mind that emetics are contraindicated. Vomiting implies violent contractions of the stomach muscles and, as these aggressive liquids may attack the stomach tissues, there is a danger of perforation

Antidotes that form insoluble compounds unable to penetrate the mucous membranes or skin have a selective effect, i.e. they counter only certain categories of chemical substances. A classical example of this type of antidote is 2,3-dimercaptopropanol which forms insoluble, chemically inert, metal sulphides. It is effective in the event of poisoning with zinc, copper, cadmium, mercury, antimony and arsenic.

Tannin (tannic acid) forms insoluble compounds with salts of alkaloids and heavy metals. Compounds of tannin with morphine, cocaine, atropine or nicotine have different degrees of stability which has to be borne in mind by the toxicologist. Tannin is used in a 0.2% solution for gastric lavage or given with a spoon by mouth in a solution of 2-3% (not more than 200 ml) in the course of 5-10 min.

Gastric lavage should be a general rule after administration of all antidotes of this group in order to eliminate the chemicals initially bound.

Antidotes with combined effects are of great interest, particularly a prescription consisting of 25 g tannin, 50 g activated charcoal and 25 g calcined magnesia. This preparation combines antidotes with both a physical and a chemical action.

In recent years increasing attention has been paid to the local application of sodium thiosulphate. This is used in cases of poisoning with arsenic, mercury, lead, hydrocyanic acid, bromine and iodine salts. It is given by mouth as a 10% solution (2-3 tablespoonfuls upon admission).

Local application of antidotes against the above types of poisoning must be combined with subcutaneous, intramuscular or intravenous injection of antidotes.

Oxidation of the toxic substance is largely made use of when opium, morphine, aconite or phosphorus have been ingested. The most widespread antidote is potassium permanganate, which is used in a 0.02-0.1% solution for gastric lavage. This preparation is inefficient in the event of poisoning by cocaine, atropine and barbiturates (luminal and its derivatives). Gastric lavage is carried out according to general practice in these cases.

Absorptive application. Absorptive antidotes with chemical action can be divided into two basic subgroups: *(a)* antidotes which react with certain intermediate products formed by the reaction of the poison with the substrate, and *(b)* antidotes which directly interfere with the reaction between the poison and certain biological systems or structures. In this case the chemical mechanism is frequently associated with the biochemical mechanism of the antidotal effect.

The following example is characteristic of the mechanism of the antitoxic effect of chemical antidotes that act *indirectly*. Cyanogen compounds are widely used in various industrial processes, and the prevention and treatment of cyanide poisoning is therefore an important problem. So far, there is no antidote capable of inhibiting the interaction between the cyanide and the vulnerable enzyme system. Once absorbed by the blood, the cyanide is carried to tissues where it reacts with the trivalent iron of oxidised cytochrome-oxidase, one of the enzymes vital for tissular respiration. The oxygen taken up by the body is thus no longer capable of reacting with the enzyme system, which develops an acute oxygen deficiency. However, the complex which the cyanide forms with the iron of cytochrome-oxidase is not stable and easily dissociates.

The antidotal treatment consequently takes three basic directions: *(a)* neutralisation of the poison in the blood immediately after its absorption; *(b)* fixation of the circulating poison to limit the amount absorbed by the tissues; *(c)* neutralisation of the poison released into the blood after dissociation of the cyanomethaemoglobin and cyano-substrate complex.

Direct neutralisation of cyanides can be achieved with glucose which reacts with hydrocyanic acid to form the only slightly toxic cyanohydrin. A much more active antidote is β-oxyethylmethylenediamine. Both these antidotes must be injected intravenously during the first few minutes or seconds following the uptake of the poison.

The second approach of antidotal treatment, which aims at the fixation of the poison circulating in the blood, is more common. Cyanides do not react with haemoglobin, but actively combine with methaemoglobin to form the complex cyanomethaemoglobin. Although this complex is not very stable, the poison is maintained in this state for a certain period of time. It is therefore important to introduce methaemoglobin-forming antidotes. This is achieved by inhalation of amylnitrite vapours or by intravenous injection of a sodium nitrite solution. The cyanide present in the blood plasma combines with methaemoglobin, thus losing considerably in toxicity. It should be borne in mind that methaemoglobin-forming antidotes may affect the arterial blood pressure: while amylnitrite causes a pronounced but short fall in pressure, sodium nitrite is known for its prolonged hypotensive effect. When applying methaemoglobin-forming agents, it should be considered that methaemoglobin does not take part in the oxygen transfer, but may by itself become the cause of oxygen deficiency. Therefore, the application of methaemoglobin-forming antidotes is subject to special rules. It is up to the physician to find the optimal relation between haemoglobin and methaemoglobin.

The third approach in antidotal treatment of cyanide poisoning consists in neutralising the cyanide released from the complex with methaemoglobin and cytochrome-oxidase. To this end, sodium thiosulphate is injected intravenously. It is a slowly acting antidote which transforms the cyanides in the presence of the enzyme rhodanase into non-toxic thiocyanates.

The specificity of chemical antidotes is limited because they do not affect the direct interaction between poison and substrate. However, the effect these antidotes produce on certain links in the mechanism of toxic action is of undoubted therapeutic value, although their administration calls for high medical skill and utmost care.

Chemical antidotes that react *directly* with a toxic substance are characterised by a high specificity, which enables them to bind the toxic compounds and eliminate them.

Complex-forming antidotes (chelating agents) form stable complex compounds with bi- and trivalent metals, which are then easily eliminated with the urine. Chelating agents can be used for both prevention and treatment of poisoning. Of great practical interest are the salts of ethylenediaminetetraacetic acid (EDTA).

Disodium calcium edetate is efficient in cases of poisoning with lead, cobalt, copper, and vanadium. The calcium contained in the antidote molecule reacts only with metals that form a more stable complex. This edetate does not react with the ions of barium, strontium and some other metals with a lower stability constant. There are a few metals with which the antidote can form toxic complexes. Its application is therefore subject to caution or contraindicated in cases of cadmium, mercury and selenium poisoning. With chronic lead poisoning this chelating agent helps to increase the urinary excretion of lead.

Penthamil (trisodium-calcium salt of diethylenetetraminopentaacetic acid) is used for acute and chronic poisoning by plutonium and radioactive yttrium, cerium, zinc, uranium and lead, and also for detecting contaminations with these isotopes. This preparation is also applied in cases of cadmium and iron poisoning. Its use is contraindicated in patients with nephritis, cardiovascular diseases and some other conditions.

Chelating agents at large also include antidotes with molecules containing free sulphydryl groups (SH-groups). Of great interest are in this respect 2,3-dimercaptopropanol (dimercaprol), generally known as BAL (British anti-lewisite), and 2,3-dimercaptopropanesulphonate (unithiol). The molecular structure of these antidotes is rather simple:

$H_2C\text{-}SH$	$H_2C\text{-}SH$
$H_2C\text{-}SH$	$H_2C\text{-}SH$
$H_2C\text{-}OH$	$H_2C\text{-}SO_3Na$
BAL (dithioglycerine)	Unithiol

There are 2 SH-groups close to each other in both these antidotes. The significance of this particular structure becomes evident in the example described hereafter, where antidotes containing SH-groups react with metals and metalloids. The reaction of dimercapto compounds with metals can be represented in the following way:

$$\text{Enzyme}\begin{matrix}\diagup SH\\ \diagdown SH\end{matrix} + \text{metal} \rightarrow \text{enzyme}\begin{matrix}\diagup S \diagdown\\ \diagdown S \diagup\end{matrix}\text{metal}$$

$$\begin{matrix}HSCH_2\\ HSCH_2\\ HOCH_2\end{matrix} + \text{enzyme}\begin{matrix}\diagup S \diagdown\\ \diagdown S \diagup\end{matrix}\text{metal} \rightarrow$$

$$\text{enzyme}\begin{matrix}\diagup SH\\ \diagdown SH\end{matrix} + \text{metal}\begin{matrix}\diagup S-CH_2\\ - S-CH\\ \diagdown HO-CH_2\end{matrix}$$

The following phases are clearly distinguishable: *(a)* reaction of the enzyme SH-groups with the metal and formation of a not-too-stable complex; *(b)* reaction of the antidote with the complex; *(c)* release of the active enzyme due to the formation of a metal-antidote complex which is eliminated with the urine. Unithiol is less toxic than BAL. Both preparations are used for the treatment of acute and chronic poisoning by arsenic, chromium, bismuth and some other metals, but not lead. In cases of chronic arsenic or mercury poisoning unithiol is given by mouth in the form of tablets. In the event of selenium poisoning it is recommended not to use BAL. There is no efficient antidote against poisoning by molybdenum, nickel and certain other metals.

These examples are far from illustrating all the various possibilities of direct absorptive action of chemical antidotes; for exhaustive details specialised manuals should be referred to.

Biochemical antidotes

They are characterised by their highly specific antidotal effect. A typical example of this class are the antidotes used for treating cases of poisoning by organophosphorus compounds, which are basic constituents of many insecticides (fenthion, malathion, paraoxone, parathion, schradan, etc.). It is known that very small doses of organophosphorus compounds inhibit the function of cholinesterase by phosphorylation, which results in the accumulation of acetylcholine in the tissues. Since acetylcholine is of vital importance for the transmission of nerve impulses in both the central and peripheral nervous systems, its excessive action leads to nervous function disorders and thus to severe pathological damage.

Enzymological research into the toxic effects of organophosphorus compounds has revealed a group of chemicals capable of restoring the cholinesterase activity inhibited by these compounds. These chemicals belong to the derivatives of hydroxamic acids and contain the oxime group R−CH−NOH. Oxime antidotes of practical importance are 2-PAM (pralidoxim), dipyroxim (TMB-4), biodoxim and isonitrosine. Under favourable conditions these cholinesterase reactivators are capable of restoring the activity of this enzyme, thus diminishing or eliminating the clinical signs of poisoning, preventing remote sequelae and contributing to a more favourable course of restoration.

However, practice has convincingly shown that the best results are obtained when biochemical antidotes are combined with physiologically acting antidotes.

Antidotes with a physiological action

The example of poisoning by organophosphorus compounds has shown that inhibition of the cholinesterase activity primarily causes acetylchòline to accumulate at the synapses. There are two ways to neutralise the toxic action of the poison: *(a)* restoration of the cholinesterase activity; *(b)* protection of the physiological systems sensitive to acetylcholine (choline-reactive systems) from the abnormally intense action of this nerve impulse mediator, resulting first in acute excitation and then in functional paralysis.

A typical example of physiological antagonism is the use of preparations depressing the sensitivity of the choline-reactive systems to acetylcholine; the most important preparation of this type is atropine. The class of physiological antidotes includes numerous drugs. The reason for this is that many authors classify these drugs in the category of functional antidotes. In the event of acute excitation of the central nervous system, which happens in many cases of poisoning, it is indicated to administer narcotics or antispasmodics. On the other hand, if the respiratory centre is acutely inhibited, analeptics are used as antidotes. There is a huge number of analogous examples. In a first approximation it can be said that antidotes with a physiological (or functional) action include all those drugs that trigger a physiological reaction countering the effect of a poison. This makes it of course difficult to draw a clear line between antidotes and drugs for symptomatic therapy.

It should be emphasised that antidotes are active drugs which must be used only when the appropriate instructions are observed.

STROIKOV, J. N.

"Clinical toxicology" (Klinična toksikologija). Monov, A. et al. (Sofia, Medicina i fizkultura, 1972). (In Bulgarian)

CIS 77-478 "Inhibitory effect of mercury on kidney glutathione peroxidase and its prevention by selenium". Wada, O.; Yamaguchi, N.; Ono, T.; Nagahashi, M.; Morimura, T. *Environmental Research* (New York), Aug. 1976, 12/1 (75-80). 12 ref.

CIS 77-172 "Cyanide intoxication: protection with chlorpromazine". Way, J. L.; Burrows, G. *Toxicology and Applied Pharmacology* (New York), Apr. 1976, 36/1 (93-97). Illus. 11 ref.

CIS 78-132 "Results of experiments in rabbits with sodium fluoride and antidotes" (Ergebnisse tierexperimenteller Untersuchungen an Kaninchen mit Natriumfluorid unter Einwirkung von Gegenmitteln). Baer, H. P.; Bech, R.; Franke, J.; Grunewald, A.; Kochmann, W.; Melson, F.; Runge, H.; Wiedner, W. *Zeitschrift für die gesamte Hygiene und ihre Grenzgebiete* (Berlin), Jan. 1977, 23/1 (14-20). Illus. 49 ref. (In German)

Antimony, alloys and compounds

Antimony (Sb)

a.w.	121.76
sp.gr.	6.7
m.p.	630 °C
b.p.	1 380 °C

a silver-white metal.

TWA OSHA	0.5 mg/m³ (and compounds as Sb)
IDLH	80 mg/m³
MAC USSR	0.5 mg/m³

Antimony is stable at room temperature but, when heated, burns brilliantly giving off dense white fumes of antimony oxide (Sb_2O_3) with a garlic-like odour. It is closely related, chemically, to arsenic. It readily forms alloys with arsenic, lead, tin, zinc, iron and bismuth.

Occurrence. In nature, it is found in combination with numerous elements, and the most common ores are stibnite (SbS_3), valentinite (Sb_2O_3), kermesite (Sb_2S_2O) and senarmontite (Sb_2O_3). World production of antimony ore in 1977 was 69 500 t (Sb content). The main sources were: Bolivia (15 156 t), South Africa (12 993 t), China (12 000 t) and USSR (7 900 t).

Extraction and production. Antimony ores are found in both narrow veins and massive bodies and they are extracted using normal metal mining techniques. The extracted ore is first sorted and crushed and then concentrated by settling and flotation. Free antimony metal is produced by the reduction of trivalent antimony sulphide with carbon or by roasting the trivalent antimony sulphide in air to obtain antimony tetroxide which is then reduced with carbon. A flux of sodium carbonate or sodium sulphate is used to cover the antimony melt during volatilisation.

Uses. High-purity antimony is employed in manufacture of semi-conductors. Normal purity antimony is used widely in the production of alloys to which it imparts increased hardness, mechanical strength, corrosion resistance and a low coefficient of friction; alloys combining tin, lead and antimony expand slightly during cooling—a valuable characteristic for the production of sharp castings, especially for printing type. Among the more important antimony alloys are babbitt, pewter, white metal, Britannia metal and bearing metal. These are used for bearing shells, printing-type metal, storage battery plates, cable sheathing, solder, ornamental castings and ammunition. The resistance of metallic antimony to acids and bases is put to effect in the manufacture of chemical plant.

HAZARDS

The principal hazard of antimony is that of intoxication by ingestion, inhalation or skin absorption. The respiratory tract is the most important route of entry since antimony is so frequently encountered as a fine airborne dust. Ingestion may occur through swallowing dust or through contamination of beverages, food or tobacco. Skin absorption is less common but may occur when antimony is in prolonged contact with skin.

The dust encountered in antimony mining may contain free silica and cases of pneumoconiosis termed "silico-antimoniosis" have been reported among antimony miners. During processing, the antimony ore, which is extremely brittle, is converted into fine dust more rapidly than the accompanying rock, leading to high atmospheric concentrations of fine dust during such operations as reduction and screening. Dust produced during crushing is relatively coarse, and the remaining operations—classification, flotation, filtration, etc.—are wet processes and, consequently, dust free. Furnacemen refining metallic antimony and producing antimony alloy, and workers setting type in the printing industry are all exposed to antimony metal dust and fume [and may present diffuse miliar opacities in the lung, with no clinical or functional signs of impairment in the absence of silica dust].

Toxicology. In its chemical properties and metabolic action, antimony has a close resemblance to arsenic and, since the two elements are sometimes found in association, the action of antimony may be blamed on arsenic, especially in foundry workers. However, experiments with high-purity metallic antimony have shown that it has a completely independent toxicology; nevertheless, it is not more than slightly toxic, and different authors have found the average lethal dose to be between 10 and 11.2 mg/100 g.

Antimony may enter the body through the skin but the principal route is through the lungs. From the lungs, antimony, and especially free antimony, is absorbed and taken up by the blood and tissues. Studies on workers and experiments with radioactive antimony have shown that the major part of the absorbed dose enters the metabolism within 48 h and is eliminated in the faeces and, to a lesser extent, the urine. The remainder stays in the blood for some considerable time with the erythrocytes containing several times more antimony than the serum. Antimony inhibits the activity of certain enzymes, binds sulphydril groups in the serum and disturbs protein and carbohydrate metabolism and the production of glycogen by the liver. Prolonged animal experiments with antimony aerosols have led to the development of distinctive endogenous lipoid pneumonia. [The therapeutic use of antimonial drugs has made it possible to detect, in particular, the cumulative myocardial toxicity of the trivalent derivatives of antimony (which are excreted more slowly than pentavalent derivatives). Reduction in amplitude of T wave, increase of QT interval and arrhythmias have been observed.] Cardiac injury and cases of sudden death have also been reported in workers exposed to antimony.

Inhalation of antimony aerosols may produce localised reactions of the mucous membrane, respiratory tract and lungs. Examination of miners and concentrator and smelter workers exposed to antimony dust and fume has revealed dermatitis, rhinitis, inflammation of upper and lower respiratory tracts, including pneumonitis and even gastritis, conjunctivitis and septal perforations.

Symptoms. The symptoms of acute poisoning include violent irritation of the mouth, nose, stomach and intestines; vomiting and bloody stools; slow shallow respiration; coma sometimes followed by death due to exhaustion and hepatic and renal complications. Those of chronic poisoning are: dryness of throat, nausea, headaches, sleeplessness, loss of appetite, and dizziness. Women seem more susceptible to the effects of antimony than are men.

Stibine (SbH_3)

ANTIMONY HYDRIDE; HYDROGEN ANTIMONIDE

m.w.	124.8
sp.gr.	4.36
m.p.	−88 °C
b.p.	−17 °C

a colourless, relatively unstable gas with a disagreeable odour.

TWA OSHA	0.1 ppm 0.5 mg/m^3
STEL ACGIH	0.3 ppm 1.5 mg/m^3
IDLH	40 ppm

Conveniently produced by dissolving zinc-antimony or magnesium-antimony alloy in dilute hydrochloric acid. However, stibine occurs frequently as a by-product in the processing of metals containing antimony with reducing acids or in overcharging storage batteries. Stibine has been used as a fumigating agent.

Stibine is an extremely hazardous gas. Like arsine it may destroy blood cells and cause haemoglobinuria, jaundice, anuria and death. Symptoms include headache, nausea, epigastric pain and passage of dark red urine following exposure.

Antimony trioxide (Sb_2O_3)

m.w.	291.5
sp.gr.	5.2
m.p.	656 °C
b.p.	1 550 °C (sublimes)

a white crystalline powder.

TLV ACGIH	0.5 mg/m^3 (handling and use) (as Sb)
ACGIH	industrial substance suspected of carcinogenic potential for man, without an assigned TLV (production)
MAC USSR	1 mg/m^3

The most important of the antimony oxides; when airborne it tends to remain suspended for an exceptionally long time. It is obtained from antimony ore by a roasting process or by oxidising metallic antimony and subsequent sublimation, and is used for the manufacture of tartar emetic, as a paint pigment, in enamels and glazes and as a flameproofing compound.

Antimony trioxide is both a systemic poison and a skin disease hazard, although its toxicity is three times less than that of the metal.

[An excess of deaths due to cancer of the lung among workers engaged in Sb smelting for more than four years, at an average Sb concentration in air of 8 mg/m^3, has been reported from Newcastle. In addition to antimony dust and fumes, the workers were exposed to zircon plant effluents and caustic soda. No other experiences were informative on the carcinogenic potential of antimony trioxide. This has been classified by the ACGIH as a chemical substance associated with industrial processes which are suspect of inducing cancer.]

Antimony pentoxide (Sb_2O_5)

m.w. 323.5
sp.gr. 3.8
m.p. 380 °C (loses oxygen above 300 °C)
an amorphous, yellowish-white powder.
MAC USSR 2 mg/m^3

Produced by the oxidation of the trioxide or the pure metal, in nitric acid under heat, it is used in the manufacture of paints and lacquers, glass and pottery and is also employed in the pharmaceutical industry. It is noted for its low degree of toxic hazard.

Antimony trisulphide (Sb_2S_3)

m.w. 339.7
sp.gr. 4.64
m.p. 550 °C
b.p. ca. 1 150 °C
grey, lustrous, crystalline masses or greyish-black powder also exists in red modification.
MAC USSR 1 mg/m^3

It is found as a natural mineral, antimonite, but can also be synthesised and is used in the pyrotechnics match and explosives industries and as a pigment and plasticiser in the rubber industry. An apparent increase in heart abnormalities has been found in persons exposed to the trisulphide.

Antimony pentasulphide (Sb_2S_5)

m.w. 403.8
sp.gr. 4.12
m.p. 75 °C (decomposes)
orange-yellow, odourless powder.
MAC USSR 2 mg/m^3

The pentasulphide has much the same uses as the trisulphide and has a low level of toxicity.

Antimony trichloride ($SbCl_3$)

ANTIMONOUS CHLORIDE; BUTTER OF ANTIMONY

m.w. 228.1
sp.gr. 3.14
m.p. 73.4 °C
b.p. 283 °C
orthorhombic, deliquescent needles fuming in air.
MAC USSR 0.3 mg/m^3

Antimony pentachloride ($SbCl_5$)

ANTIMONIC CHLORIDE; ANTIMONY PERCHLORIDE

m.w. 299
sp.gr. 2.34
m.p. 2.8 °C
b.p. 79 °C
colourless to yellow, oily liquid, fuming in air.
MAC USSR 0.3 mg/m^3

The trichloride is produced by the interaction of chlorine and antimony or by dissolving antimony trisulphide in hydrochloric acid, and the pentachloride by the action of chlorine on molten antimony trichloride. The antimony chlorides are used for blueing steel and colouring aluminium, pewter and zinc, and as catalysts in organic synthesis especially in the rubber and pharmaceutical industry. They are highly toxic substances and irritant and corrosive to the skin. The trichloride has an LD_{50} of 2.5 mg/100 g.

Antimony trifluoride (SbF_3)

m.w. 178.8
sp.gr. 4.38
m.p. 292 °C
b.p. 319 °C (sublimes)
orthorhombic deliquescent crystals.
MAC USSR 0.3 mg/m^3

Prepared by dissolving antimony trioxide in hydrofluoric acid. It is used in dyeing and pottery manufacture and organic synthesis. It is highly toxic and irritant to the skin and has an LD_{50} of 2.3 mg/100 g.

SAFETY AND HEALTH MEASURES

The essence of any safety programme for the prevention of antimony poisoning should be the control of dust and fume formation at all stages of processing.

In mining, dust prevention measures are similar to those for metal mining in general. During crushing the ore should be sprayed or the process completely enclosed and fitted with local exhaust ventilation combined with adequate general ventilation. In antimony smelting the hazards of charge preparation, furnace operation, fettling, and electrolytic cell operation should be eliminated, where possible, by isolation and process automation. Furnacemen should be provided with water sprays and effective ventilation.

Where complete elimination of exposure is not possible the hands, arms and faces of workers should be protected by gloves, dustproof clothing and goggles and, where atmospheric exposure is high, respirators should be provided. Barrier creams should also be applied especially when handling soluble antimony compounds, in which case they should be combined with the use of waterproof clothing and rubber gloves. Personal hygiene measures should be strictly observed, no food or beverages should be consumed in the workshops and suitable sanitary facilities should be provided so that workers can wash before meals and before leaving work.

Medical measures

Pre-placement and periodical medical examinations covering lungs, skin, nervous system, heart and gastro-intestinal tract are recommended.

Workers found to be suffering from impaired health should be temporarily removed from exposure. During this time a check should be kept on the patient's general condition and biologic antimony level.

[Review of the recent literature seems to indicate that workers with an Sb urinary excretion exceeding 1 mg/l exhibit signs of occupational health impairment.]

GUDZOVSKIJ, G. A.

"Metallurgy of antimony and lung function. Survey in a factory producing oxides of tin and antimony" (Métallurgie de l'antimoine et fonction respiratoire. Résultats d'une enquête menée dans une entreprise de fabrication des oxydes d'étain et d'antimoine). Cavelier, C.; Robin, H.; Mur, J. M.; Boulenguez, C.; Méreau, P.; Delebecq, J. F.; Behaguel, J. *Archives des maladies professionnelles, de médecine du travail et de sécurité sociale* (Paris), Aug.-Sep. 1979, 40/8-9 (795-803). 5 ref. (In French)

CIS 80-959 "Stibiosis, lung disease due to antimony—Arguments for considering it as an entity liable for compensation" (La stibiose, pneumopathie due à l'antimoine; arguments en faveur de son individualisation et de sa réparation). Curtes, J. P.; Develay, P.; Paumard, C. *Archives des maladies professionnelles, de médecine du travail et de sécurité sociale* (Paris), Oct. 1979, 40/10 (899-907). 14 ref. (In French)

CIS 80-1967 "Problems of toxicology and hygiene arising from combined exposure to antimony and arsenic, and damage to health observed in exposed subjects" (Príspevok k hygienicko-toxikologickej problematike kombinovanej expozície antimónu a arzénu a prejavy poškodenia zdravia u exponovaných osôb). Geist, T.; Bencko, V. *Ceskoslovenska hygiena* (Prague), Feb. 1979, 24/1 (20-26). Illus. 15 ref. (In Slovak)

CIS 79-1358 *Criteria for a recommended standard—Occupational exposure to antimony.* DHEW (NIOSH) publication No. 78-216 (National Institute for Occupational Safety and Health, 4676 Columbia Parkway, Cincinnati, Ohio) (1978), 125 p. 226 ref.

CIS 78-1349 *Antimony—Health and safety precautions.* Health and Safety Executive, London. Guidance Note EH 19 (London, HM Stationery Office, 1978), 2 p.

Health of workers engaged in antimony oxide manufacture. Davies, T. A. L. Employment Medical Advisory Service Statement (EMAS, Baynards House, Chepstow Place, London), Nov. 1973.

Apprentices

The term "apprentice" is generally applied to a person who is bound by written contract to serve an employer for a term of years while learning some handicraft, trade or profession in which the employer is reciprocally bound to instruct him. Apprenticeship is probably the earliest form of vocational training and its place is still recognised by the ILO Vocational Training Recommendation, 1962 (No. 117); at the same time it has always been a recognised way of regulating the number of skilled workers entering an occupation.

Originally, and still in many parts, skills were learnt by doing the job, watching a skilled man perform and receiving instruction from him: organised craft guilds jealously guarded the right to teach their practice. Increasing complexity of industrial processes and technological advances have accelerated the rate of change and old crafts are threatened by machine-made articles. [The Human Resources Development Convention, 1975 (No. 142) provides for the adoption and development of comprehensive policies and programmes of vocational guidance and vocational training, taking into account employment needs, opportunities and problems at both regional and national levels. Furthermore, for development purposes special schemes may be designed at the national level to enable young persons to take part in activities aimed at national economic and social development, acquiring at the same time education, skills and experience (Special Youth Schemes Recommendation, 1970 (No. 136)).]

HAZARDS AND THEIR PREVENTION

Apprentices are often subject to the same strains and hazards as young persons at work and enjoy the same legal protection, but as a select and defined group with a stable place of employment it is usually easier to ensure proper supervision of their health and safety. They tend to be more able and to have a better home background: employers, craftsmen and trade unions take more interest in them. It is, however, necessary to guard against their being used as cheap labour under the guise of training: often quality of training needs constant scrutiny. Training is usually given on the job by persons not professionally qualified as teachers. It is important that they should be responsible persons, fully skilled, safety-conscious and able to impart knowledge.

In any apprenticeship there should be an established training programme with safety and health considerations built in. Supervision of the terms and quality of apprenticeship by national or regional authorities may be necessary.

Entry into apprenticeship should be preceded by medical examination related to the requirements of the occupation and regular medical surveillance should be provided for apprentices, as indeed for other young persons (see YOUNG PERSONS).

Modern methods of training can often concentrate the basic skills of a wide range of craft apprenticeships into a shorter period than that recognised by craft regulations, but there may be resistance to change especially where possible overcrowding of the occupation has to be considered. Again technological advances render many old skills redundant and it may increasingly be necessary to change basic skills several times. It is important for young people to acquire an attitude of self-disciplined mobility which enables simple basic skills to be readily adapted to new situations and provides the groundwork for the acquisition of fresh skill.

HERFORD, M. M. E.

Vocational Training Recommendation, 1962 (No. 117) (Geneva, International Labour Office, 1962).

CIS 75-555 "Self-programmed accident prevention training for apprentices" (Selbstprogrammierte Unfallverhütung für Auszubildende). Beck, R. W. *Die Berufsgenossenschaft* (Bielefeld), Aug. 1974, 8 (322-324). (In German)

Arsenic and compounds

Arsenic (As)

ELEMENTAL ARSENIC

a.w.	74.9
sp.gr.	5.73
m.p.	817 °C at 28 atm
b.p.	613 °C (sublimes)
v.p.	1 mmHg ($0.13 \cdot 10^3$ Pa) at 372 °C

a silver-grey brittle, metallic-looking substance

a metalloid.

TWA OSHA	for inorganic As and compounds as As 0.01 mg/m^3
TLV ACGIH	for As and soluble compounds as As 0.2 mg/m^3

When arsenic is heated in air it will burn and form a white smoke consisting of arsenic trioxide As_2O_3.

Occurrence. Arsenic is found widely in nature and most abundantly in sulphide ores. Arsenopyrite (FeAsS) is the most abundant one.

Production. Arsenic "metal" is produced either by roasting the sulphide to form the oxide and then reducing the oxide with carbon or by heating arsenopyrite in the absence of air. When arsenic-containing ores are smelted the arsenic becomes gaseous and burns in air to arsenic trioxide. This is trapped by electrostatic precipitators as a crude dust, which is roasted so as to drive off arsenic trioxide. The purified As_2O_3 is collected in a cooling chamber. World annual production of arsenic and its compounds during the past few decades has been around 30 000 t with some decrease in the 1970s.

Uses. Elemental arsenic is utilised in alloys in order to increase their hardness and heat resistance, e.g. alloys with lead in shotmaking and battery grids.

There are three major groups of arsenic compounds:

(a) inorganic arsenic compounds;

(b) organic arsenic compounds;

(c) arsine gas (see separate article).

Trivalent inorganic compounds

Arsenic trichloride ($AsCl_3$)

m.w. 181.3
sp.gr. 2.16 (liquid)
m.p. −8.5 °C
b.p. 130.2 °C
v.d. 6.25
v.p. 10 mmHg (1.33·10³ Pa) at 23.5 °C

a yellowish oily liquid; dissolves in water to form As_2O_3 and HCl.

Used in the pottery industry and in the manufacturing of chlorine-containing arsenicals.

Arsenic trioxide (As_2O_3)

WHITE ARSENIC

m.w. 197.8
sp.gr. 3.86-4.09-4.15
m.p. 315 °C (sublimes)
b.p. 457 °C

soluble in water

white or transparent odourless and tasteless amorphous powder.

TWA OSHA as As 0.01 mg/m³
ACGIH arsenic trioxide production: substance associated with industrial processes suspect of carcinogenic potential for man
MAC USSR 0.3 mg/m³

Arsenic trioxide is the primary material for all arsenic compounds including the elemental form. Oral LD_{50} (rat) 15 mg/kg. Used in the manufacture of glass and as an insecticide and rodenticide.

Calcium arsenite ($Ca(As_2H_2O_4)$)

m.w. 256

white, granular powder.

Cupric acetoarsenite
(usually considered $Cu(COOCH_3)_2 \cdot 3\,Cu(AsO_2)_2$)

PARIS GREEN; VIENNA GREEN; IMPERIAL GREEN; KING'S GREEN; EMERALD GREEN; COPPER ACETATE METAARSENATE

m.w. 256

insoluble in water

a crystalline powder.

Used as a pigment, insecticide, wood preservative and in antifouling paint for ships.

Cupric arsenite ($Cu(AsO_2)_2 \cdot H_2O$)

COPPER ARSENITE; SCHEELE'S GREEN

m.w. 187.5
m.p. decomposes

insoluble in water; soluble in acids and ammonia

a yellowish green powder.

Used as a wood preservative, insecticide, fungicide and rodenticide.

Lead arsenite ($PbAs_2O_4$)

LEAD-*o*-ARSENITE; LEAD-*m*-ARSENITE

m.w. 421
sp.gr. 5.85

insoluble in water

a white powder.

On heating, it emits highly toxic fumes.

Sodium arsenite ($NaAsO_2$)

SODIUM METAARSENITE

m.w. 129.9
sp.gr. 1.87

very soluble in water; slightly soluble in ethyl alcohol

hygroscopic white or greyish-white powder.

Intraperitoneal LD_{50} (mice) 1.17 mg/kg. Used as a herbicide.

Pentavalent inorganic compounds

Arsenic acid ($H_3AsO_4 \cdot {}^1/_2\,H_2O$)

TRUE ARSENIC ACID; *o*-ARSENIC ACID

m.w. 150.9
sp.gr. 2.0-2.5
m.p. 35.5 °C
b.p. 160 °C (−H_2O)

white translucent hygroscopic crystals.

Oral LD_{50} (rat) 48 mg/kg; intravenous LD_{50} (rabbit) 8 mg/kg.

Arsenic pentoxide (As_2O_5)

ARSENIC OXIDE; ARSENIC ACID ANHYDRIDE

m.w. 229.8
m.p. 315 °C (decomposes)

white, amorphous, deliquescent powder.

TWA OSHA as As 0.01 mg/m³
TLV ACGIH as As 0.2 mg/m³
MAC USSR 0.3 mg/m³

Oral LD_{50} (rat) 8 mg/kg; oral LD_{50} (mouse) 55 mg/kg.

Calcium arsenate ($Ca_3(AsO_4)_2$)

TRICALCIUM-*o*-ARSENATE; CALCIUM-*o*-ARSENATE

m.w. 398
sp.gr. 3.6

insoluble in water

white amorphous powder.

TWA OSHA as As 0.01 mg/m³

Oral LD_{50} (rat) 20 mg/kg. Used as an insecticide and herbicide.

Lead arsenate ($PbHAsO_4$)

See separate article.

Organic arsenic compounds

Cacodylic acid ($(CH_3)_2\,AsOOH$)

DIMETHYLARSINIC ACID; HYDROXY DIMETHYL ARSINE OXIDE

m.w. 138
m.p. 195 °C

colourless crystals, hygroscopic.

Oral LD_{50} (rat) 1 350 mg/kg. Used as a herbicide and defoliant.

Arsanilic acid ($NH_2C_6H_4AsO(OH)_2$)

4-AMINOBENZENEARSONIC ACID; AMINOPHENYL ARSINE ACID; ATOXILIC ACID

m.w. 217
m.p. 232 °C

soluble in hot water

white crystalline powder.

Oral LD_{50} (rat) 216 mg/kg. Used as a grasshopper bait; a food additive permitted in the feed and drinking water of animals and/or for the treatment of food-producing animals.

Organic arsenic compound present in marine organisms

SHRIMP-ARSENIC; FISH-ARSENIC

Its chemical composition is still unknown; the molecular weight is approximately 400. It is present in marine organisms in concentrations of 1-100 mg/kg, is of low toxicity and is almost completely excreted unchanged in urine.

HAZARDS

It is possible that very small amounts of certain arsenic compounds may have beneficial effects, as indicated by some preliminary animal studies. Arsenic compounds are otherwise generally regarded as very potent poisons. Their acute toxicity, however, varies greatly as between two organic and inorganic compounds, depending on valency and solubility in biological media.

The following text will deal with adverse effects of inorganic arsenic in human beings. Organic arsenicals used as pesticides or as drugs may also give rise to adverse health effects, although such effects are incompletely documented in human beings.

Toxic effects on the nervous system have been reported in experimental animals following feeding with high doses of arsanilic acid, which is commonly used as a feed additive in poultry and swine.

Occupational exposure to inorganic arsenic compounds through inhalation, ingestion or skin contact with subsequent absorption may occur in industry. Acute effects at the point of entry may occur if exposure is excessive. Dermatitis may occur as an acute symptom but is more often the result of sensitisation (see under "long-term exposure").

Acute poisoning

Effects due to accidental *ingestion* of inorganic arsenicals, mainly arsenic trioxide, have been described in the literature. However, such incidents are very rare in industry today. Cases of poisoning are characterised by profound gastrointestinal damage, resulting in severe vomiting and diarrhoea, which may result in shock and subsequent oliguria and albuminuria. Other acute symptoms are facial oedema, muscular cramps and cardiac abnormalities. Symptoms may occur within a few minutes following exposure to the poison in solution but may be delayed for several hours if the arsenic compound is in solid form or if it is taken with a meal. When ingested as a particulate, toxicity is also dependent on solubility and particle size of the ingested compound. The fatal dose of ingested arsenic trioxide has been reported to range from 70 to 180 mg. Death may occur within 24 h, but the usual course runs from 3 to 7 days. Acute intoxication with arsenic compounds is usually accompanied by anaemia and leucopenia, especially granulocytopenia. In survivors these effects are usually reversible within 2 to 3 weeks. Reversible enlargement of the liver is also seen in acute poisoning, but liver function tests and liver enzymes are usually normal.

In individuals surviving acute poisoning peripheral nervous disturbances frequently develop a few weeks after ingestion.

Exposure to irritant arsenic compounds in air, such as arsenic trioxide, can cause acute damage to the mucous membranes of the respiratory system and can cause acute symptoms from exposed skin. Severe irritation of the nasal mucosae, larynx and bronchi as well as conjunctivitis and dermatitis occur in such cases. Perforation of the nasal septum can be observed in some individuals only after a few weeks following exposure. A certain tolerance against acute poisoning is believed to develop upon repeated exposure. This phenomenon, however, is not well documented in the scientific literature.

Long-term exposure (chronic poisoning)

Chronic arsenic poisoning may occur in workers exposed for a long time to excessive concentrations of airborne arsenic compounds. Local effects in the mucous membranes of the respiratory tract and skin effects are prominent features. Involvement of the nervous and circulatory system and the liver may also occur as well as cancer of the respiratory tract.

With long-term exposure to arsenic via ingestion in food, drinking water or medication, symptoms are partly different from those after inhalation exposure. Vague abdominal symptoms–diarrhoea or constipation, flushing of the skin, pigmentation and hyperkeratosis–dominate the clinical picture. In addition there may be vascular involvement, reported in one area to have given rise to peripheral gangrene–so-called "Blackfoot disease". Such vascular changes have not been reported in industrial arsenic exposure.

Anaemia and leucocytopenia often occur in chronic arsenic poisoning. Liver involvement has been more commonly seen in persons exposed for a long time via oral ingestion than in those exposed via inhalation, particularly in vineyard workers considered to have been exposed mainly through drinking contaminated wine. Skin cancer occurs with excess frequency in this type of poisoning.

Arsenical skin lesions, as mentioned, are somewhat different depending on the type of exposure. Eczematoid symptoms of varying degrees of severity do occur. In occupational exposure mainly to airborne arsenic, skin lesions may result from local irritation. Two types of dermatological disorders may occur:

1. an eczematous type with erythema, swelling and papules or vesicles; and
2. a follicular type with erythema and follicular swelling or follicular pustules.

Dermatitis is primarily localised on the most heavily exposed areas such as the face, back of the neck, forearms, wrists and hands. However, it may also occur on the scrotum, the inner surfaces of the thighs, the upper chest and back, the lower legs and around the ankles. Hyperpigmentation and keratoses are not prominent features of this type of arsenical lesions. Patch tests have demonstrated that the dermatitis is due to arsenic, not to impurities present in the crude arsenic trioxide. Chronic dermal lesions may follow this type of initial reaction depending on the concentration and duration of exposure. These chronic lesions may occur after many years of occupational or environmental exposure. Hyperkeratosis, warts and melanosis of the skin are the conspicuous signs. Melanosis is most commonly seen on the upper and lower eyelids, around the temples, on the neck, on the areolae of the nipples and in the folds of the axillae. In severe cases arsenomelanosis is observed on the abdomen, chest, back and scrotum along with hyperkeratosis and warts. In chronic arsenic poisoning depigmentation, i.e. leukoderma, especially on the pigmented areas, commonly called "raindrop" pigmentation also occurs. These chronic skin lesions, particularly the hyperkeratoses may develop into precancerous and cancerous lesions. Mees lines of the fingernails also occur in chronic arsenical poisoning. It should be noted that the chronic skin lesions may develop long after cessation of exposure when arsenic concentrations in skin have returned to normal.

Mucous membrane lesions in chronic arsenic exposure is most classically reported as perforation of the nasal septum after inhalation exposure. This lesion is a result of irritation of the mucous membranes of the nose. Such irritation also extends to the larynx, trachea and bronchi. Both in inhalation exposure and in poisoning caused by repeated ingestion, dermatitis of the face and eyelids sometimes extends to keratoconjunctivitis.

Peripheral neuropathy. Peripheral nervous disturbances are frequently encountered in survivors of acute poisoning. They usually start within a few weeks after the acute poisoning and recovery is slow. The neuropathy is characterised by both motor dysfunction and paresthaesia. But in less severe cases only sensory unilateral neuropathy may occur. Often the lower extremities are more affected than the upper ones. In subjects recovering from arsenical poisoning a transverse striation of the nails, so-called Mees lines, may develop. Histologic examination has revealed Wallerian degeneration, especially in the longer axons. In children exposed to arsenic, hearing loss has been reported.

Carcinogenic effects

Inorganic arsenic compounds are recognised by the International Agency for Research on Cancer (IARC) as lung and skin carcinogens. There is also some evidence to suggest that persons exposed to inorganic arsenic compounds suffer a higher incidence of angiosarcoma of the liver and possibly of stomach cancer.

Cancer of the respiratory tract has been reported in excess frequency among workers engaged in the production of insecticides containing lead arsenate and calcium arsenate, in vine-growers spraying insecticides containing inorganic copper and arsenic compounds and in smelter workers exposed to inorganic compounds of arsenic and a number of other metals. The latency time between onset of exposure and the appearance of cancer is long, usually between 15 and 30 years. For some industrial agents a synergistic action of tobacco smoking has been demonstrated. However, the role of cigarette smoking in relation to lung cancer in workers exposed to arsenic has not yet been defined. Exposure to inorganic arsenic via drinking water has been associated with a higher than normal incidence of skin cancer in Chile.

Teratogenic effects

High doses of trivalent inorganic arsenic compounds may cause malformations in hamsters when injected intravenously. With regard to human beings there is no firm evidence that arsenic compounds cause malformations under industrial conditions. Some recent evidence, however, suggests such an effect in workers in a smelting environment exposed simultaneously to a number of metals including arsenic as well as other compounds.

SAFETY AND HEALTH MEASURES

The best means of prevention is to keep exposure well below accepted exposure limits. A programme of measurement of air-concentrations of arsenic is thus of importance. In addition to inhalation exposure, oral exposure via contaminated clothes, hands, tobacco, etc., should be watched. Workers should be supplied with suitable protective clothing, protective boots and, when there is a risk that the exposure limit for airborne arsenic will be exceeded, respiratory protective equipment. Lockers should be provided with separate compartments for work and personal clothes, and adjacent sanitary facilities of a high standard should be made available. Smoking, eating and drinking at the workplace should not be allowed. Pre-employment medical examinations should be carried out. It is not recommended to employ persons with pre-existing diabetes, cardiovascular diseases, allergic or other skin diseases, neurologic, hepatic or renal lesions in arsenic work. A controversial matter at present is whether women of child-bearing age should be employed. Some companies that have women workers in shops where arsenic is present use a policy to relocate them in less exposed departments or give them a period of leave in case of pregnancy. If such a policy is to have an effect in decreasing possible risks to the fetus, exposure should obviously be decreased very early in pregnancy. Periodic medical examinations of all arsenic-exposed employees (male or female) should be performed with special attention to possible arsenic-related symptoms.

Arsenic levels in the urine are sometimes used as a means of assessing exposure. Only when inorganic arsenic and its metabolites can be specifically measured is this method useful. Total arsenic in urine may often give erroneous information about industrial exposure, since even a single meal of fish or other marine organisms (containing considerable amounts of non-toxic organic arsenic compound) may cause greatly elevated urinary arsenic concentrations for several days.

Treatment

Acute skin lesions such as contact dermatitis usually do not require other treatment than removal from exposure. If more severe symptoms from the respiratory system, the skin or the gastrointestinal tract occur, British anti-lewisite (BAL, dimercaprol) may be given . Prompt administration in such cases is vital: to obtain maximal benefit such treatment should be given within 4 h of poisoning. In addition, general treatment such as prevention of further absorption by removal from exposure and minimising absorption from the gastro-intestinal tract are mandatory. General supportive therapy such as maintenance of respiration and circulation, maintenance of water and electrolyte balance, and control of nervous system effects, as well as elimination of absorbed poison through dialysis and exchange transfusion may be used if feasible. Dimercaprol is given by deep intramuscular injection as a 5% solution in peanut oil (or a 10% solution with benzyl-benzoate in vegetable oil). It is usually given in a dose of 3 mg/kg, 4-hourly, for the first two days, 6-hourly on the third day and thereafter once or twice daily for up to seven days.

NORDBERG, G.

"Arsenic". Fowler, B. A.; Ishinishi, N.; Tsuchiya, K.; Vahter, M. *Handbook on the toxicology of metals.* Friberg, L.; Nordberg, G.; Vour, V. B. (eds.). (Amsterdam, Elsevier North Holland, 1978) (293-319).

"Arsenic, environment and health" (L'arsenic, l'environnement et la santé). Lafontaine, A. *Archives belges de médecine sociale, hygiène, médecine du travail et médecine légale* (Brussels), Apr. 1980, 38/4 (222-236). Illus. 6 ref. (In French)

"Arsenic and arsenic compounds". *IARC monographs on the evaluation of the carcinogenic risk of chemicals to humans.* Vol. 23: *Some metals and metallic compounds* (Lyons, International Agency for Research on Cancer, 1980) (39-141). 329 ref.

CIS 80-130 "Occupational exposure to inorganic arsenic: Final standard". US Department of Labor, Occupational Safety and Health Administration. *Federal Register* (Washington, DC), 5 May 1978, Part IV, 43/88 (19 584-19 630).

CIS 75-1683 *Criteria for a recommended standard: Occupational exposure to inorganic arsenic–New criteria 1975.* DHEW (NIOSH) publication No. 75-149 (Washington, DC, US Government Printing Office, Apr. 1975), 140 p. Illus. 111 ref.

CIS 80-1634 "Comparison of several methods for the determination of arsenic compounds in water and urine". Buchet, J. P.; Lauwerys, R.; Roels, H. *International Archives of Occupational and Environmental Health* (West Berlin), 1980, 46/1 (11-29). 17 ref.

Arsines

Arsine (AsH_3)

ARSENIURETTED HYDROGEN; ARSENIC HYDRIDE

m.w.	78
sp.gr.	3.48
m.p.	−116 °C
b.p.	−55 °C
v.d.	2.66

slightly soluble in water

a colourless gas with a garlic-like odour.

TWA OSHA	0.05 ppm 0.2 mg/m³
IDLH	6 ppm
MAC USSR	0.3 mg/m³

Occurrence. Arsine is produced commercially in small amounts for use in organic synthesis and in the processing of solid-state electronic components. However, arsine generation not infrequently results as a side reaction in many different processes. Every time nascent hydrogen is formed in the presence of arsenic, arsine is generated. The hydrolysis of certain metallic arsenides, such as those of sodium, zinc and aluminium, can also be a source of arsine.

Metal smelting and refining is one of the main sources of arsine poisoning. Several cases of exposure to arsine gas have occurred in England among steel bronze workers using a solution of arsenic and iron chloride in strong hydrochloric acid. Accidental exposure to arsine following the reduction of arsenical compounds may occur in processes such as galvanising, cadmium recovery and the manufacture of zinc salts. Recently, a case of acute poisoning was reported in two workers cleaning a clogged drain. The workers used a drain cleaner containing sodium hydroxide and aluminium chips, a combination that reacted, releasing hydrogen gas. The nascent hydrogen reacted with a residue of arsenic present in the drain to form arsine gas.

HAZARDS

Some 100 cases of arsine poisoning have been recorded and more than 20% of these were fatal. Arsine is one of the most powerful haemolytic agents found in industry. Experimental evidence has been obtained showing that the haemolytic activity of arsine is due to its ability to cause a fall in erythrocyte reduced glutathion (GSH) content.

Signs and symptoms of poisoning

When poisoning with arsine occurs, haemolysis develops after a latent period, whose length is inversely proportional to the extent of exposure. Inhalation of 250 ppm of arsine gas is instantly lethal; exposure to 25-50 ppm for 30 min is lethal and 10 ppm is lethal after a longer exposure. The signs and symptoms of poisoning are those characteristic of an acute and massive haemolysis. Initially there is painless haemoglobinuria, gastrointestinal disturbance such as nausea and possibly vomiting, abdominal cramp and tenderness. Acholuric jaundice accompanied by anuria and oliguria may ensue.

Evidence of marrow depression was present in some cases reported. After acute and severe exposure a peripheral neuropathy may develop and can still be present several months after poisoning.

Very little is known about repeated or chronic exposure to arsine.

The differential diagnosis should take account of acute haemolytic anaemias that could be caused by other chemical agents, such as stibine, or drugs, and secondary immunohaemolytic anaemias.

SAFETY AND HEALTH MEASURES

Care should always be taken that no arsenical compound is present whenever nascent hydrogen is generated during an industrial process. Workers should be informed of the possibility of arsine formation in all operations where arsenic is used, stored or is present as a contaminant. Since industrial exposure to arsine is mostly accidental and severe, medical prevention can only be very limited.

It is important to exclude workers presenting renal or cardiac disease or with hypersensitivity to haemolytic agents even from controlled situations of exposure, because the present permissible limits for arsine may prove harmful to them (for example subjects with congenital deficiency of GSH content in the red blood cells).

Technical prevention is, instead, of paramount importance in all the occupational settings mentioned above, in order to avoid the generation of, and consequently the exposure to, arsine.

Treatment

Upon observation of the first symptoms, like haemoglobinuria and abdominal pain, immediate removal of the individual from the contaminated environment and prompt medical attention is required. The recommended treatment, if there is any evidence of impaired renal function, consists of total replacement blood transfusion associated with prolonged artificial dialysis. Forced diuresis has proved useful in some cases, whereas, in the opinion of most authors, treatment with BAL seems to have only minimal effect.

Substituted arsines

Allied to arsine are many negative trivalent organic arsenical compounds which, depending on whether they have one, two or three alkyl or phenyl groups directly attached to the arsenic nucleus, are known as mono-, di- or tri-substituted arsines. All these compounds are readily hydrolised in water.

Dichloroethylarsine ($C_2H_5AsCl_2$)

ETHYLDICHLOROARSINE; DICK

m.w.	174.9
sp.gr.	1.74
m.p.	−65 °C
b.p.	156 °C (decomposes)
v.d.	6.03
v.p.	2.29 mmHg ($0.3 \cdot 10^3$ Pa) at 21.5 °C

a colourless liquid with an irritant odour.

This was developed as a potential chemical warfare agent.

Dichloro(2-chlorovinyl-)arsine ($ClCH{:}CHAsCl_2$)

CHLOROVINYL DICHLOROARSINE; LEWISITE

m.w.	233.3
sp.gr.	1.70
b.p.	230 °C (decomposes)
v.d.	8.05

an olive-green liquid with a geranium-like odour.

This was developed as a warfare agent but never used. The search for an antidote elucidated the mode of action of arsenic and led to the discovery of BAL.

Dimethylarsine ($(CH_3)_2AsH$)

CACODYL HYDRIDE

m.w.	106
sp.gr.	1.21
b.p.	70 °C
v.d.	3.65

a colourless liquid.

Trimethylarsine ($(CH_3)_3As$)

TRIMETHYL ARSENIC

m.w.	120
sp.gr.	1.12
b.p.	70 °C
v.d.	4.1

a colourless liquid.

These can be produced after metabolic transformation of arsenic compounds by bacteria and fungi, especially in sewage. For example, these substances were detected when a herbicide compound containing arsenic was added to the soil.

HAZARDS

These compounds do not have haemolysis of the red cells as their main effect but they do act as powerful local and pulmonary irritants and systemic poisons. This local effect on the skin is readily evident as an erythema which, in the case of dichloro(2-chlorovinyl)arsine, develops into sharply circumscribed blisters. The vapour induces marked spasmodic coughing with frothy or blood-stained sputum, the condition progressing to one of acute pulmonary oedema. BAL is an effective antidote and, especially in the early stages, may reverse the symptoms.

FOÁ, V.
BERTOLERO, F.

CIS 75-1989 "Arsine toxicity aboard the Asiafreighter". Wilkinson, S. P.; McHugh, P.; Horsley, S.; Tubbs, H.; Lewis, M.; Thould, A.; Winterton, M.; Parsons, V.; Williams, R. *British Medical Journal* (London), 6 Sep. 1975, 3/5893 (559-563). Illus. 20 ref.

CIS 77-805 "Case report–Arsine poisoning: evaluation of the acute phase". Pinto, S. S. *Journal of Occupational Medicine* (Chicago), Sep. 1976, 18/9 (633-635). 8 ref.

CIS 78-184 "Arsenic absorption in steel bronze workers". Clay, J. E.; Dale, I.; Cross, J. D. *Journal of the Society of Occupational Medicine* (Bristol), July 1977, 27/3 (102-104). 10 ref.

"Acute arsine poisoning in two workers cleaning a clogged drain". Parish, G. G.; Glass, R.; Kimbrough, R. *Archives of Environmental Health* (Washington, DC), July-Aug. 1979, 34/4 (224-227). Illus. 6 ref.

"Arsine poisoning". Kleinfeld, M. *Journal of Occupational Medicine* (Chicago), Dec. 1980, 22/12 (820-821). 6 ref.

Artists

Contemporary painters and sculptors place their emphasis on action. The modern painter is concerned with surface and textural effects. The end product of modern sculpture is generally expansive and innovative. These artists often experiment with new, untried materials, the potential hazards of which may be unknown to them. Overexposure to these materials may affect their health.

Painters and their materials

Although the traditional painting media–fresco, egg tempera, encaustic oil and watercolour–are still in use, the synthetic media are gradually taking over. The resins in use today are the polymers and copolymers of acrylics, polyvinyl chloride, and polyvinyl acetate. Other synthetic media are ethyl silicate, vinyl acetate, vinyl chloride acetate and pyroxylin. Some of the media are available commercially. Others must be formulated and/or pigmented by the artist.

The solvents, diluents and thinners used by painters are often a far greater potential health hazard than any other art materials. These liquids may produce contact dermatitis due to primary irritation, as well as hypersensitivity. High concentrations of the vapours may be irritating to the eyes, nose and throat. Painting or spraying for a considerable period of time in an enclosed room with poor ventilation may lead to excessive inhalation of solvent vapours and subsequent systemic effects.

In recent years a three-dimensional effect in paintings has been achieved by employing a variety of plastics–polyesters, acrylics, epoxies, even polyurethanes–in the creation of supports with sculptural characteristics. Liquid resins, catalysts, hardeners and plasticisers are used in the process. The major hazard is contact dermatitis which may be due to either primary irritation or allergic sensitisation. There may also be exposure to the inhalation of solvents, catalysts and some of the additives such as thickeners or binders. Other hazards are fire and explosion, since the solvents have low flash points and the resins have flash points at or below room temperature. Organic catalysts must be used to start the reaction to harden the liquid plastic to a solid. These, when not handled or stored properly, may be heat- and shock-sensitive.

Sculptors and their materials

The sculptor uses a variety of materials, tools and processes. He may carve directly in wood or stone and for modelling he may use materials such as clay, wax or plasticine, plaster, cement and polyester, acrylic and epoxy resins. He may reproduce a casting through the use of a sand mould or do shell casting and investment moulding. He may weld and braze using all sorts of scrap metal or cut, grind and form metal to desired shapes.

Each of the processes used by the sculptor may generate an environmental hazard, the severity of which depends on the character, intensity and duration of the exposure. The well known hazards of silicosis entailed in stone working and of dermatitis and allergic sensitivity in woodcarving are now overshadowed by dangers associated with the increasing use of metals and alloys.

The fundamental methods of metal casting employed by the foundry worker are used by the sculptor. Unless the workshop is equipped with adequate exhaust ventilation, the sculptor may expose himself to the fumes of toxic metals or alloys from the melting or pouring operations. In using the shell moulding process, he may expose himself to the thermosetting plastics used as binders and catalysts. The investment casting process presents the potential hazard of inhalation of dust and solvent vapours and skin contact with acids.

There are potential health hazards associated with welding and cutting of materials, particularly with arc welding. There may be inhalation of fumes and gases, exposure to ultraviolet radiation, electric shock and fire, and eye hazards from projections of hot metal. Toxic fumes can be liberated during the melting of metals and their alloys depending upon the composition of the electrode and the metal being welded. Precautions should be taken in welding on alloys containing lead, zinc, cadmium, beryllium, chromium and manganese. Excessive exposure to the fumes may cause systemic poisoning. Metal fume fever frequently occurs after inhalation of high concentrations of zinc oxide fumes although copper, iron, magnesium, cadmium and other metal oxides can also cause this condition. The controls and precautionary measures developed by industry to prevent ill effects on health of workers using these processes should also be applied by the sculptor.

The materials and processes now being used by painters and sculptors were usually developed for various industrial uses and precautionary protective measures have been adopted by the industries using them. The artist should familiarise himself with these precautionary measures before attempting to use these substances. Public health authorities through their occupational health or industrial hygiene units can assist the artist in evaluating potential hazards and recommending appropriate control measures (see also ACTORS AND PLAYERS; CINEMA INDUSTRY; ENTERTAINMENT INDUSTRY).

SIEDLECKI, J. T.

General:

"The social and working conditions of artists". Cornwell, S. C. *International Labour Review* (Geneva), Sep.-Oct. 1979, 118/5 (537-556). 27 ref.

Artistic foundry:

CIS 77-834 "Lead poisoning hazard in the pewter industry and artistic bronze foundries" (Rischio di saturnismo nella lavorazione del peltro e nelle fonderie artistiche di bronzo). Carnevale, F.; D'Andrea, F.; Grazioli, D. *Lavoro Umano* (Naples), July 1976, 28/4 (104-110). 4 ref. (In Italian)

Painting and photography:

CIS 78-860 "Health hazards to commercial artists". Foote, R. T. *Job Safety and Health* (Washington, DC), Nov. 1977, 5/11 (7-13).

Plastics:

Considerations on artist's work making use of plastics materials (Réflexions sur le travail d'artiste à partir d'un matériau : les matières plastiques). Arcier, A. F. Thesis (Université de Montpellier, Faculté de Médecine, 1981), 87 p. Illus. 49 ref.

School of art:

Cooper Union School of Art, New York, N.Y. DHEW (NIOSH) Health evaluation determination Report No. 75-12-321 (National Institute for Occupational Safety and Health, 4676 Columbia Parkway, Cincinnati) (1976).

Sculpture:

"Potential health hazards of materials used by artists and sculptors". Siedlecki, J. T. *Journal of the American Medical Association* (Chicago), 1968, 204/13 (1 176-1 180). 10 ref.

"Occupational profiles for the industrial physician. Stonemason and sculptor" (Steinmetz und Steinbildhauer). Stark, K. *Arbeitsmedizin–Sozialmedizin–Präventivmedizin* (Stuttgart), Nov. 1977, 12/11, Suppl. 128, 3 p. (In German)

Asbestos

Asbestos is a term used to describe the chain silicates which occur naturally in fibrous form and are commercially useful. The rise of production (log scale) of the three principal kinds of asbestos since the start of the industry is seen in figure 1. There has been an exponential rise, and since about 1920 the rate has been greater for the two amphiboles–amosite and crocidolite–than for chrysotile, though the actual tonnage of chrysotile far exceeds the amphiboles. The total world production now exceeds 5 million t, about half of which comes from the USSR. Most of the amphibole asbestos comes from South Africa. The diagram at the top of the figure shows the long lag before the asbestos-related diseases were first suspected and the further interval of several years before a causal relationship was generally accepted.

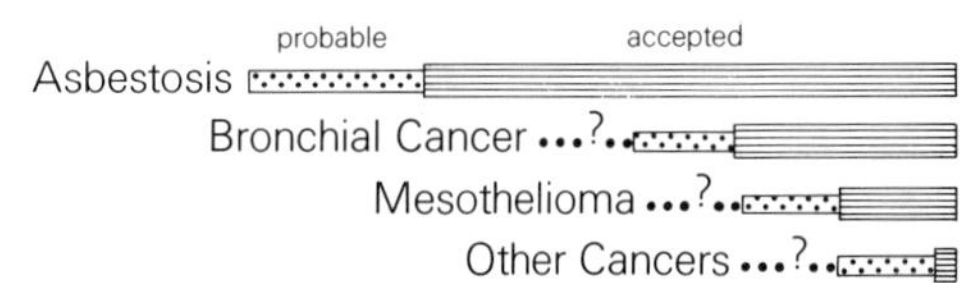

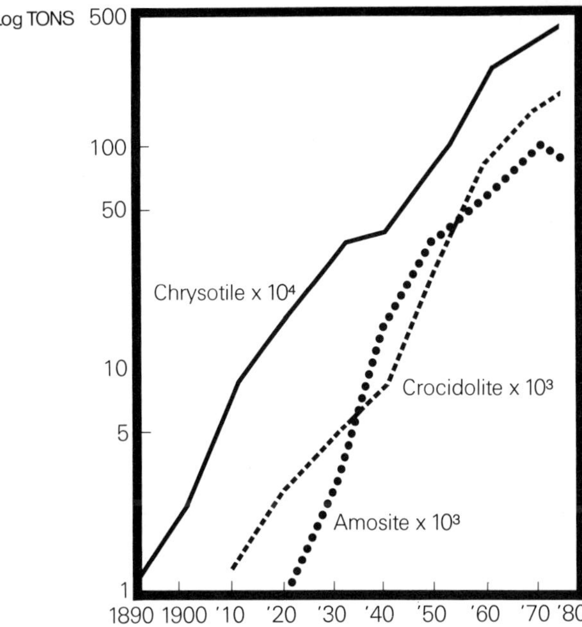

Figure 1. Rise of production of the three principal kinds of asbestos since 1890 and dates of acceptance of the causal relation between asbestos and various diseases.

There are two main types–chrysotile and the amphiboles. Table 1 shows the composition of the various types.

Chrysotile. This is a hydrated magnesium silicate which occurs in serpentine rock of the same chemical composition. It is white or greenish in colour. The fibres occur in multiple vertical seams, usually about 2 cm in width, but occasionally much thicker. The mineral is widely distributed throughout the world. The main centres of production are at Asbestos and Thetford in Quebec Province; Cassiar in British Columbia, Asbest, near Sverdlovsk in the USSR; in Southern Africa, in Swaziland and in Zimbabwe. Small quantities are also mined in Australia, Cyprus, Italy, Turkey and the United States. The large deposits in Quebec have been mined continuously since 1876, and those near Sverdlovsk since 1885. The deposits in South Africa have been mined continuously for about the last 40 years. The harshness of the fibres varies in different deposits, but chrysotile is in general the softest type of asbestos and so is most used for spinning and weaving to produce asbestos cloth and tape.

Amphiboles. There are five asbestiform amphiboles–amosite, crocidolite, anthophyllite, tremolite and actinolite. Tremolite and actinolite are of little commercial value but may be mixed with true talc (an amorphous magnesium silicate) to make commercial talc. Cosmetic talcs are in general free of fibrous silicates.

Amosite. This is an iron magnesium silicate. It is grey to brown in colour and is found only in the North-East Transvaal where it has been mined since 1916. The fibres are harsh and occur in seams up to about 30 cm wide; it

Table 1. Varieties of asbestos[1]

Variety	Colour	Major components (%)			Approximate formulae
		Si	Mg	Fe	
Chrysotile	white	40	38	2	$3MgO, 2SiO_2, 2H_2O$
Amphiboles					
Amosite	grey brown	50	2	40	$5.5FeO, 1.5MgO, 8SiO_2, H_2O$
Anthophyllite	white	58	29	6	$7MgO, 8SiO_2, H_2O$
Crocidolite	blue	50	–	40	$Na_2O, Fe_2O_3, 3FeO, 8SiO_2, H_2O$
Tremolite	white }	55	15	2	$2CaO, 5MgO, 8SiO_2, H_2O$
Actinolite	white }				$2CaO, 4MgO, FeO, 8SiO_2, H_2O$

[1] From Hodgson, A. A. (1965). *Fibrous silicates*. Lecture Series No. 4. The Royal Institute of Chemistry and the Asbestos Information Committee, London.

is the longest fibre type of asbestos. The harshness of the fibres makes it unsuitable for spinning, but when the fibres are "opened" it has large bulk and so is excellent for heat insulation.

Crocidolite. This is a sodium iron silicate, chemically very similar to amosite, but with a characteristic lavender-blue colour. Production is now mainly from a range of hills in the North-West Cape in South Africa extending over 500 km. Small quantities have been mined in Western Australia and in Bolivia. The fibres are intermediate in harshness between chrysotile and amosite, and the longer fibres can be spun. Amosite and crocidolite both occur in seams in banded ironstone.

Anthophyllite. This is a magnesium silicate containing varying quantities of iron. It occurs in fibrous masses in which the fibre bundles are short. The main production was in the Paakkila area of North East Finland where it has been mined since 1918. Mining has now ceased.

Identification and properties. In bulk, the blue colour of crocidolite and the grey-brown colour of amosite contrast with the white or greenish colour of chrysotile and anthophyllite. These differences in colour, together with the differences in harshness between the various types, made identification relatively easy. Chemical analysis can be used to confirm and, often, identify the source of the fibre.

But the more frequent need is to know which and how much of each type of fibre is present in small samples, in airborne dust or lung residues. Major advances in the last ten years now permit the identification of all minerals–fibrous and non-fibrous–in such samples (see FIBRES, NATURAL MINERAL). By using an electron microscope with an energy dispersion X-ray analysis attachment the morphology of even the smallest fibres can be recorded. Many of the finest fibres are too small to be seen with the optical microscope. The X-ray dispersion equipment provides quantitative elemental assessment even at several points on a single fibre only 0.2 µm in diameter and a few µm long. The amount of each element can at once be compared with the values for the standard mineral samples stored in the computer. X-ray diffraction is useful and quicker but cannot separate chrysotile fibres from particles of the parent serpentine rock, nor always separate the types of amphiboles.

The features of asbestos which make it so useful are the high tensile strength and flexibility of the fibres; and their resistance to heat and abrasion and to many chemicals. The fibre bundles split lengthways into very fine fibrils. In the case of amosite and crocidolite the fibrils are straight rods about 0.4 µm and 0.2 µm in diameter respectively. The fibrils of chrysotile are finer still, being about 0.16 µm in diameter. They are tubes in the form of a rolled scroll. The longer fibres are curled, which helps identification under the microscope. The longitudinal splitting of the fibres increases their respirability and biological activity.

The discovery of the importance of diameter and length rather than composition of the fibres in the aetiology of at least mesothelioma and asbestosis has directed attention to other fibrous minerals which might be expected to have similar biological effects as asbestos if inhaled in sufficient quantities. The possible implication of fibrous erionite, a type of zeolite, in the occurrence of mesotheliomas in Turkey is an example.

All types of asbestos have a large surface area so that, during the preparation of the fibre by the separation of the bundles, contamination by oils and other substances can readily occur. Virgin asbestos may also contain small amounts of oils and waxes. Trace amounts of a number of metals, including nickel, chromium, cobalt and manganese are present in many samples. The contribution of these to the biological effects of asbestos is not considered important on present evidence.

Extraction and preparation. The proportion of fibre in the deposit varies widely–the fibre content of the richest commercial seams may be as high as 30% and that of the poorest as low as about 3%. The mineral is mined or quarried with its parent rock. After preliminary sorting, the ore is crushed and passed through a rotary drying kiln or dried in the sun to facilitate the subsequent separation of the fibre. This is done by passing the crushed material over sloping, vibrating sieving trays. The fibres have a smaller falling velocity than the more compact particles and are air-lifted by suction applied to the top of the sieves as the material passes down. This process is repeated a number of times and this gradually improves the quality of the fibres and sorts them by length. The final fibre is graded to meet close specifications and filled into sacks which hold about 45 kg. These sacks were, at one time, made from hessian (burlap); however, in order to reduce the dust hazard, the sacks are now made from paper or plastic. Much of the fibre is now distributed in sealed containers. The very longest fibres used to be prepared by hand "cobbing" in which the parent rock is knocked off the ends of the fibres with a hammer.

The material from the separating mill was largely unopened bundles of fibres. For many purposes it was necessary to open the fibre by separating the bundles into their constituent fibres, which greatly increases the bulk of the material. In the asbestos textile industry the material is passed through several types of mills which separate the fibres before they are passed into the carding machines in which they undergo the first stage of asbestos yarn production. Of recent years more of the fibre is "opened" at the mills.

A major improvement to the dustiness in the first stage of textile production has been the development of a wet process for forming the thread. This is done by the chemical precipitation of the fibres from an aqueous suspension of the chrysotile.

Uses. Asbestos has a long history. It was first used in Finland about 2500 BC to strengthen clay pots. In classical times, the indestructible shrouds in which the ashes of the eminent were preserved were woven from asbestos. The word "asbestos" comes from the Greek meaning "inextinguishable" or "indestructible". Its use for lamp-wicks has continued through the ages up to the present day.

The modern asbestos industry dates from 1880 when the large chrysotile deposits in Canada and the USSR were first exploited. The initial stimulus came from the cotton industries of Lancashire and France to produce incombustible fabrics and improved materials for gland packings. The industry has expanded rapidly in the last 60 years, and especially since the Second World War.

The property which makes the mineral so valuable is its relative cheapness (approximately US$120 to $3 600 a tonne) compared to man-made mineral fibres.

Over 1 000 uses for asbestos have been described. About 70% of the material goes into asbestos-cement products—roofing sheets, wall boards, pipes, and pressure pipes (approximately 10-20% fibre). Since the Second World War, there has been a great expansion of its use as a filler for plastics for many purposes, especially floor tiling. Combined with calcium silicate and magnesia it provided an excellent heat insulant for boilers and pipes. It was widely used for fireproofing bulkheads in ships, stanchions and girders in buildings, and for improving the fire resistance of cellulose and other materials. Chrysotile and crocidolite can be spun and woven using modified cotton industry machinery. The asbestos cloth is used for fireproof clothing and curtains. The tapes are impregnated with plastics and used in brake and clutch linings, though much of the material for this purpose is now formed directly by mixing plastics, asbestos, and other fillers. The large surface area and strength of the fibre, even when wet, make it a useful material in filters, including gas masks in the Second World War.

Spraying of asbestos fibre mixed with cement and other binders started in about 1935 and was first used for insulation of railway carriages. Later it was used in greatly increased amounts for fire protection and insulation in naval ships and in storage buildings. After the Second World War its use was further expanded for encasing structural steel in buildings to prevent rapid bending in the event of fire. The process was very dusty. The type of fibre used differed, in the United Kingdom more crocidolite was used than in the United States or in Europe. In the late 1960s when the extent of the hazards from this use of asbestos became apparent spraying decreased and is now banned in many countries.

In the past ten years there has been a great reduction in many countries in the use of asbestos for insulation by substituting man-made mineral fibres and other products. In several countries the use of crocidolite has become uneconomic as a result of very stringent hygiene standards. The use of asbestos in cement products, friction materials and as a filler was increasing until about 1975 when the general state of the economy altered the trends. For some purposes such as friction materials there are at present no satisfactory substitutes, for other uses the cost of the alternatives rules out their general use.

The specific diseases associated with asbestos are: asbestosis (a form of fibrosis of the lung); cancers of the bronchi, pleura and peritoneum and probably other organs; and asbestos corns of the skin. All these, with the exception of corns, are due to the inhalation of asbestos fibres and consequently any process which gives rise to large amounts of asbestos dust may constitute a health hazard.

SAFETY AND HEALTH MEASURES

Dust control. Prevention of dust production and its effective control at the site of production is the basis of technical control. Once the dust is airborne in the general atmosphere, its elimination and control become expensive and relatively ineffective. Thus, successful technical control starts with enclosing machines and applying local exhaust ventilation at points where the equipment has to be opened—for example where bags of fibre are fed into mixers or the fibre comes out of the machine at the bagging end of the mills. Damping of the fibre before mixing with other products and during spinning and weaving can greatly assist the elimination of dust production. In many circumstances, such as removing old insulation and spraying new material, personal protection is essential. A well fitting dust mask may be adequate for some jobs where exposure is intermittent; where longer exposures occur in conditions where dust control is inadequate, such as the removal of old insulation in large amounts, full respiratory protective equipment should be used. It is now common practice to require complete isolation of the area where old insulation is being removed to protect those in the vicinity. Exhaust ventilation is required where asbestos-containing products are ground, sawn, drilled, or turned, and the cleaning up should be done by vacuum cleaners rather than brushes.

As asbestos dust in clothing has been shown to be a possible hazard, a change of clothing at the job should be provided and its use made obligatory. Laundering of the clothing will be needed. Asbestos-containing products are so widely used that complete control under all conditions is clearly impracticable. At present there is little evidence to show that occasional slight, intermittent exposure is harmful (see also FIBRES, NATURAL MINERAL).

GILSON, J. C.

CIS 79-1254 *Asbestos properties, applications and hazards*—Volume 1. Michaels, L.; Chissick, S. S. (Chichester, West Sussex, John Wiley, 1979), 553 p. Illus. 937 ref.

CIS 80-50 *Asbestos: Final report of the Advisory Committee—Papers commissioned by the Committee.* Health and Safety Commission (London, HM Stationery Office, 1979), 2 Vols., 100 + 103 p. Illus. 167 ref.

CIS 80-937 *Asbestos: An information resource.* DHEW publication No. (NIH) 79-1681 (Office of Cancer Communications, National Cancer Institute, Bethesda, Maryland), May 1978, 177 p. 367 ref.

CIS 80-686 "A review of asbestos substitute materials in industrial applications". Pye, A. M. *Journal of Hazardous Materials* (Amsterdam), Sep. 1979, 3/2 (125-147). Illus. 3 ref.

Asbestosis

Historical

Fibrosis of the lung caused by asbestos was first described with post mortem examination by Montague Murray in 1899. The word "asbestosis" was coined by Cooke in 1927. Few papers on the biological effects of asbestos appeared in the first quarter of the century (less than one per year). In the second quarter, after the recognition of asbestosis in the 1930s as a major hazard in the asbestos textile industry, the number rose to about ten per year. In the third quarter and especially after 1960 the numbers increased dramatically to over 200 per year (Dr. P. V. Pelnar, personal communication). This arose from the recognition of asbestos as an important occupational carcinogen. Since the previous edition of this encylopaedia much new information has become available. This comes principally from epidemiological surveys of asbestos-exposed groups, and optical and electron microscopy of sections of human lung to measure the amount, size distribution and type of fibre. Animal experiments have compared the fibrogenic and oncogenic action of the major types of asbestos separately. This is information difficult to obtain in man where exposure to more than one type is usual.

Definition

Asbestosis is defined as a diffuse interstitial fibrosis of the lung, the result of exposure to asbestos dust. Neither the clinical features nor the pathology are sufficiently different from other causes of interstitial fibrosis to allow confident diagnosis without evidence of significant exposure to asbestos dust in the past, or the detection of asbestos fibres or bodies in the lung tissue greatly in excess of that commonly seen in the general population. For the reasons given later, asbestosis is usually used to describe the parenchymal fibrosis but not that occurring in the parietal pleura.

Pathology

The retained fibres in the alveolar region are 3 μm or less in diameter but may be up to 200 μm long. Animal experiments strongly point to the longer fibres, 5 μm and over, as being much more fibrogenic than shorter fibres. A proportion of the longer fibres, especially amphiboles, become coated with an iron protein complex producing the drumstick appearance of asbestos bodies (figure 1). All types of asbestos cause similar fibrosis. The fibrosis starts in the respiratory bronchioles with collections of macrophages containing fibres, and others lying free. These deposits organise, collagen replacing the initial reticulin web. Initially only a few respiratory bronchioles are affected, but the fibrosis spreads centrally to the terminal bronchioles and peripherally to the acinus. The areas increase in size and coalesce causing diffuse interstitial fibrosis with shrinkage. The process starts in the bases spreading upwards as the disease progresses; in advanced disease the whole lung structure is distorted and replaced by dense fibrosis, cysts, and some areas of emphysema.

The pleura, both visceral and parietal surfaces, are affected by the fibrosis and to a degree which is much greater than in other types of pneumoconiosis. The visceral surface may be sclerosed up to 1 cm thick. In the parietal pleura thickening starts as a basket-weave pattern of fibroblasts, the sheets of fibrosis lying along the line of the ribs especially in the lower thorax and posteriorly. The edges become rolled and crenated and, after many years, calcified.

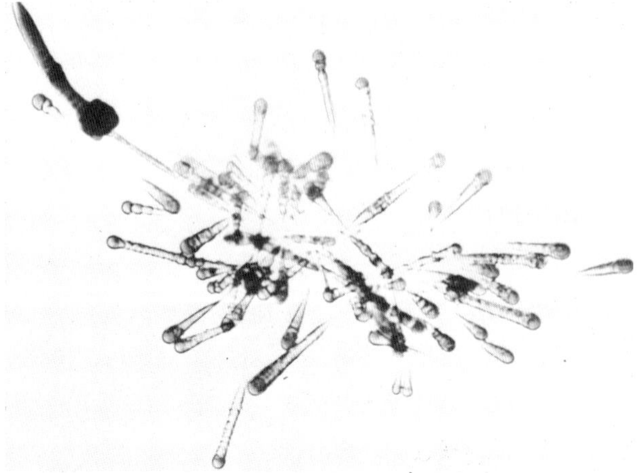

Figure 1. Asbestos bodies.

The parietal thickening may be extensive and thick with little or no parenchymal fibrosis. The reasons for this are not fully understood but indicate the need to separate, if possible, parietal and visceral pleural thickening in life.

Diagnosis and types

Table 1 lists the types of fibrosis in the lung caused by asbestos that can be partially or well separated clinically. Recent epidemiological research indicates that asbestosis and pleural plaque may have differing aetiologies, natural histories, and significance in terms of morbidity and mortality.

Table 1. Types of lung fibrosis caused by asbestos

Parenchymal		Asbestosis
Pleural: Visceral:	Acute	Asbestosis
	Chronic	Asbestosis
Parietal:	Hyaline	Pleural plaques
	Calcified	Pleural plaques

Asbestosis

The signs and symptoms of asbestosis are similar to those caused by other diffuse interstitial fibroses of the lung. Increased breathlessness on exertion is usually the first symptom, sometimes associated with aching or transient sharp pains in the chest. A cough is not usually present except in the late stages when distressing paroxysms occur. Increased sputum is not present unless there is bronchitis, the result of smoking. The onset of symptoms (except following very heavy exposure) is usually slow and the subject may have forgotten having any contact with asbestos. Persistent dull chest pain and haemoptysis indicate the need to investigate further the diagnosis of bronchial or mesothelial cancer.

The most important physical sign is the presence of high-pitched fine crepitations (crackles) at full inspiration and persisting after coughing. They occur initially in the lower axillae and extend more widely later. Agreement between skilled observers on detecting this sign is good but it may vary from day to day in the early stages. It may also be present as an isolated sign in 2-3% of otherwise normal individuals. There are now means of recording this sign on tape. Other sounds—wheezes and rhonchi—are of no help in diagnosis, but indicate associated bronchitis. Clubbing of the fingers and toes was formerly regarded as an important physical sign. There is an impression that it is now less frequently seen. Its severity does not relate well to other aspects of the diagnosis. There is poor agreement between observers except when the clubbing is very pronounced. It is possible that its presence relates to the rapidity of progression of the disease.

The chest radiograph remains the most important single piece of evidence, even though the appearances are similar to other types of interstitial fibrosis. When the radiography is classified by three or more skilled readers using the ILO 1971 scheme independently, it is found that virtually all cases of asbestosis are picked up by one or more of the readers as Category 1/0 or above. The radiographic appearances are well illustrated in the set of standard films of the ILO 1980 Classification of the radiographic appearances of the pneumoconioses (see PNEUMOCONIOSES, INTERNATIONAL CLASSIFICATION OF). The classification provides a means of recording the continuum from normality to the most advanced stages on a 12-point scale of severity (profusion) and of extent (zones) affected. The earliest changes usually occur at the bases with the appearance of small irregular (linear) opacities superimposed on the normal branching architecture of the lung. As the disease advances the extent increases and the profusion of irregular opacities progressively obscures the normal structures. Shrinkage of the lung occurs, with elevation of the diaphragm. In advanced cases distortion of the lung with cysts (honeycomb lung) and bullae occur. The hilar glands are not enlarged or calcified unless exposure has been to mixtures of silicious dusts. This may occur, for example, in making asbestos roofing shingles or pressure pipes, and in mining. The small opacities may then be rounded rather than irregular.

The pattern of lung function provides the important third component in diagnosis. The functional changes are the result of a shrunken and non-homogeneous lung, without obstruction of the larger airways (restrictive syndrome). The total lung volume is reduced and especially the forced vital capacity (FVC), but the ventilatory capacity ($FEV_{1.0}$) is only reduced in proportion to the FVC, so the ratio $FEV_{1.0}$/FVC is normal or even raised. The transfer factor for carbon monoxide is reduced in later stages, but in the early stage an increase of ventilation on a standard exercise test may be the only alteration indicating impairment of gas exchange. Although the restrictive syndrome is the commonest pattern (about 40%), in about 10% of cases airway obstruction is the main feature and in the remainder a mixed pattern is seen. This is thought to be largely due to the confounding effects of cigarette smoking.

Visceral pleurisy: chronic and acute

This occurs in two forms–chronic and acute. The former is the commoner and is a usual accompaniment of parenchymal disease, but its severity does not run parallel with the parenchymal disease. The diagnosis is radiographic. In some cases one or both of the costophrenic angles are filled in but the more specific feature is the appearance of well defined shadows running parallel to the line of the lateral chest wall and separated from it by a narrow (1-2 mm) clear zone. This is due to the thickened pleura seen "edge on". It is illustrated in the ILO 1980 standard set of films. The thickening is best seen in the middle and lower third of the lateral chest wall, the apices are usually spared. It is common in those only lightly exposed to find this pleural thickening as the *only* radiographic feature. It is readily missed when present only over a short length of the wall and if the radiographic technique does not give a clear picture of the periphery of the lung. When the visceral pleura is greatly thickened it causes veiling of the lung field, obscuring both the normal structure and parenchymal changes. This is probably the basis of the "shaggy heart" and the "ground glass" appearance described in the early accounts of asbestosis. The wide recognition that small areas of pleural thickening may be the only sign of past exposure to asbestos is recent, and it seems to be a feature of the effects of low exposure to the dust. It is likely to remain an important observation for monitoring exposure to improved conditions in the future.

Acute pleurisy affecting the bases, and costophrenic angles, with effusions, sometimes blood-stained, is now a recognised sequel to asbestos dust exposure. It is associated with pain, fever, leucocytosis and a raised blood sedimentation rate. It settles in a few weeks but leaves the costophrenic angles obscured. No precipitating factors have been identified. Its recognition is important. Firstly, the cause may be missed unless an adequate occupational history is taken; secondly, not all effusions in asbestos workers signify the onset of an asbestos-related cancer. A few weeks of observation may be necessary to confirm the aetiology.

Summary of diagnosis

The diagnosis of asbestosis therefore depends upon–

(a) a history of significant exposure to asbestos dust rarely starting less than 10 years before examination;

(b) radiological features consistent with basal fibrosis (Category 1/0 and over, ILO 1980);

(c) characteristic bilateral crepitations;

(d) lung function changes consistent with at least some features of the restrictive syndrome.

Not all the criteria need to be met in all cases but *(a)* is essential, *(b)* should be given greater weight than *(c)* or *(d)*; however, occasionally *(c)* may be the sole sign. Other investigations are not of much help. Asbestos bodies in the sputum indicate past exposure to asbestos but are not diagnostic of asbestosis. Their absence when there is much sputum and marked radiological changes of fibrosis suggest an alternative cause for the fibrosis.

Immunological tests may be positive but do not help in consistent separation of asbestosis from other types of fibrosis. Lung function results must be assessed in relation to appropriate standards allowing for ethnic, sex and age differences and for cigarette smoking.

Asbestos corns on the fingers–areas of thickened skin surrounding implanted fibres–are now much less common because much of the asbestos fibre is packed mechanically and gloves are worn. Corns do not lead to skin tumours and disappear on removal of the fibres.

Pleural plaques

Parietal pleural plaques alone rarely cause symptoms. They may occur alone or with asbestosis. The diagnosis in life is radiological and the appearances are more specific than in the case of parenchymal fibrosis. PA films will detect most cases, but because they are frequently thickest posteriorly their full extent is best seen using oblique views. The ILO 1980 standard films show their appearance and the scheme provides, for the first time, a separation of parietal (circumscribed) and visceral (diffuse) pleural thickening. The plaques lie along the line of the ribs, and when thick cast a well defined shadow over the lung field extending in from the lateral chest wall, where they may also be seen "edge on".

Separation from visceral thickening depends largely on a defined edge to the shadow. Both types may occur together. Dependent mostly on the length of time since first exposure, and age, patchy calcification occurs in the edges. This produces a bizarre pattern of dense shadows likened to "guttering candle wax" or a "holly leaf". The onset of calcification reveals many small plaques not previously visible. When calcification occurs in a crater-shaped plaque on the dome of the diaphragm a diagnosis of past exposure to asbestos or related minerals can be made with confidence.

Sources of exposure to asbestos

Formerly it was thought easy to establish past exposure to asbestos by inquiry about work in manufacturing plants, or the application of the fibre for insulation. Now it is realised that only the most detailed history of all jobs, residences and occupations of the family will reveal possible exposures to asbestos. The reasons for this change are–

(a) the much wider use of asbestos in thousands of products especially since the Second World War (see ASBESTOS);

(b) the recognition that significant exposure to asbestos occurred around mines and manufacturing plants in the past;

(c) the discovery of family exposure to the dust brought home on clothing, and also that those working in an area where lagging is in progress may be affected, even though they are not engaged in lagging;

(d) the finding that calcified pleural plaques, indistinguishable from those occupationally exposed, also occur in the general population in localised areas in several countries (Finland, Czechoslovakia, Bulgaria, Turkey and others).

With the discovery of such diversity of sources of possible exposure, but virtually no quantitative information about its severity, and few long-term follow-up studies of those exposed, it is not surprising that there is controversy about the health hazards. However, some conclusions emerge which must be subject to revision in the future.

(1) Asbestosis is primarily occupational in origin, the result of mining, milling, manufacturing, applying, removing or transporting asbestos fibre. Exposure is much less when the fibre is bound in the product (asbestos cement and asbestos plastic and paper products). Also exposure in the past was much greater than it is today with the use of the best working practices.

(2) Asbestosis may have been caused by home exposure from dusty clothing at a time when there was no dust or hygiene control in the factories.

(3) Asbestosis does not result from the very limited exposure to which the general public is or has been subject, even though asbestos fibres are detectable in the lungs of a high proportion of adults in industrialised areas. The median numbers of fibres so detected are two to three orders of magnitude less than that found in those occupationally exposed.

(4) There are and have been important differences between countries in the use of asbestos, so that exposure for the same occupation varies widely. For example, dry wall fillers (spackling) contain asbestos in the United States but not the United Kingdom; thus sanding of internal walls during construction and maintenance is a source of exposure in the former but not in the latter. On the other hand, spraying of crocidolite was much more widespread in the 1940s in the United Kingdom than elsewhere.

(5) Pleural plaques can arise at levels of exposure probably much lower than that required to produce asbestosis. In addition it is probable that other minerals can cause plaques. For example, among chryosotile miners in Quebec calcified plaques are limited to those who have worked in two out of the eight mines. The minerals causing the plaques in the general population have not been fully established. Tremolite, an amphibole often present in deposits of asbestos, may be important.

(6) Whether chrysotile and the amphiboles differ in fibrogenicity in man is uncertain, but some evidence indicates that the amphiboles may be more fibrogenic. In animals there is little difference but the amphiboles remain in the lung much longer than the chrysotile.

The relation of asbestosis to dose of dust

In only a few instances are there records of past dust sampling to relate to the prevalence or incidence of asbestosis. But the information has been exhaustively analysed for miners and millers in Quebec, a group of asbestos cement workers in the United States and asbestos textile workers in the United Kingdom, because of its relevance to setting hygiene standards. In North America the dust was measured in millions of particles/ft^3, in the United Kingdom in fibres/cm^3–the measurement now internationally used. All the data show a clear relation between estimated dose of dust (concentration x time of exposure) and the incidence or severity of disease, but are insufficiently precise to determine whether there is a threshold level below which asbestosis will not occur. A cautious conclusion from the North American studies is that at about 100 million particles/ft^3/yr there might be a threshold or that the risk of developing asbestosis would be as low as 1%. In the textile plant in the United Kingdom the conclusion was "the concentration such that 'possible' asbestosis occurs in no more than 1% of men after 40 years' exposure could be as high as 1.1 fibres/cm^3 or may have to be as low as 0.3 fibres/cm^3". More precise information will only become available when the dust sampling introduced widely after the mid-1960s is related to the incidence of disease in the future.

The relation of asbestosis to lung cancer

The important questions here are: firstly, is there an excess risk of bronchial cancer only in those who also have some degree of asbestosis? Secondly, if the dust exposures are low enough to eliminate asbestosis, will the excess lung cancer risk also be reduced to an acceptably low level? Neither question can be answered at present, and so disagreement is likely. It is known that there is a close association between asbestosis and lung cancer, about 50% of those dying from or with asbestosis have a lung cancer at post mortem. Among those knowledgeable about details of the dose-response data there would probably be agreement that dust exposures low enough to eliminate asbestosis will also reduce the excess bronchial cancer risk to a very low value. This does not extend to the risk of mesothelioma, which is not nearly so closely related to that of asbestosis (see ASBESTOS (MESOTHELIOMA AND LUNG CANCER)).

PREVENTION

This depends on successful control of dust exposure and medical surveillance to protect the *individual*, as far as is possible, and for the detection of health trends in the *group*.

Engineering control

Replacement of asbestos by other materials believed to be safer has been widespread since the mid-1970s. Man-made mineral fibres and other insulating materials are rapidly replacing asbestos for heat insulation. But for other uses, for example asbestos cement, friction materials and some felts and gaskets, substitution is not at present practicable.

Dust control has been gradually improved by partial or complete enclosure of plants and the wide use of well designed local exhaust ventilation. In the textile section a completely new wet process of forming the thread has greatly reduced dust levels, previously difficult to control. During maintenance work on old insulation much stricter control of exposures is possible by isolation of the working areas, and by training in the use of good working practices to reduce the dust, for example damping of the insulation before removal, and the use of vacuum cleaning in place of sweeping. But removal of old insulation is likely to remain for many years a major potential source of high exposure (see also DUST CONTROL, INDUSTRIAL).

Medical surveillance

The insidious onset of asbestosis and the lack of highly specific features indicate the need for well recorded and systematic, initial, and periodic examinations of asbestos workers. This ensures the best chance of detecting the earliest signs. Physical examination of the chest, full-sized, high technical quality chest radiographs and tests of FVC and FEV_{1-0} are the minimum required. The interval will vary from annually up to four times yearly, with more frequent visits when there are clinical reasons. There is increasing evidence that the radiological features of asbestosis are in part cigarette-smoking dependent, which requires the recording of smoking histories. This and the multiplicative effects of asbestos dust and cigarette smoking on the risk of bronchial cancer provide the strongest possible grounds for stopping cigarette smoking in those potentially exposed to asbestos. Personal advice on the special dangers of smoking and limiting opportunities for smoking at work are essential steps in prevention. Full personal protective equipment will be required where dust levels cannot be lowered to the hygiene standard. The system of periodic examinations also provides, if properly analysed, essential information about the effectiveness or failure of the engineering control of the dust. Tabulation, by age and years of exposure, of the results of classifying the chest films on the ILO 1980 scheme–preferably by independent readers–gives early evidence of trends in the prevalence of asbestosis. This valuable information will be missed if the *group* findings are not examined in detail.

Treatment

There is no specific treatment for asbestosis. Where the rate of progression appears unusually rapid further special investigations, including lung biopsy, may be justified if it is likely to assist in the differential diagnosis, and influence treatment–for example the use of steroids, but these are not of proved value. The severity of past exposure is the only factor known to influence progression rate. Thus, those with some evidence of asbestosis, if young or middle-aged, should be removed from further exposure. In cases where exposure has not been heavy and asbestosis is only detected late in life, progression may be very slow and the grounds for removal from work with asbestos, under good conditions, are less compelling.

The widespread and often misleading publicity given to the hazards of exposure to asbestos may cause much anxiety to those with asbestosis, both for their own health and for that of their family. Reassurance, and the putting of the likely prognosis in true perspective, are an important part of good treatment. The special risks of continuing cigarette smoking need emphasis. Mesotheliomas are a rare complication in those exposed only to chrysotile.

Compensation

The conventions on the awarding of compensation for asbestosis vary in different countries. Unusual breathlessness on exertion, as a cause of disability, may be required, even though it is not essential for a confident diagnosis of asbestosis. Compensation may be limited to those with evidence of parenchymal disease; pleural fibrosis–parietal or visceral–alone may not be accepted. Lung (bronchial) cancer is usually accepted as part of the disease provided there is at least some evidence of parenchymal fibrosis, but may be rejected if there is no radiological evidence of pleural or parenchymal fibrosis. There is plenty of opportunity for disagreement, especially when a factor for uncertainty of prognosis is included. It is now established that asbestos dust alone may cause lung cancer although the absolute risk is very small compared with that from the combined effects of cigarette smoking and asbestos dust. It has not been established that pleural plaques alone result in an increased risk of bronchial or mesothelial tumours, above that for similar exposures to asbestos dust without these pleural changes. The considerable uncertainty about the likely rate of progression of the fibrosis makes assessment on first diagnosis especially difficult. Lung biopsy is not justifiable solely for compensation assessment.

GILSON, J. C.

Biological effects of asbestos. Proceedings of a Working Conference, Lyons. Bogovski, P.; Gilson, J. C.; Timbrell, V.; Wagner, J. C. (eds.). IARC Scientific Publication, No. 8 (Lyons, International Agency for Research on Cancer, 1973), 346 p. Illus. Ref.

Biological effects of mineral fibres. Wagner, J. C. (ed.). IARC Scientific Publication, No. 30, INSERM Symposia Series, Vol. 92 (Lyons, International Agency for Research on Cancer, 1980), 2 vols., 1 007 p. Illus. Ref.

"State of the art: asbestos-related diseases of the lungs and other organs: their epidemiology and implications for clinical practice". Becklake, M. R. *American Review of Respiratory Disease* (Baltimore), July 1976, 114/1 (187-227). 231 ref.

CIS 79-1546 "Asbestosis: a study of dose-response relationships in an asbestos textile factory". Berry, G.; Gilson, J. C.; Holmes, S.; Lewinsohn, H. C.; Roach, S. A. *British Journal of Industrial Medicine* (London), May 1979, 36/2 (98-112). Illus. 17 ref.

CIS 80-50 *Asbestos: Final report of the Advisory Committee–Papers commissioned by the Committee.* Health and Safety Commission (London, HM Stationery Office, 1979), 2 vols., 100 + 103 p. Illus. 16 ref.

Health hazards of asbestos exposure. Conference Proceedings. Annals of the New York Academy of Sciences, Vol. 330 (New York, 1979), 813 p.

CIS 80-59 *Asbestos and disease.* Selikoff, I. J.; Lee, D. H. K. (New York, Academic Press Inc., 1979), 549 p. Illus. 810 ref.

Asbestos (mesothelioma and lung cancer)

While pulmonary fibrosis due to exposure to asbestos (asbestosis) has been known for decades, the first reports of individual cases of asbestosis combined with *pulmonary cancer* which appeared from time to time in various countries were accepted more as a curiosity. They did not attract much attention until in 1947 a British Chief Inspector of Factories, E. R. A. Merewether, reported that lung cancer was found to be the cause of death in 13.2% of persons known to have asbestosis who had died and been autopsied between 1923 and 1946. A similar high proportion of cancer deaths in asbestosis was found by other pathologists and the probability of a role of asbestos in pulmonary carcinogenesis was definitely established by an epidemiological study by Doll in 1955, and confirmed by further studies.

Soon afterwards a new surprising discovery was made in South Africa. An accumulation of cases of an otherwise very rare tumour of the pleura and peritoneum, the *malignant mesothelioma,* was reported by Wagner in 1959 and related to exposure to the locally mined type of asbestos, crocidolite. Soon afterwards cases were identified in non-mining occupational exposures to asbestos in England, in the United States and elsewhere. In contrast with asbestosis, and in contrast with asbestos-related pulmonary cancer, mesothelioma was found also in persons whose exposure was not necessarily occupational.

Bronchogenic carcinoma related to asbestos

Bronchogenic carcinoma of the lung. This is a disease very common in the general population. While in many countries the total mortality from cancer slowly declines, the incidence and mortality from lung cancer increases and stands as the most frequent cause of death from cancer, particularly in cigarette smokers. It begins with transformation of the mucous membrane lining the inside of the bronchus at various levels and such foci of transformation may remain at their initial spot for some time shedding at times atypical or metaplastic cells into the sputum without causing other symptoms. This is the period in which we sometimes may succeed in discovering these pre-cancerous, or the earliest cancerous, changes by sputum cytology sooner than by other diagnostic methods. Some of such early alterations of cells is reversible and may spontaneously heal when the cause disappears, e.g. when the person stops smoking. When the original focus develops definite cancer cells, the focus begins to grow, to bleed and slowly to obstruct the airway, a growing malignant tumour becomes visible on the radiogram, and unless it can be surgically removed as soon as confirmed, it tends to spread through growth and through dissemination by blood and by lymph and to lead eventually to death. Supporting treatment by chemotherapy and radiation successfully prolongs life and radical surgery can provide complete healing.

The various components of the bronchial lining may undergo malignant transformation and consequently the carcinoma may be composed of various cells and have various histological appearances such as adenocarcinoma or squamous, or oat-cell carcinoma.

There are no histological or other characteristics which would specify the individual lung cancer as cancer caused by asbestos.

In many cases of asbestos-linked pulmonary cancers the lungs also show pulmonary fibrosis-asbestosis microscopically, and often macroscopically, and on X-ray examination. Some scientists believe that so-called "asbestos lung cancer" can only develop on a pathologically changed terrain of asbestotic fibrosis. There is evidence of such a possibility in human pathology: the scar-carcinoma. Others believe that exposure to asbestos alone, particularly in a smoker, may provoke cancerous growth without also causing asbestosis. The decision between the two opinions is difficult to reach because in individual clinical cases of bronchogenic carcinoma we cannot distinguish what is an "asbestos cancer", a "cigarette cancer", or lung cancer from yet another cause. Thus, in most countries bronchogenic carcinoma is considered an occupational disease due to asbestos, e.g. for workmen's compensation, only in the presence of coexisting asbestosis. If pulmonary fibrosis were a prerequisite for development of asbestos-linked lung cancer, it would follow that lowering exposures to asbestos to levels which effectively prevent asbestosis would automatically eliminate "asbestos lung cancer".

Epidemiological data

In man the link of lung cancer with asbestos has been mainly *epidemiological*. While asbestosis cannot occur without exposure to asbestos and consequently every case of asbestosis must be linked with such exposure, with pulmonary cancer the situation is quite different. It is a rather common disease in the general population. The link with exposure to asbestos is based on finding whether in those exposed to asbestos lung cancer occurs more frequently than in those unexposed, i.e. whether in those exposed there is an excess incidence of lung cancers.

Since Doll's study a number of other epidemiological studies, of various levels of excellence, have been carried out which confirm that indeed there is an excess of bronchogenic carcinoma in persons exposed to asbestos, under certain circumstances, and thus that asbestos must be considered one of a number of carcinogenic substances.

What are the circumstances of a manifest risk of cancer in asbestos exposure? It has been established that smoking cigarettes greatly increases this risk. In fact the large majority of lung cancers attributed to asbestos exposure have occurred in smokers. A lung cancer in an asbestos-exposed non-smoker has been a rarity. Table 1 shows the effect of both exposures together, while each of the two exposures also carries a risk by itself. A particular exposure to asbestos in the reported group of workers increased the basic risk of pulmonary cancer in non-smokers. However, since the risk in non-smokers was very small, its further increase still meant only very few cases, if any at all. On the other hand, when the basic risk of exposure to asbestos was combined with the 11.8 times higher risk of a smoker, this combination necessarily produced a serious risk leading to an excess of incidence of pulmonary cancer. This experience has an important practical implication: most "asbestos cancers of the lungs" could be prevented if the workers did not smoke. In fact it was found that the risk for the asbestos workers who had stopped smoking declined after 10 years to the low level existing for non-smokers.

The bronchogenic carcinoma has a long latent period, usually 20 years or more. Consequently, what excesses of incidence of pulmonary carcinoma linked with asbestos have been found to date must be linked with exposures 20 years or more before development of the tumour. It is known that exposures in those days were generally very high. But we usually do not have any precise measurements. Thus in most existing epidemiological studies it has not been easy, and in some not possible, to establish a relation between the incidence of cancer and a certain quantitative level of exposure, other than that the exposure had been high.

Table 1.

	Asbestos exposure		
	Little	Moderate	Heavy
Non-smokers	1.0	2.0	6.9
Moderate smokers	6.3	7.5	12.9
Heavy smokers	11.8	13.3	25.0

From: McDonald, J. C. "Asbestos-related diseases: an epidemiological review" (587-601). *Biological effects of mineral fibres.* Wagner, J. C. (ed.). IARC scientific publications No. 30 (Lyons, International Agency for Research on Cancer, 1980), Vol. 2.

One quantitative measure commonly used is the duration of exposure in years. In other studies the period since first exposure is used, and in yet others, both the period since first exposure and the duration of exposure. Only a few investigations have had the additional benefit of actually measured data on past levels of exposure. An example of the latter is the series of epidemiological studies of workers of the chrysotile mines of Quebec carried out by J. C. McDonald and his collaborators. This and some other studies showed a *dose-response relationship*, i.e. the higher was the dose, in terms of level of exposure, or of periods of exposure, or of both of them combined, the higher was the excess incidence of bronchogenic cancer. In fact the excess incidence of lung cancer and statistically significantly increased relative risk was usually found only in groups of persons most severely exposed (see table 2).

Table 2. Relative risks of lung cancer in relation to accumulated dust or fibre exposure, before and after correction of work histories, with controls matched for smoking

	Accumulated dust exposure (millions of particles per cubic foot × years)				
	< 30	30 < 300	300 < 1 000	> 1 000	All
Before correction					
Cases	89	73	56	27	245
Controls	108	87	42	8	245
Relative risk	*1*	*1.02*	*1.62*	*4.10*	–
After correction					
Cases	85	73	59	27	244
Controls	101	89	44	10	244
Relative risk	*1*	*0.97*	*1.59*	*3.21*	–
	Accumulated fibre exposure (fibres per ml × years)				
	< 100	100 < 1 000	1 000 < 3 000	> 3 000	All
After correction					
Cases	86	76	56	26	244
Controls	110	87	35	12	244
Relative risk	*1*	*1.12*	*2.05*	*2.77*	–

From: McDonald, J. C.; Gibbs, G. W.; Liddell, F. D. K. "Chrysotile fibre concentration and lung cancer mortality: a preliminary report" (811-817). *Biological effects of mineral fibres.* Wagner, J. C. (ed.). IARC scientific publication No. 30 (Lyons, International Agency for Research on Cancer, 1980), Vol. 2.

The results of epidemiological studies seem to support the "threshold" hypothesis: in all studies there are lower levels of exposure to asbestos under which no excess of cancer is found. However, the supporters of the "one-hit" hypothesis argue that negative findings do not truly reflect reality; that any exposure to asbestos leads to damage and that in principle there cannot be any safe level of exposure to a carcinogen; and that the excess of cancers in low exposures is simply so small that the epidemiological methods are not sensitive enough to detect it and never will be. Therefore, those holding this view say the response (excess of cancer in low exposure) cannot be measured but must be calculated from the dose-response curve established at higher exposures, by extrapolating it into low exposure levels, provided the assumption is correct that the relation is linear and the line tends to the zero-point (see figure 1).

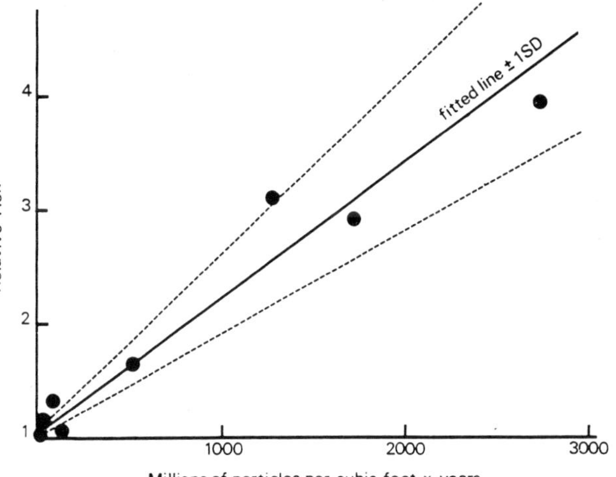

Figure 1. Relative risk of lung cancer related to accumulated dust exposure. (From: McDonald, J. C.; Gibbs, G. W.; Liddell, F. D. K. "Chrysotile fibre concentration and lung cancer mortality: a preliminary report" (811-817). *Biological effects of mineral fibres.* Wagner, J. C. (ed.). IARC scientific publications No. 30 (Lyons, International Agency for Research on Cancer, 1980), Vol. 2.)

Safe occupational exposure level

Thus the cancer risk of high past exposures to asbestos has been established. In most industries, however, progress has been made and exposures markedly lowered. The most important practical question of the last decade then is: how far down must the exposure be lowered to remove, if possible, the increased risk of cancer, or of any disease? Obviously there would be no increased risk with no exposure. But what is the level of exposure, if any, which is low enough to be also tolerated without increased incidence of cancer or any disease? Such a no-effect level of exposure seems to be reported by all epidemiological studies and demonstrated by common experience in industrial populations. Reaching some scientific decision on this question would have far-reaching practical consequences.

In consideration of this problem some questions of principle of carcinogenicity must be kept in mind. There are two schools of thinking. According to the "one-hit" hypothesis one molecule of a carcinogen, or one fibre, by damage to the biologically vulnerable material such as DNA of its nucleus, can change one cell into a cancerous cell. This one cell then multiplies and eventually produces cancer. On the other hand the "threshold" hypothesis recognises the possibility of repair of lesions in DNA and the existence of a complex defence mechanism in the body which copes with individual deviated cells, such as cancer cells, spontaneously appearing throughout the lifetime; cancer can only develop if and when the defence system is overcome by large numbers of such cells, created by a large number of "hits", i.e. if and when the "threshold of tolerance" is overstepped.

In contrast with, or in addition to calculations and extrapolations, an alternative approach to objective evaluation of the health effect of low doses of asbestos is available: to relate the finding of pulmonary health or disease to *the lung burden*, as represented by the kinds and numbers of fibres present in the lung tissue. The rationale of this approach is that in the case of a durable mineral like asbestos the fibres found in the lung should reasonably well reflect the relevant past exposure, both qualitatively and quantitatively, and that it should be possible to identify objectively past low exposures and evaluate their effect or lack of effect on health.

It was demonstrated and repeatedly confirmed in the lungs from autopsies that there are significant differences: very high numbers of asbestos fibres were found in the lungs of persons with cancer and with asbestosis (with or without cancers), much less in persons with no asbestos-linked disease but with past occupational exposure to asbestos, and still less in persons with no asbestos-linked disease and no occupational exposure to asbestos. The differences between the groups were found to be in orders of magnitude with virtually no overlapping. The strength of the evidence has been such that some scientists argue that, whatever disease is in a person with low numbers of fibres in the lungs, that this disease is unlikely to be caused by asbestos and must have had another cause. That there is no "asbestos cancer" with less than 0.5-1 million fibres/g of lung tissue (which is considered the upper limit of the low burden found in those not occupationally exposed).

Obviously, these conclusions are valid only with certain precautions: the fibres may be very unevenly distributed through the lungs, and very different readings may be obtained from various parts of the same lung. This particularly applies to results of biopsies. A standardised technique for tissue sampling, preparing and counting the fibres is required if the quantitative results even from the post-mortem lungs are to be compared between laboratories. The fibres, mainly the very small ones, migrate through the lungs and through the body. Has their presence in some locations relevance to disease or is it a simple consequence of their mobility? The various kinds of asbestos have various durability. Chrysotile seems to disappear gradually from the living lung, whereas the amphiboles mostly remain. When death occurred long after exposure had ceased, then even when past exposure was limited to mining chrysotile, more admixed tremolite than chrysotile was found in the lungs. When there had been mixed exposure to chrysotile and amosite or crocidolite, then again much more amphibole was found in the lung than chrysotile. Also, in persons not occupationally exposed to asbestos, among the few fibres found in the lungs are almost only amphiboles, although 95% of asbestos mined and used is chrysotile.

With these limitations in mind, however, the assessment and measuring of the lung burden appears to be a promising approach. It may offer an objective measure of the dose required to produce one or another asbestos-related disease, or a dose which produces no disease, and enables us to assess the levels of past exposures leading to the lung dose in question. If this is achieved, perhaps the gap in the dose-response curve left by epidemiology can be filled by data rather than by extrapolation. Then an answer may be found to the very practical question: Is there a tolerable level of exposure to asbestos, and if so, what is that level?

Somewhat similar information on lung burden can be obtained by *bronchopulmonary lavage*. The information is less direct than counting asbestos bodies or fibres in the lung tissue, but can be gained during life, and the result reflects more uniformly the situation of the whole lung than biopsy.

Animal experiments have not added much information about asbestos-linked pulmonary cancer, mainly because inhalation studies did not produce many cases of bronchogenic carcinoma.

Neither animal experiments nor epidemiology have been able to show convincingly any substantial difference in pulmonary carcinogenicity between the various kinds of asbestos, between chrysotile and the amphiboles. All kinds have proved to be carcinogenic and excess of pulmonary cancer was found even in exposure to anthophyllite, although this amphibole has never been shown to cause mesothelioma in man.

Malignant mesothelioma related to asbestos

Malignant mesothelioma is a rare tumour of the lining of the cavity of the chest or of the abdomen and occasionally of the pericardium. It presents with pain and effusion of fluid into the cavity, grows mainly locally and, being highly malignant, leads to death, within months rather than years. No effective therapy has yet been found, although surgical intervention, radiotherapy and chemotherapy seems to prolong life in some cases. Smoking plays no role in mesothelioma. There is no convincing evidence that calcified pleural plaques (see the article on ASBESTOSIS) are precursors and lead to pleural mesothelioma. Pulmonary fibrosis (asbestosis) is not necessarily present in mesothelioma.

Diagnosis

The diagnosis is not as simple as it would appear. During life pleural mesothelioma may be difficult to distinguish from subpleural pulmonary cancer, or from chest metastases of cancers originating elsewhere. In the abdominal cavity cancers of all local organs may occasionally clinically appear like mesothelioma, for example ovarial tumours. The finding of hyaluronic acid and of the tumour cells in the pleural or abdominal fluid helps in clinical diagnosis. However, the only really decisive diagnosis is by autopsy, and even then mainly by excluding other primary cancers causing pleural or abdominal metastases which may mimic the mesothelioma. The pathological picture of mesothelioma differs from one case to another: epithelial, fibrous, and mixed forms have little specificity, and so even a histological diagnosis at autopsy (or at biopsy) requires a great deal of experience. This is the reason why in some countries the definite decision on the final diagnosis of primary mesothelioma is reserved to special "mesothelioma panels" of experienced pathologists.

The difficulty of diagnosis, particularly in cases with no autopsy, may have led to underdiagnosing in the past, and so it is hard to know whether in the past the primary mesothelioma, not unknown to pathologists, really was or only appeared to be extremely rare. Following the Wagner report of the link between mesothelioma and exposure to asbestos, this disease is being found more frequently where asbestos is used. Some underdiagnosing still may exist, e.g. due to insufficient rate of autopsy and/or lack of laboratory facilities and expertise in the developing countries. However, there is also overdiagnosing owing to active searching for mesothelioma wherever there is known to have been exposure to asbestos: in the Canadian province of Quebec, where chrysotile asbestos is mined, the frequency of reported mesothelioma was twice as high as in the other Canadian provinces, but the consequent revision by the Canadian mesothelioma panel of pathologists rejected this diagnosis in half of the reported cases.

Aetiology

Following the Wagner report and further studies a belief has spread, particularly among laymen, that every case of mesothelioma had been caused by asbestos and in every case of mesothelioma an asbestos cause was automatically anticipated. This generalisation is not supported by facts. Mesothelioma was known to pathologists long before any substantial quantity of asbestos was ever mined or used. It has occurred in children; human mesotheliomas were reported to result from exposures to ionising radiation, to farming sugar cane in India, to previous pleuropulmonary disease and previous pneumothorax; and there has been a striking endemic accumulation of cases of mesothelioma in persons exposed to a non-asbestos mineral, erionite, in Turkey. In the classical animal experiments Stanton produced pleural mesothelioma also by non-asbestos fibres, including glass fibre, and mesothelioma was produced in animals by as disparate causes as avian leucosis virus and ethylene oxide. Finally, in all epidemiological studies investigating mesothelioma as related to asbestos exposure there has always been a certain proportion of cases in whom no previous exposure to asbestos was traced even if the study had been specifically designed to search for such exposures. The mesothelial tissue obviously reacts to a wide variety of stimuli and causes by formation of a tumour.

Yet asbestos, although not the only cause of mesothelioma, appears to be its most important cause in modern times. Strong evidence for this comes from epidemiological studies.

These studies are, however, complicated by several circumstances. Mesothelioma is a rare tumour. In the general population of Canada and of the United States the frequency of about 2 cases per million inhabitants per year has been established, as a background for comparison. Occupational exposures have led to higher incidence of mesothelioma, but the excess of cases varies widely. Exposure to anthophyllite, even occupational, has not been shown to cause any cases at all. Also, surprisingly, mining chrysotile in Canada, in Italy, in the USSR and in Cyprus has led to very few, if any, cases of mesothelioma, although with chrysotile mesothelioma was produced in animals. Textile and other asbestos manufacturing is linked with much higher incidence of mesothelioma, and the use of asbestos, particularly for insulation, now virtually abandoned, has led to a still higher incidence. In the last instances chrysotile has mostly been used together with amphiboles, crocidolite and/or amosite. The difference in incidence is well illustrated in table 3.

Strikingly high accumulations of cases of mesothelioma were repeatedly reported in exposures to crocidolite: in the areas of crocidolite mining in South Africa and in Australia, in the war-time manufacture of gas-mask filters in the United Kingdom and Canada, and also in asbestos factories in England where crocidolite was used, or in shipyards where it was sprayed for insulation. The above epidemiological evidence has led to the widely accepted belief that crocidolite is more dangerous than the other kinds of asbestos. This belief has led to much stricter standards of permissible exposure to crocidolite in several countries and to outright ban of crocidolite in some. It must be admitted, however, that in the situations leading to the above accumulations of cases of mesothelioma linked with crocidolite the past exposure levels are not known and may have been exceptionally high because of exceptional circumstances, or other confounding factors unknown today.

Equally unclear is the level of risk of exposure to *amosite*. In the South African Transvaal areas of amosite mining, in contrast with crocidolite mining areas of the Cape Province, extremely few cases of mesothelioma were detected. However, amosite is claimed to be the only material used in Selikoff's study of a cohort of former workers in an asbestos factory in Patterson, New Jersey, and amosite was used along with other materials, but with no crocidolite, in Selikoff's other cohort of insulation workers, in which the incidence of mesothelioma and mortality from cancers were among the highest on record. According to the *Report of ILO Meeting of Experts on the Safe Use of Asbestos, 1973*, "Evidence indicates that the risk was highest in insulation workers heavily exposed in the past. In this special section of the industry the proportion [of mesothelioma] may have been in the order of 10%. This

Table 3. Employment in occupational groups ten or more years before death for 344 male cases and their matched controls (Canada, 1960-1972; United States, 1972)

Occupational group	Jobs		Men		Relative risk
	Cases	Controls	Cases	Controls	
A. Insulation	27	1	27	1	46.0
B. Asbestos production and manufacture			25	7	6.1
Mining and milling (chrysotile)	4	2			
Manufacture	12	2			
Asbestos cement products	3	2			
Factory using asbestos	7	1			
C. Heating trades (excl. insulation)			70	27	4.4
Job necessitating heat-protective clothing	11	7			
Installing or repairing furnaces or boilers	28	9			
Steamfitter	5	2			
Boiler maker	2	1			
Plumbing and heating	23	8			
Welder	14	8			
D. Shipyards	49	17	21	13	2.8
E. Construction industry			45	30	2.6
Building trades	59	36			
Building demolition	3	2			
Painting	13	3			
Sheetrock spackling	1	0			
F. Other listed jobs (excl. men in groups A-E)			55	90	1
G. None of the above			101	176	
			344	344	

From McDonald, A. D.; McDonald, J. C. "Malignant mesothelioma in North America", *Cancer* (Philadelphia), 1980, 46/7 (1 650-1 656).

compares with over 80% in the case of some potent chemical carcinogens such as beta-naphthylamine." This part was re-adopted by the ILO Meeting of Experts on the Safe Use of Asbestos, Geneva, October 1981. We have no exact measurements of dustiness in the years long past and we only have sketchy information on what levels of exposures may have been during wartime in the factory, and at construction sites with notoriously bad housekeeping or in shipyards under pressure of wartime conditions, where the insulation workers operated. Again, in some countries stricter standards are applied to amosite than to chrysotile, but less strict than to crocidolite.

The difficulty of determining the cause of mesothelioma is also determined by the evidence that the latent period of mesothelioma is extremely long: the tumour is believed to appear 20, 30, or even up to 50 years after the first exposure. After so many years correlation to exposure is difficult and often impossible in the absence of records of dustiness decades ago, and considering the unreliability of human memory. In studies in which the relation to measured asbestos exposures is investigated there is evidence that the incidence of human mesothelioma is dose-related (table 4), and it is a common experience confirmed by all epidemiologic studies that there were levels of occupational exposures to asbestos at which no mesothelioma developed. This is true for all kinds of asbestos. The documented existence of such a level gives hope that a practical threshold for causing mesothelioma can be identified.

Table 4. Mesothelioma death rates

Exposure category and duration (years)	Pleura	Peritoneum	S years	Rate per 100 000 S years
Males				
Low to moderate				
< 2	3	1	12 031	33
> 2	3	4	7 500	93
Severe				
< 2	6	10	15 428	104
> 2	7	12	7 827	243
Laggers				
< 2	3	2	7 893	63
> 2	1	4	2 690	186
Females				
Low to moderate	1	0	2 066	48
Severe				
< 2	8	5	9 538	136
> 2	4	3	4 388	360

From: Newhouse, M. L.; Berry, G. "Patterns of mortality in asbestos factory workers in London". *Annals of the New York Academy of Sciences* (New York), 1979, 330 (53-60).

Occasional cases are brought up of mesothelioma linked with indirect or non-occupational exposures to asbestos. The assumption is made that the persons involved were exposed to low levels of dustiness, certainly less than industrial "asbestos workers". Certain cases may be regarded as cases of *para-occupational exposure*, i.e. in persons who had worked next to an asbestos worker. In shipbuilding cases are known of welders whose mesothelioma is linked with exposure to asbestos from a fellow worker applying insulation nearby. Exposure in these cases clearly cannot be considered low; in addition a possible carcinogenic effect of high levels of exposure to iron fumes was postulated by some authors, as one hypothesis to explain the higher risk of mesothelioma generally observed in shipyards. Accumulations of cases of mesothelioma have also been observed among members of the general population in the vicinity of asbestos factories and mines. It is questionable if the exposure of these *neighbourhood cases* had been low. The vicinity of the South African crocidolite mines was notoriously dry and dusty and the factory exhaust ventilation systems were said to have blown asbestos into the streets around the asbestos factories in Hamburg and London. No accumulation of any cancer could be found to be linked with dustiness around chrysotile mines in Canada or around the amosite factory in New Jersey. Perhaps most conspicuous are the *household cases* of mesothelioma in members of the family of asbestos workers who themselves had never been occupationally exposed to asbestos. Asbestos dust brought home from the job on the worker's clothes is considered to be the cause. Again it is doubtful whether exposure of the members of the family was necessarily low. Years of the bad practice of coming home in industrial work clothes could contaminate the household substantially. In fact levels of asbestos as high as 5 000 mg/m^3 of air were measured in the homes of asbestos workers, which is 1 to 3 orders of magnitude higher than levels measured at 800 m from asbestos spray sites. Less clear are cases which developed many years *after a very short period of exposure*. It seems probable that such exposures, although short, were very high.

In animal experiments all kinds of asbestos cause malignant mesothelioma in a certain number of animals, when introduced into the pleural or peritoneal cavity, and the higher the dose of asbestos introduced the more tumours resulted, i.e. a dose-response was established also in animals. There were doses with which no cases, or no excess of tumours was seen (table 5). Another important observation was made by Stanton and confirmed by others. Carcinogenic potency in animals did not depend on the chemical composition as much as on the size and shape of the fibre. Only very thin and comparatively long fibres were carcinogenic, the diameter < 1.5 microns and length > 8 microns appearing critical. Asbestos fibres occur in this size but most other natural and man-made fibres are thicker than the critical diameter. If, however, a non-asbestos fibre is made thin enough, and long enough, as is the case of some newer man-made fibres, they appear to be as carcinogenic as asbestos. An interesting conclusion is that we are dealing here with physical rather than chemical carcinogenicity. The practical conclusion is much more important: possible substitutes for asbestos may be equally carcinogenic, if we assume that their ability to replace asbestos is based on their similar technological properties, which in turn may well depend on their similar fibres sizes and shapes.

Table 5. Number of mesotheliomas in groups of 50 hamsters following an intrapleural injection

Dose mg	Chrysotile	Amosite	Anthophyllite	Crocidolite
25	9	–	–	–
10	4	3	3	10
1	0	0	–	2

From: Smith, W. E.: "Experimental studies on biological effects of tremolite talc on hamsters" (43-48). *Proceedings of Symposium on Talc, Washington (DC), May 8, 1978* (Washington, DC, Bureau of Mines, 1978).

Do the very short fibres have no carcinogenic effect then? For testing in animals the hypothesis of size and shape of the fibres critical for carcinogenesis it proved difficult to obtain adequate thin and short fibres. Objections were raised that by ball-milling not only had the length been diminished but the crystal structure of the fibres had also been lost or changed by the heat or otherwise. In other experiments the samples of short fibres contained an admixture of long fibres and the effect appeared to be proportionally related to the percentage of the long fibres in the sample. It proved technically very difficult to produce a reasonable quantity of a uniform sample of thin asbestos fibres a few microns long with all crystal and chemical properties preserved, a sample which could be used in a reasonable number of animals for injection, still less for inhalation experiments. At least in one series of experiments in which good samples were available, the short fibres produced no cancers while the long ones did. The practical importance of this kind of testing is based on the fact that, except in the vicinity of industrial sources, the public at large is exposed almost exclusively to very short asbestos fibres. Such airborne fibres come from natural outcroppings, from the various city activities such as construction, demolition and insulation. Further, very short fibres are ingested with water and beverages, and even some injection drugs were shown to contain very short asbestos fibres. Such exposures are a matter of legitimate questioning by the public. However, continuing evidence of no-effect has been reassuring.

The fact that most cases of mesothelioma are caused by asbestos, and the very malignant course of mesothelioma, for which there is so far no cure, leads to a high level of public attention to this disease. However, it is the asbestos-linked bronchogenic cancer of the lungs which must call for our attention still more urgently. No matter how tragic every single case of mesothelioma is, the absolute number of cases is fortunately small.

Statistical data

Altogether 668 cases of mesothelioma were identified in the entire population of Canada during 15 years (1960-75) and of the United States during one year (1972) after an active search among all pathologists of both countries, where there are several major mining areas of chrysotile, a large manufacturing industry using chrysotile, amosite and crocidolite, and many large shipyard operations. In contrast, in the United States alone there were 50 481 deaths from lung cancer among the male population in one year (1969). In the Federal Republic of Germany an annual average (1976-78) of approximately 150 cases of mesothelioma were reported by all the pathologists out of 63 000 autopsies from 750 000 deaths a year. Among workers exposed to asbestos a sum of 15 studies done in various countries showed 7 885 deaths from all causes over various periods up to 30 years. Of these, 882 (11.2%) were due to lung cancer and 198 (2.5%) to mesothelioma, both pleural and peritoneal.

Medical monitoring

In some countries an effort is made to identify all persons working with asbestos now or having worked with asbestos in the past and to follow up their state of health through their life. While very little can be done to prevent fatal mesothelioma by such medical monitoring, there is some hope for early detection of a curable bronchogenic cancer, and it is possible to persuade the workers examined to stop smoking, and so to reduce the risk of cancer substantially. With the above figures of the incidence of the two diseases in mind, the practical relevance of an attempt to decrease the frequency of lung cancer is obvious.

Some other cancers

Most inhaled asbestos is removed from the respiratory passages by the mucocilliary escalator and leaves the body via the gastrointestinal tract. Consequently *cancers of the oesophagus, the stomach and the intestines* were to be suspected in substantial industrial exposures. Some occupational epidemiological studies do indicate excess incidence of gastrointestinal cancers but the excess is generally less than that of the pleuropulmonary cancer. Other epidemiological studies do not show any excess, however. In steady populations with exposure to asbestos in drinking water in the vicinity of the asbestos mines in Quebec and in the city of Duluth in the United States, no excess of digestive cancers has been detected, while one study done in California indicated excess linked to asbestos in water. The population of an area with drinking water carried by asbestos-cement pipes over 30 years was studied and no excess of cancers was detected. The possible link to asbestos exposure has not yet been sufficiently clear. The main reason is that cancers of the digestive tract are very common among the general population. Their origin is very strongly influenced by nutritional circumstances and habits, and these habits in turn are influenced by climate, geographical location, and mainly by the ethnic origin of the population. Several animal experiments involving food or drinking water containing asbestos have been undertaken, one of them an unusually extensive experiment with drinking water containing amosite, but none so far has produced any significant excess of digestive cancers.

Carcinoma of the larynx has been found in excess in several groups of asbestos-exposed workers. However, the causal link with inhalation of asbestos is as yet far from clear. A much stronger link between carcinoma of the larynx and cigarette smoking, and even with consumption of alcohol, has been demonstrated, and both of these habits also appear to be high among the causes of carcinoma of the oesophagus.

PELNAR, P. V.

"Asbestosis and carcinoma of the lung". Merewether, E. R. A. (69-81). *Annual report of the Chief Inspector of Factories for the year 1947* (London, HM Stationery Office, 1949).

"Mortality from lung cancer in asbestos workers". Doll, R. *British Journal of Industrial Medicine* (London), Apr. 1955, 12/2 (81-86). 14 ref.

"Diffuse pleural mesothelioma and asbestos exposure in the North Western Cape Province". Wagner, J. C.; Sleggs, C. A.; Marchand, P. *British Journal of Industrial Medicine* (London), Oct. 1960, 17/4 (260-271). Illus. 34 ref.

"Epidemiology of mesothelioma from estimated incidence". McDonald, J. C.; McDonald, A. D. *Preventive Medicine* (New York), 1977, 6 (426-446). Illus. 185 ref.

"Carcinogenicity of amosite asbestos". Selikoff, I. J.; Hammond, E. C.; Churg, J. *Archives of Environmental Health* (Chicago), Sep. 1972, 25/3 (183-186). 15 ref.

"Mortality experience of insulation workers in the United States and Canada, 1943-1976". Selikoff, I. J.; Hammond, E. C.; Seidman, H. *Annals of the New York Academy of Sciences* (New York), 1979, 330 (91-116).

"Mechanisms of mesothelioma induction with asbestos and fibrous glass". Stanton, M. F.; Wrench, C. *Journal of the National Cancer Institute* (Bethesda), Mar. 1972, 48/3 (797-821). Illus. 28 ref.

"Asbestos-related diseases in the Federal Republic of Germany". Woitowitz, H. J.; Beierl, L.; Rathgeb, M.; Schmidt, K.; Rodelsperger, K.; Greven, U.; Woitowitz, R. H.; Lange, H. J.; Ulm, K.; *American Journal of Industrial Medicine* (New York), 1981, 2/1 (71-78). Illus. 14 ref.

Ashes and cinders

Ashes and cinders are the incombustible residue left after the burning of any substance. They are encountered most frequently and in the largest quantities in industrial processes in which fuel is converted into heat, e.g. furnaces, kilns, ovens, boilers, etc., following the disposal of waste by combustion, and in domestic waste from homes which employ solid fuels for heating purposes.

Composition. Ashes usually appear as a soft powder or flakes varying in colour from silver to black; cinders, especially of coal, are grey to black irregular lumps. In the case of coal they are composed chiefly of earthy or mineral matter. However, all ashes and cinders contain the solid residue of any non-volatile oxides or salts of metals (e.g. sodium, calcium, magnesium, iron) or non-metallic elements (e.g. silica) that may have been originally present in the fuel or waste prior to combustion. The cinders obtained from furnaces in the metallurgical industries contain mainly silicates and are dealt with in the article SLAG, BASIC.

Uses. Certain industries use the process of ashing to concentrate and recover valuable incombustibles found in organic matter. Pearl ash is obtained from the combustion of organic material such as vine shoots and lees and beetroot vinasse and is composed of impure potassium carbonate. Seawood is ashed to obtain potassium salts and iodine. Volcanic ash is used as a fertiliser and as a building material. Bones are incinerated to produce charcoal, chardust and bone ash. Ashes and cinders from refuse incineration plants are often used for waste fill for recovering land.

HAZARDS AND THEIR PREVENTION

Ashes and cinders may retain their heat for considerable lengths of time. Furnace cleaners, dustmen collecting domestic waste, workers recovering valuable residues, etc., may be exposed to severe burn hazards when touching ashes or cinders which were presumed to have cooled. Fires may also be caused in refuse collection vans or waste depots by mixing hot ashes and cinders with combustible waste.

Workers required to enter or carry out maintenance on furnaces, etc., should wait until the ashes and cinders have fully cooled or, where this is impossible, suitable heat-protective clothing, foot protection and respiratory protective equipment should be provided. Mechanical furnace emptying systems are a considerable safety aid in this respect. Refuse collectors should be supplied with hand protection and protective clothing, and publicity should be issued warning housewives of the dangers of inserting hot cinders into their domestic refuse. Residue recovery work should be arranged to ensure that ashes and cinders for treatment can be mechanically handled.

Hazardous substances present in only trace quantities in the original substance may be found in dangerous concentrations in the ashes and cinders. Firing of boilers is commonly done by crude oils, most of which contain traces of vanadium and, to a lesser extent, manganese (see BOILERS AND FURNACES, CLEANING AND MAINTENANCE OF).

Ashing processes may liberate large quantities of obnoxious fumes and, consequently, adequate ventilation should be provided around the combustion areas, and the waste gases should be filtered before release into the atmosphere.

Movement of ashes and cinders may generate large quantities of dust which, depending on the original material, may contain toxic or other hazardous substances. Fine ash has caused cases of chronic conjunctivitis. Where possible, movement should be by mechanical means and in closed containers; however, respiratory protective equipment may sometimes be necessary for intermittent intense exposure. Refuse collection vehicles should be fully enclosed to protect workers and reduce air pollution. Workers exposed to ash and cinder dust should be provided with suitable sanitary facilities and encouraged to practise a high level of personal hygiene.

REED, D. W.

"Influence of ash from coal gasification on the pharmacokinetics and toxicity of cadmium, manganese and mercury in suckling and adult rats". Kostial, K.; Kello, D.; Rabar, I.; Maljkovic, T.; Blanusa, M. *Arhiv za higijenu rada i toksikologiju* (Zagreb), 1979, 30/suppl. (319-326). Illus. 11 ref.

CIS 80-1552 "Research findings on the toxicity of quartz particles relevant to pulverized fuel ash". Raask, E.; Schilling, C. J. *Annals of Occupational Hygiene* (Oxford), 1980, 23/2 (147-157). Illus. 26 ref.

CIS 80-1553 "Clinical and experimental studies of the effects of pulverized fuel ash—A review". Bonnell, J. A.; Schilling, C. J.; Massey, P. M. O. *Annals of Occupational Hygiene* (Oxford), 1980, 23/2 (159-164) 9 ref.

Asphalt

MINERAL PITCH; BITUMEN

Asphalt

sp.gr.	0.95-1.1
b.p.	< 470 °C
f.p.	204 °C
i.t.	485 °C

soluble in oil, turpentine, petroleum, carbon disulphide; insoluble in water, alcohol and acids

a dark-brown to black cement-like material which may be solid, semi-solid or liquid, with a weak aromatic odour.

TLV ACGIH 5 ppm 10 mg/m³ fumes

The main constituents of asphalt are bitumens and small amounts of sulphur, oxygen, nitrogen and traces of other minerals. Asphalt should not be confused with tar which is chemically and physically dissimilar (see TAR AND PITCH).

Occurrence. Asphalt is a constituent of most crude petroleum, but it also occurs in natural deposit, where it is usually the residue resulting from the evaporation and oxidation of liquid petroleum. This may accumulate in extensive pits or lakes, such as Pitch Lake on the Island of Trinidad and other deposits in Venezuela, California, China, Russia and the Val-de-Travers in Switzerland.

Production. It is separated from the petroleum by refining processes that also yield gasoline, kerosene and other valuable petroleum products. By modifying the refining processes, different kinds of asphalt may be obtained, varying from sticky liquids to heavy, brittle solids. The three distinct types of asphalt made from petroleum residues are shown in table 1.

Uses. A wide variety of applications include paving streets, highways and airfields; for making roofing, waterproofing and insulating materials; for lining irrigation canals and reservoirs; and for the facing of dams and levees. Asphalt is also a valuable ingredient of some paints and varnishes.

Table 1. Types of asphalts, their properties and uses

Asphalt type and % of production	Manufacturing process	Properties	Uses
Straight-run asphalt[1] 70-75%	Distillation or solvent precipitation	Nearly viscous flow	Roads, airports, runways, hydraulic works
Air-blown asphalt[2] 25-30%	Resetting with air at 200 to 300 °C	Resilient. Viscosity less susceptible to change than straight-run asphalt	Roofing, pipe coating, paints, underbody coating, laminates
Cracked asphalt (sulphurised asphalt and sludge asphalt)[3] <5%	Heating to 400-500 °C	Nearly viscous flow. Viscosity more susceptible to high temperature changes than that of straight-run asphalt	Insulation board, saturant, dust laying

[1] This term is applied to the semi-solid to solid residues obtained from the distillation of semi-asphaltic and asphaltic petroleums and pressure-tars. The following terms are common synonyms for straight-run asphalt: residual asphalt, petroleum asphalt, petroleum pitch, petroleum residue, road binder, carpeting medium, and seal-coating material. [2] A term used to designate the product obtained by blowing air through residual oil at elevated temperatures. It is also known as oxidised asphalt, oxygenised asphalt, oxygenated asphalt, Byerlite, and condensed asphalt. [3] Sulphurised asphalt. The product obtained by heating residual oil or residual asphalt with sulphur at high temperatures. It is also called Dubbs asphalt, Pittsburgh flux, and Ventura flux. Sludge asphalt is the asphaltic product separated from the acid sludge produced in refining petroleum distillates with sulphuric acid. It is also known under the names acid-asphalt, acid-sludge asphalt, etc.

HAZARDS

Handling of hot asphalt can cause severe burns because it is sticky and is not readily removed from the skin. The principal concern from the industrial toxicological aspect is irritation of the skin and eyes by fumes of hot asphalt. These fumes may cause dermatitis and acne-like lesions as well as mild keratoses on prolonged and repeated exposure. The greenish-yellow fumes given off by boiling asphalt can also cause photosensitisation and melanosis.

Although all asphaltic materials will support combustion if heated sufficiently, the asphalt cements and oxidised asphalts will not normally burn unless their temperature is raised about 260 °C. The flammability of the liquid asphalts is influenced by the volatility and amount of petroleum solvent added to the base material. Thus, the rapid-curing liquid asphalts present the greatest fire hazard, which becomes progressively lower with the medium and slow-curing types.

Toxicity. Because of its insolubility in aqueous media and the high molecular weight of its components, asphalt has a low order of toxicity, as a result of which the oral LD_{50} (not available in the published literature) is probably indeterminate. The writer recalls chewing asphalt as a youngster whenever it was available, usually in the summertime at a building site. The asphalt was popular because of its chewiness, tang and belief that it whitened the teeth. The habit of chewing asphalt is quite common among workers in asphalt plants. Several cases of pyloric obstruction have been reported in asphalt workers who had the habit of chewing and occasionally swallowing asphalt. The indigestible mass accumulates in the stomach, forming a stony concretion there.

The effects on the tracheobronchial tree and lungs of mice inhaling an aerosol of petroleum asphalt and another group inhaling smoke from heated petroleum asphalt included congestion, acute bronchitis, pneumonitis, bronchial dilation, some peribronchiolar round cell infiltration, abscess formation, loss of cilia, epithelial atrophy and necrosis. The pathological changes were patchy and some animals were relatively refractory to treatment. It was concluded that these changes were a general phenomenon caused by breathing air polluted with aromatic hydrocarbons and that the degree of change is dose dependent. Guinea pigs and rats inhaling fumes from heated asphalt showed effects such as chronic fibrosing pneumonitis with peribronchial adenomatosis and the rats developed squamous cell metaplasia, but none of the animals had malignant lesions. However, the subcutaneous injection of a mixture of aromatic and saturated fraction of asphalt produced a variety of benign and malignant tumours in mice. From this it was concluded that the dose was more important for tumour development than the exposure time.

Studies indicate that petroleum asphalt, shale oil asphalts and coal tars show distinct variations in their relative carcinogenicity for experimental animals. It was reported that the carcinogenicity of petroleum asphalts can be reduced by dilution with natural asphalts or by converting them to air-blown asphalts. Comparison of the number of cancers in mice exposed to steam-refined asphalt and air-refined asphalt showed that the latter was more carcinogenic. This was attributed to the increased complexity of the molecules, since air refining favours polymerisation and condensation. Some tumour formation was noted in mice painted with four road asphalts but none were induced by roofing asphalt or its vapours.

Because of the lack of continuous skin contact, no significant cancer hazard is postulated for men working with molten asphalt. The hardness and high melting point of undiluted air-refined asphalt makes prolonged application to the skin impossible.

SAFETY AND HEALTH MEASURES

Since heated asphalt will cause severe skin burns, those working with it should wear loose clothing, in good condition, with the neck closed and the sleeves rolled down. Hand and arm protection should be worn. Safety shoes should be about 15 cm high and laced so that no openings are left through which hot asphalt may reach the skin. Face and eye protection is also recommended when heated asphalt is handled. Changing rooms and proper washing and bathing facilities are desirable. At crushing plants where dust is produced and at boiling pans from which fumes escape, adequate exhaust ventilation should be provided.

Asphalt kettles should be set securely and be levelled to preclude the possibility of their tipping. Workers should stand on the windward side of the kettle. The temperature of the heated asphalt should be checked frequently in order to prevent overheating and possible ignition. If the flash point is approached, the fire under the kettle must be put out at once and no open flame or other source of ignition should be permitted nearby. Where asphalt is being heated, fire-extinguishing equipment should be within easy reach. For asphalt fires, dry chemical or carbon dioxide types of extinguishers are considered as most appropriate.

Treatment. If molten asphalt strikes the exposed skin, it should be cooled immediately by quenching with cold water or by some other method recommended by medical advisers. An extensive burn should be covered with a sterile dressing and the patient taken to a hospital, while minor burns should be seen by a physician. Solvents should not be used to remove asphalt from burned flesh. No attempt should be made to remove particles of asphalt from the eyes, but the victim should be taken to a physician at once.

GERARDE, H. W.

CIS 78-432 *Criteria for a recommended standard–Occupational exposure to asphalt fumes.* DHEW (NIOSH) publication No. 78-106 (National Institute for Occupational Safety and Health, 4676 Columbia Parkway, Cincinnati, Ohio) (1977), 143 p. Illus. 80 ref.

See also ROAD MAINTENANCE.

Asphyxia

Asphyxia and suffocation are usually regarded as synonymous terms meaning severe impairment or suspension of respiratory function, although, by derivation, "asphyxia" means a state of pulselessness (Greek, *a-sphuxis*, pulse) and "suffocation" means choking (Latin, *suffocare*, to choke). Suffocation often implies an element of violence and, more correctly, is one cause of asphyxia. By convention and prolonged usage, the term asphyxia has been applied to conditions in which, as a result of interference with the respiratory functions, there is a diminished supply of oxygen to the blood and tissues and usually a diminished elimination of carbon dioxide from the lungs. The causes of such conditions include mechanical obstruction to breathing, certain gases and other chemicals, and some diseases.

A supply of oxygen and its successful utilisation are fundamental needs of every cell. Oxygen is taken into the lungs during the process of respiration which is controlled by the respiratory centre of the brain. In the pulmonary alveoli, oxygen is absorbed into the bloodstream by chemical combination with the haemoglobin contained in the red blood corpuscles. The oxygenated haemoglobin is transported to all the body through the circulatory system and subsequently the oxygen is released from the oxyhaemoglobin to the body tissues where carbon dioxide, a waste product of cellular metabolism, is produced continuously. This is taken up by the blood and carried to the lungs where it is released with the expired air. If the supply of oxygen to the brain is interrupted and falls below minimal requirements, then asphyxia, possibly fatal, will ensue, the degree of severity depending upon the extent of the oxygen deficiency. Further details will be found in the article HYPOXIA, ANOXIA.

The usual features of the state of hypoxia are: cyanosis with capillary dilation and stasis, oedema and serous effusions due to increased capillary permeability, and petechial haemorrhages (Tardieu spots). Some tissues are more sensitive to oxygen deficiency than others. Nervous tissues are first affected and their functions are disturbed even by a slight degree of oxygen lack, while sudden deprivation may cause almost immediate unconsciousness.

When asphyxia results from occlusion of the airway, a fulminating type of hypoxia and an accumulation of carbon dioxide develop together: gasping for breath and marked cyanosis occur. In asphyxia due to carbon monoxide there is no carbon dioxide retention, and the hypoxia does not produce respiratory distress or cyanosis.

Conditions producing asphyxia

A decreased supply of oxygen may result from:

(a) disease processes especially of the circulatory or respiratory systems and from an insufficiency of haemoglobin–respiration may be impaired by damage to the nervous system as in poliomyelitis or following an electric shock;

(b) breathing a rarefied or vitiated atmosphere, containing less than 17% oxygen, as in high-altitude climbing or flying or near intense conflagrations;

(c) breathing an atmosphere which contains a physiologically inert gas but which is dangerous if the gas is present in a concentration sufficient to prevent adequate oxygen from entering the lungs (simple asphyxia);

(d) inhaling a gas which reacts chemically with constituents of the body to prevent oxygen from being transported or utilised (chemical asphyxia);

(e) direct interference with the respiratory centre, as in the case of narcotic or anaesthetic poisons;

(f) mechanical interference with respiration by: impeding or blocking the passage of air through the nostrils and mouth (suffocation, smothering, gagging); impaction of foreign bodies in the glottis, larynx or main air tubes (choking); inhaling fluid (drowning); constriction of the neck (hanging, strangling); compression and mechanical fixation of the chest (traumatic asphyxia);

(g) occlusion of the air passages as from the effects of scalds or corrosive substances, acute inflammatory or infective processes, acute laryngeal spasm or oedema and tumours.

Asphyxiants in occupational medicine

Asphyxiation and suffocation are of interest to both forensic and occupational medicine. In forensic medicine, some authorities restrict the term asphyxia to those conditions of violence in which there is mechanical interference with the entry of air into the lungs. In occupational medicine, although some of the violent forms of asphyxia may occur in the course of employment, those compounds which usually comprise the two groups known as simple asphyxiants and chemical asphyxiants constitute the main problem.

Simple asphyxiants. These are gases which in general do not act as poisons but which may cause injury or even death by displacing the available oxygen in the atmosphere and reducing it to a level which will not support life. They include nitrogen, hydrogen, acetylene, methane, ethane, propane and butane, helium, neon, argon and carbon dioxide. Although the main action of the latter gas is that of a simple asphyxiant, it may cause unconsciousness when present in concentrations in excess of about 5-7%. The effect of these gases is proportional to the extent that the partial pressure of oxygen in the inspired air is reduced by their presence. They are hazardous only in high concentrations such as may occur in confined spaces.

Chemical asphyxiants. These include gases which do not exlude oxygen from the lungs but which exert a chemical action either upon the blood, thus preventing it from transporting oxygen although the lungs are well aerated, or upon the tissues, preventing them from using oxygen although it may be brought to them in ample quantity by the blood. Chemical asphyxiants of particular importance in industry are carbon monoxide and the cyanides. Hydrogen sulphide and chemicals which produce abnormal amounts of methaemoglobin are also classified as chemical asphyxiants. Signs of asphyxia may also be produced by narcotics and anaesthetic compounds such as ether, chloroform, nitrous oxide and carbon disulphide. Such compounds have effects on the nervous tissues including the respiratory centre, and excessive intake causes depression and eventual failure of respiration. Narcotics also depress tissue oxidation by interfering with dehydrogenase systems. The action of a number of lung irritants, e.g. ammonia, sulphur dioxide, chlorine, phosgene, nitrogen dioxide and bromomethane is such that an asphyxial death may result from excessive exposure to them (see GASES AND VAPOURS, IRRITANT).

SAFETY AND HEALTH MEASURES

The ideal measure in preventing hazards due to asphyxiant gases and chemicals is to replace the offending substance by one that is innocuous or less toxic. If substitution is not feasible some change in the process, or in the method of performing it, e.g. automatic handling or enclosure, may be possible.

One of the most effective measures for the control of air contamination is the application of a ventilation system, either by general (dilution) ventilation or by a local exhaust system. The former is not as satisfactory as is local exhaust ventilation and should not be used in the control of highly toxic agents. A local exhaust system implements a cardinal principle of prevention in that it removes the contaminating material at its point of origin; however, for maximum efficiency it needs proper maintenace and regular measurements of the air flow.

As a last resort, or in conditions of emergency, respiratory protective equipment may be required. This may be of the supplied-air or air-purifying type. No air-purifying device is suitable for protection in an atmosphere which is deficient in oxygen and, under such circumstances, only supplied-air devices such as hose masks or self-contained breathing apparatus should be used. Protection against gases and vapours may be afforded by an air-purifying device, such as a canister respirator or a chemical cartridge respirator, or by supplied-air equipment, depending upon the degree of contamination.

Special precautions are required and must be observed for the entry of confined spaces such as boilers, tanks, furnaces, etc., which may contain asphyxiants. The precautions include: thorough decontamination and testing of the atmosphere in the space; the wearing of an efficient safety belt or harness and a supplied-air respirator; provision of rescue and resuscitation equipment; assistance from other persons; and proper instruction in safety measures.

The wearing of suitable protective clothing is necessary when there is risk of exposure to asphyxiants which may be absorbed through the skin.

Treatment. See RESUSCITATION.

SMITH, G. C.

"Mechanical asphyxias" (Asphyxies mécaniques). Noto, R. *Encyclopédie médico-chirurgicale* 16520 A 10-7.1975 (Editions Techniques SA, 18 rue Séguier, 75006 Paris, 1975), 8 p. (In French)

"Deaths from asphyxia among fishermen". Glass, R. I.; Ford, R.; Allegra, D. T.; Markel, H. L. *Journal of the American Medical Association* (Chicago), 1980, 244/19 (2 193-2 194). 6 ref.

Asthma, occupational

Asthmas of occupational origin are being regarded with renewed interest. From the medical point of view, new techniques have resulted in increased knowledge of the physiopathological mechanisms involved, while from the occupational aspect, the introduction of new chemical products has resulted in many new hazards. It is sometimes, however, difficult to show the relationship between the occupational hazard and the health impairment involved, and this points to the desirability of amending existing legislation.

Asthma is difficult to define in spite of the fact that the mechanisms involved are known, and considerable disagreement remains among the writers on this subject. The following anatomo-clinical definition is commonly applied: asthma is a reversible reduction in the diameter of the bronchi which, by muscular contraction, hinders the passage of air. This is caused by various agents that are liberated through allergic reactions; by direct or indirect irritant phenomena; by neurogenic causes; or by pharmaco-dynamic agents. To this is added a non-specific bronchial hyper-reactivity and inflammation of the mucous membrane.

Asthmatic bronchitis is generally excluded even though borderline cases can exist. On the contrary, there is a tendency at the present time to group together coughs, hay fever, allergic rhinitis, catarrh, etc., all of which frequently precede or accompany asthma attacks. Within this group, occupational asthma only varies from the point of view of its aetiology in that it is brought about by a single or repeated exposure to an active substance present in the working environment.

Physiopathology

The bronchial tonus is a state of equilibrium between the contraction and the relaxation of the smooth muscles and a break in the latter gives rise to a bronchospasm. The maintenance of this balance calls upon the nervous system (parasympathetic vagus fibres which are joined to cholinergic or histaminic receptors). The mast cells also come into play and stock the active substances. Lastly, there are circadian variations, tonus being at a maximum during the night-time. Although an asthma attack can be started by irritant factors, it is usually exaggerated immune phenomenona that are the cause of the allergy.

Gell and Coombs have classified four types of reactions according to their physiopathological mechanism and, among these, two humoral reactions play a part in allergic asthma. These are: type I or anaphylactic (reaginic antibody) reaction which brings E immunoglobulin (IgE) into play and which in turn liberate chemical mediators from basophilic cells. This is a rapid reaction which takes place in less than 1 h after contact by the sensitised subject. Type III reaction is much more rare. This is a precipitating antibody allergy. Prolonged contact with the allergen leads to the formation of G immunoglobulin (IgG), which joins the allergen to form a large molecule called an "immune complex". Symptoms appear after a period of 4-6 h, becoming more severe with the length of the latent period; they are often preceded by type I symptoms.

Particular mention should be made of substances, such as certain pesticides, which can cause a bronchospasm through the accumulation of acetylcholine.

Clinical aspects

The clinical aspects of occupational asthmas are no different from those of common asthma.

The attack usually starts in the evening after work or, more frequently, during the night. It is sometimes preceded by a fit of sneezing, rhinorrhea and spasmodic coughing. Classic expiratory bradypnoea ensues progressively. The attack passes off slowly accompanied by viscous expectoration. An over-riding bronchial infection may appear, resulting in an irritative focus. In the long term the infection may become definitely localised in the bronchial tree, with the continuation of dyspnoeic asthma, bronchitis-emphysema and, finally, irreversible chronic respiratory insufficiency. Dramatic acute incidents may occur in the course of these developments such as severe asthmatic attacks requiring urgent resuscitation.

Diagnosis

Asthma is almost entirely a matter for clinical diagnosis although it may sometimes be helpful to carry out supplementary examinations, particularly when the sequence of events or the variety of possible causative agents does not permit a conclusion to be drawn.

Many lung function tests exist, some of which are quite simple. They include: vital capacity (VC), forced expiratory volume (FEV), peak expiratory flow rate (PF or $\sqrt{E_{max}}$). The results of these tests will show up a reduction in the ventilatory values. The Tiffeneau index $\left(\frac{FEV}{VC}\right)$ and the $\sqrt{E_{max}}$ 75 and $\sqrt{E_{max}}$ 50, which indicate the bronchial and the bronchiolar condition, are more interesting tests. The measurement of bronchial resistance, compliance and sensitivity to bronchomotor tests require the use of a specialised laboratory.

Immunological investigations usually consist of skin tests, respiratory tests and the measurement of specific antibodies.

Skin tests enable the existence of a special reactivity to a particular substance to be explored, without taking strict account of respiratory symptoms. The antigens, which are large molecular structures of animal or vegetable origin, are used directly or after having undergone preparation, when the allergen is a small molecule which acts as a half antigen. Epidermal tests consist of placing the substance on the skin: this gives a delayed result. Intracutaneous reactions make use of pure, sterilised extracts of the suspected substance and require both an immediate and a delayed reading. These tests should be performed with great care. The "prick test" consists of bringing the substance into contact by means of a light scratch, and gives an immediate result followed by a partially delayed result.

Biological examinations, although frequently used, do not always enable a precise diagnosis to be made. Blood eosinophilia is of interest but not reliable. At the present time, measurement of global immunoglobulins (IgE and IgG) is to be preferred, because they are often found to have undergone some change. It is possible in certain cases to look for the particular immunoglobulins corresponding to a suspected allergen, using specific IgE or the Radio-Allergo-Sorbent Test (RAST). This method has great significance when it is positive, but many allergens cannot be found by this means.

The Lymphoblastic Transformation Test (LTT) and the one based on basophil graining, carried out *in vitro,* have given varied results and there is little unanimity among the authors.

Asthma may be induced through the respiratory tract and there are a number of methods employed for making provocation tests on this basis which are coming increasingly into favour. One method is to inhale an aerosol containing the suspected allergen while a second method is to cause the subject to handle the suspected substance; a third method is to reproduce the occupational exposure (but this is difficult to achieve), carrying out at the same time respiratory function tests or a rhinomanometric test, or both. These provocation tests will be repeated over a period of 24 h and performed in a hospital. Both immediate and delayed reactions are observed, as well as extra-respiratory symptoms such as fever, arthralgia and anxiety, which should be treated as necessary. These tests are of interest; nevertheless any hyper-reaction of a bronchial nature must be taken into account when interpreting the results, and once again a negative result cannot lead to any conclusion.

A further test consists of alternately removing the patient from exposure and re-subjecting him to it. This is perhaps the most interesting and most searching test of all for the detection of an occupational origin for asthma. Removal of the patient may be achieved during a long holiday or on the occasion of a prolonged sickness. The development of symptoms may then be observed. A lung function test may be performed before work is recommenced. A regression of the respiratory symptoms should be seen, followed by their disappearance. If the respiratory symptoms and obstructive syndrome reappear when work is restarted, its occupational origin is definitely confirmed.

Classification of occupational asthmas

Several classifications have been proposed. From the practical point of view, the present classification is made according to the origin of the allergen concerned. This may be vegetable, animal or chemical and has the advantage for the occupational physician that it takes account of the associated allergenic factors.

There are numerous allergens of vegetable origin but it is often difficult to identify the precise causative factor, since many of these are made up of complex molecules or have a number of constituents which cannot always be identified.

The most common causes are to be found among the exotic woods and these include *Perbora granda,* mahogany, teak, Brazilian rosewood, red cedar and Lebanon cedar, sequoia or Californian redwood (and many varieties of African hardwood). Green woods, leaves such as walnut, and resinous woods are also to be included in this list. In addition, the glues, insecticides and the chemicals used for treating the wood (which include formaldehyde, pentachlorophenol and the bichromates) are often the causative agent. The asthma resulting from these causes can develop and be accompanied by an allergic extrinsic alveolitis.

Among occupations that involve exposure to flour and cereals, milling, baking and the manufacture of pasta are the most widespread. The patient suffers acutely during work, after which there are almost permanent symptoms of dyspnoea. It is difficult to identify the exact cause in these cases because there are so many allergens to take into account. These include flour; pollens–especially in the flour of buckwheat; moulds such as *Cladosporium, Penicillium, Aspergillus* and *Alternaria;* insects, among which are the granary weevil *(Sitophilus),* the flour moth *(Ephestia)* and the various arthropods. It may sometimes be necessary to make tests for asthma with the dust in the workroom, the effects of which may be compared with silo asthma. It is to be noted that the use of alkaline persulphates in bakeries is prohibited in many countries.

Among the vegetable oils the most important is castor oil, of which the oil cake is used in the manufacture of fertiliser, soap and refined oil. Workers in the refineries, dockers and those engaged in its transport are most commonly affected, but dealers and users may also be affected. Linseed oil cake (cattle food) and peanut, colza and soya oils are also allergenic.

Other allergens of vegetable origin include gum arabic and karaya gum, which are used in the printing industry. Cotton gives rise to byssinosis, which develops with the "Monday symptoms" of chest tightness or asthma, may lead to pulmonary fibrosis, and is a dust-related disease. Similar symptoms are found in occupational exposure to hemp and flax. There is some discussion about the allergenic properties of green coffee. Certain vegetable products used in the pharmaceutical industry can also produce asthma: ipecacuanha, liquorice, pyrethrum extract and natural quinine.

Asthma may also be caused by products of animal origin, and the related allergens are often of a complex nature. Hair, nails, teeth, feathers, animal debris and dried excreta are all possible sources. Mention may also be made of farmyard or laboratory animals, birds, insects or insect debris, the arthropods including the arachnida and finally the water flea used for fish food.

Chemical and industrial asthmas are caused by medicines, plastics and many different chemicals used in industry. In the case of medicines, the most important group includes the antibiotics (betalactamines, streptomycin, and precursors such as phenylglycine), sulphaguanidine, largactil, piperazine, benzalkonium chloride and others.

In the plastics industry symptoms of asthma may be traced back to the monomers and the additives used as hardeners, catalysts, etc., and only rarely to an exposure to the polymers themselves, although the latter may still contain traces of the former which can be liberated during use. Cases of asthma may thus be found both in the manufacturing stage and during the use of plastics materials.

Organic isocyanates, and particularly toluene diisocyanate (TDI), which is widely used in the manufacture of paints, lacquers, adhesives, polyurethane and other substances, may give rise to bronchospasms at the first exposure, and sensitisation can develop after exposure to concentrations at a TLV as low as 0.02 ppm.

Phthalic anydride is used in the manufacture of epoxy resins, adhesives and paints and is also responsible for what is known as "meat-packers' asthma". Formaldehyde is an irritant at high concentrations and a sensitiser at low concentrations.

Aliphatic and alicyclic amines are used in the plastics industry as hardeners and catalysts, particularly ethylenediamine and diethanolamine. The aromatic amines (paraphenylenediamine and derivatives) are found in the rubber industry, but rarely give rise to trouble because the process takes place in closed vessels. They can have an irritant and a pharmacological action through the liberation of histamine and of acetylcholine.

Proteolytic enzymes are widely used in the pharmaceutical industry and in the manufacture of washing powders, and can lead to occupational asthma. Exposure may take place during extraction or purification and at the manufacturing stage. Strict regulations concerning their use have been introduced in Sweden. Among the best known are trypsine, bromeline, papaine and *Bacillus subtilis*.

Certain metals may cause asthma and particularly the platinum salts (ammonium and sodium chloroplatinates). They may be encountered in platinum refineries, as catalysts in the chemical industry, in photography, and in the pharmaceutical industry, where they are used in the hydrogenation of streptomycin. They react with a semi-antigen to liberate histamine. Other metals concerned are nickel, chromates, cobalt, lead, vanadium and the phenylmercuric compounds.

Among other compounds that may induce symptoms of asthma are the alkaline persulphates used in the manufacture of hydrogen peroxide and hair bleaches as well as certain other hair-tinting preparations such as henna and textile dyes containing azo- and diazo-compounds and anthraquinone, which behave as semi-antigens and are accompanied by rhinitis. Sericin, found in the silk industry and in hairdressing, and diazomethane, which is a potent cause of asthma and should be used in a closed vessel, should not be overlooked. Finally, mention should be made of welding operations, where exposure to irritant vapours of chlorine and its derivatives and to allergens such as rosin, abietic acid and ethanolamine may occur.

SAFETY AND HEALTH MEASURES

The prevention of asthma by technical means should be the first consideration. This means the reduction of airborne concentrations of dust and vapours containing these substances, making use of exhaust ventilation and hoods or enclosures in well ventilated premises.

The basic preventive method consists of reallocation of the individual to different work. Where this is impossible, it may be necessary for him to be given the necessary vocational training to enable him to move to a completely new activity. In principle all asthmatics, those suffering from rhinitis or spasmodic coughing and atopic subjects should be encouraged to take up occupations where they will not be exposed to substances that are known to be a cause of asthma.

Treatment

The treatment of asthma attacks continues to be based on the use of the classic broncho-dilators derived from theophylline. In cases where there is a liberation of histamine (exposure to aliphatic and alicyclic amines, "bakers' asthma", or that resulting from exposure to persulphates, platinum salts and natural textiles), antihistaminics may be helpful. Among the sympathicomimetic products, only the betamimetics continue to be used, particularly salbutamol either as an aerosol or in the form of tablets and suppositories, or intravenously. In some cases the corticoids may help, but only in severe cases or at a later stage of the illness, and they do not always dispense with the need for ventilatory assistance.

When possible, the basic treatment of asthma should be immunological. Disodium chromoglycate may sometimes be used in cases where the following are involved: flour and cereals, insects of the *Arthropoda* phylum, animal debris, "baker's asthma", sericin, soya, gum arabic, henna, piperazine, platinum, proteolytic enzymes and *Candida tropicalis*. Desensitisation is more difficult in the case of the first three and may even be dangerous in the case of sericin.

Medico-legal aspects

Although occupational asthma is by definition caused by a single or repeated exposure to a hazard encountered during work, it is not always easy to ascertain the cause-effect relationship. The relationship must be based on two considerations: the occupational exposure and the individual factor.

The allergen may be purely occupational as, for example, a chemical substance like formaldehyde or an isocyanate, but it may also be widespread, though mainly at the workplace, as in the case of animal or vegetable allergens where cotton or sawdust is to be found; others may be dispersed generally, e.g. dust or pollen, and evidence that these were present at the workplace has to be shown.

The individual factor has to be shown up by means of an examination and careful questioning, bearing in mind the possibility of atopic polysensitisation or asthma due to a few particular sensitisers.

Table 1. Known agents of occupational asthma

Abietic acid
Acacia
Acacia gum
Acaridae
Acrylic fibres
Acrylic precursors
Actinomyces
Alicyclic amines
Alkyl phosphates
Aliphatic aldehydes
Aliphatic amines
Anthraquinone dyestuffs
Ampicillin
Arthropods
Azo dyes

Bacillus subtilis
Bakeries
Barley
Benzalkonium chloride
Benzylpenicillin
Betalactamines
Bromelain

Candida tropicalis (proteins)
Carbamates
Castor oil
Cats
Chlorine
Chlorthion
Chromium
Cobalt
Cockroaches
Colophony
Colorado beetles
Cotton
Cows

DDVP
Dogs
Diazinon
Diazomethane
Diethanolamine
Diethylene diamine
Diethylene triamine

Epoxy resins and hardeners
Ethylhexylamine
Exotic woods

Flax
Flour or meal
Formaldehyde

Grain silos
Graminaceous pollens
Green coffee
Ground nuts
Guinea-pigs
Gum arabic

Hair, horns, feathers, etc., of animals
Hamsters
Hemp
Henna
Hexamethylenetetramine

Industrial perfumes
Insecticides
Ipecacuanha

Jute

(Karaya) gum

Laboratory animals
Lead
Liquorice
Locusts

Mercury diphenyl
Mercury (organic compounds)
Metampicillin
Mice
Mites
Moulds

Nickel
Nitric oxide

Oats
Oil cake
Oleandomycin
Organic isocyanates
Organiphosphorus compounds

Panonychus ulmi
Papain
p-Dichlorobenzene
p-Formaldehyde
Penicillins
Persulphates
Pesticides
p-Phenyldiamine
Phenyl-formaldehyde resins
Phenylmercuric nitropropionate
Phenylglycine
Phenylhydrazine
Phosphoramines
Phthalic acid
Piperazine
Platinum salts
Polyamides
Polyesters
Proteolytic enzymes
Pyrethrum

Quinine

Rabbits
Rats
Red spiders
Rice
Rye

Sericin
Silk
Soya
Spiramycin

Textiles, natural
Textiles, synthetic
Thrombin
Triethylene diamine
Triethylene tetramine
Trimellitic anydrides
Trypsin

Urea-formaldehyde resins

Vanadium
Vanillin
Vegetables (pharmacological action
Viscose

Water fleas
Welding fumes
Wool

Several criteria as to the causative factor may be put forward. These may be set out schematically as follows:

(1) An allergic occupational asthma may be definitely attributable when there is only one occupational allergen in question and when the disease disappears after the work is stopped.

(2) An "occupational pseudo-asthma" may be caused where there is some irritative substance such as dust and if it disappears when the subject relinquishes the occupation concerned.

(3) Certain asthmas are found among persons who are polysensitive to allergens present both at work and at home and which disappear when the work is no longer performed. These are occupational asthmas because the occupational allergen creates an additive effect.

(4) Asthmas occurring among polysensitive individuals who are not exposed to allergenic dusts during work, and where the attacks cease when the work is no longer performed, are regarded as occupational because the work gives rise to an irritant trigger effect.

(5) Cases which arise when the individual is sensitive to an occupational allergen, but the attacks do not stop when the work is relinquished, should not be regarded as occupational asthmas.

In order to arrive at a conclusion as to the causative factors of an asthma, a lung function test and a bronchomotor test must be carried out in addition to the clinical examination and history; this should be completed by skin tests and by the determination of IgE, both total and specific, when possible. The key factor will, however, be a provocation test carried out in hospital or a test based on the stoppage and restarting of work. Allied to these there should be an exact time evaluation of the occupational exposure, to reach a conclusive decision.

FURON, D.
CANTINEAU, A.
FRIMAT, P.

CIS 80-788 *Manual of internal medicine*–Vol. 4: *Disorders of the respiratory tract*; Part 2: *Bronchitis, asthma, emphysema* (Handbuch der inneren Medizin–Bd. 4. Erkrankungen der Atmungsorgane; Teil 2. Bronchitis, Asthma, Emphysem). Ulmer, W. T. (West Berlin, Springer Verlag, 1979), 788 p. Illus. (In German)

The study of allergies (Allergologie). Charpin, J. (Paris, Flammarion-Médecine-Science, 1980), 905 p. (In French)

Occupational Asthma. Presented to Parliament by the Secretary of State for Social Services by Command of Her Majesty (London, HM Stationery Office, Jan. 1981), 21 p.

CIS 77-1400 "Bronchial asthma of occupational origin". Brooks, S. M. *Scandinavian Journal of Work, Environment and Health* (Helsinki), June 1977, 3/2 (53-72). Illus. 134 ref.

Practical guide to occupational asthma (Guide pratique de l'asthme professionnel). Gervais, P.; Diamant-Berger, O.; Rabaud, A. (Lyons, Laboratoire Fison-Gorda, 1979), 48 p. (In French)

"Evaluation of a hospital workshop of exposure in the diagnosis of occupational asthma" (Bilan d'un atelier hospitalier d'exposition dans le diagnostic de l'asthme professionnel). Gaultier, M.; Gervais, P.; Conso, F.; Dally, S.; Diamant-Berger, O. *Archives des maladies professionnelles, de médecine du travail et de sécurité sociale* (Paris), Aug.-Sep. 1979, 40/8-9 (783-793). Illus. 16 ref. (In French)

"Occupational type bronchial provocation tests: testing with soluble antigens by inhalation". Harries, M. G.; Burge, P. S.; O'Brien, I. M. *British Journal of Industrial Medicine* (London), Aug. 1980, 37/3 (248-252). Illus. 7 ref.

Audiometry

Audiometry, which here means the determination of monaural hearing threshold levels for pure tones by air conduction, should be a component of a hearing conservation programme and carried out when indicated by that programme. Audiometry alone does not prevent occupational hearing loss but is essential to determine the hearing status of the workforce, to monitor the effectiveness of noise control measures, and to identify individual workers who are adversely affected by noise.

A hearing conservation programme must also include a defined policy, a detailed noise survey, determination of noise exposure (dosimetry), reduction of excessive noise by engineering or administrative methods, provision of hearing protection, education of those exposed, and monitoring by repeated noise and noise dose measurements and by audiometry.

Indications for audiometry

Reference or baseline audiometry should be performed on all persons likely to receive at work a daily noise dose (DND) in excess of 0.1 (or an equivalent of 80 dB(A) L_{eq}.A.8). (For methods of calculating equivalent continuous sound level (L_{eq}), see the UK Department of Employment Code included in the bibliography at the end of this article.)

Monitoring or periodic audiometry should be performed on all persons likely to receive at work a DND in excess of 0.33 (or an equivalent of 85 dB(A) L_{eq}.A.8).

Another audiogram should be taken within 90 days of initial exposure for comparison with the reference audiogram. In the absence of significant change of either the audiometric pattern or the noise exposure, monitoring audiometry should be repeated every one or two years.

Consultant audiometry in specialist clinics should be performed when reference or monitoring audiograms are abnormal (as defined below), or when an assessment needs to be made for compensation purposes.

Equipment

The audiometer should be a discrete frequency, pulsed-tone, air conduction type and, whether automatic or manual, should conform with international or national standards.

An automatic (self-recording) audiometer, preferably controlled by a microprocessor, should be used whenever practicable.

There are several reasons why an automatic audiometer is preferable to a manual one: virtual elimination of systematic error due to the operator, the unambiguous nature of the subjects' responses, the provision of a permanent record without possibility of transcription error, and a visible indication of the quality of the subjects' test performance.

Manual audiometers have the advantage of being less expensive and, with standardisation of testing procedures, can give satisfactory though less precise results. Where the cost of the audiometer is likely to be a significant percentage of the total budget for the hearing conservation programme, it would be better to provide a manual audiometer than no audiometer at all. While automatic audiometry is ideal, its cost must be balanced against the likely benefits achieved by spending the equivalent amount on noise control.

Background noise levels should be sufficiently low to permit accurate results when testing hearing threshold levels down to 0 dB. Detailed tables of maximum acceptable background noise levels (as published in the (UK) National Physical Laboratory Acoustics Report AC 60, February 1973, by B. F. Berry) are available. In many circumstances the necessary levels will only be achieved by the use of a sound isolating booth. Once again, however, common sense and practical considerations are important, as the money spent on purchasing a booth may be better spent on noise reduction. As a general guide, if the audiograms of a person with known normal hearing are similar when taken under "ideal" conditions and when taken under the conditions likely to prevail in a particular occupational test setting, then those latter conditions are probably satisfactory for reference and monitoring audiometry.

Regular calibration of audiometers is essential. Where daily calibration is impracticable, objective tests of calibration (hearing levels and frequencies) should be done at least annually. New audiometers should be checked at more frequent intervals until calibration stability is established. Regular subjective checks, i.e. testing a person with known normal hearing, will disclose any gross discrepancy which might occur between formal calibrations.

The operator (audiometrician) should be fully trained and competent in basic pure-tone audiometry, and have access to the advice of an occupational physician and an otologist.

Procedures

Prerequisites. Recent exposure to loud noise may temporarily elevate hearing threshold levels. These temporary threshold shifts (TTS) will be minimised if, in the 16 h prior to reference audiometry, noise levels in excess of 80 dB(A) are avoided. For monitoring audiometry, insistence on 16 h away from noise may be impracticable. In such cases audiometry should be preceded by a period of quiet of not less than 7 h, and for this purpose the wearing of appropriate hearing protection while at work should be adequate.

Health screening by the operator or other trained person prior to audiometry should include (i) the medical and occupational history, particularly in relation to past ear disease and exposure to noise, and (ii) an examination of the ear, nose and throat.

Test frequencies should include 0.5, 1, 2, 3, 4 and 6 kHz.

Audiometry techniques will vary depending on the type of audiometer, e.g. automatic or manual, and should be performed in a standard way for each person according to the manufacturers' instructions and in conformity with international and national standards.

Interpretation of reference audiograms

Medical considerations. Many cases of hearing impairment or possible disease will be detected by initial audiometry performed as part of a hearing conservation programme. In such cases, the person should be informed of the test findings and a decision should be made regarding the need to refer to a medical practitioner for diagnosis and possible treatment. Referral should be made if any of the following criteria are met (as recommended by the National Acoustic Laboratories, Australia):

Abnormal low frequency hearing level (2 kHz and below)	hearing loss at 0.5 or 1 kHz: ⩾ 25 dB hearing loss at 2 kHz:

	≥ 30 dB at age < 40 ≥ 35 dB at age 41-45 ≥ 40 dB at age 46-50 ≥ 45 dB at age 51-60 ≥ 50 dB at age > 60
Abnormal high frequency hearing level (3 kHz and above)	no specific recommendation
Asymmetrical hearing levels	difference between L and R at 0.5, 1 or 2 kHz: ≥ 25 dB, or difference between L and R at 3, 4 or 6 kHz: ≥ 35 dB
Symptoms or signs	discharge from ear, foreign body in canal

Notes: (1) occlusion of canal by wax can usually be remedied by a suitably trained operator; (2) symptoms of pain, tinnitus or dizziness usually result in self-referral.

Abnormal hearing levels at frequencies 2 kHz and below are due to noise exposure only in the most advanced cases of noise-induced deafness, when a considerable hearing loss will already exist at the higher frequencies. Low frequencies are of limited value, therefore, for the purpose of detecting noise-induced hearing changes and monitoring the effectiveness of a hearing conservation programme. But abnormal low frequency thresholds are often due to conditions amenable to medical treatment, so the likelihood of benefit from a medical referral is reasonably high.

Hearing losses at frequencies of 3 kHz and above can be caused by a large number of factors, e.g. noise, some virus infections, certain forms of medication, blows to the head, hereditary disorders, and vascular disease. Because the damage from these causes is usually irreparable, and because most of these conditions, apart from early noise-induced deafness, also cause changes at low frequencies, it is probably not necessary to refer for medical treatment those persons who only have high frequency loss. There may, however, be a need to refer them for a hearing aid or other rehabilitation.

Rehabilitation and/or hearing aids should be considered for persons with irreversible hearing loss of such severity that they perceive their hearing loss as being serious and are keen to receive help. In general, those whose total hearing loss in each ear, at frequencies of 0.5, 1 and 2 kHz, equals or exceeds 75 dB should be referred for assessment.

Unreliable audiograms. The occasional person is unable to respond to the audiometric signals in a consistent and reliable manner, but difficulties are usually overcome by careful re-instruction by the operator and commencing the test with above hearing threshold. If re-instruction does not work, and no symptoms or signs likely to cause the difficulty can be found, then it should be assumed that the person is unable or unwilling to participate in the audiometric programme and no further action need be taken.

Interpretation of monitoring audiograms

Criteria are needed for taking action when the comparison between a monitoring audiogram and the reference audiogram indicates that a significant change in hearing threshold has occurred. While deteriorating group hearing levels can warn that a hearing conservation programme is failing, the arrest of the deterioration requires the identification and treatment of the affected individuals. The individual may be inadequately protected, or unusually susceptible to noise and in need of special protection or quieter work. Although the early stage of noise-induced hearing loss is usually manifested as a deterioration of threshold levels in the frequency range 3 to 6 kHz, other frequencies should also be included in any criteria. Changes at 2 kHz or below may exceed those at higher frequencies in cases of advanced noise-induced hearing loss, and changes in the lower frequencies may also be indicative of other pathology. The following action programme (as recommended by the National Acoustic Laboratories, Australia) should be initiated whenever there is a difference of +15 dB or more, at any test frequency from 0.5 to 6 kHz, between the monitoring and reference thresholds:

(1) Ensure that no transient condition exists that could account for the difference, e.g. a cold, earache, recent noise exposure. If such a condition exists, defer action and retest the person as soon as possible after remission of the symptom(s).

(2) If no such condition exists, remove and replace the earphones, repeat the audiogram and, at each frequency, average the results with those of the audiogram already obtained. If, at each frequency, the average hearing level does not differ by 15 dB or more from the corresponding hearing level in the reference audiogram take no further action.

(3) If at any frequency the average of the two audiograms differs by 15 dB or more from the corresponding hearing level of the reference audiogram, then:

(a) advise the person of the test results;

(b) carefully check the person's noise exposure history since the last audiogram was taken;

(c) check the suitability and condition of the person's protective equipment (which should be brought along to every hearing test), the techniques of fitting and the frequency of use;

(d) take any corrective action that appears necessary in the light of the foregoing checks.

(4) If any of the medical referral criteria are exceeded (see above), or if any change in hearing level since the most recent audiogram exceeds 20 dB (and is of known cause), suggest that the person seek medical advice.

(5) Retest in 6 months.

Records

All audiometry records should be carefully filed and retained throughout the person's employment and for at least seven years after employment has terminated.

DOUGLAS, D. B.

Audiometry in industry. Health and Safety Executive Discussion Document. (London, HM Stationery Office, 1978), 18 p.

Code of practice for reducing the exposure of employed persons to noise. Department of Employment (London, HM Stationery Office, 1972), 33 p. Illus.

Code of practice for hearing conservation. AS 1269-1979 (Sydney, Standards Association of Australia, 1979).

Criteria for assessing hearing conservation audiograms. N.A.L. Report No. 80, National Acoustic Laboratories, Commonwealth Department of Health (Canberra, Australian Government Publishing Service, 1980).

Audiometers. International Electrotechnical Commission. IEC Publication 645–First edition (Geneva, Bureau Central de la Commission Electrotechnique Internationale, 1979), 41 p.

Evaluation of the hearing loss: Notes of industrial audiology (Lineamenti e note di audiologia industriale). Rossi, G. (Turin, Minerva Medica, 1979), 133 p. Illus. 136 ref. (In Italian)

State of the art of the audiometric screening in industry (Stato attuale dello screening audiometrico nell'industria). XVII National Congress of the Italian Society of Audiology. Rossi, G. (Rome, Edizioni Luigi Pozzi, 1981), 93 p. Illus. 468 ref. (In Italian)

Audiovisual aids

Audiovisual (A/V) is a term used to describe instructional materials and equipment designed to facilitate teaching and learning by making use of both hearing and sight.

Materials commonly regarded as audiovisual include actual equipment and models, flip charts and posters, slides and other projected transparencies, recordings, filmstrips, videotapes and films. Facilities might include chalkbaords, flannel boards, hook and loop boards, bulletin boards, magnetic boards, and display cases. A/V equipment includes simulators, projectors of all kinds, tape recorders for sound and video, cameras, monitors, and the full range of television studio equipment, and cable television hardware. Activities such as demonstrations and experiments are also usually considered part of audiovisual programmes.

The following communications media can be ranked by the degree of sensory experience that each of them creates, as follows (starting from the concrete and going to the abstract):

Actual direct experience
Simulations: "hands on" devices
Demonstrations: "show and tell"
Field trips: familiarisation
Exhibits: displays
Live television
Motion pictures
Still pictures: photos and slides
Auditory aids
Graphics: charts and graphs
Words: spoken and written

A/V aids assist the communications/learning process by presenting information concisely and impressively. They involve people more than words. Mere words and sentences, whether spoken or written, not only contain relatively limited information, but can be misconstrued by learners. The more concrete the communication, the more effective it will be. Designing a communication or training programme requires more than just knowing the subject-matter. It demands an ability to use audiovisual aids effectively, and to know when each can be most productive. With properly defined objectives their uses are endless. Only imagination–and budget–define the limits.

Safety professionals are using audiovisual aids as tools in presenting information for:

(a) new employee indoctrination;

(b) training supervisors for their role in accident prevention;

(c) specific safety procedures;

(d) basic fire prevention techniques;

(e) job safety analysis programmes;

(f) fire brigade training.

Safety professionals should be aware of the range of audiovisual resources and their appropriateness both for transmitting the message content and for audience appeal.

Often a combination of aids is necessary, and for this reason many safety departments have a whole range of audiovisual aids available. Some of the most popular of these are reviewed below.

Slides. Slide presentations have become increasingly popular with the addition of sound (figures 1 and 2). Programmes can be recorded on cassettes and synchronised with slides by a device that automatically advances the slides in step with narration. Learner response is encouraged when cassette recorders are equipped with a "stop" button so that the slide presentation questions can be posed to the class, the presentation stopped to allow time for class response (oral or written) and then started again to provide the correct answers and the necessary explanation or discussion.

FOR READABILITY

-PRINT-

USE LARGE LETTERS

don't scribble like this!

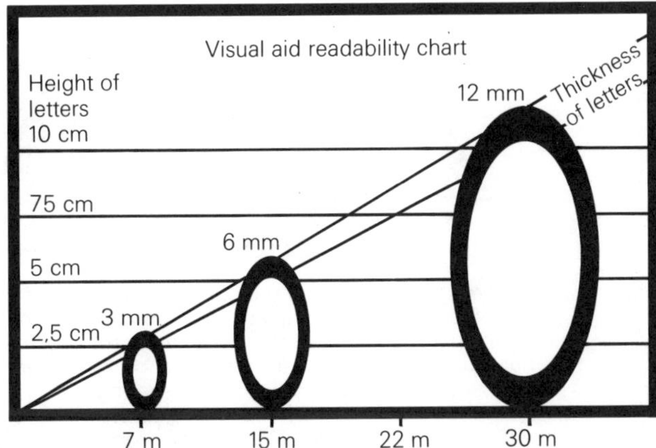

Spacing: ½ to 1 times letter height
Color: Black on white is BEST
Black on yellow is OK
Black on blue is WORSE

IF YOU CAN READ 2x2" SLIDES WITHOUT A MAGNIFIER, PEOPLE IN REAR SEATS CAN PROBABLY READ THEM ON THE SCREEN

Figure 1. *Above:* height and thickness of letters for easy viewing of projected or non-projected visual aids. *Below:* rough rule-of-thumb for the lettering on a 5 x 5 cm (2 x 2 in) slide intended for projection. *(By courtesy of the Eastman Kodak Company.)*

One international organisation has designed basic safety presentations available in half a dozen languages. Each presentation consists of a package comprising a tape cassette, slides, and a printed copy of the narration for the instructor who will conduct the safety training session. The specific advantages of slide presentations include the following:

(a) almost anyone can make a slide;

(b) slides can easily be updated;

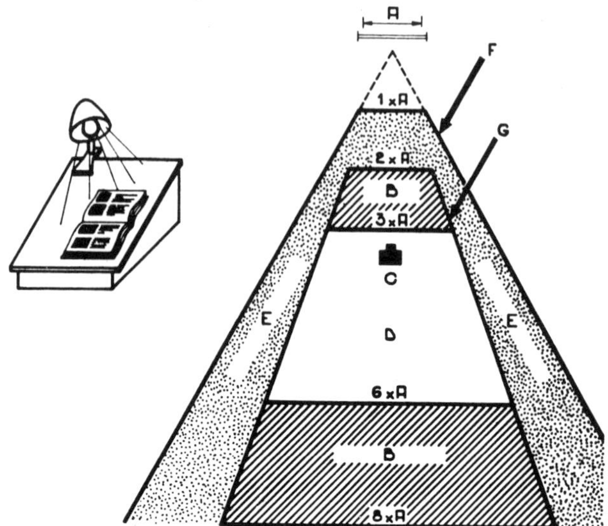

Figure 2. The best positions for viewing a projected 5 x 5 cm slide are located in the area within 20 ° of the projection axis for beaded screens and within 30 ° of the projection axis for mat screens. A. Maximum width of projected image. B. Fair seating. C. Approximate limit of 5 x 5 cm projector. D. Best seating. E. Poor seating. F. 30 ° angle from centre of screen. G. 20 ° angle from centre of screen. *Left:* the use of a lamp clipped to the lecturer's desk makes it possible to read the script when the room is in darkness.

(c) they are inexpensive;

(d) they can be geared to the desired audience;

(e) audio tape for automatic advance is available;

(f) it can be coupled to other projectors, both slide and film;

(g) it can easily be stopped for discussion.

The potential disadvantages are that:

(a) slides can easily get out of order;

(b) they can stick in the holding tray;

(c) they can easily be projected upside-down or backwards;

(d) bad slides are distracting;

(e) the simplicity of production can give a sense of "false security", resulting in poorly organised presentations.

Demonstrations. Exhibits and models make very effective three-dimensional displays for use in instruction. Such exhibits can be made for the purpose of demonstrating the safe working of a machine or process. Examples of first-aid equipment, protective clothing, rescue equipment, respiratory protective equipment and fire protection appliances can also be featured.

A small mannikin with articulated joints and spine is often used for showing the correct and safe method of lifting and carrying heavy loads. Demonstrations of fire fighting can sometimes be organised with the assistance of the local fire service. Demonstrations of good and bad lighting can easily be arranged in a lecture room. The effect of an impact on hard hats, safety shoes, and eye protection devices can be shown. The teaching of resuscitation by demonstration plays an important part in first-aid training.

Flip charts. Flip charts are a development of paper pads. Usually charts are on heavier paper, prepared in advance, and used in more formal meetings; frequently, blank pages are provided for "on-the-spot" additions. Flip charts might combine specially made material mounted on large sheets hinged in briefcase-sized easel-binders that are used for desk-top or bench-top discussion. Portable units are easy to make and to use for public or off-the-job safety.

Overhead projection. There are many ways to present visual materials, but overhead projection offers special advantages (figure 3):

(a) you can present to a group of any size in a fully lighted room;

(b) you face your audience at all times;

(c) you can reveal material point by point;

(d) the audience is not distracted by the machine;

(e) overhead projection equipment is easy to operate;

(f) overhead transparancies can be made quickly and inexpensively.

Lettering is a key to the effectiveness of transparencies for overhead projection. Careless lettering can detract from even the best illustration while neat, well planned lettering can be effective by itself. Letters should be at least ¼ in or 7.5 mm high (if the original can be read from a distance of 10 ft or 3 m, the transparency should project well). The letters can be applied to transparencies by hand, stencil, tracing, transfer letters and symbols, and lettering tape made from an imprinting machine.

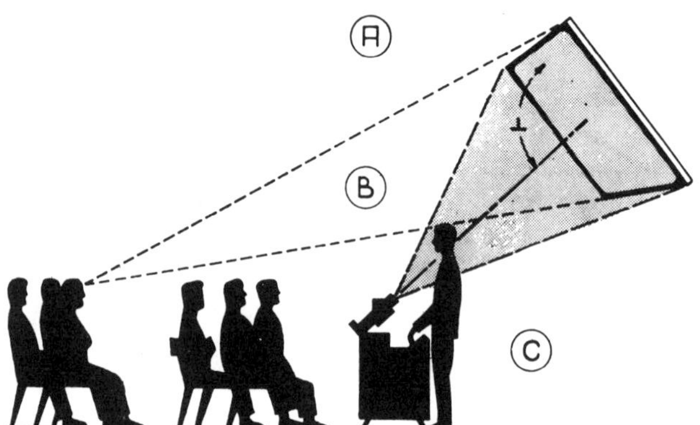

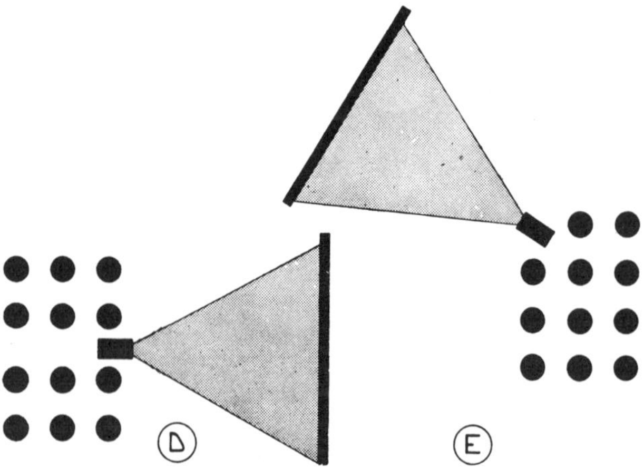

Figure 3. The use of a tilted screen for overhead projectors reduces the keystoning effect and improves visibility. A. Screen high and tilted. B. Cone of unobstructed vision. C. Projector low. D. Aisle directly in front of projector. E. Oblique viewing. *(By courtesy of Technifax Corporation.)*

It may be helpful to use colour for clarity and emphasis of certain points. Colour attracts attention, is pleasing to the eye and provides variety.

Where possible, each original should be limited to one point or comparison. Paragraphs should be broken into sentences, and sentences into phrases and key words. Use a maximum of six or seven lines, and six or seven words per line.

Reproduction equipment is available for making transparencies from almost any original material.

To add impact to pictorial transparencies overlays can be used. This method helps simplify difficult concepts and also allows the presenter to build up the story in a meaningful way. It involves the use of two or more transparencies in sequence, one over the other. Using no more than two overlays and different colours for each makes this a very effective way to present step-by-step information. Overlays are hinged to the frame on one side with tape to allow the base transparencey to be presented first, then each overlay is flapped over it to complete the message.

With an overhead projector transparencies can be projected unframed, but there are good reasons for using frames. A frame blocks light around the edge of the transparency, adds rigidity for handling and storage, and provides a convenient border for writing notes.

Chalkboards. A chalkboard is a good, straightforward, simple device that can be used very effectively in many situations. All we have to do is use its strengths and be aware of its limitations. As for its strengths–

(a) it permits easy changing of material;

(b) it is widely available;

(c) it encourages a spontaneous flow of communication between speaker and audience;

(d) coloured or fluorescent chalk can give very eye-catching effects;

(e) chalk talks are easy to tailor;

(f) mistakes are easily corrected.

The limitations of chalkboards include the following:

(a) there is loss of face-to-face contact with the audience;

(b) dust from chalk and erasing can be annoying;

(c) good lighting is necessary and ordinary white chalk may not show up well on dusty boards.

Posters are a useful part of a successful safety programme. They serve as reminders, warnings, and motivators. They deal with many accident problems, both occupational and non-occupational; and they employ many types of art styles and formats.

Posters should be changed frequently. A week is about the maximum time a poster should be displayed. Many organisations change posters every two or three days. The location of posters should be carefully considered. Put your posters up–either in poster frames, or on bulletin boards–at strategic places, i.e. where people pass, where they tarry, where they stop and read: near vending machines, smoking stations, lounges or cafeterias, tool rooms, conference or meeting rooms, near telephone booths, near water fountains, near time clocks, outside supervisors' offices. Posters should be displayed one at a time. There are exceptions, depending on the location and the particular display. But as a rule posters should not be massed. The attention factor is higher, readability is better if the number of posters in one display is kept to a minimum.

Posters should be well lighted. Flashing lights and coloured lights are effective. Frame your posters. A frame does a lot for a poster–as it does for a picture. Don't use posters that are dog-eared or dirty.

Also use your poster locations for news bulletins, press clippings of accidents, home-made posters and signs. Apart from the value that such communications have in and of themselves, they also draw people to your poster locations, condition them to check these spots for interesting, timely reading matter. For certain locations and situations jumbo safety posters and safety banners are effective.

Films. Movies are excellent for training and motivating. Because of their higher cost compared with that of other visual aids, they should be planned with special care. Usually movies are important enough to justify commercial production or at least professional advice before and during production. Some homemade movies, however, have proved to be effective.

The principal sizes of motion picture film are 16 mm, and 8 mm (both standard-8 and super-8 sizes). The 16 mm size is widely used by industrial organisations, and the 8 mm size is gaining attention because of its lower cost. The development of magnetic and optical sound, instant movie film, cassette loop projectors, improved colour stock, rear-projection screen, self-threading cartridges, single-frame viewing, and super-8 equipment has resulted in greater use of motion pictures for in-house training. Standard-8 and super-8 projectors and cameras are smaller and lighter than 16 mm equipment–an advantage where portability or size is a factor. These projectors can be used for continuous showing of safety films in such locations as cafeterias and lounges, with either a regular screen or small self-contained rear-projection unit.

A wide selection and variety of 16 mm films are available from insurance companies, local safety organisations, commercial film libraries and industrial producers.

The disadvantages of films include the fact that they:

(a) are expensive (even 8 mm, if done well);

(b) cannot be produced by the casual amateur;

(c) tend to become dated quickly (maximum life 8-10 years, usually, normally 5-7);

(d) are slow and costly to revise;

(e) are expensive to copy.

Video tape and closed circuit television (CCTV). The word video may be used to describe any type of television equipment such as cameras, tape recorders, tape players, video recording tape, monitors and a full range of television studio equipment and cable television hardware.

The constraints on top management severely restrict opportunities for personal communication with employees. Video tape and CCTV have proved to be an effective means of transmitting safety concepts when face-to-face meetings are next to impossible. Other major forces behind the accelerated growth of organisational television include: (1) improvements in video technology; (2) decreased cost of telecommunications, (3) increasing costs of travel; (4) increasing need for information.

Specific advantages of television include the following:

(a) it requires no special lighting to produce tapes;

(b) black and white television is relatively inexpensive;

(c) new, lower-cost colour tapes are becoming available;

(d) it can be stopped at any point for audience discussion;

(e) it can be replayed immediately;

(f) editing can be done on the spot;

(g) cassettes for reproduction are relatively inexpensive;

(h) good quality transfer of tape to film is now possible.

Its disadvantages are that:

(a) the initial investment required is higher than for movie cameras and projectors;

(b) editing is tedious;

(c) various types of equipment and tape are non-compatible;

(d) operator training is necessary.

Multi-image projection. Multi-image projection is the technique of using more than one projector simultaneously. Actually it is an expansion of the basic sound/slide programme technique that has gained tremendous popularity. A programmer allows you to turn slide and movie projectors on and off automatically, advance slides, and change slide projector lamp currents from off to full-on at nearly any rate of change. These functions may be performed in any sequence desired.

The advantages are that:

(a) visual information can be presented both sequentially and spatially, enabling the viewer to see not only the order but also the relationship of information;

(b) the technique compresses the time needed to create an impression;

(c) it tends to develop greater viewer involvement and even excitement.

The potential disadvantages are that:

(a) operator training is necessary;

(b) the technique requires sizeable expenditure;

(c) it takes time to develop.

In summary, it should be remembered that each audiovisual system has its own features that represent a possible advantage or disadvantage when compared with another type of audiovisual system. The decision-maker must weigh needs and objectives against all the other factors including budget and audiovisual capabilities.

ETTER, I. B.

Communications for the safety professional. Konikow, R. B.; McElroy, F. E. (Chicago, National Safety Council, 1975), 518 p. Illus. 19 ref.

Poster directory No246 (Chicago, National Safety Council, 1979), 80 p. Illus.

"Nonprojected visual aids". Data Sheet 1-564-68, National Safety Council, *National Safety News* (Chicago), Jan. 1980, (65-73). Illus. 6 ref.

Projected, still pictures (slides, strips and opaque or overhead projections). Data Sheet No. 574, Revision A (extensive). (Chicago, National Safety Council, 1970), 12 p. Illus. 2 ref.

Method for calculation and preparation of projected-image size and projection distance tables for audio-visual projectors (American National Standards Institute Inc., 1430 Broadway, New York), (1975), 12 p. Illus.

The role of films in accident prevention teaching and training methods (Rôle du film dans les méthodes d'enseignement et de formation à la politique de prévention des accidents du travail) (Brussels, Commissariat général à la promotion du travail, 1976), 69 p. Illus. (In French)

Automation

The word "automation" was coined by Harder of the Ford Motor Company to describe machinery being developed by Ford to move engine blocks automatically into and out of transfer machines in the automobile industry. The word first appeared in print during 1948 in a description of the work of Ford's newly established "automation group" and was defined as "the art of applying mechanical devices to manipulate workpieces into and out of equipment, turn parts between operations, remove scrap, and perform these tasks in timed sequence with the production equipment so that the line can be put wholly or partially under pushbutton control at strategic stations". This concept became known as "Detroit automation" and was carried a stage further by Diebold when he described automation as "the use of machines to run machines".

These definitions deal only with the engineering and productivity aspect of automation; the human element represented by the production worker is completely excluded. To remedy this omission, it is now common to talk of a production entity as a "system" which includes all the mechanical devices, equipment and human operations necessary to process materials or information into the desired product or service. The prime consideration in automation is optimisation of production by man and/or machines, and the factor which determines the degree of automation is the quantitative and qualitative relationship between the human effort on the input side and the objective results on the output side of the process and the extent to which this relationship is influenced by the environment and the comparative capabilities and limitations of men and machines.

Figure 1 contains a block diagram showing this relationship in a simplified form. Automation should here be understood as the degree to which human work is penetrated or replaced by the use of machines (whole-system mechanisation) measured at the output end of the system. However, in evaluating the system, it is necessary to measure not only the output but the stresses and demands on the workers involved.

Automation, the broken-line arrow in figure 1, here indicates the replacement of human work by mechanical equipment with the result that man is no longer compelled to participate continuously and at a preset rhythm in the production process. Where man is replaced as a

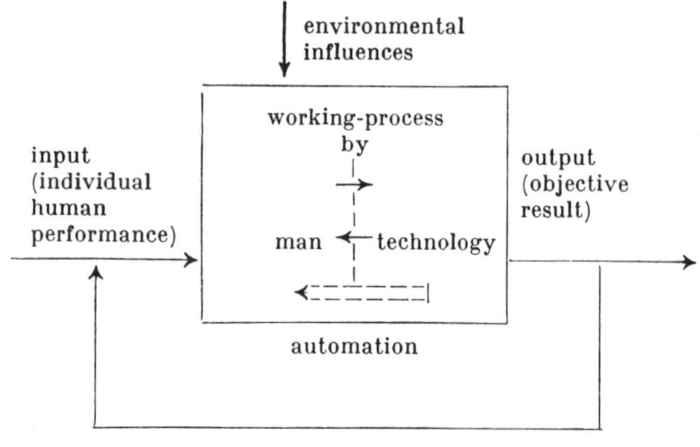

Figure 1. The man-work system.

source of power or as a means of supplying information for power modulation, the process is one of mechanisation. Automation implies a further stage in which equipment takes over the function of automatic data processing. Thus it can be seen that mechanisation forms part of automation.

There are four stages in the progression from complete manual labour to full automation; in these stages, mechanical aids are used respectively to supplement, amplify, alleviate or replace man's own efforts.

(a) The supplementary function is applied to improve or extend human ability (e.g. by the use of tools, gauges, microscopes, etc.).

(b) The amplification function is employed to overcome the limitations of human strength and endurance (the use of natural forces or fuel to drive machines, mechanisation). Electricity consumption statistics can be used to illustrate this aspect. In the industrialised countries electricity consumption per industrial worker is 13 500 kWh per annum; if it is assumed that normal human power output is 180 kWh per annum, it can be seen that each worker is backed up by 75 "robots" which amplify his power potential.

(c) The alleviation function is employed primarily in the control of complicated processes (e.g. the initial stage of true automation by automatic measurement, control, regulation, data processing).

(d) The replacement function comes into operation in the final stage when an automated process completely eliminates the need for man's participation in a given operation.

It can already be seen that automation can be viewed from numerous points of view and consequently when analysing man-work relationships, the engineer and physician will need to employ, in addition to their own specialised knowledge, disciplines such as physiology, anatomy, applied psychology and sociology.

Genesis and growth of automation

The original stimulus to mechanise and automate is probably to be found in the desire to reduce the human content of work because human labour was irksome and arduous or endangered the safety and health of the worker (safety and health incentive for automation) or because human labour was expensive (economic incentive). However, two further stimuli then came into play: automation was seen as a way of overcoming certain limitations in human capabilities (technical incentive for automation) or of compensating for shortages of labour in general or skilled workers in particular (labour market incentive).

The basis for the technical incentive is to be found in the strictly limited capacity of the human body. For example, although the human may achieve a brief peak output of 6 hp, the normal continuous rating is only about 0.1 hp; in addition, man and his performance are highly susceptible to environmental influences. However, automation is not a synonym for technical revolution in which man is to be completely replaced by the machine in industry; it is more a further step in the continuous process of technical development. This technical development is well illustrated by figure 2 which shows the levels of power that man has been able to harness and control for his service at different points in history. At the earliest stages the only power available was that provided by human and domestic animal muscles, wind, water and fire. The entry into the second stage was marked by Watt's discovery of the steam

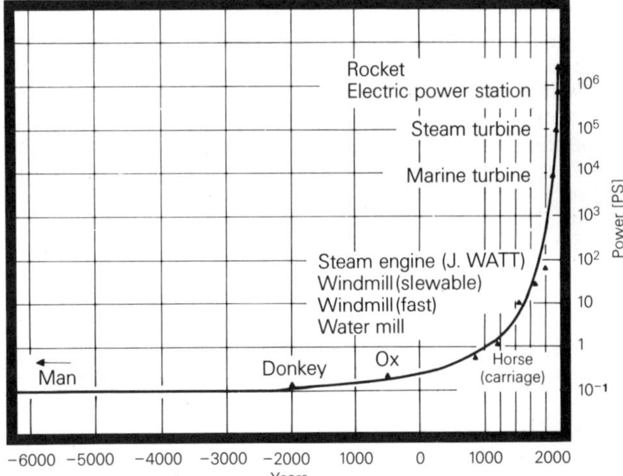

Figure 2. The increasing levels of power at man's command during the course of history.

engine in 1770, which provided a mobile and constantly available source of mechanical power; the mobility factor was furthered by the invention of the internal combustion engine. The third stage starts with the invention of the electrical generator and the development of an electricity distribution system.

However, it is not only the limited power rating of the human body that has stimulated automation on technical grounds: every single function has definite limitations. Figure 3 illustrates the limitations in the speed and accuracy of manual movements and shows that speed and accuracy are in fact reciprocal functions. At any stipulated speed of operation, there will be a corresponding degree of accuracy that can be achieved; and as the speed increases, this degree of accuracy will be reduced. Consequently, when man cannot meet the combined speed and accuracy requirements for a certain operation, then there is a technical incentive to automate. The curve in figure 3 indicates clearly the area in which human accuracy and speed combinations are adequate and the area in which it will be essential to mechanise and automate on technical grounds; in this area, technical requirements will also make man redundant.

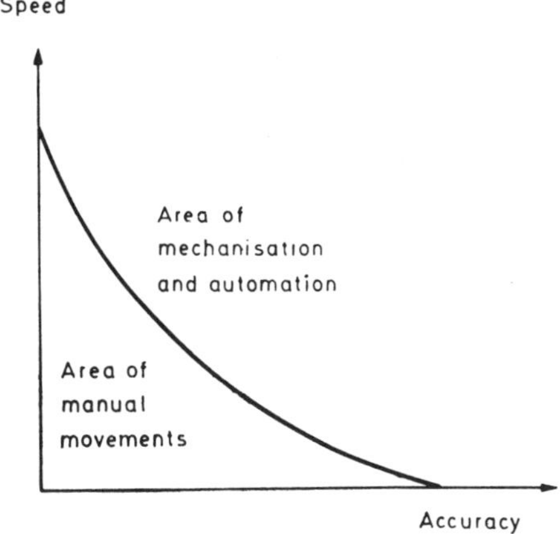

Figure 3. This diagrammatic presentation of the limitations of the speed and accuracy of manual movements indicates the range in which manual operation is adequate and the area in which man may be replaced by mechanisation or automation for technical reasons.

Conditions needed for automation

Four basic requirements have to be met before a process can be automated. Firstly, automation must be technically feasible, i.e. an adequate supply of energy must be available in a suitable form and sufficiently sensitive and reliable mechanisms must be available for collecting and processing information and for regulating the plant operation accordingly. Secondly, automation must be a profitable proposition. Thirdly, the organisational structure of the enterprise (scientific management) must be capable of dealing with the changes resulting from automation, e.g. reduced flexibility in production, increased output, etc. Finally, it must be possible to obtain sufficiently skilled personnel to operate and maintain the automated plant.

Automation and the worker

The relationship between automation and the working man may be viewed from three points of view:

(a) the influence of the operator on the efficiency of the man-machine system;

(b) the stresses imposed on the worker involved in an automated system; and

(c) the various capabilities required of the human part of the man-machine system at different levels of mechanisation and automation.

A prerequisite for a study of these three facets is a definition of the various degrees of mechanisation or automation at which the machine takes over different capacity or skill. The result of this search for a definition can be shown diagrammatically in a graph in which the level of human ability is given along the ordinate axis and the degree of automation along the abscissa axis; the result is a very uneven curve with pronounced jumps. The two most extreme but scientifically still acceptable curves have a U-shape; however, in one case, the U is upright and in the other it is inverted which means that in the one case automation requires a low level of human skill and in the other a high level.

One conclusion to be drawn from such an exercise is that the demands automation makes of the worker and the stresses to which it subjects him are not solely dependent on the level of technical complexity; a variety of other factors enter into play including work organisation, ergonomic layout of the workpoint and the safety and health hazards to which the worker is exposed. However, the analyses carried out so far have not led to a clear demarcation in the effects of these factors and, consequently, opinions on this matter are still somewhat contradictory. To rectify this situation, the scientific study of the effects of automation on the worker is being concentrated on an analysis of three factors: degree of mechanisation; work organisation; and process design.

Man's function in automated processes. The results have already indicated a number of trends. Firstly it has been found that the influence of the worker on the quantitative and qualitative output of the productive system is gradually being reduced and that his efforts are increasingly devoted to supervising and monitoring the process machinery and to the elimination of faults; nevertheless, these functions can also be automated and the worker can then also be relieved of these tasks. This is in line with the general trend in which the scope of human employment in industry is decreasing; however, since the requirement for human intervention in the event of breakdowns remains significant, the paradoxical situation may develop in which insufficient demand may result in excess demand.

There has been a qualitative change in the contribution that the worker is expected to make to the work process; this contribution has progressed from fundamental energy output, through precision co-ordination, to decision making; moreover, in many cases, reduction in frequency of intervention in the system is accompanied by an increase in the complexity of any intervention that has to be made.

The experience accumulated by the worker tends to be more and more based on contact with large-scale production plant; this experience grows only slowly since opportunities for action diminish with increased system reliability. In addition, where technology advances rapidly, acquired experience rapidly grows out of date. This aspect raises a question mark over the value of the aged in employment who, being less adaptable, are of less value to the employer as technological change accelerates.

The human contribution to the over-all system objective can thus be formulated as follows: the worker becomes less of a cost factor in production, yet his influence on system reliability is increased. Consequently, it is necessary to determine experimentally the relative advantages and disadvantages of men and machines for individual tasks (including auxiliary and control functions) and then to distribute work activities on the basis of these findings. Table 1 shows the results of such an experimental study on the relative merits of men and machines; however, an important factor that should be taken into account is the time-dependent nature of human abilities, especially in the event of prolonged inactivity.

By suitable work organisation, the various tasks can be delegated to different workers. The significance of teamwork increases since good co-operation between specialised workers is essential in the case of complicated plant highly susceptible to breakdowns, and the importance of the maintenance and servicing personnel grows.

Automation, occupational safety and health. Since mechanisation and automation entail heavy capital investment, greater recourse is had to shift work and the problems posed by shift work, particularly night work, require special consideration.

Conventional work study and job evaluation techniques employed for establishing piecework rates place considerable stress on physical capacity; consequently, the trend to the use of man for more intellectual functions inside the production system may easily lead to a formal depreciation of human work. Job evaluation procedures and pay schedules must be adapted to meet this development. As the extent of human intervention in the production process decreases, the influence of financial productivity incentives on system output falls correspondingly; therefore, if production incentives are to have any meaning at all, new criteria must be sought for calculating them.

Degree of responsibility is an important criterion in job evaluation. During the first stages of automation, the worker's responsibility increases due to the greater cost of the equipment he is operating; however, at the more advanced stages of automation, there is a fall-off in responsibility, since the worker no longer has any scope for really effective intervention in the process.

The growth of automation can lead to a steady improvement in environmental working conditions owing primarily to the spatial separation of the worker from the process plant. The result is the elimination of health hazards and the prevention of occupational diseases; however, there will also be a reduction in accident hazards. A study of the causes of fatal accidents

Table 1. Relative advantages and disadvantages of men and machines

Item	Machine	Man
Speed	Much superior	Time lag of 1 s
Power	Consistent at any level–large, constant, standard forces	2 hp for about 10 s 0.5 hp for a few mins 0.2 hp for continuous work over a day
Consistency	Ideal for routine, repetition, precision tasks	Not reliable–should be monitored by machine
Complex activities	Multi-channel	Single-channel
Memory	Best for literal reproduction and short-term storage	Large store, multiple access–better for principles and strategies
Reasoning	Good deductive	Good inductive
Computation	Fast and accurate but poor at error correction	Slow and subject to error but good at error correction
Input sensitivity	Certain extra-human senses, e.g. ability to perceive ionising radiation	Accepts stimuli over a wide energy range and possesses receptors that can deal with wide variety of stimuli, e.g. eye deals with relative location, movement and colour. Good at pattern detection. Can detect signals in high noise levels.
	Can be designed to be insensitive to extraneous stimuli	Affected by heat, cold, noise and vibration (exceeding known limits)
Overload reliability	Sudden breakdown	Gradual degradation
Intelligence	None	Can deal with the unpredicted and unpredictable–can anticipate
Manipulative abilities	Specific	Great versatility

in a railway shunting yard has shown that the introduction of automatic coupling considerably reduced the accident frequency rate since couplers were no longer required to step between the railway trucks; much the same findings apply to the operation of placing drag shoes and to the crossing of railway lines in general.

Bodily wear and tear in workers caused by strenuous muscular work will also be reduced by automation; the application of ergonomics at the design stage of automated processes will reduce the incidence of strained working postures and heavy static muscular loads. Similarly, the ergonomic problems of monitoring and supervision work can be solved by correct layout of work posts and seating arrangements, the use of well designed indicators and control panels and adequate worker selection and training.

Automation will also help in preventing the spread of infections and diseases at work since the density of workplace occupancy will be drastically reduced; this will be an important factor during influenza epidemics.

Human capacities employed in automated processes. Automation is an organic stage in development. It does, however, make greater intellectual demands, and increasing importance will have to be placed on worker selection. The features that it will be necessary to look for in candidates include personality, general intelligence and the ability to make correct decisions rapidly; the "old" virtues such as reliability will once more gain prominence. Regular reappraisal of job scope will be essential. Automation will thus entail a restructuring of leadership tasks: bureaucratic and administrative attitudes will have to make way for a more incisive and critical approach in which decision making is backed up by the availability of a wider range of data than was previously the case.

The employer must make provision for workers to improve their qualifications by further education. This is vital in view of the fact that since technology is advancing so rapidly no course of education can be considered complete; this problem may even make it necessary to modify the structure of basic education systems.

It is obviously difficult to give quantitative expression to such qualitative factors as professional position, ability and knowledge, and skill requirements. Studies have been made on the reclassification of industrial activities in relation to stages of technological advancement. The industrially employed population of a given country can be depicted as a sort of pyramid in which the base is formed of unskilled workers whilst industrial management constitutes the apex. The shape of this pyramid can be modified in two different directions by the advent of automation. Firstly, the apex of the pyramid can be widened whilst the base becomes narrower. For example the proportion of unskilled workers in the United States has fallen from 25.4% in 1900 to 14.2% in 1977. Since the end of the Second World War the demand for skilled workers has risen significantly, as has that for commercial and technical staff, plant managers and the top executives that constitute the pyramid apex. The proportion of craftsmen, foremen and kindred workers–being the most important part of the total amount of skilled workers–increased from 35.6% in 1960 to 39.6% in 1977.

Secondly, the general aspect of the pyramid may be elevated since, with the spread of automation, an enlargement of knowledge and skills will become necessary at all levels in the professional hierarchy.

Application of automation to industry

It has been suggested that industrial production can be defined as each activity which has the purpose of increasing the profit or the value of a product or of a service. However, automation has not reached the same level of development in all forms of industrial production.

The most outstanding progress has been made in the industries involved in the generation, transformation, and distribution of power. In the processing industries which are engaged mainly in producing and handling amorphous materials by flow processes, the need for automation was realised at an early date since, especially where round-the-clock working was carried out, working conditions were poor and the accident and disease rates high. The materials employed by these industries may be gaseous, liquid or solid, although the solids are usually in particulate or granular form. Consequently, materials handling and processing lend themselves readily to automation; however, in indvidual cases the stage of automation reached will depend on the quantity of the materials handled and whether processing can be made continuous or is still carried out by batch techniques. Thus automation is highly developed in the petroleum industry, the plastics production industry and the iron and steel industry.

It is in the manufacturing industries (concerned mainly with the production of piece goods and the processing of products with a well defined geometric shape) that automation presents the greatest problems and has therefore made the least progress. The finished products vary considerably in material, size, shape and tolerances and consequently a multitude of different operations have to be carried out on specific machines. The manufacturing industries are still labour intensive because the products cannot be conveyed from one process to another in a standardised manner. Therefore, individual and accurate positioning is required prior to each new operation and the automation of this type of workpiece positioning and handling presents considerable problems.

Since automation is such a complex phenomenon it is hardly justifiable to talk of its effects on the worker or man in general; it is more a question of considering the widely varied reactions that different individuals will demonstrate when they are confronted with automation.

Where there is a change-over to automation, guarantees must be given on the new working conditions and the worker must be reassured as to the demands that will be made on him and as to the demands he himself makes.

Society must be expected to create the conditions necessary for optimal use of work and effort by providing and guaranteeing employment, by measures that will help the individual in mastering his work and by assistance in all situations in which the individual wishes improvement in his working capacity or way of life (see also ROBOTS AND AUTOMATIC PRODUCTION MACHINERY).

ROHMERT, W.

Technique, work organisation and work. An empirical research into automatised production (Technik, Arbeitsorganisation und Arbeit. Eine empyrische Untersuchung in der automatisierten Produktion). Studienreihe des Soziologischen Forschungsinstituts SOFI Göttingen (Frankfurt on Main, Aspekte-Verlag, 1976). (In German)

Alienation and freedom. The factory worker and his industry. Blauner, R. (Chicago and London, University of Chicago Press, 1964), 222 p. Illus.

Collective bargaining and the challenge of new technology (Geneva, International Labour Office, 1974), 71 p.

Automation in developing countries (Geneva, International Labour Office, 1974), 246 p.

"What industrial robots can do" (Ce que savent faire les robots industriels). Defaux, M. *L'usine nouvelle* (Paris), 12 June 1980, 24 (96-102). Illus.

Automobile industry

The automobile industry is engaged in the manufacture of motorcycles, automobiles and industrial vehicles (lorries, buses, trams, trolleybuses, agricultural and industrial tractors). It forms part of the engineering industry.

Production processes. The complete cycle of production includes:

(*a*) casting (steel works, foundries) for the melting of steel, cast iron and non-ferrous metals (aluminium, copper, tin, zinc, lead) and subsequent treatment of castings and semi-finished materials (sheet, strip, wire, nuts, bolts, etc.). Supplementary products are prepared in special departments; lubricating and cutting oils, paints, rubber and plastic components;

(*b*) forging and machining, including drop forging, finishing and heat treatment, followed by machining of cylinder blocks, crankshafts, etc., up to the production of sub-assemblies and mechanical components and to engine assembly and testing. At the same time other departments are engaged in the fabrication of the bodywork by pressing sheet metal, preparing sub-assemblies up to the painting of various parts and of the body as a whole. Special sections deal with electroplating of accessories, the production of electrical components, brakes, injection pumps and upholstery (seats, interior panels, dashboards, etc.). The larger manufacturers are fully equipped to produce press dies, mould boxes, patterns, tools and other equipment;

(*c*) final assembly of the mechanical parts and the upholstery on to the body, finishing, delivery testing and subsequent distribution of the finished product complete the production cycle.

All these processes are accompanied by corresponding activities of supplying and distributing materials and power (electricity, petroleum and its derivatives, illuminating gas, compressed air), together with those involving the construction and maintenance of plant and equipment.

The scale and variety of the production cycle determine the working conditions, the materials used, the products manufactured and the arduousness, type and rhythm of work.

HAZARDS AND THEIR PREVENTION

The processes involved in automobile manufacture are so numerous and varied that it is necessary to deal with the hazards of individual processes and the relevant safety measures, according to their sequence in the production cycle.

Steel works, foundries. In iron and steel works and foundries, the basic problems of occupational hygiene are those of climatic conditions, noise and air pollution due to dust and gas. The climatic conditions are determined by two basic factors, the large size of the workplace, the need for continual replacement of considerable quantities of air in order to exhaust fumes and dust, and the presence of heat sources, especially of the radiant type such as furnaces, molten metal (about 900-1 200 °C), hot forging (ingots, billets, sheet, wire and castings–about 500-800 °C). The radiant temperature levels are highest in the vicinity of steel furnaces (up to 42-45 °C ambient temperature and 80-90 °C on the globe thermometer; relative humidity about 40%); however, the work cycle involves alternate periods of intense radiant heat exposure with long intervals of exposure to normal ambient conditions. In corresponding sections of the foundry, the radiant heat level is lower due to the smaller quantity of hot metal involved but the exposure is more continuous with considerable danger, during hot weather, of heat stroke and digestive system disorders due to excessive consumption of beverages. The subsequent processes of rolling, drawing and winding and of foundry knockout and fettling involve less intense thermal exposure.

Workers may be protected against heat by providing a flow of water over the walls and openings of furnaces, by applying insulating material or heat-reflective surfaces to heat sources (e.g. by using a polished sheet to reflect the heat away from the worker), by ventilation, by instituting suitable rest periods, by medical selection of workers and the use of acclimatisation procedures. Workers should also be provided with heat protective clothing and eye and face protection for cataract prevention. Air-conditioned rest rooms near the workshop are valuable for use during rest periods.

Atmospheric contaminants encountered in iron and steel works and foundries include:

(a) fumes and dust (iron oxide, carbonates, silicates and free silica in different crystalline forms, depending on the heat treatment processes);

(b) gases (carbon monoxide, carbon dioxide, sulphur dioxide);

(c) fumes (phosphorus, lead, fluorine, manganese, etc., depending on the composition of the molten metal).

Exhaust ventilation should be provided at source and the air drawn off should be filtered.

Pre-employment and periodic medical examinations are essential to detect early signs of pneumoconiosis, chronic bronchitis and emphysema, pulmonary tuberculosis, chronic carbon monoxide poisoning, etc.

High levels of noise and vibration are encountered in processes such as furnace loading, mechanical decoring, stripping and knockout of castings, and fettling with pneumatic tools; in forging shops, in particular, intermittent impulse noise at levels of 118-120 dB with a high upper frequency content (up to 8 000 Hz) is common. Noise control measures are difficult to apply and, consequently, personnel selection, regular audiometry and the use of hearing protection equipment (ear muffs, earplugs) are of prime importance. Nevertheless, such measures are not fully effective and occupational hearing loss is not uncommon.

The automobile industry has been subject to considerable mechanisation and automation. This has led to a reduction in muscular work and to a lack of direct contact with the process. The worker can now be located in an insulated and air-conditioned cabin to supervise operations by remote control and the incidence of accidents in mechanised foundries is significantly lower than that in traditional foundries.

Engine and transmission manufacture. In machine-tool shops, the introduction of automatic cycle or numerical control and transfer lines has, to a large extent, eliminated the common machining accidents and diseases (foreign bodies in the eyes, hand injuries, skin disorders due to cutting and grinding fluids). The machining of components such as camshafts, gear and differential pinions, brake drums, etc., which was previously carried out on individual machine tools is now done on transfer machines which may complete a cycle of up to 60 or 70 drilling, reaming, milling operations, etc., with automatic tolerance control. These machines, which may be over 200 m in length, are fitted with optical and acoustic fault warning systems and require only supervisory and maintenance personnel.

In case-hardening, tempering, nitrate salt baths and other metal heat treatment processes using furnaces and controlled atmospheres, the microclimate may be

Figure 1. Automation of engines manufacture by means of numerical control machines.

oppressive and various airborne toxic substances encountered, e.g. carbon monoxide, carbon dioxide, cyanides, etc. Atmospheric conditions may also be unpleasant where large quantities of cutting oil are used in machining axles, bushings, collars, etc., on broaching, milling, hobbing and planing machines. Machine attendants and workers handling swarf and centrifuging cutting oil prior to filtration and regeneration, are exposed to the risk of dermatitis. Where possible, machine tools should be fitted with effective splash guards and the fluids employed should be changed frequently; barrier creams may have only limited effectiveness under such conditions and the use of gloves, which may retain oil in prolonged contact with the skin, is definitely dangerous. Exposed workers should be provided with oil-resistant aprons and encouraged to wash thoroughly at the end of shifts (see GRINDING AND CUTTING FLUIDS).

Grinding and tool sharpening no longer present a pneumoconiosis hazard where silicon carbide and aluminium oxide grinding wheels are used, although there is still the danger of bursting wheels and flying particles and the possibility of fires where magnesium and aluminium alloys are being ground. Grinding wheels should be fitted with screens, and eye and face protection and respiratory protective equipment should be worn by grinders. The production of sub-assemblies and mechanical groups culminates in engine assembly on a belt conveyor, followed by engine testing and running-in; these operations are carried out in special premises fitted with equipment for removing exhaust gases (carbon monoxide, carbon dioxide, unburnt hydrocarbons, aldehydes, nitrogen oxides) and with noise control facilities (booths with sound-absorbent walls–insulated bedplates). Noise levels may be as high as 100-105 dB with peaks at 600-800 Hz (see ENGINE TESTING).

Body construction. Engine assembly is accompanied by bodywork construction. The main process is the pressing of steel sheet, strip and light sections on automatic presses ranging in capacity from 20-2 000 t.

The chief hazards in presswork are crushing and amputation injuries, especially of the hands, due to trapping in the press, and hand, foot and leg injuries, caused by scrap metal from the press. These can be prevented by the use of effective machinery guarding, safety controls, automatic feed and ejection systems, collection of press scrap and the use of personal protective equipment such as aprons, foot and leg protection and hand and arm protection. A recent development in press-tool production is the use of glass fibre reinforced plastics for pattern making, and cases of dermatitis and respiratory system disorders of mechanical and allergic origin have been encountered due to exposure to the various resins and the glass fibres.

Sheet steel pressings are assembled into body subgroups using suspended or transfer electric welding machines. The transfer machines, which operate faster and more accurately, make the work less arduous, lessen the danger of projections of hot metal and sparks and pollute the atmosphere less with the combustion products of mineral oil contaminating the sheet metal. Although suspended spot welding machines are fitted with a counterbalance system, they are still heavy and cumbersome to manipulate. Small components are welded, either by means of a gas torch or an inert-gas-shielded arc (argon or carbon dioxide). During this work, employees are exposed to intense visible and ultraviolet radiation and to the inhalation of combustion gases, vapours from pickling acids, metal fumes and gases from the electrode coatings, e.g. oxides, manganese, copper, zinc, iron, etc., silica, carbon monoxide, nitrogen oxides, carbon dioxide, ozone, etc. It is, therefore, necessary to provide local exhaust ventilation, protective screens and partitions, welding visors or goggles, gloves and aprons; welders should also be subject to periodic medical examinations (see WELDING AND THERMAL CUTTING). [Nowadays some welding operations, such as linear spot welding, are increasingly carried out by means of robots (see figure 2).]

Figure 2. Use of robots in body construction.

Some assembly techniques and body panel defect retouching processes entailed soldering with lead and tin alloys (also containing traces of antimony). Soldering and especially the grinding away of excess solder produced a severe risk of lead poisoning. It was difficult to provide adequate protection against this hazard and, consequently, the present trend is to improve the quality of pressings, thus reducing the need for retouching and at the same time, improve the medical supervision of exposed workers. Noise levels in these processes may range up to 95-98 dB, with peaks at 600-800 Hz.

Electroplating. Certain pressed steel components and castings, e.g. bumpers, mouldings, handles, etc., are electroplated with chrome, nickel, cadmium, copper, etc., and then buffed and polished. This work is normally carried out in separate workshops and involves exposure to, inhalation of or contact with vapours from the acid plating baths. These baths should be fitted with special lip ventilation, and anti-foaming surface tension agents should be added to the liquid; workers should wear eye and face protection, hand and arm protection and aprons, and should be subject to periodic health checks. The work of inserting and removing components from the vats is very hazardous and should be mechanised wherever possible. Manual buffing and polishing of plated components of felt belts or discs is strenuous since the worker is required to hold the component in his fingers pressed against the rotating felt; it also entails exposure to cotton, hemp and flax dust. This work can also be mechanised by transfer-type polishing machines (see GRINDING AND POLISHING).

Painting. Automobile bodies from the assembly line enter the paint shop where they are degreased often by the manual application of toluene, xylene, heptane, etc., phosphatised in a closed tunnel and undercoated. The undercoat is then rubbed down by hand with an oscillating tool using wet abrasive paper and the final layers of paint are applied and then stoved in an oven.

In painting shops, there is potential exposure to the inhalation of toluene, xylene, propylene, butyl and amyl acetate and methyl alcohol vapours; inhalation of paint pigments is usually of secondary importance though some paints may still contain salts of chromium and lead.

Spray painting is carried on in special booths which have water curtains on the walls and under the grilled floor and a continuously filtered air supply; the workers themselves wear suitable respiratory protective equipment. However, a relatively recent development is paint dipping for the primer/surface coat and electrostatic or electrophoretic processes for the finish coats; in such cases, personnel remain outside the booth and are required for supervision and maintenance purposes only.

The prohibition of benzene as a solvent, the elimination of lead pigments from paints and increasing mechanisation in modern plants have virtually abolished the risks of these processes in automobile manufacture; however, in the production of commercial vehicles (lorries, trams, trolleybuses), spray painting is still widely employed due to the large surfaces to be covered and the need for frequent retouching.

The painted bodywork is dried in hot air and infrared ovens fitted with exhaust ventilation and then moves on to join the mechanical components in the final assembly shop in which the body, engine and transmission are married together and the upholstery and internal trim are fitted.

Vehicle assembly. It is here that conveyor belt work is to be seen in its most highly developed version. Each worker carries out a repetitive and limited task on each vehicle; he scarcely moves from his work station whilst a conveyor system transports the bodies gradually along the assembly line. These processes demand constant vigilance and may be highly monotonous and act as stressors on certain subjects (see MECHANICAL ENGINEERING, WORK ORGANISATION IN; MONOTONOUS WORK; STRESS).

Although normally not requiring great energy expenditure, these processes are very demanding due to the postures or movements the worker is obliged to adopt, such as when installing components inside the vehicle or working under the body (with hands and forearms above head level). Nowadays the growing use of robots in body construction and vehicle assembly is substantially improving several work operations (figure 2).

After final assembly the vehicle is tested, finished and despatched. Inspection can be limited to roller tests on a roller bed (where ventilation of exhaust fumes is important) or can include track trials on different types of surface, water and dust tightness trials, and road trials outside the factory. Prior to despatch the vehicle may be sprayed with a protective coating of wax dissolved in gasoline.

Testing of prototypes. A very special operation of the automobile industry is the testing of automobile prototypes. Test drivers are exposed to a variety of physiological stresses such as violent acceleration and deceleration, jolting and vibration, carbon monoxide and exhaust fumes, noise, work spells of prolonged duration and different ambient and climatic conditions because of changes of latitude.

General hazards. Processes in the automobile industry are highly co-ordinated in a production cycle which necessitates: wide use of mechanical handling equipment such as timed belt conveyors, overhead conveyors, gantry cranes, inplant transport trucks (electric, diesel, gasoline, liquefied petroleum gas); the storage of large quantities of dangerous substances including toxic substances (cyanide salts, trichloroethylene, carbon tetrachloride), acids and other corrosive substances, flammable and explosive substances (liquefied petroleum gas, gasoline, paint solvents, mineral oils); services for the generation and distribution of electricity and the maintenance of the generators, distribution lines and electrical installations to various machines and for lighting, at voltages of from 12-500 000 V (electrical hazards are amongst the greatest hazards for the automobile industry workers); compressed-air supply from the compressor house, often the source of noise and vibration, through the distribution network into the various departments and to each job; and integrated steam heating and industrial and drinking-water systems.

Despite the tremendous complexity of work in the automobile industry, the risks of occupational disease are nevertheless limited to silicosis, solvent poisoning, lead poisoning, skin diseases due to mineral oils, and deafness. Occupational accidents are much more varied and usually include wounds and crush injuries to the arms and legs, and sometimes to the head (foreign bodies in the eyes) and the body. Cases of carbon monoxide and solvent poisoning are rare and accidents during vehicle testing are even rarer. New hazards have, however, recently been introduced in the form of X-rays and radioisotopes used for non-destructive testing.

The working force is predominantly male and the average age is less than 40-45 years in the foundries, press shops and in assembly-line work, although the age level is somewhat higher in machine shops; women workers are employed mainly in upholstery departments. Common complaints amongst these workers are digestive system disorders (gastroduodenitis) and nervous system disorders (anxiety and depressive states), which although not characteristic of the automobile industry, may be attributed to the prevalence of shift work.

[Stress at work is not uncommon, mainly among young workers. It is often evidenced by an increased absenteeism and lack of interest in the job.]

CROSETTI, L.

Safety:

CIS 74-565 "Ford–Prevention ... an American-style challenge in South-West France" (Ford–Prévention ... un défi à l'américaine dans le sud-ouest français). *Travail et sécurité* (Paris), May 1973, 5 (278-293). Illus. (In French)

Ergonomics:

CIS 77-580 *Workpost profiles—A method of analysis of working conditions* (Les profils de postes–Méthode d'analyse des conditions de travail). Service des conditions de travail de la Régie nationale des usines Renault (Paris, Editions Masson, 1976), 107 p. Illus. (In French)

CIS 77-1483 *Human work design–Examples from the German automobile industry* (Gestaltung der menschlichen Arbeit–Beispiele aus der deutschen Automobilindustrie). Arbeitskreis "Neue Arbeitsstrukturen der deutschen Automobilindustrie" (Ford-Werke AG, Postfach 21 03 69, 5000 Köln 21) (1976), 66 p. Illus. (In German)

"Practical applications of ergonomics in mechanical engineering" (Application de l'ergonomie à l'industrie mécanique).

Parmeggiani, L. (245-262). *Ergonomics in industry, agriculture and forestry.* Occupational Safety and Health Series No. 35 (Geneva, International Labour Office, 1977), 535 p. Illus. (In French)

Assembly line:

CIS 1194-1968 *Study of assembly line working* (Contribution à l'étude du travail en chaîne). Valentin, M.; Melcer, J. P. (Boulogne-Billancourt, Service médical de la Régie nationale des usines Renault, 1967), 91 p. Illus. 15 ref. (In French)

Noise:

CIS 74-390 "Noise control in a large plant" (Lärmbekämpfung in einem Grossunternehmen). Eich, J.; Hipp, P.; Hombach, W.; Seeger, O. W.; Solbach, A. *Sicher ist sicher* (West Berlin), July-Aug. 1973, 24/7-8 (338-351). Illus. (In German)

Painting:

CIS 77-204 "Reducing solvent emissions in automotive spray painting". Roberts, R. E.; Roberts, J. B. *Journal of the Air Pollution Control Association* (Pittsburgh), Apr. 1976, 26/4 (353-358). Illus.

Presses:

CIS 75-1844 "Safety with presses–Consequences of new safety regulations in the sheet-metal converting industry" (Unfallverhütung an Pressen–Auswirkungen der neuen Sicherheitsbestimmungen im Fertigungsbereich). Kärcher, W. *Bänder, Bleche, Rohre* (Düsseldorf), Oct. 1974, 10 (390-394). Illus. (In German)

Automobiles, safety and health design of

In 1977, 4.6 million traffic accidents occurred throughout the world, resulting in 6.3 million injuries and 180 000 fatalities. In the United States approximately 50 000 people are killed annually by motor vehicle accidents, and such accidents are the leading cause of death for young people under the age of 24. Fatality rates (number of fatalities per 100 million vehicle km) vary widely from country to country, depending upon the population of vehicles, the road environment, traffic safety laws and the degree of familiarity with motor vehicles (figures 1a and 1b).

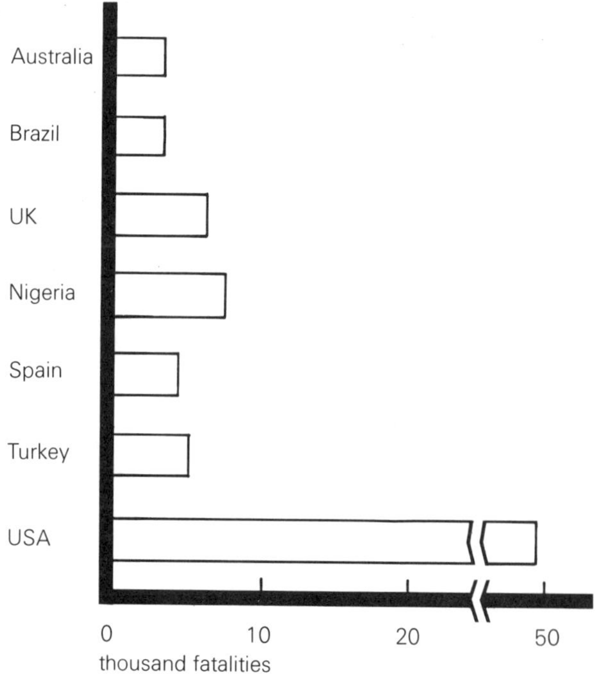

Figure 1a. Deaths resulting from traffic accidents.

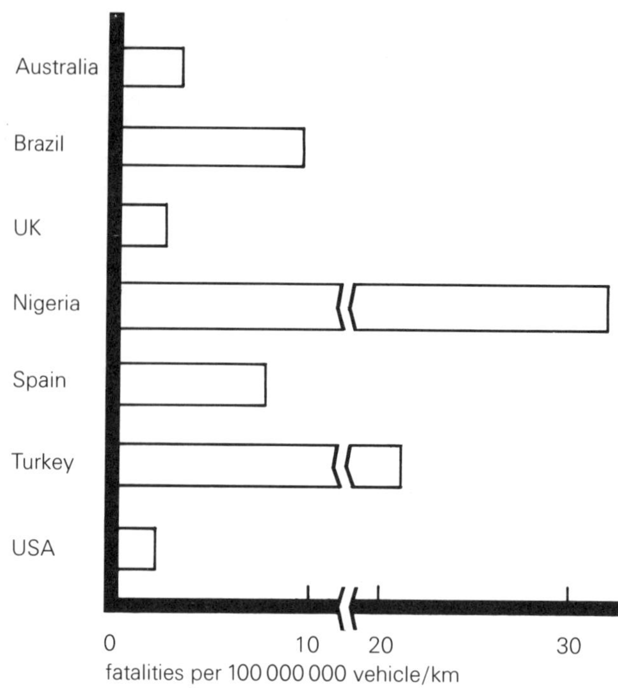

Figure 1b. Traffic accident fatality rates.

Safe design

In order to reduce casualties, many nations have enacted traffic safety measures, traffic laws and safety design requirements for motor vehicles and have improved the road environment. The Federal Motor Vehicle Safety Standards (FMVSS) started in 1966 in the United States are an example of the efforts made to regulate motor vehicle safety design criteria, both to prevent accidents and to improve the "crashworthiness" of vehicles.

A motor vehicle accident occurs when the human factor, the road environment factor or the vehicular factor shows a failure, either alone or in combination (figure 2). Accident investigations indicate that human factors are much more frequently involved than environmental or vehicular factors (figure 3). Safe driving requires ade-

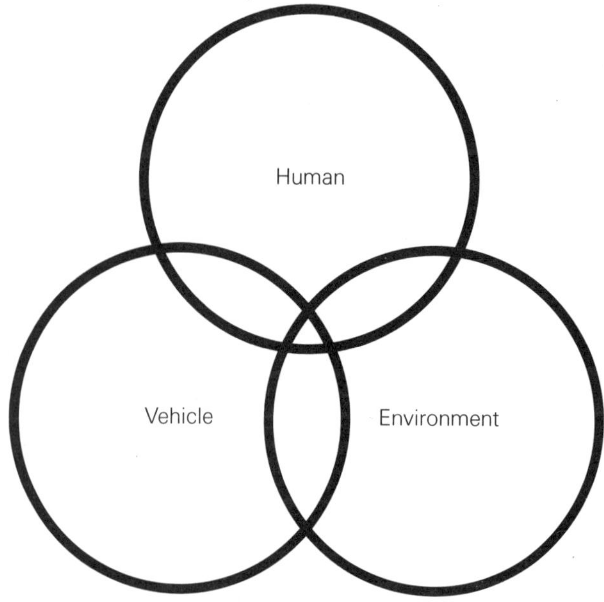

Figure 2. Accident factors in motor vehicles.

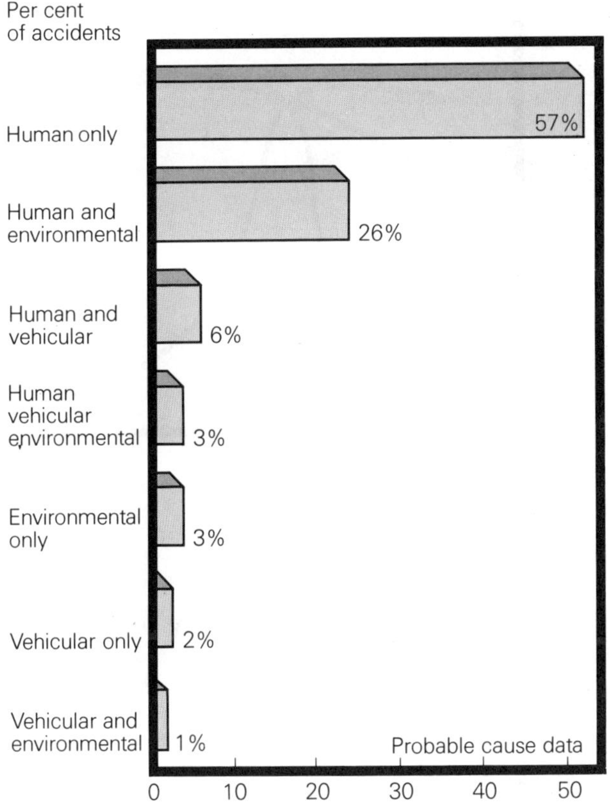

Figure 3. Factors of motor vehicle accidents (according to data collected by the Institute for Records in Public Safety, Indiana University, USA).

quate perception, sound judgement and swift and appropriate action (figure 4). Driving aptitude is determined by tests, and these are particularly important in the case of occupational drivers.

The driver obtains 90% or more of the information necessary for safe driving through visual recognition. The importance of visual information is evidenced by the fact that the night fatality rate is more than three times the day rate (US data). To provide optimum conditions, both the direct field of vision (windshield wiping area, front pillar and side windows) and the indirect field of vision (inner rear-view mirror, outside mirror and rear window) have been studied and improved. At night powerful lighting systems enable the driver to see his way and to detect obstacles. Intensity and beam distribution of head-lights are so designed as to provide adequate illumination without causing undesirably high glare in the direction of an oncoming driver; two-beam head-lights are commonly used throughout the world. To warn other road users of a motor vehicle presence and intentions, side lights, tail lights, brake lights and turn indicators are used. Their configuration, dimensions, colour and position are set in various standards.

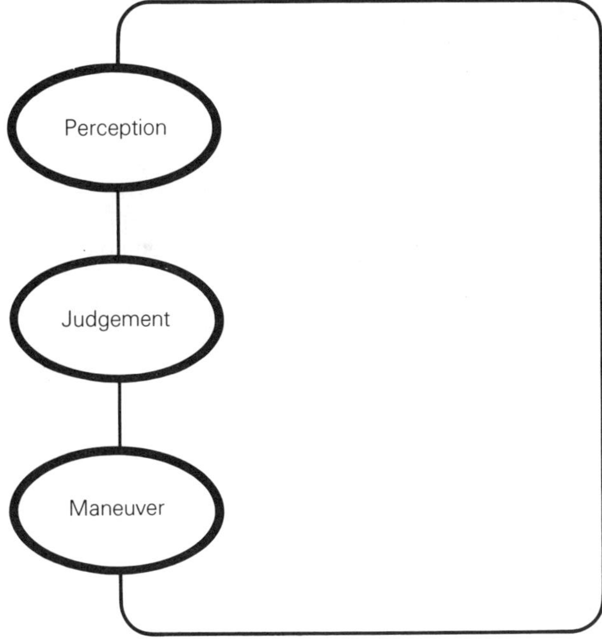

Figure 4. Driving process.

Steering control is also an important safety factor. Until recently the recirculating ball type of steering was most commonly used. However, the rack and pinion type is now popular as it ensures a more direct "road feel". Spring stiffness, damping factor, mass distribution, suspension configuration, steering geometry and various other factors are carefully balanced to provide improved directional and transitional stability. Human response to vehicle handling factors has also been studied in order to improve the capability of the driver to avoid accidents.

As vehicle average speed tends to increase, so does the importance of the braking system. This has led to the use of fade- and water-resistant disc brakes. Further improvements have been made with the incorporation of proportioning valves which regulate the brake fluid pressure between front and rear brakes, allowing quicker stops without swerving. In addition, antiskid braking systems have been developed to prevent loss of control or skidding on wet or icy road surfaces. Antiskid systems also reduce stopping distances on wet or dry road surfaces. To prevent accidents caused by failure of the brake system, dual braking systems have been widely introduced.

Another very important factor in motor vehicle safety is driver fatigue. Power steering, power brakes, improved riding comfort, air conditioning and the reduction of noise all contribute to preventing driver fatigue. Human engineering is also applied to seat and instrument panel design and the like to ensure that the driver is comfortable and can easily perform the necessary functions (see ROAD ACCIDENTS, HUMAN FACTOR).

Alcohol and drugs

Alcohol and drugs reduce the driver's perception ability, his capacity to exercise judgement and his manoeuvring skill. Alcohol is directly responsible for a significant proportion of fatal traffic accidents. In the United States, for example, alcohol consumption is a factor in at least half of the fatal motor vehicle accidents, whereas in Japan, where drunken driving penalties are very severe, less than 10% of fatal motor vehicle accidents are ascribed to alcohol. It is possible to detect drunken drivers and to interlock the vehicle's starting system; prototypes of such devices have been developed by some American motor vehicle manufacturers.

Crashworthiness

The safety of vehicle occupants has four facets: occupant ejection; interior impact; survival space; and vehicle fire safety.

According to accident statistics, the probability of being fatally injured in the case of an ejection is several times that of occupants who are retained in the vehicle. Current efforts to prevent ejection include the strengthening of door hinges and door-locks, the use of anti-inertial door-lock releases, the prevention of windshield penetration, the upgrading of windshield retention and, last but not least, the provision of three-point seat belts. A seat belt is efficient not only in the case of an interior impact but also for preventing ejection.

To upgrade interior impact safety, steering wheels and instrument panels are designed to absorb impact energy. Most modern passenger cars are equipped with collapsible steering columns. Cushioned seats, head restraints and seat belts also contribute to increased impact safety. According to the National Highway Traffic Safety Agency in the United States, the effectiveness of three-point seat-belts is approximately 50%, when they are used. But seat belts cause some inconvenience and discomfort, and some car occupants are reluctant to wear them. To solve these problems, the emergency locking retractor, seat-belt tension reliever and tongueless buckle have been developed. Considering seat-belt effectiveness, more than 20 countries impose their use and can claim significant reductions of casualties. In the United States automatic restraint systems (i.e. air bags and passive seat-belts) that require no action by the occupants have been developed; they will be made compulsory and will be introduced on the domestic market as from September 1981.

The maintenance of adequate survival space is also of fundamental importance. Integrity of the passenger compartment is achieved by impact energy absorption of the front end, rear end and side structures. Most modern vehicle-body structures provide high stiffness.

Anthropomorphic dummies have been developed to evaluate impact safety. On the other hand human tolerance research is progressing rapidly. Scientists in the United States have developed detailed mathematical criteria for the maximum acceleration (and its duration) that the human head can tolerate without fatal injury. These criteria form the basis of the recent FMVS standard on occupant crash protection.

Interior flammability is also an important factor in occupant safety. Materials are selected that will prevent the spread of fire, and measures are taken to combat fuel leakages from fuel pipes in the engine compartment or from fuel tanks. In the United States fuel-system integrity and the flammability of interior materials are strictly regulated.

Exhaust gas emission control

The major components of automobile exhaust emission are hydrocarbons (HC), carbon monoxide (CO) and nitrogen oxides (NOx). These compounds exist in the atmosphere naturally but air pollution is generally associated with advancing industrialisation. As part of the industrial air pollution is caused by motor vehicles (a small proportion of the HC and NOx, but most of the CO pollution), motor-vehicle emission control legislation has been enacted.

One method of lowering the emission of HC, CO and NOx is to suppress them in the combustion chamber where the air/gasoline mixture is burned. Exhaust gas recirculation (EGR) contributes to a reduction of NOx emission levels. Lean fuel-air mixtures also result in low concentrations of NOx (figure 5a). With EGR and/or a lean burn system, residual CO and HC are converted by thermal reactors or catalysts. However, combustion tends to be unstable and inefficient with excessive EGR and/or lean fuel-air ratios. To solve these problems, stratified engines or similar ones have been developed. Three-way catalysts (platinum-rhodium) provide a means of achieving lower NOx emission levels while maintaining good control of HC and CO emissions; this requires a very accurate control of the air-fuel ratio (figure 5b) and is accomplished by employing an improved carburettor and an electronic fuel-control system which monitors the exhaust oxygen concentration (figure 6).

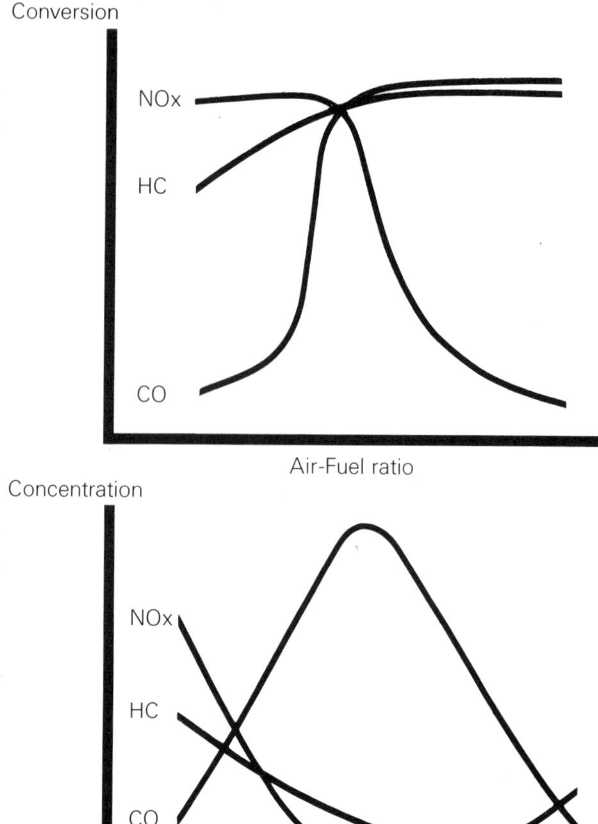

Figure 5a (above) and 5b (below). Exhaust gas emission control.

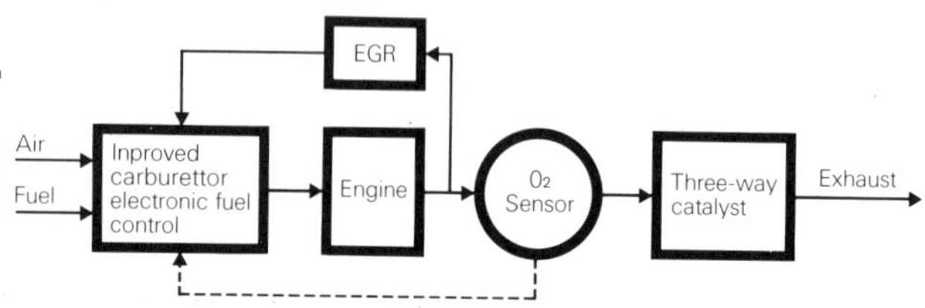

Figure 6. Exhaust gas emission control.

KASHIWAGI, M.
TANAKA, Y.

World road statistics 1974-1978 (Geneva, International Road Federation, 1979).

Accident facts 1978 (Chicago, National Safety Council, 1978), 96 p. Illus.

CIS 80-542 "Forces exerted on the driver in the event of overturning of construction equipment" (Erforschung der beim Überrollen einer Baumaschine auf den Fahrer einwirkenden Kräfte). Glöckner, H. *Die Tiefbau-Berufsgenossenschaft* (Munich), Sep. 1979, 91/9 (640-643). Illus. (In German)

Seat belts and other devices to reduce injuries from traffic accidents. Report of a WHO Technical Group. EURO Reports and Studies 40 (Copenhagen, World Health Organisation Regional Office for Europe, 1981), 53 p. Illus. Ref.

CIS 77-975 *Noise in drivers' cabs of motor vehicles* (Autojen ohjaamomelu). Kiertokirje 4/76 (Directorate of Labour Protection, Tampere, Finland) (1 Apr. 1976), 9 p. (In Finnish, Swedish)

CIS 77-1353 "Carbon monoxide pollution inside a moving car" (Pollution par le monoxyde de carbone à l'intérieur d'une voiture en circulation). Delsey, J.; Joumard, R.; Vidon, R. *Pollution atmosphérique* (Paris), Oct.-Dec. 1976, 18/72 (313-319). 6 ref. (In French)

"Catalysts for automobile emission control". Kummer, J. T. *Progress in Energy and Combustion Science* (Oxford), 1980, 6/2 (177-199). Illus. 168 ref.

Avalanches and glaciers

Persons working in high-altitude mountain regions may be exposed to the hazards of avalanches and glaciers.

Avalanches

Avalanches are masses of snow moving rapidly down a mountain slope or cliff and are responsible for a large number of injuries each year on mountain building and civil engineering sites. There are two main types of avalanche: slab avalanches and avalanches of loose snow.

Slab avalanches. These occur when a slab of packed snow is set in movement, often caused by the wind. The size of the slab is usually quite small but slabs are often dangerous because they may set off further avalanches of loose snow.

Loose-snow avalanches. Dry-snow avalanches usually occur in winter during or within several days of a heavy snowfall. Once the snow has been set in motion, it accelerates rapidly by forming an aerosol with the air. The damage is caused by a shock wave which propagates in air at the speed of sound and may produce pressure up to 800 kgf/m². After the avalanche has passed, the over-pressure phenomenon is replaced by a reciprocal under-pressure which is powerful enough to tear off the roof of a house. Finally, when an avalanche meets a stationary object in its path, the impact pressure may be many tonnes per square metre. The horizontal length of the avalanche cone is proportional to the square of the velocity and consequently the effects may be felt at a considerable distance from the avalanche slope and even on the opposite side of a valley.

Wet-snow avalanches usually move at a lower speed and they therefore more readily follow well established avalanche tracks; however, since the moving mass has greater density, the pressure it exerts on an object in its path will be significantly greater (10 000-20 000 kgf/m²) and even much higher in exceptional circumstances.

Spring avalanches of melting snow occur during the warm hours of the day, have low velocity and always follow established tracks. They present little danger, although the cone of firn debris (consolidated granular snow not yet changed to glacier ice) may considerably hamper traffic movements on mountain roads.

Avalanche release factors

Snowfall. A heavy fall of snow may produce a layer of snow that may move as an immediate avalanche within 3 days of its deposition. Immediate avalanches account for 60-80% of all avalanches and, for example, the Alpine avalanche disasters of 1951 were the consequence of exceptionally heavy snowfalls. In France it is accepted that, at an altitude of 1 500 m, following snowfalls of 50 mm water equivalent (i.e. about 50 cm of snow), there is a serious avalanche hazard. When a level of 100 mm water equivalent has been exceeded, the danger becomes widespread and avalanches are numerous.

Structure of snow layer. This is also one of the prime factors in avalanche causation. The structure of the snow layer must be examined on site by qualified personnel who carry out periodical analyses of the various snow strata and follow developments.

Wind, temperature. Avalanches usually occur during periods of high wind which cause an overloading and wind slabs. There is usually a rise in temperature but this is no more than a factor that normally accompanies heavy precipitations; this temperature rise reduces the cohesion between snow strata but at the same time promotes compaction and consolidation. Temperature rise therefore has, at first, a dangerous effect which does not, however, persist except when there is an important melting.

Slope. On slopes with an incline of over 140% there is no danger of avalanche since the snow cannot accumulate over the long term. The lower limit of dangerous incline is, however, quite low (30%) although avalanche defence construction is not built on slopes of less than 70% incline. Local conditions such as relatively smooth ground surfaces or the presence of convex slopes will increase the avalanche hazard.

SAFETY MEASURES

Protection is expensive. If a construction site is to be established in an avalanche region, documentation on avalanche occurrences should be obtained going as far back in time as possible and a detailed examination of the area should be made from the ground or air.

Protection may be either temporary or permanent. The principle behind temporary protection is to prevent the accumulation of heavy layers of snow that could become destructive avalanches. Consequently, every time meteorological conditions indicate the necessity, the hazard zone should be cleared and traffic halted so that deposited snow can be artificially released; nevertheless, such action requires considerable experience. Permanent protection may be either passive or active. With passive defence, the avalanche is allowed to start but it is not allowed to reach the protected areas. Tunnels, roofs, avalanche breakers, splitting wedges of sufficient strength to resist the pressures indicated above can be constructed to divert the moving snow, or break its acceleration, or barrier walls can be built to halt it completely. Since these structures are built in the zones endangered by the avalanche (i.e. the area around the site or its communication routes) they can be constructed with the help of heavy construction plant such as bulldozers and cranes. Active defence entails the execution of work at the snow accumulation zone, such as modification of the ground surface (e.g. reafforestation) and the construction of earthworks or structures to modify the shape or character of snow deposits or to support their weight (e.g. earth terraces, walls, snow bridges, snow rakes, etc.)

When a defence network is being devised, the following questions should be asked:

(a) Is it possible to move the object requiring protection to a safer area at a cost lower than that of providing protection?

(b) Is it necessary to provide complete and permanent protection? In this case, both active and passive protection may be necessary simultaneously.

(c) Is it sufficient to provide protection for a limited period? The reply will be affirmative in the case of construction sites for example; in such cases, active protection using light materials such as nets may be an attractive proposition and exceptional dangers can be dealt with by temporary site evacuation.

(d) Is intermittent protection sufficient, such as in the case of communication routes?

In countries where such problems exist there is an increasing number of institutions or services specialised in snow safety. It is recommended to seek their advice in any meterological condition before approaching an exposed area.

Glaciers

When a glacier snout ends in a steep slope, glacial movement may result in falls of ice blocks that constitute a considerable danger for persons working in their path. These falls may occur at any time of the day and protection should be provided by an alarm system backed up by passive defence structures such as nets around the work area.

Ice avalanches. These are the result of the fall of all or part of a glacier tongue and the amount of ice involved may be several million cubic metres. These catastrophic phenomena occur during the warm season and are due to a reduction in the coefficient of friction between the glacier and its bed following a rise in the amount of sub-glacial melt water. Typical examples occurred in Altels, Switzerland, in 1895 (6 dead); Tour, France, in 1949 (6 dead); Huascaran, Peru, in 1962 (4 000 dead); and Mattmark, Switzerland, in 1965, when 2 000 000 m^3 of ice broke off from the Allalin glacier killing 88 men working on the Mattmark dam site. In this case, the ice moved down a 60% incline and entirely covered the site huts over a length of 350 m. Around 1940 the glacier presented no danger since the snout covered the area on which the site huts were later located. The danger therefore developed due to the subsequent extensive deglaciation.

To prevent further catastrophes of this type, periodical surveys should be made of glaciers with a mean slope exceeding 45% situated above inhabited areas, since the retreat of a glacier may lead to the development of dangerous situations. The establishment of a building site in a glacier region not previously inhabited should be preceded by consultation with glaciological specialists.

Floods of glacial origin. These are the most common cause of accidents attributable to glaciers.

Glaciers often form natural dams leading to growth of a lake which empties itself periodically, flooding the neighbouring region. In this way, certain glaciers such as the Allalin have caused terrible disasters. In the Kerguelen Islands, a glacial lake containing 100 million cubic metres of water emptied itself in 3 days. Sometimes, lakes may form on the glacier and, at the Gorner glacier in Switzerland, there is a lake which empties itself every year flooding the Zermatt valley. Pockets of interglacial water may also rupture and cause flooding as in the disaster at Tête-Rousse, France, in 1892 which caused a hundred or more deaths.

The retreat of a glacier may result in a lake being formed between a deposit of moraine and the glacier snout. Such phenomena are common in the Peruvian section of the South American cordillera and have caused numerous disasters. In 1941, the rupture of the morainic wall of Lake Palcacocha, which contained 4 000 000 m^3 of water, led to the death of 6 000 people living 25 km downstream. Rupture of the morainic wall may have many different causes such as the liquefaction of and strata due to an earth tremor, regressive erosion, and ice-falls from neighbouring glaciers, which may entail several thousands of cubic metres of material and will cause waves that submerge the natural morainic dike; this phenomenon occurred in 1951 in Lake Artesoncocha which fortunately overflowed into Lake Paron which was then at its lowest level.

The only safety measure possible in such circumstances is to empty the lake when a danger level is attained. In Peru at the moment, a number of lakes behind morainic dams are in the process of being emptied; however, this very delicate operation may, if carried out incorrectly, produce a disaster such as that at Lake Jankarurish which caused 500 deaths in 1950. Consequently, lake emptying should be undertaken by specialists and only if all the necessary precautions have been taken.

GILLET, F.

"Glaciological problems set by the control of dangerous lakes in Cordillera Blanca, Peru". Lliboutry, L.; Morales, B.; Pautre, A.; Schneider, B. *Journal of Glaciology* (Cambridge), 1977, 18/79 I, II, III (239-290).

"Dangers entailed by glaciers" (Dangers présentés par les glaciers). Gillet, F. *Protection civile et sécurité industrielle* (Paris), Nov. 1969, 181 (15-19). (In French)

CIS 75-226 *Avalanche hazards to accommodation and construction sites* (Snøskredfare ved oppholds- og anleggssteder). Verneregler nr. 19 (Direktoratet for arbeidstilsynet, Postboks 8103, Oslo-Dep), (1974), 4 p. (In Norwegian)

"Injuries of buried victims" (Pathologie des ensevelis). Lapras, A. *Le Concours Médical* (Paris), 27 Dec. 1980, 47 (7289-7295). (In French)

"Medical aspects of injuries caused by avalanches: hypothermia and frostbite" (Aspects médicaux de l'accident par avalanche : hypothermie et gelures). Dubas, F. *Zeitschrift für Unfallmedizin und Berufskrankheiten* (Zurich), 1980, 73/4 (164-167). Illus. (In French)

Aviation–flying personnel

This article deals with the occupational safety and health of the crew members of civil aviation aircraft; there are separate articles on AVIATION–GROUND PERSONNEL and AIRPORTS.

Technical crew members. The technical personnel are responsible for the operation of the aircraft, and includes the pilot-in-command, the co-pilot, the flight engineer and the navigator.

The pilot-in-command is the commanding officer of the aircraft and is called "Captain" by the other crew members. He is the legal representative of the airline in relations with third parties to the extent that his authority is specified in the various regulations. The liability of the pilot-in-command to his company and third parties is fixed in his contract of employment and in governmental regulations.

The co-pilot takes his orders direct from the captain and acts as his deputy upon delegation or in the latter's absence. He is the main assistant to the pilot-in-command and acts as a monitoring crew member during all stages of the flight. He is responsible for the handling of all navigation material if there is no navigator on board, and for the collection, completion and despatch of all paper-work.

In the case of a two-engine aircraft (usually flown without a flight engineer), the co-pilot is required to have special knowledge of all technical details of the aircraft. The crew of a four-engine aircraft includes a flight engineer. He is responsible for the mechanical condition of the aircraft and equipment and has to perform certain well defined control functions prior to flight, during aircraft preparation and during and after flight.

On certain long-distance flights, the crew may be supplemented by a navigator. He is, while on flight duty, directly responsible to the pilot-in-command for the safe and efficient navigation of the aircraft, including operation of the navigation and radio equipment and instruments and all other available navigation and communications facilities.

On certain long hauls, the crew may be further supplemented by a pilot with the qualifications of a pilot-in-command, an extra flight engineer and possibly an extra navigator.

National and international law stipulates that aircraft technical personnel may fly only when in possession of a valid licence and one of the conditions for the renewal of such a licence is a periodical medical examination: every 6 months for airline transport pilots and commercial pilots over 40 years, every 12 months for commercial pilots under 40 years and for navigators and flight engineers. The minimum requirements for these examinations are specified by the International Civil Aviation Organisation (ICAO) and by national regulations. A certain number of physicians experienced in aviation medicine may be authorised for such examinations by the authorities concerned, e.g. full-time or part-time air ministry physicians, airforce flight surgeons, airline medical officers, private practitioners designated by the national air ministry.

Cabin crew members. The cabin crew, air hostesses and air stewards are in charge of passenger service. It is their task and responsibility to serve the passengers and make their journey as agreeable as possible. In emergency situations they are responsible for the organisation of emergency procedures and for the evacuation of the passengers.

The minimum cabin crew is 2 to 14 cabin attendants according to the type of aircraft. The cabin crew may be supplemented by a purser who, when on board, is in charge of cabin personnel and, at stations abroad, responsible for the whole crew.

National regulations do not usually stipulate that the cabin crew should hold licences in the same way as the technical crew; however, crew members must be certified as having received appropriate instruction in emergency procedures. Periodic medical examinations are not usually required by law but the airline's own company regulations will stipulate annual medical examinations for the purpose of health maintenance.

HAZARDS AND THEIR PREVENTION

Flying personnel are exposed to a wide variety of stress factors (both physical and psychological), to the hazard of a flight accident and to the contraction of a number of diseases.

Physical stress. Lack of oxygen, one of the main concerns of aviation medicine in the early days of flying, has now become a minor consideration in modern air transport. In the case of a jet aircraft flying at 12 000 m altitude, the equivalent altitude in the pressurised cabin is only 2 300 m and, consequently, symptoms of anoxia will not be encountered in healthy persons. Oxygen deficiency tolerance varies from individual to individual but for a healthy, non-trained subject the presumed (arbitrary) altitude threshold at which the first symptoms of anoxia occur is 3 000 m.

Motion sickness (dizziness, malaise and vomiting due to the abnormal movements and attitudes of the aircraft) was a problem for civil aviation crews and passengers for many decades; the problem still exists today in the case of small sports aircraft, military aircraft and aerial acrobatics. In the modern jet airliner it is much less serious and occurs less frequently due to higher aircraft speeds and take-off weights, higher cruising altitudes which take the aircraft above the turbulence zones and to the advent of airborne radar which enables squalls and storms to be located and circumnavigated.

Aircraft noise, while becoming a more important problem for ground personnel, is less serious for the crew members of a modern jet aircraft than was the case with the piston-engined plane, due to the efficiency of the noise-control measures such as insulation in modern aircraft. Improvements in modern radio communications equipment has minimised background noise levels in earphones.

Whereas the temperatures of the cabin air can rapidly be set at an agreeable level, the humidity of this air cannot be raised to a sufficient degree due to the large temperature difference between the aircraft interior and exterior. Consequently both crew and passengers are exposed to extremely dry air, especially on long distance flights and although the dry air is not dangerous in itself, precautions should be taken to avoid local and general exsiccosis (adequate liquids intake, use of nose ointments).

Certain environmental factors in the upper atmosphere, such as radioactivity and ozone must also be taken into account. However, world-wide studies have proved that, quantitatively, these factors are negligible at present cruising altitudes which do not exceed 13 000 m. Evaluation of the influence of these factors at higher altitudes (up to 30 000 m) is an important aspect of the preliminary studies on commercial supersonic air transport and sometimes also on other aircraft.

The problem of temporary, localised increases in radioactivity due to fallout from nuclear explosions was the subject of very careful study in the early 1960s. Although radioactivity monitoring facilities detected increased levels of radioactivity, on no occasion were aircraft subject to levels exceeding the maximum permissible established by the International Commission on Radiation Protection.

The cockpits of earlier aircraft often contained large numbers of instrument dials painted with radioactive luminous compounds. Measurements proved, however, that radiation from these dials was far below maximum permissible levels. In modern aircraft the need for luminous dials is far less important since the lighting levels in the cockpit can be adjusted over a wide range, and even brightly illuminated control panels do not inconvenience the pilot. Moreover, visual flight is now very rare at night and the runways of modern airports are very well lit.

Ergonomic problems. The main ergonomic problem as far as the cockpit crew is concerned is the need to work for many hours on end in a sitting but unsettled position and in a very limited working area; in this position, it is necessary to carry out a variety of tasks such as movements of the arms, legs and head in different directions, consulting instruments at a distance of about 1 m, above, below, to the front and to the side, scanning the far distance, reading a map or manual at close distance (30 cm), listening through earphones or talking through a microphone. Seating, instrumentation, lighting, cockpit microclimate, radio communications equipment have been and still are the object of continuous improvement. As far as the cabin crew are concerned, the problem is one of standing and moving around. During climbing, descent and in bumpy weather, the crew will be required to walk on an inclined floor and, at various times, may be required to lift and carry heavy or bulky loads in restricted spaces or whilst maintaining uncomfortable body postures, e.g. when serving plates across a three-row seat.

Workload. The workload depends both on the task, the ergonomic layout and the hours of work and many other factors. The additional factors affecting the cockpit crew include:

(a) beginning and end of last flight;

(b) duration of rest time between present and last flight;

(c) duration of sleep-time during this rest period;

(d) commencement of pre-flight briefing;

(e) problems arising during briefing;

(f) delays preceding departure;

(g) timing of flights;

(h) meteorological conditions;

(i) quality and quantity of radio communication;

(j) visibility during descent, glare and protection from sun;

(k) turbulence;

(l) technical problems.

Certain of these factors may be equally important for the cabin crew. However, the latter are subject to the following specific factors:

(a) pressure of time due to short duration of flight, high number of passengers, of extensive service programme;

(b) extra services demanded by passengers, the character of certain passengers;

(c) passengers requiring special care and attention (children, invalids);

(d) extent of preparatory work.

The measures taken by airline managements and government administrations to keep crew workload within reasonable limits include: improvement and extension of air-traffic control, execution of preparatory work by despatchers, automation of cockpit equipment; and for cabin crew, the standardisation of service procedures, the provision of airline nurses for special tasks (child care), provision of efficient and easy-to-handle galley equipment.

Psychological stress. One of the most important psychological factors (and certainly the most widely discussed) concerning crews as a group is flight fatigue and recovery. Here, biological rhythms, especially sleep intervals and duration, with all their psychological and physiological implications are of prime significance. Time shifts due either to night flights or to east/west or west/east flights across a number of time zones create the greatest problems.

Recovery depends not only upon the rest period available and on the way it is arranged but also on the advantage the individual takes of it. The better the balance between the individual professional and private life the greater this advantage will be. Two important aspects are a reasonable, healthy mode of living and

Figure 1. The cockpit of a modern aircraft.

sufficient physical exercise. Susceptibility to sleep deprivation, fatigue, nervous tension and the consequent difficulties encountered at work on the one hand, and constitutional capacity for rapid recovery on the other, vary from individual to individual and many cases of fatigue cannot be treated by general regulations but only by individual psychological understanding and guidance.

In recent years, aircraft crews have been confronted with a further mental stress factor: the likelihood of hijacking, bombs and armed attacks on aircraft. Although security measures in civil aviation have been considerably increased, the deciding factor in the prevention of air-piracy and other criminality will be the solidarity and force of world-wide public opinion.

Hours of work. Company regulations are based on compulsory international, national and local regulations and must be used for planning and execution of all company flights. These duty regulations guarantee reasonable crew utilisation as well as the fulfilment of the general policy of flight safety. They include maximum flight time per year and per month, free days, maximum duty time for day and night assignment, duration of rotations, rest time during the rotations, number of landings, regulations for time as a stand-by on the reserve list, special regulations for various sectors.

Some of the hours of work requirements of a typical European airline with average flight and duty time regulations may be given, as an example: maximum flight time per year: 850 h; maximum flight time per month: 85 h; maximum per-flight duty assignment: standard crew 10.5 h, enlarged crew 15 h.

Accidents. An aircraft accident is practically never a hazard resulting from a single well defined cause; in almost every instance a number of technical and human factors coincide in the causal process. One important, although relatively rare, type of human failure is sudden death due, for example, to myocardial infarction; other failures include sudden loss of consciousness (e.g. epileptic fit, cardiac syncope and fainting due to food poisoning or other intoxication). Human failure may also be the result of slow deterioration (e.g. gradual loss of vision or hearing capacity).

Psychological failure is often more a problem of professional skill than of neurotic or mental disorders but occupational and psychiatric aspects cannot always be strictly differentiated.

Advances in accident prevention are impossible unless former accidents are investigated and incidents evaluated. Systematic screening of all, even minor, incidents by an accident investigation board comprising technical, operational, legal and medical experts, is most valuable.

Accident prevention is the most important task of aviation medicine. Careful personnel selection, regular medical examination, survey of absence due to illness and accidents, continuous medical contact with crew working conditions and hygiene surveys can considerably decrease the danger of sudden incapacitation or slow deterioration. Psychiatric selection should be closely linked with professional selection and psychiatric examinations and psychological tests should not be considered as alternatives but as complementary one to the other.

In a well operated, modern airline of significant size, it is essential that the company should have its own medical service for these purposes. However, even the best medical service is insufficient without the continuous co-operation of all crew members, and systematic education to awaken a sense of personal responsibility in each crew member regarding his fitness for flight is an essential part of training.

Flight personnel should start flight duty in a sufficiently good mental and physical state to ensure that the fatigue occasioned by the flight itself will not affect safety. Fitness for flight duty may be impaired by mental stress and it is the responsibility of the crew member himself to decide whether or not he is fit for flight duty when he is subject to mental stress; it should be borne in mind, however, that mental stress is obviously not to be equated with mental or nervous disease.

A number of strict regulations are necessary: crew members should not consume quantities of alcohol in excess of what is compatible with professional requirements and no alcohol at all should be consumed during and for at least 8 h before flight duty. Drugs are generally not allowed during or immediately preceding flights, although exceptions may be conceded by the responsible flight physician. Other accident-prevention regulations deal with oxygen supply, crew meals and procedures in case of illness.

The deciding factor in accident prevention is, however, an excellent standard of professional skill. Commercial aviation differs from other professions in that successful graduation after training does not qualify the individual for permanent pursuit of his occupation; both the practical and theoretical aspects of the crew member's professional skill are rigorously checked every few months.

Diseases. Specific occupational diseases of crew members are not known. However, certain diseases may be more prevalent among crew members than among persons in other professions. Common colds of the upper respiratory system are frequent. This is due partly to changes of climate and temperature, dryness of the air during the flight, irregularities of schedule, etc. On the other hand, common afebrile catarrh, which is without importance for an office employee, may incapacitate a crew member if it jeopardises the clearing of the pressure on the middle ear during ascent and, particularly, during descent. Frequent travel to tropical areas entails increased exposure to infectious diseases, the most important being malaria and infections of the digestive system.

GARTMANN, H.

General:

Occupational health and safety in civil aviation. Tripartite technical meeting for civil aviation (Geneva, International Labour Office, 1977), 103 p.

Health:

"Preventive medicine aspects and health promotion programs for flight attendants". Alter, J. D.; Mohler, S. R. *Aviation, Space and Environmental Medicine* (Washington, DC), Feb. 1980, 51/2 (168-175). 50 ref.

CIS 76-1132 "Pregnant stewardess–Should she fly?" Scholten, P. *Aviation, Space and Environmental Medicine* (Washington, DC), Jan. 1976, 47/1 (77-81). 56 ref.

CIS 77-1878 "Noise hazard for civil aviation personnel in the GDR according to age and length of exposure" (Die Lärmgefährdung des fliegenden Personals der zivilen Luftfahrt der DDR in Abhängigkeit von Alter und Expositionszeit). Kressin, J. *Zeitschrift für die gesamte Hygiene und ihre Grenzgebiete* (Berlin), May 1976, 22/5 (312-318). Illus. 63 ref. (In German)

Ergonomics:

CIS 76-979 *Vibration and combined stresses in advanced systems.* Von Gierke, H. E. (National Technical Information Service, Springfield, Virginia 22161) (1975), 260 p. Illus. 448 ref.

CIS 74-1783 "Ergonomic aspects of crew seats in transport aircraft". Hawkins, F. *Aerospace Medicine* (Washington, DC), Feb. 1974, 45/2 (196-203). Illus. 16 ref.

Aviation–ground personnel

Once an aircraft has been made ready for take-off by the personnel and operations described in the article AIRPORTS, it is taken over by air-traffic control, which must–

(a) provide advice and information for the safe and efficient carrying out of flights;

(b) prevent collisions on the ground between moving aircraft and obstacles on the manoeuvring areas;

(c) prevent collisions in flight;

(d) regulate and speed up air traffic;

(e) alert and assist the right bodies when aircraft need help from the search and rescue services.

Air traffic control is based on services of three quite different types–

(a) the air traffic control services proper;

(b) the flight information services;

(c) the alerting services.

Air traffic control services

Before receiving permission to take off, the captain of an aeroplane must hand in his flight plan, which may be accepted or refused. Once it is accepted, it becomes a contract between the pilot and the air safety services.

During flight the controllers are responsible for the safety of the aeroplane, above all in respect of collision risks, and the pilot has to follow their orders.

The air traffic control services proper comprise three principal elements–

(a) regional, or *en route*, control;

(b) approach control;

(c) aerodrome control.

Regional and approach control

Regional control follows without interruption the progress of each aircraft along pre-established routes and transmits the necessary information to the pilot.

Approach control provides the same services for airborne aircraft under control in the neighbourhood of the aerodrome where the manoeuvres preparatory to landing or subsequent to take-off are carried out.

Each of these two types of control exists at present in two variants.

The non-automated, or manual, system. Radio communications between controller and pilot are supplemented by information from primary or secondary radar equipment. The trace of the aeroplane can be followed as a mobile echo on display screens formed by cathode-ray tubes.

The automated system. In this system, which is at present in full expansion, information on the aeroplane is still based on the flight plan and primary and secondary radar, but computers make it possible to present in alphanumeric form on the display screen all data concerning each aeroplane and to follow its route. Computers are also used to anticipate conflict between two or more aircraft on identical or converging routes on the basis of the flight plans and standard separations. Automation relieves the controller of many of the activities he carries out in a manual system and leaves him more time for taking decisions. Conditions of work are different in the two systems.

In the manual system the screen is horizontal or sloping and the operator leans forward in an uncomfortable position with his face between 30 and 50 cm from it.

The perception of mobile echoes in the form of spots depends on their brightness and their contrast with the illuminance of the screen. As some mobile echoes have a luminosity as low as 0.3 cd and the brightest seldom exceed 50 cd, the working environment must be very weakly illuminated to ensure the greatest possible visual sensitivity to contrast.

In the automated system the electronic data display screens are vertical or almost vertical and the operator can work in a normal sitting position with a greater reading distance. He has horizontally arranged keyboards within reach to regulate the presentation of the characters and symbols conveying the various types of information. He can alter the shape and brightness of the characters. The lighting of the room can approach the intensity of daylight, for contrast remains highly satisfactory at 160 lx.

These features of the automated system place the operator in a much better position to increase his efficiency and reduce his visual and mental fatigue.

Work is carried out in a huge artificially lighted room without windows, which is filled with display screens. This closed environment, often far from the aerodromes, allows little social contact during the work, which calls for great concentration and powers of decision. The comparative isolation is mental as well as physical and there is hardly any opportunity of diversion. All this has been held to favour exaggerated mental stress and psychoneurotic troubles.

Aerodrome control

Aerodrome control is carried out from the control tower by radio. It concerns aircraft that are airborne in the neighbourhood, landing or taking off or manoeuvring on the ground.

The walls of the room are transparent, for there must be perfect visibility. The working environment is thus completely different from that of regional or approach control. The controller has a direct view of aircraft movements and other activities. He meets some of the pilots and takes part in the life of the aerodrome. The atmosphere is no longer than that of a closed environment and it offers a greater variety of interest.

The runway staff. These guide the aircraft moving on the ground and carry out various duties in the neighbourhood of those that are stationary.

In-flight information and alert posts

In these posts the atmosphere is similar to that of aerodrome control, but problems involving human factors are not determined by the nature of the professional activities as they are for the controllers.

OCCUPATIONAL PATHOLOGY OF AIR TRAFFIC CONTROLLERS

In view of the equipment used and the mental factors governing the efficiency of automated control systems, the occupational pathology of air traffic controllers should be considered under four main headings–

(a) the hazard of X-rays from the cathode-ray tubes of the radar screen;

(b) the ocular health hazard;

(c) hearing impairment;

(d) nervous stress.

The hazard of X-rays

The screens are fed at a high voltage and are thus generators of low-power X-rays. All the measurements that have been taken show that up to 25 kV the tubes normally used discharge a negligible quantity of radiation. Moreover, the radar-receiving equipment used for air control is still operated at well under 25 kV.

The exposure limit of ionising radiation generally permitted is 5 rem in 30 years for the population at large. In view of the facts that radar reading tubes are at present operated at a maximum of 18 kV, that the front of the tube is shielded to stop the passage of X-rays, and that a controller is not exposed to radiation for more than 2 000 h in a year, it is clear that irradiation by X-rays throughout the whole career of a radar controller does not reach a dangerous value.

Ocular health hazard

Sun dazzle: eyestrain of control tower operators. The broad transparent surfaces of the control tower sometimes result in dazzling by the sun, and reflection from surrounding sand or concrete can increase the luminosity. This strain on the eyes may produce asthenopia, though often of a temporary nature. It may be prevented–

(a) by surrounding the control tower with grass and avoiding concrete, asphalt or gravel;

(b) by giving a green tint to the transparent walls of the room. If the colour is not too strong, visual acuity and colour perception remain adequate while the excess radiation that causes dazzle is absorbed.

Eyestrain due to the radar screen. Symptoms due to long and repeated reading of the radar screen are supra-, peri- and retro-orbital headaches, a feeling of heaviness in the eyeballs, prickling, lacrimation and, sometimes, congestion of the edge of the eyelids and conjunctival hyperaemia.

Until about 1960 there was a good deal of disagreement among authors on the frequency of this eyestrain, but it does seem to have been high. Since then, attention given to refractive errors in the selection of radar controllers, their correction among serving controllers and the constant improvement of working conditions at the screen have helped to lower it considerably.

This form of eyestrain is generally due to a defect or anomaly in the ocular function. The correction of various ametropias, phorias and troubles of accommodation puts an end to it, except where there is hypermetropic astigmatism of more than 1 dioptre or hypermetropia of more than 1.5 dioptres. Persons suffering from this degree of hypermetropia must not, therefore, be selected for the work.

Sometimes, however, eyestrain appears among readers with excellent sight. It may then be attributed to too low a level of lighting in the room, irregular illumination of the screen, the brightness of the echoes themselves and, in particular, flickering of the image. Progress in viewing conditions and insistence on higher technical specifications for new equipment are leading to a marked reduction in this source of eyestrain or even its elimination.

Strain in accommodation has also been considered until recently to be a possible cause of eyestrain among operators who have worked very close to the screen for an hour without interruption.

Visual pathology is becoming much less frequent and is bound to disappear or to occur only very occasionally in the automated radar system, for example when there is a fault in a scope or the rhythm of the images is badly adjusted.

Hearing impairment. This is not of a specific character among air traffic controllers. A reduction in hearing acuity over the period of the career is comparatively rare (0.24 to 0.43% of all controllers examined).

Nervous stress. The chief duty of the controller is to take decisions on the movements of aircraft in the sector he is responsible for: flight levels, routes, changes of course when there is conflict with the course of another aircraft or when congestion in one sector leads to delays, air traffic, and so on. In non-automated systems the controller must also prepare, classify and organise the information his decision is based on. The data available to him are comparatively crude and must first be digested. In highly automated systems the instruments can help the controller in taking decisions and he may then only have to analyse data produced by teamwork and presented in rational form by these instruments. Yet, when all is said and done, although his work may be greatly facilitated, the responsibility for approving the decision proposed to him remains his and his activities still give rise to stress. The responsibilities of the job, pressure of work at certain hours of dense or complex traffic, sustained concentration, awareness of the catastrophe that may result from an error, all these create a situation of continuous tension, which may lead to emotional stress in unstable subjects and so to nervous strain.

The most important factors conducive to nervous strain have nothing to do with the working environment and include family problems, financial and occupational difficulties and fatigue caused by a secondary gainful occupation. Those connected with the workplace or working conditions tend to become much less important on account of the many improvements introduced in this field. The problems include noise, heat, cramped surroundings, underground rooms and inadequate ventilation.

The fatigue of the controller may assume the three classic forms of acute fatigue, chronic fatigue or overstrain and nervous exhaustion, which can arise from a state of anxiety and develop into anxiety neurosis.

The question has been raised whether repeated emotional stress does not lead to more rapid physiological wear and tear among air traffic staff, particularly in radar control, than among persons not subjected to so much stress, and it is still the subject of study. The conclusions that can be based at present on available statistics are as follows:

(1) Illness due to mental stress in both the automated and the non-automated systems accounts for a great part of unfitness arising during service (Zetzmann gives the following percentages for diseases leading to unfitness: mental stress 45.6, cardiovascular affections 24.5, nervous affections 12.25 and gastrointestinal and metabolic affections 8.84). Permanent unfitness due to these affections, however, is not abnormally high.

(2) The psychosomatic pathology of the controller relates, in order of decreasing frequency, to–

(a) the gastrointestinal system;

(b) the cardiovascular system;

(c) the neuropsychic system.

Visual and auditory pathology has been decreasing markedly as a result of recent technical progress, particularly automation. Greater care in selecting staff must also be playing a part.

(3) Difficulties in evaluating the effects of a given form of stress arise inevitably from the effects of many individual variants, some of which have nothing to do with work.

(4) The radar controller has little or no influence on his workload and he may feel great fatigue when there is an increase in air traffic, even in highly perfected automated systems.

These considerations show the importance of—

(a) current research on the measurement of stress in the controller by objective physiological, biochemical and psycho-technical methods;

(b) the possible effects in the reduction of stress of different rhythms of work and rest.

PROTECTION OF AIR TRAFFIC CONTROL PERSONNEL

Selection of control personnel

Standards of fitness for air traffic controllers have been recommended by the International Civil Aviation Organisation (ICAO), and detailed standards are set out in national military and civil regulations, those relating to sight and hearing being particularly precise.

Visual acuity equal to unity in each eye is generally required for distant vision. If this can be achieved only with glasses, uncorrected vision must reach at least 0.2 or, under certain regulations, 0.3. Similar standards are required for near vision. The regulations provide for the rejection of persons suffering from ametropias exceeding certain narrow limits, including hypermetropias of over 1.5 dioptres.

Limits are also laid down for accommodation, convergence and phorias. Colour vision must be normal, for the lights of aeroplanes must be correctly identified and it is highly probable that coloured symbols will be used in future in the automated system to display data on the screen.

Hearing acuity must be normal in each ear. An acceptable impairment, evaluated in decibels at different frequencies, must leave hearing that ensures full understanding of spoken communications between the aeroplane and the receiving post.

Selection is followed by annual re-examinations.

Arrangement of premises and equipment

A rational arrangement of the premises is mainly one that facilitates the adaptation of the scope readers to the intensity of the ambient lighting. In a non-automated radar station adaptation to the semi-darkness of the scope room is achieved by spending 15 to 20 min in another dimly lighted room. The general lighting of the scope room, the luminous intensity of the scopes and the brightness of the spots must all be studied with care. In the automated system the signs and symbols are read under an ambient lighting of from 160 to 200 lux and the disadvantages of the dark environment of the non-automated system are avoided.

With regard to noise, despite modern sound-insulating techniques, the problem remains acute in control towers installed near the runways.

Eye protection by special glasses

Readers of radar screens and electronic display screens are sensitive to changes in the ambient lighting. In the non-automated system the controllers must wear glasses absorbing 80% of the light for between 20 and 30 min before entering their workplace. In the automated system special glasses for adaptation are no longer essential, but persons particularly sensitive to the contrast between the lighting of the symbols on the display screen and that of the working environment find that glasses of medium absorptive power add to the comfort of their eyes. There is also a reduction in eyestrain.

Runway controllers are well advised to wear glasses absorbing 80% of the light when they are exposed to strong sunlight.

Rhythm of work and rest

Air traffic control calls for an uninterrupted service 24 h a day all the year long. The conditions of work of controllers thus include shift work, an irregular rhythm of work and rest, and periods of work when most other people are enjoying holidays.

Periods of concentration and of relaxation during working hours and days of rest during a week of work are indispensable to the avoidance of operational fatigue. This principle cannot unfortunately be embodied in general rules, for the arrangement of work in shifts is much influenced by variables that may be legal (maximum number of consecutive hours of work authorised) or purely professional (workload depending on the hour of the day or the night) and by many other factors based on social or family considerations. With regard to the most suitable length for periods of sustained concentration during work, experiments show that there should be short breaks of at least a few minutes after periods of uninterrupted work of from half-an-hour to an hour-and-a-half but that there is no need to be bound by rigid patterns to achieve the desired aim: the maintenance of the level of concentration and the prevention of operational fatigue. What is essential is to be able to interrupt the periods of work at the screen with periods of rest without interrupting the continuity of the shift work. Further study is necessary to establish the most suitable length of the periods of sustained concentration and of relaxation during work and the best rhythm for weekly and annual rest periods and holidays with a view to drawing up more unified standards.

Protection of runway staff against noise

Runway staff who have to work very close to aeroplanes moving on the ground are exposed to violent noise, which is dangerous for their hearing. They must therefore be supplied with protective devices: earplugs or, better still, earmuffs or helmets. They must also be clearly informed of the danger to their hearing of exposure to loud noise, so that they can avoid unnecessary exposure and take suitable protective measures.

EVRARD, E.

"Morbidity experience of air traffic control personnel 1967-77". Booze, C. *Aviation, Space and Environmental Medicine* (Washington, DC), Jan. 1979, 50/1 (1-8). Illus. 9 ref.

Abstract of aeronautical and space medicine (Précis de médecine aéronautique et spatiale). Evrard, E. (Paris, Maloine, 1975), 706 p. (In French)

CIS 79-1457 "Review of stress in air traffic control: its measurement and effects". Crump, J. H. *Aviation, Space and Environmental Medicine* (Washington, DC), Mar. 1979, 50/3 (243-248). 64 ref.

"Stress in air traffic personnel: Low-density towers and flight service stations". Melton, C. E.; Smith, R. C.; McKenzie, J. M.; Wicks, S. M.; Saldivar, J. T. *Aviation, Space and Environmental Medicine* (Washington, DC), May 1978, 49/5 (724-728), Illus. 19 ref.

Meeting of experts on problems concerning air traffic controllers, ATC/1979/1 (Geneva, International Labour Office, 1979), 79 p.

Aziridines

Ethyleneimine

AZIRIDINE; AZACYCLOPROPANE; DIHYDRO-1H-AZIRINE; DIMETHYLENEIMINE; EI; ETHYLIMINE

m.w. 43
sp.gr. 0.83
m.p. −71.5 °C
b.p. 56 °C
v.d. 1.5
v.p. 160 mmHg ($21.28 \cdot 10^3$ Pa) at 20 °C
f.p. −11.1 °C
e.l. 3.6-4.6%
i.t. 322 °C

H_2C—NH—H_2C (ring)

miscible with water in all proportions, very soluble in ether, soluble in ethanol

a colourless volatile liquid with an ammoniacal odour.

TWA OSHA 0.5 ppm 1 mg/m³ skin
MAC USSR 0.02 mg/m³ skin

Production. Ethanolamine is esterised by sulphuric acid to ethanolamine-sulphuric acid ester, which is reacted with sodium hydroxide in a normal reaction to form ethyleneimine.

Uses. By virtue of its high reactivity, ethyleneimine is used in a large number of organic syntheses. Of prime significance are the polymerisation products, i.e. polyethyleneimines, which are used primarily as auxiliaries in the paper industry and as flocculation aids in the clarification of effluents. To a limited extent some N-substituted ethyleneimine derivatives are also used in the textile industry, e.g. for wash-resistant flameproof finishing.

HAZARDS

Ethyleneimine presents a fire and explosion hazard; it is toxic by inhalation, ingestion and skin absorption, a strong skin irritant, and may produce corneal damage.

Fire and explosion. In the absence of acids, the ring structure of ethyleneimine is comparatively stable; however, small amounts of acid will catalyse the strongly exothermic polymerisation process. In fact carbonic acid in the atmosphere would suffice for this purpose. Under favourable conditions polymerisation may proceed with explosive violence. Its flammability is akin to that of gasoline and similar petroleum products.

Skin and mucous membrane effects. After a very brief period of contact, liquid ethyleneimine gives rise to definite signs of skin irritation, i.e. inflammation and blistering, and to deep necrosis of the skin. This applies to the fullest extent for contact with any part of the body. Depending on the concentration and length of exposure, skin irritation is evident after only 5 min, but several days may expire before it is noticed. Ethyleneimine vapours severely irritate the eyes and upper respiratory system. Prolonged keratitis and conjunctivitis, which respond to therapy with extreme difficulty, have been observed, and ulcerations have been found on the septum and vocal chords. Fatal ethyleneimine intoxication caused mainly by skin absorption has been observed. Cases of skin sensitisation have not been reported in plants processing this compound.

Inhalation effects. Between 30 and 120 min after inhalation of ethyleneimine, the actual time depending on the concentration of the mixture inhaled, the patient suffers nausea, retching, and periodic vomiting. The vomiting is similar to that occurring after inhalation of ethylene oxide and can probably be attributed to attack of ethyleneimine on the vomiting centre. Headache, dizziness, pain in the vicinity of the temples, and dullness are other symptoms. Inhalation of ethyleneimine also causes severe irritation of the mucous membranes of the mouth, nose, and upper respiratory tract. Other ailments caused by inhalation of ethyleneimine are nasal secretion, laryngeal oedema, pronounced diphtheria-like mutations of the trachea and bronchi, bronchitis, shortness of breath, oedema of the lungs and secondary bronchial pneumonia.

Experimental toxicity. In animal experiments the inhalation of ethyleneimine has caused typical severe injuries to the kidneys, such as hyperaemia and tubule necrosis, although these were found to exist only in the form of a slight temporary disorder in human beings. A slight increase in the serum transaminases (SGOT and SGPT) has also been observed after prolonged contact and after repeated ethyleneimine intoxication.

Ethyleneimine is highly toxic by ingestion, percutaneous absorption and inhalation. Animal experiments have shown that the symptoms are independent of the type of application and consist of cramps, vomiting and dizziness, kidney injuries, leucopenia and blindness. Ethyleneimine penetrates animal skin so quickly that its percutaneous toxicity is not decreased if it is washed off 1 min after contact. Undiluted ethyleneimine leads to extensive degeneration of the tissues, but not to sensitisation. The toxicity and effect on the skin of the oligomers are less than those of ethyleneimine and those of polymeric ethyleneimine are less than those of the oligomers.

Carcinogenicity and mutagenicity. Ethyleneimine is carcinogenic to strains of mice by oral administration (increased incidence of liver-cell and pulmonary tumours). Subcutaneous injection in suckling mice produced an increased incidence of lung tumours in males; and in one experiment on rats tumours were produced at the injection site.

To humans ethyleneimine is a suspected carcinogen. It is mutagenic in various test systems and has been shown to cause mutations or chromosome aberrations in plants, bacteria, insects, mammals and human cultured cells.

SAFETY AND HEALTH MEASURES

Complete control of vapours by local exhaust ventilation is essential. Canister-type respiratory protective equipment for alkaline gases may be worn for protection against brief exposure to low concentrations. Prolonged exposure to levels above the recommended maximum permissible levels necessitates use of supplied-air or self-contained equipment.

Protection against skin contact should be ensured by use of hand and arm protection, eye and face protection, aprons, etc. Liquid ethyleneimine slowly penetrates rubber protective apparel and may constitute a source of skin contact; contaminated gloves, etc., should therefore be discarded.

An efficient control device for the immediate detection of the slightest leak, e.g. a gas chromatograph for periodic analysis of air at selected locations, should be installed together with an alarm system that initiates a carefully planned programme to cope with emergency situations for which personnel must be thoroughly schooled. To prevent spontaneous polymerisation, an alkaline inhibitor should be added.

Treatment. There is no specific treatment for injuries caused by ethyleneimine. Therapeutic measures are therefore determined by clinical examination of the patient. After ethyleneimine inhalation, the patient should be kept warm and transported immediately to a physician. Therapy for ethyleneimine inhalation poisoning is directed mainly at the circulatory system because frequent periodic vomiting over a period of several hours could easily lead to the collapse of the peripheral circulatory system and, in addition, the action of ethyleneimine could cause a collapse of the central circulatory system. An intensive examination of the kidney functions is also essential. In order to suppress an inflammation of mucous membranes caused by the inhalation of ethyleneimine for several hours, it is always advisable to administer hydrocortisone preparations.

Since there is also no specific treatment for skin burns caused by ethyleneimine, they should be treated like normal burns. Clothing contaminated with ethyleneimine should be removed immediately, the affected skin should be thoroughly washed with water, and a physician summoned.

Propyleneimine (C_3H_7N)

2-METHYLAZIRIDINE; 2-METHYLAZACYCLOPROPANE; METHYLETHYLENEIMINE; 1,2-PROPYLENEIMINE

m.w. 58.1
sp.gr. 0.80
b.p. 66-67 °C
v.d. 2.0
TWA OSHA 2 ppm 5 mg/m³ skin

To date, propyleneimine has hardly been used in the chemical industry. The acute toxic properties have generally been considered as similar to those of ethyleneimine. No industrial injuries have been reported other than severe eye burns.

Carcinogenicity and mutagenicity. Propyleneimine is carcinogenic in rats following oral administration, producing a variety of malignant tumours. No data are available as regards human carcinogenicity or, for the time being, on mutagenic effects.

Derivatives of ethyleneimine

Some derivates have been tested and used in the treatment of neoplastic diseases. Some derivates have also been tested as insect chemosterilants.

Tris(1-aziridinyl)phosphine oxide ($C_6H_{12}N_3OP$)

1,1',1''-PHOSPHINYLDYNETRISAZIRIDINE; APHOXIDE; APO; ENT-24915; TEF; TEPA; PHOSPHORIC ACID TRIETHYLENE IMIDE; N,N',N''-TRI-2-ETHANEDYLPHOSPHORIC TRIAMIDE

Tris(2-methyl-1-aziridinyl)phosphine oxide ($C_9H_{18}N_3OP$)

1,1',1''-PHOSPHINYLDYNETRIS(2-METHYL)AZIRIDINE; MAPO; METAPOXIDE; EMT 50,003; METEPA; N,N',N''-TRIS(1-METHYL-ETHYLENE)-PHOSPHORAMIDE

These compounds have moderate (TEPA) or suspected (MAPO) carcinogenic effects on animals, are mutagenic (MAPO more strongly) and show teratogenic effects on chick embryos. No data are available as regards carcinogenic effects in human beings.

Tris(1-aziridinyl)phosphine sulphide ($C_6H_{12}N_3PS$)

1,1',1''-PHOSPHINOTHIOYLDYNETRISAZIRIDINE; THIOPHOSPHAMIDE; THIOTEPA; N,N',N''-TRI-1,2-ETHANE-DIYLPHOSPHOROTIC TRIAMIDE; TRIS(ETHYLENEIMINO)-THIOPHOSPHATE; TESPA

This compound is carcinogenic in mice and rats and is able to induce mutations in bacterial test systems, mammals and cultured human cells; nine cases of acute leukaemia following its therapeutic use were reported.

2,4,6-Tris(1-aziridinyl)-*s*-triazine ($C_9H_{12}N_6$)

1,1',1''-*s*-TRIAZINE-2,4,6-TRIYLTRISAZIRIDINE; TEM; TET; ENT 25,296; TRIS(ETHYLENEIMINO)TRIAZINE

This compound is carcinogenic in mice and rats and has been shown to be mutagenic in insects, mammals and cultured human cells. No human data are available with regard to carcinogenicity.

Tris(aziridinyl)-para-benzoquinone ($C_{12}H_{13}N_3O_2$)

2,3,5-TRIS(1-AZIRIDINYL)-2,5-CYCLOHEXADIENE-1,4-DIONE; 1,1',1''-(3,6-DIOXO-1,4-CYCLOHEXADIENE-1,2,4-TRIYL)TRISAZIRIDINE; TEIB; TRIAZIQUINONE; 2,3,5-TRIS(ETHYLENEIMINO)PARABENZOQUINONE

The compound is carcinogenic in rats and mutagenic in mammals and cultured human cells. In some cases treatment with triaziquinone was associated with later occurrence of monocytic leukaemia and neoplastic reticulosis.

THIESS, A. M.
FLEIG, I.
STOCKER, W. G.

CIS 2588-1969 "The pathology and histology of ethyleneimine poisoning in man" (Pathologie und Histologie nach Äthyleniminwirkung beim Menschen). Birnstiel, H.; Thiess, A. M. *Internationales Archiv für Gewerbepathologie und Gewerbehygiene* (West Berlin), 7 May 1969, 25/2 (99-114). Illus. 19 ref. (In German)

"The evaluation of ethyleneimine in industrial toxicology" (Die gewerbetoxikologische Beurteilung von Äthylenimin) (365-370). 13 ref. I. *Internationales Symposium der Werksärzte der chemischen Industrie vom 27 bis 29 April 1972, Ludwigshafen.* (In German)

"Aziridines" (31-113). *IARC monographs on the evaluation of carcinogenic risk of chemicals to man.* Vol. 9: *Some aziridines, N-, S-, O-mustards and selenium* (Lyons, International Agency for Research on Cancer, 1975), 268 p.

CIS 79-456 "Hazardous chemical reactions–56. Aziridines" (Réactions chimiques dangereuses–56. Aziridines). Leleu, J. *Cahiers de notes documentaires–Sécurité et hygiène du travail* (Paris), 4th quarter 1978, 93, Note No. 1146-93-78 (571-572). (In French)

Azo and diazo dyes

Azo dye is a comprehensive term applied to a group of dyestuffs that carry the azo (–N=N–) group in the molecular structure. The group may be divided into subgroups of monoazo, diazo and triazo dye and further in accordance with the number of the azo group in the molecule.

Physico-chemical data

Physico-chemical data of some azo and diazo compounds of toxicological interest are summarised in table 1. It is important, from the toxicological viewpoint, to take into account that the commercial grade dyestuffs usually contain impurities up to 20% or even higher. The composition and quantity of the impurities are variable depending on several factors such as the purity of the starting materials for the synthesis, the process of synthesis employed and the requirements of the users.

Production. Azo dye is synthesised by diazotisation or tetrazotisation of aromatic monoamine or aromatic diamine compounds with sodium nitrite in the HCl medium, followed by coupling with dye intermediates such as various aromatic compounds or heterocyclic compounds. When the coupling component carries an amino group, it is possible to produce long-chained polyazo dye by the repetition of diazotisation and coupling.

$$R{-}NH_2 + NaNO_2 + 2HCl \rightarrow R{-}\underset{\displaystyle Cl}{\underset{|}{N}}{\equiv}N + 2H_2O + NaCl$$

$$R{-}\underset{\displaystyle Cl}{\underset{|}{N}}{\equiv}N + R'H \rightarrow R{-}N{=}N{-}R' + HCl$$

Table 1. Physico-chemical properties of some monoazo and diazo compounds

Compounds (Colour Index No.)	Molecular weight	Description	Melting point (°C)	Solubility	
				Soluble in	Insoluble in
Monoazo compounds					
Amaranth (16185)	604.5	dark red-brown crystals		water (slight)	–
Azobenzene (–)	182.2	orange-red leaflets	68	ethyl alcohol, ethyl ether, glacial acetic acid	water
Carmoisine (14720)	502.5	reddish-brown crystals		water, ethyl alcohol (slight)	–
Chrysoidine (11270)	248.7	reddish-brown crystals	118-118.5	water, ethyl alcohol, cellosolve	benzene
p-Dimethylamino-azobenzene (11020)	225.3	yellow leaflets	114-117	organic solvents	water
Orange I (14600)	350.3	reddish-brown crystals	350.3	water	organic solvents
Orange G (16230)	452.4	yellowish-red crystals		water	organic solvents
Ponceau MX (16150)	480.4	dark-red crystals		water	oils
Ponceau 3R (16155)	494.5	dark-red crystals		water	oils
Sudan I (12055)	248.3	brick-red crystals	134	ethyl alcohol, acetone, ethyl ether, benzene	water
Sudan II (12140)	276.3	brown-red crystals or red needles	166	ethyl alcohol, acetone, ethyl ether, benzene	water
Sunset Yellow FCF (12140)	452.4	orange-red crystals		water	–
Diazo compounds					
Scarlet Red (26105)	380.4	dark-reddish-brown crystals	181-188	organic solvents	water
Sudan III (26100)	352.4	reddish-brown or yellowish-red crystals	195	organic solvents	water
Sudan Red 7B (26050)	379.2		128-131.5	water	ethyl alcohol, acetone benzene

R–N=N–R′ monoazo dye
R–N=N–R′–N=N–R″ diazo dye
R–N=N–R′–N=N–R″–N=N–R‴ triazo dye

For example tetrazotisation of benzidine and coupling with naphthionic acid yield Congo Red.

Use. Azo compounds are among the most popular groups of various dyes including direct dyes, acid dyes, basic dyes, naphthol dyes, acid mordant dyes, disperse dyes, etc., and are extensively used in textiles, fabrics, leather goods, paper products, plastics and many other items.

HAZARDS

In general, azo dyes as a group represent a relatively low order of general toxicity. Many of them have an oral LD_{50} of more than 1 g/kg when tested in rats and mice, and the rodents can be given lifetime laboratory diets containing more than 1 g of the test chemical per kg of diet. A few may cause contact dermatitis but usually with only mild manifestations; in practice, it is rather difficult to determine whether the dye *per se* or co-existing material is responsible for the observed skin lesion.

In contrast, increasing attention has been focused on the carcinogenic potentials of the azo dyes. Although confirmative epidemiological observations are as yet rare, the data from long-term experiments are accumulating to show that some azo dyes are carcinogenic in laboratory animals. Cases of chronic oral administration with positive results and confirmative evaluations from an IARC monograph are summarised in table 2. The main target organ under such experimental conditions is the liver, followed by the urinary bladder. The intestine is also involved in some cases. It is, however, very problematic to extrapolate these findings to man.

Table 2. Carcinogenicity of azo dyes after long-term oral administration to animals[1]

Compounds (Colour Index No.)	Species	Tumours in
p-Aminoazobenzene (11000)	Rats	liver
o-Aminoazobenzene (11160)	Rats, mice and hamsters	liver, gall bladder, lungs and urinary bladder
Azobenzene (–)	Mice	liver
Chrysoidine (11270)	Mice	liver, blood (leukaemia), and reticulum cell sarcoma
Citrus Red No. 2 (11855)	Rats and mice	urinary bladder
p-Dimethylamino-azobenzene (11020)	Rats Dogs	liver urinary bladder
Oil Orange SS (12100)	Mice	intestine
Ponceau MX (16150)	Mice	liver and intestine
	Rats	liver
Ponceau 3R (16155)	Rats	liver

[1] From *IARC monographs on the evaluation of carcinogenic risk of chemicals to man*, Vol. 8 (1975).

Recently, bacterial tests, especially that of Ames, have been widely used to screen out, by means of mutagenicity, the carcinogenic potential of various chemicals. Accordingly, azo dyes have been subjected to the Ames test and some of them were proved to be positive mostly (but not necessarily) in the presence of the S_9 mix, a bioactivation system. Examples are shown in table 3 in comparison with animal carcinogenicity data. In other experiments, Red GTL (C.I. Basic Red 18) was demonstrated to be mutagenic (without S_9 mix) among seven monoazo dyes tested, while two diazo compounds, Brown 5R (C.I. Acid Orange 45) and Guiba Black D (C.I. Direct Black 17) were positive (without S_9 mix) among the eight.

Table 3. Comparison of mutagenicity in bacterial tests and carcinogenicity in animals[1]

Compounds	Ames test[2]		Carcinogenicity
	$-S_9$	$+S_9$	in animals[3]
Amaranth	−	−	±
Carmoisine	−	−	−
Chrysoidine	−	+	+
Orange I	−	−	±
Orange G	−	−	±
Ponceau MX	−	−	+
Ponceau 3R	−	−	+
Sudan I	−	−	+
Sudan II	−	+	±
Sunset Yellow FCF	−	−	−

[1] From Garner and Nutman (1977), with slight modifications. [2] Test with *S. typhimurium TA 1538* in the absence (−) or presence (+) of the S_9 mix. Concentration of test compounds were 50 and 100 μg/plate. [3] Evaluations given in *IARC monographs on the evaluation of carcinogenic risk of chemicals to man*, Vol. 8 (1975): +, positive; ±, inconclusive; −, negative.

Concerning the mechanism of carcinogenesis, most of the carcinogenic azo dyes are not direct carcinogens, but pre-carcinogens; they require conversion by *in vivo* metabolic activation through proximate carcinogens to be ultimate carcinogens. For example, methylaminoazobenzene first undergoes N-hydroxylation and N-demethylation at the amino moiety, and then sulphate conjugation takes place with the N-hydroxy derivative to form the ultimate carcinogen which is reactive with the nucleic acid.

The benzidine-derived diazo dye may give rise to benzidine when reduced at two azo groups *in vivo* or by intestinal bacteria: for example from the urine of hamsters given Direct Black 38 monoacetylbenzidine was isolated, a compound that is mutagenic when assayed with the Ames test in presence of the bioactivation system.

SAFETY AND HEALTH MEASURES

The workshops should preferably be separated from other plants. The floor should be flat and waterproof so that it can be washed when necessary. The entrance of personnel should be regulated to avoid unnecessary exposure. Industrial hygiene measures such as enclosure of the system and installation of local/general ventilation systems should be taken to minimise exposure to hazardous chemicals including the end product, the reagents and the by-products. Packaging should preferably be automated. A complete set of clean working clothing should be supplied and put on each day at the beginning of the shift. Personal protective clothing, e.g. gloves, aprons, and footwear, must be able to resist the chemicals. When vapour or dust is expected to rise at high concentrations, use of hoods with clean air supplies is recommended; safety masks are not always reliable. After each shift, a bath or shower with the use of good quality soap should be taken to remove the chemicals which might be remaining on the skin. Such hygienic engineering and personal protective measures are especially important when the production process suggests possible exposure to suspected carcinogens. Health examination should be carried out before employment and periodically thereafter. The examination items may include hepatic function tests and urological tests (urinary sediments and cytology with Papanicolaou's staining) depending on the chemicals handled. Urinalysis for total aromatic amino compounds or diazotisable metabolites can be employed as a measure of biological monitoring of the exposure. Such a measure has the advantage of detecting absorption both via inhalation and via skin contact, while the results should be evaluated only on a group basis. Further developments are in progress to detect specific marker metabolites rather than less specific groups of chemicals.

On DIMETHYLAMINOAZOBENZENE see separate article.

IKEDA, M.

"A field survey on the health status of workers in dye-producing factories". Ikeda, M.; Watanabe, T.; Hara, I.; Tabuchi, T.; Nakamura, S.; Kosaka, H.; Minami, M.; Sakurai, Y. *International Archives of Occupational and Environmental Health* (West Berlin), 15 Nov. 1977, 39/4 (219-235). Illus. 18 ref.

"Testing of some azo dyes and their reduction products for mutagenicity using *Salmonella typhimurium TA 1538*". Garner, R. C.; Nutman, C. A. *Mutation Research* (Amsterdam), 1977, 44 (9-19). Illus. 23 ref.

"Mutagenicity of anthraquinone and azo dyes in Ames' *Salmonella typhimurium* test". Venturini, S.; Tamaro, M. *Mutation Research* (Amsterdam), 1979, 68 (307-312). Illus. 15 ref.

IARC monographs on the evaluation of carcinogenic risk of chemicals to man. Vol. 8: *Some aromatic azo compounds* (Lyons, International Agency for Research on Cancer, 1975), 357 p. Ref.

Biological monitoring for industrial exposure control. Linch, A. L. (Cleveland, CRC Press, 1974), 188 p. Illus. 312 ref.

Direct Black 38, Direct Blue 6, and Direct Brown 96 benzidine-derived dyes. NIOSH/NCI Current Intelligence Bulletin 24. DHEW (NIOSH) publication No. 78-148 (National Institute for Occupational Safety and Health, 4676 Columbia Parkway, Cincinnati) (17 Apr. 1978), 11 p. 18 ref.

CIS 79-479 *Guide to good practice for the safe handling of diazo compounds* (Drawing Office Material Manufacturers and Dealers Association (DOMMDA), 52-55 Carnaby Street, London) (1978), 23 p. Illus.

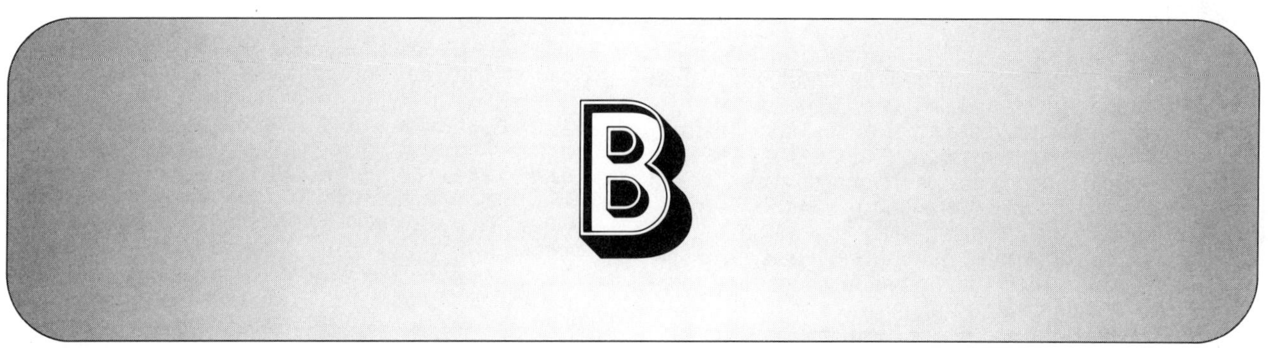

Back pain

Back pain may be defined as acute, chronic or intermittent pain occurring over the posterior neck or trunk from the occiput to the coccyx without abnormal X-ray or neurological signs. The pain may radiate anteriorly or into the nearest limb. Synonyms include lumbago, low back pain syndrome, sacroiliac strain, proximal sciatica, fibrositis, pleurodynia, torticollis, and disc syndrome.

Aetiology

The cause of the pain is unknown, though rarely (in about 0.1% of cases) it may progress to, or be followed by, a true prolapsed cervical, thoracic or lumbar intervertebral disc. The signs and symptoms of nerve root pressure will then appear, namely anaesthesia, paraesthesia, muscular wasting and loss of a tendon reflex. It must be stressed that an isolated neurological finding (e.g. 1 cm of unilateral calf-wasting, a single lost tendon reflex, or a patch of anaesthesia) is a relatively common finding in the normal population and should not in itself be regarded as the forerunner of a prolapsing disc. It must be remembered that tumours and all other causes of compression of a spinal nerve root are 10 times rarer (about 0.01% of back pain cases) than prolapsed intervertebral discs. Although the aetiology of back pain without abnormal X-ray or neurological signs is uncertain, there are more hypotheses about its origin than there are anatomical structures in the back.

Symptoms

The history of the pain may be one of gradual or sudden onset with or without a remembered incident. A very common onset is an incident or accident with an immediate sharp pain followed by a pain-free interval of, usually, a few minutes followed in turn by an ache gradually increasing in intensity and spreading in area. There is a constantly occurring segmental syndrome in almost all cases of back pain, whatever ultimate diagnosis is made, namely:

(a) dull ache which may be central, unilateral or bilateral;

(b) tender spot, usually paravertebral between 1 cm and 5 cm from the midline; there may also be a central tender spot;

(c) an area of hyperaesthesia encompassing the tender spot and the painful area as pointed out by the patient;

(d) limitation of the trunk or neck movement by pain in at least one direction—the usual movements tested are flexion, extension, lateral flexion and lateral rotation of the spine.

Incidence and course

In general about 90% of cases of back pain will clear, and of these only 10-15% will recur within one year. Cases may occur for the first time in childhood or in old age, but the modal age group is 35-40 years for the first onset of back pain; the first attack rate then decreases with age. In an average engineering works approximately half the cases occur in work and half at home: 10% of the employees each year will have an attack of pain sufficiently severe for them to seek medical help. In most works studied the condition occurred with equal frequency among males and females, and among manual and sedentary workers, but it was commoner among very heavy manual workers such as dockers or miners.

Diagnosis

This may be made on the history and physical examination alone, but it is strongly advisable to have at least an anteroposterior and a lateral X-ray of the spine. The important clues in the history that suggest a case of nerve root pressure rather than a self-limiting back pain are:

(a) paraesthesia in the relevant dermatome (this is usually in the arm or leg in cervical or lumbar nerve root compression or on the chest wall in thoracic cases); and

(b) that, though the pain may previously have been worse in the back or neck, it has now altered in quality and has become most severe in the distal part of the limb, for the dermatomes which correspond to the commonly occurring discs, namely, $C_{5,6,7}$ and $L_{3,4,5}$, are in the forearm and hand, and leg and foot respectively.

PREVENTIVE MEASURES

These may be divided into: *(a)* prevention of the first attack of back pain; and *(b)* prevention of recurrence.

The commonest cause of back pain is a twisting lift, usually with an accident or incident occurring before the completion of the task (e.g. the foot or the load slipping). However, cases may also occur with an over-reach, a sideways bend or during the extension of the neck or trunk.

Flexion is the commonest method of lifting a weight and the commonest movement in causing back pain; simple common sense prevention lies:

(a) in mechanisation and mechanical handling;

(b) in training in correct, kinetic methods of lifting and carrying;

(c) in setting a limit to the maximum weight to be carried by one worker.

The ILO Maximum Weight Recommendation, 1967 (No. 128), states that where the maximum permissible weight which may be transported manually by one adult male worker is more than 55 kg, measures should be taken as speedily as possible to reduce it to that level. It is also recommended that, where adult women or young persons are engaged in manual transport of loads, the weight of such loads should be substantially less than that permitted for adult male workers. However, there is as yet no published evidence that either lifting instruction or limitation of maximum weight does, in fact, reduce the incidence of back pain sickness absence, and there is urgent need for controlled trials in this field.

With regard to the prevention of a recurrence, there is no evidence that back strengthening exercises are of use. There is evidence, however, that recurrences tend to be caused by the same pattern of movement. If it was twisting to the right that caused the first episode, it is likely that twisting to the right will also cause the second; similarly, if the pain was caused by hyperextension originally, then a recurrence will probably occur at the next incident of severe hyperextension. Therefore, the taking of a detailed history of the attack is essential in order to point out to the patient the folly of his ways and how he can alter his habits to avoid the precipitating circumstance that may lead to a recurrence of the pain.

Ergonomics may be of help in eliminating pain-producing work situations. For although there is no such clinical entity as an "industrial back", there may be certain working postures that cause a run of similar backs (see POSTURES AND MOVEMENTS, OCCUPATIONAL). In such cases the machine or procedure should be ergonomically studied with a view to eliminating the hazard.

Pre-employment examination. In the medical management of a population whose tasks involve a back pain risk, such as dockers and miners, it has been recommended that pre-employment spinal X-rays should be taken to eliminate those radiological conditions thought to be precursors of back pain, namely *spina bifida*, asymmetrical lumbosacral facets, spondylolistheses, narrowed invertebral spaces, osteoarthritis of the joints of the vertebral arches, etc. However, apart possibly from unilateral sacralisation of the fifth lumbar vertebra, there is no X-ray abnormality more frequent among those suffering from back pain than in the general population, and a strong case cannot yet be made out for pre-employment medical examinations including spinal X-rays.

Treatment

As the cause of this commonly occurring pain is unknown, the treatment must be either none or empirical. As three-quarters of the cases which remain at work clear in the first week, there is a strong argument for taking an accurate history, for making not only a physical examination of the trunk or neck, but also a neurological examination of the relevant limb or limbs, and for reassuring the patient that, as he has no objective or subjective neurological signs, he is unlikely to have a "disc". Analgaesics and infrared radiation therapy may be given daily and the patient should be informed that, provided he keeps moving and at work (other than very heavy or "dangerous-to-himself-or-others" work), the pain should clear within a week. With regard to other possible treatments, it has been found in a clinical trial with 80 back-pain cases randomly allocated between rotational manipulation and detuned short-wave diathermy, that patients receiving manipulation had somewhat greater relief of pain on the third day but, by the seventh day, the relief of pain was the same for each series. Any case in which, with normal activity, the pain has not lessened within 14 days, should have spinal X-rays and a full neurological examination.

GLOVER, J. R.

General:

CIS 75-184 *Contribution to the study of low back pain of occupational origin* (Contribution à l'étude des lombalgies en milieu professionnel). Mauduit, M. (Paris, Université de Paris VI, Faculté de médecine Pitié-Salpêtrière, 1973), 100 p. 196 ref. (In French)

"Degenerative disease and injury of the back" (30-99). Illus. Ref.

Occupational safety and health symposia 1977. DHEW (NIOSH) publication No. 78-169 (Cincinnati, National Institute for Occupational Safety and Health, 1978), 331 p.

Prevention:

CIS 78-1986 "A study of three preventive approaches to low back injury". Snook, S. H.; Campanelli, R. A.; Hart, J. W. *Journal of Occupational Medicine* (Chicago), July 1978, 20/7 (478-481). 20 ref.

CIS 77-872 "Pre-employment back X-rays". Montgomery, C. H. *Journal of Occupational Medicine* (Downers Grove, Illinois), July 1976, 18/7 (495-498). 56 ref.

Kinetic methods of manual handling in industry. Himbury, S. Occupational Safety and Health Series No. 10 (Geneva, International Labour Office, 1967).

Bagasse

Bagasse is the fibre of sugar-cane stalks, which remains after the sugar-containing juice has been extracted. It is composed mainly of cellulose, and analysis indicates a 4% ash content and a 2% protein content; it contains virtually no silica (0.5% by weight of the ash). When dry, bagasse is rapidly broken down into dust.

Uses. In industry, bagasse is used principally in the manufacture of different types of paper and cardboard and a variety of building materials such as insulating wallboard and fillers for veneered doors. It has also been employed as a fuel, a fertiliser ingredient and poultry feed and in the manufacture of explosives and refractory bricks. More recently, it has been mixed with mud to form "sludge" used in oil-field drilling, and has been used as a filler for plastics and asphalt.

HAZARDS

The inhalation of dried bagasse dust may cause a disease of the respiratory system called "bagassosis". Bagasse is a flammable substance and airborne bagasse dust may be exposive.

Bagassosis. This disease is caused by the inhalation of airborne bagasse dust. It occurs almost exclusively in manufacturing-plant employees who are engaged in bale breaking, shredding and processing the dry fibre after it has been baled in a warm moist state, stored outdoors for long periods and allowed to become mouldy. On rare occasions, bagassosis has occurred in laboratory workers and gardeners handling old fibre. Bagassosis never occurs in workers handling fresh, moist sugar-cane fibre in the sugar industry. It has now been determined that the disease is caused by fungus-like bacteria known as thermophilic (heat loving) actinomycetes, which thrive at high temperatures (60 °C) and grow in stored bagasse. These organisms produce an allergic reaction in the lungs when inhaled. [Bagassosis is presently considered as a form of extrinsic allergic alveolitis.]

Cases of bagassosis have been reported in Louisiana, Texas, Illinois and Missouri in the United States, in Puerto Rico, the United Kingdom, Italy, Spain, India, the Philippines, Peru, Mexico and Thailand.

Symptoms include shortness of breath, cough, sputum production, fever, chills, chest pain, weakness, decreased appetite, loss of weight and occasional expectoration of blood. The period between exposure to bagasse dust and the onset of illness may vary from a day to many months. Separation of the patient from contact with bagasse dust usually results in gradual but complete recovery after a period of weeks or months, although some patients may remain symptomatic for a year or longer.

Once an individual has contracted bagassosis he should avoid all future contact with dry sugar-cane fibres since relapses almost invariably occur when contact is re-established and each episode of illness tends to be more severe and prolonged. The danger of permanent disability increases proportionately.

SAFETY AND HEALTH MEASURES

The keys to bagassosis prevention are effective dust control and the elimination of bacteria and fungi from the fibre material.

It should be re-emphasised that, whereas the fresh fibre is essentially innocuous, the handling of mouldy material which has been stored outdoors for long periods must be approached with great caution and avoided if possible. Stored bales of bagasse should never be broken, shredded or processed in a dry state. Violation of this principle has been responsible for all major outbreaks of bagassosis. Wetting the bales before they are broken or breaking them under water are helpful methods of dust control. Good ventilation and other techniques of dust elimination should be enforced. Workers should also be provided with respiratory protective equipment and required to wear it at all times when working with stored bagasse.

In Louisiana a method of producing "clean", "dehydrated" or "pasteurised" bagasse has been developed in which the fibre is dried at around 1 000 °C before being made up into bales, which are then stored indoors and protected from moisture until they are shipped to manufacturing plants for industrial use. Material produced by this process is regarded as less dusty, less likely to contain bacteria and, therefore, less likely to produce bagassosis; however, it should still be handled under the same dust-control precautions as outlined above. Treating the fibre with 1 or 2% propionic acid may also eliminate fungi and make the material safer to handle.

Most of the important methods of utilising bagasse for manufacturing purposes require that it undergo a stage in which it is cooked at high temperatures. Presumably this process effectively destroys the disease-producing fungi and any other potential allergenic properties of the fibre, since workers engaged in handling or cutting the manufactured products never develop bagassosis.

An exception to this observation relates to chemically treated press boards which are produced by an entirely dry process. There is some evidence that the dust from finished products of this kind may still be harmful. An outbreak of bagassosis at a factory in Louisiana which employed this method is the largest which has ever occurred in the history of this disease (200 cases).

The dust-control measures applied for the prevention of bagassosis will normally be sufficient to prevent the formation of explosive concentrations of bagasse dust. However, where dangerous atmospheric concentrations may occur, the normal measures for the prevention of dust explosions should be applied. All sources of ignition should be eliminated and only specially protected electrical equipment should be allowed in hazardous areas. Sources of friction or static electricity in machines, etc., should be eliminated, and welding, cutting and smoking in dusty areas should be prohibited. A full explosion venting system may be necessary in some areas. In view of the fire hazard, fire exits, fire-fighting equipment and instruction in procedures in case of explosion or fire are also necessary.

BUECHNER, H. A.

"Bagassosis: IV Precipitins against extracts of thermophilic actynomycetes in patients with bagassosis". Salvaggio, J.; Arquembourg, P.; Seabury, J.; Buechner, H. A. *American Journal of Medicine* (New York), Apr. 1969, 46/4 (538-544). Illus. 15 ref.

"Bagassosis: a histopathologic study of pulmonary biopsies from six cases". Boonpucknavig, V.; Bhamarapravati, N.; Kamtorn, P.; Sukumalchandra, Y. *American Journal of Clinical Pathology* (Philadelphia), 1973, 59/4 (461-472). Illus. 22 ref.

"Elimination of bagassosis in Louisiana paper manufacturing plant workers". Lehrer, S. B.; Turer, E.; Weill, H.; Salvaggio, J. E. *Clinical Allergy* (Oxford), 1978, 8/1 (15-20). Illus. 5 ref.

Bakeries

The industrial sector which covers the manufacture of foodstuffs from starches and sugars groups together bakeries and biscuit-, pastry- and cake-making establishments in which the safety and health hazards presented by the raw materials, the plant and equipment and the manufacturing processes are all similar (see BISCUIT MAKING). This article deals with small-scale artisanal production and covers bread and various related products.

Production. There are three main stages in breadmaking (mixing and moulding, fermentation, and baking) which are carried out in different work areas, i.e. the raw materials store, the mixing and moulding room and cold and fermentation chambers, the oven, the cooling room and the wrapping and packaging shop. The sales premises are frequently attached to the manufacturing shops.

Flour, water and salt and yeast are mixed together to make dough; hand mixing has been largely replaced by the use of mechanical mixing machines. Beating machines are used in the manufacture of other products. The dough is left to ferment in a warm, humid atmosphere after which it is divided, weighed, moulded and baked.

Small-scale production furnaces are of the fixed-hearth type with direct or indirect heat transfer. In the direct type the refractory lining is heated either intermittently or continuously before each charge. Off-gases pass to the chimney through the adjustable orifices at the rear of the chamber. In the indirect type the chamber is heated by steam passing through tubes in the chamber wall or by forced hot-air circulation. The oven may be fired by wood, coal, oil, town gas, liquefied petroleum gas or electricity. In country regions ovens with hearths heated directly by wood fires are still found.

Bread is charged into the oven on paddles or trays.

The oven interior can be illuminated so that the baking bread can be observed through the chamber windows. During baking, the air in the chamber becomes charged with water vapour given off by the product and/or introduced in the form of steam. The excess usually escapes up the chimney but the oven door may also be left open.

HAZARDS AND THEIR PREVENTION

Working conditions. The working conditions in artisanal bakehouses have the following features:

(a) night work starting at 2.00 or 3.00 h, especially in Mediterranean countries, where the dough is prepared in the evening;

(b) premises often infested with parasites such as cockroaches, mice and rats which may be carriers of pathogenic micro-organisms–suitable construction materials should be used to ensure that these premises are maintained in an adequate state of hygiene (see PEST CONTROL AT THE PLANT);

(c) house-to-house bread delivery which is not always carried out in adequate conditions of hygiene and which may entail an excess workload;

(d) low wages supplemented by board and lodging.

Premises. These are often old and dilapidated and lead to considerable safety and health problems. The problem is particularly acute in rented premises for which neither the lessor nor the lessee can afford the cost of renovation. [Floor surfaces can be very slippery when wet, although reasonably safe when dry; non-slip surfaces should be provided whenever possible.] General hygiene suffers owing to defective sanitary facilities, increased hazards of poisoning, explosions and fire, and the impossibility of modernising heavy bakehouse plant owing to the term of the lease.

Small premises cannot be suitably divided up; consequently traffic aisles are blocked or littered, equipment is inadequately spaced, handling is difficult and the danger of slips and falls, collisions with plant, burns and injuries resulting from overexertion is increased. Where premises are located on two or more storeys there is the danger of falls from a height.

Basement premises often lack emergency exits, have access stairways which are narrow, winding or steep and are fitted with poor artificial lighting. They are usually inadequately ventilated and consequently temperatures and humidity levels are excessive; the use of simple cellar ventilators at street level merely leads to the contamination of the bakehouse air by street dust and vehicle exhaust gases.

Accidents

Knives and needles are widely used in artisanal bakeries and entail a risk of cuts and puncture wounds with subsequent infection; heavy, blunt objects such as weights and trays may cause crush injuries if dropped on the worker's foot. Ovens present a number of hazards: depending on the fuel used, there is the danger of fire and explosion; flashbacks, steam, cinders, baked goods or unlagged plant may cause burns or scalds; firing equipment which is badly adjusted or has insufficient draw, or defective chimneys may lead to the accumulation of unburnt fuel vapours or gases or of combustion products, including carbon monoxide, which may cause intoxication or asphyxia; defective electrical equipment and installations, especially of the portable or mobile type, may cause electric shock; and the sawing or chopping of wood for wood-fired furnaces may result in cuts and abrasions.

Flour is delivered in sacks weighing up to 100 kg and these must often be lifted and carried by workers through tortuous gangways (steep inclines and staircases) to the storage rooms. There is the danger of falls whilst carrying heavy loads and this arduous manual handling may cause back pain and lesions of intervertebral discs. The hazards may be avoided by: providing suitable access ways to the premises; stipulating a suitable maximum weight for sacks of flour; using mechanical handling equipment of a type suitable for use in small undertakings and at a price within the range of most artisanal workers; and by wider use of bulk flour transport which is, however, suitable only when the baker has a sufficently large turnover. [In mechanised bakeries dough which is in an active state of fermentation may give off dangerous amounts of carbon dioxide; thorough ventilation should therefore be applied wherever the gas is likely to accumulate (dough chutes, etc.).]

[A wide variety of machines are used in bread manufacture, particularly in industrial bakeries. Here also, mechanisation can bring serious accidents in its wake. Modern bakery machines are usually equipped with built-in guards whose correct operation often depends upon the functioning of electrical limit switches and positive interlocks. Feed hoppers and chutes present special hazards which can be eliminated by extending the length of the feed opening beyond arm's length to prevent the operator from reaching the moving parts; hinged double gates or rotary flaps are sometimes used as feeding devices for the same purpose.

Nips on dough brakes can be protected by either fixed or automatic guards. A variety of guards (covers, grids, etc.) can be used on dough mixers to prevent access to the trapping zone while permitting insertion of additional material and scraping of the bowl. Increasing use is made of bread-slicing and wrapping machines with alternating saw blades or rotary knives; all moving parts should be completely enclosed, interlocking covers being provided where access is necessary.]

Health hazards

Bakehouse workers are usually lightly clothed and sweat profusely; they are subject to draughts and pronounced variations in ambient temperature when changing, for example, from oven charging to cooler work. Airborne flour dust may cause rhinitis, buccopharyngeal disorders, bronchial asthma and eye diseases; sugar dust may cause dental caries. Airborne vegetable dust should be controlled by suitable ventilation. Allergic dermatitis may occur in persons with special predisposition.

The above health hazards and the high incidence of pulmonary tuberculosis amongst bakers emphasise the need for medical supervision with frequent periodical examinations; in addition, strict personal hygiene is essential in the interests of both workers and the public in general.

VILLARD, R. F.

CIS 77-947 "Simultaneous exposure to airborne flour particles and thermal load as cause of respiratory impairment". Beritič-Stahuljak, D.; Valić, F.; Cigula, M.; Butković, D. *International Archives of Occupational and Environmental Health–Internationales Archiv für Arbeits- und Umweltmedizin* (West Berlin), 5 July 1976, 37/3 (193-203). 39 ref.

CIS 77-2060 "Bakers' asthma". Hendrick, D. J.; Davies, R. J.; Pepys, J. *Clinical Allergy* (Oxford), May 1976, 6/3 (241-250). 15 ref.

CIS 75-1700 "Occupational dental disease due to acids" (Berufsbedingte Säureschäden der Zähne). Reinhardt, J.; Kittner, E. *Zentralblatt für Arbeitsmedizin und Arbeitsschutz* (Heidelberg), Mar. 1975, 25/3 (72-75). Illus. (In German)

"Rest periods for bakery workers" (Timpul de odinha al cocătorilor). Marin, M. D.; Rebegebel, A. *Industria alimentară* (Bucarest), Feb. 1971, 22/2 (88-93). Illus.

Safety requirements for bakery equipment. ANSI Z50.1.1977 (American National Standards Institute, 1430 Broadway, New York) (1977), 24 p. 30 ref.

"Bread slicing machines" (Machines à couper le pain). Vacheret, J. M. Note 1212.97.79. *Cahiers de notes documentaires* (Paris), 4th quarter 1979, 97 (511-520). Illus. (In French)

Bamboo and cane

Bamboo

Botanically, bamboo includes many species of *Bambuseae* of the *Gramineae* family. It grows wild in tropical or subtropical areas, especially in South-East Asia and can be easily cultivated by planting the stalks.

Some species of bamboo are used as vegetables and may also be pickled or preserved. Bamboo stems are used for many purposes in place of wood and in certain instances may be preferable to it. In some parts they are used for many kinds of construction: houses are entirely built from bamboo with the stems as uprights and the walls and roofs made from split stems or lattice work. Boats and rafts are also made from bamboo and the material is also widely used for fencing. It is employed for furniture, containers and handicraft products of every kind, and especially for umbrellas or walking sticks; in some areas high-quality paper is made from pulped bamboo.

The harvest of bamboo shoots involves frequent visits to the forest because, to be edible, the shoots have to be cut at the right stage of growth. Stem harvesting takes place only when the stems have grown long and thick enough for the purpose for which they are to be used. The stems are cut by a hatchet, sharp knife or saw: the stumps are left in the ground.

HAZARDS AND THEIR PREVENTION

Venomous snakes (see ANIMALS, VENOMOUS) present a hazard in the forest; falls may be caused by stumbling over the stumps; tetanus is a risk associated with all cuts. During work with bamboo stems, accidental cuts can be caused by the knives used in splitting the stem, or by the sharp edges of the split bamboo itself. Skin punctures may also occur amongst handicraft workers. Hyperkeratosis of the palms and fingers has been observed in workers making bamboo containers to hold bananas for export.

First aid and medical treatment is required to deal with snake bites and to prevent tetanus. All cutting knives and saws should be carefully maintained.

Cane

Cane is sometimes confused with bamboo, but is quite distinct botanically and comes from varieties of the rattan palm, notably *Calamus margaritae*. Rattan palms grow freely in tropical and subtropical areas, particularly in South-East Asia, where harvests are usually gathered from the wild trees often in uncultivated mountainous areas. The stems of the plants are cut near the roots, dragged out from the thickets and sun-dried. The leaves and the bark are then removed and the stems sent for processing. Cane is used for making furniture, expecially chairs, baskets, containers and other handicraft products. It is very popular due to its appearance and elasticity. In manufacture it is necessary to split the stems.

Figure 1. Bamboo.

HAZARDS AND THEIR PREVENTION

In harvesting, the workers are exposed to the dangers of remote forests including snakes, venomous insects, etc. The bark of the tree has thorns which may tear the skin and workers are also exposed to cuts from knives. Gloves should be worn when the stems are handled. Cuts are also a risk during manufacture, and hyperkeratosis of the palms and fingers may often occur amongst workers owing, probably, to the friction of the material.

KO, Y.

Bananas

The banana (different varieties of the genus *Musa*) is grown widely over the tropical, and in some places subtropical areas of the world. It may be cultivated in plantations or by small proprietors, who may be united in co-operatives. The fruit is a staple item of diet in the countries in which it grows and also forms a valuable

Figure 1. Banana cutters at work.

export crop for many of these countries (e.g. Brazil, Cameroon, India, Ivory Coast, Central America and some of the Caribbean islands).

Cultivation. [Banana growing is difficult to mechanise and frequently all cultivation work is done manually. However, on very large plantations where the ground is flat mechanised cultivation using heavy tractors and equipment is being increasingly developed. Where drainage is necessary mechanical shovels may be used to make the larger ditches. Planting material is transported to the field by tractor and trailer but planted by hand, having previously been dipped in boiling water to kill nematodes.]

The buds are planted in soil prepared with basic fertilisers: tilling, banking with additional soil, weeding and irrigation are necessary during growth. Pesticides and additional fertilisers must be applied at the appropriate stage [aerial or mechanical spraying may be resorted to]. As the plant grows, supports, usually made from bamboo stems, have to be fixed to provide protection against storm damage. Fruit is harvested after about 10 months.

[Harvesting is manually done by means of a long pole with attached knife. The banana hands are detached on to the shoulder of the worker and a second removes them, attaches a nylon cord and puts them on an overhead cableway which conveys them to the tractor. There has been a radical change in the system of shipping bananas in the last ten years. Formerly bananas were shipped simply in hands in cold space, now they are usually washed and wrapped in polyethylene and packed in corrugated cardboard boxes. This work is done manually, often by women and young persons (see PLANTATIONS).]

HAZARDS AND THEIR PREVENTION

Poisoning from pesticides is the main hazard during cultivation and care and cleanliness in handling are important; in some areas there may be risk from snakes and insects (see ANIMALS, VENOMOUS). In harvesting, agricultural implements, such as machetes, may inflict severe wounds. There is no evidence that the fluid from the cut plant causes dermatitis.

Loading for export usually involves the lifting and carrying of heavy weights, often by women, and long hours of work while the banana ships are in port. The main export of bananas is to temperate countries; they are shipped unripe at low temperatures and ripened in the receiving country. Fruit ripening is often carried out by exposing the bananas to ethylene in special ripening chambers and strict precautions against explosion are necessary (see FRUIT RIPENING).

KO, Y.

Industrial and labour problems:

General Report. Committee on work on plantations. Seventh Session (Geneva, International Labour Office, 1976), 186 p.

"Occupational exposure of users of DBCP (Di-Bromo-Chloro-Propane) as a nematocide in banana plantations in Israel". Dror, K.; Lemesch, C.; Pardo, A. *Agricultural Medicine and Rural Health* (Usuda, Nagano, Japan), Summer 1981, 6/1 (11-18). 10 ref.

Band saws

Definition

The band saw is a machine with a cutting tool comprising an endless steel band with saw teeth formed on one edge, with a rectilinear, continuous cutting movement. This blade travels over the rims of two wheels in the same way as a drive belt. The first of these wheels, the drive wheel, acts like a flywheel; the other is the return wheel and is used to adjust blade tension (figure 1). The stock to be sawn is placed on a table or fixed to a carriage and forced against the blade. In certain special machines the machine itself is moved whilst the stock being sawn remains stationary.

Types of band saw

The first band saw was patented by William Newberry in London in 1808. However, it was not until around 1850 that an effective band saw blade became available. The machines of this period were of artisanal construction and operated by hand wheels or pedals. And it was not until the development of a power drive that the band saw came into wide use. The principle of the machine has remained unchanged; successive developments have related only to construction details, blade guide devices, stock feed and guarding.

The various types of band saw can be divided into four categories:

(1) vertical table band saws (the most common type);
(2) resawing band saws;
(3) band mills

(a) with a vertical blade;

(b) with a horizontal blade;

(4) portable band saws.

There are also band saws with a reciprocating movement but these are exceptional.

Machine construction

Machines should always be manufactured in accordance with good engineering practice. The chassis should be rigid and designed in such a way that it can be easily and solidly fixed to the floor, the wall or mountings. The electrical equipment should be in accordance with requirements. The machine should be fitted with all general protective devices and designed to allow the easy fitment of safety devices specific to the type of work or the location. The start and stop controls should be within easy reach of the machine operator and protected against unintentional operation. It is advisable to fit a sawdust exhaust ventilation collection device which will ensure better visibility at the cutting point and keep the machine and its surroundings free from excessive build-up of waste.

Machine location

The machines should be located in such a way that they can be operated easily for all the types of sawing for which they are intended. Their surroundings should be kept clear and well marked to ensure easy delivery and handling of the stock. The floor should be level, even, without bumps or holes and with a non-slip surface. The lighting, whether natural, artificial or a combination of both, should be adequate and glare-free.

Machine use, working procedures

A band saw, even if well guarded, may be dangerous for workers not trained in its use. The following general safety rules should be observed in the use of band saws:

(a) only persons who have been instructed and trained by skilled band saw operators should be allowed to use them;

(b) the use and maintenance of the guards should be an integral part of the instruction provided;

(c) authorisation to use the machine should be withdrawn from persons who omit to use the guards or use them incorrectly;

(d) the operating procedures taught should give priority to safety; output should not be increased at the expense of safety;

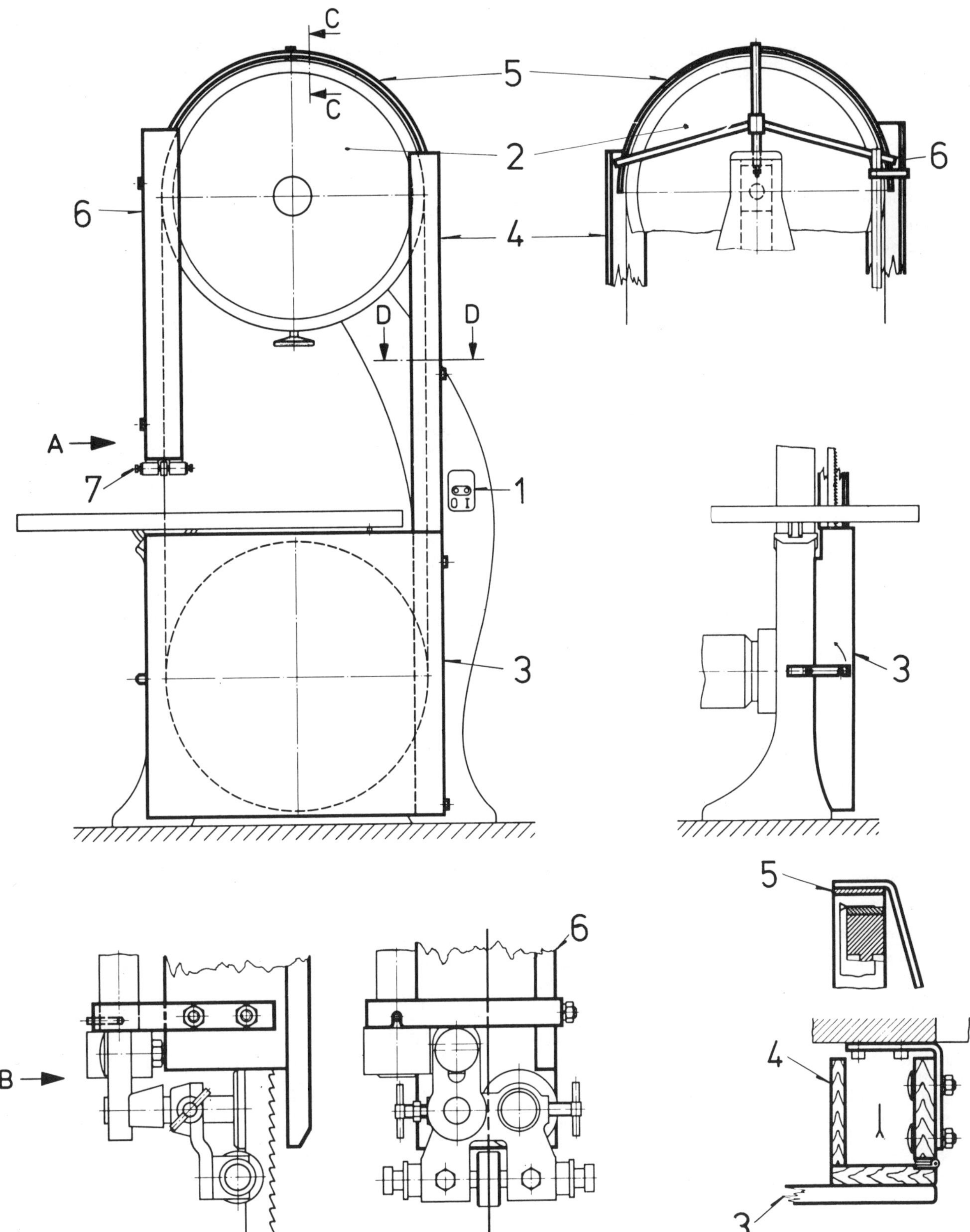

Figure 1. Band saw. 1. Switch, recessed push-button. 2. Solid upper driving wheel; in the case of a spoked wheel, a wheel guard is necessary. 3. Enclosure of the lower wheel and the rising and descending parts of the blade under the table. 4. Enclosure of the rising part of the blade above the table. 5. Guarding strip over the upper driving wheel. 6. Guard for the descending part of the blade above the table, which automatically follows the upper blade guide when it is moved vertically. 7. Band saw guide. A = side view; B = front view; C and D = cross-sections.

(e) the operator of a log band mill should, in addition to being a skilled saw operator, be aware of his responsibility and be able to exercise authority. He is required not only to operate the machine but also ensure strict discipline among his gang; the slightest relaxation of attention or false move on his part or by his colleagues may lead to a severe or fatal accident. Well outlined operating procedures and clear and simple instructions will facilitate his task.

HAZARDS

Accidents on band saws are due primarily to:

(1) contact with the working part of the blade:

(a) at the end of the saw cut when the worker's hands or body are close to the saw blade;

(b) when offcuts or sawdust are being removed from the work table;

(c) as the result of false moves, slipping, falling;

(2) contact with the non-cutting part of the blade:

(a) when handling objects or intervening close to the blade;

(b) as a result of false moves, slipping and falling;

(3) contact with the drive and return wheels:

(a) whilst adjusting the return wheel;

(b) when moving legs or feet under the table;

(c) when handling objects near to the wheels;

(d) as the result of false moves, etc.;

(4) contact with the drive or transmission system of the machine;

(5) being caught or crushed by a carriage or the automatic feed device;

(6) the blade flying off the wheels or breaking;

(7) design defects in the machine, the tool or the safety devices;

(8) incorrect working procedures.

SAFETY MEASURES

The general safety measures concerning the design, location and use of band saws have already been indicated above. Given below are the specific safety measures.

The blade wheels (drive and return wheel)

The wheels should be perfectly balanced both statically and dynamically in order to ensure their correct vibration-free operation. They should be of the solid type and have a smooth surface; if this is not the case they should be completely enclosed. The shape of the wheel close to the rim should prevent unintentional contact with the saw teeth protruding from the rim.

The casing should not inhibit evacuation of sawdust; it should be openable to allow blade changes and machine maintenance.

Rim surfacings should be examined periodically and replaced if they are torn, unstuck or markedly worn.

The blade (band)

The blade should be of first quality steel. Usually its thickness should not exceed 1/1 000 of the wheel diameter but it should not be less than 0.5 mm. The blade width will be between 5 and 250 mm depending on the machine. Blade speed is around 20-25 m/s for normal saws and up to 50 m/s for certain log band mills.

The weak point of the band is the braze. This should be scrupulously in accordance with the manufacturer's indications. Correct brazing is not possible unless use is made of equipment with precise temperature control. The blade teeth should be chosen in accordance with the type of wood and work being done. In principle, the teeth are small and straight for hard and dry woods; they should be larger and with a great set for soft or moist woods in order to prevent jamming and blade heating.

The set is obtained by bending alternate teeth to the right and to the left; it is dangerous to bend the whole of the tooth since in doing so one may produce at the base of the tooth the start of a crack which may result in the blade breaking.

Blade adjustment

Correct adjustment is decisive in preventing the blade jumping from the wheels, breaking or wearing too rapidly.

Adjustment ensures:

(1) correct tension of the band ensuring blade rigidity and adherence to the wheel rim. Adjustment is carried out by changing the distance between the wheels, i.e. moving the centre of the return wheel. Blade tension depends on blade width; for normal blade sizes it is around 10 kg/mm of blade width. In practice on normal saws it is sufficient to tension the band until it no longer chatters. There is usually an elastic component in the tensioning system to compensate for the effects of dilation and jolts during sawing;

(2) the correct positioning of the blade on the wheel rim. This adjustment is made by slightly tilting the axis of one wheel in relation to the other so that:

(a) wide blade teeth are not in contact with the wheel rim covering, in order to spare it and the saw guides from wear,

(b) narrow saw blades are, in contrast, in contact with the wheel rim covering to prevent the blade flying off.

The saw guides

The guides are designed to ensure that the blade runs true in both planes. The guides should be in contact with the whole width of the blade and should be located on both sides of the work piece and as close to it as possible. They should therefore be adjustable. In the case of table band saws, one of the guides is located directly underneath the table. Usually the rear guide is in the form of a rotating disc with its axis parallel to the direction of stock feed; the blade rests on the surface of the disc close to its periphery. Lateral guidance is obtained by rollers or by plastic, compressed wood, graphite, etc., friction pads located on each side of the blade. A good guide is essential to ensure accurate and safe sawing.

Blade maintenance

Safety depends to a great extent on the condition of the blade. It is necessary to check the blade closely at regular intervals to detect any start of a breakage, deformation, etc., and to remedy it immediately. Sharpening should be particularly accurate and regular to ensure easy sawing without jolts and in order to reduce strain on the blade.

Blade guarding

The non-working parts of the blade should be enclosed over their whole length both in front and laterally. The lateral guides of the working part of the blade should be connected to the guides so that they are brought as close as possible to the stock being cut.

A circular guard should be located around the half circumference where the band is in contact with the wheel. The distance of this guard from the wheels should be as small as possible taking into account the need to replace the blade. It should be lined with a soft material to prevent damage to the blade in case of breakage.

Guards for the working part of the blade are problematic. Numerous solutions have been proposed but none of them has proved entirely satisfactory. These guards are all fragile, affected by sawdust or off-cuts and/or a hindrance during work. We believe that in using table band saws of the normal type, the hazard of the unguarded working part is extremely small if the saw operator knows his job, works conscientiously and uses push sticks and other accessories. In the case of log band mills, on the other hand, the working part of the blade is an enormous hazard in view of the length of unguarded blade–usually located near the work station, the power of the machine, exceeding 100 hp, and the heavy work done by saw operators close to the blade. The blade guard should:

(a) be very robust;

(b) cover the blade until the stock is immediately adjacent to the saw teeth, and move away rapidly;

(c) re-cover the blade when the stock has passed through;

(d) permit the carriage to return to the start position.

Initially the guards were fixed but located at a distance from the blade or close to it and hand operated. Their effectiveness was illusory because, due to negligence, they were left in the open position or were too distant from the blade. Subsequently an attempt was made to control the opening and closure of the guard by means of mechanical devices operated by the movement of the carriage. Although this proved better it did not give satisfaction because it was complicated, vulnerable and easily knocked out of adjustment. In recent years designers have developed guards operated by servo-motors, the opening and closing being controlled by the carriage movement. They cover entirely the working part of the blade when the carriage is at a halt, i.e. during loading and unloading operations. Once the carriage begins to move, the guard slides aside and then, when the cut has been completed, it returns to its original place. This is a practical and valid solution (figures 2 and 3).

Miscellaneous guards

The carriages for log band mills are extremely heavy and move at speeds of up to 3 m/s. They are dangerous and may cause accidents:

(a) owing to workers' being crushed by the wheels. These accidents can be prevented by the installation of vertical plates almost touching the rail and located immediately in front of and behind each wheel, unless a total longitudinal guard prevents the worker's feet from being placed on the track;

(b) by workers' being caught by the projecting parts of the carriage, in particular the ends of the cross beams. Carriages with smooth sides eliminate this type of accident.

The pits for vertical log band mills should be covered completely with strong material. Any access hole should be covered by a door which is so designed as to prevent falls when in the open position. A fixed ladder, or better still, a staircase, should provide easy access to the pit bottom.

The controls of log band mills should be locked when in the "off" position.

The drive rollers of resawing band saws should be guarded by a casing, the edge of which should be at least 1 cm from the work piece.

Figure 2. Horizontal log band mill. A guard with vertical movement. 1. Blade guard. 2. Servo-control motor. 3. Sawdust exhaust ventilation collector. 4. Log being sawn.

Figure 3. Vertical log band mill. Guard with a horizontal movement. 1. Blade guard. 2. Servo-motor.

Brakes

Band saws should be fitted with an effective, progressive brake.

CHAVANEL, A.

CIS 79-955 "Safety devices on vertical band-saw machines" (Sicherheitstechnische Einrichtungen an Bandsägemaschinen). Behrens, L. *Sicherheitsingenieur* (Heidelberg), Oct. 1978, 9/10 (54-59). Illus. (In German)

CIS 74-1754 "Table band saws" (Scies à ruban à table). Tobelem, W. *Cahiers de notes documentaires–Sécurité et hygiène du travail* (Paris), 1st quarter 1974, 74, Note No. 876-74-74 (5-13). Illus. 2 ref. (In French)

CIS 75-251 "Log band saws" (Scies à ruban à grumes). Tobelem, W. *Cahiers de notes documentaires–Sécurité et hygiène du travail* (Paris), 2nd quarter 1974, 75, Note No. 893-75-74 (177-186). Illus. 3 ref. (In French)

Barium and compounds

Barium (Ba)

a.w. 137.3
sp.gr. 3.51
m.p. 725 °C
b.p. 1 640 °C

a silver-white metal which may ignite spontaneously in air in the presence of moisture, evolving hydrogen.

TWA OSHA 0.5 mg/m³ (water-soluble compounds)
IDLH 250 mg/m³

Barium is abundant in nature and accounts for 0.04% of the earth's crust. The chief sources are the minerals, barite (barium sulphate, $BaSO_4$) and witherite (barium carbonate, $BaCO_3$). Barium metal is produced in only limited quantities, by aluminium reduction of barium oxide in a retort and is little used in industry.

Barite ($BaSO_4$)

BARIUM SULPHATE

m.w. 233.4
sp.gr. 4.50
m.p. 1 580 °C
b.p. (transition point 1 149 °C, monoclinic)

insoluble in water, ethyl alcohol, dilute acids

fine, heavy, odourless powder or polymorphous crystals.

Crude barite ore is washed free of clay and other impurities, leached with sulphuric acid in lead-lined tanks, dried and ground. It is the basis for most other barium compounds, through barium sulphide, which is produced by reduction of natural barite with coal.

The greatest single use is in the manufacture of lithopone, a white powder containing 20% barium sulphate, 30% zinc sulphide and less than 8% zinc oxide, which is widely employed as a pigment in white paints. Chemically precipitated barium sulphate–*blanc fixe*–is used in high-quality paint, in X-ray diagnostic work, glassmaking and papermaking. Crude barite is used as a thixotropic mud in oil-well drilling.

Barium hydroxide ($Ba(OH)_2$)

m.w. 315.48
sp.gr. 2.18
m.p. 78 °C
b.p. 780 °C(-$8H_2O$)

soluble in water; slightly soluble in ethyl alcohol; insoluble in acetone

a white powder.

TWA OSHA see Barium

It is used in glass manufacture, synthetic rubber vulcanisation, lubricants, pesticides, the sugar industry and animal and vegetable oil refining.

Barium carbonate ($BaCO_3$)

m.w. 197.35
sp.gr. 4.43
m.p. 1 740 °C (90 atm)
b.p. decomposes

insoluble in water and ethyl alcohol; soluble in acids

a heavy, white powder.

MAC USSR 0.5 mg/m³

It is obtained as a precipitate of barite and is used in the brick and tile industry, in glazes and as a rodenticide.

Barium oxide (BaO)

It is produced by heating barium carbonate with coal. It is a white alkaline powder which is used to dry gases and solvents. At 450 °C it combines with oxygen to produce barium peroxide (BaO_2), an oxidising agent in organic synthesis and a bleaching material for animal substances and vegetable fibres.

Barium chloride ($BaCl_2.2H_2O$)

m.w. 208.25
sp.gr. 3.1
m.p. 113 °C (−$2H_2O$)
b.p. 35.7 °C

soluble in water; slightly soluble in hydrochloric acid; very slightly soluble in ethyl alcohol.

TWA OSHA see Barium

It is obtained inter alia by roasting barite with coal and calcium chloride and is used for chlorine and sodium hydroxide manufacture, as a flux for magnesium alloys, in pigment manufacture and in textile dyes and finishes. It has also been used as a medicament, in treating complete heart block, as a cardiac muscular stimulant.

Barium nitrate ($Ba(NO_3)_2$)

m.w. 261.35
sp.gr. 3.24
m.p. 592 °C
b.p. decomposes

soluble in water; slightly soluble in acid; insoluble in ethyl alcohol.

TWA OSHA see Barium

It is used in the pyrotechnics and electronics industries.

HAZARDS

Barium metal has only limited use and presents mainly an explosion hazard. The soluble compounds of barium (chloride, nitrate, hydroxide) are highly toxic, the inhalation of the insoluble compounds (sulphate) may give rise to pneumoconiosis and many of the compounds, including the sulphide, oxide and carbonate may cause local irritation to the eyes, nose, throat and skin. Certain compounds, particularly the peroxide, nitrate and chlorate, present fire hazards in use and storage.

Toxicity. The soluble compounds by the oral route are highly toxic and the fatal dose of the chloride has been stated to be 0.8-0.9 g; but, although poisoning due to the ingestion of these compounds occurs occasionally, very few cases of industrial poisoning have been reported. However, poisoning may result when workers are exposed to atmospheric concentrations of the dust of soluble compounds such as may occur during grinding. These compounds exert a strong and prolonged stimulant action on all forms of muscle, markedly increasing contractility. In the heart, irregular contractions may be followed by fibrillation, and there is evidence of a coronary constrictor action. Other effects include intestinal peristalsis, vascular constriction, bladder contraction and an increase in voluntary muscle tension. [These compounds also have irritant effects on mucous membranes and the eye.

Barium carbonate, an insoluble compound, has been found to be without pathological effects from inhalation; however, it can cause severe poisoning from oral intake and in rats it impairs the function of the male and female gonads; the fetus is very sensitive to barium carbonate during the first half of pregnancy.]

Pneumoconiosis. Barium sulphate is characterised by its extreme insolubility, and it is therefore non-toxic to humans. For this reason and due to its high radio-opacity, barium sulphate is used as an opaque medium in X-ray examination of the gastrointestinal, respiratory and urinary systems. It is also inert in the human lung, as has been demonstrated by its deliberate introduction into the bronchial tract as a contrast medium in bronchography and by industrial exposure to high concentrations of fine dust.

However, inhalation may lead to deposition in the lungs in sufficient quantities to produce baritosis—a benign pneumoconiosis. This occurs principally in the mining, grinding and bagging of barite but has been reported in the manufacture of lithopone. In the first reported case, there were symptoms and disability, but these were associated with other lung disease. Subsequent studies have contrasted the trivial nature of the clinical picture and the total absence of symptoms and abnormal physical signs with the well marked X-ray changes. These show disseminated nodular opacities throughout both lungs, which are discrete but sometimes so numerous as to overlap and appear confluent. No massive shadows have been reported. The outstanding feature of the radiographs is the marked radio-opacity of the nodules, which is understandable in view of the substance's use as a radio-opaque medium. The size of the individual elements may vary between 1 and 5 mm in diameter, although the average is about 3 mm or less, and the shape has been described variously as "rounded" and "dendritic". In some cases, a number of very dense points have been found to lie in a matrix of lower density (see the X-ray picture under PNEUMOCONIOSES, INTERNATIONAL CLASSIFICATION OF).

In one series of cases dust concentrations of up to 11 000 particles per cubic centimetre were measured at the workplace and chemical analysis showed that the total silica content lay between 0.07 and 1.96%, quartz not being detectable by X-ray diffraction. Men exposed

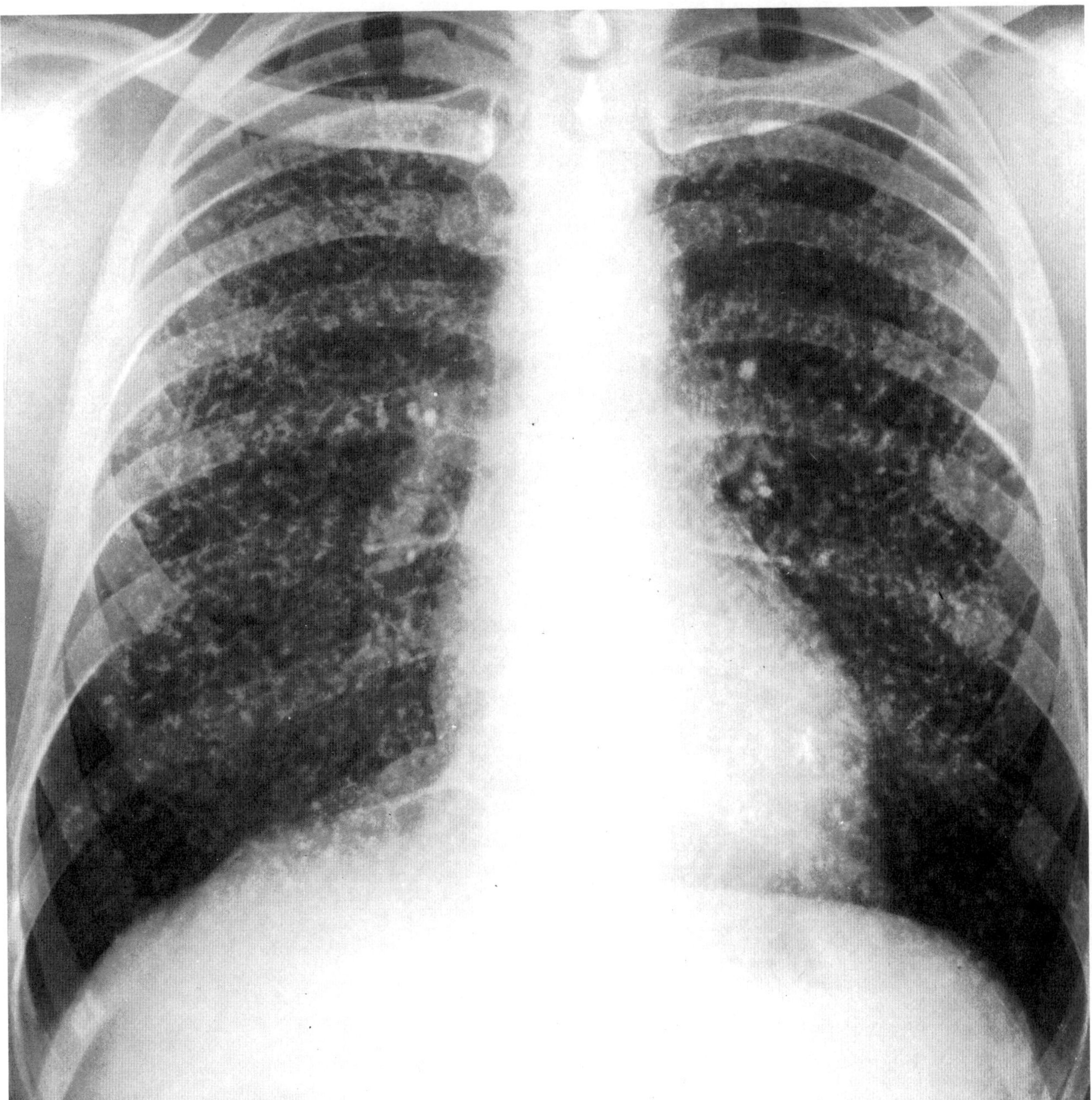

Figure 1. Baritosis in a man who had worked 13 years in a barite grinding factory and had had no other dust exposure. The man was symptomless and had excellent function.

for up to 20 years and exhibiting X-ray changes were symptomless, had excellent lung function and were capable of carrying out strenuous work. Years after the exposure has ceased, follow-up examinations show a marked clearing of X-ray abnormalities.

Reports of post-mortem findings in pure baritosis are practically non-existent. However, baritosis may be associated with silicosis in mining due to contamination of barite ore by siliceous rock, and, in grinding, if siliceous millstones are used.

SAFETY AND HEALTH MEASURES

To keep the dust concentration to below the recommended levels, processes should be enclosed and/or exhaust ventilation installed.

Adequate washing and other sanitary facilities should be provided for workers exposed to toxic soluble barium compounds and rigorous personal hygiene measures should be encouraged. Smoking and consumption of food and beverages in workshops should be prohibited. Floors in workshops should be made of impermeable materials and frequently washed down. Employees working on such processes as barite leaching with sulphuric acid should be supplied with acid-resistant clothing and suitable hand and face protection.

Although baritosis is benign, efforts should still be made to reduce atmospheric concentrations of barite dust to a minimum. In addition, particular attention should be paid to the presence of free silica in the airborne dust. Workers who are to be exposed to the inhalation of barite dust should be subject to a pre-employment examination and persons manifesting respiratory disorders should be excluded; periodical medical examinations should then follow regularly.

Treatment. Splashes of soluble barium compounds on skin and in eyes should be washed copiously with water. Persons who have ingested soluble barium compounds should be made to vomit immediately. Intoxication should be treated symptomatically.

DOIG, A. T.

Accident hazard:

CIS 76-193 "Hazardous chemical reactions–35. Calcium, strontium and barium" (Réactions chimiques dangereuses–35. Calcium, strontium et baryum). Leleu, J. *Cahiers de notes documentaires–Sécurité et hygiène du travail* (Paris), 4th quarter 1975, 81, Note No. 986-81-75 (489-491). (In French)

CIS 1453-1970 "Barium plus halogenated hydrocarbons may explode". Tracy, H. L.; Moshenrose, H. D. *American Industrial Hygiene Association Journal* (Akron, Ohio), 1969, 30/6, 562.

Pneumoconiosis:

"Baritosis: a benign pneumoconiosis". Doig, A. T. *Thorax* (London), 1976, 31/1 (30-39).

Toxicity:

CIS 79-129 "Experimental data for establishment of the maximum permissible concentration of barium fluoride in the workplace air" (Èksperimental'nye materialy k obosnovaniju predel'no dopustimoj koncentracii ftoristogo barija v vozduhe rabočej zony). Popova, O. Ja. *Gigiena truda i professional'nye zabolevanija* (Moscow), May 1978, 5 (34-37). 4 ref. (In Russian)

Bark

Bark, the protective outer covering of trees and shrubs, is made up of three layers. The outer corky layer is rough and heavy. This layer, which is analogous to human skin, is usually referred to as the epidermis. The primary duty of the outer layer is to provide a protective coat. The texture varies considerably from tree to tree; some barks, such as that of the birch, are smooth and others, such as that of the hard maple, are extremely rough. The middle layer or cortex usually contains chlorophyl. The stems of woody plants contain a middle layer which is usually replaced by outer bark tissue as growth continues. The inner layer, next to the wood, contains the phloem cells which carry food up and down the stem.

Stripping and processing. Stripping is the removal of the bark from the tree prior to industrial processing. The stripping is usually done in spring when the sap is rising. Circular incisions with a vertical cut effect the removal of the bark in large sheets. The bark may also be removed by injecting steam between the bark and the wood. Bark may also be removed from cut logs in large metal rotating containers called debarking drums. This process is used frequently in the pulp and paper industry. Here the bark is a waste product and is usually burned.

Uses. The primary use of bark is in the production of cork from the outer layer of cork oak trees. Tannic acid is made from the bark of hemlock and other trees. Quinine is produced from the bark of the cinchona tree and cough medicines are made from the bark of cherry trees. Cinnamon tree bark is used for flavouring. Bark has also been used in the past to make clothes and houses and one of its popular historical uses was in making canoes from the bark of the canoe birch.

HAZARDS AND THEIR PREVENTION

The main hazard in the industrial use of bark is the irritant property of the powdered material. Positive measures must be taken to ensure adequate ventilation in the grinding rooms and proper disposition of the bark dust. A severe lung condition called "maple bark disease" caused by the inhalation of the spores of *Cryptostroma corticale*, a black mould which grows beneath the bark of hard maple, has been described. These spores are liberated from under the bark of hard maple pulp wood during the sawing and debarking process. Inhalation of large numbers of the spores results in a hypersensitivity disease of the lung. [Maple bark disease is presently considered as a form of extrinsic allergic alveolitis.] Positive control measures such as elimination of saw operations, wetting material during the debarking process with a detergent, and installation of air-conditioned booths for the workers can adequately control the contamination of the air with the spores.

WENZEL, F. J.

"The epidemiology of maple bark disease". Wenzel, F. J; Emanuel, D. A. *Archives of Environmental Health* (Chicago), Mar. 1967, 14/3 (385-389). Illus. 8 ref.

Barrier creams and lotions

The fundamental principle in the prevention of industrial dermatitis is to prohibit skin contact with possible irritants. The ideal method in which this contact can be reduced lies in total enclosure or complete mechanisation of the dangerous process, where this is possible. Nevertheless there are many processes which cannot be totally enclosed and there are many operations in which the use of protective clothing is impractical, yet protection of the worker's skin is still essential. In these circumstances the use of barrier substances will do much to provide adequate and necessary protection. When applied to the skin, barrier creams or lotions leave a thin film, which acts as a barrier against skin irritants. Barrier creams and lotions are less effective than protective clothing but nevertheless useful. They protect the skin from various irritants which may escape into the workplace in spite of other preventive measures. The barrier substances now have an important place among the methods designed to prevent industrial dermatitis.

Types of creams and lotions

Barrier creams and lotions will be considered under four headings, as follows:

(a) water-resistant types;

(b) oil-resistant types;

(c) film-forming types; and

(d) a miscellaneous group of preparations with specific uses.

Water-resistant types. Water-resistant barrier creams are intended primarily to protect the skin from emulsified oils and solvents, liquid coolants, mist and spray from alkali baths or plating solutions, lime, cement, water, and water solutions. This type leaves a film of water-repellent substance, such as lanolin, wax, petrolatum, shellac, ethyl cellulose, or silicones, on the skin. The introduction of silicones into cream formulations has greatly widened the range of protection of these preparations.

Oil-resistant types. These creams are designed to protect the skin against dirt, dust, oil, greases, solvents, paints, lacquers and various other non-aqueous irritants. They may contain oil and solvent repellents, such as sodium alginate, methyl cellulose, sodium silicate, etc. Lanolin has some oil-repellent as well as water-repellent properties and may be used as a good emulsion base.

Film-forming types. This group of barrier substances consists of materials capable of forming protective films on the skin. These preparations, sometimes referred to as "invisible gloves", are based on the use of a solvent-resistant, film-forming materials such as vinylchloride latex, sodium polyacrylate and cellulose ether. The invisible-glove type of film also offers good protection against dermatitis of the face from the rubber edges of respiratory protective equipment.

Miscellaneous types. This fourth group includes a variety of barrier substances designed for protection against individual types of chemical or physical irritants. Barrier creams against the phototoxic or photosensitising action of the heavy coal tar distillates, oil distillation residues and other photodynamic chemicals, and also against excessive sunlight, may contain chemical light screens such as salol, methyl salicylate, aesculin, cychloform, methyl benzoate and quinine oleate. Their effectiveness as screens for radiation (290-320 nm) varies from 30 min to 3-4 h. The physical light screens block the passage of all light rays. A cream containing zinc oxides or titanium dioxide will protect the skin from sunlight. Antiflash creams are used in steel foundries and elsewhere when there is exposure to intense heat for short periods. The problem of radiation hazards has long been of great importance in industry and in some technical professions. Special barrier creams which, when applied to the body surfaces, will serve to lessen the injurious flash effects of an atomic explosion are being developed. Insect repellent creams may contain such chemicals as diethyltoluamide, dimethyl phthalate and phenyl cychlohexanol.

Selection

The development of barrier creams and lotions to prevent occupational dermatitis is in a state of flux. Today, a large number of barrier substances are available, but practice does not always prove the claims made for them by their manufacturers or advertisers. In certain circumstances, for specific purposes, they can be valuable, but only when used in conjunction with good sanitary and washing facilities. Besides, it should be emphasised that the barrier creams and lotions are not therapeutic agents but prophylactic measures to reduce the contact of the skin to certain specific harmful agents. Choosing the best barrier cream to protect the skin against an industrial irritant is a serious problem. The employer may need expert aid in making his choice as to which among these is suited to his work, and the correct choice of substance is of first importance.

In order to secure desired protective effects, a barrier cream should meet certain basic requirements and correctly formulated barrier creams should have the following properties:

(a) they should offer effectual protection from the harmful agents in question;

(b) they should be non-irritating and non-sensitising;

(c) they should have good cosmetic properties to ensure that they can be applied easily and feel pleasant to the skin;

(d) they should be easy to remove with soap and water and yet they should adhere to the skin and not rub off easily under actual working conditions;

(e) they should preserve the skin tissue in a healthy condition and hence should contain a reconditioning factor to repair surface damage;

(f) they should be bacteriostatic to prevent infection through surface damage to the skin.

Finally, it is noted that apart from meeting all these basic requirements, a cream must also be economical to use.

Application

The barrier substance should be applied before starting work and removed from the skin by washing after each spell of work. It is important that the skin should be clean and dry before applying the substance.

The need for personal cleanliness cannot be overemphasised in the prevention of industrial dermatitis. Workers should be instructed and encouraged to observe a strict personal hygiene routine including a wash before lunch and at the end of the day's work. The use of barrier cream makes such routines much more easily enforced since the wearer of barrier cream necessarily removes it with a skin cleanser at lunch time and again after work, and thus removes at the same time any irritants which are on the skin. In this regard, combination of barrier cream and efficient skin cleanser is worth while. The barrier cream prevents dirt, dust, oil, paint and other soils from becoming firmly embedded in the skin, making it much easier to remove them with a skin cleanser. Therefore, the use of suitable skin cleanser also should be encouraged.

The application of a skin conditioning cream at the end of the day's work is effective in lessening the defatting and drying action of various degreasing materials. The extensive use in industry of detergents and solvents which cause severe defatting and drying of the skin, presents the problem of dry, chapped hands and skin, and much can be done to overcome this by the application, following a day's work, of a super-fatted cream designed to temporarily substitute fatty material for natural skin oils. Most skin conditioning creams contain lanolin, glycerine or other substances that make the skin pliable.

Dispensing

Barrier creams should be placed on conveniently located shelves in the washrooms or in automatic dispensers which can be fitted to the wall. Where this is not practicable, they may be given to the worker in individual cans or tubes, which he can keep at or near his resting place or in his locker. Most workers are willing to use barrier substances when they are freely and frequently furnished by the plant medical service.

Well designed barrier creams and lotions are useful and although workers sometimes object to them they can usually be educated to accept them. However, it is not sufficient to provide workers with barrier substances. Not only must they be provided with these substances free of charge but also they must be taught to use them properly. It is important to bear in mind always that the use of barrier creams and lotions is only one of a series of preventive measures against industrial dermatitis and reliance should therefore never be placed solely on them since neglect of other preventive measures will necessarily result in the failure of dermatitis prevention.

NOMURA, S.

CIS 964-1965 *On the cutaneous effects of neutral ointments and emulsions in relation to their composition–Principles of cosmetics, dermatological protection and therapy* (Über die Wirkungsweise indifferenter Salben und Emulsionssysteme an der Haut in Abhängigkeit von ihrer Zusammensetzung–Grundlagen für Hautpflege, Hautschutz und Therapie). Tronnier, H. (Aulendorf-in-Württemberg, Editio Cantor KG, 1964), 178 p. Illus. 310 ref. (In German)

"Test concentrations and vehicles for dermatological testing of cosmetic ingredients". Maibach, H. I.; Akerson, J. M.; Marzulli, F. N.; Wenninger, J.; Greif, M.; Hjorth, N.; Andersen, K. E.; Wilkinson, D. S. *Contact Dermatitis* (Copenhagen), Oct. 1980, 6/6 (369-404).

Batteries, dry

There are three different types of dry battery:

(a) batteries with a solid depolariser made from manganese dioxide;

(b) batteries with a gaseous depolariser of atmospheric oxygen; and

(c) batteries with a solid depolariser made from mercuric oxide (see BATTERIES, DRY (MERCURY)).

Types *(a)* and *(b)* are Leclanché-type batteries and they differ solely in the composition of the depolarising mixture: activated carbon for *(b)*, manganese dioxide for *(a)*.

The solid-depolariser (manganese dioxide) Leclanché battery is the most widely used. Its manufacture commenced at the end of the 19th century in the United States, but already by 1920 industrial production was at a high level and currently annual world production is in the billions of units.

Materials and manufacture. Manganese dry batteries comprise five main units: the depolarising mixture; the gelatinous paste; the carbon element; the zinc container; and the sealing compound.

The depolarising mixture comprises 60-70% manganese dioxide, the remainder being made up of graphite, acetylene black, ammonium salts, zinc chloride and water; relative percentages vary according to the formulation used. Acetylene black is preferred to graphite for batteries subject to intermittent discharge. The manganese is delivered finely ground. The depolarising mixture forms the major part of the cell and acts both as the cathode and the depolariser.

The materials are weighed and then carried and fed into a grinder-mixer. During this process, the worker is exposed to emissions of dust; that of the manganese is relatively heavy, whilst the others are light. If no adequate local exhaust ventilation is provided at the weighing point, the maximum allowable atmospheric concentration of manganese will often be exceeded. The worker will be exposed for periods averaging 10-15 min/h for weighing and batching, and 20-25 min/h for mixer loading. For 3-4 min, the mixing is dry and the amount of dust liberated is considerable (with dust emissions often exceeding allowable concentrations when hermetically sealed apparatus is not used); after this, electrolyte or water is added and the substance is mixed wet for 15 min. The electrolyte contains ammonium chloride, zinc chloride and ammonium salts. The worker removes the moistened mixture using a small shovel.

Some mixers do not do the grinding; these are sealed machines with local exhaust ventilation. This avoids dust emission during dry grinding; however, the mixture must then be treated in a flocculator which recompresses it. As soon as the mixture is moistened, no further dust is given off. In more modern plants grinding and mixing is done in sealed equipment; only during the loading process is the worker exposed to a poisoning hazard. It has been found that a moisture content of 10-15% increases the output of the cell and significantly reduces the risks of poisoning in pressing operations. In the manufacture of certain batteries, the mixture is again dried in an oven, then sifted and remoistened. Emptying the oven and sifting release dust particles.

The prepared mixture is pressed on a hand-fed tableting press or on an agglomerating press. On the tableting press the operator feeds the machine by hand and then removes and inspects the finished tablets; dust exposure occurs only during the daily cleaning which lasts about one hour and is carried out using rags and kerosene. The agglomerating press compacts the mixture directly around the carbon element. Pressing may be done manually but the mixture is more moist and workers' hands and subsequently faces may be contaminated with it; when the mixture has dried it may give off inhalable dust. Irritant dermatitis, especially of the hands, may occur owing to the slightly corrosive action of the electrolyte salts. The eyes may be irritated by being rubbed with contaminated hands or by splashes of depolarising paste ejected from the press–although this is rare.

After a few days storage to allow the agglomerate to harden, the mixture is wrapped on a hand-fed machine. This process presents no hazards. In some cases wrapping may be done by hand and this may result in skin contamination. Not all batteries are wrapped, however.

The agglomerates are then placed in trays and soaked in electrolyte, following which they are passed on for assembly. Pressing and wrapping machines are noisy and noise levels should be checked. It should also be ensured that the working tempo of semi-automatic machines is matched to the workers' capabilities. Poisoning hazards during this production phase are related to emissions of manganese dusts (central nervous system disorders, manganese-induced Parkinson-like syndromes) and graphite dust (pneumoconiosis, allergic reactions) during the following operations: weighing, mixer loading, dry mixing and perhaps mixture drying (during oven discharge and sifting), press cleaning, and hand pressing and wrapping. Workers may be exposed to the danger of irritant dermatitis and of conjunctivitis.

The gelatinous paste is the conducting medium of the current; it absorbs most of the electrolyte. It separates the anode from the cathode of the element and is situated between the depolarising mixture and the zinc container. This paste is made up of a mixture of maize and flour starches, with zinc chloride and ammonium chloride to assist gelatinisation. The electrolyte comprises mainly water (60-70%), ammonium chloride (20-30%) and zinc chloride (7-9%). It is prepared in large vats about once per week; mixing is carried out at room temperature using compressed air; the constituents are usually poured in from sacks and there is no need for weighing. Subsequently the electrolyte is passed over zinc chips which remove metallic impurities and finally through vats containing magnesium dioxide to complete the purification process. About once a month, a small amount of electrolyte is used to produce a mixture containing a small amount of mercury (for protection of the zinc containers). A full sack weighing about 12.5 kg (no weighing required) of sublimate is tipped into 25 l of electrolyte; mixing is carried out at room temperature and, if done by hand, there may be a danger of mercury poisoning (digestive and nervous system disorders). A quantity of 250 cm^3 of this solution is added to 35 l of normal electrolyte and the gelatinous paste is mixed. This last process is carried out 8-10 times per day and, each time, lasts 10-15 min. Mixing is automatic and the worker is required only to add the necessary materials; there is no poisoning hazard during this operation. The paste is now ready to be filled into the zinc containers.

The carbon rod is located centrally in the cell; it is surrounded by depolarising mixture and collects and conducts the current from the cathode to the cell's external positive terminal. Since it is relatively porous, this rod also allows the escape of gases which would otherwise cause a pressure build-up. These rods are normally made from a petroleum coke which is calcined, ground and then mixed with coal tar pitch. After this they are heated and cooled progressively. The electrode rods are not normally manufactured on-site but delivered ready for use. There is consequently no poisoning hazard.

The zinc container forms the cell case and the anode. The zinc is virtually pure and contains only trace quantities of lead. The containers are produced from zinc blanks which are hand fed into a hot press; the pressed case is then trimmed.

Some machines carry out both operations simultaneously, producing a noise level that often exceeds 80 dB. The container may also be made from zinc sheet which is folded and soldered. There is no danger of poisoning but excessive noise levels should be avoided and the tempo of the machines should be matched to the worker's capabilities.

The cells are assembled using one of two techniques:

(a) The gelatinous paste is drawn up and the required quantity is automatically poured into the cases, after which a worker places the agglomerate in position. There is no contact with the mercury-containing paste. This work may also be done manually: the paste is fed with a pump into the cases which are heated in a water bath; once the paste has hardened, the worker inserts a star-shaped base onto which is placed the agglomerate. There is no contact with the paste.

(b) Chromate assembly: the cases have to receive a chromate finish before being filled. The machine comprises a rotating plate onto which the worker places the cases which are automatically filled with a chromate solution (hydrochloric acid and chromic acid). They are then emptied and filled with the paste onto which the agglomerate is placed. Normally, there is no contact with the chromium compound but contamination may occur accidentally when a case is placed upside down or when the machine is being cleaned. Such contact may cause dermatitis.

The sealing compound is made from an insulating material such as wax or pitch which is poured in hot and then flame heated to ensure a better bond to the zinc. After this the cells are welded together where necessary and finished.

During this final manufacturing process, workers may be exposed to welding fumes (tin) and fumes from melted bitumen (used for sealing multi-cell batteries). Where the assembly process is not mechanised, workers may be exposed to the inhalation of the paraffin wax used to render the cells leaktight and seal the bases of the various units. Assembly and finishing work is usually carried out by women and a particular check should be kept on the working tempo of the machines employed.

SAFETY AND HEALTH MEASURES

As regards the exposure to manganese dioxide and graphite, the following safety and health measures should be adopted:

(a) hermetic sealing of operations emitting dust; maximum mechanisation of manual handling; vacuum cleaning of workshops once a day;

(b) supply of clothing and gloves; headware and, if necessary, masks for special situations; compulsory showering; separate lockers for work and street clothes. Consumption of food and drink at the workplace should be prohibited;

(c) periodic medical examination of exposed workers (every six months);

(d) persons suffering from nervous disorders, liver, kidney or blood complaints or subject to hypertension, upper respiratory tract or bronchopulmonary disease or dermatitis should not be employed.

As regards exposure to sublimate, and chromating of containers, mechanisation of mixing, mechanisation of paste filling and chromating of containers, use of gloves where necessary and personal hygiene are recommended.

DUCREY, L.

CIS 77-425 "Factories and Industrial Undertakings (Dry Batteries) Regulations 1976–Legal Notice 196 of 1976" *Hong Kong Government Gazette* (Hong Kong), 6 Aug. 1976, 32, Legal Supplement No. 2, p. B607-611.

Batteries, dry (mercury)

Two different chemical forms of mercurials are used in mercury battery production: *(a)* mercuric oxide; and *(b)* metallic mercury. Mercuric oxide was first used in the dry battery known as Ruben's cell or mercury cell, and is used also at present. Metallic mercury also has been introduced in dry battery production, and this type of dry battery is now produced in large quantities. Because both types of mercury battery are small and can discharge constantly for a long time, they are suitable for use in hearing aids, radio receivers, photographic light meters, watches, calculators, electronic measuring instruments and weapons.

Materials and manufacture. The mercury battery with mercuric oxide comprises five main units; the cathode, the gelatinous substance which separates anode and cathode, the depolarising mixture, the electrolyte and the anode. The cathode is zinc ribbon or zinc powder which is pressed into tablets. The gelatinous substance contains carboxymethyl cellulose, polystyrene membrane or polyvinylalcohol membrane. Mercuric oxide is the main component of the depolarising mixture which contains also graphite as a minor component. The case acts as the anode. The electrolyte contains potassium hydroxide, zinc oxide and water.

The other type of mercury battery, in which metallic mercury is used, comprises three main units: the case, the anodic unit and the cathodic unit. The anodic unit is a mixture of silver oxide, graphite, manganese dioxide and other substances. These substances are weighed, batched, graded and mixed. The prepared mixture is pressed into tablets and put into a can, which acts as the anode. The cathodic unit is prepared as follows: zinc powder is mixed with metallic mercury, the electrolyte is added to the mixture, and a paste is made. The prepared paste is weighed and put into a can, which acts as the cathode. The two cans, the cathode and the anode, are then mechanically assembled.

HAZARDS

Mercuric oxide batteries. During the preparation of the depolarising mixture workers may be exposed to mercuric oxide and graphite. Because enclosed machines are used in modern plants, the exposure takes place at the stage of manual feeding of the machinery. Mercuric oxide by inhalation is believed to cause hazards similar to mercuric chloride or other divalent inorganic mercury compounds. Mercuric chloride is unstable and easily emits mercury vapour. Graphite powder may cause pneumoconiosis and allergic reactions. Carboxymethyl cellulose may also accumulate in the lungs. The electrolyte is very corrosive to the skin and mucosa. Machines for the preparation of the depolarising mixture may produce noise levels in excess of 80 dB.

Metallic mercury batteries. During the preparation of the anodic unit workers may be exposed to silver oxide, graphite and manganese dioxide dusts. In spite of the use of the enclosed machinery, weighing and feeding often remain manual and workers are exposed to dusts. Silver oxide may cause argyrosis. Excessive absorption of manganese dioxide dust may cause central nervous system manifestations known as the Parkinson's syndrome. Machines for the batching, grading, mixing and tableting may produce noise levels in excess of 80 dB.

The cathodic unit is produced by a separate production line. First metallic mercury is mixed with zinc powder, and the electrolyte is added to the mixture to make the paste. These processes are almost completely enclosed and exposure to mercury vapour and zinc powder dust can take place only during emptying and feeding the machinery. Thereafter, amounts of several tens of milligrammes of the paste are weighed and put into a can manually. During the weighing procedure exposure to mercury vapour is more probable than in any other phase of the job because of the manual and small-scale weighing carried out with an analytical balance. The electrolyte is alkaline and corrosive to the skin and mucosa, and irritations and/or dermatitis may occur. The two cans, one containing the anodic and the other the cathodic unit, are then assembled mechanically by manual feeding. In this job the worker may also be exposed to mercury vapour. In a modern Japanese factory the workers were exposed to mercury vapour at concentrations of 0.001-0.02 mg Hg/m^3 during the processes described above.

Because silver oxide is very expensive, it is recycled from the discharged cells, which are mainly used in watches, and mercury is also recycled. During the recycling process workers also may be exposed to mercury vapour and silver oxide.

SAFETY AND HEALTH MEASURES

In the production of cells with mercuric oxide the most hazardous substance is mercuric oxide. The content of mercury in the depolarising mixture is far higher than in the case of the cells containing metallic mercury, and there is some risk of mercury poisoning. Therefore, in periodical medical examinations of workers not only mercury concentrations in urine and blood but also clinical manifestations and complaints such as tremor, restlessness and proteinuria should be checked.

In the production of cells containing metallic mercury the most hazardous substance is the metallic mercury. Because of the low concentrations of mercury in the cathodic unit the mercury concentration in the ambient air is expected to be far lower than in the case of mercuric oxide. From pre- and post-employment examinations in a modern plant it was found after 8 months of mercury vapour exposure (0.001-0.02 mg Hg/m^3, 8 h/day, 40 h/week) that inorganic and organic mercury in plasma and organic mercury in red blood cells were increased significantly, but neither increased excretion of urinary mercury nor proteinuria were observed. The periodical medical examination carried out every six months, for about five years, in the same plant showed similar results, and the negative finding of urinary β2-microglobulin indicated the absence of renal tubular injuries. Most of the workers engaged in mercury battery production are women, and women are often engaged in the small-scale weighing. Therefore, the risk of prenatal exposure to mercury vapour should be kept in mind. In spite of the absence of complaints and symptoms of mercury vapour intoxication, biological monitoring of mercury vapour exposure should be carried out in periodical medical examinations. This requires simultaneous analysis of mercury concentrations in plasma, red blood cells and urine and the selective determination of inorganic and organic mercury.

Mercury concentration in hair also indicates the body burden of mercury. But the organic mercury concentration in hair is reduced considerably by the artificial hair waving (washing, dyeing or bleaching have no effects). In addition, in exposure to low concentrations of mercury vapour there is a metabolic interaction of inorganic and organic mercury; therefore inorganic and organic mercury must be determined separately, and the mercury concentration in samples of hair that has been artificially waved is meaningless.

Workers should have separate lockers for working and street clothes, and be forbidden to eat, drink and smoke at the workplace. Working clothes should be washed in the factory, and the waste water should be checked for the mercury concentration.

Workers exposed to manganese dioxide should be kept under medical surveillance for the central nervous system manifestations.

The electrolyte is alkaline, and precautions should be taken to avoid irritation of hands, face and eyes; in particular protective gloves and caps should be used, and eyewash facilities should be immediately available.

Workers exposed to dusts should receive periodical medical examinations and be supplied with respiratory protective equipment. Dusty or mercury vapour generating machinery should be enclosed or fitted with effective exhaust ventilation.

ISHIHARA, N.

"Inorganic and organic mercury in blood, urine and hair in low level mercury vapour exposure". Ishihara, N.; Urushiyama, K.; Suzuki, T. *International Archives of Occupational and Environmental Health* (West Berlin), Dec. 1977, 40/4 (249-253). 11 ref.

"Effects of artificial hair-waving on hair mercury values". Yamamoto, R.; Suzuki, T. *International Archives of Occupational and Environmental Health* (West Berlin), Sep. 1978, 42/1 (1-9). Illus. 14 ref.

"Interaction of inorganic to organic mercury in their metabolism in human body". Suzuki, T.; Shishido, S.; Ishihara, N. *International Archives of Occupational and Environmental Health* (West Berlin), Dec. 1976, 38/2 (103-113). Illus. 17 ref.

Batteries, secondary or rechargeable, or accumulators

These batteries consist of a series of identical cells each delivering the same voltage, so that the voltage of the battery depends on the number of cells. Different electrochemical couples give different voltages, usually in the range of 1.5 to 2 volts per cell. The lithium-sulphur dioxide couple produces 3 volts.

Essentially each cell consists of positive and negative plates in a suitable electrolyte and insulated from each other by suitable separators which either space out the plates, preventing physical contact, or are porous.

There are a large variety of potential couples. The principal commercial products are the lead-acid cell which includes about 90% of the market, and nickel-cadmium which has certain advantages in terms of long life. Silver-zinc/alkalis are used on a smaller scale and couples under serious development include sodium-sulphur and lithium-chlorine.

Other couples which have been used from time to time and may well be revived in view of the energy shortage and the importance of storing electrical energy from atomic sources include nickel-iron, nickel-hydrogen and zinc-chlorine. Further potential couples include nickel-zinc, zinc-air, iron-air, silver-iron, silver-zinc, sodium-sulphur, lithium-iron sulphide, lithium with chlorine or iodine and sodium-chlorine.

The two couples which seem to be the most promising developments are sodium-sulphur and lithium-sulphur or lithium organic batteries. If developed successfully, these batteries have considerable advantages in voltage and energy density.

Most of the potential couples at present pose unsolved problems. For instance zinc-containing batteries show problems in recharging, and other couples may have a very limited number of rechargable cycles, or may simply be too expensive.

Raw materials

Lead-acid. The principal materials include metallic lead, lead oxide, alloys containing antimony, small amounts of arsenic, selenium or calcium or cadmium, the latter two being used in maintenance-free batteries. Additives include carbon black, solka floc, amorphous silica and expanders, with sulphuric acid as the electrolyte. Composition containers, consisting of coal tar pitch and crocidolite asbestos are now rarely used. Hard rubber containers which contained natural or synthetic rubber accelerators, antioxidants, stabilisers, ebonite dust and coal dust, are being largely replaced by injection-moulded polypropylene containers.

Separators are now made from porous polyvinyl chloride, kraft paper impregnated with phenol formaldehyde resin, or in some cases porous rubber. Glass fibre mats are also used.

Nickel-cadmium. Raw materials include nickel sulphate and hydroxide, cadium oxide, cobalt sulphate, lithium hydroxide, ferrous sulphate, sodium hydroxide, graphite and paraffin. Separators are normally thin plastic rods and the containers are either welded mild steel or polypropylene.

Other couples. Basically, the raw materials for the electrodes are named in the couples. Other additives will certainly be required to make commercially produced batteries but what these will be is not established. For instance a nickel-hydrogen system requires a noble metal catalyst and also operates at high pressures, requiring a steel pressure vessel. Nickel-iron batteries may be developed using nickel foil.

In experimental lithium-chlorine batteries, a boron nitride separator has been used and the cell operates at high temperatures.

Since a wide variety of raw materials and manufacturing methods may be developed, it would seem best for any doctor or hygienist concerned with the environmental health control of future secondary battery manufacture to make a careful study of the whole process. This will enable him to determine at what points there may be a risk to health.

Manufacturing processes

Lead-acid. The process flow chart (figure 1) shows the manufacturing sequence. The lead-acid battery cell is made up from a group of plates which are based on a grid cast or wrought from lead-antimony or lead-calcium alloys. The grid supports the active material and also acts as the conducting material. The active material is made from lead oxides and suitable additives. Either during the manufacture or after assembly the plates are formed electrically to lead peroxide (positive) and spongy lead (negative).

The oxide paste is pressed into the metal grids, usually mechanically. When dried, it forms a hard plate. The plates are then cleaned and assembled with separators into cells. Each cell is connected with a lead connector to the adjacent cell, leaving a positive and negative terminal at the ends of the battery. The process flow chart shows the sequence of manufacture.

Waste lead and sulphuric acid are removed from the factory effluent prior to discharge into the sewers in accordance with local regulations.

Nickel-cadmium. The process flow chart (figure 2) shows the manufacturing sequence.

The negative active material is made from cadmium oxide into a paste. The dried paste is mixed with graphite, iron oxide and paraffin, and milled and finally compacted between rollers. The positive material is made from a slurry of nickel hydroxide, cobalt sulphate and sodium hydroxide, dried and ground with graphite flake. Sintered nickel plates impregnated with the slurry may also be used. The resulting positive and negative powders may be compressed to briquettes or fed directly into a pocket made from two perforated mild steel strips. The plates are made up from interleaving the edges of the strips and cut to length. An edge and suitable lugs are spot welded onto the plate, which is finally compressed and assembled into cells in a similar manner to the lead-acid battery. The plates are normally separated by small plastic rods which prevent electrical contact. The joining of cells is by means of bolting. The steel containers are sealed by welding at the bottom.

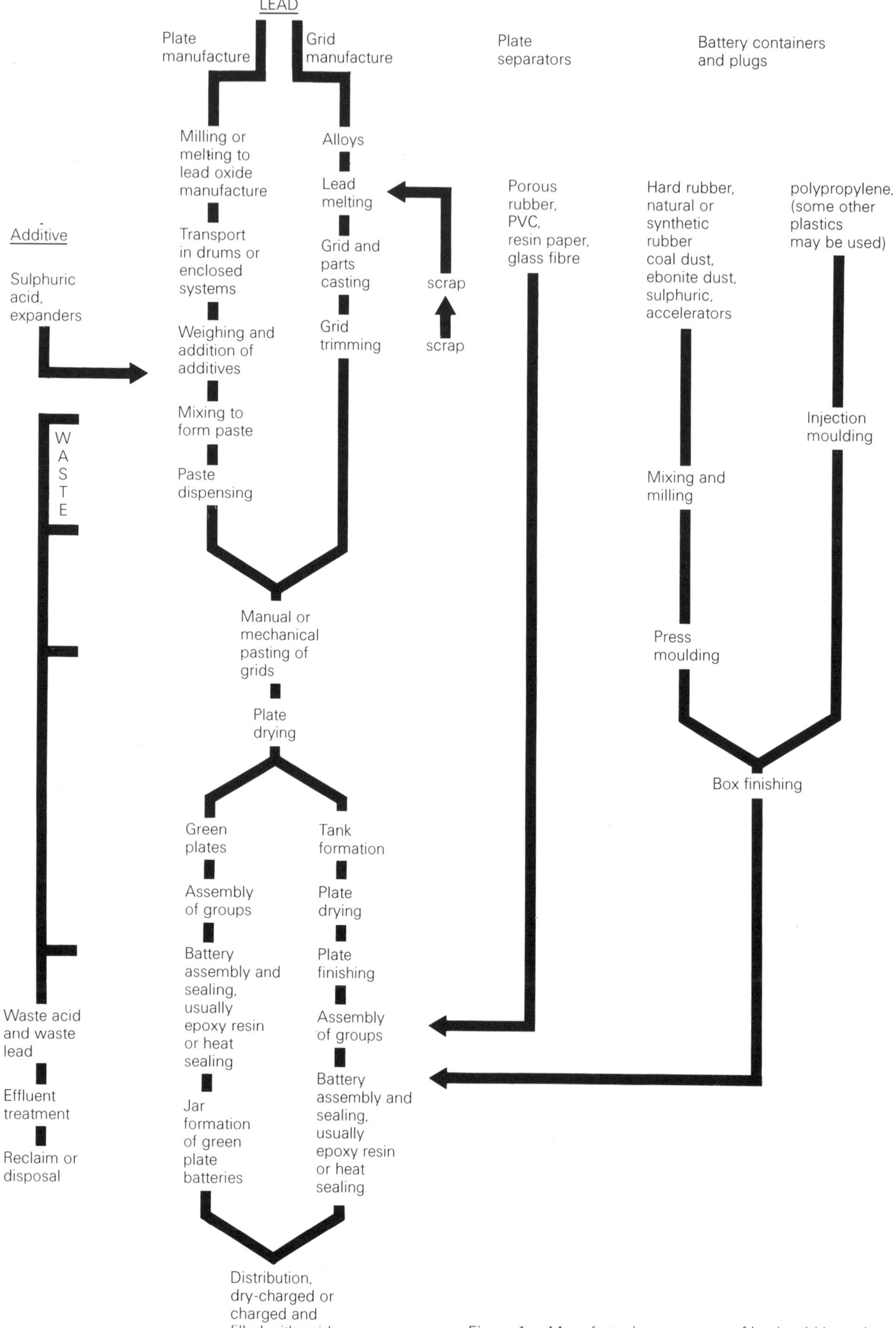

Figure 1. Manufacturing sequence of lead-acid batteries.

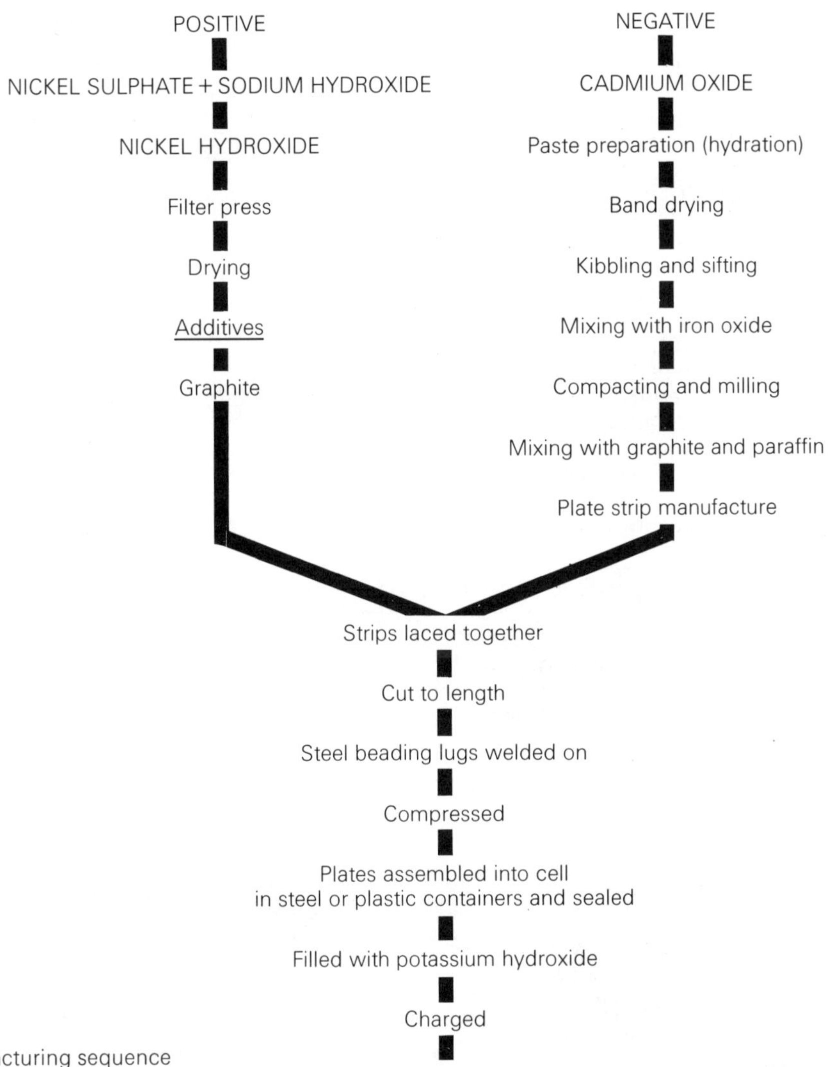

Figure 2. Manufacturing sequence of nickel-cadmium batteries.

Other couples. Sodium-sulphur batteries involve placing liquid sodium at high temperatures into a suitable container made of fused alumina. This is surrounded by a steel tube containing sulphur. Liquid sodium is handled in an enclosed process to prevent the risk of spontaneous combustion.

Other couples are either manufactured on a small scale, or are laboratory processes. It is not clear which couples may be developed commercially and how the process would be set up.

HAZARDS

Lead-acid. Most industrialised countries have regulations controlling the use of lead in industry. In countries now developing industries, some already have suitable regulations, and many international companies with factories in developing countries apply the standards used in the country of origin.

In the manufacture of lead-acid batteries, lead is still the principal hazard and occurs mainly in the pasting areas and the plate finishing areas. While pasting is a wet process, the methods of handling paste often cause splashing which results in dry paste on the clothes, aprons, tools, machinery and the hands. Movement causes the dry paste to give off fine dust.

In the plate finishing departments the handling of large numbers of dry, and hence dusty plates, also gives rise to dust. Assembly of dry plates into groups can also cause a dust hazard, although this is usually easier to ventilate satisfactorily.

Casting, which uses metallic lead, should not pose a health problem unless the dross which forms on top of molten lead is handled carelessly. The principal problem in this department is the discomfort from radiant heat and the risk of burns from splashes of molten metal. Good design and layout can minimise these hazards, especially mechanical casting, using low-temperature, well insulated lead pots.

Tank formation, providing plates are damped, with modern formation circuits which do not require connection by tacking (lead soldering), does not pose a large hazard.

Once the groups are in the battery, the soldering of intercell connectors and addition of terminal posts can be easily controlled. In many cases these processes are now fully automatic.

Oxide manufacture and handling, which used to be a major hazard, should now be controlled by total enclosure and mechanical, enclosed handling and weighing of the powder. Problems may arise in small factories which do not have enclosed handling methods and paste mixing methods. Here, good ventilation is essential and suitable respirators may need to be worn. Where control is not fully adequate these processes are best isolated from all other processes.

The basic principles of safe manufacture are to enclose totally such processes as oxide manufacture, handling and paste mixing. The pasting process requires good layout, good general ventilation (10-15 air changes per hour), suitable protective clothing and strict cleanliness of operation at all times, together with safe disposal of any spilled paste.

Formation used to cause erosion of the teeth. Modern methods of spray prevention or control by ventilation can prevent this risk. Suitable safety precautions are also required for the handling and dilution of concentrated acid. Plate cutting, group assembly and burning should be carried out under suitable exhaust ventilation at all times. In the past the risk has often been inadequately controlled because of the amount of handling of dry plates away from suitable exhaust ventilation.

The use of calcium alloys if accidentally contaminated with antimony-arsenic-lead alloys may result in a dross containing calcium arsenide or stibide. This compound, if wet with water, will form stibine and arsine gas. Strict precautions are necessary to prevent such a risk occurring.

Nickel-cadmium. Although there is current concern over the potential toxic effects of nickel and, in particular in the United States, over its possible carcinogenic effects, we have not observed any excessive cancer or other health hazards in the use of nickel in the manufacture of nickel-cadmium batteries. Cadmium continues to be the principal hazard. New methods of preparing the active raw material have considerably reduced the hazard in this area. The principal problems remain in the plate manufacturing process and in the reclaiming of cadmium from scrap plates. No excess of chronic cadmium effects have been observed in the assembly of nickel-cadmium batteries, and recent improvements in layout should preclude the development of any cases in the future.

The handling of caustic soda and of lithium requires safe control and suitable ventilation.

Other couples. Some of the raw materials which are used in the development of other couples may also produce problems. The handling of liquid sodium requires care to avoid spontaneous combustion and possible burns. Lithium compounds, particularly the hydroxide, are extremely caustic. Other lithium compounds, if absorbed, can have an effect on the central nervous system.

Chlorine gas, which may be used in some couples, needs to be carefully contained and a written safety drill worked out, should chlorine gas accidentally escape. The use of hydrogen in nickel-hydrogen batteries poses explosion and fire risks and there is a need to provide safe pressure vessels to contain the active parts of these batteries.

Some batteries operate at very high temperatures, 200-400 °C, where burns may be a problem. There are as yet undetermined additives which may be required in the production of other couples, but these will largely depend on the technical need for such materials in the future development of these batteries.

HEALTH AND SAFETY MEASURES

Lead-acid. Over the past 15 years the increase in biological monitoring, and particularly the estimation of lead in blood by a reliable laboratory, has enabled doctors, hygienists and engineers not only to ensure increased safety for individuals but, by suitable use of

Figure 3. Ventilation system in the manufacturing of nickel-cadmium battery plates.

group results, to improve the environmental controls. Lead in air estimations are useful for the improvement of engineering design, together with Tyndall beam photography of dust clouds. Individual lead-in-air estimations are very variable and so large numbers of samples are required to give as much information as a blood lead sample gives about individual exposure. For this reason, lead-in-air concentrations are not really suitable as legal control limits but they have the advantage over blood lead measurements in that, properly planned, they can indicate sources of lead.

Other measures include provision of suitable protective clothing and good washing and changing facilities. Eating and smoking should be prohibited in any lead-using workshop or before washing and, in the case of eating, the removal of any contaminated protective clothing. In order to prevent the transport of lead or other toxic materials home, protective clothing should be kept at work and laundered by the company. The chief route of lead transport to the home is on shoes and socks.

Strict housekeeping and the design of factories to avoid dust or dirt accumulation is important. Training of management, supervisers and all operatives is vital and needs to be kept up to date.

Medical supervision and the temporary removal from lead work when blood lead levels indicate excessive absorption should prevent any cases of clinical lead poisoning or long-term harmful effects.

Nickel-cadmium. Here the development of enclosed manufacturing and handling of raw materials has considerably improved the environmental control and reduced the incidence of proteinuria amongst employees. The basic principles of control are similar to those for lead-acid batteries, including the prohibition of eating and smoking in workshops.

Tests are now being carried out to see if cadmium in blood can be used as a reliable control of absorption. However, since cadmium does not have an affinity for red cells, normal cadmium-in-blood levels are low, and cadmium workers do not normally show more than 3-4 μg/100 ml of cadmium in blood, it is therefore necessary to have a laboratory accuracy to about 0.2 μg/100 ml.

At present the monitoring includes annual or six-monthly estimations for proteinuria, β2-microglobulin, lung function tests and examination to indicate any developing emphysema, together with tests for anosmia. Since, under present-day conditions, adverse effects are not normally apparent in less than 10 to 20 years, environmental control largely depends on cadmium-in-air estimations and Tyndall beam dust photography. The former, of course, has all the problems of lead-in-air estimations.

Recent concern and legislation has increased the responsibility of manufacturers for environmental pollution, with the result that discharge of exhaust air and of lead, cadmium, acid, alkali and other potentially toxic materials to the atmosphere must be monitored and conform to local regulations or recognised standards. It has not been found necessary to filter all exhausts, but exhaust from the following processes are filtered or scrubbed; oxide manufacture, paste mixing, mechanical plate cutting and the powder filling of tubular plates, as well as recovery of batter scrap.

In two studies (Potts 1965, and Kipling and Waterhouse 1967) an excess of prostatic cancer was shown, which it was thought might be related to cadmium. However, in a follow-up study of the population, no excess of prostatic cancer was observed in employees first exposed after 1947, and three series of animal experiments have failed to show that cadmium is related to prostatic cancer (Roe et al. 1973, Levy et al. 1973, and Chandler 1979).

Other couples. In the present climate of concern for health and safety, potential hazards should be recognised in the pilot plant production stage, so that any commercially developed processes should be of safe design and adequately controlled from the start. Some of the potential hazards are mentioned in the previous section, and other hazards may need to be considered and controlled depending on the additional materials used and the method of manufacture.

MALCOLM, D.

CIS 79-1923 "Experimental analysis of lead-in-air sources in lead-acid battery manufacture". Caplan, K. J.; Knutson, G. W. *American Industrial Hygiene Association Journal* (Akron, Ohio), July 1979, 40/7 (637-643). Illus.

CIS 79-133 "Lead dust control in the accumulator industry: Good ventilation and personal hygiene" (Ackumulatorindustri bekämpar blydamm: God ventilation och personlig renlighet). Saarinen, J. *Arbetsmiljö* (Stockholm), 1978, 10 (11-13). Illus. (In Swedish)

CIS 77-1626 "Dose-response analysis of cadmium-induced tubular proteinuria—A study of urinary β_2-microglobulin excretion among workers in a battery factory". Kjellström, T.; Evrin, P. E.; Rahnster, B. *Environmental Research* (New York), Apr. 1977, 13/2 (303-317). 21 ref.

"Further studies on the effect of cadmium on the prostate gland—II. Absence of prostatic changes in mice given oral cadmium sulphate for eighteen months". Levy, L. S.; Clack, J.; Roe, F. J. C. *Annals of Occupational Hygiene* (Oxford), Feb. 1975, 17/3-4 (213-220). 31 ref.

CIS 78-544 *Health and safety guide for storage battery manufacturers.* DHEW (NIOSH) publication No. 77-190 (National Institute for Occupational Safety and Health, 4676 Columbia Parkway, Cincinnati) (July 1977), 110 p. 14 ref.

BCG

BCG *(Bacillus Calmette-Guérin)* is a vaccine made from bovine-type tubercle bacilli attenuated by repeated culture on a bile-potato culture medium. It continues to be a major weapon in tuberculosis control.

However, interest in this vaccine has been somewhat reduced recently in a number of countries where active tuberculosis control programmes employing modern means of detection and treatment already have led to a spectacular fall in tuberculosis morbidity and mortality. On the other hand, BCG remains important for other regions such as Africa, Asia and Oceania, where tuberculosis continues to be a significant disease entity with some 7 million contagious cases throughout the world.

In principle, BCG vaccination is intended for persons with a negative tuberculin reaction. However, if vaccination were given to a person who has already been infected with tubercle bacilli, it would produce only a somewhat accelerated local reaction. Direct BCG vaccination (without a preceding tuberculin test) is therefore commonly used in the lower age groups.

The methods of testing for tuberculin sensitivity include: cuti-test, stamp, ring, percuti-test and intradermal test. The latter is the most reliable and precise and should therefore be given preference. It comprises an intradermal injection of 0.1 ml of tuberculin (5 I.U.). It is generally agreed that the reaction is "positive" if, on the third day, there is an induration of 6 mm diameter or more. However, this criterion does not apply in countries where so-called low-grade tuberculin sensitivity is prevalent.

Any uninfected person can receive BCG vaccination. There are certain contraindications, which may be absolute (solid malignant tumours and leukemias, chronic nephritis, lipoid nephrosis, poorly compensated heart disease, silicosis) or relative (in particular evolutive and convalescent acute disease and extensive dermatitis).

Indications for BCG vaccination programmes depend on various factors: tuberculosis morbidity in the region, level of development, detection and treatment facilities, and appreciation of social, pathological or occupational risk factors. In countries where BCG vaccination is compulsory, the lists of persons included are relatively long and, from the occupational health point of view, include for example:

(a) students taking various courses, in particular physics, chemistry and biology, medical and dental students, students in nursing, social work and midwifery schools;

(b) teachers;

(c) hospital workers, civil servants, industrial and commercial workers;

(d) persons working in dirty environments or handling foodstuffs, in particular;

(e) miners;

(f) public transport workers.

In these cases, persons who have not acquired tuberculin sensitivity following two successive vaccinations are considered to have met their legal obligations. On the other hand, for persons who have lost their tuberculin sensitivity renewed BCG vaccination is sometimes offered. Even if adequate case-finding and treatment facilities are available, serious forms of tuberculosis such as miliary disease and meningitis entail a relatively high fatality. These forms are mostly observed in infants. Consequently, in countries where tuberculosis morbidity is still high, or under high risk conditions, it is advisable to vaccinate all children as early as possible shortly after birth or during the first years of life.

Most vaccination methods are satisfactory provided a good quality vaccine is used with a rigorous technique. The response to BCG vaccination is dose-dependent, and experience has shown that the intradermal route is therefore the most effective. Under no circumstances can BCG vaccination cause or aggravate any primary infection.

Complications of BCG vaccination are rare and usually benign. However, exceptional cases have been reported of fatal generalised infection ("BCG-itis") in children with immunodepression; the incidence of this type of complication is not more than one per million. In most cases complications are restricted to local or general occurrences:

(a) local occurrences such as prolonged ulceration or violent local reaction which may require *in situ* chemotherapy: the incidence is around 1%;

(b) regional occurrences in the form of adenitis in the lympth nodes draining the region around the point of injection. Reabsorption is usually spontaneous but local injection of antituberculosis drugs and corticoids may be helpful where there is fistulisation. The incidence of this type of complication is highest in the newborn, who for this reason may be given a reduced dose; it is virtually zero in adolescents and young adults.

These complications should not give rise to consideration in countries where tuberculosis is still highly prevalent, in view of the considerable advantages of vaccination. In other countries they should probably be taken into account, however, and the indications for vaccination should be more restrictive and matched to the circumstances.

Adequacy of vaccination is shown by the appearance of tuberculin sensitivity, which can be checked by a low-dose tuberculin test 2-3 months after vaccination.

Even a "successful" vaccination does not totally protect the subject from a subsequent tubercular disease but the risk may be as much as five times less than that for a non-vaccinated person. This immunity may last for at least 15 years.

To conclude:

(1) BCG vaccination is one of the most effective weapons against tuberculosis.

(2) Its contraindications are very limited.

(3) It must be applied under rigorous conditions. Preference should be given, where possible, to the intradermal technique, both for evaluating tuberculin sensitivity and for the vaccination itself.

(4) The complications are rare and usually benign and, consequently, under most current circumstances should not prohibit use of BCG vaccination. However, certain factors, in particular of epidemiological, social or occupational nature, may warrant an extension or a reduction in the scope of the vaccination programme.

(5) The protection given by vaccination is neither absolute nor definitive. Revaccination after a period of 10-15 years may be indicated. This applies in particular if the first vaccination was given shortly after birth.

MARCHAND, M.

WHO Expert Committee on Tuberculosis. Ninth report. Technical report series No. 552 (Geneva, World Health Organisation, 1974), 40 p.

"Prevention of tuberculosis in mining areas. The experience of the Nord and the Pas-de-Calais Basin" (Lutte antituberculeuse en milieu minier. L'expérience du bassin du Nord et du Pas-de-Calais). Amoudru, C.; Michot, R.; Azencott, A.; Deflandre, J. *Acta Tuberculosea et Pneumologica Belgica* (Brussels), 1975, 66/1-2 (128-137). (In French)

"Control of tuberculosis in hospital personnel". Rosenthal, S. R. *Hospital Practice* (New York), 1974, 9/6 (85-88).

"Skin allergy to tuberculin in medical students and student nurses. Importance of the age of primary BCG vaccination" (Allergie cutanée à la tuberculine chez les étudiants en médecine et les élèves infirmières. Importance de l'âge de la primo-vaccination par le BCG). Lewi, S. *Semaine des Hôpitaux de Paris* (Paris), 1976, 52/12 (745-750). (In French)

"Revised requirements for dried BCG vaccine" (116-147). 11 ref. *WHO Expert Committee on Biological Standardisation. Thirtieth report.* Technical report series (Geneva, World Health Organisation, 1979), 199 p.

Beat diseases

Beat diseases are statutorily defined in the United Kingdom legislation as shown in table 1.

These diseases have been an affliction of miners from earliest times and are still widely prevalent, as is shown in table 2. It will be seen that beat knee is the most widespread condition.

Aetiology

Beat knee. The dominant environmental factor in the causation of the beat diseases, in particular beat knee, is the height of the working face since, when a miner is compelled to work in a kneeling position, to crawl or

Table 1. The statutory definition of beat diseases in the United Kingdom

Description of disease or injury	Nature of occupation
Beat hand	
Subcutaneous cellulitis of the hand	Manual labour causing severe or prolonged friction or pressure on the hand
Beat knee	
Bursitis or subcutaneous cellulitis at or about the knee due to severe or prolonged external friction or pressure at or about the knee	Manual labour causing severe or prolonged external friction or pressure at or about the knee
Beat elbow	
Bursitis or subcutaneous cellulitis at or about the elbow due to severe or prolonged external friction or pressure at or about the elbow	Manual labour causing severe or prolonged external friction or pressure at or about the elbow

Table 2. Number of new cases of beat diseases for which new claims were made under the United Kingdom Social Security Act, 1971-79[1]

Type of beat disease	1971	1973	1975	1977	1979[2]
Beat hand	13	6	6	4	1
Beat knee	1 010	641	480	350	284
Beat elbow	163	107	73	68	56
Total	1 186	754	559	422	341

[1] From *Health and safety–Mines and quarries* (London, Health and Safety Executive, various years). [2] Provisional.

work lying on his side, the repeated pressures on the weight-bearing areas of the body devitalise the tissues in these areas, the knees, the elbows and the hands. It has been shown that during the performance of routine tasks in the kneeling position, the surface of the knee may be subject to pressures of up to 14 kgf/cm^2. In addition, infection may intervene and play a significant role.

The condition is also seen in carpet layers, parquet floor workers, asphalt layers and joiners.

Figure 1. Coalface mineworkers waiting to go below ground. Characteristic work postures while at rest are displayed. Attention is drawn to the relationship of the body to the knees and ankles.

Beat elbow. This is more commonly associated with lower working heights than those which predispose to beat knee. In thinner seams the miner works on his side rather than on his knees and the number of notified cases of beat elbow is higher than that of either of the other beat conditions.

Beat hand. Beat hand is due to the infection of the subcutaneous tissues of the palmar aspects of the hand, thumb or fingers via a break in the integrity of the skin. Local cellulitis develops and may extend to the fascial spaces and tendon sheaths. Predisposition to this condition may be produced by any environmental factor which reduces the resistance of the healthy tissues to infection, e.g. roughness of hand tool shafts, or wet conditions leading to maceration of the skin surface. Bursae do not form part of the beat hand syndrome, whereas they are usually involved in beat knee and elbow. Beat hand can also affect joiners, navvies and caulkers.

Symptoms and signs

Beat knee. The dominant symptom is pain. The clinical signs are determined by the relative extent of the bursitic element and the infective element. There are definite clinical patterns related to the arrangement of the bursal anterior to the knee joint. These patterns fall into three main groups:

(a) simple bursitis;

(b) cellulitis;

(c) combinations of bursitis and cellulitis.

A sympathetic effusion into the knee joint may occur. Constitutional disturbances are unusual. Repeated attacks may cause a chronic condition characterised by a painless enlargement of the bursa, usually the prepatellar. The skin over the affected bursa is thickened and recurrences of the acute state are common.

Beat elbow. The tissues around the elbow are swollen and painful, displaying all the signs of acute infection. There is frequently an associated inflammatory reaction in the bursa over the olecranon process. The condition is often associated with a history of an injury to the olecranon area. Evidence in the form of an abrasion of the skin in that part may be seen. Chronic, painless enlargement of the olecranon bursa is often found in miners working in relatively thin seams. This is a potential beat elbow.

Beat hand. In the typical clinical picture of beat hand the affected hand becomes swollen and painful, displaying the classical signs of inflammation. The cellulitis may be local and confined to the skin and subcutaneous tissues. The infection may extend to involve the fascial spaces and the tendon sheaths. Constitutional reactions are more common in beat hand than in other beat conditions, and they may be severe.

PREVENTIVE MEASURES

The ultimate solution to the prevention of beat diseases in miners lies in the wider application of mechanised mining techniques, and increased mechanisation in the coalmining industry in the United Kingdom has already proved an important factor in reducing the incidence of these diseases. In addition, the establishment of a medical service staffed by skilled personnel at the mines themselves, to ensure early diagnosis and treatment of injuries, has also played an important part in lessening the impact of the beat diseases.

The intelligent use of suitable knee pads, elbow pads and hand protection by men at risk is of value in protecting the weight-bearing areas, and the miner himself has his part to play by maintaining in good health the skin covering the areas liable to injury.

Treatment

Beat knee. When pain is acute, kneeling is impossible and rest is necessary. In the acute, simple state aspiration of the bursa under strict aseptic conditions may be considered but this should not be undertaken lightly since a bursa may communicate with the knee joint. In the acute phase good results have been obtained from a combination of careful skin cleansing of the affected area, quadriceps exercises with the knee in the right-angle position and high-dosage ultraviolet radiation. Resolution and return to kneeling work is hastened by the firm application of a mines-type dressing to the front of the affected knee.

When severe infection is present, appropriate antibiotic therapy is indicated. Surgery should be avoided even in the chronic state since excision of a diseased bursa may leave a painful scar which prevents kneeling.

Beat elbow. This condition usually resolves with rest and local applications of ultraviolet radiation. A suitable antibiotic will be necessary where the infection is severe or in the presence of constitutional reactions.

Beat hand. The treatment of beat hand is the same as that for any severe infection of the hand from whatever cause. Rest and adequate antibiotic therapy is essential. Surgical intervention calls for the highest skills and clinical judgement. Planned rehabilitation to restore full function of the hand is an integral part of efficient treatment.

WATKINS, J. T.

General:

CIS 78-782 *Beat conditions, tenosynovitis.* Guidance Note MS 10, Health and Safety Executive (London, HM Stationery Office, 1977), 2 p. 7 ref.

CIS 78-794 *Occupational beat conditions (excluding beat knee)* (Les bursites d'origine professionnelle (à l'exclusion de celles du genou)). Beaucousin, M. (Université de Paris VI, Faculté de médecine Broussais–Hôtel-Dieu, Paris) (1977), 57 p. 25 ref. (In French)

Beat elbow:

CIS 462-1967 "A study of beat elbow and related conditions at one colliery". Archibald, R. McL.; Kay, D. G.; Scott, E. *Occupational Health* (London), 1966, 18/3 (118-123). Illus.

Beat knee:

"Evaluation of occupational stresses in parquet floor workers" (Ocena narazenia zawodowego parkieciarzy). Wiercioch, B. *Medycyna Pracy* (Lodz), 1973, 24/2 (225-228). (In Polish)

CIS 475-1963 "Aetiology and pathology of beat knee". Sharrard, W. J. W. *British Journal of Industrial Medicine* (London), Jan. 1963, 20/1 (24-31). Illus. 6 ref.

Benzanthrone

Benzanthrone ($C_{17}H_{10}O$)

1,9-BENZANTHRONE; MESOBENZANTHRONE

m.w.	230.25
m.p.	170-171 °C
v.p.	1 mmHg (0.13·10³ Pa) at 225 °C

insoluble in water; slightly soluble in ethyl alcohol and other organic solvents

pale yellow needles.

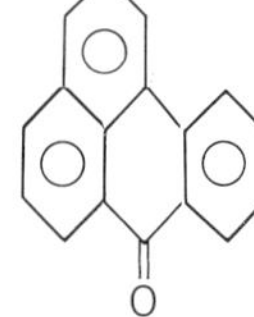

Production. Of the several known methods of synthesis, only the anthraquinone/glycerol process is used industrially. The anthraquinone is usually first reduced with powdered copper, and the glycerol is converted to acrolein before condensation with anthrone.

$CH_2-CH-CH_2$ (OH OH OH) glycerol; $CH_2=CH-CHO$ acrolein; anthraquinone; anthrone → benzanthrone

In the industrial process, anthraquinone is dissolved in concentrated sulphuric acid and copper powder is added. After the copper has dissolved, a glycerol/water mixture is slowly added and the bath is heated to 120 °C and stirred. The bath is then cooled, drowned in water, and the crude product filtered off and washed with water. It is purified by boiling with dilute caustic and filtering.

Uses. Benzanthrone is the starting product for an important group of vat dyes. Fusion of benzanthrone with potassium hydroxide produces dibenzanthrone (violanthrone–$C_{34}H_{16}O_2$) or Indanthrene Dark Blue 30.

Nitration of violanthrone in a solvent and reduction of the nitrocompound produces Vat Green B, also known as Indanthrene Black BB. Oxidation of violanthrone results in a dihydroxy derivative which is an intermediate in the production of Vat Jade Green. Other important dyes in the benzanthrone series are Vat Olive Green B, Vat Olive T and Indanthrene Brilliant Violet 2R.

Benzanthrone and its derivatives are used as light sensitisers for degradation of plastics, daytime fluorescent pigments, and in coloured chemical smokes, too.

HAZARDS

The most common hazard of benzanthrone is skin sensitisation due to exposure to benzanthrone dust. Sensitivity varies from person to person, but after exposure of between a few months and several years, sensitive persons, especially those who are blond or red-headed develop an eczema which may be intense in its course and the acute phase of which may leave a hazel or slate-grey pigmentation, especially around the eyes. Microscopically, atrophy of the skin has been found. Skin disorders due to benzanthrone are more frequent in the warm season and are significantly aggravated by heat and light.

The general effects ascribed to benzanthrone are manifold and relatively vague. Persons with periods of exposure to benzanthrone varying between months and years complain of loss of appetite, intolerance of fatty foods, fatigue and weakness. Complaints of a decrease in sexual activity are strikingly frequent. There is often an objective weight loss and there have been reports of neurasthenic syndrome, changes in corneal and cremaster reflexes, neurocirculatory disorders with accelerated pulse and reduced blood pressure. Studies in Czechoslovakia indicate that the reduction in sexual activity has a factual basis. There have also been reports of liver function impairment and gastritis with decreased acidity.

Experimentally, a lowering of the coagulation time, increase in plasma fibrinogen and in the number of platelets, decrease of bleeding time, and a decrease in the number of red blood cells as well as of haemoglobin content has been observed in rats treated with benzanthrone. In rabbits and in guinea-pigs the intraperitoneal application of benzanthrone induced vascular congestion of the lamina propria and submucosa and localised damage of the epithelial layer of the urinary bladder; oral and topical administration, however, did not alter the appearance of the bladder. A lowered body ascorbic acid level combined with the effect of benzanthrone or its metabolites excreted in the urine may be responsible for the epithelial damage. The impairment of the gametogenic function by benzanthrone has also been proved.

SAFETY AND HEALTH MEASURES

Adequate dust-control measures, including exhaust ventilation, are essential in benzanthrone production. Personal protective equipment such as respiratory protective devices and skin protection are not to be recommended since the friction and the increase in sweating that they provoke may make the skin more vulnerable to the action of benzanthrone. It has been found possible to decrease the incidence of skin disease by limiting the use of benzanthrone to the colder seasons. A special diet (150 mg of vitamin C daily) is provided in the USSR to exposed persons.

Persons with fair skin are less suitable for work with benzanthrone and persons with skin diseases, especially eczema, should not be exposed to this substance. The employment of persons with liver, digestive system, and especially gastric, diseases, blood coagulation disorders and urinary bladder diseases, as well as of persons with an outstanding neurasthenic syndrome or vegetative disorders is not advisable in benzanthrone plants. Should serious skin disease develop during work with benzanthrone, exposure should be terminated.

MARHOLD, J. V.

CIS 78-130 "Effect of benzanthrone on testis and male accessory sex glands". Singh, G. B.; Khanna, S. K. *Environmental Research* (New York), Dec. 1976, 12/3 (327-333). Illus. 22 ref.

CIS 1489-1971 "Toxicity of dyes with special reference to benzanthrone". Singh, G. B. *Indian Journal of Industrial Medicine* (Calcutta), Sep. 1970, 16/3 (122-129). 16 ref.

"Our experience from investigation of the state of health of people employed in benzanthrone production" (Naše zkušenosti se sledovǎnim zdravotniho stavu zaměstnancû, pracujicich přivýrobě benzanthronu). Horáková, E.; Merhaut, J. *Pracovni Lékařstvi* (Prague), Feb. 1966, 18/2 (78-81). 3 ref. (In Czech).

"Health status of workers of benzanthrone manufacturing plants" (Sostojanie edorov'ya rabočih sovremennogo proizvodstva benzantrona). Kleiner, A. I.; Sonkin, I. S.; Nestrugina, Z. F.; Krylova, E. V.; Rezenkina, L. D.; Ermilova, I. I. *Gigiena Truda i professional'nye zabolevanija* (Moscow), Oct. 1979, 10 (43-45). 4 ref. (In Russian)

Benzene

Benzene (C_6H_6)

BENZOL; COAL NAPHTHA

m.w.	78
sp.gr.	0.88
m.p.	5.5 °C
b.p.	80.1 °C
v.d.	2.8
v.p.	75 mmHg (9.97·10^3 Pa) at 20 °C
f.p.	−11 °C
e.l.	1.3-7.1%
i.t.	562 °C

slightly soluble in water; very soluble in organic solvents and oils

a colourless volatile liquid with a specific odour.

TWA OSHA	1 ppm 3 mg/m^3 5 ppm 15 mg/m^3 ceil
TLV ACGIH	10 ppm 30 mg/m^3: industrial substance suspected of carcinogenic potential for man
STEL ACGIH	25 ppm 75 mg/m^3
IDLH	2 000 ppm
MAC USSR	5 mg/m^3 skin

Benzene is often referred to as "benzol" in its commercial form (which is a mixture of benzene and its homologues) and should not be confused with benzine, a commercial solvent which consists of a mixture of aliphatic hydrocarbons.

Production. Benzene is a constituent of coal tar from which benzol is obtained by distillation. The designation benzol 90/100 indicates a substance containing 90% of hydrocarbons distilling below 100 °C. Benzol is extracted from coal gas and coke oven gas by a stripping or scrubbing operation in a tower up which the gas is passed. In the petroleum industry large quantities of benzene are produced by catalytic reforming, dealkylation and dehydrogenation processes or by cyclisation and aromatisation of paraffin hydrocarbons.

Uses. In industry, benzene is used as a fuel, as a chemical reagent and as a solvent.

In certain parts of the world a major use of benzene is as an additive of motor fuel and large quantities are used for this purpose.

Benzene is chemically reactive and serves as a raw material for a great number of chemical syntheses. Nitration of benzene produces nitrobenzene and the aromatic nitro compounds which can be used directly in finished products or be reduced to amino substances such as aniline and phenylenediamine which can, in turn, be converted to nitroso compounds and coupled with other compounds to form complicated dyes and dyestuffs. Benzene can be sulphonated or chlorinated as part of an elaborate synthesis or it can be alkylated by a Friedel-Crafts reaction. The halogenated pesticides (DDT, chlorinated diphenyls) account for many of the chloro-derivatives. Benzene is used extensively in the manufacture of styrene, phenols, maleic anhydride and a number of detergents, explosives, pharmaceuticals and dyestuffs.

Benzene is a good solvent for a large number of materials such as rubber, plastics, paints, inks, oils and fats, and is very effective as an extracting agent for seeds and nuts. The volatility is a valuable property when the rapid drying of the solvent is an important part of a process as in photogravure printing or paint spraying. In spite of these valuable solvent properties, the hazards associated with the use of benzene are such that less effective solvents are to be preferred. Because of these hazards, the use of benzene as a dry-cleaning liquid is now very limited.

Metabolism. The fate of benzene in the body has been the subject of considerable study. A large proportion is eliminated in the exhaled air, but 15-60%, depending on the circumstances, is metabolised. The main metabolic transformation is oxidation by microsomal mixed function oxidases to benzene epoxide, a highly reactive substance which can react with cellular constituents, e.g. proteins and nucleic acids. Benzene epoxide can be transformed by a non-enzymatic rearrangement into phenol, or can be hydrated by an epoxide hydratase and then be reduced to catechol, or can be condensed with glutathione to form mercapturic acid. The oxidation phase occurs mainly in the liver and is followed by a conjugation phase of phenol and catechol with sulphate or glucuronic acid, and then conjugates are excreted in the urine.

Mechanism of action. From the biological point of view, it seems that the bone marrow and blood disorders found in chronic benzene poisoning can be attributed to the conversion of benzene to benzene epoxide. It has been suggested that benzene might be oxidised to epoxide directly in bone marrow cells, such as erythroblasts.

As far as the toxic mechanism is concerned, benzene metabolites seem to interfere with nucleic acids. Increased rates of chromosome aberrations have been observed both in man and in animals exposed to benzene (see CHROMOSOME ABERRATIONS).

Any condition likely to inhibit further metabolism of benzene epoxide and conjugation reactions, especially hepatic disorders, tends to potentiate the toxic action of benzene. These factors are of importance when considering differences in individual susceptibility to this toxic agent.

To see the very special toxic nature of benzene in its true perspective, comparison with the benzene homologues (toluene, xylenes) is imperative. Both toluene and xylenes have the same acute and subacute toxic potential as that described for benzene. Yet, in chronic exposure their action on the bone marrow seems to entail essentially different mechanisms and is considerably less pronounced. Human and animal research has shown that the major metabolic degradation process to which benzene homologues are subject is not oxidation of the benzene ring, but predominantly oxidation of the substituted methyl group with the production of aromatic acids devoid of antimitotic and cytotoxic properties.

HAZARDS

Benzene evolves very toxic, flammable vapours and grave risks are thus associated with its industrial use.

Health hazards

Acute poisoning. Benzene exerts the acute narcotic action common to many other hydrocarbons and has a local irritant effect on the skin and mucous membranes.

Chronic poisoning. The outstanding feature of benzene is its ability to damage blood-forming tissues of chronically exposed persons, with resulting *hyporegenerative anaemia* of various degree. The onset of chronic benzene poisoning is extremely insidious and its ultimate injury potentially incurable. The early symptoms are by no means specific to benzene exposure—vague complaints of fatigue, loss of appetite, headache, dizziness and an anaemic appearance, all of which may be common to quite other causes. Blood examinations at this stage may reveal only slight abnormality; the initial response may, in fact, be an actual rise in the number of erythrocytes instead of the expected fall. Later, however, the most significant change in the blood picture appears—a fall in the total polymorphonuclear leucocyte count to below 4 000/mm^3 with relative lymphocytosis, macrocytic, normochromic or slightly hyperchromic anaemia and thrombocytopoenia. For some time the appearance of bone marrow remains virtually unchanged except for a decrease in polychromatic erythroblasts, myelocytes and metamyelocytes. In severe cases, especially if exposure is not terminated, *true aplastic anaemia* may develop with partial or total destruction of all elements of the bone marrow. Haemorrhagic manifestations are relatively common. For some considerable time the patient's general condition remains quite good although there is pallor, hypotension and slight increases in temperature in some cases. Even if the subject is immediately removed from exposure, recovery is always protracted and there is frequently permanent impairment of blood function; relapses are common and severe cases prove fatal.

Even though hyporegenerative anaemia resulting from chronic exposure to benzene has been known for over a century, it was not until 1928 that the action of benzene was shown to be a cause of *leukaemia*. Since then, approximately 200 cases of benzene leukaemia have been reported, either as single cases or as outbreaks.

In Italy, in the provinces of Milan and Pavia, outbreaks of benzene blood diseases, including acute leukaemia, occurred years ago in shoe factories and in rotogravure plants where benzene was used as a solvent, and many such cases were observed at the institutes of occupational health of those cities (table 1). In the factories concerned a high concentration of benzene vapours at the workplace was demonstrated, and the incidence of acute leukaemia among subjects occupationally exposed to benzene appeared to be about 20 times higher than expected.

Table 1. Fatal cases of blood disease due to chronic benzene poisoning seen at the institutes of occupational health of Milan (1943-74) and Pavia (1960-74)

Total deaths	34
Aplastic anaemia	10
Acute leukaemia	19
Erythroleukaemia	4
Acute erythraemia	1

An outbreak of acute leukaemias from benzene exposure has recently been reported among Turkish workers chronically exposed to high concentrations of benzene.

Recent epidemiological studies carried out in the United States and in Japan confirm an increased risk of leukaemias in workers with chronic exposure to benzene. In some French and American reports also, cases of chronic leukaemias have been attributed to benzene. However, experience has shown that the leukaemia cases which occurred in factories where cases of aplastic anaemia had also been observed and, where exposure to benzene had been heavy, were all acute.

Negative epidemiological data for increased risk of leukaemia were, however, obtained in mortality surveys of workers of the chemical or petrochemical industry exposed to low concentrations of benzene, never exceeding 25 ppm. In our experience, no further cases of benzene leukaemias have been observed since benzene was replaced by its homologues, containing only traces of the substance.

Benzene leukaemias are mostly acute myeloblastic leukaemias; a few cases of acute erythroleukaemia have also been reported. Most cases are leucopoenic or shown only moderate leucocytosis, except in the terminal phase. Moderate or no splenomegaly is present. The bone marrow shows extensive infiltration by undifferentiated cells, mostly micromyeloblasts or paramyeloblasts. Leukaemia occurs most frequently in subjects with hyporegenerative anaemia; in such cases the bone marrow changes from a hypoplastic to a leukaemic pattern during the evolution of the disease. Generally, benzene haemopathy occurs in subjects exposed to concentrations of benzene much greater than the threshold limit value; in some cases a long latent period (up to 12 years) between cessation of exposure and occurrence of leukaemia has been observed.

As far as the pathogenesis is concerned, we have suggested that stable chromosome aberrations induced by benzene, persisting in bone marrow cells, might give rise to abnormal clones, which may eventually cause the development of leukaemia even after long latent periods. The treatment of benzene leukaemia is the same as that prescribed for acute leukaemia of unknown origin. In addition to supportive treatment (such as blood transfusions, antibiotics, etc.), corticosteroids, 6-mercaptopurine and folic acid antagonists are used; these may induce temporary remissions, thus prolonging the survival time. It is important to follow up the bone-marrow pattern rather than that of the peripheral blood, since an infiltration of the bone marrow by immature cells may call for treatment with antimetabolites even in the absence of leucocytosis in the peripheral blood.

Fire and explosion

Benzene is a flammable liquid, the vapour of which forms flammable or explosive mixtures in air over a large range of concentrations; the liquid will evolve vapour concentrations in this range at temperatures as low as −11 °C. In the absence of precautions, therefore, at all normal working temperatures flammable concentrations are liable to be present where the liquid is being stored, handled or used. The risk becomes more pronounced when accidental spillage or escape of liquid occurs.

SAFETY AND HEALTH MEASURES

From the above description of the dangers associated with benzene, it is apparent that exceptional precautionary measures are justifiable for the prevention of injury to health and that reliable safeguards must be provided to prevent injury from fire and explosion.

Health precautions

It is now recognised that the use of benzene should be abandoned for any industrial purpose where an effective, less harmful substitute is available. It is seldom the case that a substitute is available when the benzene is being used as a reactant in a chemical synthesis. In such cases it is the particular molecular configuration of benzene that makes the reaction possible. In some few cases, however, it may be possible to adopt a different synthesis and reach the final product by another synthetic route that does not involve benzene, such as in the production of phenol from toluene or cumene instead of from benzene; nevertheless, possibilities of this type are rare. On the other hand it has proved possible to adopt substitutes in almost all the very numerous operations where benzene has been used as a solvent. The substitute is not always as good a solvent as benzene, but it may still prove the preferable solvent because less onerous precautions are required. Such substitutes include:

(a) benzene homologues, especially toluene and xylene;

(b) cyclohexane;

(c) aliphatic hydrocarbons, either pure as is the case with hexane, or as mixtures as is the case with the wide range of petroleum solvents;

(d) solvent naphthas which are relatively complex mixtures of variable composition obtained from coal or certain petroleum products. They contain virtually no benzene and very little toluene; the main constituents are homologues of these two hydrocarbons in proportions that vary depending on the origin of the mixture;

(e) various other solvents chosen to suit the material to be dissolved and the relevant industrial processes. They include alcohols, ketones, esters and chlorinated derivatives of ethylene.

In modern petrochemical plants, it is possible to eliminate benzene almost completely from the finished product; however, in coal-based chemical production plants, which are usually older, the elimination of benzene is not always possible. Consequently, it has been found necessary to accept, at least for the time being, a certain proportion of benzene in the above-mentioned solvents.

The ILO's Benzene Convention, 1971 (No. 136) defines as products containing benzene products the benzene content of which exceeds 1%. The Convention stipulates in Article 4 that the use of benzene and products containing benzene shall be prohibited in certain work processes to be specified by national laws or regulations; this prohibition shall at least include the use of benzene and products containing benzene as a solvent or diluent, except where the process is carried out in an enclosed system or where there are other equally safe methods of work.

When benzene must be used, for chemical synthesis, in motor fuel, in analytical or research work carried out in laboratories, or in the production of benzene itself, all the precautions to prevent the escape of vapour into the workroom atmosphere must be taken with meticulous thoroughness. Wherever possible, the plant should be totally enclosed; whether the enclosure is complete or whether openings are provided for the purposes of manipulation, enclosures should be supplemented by exhaust ventilation, so that, even if the process has ceased and the plant is open for maintenance or repair, the vapour will be prevented from contaminating the air of the workroom. When ventilation systems are being installed, the high density of benzene vapours should be borne in mind. The atmosphere of the workroom should be tested periodically for benzene to ensure that the precautions taken are effective.

Experience in the years following the first recommendation of maximum permissible atmospheric concentrations has resulted in a lowering of these suggested levels. From the original figure of 100 ppm postulated in 1937 by the American Conference of Governmental Industrial Hygienists, the threshold limit in the United States was lowered to 35 ppm in the period 1951 to 1957, and to 10 ppm or 30 mg/m^3 (TWA) at the present time; the maximal excursion allowed is 20 ppm. When considering these values one must take into account the possible absorption by skin and mucous membranes. The same agency has included benzene among the substances suspected of carcinogenic potential for man. On this basis the NIOSH in 1976 recommended that the exposure to benzene be kept "as low as possible"; since 1 ppm is considered "the lowest level at which a reliable estimate of occupational exposure to benzene can be determined, NIOSH recommends that ... no worker be exposed to benzene in excess of 1 ppm (3.2 mg/m^3) in air, as determined by an air sample collected at 1 l/min for 2 h". The MAC in the USSR is now 1.5 ppm (5 mg/m^3).

Article 6(2) of ILO Convention No. 136 stipulates that where workers are exposed to benzene or to products containing benzene the employer shall ensure that the concentration of benzene in the air of the places of employment does not exceed a maximum which shall be fixed by the competent authority at a level not exceeding a ceiling value of 25 ppm (80 mg/m^3). Convention 136 has been ratified so far by 21 member States.

Efforts should be made to encourage the monitoring of the atmosphere in small undertakings where the hazards are generally higher and medical supervision seldom adequate. Although not very specific, danger-level warning systems such as those using colour reactions of benzene and its homologues may be used as a first step. The recommendation already adopted by the ILO and the Council of Europe to use, on the labels of benzene containers, the skull and crossbones symbol for toxic substances is fully justified. It is recommended that where solvents and thinners contain benzene, the label should indicate both the designation of the product and its benzene content by volume.

Whenever it is necessary to enter a vessel that has contained benzene, the safety and health measures applicable to work in confined spaces should be implemented.

Medical supervision

No person should be employed for work in a job involving exposure to benzene unless he is in a good state of health. The assessment of fitness should include consideration of previous medical history and occupational history.

The occupational history should take into account any previous exposure to benzene, other radiomimetic substances or ionising radiations. The medical examination should include a thorough physical examination and a haematological examination. The latter is of prime importance and should cover haemoglobin determination, red cell, white cell and platelet counts, white cell differential count and red cell and leucocyte morphology.

Young persons of either sex under 18 years of age should not be exposed to benzene since haematologists recognise that adolescents have a lower resistance to bone-marrow poisons. Pregnant women and nursing mothers should not be exposed to benzene and, since pregnancy is often not diagnosed until some considerable time after conception, special precautions are necessary where women of childbearing age are exposed to a benzene hazard. Also subjects with liver diseases and subjects with microcytaemia should not be exposed to benzene.

Periodic examinations should be carried out in the same way as pre-employment examinations. Individual susceptibility to benzene varies considerably and, consequently, particular attention should be paid to any haematological abnormalities found during the first periodic examination since these may be manifestations of special sensitivity to benzene. Persons in whom haematological abnormalities are found during the first periodic examination should be removed from exposure to benzene. Particular attention must be paid to the differential count of white blood cells; in fact, in leucopoenic cases of acute myeloblastic leukaemia only a few micromyeloblasts may be present in the peripheral blood and these might at first glance be taken for lymphocytes. Whenever there is the slightest suspicion of leukaemia, a bone-marrow biopsy is warranted. This may also reveal a hyperplastic process in cases with peripheral pancytopoenia. It is probable that in the past several cases of benzene leukaemia have been overlooked and classified as aplastic anaemia on the basis of the peripheral blood findings only.

Biological monitoring

Exposure tests are usually employed as a counterpart to environmental monitoring since they indicate the uptake of benzene. In particular, estimation of urinary phenols by specific and sensitive techniques is recommended and should be carried out on a group basis, possibly with the knowledge of a background value for the subjects on study. A group of experts, who met in Paris in November 1976 to discuss the toxicology of benzene, suggested as guidelines the following values for individual results of urinary phenols determined immediately after shift exposure:

(a) about 100 mg/l indicates a shift exposure of approximately 200 ppm-hours (i.e. 25 ppm for 8 hours);

(b) about 50 mg/l indicates exposure to approximately 80 ppm-hours (i.e. 10 ppm for 8 hours);

(c) over 25 mg/l indicates some benzene exposure;

(d) less than 10 mg/l probably shows an absence of significant exposure.

The results should be corrected by creatinine or specific gravity determinations, and results of samples with specific gravity less than 1 010 or creatinine less than 0.5 g/l should be discarded.

The determination of urinary sulphoconjugates is not recommended due to the scanty sensitivity; in fact they increase only for air concentrations of benzene higher than 40 ppm.

Breath sampling of benzene is more sensitive and specific than phenol determination in urine and deserves further investigation. Samples taken immediately before a new shift (i.e. 16 hours after the end of a shift) reflect exposure during previous shifts. Samples taken at the end of a shift are greatly influenced by exposure toward the end of the shift. There is no accumulation during the week.

Fire prevention

The precautions described above to ensure that a toxic concentration of benzene vapour is not present in the working atmosphere are fully adequate to ensure that flammable mixtures will not form with the air in normal circumstances. Additional precautions are needed, however, to cover the risk of accidental spillage, leakage or overflow of the liquid from storage or process vessels. To prevent escaping liquid from flowing from a place of safety to a place where it could become ignited and to prevent it from spreading over an extensive surface from which large evolution of vapour could occur, mounds should be provided round storage tanks in the open; a similar effect is achieved inside the factory by the design of the floor, by the provision of sills at doorways and, in certain instances, by the provision of curbs round the process vessels.

Open flames and other sources of ignition should be excluded where benzene is stored and used.

Benzene derivatives

A large number of benzene derivatives are formed by processes of halogenation, nitration, reduction, diazotisation and alkylation. Those most commonly produced by and used in industry are dealt with in the following separate articles in this encyclopaedia: CHLOROBENZENE AND DERIVATIVES; NITROBENZENE; NITROCOMPOUNDS, AROMATIC; DINITROPHENOL; ANILINE; AMINES, AROMATIC; AZO AND DIAZO DYES.

FORNI, A.
VIGLIANI, E. C.
TRUHAUT, R.

"International Seminar on Interpretation of Data and Evaluation of Current Knowledge of Benzene. Vienna, June 10-11 1980". *International Archives of Occupational and Environmental Health* (West Berlin), Feb. 1981, 48/1 (107-111).

"International Workshop on Toxicology of Benzene. Paris, 9th-11th November 1976". Truhaut, R.; Murray, R. *International Archives of Occupational and Environmental Health* (West Berlin), 1978, 41/1 (65-76). 4 ref.

CIS 80-1358 *Assessment of health effects of benzene germane to low-level exposure.* Publication No. EPA-600/1-78-061 (US Environmental Protection Agency, Office of Research and Development, Environmental Research Information Center, Cincinnati) (1978), 112 p. 313 ref.

Human biological monitoring of industrial chemicals. 1: *Benzene.* Lauwerys, R. Commission of the European Communities, EUR 6570 EN (Brussels-Luxembourg, ECSC-EEC-EAEC, 1979), 46 p. 84 ref.

"Benzene and leukemia". Vigliani, E. C.; Forni, A. *Environmental Research* (New York), 1976, 11 (122-127). 19 ref.

Criteria for a recommended standard. Occupational exposure to benzene. DHEW (NIOSH) publication No. 74-137 (National Institute for Occupational Safety and Health, 4676 Columbia Parkway, Cincinnati) (1974), 137 p. Illus. 142 ref.

Convention concerning protection against hazards of poisoning arising from benzene, 1971 (Geneva, International Labour Office, 1971).

Recommendation concerning protection against hazards of poisoning arising from benzene, 1971 (Geneva, International Labour Office, 1971).

Benzoyl peroxide

Benzoyl peroxide ($(C_6H_5CO)_2O_2$)

DIBENZOYL PEROXIDE

m.w.	242.22	⟨O⟩-COOOOC-⟨O⟩
sp.gr.	1.33	
m.p.	103-105 °C	
b.p.	explosive decomposition just above m.p.	

slightly soluble in water; soluble in many organic solvents

white, granular, tasteless, odourless solid.

TWA OSHA 5 mg/m³

IDLH 1 000 mg/m³

Maximum permissible concentration Bulgaria 0.05 mg/m³

Benzoyl peroxide is highly flammable and explosion-sensitive to heat, friction, shock and certain chemicals.

Production. Benzoyl peroxide is formed by the reaction of benzoyl chloride with a solution of sodium peroxide in water or with an alkaline solution of hydrogen peroxide. Dispersion agents such as water, plasticisers, silicone oil, wheat starch or gypsum are commonly added in commercial formulations. Annual United States production of benzoyl peroxide is on the order of 4 000 000 kg.

Uses. Most benzoyl peroxide is used in the polymer industry to initiate free-radical polymerisations and copolymerisations of vinyl chloride, styrene, vinyl acetate and acrylics, and to cure thermoset polyester resins and silicone rubbers. These reactions result from the thermal decomposition of benzoyl peroxide into useful free radicals in the temperature range of about 80 °C to 120 °C. Decomposition at lower temperatures is achieved by adding a promoter, usually dimethyl aniline. This permits room temperature curing of thermoset polyester resins and polymerisation of acrylics, making possible their widespread use in such applications as auto body putty, mine roof bolts, furniture, bowling balls, acrylic dental and optical castings, and many others.

Benzoyl peroxide is used in medicine for the treatment of acne. It is the preferred bleaching agent for flour, and has been used for bleaching cheese, vegetable oils, waxes, fats, etc. It is used as a free-radical source in many organic syntheses and has been used as a burn-out agent for acetate yarns.

SAFETY AND HEALTH HAZARDS

The major hazards of benzoyl peroxide are fire and explosion. Health hazards are relatively minor but not negligible.

Safety. Dry benzoyl peroxide ignites instantly and burns furiously. Explosive decomposition can be initiated by heat, friction, shock or contamination. Benzoyl peroxide is stable at room temperature but decomposes exothermically when heated. The reaction can accelerate into a violent spontaneous decomposition or explosion if the heat of decomposition is not carried away quickly enough. Decomposition is instantaneous and explosive just above the melting point. Heat from friction or heavy shock can cause vigorous decomposition. Benzoyl peroxide reacts exothermically with strong acids or bases, sulphur compounds, reducing agents and polymerisation promoters and accelerators, such as dimethyl aniline, amines, or metallic naphthenates. Contamination with these can initiate violent decomposition and fire. The violence of decomposition is greatly affected by the degree of confinement of the peroxide. While decomposition may occur without ignition, the products are highly flammable and can form explosive mixtures in air.

These hazards of dry benzoyl peroxide are greatly reduced by dispersing it in non-solvent diluents that absorb any heat of decomposition and provide other benefits. Benzoyl peroxide is commonly produced in hydrated granular form with 20 or 30% water and in various pastes, usually containing about 50% of a plasticiser or other diluents. These formulations have greatly reduced flammability and shock sensitivity compared to dry benzoyl peroxide. Some are fire-resistant. The hardeners used with plastic resin fillers, such as auto body putty, typically contain 50% benzoyl peroxide in a paste formulation. Flour bleach contains 32% benzoyl peroxide with 68% grain starch, calcium sulphate dihydrate or dicalcium phosphate dihydrate and is considered non-flammable. Acne creams, also non-flammable, contain 5 or 10% benzoyl peroxide.

Health. The known toxic effects of benzoyl peroxide are limited to a potential for irritation of skin, mucous membranes and eyes with prolonged contact, and a low incidence of allergic contact dermatitis (1-2.5% of users of acne creams). The health hazards of diluents should be considered, as they may be more serious.

SAFETY AND HEALTH MEASURES

The properties of and safe handling procedures appropriate for the benzoyl peroxide composition or formulation in use must be thoroughly understood by all persons using it. Benzoyl peroxide should be stored in its original containers in a cool, ventilated place apart from other flammable or reactive materials. It should be protected from flame, static electricity, sparks or sources of heat such as steam-pipes, radiators or direct sunlight. Dry benzoyl peroxide must be protected from shock or friction. No more than is needed for a single work shift should be removed from storage at one time. Great care must be taken to prevent direct mixing or contamination of benzoyl peroxide with accelerators, promoters and reducing agents. It should never be added to hot solvents, monomers or reaction vessels. Wet or dilute formulations must not be allowed to separate or dry out. Spills should be cleaned up immediately, using non-sparking tools and an inert moist diluent such as vermiculite or sand if necessary. Waste containing benzoyl peroxide can be added slowly ten times its weight of 10% sodium hydroxide in water and stirred vigorously until the peroxide is decomposed. This may take 3 h at 20 °C. Small amounts of waste can be spread in an open incinerator or trench and ignited by a torch on a long pole. Rigid containers of uncertain age or condition should not be opened but carefully broken and burned from a safe distance. Empty benzoyl peroxide containers should be destroyed.

Persons handling benzoyl peroxide should use safety glasses or face shields, gloves, aprons and other protective clothing as necessary to prevent skin contact and, especially, to protect against burns in the event the benzoyl peroxide flashes. Clothing and equipment that generate static electricity must be avoided.

Storage and handling areas should be protected by a deluge system or sprinklers. In case of fire, water should be applied by the sprinkler system or by hose from a safe distance, preferably with a fog nozzle. Portable extinguishers should not be used. Benzoyl peroxide threatened by fire should be wetted from a safe distance for cooling.

Treatment. Benzoyl peroxide should be washed from the skin after handling, or promptly if there is irritation. In case of eye contact, the eyes should be flushed with large amounts of water and medical attention should be obtained. Sensitised individuals should avoid further contact with benzoyl peroxide.

WOODCOCK, R. C.

"Peroxides and peroxy compounds, organic". Mageli, O. L.; Sheppard, C. S. (766-820). 168 ref. *Encyclopedia of Chemical Technology,* Vol. 14. Kirk, R. E.; Othmer, D. F. (eds.). (New York, John Wiley and Sons, 2nd ed., 1967).

Criteria for a recommended standard. Occupational exposure to benzoyl peroxide. DHEW (NIOSH) publication No. 77-166 (National Institute for Occupational Safety and Health, 4676 Columbia Parkway, Cincinnati) (1977), 117 p. 100 ref.

"Acne vulgaris: Current concepts of pathogenesis and treatment". Hurwitz, S. *American Journal of Diseases of Children* (Chicago), May 1979, 133/5 (536-544). Illus. 39 ref.

Loss prevention data–Organic peroxides, Report No. 7-80 (Factory Mutual Research Corporation, Public Information Division, 1151 Boston-Providence Turnpike, Norwood, Massachusetts) (Mar. 1972), 8 p.

Loss prevention data–Organic peroxides hazard classification (tentative), Report No. 7-81 (Factory Mutual Research Corporation, Public Information Division, 1151 Boston-Providence Turnpike, Norwood, Massachusetts) (Feb. 1974), 10 p.

Benzyl chloride

Benzyl chloride ($C_6H_5CH_2Cl$)

α-CHLOROTOLUENE

C_6H_5–CH_2Cl

m.w.	126.6
sp.gr.	1.1
m.p.	−39 °C
b.p.	179.3 °C
v.d.	4.4
v.p.	0.9 mmHg ($0.12 \cdot 10^3$ Pa) at 20 °C
f.p.	67.2 °C
e.l.	1.1%
i.t.	584 °C

insoluble in water; miscible with ethanol and ether

a colourless liquid with an unpleasant irritating odour.

TWA OSHA	1 ppm 5 mg/m³
NIOSH	5 mg/m³/15 min ceil
IDLH	10 ppm
MAC USSR	0.5 mg/m³

Production. It is obtained commercially by chlorinating toluol at a high temperature (60-80 °C) in the presence of a catalyst.

Uses. Benzyl chloride serves as an intermediate in the manufacture of butyl benzyl phthalate, and as the basic product in the manufacture of benzal chloride, benzyl alcohol and benzaldehyde. It is used in the manufacture of quaternary ammonium chlorides, dyes, tanning materials and in pharmaceutical and perfume preparations.

HAZARDS

As a result of its strong irritant properties benzyl chloride concentrations of 6-8 mg/m³ cause a light conjunctivitis after 5 min of exposure. Airborne concentrations of 50-100 mg/m³ immediately cause weeping and twitching of the eyelids, and in concentrations of 160 mg/m³ it is unbearably irritating to the eyes and mucous membrane of the nose. The complaints of workers exposed to 10 mg/m³ and more of benzyl chloride included weakness, rapid fatigue, persistent headaches, increased irritability, feeling hot, loss of sleep and appetite, and in some, itching of the skin. Medical examinations of workers revealed asthenia, dystonia of the autonomic nervous system (hyperhidrosis, tremors in the eyelids and fingers, unsteadiness in Romberg's test, dermatographic changes, etc.). There may also be disturbances of liver function, such as increased bilirubin content of the blood and positive Takata-Ara and Weltmann tests, a decrease in the number of leucocytes, and a tendency to illness similar to colds and allergic rhinitis. Cases of acute poisoning have not been reported. Benzyl chloride can cause dermatitis and, if it enters the eyes, the result is intense burning, weeping and conjunctivitis.

Benzyl chloride has been found to be carcinogenic in rats, in which it produced local sarcoma by subcutaneous injection. No epidemiological or case studies are available as regards carcinogenic effects in man.

SAFETY AND HEALTH MEASURES

The manufacturing process should be enclosed as completely as possible. Effective ventilation should be provided together with local exhaust equipment at the main sources of benzyl chloride. Personal protective equipment should include industrial filter respirators, eye and face protection as well as hand and arm protection. Workers who are exposed to this substance are advised to have a special diet (milk, curds). Periodic medical examinations, given at least once per year, should include investigations of the blood, and the function of liver and kidneys.

Treatment

Benzyl chloride on the skin should be washed off with alcohol first and then with soap and water. Should it enter the eyes, they should be profusely irrigated with a 2% solution of soda containing a few drops of atropin and cocaine.

MIHAJLOVA, T. V.

CIS 75-1670 *Properties and essential information for safe handling and use of benzyl chloride.* Chemical Safety Data Sheet SD-69 (Manufacturing Chemists Association, 1825 Connecticut Avenue, NW, Washington, DC) (Revised 1974), 15 p.

Benzyl chloride (217-223) 18 ref. *IARC monographs on the evaluation of carcinogenic risk of chemicals to man.* Vol. 11: *Cadmium, nickel, some epoxides, miscellaneous industrial chemicals and general considerations on volatile anaesthetics* (Lyons, International Agency for Research on Cancer, 1976), 306 p. Ref.

CIS 79-435 *Criteria for a recommended standard: Occupational exposure to benzyl chloride.* DHEW (NIOSH) publication No. 78-182 (National Institute for Occupational Safety and Health, 4676 Columbia Parkway, Cincinnati) (1978), 90 p. 100 ref.

Beryllium, alloys and compounds

Beryllium (Be)

a.w.	9.01
sp.gr.	1.85
m.p.	1 278 ± 5 °C
b.p.	2 970 °C

a grey to silver metal.

TWA NIOSH	0.0005 mg/m³/130 min ceil
TLV ACGIH	0.002 mg/m³, industrial substance suspect of carcinogenic potential
MAC USSR	0.001 mg/m³

Beryllium is notable for its lightness in weight, high tensile strength and corrosion resistance.

Sources. Beryl ($3BeO.Al_2O_3.6SiO_2$) is the chief commercial source of beryllium, the most abundant of the minerals containing high concentrations of beryllium oxide (10-13%). Major sources of beryl are Argentina, Brazil, India, Zimbabwe and South Africa. In the United States, beryl is found in Colorado, South Dakota, New Mexico and Utah. Bertrandite, a low grade ore (0.1-3%) with an acid-soluble beryllium content is now being mined and processed in Utah, United States.

Production. The two most important methods of extracting beryllium from the ore are the sulphate process and the fluoride process.

In the sulphate process, crushed beryl is melted in an arc furnace at 1 650 °C and poured through a high velocity water stream to form frit. After heat treatment, the frit is ground in a ball mill and mixed with concentrated sulphuric acid to form a slurry which is sprayed through a jet into a direct-heated, rotating, sulphating mill. The beryllium, now in a water-soluble form, is leached from the sludge, and ammonium hydroxide is added to the leach liquor which is then fed to a crystalliser where ammonium alum is crystallised out. Chelating agents are added to the liquor to hold iron and nickel in solution, sodium hydroxide is then added and the sodium beryllate thus formed is hydrolysed to precipitate beryllium hydroxide, which may be converted to beryllium fluoride for reduction by magnesium to metallic beryllium, or to beryllium chloride for electrolytic reduction.

In the fluoride process (figure 1) a briquetted mixture of ground ore, sodium silicofluoride and soda ash is sintered in a rotating hearth furnace. The sintered material is crushed, milled and leached. Sodium hydroxide is added to the solution of beryllium fluoride thus obtained and the precipitate of beryllium hydroxide is filtered in a rotary filter. Metallic beryllium is obtained as in the previous process by the magnesium reduction of beryllium fluoride or by electrolysis of beryllium chloride.

Uses. Beryllium is used in alloys with a number of metals including steel, nickel, magnesium, zinc and aluminium, the most widely used alloy being beryllium-copper which has a high tensile strength and a capacity for being hardened by heat treatment. Beryllium bronzes are used in non-spark tools, electrical switch parts, watch springs, diaphragms, shims, cams and bushings.

One of the largest uses of the metal is as a moderator of thermal neutrons in nuclear reactors and as a reflector to reduce the leakage of neutrons from the reactor core. A mixed uranium-beryllium source is often used as a neutron source. As a foil beryllium is used as window material in X-ray tubes. Its lightness, high elastic modulus and heat stability make it an attractive material for the aircraft and aerospace industry.

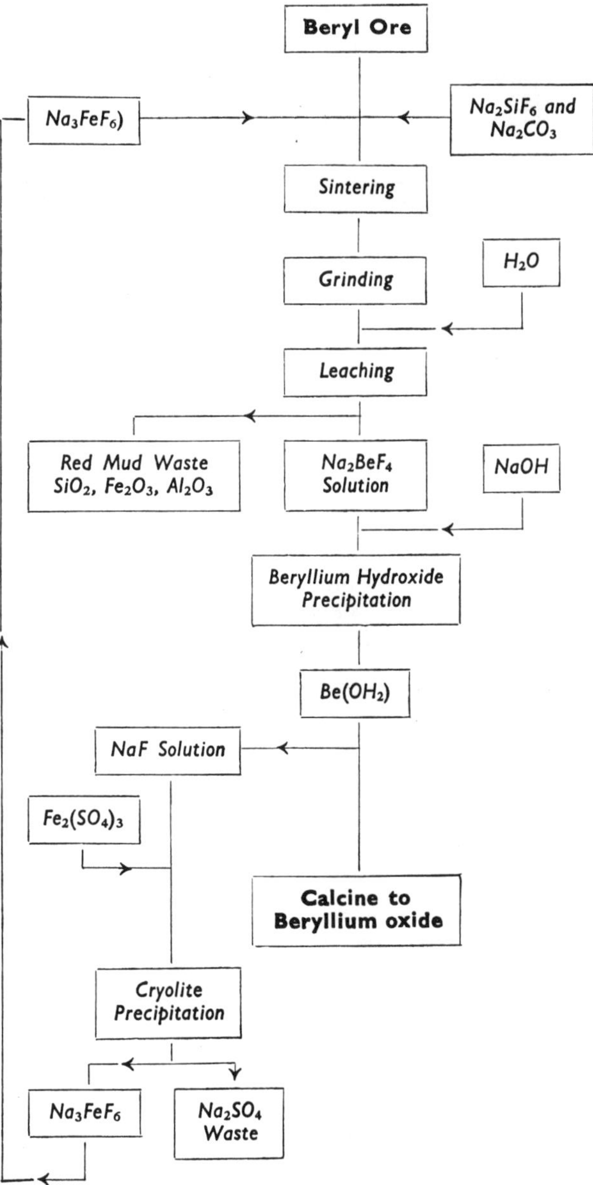

Figure 1. Production of beryllium oxide (fluoride process). *(From White, D.W. and Burke, J.E.* The metal beryllium *(American Society for Metals, Cleveland, Ohio, 1955).)*

Beryllium oxide (BeO)

BERYLLIA

m.w.	25
sp.gr.	3.02
m.p.	2 530 ± 30 °C

soluble in acids and alkalis; insoluble in water

a white amorphous powder.

TWA USA, MAC USSR } see Beryllium

Beryllium oxide is made by heating beryllium nitrate or hydroxide.

It is used in the manufacture of ceramics, refractories and other beryllium compounds. It was used for the manufacture of phosphors for fluorescent lamps until the incidence of beryllium disease in the industry caused its use for this purpose to be abandoned (1949 in the United States).

Beryllium fluoride (BeF_2)

m.w.	47.02
sp.gr.	1.99

m.p. sublimes 800 °C
readily soluble in water; sparingly soluble in ethyl alcohol
a hygroscopic solid.
TWA USA } see Beryllium
MAC USSR

It is made by the decomposition at 900-950 °C of ammonium beryllium fluoride. Its main use is in the production of beryllium metal by reduction with magnesium.

Beryllium chloride ($BeCl_2$)
m.w. 79.9
sp.gr. 1.90
m.p. 405 °C
b.p. 520 °C
very soluble in water; soluble in ethyl alcohol, benzene, ethyl ether and carbon disulphide
white or slightly yellow deliquescent crystals.
TWA USA } see Beryllium
MAC USSR

This compound is manufactured by passing chlorine over a mixture of beryllium oxide and carbon.

Beryllium nitrate ($Be(NO_3)_2.3H_2O$)
m.w. 187.08
sp.gr. 1.56
m.p. 60 °C
b.p. 142 °C
soluble in water and ethyl alcohol white to faintly yellow deliquescent crystals.
TWA USA } see Beryllium
MAC USSR

It is produced by the action of nitric acid on beryllium oxide. It is used as a chemical reagent and as a gas mantle hardener.

Beryllium nitride (Be_3N_2)
m.w. 55.06
m.p. 2 200 ± 100 °C
hard, refractory white crystals.
TWA USA } see Beryllium
MAC USSR

This compound is prepared by heating beryllium metal powder in an oxygen-free, nitrogen atmosphere at 700-1 400 °C. It is used in atomic energy reactions including the production of radioactive carbon isotope, ^{14}C.

Beryllium sulphate hydrate ($BeSO_4.4H_2O$)
m.w. 177.2
sp.gr. 1.71
m.p. 100 °C
soluble in water; insoluble in ethyl alcohol
colourless crystals.
TWA USA } see Beryllium
MAC USSR

It is produced by treating the fritted ore with concentrated sulphuric acid. It is used in the production of metallic beryllium by the sulphate process.

HAZARDS

Fire and health hazards are associated with processes involving beryllium.

Fire hazards. Finely divided beryllium powder will burn, the degree of combustibility being a function of particle size. Fires have occurred in dust filtration units and during the welding of ventilation ducting in which finely divided beryllium was present.

Health hazards. Beryllium and its compounds are highly toxic substances. The toxic reaction is body-wide rather than in the lung alone, in much the same way that lead may enter the body by inhalation and cause systemic disease.

Beryllium enters the body almost entirely by inhalation. Experimental evidence suggests that little beryllium is absorbed through the intestinal wall. Traumatic introduction of beryllium and its compounds subcutaneously can give rise to local damage, but beryllium does not enter the body through the unbroken skin.

Metabolism. Beryllium is poorly absorbed from the gastrointestinal tract, reflecting, in part, the precipitation of beryllium by ingested phosphate. In man, beryllium absorbed from the lung is widely distributed, although the rate of uptake, distribution, and excretion is invariably influenced by the chemical and physical characteristics of the inhaled material. Highly refractory oxides would be expected to persist in the lung for an extended period. Although beryllium in the urine indicates that it is excreted, the biological half-life is prolonged, the element being detectable longer than 20 years after the last exposure.

Immunology. Early speculation that the pathogenesis of chronic beryllium disease involves the immune mechanism was based upon the inconsistent incidence among similarly exposed groups of people, the delay in onset, the delayed cutaneous response to skin testing, the characteristic granulomatous lesion, and the response to steroid treatment. Current thinking continues to implicate immune mechanisms, although no "beryllium antigen" has been identified. Presumably the beryllium ion, too small itself to serve as a complete antigen, becomes associated with a protein or other macromolecule to acquire antigenic potency.

The repeated observation that circulating gammaglobulin levels are increased in chronic beryllium disease has led to searches for specific humoral antibodies; however, none have been convincingly demonstrated. On the other hand, evidence for delayed hypersensitivity mechanisms continues to accumulate. Lymphocytes from persons with beryllium hypersensitivity are highly reactive to beryllium compounds and undergo "blastogenic transformation" in their presence. Lymphocytes from non-sensitive donors do not transform in this way.

Lymphocytes from beryllium-sensitised donors are stimulated to transform by contact with macrophages containing phagocytised beryllium oxide. The beryllium content of these cells is much lower than that quantity which would be required to stimulate transformation directly, i.e. without the initial *in vitro* processing. Cell-free supernatants from positive lymphocyte transformation preparations have been shown to contain migration inhibition factor (MIF), which supports the interpretation that immune mechanisms are fundamental in chronic beryllium disease. Clinical studies of MIF in chronic beryllium disease have shown a reasonably consistent positive response in small studies.

Skin injuries. Acid salts of beryllium cause allergic contact dermatitis. Such lesions may be erythematous, papular or papulovesicular, are commonly pruritic, and are found on exposed parts of the body. There is usually a delay of 2 weeks from first exposure to occurrence of the dermatitis except in heavy exposures when an irritant reaction may be immediate. This delay is regarded as the time required to develop the hypersensitive state. The reaction in sensitised persons is prompt.

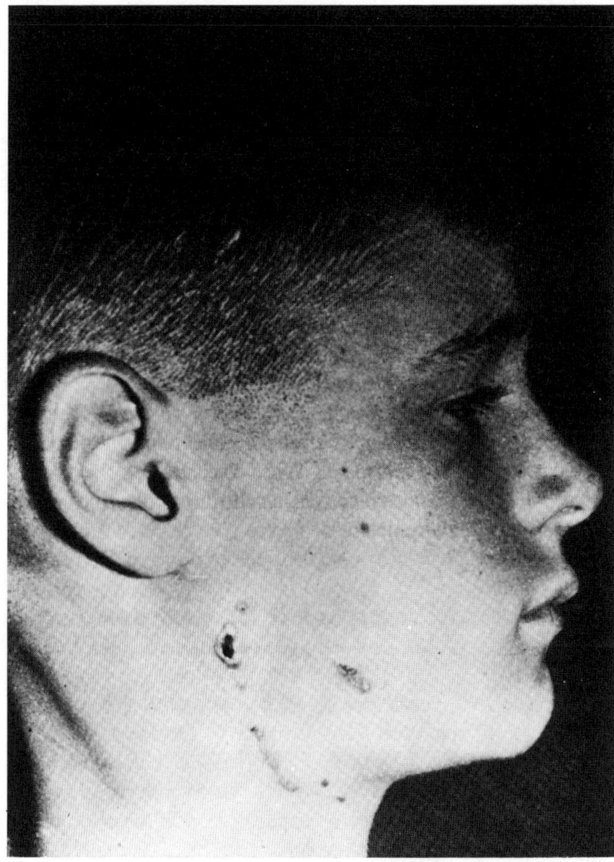

Figure 2. Beryllium granulomata. *(By courtesy of Grier, Nash and Freiman.)*

Accidental implantation of beryllium metal or crystals of a soluble beryllium compound in an abrasion, a crack in the skin or under the nail may cause an indurated area with central suppuration (figure 2). Healing of an ulcer is often followed by repeated breakdown of the scar until the beryllium compound has been completely removed mechanically. Rarely have such lesions been reported in recent years.

Conjunctivitis and dermatitis may occur alone or together. In cases of conjunctivitis, periorbital oedema may be severe.

Acute effects

Beryllium nasopharyngitis is characterised by swollen and hyperaemic mucous membranes, bleeding points, fissures and ulceration. Perforation of the nasal septum has been described. Removal from exposure results in reversal of this inflammatory process within 3-6 weeks.

Involvement of the trachea and bronchial tree following exposure to higher levels of beryllium causes non-productive cough, substernal pain and moderate shortness of breath. Rhonci and/or rales may be audible, and the X-ray of the chest may show increased bronchovascular markings. The character and speed of onset, and severity of these signs and symptoms depend on the quality and quantity of exposure. Recovery is to be expected within 1-4 weeks if the worker is removed from further exposure.

Following even brief, intense exposures, a severe chemical pneumonia may ensue with pulmonary oedema. There have been 18 such cases ending in death. These fatalities predate clear recognition of the hazard, and steroid therapy was not available. Such cases follow a clinical pattern similar to that of any penumonia due to the inhalation of a heavy dose of a pulmonary irritant.

In addition, there have been a variety of clinical pictures suitably brought together by the following criteria: significant beryllium exposure, non-productive cough, dyspnoea, bilateral widespread chest X-ray changes, weight loss in most cases, and reversal of all abnormal signs and symptoms within a few weeks to 6 months (in a few instances 1 year). Clinical reports describe considerable variation in onset, chest X-ray picture, and remission of symptoms. The non-specificity of the syndrome requires care in assessing the quality and quantity of exposure in order to make a correct diagnosis. Such cases should be increasingly rare, but the variety of new uses for beryllium compounds makes it possible that acute beryllium pneumonitis will occur through ignorance or urgency of research, and difficulty in engineering control of unusual operations.

Chronic beryllium disease

This is an intoxication arising from the inhalation of any beryllium compound (beryl excepted). The precise quality and quantity of the disease-producing dose is at present unknown, although there is some knowledge of both certainly harmful and probably safe dose levels gained from a registry of case records established in 1952 in the United States.

Signs and symptoms. Chronic beryllium poisoning in most cases affects the respiratory tract, although the onset may be marked by weakness, easy fatigue, and weight loss without cough or dyspnoea. The usual story is that, following an illness, elective surgery or uneventful pregnancy, the patient becomes aware of non-productive coughing and shortness of breath on effort. Routine chest X-ray surveys have discovered a number of cases of asymptomatic radiologic beryllium disease. A remarkable number of case histories reveal a delay from the last beryllium exposure to detectable evidence of disease of periods of 5-10 years and, in some cases, 30 years.

Chronic beryllium disease proves to be of long duration in most cases with exacerbation and remission in the majority while a few have a static course over a number of years. The clinical variants are many. Chronic beryllium intoxication may be asymptomatic, productive only of chest X-ray changes. Mildly disabling disease results in some non-productive cough, dyspnoea on unusual effort in addition to chest X-ray changes. Joint pains and weakness are common complaints. As complications, patients in this group may have renal calculi, and periods of hypercalciuria. The clinical course in this group varies but usually is stable for years with eventual evidence of pulmonary and/or myocardial failure.

Moderately severe beryllium poisoning causes distressing cough, shortness of breath, with marked chest X-ray changes. Serum proteins are at some time found to be elevated with increased γ-globulin fraction. Liver and/or spleen may be enlarged with laboratory evidence of disturbed liver problem. Depending on when the patient is seen in the course of the disease, there may be weight loss, osteoarthropathy, increase in hematocrit, disturbed liver function, elevated serum uric acid, hypercalciuria with and without stones, spontaneous skin lesions mimicking those of Boeck's disease. Lung function tests show such cases to have little or no ventilatory defect but measurable difficulty in diffusing gases. In cases of long duration, obstructive lung disease is established in some patients who have never smoked cigarettes. With steroid therapy, these patients survive many years and a few have had long remissions.

Severely disabled patients with beryllium disease may show, in addition, a striking cachexia and, after some time, signs of right heart impairment with a severe non-productive cough. Spontaneous pneumothorax may be a serious complication in these cases. There are a few cases of chronic beryllium poisoning with bouts of chills and fever in the course of moderately serious disease, a complication which carries a bad prognosis.

Neighbourhood cases of chronic beryllium disease need special mention. There are 60 individuals known who were exposed by proximity (less than 1 km) and/or contaminated clothes of workers, to toxic beryllium compounds and became ill.

Roentgenologic features. The X-ray pattern in chronic beryllium disease is non-specific and is similar to that which may be observed in sarcoidosis, Hamman-Rich syndrome, tuberculosis, mycoses, and dust disease (figure 3). Early in the course of the disease films may show granular, nodular or linear densities or a combination of them. With evolution of the disease the densities may increase, decrease, or remain unchanged, with or without fibrosis or emphysema. Gross structural change is almost always bilateral and symmetrical, and relative emphysema may appear in adjacent spaces. Hilar adenopathy, seen in approximately one-third of patients, is usually bilateral and accompanied by mottling of the lung fields. The absence of lung changes in the presence of adenopathy is a relative but not an absolute differential consideration in favour of sarcoidosis, as opposed to chronic beryllium disease.

The X-ray picture does not correlate well with clinical status and does not reflect particular qualitative or quantitative aspects of the causal exposure. Although animal experiments show beryllium to be a carcinogen, there is still uncertainty as to whether or not beryllium is a human carcinogen. There is active inquiry as to the likelihood that lung cancer is increased in either beryllium-exposed populations or in patients with beryllium disease. Death is caused in most cases by

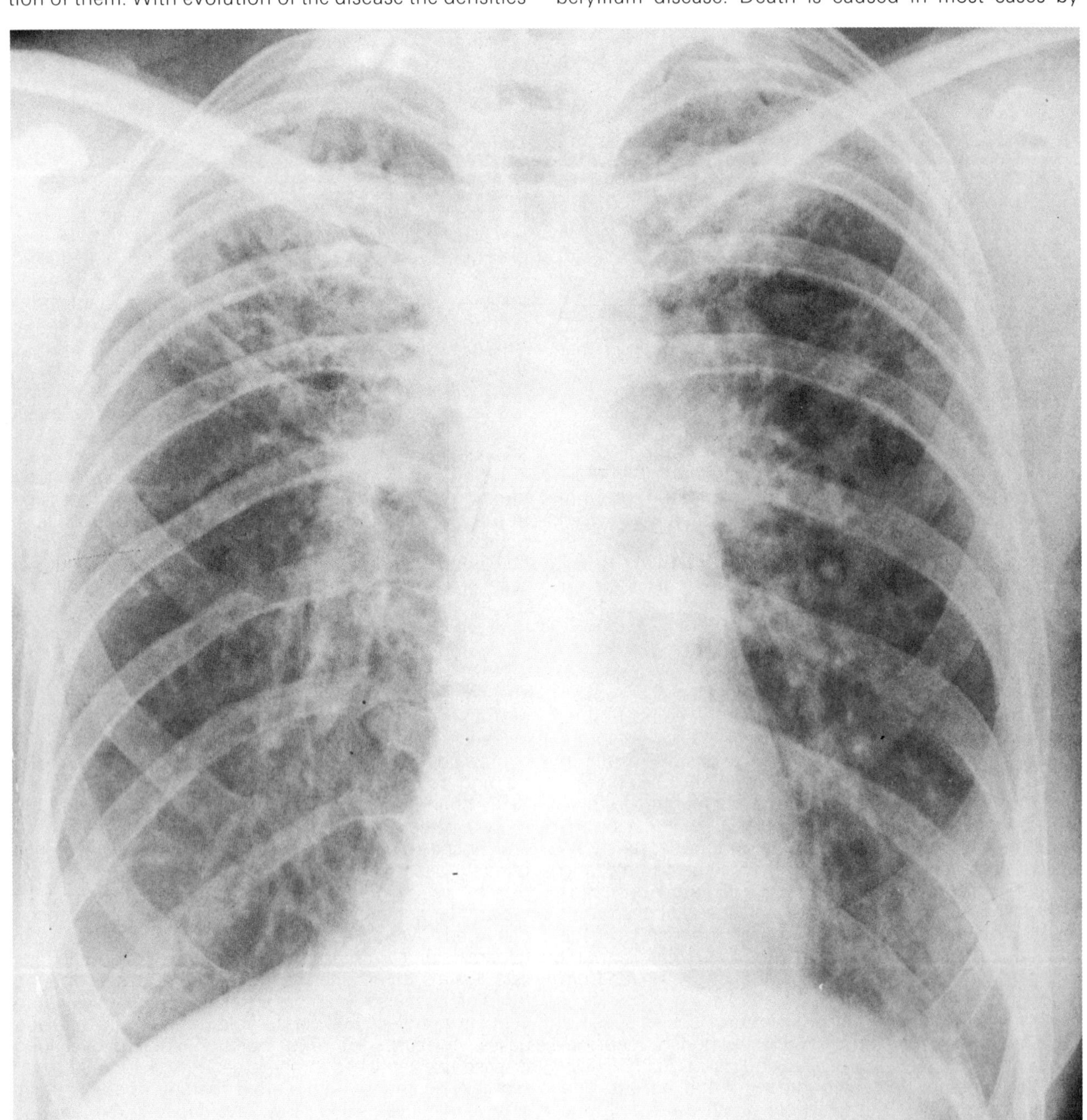

Figure 3. Berylliosis.

pulmonary insufficiency or right heart failure, sometimes by inanition with little pulmonary insufficiency, by uncontrollable pneumothorax, and, in rare instances, by renal failure or myocarditis.

Pathology. The pathologic reaction in chronic beryllium disease is characterised by formation of pulmonary non-caseating granulomata. In addition, there is marked round-cell interstitial infiltration, which best explains functional disability. Fluctuation in the extent of such inflammation may parallel clinical activity. Necrosis may be seen. As in sarcoidosis, asteroid and Schaumann's bodies, hyalinisation of granulomata are present in varying degree. In many cases, blebs and bullae form and may cause pneumothorax. Rapidly progressive disease may result in confluent bullae replacing large areas of lung. Pleural thickening is usual, though not great enough to be visible on X-ray.

Diagnosis. Diagnosis of chronic beryllium disease is based upon demonstration of consistent clinical findings and a history of exposure. Exposure is established on the basis of epidemiological evidence or tissue assays of beryllium. In lung tissues of individuals who have had no occupational or "neighbourhood" exposure to beryllium it would be unusual to find concentrations of the element as high as 0.05 μg/100 g.

Sarcoidosis is the disorder most closely resembling chronic beryllium disease, and the differentiation may be difficult. Thus far no cystic bone disease or involvement of the eye, parotid, or tonsil has appeared in chronic beryllium disease. Similarly the Kveim test is negative. Assessment of the histopathologic features of chronic beryllium disease shows interstitial pneumonitis to be relatively more prominent than in sarcoidosis, although granulomata in both may be similar.

Skin testing to demonstrate beryllium sensitisation is not recommended, in that the test itself is sensitising, may possibly trigger systemic reactions in sensitised people, and does not of itself establish that the presenting disease is necessarily beryllium-related.

More sophisticated immunological approaches in differential diagnosis are not generally applied at the present time.

Prognosis. The prognosis of chronic beryllium disease has altered favourably during the years, suggesting that longer delay in onset may reflect lower exposure or lower beryllium body burden, resulting in milder clinical course. Clinical evidence is that steroid therapy, if used when measurable disability first appears, in adequate doses for long enough, has improved the clinical status of many patients allowing some of them to return to useful jobs. Neither lung function study nor X-ray picture consistently reflect this improvement, which may be dramatic. There is no clear evidence that steroids have cured chronic beryllium poisoning and there is as yet no documented case of spontaneous cure, although there are a number of cases of improvement without therapy. Current (1980) opinion is that a number of patients have long remissions or reach a stage of "burned-out" disease with considerable residual damage.

SAFETY AND HEALTH MEASURES

The precautions must cover the fire hazard as well as the much more serious toxic danger.

Fire prevention. Arrangements must be made to prevent possible sources of ignition, such as the sparking or arcing of electrical apparatus, friction, etc., in the vicinity of finely divided beryllium powder. Equipment in which this powder has been present should be emptied and cleaned before acetylene or electrical welding apparatus is used on it. Oxide-free, ultrafine beryllium powder that has been prepared in inert gas is liable to ignite spontaneously on exposure to air.

Suitable dry powder—not water—should be used to extinguish a beryllium fire. Full personal protective equipment, including respiratory protective equipment, should be worn and firemen should bathe afterwards and arrange for clothing to be laundered separately.

Health protection. Beryllium processes must be conducted in a carefully controlled manner to protect both the worker and the general population. The main risk arises from airborne contaminant and the process and plant should be designed to give rise to as little dust or fume as possible. Wet processes should be used instead of dry processes, and the ingredients of beryllium-containing preparations should be unified as aqueous suspensions instead of as dry powders; whenever possible the plant should be designed as enclosed units. The permissible concentration of beryllium in the atmosphere is so low that enclosure must be applied even to wet processes, otherwise escaping splashes and spills can dry out and the dust can enter the atmosphere.

Operations from which dust may be evolved should be conducted in the maximum degree of enclosure consistent with the needs of manipulation. Some operations are performed in glove boxes, but many more are conducted in enclosures provided with exhaust ventilation similar to chemical fume cupboards. Machining operations may be ventilated by high-velocity low-volume local exhaust systems or by hooded enclosures with exhaust ventilation.

To check the effectiveness of these precautionary measures, atmosphere monitoring should be done in such a manner that the daily average exposure of workers to respirable beryllium can be calculated.

Maintenance and good housekeeping is particularly important in a beryllium processing area. The area should be cleaned regularly by means of a proper vacuum cleaner or a wet mop. Beryllium processes should be segregated from the other operations in the factory.

Personal protective equipment should be provided for workers engaged in beryllium processes. Where they are fully employed in processes involving the manipulation of beryllium compounds or in those associated with the extraction of the metal from the ore, provision should be made for a complete change of clothing so that the workers do not go home wearing clothing in which they have been working. Arrangements should be made for the safe laundering of such working clothes as well as protective overalls to ensure that laundry workers are not exposed to risk. These arrangements should not be left to the normal home laundering procedures; cases of beryllium poisoning in the families of workers have been attributed to contaminated clothing taken home or worn in the home by the workers.

An occupational health standard of 2 μg/m^3, proposed in 1948 by a committee operating under the auspices of the United States Atomic Energy Commission, continues to be widely observed. Existing interpretations generally permit fluctuations to a "ceiling" of 5 μg/m^3 as long as the time-weighted average is not exceeded. Additionally, an "acceptable maximum peak above the ceiling concentration for an 8-hour shift" of 25 μg/m^3 for up to 30 min is also permissible. These operational levels are achievable in current industrial practice, and there is no evidence of adverse health experience among persons working in an environment thus controlled. Although there had previously been discussion that these levels might be raised, the possibility that beryllium may be a human carcinogen has prompted considerations of whether or not a reduction in current levels should be recommended.

Pre-employment and periodical medical examinations of workers exposed to beryllium and its compounds are made compulsory in a number of countries. Because of difficulty in diagnosis of chronic beryllium disease, it is unwise to allow a worker with abnormal chest X-ray changes to work with beryllium or its compounds.

Treatment. It is necessary to have prompt and competent medical care available where accidental over-exposure may occur. Oxygen, judicious use of steroids, and absolute bed rest are necessary for beryllium pneumonitis and the more serious complication of acute poisoning. Any worker who has had acute beryllium poisoning should be excluded from further beryllium exposure.

In chronic beryllium disease, with objective signs (not X-ray changes alone) of pathologic change, doses of steroids in the range of 15 mg, to as high as 80 mg of prednisone a day for short periods, have proved effective if continued, in most cases, for years. Control of side effects has been improved by using the method of Harter, which is to give the steroid in one dose every other day.

HARDY, H. L.
TEPPER, L. B.
CHAMBERLIN, R. I.

"Beryllium". Tepper, L. B. *Metals in the environment.* Waldron, H. A. (ed.). (London, Academic Press, 1980).

"Beryllium disease. A clinical perspective". Hardy, H. L. *Environmental Research* (New York), Feb. 1980, 21/1 (1-9). 34 ref.

"Modern views on beryllium toxicity" (Zur Gesundheitsgefährdung durch Beryllium aus heutiger Sicht). Preuss, O.; Oster, H. *Arbeitsmedizin, Sozialmedizin, Präventivmedizin* (Stuttgart), Nov. 1980, 15/11 (270-275). 29 ref. (In German)

CIS 79-40 "Reversible respiratory disease in beryllium workers". Sprince, N. L.; Kanarek, D. J.; Weber, A. L.; Richard, I.; Chamberlin, R. I.; Kazemi, H. *American Review of Respiratory Disease* (New York), June 1978, 118/6 (1 011-1 017). Illus. 12 ref.

CIS 78-1077 "Specific humoral and cellular reactions in berylliosis" (Specifičeskie gumoral'nye i kletočnye reakcii pri berillioze). Vasil'eva, E. V.; Ermakova, N. G.; Orlova, A. A. *Gigiena truda i professional'nye zabolevanija* (Moscow), July 1977, 7 (8-12). 19 ref. (In Russian)

CIS 76-165 "Hazardous chemical reactions–33. Beryllium" (Réactions chimiques dangereuses–33. Béryllium). Leleu, J. *Cahiers de notes documentaires–Sécurité et hygiène du travail* (Paris), 3rd quarter 1975, 80, Note No. 976-80-75, 395 p. (In French)

Beverage or soft drink industry

Soft drinks (non-alcoholic beverages) include a variety of beverages which may be divided into two basic types: carbonated and non-carbonated, or still, drinks. Carbonated beverages include sparkling sodas, colas and flavoured effervescent drinks which had their origins in early attempts to duplicate the aerated water of famous European springs.

The first commercial production of carbonated drinks began with the opening of bottling plants in Europe and the United States between 1789 and 1821.

[Because of the seasonal variation in demand, there are fluctuations in the production of soft drink industries. The following example indicates the percentage of output during the year in the United Kingdom's soft drink industry over the period 1968-76: Jan.-Mar. 19.6; Apr.-June 26.9; July-Sep. 29.5; Oct.-Dec. 23.9.]

Processing. The water supply for soft drinks manufacture is critical in that it affects the taste of the finished product. Even with a good water supply, the water is usually treated to ensure the desired mineral content and chemical balance. The treated water is then mixed with the other ingredients and, in sweetened beverages, with sugar or artifical sweeteners. The mixture is then piped to the filling apparatus where it flows into bottles, cans or tanks. Carbon dioxide is usually added prior to the filling operation in beverages to be carbonated. Carbonated beverages are normally filled at temperatures slightly above freezing. The filled containers are then placed in cardboard, wood or plastic cases and usually stacked on pallets ready for distribution.

HAZARDS AND THEIR PREVENTION

[The seasonal fluctuations in production inevitably have effects on the occupational health and safety problems associated with this industry. Seasonal work is characterised both by exceptional effort, which can cause additional physical and nervous fatigue, and by the use of temporary workers, who often have little experience. Moreover, safety measures and machine maintenance are often neglected in times of peak activity.]

The machinery guarding of high speed bottling and canning equipment is of prime importance. Normally, guards are provided by the manufacturers of this equipment. However, often the user must provide additional guarding to protect his employees from injury. All transmissions such as chain and sprocket assemblies, gear drives and belt and pulley drives must be completely enclosed. Many moving elements on fillers, crowners and sealers are difficult to guard.

Overhead product conveyors can create hazards from falling cases and bottles. Nets or wire mesh screening should be installed under these conveyors to prevent injury to persons working below them. Nips between conveyor belts and drums and pulleys should be guarded against trapping risk: bridges with walkways will obviate the risk of persons stepping over conveyors. Frequent stop-buttons should be provided for the conveyors.

Ammonia is frequently used as a refrigerant in bottling plants. Emergency procedures must be established in case of a leak in the system. Responsibility should be assigned for shutting off the supply of ammonia to prevent its spread. Ammonia respirators strategically located are required for immediate availability to persons working in ammonia hazard areas. Ammonia is flammable and highly soluble in water. Therefore, water is a choice extinguishing agent for controlling ammonia fires. Similar precautions are necessary with other refrigerants, such as the flammable halogenated hydrocarbons.

Broken glass is a constant source of injury in any bottling operation. Bottling machines should, as far as practicable, incorporate fixed shields strong enough to contain fragments of bursting bottles or syphons. All employees working on bottling lines, and especially with broken glass, must be provided with eye, face and hand protection. Gauntlets to protect the arm are also desirable. Spectacle-type safety glasses with side shields are generally sufficient protection for employees working around bottling lines. Enclosure-type goggles should be used by employees disposing of and handling broken glass. In areas where bottles burst with regularity, full face shields are recommended. Special containers for the disposal of broken glass must be provided to prevent the "fly back" of glass fragments when bottles are thrown into them. The containers should be equipped with either a hinged cover or a chute of sufficient length to eliminate "fly back".

Mechanisation of product handling is recommended where economically feasible; i.e. case packers, palletisers, conveyors, etc. Employees must be thoroughly trained in correct lifting and setting down techniques in order to prevent strains, sprains and back injuries wherever it is not possible to mechanise the handling of products. Where hand stacking is necessary, the work should be arranged in such a way as to minimise the need for twisting or turning. Fork-lift trucks used for palletised handling should be equipped with load safety racks and overhead guards for driver protection.

Delivery men must be trained in handling products to and from trucks to avoid slips and falls. The trucks should be equipped with convenient handholds and footholds to facilitate climbing up and down in order to gain access to the product. Footholds and running boards should be made of open grating to prevent the accumulation of mud and snow.

Wet floors in a bottling or canning operation are a source of slips and falls. Floors should be well maintained with special arrangements for draining. Employees should be encouraged to wear safety shoes with tread-type soles. Tennis shoes and sandals should not be permitted. Where employees are required to work in excessive amounts of water, rubber safety boots should be provided. Employees at fillers, soakers, case packers and other equipment located in wet areas should be provided with raised platforms with open gratings. In general, waterproof aprons should be provided to protect against damp and splashes and also to give some protection to the body against flying glass.

To gain access to mixing tanks and other locations above floor level, the use of portable ladders should be avoided. Instead, these areas should be equipped with fixed platforms, catwalks and stairways made of open grating. Platforms and catwalks should be equipped with standard hand rails and toe boards.

Caustic substances in dry or liquid form are widely used for cleaning bottles, mixing tanks, etc. Employees handling caustics in either form must be required to wear eye and/or face protection, chemical gloves and aprons. Emergency showers or face-wash fountains should be installed in areas where caustics are handled. When repairing or cleaning the inside of soakers, employees are frequently subject to caustic burns.

Due to the prevailing damp conditions all electrical apparatus should be especially protected and efficiently earthed. Portable electric apparatus, in particular, needs care.

[In bottling plants noise in excess of 90 dBA is a common problem and persons who have been working there for several years often show hearing damage. The main source of noise are the conveyors transporting the bottles between the different processing points and the machinery (bottle washing, filling and capping, packing, palletting). Noise control measures should mainly aim at reducing the number of impacts and the speed of travel of the bottles. Acoustic barriers and enclosures can be also of use.]

Good washing and sanitary facilities are required for all employees. First-aid provisions should include waterproof plasters and dressings.

ESPINOSA, D. P.

General:

General report. Second Tripartite Technical Meeting for the Food Products and Drink Industries (Geneva, International Labour Office, 1978), 121 p.

Labour and social problems arising out of seasonal fluctuations of the food products and drink industries. Report II, Second Tripartite Technical Meeting for the Food Products and Drink Industries (Geneva, International Labour Office, 1978), 51 p.

Safety standards:

CIS 77-1743 *Occupational safety in the distilling and soft drink industries* (Arbeitsschutz in der Gärungs- und Getränkeindustrie). (Berlin, Verlag Tribüne, 1976), 48 p. Illus. (In German)

Guidelines for the control of industrial wastes. 11: Soft drinks. Isaac, P. C. G.; Partner, J. D.; Watson, D. M. WHO/WD/77.18 (Geneva, World Health Organisation, 1977), 12 p. Illus. 11 ref.

Noise:

CIS 78-89 *Noise emission of bottling plant and noise control measures* (Geräuschemission von Getränkeabfüllanlagen und Massnahmen zur Lärmminderung). Probst, W. Forschungsbericht nr. 172 (Bundesanstalt für Arbeitsschutz und Unfallforschung, Martener Strasse 435, 4600 Dortmund-Marten) (1977), 255 p. Illus. 31 ref. (In German)

Beverages at work

On the average, water accounts for 55-60% of body weight; three-quarters of this water is to be found in the body cells and the remaining quarter in the extra-cellular liquid (plasmatic and interstitial), the volume of which may vary. Water is the most mobile of the important body constituents and it takes part in the functions of absorption, diffusion, secretion, excretion and thermoregulation. Its metabolism is closely linked with that of the electrolytes in the body (primarily Na^+ and Cl^- in the extra-cellular liquid and K^+ in the cells); together with these elctrolytes, it constitutes an isotonic liquid of relatively constant composition.

The body may absorb water in a variety of forms: it is estimated that, for a 70 kg man at rest in a temperate climate, average daily water intake is 1 500 cm^3 in the form of beverages, 700 cm^3 in the form of foodstuffs and 300 cm^3 by means of metabolisation (oxidation water). Water intake and the sodium concentration in foodstuffs (which is around 125 mEq/l in the case of light work in a temperate climate) are regulated, in the normal subject, by thirst which may become manifest as soon as the water deficit reaches 0.5% of body weight.

Water is eliminated by various routes. Over a 24-h period, the normal subject in the above conditions will eliminate 1 500 cm^3 of water in the urine, 500 cm^3 through the skin, 400 cm^3 through the lungs and 100 cm^3 in the stools. If the skin temperature of a person at rest does not exceed 32 °C, only small quantities of water are eliminated through the skin; elimination is imperceptible and is called "insensible perspiration". During physical work, or when the environmental temperature rises, sweat becomes visible and may be eliminated from the body at a rate of up to 1 l/h and several litres per day. Sweating also entails the elimination of electrolytes, metabolic products such as lactic acid, urea, ammonia, etc. (which increase in quantity during physical exertion), and vitamins, especially those of group B. Urinary elimination of phosphates also increases significantly as a result of physical effort.

Hot work and arduous physical work

In general it is not until there has been non-compensated elimination of liquids equivalent to about 5% of body weight that the first disorders related to body dehydration become manifest. Dehydration is a hazard in desert climates where a worker may eliminate up to 10-12 l of sweat in an 8-h shift; in such cases, if an appropriate quantity of liquid is not absorbed, the body may come dangerously close to what is widely considered the fatal dehydration threshold, i.e. the point at which body weight has been reduced by 10-15% due to loss of liquid. In a temperate climate dehydration in workers does not give rise to clinical manifestations but it does increase fatigue and the incidence of accidents and also reduces output; moreover, in the case of non-acclimatised workers who drink large quantities of water without compensating for the electrolyte loss, dehydration may produce heat cramp, a condition that may occur when blood sodium and blood chloride levels are low. If, during heat exposure, large quantities of pure water are ingested, the body sweats more intensely than if only moderate quantities of water or salted drinks are absorbed. On the other hand, beverages containing excessive quantities of sodium chloride may reduce output and, in the long run, lead to kidney disorders.

Even before the First World War, it had been observed that, in soldiers who were required to march very long distances, the urinary elimination of chlorides decreased significantly; it was consequently recommended that, in such cases, soldiers should be given isotonic beverages. Starting in the 1930s, in the iron and steel industry and in building and civil engineering work in tropical countries, wide use was made of sodium chloride tablets and powdered salt dissolved in the drinking water. However, the gastrointestinal disorders resulting from unduly high salt concentrations led to a reduction in sodium chloride levels between 0.1% and 0.04%.

The composition of beverages should be related to the amount of salt the worker normally consumes in his food. The saltiness of food varies considerably from country to country; in certain countries the normal dietary salt is sufficient to protect workers against excessive electrolyte loss, whilst in others workers require salted drinks if they are to maintain their output. In temperate climates it is not normally necessary for these salted beverages to contain more than 500 mg/l of sodium and 250 mg/l of potassium; these concentrations can be tolerated by the digestive system and are acceptable to the taste, especially if the beverage is flavoured. Use of sodium chloride tablets is indicated only if urinary elimination of chlorides falls to 2 g or less in 24 h. It is useful to add to these beverages vitamins especially thiamine (1-2 mg/l), vitamin C and phosphates. The inclusion of a small amount of carbon dioxide makes the beverage more pleasant to drink; however, this practice should not be abused or gastrointestinal disorders may be the result.

Body tolerance to water deficiency is quite high and workers tend to drink large quantities at long intervals. Iron and steel workers drink during tapping and miners drink only after 3-4 h at work, by which time they may have excreted up to a litre of sweat. Beverages should also be freely accessible and palatable and workers should be encouraged to drink small quantities at relatively short intervals.

To aid digestion, beverages should not be too cold and beverages for persons doing heavy physical work in the open air (forestry and agricultural workers, etc.) or exposed to the cold, should preferably be warm.

Beverages for workers who are not exposed to high temperature may reflect more, in their composition, the individual habits. However, it is advisable to avoid excessive consumption of carbonated water, coffee, cola beverages and milk. Tea is preferable and it is usually well accepted especially when available from an automatic machine.

Beverages in the prevention of poisoning

The use of milk in the prevention of lead poisoning has long been a topic of debate. This practice dates from the time when the treatment of lead poisoning was based, to a large degree, on the finding that lead has the same metabolic pathway as calcium; the administration of milk was therefore indicated in the treatment of lead colic and of cases of excessive lead mobilisation. However, in the use of milk for lead poisoning prophylaxis, the following problems are encountered: *(a)* the prophylactic efficiency of milk has never been demonstrated unequivocally—and some studies even indicate that milk may facilitate lead absorption; *(b)* a quantity of over half-a-litre of milk consumed during work each day is difficult to digest; *(c)* it is difficult to ensure that all lead-exposed workers consume sufficient milk; *(d)* if reliance is placed on this measure, other safety measures may be neglected. Milk is not an antidote for toxic agents in industry and its prophylactic use is not recommended; however, there is no reason for not recommending its use as a refreshing beverage, provided it is well tolerated and consumed in moderate quantities.

It is a general principle that, in the prevention of occupational poisoning, no beverage or medicament (e.g. vitamin C for benzene exposure, sulphurated water for mercury exposure, EDTA (edetic acid) for lead exposure, etc.) should ever be considered an adequate substitute for effective engineering control of the hazard.

Beverages are also used to control certain acute, reversible effects of occupational origin such as headache and arterial hypotension resulting from exposure to nitrates in the explosives and pharmaceutical industries, which are readily eliminated or corrected by a cup of tea or coffee.

Alcoholic beverages

The toxic action of alcoholic beverages is dealt with in the article ALCOHOLISM. It should be pointed out, however, that certain types of work, or exposure to certain agents, may potentiate the effects of alcohol. In persons doing heavy physical work alcohol has a deleterious effect due to its action on the cardiovascular system; contrary to common belief, this effect is even more dangerous in persons doing hot work or cold work. Various chemicals (calcium cyanamide, thiocarbamates, furfural and related compounds) may provoke violent reactions of the type encountered in disulphiram exposure. Persons doing intellectual work or persons whose work necessitates a high degree of psychomotor co-ordination (precision work or work demanding considerable balance control) would also be very adversely affected in their performance by alcoholic beverages. The consumption of all alcoholic beverages during working hours should, therefore, be forbidden and the amount of alcoholic drinks taken during meal breaks should not exceed 250 cm³ of wine, 500 cm³ of beer or the alcohol equivalent in other beverages. However, in certain countries measures of this type will be successful only if a varied range of non-alcoholic drinks is made available to satisfy all tastes, if building sites are equipped with drinking water from the very start and if workers are well informed of the hazards of alcoholic beverages. The distribution of beverages at the workplace has been considerably simplified by the use of special mobile distribution units run from central catering facilities, and by the introduction of automatic distribution or vending machines.

PARMEGGIANI, L.

Drinking habits:

CIS 1191-1973 "Drinking habits in heavy industry" (Trinkgewohnheiten in der Schwerindustrie). Thielen, R. G. *Arbeitsmedizin-Sozialmedizin-Arbeitshygiene* (Stuttgart), Nov. 1972, 7/11 (317-321). Illus. (In German)

Drinking fountains:

CIS 75-850 *Standard for drinking-fountains and self-contained mechanically refrigerated drinking-water coolers.* ARI 1010-73 (Air-conditioning and Refrigeration Institute, 1815 North Fort Myer Drive, Arlington, Virginia 22209), 1973, 7 p.

Beverage distribution:

"Automatic machines for distribution of beverages at workplaces" (Les distributeurs automatiques de boissons sur les lieux de travail). Chesnay, C. *Revue de la sécurité* (Paris), Mar. 1980, 16/166 (29-32; 53-54). Illus. (In French)

Biological monitoring

The determination of the concentration of toxic substances in the air of working environments–environmental monitoring–is essential for the surveillance of hygiene conditions and for a preliminary identification of risk, but it does not permit an accurate evaluation of the occupational exposure of individual workers. This method has been influenced by the pragmatism of the industrial hygiene schools which, in the 1940s and 1950s, were composed almost exclusively of scientists from non-biological disciplines. The method was called into question when more detailed studies began to be made on the metabolic fate of exogenous substances in the light of the results of experimental studies on animals, and later on human volunteers. These studies have made it possible to define, with increasing accuracy, the relationships between exposure, absorption, biotransformation, retention and excretion of exogenous substances.

From the study of the interaction between exogenous noxae and the living organism, it was seen that the response of the organism depends ultimately on the concentration reached by the exogenous substances in the sites where a damaging action may develop. Such concentration is related not only to the physical and chemical properties of the individual substances and the environmental conditions in which they are used, but also to the mode of impact of these substances with the organism and the host's own biological factors.

The results of these studies have been to attribute a defined role to health surveillance of workers exposed to industrial poisons, a role which went beyond simple confirmation of the development of the impairment. Instead, it was possible to use methods that could reveal both an excessive absorption as early as possible (before alterations of any biological significance occurred) and initial biological effects (while they were still reversible and such as to cause no impairment of health).

Thus, environmental monitoring was joined by *biological monitoring*, which may be defined as the indirect identification and quantification of exposure to exogenous agents.

Biological indicators

Biological monitoring is performed by the determination, on biological samples of exposed organism, of the exogenous agents, their metabolites and the metabolic effects they produce. These determinations are used as *biological indicators*.

The biological samples where the indicators may be determined consist of:

(a) blood, urine, saliva, sweat, faeces;

(b) hair, nails;

(c) expired air.

With biological monitoring, information can be obtained that would not otherwise be available, i.e.:

(a) on the evaluation of absorption and/or exposure over a prolonged period of time (not only of the amount of substances present in the working environment at a given time, when the analysis of environmental pollutants is carried out);

(b) on the amount of a substance absorbed as a result of movements within the working environment or of accidental causes, which often cannot be checked (not only on the amount normally present in the assigned workplace);

(c) on the amount absorbed by the organism via various routes (not only via the respiratory route, as is presumed in environmental monitoring);

(d) on the evaluation of the over-all exposure, as the sum of different sources of contamination, which may also exist outside the working environment;

(e) on the amount absorbed by the subject, taken as an individual, as related not only to his workplace, but also to climatic factors, the subject's particular way of withstanding physical effort, age, sex, individual genetic characteristics, the functional condition of the organs responsible for biotransformation and elimination processes, etc.;

(f) on whether the subject has been exposed to a risk which could not be proven in any other way and, in some cases, when.

Biological indicators may be indicators of internal dose or indicators of effect.

The indicators of internal dose can be further divided into:

(a) true indicators of dose, that is, capable of indicating the quantity of the substance at the sites where it exerts its effect (this is still a theoretical problem in industrial toxicology);

(b) indicators of exposure which can provide an indirect estimate of the degree of exposure, since the level of the substance in biological samples is closely correlated with the levels of environmental pollution;

(c) indicators of accumulation that can provide an evaluation of the concentration of the substance in organs and/or tissues from which the substance, once deposited, is only slowly released.

The indicators of effect are those that can identify early and reversible alterations. The aim is to be able to assess alterations that develop in the *critical organ*, that is, in the organ where the concentration of the poison that can cause functional alterations is first reached.

The study of the concentration of a substance in the working environment and the simultaneous determination of the indicators of dose and effect in exposed subjects allows information to be obtained on the relation between occupational exposure and the concentration of the substance in biological samples, and between the latter and early effects.

Knowledge of the relationships between the dose of a substance and the effect it causes is an essential requirement if a programme of biological monitoring is to be put into effect. It is therefore worth while to try and define these relationships.

Dose-effect relationship. The evaluation of the relationship between dose and effect is based on analysis of the degree of association existing between indicator of dose and indicator of effect and on the study of the quantitative variations of the indicator of effect with every variation of indicator of dose. The degree of association–positive or negative–is expressed by a correlation coefficient (r); the quantitative variation of an effect according to the variations in dose is calculated and is represented by the regression curve that best fits the data under analysis.

With the study of the dose-effect relationship it is possible to identify at which concentration of the toxic substance the indicator of effect exceeds the values currently accepted as "normal".

Furthermore, in this way it may also be possible to examine what might be the no-effect level, if such exists, by which is meant a non-adverse effect. With such data it is possible to identify the subjects in a group who have abnormal levels compared to those obtained from the study of a reference group considered as "normal".

Dose-response relationship. Since not all the individuals of a group react in the same manner, it is necessary to study how the group responds to exposure by evaluating the appearance of the effect compared to the internal dose. This is what is meant by *response*, which is the percentage of subjects in the group who show a specific quantitative variation of an indicator at each dose level.

By studying these relationships it is also possible to ascertain whether, following exposure to a given substance, different groups of a working population show the same effect. For example it is now common knowledge that erythrocyte proptoporphyrin increases in female subjects more markedly than in male subjects at the same internal lead dose.

Choice of indicators

Knowledge of the metabolism of an exogenous substance in the human organism and of the alterations that occur in the critical organ is essential for the selection of the parameters that can serve as indicators of dose and effect. Unfortunately, such knowledge is insufficient and thus sets a limit on any biological monitoring programme.

The conditions necessary for successful biological monitoring can be thus summarised as follows:

(a) the existence of indicators;

(b) the existence of sufficiently accurate, sensitive and specific analytical methods that will guarantee technical reliability in the use of these indicators;

(c) the possibility of using readily obtainable biological samples on which the indicators can be measured;

(d) the existence and knowledge of dose-effect and dose-response relationships.

It must also be emphasised that biological monitoring cannot be used as a means of worker surveillance when acute exposure conditions exist or when biologically inert substances, local irritants, infectious agents, allergens, mutagenic, teratogenic or carcinogenic substances are involved.

Once the said conditions can be considered satisfactory, it is then essential to use indicators that have sufficient *predictive validity*.

In this context the *validity* of a test is the degree to which the parameter under consideration predicts the situation as it really is or as it would be using more accurate measuring instruments. Validity is given by the combination of two properties: sensitivity and specificity. If a test possesses a high sensitivity, this means that it will give few false negatives; if it possesses high specificity, it will give few false positives.

Time of sampling and other interfering factors

In carrying out a biological monitoring programme, it is indispensable to know exactly what the characteristics and behaviour of the indicators under study are in relation to length of exposure, time elapsed since the beginning and end of exposure, and other factors that may give a false interpretation of the results obtained.

For example, in the exposure to metals the levels of indicators of internal dose vary with the time interval after the end of exposure. Subjects with a clinical picture of lead poisoning may show blood lead levels very close to those of the non-occupationally exposed, when the measurement of lead in blood is made some time after the end of the exposure; however, the urinary excretion of lead provoked by chelating agents will confirm the existence of a high accumulation of lead in the organism. For other metals, like cadmium and mercury, the indicator of dose may be significantly correlated with exposure only several months after the start of work (figure 1).

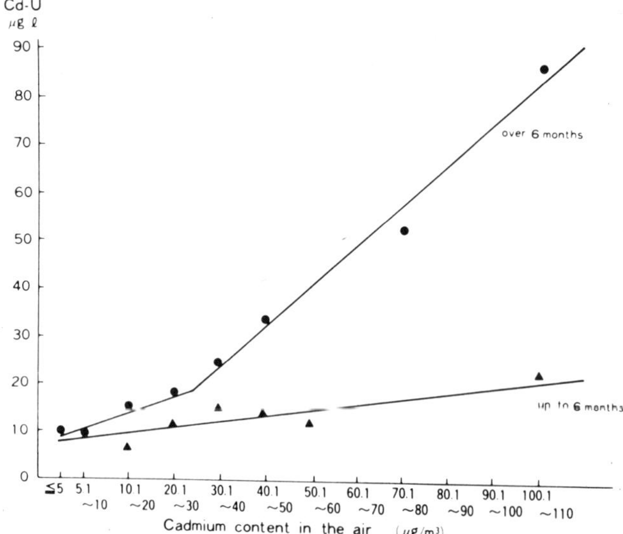

Figure 1. Relationship between cadmium concentration in air and cadmium concentration in urine (CdU) of workers exposed in battery-making plants. *(From A. Harada in Nordberg (ed.), 1976.)*

The use of hair as a biological specimen for the verification of exposure to metals, while useful in monitoring the general population, is not recommended for occupational exposure since the sample may have undergone external contamination and the values obtained would not represent the internal dose.

Other factors can influence the level of metal in the blood independently of exposure and body-burden: for example blood lead and blood cadmium levels may vary according to the number of circulating red cells.

In exposure to organic substances, the collection time of the biological samples becomes all the more important in view of the different speed of metabolic processes and consequently of the more or less rapid excretion of the absorbed dose. For example the biological monitoring of xylenes is carried out by measuring the concentration of methylhippuric acid in urine. The excretion of this urinary metabolite is practically completed within 16 h after the end of exposure and the highest levels of methylhippuric acid are found at the end of exposure, when exposure is constant over time. In this case the current practice of collecting the urine samples at the end of the shift is correct. However, because of the rapid metabolism of xylenes, if exposure is not stable over time, ample variation in excretion of methylhippuric acid may occur during the working day (figure 2).

In the biological monitoring of trichloroethylene, the determination of trichloroacetic acid in the urine must be made at the end of the working week since this solvent is metabolised rather slowly, with a consequent accumulation process (figure 3). In addition, when the analysis is made on spot samples, the levels of the

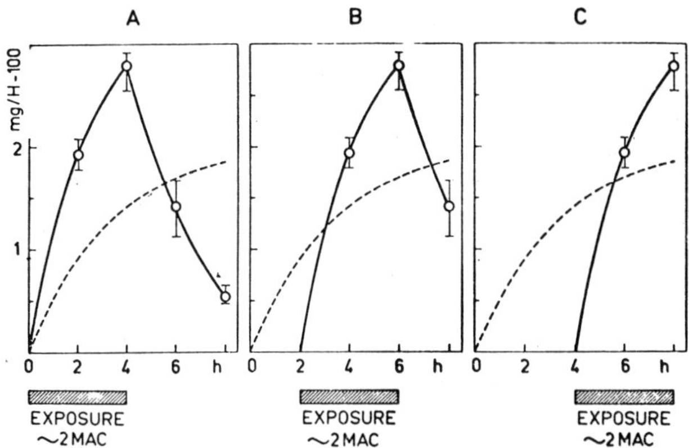

Figure 2. Methylhippuric acid levels in urine collected at the end of the work shift in subjects exposed to xylenes mainly at the beginning (A), half way through (B) and at the end of the work shift (C). (From "Exposure test for xylenes". Sedivec, V.; Flek, J. *International Archives of Occupational and Environmental Health* (West Berlin), 1976, 37/3 (219-232). 16 ref. *By courtesy of Springer Verlag*).

substances under study must be corrected to an urinary standard specific weight, for example 1 016 or 1 024, or according to the creatinine content, so as to reduce the possible wide variations in values. In particular, urine with a specific weight below 1 010 or higher than 1 030 or with a creatine concentration lower than 0.5 g/l or greater than 3 g/l should be discarded.

The concentration of metabolites of organic substances is a useful test when the exposure of a homogeneous group of workers has to be evaluated, but is more difficult to interpret if the exposure of an individual worker is to be assessed. In this case the levels of some urinary metabolites or organic substances may be influenced by exposure other than occupational. For example in the case of toluene, eating prunes, which contain benzoic acid, can give rise to a high excretion of hippuric acid. And increased amounts of urinary

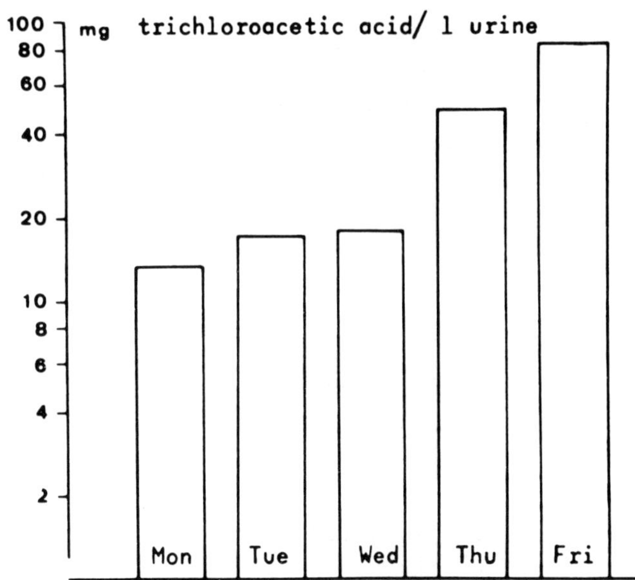

Figure 3. Average concentration of trichloroacetic acid in 1 071 urine samples of workers exposed to trichloroethylene plotted over the working days of the week (from G. Lehnert, R. D. Landendorf, D. Szadkowski, *International Archives of Occupational and Environmental Health*, 41/95, 1978).

phenols are found after taking drugs containing phenylic groups.

Such remarks are intended as a reminder to occupational health physicians that the test they use may give false indications, of which false negatives cause the most concern. This problem has until now been dealt with mainly at an experimental level. The biotransformation of benzene in rats is accelerated by administration of phenobarbital, probably by induction of the microsomial enzymes responsible for metabolising the solvent; as a consequence the elimination of urinary phenols is lower than would be expected from the degree of exposure.

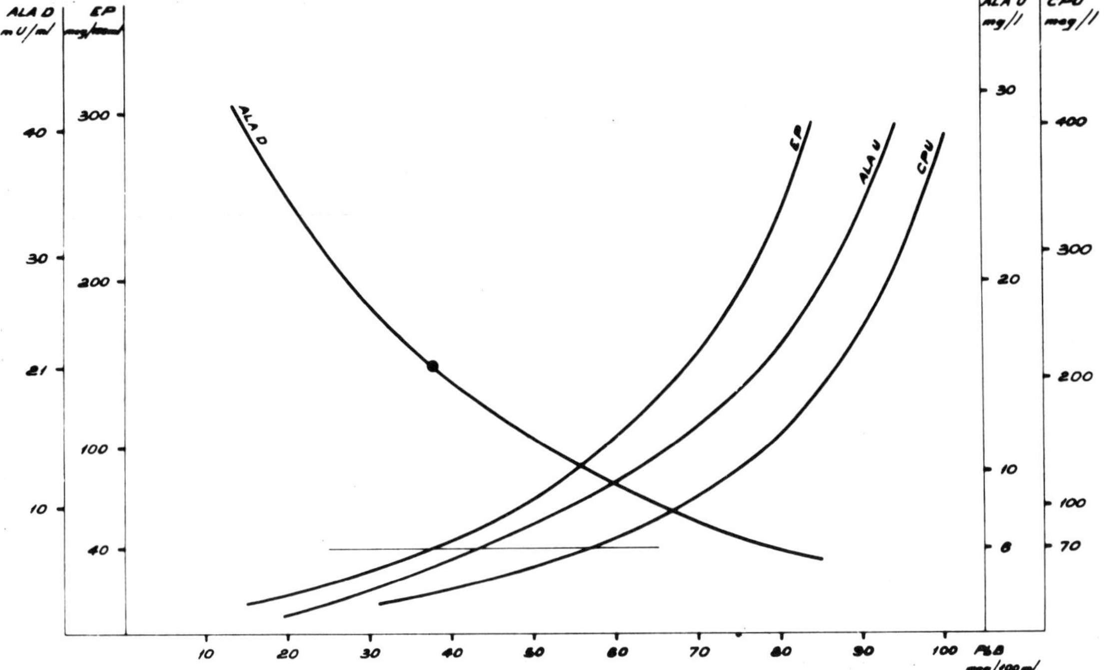

Figure 4. Relationship between PbB and indicators of effect in adult males currently exposed to lead (from L. Alessio and G. Camerini in *L'intossicazione professionale da piombo inorganico e la sua prevenzione* (Milan, Cortina, 1975)).

Retention of selenium can be influenced by the simultaneous absorption of mercury. These interactions in the human organism have not been extensively studied yet, but should be paid particular attention in combined exposure to two or more exogenous substances or in consumption of drugs that can interfere with the metabolism of the substance under biological monitoring.

For the study of early biological effects resulting from the absorption of a toxic substance, it is not only necessary to study the alterations that occur in the critical organ, but also to know which indicator of effect gives the earliest possible information on such alterations.

For example, in cadmium-exposed subjects the determination of proteinuria with trichloroacetic acid can reveal tubular damage; however, the detection of low molecular weight proteins in urine can reveal the same effect earlier. The appearance of increased urinary excretion of coproporphyrins in lead-exposed subjects is evidence of an alteration in heme synthesis, which, however, can be revealed by the determination of erythrocyte protoporphyrins and urinary ALA at lower internal dose levels than those capable of causing an increased coproporphyrin excretion (figure 4).

The existence of situations, other than exposure, that can influence the indicator levels must also be taken into consideration in assessing the indicators of effect. For example, an increase in erythrocyte protoporphyrins can be detected in sideropoenic subjects who show no abnormal lead absorption; in thalassemic subjects with high recticulocytosis even after an abnormal lead absorption, erythrocyte ALA D levels can be uninhibited.

To sum up, biological monitoring can be recommended for those substances that comply with the requirements listed above. In view of the uncertain relationship existing between external and internal dose and between internal dose and effect (or response) for the majority of industrial agents it is difficult to propose indicator values that can be taken as threshold values, i.e. values below which we can be sure that the subject will be safe even from long-term effects. There are only a few substances for which indicator levels can be indicated with sufficient accuracy to be taken as a reference for a correct health surveillance of exposed populations. The aim of good industrial hygiene practice should be to bring exposure levels in the working environment closer and closer to the levels in normal everyday exposure outside the plant. The time schedule for carrying it into effect cannot, of course, be established by the physician alone, because of the effects of social, political and economic conditions. With these in mind a "threshold of acceptability" will be established; this certainly is not a biological parameter, since by "acceptability" is meant the acceptability of a risk.

Lastly, a number of practical suggestions may be made. For each subject undergoing pre-employment examination, the parameters to be used as indicators of dose and/or effect in subsequent periodic examinations should be assessed, so as to have a reference point for each worker. In addition, in periodical examinations, whenever possible an indicator of dose should be measured at the same time as an indicator of effect. Reference values for the general population, broken down into age and sex groups as a minimum, should be available. Differences among countries and areas within the same country as regards regulations and habits in collecting, analysing and storing human samples should also be borne in mind.

FOÁ, V.
ALESSIO, L.

The use of biological specimens for the assessment of human exposure to environmental pollutants. Berlin, A.; Wolff, A. H.; Hasegawa, Y. (eds.). (The Hague, Boston, London, Martinus Nijhoff for the Commission of the European Communities, 1979), 368 p. Ref.

"Biological criteria for selected industrial toxic chemicals: a review". Lauwerys, R. *Scandinavian Journal of Work, Environment and Health* (Helsinki), Sep. 1975, 1/3 (139-172). Illus. 373 ref.

Effects and dose-response relationship of toxic metals. Nordberg, G. F. (ed.). (Amsterdam, Elsevier Scientific Publishing Company, 1976), 558 p. Illus. Ref.

"Applicability and limits of biological monitoring in occupational health and epidemiology" (Anwendbarkeit und Grenzen des "Biological Monitoring" in der Arbeitsmedizin und Epidemiologie). Schiele, R. *Zentralblatt für Arbeitsmedizin, Arbeitsschutz und Prophylaxe* (Dortmund-Marten), Jan. 1978, 28/1 (1-8). 21 ref.

"Biological monitoring". Zielhuis, R. L. *Scandinavian Journal of Work, Environment and Health* (Helsinki), Mar. 1978, 4/1 (1-18). 30 ref.

Biological rhythms

Human beings live according to statistically validated physiological changes recurring with a reproducible waveform–their biological rhythm.

Rhythms can be viewed macroscopically from the inspection of original data or of averages in displays–as impressions, abstractions or intuitive inferences–without the provision of inferential statistical point-and-interval estimates of rhythm characteristics. This approach has to be distinguished from a microscopic approach, which is complementary to the macroscopic one in that it relies on the objective quantification of temporal characteristics in biological data–for instance, by testing the fit of mathematical models to time series and then estimating the temporal parameters on the basis of the best-fitting curve. Figure 1 provides on top an early macroscopic spectral view, whereas the bottom of figure 1 summarises results of a microscopic approach to a variable such as urinary 17-ketosteroid. Table 1 describes the period ranges for different rhythmic components of time series.

Chronobiology

By appropriate methods, chronobiology, a section of biology, is achieving the objective description of biological time structure–the sum total of non-random, and thus predictable, temporal aspects of organismic behaviour, including, with the spectrum of biologic rhythms, developmental changes, growth and age trends. It is now recognised that biological time structure characterises individuals as well as groups of populations of organisms or of their subdivisions: organ systems, organs, tissues, cells and intracellular elements (including electron-microscopic ultrastructure). For a number of rhythms validated as a bioperiodic aspect of data, a few basic characteristics have been measured: the period, the acrophase (measure of the timing of predictable rhythmic changes), the amplitude (measure of the extent of predictable rhythmic changes), the waveform and the (rhythm-adjusted) level and range. Table 2, complemented by table 3, summarises some major features of rhythms.

Circadian rhythms

The term circadian (derived from Latin *circa dies*: about a day) refers to an average period of precisely 24 h (also referred to as dian, if the period is longer than 23.8 h but shorter than 24.2 h) or any other duration between 20

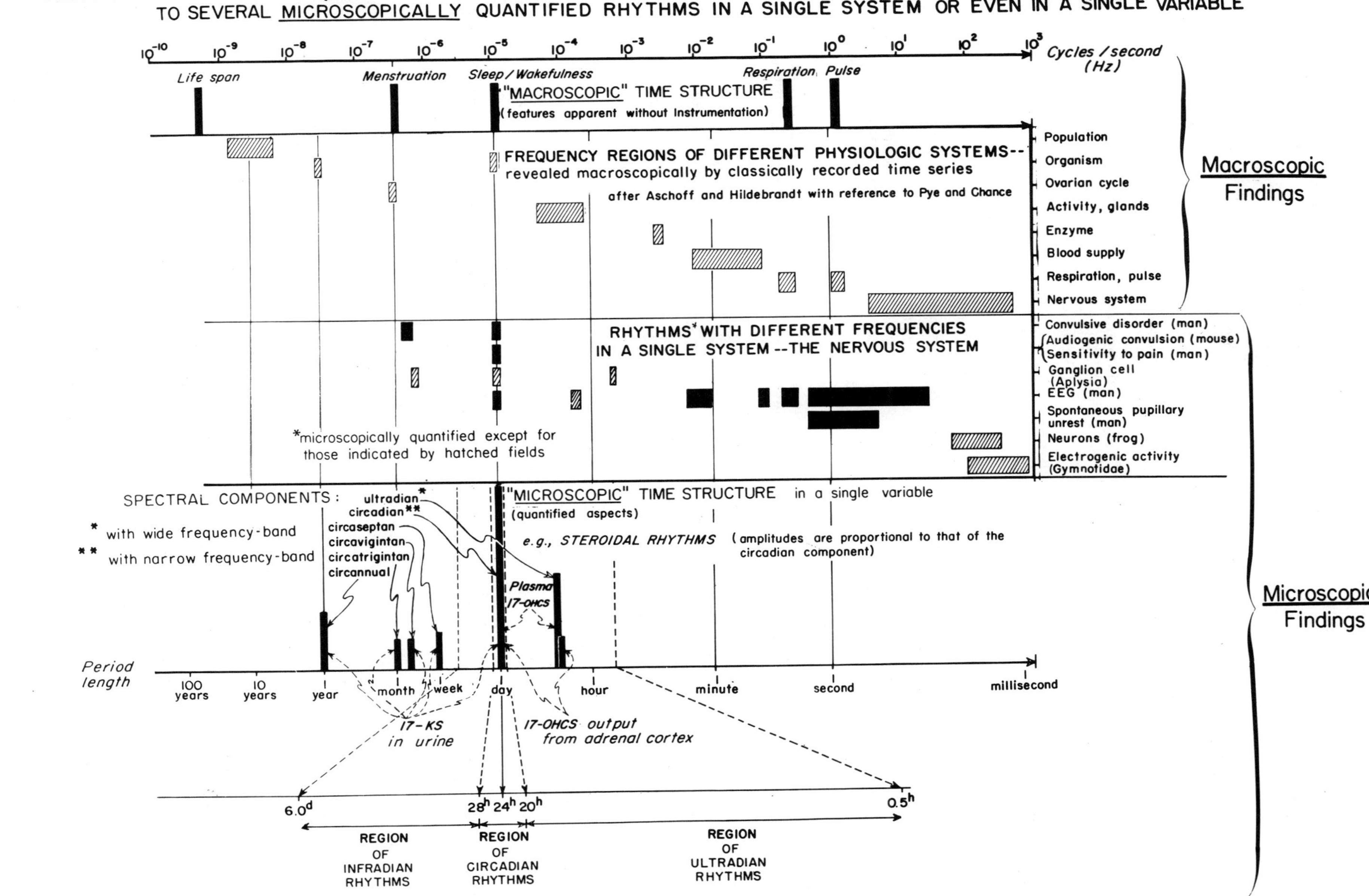

Figure 1. Physiological spectral domains (D.).

and 28 h. A rhythm of such period is well evident in a number of functions relating to:

(1) the exogenous influences of the day/night cycle with its social exigencies as well as effects of lighting, electromagnetic phenomena, temperature, etc. These synchronisers act upon the second category of functions;

(2) a multifrequency web of endogenous cycles, some of them *(a)* cellular, such as a sequence consisting of the circadian labelling of phospholipid (as a membrane phenomenon) followed by RNA labelling, DNA formation and mitosis in liver and some other tissues, others *(b)* hormonal (corticotropin-releasing factor, ACTH, 17-OH-corticosteroid (see figure 2), and yet others *(c)* neural and neuroendocrine. The circadian amplitude- and pace-resetting role of the suprachiasmatic nucleus at several levels of organisation *(a)-(c)* can be viewed in the context of a broader network involved in the modulation of several frequencies—ultradian to infradian (figure 3).

Table 1. Period ranges for terms describing biological rhythms[1]

Domain region	Range
ultradian	$\tau < 20$ h
circadian	20 h $\leqslant \tau \leqslant$ 28 h
dian	23.8 h $\leqslant \tau \leqslant$ 24.2 h
infradian	$\tau > 28$ h
circaseptan	$\tau = 7 \pm 3$ d
circadiseptan	$\tau = 14 \pm 3$ d
circavigintan	$\tau = 21 \pm 3$ d
circatrigintan	$\tau = 30 \pm 5$ d
circannual	$\tau = 1$ yr $\pm$ 2 mth

[1] τ = period; h = hour; d = day; mth = month; yr = year. Terms coined by analogy to usage in physics.

Just as frequencies higher than those audible or visible are called ultrasound and ultraviolet, frequencies higher than 1 cycle per 20 h are designated as ultradian. By the same token, as frequencies lower than audible or visible are called infrasound and infrared, rhythms with a frequency lower than 1 cycle per 28 h are designated as infradian.

Table 2. Some major features of biological rhythms

During most of our life most components in a spectrum of physiological rhythms—

(1) are locked into a spectrum of environmental cycles
with synchroniser frequencies "acceptable" to ("resonant" with) the organism,

(2) yet modulated and influenced
from within and without (by other spectral components or non-cyclic factors).

Environmental factors (synchronisers) may be manipulated so that rhythms—

(3) are shifted (in response to schedule shifts) with:
(a) asymmetries (e.g. delay of rhythm faster than advance or vice versa) (see table 3 and figure 4),
(b) polarities (i.e. some rhythms advance while others delay in the same organism, responding to the same synchroniser shift),

(4) or persist under certain conditions of isolation
with transients and eventually new stable internal time relations, in the absence of known external synchronisers, modulators and influencers—usually with "free-running" periods, different from known environmental ones.

The organism may be manipulated by various means including surgery or disease, so that rhythms—

(5) change characteristics (alter amplitude, mesor or acrophase), desynchronise, transfer variance, frequency-multiply and/or frequency-demultiply or, more broadly, spectrally compromise as does thermovariance in cancer,

(6) or disappear
as does the circadian rhythm in blood eosinophils in Addison's disease.

Sleep-wakefulness, body temperature, pulse and blood pressure, the composition of blood, the excretion

Table 3. Comparative physiology of circadian acrophase-shifts reveals structured adjustment in the form of differences in rate (asymmetry) or direction (polarity) of rhythm adjustment after schedule shift

Organism (Investigators)	Variable	Extent of synchroniser-shift: Hours (on a 24-h cycle)	Extent of synchroniser-shift: Degrees	Hours to adjust to regimen: Advance	Hours to adjust to regimen: Delay	Difference (asymmetry)[1]
					Faster advance	
Albizzia julibrissin: silk tree (Koukkari and Halberg)	Pinnule angle	4[2]	60	~ 24	~ 52	28
Fringilla coelebs: Chaffinch (Aschoff and Wever)	Jumping activity[3]	6	90	~ 60	~ 120	60
					Faster delay	
Tribolium confusum: flour beetle (Chiba, Cutkomp and Halberg)	O_2 consumption	6	90	> 48	> 24	24
Rattus norvegicus: Sprague-Dawley rat (Halberg)	Intraperitoneal temperature	6	90	~ 312	~ 120	192
Macacus rhoesus, M cyanomolgus: monkeys (Halberg)	Axillary temperature	6	90	> 120	> 72	48
Homo sapiens (Halberg)	Oral temperature[4]	~ 10	~ 150	~ 312	~ 192	120
Homo sapiens	Urinary variables[4]	~ 10	~ 150		P	

[1] Unless polarity *(P)* prevails. [2] 4 h = half the dark span in Albizzia; 6 h = half the dark span in rat, monkey and chaffinch. [3] Macroscopic study by inspection of plots; all other studies microscopic, i.e. validated by inferential statistical parameter estimation. [4] In social setting. Studies by Aschoff on men isolated in a bunker reveal faster advance than delay.

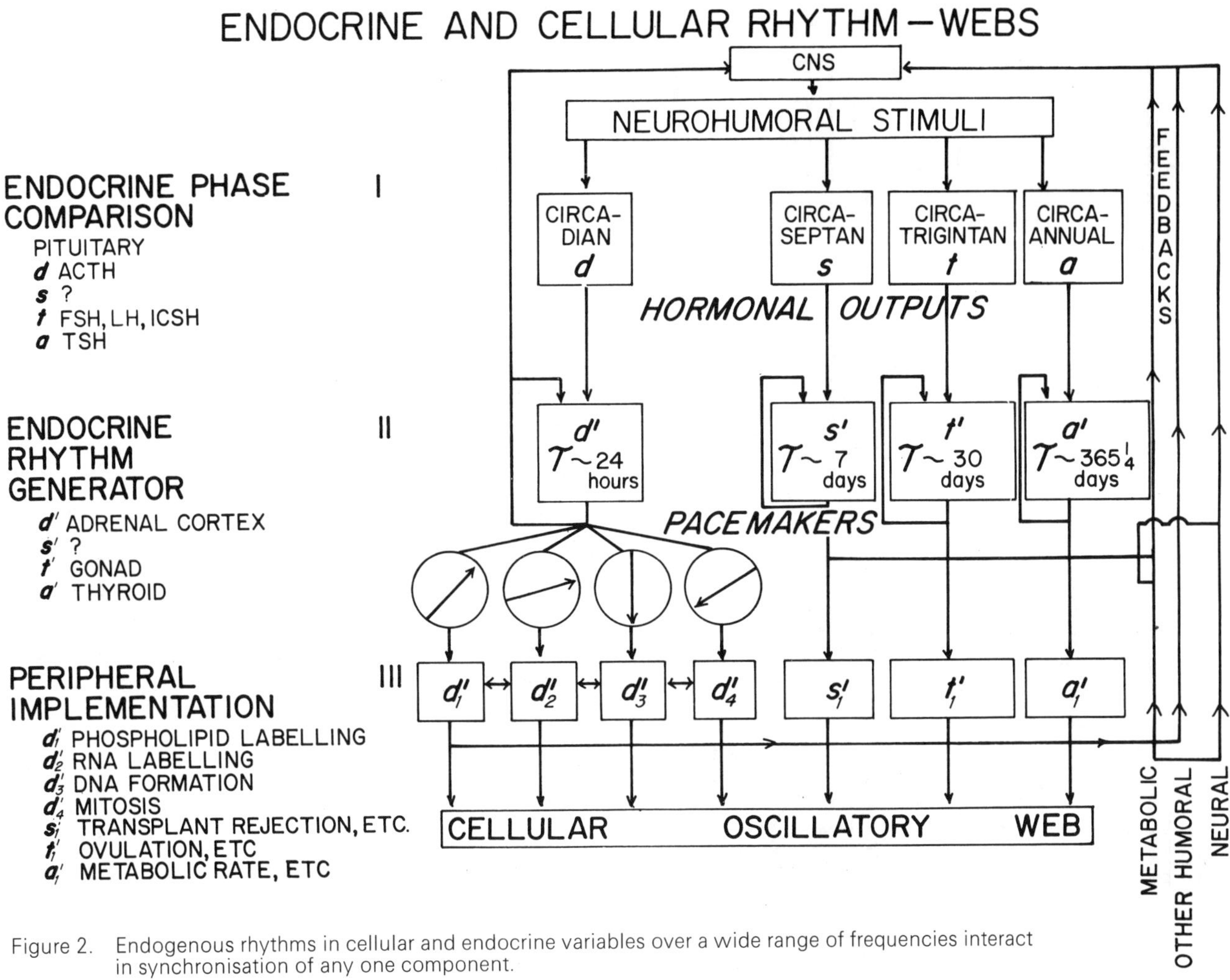

Figure 2. Endogenous rhythms in cellular and endocrine variables over a wide range of frequencies interact in synchronisation of any one component.

rate of substances in urine, competence at work, susceptibility to drugs and even the probability of being born or dying, all depend to some extent on several components in the spectrum of rhythms. Circatrigintan rhythms include not only the well known menstrual cycle but rhythms with periods of about 30 days found in women before menarche and after menopause and characterising the spectrum of male core temperature variability as well.

A considerable number of studies indicates that human beings and other animals (as well as plants) continue to show certain about-daily, about-monthly and about-yearly variations in the apparent absence of known environmental periodicity. A man or a woman isolated in a cave for 4-6 months without time information continues to exhibit an about-24-h rhythm in body temperature, pulse, adrenocortical and other variables. Circadian rhythms also persist during flight in extraterrestrial space, for the few weeks of experience studied thus far.

Endogenous time structure

Endogenicity is suggested by the desynchronisation of rhythms from precise environmental schedules, e.g. of circadian rhythms from precise 24.0-h (solar day) and 24.8-h (lunar day) schedules or circaseptan rhythms from precise 7-day schedules and of circannual rhythms from a precisely 1-yr period. As an example of circadian desynchronisation the rhythm in intraperitoneal temperature, telemetered at intervals of approximately 10 min for 116 days from 6 blinded, mature, inbred Minnesota Sprague-Dawley rats, kept singly housed with food freely available at 24 °C environmental temperature, in light and darkness alternating at 12-h intervals, persisted with a 24.3-h period (rather than a 24-h period) and a 0.57 °C amplitude. Similar amplitude (of 0.61 °C) at a precise 24-h average period was found in five concomitantly evaluated intact (rather than blinded) rats kept under the same conditions.

Acrophase charts

Circadian, circaseptan, circatrigintan and circannual acrophase charts already are available to show the timing of rhythms in relation to a schedule consisting of about 16 h of activity alternating with about 8 h of rest each day. For example relatively high values of oral temperature tend to occur shortly after the "middle" of the activity span, that is at about 1600 h for someone who gets up at 0700 h and retires at 2300 h.

Schedule shifts

The temporal placement along the 24-h scale of the human core temperature rhythm can be shifted. Such a schedule-shift-induced change in timing of the human oral temperature rhythm usually, but not invariably, occurs more rapidly when the schedule of living is delayed, as in a flight from Minnesota to Japan, than when the schedule is advanced, as in a flight from Japan to Minnesota. Figure 4 shows that internal timing, namely the relation between two rhythms themselves, may differ during schedule shifts, and the performance of a healthy man adjusts much more slowly to an eastward

I. SYNCHRONIZERS
   SOCIAL SCHEDULES,
   ENVIRONMENTAL CYCLES

II. TRANSDUCERS
   SENSE ORGANS; MUSCLE
   STRETCH RECEPTORS;
   CELLULAR RECEPTOR SITES;
   MEDULLARY NUCLEI; ETC.

III. MODIFIERS
   RETICULAR FORMATION;
   THALAMUS, TECTUM,
   AND SENSORY CORTEX (CODING);
   LIMBIC SYSTEM

IV. LOGIC CONTROLLERS
   ASSOCIATION CORTEX;
   FRONTAL CORTEX

V. PACE-RESETTERS
   SUPRACHIASMATIC NUCLEUS;
   OTHER POSSIBLE RELATED
   AREAS

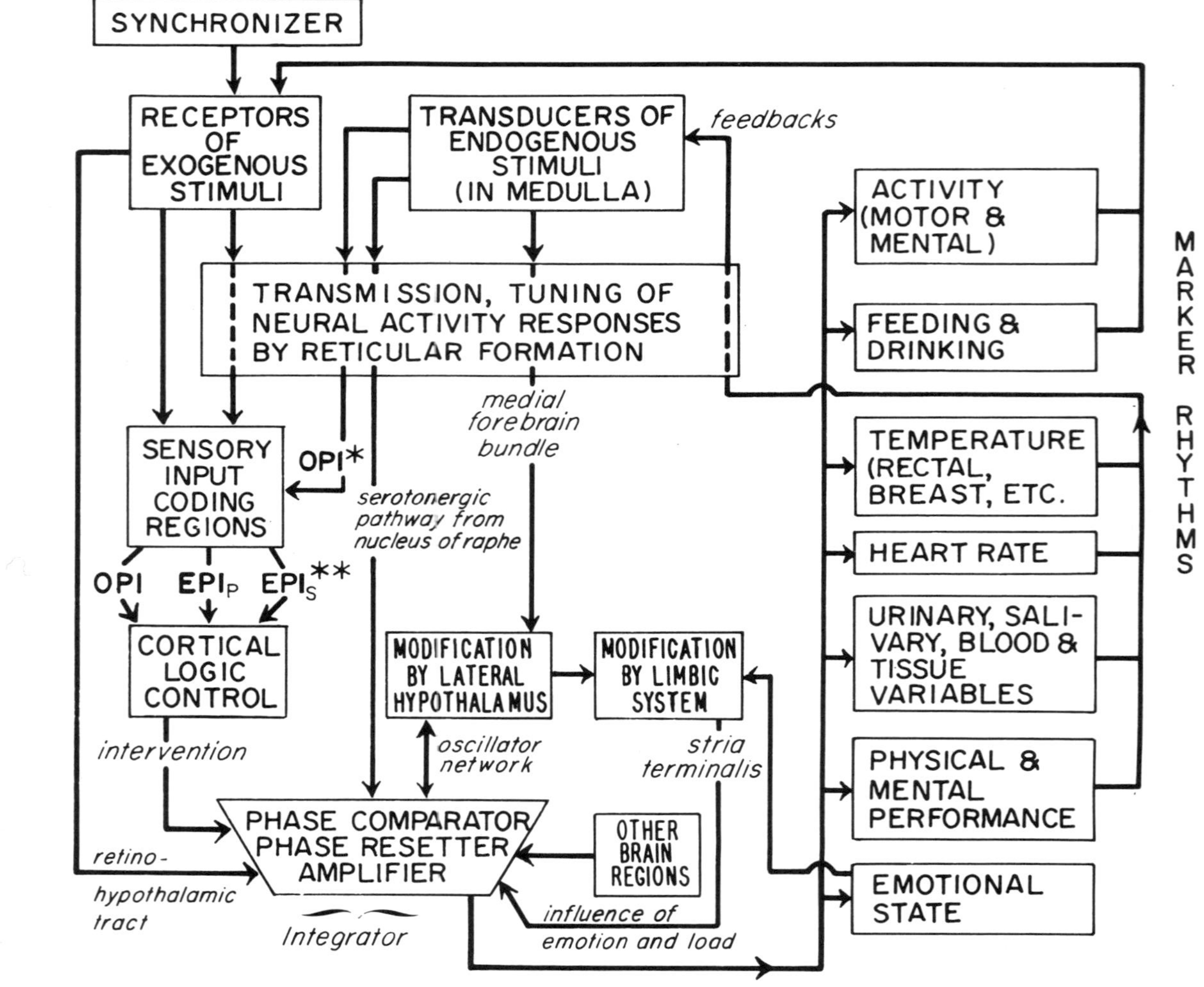

* *OPI = Organismic Phase Information (Composite)*

** *EPI = Environmental Phase Information (Composite); EPI_P denotes physico-chemical and EPI_S denotes socioecological. See Halberg, F., Halberg, E., Barnum, C.P., Bittner, J.J.: Physiologic 24-hour periodicity in human beings and mice, the lighting regimen and daily routine. In: Photoperiodism and Related Phenoma in Plants and Animals, R.B. Withrow, Ed., Ed. Publ. No. 55, Amer. Assoc. Adv. Sci., 1959, pp. 803-878.*

Figure 3. Synchronisation of neural rhythms.

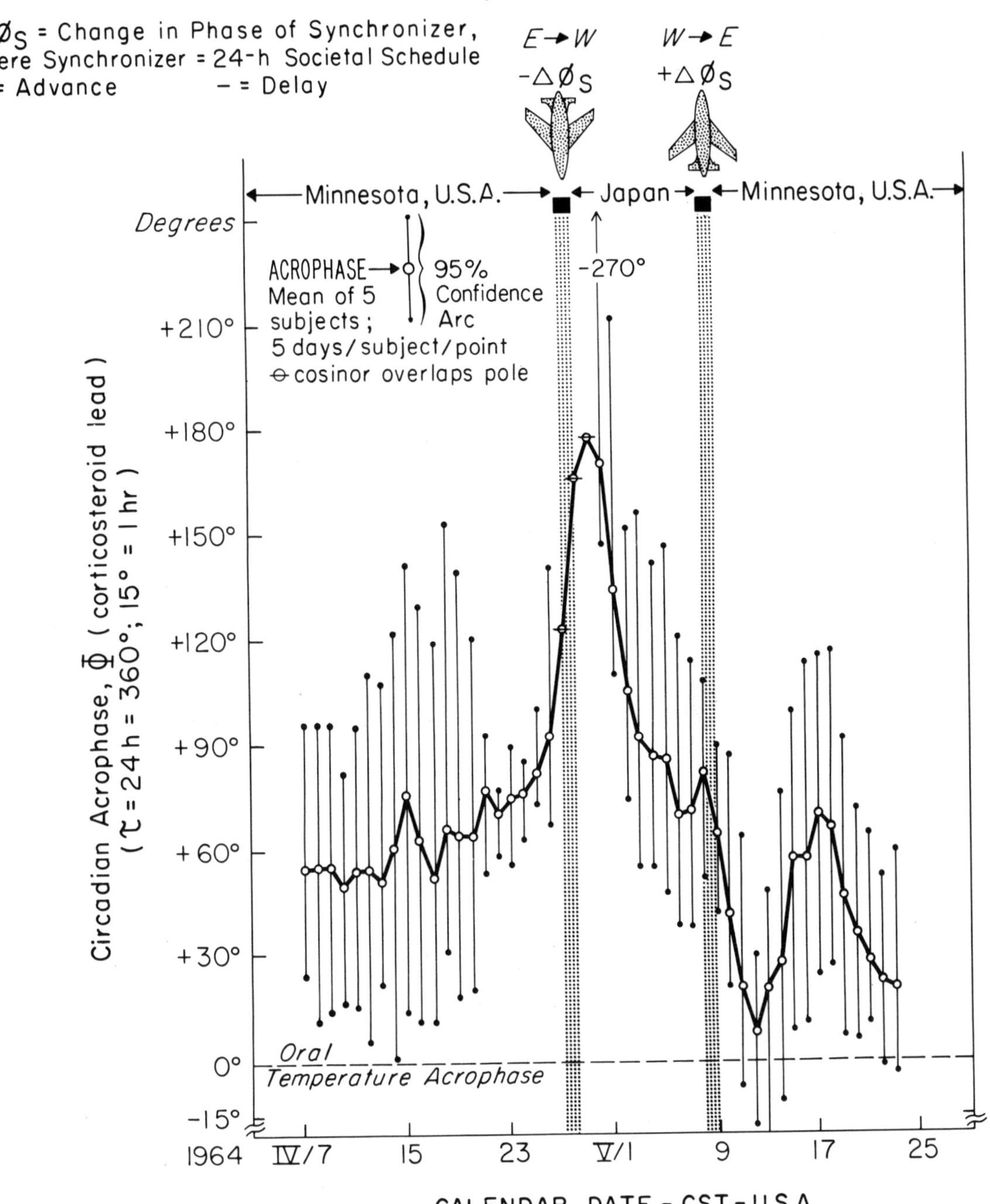

Figure 4. Time relation between 2 circadian rhythms temporarily changed after shifts in schedule (transmeridian flight).

flight than does either oral temperature or heart rate. Figure 5 in turn shows a polarity of the system, in that during an adjustment following a flight from Minnesota to Japan, 17-hydroxy-corticosteroids, 17-ketosteroids and calcium in the urine may delay (incidentally, at different rates–the phenomenon of intra-individual within-shift asymmetry), whereas other functions, namely sodium and potassium excretion, advance. Such a polarity may even be seen during a change in schedule without geographic displacement.

Schedule shifts can involve a deficit in performance. By treatment at different circadian times with a corticosteroid the human circadian system can be manipulated and the acrophase displaced in different directions.

Basic concept

There are three major dimensions of organismic complexity:

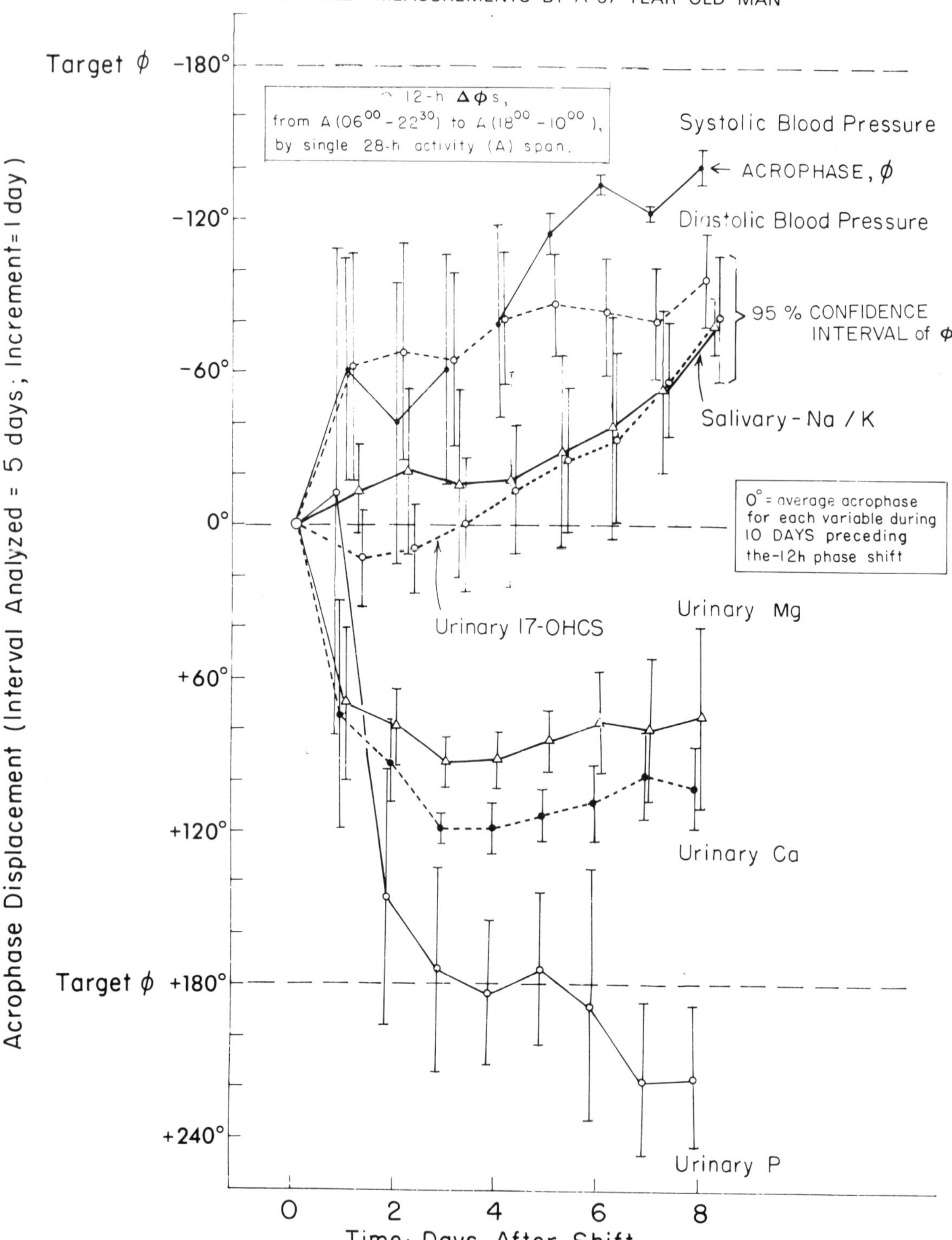

Figure 5. Differences in polarity of shift for different circadian rhythms in same subject following delaying-shift of schedule (data of H. Levine).

(a) the different physiological systems—neural systems, glandular secretions, and intracellular variations;

(b) the spectrum of frequencies with which the bioperiodicities occur in all these systems;

(c) the evolution of this entire network of rhythmic variation with the growth, development, maturation and ageing of the individual.

Multiple influences and linkages exist among the different physiological systems and among rhythms of different frequencies. Such a dynamic web of physiological interaction among biochemical compounds in anatomical locations cannot be replaced by a simple environmental control and unitary or isolated biological clocks. Simplifications all too often imply exclusive control by the environment *or* by the genetic code, when in fact both are pertinent.

Application

Chronobiological information is of applied as well as basic interest. Rhythms are found everywhere as characteristics of living things. Some of them are important to agriculture, others to health care, and still others to work efficiency or education. Every student's body can be used as a free yet immensely useful laboratory aid. A few self-measurement tools may be of much greater educational value than rather expensive pieces of other equipment. This realisation should be capitalised on as part of a broader approach towards a chronobiological literacy. In the context of its broader educative value, such literacy may also be the best possible way to effect the entry of self-help for health care into primary and secondary schools. Rhythmometry provides means for early detection of disease by individualised and time-specified ranges of normal values. These same reference intervals that define health also allow us to detect harbingers of risk—as rhythm alterations—before disease becomes established.

The study of ageing also reveals rhythm alterations. These may provide a better understanding of primary ageing and a better treatment of certain secondary diseases of the elderly. Blood pressure elevation may be recognised earlier by systematic self-measurements. The effects of antihypertensive or antiasthmatic treatments also are thus better gauged. The result is better compliance with a treatment. The same self-measurement of blood pressure may have added merits for specifying the timing of treatment—for better results. Life styles may be optimised by timing nutrition and exercise. Optimisation is particularly pertinent in relation to safety, health and productivity of shift workers.

OCCUPATIONAL SAFETY AND HEALTH ASPECTS

Information on rhythms has a bearing on occupational health and preventive medicine. Thus, autorhythmometry (computer analyses of self-measurements) on circadian variables such as grip strength, eye-hand skill and arm-hand steadiness, as well as on indices of mental performance (random-number addition, short-term memory) serves to monitor the worker's productivity by his performance (which represents an unspecific yet non-trivial aspect of health as well). In the short run an imposed rhythm alteration, resulting from changes in schedules, is associated with decreased performance (see SHIFT WORK). Moreover, rhythms should also not be overlooked in studying the distribution of accidents during the day and the year. For example, many studies of accidents have shown that the accident frequency rate for night shifts is lower than that for day shifts and that, during work on either day or night shifts, the accident-incidence curve rises between the third and sixth hour of the shift, falling again at the end of the shift. In the morning the accident curve reaches a minimum soon after the beginning of the shift and again just before the end. A classical psychological account of the accidents before the first minimum is that workers starting a morning shift are still drowsy after awakening. During the morning, feeling fresh and fit, they concentrate more on their work; but then in the second half of the shift they become negligent and careless, and this leads to another increase in accidents until the "end-spurt". Night-workers, however, have sufficient time to prepare themselves, getting up 2-3 h before starting work, and arrive calmer and more relaxed at work, with an attitude more beneficial to safety. The role of ultradian rhythms and of their changes in frequency along the 24-h scale remains to be investigated in a chronopsychological endeavour toward optimisation of the working time.

A systematically unequal distribution of accidents over different months has led to the expression "seasonal cycle" to indicate an increasing number of accidents up to summer, followed by a decrease in the winter. Circannual rhythms have been documented for many body functions in human beings and other life forms. A circannual rhythm structure contributes to the aforementioned unequal distribution of accidents along the one-year scale, just as circadian rhythms contribute to differences in health and performance along the one-day scale.

HALBERG, F.

Glossary of chronobiology. Halberg, F.; Carandente, F.; Corélissen, G.; Katinas, G. S. (Milan, Il Ponte, 1977), 189 p. Illus. Bilingual edition (English and Italian)

"Chronobiologic optimization of aging" (5-56). Illus. 87 ref. Halberg, F.; Nelson, W. *Aging and biological rhythms.* Samis, H. V.; Capobianco, S. (eds.). (New York, Plenum Press, 1978).

Biological rhythms in psychiatry and medicine. Luce, G. G. (Washington, DC, US Government Printing Office, 1970), 183 p.

Shift work and health. Rentos, P. G.; Shepard, R. D. (eds.). (Washington, DC, US Government Printing Office, 1976), 283 p. Illus.

CIS 78-1178 *Recommendations concerning the recording of workers' biological rhythms for the purposes of work organisation and occupational safety* (Metodičeskie rekomendacii po učetu bioritmov čeloveka v organizacii i ohrane truda). Smirnov, K. M. (Leningrad, Vsesojuznyj naučno-issledovatel'skij institut ohrany truda VCSPS, 1976), 32 p. Illus. 44 ref. (In Russian)

CIS 79-267 "Daily fluctuations in worker performance" (Tagesschwankungen menschlicher Arbeitsergiebigkeit). Müller-Seitz, P. *Wt–Zeitschrift für industrielle Fertigung* (West Berlin, 1978), 68/1 (23-26). Illus. 20 ref. (In German)

Biomechanics

Biomechanics is concerned with the study of mechanical forces acting upon anatomical structures during human movement *per se* or as a result of the interaction between man and the physical environment. The study and practice of biomechanics require knowledge of functional anatomy, kinesiology, work physiology, mechanical engineering and electronic instrumentation. Laboratory methods commonly employed in research and field studies include, among others: kinematography; the measurement of inertial properties of the whole body and its segments; quantification of range, strength, velocity and acceleration of joint movement; work physiological measurement (metabolism, heart rate, etc.); and electrophysiological kinesiology. The latter includes the recording of bioelectric signals generated by muscles and other organs involved in a specific activity producing indices of physical effort and degrees of co-ordination necessary to perform a given task.

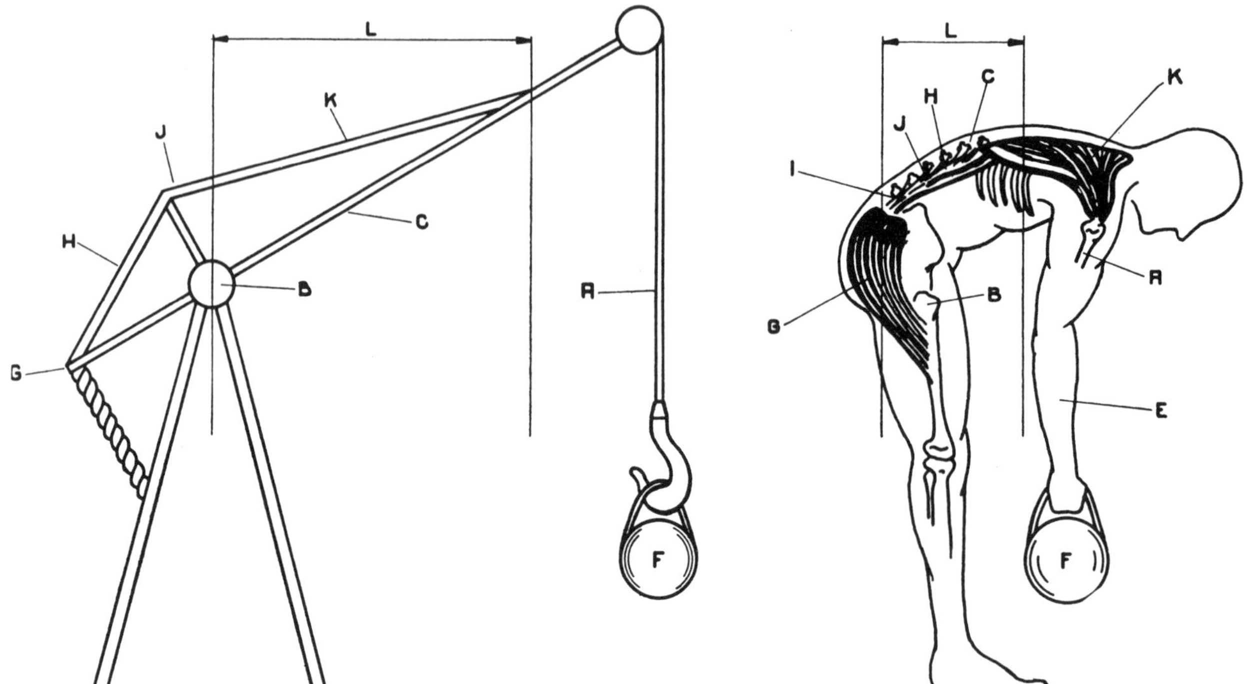

Figure 1. In lifting the human body can be compared to a crane; identical computational procedures can be used to predict mechanical failure of analogue structures of either. A. Humerus. B. Socket of hip joint. C. Vertebral column. E. Arm. F. Load. G. Muscles–buttock *(gluteus maximus)*. H. Muscles–back *(sacro spinalis)*. I. Fifth lumbar vertebra. J. Spinous process of a vertebra. K. Trapezius muscle. L. Distance from the centre of mass of combined body-load aggregate to lumbosacral joint.

Application to occupational health and safety

Within this context, biomechanics is concerned with the prevention of pathological responses to mechanical noxae or traumatogenic movements encountered in work situations. Such responses may be instantaneous and, thus, come under the heading of "accident" or they may be evoked by repetitive stress experienced over a period of time. In the latter case the resulting impairment will be classified as "disease" (e.g. tenosynovitis, carpal tunnel syndrome, etc.). Sometimes, a lengthy interval of time–weeks, months, or even years–may elapse before ergogenic diseases become manifest (e.g. traumatic arthritis, thromboses of the small blood vessels of the hand, or the various types of vibration-generated disease).

Some causes of disability of biomechanical origin

Man at work, whether manipulating a simple hand tool or operating a complex machine, forms part of a man-equipment-task system, where mechanical implements and anatomical structures interact. Should the design of the external "mechanical" environment subject the internal "biomechanical" environment to excessive forces, pressures and moments, neuro-musculo-skeletal or vascular disease may result. The following may serve as examples.

Example 1. Frequently the application of theoretical considerations may suffice for estimating the relative stressfulness of a task. When engaged in lifting or carrying, the human body becomes the analogue of a crane (figure 1). Mechanical failure of analogue components of either can be reliably predicted by the computational procedures of engineering mechanics. The critical dimension is *L*–in man, the distance from the lumbosacral joint to the centre of mass of an aggregate comprising the load and those body segments involved in the task. In near-static situations distance *L*, multiplied by the weight of the aforementioned aggregate, is known as the "biomechanical lifting equivalent". Therefore, the magnitude of a near-static lifting task should never be expressed in terms of "weight" alone, but should always be assigned the dimensions of "torque" (i.e. the product of force multiplied by the distance from its point of application). For this reason light and fragile objects packed into bulky protective containers may be biomechanically much more stressful and hazardous to handle than small and compact ones of greater weight (figure 2).

Example 2. Experimental methods of analysis may be required in other circumstances. Some of the stronger muscles involved in the gripping of a tool originate from

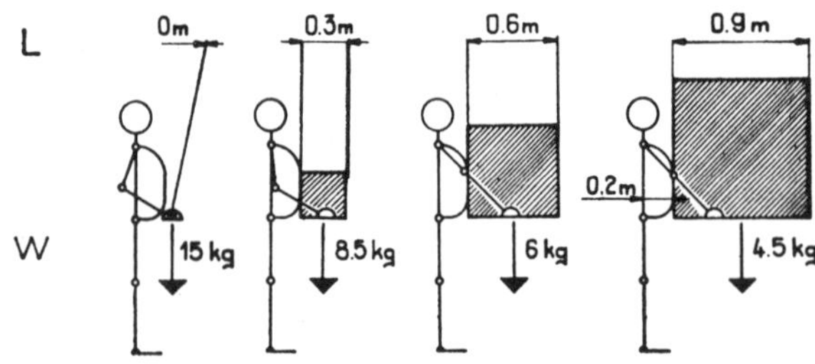

$$(0.2\,m + \tfrac{1}{2}L)\,(W) = M_\varepsilon \approx 3\,kgf.m \approx 29\,N.m$$

Figure 2. The "moment concept" applied to the derivation of biomechanical lifting equivalents. All the loads represented in the figure produce approximately equal bending moments on the sacrolumbar joint. 0.2 m = approximate distance from the joints of the lumbar spine to the front of abdomen (a constant for each individual). L = length of one side of a cube of uniform density lifted during the standard task. W = weight of the cube handled. M_ε = the biochemical lifting equivalent.

the elbow and are connected to the fingers by tendons passing through a narrow passage in the wrist–the carpal tunnel. Whenever the design of workplace or equipment demands deflexion of the wrist towards the ulna, simultaneously with strong gripping, then the tendons in the carpal tunnel bunch up, press against each other, and may squeeze the median nerve. This is known to be conducive to one of the most common wrist diseases–carpal tunnel syndrome. Prior to the introduction of a new work method, or hand tool, likely to be conducive to ulnar deflexion, a "biomechanical profile" of the operation should be constructed in order to forestall any exposure of the working population to a potentially higher occupational hazard. In wire cutting (figure 3) such a "profile" consists of a simultaneous tracing of the force exerted by the tool while cutting and the muscular effort necessary to perform the task employing various degrees of ulnar deflexion. An analogue of the former is obtained from a strain gauge embodied in the tool handle, while the latter is represented by the integrated surface myogram. Inasmuch as in this specific work situation the area under the myogram is proportional to isometric muscular work, the "biomechanical profile" shows that individuals working in ulnar deviation have to make an effort about six times as large as if the wrist were kept straight while it takes eight times as long to perform each cut. Comparison with clinical findings shows that the "biomechanical profile" is of highly predictive value. Over 60% of all people using a wrist deflecting tool suffer, after some time, from discomfort or disease of hand or wrist (table 1).

Table 1. Difference in incidence of carpal tunnel syndrome (subsequent to 3 months of work under high incentive conditions), between two independent samples of assembly workers trained to hold wire cutters in different ways

	No complaint	Carpal tunnel syndrome	Total
Sample I *Wrist straight*	*52*	*4*	*56*
Sample II *Ulnar deviation*	*16*	*25*	*41*
Total	*68*	*29*	*97*

(Chi-square ≈ 30. Significance level better than 0.001).

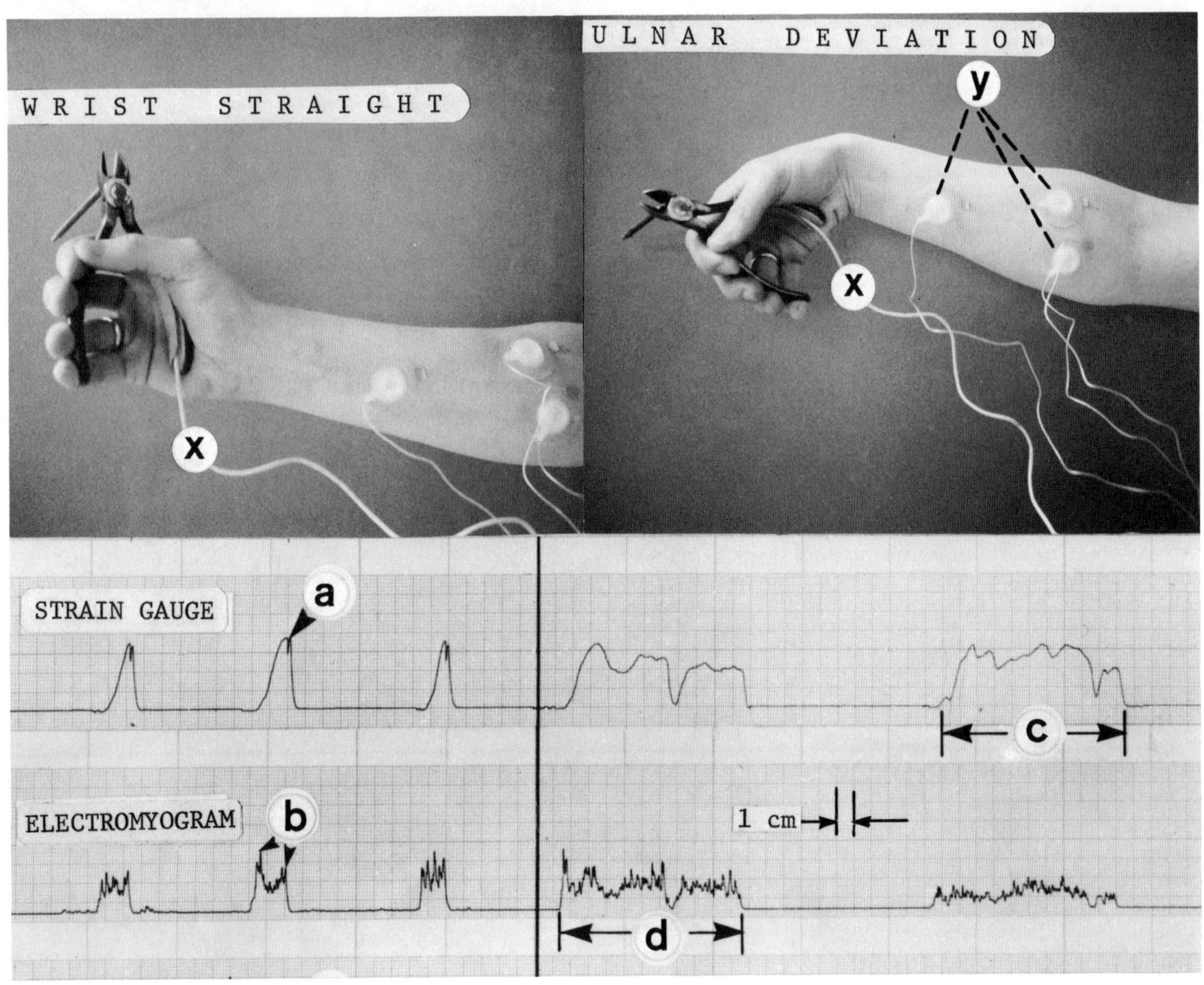

Figure 3. Biomechanical profile of wire cutting. The "profile" shows that ulnar deviation increases the effort required for cutting sixfold. Paper speed 1 cm/s. x. Cable from strain gauge to recorder. y. Electrodes to gather electrical signals produced by flexor muscles to the fingers during exertion. a. Readout from strain gauge (height of tracing proportional to force exerted by tool). b. "Integrated" electromyogram of finger flexors (area under myogram is proportional to muscular effort). c. Readout from strain gauge. d. "Integrated" electromyogram of finger flexors.

Figure 4. At normal work surface height displacement of the wrist can be performed by simple rotation of the upper arm and with little effort. When work surface is 10 cm too high then identical wrist displacement requires full shoulder swing with arm in abduction. Resulting discomfort in muscles of shoulder and chest may mimic symptoms of heart disease.

Example 3. Light-path analysis (chronocyclography), one of the oldest and most reliable procedures in occupational biomechanics, is performed by attaching small flashing light bulbs to select anatomical reference points and, subsequently, photographing the operation. Chronocyclography is of particular advantage in the analysis of seated work situations and the optimisation of chair heights in accordance with the needs of individual workers. A chair only 10 cm too low may force an individual to keep one arm abducted while performing an assembly operation. In this posture displacement of the wrist demands a full shoulder swing, which requires much effort. After a while the individual is likely to experience a sensation of discomfort in those muscles of the shoulder and the chest involved in keeping the arm abducted. This may trigger, especially in overweight individuals, an unwarranted fear of heart disease (figure 4).

Biomechanical manipulators

Knowledge of biomechanics is also applied to the design of manipulators which handle radioactive, infectious or dangerous materials (figure 5). Such apparatus consists of a mechanical "hand" connected to the real hand of a human operator by means of mechanical linkages which pass through protective translucent shielding. Whenever the human hand grasps, manipulates or positions an object outside of the protective enclosure, then the mechanical hand inside performs the same manoeuvre at identical displacement velocity and acceleration. Thus, the operator receives visual feedback identical with the one experienced in normal eye-hand co-ordination. This master-slave technology, while developed from biomechanics, is now generally considered to be in the field proper of "bionics". These devices have made the medical use of potentially dangerous materials, such as

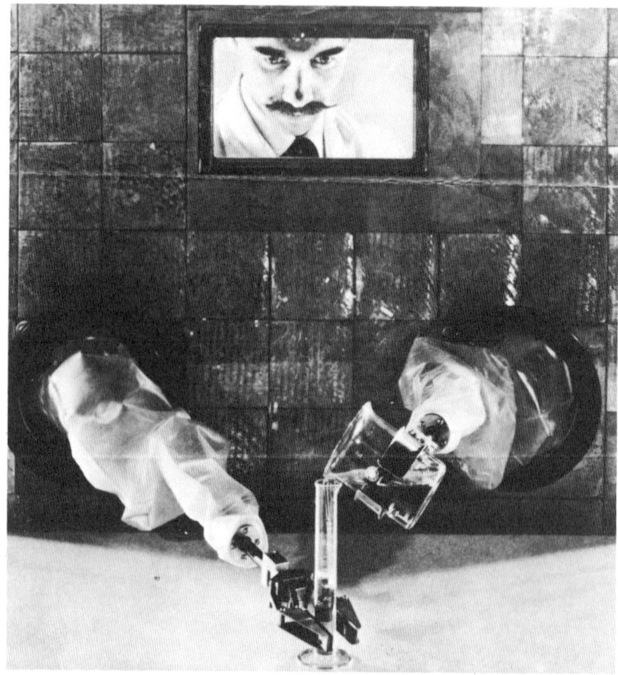

Figure 5. Manipulators for handling radioactive materials. *(By courtesy of the United Kingdom Atomic Energy Authority.)*

radioactive isotopes, a practical feasibility and are now commonly used in medical institutions.

TICHAUER, E. R.

Proceedings of the first international congress of electromyographic kinesiology. Rosselle, N. (ed.). (Louvain, Nauwelaerts Publishing House, 1968), 225 p. Illus. Ref.

Proceedings of the fifth international congress of biomechanics. Komi, P. V. (ed.). (Baltimore, University Park Press, 1976), 549 p. Illus. Ref.

Your posture at work–afflictions of arms, back and legs (Din stallning i jobbet–Belastningar på armar, rygg och ben). Tufuesson, B. (Stockholm, Almqvist and Wiksell Forlag AB, 1977), 102 p. Illus. 87 ref. (In Swedish)

The biomechanical basis of ergonomics–Anatomy applied to the design of work situations. Tichauer, E. R. (New York, John Wiley and Sons, 1978), 99 p. Illus. 86 ref.

Biometrics

Biometrics (Greek *bios* life, *metron* measure) is a widely useful but specialised branch of statistics dealing with living things, especially micro-organisms, plants, animals, and man (see STATISTICS). It uses mathematical and statistical methods to describe and analyse data concerning the variation of biological characteristics obtained either from observation or experiment. It is of particular value and importance in reporting and comparing:

(a) population statistics;

(b) environmental effects;

(c) feeding, nutrition, and growth; and

(d) individual requirements for oxygen and water.

Established some 80 years ago in rather narrow agricultural and genetic fields, biometrics has become an applied science of influential scientific deployment because it allows significant country-to-country, area-to-area, and person-to-person comparisons to be made. It currently takes in such large and specialised subtopics as medical statistics, biostatistics, mathematical genetics, toxicological/potency assays and demography. It also covers important although small segments of agronomy, biochemistry, biophysics, botany, entomology, horticulture, nutrition, pathology, pharmacy, pharmacology, veterinary medicine, virology, zoology and their related branches of the life sciences (table 1).

By contrast, biometrics is relatively underdeveloped in certain important fields such as ergonomics, industrial hygiene, ground-level meteorology, human factors engineering, paediatrics, psychology and geriatrics. These are occupationally important studies where there is scope and requirement for biometrics to be applied, and new information to be generated in ways more sophisticated than variance studies and distribution analysis alone. In the remainder of this article certain key uses of biometrics in occupational health and safety will be described. These are examples of the diffuse larger field: they will convey selected principles of utility and significance whenever further industrial development is planned, and working populations receive industrial health care.

Parameters and indices

The specialised parameters, indices and techniques of principal utility in occupational biometrics are shown in table 2. Most computer centres possess or have access to written programmes for all standard procedures, and it is good professional practice to consult with their programme librarians on availability, size, and applicability of pre-programmed procedures. Effective use of available computer help should be sought:

(a) before the planning of all major biometric surveys, and

(b) on every international project in which country-to-country comparisons are sought or may later be desired.

Table 1. Biometrics as a specialisation in the life sciences

Biometrics	
Class I	Class II
Biostatistics	Agronomy
Demography	Biochemistry
Mathematical genetics	Biophysics
Medical statistics	Botany
Toxicology/potency bioassays	Entomology
Veterinary statistics	Horticulture
Vital statistics	Nutrition
	Pathology
Other numerically based life sciences	Pharmacology
	Veterinary medicine
	Virology
	Zoology
	Other life sciences having quantitation as subsidiary component

Development of unique specialisation ← in computer and statistical techniques

Table 2. Biometric parameters, indices and techniques

Applicable techniques		
Class I	Class II	Class III
Frequencies	All-or-none techniques	Normalisation techniques: Per unit population per square area* per surface area* per unit body per unit muscle mass
Distributions	Dose response curves	
Ranking	Graded response techniques	
Variance weight*	Biological confidence limits	Power law techniques: for temperature* for age*
Significance	Biological modelling	
Regression	Genetic probability	Productivity indices: allowances standard elements* standard times*
Correlation	Growth analysis	
Co-variance	Lethal dose analysis	
Factorisation	Logit analysis	Performance indices: learning curves tracking and response times on target
Matrices	Probit analysis	
Curve fitting	Event analysis	
Discriminants	Mortality techniques	
Transformations		
Other basic procedures statistics	Other standard biometric procedures	Other specialised parameters and indices unique to biometry

* Key indices for normalising data.

Correct selection and use of appropriate parameters and indices greatly increases the biometric value and acceptability of survey findings. Examples of key indices are indicated by asterisks in table 2.

Non-parametric biometrics

An important segment of biometrics involves growth, form, and the presence or absence of specific features or phenomena. Although valuable biometric data can be obtained from dimensional photographs and displays, and from assessing age or maturity, normality or abnormality, the findings so obtained are usually quantitated only in their frequencies, appearances, categorisation or ranking (e.g. specimen x is the eighth largest in a sample of 18). These are non-parametric data. Important examples of non-parametric biometry are in displays of:

(a) body growth;

(b) size and shape;

(c) motion and motion-study including work study;

(d) productivity studies including time study.

Two-dimensional and three-dimensional plots/graphs and conventional, stereo- or time-lapse photographs are specific examples of non-parametric biometric tools from which useful quantitation may result.

Biometric methods

These vary from routine use of mean, range, standard deviation, standard error and fiducial (e.g. 98%) limits to simple and complex analysis of variance, and from calculations of regression and correlation to use of matrices and calculations of percentiles and ED_xs (ED_x is that dose or concentration of a drug, toxin or gas which produces some specified effect in x% of a large test animal or cell population). Human population studies usually embrace tabulation techniques, calculations of distribution, predictions of future population composition, and trend setting by multiple regression techniques.

Clinical trials involve such special techniques as group comparisons, blank (or placebo) treatments and single-blind or double-blind trials. In single-blind trials, doses are known to those administering drugs to patients but, in double-blind trials, neither those in charge of administration nor those examining the patients are aware which person received which dose and consequently reporting bias is reduced. There are many uniquely valuable techniques in medical statistics.

Growth and maturity studies include such techniques as calculation of standards, single and multiple correlations (e.g. height *vs.* age; height *vs.* nutritional level and age; height *vs.* race, nutrition, age and sex), and tests of significant deviation from norms and predicted values. Human environment studies usually involve the statistics of stress response (e.g. rise in heart rate, correlation with cardiac output, regression on body temperature and rate of work, etc.). Graphical, population-to-population displays of stress-strain relationship are based on:

(a) the "ideal" or "average" man or woman in a group;

(b) levels of stress;

(c) adjustment or "normalisation" for different body size (e.g. cubic centimetres of oxygen consumed per minute per kilogram of body weight) and so on.

Likewise, effects of heat and cold are often normalised in terms of increment of temperature shift (e.g. litres per minute pulmonary ventilation per 10 °C shift in air temperature).

Computerisation of biometric data

The scope, data availability and cost feasibility of many biometric programmes have been dramatically improved by use of computers. The modern electronic computer is of third generation type (first generation = thermionic tubes; second generation = printed circuits/transistors; third generation = solid state) and is either digital, analogue, or true analogue-digital hybrid in function. Conventional second generation types, some of which include A-D (analogue to digital) converters, are in use in many countries and are contributing to the speedy handling of biometric data. They are commonly programmed for handling business-type tasks such as record keeping, tabulation and routine costing, and are effective for hospital record keeping, population studies and the like. Larger computers are usually furnished with satellite and peripheral equipment, and can be shared in their use by remote access input and output devices, often using long distance telephone circuitry. In such cases it is common to find biometric data displayed visually as well as by printed table, and such devices as graphical x-y recorders and prepared co-ordinate axis displays on cathode-ray tubes are coming into use by bio-scientists. In general, researchers use analogue or true hybrid computers in preference to digital types. They often plan experimental programmes by mathematical modelling. Controlled test programming and biometric data outputs are sometimes processed on the same computer.

FLETCHER, J.

Biostatistical analysis. Zarr, H. H. (Englewood Cliffs, New Jersey, Prentice-Hall, 1974), 620 p. Illus. Ref.

Articles on biometrics are currently issued in:

Biometrika (University College, Gower Street, London), *Biometrics–Journal of the Biometric Society* (PO Box 5962, Raleigh, North Carolina 2765).

Biotransformation of toxic substances

Within the organism most foreign organic substances and also certain inorganic substances undergo metabolic transformations through catalysis by intracellular and extracellular enzymes. These reactions generally give rise to derivatives that are more polar than the substance originally absorbed, which are thus more easily eliminated from the organism. All tissues (including those of liver, kidneys, intestines, skin and placenta) have the capacity, varying from organ to organ, to metabolise foreign substances, but the main site of biotransformation is the liver, and in particular the parenchymatous cells.

The biotransformation is catalysed, in accordance with the chemical structure of the foreign substance, by enzymes located in different components of the cell (soluble fraction of the cytoplasm, endoplasmic reticulum, mitochondria, lysosome, nucleus, etc.). Many reactions are catalysed by enzymes of the smooth endoplasmic reticulum still known as microsomes when this reticulum is isolated after the tissue has been homogenised and the organelles of the cell fractioned centrifugally.

Principal biotransformation reactions

Reactions of biotransformation undergone by foreign substances in the organism can be placed in two classes: what are known as phase I reactions, mainly oxidation, reduction and hydrolysis, and what are known as phase II reactions, comprising biosynthetic conjugation reactions through which foreign substances or their metabolites resulting from phase I metabolic reactions are coupled with endogenous substrates.

Phase I reactions may themselves be subdivided into two classes: reactions catalysed by microsomic enzymes (enzymes of the endoplasmic reticulum) and those catalysed by non-microsomic enzymes.

Oxidation catalysed by microsomic enzymes

Reactions of oxidation may be presented schematically in the following way:

$$RH + 2e^- + 2H^+ + O_2 \rightarrow ROH + H_2O$$

foreign substance → metabolite

Molecular oxygen and an electron donor are needed for the reaction. One of the atoms of the oxygen molecule is incorporated in the foreign substance and the other is reduced to water. It is for this reason that the enzymes catalysing these reactions are known as mono-oxygenases (MO) or mixed function oxidases.

The detailed mechanism of this reaction has not yet been fully established, but it is known that a haemoprotein inhibited by carbon monoxide known as cytochrome P450 has a central part in the oxidation reactions. The foreign substance to be oxidised attaches itself to the oxidised cytochrome P450 to produce a complex that is progressively reduced and is capable of fixing molecular oxygen. The electrons come mainly from NADPH and are transferred to the cytochrome P450-substrate complex through various intermediates including the cytochrome C reductase still known as NADPH cytochrome P450 oxidoreductase; it is possible that they also come from NADH via cytochrome b5. The dissociation of the activated complex gives rise to an oxidised substrate with regeneration of the oxidised cytochrome P450. The fixation of the substrate on the cytochrome P450 entails spectral modifications. Oxidisable foreign substances have thus been arranged in three classes in accordance with the type of spectrum produced (type I: for example phenobarbital; type II: for example aniline; type III, known as inverted type I: for example acetanilide).

The concentration of a cytochrome P450 varies from one tissue to another, and is generally highest in the liver, which explains, partly at any rate, the great metabolic activity of the liver in respect of many foreign substances.

Cytochrome P450 can itself be inhibited by various foreign substances, with the production of cytochrome P420, a denatured form of cytochrome P450 that is not functional in oxidation reactions.

A few examples follow of oxidation reactions catalysed by microsomic enzymes:

- hydroxylation of aromatic compounds, for example the hydroxylation of benzene into phenol with the formation of benzene epoxide as a highly reactive intermediate metabolite

benzene —MO→ benzene epoxide —non-enzymatic→ phenol

- hydroxylation of aliphatic chains of aromatic compounds, for example the hydroxylation of *n*-propylbenzene into ethylphenylcarbinol:

$$C_6H_5CH_2CH_2CH_3 \xrightarrow{MO} C_6H_5CHOHCH_2CH_3$$

n-propylbenzene → ethylphenylcarbinol

- hydroxylation of alicyclic compounds, for example the production of cyclohexanol from cyclohexane

cyclohexane —MO→ cyclohexanol

- oxidative O-dealkylation, for example the dealkylation of 4-nitroanisole into 4-nitrophenol

4-nitroanisole —MO→ 4-nitrophenol $+ CH_2O$

- oxidative N-dealkylation, for example the demethylation of dimethylformamide into monomethylformamide

$$(CH_3)_2N{-}CHO \xrightarrow{MO} CH_3NH{-}CHO + CH_2O$$

dimethylformamide → monomethylformamide + formaldehyde

- S-dealkylation, for example the demethylation of thioethers into thiols

$$R\,S\,CH_3 \xrightarrow{MO} RSH + CH_2O$$

thioether → thiol + formaldehyde

- oxidative desulphurisation, for example the transformation of parathion (*o,o*-diethyl *o*-(4-nitrophenyl) thiophosphate) into paraoxon

$$O_2N{-}C_6H_4{-}O{-}P(S)(OC_2H_5)_2 \xrightarrow{MO} O_2N{-}C_6H_4{-}O{-}P(O)(OC_2H_5)_2$$

parathion → paraoxon

- dehalogenation, for example the dechlorination of carbon tetrachloride into chloroform.

Oxidation catalysed by non-microsomic enzymes

Reactions of oxidation can be catalysed by enzymes present in mitochondria, the soluble fraction of the cytoplasm or the plasma.

Examples are the oxidation of ethanol (and other alcohols) by alcohol dehydrogenase (ADH) in the presence of NAD and the oxidation of primary amines into aldehydes by an amine oxidase.

$$C_2H_5OH + NAD \xrightarrow{ADH} CH_3CHO + NADH_2$$

ethanol → acetaldehyde

$$RCH_2NH_2 \rightarrow RCH = NH \rightarrow RCHO + NH_3$$

primary amine → aldehyde

Reduction reactions catalysed by microsomic enzymes

Nitro-aromatic compounds are reduced to amines through the formation of hydroxylamine by microsomic enzymes in the presence of NADPH and the absence of oxygen.

Nitrobenzene is thus reduced to aniline

$$C_6H_5NO_2 \rightarrow C_6H_5HN_2$$

Microsomic azoreductases catalyse the reduction of azo-derivatives to amines.

p-Dimethylaminoazobenzene, for example, is reduced to aniline and dimethylphenylenediamine

$$(CH_3)_2N\text{-}C_6H_4\text{-}N{=}N\text{-}C_6H_5 \longrightarrow N(CH_3)_2\text{-}C_6H_4\text{-} + NH_2\text{-}C_6H_5$$

Microsomic enzymes can also dehalogenate foreign substances through a reducing mechanism. For example DDT (1,1,1-trichloro-2,2-*bis*(4-chlorophenyl) ethane) is transformed into DDE (1,1-dichloro-2,2-*bis*(4-chlorophenyl) ethylene)

$$Cl\text{-}C_6H_4\text{-}CH(CCl_3)\text{-}C_6H_4\text{-}Cl \longrightarrow Cl\text{-}C_6H_4\text{-}C(=CCl_2)\text{-}C_6H_4\text{-}Cl$$

DDT DDE

Reduction reactions catalysed by non-microsomic enzymes

The reactions of reduction catalysed by non-microsomic enzymes include:

- the reduction of disulphides to mercaptans;
- the reduction of hydroxamic acids to amides;
- the reduction of sulphoxides to sulphides;
- the reduction of N-oxides to amines;
- the dehydroxylation of hydroxylated aliphatic or aromatic compounds.

Hydrolysis catalysed by microsomic and non-microsomic enzymes

Esterases and amidases present in different components of the cell and in the plasma catalyse the hydrolysis of many esters and amides.

The aliphatic nitriles can be hydrolysed with liberation of the CN^- group.

CH_3CN	→	$HCOOH$	+ CN^-
acetonitrile		formic acid	cyanide

Conjugation

The conjugation of foreign substances or their metabolites with various endogenous substrates generally forms polar derivatives that are more easily eliminated from the organism by the urinary passages or through the action of the bile. The principal reactions of conjugation are the following:

(a) Conjugation with glucuronic acid. Like most reactions of conjugation, glucuronoconjugation takes place in two stages. A glucuronic acid donor (uridinediphosphate-α-D-glucosiduronic acid) is synthetised first. This reaction is followed by the transfer of the glucuronic acid to the foreign substance or its metabolite. The synthesis of the donor cofactor is catalysed by enzymes of the soluble fraction of the cytoplasm and the transfer of the glucuronic acid is catalysed by microsomic enzymes.

Glucuronoconjugates are generally classed as:

(i) *o*-glucuronides formed from phenol derivatives, alcohols (for example dichloroethanol), carboxylic acids and hydroxylamines;

(ii) *n*-glucuronides formed from aromatic amino-derivatives (for example aniline), amides and nitrogen-containing heterocyclic compounds;

(iii) *s*-glucuronides formed from thiol derivatives.

(b) Sulphoconjugation. The donor cofactor is adenosine-3′-phosphate-5′-phosphosulphate, which transfers its sulphate group, when acted on by a sulphotransferase, to a phenol derivative, an alcohol or an amine.

(c) Methylation. A methyl group is transferred from the cofactor S-adenosylmethionine to an amino-derivative, phenol or a thiol group. Certain inorganic bodies, such as selenium, can also be transformed into methyl derivatives *in vivo*.

(d) Acetylation. Coenzyme A is the intermediate in this reaction. Acetylation affects mainly the aromatic amines, the sulphonamides and the derivatives of hydrazine.

(e) Conjugation with glycine. The aromatic acids, such as benzoic acid, conjugate with glycine to form hippuric acid or its homologues. ATP and coenzyme A are the intermediates in this reaction.

(f) Conjugation with glutathione. Various aromatic compounds, such as benzene, naphthalene, polycyclic hydrocarbons, the derivatives of monohalogenated benzene and certain halogenated aliphatic hydrocarbons, conjugate with glutathione to form premercapturic acids, which are conjugates with the *L*-acetylcysteine fraction of glutathione.

(g) Conjugation with sulphur. This reaction concerns quasi-selectively the cyanide radical, which conjugates with the sulphur of thiosulphate to form thiocyanate

$$CN^- + S_2O_3^- \rightarrow CNS + SO_3^-$$

Various biotransformations

Metabolic transformations that cannot be placed in one of the foregoing classes may also take place, for example the opening of rings of heterocyclic compounds or, conversely, the cyclisation of other substances.

Combination of several reactions

The same chemical substances can naturally undergo a series of metabolic transformations in parallel or in series. The metabolism of benzene is an example (figure 1).

It should also be mentioned that in man the intestinal flora in the colon or rectum can also metabolise foreign substances, chiefly those reaching the large intestine (substances that are hardly lipide-soluble and therefore not easily absorbed) or those undergoing an entero-hepatic cycle.

Factors affecting the biotransformation of foreign substances in the organism

The fate of a foreign substance in the organism varies greatly from one individual to another and even, over time, in the same individual. Different factors, both endogenous and exogenous, can affect the metabolic capacity of the organism to deal with foreign substances.

Endogenous factors

These may be divided into genetic factors and physio-pathological factors.

Genetic factors. When they are subjected to similar environmental influences, identical twins metabolise foreign substances in a more or less identical way whereas dissimilar twins show marked differences in

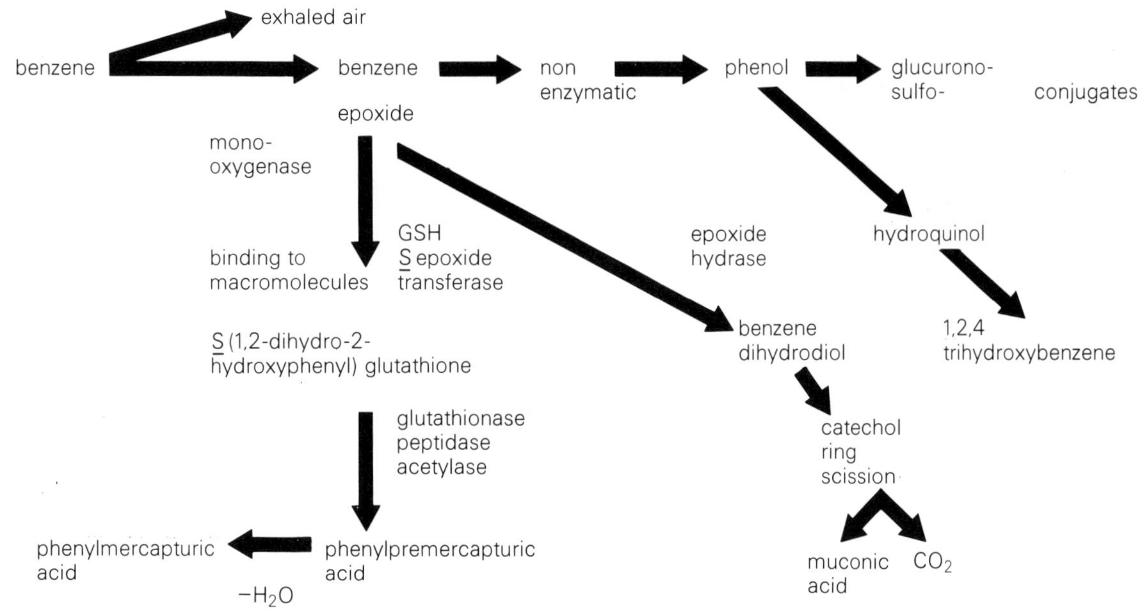

Figure 1. Metabolic biotransformation of benzene *in vivo*.

their capacity to metabolise foreign substances. This shows the importance of genetics in the metabolism of foreign substances.

Physiopathological factors. It has been clearly shown that in animals, and sometimes in human beings too, various physiopathological factors (age, sex, hormones, pregnancy, nutritional state, disease, etc.) affect the biotransformation of foreign substances. For example the microsomic oxidation of foreign substances is reduced among the cirrhotic.

Exogenous factors

It is well known that exposure to various chemical substances can inhibit or stimulate the activity of the enzymes that metabolise foreign substances. This explains certain forms of interaction between toxic substances (antagonism or synergism).

Certain organophosphorus pesticides, for example, inhibit the esterases that inactivate malathion, whose anticholinesterase effect is thus increased *in vivo*. Many chemical substances are inductors of microsomic enzymes: that is to say, they can stimulate the synthesis of new molecules of microsomic enzymes. Induction occurs mainly in the liver, the microscopic examination of which shows a proliferation of the membranes of the smooth endoplasmic reticulum of the cells. Very many substances can act in this way, and their structures vary greatly. They include medicines (barbituric hypnotics, phenylbutazone, etc.), organochlorine pesticides (such as DDT), polycyclic hydrocarbons and certain components of tobacco smoke. Stimulation of the activity of the microsomic enzymes accelerates the biotransformation of various chemical bodies. The intensity of the action is increased or reduced, according as the metabolite produced is more or less active than the original substance.

Consequences of metabolic transformations

It has already been stated that most metabolic transformations tend to make the chemical substance more polar, and so more easily eliminated from the organism (mainly by renal or biliary excretion or both). Generally speaking, metabolic transformations tend to reduce the toxicity of the foreign substance. Examples are the hydrolysis of the organophosphorus ester paraoxon into 4-nitrophenol and diethylphosphate and the conjugation of phenol into phenolglucuronide. It can happen, however, that the biotransformation of a foreign substance gives rise to an intermediate that is more poisonous, the reaction then being known as one of activation. Examples are the transformation of methanol into formaldehyde and the oxidation of parathion into paraoxon, a powerful inhibitor of the cholinesterases.

Importance of the study of metabolic transformations

The study of the biotransformations undergone by foreign substances can be of practical importance in protecting the health of the workers.

The discovery of a highly reactive intermediate in the metabolic chain of a foreign substance may point to a certain hitherto unsuspected toxic activity.

For example the discovery that a highly reactive epoxide can be produced from molecules such as those of vinylidene chloride, 2-chlorobutadiene or even ethylene makes it reasonable to suspect these substances of a mutagenic, and even a carcinogenic, action. Another very important practical application of metabolic studies is to suggest methods for the biological monitoring of workers, though these methods must, of course, first be confirmed by clinical studies. The discovery, for example, of the demethylation of dimethylformamide, first in animals and then in human beings, has led to the suggestion that the excretion of *N*-monomethylformamide in the urine should be measured as a biological method of assessing the intensity of the exposure of workers to dimethylformamide. Clinical studies have confirmed the value of this method, which has the advantage over atmospheric analysis of taking into account the exposure of the workers through all channels and in particular through the lungs and the skin.

LAUWERYS, R.

The biochemistry of foreign compounds. Parke, D. V. (Oxford, Pergamon Press, 1968), 269 p.

Fundamentals of drug metabolism and drug disposition. LaDu, B. N.; Mandel, H. G.; Way, E. L. (Baltimore, Williams and Wilkins, 1972), 615 p.

"Biological criteria for selected industrial toxic chemicals. A review". Lauwerys, R. *Scandinavian Journal of Work, En-*

vironment and Health (Helsinki), Sep. 1975, 1/3 (139-172). 373 ref.

"Drug metabolism under pathological and abnormal physiological states in animals and man". Kato, R. *Xenobiotica* (London), 1977, 7/1 (25-92).

"Biotransformation of organic solvents. A review". Toftgoard, R.; Gustafsson, J. A. Scandinavian Journal of Work, Environment and Health *(Helsinki), Mar. 1980, 6/1 (1-18). Illus. 109 ref.*

Biscuit making

The various undertakings engaged in the industrial production of foodstuffs made from flour and sugar (i.e. bakeries and factories manufacturing rusks, biscuits, wafers, cakes and pastry) have a number of points in common since the raw materials, the manufacturing processes and the production plant are similar from the point of view of occupational safety and health.

The manufacture of sweets or candies is dealt with in the article CONFECTIONERY INDUSTRY and the production of bread is dealt with in the article BAKERIES.

Manufacture. The complexity of the product and the manufacturing processes increases as one moves from bread to pastry goods. Figure 1 shows the various combinations of processes possible.

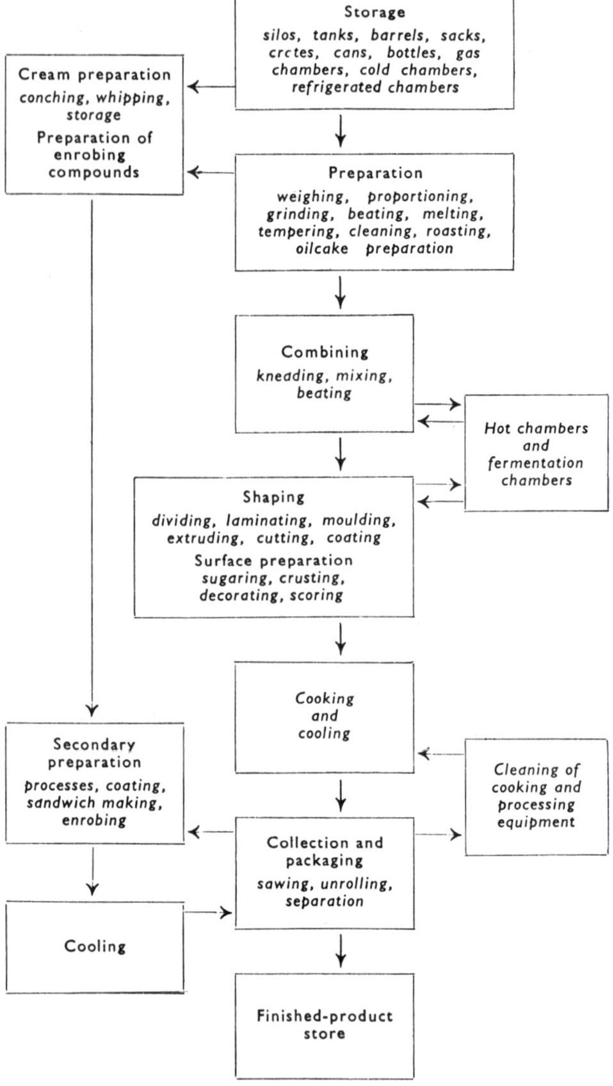

Figure 1. Industrial biscuit manufacture.

HAZARDS AND THEIR PREVENTION

Work premises. The high degree of industrialisation in this sector, demonstrated by a strict division of work processes and a considerable degree of mechanisation, does not mean that the undertakings involved are large in size. Certain factories or factory departments are highly automated.

The various workshops may be classified by their predominant environmental features: cold to very cold; cool and dry; dusty (preparation of raw materials, mixing); normal but tendency to humidity (manufacture); noisy (sugar grinding, cream preparation); and air-conditioned.

Raw materials. In certain cases, some of the raw materials are stored in silos (flour, sugar) or in tanks (fats) from which they are despatched by pneumatic conveyors or pumps to the point of use. In other cases, all or a large part of the raw materials are stored and handled in sacks, crates, drums, cans, cartons, or glass or plastic bottles. The weight of a single container and its contents may be as high as 200 kg.

The raw materials may have specific hazards or the hazard may be related to the processing equipment or to the lifting and carrying operations.

Manufacturing stages. The individual ingredients are first prepared separately and, subsequently, brought together in the finished product.

In the preparation and proportioning of ingredients, the accident hazards are usually those of bruises, cuts and abrasions caused by blunt, sharp-edged and pointed tools or open containers. The health hazards are due to airborne flour and sugar dust (rhinitis, asthma, eye diseases, dental caries) and to allergies resulting from contact with certain spices and essences. Crates and boxes made from exotic woods or arriving from tropical countries may be the cause of various health disorders.

Sugar is ground down to a fine powder using sugar mills which may function according to different principles but which all produce a large amount of noise with a considerable high-frequency component. Mill tools, discs, wheels and gear trains must be fully enclosed or guarded. The in-feed and out-feed points of certain machines may be the source of large amounts of airborne dust. Lifting and carrying of heavy sacks may be arduous.

Fats are plastified in fat mills, in heat exchangers or directly in a mixer (creaming). Fruit may be received whole, pulped, puréed or dry; dry fruit is sorted, washed, dried in a centrifuge and, where necessary, chopped. Nuts are usually received ready-husked and are then gassed immediately or periodically in special chambers, sorted on machines of which the needle-type may be hazardous, scalded and, in the case of almonds, blanched on grooved-roller machines, knife-grated or machine-grated, roasted (in the case of hazel nuts), ground in roller-, plate- or knife-type attrition mills, and cooked in open-flame or water-jacketed or steam-jacketed boilers, ready for the making of marzipan or nougat. These mixtures are then ground once again unless they are to be used in confectionery preparations, and the handling of these hot mixtures entails a considerable hazard of burns and scalds.

The prepared materials may undergo many further processes such as kneading, mixing, beating, moulding, fermentation, laminating, surface finishing—all of which may be preceded or followed by cooking or refrigeration.

These processes may be accompanied by the appearance of pests such as cockroaches and rodents and, in summer, workers may be exposed to bites from venomous insects attracted by the foodstuffs.

Kneading machines, mixing machines and beating machines may cause severe injuries; the first two types are particularly dangerous when the dough bowl is being scraped. Kneading machines should be fitted with guards that allow additional materials to be inserted or dough samples to be removed during mixing. The dough bowls are emptied by means of spatulas and the contents tipped by hand into feed hoppers. Although it presents no postural problems, this work may be quite arduous.

The hoppers feed the dough into dough-rolling and dividing machines which produce either a continuous ribbon of dough or individual dough pieces. Safe working practices should be strictly enforced for dough-rolling and dough-dividing machines, otherwise workers' hands may easily be crushed or mutilated.

Fermenting chambers and some hot or cold chambers are equipped with ultraviolet bactericidal lamps which may produce eye disorders. The loading of scales is controlled by synchronised systems and entails the danger of nipped or broken fingers.

Rolling machines may be continuous, reversible, alternating or linked in series, and are normally protected by interlocked guards, as are dough dividers. Nevertheless, workers may inactivate or scotch this protective equipment and thus introduce an increased danger of crushed hands or fingers.

Hoppers for liquid and semi-liquid batters are fitted with piston-type volumetric proportioning pumps which may cause finger cuts.

Wire-type cutters, slicers, circular saws for wafers and rusk cutters should be suitably guarded. When the stock is hand-fed, a push stick should be employed. Brushes for cleaning grills, baking trays, belts and moulds have a very vigorous action and should, therefore, have no sharp or projecting edges. Brushes made from animal bristle or goose feathers may cause puncture wounds which readily become infected.

Biscuit-making firms often build their own machines such as rotary tart machines, forming machines, stripping machines, etc., to overcome specific production problems. In these prototype machines, safety devices or guarding are often inadequate or completely non-existent.

Products on trays or waffle plates or in moulds are baked in a continuous oven which is gas- or oil-fired or electrically heated, either directly or indirectly. Each of these heating systems has its own specific hazards. In the event of an electricity failure, there is a fire hazard since the feed belts will come to a halt, the products in the oven will overheat and may burst into flames when they exit from the furnace after the belt has been restarted. Even when ovens are well lagged, they still give off enough heat to raise the workplace temperature to 40 °C. There is a definite burn hazard with certain types of baking equipment or operation. During baking, the dough or batter gives off large quantities of steam which is used to obtain certain special effects in the product. Overcooked items may have a tendency to burst into flames spontaneously.

Mould-stripping and the handling of trays and moulds necessitates the wearing of gloves which may cause maceration of the hands. The hand-scoring of loaves before they enter the oven may result in cuts to the hands. Considerable amounts of pressurised fluids such as steam and hot water are employed, which entails dangers of burns and scalds.

The most important hazard associated with cleaning equipment and operations is that of falls. Workers employed in equipment washing rooms are exposed to hot, humid atmospheres. Firms which sell their products in returnable metal boxes or use metal boxes for internal conveying are often equipped with special washing equipment and metal-working presses for repair work.

In the wrapping and packaging department, many of the hazards are similar to those in the paper industry. Materials such as waxed paper, corrugated paper, laminated aluminium foil, cellulose, acetate foil, etc., are usually supplied in rolls weighing up to 50 kg. The handling of these rolls may be a difficult task, especially for women. Other hazards are presented by heat-sealing equipment, knives and pushers, loaders and chains, in addition to all the other moving parts of an automatic machine. Boxes may also be crimped.

Many jobs which are done by women entail tiring postures. Some of these jobs can be automated; in many cases, the chassis of conveyor belts can be redesigned to allow persons to adopt a standing, sitting or leaning position as desired.

Team work on production or packaging belts necessitates adequate equipment control if accidents are to be avoided. Biscuit-making firms usually operate one or two day shifts, except for certain jobs done by men.

Workers should wear close-fitting garments and adequate foot and leg protection; jewelry should be removed during work and long hair should be retained in a cap or scarf.

Manual lifting and carrying of materials is a frequent cause of back pain; in-plant transport equipment (such as fork-lift trucks) presents traffic problems especially where doors are concerned. Correct stacking techniques should be employed.

This sector of industry should be subject to the strict general hygiene requirements applicable to all food processes.

VILLARD, R. F.

CIS 76-1747 *Occupational safety and health in bakeries and confectioneries* (Arbeitsschutz in Bäckereien und Konditoreien). Schall, J.; Kriems, P. (West Berlin, Verlag Tribüne, 1975), 62 p. Illus. (In German)

CIS 78-2054 *Safety requirements for bakery equipment.* ANSI Z50.1-1977 (American National Standards Institute, 1430 Broadway, New York) (1977), 24 p. 30 ref.

CIS 77-46 "Guard for kneading machines" (Protecteur pour pétrins). *Travail et sécurité* (Paris), Sep. 1976, 9 (410-412). Illus. (In French)

CIS 77-1837 *Dough dividers.* Guidance Note PM6, Health and Safety Executive (London, HM Stationery Office, 1976), 10 p. Illus.

CIS 77-1250 *Directives for baking ovens* (Richtlinien für Backöfen). ZH 1/553, Hauptverband der gewerblichen Berufsgenossenschaften (Cologne, Carl Heymanns Verlag, 1976), 12 p. (In German)

Bismuth and compounds

Bismuth (Bi)

a.w.	209
sp.gr.	9.8
m.p.	271 °C
b.p.	1 560 °C (760 mmHg)

a pinkish-silver, hard, brittle metal, superficially oxidised by air.

Bismuth telluride:

TLV ACGIH	10 mg/m^3
STEL ACGIH	20 mg/m^3

Se-doped bismuth telluride:

TLV ACGIH	5 mg/m^3
STEL ACGIH	10 mg/m^3

Occurrence. In nature, bismuth occurs as the free metal and in ores such as bismutite (carbonate), bismuthinite (double bismuth and tellurium sulphide), where it is accompanied by other elements–mainly lead and antimony.

Production. Bismuth is a by-product of lead, copper and tin ore refining. It is removed from the other metals by crystallisation, by the Betterton-Knoll process or by electrolysis.

Uses. Bismuth is used in metallurgy for the manufacture of numerous alloys, especially alloys with a low melting point (Wood, Darcet, etc.). Some of these alloys are used for welding.

Bismuth telluride is used as a semiconductor. Bismuth oxide, hydroxide, oxychloride, trichloride and nitrate are employed in cosmetics. Other salts are used medicinally: the organic salts (succinate, orthoxyquinoleate, etc.) possess bactericidal properties that are used in the treatment of anginas; butylthiolamate is an antisyphilitic derivative; the slightly soluble mineral salts (subnitrate, carbonate, phosphate) are employed in gastroenterology as antacids, adsorbents and for protecting the mucous membranes.

HAZARDS

The toxicity of bismuth depends on the way in which it enters the organism. The organic derivatives attack the kidneys and, to a lesser degree, the liver. They cause alterations of the convoluted tubules and may result in serious, and sometimes fatal, nephrosis.

The insoluble mineral salts, taken orally over prolonged periods in doses generally exceeding 1 gram per day, may provoke brain disease characterised by mental disorders (confusional state), muscular disorders (myoclonia), motor co-ordination disorders (loss of balance, unsteadiness) and dysarthria. These disorders stem from an accumulation of bismuth in the nerve centres which manifests itself when bismuthaemia exceeds a certain level–estimated at around 50 μg/l. In most cases, bismuth-linked encephalopathy gradually disappears without medication within a period of from ten days to two months, during which time the bismuth is eliminated by the urinary passages. Fatal cases of encephalopathy have, however, been recorded.

These neuropsychic effects have been observed in France and Australia since 1973. They are caused by a factor not yet fully investigated which encourages the absorbtion of bismuth through the intestinal mucous membrane and leads to an increase in bismuthaemia to a level as high as several hundred μg/l.

[Inhalation in animals of insoluble compounds such as bismuth telluride provoke the usual lung response of an inert dust. However, long-term exposure to bismuth telluride "doped" with selenium sulphide can produce in various species a mild reversible granulomatous reaction of the lung. Therefore for the "doped" compound the ACGIH has recommended a TLV of 5 mg/m^3.

Some bismuth compounds may be the origin of dangerous chemical reactions. Bismuth pentafluoride decomposes on heating and emits highly toxic fumes.]

Bismuth and its compounds do not appear to have been responsible for poisoning associated with work. It is therefore looked upon as the least toxic of the heavy metals currently used in industry. The danger of encephalopathy caused by breathing in metallic dust or oxide smoke is remote in the extreme. The poor solubility of bismuth and bismuth oxide in blood plasma and its fairly rapid elimination by the urinary passages (its half-life is about 6 days) argue against the likelihood of a sufficiently acute impregnation of the nerve centres to reach pathological proportions.

SAFETY AND HEALTH MEASURES

Classic preventive measures (extraction of dust and smoke at source and ventilation of places of work) combined with elementary hygiene (washing of hands before meals and a ban on the eating of food in the place of work) should suffice virtually to eliminate the risk of poisoning at work.

In the event of serious exposure caused by the technical failure of preventive measures, the level of bismuth in the blood and urine indicates the degree of toxic impregnation of the personnel so exposed.

BOITEAU, H. L.

CIS 75-1340 "Hazardous chemical reactions–28. Arsenic, antimony and bismuth" (Réactions chimiques dangereuses–28. Arsenic, antimoine, bismuth). Leleu, J. *Cahiers de notes documentaires–Sécurité et hygiène du travail* (Paris), 1st quarter 1975, 78, Note No. 950-78-75 (129-130). (In French)

"Bismuth and encephalopathy" (Bismuth et encéphalopathie). Galland, M. C.; Rodor, F.; Jougland, J. *Médecine et Hygiène* (Geneva), 15 Aug. 1979, 37/1342 (2579-2582). 52 ref. (In French)

Bladder cancer

Rehn, in 1895, reported that three men reporting at his clinic with bladder cancer had each worked at the same factory producing magenta (fuchsin) from commercial aniline. A fourth bladder cancer patient had worked on the same process elsewhere. Because bladder cancer is a relatively rare disease, Rehn deduced that these tumours were job-related and called the disease "aniline cancer", a misnomer since it is now known that aniline is not a causative factor. During the succeeding 50 years, there were many reports of occupational bladder cancer in the chemical industry in many different countries. Those directly concerned with the disease, such as factory medical officers, came to the conclusion that aniline, benzidine, and 1- and 2-naphthylamine were the likely causative agents.

The first definitive epidemiological survey was conducted by Case and his colleagues (1954) into bladder cancer mortality in selected factories in the British chemical industry. By comparing exposure to specific aromatic amines (job histories) with mortality from the disease, they showed that the manufacture and use of benzidine, and 1- and 2-naphthylamine, but not aniline, were responsible for the occupational cancers. Additional studies demonstrated that the manufacture of auramine and magenta from aniline were associated with bladder cancer mortality. At the same time, the induction of bladder tumours in those working with 4-aminobiphenyl was described. These chemicals are the only ones clearly associated with occupational bladder cancer at this time. These observations have been confirmed in 1956 and 1967. It is also firmly established that human bladder cancer is associated with cigarette smoking and certain drugs, such as 2-naphthylamine mustard (Chlornaphazin), cyclophosphamide, and probably phenacetin.

Human infection by *Schistosomum hematobium*, a bigenetic trematode, leads to the desposition of ova in the urothelium. The consequent disruption of the urothelium may be associated with the development of bladder cancer. As, particularly in Egypt, *S. hematobium* infection is prevalent in agricultural workers, these tumours may be regarded as occupational in origin (see SCHISTOSOMIASIS).

Occupational exposure

Case and his colleagues assembled a population of 4 622 men who had worked between 1920 and 1950 for more than six months in selected factories in the British chemical industry. There were 127 death certificates mentioning cancer of the urinary bladder among these men, whereas only 3-5 would have been expected from mortality statistics for an age-matched segment of the male population of England and Wales. Working in the chemical industry increased the risk of death from bladder cancer thirty- to fortyfold. Examination of job histories revealed subpopulations that had worked with only one of the suspected chemicals. Working with aniline alone revealed a slight, but statistically significant, excess of risk from bladder cancer that disappeared when those using aniline to manufacture auramine and magenta were excluded. The manufacture of auramine and magenta was independently shown to present excess risk. Using men exposed to only one of the other suspect chemicals, with or without exposure to aniline, it was demonstrated that 2-naphthylamine and benzidine were potently carcinogenic with a latency from the time of entry into the industry to tumour diagnosis of about 18 yr. 1-Naphthylamine presented a lesser risk and a latency of about 22 yr. Five years' exposure was needed for tumour induction by 1-naphthylamine, compared to 6 mth with 2-naphthylamine. However, the 1-isomer as manufactured at that time in Britain contained 4-10% of the 2-isomer, so the carcinogenicity of "pure" 1-naphthylamine cannot be considered established.

The only other established occupational bladder carcinogen is 4-aminobiphenyl. The substance is clearly a potent bladder carcinogen since 55 of 315 workers developed tumours. Tumour latency was not established. The natural history of 4-aminobiphenyl-induced occupational tumours was also followed cytologically and it was clearly demonstrated that in some cases the precancerous stage (cytologically positive; cytoscopically negative) lasts for many months in some patients.

The chemical industry makes chemicals to be used by other industries. The carcinogenic aromatic amines have induced bladder tumours in workmen in several user industries. The first to be identified was the United Kingdom rubber industry, which used as a rubber compounding ingredient Nonox S, a complex of 1- and 2-naphthylamine and acetaldehyde. The first indication of hazard came from Case's study of the British chemical industry. Case originally intended to use morbidity rather than mortality data and to match the bladder cancer incidence in the chemical industry against the morbidity in males in a large industrial city without a chemical industry. Choice of the city of Birmingham, England, demonstrated that one area of the city had a large excess of bladder cancer that was traced to the use of Nonox S by the local rubber industry, and the epidemiology of bladder cancer in this industry was repeated in 1954. British electric cable makers used rubber for insulation and also demonstrated an excess risk from bladder cancer. Nonox S appears only to have been used in Britain. The United States rubber industry, for example, does not present as well defined a bladder cancer hazard, although it is believed that cases of the occupational disease have occurred.

Demonstration of the increased bladder cancer risk in the British rubber industry led in 1959 to the abandonment of the use of the then suspect human bladder carcinogens by the industry. A retrospective-prospective survey of all workers in the British rubber and electric cable making industries was set up with the primary objective of determining whether elimination of the hazardous materials had, in fact, removed the excess bladder cancer risk. The names of all persons making rubber, with the date of entry into the industry and job descriptions, were determined and death certificates for each member of this population were collected when they died. Mortality data for those men exposed before the abandonment of the hazardous aromatic amines were compared with those for men who entered the industry after 1949 and with mortality in the general population of England and Wales. It is still too early to be sure that the excess risk from bladder cancer has been completely removed. Further analysis of this data base should give useful information.

Textile dyers and other textile processers have been shown to present an excess risk from bladder cancer. Many textile dyestuffs are made by diazotising aromatic amines, such as benzidine or its congeners, and coupling with suitable phenols or amines. Thus, textile dyers may be exposed to carcinogenic aromatic amines through their presence as impurities in the dyestuffs or through the practice of removing the dyestuff reductively from misdyed fabric, leading to the regeneration of the amine. Much more important, however, is the ability of human gut flora and the widely distributed tissue enzyme, azoreductase, to split the dyestuff reductively to liberate two aromatic amino-groups from each azo-group. Reduction by the gut flora is of special importance since it may split highly polar dyestuffs that otherwise would not be significantly absorbed from the gastrointestinal tract. The widely used direct dyes are made, for example, from benzidine or its congeners. Liberation of benzidine in the gastrointestinal tract means that ingestion of these dyes results in the same consequences as the ingestion of benzidine itself. Bladder cancer among Japanese kimono painters who now use benzidine-based dyestuffs instead of the traditional natural pigments has been described. The practice of pointing their brushes between the lips leads to the ingestion of considerable quantities of dyestuffs, liberation of free benzidine, and the consequent incidence of bladder cancer.

Other examples of occupational bladder cancer have been sought by reviewing, as far as is possible, all cases of bladder cancer in large city populations. These populations are then matched for age, sex and various other characteristics with members of a control (non-bladder cancer) population. The frequency with which specific occupations occur in cancer or control patients gives some feeling for possible risk inherent in such occupations. The small numbers working in a given occupation, even when a population of 1 000 or more bladder cancer patients is assembled, means that such studies can do little more than indicate the need for proper epidemiologic surveys of particular industrial environments. Moreover, differences in industrial practice and employment patterns in different localities means that results obtained in one city may not be comparable with those found in another. Among occupations suspected to be at risk as a result of such surveys were those engaged in hair care (exposed to aromatic amines in hair dyes) and medical personnel (who may have previously used benzidine for occult blood in urine or stools).

It has also been suggested that rodent operatives using alpha-naphthylthiourea (ANTU) contaminated with the beta-isomer may be at risk from bladder cancer. This needs to be substantiated.

Exposure levels and latency

There is little information about the level of exposure of workpeople to 4-aminobiphenyl, benzidine or 2-naphthylamine in the occupational environment. In some processes, such as the flaking (powdering) of freshly distilled 2-naphthylamine, gross exposures must have occurred and 95% of workmen engaged in this process developed bladder cancer. An appreciation of the level of tumour response at lower levels of exposure would be valuable, but is not available. Most epidemiological surveys followed the example of Case and his colleagues and have taken the number of years worked in the hazardous occupation as the most meaningful index of exposure.

Case and his colleagues determined the mean latency from time of entry into the industry to clinical recognition of bladder cancer to be 18 yr for 2-naphthylamine and benzidine, and 22 yr for 1-naphthylamine. However, they noted that from the mortality and morbidity trends they calculated the number of instances of the disease would approximately double before the lifespan of this population of 4 622 men was complete. The extra cases would occur in an ageing population and consequently lengthen the latencies reported in 1954. The 262 instances of the disease included bladder cancer patients that were still living, bladder cancer deaths that were certified, and bladder cancer patients that had died from other causes and did not have bladder cancer recorded on their death certificates.

A firmer appreciation of the latency of induced bladder cancer may be derived from the use of the drug, 2-naphthylamine mustard, to treat polycythaemia. Doses of from 200-350 g led to the dose-related induction of bladder carcinomas in 2.5-11 yr after the start of treatment. It is not completely clear, however, whether these short latencies are due to the high doses of aromatic amines used, to the cytotoxic mustard moiety, or to other treatment modalities used in the management of polycythaemia.

PREVENTIVE MEASURES

The most satisfactory form of prevention of aromatic-amine-induced bladder cancer is to seek safer alternatives for 4-aminobiphenyl, benzidine, 2-naphthylamine, and other carcinogenic amines. Much progress has been made in this area since the high potency of these chemicals to the human urinary bladder was firmly established. Much 2-naphthylamine, for example, was formerly used to prepare 2-naphthylamine sulphonic acids for dyestuff manufacture. The alternative procedure is to sulphonate 2-naphthol, which may then be converted to 2-naphthylamine sulphonic acids without use of the potent bladder carcinogen, 2-naphthylamine as an intermediate. 2-Naphthylamine sulphonic acids are generally regarded as not being carcinogenic, although there is a lack of data to support this.

If a hazardous aromatic amine is to be manufactured or used, the basic principle must be its complete containment by the most advanced engineering techniques. Special attention must be paid to protecting maintenance personnel who have to enter the plant if it malfunctions. Such manufacture should, if possible, be limited to processes involving the complete use of the carcinogen within the closed environment. The ultimate product should contain as little as possible of the free carcinogenic aromatic amine, and should not be capable of metabolism to the aromatic amine itself, or its proximate carcinogen derivatives.

Clinical aspects

Occupational bladder cancer is generally considered to be clinically and pathologically indistinguishable from the naturally occurring disease. Therefore, its management will not differ from that of other bladder cancer patients having an equivalent tumour.

Those exposed to 4-aminobiphenyl, benzidine or 2-naphthylamine differ from the general population in being at high risk of bladder cancer. The need for early diagnosis to give therapy the best chance of success has led to the introduction of screening techniques. Originally, visible or occult haematuria was the screen of choice, but blood in the urine may signify the presence of a well established tumour. Periodic cytoscopy is traumatic to the patient who will often default from the programme. Urinary cytology, the examination of cells that have desquamated from the urothelium into the urine, represents one of the best available approaches for populations at high risk of bladder cancer. It is not entirely satisfactory, however, for screening low risk populations (e.g. with naturally occurring tumours) because the proportion of false positive results that leads to a demand for cytoscopy is too high. There is a need for new and more exact screening techniques.

In a follow-up of 503 workers who had been exposed to 4-aminobiphenyl, there were 435 with negative urinary cytology, 59 with positive cytology, and 9 with cytology of doubtful significance. Thirty-five of 59 patients with positive cytological findings developed bladder carcinoma, 7 retained positive cytology without tumour at the time of the report, 7 died from other causes and 10 were lost to follow-up. Negative cytology concealed two cases with bladder carcinoma and one with papilloma.

Legislation

Some nations have prohibited the manufacture and use of one or more of the aromatic amine human bladder carcinogens; others have attempted to regulate their manufacture and use. These two strategies are well illustrated by the approach to legislation in the United Kingdom and the United States.

The United Kingdom made occupational bladder cancer a prescribed industrial disease in 1953 (Statutory Instrument, 1953). This meant that workpeople who knew and could prove that they had been exposed to the noxious aromatic amines could claim industrial compensation. These people were also able to obtain compensatory damages in court if they could demonstrate that their employer knew, or should have known, that the claimant had been exposed to the specific human bladder carcinogens. In 1967 the United Kingdom Government proceeded to make an order to prohibit the manufacture and use of 4-aminobiphenyl, 4-nitrobiphenyl, 2-naphthylamine and, except under carefully defined conditions, benzidine, or their salts. The use of substances containing 1% or less of the prohibited substances was strictly controlled with mandatory medical surveillance, including urinary cytology for those employed (Statutory Instrument, 1967).

The United States approached the problem quite differently. It was decided to regulate use of carcinogens in order to minimise the exposure of the workmen. Of 14 substances declared to be human carcinogens, 7 were aromatic amines or related compounds, namely: 4-aminobiphenyl, 4-nitrobiphenyl, benzidine, 3,3'-dichlorobenzidine, 4,4'-methylene-*bis*(2-chloraniline), 2-naphthylamine, and 2-acetylaminofluorene (Department of Labor, 1974). 2-Acetylaminofluorene is, in fact, only an experimental carcinogen as its use as a pesticide was prevented by the timely demonstration of its carcinogenicity (Wilson et al., 1941). The US Department of Labor is now moving to regulate all human and animal carcinogens in the workplace (Department of Labor, 1980).

It is clear that both the UK and the US approaches to occupational cancer are needed. While all carcinogens in the workplace may be regulated to reduce worker exposure to the minimum feasible, all carcinogens cannot be prohibited. On the other hand, the difficulty of restraining chemicals such as 2-naphthylamine or 4-aminobiphenyl from entering the occupational environment is such as to make their continued use of questionable merit.

Conclusions

It cannot reasonably be doubted that exposure to 4-aminobiphenyl, benzidine or 2-naphthylamine leads to bladder cancer in exposed workpeople. The manufacture of auramine and magenta presents a bladder cancer hazard. While it is clear that 1-naphthylamine, containing 2-naphthylamine as an impurity, is carcinogenic to the human bladder, there is no convincing evidence that 1-naphthylamine free from the 2-isomer is hazardous. The relatively few human bladder carcinogens identified among the aromatic amines contrasts markedly with the large number of these chemicals that have been shown to be carcinogenic to the bladder or other tissues in laboratory animals. It is pertinent to consider that most aromatic amines have been studied in man only for their ability to give bladder cancer. There is a need for wider ranging epidemiologic studies on worker populations exposed to other aromatic amines.

CLAYSON, D. B.

"The epidemiology of bladder cancer: a second look". Wynder, E. L.; Goldsmith, R. *Cancer* (Philadelphia), 1977, 40 (1246-1268).

"Carcinogenic aromatic amines and related compounds" (366-461). Clayson, D. B.; Gardner, R. C. *Chemical Carcinogens.* ACS monograph No. 173. Searle (ed.) (Washington, DC, American Chemical Society, 1976).

"Cancer of the urinary bladder in Finland. Association with occupation". Tola, S.; Tenho, M.; Korkala, M. L.; Järvinen, E. *International Archives of Occupational and Environmental Health* (West Berlin), Apr. 1980, 46/1 (43-51). 11 ref.

"A survey of occupational cancer in the rubber and cable-making industries: results of five-year analysis". Fox, A. J.; Lindars, D. B.; Owen, R. *British Journal of Industrial Medicine* (London), Apr. 1974, 31/2 (140-151). Illus. 41 ref.

"Bladder cancer due to exposure to para-aminobiphenyl: a 17-year followup". Melick, W. F.; Naryka, J. J.; Kelly, R. E. *Journal of Urology* (Baltimore), 1971, 106 (220-226).

Etiology of bladder cancer metabolic agents (31-39). Yoshida, O.; Miyakawa, M. *Analytic and experimental epidemiology of cancer.* Nakahara, W. et al. (eds.) (Baltimore, University Park Press, 1973).

Blasting and shotfiring

Blasting and shotfiring are processes of fragmenting or loosening solid materials such as rock, earth or masonry be means of an explosive charge. The normal sequence consists of drilling a hole, inserting a charge, stemming (covering the charge with a dense material to prevent dissipation of the explosive force) and firing by means of a detonator or fuse.

Mine and quarry blasting. Many types of explosive are used in various ways for breaking up minerals or rock either on the surface or underground to permit their excavation. The principal types of explosives used are black blasting powder (gunpowder), special explosives for safe use in coal mines (permitted explosives), gelignites, dynamites, liquid oxygen and ammonium nitrate/fuel oil mixtures (AN/FO) either in powder or slurry form. Accessories used in connection with initiation of explosives are safety fuse (having a regular burning speed), plain detonators, electric instantaneous and delay detonators, detonating relays, detonating fuse (does not burn but detonates at high velocity), igniter cord and connectors, exploders (blasting machines), circuit testers, shotfiring cables, detonator crimpers and fuse igniters.

Most explosives are initiated in one of two ways: gunpowder is exploded by lighting a safety fuse in contact with it while most other explosives are fired by means of a detonator. Plain detonators are initiated by a safety fuse, electric detonators by an electric current from an exploder, and a detonating fuse by either a plain detonator and safety fuse or an electric detonator. Igniter cord is used for lighting unlimited numbers of safety fuses in desired rotation. Delay detonators are used for sequence firing, for giving good fragmentation and for minimising ground vibrations. Detonating relays are delay devices for use with detonating fuse.

Blasting is basically either primary or secondary. Primary blasting breaks solid rock or mineral *insitu* while secondary blasting further reduces in size material from primary blasting, which is too large for crushing plants. The three ways in which explosives are used are:

(a) for primary blasting in boreholes (figure 2);

(b) for primary blasting of bulk charges in underground chambers–heading or coyote blasting (figure 1); and

(c) for secondary blasting either in boreholes or by surface charges–plaster shooting or mudcapping (figure 3).

Most blasting work is by borehole charges. Boreholes may range from about 2.5 to 38 cm in diameter and up to

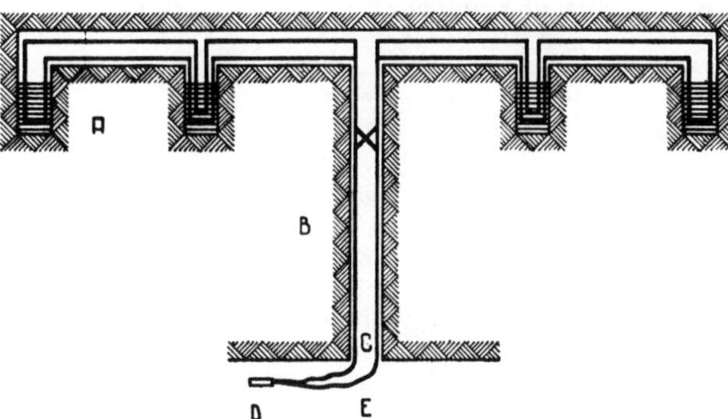

Figure 1. Heading (primary) blasting: plan. A. Chamber containing explosive. B. Stemming. C. Detonating fuse. D. Detonator. E. Entry from quarry floor.

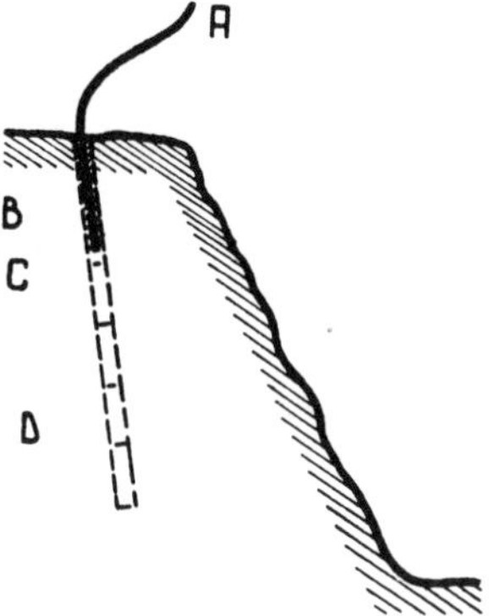

Figure 2. Primary blasting: simple borehole charge. A. Fuse. B. Stemming. C. Detonator inserted in explosive (primer cartridge). D. Explosive cartridges.

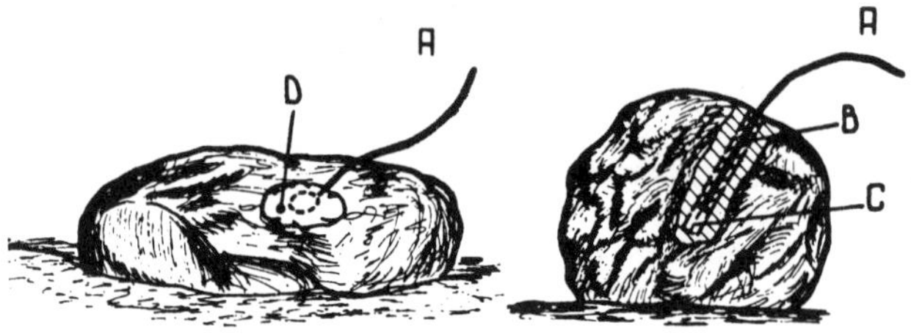

Figure 3. Secondary blasting. *Above:* surface charge; *below:* borehole charge. A. Fuse. B. Stemming. C. Charge. D. Charge covered with clay.

30 m in depth. Some explosives are specially packaged to suit the more common borehole dimensions. Both borehole charges and heading blasts require stemming. Stemming is the name given to materials packed into boreholes or tunnels after the explosive has been loaded. Normally rock fines or sand or mixtures of sand and clay are used to fill boreholes, while stones are used for heading blast tunnels. Stemming when well tamped (closely packed) ensures maximum efficiency from the explosive. [Water stemming for shotholes presents some advantages because it can reduce blasting fumes and dust concentration provided that the plastic ampoules filled with water do not leave free space for the escape of gases along the walls of the shothole. Stemming with sodium chloride fed in with compressed air is also recommended instead of clay stemming.]

An experienced engineer should be responsible for the design of heading blasts or the more complicated borehole patterns, including size, depth, angle and spacing of boreholes and the types and quantities of explosives required. Shotfirers should follow any expert guidance available and aim at safety and economy in the use of explosives.

Other blasting. Explosives are used extensively for civil engineering works such as dam construction, railway, road and water supply tunnels, reservoirs, road cuttings and drainage ditches, well blasting, footings for foundations and rock excavation for levelling construction sites. They are used for the breaking up of blast furnace hearths, breaking and cutting of steel plate and girders, metal forming, demolition of buildings, factory chimneys, bridges, etc. Explosives are used underwater for blasting channels, removing sandbanks and rocks and dispersal of wrecks which are a hazard to navigation. They are necessary for seismic prospecting on land or underwater and are used in agriculture for general land clearance and removal of tree stumps, boulder breaking and stream diversions.

HAZARDS AND THEIR PREVENTION

The hazards represented by using explosives in mines and quarries depends on their consumer use in the extraction of minerals and their potential danger of detonation due to the chemical compound or construction of explosives. The danger of accidental detonation can, in principle, arise in every aspect connected with the application of explosives, their means of transport, storage, distribution and handling, and while in use. The hazards resulting from these actions and safety measures are contained in national safety regulations for mining enterprises. Although these regulations differ from country to country, their common goal is to make the use of explosives as safe as possible. The most common causes of accidents and respective safety measures are as follows.

During transportation

(a) Use of transport vehicles which do not sufficiently protect the explosives against shocks, friction, collisions, light or sparking;

(b) the failure to clear roads of other traffic while transporting explosives;

(c) employment of untrained or undisciplined workers to load, transport and unload explosives.

Safety regulations relating to the transport of explosives in most of the countries exploiting minerals are strict. Among these is the provision that explosives should always be transported in specially designed vehicles bearing a special sign or inscription entitled "DANGER–EXPLOSIVES". Detonators should be separated from other explosives by transporting them in separate compartments or, in the case of small quantities, an effective barrier (e.g. sandbags) should be erected to separate them from other explosives. The route followed, either on the surface or underground, by transport carrying explosives should be cleared of all other traffic and any other works obstruction. The distance between vehicles carrying explosives and the speed of vehicles or trains should be limited to safety values (e.g. in underground mines, not less than 10 m between two manually operated wagons, and the speed of underground trains not greater than 10 km/h).

When storing, distributing and handling

Accidental detonation in this case is caused by:

(a) use of unsuitable rooms for storing explosives;

(b) lack of systematic control of wiring systems, fire-alarm systems, store closing devices and stray currents;

(c) lack or bad state of safety dams in surface stores, and of fire-bridges in underground storage;

(d) careless handling, unwrapping and distribution of explosives.

To avoid the hazard of accidental detonation, the greatest care is required when storing, distributing or handling explosives. Safety regulations state, therefore, that stores of explosives should be placed in dried and intermediate layers, well ventilated and protected from water, not nearer than 100 m from shafts and at least 20 m from any mechanical or electrical devices. The area surrounding stores of explosives should be kept clear of inflammable debris. Separate compartments should be provided for detonators. Boxes with explosives should never be opened inside a store but at least 15 m distant from it. All activities in the store should be carried out by authorised workers, and entry to the store should be restricted to the storekeeper and supervisors.

During use

Accidents occur most often during the use of explosives:

(a) when preparing and loading the blast holes;

(b) as a result of failure to withdraw fast enough from the blast area;

(c) when returning to blasting points too soon after firing;

(d) as a result of entry by strangers to the blasting point.

Many injuries are caused by the premature detonation of explosives or by misfires. Premature detonation may be caused by over-vigorous tamping of charges or friction of metal on a tamping rod (only wooden tamping rods should be used); it may also result from using too short a fuse. In electrical shotfiring, premature detonation can be caused by making connections to electrically charged lines, by accidental firing while persons are still at blasting points, by electric storms, stray currents, static electricity and even by electromagnetic radiation from nearby radio, television or radar installations. To ensure properly managed shooting works, the most conscientious and properly selected personnel should be employed. The selection of candidates for shotfirers should be the responsibility of the mines supervising body.

Care should be taken that the blasting point is free of detonating gas, inflammable objects, sparking or damaged wiring systems, stray currents and static electricity, and is well protected from the unexpected entry of incompetent personnel. Adequate shooting technology and composition of explosives (e.g. the right amount required to charge the explosive holes, the use of proper fuses) must conform to local mines' geological conditions and applied mining technology. In the case of blasting underground, no person should return to a blasting point until all smoke and fumes have dispersed. [The composition and quantity of poison gases produced during shotfiring underground depends upon the composition of the explosives, the consistency and composition of the rocks and other conditions. When ventilation is poor the rocks can adsorb a part of gases. The most dangerous gases produced are nitrogen oxides and carbon monoxide; large quantities of nitrogen, carbon dioxide and water are present. The ventilation should therefore be adapted to the nature of the explosive and may vary, for instance, between 1 500 m^3 of air per kg of nitroglycerine to 100 m^3 of air per kg of potassium chloride.]

Both below and above ground, no work should be resumed after firing until all blasting points have been carefully checked to ascertain if there have been any misfires or if the surroundings have become unsafe as a result of blasting. If there is any doubt that not all of the blasting charges have exploded, return to blasting points should not be authorised for 15 to 30 min (the period varies and depends on national safety regulations). Misfires often provoke unexpected blow-ups and account for about 30% of all fatal accidents and injuries resulting from explosives.

Training and qualification of shotfirers

In many countries, government authorities require persons using explosives either:

(a) to hold a qualification of a recognised professional or technical institution which covers training in the handling, storage and use of explosives; or

(b) to possess a shotfirer's certificate granted by the authorities on successful completion of an examination.

Government organisations sometimes conduct shotfirers' training courses while some large commercial undertakings arrange such courses for their own employees. In a few countries, knowledge is simply passed on individually by older to new employees but this is not to be recommended because unsafe practice may be perpetuated in this way. Shotfirers' certificates are in some countries valid for a fixed period of years after which they may be renewed at the discretion of the authorities, which also have the power to revoke or suspend certificates for breaches of safety regulations or other good reasons. Only mature, intelligent and stable persons should be selected for shotfiring work.

Legal requirements

Laws relating to explosives have as their object the minimising of injury to persons and of damage to property. Many countries have laid down legal requirements concerning the following matters:

(a) safety conditions for explosives manufacture (including on-site mixing of AN/FO);

(b) import and export controls;

(c) provision of special port-handling areas for cargoes of explosives (to minimise the effects of any accidental explosion);

(d) packing and labelling (see DANGEROUS SUBSTANCES, LABELLING AND MARKING OF);

(e) construction and labelling of vehicles for carriage of explosives by road;

(f) carriage of explosives through road tunnels, by rail or on inland waterways;

(g) location and construction of storage magazines (e.g. the United States and the United Kingdom have standard tables of distances from buildings, etc., recommended for storage of various quantities of explosives);

(h) keeping of accurate stock books and maintenance of security at storage magazines;

(i) qualifications for employment as a shotfirer;

(j) minimum length of safety fuse and maximum number of fuses to be lit by one person;

(k) procedures for dealing with misfires;

(l) control over ground vibrations for protection of property.

KNAPP, J. H.
ZAJACZKIEWICZ, R.

CIS 79-1443 "Stemming of explosive charges" (Verdämmung von Sprengladungen). Dubsky, M.; Vogel, H. *Schweizer Bauwirtschaft–Journal suisse des entrepreneurs* (Zurich), 18 May 1979, 78/25 (13-15 German) (50-53 French). Illus. (In French, German)

CIS 78-2014 *Safety recommendations for ammonium nitrate based blasting agents–Revision of Information Circular 8179.* Damon, G. H.; Mason, C. M.; Hanna, N. E.; Forshey, D. R. Information Circular 8746 (Bureau of Mines, 4800 Forbes Avenue, Pittsburgh) (1977), 31 p. 48 ref.

"Two potential safety problems which may occur when blasting big boreholes underground". Day, P. R.; Webster, W. *Canadian Mining Journal* (Westmount), July 1980, 101/7 (22-26). Illus.

CIS 75-1442 "Blasting fumes in coal mines" (Sprengschwaden im Steinkohlenbergbau). Graefe G. *Glückauf* (Essen), Apr. 1975, 111/8 (368-374). Illus. 18 ref. (In German)

CIS 76-1088 "Neurological and EEG signs of acute poisoning from shotfiring fumes in uranium mines" (Akutní oitrava odstřelovými zplodinami v uranových dolech v neurologickém a elektroencefalografickém obraze). Kůel, V. *Pracovní lékařství* (Prague), May 1975, 27/5 (156-160). Illus. 31 ref. (In Czech)

CIS 78-1861 "Physical air foam for control of dust and harmful gases" (Vozdušno–mehaničeskaja pena dlja pogloščenija pyli i vrednyh gazov). Abramov, F. A.; Zadara, V. M. *Bezopasnost' truda v promyšlennosti* (Moscow), Mar. 1978, 3 (28-29). Illus. (In Russian)

"Keep your blasting under control". Dick, R. A.; Siskind, D. E. *Coal Age* (New York), Nov. 1979, 84/11 (104-108). Illus. 3 ref.

Blastomycosis

Definition. Blastomycosis is an infection of humans and lower animals caused by a dimorphic fungal agent, resulting in pulmonary, systemic, or cutaneous lesions.

There have been two different types described: North American blastomycosis and South American blastomycosis. South American blastomycosis is caused by the fungal agent *Paracoccidioides brasiliensis*, which exists primarily in the tropical and subtropical forests of Brazil, but also may be seen in other South and Central American countries. South American blastomycosis is primarily a cutaneous disease of the oral mucosa, facial skin, and perianal mucosa. This disease seems to be primarily the result of using leaves and twigs which are naturally contaminated with the mould conidia to clean teeth and gums, and for other hygienic practices. Although most of these cases are in the rural population, the disease seems to be more of a cultural and environmental disease than an occupational disease, and thus will not be considered further here.

North American blastomycosis, also called Gilchrist's disease, is primarily a systemic and sometimes cutaneous disease of mammals caused by the fungus *Blastomyces dermatitidis*. This disease exists primarily in the United States and Canada, but a few cases have been reported in Mexico, Central America, and in Africa. In many ways it resembles histoplasmosis and coccidioidomycosis, the two major systemic mycotic diseases in the United States. Infection can range from a mild self-limiting disease to an acute or chronic infection localised in the lungs, to an acute or chronic systemic infection with symptoms varying according to the organs affected. Lungs and skin are affected most frequently but any organ system may be involved.

Occurrence. The fungal agent *B. dermatitidis* is a soil organism with very limiting environmental parameters for growth. Although the specific growth requirements are not known, growth limitations include factors of altitude, soil type, forest type, and general climate. The organism is often isolated from wood or soil where animals had been kept in previous years. The amount of manure or bark in the soil seems to be important to growth of the organism. Even though *B. dermatitidis* has been found to grow well in avian or bat guano, this does not seem to be an essential growth factor as it is for *Histoplasma capsulatum*. The geographic distribution of *B. dermatitidis* is similar to *H. capsulatum*, i.e. the greatest incidence is in the Ohio and Mississippi River basins.

The source of infection for agricultural workers is through soil contact. Inhalation of conidia-laden dust is the major route of transmission. Preparing soil for planting of crops is a major activity where close contact with soil is present. The cleaning or remodelling of old animal structures is another activity with exposure potential. Clearing of wooded areas for crop production may also involve exposure. Several mammalian species (including dogs, horses, and cats) can be infected in addition to man, but they are dead-end hosts.

The majority of reported cases have been a widely disseminated chronic granulomatous disease with pulmonary and cutaneous involvement. Most cases of blastomycosis in man occur sporadically. However, one outbreak reported in the state of Minnesota involved members of four families who were engaged in building a cabin in a wooded site. Of 21 persons exposed, three were severely ill, and four had a clinically mild disease of short duration. An additional 11 persons had radiographic and serologic evidence of infection, but were asymptomatic. This report suggests that many sporadic cases may be of the mild self-limiting form. Many of these latter cases could easily go undetected, thus providing false information on prevalence of the disease.

Symptoms. The lesions in man are primarily associated with skin and lungs. It was previously thought that skin was a primary route of infection, but more recent studies indicate that the majority of infections are by inhalation of conidia, and that cutaneous infections are secondary. On rare occasions cases have been reported of subcutaneous inoculation of spores.

The illness is similar to histoplasmosis in that it is primarily a respiratory infection, which may disseminate, and the symptoms have a wide range of severity depending on dose and host immunity. Many infections are subclinical. In acute respiratory infections, fever, cough, bloody or purulent sputum and chest pain are the major symptoms. A secondary disseminated infection may occur, involving skin, bones, genito-urinary tract and reticuloendothelial system. The specificity and degree of symptoms depend on degree of infection and organs involved.

Diagnosis. Diagnosis of blastomycosis is made on the basis of compatible lung and skin lesions, a productive cough, history of exposure to soil in an area endemic to the disease, and laboratory indications of the presence of *Blastomyces*. The diagnosis is confirmed by isolation of the organism from sputum culture.

Serology may also be used to help confirm the diagnosis. Agar-gel precipitation and complement fixation tests are most commonly used. A titer rise over a ten-day to two-week period is indicative of a recent infection. Skin testing with blastomycin has limited diagnostic value.

PREVENTIVE MEASURES

Since humans are infected through inhalation of conidia-laden dust, preventive measures include protection against potentially contaminated dusts. For example, respiratory protection should be used when cleaning out old farm structures.

Treatment. Amphotericin-B is the drug of choice for treatment. In certain instances surgical resection of chronically infected tissues may be indicated.

DONHAM, K. J.

"Blastomycosis I. A review of 198 collected cases in veterans administration hospitals". Busey, J. F. et al. *American Review of Respiratory Disease* (New York), May 1964, 89/5 (659-672). Illus. 23 ref.

"Pulmonary blastomycosis". Wasrath, S. P.; Atal, P. R.; Kishare, B.; Arora, R. C. *Journal of the Indian Medical Association* (Calcutta), 16 Aug. 1971, 57/4 (144-145). 5 ref.

"Occupational hazards from deep mycoses". Schwarz, J.; Kauffman, C. A. *Archives of Dermatology* (Chicago), Sep. 1977, 113/9 (1270-1275). 96 ref.

CIS 77-825 "Isolation of blastomyces dermatitidis from pigeon manure". Sarosi, G. A.; Serstock, D. S. *American Review of Respiratory Disease* (New York), Dec. 1976, 114/6 (1179-1183). Illus. 7 ref.

"The gamut of paracoccidioidomycosis". Restrepo, A.; Robledo, M.; Giraldo, R.; Hernandez, H.; Sierra, F.; Gutierrez, F.; Lopez, R.; Calle, C. *American Journal of Medicine* (New York), July 1976, 61/1 (33-42). Illus. 30 ref.

Bleaching and bleaching agents

Bleaching agents are substances or mixtures which have the ability chemically to remove dyes or pigments that exist naturally in a material or that have been added to it in an industrial process. Bleaching agents are widely used in the treatment of cellulose, in the paper and paper pulp industry and the textile industry.

Their action is based on the principle of oxidation and the main active agents are chlorine, calcium hypochlorite, sodium hypochlorite, Javel water (potassium hypochlorite solution), chlorine dioxide, sodium chlorate, sodium peroxide, detergents, etc. In certain circumstances, use is made of chemical compounds which give the bleaching agent the specific degree of acidity or alkalinity needed for the chemical reaction, or which themselves take part in the reaction during which active oxygen is released.

[Cotton bleaching is effected by means of sodium or calcium hypochlorites, sometimes sodium chlorite (for quick action) or hydrogen peroxide at 0.5-1 volume %. Silk and wool bleaching is usually effected by exposure to vapours of sulphur dioxide or by immersion in hydrogen peroxide at 1-2% volumes for 12 h.

Optical bleaching is based on the property of some substances (usually toluylene derivatives) to absorb ultraviolet radiations of the visible light and to emit a bluish light, masking the yellow colour of the textile fibres. Such substances are more and more currently used in commercial detergents.]

Chlorine

This is the most important of the bleaching agents. The hazards and safety and health measures relating to this element are dealt with in the article CHLORINE AND INORGANIC COMPOUNDS.

Chlorinated lime

CHLORIDE OF LIME, CALCIUM OXYCHLORIDE, BLEACHING POWDER

A relatively unstable chlorine carrier in solid form, it is a complex chemical compound of indefinite composition, presumably consisting of varying proportions of calcium hypochlorite, calcium chlorite, calcium oxychlorite, calcium chloride and free calcium hydroxides. Commercial chlorinated lime usually has a content of 24-37% available chlorine. It is a white or greyish-white powder with a strong odour of chlorine. It is produced by reacting lime with chlorine.

Chlorinated lime is extremely hygroscopic and, on exposure to moisture, it rapidly decomposes with a very intense evolvement of active chlorine. When mixed with flammable substances and heated to a temperature of 100 °C, it may react explosively. The presence of catalysts such as iron, manganese, etc., may lead to the development of dangerously high pressures in closed vessels, especially if the temperature of the substances is high. The toxic effect of chlorinated lime is due to its chlorine content.

During the production, storage and use of chlorinated lime, steps should be taken to ensure that this substance is not mixed with organic materials or metallic catalysts; rapid heating and storage in moist areas should also be avoided. During the use of chlorinated lime, adequate exhaust and general ventilation is necessary to prevent the formation of dangerous concentrations of chlorine gas.

Sodium hypochlorite (NaClO)

Potassium hypochlorite (KClO)

Both these substances are highly unstable; however, the hypochlorite ion in aqueous solutions is remarkably stable and these aqueous solutions are available under such names as Javel water, chlorosol, chlorox, etc.

Sodium hypochlorite is produced by passing chlorine through a solution of sodium or, electrolytically, from sodium chloride. Apart from its use as a bleaching agent in the textile industry, Javel water is used in the manufacture of synthetic fibres, aniline dyes and as an oxidising agent.

Sodium hypochlorite is an extremely strong oxidising agent which releases chlorine when heated to a temperature above 35 °C. Its toxic effect is due to the presence of chlorine. It has a pronounced irritant effect on the skin and the use of Javel water for hand cleaning may produce skin disorders, and systematic use of this product may produce a generalised toxic effect.

Chlorine dioxide (ClO_2)

See CHLORINE AND INORGANIC COMPOUNDS.

Miscellaneous bleaching agents

A further important group of bleaching agents comprises substances which liberate oxygen. The most common are sodium peroxide (Na_2O_2), hydrogen peroxide (H_2O_2), sodium perborate ($NaBO_3.4H_2O$), sodium percarbonate (CO_4Na_2 or $C_2O_6Na_2$), sodium orthophosphate, sodium metaphosphate and sodium polyphosphate, etc.

Among the bleaching agents used domestically are alkaline salts, especially sodium salts, which are marketed in mixtures with inert fillers. The most important material is sodium carbonate (Na_2CO_3).

Sodium peroxide and hydrogen peroxide are the most dangerous of these substances. When mixed with organic substances they may be subject to explosive reactions; they may cause chemical burns of skin and mucous membranes, although burns caused by hydrogen peroxide are less serious than those caused by sodium peroxide.

Sodium carbonate is an alkaline substance which is an irritant to the skin, respiratory system, eyes and even the digestive system. It may produce eczema and skin necrosis.

SAFETY AND HEALTH MEASURES

See ALKALINE MATERIALS; CHLORINE AND INORGANIC COMPOUNDS; HYDROGEN PEROXIDE; OXIDISING SUBSTANCES.

LUKANOV, M.

General:

CIS 80-2056 "Burns due to caustic droplets in the textile industry" (Brûlures dues aux gouttelettes caustiques dans l'industrie textile). Caisse nationale française de l'assurance-maladie. *Cahiers de notes documentaires–Sécurité et hygiène du travail* (Paris), 3rd quarter 1980, 100, Note No. 1276-100-80 (Recommendation No. 175) (439-440). (In French)

CIS 79-1462 "Safety and health in the textile industry–1. Occupational safety and health in general–2. Machines and installations for bleaching, dyeing, printing, finishing" (Hygiène et sécurité dans l'industrie textile: 1. Mesures générales de prévention–2. Machines et installations de blanchiment, teinture, impression, apprêt). Editions INRS No. 575 et 580 (Institut national de recherche et de sécurité, 30 rue Olivier-Noyer, Paris) (1979), 74 and 142 p. Illus. 18 ref. (In French)

Hypochlorites:

CIS 74-1627 "Hazardous chemical reactions–13. Hypochlorous acid and hypochlorites" (Réactions chimiques dangereuses–13. Acide hypochloreux et hypochlorites). Leleu, J. *Cahiers de notes documentaires–Sécurité et hygiène du travail* (Paris), 1st quarter 1974, 74, Note No. 887-74-74 (139-141). (In French)

CIS 76-26 "Fire hazards of calcium hypochlorite". Clancey, V. J. *Journal of Hazardous Materials* (Amsterdam), Sep. 1975, 1/1 (83-94).

CIS 77-1225 *Sodium hypochlorite.* H52, Information sheets on hazardous materials (Fire Protection Association, Aldermary House, Queen Street, London EC4N 1TJ) (1976), 2 p.

Blind workers

Rehabilitation of newly blinded persons is one of the prime functions of welfare services for the blind and is an essential factor in ensuring that a person who loses his sight due to an accident or disease is rapidly reintegrated into normal social and occupational life.

Rehabilitation programmes

These commence by restoring the patient to full physical and psychological health and then utilise past educational and professional capacities and skills to return to society a person who is well balanced, able and resourceful and who can continue his life with dignity, usefulness and happiness within his social framework.

Basic training consists primarily in instruction in the use of Braille, in typewriting, handling of special machines and appliances designed for the blind and in learning to become independently mobile once again. At the end of this basic training period, those blind who cannot return to their former jobs or similar employment are given the opportunity of undergoing vocational training in the workshops of rehabilitation centres or of studying in special institutes for the blind or in normal educational establishments.

Sound basic and vocational training must be backed up by first-class representation to an employer and by job placement by a competent vocational guidance officer, especially where there is built-in resistance amongst employers to the employment of handicapped persons; regular visits from this officer may be needed subsequently to ensure that the worker is consolidated in his job. The resettlement of the blind is easier in industrial than in rural areas.

Work for the blind

The blind can be trained for a wide variety of jobs, and progressive thinking in this field is based on the concept that a blind person can, in broad terms, be trained to do the type of work he might have been expected to undertake before losing his sight.

In predominantly agricultural areas, such as in the developing countries, blind persons are frequently trained for agricultural work such as animal husbandry (pigs, poultry), market gardening (fruit, vegetables, flowers) and tree nursery work. The traditional craftsman and handicraft work for the blind such as chair caning, carpet weaving, basketware, knitting, etc., is still taught but is nowadays reserved primarily for blind workers who, due to old age or additional handicaps, are unable to do other work.

Figure 1. These blind persons are doing electrical work under the supervision of an ILO expert who is himself blind.

Workshops for the blind, which now have a long history, still have an important part to play. In former times, workshop organisers or agencies for the blind were required to provide premises, tools, materials, markets and even supplementary financial assistance for the support of blind workers; modern workshops, however, provide advanced training in an industrial setting for blind persons who will ultimately be transferred to open industry and, at the same time, they offer permanent employment to those who cannot be employed outside the workshop's "sheltered" environment. Many of these workshops are, in fact, factories equipped with modern machinery, manufacturing basic products and handling subcontract work along strictly industrial lines.

In countries where rehabilitation of the blind has made the greatest progress, many blind persons are employed in open industry along with sighted colleagues and earn equal wages. Increasing numbers of blind persons are now finding work in commerce and offices doing clerical, administrative or professional work. Rehabilitation centres and vocational training establishments for the blind offer courses in shorthand and typing, switchboard operation, office administration, shopkeeping, physiotherapy, piano tuning and music. Recent statistics have shown that, in certain countries, up to 10% of the blind population is employed in commercial, administrative and professional occupations.

The list of occupations in which blind persons can be successfully employed is impressive and includes: writing journalism, social welfare work generally and social work for the blind in particular, engineering, general management, personnel administration, shopkeeping, administration, architecture, music, teaching and lecturing, the law, public relations, the Church, medicine, librarianship, secretarial work, accountancy, interpreting and computer programming. New employment opportunities are still being developed.

Safety of blind workers

The basic requirements for the safe employment of blind workers are suitable training and the avoidance of heavy and crude work which would hinder the development of compensatory faculties; low background noise at the workplace is also important since the blind worker relies, to a large extent, on his ears to obtain much of the information other people receive through their eyes. Where manual assembly work or work on machines is done, finger and hand protection is essential, the workbench should be of limited size, and controls and tools should be ergonomically positioned and distinguishable by their shape.

Where possible, visual warning signals should be replaced by acoustical signals, sharp or projecting parts on machines should be rectified and, above all, efficient maintenance and good housekeeping should be practised to ensure that falls and collisions are avoided.

The adaptation of a metal-working lathe for use by a blind worker in an automobile plant showed that the following factors were of prime importance:

(a) the chuck should be readily accessible, easy to operate and should not harbour swarf;

(b) the parts to be produced by the blind worker should be small, distinctively shaped and easy to insert and remove;

(c) the danger area around moving parts should be protected by interlocked machinery guarding;

(d) protection should be provided against flying swarf;

(e) all-enclosed machine design with easy-to-reach and foolproof controls should be adopted;

(f) shielding should be installed against cutting-oil spray.

It can, in fact, be seen that the majority of these factors are applicable to any safe and ergonomically designed operation.

International and national legislation. Statistics show a remarkable progress in the employment of the blind in countries where special laws or regulations for the occupation of disabled persons are in existence. The ILO Vocational Rehabilitation (Disabled) Recommendation, 1955 (No. 99), deals with the principles and methods of vocational guidance, vocational training and placement of disabled persons and mentions three main methods of promoting the employment of the disabled, including the blind:

(a) the quota system by which undertakings of a certain size are required to employ a given number of disabled persons;

(b) reservation of certain designated occupations for disabled persons;

(c) preference in certain occupations to seriously and multiple handicapped persons.

Legal measures of this kind have been introduced in many countries and have proved most valuable in the placement of the blind, especially in times of unemployment. In addition to stipulating preferential placement of the disabled, these laws also provide for special equipment of the workplace by the employer, the maintenance and restoration of working capacity, protection from dismissal, supplementary holidays and further privileges.

Partially sighted workers

The above principles apply in almost their totality to the partially sighted worker, i.e. persons suffering from cataract, glaucoma, macular lesions, myopia, optic atrophy, *retinitis pigmentosa*, diabetic retinopathy and amblyopia.

In many countries special institutions separate from the centres for the blind cater for training of the partially sighted, and the job-placement officers dealing with disabled persons also assist in the employment, follow-up and after-care of the visually handicapped.

GEISSLER, H.

"Technical development and its part in the rehabilitation of the blind". Garland, C. W. *Rehabilitation* (London), Jan.-Mar. 1976, 96 (8-11).

"Workshop on low vision mobility". Apple, L. E.; Blasch, B. B. *Bulletin of Prosthetics Research* (Washington, DC), 1976, 10/26 (46-138).

The prevention of blindness. Report of a WHO study group. Technical report series No. 518 (Geneva, World Health Organisation, 1973), 18 p. Ref.

Blood diseases, occupational

Although benzene, ionising radiation and lead are the most common causal agents in occupational blood diseases, technological development has led to a significant increase in the number of occupational haemotoxic agents; moreover, the list of these agents is steadily growing. Since these agents are of varied nature (chemical, physical or biological), their haemotoxic activity differs.

Aetiology

When evaluating the effects on the blood of occupational exposure, it is necessary to consider:

(a) the haemotropic properties of the occupational agents, and the personal factors;

(b) the specific features of the work station.

The haemotropic property depends on the specific characteristics of a toxic substance, its physicochemical properties and its route of entry into the body. In addition, one should not overlook the significance of age, sex, race and, in particular, acquired or hereditary personal predisposition.

The specific features of the work station are related to the presence and concentration of haemotoxic substances in the environment, time of exposure, effectiveness of the ventilation system or even use of personal protective equipment.

The hazard is particularly increased during work such as the degreasing of metal components, dissolving and reprocessing rubber, the manufacture of solvents for glues, paints and varnishes, the use of printing inks (heliogravure and photogravure), dry cleaning, etc. Other activities which may be the source of agents producing occupational blood diseases include: the leather industry, textile industry, woodworking industry, battery manufacture, work in the vicinity of sources of ionising radiation or carbon monoxide, etc.

Types of occupational blood impairment

The blood disease may present itself as a reduction or, in certain rarer cases, an increase in the number of blood cells in the peripheral blood. In the former instance the cause may be of central origin and in this case the pathological process is located in the bone marrow (central or medullar cytopoenia), or if cell lysis occurs in the blood, the origin of this process is peripheral (peripheral cytopoenia).

Another aspect of the effect of haemotoxic agents may be seen indirectly in blood coagulation disorders and result in either haemorrhagic syndromes or, on the other hand, a tendency to thrombosis.

If carbon monoxide or nitro or amino compounds are found in the work environment, there may be denaturation of haemoglobin and the formation of new compounds (carboxyhaemoglobin, methaemoglobin).

Immunoallergic cytopoenia may also occur (immune thrombopoenia and leukopoenia and, more rarely, immune haemolytic anaemia). In spite of the occupational significance of this type of blood disorder, the lack of adequate documentation makes it impossible to make any precise interpretation.

Bone marrow reaction due to haemotoxic agents: myelopathies

Benzene and ionising radiation head the list of occupational agents having selective action on bone marrow.

Hypoplastic or aplastic myelopathies (medullar or central cytopoenia). The essential feature of this type of myelopathy is the dominant suppression of one blood cell series whereas the others may be more or less unaffected. In the latter case, the clinical picture is of septic or febrile haemorrhagic syndrome which leads rapidly to generalised debility. The course of hypoplastic conditions of the haematopoietic system is slow and may be characterised by repair, followed by stabilisation or, in some cases, aggravation.

Hyperplastic myelopathies. In the case of bone marrow hyperplasia, it is the white cells that are often affected;

red cell involvement is exceptional. Persistent leukocytosis may provide valuable information for early detection. Hyperplasia which is of a transitory nature has a favourable prognosis whereas persistent hyperplasia, especially where there is transformation towards a leukaemic state, tends to be fatal in spite of remissions of varying duration.

Effect of haemotoxic agents on the blood (peripheral cytopoenia)

When a haemotoxic agent acts directly on components circulating in the blood, the number of blood components may be reduced as a result of direct cell lysis. The dominant feature of this type of cytopoenia is that it affects only one series of blood cell (usually red cells) but never two or three at the same time. Onset is sudden and is accompanied by a dramatic clinical picture, and the course of the disease is generally short.

Haemolytic syndrome. Intravascular haemolysis is a consequence of the direct action of haemotoxic substances on red blood cells and leads to their degradation and the transfer of haemoglobin into the plasma. In industry the typical example of this process is seen in the haemolytic crisis caused by arsine; aromatic amines and nitro compounds (nitrobenzene, aniline and trinitrotoluene), lead, and alkaline chlorates also have haemolytic properties but of lower intensity.

Persons with an enzyme deficit are particularly prone to haemolytic crisis; this deficit usually relates to red cell enzymes (glucose-6-phosphate dehydrogenase deficiency) or in some cases, haemoglobin constitution.

Coagulation disorders (coagulopathies)

These diseases may take the following forms.

Hypocoagulability. Usually manifested by signs of a haemorrhagic syndrome (ecchymosis, purpura, petechiae, epistaxis, haemorrhage, etc.). It results from a reduction in the activity of certain coagulation factors (V, VII, VIII, IX and others) or a lack of fibrinogen (hypofibrinogenia), or an insufficiency of thromboplastin production, or a rise in the level of factors which prevent coagulation (coagulation inhibitors).

Hypercoagulability. This may be seen in the formation of localised coagulation (thrombosis, a tendency to emboli) or generalised coagulation (disseminated intravascular coagulation).

Among the factors that may affect the coagulation process, mention should be made of the liver which has a complex role to play in the synthesis of certain coagulation proteins. Severe toxic involvement of liver tissue by carbon tetrachloride, carbon disulphide or certain medicinal products may lead to coagulation disorders. Factors dependent on vitamin K (II, V, IX, X) may be reduced and fibrinogen synthesis inhibited. Parallel to this, haemopathies which are accompanied by quantitative or qualitative changes in platelets may contribute to coagulation disorders and predisposition to a haemorrhagic syndrome. For example in the case of toxic bone marrow effects due to benzene or toluene diisocyanate, it is possible that haemorrhagic thrombocytopoenia may appear as a dominant haematologic sign, and occur during the process of haemostasis. It has also been shown that DDT, ethanol and ionising radiation (in the final stage of changes) may also be involved in the onset of haemorrhagic syndrome.

Qualitative changes in haemoglobin

This blood component has the property of moving readily from one form to another (oxyhaemoglobin ←→ haemoglobin ←→ reduced haemoglobin) and has a strong affinity for carbon monoxide (CO). This toxic gas binds virtually irreversibly with haemoglobin in the blood to form carboxyhaemoglobin, thus preventing the haemoglobin from taking up oxygen and producing severe generalised hypoxia with a possible fatal outcome.

Haemoglobin may undergo another qualitative modification resulting in methaemoglobin. Two groups of products may be responsible for this process: mineral substances such as nitrites, nitrates and chlorates, and secondly, organic substances such as aromatic amines and nitro-compounds (nitrobenzene, nitrophenol, nitrotoluene, nitroglycerine, aniline, etc.).

Methaemoglobin is not able to transport oxygen and the result is inadequate oxidation of body tissues and generalised cyanosis of the skin when the level reaches 1.5 g/l.

Agents

Benzene

This is an important aromatic hydrocarbon solvent which may induce polymorphic changes affecting one or more types of blood corpuscle.

Latent benzene poisoning (incipient benzene poisoning). At this stage there may be discrete bone marrow hypoplasia. Examination of peripheral blood may reveal moderate anaemia (3.5-4 million RBC/mm^3), slight leukopoenia (3 000 WBC/mm^3) or, on the contrary, moderate leukocytosis, which may be persistent. Thrombocytopoenia is of indiscutable significance in the early diagnosis of benzene poisoning.

Manifest benzene poisoning. At this stage, detailed haematological examination may show, in certain cases, manifest bone marrow hypoplasia or aplasia:

(a) white blood cell anomalies are the most frequent and show leukopoenia accompanied by neutropoenia. The white blood cell count may fall to 2 000, 800 or even less;

(b) red blood cell anomalies may show a sharp fall in the number of red blood cells (to below 1 million). The anaemia may display slight hyperchromia and be accompanied by anisopoikilocytosis and anisochromasia;

(c) thrombocyte series anomalies may be pronounced and the number of thrombocytes may fall to 50 000, and sometimes 20 000/mm^3 or less. Bleeding time, coagulation time and clot-retraction time may be markedly increased.

The clinical picture of manifest benzene poisoning is that of a febrile, septic, haemorrhagic and asthenic syndrome.

In the benign forms the course to cure is usually slow and may be punctuated by recurrences. The malignant forms may have a fatal outcome.

Special forms of benzene-induced blood diseases. These are most frequently found in the white blood cells. The presence of leukopoenia or, almost as frequently, lymphocytosis may help in the diagnosis of latent benzene poisoning. Polynucleosis may sometimes be the only sign of the haemotoxic effect of benzene. Persistent leukocytosis, sometimes linked with polynucleosis, may be considered the warning sign of bone marrow hyperplasia.

It seems that a certain number of benzene leukaemias may be masked by anaemia or haemorrhagic purpura; cytogenetic studies show that benzene causes chromosome aberrations and attempts have been made to prove that leukaemia occurs as a consequence of somatic mutation. They may also appear following persistent leukocytosis or, occasionally, following an aplastic phase in the bone marrow.

As far as special red blood cell effects are concerned, there have been reports, although rare, of a polyglobulia and discrete haemolysis.

Toluene, xylene

If these benzene homologues contain benzene impurities, they may affect, although to a lesser degree, the red or white cells, thus producing leukopoenia or anaemia. Toluene diisocyanate may produce a true haemorrhagic syndrome affecting the bone marrow and producing primarily thrombocyte series suppression.

Turpentine

This may sometimes have the effect of producing slight thrombopoenia.

Lead

This metal may have a peripheral effect (cytolysis) and a central effect (bone marrow inhibition) affecting mainly the red cells. It also impedes various stages of haemoglobin synthesis thus producing secondary porphyria (coproporphyrinuria).

In the case of manifest lead poisoning, one often finds moderate normochromic or slightly hypersideraemic anaemia. This is often accompanied by anisopoikilocytosis, basophilic stippling and, in the regenerative phase, pronounced reticulocytosis. Haematologic detection of lead poisoning is based on these two latter phenomena. Slight sideraemia is found particularly during attacks of colic.

Mercury, manganese and cobalt

At the onset of poisoning due to one of these metals, it is usually possible to observe transient polyglobulia.

Arsenic

The haematological changes in arsenic poisoning may be in the form of polyglobulia, anaemia and, sometimes, slight leukopoenia.

Carbon monoxide

The reaction between this substance and haemoglobin produces carboxyhaemoglobin which prevents haemoglobin from taking up oxygen and releasing it in the tissues; this leads to generalised hypoxia.

Aromatic amines and nitro compounds

Exposure to these substances leads to the production of methaemoglobin which prevents oxygen uptake by haemoglobin followed by hypoxia and generalised cyanosis of the skin. Haematological examination will often show the presence of Heinz bodies and, sometimes, slight anaemia.

Arsine (hydrogen arsenite)

Exposure to this substance may produce intravascular haemolysis, which is marked by signs of pronounced general weakness accompanied by dyspnoea, cyanosis, fever and haemoglobinuria. Haematological examination shows a rapid fall in red blood cells and haemoglobin and a fall in haematocrit values.

Insecticides

Prolonged exposure to DDT may produce agranulocytopoenia or, in certain cases, anaemia or thrombocytopoenia (although this is not unanimously accepted). Phosphoric acid esters may sometimes be the cause of slight anaemia or neutropoenia.

Vinyl chloride

Apart from a possible carcinogenic action on the liver, this substance has a haemotoxic activity in white blood cells, frequently causing lymphocytosis. Moderate thrombopoenia may also be observed.

Naphthalene

The rare observations of naphthalene poisoning reveal leukopoiesis inhibition (leukopoenia) and a haemolytic effect.

Physical agents

Ionising radiations

Their haematological effect can be seen by central action on the bone marrow which affects, in particular, the leukocytes series. The changes that follow may be either acute or chronic depending on the type of exposure.

Blood changes due to acute exposure. These are due to sudden whole-body irradiation or irradiation of radiosensitive body tissues to a dose greater than 100 rad (accident in a nuclear reactor).

Hypoplastic states affect in particular the leukocyte series and produce leukopoenia followed by neutropoenia, which occurs prior to other haematological changes. In this case, following an initial and temporary fall in the white blood cell count during the first days, a significant white blood cell count reduction takes place between the fourteenth and twenty-eighth day. A return of the leukocyte count and, in particular, the neutrophile count to normal levels is a favourable sign from the prognostic point of view.

Anomalies in the red cell series are displayed by an anaemia which occurs from the thirtieth day following radiation. The appearance of reticulocytosis is considered a favourable prognostic sign.

Changes due to chronic exposure. These are of greater significance from the occupational point of view. Such exposure is the result of fractionated exposure over a number of days, weeks or even years (therapeutic radiation, radiological diagnosis).

The leukocyte series—which is the most frequently affected—displays leukopoenia accompanied by slight lymphocytosis. In some cases anomalies other than lymphocytosis may occur: for example lymphopoenia or an increase in the number of hypersegmented neutrophils, or a slight increase in the number of binucleated lymphocytes. These morphological disorders are an extremely sensitive haematological indicator and also an indication of the hazard of chronic exposure to ionising radiation.

The effects of ionising radiation may also lead to other late haematological changes that appear following exposure possibly lasting several decades. The result is primarily a hyperplastic proliferation of the leukocyte series which, in certain cases, may undergo malignant transformation. Chronic myeloid leukaemia in radiologists is, from the occupational point of view, the most common form. It is considered that brief repeated exposures may have a leukaemogenic effect in the same way as massive whole-body irradiation.

The red cell series may also be affected, resulting in normochromic or slightly hyperchromic anaemia.

The platelet series may undergo significant changes (thrombocytopoenia) in the final phase.

Biological factors

North-south migrational movements of immigrant workers from various countries have, over recent decades, presented the problem of major parasitic eosinophilias and their recognition within the framework of occupational diseases. When these have been initiated during work under special conditions or in tropical regions, they may be recognised as occupational diseases.

Ancylostomiasis. This may produce anaemias of varying severity, accompanied by leukocytosis. Eosinophilia is not very pronounced during the first year.

Brucellosis. In the chronic form of this disease, there is a certain degree of anaemia, normally slightly hypochromic.

Schistosomiasis. The haematological effects of this parasitic disease are the development of major eosinophilia.

Malaria. The haematological findings in this disease are the presence of parasites in the blood and the development of haemolytic anaemia followed by haemoglobinuria.

Blood changes occurring in certain occupational diseases and under certain special working conditions

In summary, these may be in the form of polycythaemia occurring in cases of pulmonary disease (pulmonary silicosis, chronic bronchitis, emphysema). Transitory polycythaemia may also occur during work in overheated premises (haemoconcentration phenomenon) or physical effort ("work erythrocytosis") accompanied by transitory leukocytosis.

High-altitude work at low barometric pressure produces a significant stimulation of erythropoiesis followed by polycythaemia (7-8 million/mm^3). A significant increase in the platelet count may also be frequently seen.

Haemorrhagic syndrome in carbon tetrachloride poisoning is mainly the result of a toxic lesion of the liver parencyma.

Haemoglobinuria resulting from haemolysis has been reported in cases of burns to large skin areas, during long marches, and during work in low-temperature premises ("refrigeration haemoglobinuria"), etc.

Diagnosis

There is no specific diagnosis for occupational blood disorders (with the exception of spectrophotometric analysis of denaturated haemoglobin). The diagnosis will be based primarily on a thorough occupational history, clinical and haematological examinations and the specific toxicological analyses related to a given occupational disease.

Haematological analysis is based on the results of a complete examination of peripheral blood involving a number of physiological constants and often on myelography (and in rarer cases on adenography and splenography). Other haematological indicators may be of value in the diagnosis, including the appearance of basophilic stippling of the erythrocytes in lead poisoning, the presence of Heinz bodies (due to methaemoglobin precipitation in the erythrocytes) in aromatic amine or nitro compound poisoning, the occurrence of reticulocytosis in the regenerative phase of haemolytic anaemia, an increase in binuclear lymphocytes in chronic exposure to ionising radiation, etc. The presence of granulations in the leukocytes, although not characteristic of the action of any specific toxic substances, gives an idea of the degree of severity of their haemotoxicity.

These haematological examinations can be used to determine the type of occupational haemopathy and the degree of change that has taken place. They are also valuable from the point of view of prognosis.

PREVENTION

Preventive measures should cover, in particular, systematic monitoring of haemotoxic substances in the workplace environment and dosimetry where there is a danger of exposure to ionising radiations, etc. Medical examinations of persons exposed to haemotoxic hazards should be carried out on a regular basis and include complete haematological examinations and the toxicological analyses relevant to the specific occupational disease in question. Workers already suffering from a blood disease should be carefully examined and they should be advised against any work with haemotoxic substances.

All the above procedures together are essential in determining the true nature and extent of the occupational blood disease hazards and their effective prevention.

Treatment

The first step is to remove the victim from exposure to the substances in question and remove all trace of poison from his body and clothing. Another emergency treatment measure is to administer substances which absorb, neutralise or eliminate the toxic substance. Specific types of diseases require specific treatment as summarised below:

Agranulocytosis: isolation of patient, antibiotics, transfusions, tonics;
Bone marrow hyperplasia: cytostatics, transfusions, hormone therapy;
Haemolytic anaemia: transfusions, where necessary—total blood replacement, vitamins K and C;
Lead poisoning anaemia: in severe cases, transfusions, EDTA;
Thrombopoenia: siliconised blood, platelet transfusions, vitamins K and C;
Carbon monoxide: oxygen inhalation (analeptics, cardiotonics, and antibiotics where necessary);
Aromatic amines and nitro-compounds: 10 ml of methylene blue 1% in 500 ml of glucosed serum by slow i.v. injection, oxygen, washing of contaminated skin.

VALCIC, I.

CIS 75-782 "Blood picture and occupational medicine" (Hémogramme et médecine du travail). Housset, H.; Cohen, A.; Boucker, M.; Legrand, G. *Cahiers de médecine interprofessionnelle* (Paris), 3rd quarter 1973, 13/51 (39-58). Illus. (In French)

"Blood findings reported to the Marseilles Poison Centre 1973-1976" (Les manifestations hématologiques signalées au Centre anti-poisons de Marseille de 1973 à 1976) (17-23). Jouglard, J. (ed.). *Sang et toxiques.* XVes Journées du Groupement français des Centres anti-poisons (Paris, Masson, 1976). (In French)

CIS 75-781 "The blood, an index of aptitude and exposure" (Le sang, index d'aptitude et d'exposition). *Cahiers de médecine interprofessionnelle* (Paris), 3rd quarter 1973, 13/51 (5-100). Illus. (In French)

CIS 78-789 *Thalassaemia and the occupational physician* (Les thalassémies et la médecine du travail). Dardel, A. (Bordeaux, Université de Bordeaux II, 1977), 150 p. 91 ref. (In French)

Body reference man

A quantitative knowledge of the human body is necessary for the study of physiopathology and industrial toxicology for preventive purposes. Many biological data are needed to enable the assessment of the absorbed doses of toxic substances or physical agents to which the body has been exposed for a given time, the amounts metabolised and retained and hence exposure limits and hygienic standards. In the last 30 years a great many such data have therefore been gathered and studied, and a number of anatomic, metabolic and functional data have been defined for use in prevention of occupational risks.

The most outstanding and co-ordinated effort in this field has been carried out by the International Commission on Radiological Protection (ICRP), which is especially interested in such data for the purpose of dosimetry and radiation protection. The ICRP has defined a "reference man" as "a typical occupational individual" for purposes of radiation protection, with some indication of variability of the occupational group about the norm. Such a "reference man" does not correspond to a statistical definition and is not representative of a defined population, although his parameters are in essence those of what is believed to be a typical individual of Western Europe or Northern America in habitat and custom, Caucasian, between 20 and 30 years of age, who lives in a temperate climate (average temperature between 10 and 20 °C). Being quantitatively defined, his values can easily be adapted to any individual under consideration provided that specific differences are taken into account.

The data shown in table 1 were taken from the report adopted in 1974 by a Task Group of Committee 2 of the ICRP, under the chairmanship of Dr. Snyder. They have been selected as being those of greatest interest for occupational health.

PARMEGGIANI, L.

Table 1. Data of reference man

	Adult male	Adult female
Weight of total body	70 kg	58 kg
Length of total body	170 cm	160 cm
Surface area of total body	18 000 cm²	16 000 cm²
Total body water	600 ml/kg	500 ml/kg
Total blood volume	5 200 ml	3 900 ml
Weight of total blood	5 500 g	4 100 g
Volume of plasma	3 000 ml	2 500 ml
Weight of plasma	3 100 g	2 600 g
Surface area of alveoli	72 m²	66 m²
Metabolic rate	17 cal/min-kg	16 cal/min-kg
Total lung capacity	5.6 l	4.4 l
Functional residual capacity	2.2 l	1.8 l
Vital capacity	4.3 l	3.3 l
Dead space	160 ml	130 ml
Minute volume, resting	7.5 l/min	6.0 l/min
Minute volume, light activity	20.0 l/min	19.0 l/min
Air breathed, 8 h working light activity	9 600 l	9 100 l
Air breathed, 8 h non-occupational activity	9 600 l	9 100 l
Air breathed, 8 h resting	3 600 l	2 900 l
Air breathed, 24 h total	2.3×10^4 l	2.1×10^4 l
% of total air breathed at work	42	43
Urine volume	1 400 ml/d	1 000 ml/d
Urine specific gravity	1.02	
Urine pH	6.2	
Urine solids	60 g/d	50 g/d
Urine urea	22 g/d	
Urine "sugars"	1 g/d	
Urine bicarbonates	0.14 g/d	0.12 g/d
Total body content:	*amount (g)*	*% of total body weight*
Oxygen	43 000	61
Carbon	16 000	23
Hydrogen	7 000	10
Nitrogen	1 800	2.6
Calcium	1 000	1.4
Phosphorus	780	1.1
Sulphur	140	0.20
Potassium	140	0.20
Sodium	100	0.14
Chlorine	95	0.12

Report of the Task Group on Reference Man. ICRP Publication 23, International Commission on Radiological Protection (Oxford, New York, Toronto, Sydney, Braunschwig, Pergamon Press, 1975), 480 p. Illus. Ref.

Boiler making

Boiler making is a branch of mechanical engineering dealing with the manufacture of stationary pressure vessels and steam boilers and of their ancillary equipment.

There are two basic boiler manufacturing techniques based on the use of cast iron and steel respectively.

Manufacture of cast-iron boilers

The boiler sections (shell rings, shell halves, etc.) are obtained by pouring molten cast iron into non-permanent sand moulds. The main processes of this manufacturing technique are the preparation of pig iron and charging it into the furnace, melting, tapping the molten metal and pouring it into the moulds, preparation of the moulds and cores, knocking-out, fettling and partial machining of the castings.

Harmful factors characterising the conditions of work in foundries are dust, microclimate, gases and noise.

The greatest dust concentrations are found in sand preparation, knocking-out and fettling shops, where they usually vary from 10 to 50 mg/m³.

Unfavourable microclimatic conditions are created by high air temperatures and radiant heat during melting and pouring, to a lesser degree when knocking out and leaving the casting to cool, and also during core baking and mould drying. The energy of radiant heat to which melting-furnace operators are exposed amounts to 8-20 mW/cm²; pourers are exposed to 2.8-7.2 mW/cm², and knockers-out to 1.2-8 mW/cm².

The air in foundries where production goes on without interruption is never free from carbon monoxide. Mean CO concentrations of 18-20 mg/m³ are measured in modern iron foundries. Higher CO concentrations may be observed in shops enclosed by buildings, especially in bays where hot operations are performed (pouring, moulding, knocking out) and where the air intake zones are polluted by carbon monoxide emissions from cupolas, melting furnaces, untight baking ovens, etc.

Other gases polluting the air of foundry shops are:

(a) ammonia and formaldehyde, which are released from urea-formaldehyde resins used as binders for core sands;

(b) acrolein which is set free during the baking of cores made of sand containing up to 2% vegetable oils

which decompose at the baking temperature of 220-250 °C and release acrolein.

Noise in foundries is produced mainly by the knocking-out grate. This noise is of the impulse type and has a broad frequency spectrum. A less intensive low-frequency noise is produced by moulding machines (90-95 dBA). Extensive mechanisation of the various manufacturing processes has done away with heavy physical work and considerably reduced the accident rate.

Dust control measures in knocking-out bays consist in enclosing the knocking-out grate as far as this is compatible with technological requirements, and in providing for exhaust ventilation from under the enclosure. The best solution for such enclosures is to equip the grates with mobile covers which remain interlocked for a few minutes after the shaking or vibrating mechanism of the grate has stopped, and then move back to give access to the casting, i.e. when the dust-laden air under the cover has been exhausted. Effective mechanical ventilation must be provided for the moulding and core sand processing and conveying equipment (grinding mills, sieves, mixers). Despite the progress made in dust control in foundries, the dust problem cannot be solved without correctly planning the various shops and in particular by segregating dusty operations and automating them.

Local mechanical ventilation is used to prevent heat stress and thermal burns, and during the cold season the ventilation air is heated. The location of the air intake openings of these ventilation systems must be carefully chosen to avoid sucking in gas-polluted air. All equipment releasing heat and gases should be enclosed as far as possible and provided with local exhaust devices for mechanical or natural ventilation (melting furnaces, conveyors between the pouring bay and knocking-out grate, used-sand tunnels, etc.). It is advisable to use barrel-type ladles with a small opening in order to avoid splashes of hot metal during pouring.

The pourers must be equipped with protective clothing of wool or canvas, having neither pleats nor pockets. To prevent burns to the feet and legs (which are involved in 40% of all burns notified in melting and pouring shops), the workers should wear boots made of thick smooth leather, with leather soles and a thin asbestos lining at the rear, or felt boots; the trouser legs should be worn over the boots.

To protect the face and eyes against the heat radiated by the molten metal, it is advisable for pourers to wear canvas masks with a leather frame holding dark-violet glass eyepieces shielding against excessive brightness and infrared radiation.

Manufacture of steel boilers

In this case the boiler elements are made from rolled stock, i.e. steel plate with thickness varying from 10 to 200 mm according to the operating pressure requirements to be met by the boiler. The plate thickness determines the manufacturing technique to be adopted.

Thinner plate is cold-formed on plate bending machines into cylinders which are assembled by riveting (for pressures of up to 10-12 kgf/cm² or 140-170 psig). Boilers for higher pressures (up to 30 kgf/cm² or 420 psig) are constructed from hot-rolled shells and hot-forged ends (forging temperatures of 1 000-1 100 °C) which are welded to each other. However, the wall thickness required for pressures of 100 kgf/cm² (1 400 psig) and more do not permit assembly by welding, and the shells are therefore hot-forged on hydraulic presses by piercing a steel ingot and rolling it subsequently to the required thickness. The steels used for this purpose are chrome-nickel or chrome-nickel-molybdenum alloyed grades of the necessary strength and reliability. The forged ends are machined on vertical turning and boring mills.

HAZARDS AND THEIR PREVENTION

Harmful factors in the manufacture of steel boilers are microclimate, noise, vibration, gases, the risk of burns and electrical accidents. Although the steel is only heated for hot forming (up to 1 100 °C) and not molten, convection heat and radiant heat are determining factors for the conditions of work in this industry. Apart from incandescent workpieces, the surfaces of reheating furnaces, the flooring (especially when covered with steel plates) and the mechanical equipment are sources of heat (40-200 °C).

The radiant heat energy varies from 4-16 mW/cm² at a distance of 1-2 m from the hot forgings to 20 mW/cm² in front of open furnace doors. The air temperature in forging shops is therefore 10-20 °C higher than that outdoors, especially when the natural ventilation is insufficient.

Intense high-frequency noise of up to 120 dBA is produced when low-pressure boiler shells are riveted with pneumatic hammers, but this technique is very rarely used nowadays. Recurrent noise of the impulse type of up to 100-110 dBA is experienced near some bending rolls when steel plate is cold-formed. Low- to medium-frequency noise of 90-95 dBA is produced by the burners of gas-fired furnaces, presses, fans and other equipment. Vibration is transmitted to the hands of workers holding pneumatic riveting hammers. The replacement of riveting by welding in boiler making has done away with noise and vibration, but introduced the hazard of ultraviolet radiation and respirable dust (particles of less than 2 μm). Depending on the electrodes used, the dust may contain up to 20% silicon compounds, up to 8% manganese, up to 0.15% chromium and even fluorine. The presence of these toxic substances in the dust constitutes a great health hazard, and special measures must be taken to prevent their inhalation.

Apart from dust, welding gives rise to small concentrations of carbon monoxide and nitrogen oxides. When welding is done in confined spaces, these gases may accumulate in dangerous concentrations.

When hot plate is rolled or bent and when boiler parts are welded, there is a hazard of burns from flying scale particles or welding sparks.

Efficient natural ventilation is of the utmost importance for preventing excessive air temperatures in boiler-making shops. Additional measures are water curtains in front of furnace doors, thermal insulation of furnace surfaces and of the floor portions which come in contact with hot metal by putting heat-insulation joints between the floor plates and the surrounding flooring.

The installation of air showers at the workplaces of hot rollers, reheating furnace attendants and other heat-exposed workers effectively relieves heat stress. When powerful fans and air ducts are installed, attention should be paid to noise absorption and vibration isolation to reduce ventilation-system noise. If noise cannot be reduced to acceptable levels by technical control measures during particularly noisy operations (riveting, fettling), the exposed workers must wear ear protectors.

Work with pneumatic hammers presents the hazard of neurovascular disorders, which may be induced by the transmission of intense vibration to the hands. These disorders are detrimental to working capacity. Riveting should be discontinued in boiler making, but when pneumatic or vibrating tools have to be used, they should be equipped with vibration dampers, and special work and rest schedules should be established for riveters to limit their exposure.

Conditions of welding work can be fundamentally improved by changing over to machine welding with automatic welding heads and by replacing arc welding by resistance welding. When welding is done in confined spaces, effective local exhaust ventilation must be provided for.

Electric welders should use screens or wear masks with special eyepieces to protect their eyes against harmful radiation. In manual arc welding, handy shape and simple construction of the electrode holder is very important because it should enable the electrode to be held at any angle. Its handle is made of an electrically and thermally insulating material. A self-locking switch is provided for electrical safety; it breaks the welding-transformer circuit as soon as welding is finished, and remakes the circuit when the electrode strikes the workpiece. Welders working inside boilers must be protected with rubber insulating mats laid on heat-insulating linings.

All workers handling hot metals and all welders must be equipped with protective clothing: canvas or linen suits impregnated with special flame-retarding compounds. The trouser legs are worn over the boots which should have smooth tops without laces.

ŠKARINOV, L. N.

Handbook of occupational hygiene (Spravočnik po gigiene truda). Karpov, B. D.; Kovšilo, V. E. (Leningrad, "Medicina", 1979), 446 p.

CIS 79-1895 *Vibration absorbers for noise reduction in boilermaking* (Lärmminderung im Behälterbau durch Schwingungsabsorber). Albrecht, H. Forschungsbericht Nr. 192 (Bundesanstalt für Arbeitsschutz und Unfallforschung, Postfach 170202, 4600 Dortmund 17) (1978), 23 p. Illus. 3 ref. (In German)

Boilers and furnaces, cleaning and maintenance of

Steam boilers are pressure vessels in which water is converted to steam for heating, processing or electric power generation purposes. The generating capacity of such boilers can vary from a few kilogrammes to several thousand tonnes of steam per hour. Steam may be produced at pressures ranging from less than 200 kPa to over 15 MPa and may be saturated or superheated. The temperature of superheated steam generated in a boiler may be as high as 600 °C. Boilers may be fired by solid fuel, oil, gas or electricity.

The types of boiler likely to be encountered are shell boilers (also known as fire-tube or return-tube boilers), water-tube boilers, coil boilers (sometimes referred to as steam generators) and electrode boilers. In shell boilers the water is contained within the cylindrical boiler shell and is heated by tubes through which pass the hot combustion gases. Water-tube and coil boilers contain the water in tubes which are heated by burners firing into a furnace. Electrode boilers heat the water by passing an alternating electric current through the water (see BOILERS AND PRESSURE VESSELS).

All types of boiler may become fouled on the water side by deposits resulting from mineral matter in the feedwater. Such deposits may range from calcium carbonate in low-pressure boilers to hard silicates which have very low thermal conductivity. These drastically reduce efficiency and often lead to overheating. Accumulation of deposits in shell boilers may result in choking

Figure 1. Scale deposition in a water-tube boiler.

of water-gauge glass connections and those to water-level controls. This may lead to furnace collapse either directly or as a result of incorrect water levels being indicated by the gauge glasses or maintained in the control chambers (figure 1). Deposition often leads to on-load corrosion in high-pressure boilers.

Fire-side fouling occurs mainly in solid fuel and oil boilers and arises from the presence of impurities in the fuel. Such impurities include ash, metal oxides, sulphur oxides and vanadium compounds. Such deposits reduce boiler efficiency and may give rise to problems of corrosion and/or erosion.

Vanadium compounds can promote serious fire-side corrosion at temperatures in excess of 600 °C.

Methods of cleaning

Boilers may be cleaned on or off load (with or without taking the boiler out of service), the former being preferred wherever possible. Cleaning methods for fire and water sides may be both physical and chemical.

Water-side cleaning

On-load cleaning. Cleanliness may be achieved and maintained by a carefully controlled programme of water treatment and conditioning which must be appropriate to the feedwater available. Such treatment is designed to remove harmful impurities such as calcium and magnesium, silica, iron, copper and oxygen. Most of these impurities may be removed by ion-exchange but chemicals are still used to take care of residual amounts and to remove oxygen. These chemicals include sodium hydroxide, sodium phosphate, chelating agents (e.g. EDTA), hydrazine, sodium sulphite, ammonia and certain amines. Precipitated impurities are removed by blow-down.

Off-load cleaning–physical methods. Thin soft deposits are often removed by brushing. In water-tube boilers this is accomplished by means of rotary brushes, scatter scalers or bullet brushes, the latter being fired through the tubes by compressed air. Another method sometimes used for removing soft deposits is high-pressure water jetting; this is used both in shell boilers and water-tube boilers.

Off-load cleaning–chemical methods. Here chemicals are added to the boiler which react with the scale and render it soluble. The chemicals most commonly used are inhibited acid (usually hydrochloric acid), inhibited hydrochloric acid with ammonium bifluoride (when silicate scales are present) and ammonia/citric acid/sodium bromate solutions (for removing copper).

Fire-side cleaning

On-load cleaning–physical methods. The most widely used method of on-load fire-side cleaning is soot blowing using steam or air jets with the aim of removing soot, dust and grit deposits which have been deposited on tubes and in gas paths where they could seriously impede gas flow and may elevate flue gas temperature to a dangerously high level. Smoke tubes in shell boilers are cleaned on-load by steam lancing individual tubes. On-load water lancing of boilers should not be attempted without consulting the boilermakers and other competent specialists.

On-load cleaning–chemical methods. Several chemical additives have been developed to prevent the build-up of deposits on the fire-side. These include magnesium oxide and various proprietary formulations.

Off-load cleaning. Physical methods are of prime importance in off-load fire-side cleaning. Dust is usually removed by air blowing or vacuum cleaning. Adherent deposits are removed by air lancing or by soaking with detergent and water washing. Washing should be carried out at 70 °C and the surfaces washed down afterwards with an alkaline solution to minimise corrosion. Adherent deposits are sometimes removed by shot blasting using trays to collect spent shot and debris.

HAZARDS

Boiler maintenance personnel can be presented with hazards both in on-load and off-load situations.

Most on-load hazards are associated with scales and deposits either on the water or on the fire-side. As mentioned above, water-side deposits may cause blockage of feed lines or automatic controls and may render gauge glasses inoperative. These failures can lead to shortage of water and collapse of furnaces or water tubes. On-load water-side cleaning may cause large pieces of scale to break away. In shell boilers, such detached pieces may block gauge glasses or automatic level controls. In water-tube boilers, tubes may become completely blocked, leading to rupture of the tube. In modern boilers with high rates of heat transfer, thin layers of scale (often less than 0.5 mm in the case of silicates) can cause overheating and failure. Scale formation in tubes can lead to "on-load corrosion" in high-pressure water-tube boilers. This is often associated with the liberation of hydrogen, giving rise to hydrogen embrittlement. Cases have been known where complete pieces of tube have blown out and have been ejected through the boiler casing; one such occurrence caused the death of a maintenance operative. Conditioning chemicals used to minimise the formation of water-side deposits present their own hazards. Sodium hydroxide (caustic soda) causes severe skin burns, hydrazine is highly toxic and certain amines used in boiler water treatment also possess toxic properties. The long-term effects of chelating agents are unknown, but care must be taken when they are handled. Proper caution should be exercised in the storage of chemicals. Sodium sulphite must not be allowed to come into contact with acids, as poisonous fumes are liberated. Strict personal hygiene must be observed at all times and food must not be eaten in areas where chemicals are stored. When chemicals are used or handled, protective clothing must be worn. On the fire side, blockages or partial blockages of gas paths may lead to overheating. Solid matter in gas streams erodes furnaces and water tubes and can cause a dangerous reduction in metal thickness.

When a boiler is opened for inspection, there are potential hazards in the removal of lagging. Materials used for lagging may contain asbestos or other materials having harmful effects on the lungs. Hazards during off-load cleaning arise from the cleaning methods themselves. Acids used in the chemical descaling of boilers can cause skin burns and may result in loss of sight if they get into the eyes. Particular care must be taken where acid mixtures containing fluorides are used; these mixtures generate hydrofluoric acid, which causes very serious skin burns that heal only with difficulty. When water-side or fire-side deposits are removed by mechanical means, there is always the problem of dust. This dust may contain among other things silica, asbestos, vanadium and arsenic compounds. These and other substances found in fire-side residues may be toxic, corrosive or carcinogenic and may give rise to kidney, liver, lung and bladder damage. Wetting agents are sometimes used in cleaning and this can produce allergic-type reactions in some workers. Deaths and serious injuries have occurred as a result of accidental lighting of burners while men were still inside boiler furnaces. Accidents have been reported resulting from entry into boilers which had not been properly ventilated or which had not had sufficient time to cool; the former may lead to asphyxia and the latter to thermoregulatory disorders, heat cramps and heat strokes (see CONFINED SPACES). Dangers may also arise from missiles and falling objects. Shot used for cleaning the fire-side surfaces of water-tube boilers may ricochet, and bullet brushes may be ejected from tubes into drums at high speed. Pieces of bonded ash may fall from a height (figure 2).

Figure 2. A piece of ash weighing 900 kg that has fallen to the bottom of a boiler. (*By courtesy of the Regional Director, Central Generating Board, Midlands Region, Solihull, Birmingham, England.*)

HEALTH AND SAFETY MEASURES

Entry into the furnaces, drums or shell of any boiler should not be made until a permit-to-work has been issued. Such a permit should only be issued after a competent and responsible person has ascertained that the boiler has cooled to below 40 °C. It should be well ventilated to ensure that the internal and external pressures are equal, thus minimising the risk of a manhole door being blown out or maintenance personnel being sucked in by the vacuum. There should be an adequate supply of air and no toxic, flammable or explosive gases or vapours should be present. All valves and dampers controlling the entry of steam, water and flue gases should be shut and locked or the pipes blanked off. Where two or more boilers share common blowdown arrangements, the blowdown valve of the boiler being inspected should be shut and locked and the key removed. Before inspecting electrode boilers, the boiler should be completely isolated from the mains supply.

Workers should wear waterproof clothing and garments to protect their hands, face and eyes from acids and other chemicals. Safety helmets should be worn. Where dust is a hazard, dust masks fitted with suitable filters should be worn and full breathing apparatus used if toxic or noxious fumes are present or if an oxygen deficiency can arise. When a man is working inside a boiler or furnace, he should be equipped with a safety belt and lifeline and be accompanied by a responsible person who should remain on the outside. Tools should be prevented from falling on to workers below.

After chemical cleaning of the water side has been completed it is essential that the boiler be completely flushed to remove any harmful chemicals and then force-draught ventilated to ensure that no noxious fumes, or flammable or explosive gases remain. The atmosphere should be tested before entry. Scaffolding or staging used by workers should be inspected before use. Fuses controlling electrical installations connected to the plant should be removed and switches locked in the "off" position; warning notices should be prominently displayed indicating that men are at work inside. Portable lighting should have a low-voltage supply; in some countries 25-50 V is considered suitable, in others a 12-24 V supply is required. When bullet brushing is in progress, the receiving drum should be mattressed to prevent brushes flying into other tubes. Manhole openings should be blanked off to prevent brushes flying out and injuring personnel. Vacuum cleaners should be used to remove dust, thus limiting the formation of dust clouds.

First-aid equipment–including oxygen-therapy apparatus–should be provided and persons with special training in resuscitation techniques should be available at all times while cleaning and maintenance operations are in progress. Washing and sanitary facilities should be provided and the need for personal hygiene emphasised.

Pre-employment and periodical medical examination of boiler and furnace maintenance workers is advisable. These examinations should include, in particular, chest X-rays, especially for persons who may be exposed to silica or asbestos dust, and careful skin examination. Where there is a likelihood of bladder tumours due to asbestos ingestion, special examinations should be carried out.

Boiler and furnace inspection

After any type of steam boiler has been cleaned, it should be inspected by a competent person. This person should examine both the structural condition and cleanliness. Depending on the condition and age of the equipment, it may be necessary to have portions of lagging or seating brickwork removed. Selected critical tubes may be subjected to non-destructive methods of examination (such as ultrasonics or radiography) to check satisfactory thickness and freedom from defects. In doubtful cases, tubes may be removed and sectioned to check the internal surfaces for scale and/or pitting. In the absence of non-destructive test facilities, thicknesses may be checked by removal of tubes and drilling of plates. If riveted seams have leaked at intermittent intervals, metallurgical and chemical analyses may be necessary to investigate the possible presence of stress corrosion (caustic cracking). In the case of internal inspection, full accessibility must be ensured to the extent that construction permits and steam-drum baffles and other impedimenta should be removed. If internal access is limited, a hydraulic test may be applied and deflections measured at critical points.

Mountings should be examined separately. Particular attention should be given to the correct functioning of safety valves and boiler control-float chambers, where fitted. Finally, the boiler should be examined under working conditions; operational checks should be made of any automatic control or alarm systems and a steam test carried out on the safety valves.

PHILLIPS, J. D.

CIS 78-1234 "Corrosion" (corrosion). *APAVE* (Paris), Oct.-Nov.-Dec. 1976, 57/196 (35-89). Illus. 21 ref. (In French)

CIS 78-636 "Cause of material damage in the bottom zone of sandlime brick hardening autoclaves" (Über die Ursache von Werkstoffschädigungen im Sohlenbereich von Steinhärtekesseln). Schlegel, D.; Gerken, G. *TÜ* (Düsseldorf), June 1977, 18/6 (197-201). Illus. 1 ref. (In German)

CIS 80-656 *Abnormal corrosion of boiler tube welds during chemical cleaning.* Christopher, N. S. J.; Nelson, K.; Ashworth, V. (90-103). Illus. Technical Report, Vol. XIII (British Engine Insurance, Longridge House, Manchester M60 4DT) (1978).

CIS 76-1041 "Occupational poisoning due to vanadium compounds inhaled by workers cleaning out fuel oil-fired heaters" (Profesionální otravy sloučeninami vanadu u pracovníků v kotelnách na mazut). Kůelová, M.; Šírl, J.; Popler, A. *Pracovní lékařství* (Prague), May 1975, 27/5 (170-173). 20 ref. (In Czech)

Boilers and pressure vessels

Steam boilers opened the era of "industrial revolution" i.e. growing mechanisation. At the same time they opened the era of technological risks (both to workers and the public at large) and of related preventive measures.

The number of boilers grew very rapidly, and explosions of boilers were for some time a common plague. In about the middle of the 19th century, however, a number of boiler-owners' associations were set up and safety codes and regulations issued in Europe and America.

Boilers, as well as fired or unfired positive or negative internal pressure vessels (for short: BPV) and related piping may be damaged or destroyed by explosions or implosions following: either an excessive rise or loss of internal pressure beyond rated working values, or failure of walls or components at any pressure.

Explosions and implosions cause damage and injury as a result of: propagated pressure waves; flames, fumes, escaping fluids; and flying fragments. Exposed hot surfaces may cause burns, fires and further explosions. Electrical heating adds electrical risks.

Such events may arise from a number of causes. They can be prevented by careful design, installation, operation, attendance, maintenance and inspection.

Guidance on safe *design* is given by the regulations and codes promulgated in many countries; although they have common fundamental criteria, they may differ in details. They normally classify BPV according to these parameters: operating pressure range (low, medium, high), temperature, kind of fluid, capacity and purpose (from nuclear and fossil fuel electricity generating plants to espresso coffee making machines), and installation.

Especially stringent regulations often apply to BPV installed in road or rail vehicles, aircraft and ships, because of the risk of unreliable operation they entail and other specific conditions, as well as in view of the extent of the damage that BPV explosions may produce to the vehicles themselves and to the public.

Special regulations apply also to pressure systems of nuclear plants, which include systems relying on liquid and vaporised metals as well as gases and steam.

BPV should be *installed* in such places as will minimise the consequences of malfunctions and explosions or implosions. Larger BPV should be installed in detached rooms or areas, to which only their attendants and other authorised persons should have access. Fire and combustion gases containing particulate matter and noxious gases should be treated so as to minimise pollution. The frame structure should have the necessary strength and stiffness not only to bear the dead loads but also to ensure reliable operation of the boilers in the event of earthquakes.

BPV should be *operated* in accordance with the manufacturers' instructions and attended by fully competent persons. The BPV user should ascertain that such persons have received and assimilated appropriate training and have been licensed as may be required by law.

A *preventive maintenance system* of all BPV, including unattended, automatically controlled ones, should be set up and strictly followed. Cleaning implies hazards to health (see BOILERS AND FURNACES, CLEANING AND MAINTENANCE OF). Safe access to all external and, where necessary, internal parts should be provided by means of staircases or fixed ladders, platforms, openings and manholes.

From time to time *special testing, and inspection* should be carried out by a competent person. For some categories of BPV it is required that besides periodical inspection by a competent person designated by the owner, further examinations be made by independent inspectors belonging to a state authority or to a recognised technical association. During inspections, besides the risks encountered in maintenance, account should be taken of risks arising from certain non-destructive testing methods, mainly ultrasound and ionising radiation (see ULTRASOUND; RADIATION, IONISING, INDUSTRIAL USES OF).

Boilers

A boiler is an installation in which heat is used to produce steam or superheated water for utilisation outside the installation. The heat is obtained from burning organic fuels, technological processes, electrical energy or hot waste gases.

Parts belonging integrally or partially to a boiler are the furnace, superheater, economiser, air heater, shell, brick-lining, thermal insulation, enclosure, etc.

A boiler plant groups the boiler and the auxiliary equipment ensuring normal continuous operation of the boiler. Apart from the latter, a boiler plant can comprise blowers, heating-surface cleaning equipment, fuel feed and processing systems within the boiler-plant boundaries, slag and ash removal equipment, fly ash collectors and other flue gas cleaning devices, gas and air ducts not directly connected to the boiler, water and steam piping, fuel supply systems, valves and fittings, automatic controls, monitoring and safety apparatus, feed-water treating equipment and a stack.

A boiler designed for raising steam is a steam generator, and one designed for producing hot water is a hot-water boiler.

Types of boiler

Boilers may be classified from many points of view. From the point of view of construction a distinction must be made between fire-tube and water-tube boilers.

In a fire-tube boiler the fuel combustion products are led through heating-surface tubes or flues, while the water or steam/water mixture surrounds the tubes. In a water-tube boiler the water, the steam/water mixture and steam pass through heating-surface tubes surrounded by the fuel combustion products.

Another major difference between these two types of boiler is their storage capacity which can be expressed as the relationship of the boiler volume to the heating-surface area. For fire-tube boilers this relationship is considerably higher than for water-tube boilers, and the former therefore present certain advantages: greater ease of tending, possibility of using simpler regulating systems and greater versatility of operation. On the other hand, fire-tube boilers present a greater explosion hazard than water-tube boilers since a leak forms suddenly owing to deterioration of the shell-wall strength or of other boiler elements, the enormous energy stored in the superheated water is set free and may cause the destruction of the building in which the boiler is installed. Other drawbacks of this boiler type are its limited steam-raising capacity and narrow range of steam parameters. Fire-tube boilers are used for recovering waste heat, for agricultural purposes and in cases where it is impossible or undesirable to use electric waterheaters or electric boilers.

From the viewpoint of reliability in operation it is interesting to classify boilers according to steam pressure. However, this basis of classification is very relative as it changes with the evolution of boiler construction, and the criteria determining high and low pressure limits show an ever-increasing trend in terms of absolute steam pressure. The requirements to be met by the construction, materials used and by safety and regulating devices increase as a function of pressure. The same applies to testing and control procedures.

Fire-tube boilers may be of either the low-pressure or the medium-pressure type, as they are generally built for pressures of up to 20-24 kgf/cm^2 or 140-170 psig (1 960-2 350 KPa), and only waste-heat boilers heated by waste gases from industrial plants are designed for higher pressures which may in certain cases attain 100 kgf/cm^2 or 1 400 psig (9 800 KPa).

Water-tube boilers are widely used on account of their greater flexibility in operation and their larger steam production. In these boilers the water and steam/water mixture move in a closed circuit consisting of riser tubes through which the steam/water mixture is led into the steam drum, and of downcomer tubes through which non-evaporated water returns to the inlet header. If this circulation is induced by the difference in the densities of the water in the downcomer tubes and the steam/water mixture in the riser tubes, one speaks of a natural-circulation boiler; if the circulation is driven by a pump, the boiler is called a forced-circulation boiler; and if the water is entirely converted to steam after one single passage through the bundle of evaporator tubes, one speaks of a once-through boiler. The latter type of boiler requires less metal per unit of steam raised, but this advantage is outweighed by the more complicated automatic control system which is needed to maintain optimal conditions of operation.

A very important element of boilers with natural or forced water circulation is the drum in which steam and water are separated; therefore, these boilers may also be referred to as drum-type boilers. Once-through boilers have no drum.

In terms of the circulation of the combustion products in the boiler and the pressure applied to the boiler furnace, it is possible to distinguish between natural-draught boilers, in which the flue resistance is overcome by the difference in density between the atmospheric air and the gases in the stack, and forced-draught boilers, in which the flue resistance is overcome by a blower. There are also balanced-draught boilers which occupy an intermediate position between the above two types: the pressure in the furnace or at the flue inlet is maintained close to atmospheric by the combined action of smoke exhausters and blowers. Pressurised-furnace boilers have supercharged furnaces from which the gases pass into the flue at a pressure of 1 kgf/cm^2 or 14 psig (98 KPa) or more above atmospheric pressure.

Depending on the fuel used, boilers may be classified as solid-fuel boilers (fired with coal, lignite, wood scraps, peat, solid wastes, etc.), liquid-fuel boilers (mineral oil), gaseous-fuel boilers (natural gas, waste gases from metallurgical, chemical or petroleum-converting processes), and multiple-fuel boilers.

Boiler defects and failures

Faulty design. Nowadays defects due to faulty design are extremely rare, for boilers are designed to specifications that are closely controlled by technical inspectorates at the design stage. These specifications cover both material strength and reliability of operation, including accessibility for inspection of all sub-assemblies that are vital for safe operation; they also cover any replacements that may be required. Defects are often found in materials of boiler elements and in welded, flange, rolled and other joints, since manufacturing techniques cannot guarantee parts that are absolutely free of flaws. The problem here is to establish, for every component, what defects can be tolerated since they constitute no risk for given modes of operation, and to institute inspection techniques capable of detecting inadmissible defects in time.

Water leakage. This results in the water level falling short of the lower limit and is a frequent cause of boiler failure, although every boiler should be fitted with not less than two independent water gauges. If there is not enough water in water-tube boilers, the heating-surface tubes burn through, and in fire-tube boilers the boiler shell may burst. In the event of a water shortage in a boiler the combustion process in the furnace must be stopped immediately. If this is not done, or if an attempt is made to complete the water level in the boiler by admitting more feed water, fire-tube boilers will inevitably explode, and in water-tube boilers the steam drum will suffer permanent deformations, which represent a considerably greater prejudice than damage to the heating-surface tubes. As a rule, the latter cannot be saved by increasing the feed-water supply.

Loss of wall thickness. This is due to external or internal corrosion and erosion. The degree of chemical attack (corrosion) of the inside and outside tube surfaces and the extent of mechanical abrasion (erosion) of the tubes by gases containing incombustible particles depend on the quality of the feed water and composition of the combustion products. As a rule, losses in wall thickness are localised phenomena; they give rise to flaws through which small quantities of steam may escape. If this is noticed in time, the boiler should be shut down to replace the tubes in which the walls have become thinner than admitted by standard specifications. However, if this is not recognised in time, the damage will rapidly become more serious, and there will be a risk of explosion in the case of fire-tube boilers.

Gas explosions. In boilers fired with liquid fuel or gas there is a danger that fuel vapours or unburnt gases accumulate in the furnace and mix with the oxygen from air. These explosive mixtures may be ignited by the hot furnace walls or by the burner flames and cause an explosion. The result may be damage to small furnaces or, if the furnace walls are not tight or if inspection openings have been left open, blazing flames may cause burns to the attending personnel. To prevent such explosions, furnace ventilation should be activated for some time after the burners have been shut and before they are lit.

To limit damage from explosions, special explosion vents are provided which discharge the explosion gases to a safe place outside the plant. With large boiler plants with considerably greater furnace volumes such explosions may present little or no danger, because it is possible to provide for adequately reinforced furnace walls. Gas explosions in furnaces burning pulverised coal are rare occurrences. Gas explosions may also occur in the fuel processing and feeding systems, and adequate safety measures should be adopted, such as atmosphere monitoring, vents and reinforced construction.

Impurities in water supply. The boiler manufacturer establishes the optimal composition of the boiler feed water to ensure minimum corrosive effects and prevent deposits of scale on the inside of the heating-surface tubes. Such deposits are detrimental to the efficiency of the boiler and may cause overheating and deterioration of the metal. Important scale deposits on the heating surface of fire-tube boilers may give rise to leakage and, if larger portions become leaky, to a boiler explosion. Water should be chemically treated if necessary and scale deposits eliminated by mechanical or chemical means (see BOILERS AND FURNACES, CLEANING AND MAINTENANCE OF).

Excess pressure in the boiler. In the event of a rapid rise in pressure inside the boiler due to a failure of safety devices or insufficient release of excess steam from the boiler, boiler parts under pressure may become leaky. If the leakage is important, a boiler explosion causing serious damage may ensue.

Failure of automatic regulating systems and safety devices. Modern boilers are fitted with systems which protect against breakdown and damage, and practically exclude the possibility of explosions or emergencies provided that they are operating reliably. It is therefore of the utmost importance to ensure that these systems are kept in perfect condition and frequently tested.

Pressure vessels

Many vessels contain fluids under positive (i.e. higher than atmospheric) pressure; they are designed for thermal, physical, chemical or other technological processes or used for the storage and transport of compressed, liquefied or dissolved gases and liquids. Other vessels operate at negative pressure, i.e. less than atmospheric pressure. Positive pressure vessels may accidentally be subjected to negative pressure and vice versa.

Although there are many different types of pressure vessel with all sorts of mountings and for various purposes, it is possible to distinguish two basic parts in these vessels: the shell and the external mountings. From the safety viewpoint the vessel shell is of fundamental importance and must afford the necessary strength and tightness under all the operating conditions for which the vessel has been designed. The great majority of pressure-vessel shells consist of a vertical or horizontal cylinder with convex, concave or flat ends which either are welded to the cylinder or may be dismountable. The shells can also be spherical or have flat walls forming a cube or other geometrical solid. In some cases inert gas injection is used either to purge vessels or to slow chemical reactions or to bring them to an emergency stop. Monitoring and safety devices should then be fitted in order to avoid dangerous excess velocity, flow, pressure or temperature.

The design features and measures to ensure the safety of some of the most widely used positive pressure vessels are discussed below.

Autoclaves

These vessels are designed for the steam treatment of various products of the building materials, chemical and food industries. A special design feature of autoclaves is that one of the ends opens like a door and is closed by a quick-opening slide lock or bayonet lock. The hinged end is rapidly sealed by turning a rotary lid having lugs with tapered locking surfaces. As the latter have a relatively small area, the cross section of the autoclave must be perfectly circular at the door end, which is particularly important for autoclaves of large diameter (2 m and more). From the safety viewpoint it is therefore necessary to choose the correct wall thickness not only to withstand the inside pressure and steam temperature but also to maintain the circular shape of the shell throughout its service life. Should the cross section become oval (e.g. due to non-uniform heat distribution over the shell circumference), local leakage may result at the bayonet lugs attached to the lid which remains circular. Such a leakage may give rise to a spontaneous shift in the angular position of the lid or to the distortion of one or several lugs, and the lid may be blown open with an energy equal to that of an autoclave explosion. In this case the autoclave end and shell are generally torn apart and travel at high speed in opposite directions. Autoclaves should therefore be periodically checked for ovalisation of the cross section and for tightness of the slide lock, i.e. for inadmissible distortions of the locking lugs, gasket wear, etc. The maximum permissible ovalisation of the cross section (as a rule less than 1%), admissible distortions of the lugs and periodicity of inspection must be specified by the manufacturer in the documentation supplied with the autoclave.

Furthermore, every autoclave must be equipped with an interlocking system which prevents the slide lock from being opened as long as the autoclave is under pressure.

Digesters

As regards the technical process–steam treatment of a product–digesters are like autoclaves. The difference lies in the vessel construction: the product to be processed (generally a liquid suspension) is charged through a pipe at the upper end of the upright digester and discharged through another pipe at the bottom. As digesters have no open end, they are considerably less dangerous than autoclaves, though their diameter may attain 5-6 m or more and their height up to 15-20 m. Digesters are used in the paper industry for pulp production. The inside surface of digesters is covered with a protective lining or cladded with a corrosion-resistant metal. The protective lining must be periodically inspected because defects in it would rapidly bring about a deterioration of the strength of the shell leading to its failure.

Fired stills

These are vessels for distilling petroleum or mineral oil, for coking petroleum and other products, in which the process relies on combustion heat. The temperature inside a still may attain 500 °C at low absolute pressures of up to 0.2 kgf/cm² or 3 psig (20 KPa). Stills can have diameters of 3 m and more, and lengths of up to 20 m and more. The safety measures to be taken during the operation of stills are basically directed at the protection of the personnel from burns and at the prevention of fires.

Heat exchangers

They are designed for heating or cooling one fluid by another or for condensing steam or vapours. The heat exchange between the two fluids takes place through tubes generally arranged inside a cylindrical shell which may be rigid, horizontal or vertical, and may have a compensator or floating head. Heat-exchanger tubes are either straight or bent into loops or coils. The heating surface of the tubes may attain 2 000 m² and the length of straight tubes 18 m. The characteristics of the two fluids may be very different from each other. Heat exchangers for production processes in the synthetic fuel industry, for instance, operate under high pressure. They are of the tube-in-tube type, have a heating surface of up to 500 m² and smooth or finned tubes. Heat exchangers are widely used in practically all branches of industry, but in particular in chemical, gas and power generating plants.

The safety measures to be taken depend on the characteristics of the fluids passing through heat exchangers, i.e. working pressure and temperature. As the inside surface of heat-exchanger shells is generally inaccessible for inspection, the necessary examinations of their metallic elements and welded joints must be carried out with ultrasonic testing instruments from the outside of the shell. To check the intensity of the corrosive attack on the inside surfaces of heat exchanges and in particular of their shells, blank test pieces kept outside the vessel are used for comparison; these blanks are made from the same material as the heat-exchanger shell.

Reactors, regenerators, contactors

These vessels found in the chemical, petroleum and gas industries can be horizontal or vertical, have diameters of up to 5-6 m and lengths or heights of up to 25-30 m, with inside pressures of up to 20 kgf/cm² or 280 psig (1 960 KPa). The safety measures to be taken are determined by the pressure, temperature and properties of the fluids processed in the equipment, which may be harmful (toxic) or flammable. All details decisive for the safety of these vessels must be stated in the instructions to be provided by the manufacturer and strictly observed by the equipment owner who should take into account the conditions of operation, which may entail less frequent servicing. Manufacturers' instructions regarding safety have to be considered as minimum requirements that can always be improved upon.

Column-type vessels

These are used for various purposes, such as fractionating columns for the separation of butane, propane, methane and other gases, separators for synthetic fuel, extraction columns, etc. There are columns with diameters of up to 7 m and heights of up to 60 m; the pressure of the fluid processed does not, as a rule, exceed 40 kgf/cm² or 560 psig (3 920 KPa). Columns are generally installed outdoors, and if they are high they must be designed to withstand calculated wind loads which should be specified in the documentation supplied with the equipment. The higher the centre of gravity of columns is above the ground level, the more stringent should be the calculations regarding resistance to seismic shocks if the equipment is to be installed in a seismically active region. Ladders and platforms should be provided for safe access to the top and intermediate heights.

Storage and transport vessels for gases, liquefied gases and liquids

Stationary vessels or tanks of cylindrical or spherical shape may attain volumes of 800 m³; the pressure of their contents can be expressed in terms of liquid column height and may attain 15-20 kgf/cm² or 210-280 psig (1 370-1 960 KPa). Vessels used for the transport of gases and liquids are subdivided in tanks, barrels and cylinders. Transport tanks are mounted on railway wagon underframes, truck or trailer chassis, and also on other means of transport. Barrels are vessels with a capacity of more than 100 l (0.1 m^3) and have rims for rolling them over. Cylinders are vessels with a capacity of less than 100 l (0.1 m^3) and have one or two necks with openings for screwing in valves or pipe connections.

The classification of transport vessels is only approximate, especially as regards their volume and, since they present greater hazards than stationary vessels, it is advisable to classify them by categories which take into account specific requirements to be met by their construction and use. When determining their wall thickness, for instance, internal impact loads occurring during transport must be taken into account, i.e. jerks produced when tank wagons and tank trucks start and stop, and also stresses occurring in barrels and cylinders when they are filled or emptied. Transport tanks are equipped with manholes or inspection covers and safety valves, whereas barrels and gas cylinders have none.

It is very important that the colour codes identifying contained fluids be the same for all transport vessels. An effort is being made at the international level to harmonise colour codes for transport on roads, railways, ships and planes. Filling rules should be established in accordance with the type of vessel and degree of hazard presented by the fluid to be transported. Every vessel must be clearly marked with the characteristics of the fluid, the names of the manufacturer and the owner, and the periodicity of inspection (see GAS CYLINDERS).

Storage and transport vessel defects and failures

Storage and transport pressure vessels may become defective or fail during operation or use in a similar manner as steam and hot-water boilers. Therefore all remarks made with regard to failures in boiler construction and operation also apply to them. Certain safety measures to be taken are determined by the properties of the fluid they contain. In the case of oxygen, for instance, everything must be done already at the construction stage to avoid contamination of the inside of the vessel by a lubricant, because a chemical reaction takes place when oxygen comes in contact with oil or grease. This reaction causes a rise in temperature which, in turn, determines a rise in pressure and may result in an explosion. Another hazard is the ignition of oil vapours in oxygen containers and air receivers. If pressure vessels are intended to receive flammable or harmful (toxic) gases, vessel tightness, appropriate design of sealing elements (flanges, shut-off devices, etc.) and correct choice of sealing materials are of great importance, because leakage of these gases may give rise to fires or poisoning. All precautions to be taken in accordance with the properties of the fluids which pressure vessels may receive must be laid down in the manufacturer's instructions and be strictly observed by the vessel user.

The commonest defects and failures which may arise during the operation and use of storage and transport pressure vessels are as follows.

Rise in pressure. In stationary industrial vessels the pressure may rise as a result of a processing problem or jammed safety device. Overpressure may also be caused by faulty installation or unsuitable colour of paint in the case of vessels intended for operation under ambient temperatures and having no insulation cover. If such vessels are heated by a nearby source of heat or solar radiation, the inside pressure will inevitably rise. In the case of gas cylinders and barrels, which have no safety devices, the probability of an explosion merely depends on the degree of the unforeseen rise in temperature or on vessel leakage. Overpressure in gas cylinders and barrels may also be due to overfilling because of an incorrect indication read on a level gauge or pressure gauge. Danger also occurs when a vessel is filled with a substance for which it has not been designed.

Loss in wall thickness. This phenomenon may be due to corrosion or erosion. It is similar to that described for boilers. However, in storage and transport vessels, where the inside surface is inaccessible for periodical inspection, wall thinning as a cause of failure is of great importance and requires the application of special methods (use of blank test pieces with ultrasonic flaw detection or equivalent non-destructive tests).

Failure or incorrect adjustment of safety, automatic shut-off and alarm devices. This applies in particular to vessels with quick-opening lids. Interlocking safety devices must be correctly adjusted to ensure that the hinged lid stays closed and may only be opened when the vessel is no longer under pressure.

Special attention should be paid to the safety of cylinders for the storage and transport of compressed, liquefied or dissolved gases, because (1) these cylinders may be found anywhere in industry, (2) there are always people working near oxygen, carbon dioxide or acetylene cylinders, and (3) propane or butane cylinders are kept near or in living quarters (e.g. kitchens). Failures of these cylinders may therefore have particularly serious consequences.

Valves and gauges

Boiler plants and pressure vessel installations comprise many different valves, i.e. devices for opening, shutting off and regulating the flow of working fluids and for protecting BPV from pressures beyond fixed values. Additional pressure relief devices are breaking discs and fusible plugs.

Gauges are intended for measuring and monitoring the pressure and temperature of the working fluid, its level (if the fluid in the boiler or vessel must be present in both its liquid and gaseous phases), velocity and rate of flow, etc. They may be used as warning devices and control actuators.

Valves are characterised by their throughput capacity or diameter, by the parameters and properties of the working fluid (pressure, temperature, acidity, alkalinity, etc.), by design type (gate valve, butterfly valve, cock, etc.) and by purpose (shut-off, regulating, safety valve).

Valves must meet stringent requirements regarding the strength and other characteristics of the material of which their main parts are made, as well as the tightness of their closing and packing elements, and adequately tested.

Shut-off and regulating valves can be manually operated by turning a handwheel on the bonnet or actuated by means of linkage rods 6-8 m away or remote-controlled with the aid of electric, hydraulic or pneumatic actuators, capable of signalling their actual state (open, partially closed, closed).

Safety valves must meet most stringent requirements in design and construction in order to ensure the highest reliability in operation. They are used to protect BPV from unwanted pressure by automatically opening (or actuating) a working fluid discharger into the atmosphere or transferring it into a receiver of lower pressure. The admissible excess pressure (3-10% above the working pressure according to class and purpose of the equipment) at which safety valves must open, and the amount of reduction in pressure after which it should close again, are laid down by existing rules. To check whether safety valves operate reliably, they are periodically, i.e. several times a day, opened by hand. The number of safety valves equipping an installation depends on the volume of working fluid and discharge capacity of the valves. Boilers should be fitted with not less than two safety valves. The nominal discharge capacity of safety valves is determined by testing at least three valves and rated to be 90% of the actual average discharge capacity measured. Testing, removal or installation of safety valves may involve fluids at extreme high pressure or temperature. Ear drum protection, eye protection, gloves and protective clothing are some of the measures necessary in and around the area of testing, removal and installation.

On account of the considerable noise produced by the blowdown of safety valves, design measures are taken to reduce the noise level to admissible limits. Escaping fluids should be collected and guided to a location where they can be treated in order to be made harmless.

In pressure vessels for fluids which are detrimental to the closing elements of valves or cause them to stick (e.g. gumming by resinifying substances), *bursting discs* are used instead of safety valves. These one-time membranes or discs break if the pressure exceeds a predetermined limit and discharge the working fluid from the vessel.

Fire actuated or steam actuated *fusible plugs* are additional safety devices. Gauges are of many kinds for many purposes. The operation of BPV depends in particular on pressure gauges and liquid-level gauges. The most widely used type of *pressure gauge* is one with an elastic measuring element, a tube with oval cross section bent into a ring: pressure changes inside the tube modify the cross-sectional shape so that the tube tends to straighten or curl; this motion is multiplied by a quadrant and pinion mechanism and transmitted to a pointer on a circular scale graduated in units of pressure (pascal, N/cm^2, kgp/cm^2, kgf/cm^2, bars, psig). The pressure-gauge connection pipe must be fitted with a three-way cock for purging the connection pipe and connecting a test gauge. To ensure that the elastic measuring tube always contains the liquid phase of the working fluid, the connection pipe must have a liquid seal (bending the pipe into 2-3 coil helices generally suffices). To maintain the reliability of the gauge, the working pressure in the equipment should not exceed two-thirds of the gauge scale. A control pointer fitted coaxially to the gauge pointer is provided to monitor the maximum pressure at which the equipment may be operated; this control pointer only moves in the direction of the pressure rise. On analogue scales the maximum permissible pressure is generally marked by a red line thus giving a warning of approaching danger. Such a warning cannot be given by digital scales; therefore analogue scales should be preferred. Pressure gauges are periodically (as a rule once a year) examined on a test bench or *in situ* by means of a test gauge. The number of pressure gauges equipping an installation, their location and their degree of precision are laid down by existing rules in accordance with the working-fluid pressure and temperature and the type and capacity of the equipment.

With BPV in which the working fluid is present in both its liquid and gaseous phases, safety in operation depends on the position of the interface between the two phases. This position is monitored by means of a *liquid level gauge*. The simplest level gauge consists of a vertical glass tube connected by pipes to the two phases of the working fluid. The connection pipes are fitted with cocks enabling the pipes to be periodically scavenged (to prevent obstructions). In the case of high working-fluid pressures and temperatures the glass-tube gauges are replaced by sight glasses made of special glass or mica which are tightly held in their frames by bolted sealing strips. For better visibility of the liquid level the sight glasses are illuminated. The scale of the gauge must bear clearly visible marks corresponding to the admissible "lower" and "upper" levels; these marks should be at a distance of not less than 25 mm from the boundaries of the visible gauge portion. The number and locations of level gauges are laid down by inspectorate rules. Steam boilers should have at least two level gauges fitted to each steam drum or to the boiler shell.

Monitoring and control devices

In addition to safety valves and gauges, BPV may be fitted with a number of control systems: automatic regulation, overheat protection, automatic shutdown, monitoring and signalling devices, remote controls, automatic control of periodic and starting-up operations. The automatic regulation system consists of a number of independent regulators for stabilising the different parameters of continuous processes, and of groups of interconnected regulators for stabilising certain parameters associated with a single process. The overheat protection system is to prevent, locate and eliminate emergency conditions which may arise when limit parameters have been exceeded; the system consists of alarms and circuits triggering safety devices. The automatic shutdown system is designed to prevent damage which may result from the spontaneous arrest or start of boiler mechanisms which are connected to common control circuits under normal operating conditions, and also to simplify manual or remote-control operations and to exclude errors which may be made by attendants when manually controlling certain processes or emergency conditions. The electronic control gear is assembled in compact blocks and installed by groups in cabinets. The recording and display instruments and the control switches are mounted on the boiler or power control panel; safe ergonomic criteria should be followed in order to avoid excess fatigue, errors and accidents.

ZVER'KOV, B. V.

CIS 79-1852 "Technical rules for steam boilers" (Technische Regeln für Dampfkessel (TRD)). Bekanntmachung. *Bundesarbeitsblatt* (Stuttgart), Sep. 1979, 9 (99-128). Illus. (In German)

CIS 80-1829 *Pressure relief devices* (Institution of Mechanical Engineers, Holt Saunders Limited, 1 St. Anne's Road, Eastbourne, United Kingdom) (1979), 102 p. Illus. 74 ref.

CIS 78-1237 *Safety standard for pressure vessels for human occupancy*. ANSI/ASME PVHO 1-1977 (American Society of Mechanical Engineers, 345 East 47th Street, New York, NY) (15 Oct. 1977), 67 p. Illus. 21 ref.

ASME boiler and pressure vessel code (American Society of Mechanical Engineers, 345 East 47th Street, New York, NY) (last revision).

Model code of safety regulations for industrial establishments for the guidance of governments and industry (Geneva, International Labour Office, 3rd impression, 1962), 523 p.

Bolt guns

Bolt guns are tools which use an explosive charge to force a nail or stud into a structure made of materials such as brick, concrete or steel; they are widely used in building and civil engineering. In the pistol type, the explosive force acts directly on the projectile (figure 1), in the hammer type, it propels a hammer on to the projectile (figure 2). The characteristic common to all these tools is that they force the nail or stud instantaneously into the material, where it remains fixed by pressure without the need for any binding material or support.

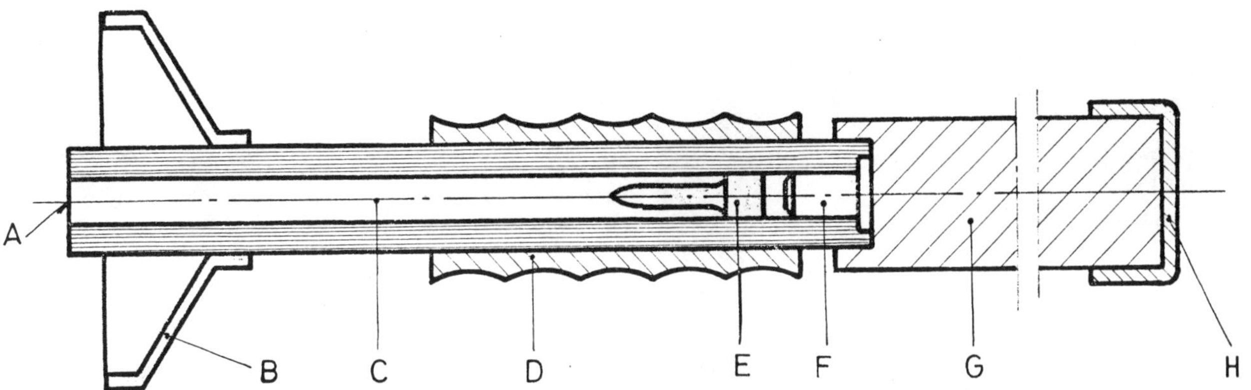

Figure 1. Cross-sectional diagram of a direct-firing bolt gun. A. Muzzle. B. Flash guard. C. Barrel. D. Hand grip. E. Projectile. F. Breech with cartridge. G. Firing mechanism including firing pin and safety catch. H. Hammer.

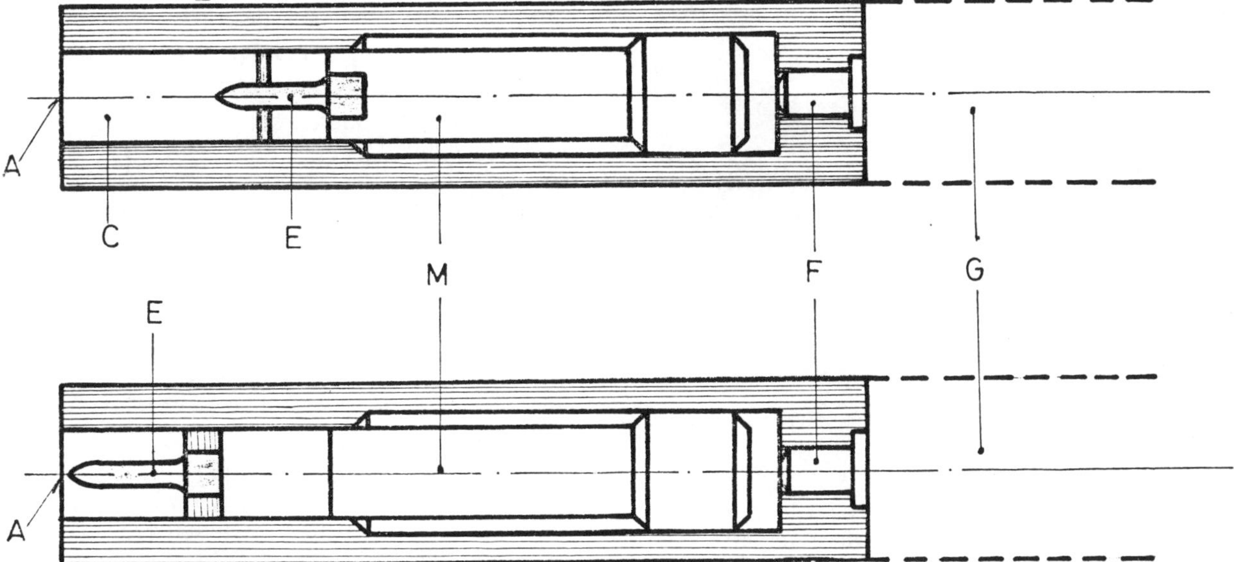

Figure 2. Cross-sectional diagram of a hammer-type (indirect-firing) bolt gun. A. Muzzle. C. Barrel. E. Projectile. F. Breech with cartridge. G. Firing mechanism including firing pin and safety catch. M. Hammer.

A bolt gun consists essentially of the barrel, the breech, the firing pin and a hand grip to enable the gun to be held securely. It is also fitted with a flash guard and some type of safety catch to prevent accidental firing.

Loading and firing

Explosive charge. The cartridge and the projectile do not form a single unit. Cartridges of different strengths can be used and should be clearly marked so as to indicate the charge contained. This is usually shown by means of a recognised colour code which must be clearly visible at the nose and on the base of the cartridge. The projectiles (nails, pins, etc.) are especially designed, both as regards their shape and the material used, according to the type of gun with which they are intended to be fired. The cartridge is placed into the breech between the firing pin and the barrel. The projectile may be introduced through the muzzle or through the breech. Some guns have a cylinder or an arrangement for belt feed enabling a number of cartridges to be loaded at one time. The cartridge is usually fired by means of a lever- or twist-type trigger or, in rare cases, by means of a blow from a hammer. It must not be possible to fire the gun unless the various safety catches have been freed.

HAZARDS

The main causes of accidents in the use of bolt guns are:

(a) careless handling of the tool;

(b) firing of projectiles into unsuitable target materials;

(c) ricochets of the bolt from the surface or the interior of the target material;

(d) projection of debris of the target material or of parts of the bolt or cartridge case;

(e) firing of projectiles whilst unauthorised persons are in the firing zone.

Accidents may also occur before or after firing due to incorrect loading and unloading procedures, attempts to rectify faults in a loaded tool, etc., or at the moment of firing, e.g. by firing into a material which is too brittle or which does not have uniform density. An accident may involve the user and/or any other person in the vicinity. When the user himself is injured, it is frequently either because he has pointed the tool towards his own body, has tried to rectify a fault on his loaded tool, has used the tool without a guard or with an unsuitable guard, has inserted the stud at a point too close to the corner of a structure, or because the projectile has ricocheted. Other persons may be injured by projected fragments of the target material, by shrapnel or by the complete projectile if it is fired accidentally or if it is aimed at, for example, a wall which is of inadequate strength and allows the projectile to pass straight through and strike a person on the other side.

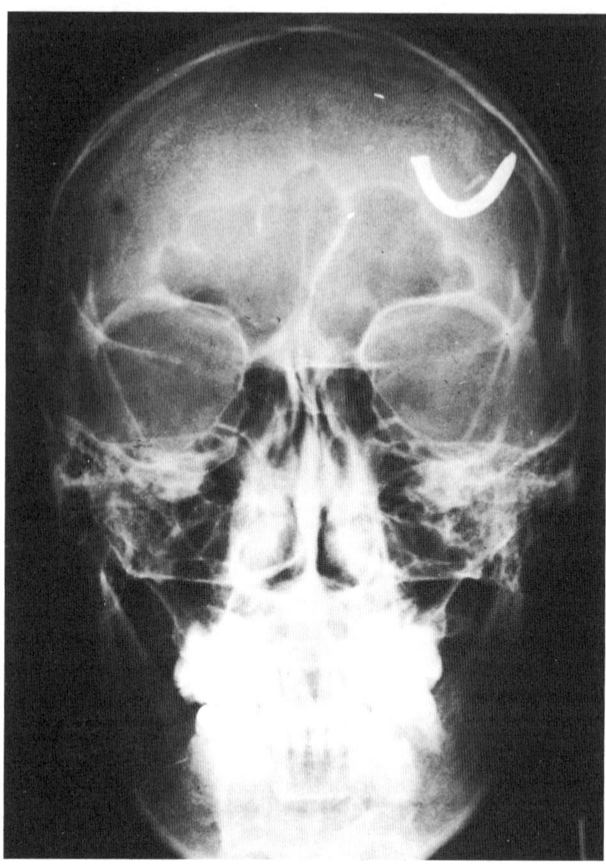

Figure 3. Skull injury caused by the ricochet of a pin from a bolt gun.

SAFETY MEASURES

Safety precautions must be taken both in regard to the constructional details and to the use of the tool.

Constructional details. Bolt guns must be constructed according to the highest standards and with best quality materials only. The firing mechanism of the gun and the safety devices must be such that the gun will only fire when the following conditions are fulfilled:

(*a*) the flash guard must be fitted;

(*b*) the muzzle and the periphery of the flash guard must be firmly pressed against the working surface with a pressure of not less than 5 kg;

(*c*) the angle between the axis of the gun and the perpendicular to the working surface must not exceed 7 °.

The safety devices must prevent the gun from being fired during normal handling and when dropped on to a hard surface from a height of 3 m. The flash guard must be made of strong material and its outer rim must extend to at least 50 mm from the axis of the barrel. Suitably shaped flash guards must be provided in order to enable the gun to be fired with safety in corners, around or close to projections such as in the case of steel sections or wooden mouldings, etc.

Use. Only approved types of bolt gun which conform to safety regulations should be used. Cartridges, nails, pins or other projectiles should be of a type for which the tool was designed. Bolt guns should only be used by a responsible person who has been adequately instructed in the correct procedures. The user should wear protective clothing including safety spectacles, a hard hat in the case of overhead firing, and, in certain special cases, ear protection and some protection for the body such as a leather apron. After use, the gun, cartridges, bolts and safety equipment should be stored in a suitable box.

Bolt guns should not be used in explosive or flammable atmospheres. Guns should only be loaded immediately before firing is to take place; they should not be used for firing:

(*a*) into a target material of insufficient strength which might allow the projectile to pass straight through;

(*b*) in the vicinity of gas pipes or electrical conduits;

(*c*) into materials such as cast iron, pottery, earthenware or other brittle materials;

(*d*) into hardened steel or hard or closely compacted stone;

(*e*) into any material which is insufficiently rigid;

(*f*) across or through a pre-existing hole unless a special guide is provided for the projectile;

(*g*) into masonry or concrete at a point less than 10 cm from the edge of this material;

(*h*) at any point within 5 cm from another bolt which is badly fixed, bent, damaged or wedged or from a zone where the target material may be fractured or otherwise damaged.

Care and maintenance. It is essential for safety reasons that bolt guns are carefully cleaned and checked after each day's use. With intensive use, they should be cleaned and checked more frequently. Any defective gun should be taken out of service immediately.

Cartridge-assisted hammers

This type of tool in which the cartridge is detonated by a heavy blow from a club hammer is safer than the pistol-type gun and is, therefore, to be preferred. However, the safe practices described above are also applicable to this tool except that:

(*a*) the flash guard may be dispensed with, providing the projectile can be placed directly in contact with the target material before firing;

(*b*) the user need not wear a safety helmet;

(*c*) the distance to be left free from the edge of the material (10 cm) or from a previously fired and damaged bolt (5 cm) may be reduced;

(*d*) permission may be given to fire across or through an existing hole. The special conditions to be fulfilled in such cases should be set out precisely.

CHAVANEL, A.

CIS 77-1532 *Explosive actuated fastening tools.* Data Sheet–Occupational Safety and Health No. E-1 (Canada Safety Council, 1765 St. Laurent Boulevard, Ottawa) (1977), 14 p. Illus. 11 ref.

CIS 77-1441 *Bolt guns* (Pistolets de scellement). Editions INRS No. 549 (Institut national de recherche et de sécurité, 30 rue Olivier-Noyer, Paris) (Mar. 1976), 29 p. Illus. 6 ref. (In French)

CIS 78-2044 *Safety in the use of cartridge operated fixing tools.* Health and Safety Executive, London. Guidance Note Plant and Machinery/14 (London, HM Stationery Office, May 1978), 7 p. 4 ref.

CIS 79-1842 *Use of explosive-actuated tools* (Arbeiten mit Schussapparaten). VBG 45, Hauptverband der gewerblichen Berufsgenossenschaften (Cologne, Carl Heymanns Verlag, 1979), 16+7 p. (In German)

Bone and bone meal

Bone is the hard connective tissue which forms the basic component of the skeleton of vertebrates. On the basis of shape and structure, bones can be divided into three different types: long or tubular; flat or broad; and short.

Composition. Bones comprise 20-25% water and 75-80% solid materials of which 67-70% are organic substances and 30-33% inorganic substances. However, this chemical composition will vary considerably as a function of age and diet. The mineral composition of bone is approximately 21-25% calcium, 9-13% phosphorus, nearly 1% magnesium and up to 5% carbonic acid together with traces of potassium, sodium and fluorine. In domestic animal bones, the main chemical compounds are $Ca_3(PO_4)_2$, $CaCO_3$, $CaF_2.3Ca_3(PO_4)_2$ and $CaCO_3.3Ca_3(PO_4)_2$ which form complex crystals similar to those of the apatite group. The collagenous component of bone–ossein–is a mixture of proteic substances, mainly collagen; in fact, collagen accounts for 93% of the total protein content of the long bones, the remainder comprising elastin (1.2%) and alkali-soluble proteins–mucins and mucoids (5.7%). Bone also contains glycogen, various enzymes (proteinases, peptidases, phosphorylases, alkaline phosphatases) and vitamins A and C.

Uses. Bone is a valuable raw material for the industrial production of edible and non-edible fats; after fat extraction, the limb bones are also used in the manufacture of haberdashery, musical instruments, etc. However, the prime uses of bone are adhesives manufacture (gelatine and glue) and the production of bone meal for use in fodder and fertilisers.

Bone meal production. The main processes in the conversion of bone to bone meal are: sorting, crushing, degreasing, cleaning and secondary crushing, separation of fully crushed bone from nitrogenous waste, washing and steam and hot-water treatment for glue extraction, drying and milling. Bone-meal production has become centralised and is often highly mechanised; however, a certain amount of small-scale production is still carried on and it is here that the poorest conditions in this industry are usually found.

HAZARDS

The processes of bone crushing and milling give off considerable quantities of dust, comprising mainly organic material, and prolonged exposure to high concentrations of this dust may give rise to eye and respiratory system disorders. Chemical treatment of bones such as fat extraction using organic solvents or sulphuric acid may result in the release of dangerous gases and vapours (gasoline, toluene, dichloroethanes, sulphur dioxide, etc.).

Workers handling raw bones are in danger of contracting zoonoses such as anthrax, glanders, foot and mouth disease, aphthous fever, brucellosis, Q fever, etc., from infected material. Finally, the decomposition of organic matter in areas used for the storage of raw bones, especially where standards of cleanliness and good housekeeping are poor, will be accompanied by the generation of unsavoury odours and gases such as hydrogen sulphide, ammonia, indole, etc., which may cause air pollution and incommode people in neighbouring residential areas as well as having an undesirable effect on the health and well-being of workers.

SAFETY AND HEALTH MEASURES

Raw bones should be degreased as early as possible after reception and stored in a well ventilated, covered area to reduce the evolution of unpleasant odours and gases from rotting organic material. Incoming material should be subject to veterinary or public health service supervision to ensure that infected material is detected at an early stage, and untreated bones should be carefully separated from the finished products to limit the danger of contamination.

Manual lifting and carrying should be replaced by mechanical handling, and bone processing should be mechanised and carried out in enclosed equipment to prevent the escape of dangerous or obnoxious substances into the workplace air. Dusty processes and conveyors carrying fine bone meal should be fitted with exhaust ventilation equipment. Dust-laden air should be filtered before being released into the atmosphere. Air exhausted from workshops housing malodorous processes should not be recycled. For the final processes in the preparation of animal feed and industrial products, it is advisable to have a ventilation system completely separate from that used for the more obnoxious processes.

In degreasing and fat-extraction processes, the least dangerous solvent should be used, and under no circumstances should benzene be employed. Exhaust ventilation should be installed to collect any toxic vapours or gases evolved and, where flammable solvents are employed, appropriate fire-protection and prevention measures should be taken.

High standards of personal hygiene are essential for persons handling raw bones, and particular care is essential to ensure that any small cuts or abrasions do not provide a portal of entry for pathogenic micro-organisms; first-aid facilities should be available for the immediate disinfection of minor wounds. Adequate sanitary facilities and work-clothing laundry services should be provided and workers should be subject to periodic medical examinations. Hand and arm protection should be worn for the manipulation of raw bones and eye and face protection should be supplied to persons exposed to splashes from solvents or other chemicals.

In certain countries, work on the processing of bones and bone products is considered unsuitable employment for women and young persons.

GALKINA, K. A.

Bone charcoal:

CIS 1609-1968 "Pulmonary anthrax caused by contaminated sacks". Enticknap, J. B.; Galbraith, N. S.; Tomlinson, A. J. H.; Elias-Jones, T. F. *British Journal of Industrial Medicine* (London), Jan. 1968, 25/1 (72-74). 6 ref.

Boranes

BORON HYDRIDES; HYDROGEN BORIDES

Boranes are highly toxic chemical compounds composed of boron and hydrogen. They easily oxidise and are strong reducing agents. Many boranes are highly flammable and may ignite spontaneously or explode in moist air or in the presence of oxygen. They hydrolise at room temperature and release hydrogen and boric acid giving off heat. The boron hydrides include diborane (B_2H_6), tetraborane (B_4H_{10}), pentaborane (9) (B_5H_9), pentaborane (11) (B_5H_{11}), hexaborane (B_6H_{10}), decaborane ($B_{10}H_{14}$) and others. The three most frequently encountered in industry are diborane, pentaborane (9) and decaborane.

Production. Diborane is obtained by reaction of lithium aluminium hydride with BF_3 and from boron trialkyls and hydrogen; the remaining boranes are produced mainly by pyrolysis of diborane under various conditions.

Uses. The boron hydrides are used mainly in high-energy fuels. Other uses include welding, the pharmaceutical and perfume industry boronising metals, as reducing agents in organic synthesis and as a vulcanising agent in the rubber industry (decaborane) (see ROCKET PROPELLANTS).

HAZARDS

Boranes are highly toxic by inhalation, skin absorption or ingestion. They may produce both acute and chronic poisoning. The toxicity of the hydrides is many times greater than that of other boron compounds and pentaborane is considered more toxic than diborane and decaborane. It is more toxic than phosgene and HCN. The relationship between reducing capacity and toxicity has been reported. Experiments have indicated that repeated exposure to boranes produces cumulative effects. Vapours have an irritating effect on skin and mucous membranes. In particular pentaborane and decaborane produce marked irritation of the skin and mucous membranes, causing necrotic changes, serious kerato-conjunctivitis with ulceration and corneal opacities.

The boron hydrides present a considerable fire and explosion hazard. They undergo an explosive reaction with most oxidising agents, including the halogenated hydrocarbons.

Diborane (B_2H_6)

BORON HYDRIDE; BOROETHANE

m.w.	27.7
sp.gr.	liquid 0.45, solid 0.58
m.p.	−165.5 °C
b.p.	−92.5 °C
v.d.	0.96
v.p.	224 mmHg (29.8·10³ Pa) at −112 °C
e.l.	0.9-98%
f.p.	−90 °C
i.t.	38-52 °C

a colourless gas with sickly-sweet, nauseating smell.

TWA OSHA	0.1 ppm 0.1 mg/m³
IDLH	40 ppm

It hydrolises exothermically in water to hydrogen and boric acid. It ignites spontaneously in moist air.

Health hazard. Animal experiments have given an LD_{50} of 30-90 mg/m³ for 4-h exposure, and the symptoms of acute poisoning are respiratory distress, hypoxia, pulmonary haemorrhage and oedema followed by death. Experiments on chronic poisoning with concentrations of 1.1-6.8 mg/m³ led to death after 7-10 exposures. Autopsy revealed rhinitis, pneumonia and structural lung damage. Animals surviving after exposure to low concentrations manifested no pathological changes.

Acute poisoning in man gives rise to a syndrome like that of metal fume fever including the following symptoms: tightness, heaviness and burning in chest, coughing, shortness of breath, pericardial pain, nausea, shivering and drowsiness. The earliest symptoms are those indicating respiratory damage. Signs of intoxication may occur soon after exposure or following a latent period of up to 24 h and then persist for from 1-3 days or more. Long-term exposure to low concentrations produced pulmonary irritation, headaches, dizziness, muscular fatigue and weakness and occasional, transient tremors. An increase in non-protein blood nitrogen and positive cephalin flocculation tests also occurred after exposure.

Pentaborane (9) (B_5H_9)

PENTABORANE STABLE; PENTABORON ENNEAHYDRIDE

m.w.	63.2
sp.gr.	0.66
m.p.	−46.8 °C
b.p.	58.4 °C
v.d.	2.2
v.p.	66 mmHg (8.78·10³ Pa) at 0 °C
e.l.	0.42%

a colourless, volatile liquid with a sweetish, unpleasant odour.

TWA OSHA	0.005 ppm 0.01 mg/m³
STEL ACGIH	0.015 ppm 0.03 mg/m³
IDLH	3 ppm

It decomposes at temperatures above 150 °C and hydrolises in water. It ignites spontaneously in air.

Health hazard. Pentaborane is considered to be the most toxic of the boron hydrides when inhaled. It is also readily absorbed through undamaged skin, although to a lesser extent than decaborane. Inhalation experiments with small animals gave an LD_{50} of 9-46 mg/m³. The symptoms were those of central nervous system damage. Acute poisoning was accompanied by excitation, tremor, spasms. Death occurred due to cardiac insufficiency. Prolonged exposure to a vapour concentration of 2.5 mg/m³ for 6 months lead to loss of weight and appetite, decreased activity, listlessness and tremors. Autopsy revealed no serious pathological changes.

In man, the initial symptoms of slight intoxication may be nausea and drowsiness. Exposure of moderate severity leads rapidly to headaches, dizziness, nervous excitation and hiccups. There may also be muscular pains and cramps and spasms of the muscles of the face and extremities. In more serious cases, after 40-48 h, the symptoms are loss of mental concentration, incoordination, disorientation, cramps and convulsions, semi-coma and persistent leucocytosis. Functional tests suggested liver and kidney damage.

Decaborane ($B_{10}H_{14}$)

DECABORON TETRADECAHYDRIDE

m.w.	122.3
sp.gr.	0.94
m.p.	99.5 °C
b.p.	213 °C (decomposes above 170 °C)
v.p.	100 mmHg (13.3·10³ Pa) at 142 °C
f.p.	80 °C

soluble in ethyl alcohol, ethyl ether, benzene and chloroform; very soluble in carbon disulphide

white, orthorhombic crystals with an intense, bitter odour.

TWA OSHA	0.05 ppm 0.03 mg/m³ (skin)
STEL ACGIH	0.15 ppm 0.9 mg/m³
IDLH	20 ppm

It hydrolises slowly in water.

Health hazard. The LC_{50} for 4-h exposure to decaborane was 122-230 mg/m³ for small animals. Cutaneous application of doses of 47-440 mg/kg had the same effect. The main acute toxic symptoms were restlessness, depressed breathing, incoordination, general weakness, spasmodic movements and convulsions. Corneal opacity was also observed. In experiments with other modes of entry, cardiovascular effects were observed including bradycardia and periods of moderate hypertension preceding the terminal fall in blood pressure. Symptoms of chronic poisoning are similar to those of acute poisoning, but with more pronounced neurotoxic activity. Marked restlessness and excitement may be accompanied by aggressiveness, convulsions and violent tremor. Erythrocyte count rises and haematocrit readings are higher. Preterminal symptoms include: haematuria, albuminuria, increased bromsulphthalein retention and a rise in urine creatinine.

In man, mild exposure to decaborane produced headaches, dizziness, drowsiness and nausea. Symptoms of central nervous system damage predominated but were less marked than with pentaborane. Other symptoms, similar to those described in connection with diborane and pentaborane have been observed, although less frequently.

SAFETY AND HEALTH MEASURES

These must be directed at preventing toxic exposure and fire and explosion hazards in storage, handling and processing.

Storage. It is advisable to locate storage tanks in the open in areas clearly marked as danger zones with warning signs and "No Smoking" notices. Pentaborane should be kept under a dry, nitrogen atmosphere and diborane should be refrigerated to prevent decomposition. Containers should be shielded from direct sunlight. All containers should be thoroughly flushed before maintenance work is carried out on them and, in the event of leakage, only persons with adequate personal protective equipment should be allowed in the vicinity of containers. Any containers situated indoors must be adequately ventilated and away from any source of ignition.

Handling and processing. All equipment should be fully enclosed and carefully designed to eliminate any likelihood of auto-ignition. All equipment must be purged before opening. All possible sources of ignition should be eliminated and processes in which boron hydrides are heated must be minutely controlled. Automatic leak detectors with visual and acoustic warning signals should be fitted. Where possible, processes should be automated or operated by remote control.

The threshold limit concentrations recommended by the ACGIH are below the odour threshold (3.3 ppm for diborane, 0.8 ppm for pentaborane, 0.7 ppm for decaborane). Consequently vigilant monitoring of workplace concentrations is essential. Workers should be informed of the high toxicity of these substances and instructed on safety and emergency procedures in the case of leakage or personal exposure.

The extremely hazardous nature of the boron hydrides makes the use of personal protective equipment necessary even where maximum engineering safety measures are applied. Workers should be supplied with respiratory protective equipment either of the cartridge-type respirator or canister type with a filter comprising silica gel, hopcalite and activated charcoal, the combined canister-supplied air type or the self-contained type depending on the foreseen concentrations. Eye and face protection, impervious overalls, hand and arm and foot and leg protection, should also be worn. Adequate sanitary facilities should be provided and workers required to wash thoroughly at the end of a shift and before meals. Smoking, eating and drinking in workshops should be prohibited.

Fire fighting. Boron hydride fires are difficult to extinguish. For fighting diborane fires, it is recommended to apply an inert substance such as liquid nitrogen; calcined silica gel or sand should be applied to extinguish penta- and decaborane flames. Highly halogenated hydrocarbons (CCl_4) and also ketones, aldehydes and certain organic acid esters combine with boranes to form explosive solutions which are sensitive to shock, friction and heat. Such compounds must therefore never be used as extinguishing agents.

First aid. Persons exposed to boron hydrides should be immediately moved from exposure; after contaminated clothing has been removed, the victim's skin should be thoroughly washed with water. Treatment is symptomatic and supportive.

KASPAROV, A. A.
BALYNINA, E. S.

Boron hydride chemistry. Muetterties, E.; Earl, L. (New York, San Francisco, London, Academic Press, 1975), Ch. 2, 39 p.

Harmful substances in industry (Vrednye vescestva v promyslennosti). Lazarev, N. V.; Gadaskina, I. D. (Leningrad, Himija, 1977), 3, 315 p. (In Russian)

CIS 75-432 "Studies on the interaction of several boron hydrides with liver microsomal enzymes". Valerino, D. M.; Soliman, M. R. I.; Aurori, K. C.; Tripp, S. L.; Wykes, A. A.; Vesell, E. S. *Toxicology and Applied Pharmacology* (New York), Sep. 1974, 29/3 (358-366). Illus. 32 ref.

"Mechanisms of decaborane toxicity". Naeger, L. L.; Leibman, K. *Toxicology and Applied Pharmacology* (New York), 1972, 22/4 (517-527). Illus. 29 ref.

Boron, alloys and compounds

Boron (B)

a.w.	10.8
sp.gr.	2.34 (crystalline), 2.37 (amorphous)
m.p.	2 300 °C
b.p.	2 550 °C (sublimes)

an amorphous or crystalline, insoluble, brownish-black powder which is fairly stable at normal temperature, and combines with metals, oxygen, nitrogen, carbon and other substances.

Boron is a component of all animal tissues but its role in the body is not clear. Normal boron content of the blood is 9.8 μg/100 g with a range of 3.9-36.5, and of the urine, 715 μg/l, with a range of 40-6 600. Most boron compounds are toxic. Ores and some refractory compounds (titanium and niobium borides) are non-toxic. Boron trifluoride is also corrosive to skin and boride dust has caused pneumoconiosis.

Occurrence and production. Boron accounts for 0.001% of the earth's crust. It is not found free in nature but occurs as
borax ($Na_2B_4O_710H_2O$),
colemanite ($Ca_2B_6O_{11}5H_2O$),
boronatrocalcite ($CaB_4O_7NaBO_28H_2O$), and
boracite ($Mg_7Cl_2B_{16}O_{30}$).
The most important sources of boron are the minerals rasorite (kernite) and borax, and the largest known deposits have been found in Turkey, the USSR, the United States, Tibet, Italy, Argentina and Bolivia.

Depending on the situation of the deposit, the ore is extracted by underground or opencast mining. Boron is obtained by the reduction of the trioxide with magnesium or aluminium.

Uses. Elemental boron is used in metallurgy as a degasifying agent and is alloyed with aluminium, malleable iron and steel. It is also finding increasing use in the nuclear industries as a constituent of neutron shielding materials, e.g. in reactors.

Boric anhydride (B_2O_3)

BORON OXIDE

m.w.	69.6
sp.gr.	2.46
m.p.	about 450 °C
b.p.	1 860 °C

slightly soluble in water; soluble in ethyl alcohol
a vitreous, colourless, crystalline, hygroscopic solid.

TWA OSHA 15 mg/m³
TLV ACGIH 10 mg/m³
STEL ACGIH 20 mg/m³
MAC USSR 5 mg/m³

It forms boric acid when dissolved in water; it is used in the manufacture of glass, enamels and glazes.

Health hazard. This has a lower order of toxicity either by inhalation or ingestion, with an LD_{50} for laboratory animals of 3.16 g/kg. It has an irritant effect on the skin and mucous membranes of the eyes of rabbits. Prolonged absorption leads to loss of weight, dysproteinaemia, carbohydrate metabolism disorders, moderate changes of the liver and kidneys, and vascular disorders.

Boric acid (H_3BO_3)

BORACIC ACID

m.w. 61.8
sp.gr. 1.43
m.p. 169 ± 1 °C (transition point to HBO_2)
b.p. 300 °C

at 0 °C, a saturated aqueous solution contains 2.6% acid; at 100 °C, 39.7%; soluble in organic solvents

boric acid starts to dehydrate when heated to 70°C and metaboric acid (HBO_2) forms; complete dehydration results in boric anhydride (B_2O_3); the salts of boric acid are derivatives of metaboric or tetraboric acid ($H_2B_4O_7$); boric acid forms esters with methanol and ethanol in the presence of sulphuric acid

scale-like colourless crystals.

MAC USSR 10 mg/m³

Occurrence. Boric acid exists in nature as sassolite, a mineral, and also occurs in hot mineral-water sources.

Production. Boric acid extraction techniques depend on the type of deposit. The minerals are industrially processed with sulphuric acid. After separation of the sludge, boric acid crystals form while the filtrate is cooled.

Uses. Industrial uses of boric acid include the preparation of enamels and glazes, the manufacture of optical and stained glass, welding and soldering ("jeweller's borax"), metallurgy, electroplating, and the manufacture of dyes, paper, leather and pharmaceutical products. Boric acid is also used as a disinfectant, preservative and fertiliser. In laboratory practice boric acid serves for the preparation of buffer solutions, and in medicine as a disinfectant.

Borax ($Na_2B_4O_7.10H_2O$)

SODIUM TETRABORATE DECAHYDRATE; TINCAL

m.w. 381.4
sp.gr. 1.69-1.72
m.p. 75 °C (60 °C $-8H_2O$)
b.p. 320 °C ($-10H_2O$)

slightly soluble in water with alkaline reaction; insoluble in ethanol

large colourless prismatic crystals.

TLV ACGIH anhydrous 1 mg/m³
decahydrate 5 mg/m³
pentahydrate 1 mg/m³

Occurrence. Borax occurs in nature in the form of minerals, i.e. tincal (or crude borax) and kernite (or rasorite). It is contained in many mineral waters and salt lakes.

Production. Borax is obtained from boric acid, tincal, kernite, and also from the water of salt lakes (fractional crystallisation). It can be produced by reaction of boric acid with soda. Anhydrous borax is obtained by heating common borax to 350-400 °C.

Uses. Borax is used in the manufacture of special kinds of glass, enamels and glazes, as soldering and welding fluxes, for dissolving fats, for the fixation of mordants on textiles, in silk spinning, and in the soap, leather and cosmetics industries.

Health hazard. These compounds may enter the body by inhalation, ingestion or by absorption through the mucous membranes or skin burns. Absorption through damaged skin is rapid and almost complete; absorption also occurs through undamaged skin but not to a sufficient extent to cause poisoning.

Following absorption, there is a rise in the concentration of boron in the cerebrospinal fluid, but the highest concentrations are found in the brain tissues, the liver and adipose tissue. Repeated doses have a cumulative effect, retention being greatest in bone tissue. Retention in the bodies of growing animals is more pronounced. Elimination is mainly in the urine but also to a lesser extent in faeces, milk and sweat. In laboratory animals the LD_{50} for ingested boric acid was 2.66-3.45 g/kg and for borax, 2-5.33 g/kg. Symptoms include depression, cramps, ataxia, fall in body temperature and violent colouring of the skin and mucous membranes. Shock-like syndrome may develop and histological damage to the renal glomeruli and tubules and hyperchromatism and atrophy of certain central nervous system cells have been reported. Experiments have shown the ability of boric acid to penetrate the placental envelope affecting both sexual cycle and reproductive function in females, with large doses producing sterility.

Workers industrially exposed to borax often suffer from chronic eczema. Long-term exposure to borax dust may lead to inflammations of the mucous membranes of the airways (bronchitis, laryngitis) and to conjunctivitis.

Boron trifluoride (BF_3)

m.w. 67.8
sp.gr. 2.99
m.p. −126.7 °C
b.p. −99.9 °C

very soluble in concentrated acids and organic solvents

a colourless gas with a pungent, suffocating odour.

TWA OSHA 1 ppm 3 mg/m³ ceil
MAC USSR 1 mg/m³

It hydrolises in air, forming dense, white fumes and decomposes in water to form boric acid and fluoboric acid. It is used as a catalyst, as a flux for magnesium and as a fumigant.

Health hazard. The principal feature in the acute action of this substance is the irritation of the mucous membranes of the respiratory tract and eyes. In animal acute experiments, a concentration of 42 mg/m³ proved fatal in some cases. Examination revealed a fall in the inorganic phosphorus level in the blood and autopsy showed pneumonia and degenerative changes in the renal tubules. Long-term (4 months) exposure to 3 and 10 mg/m³ of boron fluoride produced irritation of the respiratory tract, dysproteinaemia, reduction in cholinesterase activity and increased nervous system lability. Exposure to high concentrations results in a reduction of pyruvic acid and inorganic phosphorus levels in the blood, and dental fluorosis. Examination of workers exposed to high concentrations of boron fluoride for 10-15 years revealed dryness and bleeding of the nasal mucous membrane.

Boron tribromide (BBr_3)

BORON BROMIDE

m.w. 250.6
sp.gr. 2.65
m.p. −45 °C
b.p. 91.7 °C
v.p. 40 mmHg ($5.32 \cdot 10^3$ Pa) at 14 °C, 100 mmHg ($13.3 \cdot 10^3$ Pa) at 33.5 °C

colourless fuming liquid.

TLV ACGIH 1 ppm 10 mg/m³

STEL ACGIH 3 ppm 30 mg/m³

It is used in radioelectronics. Boron bromide emits toxic fumes; it can explode when heated; it reacts with water as well as with organic matter, emitting corrosive fumes.

Boron trichloride (BCl_3)

m.w. 117.2
sp.gr. 1.35
m.p. −107.3 °C
b.p. 12.5 °C

a colourless fuming liquid at low temperature; decomposed by water or ethyl alcohol.

It is used as a catalyst for organic reactions.

Health hazard. This substance fumes and hydrolises in moist air forming hydrochloric acid and various decomposition products, which include oily liquids with powerful irritant and corrosive action. Exposure to 100 ppm has proved fatal to certain laboratory animals; symptoms included pulmonary lesions and irritation of the mucous membranes. The toxic action of the halogenated borons is considerably influenced by their decomposition products (hydrofluoric acid−HF, fluoboric acid−HBF_4, hydrochloric acid−HCl).

Boron hydrides

See separate article.

Boron carbide (B_4C)

m.w. 55.3
sp.gr. 2.52
m.p. 2 350 °C
b.p. > 3 500 °C
MAC USSR 6 mg/m³

black, hard crystals used as an abrasive.

Boron nitride (BN)

sp.gr. 2.34
m.p. 3 000 °C (under pressure)

white crystalline powder; slightly soluble in water, chemically stable; very hard.

MAC USSR 6 mg/m³ (hexagonal and cubic crystals)

It is used as an abrasive, for sockets of electronic lamps, and in rocket and nuclear engineering.

Health hazards. Boron-carbide and boron-nitride dust are fibrogenic. Chronic exposure to BN dust (100-200 mg/m³ four times a week over 6 months) and to B_4C dust (300-350 mg/m³ over 12 months) causes catarrhal-desquamative bronchitis, emphysema and moderately pronounced diffuse lung fibrosis in rats. Acute and chronic inflammatory diseases of the upper airways have been observed in boron-carbide workers exposed for 12-14 years.

Borides

sp.gr. 2.49-8.01
m.p. 2 100-3 040 °C

MAC USSR 1 mg/m³ for CrB_2, 2-5 mg/m³ for ZrB_2, 4 mg/m³ for CaB_6, 4 mg/m³ for Mo_2B_5, 10 mg/m³ for TiB_2, 10 mg/m³ for NbB_2

Boron forms borides with a number of metals: calcium (CaB_6), molybdenum (Mo_2B_5), chromium (CrB_2), titanium (TiB_2), zirconium (ZrB_2) and others.

They are obtained by reaction of the metal oxide with boron carbide, by electrolysis of molten mixtures of alkaline and alkali earth metal borates with oxides of high-melting point metals, by thermometallurgical reduction of mixtures of metal oxides and boron.

They are used for coating steel and other metals, as alloys, catalysts and semiconductors.

Health hazards. The toxicity of these compounds depends to a large extent on the metals with which the boron atoms are combined−molybdenum, chromium, tungsten, niobium, etc. Boride dust causes experimental pneumoconiosis.

Organoboron compounds

They are generally liquids in which boron is combined with different organic radicals of the R_3B type, e.g. ethyl, propyl, butyl, phenyl, etc.

Tripropylborate ($(C_3H_7)_3B$)

sp.gr. 0.72 (25 °C)
m.p. 65.5 °C
b.p. 164.5 °C

Triethylborate and tripropylborate are obtained by reaction of ethyl- or propylmagnesium bromide with boron fluoride; they ignite in contact with air.

They are used as raw materials for the production of boron hydrides and as polymerisation catalysts.

Health hazards. The toxicity of organoboron compounds has been studied experimentally. The LD_{50} of the most toxic ones, tricresylborate and triphenylborate, is 200-400 mg/kg; triethylborate, 230 mg/kg; tripropylborate, 1 200 mg/kg (rats).

Symptoms of poisoning are respiratory disorders, diarrhoea and shock syndrome after absorption of larger doses. Organoboron compounds exert localised irritating effects. Chronic poisoning with tripropylborate vapours in concentrations of 1-5 mg/m³ cause an increase in the blood carboxyhaemoglobin level and temporary reticulocytosis. The toxic mechanism is due to the fact that organoboron compounds oxidise in contact with air and form CO.

The MAC recommended in the USSR for tripropylborate is 3 mg/m³.

The safe concentration recommended for meta-carboranes ($C_2B_{10}H_{12}$), organoboron compounds used in the synthesis of heat-stable polymers, is 2-4 mg/m³. After exposure to nitrogen derivatives of *o*-carborane and 1,2-dicarbaundecaborane neurotoxic effects have been remarked.

SAFETY AND HEALTH MEASURES

It is important to prevent the air of industrial premises from becoming polluted with vapours and aerosols of boric acid, boric anhydride, borax, boron fluoride and boron chloride by tightly enclosing machinery and equipment, by local exhaust ventilation, and by mechanisation of packaging and handling. Gloves, aprons and protective footwear must be worn during work where there is a risk of thermal or chemical burns by hot solutions, vapours and concentrated sulphuric acid. The workers must observe personal hygiene rules and take a shower at the end of the shift. Smoking, drinking and eating at the workplace should be forbidden.

Treatment. In case of boric acid poisoning, vomiting is to be artificially induced or a gastric lavage is to be carried out. An exchange transfusion, intravenous injection of an electrolyte solution and cortisone application against dermatitis may be indicated. If boron chloride or boron fluoride is involved, the main hazards are hydrolysis products such as hydrochloric acid, hydrofluoric and borohydrofluoric acids; therefore, a therapy to neutralise these acids is to be applied.

In the event of splashes into the eyes, they should be immediately washed with clean water. In case of illness, the physician should be consulted.

KASPAROV, A. A.
BALYNINA, E. S.

Harmful substances in industry. III: *Inorganic and hetero-organic compounds* (Vrednye vescestva v promyslennosti–III: Neorganiceskie i elementorganiceskie soedinenija). Lazarev, N. V.; Gadaskina, I. D. (Leningrad, Izdatel'stvo "Himija", 7th ed., 1977), 608 p. 58 ref. (In Russian)

"Toxicologic studies on borax and boric acid". Weir Jr., R. J.; Fisher, R. S. *Toxicology and Applied Pharmacology* (New York), 1972, 23/3 (351-364).

CIS 78-1968 "Boric acid" (Acide borique). Morel, C.; Cavigneaux, A.; Protois, J. C. Fiche toxicologique nº 138. *Cahiers de notes documentaires–Sécurité et hygiène du travail* (Paris), 3rd quarter 1978, 92, Note No. 1138-92-78 (469-472). 24 ref. (In French)

CIS 75-410 *Harmful effects of barium and boron on man* (Action nocive du baryum et du bore sur l'organisme humain). Šejbl, J.; Sejblová, S. Translation INRS 32 B-74 (Institut national de recherche et de sécurité, 30 rue Olivier-Noyer, Paris) (1974), 8 p. (In French)

CIS 77-1031 *Criteria for a recommended standard–Occupational exposure to boron trifluoride.* DHEW (NIOSH) publication No. 77-122 (National Institute for Occupational Safety and Health, 4676 Columbia Parkway, Cincinnati, Ohio) (Dec. 1976), 83 p. Illus. 73 ref.

Brewing industry

Brewing is one of the oldest industries: beer in different varieties was drunk in the ancient world and the Romans introduced it to all their colonies. Today it is brewed and consumed in almost every country, particularly in Europe and areas of European settlement (table 1).

Table 1. International per capita consumption of beer, based on total population, 1977

Country	Litres of bulk beer	Country	Litres of bulk beer
Australia	136.2	Italy	14.0
Austria	103.1	Luxembourg	127.0
Belgium	130.1	Netherlands	83.9
Bulgaria	45.0	New Zealand	128.5
Canada	86.0	Norway	45.47
Czechoslovakia	140.0	Poland	33.9
Denmark	116.06	Portugal	27.0
Finland	55.3	Romania	35.20
France	46.21	Spain	46.90
Germany (Dem. Rep.)	126.4	Sweden	53.6
Germany (Fed. Rep.)	148.7	Switzerland	68.3
Hungary	80.0	United Kingdom	119.5
Iceland	16.0	United States	85.6
Ireland	126.2	USSR	24.0
		Yugoslavia	40.4

The brewing process. The grain used as the raw material is usually barley, but rye, maize, rice and oatmeal are also employed. In the first stage the grain is malted either by causing it to germinate or by artificial means. This converts the carbohydrates to dextrin and maltose and these sugars are then extracted from the grain by soaking in a mash tun and then agitating in a lauter tun. The resulting liquor, known as sweet wort, is then boiled in a copper with hops (figure 1) which give a bitter flavour and help to preserve the beer. The hops are then separated from the wort and it is passed through chillers into fermenting vessels where the yeast is added–a process known as pitching–and the main process of converting sugar into alcohol is carried out. The beer is then chilled to 0 °C, centrifuged and filtered to clarify it and is then ready for dispatch by keg, bottle or bulk transport (see FERMENTATION, INDUSTRIAL).

Figure 1. Hops.

HAZARDS AND THEIR PREVENTION

Manual handling. This accounts for most of the injuries in breweries: hands are bruised, cut or punctured by jagged hoops, splinters of wood and broken glass. Feet are bruised and crushed by falling or rolling barrels. Much can be done to prevent these injuries by suitable hand and foot protection. Increase in automation and standardisation of barrel size (say at 50 l) can reduce the lifting risks. The back pain caused by lifting and carrying of barrels, etc., can be dramatically reduced by training in sound lifting techniques such as the kinetic method. Falls on wet and slippery floors are common. Non-slip surfaces and footwear and a regular system of cleaning are the best precaution.

Machinery. Where malt is stored in silos, the opening should be protected and strict rules enforced regarding entry of personnel. Conveyors are much used in bottling plants; traps between belts and drums in the gearing can be avoided by efficient machinery guarding. Where there are walkways across or above conveyors, frequent stop buttons should also be provided. In the filling process, very serious lesions can be caused by bursting bottles: stout guards on the machinery and face guards, rubber gloves, rubberised aprons and non-slip boots for the workers can prevent injury.

Electricity. Owing to the prevailing damp conditions, electrical installations and equipment need special protection, and this applies particularly to portable apparatus. Wherever possible, low voltages should be used, especially for portable inspection lamps. Steam is used extensively and burns and scalds occur: lagging and protection of pipes should be provided, and safety locks on steam valves will prevent accidental release of scalding steam.

Health hazards. Handling of grain can produce barley itch caused by a mite infesting the grain. Mill-workers'

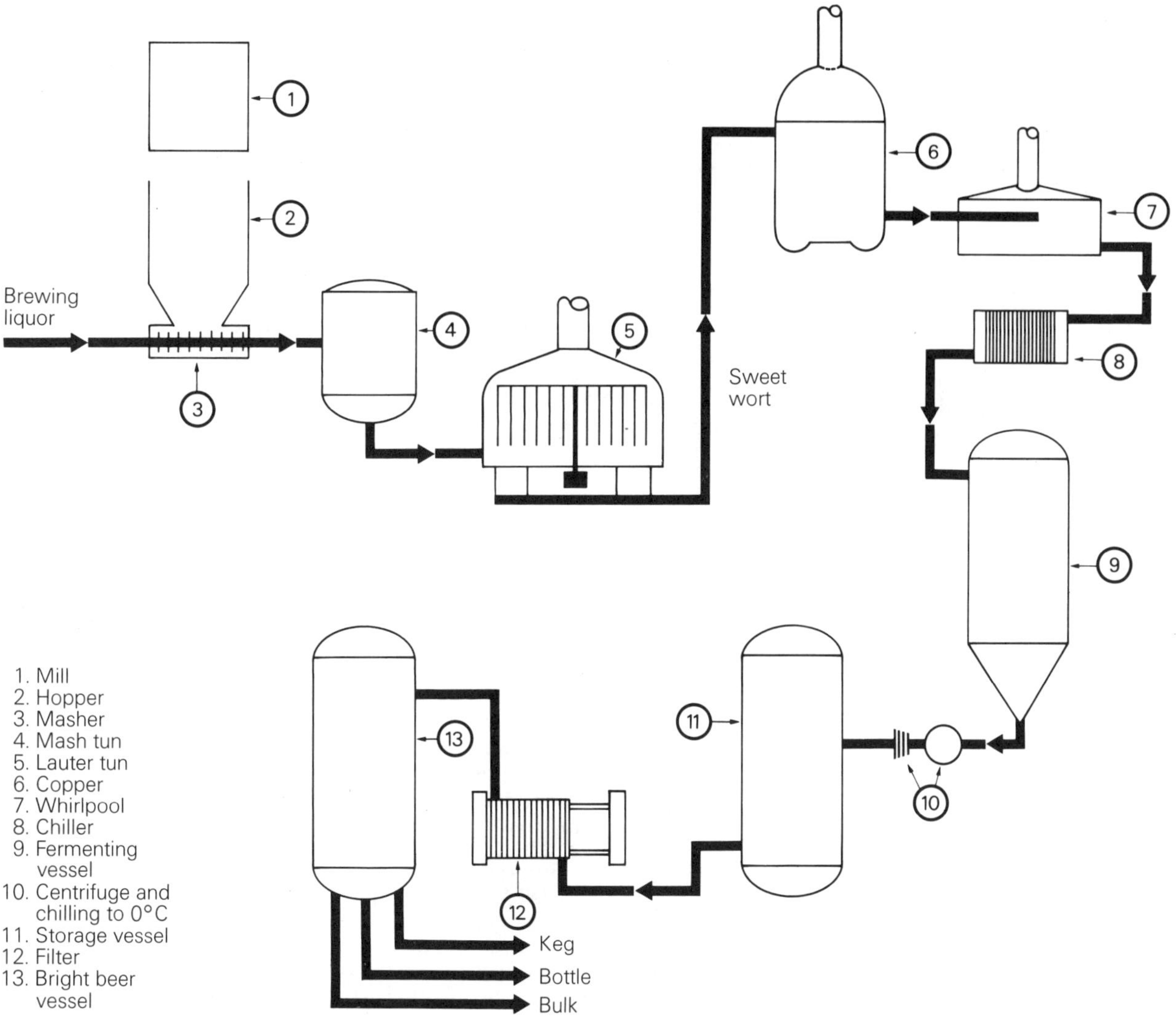

Brewing liquor

Sweet wort

Keg
Bottle
Bulk

1. Mill
2. Hopper
3. Masher
4. Mash tun
5. Lauter tun
6. Copper
7. Whirlpool
8. Chiller
9. Fermenting vessel
10. Centrifuge and chilling to 0°C
11. Storage vessel
12. Filter
13. Bright beer vessel

Figure 2. A flow chart of the brewing process.

Figure 3. Bottling plant in a brewery. *(By courtesy of Institut national de recherche et de sécurité.)*

asthma, sometimes called malt fever, has been recorded in grain handlers and has been shown to be an allergic response to the grain weevil *(Sitophilus granarius)*. Manual handling of hops can produce a dermatitis due to the absorption of the resinous essences through broken or chapped skin. Preventive measures include good washing and sanitary facilities, efficient ventilation of the workrooms and medical supervision of the workers.

When barley is malted by the traditional method of steeping it and then spreading it on floors to produce germination, it may become contaminated by *Aspergillus clavatus*, which can produce growth and spore formation. When the barley is turned to prevent root matting of the shoots or when it is loaded into kilns, the spores may be inhaled by the workers. This may produce extrinsic allergic alveolitis, which in symptomatology is indistinguishable from farmers' lung; exposure in a sensitised subject is followed in about 6 h by a rise in body temperature and dyspnoea. There is also a fall in vital air capacity and a decrease in the carbon monoxide transfer factor.

In an exposed population, the incidence of the disease is about 5% and continued exposure produces severe respiratory incapacity.

With the introduction of automated malting, where man is not exposed, this disease has largely been eliminated.

Carbon dioxide. This is formed during fermentation and is present in fermenting tuns and vats and vessels that have contained beer. Concentrations of 10%, even if breathed only for a short time, produce unconsciousness, asphyxia, and eventual death. Carbon dioxide is heavier than air and efficient ventilation with extraction at low level is essential in all fermentation chambers where open vats are used. As the gas is imperceptible to the senses, there should be an acoustic warning system which will operate immediately if the ventilation breaks down. Cleaning of confined spaces presents serious hazards: the gas should be dispelled by mobile ventilators before men are permitted to enter, safety belts and lifelines and respiratory protective equipment of the self-contained or supplied-air type should be available, and another workman should be posted at hand for supervision and rescue if necessary.

Gassing. This has occurred during relining of vats with protective coatings containing toxic substances such as trichloroethylene. Similar precautions should be taken as those listed above against carbon dioxide.

Refrigerant gases. Chilling is used to cool the hot wort before fermentation and for storage purposes. Accidental discharge of refrigerants can produce serious toxic and irritant effects. In the past, chloromethane, bromomethane, sulphur dioxide and ammonia were mainly used but recently these have been replaced by chlorofluoro-derivatives of methane, the least toxic being dichlorodifluoromethane (freon 12). The gas is safe unless heated, when it decomposes, emitting highly toxic phosgene and fluoride fumes. Adequate ventilation and careful maintenance will prevent most risks but breathing apparatus should be provided for emergencies and frequently tested. Precautions against explosive risks may also be necessary, e.g. flameproof electrical fittings, elimination of naked flames.

Hot work. In some processes, such as cleaning out mash tuns, men are exposed to hot, humid conditions while performing heavy work; cases of heat stroke and heat cramps can occur especially in men new to the work. These conditions can be prevented by an increased salt intake, adequate rest periods and the provision and use of shower baths. Medical supervision is necessary to prevent mycoses of the feet (athlete's foot), which spread rapidly in hot, humid conditions.

Throughout the industry, temperature and ventilation control, with special attention to the elimination of steam vapour, and the provision of personal protective clothing are important precautions, not only against accident and injury but also against more general hazards of damp, heat and cold, e.g. warm working clothes for workers in cold rooms.

Control should be exercised to prevent excessive consumption of the product by the persons employed and alternative hot beverages should be available at meal breaks.

Noise. When metal barrels replaced wooden casks breweries were faced with a severe noise problem. Wooden casks made little or no noise during loading, handling or rolling but metal casks when empty create high noise levels. Modern automated bottling plants generate a considerable volume of noise.

The noise produced in handling metal casks can be reduced by the introduction of mechanical handling on pallets. In the bottling plants the substitution of nylon or neoprene for metal rollers and guides can substantially reduce the noise level.

EUSTACE, J. F.

CIS 74-1082 "Prevalence of respiratory symptoms and sensitization by mould antigens among a group of maltworkers". Riddle, H. F. V. *British Journal of Industrial Medicine* (London), Jan. 1974, 31/1 (31-35). 12 ref.

Dust and accidents in malthouses. Guidance note EH 24 from the Health and Safety Executive (London, HM Stationery Office, 1979), 2 p.

"Practical applications of ergonomics in malthouses and breweries" (Applications pratiques de l'ergonomie dans les malteries et brasseries). Gaeng, E. *Bulletin de l'AISS.* Kolloquium "Mensch Maschine Umwelt" (48-53) (Geneva, International Social Security Association, 1975).

"Cancer morbidity and causes of death among Danish brewery workers". *International Journal of Cancer* (Geneva), 1979, 23/4 (454-463).

Brick and tile manufacture

This article deals with bricks and tiles used as building materials: refractory bricks and ceramic tiles are dealt with in the articles REFRACTORIES and POTTERY INDUSTRY, respectively. Bricks and tiles made from clay have been used as building material since the earliest times in many parts of the world. When properly made and fired they are more durable than some stones, resistant to weather and great changes of temperature and moisture. The brick is an oblong of standard size, varying slightly from region to region but essentially convenient for handling with one hand by a bricklayer; roofing tiles are thin slabs, either flat or curved; clay tiles may also be used for floors.

Materials and processing. The basic material is clay of various kinds with mixtures of loams, shales and sand according to local supply and needs, to give the required properties of texture, plasticity, regularity and shrinkage and colour (see CLAY; SILICA AND SILICATES).

Extraction of the clay is nowadays often fully mechanised; manufacture usually takes place alongside the extraction hole but in large works the clay is sometimes conveyed in skips on ropeways. The subsequent processing of the clay varies according to its constitution and to the end product but in general includes crushing, grinding, screening and mixing.

Clay for wire-cut bricks is broken up by rollers, water is added in a mixer, the mixture rolled again and then fed through a horizontal pugmill. The plastic clay extruded is then cut to size on a wire-cutting table. Semi-dry and stiff plastic material is produced by rolling and screening and is then fed to mechanical presses. Some bricks are still hand moulded.

Where plastic material is used, the bricks have to be dried either by sun and air or more frequently in regulated kilns, before firing; bricks made from semi-dry or stiff plastic may be fired immediately. Firing may take place in ring kilns, often hand fed, or in tunnel kilns, mechanically fed; the fuels used will vary according to local availability. A finishing glaze is applied to some decorative bricks.

HAZARDS AND THEIR PREVENTION

Mechanical hazards. The machinery used is heavy and can give rise to serious injuries unless secure machinery guarding is maintained. Chute feeds should be protected against falls. The rollers and the edge runner pans, the worms of the pugmills, the tools and dies of the presses in particular need protection. At presses where a taker-off works in conjunction with the presser, protection at the taking-off point should not be overlooked. The rough and heavy nature of the material means that all machinery and guards are exposed to frequent damage, and maintenance is therefore of cardinal importance. In many older brick works there will be a great deal of transmission working often at low level, and the maintenance of guarding here is most important. Oiling, greasing or minor repairs should be carried out only when the machinery is at rest.

In the rough and often damp conditions, all electrical installations and fittings should be especially protected

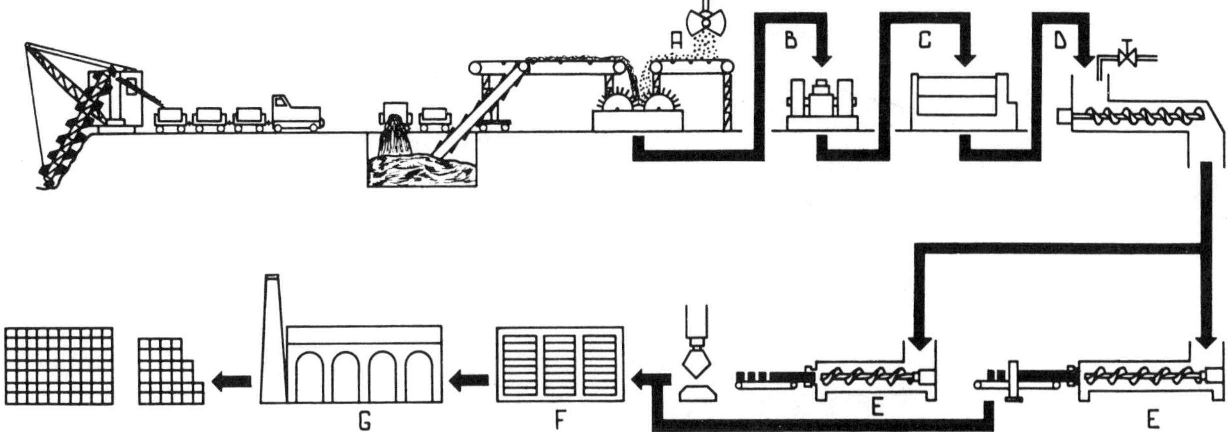

Figure 1. The manufacture of bricks and tiles. A. Clay proportioning. B. Pugging. C. Rolling. D. Mixing. E. Moulding. F. Drying. G. Kilning.

to obviate risk of electric shock. Floors, runways and yards should be of solid construction, easily cleaned to guard against falls, overturning of barrows and accumulation of dirt and wet. Good lighting at all points where persons have to work is necessary, including yards, access ways and kiln tops.

Health hazards. The materials used for building bricks and tiles do not usually produce highly dangerous dust unless the content of free silica in the clay is relatively high (fire clay, see REFRACTORIES). To prevent hazards, dust-producing equipment such as crushers, conveyor belts and vibrators should be enclosed and local exhaust ventilation may be necessary. Raw materials should be moistened during working and premises should be cleaned by methods that do not raise dust.

If sand is applied to green bricks by compressed air, efficient local exhaust ventilation is necessary to prevent inhalation of silica: a more satisfactory method of obviating the health risk is to apply sand in a wet state by mechanical means.

Where bricks or tiles are glazed, only non-lead glazes should be used.

There is some risk of carbon monoxide poisoning from the kilns: it is essential that correct draught conditions be maintained.

The work is usually arduous and may involve much exposure to dust and to lubricants, to the vagaries of climate and the heat of furnaces. Suitable protective clothing is necessary together with changing and drying rooms. Washing facilities should include showers. Messrooms and cabin shelters are also necessary.

Mechanisation is gradually removing many heavy manual operations, such as the movement of bricks from presses to dry wagons, from dryers to kilns, and the filling and emptying of kilns, but pre-employment medical examination should still be directed to fitness for lifting and carrying in many jobs: training in lifting and handling is desirable.

ROSCINA, T. A.

CIS 1460-1971 "A contribution to the study of the silicosis hazard in brickworks in the province of Rome" (Contributo allo studio del rischio di silicosi nell'industria dei laterizi in provincia di Roma). De Luca, F.; Benvenuti, F.; Cisbani, B. *Securitas* (Rome), Aug. 1970, 55/8 (669-682). Illus. 9 ref. (In Italian)

"Kangri cancer in the brick industry". Svindland, H. B. *Contact Dermatitis* (Copenhagen), Jan. 1980, 6/1 (24-26). Illus. 3 ref.

CIS 74-905 "Physiological assessment of the strenuousness of work done by women employed in a factory producing ceramic products for the building industry" (Ocena stopnia uciazliwości pracy kobiet zatrudnionych w zakladach wytwórczych ceramiki budowlanej, na podstawie wskázników fizjologicznych). Lotach, H.; Puchalska, H. *Prace centralnego instytutu ochrony pracy* (Warsaw), 1973, 123/77 (121-140). 24 ref. (In Polish)

CIS 80-862 "Safety in the brickyard". Bennett, L. G. *Safety Management–Veiligheidsbestuur.* Arcadria (S.A.), Sep. 1979, 5/9 (21-23). Illus. 1 ref.

Briquette manufacture

Briquettes, which are also known as "patent fuel", are made from a combustible, usually coal, coal dust, coke, slurry lignite or peat which is compacted into a shape convenient for transport and use. It may be in block, ovoid (boulet) or cylindrical form varying in weight from 1-12 kg in blockform, and from 40-170 g in the smaller sizes. Production in some countries is very large, e.g. in the Republic of Korea, a single factory has an annual production of 700 million briquettes ranging in size from 4.5-12 kg and consuming 3 million tonnes of coal per year.

Production. The raw material is crushed and powdered and then passed through dryers. Crushed coal-tar pitch is usually added as a binding agent, though in some cases the binding agent may be petroleum pitch, natural bitumen, starch, clay or lime. In some cases hot liquid pitch is sprayed on the coal briquetting material in a mixing drum. The mixture of the raw material and binding agent is then compressed into shape at the briquetting presses. The pressing process generates great heat and the freshly formed briquettes are given time to cool before being loaded into transport vehicles.

Uses. Briquettes are used as domestic fuel, as boiler fuel, in gas generators, and for coking.

HAZARDS

Accidents. The most severe accidents in briquette production are those from fires and coal-dust explosions in the rotary driers: injuries also occur from falls and from being caught in conveyors or presses. In the case of lignite, the steam-saturated waste gases of the drying plant may cause fog or a fine precipitation of the vapour in the plant area leading to slippery floors.

Diseases. The most important disease is that caused by polycyclic aromatic hydrocarbons. The type of disease depends on the particle size of the pitch aerosol, the air

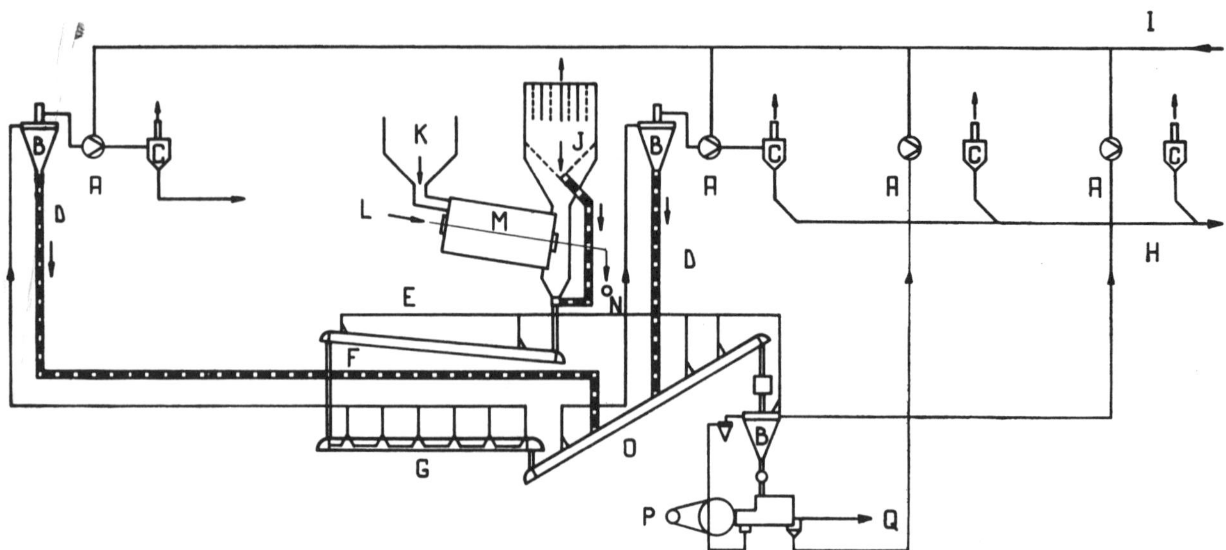

Figure 1. Briquette manufacture. A. Ventilator. B. Dry separator. C. Wet separator. D. Dust pipe. E. Collecting chain conveyor. F. Chain conveyor. G. Cooling chain conveyor. H. Coal sludge to the purification plant. I. Injected water. J. Dust collector. K. Lignite. L. Hot air. M. Rotary drier. N. Condenser. O. Dry precipitator for the presses. P. Briquette extruder. Q. Transportation.

concentration, the exposure period and personal hygiene. High boiling point tar fractions contain carcinogens which are primarily effective on the epidermis and seldom on the bronchial mucous membrane. The low boiling point tar fractions contain 2-naphthylamine and may give rise to erythema, pigmentation, burns and excoriations, hyperkeratosis, papilloma, warts and spinocellular carcinoma on the scrotum; the mortality rate with scrotal carcinoma is high. Carcinoma of the respiratory and urinary passages are relatively rare. The concentration of potassium and sodium in lignite may be high and its alkaline content may have a detrimental effect on the skin and mucous membrane. The presence, in lignite, of *Streptococcus faecalis* and *Actinomyces* may cause eczema, epidermomycosis, pyodermatitis, allergic bronchial asthma and superinfection of non-occupational bronchitis. Work on driers involves considerable heat stress. At hammer mills and high performance roller presses there is a risk of hearing loss from noise (see TAR AND PITCH; PEAT AND LIGNITE).

SAFETY AND HEALTH MEASURES

The prevention of health hazards is ensured by regular medical examinations. In pre-employment examinations, the condition of the skin and circulatory, respiratory and urinary systems must be examined with regard to future hazards. Skin inspection and cytological control of the bronchial secreta and urine is recommended at intervals of 6-12 months; chest X-ray at intervals of 1-2 years is also necessary.

Instruction in hygiene and accident prevention is essential. The production of dust should be kept to a minimum. Skin protection can be obtained with kaolin, which should be provided. Workpeople should immediately inform the proper authorities of any unusual symptoms. At workplaces exposed to heat, reasonable drinking habits should be observed. Hearing protection should be worn at noisy sites. Adequate first-aid knowledge and facilities should be given to workers. Messrooms, washing and sanitary facilities and first-aid rooms should be sited as far as practicable from any dust-producing plant.

Hazards of dust and dust explosion can be prevented by dust exhaust installations. The site of each operation should have a separate ventilation system. The largest quantity of dust (~90%) comes from gases expelled from lignite at the driers. Optimal collection is obtained by separate electrostatic precipitators. For the presses, wet and dry cyclones are usually used; for dedusting rooms and cooling, air cyclones or electrostatic precipitators are used. A continuous supply of material at the driers avoids ignition and explosion. It is obtained by electronically controlled vibrators and alarm systems, as well as by automatic control of the circulating and ejection speed of the driers. In briquetting, rich in binding agents, hazardous pitch dust occurs at delivery, during crushing and when turning up the pitch. Dust-free processing using liquid pitch is preferable although enclosure of the plants is still necessary. In these plants the pressure at every exit for the pitch should be maintained below atmospheric. Pitch-free processing is the best technical prevention.

Some countries have detailed regulations concerning the safety and health measures to be observed in briquette manufacture.

ZORN, H.

Bromine and compounds

Bromine (Br_2)

m.w.	159.8
sp.gr.	2.93
m.p.	−7.2 °C
b.p.	58.8 °C
v.d.	5.5
v.p.	175 mmHg ($23.27 \cdot 10^3$ Pa) at 21 °C

slightly soluble in water; very soluble in ethyl alcohol, chloroform, ethyl ether, carbon disulphide, carbon tetrachloride

a dark reddish-brown, fuming, volatile liquid with a suffocating odour.

TWA OSHA	0.1 ppm 0.7 mg/m³
STEL ACGIH	0.3 ppm 2 mg/m³
IDLH	10 ppm
MAC USSR	0.5 mg/m³ skin

It attacks all metals and organic tissues and rapidly vaporises at room temperature.

Occurrence. Bromine is widely distributed in nature in the form of inorganic compounds such as minerals, in seawater and in salt lakes. Small amounts of bromine are also contained in animal and vegetable tissues. It is obtained from salt lakes or boreholes, from seawater and from the mother liquor remaining after the treatment of potassium salts (sylvinite, carnallite).

Production. Bromine can be produced industrially by chlorine displacement; electrolysis or solvent extraction are also used. The principle of the chlorine displacement technique is that bromide ions are oxidised in solution by chlorine to free elemental bromine according to the following reaction: $Cl_2 + Br^- \rightarrow Cl^- + Br_2$. Vaporisation by steam results in the recovery of free bromine directly. The main adaptations of the chlorine displacement technique are the Koubirschsky process (for raw material with high bromine content), and the Daw process (for raw material with low bromine content).

Uses. Bromine is used for gold extraction, bleaching, pharmaceuticals, dyestuffs and for fuel additive production; this latest use accounts for the majority of industrial production.

HAZARDS

Bromine is a highly corrosive, liquid halogen, the vapours of which are extremely irritating to the eyes, skin and mucous membranes. On prolonged contact with tissue, bromine may cause deep burns which are long in healing and subject to ulceration; bromine is also toxic by ingestion, inhalation and skin absorption. Bromine will cause the ignition of organic materials such as sawdust. It combines readily with potassium, phosphorus and tin, and the reaction may be accompanied by spontaneous ignition.

Health hazards. Escape of bromine into the the workplace air is the main toxic hazard during bromine production. This escape is caused by leaks in the columns as a result of malfunction, excess of chlorine introduced into the column, disorders of thermal equilibrium or blockage of the ducting; it can also arise from cracks in the ceramic components of the bromine conduits, following increase in temperature of the reaction mass, from faulty joints or from unsafe bottling techniques.

A bromine concentration of 0.5 mg/m³ should not be exceeded in case of prolonged exposure; in a bromine concentration of 3-4 mg/m³, work without a respirator is impossible. A concentration of 11-23 mg/m³ produces severe choking and it is widely considered that 30-60 mg/m³ is extremely dangerous for man and that 200 mg/m³ would prove fatal in a very short time. Bromine vapours can cause acute as well as chronic poisoning. They enter the body by the respiratory system, the skin and the digestive system; however, vapour inhalation constitutes the most dangerous mode of entry.

Bromine is very irritating for the mucous membranes. It has cumulative properties, being deposited in the tissues as bromides and displacing other halogens (iodine and chlorine). Long-term effects include disorders of the nervous system.

Exposure to rather low concentrations results in copious mucous secretion in the upper airways, inflammation of the eyelids, lacrimation, coughing, epistaxis, respiratory difficulties, vertigo and headache. Occasionally, these symptoms are followed a few hours later by nausea, diarrhoea accompanied by stomach pains, hoarseness and respiratory difficulty with symptoms of asthma; crepitations are heard in the lungs. In some cases, a generalised vesicular or morbilliform rash has been seen.

Inhalation of high bromine concentrations causes inflammatory lesions to the mucous membranes of the upper airways. The tongue and palate look inflamed and become oedematous and spasm of the glottis occurs; there is an asthmatic bronchitis and the expired air may have a characteristic odour. Photophobia and blepharospasm occur. High concentrations can produce fatal chemical burns of the lungs.

Persons exposed regularly to concentrations 3-6 times higher than the exposure limit for one year complain of headache, pain in the region of the heart, increasing irritability, loss of appetite, joint pains and dyspepsia. During the fifth or sixth year of work there may be loss of corneal reflexes, pharyngitis, vegetative disorders and thyroid hyperplasia accompanied by thyroid dysfunction. Cardiovascular disorders also occur in the form of myocardial degeneration and hypotension; functional and secretory disorders of the digestive tract may also occur. Signs of inhibition of leucopoiesis and leucocytosis are seen in the blood. The blood concentration of bromine varies between 0.15 mg/100 cm³ to 1.5 mg/100 cm³ independently of the degree of intoxication.

When bromine comes into contact with the skin it causes strong chemical burns which take long to heal completely and leave deep scars.

Hydrogen bromide (HBr)

m.w.	80.9
sp.gr.	3.5
m.p.	−88.5 °C
b.p.	−67

soluble in water and ethyl alcohol

a corrosive colourless gas.

TWA OSHA	3 ppm 10 mg/m³
IDLH	50 ppm
MAC USSR	2 mg/m³

Hydrobromic acid (HBr 47% + H_2O)

m.w.	80.9
sp.gr.	1.49
m.p.	−11 °C
b.p.	126 °C

a corrosive, faintly yellow liquid with a pungent smell, which darkens on exposure to air and light.

The same exposure limits can apply as for HBr. Hydrogen bromide is obtained by the combination of bromine vapour and hydrogen at a high temperature and in the presence of platinised asbestos or pumice stone; hydrobromic acid is obtained by dissolving HBr in water. Hydrogen bromide and its aqueous solutions are used for manufacturing organic and inorganic bromides, as reducing agents and catalysis and in the alkylation of aromatic compounds.

The toxic action of hydrobromic acid is 2-3 times weaker than that of bromine. Both the gaseous and aqueous forms irritate the mucous membranes of the upper respiratory tract. Chronic poisoning is characterised by upper respiratory catarrh and dyspepsia, slight reflex modifications and diminished erythrocyte counts. Olfactory sensitivity may be reduced. Contact with the skin or mucous membranes may cause burns.

Bromic acid ($HBrO_3$)

Hypobromous acid (HBrO)

The oxygenated acids of bromine are found only in solutions or as salts. Their action on the body is similar to that of hydrobromic acid.

Ferroso-ferric bromide ($FeBr_2-FeBr_3$)

Ferroso-ferric bromides are solid substances used in the chemical and pharmaceutical industries and in the manufacture of photographic products. They are produced by passing a mixture of bromine and steam over iron filings. The resultant hot syrupy brome salt is tipped into iron containers where it solidifies. Wet bromine (i.e. bromine containing more than about 20 ppm of water) is corrosive to most metals, and elemental bromine has to be transported dry in hermetically sealed monel, nickel or lead containers. To overcome the corrosion problem, bromine is frequently transported in the form of ferroso-ferric salt.

Potassium bromide (KBr)

m.w. 119
sp.gr. 2.75
m.p. 730 °C
b.p. 1 435 °C

soluble in water, slightly soluble in ethyl ether

colourless crystals or white granules or powder.

It is produced by the reaction of ferroso-ferric bromide with potassium carbonate in the presence of dry steam. It is used in the manufacture of photographic papers and plates and in process engraving. It has no pronounced toxic properties.

Bromophosgene ($COBr_2$)

This is a decomposition product of bromochloromethane and is encountered in the production of gentian violet. It results from the combination of carbon monoxide with bromine in the presence of anhydrous ammonium chloride.

The toxic action of bromophosgene is similar to that of phosgene.

Cyanogen bromide (BrCN)

This is used for gold extraction and as a pesticide. Its toxic action resembles that of hydrocyanic acid; cyanogen bromide also has a pronounced irritant effect and high concentrations may cause pulmonary oedema and lung haemorrhages.

Bromomethane

See separate article.

Bromoethane (CH_3CH_2Br)

ETHYL BROMIDE; MONOBROMOETHANE

m.w. 109
sp.gr. 1.46
m.p. −118.9 °C
b.p. 38.4 °C
v.d. 3.8
v.p. 400 mmHg ($53.2 \cdot 10^3$ Pa) at 21 °C
e.l. 6.7-11.3%
i.t. 511 °C

soluble in water; very soluble in ethyl alcohol, ethyl ether, chloroform and in other organic solvents

a colourless, volatile, flammable liquid with an ethereal odour and burning taste; it becomes yellowish on exposure to air.

TWA OSHA 200 ppm 890 mg/m³
STEL ACGIH 250 ppm 1 110 mg/m³
IDLH 3 500 ppm
MAC USSR 5 mg/m³

It is produced by the reaction of potassium bromide with refrigerated sulphuric acid and ethyl alcohol. It is used for the ethylation of gasoline, as a refrigerant and an anaesthetic, and also for extinguishing fires.

Bromoethane may enter the body through the skin or respiratory tract. Acute or chronic symptoms of poisoning affect, principally, the nervous system. It has a narcotic effect at concentrations of 3-10%. Acute poisoning is characterised by redness of the face, dilation of pupils, a rapid pulse and tremors of the extremities. In severe cases with a fatal outcome, respiratory disorders, cyanosis, collapse and respiratory paralysis are seen.

Chronic occupational exposure to bromoethane gives rise to nervous disorders and impaired sensitivity. If the workplace concentrations are relatively high, the symptoms of poisoning appear within a few days. The exposed workers complain of headache, dizziness, weakness and numbness of the extremities, and a sensation of heaviness in the body. Exhalation from the mouth has a characteristic ether smell. The disease evolves towards intensified weakness of the extremities, disturbances of gait and increased tendon reflexes. Speech disturbances, systagmous, tremor of fingers, salivation reflex modifications may be observed. Recovery is slow.

Animal experiments have shown that bromoethane may be absorbed through the skin.

Bromoethane forms explosive concentrations with air and is a dangerous fire hazard. It readily decomposes into volatile toxic products such as hydrobromic acid, particularly in the presence of hot surfaces or open flames.

Dibromoethane (CH_2BrCH_2Br)

See separate article.

Bromoform ($CHBr_3$)

TRIBROMOMETHANE

m.w. 252.8
sp.gr. 2.89
m.p. 8.3 °C
b.p. 149.5 °C

slightly soluble in water; very soluble in ethyl alcohol, benzene, chloroform

a heavy liquid with a chloroform-like odour and sweetish taste.

TWA OSHA 0.5 ppm 5.0 mg/m³ (skin)
MAC USSR 5.0 mg/m³

It is produced by adding bromine to a heated solution of acetone and sodium carbonate, by treating chloroform with aluminium bromide or by the electrolysis of potassium bromide in ethyl alcohol. It is used for the synthesis of pharmaceuticals, in shipbuilding and the aircraft and aerospace industry, and in fire extinguishers.

Bromoform has narcotic and irritant effects and may be absorbed through the skin. Workers who were exposed to bromoform and bromoethane in concentrations of up to 100 mg/m³ for longer periods complained of headaches, dizziness, pain in the chest and digestive disorders; other symptoms observed were central nervous system disorders, hepatic dysfunction and liver disease. Cases of heavy poisoning have not been reported in industry.

Dibromomethane (CH_2Br_2)

METHYLENE BROMIDE

sp.gr. 2.49
m.p. −52 °C
b.p. 97 °C

very soluble in ethyl alcohol, ethyl ether, acetone.

MAC USSR 10 mg/m³

It exerts narcotic and irritant effects; it causes liver and kidney disorders and produces blood changes including neutrophil leucocytosis with relative lymphocytosis and vitamin C deficiency. Its toxic effects are in many respects similar to those of bromoform; however, bromoform is more toxic than dibromomethane.

Bromobenzene (C_6H_5Br)

PHENYL BROMIDE

m.w.	157
sp.gr.	1.43
m.p.	30.6 °C
b.p.	156 °C
v.d.	5.4
v.p.	10 mmHg ($1.33 \cdot 10^3$ Pa) at 40 °C
f.p.	51 °C
i.t.	621 °C

soluble in organic solvents slightly soluble in water; very soluble in organic solvents

a colourless liquid.

MAC USSR 3 mg/m³

It is produced by the bromination of benzene in the presence of iron or aluminium powder or by the action of ferric bromide on benzene; it is used in organic synthesis.

This substance is a flammable liquid and constitutes a moderate fire hazard.

Like other halogenated hydrocarbons, bromobenzene has narcotic properties. It affects the mucous membranes of the upper airways and brings about changes in the liver, blood and testicular epithelium.

The chronic effect of bromobenzene has been studied in rats exposed to concentrations of 20 mg/m³ and 3 mg/m³ for 4.5 months. There was no effect at 3 mg/m³; 20 mg/m³ arrested growth, caused inhibition of the nervous system, disorders of liver function and reduction of sulphydryl groups in serum and liver homogenate. There was also a reduction of the serum albumin concentration. The addition of methionine and cystine to the diet of the animals prevented liver damage and restored normal growth.

Other brominated aromatic hydrocarbons, such as bromotoluene and bromoxylene, exert similar effects.

Brominated ketones

Bromacetone, bromomethyl ethyl ketone and dibromomethyl ethyl ketone are highly irritant, especially for the eyes. During the First World War brominated ketones were used as war gases.

GENERAL SAFETY AND HEALTH MEASURES

The prime objective in preventing health injury in the use of bromine and its compounds is to ensure that the vapour and gases of bromine and its volatile compounds cannot escape into the workplace atmosphere.

Steps should be taken to ensure the elimination of leaks in equipment and ducting. Equipment for the manufacture of bromine and its compounds must be made of materials that will resist the corrosive action of these substances. Ducts should be made of ceramic material or special glass and the substances should be conveyed by means of inert gas under pressure, by vacuum or by the use of glandless pumps. Continuous supervision of the gas-tightness of equipment and ducting and of the atmospheric concentration of toxic substances is essential. The workplaces should have good general ventilation and places where gas may escape should be provided with local exhaust ventilation. Reactors in which the bromination of organic compounds takes place should either be isolated or protected by special hoods. The industrial buildings in which bromomethane or bromoethane are processed should be constructed of materials that do not absorb these substances, i.e. of ceramic or similar materials with special linings. Production premises where toxic compounds are processed are generally equipped with remote control systems of the most dangerous processes. All persons handling bromine and its compounds should be familiar with the safety rules and precautions to be observed when working with these substances.

The filling of bromine into carboys or other containers should be mechanised and the filling premises must be equipped with good exhaust ventilation. Bromine is generally stored and transported in dark glass carboys with well ground stoppers. During transport the stoppers should be sealed with clay and wrapped in vegetable parchment or resistant film.

The maintenance of machinery and equipment is the most hazardous aspect of operations employing bromine and its compounds. Before undertaking any such work, reaction vessels and ducting must be completely empty, carefully rinsed with water, steamed out and purged with inert gas. Where there is spillage, water or steam should be used to clean contaminated areas. It should be strictly forbidden to have repair work carried out by one man without a standby worker. The air in the damaged equipment must be checked for the presence of toxic substances before the repair work is started. If necessary the workers have to wear gas masks after having been especially trained in their use.

Due to the fire and explosion hazard, bromine must not be stored, transported or treated along with acetylene, butane, methane, benzene, turpentine or metallic powders. Vessels containing bromine and its compounds should be kept tightly closed. Large quantities should be stored out of doors, shielded from direct exposure to sunlight and away from areas of acute fire hazard, high-temperature processes and readily oxidisable materials.

Persons working with bromine and its compounds must be supplied with (and trained in the use of) personal protective equipment, including respiratory protective equipment, of the canister or self-contained type (depending on the nature of the work), hand protection, eye and face protection and special chemical resistant working clothes; these working clothes must be regularly washed and decontaminated. Pre-employment and periodic medical examination of workers exposed to bromine and its compounds is necessary to make sure that there are no contraindications, e.g. liver or kidney disorders.

Treatment. First aid in case of poisoning includes transportation to fresh air, rest, warmth, oxygen therapy and hospitalisation. Skin contamination with bromine or bromine compounds must be quickly removed with water and sodium bicarbonate solution.

ALEXANDROV, D. D.

"Toxicology of irritant poisons and their role in occupational pathology" (Toksikologija razdrăajuščih jadov i ih rol' v razvitii professional'noj patologii). Ivanov, N. G. *Žurnal vsesojuznogo himičeskogo obščestva im. D. I. Mendeleeva* (Moscow), Feb. 1974, 19/2 (213-217). (In Russian)

CIS 75-1048 "Hazardous chemical reactions–20. Bromine" (Réactions chimiques dangereuses–20. Brome). Leleu, J. *Cahiers de notes documentaires–Sécurité et hygiène du travail* (Paris), 4th quarter 1974, 77 Note No. 931-77-74 (599-601). (In French)

CIS 75-138 "Hazardous chemical reactions–14. Hydrobromic acid. Bromides. Bromates" (Réactions chimiques dangereuses–14. Acide bromhydrique. Bromures. Bromates) Leleu, J. *Cahiers de notes documentaires–Sécurité et hygiène du travail* (Paris), 2nd quarter 1974, 75, Note No. 901-75-74 (265-269). (In French)

CIS 77-1037 "Experimental data for establishment of a hygiene standard for bromine and hydrogen bromide concentrations in the workplace air" (Ėksperimental'nye materialy k gigieničeskomu normirovaniju soderžanija broma i bromistogo vodoroda v vozduhe rabočej zony). Ivanov, N. G.; Kljačkina, A. M.; Germanova, A. L. *Gigiena truda i professional'nye zabolevanija* (Moscow), Mar. 1976, 3 (36-39). 5 ref. (In Russian)

Bromomethane (CH_3Br)

METHYL BROMIDE; MONOBROMOMETHANE

m.w.	95
sp.gr.	1.73
m.p.	−95 °C
b.p.	3.6 °C
v.d.	3.3
v.p.	1 824 mmHg (242.6·10³ Pa) at 25 °C
e.l.	10-16%
i.t.	537 °C

a colourless gas with a faint chloroform-like odour.

TWA OSHA	20 ppm 80 mg/m³
TLV ACGIH	5 ppm 20 mg/m³ skin
STEL ACGIH	15 ppm 60 mg/m³
IDLH	2 000 ppm
MAC USSR	1 mg/m³

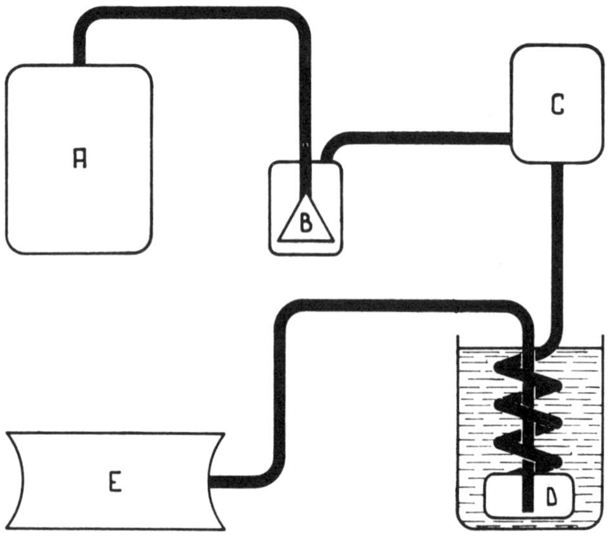

Figure 1. Bromomethane manufacture. A. Still. B. Acid scrubber. C. Alkali scrubber. D. Brine-cooled condenser receiver. E. Storage cylinder. Reaction: sodium bromide + methyl alcohol + sulphuric acid → bromomethane + sodium sulphate + water.

Production. This is shown in figure 1.

Uses. Bromomethane is used as a low boiling solvent in the manufacture of aniline dyes, as a refrigerant, fumigant or insecticide, soil fungicide and rodenticide, and a fire-extinguishing agent. In view of its extreme toxicity, especially when concentrated in confined spaces, its use as a fire extinguisher must be limited. In Australia it is used for total herbage destruction (see HERBICIDES).

HAZARDS

The gas gives no warning of its presence and disperses slowly, in addition to being one of the most toxic organic halides. For these reasons it is among the most dangerous materials encountered in industry. Entry to the body is mainly by inhalation, whereas the degree of skin absorption is probably insignificant. Unless severe narcosis results, it is typical for the onset of symptoms to be delayed by hours or even days. A few deaths have resulted from fumigation. A number have occurred due to leakage from refrigerating plants, or from the use of fire extinguishers. Lengthy skin contact with clothing contaminated by splashes can cause second-degree burns.

Exposures to 35 ppm bromomethane caused symptoms in 31 of 90 workers after a period of 2 weeks. Symptoms included a loss of appetite, nausea, vomiting, headache, giddiness, visual disturbances, lethargy, and faintness, but there was no indication of pulmonary involvement. The above symptoms are likely to occur when blood bromine, measured as bromine ion, is about 10-15 mg/100 cm³. After inhalation of the gas, convulsions may occur without warning, but they may be preceded by giddiness, numbness of the arms and hands, staggering, weakness, drowsiness or headache, vomiting and delirium. Convulsions are often rapidly followed by coma and massive pulmonary congestion, after which the prognosis is extremely poor. Many patients have survived the convulsions, but suffered permanent brain damage. In one clinical case, an intensive exposure was followed by convulsions after a latent period of 5 h, but without pulmonary symptoms. Clinical examination suggested damage predominantly in the cerebellum and the pyramidal tracts of the brain. An electroencephalogram was grossly abnormal, showing slow waves at 5-7 Hz in brief bursts, with a suggestion of periodicity. In this case, serum bromide was 55 mg/100 cm³ on admission. The patient developed spasticity of his right limbs, and this, together with convulsions, was still present 9 months after injury. In cases with prominent pulmonary oedema, high temperatures, up to 41.9 °C, have been recorded, probably due to central nervous system damage. The pulse rate may be greatly elevated and the respiratory rate is commonly 50/min. Two cases of poisoning have been described which were characterised by complete absence of damage to the nervous system. Renal damage was prominent in one and hepatic damage in the other.

Bromomethane may damage the brain, heart, lungs, spleen, liver, adrenals, and kidneys. Both methyl alcohol and formaldehyde have been recovered from these organs, and bromide in amounts varying from 32-62 mg/300 g of tissue. The brain may be acutely congested with oedema and cortical degeneration. Pulmonary congestion may be absent or extreme. Degeneration of the kidney tubules leads to uraemia. Damage to the vascular system is indicated by haemorrhage in the lungs and brain. Bromomethane is said to be hydrolysed in the body with the formation of inorganic bromide. The systemic effects of bromomethane may be an unusual form of bromidism with intracellular penetration by bromide.

No reasons have been advanced for the complete sparing of the lungs in some cases where brain damage has been both severe and permanent. Severe neurological signs seem to be dependent on a sudden exposure to high concentrations following continuous slight exposure. Pulmonary involvement in such cases is less severe.

[An acneiform dermatitis has been observed in persons repeatedly exposed.

Cumulative effects have been reported after repeated inhalation of moderate concentrations of bromomethane, more often with disturbances of the central nervous system.]

SAFETY AND HEALTH MEASURES

[First of all, as far as practicable, less hazardous substances should be used instead of bromomethane, such as chloromethane in refrigeration and fluorocarbons also in refrigeration and as fire extinguishers.]

In addition to safety and health measures applicable to BROMINE AND COMPOUNDS, the following are recommended.

A pungent indicator such as chloropicrin can be added to bromomethane to give warning of its presence. Continuous monitoring for its presence is not practical and laboratory techniques are time-consuming and unsuitable; however, periodical monitoring with a halide lamp leak detector is an advisable precaution, provided the sensitivity is at least at the level of the TLV.

If possible, any process involving bromomethane should be isolated from the operator or carried out under efficient local exhaust ventilation. No one should work alone when exposed to bromomethane infumigation. Personal protection should consist of airline masks supplied by plant-compressed air, positive pressure hose masks supplied by externally lubricated blowers, or self-contained respiratory protective apparatus. Canister-type gas masks are also useful except in emergency situations, during which the vapour concentration may exceed 20 000 ppm. Impervious rubber or PVC hand protection, overalls which can be buttoned to the neck, and rubber foot and leg protection should be supplied. Emergency showers or washing facilities should be available immediately.

Treatment. The patient should immediately go or be helped to fresh air. All contaminated clothing, shoes and socks should be removed and the contaminated skin thoroughly washed gently with a solution of bicarbonate of soda or, if not available, with soap and water. If the inhalation of high concentrations of bromomethane is suspected, the patient should remain under medical observation for at least 48 h. Should pulmonary oedema develop, the administration of 100% oxygen and bed rest are essential. Convulsions require the administration of barbiturates and anti-convulsants. In many cases the period of convalescence is long and prolonged rehabilitation will be required.

LONGLEY, E. O.

Acute intoxication:

CIS 75-1032 "Collective acute methyl bromide poisoning" (Intoxicación aguda colectiva por bromuro de metilo). García Rico, A. M.; Carcés, J.; Más, J.; Nolla, R. *Medicina clínica* (Barcelona), Oct. 1974, 63/6 (291-296). 14 ref. (In Spanish)

"Occupational methyl bromide intoxication" (Professionele intoxicatie door methylbromide). Van Den Oever, R.; Van De Mierop, L.; Lahaye, D. *Archives belges de médecine sociale, hygiène, médecine du travail et médecine légale–Belgisch Archief van Sociale Geneeskunde, Hygïëne, Arbeidsgeneeskunde en Gerechtelijke Geneeskunde* (Brussels), June 1978, 36/6 (353-369). 28 ref. (In Flemish)

Chronic exposure:

"Bromine in blood, EEG and transaminases in methyl bromide workers". Verberk, M. M.; Rooyakkers–Beemster, T.; De Vlieger, M.; Van Vliet, A. G. M. *British Journal of Industrial Medicine* (London), Feb. 1979, 36/1 (59-62). Illus. 11 ref.

Fumigation:

CIS 76-1392 *Guidance note: fumigation using methyl bromide (bromomethane).* Health and Safety Executive, London. General Series/1 (London, HM Stationery Office, June 1976), 12 p. 16 ref.

Safety:

CIS 1642-1968 *Properties and essential information for safe handling and use of methyl bromide.* Chemical safety data sheet SD-35 (Manufacturing Chemists' Association, 1825 Connecticut Avenue NW, Washington, DC) (revised 1968), 16 p.

Bronchitis, chronic, and emphysema

In 1819 Laennec described "pulmonary catarrh" as an inflammation of the mucous membrane of the bronchi causing abundant secretion of mucus. Laennec also said that emphysema was "an increase in the size of the air spaces in the lungs" and this anatomical description is similar to the modern definition. It is now obvious that chronic bronchitis and emphysema are separate conditions that occur more commonly in the same persons than independently. Diagnostic confusion of emphysema in life with chronic bronchitis chiefly occurs when the latter is accompanied by generalised obstruction of the airways causing the lungs to become overfilled with air. This condition is defined below.

Definitions

The Ciba Symposium (1959) defined chronic bronchitis as chronic or recurrent cough together with expectoration which occurred "on most days for at least three months in the year during at least two years". When used for clinical or epidemiological purposes, it is necessary to exclude localised chronic disease such as pulmonary tuberculosis or bronchial carcinoma and generalised lung disease due to specific causes such as pneumoconiosis. The British Medical Research Council's Committee on Chronic Bronchitis (1965) distinguished simple from chronic mucopurulent bronchitis and used the term chronic obstructive bronchitis when, in addition to the symptoms indicating excessive mucus secretion, there was evidence of irreversible generalised obstruction of the airways. The latter, recognised clinically because of dyspnoea, was measured by ventilatory function tests such as the forced expiratory volume in 1 ($FEV_{1.0}$). This volume and its ratio to the forced vital capacity are both reduced in airways obstruction.

Emphysema was defined by the Ciba Symposium as the pathological state of the lung "characterised by increase beyond the normal in the size of air spaces distalt to the terminal bronchiole either from dilation or from destruction of their walls". However, common usage restricts emphysema to lungs with actual destructive changes as shown by reduced numbers of alveolar wall intercepts in the inflated lung. In life this destruction causes loss of elastic recoil, thus reducing the efficiency of expiration and the emptying of the lungs, even in the absence of airways obstruction from bronchial constriction. The difficulty in precise differentiation of emphysema from chronic obstructive bronchitis in life has led to the increasing use of the latter term. In the United States there has been a transfer of certification at death so that bronchitis and emphysema have diminished whereas chronic obstructive pulmonary disease has increased and the combined total is still increasing yearly, though it is still proportionately only half that of chronic bronchitis in the United Kingdom.

The natural history of chronic obstructive bronchitis

As seen in patients attending clinics and in hospitals, chronic productive cough followed after some years by the onset of breathlessness is initially episodic and may follow an infection such as influenza or pneumonia. But it later becomes persistent and after 20 or 30 years with recurrent acute illnesses or exacerbations, respiratory failure ensues and progresses until death with or without congestive heart failure *(cor pulmonale).*

An eight-year prospective study of working men, 30 to 59 years of age, in London by Fletcher et al. (1976) was designed to witness the inception of chronic bronchitis. After eight years of study few persons had acquired symptoms of bronchitis for the first time and the rates of deterioration of ventilatory function as shown by the $FEV_{1.0}$, though differing in trend from one person to another, showed few sudden alterations. The study pinpointed the fact that some persons are particularly susceptible to the harmful effects of cigarette smoking as shown by productive cough or by airways obstruction. Others appear to be able to smoke with impunity.

But the probability exists that even before smoking has begun, children may already have suffered damage as a result of respiratory infections such as bronchiolotis or pneumonia and thus have become susceptible to environmental irritants. Children under 5 years of age in Britain who were subject to chronic exposure to air pollution were found to suffer from an increased frequency of acute bronchitis or pneumonia. A follow-up of these children when aged 20 showed that those with a history of previous lower respiratory tract illnesses were more likely to become adults with chronic cough, particularly when they became cigarette smokers.

The evolution of airways obstruction remains, however, largely unknown. It exists in children with a past history of asthma but also in those with past attacks of bronchitis. In adults the normal rate of decline of the $FEV_{1.0}$ with age is accelerated in those who are smokers compared with non-smokers. Sputum production is also correlated with a lower than normal FEV. Yet acute episodes of chest illness in persons with chronic obstructive bronchitis are not necessarily followed by change in the over-all downward trend.

The absence of an effect from infection in adults is surprising, for attacks of uncomplicated influenza are associated with mild airways obstruction and an increase in total pulmonary resistance. Even a greater bronchial response to the inhalation of histamine or carbachol occurs during colds and influenza. There is also evidence that the small peripheral airways become diseased and obstructed in a manner suggestive of infection before the onset of disease of the larger bronchi. These small airways may be the source and antecedent of emphysema. The progress of the natural history of chronic obstructive pulmonary disease may thus involve many factors which are both personal and environmental and operate at different periods of childhood and adult life.

Epidemiology and causation

The many studies of mortality and of morbidity carried out in different countries have begun to indicate causative factors operative in both developed and developing countries. The use of symptomatic questionnaires, from the first edition of the MRC's questionnaire (1960) to its most recent version in 1976 and that of the American Thoracic Society in the United States (Comstock et al., 1979), has been essential to field studies of chronic bronchitis. Spirometry has aided the problem of measuring generalised airways obstruction. Examination of the lungs at post-mortem by thin lung slices has made it possible to compare emphysema in different countries but has not progressed beyond the confirmation of male preponderance and a relationship to smoking.

In relation to chronic obstructive bronchitis five factors have been identified, three which can be partly quantified and two which do not lend themselves to measurement. The three most significant factors are smoking, air pollution and occupation which will now be considered.

Smoking. Each and every study has shown the harmful effect of tobacco smoking, whether this occurs in towns or in rural areas. A graded relationship exists between the quantity of cigarettes smoked daily and symptoms and also between smoking and the rate of decline of the FEV with time. Giving up smoking reverses such relations but requires time before this effect is manifested. Pipe and cigar smokers are much less affected than cigarette smokers yet have a higher incidence of bronchitis than non-smokers. The effect of cigarette smoking is massive and obscures the effect of other environmental agents such as dust. It also accounts for the excess of chronic obstructive bronchitis in men compared with women.

Air pollution. There are many forms of air pollution, from that of wood smoke in the air of huts, as in New Guinea, to the increased concentrations of smoke particles and SO_2 in the ambient air of coal-burning cities. The effect of the lacrimatory smogs of California on the respiratory tract must also be considered. Smoke from untreated open coal fires is the most harmful pollutant in high enough concentrations and, when combined with anticyclonic weather with temperature inversions, it can cause lethal effects. Restriction of coal burning which emits dark smoke in Britain has dispelled this hazard. But the effect of chronic exposure to much lower levels of pollution remains, even though it can no longer be demonstrated as readily as the effects of high concentrations. The effect of pollution on the health of young children has already been mentioned above.

The contribution to respiratory disease of pollutants such as the oxides of nitrogen, ozone, and the oxidants resulting from the effect of bright sunlight on exhaust gases from cars has not been evaluated to the same extent as pollutants from coal. All such do have an adverse effect on lung function.

Occupational and social factors. These factors are considered together because of the difficulty of separating domestic and home environment from that of occupation. Indeed, though miners and foundry workers have a high standardised mortality ratio from bronchitis compared with that of men in other occupations, so also do their wives, even though the mortality rates in wives are so much lower than those of their husbands. As already discussed, mortality rates are not necessarily the best method of discerning differences, particularly in the case of chronic disease. A sample study of sickness absence directed to the relationships between types of illness and occupation was carried out in 1961-62 in the United Kingdom by the Ministry of Pensions and National Insurance. The sample was one of 5% among men aged 18 to 63 and 2.5% among women. Higgins (1974), whose review of the epidemiology of chronic respiratory disease literature should be consulted, has extracted certain tables reproduced below in modified form.

Table 1 shows the interplay of social class as defined by occupational groups and the rates of bronchitis (incidence is expressed as inception-rates). The table also shows the effect of sheer physical exertion in Classes III to V. It is a fact that, apart from agricultural workers, those in occupations where the air is dusty are often performing heavy work. Table 2 shows the bronchitis rates and ratios in certain individual occupational groups.

It is difficult to discern the precise meaning of these differences. In the case of miners, for instance, nearly every cause of sickness-absence has a higher rate than the same illness in other workers. Migraine as a cause of absence has a rate nine times higher in miners than in non-miners. In order to show that bronchitis in particular occupations is related to the conditions of work, there should be a discernible effect from degree of pollution of the air and of the duration of exposure. These points are considered in other surveys.

Higgins has related the results of field surveys in different countries of the incidence of bronchitis symptoms and the mean ventilatory capacity for those heavily exposed to dust and those not exposed. There is evidence from table 3 that dust-exposed men had a higher rate of bronchitis and a lower ventilatory capacity than those not so exposed in most of the groups. In two integrated steel works in South Wales it was found impossible to relate the rates of bronchitis to the measured atmospheric pollution to which the men were

Table 1. Bronchitis inception rates and incapacity for work caused by bronchitis (men aged 18-63 in the UK survey of 1961-62)[1]

Occupational category	Rates/1 000 men standardised for age	Work incapacity (days/1 000 men)
All categories	*36.9*	*1 235*
Professional and intermediate I and II	*15.4*	*354*
Skilled III	*35.1*	*1 040*
Heavy	50.0	1 442
Medium	35.9	1 106
Light	25.8	679
Partly skilled IV	*47.7*	*1 573*
Agricultural	19.3	506
Non-agricultural: heavy	60.1	2 196
medium	47.1	1 540
light	38.9	1 429
Unskilled V		
Heavy	57.1	2 384
Medium	50.1	1 732

[1] From Ministry of Pensions and National Insurance (1965), after Higgins (1974).

Table 2. Bronchitis inception rates[1] in certain occupations, United Kingdom, 1961-62

Occupation	Rate: (inception/ 100 men[2])	Ratio: (rate for occupation/ rate for all occupations)
All	3.69	100
Miners and quarrymen	7.24	196
Furnace, forge, foundry and rolling mill workers	4.71	128
Construction workers	4.1	111
Engineering workers	3.8	103
Textile workers	3.53	96
Electrical and electronic workers	3.43	93
Woodworkers	3.27	89
Farmers, foresters and fishermen	2.25	61
Administration and managers	1.08	29

[1] Modified from Higgins (1974). [2] Standardised for age.

exposed. This negative result was attributed to the blanketing effect of smoking. The role of dust in the development of bronchitis in coal miners was studied and an increased prevalence in men as their cumulative dust exposures increased was reported. Rogan et al. (1973) added to this from the same study in 20 collieries by showing a progressive reduction in $FEV_{1.0}$ with increasing cumulative exposure to airborne dust. A series of studies in the Federal Republic of Germany among miners, iron and steel workers, and cement, ceramic and engineering workers sought to define the relation between symptoms, pulmonary function and the burden of inhaled dust. The principal results were that age and smoking exerted a greater influence on the lung than dust exposure. But where non-smokers could be studied, there was evidence, particularly in iron and steel works, that dust had an independent deleterious effect by itself (see AIR POLLUTION).

It is impossible to draw firm conclusions from all these studies. The problem of the relation of occupation to chronic obstructive bronchitis is that there is no identifiable specific agent or pathological effect such as is operative in the pneumonconioses due to silica, asbestos or other noxious chemicals. Even in the case of byssinosis occurring in cotton operatives the occurrence of symptoms on the first day of the week with gradual decrease during the week is unlike that of chronic bronchitis itself. The symptoms of byssinosis resemble those of chronic bronchitis more closely than do those of the pneumoconioses, but the finding of a decline in ventilatory function in byssinosis during the working day contrasts with the relatively unvarying obstruction associated with bronchitis (see BYSSINOSIS). As to the effect of occupation on the rate of progression of chronic obstructive bronchitis, it is probable that once cough and sputum and dyspneoa have developed, those concerned are forced to leave occupations where the air is smoky or dusty. The loss of earnings by men formerly performing skilled work is one reason why men may tend to drift down the social scale until they become relatively unemployable.

Infective and familial factors. Comment concerning infection has been made above, but as there is doubt regarding the contribution both of bacterial and of virus infection to the natural history of chronic obstructive lung disease it is impossible to analyse the subject further. It is important to draw a distinction, however, between the probable importance of infection in relation to chronic bronchitis itself and the absence of any

Table 3. Population surveys of bronchitis and ventilatory capacity in relation to high dust exposure (HDE)[1]

Country and locality	Author	Industry	Age	Incidence of bronchitis		Ventilatory capacity	
				HDE	Control	HDE	Control
England							
Leigh	Higgins (1956)	Coal mining	55-64	23.5	10.7	2.17	2.39
Staveley	Higgins (1959)	Mining	55-64	20.8	14.8	2.44	2.52
		Foundry	55-64	7.8	14.8	2.36	2.52
United States (West Virginia)							
Mullens	Enterline (1967)	Coal mining	21-64	13.0	6.5	3.01	3.18
Richwood	Enterline (1967)	Coal mining	21-64	8.9	9.1	3.29	3.28
Canada	Parsons (1964)	Fluorspar	20-70	21.3	4.0	2.86	3.09
South Africa	Sluis-Cremer (1967)	Gold mining	⩾35	10.6	3.8	3.12	3.27
Northern Ireland	Pemberton (1968)	Flax	⩾35	12.6	5.6	–	–

[1] Modified from Higgins (1974).

relation to generalised airways obstruction. Thus the quantity of sputum is unquestionably increased during respiratory virus infections such as common colds and the purulence of the sputum seen during acute illnesses in those with chronic bronchitis is related to quantitative increase in the bacteria. It is also true that influenza epidemics are accompanied by a period of excess mortality among those with pre-existing chronic bronchitis and emphysema.

Familial factors include heredity and, though no proof has yet been offered in favour of a true genetic susceptibility to chronic obstructive bronchitis, there is evidence of this in the case of emphysema. The deficiency of serum alpha$_1$-antitrypsin protein was shown to be related to emphysema soon after its original description. It is now accepted that only the homozygous persons with the PiZ phenotype have a high risk of emphysema, the suggestion of an increased frequency of chronic obstructive lung disease, particularly in those with the PiMZ phenotype, being unproven.

Apart from the question of genetic abnormality there is evidence that familial factors are important in chronic bronchitis and these may be environmental in origin because they may be concerned with exposure to adverse circumstances of children in the home. The suggestion made above that there is a factor of individual susceptibility to the inimical effects of cigarette smoking is yet to be proven. Nor is it known whether such susceptibility is also manifest with other inhaled irritants.

Prevention and rehabilitation

As it is at present impossible to identify persons who are particularly susceptible to environmental irritants such as dust and tobacco smoke, prevention has to be directed to those already afflicted by symptoms of chronic obstructive bronchitis. As cough and sputum occur in healthy persons during and shortly after acute intercurrent respiratory infections, they cannot serve as danger signals indicating chronic bronchitis unless they are experienced over prolonged periods of time. But when either dyspnoea or wheezing occur alone or added to cough and sputum, the subjects should be investigated in order to determine whether or not there is evidence of airways obstruction. An abnormally low $FEV_{1.0}$ is often found in those with a history of asthma even when symptom-free, but these persons will already know something about their individual precipitating circumstances. When there is no history of attacks of asthma in childhood or early adult life, a low $FEV_{1.0}$ should be regarded as an indicator of the need for preventive action.

Such action can be partly advisory and partly administrative. The advice to give up smoking and to avoid exposure to smoky or dusty environment may be resented, particularly if the occupation being followed is one with high earning capacity. Men with chronic obstructive bronchitis are often good workers whose skill should be preserved and, wherever possible, they should be offered retraining for work not involving harmful environmental exposure. A second but important corollary is to insist that work should not be resumed after acute infective attacks causing increased cough and sputum, until adequate time has elapsed for recovery. A convalescent period away from home in a better environment may be suitable for those who have had attacks with signs suggesting pneumonia. The prompt treatment of acute respiratory illnesses with antibiotics is better than attempts to use antibiotics as prophylactic agents. The scope of medical treatment for those with forced expiratory volumes less than 1.5 l in the first second of expiration requires frequent monitoring of the pulmonary function. Regular use of antispasmodic drugs such as isoprenaline or analogous drugs by inhalation or orally is practised by many physicians but it has yet to be shown that it will reverse the obstruction of chronic bronchitis or of emphysema. Breathing exercises improve morale but are not capable of reversing airways obstruction.

STUART-HARRIS, C.

"Definition and classification of chronic bronchitis for clinical and epidemiological purposes". Medical Research Council's Committee on the Aetiology of Chronic Bronchitis. *Lancet* (London), 10 Apr. 1965, 1 (775-779). 24 ref.

The natural history of chronic bronchitis and emphysema. Fletcher, C.; Peto, R.; Tinker, C.; Speiser, F. E. (Oxford, University Press, 1976), 272 p. 208 ref.

"The relations between structural changes in small airways and pulmonary function tests". Cosio, M.; Ghezzo, H.; Hogg, J. C.; Corbin, R.; Loveland, M.; Dosman, J.; Macklem, P. T. *New England Journal of Medicine* (Boston), 8 June 1978, 298 (1277-1281). 22 ref.

"Standardized respiratory questionnaires; comparison of the old with the new". Comstock, G. W.; Tockman, M. S.; Helsing, K. J.; Hennessy, K. M. *American Review of Respiratory Disease* (New York), Jan. 1979, 119/1 (45-53). 11 ref.

Report on an enquiry into the incidence of incapacity for work. Ministry of Pensions and National Insurance (London, HM Stationery Office, 1965).

Epidemiology of chronic respiratory diseases; a literature review. Higgins, I. T. T. (Washington, DC, Office of Research and Development, Environmental Protection Agency, Aug. 1974). 590 ref.

"Role of dust in the working environment in development of chronic bronchitis in British coal miners". Rogan, J. M.; Attfield, M. D.; Jacobsen, M.; Rae, S.; Walker, D. D.; Walton, W. H. *British Journal of Industrial Medicine* (London), 1973, 30/3 (217-226). 26 ref.

"Epidemiology of chronic bronchitis" (Epidemiologie der chronischen Bronchitis). Minette, A. *Revue de l'Institut d'Hygiène des Mines* (Hasselt, Belgium), 1978, 33/2 (56-99). Illus. 108 ref. (In German)

Brucellosis

Brucellosis is an infectious disease caused by bacteria of the *Brucella* genus. The disease is transmitted to man from animals and, over the centuries, has enjoyed a variety of names in different places, such as: undulant fever, Gibraltar fever, Malta fever, Mediterranean fever, Neopolitan fever, infectious abortion, Bang's disease. The vulnerability of most domestic and wild animals to *Brucella* explains the prevalence of the disease in the major animal husbandry regions in the world. Very few countries seem to escape the disease altogether, and those that appear untouched are in fact countries where no systematic research has been conducted into the infection. The most severly hit areas are Europe and North and South America. The occurrence of the disease in Africa and Asia is extremely variable, though there is a clear predominance in the countries of the Mediterranean basin.

The principal features of the disease in animals, whose symptomatology varies, are abortion in the female and infection of the genital glands in the male. It is often not clinically apparent.

In man the disease develops in several stages. The two- to four-week incubation period corresponds to the primary focal infection: contamination is followed by the spread of the germ through the lymphatic vessels and constitution of the primary (and generally glandular) focus of infection. This stage may be followed by acute brucellosis resulting from septicaemic dissemination, which is characterised by a feverish state lasting from two to four months in the absence of any active treatment. The fever is sometimes relatively mild and may be either stationary or undulant. The septicaemic stage may be accompanied by the formation of secondary focuses of infection that continue to develop after the septicaemic stage: this is a localised, sub-acute form of brucellosis. The main focuses of infection are the joints, bowels, genitals and meninges.

Chronic brucellosis may develop very quickly or only months or years after an acute infection; it may even occur when an acute phase of brucellosis has passed unnoticed. Chronic injuries to the joints and bowels caused by brucellosis are extremely varied. People suffering from chronic brucellosis frequently present no precise local symptoms, the major manifestations of the infection being physical and mental weakness.

The disease is very long in developing and prognosis varies widely from one case to another according to the localisation of the infection. Brucellosis confers definitive cross-immunity to the various species of the *Brucella* genus.

Aetiology

Several species of *Brucella* and numerous sub-types have been recorded. *B. melitensis*, of which there are three known sub-types, is principally found in sheep and goats, and also in camels and dromedaries. *B. abortus*, which is responsible for brucellosis in cattle, is the most commonly found species throughout the world, of which there are nine known sub-types. *B. suis* is encountered among pigs (bio-types 1, 2 and 3), hares (bio-type 2), caribou and reindeer (bio-type 4). Other less common species include *B. ovis*, *B. canis*, *B. neotomae*; except for *B. ovis* and *B. neotomae*, all these species of *Brucella* are pathogenic in man and have a virtually identical symptomatology, although *B. melitensis* is responsible for the more severe form of the disease.

Epidemiology

Although a few individual cases have been described, man-to-man contamination is virtually unknown. Contamination therefore originates almost exclusively in infected animals.

Focuses of infection. Brucellosis is essentially connected with the breeding of cattle *(B. abortus)*, sheep and goats *(B. melitensis)* and pigs *(B. suis)*. There is, however, no absolutely specific connotation of the various species of *Brucella* with these species of animal.

Once focuses of infection have declared themselves, bacterial contamination is not restricted to the infected animals. The latter excrete *Brucella* through their milk, urine, placenta after giving birth and foetal matter after aborting. *Brucella* germs, in fact, spread extremely rapidly throughout the infected farming area. The duration of excretion may be prolonged, particularly in goats, where a single animal has been known to produce a vaginal excretion of *B. melitensis* for 33 weeks. Cattle as a rule remain infected throughout their lives. Although sheep have a natural tendency to sterilise themselves on average within a space of six months, some 20% of the infected animals continue to be carriers of the germ for much longer.

Once the *Brucella* have been excreted by the infected animals, the survival of the germ may be as long as 70 to 80 days in damp ground. The *Brucella* may also infect other domesticated animals (dogs, cats, poultry) or wild animals (rodents), which in turn become carriers of the germ. In an infected farming area the animals' quarters, milk and water containers and other instruments may all be contaminated by *Brucella*. For all these reasons, infected farms are centres of infection where brucellosis persists for an extremely long time and from which it spreads to other areas.

About two-thirds of the cases of infection in man are due to direct contamination, generally of an occupational nature; other instances are attributable to contamination of food. Whereas brucellosis caused by food contamination is just as prevalent in urban as in rural areas, the form of the disease transmitted by contact is almost exclusively rural. People employed in agricultural occupations–cattle breeders and those coming into contact with them, shepherds and, in general, anybody who works in areas infested by sick animals–are therefore the chief victims. Veterinarians and abattoir workers are particularly susceptible to the disease.

Direct contamination. More than 70% of cases of brucellosis are caused by direct contamination. This may occur through the hands, by contact with sick animals (milking and handling of animals), from contact with manure and from dead animals, especially aborted fetuses. As an occupational disease, therefore, brucellosis is most commonly found among farmers, shepherds, veterinarians, butchers and abattoir workers. However, because the *Brucella* tend to spread throughout the infected areas, most of the inhabitants of a farm frequently become contaminated.

Contamination also occurs in people who have spent only a few hours in a farm, and in bricklayers, electricians and other workers brought in to repair contaminated sheds.

Contamination in food. If meat is cooked for a sufficiently long time to kill off the *Brucella*, the slaughtering of infected animals for consumption may be authorised. The principal food carriers of *Brucella* are milk and, above all, fresh cheese from cows, sheep and goats (the survival of the *Brucella* varies from 3 to 20 days, depending on the type of cheese). Cooked pork meats are something of a hazard in areas where *B. suis* is prevalent.

To a lesser degree, contamination has been caused by water from wells contaminated by liquid manure and by vegetables that have been treated with sheep manure.

Occupational disease

Anybody coming into contact with infected animals in the course of their work or employed in businesses and industries dealing in meat products may be contaminated. They include stockbreeders, shepherds, veterinarians, abattoir workers, butchers, dairy workers, canning factory employees and the drivers of motor vehicles used to transport the animals and meat. Persons working with hides and wool and goats' hair are also exposed to the disease. Finally, it must be borne in mind that the *Brucella* are highly contagious bacteria in the laboratory, where they represent a health hazard for the staff of medical or veterinary analysis laboratories and pharmaceutical centres where brucellosis vaccines and the antigens used in diagnosis are prepared.

Doctors treating persons in the categories listed above must always carry out tests for brucellosis whenever confronted with a suspicious disease. The symptoms of the disease are not characteristic and laboratory tests are necessary for diagnosis. *Brucella* in the blood may be detected by haemoculture and serum tests (agglutination reaction), which must be repeated after a month if initial tests are negative. Diagnosis of chronic brucellosis is particularly difficult when the acute stage of the disease has not been identified. In such cases, serum reactions may be negative. A positive intradermal test is indicative of previous invasion of the body by *Brucellae* and not of active disease.

In the undertakings referred to the periodical tests carried out by the works physicians must include serum reaction tests for brucellosis.

Finally, people who have suffered from brucellosis even in the distant past may have hypersensitive reactions when brought into contact with *Brucella* antigen again. These allergic reactions, which may affect the skin, respiration or joints, occur suddenly in the course of work. The prognosis is not generally serious but they are most unpleasant and may disrupt a person's occupational activities considerably.

Prevention

Clearly the only truly effective means of prevention is animal prophylaxis by elimination of infected animals and vaccination of healthy animals. This, however, is a difficult and expensive process and in many countries comes up against economic obstacles.

Individual hygiene is obviously very much to be encouraged, including the wearing of protective clothing, gloves and boots and the careful washing of hands. Other indispensable measures include the cleaning of the working area, disinfection of utensils and a ban on the consumption of food and drink and on smoking on the premises. It is, however, difficult to enforce strict observance of these measures at all times, especially in rural areas, and the vaccination of persons employed in exposed occupations may therefore be advisable.

Vaccination by live vaccine (strain 19-BA) is effective and has been very successful in reducing brucellosis in man in countries where it has been used. The vaccine generally causes sensitisation conducive to allergic reactions, however, and its repetition has to be decided on with great caution.

Dead vaccines are ineffective in man as they can not be used with the additives employed in animal vaccines. The trend nowadays is therefore towards vaccines using antigenic extracts of *Brucella* (glyco-peptidic complexes), which do not have the disadvantages of live vaccines and provide effective, if short-lived, immunity. In France success in immunisation of occupationally exposed persons has been reported following the use of a phenol-insoluble, antigenic fraction designated "PI". Use of such a non-living vaccine has decided advantages over a live attenuated proportion. Vaccination should be repeated about every two years.

Treatment

The acute and subacute forms of the disease must be considered separately from the chronic form. Supportive therapy is recommended, including bed rest and adequate diet. Patients can recover without treatment and this should be recognised in evaluating any new treatment.

The most widely recommended antibiotic is tetracycline, at a daily dose of 1-2 g orally, for three weeks, which may be prolonged if necessary. If a relapse occurs, tetracycline therapy should be repeated for 2-3 weeks. In severe cases it may be necessary to adminster tetracycline parenterally. In severe cases with demonstrable localised lesions such as those due to *B. suis* or *B. melitensis*, it may be necessary to combine the tetracycline treatment with streptomycin (1 g daily for 2 weeks).

In some critically ill patients the initial dose of antibiotic may be followed by a severe reaction including shock. The antibiotics should be used with caution. With prolonged use of the antibiotics, it is necessary to guard against consequent alterations in the intestinal flora by the use of lactobacillus preparations and replacement of depleted vitamin B component. Failure to provide for this replacement therapy may allow the development of serious disturbances of intestinal function, stomatitis and *pruritus ani*, which may persist for months after therapy has been discontinued.

Corticosteroids may be justified in severe septicaemic and certain visceral forms of brucellosis, but should not be administered for more than a few days because of their immuno-suppressive activity.

In chronic brucellosis treatment with antibiotics and even surgical intervention may be useful. However, in the chronic forms of the disease that are characterised by physical and mental weakness, antibiotics are often ineffective. In such cases the infected person, who is hyperergic, must possibly be desensitised by antigenic therapy (subcutaneous injections of increasing doses of vaccine). Physiotherapy may be a useful adjunct.

ROUX, J.

Joint FAO/WHO Expert Committee on Brucellosis. Fifth Report. WHO Technical report series No. 464 (Geneva, World Health Organisation, 1971), 76 p.

CIS 74-791 "Symptomatology of chronic brucellosis". McDevitt, D. G. *British Journal of Industrial Medicine* (London), Oct. 1973, 30/4 (385-389). 8 ref.

CIS 79-794 "Passive haemoagglutination in the individual or epidemiological diagnosis of human brucellosis" (Hémagglutination passive appliquée au diagnostic individuel ou épidémiologique de la brucellose humaine). Renoux, G.; Renoux, M. *Semaine des hôpitaux* (Paris), 8-15 Dec. 1978, 54/43-44 (1337-1342). 10 ref. (In French)

CIS 79-192 "Brucellosis (Bang's disease) considered as an occupational disease" (Die Brucellose (Morbus Bang) als Berufskrankheit). Parnas, J. *Bundesgesundheitsblatt* (Cologne), 16 May 1978, 21/11 (161-166). 11 ref. (In German)

"Brucellosis in the United States 1960-1972. An abattoir-associated disease. Part III. Epidemiology and evidence for acquired immunity". Buchanan, T. M.; Hendricks, S. L.; Patton, C. M.; Feldman, R. A. *Medicine* (Baltimore), 1974, 53/6 (427-439).

Building construction

The building and civil engineering sector includes activities such as the construction, upkeep, repair and demolition of small, medium and large permanent or temporary buildings erected for a large variety of purposes, using all sorts of materials and techniques which may be simple and manual or assisted by powerful computer-programmed machinery.

The main differences from other industries are that the worksite is not fixed but changes from building to building, and is not protected but exposed to all weather conditions, that the nature of the work is not repetitive but changes gradually with time, and that the composition of the workforce is not permanent. Except for a small hard core, labour is generally recruited in the site area or includes a large proportion of migrant workers and is therefore composed of people with little or no specialisation. According to the site environment, and in particular to the ground, it is often necessary to change work methods and tools rapidly. During many phases of the work, and especially as it nears completion, workers and equipment belonging to various contractors are found on the same site.

As building and civil engineering activities have many things in commmon, the information given in CIVIL ENGINEERING should be borne in mind where applicable.

Site layout and work planning

Safe and healthy conditions of work can—and should—be provided for at the site planning stage. The first step to be taken in planning is a close examination of the site area.

It should include: a study of the ways of access for mechanical equipment, and people, and of traffic ways on the site itself; the choice of the means of transport together with the equipment needed for their maintenance; the construction of roads, sewerage systems, water mains, electric power supply lines–not to forget the installations which will be required after completion of the building. When laying out the areas and premises for storing materials and products, the need to isolate and protect those intended for dangerous and harmful substances should be borne in mind. Accommodation planning should include the provision of parking space for private transport, locker rooms, canteens, bad-weather shelters and first-aid rooms. Special attention should be paid to the location of concrete-making plant and fixed machinery releasing dust or producing noise.

The safety of the workforce and public and the protection of the environment should be integrated into the planning of all phases of the work. The safety of the people dwelling or working in the planned building and of the workers who will have to maintain, repair or demolish it can be considerably enhanced when these requirements are borne in mind at the planning stage.

The work programmes should be drawn up jointly by the designers and contractors. Check lists should be used to help in preventing hazards.

Personnel, time and equipment should be allotted for periodical checks of the state of plant, machinery, storage premises and personal protective equipment. The building under construction and the state of temporary structures (scaffolds, etc.) should also be checked. Persons working in isolation or in a potentially hazardous environment must be supervised. Constant housekeeping helps to prevent accidents of the type caused by nails protruding from boards lying on the ground (danger of tetanus), fires, and infestation by rodents and insects.

For large construction sites safety committees should be organised in which all those engaged in construction work should participate: managers, supervisors and workers.

Training of the workforce is an important factor in work organisation. The workers should be given adequate vocational training and be familiar with the basic principles of safety and hygiene. Even if such training has been given outside the construction site, each worker should be made familiar with the particular features of his task, the hazards associated with it and the corresponding collective and personal safety measures before he is assigned to his job. This is particularly important for newcomers to the building trade and for migrant workers.

Manual lifting and carrying of heavy or cumbersome loads should be taught to prevent injuries and strains during both individual and group work.

Medical pre-employment examination is useful to direct each worker to the type of job which is most suitable for him and also to avoid his assignment to a task which could be particularly dangerous for him or endanger other workers. This is of special importance for persons applying for jobs such as drivers of cranes, trucks or locomotives.

Since the activities and the work environment change continuously, the man on the spot is best suited to recognise hazards; therefore, each member of the site workforce should be encouraged and trained immediately to communicate such hazards, thus contributing efficiently to general safety.

In view of the hazards associated with certain types of work and work environment, it is often necessary to provide the workers with protective clothing, e.g. against water, humidity, bad weather. Personal protection also includes equipment to protect the head, eyes, ears, face, hands, respiratory tract, feet and entire body against various hazards such as falls, dangerous or harmful substances, welding fumes, paint mists, cement dust. Hard hats for head protection are a must for everybody on the site, including supervisors and visitors. Safety footwear of various types is required to protect against falling objects or for work on inclined surfaces (see ROOFING), as well as against injuries and infections (see TETANUS).

There should not only be a sufficient supply of protective clothing and equipment but provision should also be made for their regular cleaning and upkeep, which is particularly important for respirators, safety belts and harnesses.

Gravity and falls

It should be borne in mind that the force of gravity is encountered everywhere and at any time in building construction, that it acts on any lifeless and live mass, and that it may cause falls of ground, fixed or temporary structures, equipment, materials and also persons. Constant attention must therefore be directed at preventing falls from the planning stage through the various work phases (see FALLS; FALLS FROM HEIGHTS, PERSONAL PROTECTION AGAINST; ROOFING; SAFETY BELTS; SAFETY NETS).

Falls of persons moving on level ground sometimes result in injuries as serious as those caused by falls from heights. They may be due to slipping or stumbling and can be avoided by clearing away or at least by drawing attention to obstacles or irregularities in passageways, and by providing for good lighting and housekeeping (immediate elimination of spills, fallen objects, etc.).

Falls from heights can be prevented by installing guard rails, which should include at least a toeboard and be made of materials strong enough to withstand a stumbling person or the shock of a wheelbarrow. Guard rails should be at least 1 m high.

If there is no protection against falls from heights, fall-arresting devices should be used. A single person may be secured by a lifeline which is safely anchored at one end and fastened at the other to a harness or safety belt worn by the person. The impact caused by the sudden arrest is extremely dangerous for the human body; therefore, belts should be worn only when there is practically no height of fall (work on an elevated level), whereas harnesses are normally designed for heights of fall of 1-1.5 m. Beyond these heights, it is advisable to use gradually braking fall arresters. Another system which permits anchorage points to be used at greater distances is that of automatic rope tightening drums which keep the lifeline taut while the secured worker moves. However, before deciding to use any such fall arresters, it should be ensured that there are no obstacles within the potential trajectory of fall.

Collective fall-arresting devices (e.g. safety nets) may be installed at considerably greater distances than those admitted for individual fall arresters, but they are effective only if the trajectory of fall is intercepted by the collecting surface. When calculating the minimum width of this surface at a given level, it should be borne in mind that this trajectory is a parabolic curve which is the more distant from the vertical the higher the horizontal speed at which the person or object is moving at the start of the fall. This means that the horizontal distance between the trajectory and the vertical of the point of fall increases as a function of the vertical distance. Collecting surfaces may be rigid if they are intended to catch falling objects, but should be strong enough. If they are intended to catch falling persons, rigid materials may only be used for rather moderate heights; for greater heights not exceeding 6 m elastic surfaces such as safety nets should be used.

Machines and tools

The machinery, plant and tools used on construction sites must not only meet the same safety requirements as those used in industry (see MACHINERY GUARDING; ELECTRIC POWER TOOLS, PORTABLE; PNEUMATIC TOOLS; WELDING AND THERMAL CUTTING; BOILERS AND PRESSURE VESSELS), but should also be reliable, sturdy and safe to operate under rough conditions on sites where they may be exposed to mud, dust, rain, extreme temperatures, wind and so on. All these factors should be borne in mind when machinery and plant are chosen, when testing them before acceptance and when installing, operating and maintaining them.

Motor-driven plant and machinery are sources of harmful noise and vibration (see NOISE MEASUREMENT AND CONTROL; VIBRATION) and may also produce and spread harmful dust. Internal combustion engines release exhaust gases which may cause poisoning and asphyxia; therefore, they must not be used in wells, tunnels, trenches and other confined spaces if there is not abundant mechanical ventilation.

Mobile track-bound, wheeled or crawler-type plant with drivers' cabs must be equipped with seats which absorb vibration to the highest possible degree and are adjustable. The drivers should also be protected by falling object protective structures (FOPS). All wheeled and crawler-type plant moving on rough ground should be fitted with roll-over protective structures (ROPS) and other guards described under TRACTORS.

Lifting equipment

Lifting equipment of various types and sizes is used on construction sites. The safety of this equipment depends in general on the choice of types of crane or hoist which are adequate for the task to be performed (heavy-duty or precision work), on the choice of a qualified operator, and on periodic checks of the good condition and correct operation of all mechanical, electrical and hydraulic elements, load-carrying appliances, controls and safety devices. Other important factors of safety are anchorage to strong fixed structures, the stability and horizontal adjustment of runways, rails and ground, the observance of the rated loads and jib radius (which have been specified by the manufacturer but should be reduced in certain cases), the measures taken to ensure that loads being lifted do not fall or that the jib and load do not collide with neighbouring equipment, and the precautions adopted to protect the site personnel and people moving outside the site from falling objects (see CRANES AND LIFTING APPLIANCES).

As regards builders' hoists, the main hazards are those of being struck by the hoist platform or by other moving parts and of falls from the unloading landing. The hazards of these hoists are identical to those of lifts in factories or blocks of flats, and all safety measures should therefore meet the criteria given for LIFTS, ESCALATORS AND HOISTS, whereas the safety devices should be of a more rugged design and construction, since they are liable to be soiled by cement and dust and exposed to moisture or rain and snow when installed outdoors.

Electric installations and appliances

From the point of view of electrical safety, construction sites are classified as wet or damp locations. Unless more stringent precautions have to be taken when required by certain site characteristics or by national legislation, they should meet the requirements set out under ELECTRIC INSTALLATIONS, TEMPORARY, i.e. use of portable distribution panels and fault-current protection relays. Portable electric tools should correspond to the details given under ELECTRIC POWER TOOLS, PORTABLE.

Overhead electric lines should be installed and protected in such a manner that no mobile parts of cranes or other equipment moving near or under the lines can come in contact with the conductors.

Electric installations, switchgear and safety devices are easily damaged by the rough work and severe conditions prevailing on sites. They should therefore be frequently checked by a qualified electrician. Visual inspection should be supplemented by resistance and insulation tests. The same applies to earth connections. It is convenient to install the earth electrodes required for the structure to be erected when opening the site and to use them for the temporary site network.

Dangerous and harmful materials, substances and products

Construction work involves the handling and use of many different materials, products and substances which are flammable, explosive or harmful to health.

Apart from timber, there are plastics and elastomers which are combustible and give off heat, flames, smoke and asphyxiating or toxic gases when burning. Engine fuels and certain types of lubricant are not only combustible but may also release vapours which, when accumulating, form explosive gas/air mixtures. Acetylene and oxygen, as well as hydrogen and liquefied petroleum gases (LPG) used for welding, cutting, heating and lighting constitute ignition hazards; one of the LPG (propane) is heavier than air and may accumulate in basements, underground premises and bottom portions of premises, thus creating a latent hazard of asphyxia and explosion. Organic solvents, which are also used as paint thinners and with certain plastics, and most of the vapours released by the substances and products mentioned above form explosive mixtures with air and present health hazards. Formation of explosive mixtures is countered by reducing the gas or vapour concentration, by exhausting the gases or vapours, or by dilution ventilation. Ignition of explosive mixtures can be avoided by suppressing all possible sources of heat or sparks.

The main health hazards are acute or chronic forms of skin disease, respiratory or digestive system impairment, and cancer. Pneumoconiosis and silicosis are hazards associated with the extraction and processing of rock, gravel and sand, with mechanical cutting of stone, bricks and ceramic materials, with sandblasting, and with exposure to inorganic or organic dusts and fibres such as talc, asbestos, glass wool and plastics.

Other health impairment may be caused by a large number of materials, substances and products such as lime; cement containing chromium, nickel or cobalt compounds; mould oils for concrete formwork; accelerators or retarders for concrete curing; natural or artificial asphalts and bitumens; various products used for sealing, thermal insulation, sound insulation and flame retardant treatment; lead used for pipework and roofing; rubber or plastics-based materials; paints and varnishes for concrete, metal or timber; adhesives, putties and solvents; cleansing, degreasing and polishing products; wood preservatives; etc.

The severity of health impairment depends largely on the types of material or products and on the substances they contain, on the conditions and environment in which they have been used, and on the individual susceptibility of the exposed worker.

It is necessary to obtain thorough information on the properties of the products it is intended to use. The danger symbols and instructions given on the containers should be strictly observed. If there are no instructions, the manufacturer should be contacted to provide them, and they should be compared with information obtainable from competent sources.

When harmful substances are used, their quantities should be limited as far as possible; cold application methods should be given priority over hot ones, and it is preferable to use brushes or knives instead of spray guns. In confined spaces or basements where concentrations exceeding the exposure limits may easily build up, efficient ventilation should be provided by exhausting the polluted air or blowing in air which is known to be clean.

Individual susceptibility to a substance generally presents an insuperable difficulty. To avoid aggravation and severe impairment, hypersensitive workers should be employed on other tasks where they do not come in contact with the substance, or the use of the substance should be discontinued.

Wearing of personal protective equipment such as gloves, goggles, face screens, respirators and protective clothing is imperative in confined spaces, but may also be necessary for work outdoors. To avoid prolonged exposure to harmful agents, certain tasks should be interrupted by spells of work or rest in fresh air.

Pre-employment and periodic medical examinations are required by national legislation for workers exposed to many categories of substances known to be harmful to health, but may also be useful with other less-known substances.

Scaffolds, formwork and falsework

Scaffolds are temporary structures of steelwork, timber or bamboo. The criteria for their erection are the same as those for permanent structures as regards: stability to vertically or horizontally applied dynamic loads due to people, materials, lifting equipment, wind, snow and ice; safe supports and solid anchorage; measures against falls of persons and objects.

Certain types of scaffold rest directly on the ground (see SCAFFOLDING). Prefabricated frames of an approved type offer the advantage that the distances between vertical and horizontal elements, required by the laws of stability, are automatically respected. Other scaffolds are of the cantilever type (outrigger and jib scaffolds) and must always be secured to stable, sturdy structures; they should never be held in position by counterweights.

Formwork to receive poured concrete and falsework to support brickwork, concrete or steelwork structure during erection are also temporary structures which should be designed and constructed according to engineering rules. When these structures also serve as working platforms or walkways, they should be fitted with guards against falls.

Climbing or rising formwork combined with scaffolds is used for the erection of high structures of constant or variable cross section. A frequently encountered type of this formwork for reinforced concrete is that comprising three or four levels at which the following tasks are performed: placing of the steel reinforcement at the upper level, pouring of concrete at the next one, and checks and finishing operations at the lower one. This type of formwork affords protection against falls and bad weather; a special hoist is provided for workers and materials.

The generic term "scaffolding" also covers vertically and/or horizontally moving assemblies. Suspended scaffolds with hand-operated or powered platforms are particularly dangerous. The people working on these platforms should be equipped with safety harnesses and lifelines attached to a reliable anchorage independent of the scaffold suspension system. Mobile scaffolds moving on the ground should be provided with: levelling devices to ensure the horizontal position of the wheel base and the vertical position of the fixed or extending structure supporting the work platform; wheel locking devices; and brakes for the various movements.

To lift and maintain workers in elevated positions for assembly, finishing, repair or maintenance tasks, aerial baskets attached to the end of a truck-mounted, hinged swinging boom are generally used. Particular safety measures in this connection are a switch operated by the man in the basket to lock the controls and the possibility of manually lowering the basket in the event of an emergency or breakdown of the control system.

Trenches

For large-size excavations it is usual to envisage special methods of integrated safety to make sure that there will be no collapse during the entire work period. This is not always the case when trenches of modest cross section are dug for short durations, e.g. for making foundations or for laying pipelines or cables. If the ground is not exceptionally stable, trench sides may collapse and crush and/or suffocate people working on the trench bottom. Therefore, when trenches are dug at depths beyond those set by national legislation (generally between 1.0 and 1.5 m), the sides must either be sloped (at least 45 °) or be sturdily timbered or shored. As shoring from inside a trench is dangerous and time-consuming, it is preferable to use prefabricated shoring cages which are placed and removed by crane. The condition of trench sides and shoring should be checked periodically, especially after rainfalls, periods of dryness, frost and inrushes of water. Ladders should be installed at regular distances to permit quick evacuation. Sources of vibration and loads such as excavated ground should not be allowed near the trench edges. Guards should be installed to prevent falls of people or objects into trenches (see TRENCHING).

Concrete and reinforced concrete work

The materials used for concrete making and for the reinforcement, the methods applied for mixing, placing, pouring and curing, the design, construction and removal of formwork and the supervision procedures, including frequent tests of the concrete strength, checks of the site records and stress analyses, must conform to the provisions of national legislation and/or international engineering standards.

Reinforced concrete structures may be prestressed by tensioning steel tendons embedded in the concrete before or after pouring. The tensioning operation by means of hydraulic jacks involves great risks on account of the enormous energy applied to the tendons and accumulated in the structure. Breakage of tendon wires or cables, failure of concrete elements, breakdowns or errors in jack operation may bring about the sudden release of the stored energy and cause serious damage and injury. Therefore, this operation should be effected when there are no workers on the site, and the jack operators, who are protected by sturdy screens, should carefully watch the gauges and tendon attachments. Post-tensioning in the factory or on the site should not be carried out before the concrete has reached the necessary strength (see CONCRETE AND REINFORCED CONCRETE WORK).

Assembly of prefabricated elements

This technique makes use of prefabricated reinforced concrete parts or of steel frames. Its application presupposes programming methods similar to those used in industry and relies on a highly skilled workforce. From the safety point of view it offers some advantages: small number of workers exposed to hazards on the site, short hazard exposure times thanks to the rapid progress of assembly; possibility of attaching guards against falls of people or objects to the elements before they are lifted off the ground.

Simple prefabricated reinforced concrete elements such as columns, beams, panels, slabs and roof trusses, or also some larger and more complex blocks, may be prepared on the site ground or in remote factories. All elements to be assembled must be designed and constructed to resist not only the dynamic forces which will act on them when they are in place, but also the stresses acting on them during transport, storage and lifting. They must be equipped with devices for safe transport and lifting and for the attachment of guard rails or safety nets and service platforms. The assembly should thoroughly respect the programme previously established in common by the manufacturer and user. The lifting appliances should be designed for remote unhooking of the elements to avoid workers' having to climb on top of them in order to take off the slings and hooks. Large-surface panels should not be lifted while strong wind is blowing. Falls to the outside are prevented by installing guard rails and nets or by platforms installed with the aid of cranes. The workers fitting the elements to each other should be equipped with safety harnesses and lifelines with fall arresters which resist heat and sparks released during welding.

When steel structures are erected, it is advisable to assemble on the ground a maximum of elements to reduce the number of junctions to be made at height. Floors in multi-storey buildings should be laid only when there are no hazards due to work going on at higher levels.

When assembly programmes involving special systems and equipment are being established, it is important to make sure that the necessary stability will be guaranteed at all intermediate phases and positions, to bear in mind the amount and direction of the forces applied—either directly or through ropes—by jacks or winches, and to anticipate dynamic effects.

Repair and maintenance

This type of work is frequently associated with conditions inferior to those prevailing during the erection of new buildings, because the structural elements on which work is to be done or on which the workers have to move are generally in a bad state of conservation, because repairs are urgent in order to protect premises against bad weather, and because work may have to be carried out near electrical installations or gas mains. Repair and maintenance work also involves hazards to strangers such as passengers, employees or customers in public or private buildings (falls of objects, stacked materials, lifting equipment).

Work in confined spaces, such as tanks, underground chambers (cable manholes), galleries for cables and mains, is particularly dangerous because they may contain oxygen-deficient atmospheres, toxic vapours or gases, or explosive gas-air mixtures. Sewers, in addition to these hazards, present those of drowning, rat bites, cuts and punctures, infections, etc. (see SEWERS).

Before starting such work, all possible hazards should be carefully mapped in order to devise the corresponding safety measures. All work phases should be thoroughly programmed and supervised. After completion of work, all sources of potential accidents—such as means of access or tools abandoned in dangerous positions—should be eliminated (see CONFINED SPACES).

For work on roofs see ROOFING.

Demolition and dismantling

These activities are associated with hazards inherent in the structures to be demolished and hazards due to the methods used. It is important to explore these hazards before starting work, i.e. to find out about the techniques used for erecting structures and for modifying them in the course of time, to ascertain whether or not there are internal stress factors, to check the state of conservation of the materials, and to locate electrical installations, gas piping, etc.

When choosing the appropriate demolition or dismantling methods, a programme of successive work phases should be elaborated, bearing in mind that certain changes will be introduced into the interaction of forces: thrust no longer countered by arches or vaults, cancellation of prestressing in reinforced concrete or steel elements, strains caused by falling or striking masses, shock waves from demolition blasts.

Indoor installations have to be shut off. Structural parts liable to yield have to be reinforced. Access to the structure to be demolished has to be blocked up except for a few controlled and protected passages. Demolished materials should be carried away immediately and not left as dead weight on floors to be demolished. Demolition of certain structural parts must not cause collapse of others.

Demolition personnel should be carefully selected and trained in the techniques to be used. Unsafe work stations on the structure being demolished must be avoided. Good visibility should be maintained by controlling the spread of dust, i.e. by avoiding the free fall of demolished materials or by wetting them. The progress of work and safety of the demolition workers should be continuously supervised by qualified persons. To ensure safe conditions of work, it is sometimes necessary to erect scaffolds or other temporary structures and to use special equipment such as aerial baskets which must, of course, not be subjected to inadmissible static or dynamic loads.

ANDREONI, D.

CIS 80-844 *Building work—A compendium of occupational safety and health practice.* Occupational Safety and Health Series, No. 42 (Geneva, International Labour Office, 1979), 256 p. Illus.

Safety and health in building and civil engineering work. An ILO code of practice (Geneva, International Labour Office, 1972), 386 p.

Proceedings of International Symposia on the Prevention of Occupational Risks. ISSA International Section on the Prevention of Occupational Risks in the Construction Industry (General Secretariat c/o OPPBTP, Issy-les-Moulineaux, France).

Construction industry. OSHA 2207 Revised, Oct. 1, 1979. US Department of Labor, Occupational Safety and Health Administration (Washington, DC, US Government Printing Office, 1980), 618 p. Illus.

Safety in building construction (La sicurezza nelle costruzioni edili). Andreoni, D. Serie C, No. 33 (Rome, ENPI, 1977), 692 p. Illus. (In Italian)

Check lists for occupational safety (Prüflisten für Arbeitssicherheit). No. 1-37. Arbeitsgemeinschaft der Bau-Berufsgenossenschaften, Tiefbau Berufsgenossenschaft (Frankfurt am Main). (In German)

"Morbidity study in building and public works" (Etude de la morbidité dans l'industrie du bâtiment et des travaux publics). Halewick de Heusch, J. M. *Revue de médecine du travail* (Paris), 1980, 8/2 (87-100). (In French)

Burns and scalds

Burns and scalds are thermal injuries in which a portion of the body surface is exposed to either dry or moist heat of sufficiently high temperature to cause local and systemic reactions. Burns may also be caused by chemical substances (see BURNS, CHEMICAL), electricity (see ELECTRIC CURRENT, PHYSIOLOGY AND PATHOLOGY OF) and ionising radiations.

Causes. [Statistics from France show that in 1977 burns accounted for 34 366 out of a total of 1 025 968 accidents with absence from work (3.3%), for 1 498 out of a total of 112 146 accidents with permanent disability (1.3%) and for 717 331 out of a total of 28 496 598 working days lost for temporary disability for all accidents (2.5%).

As for the United Kingdom, table 1 shows the number of accidents in all premises covered by the Factory Act and the number caused by burns and scalds in 1971-75.]

Table 1. Accidents from all causes and accidents caused by burns and scalds in premises covered by the Factory Act in the United Kingdom, 1971-75

Type of accident	1971-73 (average)	1974 (average)	1975 (average)
Serious accidents	47 600	44 600	38 620
Including:			
Deeply penetrating burns and scalds	1 340 (2.8%)	1 620 (3.6%)	1 220 (3.2%)
Burns and scalds covering more than 1 ft²	540 (1.1%)	360 (0.8%)	340 (0.9%)
Minor accidents (groups 2 and 3)	266 500	256 920	243 140
Including minor burns and scalds	7 600 (2.9%)	7 200 (2.8%)	7 060 (2.9%)

The causes of burns are many and varied: contact with the surfaces of hot vessels or pipes, contact with workpieces that are heated during processing, projections of molten metal and splashes of hot liquid, falls into vats of hot liquid, leakage of steam under pressure, fires and explosions including the ignition of flammable working clothes, friction due to contact with moving parts, etc. See tables 2 and 3.

Table 2. Breakdown by cause of cases of burns hospitalised in a French specialist burn clinic

Cause	Percentage of all hospitalised burn cases	Percentage of cases of occupational origin
Hot liquid	47	5
Flame	33	33
Explosion	13	29
Electricity	2	2
Molten metal	1	11
Chemical	2	20
Hot solid	2	0

Table 3. Breakdown by site of burns of occupational origin hospitalised in a French specialist burn clinic (Percentages)

Site	Percentage
Head	81
Trunk	51
Lower limbs	64
Upper limbs (except hands)	52
Hands	65

Occupations at risk. Burn injuries may be sustained in virtually any industry, although those in which the risk is greatest are: those employing furnaces, kilns and ovens such as the iron and steel industry and smelting and refining, the cement industry, pottery industry, etc.; those involving heat processing or molten metal processing, e.g. galvanising, heat treatment of metals, wire drawing, welding and cutting, etc.; those employing hot liquids or steam for heating, processing or cooking, e.g. the food industries, chemical industry, plastics industry, pulp and paper industry, etc. However, even those industries which seem at first sign to be devoid of hot processes, such as agriculture, clothing manufacture, etc., may still experience burn accidents even if only as a result of inadequately guarded space-heating equipment.

Burn classification

The severity of a burn is dependent on the depth to which the injury penetrates, commonly expressed as "degree of burn" and the extent of the body affected, expressed as a percentage of body surface area; other factors include age and health condition of the victim, contamination of the wound by the burn agent, e.g. in the case of a burn by a hot toxic substance.

Degree. Different systems of classification exist, some being based on three degrees of burn and some on four.

First-degree burns result in erythema and swelling of the wounded area. Pain may be intense, but there is no necrosis or blistering; pain soon diminishes and healing occurs in a few days without scarring.

Second-degree burns still affect only the epidermis with some tissue destruction and blistering. After the blisters break or are pierced, the skin dries and heals by regeneration.

Third-degree burns involve complete destruction of the epidermis and necrosis of hair follicles and sweat glands. There may be large blisters or brownish-black eschar marks. Complications are not uncommon and skin grafting may be necessary.

Fourth-degree burns may penetrate to areas below subcutaneous fat and involve muscle, bone and even whole organs. In certain classifications, the third and fourth degree are combined to form a single category.

Extent. The extent of the burn is expressed as a percentage of total body area; for this purpose the body may be divided into sections representing the fraction of skin area indicated below:

Head, 9%
Upper extremities, each 9%
Front or back of trunk, each 18%
Lower extremities, each 18%
Genital organs, 1%

Types of burn

Although all burns have essentially the same features, the specific nature of the lesion may be influenced by the type of causative agent, e.g. steam, hot water, other hot liquids, molten metals, hot solids, flames, etc.

Explosion burns. These usually affect the exposed parts of the body, i.e. the hands and face. In certain cases, the victim's clothing may protect the rest of his body from being affected.

Steam burns. These are usually superficial and affect mainly the exposed parts of the body.

Hot-water burns. The depth of the burn depends on the temperature of the water; there is blistering (broken or intact blisters) and the colour of the subjacent skin varies from pink to dark red depending on the severity. This type of burn is often more severe if the victim was heavily clothed, since clothing prolongs the contact between the hot water and the skin.

Molten-metal burns. These more frequently affect the lower limbs and they are extremely deep since the metal becomes literally incrusted in the skin.

Hot-solid burns. Although normally not extensive, they are often very deep.

Flame burns. These are always deep and their severity depends on the depth. They may have the appearance of large brownish or livid white patches below a thin layer of carbonised epidermis.

Systemic reactions

These depend on the degree, extent and site of injury and individual factors in the victim and may include: modification of blood flow, oedema, shock, acid-base imbalance. Specific injury to individual organs may also result in cases of severe burn, e.g. to the stomach (Curling's ulcer), kidneys (lower nephron nephrosis) and liver.

Prognosis

This is based on: the relationship between the degree of burn and its extent; the integrity of important functional zones such as the face, hands, genital organs, and the respiratory tract; the presence of traumatic lesions or absorption of toxic substances; age and health of victim.

Burns affecting the aged in employment are always more problematic and burns of limited extent may often be fatal. Existing conditions such as alcoholism, renal disorders, mental disorders, epilepsy, etc., or concomitant trauma will always result in less favourable prognosis. However, no matter what the physiological condition of the victim, it is possible that burns of apparently moderate severity may take a dramatic course following body fluid imbalance during the first hours: the burn may pump large quantities of liquid from the body to such an extent that the blood is concentrated and should be diluted immediately by appropriate liquid injections.

First aid

The first-aid treatment of burns should be aimed at keeping the burnt area sterile. Carefully remove clothing; however, should the fabric stick to the burnt area, do not pull it off but cut round it. Do not break blisters. Cover the wound with a strip of lint or with the cleanest material available. Bind the lint lightly in position with cotton wool.

Experience over the last decade has shown the value of first-aid treatment of superficial burns by immersion in iced or cold water for about 10 min, or the application of ice as soon as possible after the accident. This treatment immediately cools the wound, extinguishes burning clothes, etc., and limits the ensuing oedema; cold water also removes acids or bases. This treatment should be followed by soap and water washing (taking precautions not to burst blisters) and application of a sterile dressing. The wound should not be touched for 8-12 days to allow complete skin regeneration.

In the event of a burn which covers 10% or more of the body area, immediate hospitalisation is essential; where the face, hands or perineal region are involved, the threshold figure is approximately 5%. If the victim can be transported to a hospital specialising in burns within an hour of the accident, wrapping in a clean sheet will suffice; if this is not possible the victim should be sent to the nearest hospital or the nearest physician immediately. In all cases, except electrical burns, victims should not be given liquids to drink.

Whenever a victim is hospitalised he should be accompanied by a note indicating the circumstances of the accident, especially the causal agent, and any treatment given. If transport of the victim is to be a lengthy process (e.g. in the case of isolated workplaces, or on board ship), relief of pain is essential.

Specialist hospital treatment

On arrival, the patient is examined and, on the basis of burn degree and extent, an initial prognosis is arrived at; special attention is paid to the pain factor and respiratory and cardiovascular disorders. Urinalysis, antitetanus injection and analgesic treatment is carried out. Reconstitution of normal blood composition is ensured by administration of macromolecular and electrolytic solutions which re-establish physiological balance.

Where skin has been destroyed over more than 30% of the body area, plasma loss and bacteria contamination may be fatal. Recent research has concentrated on reducing the danger of bacteriological infection especially by *Pseudomonas pyocyanea*, and sulphamylon, colistine gentamyacine or acerbine ointments, or 0.5% silver nitrate solution have been used to this end.

Widespread or deep burns are treated by skin grafting, the grafts being taken from the patient himself if his condition permits or from volunteers (homografts).

COLSON, P.
BIRON, G.

CIS 79-8 *Occupations at high burn risk–A summary report* (National Institute for Occupational Safety and Health, 4676 Columbia Parkway Cincinnati, Ohio) (1978), 152 p. Illus. 10 ref.

CIS 76-2065 "Causes, diagnosis and immediate treatment of burns" (Ursachen, Diagnose und erste Versorgung von Verbrennungen). Müller, F. E. *Zentralblatt für Arbeitsmedizin, Arbeitsschutz und Prophylaxe* (Heidelberg), Apr. 1976, 26 (82-88). Illus. 6 ref. (In German)

CIS 78-1683 "Occupational burns" (Les brûlures d'origine professionnelle) Philbert, M.; Rieu, M.; Dalloz, S. *Archives des maladies professionnelles* (Paris), Oct.-Nov. 1977, 38/10-11 (879-893). (In French)

CIS 76-623 "Fire and burns at work" (Incendie et brûlures au travail). Pardon, N.; Pommier-Perrot, M. P.; de Frémont, H.; Ogier, J.; François, R. C.; Lebeaupin, R.; Monsaingeon, A.; Bansillon. *Cahiers de médecine interprofessionnelle* (Paris), 2nd quarter 1975, 15/58 (5-75). 99 ref. (In French)

CIS 79-2070 "Burn first aid". *National Safety News* (Chicago), June 1979, 119/6 (83-87). Illus. 8 ref.

Burns, chemical

Most chemical burns result from the action of corrosive substances which destroy tissue at the point of contact. The skin, eyes and digestive system are the most commonly affected parts of the body. The corrosives may be either acid or alkali, the main feature being the hydrogen or hydroxyl concentration. Extremes above pH 11.5 or below 2.5 are not tolerated by the body and will almost always result in irreversible tissue damage. An outstanding feature of chemical burns is the fact that tissue destruction is progressive: acids tend to be neutralised by the available or exposed tissue whereas alkalis continue to cause damage unless neutralised by other means. Protein breakdown is most prominent, being marked by coagulation, precipitation and actual dissolution of tissue constituents (colour plate 7).

The number of serious chemical burns is lower than that of thermal burns and scalds–chemical burns account for only 2% of the cases dealt with by specialised burn clinics.

Causes. Common causes of chemical burns are splashes produced during decanting, spills occurring whilst carrying open containers of chemicals, splashes from violent chemical reaction, or jets from leaking vessel, etc. Chemicals widely used in industry which are frequent causes of burns are sulphuric, hydrochloric, nitric, acetic, formic and hydrofluoric acids and ammonia, potassium hydroxide and phosphorus.

Severity. The severity of a chemical burn depends on *(a)* the concentration of the chemical; and *(b)* the duration of contact.

The first factor cannot usually be controlled but correct first-aid treatment may reduce the duration of contact.

In the same way as thermal burns and scalds, chemical burns may be classified according to the degree of tissue injury. See BURNS AND SCALDS.

HEALTH HAZARDS

In the case of most minor burns, the health hazard is limited to slight superficial tissue destruction and moderate pain. However, in the case of severe or extensive burns, the health hazard may be threefold:

(a) severe pain and shock;

(b) loss of body fluid;

(c) absorption by the bloodstream of the injurious chemical and the poisonous products of tissue decomposition.

General first aid

Immediate and competent first aid can limit substantially the effects of contact with a corrosive substance.

Skin burns. Where working clothes have been contaminated, these should be rapidly removed and the affected area of the body should be flushed copiously with running water to wash away the chemical. Any delay in this treatment will allow the chemical to attack the tissue and no secondary treatment, such as the use of a neutralising substance, will compensate for this loss of time. In addition, the reaction between the injurious chemical and the neutralising agent may produce sufficient heat to aggravate the primary lesion.

After the chemical has been removed by irrigation, the wound should be covered with a sterile dressing and the victim despatched to a specialised burns treatment centre. The victim should always be accompanied by a note specifying the circumstances of the accident, i.e. the name of the chemical, its concentration, degree of purity and temperature at the time of the accident.

Eye burns. In the case of chemical contact with the eye, the need for speed in first-aid treatment is even greater since if damage occurs to this vital organ it may lead to permanent blindness. The eye should be irrigated with copious amounts of water for 15-20 min; the water should be at low pressure and the eye should be kept exposed by pulling back the eyelids. A suitable technique may be to hold the victim's face under water whilst he blinks vigorously for 10 min or so, with brief pauses for catching breath. Without specialised supervision, neutralising treatment should not be attempted since further damage to the eye may be occasioned. The victim should then be transported to an ophthalmologist, since the severity of chemical eye injuries may not be fully apparent until some days after the accident.

Ingestion of corrosive substances. Vomiting must not be induced since this may aggravate injury. If the victim is conscious and can swallow, he should be made to drink large quantities of water.

Specific first aid

Immediate irrigation is the most important first-aid measure in all cases of chemical burns; however, different chemicals may call for slight variations in approach.

Acid burns. The lesions rapidly attain their maximum severity. Neutralisation with 2-5% sodium bicarbonate solution is less effective than water flushing and there is a danger of additional aggression. Washing with a 5% triethanolamine solution is now being recommended.

Alkali burns. These will usually be more serious than acid burns. Where a water hose is available and clothes are contaminated, time may be saved by inserting the hose under the clothing in order to flood the injured zone.

Phosphorus burns. Lesions should be immediately covered with clean, or, if possible, sterile, wet towels; ointments should not be applied.

First-aid equipment

Industries in which there is a danger of chemical burns should be fitted with chemical emergency showers and eye irrigation fountains. They should be placed in a convenient position near to danger zones, clearly marked, and personnel should be fully instructed in their operation. See also separate articles on individual substances.

PREVENTIVE MEASURES

Wherever possible the least corrosive substance should be employed in a process. Containers should be clearly labelled and the workers should be informed of the hazards, especially of those involved in using unsuitable containers for the decanting of corrosive substances. Handling aids should be used when carrying or decanting containers, and workers exposed to splashes, etc., should wear chemical-resistant eye and face protection, hand and arm protection and foot and leg protection; work clothing should also be of the chemical-resistant type.

Where practical, mechanisation or automation of corrosive substance handling should be instituted.

COLSON, P.
BIRON, G.

CIS 77-1102 *Study of the causes and consequences of occupational burns* (Etude des causes et conséquences des brûlures d'origine professionnelle). Dalaloz, S. (Université de Paris VI, Faculté de médecine Broussais, Hôtel-Dieu, Paris) (1977), 89 p. 34 ref. (In French)

"State of the art of the treatment of severe thermal and chemical burns and inhalation injuries" (Der gegenwartige Stand der Behandlung von schweren thermischen und chemischen Hautschädigungen sowie Inhalationsschäden). Fiege, U. *Zentralblatt für Arbeitsmedizin, Arbeitsschutz, Prophylaxe und Ergonomie* (Heidelberg), Apr. 1981, 31/4 (130-141). Illus. 31 ref. (In German)

"Possibilities of medical rehabilitation in chemical burns and injuries of the eyes" (Möglichkeiten der medizinischer Rehabilitation von Patienten mit Augenveratzungen und -verletzungen). Kruger, K. E. *Zeitschrift für die gesamte Hygiene und ihre Grenzgebiete* (Berlin), 1974, 20/2 (129-131). (In German)

Bursitis and tenosynovitis

Bursitis

This is an inflammatory condition affecting a bursa, in particular the bursae of the elbow and knee joints.

Occupational causes

Bursitis is usually caused by prolonged, repeated pressure or by repeated jolts to the joint in question. While the classical types of bursitis such as "parson's knee", "housemaid's knee" or "weaver's bottom" are seldom seen today, their place has been taken by others, such as "carpet layer's knee" caused by pushing a carpet-stretching block with the knee and "conveyor belt shoulder" in which bursitis occurs as the result of repeated jolts when heavy objects are removed from an overhead conveyor. Bursitis in the elbow is the most common form met with in industry and is usually the result of prolonged pressure on the elbow; using the elbow as a fulcrum when lifting heavy objects, such as the forms from a small printing press, will rapidly cause bursitis.

Symptoms. In the case of bursitis in the knee or elbow, the symptoms are usually pain and swelling; with bursitis of the shoulder, the pain is usually widespread but swelling may not be apparent.

PREVENTIVE MEASURES

Tools or processes which produce prolonged or repeated pressure to be exerted on a joint should, where possible, be redesigned to eliminate this cause of injury. For example a mechanical carpet-stretcher should be used in place of the spiked block (knee-kicker) that has to be pushed or struck with the knee. The height of conveyors should be such that it is not necessary to unload goods from them on to the shoulder, or better still the process should be mechanised. Where such measures are not feasible, the work should be done, for example, by tall men and not by short women.

Where equipment redesign is not practicable, localised pressure on the joint should be alleviated by using a soft pad which distributes the pressure over a larger area. However, very soft materials such as low-density foamed plastics are not suitable for these pads.

Treatment. Bursitis in the early stages can be treated by avoiding further pressure on the joint and prescribing that the joint is rested until the symptoms abate. In advanced cases treatment will be medical or surgical.

Tenosynovitis

This term is applied loosely to conditions in which the free movement of a tendon is restricted, usually by a swelling of the tendon or its sheath. While any tendon may be involved, in industry it is the hand, wrist and forearm tendons that are more frequently affected. Many clinical varieties of the disease may be observed, including: stenosing tenosynovitis; de Quervain's disease; carpal tunnel syndrome; and rheumatoid and tubercular tensynovitis. In carpal tunnel syndrome, the swollen tendon presses on the median nerve in the carpal tunnel.

Occupational causes

The most common causes in industry are processes which entail very rapid, repetitive finger and hand movements in manipulating small workpieces. Typical examples of industries having a high incidence of tenosynovitis are telephone switchgear, electronics, printing (stripping and interleaving), domestic electrical equipment and automobile (small-component assembly [and press operations) industries, pottery (glaze dipping), brick making, industrial poultry (evisceration and trussing chicken), etc.]

The incidence of the disease may be increased by the existence of one or more of the following factors:

(*a*) productivity incentives and overtime which encourage excessive use of the muscles. Work is often continued until the wrist is so sore that further work is impossible;

(*b*) unsuitable hand tools, poor workplace design, physiologically unsound work posture—all resulting in excessive muscular work;

(*c*) faulty components such as tight-fitting screws or nuts, tight-fitting or stiff plastics components which have to be forced into holes—thus increasing muscular work;

(*d*) use of excessive force or an unnecessarily tight grip on hand tools (figures 1 and 2);

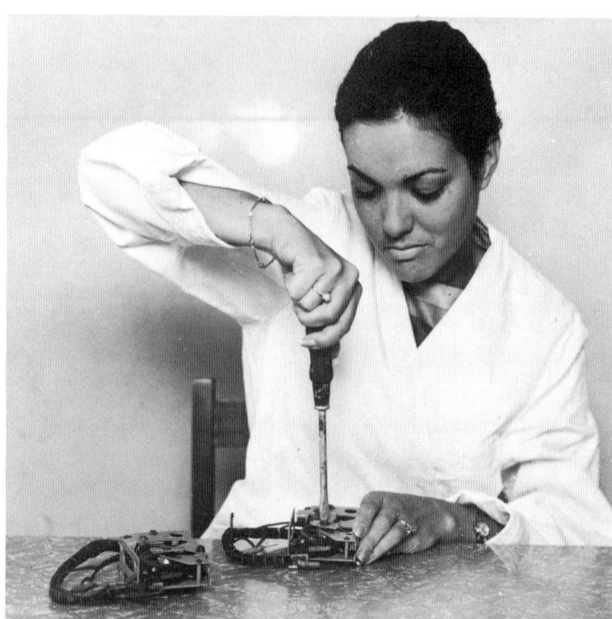

Figure 1. This method of using a screwdriver has frequently caused tenosynovitis after a few days of work.

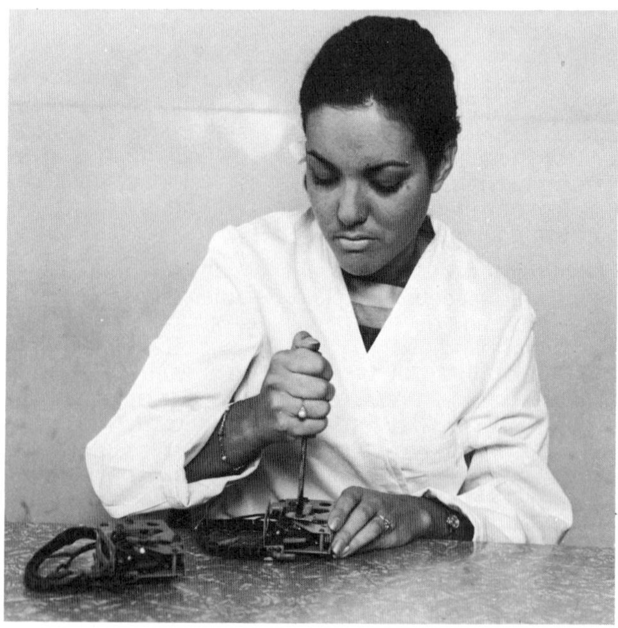

Figure 2. Work posture has been improved with regard to figure 1 by using a shorter screwdriver and proper training of worker. Further improvement in work posture would result if the operator was raised or the bench lowered.

(e) jerky work movements instead of smooth working action;

(f) return to repetitive work after holiday, sickness or other prolonged absence;

(g) full-time employment of inexperienced operator on repetitive work.

An investigation of over 500 cases of industrial tenosynovitis showed that the cause was usually a combination of two or more of these factors; in over a third of these cases the major contributing factor was the use of excessive force or too tight a grip on the tool.

Symptoms. In the early stages the subject will complain of a sore wrist or forearm; in later stages the pain will be acute and movement of the wrist or fingers limited. Pressure on the tendon sheath will be painful and there may be a grating sensation (crepitus). In advanced stages only jerky movement of the finger may be possible, and in the final stage the thumb or finger may become locked in flexion.

PREVENTIVE MEASURES

Workstations, procedures and tools should be evaluated to detect potential or actual causes of tenosynovitis. Remedial action may include redesign of the workstation, provision of special tools or modification of existing tools, elimination of badly fitting components, worker training, job rotation or job enlargement.

New employees should be properly trained, introduced gradually to repetitive work and carefully supervised. Persons who work with a jerky action or with rapid and unnecessary arm and body movements should not be employed on work where there is a possibility of tenosynovitis. While much can be done to prevent tenosynovitis, it appears that some persons are incapable of doing rapid repetitive work.

Treatment. In the early stages of tenosynovitis, treatment is simple; if the employees are instructed to report at the medical centre at the first sign of a sore wrist or forearm, infrared-ray treatment, light bandaging and rest are usually effective. A change of work or job rotation is indicated to prevent recurrence. If a large productivity incentive is paid for high work output, it is probable that the employee will not report the injury until the pain is intolerable and in these cases medical or even surgical treatment is usually necessary.

WELCH, R.

Bursitis:

CIS 78-794 *Occupational beat conditions (excluding beat knee)* (Les bursites d'origine professionnelle (à l'exclusion de celles du genou)). Beaucousin, M. (Université de Paris VI, Faculté de médecine Broussais, Hôtel-Dieu, Paris) (1977), 57 p. 25 ref. (In French)

Bursitis and epicondylitis in glass cutters in East Bohemia (Bursitidy a epikondylitidy u brusičů skla ve Východočeském kraji). Kuželová, M.; Šimko, A.; Kovařik, J.; Salandová, J.; Malá, H. *Pracovnilekarstvi* (Prague, 1980), 32/1 (23-25). 5 ref. (In Czech)

Tenosynovitis:

CIS 77-803 *Occupational disease due to overstrain of the tendon sheaths, peritendinous tissue, and muscular and tendinous insertions* (Les affections professionnelles provoquées par le surmenage des gaines tendineuses, du tissu péritendineux, des insertions musculaires et tendineuses). Dupont, M. P. (Université de Paris VI, Faculté de médecine Saint-Antoine, Paris) (1976), 91 p. 93 ref. (In French)

CIS 74-1994 "Affections of the tendons in the region of the shoulder and elbow in moulders and fettlers" (Tendopathien im Schulter- und Ellbogenbereich bei Formern und Gussputzern). Gallitz, T. *Arbeitsmedizin–Sozialmedizin–Präventivmedizin* (Stuttgart), Dec. 1973, 8/12 (282-285). Illus. 14 ref. (In German)

"A study of painful shoulder in welders". Herberts, P.; Kadefors, R. *Acta orthopedica scandinavica* (Stockholm), 1976, 47 (381-387). Illus. 23 ref.

"Peritendinitis crepitans and simple tenosynovitis: a clinical study of 544 cases in industry". Thompson, A. R.; Plewes, L. W.; Shaw, E. G. *British Journal of Industrial Medicine* (London), 1951, 8/2 (150-158). Illus. 23 ref.

See also the *Scandinavian Journal of Work, Environment and Health* (Helsinki), 1980, 6/suppl. 3.

Business machine operation

Business machines may be classified into two basic groups. First, there are small individual pieces of office equipment, usually operated by one person, such as electric typewriters, duplicating and dictating machines. The second group consists of more complex machines and machine combinations, such as large computer systems. There is, however, an overlap between the two groups. For example, input-output terminals for a computer may be located in a small office, and the computer located many miles distant (see COMPUTERS).

In recent years there has been a tremendous increase in the use of all types of business machines, particularly in the more technologically advanced countries, and it is considered that this increase in the number, variety and complexity of the machines used in offices will continue.

HAZARDS AND THEIR PREVENTION

Considering the very large numbers of persons who use business machines and the fact that machine operators such as typists, photocopying clerks, computer attendants, etc., are in the majority of cases largely unfamiliar with the technical functioning of the machines they use, the low incidence of occupational accidents and diseases in business machine operation indicates that these machines must be classed low on the list of occupational hazards.

Nevertheless, there are a number of potential safety and health hazards which include: exposure to noise, chemicals and fumes; photosensitive or other treated papers, laser beams and X-radiation; psychological problems; problems related to ergonomics; electrical hazards; fire; high-intensity light; ultraviolet and infrared radiation.

Health hazards. Most business machines are designed in such a way that the noise they produce will not endanger the operator's hearing. However, when a machine is installed without due consideration of room acoustics, or where a large number of machines or batteries of equipment are installed in a single room, the noise produced may constitute a hazard. Moreover, as the speed at which business machines operate increases, the level of noise produced rises and the potential noise hazard becomes more acute. Noise control is essentially a task for the manufacturer who should ensure that the machines are designed for quiet operation and should fit adequate damping or enclosure systems. However, where it is found that the noise levels from existing machines exceed acceptable levels, much can be done by the installation of noise absorbant partitions, vibration damping mountings, wall, ceiling and floor surfaces of absorbent material.

Operators frequently use earphones of the insert type when using dictating equipment. Improper fit of the earphones can cause an irritation of the external ear canal. Even a properly fitting earphone may cause complications when placed in an already infected ear. Prevention depends upon proper design and maintenance of the earphone and education of the personnel as to the potential hazard. Arrangements should be made for cleaning of the earphones.

Mixtures of potentially hazardous organic solvents are employed in cleaning business machines. The cleaning solutions formulated by some manufacturers contain only solvents having a low order of toxicity. Nevertheless, a potential hazard does exist and, consequently, equipment should be cleaned strictly in accordance with the manufacturer's instructions; cleaning solutions should always be used in well ventilated areas to avoid the build-up of hazardous atmospheric concentrations and should be kept covered when not in use to prevent unnecessary evaporation. Some of the solvents employed may be absorbed through intact skin and most solvents will have a defatting action on body tissues. Skin contact should therefore be minimised by use of special cleaning tools or hand protection. Highly toxic substances such as benzene or carbon tetrachloride should never be used as substitutes for the manufacturer's standard solutions. Personnel using cleaning solutions should be fully informed of the potential hazards involved and the necessary precautions. Many business equipment manufacturers have equipment-maintenance services which carry out client servicing under contract and employ personnel who have been trained in safe working procedures.

Some of the inks used in printing contain azo dyes. Studies have indicated that some of these dyes may be mutagenic and/or carcinogenic. Environmental emission is usually quite low. The primary hazard is related to skin contamination with secondary absorption through the skin. Exposure is primarily of maintenance personnel and not the user.

Some types of duplicating processes require the use of volatile liquids such as methyl alcohol and, during refilling operations or where machines are sited in small enclosed premises, potentially hazardous atmospheric concentrations of vapour may accumulate. It is also possible for toner dust to be released if the duplicating equipment is malfunctioning. Control depends upon proper installation and maintenance of equipment and provision of adequate ventilation. Persons with chronic respiratory disorders require selective job placement avoiding processes utilising irritating volatile solvents.

Video display terminals are frequently used as input-output devices. They contain a cathode ray tube that is illuminated when an electron beam excites the phosphor coating on the inner surface of the glass faceplate. At operating conditions below 20 kV no measurable X-radiation above background is produced. Clamping diodes may be added to the circuitry to turn off the power should voltage levels be reached at which X-radiation could be produced. The electromagnetic emissions from the glowing phosphor are relatively small and pose no health problems. The United States National Institute for Occupational Safety and Health has investigated and found that there is no cause and effect relationship between cataracts and the use of visual display terminals.

The primary potential problems resulting from the use of visual display units are eye strain and operator fatigue. Symptoms of eye strain are usually the result of excessive ocular muscle activity. These muscles will recover with adequate rest. Appropriate contrast between the visual object and surrounding area enhances perception and decreases eye strain. An optimal distance between the operator and the display unit will help maintain relatively constant pupil size and accommodation and lessen the possibility of eye fatigue. Even a person without noticeable refractive errors may find long-term close-up tasks more comfortable when wearing glasses that permit him or her to stay in the range of comfortable viewing. Inadequate workstation designs may force operators to compromise between efficiency and comfort and can lead to fatigue. Chair and work surfaces must be appropriate for each specific operator.

The treated papers used in duplicating equipment may contain compounds which can cause skin irritation and sensitisation. Though the occurrence of dermatitis is infrequent, it may be necessary to ensure that persons with a history of skin disease are not subject to exposure.

Where large numbers of machines or very large machines, such as computers, are located in relatively small premises, the amount of heat given off may be considerable. Air-conditioning is the ideal solution to this problem but efficient mechanical ventilation may also provide adequate control.

The high-voltage power supplies required for the operation of both computers and the cathode-ray tubes sometimes used as computer input-output devices may be a source of X-radiation. Control of potential exposure depends upon the circuit design and the installation. Many office copying machines employ ultraviolet, infrared and high-intensity visible radiation; shielding is required to eliminate operator exposure. In addition, ultraviolet radiation can, on contact with the oxygen in air, produce ozone, a gas which is a respiratory irritant. Prolonged use of duplicating equipment in enclosed spaces without ventilation can lead to levels of ozone which will produce respiratory irritation. Special applications require the use of laser optical scanners in conjunction with computer systems. The energy level is, however, usually too low to present a health hazard.

A number of potential psychological problems are associated with the use of computer systems. The worker's fear of automation, particularly where job security is concerned, may necessitate education of personnel prior to installation of computers. Many older employees may find it difficult to change work habits developed over years of employment. Computers allow the individual worker to increase his productivity greatly but, with this, there is an increase in responsibility. This, too, may lead to psychological problems of adjustment.

Human factors engineering is of particular importance for business machine operators. Reference has already been made to eye strain in relation to the use of visual display units. Prolonged operation of business machines, especially from a sitting position, may produce postural problems in persons such as typists, addressing-machine operators, card punchers and sorters, accounting-machine operators, etc. An ergonomic study of the workpost is desirable to ensure that the operator's posture and movements do not produce undesirable stress on the muscular and skeletal systems. Tables and benches should be of a suitable height, chairs should have adjustable-height seats, and adequate back rests and foot rests may also be desirable. Back supports must take into consideration the normal curvature of the spine. The shape of the chair should permit most of the body weight to be transferred to the seat through the buttocks. A number of companies arrange rest periods for business machine operators, during which light physical fitness exercises may be carried out (see CERVICO-BRACHIAL DISORDERS, OCCUPATIONAL).

Electrical hazards. Business machines are nearly all electrically operated and some are supplied at very high voltages; there is, consequently, a potential hazard of electric shock. Machines should be well maintained and regularly inspected. Sufficient electricity points should be strategically located throughout the office to minimise the need for multiway connectors and long flexible leads. Operators should be warned of the electrical hazards of makeshift electrical connections, of tampering with machines or of attempting to carry out running repairs. Any defects should be reported immediately to the supervisor or the maintenance staff.

Fire. Electrical faults may be the cause of fires in business machines and consequently fire extinguishers should be available on the premises and staff should receive instruction in fire drills. The magnetic belts and tapes used with computers and office equipment contain complex high molecular weight polymers. Though not a hazard in themselves, in the event of fire these polymers will undergo thermal decomposition releasing irritating and toxic vapours. Operating personnel and those responsible for fire fighting should be aware of the potential hazard. Many of the solvents used may also constitute a fire hazard.

KAMINER, A. J.

Office equipment:

CIS 77-1162 *Standard for office appliances and business equipment.* UL 114 (Underwriters' Laboratories, 207 E Ohio Street, Chicago) (14 Oct. 1976), 40 p. Illus.

CIS 79-566 "What you should expect from a good typewriter" (Det här ska du kräva av en bra skrivsmaskin). Berns, T.; Carlsson, L.; Juhl, B. *Arbetsmiljö* (Stockholm), 25 Sep. 1978, 11 (12-19). Illus. (In Swedish)

CIS 79-2073 *Ergonomic office design—Examples from practice* (Ergonomische Bürogestaltung—Beispiele aus der Praxis). Krafft, H. G. Nr. 19, Schriftenreihe Arbeitsschutz, Moderne Arbeitsstätten (Bundesanstalt für Arbeitsschutz und Unfallforschung, Postfach 17 02 02, Dortmund) (1978). Illus. (In German)

CIS 78-1477 *Ergonomic survey of office and data-processing machine repairwork* (Konttorikonehuolto—ja tietokonetekniikkojen työn ergonominen kartoitus). Järvinen, T.; Cedercreutz, G.; Koskela, A. (Työterveyslaitos, Työolosuhteet 10, Helsinki) (1977), 25 p. Illus. 3 ref. (In Finnish)

Photocopying:

CIS 79-479 *Guide to good practice for the safe handling of diazo compounds* (Drawing Office Material Manufacturers and Dealers Association (DOMMDA), 52-55 Carnaby Street, London) (1978), 23 p. Illus.

CIS 80-489 "Itching erythema among post office workers caused by a photocopying machine with wet toner". Jensen, M.; Roed-Petersen, J. *Contact Dermatitis* (Copenhagen), Dec. 1979, 5/6 (389-391). 1 ref.

CIS 78-1925 "Characterization of potential indoor sources of ozone". Allen, R. J.; Wadden, R. A.; Ross, E. D. *American Industrial Hygiene Association Journal* (Akron, Ohio), June 1978, 39/6 (466-471). 12 ref.

Video display terminals:

Human factors aspects of visual display unit operation. MacKay, C. Research paper 10 (Health and Safety Executive, Baynards House, 1 Chepstow Place, London) (1980), 12 p. 57 ref.

A report on electromagnetic radiation surveys of video display terminals. DHEW (NIOSH) publication No. 78-129 (National Institute for Occupational Safety and Health, 4676 Columbia Parkway, Cincinnati) (Dec. 1977), 20 p. 14 ref.

Work with visual display units. CIS bibliography N.15 (1981-1964) (Geneva, International Labour Office, 1981), 12 p.

Butadiene

Butadiene (CH_2:CHCH:CH_2)

1,3-BUTADIENE; BIETHYLENE; BIVINYL; DIVINYL; VINYL ETHYLENE; DIETHYLENE; ERYTHRENE

m.w.	54.1
sp.gr.	(liquid at −6 °C) 0.65
m.p.	−108.9 °C
b.p.	−4.4 °C
v.d.	1.9
v.p.	760 mmHg (101.08·10³ Pa) at −4.5 °C
f.p.	< 7 °C
i.t.	428.9 °C
e.l.	2.0-11.5%

insoluble in water; soluble in ethyl alcohol

a colourless, flammable, non-corrosive gas with a pungent, aromatic odour.

TWA OSHA	1 000 ppm 2 200 mg/m³
STEL ACGIH	1 250 ppm 2 750 mg/m³
IDLH	20 000 ppm
MAC USSR	100 mg/m³

Butadiene is highly reactive and tends to polymerise and form oxidation products.

Production. While hundreds of methods of obtaining butadiene have been described, its commercial production is generally by steam cracking naphtha and light oil, or by dehydrogenation of butane or butene over a catalyst. Concentration is accomplished by extractive distillation with acetonitrile or dimethylacetamide, or by complexing agents such as cuprous ammonium acetate from which butadiene is released in a relatively pure state by heat. Conversion of ethyl alcohol to butadiene by reaction with acetaldehyde was developed during the Second World War.

Uses. The chief use of 1,3-butadiene is in the manufacture of synthetic rubber by polymerisation and copolymerisation under the influence of sodium and other activators. Dimerisation to 4-vinyl-1,3-cyclohexene also occurs. It is finding increased usage in the plastics and resin field. Useful properties here include numerous addition reactions and combination with activated olefins in the Diels-Alder reaction to give hydro-aromatic compounds. With these chemicals, butadiene forms resins which serve as reinforcing and stiffening agents for rubbers, components of water and solvent-based paints, and high impact plastics with good chemical and heat-resistant properties.

HAZARDS

Physico-chemical hazards associated with butadiene result from its high flammability and extreme reactivity. Since a flammable mixture of 2-11.5% butadiene in air is easily reached, it constitutes a dangerous fire and explosion hazard when exposed to heat, sparks, flame or oxidisers. On exposure to air or oxygen, butadiene readily forms peroxides, which may undergo spontaneous combustion.

The experience of workers with occupational exposure to butadiene, and laboratory experiments on humans and animals, indicate that its toxicity is of a low order. Exposure to very high levels of gas may result in primary irritant and anaesthetic effects. However, human subjects tolerated concentrations up to 8 000 ppm for 8 h with no ill effects other than slight irritation of the eyes, nose and throat. Dermatitis (including frostbite due to cold injury) may result from exposure to liquid butadiene and its evaporating gas. Inhalation of excessive levels, which might produce anaesthesia, respiratory paralysis and death can occur from spills and leaks from pressure vessels, valves and pumps in areas with inadequate ventilation. Such very high levels are unlikely to be reached in the industry except by accident. Evidence of significant cumulative effect is lacking. While excess cases of leukaemia and gastrointestinal tract, circulatory and nervous sytem disorders have been reported in the styrene-butadiene rubber industry, worker exposures were to multiple agents and a causal relationship to butadiene has not been established. Cases of serious illness due to the industrial use of butadiene are rare or non-existent.

SAFETY AND HEALTH MEASURES

Adequate ventilation should be maintained to prevent exposure of workers to a concentration above the recommended safe limits. Workers should be instructed that smarting of the eyes, respiratory irritation, headache and vertigo may indicate that the concentration in the atmosphere is unsafe. Cylinders of butadiene should be stored upright in a cool, dry, well ventilated location away from sources of heat, open flames and sparks. The storage area should be segregated from supplies of oxygen, chlorine, other oxidising chemicals and gases, and combustible materials. Since butadiene is heavier than air and any leaking gas will tend to collect in the depressions, storage in pits and basements should be avoided. Containers of butadiene should be clearly labelled and coded appropriately as an explosive gas. Cylinders should be suitably constructed to withstand pressure and minimise leaks, and should be handled so as to avoid shock. A safety relief valve is usually incorporated in the cylinder valve. A cylinder should not be subjected to temperatures above 55 °C. Leaks are best detected by painting the suspected area with a soap solution, so that any escaping gas will form visible bubbles; under no circumstances should a match or flame be used to check for leaks.

Both in its manufacture and usage, butadiene should be handled in a properly designed, closed system. Anti-oxidants and inhibitors (such as tertiary-butylcatechol at about 0.02 weight %) are commonly added to prevent the formation of dangerous polymers and peroxides. Butadiene fires are difficult and dangerous to extinguish. Small fires may be extinguished by carbon dioxide or dry chemical fire extinguishers. Water may be sprayed over large fires and adjacent areas. Wherever possible, a fire should be controlled by shutting off all sources of fuel. No specific preplacement or periodic examinations are needed for employees working with butadiene.

Treatment. If a patient who is deeply anaesthetised by butadiene is removed from the contaminated atmosphere, he should recover completely as long as effective respiration and heart action persist. If respiratory and/or cardiac activity is suppressed, appropriate resuscitative measures (cardiopulmonary resuscitation, mouth-to-mouth, assisted respiration, etc.) should be carried out immediately after moving the patient to fresh air. Inhalation of oxygen may be beneficial; other symptomatic and therapeutic measures may be used as indicated.

WEAVER, N. K.

Investigation of selected potential environmental contaminants: butadiene and its oligomers. Miller, L. M. Report EPA/650/2-78/008, 1978; NTIS Order No. PB-291684, 195 p.

CIS 79-1967 "Inhalation toxicity studies with 1,3-butadiene–1. Atmosphere generation and control–2. Three-month toxicity studies in rats". Pullinger, D. H.; Crouch, C. N.; Dare, P. R.; Gaunt, I. F. *American Industrial Hygiene Association Journal* (Akron, Ohio), Sep. 1979, 40/9 (789-795) (796-802). Illus. 23 ref.

CIS 79-1988 "Hepatic function in butadiene-styrene rubber manufacturing workers" (O sostojenii funkcii pečeni u rabočih različnyh proizvodstv divinilmetilstirol'nogo sintetičeskogo kaučuka). Putalova, T. V. *Gigiena truda i professional'nye zabolevanija* (Moscow), June 1979, 6 (21-24). 9 ref. (In Russian)

CIS 78-25 "Butadiene". H63, Information sheets on hazardous materials, Fire Protection Association, London. *Fire Prevention* (London), Dec. 1977, 122 (45-46).

Butchery trade

Butchery covers the preparation for sale of dressed carcasses from an abattoir: sometimes the same people will be involved in slaughtering in adjacent premises, sometimes the butchery will be in small premises at the rear of a retail shop, sometimes it will be a large establishment selling meat wholesale or supplying a chain of shops.

Preparation for sale includes cutting up dressed carcasses into suitable joints and pieces, some of which are boned, tied or stuffed and sometimes wrapped in a plastics material. Some meat may be tenderised by crushing the fibres, some may be minced or mixed with other foodstuffs to make sausages, some meat may be cooked and thereafter sliced. Local or national regulations may require the addition of preservatives (usually sulphur dioxide) to some products. Until sale, many of the products may be kept under refrigeration.

HAZARDS AND THEIR PREVENTION

Accidents. Movement of heavy and awkward carcasses may cause strains of the back and shoulders: the use of rails and hooks can prevent many such strains provided that the means of attaching the carcasses are well worked out. Training in lifting and handling is also necessary.

Wet and greasy floors make slips and falls a hazard which may be prevented by floors constructed of non-slip material, such as grooved, ceramic tiling or mixes containing carborundum. Rigorous cleanliness, sloping and drainage of the floors to prevent accumulation of water are also necessary; footwear should have non-slip soles.

Knives are the cause of many accidents: most cuts are on the hand or arm opposite to that holding the knife but very serious injuries may occur, usually during boning, by the stabbing of the knife into the upper thigh, dividing the femoral vessels. Choice and care of knives and training in their use are essential; handles should be well designed and maintained and blades kept sharp (if the knife edge is dull, extra force is applied to points of resistance and accidents may follow). Suitable protective clothing is also necessary. An apron of thick leather or plastic material over the femoral area, as well as hand and arm protection such as finger stalls or gloves of chain mail may be used (figure 1).

Power saws used on the larger carcasses, slicing, mixing, mincing and tenderising machines may cause very serious injuries, sometimes involving loss of a hand, unless efficient machinery guarding to prevent access to the dangerous parts is maintained at all times. Operators should be trained in the use of the guards and it is particularly important that cleaning in motion should be prohibitied.

Steam and refrigerating plants present the same hazards as in other industries: regular examination, checking of leaks, availability of emergency respiratory equipment are necessary.

Diseases. Workers in the butchery trade are subject to the same zoonotic diseases as abattoir workers but usually to a lesser degree. Dermatitis of the primary type is often associated with abrasion of the skin by fat-soiled clothing, or exposure to brine solutions. Erysipeloid and orf are also not uncommon, the former more often on the

Figure 1. Hand protection made from wire mesh for meat cutting.

hands and the latter on the forearms. A cutaneous form of tuberculosis involving the skin of the fingers has been described.

Hazards from potentially toxic chemicals are usually limited to light and intermittent exposures to concentrated forms of chemicals used, legally or illegally, as preservatives. These include borax, boracic acid, sodium sulphite, nitrate and nitrite.

Regular medical examination, possibly including yearly X-ray examination, of butchers is desirable both for the protection of the worker's own health and for the protection of the public to ensure that communicable diseases are not transmitted. Periodic vaccination and blood tests may be necessary for some workers.

There is a higher incidence of inflammation in wounds than in ordinary industries and this is more likely to occur following a small, almost trivial injury, such as a bone prick than following a larger incised wound. Watch therefore is to be kept for such septic wounds. Full first-aid facilities should be provided and used by trained first-aiders. Any antiseptic used must not taint the meat and cetrimide is therefore the antiseptic of choice. Waterproof dressing should be applied during working hours and replaced by porous dressings at home.

In pre-employment examinations the physique of the worker is important for work which involves considerable energy expenditure including the movement of large unwieldy weights.

Premises. For public health reasons and for the protection of the workers, a high standard of hygiene is required and the premises should be designed and constructed to facilitate this. Construction materials should not readily absorb fat or water and should be easy to clean. All premises should be well lit and well ventilated. Rails for transport of carcasses should not endanger passageways or means of access. Bench space should be adequate to prevent men working too closely together, especially when knives are in use.

Well maintained washing and sanitary facilities (including showers where practicable) should be adjacent to the workrooms but separated by a ventilated space. A changing room and locker space is also required as a complete change of outdoor clothing for work is necessary. Protective clothing should be light coloured and frequently laundered.

COPPLESTONE, J. F.

Occupational safety:

"Occupational injuries affecting slaughterhouse workers and butchers" (Lesioni professionali negli squartatori dei macelli e nei macellai). Alessandri, M. *Securitas* (Rome), 1969, 54/5-6 (481-493). (In Italian)

CIS 913-1970 *Butchers' utensils* (Werkzeuge für Fleisch und Wurst). Merkblatt 30/16 (Berufsgenossenschaft für den Einzelhandel, Niebuhrstrasse 5, 53 Bonn) (1969), 8 p. Illus. (In German)

CIS 78-2004 *Data sheet on protective aprons against knife injury* (Stechschutzschürzen–Merkblatt). ZH 1/572 Hauptverband der gewerblichen Berufsgenossenschaften (Cologne, Carl Heymanns Verlag, 1977), 19 p. Illus. (In German)

CIS 74-1093 "Personal protective equipment in the butcher's trade" (Persönliche Schutzausrüstungen für den Fleischer). Neitmann, H. *Die Berufsgenossenschaft* (Bielefeld), Sep. 1973, 9 (356-359). Illus. (In German)

For occupational health see ABATTOIRS AND PACKAGING.

Button manufacture

Button manufacture is an important industry in many parts of the world. Modern button factories are often highly mechanised but, in some places, buttons are still produced at a handicrafts level.

Raw materials. It would be impossible to draw up a complete list of the materials used in button-making: they can be of animal, vegetable or mineral origin, natural or synthetic, and in recent years plastics have been increasingly used. Among the materials at present in use are the following:

(a) materials of animal origin–bones of large animals, ivory, horn, leather, blood, milk, mother-of-pearl;

(b) materials of vegetable origin–various woods, seeds of the ivory palm, cellulose, linen and other textiles;

(c) minerals–metals (iron, zinc, aluminium, brass, silver, gold) and materials of industrial or synthetic origin such as porcelain, glass, synthetic polymers (polystyrene, acrylic resins, phenol formaldehyde, polyamides, vinyl resins, etc.).

Production. Animal horns and hooves for button manufacture are autoclaved, sawn, pressed and punched into cylinders from which discs of uniform thickness are cut. The discs are then stamped, turned, bored and polished. Horns and hooves may be pulverised, mixed with urea-formaldehyde adhesive and moulded in a compression moulding press. Bones for button manufacture are first washed, degreased and bleached and then undergo much the same processing as horn and hoof.

Casein and albumin are ground, moulded, dyed in formaldehyde and machined. Leather buttons are cut from tanned and embossed leather. In the manufacture of mother-of-pearl buttons, the shells of various molluscs are broken, cut into circles, pierced, flattened, ground and polished. Wooden buttons are impregnated with various chemicals such as wood preservatives and dyes. Imitation ivory buttons are made from corozo nuts which are shelled, sawn up, ground, bleached, polished and sometimes ornamented.

Metal buttons may be pressed from steel or non-ferrous metal sheet or cast in a foundry; machining, grinding and polishing, pickling, painting, enamelling, electroplating or anodising may follow. Glass buttons are manufactured from crushed, sieved glass, sometimes mixed with clay, which is press-moulded and kilned. In the plastics processing industry, buttons are made by compression moulding or extrusion, followed by machining, coating, etc. Imitation mother-of-pearl is obtained by use of bismuth or lead salts or aluminium powder, or a product made from fish scales. Plastic buttons with a sandwich construction are also used.

Buttons are either shanked or unshanked: if unshanked the button has to be perforated to take the thread, if shanked, the shank may be added to the button form by soldering, welding or glueing. Linen or cloth buttons are made by covering button forms with fabric, usually by hand; in the high-quality clothing industry, matching buttons are often made in small undertakings or by homeworkers. Buttons are often stuck or fastened on to cards as a final operation at the factory.

HAZARDS AND THEIR PREVENTION

During the various types of button manufacturing, process workers may be subject to accident and explosion hazards, and health hazards such as toxic and irritant substances, harmful dusts, zoonoses and noise and vibration.

Accidents. Many different machines are used in the large-scale production of buttons and the risks of accident are varied. Saws, presses, boring machines and turning lathes in particular require secure machinery guarding the wheels, mops and spindles, as do grinding and polishing machines. Cuts and abrasions are common in workers holding small pieces against grinding and polishing machines; these may be infected if contaminated animal products are being used. Chips, flakes and dust may cause eye injuries unless efficient eye protection is provided and worn.

There may be a risk of dust explosion in some of the mixing processes unless the appropriate precautions are taken. It is to be hoped that celluloid is no longer used as this involves serious fire hazards both in the factory and to the eventual wearer.

Health hazards. The diversity of materials and methods of production involve an equal diversity of health hazards in the button-making industry. Only some of the more common risks can be mentioned here.

During the processing of plastic buttons, workers may be exposed to the toxic or irritant gases or vapours of the raw materials or their decomposition products, for example styrene, phenol, formaldehyde, acrylic acid, etc. A wide range of dangerous solvents may also be used including benzene, carbon disulphide and carbon tetrachloride, together with a variety of organic and inorganic acids, alkaline materials or oxidising substances which may cause chemical burns of the skin or mucous membranes. The pigments, dyes and impregnating agents used may be toxic or give rise to serious skin disease or allergic manifestations. It is essential that, whenever it is technically possible, the most harmful chemical agents should be replaced by less harmful substitutes. In particular, solvents such as benzene, carbon disulphide and carbon tetrachloride should be excluded.

Button making entails many dusty processes, in particular grinding and polishing. These dusts may be toxic (e.g. certain metals or polymers) or may result in inflammation of the upper and lower respiratory tracts, allergy or predisposition to other pulmonary conditions (e.g. wood, bone, horn, metal, glass, mother-of-pearl or plastics).

All sources of hazardous gases, vapours or dusts should be enclosed and fitted with exhaust systems. Workplaces must be thoroughly ventilated by fresh air from outside. When repairs or maintenance work have to be done in an enclosed area where the concentration of dangerous substances is above the tolerable limits, workers should be supplied with personal protective equipment including special working clothes, eye and face protection and respiratory protective equipment, depending on the nature of the working conditions.

Bones, tusks, ivory and hooves may present a risk of anthrax if the raw material comes from an infected animal and has not been thoroughly disinfected. Particular attention should be paid to the preliminary cleansing and disinfection of animal raw materials as well as immediate neutralisation of waste products liable to decomposition. Arrangements should be such as to protect those living near from unpleasant smells.

Workers engaged in the making of mother-of-pearl buttons may suffer from various diseases of infectious, allergic, or distrophic nature (see CORAL AND SHELL). These occur mainly where dry polishing gives rise to high dust concentrations.

Many manufacturing processes produce very high levels of noise and vibration. All possible measures to deal with noise and vibration should be taken when the equipment is installed, especially by isolation and absorption.

Processing on primitive, old-fashioned machine tools may entail rapid repetitive movements of hands, and, in certain cases, feet; fatigue and neuromuscular disorders of the hands are not uncommon consequences. Certain operations involve close work and where the lighting is inadequate may cause ocular fatigue.

Heating of some of the raw materials may lead, either by convection or radiation, to unpleasantly high temperatures in the factory premises and workplaces.

Mechanisation, and eventually full automation, will remove many of these hazards. Much success has been achieved in this field where buttons are made from resins and plastic materials. One such machine, in the space of 24 h, produces tens of thousands of buttons. Methods are, at present, being examined for producing buttons from synthetic materials directly in the workrooms. In such a way, the manufacture and the use of buttons will take place at the same time and reduce the work involved to a minimum.

To avoid fatigue among workers, any monotonous work involving repetitive or rapid movement should be interrupted from time to time for short rest periods. In certain cases alternation is advised for operations liable to give rise to strain of certain groups of muscles.

The lighting of workplaces should be strong enough to illuminate both workplace and work, and be free of reflection.

Periodical medical examination is desirable. The type and frequency of the examination will depend upon the specific risks attached to the type of button made.

VOLKOVA, Z.

Byssinosis

Byssinosis is a chronic respiratory disease of cotton, flax and soft hemp workers. Recently it has been found to occur in sisal workers preparing fibres prior to processing them. Byssinosis is characterised by chest tightness and breathlessness at work after the weekend break or other absence. In its late stages, which usually occur after many years of exposure to dust, the worker is severely disabled with symptoms of chronic bronchitis and emphysema. It occurs principally among those who clean and prepare fibres for spinning.

In the cotton industry byssinosis may affect workers in the ginneries where the seeds are removed, the bale pressing plants and the mixing and card rooms where the fibres are cleaned and combed. As a result of the introduction of mechanical picking, which has increased the contamination of cotton with plant debris, and the speeding up of all processes, dust concentrations in workrooms have risen and the disease has recently been found in spinners, winders and weavers.

Among flax workers in factories making linen, ropes and twines, byssinosis occurs only in the preparatory processes. In an investigation of the disease in an Egyptian village where flax is processed in the homes, it was found that workers and other members of the family were affected by the dust generated in and around their houses. Young children, old enough to express their complaint, had the characteristic symptoms of chest tightness.

Byssinosis has been described among hemp workers in Spain, where it is called cannabosis. The risk of the disease among hemp workers appears to be confined to the processing of soft hemp, which is a fibre from the stem of the plant used for making ropes and twines. Other recent surveys indicate that, in spite of high dust exposures, there is no specific hazard of byssinosis from St. Helena hemp and manila which are leaf fibres. Their dusts, however, may cause persistent cough and expectoration. The gradual changeover from natural to synthetic fibres will reduce the risk of occupational respiratory disease, since synthetic fibres do not give rise to byssinosis. Nevertheless, in developing countries the risk of byssinosis is likely to increase. The building of new textile factories to process the natural fibres which they grow in abundance is an important part of their economic expansion.

Aetiology

The world-wide prevalence of byssinosis in the cotton industry, brought to light by surveys in Europe, Asia, Africa and North and South America, has discounted old concepts of the disease, namely that it affects only the relatively few male workers who are susceptible to the dust and that it is caused by the combined effects of exposure to cotton dust and heavy atmospheric pollution.

Epidemiological studies in flax and soft hemp as well as cotton factories have revealed that at least 40% of workers in dusty processes may be affected; and in factories where dust-control measures are minimal, the prevalence may be much higher. Many of the women employed in dusty processes have recently been found to suffer from the disease. There are few climates less polluted than those in Egyptian villages and the small mill towns of the Southern States in the United States, but abundant disabling disease has been shown to occur in these places.

Recent evidence from a number of independent inquiries confirms that in cotton workers the disease is caused by some broncho-constricting agent contained in the leaves of the cotton plant but not in the fibres or seeds. There is also evidence that endotoxins from Gram-negative bacteria, which are found in cotton plants and constitute part of the normal plant flora, may be aetiologically important. While the prevalence and severity of byssinosis is determined primarily by the amount of dust in the workrooms and the length of exposure, other factors such as air pollution outside the factory, cigarette smoking and respiratory infection also appear to be aetiologically important.

Symptoms

In the early stages there are characteristic symptoms of chest tightness on the first day of work after the weekend break. In western countries this occurs on Mondays, usually towards the end of the shift, and disappears shortly after leaving the workplace. On Tuesdays the worker has no symptoms. As the disease progresses, the chest tightness accompanied by breathlessness worsens and extends to Tuesdays and then to other days. At this stage, symptoms become less pronounced as the week goes on. Eventually the worker may become severely affected on every working day with permanent and severe effort intolerance, which is not noticeably improved by giving up his dusty occupation.

In its final stages the disease cannot be distinguished from chronic bronchitis and emphysema due to non-occupational causes, except for the past history of chest tightness characteristically worse at the beginning of the week. The patient often forgets his early symptoms and is diagnosed as suffering from a non-occupational chronic respiratory disease. Chest X-rays do not show changes specific for byssinosis, nor has any specific pathology been identified in the lungs of workers who have died of this disease. Any X-ray or pathological changes in the lungs are the same as those also found in chronic bronchitis and emphysema due to non-occupational causes.

For assessing its prevalence and severity in epidemiological surveys, byssinosis may be graded as follows:

Grade ½	Occasional chest tightness or respiratory irritation on the first day of the working week.
Grade 1	Chest tightness and/or shortness of breath on every first day of the working week (Mondays in Europe and Saturdays in Arab countries).
Grade 2	Chest tightness and/or shortness of breath on the first and other days of the working week.
Grade 3	Grade 2 symptoms accompanied by evidence of permanent incapacity from diminished effort intolerance and/or reduced ventilatory capacity.

For the assessment of lung function changes in textile workers, a functional grading system based on the forced expiratory volume in the first second ($FEV_{1.0}$) of a forced expiration after maximum inspiration has been used as follows:

Grade	Acute change (Δ $FEV_{1.0}$* as a percentage of pre-shift value)	Basic value ($FEV_{1.0}$** as a percentage of predicted value)
F0	less than 5% loss	80
F½	from 5% to less than 10% loss	80
F1	10% loss or more	80
F2	10% loss or more	60-79
F3	10% loss or more	60 or less

* Difference between $FEV_{1.0}$ before and after work shift on the first working day of the week, expressed as a percentage of pre-shift value.
** $FEV_{1.0}$ in the absence of dust exposure (two days or longer); use value after bronchodilator treatment when feasible.
F0 no demonstrable acute effect; no evidence of chronic ventilatory impairment
F½ only slight acute effect
F1 definite acute effect only
F2 evidence of a slight to moderate irreversible impairment of ventilatory capacity
F3 evidence of moderate to severe impairment of ventilatory capacity

Many of the subjects in grades F2 and F3 will also show an acute fall of $FEV_{1.0}$ following exposure to dust, but for the purposes of this scheme of functional grading, only the permanent (irreversible) impairment of ventilatory capacity is necessary for inclusion. The choice of 80% of predicted normal for grade F2 is arbitrary; about 10% of "normal" persons have a $FEV_{1.0}$ under this limit. Individual values need to be compared with those expected (predicted values) for healthy individuals of the same sex, age, stature and race. Equations for calculating these values are available for Caucasians, Africans and Asians (see World Health Organisation: *Environmental health criteria for vegetable dusts*). Predicted values may also be obtained from healthy non-smokers in the population under surveillance.

Diagnosis

The diagnosis of byssinosis can be made on the characteristic history of chest tightness, which in the early stages of the disease is always worse at the beginning of the week. These symptoms are usually associated with a marked fall in ventilatory capacity during the work shift. Thus measurement of $FEV_{1.0}$ or airway resistance at the beginning and end of the work shift on Monday may help not only to confirm the diagnosis but also to identify workers who are susceptible to the dust but without symptoms.

Mode of action of the dust

A reasonable hypothesis is that the acute changes in ventilatory capacity eventually lead to an irreversible loss of lung function, and that these changes in lung function are not primarily caused by mechanical irritation but by pharmacologically active substances producing broncho-constriction. There is some conflict of opinion as to the mechanism by which inhaled dust has its acute effect on pulmonary ventilation. There are, for example, a number of agents known to be in cotton dust that could cause narrowing of the bronchioles by muscle constriction and/or oedema of the mucosa. They are histamine, antigenic and non-antigenic histamine-releasing factors and 5-hydroxytryptamine. Histamine and 5-hydroxytryptamine are usually present in such small quantities that they are unlikely to be important.

There has been a report of an antibody directed against an antigen present in the cotton plant. The titre of this antibody is higher in card room workers than in normal people and is highest in those with byssinosis. The evidence in favour of non-antigenic histamine release as a mechanism in the broncho-constrictor response to dusts causing byssinosis is as follows:

(a) cotton and certain other textile dusts cause airway constriction even in healthy persons exposed to concentrations found in workplaces;

(b) the time course of action of the dust and of dust extracts *in vivo* is consistent with an effect involving release of a stored mediator;

(c) textile dusts and bract extracts release histamine from human lung tissue without previous sensitisation; the time course of histamine release is consistent with the time course of the *in vivo* responses;

(d) the airway constrictor action of textile dusts in man is associated with increased histamine levels in blood and excretion of a major histamine metabolite; the effect can be prevented by a low dose of antihistamine drug.

The causative agent or agents have not yet been identified. Identification is complicated by the presence of a large number of chemical compounds and biological material (bacteria, fungi, viruses and others) which constitute part of the normal flora of the cotton or other plant, or have been added as contaminants during storage and processing.

In addition to this specific reaction, it seems that the dusts causing byssinosis also give rise to a non-specific irritation like other vegetable dusts.

PREVENTIVE MEASURES

These consist essentially of dust control, medical supervision and, in certain cases, personal protective equipment.

Dust control. Since the active agents of the disease are in the plant debris but not in the cotton fibres, the risk of disease could be prevented if the plant debris were extracted from the fibres immediately after picking. Less is known about the source of the causal agent of byssinosis in flax and soft hemp. Attempts have been made to pretreat cotton by steaming it to remove the active agent(s). At present there is no clear evidence that such steaming results in a degree of prevention of respiratory effects that would make it worth while to pursue this practice. Methods of removing the pharmacologically active agents from textile dusts may become feasible once these agents have been identified.

Until a more fundamental approach to the removal of dust from the fibres is possible, airborne-dust control will depend on various measures such as exhaust ventilation and general housekeeping in workrooms. Environmental monitoring by gravimetric dust sampling is necessary as a means of detecting hazardous processes and maintaining dust control at permissible levels.

Permissible levels for textile dusts vary from country to country. The most detailed and fully documented data for cotton dust are those of the United States. The US Occupational Safety and Health Administration (OSHA) sets permissible levels of respirable dust at 0.2 mg/m^3 for yarn manufacturing, 0.75 mg/m^3 for slashing and weaving and 0.5 mg/m^3 for other textile operations. These concentrations are for samples averaged over the 8-h period collected by the vertical elutriator or equivalent method. In other countries hygiene standards are set at higher levels. Dust control at the levels set by OSHA do not provide complete protection for susceptible workers.

Medical control. Medical surveillance will continue to be important until harmful dusts can be removed. It should include initial and periodic examinations.

The initial medical examination should aim at exluding from the dusty processes persons of all ages who have evidence of chronic non-specific lung disease, pulmonary tuberculosis, allergic asthma, or any other lung disease which may cause respiratory disability. This examination should include a test of ventilatory capacity–preferably the forced expiratory volume ($FEV_{1.0}$).

In the initial and periodic examinations the chest film is of no use for diagnosing byssinosis but in some countries routine X-rays are needed to detect other diseases, such as tuberculosis. The periodic examination should if possible be at yearly intervals and include the following:

(a) a standardised questionnaire relating to the characteristic symptoms occurring in byssinosis, as well as the more common symptoms of chronic non-specific chest disease;

(b) the $FEV_{1.0}$ and its change during a work shift. This will detect those who are adversely affected by the

dust. This test should be done on the first day back at work after the weekend break.

Workers who have chest tightness and/or shortness of breath on most first working days, those who exhibit a fall in $FEV_{1.0}$ over the work shift of more than 10% of the pre-shift value and those with a pre-shift $FEV_{1.0}$ of less than 60% of the predicted value, should be removed from further dust exposure. Workers who show less severe effects, such as pre-shift $FEV_{1.0}$ of more than 60% but less than 80% of predicted value or a work shift decrement in $FEV_{1.0}$ of more than 5% of the pre-shift value should be re-examined in 6 months.

Personal protection. There are certain short-term jobs–for example, in maintenance work–where exposure to high dust concentrations is inevitable. Every effort should be made to protect such people individually by efficient face masks or fresh-air hoods.

Antihistamine drugs reduce the effects of dust inhalation on the lungs, but the regular use of such drugs for this purpose is not recommended.

SCHILLING, R. S. F.

Respiratory diseases resulting from occupational exposure to vegetable dusts. Pathogenesis, early detection and control. OCH/80.1 (Geneva, World Health Organisation, 1980), 30 p. 92 ref.

"Byssinosis: Scheduled asthma in the textile industry–A review". Bouhuys, A. *Lung* (Heidelberg), 1976, 154 (3-16). 29 ref.

"Byssinosis, respiratory symptoms and spirometric lung function tests in Tanzanian sisal workers". Mustafa, K. Y.; Lakha, A. S.; Milla, M. H.; Dahoma, U. *British Journal of Industrial Medicine* (London), May 1978, 35/2 (123-128). 22 ref.

"Mill effect and dose-response relationships in byssinosis". Jones, R. N.; Diem, J. E.; Glindmeyer, H.; Dharmarajan, V.; Hammad, Y. Y.; Carr, J.; Weill, H. *British Journal of Industrial Medicine* (London), 1979, 36/4 (305-313). 15 ref.

Environmental health criteria for vegetable dusts. I: Dusts from vegetable fibers (Geneva, World Health Organisation, in preparation).

Criteria for a recommended standard...Occupational exposure to cotton dust. DHEW (NIOSH) publication No. 75-118 (National Institute for Occupational Safety and Health, 4676 Columbia Parkway, Cincinnati, Ohio) (1974), 170 p.

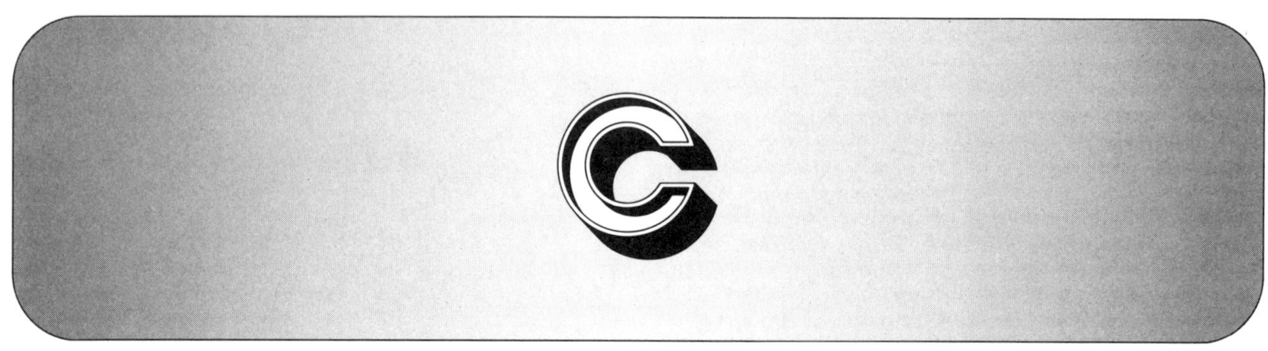

Cable transport

Ropeways have been used from earliest times as a means of transporting men and materials over difficult terrain such as rivers and ravines. However, it was only with general progress in engineering, and in particular with the development of the wire rope, that cable transport became commercially viable. Modern cable transport systems can be operated in safety, have low installation and running costs, are independent of adverse topographical conditions and have good conveying capacity. New and improved equipment is being constantly developed and special attention is being paid to higher levels of safety, automation of operation and increased capacity.

Cable transport systems can be broken down into four main categories: aerial ropeways, cable railways, cable cranes, surface lifts.

Aerial ropeways. With this system, a trolley or carrier suspended from an overhead rope travels between two points, along the line described by the rope, which may be supported at intervals by pylons, etc. In the twin-rope type, the carrier is suspended from a supporting cable and moved by a separate traction cable (see figure 1); in the single-rope type, the functions of suspension and traction are fulfilled by the same rope. The carriers may operate on a shuttle service or travel in an endless loop; provision may also be made for engaging and disengaging the carriers without halting the traction rope.

Cable railways. Here the carriers run along the ground on rails and are pulled by a traction rope or ropes.

Cable cranes. The cable crane is a combination of ropeway and crane in which the carrier or trolley is equipped with a lifting mechanism; such systems are frequently equipped with multiple suspension and traction ropes.

Surface lifts. The most widespread example of this system is the ski-lift or ski-tow in which skiers standing on their skis are pulled across the snow by holding on to or by attaching themselves to a bar or rope fixed to a combined suspension and traction cable.

Uses. Cable transport systems may be used for conveying persons or materials. They are of particular advantage in mountainous areas as a substitute for roads or railways. A good example of this suitability for difficult terrain is a 42 km aerial ropeway constructed in Norway

Figure 1. Turn-rope aerial ropeway.

between two points which would have required a road link of 180 km; in this case the aerial ropeway not only provided direct communication, it was cheaper to run and maintain than a road and could be operated during the winter when the road would have been blocked by snow.

Cable transport systems have been built with over-all conveying capacities of 600 t/h and individual trolley capacities of 20 t.

Systems designed for passenger transport are used as the basic means of conveyance in certain areas of difficult terrain; they are also used to provide access to areas of scenic beauty, sport and recreation.

HAZARDS AND THEIR PREVENTION

Improvements in materials and continuous progress in design and construction techniques have made ropeways and similar installations one of the safest means of transport. However, certain hazards continue to exist and are the result of environmental conditions, failure of the human factor in operators, and design defects or shortcomings.

Cable and carrier speed

Shuttle system. For the time being, carriers operate at speeds not exceeding 10 m/s; speeds of 12 m/s are being experimented. An increase of this speed is not necessarily a factor of safety, but of economics, as the carriers have to pick up speed when leaving and slow down when approaching a terminal. These procedures reduce the over-all distance of high travel speed along the cable line, thus only long tramways really call for high travel speeds.

Continuous system, detachable carrier. The carriers are hauled at a speed not exceeding 4 m/s (speeds of 5 m/s are being experimented). For this system as well as for the shuttle system, high speeds can only be obtained by automatic operation. As long as the equipment is of adequate design, safety does not limit the above-mentioned speed.

Continuous system, fixed carrier (chair lifts). The cable and its carriers operate at a maximum speed of 2.5 m/s (speeds of 3.0 m/s are being experimented). With fixed carriers, loading and unloading present certain hazards and set a limit to the speed.

Environmental factors

Cable transport systems may be damaged and their safe operation imperilled by high winds, cold, snow, ice, lightning and earthquakes. When installations are erected in areas subject to these natural phenomena, the necessary safety factor should be designed into the system and its components, e.g. allowance should be made for the weight of ice and snow deposits on components and for variations in the mechanical properties of materials due to severe cold. Lightning protection should be provided for pylons and stations; ropes, pylons and their foundations, in particular, should be suitably dimensioned to withstand high wind pressures or earthquakes. The influence of wind velocity differs according to the system, the exposure and shape of surfaces. As a rule, the operation of aerial tramways should be stopped when the wind velocity exceeds 16 m/s (when the wind is blowing transversally) or 20 m/s (when the wind direction is parallel to the axis of the line).

Human factors

Operations which depend for their safety on human factors such as vigilance, estimation of speeds and distances, etc., should, where possible, be automated. Control tasks in which human intervention cannot be excluded should be so designed as to minimise the stress to which the operator is subject by the ergonomic layout of indicators and control panels, rational design of visual displays for indicating instruments, use of closed-circuit television, etc.

Design factors

Stringent and comprehensive national and international requirements have been laid down concerning the safe design of cable transport systems. Modern wire ropes are subject to such high standards of quality control that complete breakage of a suitably dimensioned rope is virtually excluded. If the trolley equipment of the carrier is correctly designed, adequately strong and properly maintained, it will be impossible under normal operating conditions for the runner wheels to become dislodged from the suspension wire and for the carrier to fall; danger may arise in the event of high winds and consequently it should be stipulated that the system is halted when the wind velocity exceeds a certain predetermined level. Cabins of passenger carriers should preferably be totally enclosed to prevent falls of persons; cabin doors should have special interlocks or there should be an attendant to ensure that the door is not opened when the cabin is moving. Where it is necessary to transport passengers in carriers designed for the transport of materials, strict safety regulations should apply to the behaviour of transported persons, and passengers should be accompanied by a responsible supervisor.

Carriers, even when swinging in moderate winds, must be able to pass by pylons and stations without the risk of contact with either their structure or persons working on them. Pylons should be fitted with climbing rungs or other safe means of access and small working platforms for the use of maintenance workers. Stations should have control desks fitted with carrier-speed and wind-speed indicators, electrical switch gear and an emergency switch. Emergency lighting and a complete tool kit should also be available. Prime movers and transmission equipment should be fitted with adequate guarding.

Maintenance

Cable transport systems, especially modern automated ones, require only a small operating and maintenance staff to ensure adequate lubrication and to check for signs of wear and tear. These workers should be fully trained and well aware of the dangers of their work, especially of the dangers of cold and bad weather in the case of mountain locations. Where the ropeway is situated in an isolated region, it may be necessary to submit workers to special examinations to ensure that they are mentally suited to work in isolation.

During maintenance work on pylons workers should wear safety belts and lifelines; in bad weather they should wear suitable working clothes. Whenever practicable, visual or radio communication should be maintained between maintenance workers on pylons and the main operating station.

Cable transport systems should be regularly inspected by a competent person.

BRÜGGEMAN, H.

CIS 79-1808 "Non-destructive testing of wire ropes: Non-destructive electromagnetic testing of the Aiguille du Midi cable-car ropes" (Contrôle non destructif de câbles métalliques: le contrôle électromagnétique des câbles du téléphérique de l'Aiguille du Midi). Chesnay, C. *Revue de la sécurité* (Paris), July-Aug. 1979, 15/159 (21-29). Illus. (In French)

Cadmium and compounds

Cadmium (Cd)

a.w.	112.4
sp.gr.	8.64
m.p.	320.9 °C
b.p.	765 °C

insoluble in water; soluble in acids

a white metallic element with a bluish tinge.

Cadmium dust:

TWA OSHA	0.2 mg/m³ 0.6 mg/m³ ceil
TWA NIOSH	40 µg/m³/10 h 200 µg/m³ 5 min ceil
TLV ACGIH	0.05 mg/m³ (including salts as Cd)
STEL ACGIH	0.2 mg/m³ (including salts as Cd)
IDLH	40 mg/m³

Cadmium oxide fume:

TWA OSHA	0.1 mg/m³ 3 mg/m³ ceil
TWA NIOSH	40 µg/m³/10 h 200 µg/m³ 5 min ceil
TLV ACGIH	0.05 mg/m³ ceil
IDLH	40 mg/m³
MAC USSR	0.1 mg/m³

Cadmium oxide production:

TLV ACGIH	0.05 mg/m³

Occurrence. Cadmium has many chemical and physical similarities to zinc and occurs together with zinc in nature. In minerals and ores cadmium and zinc generally have a ratio of 1 : 100 to 1 : 1 000. World production of cadmium has increased heavily during this century. Only marginal amounts of cadmium were refined before 1920, but now, in the late 1970s, world production is in the order of 15 000 t/yr.

Production. Cadmium is obtained as a by-product in the refining of zinc and other metals, particularly copper and lead. It is obtained by precipitation from zinc electrolyte in electrolytic zinc refining, recovery from the fume of zinc calcine sintering plants, the fume of lead and copper smelters and during distillation and refining of zinc. It may be in the form of a chloride, oxide or sulphate, which is then leached, electrolysed, precipitated, etc., and cast into bars, balls or anodes for electroplating.

Uses. Cadmium is highly resistant to corrosion and is widely used for electroplating of other metals, mainly steel and iron. Almost 50% of all cadmium is used for this purpose. Screws, screwnuts, locks and various parts for aircraft and motor vehicles are frequently cadmiated in order to withstand corrosion. Large amounts of cadmium compounds are used as pigments and stabilisers in plastics. Cadmium is also used in certain alloys, and as one of the electrodes in nickel-cadmium alkaline batteries.

Compounds. Cadmium occurs in various inorganic salts. The most important are cadmium stearate, which is used as a heat stabiliser in PVC plastics, and cadmium sulphide and cadmium sulphoselenide, used as yellow and red pigments in plastics and colours. Cadmium sulphide is also used in photo and solar cells. Cadmium chloride is used in the production of certain photographic films.

Metabolism and accumulation

Gastrointestinal absorption of ingested cadmium is about 2% to 6% under normal conditions. Individuals with low iron body stores, reflected by low concentrations of serum ferritin, may have considerably higher absorption of cadmium, up to 20% of a given dose of cadmium. Significant amounts of cadmium may also be absorbed via the lung due to inhalation of tobacco smoke or occupational exposure to atmospheric cadmium dust. Pulmonary absorption of inhaled respirable cadmium dust is estimated at 20% to 50%. After absorption via the gastrointestinal tract or the lung, cadmium is transported to the liver where production of a cadmium-binding low molecular weight protein, metallothionein, is initiated. About 80% to 90% of the total amount of cadmium in the body is considered to be bound to metallothionein. This prevents the free cadmium ions from exerting their toxic effects.

It is likely that small amounts of metallothionein-bound cadmium are constantly leaving the liver and transported via the blood to the kidney. The metallothionein with the cadmium bound to it is filtered through the glomeruli into the primary urine. Like other low molecular weight proteins and amino acids, the metallothionein cadmium complex is subsequently reabsorbed from the primary urine into the proximal tubular cells, where digestive enzymes degrade the engulfed proteins into lesser peptides and amino acids. Free cadmium ions occurring in the cells as a result of degradation of metallothionein initiates a new synthesis of metallothionein binding the cadmium and thus protecting the cell from the highly toxic free cadmium ions. Kidney dysfunction is considered to occur when the metallothionein-producing capacity of the tubular cells is surpassed.

The kidney and liver have the highest concentrations of cadmium, together containing about 50% of the body burden of cadmium. The cadmium concentration in kidney cortex, before cadmium-induced kidney damage occurs, is generally about 15 times the concentration in liver. Elimination of cadmium is very slow. As a result of this, cadmium accumulates in the body with increasing age and time of exposure. Based on organ concentration at different ages the biological half-time of cadmium has been estimated in the range of 7 to 30 years.

Acute toxicity

Inhalation of cadmium compounds at concentrations above 1 mgCd/m³ in air for 8 h, or at higher concentrations for shorter periods, may lead to chemical pneumonitis, and in severe cases pulmonary oedema. Symptoms generally occur within 1 to 8 h after exposure. They are influenza-like and similar to those in metal fume fever. The more severe symptoms of chemical pneumonitis and pulmonary oedema may have a latency period up to 24 h. Death may occur after 4 to 7 days. Exposure to cadmium in the air at concentrations exceeding 5 mgCd/m³ is most likely to occur where cadmium alloys are smelted, welded or soldered.

Ingestion of drinks contaminated with cadmium at concentrations exceeding 15 mgCd/l gives rise to symptoms of food poisoning. Symptoms are nausea, vomiting, abdominal pains and sometimes diarrhoea. Sources of food contamination may be pots and pans with cadmium-containing glazing and cadmium solderings used in vending machines for hot and cold drinks.

In animals parenteral administration of cadmium at doses exceeding 2 mgCd/kg body weight causes necrosis of the testis. No such effect has been reported in humans.

Chronic toxicity

Chronic cadmium poisoning has been reported after prolonged occupational exposure to cadmium oxide fumes, cadmium oxide dust and cadmium stearates. Changes associated with chronic cadmium poisoning may be local, in which case they involve the respiratory tract, or systemic, resulting from absorption of cadmium. Systemic changes include kidney damage with proteinuria and anaemia.

Obstructive lung disease in the form of emphysema is the main symptom at heavy exposure to cadmium in air, whereas kidney dysfunction and damage are the most prominent findings after long-term exposure to lower levels of cadmium in workroom air or via cadmium-contaminated food. Mild hypochromic anaemia is frequently found among workers exposed to high levels of cadmium. This may be due to both increased destruction of red blood cells and iron deficiency. Yellow discolouration of the necks of teeth and anosmia may also be seen in cases of exposure to very high cadmium concentrations.

Pulmonary emphysema is considered a possible effect of prolonged exposure to cadmium in air at concentrations exceeding 0.1 mgCd/m^3. It has been reported that exposure to concentrations of about 0.02 mgCd/m^3 for more than 20 years can cause certain pulmonary effects. Cadmium-induced pulmonary emphysema can reduce working capacity and may be the cause of invalidity and life shortening.

With long-term low-level cadmium exposure the kidney is the critical organ, i.e. the organ first affected. Cadmium accumulates in renal cortex. Concentrations exceeding 100-300 μgCd/g wet weight lead to tubular cell dysfunction with decreased reabsorption of proteins from the urine. This causes tubular proteinuria with increased excretion of low molecular protein such as β_2-microglobulin. As the kidney dysfunction progresses, aminoacids, glucose and minerals, such as calcium and phosphorus, are also lost into the urine. Increased excretion of calcium and phosphorus may disturb the bone metabolism, and kidney stones are frequently found in exposed workers. Osteomalacia has been reported in cases of severe chronic cadmium poisoning.

In order to prevent kidney dysfunction, as manifested by β_2-microglobulinuria, also after 25 years of occupational exposure (8-h workday, 224 workdays/year) to cadmium fumes and dust, it is recommended that the average workroom concentration of respirable cadmium should be below 0.01 mg/m^3.

Excessive cadmium exposure has occurred in the general population through ingestion of contaminated rice and other foodstuffs, and possible drinking water. The itai-itai disease, a painful type of osteomalacia, with multiple fractures appearing together with kidney dysfunction has occurred in Japan in areas with high cadmium exposure. Though the pathogenesis of itai-itai disease is still under dispute it is generally accepted that cadmium is a necessary aetiological factor.

It should be stressed that cadmium-induced kidney damage is irreversible and may grow worse even after exposure has ceased.

Cadmium and cancer

Occupational cadmium exposure has been associated with an increased incidence of prostatic cancer in three different epidemiological studies. However, these studies are not totally conclusive. The International Agency for Research on Cancer (IARC) in 1976 summarised the matter as follows: "Available studies indicate that occupational exposure to cadmium in some form (possibly the oxide) increases the risk of prostate cancer in man. In addition, one of these studies suggests an increased risk of respiratory tract cancer."

SAFETY AND HEALTH MEASURES

Kidney cortex is the critical organ with long-term cadmium exposure via air or food. The critical concentration is estimated at about 200 μgCd/g wet weight. In order to keep the kidney cortex concentration below this level even after lifelong exposure, the average cadmium concentration in workroom air (8 h per day) should not exceed 0.01 mgCd/m^3. FAO and WHO in 1972 recommended that the intake of cadmium via food should not exceed 70 μgCd/day. This recommendation was based on knowledge available in 1972. More recent reports on the chronic toxicity of cadmium will probably necessitate a reduction of this figure.

Work processes and operations which may release cadmium fumes or dust into the atmosphere should be designed to keep concentration levels to a minimum and if practicable be enclosed and fitted with exhaust ventilation. When adequate ventilation is impossible to maintain, e.g. during welding and cutting, respirators should be carried and air sampled to determine the cadmium concentration.

In areas with hazards of flying particles, chemical splashes, radiant heat, etc., e.g. near electroplating tanks and furnaces, workers should wear appropriate safety equipment, such as eye, face, hand and arm protection and impermeable clothing.

Adequate sanitary facilities should be supplied and workers should be encouraged to wash before meals and to wash thoroughly and change clothes before leaving work. Smoking, eating and drinking in work areas should be prohibited. Tobacco contaminated with cadmium dust from workrooms can be an important exposure route. Cigarettes and pipe tobacco should not be carried in pockets, etc., in the workroom. Smokers should be required to wash their hands before smoking cigarettes or filling their pipe with tobacco. Contaminated exhaust air should be filtered, and persons in charge of dust collectors and filters should wear respirators while working on the equipment.

To ensure that excessive accumulation of cadmium in the kidney does not occur, cadmium levels in blood and in urine should be checked regularly.

Cadmium levels in blood are mainly an indication of the last few months' exposure. A value of 10 ngCd/ml whole blood is an approximate critical level if exposure is regular for long periods.

Cadmium values in urine can be used to estimate the cadmium body burden. Its concentration in individuals should not be allowed to reach 10 μgCd/g urinary creatinine, since when this concentration is reached there is some risk that kidney dysfunction develops.

Since the mentioned blood and urinary levels are levels at which action of cadmium on kidney has been observed, it is recommended that control measures be applied whenever the individual concentrations of cadmium in urine and/or in blood exceed 5 ng/ml whole blood or 5 μg/g creatinine respectively.

Pre-employment medical examination should be given to workers who will be exposed to cadmium dust or fumes. Persons with respiratory or kidney disorders should be excluded from such work. Medical examination of cadmium-exposed workers should be carried out at least once every year. In workers exposed to cadmium for longer periods quantitative measurements of β_2-microglobulin in urine should be made regularly. Concentrations of β_2-microglobulin in urine should normally not exceed 0.5 mg/l.

Treatment of cadmium poisoning. Persons who have ingested cadmium salts should be made to vomit or given gastric lavage; persons exposed to acute inhalation should be removed from exposure and given oxygen therapy if necessary. No specific treatment for chronic cadmium poisoning is available and symptomatic treatment has to be relied upon. As a rule the administration of chelating agents such as BAL and EDTA is contra-indicated since they are nephrotoxic in combination with cadmium.

FRIBERG, L.
ELINDER, C. G.

Occupational exposure to cadmium. Criteria for a recommended standard. DHEW (NIOSH) publication No. 76-192 (National Institute for Occupational Safety and Health, 4676 Columbia Parkway, Cincinnati, Ohio) (Aug. 1976), 86 p. Illus. 287 ref.

"Mortality and cancer morbidity among cadmium exposed workers". Kjellström, T.; Friberg, L.; Rahnster, B. (199-204). Illus. 32 ref. "Proceedings of an International Conference on Environmental Cadmium. Bethesda, June 7-9, 1978". *Environmental Health Perspectives* (Research Triangle Park, North Carolina), 1979, 28.

"Cadmium-induced osteomalacia". Blainey, J. D.; Adams, R. G.; Brewer, D. B. *British Journal of Industrial Medicine* (London), Aug. 1980, 37/3 (278-284). Illus. 23 ref.

Cadmium in the environment and its significance to man. Department of the Environment, Central Directorate on Environmental Pollution, Pollution paper No. 17 (London, HM Stationery Office, 1980), 64 p. 179 ref.

Calcium and compounds

Calcium (Ca)

a.w. 40
sp.gr. 1.54
m.p. 842.8 °C
b.p. 1 487 °C

a lustrous, silver-white metal

when finely divided, calcium ignites in air and burns with a crimson flame. It displaces hydrogen from water and may cause detonation in the presence of alkali hydroxides or carbonates.

Calcium is the fifth most abundant element and the third most abundant metal; it is widespread in nature as calcium carbonate (limestone and marble), calcium sulphate (gypsum), calcium fluoride (fluorspar) and calcium phosphate (apatite). Calcium minerals are quarried or mined; the metallic calcium is obtained by the electrolysis of molten calcium chloride or fluoride. Metallic calcium is used in the production of uranium and thorium, as a deoxidiser for copper, beryllium and steel, as a hardener for lead bearings, and in the electronics industry.

Calcium is a well known essential constituent of the human body and its metabolism–alone or in association with phosphorus–has been widely studied with special reference to the musculoskeletal system and cellular membranes. Several conditions may lead to calcium losses such as immobilisation, gastrointestinal disturbances, low temperature, weightlessness in space flights and so on. The absorption of calcium from the work environment by inhalation of calcium compounds dust does not increase significantly the calcium daily intake from vegetables and other food (usually > 0.5 g). On the other hand, metallic calcium has alkaline properties and it reacts with moisture causing eye and skin burns. Exposed to air it may present an explosion hazard.

Calcium chloride ($CaCl_2$)

m.w. 111
sp.gr. 2.15
m.p. 772 °C
b.p. 1 600 °C

soluble in water and ethyl alcohol

colourless, hygroscopic crystals.

Calcium chloride is obtained as a waste product in the Solvay ammonia-soda process. It is used in the production of barium chloride, metallic calcium and various dyes; in air-conditioning systems as a drying agent; as a refrigerant; to prevent dust formation during road construction [and in coal mines, where it also inhibits spontaneous combustion of the coal]; and as an agent for accelerating concrete curing times.

Calcium chloride has a powerful irritant action on the skin and mucous membranes and cases have been reported, amongst workers packing dry calcium chloride, of irritation accompanied by erythema and peeling of facial skin, lacrimation, eye discharge, burning sensation and pain in the nasal cavities, occasional nose bleeding and tickling in the throat. Cases of perforation of the nasal septum have also been reported.

Where possible, processes should be mechanised and enclosed to prevent dust formation, and under all circumstances adequate ventilation should be installed to collect airborne dust. Suitable protective clothing including overalls, gloves and headwear should be worn to prevent dust and splashes of calcium chloride solution coming in contact with skin and mucous membranes. Where concentrations are elevated, goggles and respiratory protective equipment are necessary. Adequate sanitary facilities should be provided and smoking, drinking and eating prohibited at the workplace.

Calcium fluoride (CaF_2)

FLUORITE; FLUORSPAR

m.w. 78.1
sp.gr. 3.18
m.p. 1 360 °C
b.p. about 2 500 °C

soluble in ammonium salts

transparent crystals.

Calcium fluoride is obtained from fluorspar, a mineral comprising 90-95% calcium fluoride and 3.5-8% silica, which is extracted by drilling and blasting. It may also be produced by the action of hydrofluoric acid on aqueous calcium salt solutions. It is used to etch glass, in the manufacture of pottery, as a flux in the iron and steel industry and as a raw material for metallic calcium and hydrofluoric acid production. Pure, colourless fluorspar is used in the optical industry.

The hazards of fluorspar are due primarily to the harmful effects of the fluorine content, and chronic effects include diseases of teeth, bones and other organs. Pulmonary lesions have been reported among persons inhaling dust containing 92-96% calcium fluoride and 3.5% silica. It was concluded that calcium fluoride intensifies the fibrogenic action of silica in the lungs. Cases of bronchitis and silicosis have been reported amongst fluorspar miners.

In the mining of fluorspar, dust control should be carefully enforced, including: wet drilling, watering of loose rock, exhaust and general ventilation. In the use of fluorspar, there is also the hazard of hydrofluoric acid being formed, and the relevant safety measures should be applied.

Calcium nitrate ($Ca(NO_3)_2.4H_2O$)

m.w. 236.2
sp.gr. 2.50
m.p. 561 °C

highly soluble in water

colourless, deliquescent crystals.

When heated to incandescence, calcium nitrate decomposes into NO_2, CaO and O_2. The commercial product may contain an admixture of lime. Calcium nitrate is obtained by dissolving limestone in nitric acid or through the absorption of nitrous gases by milk of lime. It is used as a fertiliser, in the explosives and pyrotechnics industries and in match manufacture as an oxidising agent.

Calcium nitrate has an irritant and cauterising action on skin and mucous membranes. It is a powerful oxidising agent and presents a dangerous fire and explosion hazard.

Processes employing calcium nitrate should be mechanised to the greatest possible extent. Exhaust ventilation should be installed to collect airborne dust, but some operations may necessitate the use of respiratory protective equipment and goggles. Protective clothing including overalls, gloves and footwear should be worn. It should not be stored near organic or other easily oxidisable materials.

Calcium sulphite ($CaSO_3$)

A white powder obtained by the action of sulphurous acid on calcium carbonate. It is used as a reducing agent in the production of cellulose. Occupational cases of calcium sulphite poisoning have not been reported. Accidental ingestion of a few grammes may produce repeated vomiting, violent diarrhoea, circulatory disorders and methaemoglobinaemia.

Other calcium compounds are dealt with in the following articles: ARSENIC AND COMPOUNDS (calcium arsenate and arsenite); BLEACHING AND BLEACHING AGENTS (chlorinated lime); CALCIUM CARBIDE; CALCIUM CYANAMIDE; CYANOGEN, HYDROCYANIC ACID AND CYANIDES (calcium cyanide); GYPSUM (calcium sulphate); LIMESTONE AND LIME (calcium hydroxide, calcium oxide); PHOSPHATES AND SUPERPHOSPHATES (calcium phosphate).

SADKOVSKAJA, N. I.

Calcium:

CIS 76-193 "Hazardous chemical reactions–35. Calcium, strontium and barium" (Réactions chimiques dangereuses–35. Calcium, strontium et baryum). Leleu, J. *Cahiers de notes documentaires–Sécurité et hygiène du travail* (Paris), 4th quarter 1975, 81, Note No. 986-81-75 (489-491). (In French)

Calcium chloride:

CIS 151-1972 "The favourable effect on spontaneous coal combustion of calcium chloride powder used for consolidating coal dust deposits" (Geringere Gefahr der Selbstenzündung von Kohle durch Chlorcalcium-Montan-Pulver). Externbrink, W.; Lewer, H. *Glückauf* (Essen), Aug. 1971, 107/17 (652-653). Illus. 9 ref. (In German)

Calcium fluoride:

CIS 1059-1964 *Lung cancer in a fluorspar mining community. I: Radiation, dust and mortality experience.* De Villiers, A. G.; Windish, J. P. *British Journal of Industrial Medicine* (London), Apr. 1964, 21/2 (94-109). Illus. 20 ref.

CIS 2481-1964 *Lung cancer in a fluorspar mining community. II: Prevalence of respiratory symptoms and disability.* Parsons, W. D.; De Villiers, A. G.; Bartlett, L. S.; Becklake, M. R. *British Journal of Industrial Medicine* (London), Apr. 1964, 21/2 (110-116). Illus. 28 ref.

CIS 925-1973 "Improving the occupational health characteristics of calcium fluoride coated electrodes" (Puti ulučšenija gigieničeskih harakteristic elektrodov karbonatno-fljuoritnogo tipa). Naumenko, I. M.; Marčenko, A. E. *Gigiena truda i professional'nye zabolevanija* (Moscow), Sep. 1972, 16/9 (9-12). Illus. 5 ref. (In Russian)

Calcium carbide

Calcium carbide (CaC_2)

m.w.	54.1
sp.gr.	2.22
m.p.	about 2 300 °C

greyish-black, irregular lumps, the industrial product is coloured by admixtures of coal

calcium carbide decomposes in water with the evolution of acetylene, leaving a residue of lime. The commercial grade usually contains phosphorus, sulphur and lime as impurities.

Production. Calcium carbide is manufactured by heating a mixture of crushed quicklime and coke or anthracite at a temperature of approximately 2 000 °C in an open or closed electric furnace. The reaction is as follows: $CaO + 3\ C \rightarrow CaC_2 + CO$. The process is continuous with the mixture being charged from a platform over the furnace or through piping. The molten carbide is tapped off from the bottom of the furnace and run into moulds at a temperature of 1 900-2 000 °C. After cooling for 20-24 h, the blocks of carbide are removed from the moulds and ground in a drum grinder. In technologically more advanced processes, cooling and grinding may be done simultaneously. The finished product is packed in hermetically sealed steel drums.

The principal harmful factors of this production are heat and air pollution by calcium carbide, quicklime, anthracite dust, carbon monoxide and acetylene.

Uses. Calcium carbide is used for the industrial production of acetylene, and employed in acetylene generators for acetylene lamps and oxyacetylene welding and cutting. It is also used in the manufacture of calcium cyanamide and in the pyrotechnics industry.

HAZARDS

Calcium carbide exerts a pronounced irritant effect due to the formation of calcium hydroxide upon reaction with moist air or sweat. Dry carbide in contact with skin may cause dermatitis. Contact with moist skin and mucous membranes leads to ulceration and scarring. Calcium carbide is particularly hazardous to the eyes. A peculiar type of melanoderma with strong hyperpigmentation and numerous telangiectases is often observed. Burns caused by hot calcium carbide are the most frequent accidents in this industry. The tissues are generally damaged in depths of 1 to 5 mm; the burns evolve very slowly, are difficult to treat, and often require excision. Injured workers may resume work only after the burnt skin surface is completely scarred. Persons exposed to calcium carbide frequently suffer from cheilitis characterised by dryness, swelling and hyperaemia of the lips, intense desquamation, deep radial fissures; erosive lesions with a tendency to suppuration can be observed in the mouth angles. Workers with a long professional history often suffer from nail lesions, i.e. occupational onychia and paronychia. Eye lesions with pronounced hyperaemia of the lids and conjunctiva, often accompanied by mucopurulent secretions are also observed. In heavy cases the sensitivity of the conjunctiva and cornea is strongly reduced. While the keratitis and keratoconjunctivitis evolve first without symptoms, they may later degenerate into corneal opacities.

In calcium carbide production, impurities may produce additional hazards. Calcium carbide contaminated with calcium phosphate or calcium arsenate may, when moistened, give off phosphine or arsine both of which are extremely toxic. Calcium carbide itself, when exposed to damp air, gives off acetylene which is a moderate anaesthetic and asphyxiant, and a considerable fire and explosion hazard (see ACETYLENE). On the other hand, if the production furnaces are correctly operated and the furnace rooms well ventilated, carbon dioxide and sulphur dioxide emission should present no problems.

SAFETY AND HEALTH MEASURES

Equipment which is a source of airborne calcium carbide dust should be fully enclosed and fitted with exhaust ventilation. Air ducts should be adequately inclined to prevent accumulation of dust, and dust-laden air should be cleaned in cyclones and bag filters before release into the atmosphere. Workers in furnace rooms should be protected from excessive heat and gases by good general ventilation and radiant-heat protection measures; the walls of these rooms should be equipped with windows at both the top and bottom to promote natural ventilation by convection and the roofs provided with ventilation openings.

Exhaust hoods should be installed over the furnaces; the edges of those hoods should be head height above ground and from them should be suspended fire- and heat-resistant screens or curtains. The exhaust ducts from the hoods should be connected to the furnace flues. Settled dust in the work premises should be collected using industrial vacuum cleaners or permanent exhaust systems.

The air of premises where calcium carbide is processed must be monitored for the presence of calcium hydroxide, carbon monoxide, and if the calcium carbide contains impurities (calcium phosphate, calcium arsenate), also for the presence of phosphine and arsine.

Drums and barrels of calcium carbide should be stored in dry, well ventilated premises. Fire prevention and protection measures should be taken and fire-fighting equipment installed. A notice should be displayed indicating that water should not be used in case of fire. Containers should be opened with non-spark tools.

Personal protective measures include: protective clothing in a resistant material, waterproof gloves (worn on hands protected with vaseline), head protection, goggles and respiratory protective equipment. To protect workers against radiant heat, flying sparks and hot particles, workers are provided also with foot and leg protection lined with asbestos, and aprons; the helmet is fitted with a metal shield with an insert of safety glass or fine wire mesh for eye and face protection.

Carbide acetylene generators should be located in well ventilated areas and at a distance from combustible material. Refilling and cleaning should be carried out by competent and adequately protected persons. Generators should be fitted with pressure-regulating mechanisms, a pressure gauge, safety relief or overflow pipe and a quick-acting shut-off valve.

SADKOVSKAJA, N. I.
IVANOV, N. G.

Health:

CIS 80-728 "Calcium carbide" (Carbure de calcium). Association interprofessionnelle des centres médicaux et sociaux de la région parisienne. *Cahiers de médecine interprofessionnelle*, 3rd quarter 1979, 75 (detachable insert), 1 p. 4 ref. (In French)

"Effects of calcium carbide on the organ of vision" (Vlijanic karbida kal'cija na sostojanie organa zrenija) Nazarov, P. M. *Naučnye trudy Irkutskogo medicinskogo Instituta* (Irkutsk), 1977, Series 140 (68-70) (In Russian)

CIS 75-1100 "Occupational skin diseases in the production of calcium carbide" (Professional'nye zabolevanija kŏi v proizvodstve karbida kal'cija). Anton'ev, A. A.; Ajrapetjan, M. A.; Hačatrjan, V. A. *Gigiena truda i professional'nye zabolevanija* (Moscow), June 1974, 6 (15-18). 7 ref. (In Russian)

Safety:

CIS 2399-1967 *Properties and essential information for safe handling and use of calcium carbide. (1967).* Chemical Safety Data Sheet SD-23 (Manufacturing Chemists' Association 1825 Connecticut Avenue NW, Washington DC) (revised July 1967), 12 p.

Calcium cyanamide

Calcium cyanamide ($CaCN_2$)
CALCIUM CARBIMIDE

m.w. 80.1
sp.gr. 2.3
m.p. 1 300 °C (sublimes)

a blackish-grey, shiny powder, greasy to the touch.

TLV ACGIH 0.5 mg/m^3
STEL ACGIH 1 mg/m^3

Calcium cyanamide decomposes in water, liberating ammonia.

Production. Calcium cyanamide is produced by the Frank-Caro batch process, in which finely powdered calcium carbide is nitrogenated at approximately 2 000 °C in the presence of alkaline earth halides in a vertical electrical furnace. The reaction is $CaC_2 + N_2 = CaCN_2 + C$, and is highly exothermic. Crude calcium cyanamide, in the form of dark compact masses, is mixed with gravel, powdered and then combined with oil or water to give it a certain degree of agglomeration. The final nitrogen content is around 20-21%.

Uses. Calcium cyanamide is mainly used as a fertiliser but also, in high concentrations, as a herbicide and a defoliant for cotton plants and as a pesticide for mole crickets and other parasites. In industry calcium cyanamide is used for the manufacture of dicyandiamide (the raw material for melamine) and calcium cyanide. It is also used as a desulphuriser in the iron and steel industry and for steel hardening processes.

HAZARDS

Exposure to calcium cyanamide, which occurs primarily during manufacture and agricultural application, entails a danger of irritation of the skin and mucous membranes and systemic intoxication.

Calcium cyanamide has an irritant contact effect on the exposed skin and mucosae (conjunctivae, upper respiratory tract). It is dissolved by perspiration or mucosal secretions, giving rise to free cyanamide ($CN-NH_2$), ammonia and, through reaction with carbon dioxide, calcium carbonate which has a dehydrating effect. Erythematous dermatitis has been reported in individuals exposed to calcium cyanamide in both industry and agriculture; it is of a highly itchy nature, localised mainly to the exposed parts of the body, face and legs, but also at the axillary and inguinal folds, and at the scrotum. Vesicular or bullous formations were seen in the more severe cases, with characteristics which may correspond to a second-degree burn. Continuous contact with calcium cyanamide causes the formation of ulcerations (atonic ulcer) covered with indolent, slowly healing crusts, on the palms of hands and the spaces between the fingers. Eye lesions consist of conjunctivitis and keratitis which, in the more severe cases, may lead to corneal ulceration. Inhalation of calcium cyanamide dust is the cause of rhinitis, pharyngitis, laryngitis and tracheobronchitis. Ingestion of dust deposited on the pharynx may lead to gastritis. Chronic rhinitis with perforation of the nasal septum has been reported in subjects exposed for many years to calcium cyanamide dust.

Absorption of cyanamide, liberated by calcium cyanamide, and entering the body through the digestive tract or respiratory system, causes a characteristic vasomotor reaction (erysipelas-type) which appears with intense localised erythema at the upper portions of the body, face, chest and arms. This reaction is accompanied by headache, congestion of the mucosae, cold sensation, nausea, vomiting, tachycardia, and hypotension. Circulatory collapse may follow in the more serious cases. The vasomotor attack generally regresses slowly over several hours. The effect of calcium cyanamide is potentiated by alcohol, which may trigger a vasomotor attack. Apparently cyanamide causes inhibition of the enzyme aldehydedehydrogenase, with consequent accumulation of acetaldehyde, which reaches a particularly high degree after the intake of alcohol. Thus, the phenomenon appears to be similar to that caused by disulphiram (Antabuse). The possibility of cyanic radical being liberated from calcium cyanamide is a very debatable point; it appears that metabolic degradation of cyanamide leads to the formation of ammonia without liberation of cyanides. The acute toxicity of calcium cyanamide is very low: the oral lethal dose in adults is said to be 40-50 g.

SAFETY AND HEALTH MEASURES

In plants working with calcium cyanamide, dust control measures should be implemented, including adequate exhaust ventilation. When appropriate, the skin, eyes and respiratory system of workers exposed to calcium cyanamide should be protected by the use of dustproof overalls, a hood fastened tightly round the neck, protective eyewear, respiratory protective equipment (dust masks) and cloth hand protection. Skin and face protection can be further improved by the application of waterproof barrier creams. Emergency eye-flushing installations should be provided for workers.

Good personal hygiene should be followed, with the provision of adequate sanitary facilities and encouragement of workers to shower at the end of shifts and before meals. The consumption of beverages and food in workshops should be prohibited and, in view of the potentiating effect of alcohol, all alcoholic drinks should be forbidden.

Workers should be subject to periodical medical examinations and occupational hygiene measures should be observed.

Treatment. Skin lesions should be carefully cleansed to remove residual calcium cyanamide. Healing of skin ulcers may be accelerated by application of a 1% resorcinol solution followed by a decongestive ointment.

In cases of eye contact, the eyes should be washed thoroughly in running water for at least 15 min, followed by symptomatic medication, using anaesthetic eye drops. If severe exposure has occurred, an eye specialist should be consulted.

Bronchitis should be treated symptomatically. Slight vasomotor symptoms regress spontaneously; however, administration of analeptics is advisable. In severe cases with signs of cardiovascular collapse the treatment is: intravenous hydrocortisone (100 mg) followed by venous infusion of noradrenaline (0.1% solution) until arterial pressure returns to normal.

BARTALINI, E.

Calibration

The proper interpretation of any environmental measurement is dependent on an appreciation of its accuracy, precision, and whether it is representative of the condition or exposure of interest. This article is concerned primarily with the accuracy and precision of industrial hygiene measurements. Although considerations of the location, duration, and frequency of measurements may be equally or even more important in the evaluation of potential hazards, they require knowledge of the process variables, the kinds of hazard and/or toxic effect that may result from exposures, and their temporal variations. Such considerations, which require the exercise of professional judgement, are beyond the scope of this article.

The accuracy of a given measurement is dependent on a variety of different factors including the sensitivity of the analytical method, its specificity for the agent or energy being measured, the interferences introduced by cocontaminants or other radiant energies, and the changes in response resulting from variations in ambient conditions or instrument power levels. In some cases the influence of these variables can be defined by laboratory calibration, providing a basis for correcting a field sample or instrument reading response. In cases such as the effect of variable line voltage on an instrument's response, they can be avoided by modifications in the circuitry, or the addition of a constant-voltage transformer. When the effects of the interferences cannot be controlled or well defined, it may still be desirable to make field measurements, especially in range-finding and exploratory surveys. The interpretation of any such measurements is greatly aided by an appreciation of the extent of the uncertainties. Laboratory calibrations can provide the basis for such an appreciation.

Types of calibration

Occupational health problems can arise from exposure to airborne contaminants, heat stress, excessive noise, vibration, ionising radiations, and non-ionising electromagnetic radiations. Each of these types of exposure involves a different set of measurement variables and calibration considerations, and they are considered separately. Other types of measurement requiring calibration are associated with the evaluation of ventilation systems used to control exposure to airborne contaminants.

Air sampling instruments

Air samples are collected to determine the concentrations of one or more airborne contaminants. To define a concentration, the quantity of the contaminant of interest per unit volume of air must be ascertained. In some cases the contaminant is not extracted from the air (i.e. it may simply alter the response of a defined physical system). An example is the mercury vapour detector, where mercury atoms absorb the characteristic ultraviolet radiation from a mercury lamp, reducing the intensity incident on a photocell. In this case the response is proportional to the mercury concentration, not to the mass flow-rate through the sensing zone; hence concentration is measured directly.

In most cases, however, the contaminant is either recovered from the sampled air for subsequent analysis or is altered by its passage through a sensor within the sampling train, and the sampling flow-rate must be known to be able to ultimately determine airborne concentrations. When the contaminant is collected for subsequent analysis, the collection efficiency must also be known, and ideally it should be constant. The measurements of sample mass, of collection efficiency, and of sample volume are usually done independently. Each measurement has its own associated errors, and each contributes to the over-all uncertainty in the reported concentration.

The sample volume measurement error is often greater than that of the sample mass measurement. The usual reason is that the volume measurement is made in the field with devices designed more for portability and light weight than for precision and accuracy. Flow-rate measurement errors can further affect the determination if the collection efficiency is dependent on the flow-rate.

Each element of the sampling system should be calibrated accurately before initial field use. Protocols should also be established for periodic recalibration, since the performance of many transducers and meters changes with the accumulation of dirt, as well as with corrosion, leaks, and misalignment due to vibration or shocks in handling, and so on. The frequency of such recalibration checks should be high initially, until experience indicates that it can be reduced safely.

Flow and/or volume. If the contaminant of interest is removed quantitatively by a sample collector at all flow-rates, the sampled volume may be the only airflow parameter that need be recorded. On the other hand, when the detector response is dependent on both the flow-rate and sample mass, as in many length-of-stain detector tubes, both quantities must be determined and controlled. Finally, in many direct reading instruments, the response is dependent on flow-rate but not on integrated volume.

In most sampling situations the flow-rates are, or are assumed to be, constant. When this is so, and the sampling interval is known, it is possible to convert flow-rates to integrated volumes, and vice versa. Therefore flow-rate meters, which are usually smaller, more portable, and less expensive than integrated volume meters, are generally used on sampling equipment even when the sample volume is the parameter of primary interest. Normally little additional error is introduced in converting a constant flow-rate into an integrated volume, since the measurement and recording of elapsed time generally can be performed with good accuracy and precision.

Flow meters can be divided into three groups on the basis of the type of measurement made: integrated volume meters, flow-rate meters, and velocity meters.

The response of volume meters such as the spirometer and wet test meter, and flow-rate meters such as the rotameter and the orifice meter, are determined by the entire sampler flow. In this respect they differ from velocity meters such as the thermoanemometer and the Pitot tube, which measure the velocity at a particular point of the flow cross section. Since the flow profile is rarely uniform across the channel, the measured velocity invariably differs from the average velocity. Furthermore, since the shape of the flow profile usually changes with changes in flow-rate, the ratio of point-to-average velocity also changes. Thus when a point velocity is used as an index of flow-rate, there is an additional potential source of error, which should be evaluated in laboratory calibrations that simulate the conditions of use. Despite their disadvantages, velocity sensors are sometimes the best indicators available–for example, in some electrostatic precipitators, where the flow resistance of other types of meters cannot be tolerated. Velocity sensors are also used in measurements of ventilation airflow and to measure one of the components in the determination of heat stress.

Calibration of collection efficiency. A sample collector need not be 100% efficient to be useful, provided its efficiency is known and consistent and is taken into account in the calculation of concentration. In practice, acceptance of a low but known collection efficiency is reasonable procedure for most types of gas and vapour sampling, but it is seldom if ever appropriate for aerosol sampling. All the molecules of a given chemical contaminant in the vapour phase are essentially the same size, and if the temperature, flow-rate, and other critical parameters are kept constant, the molecules will all have the same probability of capture. Aerosols, on the other hand, are rarely monodisperse. Since most particle capture mechanisms are size dependent, the collection characteristics of a given sampler are likely to vary with particle size. Furthermore, the efficiency will tend to change with time because of loading; for example a filter's efficiency increases as dust collects on it, and electrostatic precipitator efficiency may drop if a resistive layer accumulates on the collecting electrode. Thus aerosol samplers should not be used unless their collection is essentially complete for all particle sizes of interest.

Recovery from sampling substrate. The collection efficiency of a sampler can be defined by the fraction removed from the air passing through it. However, the material collected cannot always be completely recovered from the sampling substrate for analysis. In addition, the material sometimes is degraded or otherwise lost between the time of collection in the field and recovery in the laboratory. Deterioration of the sample is particularly severe for chemically reactive materials. Sample losses may also be due to high vapour pressures in the sampled material, exposure to elevated temperatures, or reactions between the sample and substrate or between different components in the sample.

Laboratory calibrations using blank and spiked samples should be performed whenever possible to determine the conditions under which such losses are likely to affect the determinations desired. When it is expected that the losses would be excessive, the sampling equipment or procedures should be modified as much as feasible to minimise the losses and the need for calibration corrections.

Sensor response. When calibrating direct reading instruments, the objective is to determine the relation between the scale readings and the actual concentration of contaminant present. In such tests the basic response for the contaminant of interest is obtained by operating the instrument in known concentrations of the pure material over an appropriate range of concentrations. In many cases it is also necessary to determine the effect of such environmental cofactors as temperature, pressure and humidity on the instrument response. Also, many sensors are non-specific, and atmospheric cocontaminants may either elevate or depress the signal produced by the contaminant of interest. If reliable data on the effect of such interferences are not available, they should be obtained in calibration tests. Procedures for establishing known concentrations for such calibration tests are discussed in detail in ACGIH (1978) and in Lippmann (1978) cited in the bibliography.

Ventilation system measurements

Air velocity measurements. Most ventilation performance measurements are made with anemometers (i.e. instruments that measure air velocities), and with the exception of the Pitot tube, all require periodic calibration. Instruments based on mechanical or electrical sensors are sensitive to mechanical shocks and/or may be affected by dust accumulations and corrosion.

Anemometers are usually calibrated in a well defined flow field that is relatively large in comparison to the size of meter being calibrated. Such flow fields can be produced in wind tunnels.

Pressure measurements. Although the standard Pitot tube and a water-filled U-tube manometer may not require calibration, Pitot tubes and other flow meters may be used with pressure gauges that do. Many direct reading gauges can give false readings because of the effects of mechanical shocks and/or leakage in connecting tubes or internal diaphragms. For pressures of approximately 2.5 cmH_2O or greater, a liquid-filled laboratory manometer should be an adequate reference standard. For lower pressures, it may be necessary to use a reference whose calibration is traceable to a National

Bureau of Standards (NBS) standard in the United States or from comparable organisations in other countries.

Heat stress measurements

The four environmental variables used in heat stress evaluations are air temperature, humidity, radiant temperature, and air velocity. The measurement of air velocity is discussed in detail below. Liquid-in-glass thermometers used to measure dry-bulb, wet-bulb, or globe thermometer temperatures should have calibrations traceable to certified national standards, but should not need recalibration. Other temperature sensors will require periodic recalibration.

Electromagnetic radiation measurements

The electromagnetic spectrum is a continuum of frequencies whose effects on human health are frequency dependent, and some are attributable to narrow bands of frequency. Thus many of the instruments used to measure frequency and/or intensity are designed to operate over specific frequency regions. Other instruments, known as bolometers, measure the total incident radiant flux over a wide range of frequencies.

Noise-measuring instruments

The accuracy of sound-measuring equipment may be checked by using an acoustical calibrator consisting of a small, stable sound source that fits over a microphone and generates a predetermined sound level within a fraction of a decibel. The acoustical calibration provides a check of the performance of the entire instrument, including microphone and electronics. Some sound level meters have internal means for calibration of electronic components only. Sound level calibrators should be used only with the microphones for which they are intended. Manufacturers' instructions should be followed regarding the use of calibrators and indications of malfunction of instruments.

Calibration standards

Calibration procedures generally involve a comparison of instrument response to a standardised atmosphere or to the response of a reference instrument. Hence the calibration can be no better than the standards used. Reliability and proper use of standards are critical to accurate calibrations. Reference materials and instruments available from or calibrated by an official bureau of standards should be used whenever possible.

Test atmospheres generated for purposes of calibrating collection efficiency or instrument response should be checked for concentration using reference instruments or sampling and analytical procedures whose reliability and accuracy are well documented. The best procedures to use are those which have been referee or panel tested, that is, methods that have demonstrably yielded comparable results on blind samples analysed by different laboratories.

Instrument calibration procedures

Calibration of sampler's collection efficiency

Use of well characterised test atmospheres. To test the collection efficiency of a sampler for a given contaminant, it is necessary either (1) to conduct the test in the field using a proved reference instrument or technique as a reference standard, or (2) to reproduce the atmosphere in a laboratory chamber or flow system. Techniques and equipment for producing such atmospheres are discussed in ACGIH (1978) and Lippmann (1978).

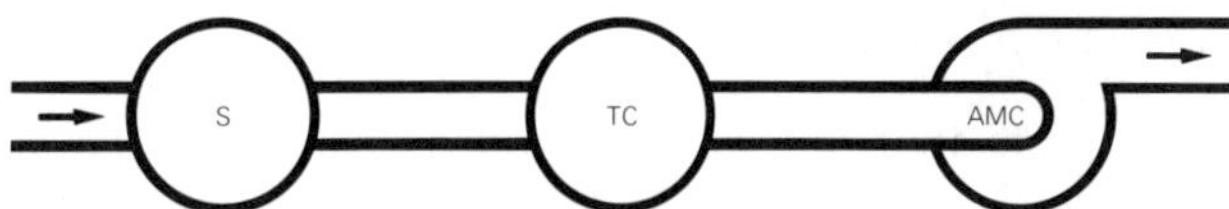

Figure 1. Sampler efficiency evaluation with downstream total collector: analysis of collections in sample under test *S* and total collector *TC*; *AMC*, air mover, flow meter, and flow control.

Analysis of sampler's collection and downstream total collector. The best approach to use in the analysis of a sampler's collection is to operate the sampler under test in series with a downstream total collector, as illustrated in figure 1. The sampler's efficiency is then determined by the ratio of the sampler's retention to the retention in the sampler and downstream collector combined. This approach is not always feasible, however. When the penetration is estimated from downstream samples there may be additional errors if the samples are not representative.

Analysis of sampler's collection and downstream samples. It is not always possible or feasible to quantitatively collect all the test material that penetrates the sampler being evaluated. For example a total collector might add too much flow resistance to the system, or be too bulky for efficient analysis. In this case the degree of penetration can be estimated from an analysis of a sample of the downstream atmosphere, as illustrated in figure 2. When this approach is used it may be necessary to collect a series of samples across the flow profile rather than a single sample, to obtain a true average concentration of the penetrating atmosphere.

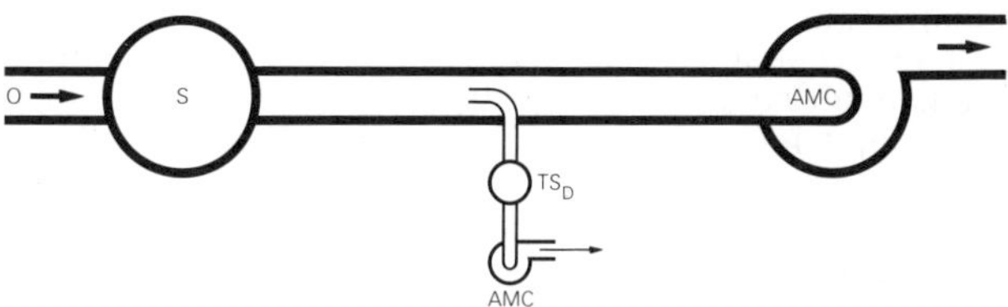

Figure 2. Sampler efficiency evaluation with downstream concentration sampler: analysis of collections in sample under test *S* and downstream sampler total collector TS_D.

Analysis of upstream and downstream samples. In some cases it is not possible to recover or otherwise measure the material trapped within elements of the sampling train such as sampling probes. The magnitude of such losses can be determined by comparing the concentrations upstream and downstream of the elements in question, as schematised in figure 3.

Determination of sample stability and/or recovery

The stability and the recovery of trace contaminants from sampling substrates are difficult to predict or control. Thus these factors are best explored by realistic calibration tests.

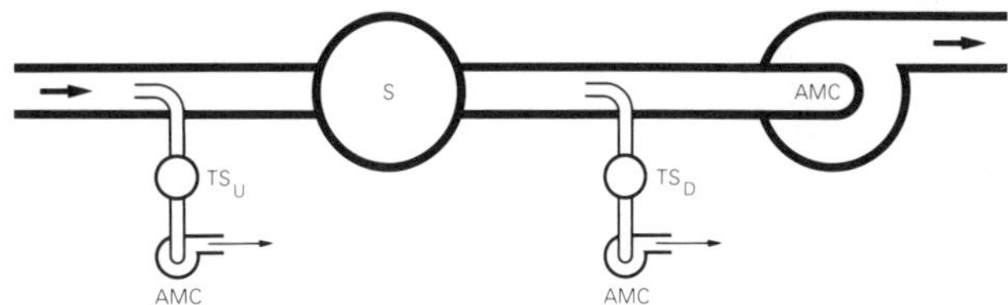

Figure 3. Sampler efficiency evaluation with upstream and downstream concentration samplers: analysis of collections in upstream and downstream samplers, total collection TS_U and TS_D.

Analysis of sample aliquots at periodic intervals after sample collection. If the sample is divided into a number of aliquots that are analysed individually at periodic intervals, it is possible to determine the long-term rate of sample degradation, or any tendency for reduced recovery efficiencies with time. These analyses would not, however, provide any information on losses that may have occurred during or immediately after collection because they had different rate constants. Such losses should be investigated using spiked samples.

Analysis of spiked samples. If known amounts of the contaminants of interest are intentionally added to the sample substrate, subsequent analysis of sample aliquots will permit calculation of sample recovery efficiency and rate of deterioration. These results will be valid only in so far as the added material is equivalent in all respects to the material in the ambient air. There are two basic approaches to spiked sample analyses: (1) the addition of known quantities to blank samples, and (2) the addition of radioactive isotopes to either blank or actual field collected samples.

When the material being analysed is available in tagged form, the tag can be added to the sample in negligible or at least known low concentrations. If there are losses in sample processing or analysis, the fractional recovery of the tagged molecules will provide a basis for estimating the comparable loss that took place in the untagged molecules of the same species.

Calibration or sensor response

Direct reading instruments are generally delivered with a direct reading panel meter, a set of calibration curves, or both. The unwary and inexperienced user tends to believe the manufacturer's calibration, and this often leads to grief and error. Any instrument with calibration adjustment screws should of course be suspect, since such adjustments can easily be changed intentionally or accidentally—for example in shipment.

All instruments should be checked against appropriate calibration standards and atmospheres immediately upon receipt and periodically thereafter. Procedures for establishing test atmospheres are discussed earlier in this article. Verification of the concentrations of such test atmospheres should be performed whenever possible using analytical techniques that are referee tested or otherwise known to be reliable.

With these techniques, calibration curves for direct-reading instruments can be tested or generated. When environmental factors such as temperature, ambient pressure, and radiant energy may be expected to influence the results, these effects should be explored with appropriate tests whenever possible. Similarly, the effects of cocontaminants and water vapour on instrument response should also be explored.

Estimation of errors

Sources of sampling and analytical errors

The difference between the air concentration reported for an air contaminant on the basis of a meter reading or laboratory analysis, and true concentration at that time and place, represents the error of the measurement. The over-all error is often due to a number of smaller component errors rather than to a single cause. To minimise the over-all error, it is usually necessary to analyse each of its potential components, concentrating one's efforts on reducing the component error that is largest. It would not be productive to reduce the uncertainty in the analytical procedure from 10 to 1.0% when the error associated with the sample volume measurement is +15%.

Sampling problems are so varied in practice that it is possible only to generalise on the likely sources of error to be encountered in typical sampling situations. In analysing a particular sampling problem, consideration should be given to each of the following:

(a) flow-rate and sample volume;

(b) collection efficiency;

(c) sample stability under conditions anticipated for sampling, storage, and transport;

(d) efficiency of recovery from sampling substrate;

(e) analytical background and interferences introduced by sampling substrate;

(f) effect of atmospheric cocontaminants on samples during collection, storage, and analyses.

Cumulative statistical error

The most probable value of the cumulative error E_e can be calculated from the following equation:

$$E_e = [E_1^2 + E_2^2 + E_3^2 + \cdots + E_n^2]^{1/2}$$

For example, if accuracies of the flow-rate measurement, sampling time, recovery, and analysis are ±15, 2, 10, and 10%, respectively, and there are no other significant sources of error, the cumulative error would be

$$E_n = [15^2 + 2^2 + 10^2 + 10^2] = [429]^{1/2} = +20.7\%$$

It should be remembered that this provides an estimate of the deviation of the measured concentration from the true concentration at the time and place the sample was collected. As an estimate of the average concentration to which a workman was exposed in performing a given operation, it would have additional uncertainty, dependent on the variability of concentration with time and space at the workstation.

Conclusions

Determinations of the concentrations of trace level contaminants in air and of heat stress, noise and radiant energies are subject to numerous variables, many of them difficult to control. Thus it is prudent to perform frequent calibration checks on all industrial hygiene instruments. Such calibrations should be based on realistic simulations of the conditions encountered in the field.

The production of test atmospheres in the range of occupational threshold limits is often difficult. This article provides a review of available techniques for the production of test atmospheres of gases, vapours and aerosols, with sketches of many of the more useful techniques.

Extreme care should be exercised in performing all calibration procedures. The following guidelines should be followed:

1. Use standard or reference atmospheres, instruments and devices with care and attention to detail.
2. Check all standard materials and instruments and procedures periodically to determine their stability and/or operating condition.
3. Perform calibrations whenever a device has been changed, repaired, received from a manufacturer, subjected to use, mishandled or damaged and at any time when a question arises with respect to its accuracy.
4. Understand the operation of an instrument before attempting to calibrate it, and use a procedure or set-up that will not change the characteristics of the instrument or standard within the operating range required.
5. When in doubt about procedures or data, make certain of their validity before proceeding to the next operation.
6. Keep all sampling and calibration train connections as short and free of constrictions and resistance as possible.

7. Exercise extreme care in reading scales, timing, adjusting and levelling if needed, and in all other operations involved.

8. Allow sufficient time for equilibrium to be established, inertia to be overcome and conditions to stabilise.

9. Obtain enough points or different rates of flow on a calibration curve to give confidence in the plot obtained. Each point should be made up of more than one reading whenever practical.

10. Maintain a complete permanent record of all procedures, data and results. This should include trial runs, known faulty data with appropriate comments, instrument identification, connection sizes, barometric pressure and temperature.

11. When a calibration differs from previous records, determine the cause of change before accepting the new data or repeating the procedure.

12. Identify calibration curves and factors properly with respect to conditions of calibration, device calibrated and what it was calibrated against, units involved, range and precision of calibration, date and name of the person who performed the actual procedure. Often it is convenient to indicate where the original data are filed and to attach a tag to the instrument indicating the items just listed.

(*Reproduced in part from* Patty's Industrial Hygiene and Toxicology*; see bibliography.*)

LIPPMANN, M.

Air sampling instruments. American Conference of Governmental Industrial Hygienists (Cincinnati, ACGIH, 5th ed., 1978).

"Calibration". Lippmann, M. Ch. 25 (1 263-1 331). 88 ref. *Patty's Industrial Hygiene and Toxicology.* Vol. 1: *General principles.* Clayton, G. D.; Clayton, F. E. (eds.). (New York, Wiley, 3rd revised ed., 1978).

Camphor

A bicyclic saturated terpene ketone. It exists in the optically active dextro- and levo-forms and as a racemic mixture of these two forms.

Camphor ($C_{10}H_{16}O$)

m.w.	152
sp.gr.	0.99
m.p.	176 °C
b.p.	209 °C
v.d.	5.2
v.p.	0.18 mmHg (20 Pa) at 20 °C
f.p.	65.6 °C
e.l.	0.6-3.5%
i.t.	466 °C

CH_3 / =O / $H_3C-C-CH_3$

soluble in ethyl alcohol, ethyl ether and most other organic solvents

a white, tough crystalline mass with a characteristic odour.

TWA OSHA	2 ppm 12 mg/m³
STEL ACGIH	3 ppm 18 mg/m³
IDLH	200 mg/m³
MAC USSR	3 mg/m³

Camphor is obtained as a natural product from wood distillation or by synthesis.

Natural camphor

This is obtained primarily from the camphor laurel tree *Cinnamomum camphora* grown in Japan, South West China and neighbouring countries; however, the industry is only a small-scale one. The chopped branches of the camphor laurel, when steam-distilled, produce a mixture containing 3% of crystalline camphor and camphor oil. The solid camphor is removed by distillation and the liquid camphor fractionally distilled. From the fractions, more camphor is obtained by freezing. The crude camphor so obtained is refined by sublimation.

Figure 1. The camphor plant *(Cinnamomum camphora).* *(By courtesy of Dr. D. K. Ferguson.)*

Synthetic camphor. Since the natural product is limited in quantity, synthetic camphor is now manufactured on a considerable scale chiefly from pinene which is obtained from the oleoresin of pine trees. The oil of turpentine from these trees consists principally of α- and β-pinene. Another source of pinene is from steam-distilled wood turpentine. In the process of manufacture the pinene is converted to bornyl chloride and then, by hydrolysis, to isolborneol and, finally, oxidised to camphor.

Uses. Natural and synthetic camphor may both be put to the same uses since the natural dextro form does not differ substantially from the optically inactive synthetic form. Camphor is used in the manufacture of plastics and films, some types of lacquers, smokeless powder explosives, perfumes, disinfectants and moth repellents. Celluloid was produced by first gelatinising pyroxylin with camphor and ethyl alcohol.

There is a steady demand for camphor in India for use in religious ceremonies in temples and for ceremonial cremation. It also acts as a weak local anaesthetic and antiseptic and is used in liniments for lumbago and related disorders and as a respiratory and circulatory stimulant.

HAZARDS AND THEIR PREVENTION

The harvesting of the camphor laurel plant, the cutting of the wood into pieces and the steam distillation for the production of natural camphor is usually done in remote mountainous areas where the plants grow wild. The natural hazards of such places are snakes and malaria-carrying insects.

Camphor, being rather volatile, may give off flammable vapours and should be packed in absolutely airtight packages or cases for transport and, on board ship, it should be stored in a well ventilated space, away from any source of heat and oxidising agents. There is a moderate fire risk if camphor is exposed to heat or flame but there is no spontaneous combustion danger.

Camphor is a powerful irritant and stimulant, diaphoretic and narcotic, poisonous in overdoses and sedative in small doses. Workers employed in storerooms or on removing camphor from the sublimation chambers may be exposed to high concentrations of camphor vapour and are liable to suffer ill effects leading, in the most severe cases, to nausea, vomiting, dizziness, excitation and convulsions. Foaming at the mouth and fainting may ensue. The first symptoms are a sweet taste in the mouth then an irritation of the throat, followed by headache, and the final state may resemble an epileptic spasm. Exposure to camphor vapour results in a decrease in the frequency of respiration and there is also a tendency for the pulse rate to be lowered.

To reduce the risk of intoxication from camphor vapour, workplaces in which this material is processed or handled should be fitted with adequate ventilation since a concentration of more than 2 ppm of camphor vapour can cause irritation and have adverse effects in the case of prolonged exposure. Workers should be fully informed of the dangers and the possible symptoms of overexposure.

KO, Yuan-ching

"Camphor exposure in a packaging plant". Gronka, P. A.; Robkoskie, R. L.; Tomchick, G. J.; Rakow, A. B. *American Industrial Hygiene Association Journal* (Akron), May-June 1969, 30/3 (276-279). 3 ref.

Campylobacter infections

Definition

A variety of animal diseases is caused by several species of the bacterium genus *Campylobacter* (formerly *Vibrio*). *C. fetus* is a venereally transmitted infection of cattle causing abortion and other types of reproductive failure. The organism *C. fetus* subspecies (ssp.) *intestinalis* causes reproductive failure in sheep. Transmission in this instance is via the oral route. The organism can be found in high numbers in fetal and placental tissue of aborted fetuses of cattle and sheep. The disease often results in chronic vaginal infections and septicaemia in cattle.

Specific types of dysentery are caused by ssp. *coli* and ssp. *jejuni* respectively in swine and cattle. Campylobacter infections are quite common in animal populations, and are among the most important economic diseases of the cattle industry in some areas.

Farmers may be exposed to the organism through direct or indirect contact with infected cattle or sheep, such as while attending sick animals or working in the environment where infected animals have been kept. Environmental exposure beyond the agricultural workplace has been documented recently, as this organism has been found to be a contaminant of streams and other bodies of water. Also, it has been found as an occasional contaminant of municipal water supplies. Outbreaks of diarrhoea in human populations have been attributed to consumption of contaminated water.

Occurrence and symptoms

Organisms similar to those causing disease in animals may also affect man. Early reports have indicated human infections to be a rare but severe disease of man. Recent reports suggest that occurrence in man is much more frequent but milder than many previously reported cases. The diagnosis has been dependent on a positive blood culture, but due to difficulties in isolation, many diagnoses may have been missed.

The first human case was reported in 1947. Since that time many cases have been reported, and the clinical manifestations associated with infection can be classified into five forms: (1) febrile illness without localisation; (2) febrile illness with localisation; (3) severe enteritis with haemorrhagic diarrhoea; (4) mild enteritis with mucoid diarrhoea; and (5) abortion or neonatal death. The localised infections have resulted in valvular endocarditis, septic arthritis, meningoencephalitis, thrombophlebitis, placentitis, and jejunitis. Enteritis and haemorrhagic diarrhoea may occur with the febrile illness regardless of whether or not the illness localised. In recent years several outbreaks of mild mucoid diarrhoea without a severe febrile illness have been related to consumption of contaminated water.

Infections in pregnant women have been recognised in the fifth to sixth month of pregnancy. The most common signs and symptoms seen were chills, fever, headache, malaise, uterine contractions, and diarrhoea. Fever may fluctuate, but persist for weeks to months. Weight loss, muscle and joint stiffness and pneumonia may be seen. Pregnancy may terminate in abortion 30-40 days following onset. Premature birth may also occur. Congenital infections in infants have resulted in severe enteritis and encephalitis, or hepatitis.

Many of the reported human cases have been associated with concurrent chronic disease. No point source was identified in many of the cases. It was suggested that the infection may be latent (as it often is in cattle) then become clinical with stress.

Epidemiology

Very few epidemiological data are available on campylobacter infections in man. Attempts to relate contraction of the disease from an animal source have not been rewarding. In most cases a point source has not been determined, and no information is available to indicate previous contact with farm animals, consumption of unpasteurised milk or consumption of meat. Other possible sources have been suggested. Several cases have followed tooth extraction, suggesting that the organism is apparently part of the oral flora. A contamination of the intestinal tract of the mother was thought to be responsible for congenital neonatal encephalitis in three infants. Recent evidence suggests the organism is a common contaminant of surface waters, and that individual cases or outbreaks of diarrhoea may result from consumption of this water. The epidemiology of this disease is in the very early stages of development. Much more information should be revealed on this disease in future years.

Diagnosis

Diagnosis is at present entirely dependent on isolation and identification of the organism in blood, spinal fluid, joint fluid, body tissues, faeces, or vaginal discharge. Multiple isolation attempts are usually necessary, and isolation attempts should correlate with symptoms, i.e. positive blood culture during fever spike, positive cerebral spinal fluid with symptoms of meningitis, etc.

PREVENTION

Prompt and sanitary disposal of placentas and aborted fetuses of sheep and cattle is essential. Control in animal populations includes vaccination of bulls and excellent environmental sanitation to prevent contamination of animal feed and water. Other general sanitary measures pertaining to enteric pathogens should be observed, i.e. proper treatment of potentially contaminated drinking water, washing of fresh vegetables, hand washing when handling potentially contaminated materials, thorough sanitary handling and thorough cooking of meat, and pasteurisation of milk.

Treatment

Streptomycin, aureomycin, tetracycline, and chloramphenicol have all been reported to result in clinical improvement of infected patients. Treatment should continue for three weeks to avoid possible relapse.

DONHAM, K. J.

"Human vibriosis (Leading articles)". *British Medical Journal* (London), 29 Apr. 1967, 2 (260-261). 8 ref.

"Human vibriosis: indigenous cases in England". White, W. D. *British Medical Journal* (London), 29 Apr. 1967, 2 (283-287). Illus. 43 ref.

"Waterborne campylobacter–Vermont". *Morbidity, Mortality Weekly Report* (Atlanta), 1978, 27/207.

"Campylobacter fetus Septicemia and hepatitis in a child with agammaglobulinemia". Wyatt, R. A.; Younoszai, K.; Anaras, S.; Myers, M. G. *Journal of Pediatrics* (St. Louis), Sep. 1977, 91/3 (441-442). Illus. 9 ref.

Cancer, environmental

Cancer is a common disease in all countries of the world. The probability that a person will develop it by the age of 75 years, given survival to that age, varies between about 9 and 32% in females and 9 and 42% in males. On average, in developed countries one person in four will die from cancer.

Geographical variation

Variation between geographic areas in the rates of particular types of cancer may be much greater than that for cancer as a whole. Known variation in the incidence of the commoner cancers is summarised in table 1. The incidence of cancer of the oesophagus, for example, varies some 300-fold between the North East of Iran (similar high rates are probably also found in adjacent parts of the Soviet Union and in central China) and Nigeria. This wide variation in frequency of the various cancers has led to the view that much of human cancer is caused by factors in the environment. In this context, the environment will be considered to encompass human behaviour, habits and life style as well as external factors over which the individual has no control.

It has been argued that the lowest rate of a cancer observed in any population is indicative of the minimum, possibly spontaneous, rate occurring in the absence of causative factors. Thus the difference between the rate of a cancer in a given population and the minimum rate observed in any population is an estimate of the rate of the cancer in the first population which is attributable to environmental factors. On this basis it has been estimated, very approximately, that some 80 to 90% of all human cancers are environmentally determined.

There are, of course, other explanations for geographical variation in cancer rates. Under-recording of cancer in some populations may exaggerate the range of variation, but certainly cannot explain differences of the size shown for some cancers in table 1. Genetic factors may also be important. It has been observed, however, that when populations migrate along a gradient of cancer incidence they often acquire a rate of cancer which is intermediate between that of their home country and that of the host country. This suggests that a change in environment, without genetic change, has changed the cancer incidence. For example when Japanese migrate to the United States their rates of colon and breast cancer, which are low in Japan, rise and their rate of stomach cancer, which is high in Japan, falls. These changes may be delayed until the first post-migration generation but they still occur without genetic change. For some cancers, change with migration does not occur. For example the Southern Chinese retain their high rate of cancer of the nasopharynx wherever they live, thus suggesting that genetic factors, or some cultural habit which changes little with migration, are responsible for this disease.

Time trends

Further evidence of the role of environmental factors in

Table 1. Variation between populations in the reported incidence of commoner cancers[1]

Type of cancer	High incidence area	Cumulative risk (%) to age of 75 years	Low incidence area	Cumulative risk (%) to age of 75 years	Range of variation
Skin	Australia[2]	> 20	India[2]	0.1	> 200
Oesophagus	Iran[2]	20	Nigeria[2]	< 0.1	300
Bronchus	England[2]	11	Nigeria[2]	0.1	100
Stomach	Japan[2]	11	Uganda[2]	0.2	50
Cervix uteri	Colombia[2]	10	Israel (Jews)	0.6	20
Liver	Mozambique[2]	8	Norway	0.1	70
Prostate	United States (black)[2]	7	Japan[2]	0.3	20
Breast	United States[2]	7	Uganda[2]	0.9	7
Colon	United States[2]	3	Nigeria[2]	< 0.1	30
Mouth	India[2]	> 2	Denmark	0.1	> 20
Rectum	Denmark	2	Nigeria[2]	< 0.1	20
Bladder	United States[2]	2	Japan[2]	0.6	4
Ovary	Denmark	2	Japan[2]	0.2	8
Corpus uteri	United States[2]	2	Japan[2]	0.2	10
Nasopharynx	Singapore (Chinese)	2	England[2]	< 0.1	40
Pancreas	New Zealand (Maori)	2	Uganda[2]	0.1	10
Penis	Uganda[2]	1	Israel (Jews)	⩽ 0.1	250

[1] All cancers with cumulative risk of 1% by the age of 75 years. Male rates are given except for female sex organs. Modified from R. Doll: "Epidemiological enquiry: Power and limitations" in P. Emmelot and E. Kriek (eds.): *Environmental carcinogenesis* (Amsterdam, Elsevier/North Holland Biomedical Press, 1979), (381-400). [2] Data used refer only to part of the country.

cancer incidence has come from the observation of time trends. The most dramatic and well known change has been the rise in lung cancer rates in parallel with but occurring some 20 to 30 years after the adoption of cigarette usage. Similarly a rise in incidence of endometrial cancer has been seen in the United States following widespread use of oestrogens for relief of menopausal symptoms with, more recently, evidence of a fall in rates following reduction in their usage. Less well understood are the substantial falls in incidence of cancers of the stomach, oesophagus, cervix and others which have paralleled economic development in many countries. It would be difficult to explain these falls, however, except in terms of reduction in exposure to causal factors in the environment or, perhaps, increasing exposure to protective factors–again environmental.

Main environmental carcinogenic agents

The importance of environmental factors as causes of human cancer has been further demonstrated by epidemiological studies relating particular agents to particular cancers. The main agents which have been identified are summarised in table 2. The relative importance of these agents varies widely. At one extreme there may be very heavy exposure of a small group to high concentrations of carcinogen–such as occurred in distillers of 2-naphthylamine all of whom, in some studies, developed bladder cancer. The contribution, however, of 2-naphthylamine to the total problem of bladder cancer was small because of the small number of people with appreciable exposure. At the other, extreme exposure of large numbers of people to low levels of carcinogen may give rise to many more cancers simply because of the numbers involved.

Table 2. Environmental agents thought to cause cancer in humans[1]

Environmental agents	Cancer sites
Radiation	
Radon and its daughter isotopes	Lung
Radium, thorium	Bone
Other X-irradiation	Leukaemia, breast, thyroid, other solid tumours
Ultraviolet light	Skin
Medicines	
Alkylating agents–melphalan, cyclophosphamide(?), thiotepa(?)	Leukaemia
Androgens	Liver
Coal tar ointments	Skin
Compound analgaesics	Kidney, ureter(?), bladder(?)
Immunosuppressive drugs	Non-Hodgkins lymphoma
Methoxsalen(?)	Skin
Steroid contraceptives	Liver, breast(?)
Stilbestrol (transplacental)	Vagina
Other oestrogens	Corpus uteri, breast(?)
Thorotrast	Liver, spleen, leukaemia
Tobacco, alcohol and related substances	
Alcohol drinking	Mouth, pharynx, oesophagus, larynx, liver
Betel chewing (with tobacco and lime)	Mouth
Tobacco smoking	Lip, mouth, pharynx, larynx, oesophagus, lung (bronchus), kidney, bladder, pancreas(?)
Other products and by-products of industry	
4-Aminobiphenyl	Bladder
Arsenic	Skin, lung
Asbestos	Lung, pleura, peritoneum
Benzene	Leukaemia
Benzidine	Bladder
Bis(chloromethyl)ether	Lung
Cadmium, some cadmium compounds(?)	Prostate, kidney
Chlorophenols(?)	Lymphoma, soft-tissue sarcoma
Chromium, some chromium compounds	Lung
Hardwood dust	Nasal sinuses
Isopropyl oil(?)	Nasal sinuses
Mustard gas	Lung, larynx, nasal sinuses
1- and 2-Naphthylamine	Bladder
Nickel, some nickel compounds(?)	Lung, nasal sinuses
Phenoxyacetic acids(?)	Lymphoma, soft-tissue sarcoma
Soot, tar and mineral oils	Skin, lung
Vinyl chloride (monomer)	Liver
Infections	
BCG vaccination(?)	Lymphoma
Epstein-Barr virus(?)	Burkitt's lymphoma, nasopharyngeal carcinoma, Hodgkins disease
Hepatitis-B virus	Liver
Herpes simplex virus type 2(?)	Cervix
Schistosomiasis	Bladder, colon and rectum(?)
Diet	
Aflatoxins	Liver
Bracken fern(?)	Oesophagus
Coffee(?)	Bladder
Fat(?)	Breast, corpus uteri, colon and rectum
Fibre deficiency	Colon and rectum
Iodine deficiency	Breast, thyroid
Meat(?)	Colon and rectum
Nitrate and nitrite(?)	Stomach
Obesity	Corpus uteri, breast(?), gallbladder(?)
Brassica vegetables (protective?)	Colon and rectum
Vitamin A (protective)	Lung, bladder, stomach(?), cervix(?)
Vitamin C (protective?)	Stomach
Reproductive and sexual behaviour	
Late first childbirth	Breast
Nulliparity or low total parity	Breast, ovary, corpus uteri
Promiscuity	Cervix
Sexual continence(?)	Prostate
Personal hygiene	
Poor oral hygiene	Mouth
Poor sexual hygiene	Penis, cervix(?)
Uncircumcision	Penis

[1] (?) signifies less well established associations; some protective factors are listed and identified specifically as such.

The most important agents among those listed in table 2 are those to which a substantial proportion of the population is exposed in relatively large amounts. They include particularly ultraviolet light; tobacco; alcohol; in some populations the betel "quid"; hepatitis-B virus and/or aflatoxins; possibly dietary fat and meat or dietary fibre deficiency; reproductive delay; and promiscuity. It has been suggested that asbestos should be in this group, but this is debatable on present evidence although the existence of widespread, low-level exposure to it cannot be denied. Deficiency of nutrients, other than fibre–vitamins A and C and perhaps others–may also contribute to an appreciable proportion of human cancers. Available data, however, are insufficient to establish this with any certainty.

Attempts have been made to estimate numerically the relative contributions of these factors to the 80 or 90% of cancers which might be attributed to environmental factors. The pattern varies, of course, from population to population and two populations are compared in table 3. Both have comparatively high rates of cancer as a whole. Apart from the substantial contribution of tobacco, either alone or in combination with alcohol, to cancer production in males in Birmingham (and indeed in most other economically developed populations) most of the cancers in both populations are attributed to diet and life style. The combinations of factors summarised in the latter category differ between the two populations. On the one hand, in the English population, the responsible factors are those which go with affluence (dietary excess, controlled reproduction, etc.) while on the other, in the African population, they are those which accompany relative poverty (specific dietary deficiencies, mould-contaminated food, poor hygiene, etc.). The net results, however, are problems of similar size.

Table 3. Estimates of the proportions (%) of human cancers attributable to various environmental factors in two populations[1]

Environmental factors	White population of Birmingham, England		Black population of Bulawayo, Zimbabwe	
	Males	Females	Males	Females
Tobacco alone	30	7	14	2
Tobacco and alcohol together	5	3	2	< 1
Sunlight	10	10	< 1	< 1
Occupational hazards	6	2	2	< 1
Radiation	1	1	< 1	< 1
Medicines	1	1	0	0
Infections	0	0	6	4
Diet and life style	30	63	57	64
Non-environmental factors				
Congenital	2	2	2	2
Other and unknown	15	11	15	24

[1] Modified from J. Higginson: "Environmental carcinogenesis: a global perspective"; in *Environmental carcinogenesis* (2-9). P. Emmelot and E. Kriek (eds.). (Amsterdam, Elsevier/North Holland Biomedical Press, 1979).

The cancers attributed, in table 3, jointly to tobacco and alcohol reflect the increasing awareness of the importance of interactions between carcinogens. Thus for example, in the case of alcohol, tobacco and cancer of the oesophagus it has been shown that an increasing consumption of alcohol multiplies many times over the rate of cancer produced by a given level of tobacco consumption. Alcohol, by itself, may not be carcinogenic but may facilitate transport of tobacco, or other carcinogens into the cells of susceptible tissues. Multiplicative interaction may also be seen between initiating carcinogens, as between radon and its daughters and tobacco smoking in miners of uranium and some other minerals. Some environmental agents may act by promoting cancers which have been initiated by another agent–this is the likely mechanism for an effect of dietary fat on development of breast cancer (through, probably, increased production of the hormones which stimulate the breast). The reverse may also occur, as, for example, in the case of vitamin A which probably has an anti-promoting effect on lung and possibly other cancers initiated by tobacco. Similar interactions may also occur between environmental and constitutional factors although they are currently less well understood.

The significance of interactions between carcinogens, from the point of view of cancer control, is that withdrawal of exposure to one of two (or more) interacting factors may give rise to a greater reduction in cancer incidence than would be predicted from consideration of the effect of the agent when acting alone. Thus, for example, withdrawal of cigarettes may eliminate almost entirely the excess rate of lung (bronchogenic) cancer in asbestos workers (although rates of mesothelioma would be unaffected).

The realisation that environmental factors are responsible for a large proportion of human cancers has laid the basis for primary prevention of cancer by modification of exposure to the factors identified. Such modification may comprise removal of a single major carcinogen; reduction, as discussed above, in exposure to one of several interacting carcinogens; increasing exposure to protective agents; or combinations of these approaches. While some of this may be achieved by community-wide regulation of the environment through, for example, environmental legislation, the apparent importance of life style factors suggests that much of primary prevention will remain the responsibility of individuals. Governments, however, may still create a climate in which individuals find it easier to make the right decision. More research is also required to ensure that advice for life style change is soundly based.

ARMSTRONG, B. K.

IARC monographs on the evaluation of the carcinogenic risk of chemicals to humans. Vols. 1-20, Supplement 1: *Chemicals and industrial processes associated with cancer in humans* (Lyons, International Agency for Research on Cancer, Sep. 1979).

"The epidemiology of cancer". Doll, R. *Cancer* (Philadelphia), 15 May 1980, 45/10 (2475-2485). 37 ref.

Cancer incidence in five continents. Vol. III. Waterhouse, J.; Muir, C.; Correa, P.; Powell, J.; Davis, W. (eds.). IARC scientific publications No. 15 (Lyons, International Agency for Research on Cancer, 1976), 584 p.

Cancer, occupational

Occupational cancer is a form of delayed toxicity, usually serious in clinical course and outcome, due to exposure to chemical and physical agents (carcinogens) in the workplace. Scientifically, occupational cancer holds a central position in the history of cancer research, as it provided the first examples of cancers for which an aetiology could be identified, first in terms of occupational exposures and subsequently in terms of the specific causative agents. Thus, to quote once more the most famous instance, cancer of the scrotum was initially linked to exposure to soot in chimney-sweeps by Percival Pott in 1775 and much later (in the 1930s) several polycyclic hydrocarbons were identified as the active compounds. From a public health viewpoint the chief relevance of occupational cancer lies in the fact that occupational determinants of cancer once identified can be removed or controlled more easily than causal factors related to personal habits under cultural influence, such as smoking and drinking.

Carcinogenesis

Biologically and clinically, cancer due to occupation is at present undistinguishable from cancer due to other causes. Cancer is not a single disease but a family of different diseases (well over a hundred), each characterised by its own aetiological (often unknown), morphological, patho-physiological and clinical profile. What makes the family resemblance of these diseases are certain broad qualitative attributes–

(a) cell neoplasia, anaplasia and heterotopia, namely new and progressive growth, often infiltrative and destructive; structural differences (tissue, cell and nuclei atypism and polymorphism) between the normal and the new growth; migration and colonisation (metastasisation) of cells to distant sites via blood and lymph vessel penetration;

(b) the carcinogenic process, namely the way in which a normal cell is transformed into a cancer (malignant) cell. The carcinogenic process (or processes) is still very incompletely understood, but some aspects deserve mention:

(1) Experimental and epidemiological investigations suggest that the process may be represented as a succession of discrete stages, taking place in the time span from first exposure to a carcinogenic agent to the appearance of the clinical cancer. The substantial length of this period, which in man may vary from a few years to several decades and does not seem related to the intensity of exposure, is in itself a striking and unexplained feature of the process.

(2) In its simplest form, as first brought out in mouse skin carcinogenesis experiments now regarded as classic, the multistage process reduced to two stages: an irreversible "initiation" stage inducing latent malignant cells and a "promotion" stage which propagates these cells into a malignant growth. By reference to this model, one can try to divide carcinogens into complete carcinogens, capable of both initiating and promoting activity (e.g. polycyclic hydrocarbons) and incomplete carcinogens, only capable of initiation (e.g. urethane) or of promotion (e.g. phorbol esters). The distinction, however, proves not at all clear-cut as the activity exhibited depends not only on the intrinsic nature of the compound but also on the dose and mode of administration.

(3) A current hypothesis equates the initiation stage with a change in the desoxyribonucleic acid (DNA) macromolecule, which carries the genetic information. The DNA lesion may be produced, for example, by reaction with a molecule of a chemical carcinogen or of one of its reactive metabolites. If it goes unrepaired by enzymatic mechanisms, it leads to "somatic mutation" at the moment of the DNA and cell replication. By definition this mutation must involve genes directly or indirectly dealing with the control of cell proliferation. The hypothesis of somatic mutation is supported by the finding that a large number of known carcinogens have also been shown to be mutagens; it also conforms more than alternative hypotheses to the soundest fact in cancer biology, namely that cancerous characteristics are transmitted from one cell generation to the next, i.e. they behave as inheritable.

(4) The nature of promotion is still very obscure. Promotion is one mechanism of what has been labelled as epigenetic carcinogenesis, in contrast to genetic carcinogenesis involving gene mutation in somatic cells. Epigenetic carcinogenesis embraces other postulated mechanisms, whose actual role remains to be established, such as inhibition of DNA repair, suppression of immune response, hormone inbalance, and alterations in the homeostasis of cell turnover. See CARCINOGENIC SUBSTANCES, EPIGENETIC.

It may be that carcinogenic substances which have not been found mutagenic in experimental test systems (e.g. saccharin) act through some epigenetic mechanisms. One relevant consequence of regarding carcinogenesis as a multistage process is (to quote Boyland) that "the concept of threshold for carcinogenic compounds becomes difficult, because the threshold must be dependent on the presence of other essential factors. ... From the point of view of carcinogenic risk any factor involved in any stage in the development of neoplasia should be considered hazardous".

Epidemiology

While current research in chemical carcinogenesis aims at investigating and characterising the role of exogenous chemicals in the different steps of the carcinogenic process, the epidemiological concept of a carcinogen is somewhat simpler and more direct. Essentially any unequivocally identified exposure which is causally associated with an increased occurrence of a cancer is regarded as a "carcinogen".

The two situations which, particularly in the past, led to the discovery of a carcinogenic risk are an unusual occurrence of a rare tumour in a working population (liver angiosarcoma in polyvinyl chloride workers, pleural mesothelioma in asbestos workers, etc.) or an excess of a "common" neoplasia in a specific industrial process (lung cancer among employees in *bis*-chloromethyl ether and chloromethyl methyl ether production in the chemical industry, lung cancer among workers involved in mustard gas production, etc.).

The period intercurring between the start of exposure and the detection of the cancer related to that exposure is, in epidemiology, considered as a whole and often–but not properly–called latency period. This period includes in fact a process with several steps, previously alluded to, of which the "latent" phase is probably only the last one: thus the period is better defined as "time since first exposure".

A considerable number of substances for which occupational exposure occurs have been proved to be carcinogenic to man (see CARCINOGENIC SUBSTANCES). In a few cases, it has not yet been possible to identify the single chemical agent(s) within an industrial process responsible for the carcinogenic effect.

It should also be emphasised that the carcinogenic risk for man in several industrial processes has never been investigated and therefore it is unknown. Moreover, the carcinogenic risks from industrial processes may not necessarily be confined to the occupational environment but may also involve the general population by air and water pollution, industrial accidents, and sometimes, as in the case of relatives of asbestos workers, by a person-to-person contamination.

Occupations involving a carcinogenic risk. While it is relatively simple to list the substances recognised as carcinogenic in the working environment, a similar classification by occupation presents several problems, and is limited by poor information about the industrial processes, which unfortunately is not rare. Moreover, in order to avoid misleading interpretation, such a list should ideally be viewed against a list including the industrial processes or occupations which have been somehow investigated with respect to the carcinogenic risk.

Another important point is the change in exposures with time within a given occupational situation. For example vintners are no longer exposed to arsenic in developed countries, and should not form a separate category from other workers with common exposure to new types of pesticides in current use. This points out the most critical limit of a classification of this type: the presence of a carcinogen in an occupational situation does not necessarily mean exposure to it and, on the other hand, the absence of a known carcinogen does not exclude the presence of an unknown carcinogenic risk.

Despite all these problems, a tentative list of occupations or industrial processes for which a carcinogenic risk has been reported is shown in table 1. The gathered evidence is limited to epidemiological studies which point out an excess of cancer in a defined occupational group. Case reports, ecological studies and studies without a precise definition of the occupation have been excluded, as well as occupations of historical interest, but virtually not existing at the present time. Table 1 *(a)* presents the occupations in which a carcinogenic risk has been demonstrated, according to the IARC Monographs Programme (see also table 1 in CARCINOGENIC SUBSTANCES) and to other available sources. Table 1 *(b)* lists the occupations reported to present a cancer excess, but for which the assessment cannot be considered definitive. There is an obvious dishomogeneity as to the degree of evidence in this group, which contains occupations reported as carrying a carcinogenic risk by only one study, often involving small numbers, or by several studies but whose limitations do not yet allow one to draw a clear-cut conclusion. Also included in table 1 *(b)* are some occupations, already listed in table 1 *(a)*, which present inconclusive evidence of carcinogenicity for sites other than those reported in table 1 *(a)*. The occupational groups listed in the table are those described in the epidemiological studies which contributed to the evidence, to whatever degree, of the

Table 1. Occupations and cancer[1]

Industry	Occupation	Site	Reported or suspected causative agent
(a) Occupations recognised to present a carcinogenic risk			
Agriculture, forestry and fishing	vineyard workers using arsenical insecticides	lung, skin	arsenic
Extractive	arsenic mining	lung, skin	arsenic
	iron-ore mining	lung	causative agent not identified
	asbestos mining	lung, pleural and peritonea mesothelioma	asbestos
	uranium mining	lung	radon
Asbestos production industry	insulated material production (pipes, sheeting, textile, clothes, masks, asbestos cement manufacts)	lung, pleural and peritoneal mesothelioma	asbestos
Petroleum industry	wax pressmen	scrotum	polycyclic hydrocarbons
Metal industry	copper smelting	lung	arsenic
	chromate producing	lung	chromium
	chromium plating	lung	chromium
	ferrochromium producing	lung	chromium
	steel production	lung	benzo(a)pyrene
	nickel refining	nasal sinuses, lung	nickel
Shipbuilding, motor vehicles and transport	shipyard and dockyard workers	lung, pleural and peritoneal mesothelioma	asbestos
Chemical industry	BCME and CMME products and users	lung (oat cell carcinoma)	BCME, CMME
	vinyl chloride producers	liver angiosarcoma	vinyl chloride monomer
	isopropyl alcohol manufacturing (strong acid process) workers	paranasal sinuses	causative agent not identified
	pigment chromate producing	lung	chromium
	dye manufacturers and users	bladder	benzidine, 2-naphthylamine, 4-aminodiphenyl
	auramine manufacture	bladder	auramine (together with the other aromatic amines used in the process)
Pesticides and herbicides production industry	arsenical insecticides production and packaging	lung	arsenic
Gas industry	coke plant workers	lung	benzo(a)pyrene
	gas workers	lung, bladder, scrotum	coal carbonisation products, β-naphthylamine
	gas-retort house workers	bladder	α/β-naphthylamine

Industry	Occupation	Site	Reported or suspected causative agent
Rubber industry	rubber manufacture	lymphatic and haematopoietic system (leukaemia)	benzene
		bladder	aromatic amines
	calendering, tyre curing, tyre building	lymphatic and haematopoietic system (leukaemia)	benzene
	millers, mixers	bladder	aromatic amines
	synthetic latex producers, tyre curing, calender operatives, reclaim, cable makers	bladder	aromatic amines
Construction industry	insulators and pipe coverers	lung, pleural and peritoneal mesothelioma	asbestos
Leather industry	boot and shoe manufacturers, repairers	nose, marrow (leukaemia)	leather dust, benzene
Wood pulp and paper industry	furniture and cabinet makers	nose (adenocarcinoma)	wood dust
Other	roofers, asphalt workers	lung	BAP

(b) Occupations reported to present a cancer excess, but for which the assessment of the carcinogenic risk is not definitive

Industry	Occupation	Site	Reported or suspected causative agent
Agriculture, forestry and fishing	fishermen	skin, lip	pitch, ultraviolet radiation
	farmers	lymphatic and haematopoietic system (leukaemia, lymphoma)	undefined
	basal bark spraying	lymphatic and haematopoietic system (lymphoma), soft tissue sarcomas	phenoxyacetic acids, chlorophenols (presumably contaminated with PCDF, PCDD, and polychlorinated benzodioxins)
	railway embankment spraying	lymphatic and haematopoietic system (lymphoma), lung cancer	phenoxyacetic acids, amitrol, monuron, durion
	pesticides appliers	lung	hexachlorocyclohexane combined and other pesticides
Extractive	zinc-lead mining	lung	radiations
	coal	stomach	coal dust
	talc	lung, pleura	talc (contaminated with asbestos?)
	asbestos mining	gastrointestinal tract	asbestos
Asbestos production industry	insulation material production (pipes, sheeting, textiles, clothes, masks, asbestos cement manufacts)	larynx, gastrointestinal tract	asbestos
Petroleum industry	oil refining	oesophagus, stomach, lung	polycyclic hydrocarbons
	boilermakers, painters, welders, oilfield workers	lung	polycyclic hydrocarbons
	petrochemical plant workers	brain, stomach	polycyclic hydrocarbons
	petroleum refining	marrow (leukaemia)	benzene
Metal industry	aluminium production	lung	benzo(a)pyrene
	beryllium refining	lung	beryllium
	smelters	respiratory and digestive system	lead
	nickel refining	larynx	nickel
	battery plant workers, cadmium alloy producers, electroplating workers	prostate, kidney	cadmium
	cadmium smelters	prostate, lung	cadmium

Industry	Occupation	Site	Reported or suspected causative agent
Shipbuilding, motor vehicles and transport	filling station, bus and truck drivers, operators of excavating machines	marrow (leukaemia)	petroleum products and combustion residues containing benzene
	hauliers	lung	polycyclic aromatic hydrocarbons
	shipyard and dockyard workers	larynx, digestive system	asbestos
Chemical industry	acrylonitrile production	lung, colon	acrylonitrile
	vinylidene chloride producers	lung	vinylidene chloride (mixed exposure to VC and acrylonitrile)
	isopropyl alcohol manufacturing (strong acid process) workers	larynx	undefined
	polychloroprene producers	lung	chloroprene
	dimethylsulphate producers	lung	dimethylsulphate
	epichlorohydrin producers	lung, lymphatic and haematopoietic system (leukaemia)	epichlorohydrin
	ethylene oxide producers	lymphatic and haematopoietic system (leukaemia), stomach	ethylene oxide
	ethylene dibromide producers	digestive system	ethylene dibromide
	flame retardant and plasticiser users	skin (melanoma)	polychlorinated biphenyls
	styrene and polystyrene producers	lymphatic and haematopoietic system (leukaemia)	styrene
	ortho- and *para-*toluidine producers	bladder	*ortho/para-*toluidine
	benzoylchloride producers	lung	benzoylchloride
	magenta producers	bladder	aniline, *o-*toluidine
Pesticides and herbicides production industry	tetrachlorodibenzodioxin producers and those exposed after accidents	lung, stomach	DCDD and TCDD dichloridibenzodioxin, trichlorodibenzodioxin
Rubber industry	rubber manufacturing	lymphopoietic system, stomach, brain, pancreas	undefined
	processors, composers, cementing synthetic plant	stomach	undefined
	general service	lymphatic and haematopoietic system (leukaemia), lymphatic and haemopoietic tissue	undefined
	synthetic latex producers and tyre curing	lung	undefined
	calender operatives and reclaim	prostate, lung	undefined
	compounding, mixing and calendering	prostate	undefined
	styrene butadione rubber producers	lymphatic and haematopoietic system (lymphomas)	styrene
	pliofilm producers	lymphatic and haematopoietic system (leukaemia)	benzene
	rubber compounding, extruding, milling	stomach	undefined
	tyre assembly	skin	mineral extender oil
		brain	undefined
Construction industry	insulators and pipe coverers	larynx, gastrointestinal tract	asbestos
Printing industry	rotogravure workers, binders	marrow (leukaemia)	benzene
	printing pressmen	buccal cavity, rectum, pancreas, lung, prostate, kidney	oil mist, solvents, dyes, cadmium, lead

Industry	Occupation	Site	Reported or suspected causative agent
	newspaper pressmen	buccal cavity	oil mist, solvents, dyes, cadmium, lead
	commercial pressmen	pancreas, rectum	oil mist, solvents, dyes, cadmium, lead
	compositors	multiple myeloma	solvents
	machine room workers	lung	oil mist
Leather industry	tanners and processors	bladder, nasal, lung	leather dust, other chemicals, chromium
	leather workers, unspecified	nose, larynx, lung, bladder, lymphatic and haematopoietic system (lymphomas)	undefined
	boot and shoe manufacturers and repairers	buccal cavity	undefined
	other leather goods manufacturers	marrow (leukaemia)	benzene
Textile industry	cotton and wool textile workers	mouth, pharynx	cotton and wool dust
Wood pulp and paper industry	lumbermen and sawmill workers	nose, Hodgkins lymphoma	wood dust, chlorophenols
	pulp and papermill workers	lymphopoietic tissue	undefined
	carpenters, joiners	nose, Hodgkins	wood dust, solvents
	wood workers, unspecified	lymphomas	undefined
Other	radium dial workers	breast	radon
	laundry and dry cleaners	lung, skin, cervix uteri	tritetrachloroethylene and carbon tetrachloride
	roofers, asphalt workers	mouth, pharynx, larynx, oesophagus, stomach	benzo(a)pyrene, other pitch volatile agents

[1] See also table 3 of the article CARCINOGENIC SUBSTANCES.

presence of a carcinogenic risk. They must *not* be regarded as the occupational groups in which such a carcinogenic exposure, likely of larger extent, might occur.

The size of the problem. In the most recent years attempts have been made to measure the magnitude of occupationally induced cancers. These estimates, based on the epidemiological data today available, show a remarkable variability ranging, according to different scientists, from less than 1 to as much as 40% of all cancers, most of the less unreliable estimates being between 1 and 10%. A single summary estimate of the fraction of cancers due to occupation appears consequently to be affected by a considerable uncertainty. The proportion of occupational cancers is in fact a relative measure, which depends on the effect of any other carcinogenic factor (e.g. smoking, alcohol, drugs, diet, etc.) and is therefore specific to a given population, in a particular country and in a specific time (see CANCER, ENVIRONMENTAL). Considering that occupational exposures are very different from country to country and often within the same country, any overall estimate of occupationally related cancers appears to be incompletely informative. It would thus be preferable to avoid a full identification between "size of the problem" and "proportion of cancers attributable to occupational exposure". Furthermore, only a few among the developed countries can provide reliable data at the present time, while the geography of the occupational cancer risk has been changing in the most recent years. There is a tendency to transfer industrial production and use of well known or suspected carcinogens from countries in which regulations have been established for carcinogenic substances, to developing countries where environmental standards and actual controls for carcinogens either do not yet exist or are ineffective.

Some methodological problems. One of the major problems in establishing a cause-effect relationship between an excess of cancer in a group of workers and their occupational exposure is to determine the role played by other non-occupational factors. This is mainly due to the frequently poor information on levels of exposure in industry and to the unreliability or entire absence of information, as in historical cohort studies on voluptuary habits. When sufficient information has been available, it has been possible to document in certain situations a synergism between occupational and non-occupational carcinogenic factors. For example asbestos and smoking have been shown to have a synergistic (multiplicative) effect in the production of lung cancer, although each of the agents is also independently capable of causing the disease. Also, epidemiological research is dealing more and more with "border-line" situations involving low exposure to a known carcinogen or exposure to a "weak" carcinogen, or, finally, exposure to a non-carcinogenic compound that for some reasons has come under suspicion. In these situations epidemiological methods may not be sensitive enough to analyse the separate effect of different carcinogens and therefore to establish with persuasiveness a causal association with occupational exposure.

Dose-response as measurable in human populations. Generally, when data on occupational exposure levels are available, the risk of developing a cancer appears to increase with the level of exposure. It has been possible to measure this pattern, for example in asbestos exposure, which, particularly in the past, presented very high levels. In other situations the results are not clear-cut, depending on the interplay with other variables like time since first exposure, length of exposure and "potency" of the carcinogen. It seems, however, acceptable that the lower the level of exposure, the lower should be the risk of developing a cancer. The problem arises when one tries to determine a "safe" level of exposure to a carcinogen.

If we assume that occupational cancer, as any other cancer, is the result of a multistage process in which more factors, both exogenous and individual, are involved, the possibility of determining a dose of a carcinogen so low as to exclude a risk for the exposed working population appears to be only theoretical at the present time.

Diagnosis

Occupational history. As previously mentioned, occupational cancer does not differ clinically and histopathologically from other cancers, although some histological types seem occasionally predominant, e.g. oat cell carcinoma for lung cancer. Therefore, the main diagnostic assessment remains a detailed occupational history of the cancer patient. The first step is to investigate whether the patient has been employed in the past in a job in which exposure to a carcinogen could occur. If at all possible, this information should be collected by interviewing the patient himself. In this case it should be remembered that, unfortunately, the worker is rarely aware of the chemical compounds present in the occupational setting in which he is or was employed. Another source of information are factory records in which, particularly for large firms, detailed information is usually available on the different jobs within the factory in which the worker has been employed.

Other sources of information which have been successfully used are health insurance records and trade union nominal rolls (two sources of information very often used in epidemiological studies). Even if the presence of a past occupational exposure to a recognised carcinogen does not in itself prove, in an individual cancer patient, a causal relationship, the detailed information on the exposure (levels, length of exposure, time since first exposure, etc.) can provide convincing evidence of its aetiological role, e.g. for compensation purposes.

Questionnaires. Interviewing a subject on his occupational history presents specific problems which merit brief attention. First of all, the most important part of the interview concerns–as opposed to the great majority of clinical histories–the remote past with the consequence that short periods of exposure can easily be forgotten. As already pointed out, another major problem arises from the fact that workers are rarely informed about chemical substances occurring in the workplace. Finally, it should be emphasised that a poor background of occupational medicine in a general clinician can easily lead to an incomplete occupational history.

The routine use of standard questionnaires for occupational history could, at least in part, confront some of these problems. Efforts should also be made to develop a standard classification and coding system for occupations based on known or possible exposures instead of social status or general industrial branch, as are the great majority of occupational codes which are available at the present time.

Screening. It is not the purpose of this section to review in detail the criteria for assessing the validity of a screening programme. A screening programme aimed at early diagnosis of a cancer (occupational or not) can only be regarded as beneficial if its application reduces the mortality in the screened population group with respect to a comparable group which was not screened. There is no evidence that screening programmes for occupationally induced cancers (e.g. lung cancer, bladder cancer, liver cancer, etc.), if evaluated against mortality reduction yardsticks, are at all beneficial, with a possible exception for skin cancer (screening by inspection). This sobering consideration should be borne in mind before embarking on the development of any screening programme for occupational cancer which may subtract valuable resources from other more effective forms of cancer control in the workplace.

Prevention

The most effective measure to prevent occupational cancer is undoubtedly to prohibit the presence of carcinogenic substances in industrial processes. Unfortunately, only a few countries have legal restrictions on the production and use of carcinogenic substances and there are often discrepancies from country to country. See CANCER, OCCUPATIONAL (LEGISLATION). Together with the removal of carcinogens from industrial processes, industrial research on non-carcinogenic substances to be used as substitutes should be developed.

A second major option is to eliminate the contacts between workers and carcinogenic substances when present in the workplace. This includes three aspects:

(a) production and transportation of carcinogens in closed systems;

(b) control of the working environment by monitoring levels of exposure and installation of air-conditioning systems effective even in cases of emergency;

(c) personal protective equipment for those workers at a higher risk of coming into contact with carcinogenic substances.

A major point concerning prevention is the need for national and international regulations on the production, use and importation of carcinogenic substances, including waste disposal procedures and control in order to avoid contamination of the general environment by active carcinogenic compounds.

As previously noted, the definition of a threshold "safe" level of exposure to a carcinogen seems far from having scientific solidity in the present state of knowledge on carcinogenic mechanisms. In the absence of other preventive measures one should take into account the simple fact that any appreciable reduction in the levels of exposure to a carcinogen goes in the direction of diminishing the risk for the exposed population.

Finally, there are other preventive measures which are in principle possible, based on preselection of workers according to personal characteristics of "susceptibility" or on frequent rotation of workers exposed to carcinogens. For the time being, most of these measures stand on uncertain scientific grounds, and they are likely to be socially controversial as well as difficult to apply, so that they cannot be regarded as general practical propositions. As a general beneficial measure, in itself and in view of the possible potentiating effect on occupational carcinogens, stopping smoking should, however, be recommended.

SIMONATO, L.
SARACCI, R.

IARC monographs on the evaluation of carcinogenic risk of chemicals to humans. Vols. 1 to 26 (Lyons, International Agency for Research on Cancer, 1972-1981).

"Chemical carcinogenesis". *British Medical Bulletin* (London), 1980, 36/1.

"Chemical agents and occupational cancer". *Journal of Environmental Pathology and Toxicology* (Park Forest South, Illinois), 1980, 3 (399-417).

Cancer, occupational (legislation)

Most industrialised countries have general laws dealing with toxic substances, but in only a few of these does legislation include special provisions for carcinogens. The first country to adopt regulations prohibiting the manufacture of certain chemicals specifically because of their carcinogenicity was the United Kingdom with regard to aromatic amines. Following unsuccessful attempts in the 1930s at its inclusion in the Workmen's Compensation Act, bladder cancer due to exposure to aromatic amines was included in the 1962 Prescribed Industrial Diseases Regulation. Finally, in 1967 specific legislation prohibiting the manufacture of 2-naphthylamine, benzidine, 4-aminodiphenyl, 4-nitrobiphenyl and their salts was enacted. It is worth noting that in 1921 the ILO had already recognised that 2-naphthylamine and benzidine were, on epidemiological evidence, causes of bladder cancer in humans. A few industrialised countries followed the example of the United Kingdom, namely Ireland, Japan, the USSR and the United States, but only the first four of these give consideration to banning the importation of carcinogens whose manufacture is prohibited.

It appears that all existing laws regulating exposure to carcinogens were formulated in the 1960s and that to date no new laws have been adopted thereafter. However, many countries, including the member States of the EEC, are at present preparing new laws which, in some cases, are at an advanced stage of formulation. It must also be noted that many nations, in the absence of specific legislation, have put various temporary mechanisms into effect following a general trend to avoid, if possible, the use of or reduce exposure to carcinogenic substances. Examples of such mechanisms are the code of practice regarding vinyl chloride in the United Kingdom and a contract concerning working conditions between the unions and the Association of Industries in Italy. In addition, various countries have dealt with specific problems by ad hoc regulations; such is the case, for instance, of the laws regulating the use of asbestos in Denmark and Belgium and that of propane sultone in the Netherlands. Most States recognise a limited right to compensation for occupational cancer.

From a survey of existing legislation and regulations dealing with exposure to chemical carcinogens at the workplace, the following observations can be made:

(a) only a limited number of countries have legislation dealing specifically with chemical carcinogens;

(b) legislation and/or regulations do not cover the same chemicals in each country. This is true even for the few countries in which legislation dealing specifically with chemical carcinogens exists;

(c) among the chemicals whose manufacture is prohibited or regulated, there are some which are no longer in use (such as 4-aminobiphenyl), or have probably never been used (like acetylaminofluorene), while compounds or industrial processes which have been proved to be carcinogenic to humans are not included;

(d) the criteria for selecting chemicals, for which the evidence of carcinogenicity relies on experimental results only, as possible human carcinogenic hazards have never been clearly formulated and are inconsistent;

(e) some of the chemicals for which the possible human hazard is recognised on the basis of experimental results have concentration limits which appear to be exceedingly high.

Table 1. Chemicals for which in 1980 the Council of European Communities proposed a directive on the protection of workers

Acrylonitrile
Arsenic and arsenic compounds
Asbestos
Benzene
Cadmium and cadmium compounds
Chlorinated hydrocarbons:
carbon tetrachloride
chloroform
para-dichlorobenzene
Lead and lead compounds
Mercury and mercury compounds
Nickel and nickel compounds

In 1974 the International Labour Conference adopted the Occupational Cancer Convention, 1974 (No. 139), Article 1 of which states: "Each member which ratifies the Convention shall periodically determine the carcinogenic substances and agents to which occupational exposure shall be prohibited or made subject to authorisation or control, and those to which other provisions of this Convention shall apply." Other provisions deal with the following essential principles of prevention: replacement of carcinogenic substances by less dangerous ones, recording of data concerning exposure and exposed workers, medical surveillance, information and education. To date the Convention has been ratified by 16 member States. The Convention, and Recommendation No. 147, which supplements it, do not specify a list of carcinogenic substances or agents to which protective measures shall apply. However, an ILO Panel of Consultants on Occupational Cancer (1976-81) drew up an indicative list of carcinogenic substances and agents to provide guidance for the implementation of the principles set forth in the ILO instruments (see table 3).

At present, most countries are preparing new laws to control possible hazards in the workplace and the EEC is trying to bring about a unified approach within its member States, with the intention of transforming all existing laws and regulations into a single, internationally valid law. More specifically, it foresees the notification of all new substances to be placed on the market in the member States, which should include information on short-term and delayed effects. Within this proposal, the carcinogenic effect is not singled out specifically, since it is listed among the various characteristics which render a substance "dangerous"; carci-

Table 2. Carcinogens for which regulatory standards were established in the United States in 1974[1]

4-Aminobiphenyl
4-Nitrobiphenyl
1-Naphthylamine
2-Naphthylamine
4,4'-Methylene bis-(2-dichloroaniline)
Methyl (chloromethyl) ether
bis (Chloromethyl) ether
Benzidine
3,3'-Dichlorobenzidine and its salts
Dimethylnitrosamine
Ethyleneimine
2-Acetylaminofluorene
4-Dimethylaminoazobenzene

[1] Ad hoc regulations have been adopted for three further substances, or groups of substances: arsenic compounds (1978), asbestos (1976) and benzene (1978).

Table 3. ILO Panel of Consultants on Occupational Cancer: indicative list[1] of carcinogenic substances[2] and agents

Group 1 – Contact should be avoided
- 2-naphthylamine
- nitrosamines (dialkyl)
- benzidine
- 4-aminodiphenyl
- 2-acetylaminofluorene
- 2-nitronaphthylamine
- 4-dimethylaminoazobenzene
- 4-nitrodiphenyl
- methylnitrosourea (MNU)
- bis(chloromethyl) ether

Group 2 – Exposure should be limited through the application of stringent protective measures
- 1-naphthylamine[3]
- propane sultone
- asbestos
- vinyl chloride
- ionising radiation and radioactive substances
- methylchloromethyl ether[3]
- diazomethane
- 1,1-dimethylhydrazine
- benzene
- β-propiolactone

Group 3 – Exposure should be kept to a minimum through the use of the most feasible and applicable controls
- inorganic arsenic
- nickel carbonyl
- 4,4'-methylene-*bis-o*-chloroaniline (MOCA)
- dimethyl sulphate
- 3,3'-dichlorobenzidine
- *o*-toluidine
- dianisidine
- ethyleneimine
- ethylene thiourea

Materials of complex composition[4] whose use represents a significant carcinogenic risk: Exposure should be kept at a minimum through the use of technical and personal protective measures
- coal tar
- high boiling petroleum residues
- cutting mineral oils
- shale oil
- creosote oil
- coal pitch
- soot

Industrial processes involving significant carcinogenic risk: Exposure should be kept to a minimum through the use of the most feasible and applicable controls
- treatment of chromium ores
- treatment of nickel ores
- auramine manufacture
- magenta manufacture
- hematite mining
- coke-oven operations
- manufacture of isopropyl alcohol
- pressing of paraffin wax from petroleum
- use of antioxidants and accelerators in the rubber and cablemaking industry

[1] A number of national lists of toxic substances for which exposure limits are prescribed make reference to carcinogenic substances. The lists will be found together with complementary information in *Occupational exposure limits for airborne toxic substances.* Occupational Safety and Health Series No. 37 (Geneva, ILO, 1977). [2] Use caution concerning any derivatives of substances possessing a carcinogenic risk. Although some are considered to be non-carcinogenic, such as the sulphonated derivatives of aromatic amines, extreme care should be exercised until results are demonstrated. [3] With these compounds, as with many others, there is difficulty in determining whether the basic chemical or its impurities (or both) is or are the active agent(s). Until such information is available, both the chemical and its contaminant(s), i.e. the mixture, must be considered to possess carcinogenic risk. [4] Using common terminology. All of these materials are known to have caused cancer in man.

nogenicity will thus enter into the over-all regulation against harmful effects. It must be noted, however, that substances already on the market will not come under this regulation, since some of them are already considered, or will be considered, under individual directives. In 1980 the Council of the European Community issued a directive on the protection of workers from the risks related to exposure to chemical agents at work; the list of chemicals considered in the directive is given in table 1.

In the United States regulatory standards were established in 1974 for 13 carcinogens, which are listed in table 2. In addition, ad hoc regulations have been adopted for the following substances, which are recognised as being human carcinogens (the dates of the latest deliberations of the US Occupational Safety and Health Administration (OSHA) are given in parentheses): arsenic compounds (1978); asbestos (1976); benzene (1978). The Toxic Substances Act was adopted in October 1976 in order to facilitate the control of toxic substances in the human environment. This Act is at present undergoing several amendments, but basically will require the notification of all new substances before marketing.

TOMATIS, L.

"Legislation concerning chemical carcinogens in several industrialized countries". Montesano, R.; Tomatis, L. *Cancer Research* (New York), Jan. 1977, 37/1 (310-316). 27 ref.

IARC monographs on the evaluation of the carcinogenic risk of chemicals to humans. Vols. 1-20, Supplement 1: *Chemicals and industrial processes associated with cancer in humans* (Lyons, International Agency for Research on Cancer, Sep. 1979).

"An evaluation of chemicals and industrial processes associated with cancer in humans based on human and animal data: IARC Monographs volumes 1-20". IARC Working Group. *Cancer Research* (New York), Jan. 1980, 40/1 (1-12). Ref.

Occupational cancer – Prevention and control. Occupational Safety and Health Series 39 (Geneva, International Labour Office, 1979), 36 p.

Cancer, occupational (statistics and registration)

There is increasing awareness, not only on the part of the individual worker and his trade union but also on that of industry and government, that some exposures at work may increase the risk of cancer. As many of the workplace exposures demonstrated or alleged to be carcinogenic may also affect the general public to a greater or lesser degree, the man in the street may also be involved. The employee is worried about his own safety, industry about the cost of any preventive control measures that may be needed and government about the health of the citizen and the competitiveness of national manufacturing industry.

While the proportion of all cancers due to exposures at work is small, these cancers are preventable. In the past industrial and occupational risks have been identified by the alert clinician (e.g. adenocarcinoma of the sinuses and nasal passages in part of the furniture and leather industries, liver tumours in vineyard workers spraying arsenic-containing solutions, liver tumours in vinyl chloride kettle cleaners and leukaemia in artisan boot and shoe repairers using benzene-containing solvents), it is very unlikely that this partly intuitive clinical approach, essentially based on the recognition of the coincidence of unusual exposure and rare tumour, could uncover more than a fraction of all risks. This being so, it is necessary to develop methods that will reveal cancer due to exposure at work systematically.

Detection and quantification of risk

There are several standard methods for assessing occupational cancer risk—each with its advantages and limitations. These fall into three broad groups, the first of which is suitable for routine exploratory studies. The second and third have not hitherto usually been used until there was a suggestion of increased risk.

Group correlation. This involves sytematic, or ad hoc, linkage of data already collected routinely, such as linkage of the statements of occupation on census records with the information on occupation and cause of death given on the death certificate, to determine whether the mortality for a given occupation is significantly higher or lower than the national average.

Cohort. In this form of investigation the number of cancers which occur in those working in a given workplace or occupation over a defined period of time is assessed and compared with the numbers that would have been expected if these individuals had experienced the national average risk.

Case-control. Individuals with a given cancer are asked questions, among others, about occupation and possibly workplace exposure; similar questions are posed to persons, usually of the same age and sex, without this cancer. The results are compared to see whether there are more cancer cases with a particular occupation or exposure than controls.

The first category of study deals with the risk of an occupational group as a whole, not with the individuals in the group; the second and third assess the experience of a series of individuals. Before looking at these methods more closely it is essential that the reader, when considering occupational risk, constantly reminds himself of the following facts:

(1) Occupation is much more than a series of exposures at work. Occupation determines income and income to a major degree influences place of residence, the type and quantity of food taken, choice of friends and of leisure activities, and frequently personal habits such as alcohol and tobacco consumption, factors which themselves often influence the risk of cancer. Demonstration of an increased risk in an occupation does not necessarily mean that the only cause was in the workplace.

(2) While an occupation is frequently accompanied by a common set of exposures, this is not inevitably so. A plumber employed by chemical industry may have a different series of exposures to one dealing with the plumbing of buildings. The nature of the exposures associated with a particular occupation may change over time and levels of exposure may vary. Many occupations may share several exposures.

(3) Although there are exceptions (vinyl chloride monomer for example), it usually requires 20 or more years' exposure before the effect of any carcinogen, occupational or otherwise, can be detected. This implies that it is of little value to correlate the exposures of today and current cancer patterns, unless the nature of the exposures can be assumed to be little changed.

(4) Not all persons exposed to a carcinogen will develop cancer in their lifetime. It is well known that only one in ten or so heavy smokers die from lung cancer, although many more die from heart and lung disease. Thus carcinogenic exposures do not inevitably result in cancer—however, they increase the probability of contracting cancer.

(5) Apart from the example of unusual microscopical types of cancer quoted in the second paragraph of the introduction, cancers caused by workplace exposure are not readily distinguished from those provoked by other agents. Exposure to the solvent *bis*-chloromethyl ether increases the risk of lung cancer; so does cigarette smoking. Skin cancer may be caused by cutting oil, and by sunlight. It may require several epidemiological studies to separate the effects of the two exposures.

Group correlations

Proportionate Mortality Ratio. In many countries the death certificate gives, in addition to the cause of death, the occupation of the deceased. If occupation has no influence on a particular cause of death then the proportion of persons dying with this disease should be the same in all occupations, e.g. 10% of all males in England and Wales should die from lung cancer whether they were schoolteachers, electricians or leather workers. If it were found that 18% of leather workers died from this particular cancer the existence of an occupational hazard in the leather industry could be suspected. In practice, the average risk of death due to a particular cause for the whole population is considered to be 100 and risks expressed as Proportionate Mortality Ratio (PMR). Thus schoolteachers with a PMR of 32 clearly appear to have less cancer than the national average, steelworkers and bricklayers with PMRs of 126 and 133 respectively much more. When calculating the PMR account should be taken of the age-structure of those employed in various occupations. The risk for nearly all forms of cancer increases with age and, should a particular occupation have either an excess or a deficit of older employees, unless care be taken to adjust for age, biased results may occur.

The advantage of the PMR is that a knowledge of the distribution of or numbers following a particular occupation is not needed. The disadvantage lies in the fact that if the proportion of one disease in an occupation is high the proportion of all other diseases, as for such analyses one can only die from one cause, will be automatically reduced: a high value could thus be because the disease is common in the occupation in question or because other major causes were uncommon.

Standardised Mortality Ratio. When the numbers employed in the various occupational groups are known, it becomes possible to calculate death rates for each group separately. Eventually everybody dies, no matter what the occupation, and in this sense the death rate for any occupation is 100%. However, the ages at which people die and their causes of death vary: it is usual to express the national rate as 100 and to compare the experience of each occupational group against this as a Standardised Mortality Ratio, or SMR, which takes account of the number and age distribution of the occupational group in question.

It is usual to consider the age-span 15-64 years when calculating the SMR for the following reasons: *(a)* these are the ages of employment and a usable statement of occupation for any death occurring before 65 is likely to appear on the death certificate; after 65 "retired" may be the only information given; *(b)* although the long latent period for cancer makes it likely that some cancers due to occupational exposures will occur after retirement, and in England and Wales (and elsewhere) analyses for the age-group 65-74 have been undertaken, they are often held to be less reliable.

Bias. The statement of occupation on the death certificate is not always reliable and the accuracy often depends on who gives the information. Hospital staff or a relative probably have a less exact idea of what a person did than the individual himself or the employer. Some occupations, such as coal miner, have a certain prestige in the community and may be reported even if this was not the person's last occupation. Fishermen and seamen absent from home at census time may be undercounted in the census, but fully represented on death certificates, giving a false impression of a very high mortality. A bias in the same direction is also noted for occupations such as firemen, policemen and aircraft pilots whose early retirement results in the under-reporting of these occupations at the census in comparison with death registration.

Information on employment obtained at the time of national censuses generally relates to current full-time occupation, which is not necessarily that followed for all or most of the working life.

The classification of occupations used in the United Kingdom is not identical with the International Standard Classification of Occupations (ISCO) of the ILO which generally permits more detailed coding. Frequently countries use their own adaptation of this classification. The traditional classification may not be appropriate to bring out new risks and may be slow to recognise new technologies. Thus insulation workers, known to have a high risk of lung cancer and pleural mesothelioma by virtue of exposure to asbestos, are included in the UK classification in the large group "Construction workers nec" and the very large increase in risk could well remain undetectable. The ISCO classification provides for separate coding for this group.

Some results. As noted above the computation of SMRs depends on linking at a given time the number of persons in an occupation dying from a particular disease and the number of persons then following that occupation. This technique has been used in England and Wales since 1911, the information about the deceased's occupation being derived from the death certificate and for the population from census data. Analyses are performed for a period of years, usually three (the year of the census, the year before and the year after), so that numerator and denominator are as closely in step as possible. Three years are chosen to obtain sufficient deaths to calculate SMRs for smaller occupational groups and for the rarer causes of death.

There are often difficulties in defining jobs. The Classification of Occupations used in the United Kingdom on census schedules included 27 occupation orders, divided into 223 occupation units. Since 1921 occupations have been grouped into five social classes, but in 1951 a new system was developed whereby people were classified by socioeconomic group. This was obtained by cross-tabulation of occupation, industry, employment status and economic position. Despite these problems the technique has shown that membership of certain occupations is associated with greater or lesser risk for certain diseases, including particular cancers, than experienced by the population taken as a whole. A selection of these is given in table 1.

The risk for cancer of all sites varies enormously, apparently being three times commoner in "other labourers" than in schoolteachers. Much of the variation is due to stomach and lung tumours. There is a 25% difference in over-all mortality between farm owners and managers (002) and agricultural workers (003). The "all sites" SMR for chemical production workers (012) is 116 whereas it is 89 for professional chemists (204); electricians (027) with an SMR close to the national average have none the less much more cancer than professional electrical engineers (197) whose SMR is 44. While labourers in the engineering trade (108) have an SMR of 99, that for other miscellaneous labourers

Table 1. Standardised mortality ratios in males aged 15-64 in England and Wales 1970-72 for selected cancer sites and occupational units[1]

Occupational unit		Cancer site						
Code no.	Title	Stomach	Large bowel	Rectum	Lung	Bladder	Brain	All
(002)	Farmers, farm managers, market gardeners	85	109	92	57	48	110	78
(003)	Agricultural workers nec	91	136	112	105	71	95	103
(007)	Coal-mine workers underground	171	112	147	114	101	101	119
(012)	Chemical production process workers nec	150	104	116	118	115	107	116
(027)	Electricians	110	130	118	100	141	89	103
(039)	Machine tool operators	182	160	153	164	115	145	163
(055)	Carpenters and joiners	108	76	103	120	160	133	111
(093)	Bricklayers, tile setters	156	107	136	147	138	112	131
(100)	Painters, decorators nec	116	101	99	136	147	116	121
(108)	Labourers in engineering and allied trades	112	90	83	109	85	65	99
(114)	Other labourers	207	151	159	199	158	138	182
(122)	Drivers of road goods vehicles	132	92	103	145	119	95	126
(148)	Commercial travellers, manufacturers' agents	61	125	92	77	68[2]	104	85
(154)	Publicans, innkeepers	95	119	104[2]	153	135[2]	—	146
(177)	Managers[3]	66	79	79	60	96	115	68
(181)	Medical practitioners (qualified)	43[2]	92[2]	—	32	—	—	61
(193)	Primary and secondary schoolteachers	37	76	114	28	88	72[2]	57
(213)	Clergy, ministers, members of religious orders	29	102[2]	—	33	—	—	60

[1] Abstracted from *Occupational Mortality. The Registrar General's decennial supplement for England and Wales, 1970-72* (Office of Population Censuses and Surveys (OPCS), 1978) and from related microfiche tables. [2] Figures based on fewer than 20 deaths. [3] This unit includes managers in a variety of production industries such as mining, chemicals, glass, steel, paper, rubber, plastics, construction, painting and transport.

(114) is very high at 182. One might suspect this latter group to be over-represented on death certificates as it is likely persons already ill might be forced to take such work. The reasons for some of the raised SMRs are not apparent and require investigation.

It will be noted that there is very wide variation in the SMR for both lung and stomach cancer. Very little is known about the causes of the latter, whereas most lung cancers are due to smoking. In table 2 the SMR for lung cancer is compared with a smoking index for the occupation in question in 1972. This index, which takes account of age, indicates the current smoking habits of the occupation in relation to the national average which is considered to be 100.

Table 2. Relationship between lung cancer and smoking habits for males aged 15-64 in England and Wales in some occupation units

Occupational unit		Lung cancer SMR	Smoking index
Code no.	Title		
(002)	Farmers, farm managers, market gardeners	57	65
(007)	Coal-mine workers underground	114	131
(027)	Electricians	100	105
(039)	Machine tool operators	164	116
(177)	Managers in mining and production nec	60	67
(181)	Medical practitioners (qualified)	32	33

There is a concordance between the proportion smoking and the lung cancer risk. The very high risk in machine tool operators in relation to the amount smoked may be due to workplace exposures or to very heavy smoking in the past, the discrepancy suggesting to the epidemiologist that this might be an occupation worth future study.

The significance of a raised SMR

Coal-miners in England and Wales have a very high SMR for stomach cancer (171), and indeed for all cancers combined their risk is 19% above the national average (SMR: 119). As might be expected from the nature of their occupation, SMRs from chest diseases (e.g. pneumoconiosis) and accidents, 252 and 156 respectively, are also raised. The question arises whether the large excess risk of stomach cancer is due to exposures at work. The observation that there is a marked social class gradient for stomach cancer, in other words that a large proportion of those belonging to the same social class as coal-miners also have a raised SMR for this cancer, casts doubts on the occupational specificity of the association. Indeed when the various occupations within this social class are compared, coal-miners are at no more risk from stomach cancer than any other group within the social class, although their over-all mortality from all causes (SMR: 132) is considerably greater. Examination of the risk in miners' wives is most instructive in that they too have very high stomach cancer risks, although not working in coal-mines. This evidence suggests that it is the social class, and all that goes with it in terms of life-style (diet, personal habits, etc.) that is important, rather than the exposures at work. Such a conclusion is valid where women do not work or are not exposed to occupational hazards. While it may be true that occupational influences were not present in married women 20 years ago, it is not so today, when a considerable proportion of married women are in active employment. Indirect influences, such as way of life, are assumed to be the same for husband and wife, but there are nevertheless large sex differences in, for example, cigarette and alcohol consumption, which cause certain cancers. Fox and Adelstein summarise this type of evidence thus: "... surprisingly, only 12% of cancer variation appeared to be associated with work. For other causes, such as circulatory and respiratory diseases, the proportion was nearer 30 per cent."

Table 3. Standardised mortality ratios before and after standardisation for social class for men aged 15-64

	Occupation order	Cancer		All causes	
		Before	After	Before	After
I	Farmers, foresters, fishermen	92	92	91	91
II	Miners and quarrymen	120	105	144	133
III	Gas, coke and chemical makers	118	102	107	95
IV	Glass and ceramic makers	119	105	109	102
V	Furnace, forge, foundry, rolling mill workers	135	119	122	114
VI	Electrical and electronic workers	107	96	104	100
VII	Engineering and allied trades workers nec	113	100	104	97
VIII	Woodworkers	107	95	96	91
IX	Leather workers	107	94	114	107
X	Textile workers	101	88	110	101
XI	Clothing workers	97	86	103	96
XII	Food, drink and tobacco workers	125	111	110	103
XIII	Paper and printing workers	96	85	91	86
XIV	Makers of other products	92	81	84	77
XV	Construction workers	126	110	111	102
XVI	Painters and decorators	123	108	111	104
XVII	Drivers of stationary engines, cranes, etc.	105	93	103	96
XVIII	Labourers nec	133	102	141	104
XIX	Transport and communications workers	120	106	111	102
XX	Warehousemen, storekeepers, packers, bottlers	110	95	108	96
XXI	Clerical workers	87	97	99	102
XXII	Sales workers	89	104	90	100
XXIII	Service, sport and recreation workers	114	106	116	109
XXIV	Administrators and managers	74	94	73	90
XXV	Professional, technical workers, artists	72	92	75	94
	All men	*100*	*100*	*100*	*100*

It will have become apparent that as a method for the routine detection of increased risk in an occupation, let alone for determining whether the increase was due to exposure at work, the group correlation approach is full of pitfalls and uncertainties. The finer the division of the occupations the more likely associations are to be due to statistical chance; the broader the categories the greater the chance of missing an increased risk in a small group.

Any association found should none the less be examined for biological plausibility and if an association seems possible the hypothesis should be tested by a cohort or case-control study. Mistakes will inevitably occur. For example the analysis by the Office of Population Censuses and Surveys (OPCS) in the United Kingdom shows an increased risk of cancer of the nose and nasal sinuses in butchers. As an unsupported finding this would probably be dismissed as a chance association: once it is known that furniture workers exposed to wood dust also have a high risk of these cancers a common exposure–sawdust–seems a plausible explanation.

In summary, this approach is useful to confirm known or suspected associations and to suggest new ones.

Occupation and the cancer registry

Ideally one would like to know not only about the fatal cancers associated with an occupation but about all those diagnosed. For some sites of cancer, such as those arising in the oesophagus, there are very few survivors even after treatment and death rates give a very good indication of the importance of the disease. The outlook in lung and stomach cancer is also poor, whereas about half of those with a colon cancer will survive and, for many kinds of skin cancer, around 98%. For cancers in which treatment results in a considerable proportion of cures, incidence or the number of newly diagnosed cancers is clearly of much greater interest than mortality. Incidence statistics are collected by cancer registries. Unlike death statistics which are national in coverage, the population covered by a cancer registry may range from a nation such as the German Democratic Republic (13 million), to a region such as Birmingham and the West Midlands in England (5 million), to quite small areas such as Iceland (200 000). Data from reliable registries are published in monographs of the *Cancer Incidence in Five Continents* series. Of these, 46, or 74%, record information on occupation. However, unlike occupation which appears on all death certificates in many countries and which is one of a series of standard questions asked by the official registering death, the cancer registry has to depend on a series of persons to enter this item of information on the case notes, and eventually the cancer registry notification forms. This may be the hospital admission clerk, a ward sister or a medical officer and the information may well be missing. It may also be difficult if not impossible to obtain information on the numbers and ages of the various occupational groups for the district covered by the cancer registry. For these reasons there has been comparatively little use of the occupational information stored in cancer registry files. None the less, in Los Angeles such information has been systematically collected for several years and incidence rates by occupation computed.

Group and individual comparison

Occupational cancer analyses of the type conducted by OPCS are cross-sectional and do not take into account the fact that the occupation at the time of death may not have been that followed for most of the deceased's working life. Indeed the illness leading to death may have resulted in a change of occupation. It must also be stressed that the OPCS approach so far deals with the average experience of groups of persons, not with individuals, and that no attempt is made to link an individual's death certificate with *his* census returns.

The OPCS is now linking individuals in a 1% sample of the population at each decennial census to assess changes in employment. In Norway and Denmark annual statements of occupation are required for all employees and, as these are identified by the unique personal identification number of the employee, can be readily linked with cancer registry notifications and death certificates. This resource enables estimates to be made for an individual of the years-at-risk ensuing from each employment and in effect makes a cohort study (see below) of the entire workforce possible. In Sweden it is possible, subject to the approval of the Data Inspection Board, to link census data with death certificate data for individuals and hence assess the level of risk associated with a given employment, taking into account factors such as residence, income, education, etc., but publications using this resources have not yet appeared.

Cohort studies

In a cohort study a series (cohort) of *individuals* with a particular exposure, or set of exposures, is identified and each individual is followed to a predetermined date to see whether he develops cancer or some other disease. To detect a carcinogenic effect there must normally be at least 15 years exposure and it is therefore useful to identify the cohort from existing records and follow the members until the present. The cohort can also be characterised and followed from the present, but this implies a long wait for the results (see the discussion of monitoring below). In both approaches the number of cancers occurring in the cohort members is compared with the number that would be expected to occur in a national average population of the same size and age composition as the study cohort followed over the same period of time. Internal comparison groups are frequently used, the experience of, say, production workers, craftsmen, storemen and maintenance personnel being compared. The technical epidemiological problems inherent in cohort studies are discussed under EPIDEMIOLOGY: two are of pertinence here.

The first requirement is to be sure that if a member of the study cohort develops cancer this can be discovered. In practice the epidemiologist gives a list of the cohort members to the vital statistics office and/or the cancer registry, which should inform him if and when a cohort member dies or is reported to the cancer registry as having malignant disease. This implies that the vital statistics office and the cancer registry be given sufficient identifying information about each cohort member to make an unequivocal match in their records. Clearly, in countries where such linkages are impossible, due to confidentiality and other restrictions, such cohort studies cannot be done and carcinogens will remain undetected. Where death certificates are not stored in one place, as in the United States where they are currently held by each state, follow-up becomes much more complicated and expensive.

The second requirement is that all persons ever employed be followed even though they may have retired or taken other employment, again necessitating good records.

In the cohort approach, risk in a particular occupation or workplace is generally compared with that of the national average. While the choice of a comparison population is discussed under EPIDEMIOLOGY, those concerned with the health of the employed should possibly choose as a basis for comparison the experience of those occupations assigned to social classes I and II by the OPCS, i.e. the administrative, managerial and professional. This in turn implies that before the at-work or on-job portion of a cancer risk can be assessed it will be necessary to have a much better understanding of the other causes of human cancer.

Monitoring. There are industries which by their very nature are more likely than others to result in the exposure of the workforce to possibly hazardous substances. Under such circumstances it becomes highly desirable to identify not only each employee but also his work stations and the exposure associated with each work station. As processes and equipment change over time it is essential to keep a record of the exposure levels associated with a particular task. Several large companies have instituted such schemes, notably for petrochemicals. Such environmental measurements need to be linked to possible health effects including cancer: the most economical and effective way is currently being studied by computer simulation in several countries.

Such large-scale monitoring of exposure and health effect, in other words a continuing cohort study, is comparatively easy for large companies. However, even within very large companies the number of persons exposed to a particular chemical may be small and to obtain a valid estimate of the effect of exposure it may be necessary to cumulate the exposure of several companies. A similar problem arises for industries characterised by small production units. For reasons of commercial secrecy it becomes difficult for any one company to direct the study and under such circumstances an outside independent organisation, such as a university or research institute, should undertake the work. Independent investigation of a potential hazard is desirable as the results, particularly if negative, are less likely to be impugned by third parties. Thus the man-made mineral fibre industry in Europe requested the International Agency for Research on Cancer to carry out a multinational, multifactory study on the risk, if any, associated with exposure to these widely used substitutes for the carcinogenic insulating material asbestos.

Case-control studies

It will be recalled that in the case-control approach persons with a given cancer are interviewed and asked about their past exposures, including those at work, the answers being compared with those to the same questions posed to "normal" controls, who do not suffer from the cancer in question. This technique has been used for bladder and lung cancer but the results are usually disappointing in that the number of persons with a given cancer who follow a specific occupation is likely to be rather small at any one time.

However, in a region where there is a relatively high prevalence of the occupation under study the case-control approach may be rewarding. Extending this concept, if cases arising within a cohort study population (which by definition is likely to have considerable numbers in one occupation) and appropriate controls in the same cohort can be interviewed, this constitutes a powerful and effective method of analysis, as the effect of confounding factors such as smoking can be taken into account. Obviously cases arising in cohorts followed from some time in the past may have died and under such circumstances information may be limited to that obtainable from surviving relatives.

Industrial records

The foregoing discussion of cohort studies, and the long induction period for cancer, implies that records must be kept for at least 60 years for all employees *no matter how short the period of employment.* While employment for less than 6 months might be regarded as of little import, 6 months' exposure to asbestos in high concentration has been shown to increase risk of lung cancer and mesothelioma.

Management. From time to time some employee records may be removed from the main files for a particular purpose, e.g. compensation claims, frequent sickness or absenteeism or on retirement. It is such persons who are of the greatest interest from the point of view of health effects and in consequence a system must be devised to ensure that such records are not "lost" when needed. In this connection the old-fashioned bound ledger with serial entries of the name of each new recruit to the company, date of commencement, etc., is invaluable as this is much less likely to be lost or misplaced than an index card or file and can serve to check that other files are complete. For cohort studies changes of job station within a factory are of great importance. If such changes entail a change in salary they are usually noted, otherwise regrettably they are not.

Labour. Trade unions can contribute significantly by maintaining and conserving membership records. A recent study of mortality in Danish brewery workers was based entirely on union records. This particular study was technically elegant in that union members either worked in a brewery or bottled mineral water, and as only the former had access to a free beer ration these two groups served to control each other.

In many countries the pattern of employment is changing. Fewer gain their livelihood from heavy manufacturing industry, there being a substantial movement into what are termed "service industries", frequently composed of small dispersed units. Although exposures may be substantial and similar in, say, the hairdressing salon, the detection of occupational risk under such circumstances becomes much more difficult and is probably most easily accomplished by the trade union for the occupation in question.

Administrative lists. When a professional licence is required to exercise a particular activity, or where persons following a particular activity need to be registered, e.g. beauticians, the task is much easier. Employment of a seasonal nature associated with a high turnover of the workforce and employing a high proportion of migrant labour, frequently illegal, is likely to result in failure to detect risk.

It is likely that in the future industrial cancer risk will be increasingly detected using records and unless these are meticulously preserved carcinogens will remain unrecognised.

The role of the occupational physician

It is the occupational physician with a knowledge of possible hazards and an insight into the records systems of the company who can best exert pressure on management to have the necessary records maintained. An increase in lung cancer risk is a not uncommon finding from industrial studies. To assess the significance of this finding it is necessary to have knowledge of other factors, notably smoking, which could give rise to the same cancer. The wise occupational physician will have made sure that his medical records for each employee contain information on smoking and alcohol consumption, so that such questions can be answered rapidly and synergistic effects, if any, detected. The solution of the Swedish building trades is worth careful examination. Under this scheme management and labour contribute to a joint nation-wide environmental health service, which undertakes the measurement of potential hazards in the workplace and provides a health service. Such arrangements for industries with a widely dispersed workforce are of capital importance in that users of chemicals and insulating material are frequently much more heavily exposed than those producing them, where the processes may be largely closed.

Comment

There is as yet no universally applicable routine method for the detection of occupational risk or for the assessment of whether a raised risk in an occupation is due to exposure at work. The systematic continued monitoring of cohorts now technically feasible in several Nordic countries, although promising, has yet to prove its worth. The detection and prevention of occupation-induced cancer will thus continue to need collaboration of both sides of industry, vital statistics offices, cancer registries, the factory physician and the epidemiologist. The key element for such work is the linkage of an individual's exposures at work and elsewhere to their health consequences, which implies the maintenance and preservation of good linkable records. Despite this, in several countries confidentiality, medical and industrial, will continue to make the detection of carcinogenic workplace exposures impossible. Under such circumstances whom does confidentiality benefit–the exposer or the exposed?

MUIR, C. S.
DEMARET, E.

Cancer incidence in five continents. Vol. III. Waterhouse, J.; Muir, C.; Correa, P.; Powell, J.; Davis, W. (eds.). IARC scientific publications No. 15 (Lyons, International Agency for Research on Cancer, 1976), 584 p.

"A critique of the standardized mortality ratio". Gaffey, W. R. *Journal of Occupational Medicine* (Chicago), Mar. 1976, 18/3 (157-160). 8 ref.

"Cancer epidemiologic surveillance in the Du Pont company". Pell, S.; O'Berg, M. T.; Karrh, B. W. *Journal of Occupational Medicine* (Chicago), Nov. 1978, 20/11 (725-740). Illus. 16 ref.

"Strategies for the development of a coherent cancer statistics system". Waterhouse, J. *World Health Statistics Quarterly* (Geneva), 1980, 33/3 (185-196). 8 ref.

Candle manufacture

A candle comprises a wick surrounded by a compact mass of material which melts readily and is moderately combustible.

Raw materials. The candle body is made from wax and/or fatty acids, and the eventual use will determine the choice and blend of materials. The quality of the candle light depends mainly on the proportioning of the wick and body of the candle. Candles are now commonly made from: vegetable waxes such as carnauba, candella and japan wax; animal waxes such as beeswax, spermaceti and shellac; mineral waxes such as ozokerite, paraffin wax, ceresin and montan wax; refined waxes which have been chemically processed or synthesised; fatty acids such as stearic or palmitic acid and greases such as candle grease. These raw materials are blended, and opacifiers, hardeners, dyes and pigments may be added.

Wicks are usually made from braided cotton which is impregnated with chemicals such as borax, ammonium sulphate, ammonium nitrate or ammonium phosphate, to ensure it is sufficiently absorbant.

Manufacturing processes. The original tallow candle was made by home workers or in small handicraft workshops from melted animal fats and continues to be made in this way in many parts of the world. In the more industrialised countries, manufacture has been concentrated in factories and is largely mechanised. The common domestic candle is machine-made from a mixture of refined paraffin wax and stearic acid; however, the ornamental candles which may contain a wide combination of materials may be machine- or hand-made. Church candles and most high-quality candles are still hand-moulded.

The raw materials are first melted in steam-heated vats which are made of metal other than copper or iron due to possible reactions between the bath and the vessel. Agitation is essential to prevent localised overheating and decomposition leading, in particular, to the release of acrolein. After being melted down the wax is formed around the wick by drawing, dipping, casting or compression moulding.

Drawing is a continuous process where an endless wick is passed through vats containing molten wax; as the wax-coated wick leaves the vat, it passes through a die which regulates the size and gradually the candle is built up to the required diameter. Finally the candle is cut to the required length. In dip production, the wick is suspended overhead and repeatedly dipped into molten wax until the required diameter has been achieved. Casting is done in separate moulds or in semi- or fully automatic casting machines; a common hand-operated casting machine will produce some 3 000 candles a day. Compression moulding using hydraulic processes and powdered wax is often employed for night lights or heating candles. Candles can also be extruded, cut to length and milled to shape.

HAZARDS AND THEIR PREVENTION

The main hazards in candle production are moving machinery, burns, falls on slippery floors and fires; skin affections and nasal irritation occur occasionally. There are no occupational hazards which are particular to the candle industry.

Machinery such as presses, extruders and automatic casting machines must be fitted with adequate machinery guarding, where possible of the interlocking type. Gears, inrunning nips and shear points should be fenced and screens should be provided where there is a risk of projections of hot wax. Packaging machines also require a high standard of guarding and all machines should be halted before cleaning and maintenance is undertaken.

With the large quantities of molten wax employed it is inevitable that some spillage occurs and that floors become slippery. Every effort should be made to reduce spillage by the provision of screens and sills, and maintenance and good housekeeping should be of a

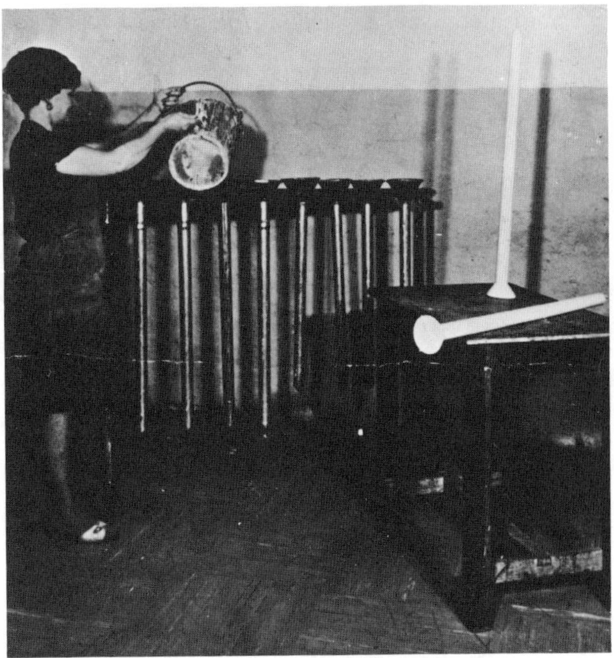

Figure 1. Hand casting of large candles.

Figure 2. Automatic candle-casting machine.

high standard to keep floors clean. Wooden floors and certain asphalt floor coverings have proved to be less slippery than other floor surfaces in this industry; concrete floors should be avoided since wax is difficult to remove from them. The wax itself is not a fire hazard at the temperature normally encountered in candle manufacture; however, suitable fire protection and prevention measures, including fire alarms, fire extinguishers and instruction in fire drills should be provided.

Molten wax is usually kept at a temperature of below 70 °C and hot-wax containers should be adequately guarded to avoid persons falling into them. On the other hand, splashes of heated wax do not usually lead to severe burns because of the relatively small amount of heat transferred.

Cuts and abrasions of the fingers and hands are common during the sawing and milling of candles and during the operation of casting machines, especially of the automatic type. Machines should be adequately guarded and workers should wear hand protection.

Vapours of paraffin are considered as non-toxic; however, they may be unpleasant and lead to some irritation of the airways. The ACGIH has recommended a TLV of 2 mg/m³ and a STEL of 6 mg/m³.

Overheated wax is likely to decompose with the release of acrolein vapours which may irritate the mucous membranes of the upper respiratory tract; where such a danger exists, wax boilers should be equipped with local exhaust ventilation and the workplaces should be of adequate height and provided with good general ventilation.

Prolonged contact with certain candle ingredients may produce skin disorders and, consequently, exposed workers should wear personal protective clothing including gloves and aprons.

TANNE, C.

Simple methods of candle manufacture. Industrial Liaison Unit of the Intermediate Technology Development Group (Intermediate Technology Publications Ltd., 9 King Street, London) (1978), 19 p. Illus.

Canning and food preserving

There are six basic methods of food preservation:

(a) heating;

(b) radiation sterilisation;

(c) antibiotic sterilisation;

(d) chemical action;

(e) dehydration;

(f) refrigeration (see also FROZEN FOOD INDUSTRY).

Briefly, the first three methods destroy microbial life whilst the latter merely inhibit growth.

Raw ingredients such as fish and meat, fruit or vegetables are taken fresh and preserved by one of the above methods or a mixture of different foods are processed to form a product or dish which is then preserved. Such products include soups, meat dishes and puddings.

Food preservation goes back to the last Ice Age, about 15 000 BC, when Cro-Magnon man discovered for the first time a way of preserving his food by smoking it. The evidence for this lies in the caves at Les Eyzies in the Dordogne in France where his way of life is so well portrayed in carvings, engravings and paintings. From then to the present day, although many methods have been used and still are, heat remains one of the principal cornerstones of food preservation.

Canning industry. The conventional method of canning is based on the original work of Appert in France for which in 1810 the French Government awarded him a prize of 12 000 francs. He preserved food in glass containers and it was in Dartford, England, in 1812 that Donkin and Hall set up the first cannery using tinned iron containers.

Today the world uses 6 million tonnes of tinplate annually for the canning industry; additionally, a substantial amount of preserved food is packed into glass jars. The process of canning consists of taking cleaned food, raw or partly cooked but not intentionally sterilised, and filling it into a can which is sealed with a lid—the can is then heated, usually by steam under pressure, to a temperature and for a period of time to allow penetration of the heat to the centre of the can, thereby destroying the microbial life. The can is then cooled in air or chlorinated water after which it is labelled and packed. Recent years have seen the introduction of continuous sterilisers which cause less damage to cans by impact and allow cooling and drying in a closed atmosphere.

Some parts of the world—particularly Japan—are now consuming foods that have been heat preserved in retortable pouches. These are bags of small cross-sectional area made from laminates of aluminium and heat-sealable plastics. The process is the same as for conventional canning but better organoleptic properties are claimed for the products because sterilisation times can be reduced. Very careful control of the retorting process is essential to avoid damage to the heat seals with subsequent bacteriological spoilage.

There have been recent developments in the aseptic packaging of food. The process is fundamentally different from conventional canning. In the aseptic method the food container and closure are separately sterilised and the filling and closing are done in a sterile atmosphere. Product quality is optimum because heat treatment of the foodstuff can be controlled precisely and is independent of the size or material of the container. It is likely that the method will become more widely used because over-all it should result in energy savings. To date most progress has been made with liquids and purées sterilised by the so-called HTST process in which the product is heated to a high temperature for a few seconds. Developments on particulate foodstuffs will follow. One likely benefit in food factories will be the reduction of noise if rigid metallic containers are replaced. Such containers may also cause problems by contaminating preserved food with lead and tin. These are minimised by new-type two-piece containers drawn from lacquered tinplate or three-piece containers with welded instead of soldered side seams.

Ionising radiation sterilisation. Ionising radiations offer hopes of providing better methods of food preservation to reduce wastage and spoilage. Some promising results have been achieved mainly using γ-emitting radio-isotopes such as cobalt-60.

"Radiation pasteurisation" using much lower doses (0.5 Mrad as compared with 4.8 Mrad necessary to kill off *Clostridium botulinum* in meat) enables the refrigerated shelf life of many foods to be considerably extended. The use, however, of radiation for sterilising canned foods requires such high radiation dosage that unacceptable flavours and odours result.

Ionising radiation has two other well recognised uses in the food industry—the radiological screening of food packs for foreign matter and the use of isotopes and X-rays for package monitoring to detect underfilling of packs.

Microwave sterilisation. Another type of electromagnetic emission that is currently finding use in the food industry is microwave energy. It is used for rapidly thawing raw frozen ingredients before further processing as well as for heating frozen cooked foods in 2-3 min. Such a method with its low moisture content loss preserves the appearance and flavour of the food.

Drying. Sun drying is man's oldest and most widely used method of food preservation. Today foodstuffs may be dried in air, superheated steam, in vacuum, in inert gas and by direct application of heat. Many types of dryers exist, the particular type being dependent on the nature of the material, the desired form of finished product, etc. Dehydration is a process in which heat is transferred into the water in the food, which is vaporised. The water vapour is then removed.

Freeze drying. The material to be dried is frozen, and placed in a sealed chamber. The chamber pressure is reduced and maintained at a value below 1 mmHg. Heat is applied to the material, the ice heats up and sublimes from the surface, the resultant water vapour being drawn off by the vacuum system. As the ice boundary recedes into the material, the ice sublimes *insitu* and percolates to the surface through the pore structure of the material.

Intermediate moisture foods. These are foodstuffs that contain relatively large amounts of water (5-30%) and yet do not support microbial growth. The technology, which is difficult, is a spin-off from space travel. Open shelf stability is achieved by suitable control of acidity, redox potential, humectants and preservatives. Most developments to date have been in foods for pet animals.

General processes. Whatever the preservation process, the food to be preserved has first to be prepared. Meat preservation involves a butchery department, fish needs cleaning and gutting, filleting, curing, etc. Before fruit and vegetables can be preserved they have to be washed, cleaned, blanched, perhaps graded, peeled, stalked, shelled and stoned. Many of the ingredients have to be chopped, sliced, minced or pressed.

A can-making department is a frequent adjunct to many canneries, where can bodies and lids are made from tin plates on power presses and, after cans or jars have been filled or sealed, they are labelled (unless the can itself proclaims its contents) and packaged for despatch.

In many places, fruit and vegetable preserving is very much a seasonal activity depending on the local crops, although this is often rectified by preserving imported fruits or vegetables in the off-season or by putting aside the local produce (in sulphur dioxide or quick deep freeze) for treatment later as convenient. Even so, large numbers of seasonal or migrant workers may be recruited at peak periods and hours of work may be long.

HAZARDS AND THEIR PREVENTION

Accidents. Where traditional methods of food preservation are still carried on, much manual effort is still required, bringing with it the usual hazards of manual work, the strains of heavy lifting and carrying of materials and equipment and handling accidents and injuries from falls of materials. Mechanisation removes some of these hazards but training in methods of lifting, safe stacking, maintenance and good housekeeping are also important.

Contamination of floors by oils, fats and water leads to falls of workers. Floors should have impervious surfaces and be well drained: attention to water-containment in the design of machines will reduce spillage. Regular and efficient cleansing is essential.

Burns and scalds from hot liquors and cooking equipment are common and similarly from steam and water used in equipment cleaning. Alkalis, acids and other chemicals used in cleaning may give rise to chemical burns of the skin and eyes. All steam equipment needs regular and careful maintenance to prevent major explosion or minor leaks. Refrigerating plant requires careful maintenance to provide against escape of toxic and/or explosive fumes.

Much of the machinery used in the preparation departments is dangerous unless efficient machinery guarding is provided and maintained. Filling and closing machines should be totally enclosed except for the intake and discharge openings. The intakes of conveyor belts and drums, pulleys and gearing should be securely protected. To prevent cuts, effective arrangements for clearing up sharp tin or broken glass are required.

Diseases. The commonest infections are occupational dermatoses, the primary dermatitis caused by irritants such as acids, alkalis, detergents and water used in cleaning; friction from fruit picking and packing and the handling of sugar much used in food manufacture. Secondary sensitisation results from the handling of many fruits and vegetables.

Some food handlers may be subject to a wide variety of skin infections including anthrax, actinomycosis, erysipeloid and tuberculous infections. Certain dried fruits are infested with mites which give trouble in sorting operations. Warts due to a virus in the fish slime are seen in fish filleters. Many infections of the skin also produce systemic effects. Many of the diseases of cattle may affect food handlers working with meat. In the most modern canneries, very little contact occurs between the worker and the materials handled after the initial unloading of the raw ingredients; most of the processes are totally enclosed and the traditional hazards are not encountered.

Apart from specific prophylactic vaccination against infectious diseases, good personal hygiene and the sanitary facilities to enable this, which are a prerequisite of any food industry as a protection to the product, are the most valuable preventive measures. Good washing facilities, including showers, and appropriate protective clothing are essential. Efficient medical care, especially for treatment of minor injuries, is an equally important requirement.

A serious health hazard in modern canning is exposure to noise. The manufacture, conveying and filling of cans at speeds of up to 1 000 per minute, leads to exposure of operators to a noise level of up to 100 dB at frequencies ranging from 500 to 4 000 Hz, an L_{eq} of about 96 dB(A) which if uncontrolled must lead in many cases to noise-induced deafness in a working lifetime. Certain engineering techniques undoubtedly lead to some noise reduction–these include sound-absorbent mounting, magnetic elevators, nylon-coated cables and speed-matching in can conveyor systems. Some radical change in the industry, such as the use of plastics containers, however, is the only hope for the future of producing a reasonably noise-free environment. At present, a hearing conservation programme based on hearing-protection equipment should be instituted.

The use of enclosures for noisy equipment in the food industry produces many hygiene problems and currently, in the absence of radical change such as plastic containers, the provision of noise refuges and personal ear protection are the only effective means of avoiding noise-induced hearing loss. The debate on the value of audiometry for such exposed workers continues and is summarised in the UK Health and Safety Executive discussion document *Audiometry in Industry*, which states: "Routine audiometric examination of groups of workers whose exposure does not exceed 85 dB(A) L_{eq} is not normally necessary. An audiometry programme should be instituted for all those working in a noisy environment of 105 dB(A) L_{eq} and above. As noise levels increase between these two values there is a corresponding increase in the desirability for the institution of an audiometric programme."

The canning industry also exposes workers to extremes of temperature. Exposure to cold can range from handling and storage of raw materials in winter or in "still air" cooled store rooms to extremes of cold in air-blast refrigeration of raw materials, similar to the ice cream and frozen foods industry.

Heat, often combined with high humidity, in cooking and sterilising can produce an equally intolerable physical environment so that heat stroke and heat exhaustion are not unknown. These conditions are found especially in processing which entails evaporation of solutions, such as tomato paste production, often in countries where hot conditions already prevail. Effective ventilation systems are essential with special attention to condensation problems. Air conditioning may be necessary in some areas.

Where ionising radiations are used, the full precautions applicable to the work, radiation protection, hazard monitoring, health screening and periodic medical examinations are necessary.

The traditional tin/lead soldering of the side seam of a food can and the awareness of the problem of lead levels in food products in the United Kingdom and elsewhere has resulted in studies of environmental lead levels in can-making units and blood lead levels in workers. Evidence has shown both to be raised but neither the environmental TLV nor the currently acceptable blood lead levels have ever been found to be exceeded. Thus the results are consistent with a "low risk lead process". The findings of this survey should exclude can-making using tin/lead solder from many of the requirements of the proposed new regulations and code of practice of the Health and Safety Commission in the United Kingdom.

GRAHAM, J. C.

Sterilisation:

Wholesomeness of irradiated food. Report of a joint FAO/IAEA/WHO Expert Committee. Technical Report Series No. 604 (Geneva, World Health Organisation, 1977), 44 p. 38 ref.

"Effects of exposure to microwaves–problems and perspectives". Michaelson, S. M. (133-156). Illus. 143 ref. *Environmental Health Perspectives* (Research Triangle Park), Vol. 8. Department of Health, Education, and Welfare (Washington, DC, Government Printing Office, 1974).

International acceptance of irradiated food. Legal aspects. Report of a joint FAO/IAEA/WHO advisory group. Wageningen, 28 Nov.-1 Dec. 1977. International Atomic Energy Agency, Legal Series No. 11 (Vienna, 1979), 70 p.

Occupational diseases:

CIS 78-855 *Workplaces and occupational disease in a meat preserving and curing plant, with reference to Finistère fish and vegetable preserving plants* (Postes de travail et pathologie professionnelle dans une conserverie de viande et salaison et extension aux conserveries de poissons et de légumes dans le Finistère). Guézénoc, L. (Université de Bordeaux II, Unités d'enseignement et de recherche des sciences médicales, Bordeaux) (1977), 110 p. 21 ref. (In French)

CIS 78-1450 *Working conditions and occupational disease in codfish drying establishments* (Les conditions de travail et la pathologie professionnelle dans les sécheries de morues). Locquet, J. (Université de Bordeaux II, Unités d'Enseignement et de Recherche des Sciences médicales, Bordeaux) (1977), 94 p. 15 ref. (In French)

CIS 77-1099 "Dermatitis in the fruit and vegetable canning industry" (La dermatitis en las industrias de conservas de vegetales). González Sánchez, A. *Salud y trabajo* (Madrid), Apr. 1977, No. 6 (12-19). Illus. (In Spanish)

Accidents:

CIS 75-247 "Vegetable canning plants" (Les conserveries de légumes). Conynck. *Prévention et sécurité du travail* (Lille), 3rd quarter 1973, No. 97 (16-25). Illus. (In French)

Caplan's syndrome

Rheumatoid pneumoconiosis (Caplan's syndrome) is a specific type of pneumoconiosis which occurs when a worker exposed to a dust hazard suffers from, or is liable to suffer from, rheumatoid disease.

Aetiology and incidence

Rheumatoid pneumoconiotic nodules develop in the lungs presumably as a result of a chemical reaction between the mineral dust in the lung and rheumatoid factor. The mechanism by which the dust foci become vulnerable is not yet understood.

Cases of Caplan's syndrome have been reported in many countries throughout the world. The majority have been found in coal workers, mainly from the United Kingdom, France, Belgium, Poland and Czechoslovakia; a few cases have been reported from the United States and Australia. The syndrome has also been recorded in workers exposed to free silica and mixed dust hazards in a wide variety of occupations in Europe, South Africa, Australia and the United States. Cases have been found in the pottery industry, brass and iron and steel foundries, gold mining, boiler scaling, sandblasting, brick and tile manufacturing, quarrying, limestone mining and metal polishing. Three cases have been reported in asbestos workers.

Pathology

Rheumatoid pneumoconiosis has consistent macroscopic and microscopic appearances and can be distinguished histologically from progressive massive fibrosis and classical silicosis. The lesions are at first discrete, measuring up to 1.5 cm in diameter, but in the majority of cases they conglomerate into larger masses which may occupy large areas of the lung tissue. Macroscopically, the lesions show a characteristic concentric arrangement of lighter and darker layers. The pale areas are grey in some instances and yellow in others. Liquefaction tends to occur in the pale areas leaving clefts.

It has been suggested that the histological criterion upon which diagnosis should be made is the presence of a central area of necrotic collagen, outside which is a zone of active inflammation consisting of a cellular infiltration of macrophages and frequently, also, of polymorphonuclear leucocytes. In this zone, collagen is being destroyed. Some of the macrophages in the inflammatory zone contain dust. When these macrophages die and disintegrate the dust is deposited and this accounts for the dark concentric rings seen in the nodules. The inflammatory zone may involve the whole or only part of the circumference of a nodule. Multinucleated giant cells are present in some instances, and these lie in a zone of fibroblasts outside the zone of inflammatory cells. The fibroblasts are orientated in a palisade manner. Outside the palisade is a zone of collagen arranged circumferentially and not necrotic.

Rheumatoid pneumoconiotic lesions have been studied by means of immunofluorescence and the main findings were that, at the periphery of the necrotic ring present in the outer zone, deposits of γ-globulin are regularly demonstrable.

Immunological studies have been carried out on miners without clinical evidence of rheumatoid disease but with *(a)* characteristic radiographic appearances of the rheumatoid syndrome, *(b)* nodular type of simple pneumoconiosis and *(c)* mixed nodular and irregular radiographic opacities. There was a high proportion of positive rheumatoid factor tests in all cases and particularly in the group with characteristic appearances.

Radiographic appearances

The radiographic appearances are characterised by the presence of multiple, well defined, round opacities 0.5-5 cm in diameter distributed throughout both lung fields (see figure 1). In many cases the background of simple pneumoconiosis is slight or absent and the opacities often appear with a suddenness that is

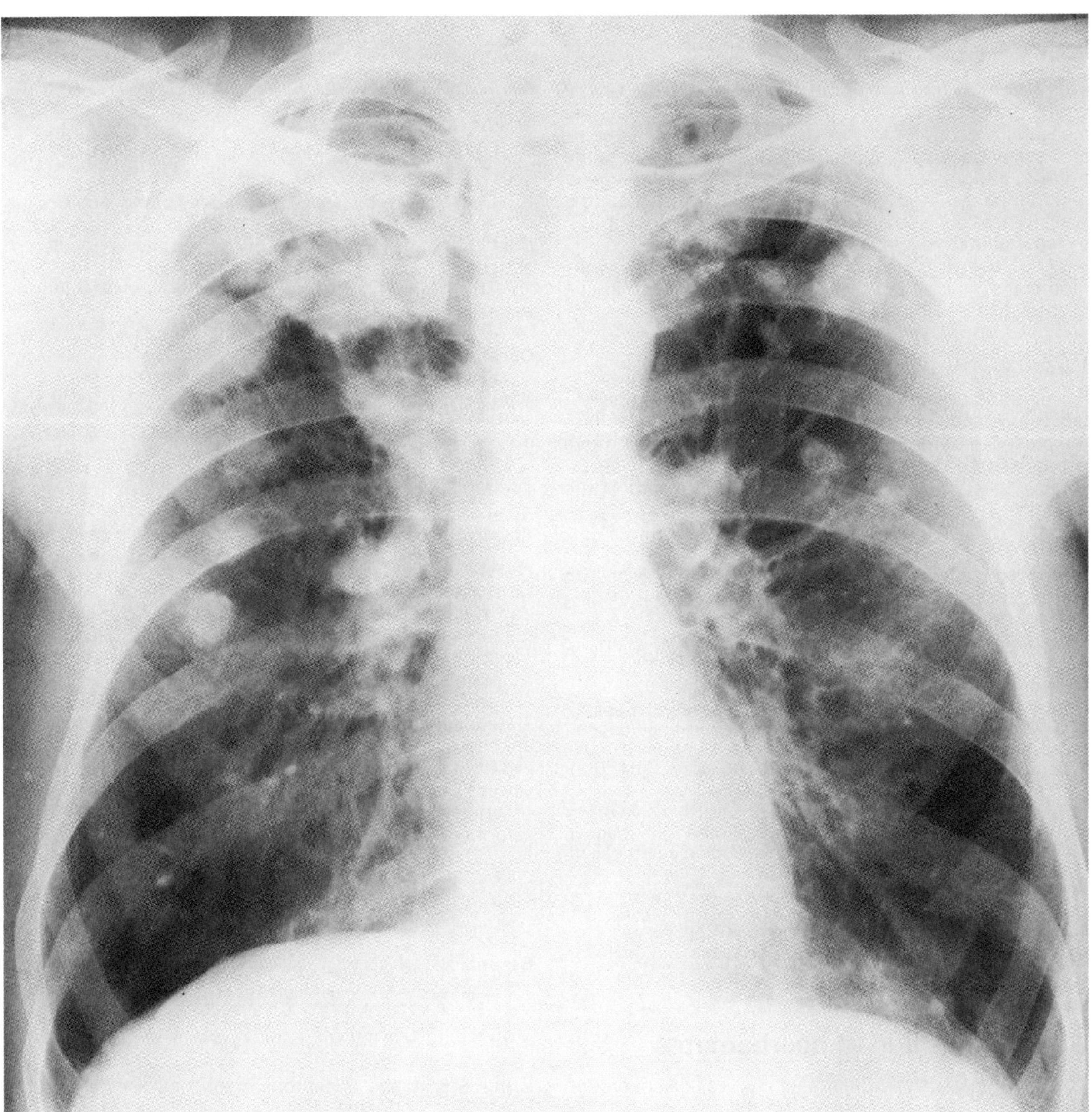

Figure 1. Caplan's syndrome.

not usually observed in the development of progressive massive fibrosis. Most frequently, the development of the opacities coincides approximately with the onset of arthritis, but cases have been seen where the arthritis preceded the lung lesions by periods of up to 6 years and, in other cases, the lung lesions have preceded the arthritis by periods of up to 10 years.

The tendency is for the opacities to increase in size and number, and crops of fresh lesions may appear at intervals of a few months. In a minority of cases a few round opacities are localised to one or more areas of the lung fields and may remain stationary. It is not uncommon for the lesions to cavitate and, when the cavitation is extensive, the radiological appearances are quite striking and almost pathognomonic of the syndrome. Following cavitation, the lesion may disappear completely or contract to a smaller irregular opacity. Calcification of the lesions is common. In many cases, after a period of years the lesions become incorporated in a mass indistinguishable radiographically from progressive massive fibrosis.

More cases would probably be recognised if it were realised that the radiographic appearances will depend on the number and size of the nodules and the extent of conglomeration. When the nodules are discrete and measure about 0.5-1.5 cm, the radiographic appearances may be indistinguishable from silicosis. When the nodules conglomerate into masses 1.5-3 cm in diameter and are distributed throughout the lung fields, the appearances are characteristic of the original description of the rheumatoid syndrome.

When the discrete nodules are few and the conglomeration large and irregular, progressive massive fibrosis may be simulated.

PREVENTIVE MEASURES

Rheumatoid pneumoconiosis might possibly be prevented if all workers exposed to a dust hazard were tested for the presence of rheumatoid factor in the blood. Positive reactors should be advised not to work in a dusty atmosphere. It is unlikely, however, that such a procedure would be practicable.

Treatment

A series of cases have been adequately treated with corticosteroids and antituberculous chemotherapy without any obvious benefit.

CAPLAN, A.

CIS 79-676 "Pulmonary function in coal workers with Caplan's syndrome and non-rheumatoid complicated pneumoconiosis". Constantinidis, K.; Musk, A. W.; Jenkins, J. P. R.; Berry, G. *Thorax* (London), Dec. 1978, 33/6 (764-768). 15 ref.

"HLA-A and B antigen frequencies in Welsh coalworkers with pneumoconiosis and Caplan's syndrome". Wagner, M. M. F.; Darke, C. *Tissue antigens* (Copenhagen), 1979, 14/2 (165-168). 8 ref.

"Rheumatoid pneumoconiosis (Caplan's syndrome) in an asbestos worker: A 17 years' follow-up". Greaves, I. A. *Thorax* (London), June 1979, 34/3 (404-405). Illus. 2 ref.

Carbamates and thiocarbamates

The biological activity of carbamates was discovered in 1923 when the structure of the alkaloid eserine (or physostigmine) contained in the seeds of Calabar beans was first described. In 1929 physostigmine analogues were synthesised, and soon such derivatives of dithiocarbamic acid as thiram and ziram were available. The study of carbamic compounds began in the same year, and now more than 1 000 carbamic acid derivatives are known. More than 50 of them are used as pesticides, herbicides, fungicides and nematocides. In 1947 the first carbamic acid derivatives having insecticide properties were synthesised. Some thiocarbamates have proved effective as vulcanisation accelerators, and rather recently derivatives of dithiocarbamic acid have been obtained for the treatment of malignant tumours, hypoxia, neuropathies, radiation injuries and other diseases.

Carbamic acid esters

Aryl esters of alkylcarbamic acid and alkyl esters of arylcarbamic acid are used as pesticides.

Baygon ($C_{11}H_{15}NO_3$)

1-ISOPROPOXYPHENYL-N-METHYLCARBAMATE; BAYER 39007

m.w. 209.2
m.p. 91.5 °C
white crystalline solid, slightly soluble in water.
TLV ACGIH 0.5 mg/m³
STEL ACGIH 2 mg/m³

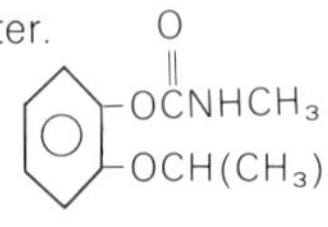

It is produced by reaction of alkyl isocyanate with phenols and used as an insecticide. Baygon is a systemic poison. It causes inhibition of the serum cholinesterase activity up to 60% after oral administration of 0.75-1 mg/kg. This highly toxic substance exerts a weak effect on the skin.

Carbaryl ($C_{12}H_{11}O_2N$)

1-NAPHTHYL-N-METHYLCARBAMATE; SEVIN

m.w. 201
sp.gr. 1.23
m.p. 142 °C
white crystals.
TLV ACGIH 5 mg/m³
STEL ACGIH 10 mg/m³
MAC USSR 1 mg/m³

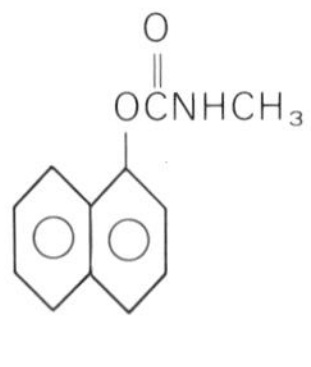

Carbaryl may be synthesised directly from 1-naphthol and methyl isocyanate, or it can be produced by the reaction of naphthyl chloroformate (obtained from 1-naphthol and phosgene) with methylamine. It is used as an insecticide.

Carbaryl is a systemic poison which produces moderately severe acute effects when ingested, inhaled or absorbed through the skin. It may cause local skin irritation. Being a cholinesterase inhibitor it is much more active in insects than in mammals. Medical examinations of workers exposed to concentrations of 0.2-0.3 mg/m³ seldom reveal a fall in cholinesterase activity.

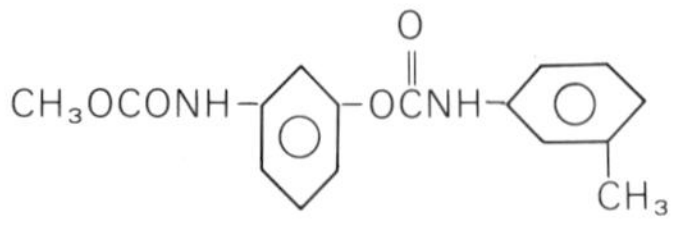

Betanal ($C_{16}H_{16}N_2O_4$)

3-(METHOXYCARBONYL)AMINOPHENYL-N-(3-METHYLPHENYL) CARBAMATE; SCHERING 38584; N-METHYLCARBANILATE; EP 452

MAC USSR 0.5 mg/m³

It belongs to the arylcarbamic acid alkyl esters and is used as a herbicide. Betanal is slightly toxic for the gastrointestinal and respiratory tracts. Its dermal toxicity and local irritation are insignificant.

Propham ($C_{10}H_{13}O_2N$)

ISOPROPYL-N-PHENYLCARBAMATE; IPC

m.w. 179

m.p. 90 °C

white crystalline solid.

MAC USSR 2 mg/m³

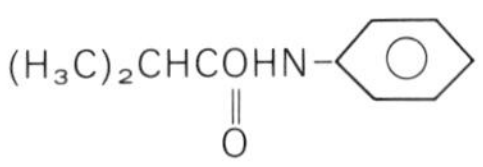

$(H_3C)_2CHCOHN-C_6H_5$ (C=O)

This compound is synthesised by reaction of the corresponding arylisocyanate with isopropyl alcohol. It is used as a root pesticide in seedbed preparation.

Isolan ($C_{10}H_{17}N_3O_2$)

1-ISOPROPYL-3-METHYLPYRAZOLYL-5-DIMETHYLCARBAMATE

m.w. 211.3

sp.gr. 1.07

b.p. 103 °C (0.7 mmHg or 90 Pa)

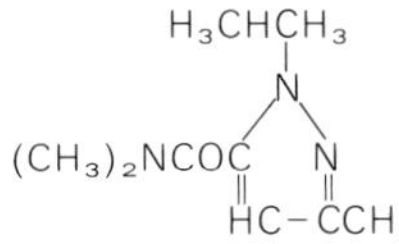

This compound is prepared by treating 1-isopropyl-3-methyl-5-pyrazolone with dimethylcarbamoyl chloride. Isolan is used as an insecticide.

This carbamate is a highly toxic member of the group, its action, like that of Sevin and others, being characterised by the inhibition of acetylcholinesterase activity.

Pyrimor ($C_{11}H_{18}N_4O_2$)

5,6-DIMETHYL-2-DIMETHYLAMINO-4-PYRIMIDINYL METHYLCARBAMATE

m.w. 238

TSRAL USSR 0.05 mg/m³

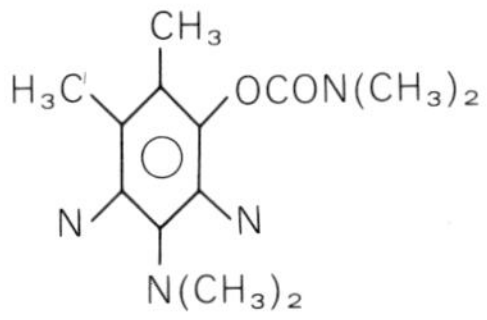

This compound is a derivative of arylcarbamic acid alkyl esters. It is highly toxic for the gastrointestinal tract. Its general absorption and local irritative effect are not very pronounced.

Thiocarbamic acid esters

Of this group of compounds the following substances are worth mentioning:

Ronite ($C_{11}H_{21}NOS$)

S-ETHYLCYCLOHEXYLETHYL THIOCARBAMATE; EUREX

m.w. 187.3

b.p. 140 °C

MAC USSR 1 mg/m³

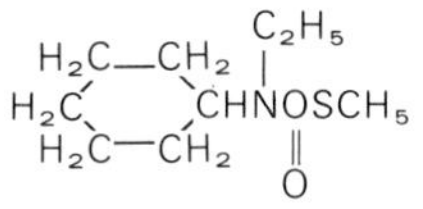

Eptam ($C_9H_{19}SNO$)

S-ETHYL-N,N-DIPROPYL THIOCARBAMATE

m.w. 189

b.p. 127 °C

MAC USSR 2 mg/m³

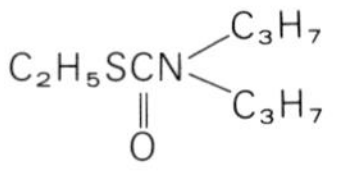

$C_2H_5SCN(C_3H_7)_2$ (C=O)

Tillam ($C_{10}H_{21}SNO$)

S-PROPYL-N-ETHYL-N-BUTYL THIOCARBAMATE

m.w. 203.3

MAC USSR 1 mg/m³

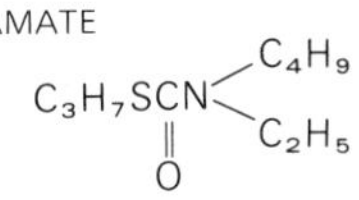

$C_3H_7SCN(C_4H_9)(C_2H_5)$ (C=O)

These esters are synthesised by reaction of alkyl thiocarbamates with amines and of alkaline mercaptides with carbamoyl chlorides. They are effective herbicides of selective action.

The compounds of this group are slightly to moderately toxic. Their toxicity is insignificant when they are absorbed through the skin. They can affect the oxidative processes as well as the nervous and endocrine systems.

Dithiocarbamates and bisdithiocarbamates

This group includes the following products which have much in common as regards their use and their biological effects:

Ziram ($C_6H_{12}N_2S_4Zm$)

ZINC DIMETHYL DITHIOCARBAMATE

m.w. 305.8

sp.gr. 1.71

m.p. 246 °C

white crystals.

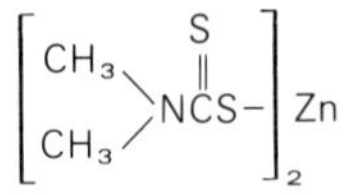

$[(CH_3)_2NC(=S)S-]_2Zn$

This zinc salt is produced by reaction of sodium dimethyl-thiocarbamate with a water-soluble zinc salt. It is used as a vulcanisation accelerator for synthetic rubbers, and, in agriculture, as a fungicide and seed fumigant.

This compound is very irritant to the conjunctiva and upper airway mucous membranes. It can cause extreme pain in the eyes, skin irritation and liver function disorders. It has embryotoxic and teratogenic effects.

TTD ($C_6H_{12}N_2S_4$)

TETRAMETHYLTHIURAM DISULPHIDE; THIRAM; DISULPHIRAM; BIS-DIMETHYLTHIOCARBAMYL DISULPHIDE

m.w. 240

sp.gr. 1.3

m.p. 156 °C

b.p. 129 °C at 20 mmHg

grey-yellow powder.

TWA OSHA 5 mg/m³

STEL ACGIH 10 mg/m³

IDLH 1 500 mg/m³

MAC USSR 0.5 mg/m³

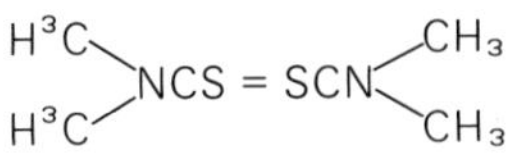

$(H^3C)_2NCS = SCN(CH_3)_2$

This compound is obtained by oxidation of dimethyl-dithiocarbamate produced by reaction of dimethylamine with carbon disulphide in an alkaline medium. It is used as a seed fumigant.

It irritates the skin, causes dermatitis and affects the conjunctiva. It increases sensitivity to alcohol.

Nabam ($C_4H_6N_2Na_2S_4$)

DISODIUM ETHYLENEBIS- (DITHIOCARBAMATE)

m.w. 256.4

colourless crystals.

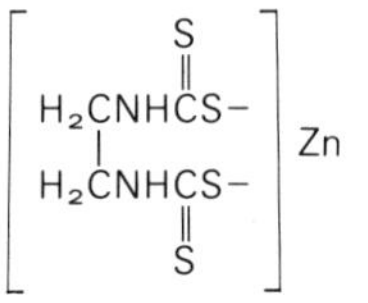

This compound is prepared by treating ethylenediamine with carbon disulphide in the presence of sodium hydroxide. Nabam is a plant fungicide and serves as an intermediate in the production of other pesticides.

It is irritating to the skin and mucous membranes and it is a narcotic in high concentrations. In the presence of alcohol it can cause violent vomiting.

Ferbam ($C_9H_{18}N_3S_6Fe$)

FERRIC DIMETHYLDITHIOCARBAMATE

m.w. 416.5

m.p. 180 °C (decomposes)

dark-brown powder, slightly soluble in water.

TWA OSHA 15 mg/m³

TLV ACGIH 10 mg/m³

STEL ACGIH 20 mg/m³

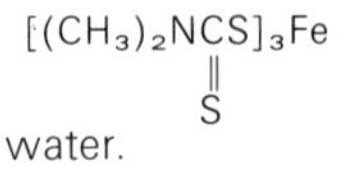

$[(CH_3)_2NC(=S)S]_3Fe$

Ferbam is obtained by adding carbon disulphide to dimethylamine dissolved in alcohol and by subsequent precipitation with iron salts. It is used as a fungicide.

It is relatively little toxic, but may cause renal function disorders. It irritates the conjunctiva, the mucous membranes of the nose and upper airways, and the skin.

Zineb ($C_4H_6N_2S_4Zn$)

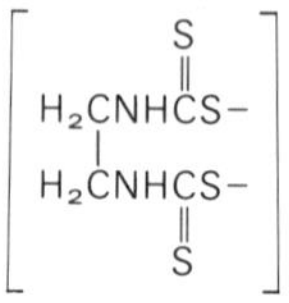

ZINC ETHYLENEBIS-DITHIOCARBAMATE

m.w. 275.8
m.p. 240 °C

white powder or crystals, insoluble in water.

MAC USSR 0.5 mg/m³

$$\left[\begin{array}{l} H_2CNHC(=S)S- \\ | \\ H_2CNHC(=S)S- \end{array}\right] Zn$$

This compound is obtained by reaction of sodium ethylenebis-dithiocarbamate with a zinc salt. It is used as an insecticide and fungicide.

Zineb can cause irritation of the eyes, nose and larynx, and is harmful if inhaled or swallowed.

Maneb ($C_4H_6N_2S_4Mn$)

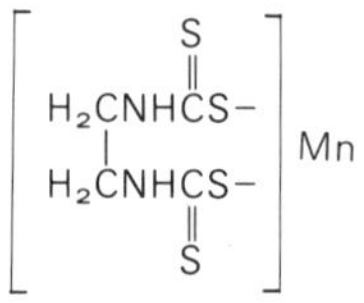

MANGANESE ETHYLENEBIS-DITHIOCARBAMATE; MANGATE

m.w. 265.3
m.p. 120 °C

yellow or brown powder which decomposes when heated.

MAC USSR 0.5 mg/m³

$$\left[\begin{array}{l} H_2CNHC(=S)S- \\ | \\ H_2CNHC(=S)S- \end{array}\right] Mn$$

Maneb is obtained by neutralising an aqueous solution of acetic acid and adding a solution of manganese chloride. It is used as a fungicide.

This compound can cause irritation of the eyes, nose and larynx, and is harmful if inhaled or swallowed.

Vapam ($CH_3NHC(=S)SNa$)

SODIUM METHYLDITHIOCARBAMATE; CARBATION

m.w. 130.2

white crystalline powder of unpleasant smell similar to that of carbon disulphide.

Vapam is an effective soil fumigant which destroys weed seeds, fungi and insects.

This substance irritates the skin and mucous membranes.

CHRONIC POISONING HAZARDS

The specific effects produced by acute poisoning have been described for each substance listed. A review of the specific effects gained from an analysis of published data makes it possible to distinguish similar features in the chronic action of the different carbamates. Some authors believe that the main toxic effect of carbamic acid esters is the involvement of the endocrine system. One of the peculiarities of carbamate poisoning is the possible allergic reaction of the organisms. The data so far available suggest that the development of delayed effects may be the main potential hazard. Results from animal experiments are indicative of embryotoxic, teratogenic, mutagenic and carcinogenic effects of some carbamates.

SAFETY AND HEALTH MEASURES

Our knowledge of the harmful effects of carbamates has considerably widened during the past years, so that it is possible to elaborate appropriate measures to prevent exposure during carbamate production and use. Production equipment must be enclosed and provided with exhaust ventilation. No worker must be authorised to enter a reactor for the synthesis of these compounds before it has been purged, cleaned, checked and certified safe for entry.

Personal protective equipment should be issued to the workers engaged in the production of carbamates or in their agricultural use, in particular eye protectors and respirators to those who spray crops. Showers should be provided, and their regular use should be encouraged.

KUNDIEV, Ju. I.
DOBROVOL'SKIJ, L. A.

Skin absorption of pesticides and prevention of poisoning (Vsasyvanie pesticidov čerez kŏu i profilaktika otravlenij). Kundiev, Ju. I. (Kiev, Zdorov'ja, 1975), 189 p. (In Russian)

Handbook on pesticides (hygiene in use and toxicology) (Spravočnik po pesticidam (gigiena primenenija i toksikologija)). Medved', L. I. (ed.) (Kiev, Urŏaj, 2nd ed. 1977), 374 p. (In Russian)

Pesticides–toxic effects and prevention (Pesticidi. Toksično dejstvie i profilaktika). Kalojanova-Simeonova, F. (Sofia, Bălgarskata akademija na naukitem, 1977), 307 p. (In Bulgarian)

Pesticides and the environment (Pesticidy i okrŭajuščaja sreda). Mel'nikov, N. N. (Moscow, Himija, 1977), 240 p. (In Russian)

CIS 77-414 *Some carbamates, thiocarbamates and carbazides.* Bates, R. R.; Boyland, E.; Van Esch, G. J.; Hathway, D. E.; Lijinsky, W.; Mohr, U.; Murphy, S. D.; Okulov, V. B.; Ramel, C.; Teichman, B.; Terracini, B. *IARC monographs on the evaluation of carcinogenic risk of chemicals to man.* Vol. 12 (Lyons, International Agency for Research on Cancer, 1976), 282 p. Illus. 775 ref.

CIS 77-1698 *Carbaryl.* Data sheets on pesticides No. 3 (Geneva, World Health Organisation; and Rome, United Nations Food and Agriculture Organisation, Jan. 1975), 13 p. 6 ref.

CIS 77-511 *Criteria for a recommended standard–Occupational exposure to carbaryl.* DHEW (NIOSH) publication No. 77-107 (National Institute for Occupational Safety and Health, 4676 Columbia Parkway, Cincinnati) (Sep. 1976), 192 p. Illus. 162 ref.

Carbon black

Carbon black

m.w. 12

a black, odourless, powdery material.

TWA OSHA 3.5 mg/m³
STEL ACGIH 7 mg/m³

Carbon black is a general term used to designate a finely divided form of carbon made by thermal decomposition of natural gas or oil or a mixture of both. It should not be confused with animal and mineral blacks, which possess the nature of chars. For practical purposes, this article does not include lamp black, generally made of coal-tar-creosote, nor acetylene black which is produced from this gas.

The history of the use of blacks could be traced to primitive times, but the first inks of this kind were used in the early Chinese and Egyptian civilisations, as far back as 2 500 BC.

Depending on the process of manufacture, there are variations in the chemical composition of carbon black. It contains 88-99.5% of carbon; 0.3-11% of oxygen; 0.1-1% of hydrogen; up to 1% of inorganic materials; small amounts of tarry matter and traces of sulphur. The size of the particles in carbon black varies from 5-500 nm. Under the electron microscope they appear spherical. X-ray diffraction reveals a structure similar to, but less regular than, graphite.

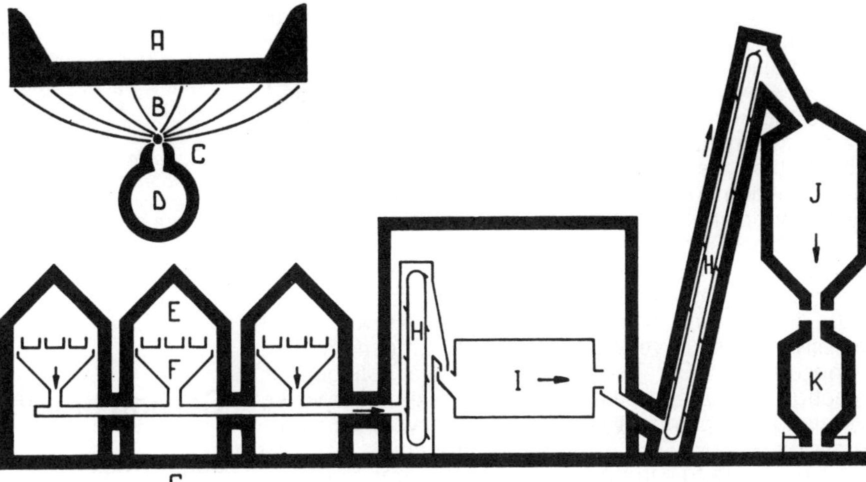

Figure 1.
Manufacture of carbon black: channel process. A. Channel. B. Flame. C. Lava tip. D. Burner pipe. E. Channels. F. Hoppers. G. Burner buildings. H. Elevator. I. Dustless process. J. Carbon-black storage tank. K. Hopper car. *(By courtesy of Ashland Chemical Co.)*

Production. Three main processes are in use: channel, furnace and thermal. Only the first two are described, since the last one is of less commercial importance.

(a) The channel process involves the burning of natural gas, or natural gas enriched with oil. The impingement of flames against structural steel elements, called "channel irons", which move slowly to and fro, leaves a layer of carbon black. This is scraped off by a stationary blade and dropped into a hopper. From the hopper it is screw-conveyed out of the burner house for further processing, such as pelleting and packaging (see figure 1).

(b) The furnace process yields carbon black from only one large flame in a combustion chamber. The black is carried out of the furnace by the hot combustion gases, quenched by a direct water spray and directed to another water spray cooler. The stream continues to precipitators, cyclones, and bag filters where the black is collected and dropped into attached hoppers and the conveying system (see figure 2).

The most carbon black is produced in the United States mainly in Texas. Other important producing countries are the United Kingdom, Canada and France.

Uses. The rubber industry is the principal user of carbon black because it increases the resistance of rubber to abrasion and wear. Its production was developed tremendously for the tyre industry as a reinforcing filler. It is estimated that each tyre contains more than 2 kg of carbon black. The second most important user of carbon black is the newspaper industry, which utilises it in printing and other types of inks. The fabrication of phonograph records, protective coatings, carbon paper and batteries also consumes carbon black. Its principal functions in the above uses are as a filler, reinforcing agent, pigment, electric conductor and chemical reducing agent.

HAZARDS

The epidemiological investigations of the hazards of carbon black were initiated in 1950, when the incidence of cancer in this industry was studied. It has now been demonstrated that the cancer incidence of workers in the carbon black industry is similar to or lower than that of the general population. The last fact seems to have an explanation in the adsorption qualities of the carbon black for the aromatic chemicals compounds. The original incrimination of carbon black as a carcinogenic agent

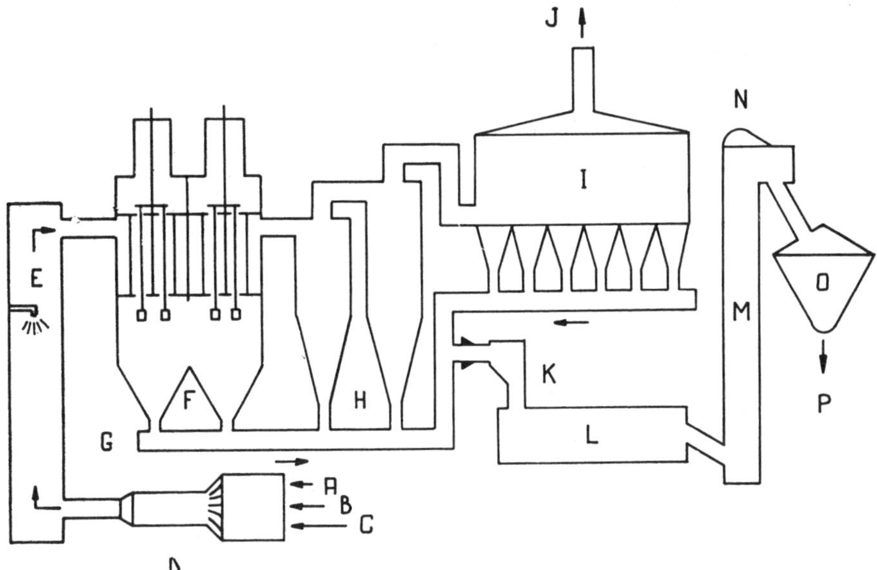

Figure 2.
Manufacture of carbon black: furnace process. A. Air. B. Oil. C. Gas. D. Reactor. E. Water. F. Precipitator. G. Conveyor. H. Cyclones. I. Bag filter. J. Clean gases. K. Pulveriser. L. Dustless process. M. Elevator. N. Screens and magnet remove dust, oversize and contaminants. O. Storage tank. P. Product to bags and hopper cars. *(By courtesy of Ashland Chemical Co.)*

is due to the presence of impurities. For instance it has been demonstrated that the composition of European carbon blacks is different from the American. In the first up to 1% by weight of 3,4 benzpyrene has been found, while the second is practically free of this substance. Some information presented in the literature blaming carbon black as a cancer producing agent of the salivary glands in humans is not conclusive. Pneumoconiosis has been found in workers engaged in the production of carbon black and related products, especially of channel black, with a radiological appearance of small round shadows (forms *p* and *q* of the international classification) and marking of bronchi and vessels. A few cases of massive fibrosis have occurred, probably after combined exposure to SiO_2. There is a moderate collagenous reaction and emphysema may be present. However, ventilatory impairment is not usually severe and progression is slow. The evidence of bronchitis resulting from exposure to carbon black is not fully convincing.

Oral mucosal lesions, including leucoplakia and keratosis, have sometimes been reported, as well as skin conditions such as follicular coniosis.

Carbon black can be ignited in the presence of open flames, and burns slowly with production of carbon monoxide. When carbon black exceeds 8% volatile material it should be considered as an explosion hazard.

Animal experiments for the determination of carcinogenicity of carbon black showed that the feeding of this substance to mice leads to no definite changes and that it can adsorb a carcinogen and make it ineffective. The inhalation of high concentrations of channel and furnace carbon black for prolonged periods of time was without significant effects in laboratory animals. In the case of skin contact, harmful effects were observed and effective adsorption of known carcinogens occurred. Investigations on the peritoneal reactions of carbon black in rats showed that apart from the types composed of large spherical particles, carbon black is not completely inert.

SAFETY AND HEALTH MEASURES

There are no well demonstrated health hazards in the carbon black industry at present. In some factories in Texas the workers cover their faces with talcum powder in order to prevent the penetration of carbon black into the skin.

A time weighted average threshold limit value of 3.5 mg/m³ has been adopted in various countries and recently confirmed by NIOSH pending further investigations on carbon black exposure. Because of the fineness of dust (80% of particles usually less than 0.5 µm in diameter) highly efficient collection systems should be used with a relatively low velocity of air stream.

RISQUEZ-IRIBARREN, R.

Health impairment:

CIS 76-957 "Carbon-black pneumoconiosis" (Pneumoconiose au noir de fumée). Cocarla, A.; Cornea, G.; Dengel, H.; Gabor, S.; Milea, M.; Papilian, V. V. *International Archives of Occupational and Environmental Health* (West Berlin), 26 Jan. 1976, 36/3 (217-228). Illus. 17 ref. (In French)

"A mortality study of carbon black workers in the United States from 1935 to 1974". Robertson, J. McD.; Ingalls, T. H. *Archives of Environmental Health* (Chicago), 1980, 35/3 (181-186).

Exposure limits:

CIS 76-653 "Establishment of a maximum admissible concentration for airborne carbon black in industrial workplaces" (Obosnovanie predel'no dopustimoj koncentracii pylej černyh promyšlennyh să v vozduhe proizvodstvennyh pomeščenij). Troickaja, N. A.; Veličkovskij, B. T.; Bikmulina, S. K.; Săina, T. G.; Gorodnova, N. V.; Andreeva, T. D. *Gigiena truda i professional'nye zabolevanija* (Moscow), Mar. 1975, 3 (32-36). Illus. 16 ref. (In Russian)

Criteria for a recommended standard–Occupational exposure to carbon black. DHEW (NIOSH) publication No. 78-204 (National Institute for Occupational Safety and Health, 4676 Columbia Parkway, Cincinnati) (Sep. 1978), 99 p. Illus. 86 ref.

Carbon dioxide

Carbon dioxide (CO_2)

m.w.	44
sp.gr.	1.98
m.p.	−56.6 °C (at 5.2 atm)
b.p.	−78.5 (sublimes)
v.d.	1.53

a colourless, odourless gas with a faint acid taste.

TWA OSHA	5 000 ppm 9 000 mg/m³
TLV NIOSH	10 000 ppm/10 h
NIOSH	30 000 ppm/10 min ceil
STEL ACGIH	15 000 ppm 18 000 mg/m³
IDLH	50 000 ppm

Occurrence. Carbon dioxide is present in the normal atmosphere in concentrations varying from 0.03% to 0.06%. It is also found in solution in spring water which is sometimes so charged with the gas under pressure that it is effervescent. It is evolved in large quantities from vents and fissures in the earth in volcanic regions. The gas is also present in exhaled air and its concentration rises in the atmosphere of a crowded room.

The gas is produced when carbonaceous substances burn in an excess of air or oxygen. It is a product of fermentation processes, synthetic ammonia production, limestone calcination, the reaction of sulphuric acid with dolomite and numerous other reactions in industrial chemistry.

Production. Flue gases from furnaces are cooled and cleaned in water scrubbers and are then passed into absorbers in which the carbon dioxide is absorbed by ethanolamine in the Girbotol process or by alkali in the alkali-carbonate process. When a fermentation process is the source of the commercial carbon dioxide, the gas is treated in water scrubbers and by activated carbon before being passed to the compressors. The gases from lime kilns contain 40% of carbon dioxide which is separated by alkali absorption and obtained as sodium carbonate. In addition, carbon dioxide is the by-product of one operation for the synthesis of ammonia.

Uses. Carbon dioxide is used in industry in the form of gas, liquid or solid. As a gas it is used for carbonating beverages such as soft drinks, to neutralise excess alkali in the chemical industry, in shielded-arc welding and, in market gardening, as an addition to the atmosphere of greenhouses to increase growth rate. As a liquid it is used in fire-extinguishing equipment, in cylinders for inflating life rafts, for refrigerated storage and for the manufacture of solid carbon dioxide. In the solid form (dry ice), it is used for refrigeration purposes in the production of ice cream, meat products and frozen foods. In this form it is used to embrittle the flash of moulded rubber articles so that it can easily be broken and removed. It is used to chill golf ball centres before winding and for many purposes in laboratories and hospitals. Carbon dioxide in its solid or gaseous form is commonly used to produce an inert atmosphere in vessels or plant where there would otherwise be an explosive risk.

HAZARDS

A concentration of 5% of carbon dioxide in the atmosphere (i.e. 50 000 ppm) may produce shortness of breath and headache. At a concentration of 10%, carbon dioxide can produce unconsciousness in an exposed person, who will die from oxygen deficiency unless he is removed to a normal atmosphere or is given oxygen resuscitation. Carbon dioxide does not give a warning of its presence in an asphyxiating concentration, and a person may unwittingly enter a confined space or descend into a tank or vessel and be overcome before he becomes aware of the danger and can make his escape.

[In addition to producing sudden oxygen deficiency at high concentrations, carbon dioxide is now recognised as being potentially toxic at low concentrations in consequence of cellular membrane effects and biochemical alterations such as increased $P(CO_2)$, increased concentrations of bicarbonate ions, acidosis (see ACID-BASE BALANCE). Long-term exposure to levels of carbon dioxide between 0.5 and 1%, as may occur in submarines, is likely to involve increased calcium deposition in body tissues, including kidney; concentrations of 1 to 2% appear to be dangerous after exposure for some hours even if there is no lack of oxygen.]

SAFETY AND HEALTH MEASURES

When the gas is extracted from the process as a commercial by-product, it is unlikely that the workroom atmosphere will be seriously contaminated. When the gas is not collected for commercial purposes it may well be that the most viable means of preventing the contamination of the workroom atmosphere is the provision of efficient local exhaust ventilation.

MATHESON, D.

Ecology:

"Man's global redistribution of carbon". Björkström, A. *Ambio* (Oslo), 1979, 8/6 (254-259). Illus. 22 ref.

Occupational exposure:

"Poisoning due to prolonged exposure to carbon dioxide" (Die protrahierte CO_2-Vergiftung). Zink, P.; Reinhardt, G. *Beiträge zur gerichtlichen Medizin* (Vienna), 1975, 33 (211-213). (In German)

CIS 77-442 *Criteria for a recommended standard–Occupational exposure to carbon dioxide.* DHEW (NIOSH) publication No. 76-194 (National Institute for Occupational Safety and Health, 4676 Columbia Parkway, Cincinnati) (Aug. 1976), 169 p. 151 ref.

Fire extinguishers:

CIS 78-1506 *Carbon dioxide extinguishing systems.* NFPA No. 12-1977 (National Fire Protection Association, 470 Atlantic Avenue, Boston) (1977), 93 p. Illus. 12 ref.

Carbon disulphide

Carbon disulphide (CS_2)

CARBON BISULPHIDE

m.w.	76
sp.gr.	1.26
m.p.	−110.8 °C
b.p.	46.3 °C
v.d.	2.6
v.p.	360 mmHg ($46.8 \cdot 10^3$ Pa) at 25 °C
f.p.	−30 °C
e.l.	1.3-50%
i.t.	100 °C

slightly soluble in water; soluble in ethyl alcohol, ethyl ether

a clear, colourless liquid with a sweet, etheral odour when pure, while the technical product is foul smelling.

TWA OSHA	20 ppm
OSHA	30 ppm ceil
	100 ppm/30 min peak
TLV NIOSH	1 ppm
NIOSH	10 ppm ceil
TLV ACGIH	10 ppm 30 mg/m³
IDLH	500 ppm
MAC USSR	10 mg/m³

Occurrence. Minute amounts may be present in coal tar and crude petroleum.

Production. By heating charcoal with vaporised sulphur and also by reacting sulphur with petroleum hydrocarbons.

Uses. It is an important solvent of alkali cellulose, fats, oils, resins and waxes, in addition to being used in the production of viscose rayon, in the manufacture of optical glass, as a pesticide and in oil extraction.

HAZARDS

The first cases of poisoning were observed during the 19th century in France and Germany in connection with the vulcanisation of rubber. After the First World War, the production of viscose rayon expanded, and with it the incidence of acute and chronic poisoning from carbon disulphide, which has remained a serious problem in some countries. Nowadays, acute and, more often, chronic poisoning occurs particularly in the viscose rayon industry, although improvements in technology and hygienic conditions in plants have virtually eliminated such problems in a number of countries.

Carbon disulphide is primarily a neurotoxic poison, therefore those symptoms indicating central and peripheral nervous system damage are the most important. It was reported that concentrations of 0.5-0.7 mg/l (160-230 ppm) caused no acute symptoms in man, 1-1.2 mg/l (320-390 ppm) were bearable for several hours, with the appearance of headaches and unpleasant feelings after 8 h of exposure; at 3.6 mg/l (1 150 ppm) giddiness set in; at 6.4-10 mg/l (2 000-3 200 ppm) light intoxication, paraesthesia, and irregular breathing occurred within 1/2-1 h. At concentrations of 15 mg/l (4 800 ppm), the dose was lethal after 30 min and at even higher concentrations, unconsciousness occurred after several inhalations.

Acute poisoning occurs mainly after accidental exposures to very high concentrations. Unconsciousness, frequently rather deep, with extinction of cornea and tendon reflexes, occurs after only a short time. Death sets in by a blockage of the respiratory centre. If the patient regains consciousness, motor agitation and disorientation follow. If he recovers, frequently late sequelae include psychic disturbances as well as permanent damage to the central and peripheral nervous systems. Subacute cases of poisoning usually occur from exposure to concentrations of more than 2 mg/l. They are manifested mainly in mental disorders of the maniac-depressive type; more frequent at lower concentrations, however, are cases of polyneuritis.

Chronic poisoning. It begins with weakness, fatigue, headache, sleep disturbances, often with frightening dreams, paraesthesia and weakness in the lower extremities, loss of appetite and stomach ailment. Neurovegetative symptoms are also seen and impotence is rather frequent. Continued exposure may give rise to polyneuritis, which is said to appear after working in concentrations of 0.3-0.5 mg/l for several years; an early sign is the dissociation of tendon reflexes in lower extremities. Damage to the brain nerves is less frequent, but *neuritis n. optici*, and vestibular and sense-of-smell disturbances were observed.

In exposed workers, the examination of nerve conduction velocity is of advantage because this is reduced even earlier than the impairment becomes demonstrable by routine neurological examinations. (The normal bottom limit is approximately 52 m/s.) The EEG shows diffuse abnormalities in more serious cases of poisoning; however, these are not specific for carbon disulphide only. Epigastric pain, indisposition and vomiting, all due to atrophic gastritis, as shown by biopsy examinations, may be present. A slight anaemia and a moderate leucocytosis have sometimes been recorded. An increase in coagulation time and decreased plasmin and plasminogen activity have also been reported.

In exposed workers, disorders occur in the sexual sphere (hypo, asthenospermia), and excretion of 17-ketosteroids, 17-hydroxycorticosteroids, and androsteron decreases during exposure. In women menstrual disturbances, metrorrhagia and more frequent abortions are described. Carbon disulphide passes the placenta, so that concentration in the tissues of the fetus is the same as in the mother.

The relationship between carbon disulphide and atherosclerosis is a topic of special interest. Prior to the Second World War, not much attention was paid to this pattern, but thereafter, when classic carbon disulphide poisoning ceased to occur in many countries, several authors noted the development of atherosclerosis of the brain vessels in workers of lower age groups in viscose rayon plants.

Ophthalmodynamographic studies in young workers who were exposed to carbon disulphide concentrations of 0.2-0.5 mg/l for several years, showed that the retinal systolic and diastolic blood pressure was higher than that of the brachial artery. This increase was due to arterial hypertension in the brain and it was reported that arterial spasms appeared before subjective complaints. Rheoencephalography has been recommended for assessment of brain vessel function. Changes in resistance are caused by arterial pulsation, especially of intracranial vessels and could therefore lead to the discovery of possible increased rigidity or spasms of cranial vessels. In Japanese workers a higher incidence of small, round, retinal haemorrhages and microaneurysms was observed.

In chronically exposed men, arteriolocapillary hyalinosis was found, which represents a special type of carbon disulphide arteriosclerosis. Therefore, carbon disulphide may be assumed to be a contributing factor to the origin of this sclerosis, but not a direct cause. This hypothesis, as well as the results of biochemical examinations, seems to be supported further by reports about the significant increase of atherosclerosis, frequently in younger persons who were exposed to carbon disulphide. With regard to the kidneys, it seems that glomerulosclerosis of the Kimmelstiel-Wilson type is more frequent in persons exposed to carbon disulphide than in others. Recently, British, Finnish and other investigators have shown that there is increased mortality from coronary heart disease in male workers exposed for many years to relatively low carbon disulphide concentrations.

The absorption of carbon disulphide through the respiratory tract is rather high and about 30% of the inhaled quantity is retained when a steady state of inhalation is reached. The time required for the establishment of this state varies in length from rather short, to several hours if light physical work is done. After termination of exposure, part of the carbon disulphide is rapidly excreted through the respiratory tract. The length of the desaturation period depends on the degree of exposure. Approximately 80-90% of the absorbed carbon disulphide is metabolised in the body with the formation of dithiocarbamates and possible further cyclisation to thiazolidane. Owing to the nucleophilic character of carbon disulphide, which reacts especially with $-SH$, $-CH$, and $-NH_2$ groups, perhaps other metabolites are formed too.

Carbon disulphide is also absorbed through the skin in considerable amounts, but less than through the respiratory tract. Dithiocarbamates easily chelate many metals such as copper, zinc, manganese, cobalt, iron. In fact an increased zinc content has been demonstrated in the urine of animals and men exposed to carbon disulphide. It is also believed that a direct reaction takes place with some of the metals contained in metalloenzymes.

Liver microsome tests have demonstrated the formation of carbon oxysulphide (COS) and atomic sulphur which is bound covalently to microsomal membranes. Other authors have found in rats that carbon disulphide after oxidative decomposition binds primarily to protein P-450. In urine it is excreted in a fraction of 1% as carbon disulphide, of the retained amount it is excreted to about 30% as inorganic sulphates, the remainder as organic sulphates and some unknown metabolites, one of which is thiourea according to Yugoslav authors.

It is assumed that the reaction of carbon disulphide with vitamin B_6 is very important; its metabolism is impaired, which is manifested by enhanced excretion of xanthurenic acid and decreased excretion of 4-pyridoxine acid, and further in a reduced serum pyridoxine level. It appears that copper economy is disturbed as indicated by the reduced level of ceruloplasmin in exposed animals and humans. Carbon disulphide interferes with serotonine metabolism in the brain by inhibiting certain enzymes. Furthermore, it has been reported that it inhibits the clearing factor (lipase activated by heparin in the presence of α-lipoproteins), thus interfering with the clearing of fat from blood plasma. This may result in the accumulation of cholesterol and lipoid substances in vessel walls and stimulate the atherosclerotic process. However, not all reports about the inhibition of the clearing factor are so convincing. There are many, although often contradictory, reports about the behaviour of lipoproteins and cholesterol in the blood and organs of animals and men exposed to carbon disulphide for a long time, or poisoned by it.

Impaired glucose tolerance of the chemical diabetes type was found. It is elicited also by the elevated level of xanthurenic acid in serum which, as was demonstrated in experiments, forms with insulin a complex and reduces its biological activity. Neurochemical studies have demonstrated changed catecholamine supplies in the brain as well as in other nervous tissues. These findings show that carbon disulphide changes the biosynthesis of catecholamines probably by inhibiting dopamine hydroxylase by chelating enzymatic copper.

Examination of animals poisoned by carbon disulphide revealed a variety of neurologic changes. In man the changes included serious degeneration of the grey matter in the brain and cerebellum, changes in the pyramid system of pons and spinal cord, degenerative changes of peripheral nerves and disintegration of their sheaths. Also described were atrophy, hypertrophy and hyalin degeneration of muscle fibres.

Fire and explosion. Carbon disulphide is also highly flammable and explosive.

SAFETY AND HEALTH MEASURES

Of special importance in the prevention of health problems by carbon disulphide is the enclosure of processes involving it, as well as local exhaust and general ventilation. It is also advisable to replace carbon disulphide by less toxic substances whenever possible. The workers should be instructed about its toxicity and fire hazard. Wearing of respiratory protective equipment should be left to cases of emergency.

Frequent control of carbon disulphide concentrations in the air, such as in viscose rayon plants, is essential. This can be done by colorimetric indicators, with continual analyses, and portable air samplers. The carbon disulphide content in the air can also be measured by exposure tests such as determining the amount in urine, blood, or expired air. However, these tests are generally useful only for a rough estimation of exposure. A widely used test in recent years is the iodine-azide test. It is based on the presence of carbon disulphide sulphur metabolites in urine. The level of exposure can be estimated by the rate of iodine reduction by sodium azide. Persons still producing a positive test on the second day after terminating work should be placed under careful medical control.

On the basis of pre-employment and periodic medical examinations, persons should be considered unsuitable for work with carbon disulphide if they are neurotic or have serious nervous and mental afflictions, or if they have diseases of the liver or kidneys, and atherosclerotic changes. Young persons of either sex should not be employed if below the age of 18. There is no evidence that women are more sensitive to this material but, since it is highly toxic, their work with it is not recommended, or only at low exposures, because of their maternal functions and their labile condition in climacterium.

The frequency of periodic medical examinations depends upon the hygienic conditions at the workplace. If carbon disulphide concentrations exceed recommended limits, two or three medical examinations annually are advisable. A neurologist and, if possible, a psychiatrist should participate in such examinations. Whenever symptoms have been noted, the worker should be removed from exposure and, if a diagnosis of poisoning has been established, he should not be permitted to continue such work.

Treatment. In acute cases the patient should be removed from exposure, made comfortable, kept warm, but not hot, and given resuscitation if necessary.

In chronic poisoning, the prognosis is generally good if the patient has been removed from exposure in due time. If, however, the peripheral or central nervous systems have been affected, the prognosis in regard to complete recovery is very dubious; this has been confirmed over years of observations.

TEISINGER, J.

Biological effects:

"Specific binding of CS_2 metabolites to microsomal proteins in the rat liver". Savolainen, H.; Javisato, J.; Vainio, H. *Acta pharmacologica et toxicologica* (Copenhagen), 1977, 41/1 (94-96). Illus. 3 ref.

"The mechanism of chronic effect of carbon disulphide on carbohydrate metabolism". Kujalova, V.; Sperlingova, I.; Frantik, E. (478-479). 3 ref. *XIX International Congress on Occupational Health: Abstracts* (Zagreb, 1978).

"The effect of carbon disulphide on development of atherosclerotic changes in rabbits fed on standard and atherogenic diet". Wronska-Nofer, T.; Szendzikowski, S.; Laurman, W. (492). 4 ref. *XIX International Congress on Occupational Health: Abstracts* (Zagreb, 1978).

"Carbon disulfide metabolites excreted in the urine of the exposed workers. II. Isolation and identification of thiocarbamide". Pergal, M.; Vukojevic, N.; Djuric, D. *Archives of Environmental Health* (Chicago), July 1972, 25/1 (42-44). Illus. 4 ref.

Health impairment:

CIS 77-1066 *Behavioral and neurological effects of carbon disulfide.* Tuttle, T. C.; Wood, G. D.; Grether, C. B. DHEW (NIOSH) publication No. 77-128 (National Institute for Occupational Safety and Health, 4676 Columbia Parkway, Cincinnati) (Dec. 1976), 156 p. Illus. 24 ref.

"Ten-year coronary mortality of workers exposed to carbon disulfide". Tolonen, M.; Nurminen, M.; Hernberg, S. *Scandinavian Journal of Work, Environment and Health* (Helsinki), June 1979, 5/2 (109-114). Illus. 12 ref.

Exposure limits:

Carbon disulphide. Environmental health criteria No. 10 (Geneva, World Health Organisation, 1979), 100 p. 373 ref.

Safety:

CIS 78-454 *Carbon bisulfide (carbon disulfide).* Data Sheet 341, Revision B (Extensive) (National Safety Council, 444 North Michigan Avenue, Chicago) (1977), 4 p. 7 ref.

Carbon monoxide

Carbon monoxide (CO)

m.w.	28.01
sp.gr.	1.25
m.p.	−205.1 °C
b.p.	−191.5 °C
e.l.	12.5-74%
i.t.	608.9 °C

a colourless, tasteless and almost odourless gas which is lighter than air and burns in air with a blue flame.

TWA OSHA	50 ppm 55 mg/m³
TLV NIOSH	35 ppm/10 h
	200 ppm ceil
STEL ACGIH	400 ppm 440 mg/m³
IDLH	1 500 ppm
MAC USSR	20 mg/m³

Occurrence. Carbon monoxide is produced when organic material, such as coal, wood, paper, oil, gasoline, gas, explosives or any other carbonaceous material, is burned in a limited supply of air or oxygen. When the combustion process takes place in an abundant supply of air without the flame contacting any surface, carbon monoxide emission is not likely to result. CO is produced if the flame contacts a surface which is cooler than the ignition temperature of the gaseous part of the flame. The exhaust gas of gasoline-fuelled combustion engine (spark ignition) is the most important source of ambient CO and contains 1 to 10% of CO depending on the mode of operation of the engine. The diesel engine (compression ignition) exhaust gas contains about 0.1% of

CO when the engine is operating properly, but maladjusted, overloaded or badly maintained diesel engines may emit considerable amounts of CO. Motor vehicles account for about 55 to 60% of global man-made emissions of CO. Thermal or catalytic afterburners in the exhaust pipes considerably reduce the amount of CO emitted. Other major sources of CO are cupolas in foundries, catalytic cracking units in petroleum refineries, distillation of coal and wood, lime kilns and the kraft recovery furnaces in kraft paper mills, manufacture of synthetic methanol and other organic compounds from carbon monoxide, the sintering of blast furnace feed, carbide manufacture, formaldehyde manufacture, carbon black plants, coke works, gas works, and refuse plants.

Any process where incomplete burning of organic material may occur is a potential source of carbon monoxide emission. Thus, sources of CO exposure are quite ubiquitous.

Production and uses. Carbon monoxide is produced on an industrial scale by the partial oxidation of hydrocarbon gases from natural gas or by the gasification of coal or coke. It is used as a reducing agent in metallurgy especially for the Mond nickel recovery process, in organic syntheses such as the Fischer-Tropsch and oxo processes, and in the manufacture of metal carbonyls. Several industrial gases that are used for heating boilers and furnaces and driving gas engines contain carbon monoxide. Water gas contains about 40%, blast-furnace gas about 30%, producer gas about 25%, and coal gas about 5% of CO.

HAZARDS

Carbon monoxide is thought to be by far the most common single cause of poisoning both in industry and in homes. Thousands of persons succumb annually as a result of CO intoxication. The number of victims of non-fatal poisoning that suffer from permanent central nervous system damage can be estimated to be even larger. The magnitude of the health hazard due to carbon monoxide, both fatal and non-fatal, is huge and poisonings are probably more prevalent than is generally recognised.

A sizeable proportion of the workforce in any country has a significant occupational CO exposure. CO is an ever-present hazard in the automotive industry, garages and service stations. Road transport drivers may be endangered if there is a leak of engine exhaust gas into the driving cab. Occupations with potential exposure to CO are numerous, e.g. garage mechanics, charcoal burners, coke oven workers, cupola workers, blast furnace workers, blacksmiths, miners, tunnel workers, Mond process workers, gas workers, boiler workers, pottery kiln workers, wood distillers, cooks, bakers, firemen, formaldehyde workers, and many others. Welding in vats, tanks or other enclosures may result in production of dangerous amounts of CO if ventilation is not efficient. The explosions of methane and coal dust in coal mines produce "afterdamp" which contains considerable amounts of CO and carbon dioxide. If ventilation is decreased or CO emission increases owing to leaks or disturbances in process, unexpected CO poisonings may occur in industrial operations that usually do not create CO problems.

Toxic action. Small quantities of CO are produced within the human body from the catabolism of haemoglobin and other haemo-containing pigments leading to an endogenous carboxyhaemoglobin (COHb) saturation of about 0.3-0.8% in the blood. Endogenous COHb concentration is increased in haemolytic anaemias and after bruises or haematomas which result in increased haemoglobin catabolism.

CO is easily absorbed through the lungs into the blood. The best understood biological effect of CO is its combination with haemoglobin to form carboxyhaemoglobin. Carbon monoxide competes with oxygen for the binding sites of the haemoglobin molecules. The affinity of human haemoglobin for CO is about 240 times that of its affinity for oxygen. The formation of COHb has two undesirable effects: it blocks oxygen carriage by inactivating haemoglobin and its presence in the blood shifts the dissociation curve of oxyhaemoglobin to the left so that the release of remaining oxygen to tissues is impaired. Because of this latter effect the presence of any percentage level of COHb in the blood interferes with tissue oxygenation considerably more than an equivalent reduction of haemoglobin concentration, e.g. through bleeding. Carbon monoxide also binds with myoglobin to form carboxymyoglobin, which may disturb muscle metabolism, especially in the heart. The biological significance of CO combination with other haem proteins, such as cytochrome oxidase and cytochrome P-450, is inadequately investigated for the time being.

The approximate relation of carboxyhaemoglobin (COHb) and oxyhaemoglobin (O_2Hb) in blood can be calculated from the Haldane's equation. The ratio of COHb and O_2Hb is proportional to the ratio of the partial pressures of CO and oxygen in alveolar air:

$$\frac{COHb}{O_2Hb} = 240\,\frac{pCO}{pO_2}$$

The equation is applicable for most practical purposes to approximate the actual relationship in equilibrium state. For any given CO concentration in the ambient air, the COHb concentration increases or decreases towards the equilibrium state according to the equation. The direction of the change in COHb depends on its starting level. For example continuous exposure to ambient air containing 35 ppm of CO would result in equilibrium state of about 5% COHb in blood. After that, if the air concentration remains unchanged there will be no change in COHb level. If the air concentration increases or decreases, the COHb also changes towards the new equilibrium. A heavy smoker may have a COHb concentration of 8% in his blood at the beginning of a work shift. If he is continuously exposed to a 35 ppm CO concentration during the shift, but is not allowed to smoke, his COHb level gradually decreases towards the 5% COHb equilibrium. At the same time, the COHb level of non-smoking workers gradually increases from the starting level of about 0.8% endogenous COHb towards the 5% level. Thus, absorption of CO and build up of COHb is determined by gas laws and the solution of Haldane's equation will give the approximate maximum value of COHb for any ambient air CO concentration. It should be remembered, however, that the equilibrium time for man is several hours for air concentrations of CO usually encountered at worksites. Therefore, when judging the potential health risk of exposure to CO it is important that the exposure time is taken into account in addition to CO concentration in the air. Alveolar ventilation is also a major variable in the rate of CO absorption. When alveolar ventilation increases, for example during heavy physical work, the rate of absorption increases and the equilibrium state is approached more rapidly compared to the situation with normal ventilation.

The biological half-life of COHb concentration in the blood of sedentary adults is about 2-5 h. The elimination of CO becomes slower with time and the lower the initial level of COHb, the slower the rate of excretion.

Acute poisoning. The appearence of symptoms depends on the concentration of CO in the air, the exposure time, the degree of exertion and individual susceptibility. If the exposure is massive, loss of consciousness may take place almost instantaneously with few or no premonitory signs and symptoms. Exposure to concentrations of 10 000-40 000 ppm leads to death within a few minutes. Levels between 1 000 and 10 000 ppm cause symptoms of headache, dizziness and nausea in 13-15 min and unconsciousness and death if exposure continues for 10-45 min, the rapidity of onset depending on the concentrations. Below these levels the time before the onset of symptoms is longer: levels of 500 ppm cause headache after 20 min and levels of 200 ppm after about 50 min. The relation between carboxyhaemoglobin concentrations and the main signs and symptoms is shown in table 1.

The victim of poisoning is classically described as being cherry red. In the early stages of poisoning, the patient may appear pale. Later, the skin, nailbeds and mucous membranes may become cherry red due to a high concentration of carboxyhaemoglobin and a low concentration of reduced haemoglobin in the blood. This sign may be detectable above 30% COHb concentration, but it is not a reliable and regular sign of CO poisoning. The patient's pulse is rapid and bounding. Little or no hyperpnoea is noticed unless COHb level is very high.

Where the symptoms or signs described above occur in a person whose work may expose him to carbon monoxide, poisoning due to this gas should be immediately suspected. Differential diagnosis from drug poisoning, acute alcohol poisoning, cerebral or cardiac accident, diabetic or uraemic coma may be difficult, and the possibility of carbon monoxide exposure is often unrecognised or simply overlooked. Diagnosis of carbon monoxide poisoning should not be considered established until it is ascertained that the body contains abnormal quantities of CO. Carbon monoxide is readily detectable from blood samples or, if a person has healthy lungs, an estimate of blood COHb concentration can be rapidly made from samples of exhaled end-alveolar air which is in equilibrium with blood COHb concentration.

Critical organs in respect to CO action are the brain and the heart, both of which are dependent on continuous uninterrupted supply of oxygen. The heart is burdened by two mechanisms by carbon monoxide, because its work is increased in order to provide the peripheral oxygen demand while its own oxygen supply is reduced by CO. Myocardial infarction may be precipitated by carbon monoxide.

Acute poisoning may result in neurological or cardiovascular complications which are evident as soon as the patient recovers from the initial coma. In severe poisoning, pulmonary oedema (excess fluid in the lung tissues) may emerge. Pneumonia, sometimes due to aspiration, may develop after a few hours or days. Temporary glycosuria or albuminuria may also occur. In rare cases acute renal failure complicates the recovery from poisoning. Various cutaneous manifestations are occasionally encountered.

After severe CO intoxication the patient may suffer from cerebral oedema with irreversible brain damage of varying extent. The primary recovery may be followed by a subsequent neuropsychiatric relapse days or even weeks after poisoning. Pathological studies of fatal cases show the predominant nervous system lesion in white matter rather than in neurones in those victims who survive a few days after the actual poisoning. The degree of brain damage after CO poisoning is determined by the intensity and duration of exposure.

Until recently, neurological sequelae were thought to be rare and to occur only after extremely severe poisoning. However, on regaining consciousness after severe CO poisoning, 50% of the victims have been reported as presenting an abnormal mental state manifested as irritability, restlessness, prolonged delirium, depression or anxiety. A 3-year follow-up of these patients revealed that 33% had personality deterioration and 43% had persistent memory impairment.

Table 1. Principal signs and symptoms with various concentrations of carboxyhaemoglobin[1]

Carboxyhaemoglobin concentration (%)	Principal signs and symptoms
0.3-0.7	No signs or symptoms. Normal endogenous level.
2.5-5	No symptoms. Compensatory increase in blood flow to certain vital organs. Patients with severe cardiovascular disease may lack compensatory reserve. Chest pain of angina pectoris patients is provoked by less exertion.
5-10	Visual light threshold slightly increased.
10-20	Tightness across the forehead. Slight headache. Visual evoked response abnormal. Possibly slight breathlessness on exertion. May be lethal to fetus. May be lethal for patients with severe heart disease.
20-30	Slight or moderate headache and throbbing in the temples. Flushing. Nausea. Fine manual dexterity abnormal.
30-40	Severe headache, vertigo, nausea and vomiting. Weakness. Irritability and impaired judgement. Syncope on exertion.
40-50	Same as above, but more severe with greater possibility of collapse and syncope.
50-60	Possibly coma with intermittent convulsions and Cheyne-Stokes respiration.
60-70	Coma with intermittent convulsions. Depressed respiration and heart action. Possibly death.
70-80	Weak pulse and slow respiration. Depression of respiratory centre leading to death.

[1] There is considerable individual variation in the occurrence of symptoms.

Repeated exposures. Carbon monoxide does not accumulate in the body. It is completely excreted after each exposure if suffcient time in fresh air is allowed. It is possible, however, that repeated mild or moderate

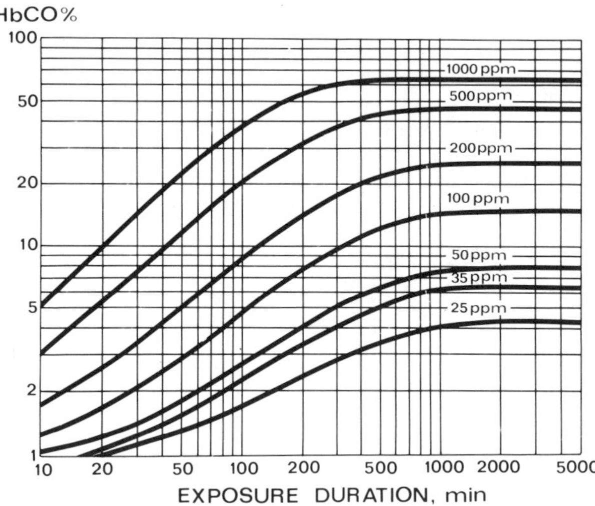

Figure 1. Accumulation of carbon monoxide in human blood at various concentrations of carbon monoxide in the air, according to Stewart et al. (1970).

poisonings which do not lead to unconsciousness would result in death of brain cells and ultimately lead to permanent central nervous system damage with a multitude of possible symptoms such as headache, dizziness, irritability, impairment of memory, personality changes and a state of weakness of the limbs.

Individuals repeatedly exposed to moderate concentrations of CO are possibly adapted to some extent against the action of CO. Mechanisms of adaptation are thought to be similar to the development of tolerance against hypoxia in high altitudes. An increase in the haemoglobin concentration and in a haematocrit has been found to occur in exposed animals, but neither the time course nor the threshold of similar changes in exposed men have been accurately quantified.

Altitudes. At high altitudes the possibility of incomplete burning and greater CO production increases because there is less oxygen per unit of air than at sea level. The adverse body responses also increase due to reduced oxygen partial pressures in breathed air. The oxygen deficiency present at high altitudes and the effects of CO apparently are additive.

Methane-derived halogenated hydrocarbons. Dichloromethane (methylene chloride) which is a major component of many paint strippers and other solvents of this group are metabolised in the liver with the production of CO. COHb concentration may increase up to moderate poisoning level by this mechanism.

Effects of low level exposure to carbon monoxide. In recent years considerable efforts of investigation have been focused on biological effects of COHb concentrations below 10% upon both healthy persons and patients with cardiovascular diseases. Patients with severe cardiovascular disease may lack compensatory reserve at about 3% COHb level, so that the chest pain of angina pectoris patients is provoked by less exertion. As to the central nervous system, the ability to perform complex tasks requiring both judgement and motor co-ordination seems not to be affected adversely by COHb concentrations below 10%. Carbon monoxide readily crosses the placenta to expose the fetus, which is sensitive to any extra hypoxic burden in such a way that its normal development may be endangered.

Susceptible groups. Particularly sensitive to the action of CO are individuals whose oxygen carriage capacity is decreased due to anaemia or haemoglobinopathias, those with increased oxygen needs due to fever, hyperthyroidism or pregnancy, patients with systemic hypoxia due to respiratory insufficiency, and patients with ischemic heart disease and cerebral or generalised arteriosclerosis. Children and young individuals whose ventilation is more rapid than that of adults attain the intoxication level of COHb sooner than healthy adults. Also, smokers whose starting COHb level is higher than that of non-smokers would more rapidly approach dangerous COHb concentrations at high exposures.

SAFETY AND HEALTH MEASURES

Processes should, where possible, be controlled to prevent or minimise carbon monoxide formation, or arrangements should be made to destroy the gas as it is formed. In addition, CO concentrations can be reduced by general dilution ventilation, by installing local exhaust systems and by enclosing the CO sources. Ventilation should be designed to accommodate the largest levels of emission anticipated. Ventilation ducts and pipes designed to disperse fumes must be regularly checked for possible obstruction. Careful attention must be given to metering equipment, piping and burners to prevent leaks. All industrial heating devices should be checked at regular intervals during operation to guarantee that no unexpected CO emission takes place due to progressive damage in the devices. It is essential that properly trained personnel be made responsible for the installation and maintenance of all fuel burning equipment. During the operation of pressure gas producers, unless special precautions are taken there is always a copious escape of gas from poke-holes as soon as the plugs are withdrawn, and gassing accidents during poking and fuel-bed gauging are frequent. Poke-holes should therefore be fitted with a device which creates, in the mouths of the poke-hole, a curtain of steam or compressed air directed downwards so as to balance the pressure of the gas.

In gas manufacture, distribution and use, carbon monoxide may occur during the charging and discharging of retorts, the charging of producers, the emptying of purifiers, and also as a result of leakage from plant, gas taps being left on with gas not ignited, or water-seal failure. In gas manufacture, arrangements should be made, where possible, to ignite gas escaping when feed openings are uncovered, and lids or cover plates should be well fitting to minimise leakage. Water seals should be regularly maintained and frequently inspected to see that they function correctly.

Acute carbon monoxide poisoning may easily occur in the coke industry. Fitting blank flanges, filling, pricking up or measuring the temperatures of the ovens and repairing transmission lines or gas-blowers are operations where exposure to high concentrations of carbon monoxide in the air may occur. When welding gas pipes the gas supply must be cut off and the pipes purged with steam or an inert gas. The most important sources of CO in foundries are the cupolas and the iron casting operation. About 20 to 30% of the gases emitted from the cupolas is carbon monoxide, which can seep into the work environment if a leak occurs, exposing workers in the area to high concentrations of CO. Electric furnaces and oil-heated crucibles produce relatively little CO. During and after casting, the carbonaceous materials in the moulding sand are burned by the hot metal, and CO evolves.

Where it is necessary to run a petrol engine in an enclosed space such as in a repair garage, the vehicle exhaust pipe should be connected directly to an exhaust ventilator manifold. Opening of large service doors increases natural dilution ventilation and can be used as a measure to reduce CO concentration in the air, although the effectiveness of this method varies. Substitution of propane-fuelled or electric-powered lift trucks for petrol-powered equipment can reduce or eliminate exposure in warehouses and factories.

Many cases of poisoning occur during maintenance work on a plant that contains carbon monoxide. Where possible, plant should be shut down, isolated and purged before maintenance work is undertaken. Wherever it is necessary to enter a plant that may contain carbon monoxide, the precautions applicable to entry into confined spaces should be observed.

Containers of carbon monoxide must be checked for leaks upon their arrival, upon filling and at regular intervals thereafter at least every three months. Automated alarms should preferably be employed in premises where CO is stored. Containers of CO and worksites with risk of high exposure to CO emissions should be provided with labels and warning signs. All workers employed in areas where there is a hazard of high concentrations of CO in the air should be adequately instructed about its hazardous properties, symptoms of poisoning, location of masks, and appropriate emergency procedures. Adherence to strict codes of safe working practices is obligatory to prevent gassing accidents.

Respiratory protective equipment may be necessary in emergencies for evacuation purposes, during the clean-up of the area, or in fire fighting. A pressure-demand type self-contained breathing apparatus should be used in high CO concentrations. Satisfactory facepiece fit of respiratory equipment must be ensured. It should be stressed that activated charcoal filters do not provide adequate protection. A filter containing hopcalite may be used for a short period in an emergency. Nobody should be allowed to work alone in areas where exposure to high concentrations of carbon monoxide is anticipated.

Where the CO hazard is severe and regular an effective monitoring system is desirable. Static continuous multipoint samplers can be fitted with automatic alarms and with a recorder to provide permanent records of fluctuations in carbon monoxide concentrations at the various sampling points. Study of these records may provide an indication of leaks or malfunctions in a system, enabling maintenance work to be done before there is a major risk to the workers. Portable continuous samplers can be used to trace the source of CO emission and to sample a suspect area at a representative point. They can also be used in an area where men have to work in low concentrations of CO to monitor levels so that the ceiling level is not exeeded. Traditional monitoring methods should not be completely forgotten. Caged canaries have been used in mines and steel works as an effective and rapid monitoring form. The canary absorbs carbon monoxide much more rapidly than man and therefore falls off its perch before workers begin to be affected.

Trained rescue men must be available on each shift in plants where there is a possibility of dangerous CO emissions into the workplace. Rescue teams should be supplied with self-contained respirators and oxygen resuscitation apparatus.

Biological monitoring. Measurement of blood carboxyhaemoglobin concentration or end-alveolar air concentration of CO after a working shift of a group of exposed workers is a powerful means in evaluating the magnitude of occupational exposure. It is important, however, that the results of non-smokers and smokers be judged separately and, therefore, smoking habits must be registered. The carboxyhaemoglobin concentrations of cigarette smokers usually are between 3 and 10%, and may be as high as 20% in inhaling cigar smokers.

Medical prevention. Pre-employment medical examinations should be performed with special attention to cardiovascular diseases, anaemia, respiratory insufficiency or any other disease or medical condition which could be exacerbated by the hypoxic effect of carbon monoxide. Pregnant women should not be employed in tasks where there is a risk of dangerous concentrations of CO in the air.

Treatment. Biochemical repair starts as soon as the worker is exposed to fresh air, if the victim is still breathing. Therefore, the victim should immediately be moved from the exposure to uncontaminated air. If breathing has stopped, resuscitation should be started without delay, using mouth-to-mouth respiration and, if necessary, external cardiac massage. Pure oxygen by the best method available should be administered to hasten the carbon monoxide excretion. Hyperbaric oxygen treatment at three atmospheres of pressure reduces the biological half-life of carboxyhaemoglobin to 24 min. The toxicity of oxygen itself should not be forgotten, however. The addition of 5% of CO_2 to oxygen, to serve as a respiratory stimulant, entails some risk of compounding the mild metabolic acidosis that arises from tissue hypoxia. Stimulant drugs should be avoided and methylene blue must not be injected. Absolute rest in bed for at least 48 h should be ensured. Avoiding early physical activity may be prophylactic as to possible later complications of CO poisoning. After recovery, the patient must be watched for late neurological and cardiac complications.

KURPPA, K.
RANTANEN, J.[1]

[1] The authors wish to acknowledge the contribution of Dr. G. O. Lindgren, whose review in the previous edition served as an excellent background for this article and has been partly quoted as such.

"The effect of carbon monoxide on humans". Stewart, R. D. *Journal of Occupational Medicine* (Chicago), May 1976, 18/5 (304-309). Illus. 42 ref.

Carbon monoxide. Environmental health criteria 13 (Geneva, World Health Organisation, 1979), 125 p. Illus. 356 ref.

"Carbon monoxide criteria. With reference to effects on the heart, central nervous system and fetus". Rylander, R.; Vesterlund, J. *Scandinavian Journal of Work, Environment and Health* (Helsinki), 1981, 7/suppl. 1, 39 p. Illus. 94 ref.

"Warning levels and alarm instrumentation for carbon monoxide in the atmosphere". Jones, J. G. *Annals of Occupational Hygiene* (Oxford), Aug. 1975, 18/1 (79-82). 1 ref.

CIS 78-452 "End-expired air technic for determining occupational carbon monoxide exposure". Smith, N. J. *Journal of Occupational Medicine* (Chicago), Nov. 1977, 19/11 (766-769). 8 ref.

"Methods to reduce carbon monoxide levels at the workplace". Haag, W. M. *Preventive Medicine* (New York), May 1979, 8/3 (369-378). 25 ref.

Occupational exposure to carbon monoxide–Criteria for a recommended standard. US Department of Health, Education and Welfare. HSM 73-11000 (National Institute for Occupational Safety and Health, 4676 Columbia Parkway, Cincinnati) (1972). Illus. 129 ref.

"Experimental human exposure to carbon monoxide". Stewart, R. L.; Petersen, J. E.; Baretta, E. D.; Backmand, R. T.; Hosko, M. J.; Herrmann, A. A. *Archives of Environmental Health* (Chicago), Aug. 1970, 21/2 (154-164). 9 ref.

Carbon tetrachloride

Carbon tetrachloride (CCl_4)

TETRACHLOROMETHANE; PERCHLOROMETHANE

m.w.	153.84
sp.gr.	1.59
m.p.	−22.9 °C
b.p.	76.7 °C
v.d.	5.3
v.p.	113 mmHg ($14.7 \cdot 10^3$ Pa) at 25 °C

insoluble in water; highly soluble in ethyl ether and benzene

a clear, colourless liquid, with an ethereal odour.

TWA OSHA	10 ppm
OSHA	25 ppm ceil
OSHA	200 ppm/5 min/4 h peak
NIOSH	2 ppm/1 h ceil
TLV ACGIH	5 ppm 30 mg/m³ skin. Industrial substance suspected of carcinogenic potential for man
STEL ACGIH	20 ppm 125 mg/m³
IDLH	300 ppm
MAC USSR	20 mg/m³ skin

Production. It is obtained from carbon disulphide and chlorine in the presence of a catalyst, or by the chlorination of aliphatic hydrocarbons.

Uses. As a result of its easy availability and low cost, carbon tetrachloride has been used extensively as a chemical intermediate, fire extinguishing agent, fumigant, suppressant of the flammability of mixtures of more flammable materials, and as a solvent. This latter application has been the cause of most problems due to improper handling at home and in industry and this use has been essentially terminated.

HAZARDS

Most carbon tetrachloride intoxications have resulted from the inhalation of the vapour; however, the substance is also readily absorbed from the gastrointestinal tract. Being a good fat solvent, carbon tetrachloride removes fat from the skin on contact, which may lead to development of a secondary septic dermatitis. Since it is absorbed through the skin, care should be taken to avoid prolonged and repeated skin contact. Contact with the eyes may cause a transient irritation, but does not lead to serious injury.

Carbon tetrachloride has anaesthetic properties and exposures to high vapour concentrations can lead to the rapid loss of consciousness. Individuals exposed to less than anaesthetic concentrations of carbon tetrachloride vapour frequently exhibit other nervous system effects such as dizziness, vertigo, headache, depression, mental confusion, and incoordination. It may cause cardiac arrhythmias and ventricular fibrillation at higher concentrations. At surprisingly low vapour concentrations, gastrointestinal disturbances such as nausea, vomiting, abdominal pain and diarrhoea are manifested by some individuals.

The effects of carbon tetrachloride on the liver and kidney must be given primary consideration in evaluating the potential hazard incurred by individuals working with this compound. It should be noted that the consumption of alcohol augments the injurious effects of this substance. Anuria or oliguria is the initial response, which is followed in a few days by a diuresis. The urine obtained during the period of diuresis has a low specific gravity, and usually contains protein, albumin, pigmented casts and red blood cells. Renal clearance of inulin, diodrast, and *p*-aminohippuric acid are reduced, indicating a decrease in blood flow through the kidney as well as glomerular and tubular damage. The function of the kidney gradually returns to normal, and within 100 to 200 days after exposure, the kidney function is in the low normal range. Histopathological examination of the kidneys reveals varying degrees of damage to the tubular epithelium.

Experimental toxicity. In animals carbon tetrachloride was demonstrated to be a more specific hepatotoxin than six other aliphatic chlorinated hydrocarbons, and after a single intraperitoneal injection or after a single exposure to its vapour, it was found that the dose or exposure necessary to cause anaesthesia was approximately 200 times greater than that needed to cause liver damage. Liver damage induced by single exposure to carbon tetrachloride is manifested grossly by the swollen fatty appearance of the liver; focal areas of necrosis are seen in the liver of animals after large exposures.

While numerous references describe alterations in blood and liver enzyme levels and induction of liver enzymes and suggest possible routes of metabolism of carbon tetrachloride, these are of limited value in therapy. Prevention of illness by eliminating excessive exposure is essential.

In human beings, kidney damage is apparently more prominent following exposures to carbon tetrachloride than in animals. The effects seen in animals exposed repeatedly over a prolonged period of time are even more disturbing. The structure and function of the liver is most markedly altered by repeated low magnitude exposures, whereas alterations in the structure and function of the kidney are less prominent.

Undesirable effects including fatty metamorphosis and cirrhosis of the liver, parenchymatous degeneration of the tubular epithelium of the kidneys, and alterations in the behaviour, gross appearance and growth, as well as death, were observed in all of the animal species exposed to vapour concentrations of 200 and 400 ppm for long periods of time.

At vapour concentrations of 5 ppm, no significant changes were detectable in any of the species.

Although studies in aminals fed hepatotoxic levels of carbon tetrachloride have produced cancer in rats and mice, the effects at non-hepatotoxic levels have not been adequately evaluated nor have epidemiological studies been conducted in exposed human populations.

The over-all metabolism of carbon tetrachloride including absorption, distribution, and excretion by monkeys exposed to its vapour has been reported. Using the ^{14}C-labelled compound, it was found that approximately 30% was absorbed. At least 51% of the amount absorbed during an inhalation period was eliminated in the expired air within 1 800 h. The remainder was excreted in the urine and faeces. Approximately 4.4% was eliminated as carbon dioxide. Of the radioactivity excreted in the urine 94% was a non-volatile unidentified metabolite, while small amounts were contained in ureas and carbonates.

SAFETY AND HEALTH MEASURES

With the recognition of the health hazards incurred by exposures to carbon tetrachloride and the consequent development of the methodology for minimising such exposures, the incidence of such intoxications in industry has been reduced. Industrial experience has been excellent when prescribed limits have not been exceeded and the converse has likewise been true.

Measures should be aimed at preventing the inhalation of vapour and avoiding prolonged or repeated skin contact. When it is used in larger quantities, carbon tetrachloride should not be kept in open containers. Ventilation should be provided, particularly at floor level since the vapours are heavier than air. Employees who must be exposed to it should wear adequate respiratory protective equipment as well as hand protection to prevent dermatitis. Anyone who enters tanks or other enclosed areas where high concentration of the vapour may be present should do so only with adequate respiratory protection, a lifeline and harness, while being observed by an attendant on the outside. Carbon tetrachloride should be stored in a cool ventilated place and labelled properly.

Treatment. Persons suspected of having been overexposed to this substance should be removed to fresh air immediately and a physician called. Artificial respiration should be started if breathing has stopped. In case of spillage on the person, contaminated clothing must be removed, and if carbon tetrachloride has been splashed into the eyes, they should be irrigated with copious amounts of warm water for 15 min. If the material has been swallowed, vomiting should be induced. Therapy should be supportive with particular attention to the liver and kidneys. In cases of suspected recent exposure, catecholamines such as adrenalin should be avoided since carbon tetrachloride is known to sensitise the heart to their action.

GEHRING, P. J.
ROWE, V. K.
TORKELSON, T. R.

CIS 77-735 "Carbon tetrachloride toxicity potentiated by isopropyl alcohol". Folland, D. S.; Schaffner, W.; Ginn, H. E.; Crofford, O. B.; McMurray, D. R. *Journal of the American Medical Association* (Chicago), Oct. 1976, 236/16 (1853-1856). Illus. 20 ref.

CIS 78-1088 "Cochleovestibular findings in carbon tetrachloride exposure" (Cochleo-vestibuläre Befunde bei Tetrachlorkohlenstoffexposition). Wenkebach, P.; Skurczynski, W. *Zeitschrift für die gesamte Hygiene und ihre Grenzgebiete* (Berlin), 1977, 23/3 (145-147). 9 ref. (In German)

"Carbon tetrachloride" (53-60). 28 ref. *IARC monographs on the evaluation of carcinogenic risk of chemicals to man.* Vol. I (Lyons, International Agency for Research on Cancer, 1972), 184 p. Ref.

CIS 74-1678 "Prevention and treatment of carbon tetrachloride hepatotoxicity by cysteine: Studies about its mechanism". De Ferreyra, E. C.; Castro, J. A.; Díaz Gómez, M. I.; D'Acosta, N.; De Castro, C. R.; De Fenos, O. M. *Toxicicology and Applied Pharmacology* (New York), Mar. 1974, 27/3 (558-568). Illus. 25 ref.

Carcinogenic substances

Before the Industrial Revolution began life centred on the village, and the main occupations were those closely related to agriculture. The Industrial Revolution resulted in a significant change in the social and economic structure of society. Workers were drawn from the land to the new factories which were springing up. These changes, coupled with those resulting from an increasing use and discovery of new chemicals and processes, resulted in both an increase in the numbers of persons exposed and an increase in the levels of exposure.

The high levels of exposure have resulted in the fact that almost all agents that have been recognised as being carcinogenic in humans were first identified as such in the occupational setting. Although the concentration of a carcinogen at work is usually greater than in the various environments outside the workplace, technological advances and improvements are such that jobs and exposures are constantly changing. This means that new jobs and exposures come into being as others cease to exist. Most cancers have a long latency period of 10 to 25 years or more. Thus an epidemiological study carried out today is in a sense historical, since it is demonstrating the results of exposures 10 to 25 years or more ago. Unless careful measurements were made at that time, it is not possible to get an accurate idea of the exposures then involved, resulting in today's tumours. In a similar way, measurements of exposure made at work today may only be of value 10 to 25 years or more in the future.

As other causes of death have declined in frequency, in particular the infectious diseases, so cancer has increased in importance both in relative and in absolute terms. In several countries it now accounts for about 20% of deaths, the vast majority of fatal human cancers being carcinomas. The estimations of the percentage of cancer cases due to occupational exposure among the total human cancer cases have varied from 1% to 40%. It would be rather difficult at present to state to what extent the contribution of occupational exposure to the total human cancer burden has been gossly under- or over-estimated. While there is uncontrovertible evidence that a certain proportion of human cancers are due to identified occupational exposures, it is not possible in the present state of our knowledge to quantify exactly how many human cancers could be prevented by the improvement of working conditions. The main reasons for this are that:

(a) we do not know how many chemicals are actually carcinogenic to humans;

(b) we do not know with any precision how many cancer cases are actually due to exposure to recognised carcinogens. In fact in most cases the counting of occupational cancers is limited to those which have had the doubtful privilege of being the object of a scientific publication;

(c) we do not know enough about the effects of low levels of exposure and of exposure to multiple carcinogenic or promoting agents.

Definitions

The existence of numerous and varied definitions illustrates the difficulty in drawing up an exact and uniform definition of a carcinogen. Examples of definitions are as follows:

From the International Agency for Research on Cancer (IARC) in its monographs:

"The induction by chemicals of neoplasms that are usually observed, and/or the induction by chemicals of more neoplasms than are usually found, although fundamental differences in the mechanisms may be involved. Etymologically, the term 'carcinogenesis' means the induction of cancer, that is, of malignant neoplasms; however, the commonly accepted meaning is the induction of various types of neoplasms or of a combination of malignant and benign tumours. Within the monographs, the words 'tumour' and 'neoplasm' are used interchangeably. (In the scientific literature the terms 'tumourigen', 'oncogen' and 'blastomogen' have all been used synonymously with 'carcinogen', although occasionally 'tumourigen' has been used specifically to denote the induction of benign tumours)."

From the United States Occupational Safety and Health Administration (OSHA):

"'Potential occupational carcinogen' means any substance, or combination or mixture of substances, which causes an increased incidence of benign and/or malignant neoplasms, or a substantial decrease in the latency period between exposure and onset of neoplasms in humans or in one or more experimental mammalian species as the result of any oral, respiratory or dermal exposure, or any other exposure which results in the induction of tumors at a site other than the site of administration. This definition also includes any substance which is metabolised into one or more potential occupational carcinogens by mammals."

From the Commission of the European Communities:

"Substances or preparations which, by inhalation, ingestion or cutaneous penetration can induce cancer or increase its frequency."

Concepts concerning the origins of cancer

There is a considerable amount of information available on the geographical distribution of cancer. Particularly useful in this context are the data collected by cancer registers. These data show enormous variations in the frequency of various types of cancer from country to country as well as demonstrating marked differences within a country. As examples, cancer of the prostate is rare in Japan, but common in Canada; and in England and Wales cancer of the bladder occurs most frequently in Yorkshire and Humberside.

Geographical differences of this kind have led to the concept that the majority of human cancers are due to external factors. The evidence includes case reports, descriptive and analytical epidemiological studies (including studies on emigrants) and is backed up by studies on the induction of tumours in animals. External factors include agents associated with air, water, food, life-style, exposure to agents at work, as well as changes in body metabolism, which may be caused by diet, disease, etc. In addition, the effect of such factors may be influenced by genetic constitution. See CANCER, ENVIRONMENTAL.

These various factors may interact with each other in such a way that they influence the probability of developing cancer. The possibility of interaction between various agents and/or risk factors may represent a serious difficulty in singling out a particular agent as the specific cause of cancer. However, some factors have been positively identified as carcinogenic for humans and others have been suggested as carcinogens based on experimental evidence. As a particular example, it has been shown that both alpha radiation and asbestos act synergistically with cigarette smoking in causing lung cancer in exposed workers.

It is possible that chemicals can lead to the development of cancer in one or both of two ways:

(a) by initiating changes in cells;

(b) by permitting selective growth of populations of changed cells.

Some chemicals may act in both ways. See CANCER, OCCUPATIONAL.

Tests for determining carcinogenic potential

Occupational cancer is related to previous exposure at work to a carcinogen, which may involve the interaction of a number of other factors. The existence of such an exposure indicates that either the agent has not been recognised as a carcinogen, or that inadequate protective measures were taken to reduce the exposure sufficiently.

Since relatively few substances have been identified as carcinogenic for humans, a great deal of research effort has been put into tests aimed at detecting agents with a carcinogenic potential. Classically this has been carried out on test animals exposed for the major portion of their life span, but the recognition in recent years that carcinogens have mutagenic effects has led to the development of a number of short-term mutagenicity tests, which can be used, in particular, for assessing new chemicals for which there is a likelihood of significant human exposure. It is therefore increasingly likely that tests on animals will be reserved for those agents that are shown to be mutagenic.

Mutagenicity tests. A mutagen produces alterations in the genetic apparatus of an organism (see MUTAGENIC EFFECTS). Individual mutagenicity tests have limited predictability since they may produce both false positive results and false negative results. To minimise these false results more than one test is normally used.

The mutagenicity test systems selected should be relatively easy to perform, sensitive to the class of chemical under investigation and reasonable in demands on resources, and should have demonstrated utility in screening a number of classes of compounds.

It should be recognised that there does not exist a minimum set of *in vitro* mutagenicity tests which can presently detect all genetically active agents. Nevertheless an extensive data base exists for a correlation between mutagenicity as demonstrated by *in vitro* tests and carcinogenicity in animals. The best correlation has been established using results from bacterial tests. However, any positive results from any mutagenicity test should be taken as an alert for potential carcinogenicity.

For screening purposes it will generally be necessary to obtain information on mutations in:

(a) bacterial cells, which in most instances would be information on the potential of a chemical to produce gene (point) mutations. Experimental evidence indicates that bacteria such as *Salmonella typhimurium* and *Eschericia coli* are presently suitable;

(b) non-bacterial cells, which generally involves the determination of the potential of a chemical to produce chromosomal changes. The micronucleus test, *in vitro* chromosomal aberrations and *in vivo* cytogenetics in bone marrow are presently suitable tests.

Long-term studies in test animals. There is a considerable amount of evidence that shows that chemicals which cause cancer in humans are known to cause cancer in animals, although they often act at different sites. At present the only known exception is arsenic. It would therefore be reasonable to regard chemicals which are causally associated with cancer in animals as if they represented a carcinogenic risk for man. However, no scientifically unexceptionable methods presently exist for directly extrapolating from experimental data to humans which permit an accurate quantitative assessment to be made of human risks solely on the basis of experimental results. The good empirical correlation which exists between experimental and human data for the limited number of identified human carcinogens strongly supports the view that experimental data can instead predict a qualitatively similar response in humans.

A long-term carcinogenicity study is performed so that test animals can be observed for a major portion of their life-span for the development of neoplastic lesions during or after exposure to various doses of a test substance by an appropriate route.

Normally two species are used and, because of the long latent period required for induction and manifestation of tumours, it is generally agreed that treatment of test animals should be started in young animals and continued for the duration of the experiment. The choice of the route of administration depends upon the physical and chemical characteristics of the test substance and the form typifying human exposure. It is generally advised that more than one dose level be used, the highest being the one nearest to the maximum tolerated dose, that is the highest dose permitting a normal life-span and producing only minimal toxic effects. A lower dose is generally selected as ½ or ¼ of the high dose. If a dose-response relationship is desired, then 3 or more dose levels should be used scaled by a factor of 3 or 5, or even 10 from the high test dose.

As for the duration of observation, a long finite time is generally preferable to the entire life-span. As a general guideline, it is advised that survivors in all groups are killed and the whole experiment is terminated if mortality in the control or low-dose group ever reaches 75% and that, in any case, the study is not continued beyond 130 weeks for rats, 120 weeks for mice and 100 weeks for hamsters, irrespective of mortality.

Gross pathology should be performed in all animals by a trained laboratory animal pathologist or by experienced technicians under the guidance and responsibility of the pathologist and detailed histological examination should be conducted, as a minimal requirement, in the control and the highest effective dose groups. For more details on guidelines for long-term as well as for short-term carcinogenicity tests, see: "Long-term and short-term screening assays for carcinogens: a critical appraisal". IARC supplement 2, IARC, Lyons, 1980.

With regard to the assessment of evidence for carcinogenicity from experimental animal studies, the IARC has classified such assessments into four groups:

(1) *Sufficient evidence* of carcinogenicity indicates that there is an increased incidence of malignant

tumours: *(a)* in multiple species or strains, or *(b)* in multiple experiments (preferably with different routes of administration or using different dose levels), or *(c)* to an unusual degree with regard to incidence, site or type of tumour, or age at onset. Additional evidence may be provided by data concerning dose-response effects, as well as information on mutagenicity or chemical structure.

(2) *Limited evidence* of carcinogenicity means that the data suggest a carcinogenic effect but are limited because *(a)* the studies involve a single species, strain or experiment; or *(b)* the experiments are restricted by inadequate dosage levels, inadequate duration of exposure to the agent, inadequate period of follow-up, poor survival, too few animals, or inadequate reporting; or *(c)* the neoplasms produced often occur spontaneously or are difficult to classify as malignant by histological criteria alone (e.g. lung and liver tumours in mice).

Table 1. Chemicals, groups of chemicals and industrial processes carcinogenic to humans

Chemicals and groups of chemicals	Main target sites
4-Aminodiphenyl	Bladder
Arsenic and certain arsenic compounds	Skin, lung
Asbestos	Lung, pleura, peritoneum, gastrointestinal tract, larynx
Benzene	Haematolymphopoietic system
Benzidine	Bladder
N,N-*bis*-2-chloroethyl-2-naphthylamine (chlornaphazine)	Bladder
bis-Chloromethyl ether and technical grade chloromethyl methyl ether	Lung
Chromium and certain chromium compounds	Lung
Diethylstilboestrol	Female genital tract, breast
Melphalan	Haematolymphopoietic system
Mustard gas	Lung, pharynx
2-Naphthylamine	Bladder
Soots, tars and mineral oils[1]	Skin, lung, bladder, gastrointestinal tract
Vinyl chloride	Liver, lung, central nervous system, haematolymphopoietic system
Conjugated oestrogens	Uterus
Cyclophosphamide	Bladder
Industrial processes	**Main target sites**
Manufacture of auramine[1]	Bladder
Underground haematite mining	Lung
Manufacture of isopropyl alcohol by the strong acid process[1]	Nose and nasal sinus
Nickel refining[1]	Lung, nose and nasal sinus
Boot and shoe manufacture and repair (certain occupations)	Nose
Furniture and cabinet-making industry (certain occupations)	Nose

[1] The specific compound(s) which may be responsible for a carcinogenic effect in humans has (have) not been identified.

(3) *Inadequate evidence* indicates that because of major qualitative or quantitative limitations, the studies cannot be interpreted as showing either the presence or absence of a carcinogenic effect.

(4) *Negative evidence* means that, within the limits of the tests used, the chemical is not carcinogenic. The number of negative studies is small since, in general, studies that show no effect are less likely to be published than those suggesting carcinogenicity.

The classification of *sufficient evidence* and *limited evidence* only refer to the strength of the experimental evidence that these chemicals are or are not carcinogenic on the basis of the available data, and do not refer to the extent of their carcinogenic activity.

Human carcinogens

The ultimate "proof" of whether or not an agent is carcinogenic for humans is demonstrated by human

Table 2. Probable human carcinogens

Higher degree of evidence	Lower degree of evidence
Aflatoxins	Acrylonitrile
Cadmium and certain cadmium compounds[1]	Aminotriazole (amitrole)
Chlorambucil	Auramine
Azathioprine	Beryllium and certain beryllium compounds[1]
Nickel and certain nickel compounds[1]	Carbon tetrachloride
tris-1-Aziridinyl-phosphine sulphide (Thiotepa)	Dimethylcarbamoyl chloride
	Dimethylsulphate
	Ethylene oxide
	Iron dextran
	Oxymetholone
	Phenacetin
	Polychlorinated biphenyls

[1] The specific compound(s) which may be responsible for a carcinogenic effect in humans has (have) not been identified.

Table 3. Excess cancer incidence in occupational groups

Occupational group	Cancer site(s)	Percentage excess reported
Coal miners	Stomach	40
Chemists	Pancreas	64
	Lymphatic system (lymphomas)	79
Foundry workers	Lung	50-150
Textile workers	Mouth and pharynx	77
Printing pressmen (newspapers)	Mouth and pharynx	125
Metal miners	Lung	200
Coke by-product workers	Large intestine	181
	Pancreas	312
Cadmium production workers	Lung	135
	Prostate	248
Rubber industry:		
Processing	Stomach	80
	Leukaemia	140
Tyre building	Bladder	88
	Brain	90
Tyre curing	Lung	61
Furniture workers	Nasal cavity and sinuses	300-400
Shoe workers	Nasal cavity and sinuses	700
	Blood marrow (leukaemia)	100
Leather workers	Bladder	150

studies which provide adequate evidence of carcinogenicity and indicate a causal relationship between exposure and human cancer. Such studies are of considerable value today in assessing human carcinogenicity, but it is likely that their value will in the future become reduced as more and more carcinogens are identified and controlled.

Evidence of carcinogenicity for humans is obtained from three main sources:

(a) case reports;

(b) descriptive epidemiological studies which observe the distribution of cancer in human populations;

(c) analytical epidemiological studies which investigate the hypotheses suggested by the descriptive studies aimed at examining individual exposure to an agent or group of agents and its association with an increased risk of cancer.

Although in some cases a single epidemiological study may be sufficient to indicate a cause-effect relationship, the most convincing evidence of causality is given when several studies performed independently of one another produce similar results.

In the cases of agents where there is insufficient evidence available for determining a causal relationship, reliance has to be placed on the available human data together with non-human data such as those obtained from animal studies.

Probably the best review of chemical carcinogens is contained in the IARC monographs. From these monographs a total of 39 chemicals, groups of chemicals or industrial processes for which data of carcinogenicity from human and animal studies were available, were classified as either carcinogenic to humans (group 1, see table 1) or probably carcinogenic to humans (group 2, see table 2). The chemicals in group 2 were further subdivided into subgroups A and B indicating respectively higher or lower degree of evidence.

It should be noted, however, that this list is not exhaustive, and does not include agents, such as ionising radiations, which can produce a wide variety of cancers, in particular leukaemia. (As regards occupation, a more detailed analysis is provided in the article CANCER, OCCUPATIONAL).

The evidence available from short-term mutagenicity tests and from long-term studies in test animals indicates that there are many more agents than those in tables 1 and 2 which may play a role in the development of occupational cancer. This tends to be supported by the observed excess incidence of cancer in various groups of workers (table 3) for which no specific causative agent has yet been identified.

It is therefore evident that although a considerable amount of data are currently available on occupational cancer, there is a need for more research to be carried out to answer some of the problems still outstanding. It is also evident that the toll of occupational cancer can be considerably reduced by limiting or eliminating exposure of workers to known carcinogens.

HUNTER, W. J.

"Evaluation of the carcinogenicity of chemicals: a review of the monograph program of the International Agency for Research on Cancer". Tomatis, L.; Agthe, C.; Bartsch, H.; Huff, J.; Montesano, R.; Saracci, R.; Walker, E.; Wilbourn, J. *Cancer Research* (Chicago), Apr. 1978, 38 (877-885). 50 ref.

Principles and methods for evaluating the toxicity of chemicals. Part 1. Environmental health critera No. 6 (Geneva, World Health Organisation, 1978), 272 p. Illus. Ref.

Convention 139. Convention concerning prevention and control of occupational hazards caused by carcinogenic substances and agents. International Labour Conference (Geneva, International Labour Office, 1974).

Recommendation 147. Recommendation concerning prevention and control of occupational hazards caused by carcinogenic substances and agents. International Labour Conference (Geneva, International Labour Office, 1974).

"Identification, classification and regulation of potential occupational carcinogens". Department of Labor, Occupational Safety and Health Administration. *Federal Register* (Washington, DC), 22 Jan. 1980, 45/15 (5 002-5 295).

Short term test systems for detecting carcinogens. Norpoth, K. H.; Garner, R. C. (eds.). (West Berlin, Heidelberg, Springer, 1980), 417 p. Illus. Ref.

"Occupational chemical pollution and cancer" (Pollution chimique professionnelle et cancer). Lauwerys, R. *Archives belges de médecine sociale, hygiène, médecine du travail et médecine légale* (Brussels), June-July 1979, 37/6 (337-384). 257 ref. (In French)

Long-term and short-term screening assays for carcinogens: a critical appraisal. IARC monographs on the evaluation of carcinogenic risk of chemicals to humans. Supplement 2 (Lyons, International Agency for Research on Cancer, 1980), 426 p.

Carcinogenic substances, epigenetic

It is generally recognised that carcinogenesis is a multistep process which may involve different kinds of causative agents whose distinctive characteristics are still unclear. At present, chemical carcinogens can be tentatively and broadly grouped in two categories, the first including substances that alter in a still unspecified manner the cellular genome through either a direct chemical reaction with DNA or other indirect mechanisms *(genotoxic carcinogens)*, and the second including substances that are believed to act by altering gene expression without permanent damage to the genetic material *(epigenetic carcinogens)*. It must be emphasised that the mechanism of action of the latter category is not well understood and that at present epigenetic carcinogens are operationally defined as such by the absence of mutagenicity as demonstrated by short-term assays and by their role in the whole process of carcinogenesis, even though some of them may be unable to produce tumours by themselves. It is thought that even a tentative classification of carcinogens according to their mechanism of action may be important in defining the relative risk involved in human exposure.

The classical two-stage mouse skin model has shown that epidermal cells modified by a subcarcinogenic dose of a carcinogen (a process called *initiation*) continue to behave normally unless a second stimulus *(promotion)*, consisting of repeated topical applications of croton oil, will transform initiation cells into tumours. Whereas initiation is caused by genotoxic carcinogens and is irreversible, promotion is dose dependent and reversible. Promoting agents may be considered a good example of epigenetic carcinogens. In the skin model, the potent principle of croton oil, the phorbol ester 12-*o*-tetradecanoyl-phorbol-13-acetate (TPA), is not carcinogenic by itself, is not mutagenic, and its promoting action is accompanied by inhibition of cell differentiation and increased proliferation, and by alteration of several enzymes and cell surface properties.

In other experimental models it is more difficult to understand whether a chemical is a weak complete carcinogen or has only an enhancing effect on cells already initiated by either a genotoxic carcinogen or an unknown factor. In liver carcinogenesis it has been demonstrated that phenobarbitol given in the diet after a small dose of a carcinogen greatly enhances the production of liver tumours in the rat. Several other chemicals such as DDT, chloroform, carbon tetrachloride, TCDD, and polychlorinated byphenyls, exert similar enhancing effects, but they can induce the same type of tumours in mice and rats when administered alone at high doses continuously for life. Since they have not been found to be mutagenic in short-term tests, it can be hypothesised that their carcinogenic potentiality is mostly related to a promoting activity on cells initiated by other carcinogens unintentionally contaminating the animal environment.

Also the artificial sweeteners saccharin and sodium cyclamate may be thought to act as epigenetic rather than genotoxic carcinogens because they are not mutagens, produce tumours in animals only after exceedingly high doses, and greatly enhance the incidence of urinary bladder tumours in rats receiving a sub-threshold dose of a carcinogen.

In other experimental models it has been found that chemicals with a specific action on endocrine organs may exert carcinogenic activity on the same organs. In particular, thiourea, thiouracil and derivatives are anti-thyroid substances which produce thyroid tumours in the experimental animals by an indirect mechanism of hormonal imbalance; however, they increase the incidence of tumours also in organs which are not under specific hormonal control. The same reasoning can be applied to synthetic oestrogens which have been shown to produce tumours principally in oestrogen-responsive tissues.

On the basis of many experimental results, it is reasonable to hypothesise that certain carcinogens are endowed with both initiating and promoting capacities. However, particularly when the carcinogenic potentiality is achieved after a long-term treatment at high doses, the biological effects of impurities should be investigated. Moreover, various types of interactions may occur when mixtures of chemicals are considered. This is of particularly frequent occurrence in the case of occupational exposure, and is also well illustrated by cigarette smoking. Chemical analysis has shown that cigarette smoke contains several known carcinogens; however, biological studies and epidemiological observations are consistent with the presence of promoting chemicals. Also, cigarette smoking increases in a multiplicative way lung cancer incidence in workers exposed to other carcinogens, particularly to asbestos fibres.

It is evident from the examples reported that epigenetic carcinogens are not readily defined and include chemicals which may act in different ways. Clearly, much more research is urgently needed on this group of carcinogens because of their potential importance in the human environment. On the one hand, if not tested in a properly designed experiment, a pure promoter, even of considerable potency in enhancing tumour formation, may not be recognised. On the other hand, it is of importance to identify epigenetic carcinogens as such, because they appear to have threshold levels of action and because their effects seem, to a certain extent, to be reversible. Moreover, recent studies have led to the discovery of substances which can inhibit the process of promotion through different mechanisms. If these studies are confirmed, it may become feasible to counteract the action of at least some of the epigenetic carcinogens and in this way to block the development of tumours from cells transformed by genotoxic carcinogens.

DELLA PORTA, G.

Cardiacs at work

An acquired heart condition, whether of occupational or non-occupational origin, presents the occupational physician with a number of complex problems, which usually necessitate solutions varying from case to case, and which require close supervision of the worker in question.

The conditions may be disparate, functional or organic with variable consequences. The lesions may involve the coronary system (i.e. the heart vessels), the endocardium (the internal membrane of the heart), the myocardium (the cardiac muscle) and pericardium (the serous envelope of the cardiac muscle). These lesions, or simple functional disorders, may give rise to a variety of symptoms: palpitations, cardiac rhythm disorders, cardiac insufficiency, precordial pain, angina pectoris, fainting fits. Finally, injuries may result in heart valve ruptures or heart wounds.

Heart disease and work

About 6% of all workers suffer from a heart condition although, among elderly workers, the incidence is greater. In general, morbidity and absenteeism among cardiacs is similar to that for other groups of workers; however, in France for example, cardiacs come immediately behind tubercular patients in the list of persons receiving insurance payments for "long-term illness" or "disability".

Living and working conditions (hours of work, speed of work, nutrition, smoking) are factors at least as important in the aetiology of heart disease as physical work itself. Coronary diseases are found more frequently amongst sedentary workers or persons with heavy responsibility, e.g. manager's or executive's coronary disease.

It has been asked whether cardiovascular conditions raise the accident frequency rate. It was at one time believed that hypertensive persons were more liable to accidents; however, statistics have disproved this and precise and detailed studies have shown that persons with cardiovascular disorders are no more accident-prone than other workers.

From the physiology of the heart and the phenomena that allow the heart to adapt to effort, it can be seen that arduous work, i.e. work which places high demand on the heart pump, has an undesirable effect on the myocardium and may aggravate existing lesions. To such effects can be related the majority of contraindications for the employment of cardiacs.

Work unsuitable for cardiacs

Although a cardiac condition may have been diagnosed in a worker, it does not necessarily follow that his job should be changed, or that he should be classified as a "handicapped worker", or that he should no longer continue to work. Only a detailed medical examination can provide the necessary information for evaluating the individual patient's work capacity. Such an evaluation should be carried out on each worker with a suspected cardiovascular disorder, who is to be employed on a job involving:

(a) heavy lifting and carrying entailing high levels of energy expenditure;

(b) uncomfortable postures (e.g. standing, lying, crouching, etc.), or dangerous positions (e.g. work at heights);

(c) repeated or prolonged movements (e.g. movements round a machine, painting a large surface, climbing stairs, inclines or scaffolds, etc.);

(d) inclement climatic conditions (e.g. exposure to heat, hot work, work outdoors, work in humid atmospheres, at high altitudes, in compressed-air, or in low atmospheric pressures);

(e) exposure to vibrations (road drills, percussion drills, heavy contractors' plant, etc.);

(f) work entailing the safety of others.

This list does not claim to be exhaustive; however, it will have little value at all if it is not adapted to each individual case. An absolute contraindication for one cardiac may be only relative for another, for example:

(a) Persons suffering from neurocirculatory asthenia have almost complete working capacity provided they are kept under treatment–they can do most jobs although it is preferable not to employ them on work which entails the safety of others or which necessitates high rates of effort.

(b) In the case of persons with isolated, benign or severe hypertension, it is necessary to avoid sudden bursts of effort, exposure to inclement weather and factors such as noise that may produce nervous fatigue.

(c) Workers with valvular heart disease should not be exposed to an infectious disease hazard, toxic substances or vibrations–patients with valvular heart disease which is likely to decompensate easily should do no heavy work at all and the general contraindications listed above must be applied more rigorously.

(d) In the case of persons suffering from angina pectoris the problems are more complex. The onset of an attack is not the result of the normal routine effort of the individual, or even of heavy effort; it usually follows some special exertion such as walking, riding a bicycle or climbing stairs. A precise evaluation should be made of the resistance of these patients to various forms of exertion.

(e) Cardiologists consider that myocardial infarction is not, in itself, a contraindication for most types of work–however, persons recovering from an infarction must be closely supervised, and work entailing the safety of others, exposure to toxic substance and burn or trauma hazards should be avoided.

(f) In persons who have undergone heart surgery, it seems that in most cases the operation results in some increase in work capacity–however, little is known of the long-term outcome of such operations in any type of heart condition.

(g) A good half of male patients who carry pace-makers recover their former capacity for work; the results are less favourable in the case of female patients.

(h) Exposure to the inhalation of toxic substances such as the chlorine or fluorine derivatives of hydrocarbons, carbon disulphide, carbon monoxide, and certain glycol derivatives is particularly to be avoided in the case of patients suffering from arrhythmias or extrasystolic disease since this can result in fatal ventricular fibrillation in such cases.

(i) The employment of cardiacs in key safety jobs or in jobs on which the safety of others may depend requires the particular attention of the industrial medical officer–one cannot be over-careful if there is the slightest danger of the cardiac worker undergoing a sudden breakdown.

Screening for heart disease

It is widely held that many heart disorders could and should be detected by the family doctor or the school doctor before vocational training or employment; in this way, suitable vocational guidance of young cardiacs would be possible before or after treatment.

The detection by the medical officer of heart disorders in persons already at work is possible only by means of rigorous clinical examination, in which medical history (personal and hereditary antecedents) is of considerable significance. This examination should include auscultation, measurement of arterial blood pressure and a peripheral vascular examination together with a chest X-ray.

Any indication of the slightest organic or functional cardiovascular anomaly should be followed up by a complete examination which will include functional exploration by a specialist and an electrocardiogram.

Some countries have "occupational cardiology centres" to which the industrial medical officer may send his patients for consultation and to obtain all necessary data on cardiovascular condition and work capacity. Elsewhere, occupational medical services themselves organise this type of consultation with the same objective.

Rehabilitation

Although it is recognised as essential for handicapped persons with motorial disorders, and has long been practised under the name of "post-cure treatment" for persons who have had pulmonary tuberculosis, functional rehabilitation is still not adequately organised for cardiacs in some countries. The value of progressively rehabilitating cardiacs to effort is, nevertheless, well known although a distinction is often not made between the effort required to resume normal everyday tasks and that needed as a preparation for a return to work.

Supervised functional rehabilitation is especially indicated during acute disease or acute episodes in chronic conditions which have halted the patient's normal activity for a period of time (e.g. infarction, cerebrovascular injury, infectious valvulitis, etc.).

In the absence of practical functional rehabilitation facilities, rehabilitation to effort by means of early but progressive return to work in the worker's previous employment or in a more suitable job may be recommended.

Occupational rehabilitation

Persons with coronary disease require special supervision and, where this is possible, return to work is generally possible following early psychological preparation and physical re-education. It is advisable for the worker to return to some sort of suitable work–at least part time; careful placement in a new job is preferable to inactivity, since regular physical exercise has a favourable effect on the health of the cardiovascular patient. Where possible, it is preferable to re-employ the patient on his original job or in the same service, and at an equivalent wage. Where difficulties in readaptation occur, on-site supervision may be necessary.

This type of rehabilitation is excellent provided the progressive increase in hours of work and workload is carefully planned by the industrial medical officer in collaboration with the cardiologist–if the plan is strictly observed and medical supervision is well carried out the results are even better. However, danger may arise if these conditions are not met and, all in all, it is only in large undertakings that this type of procedure is possible; not all firms have special rehabilitation workshops.

Retraining

Since it is often impossible to provide cardiacs with the benefit of functional or occupational rehabilitation, it is necessary to advise an apprenticeship in a new trade. However, this is required only where there is incompatibility for the original job and where no other suitable job can be found in the firm.

Retraining is most suitable in the case of young workers who received dubious vocational guidance at the beginning of their career or in the case of persons under the age of 45 years who cannot return to their previous job; in other cases it may prove the wrong solution. Most of the numerous modern trades taught in occupational retraining centres are suitable for cardiacs. However, this type of apprenticeship requires educational qualifications usually beyond the capabilities of manual workers.

When the time comes to place a cardiac in an undertaking, he has no easier task than other handicapped workers. In fact, where the disease is no more than early senescence (a frequent occurrence), the obstacles are usually great. Consequently, it is desirable for the industrial medical officer to detect signs of ageing in workers and arrange, in advance, the change of post which will finally prove necessary; this is the most satisfactory form of job transfer.

Supervision

Periodical examinations for cardiacs are the same as those used for screening; they require close collaboration between the cardiologist and the medical officer: the first has the knowledge of the organic and functional condition and the second, the knowledge of the workload entailed by the job; together, they can evaluate the suitability of the work and decide on the type and frequency of the periodical examinations. The medical officer should pay particular attention to the supervision of young persons and pregnant women and, at regular intervals, check that working conditions have not changed. It is advisable that no cardiac should be transferred to another job, for any reason whatsoever, without the approval of the medical officer. It is important that a cardiac patient avoid overwork and dangerous excess and that he follows a careful diet. The industrial physician should play an important role in this respect: while not interfering with any treatment of a patient that may have been initiated by another doctor, he can and should assert his role as an adviser to the workers under his care, paying particular attention to those with heart complaints so that they may organise their lives smoothly, observing their correct diet, avoiding overexertion and exposure to toxic substances including alcohol and tobacco. In this way he will play an important part in the life of the workers by teaching them the best way to protect their health.

FOURCADE, J.
MOURET, A.

CIS 78-1781 *Continuous dynamic electrocardiography in joint occupational medical services* (Electrocardiographie dynamique continue en médecine du travail interentreprises). Labriffe, H. (Laboratories Ames, Département des Laboratoires Miles, Tour Maine-Montparnasse, Paris) (1978), 111 p. Illus. 69 ref. (In French)

CIS 76-875 "Anticoagulants and work" (Anticoagulants et travail). Slim-Kerriou, V.; Bremond, P.; Cargill, C.; Dupret-Lagrange, F.; Goumot, H.; Horville, R.; Miroux, M.; Auzoux, L.; Borel, P.; Voulfow, J.; Houdoy, A. M.; Lehmann, M.; Morel, J.; Rossel, G.; Simon-Vie, A. *Cahiers de médecine interprofessionnelle* (Paris), 2nd quarter 1975, 15/58 (77-95). 3 ref. (In French)

"Study of the socio-occupational fate of persons carrying a pace-maker" (Etude sur le devenir socio-professionnel des sujets porteurs de pace-makers cardiaques). Lazarini, H. J.; Besse, P.; Doignon, J.; Peres, M.; Serise, A.; L'Epée, P. *Archives des maladies professionnelles, de médecine du travail et de sécurité sociale* (Paris), Oct.-Nov. 1975, 36/10-11 (633-639).

CIS 75-282 "Psychological suggestions for the prevention of myocardial infarction in industrial medicine" (Psychologische suggesties voor een bedrijfsgeneeskundige preventie van het hartinfarct). Appels, A. *Tijdschrift voor sociale geneeskunde* (Bussum, Netherlands), 5 Apr. 1974, 52/7 (225-226, 233). (In Dutch)

"Role of industry in preventive cardiology" (135-178). Ref. *Occupational safety and health symposia 1977.* DHEW (NIOSH) publication No. 78-169 (National Institute for Occupational Safety and Health, 4676 Columbia Parkway, Cincinnati) (June 1978).

"The work evaluation of the cardiac patient". DeBusk, R.; Davidson, D. M. *Journal of Occupational Medicine* (Chicago), Nov. 1980, 22/11 (715-721). Illus. 20 ref.

Cardiovascular diseases

Under the present heading it is intended to take into account those occupational factors which have a bearing on the development of cardiovascular diseases (see also SILICOSIS). Based on present-day knowledge, the part played by occupation in the aetiology of such diseases appears to be of limited importance. However, several physical and chemical factors have to be taken into consideration because of their potential cardiovascular damaging effects.

Physical factors

High temperatures may give rise to acute cardiac attacks. There appears, however, to be no special cardiovascular pathology related to repeated exposure to high temperatures.

Noise will bring about a small increase in heart rate and a slightly higher blood pressure while at rest; under effort, these effects are no longer perceptible. Noise cannot be included among those factors leading to cardiovascular diseases.

Repeated microtraumata of the extremities are said to be one of the aetiological factors found with Raynaud's disease among workers using hand-held pneumatic percussion tools.

Electrocution can be a cause of death following ventricular fibrillation. Cases have been described of myocardial infarction having lasted several hours or even days after an electrocution; little is known about their physiology or pathology.

High levels of energy expenditure are sometimes blamed as a cause of an infarction or of sudden death due to an upset in the ventricular rhythm. An examination of the complications related to exercise tests does not however bring to light any apparent relationship: ventricular arrhythmias, which rarely appear during these exercise tests (less than 1/10 000), make their appearance just as much during light exercise as during maximum exertion. Cases of cardiac arrest observed during physical training sessions occur haphazardly and do not appear to be linked with particularly strenuous effort. Similarly, cases of infarction or sudden death are spread uniformly over all hours of day and night.

Regular physical exercise is often considered to furnish protection against coronary atherosclerosis, just as excessively sedentary habits are considered to be a risk factor. Based on the numerous works covering this subject, it would seem that regular physical activity is probably beneficial, but no study has yet provided positive proof of this beneficial action.

Occupational stress is often referred to as a risk factor in coronary disease. This factor is particularly difficult to isolate, and no serious study has yet been put forward which clearly demonstrates that stress as such plays an unfavourable role.

Occupational poisoning

Among the substances which have been called into question in connection with the development of cardiovascular diseases, the most important are cobalt, cadmium, the halogenated hydrocarbons, carbon disulphide, the aliphatic nitrates and carbon monoxide. In certain cases suspicion based on the results of experimental conditions with laboratory animals has not been confirmed in man (cadmium); some of the substances are only toxic under very heavy conditions of exposure, so that the chronic occupational risk would seem of little importance (halogenated hydrocarbons); finally, certain toxic substances are so commonly encountered that the occupational hazard is less than that found generally (carbon monoxide).

Cobalt. In about 1966, in Belgium, Canada and the United States, over 150 cases of cardiomyopathy among beer drinkers were attributed to the use of cobalt as a foaming agent in the manufacture of beer. This epidemic disappeared when the use of cobalt in the manufacture of beer was stopped.

The toxic effect of cobalt on the heart has been confirmed by animal studies. These have shown alterations in the heart myofibrils and mitochondria which were non-specific and identical to those seen in man and were similar to those seen in cases of primary cardiomyopathy. These lesions are due to the cobalt interfering with certain enzymes occurring in the Krebs cycle (a secondary decrease in the proportions of zinc and iron).

It might be expected that cardiomyopathies would be encountered among workers exposed to cobalt, but only one case has been described by Barbouki and Daseki in 1972. The extreme rarity of these cases is probably related to the fact that the toxic effect of cobalt upon the heart only makes its appearance in the presence of favourable factors such as malnutrition and alcoholism.

Cadmium. In animal experimentation repeated doses of cadmium can give rise to arterial hypertension whose pathogenesis remains debatable. It has been suggested that the vascular reactivity is modified, with an interference in the action of the vasopressin and an excessive secretion of renin; in such cases the level of renal cadmium is excessively high with a relative fall in the level of zinc.

In man cadmium has in consequence been associated with the aetiology of arterial hypertension. Schroeder has demonstrated that the kidneys of hypertension cases contained more cadmium than those of persons who died of other causes; in the United States it has been suggested that there is a relationship between the incidence of arterial hypertension and the amount of cadmium pollution in the atmosphere. These results have not, however, been confirmed; furthermore, workers exposed to cadmium, and whose renal cadmium level is high, have shown no tendency towards arterial hypertension.

At the present time there is thus no convincing argument to the effect that cadmium influences the development of hypertension in man; this example shows that it is only with the greatest caution that the results of animal experiments should be extrapolated to man.

Halogenated hydrocarbons. The freons or fluoroalkenes and the fluorochlorine derivatives may also be included in this group.

In man the toxic effect of these chemicals on the heart has been known from the earliest use of chloroform, which is no longer used as an anaesthetic because of its upsetting effect on the ventricular rhythm.

The freons and the propellant gases in general used in aerosols were considered as having very little toxic effect, but this point of view has had to be reviewed following the fatal effect of the voluntary inhalation of commercial aerosols by young persons, with the object of obtaining similar effects to those produced by psychotropic drugs; in view of their youth, the sudden nature of the death and the negative results obtained from autopsies lead to the suspicion that death was the result of ventricular fibrillation.

In animal experiments, exposure to the freon gases lowers the electrical activity of the heart and slows down the conduction of the impulse; this leads to bradycardia with escape jonctional or ventricular rhythms and finally to asystole. These effects are dependent on the concentration of freon and will be reinforced by hypoxia and acidosis. When the latter are foreseen and corrected by the administration of oxygen, the animal will survive. Freons cause the heart muscle to become particularly sensitive to the arrhythmic effects of the catecholamines as may be shown by the administration of epinephrine to animals with freon poisoning. This leads to the appearance of tachyarrhythmia followed by death from ventricular fibrillation. This "sensitisation of the heart muscle" is linked with an inhomogeneous depression of the conductive tissue thereby permitting serious ventricular arrhythmia to appear under the effect of the catecholamines; a similar mechanism has been suggested in the case of chloroform and trichlorethane. The halogenated hydrocarbons also reduce the contractility of the heart muscle with a decrease of the cardiac flow and of arterial pressure.

Freon gases used in aerosol sprays have been implicated as a contributory factor in the death of asthmatic patients who have made excessive use of bronchodilators. In such cases all the necessary factors for increasing their toxicity would be present–hypoxia, acidosis and stress, as well as the sympathomimetic bronchodilators. Nevertheless, the contribution of the aerosol spray propellant to the death of the patient has yet to be established.

Chronic poisoning by halogenated hydrocarbons has never been considered to have been the cause of a cardiovascular pathology.

Carbon disulphide. Chronic poisoning from carbon disulphide can result in arterial hypertension and an increased incidence of ischemic cardiac disease. The theory has been put forward that carbon disulphide has a toxic effect on the walls of the arteries. The toxic effects of carbon disulphide may be explained indirectly in terms of an arteriosclerosis of the kidney, the brain and the heart.

In animal experiments carbon disulphide impairs the lipid metabolism so that the fat-clearing capacity of the arterial wall is reduced and the synthesis of cholesterol by the liver is increased.

In man there is no significant increase in cholesterol from chronic exposure to carbon disulphide. It has been suggested that cerebral and renal atherosclerosis may be provoked, but this has never been established.

A secondary effect of carbon disulphide that has been well demonstrated is arterial hypertension. Epidemiological inquiries have shown, however, that this effect is of little importance, since the blood pressure of an exposed group was found to be 149/90 against 133/82 in a group of workers that had no exposure.

Coronary atherosclerosis is distinctly more common among workers exposed to carbon disulphide. This has now been confirmed by a Finnish study which showed an incidence of coronary disease 2.5 to 5 times higher among exposed subjects. The mechanism of this harmful effect of carbon disulphide has not yet been made clear; it has not been found among Japanese workers, among whom the effects are more often of a cerebral vascular nature. This last finding lends weight to the opinion that carbon disulphide speeds up the development of a latent atherosclerosis; the Japanese in fact seem to have a defence as far as their coronary circulation is concerned but pay heavily in cerebral vascular diseases.

Detailed investigations have been made into the retinal circulation following research into early signs of vascular damage; Japanese studies demonstrated the appearance of microaneurysms in exposed subjects, but this has not been confirmed by the Finnish research workers. Nevertheless, the Finnish research workers have, with the aid of retinal angiography, been able to detect signs of alterations in the retinal circulation which, although barely specific, are constant and reproducible; in every case these signs precede the appearance of other secondary effects of carbon disulphide and thus provide a means for the early detection of subjects who are most at risk.

Aliphatic nitro compounds. The symptoms of acute nitroglycerine poisoning are related to its vasodilator effects which include headaches, tachycardia, hypotension, fainting fits and possibly orthostatic fainting. These effects are benign and related mainly to the effect of the nitroglycerine on the venous wall including venous ectasia with peripheral accumulation of venous blood and failure of the heart pump.

Chronic exposure to aliphatic nitro compounds is accompanied by a progressive disappearance of the secondary effects; the vasodilatation appears to be offset by the sympathetic tonus. Among workers subject to repeated exposure, an increase is to be seen in the diastolic arterial pressure together with a reduction in systolic arterial pressure.

If exposure to nitroglycerine is discontinued, 5% of patients who have been exposed over several months develop symptoms of myocardiac ischemia ranging from spontaneous angina through myocardial infarction up to sudden death. The symptoms appear 48 to 72 hours after cessation of exposure; they may be reversed by the sublingual application of nitroglycerine or possibly by returning the patient to work.

Coronary incidents occur frequently among young persons and among women who are generally free from coronary atherosclerosis. As a possible explanation of these cases, the occurrence has been suggested of coronary spasms linked with generalised vasoconstriction originating in the sympathetic system, which gradually takes over as the dilating effect of the aliphatic nitro compounds disappears; Lange and collaborators have provided support for this hypothesis by showing the reversibility of a coronary spasm treated sublingually with nitroglycerine. Patients that have come forward with an angina pectoris without infarction and have been withdrawn from exposure to nitroglycerine have a normal life prognosis.

The relative scarcity of cases described leads to the supposition that an individual predisposition exists which is as yet unidentified for preventive purposes. In the light of the presumed physiopathology of this syndrome, it seems likely that coronary patients are at greater risk than normal subjects and should not be subjected to exposure to these compounds.

Carbon monoxide. In the case of acute carbon monoxide poisoning, the electrogram will frequently show alterations of the type associated with subepicardiac ischemia (reversal of "T" waves) sometimes accompanied by an increase in blood enzymes; anatomo-pathological studies subsequently reveal diffuse small necrotic areas in the heart muscle, which cannot be described as a systematic infarction topography. When the poisoning does not lead to death, these anomalies on the electrocardiogram disappear progressively.

The effects of moderate exposure to carbon monoxide are similar to those of hypoxia. In a normal subject the adaptation of the cardiac circulation to a sub-maximal effort rests unchanged so long as the maximum oxygen consumption decreases in line with the level of carboxyhaemoglobin; this results simply in a reduction in the maximal arteriovenous difference.

In the case of coronary subjects electrocardiographic and clinical signs of myocardiac ischemia appear after less strenuous effort; after smoking several cigarettes not containing nicotine (HbCO = 7.8%), the appearance of angina pectoris is to be seen after slight effort and with a lower pressure/time index. Carbon monoxide does not, however, appear to have direct toxic effects on the heart muscle, its action being the result of the anoxia.

Chronic exposure to carbon monoxide in animal experiments gives rise to modifications in the arterial wall: an increased permeability of the endothelium with subintimal oedema accompanied by the accumulation of lipids at the level of the arterial wall. Both macroscopically and microscopically the lesions are similar to those found in human atherosclerosis. In animals isolated hypoxia gives rise to similar lesions and the toxicity of carbon monoxide appears to reside entirely in the lack of oxygen which it engenders.

Cigarette smokers who inhale the smoke arrive at a carboxyhaemoglobin level varying from 5 to 15%. It has been clearly established that the incidence of coronary disease is twice to three times as high among smokers as among non-smokers. This effect, which is more pronounced among young people, is directly proportional to the number of cigarettes consumed. It seems that the toxic effect of tobacco is due neither to the nicotine, nor to the tars, but to the atherogenic action of the carbon monoxide, which itself is no doubt linked with anoxia (see SMOKING).

Workers subject to chronic exposure to carbon monoxide are thus susceptible to early onset of coronary disease. An exposure limit for carbon monoxide of 50 ppm is too high because after five hours such a concentration will give rise to a HbCO level of 5% or more; it should be remembered, however, that smoking plays a role that is just as important as that of occupational exposure to carbon monoxide.

DETRY, J. M.
LAVENNE, F.

General:

"Cardiovascular disease and environmental exposure". Rosenman, K. D. *British Journal of Industrial Medicine* (London), May 1979, 36/2 (85-97). 116 ref.

"On the causal relationship between myocardial infarction and occupation" (Zur Frage des Ursachezusammenhanges zwischen Myokardinfarkt und Beruf). Hartung, M.; Kentner, M.; Raithel, H. *Arbeitsmedizin, Sozialmedizin, Präventivmedizin* (Stuttgart), Oct. 1979, 14/10 (240-244). Illus. 15 ref. (In German)

Special:

"Cadmium as a factor in hypertension". Schroeder, H. A. *Journal of Chronic Diseases* (New York), July 1965, 18 (647-656). 42 ref.

"Cardiac arrhythmias and aerosol sniffing". Reinhardt, C. F.; Azar, A.; Maxfield, M. E.; Smith, P. E. Jr.; Mullin, L. S. *Archives of Environmental Health* (Chicago), Feb. 1971, 22 (265-279). Illus. 37 ref.

"Cobalt-beer cardiomyopathy". Alexander, C. S. *American Journal of Medicine* (New York), Oct. 1972, 53 (395-417). Illus. 105 ref.

"Nonatheromatous ischemic heart disease following withdrawal from chronic industrial nitroglycerin exposure". Lange, R. L.; Reid, M. S.; Tresch, D. D.; Keelan, M. H.; Bernhard, V. M.; Coolidge, G. *Circulation* (New York), Oct. 1972, 46 (666-678). Illus. 11 ref.

"Carbon monoxide, smoking and atherosclerosis". Astrup, P.; Kjeldsen, K. *Medical Clinics of North America* (Philadelphia), Mar. 1974, 58/2 (323-350). Illus. 70 ref.

"Angina pectoris, electrocardiographic findings and blood pressure in Finnish and Japanese workers exposed to carbon disulfide". Tolonen, M.; Hernberg, S.; Nordman, C. H.; Goto, S.; Sugimoto, K.; Baba, T. *International Archives of Occupational and Environmental Health* (West Berlin), 1976, 37/4 (249-264). Illus. 34 ref.

Cardiovascular exercise tests

Exercise tests are widely used as a physiological load in studying the functional capacities of the cardiorespiratory system. A prime application of such tests is to determine the exercise capacity of healthy persons: workers, sportsmen and others. In clinical medicine exercise tests may be used also as an aid in diagnosis since some functional abnormalities, e.g. in the electrical activity of the heart, are detectable during exercise even before the appearance of related objective signs or subjective symptoms at rest. Another broad field for exercise tests in medicine is in the assessment of the functional capacity, both for judging the disability of persons known to be suffering from cardiovascular disease, and also more generally for observing the progress of rehabilitation after illness or injury.

Functions tested

Exercise requires energy. For very short performances this may be obtained from the stored chemical energy of organic phosphate bonds and from the anaerobic conversion of sugar to lactic acid, but ultimately energy is derived from the oxidation of nutrients. The "oxygen supply line" goes through the lungs to the blood, and further with the aid of the pumping action of the heart to the muscles and other organs. The capacity of the oxygen supply can be measured by determining the maximum oxygen intake ($\dot{V}O_2$ max expressed in l/min), i.e. the largest intake of oxygen which can be attained by working against increasing workloads. It has been shown that healthy lungs do not place a limit to the maximum oxygen intake. Thus, the oxygen supply to the consumer organs depends on the pumping action of the heart, on the amount of oxygen contained in the arterial blood that is delivered to the tissues, and on the ability of the tissues to extract oxygen from blood. The pumping action is determined by the heart rate and the stroke volume, while the mean arteriovenous oxygen difference indicates the delivery of oxygen from blood to the organs. Whereas all the components of this "oxygen supply line", i.e. heart rate, stroke volume, and mean arteriovenous oxygen difference can be measured at rest and during exercise in the laboratory, only the heart rate can be readily measured without special equipment.

The heart rate is of prime interest. It is easily counted during exercise by placing a stethoscope over the heart or brachial artery or by palpating the carotid artery in the neck. Heart rate may also be taken from the electrocardiogram, and both portable and telemetric heart rate counters (cardiotachometers) are available. After about 4 min of work, the heart rate rises to a stable level, provided it is possible to obtain a steady-state response to the imposed load.

It is necessary to carry out electrocardiography if the cardiovascular function test is performed for the detection of heart disease. Special leads must be used to avoid muscle noise and it has been found that an arrangement with one electrode at the *manubrium sterni* or forehead and the other at the fifth intercostal space in the left anterior axillary line is suitable for monitoring during exercise.

Blood pressure measurement is often included in exercise tests. It should be noted, however, that its response to exercise is relatively independent of the exercise capacity. The usual auscultatory measurement of systolic pressure is applicable during exercise, but the diastolic readings are, in most cases, not valid. With increasing workload, the systolic pressure rises: readings of 300 mmHg ($39.9 \cdot 10^3$ Pa) and more have been recorded without apparent discomfort or hazard to the subject. More important a warning signal may be hypotension on exercise, particularly a decline of the systolic pressure with increasing loads, after a peak pressure has been reached. In testing cardiovascular patients, blood pressure should be measured at 2-min intervals or at the end of each step of increasing the workload. The rate-pressure product (heart rate × systolic pressure) gives a measure of the work of the heart.

The exercise capacity is mostly expressed in terms of oxygen intake. The oxygen intake of the standard test exercises is well known and in most cases can be estimated with sufficient accuracy without actually determining the oxygen intake during the test. This "shortcut" rests on the assumption that all subjects perform the test exercise with the same mechanical efficiency. Obviously, this hypothesis cannot be valid for subjects with locomotor or neurological disabilities; when testing such subjects, the oxygen uptake should be directly determined.

Exercise type and intensity

Maximal or submaximal loading is used in exercise testing. Maximum aerobic power ($\dot{W}$ max) is the rate of work at which the oxygen intake reaches a plateau, $\dot{V}O_2$ max. For a short time work can be carried out at an even higher rate, while the extra energy is obtained anaerobically. When a subject is required to work against a series of increasingly heavy loads, both oxygen intake and heart rate show an approximately linear relation to the level of external work. However, these relations often deviate from linearity at both ends, for the heart rate below 100 and above 175 beats per min. Where there is no interference from factors such as emotion or high environmental temperature, which tend to increase heart rate, the load at which the maximum heart rate is attained is reasonably close to that at which the maximum oxygen intake is reached. The linearity and the approximate coincidence of the two maxima make it possible to assess the maximum oxygen intake from the heart rate alone, and further to extrapolate from heart rate in submaximal work to the predicted maximum heart rate, taken as the mean for the age group. Thus aerobic power may be:

(a) directly determined from the measurement of oxygen intakes at increasing loads up to maximum;

(b) indirectly indicated by the load at which the heart rate reaches its maximum;

(c) extrapolated, more indirectly, from the heart rate at a series of submaximal loads; or even

(d) extrapolated from a single submaximal load only, assuming an average slope of the heart rate against the workload.

Obviously, the accuracy of the prediction falls as the method becomes more indirect, and if the extrapolation is made from a heart rate observed in light work far below the maximum.

The intensity of work required to raise the heart rate to a predetermined level may be used as a measure of physical working capacity, without calculating the maximum oxygen intake. In young adults, 170 beats per minute have been widely used as such a level: the corresponding rate of work ($\dot{W}_{170}$) is expressed in watts or in kgfm/min. If a heart rate of 170 beats per minute is not reached with the loads used, $\dot{W}_{170}$ may still be estimated by linear extrapolation from a series of lower workloads and from the corresponding heart rates. It should be pointed out that the $\dot{W}_{170}$ is generally not identical with the maximum aerobic power. In subjects with a maximum heart rate above 170, $\dot{W}_{170}$ is lower, and in those with a maximum heart rate below 170 beats per minute higher, than the aerobic power. With age the maximum heart rate decreases: a practical estimate is 220 − age in years, which gives 200 beats per minute for a person of 20 years and 160 for one of 60 years. In aged subjects it will thus be more realistic to determine, for example, $\dot{W}_{150}$ or even $\dot{W}_{130}$.

Common test exercises

Any exercise in which large muscle groups are used dynamically may, in principle, be used for increasing energy expenditure and hence the heart output. The need to achieve standardisation and to obtain ease of performance limits the choice to such activities as running, walking, pedalling an ergometer, stepping, or hand cranking. Little learning is needed from the subject for any of those forms of exercise, and learning does not essentially affect the energy needed for performing a unit of work, i.e. the mechanical efficiency.

Ergometers. The bicycle ergometer consists of a pedal mechanism in which work is done against a friction belt or electromagnetic brake. The calibration of the brake resistance must be checked regularly. This is easily done for mechanically braked ergometers, but may be quite complicated with the other types. The saddle should be adjusted so that the knees extend to approximately 170 °. The subject should be required to work at a rate of 50-60 pedal revolutions/min. The load is expressed as watts (1 W = 1 newton·metre per second ≃ 6 kgfm/min). Mechanical efficiency is relatively constant and the energy requirement is 4.4 times the mechanical work (mechanical efficiency 23%). A metronome seen and heard by the subject is useful for pacing: however, the subject may also pace himself by means of speedometer. Most subjects are able to start ergometer work at a level of 50 W which is subsequently increased by increments of 25 or 50 W.

The cranking ergometer is designed for arm work and consists of a crank (optimum radius 30 cm), the axis of which is located 1 m above floor level. Cranking, however, does not mobilise sufficiently large muscle groups and increases the heart rate relatively more than leg work does. Therefore cranking ergometers are useful only for special purposes (e.g. for leg invalids).

Stepping tests. In these tests, the subject performs a constant amount of external work whilst stepping on and off a bench. The work done is 5.7 times the calculated requirement for lifting the body weight, independent of marked variations in step height or frequency; increments of 25 or 50 W are commonly used. In order to attain the maximum work rate in young subjects the total height of the steps should be 40-50 cm. A series of benches from 20 cm to 50 cm high, differing by 5 cm each, may prove useful in testing subjects with a large range of stature. The stepping frequency may be varied between 15 and 45 cycles (60 to 180 paces) per minute. The test should be supervised in such a way that the lift of the body's centre of gravity corresponds to the step height. Stepping frequencies of 15 and 25 cycles per minute and a stepping height of 40 cm have been found to raise the heart rate to around 110 and 150 beats per minute, respectively. The double (Masters' type) step with a height of 23 cm results in slightly lower mechanical efficiency; the energy requirement is 6.1 times the mechanical work of lifting.

Treadmills. These consist of a continuous belt which is moved across a smooth platform by an electric motor; the platform is provided with handrails and the speed of the belt and the angle of incline of the platform are adjustable. On exercise, the belt moves towards the subject who is required to walk forward in order to maintain his relative position. Treadmills have no major advantages over simple bicycle ergometers and are most expensive; in addition, with inexperienced subjects and/or high workloads, special care is required if accidents are to be avoided.

The maximum oxygen uptake on a treadmill is some 7-12% higher than on a bicycle ergometer and about 3% higher than in a step test. The energy expenditure of a subject walking on a treadmill at the speed of 80 m/min is approximately three times that of the same subject at rest. Each 2.5% increment in the angle of incline increases the energy cost by an amount equal to that of the metabolism at rest. The maximum oxygen intake may be reached also on a horizontal treadmill, by increasing running speed.

Running. Maximum oxygen intake is an important determinant of maximum capacity particularly in such performances, which last from 1 min to 1 h. The distance walked or run in 12 min by motivated men has been shown to correlate satisfactorily with the aerobic power. This Balke-Cooper test is a maximal test. The result can be simply expressed as the distance covered: it is also possible to make an estimate of the maximum oxygen intake by using the following equation (validated on US men): maximum oxygen intake ml/kg/min = 22.36 × distance km−11.29. The Balke-Cooper test should be used only in testing young, healthy subjects with some running experience.

Standardisation of test conditions

The load may be constant, increased in steps or continuously (figure 1). A test comprising a series of increasing loads, with an almost steady state at each level, no intermittent rest periods, and with 4-6 min for

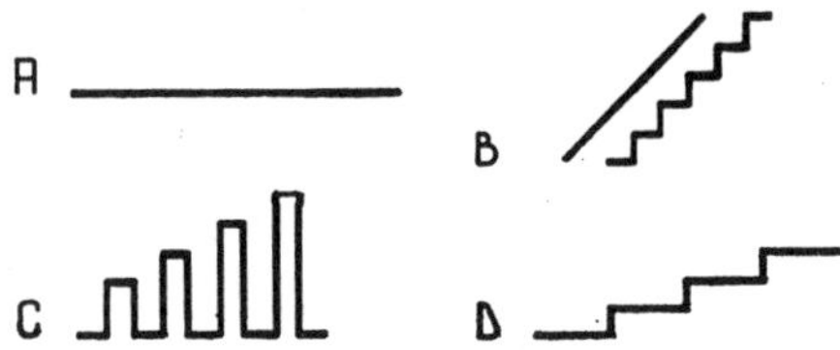

Figure 1. Types of load used in exercise tests. A. Single-level load. B. Continuous or nearly continuous increase in load. C. Discontinuous series of increasing loads with intermittent rest periods. D. Continuous series of increasing loads with an almost steady state at each level. *(Adapted from World Health Organisation, 1971.)*

each step, is satisfactory for most purposes and can be carried out in a relatively short time. For determining $\dot{W}_{170}$ of young people the first load should raise the heart rate to 120-140, the second to 140-160 and the third to 160-180 beats per min. If the load is increased continuously, the oxygen intake and heart rate will at each level of load be somewhat lower than they would be at the end of each step of load if increased stepwise. This will cause a small systematic difference in the prediction of maximum oxygen intake from the heart rate at submaximal loads: a continuous increase of loads will result in a slightly higher estimate than steps of load at least 4 min long would give. For screening purposes, a single load test may be employed.

Tests should be arranged in such a way that the material test conditions remain constant and that the subject is not affected by factors which might influence the results. For example, heart rate at rest and during exercise rises whilst food is being digested; consequently a minimum of 2 h should be allowed to elapse after a meal, or after the consumption of coffee or alcoholic beverages. The subject is closest to the standard state in the morning and he should have a rest of 1 h without smoking before the test starts. After heavy physical work, 24 h are required before the response to exercise returns to normal. Loss of sleep during the preceding night is without effect in healthy subjects. Ambient temperature has a major effect on the cardiovascular system. Blood pressure rises in a cool environment, and heat stress leads to an increase in heart rate at temperatures above 25 °C. The testing room should be at a temperature of 15-25 °C; if very accurate results are required, ambient temperature should not vary more than 3 °C. The relative humidity should preferably be less than 65%.

Precautions and contraindications

Healthy, young persons may be subjected to submaximal exercise tests without special precautions since the test conditions do not impose a greater load on the body than many everyday activities. Even in the case of elderly or sick subjects, and persons with suspected cardiorespiratory disease, the risk associated with cardiovascular function tests is quite low; nevertheless, subjects of this type must be medically examined prior to testing. When maximal testing is employed, medical examination is advisable in all cases.

When patients are tested, a physician must for safety reasons be present and observe the subject throughout the test: behaviour and general condition often give at least as important information as the numerical results. Continuous ECG monitoring adds to the safety. Also when healthy subjects are tested, a physician should be at hand for consultation and possible emergency situations. Appropriate equipment to deal with medical emergencies should be available. The subject should always be allowed to discontinue the test at will.

At the end of the test exercise, postural hypotension and excessive cooling should be avoided; it is preferable to taper off the exercise than halt it abruptly.

Factors affecting aerobic power

When the body or some part of it is moved against gravity, the energy required is proportional to body weight; this is the situation in a step test. On the other hand, energy expenditure in pedalling a bicycle is practically unaffected by the body size. In many occupations, a large part of the physical work consists of moving the body and, where this is the case, working capacity may be indicated by maximum oxygen uptake expressed as cubic centimetres of oxygen per kilogramme of body weight. However, where it is necessary to assess capacity to move external loads, it is preferable to use measurements based on an absolute scale. Maximum heart rate decreases with age; this affects the prediction of the maximum aerobic power from the heart rate in

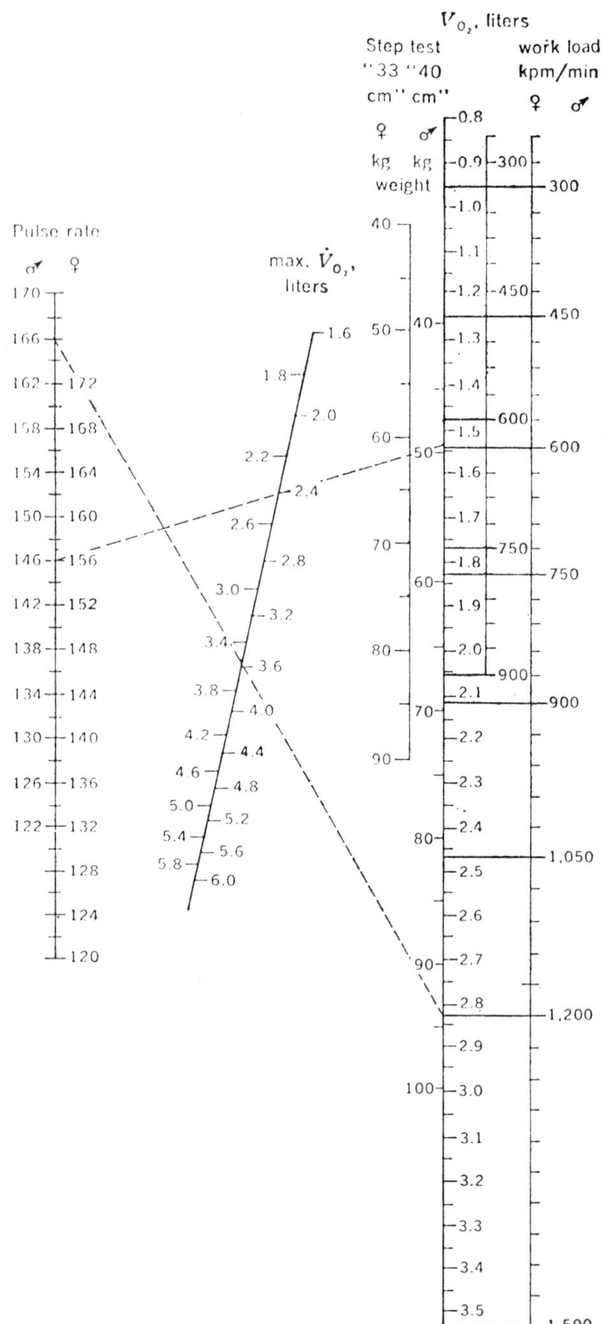

Figure 2. The adjusted nomogram for calculation of maximal oxygen intake from submaximal heart rate and O_2-intake values (cycling and step test, 22 cycles/min). In tests without direct O_2-intake measurement it can be estimated by reading horizontally from the "weight" scale (step test) or "workload" scale (cycle test) to the oxygen intake ($\dot{V}O_2$ litres) scale. The point on the oxygen-intake scale should be connected with the corresponding point on the "pulse rate" scale, and the predicted maximal O_2-intake read on the middle scale (max. $\dot{V}O_2$ litres). A female subject (61 kg) reaches a heart rate of 156 at step test; predicted max $\dot{V}O_2$ = 2.4 l/min. A male subject reaches a heart rate of 166 at cycling test on a workload of 1 200 kpm/min (= 200 W): predicted max $\dot{V}O_2$ = 3.6 l/min (exemplified by dotted lines). *(From Åstrand and Rodahl, 1977.)*

submaximal work. Two slightly different alternatives for the prediction are available, the Åstrand nomogram (figure 2) with an age correction, or the equation proposed by von Döbeln et al. (1967) where the age effect is included. When using the Åstrand nomogram, the value obtained has to be multiplied by the following correction factor:

Age in years	Correction factor for Åstrand nomogram
15-24	1.10
25-34	1.00
35-39	0.87
40-44	0.83
45-49	0.78
50-54	0.75
55-59	0.71
60-64	0.68
⩾ 65	0.65

The equation of von Döbeln et al. has the following form:

Estimated

$$\dot{V}O_2\text{max l/min} = 1.29 \times \sqrt{\frac{L}{f_h - 60}}\, e^{-0.00884 \times T},$$

where

L = load of bicycle ergometer kgfm/min,
f_h = heart rate at the end of 6 min at the load L (beats/min),
T = age (years), and
e = the base of natural logarithms (2.718..).

The maximum aerobic power of men gradually decreases from the age of 25-30 years onwards and at 70 years of age it is only half of that at the age of 20. In women the peak level is reached after puberty, but the decline starts later, in the menopause. The mean aerobic power per kilogramme of body weight of young women is 70% of that of young men. The maximum aerobic power and the maximum oxygen intake are proportional to body size, and thus vary widely in different ethnic groups. The maximum oxygen intake per kg body weight shows much smaller international differences: mean values from 34 to 67 ml/kg/min have been published for men and from 27 to 43 ml/kg/min for women of 20-30 years.

In all communities, the maximum aerobic power shows wide inter-individual variations. The highest values are observed in sportsmen such as middle- and long-distance runners. Occupational work has much less effect than sport: only in occupations requiring extremely hard effort, such as woodcutting, is the mean aerobic power higher than among the general population. In industrialised communities, leisure-time physical activity appears to have more training effect on the maximum aerobic power than does occupation.

When a subject starts to train, his heart rate both at rest and at submaximal exercise may decrease considerably in as little as a few weeks. The decrease of work heart rate over the first days is largely due to habituation to the test situation. Hard training over a period of several months may increase the maximum aerobic power by 10-20% (figure 3).

Testing for work capacity

When workers are recruited for hard physical work, the cardiovascular function test can be used to indicate the work rate the subject will be able to tolerate (see MUSCULAR WORK). The acceptable levels of workload are discussed in the article on ENERGY EXPENDITURE. Taking heart rate as an indicator of the physiological load, a figure of 50% of the available range from 60 beats per minute at rest to 180 beats per minute at maximum exercise (for an average subject of 40 years) will set the limit for work heart rate at 120 beats per min, a figure 33% of the same range correspondingly at 100 beats per minute. The former limit, 50%, is generally considered as acceptable for top workloads, the latter, 33%, corresponds to a maximum allowable level of energy expenditure for an 8-h day.

Diagnostic applications

Cardiovascular function tests as described can be used for the diagnosis of ischaemic heart disease. In subjects with a decreased reserve, various means can be used for raising the heart muscle oxygen supply to the critical level. Poorly oxygenated heart muscle regions may give rise to a horizontal or downward-sloping ST depression or other electrocardiogram changes. Although such changes are associated statistically with decreased life expectancy, they are not in all cases signs of heart disease. An orthostatic test is helpful for the differentiation of such cases in which the ST-T changes are due to poor control of blood distribution and those due to ischaemic heart disease. The orthostatic test is best carried out 5-15 min after the exercise test. If ST depression is accompanied by chest pain, the patient should be adequately treated; however, an ST depression in an ostensibly healthy person can be considered more as a risk factor than as a manifestation of ischaemic heart disease.

Application to rehabilitation

Cardiovascular function tests have several uses in rehabilitation. Firstly, in combination with electrocardiogram monitoring they may be used to demonstrate to the patient his safe working capacity. Secondly, repeated tests are useful for measuring the effect of training on the heart rate response to exercise. Thirdly, in rehabilitating patients with ischaemic heart disease, repeated exercise electrocardiograms at different loads show whether the ischaemic ST depression undergoes any change.

KARVONEN, M. J.

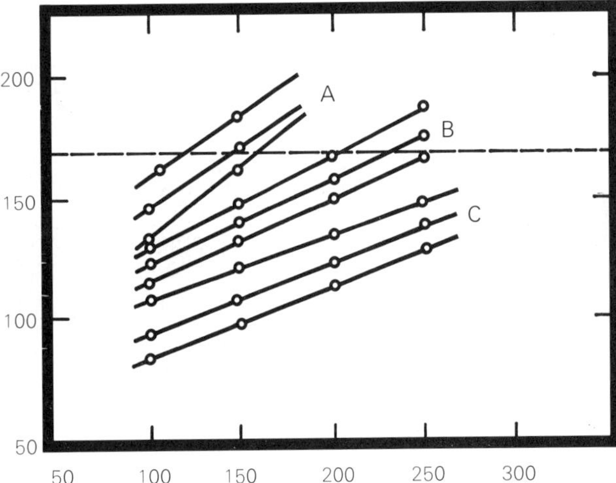

Figure 3. Dependence of heart rate response on degree of physical fitness. Ordinate: heart rate per minute; abscissa: workload in watts. A. Cardiacs (valvular defects, circulatory insufficiency). B. Normal cases. C. Athletes. *(Adapted from E. W. Banister and S. R. Brown, 1968.)*

Cardiovascular survey methods. Rose, G. A.; Blackburn, H. (Geneva, World Health Organisation, 1968), 188 p. Illus. 228 ref.

Fundamentals of exercise testing. Lange Andersen, K.; Shephard, R. J.; Denolin, H.; Varnauskas, E.; Masironi, R. (Geneva, World Health Organisation, 1971), 133 p. Illus. 158 ref.

Habitual physical activity and health. Lange Andersen, K.; Masironi, R.; Rutenfranz, J.; Seliger, V. (Copenhagen, World Health Organisation, Regional Office for Europe, 1978), 188 p. Illus. 191 ref.

Human physiological work capacity. Shephard, R. J. (Cambridge, University Press, 1978), 303 p. Illus. 740 ref.

Stress testing. Principles and practice. Ellestad, M. H. (Philadelphia, F. A. Davis, 1975), 296 p. Illus. 355 ref.

A textbook of work physiology. Åstrand, P. O.; Rodahl, K. (New York, McGraw-Hill, 2nd ed., 1977), 681 p. Illus.

"The relative energy requirements of physical activity". Banister, E. W.; Brown, S. R. (267-322). *Exercise physiology.* Falls, H. B. (New York and London, Academic Press, 1968).

"An analysis of age and other factors related to maximal oxygen uptake". Von Döbeln, W.; Åstrand, I.; Bergström, A. *Journal of Applied Physiology* (Bethesda, Maryland), May 1967, 22/5 (934-938). Illus. 15 ref.

Cardiovascular system

The cardiovascular system comprises a closed canalicular network made up of arteries, capillaries, veins and a central pumping organ: the heart. Following a cardiac contraction, the blood is distributed to the arteries and then to the capillary areas, returning through the veins back to the heart. The arteries and veins are, in fact, no more than simple transit vessels whereas the capillaries have an important functional significance since they are involved in the vital exchange of substances between the blood and the intra-cellular spaces, resulting in important modifications in the blood's chemical composition and physical properties.

The heart is made up of four chambers, two atria and two ventricles. Venous blood arrives in the right atrium (RA), passes into the right ventricle (RV), and from there via the pulmonary artery (PA) into the pulmonary circulation where haematosis occurs through the capillaries. The arterial blood flows back to the left atrium (LA) through the pulmonary veins and from there into the LV where the systemic circulatory cycle recommences. There are thus two clearly separate circulations which start and finish at the heart: the lesser or pulmonary circulation which starts at the RV and finishes at the LA and the greater or systemic circulation which starts from the LV and finishes at the RA.

The four cardiac chambers are separated by a system of valves, the tricuspid valve separating the RA from the RV and the mitral separating the LA from the LV; these are membranous folds made entirely of conjunctive, elastic tissue and are designed to direct the blood flow inside the heart, their movement being essentially passive. They open when the pressure in the atrium is greater than that in the ventricle and close when the pressure is higher in the ventricles. The sigmoid valves at the start of the large vessels open when the pressure is higher in the ventricles and close when the pressure in the aorta (A) or the pulmonary artery exceeds the ventricular pressure (diastole). There are thus two phases in the cardiac cycle: the phase during which the ventricle is filled–the diastole, and the phase of ventricular injection on contraction of the myocardium–or systole. The cycle occurs around 75 times per minute but this figure will vary considerably depending on the age and physiological condition of the subject.

General properties of the heart

The myocardial fibre (striated muscle cell of special structure) has four basic properties controlling cardiac function:

(a) *rhythmicity* which is the power of itself creating the impulses which cause its contraction thus imposing a rhythmic activity on the heart;

(b) *conductivity* which permits the propagation of the excitation wave from the atria to the ventricles via the conductor system (made up of special cells known as the bundle of His and the network of Purkinje). The total cardiac muscle is thus affected by this process of excitation, which is propagated by a mechanism similar to that of the propagation of nervous influx; however, the conductivity is common to the whole cardiac muscle mass in the same way as rhythmicity, thus giving the myocardial cell an important role in certain pathological situations (rhythm disorders, extrasystoles, etc.);

(c) *irritability* which is the property of the myocardial fibre to respond by a contraction to any excitation or stimulus (electric current, chemical or physical agent, etc.). However, the heart is not irritable to the same degree throughout the whole phase of its activity. During systole, the muscle becomes completely refractory to any stimulus whatsoever (absolute refractory phase). The cell recovers its irritability progressively during diastole to reach its maximal level at the end of diastole;

(d) *contractility* which ensures myocardial response to intrinsic or extrinsic stimuli and controls cardiac flow (normally 4-5 l/min). The latter is dependent on two main factors: heart rate and stroke volume, which is the difference between the ventricular blood volume at the end of diastole and the ventricular volume at the end of systolic ejection.

The control of systolic flow has been studied extensively on the basis of two basic concepts:

(a) diastolic ventricular volume is dependent on effective filling pressure; the energy produced is proportional to the initial length of the cardiac muscle fibre (Starling's Law of the Heart). Thus the larger the diastolic volume (the longer the muscle cell at the end of diastole), the greater the force of myocardial contraction (and, consequently, the energy produced) as a result of the increase in the tension of the muscle fibres under the effect of better filling;

(b) heart rate also plays an important role in determining systolic ejection volume, systolic and diastolic ventricular volume and cardiac output.

The various factors which tend to increase heart rate may act simultaneously by increasing contractility so that the effective filling pressure is maintained at a relatively constant level, unless heart output does not meet peripheral demand; in this case, diastolic filling may be increased by raising diastolic pressure (cardiac insufficiency–CI). Neurovegetative and hormonal mechanisms (catecholamines) are capable of controlling and modifying the very properties of myocardial contractility.

The stroke volume has an effect on the arterial wall, the tension of which varies depending on blood pressure. Blood pressure itself is directly proportional to the volume of blood ejected per minute (minute volume–Vm) and peripheral resistance (R). Any change in these factors will lead to a similar variation in arterial blood pressure. If one of the two is cancelled, arterial pressure falls to zero. Consequently all the parameters influencing cardiac output (systolic output, heart frequency, the state of myocardial contractility, total volume and blood viscosity, etc.) or the R (the cross-sectional area of the vessels subject to neuro-humoral influences and control) will have an effect on blood pressure. Thus, resistance will vary inversely to the square of the transverse cross-sectional area, i.e. small reductions in cross-sectional area (stimulation of the vaso-constrictor nerves) may produce considerable increases in resistance and consequently of the arterial blood pressure. The arrival of blood in the capillaries and its constant renewal are the basic objective of circulatory system function.

It is here that the exchanges occur between the blood and the interstitial liquids—a sort of transmitter between the blood and the various specific organ cells. The composition and physical properties of these liquids, i.e. the conditions for cell life, depend on these exchanges. The capillaries are therefore of prime functional importance.

On leaving the capillaries, the blood flows into the venous system (venules, veins), true systems of confluent tubes culminating via the two general collector trunks (superior vena cava and inferior vena cava) into the right atrium. In view of its large capacity and muscular activity, this is an important reservoir of variable capacity and is very useful for the adaptation of blood circulation. The pressure of blood on the walls, skeletal striated muscle, acts on the blood column by facilitating centripetal movement. However, it is the pressure gradient between mean systemic pressure and the pressure of the right atrium (normal central venous pressure = 0-3 mmHg) which has the main effect on the venous flow.

The increase in blood volume, vascular tonus and dilation of the small systemic vessels tend to increase venous flow. Any change in these various factors in the opposite sense will tend to reduce systemic venous flow and, consequently, central venous pressure.

In the normal lung the pulmonary arteries are the terminal arteries. They extend to the first alveoli and follow the route of the bronchi. On reaching the alveoli, the precapillary vessels divide and form a large network of capillaries between the alveolar walls. It is here that gaseous exchange takes place. The pulmonary veins are also terminal and are located at the periphery of the lobule. Pulmonary and systemic circulation are connected in series forming a continuous circuit; their haemodynamic difference has important variants: systemic circulation is a circuit with high resistances and a large pressure differential between the arterial and venous system, whereas pulmonary circulation has very low resistances. Mean pulmonary arterial pressure varies between one-fifth and one-sixth of systemic pressure.

Coronary circulation. This is ensured by two arteries, the right and left coronary artery. Although anatomically anastomosised, they act physiologically as terminal arteries. Venous return is centred on the coronary sinus but the left coronary accounts for 75-95% of the flow. To summarise, during effort, coronary flow increases considerably whereas myocardial oxygen extraction remains largely the same as at rest. Aortic pressure plays a major part in the regulation of coronary circulation; when it increases it raises the flow and vice versa. Tachycardia also increases it, paradoxically, as a result of the fall in coronary resistance. Myocardial oxygen consumption is the ultimate factor by which all the others act.

Cardiac work. This is the quantity of energy that the heart transmits to the volume of blood to propel it through the vessels. This energy is produced by the oxidation of organic substances such as glucose, glycogen, lactic acid, etc., partially converted to mechanical energy during myocardial contraction. There are two types of energy. First *potential* energy (the more important) is used to move the mass of blood from a low venous pressure to a high arterial pressure. This work, carried out by the left ventricle to raise the blood pressure at each systole, is equal to the product of stroke volume multiplied by mean ventricular ejection pressure minus the left atrial pressure. The same formula is applicable to the right ventricle, the work of which is one-seventh of that of the left ventricle. The second form of energy, *kinetic* energy, is used to increase the velocity of the blood mass through the aortic and pulmonary valves. This energy is proportional to the quantity of blood ejected and to the square of the ejection velocity. Normally, the work that the left ventricle requires to create the kinetic energy accounts for 2-4% of the total work. Under certain abnormal conditions, such as aortic stenosis, kinetic energy produced by the left ventricle accounts for 50% of the total work.

The heart's work and its distribution during systole can be broken down into three components:

(a) the *work of contractile components* indicates the total mechanical energy supplied by the energy generators—the fibre contractile components;

(b) the *work of the fibres that is released during ejection*, which is the energy consumed in the propagation of the systolic wave—a part of which comes from the contractile components still in operation, whilst the other comes from the elastic components;

(c) the *work of tensioning the elastic components*, which is the energy stored in the elastic components during the isometric contraction and returned during the final stage of ejection.

In a normal subject, an average of 90% of the contractile work is used in the propagation of the systolic wave; 25-45% of this work is stored in the elastic components; only 10%, on average, is not returned and is probably released during the period of isometric relaxation.

The concept of cardiac reserve is a basic one: the normal subject can, in effect, during physical effort increase his output by up to five times. At this point, the myocardium will increase its work since the ventricular work is proportional to the output and to the pressure. This margin of adaptability to an increase in work is the cardiac reserve.

Cardiac insufficiency may be defined as the inability of myocardial function to ensure an output that meets the body's requirements. In cardiac insufficiency, the output is not necessarily reduced in absolute figures. It may, in fact, remain high in certain pathological conditions (anaemia, Paget's disease, beri-beri, etc.), even when cardiac insufficiency occurs. Here one speaks of *high-output failure* in contrast to the *low-output failure* that one observes, for example, in the case of valve lesions or uncompensated coronary-artery disease. The start of cardiac insufficiency may be situated at the point where cardiac reserve is compromised. The patient in a state of insufficiency is no longer capable of increasing his cardiac work to the same degree, as the defective ventricle has been described as having a slower increase in isometric contraction, a shortening of ejection time, a lower systolic pressure, a greater diastolic volume, a higher filling pressure and a reduction in the mechanical efficiency of myocardial contraction. All in all, there is a reduction in contractility. The defective heart does not fully utilise the energy obtained from glucose degradation. The result is a reduction in cardiac output which gives rise to a greater arterio-venous difference in haemoglobin oxygen saturation. Left cardiac insufficiency is characterised by symptoms resulting from pulmonary congestion due to high pressures in the pulmonary vascular bed especially at the post-capillary level. The predominant symptom is dyspnoea, which apparently results from the increase in respiratory work due to the rigidity and the oedematous infiltration of the pulmonary parenchyma. Congestion oedema of the walls of the respiratory network increases the resistance of respiratory flow. Extravasation of blood from the bronchial capillaries into the alveoli may produce haemophthysis. The symptoms, such as fatigue, gastrointestinal disorders and renal dysfunction are attributable, in part, to a reduction in cardiac output, which affects the normal functioning of the various organs. Right cardiac insufficiency produces systemic venous congestion associated with an increase in the central venous pressure. Peripheral oedema comprises an accumulation of liquid in the interstitial space occurring, first of all, in the extremities of the lower limbs and rising progressively to the junction of the limbs, depending on the extent of liquid retention. The serous membranes may be the site of large accumulation of liquids (ascites, pleural or pericardial effusions) containing large quantities of protein (3-6%). Generally, venous congestion, oedema and effusions occur only when the total quantity of body liquid increases. Renal water and salt retention play an important role. The exact mechanisms of the symptoms of cardiac insufficiency are extremely complex. In addition to the haemodynamic theories (increase in the pressure in the vascular bed upstream from the defective ventricle), there are other factors (hypoxia, reduction in cardiac output, changes in capillary permeability, hormonal factors, salt and water retention).

HEALTH MEASURES

Hygiene of the cardiovascular system is limited basically to the methods for the prevention of cardiovascular disease. These relate to the prevention of congenital diseases in which many of the exogenous factors are recognised as teratogenic when they occur during the first months of gestation (pharmacodynamic agents such as thalidomide; infections such as rubella, mumps, etc., nutritional disorders and hypoxia).

As far as rheumatic heart disease is concerned, preventive measures are related primarily to the treatment of β-haemolytic streptococcal angina. Since around half of all anginas are due to this micro-organism, it is advisable to treat all febrile anginas as streptococcal anginas, and the only effective therapy is the use of antibiotics (penicillin, which has a success rate of around 100%). Other antibiotics in the erythromycin and tetracyclin group have a very good level of effectiveness. Unfortunately, a third of patients with valvular cardiopathies do not have a history of characterised angina but suffer from latent or larval streptococcal infection. The only hope of eradication of rheumatic heart disease at present seems to be a vaccination against group-A haemolytic streptococcus, but the enormous difficulties encountered with this vaccination are widely known. In patients already suffering from rheumatic heart disease, the prevention of new rheumatic flare-ups is the main objective and is achieved by long-term antibiotic treatment which may even last years.

In young adults efforts should be made to detect risk factors for the development of ischaemic myocardial disease or atherosclerotic disease (hypercholesterolaemia, arterial hypertension, obesity, diabetes, gout, smoking, physical inactivity, stress) and to combat these various factors (which are often associated) effectively, using medical methods, diet, physical exercise, occupational health measures, etc. These measures are all the more important when the risk factors are encountered in a subject already at particularly high risk due to hereditary and familial factors (familial hypercholesterolaemia, arterial hypertension, diabetes, etc.).

In patients already suffering from coronary artery disease, physical exercise is a powerful therapeutic agent. The functional results of physical training in these patients has been clearly demonstrated. A seated position after myocardial infarction reduces the loss of "posture" and tachycardia. Early passive and then active mobilisation reduces ankylosis of the joints and muscular atrophy. Regular exercise of the legs and the respiration help to reduce respiratory and embolic complications. Later, progressive mobilisation of larger muscle groups (back, arms, legs) makes it possible to increase cardiac efficiency. This raises the symptom threshold that results in incapacity in the patient. Heart rate and heart work for a given task decrease and the systolic volume increases. Contractions are stronger and better synchronised.

It is currently too early to say whether long-term prognosis is improved by physical training. The outcome of all this is that every man over the age of 30 should have an annual medical examination (arterial blood pressure, electrocardiogram at rest and during effort, chest X-ray, blood cholesterol level), for the early detection of risk factors for the development of cardiovascular disease. In older subjects, this examination becomes increasingly essential in view of the much higher probability of pre-existing stigmata of a cardiovascular disorder.

HAENNI, B.

Carpets, handwoven

Hand-knotted carpets

All oriental carpets are handwoven. Many of them are made in family workplaces, all the members of the household working from time to time on the loom. In some cases it is only a part-time occupation of the family. In some districts the manufacture is no longer home work but a factory industry.

The processes involved in the manufacture of a carpet are yarn preparation, consisting of wool sorting, washing, spinning and dyeing, and the actual weaving.

Yarn preparation. In some cases the yarn is received at the weaving place already spun and dyed. In other cases the raw fibre, usually wool, is prepared, spun and dyed at the weaving place. The first process of sorting into grades is done by women who usually sit on the floor. Wool washing and hand spinning follow. The dyeing is carried out in open vessels using mostly aniline or alizarine dyestuffs, the natural dyestuffs no longer being used.

Designing. In the handicrafts weaving or tribal weaving, as it is sometimes called, the designs are traditional and no new designs need to be made, but in undertakings employing a number of workers there may be a designer who first sketches the design of a new carpet on a sheet of paper and then transfers it in colours on to squared paper from which the weaver can ascertain the number and arrangement of the various colours of knots to be woven in the carpet.

Figure 1. An old type of loom at which the weaver is required to squat on the floor.

Weaving. In most cases the loom consists of two horizontal wooden rollers supported on uprights. One roller is about 10 to 30 cm above floor level and the other roller about 3 m above it. The warp yarn passes from the top roller to the bottom roller in a vertical plane. The number of weavers per loom is usually one but for wide carpets there may be as many as six. In about 50% of the cases the weaver squats on the floor in front of the bottom roller. In other cases he may have a horizontal plank to sit on. The plank is raised as weaving proceeds. The weaver has to tie short pieces of woollen or silk yarn into the form of knots round pairs of warp threads and then thread by hand across the whole length of the carpet one or two shots or picks of weft. The picks of weft are beaten up into the fibre of the carpet by means of a beater or hand comb. The tufts of yarn protruding from the fibre are trimmed or cut down by scissors.

As the carpet is woven it is in many cases wound on to the lower roller which increases in diameter. Where the worker squats on the floor the position of the lower roller prevents him from stretching his legs and as the diameter of the roller increases with the build-up of the carpet he has to sit further back but must still lean forward to reach the position in which he ties in the knots of yarn. Figures 1 and 2 show old types of looms at which the weaver must squat on the floor or on a plank. An improved type of loom at which the weaver may sit on a chair is shown in figure 3.

In some parts of Iran the warp in the carpet loom is horizontal instead of being vertical, and the worker sits on the carpet itself whilst working–which makes the task more difficult.

HAZARDS

Workers are subject to skeletal deformations, eyesight disorders and mechanical and toxic hazards.

Skeletal deformation. The squatting position that the weaver must occupy on the old type of loom, and the need for him to lean forward to reach the place into which he knots the yarn, causes in time very serious troubles. If a worker is employed from a young age, the legs may become deformed *(genu valgum)*, or serious crippling arthritis, or water on the knee may develop. The deformation of the pelvis, often in a restricted form, sometimes occurs; it is particularly serious for women as it may necessitate a Caesarian operation in the case of pregnancy. The lateral curvature of the spinal column (scoliosis) and lordosis are other common maladies.

Eyesight disorders. The constant close attention that the weaver must give to the point of weaving or knotting may cause considerable eyestrain, particularly if the lighting is inadequate. It should be mentioned that in some family concerns there is no electric light available and only oil lamps are used for work after dark. There have been cases of almost total blindness occurring after only about 12 years of employment at this work.

Figure 2. In the old type of loom the weaver sits on a plank which is raised from time to time as the work proceeds and, sometimes, goes up to 4 m from the floor. The worker does not have sufficient room for his legs and is forced into an uncomfortable position.

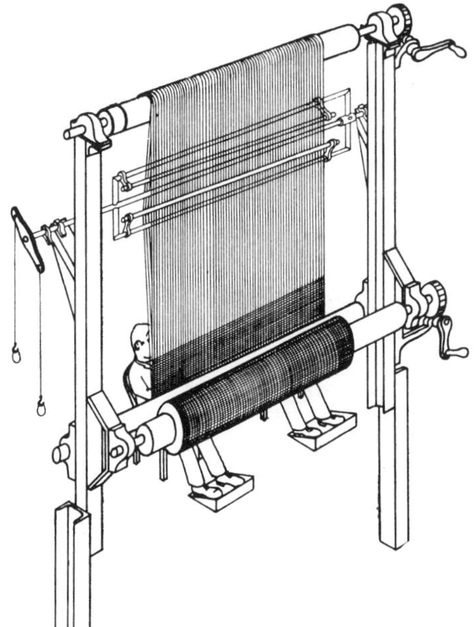

Figure 3. A new type of loom developed by Mr. Radjabi, which allows the worker to stretch his legs and to adjust the height and inclination of the carpet to suit his requirements.

Hand and finger disorders. The constant tying of knots and the threading of the weft yarn through the warp threads may result in swollen finger joints, arthritis and neuralgia causing permanent deformation of the fingers.

Mental troubles. The very fine work requiring a high degree of skill and constant attention leads in some cases to nervous illness manifesting itself in hand trembling and sometimes mental troubles.

Mechanical hazards. As no power machinery is used, there are practically none. If the looms are not kept in a good state of maintenance the wooden lever tensioning the warp may break and strike the weaver in falling. This danger may be avoided by using special thread tensioning gears.

Toxic hazards. The dyestuffs used, particularly if they are used with potassium or sodium bichromate, may cause skin infections or dermatitis. There is also the risk from the use of ammonia, strong acids and alkalies. Lead colours are sometimes used by designers and there have been cases of lead poisoning due to their practice of smoothing the tip of the paintbrush by placing it between the lips.

Infection. There is a danger of anthrax infection from raw wool.

SAFETY AND HEALTH MEASURES

At the sorting process for the raw material, all wool or camel hair, goat hair, etc., should be sorted over a metal grid provided with slight exhaust ventilation to draw any dust through the grid to a dust collector situated outside the factory. Where it is known that any such wool or hair, whether imported, or locally produced, is likely to be anthrax infected, the appropriate government authority should ensure that it is properly disinfected before being delivered to any workshop or factory.

In the wool-washing and dyeing processes, rubber gloves and waterproof aprons should be provided for the workers and all waste liquors should be neutralised before being discharged into waterways or sewers. Adequate ventilation is a necessity where wool-washing and dyeing processes are being carried on.

Very good lighting is required for the designing room. The use of leaded paints should be prohibited and harmless waterpaints substituted. Exceptionally good lighting is required for weaving work. The most important improvement required in many handweaving workshops is the provision of comfortable seats and the abolition of the necessity to squat on the floor in an unhealthy and uncomfortable fashion. This will require the raising of the lower carpet roller of the loom in many cases. It is really a problem in ergonomics. Such an improvement will not only improve the health of the worker, but will increase his efficiency and most likely lead to higher productivity after adaptation.

The workrooms should be kept clean and well ventilated, and properly boarded or covered floors substituted for earth floors. Adequate heating is required during cold weather. Manual manipulation of the warp places great strain on the fingers and may cause arthritis; wherever possible, hooked knives should be used for holding and weaving operations. Pre-employment and annual medical examination of all workers is highly desirable.

Hand-tufted carpets

The manufacture of carpets by the tying of knots of yarn by hand is a very slow process. The number of knots varies from 2 to 360 per square centimetre according to the quality of the carpet. A very large carpet with an intricate design may take over a year to make and involve the tying of hundreds of thousands of knots.

An alternative method of manufacture which eliminates the need for tying the very large number of knots is to use a special kind of handtool fitted with a needle through which the yarn is threaded. A sheet of coarse cotton cloth is suspended vertically and the design of the carpet traced on it in outline. The weaver places the handtool against the cloth and, by pressing a button, the needle is forced through the cloth leaving a loop of yarn of about 10 mm depth on the reverse side. The tool is then moved about 2 or 3 mm horizontally and the button pressed to make another loop on the reverse side, leaving also a loop on the face of the cloth. As many as 30 loops on each side can be made in one minute, but the weaver has to stop to change the colour of yarn for different parts of the pattern from time to time. When the looping operation has been finished, the carpet is taken down and placed reverse side up on the floor. A rubber solution is applied to the back and a covering or backing of stout jute canvas placed over it. The carpet is then placed face upwards and the protruding loops of yarn are trimmed by portable electric clippers. In some cases the design of the carpet is made by cutting or trimming the loops to varying depths.

HAZARDS AND THEIR PREVENTION

These are considerably less than in the case of the hand-knotted carpets. The operator usually sits on a plank in front of the canvas and has plenty of leg room. The plank is raised as the work proceeds. It would be better to have a backrest and a cushioned seat which could be moved along the plank as work proceeds horizontally. There is less eyesight strain, and no hand or finger operations are required likely to cause trouble. The rubber solution used in the carpet usually contains a solvent which is both toxic and highly flammable. The backing process should be carried out in a separate workroom with good exhaust ventilation, at least two fire exists, and with no open flames or lights. Any electrical equipment in this room should be of certified flameproof construction. The minimum amount of flammable solution should be kept in this room and fire extinguishers provided. A fire-resisting store for the flammable solutions not in use should not be situated inside any occupied building but preferably in an open yard.

Legislation

In most countries the general provisions of factory legislation cover the necessary standards required for safety and health in this industry; however, they may not apply to family undertakings and/or home work. Special regulations exist in some countries regarding the sorting and washing of wool, or other animal fibres likely to be infected with anthrax.

RADJABI, M. E.

Ergonomics:

"Ergonomics in traditional Iranian industries". Kavoussi, N. *Journal of Human Ergology* (Tokyo), 1976, 5/2 (145-147).

CIS 493-1968 "Physiological study of the manufacture of Persian carpets" (Fiziologičniproučvanija pri proizvodstvoto na persijshi kilimi). Enčev, M. *Trudove na Naučnoizsledovatelskija institut po ohrana na truda i professionalni zabolijavania* (Sofia), Oct. 1967, 14 (55-65). 4 ref. (In Bulgarian)

Visual impairment:

"An occupational health study in the carpet industry in Kerman, Iran". Kavoussi, N. *Work–Environment–Health* (Helsinki), 1973, 10/1 (48-51). Illus. 6 ref.

Carpets, machine-made

The art of carpet weaving is believed to have originated in Southern Persia about 4 000 BC. It was not until 1840-1845 that power was first applied to the weaving of carpets by E. P. Bigelow in the United States.

Today the main machine-made carpet manufacturing countries are the United States, United Kingdom, Canada, Belgium and Japan; however, most European and South American countries have their own carpet-making centres.

Types of carpet. There are two main types: woven carpets and non-woven carpets.

Woven carpets have a surface of pile yarn woven into a backing of warp and weft threads. Woollen and worsted, animal hair and man-made fibres such as rayon, acrylic and polyamide fibres as well as blends of all these are used to produce pile yarn. The backing yarns are usually of cotton or jute. The types of non-woven carpets are more varied.

Tufted carpets are made by inserting loops of pile yarn, usually artificial fibre, into a pre-woven backing, e.g. hessian, and the tufts of pile yarn are secured in position by latex applied to this backing. Needleloom carpets are made by inserting tufts of loose fibrous material into the base fabric, usually of jute. Needle Axminster carpets are similar to tufted carpet but the tufts of pile yarn are secured by a cotton locking thread passing through the loop of the tuft beneath the carpet. Electrostatically flocked carpets are made from chopped fibres, electrically charged, which are projected on to a backing fabric coated with adhesive, the backing also being electrically charged. Knitted carpets are made on knitting looms. Finally, the principle of bonding fibre or yarn to a backing material is the basis of many other methods of carpet manufacture such as the Neko, Giroud and Brandon processes.

Manufacturing techniques. Manufacture begins with the preparation of the yarns and although some of the large firms start with the raw wool, cotton, jute or other untreated fibres, it is more common for the carpet manufacturer to buy both the pile yarn in hank form ready for dyeing and the backing materials ready for use.

Woven carpets are produced on looms to which have been fitted special mechanisms for forming the pile. In Brussels and Wilton carpeting, pile yarns are led into the loom in continuous lengths parallel to the warp-backing threads and, as weaving proceeds, are drawn up in rows of loops above the backing by a pile-forming mechanism. For Brussels carpets, the pile loops are left uncut but, for Wilton, the loops are cut by a small knife carried on the pile-forming wire. In both spool and gripper Axminster weaving, the pile is formed by the mechanical insertion of individual tufts of yarn between the warp threads. These are bound into the backing in rows by two or more shots of weft. The tufts are formed from short lengths of yarn cut by automatic knives from a series of pile ends brought down over the weaving face.

The method of manufacture for tapestry carpets is the same as for Brussels and Wilton but the pattern is obtained by printing each pile yarn end separately. The pile yarns are then wound on to a beam in the correct order and the carpet is produced as described above. Two separate processes are involved in the manufacture of Chenille Axminster carpets. The pile is first woven separately on a Chenille weft loom on which the pile yarn forms the weft. The weave is then cut into strips and steamed to form V-shaped pile tufts known as "fur". The second process consists of weaving the Chenille fur into carpets on a setting loom in which the ropes of fur form

Figure 1. Gripper Axminster loom.

Figure 2. Carpet-tufting machine.

the weft and are bound into the backing as it is woven to form the pile of the carpet.

In the manufacture of tufted carpets, the pile yarn is led from cones located on a creel and each pile end is led to a needle in the tufting machine, the number of needles varying according to the width of carpet being produced. A backing cloth is passed under the needle by feed rollers and the pile is inserted through the backing cloth by the motion of the needles. Loop and cut-pile carpet can be manufactured. Tufted carpets are usually made from artifical or synthetic fibres, but recently wool has been used. Tufted carpets are normally dyed after making but pile yarn may be dyed before the tufting process. The backing of the carpet is coated with latex in order to secure the tufts and give the carpet body. [The most recent technical change in carpet production in developed countries is the printing of tufted carpets. This technique is now so developed that good printed tufted carpets are taking over much of the Axminster trade. The main advantage is that colours are applied after the carpet is made and this gives greater flexibility in changing designs.]

The process for needleloom carpets consists of the attachment of a lap or batt of loose fibrous material to a base fabric of jute. A beam having needles pointing vertically downwards rises and falls continuously. The batt with the base fabric underneath is passed under the beam and the needles penetrate the material, inserting tufts of pile material into the base fabric.

HAZARDS AND THEIR PREVENTION

Carpet-making machines present considerable mechanical hazards; there is also a danger of anthrax infection from infected wool and a number of the chemicals used in carpet dyeing and finishing may prove toxic.

Accidents. Carpet weaving machinery is complex and involves a multiplicity of moving parts which are potentially dangerous. Gear wheels, driving belts and other moving parts should be guarded. It is difficult to fence some operative parts, e.g. the trap between the grippers and the fixed parts at Axminister looms but photoelectrically operated stopping devices have been tried. Shuttles sometimes fly out of looms and cause injuries to weavers and suitable shuttle guards should be provided. In view of the risks, great reliance must be placed on the correct training of the operatives and the introduction of safe methods of working. The finished rolls of carpet are extremely heavy and suitable mechanical handling and inplant transport equipment should be employed for moving them; suitable stacking techniques or storage stands are also necessary.

Diseases. Hardwearing low grade wools are often used in carpet manufacture and with some wools of this type there will be a risk of anthrax. Once the wool starts on the manufacturing process, complete control of the anthrax spore is difficult. Application of mechanical ventilation at all handling and dust-producing points will mitigate the hazard, but the only complete solution is to disinfect all infected wool before use. Precautions against anthrax in carpet manufacture should include the use of warning notices and the issue of personal cards to all workers to make sure that the early symptoms are at once recognised.

In carpet dyeing, there is a risk of chrome ulceration where chrome salts are used. To combat this, personal protective equipment, including hand and arm protection, impervious aprons, etc., should be provided.

Certain pesticides used as mothproofing agents for wool (e.g. dieldrin) are toxic when inhaled, ingested or absorbed through the skin. Protective clothing (rubber gloves and aprons) should be worn and great care taken to avoid spillage. A good standard of general ventilation and, where necessary, local exhaust ventilation should be provided.

DAVIS, J. A.

Castor oil and bean

The castor plant, *Ricinus communis*, a member of the spurge family, is a coarse, erect annual herb in temperate climates and a tree-like perennial growing to over 12 m tall in warmer regions. When ripe, the prickly seed pods separate into three parts, each containing a seed, the castor bean, which contains a thick, colourless or greenish oil.

The castor plant is a native of tropical Africa and Asia but it is now grown widely in tropical, subtropical and temperate climates. Brazil and India are the major producers. The USSR, Thailand, the United States, Ecuador, South Africa, Tanzania, Paraguay, Ethiopia, Romania and the Sudan also produce large amounts of castor beans. About 50% by weight of the beans is extractable oil.

Processing. Pressing removes between 70 and 90% of the oil from the bean, and this is usually followed by extraction with fat solvents. Solvent extraction, by decreasing the oil content, leads to a more finely divided and dispersible pomace. Steam treatment, which is usually used to remove residual solvent and oil, also has the effect of reducing the toxic potential of the pomace.

Uses. Some varieties of castor plant are ornamental but the majority are cultivated for the production of castor bean from which castor oil is extracted.

Although castor oil has medicinal value as a cathartic, its major use is as an industrial raw material. Dehydrated castor oil is used as a fast drying agent for paints and varnishes, in the manufacture of lubricants resistant to high temperatures, plastics, viscose, fungicides, rubber products, soaps and plasticisers. Castor pomace, the residue left after pressing, is used as a fertiliser and, if adequately detoxified, as an animal food.

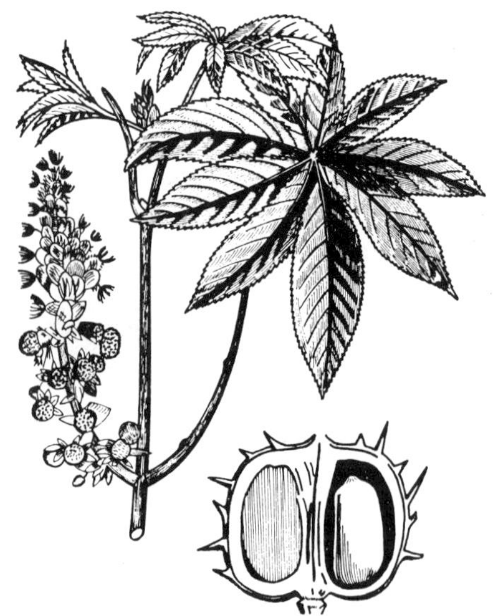

Figure 1. Castor-oil plant.

HAZARDS AND THEIR PREVENTION

Castor beans contain a toxic protein, ricin. The toxic intravenous dose of ricin for guinea-pigs is as low as 1 mg/kg, and in man ingestion of this substance produces nausea, diarrhoea, tenesmus, abdominal cramps and haemorrhagic changes in the gastrointestinal tract. Ricin is also capable of agglutinating red blood cells. The toxic properties of ricin are destroyed by heat; the power to agglutinate red blood cells is, however, destroyed at a lower temperature and consequently the agglutination test is not always adequate to rule out toxicity. Animals and man can develop immunity to ricin.

Castor beans also contain several potent allergens capable of causing severe hay fever, asthma and urticaria in sensitised humans. Cross sensitisation between castor allergens and materials in closely related plants may occur.

The ricin and allergens in castor beans remain in the residue or castor pomace when the oil is extracted so that the castor oil of commerce is not hazardous. Sensitivity to castor oil when used medicinally is extremely rare, only one case having been reported in the literature.

Castor pomace is both toxic and allergenic and can be a hazard to those who handle it, live near processing plants or are exposed to fertilisers which contain it. Its toxic properties vary with the method of extraction and whether or not it has been treated to destroy the ricin or to reduce allergenicity.

Occupational illness from castor pomace exposure was first reported in laboratory workers in 1913; occasional cases have occurred in agricultural workers; many cases have been reported in castor-oil mills, in dock workers, railway workers, fertiliser blenders, gardeners and other occupationaly exposed persons. Community outbreaks of severe asthma associated with airborne castor pomace have been reported from the United States, Hungary, Germany, Brazil, South Africa, Italy and France. The potency of castor allergens is such that bags used to carry castor pomace have been shown to contaminate green coffee sufficiently to affect coffee handlers. In most cases, allergic sensitivity is responsible for these outbreaks but occasionally ricin toxicity has also played a part.

Skin testing with castor products is not recommended as a routine diagnostic procedure because of the violent reactions that may occur unless done with great care and very dilute preparations. The passive transfer of reagins to the skin of monkeys and other animals has been shown to be useful in diagnosis.

Castor pomace that has not been treated to reduce the ricin and allergen content is so hazardous to health that it should not be an article of commerce. When it must be handled, extreme precautions are recommended, including the use of respiratory protective equipment, eye protection and overalls.

Pomace in which the ricin and allergen content have been reduced can be handled safely if proper measures are taken to minimise dust dissemination. Local exhaust ventilation, the use of properly labelled and dust-proof containers and maintenance and good housekeeping are essential. Personal protection, including respirators, should be used for temporarily dusty situations. Persons subject to allergic asthma should not work with castor pomace. Plant effluents containing castor dust should not be discharged into community air.

Ultimately, both the protection of employees and the general public and the maximum use of castor pomace in fertilisers and animal feeds will depend upon the development and application of new techniques for detoxification and deallergenisation.

COOPER, W. C.

"Poisoning by castor beans" (Otrovanje sjemenkama ricinusa). Maretic, Z. *Arhiv za higijenu rada i toksikologiju* (Zagreb), Sep. 1980, 31/3 (251-257). Illus. 9 ref. (In Serbocroatian)

The harmful effects on health of Ricinus communis *seeds* (De schadelijke werking van het zaad de *Ricinus communis* op de gezondheid). Vroege, D. (Rotterdam, Veen en Scheffers, 1971), 171 p. Illus. 313 ref. (In Dutch)

"Collective asthma due to castor bean allergy" in Ourinhos, S.P.: "Follow up study after industrial processing of castor bean was stopped". Strauss, A. *Revista do Instituto de Medicina Tropical de São Paulo* (São Paulo), 1975, 17/2 (79-82).

"Castor bean allergy in the upholstery department of a furniture factory". Topping, M. D.; Tyrer, F. H.; Lowing, R. K. *British Journal of Industrial Medicine* (London), Aug. 1981, 38/3 (293-296). Illus. 15 ref.

Catalysts

When the rate of a chemical process is increased by the presence of a substance which itself is chemically unchanged at the end of the reaction, that substance is called a catalyst and the mechanism by which the rate is increased is called catalysis.

Since the catalyst is not consumed during reaction it is possible to form large amounts of products using a relatively small quantity of catalyst. It is further possible, by careful choice of catalyst, to increase the yield of the required product by selective promotion of one reaction relative to unwanted side-reactions. Thus by the use of catalysts it is possible to increase the yield of a product to such an extent that a process becomes economically suitable for commercial manufacture.

Industrial catalysts may be classified in two basic groups: homogeneous catalysts, where the catalyst is in the same phase (generally solution) as the reactants, and heterogeneous, where the catalyst is in a different phase (generally solid) from the reactants (liquids or vapours).

Industrial history. Industrial catalysis dates from 1746 with the use of nitrogen oxides to accelerate the oxidation of sulphur dioxide in the lead chamber process. This was followed by the use of platinum as an oxidation catalyst for several inorganic processes: the oxidation of sulphur dioxide, the oxidation of hydrogen chloride to chlorine (Deacon, 1860) and the oxidation of ammonia (Ostwald, 1905). The synthesis of ammonia was added to the list in 1914 by Haber and Mittasch.

The first industrial application of catalysis to organic chemistry was made in 1902 with the hardening of fats catalysed by nickel. This was followed by the synthesis of methanol from carbon monoxide and hydrogen and the production of phthalic anhydride by the partial oxidation of aromatic hydrocarbons.

The enormous growth of the petrochemical industry over the last half-century, and the very existence of some common materials (plastics and synthetic rubber) are due to developments in catalyst technology. Each process in oil refining and petrochemical industries has seen major advances in catalyst technology.

Many modern catalysts for heterogeneous processes are extremely complex. Most consist of one, two or three principal components often carried on a support material which usually plays no part in the chemical process and is often termed "inert". It is usual, however, for other components to be added to modify the performance of the catalyst. These extra components can be "activators" or "promotors" which are added to enhance the rate of particular reactions. They can also be "inhibitors", "dopants" or "partial poisoning agents" which are added to prevent the occurrence of unwanted reactions. The activating or inhibiting processes are sometimes achieved by adding a component to the chemical feedstock so that they interact with the catalyst during the operation of the chemical process.

It should be borne in mind that the chemical and hence toxicological behaviour of some catalyst components, especially when alloyed or in some form of chemical combination with other components, will not necessarily be the same as when that component is in the free state.

HAZARDS

Certain chemical processes may be accelerated by a number of catalysts. For this reason certain reactions occur repeatedly in the following catalogue of some industrial catalysts.

Aluminium compounds

Aluminium oxide (Al_2O_3). This compound is widely used as a catalyst support. It is also used in bifunctional catalysts in petroleum reforming reactions and as a promotor of iron catalysts in the synthesis of ammonia. Silica-alumina has acidic properties which are used in petroleum cracking and olefin polymerisation reactions.

There is no evidence that aluminium oxide has been a cause of serious health injury. It is, however, listed as a nuisance particulate with a TLV of 10 mg/m^3 (total dust) or 5 mg/m^3 (respirable dust).

Aluminium chloride ($AlCl_3$). This is used in catalytic alkylation to produce octane isomers and in the Friedel-Crafts' synthesis of alkylated aromatic compounds.

Anhydrous aluminium chloride can react explosively with water and forms hydrochloric acid fumes in moist air. Where the material is stored in metal containers there is thus a risk of hydrogen generation and attendant fire and explosion risk. Aluminium chloride dust is an oral and inhalation irritant.

Aluminium alkyls. Compounds such as $Al(C_2H_5)_3$ are used in conjunction with titanium chloride in the Ziegler-Natta stereospecific polymerisation of olefins to produce synthetic rubber.

Aluminium alkyls are spontaneously flammable when exposed to air, and violently explosive when in contact with water. In an undiluted state they must be stored under inert gas (N_2 or A) and all possibility of contact with water avoided. Solutions containing up to 20% alkyls in inert solvents can be used without risk of spontaneous ignition.

Fumes from spillages can cause death from pulmonary haemorrhage. Aluminium alkyls have listed TWA values of 2 mg/m^3.

Antimony compounds. Antimony oxide (Sb_2O_3) may be a component of complex oxide catalysts for the oxidation and ammoxidation of propylene, the oxidation of butanes to buta-1,3-diene and the oxidation of hydrocarbons to acetic acid.

Most antimony compounds are highly toxic by inhalation. They are also skin and mucous membrane irritants.

Antimony trioxide has a listed TWA of 0.5 mg/m^3 and antimony trioxide production is suspected of carcinogenic risk.

Barium compounds. Barium compounds may be added to silver catalysts for the oxidation of ethylene.

Metallic barium can be explosive in the form of dust when exposed to heat or by chemical reaction. There is a violent reaction with water and some halogenated hydrocarbons.

Soluble barium salts are poisonous taken orally and also irritate the eyes, nose, mouth and skin. Soluble barium compounds have a listed TWA of 0.5 mg/m^3.

Biological catalysts, enzymes

Enzymes are protein catalysts found in all living cells and are responsible for catalysing chemical reactions in biological processes. Their use is, of course, commonplace in the brewing and food industries. Yeasts induce fermentation through the agency of the enzyme zymase, which converts glucose and some other carbohydrates to carbon dioxide and water in the presence of oxygen or into alcohol and carbon dioxide in the absence of oxygen.

Bound or immobilised enzymes, in which the enzyme is attached to a water insoluble matrix, have found use as catalysts in a number of industrial processes, e.g. the enzyme amino acid actylase has been used in the production of optically pure amino acids, glucose isomerase has been used in the production of fructose syrup and β-galactosidase has been used in the production of glucose and galactose from lactose. The industrial use of other bound enzymes is anticipated (see Royer, 1980).

Bismuth compounds

Bismuth oxides. Bismuth compounds, in the form of complex mixed oxides, are used in selective oxidation reactions, for example propylene to acrolein; benzene, toluene and naphthalene to maleic anhydride, benzoic acid and phthalic anhydride respectively. Similar catalysts containing bismuth oxide are used in the ammoxidation of propylene to acrylonitrile.

Bismuth and its salts can produce kidney damage although the degree of damage is usually mild.

Calcium compounds

Calcium oxide (CaO). It is used as a promoter of iron catalysts in the ammonia synthesis.

Calcium oxide is a powerful caustic to living tissue. It has a listed TWA of 2 mg/m^3.

Carbon. Carbon in various forms is commonly used as a catalyst support. Activated charcoal is used as a catalyst for the production of phosgene from carbon monoxide and chlorine.

Carbon dusts are considered to be nuisance dusts. Some forms of carbon dust can cause irritation of the eyes and mucous membrane.

Chromium compounds. Chromium compounds, usually complex oxides, are used in many catalytic reactions. Different formulations are used as catalysts in the following processes: the production of butadiene and butene from butane, the reduction of nitrobenzene to aniline, the production of saturated and unsaturated alcohols from fatty acids and esters, the conversion of aliphatic hydrocarbons to aromatics, the Ziegler olefin polymerisation and the synthesis of methanol from carbon monoxide and hydrogen. Chromium compounds are also used as activators of iron oxide catalysts in the shift process for the conversion of carbon monoxide.

Although some hexavalent chromium compounds are carcinogenic, cause chrome ulceration of the skin and nasal septum and a sensitisation dermatitis, the compounds of trivalent chromium—e.g. Cr_2O_3—are relatively less toxic. Chromium and soluble chromic and chromous salts have listed TWA values of 0.5 mg/m^3. Chromates have a value of 0.05 mg/m^3.

Cobalt compounds

Metallic cobalt (Co). Cobalt is used as a catalyst in the Fischer-Tropsch synthesis of paraffins from hydrogen and carbon monoxide and in the shift process for the conversion of carbon monoxide.

Cobalt oxides. Cobalt molybdates are widely used catalysts in hydrodesulphurisation reactions, e.g. the conversion of thiophene to butane. Cobalt is also a component of molybdenum-based catalysts for the oxidation and ammoxidation of olefins and alcohols.

Cobaltous-cobaltic oxide Co_3O_4 has been used as a catalyst for the oxidation of ammonia.

Cobalt hydrocarbonyl ($HCo(CO)_4$) and cobalt tetracarbonyl ($Co_2(CO)_8$). They are used in the oxo process for the synthesis of aldehydes from olefins, carbon monoxide and hydrogen, and in hydrogenation reactions.

Industrial experience has indicated that cobalt dust and fumes may cause dermatitis and pneumoconiosis. [In addition a cobalt phthalocyanine catalyst has been recently suspected of carcinogenic effects.] The metal, dust and fume has a proposed TWA of 0.05 mg/m^3.

The toxicity of carbonyls is related to their ready decomposition to release carbon monoxide. Symptoms are due in part to carbon monoxide and in part to the direct irritating action of the carbonyl.

Copper compounds

Metallic copper (Cu). Copper is used to catalyse the synthesis of phenol, the production of formaldehyde from methane and the dehydrogenation of alcohols to aldehydes and ketones. Copper alloyed with other metals (e.g. Au, Ni, Os, Ru, Pt) in the form of bimetallic catalysts is used in reforming reactions such as the conversion of naphthas to high octane compounds.

Cupric oxide (CuO). This may be a component of mixed oxide catalysts for partial oxidation reactions, the reduction of fatty acids and esters, the synthesis of methanol and the low temperature water shift reaction.

Copper oleate $Cu(C_{12}H_{33}O_2)_2$. This compound is a catalyst for the synthesis of phenol.

Cuprous chloride (CuCl). This compound is used to catalyse the dimerisation of acetylene to vinyl acetylene and aniline to azobenzene.

Cupric chloride ($CuCl_2$). This catalyst is used in the oxychlorination of ethylene to ethylene dichloride, a precursor of vinyl chloride. It may also find use in the production of chlorine from hydrogen chloride, acetaldehyde from ethylene (Wacker process), and in the vinyl acetate synthesis.

Copper pyrophosphate (CuHPO4). This compound catalyses the dimerisation of isobutylene to diisobutylene.

Inhalation of copper dust has been reported to cause damage to lung cells, liver and pancreas in animals. Copper fume, dust and mist has a TWA of 1 mg/m^3.

Sublimed copper oxide has been reported as being a possible cause of one form of metal fume fever.

Gallium, germanium, indium, iridium

See platinum.

Gallium and indium are highly toxic by subcutaneous routes. Germanium and iridium are reported to be less toxic.

Iron compounds

Metallic iron (Fe). Finely divided iron is one of the catalysts used in the synthesis of ammonia and in the vapour phase hydrolysis of benzene to benzyl alcohol. Iron may also be present in bimetallic catalysts for petroleum reforming.

Iron oxides. Catalysts containing iron oxides are used in the shift reaction for the production of hydrogen from carbon monoxide and steam. Iron oxides may also be components of mixed oxide partial oxidation catalysts, e.g. the molybdenum-based catalysts for the oxidation and ammoxidation of olefins and alcohols.

Ferrous chloride ($FeCl_2$). This may be a component in complex catalysts for the oxidation of ethylene to acetic acid.

Iron dust can form conjunctivitis, choroiditis, retinitis and siderosis of tissues. Iron oxide fume (associated with welding) has a TWA of 5 mg/m^3.

Lithium compounds

Metallic lithium (Li). This metal is used in stereospecific polymerisation reactions.

Lithium alkyls. These compounds are used as anionic polymerisation catalysts in reactions such as styrene polymerisation.

Metallic lithium is a highly reactive metal which can constitute a fire hazard when exposed to heat. Lithium oxide and hydroxide form extremely caustic solutions and the lithium ion has central nervous system toxicity.

Magnesium compounds

Magnesium alkyls and dialkyls are used as catalysts in the polymerisation of styrene.

Manganese compounds

Manganese dioxide (MnO_2). This is one of the catalysts used in the synthesis of methanol from carbon monoxide and hydrogen.

Industrial exposure to manganese dusts and fumes occurs during the mining, crushing and sieving of the ore and in the vicinity of the reduction furnaces; reported cases of poisoning (which can consist of damage to the upper respiratory tract and central nervous system) are associated with these operations rather than with its use as a catalyst.

Mercury compounds

Mercuric sulphate ($HgSO_4$). In the presence of sulphuric acid this salt is used to catalyse the hydrolysis of acetylene to acetaldehyde.

Mercuric chloride ($HgCl_2$). This is used as a catalyst in the production of vinyl chloride from acetylene and hydrogen chloride.

Mercury and its salts are general systemic poisons. The chief effects are on the central nervous system and on the teeth and gums. Mercury compounds have listed TWA values of 0.05 mg/m^3.

Molybdenum compounds

Molybdenum oxides. Molybdenum trioxide is used to catalyse the polymerisation of olefins.

Following the success of bismuth molybdate, a range of molybdenum-based catalysts are used for the partial oxidation of aromatic hydrocarbons and for the oxidation and ammoxidation of olefins and alcohols.

Sulphided cobalt molybdenum oxide catalysts are widely used for hydrodesulphurisation reactions using all kinds of petroleum feedstocks.

There are listed TWA values for molybdenum compounds of 10 mg/m^3 (insoluble compounds) and 5 mg/m^3 (soluble compounds).

Nickel compounds

Metallic nickel (Ni). Metallic nickel is used as a catalyst often alloyed with copper, cobalt or iron for hydrogenation and reforming processes. It is also used as a catalyst for the methane conversion reaction and for the Fischer-Tropsch (methanation) reaction.

Nickel oxides. Mixed, nickel-containing oxides are used as partial oxidation catalysts and also as hydrodesulphuration catalysts (cobalt nickel molybdate).

Some nickel compounds cause dermatitis and some are reported to be carcinogenic by inhalation. Metallic nickel has a listed TWA value of 1 mg/m^3 and soluble nickel compounds have a value of 0.1 mg/m^3.

Nitrogen oxides

Nitric oxide (NO) and nitrogen dioxide (NO_2). These nitrogen oxides are used to catalyse oxidation in the manufacture of sulphuric acid. Although most of the production of this acid is by the contact process, in which other catalysts are used, a significant amount of the acid is still produced by the earlier process and precautions in respect of these nitrous fumes must be considered. Nitrous fumes belong to a group of lung tissue irritants which have a delayed effect that enhances their danger. Nitric oxide has a listed TWA value of 30 mg/m^3 and the value for nitrogen dioxide is 9 mg/m^3.

Palladium compounds

Metallic palladium (Pd). This metal is used as a catalyst with ethyl anthraquinone, which is treated first with hydrogen and then with oxygen to produce hydrogen peroxide, the ethyl anthraquinone being regenerated in the second stage of the process.

The metal is also used as an oxidation catalyst in automobile exhaust and, alloyed with gold, as a reforming catalyst.

Palladium oxide (PdO). This oxide catalyses the reduction of furan to tetrahydrofuran.

Palladium chloride ($PdCl_2$). This salt is associated with copper chloride in the Wacker process for acetaldehyde formation, and vinyl acetate synthesis. No cases of industrial poisoning by palladium or its compounds have been reported.

Peroxides

Lithium peroxide (Li_2O_2) and sodium peroxide (Na_2O_2). These inorganic peroxides catalyse the polymerisation of styrene to polystyrene.

Benzoyl peroxide ($(C_6H_5CO)_2O_2$). Lauroyl peroxide ($[CH_3(CH_2)_{10}CO]_2O_2$). Methyl ethyl ketone peroxide ($(CH_3COC_2H_5)_2O_2$). These organic peroxides (and others) give rise to free radicals under certain conditions and initiate a number of polymerisation reactions.

In general, peroxides decompose with the liberation of oxygen and thus constitute a fire hazard. Organic peroxides are particularly dangerous, and benzoyl peroxide has caused disastrous explosions in factories where it is manufactured and used. It is now customary to supply most organic peroxides already mixed with a plasticiser or phlegmatiser, which reduces the risk from one of explosion to one of fire.

Inorganic peroxides are of variable toxicity and may cause injury by skin contact or to the mucous membrane. Organic peroxides cause irritation to the skin, eyes and mucous membrane.

Phosphorus

Phosphorus oxides. Phosphorus is used as a component of complex oxides used as partial oxidation catalysts in processes such as the oxidation of butanes to maleic anydride.

Inorganic phosphorus compounds are of variable toxicity. Large doses may cause serious disturbances particularly in calcium metabolism. Metaphosphates may be highly toxic causing irritation and haemorrhages in the stomach as well as liver and kidney damage.

Elemental phosphorus is extremely reactive and highly toxic.

Platinum

Metallic platinum (Pt). This metal is used as a catalyst often alloyed with rhodium and palladium in the oxidation of ammonia. It is used as a catalyst in the contact process for production of sulphuric acid. It catalyses reforming and isomerisation reactions in the petroleum industry and the production of formaldehyde from methyl alcohol or methane. It is used as a catalyst in fuel cells in which electrical energy is generated by the reaction of fuel and oxygen.

It is a common component in bimetallic reforming catalysts (with, e.g., gallium, germanium, indium, iridium) and the metal is also used in automobile exhaust catalysts.

Platinum on porous graphite is used as a catalyst in the synthesis of hydroxylamine from nitric oxide and hydrogen in aqueous sulphuric acid.

Platinic oxide (PtO_2). This catalyses the reduction of benzene to cyclohexane.

Platinum complex salts cause irritation to the skin, eyes and respiratory system. Soluble platinum salts have listed TWA values of 0.002 mg/m^3.

Potassium

Potassium oxide (K_2O). This compound is used as a promotor of iron catalysts for the Fischer-Tropsch process and for the synthesis of ammonia.

Potassium alkyls. These are used as catalysts in the polymerisation of styrene.

Potassium oxide and potassium alkyls react violently with water to produce caustic potassium hydroxide.

Rhenium compounds

Metallic rhenium (Re). This metal is used, alloyed with platinum, in catalysts for petroleum reforming.

Rhodium compounds

Metallic rhodium (Rh). This metal is used with platinum, in ammonia oxidation catalysts. It is also used in some automobile exhaust catalysts.

Rhodium carbonyls (e.g. $RhClCO[(C_6H_5)_3]_2$). These compounds are used in carbonylation reactions such as the production of acetic acid from methanol and carbon monoxide, and in the production of polyhydric alcohols (e.g. glycol) directly from carbon monoxide and hydrogen. (See cobalt hydrocarbonyl for the effects of carbonyl exposure).

Rhodium metal fume and dust has a listed TWA value of 0.1 mg/m^3. The soluble salts have TWA values of 0.001 mg/m^3.

Ruthenium compounds

Ruthenium metal (Ru) may be alloyed with copper in catalysts for petroleum reforming.

Silicon compounds

Silicon oxide, silica (SiO_2). Silica and alumina-silicas are used extensively as cracking catalysts and in dual function catalysts for hydrocracking and similar processes. They are also used as supports for a wide range of catalysts. Inhalation of silica dust may cause silicosis.

Silver

Metallic silver (Ag). This metal is used as catalyst in the oxidation of ethylene to ethylene oxide.

Metallic silver causes argyria. It has a listed TWA of 0.1 mg/m^3.

Sodium compounds

Sodium alkyls are used as catalysts in the anionic polymerisation process in the production of synthetic rubber.

Sodium alkyls are very reactive, producing caustic sodium hydroxide on contact with water.

Thorium compounds

Thorium oxide (ThO_2). This is one of the catalytic agents used in the Fischer-Tropsch synthesis for the production of paraffins from carbon monoxide and hydrogen.

Industrial experience with thorium oxide has extended over a very considerable period of years and has been good. However, thorium oxide is a radioactive substance and has proved to be carcinogenic when taken internally. No toxicity other than by radiation has been described.

Tin compounds

Metallic tin (Sn). This metal may be alloyed with platinum in catalysts for the reforming of petroleum.

Tin oxides. Mixed tin oxides, principally tin-molybdenum and tin-antimony, are used as catalysts for various selective oxidations of propylene.

Stannous octoate ($(C_7H_{13}COO)_2Sn$). This is used in conjunction with triethylenediamine in the production of polyurethane.

Metallic tin is regarded as non-toxic. Some inorganic tin salts are irritants and alkyl tin compounds may be highly toxic and produce skin rashes. Organic tin compounds are absorbed by the skin. Inorganic tin compounds (except SnH_4 and SnO_2) have a listed TWA value of 2 mg/m^3. Organic tin compounds have a TWA value of 0.1 mg/m^3.

Titanium compounds

Titanium chloride ($TiCl_4$). This compound is used with aluminium alkyls to catalyse Ziegler-Natta type stereospecific polymerisation reactions. This compound reacts with water with the liberation of hydrochloric acid vapour.

Triethylenediamine ($(C_3H_4)_3N_2$). This compound together with stannous octoate catalyses the combined condensation and polymerisation by which the polyurethanes are produced from diisocyanates.

The vapours of aliphatic amines and polyamines cause irritation of the eye and of the mucous membranes of the nose and throat. Exposure to the vapour may cause lung irritation, respiratory difficulty and coughing. There may be sensitisation with asthmatic symptoms. These substances also cause dermatitis.

Uranium compounds

Uranium oxides. Mixed uranium oxide catalysts are used for selective oxidation reactions, for example uranium-antimony oxide is used for the oxidation and ammoxidation of propylene. Uranium compounds are highly toxic and permissible levels are based on both chemical and radiotoxicity. Uranium oxides are recognised carcinogens. The soluble and insoluble compounds of uranium have listed TWA values of 0.2 mg/m^3.

Vanadium compounds

Vanadium pentoxide (V_2O_5). This is a most effective and versatile oxidation catalyst. It is used for the oxidation of sulphur dioxide to sulphur trioxide in the manufacture of sulphuric acid. It is also used in the oxidation of ethylene to acetic acid and acetaldehyde, in the oxidation of naphthalene to phthalic anhydride, and in the oxidation of benzene or C hydrocarbons to maleic anhydride. Vanadium compounds act chiefly as irritants to the conjunctivae and respiratory tract. Most attention has been focused on the hazard associated with the dust and fume. Vanadium pentoxide has a proposed TWA value of 0.05/m^3 (dust).

Zeolites

These compounds are essentially crystalline aluminosilicates with large internal pore volumes which are capable of holding active ionic and metallic species. They are extensively used in the petrochemical industries for such reactions as shape selective cracking, hydrogenation and reforming.

Zeolite dust is regarded as a nuisance dust.

Zinc compounds

Zinc oxides. Mixed zinc-containing oxides are used for a variety of catalysts, for example zinc-chromium catalysts are used for the oxidation of butane to butadiene, the synthesis of methanol and the reduction of unsaturated fatty acids to unsaturated alcohols. Zinc-copper oxides are used in the low temperature water shift reaction.

Zinc compounds are generally of low toxicity. The oxide is virtually innocuous although there are hazards associated with the freshly formed oxide. Zinc oxide fume has a listed TWA value of 5 mg/m^3.

[**Miscellaneous**

DMPAN($(CH_3)_2NHC_2CH_2CN$). Dimethylaminopropionitrile, a catalyst used in the manufacture of foam resins, has recently been shown to provoke urinary retention and to have other neurotoxic effects in workers exposed to it.

4,4'-Methylene-bis-*2-chloroaniline* ($C_{13}H_{12}N_2Cl_2$) (3-3'-DICHLORO-4,4'-DIAMINODIPHENYLMETHANE; MOCA; DACPM). MOCA is used as a catalyser in the polyaddition of polyurethanes. It has been heavily suspected of being a human carcinogen on the ground of evidence of its mutagenic properties, carcinogenic effects in rat, mouse and dog (with adenocarcinomas of the lungs, hepatocellular carcinomas, mammary adenocarcinomas, haemangiosarcomas), and its presence in the urine of exposed workers. In the absence of conclusive epidemiological evidence in humans, it is recommended to treat MOCA as a potential occupational carcinogen and to prevent any skin contact with it.

Niax catalyst ESN. This is composed of DMPAN (95%) and of *bis*-2-dimethylaminoethylether. It has shown the same urinary effects as DMPAN in addition to liver dysfunction. It is also used in the manufacture of polyurethane foam.]

SAFETY AND HEALTH MEASURES

Detailed indications of hazards have been given for some of the most dangerous substances used as catalysts, e.g. aluminium alkyls, cobalt carbonyl, mercury compounds, nitrogen oxides and organic peroxides. In addition to the special treatment given to these particularly hazardous substances the harmful properties of many other catalysts have been described. These substances have to be introduced into the plant at the beginning of a process, and require to be withdrawn, regenerated and replaced. In the course of this work precautions must be taken to prevent injury to the operators.

When substances that may cause injury to health are used as catalysts, precautions must be taken to prevent or reduce personal contact with them. Often the route of absorption of injurious solid material is by inhalation of the dust. If possible, the catalyst should be used in a granular or pelleted form to avoid the risk of airborne dust when it is being introduced into the plant, removed or regenerated. When the catalyst can be used in the form of a fluidised bed it can be moved mechanically within enclosed plant from the reaction chambers to regenerating chambers and back again. By such an arrangement the process worker is protected from contact with a harmful catalyst, but a maintenance worker who has to enter the plant would still be exposed. In such circumstances the maintenance worker should wear protective clothing, personal protection equipment, including eye protection, and suitable respiratory protective equipment according to the nature of the risk. These considerations are concerned with solid substances associated with heterogeneous catalysts. When the catalyst takes the form of a harmful gas or vapour, exhaust ventilation, breathing apparatus and protective clothing should be provided.

GENTRY, S. J.
HOWARTH, S. R.
JONES, A.

General:

"Immobilized enzymes". Royer, G. P. (29-74). *Catalysis Reviews: Science and Engineering.* Vol. 22 (New York, Manel Dekker Journals, 1980).

Selective oxidation of hydrocarbons. Hucknall, D. J. (London and New York, Academic Press, 1974), 220 p.

Scientific bases for the preparation of heterogeneous catalysts. Delmon, B.; Jacobs, P.; Poncelet, G. (eds.). (Amsterdam, Oxford, New York, Elsevier Scientific Publishing Co., 1976), 706 p.

"The kinetics of some industrial heterogeneous catalytic reactions". Tempkin, M. I. *Advances in Catalysis.* Vol. 28. Eley, D. D.; Pines, H.; Weisz, P. B. (eds.) (New York, Academic Press, 1979).

Occupational health:

CIS 79-1273 "Giant-cell tumours due to the effects of cobalt dust (cobalt phthalocyanine, Merox catalyst)" (Riesenzelltumoren nach Einwirkung von kobalthaltigem Staub–Kobalt-Phthalocyanin (Merox-Katalysator)). Schulz, G. *Staub* (Dusseldorf), Dec. 1978, 38/12 (480-481). 7 ref. (In German)

CIS 75-129 "Health status of workers engaged in platinum catalyst production" (O sostojanii zdorov'ja rabočih proizvodstva platinovogo katalizatora). Gladkova, E. V.; Odincova, F. P.; Volkova, I. D.; Vinogradova, V. K. *Gigiena truda i professional'nye zabolevanija* (Moscow), Mar. 1974, 3 (10-13). 7 ref. (In Russian)

CIS 79-1044 *Special hazard review with control recommendations for 4,4'-methylenebis(2-chloroaniline).* DHEW (NIOSH) publication No. 78-188 (National Institute for Occupational Safety and Health, 4676 Columbia Parkway, Cincinnati) (Sep. 1978), 67 p. 30 ref.

Experimental alert No. 4 concerning the occupational hazard due to Niax catalyst ESN. International Occupational Safety and Health Hazard Alert System (Geneva, International Labour Office, 1980), 5 p.

Cataract, occupational

Cataract is a disease of the crystalline lens with the clinical appearance of clouding. Cataracts are caused by the action of chemical or physical factors disturbing the internal respiration and the metabolic processes in the lens, thus inhibiting its nutrition and giving rise to the atrophy of epithelial elements of the capsule. Cataract reduces the visual capacity.

Types of cataract. Congenital and acquired cataracts are characterised by grey-white opacities in the lens substance or on its capsule. They are classified according to their location and shape as capsular, anteriopolar or posteriopolar, central and fusiform, zonary or lamellar, nuclear, cortical, total; according to consistency as soft, anuclear and hard; according to evolution as stationary and progressive.

Acquired cataracts also include lens opacities caused by occupational factors such as electricity, ionising and non-ionising radiation, toxic substances.

Microwave cataract. Microwave radiations in the decimetre and centimetre range can cause lens opacities if protective measures are not thoroughly observed. High-energy densities (hundreds of mW/cm^2) and prolonged exposure can lead to rapid development of cataract. It appears that cataracts are brought about by an increase in eyeball temperature and by disturbed metabolic processes in the lens. During slit-lamp examinations, opacities can be recognised as numerous points of various sizes and non-homogeneous density in different layers of the lens. In the initial stage there may be indeterminate cloud-like or plumose opacities round the posterior pole of the lens, or also grey-white punctiform accumulations of various sizes which are disseminated in the cortical layers and equatorial zone of the lens. These opacities are liable to progress with regard to both size and density.

Microwave cataracts can be distinguished from infrared cataracts by the type and location of the opacities.

To protect the eyes from microwave radiation, filter-glass spectacles have been developed; they may be part of the safety helmet or be worn separately. The glasses are covered with tin dioxide and mounted in frames of radiation-absorbent or reflecting material. Workers employed on very high-frequency generators and radar installations should regularly undergo eye examinations. Special precautions have to be taken when using microwaves for therapeutic purposes in the face and eye region.

Radiation cataract. Ionising radiation (X-rays, gamma radiation, neutrons) is a serious cataractogenic factor. Cataract develops after exposure to both high doses and repeated small doses. The latency varies between 6 months and 2 years; it may extend to 8-12 years in certain cases. The higher the dose, the shorter is the latent period of cataract. After a strong radiation exposure the opacities in the lens rapidly progress and grow to a total cataract. The hazard of cataract development is greater after repeated exposures to small doses of hard rays than to X-rays. Radiation cataract is encountered in persons employed in X-ray services, radiographic testing, nuclear power plants, and also in workers handling radioactive isotopes. With this type of cataract, opacification appears first near the posterior pole of the lens under the capsule, taking the shape of tiny granules or vacuoles. These granules gradually grow to a disc which sharply contrasts with the transparent part of the lens.

Cataracts caused by gamma rays and neutrons are very similar to X-ray cataracts. The disc-shaped opacity in the posterior polar region of the lens is generally surrounded by dot-like opacities. At a later stage vacuoles and belt-shaped clouding appear under the anterior capsules, epithelisation takes place, and a thick film forms. The complete opacification resembles asbestos; the visual capacity is reduced to mere light sensation. Ionising radiation also causes retinal damage. Radiation cataracts generally progress slowly. The initial opacities remain unchanged for years without noteworthy loss of sight. Signs of radiation disease are not necessarily evident.

Preventive measures to protect workers from ionising radiation include lead shields or containers, spectacles with leaded glasses and observance of safe distances. During X-ray or gamma-ray treatment in the face region, the eyes are protected by lead masks.

Toxic cataract. This is encountered rather frequently after long-term exposure to certain nitro compounds such as trinitrotoluene (TNT, trotyl) and results from systemic poisoning. Development of cataract is observed 3-4 years after starting work involving contact with the toxic substance. Opacification progresses slowly. The initial stage is characterised by clouding foci appearing like a ring in the equatorial belt of the lens when examining it with the slit lamp. The second stage is observed after exposure for more than 10 years: a multitude of dot-like opacities under the capsule form a second ring in the pupillary region. The cuneiform or wedge-shaped stage observable 3-6 years later is characterised by tapered opacities pointing towards the centre of the lens. During the subsequent stage the clouding becomes denser and migrates to the centre. Eye examination becomes more difficult, the visual capacity is severely impaired, and the lens becomes opaque (mature cataract). Cessation of exposure to the toxic substance during cataract development (first and second stages) stops the pathological progress, but if the changes are pronounced (third and fourth stages) the cataract progresses.

Treatment. In the initial stages general and local vitamin therapy is indicated (riboflavin or vitamin B_2, thiamin, ascorbic acid or vitamin C). To stabilise the clouding process and reverse opacification, iodine preparations, potassium chloride, methyluracil and ATF are used. Mature forms of cataract must be treated by operation.

KALYADA, T. V.

"Occupational cataract" (Katarakta professional'naja). Larionov, L. N.; Kudrjavceva, S. V. *Spravočnik profpatologa* (Leningrad, "Medicina", 1977) (132-134). (In Russian)

CIS 79-491 "Is cataract an occupational disease?" (Cataract, een beroepsziekte?). Dalderup, L. M. *Tijdschrift voor sociale geneeskunde* (Amstelveen, Netherlands), 8 Nov. 1978, 56/22 (748-751). 49 ref. (In Dutch)

CIS 74-470 "The cataractogenic activity of chemical agents". Gehring, P. J. *Critical Reviews on Toxicology* (Cleveland), Sep. 1971, 1/1 (93-118). Illus. 224 ref.

Catering, industrial (operation)

The type of catering system adopted will depend on the number of workers to be catered for, the length of the meal break and local customs, and may be characterised by food prepared on-site or delivered from outside (hot, refrigerated or deep-frozen cooked meals); and table service or self-service.

Canteen premises

The design of the canteen premises will depend on: the number of meals to be prepared or heated, and on how meals are served. However, the rules applicable to the various functions are the same. The canteen location should be selected to facilitate the reception of supplies and food distribution. Its layout should be in line with food hygiene requirements, and should facilitate product flow and the movement of staff.

Food should be received into a special room in which weight and quality are checked. Perishable foodstuffs, including beverages other than wine should be kept in separate cold-storage rooms: for meat, eggs and dairy products, vegetables, and frozen or deep-frozen products; non-perishable foodstuffs will be kept in a constant-temperature, dry, dust-free room.

Food preparation. Food preparation comprises:

(a) preparation itself, including (i) preparation of vegetables, which should be located at a distance from the main area, preferably in adjoining premises containing fixed or mobile sinks alongside the vegetable peeling machines and in front of the preparation tables equipped with vegetable cutters and salad spinners; (ii) meat preparation: a table close to the cooking installation in small canteens or a special butchery room in large ones; and (iii) fish preparation: a table and two sinks for scale removal;

(b) cooking: depending on the size of the canteen these areas should be equipped with a cooker, an oven, and saucepans in small establishments, or with a cooking unit with ovens, grills, a sauteuse, simmer plates, deep fryer with automatic filtration, autoclave, tipping cauldrons in big ones. Mass production of food demands ever-increasing automation comprising "transfer machines" or "automats" in which the meals being prepared travel continuously along a conveyor specially designed for the product, such as continuous cookers, roasters, rotary sauteuse, etc. In addition to this, there will be general kitchen machines such as: beaters, mixers, meat grinders, ham slicers, etc. A refrigerator should be provided for storage of ready-to-serve dishes (hors-d'oeuvre and similar products, pastries).

(c) service: this unit is the essential link between preparation and consumption.

Washing up. Washing of kitchen equipment should, if possible, be kept separate from the cooking area and from the dish washing unit. It will require a three-sink unit (soaking, washing, rinsing) and a draining board. The dish washing unit is designed for dishes, plates, and cutlery from the dining room. It should be equipped with two sinks, one for soaking and washing and the other for rinsing, and with a draining board. If the number of persons catered for exceeds 50, a dish-washing machine becomes essential.

Waste recuperation. An independent room should be provided to collect waste food, waste water, and packaging before they are disposed of. Dustbins should be stored at a temperature not exceeding 15 °C.

Kitchen staff. The kitchen staff should have their own cloakrooms, washrooms, toilets, showers and, where necessary, their own dining room close to the working area but independent from the kitchen itself.

Dining room. Cloakrooms and sanitary installations should be provided for the diners prior to their entry into the dining room. The entrance should be sufficiently wide and, if possible, there should be a separate exit.

The environment should be attractive (with pleasant colours, decoration and flowers). Lighting should preferably be natural, well distributed and without violent contrasts or glare and should ensure good colour rendering; a level of illumination of 300 lx is recommended. Sound insulation should be sufficient for the diners to hear each other without difficulty. An air volume of at least 10 m^3 per person should be provided, with sufficient ventilation to eliminate odours (slight over pressure), at a rate of about 6 l/s per occupant. A comfortable level of heating should be maintained if possible by air conditioning with blown air ensuring suitable temperature and humidity. The floor should be covered with plastic tiles or stone tiles where there is fear of soiling. Gangways should be sufficiently wide to ensure easy movement of persons, trays and trolleys. Tables should be made of materials resistant to impact, heat, humidity, and staining and which are easy to maintain. Chairs should be solid, stable and should not be noisy. The crockery should be unbreakable and the cutlery in stainless steel.

Cafeteria. This should have a comfortable environment and should contain, in particular, in addition to tables and chairs, a counter with coffee machines, percolators, chocolate serving machines, shelving for beverages, sinks, and various display stands.

Table service. Table service can take the form of group service at table (virtually abandoned), individual service, trolley service, or rapid service. In the latter case the diner, on entering the dining room, receives a menu on which he marks the items he wants; this reduces waiting time.

Service equipment is reduced to a minimum: two serving buffets and hot table for dishes waiting to be served. The kitchen is separated from the dining room by a wall with a serving hatch; an automatic or semi-automatic dumb waiter is provided if the kitchen is situated on another floor.

Snack bar. The snack bar comprises basically a *serving counter* in front of which are fixed stools. On the other side of the counter there is a serving shelf and at each end trays for cutlery. A drinks compartment is provided and display units for fruit and pastries are located on the free areas of the counter.

Refrigerated and water-heated serving equipment, and wheeled tables for the plates are located close to the serving counter. A limited variety of dishes, pre-prepared for rapid cooking, is provided in the interests of rapid service.

Self-service. Self-service makes it possible to reduce the number of serving personnel and the time taken for the meal.

On entering the restaurant, the diner takes a tray and serves himself at one or more serving counters containing various compartments: for crockery and cutlery, glasses and bread; refrigerated units for hors-d'oeuvre, salads, entrées, ice creams, desserts, beverages; heated units for hot dishes. These counters may be linear, rotary (rotating installations on which the individual components of the meal are displayed) or scramble-line or free-flow system (comprising specialised stands for hors-d'oeuvre, hot dishes, desserts, etc.).

The semi-self-service system is a variant of the linear self-service system. The hot main dish is served at table by a waitress and so it is not cold by the time the diner has finished his first course. This disadvantage of self-service can, however, be avoided by using heat-retaining plates, plate covers, plate heaters on the table, which nullifies all the advantages of this service system.

Prepared meal trays. Prepared meal trays can be served at a counter and a different menu can be served at each of several counters. The advantage of this sytem is to offer a meal designed on dietetic principles.

Automatic restaurant. The automatic restaurant provides rapid service of a complete meal at regular hours with a reduced staff or at irregular hours (without service) whilst eliminating kitchen equipment. It consists of a self-service panel made up of distributors, the number of which will depend on the number of diners, comprising a soft drinks and bottled drinks distributor; a multi-stage refrigerated distributor with rotating sections containing hors-d'oeuvre; cooked dishes; desserts, fruit or pastries; and a one or two magnetron microwave oven for reheating dishes in a few seconds or cooking them in a few minutes.

Serving trolleys for cutlery, bread, trays and napkins are also provided.

Automats providing goblet service of drinks, packaged drinks, cooked dishes, refrigerated foods or canteen tickets can also be used to modernise a restaurant or mess room or improve service in a self-service restaurant or snack bar.

They are used increasingly for distribution, outside meal breaks, of snacks such as sandwiches, pastries, confectionery, fruit, yoghourts, desserts, salted products and dried fruits.

Mess room. The mess room, although it is hardly a satisfactory solution, still exists in many firms; it serves the dual function of a kitchen and dining room. The room should be well ventilated and light, and adequately heated during the cold season. The walls and floors should be made from impermeable materials. The mess room should be cleaned after each meal and access should be forbidden outside of stipulated hours.

Some means of heating meals (lunch pail heaters, water heaters, hotplates or cookers) should be provided, as well as a refrigerator and one cold drinking water fountain and one hot water outlet for every ten users. An adequate number of tables and seats should be provided.

Food hygiene

The task of ensuring food hygiene entails correct selection of raw materials and a knowledge of the specific precautions involved; purchasing goods which offer every guarantee of cleanliness and checking them after reception; ensuring hygiene during conservation, preparation and serving; maintaining the hygiene of premises and equipment; checking on personnel hygiene and ensuring good training.

Choice of foods. Meat always involves the risk of contamination and the smaller the cuts the greater the hazard; individual portions should be cut as shortly as possible before consumption. In very large restaurants purchase of carcasses or quarters involves skilled personnel working in specially equipped premises. In a medium-sized restaurant quarters that are purchased ready dressed should be cut up in a separate part of the kitchen at a distance from any heat source and from all passageways. In a small restaurant a check should be kept on the supplying butcher's installations, handling and transport procedures and it should be ensured, in particular, that final cutting takes place immediately prior to delivery.

Minced meat or poultry and products containing mince demand strict standards of hygiene in both their preparation and transportation. The use of minced horsemeat or poultry should be prohibited. Offal, which is highly perishable, and in particular ox tongue, require especially strict hygiene conditions in transport and delivery. Hung meat should be excluded from group catering practice.

Eggs (chicken eggs in shells only) should be fresh or extra-fresh, unwashed with a minimum unit weight of 50 g. Milk should be pasteurised or, preferably, sterilised. A check should be made on the hygiene of the conditions under which fresh cheeses and yoghourts are prepared and their freshness should be verified. Unpasteurised fresh goat's cheese should be forbidden. For pasteurised or sterilised cream, it is necessary to require the same guarantees from suppliers as for milk and cheese.

Preserves and semi-preserves should have indicated on their packaging the date of manufacture or preparation; this is particularly important in the case of semi-preserves with a limited storage life. Cured products, sausages and similar meats should bear an indication of their composition, their date of manufacture and the date by which they should be consumed; choice of suppliers and checks on their production are very important. The freshness and presentation of fresh fish should be checked. Mussels should not be used for group catering. Preference should be given to fish, deep frozen at sea. The use of green vegetables grown on sewage farms should be prohibited; the origin of watercress in particular should be ascertained.

Frozen and deep-frozen products (meat, poultry, offal, butter, animal fats, ices and ice creams and other foodstuffs) should be subject to precise requirements for manufacture, warehousing, transport, distribution, storage, and defreezing; these may be laid down in national legislation.

Drawing up supply contracts. When drawing up contracts with suppliers, it is necessary to lay down technical specifications (in particular hygienic and bacteriological) for the supply of foodstuffs and also to stipulate access for inspection of manufacturing premises (bread, meat, prepared meats).

Foodstuff reception. The conditions under which dairy products, meat, fish, prepared meats, and frozen and deep-frozen products are transported and the packaging in which products are supplied (eggs, butter, preserves) should be checked on reception. An even closer check should be made on the cleanliness of frozen tongue, fresh fish, fresh cheese and yoghourts, ice creams, cans of preserves; any tins that are blown or leaking should be refused. Random checks should be carried out and it is advisable periodically to call upon the services of a specialist (veterinarian, etc.).

Conservation of foodstuffs. Premises in which foodstuffs are stored should be well lit and ventilated. The walls and the ceilings should be washed regularly or whitewashed at least once a year.

The floor should be made from hard, smooth material or have an impermeable surfacing; it should be well drained through a proper drainage system and washed down regularly. Foodstuffs should not be stored on the floor but on staging, racks, grids or pallets, shelving or boxes at a distance from non-edible materials. If they are not wrapped they should be protected by transparent partitions or fine mesh.

Cold stores should be specialised if possible, their temperature checked each day and their operation checked at least once per year, preferably during the warm season.

Meats and meat products should be preserved at between 0 and 3 °C with a relative humidity of 80-90%, and non-cut-up meat should be hung up. Dairy products and eggs should be stored at between 4 and 6 °C; fresh vegetables at 8 °C; and deep frozen foodstuffs at −18 °C. Semi-preserves should be kept in cold storage since their shelf-life depends on the storage temperature. Fresh fish should be placed in icy water and stored in a refrigerated

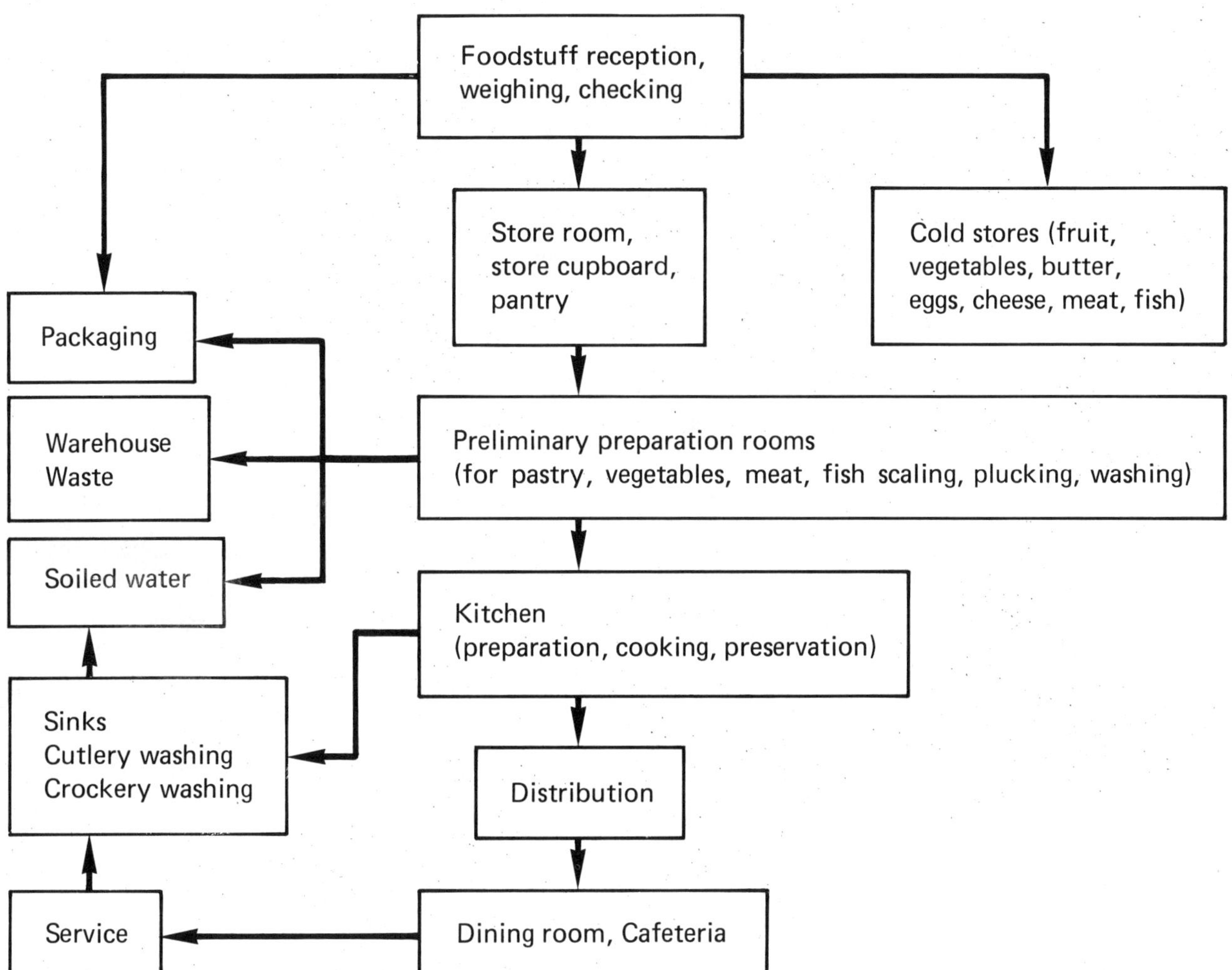

Figure 1. Flow sheet for a restaurant. Path followed by foodstuffs ("step by step" principle).

area. Animals should not be allowed into these storage areas and measures should be taken to prevent the entry of flies, insects and rodents and to ensure their extermination (using techniques which will not affect the safety of the foodstuffs).

Hygiene during preparation. Foodstuffs should be checked once again at the preparation stage.

The kitchen work stations should be laid out in such a way as to follow the "step by step" procedure with the raw materials being cleaned at each step along their progression (figure 1). Dustbins should not be allowed in the proximity of cooking stoves.

Vegetables require extremely careful cleaning, copious washing, immersion in water containing a small amount of Javel water (5-6 drops per litre of water), followed by washing once again and removal of excess water. Cutting and slicing should be carried out with perfectly clean equipment. Where vegetables are eaten raw, seasoning (parsley, mayonnaise, etc.) and the preparation of plate portions should be done with great care; until the moment of consumption, the portions should be kept in a refrigerated place.

Rice should have been recently boiled or kept cold immediately after boiling.

Prepared meats should be made up into portions just before serving, using well cleaned and disinfected utensils. Minced meat requires special precautions. Meat should be fresh and taken from carcasses that have been prepared and transported under good hygienic conditions. Mincing should be carried out in the butchery room where this exists, or on a work table reserved especially for the purpose. Mincing machines should have a minimum of handling; the parts that are in contact with the meat should be cleaned immediately after use and stored away from possible soiling, or kept immersed in an approved antiseptic solution (chlorinated water) and rinsed before re-use. Kitchen staff should be provided with head coverings and clean working clothes; the use of gloves and a face mask is recommended. Mincing should be carried out in a single operation and the minced meat should be kept at a temperature of between 0 and 2 °C whilst awaiting cooking, which should take place within a period of no more than 2 hours; leftovers should not be re-used. Deep-frozen minced meat should be eaten immediately after defreezing.

Poultry represents considerable bacteriological problems. Cleaning and cutting up should be carried out on surfaces that are cleaned and disinfected after each use. Fresh fish require the same precautions. Eggs which are cracked or soiled should not be used. Before breaking, eggs may if necessary be dipped in a quaternary ammonium solution. Personnel should wash their hands after each series of eggs is broken, which should be only shortly before cooking.

Pressed and cooked cheeses should be grated with disinfected equipment and placed on plates shortly before cooking. It is necessary to avoid grinding too close to the rind. Spices should be taken from their containers with spoons; these containers should be emptied, washed and disinfected regularly. Spices should be boiled in the sauces. It is advisable to use "debacterialised" products.

Pastry creams and butter-based creams require considerable precautions in addition to those essential for eggs; bottled milk should preferably be boiled with the sugar and the vanilla; metal containers should undergo prior sterilisation in an oven for 10 minutes at 120 °C; as soon as it has finished boiling, the cream should be placed in a cold room; forcing bags should be disinfected each day by boiling for 20 minutes; utensils should be dipped in water containing a small amount of Javel water or in an antiseptic solution which may be used in contact with food products.

Creams and fresh pastry goods containing cream should be consumed on the day they are prepared; leftovers should never be re-used. The same precautions should be taken for ice creams.

Cooking should be sufficiently long to ensure the destruction of germs. Meat and offal should be cooked in pieces weighing less than 3 kg and should be kept after cooking at 70 °C or stored in a cold room at +2 °C. The same applies to vegetable broths. Boiled meat in sauces containing vegetables and herbs and fish cooked in court bouillon should not be left in the liquid.

Table oils should not be used for frying; overheating should be avoided (by use of a thermostat) and frying pans should be cleaned of carbon encrustation.

Frozen or deep-frozen foodstuffs that have to be defrozen before cooking should be protected from soiling and defrozen at a temperature of 0 to 4 °C. Frozen tongues should be delivered at the earliest on the day before they are to be consumed, stored at +2 °C, blanched in boiling water without defreezing and eaten within 2 hours of cooking. Skinning and portioning should be carried out whilst observing a maximum of personnel and equipment hygiene. Leftovers should not be re-used.

Cooked dishes prepared externally should, if they are hot, be kept at a temperature of 65 °C from cooking to consumption and eaten on the day of preparation. If they are refrigerated they should be kept at a temperature of +3 °C; if frozen or deep-frozen they should be kept at a temperature of −18 °C and reheated so that their temperature is raised to 65 °C in less than 1 hour; and this temperature should be maintained until consumption. Refrigerated dishes should be consumed within 6 hours of preparation.

The use of leftovers should be prohibited as a rule. Under exceptional circumstances, good quality foodstuffs may be kept in cold storage at +2 °C and eaten the next day. Consommés, meat gravy, sauces, stuffing, and minced meat should never be re-used.

Hygiene of kitchen premises and equipment. Floors should be of hard material (tiles, cement); they should be smooth or covered with an impermeable surfacing and kept clean; a slope should be provided to ensure the run-off of washing water, with the water running out through a grilled orifice fitted with a syphon. A holding tank may be acceptable provided it has a smooth surface with rounded angles and is covered with an airtight lid and is emptied at least once per day (by a mechanical process) and subsequently disinfected.

Walls should be painted or have a special smooth covering, and light in colour. Hard materials in the butchery room should be rot proof and have a smooth surface up to a height of at least 2 m; the rest of the walls and the ceiling should be covered with washable paint.

Pantries, cold rooms and the kitchen itself should be well ventilated. Normal ventilation should be sufficient in small kitchens but air conditioning is recommended for large kitchens and the air flow should be 20-30 times the volume of the premises each hour. Cookers should be fitted with exhaust ventilation hoods linked up to a single ventilation conduit; conduits should be fitted with grease filters that are cleaned regularly and, if necessary, an activated carbon filter to deodorise the vapours. In large kitchens, hood ventilation is tending to be replaced by "induction" ventilation (a narrow flow of air at ceiling level).

Premises should be supplied with hot and cold drinking water for the preparation of foodstuffs and for the various cleaning processes.

Working tables, utensils and equipment should be in impermeable materials, smooth, rot resistant, corrosion resistant and easy to clean.

Lighting should be glare and shadow free, should not alter colour rendition and should have a level of at least 300 lx at the work posts; general lighting should be 250 lx for the kitchen and 150 lx for the other premises. Light fittings should be sealed, heat resistant and easy to clean.

Noise control in the kitchen should be achieved by providing non-resonant working surfaces and walls and ceilings of materials that do not reflect sound.

Premises and equipment should be kept clean. Every day the floors should be washed (and not dry swept); pots, pans and other cooking equipment should be washed using an approved detergent followed by copious rinsing, disposable paper towels being used for drying. The equipment should be stored in a closed cupboard, or where this is not possible, it should be covered by a clean towel. Ovens, sinks, and service trolleys should also be cleaned each day. Weekly cleanings should include the walls up to a height of 1.5 m, the underside of furniture, hoods, windows, metal surfaces, and the floors of adjoining premises. Periodic cleaning should include the light fittings, windows, doors, and ceilings and walls of storage cupboards. Regular chimney sweeping should not be forgotten. Cutting and preparation tables should be kept constantly clean; kitchen utensils and equipment used for food preparation should be cleaned as and when they are used. Moving parts of food preparation machines in contact with the foodstuffs should be stripped down, washed (and brushed if necessary), soaked in a disinfectant solution, rinsed, dried and stored away from likelihood of contamination. The same applies to knives. Cleaning products should not present a danger for the foodstuffs. Waste and garbage should be placed in a container with a close-fitting lid which should be emptied, cleaned and disinfected each day. Suitable measures should be taken to prevent the entry of flies, insects and rodents. Rodent and insect extermination should be carried out annually.

A bacteriological inspection should be carried out periodically, culture papers being used if necessary.

Automatic food distributors should be checked and cleaned periodically (pigeon holes, openings, tanks, piping). The freshness, quality and correct temperature of the foods should also be checked.

Hygiene in the dining room and cafeteria. The walls should have a smooth covering or be in washable paint, they should be maintained in good condition and cleaned periodically. The floor should be made of a hard material or have an impermeable covering; it should be washed after each meal and not dry brushed. The dining room should be well ventilated.

Soap, disposable towels and toilets which do not connect directly with the dining room should be made available for diners.

Tables, made from washable materials, should be cleaned after each diner has finished and should be washed after each service with an approved detergent. Table cloths, in plastic or waxed cloth, should have the same maintenance; fabric table cloths should be covered with paper place settings, which should be changed for each diner.

The equipment should be in good condition: glasses and cups should be intact; water jugs should be emptied after each meal. The normal preventive and disinfection measures should be taken to deal with rodents and insects.

Information and education of staff. The occupational medical officer or a dietician should provide information and education for the staff, who are often ignorant of the rules of personal and food hygiene. There should be regular supervision to ensure personal and clothing hygiene.

The hygiene of staff and sanitary installations. Staff should be subject to pre-employment and periodic medical examinations and examinations after absence due to sickness or accidents (anti-tuberculosis vaccinations, follow-up for post-vaccination allergy, rhino-pharyngeal swabs, examination of stools, etc.).

Workers should wear working clothes or full aprons that are white or light coloured, non-flammable and kept clean. Headwear should be compulsory; it should cover the hair totally. Sanitary gloves, used for the production of minced meat and for slicing should, after use, be soaked and disinfected with an approved antiseptic solution, dried with a clean towel, and hung in a closed cupboard (with, if necessary, exposure to ultraviolet radiation).

The workers should keep their hands constantly clean and should wash them after using the toilet; hand washing installations with non-manual controls, disposable towels and a foot operated covered waste bin should be provided in the vicinity of the workstations. Sanitary installations located near the workplace should comprise wash basins and toilets (see SANITARY FACILITIES).

Workers' safety

Reception and storage of goods. The hazards here are basically those of manual handling. If necessary an unloading bay with ramps and raised quays should be installed. Traffic areas should be free of obstacles (steps, narrow or low doors, sharp inclines). Suitable, stable mechanical handling equipment of the multi- or single-purpose type (e.g. drum rollers) should be used. Bottle crates should be made from wood with special nails which do not come loose or catch; better still the crates should be made from plastics. Storage racks and shelving should be of a suitable height and adequately spaced.

Cold and chilled rooms should be fitted with an opening mechanism fitted on the inside of the door and with an acoustic or optical warning signal that can be actuated by any person accidentally locked inside.

Kitchen and annexes. Individual hazards include: falls, cuts, burns and electrocutions, while group hazards include fires and explosions.

The floor should have a non-slip surface and be cleaned with an emulsion that has a good degreasing effect (sodium carbonate detergent); one may also use a mineral product which absorbs grease instead of the usual sawdust. Traffic areas should be adequately dimensioned so that movement and handling are unhindered.

Knives for meat should be fitted with a ferule, grooved handles or, better still, formed to the shape of the hand. They should be kept on knife racks with a permanent magnet or in individual sheaths with a protective sleeve. Three-fingered gloves in chain mail should be worn on the left hand, and a leather wrist band with steel mesh or a strong plastics sleeve. Knives for cutting fish should be of the same standard as those for meat whether for medium-sized or frozen fish. It is advisable to wear a safety apron.

Butchery machines should be well guarded. Circular saws should be fitted with a blade guard. A device should be fitted on screw mincers to prevent fingers touching the screw and it should be possible to lock the controls whilst the machine is being assembled and stripped down. Mixers, mixer-beaters, peelers, vegetable cutters, etc., should be fitted with guards for the drive transmissions (sheet-metal guards or grills) and mixer arms (cover, grill, etc.) and with protection against electrical hazards (earthing, circuit breakers). Autoclaves should be checked regularly.

Ovens should be fitted with guard rails to avoid burns and contact with sharp edges; the use of heat resistant gloves is advisable when working with pastry ovens having sheet metal trays; clothing made of flammable fabrics should be prohibited.

Gas cooking equipment should be fitted with a thermostat and a flame-failure protection device; electric cookers should be fitted with thermostats and temperature limiting devices.

Operating, cleaning and maintenance instructions should be displayed and systematic checks carried out once per year.

Hand-operated dumb waiters should be fitted with devices preventing access to the underside of the platform while the apparatus is in use and preventing falls into the shaft; the safety devices for electric dumb waiters should be checked daily.

For fire protection purposes, an adequate number of fire extinguishers should be provided and maintained in good order; fire hydrants should be provided where necessary and fire instructions displayed.

COURAU, P. J.

Microbiological aspects of food hygiene. Report of a WHO expert committee with the participation of FAO. Technical report series No. 598 (Geneva, World Health Organisation, 1976), 103 p. Illus. 257 ref.

"Hygiene problems in large kitchens" (Hygiene-Probleme in der Grossküche). Bansemir, K. *Arbeitsmedizin–Sozialmedizin–Präventivmedizin* (Stuttgart), July 1979, 14/7 (161-163). 8 ref.

CIS 78-1763 *Dietary problems of workers lunching in staff canteens* (Problèmes d'hygiène alimentaire du repas de midi dans les restaurants d'entreprise). Allo, J. J. (Université de Paris VI, Faculté Broussais–Hôtel-Dieu, Paris) (1978), 108 p. 34 ref. (In French)

"Dietary mistakes in the work environment (or some observations concerning dietary mistakes in the work environment and industrial catering)" (A propos des erreurs nutritionnelles en milieu de travail (ou quelques constatations concernant les erreurs nutritionnelles en milieu de travail et en restauration collective). Penneau, D.; Domont, A.; Loriot, J.; Cousteau, J.; Proteau, J.; Vie, E.; Allo, J. J. *Archives des maladies professionnelles, de médecine du travail et de sécurité sociale* (Paris), 1980, 41/1 (26-28). (In French)

"Food hygiene–an educational problem?". Oakley, B. *Health and Safety at Work* (Croydon), Jan. 1982, 4/5 (23-25). Illus.

Catering, industrial (organisation)

Industrial conditions demand that the worker should be able to obtain a regular and sufficient supply of food and drink during his working hours and it is increasingly appreciated that the employer has a responsibility to ensure that facilities are available. The type of provision will vary according to the circumstances, national habits, size and location of the enterprise, type of work carried out, arrangement of hours.

Where workers have to travel long distances in congested transport services or, as in many developing countries, on foot through the countryside, a catering service at the place of work becomes a necessity. Organising an efficient service to the satisfaction of all concerned presents many problems: the type of service, quality, price, obtaining of meals and snacks have to be tailored to suit local requirements.

Types of service that can be organised include centralised canteens, sectional canteens, kiosks or serving counters, mobile services, automatic vending machines.

Centralised canteens

Under this system the entire service is located at one place in the factory. Such a place should be within easy reach of the workers but this is not always simple to arrange. For various administrative reasons, canteens in new factories are often located near the main gate or in some quiet corner. It is generally considered that the dining hall should provide accommodation for about 30% of the working force on the premises at any one time. Both midday meals as well as tea and snacks are served at fixed hours in the dining hall.

Sectional canteens and kiosks

In bigger establishments employing 3 000 workers and above, or where different sections of the factory are spread out, the distance from the workroom to a canteen may take away a sizeable part of the lunch or other interval allowed to the workers. Such factories need decentralised services, viz. sectional canteens, kiosks or a mobile service, run as branches of the central canteen. Their number and size will depend on the employee population as well as the area of the factory. The arrangements vary from mere selling counters near the plants (kiosks) to a permanent structure having all facilities of the central canteen except the kitchen (sectional canteens). All cooking is done at the central kitchen and great care is needed to ensure that food is delivered in an appetising condition.

Mobile services

Under this system, beverages and snacks are transported on mobile catering trolleys which go round different departments of the factory at fixed hours. The equipment consists of a stainless steel or aluminium cabinet mounted on a trolley with compartments or drawers for keeping snacks wrapped in unit packings. Additional fittings for keeping soft drink bottles, cups and saucers and a small water tank are provided on either side of this cabinet. Small trolleys are convenient for one person to handle and also attend to the service. Larger trolleys may be subject to frequent breakdowns especially if the roads inside the factory are not well maintained. Power-driven vehicles can take many times more material but practice has shown that they are usually not suitable.

Group canteens

Small undertakings sometimes find it difficult to provide canteen facilities on their own. Two or more such small units can combine their efforts to establish a central canteen at a convenient spot with kiosks or sectional canteens. Small factories may often find it easier to provide simple meals in an unpretentious canteen if a capable person can be found to manage it.

Messrooms

Where such provision is not practicable the importance of providing a messroom should not be ignored. An attendant is usually necessary if standards are to be maintained.

Special services

Some types of work present special catering problems, e.g. in particular, building and construction work, dock work. Arduous outdoor labour in all climatic conditions demands good feeding arrangements but sites are often isolated and frequently on the move and the labour force varies greatly in number and content during the course of the work; migrant and seasonal workers with different food habits may be employed. On a small site the minimum requirement is a well kept messroom with facilities for heating food, boiling water and for washing up, tables and seats. On a large site of longer duration a full hot meal service should be provided wherever possible. Fitted service vans may be used to convey food to workers at distant sites.

Service

In most factories, the day shift has the highest complement and arranging for their midday meal needs careful planning. The canteen manager must be able to estimate the number of meals required to prevent unnecessary wastage of food and avoid complaints regarding shortages. A system whereby workers buy a lunch token as soon as they enter the factory or give advance notice enables the canteen manager to cook for the required number.

Staggering of the meal hours and the mode of service helps to ensure efficiency in the dining hall. Here, sectional canteens will be of great help. The system of self-service has many advantages, the most important being the cost. Waiter service, besides being expensive, is time-consuming. A practice, wherein ready lunch plates properly covered are laid down on the table just a few minutes before the scheduled time, has been successfully tried in some factories. This eliminates workers rushing round with plates or trays in hand and ensures speedy and yet orderly handling of the clientele. The menu, of course, has to be common for everybody.

Rest periods for refreshing warm or cold beverages such as tea, coffee, mineral waters can also be staggered if the dining hall is to be used for this purpose. But kiosk service is more suitable for such breaks particularly in big plants. Trolley services may be useful where there is no health or production reason against partaking of food on the job. In many places, coin-operated dispensing machines are increasingly popular, particularly for hot or cold beverages, which can thus be taken as required.

For the workers of the second and night shift and for those on overtime, it may be convenient to run the central canteen at fixed hours with a skeleton staff or run one or more kiosks depending on the number employed. Provided sufficient numbers come forward, an evening meal can also be economically arranged. Planning of menus and timing of meal breaks for night shift workers requires special care; a mere duplication of daytime menus does not usually meet the dietary needs or personal tastes of night workers.

Menu and portions

Selecting a standard menu acceptable to all tastes and yet meeting the nutritional requirements of workers and priced at an economic level is a difficult proposition in the context of rising costs. A good midday meal menu should provide adequate nutrition and should be well balanced. Items which are acceptable to the majority must necessarily find a place in the menu. Most workers in all countries are conservative in their tastes but, while due regard must be paid to habits (and prejudices), a canteen can serve a useful educative purpose in encouraging good dietary practices. Sometimes foods can be introduced to correct deficiencies and this is particularly valuable for young workers, but the introduction must be tactful and the co-operation of the workers ensured. In some places, it is possible to make special dietary arrangements for workers suffering from ailments which might otherwise involve hardship or even absence from work; some places also arrange to supplement the diet of workers engaged in specially hazardous occupations.

Portions—the quantitative unit of the items served either as snacks or as items of lunch—need to be clearly indicated and so also their prices. Items like tea, coffee and snacks can fetch a comparatively higher price and part of the extra cost incurred on the midday meal can to some extent be compensated in this way.

Operating costs

To ensure that the benefits of a catering scheme are available to those who most need it, it is essential that prices should be kept low. Many employers recognise that good feeding is not only a welfare matter but also has an effect on human relations and on production; they are therefore ready to subsidise the catering costs to a greater or lesser degree. At least, they are often prepared to carry all overhead costs, provision and maintenance of buildings and equipment, heating, lighting and ventilation, leaving only the food and labour costs to the catering management and not expecting any financial profit. In many places meals are available for young workers well below cost-prices.

Methods of management are often the subject of discussion: industrial catering contractors are sometimes favoured for their expert knowledge but this means that management has less control over quality and price of food and workers are often suspicious of the profit motive involved. In many undertakings it has been found advantageous to associate employees with management in a canteen committee to advise on the running of the service and to deal with complaints. Often the catering services are the focus of general malaise in an establishment and a genuine collaboration in dealing with complaints may have wider effects than improvement of the food or beverages served. In some places a joint committee is directly responsible for the running of the canteen.

Legislation

Many countries have legislation making it obligatory for the employers to provide a catering service or, at least, messrooms. These requirements may be limited to premises of a certain size and to those with special health hazards. It is true to say, however, that the will to provide an efficient catering service is of more importance than legislative compulsion. Many countries also have strict hygiene laws which apply to the storage, preparation and service of food.

THACKER, P. V.

Recommendation 102. Recommendation concerning welfare facilities for workers. International Labour Conference (Geneva, International Labour Office, 1956).

Cellulose and derivatives

Cellulose

sp.gr.	1.27-1.61
m.p.	260-270 °C (decomposes)
TLV ACGIH	10 mg/m^3 of total dust; 5 mg/m^3 of respirable dust
STEL ACGIH	20 mg/m^3

Cellulose is a polymeric carbohydrate containing 44.4% carbon, 6.2% hydrogen and 49.4% oxygen. The size of the cellulose molecule or the degree of polymerisation (DP) varies, even in the same plant. The DP of native cellulose is more than 3 000. The DP of technical cellulose varies between 500 and 2 000. If the DP is 1 000, the molecular weight of cellulose is about 162 000. The DP of cellulose is of great technical importance. The higher it is, the stronger is the cellulose. The structural formula of cellulose is:

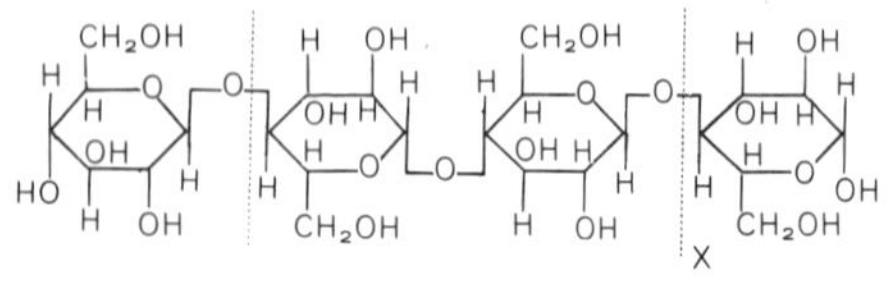

Cellulose is insoluble in alcohol and ether and also resists fairly well the action of acid and alkaline solutions. Treated with strong acids under special conditions or with the enzyme cellulose, it is converted into glucose. Treated with concentrated alkaline solutions, it is converted into alkali cellulose, which, after removal of the alkali, has a greater reactivity than natural cellulose.

Sources. Cellulose forms the major part of the cell walls in plants. In cotton, cellulose is almost pure, but in wood it is combined with other substances such as hemicellulose, lignin, resins, terpenes, fats, starches, proteins and minerals. The most important sources of cellulose are wood, jute, linen, cotton, hemp, straw and fast growing reedy plants.

Production. The aims of manufacturing pulp are to separate the cellulose-containing constituents of plants from each other and from other substances such as hemicellulose and lignin. This can be achieved with three principal methods: the mechanical, chemimechanical and chemical methods. In industry the term cellulose is given to the purified end product, the chemical pulp.

Mechanical pulping. Logs are passed through a pulp grinder, which separates fibres from the log's surface. This mechanical separation of fibres also causes disruption or splitting of fibres. Therefore, the end product known as ground pulp is rather feeble, but the fibre yield is almost 100%. Wood chips pretreated by water or steam and then mechanically ground are used for so-called thermomechanical pulp.

Chemimechanical pulping. The purpose of these methods is to produce a pulp with less split and broken fibres than is possible with only the conventional mechanical treatment. The coarse pulp required as a starting material is first obtained by impregnating wood chips or sawdust with an aqueous cooking liquor having a pH of 4.0 to 12.7, and then by mechanically separating the fibres from the impregnated chips at a temperature of about 140 to 180 °C. A preferred cooking liquor contains an alkali metal sulphite and/or an alkali metal bisulphite.

Chemical pulping. Through chemical methods the lignin which attaches the fibres to each other is soaked away by certain chemicals. This leads to liberating the fibres without injuring them, and yields a strong pulp. Chemical reactions with the cellulose and other constituents of wood cannot, however, be avoided, and the yield is thus only 35-55%, depending on the method used.

The wood used as raw material is peeled and chipped. The chips are treated with a cooking liquor while under high pressure and temperature. After cooking, the cooking liquor is washed out and impurities removed. The pulp yielded can be further treated with various bleaching processes.

Depending on the chemical used, the chemical methods can be subdivided into alkaline or acidic methods. Of the alkaline methods the kraft or sulphate process is the most common. Any kind of wood can be used as a raw material, but pine is the most common. The chipped wood or other lignocellulosic material is cooked in an aqueous solution containing sodium hydroxide and sodium sulphide at a temperature of about 170 °C for the time required to produce a pulp of the required yield. The resulting pulp contains lignin and carbohydrates in a ratio which is determined by the specific conditions of the process, such as temperature, time and quantity of liquor. Because of the presence of lignin and other impurities, the pulp yielded is brown in colour, and can be used as such for the production of kraft paper and packing cardboard. The cooking liquor spent is usually concentrated and burned in a steam boiler, and the reagent can be recycled into the process. Turpentine and pine oil are by-products of this process.

The most important acid pulping method is the sulphite process. Spruce is preferably used as a raw material, because woods containing significant amounts of resins are less suitable for the process. The digesting of wood chips is achieved by cooking them in an aqueous sulphite digesting liquor at elevated temperatures and pressures. After cooking, the substantially lignin-free cellulosic fibre is removed from the liquor and is subjected to various stages of washing and purification. The cooking liquor usually contains an excess of sulphur dioxide and calcium bisulphite. The waste liquor is a source of carbohydrates, and can be used for the production of industrial alcohol and fodder yeast as by-products of the process. The bisulphite process is a variation of the conventional sulphite process. The cooking liquor contains magnesium-, ammonium- or sodium-bisulphite without an excess of sulphur dioxide. The pH of the liquor is 4-6. In this process also pine can be used as a raw material.

In addition to the processes already mentioned, there are numerous less important variations and combinations of chemical pulping processes. In the cooking solutions such chemicals as ammonia, acetic acid, sulphonates, phosphoric and nitric acid, chlorates, nitrogen dioxide, water-ethanol mixture and oxygen have been used.

Unbleached pulp is brownish in colour, the sulphite pulp being brighter than the sulphate pulp. The discolouring compounds in the pulp, such as lignin, tannin, carbohydrates and resins, are removed with bleaching processes. For these processes such compounds as chlorine, chlorine dioxide, hypochlorites, hydrosulphites, peroxides, oxygen and ozone can be used as bleaching agents.

In the 1970s the leading producers of cellulose and paper products were the United States, Canada, Japan, the USSR, Sweden and Finland.

Uses. Cellulose is used in the manufacture of paper, fillers, cellophane, celluloid, photographic film, rayon, viscose and other chemical derivatives such as certain explosives, etc. The kraft or sulphate cellulose is strong, and unbleached it is suitable for linerboards, wrapping and bag paper. Bleached draft or sulphate cellulose is used for high quality paper products. Sulphite cellulose has less strength and intrinsic viscosity than sulphate cellulose. Unbleached sulphite cellulose is suitable for the manufacture of newsprint, wrapping paper and paperboards. Bleached sulphite cellulose is used in the manufacture of printing and tissue papers.

HAZARDS

There are special hazards in the cellulose industry because of the many gases used in the production and bleaching processes and the waste gases which are evolved during the pulping process. The gases used or produced in the sulphate process are hydrogen sulphide, mercaptans (including methyl mercaptane), and terpenes and lye vapours. The substances used or produced in the sulphite process are sulphur dioxide, carbon dioxide, methyl alcohol, furfural and possibly carbon monoxide and hydrogen sulphide. All these substances, like the gases in the bleaching process (chlorine, chlorine dioxide, etc.), may cause light or severe intoxication depending on the concentration, and the symptoms may vary from light irritation of the eyes and the upper respiratory system to pulmonary oedema. Hydrogen sulphide, furfural and the mercaptans may in high concentrations also cause acute intoxications with symptoms from the central nervous system. In the pulp mill processes most of the acute intoxications are, however, due to chlorine, chlorine dioxide or sulphur dioxide.

Chemicals used in the cellulose industry such as acids, alkalis and bleaching agents may cause skin irritation and dermatitis.

During the peeling and chipping of wood, there may be a harmful degree of noise. Special problems in the cellulose industry include water pollution caused by mill effluents, and air pollution caused especially by mercaptans from the sulphate process.

SAFETY AND HEALTH MEASURES

Epidemiologic studies have shown that workers involved in pulp mill processes and exposed to irritant gases (especially chlorine, chlorine dioxide) appear to have symptoms of chronic bronchitis and poorer results in some ventilatory function tests than non-exposed reference groups. Pre-employment examinations of persons exposed to irritants should be performed to exclude persons with chronic cardiorespiratory diseases from these jobs. Further health measures should concentrate upon the prevention of acute intoxications.

Many of the dangers associated with pulping processes have been removed by automation. This concerns especially the corrosive and irritating or otherwise dangerous liquids, the flow of which through pipelines can safely be controlled by modern techniques. The risks of harmful exposure are usually connected with disturbances in the industrial process, leaks in the apparatus or maintenance and repair work. Other factors connected with exposure from leaks and repair work are mostly poor ventilation, non-use of personal protective equipment and defective or ineffective protection equipment.

To prevent exposure to toxic gases, the following safety measures should be performed:

(a) periodic inspection of apparatus and equipment should be performed to prevent unexpected leaks;

(b) good general ventilation should be provided at the workplaces and a high standard of local exhaust ventilation should be provided at points of risk in the process. Suitable respiratory protective equipment should be available for use in emergencies;

(c) the use of personal protective equipment should be mandatory in all risk situations;

(d) process workers should be trained to follow the proper procedures;

(e) maintenance and repair workers should be provided with special instructions and regulations. Work in autoclaves, digesters, tank chambers, or any other confined space where dangerous gas or vapour is liable to be present requires the precautions described under CONFINED SPACES;

(f) ambient air concentrations of harmful gases should be monitored continually for the registration of peak concentrations of gases caused by disturbances in the process. Because man's olfactory organ may develop a tolerance to increasing quantities of irritating gases, the sense of smell is not to be relied upon in these situations.

Contamination of the atmosphere outside the factory can be avoided by recirculating certain gases or by passing them through scrubbers or chemical absorbers before the exhaust air is emitted into the open air.

Water pollution can be prevented by burning the black cooking liquor from the sulphate process or by using the sulphite cooking liquor for the production of fodder yeasts or as a fuel in steam boilers.

Derivatives

Cellulose acetate

See FIBRES, MAN-MADE

Cellulose triacetate

See FIBRES, MAN-MADE

Cellulose nitrate

NITROCELLULOSE; CELLOIDIN; PYROXYLIN; GUNCOTTON; NITROCOTTON

m.w. 594.28-459.28
sp.gr. 1.66
m.p. ignites at 169-170 °C
f.p. 12.8 °C

a pulpy, cotton-like amorphous solid (dry), or colourless liquid to semi-solid solution, depending on the degree of nitration.

It is produced by treating cellulose with a mixture of concentrated nitric and sulphuric acids, the excess of which is removed by washing, digesting and boiling procedures.

Cellulose nitrates are used for fast-drying automobile lacquers, high explosives, colloidin, rocket propellants, medicine, printing-ink bases, flashless propellant powders, coating book-binding cloths and leather finishing.

Cellulose nitrate constitutes a serious fire hazard in most of its industrial forms. When the nitrogen content of its chemical composition is high, it is classified as an explosive. When damp nitrocellulose with a lower nitrogen content is used in normal industrial processes precautions for a dangerously flammable substance should be observed. The workrooms should be of fire-resisting structure, open lights and other sources of ignition should be prohibited and adequate means of escape should be provided. A copious supply of water should be available for extinguishing purposes.

Celluloid

sp.gr. 1.35-1.60

Celluloid is prepared by plasticising cellulose nitrate with camphor. Pigments are generally added at this stage. It is thermoplastic and is used in the manufacture of innumerable articles such as toys, combs, photographic film and pen barrels. Incorporated with solvent it is used in a number of adhesives.

Celluloid is a very flammable substance and, for this reason, is being replaced by cellulose acetate for many purposes. Precautions similar to those recommended for cellulose nitrate should be adopted in workrooms where celluloid is used.

Ethyl cellulose

sp.gr. 1.07-1.18
white, granular, thermoplastic solid, softening at 100-130 °C.

Ethyl cellulose is made by treating alkali cellulose with chloroethane or ethyl sulphate. It may also be prepared from cellulose and ethyl alcohol in the presence of a dehydrating agent. It is used for making plastics, lacquers, flashlight cases, furniture trim, coatings, adhesives and wire insulation.

Precautions for a moderate fire hazard should be adopted for the solid material. It is frequently handled as a paste or a solution in a flammable solvent, when precautions for flammable liquids and vapours may have to be observed.

Methyl cellulose

insoluble in alcohol, ether and chloroform; soluble in glacial acetic acid; it swells in water to a viscous colloidal solution
a greyish-white, fibrous powder.

Methyl cellulose is produced by treating alkali cellulose with chloromethane or dimethyl sulphate. It is also made by the reaction of cellulose, methyl alcohol and a dehydrating agent. It is used in the pharmaceutical industry, cosmetics and the textile industry; it acts as a dispersing, thickening, emulsifying and sizing agent. It is also used in adhesive preparations, to prevent flocculation of pigments, for films and sheetings, binders in ceramic glazes, leather tanning, and as a food additive.

Methyl cellulose is stable up to about 300 °C and is flammable when ignited. A plentiful supply of water should be available for extinguishing fires of the solid material. Fire-resisting structure and means of escape should be such as are required for a moderate fire risk.

Sodium carboxymethyl cellulose

SODIUM CELLULOSE GLYCOLATE; CELLULOSE GUM; CM CELLULOSE; CMC
sp.gr. 1.59
a colourless, odourless, tasteless powder.

Sodium carboxymethyl cellulose is produced by the reaction between alkali cellulose and sodium chloroacetate. It is soluble in water and insoluble in organic liquids. It reacts with heavy-metal salts to form films that are insoluble in water, transparent and unaffected by organic materials.

Sodium carboxymethyl cellulose is used for detergents, soaps, food additives, textile manufacturing (sizing), coating of paper and paper board to lower porosity, for drilling muds, emulsion paints, pharmaceuticals and cosmetics.

This substance is considered non-toxic, and does not present a serious fire risk. It is, however, listed as a suspected carcinogen.

GRENQUIST, B.

"Pulp bleaching and purification". Partridge, H. D. (273-309). *Chlorine–its manufacture, properties and uses.* Sconce, J. S. (ed.). (Huntington, New York, Robert E. Krieger, 1972).

Pulp mill processes. Pulping, bleaching, recycling. Halpern, M. G. (Park Ridge, New Jersey, Noyes Data Corporation, 1975), 403 p.

Chlorine accidents in the pulpmill industry–incidence and causes. Grenquist, B. (Helsinki, Institute of Occupational Health, 1978), 43 p. (In Finnish with English summary)

CIS 175-1971 "Evaluation of the health hazard from vegetable dust in the cellulose and cardboard industry" (Gigieniceskaja ocenka rastitel'noj pyli celliulozno-kartonnogo proizvodstva). Cyganovskaja, L. H. *Gigiena i sanitaria* (Moscow), June 1970, 35/6 (26-31). Illus. 9 ref.

"Autoignition of cellulose nitrate. Account of an investigation into the cause of a disastrous fire in a Dutch paint works". Groothuizen, T. M. et al. *TNO-nieuws* (Bussum, Netherlands), Dec. 1971, 26/12 (620-628). Illus. 6 ref.

CIS 77-1087 "Safety standards in the use of nitrocellulose" (Normas de seguridad en el uso de la nitrocelulosa). Villanueva Muñoz, J. L. *Salud y trabajo* (Madrid), Dec. 1976-Jan. 1977, 5 (20-22). Illus. 4 ref. (In Spanish)

Organic coatings manufacture (Boston, National Fire Protection Association, 1976), 41 p.

"A case of allergic contact dermatitis due to sodium carboxymethyl cellulose". Hamada, T. *Japanese Journal of Industrial Health* (Tokyo), July 1978, 20/5 (207-211).

Cement

Cement dust without free silica MAC USSR 6 mg/m³.
Cement dust containing less than 10% free silica MAC USSR 5 mg/m³.
Asbestos cement dust containing more than 10% asbestos MAC USSR 5 mg/m³.

Cement is a hydraulic bonding agent used in building construction and civil engineering. It is a fine powder obtained by grinding the clinker of a clay and limestone mixture calcined at high temperatures. When water is added to cement it becomes a slurry that gradually hardens to a stone-like consistency. It can be mixed with sand and gravel (coarse aggregates) to form mortar and concrete.

There are two types of cement: the natural cements and the artificial cements. The natural cements are obtained from natural materials having a cement-like structure and require only calcining and grinding to yield hydraulic cement powder. Artificial cements are available in large and increasing numbers. Each type has a different composition and mechanical structure and has specific merits and uses. Artificial cements may be classified as portland cement (named after the town of Portland in England) and aluminous cement.

The world production of cement has attained 759 million tonnes in 1977, 261 million of which were produced in Europe, 199 million in Asia and 127 million in the USSR. There has been a steady increase in the amount produced in the last 30 years, except in the period 1973-75.

Production. The portland process, which accounts for by far the largest part of world cement production, is illustrated in figure 1 and comprises two stages: clinker manufacture and clinker grinding. The raw materials used for clinker manufacture are calcareous materials such as limestone and argillaceous materials such as clay. The raw materials are blended and ground either dry (dry process) or in water (wet process). The pulverised mixture is calcined either in vertical or rotary-inclined kilns at a temperature ranging from 1 400 to 1 450 °C. On leaving the kiln, the clinker is cooled rapidly to prevent the conversion of tricalcium silicate, the main ingredient of portland cement, into bicalcium silicate and calcium oxide.

The lumps of cooled clinker are often mixed with gypsum and various other additives which control the setting time and other properties of the mixture in use. In this way it is possible to obtain a wide range of different cements such as normal portland cement, rapid-setting cement, hydraulic cement, metallurgical cement, trass cement, hydrophobic cement, maritime cement, cements for oil and gas wells, for highways or dams, expansive cement, magnesium cement, etc. Finally, the clinker is ground in a mill, screened and stored in silos ready for packaging and despatch. The chemical composition of normal portland cement is:

calcium oxide (CaO): 60-70%
silicon dioxide (including about 5% free SiO_2) (SiO_2): 19-24%
aluminium trioxide (Al_3O_3): 4-7%
ferric oxide (Fe_2O_3): 2-6%
magnesium oxide (MgO): < 5%

Aluminous cement produces mortar or concrete with high initial strength. It is made from a mixture of limestone and clay with a high aluminium oxide content and without extenders, which is calcined at about 1 400 °C. The chemical composition of aluminous cement is approximately:

aluminium oxide (Al_2O_3): 50%
calcium oxide (CaO): 40%
ferric oxide (Fe_2O_3): 6%
silicon dioxide (SiO_2): 4%

Recently, as a consequence of fuel shortage, there has been increased production of natural cements, especially of the volcanic tuff. If necessary, this is calcined at 1 200 °C, instead of 1 400-1 450 °C as required for portland. The tuff may contain 70-80% amorphous free silica and 5-10% quartz. When calcination is necessary, amorphous silica is partially transformed in tridimite and crystobalite. See SILICA AND SILICATES; SILICOSIS.

Uses. Cement is used as a binding agent in mortar and concrete—a mixture of cement, gravel and sand. By varying the processing method or by including additives, different types of concrete may be obtained using a single type of cement, e.g. normal, clay, bituminous, asphalt tar, rapid-setting, foamed, waterproof, microporous, reinforced, stressed, centrifuged concrete, etc.

HAZARDS

In the quarries from which the clay, limestone and gypsum for cement are extracted, workers are exposed to the hazards of climatic conditions, dusts produced during drilling and crushing, explosions and falls of rock and earth. Road transport accidents occur during haulage to the cement works.

During cement processing the main hazard is dust. Dust levels ranging from 26 to 114 mg/m^3 were in the past measured in quarries and cement works. In individual processes the following dust levels were reported: clay extraction—41.4 mg/m^3; raw materials crushing and milling—79.8 mg/m^3; sieving—384 mg/m^3; clinker grinding—140 mg/m^3; cement packing—256.6 mg/m^3; and loading, etc.—179 mg/m^3. In modern factories using the wet process 15-20 mg dust/m^3 air are occasionally the upper short-time values. Also the air pollution in the neighbourhood of cement factories is around 5-10% of the old values thanks in particular to the widespread use of electrostatic filters. The free silica content of the dust usually varies between the raw material (clay may contain fine particulate quartz, and sand may be added) and the clinker or the cement from which all the free silica will normally have been eliminated.

Other hazards encountered in cement works include high ambient temperatures, especially near furnace doors and on furnace platforms, radiant heat and high noise levels (120 dB) in the vicinity of the ball mills. Carbon monoxide concentrations ranging from trace quantities up to 50 ppm have been found near limestone kilns.

Pathological conditions encountered in cement industry workers include diseases of the respiratory system, digestive disorders, skin diseases, rheumatic and nervous conditions, hearing and visual disorders.

Respiratory tract diseases. These are the most important group of occupational diseases in the cement industry and are the result of inhalation of airborne dust and the effects of macroclimatic and microclimatic conditions in the workplace environment.

Chronic bronchitis, often associated with emphysema, has been reported as the most frequent respiratory disease, although sickness absence data show a lower morbidity rate for such disease than medical examinations.

Normal portland cement does not cause silicosis because of the absence of free silica. However, workers engaged in cement production may be exposed to raw materials which present great variations in free silica content. Acid-resistant cements, used for refractory plates, bricks and dust contain high amounts of free silica and exposure to them involves a definite risk of silicosis (see REFRACTORIES).

Cement pneumoconiosis has been described as a benign pin-head or reticular pneumoconiosis, which may appear after prolonged exposure and presents a very slow progression. However, a few cases of severe pneumoconiosis have also been observed, most likely following exposure to materials other than clay and portland cement.

Some cements also contain varying amounts of diatomaceous earth and tuff (volcanic ash). It is reported that when heated, diatomaceous earth becomes more aggressive due to the fact that the amorphous silica is transformed into cristobalite, a crystalline substance even more pathogenic than quartz. Under such conditions, concomitant tuberculosis may make the course of the cement pneumoconiosis much more severe. In addition, workers employed on the manufacture of asbestos-cement products are exposed to a definite asbestosis hazard.

Digestive disorders. Attention was drawn to the apparently high incidence of gastroduodenal ulcers in the cement industry. Examination of 269 cement plant workers revealed 13 cases of gastroduodenal ulcer (4.8%). Subsequently, gastric ulcers were induced in both guinea-pigs and a dog fed on cement dust. However, a study at a cement works showed a sickness absence rate of 1.48-2.69% due to gastroduodenal ulcers. Since ulcers may pass through an acute phase several times a year, these figures are not excessive when compared with those for other occupations.

Skin diseases. These are widely reported in the literature and have been said to account for about 25% and more of all the occupational skin diseases. Various forms have been observed, including inclusions in the skin, periungal erosions, diffuse eczematous lesions, cutaneous infections (furuncles, abscesses and panaritiums). However, these are more frequent among cement users (e.g. bricklayers, masons, etc.) than among cement manufacturing plant workers.

As early as 1947 it was suggested that cement eczema might be due to the presence in the cement of hexavalent chromium (evidenced by the chromium solution test). Probably the chromium salts enter the dermal papillae, combine with proteins and produce a sensitisation of an allergic nature. Since the raw materials used for cement manufacture do not usually contain chromium, the following have been listed as the possible sources of the chromium in cement: volcanic rock, the abrasion of the refractory lining of the kiln, the steel balls used in the grinding mills and different tools used for crushing and grinding the raw materials and the clinker. It has been stated that sensitisation to chromium may be the leading cause of nickel and cobalt sensitivity and that the high alkalinity of cement is an important factor in cement dermatoses. See ECZEMA, OCCUPATIONAL.

Rheumatic and nervous disorders. The wide variations in macroclimatic and microclimatic conditions encountered in the cement industry have been said to favour the appearance of various disorders of the locomotor system (arthritis, rheumatism, spondylitis and various muscular pains) and the peripheral nervous system (back pain, neuralgia, radiculitis of the sciatic nerves).

Hearing and vision disorders. Moderate cochlear hypoacusia in the workers of a cement mill has been reported. The main eye disease is conjunctivitis, which normally requires only ambulatory medical care.

Accidents. Accidents in quarries are due in most cases to falls of earth or rock, or occur during transportation. In cement works the main types of accident injuries are bruises, cuts and abrasions which occur during manual handling work; serious accidents are rare.

SAFETY AND HEALTH MEASURES

A basic requirement in the prevention of dust hazards in the cement industry is a precise knowledge of the composition and, especially, of the free silica content of all the materials used. Knowledge of the exact composition of newly developed types of cement is particularly important.

In quarries, excavators should be equipped with closed cabins and ventilation to ensure a pure air supply, and dust suppression measures should be implemented during drilling and crushing. The possibility of poisoning due to carbon monoxide and nitrous gases released during blasting may be countered by ensuring that workers are at a suitable distance during shotfiring and do not return to the blasting point until all fumes have cleared. Suitable protective clothing may be necessary to protect workers against inclement weather.

All dusty processes in cement works (grinding, sieving, transfer by conveyor belts) should be equipped with adequate ventilation systems and conveyor belts carrying cement or raw materials should be enclosed, special precautions being taken at conveyor transfer points. Good ventilation is also required on the clinker cooling platform, for clinker grinding and in cement packing plants.

The most difficult dust control problem is that of the clinker kiln stacks which are usually fitted with electrostatic filters, preceded by bag or other filters. Electrostatic filters may be used also for the sieving and packing processes, where they must be combined with other methods for air pollution control. Ground clinker should be conveyed in enclosed screw conveyors.

Hot work points should be equipped with cold air showers and adequate thermal screening should be provided. Repairs on clinker kilns should not be undertaken until the kiln has cooled adequately and then only by young, healthy workers. These workers should be kept under medical supervision to check their cardiac, respiratory and sweat function and prevent the occurrence of thermal shock. Persons working in hot environments should be supplied with salted drinks when appropriate.

Skin disease prevention measures should include the provision of shower baths and barrier creams for use after showering. Desensitisation treatment may be applied in cases of eczema: starting with removal from cement exposure for 3-6 months to allow healing; desensitisation consists of application of 2 drops of 1 : 10 000 aqueous potassium dichromate solution for 5 min, 2-3 times per week. In the absence of local or general reaction contact time is normally increased to 15 min, followed by an increase in the strength of the solution. This desensitisation procedure can also be applied in the case of cobalt, nickel and manganese. It has been found that chrome dermatitis—and even chrome poisoning—may be prevented and treated with ascorbic acid. The mechanism for the inactivation of hexavalent chromium by ascorbic acid involves reduction to trivalent chromium, which has a low toxicity, and subsequent complex formation of the trivalent species.

PRODAN, L.

Respiratory diseases:

CIS 75-58 "Occupational respiratory diseases in modern cement production" (Professional'nye zabolevanija organov dyhanija v uslovijah sovremennogo cementnogo proizvodstva). Mal'ceva, L. M.; Tatanov, Ju. A. *Gigiena truda i*

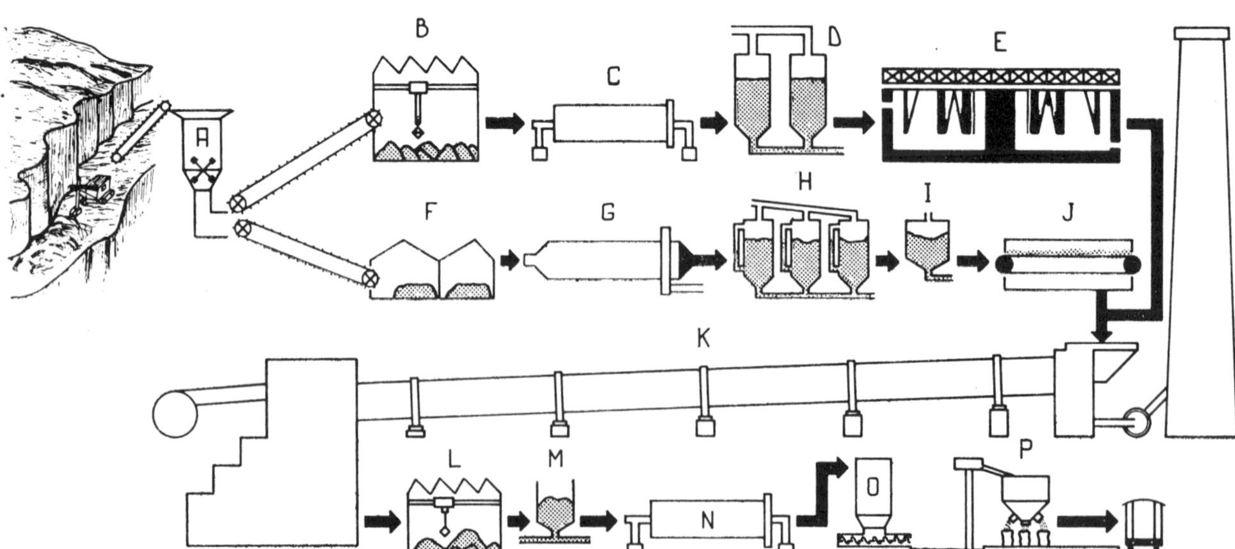

Figure 1. The manufacture of cement. A. Crushing. B. Storage-wetting (wet process). C. Grinding. D. Proportioning. E. Slurry storage basin. F. Pre-blending (dry process). G. Grinding-drying. H. Blending. I. Granulation. J. Drying. K. Rotary kiln. L. Clinker. M. Additives. N. Grinding. O. Storage. P. Packaging.

professional'nye zabolevanija *(Moscow), Mar. 1974, 3 (14-17). 17 ref. (In Russian)*

CIS 76-207 "Cement, asbestos, and cement-asbestos pneumoconioses—A comparative clinical-roentgenographic study". Scansetti, G.; Coscia, G. C.; Pisani, W.; Rubino, G. F. Archives of Environmental Health *(Chicago), June 1975, 30/6 (272-275). Illus. 15 ref.*

Skin diseases:

CIS 76-1320 "Skin sensitivity to cement in dermatitis sufferers" (Sensibilité cutanée au ciment chez les porteurs de dermite). Amphoux, M.; Bensoussan, M.; Robin, J. Revue de médecine du travail *(Paris), 1975, 3/5 (361-368). Illus. 17 ref. (In French)*

Safety:

CIS 75-2048 Safety and health in cement works *(L'hygiène et la sécurité dans les cimenteries). Cambie, R. (Caisse régionale d'assurance maladie du Nord de la France, 11 boulevard Vauban, Lille) (1974), 11 p. (In French)*

Hygiene:

CIS 79-376 "Prevention of dust hazard in cement works" (Prévention du risque des poussières dans les cimenteries). Comité technique national des pierres et terres à feu, Caisse nationale de l'assurance maladie. Cahiers de notes documentaires—Sécurité et hygiène du travail *(Paris), 4th quarter 1978, 93, Note No. 1152-93-78 (Recommendation No. 142) (615-616). (In French)*

CIS 79-389 Reduction of concrete block machine noise with accessory soundproofing attachments *(Lärmminderung an Betonsteinmaschinen durch schallschluckende Nachrüstsätze). Seiler, H. Forschungsbericht Nr. 177 (Bundesanstalt für Arbeitsschutz und Unfallforschung, Postfach 4600, Dortmund-Dorstfeld) (1978), 33 p. Illus. 33 ref. (In German)*

Centrifuges

A centrifuge is a machine in which centrifugal force is used to separate solids from liquids, to fractionate liquids of different densities, and to remove excess moisture from bulky materials such as textiles.

Uses. These machines are in common use in industry. They are found in engineering works for the removal of oil from swarf, in many chemical processes for the batch separation of solids from liquids; in laundries and other factories where wet textiles are handled, they are usually referred to as "hydroextractors". The domestic spin-dryer is a modified form of hydroextractor.

Design and construction

There are two basic types of centrifuge: swinging bucket machines and perforated basket centrifuges.

Swinging bucket machines. These are usually small in size, revolve at a very high speed and are, in the main, used in research and in process control laboratories (figure 1).

Essentially the machine consists of a vertical driving spindle on the top of which a disc is horizontally mounted; round the periphery of the disc are hinged a number of arms and on the free end of each arm is a small bucket for holding the materials to be centrifuged. At high speed, the arms are thrown out to a horizontal position and, as a consequence, the more dense parts of the material tend to migrate towards the periphery, i.e. the bottoms of the buckets.

The speed of rotation may be such that it is not unknown for a machine to disintegrate. This problem, and the risk of physical contact with the rotating parts, makes it essential that, when in motion, the moving parts be enclosed. The enclosure should be so designed that it will contain the flying pieces should the machine break

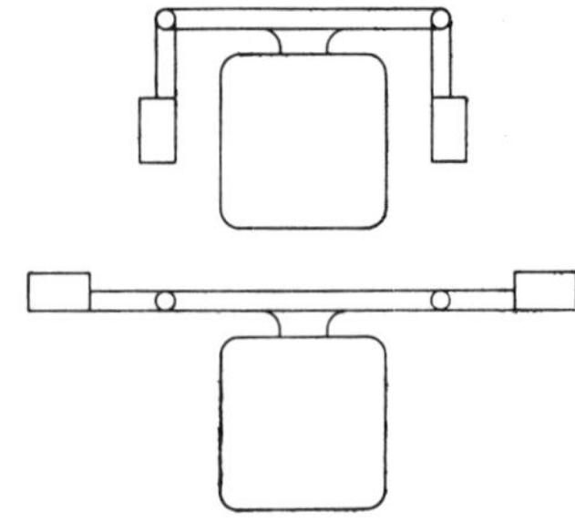

Figure 1. Swinging-bucket type of centrifuge.

up. If it is necessary to remove the enclosure for loading, it must be interlocked with the driving mechanism.

Perforated basket centrifuges. These are commonly used in factories as production units. Their main feature is a perforated basket which rotates about its axis. The basket is enclosed by a casing, a "monitor casing", in which provision is made for continuous draining (figure 2).

The method of loading depends on the type of material being processed; slurries and liquids containing fine solids are generally pumped or mechanically fed into the baskets, whilst bulky goods are loaded by hand.

HAZARDS AND THEIR PREVENTION

The presence of a component, perhaps of considerable diameter, revolving at a very high speed constitutes a severe hazard.

Contact with basket or contents. This is particularly dangerous if the interior of the basket is fitted with ribs to prevent surge.

When loading a centrifuge, it is almost impossible to exactly balance the contents of a basket. Any out-of-balance results in an oscillation of the basket as it spins. As the basket oscillates, the gap between the top of the basket and the underside of the opening in the monitor casing varies in width; hands or fingers inserted into the gap are liable to be caught between the basket and the casing and amputated.

Entanglement in the materials being centrifuged may occur, especially in laundries and places where textiles are processed. There have been instances of employees getting their hands or arms caught in a length of material or by a garment in a moving basket and amputation has occurred.

This hazard of contact with the basket or its contents can be eliminated by the provision of an efficient

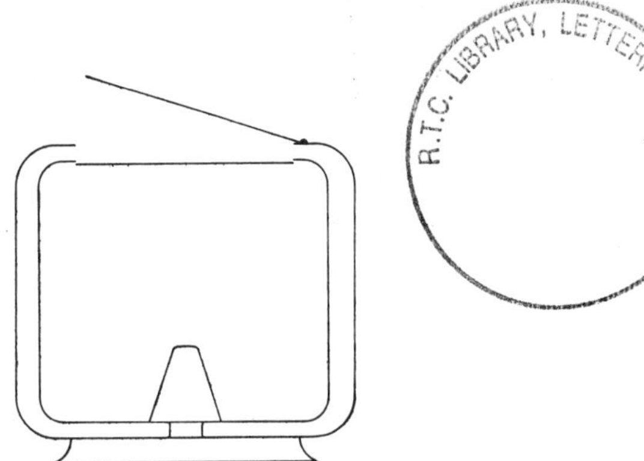

Figure 2. Centrifuge (hydroextractor) of the type used in laundries. The outside shell is the monitor casing, the inside is the perforated basket. At the top, the interlocked cover feed opening.

interlocked cover on the feed opening of the monitor casing. The interlocking arrangements must ensure that the machine cannot be started if the cover is in the open position, and once the machine is in motion it should be impossible to open the cover until the basket has stopped completely.

Bursting of basket. This is not a common occurrence but when it does happen, one of the following factors are usually present: *(a)* over-speeding; *(b)* corrosion; *(c)* oscillation.

It is vital that the safe maximum speed of the basket should never be exceeded. If the driving mechanism is such that the safe speed of the basket can be exceeded, a suitable governor is required, which must be so set as to keep the speed of rotation within safe limits. All centrifuges should be marked with the safe maximum speed of the basket.

Corrosion is a problem in factories where centrifuges are used for the processing of chemicals. It results in a general deterioration in the structural strength of the machine. When possible the basket, and indeed the whole of the interior of the machine, should be constructed of a material which is resistant to the materials handled. If this is not practicable, the inside of the centrifuge requires an anticorrosive lining made of a suitable material such as lead, plastics, etc.

Severe oscillation of the basket is brought about by out-of-balance loading or by excessive liquor surge. In the case of bulky articles, such as textiles, the provision of a spacing divider in the basket assists the operator in balancing the load.

Liquor surge arises when too much free liquid is present and its movement creates an increasing state of unbalance. The surge can be controlled by installing in the basket a number of longitudinal ribs; these have the effect of preventing the free movement of the liquid. Buffers are occasionally fitted between the outside of the basket and the inside of the monitor casing. When excessive oscillation occurs, the basket strikes the buffers and the speed of the basket is reduced. Trip devices are available which will automatically cut off the power if the oscillation exceeds a predetermined limit.

Displacement of basket cover. At high speeds, bulky goods tend to rise in the basket as it revolves. To counteract this, a removable cover or partial cover is fitted to the basket; this cover often takes the form of an annular ring. Unless this cover or ring is securely locked in position before the machine is started, there is a danger that it will become detached when the basket is at its maximum speed. Rings have been known to cut through the side of the monitor casing and to be projected into the workroom with fatal results. Interlocking devices are quite practicable.

Brakes. All centrifuges must be fitted with an efficient brake. The brake should be interlocked with the power supply in such a way that it cannot be applied until the power has been cut off.

Inspection. It is desirable that centrifuges should be the subject of careful maintenance and should be inspected by a competent person once during each period of 12 months, and records should be kept. If the machine is used for corrosive materials, frequent routine inspections should be made at regular intervals between each major inspection.

Operator training. Before a person is left in charge of a centrifuge, he should be trained in its use; this training should make special reference to the dangers of these machines.

HALL, S.

Industrial centrifuges:

Construction and use of basket centrifugal separators for industrial purposes. Resolution AP (71) 1, Council of Europe Committee of Ministers (Partial agreement in the social and public health field) 19.2.1971 (Strasbourg).

CIS 75-1257 "Centrifugal separators" (Centrifugeuses). Lacore, J. P. Fiche technique de sécurité 9 (Institut national de recherche et de sécurité). *Cahiers de notes documentaires–Sécurité et hygiène du travail* (Paris), 1st quarter 1975, 78, Note No. 939-78-75 (5-23). Illus. 9 ref. (In French)

Explosion hazards:

CIS 77-1525 "User guide for the safe operation of centrifuges" (Institution of Chemical Engineers, 165-171 Railway Terrace, Rugby) (No date), 36 p. Illus. 9 ref.

"Inertising techniques and design modifications for the prevention of explosions and fire in filter centrifuges" (Inertisierungstechnik und Konstruktionsmassanhmen zur Vermeidung von Explosion und Brand an Filter-Schleudern). Simon, D. I. E. (367-396). *Report of the 3rd International Symposium on the Prevention of Occupational Risks in the Chemical Industry* (Berufsgenossenschaft der chemischen Industrie, Gaisbergstrasse 11, 69 Heidelberg). Illus. 8 ref. (In German)

Laboratory centrifuges:

CIS 78-1519 *Directives concerning safety measures for laboratory centrifuges* (Anvisningar om skyddsagårder vid laboratoriecentrifuger). Anvisningar nr 122 (Arbetarskyddsstyrelsen, Fack, 100 26 Stockholm) (Apr. 1978), 11 p. Illus. (In Swedish)

Cervicobrachial disorder, occupational

Since the first case of an occupational cervicobrachial disorder was found among computer key-punchers in 1958, the same kind of disorder has been observed among typists, telephone operators, cash-register operators, office workers using ball-point pens and others, and many young female workers have been suffering from discomfort of the neck, shoulders and arms in many industries where a rapid development of mechanisation and automation had required simple and repetitive tasks.

From 1972 to 1975 a Japanese committee demonstrated that a broad cause/effect relationship exists between working conditions and the disorder in the cervicobrachial areas as related to chronic fatigue of the shoulders, neck, back and other parts of the body. Later on four working groups studied the aetiology of the disorder, its natural history in relation to workload, the screening techniques and the health services to be provided for the workers. Finally, the syndrome was recognised as a new chronic occupational hazard arising from a combination of technological changes and increased working speed. Previously, great interest was taken in health effects of heavy workload but not much attention was devoted to the potential hazards of light mechanical activities. In future, it will be necessary to take into consideration possible adverse health effects whenever new occupations or working conditions are introduced, even if they are physically light and mechanised.

Aetiology. Mechanised and automated processes imply two main problems: first, they have apparently relieved physical workloads but have increased the consistency of physical movements by concentrating the workload in bodily parts such as arms, shoulders, neck, eyes, etc.; second, they generally produce an elevated rate of work with monotonous tasks and requires increased attention and patience. Such working methods easily result in increasing mental stress. Moreover, a combination of several factors which seem to be tolerable by themselves often leads to an occupational hazard under the stress of a high working speed.

The study of the disorder among conveyer system workers at a cigarette company and cash-register operators at supermarkets has pointed out the following aetiological factors: the muscle load on the upper limb consisting of a dynamic fraction in repetitive activity and of a static fraction (such as raising an arm); the muscle load on the whole body resulting from an uncomfortable working posture (such as working at a narrow cash-register counter); the mental stress resulting from high speed work and/or tension in the service trades; environmental factors such as noise, low temperature, unsuitable ventilation, poor lighting, etc.; and unfavourable working conditions due to poor personnel management.

Natural history and clinical findings. At an early stage of the disorder, subjective complaints and/or clinical findings are different according to the characteristics of the workload. However, there are no clinical differences among patients with serious illness, even if they perform different jobs.

Natural history of the disorder is classified into five grades as follows.

1st grade: subjective complaints without clinical findings.

2nd grade: subjective complaints with induration and tenderness of neck, shoulder and arm muscles.

3rd grade: includes 2nd grade plus any of the following:

(1) increasing tenderness and/or enlargement of affected muscles;

(2) positive neurologic test such as Addison's test, Morley's test and others;

(3) paraesthesia;

(4) decrease of muscle strength;

(5) tenderness of spinous processes of the vertebrae;

(6) tenderness of paravertebral muscles;

(7) tenderness of nerve plexes;

(8) tremor of hand and/or eyelid;

(9) kinesalgia of neck, shoulder and upper extremity;

(10) functional disturbance of peripheral circulation;

(11) severe pain or subjective complaints of the neck, shoulders, or upper extremities.

4th grade: divided into two groups:

(a) severe type of 3rd grade;

(b) directly develop from 2nd grade without passing through 3rd grade, but having specific findings as follows:

(1) orthopaedic diagnosis of neck-shoulder-arm syndrome;

(2) organic disturbance such as tendinitis or tenosynovitis;

(3) autonomic nervous disturbance such as Raynauld's phenomenon or passive hyperaemia;

(4) mental disturbance such as anxiety, sleeplessness, thinking disfunction, hysteria or depression.

5th grade: disturbance not only in work but also in daily life.

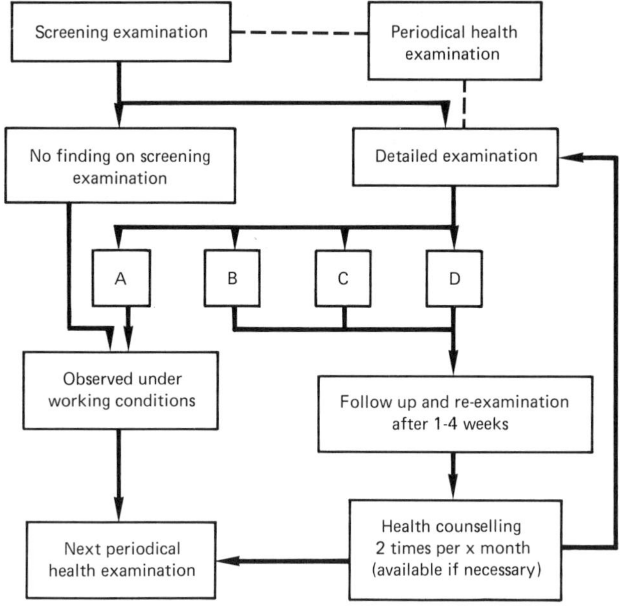

Figure 1. Medical surveillance system.

Diagnosis. The study of telephone operators and cash-register operators as workers at risk suffering from the disorder as compared with a control group of nursing students led to the conclusion that examination techniques can be classified into four categories as follows:

(1) Useful techniques for finding an aetiologic relationship between workload and clinical findings.

(2) Effective techniques for early case finding.

(3) Diagnostic techniques for classification of the grade.

(4) Useful techniques for the diagnosis of clinical type.

The most important information about the disorder relating to workload and working conditions was often given by the workers' past and present histories.

Medical surveillance. The occupational health services should supervise workers' health and working conditions. Many instructions for typists, key-punchers, cash-register operators, telephone operators and stenographers need extensive revision. Training courses and health education concerning operating method and preventive measures should be provided for the workers before they are engaged. A medical surveillance system should be provided for the workers at risk of the disorder as shown in figure 1. Periodical health examinations should be available for the workers more than twice per year; health counselling should be provided even more frequently up to twice a month. Each examination will have a score, including an indication of health needs (scores from 1 to 3) and of working environment improvement (scores from A to D) as follows.

Indication for health care: (1) Needs no specific treatment; (2) Needs specific treatment; (3) Needs medical care.

Indication for working conditions: A. Needs no specific measures; B. Needs measures to improve working conditions, however the current job can be carried out; C. Needs reduction of work or change of work; D. Should stop working.

The check list for the control of working conditions designed by the Japanese committee consists of four categories as follows:

(1) Working conditions: this includes working hours and rest, the quality and quantity of work, working speed and consistency of work.

(2) Physical posture necessary to work with various kinds of instruments and machines.

(3) Working environment: temperature, lighting, noise, etc.

(4) Human relations among workers, and between workers and supervisors.

In Japan some regulations for the improvement of working conditions such as shortening of working hours, introduction of new business machines with lighter key touch, air temperature and lighting control and health examination for the workers were enforced. They have reduced the number of workers suffering from the disorder and the new working conditions have changed the natural history of the disorder, although they have not completely solved the relevant problems.

AOYAMA, H.

"Health hazard among cash register operators and the effects of improved working conditions". Ohara, H.; Aoyama, H.; Itani, T. *Journal of Human Ergology* (Tokyo), Sep. 1976, 5/1 (31-40). Illus. 6 ref.

"Shoulder muscle tenderness and physical features of female industrial workers". Onishi, N.; Nomura, H.; Sakai, K.; Yamamoto, T.; Hirayama, K.; Itani, T. *Journal of Human Ergology* (Tokyo), Dec. 1976, 5/2 (87-102). Illus. 10 ref.

"Recent trends in research on occupational cervicobrachial disorder". Aoyama, H.; Ohara, H.; Oze, Y.; Itani, T. *Journal of Human Ergology* (Tokyo), Sep. 1979. 8/1 (39-45). 19 ref.

"Localized fatigue in accounting-machine operators". Maeda, K.; Hünting, W.; Grandjean, E. *Journal of Occupational Medicine* (Chicago), Dec. 1980, 22/12 (810-816). Illus. 24 ref.

"Neurological and neurophysiological aspects of the cervical syndrome and related painful conditions with regard to work and especially monotonous postures" (Neurologische und neurophysiologische Aspekte des Cervicalsyndroms und verwandter Schmerzzustände unter Berücksichtigung pathogener Arbeitsbedingungen, insbesondere monotoner Sitzhaltungen). Kropp, H. *Arbeitsmedizin–Sozialmedizin–Präventivmedizin* (Stuttgart), June 1979, 14/6 (137-141). (In German)

"Occupational cervicobrachial syndromes. A review". Waris, P. *Scandinavian Journal of Work, Environment and Health* (Helsinki), 1979, suppl. 3 (3-14). 115 ref.

Occupational cervicobrachial disorder–Preventive measures in the field of occupational health (Keikenwan syogai–Shyokuba ni okeru sono taisaku). Aoyama, H. (ed.). (Tokyo, Rodokijunchosakai, 1979), 424 p. Illus. 121 ref. (In Japanese)

Charcoal burning

Wood charcoal is the residue obtained when wooden material is partially burned or heated in the absence of air at a temperature exceeding 400 °C. It has a high calorific value and can be burned almost smokelessly. Charcoal is therefore widely used in many developing countries as domestic and also as industrial fuel (e.g. for the reduction of minerals). Because of its absorbtive properties, active charcoal is also used for a variety of purposes by the chemical industry (e.g. for purification of air and water).

Production. Traditionally charcoal is made in pits or heaps covered with earth. However, with portable steel kilns and fixed installations such as masonry kilns the output of charcoal can be increased from 30 to 70-80 kg per stacked m^3 of wood and its quality can be improved at the same time. Large-scale charcoal production can be carried out by a continuous process in a retort, permitting the recovery by distillation of by-products such as acetic acid, tar and methanol.

HAZARDS

When portable kilns are used the main risk is burns when handling or touching hot parts. In fixed installations there is a danger of carbon monoxide poisoning and inhalation of dust, which may lead to diseases of the respiratory system. When using retorts and recovering chemical by-products there are safety and health risks similar to those that are common in the chemical industries.

SAFETY AND HEALTH MEASURES

Portable steel kilns should be fenced off during burning and cooling. Protective gloves must be used when handling hot kiln parts. Special care must be taken to stand clear of the air conduits when reversing the draught because of the possibility of blow-outs of hot gases.

Masonry kilns should be of heat-resistant construction to avoid collapsing during charcoal removing. They should only be entered when there is no further danger of carbon monoxide poisoning. Lift trucks or conveyor belts may render unloading easier and safer.

Water should be kept readily available during unloading in case of spontaneous combustion. Face shields should be worn during sieving. Charcoal should be conditioned for 24 h before stacking in a confined space.

STREHLKE, B.

Chelating agents

Chelating agents, also known as chelants or chelators, are organic compounds capable of binding metal ions to form ring structure complexes, named chelates. The term chelate is derived from the Greek word χηλή, meaning "crab's claw".

The metal atoms which more readily bind to the electron-donor groups (ligands) of polydentate chelants are those of the transition series. These metals, with the d-orbitals open for sharing electron pairs with the ligand atoms or groups, offer different co-ordination numbers, ranging from 2 to 10, but most commonly 4 and 6, which determine the geometrical structure and the physico-chemical characteristics of the chelates.

The co-ordination number of the metal (M) is not necessarily fixed, but it may vary depending on its oxidation state and on the characteristics of the ligand (L). The co-ordinate bond formed between M and L is covalent in nature and its stability depends on the electronic structure of M and L and on the allowed stereochemistry of the chelate; e.g. if the steric configuration of the chelating molecule allows a perfect fit of the metal atom with its orbitals and co-ordination numbers matching the positions of the donor atoms, a very stable chelate, with a nearly perfect geometrical configuration, is predicted.

In order to function as a chelating agent an organic molecule must have at least two functional groups with electron-donor atoms which are capable of donating a pair of electrons to the metal atom. These groups must be situated in such a way as to permit the formation of a ring structure which includes the metal atom.

The functional groups found in chelating agents are either the protonated ligands, which exchange their protons for metallic cations, or the non-protonated basic co-ordinating groups, which donate their unshared pair of electrons to the metal atom (see table 1).

Table 1. Protonated and non-protonated ligands

Protonated	$-C(O)OH$ $-SH$	$-SO_3H$ $-N(O)H$	$-P(O)(OH)_2$ $-OH$ (phenolic or enolic)	$-C(S)SH$
Non-protonated	$-NH_2$ $=O$	$=NH$ $-O-R$	$=N-$ $-S-$ $=NOH$	$-AsR_2$ $-OH$ (alcoholic)

Chelants in medicine

A great variety of chelating agents, either of natural or synthetic origin, have been described, but for therapeutic purpose only a few of them have been extensively tested *in vitro* and *in vivo*. Chelating agents have been widely used for the chelation therapy of heavy-metal poisoning. Other uses of chelants in medicine include diagnostic techniques with radionuclides, the selective staining of tissues for microscopy, the treatment of hypertension, rheumatism and cancer, as well as endodontic therapy.

Since heavy metal toxicity continues to plague the industrial societies, an approach to this problem has been sought through the use of chelants which either enhance the excretion of the toxic metals, and thus reduce their body burden, or decrease their gastrointestinal absorption, if the ingested metal is still in the gastrointestinal tract, by forming insoluble, non-absorbable chelates. See ANTIDOTES.

The most used chelants in the therapy of heavy-metal poisoning are ethylenediaminetetraacetic acid (EDTA), diethylenetriaminepentaacetic acid (DTPA), 2,3-dimercaptopropanol (BAL), penicillamine (PA), desferrioxamine B (DFOA), and dimercaptosuccinic acid (DMS).

Ethylenediaminetetraacetic acid

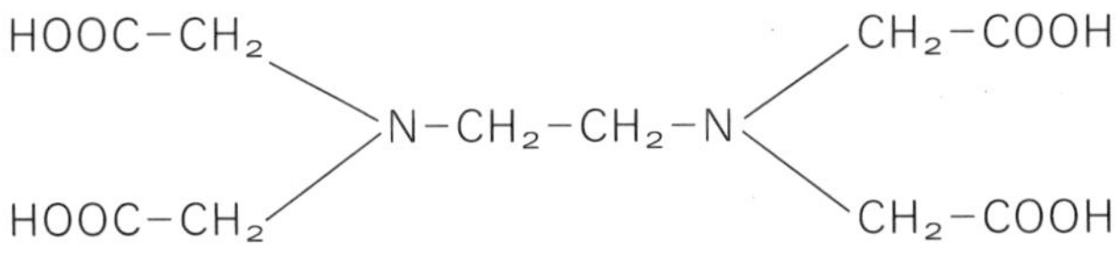

$$\begin{array}{l} HOOC-CH_2 \\ \quad\quad\quad\quad\; \searrow \\ \quad\quad\quad\quad\quad N-CH_2-CH_2-N \\ \quad\quad\quad\quad\; \nearrow \\ HOOC-CH_2 \end{array} \begin{array}{l} CH_2-COOH \\ \swarrow \\ \\ \nwarrow \\ CH_2-COOH \end{array}$$

Ethylenediaminetetraacetic acid, also known as edetic acid, or versenate, or EDTA, is a synthetic polyaminopolycarboxylic acid largely used in chelation therapy.

Since the free acid or the Na salt of this compound are rather toxic (LD_{50} of about 0.8 mM/kg when administered parenterally), the less toxic calcium disodium form ($CaNa_2EDTA$) is more frequently used. $CaNa_2EDTA$, herein simply referred to as EDTA, prevents the onset of a dangerous hypocalcaemic state. Moreover, the calcium disodium salt has shown similar effectiveness in the removal of toxic metals as the parent compound or its sodium salt.

EDTA is essentially not metabolised by the human body and it is rapidly excreted in the urine. About 50% of EDTA administered intravenously is excreted within 1 h and 90% within 7 h. EDTA and its metal chelates do not permeate the cellular membrane to a significant extent; thus most of the EDTA remains in the extracellular fluids until excreted into the urine. Due to this property, orally administered EDTA is capable of binding metals in the gastrointestinal tract, thus preventing their absorption. However, there is some indication that chelates of EDTA with trivalent ions, such as Y^{3+} and Cr^{3+}, may penetrate to a limited extent the cellular membrane.

EDTA has been extensively employed in lead-poisoned adults and children for both therapeutic and diagnostic purposes. It is also used externally in solution or paste to treat skin lesions in chromium-exposed electroplating workers. It has not been found effective in the treatment of workers exposed to iron, nickel, manganese, vanadium or cobalt. EDTA in the acid form has also been tested in endodontic therapy, but its effectiveness in enlarging the root canal for a better and cleaner preparation has not been fully established.

Although chelation therapy with EDTA has been found effective in chronic lead intoxication, there are some doubts whether it is valid for the treatment of acute lead poisoning, especially in the case of lead encephalopathy, where the administration of the chelant has occasionally triggered the deterioration of clinical conditions.

Administration. $CaNa_2EDTA$ is generally administered by a daily intravenous infusion of 1 g dissolved in 500 ml of dextrose solution for the period of 3 to 5 days. A second course of treatment, if required, should not be started until after an interval of 3 or more days. EDTA is also available in 50 mg tablets to prevent absorption of lead from the gastrointestinal tract.

Side effects. $CaNa_2EDTA$ is the most widely used chelant for its relatively low toxicity (LD_{50} by i.v. administration between 10 and 20 mM/kg). However, treatment of patients with high doses has been known to produce severe renal lesions including haematuria, proteinuria, elevated BUN and histopathological changes of the tubuli. Thus, chelation therapy should be avoided in patients suffering from renal diseases.

There is a possibility that most side effects of EDTA are due to the chelation of endogenous essential metals, especially of zinc, and that supplementation of zinc during chelation therapy may be effective in reducing the toxicity of this chelant.

Related chelants. Other polyaminopolycarboxylic acids with known chelating properties are: diethylenetriaminepentaacetate (DTPA), triethylenetetraaminehexaacetate (TTHA), 2,2'-*bis*[di-carboxymethyl-amino] diethylether (BADE) and ethylenediaminedi-(*o*-hydroxyphenylacetate) (EDHPA). While the metabolism and toxicity of DTPA, TTHA and BADE are comparable to those of EDTA, EDHPA is far more toxic than its related compounds and it is prevalently excreted in the bile.

Penicillamine

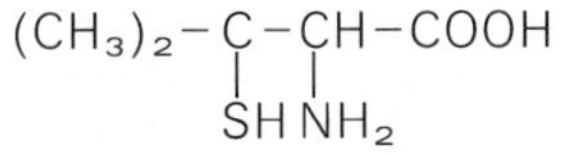

$$(CH_3)_2-\underset{SH}{\underset{|}{C}}-\underset{NH_2}{\underset{|}{CH}}-COOH$$

β,β-Dimethylcysteine, also known as penicillamine (PA), is available in two optical forms, *d* and *l*, or as a *dl*-racemic mixture. The metabolic behaviour and the toxicity of the two stereoisomers differ significantly as a result of a greater participation of *l*-PA in metabolic reactions. The distinctly lower toxicity of *d*-PA (LD_{50} about 17 mM/kg, by oral route in rats) has favoured the use of this isomer over the *dl*-racemic PA (LD_{50} about 2.4 mM/kg, p.o. or i.p. in rats).

The principal use of *d*-PA has been in the therapy of Wilson's disease, which is characterised by abnormal deposition of copper in liver, brain, kidneys and cornea. *d*-PA has also been used in the chelation therapy of plumbism; the main advantage of *d*-PA over EDTA is that the former is absorbed through the gastrointestinal tract and can be given orally, while EDTA must be given by intravenous infusion. Treatment of chronic mercury poisoning with *d*-PA has also been successful. Although *d*-PA chelates other heavy metals, the therapeutic effectiveness in poisoning from such metals has not been fully validated.

Recently *d*-PA has been used in the treatment of rheumatoid arthritis with some success, but the side effects have been so severe as to discourage the prolonged administration of the drug.

Administration. In patients with Wilson's disease *d*-PA is given in capsules of 250 mg for a dose of 1 capsule 3-4 times per day for a prolonged period. In saturnism or hydrargyrism 4-6 capsules daily are given for 4-10 days.

Side effects. Like penicillin, its parent compound, penicillamine may cause allergic reactions. The side effects of *dl*-PA are much more pronounced than those of *d*-PA. The most frequently encountered complaints in the *d*-PA therapy are fever, joint pains, maculo-papular skin eruptions, and eosinophilia.

Long-term therapy with *d*-PA can also result in nephrotic syndrome, interference with haematopoiesis, lupus erythematosus and neurological disorders. These side effects may be caused by interference of PA with vitamin B_6 and the endogenous essential trace metal metabolism. Administration of pyridoxine, 25-50 mg/day, as well as a supplement of essential metals, are advisable during long-term PA therapy.

Related chelants. Penilloic and penicilloic acids are degraded in the gastrointestinal tract to *d*-PA; thus they might be used as "prechelators" in occupational medicine. The N-acetyl derivative of *dl*-PA as well as the mercaptoethyliminodiacetate (MEIDA) are chelants related to PA, but their application in chelation therapy has been purely of academic nature.

Desferrioxamine B

$$NH_2-(CH_2)_5-\underset{\displaystyle OH}{\underset{|}{N}}-CO-(CH_2)_2-CONH-(CH_2)_5-$$

$$\underset{\displaystyle OH}{\underset{|}{N}}-CO-(CH_2)_2-CO-NH-(CH_2)_5-\underset{\displaystyle OH}{\underset{|}{N}}-CO-CH_3$$

Desferrioxamine B (DFOA), also known as deferoxamine, is a natural chelant of iron prepared by the removal of Fe^{3+} from a siderophore, ferrioxamine B (FOA), produced by *Streptomyces pilosus*.

DFOA can be virtually considered a selective chelant of Fe^{3+} since the stability constant of this chelate (FOA) is about 10^{31}, many orders of magnitude higher than that of chelates formed with other metal ions. The iron-binding capacity of DFOA is 8.5 mg Fe per 100 mg DFOA. Due to these properties, DFOA is effective in mobilising iron from storage and enhancing its excretion.

Peroral administration of this chelant is ineffective due to its poor intestinal absorption, but subcutaneous or intramuscular injections are as effective in Fe-chelation as intravenous infusion.

DFOA is rapidly metabolised in the mammalian organism. In dogs 70% of intravenously injected DFOA is excreted in the urine within 72 h, partly unchanged (28%) and the remainder as degradation products, some of which still possess Fe^{3+}-chelating properties.

DFOA is used in the treatment of idiopathic haemochromatosis and other Fe-storage diseases as well as in the therapy of occupational siderosis, where iron is prevalently deposited in the lungs. Although this chelant has been proven effective in mobilising iron from stores, the therapeutic value of Fe-chelation in iron storage diseases remains controversial. However, the use of DFOA or other chelants for diagnostic purposes is generally accepted.

Administration. DFOA is usually administered intramuscularly. Intravenous administration should if possible be avoided, since the side effects are principally targeted to the vascular system.

The recommended dose is 30 μM/kg (about 18 mg/kg) per day, which can be given intramuscularly in two fractions. For long-term therapy the daily dose is reduced to 15 μM/kg (about 9 mg/kg). In cases of acute Fe intoxication caused by oral intake of large doses of inorganic iron the recommended treatment consists of 5-10 g DFOA given via nasogastric tube following gastric lavage. In severe cases 1 g DFOA is administered slowly by intravenous infusion at a rate not higher than 15 mg/kg per hour.

Side effects. Toxic effects of DFOA at the recommended doses are generally rare, transient and inconsequential. They are more frequent when the administration is intravenous rather than intramuscular. The symptoms are nausea, headaches, flickering of the eyes and muscular cramps. Rapid intravenous infusion may also cause marked decrease in blood pressure, tachycardia, and occasionally a shock-like state, convulsions and erythema. Long-term treatment with DFOA produced cataracts in experimental animals, but this effect is uncommon in humans.

BAL

$$\underset{\displaystyle SH}{\underset{|}{CH_2}}-\underset{\displaystyle SH}{\underset{|}{CH}}-CH_2OH$$

BAL (the abbreviation for British anti-Lewisite) is 2,3-dimercaptopropanol or dimercaprol. The main use of this chelant has been in the treatment of arsenical poisoning. Intoxications with arsenic, which previously would have been fatal, may now be successfully treated with BAL. In cases requiring artificial kidney or peritoneal lavage, considerable amounts of arsenic are found in the dialysate. However, it should be pointed out that the effectiveness of BAL is not the same for all arsenical compounds.

Mercury, gold, antimony and bismuth poisoning may also respond to BAL treatment.

Administration. Because of its poor solubility in water BAL is, as a rule, dissolved in lipid solvents and administered intramuscularly.

The recommended doses for arsenic poisoning are as follows: first day, 3 mg/kg every 4 h; second and third day, 3 mg/kg every 6 h; following days, 3 mg/kg twice per day until recovery.

A glucoside of BAL, known as BAL intrav, did not prove to be practical due to the poor stability of the available preparations.

In the case of topical application, 5-10% solutions or 3-5% ointments are employed.

Side effects. The LD_{50} in rats and rabbits by intramuscular or subcutaneous injection is about 100 mg/kg. In man the administration of 5 mg/kg may induce secondary reactions including vomiting, headache, body pains, burning sensations in the mucosae, tachycardia and hypertension.

Dimercaptosuccinic acid

$$HOOC-\underset{\displaystyle SH}{\underset{|}{CH}}-\underset{\displaystyle SH}{\underset{|}{CH}}-COOH$$

The meso-2,3-dimercaptosuccinic acid (DMS) is a chelating agent recently tested for the mobilisation of metals in experimental animals.

Treatment with DMS has been shown to be more effective than penicillamine or EDTA in removing lead from lead-exposed mice. At physiological pH DMS has a higher affinity for mercury than BAL. Treatment with DMS has been found effective in reducing mercury body-burden in mice, rats or guinea-pigs exposed to $HgCl_2$ or MeHgCl.

The advantages of DMS are: first, it is effective by either parenteral or oral route of administration; second, its toxicity is in the same range as that of D-penicillamine (the LD_{50} for intraperitoneal DMS in mice is between 3.0 and 4.0 g/kg) and it is approximately 35 times less toxic than BAL; third, compared with penicillamine the efficiency of DMS on the basis of mg/kg is four times higher in the mobilisation of Hg.

As far as it is known, this compound has been clinically tested only in chronic saturnism. In such patients DMS, when administered orally for six days at dosages ranging from 8.4 to 42.2 mg/kg/day, induced a significant decrease of blood-lead concentration and an increase of urinary lead excretion. The existing data from animal studies would predict a successful application also in the therapy of hydrargyrism.

Mixed ligand chelate therapy

A therapy for heavy metal poisonings using a combination of chelants has been proposed. For acute lead intoxication a combination of EDTA and BAL has been recommended. This therapy has been found effective especially in lead-poisoned children.

Other cocktails of ligands utilised in experimental animals have recently been reported to be effective in the chelation of cadmium, plutonium and other metals. These cocktails include two or more of the following chelants: EDTA, DTPA, PA, salicylic acid, sulphosalicyclic acid, nitrilotriacetic acid and many others. Although certain data indicate a high effectiveness in the mobilisation of metals, the mechanism of action and the therapeutic usefulness remain controversial. Moreover, further studies are needed to assess the side effects of each combination of ligands.

Adverse effects of chelation therapy

Chelating agents, with the possible exception of some natural ones, are not selective in binding metals. Thus, in biological systems they bind not only the target metal but also essential metals which play major roles in enzymatic reactions.

Chelation may alter excretion and/or distribution of essential trace metals producing a functional, albeit transitory, deficit of these metals resulting in enzyme inhibition and altered metabolic pathways. Other reactions may also occur between these drugs and biologically important compounds (e.g. vitamin B_6 depletion by penicillamine). Supplementation of one or more essential nutrients, known to be affected by the administered chelant, as well as continuous medical supervision, are indicated in long-term and/or large dose therapy.

IANNACCONE, A.

"D-Penicillamine in the treatment of occupational mercurialism" (La D-penicillamina nel trattamento del mercurialismo professionalle). Morelli, A.; Iannaccone, A.; Cicchella, G. *Securitas* (Rome), Sep. 1965, 50/9 (71-84). Illus. 45 ref. (In Italian)

"Treatment of lead poisoning by 2,3-dimercaptosuccinic acid". Friedheim, E.; Graziano, J. H.; Popovac, D.; Dragovic, D.; Kaul, B. *Lancet* (London), 9 Dec. 1978, II/8102 (1234-1236). Illus. 15 ref.

The chelation of heavy metals. Catsch, A.; Harmuth-Hoene, A.-E.; Mellor, D. P. (ed.); Levine, W. G. (Oxford, Pergamon Press, 1979), 239 p. Illus. 763 ref.

Mixed ligand chelation therapy. Bulman, R. A.; Crawley, F. E. H.; Geden, D. A. *Nature* (London), 4 Oct. 1979, 281/5730 (406). 4 ref.

"Effect of zinc supplementation during chelation therapy in plumbism: a case report". Finelli, V. N.; Lerner, S.; Hong, C.; Lohiya, G. *Medicina del Lavoro* (Milan), Mar.-Apr. 1980, 71/2 (149-156). Illus. 16 ref.

Chemical industry

The business of the chemical industry is to change the chemical structure of natural materials in order to derive products of value to other industries or in daily life. Chemicals are produced from these raw materials—principally minerals, metals and hydrocarbons—in a series of processing steps. Further treatment, such as mixing and blending, is often required to convert them into end-products, for example paints, adhesives, medicines and cosmetics. So the chemical industry covers a much wider field than what is usually called "chemicals" since it also includes such products as artificial fibres, resins, soaps, paints, photographic films, etc.

Chemicals fall into two main classes: organic and inorganic. Organic chemicals have a basic structure of carbon atoms, combined with hydrogen and other elements. Oil and gas are today the source of 90% of world organic chemical production, having largely replaced coal and vegetable and animal matter, the earlier raw materials. Inorganic chemicals are derived chiefly from mineral sources. Examples are sulphur, which is mined as such or extracted from ores, and chlorine, which is made from common salt.

The products of the chemical industry can be broadly divided into three groups, which correspond to the principal steps in manufacture:

(a) base chemicals (organic and inorganic) are normally manufactured on a large scale up to several million tonnes per year, and are normally converted to other chemicals;

(b) intermediates are derived from base chemicals. Most intermediates require further processing in the chemical industry, but some, solvents for example, are used as they are;

(c) finished chemical products are made by further chemical processing. Some of these (drugs, cosmetics, soaps) are consumed as such; others, such as fibres and plastics, dyes and pigments, are processed still further.

Thus the main sectors of the chemical industry are as follows:

(1) Basic inorganics: acids, alkalis, and salts, mainly used elsewhere in industry; industrial gases, such as oxygen, nitrogen and acetylene.

(2) Basic organics: feedstocks for plastics, resins, synthetic rubbers, and synthetic fibres; solvents and detergent raw materials; dyestuffs and pigments.

(3) Fertilisers and pesticides (including herbicides, fungicides, insecticides, etc).

(4) Plastics, resins, synthetic rubbers, cellulosic and synthetic fibres.

(5) Pharmaceuticals (drugs and medicines).

(6) Paints, varnishes and lacquers.

(7) Soaps, detergents, cleaning preparations, perfumes, cosmetics and other toiletries.

(8) Miscellaneous chemicals, such as polishes, explosives, adhesives, inks, photographic film and chemicals.

It would be quite impossible, with the enormous number of products involved, to design a detailed flow diagram or genealogical tree of raw materials, processes, intermediates and finished products of the chemical industry. Figure 1 does, however, give a condensed summary of the essential information and demonstrates the complex inter-relationship which inevitably exists between the various sectors of the industry.

In the ISIC (International Standard Industrial Classification of All Economic Activities) system, used by the United Nations to classify economic activity into ten major divisions, the chemical industry is classified as Division 35, one of the nine subdivisions of Major Division 3: Manufacturing. Division 35 is further subdivided into industrial chemicals (351), other chemicals (352), petroleum refineries (353), miscellaneous coal and petroleum products, e.g. asphalt (354), rubber products including tyres (355), and plastics processing (356).

In reporting chemical industry statistics each country normally uses its own classification system, and this can be misleading. Thus comparison between countries of total chemical industry performance cannot be based on national sources. However, international bodies like the OECD (Organisation for Economic Co-operation and Development) and the United Nations normally supply data on the ISIC basis, though with a delay of about two years.

Trade statistics are published internationally under the Standard International Trade Classification (SITC) which differs from the ISIC system. Trade statistics by individual countries nearly always refer to SITC section 5, which covers about 90% of total chemicals reported in the ISIC system.

The chemical industry has grown much more rapidly in the past 30 years than industry as a whole, and now, in the early 1980s, it accounts for about 10% of the output of manufacturing industry in the OECD countries, or just under 9% of total industrial output in the area. Table 1 shows average growth rates for nine major OECD countries for the period 1970 to 1978.

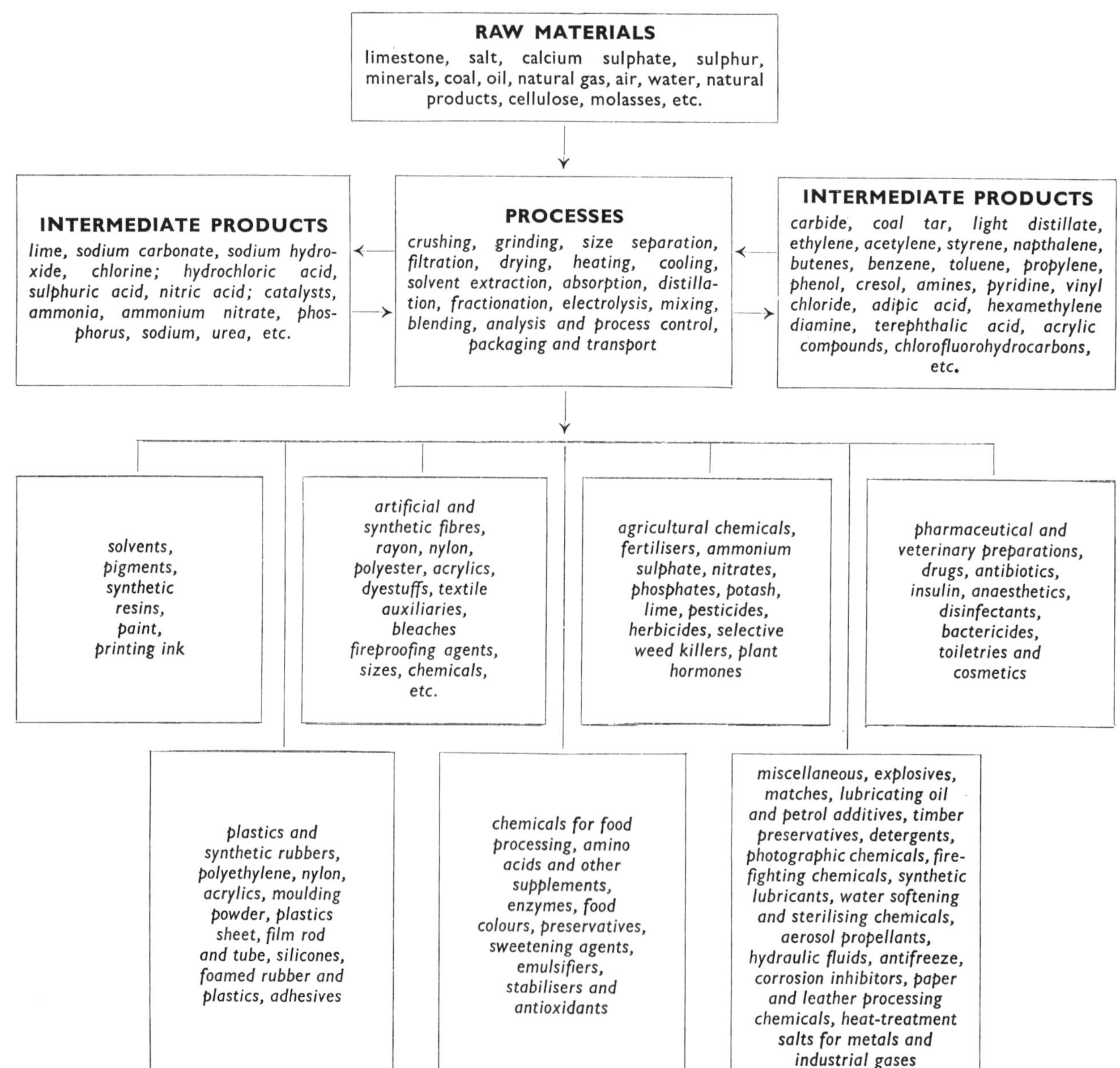

Figure 1. The chemical industry and its raw materials, processes and products, with the exclusion of petroleum oil, gas, glass and metals.

Table 1. Comparative growth rates (% per year) of chemical and industrial production in nine OECD countries, 1970-1980.

Country	Chemical production growth rate (A)	Industrial production growth rate (B)	Difference (A–B)
United States	6.1	3.8	2.3
Germany (Fed. Rep.)	4.2	1.4	2.8
United Kingdom	3.4	1.1	2.3
France	5.3	3.2	2.1
Italy	5.7	2.8	2.9
Spain	17.1	7.0	10.1
Netherlands	5.8	3.2	2.6
Belgium	8.0	2.3	5.7
Japan	4.8	3.8	1.0

Much of the chemical industry is highly capital-intensive (for example the first four major sectors listed above) and is also strongly dependent on research and development (e.g. pharmaceuticals). The combined result of these two factors is that the industry employs an abnormally low number of unskilled manual workers for its size, in comparison with manufacturing industry in general.

Total employment in the industry rose slightly during the period of rapid growth prior to 1970, but since then the lower growth of industrial production and improved productivity of chemical output has resulted in a slow decline in employment of perhaps 1% per year in the major OECD countries. Table 2 illustrates this decline.

Table 2. Chemical industry employment in nine OECD countries, 1970-77

Country	Employment in 1970	Employment in 1977
United States	1 171 200	1 071 000
Germany (Fed. Rep.)	590 340	551 000
United Kingdom	519 500	450 000
France	300 000	320 500
Italy	264 000	293 000
Spain	120 970	148 500
Netherlands	110 000	87 500
Belgium	59 720	64 775
Japan	542 000	491 000

In regard to investments there are large fluctuations from year to year in individual countries. In 1976 it would appear that chemical industry investment in Western Europe increased over the previous year's figure by about 2%. Table 3 shows chemical industry investment by country.

Table 3. Chemical industry investment in eight Western European countries and the United States, 1974, 1975 and 1976 ($000 000)

Country	1976	1975	1974
United States	6 680	6 250	5 690
Germany (Fed. Rep.)	2 183	2 276	2 169
Italy	1 310	1 302	999
United Kingdom	1 227	1 298	984
France	1 046	1 166	1 038
Netherlands	692	537	469
Belgium	684	477	295
Spain	426	440	311
Sweden	234	179	149
Total of eight Western European countries	7 802	7 675	6 384

SAFETY AND HEALTH

Industrial accidents

An examination of the statistical records of accidents in the industry demonstrates clearly that the majority of accidents are of a kind that are common to all industries: falls of people or materials, contact with moving or stationary objects, hand tools, electricity, fire and explosion. The acute injuries resulting from the risks of material peculiar to the chemical industry form a relatively small percentage of the whole, the only exception being eye injuries by chemicals, which form a high proportion (up to 40%) of total industrial eye injuries. On the other hand, gassing accidents form only a few per cent of all accidents in industry, yet this is a special hazard which might be expected to feature high in the list of causes of injury. Fatalities by accidents in the chemical industry form only a very small proportion of total fatalities by accidents. In the Federal Republic of Germany for example there were in 1977: 15 000 fatalities by traffic, 8 300 by accidents in or around home and 3 000 by industry, of which 41 in the chemical industry.

In the United Kingdom in 1978 there were 651 fatal accidents in industry of which only 15 in the chemical industry (for comparison: mining accidents resulted in 88 fatalities; in agriculture, forestry and fishing the figure was 57; and in the construction industry it was 157).

Safety of the environment

Though the number of casualties and fatalities in the chemical industry may not be alarming, accidents in this industry and in the transport of its products by road or rail may pose risk to the public: the physical and toxicological consequences of spilling and fires and explosions which occur from time to time. One of the most severe was the "Seveso" accident in Italy 1976, where due to a reaction which ran out of control a large cloud of an extremely toxic and persistent substance (tetrachloro-dibenzo-para-dioxine) was spread and settled on the earth for miles around the factory. Though fortunately the toxicological effects on the people living in that area were remarkably small, taking into consideration the extent of contamination and toxicity of the compound, the problems posed by the necessity of evacuating a large number of inhabitants of the region and of decontaminating the area were enormous. These and similar accidents have resulted in a call for stricter control of technical safety measures by industry to protect the general population outside the factory and for better organisation at national and international level to meet and limit the consequences of such accidents. The same applies to hazards to the public by air and water pollution and by waste disposal, for which stricter national and (though more difficult) international requirements have been made and still more may be necessary.

Occupational diseases

Due to the nature of the materials handled in the chemical industry, many of which are used because of their reactive abilities, and the fact that chemical reactivity and biological activity very often run parallel, special hazards may be expected to exist for those exposed to these materials. Thus constant alertness is required in regard to possible short- or long-term toxicological effects and in order to keep exposures below hazardous levels. Though apparently, taking into consideration the variety and the quantity of the relevant chemicals, industry in general succeeds in these efforts, unfortunate incidents still occur, partly by acute unexpected exposures (comparable with accidents that occur in any industry), partly because of insufficient knowledge of toxicological properties, especially in regard to long-term effects (e.g. β-naphthylamine, vinyl chloride monomer).

Poisoning as a result of absorption of lead or mercury compounds has been recognised for generations, yet, in the 20th century, in any highly industrialised society, cases of lead poisoning or mercury intoxication (or at least effects relatable to excessive exposure) still occur from time to time.

Skin diseases caused by contact of the skin with primary irritants or sensitising agents also occur, though they are always preventable by proper hygienic working methods. A greater problem is posed by the control of long-term effects, especially those leading to malignancies like cancer. In the first place they are harder to detect (because of the usually long latency time between exposure and effect and the many other non-industrial, possible causes of similar clinical conditions); and secondly adequate toxicological testing for these effects is extremely difficult. Epidemiological surveys (relating group health data to group exposure data over many years) have contributed much to the detection of relations between certain exposures and the occurrence of certain malignancies, but this too is not a very sensitive method of detection and, moreover, results come too late to save those already exposed and affected.

Thus occupational cancers and ways to prevent them are very much in the limelight, the more so because, apart from the workers in industry, large populations are brought into contact with many chemical products as consumers. As scientific evidence is often insufficient and conflicting and no agreement exists yet as regards the existence of certain levels of exposure below which the risk is practically nil, the principles and policies of control differ widely, ranging from a total ban on all compounds considered to have carcinogenic effects on man at one extreme to excessive reliance on what is technically and economically feasible in the relevant industry at the other.

Still, apart from some unfortunate incidents in industry, cancer statistics over the past 20 years (a period which allows for a reasonable latency time) do not give reason for alarm, since they show that total deaths from cancer (after deduction of lung cancer) have not increased during the past 15-20 years, which of course does not prove that no individual cases caused by chemical products have occurred which may be undetectable in the large total figures.

Anyhow, there is a potential cancer problem in industry needing constant alertness, as shown by various unfortunate incidents: bladder cancer caused by β-naphthylamine, pleural mesothelioma caused by asbestos, nasal carcinoma in the shoe and wood industries, scrotal cancer caused by cutting oils, leukaemia caused by benzene and, rather recently, angiosarcomas (a rare liver malignancy) in vinyl chloride production and polymerisation workers.

The vinyl chloride case, in particular, rightly caused much alarm. Firstly vinyl chloride had always been considered to be a relatively harmless compound (except for some acute narcotic effects) and had even been considered by the medical profession for use as an anaesthetic. Secondly, the discovery of the malignant effects (almost at the same time by laboratory testing by Viola and Maltoni and by industrial epidemiological surveys in the United States) came, as usual, many years after many people engaged in vinyl chloride polymerisation, especially the reactor cleaners, had been exposed to high levels of vinyl chloride. So although, as a result of considerable efforts made by the vinyl chloride industry, conditions now seem safe, cases still occur in this previously high exposure group and there is no way to prevent this.

As this is the usual pattern in these incidents, extensive toxicological testing, especially on these long-term effects, is now routine and in many countries obligatory before workers or the public are exposed to the compound. Also on account of great advances in technology, industrial and personal hygiene, and improvement in toxicological investigation techniques and in medical supervision of workers, it is expected that such incidents can be better prevented now. In addition specialised methods of risk analysis and assessment are being developed by many scientists. Nevertheless, especially in regard to carcinogenicity, there is still no general agreement. Weighing the risks against the benefits (risk/benefit assessment) will always also be a matter of politics and economics and thus of much debate in which, apart from national concepts, psycho-social factors play a role: some (high) risks are accepted by society while other (much smaller) risks are not or not yet accepted.

International co-operation

In this respect the awareness of the importance of international co-operation both by industry and governments has grown, leading to pooling of experience, and collaboration in testing programmes and in fixing standards. International bodies like the ILO and WHO have long contributed in this effort by providing guidelines and information on hazards and ways to avoid them and by stimulating and reviewing scientific work. In specific fields the ILO has published several recommendations on protection against hazards from exposure (e.g. on benzene).

The International Agency for Research on Cancer (IARC), established by the WHO in 1965, is especially active in this field and has published several monographs in addition to organising symposia and keeping research registers.

In 1972 occupational medical doctors, who have long co-operated in the Permanent Commission and the International Association on Occupational Health through groups on lead, asbestos, artificial fibres, etc., set up a special section (MEDICHEM) dealing with problems in the chemical industry.

The chemical industry in Europe has founded a Bureau in Brussels (European Council of Chemical Manufacturers' Federations–CEFIC), which has a medical advisory committee that contributes to the exchange of experience and opinion, and a scientific bureau on toxicology (the European Chemical Industry Ecology and Toxicology Centre–ECETUC) dealing with co-ordinated toxicological research efforts and the assessment of results. In the United States a more or less similar research institute was established by industry (the Chemical Industry Institute of Toxicology–CIIT).

At the governmental level many countries now maintain close contact, for example in the European Economic Community (EEC), which has an action programme in this field and has issued directives in regard to special compounds such as vinyl chloride, lead, cadmium, etc., and through the EEC with the United States. As regards occupational health care the ILO adopted its Occupational Health Services Recommendation in 1959 which served as a basis for the EEC recommendations published in 1962. In many countries this occupational health care is now obligatory and in others rapidly developing.

Although it will probably never be possible to achieve absolute safety in the chemical industry, awareness of the potential hazards, the combined efforts of industry, workers (and their unions) and governments as well as the many improvements in toxicological testing will probably prevent a recurrence of the unfortunate incidents of the past. Legislation in certain respects will be necessary, but the main effort must come from industry at all levels. Good training and education of all workers, supervisors and management is essential in this, as is also adequate occupational health care, including industrial hygiene, in factories.

Thus the chemical industry can continue to make a substantial contribution to economic progress with its constant efforts not only to develop new products but also to introduce improved methods of making existing products better and in greater safety.

DE BOER, L.

General:

The chemical industries and the working environment. Chemical Industries Committee, Eighth Session, Geneva 1976 (Geneva, International Labour Office, 1976), 78 p.

"Health and safety trends in the UK chemical industry". Bean, N. W. (411-427). Illus. 9 ref. *Third International Symposium on the Prevention of Occupational Risks in the Chemical Industry. Frankfurt am Main, 21-23.6.1976* (Geneva, International Social Security Association, ISSA, 1976), 598 p.

Health:

CIS 78-1389 *Occupational pathology of the upper airways in the chemical industry* (Professional'naja patologija verhnih dyhatel'nyh putej v himičeskoj promyšlennosti). Soldatov, I. B.; Danilin, V. A.; Mitin, Ju. V. (Moscow, "Medicina", 1976), 187 p. Illus. 269 ref.

CIS 79-164 "Cancer epidemiologic surveillance in the Du Pont Company". Pell, S.; O'Berg, M. T.; Karrh, B. W. *Journal of Occupational Medicine* (Chicago), Nov. 1978, 20/11 (725-740). 16 ref.

Safety:

CIS 78-1152 *Flowsheeting for safety–A guide on safety measures to consider during the design of chemical plant.* Wells, G. L.; Seagrave, C. J.; Whiteway, R. M. C. (Institution of Chemical Engineers, George E. Davis Building, 165-171 Railway Terrace, Rugby, Warwickshire CV21 3HQ) (undated), 60 p. Illus. 144 ref.

CIS 78-1150 *Chemical process hazards with special reference to plant design–VI.* Symposium series No. 49 (Institution of Chemical Engineers, George E. Davis Building, 165-171 Railway Terrace, Rugby, Warwickshire CV21 3HQ) (1977), 168 p. Illus. 205 ref.

CIS 78-1151 *Process industry hazards–Accidental release, assessment, containment and control.* Symposium series No. 47 (Institution of Chemical Engineers, George E. Davis Building, 165-171 Railway Terrace, Rugby, Warwickshire CV21 3HQ) (1976), 224 p. Illus. 175 ref.

"Risk assessment and safety analysis in chemical engineering" (Risikoermittlung und Sicherheitsanalysen in der chemischen Technik). Pilz, V. *Chemie-Ingenieur-Technik* (Weinheim, Federal Republic of Germany), 1980, 52/9 (703-711). 30 ref. (In German)

Chemical reactions, dangerous

Work in chemical plants is characterised by multifarious technical processes involving the use of a wide range of raw materials and finished products with flammable, explosive or toxic properties.

In the basic chemical reactions certain substances are chemically converted into others, with the absorption or release of heat. These thermal effects require special cooling systems (exothermic reactions) or heating systems (endothermic reactions) to ensure optimal operating conditions and the safe operation of the process equipment. One of the prerequisites for these conditions is the use of catalysts or initiators. Any perturbation of the parameters of a chemical reaction may result in a dangerous situation.

Dangerous chemical reactions are reactions involving toxic or flammable substances or mixtures and proceeding very rapidly. This type of reaction with toxic, flammable or explosive substances (acids, alkalis, aldehydes, esters, hydrocarbons, metals and their compounds, nitrates, peroxides) includes processes such as oxidation, sulphonation, chlorination, nitration, hydration, polymerisation, polycondensation, etc.

Dangerous chemical reactions may therefore be classified according to the types of substances and compounds involved, as follows:

1. reactions with toxic substances;
2. reactions with flammable substances;
3. processes with high-speed reactions;
4. mixed reactions.

Although this classification is arbitrary, it enables the hazard of a chemical reaction to be assessed according to certain parameters, such as the exposure limits in the case of toxic substances and the limits of flammability, auto-ignition temperature, flash point, etc., in the case of flammable and explosive substances or mixtures.

Dangerous chemical reactions of hazard class 1 feature the use of toxic substances as raw materials or final products formed at various stages in the course of the basic reaction or as a result of side reactions. In nitration processes, for instance, the basic nitrating agent is nitric acid, the vapours of which are both toxic and strongly oxidising. If the normal course of the process is disturbed, for example by a malfunction or emergency, the dual nature of the nitrating agents gives rise as a side reaction to nitrogen oxides, which are highly toxic.

Dangerous chemical reactions of hazard class 2 involve the use of explosive substances as raw materials, intermediate, final and secondary products, as well as physical and chemical interactions that may give rise to the formation of explosive mixtures of such substances. In certain cases chemical reactions of this hazard class associated with cooling and condensation may be even more dangerous than processes in which the product is heated. The danger results from the change in state of aggregation of the mixture, since flammable mixtures are liable to form in the gas or vapour phase. Benzene, for instance, with its thermal limits of flammability becomes more explosive during cooling than during heating.

Dangerous chemical reactions of hazard class 3 are reactions accompanied by an intensive release of gases or vapours, and of heat (exothermic reactions). The rise in temperature accelerates the basic and side reactions, the latter causing the formation of products with toxic, flammable or explosive products or of explosive mixtures, and the release of large amounts of gases. All nitration reactions are exothermic, and the rise in temperature is associated with side reactions accompanied by an intensive release of gases (nitrogen oxides).

Reactions of hazard class 4 combine the features of those of classes 1, 2 and 3, or parts of them. A large proportion of dangerous chemical reactions belongs to this class of mixed reactions.

Basic safety and health measures in chemical laboratories and plants

Protective goggles and gloves must be worn when concentrated acids and alkalis are being handled. These and other corrosive liquids should be transferred from one container into another with the aid of glass pipettes with rubber bulbs under an exhaust hood.

All reactions involving highly flammable and ignitable liquids should be carried out in exhaust hoods with their fans running and screens closed. Liquids with a low boiling point (acetone, benzene, ethers) must be distilled and heated in round-bottomed flasks of heat-resistant glass in baths filled with a heat-transfer agent (water, oil, liquid silicone, sand) to be chosen according to the boiling point of the substance. When ethers are distilled, hydroxides may form which are liable to decompose explosively under the action of a shock, friction or slight heating. To prevent explosions, liquids which are likely to form peroxides should not be completely distilled: at least 10% by volume of the liquid distilled should remain in the flask. In certain cases stabilisers, i.e. oxidation inhibitors, are used to prevent the formation of peroxides. When peroxides are to be handled, hermetically sealed boxes (chambers) must be used.

Work with highly volatile organometallic compounds (with boiling points between 20 and 25 °C) and with spontaneously flammable volatile substances must be carried out under vacuum or inert gas. Metallic potassium or sodium must not be allowed to come into contact with water.

Before technical processes involving dangerous chemical reactions are elaborated, safe formulations, the physical and chemical properties of the products, reagents and their mixtures, the laboratory procedure and the type of equipment used must be studied in detail. Dangerous processes should be carried out in isolated premises meeting the requirements regarding explosion and fire safety. Work with volatile toxic substances must be carried out in enclosed equipment under negative pressure. Processes which may give rise to explosion or release of toxic substances should be controlled and monitored from a remote location.

BOBKOV, A. S.

CIS 79-1377 *Handbook of reactive chemical hazards.* Bretherick, L. (London, Butterworth Group, 1979), 1 281 p.

CIS 78-1964 "Chemical reactions–Analysis of mechanisms and attempt at hazard prevention" (Les réactions chimiques–Analyse des mécanismes et essai de prévision des risques). Leleu, J. *Cahiers de notes documentaires–Sécurité et hygiène du travail* (Paris), 3rd quarter 1978, 92, Note No. 1130-92-78 (403-411). 5 ref. (In French)

Third international symposium on the prevention of occupational risks in the chemical industry. Frankfurt am Main 23.6.1976. Report (Geneva, International Social Security Association, 1976), 598 p. Illus. Ref.

Protection system against potentially dangerous processes in chemical technology (Systema zaščity potencyal'no opasnyh processov v himičeskoj tehnologii). Obnovlenski, P.; Musiakov, L. A.; Cheltsov, A. V. (Leningrad, Himia, 1978), 224 p. (In Russian)

Occupational safety and health in the USSR scientific institutions (Ohrana truda v naučnyh ucreždeniah Akademii nauk USSR). Cehmacev, G. G. (ed.). (Moscow, Nauka, 1972), 575 p. (In Russian)

See also the series "Hazardous chemical reactions" (Réactions chimiques dangereuses) published in *Cahiers de notes documentaires–Sécurité et hygiène du travail* (Paris). (In French)

Chemicals, new

The world around us is filled with and made up of chemicals, and virtually everything that nurtures our civilisation is made possible by the use of chemicals. To differentiate accurately between "new" and other chemicals is impossible today, but at some time in the not too distant future it may become possible to keep a running inventory of new chemicals that are produced in sufficient quantities to enter world commerce. It is difficult to establish how many chemicals are in existence, but a 1978 estimate based upon the American Chemical Society's Chemical Abstract Service (CAS) identified 4 039 907 distinct chemical compounds and it has been reported that the total increases at an average rate of about 6 000 per week. Most of the chemicals listed by the CAS have either been isolated from natural products or synthesised for research purposes, and can hardly be considered as chemicals in "everyday use". Current estimates from the Environmental Protection Agency (EPA) in the United States indicate that there may be as many as 63 000 of the latter type. The average person thinks of chemicals in everyday use as those around the home, the workshop and the garden (perhaps also on farms and in factories); but even without pesticides, food additives, and pharmaceuticals there are some 50 000 chemicals on the ever-growing common use list including agricultural chemicals (other than pesticides), fuels for power production, and chemical consumer products. It is estimated that the total trade in chemicals among countries of the Organisation for Economic Co-operation and Development (OECD) exceeded $50 000 million per annum for several years in the late 1970s, and the world figure is probably three times higher. The total number of chemicals in commerce throughout the world has been put as high as 100 000 by EPA, and new chemicals placed on the market each year have been estimated by EPA, the World Health Organisation (WHO), OECD, and others to range from 200 to 1 000. In the United States alone production of the 50 largest-volume chemicals has been increasing by about 8% per year over the last few years (during the 1970s) and should be in excess of 225 million tonnes by the time this encyclopaedia is published.

Approaches to the control of toxic chemicals

Before new chemical hazards can be controlled they must be identified and evaluated. The ultimate objective of testing chemicals is to obtain the information needed to form a sound basis for recommending "safe" levels of exposure for humans in contact with them. Safe is a relative term and no matter what adverse effect is associated with a chemical, absolute safety for that chemical can never be assured–except by its elimination. This cannot often be accomplished, and perhaps it should not be except in a few instances of particularly hazardous chemicals and/or chemicals easily replaced. The hazard associated with a chemical is a function of its toxicity and risk, and is dependent also on its conditions of use–which in turn determine the exposure of individuals and of populations. Evaluation of the toxicity of a chemical requires information on many parameters and the process by which toxicological and/or epidemiological studies are assessed (what level of chemical produces what adverse effect with what probability) is referred to as risk assessment.

Information on hazard and risk assessment of new chemicals (or those that have only recently been introduced) is usually non-existent, and exposure information can therefore be derived only indirectly from data on production and use. Since these are difficult to obtain, there is a severe limitation on any sort of evaluation of exposure effects. Generally the necessary information is available only to the industries that manufacture, use and/or sell the chemicals. Industries can therefore be an important source for information on hazards associated with particular chemicals, but many are unwilling to reveal it.

Legislative approaches to societal needs. Societal concern with protecting the public against perceived harm, real or imaginary, is increasing in virtually every country, and many have enacted legislation empowering ministries, agencies or boards to issue or amend detailed regulations to control drugs, narcotics, pesticides, household chemicals, etc. In addition, food additives, chemical additives in gasoline, cosmetic ingredients, coatings of cooking utensils, paints containing toxic agents, and other substances have been made subject to legislative control. Many countries regulate air and water quality by the adoption of emission standards applicable to industries, domestic fuel-burning installations, motor vehicles, etc. Standards have also played a prominent role in legislation dealing with chemical hazards in the work environment, although sometimes the standards adopted are not strictly enforceable, since they have no official status (for example, a number of threshold limit values).

Several countries have recently adopted (or are in the process of drafting) legislation that reflects a more comprehensive approach to the control of chemicals which may adversely affect human health and the environment. These countries, which include Canada, Colombia, France, Japan, Norway, Sweden, Switzerland, the United Kingdom and the United States, have or will have legislation empowering their authorities to request information from industry about any chemical, irrespective of its intended purpose, and in some countries information concerning toxicity testing performed on certain agents.

Control of toxic substances. An example of such legislation is the Toxic Substances Control Act of 1976, or Public Law 94-469 of the United States, better known as the TSCA. The four major activities provided for by this Act are:

(a) to gather information on chemicals;
(b) to require testing of chemicals identified as possible risks;
(c) to screen new chemicals proven to present a risk;
(d) to control chemicals proven to present a risk.

The basic aim of TSCA is to prevent unreasonable chemical risks; in this regard it is unlike most other laws in that the Government will be empowered to take action *before* widespread and possible serious harm has occurred. Eight product categories, namely tobacco, nuclear material, firearms and ammunition, and substances used solely as pesticides, food, food additives, drugs, and cosmetics, are exempt from TSCA's authority.

To achieve its first major purpose the TSCA empowers the Government to monitor the entire life cycle of chemical substances–through product development, testing, manufacturing, processing, distribution, use, and disposal. Because of this unusual "cradle to grave" authority there had to be an inventory of all chemicals already existing in US commerce. Once this inventory had been made, then all chemicals not included in it would be considered "new". New chemicals are subject to review by the EPA before being manufactured or imported for a commercial purpose. The Initial Chemical Substances Inventory was published in 1979 and contained a listing of 44 000 chemicals; a Revised Inventory came out in 1980. Security procedures to safeguard confidential business information are necessary and are applied.

The second major purpose–the testing of chemicals–is addressed in greater detail later. In determining the testing rules, the EPA is required to consider the relative cost of various protocols that could be used to develop the needed data, and also the foreseeable availability of test facilities and personnel for performing the tests. The TSCA provided for the establishment of an Interagency Testing Committee (ITC) to recommend chemicals to the EPA for testing. The ITC, formed in 1977, includes representatives of eight agencies concerned with some aspects of research and/or standards development and/or regulatory action for chemicals. The ITC, which can designate a priority list of up to 50 chemicals, published its first list in the Federal Register 1977. This included four individual chemicals and six chemical categories, recommended for testing on the basis of knowledge regarding use, exposure, toxicity, environmental release, etc.; special attention was given to those suspected of causing cancer, gene mutations, or birth defects. Within 12 months after the ITC has designated its priority chemicals the EPA must either initiate rule-making proceedings to require testing, or publish reasons for not doing so. A general policy statement was issued in 1978 concerning TSCA's "substantial risk notification procedures". These procedures require that anyone who has information providing reasonable grounds for concluding that a chemical substance presents a substantial risk of injury to health or the environment must immediately notify the EPA. Several months later, the EPA published its first ruling regarding the provision of copies or lists of health and safety studies from any person who manufactures, processes, or distributes certain chemical substances identified in the ITC priority list. In mid-1980 rules were published requiring these same groups to keep records of "significant adverse reactions" to health and the environment alleged to have been caused by a chemical substance. The law already requires that records of alleged employee health effects must be kept for 30 years. All other records of alleged adverse reactions (including consumer and environmental effects) must be retained for 5 years.

The third major activity under TSCA concerns the important premanufacture notification programme. This screening process requires some fundamental changes in the way the chemical industry has operated in the past, and concerns *new* chemicals only. New chemicals are those not included in the inventory (previously mentioned). Once this aspect of the law had been put into effect in 1980, manufacturers and importers of chemicals could not knowingly process or use for commercial purposes a "new" substance which has not first passed through premanufacture notification. Certain information is required to be submitted to the EPA with the premanufacture notice: the common or trade name; the chemical identity and molecular structure; estimated output; proposed use categories; methods of disposal; workplace exposure levels; and a description of by-products, impurities and other related products. Data on the health and environmental effects of the chemical (known to and within the control of the submitter) must also be supplied.

The fourth major activity under TSCA is to control chemical substances found to pose an unreasonable risk to human health or the environment. The EPA may prohibit or limit the manufacture, processing, distribution in commerce, use, or disposal of a chemical. Action towards this can range from a complete ban to a simple labelling requirement. Although there have been many chemical candidates for attention, the only chemicals actually named thus far for regulation and eventual banning are the polychlorinated biphenyls.

Some issues on testing and evaluation

It has been estimated that only 6%, or 6 000, of the chemicals in present-day commerce have been laboratory-tested for toxicity. This is, in itself, difficult to believe when one considers the large numbers of chemicals, and it becomes more serious when the lack of quality and uniformity of testing in years gone by are taken into account. Moreover, an overwhelming proportion of the tests were not adequately designed to assess chronic toxicity even roughly (though admittedly that was not always the intent).

We already know from experience that the chronic effects (if any) of a new chemical may take years from the time of its introduction until they manifest themselves, and that the establishment of causal relationships will be difficult as well as costly. Consequently, suspect chemicals must be identified and tested before entering into large-scale commercial circulation. Chemicals with chronic health effect can be identified by epidemiological studies, short-term tests for predicting oncogenicity, and long-term chronic toxicity tests.

Epidemiology. Epidemiologic studies may provide the most satisfactory data for the investigation of human health effects resulting from exposure to environmental chemicals, since they avoid the uncertainties of extrapolating experimental animal or *in vitro* system data to humans, i.e. the high doses of the bioassay to the probable low-dose of human exposure. However, epidemiology has inherent limitations, the chief among these being the long latency period for the development of most cancers and other chronic diseases in humans; thus, disease detection does not occur until after a population has been exposed and a significant number of its members contract the disease. To prevent chemically induced chronic disease, timely and reliable data on the chronic health effects of chemical substances are essential and, while epidemiologic studies are extremely useful, they cannot be relied upon as the primary means of obtaining the data needed for the regulation of toxic substances.

Short-term tests. It is generally agreed among toxicologists that the goal of an effective testing programme is to develop a battery of reliable short-term tests that will predict carcinogenicity accurately. As noted above, human beings risk potential exposure to thousands of chemicals, most of which have not been tested adequately, if at all, for oncogenic effects. Given this, in addition to the reported flow of newly produced chemicals, it is obvious that there is need for quick, inexpensive short-term tests with high predictive value in identifying possible oncogenic chemicals and helping to set priorities for testing in long-term bioassay.

Although a number of short-term *in vitro* tests are in use or under development, the Salmonella/microsome/mutation (S/m/m) test developed by Ames and McCann in the United States appears to be the most popular. It is relatively inexpensive, gives quick results, and shows sensitivity to very small doses of chemicals. Moreover, correlations with the results of long-term animal bioassays have been good, with about 90% accuracy in predicting carcinogens and non-carcinogens. But this still means that a significant number of chemicals could be misevaluated, since there is such a large and varied universe of chemicals currently in use or under development. Moreover, short-term tests detect the mutagenicity of a chemical rather than its capacity to produce tumours, which is the "endpoint" of a long-term bioassay. In so far as chemically induced carcinogenicity and mutagenicity correlate, the test may be a satisfactory indicator, but the extent of this correlation is not conclusively known; not all mutagens are demonstrable carcinogens and vice versa. There are other valid reasons why short-term tests are not capable of replacing long-term ones, although they certainly have a valuable place in bioassay. However, until there is some good method for validating the systems and the laboratories using them, there will always be doubt regarding their accuracy and hence their predictability (and perhaps the same could be said about much long-term testing).

Long-term chronic toxicity tests (oncogenicity). A long-term oncogenicity test that is properly designed and conducted is at present the definitive test model for estimating the cancer risk of chemicals for humans. Having mammalian tumour-induction as its endpoint, the oncogenicity bioassay is the only source of direct evidence (other than in the human) of chemically induced tumours in the mammalian species. Moreover, of all test systems, it comes closest to mimicking human routes of exposure and metabolic/pharmacologic processes which activate and distribute chemicals.

The IARC recently revised its assessment criteria for evaluating chemically induced oncogenicity in humans and/or experimental animals and concluded: "In the presence of appropriate experimental carcinogenicity data and in the absence of adequate human data, it is reasonable to regard chemicals for which there is 'strong evidence' of carcinogenicity (i.e. unquestionable production of malignant tumours in animals) as if they were carcinogenic to humans." It is generally agreed, therefore, that long-term bioassays can predict human risk and should be used for the identification and removal of cancer-causing chemicals.

Long-term chronic toxicity tests (non-oncogenicity). Ideally, toxicity testing for non-oncogenic effects in experimental animals would be accomplished by short-term exposures, either acute or subacute. The former involves exposures lasting up to 24 h, and the latter can last up to a tenth of the life-span of the experimental species. These (especially the first) are not as reliable as long-term, chronic toxicity experiments, which may involve exposures lasting as long as the whole life-span of the test species, but the expense of chronic testing can be prohibitive. Scientists have attempted to design an optimal chronic toxicity test system that will provide adequate toxicological information while reducing the investments of time and money for each chemical tested. Short-term (3-4 month) chronic tests are often used, and some toxicologists consider them to have a number of advantages over the longer-term chronic studies, apart from decreased time and expense.

However, it is generally agreed that, since humans are exposed to many synthetic chemicals in their environment for most of a lifetime, chronic toxicity testing for assessment of health effects in humans should be for a comparable period of time, i.e. lifetime or near-lifetime of laboratory animals. In this way, age-related factors such as altered tissue sensitivity, changing metabolic and physiological capability, and spontaneous disease, which can influence the degree and nature of toxic responses, can be assessed. Moreover, with long-term testing, chemicals may produce different toxic responses with repeated dosing; different metabolic pathways may become involved; bioaccumulation of the chemical in the host tissue may occur.

In brief, then, the non-oncogenic chronic toxicity test is considered to be the most reliable means of establishing a no-observed-adverse-effects level which, in turn, can be used to define the lifetime "acceptable daily intake" (ADI) of a chemical for humans who have been exposed to that chemical over a major portion of their lifetime.

Readers interested in the various scientific aspects of toxicity testing procedures, e.g. study design issues, test species, data collection, good laboratory practice, etc., are referred to the literature cited (see also ANIMAL

EXPERIMENTATION). Details concerning economic aspects are also of obvious importance; costs can vary considerably, depending upon the length of testing, species, number of species and animals per species, endpoint effects (which determine evaluation techniques), etc. Requirements by the TSCA call for two species to be used for each chemical in evaluation of chronic effects, whether oncogenic (mice and rats) or non-oncogenic (one rodent and one non-rodent species). Estimates indicate that pre-chronic testing, i.e. any short-term testing that may be needed prior to chronic testing, would cost approximately $50 900 and $100 500 per chemical for the oncogenic and non-oncogenic testing; and, that full-blown, lifetime tests for chronic effects of a chemical cost $387 700 and $528 700, respectively. On the basis of these estimates, the cost of a testing programme to meet the requirements of the TSCA, would range from approximately $400 000 per chemical for an oncogenic study where pre-chronic testing information is already available, to $900 000 per chemical for a combined study for which there is no available information from pre-chronic tests. It is seen, then, that a toxicity testing programme designed to test 15 to 30 *new chemicals* per year, using test rules and standards similar to those currently required by law in the United States, would cost about $900 000 per chemical (because pre-chronic testing and certain oncogenic tests would be required); this means a total annual cost ranging from a minimum of $6 million to a staggering peak of $24 million.

From the above, one can readily appreciate the drive and motivation in several countries to develop and standardise short-term tests by which the health effects and risks for humans can be satisfactorily determined. Furthermore, such expense dramatically points to the need for international co-operation in testing and information exchange (see below).

Closing considerations

Responsible societies are realising more and more that potentially hazardous chemicals must be laboratory tested so that suspected harmful products can be identified, and either controlled or eliminated, before being introduced into commerce; only in this manner can the risk of exposing unsuspecting populations be minimised. It is one thing to introduce chemicals of a known hazardous entity, e.g. to workers who are trained and equipped to handle them, but quite another to expose whole populations (small or large) who are unwary and uneducated with regard to the risks involved and without means to counter them.

Although decisions to ban or to restrict the use of any particular chemical or class of chemical substances for one or more uses must take into account the benefits and the liabilities that will result, in many cases such decisions will have to be considered against the background of the complex socio-economic picture as a whole. Among factors of particular relevance are the size and scope of the industry affected, its impact on national income and gross national product, its impact on employment, and its effect on other national problems such as balance of trade, preservation of natural resources and impact on related industries.

Legislation for the control of toxic chemicals will only function effectively if there is a certain amount of co-operation between industry, government and the public. An important aspect of this co-operation is information exchange, which is not only needed at the local level where industry impinges on government and the public, and government impinges on both, but also at the international level. A world-wide co-operative effort will eventually be essential since information generated can be exchanged between countries thus reducing the high costs of testing. Such an international effort, in order to succeed, would have to be based on the use of standardised procedures for testing and the exchange of information. Obviously, it would be a boon to developing countries unable to afford the high costs of implementing toxic substances legislation but undergoing industrialisation with heavy reliance on international trade in chemicals. Hopes for an international co-operative effort currently lie with the International Programme on Chemical Safety (IPCS), which is a joint programme of the United Nations Environment Programme (UNEP), the International Labour Organisation (ILO) and the WHO.

FAIRCHILD, E. J.II

"Chemicals: How many are there?" Maugh, T. J. *Science* (Washington, DC), 13 Jan. 1978, 199/4 (162).

Registry of toxic effects of chemical substances. Lewis, R. J., Sr.; Tatken, R. L. (eds.). DHEW (NIOSH) publication No. 79-100 (National Institute for Occupational Safety and Health, 4676 Columbia Parkway, Cincinnati) (Jan. 1979), 1 362 p.

Toxic Substances Control Act (TSCA) Chemical Substance Inventory (US Environmental Protection Agency, Office of Toxic Substances, Washington, DC) (May 1979), 4 vols. Supplement I (Oct. 1979), 90+50+15 p.

CIS 80-1321 *Principles and methods for evaluating the toxicity of chemicals: PartI.* Environmental health criteria 6 (Geneva, World Health Organisation, 1978), 272 p. Illus. 546 ref.

Principles for evaluating chemicals in the environment. A report prepared for the EPA by the Environmental Studies Board and the Committee on Toxicology of the National Academy of Sciences–National Research Council (Washington, DC, 1975).

"Evaluation of the carcinogenicity of chemicals: A review of the Monograph Program of the International Agency for Research on Cancer". Tomatis, L.; Agthe, C.; Bartsch, H.; Huff, J.; Montesano, R.; Saracci, R.; Walker, E.; Wilbourn, J. *Cancer Research* (Baltimore), Apr. 1978, 38/4 (877-885). 50 ref.

"Proposed health effects test standard for Toxic Substances Control Act test rules". Environmental Protection Agency. *Federal Register* (Washington, DC), May 1979, 44/91 (27-34).

Regulations or other provisions applicable to the introduction of new substances. Law and practice report. ILO-MELE/1977/IV (Geneva, International Labour Office, 1977), 10+12 p.

Chlorinated nitroparaffins

Uses. They are used most frequently as solvents and intermediates in the chemical and synthetic rubber industries, but also as pesticides, especially fumigants, fungicides, and mosquito ovicides.

HAZARDS

When exposed to heat or flame, chlorinated nitroparaffins are easily decomposed into dangerous fumes such as phosgene and nitrogen oxides. These highly toxic fumes may result in the irritation of mucous membranes and pulmonary damage with varying degrees of acute oedema and death. However, no information about accidental exposures of humans has been reported.

The toxicity of some of the substances shown in the above list has not been clearly elucidated. In general, however, experimental exposures to high concentrations

	m.w.	sp.gr.	b.p. (°C)	v.p. mmHg (25 °C)	v.d.	f.p. (°C)	solubility	Exposure limits TWA OSHA	TLV ACGIH
chloronitromethane ($ClCH_2NO_2$)	95.5	1.46	122-123				water		
dichloronitromethane (Cl_2CHNO_2)	130		107						
1-chloro-1-nitroethane ($CH_3CHClNO_2$)	109.6	1.26	129	11.9	3.8	56.1 (oc)	ethyl alcohol		
1,1-dichloro-1-nitro-ethane ($CH_3CCl_2NO_2$)	144	1.42	124	16.0	5.0	75.5 (oc)	water	10 ppm 60 mg/m³ ceiling	10 ppm 60 mg/m³ ceiling
1-chloro-1-nitropropane ($CH_3CH_2CHClNO_2$)	123.5	1.21	139.5	5.8	4.3	62.2 (oc)	water, ethyl ether, ethyl alcohol	20 ppm 100 mg/m³	2 ppm 10 mg/m³
1-chloro-2-nitropropane ($CH_3CH(NO_2)CH_2Cl$)	123.5	1.20	172-173						
2-chloro-1-nitropropane ($CH_3CHClCH_2NO_2$)	123.5	1.23	172				water, ethyl ether, ethyl alcohol		
2-chloro-2-nitropropane ($CH_3CCl(NO_2)CH_3$)	123.5	1.20	133.6	8.5	4.3	57.2 (oc)	water, ethyl ether, ethyl alcohol		

produced damage not only to the respiratory system but also possibly to the liver, kidneys, and cardiovascular system. In addition, ingestion has caused congestion of the gastrointestinal tract and skin irritation resulted from contact with large amounts. No significant reports about chronic local or systemic cases of poisoning in industrial workers have been recorded.

SAFETY AND HEALTH MEASURES

The most important methods of technical control to prevent hazards are general or local exhaust ventilation. General ventilation entails dilution of contaminated air with fresh air by fans or blowers in the working environment. Local exhaust ventilation usually means removal of the contaminants from the environments where harmful fumes are generated. The working room concentration should be maintained below the exposure limits by using both of these methods. If it is not possible to reduce excessive amounts of contaminants in the air by only the ventilation methods, enclosure of a process or segregation of personnel is recommended.

Preplacement examinations including chest X-rays of personnel should be performed in order to discover individuals with pre-existing pulmonary disease or those highly susceptible to skin irritation. Employment should be restricted to persons free of a history of pulmonary disease. Workers repeatedly exposed to low concentrations of chlorinated nitroparaffins should have periodic medical examinations, including lung function test, although the effect of repeated small doses has not been determined. At high concentrations, respiratory protective equipment should be worn.

Treatment. In case of eye or skin contact, washing with large amounts of water is necessary. For the relief of pulmonary distress after inhalation exposure, there is no specific treatment, only symptomatic, which also applies to chronic effects, including dermatoses.

TOYAMA, T.
KONDO, H.

Documentation of the Threshold Limit Values Fourth Edition 1980 (American Conference of Governmental Industrial Hygienists Inc., 6500 Glenway, Cincinnati, Ohio) (1980), 486 p. Ref.

Chlorine and inorganic compounds

Chlorine (Cl_2)

m.w. 70.91
sp.gr. 1.56 (liquid, −34.6 °C)
m.p. −101 °C
b.p. −34.6 °C
v.d. 2.5
v.p. 4 800 mmHg (638.4·10³ Pa) at 20 °C

slightly soluble in water; soluble in alkalis

a gas, with a pungent odour and a greenish-yellow colour.

TWA OSHA 1 ppm 3 mg/m³ ceil
NIOSH 0.5 ppm/15 min ceil
STEL ACGIH 3 ppm 9 mg/m³
IDLH 25 ppm
MAC USSR 1 mg/m³

Occurrence. Chlorine is widely distributed in nature although not as a free gas. It forms 0.15% of the Earth's crust, mostly as sodium chloride in sea water and natural deposits, as carnallite ($MgCl_2.KCl.6H_2O$) and as sylvite (KCl).

Production. On a commercial scale chlorine is produced mainly by the electrolysis of sodium chloride (common salt) in the form of a brine solution. Chlorine, hydrogen and sodium hydroxide (caustic soda) are produced according to the following equation:

$$\underset{\text{salt}}{2\,NaCl} + \underset{\text{water}}{2\,H_2O} = \underset{\text{chlorine}}{Cl_2} + \underset{\text{sodium hydroxide}}{2\,NaOH} + \underset{\text{hydrogen}}{H_2}$$

Two types of cell are traditionally used for the electrolytic process—the diaphragm cell and the mercury cell (figure 1). In the former, a diaphragm composed of asbestos material separates the anode from the cathode solutions to isolate the chlorine which forms at the cathode, since mixing would lead to the formation of sodium hypochlorite. In the mercury cell sodium is liberated at the mercury cathode at the bottom of the cell. The sodium-mercury amalgam is led away to a separate part of the cell where it acts as an anode and causes sodium to return to solution as sodium hydroxide while hydrogen is given off from a cathode above. The advantage of the mercury cell over the diaphragm cell lies in the purity and strength of the caustic soda solution

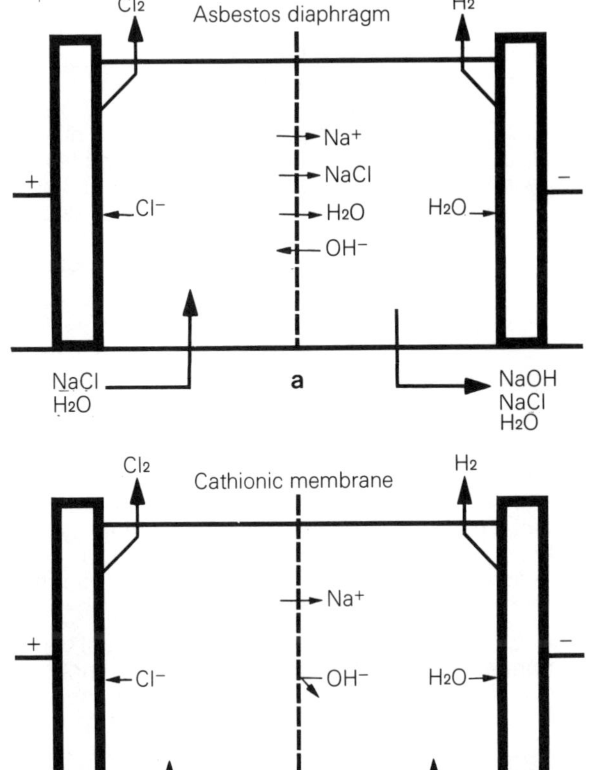

Anode (+) $Cl^- = \frac{1}{2}Cl_2 + e^-$

Cathode (−) $H_2O + e^- = \frac{1}{2}H_2 + OH^-$

Cell $NaCl + H_2O = \frac{1}{2}Cl_2 + \frac{1}{2}H_2 + NaOH$

Cl_2+ NaCl dil. H_2+ NaOH conc.

Cathionic membrane

Anode Cathode   Anode Cathode

c

NaCl conc. H_2O

Figure 1. Diagrammatic presentation of a diaphragm cell *(a)*, a cathionic membrane cell *(b)*, and a section of a bipolar membrane cell *(c)*. From the Report to the National Congress (1980): *Petrochemicals–technology, working environments, prevention and pathology* (Petrochimica–Tecnologia, ambienti di lavoro, prevenzione e patologia). *(By courtesy of the Italian Society of Occupational Health and Industrial Hygiene.)*

which can be obtained in a concentration as high as 50%. The mercury cell method is expensive, however, and presents a health hazard; exposure to mercury is usually a greater health problem during production than exposure to the chlorine.

Uses. The first important use of chlorine was in the production of bleaching powder for the paper and textile industries. However, with the development of techniques for transporting liquid chlorine over long distances, gaseous chlorine came to be increasingly used as a replacement for bleaching powder in bleaching processes, and the pulp and paper industry and textile industry are still heavy consumers of chlorine. It is also used for bleaching cellulose for artificial fibres.

Chlorine is a powerful disinfectant and is used in water supply and treatment for the chlorination of drinking water, swimming pools, etc. and as a disinfectant for refuse.

The chemical industry now consumes large quantities of chlorine for the manufacture of inorganic and organic chlorine compounds including metallic chlorides (aluminium chloride, iron chloride, etc.), solvents (carbon tetrachloride, trichloroethane, chloroform, etc.), refrigerants (chloromethane, dichloromethane, etc.), pesticides (aldrin, dieldrin, chlordane, DDT, etc.), polymers (polyvinyl chloride, synthetic rubbers, etc.).

HAZARDS

Molecular chlorine is notable as a danger to health and a certain limited risk of explosion is associated with its manufacture and use, because of its high oxidising power. Chlorine ion, contrary to fluorine ion, is practically harmless.

Health hazard. Chlorine is a mucous membrane and respiratory system irritant. It will react with body moisture to form acids and, at high concentrations, it will act as an asphyxiant by causing cramps in the muscles of the larynx and swelling of the mucous membranes. The presence of chlorine in the atmosphere is, to some extent, detectable by its characteristic odour and irritant properties. Consequently, in the event of leakage, workers usually have sufficient warning to escape and avoid excessive exposure. Table 1 indicates the hazard of chlorine at different atmospheric concentrations.

Table 1. Hazard of chlorine at different atmospheric concentrations

Chlorine concentration in air		Degree of hazard
ppm	mg/m³	
0.2-0.5	0.3-1.5	No noxious long-term effect
0.5	1.5	Slight odour (tentative limit proposed by author)
1-3	3-6	Definite odour, irritation of eyes and nose
6	15	Irritation of the throat
30	90	Intense coughing fits
40-60	120-180	Exposure without effective respirator for 30-60 min or more may cause serious damage
100	300	May cause lethal damage
1 000	3 000	Danger to life even after a few deep inhalations
10 000	30 000	Filter respirator inadequate. Self-contained or supplied-air apparatus needed

Acute exposure. The first symptoms of exposure to chlorine are irritation to the mucous membranes of the eyes, nose and throat, which increases to smarting and burning pain: this irritation spreads to the chest. A reflex cough develops, which may be intense and often associated with pain behind the breastbone; the cough may lead to vomiting. Cellular damage may occur with excretion of fluid in the alveoli. This may prove fatal if adequate treatment is not given immediately (complete rest, oxygen therapy, immediate transfer to hospital). Vomit frequently contains blood due to lesions of the mucous membrane caused by the gas. Other common symptoms include headaches, general indisposition, anxiety and feeling of suffocation.

Most cases of fatal acute chlorine poisoning have been due to massive leakages or explosions of gas containers from which rapid escape has not been possible. Massive inhalation produces pulmonary oedema, fall of blood pressure, and in a few minutes cardiac arrest.

Chronic exposure. Chlorine concentrations considerably higher than current threshold values may occur without being immediately noticeable; men rapidly lose their ability to detect the odour of chlorine in small concentrations. It has been observed that prolonged exposure to atmospheric chlorine concentrations of 5 ppm results in disease of the bronchi and a predisposition to tuberculosis, while lung studies have indicated that concentrations of 0.8-1 ppm cause permanent, although moderate, reduction in pulmonary function. Acne is not unusual in persons exposed for long periods of time to low concentrations of chlorine and is commonly known as "chloracne". Tooth enamel damage may also occur.

Fire and explosion. Chlorine forms flammable and explosive mixtures with hydrogen, and its reactions with some organic compounds, such as hydrocarbons, alcohols and ethers can become explosively exothermic unless they are properly controlled. See OXIDISING SUBSTANCES.

SAFETY AND HEALTH MEASURES

First of all it is essential to keep the concentration of chlorine in air below the exposure limits. Secondly, for all operations involving chlorine, precautions must be provided to cope with the risk of injury to health; for some processes there must be additional precautions against the dangers of fire and explosion.

Health protection. Leaks in chlorine lines, equipment or containers should be given prompt attention; emergency repair kits should be available for use by experienced maintenance men. Before normal repair work is undertaken the tanks, pipelines and equipment affected should be purged with dry air and isolated from any other source of chlorine. A major dangerous escape of chlorine is generally due to some disturbance in a reaction associated with the process or to some irregularity in the operation of the plant. Such incidents should be prevented or reduced to a minimum by the arrangements made in the planning of the process and in the design of the plant.

Because of the widespread dangers that could result from a serious escape of chlorine, special care should be devoted to the training of the process workers who have to operate the controls; this training should include instruction in the action to be taken in an emergency. The co-operation of the chlorine suppliers should be sought for the education of workers in the proper handling and operation of chlorine cylinders, containers and tank wagons which are often used directly on the process. Some of the men engaged in the process should be given instruction and training in first aid and in the use of the oxygen resuscitation apparatus which should always be available. The workers should be made familiar with the use of respiratory protective equipment. Gas masks and filters should be regularly examined by a competent person; they should be kept in a suitable and accessible place, and each worker should know where to find his own specially fitted mask.

Medical examination of workers exposed to chlorine should take place at least once a year, if possible more often, especially at the outset of the employment. People who are susceptible to infections of the air passages, who have had serious lung diseases, or who suffer from heart disorders, should not be employed in industries involving exposure to chlorine.

Environmental protection. Free chlorine destroys vegetation and, as it may occur in concentrations causing such damage under unfavourable climatic conditions, its release into the surrounding atmosphere should be prohibited. If it is not possible to utilise the liberated chlorine for the production of hydrochloric acid or the like, every precaution must be taken to bind the chlorine, for instance by means of a lime scrubber. Special technical safety measures with automatic warning systems should be installed, in the factories and in the surroundings, wherever there is a risk that appreciable quantities of chlorine may escape to the surrounding atmosphere.

Technical preventive measures. From the point of view of environmental pollution, particular attention should be paid to cylinders or other vessels used for the transport of chlorine or its compounds, to measures for the control of possible hazards, and to steps to be taken in case of emergency. As regards cylinders:

(*a*) protective covers for valves should always be secured, even when the cylinders are empty;

(*b*) fusible plugs should not be interfered with nor exposed to heat sources;

(*c*) cylinders should always be kept away from flammable materials in a clean, well ventilated place and in fireproof surroundings, protected from strong sunlight and rain;

(*d*) badly fitting connections should not be forced and the correct tool should always be used for opening and closing valves: they should never be hammered;

(*e*) when being emptied, the key should be opened fully: it should not be used at any time to regulate the flow of chlorine;

(*f*) implements and other equipment used for emptying gas cylinders should be clean and free of grease, dust or grit. See GAS CYLINDERS; and GAS AND AIR, COMPRESSED.

Measures for the control of hazards include the following:

(*a*) all equipment should be sufficiently strong, both mechanically and chemically, to withstand the working temperatures and pressures;

(*b*) new or freshly repaired equipment should be thoroughly cleaned before being put into service: neither hydrocarbons nor alcohol should be used for this purpose;

(*c*) all equipment and tubing that have been used for chlorine should be thoroughly purged before any repairs are carried out on them;

(*d*) checks should be made at least daily for any possible leakage of chlorine;

(*e*) cylinders should be stored on a cement floor sloping towards a drain capable of collecting all the liquid in the cylinders; this drain should be about 2 x 2 x 5 m deep. Under no circumstances should water be allowed to run on to the chlorine in this drain.

The following steps should be taken in emergencies:

(a) the most important consideration is to ensure that all emergency equipment that may be required is kept readily available. Provision should be made for first aid in case of need;

(b) adequate protective clothing should be donned before entering an emergency zone, or other appropriate safety measures should be established;

(c) in the event of an escape of chlorine, repairs or investigations should only be performed by qualified persons having the correct equipment;

(d) any leaking container should be placed in such a position that only gas and not liquid can escape;

(e) approach from the windward side so that escaping gas is carried downwind to a lower level;

(f) in no circumstances should water or other liquid be directed towards leaking containers. Water may be used on containers that are not leaking in order to keep them cool in the case of a fire;

(g) if a container commences to leak during transport, it should be carried on to its destination;

(h) personal protective equipment should be kept in a prominent position.

Fire and explosion prevention. Since hydrogen and chlorine can form explosive mixtures, care must be taken to prevent accidental mixing by keeping the two gases separate. When they are deliberately made to react to form hydrogen chloride, the reaction must take place under suitably controlled conditions. The temperature of highly exothermic chlorinations must be carefully regulated to prevent the reaction from getting out of control since the temperature and pressure would then rise with explosive violence.

First aid. First aid measures to be taken are as follows:

(a) evacuate the contaminated zone;

(b) call a doctor;

(c) never attempt to neutralise chlorine with other chemicals;

(d) splashes of liquid chlorine or chlorine water will attack and destroy clothing and if such clothing is in contact with the skin, it will lead to irritation and burning. The affected clothing should be removed and the skin should be washed copiously with soap and water;

(e) *in case of inhalation:* if the victim is conscious, transport him to a quiet place and lay him down with the upper part of the body elevated, loosen his clothing, particularly a tight collar or belt and cover him with a blanket;

(f) if the victim is *unconscious but breathing,* in addition to the foregoing, let him inhale oxygen at low pressure until the arrival of the doctor;

(g) if the victim is *not breathing,* quickly stretch him out on the ground, on a blanket if available, loosen his collar and belt, and start artificial respiration without delay, with administration of oxygen, and continue until the arrival of the doctor.

Oxides

In all, there are five oxides of chlorine. They are dichlorine monoxide (Cl_2O), chlorine monoxide (ClO), chlorine dioxide (ClO_2), chlorine hexoxide (Cl_2O_6) and chlorine heptoxide (Cl_2O_7); they have mainly the same effect on the human organism and require the same safety measures as chlorine. The one most used in industry is chlorine dioxide.

Chlorine dioxide (ClO_2)

m.w. 67.5
sp.gr. 3.09 (11 °C) (liquid)
m.p. −59.5 °C
b.p. explodes
v.d. 2.4

explosive, green gas or red liquid with a characteristic pungent odour.

TWA OSHA 0.1 ppm 0.3 mg/m³
STEL ACGIH 0.3 ppm 0.9 mg/m³
IDLH 10 ppm

Production. This gas is prepared from sodium chlorate, sulphuric acid and methyl alcohol or from sodium chlorate and sulphur dioxide. Smaller amounts may be made at the point where it is to be used by the oxidation of sodium chlorite with chlorine, hydrochloric acid or hypochlorite.

Uses. It is employed as a bleaching agent for wood pulp, textiles, fats, oils and flour. It is used in the treatment of municipal water supplies and bathing pools, in the control of industrial micro-organisms and in a variety of chemical processes.

HAZARDS AND THEIR PREVENTION

Chlorine dioxide is a toxic, corrosive and explosive substance. Its behaviour as a respiratory and eye irritant is similar to that of chlorine but more severe in degree. Acute exposures by inhalation cause bronchitis and pulmonary oedema, the symptoms observed in affected workers being coughing, wheezing, respiratory distress, nasal discharge, eye and throat irritation. Chlorine dioxide is a powerful oxidising agent and a powerful supporter of combustion. In addition, the gas is explosive when its concentration in air exceeds 10%. Concentrations above this figure may ignite at 130 °C. Oxidisable organic dusts can lower the decomposition temperature.

Precautions must take account of respiratory dangers, corrosive properties, risks of fire and explosion.

The safeguards that should be provided for the dangers associated with the inhalation of the gas are similar to those required for chlorine. Safety showers and eyewashing facilities should be readily available to counter its corrosive action in contact with skin or eyes.

To avoid dangerously high concentrations of the gas, it is swept by air or nitrogen out of the reaction vessel in which it is prepared in concentrations below 10% and it is absorbed in chilled water, the flow of which is adjusted to produce a solution containing 6-10 g/l. Oxidisable dust is stripped from the air or nitrogen streams by means of suitable filters.

Chlorates

These are the salts of chloric acid ($HClO_3$). They are strong supporters of combustion and their main hazard is associated with this property. The potassium and sodium salts are typical of the group and are those most commonly used in industry.

Potassium chlorate ($KClO_3$)

m.w. 122.55
sp.gr. 2.32
m.p. 356 °C
b.p. 400 °C (decomposes)

transparent, colourless crystals or white powder.

Sodium chlorate ($NaClO_3$)

m.w. 104.65
sp.gr. 2.49
m.p. 248-261 °C

colourless, odourless crystals.

Production. These chlorates are produced by electrolysing hot, concentrated solutions of the corresponding chlorides. The potassium salt is commonly prepared by the interaction of solutions of potassium chloride and sodium chlorate.

Uses. Potassium chlorate is used in the manufacture of explosives, matches, pyrotechnics, percussion caps, pharmaceutical products, dyes; it is used in bleaching, in the printing of textile fabrics, in pulp and paper manufacture and as a disinfectant.

Sodium chlorate is also used in the manufacture of matches, explosives, flares and pyrotechnics and pharmaceutical products; and as a herbicide and defoliant, a textile mordant, in tanning and leather finishing and in the manufacture of other chemicals.

HAZARDS AND THEIR PREVENTION

Chlorates are powerful oxidising agents and the main dangers are those of fire and explosion. They are not themselves explosive but they form flammable or explosive mixtures with organic matter, sulphur, sulphides, powdered metals and ammonium compounds. Cloth, leather, wood and paper are extremely flammable when impregnated by these chlorates. Special consideration must be given, therefore, to personal protective equipment; working clothes worn by chlorate workers should be washable and should not be worn away from the immediate work area; they should be washed and rinsed each day. They should include overalls, hand and arm protection such as plastic gloves, a washable head covering and foot protection such as rubber boots. Wood should be avoided in the structure and floors of the buildings and in the plant or equipment where chlorates are handled, and a copious water supply should be available for extinguishing fire. Chlorates are harmful if absorbed by ingestion or by inhalation of the dust. This can provoke sore throat, coughing, methaemoglobinaemia with bluish skin, dizziness and faintness, and anaemia. In case of large absorption of sodium chlorate an increased sodium content in the serum will be seen. The dust should be reduced to a minimum; dust respirators may be necessary.

Perchlorates. Perchlorates may enter the body either by inhalation as dust or by ingestion. They are irritant to skin, eyes and mucous membranes. They cause haemolitic anaemia with methaemoglobinaemia, Heinz bodies in the red cells, liver and kidney injuries. Precautions must be the same as with chlorates.

Perchlorates

See OXIDISING SUBSTANCES.

Sulphur compounds

Sulphur chloride (S_2Cl_2)

SULPHUR MONOCHLORIDE

m.w. 135
sp.gr. 1.68
m.p. −80 °C
b.p. 135.6 °C
v.d. 4.66
v.p. 10 mmHg ($1.33 \cdot 10^3$ Pa) at 27.5 °C
f.p. 118.3 °C
i.t. 233.9 °C

yellow, oily, fuming liquid with a penetrating odour.

TWA OSHA 1 ppm 6 mg/m³
STEL ACGIH 3 ppm 18 mg/m³

Sulphur chloride is made by passing chlorine over molten sulphur and purifying by distillation. It is used in the manufacture of other chemicals and in the vulcanising of rubber.

Sulphur chloride is a flammable liquid which gives rise to a moderate fire hazard associated with the evolution of the dangerous decomposition products, sulphur dioxide and hydrogen chloride. It is a fuming, corrosive liquid which is dangerous to the eyes; the vapour is irritating to the lungs and mucous membrane. In contact with the skin, the liquid can cause chemical burns. It should be handled under the maximum degree of enclosure and workers should be provided with personal protective equipment including eye protective equipment and respiratory protective equipment.

Sulphur dichloride (SCl_2)

m.w. 103
sp.gr. 1.62 (15 °C)
m.p. −78 °C
b.p. 59 °C (decomposes)
v.d. 3.55

reddish-brown, fuming liquid with pungent chlorine odour.

Sulphur dichloride is made by passing excess chlorine into sulphur chloride at 6-10 °C followed by carbon dioxide to drive off the excess of chlorine. It is used in the manufacture of insecticides and other chemicals and for vulcanising in the rubber industry.

The toxic and corrosive properties of this substance are similar to those of sulphur chloride and similar precautions must be adopted.

Thionyl chloride ($SOCl_2$)

SULPHUR OXYCHLORIDE

m.w. 119
sp.gr. 1.65
m.p. −105 °C
b.p. 78.8 °C (746 mmHg)
v.p. 100 mmHg ($13.3 \cdot 10^3$ Pa) at 21.4 °C

clear, yellow liquid with a sharp odour.

Thionyl chloride is manufactured by adding a sulphur trioxide to sulphur chloride at 75-80 °C and passing a stream of chlorine into the mixture to reconvert the separated sulphur into sulphur chloride. It is used as a reagent and as a catalyst in organic synthesis and in the manufacture of acid chlorides and anhydrides.

Thionyl chloride possesses toxic and corrosive properties; the vapour is a respiratory irritant. For these health hazards, preventive measures should be adopted that are similar to those recommended for sulphur chloride.

Sulphuryl chloride (SO_2Cl_2)

SULPHURIC OXYCHLORIDE; CHLOROSULPHURIC ACID

m.w. 135
sp.gr. 1.67
m.p. −54.1 °C
b.p. 69.1 °C
v.d. 4.65
v.p. 100 mmHg ($13.3 \cdot 10^3$ Pa) at 17.8 °C

colourless liquid with a pungent odour.

Sulphuryl chloride is formed by the direct combination of sulphur dioxide and chlorine in the presence of a catalyst which may be charcoal, camphor or acetic anhydride. It is also obtained by heating chlorosulphonic acid, with mercuric sulphate, antimony or tin as catalyst. It is used in the manufacture of pharmaceuticals and dyestuffs and generally in organic synthesis as a chlorinating, dehydrating or acylating agent.

Sulphuryl chloride is a corrosive liquid which, in contact with the body, can cause burns; the vapour is a respiratory irritant. The precautions are similar to those recommended for sulphur chloride.

Chlorsulphonic acid ($ClSO_2OH$)

SULPHURIC CHLOROHYDRIN

m.w.	116.5
sp.gr.	1.76
m.p.	−80 °C
b.p.	158 °C
v.d.	4.02
v.p.	1 mmHg (0.13·10³ Pa) at 32 °C

colourless to light yellow liquid with a pungent odour.

Chlorsulphonic acid is made by treating sulphur trioxide or fuming sulphuric acid with hydrochloric acid. It is used in the manufacture of detergents, pharmaceuticals, dyes, pesticides and ion-exchange resins.

Chlorsulphonic acid is a very corrosive liquid which can cause severe burns in contact with the skin; the vapour can cause conjunctivitis and is a respiratory irritant. The precautions that should be adopted are similar to those recommended for sulphur chloride.

CONRADI FERNANDEZ, L.
INCLAN CUESTA, M.

CIS 79-475 *Chlorine* (Le chlore). Dandres, R. Edition INRS nº 231 (Institut national de recherche et de sécurité, 30 rue Olivier-Noyer, Paris) (3rd ed., May 1978), 297 p. Illus. (In French)

CIS 77-771 *Chlorine–Codes of practice for chemicals with major hazards* (Chemical Industries Association Ltd., Alembic House, 93 Albert Embankment, London) (Jan. 1975), 23 p. Illus.

CIS 77-432 *Criteria for a recommended standard–Occupational exposure to chlorine.* DHEW (NIOSH) publication No. 76-170 (National Institute for Occupational Safety and Health, 4676 Columbia Parkway, Cincinnati) (May 1976), 155 p. 159 ref.

CIS 80-727 "Chlorine dioxide" (Bioxyde de chlore). Association interprofessionnelle des centres médicaux et sociaux de la région parisienne. *Cahiers de médecine interprofessionnelle* (Paris), 3rd quarter 1979, 75 (detachable insert), 1 p. 4 ref. (In French)

CIS 74-1677 "Toxic properties of chlorine trifluoride". Dost, F. N.; Reed, D. J.; Smith, V. N.; Wang, C. H. *Toxicology and Applied Pharmacology* (New York), Mar. 1974, 27/3 (527-536). 25 ref.

Storage and use of sodium chlorate. Guidance note CS 3 (Health and Safety Executive, Baynards House, 1 Chepstow Place, London) (1980).

"Hazardous chemical reactions–4. Perchlorates" (Réactions chimiques dangereuses–4. Perchlorates). Leleu, J. *Cahiers de notes documentaires–Sécurité et hygiène du travail* (Paris), 4th quarter 1972, 69, Note No. 816-69-72 (405-408). (In French)

CIS 74-1690 "Hazardous chemical reactions–12. Chlorites" (Réactions chimiques dangereuses–12. Chlorites). Leleu, J. *Cahiers de notes documentaires–Sécurité et hygiène du travail* (Paris), 1st quarter 1974, 74, Note No. 886-74-74 (135-138). (In French)

Chlorobenzene and derivatives

Chlorobenzene (C_6H_5Cl)

PHENYL CHLORIDE; MONOCHLOROBENZENE

m.w.	112.56
sp.gr.	1.11
m.p.	−45 °C
b.p.	131-132 °C
v.d.	3.88
v.p.	10 mmHg (1.33·10³ Pa) at 22.2 °C
	11.8 mmHg (1.57·10³ Pa) at 25 °C
f.p.	28 °C
e.l.	1.3-7.1%

insoluble in water; very soluble in ethyl alcohol, ethyl ether, benzene, chloroform

colourless liquid.

TWA OSHA	75 ppm 350 mg/m³
IDLH	2 400 ppm
MAC USSR	50 mg/m³

1,2-Dichlorobenzene ($C_6H_4Cl_2$)

o-DICHLOROBENZENE

m.w.	147.01
sp.gr.	1.31
m.p.	−17 °C
b.p.	180.5 °C
v.d.	5.05
v.p.	1.56 mmHg (0.21·10³ Pa) at 25 °C
f.p.	66.1 °C
e.l.	2.2-9.2%

practically insoluble in water; miscible with ethyl alcohol, ethyl ether, benzene

colourless liquid.

TWA OSHA	50 ppm 300 mg/m³ ceil
IDLH	1 700 ppm
MAC USSR	20 mg/m³ skin

1,3-Dichlorobenzene ($C_6H_4Cl_2$)

m-DICHLOROBENZENE

m.w.	147.01
sp.gr.	1.29
m.p.	−24.76 °C
b.p.	173 °C
v.d.	5.08
v.p.	1 mmHg (0.13·10³ Pa) at 12.1 °C

practically insoluble in water; soluble in ethyl alcohol, ethyl ether

colourless liquid.

MAC USSR	20 mg/m³ skin

1,4-Dichlorobenzene ($C_6H_4Cl_2$)

p-DICHLOROBENZENE

m.w.	147.01
sp.gr.	1.46
m.p.	53.5 °C (α-modification)
	54 °C (β-modification)
b.p.	174.1 °C
v.d.	5.08
v.p.	10 mmHg (1.33·10³ Pa) at 54.8 °C
f.p.	65.6 °C

practically insoluble in water; slightly soluble in ethyl alcohol; soluble in ethyl ether, benzene, chloroform, carbon disulphide

colourless or white crystals.

TWA OSHA	75 ppm 450 mg/m³
STEL ACGIH	110 ppm 675 mg/m³
IDLH	1 000 ppm
MAC USSR	20 mg/m³ skin

1,2,3-Trichlorobenzene ($C_6H_3Cl_3$)

vic-TRICHLOROBENZENE

m.w.	181.46
sp.gr.	1.69
m.p.	52.6 °C
b.p.	221 °C
v.d.	6.26
v.p.	1 mmHg (0.13·10³ Pa) at 40.0 °C
f.p.	113 °C

insoluble in water; sparingly soluble in ethyl alcohol; freely soluble in benzene, carbon disulphide

platelets.

MAC USSR 10 mg/m³

1,3,5-Trichlorobenzene ($C_6H_3Cl_3$)

sym-TRICHLOROBENZENE

m.w. 181.46
m.p. 63.4 °C
b.p. 208.4 °C
v.d. 6.26
v.p. 10 mmHg (1.33·10³ Pa) at 78.0 °C
f.p. 107 °C

insoluble in water; sparingly soluble in ethyl alcohol; freely soluble in ethyl ether, benzene, petroleum ether, carbon disulphide, glacial acetic acid

white crystals.

MAC USSR 10 mg/m³

1,2,4-Trichlorobenzene ($C_6H_3Cl_3$)

unsym-TRICHLOROBENZENE

m.w. 181.46
sp.gr. 1.46
m.p. 17 °C
b.p. 213 °C
v.d. 6.26
v.p. 1 mmHg (0.13·10³ Pa) at 38.4 °C
f.p. 110 °C

insoluble in water; sparingly soluble in ethyl alcohol; miscible with ethyl ether, benzene, petroleum ether, carbon disulphide

colourless liquid.

TLV ACGIH 5 ppm 40 mg/m³
MAC USSR 10 mg/m³

Hexachlorobenzene (C_6Cl_6)

PERCHLOROBENZENE

m.w. 284.80
sp.gr. 2.04
m.p. 231 °C
b.p. 323-326 °C, sublimable
v.d. 9.8
v.p. 1 mmHg (0.13·10³ Pa) at 114.4 °C
f.p. 242.2 °C

monoclinic prisms.

MAC USSR 0.9 mg/m³ skin

1-Chloro-3-nitrobenzene ($C_6H_4ClNO_2$)

m-CHLORONITROBENZENE

m.w. 157.56
sp.gr. 1.53
m.p. 46 °C
b.p. 236 °C

insoluble in water; sparingly soluble in cold ethyl alcohol; freely soluble in hot ethyl alcohol, chloroform, ethyl ether, carbon disulphide, glacial acetic acid

pale-yellow crystals.

MAC USSR 1 mg/m³ skin

1-Bromo-4-chlorobenzene (C_6H_4BrCl)

m.w. 191.5
m.p. 67.4 °C
b.p. 196.3 °C
f.p. 64.5 °C

white crystals.

Production. Monochlorobenzene and dichlorobenzenes are produced by the chlorination of benzene in the presence of an iron catalyst. Monochlorobenzene is first isolated from dichlorobenzenes by distillation, while separation of three isomers of dichlorobenzene is achieved by distillation and crystallisation. Chlorination of nitrobenzene gives *m*-chloronitrobenzene. Heating of tetrachloroquinone with phosphorus trichloride and phosphorus pentachloride yields hexachlorobenzene.

Uses. Monochlorobenzene and dichlorobenzenes are widely used as solvent and chemical intermediates. Dichlorobenzenes, especially the *p*-isomer, are employed also as fumigants, insecticides and disinfectants. A mixture of trichlorobenzene isomers is applied to combat termites. Hexachlorobenzene is used as a fungicide and for organic synthesis.

HAZARDS

Chlorobenzene. In humans eye and nasal irritation may occur at 200 ppm, followed by the depression of the central nervous system at higher concentrations. Repeated exposure of animals at 1 000 ppm for 7 h/day, 5 days/week for 32 times resulted in injury to the lungs, liver and kidneys. Blood dyscrasia as observed with benzene exposure will not be caused by chlorobenzene.

Dichlorobenzenes. Two isomers, *o*- and *p*-, are primarily irritants of eyes, nose and skin. At higher concentrations, effects on the central nervous system such as severe headache may occur. Animal experiments revealed toxicity to liver and kidneys, and the effects were more prominent with the *o*-isomer. Hepatic damage due to the *p*-isomer was also reported among those who used *p*-isomer as a mothproofing agent at home. Solid particles of *p*-dichlorobenzene cause pain when dropped in the eye and produces a burning sensation when kept in contact with skin. The toxicity of *m*-isomer is not well elucidated.

Trichlorobenzene. 1,2,4-Trichlorobenzene causes transient erythema by single painting on mouse skin, and acanthosis and hyperkeratosis (but not chloracne) by repeated application to rabbit ears. Findings in rats, rabbits and monkeys after vapour exposure at 25, 50 and 100 ppm, 7 h/day for 6 months were essentially negative.

Hexachlorobenzene. An epidemic took place after consumption as food of seed wheat treated with hexachlorobenzene as fungicide. The symptoms in adults, diagnosed as *porphyria cutanea tarda,* began with bullous lesions which progressed to ulceration healing with pigmented scars. In children the initial lesions resembled comedones and milia. In severe cases the peculiar appearance was called "monkey disease" because of the resemblance. Massive discharges of porphyrins were detected in urine and faeces of the patients. The high cumulative tendency of this compound coupled with chemical porphyria was demonstrated in animal experiments. In Syrian golden hamsters hexachlorobenzene produced hepatocellular carcinoma and other malignant tumours when the animals were given for life a diet containing hexachlorobenzene at 50, 100 and 200 ppm. Animal studies for placental transfer were positive, while teratogenicity and dominant lethal studies were negative.

1-Chloro-3-nitrobenzene. Analogous to nitrobenzene, this compound is capable of producing methaemoglobinaemia when absorbed.

SAFETY AND HEALTH MEASURES

General occupational hygiene measures, especially those for organic solvents, are applicable to cases of chlorinated benzenes to minimise vapour inhalation and skin contact. Urinary excretion of 2,5-dichlorophenol can serve as an index of exposure in work environments, while blood level of hexachlorobenzene is taken as an indicator of exposure in occupational and general environments.

Treatment. Treatment is generally symptomatic. EDTA is considered effective to accelerate the excretion of persistent hexachlorobenzene.

IKEDA, M.

CIS 77-177 "Effect of chlorinated benzenes on the metabolism of foreign organic compounds". Carson, G. P.; Tardiff, R. G.

Toxicology and Applied Pharmacology (New York), May 1976, 36/2 (383-394). 22 ref.

"Evaluation of chlorobenzene exposure based on 4-chlorocatechol determination in urine" (Ocena narazenia na chlorobenzen na podstawie oznaczania 4-chlocatecholy w moczu). Dutkiewicz, T.; Pacholuk, B. *Medycyna Pracy* (Lodz), 1980, 31/4 (289-295). Illus. 11 ref. (In Polish)

CIS 79-499 "Aplastic anemia following exposure to paradichlorobenzene and naphthalene". Harden, R. A.; Baetjer, A. M. *Journal of Occupational Medicine* (Chicago), Dec. 1978, 20/12 (820-822). 16 ref.

CIS 78-763 "Chronic inhalation exposure of rats, rabbits, and monkeys to 1,2,4-trichlorobenzene". Coate, W. B.; Schoenfisch, W. H.; Lewis, T. R.; Busey, W. M. *Archives of Environmental Health* (Washington, DC), Nov.-Dec. 1977, 32/6 (249-255). 19 ref.

Hexachlorobenzene:

CIS 77-2018 *Hexachlorobenzene.* Data sheets on pesticides No. 26 (Geneva, World Health Organisation; Rome, United Nations Food and Agriculture Organisation, 1977), 9 p. 8 ref.

"Carcinogenic activity of hexachlorobenzene in hamsters". Cabral, J. R. P.; Shubik, P.; Mollner, T.; Raitano, F. *Nature* (London), 1977, 269/5628 (510-511). Illus. 8 ref.

CIS 76-1909 "The effects of pentachloronitrobenzene, hexachlorobenzene, and related compounds on fetal development". Courtney, K. D.; Copeland, M. F.; Robbins, A. *Toxicology and Applied Pharmacology* (New York), Feb. 1976, 35/2 (239-256). 19 ref.

"Human blood samples as indicators of occupational exposure to persistent chlorinated hydrocarbons". Lund, E. G.; Bjørseth, A. *Science of Total Environment* (Amsterdam, 1977), 8/3 (241-246). Illus. 14 ref.

Chloroethylamines

The chloroethylamines form a group of biologically highly active substances with a very dangerous local as well as general action.

In their chemical structure as well as in their effects, the *bis*-2-chloroethylamines $R-N(CH_2-CH_2Cl)_2$ are similar to *bis*-2-chloroethylsulphide (sulphur mustard, mustard gas) $S(CH_2-CH_2Cl)_2$ and they have been called, therefore, nitrogen mustards.

They are mostly high-boiling, oily liquids with very limited water solubility. Colourless when pure and freshly prepared, they turn yellow on storage and decompose on standing to form solid polymeric compounds. They form salts with halogen and other acids (hydrochlorides are usually prepared) which are stable, colourless, crystalline solids, less active than the free bases.

Several thousands of compounds which may be considered as nitrogen mustards have been synthesised and their biological activity tested. The most important members of the group are:

1. N-(2-chloroethyl)diethylamine

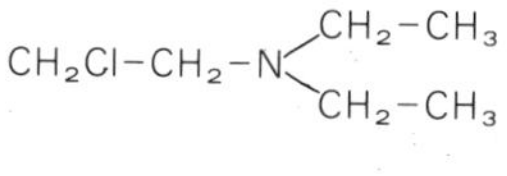

   $CH_2Cl-CH_2-N(CH_2-CH_3)_2$

2. N-(2-chloroethyl)dibenzylamine
   DIBENAMINE; SYMPATHOLYTIN

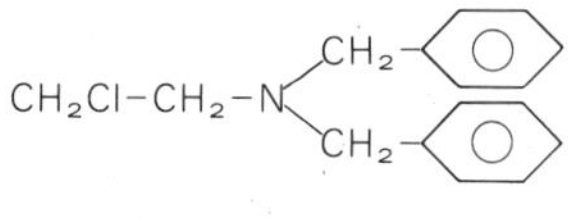

   $CH_2Cl-CH_2-N(CH_2-C_6H_5)_2$

3. N-(2-chloroethyl)-N-(1-methyl-2-phenoxyethyl)-benzylamine
   PHENOXYBENZAMINE; DIBENYLINE

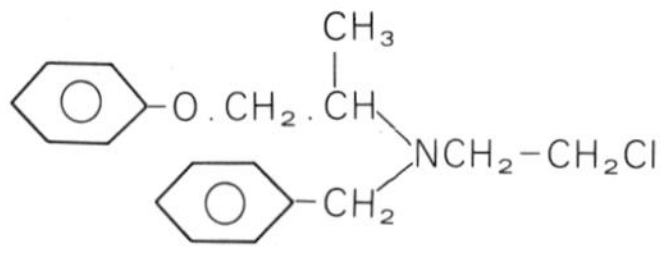

   $C_6H_5-O.CH_2.CH(CH_3)-N(CH_2-C_6H_5)-CH_2-CH_2Cl$

4. 1,6-*bis*-((2-chloroethyl)amino)-1,6-dideoxy-D-mannitol
   MANNITOL MUSTARD; MANNOMUSTINE

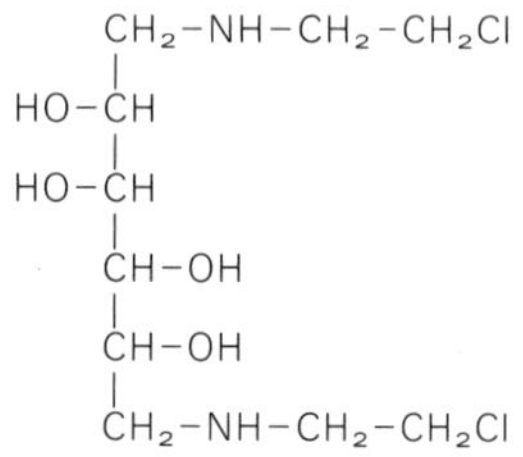

   $CH_2-NH-CH_2-CH_2Cl$
   $HO-CH$
   $HO-CH$
   $CH-OH$
   $CH-OH$
   $CH_2-NH-CH_2-CH_2Cl$

5. *bis*-(2-chloroethyl)amine

   $H-N(CH_2.CH_2Cl)_2$

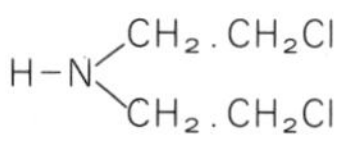

6. N,N-*bis*-(2-chloroethyl)methylamine
   NITROGEN MUSTARD; MBA; HN-2; CHLORMETHINE

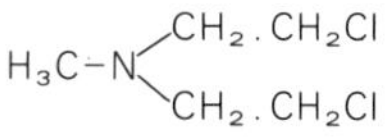

   $H_3C-N(CH_2.CH_2Cl)_2$

7. 4-(4-(*bis*(2-chloroethyl)amino)phenyl)butyric acid
   CHLORAMBUCIL; LEUKERAN

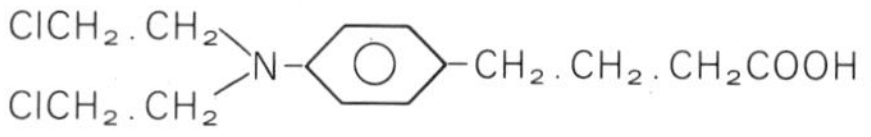

   $(ClCH_2.CH_2)_2N-C_6H_4-CH_2.CH_2.CH_2COOH$

8. 3-(4-(*bis*-(2-chloroethyl)amino)phenyl)alanin
   SARCOLYSIN; *d*-stereoisomer: MEDPHALAN; *l*-s.: MELPHALAN; *dl*-s.: MERPHALAN

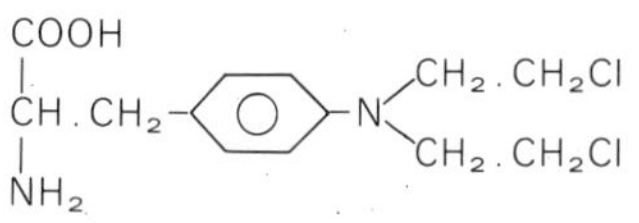

   $HOOC-CH(NH_2).CH_2-C_6H_4-N(CH_2.CH_2Cl)_2$

9. 5-(*bis*-(2-chloroethyl)amino)uracil
   URACIL MUSTARD; URAMUSTINE

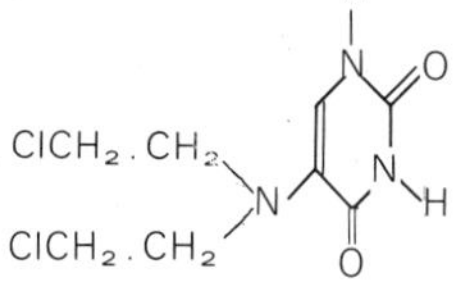

   $(ClCH_2.CH_2)_2N-$ (uracil ring, N–H, C=O)

10. 2-(*bis*-(2-chloroethyl)amino)tetrahydro-2H-1,3,2-oxazaphosphorine-2-oxide
    CYCLOPHOSPHAMIDE; ENDOXANE

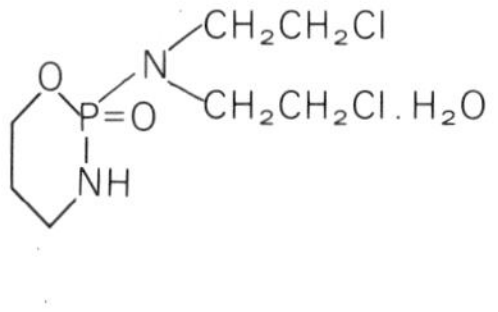

    $N(CH_2CH_2Cl)_2$ on P=O of ring (O, NH) $.H_2O$

11. *bis*-((4-(*bis*-(2-chloroethyl)amino)phenyl)-acetate)oestradiol
OESTRADIOL MUSTARD

12. *tris*-(2-chloroethyl)amine
HN-3

$ClCH_2 . CH_2 . N(CH_2 . CH_2Cl)_2$

The structural configuration necessary for the activity is the presence of a halogen (most frequently chlorine, although bromine and iodine compounds are also active) in the β-position. The β-halopropyl derivatives are active too. The amines with a single haloethyl group (1,2,3) are less potent. For the full characteristic action, the presence of at least two haloethyl groups is necessary. The third substituent on the tertiary amine may be a wide variety of aliphatic (6), aromatic (7,8) or heterocyclic (9,10) groups without abolishing activity, even though influencing the potency and, to a certain extent, the site of action. The secondary amine compounds (5) are less effective than the tertiary ones, still possessing, however, a rather strong mustard activity whereas the quaternary amines do not.

Production. Chloroethylamines are prepared by the action of thionyl chloride on the appropriate ethanolamine:

$$R-N(CH_2-CH_2OH)_2 + 2\,SOCl_2 \rightarrow R-N(CH_2-CH_2Cl)_2.HCl + 2SO_2 + HCl$$

The free base is liberated by treating the hydrochloride with sodium hydroxide or carbonate.

Uses. The vesicant properties of this type of compound were first noted in 1935, and intensive study of their biological effects was stimulated by interest in their potential military application at the advent of the Second World War. They have never been used in this way, fortunately, but these studies led to their early application in the treatment of neoplastic disease in man and in the cancer research. After the primitive nitrogen mustard (6), which must be injected intravenously because the compound is very highly reactive, further compounds have been introduced in the therapy of malignant diseases (7,8,9,10,11). The main advantage of the newer members of this group is that they are well absorbed when administered orally. Their margin of safety is greater too. Tertiary amines with a single 2-chloroethyl group, as exemplified by dibenamine (2) and phenoxybenzamine (3), have been used as alpha adrenergic blocking agents. Other uses, such as for fungicides and industrial intermediates, are also very limited.

HAZARDS

The intrinsic effect of chloroethylamines and other nitrogen mustards consists in damaging living matter, particularly dividing cells (cytotoxic effect). The manifestations of this feature include local irritation, and teratogenic, mutagenic, and carcinogenic activity. These effects resemble, on the whole, those of ionising radiation and the term "radiomimetic effect" is frequently used. However, the reason for the biological activity of nitrogen mustards is very different from that of ionising radiation, i.e. the chloroethylamines (or their metabolites) react readily with important groups of proteins and nucleic acids. They transfer alkyl groups to essential cell constituents by combining with amino, sulphydryl, carboxyl, and phosphate groups, belonging, consequently, in the class of biological alkylating agents, together with sulphur mustard, haloalkylethers, derivatives of ethylenimine, the epoxides, the beta-lactones, and other substances. The alkylation of DNA and, more specifically, guanine, results in the formation of single strand breaks or interstrand crosslinks and, ultimately, in an inhibition of DNA synthesis. A marked decrease of the RNA chain length in both nucleoplasm and nucleolus of some cells under the influence of nitrogen mustard has been proved. Lesions can be produced which affect base pairing during DNA replication and cause permanent genetic alterations or tumour initiation.

In contact with the skin and eye, the nitrogen mustards are severely vesicant and damaging, their action being similar to that of the sulphur variety of mustard gas but with more immediate effect and deeper injury. Accidental burns of human skin leading to deep and slow-to-heal ulcers, and accidental corneal damage, including the injury of iris and lens, have been reported. Industrially, irritation of the eyes, nose, skin and respiratory system occurs following exposure to vapours. Delayed pulmonary oedema following accidental vapour exposure has been observed.

Nitrogen mustards are absorbed through the skin, respiratory and gastrointestinal tract in sufficient amounts to produce systemic intoxication. It has been observed that workers briefly exposed to nitrogen mustard in concentrations estimated to be between 10 and 100 ppm became severely ill with nausea, vomiting and dilated pupils. In normal human volunteers single oral doses of some milligrams resulted in nausea, vomiting and diarrhoea. Larger doses produce neurotoxic effects such as prolonged tremor, uncoordinated movements, ataxia, derangement of positional reflexes, and convulsions.

The basic action of nitrogen mustards may explain their preferential toxicity for rapidly multiplying cells and the predominant damage to organs which have a high rate of cell turnover. Bone marrow depression may be observed following their application, with leucopoenia and bleeding as an early sign. Alopecia may occur. They can cause impairment of menses and spermatogenesis; mutagenic activity has been proved. They can exert cytotoxic effects on the embryo, the resultant teratogenicity depending largely on the stage of fetal development and the extent to which the compound or its active metabolites pass the placental barrier. Lymph node damage, involution of lymphatic tissues and the thymus, and immunosuppression may occur; intercurrent infections may come out. The development of tumours following exposure to some of these substances could be due to a direct chemical action or to impairment of the immune system or to both. Most of the very dangerous effects have been proved in animals only, or, to a lesser extent, have been observed during therapeutic use. There have been very few reports of severe and lasting general injury from industrial exposure.

SAFETY AND HEALTH MEASURES

In view of the pronounced local damaging action of the chloroethylamines, any possible contact with the skin and eyes and any exposure to high vapour concentrations should be avoided.

Although the danger of a severe general effect seems to be relatively small in industry, the possibility of bone marrow damage, reduction in fertility, teratogenic and mutagenic activity and carcinogenicity must be considered.

Only mature, experienced and responsible men should be employed in work with chloroethylamines. High standards of maintenance, good housekeeping and cleanliness should be observed in the work premises. Personal hygiene should be emphasised and personal protective equipment worn. Even if not expressly mentioned in actual official carcinogen lists, analogous safety and health measures and regulations can be recommended for chloroethylamines as for the carcinogens listed.

MARHOLD, J.

"The toxicity and pharmacological action of the nitrogen mustards and certain related compounds". Anslow, W. P.; Karnovsky, D. A.; Jager, B. V.; Smith, H. W. *Journal of Pharmacology and Experimental Therapeutics* (Baltimore), 1947, 91 (224-235). 10 ref.

"Biological effects of alkylating agents". Van Duuren, B. L. *Annals of the New York Academy of Sciences* (New York), 1969, 163 (589-1029).

"Mustards" (117-241). Ref. *IARC monographs on the evaluation of carcinogenic risk of chemicals to man.* Vol. 9 (Lyons, International Association for Research on Cancer, 1975).

Chloroform

Chloroform ($CHCl_3$)

TRICHLOROMETHANE

m.w.	119.4
sp.gr.	1.48
m.p.	−63.5 °C
b.p.	61.2 °C
v.d.	4.1
v.p.	200 mmHg (26.6·10³ Pa) at 25.9 °C

miscible with most organic solvents; very slightly soluble in water (0.822 g/100 g)

a colourless non-flammable liquid with a characteristic odour.

TWA OSHA	50 ppm 240 mg/m³
NIOSH	2 ppm/1 h ceil
TLV ACGIH	10 ppm 50 mg/m³: industrial substance suspected of carcinogenic potential for man
IDLH	1 000 ppm

In sunlight it decomposes slowly to phosgene, chlorine and hydrogen chloride. Phosgene is also formed by the action of strong oxidants on chloroform.

Production. Chloroform is obtained by the chlorination of methane and by the hydrochlorination and chlorination of methanol. Ethanol and acetone were also used to synthesise chloroform; after chlorination to chloral hydrate and 1,1,1-trichloroacetone the trichloro derivatives were decomposed to chloroform.

Uses. Chloroform was one of the first inhalation anaesthetics. It is now used mainly for synthesis of fluorocarbon 22 ($CHClF_2$, refrigerant, propellant and material for the production of tetrafluoroethylene and polytetrafluoroethylene) and other chemicals; it was also used as a solvent, extractant, insecticidal fumigant, preservative and sweetener of pharmaceutical preparations. Its use in anaesthesiology and pharmacy has been discontinued.

Occurrence. Chloroform is found in drinking water and in the ambient atmosphere. Most of the chloroform in tap water is formed through chlorination of organic substances. Its level is usually below 1 mg/l water and the biological significance of these low-level concentrations can hardly be evaluated at present. Chloroform in air may result at least partly from photochemical degradation of trichloroethylene.

HAZARDS

Chloroform is one of the most dangerous volatile chlorinated hydrocarbons. It may be harmful by inhalation, ingestion and skin contact. It may cause narcosis, respiratory paralysis, cardiac arrest or delayed death due to liver and kidney damage. It may be misused by sniffers. Liquid chloroform may cause defatting of the skin and chemical burns. It is teratogenic and carcinogenic for mice and rats.

Experimental toxicity. The oral LD_{50} for dogs and rats is about 1 g/kg; 14-day-old rats are twice as susceptible as adult rats. Mice are more susceptible than rats. Liver damage is the cause of death. Histopathological changes in the liver and kidney were observed in rats, guinea-pigs and dogs exposed for 6 months (7 h/day, 5 days/week) to 25 ppm in air. Fatty infiltration, granular centrilobular degeneration with necrotic areas in the liver and changes in serum enzyme activities as well as swelling of tubular epithelium, proteinuria, glucosuria and decreased phenolsulphonephtalein excretion were reported. Chloroform was found to be fetotoxic in inhalation experiments in rats and caused malignant kidney tumours in male rats, tumours of the thyroid gland in female rats and hepatomas and hepatocellular carcinomas in mice.

Biotransformation. The fate of chloroform in the mammalian body is not fully known; chloroform is partly exhaled unchanged, partly as carbon dioxide; three metabolites were detected in urine and one of them was identified as urea. Experiments with ^{14}C-chloroform revealed ^{14}C covalent binding in the liver and kidney tissues. The first product of chloroform biotransformation in the body may be trichloromethanol which can be converted to phosgene by dehydrochlorination. $CHCl_2$ and CCl_3 radicals as well as formaldehyde were also proposed as possible intermediates. The toxic compound is supposed to alkylate the sulphydryl group containing derivatives and macromolecules of tissues.

Acute exposure. Persons exposed to chloroform vapour in air may develop different symptoms depending on the concentration and duration of exposure: headache, drowsiness, feeling of drunkness, lassitude, dizziness, nausea, excitation, unconsciousness, respiratory depression, coma and death in narcosis. Death may occur due to respiratory paralysis or as a result of cardiac arrest. Chloroform sensitises the myocardium to catecholamines. A concentration of 10 000-15 000 ppm of chloroform in inhaled air causes anaesthesia and 15 000-18 000 ppm may be lethal. Narcotic concentrations in blood are 30-50 mg/100 ml; levels of 50-70 mg/100 ml blood are lethal. After transient recovery from heavy exposure failure of liver functions and kidney damage may cause death. Effects on heart muscle have been described. Inhalation of very high concentrations may cause sudden arrest of heart's action (shock death).

Long-term exposure. Workers exposed to low concentrations in air and persons with developed dependence on chloroform may suffer from neurological and gastrointestinal symptoms resembling chronic alcoholism; hepatomegaly, toxic hepatitis and fatty liver degeneration were also observed.

Late effects. Because of its recognised carcinogenicity for mice and rats, chloroform is suspected of being carcinogenic for humans. For that very reason its exposure limit in workplace air was recently lowered. For the same reason, all cases of tumour occurrence where connection to chloroform exposure is probable or possible should be published in relevant periodicals or reported to international authorities such as the ILO, WHO or the International Agency for Research on Cancer.

Analysis. Determination of chloroform in air and in drinking water can be performed by gas chromatographic methods. Methods formerly used included spectrophotometry (Fujiwara reaction: heating with pyridine and alkali and photometry at 530 nm) and methods based on the determination of thermally liberated chloride ions. No suitable exposure test in biological material has been developed so far for a quantitative monitoring of chloroform hazard.

Table 1. Dose-effect relationship in man

Atmospheric chloroform concentration		Response and symptoms
ppm	mg/m³	
20-200	100-1 000	Distinct characteristic sweet odour; olfactory fatigue follows
20-70	100-350	Slight subjective complaints may occur
20-360	100-1 800	Fatigue, lassitude, feeling of drunkness, somnolence, irritability, psychotic unbalance, loss of appetite, digestive disturbances, hepatomegaly and toxic hepatitis in some after 1-4 years of exposure
1 000	5 000	Dizziness, nausea, etc.
4 000	20 000	Disorientation which may result in falling
15 000	75 000	Rapid loss of consciousness

SAFETY AND HEALTH MEASURES

Chloroform should not be used as an industrial solvent but should be replaced by non-carcinogenic solvents (carbon tetrachloride, trichloroethylene, tetrachloroethylene, 1,2-dichloroethane and 1,1,2,2-tetrachloroethane are also believed to be carcinogenic). The use of chloroform should also be avoided in laboratories, where it has been used extensively for extractions.

All shipping and storage containers for chloroform should bear a label giving a warning that the substance is toxic and carcinogenic and thus highly dangerous. Containers must be closed when not in use and the work with chloroform must be carried out with adequate ventilation; operations involving the use of chloroform should be isolated. In areas where chloroform is used, warnings should be posted at entrances as well as at workplaces. All workers should be given instruction in the safe handling of chloroform and be well acquainted with the hazards and the suspected carcinogenic potential of chloroform. Breathing of vapours and skin contact should be avoided. Protective clothing, gloves, eye protective devices, shields and respirators should be provided and used when indicated.

The concentration of chloroform in the workplace air must be measured by methods sensitive enough to estimate the level within the hygienic standard.

Medical prevention. Only persons in a good state of health should be employed for jobs involving the risk of chloroform exposure. Persons with liver dysfunction, alcoholics and persons with drug addiction in their medical history, and pregnant women and nursing mothers should not come in contact with chloroform.

Emergency treatment. The victim should be removed immediately to fresh air and artificial respiration should be started if breathing stops. If chloroform has entered the eyes, they should be irrigated with plenty of water (for about 15 min). In the case of chloroform ingestion, vomiting should be induced. Clothing or shoes should be removed when wet with chloroform. A physician should be called without delay. Adrenalin should not be applied to persons suffering from acute chloroform poisoning. Pending admission to hospital the patient should be held under suitable thermal and rest conditions. Victims should be hospitalised for some days to ensure detection of possible delayed alterations in liver and kidney function.

BARDODĚJ, Z.

CIS 77-1021 "The toxicity of chloroform as determined by single and repeated exposure of laboratory animals". Torkelson, T. R.; Rowe, V. K. *American Industrial Hygiene Association Journal* (Akron, Ohio), Dec. 1976, 37/12 (697-705). 14 ref.

Chloroform. Current Intelligence Bulletin No. 9 (National Institute for Occupational Safety and Health, Rockville, Maryland) (15 Mar. 1976), 9 p. 15 ref.

"Hepatic functions in workers exposed by inhalation to chloroform vapours" (Funzionalità epatica in operai esposti all'inalazione di vapori di chloroformio). Gambini, G.; Farina, G. *Medicina del Lavoro* (Milan), Nov.-Dec. 1973, 64/11-12 (432-436). 13 ref. (In Italian)

"Chloroform" (401-427). 121 ref. *IARC monographs on the evaluation of the carcinogenic risk of chemicals in humans.* Vol. 20: *Some halogenated hydrocarbons* (Lyons, International Agency for Research on Cancer, 1979).

CIS 75-1353 *Criteria for a recommended standard–Occupational exposure to chloroform.* DHEW (NIOSH) publication No. 75-114 (Washington, DC, US Government Printing Office, 1974), 120 p. 124 ref.

Chloromethane

Chloromethane (CH_3Cl)

METHYL CHLORIDE; MONOCHLOROMETHANE

m.w.	50.5
sp.gr.	0.92
m.p.	−97 °C
b.p.	−23.7 °C
v.d.	1.74
v.p.	760 mmHg (101.32·10³ Pa) at −24 °C
f.p.	0 °C
e.l.	8.1-17.4%
i.t.	632 °C

soluble in water, ethyl alcohol; very soluble in ethyl ether

a colourless and practically odourless gas; at temperatures above 400 °C or in strong ultraviolet light in the presence of air and moisture it decomposes with the emission of hydrogen chloride, carbon dioxide, carbon monoxide and phosgene.

TWA OSHA	100 ppm
OSHA	200 ppm ceil 300 ppm/5 min peak
TLV ACGIH	50 ppm 105 mg/m³
STEL ACGIH	100 ppm 205 mg/m³
IDLH	10 000 ppm
MAC USSR	5 mg/m³

Production. Chloromethane is manufactured by the interaction of methyl alcohol and hydrochloric acid at atmospheric pressure and temperatures of 100 to 150 °C, in the presence of catalysts.

Uses. It is used as a refrigerating agent, though not as widely as in earlier times [because of the prohibition in some countries, for safety reasons, of its use in refrigerators]; as a methylating agent, as a solvent in the synthetic rubber industry; in petroleum refining and as an extractant for oils, fats and resins.

HAZARDS

Since chloromethane is an odourless gas and therefore gives no warning, it is possible for considerable exposure to occur without those concerned becoming aware of it. There is also the risk of individual susceptibility to even mild exposure. In animals it has shown markedly differing effects in different species, with greater susceptibility in animals with more highly developed central nervous systems, and it has been suggested that human subjects may show an even greater degree of individual susceptibility. A hazard pertaining to mild chronic exposure is the possibility that the "drunkenness", dizziness and slow recovery from slight intoxication may cause failure to recognise the cause and that leaks may go unsuspected. This could result in further prolonged exposure and accidents. The majority of fatal cases recorded have been caused by leakage from domestic refrigerators or defects in refrigeration plants. It is also a dangerous fire and explosion hazard.

Toxicity. Severe intoxication is characterised by a latent period of several hours before the onset of symptoms such as headache, fatigue, nausea, vomiting and abdominal pain. Dizziness and drowsiness may have existed for some time before the more acute attack was precipitated by a sudden accident. Chronic intoxication from milder exposure has been less frequently reported, possibly because the symptoms may disappear rapidly with cessation of exposure. The complaints during mild cases include dizziness, difficulty in walking, headache, nausea and vomiting. [The most frequent objective symptoms are a staggering gait, nystagmus, speech disorders, arterial hypotension, and reduced and disturbed cerebral electrical activity.] In this respect, chloromethane is regarded as more toxic than dichloromethane, since mild prolonged intoxication is liable to cause permanent injury of the heart muscle and the central nervous system, with a change of personality, depression, irritability and occasionally visual and auditory hallucinations [increased albumen content in the cerebrospinal fluid, with possible extrapyramidal and pyramidal lesions, suggesting a diagnosis of meningo-encephalitis.] In fatal cases, autopsy has shown congestion of lungs, liver and kidneys.

SAFETY AND HEALTH MEASURES

All employees working with apparatus in which chloromethane is used should be instructed about its potential toxic effects, and employers warned that it should be filled only in gas cylinders, moved carefully and safeguarded by holding devices. These cylinders should never be stored near open flames since its ignition temperature is 632 °C. All cylinders and equipment should be frequently checked for leaks and provided with exhaust ventilation in order to keep the air concentration below the exposure limits. The air intake of the exhaust apparatus should be as low as possible, since chloromethane, with its high specific gravity, has a tendency to accumulate at a low level. If enclosures have to be entered which are heavily contaminated with the gas, air or oxygen masks should be provided, together with a safety belt and lifeline under the supervision of a helper on the outside. Estimations of the air concentrations should be made periodically [and if any risk of leakage exists continuous sampling systems with alarm devices are recommended]. Persons suffering from disorders of the central nervous system, anaemia, alcoholism or liver or kidney disease should be prohibited from working with chloromethane. [Chloromethane is present in the expired air in significant quantities, therefore breath analysis has been suggested as an exposure test; more recently the presence of S-methylcysteine in the urine of workers exposed to chloromethane has been reported.]

Treatment. Workers complaining of any of the symptoms of mild intoxication should be removed from exposure for several weeks. In cases of acute intoxication, the patient should be removed into fresh air [and be given oxygen and artificial respiration in the event of respiratory difficulty]. If convulsions are present, sedatives may be given, excluding those such as chloroform or chloral hydrate which are known to have a potential toxic effect on the liver. [Epinephrine is also contraindicated.]

BROWNING, E.

General:

CIS 1245-1971 *Properties and essential information for safe handling and use of methyl chloride.* Chemical Safety Data Sheet SD-40 (revised) (Manufacturing Chemists' Association, 1825 Connecticut Avenue, NW, Washington, DC) (1970), 18 p.

Chronic intoxication:

CIS 74-1390 "Chronic methyl chloride intoxication in six industrial workers". Scharnweber, H. C.; Spears, G. N.; Cowles, S. R. *Journal of Occupational Medicine* (Chicago), Feb. 1974, 16/2 (112-113). 5 ref.

Medical surveillance:

CIS 76-1769 "Principles of preventive medical examinations–Hazards of monochloromethane (methyl chloride)" (Berufsgenossenschaftliche Grundsätze für arbeitsmedizinische Vorsorgeuntersuchungen–Gefährdung durch Monochlormethan (Methylchlorid)). Hauptverband der gewerblichen Berufsgenossenschaften. *Arbeitsmedizin–Sozialmedizin–Präventivmedizin* (Stuttgart), Dec. 1975, 10/12 (248-250). 9 ref. (In German)

"State of health in apprentices–future workers with methylchloride" (Zdravotní stav učňu-budoucích pracovníku s methylchloridem). Hassmanova, V.; Hassman, P.; Borovskà, D.; Fiedlerovà, D.; Nyplovà, H.; Malà, H. *Pracovni Lékarstvi* (Prague), 1978, 30/9 (340-346). 22 ref. (In Czech)

"Detection and identification of S-Methylcysteine in urine of workers exposed to methyl chloride". Van Doorn, R.; Borm, P. J. A.; Leijdekkers, Ch. M.; Henderson, P. Th.; Reuvers, J.; Van Bergen, T. J. *International Archives of Occupational and Environmental Health* (West Berlin, Heidelberg), June 1980, 46/2 (99-109). Illus. 20 ref.

Chloronaphthalenes

The specific gravity of commercial chloronaphthalenes ranges from 1.5-1.7, the melting point from 85-130 °C, the boiling point from 288-371 °C, and the chlorine content from 43-70%. They are soluble in many solvents and oils. The physical states vary from a mobile liquid to a waxy solid and their properties differ according to the degree of chlorination. For example the greater the chlorine content, the more solid the material. The TWA OSHA for trichloronaphthalene (skin) is 5 mg/m³ and for pentachloronaphthalene (skin) 0.5 mg/m³. The MAC USSR for trichloronaphthalene and mixtures of tetra- and pentachloronaphthalenes is 1 mg/m³, and that for the higher chlorinated naphthalenes, 0.5 mg/m³.

Production. The chloronaphthalenes are made by the chlorination of boiling naphthalene ($C_{10}H_8$), the hydrogen of the naphthalene being replaced by chlorine.

Uses. The chloronaphthalenes in industrial use are mixtures of tri-, tetra-, penta- and hexachloronaphthalenes. [For the last 20 years plastics have been substituted for chlorinated naphthalenes; however, these are still used in industry owing to their chemical stability, their dielectric and water repellant properties, their thermoplasticity, their non-flammability and in particular their low cost.] Chloronaphthalenes are used in the manufacture of electric condensers, in the insulation of electric cables and wires, and as additives for extreme pressure lubricants and crayons for foundry use.

HAZARDS

The toxicity of chloronaphthalene increases with a higher degree of chlorination, but this is accompanied by a decrease in solubility and volatility. [All chloronaphthalenes are capable of being absorbed through the intact skin.]

Workers engaged in manufacturing chloronaphthalenes, or coating wires or condensers with it, and those stripping wires insulated with it have been found to be affected by associated health hazards. Acne-form dermatitis and toxic hepatitis are the primary problems resulting from contact with chloronaphthalenes. The acne-form lesions caused by chlorinated hydrocarbons are more striking than those produced by other acne-forming substances. The acnes caused by chlorinated hydrocarbons such as chloronaphthalenes, chlorophenols and chlorobiphenyls have been called "chloracne".

The fumes of chloronaphthalenes, as well as repeated contact with solid substances, cause skin lesions on the exposed parts and when the substances penetrate the clothing, they also cause skin lesions underneath. Eruptions are usually located on the face, surface of ear lobes, behind the ears, on the neck, shoulders, arms, chest, abdomen and even the scrotum. The eruptions begin as small comedones or as blind cysts and develop into hard cyst-like elevations, some of which may suppurate. The lesions occur in follicles and resemble the comedones and pustules of *acne vulgaris*. However, chloracne differs is some respects from *acne vulgaris*. Whereas in *acne vulgaris* the skin is usually greasy, in chloracne it may be dry. The comedones of chloracne are non-inflammatory types in contrast to those of *acne vulgaris*, which are usually inflammatory lesions. The duration of exposure necessary for the formation of chloracne may vary from several weeks to about one year, depending mostly upon working conditions.

Chloronaphthalenes have a photosensitising action on the skin.

[Pruritus is frequently complained of among other symptoms, the most frequent are eye irritation, headache, fatigue, vertigo and nausea.] The higher chlorinated naphthalenes may cause severe injury to the liver, characterised by acute yellow atrophy or by subacute necrosis. Systemic injury may occur from inhaling vapours or from ingestion.

SAFETY AND HEALTH MEASURES

For the prevention of skin hazards and systemic poisoning, cleanliness of the operation and especially adequate ventilation are of paramount importance. Condenser impregnation and other operations involving melting of chloronaphthalene should be enclosed or provided with effective local exhaust ventilation. Contact of the skin with chloronaphthalenes or with materials impregnated with these chemicals should be avoided as far as possible. Work clothes should be frequently inspected and laundered daily. Barrier creams have been found effective if used properly. Good personal hygiene, including a daily shower, is essential for workers handling chloronaphthalenes. [Pre-employment and regular periodical medical examinations should be carried out.]

NOMURA, S.

CIS 1465-1972 "Chloronaphthalenes–$C_{10}H_{(8-n)}Cl_n$" (Chloronaphtalènes). Morel, C.; Cavigneaux, A. Fiche toxicologique n° 93. *Cahiers de notes documentaires–Sécurité et hygiène du travail* (Paris), 1971, 65, Note No. 771-65-71 (475-478). 17 ref. (In French)

CIS 2641-1972 "Clinical effects of chlorinated naphthalene exposure". Kleinfeld, M.; Messite, J.; Swencicki, R. *Journal of Occupational Medicine* (Chicago), May 1972, 14/5 (377-379). Illus. 6 ref.

"Analytical determination of chlorinated naphthalenes occurring in concentrations of industrial hygienic importance" (Die analytische Bestimmung chlorierter Naphthaline in arbeitshygienisch bedeutsamen Konzentrationsbereichen). Schaffernicht, H. *Zeitschrift für die gesamte Hygiene und ihre Grenzgebiete* (Berlin), 1978, 24/12 (893-895). 3 ref. (In German)

Chloropicrin

Chloropicrin (CCl_3NO_2)

NITROTRICHLOROMETHANE; NITROCHLOROFORM

m.w.	164
sp.gr.	1.66
m.p.	−64 °C
b.p.	112 °C
v.d.	5.7
v.p.	20 mmHg (2.66·10³ Pa) at 20 °C

soluble in ethyl alcohol, benzene, carbon disulphide; slightly soluble in ethyl ether; insoluble in water

a slightly oily, colourless, refractive liquid with a very intense odour.

TWA OSHA	0.1 ppm 0.7 mg/m³
STEL ACGIH	0.3 ppm 3.2 mg/m³
IDLH	4 ppm

Production. Chloropicrin is produced by the action of picric acid on calcium hypochlorite or the nitrification of chlorinated hydrocarbons.

Uses. It has been used as a chemical warfare agent, in dye manufacture and as a pesticide, rodenticide and soil fumigant.

HAZARDS

Chloropicrin vapours are highly irritant to the eyes, causing intense lacrimation, and to the skin and respiratory tract. Chloropicrin causes nausea, vomiting, colic and diarrhoea if it enters the stomach.

Data on the effects of chloropicrin are derived mainly from First World War experience with chemical warfare agents. It is a pulmonary irritant with a toxicity greater than chlorine but less than phosgene. Military data indicate that exposure to 4 ppm for a few seconds is sufficient to render a man unfit for action and 15 ppm for 60 s causes marked bronchial or pulmonary lesions. It causes injury particularly to the small and medium bronchi and oedema is frequently the cause of death. [Because of its reaction with sulphidryl groups, it interferes with oxygen transport and can produce weak and irregular heart beats, recurrent asthmatic attacks and anaemia.]

A concentration of around 1 ppm causes severe lacrimation and provides good warning of exposure; at higher concentrations, skin irritation is evident. Ingestion may occur due to the swallowing of saliva containing dissolved chloropicrin and produce vomiting and diarrhoea.

[Chloropicrin is non-combustible; however, when heated it can detonate and can also be shock detonated above a critical volume.]

SAFETY AND HEALTH MEASURES

Where possible, chloropicrin should be replaced by a less toxic chemical. Where there is a risk of exposure, e.g. in soil fumigation, workers should be adequately protected by wearing suitable chemical eye protection, respiratory protective equipment preferably of the supplied-air type and, in the case of high concentrations, protective clothing to prevent skin exposure. Particular care should be taken during mixing and dilution of chloropicrin; greenhouses in which soil has been treated should be clearly labelled and entry of unprotected persons prevented.

Treatment. Eye contact should be treated by copious irrigation with water or a 0.14% solution of cooking salt in water. Skin contamination should also be removed by the use of large quantities of water. Persons poisoned with chloropicrin should be carried to the fresh air. They should not be allowed to walk since absolute rest is essential. In serious cases, oxygen may be administered, but artificial resuscitation should not be practised.

REED, D. W.

General:

Chloropicrin (Chlorpikrin). Gesundheitsschädliche Arbeitsstoffe. Toxikologisch-arbeitsmedizinische Begründung von MAK-Werten (Weinheim, Verlag Chemie, 1974), 5 p. 19 ref. (In German)

Handling:

CIS 2818-1970 *Chloropicrin–Safe handling in docks* (Chloorpicrine–Veilige behandeling in de haven), P.No99 (Arbeidsinspectie, Directoraat-Generaal van de Arbeid, Voorburg, Netherlands) (1970), 8 p. (In Dutch)

Use:

CIS 2400-1971 *Chloropicrin–Soil disinfection in horticulture* (Chloorpicrine–Grondontsmetting in de tuinbouw), P.No51 (Arbeidsinspectie, Directoraat-Generaal van de Arbeid, Voorburg, Netherlands) (2nd ed. 1971), 10 p. Illus. (In Dutch)

Chloroprene

Chloroprene (CH_2:CCICH:CH_2)

2-CHLORO-1,3-BUTADIENE; β-CHLOROPRENE

m.w.	88.5
sp.gr.	0.96 at 20 °C
b.p.	59.4 °C
fr.p.	−130 °C
v.d.	3.0
v.p.	188 mmHg ($25.0 \cdot 10^3$ Pa) at 20 °C
f.p.	−20 °C (oc)
e.l.	1.9-20%

very slightly soluble in water; soluble in ethyl alcohol, diethyl ether, acetone, benzene and organic solvents

a colourless liquid the vapour of which has a pungent, ethereal odour.

TWA OSHA	25 ppm 90 mg/m³ skin
NIOSH	1 ppm/15 min ceil
TLV ACGIH	10 ppm 45 mg/m³ skin
IDLH	400 ppm
MAC USSR	0.0138 ppm 0.05 mg/m³

Chloroprene polymerises readily under the influence of light and catalysts, and rapidly oxidises to form peroxides and acidic materials. The latter reactions are inhibited by storage at less than −15 °C and/or by the addition of antioxidants to the fresh distillate.

Production. Chloroprene is manufactured by either of two processes. The acetylene process involves the addition of hydrogen chloride to vinyl acetylene in the presence of cuprous chloride to form 2-chloro-1,3-butadiene (chloroprene), which is subsequently distilled off and repurified by distillation. The butadiene process, which is more commonly used today, involves the chlorination of butadiene in gas phase to 1,4-dichloro-2-butene and 3,4-dichloro-1-butene; the latter, which is the precursor of chloroprene is distilled off and the 1,4-dichloro-2-butene is isomerised to the 3,4-isomer by distillation over copper and cuprous chloride. The 3,4-dichloro-1-butene is then dehydrohalogenated with aqueous sodium hydroxide to yield chloroprene, which is removed from the mixture by vacuum distillation.

Uses. Chloroprene is used as a chemical intermediate in the manufacture of artificial rubber.

HAZARDS

Chloroprene is flammable and should not be exposed to flame, sparks, or heated metal. It produces hydrogen chloride on burning.

Experimental work with animals and clinical observations on exposed workers indicate that chloroprene is low to moderate in toxicity on an acute exposure basis. The principal effects at high vapour concentrations include central nervous system depression, skin and eye irritation, and damage to lungs, liver, and kidneys. Its Approximate Lethal Concentration (ALC) in rats after 4 h of exposure is 2 280 ppm. Several investigators have also shown that chloroprene is mutagenic to bacteria (Ames test).

On a chronic exposure basis, chloroprene does appear to have a cumulative toxicity potential. In 4-week studies in rats and hamsters exposures at about 160 ppm and above produced growth retardation, alopecia, respiratory tract irritation, and damage to various body organs including the lungs, liver and kidneys. In the earlier literature differences in toxic effects exhibited by intermediate polymers, oxidised materials, and stabilised chloroprene have led to confusion in the interpretation of certain toxicological observations. However, recent experimental animal studies conducted in the Netherlands have shown that chloroprene is not carcinogenic, mutagenic or teratogenic, and has no adverse effects in male or female reproduction. Unlike earlier work the recent studies involved much effort to maintain the integrity of the chloroprene test samples so that the animals were exposed to pure chloroprene similar to the industrial use situation. In humans subject to chronic exposure dermatitis may result from the action of both chloroprene and related intermediates. Alopecia has also been reported in the earlier literature, but the lost hair regrew after cessation of exposure. Central nervous system effects and irritation of mucous membranes, which are readily reversible upon removal from exposure, have also occurred in humans at high inhaled levels.

SAFETY AND HEALTH MEASURES

Workmen handling chloroprene should be educated about its hazards and potential dangers. The vapour concentration in work areas should be kept below the ACGIH TLV of 10 ppm (8-hour TWA) by limiting vapour emissions or by ventilation. In the event of potential exposure to liquid or high vapour levels, the employee should wear protective garments, chemical safety goggles, an air-supplied respiratory or self-contained oxygen mask.

Dangerous chloroprene fires are best extinguished by shutting off the source of fuel. Carbon dioxide, dry chemicals, and water spray (fog nozzle) may be used as control measures. Storage or transport of chloroprene can only be conducted under special conditions and could involve dilution with appropriate solvents like xylene, heavy inhibition, and cooling ($\leqslant -15$ °C).

Treatment. The first consideration in treatment of toxic effects from exposure to chloroprene is prompt removal from exposure. In case of liquid contact, contaminated clothing should be removed and the skin thoroughly cleaned with plenty of soap and water; eyes should be copiously flushed with water and medical attention solicited. Oxygen should be administered for respiratory distress. Management of dysfunction of liver, kidneys, and other organs is supportive and symptomatic.

WEAVER, N. K.
REINHARDT, C. F.

"Biochemical and hematological evaluation of chloroprene workers". Gooch, J. J.; Hawn, W. F. *Journal of Occupational Medicine* (Chicago), Apr. 1981, 23/4 (268-272). 17 ref.

"Toxicity of β-chloroprene (2-chlorobutadiene-1,3): Acute and subacute toxicity". Clary, J. J.; Feron, V. J.; Reuzel, P. G. J. *Toxicology and Applied Pharmacology* (New York), 1978, 46/2 (375-384). 19 ref.

Unpublished studies conducted at CIVO Laboratories, Zeist, Netherlands, under the sponsorship of the Joint Industry Group on Chloroprene (Bayer AG, British Petroleum Company Limited, Denki Kagaku Kogyo K. K., and E. I. Du Pont de Nemours and Company), 1978-1980 (to be published).

CIS 79-1407 "Chloroprene (2-chloro-1,3-butadiene)–What is the evidence for its carcinogenicity?". Haley, T. J. *Clinical Toxicology* (New York), Sep. 1978, 13/2 (153-170). 149 ref.

"Study of the reproductive function in men exposed to chemicals" (Izučenie mužskoj re roduktivnoj funkcii pri dejstvii nekotoryx veščestv). Sanotsky, I. V.; Davtian, R. M.; Glushchenko, V. I. *Gigiena truda i professional'nye zabolevanija* (Moscow), May 1980, 5 (28-32). 11 ref. (In Russian)

CIS 78-413 *Criteria for a recommended standard–Occupational exposure to chloroprene.* DHEW (NIOSH) publication No. 77-210 (National Institute for Occupational Safety and Health, 4676 Columbia Parkway, Cincinnati) (Aug. 1977), 176 p. Illus. 104 ref.

Chromium, alloys and compounds

Chromium (Cr)

a.w.	52
sp.gr.	7.20
m.p.	1 890 °C
b.p.	2 482 °C

insoluble in water, soluble in dilute sulphuric and hydrochloric acids

a hard, brittle, lustrous, steel-grey metal which is very resistant to corrosion.

Cr metal and insoluble salts:

TWA OSHA	1 mg/m³
IDLH	500 mg/m³

Cr as soluble chromic and chromous salts:

TWA OSHA	0.5 mg/m³
IDLH	250 mg/m³

Chromite ore processing (chromate) as Cr:

TLV ACGIH	0.05 mg/m³ human carcinogen

Occurrence. Elemental chromium is not found free in nature, and the only ore of any importance is the spinel ore, chromite or chrome iron stone, which is ferrous chromite ($FeO\ Cr_2O_3$), widely distributed over the Earth's surface. In addition to chromic oxide, this ore contains variable quantities of iron oxide, aluminium oxide, magnesium oxide and silica, as well as trace quantities of other substances. Only ores or concentrates containing more than 40% Cr_2O_3 are used commercially and countries having the most suitable deposits are the USSR, South Africa, Zimbabwe, Turkey, the Philippines and India. The chief consumers of chromite are the United States, the USSR, the Federal Republic of Germany, Japan, France and the United Kingdom.

Extraction. Chromite may be obtained from both underground and opencast mines. The ore is crushed and, if necessary, concentrated.

Production. Ferrochrome, a chromium iron alloy, is usually produced by reduction of chromite ores with carbon in an electric furnace. The process leads to production of a high carbon ferrochrome whose use was restricted to high carbon steels. With increasing use of argon-oxygen decarburisation and similar processes in steel manufacture, this limitation is less severe. Sulphur must be kept low as it embrittles both Cr metal and nickel base alloys to which Cr is added. The principal source of chromium metal in a purer form is by the reduction of chromic oxide (Cr_2O_3) with aluminium powder by the alumino-thermit process or by electrolysis of various chromium-containing solutions.

Uses. The most important use of pure chromium is for chromium electroplating a wide range of equipment such as automobile parts and electrical equipment. Chromium is used extensively for alloying with iron and nickel to form stainless steels and with nickel, titanium, niobium, cobalt, copper and other metals, to form special purpose alloys.

Chromium compounds

Chromium forms a number of compounds in various oxidation states. Those of +2 (chromous), +3 (chromic) and +6 (chromate) are the most important; the +2 state is basic, the +3 state amphoteric and the +6 state is acidic. Commercial applications mainly concern compounds in the +6 state, with some interest in +3 state chromium compounds.

The chromous state (+2) is very unstable since it is readily oxidised to the chromic state (+3); this therefore limits the use of chromous compounds. The chromic compounds are very stable and form many compounds which have commercial applications, the principal of which are chromic oxide and basic chromium sulphate.

Chromium in the +6 oxidation state has its greatest industrial application as a consequence of its acidic and oxidant properties, and its ability to form strongly coloured and insoluble salts. The most important compounds containing chromium in the +6 state are sodium dichromate, potassium dichromate and chromic acid. Most other chromate compounds are produced industrially using sodium dichromate as the source of +6 chromium.

Production. Sodium chromate and dichromate are the starting materials from which most of the chromium compounds are manufactured. Sodium chromate and dichromate are prepared directly from the chrome ore; the sequence of operations is shown in figure 1. Chrome ore is crushed, dried and ground and soda ash is added and lime or leached calcine may also be added. After thorough mixing the mixture is roasted in a rotary furnace at an optimum temperature of about 1 100 °C; an oxidising atmosphere is essential to convert the chromium to the +6 state. The melt from the furnace is cooled and leached and the sodium chromate or dichromate isolated by conventional processes from the solution.

Chromium^{III} compounds

Chromic oxide (Cr_2O_3)
CHROMIUM SESQUIOXIDE

m.w. 152
sp.gr. 5.21
m.p. 2 435 °C
b.p. 4 000 °C

insoluble in water, acid, alkali and ethyl alcohol

a green powder which is exceptionally stable and hence used as a pigment.

TLV ACGIH (Cr^{III} compounds as Cr) 0.5 mg/m^3
MAC USSR 0.01 mg/m^3

Technically, chromic oxide is made by reducing sodium dichromate either with charcoal or with sulphur. Reduction with sulphur is usually employed when the chromic oxide is to be used as a pigment. For metallurgical purposes carbon reduction is normally employed.

Chromic sulphate ($Cr_2(SO_4)_3$)

m.w. 392.2
sp.gr. 3.01

practically insoluble in water and acids

a peach-coloured solid.

TWA OSHA 1 mg/m^3
TLV ACGIH (Cr^{III} compounds as Cr) 0.5 mg/m^3

The commercial material is normally basic chromic sulphate $[Cr(OH)(H_2O)_5]SO_4$, which is prepared from sodium dichromate by reduction with carbohydrate in the presence of sulphuric acid; the reaction is vigorously exothermic. Alternatively, sulphur dioxide reduction of a solution of sodium dichromate will yield basic chromic sulphate. It is used in the tanning of leather and the material is sold on the basis of Cr_2O_3 content which ranges from 20.5 to 25%.

Chromium^{VI} compounds

Sodium dichromate ($Na_2Cr_2O_7.2H_2O$)

m.w. 298
sp.gr. 2.52
m.p. 256.7 °C (anhydrous)
b.p. decomposes 400 °C

very soluble in water; insoluble in ethyl alcohol

a bright-orange material.

TWA NIOSH 0.025 mg/m^3/10 h day
0.05 mg/m^3/15 min ceil
TLV ACGIH (water soluble Cr^{VI} compounds) 0.05 mg/m^3
MAC USSR 0.01 mg/m^3

Sodium dichromate can be converted into the anydrous salt. It is the starting point for the preparation of chromium compounds.

Potassium dichromate ($K_2Cr_2O_7$)

m.w. 294.2
sp.gr. 2.67
m.p. 398 °C (transition from triclinic to monoclinic at 241.6 °C)

soluble in water; insoluble in ethyl alcohol

an orange-red crystalline material, more stable than sodium dichromate in humid conditions.

TWA NIOSH 0.025 mg/m^3/10 h day
0.05 mg/m^3/15 min ceil

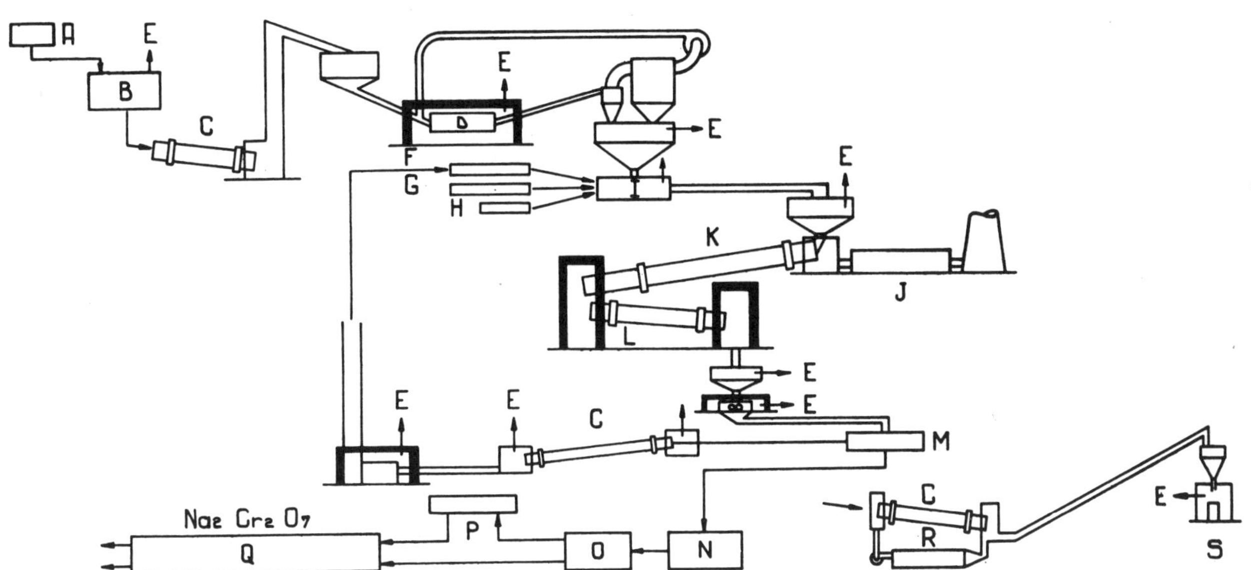

Figure 1. Production of sodium dichromate. A. Ore. B. Crusher. C. Dryer. D. Ball mill. E. Exhaust. F. Residue. G. Soda ash. H. Lime. I. Weighers and mixers. J. Gas cleaning equipment. K. Rotary kiln. L. Cooler. M. Leach. N. Liquor storage. O. Treatment. P. Granulator. Q. Covered exhausted baskets. R. Air conditioner. S. Bagging-barrelling booth.

TLV ACGIH (water soluble Cr^{VI} compounds) 0.05 mg/m³
MAC USSR 0.01 mg/m³

Chromium trioxide (CrO_3)

CHROMIC ANHYDRIDE (sometimes referred to as "chromic acid" although true chromic acid cannot be isolated from solution)

m.w. 99.99
sp.gr. 2.70
m.p. 196 °C
b.p. decomposes

very soluble in water

dark-red prismatic crystals or granular powder.

TWA OSHA 0.1 mg/m³ ceil
TWA NIOSH 0.025 mg/m³/10 h day
0.05 mg/m³/15 min ceil
TLV ACGIH (water soluble Cr^{VI} compounds) 0.05 mg/m³
IDLH 30 mg/m³
MAC USSR 0.01 mg/m³

Chromic acid anhydride is formed by treating a concentrated solution of a dichromate with strong sulphuric acid in excess. It is a violent oxidising agent, and in solution is the principal constituent for chromium plating.

Insoluble chromates

Chromates of weak bases are of limited solubility and more deeply coloured than the oxides: hence their use as pigments. These are not always distinct compounds and may contain mixtures of other materials to provide the correct pigment colour. They are prepared by the addition of sodium or potassium dichromate to a solution of the appropriate salt.

Lead chromate ($PbCrO_4$)

m.w. 323.2
sp.gr. 6.3
m.p. 844 °C
b.p. decomposes

TWA OSHA 0.1 mg/m³ ceil
TWA NIOSH 0.001 mg/m³ occupational carcinogen (certain water insoluble Cr^{VI} compounds)
TLV ACGIH 0.05 mg/m³ human carcinogen
IDLH 30 mg/m³
MAC USSR 0.001 mg/m³

This is trimorphic; the stable monoclinic form is orange-yellow, "chrome yellow", and the unstable orthorhombic form is yellow, isomorphous with lead sulphate and stabilised by it. An orange-red tetragonal form is similar and isomorphous with lead molybdate (VI) $PbMoO_4$ and stabilised by it. On these properties depends the versatility of lead chromate as a pigment in producing a variety of yellow-orange pigments.

Calcium chromate ($CaCrO_4.2H_2O$)

m.w. 158.1

sparingly soluble in water

yellow monoclinic crystals.

TLVs as for lead chromate

This compound is used as a corrosion inhibitor and in the depolarisation of batteries.

Zinc chromate ($ZnCrO_4$)

m.w. 298.8

slightly soluble in water; soluble in dilute acids

a yellow powder.

TLVs as for lead chromate

It is used as a corrosion inhibitor.

Uses. Compounds containing +6 chromium are used in many industrial operations among which may be mentioned: the manufacture of important inorganic pigments such as lead chromes (themselves used to prepare chrome greens), molybdate oranges, zinc chromate and chromium oxide green; wood preservation; corrosion inhibition; and coloured glasses and glazes. Basic chromic sulphates are widely used for tanning.

The dyeing of textiles, the preparation of many important catalysts containing chromic oxide and the production of light-sensitive dichromated celloids for use in lithography are also well known industrial outlets for chromium chemicals.

Chromic acid is used not only for "decorative" chromium plating but also for "hard" chromium plating where it is deposited in much thicker layers to give an extremely hard surface with a low coefficient of friction.

Because of the strong oxidising action of chromates in acid solution, there are many industrial applications particularly involving organic materials, such as the oxidation of anthracene; for production of anthraquinone; the oxidation of trinitrotoluene (TNT) to give phloroglucinol; and the oxidation of picoline to give nicotinic acid.

Chromium oxide is also used for the production of pure chromium metal suitable for incorporation in creep-resistant, high-temperature alloys, and as a refractory oxide. It may be included in a number of refractory compositions with advantage, for example in magnesite and magnesite-chromite mixtures.

HAZARDS

It is evident that chromium in the +3 oxidation state is considerably less hazardous. It is not readily absorbed from the digestive system and combines with proteins in the superficial layers of skin to form stable complexes.

This property probably accounts for the fact that chromic compounds do not cause dermatitis or chrome ulceration. In the +6 oxidation state, chromium compounds are irritant and corrosive and are absorbed by ingestion, through the skin and by inhalation.

The effects of chromium encountered in occupations where chromium compounds are manufactured and used are due to +6 chromium and involve mainly the skin and respiratory system. Occupational exposure to chromium and its compounds has been found to cause skin and mucous membrane irritation and corrosion, dermatitis and chrome ulceration, and has been related to an increase in the incidence of lung cancer.

Typical industrial hazards are inhalation of the dust and fume arising during the manufacture of dichromate from chromite ore and the manufacture of lead and zinc chromates, inhalation of chromic acid mists during the electroplating and surface treatment of metals and skin contact with +6 chromium compounds in manufacture or use.

Chromium +6 may arise as dust and fume in the welding of stainless steels.

Chrome ulceration. This is the commonest lesion resulting from occupational exposure to chromium. It is due to the corrosive action of +6 chromium which penetrates the skin through cuts and abrasions.

The lesion begins as a painless papule, commonly on the hands, forearms and feet, which may be ignored until the surface ulcerates. Unchecked, the ulcer penetrates deeply into soft tissues and may reach underlying bone. Usually the ulcers are circular, the edge is raised and hard and the base is covered with exudate or firmly adherent crusting (figure 2). Unless treated at an early stage,

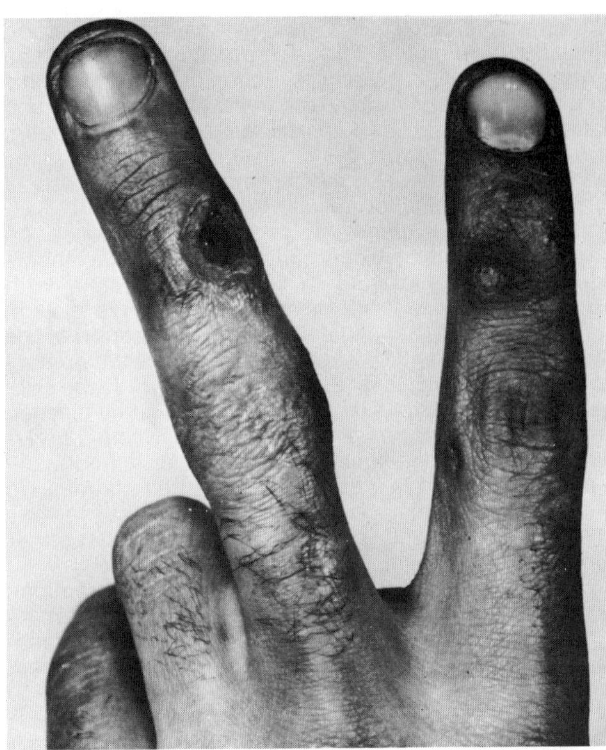

Figure 2. Chrome ulceration of the skin of a fitter employed in a chromium-plating works. *(By courtesy of Dr. R. I. McCallum.)*

healing is slow. An atrophic scar remains. Neoplastic change never occurs.

Multiple ulcers, often the site of secondary infection, were at one time common and penetration to underlying bone occurred occasionally making amputation of a terminal joint or finger necessary.

Dermatitis. The +6 compounds of chromium cause both primary skin irritation and sensitisation. In the chromates-producing industry some persons develop skin irritation, particularly at points of contact with clothing such as the neck or wrists, soon after starting work with chromates. In the majority of cases, this clears rapidly and does not recur; it is seldom necessary to recommend a change of occupation.

Despite experience in the chromates-producing industry, many cases of occupational dermatitis are attributed to sensitisation to +6 chromium which is present, at least in trace amounts, in many industrial processes. Numerous sources of exposure have been listed and there are reports on chromate dermatitis in many occupations including automobile factories, the servicing of diesel locomotives and among persons working with cement and oil. Persons affected react positively to patch testing with 0.5% dichromate and show, in most cases, no reaction to other substances used in the standard series of patch-testing materials.

One report refers to the variability in the type of lesion present in men employed on wet sandpapering of car bodies. Some had only erythema or scattered papules and in others the lesions resembled dyshidriotic pompholyx, nummular eczema or follicular irritative eczema. It has been suggested that this similarity of chromate dermatitis to types of constitutional eczema may lead to misdiagnosis of genuine cases of occupational dermatitis. It is a feature of chromate dermatitis that recovery may be slow and relapse may occur.

It has been shown that +6 chromium penetrates the skin through the sweat glands and is reduced to +3 chromium in the corium. It is postulated that the +3 chromium then reacts with protein to form the antigen-antibody complex. This theory explains the localisation of lesions around sweat glands and why very small amounts of bichromate can cause sensitisation. The chronic character of the dermatitis may be due to the fact that the primary inflammatory lesion is in the corium and that the antigen-antibody complex is removed more slowly than would be the case if the reaction had occurred in the epidermis.

Acute respiratory effects. Inhalation of dust or mist containing +6 chromium is irritating to mucous membranes. Sneezing, rhinorrhoea, lesions of the nasal septum, irritation and redness of the throat and generalised bronchospasm are well documented effects. In a few cases sensitisation occurs, resulting in typical asthmatic attacks, which recur on subsequent exposure.

Cough, headache, dyspnoea and substernal pain have been described in a man exposed for several days to concentrations of chromic acid mist of about 20-30 mg/m^3. Generalised rhonchi and moist rales were heard throughout the lungs. The signs persisted for two weeks. Another man working on the same process was similarly but less severely affected. The occurrence of bronchospasm in a man working with chromates should suggest chemical irritation of the lungs. Treatment is symptomatic.

Ulceration of the nasal septum. The commonest cause of ulceration of the nasal septum is trauma from nose-picking, surgical operation or following haematoma sustained by a blow on the nose.

Conditions in which dust or mist containing chromates can be deposited on the nasal septum, often enhanced in persons who have the habit of nose-picking, will lead to ulceration of the cartilaginous portion followed, in most cases, by perforation at the site of ulceration. The mucosa covering the lower anterior part of the septum, known as Kiesselbach's or Little's area, is relatively avascular and closely adherent to the underlying cartilage. Incoming streams of air impinge first on this site. Redness or irritation of the mucosa, accompanied by rhinorrhoea, is followed by blanching and formation of tough, adherent crusts. These cause discomfort and a strong temptation to remove them by nose-picking which often causes abrasion and further contamination of the affected area. Crusts which contain necrotic debris from the cartilage of the septum continue to form and within a week or two the septum becomes

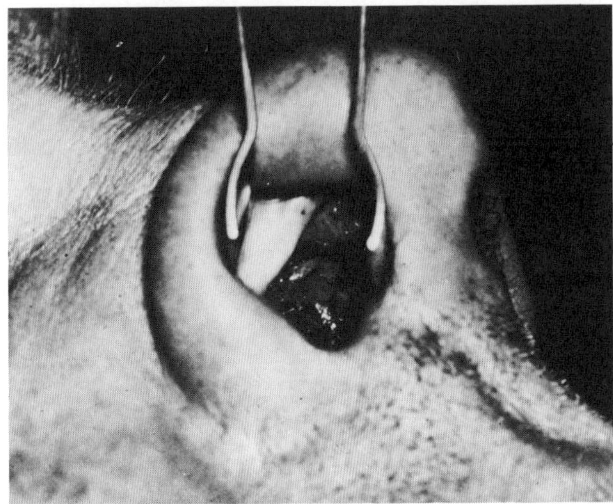

Figure 3. Ulceration and perforation of the nasal septum in a chrome plater. *(By courtesy of Dr. R. I. McCallum.)*

perforated (figure 3). The periphery remains in a state of active ulceration for up to three months, during which time the perforation may increase in size. It heals by the formation of vascular scar tissue. The bony structure is never involved and neoplastic change does not occur. Sense of smell is almost never impaired.

During the active phase, rhinorrhoea and nose-bleeding may be troublesome symptoms. When soundly healed, symptoms are rare and many persons are unaware that the septum is perforated. Whistling on inspiration may be a source of annoyance when the perforation is of small diameter and sometimes the tendency to crusting and "wet nose" persists.

Delayed effects. The confirmed long-term effect of occupational exposure to chromium is an increase in the incidence of lung cancer of men employed in the manufacture of dichromates from the chromite ore. The risk has been established for this industry in the Federal Republic of Germany, the United States and the United Kingdom. There is considerable evidence of increased risk of lung cancer among workers employed in the colour pigment industry. Preliminary findings of a study conducted by Tabershaw-Cooper Associates (United States) for the Dry Colour Manufacturers Association imply that there is an unusually high risk of lung cancer among workers exposed to lead chromate pigments. Sverre Langard and Tor Norseth (Norway) also investigated a small cohort of workers engaged in the manufacture of chromate pigments, principally zinc chromate pigment; an increased incidence of lung carcinoma is recorded. There has also been other evidence of lung cancer reported among pigment works in the Federal Republic of Germany and in the United States.

The symptoms, signs, course, X-ray appearance, method of diagnosis and prognosis differ in no way from those of cancer of the lung due to other causes. It has been found that the tumours often originate peripherally and that the lungs rarely show fibrotic change. The tumours are of all histological types, the majority being anaplastic oat-celled tumours. Water-soluble, acid-soluble and insoluble chromium is found in the lungs of chromates workers in varying amounts which bear no relation to the cause of death. Thus there is no method of determining beyond doubt that lung cancer in a chromate worker is the result of his occupation.

SAFETY AND HEALTH MEASURES

On the technical side, prevention depends on proper design of processes including adequate and appropriate exhaust ventilation and the suppression of dust or mist containing chromium in the +6 state, supplemented where necessary by built-in accessory control measures, and requiring the least possible action by either process operators or maintenance staff.

Where possible, wet methods of cleaning should be used; at other sites, the only acceptable alternative is by vacuum cleaning. Spills of liquid or solid must be removed immediately to prevent dispersion as airborne dust. The concentration in the working environment must be measured at regular intervals by positional sampling and personal monitoring. Where concentrations above recommended levels are found by either method, the sources of atmospheric contamination must be identified and controlled. Dust masks of an approved type (e.g. which have an efficiency of more than 90% in retaining particles of 0.5 μm size), must be worn in situations where atmospheric sampling has shown concentrations above the recommended levels and it may be necessary to provide air-supplied respiratory protective equipment for certain jobs, such as repair and inspection of process equipment. Surface contamination should be removed by washing down or suction before work of this type begins. It is necessary also to provide overalls which are laundered daily, hand protection and eye protection, and to ensure that a system is established for daily inspection, repair and replacement of all personal protective equipment.

The medical supervision of persons employed on processes in which +6 chromium is present must include education in the nature of the risks to health with particular emphasis on the need to observe a high standard of personal hygiene. Chrome ulceration of the skin can be prevented by eliminating sources of contact and by preventing, as far as is possible, injury to the skin. Skin cuts or abrasions, however slight, must be cleaned immediately and treated with a 10% $CaNa_2$ EDTA ointment. This, together with the use of a frequently renewed impervious dressing, will ensure rapid healing of any ulcer that may develop.

Although EDTA does not chelate +6 chromium at room temperature, it reduces the +6 form to the +3 rapidly and the excess EDTA chelates the +3 chromium. Thus both the direct irritant and corrosive action of +6 chromium and the formation of protein/+3 chromium complexes, which might act as allergents, are prevented.

Careful washing of the skin after contact and care to avoid friction and sweating are important in the prevention and control of primary irritation due to chromates.

An ointment containing 10% $CaNa_2$ EDTA applied regularly to the septum before exposure would assist in keeping the septum intact, and when soreness of the nose or early ulceration is treated by regular application of this ointment, healing occurs without perforation.

For persons engaged in the production of bichromates from the ore, close supervision of health by means of routine medical examination and chest radiography is essential. The value of sputum cytology has yet to be established; examination of several hundreds of specimens over several years in a group of men so employed in the United Kingdom proved unrewarding, but with improved technique for the collection of suitable specimens and greater experience in interpretation this may prove to be a useful ancillary method of supervision in the future. The excretion of chromium in the urine and the concentration in the blood have been measured. The results may indicate absorption of chromium but are not diagnostic nor confirmative of disease.

BIDSTRUP, P. L.
WAGG, R.

CIS 79-755 *Proceedings of the Symposium on health aspects of chromium containing materials* (Industrial Health Foundation, Inc., 5231 Centre Avenue, Pittsburgh) (1978), 61 p. Illus. 7 ref.

CIS 77-1975 "A review of the carcinogenicities of nickel, chromium and arsenic compounds in man and animals". Sunderman, F. W. *Preventive Medicine* (New York), June 1976, 5/2 (279-294). 167 ref.

CIS 79-1106 "Experimental development of allergological skin tests with chromium compounds and their application during examination of industrial workers" (Eksperimental'noe obosnovanie kõnyh allergologičeskih prob s soedimenijami hroma i primenenie ih pri obsledovanii promyšlennyh rabočih). Gončarov, A. T.; Ul'janov, A. D.; Ovruckij, G. D. *Gigiena truda i professional'nye zabolevanija* (Moscow), Dec. 1978, 12 (46-47). 10 ref. (In Russian)

CIS 79-461 *Manufacture of inorganic pigments–Special case of chromium derivatives* (La fabrication des pigments minéraux–Cas particulier des dérivés du chrome). Deroo, J. L. (Caisse régionale d'assurance maladie du Nord de la France, 11 boulevard Vauban, Lille) (May 1978), 13 p. Illus. (In French)

CIS 78-114 *Chromium–Health and safety precautions.* Health and Safety Executive, London, Environmental Hygiene/2 (London, HM Stationery Office, July 1976), 2 p. 5 ref.

Criteria for a recommended standard–Occupational exposure to chromium[VI]. DHEW (NIOSH) publication No. 76-129 (National Institute for Occupational Safety and Health, 4676 Columbia Parkway, Cincinnati) (1975), 200 p.

Chromosome aberrations

Chromosomes in higher organisms are complex structures, consisting of deoxyribonucleic acid (DNA), ribonucleic acid (RNA) and proteins. The component responsible for genetic information is DNA.

DNA is a double-stranded helical molecule, composed of two polynucleotides running in an antiparallel way. The nucleotides, each comprising a base, a sugar (deoxyribose) and phosphoric acid, are linked to each other by phospate. The bases present in DNA are four, two purines (adenine and guanine) and two pyrimidines (thymine and cytosine); each purine is complementary to one pyrimidine (i.e. adenine to thymine and guanine to cytosine) and base pairing by hydrogen bonding is essential in the formation of the double-stranded DNA molecule. The sequence of the bases in the various DNA molecules constitutes the genetic code, which is transferred from cell to cell during cell division after an exact replication of the DNA molecules, which occurs during the synthetic phase (S) of the cell cycle. The information contained in the genetic code is transcribed into RNA molecules and translated through the RNA during the synthesis of proteins. The protein backbone of the chromosomes is important in regulating the activity of the various parts of the genome.

The chromosomes are visible only during the active phase of cell division, i.e. mitosis for somatic cells and meiosis for germinal cells, when the submicroscopical nucleic acid-protein filaments fold up in a complex manner to give rise to short rods, strongly stainable by basic dyes, which are fixed in number for all individuals of the same species.

The chromosomes most suitable for cytogenetic analysis are those visible during the metaphase of the mitosis, when they appear as made up by two equal filaments, the chromatids, joined at the centromere (or primary constriction). The position of the centromere divides the chromosome into two parts, the short (p) and the long (q) arm. The length of the chromosomes and the ratio between the arms are used for classifying the chromosomes and studying the chromosome complement (karyotype).

To obtain metaphase chromosomes one has to deal with spontaneously dividing cell populations (e.g. bone marrow cells) or with cells put in culture and induced to divide, like peripheral blood lymphocytes. Dividing cells can be blocked in metaphase by spindle poisons, such as colchicine or vinblastine. Cells are induced to swell by a treatment with a hypotonic solution, then are fixed and spread on slides by special techniques, so that one can obtain metaphase spreads with no overlapping of the chromosomes. Stained chromosome preparations can be observed and counted directly at the microscope under high power and selected metaphases are photographed. From photographic enlargements, the single chromosomes are cut and arranged in groups and pairs according to a conventional classification which takes into account the whole length of the chromosome and the ratio between arms; the result is a karyogram. The chromosome complement of an individual is called the karyotype. In man the normal karyotype is composed of 46 chromosomes, 22 pairs of autosomes and one pair of

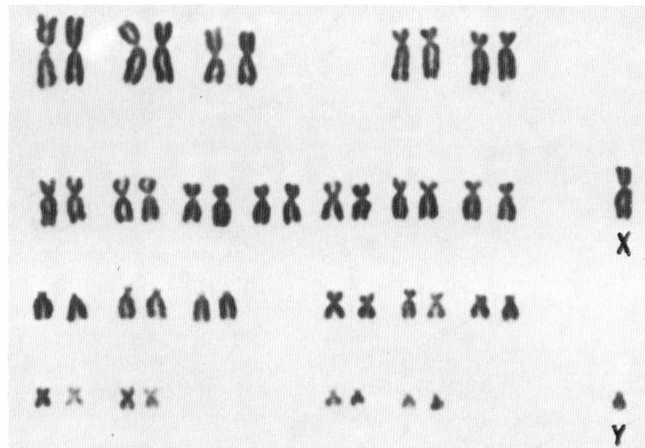

Figure 1. Normal male karyotype (46, XY).

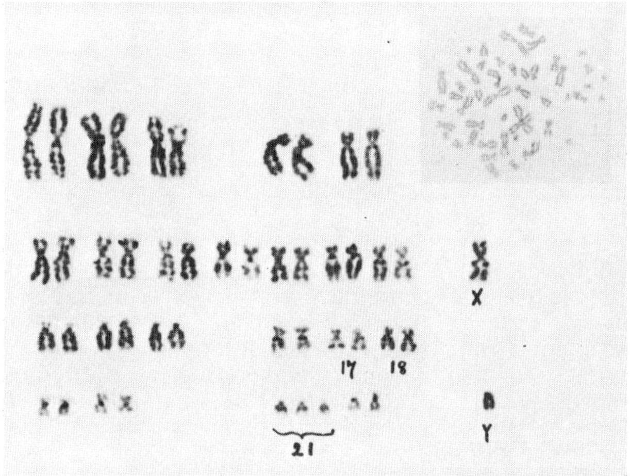

Figure 2. Giemsa-banded metaphase and karyogram of male child with Down's syndrome (trisomy 21). Karyotype: 47, XY, 21+.

sex chromosomes (XY in the male, XX in the female) (figure 1). Recent staining techniques which induce a reproducible banding in the individual chromosomes permit a more precise classification of chromosomes within groups and make it possible to recognise some normal variants in individual chromosomes (figure 2).

Abnormalities of the chromosome complement (chromosome aberrations) can be numerical or structural. The first derive from non-disjunction of chromatids during mitosis, or from non-disjunction of homologous chromosomes during meiosis of germinal cells, resulting in two abnormal cells, one with a missing and one with an extra chromosome. If the abnormal complement is compatible with survival, the abnormality is transmitted to daughter cells, resulting in an abnormal cell clone. For germinal cells, if the mature abnormal gamete is involved in fertilisation and the resulting embryo is viable, the result is an individual with an abnormal unbalanced karyotype (monosomy or trisomy for a chromosome, if a chromosome of one pair is missing or is supernumerary, respectively). Structural aberrations derive from breakage, with loss of a part of a chromosome (deletion), and eventual rearrangements within and between chromosomes, giving rise to abnormal chromosomes; in this case too, eventual aberrations can be transmitted from cell to cell, or from individual to individual if germinal cells are involved.

Chromosomal aberrations genetically transmitted are the base of numerous syndromes, characterised by multiple malformations. For example numerical aberrations are responsible for Turner's syndrome (karyotype 45, X, with a missing sex chromosome), Klinefelter's syndrome (karyotype 47, XXY), Down's syndrome (trisomy of chromosome 21) (figure 2), Edward's syndrome (trisomy of chromosome 18), Patau's syndrome (trisomy of chromosome 13) and many others; some important numerical aberrations are incompatible with life, but are described in spontaneous abortions. An example of genetically transmitted structural aberration is the "cri du chat" syndrome, with a partial deletion of the short arm of one chromosome 5. More and more syndromes associated with structural aberrations are being described now that the banding techniques have become available. Some balanced aberrations in phenytypically normal parents can become unbalanced during gametogenesis, and give rise to children with unbalanced karyotypes. Aberrations occurring during embryogenesis are responsible for the formation of "mosaics", i.e. individuals whose karyotype is made up of cells partially with normal and partially with abnormal chromosome complement.

Chromosome aberrations, both numerical and structural, can often be found in cells from solid tumours and in leukaemic cells. Except for some occasions where a specific chromosome abnormality is present (e.g. the Philadelphia chromosome in chronic myeloid leukaemia), the changes are variable and tend to increase with tumour progression.

Environmental factors, such as ionising radiations, numerous chemicals, and some biological agents (viruses) are able to induce chromosome aberrations, mainly structural, by either inducing DNA breakage or interfering with DNA synthesis or repair. The effect of numerous environmental factors on somatic chromosomes can be demonstrated *in vivo* and/or *in vitro*, in humans as well as in experimental animals or in plants, while the study of the effect on germinal cells is practically impossible in man and difficult in experimental animals.

In vivo methods consist in determining chromosome aberration rates in cells of subjects (humans or experimental animals) exposed to the suspected or known damaging agent, in comparison to unexposed controls. *In vitro* methods consist in culturing cells, either human or of other species, in the presence of the agent under study, and determining aberration rates in comparison to untreated cultures. Therefore methods *in vitro*, or *in vivo* in experimental animals can be used to screen substances suspected to be mutagenic and/or carcinogenic, while *in vivo* methods in man serve to monitor the exposure to a known damaging agent, or to check whether a suspected agent has induced a chromosome damage. The results of *in vivo* and *in vitro* methods may not be in agreement, since in the first case damaged cells may have been eliminated, and in the second case the active metabolite may not be formed.

The cell system most used in the study of chromosome aberrations from exogenous agents in man are short-term (48-72 h) cultures of peripheral blood lymphocytes, stimulated to divide by a mitogen, like phytohaemagglutinin (PHA); a part of lymphycytes, which are long-lived and rarely undergo mitosis *in vivo*, when stimulated to divide *in vitro* can show at first or second division chromosome damage that has taken place *in vivo*. This system can be used to evaluate even damage that occurred months or years before, and is a good indicator of the *in vivo* situation; however, the possibility of culture-borne aberrations should be taken into account. In experimental animals the most used system is direct preparations of bone marrow, where spontaneously dividing cells are present; this method is a good indicator of the *in vivo* situation, but it is not useful for the evaluation of past damage, owing to the short life cycle of bone marrow cells. In man the study of bone marrow cells can be used to detect a recent cytotoxic effect or to evaluate whether a past exposure has induced the formation of abnormal cell clones.

Since the rates of spontaneous and induced chromosome aberrations are generally low, numerous metaphases (generally 100) from each experimental and control sample should be counted and scored for visible aberrations. These can involve one or both chromatids, and are called respectively "chromatid" and "chromosome" aberrations. The former appear when the damage has occurred either during or after DNA replication, the latter represent damage that has occurred before DNA synthesis; therefore the significance of the various aberration types is different depending on whether one studies *in vitro* systems, where the agent is applied directly to cells, or *in vivo* systems, where the agent under study may have acted in the past. Some of the principal abnormalities are illustrated in figure 3. Among chromosome abnormalities some authors distinguish unstable changes (like acentric fragments, dicentric and ring chromosomes) from stable changes (like translocations, inversions, etc.); the former are more easily detected by direct observation, and are therefore better indicators of damage than the latter, even though stable changes can be more relevant from a biological point of view, since they can more easily give rise to abnormal clones. Even though some agents tend to induce mainly chromosome abnormalities and other chromatid abnormalities, the observed aberrations are generally aspecific and cannot be attributed unequivocally to a certain agent, unless the exposure to other damaging agents is ruled out.

The study of chromosome aberrations from physical and chemical environmental agents has become of interest in occupational medicine. Increased rates of

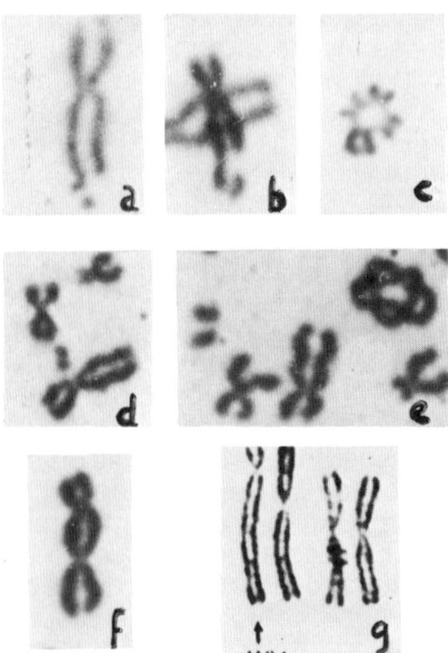

Figure 3. Some structural chromatid *(a)*-*(c)* and chromosome *(d)*-*(g)* aberrations. *(a)* Chromatid break with dislocation; *(b)* Chromatid break without dislocation; *(c)* Quadriradial (chromatid exchange); *(d)* Minute fragment; *(e)* Acentric fragment and ring chromosome; *(f)* Dicentric chromosome; *(g)* Partial karyotype showing an abnormal monocentric chromosome from pericentric inversion in chromosome No. 2.

chromosome aberrations in peripheral blood lymphocytes have been described after acute accidental exposure to *ionising radiations*, as well as in subjects with chronic occupational exposure to low radiation doses, both in the medical practice and in industry. In chronic exposure, although no strict relationship exists between cumulated dose and chromosome aberration rates in single individuals, groups with higher exposures tend to have higher rates of chromosome aberrations.

Benzene, a known myelotoxic and leukaemogenic agent, has been shown to induce chromosome aberrations. Subjects with benzene haemopathy, subjects who have recovered from benzene haemopathy, as well as subjects with past heavy exposure without haematological signs of disease, present increased rates of chromosome aberrations, in some cases persisting for years and decades after cessation of exposure in peripheral blood lymphocytes. At the time of benzene haemopathy, aberrations in bone marrow cells have also been described.

Other chemicals of environmental and/or occupational relevance like arsenic, lead, inorganic and organic mercury, cadmium, vinyl chloride and some pesticides have been demonstrated to induce chromosome aberrations in man and in experimental animals.

The use of the study of chromosome aberration rates in biological monitoring of occupational exposure to chromosome-damaging agents, however, is largely limited by the aspecificity of the chromosome damage and by the different individual susceptibility to the agent under study, which is probably the cause of the dispersion of data in even relatively homogeneous groups. Moreover, since well conducted studies require considerable time of skilled personnel, this kind of investigation cannot be extended to large, unselected population groups, and should be limited to small groups selected on the basis of known exposure to a certain agent, in which exposures to other chromosome-damaging agents can be excluded.

The biological implications of findings of increased rates of chromosome aberrations for the processes of carcinogenesis and mutagenesis are still largely unknown. Gross aberrations like breaks, dicentric chromosomes, etc., probably are responsible for cell death after a few divisions, and are therefore non-transmissible. Some type of rearrangement giving rise to abnormal monocentric chromosomes can give rise to abnormal cell clones. On the basis of various observations the hypothesis can be raised that in subjects exposed to radiation or benzene, chromosomal aberrations in bone marrow cells might give rise to a leukaemic clone, as suggested also from serial chromosome studies in a woman with benzene haemopathy evolving into acute myeloblastic leukaemia. On the other hand, cells with an unbalanced or unstable chromosome complement might be more susceptible to neoplastic transformation by other agents like viruses; moreover, chromosome aberrations in immunocompetent cells might favour the development of leukaemia by impairing the immune surveillance of abnormal clones.

Even more difficult is to evaluate the significance of increased rates of chromosomal aberrations in somatic cells for the processes of mutagenesis and teratogenesis, which involve germinal and/or embryonal cells. The study of chromosome aberrations in germinal cells is practically impossible in man; it is now known how much of the agent under study or of its active metabolite reaches the germinal cells, whether these cells have the same or a different susceptibility to chromosome damage of somatic cells, or whether damaged cells are still viable and can give rise to a vital embryo, when involved in fertilisation. Some answers to these questions might eventually be extrapolated from animal studies.

FORNI, A.

Paris Conference 1971: Standardisation in human cytogenetics. Hamerton, J. L.; Jacobs, P. A.; Klinger, H. P. (eds.). *Birth defects: original articles series*, Vol. 8/7 (New York, National Foundation, 1972), 46 p. Illus. 15 ref.

Human chromosomes. Ford, E. H. R. (New York, Academic Press, 1973), 381 p. Illus. Ref.

"Human chromosome damage by chemical agents". Shaw, M. *Annual Review of Medicine* (Palo Alto), 1970, 21 (409-432). Illus. 16 ref.

"Etiology, role and detection of chromosomal aberrations in man". Fabricant, J. D.; Legator, M. S. *Journal of Occupational Medicine* (Chicago), Sep. 1981, 23/9 (617-625). Illus. 67 ref.

"The role of human genetic monitoring in the workplace". Dabney, B. J. *Journal of Occupational Medicine* (Chicago), Sep. 1981, 23/9 (626-631). 44 ref.

Methods for the analysis of human chromosome aberrations. Buckton, K. E.; Evans, H. J. (eds.). (Geneva, World Health Organisation, 1973), 66 p. Illus. 11 ref.

"Chromosome changes and benzene exposure. A review". Forni, A. *Reviews on Environmental Health* (Tel-Aviv), 1978, III/1 (5-17). 33 ref.

Cinema industry

A large modern film studio is very like a self-contained industrial estate with most kinds of engineering, woodworking, electrical and allied crafts represented, together with many more specialised units. It may have its own power station to provide direct current electricity for arc lights and its own water supply. Up to 50 different trades, crafts and professions may be employed in the making of a film.

The film producer is responsible for finding his story; this may be an idea of his own or more usually he buys the film rights of a book or an original screenplay. He then works with a script writer to prepare first a treatment and then a detailed script. The production manager examines the script in detail, working out the practical requirements in terms of locations, studio space and estimated time to completion. The art director begins to design the "look" of the film, scene by scene, and the costume designer, in consultation with him, starts to design clothes. Models are made from the art director's sketches, and from both of these the director plans his action and camera angles and the lighting cameraman prepares his lighting plots.

The production manager has now booked the necessary studio space and has worked out a shooting schedule. From the set drawings, the construction manager dovetails the sets to use the space most economically. The actual making of the film now begins with the construction of the stage and sets, the arrangement of the lighting, the provision and issue of properties, and the filming of the scenes with the actors, stuntmen and any necessary special effects.

HAZARDS AND THEIR PREVENTION

Most safety and health hazards are those associated with the individual trades employed or those found in industry in general; there are relatively few medical problems specifically related to film work.

The stage and set. A film stage is a very large lofty "room" with a floor measuring perhaps 50 × 35 m and a height

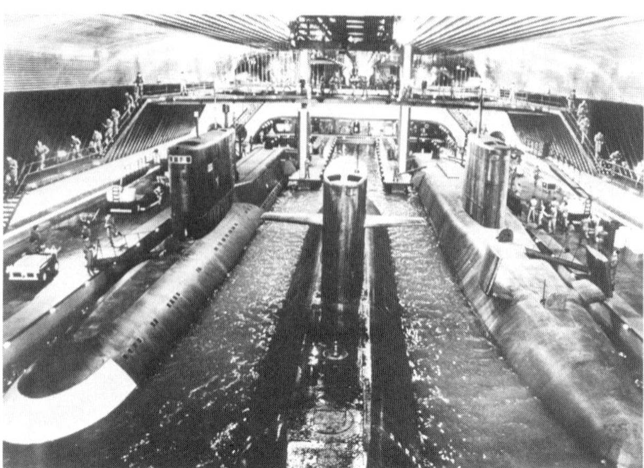

Figure 1. Pinewood Studios. The 007 Stage is the largest film stage in the world, with its size of 336' × 139' × 40'8" (103.38 m × 42.77 m × 17.54 m). Its lower elevation reservoir can hold up to 1 500 000 gallons (6 825 000 litres) of water. Here it is shown with a gigantic set from the 007 film "The Spy Who Loved Me". The three submarines in picture are all 5/8 scale models of Polaris-type fighting ships. *(By courtesy of the Rank Organisation and United Artists Corporation.)*

Figure 2. 007 Stage shown empty. *(By courtesy of the Rank Organisation and United Artists Corporation.)*

of 10 m (figures 1 and 2); this room must be proofed against light and sound. It is important to maintain good ventilation, preferably by air conditioning from above, via telescopic ducting. Older stages have wall vents but these are unsatisfactory since they are remote from the scene of activity and tend to be obstructed by pieces of scenery and the large canvas backing screens.

To provide access for large pieces of scenery and bulky equipment all the stages have very large sliding doors (moving vertically or horizontally). These heavy doors are provided with special safety devices to prevent crushing injuries. During hot weather it is usual, when film shooting is not actually in progress, to increase ventilation by keeping these doors open.

Outside scenes may be shot on location (anywhere in the world) or on the studio lot (open ground) where whole villages, castles, railway stations or even volcanoes may be built.

It is usual for everyone going abroad to have a medical examination and for all necessary inoculations to be completed before departure. In certain cases, a doctor and nurse may accompany a film unit on location but, more often, arrangements for medical care are made locally. A fully equipped first-aid box should be taken on all locations at home or abroad.

Riggers. These craftsmen erect metal scaffolding (usually tubular steel) which forms the skeleton structure of most film sets. They work at heights and the most common accidents are falls from heights and eye injuries caused by rust particles. They should wear head protection, eye and face protection and, where the nature of the work permits, safety belts and lifelines.

Underwater construction work frequently has to be carried out by rigger/divers, who should be medically examined every 6 months. A pre-employment chest radiograph is mandatory and it is wise similarly to examine the sinuses.

Stagehands. Once the carpenters and riggers have done their work, the stagehands are responsible for the completion of the set. Large set pieces will have to be moved, very often raised on rails and chains, assembled and finally "cleaned off" before other trades—plasterers and painters—can complete the operation.

Moving platforms on wheels are used for many sets. Care must be taken to avoid crush injuries to the foot or boots with reinforced toecaps should be worn. When the rostrum is moved, each wheel is steered by a jack and there is a risk of injury to the hands when the rostrum is less than 1 m high.

On film sets, it is always important to distinguish between steps, staircases, balconies, etc., which are made to be walked upon and those which are designed only for visual effect.

Problems arise when sets are "struck" after the completion of filming. All wood sections containing nails should be stacked straight on to trailers for rapid clearance. Flats (large sections of scenery) should have all protruding nails removed before being returned to the scene dock. Trailers are not always immediately available, and boards with protruding nails form a major hazard. As always in the cinema industry, the main difficulty is lack of time; a film may have over-run and the area must be quickly cleared for new building to start.

One stagehand referred to as the "camera grip" is responsible for movements of the "dolly", a small platform on wheels, on which the camera can be moved backwards and forwards. This requires the laying of rails as the movement must be perfectly smooth, without vibration.

Carpenters. Virtually anything that a film director may require can be built from wood. The carpenters' shop contains all types of woodworking machinery with the attendant hazards. All machines must have suitable machinery guarding.

Plasterers. Extensive use is made of the plasterers' art in the film industry. Not only are statues, ornaments, etc., reproduced but models of individual artists (dummies) may need to be made for special scenes. Special moulding processes are used, some of which are trade secrets. Calcium sulphate hemihydrate (plaster of Paris) is the most widely used material; it is not a noxious substance and no particular hazards to health accompany its use.

In scenes where broken glass predominates, as in bomb-blast effects or where bottles are broken in a tavern brawl, a special material, Dow's resin, is used. This is a crystal substance to which a plasticiser is added; the resultant mixture is heated in a warm oven and poured to the required shape in a mould. Good exhaust ventilation is necessary to remove the fumes which are unpleasant, although not dangerous.

Glass fibre is used extensively and may cause some direct irritation of the skin, the hands, wrists and ankles. The preparation of certain resins used in glass-fibre reinforced plastics involves the use of an accelerator and a catalyst, which are added at different stages. If they are accidentally added simultaneously, spontaneous combustion may occur.

Many surfaces to represent brickwork, roofing tiles, etc., are now rapidly made on a plastics vacuum forming machine. A roll of thin plastics, a length at a time, is pressed, when hot, on to a plaster mould. The operator of this press should be fully protected by an automatic gate whilst a sensing device should cut off the electricity if overheating occurs.

Electricians. Arc lamps require a direct current electricity supply but incandescent lamps normally use alternating current electricity taken from the public supply network.

Apart from the dangers of electric shock, there are certain special electrical hazards in the cinema industry. Many of the lights are mounted on gantries overlooking the set and a lamp which has been insecurely clamped may fall on to the people below. Electricians work at heights on these gantries (figure 3) which should be provided with safety rails at waist level. The concentration of lamps can produce environmental temperatures of around 35 °C or more and there is a risk of electricians, between bouts of activity, becoming drowsy and falling.

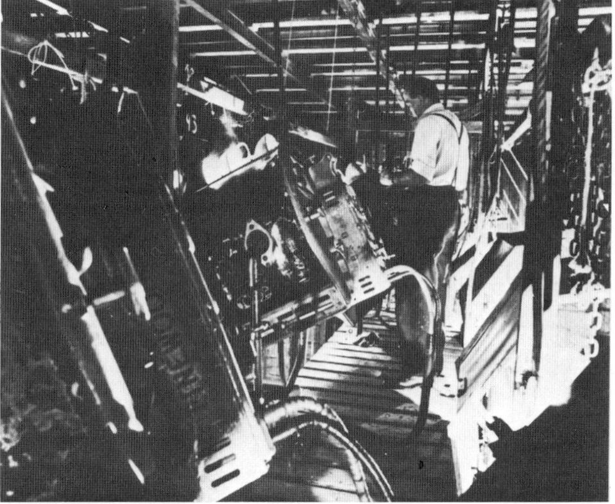

Figure 3. An electrician operating the lighting system in a film studio. He is standing on a gantry at a height of 4 m above floor level.

From time to time a lamp shatters, but usually the thick glass of the casing retains the broken glass of the bulb. If some of the newer-type quartz halogen lamps are used at an angle greater than that for which they have been designed (45 °C), a bubble appears in the glass which soon develops a pinhole and leak. The amount of halogen gas (usually iodine) which escapes is too small to constitute any danger.

Cables leading to lights, cameras and sound equipment trail on the floor and are a constant hazard, especially in the less well lit areas. New stages have an overhead system to obviate this danger.

Propertymen. These members of the team are responsible for providing various articles on the set and for moving furniture as and when required. There are no hazards other than those of lifting and carrying and the risk of dermatitis from handling dirty objects.

Animals used in films may occasionally be the concern of the property department, but fully trained animal handlers are usually employed. There is always the risk of animal bites and scratches, and appropriate treatment, with protection against tetanus, must be available.

The members of the drapes department are responsible for all soft furnishings and have no dangers to contend with other than those of wielding hammers and tacks, needles and scissors.

Film vaults. In most studios, film has to be stored. Specially designed vaults are built for this purpose. It must be remembered that before the early 1950s, nitrate film, which is highly flammable, was used exclusively. Since the invention of safety film no other type is used for film-making, but a great deal of the nitrate film is stored and exhibited all over the world. Not only is the film flammable but when it is stored in a tightly sealed can, subject to heat, an explosion may occur with the effects of a hand-grenade.

The vaults are small rooms, brick built, with teak, blast-proof doors and an explosion relief vent in the far wall leading to a vertical blast shaft. Sprinkler fire prevention systems are installed throughout the vaults. The film is marked "nitrate" or "safety" along its edge and it is wise, in addition, to mark the cans and to ensure separate storage.

Film editing. The film editor is responsible for putting the film together in its final form. He works, usually with two assistants, in a small room. There are no hazards to health particular to this occupation.

Film processing. See article PHOTOGRAPHIC PROCESSING.

Cameramen. There are no dangers to health involved in camera work other than those of the situation in which the operator may have to work. Accidents have occurred to cameramen filming from aircraft, and newsreel cameramen may of course be exposed to many hazards not usually considered occupational. Cameras (with operators) may be mounted on tall jibs (figure 4) and the normal precautions against falls from heights must be taken.

Figure 4. A mobile camera crane used in a film studio.

Stuntmen. Many scenes require dangerous stunts to be performed. The stuntmen and women who are employed for this work are specially trained and always work on a freelance basis. The work methods and the safety measures employed by these people are so specialised that it would be impracticable for the safety and health organisation of the film unit to supervise them. It is considered that their specialisation and the nature of their training is the best guarantee of their safety and even the film director does not interfere with their methods. However, accidents may occur, and the medical department should always stand by during stuntwork to provide first aid should it be necessary.

Special effects. Explosions, fires, smokes to simulate fogs, volcano eruptions may all form part of a film and are all potentially dangerous.

Black-smoke and white-smoke generators are used and are commercially available in a variety of sizes. The black smokes deposit soot and are therefore a possible cause of air pollution; the white smokes are mostly resinous and contain sulphur. Explosives are frequently used; gunpowder is made up in different containers to produce various effects, and high explosives such as dynamite are employed for actual destructive explosions. To obtain a spectacular flash, a high-velocity firework, containing naphtha in crystal form, is used.

The special-effects expert must always work in liaison with stuntmen who perform in close proximity to explosions, and a very careful timing drill has to be followed. Permits must be obtained from the authorities before explosives may be purchased and every item of explosive must be accounted for at the end of the working period. Accidents have occurred at a later date, because detonators have been overlooked and, for example, stored away with a coil of rope. Explosives should be stored only in a safe repository, built in accordance with local regulations and located in an open space.

Smokes may be irritant but, since the exposure is always of very short duration, the danger to health is minimal. Explosives can never be entirely safe and reliance must be placed on the expertise of the special effects man.

The invention of the laser, which made holography practical, has opened up new prospects for the use of light in the field of special effects. Many tried and trusted methods of filming could be made redundant by the development of holography. Travelling matte shots, glass shots and backcloths could be made more easily and the finished product vastly improved by the introduction of holograms. At present the two-dimensional backcloths commit directors to keep their cameras stationary. Should backcloths be superseded by holograms craning, panning and tracking could be employed in these scenes for the first time, since the three-dimensional backgrounds would change perspective with the movement of the camera.

Accidental exposure to laser radiation can result in damage to the skin and eye but lasers can be used with safety provided certain precautions are taken (see LASERS). The well known metal-cutting carbon dioxide lasers have no part to play in the field of lighting effects. The types used are likely to be krypton-argon ion, argon ion and helium continuous wave gas lasers. In the field of special effects it is the diffused laser light which is being filmed rather than the laser beams, so that unless a reckless person actually looks into the light source there is no risk of damage to the eyes of studio personnel. The apparatus should not be left unattended unless locked and disconnected from the power source. Beryllium-coated glass tubes may break, particles being driven into the skin with the risk of development of granulomata (see BERYLLIUM, ALLOYS AND COMPOUNDS). The risk of such accidents is reduced by built-in sensors which cut off the source of power if the water pressure drops or the water outlet is blocked. Ultrasound may be produced by galvanometer activity, but this can be tuned out. The international laser warning symbol should be exhibited where lasers are in use.

Film actors, extras and stand-ins. The hazards which film actors face are similar to those of stage actors dealt with in the article ACTORS AND PLAYERS.

It must be remembered that the film actor's day is a long one and that, after filming, he may be appearing in a stage performance in the evening. Nervous tension is always high and sympathetic consideration should be given to problems such as insomnia. In the past, there have been cases of skin trouble due to makeup but, with the high standards of manufacture and cleanliness maintained today, such cases are extremely rare. Rubber devices for altering the shape of the face occasionally cause an allergic reaction. By the time these cases come to the attention of the medical department, the need for the appliance is usually over and the lesions heal quickly. On rare occasions, a wig-cleaning solvent may cause trouble and substitution of absolute alcohol for the solvent will solve the problem. The eyes of certain actors or actresses prove sensitive to the powerful lights now necessary for colour photography. Eye drops containing an antihistamine have been found useful in these cases.

The stand-in takes the place of the actor while lighting arrangements are being tested and positions are being worked out. This is simply to enable the actor to relax. Like film extras who provide the "crowd", stand-ins have no particular problems caused by the nature of their work.

Safety and health organisation. It is advisable to employ a full-time safety officer whose job it is to analyse working procedures, detect hazards, advise suitable safety measures and supervise their implementation. Safety committee meetings should be held at regular intervals, perhaps under the chairmanship of the chief engineer, to discuss all matters of safety in the studios and plan the necessary action. Whenever new chemicals, smokes, etc., are to be used, every effort should be made to find out their composition and to anticipate any health problem.

A particular problem of safety organisation is that it is now common for film studios to provide their space and services to "renters", i.e. film companies which move into the studios to make one film or sometimes a series of films. Consequently, in many respects, the studio management has limited control over many of the activities of the production teams, while having over-all responsibility for the health and safety of all employees on its premises. It is therefore advisable that special arrangements should be made concerning the authority of the studio safety and health personnel over the employees of "renters" to ensure that the safety officer can carry out his work effectively.

A film studio should also have a medical department comprising a physician, fully trained nurses and first-aid workers. The method of operation must be flexible, but as the nursing sisters and doctor can operate most effectively in the medical department, the normal system in case of an accident is for the first-aid team to be summoned by telephone to deal with the situation at the site of the incident and then to bring the casualty to the medical department. Depending on the terrain involved, and the distance to be covered, a non-walking patient will best be conveyed on a carry stretcher, wheeled stretcher or stretcher all-terrain vehicle (combined fire-fighting and first-aid vehicle).

At various strategic locations in the studio, there should be first-aid points equipped with stretcher, blankets and first-aid boxes. The first-aiders must be prepared to bring down an injured patient from a height and, in many studios, a Neil Robertson or similar stretcher is provided for this purpose.

The amount of flammable and explosive material employed in a film studio is enormous, and a well organised and fully equipped fire service should be provided. A fireman is always asked to stand by when "live" fires are being filmed, for instance in the interior of a burning aeroplane fuselage; the special-effects technician will prepare and start the fire, the scene will be filmed and the fireman asked to extinguish the fire. This will be done with an appropriate extinguisher, avoiding those likely to create toxic risks in an enclosed space.

It should be noted that the director may wish to reshoot the scene several times. If fibre glass objects are ignited, e.g. in the case of mock-up bombings, an acrid fume is emitted and firemen should wear the necessary respiratory protective equipment. A wet gauze mask will be satisfactory for very short exposures, but self-contained or airline equipment is necessary for exposures of more than 2-3 min.

SEAGER, F. G. M.

CIS 76-547 *Stages and studios* (Bühnen und Studios). VBG 70, Hauptverband der gewerblichen Berufsgenossenschaften (Cologne, Carl Heymanns Verlag, 1974), 11+16 p. (In German)

CIS 76-1456 *Stage and film studio sets* (Bühnen- und Studioaufbauten). DIN 15920, Teil 11, Deutsches Institut für Normung (West Berlin, Beuth Verlag, 1975), 2 p. (In German)

Civil engineering

Civil engineering deals mainly with the construction of infrastructures and has an ill defined borderline with building construction. The erection of large buildings is preceded by important work pertaining to civil engineering, and the basic technical and human problems of building construction are also met with in civil engineering—often on a large scale.

Occupational safety and health conditions in civil engineering are dictated by certain fundamental but variable characteristics, viz.:

(a) work on a fixed site (bridges, harbours, dams) or on a continually moving site (roads, railways, canals, power distribution lines, pipelines);

(b) highly mechanised work or preponderant use of manual labour according to which is preferable or required by the tender;

(c) work in the country of the contractor and workforce or in a foreign country, often in a hostile natural environment.

As building and civil engineering activities have many things in common, the information given in the article on BUILDING CONSTRUCTION should be borne in mind where applicable.

Workforce

Particular attention should be paid to the recruitment, training, accommodation and conditions of life of the workforce, who may vary from a few dozen to many hundreds of workers, and among whom there are some specialists coming from industrialised countries and a large number of labourers of various nationalities or ethnic origins, sometimes accompanied by their families.

It is important to establish the camp not too far away from the construction site, in a safe and salubrious location, and to provide for everything that is needed by a community: an access road connected to the local network, roads inside the camp; means of transport; supply of water, gas and electricity; sewers; heating or cooling and air conditioning; housing for single persons and families; refuse collection and street cleaning; fire protection; health care and hospitals taking into account local diseases and potential epidemics; shops; postal services and travel agency; police services; sports and leisure-time occupations; schools for children; vocational training and language courses for adults. All this has to be co-ordinated with or authorised by the local authorities, organisations and services.

To avoid quarrels and disorders it is necessary to provide for separate accommodation, cooking facilities and places of worship for the different nationalities, ethnic groups and religions. For this purpose it is useful to consult representatives of the various groups in order to improve programmes, to facilitate the observance of the rules governing camp life and to adjust vocational training to site requirements.

Siting

A prerequisite for the proper planning and layout of construction sites is a thorough knowledge of natural hazards which may cause events resulting in the loss of human lives, material damage or destruction of structures and site equipment. Apart from information and statistics which may be supplied by local services, it is necessary to have studies made and research conducted *in situ* by experts of the technical service planning the operations or of the contractor charged to carry out the work. The frequency and probability of certain natural events can be established in this way.

For sites in the mountains it is thus possible to locate the points where there is a great danger of landslides or falls of rock or ice, to choose the site accordingly and to provide for the necessary ground consolidation and/or protective structures. The same applies to measures against avalanches which cause damage not only by the huge masses of snow they displace but also by the violent blast they generate over considerable distances. It is possible to deviate or arrest avalanches by special structures and to loosen them deliberately by explosives (see AVALANCHES AND GLACIERS; GEOLOGY AND SAFETY). Finally, a surveillance and alarm service must be set up and connected with existing or specially created meteorological stations.

As regards the hazard of periodical or sudden floods, surveillance and alarm systems are to be organised and by-channels excavated to keep the site safe.

The structures should be sufficiently strong to resist heavy storms; plant and equipment which might be overturned should be secured to anchor points. If sandstorms are to be feared, the plant and equipment chosen should either be protected against the penetration of fine sand or be so sturdy and reliable that it does not break down when exposed to sand.

Certain areas are the natural habitat of big animals hostile to man or of insects, reptiles, and poisonous or otherwise dangerous vegetation. There may be waters infested with parasites capable of transmitting disease to man (see SCHISTOSOMIASIS; LEPTOSPIROSIS), or there may be endemics which may give rise to epidemics (malaria, tropical diseases). It is therefore necessary to provide complete information to the personnel, especially workers coming from other regions, to organise a service to control these dangers, to provide for medical prevention by way of disinfestation and vaccination, to set up a first-aid service with a programmed supply of the necessary material and equipped with means of road or air transport for establishing a liaison with medical care centres, to supply the workers with protective clothing and footwear, and to distribute first-aid equipment needed in the event of snake bite to isolated camps and workers (who have to be trained in the correct use of the equipment).

In regions with extreme (hot or cold) climatic conditions, appropriate protective clothing and shelters for use during workbreaks must be provided.

There are numerous areas where work can be carried out during one or two seasons only. In these areas the work day will comprise more than one shift, and a work shift may have more hours than in normal circumstances. In hot climates the hours of work may have to be shifted partly towards the beginning of the day and partly towards the evening. Continuous shift work and night work should be associated with supplementary health and safety measures aimed in particular at ensuring visibility and at the presence of supervisors with safety qualifications.

Organisation of work and safety

The complexity and large scale of civil engineering work requires, even more than building construction, the integration of safety into the work programmes, starting with the choice of methods and techniques, the choice of materials and equipment, and the training of workers.

On large sites it is advantageous, and sometimes even required by national legislation, that the body placing the order should either directly assume, or entrust the main contractor with, the task of co-ordinating the health and safety programmes presented by the various contractors and of instituting an inter-contractor committee to ensure that the community-established rules are adhered to. A well organised system of permanent safety supervision with management and contractors' representatives is useful and advantageous even if there is no committee and if no safety programmes have been submitted by the various undertakings.

When the site is being laid out, special attention should be paid to communication roads, particularly with the workers' camp, to roads for rapid evacuation in the event of emergencies, to the supervision of the access ways, to the location of machinery and plant taking into account sources of dust and noise, to the routes taken by mobile equipment and trucks, to means of rail and air transport, etc. Particular hazards must be taken into consideration when carrying out special types of work dealt with hereafter.

Preparation of the ground

The first steps in preparing virgin ground are the elimination of trees and scrub, of rocks and stones and of the surface vegetation, using hand tools, contractors' plant or explosives. This work involves hazards typical of logging (see FORESTRY INDUSTRY) such as falls of trees, injuries by splinters and thorns, contact with poisonous, allergenic or biologically strongly active varieties of wood. The use of chain saws may give rise to accidents if they are not properly guarded, and cause impairment to the hand and arm bones and joints. Tree-felling and log-haulage equipment should be of the appropriate type, be guarded against falling trees and have roll-over protective structures (see TRACTORS).

Excavation and grading are done with the aid of earth-moving equipment such as power shovels, bulldozers, trucks, etc. The choice of the type of equipment to be purchased or hired should be governed by the requirements of the site location and the distance from maintenance and repair shops; it should be sturdy, reliable, easy to maintain and protect the operator from heat and cold, against dust and noise. Another very important safety factor is good visibility of the ground both during the day and at night. Protective structures for falling objects (FOPS) and to prevent roll-over (ROPS) of the utmost sturdiness and detailed operating and maintenance instructions are an absolute necessity (see BUILDING CONSTRUCTION).

On sites with naturally inclined ground or artificial slopes there is a potential risk of landslides or falls of rock (see GEOLOGY AND SAFETY). It is therefore important to explore as soon as possible the characteristics of the various ground layers either *in situ* or in a laboratory, to examine all the geological, hydrological, climatic and meteorological factors which may play a part, and to set up a work programme for creating a declivity that guarantees the stability of the ground or for taking other measures of ground stabilisation (e.g. protection with plastic film against rainfalls, drainage at the surface and in depth, retaining walls, front injections, rock bolting, etc.). Furthermore, the slopes have to be continuously monitored for any swelling of the ground, fissures, abnormal yieldings or other premonitory signs in order to evacuate the workforce in time or to post caution signs as long as the necessary consolidation is not completed.

When several persons have to work at the same time at different heights on sloping ground, they should be staggered at horizontal distances so that they cannot be hit by stones or objects falling from a workplace above them. As the trajectories of falling objects are often unforeseeable, it is preferable to avoid this type of simultaneous work.

Processing of materials

The materials are extracted on auxiliary sites, such as quarries, and gravel or sandpits equipped with crushing, screening, washing, bituminous material processing and concrete-making plant. The following hazards must be taken into account when the processes are programmed and work instructions elaborated: natural or work-related instability of ground; extraction plant and conveying equipment; bins; dust and noise; cement and additives; bituminous binders (fire and explosion hazards, burns, health impairment; see ROAD MAINTENANCE).

Sand is sometimes extracted in depths below the ground-water level which leads to the formation of ponds, and when extraction is continued using floating dredgers there are dangers of burial in quicksand and of drowning. Not only extraction workers are exposed to these dangers but also other people, especially children, when the pit is abandoned. It would be preferable to refill these pits at least up to the level that can be reached by the water.

Road building

All the remarks made under "Preparation of the ground" regarding the stability of natural or built-up ground and the safe use of contractors' plant near edges, on slopes and in curves are applicable to the building of roads on level ground, in cuttings and on embankments. Detailed instructions must be worked out for plant operators and workers on the ground, taking into account whether the plant is operated separately or in conjunction.

The use of bituminous material is associated with dangers already mentioned under "Processing of materials", particularly when hot asphalt is applied: burns, fires, short-term and long-term health hazards.

The traffic of plant trucks must be carefully controlled using standard traffic signals. This is even more important when road work is performed without interruption of public traffic, in which case the personnel should be supplied with highly visible clothing, and special arrangements should be made with the competent authorities.

Analogous precautions should be taken when permanent ways are being built, repaired or maintained: optical and acoustical warning signals, and written instructions taking into account the schedules of ordinary and extraordinary (service) trains.

Bridge building

A great variety of bridges are built for pedestrians, road vehicles and railways: arch, beam and suspension bridges of various dimensions and types. The techniques used in bridge building differ greatly from each other: masonry or brickwork; concrete poured on falsework; preparation of the lateral and intermediate supports followed by launching the prefabricated steel or reinforced-concrete beams, which can be pulled, pushed or lifted in place by cranes; pushing of large prefabricated beam elements, cantilevered from the supports, towards the centre, of the bay; stranding of the cables of a suspension bridge by special machines that travel from one support to the other; construction of the foundations partly by open excavation and partly by the use of compressed-air caissons.

All these operations require a large proportion of programmed safety at the planning stage or safety built into the equipment. Factors to be taken into consideration are the risk of collapse of the bridge under construction (to which natural phenomena such as wind and floods may contribute), hazards to the workers on the bridge, hazards to people below the bridge (roads, railways, river craft) and hazards to neighbouring structures (in particular electric lines).

Safety measures to be taken are the same as described earlier in this article and in other articles such as FALLS; FALLS FROM HEIGHTS, PERSONAL PROTECTION AGAINST; SCAFFOLDING; CRANES AND LIFTING APPLIANCES; DANGEROUS SUBSTANCES. Where appropriate, see also under CONCRETE AND REINFORCED CONCRETE WORK; COMPRESSED-AIR WORK; DECOMPRESSION SICKNESS.

Work near or in water

Work above or in rivers, canals, sluices, natural or artificial lakes, or the sea, is associated with the permanent and common risk of drowning, as well as other hazards which depend on the type of work and the working methods used.

When work has to be carried out at a certain distance above the water level, falls of persons may be prevented by fitting guard rails to the structure, scaffolds, formwork, falsework, floating plant and access ways. Safety nets should be used when there are no guard rails or as a supplementary safeguard.

When permanent or temporary structures are to be built in water, all the factors affecting the water level must be thoroughly studied: the regulation of canals and artificial lakes, periods of rainfall which may cause floods, periodicity of diurnal and seasonal high tides, direction of currents and waves, etc. Instructions must be worked out for the traffic of persons, vehicles and self-propelled plant on moles and embankments under construction, for the use of land-based or floating pile-driving equipment and for preventing the overturning of land-based or floating cranes or concrete placers. Coffer-dams should be provided with rescue equipment to be used in the event of exceptionally strong seepage of water from below.

When guard rails and safety nets are being installed, or when they cannot be used, self-inflating life preservers should be worn. If work is performed at night, the points where there is a risk of falling into water must be abundantly illuminated. An alarm system (e.g. a code of acoustic signals known by all those working on the site) should be established and a rescue service organised. The latter should comprise a well trained team equipped with buoys, rafts, lifelines and boats for rescuing workers from drowning and for giving them the essential first aid.

When work is performed below the water level, either in the water or in the ground below, divers and caisson workers may be exposed to the hazards described in the articles on COMPRESSED-AIR WORK; DECOMPRESSION SICKNESS; and UNDERWATER WORK. The safety measures detailed in those articles must be scrupulously observed, in particular those concerning compression and decompression times, the availability of a decompression chamber on the site or nearby, the existence of a standby compressor to be used in the event of a breakdown of the normal compressed-air supply, and the monitoring of the purity of the breathing air supplied through pipes or from cylinders.

Dam construction

The construction of earth or concrete dams (figure 1) involves setting up large but very isolated camps in remote locations where work can generally be carried on during part of the year only, but has to be continued for several years.

Figure 1. Building a dam.

The stability of dams and, at the same time, the safety of the workers building them and of the population living downstream depend on the method of construction adopted on the basis of preliminary studies concerning the geology, hydrology, meteorology and seismicity of the area, and also on the efficiency of the monitoring and alarm systems created in view of these factors.

The isolated location of the site requires particular efforts to be made for the accommodation of the workforce, the access by road and aircraft (if possible) and the site layout. The operations involve moving considerable quantities of earth, underground work, work near or in water, auxiliary sites for preparing materials, night and shift work, and entail the hazards and safety measures dealt with in relevant sections above.

As regards mechanisation, an important part is sometimes played by aerial ropeways for materials or personnel, which must meet the requirements of national legislation or, if there is none, the recommendations of international codes of practice. Other typical installations are cableways with a travelling load carriage and fixed or mobile towers or other supporting structures. Overturning of these supports must be rendered impossible during all phases of operation or at standstill, and the dynamic behaviour of the loads, as well as the effects of ice deposits on the cables and wind load on the supports, cables, carriage and suspended loads, have to be taken into account.

If the equipment is remote-controlled by radio, it has to be ensured that the frequency used is not the same as that used by others and cannot be disturbed.

Underground work

This type of work involves mainly the construction of tunnels for roads or railways, of tunnels and vertical or inclined shafts for conveying free water or water under pressure, and of caverns for power stations, workshops, military purposes, shelters, and the storage of water, petroleum products, compressed air or gases, and radioactive waste. The hazards encountered depend on the nature of the ground and the type of work.

The ground may be loose, slightly cohesive or very cohesive. Excavation in the latter gives rise to changes in water and gas circulation and modifies the stress distribution. Energy is set free both gradually and suddenly round the cavity, causing falls of rocks, creep of the walls and collapse of the temporary supports and definite lining; there may be inrushes of water and sudden outbursts of flammable or harmful gas. The contact of the ground with water and air from the surface leads to chemical reactions which contribute to changes in rock strength, rises in temperature and the formation of flammable and harmful gaseous compounds.

To prevent and control these hazards, thorough geological prospection is required. This prospection should be even more thorough when full-face tunnelling machines are used, and should be continued over the whole period of construction.

The excavating, supporting, consolidating and lining operations are associated with dangers resulting from the use of explosives, drills, materials handling and transport equipment, electrical installations, compressed-air systems and lasers; other hazards are poor lighting and visibility reduced by dust or mist, noise, humidity, heat, rock and cement dust, liquid concrete, plastics and their solvents. Poor cleanliness on the site and insufficient personal hygiene may result in the appearance and dissemination of ancylostomiasis and other contagious diseases.

Excavation may be carried out by blasting, for which purpose shot holes have to be prepared either separately by rock drills or collectively by drill carriages with several rock drills on one or more mobile stages (associated with the risk of falls from the stages); wet drilling or exhaust devices are used for dust control (see DRILLING, ROCK; BLASTING AND SHOTFIRING). In other cases full-face tunnellers excavate the full tunnel section by means of cutting bits; the length of these machines may sometimes reach several tens of metres when they are made up of sections which continuously evacuate the excavated material towards belt conveyors or trucks and prepare the concrete lining with or without preliminary consolidation grouting. When tunnellers are used, the number of workers exposed to hazards (including the presence of compressed air which may be required to control the water pressure in the ground) is smaller, but the intensity of the hazards may be greater than that associated with other working methods and equipment. Monitoring ground behaviour, the atmosphere and the workers' health must therefore be intensified. Personal protective equipment is a necessity.

Mechanical ventilation may be of the forced or exhausting type; a combination of the two is often preferable. The air quality must be continuously monitored, and standby sources of energy must be immediately ready to replace the power usually driving the fans in the event of a breakdown.

In addition to all the safety measures required to counter the above-mentioned hazards, a number of intervention and rescue teams (at least one per shift) should be formed to take action in the event of accidents and cases of poisoning, collapse, fire, explosion or flooding; they should be equipped with the necessary rescue material (which varies according to the progress of construction and type of work) and be permanently trained (see TUNNELLING).

Pipelines for gases, liquids and slurries

Pipelines for conveying water, mineral oil and other liquids, gases and coal or mineral slurries may be laid under ground or above ground level or under water (see BUILDING CONSTRUCTION; PIPELINES).

As in the case of electric line construction, pipeline construction sites are continuously moving and often pass through hostile areas. It is of paramount importance to make sure that there are roads and means of transport for materials and personnel and, in particular, reliable means of telecommunication for normal and emergency use.

ANDREONI, D.

Safety and health in building and civil engineering work. An ILO code of practice (Geneva, International Labour Office, 1972), 386 p. Illus.

Civil Engineering Work. A compendium of occupational safety practice. Occupational Safety and Health Series No. 45 (Geneva, International Labour Office, 1981), 155 p.

CIS 78-216 *Underground work* (Travaux souterrains). Mouton, P.; Andreoni, D.; Gonner, D.; Hallen, M.; Martins, C.; Stollenz, E.; Baudu, M.; Kaufmann, E.; Rouhier, F.; Bourgonnier, F.; Dubois, E.; Regnery, E. Editions OPPBTP n° 173 A 75 (Organisme professionnel de prévention du bâtiment et des travaux publics, 2 bis rue Michelet, Issy-les-Moulineaux, France) (Jan. 1977), 166 p. Illus. (In French)

CIS 78-217 *Professional diving in civil engineering work* (La plongée professionnelle dans les travaux publics). Editions OPPBTP n° 180 A 77 (Organisme professionnel de prévention du bâtiment et des travaux publics, 2 bis rue Michelet, Issy-les-Moulineaux, France) (Feb. 1977), 161 p. Illus. (In French)

CIS 77-1442 *Stability of unretained slopes and embankments* (Stabilité des pentes et talus non soutenus). Edition INRS n° 514 (Institut national de recherche et de sécurité, 30 rue Olivier-Noyer, Paris) (July 1976), 98 p. Illus. 15 ref. (In French)

CIS 80-2032 *Safety engineering in land reclamation and soil improvement–Manual* (Tehnika bezopasnosti v meliorativnom stroitel'stve–Spravočnik). Furman, I. V. (Moscow, "Kolos", 1979), 183 p. Illus. 15 ref. (In Russian)

CIS 79-1454 *Health and safety guide for highway and street construction.* DHEW (NIOSH) publication No. 78-196 (National Institute for Occupational Safety and Health,

4676 Columbia Parkway, Cincinnati) (Oct. 1978), 137 p. Illus.

"Application of anthropometry in civil engineering" (Aplicação da Antropometria na Construção Civil). *Revista Brasileira de Saúde Ocupacional* (São Paulo), Apr.-May-June 1981, 9/34 (48-73). Illus. 17 ref. (In Portuguese)

Clay

Clay is a malleable plastic material formed by the weathered disintegration residues of argillaceous silicate rock; it usually contains 15-20% water and is hygroscopic. It occurs as a sediment in many geological formations in all parts of the world and contains in varying amounts feldspars, mica, and admixtures of quartz, calcspar and iron oxide.

The quality of clay depends on the amount of alumina in it: for example a good porcelain clay contains about 40% alumina and the silica content is as low as 3-6%. On average the quartz content of clay deposits is between 10-20% but, at worst, where there is less alumina than usual, the quartz content may be as high as 35%. Content may vary in a deposit and separation of grades may take place in the pit. Particle size varies but in general individual particles are less than 20 μm in size. In its plastic state, clay can be moulded or pressed but when fired it becomes hard and retains the shape into which it has been formed.

Extraction. Clay is often extracted in opencast pits but sometimes in underground mines. In opencast pits the method of getting depends on the quality of the material and the depth of the deposit: sometimes the conditions require the use of hand-operated pneumatic tools but, wherever possible, mining is mechanised, using excavators, power shovels, clay cutters, deep digging machines, etc. The clay is taken to the surface by truck or cable transport.

In underground mining, as a rule, the workings branch off from a main road which, because of the ground pressure, is circular in cross-section and lined with concrete blocks. At predetermined intervals, fan-shaped cross cuts are driven. These are timbered, driven not more than 30 m because of the ground pressure and then stripped of the timbering. In this method of mining, the clay is extracted by pneumatic spades or cutter loaders. In a short time the detimbered roads cave in under the pressure of the invading clay. Extraction can continue in this way until all the clay within reach has been removed. Then the same series of extraction operations can be repeated at a lower level. Formerly, extraction was facilitated by blasting but this is rarely used nowadays.

The clay brought to the surface may be subjected to preliminary processing before despatch (drying, crushing, pugging, mixing, etc.) or it may be sold whole. Sometimes, as in many brickyards, the clay pit may be adjacent to the factory where the finished articles are made.

Uses. Different types of clay form the basic material in the manufacture of pottery, bricks and tiles, and refractories. Clay may be used without any processing in dam construction; *in situ* it sometimes serves as a cover for gas stored in a lower stratum of sandstone.

HAZARDS AND THEIR PREVENTION

During the extraction and preliminary processing of clay, workers may be subject to a variety of machinery and mining accident hazards and to a silicosis hazard where the silica content of the clay is high.

Accidents. Accidents are more frequent in underground than in surface working. Falls and slides of clay, collapse of supporting timbers and of undermined walls may cause fatal or serious injury. Mechanical handling and equipment such as conveyors, especially belts or pulleys, runaway vehicles on inclined light railways, falls (into shafts), slips on slippery ground, stepping across conveyor belts, repairing and cleaning machinery in motion, and misunderstanding about switching on of electric power, are common causes.

Many accidents from falling clay can be prevented; the working slope should not exceed 60-70°, overhangs should be avoided. All dangerous parts of machinery should be securely guarded: in particular conveyor pulleys, roll crushers and pug mills. Bridges over conveyor belts should be provided at suitable intervals. Cleaning and repair of machinery in motion should be prohibited.

Slippery ground on the surface should be strewn with sawdust or similar material; underground it should be covered by boards. Shaft gates should be closed immediately after use and turntables should be protected. All haulage should be stopped while persons are on the haulage incline. Safety rules for wagon coupling and the prevention of runaway wagons should be closely followed. Personal protective equipment, overalls, foot and leg protection and head protection should be provided and worn.

Health hazards. There is a silicosis risk to underground workers where there is mechanised mining of clay with a high quartz content (over 20% and possibly as much as 35%) and little natural moisture. Here the decisive factor is not merely the quartz content but also the natural dampness: if the moisture level is less than 12%, much fine dust must be expected in mechanical getting. Mechanisation increases the volume of dust at all stages. Dust-control measures including local exhaust systems should be provided wherever practicable. Periodical medical examinations, with chest X-rays, are necessary for workers at risk: at the first sign of lung changes personnel should be transferred to dust-free work.

LANDWEHR, M.

General:

Annual Reports of the Mutual Accident Insurance Association of the Pottery and Glass Industry (Jahresberichte der Berufsgenossenschaft der Keramischen und Glas-Industrie) (Würzburg). (In German)

Health effects, clay:

CIS 74-976 "The evolution of functional and radiological lung changes due to clay dust exposure" (Razvoj funkcionalnih i rendgenskih promjena pluća u toku izloženosti aerosolima gline). Beritic-Stahuljak, D.; Valić, F.; Mark, B. *Arhiv za higijenu rada i toksikologiju* (Zagreb), 1972, 23/3 (183-197). Illus. 6 ref. (In Serbo-Croatian)

CIS 79-1265 "New data concerning lung damage due to silica dust" (Données nouvelles concernant la nocivité pulmonaire des poussières de silice). Le Bouffant, L.; Daniel, H.; Martin, J. C. *Archives des maladies professionnelles* (Paris), Jan.-Feb. 1979, 40/1-2 (59-62). Illus. 3 ref. (In French)

Health effects, bentonite:

CIS 1380-1971 "Silicosis in Wyoming bentonite workers". Phibbs, B. P.; Sundin, R. E.; Mitchell, R. S. *American Review of Respiratory Disease* (New York), Jan. 1971, 103/1 (1-17). Illus. 8 ref.

"Studies on the effect of quartz, bentonite and coal dust mixtures on macrophages in vitro". Adamis, Z.; Timar, M. *British Journal of Experimental Pathology* (London), 1978, 59/4 (411-415). Illus. 13 ref.

Health effects, fuller's earth:

"Pneumoconiosis due to fuller's earth". Sakula, A. *Thorax* (London), 1961, 16 (176-179). Illus. 5 ref.

Climate and meteorology

Climate is the total condition of the atmosphere at a locality or over an area during a long period. The factors which determine climate are related to the geographical conditions of latitude, altitude and relief; they result from physical processes involving: radiant heat exchange (solar and terrestrial radiation), exchanges of heat and humidity within the atmosphere (movements of air masses, wind currents, evaporation, precipitation), variations in atmospheric electricity (lightning) and ozone, and, more generally, the chemical composition and physical state of the atmosphere.

Meteorology is the science of the study of the laws of atmospheric phenomena. Climatology, the common sector of meteorology and physical geography, is the science of climate. A number of special sectors of climatology have acquired particular importance: medical climatology, agricultural climatology, technological climatology, etc. Medical climatology studies the effect of climate, the seasons and the weather on man's health, and the methods of using climatic factors for prophylactic and therapeutic purposes.

Climatic zones and factors. The Earth can be divided by latitude into equatorial, tropical, temperate and polar zones; by geographical conditions into maritime (stimulating or sedative), plain, mountain (mostly stimulating) and desert. In addition, a distinction is made between macroclimate, specific to a relatively large geographical zone and the microclimates of restricted areas (microclimate of a forest, a valley working zone). The state of the atmosphere and the climate are characterised by combinations and variations (hourly, daily, monthly, by season or mean) of a series of parameters, certain of which are the elements of the weather: air temperature (maximum, minimum and mean); barometric pressure, relative humidity; wind currents (direction, speed, distribution); sand or dust content of air; precipitation (rain, snow); type and extent of other phenomena such as cloud, fog, frost, ice, storms; evaporation from land and water surfaces; thermal and electromagnetic radiation, etc. These various parameters have, in turn, an overall effect on the human body and on occupational safety and health, especially with regard to certain permanent or temporary physiopathological states.

Climate and man

A distinction can be made between the effects on man firstly, of the climate and its various parameters and, secondly, of the microclimate at the workplace. The effects of climate include: that of solar radiation (positive effects on general well-being and on certain cicatrising processes; danger of skin disorders) and that of darkness (disorders encountered during polar night), solar activity (on certain epidemics, psychoses and mortality in general), night-day sequence (biological rhythms: pulse, body temperature, arterial pressure, metabolism, urinary excretion). There are also the effects of the seasons, especially on certain epidemic diseases (e.g. influenza) and, finally, the effects of variations in temperature, humidity, wind, radiation, atmospheric pressure, fog and atmospheric electricity (lightning).

Effects of temperature. From the point of view of medical climatology it is particularly important to have available data on the number of days in which the mean temperature is between 15 and 20 °C, which is the most suitable temperature for outdoor activity and rest. The environmental temperature acts on the body through the nerve receptors and activates the physiological mechanisms which reduce or increase the elimination or production of heat in relation to the character of the stimuli (heat or cold).

Effects of humidity. To evaluate the effect of humidity on health, one must know the relative "physiological" humidity and the "physiological" saturation deficit. Evaluation is based on a skin temperature of 33 °C.

In many industrial operations, the relative humidity in the work premises is high; these operations may also entail contact with water or wet products, e.g. degreasing, galvanising, metal-smelting, textile industry, tanning, woodworking, sugar industry, etc.

Effects of atmospheric pressure. Most people pass their lives under conditions in which the gaseous composition of the atmosphere and the atmospheric pressure vary within only a very small range; however, modern technology has led to an increase in the number of persons who work in low atmospheric pressures; such activities include geological expeditions, work at altitude in high mountain ranges, sports aviation and aviation in general, parachuting, etc. (see ALTITUDE).

Effects of moving air masses (winds). The sensation produced by moving air varies depending on the air speed. Walking against the wind or prolonged and continuous exposure to an "air curtain" at a workplace where the temperature is high produces an unpleasant sensation due to the irritation of the tactile receptors. The effect of moving air varies; in hot weather it promotes evaporation from the skin, whilst during cold weather it accelerates heat loss; wind may also carry dust, germs, pollutants and a certain quantity of moisture; finally although soft winds may produce a pleasant sensation, high winds may cause destruction and have been at the origin of many occupational accidents on building sites (cranes, scaffolding, etc.) through disorders of neurovegetative system regulation.

Effects of caloric radiation. The periodical changes in the physical factors comprising the climatic environment and the consequent physiological repercussions are due essentially to astronomical factors: length of the day and night, alternation of seasons, maximum height of sun during the year in relation to latitude. Sunlight has a fundamental effect on physiological processes. Persons who are constantly exposed to sunlight are better able to tolerate variations in ambient temperature and resist various diseases. Nevertheless, even brief exposure of unacclimatised skin may produce sunburn, high air temperatures accompanied by intense radiation may result in heat stroke and it has also been found that prolonged exposure to very high levels (e.g. in sailors) may lead to a high incidence of skin cancer (see RADIATION, ULTRAVIOLET, VISIBLE AND INFRARED).

Consequently, workers required to work in the open and exposed to intense sunlight should be adequately protected (ultraviolet filter barrier creams, hats, clothing, tents, showers, beverages, suitable rest periods, etc.) (see HEAT AND HOT WORK; BEVERAGES AT WORK).

Effects of atmospheric electricity. Atmospheric electricity is an important factor in climatic conditions and encompasses ionisation, voltage of the electric field, and air conductivity, etc.

The main cause of ion formation in the lower layers of the atmosphere is gaseous ionisation due to radiation from radioactive materials in the earth and air, and to cosmic radiation. The effects of atmospheric ionisation have received little study. Significant ion concentrations (both polarities) have been found in work premises, with levels up to several thousand per cubic centimetre being measured. In many cases, the polarity and number of ions is a function of the industrial process in question (see AIR

IONISATION). Lightning, the most violent manifestation of atmospheric electricity, may have catastrophic consequences.

The question of the value of supplementary atmospheric ionisation for hygiene purposes in industrial premises where the ionisation level is low has not received a definitive answer. Brief therapeutic inhalation of artificially ionised negative polarity air has been applied for certain occupational diseases, such as the pneumoconioses, to treat neurotoxicosis, etc.

Microclimate. Man's protection against the deleterious effects of climate has improved as a result of higher standards of living, technological research and material welfare (nutrition, clothing, housing). The result has been the creation of controlled microclimates in closed premises: private dwellings, public buildings, hospitals, schools, etc. However, these microclimates are still subject to the external macroclimate and continue to vary from season to season.

In addition, the processes operating in industrial premises influence the environmental conditions and thus produce a microclimate with specific characteristics and create problems of relationship between work environment and meteorological conditions.

The factors that require control in dwellings, public buildings and work premises are: ventilation, heating, cooling and humidity (air conditioning), lighting, radiant heat sources, atmospheric dust and other pollutants and ionising radiation (see articles dealing with these specific subjects).

SAFETY AND HEALTH MEASURES

Hot climates subject the human body to considerable stress, especially during the course of building or agricultural work, and rapidly produce a state of fatigue. Pronounced signs of intolerance appear in persons working in air temperatures higher than 28 °C. Consequently, well planned work schedules are required under such conditions with work breaks calculated as a function of energy expenditure and climatic conditions. It is preferable to stop work during the hottest period of the day and provide an adequate supply of beverages, preferably salted, to prevent dehydration; in addition, tents or other types of shelter should be erected, fitted where possible with showers, etc. In agriculture, particular attention should be paid to synergistic effects of heat on pesticide action, and the physical problem of wearing personal protective equipment in very hot weather.

A variety of measures may be applied to combat high temperatures and infrared radiation in factories, including mechanisation and automation of hot work.

During the cold season measures should be taken to raise temperatures to a comfortable level in closed premises (see HEATING OF WORKPLACES). In addition, work should be arranged to ensure that workers do not have to move frequently from heated premises to the cold open air, and warm restrooms should also be provided either in the same building or linked to the workshop by closed corridors.

Open-air work during cold weather also requires special protective measures. Cold hands lack dexterity leading to mishandling of handtools and resulting in cuts and abrasions. The presence of ice on walkways, scaffolding or ladders may cause falls and the general mental numbness caused by cold may result in miscalculations of distance or loads with catastrophic consequences. In addition, chills will be common and exposed parts of the body may suffer from frostbite in extremely cold conditions.

Wherever possible, work on building sites should be planned to ensure that external work is done during the warm season and that workers can be employed indoors during very cold weather. Where outdoor work is unavoidable, the creation of an amenable microclimate by inflatable tents covering the whole worksite should be considered: experiments with these in countries such as Finland have shown that work can be continued under virtually normal conditions throughout the year. In all other cases, workers should be provided with protective clothing suitable for the cold and cold work, and huts equipped with heating and facilities for warming food and drinks should be provided.

Tolerance to extremes of both heat and cold can be improved by suitable diet since different climatic conditions entail different nutritional requirements, e.g. cold work requires a larger intake of carbohydrate than hot work.

MALYŠEVA, A. E.

General:

Proceedings of the World Climate Conference, Geneva 1979. WMO No. 537 (Geneva, World Meteorological Organisation, 1979), 791 p. Illus. Ref.

CIS 76-1775 *Man, climate and architecture.* Givoni, B. Architectural science series (Applied Science Publishers Ltd., Ripple Road, Barking, Essex) (2nd ed., 1976), 483 p. Illus. 269 ref.

The atmospheric environment. Frisken, W. A. (Resources for the future Inc., 1755 Massachusetts Avenue NW, Washington, DC), 1973, 68 p. Illus. Ref.

Clothing:

CIS 1056-1969 *Physiology of heat regulation and the science of clothing.* Newburgh, L. H. (ed.) (New York, Hafner Publishing Co., 1968), 457 p. Illus. 78 ref.

Weather:

CIS 77-277 "Industrial accidents and the weather" (Betriebsunfälle und Wetter). Harlfinger, O.; Jendritzky, G. *Münchener medizinische Wochenschrift* (Munich), 1976, 118/3 (69-72). Illus. 11 ref. (In German)

Clothing industry

Garment-making is one of the oldest and most universal activities. It began in the home and even children contributed to the work. The industry covers the making-up of all types of garments in textiles made from natural, artificial or synthetic fibres. Following the Industrial Revolution, the industry expanded rapidly, although workshops were small and a great deal of production was on a home work basis. The development of artificial and, later, synthetic textiles has led to great expansion of mass production.

Technical equipment. Compared with many other industries, the technical equipment of the clothing industry remains comparatively simple. Cloth is moved in bulk by mechanical-handling equipment such as power trucks and lifting machines. The power-driven band knife cuts many garments at a time where the hand shears cut only one; the long benches of sewing machines, often conveyor-fed, merely multiply the product of one machine; the steam presses replace the bank of irons heating on the stove. However, the use of bonded interlinings and the application of plastics processing techniques such as welding have introduced new types of machinery, e.g. high-frequency, compressed-air powered welding machines and power presses.

HAZARDS AND THEIR PREVENTION

Accidents in this industry are to a large extent limited to hand injuries, either cuts, burns or crushing; however, the danger of fire is ever present. The health hazards are related mainly to general working conditions although exposure to chemicals may occur in modern working processes.

Accidents. Small undertakings in unsuitable domestic premises often present a serious fire hazard. In any workroom, large or small, there is much combustible material, and combustible waste will accumulate unless very strict control is exercised. Some of the materials used nowadays are particularly flammable, e.g. foam resins used for lining and padding, and fine particulate coir. Adequate means of escape, adequate fire extinguishers and training in procedures in case of fire are necessary. Maintenance and good housekeeping not only assist in fire prevention or fire spread but are essential where goods are transported mechanically.

In general, the accident frequency and severity rates are low, but the trade produces a multiplicity of minor injuries that can be prevented from becoming more serious by immediate first aid. Band knives can cause serious wounds unless effectively protected: only that part of the knife necessarily exposed for cutting should be left unguarded; the circular knives of portable cutting machines should be similarly protected. If power presses are used, adequate machinery guarding, preferably fixed, is necessary to prevent access of the hand to the danger area. The sewing machine presents two main hazards—the driving mechanisms and the needle. In many places, long lines of machines are still driven by underbench shafting and it is essential that this shafting should be effectively guarded by enclosure or close railing: many entanglement accidents have occurred when workers stooped under benches to retrieve materials or to replace belts. Several different types of needle-guard are available, which prevent access of the finger to the area of risk.

The use of garment presses involves a serious risk of crushing and burning. Two-handed controls are widely used but are not entirely satisfactory: they may be subject to misuse (e.g. operation by the knee) and should always be set to make this impossible, as also to prevent operation by one hand; they provide no protection for anyone other than the operator. Guards are being developed which prevent the pressure-head closing on the buck if anything (e.g. most importantly the hand) comes within the area. All presses with their steam and pneumatic supplies require frequent examination. All portable electrical power tools require careful maintenance of the earthing arrangements.

Recent developments in plastics welding (to replace seaming, etc.) and in the making of foam backs usually involve the use of an electric press, sometimes operated by treadle, sometimes by compressed air. There is a risk of physical trapping between the electrodes and also of electrical burns from high-frequency current. The only sure safety measure is to enclose the dangerous parts so that the electrode cannot operate when the hand is in the danger area: double-handed control has not proved satisfactory. There is a pressing need to redesign seaming machines to ensure built-in safety.

Health hazards. The high proportion of women and girls employed provided an ideal field for exploitation and the working conditions were indicated as being the cause of an unduly high incidence of tuberculosis, anaemia and general ill health. Many of the worst conditions have now been removed but, nevertheless, vigilance is still required especially in small workshops outside the legislation control, where there may still be health risks such as eyestrain, throat irritation due to dust inhalation in those handling cloth in bulk, posture defects amounting sometimes to deformities, especially in those taken into employment in adolescence, cramps and callosities in the craftsmen cutters.

Figure 1. Use of a portable cutting machine.

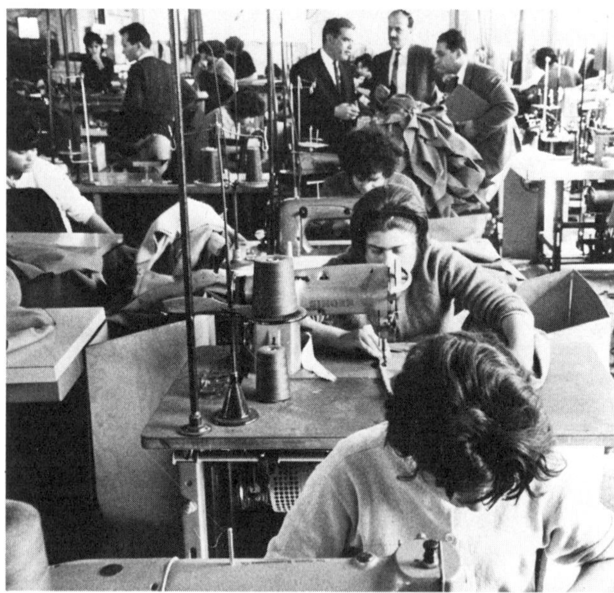

Figure 2. The sewing section of a menswear factory.

[Resin-treated fabrics for permanent pressed clothing may release formaldehyde mainly during cutting, finishing (heating promotes the liberation of formaldehyde from residual amounts of resin) and in retail shops. Exposure can cause respiratory impairment with inflammation of the mucous membranes of the upper airways and eye, skin rashes and after some time intestinal troubles and kidney disorders. Formaldehyde hypersensitivity is usually found in combination with hypersensitivity to other textile finishing agents such as ethylene urea, dimethylolurea, urea formaldehyde, melamine formaldehyde and dimethyltriazone.]

Especially in the waterproof clothing section (e.g. in binding seams), toxic solvents may be used. Benzene should be prohibited and the safest possible solvent used in well ventilated conditions; locally applied exhaust may be necessary. Toxic solvents are also to be avoided in finishing departments where they may be used for removing stains, etc.

General environmental conditions are important: ample air space, good ventilation, comfortable temperature, high standards of illumination. Where so much sedentary work is done, well designed seats are necessary [to ensure normal posture and reduce fatigue of the scapular, dorsal and lumbar muscles. Adjustable seats and backs are necessary where more than one shift is operating. Much remains to be done in the application of ergonomics to the whole industry, and in particular for the improvement of various types of machines such as high-speed sewing machines, buttonholers and button-stitchers, looping machines and others. Lighting is of paramount importance: good general illumination is to be preferred to local lighting and glare from unshaded lights is a common fault to be avoided. Noise has been found excessively high in an ancillary workshop: the zip fastener chain production plant. The extreme division of labour accompanied by narrow specialisation and quick pace may have negative effects on job satisfaction and in the long run on mental health, absenteeism and accident rates.]

Pre-employment medical examination is particularly important because the garment trade is often chosen by those not considered fit enough for heavier industry; special attention should be paid to the eyes. Periodical medical examination is desirable.

Messrooms and adequate washing and sanitary facilities should be provided.

BETTENSON, A. S.

Social problems:

Contract labour in the clothing industry. Second Tripartite Technical Meeting for the Clothing Industry (Geneva, International Labour Office, 1980), 75 p.

Safety and health:

CIS 75-246 *Safety rules for sewing machines used in industry and handicrafts* (Sicherheitsregeln für Industrie-Nähmaschinen und Handwerk-Nähmaschinen). ZH 1/437, Hauptverband der gewerblichen Berufsgenossenschaften (Cologne, Carl Heymanns Verlag, 1974), 11 p. Illus. (In German)

"Prevention of accidents in the use of sewing machines in clothing industry" (La prévention des accidents dus aux machines à coudre). Courcel, M. *Prévention et sécurité du travail* (Lille), 4th quarter 1979, 122 (25-29). Illus. 5 ref. (In French)

CIS 75-245 *Safety rules for garment ironing and pressing appliances* (Sicherheitsregeln für Bügelmaschinen und Bügelpressen). ZH 1/504, Hauptverband der gewerblichen Berufsgenossenschaften (Cologne, Carl Heymanns Verlag, 1974), 8 p. (In German)

Use of solvents in the clothing industry. Hazards and their prevention (Utilisation de solvants dans l'industrie du vêtement. Risques et moyens de prévention). Deroo, J. L. (Caisse régionale d'assurance maladie du Nord de la France, Lille), CDU 66.062.687 (21 June 1979), 15 p. Illus. 7 ref. (In French)

CIS 81-1320 "Criteria for the evaluation of toxic chemicals—Studies in the clothing industry" (Criteri di valutazione di pericolosità di agenti chimici—Ricerche nell'industria dell'abbigliamento). Granati, A.; Calsini, P.; Lenzi, R. *Medicina del lavoro* (Milan), Jan.-Feb. 1981, 72/1 (22-32). Illus. 19 ref. (In Italian)

Coal and derivatives

Coal is the general name for the residues of fossilised vegetable substances which, after having been covered by a layer of sedimentary rock which sealed them off from air and after having been subjected to pressure and temperature over geological eras, have been converted to carbon-rich compounds (carbonification).

Conditions suitable for the natural formation of coal occurred between 40 and 60 million years ago in the Tertiary Age (brown-coal formation) and over 250 million years ago in the Carboniferous Age (bituminous-coal formation) when swampland forests thrived in a hot climate and then gradually subsided during ensuing geological movements.

Classification and geographical occurrence

Brown coals. It is possible to distinguish, by their appearance, the soft brown coals (kylite, lignite) from the older, hard brown coals, which are composed completely of wood. The recent International Brown Coal Classification provides for six classes depending on the total water content (ash-free), and attributes to each class five groups depending on the tar content (moisture- and ash-free).

World reserves of brown coal are estimated at 265 billion (billion = 1 000 million) tonnes. In Europe the main deposits are in the USSR, followed by the Federal Republic of Germany and the German Democratic Republic. Other important reserves are to be found in Australia, Czechoslovakia, Poland, the United States and Yugoslavia.

Bituminous coals. These can be classified according to their degree of carbonification into: flame coal, flame-gas coal, gas coal, fat coal, steam coal, lean coal and anthracite.

An international three-figure code system is used for the description of bituminous coals; the first figure (0 to 9) denotes the volatile matter content, the second (0 to 3) the caking properties and the third (0 to 5) the coking properties.

The proven recoverable reserves of bituminous coal are 515 billion tonnes and total reserves are estimated at more than 920 billion tonnes. Out of these, 48% are located in the USSR, 28% in the United States and 12% in the People's Republic of China.

The larger European bituminous coal deposits are found in the Rhein-Westphalia and Aix-la-Chapelle-Liège coalfield, the Danube Basin, the Katovice and Ostrava coalfields, the Saar-Lothringen area and the British coalfields. There are important deposits in the Asiatic part of the USSR (Kuznetsk, Bureja and Karaganda basins), in the People's Republic of China, the United States (West Virginia, Pennsylvania, Missouri, Ohio), and in India, Australia and Japan.

Composition and structure

In the sequence—wood, peat, brown coal, bituminous coal, anthracite—the carbonaceous content rises from 50 to 91.5%, whereas the oxygen content falls from 44 to less than 2.5%, and the moisture content from 6 to less than 3.8%.

Soft brown coal (1 800-3 000 kcal/kg) contains 30-65% moisture and up to 15% mineral ballast; hard brown coal (4 800-7 200 kcal/kg) contains 20 to 30% moisture and up to 40% mineral ballast. The organic components of brown coal are bitumen and humic material—the organic decomposition products of plants and animals.

Bituminous coals may contain up to 7% moisture and up to 30% ash. In the sequence from flame coal to anthracite, the content of volatile materials (tars and gases) falls from 45 or 40 to 10 or 6% and the calorific value increases from 7 800 to 8 600 kcal/kg. Coal dust/air mixtures containing more than 70 g/m^3 of coal with a 14% content of volatile matter are explosive and can be ignited by high-temperature flames such as are found in a firedamp explosion. Coal-dust explosions consume enormous quantities of oxygen from the atmosphere and may have a secondary asphyxiating action.

Studies on the harmful effects of dust in bituminous coal mines have shown that coal mitigates the effect of silica on the pulmonary tissue; the action of flame-gas coal is more pronounced than that of anthracite in this respect. Bituminous coal consists of three maceral groups:

(a) vitrinite with the macerals, telinite and collinite;

(b) exinite with the macerals, sporinite (macrospores and microspores), kutinite, resinite and alginite;

(c) inertinite with the macerals, mikrinite, semifusinite, fusinite and sclerotinite.

Bituminous coals may be composed of various macerals and these microscopically recognisable maceral combinations are divided into seven microlithotypes (strata types): vitrite, clarite, durite, fusite, vitrinertite, trimacerite and carbominerite. A macroscopic classification by appearance (surface polish) is also employed and contains the following classes: bright coal, dull coal, semi-dull coal and fibrous coal.

The bituminous coal macerals and microlithotypes have different chemical and physical properties which can be used for determining coal type and grade, for example in formulating good coking-coal mixtures.

An external index of carbonification is the rise in reflectance (under oil) from about 0.6% for flame coal to 5% for anthracite. Determination of carbonification grade in vitrinite by reflection measurements gives a more accurate result than does the quantification of volatile matter.

The chemical composition of bituminous coal is extremely difficult to analyse due to the insolubility of the material, the multitude of aromatic constituents, and the lack of knowledge about their combination. Spectroscopic and chemical analyses show that bituminous coal is a mixture of organic compounds of a predominantly aromatic character. The molecules have a mean molecular weight of 2 000 or more. In low-carbonification coals (with a carbon content of up to 85%), polycondensed aromatic ring systems are cross-linked through aliphatic side chains. Aromatisation and polycondensation increase with rising carbonification grade, whereas the linked side chains fall in number and are absent at above a 91% carbon content.

Uses of coal

Brown coal is used primarily for the thermal generation of electricity. Large quantities are formed into briquettes under high pressure, mainly for heating purposes, but also for coke production and gas manufacture in countries with little bituminous coal. Bitumen-rich brown coal is also a valuable raw material for basic chemicals.

In 1978, 31% of the world's primary energy requirements were met from bituminous coal. The utilisation of raw coal for space and process heating, which was formerly its principal application, is now falling off. Coal can, however, be converted in many different ways into valuable and marketable sources of secondary energy. In this form it is of importance for the generation of electricity in thermal power stations. Large amounts of bituminous coal are turned into coke for metallurgical purposes and the organic compounds that occur as by-products are used as feedstock in coal chemistry. Advanced combustion techniques are being increasingly used in industry, e.g. fluidised-bed combustion. Gasification and liquefaction processes are being used to produce Substitute Natural Gas (SNG), transport fuels, hydrogen and other sources of energy as well as raw material for the chemical industry.

The basic process for the manufacture of liquid transport fuel was developed in Germany during the 1920s and 1930s. There were, at one time, 12 coal hydrogenation plants in operation (production in 1944 was 4 million tonnes). At the present time (1979), a coal liquefaction plant is in operation with a commercial output of 200 000 tonnes annually, a further plant for 2 million tonnes a year is under construction, and a similar one is being planned. Hydrogenation and gasification processes are being successfully developed elsewhere in the world, particularly in the United States, where there are large reserves of cheaply mined coal, and in the United Kingdom, Japan, Australia and South Africa.

In the case of gasification, it is necessary to distinguish between gasifiers of the fixed bed, fluidised bed and fully entrained (suspension) types through which one of two processes can be followed, either where heat is produced by partial burning of the coal, or where heat is furnished from an outside source, and which yield low-calorie gas (used for heating the coal from below), synthesis gas (for chemical processes) or methane (SNG) as products. The question of producing gas directly from coal making use of the process gas from high-temperature nuclear reactors may become an economic proposition in the future.

Research is being carried on into the generation of electricity from coal by means of new direct conversion techniques:

(a) the magnetohydromechanical process (direct generation in gas plasma);

(b) electrogasdynamic conversion (ionisation within coronary discharge); and,

(c) the use of fuel cells with a hydrogen/carbon monoxide mixture.

Coal derivatives and their uses

The chart in figure 1 shows the immense variety of coal derivatives and their wide range of uses. These products are obtained partly by mechanical conversion (lump coal, briquettes), partly by conversion through combustion, distillation or chemical transformation to different energy forms (electricity, heat, gas, coke, fuels), and to other basic chemicals (such as hydrogen, carbon monoxide, carbon dioxide, methane, acetylene, ethylene, benzene, ammonia) or, by further processing, to a broad range of chemical products (dyes and dyestuffs, pharmaceuticals, plastics, etc.).

Changing role of coal as fuel

Throughout the world, coal is being replaced increasingly by liquid and gaseous fuels. In 1900, 94% of the world's energy requirements were met by coal and only 5% by petroleum and natural gas. As late as 1940, coal still accounted for 75% of consumption and petroleum and natural gas 23%; however, by 1967, the figure for coal had fallen to 40% whereas that for petroleum and natural gas had risen to 58%, and in 1978 these figures had become 31% for coal and 63% for petroleum and natural gas. Coal, however, represents 81% of the world's fossil fuel reserves against 19% for petroleum and natural gas. Since the so-called oil crisis of 1973-74, it has become increasingly apparent that, as the end of the 20th century approaches, this disparity between reserves and consumption must be reduced, and that in order to meet the requirements other additional energy sources will have to be drawn upon, such as nuclear power, energy from the sun, geothermal power and nuclear fusion. Energy consumption forecasts indicate that the figure for 1978, which was 9 000 million tonnes of bituminous coal equivalent (BCE), will, on the basis of an average annual growth rate of 3%, have doubled by the year 2000.

Replacement of coal chemistry by oil/gas chemistry

For over a century, coal has been the dominant feedstock for the organic chemicals industry; however, since 1945

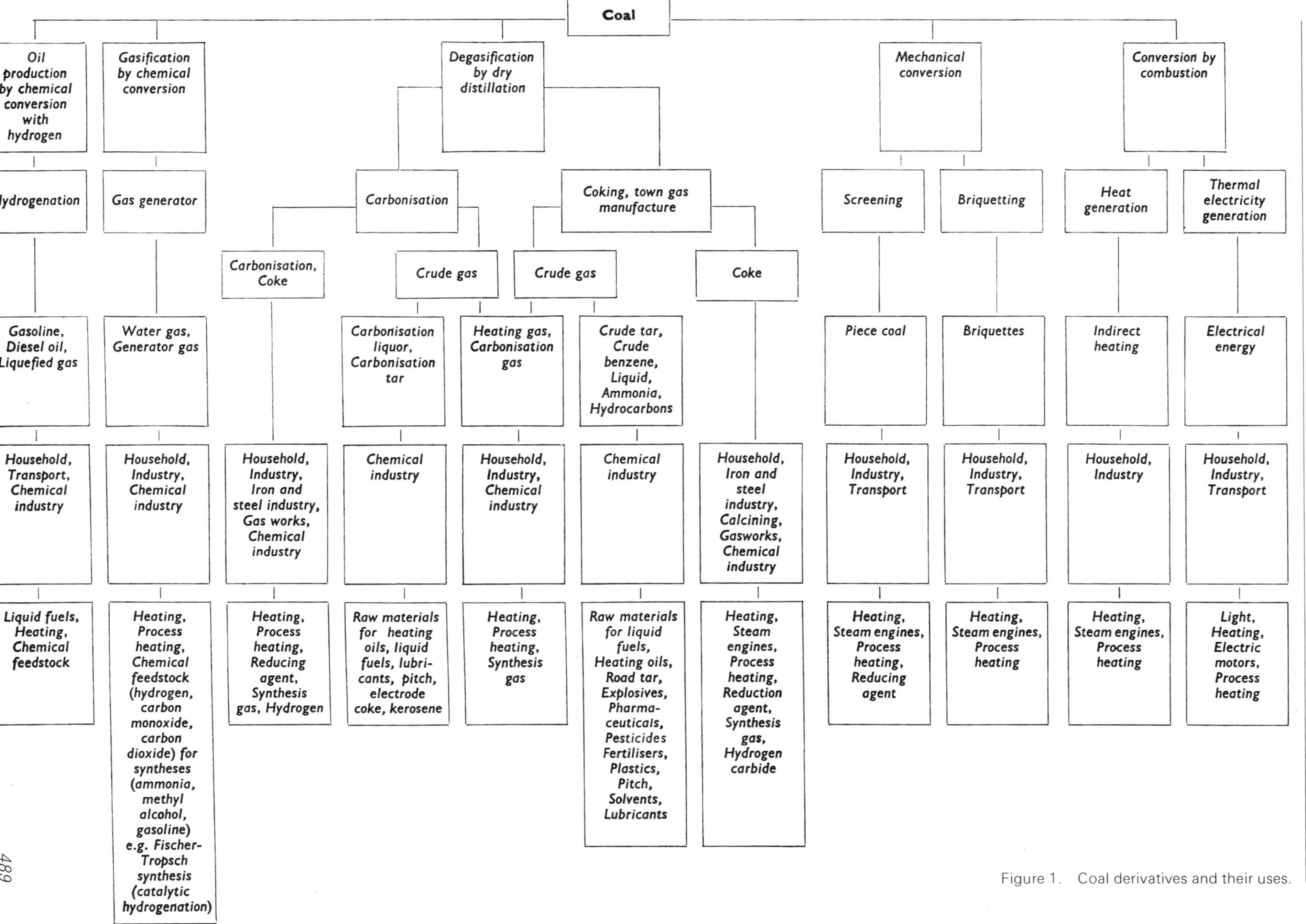

Figure 1. Coal derivatives and their uses.

its dominance has been increasingly reduced by petroleum and natural gas in most industrial countries. The Federal Republic of Germany is widely considered the stronghold of coal chemistry and here, in 1957, 76% of primary chemicals were still obtained from coal and only 24% from petroleum and natural gas. By 1968, total consumption had risen fivefold but the proportion of output accounted for by coal had fallen back to 13% and by 1978 totalled only 5%. Growth in aliphatic hydrocarbons is faster than in aromatics of which 50% were still obtained from coal in 1966 in Europe. By 1974 this figure had fallen to 15%. Since 1974, following the steep rise in the price of crude oil, the return to coal chemistry as a replacement for oil/gas chemistry is being discussed and research is being instituted into the further development of known coal chemistry processes.

Coal chemistry processes and products

Degasification (carbonisation, coking), gasification and hydrogenation of coal give rise to numerous basic organic chemicals used in the chemical industry.

In the carbonisation of brown coal, the carbonisation liquor yields mainly phenol (feedstock for dyes and plastics) and the carbonisation tar yields paraffins (for detergents and surface active agents). In the coking of bituminous coal, it is possible to obtain crude tar, crude benzene, ammonia, methane, ethane and ethylene. Distillation of crude tar and crude benzene yields:

(a) benzene, toluene, xylene, phenol and cresols that can be used for the production of plastics, synthetic fibres, pesticides, perfumes and essences, pharmaceuticals, sweetening agents, dyes, explosives and solvents;

(b) naphthalene for naphthols, phthalic acid and N-sulphonic acids for dyes; and

(c) unsaturated liquid hydrocarbons such as styrene, coumarone, indene and pyridine.

Crude tar also yields heavy oils (fuels, heating, detergent and impregnating oils), anthracene oil (anthraquinone dyes, phenanthrene, carbazol) and pitch residues (roofing felt, road tar).

Ammonia is processed to ammonium salts, nitric acid, liquid ammonia and soda. Methane is used as a heating and fuel gas, for the production of carbon tetrachloride (solvent, fire-extinguishing agent, and raw material for dye and pharmaceutical manufacture) and for the production of acetylene which is itself employed as a fuel or welding gas and for the production of acetaldehyde and its derivatives—acetic acid, acetone, ethyl alcohol and butanol. In Reppe synthesis, using pressures up to 30 kgf/cm^2 (29.4·10^5 Pa) and catalysts such as metal carbonyls, acetylene is chlorinated to trichloroethane and used to produce plastics such as polyvinyl chloride, synthetic fibre such as polyacrylonitrile fibre and aromatic compounds such as benzene, styrene and azulene. Ethane is the main feedstock for the production of ethylene which may be polymerised to polyethylene, chlorinated to dichloroethane and trichloroethane or hydrogenated to ethyl alcohol.

Synthesis gas (carbon monoxide and hydrogen) obtained from coal gasification is used as the starting point for Fischer-Tropsch synthesis which, using normal pressures and a cobalt catalyst at 160-205 °C, produces crude kerosene, benzene, gasoline and gas oil; medium pressure syntheses use cheaper iron catalysts. The Lurgi-Ruhrchemie process operates at 220-250 °C and 25 kgf/cm^2 (24.5·10^5 Pa) and produces gasoline, diesel oil and kerosene; the Kellog process uses pulverised catalysts at 300-340 °C at 30-45 kgf/cm^2 (29.4·10^5-44.1·10^5 Pa) and produces mainly gasoline and diesel oil. A process similar to Fischer-Tropsch synthesis produces methyl alcohol (solvent, dry-cleaning agent, antifreeze agent) from synthesis gas at 380-400 °C and 200-225 kgf/cm^2 (196.1·10^5-245.1·10^5 Pa) using a chromium oxide-zinc oxide mixed contact process.

Coal or coal products (tars, extracts) can be hydrogenated by the Bergius-Pier process (10-1 000 kgf/cm^2 (9.8·10^5-980.6·10^5 Pa) and 200-250 °C) to liquid fuels, light oils, lubricant oils, heating oils, kerosene and various chemical by-products.

Large-scale hydrogenation is a two-stage process: in the slurry phase, the high-molecular feedstock (coal pitch treated with oil, tar or petroleum residues) is primarily cracked in the presence of a fine particulate contact catalyst; in the gas phase, there is further cracking and then hydrogenation of the medium and heavy oils obtained in the slurry phase, using, for example, a tungsten or molybdenum catalyst.

STAHL, R.

"Coal gasification and occupational health". Young, R. J.; McKay, W. J.; Evans, J. M. *American Industrial Hygiene Association Journal* (Akron, Ohio), Dec. 1978, 39/12 (985-997). Illus. 102 ref.

"Occupational health aspects related to coal liquefaction". Shepard, H. G. *Medical Bulletin* (New York), Summer 1981, 41/2 (119-131). Illus. 12 ref.

Energy and ergonomic problems:

"Coal scenario AD 2000". Atkinson, T. *Mining Engineer* (London), Jan. 1981, 140/232 (489-499). Illus. 12 ref.

"Hard coal mining as a management task" (Steinkohlenbergbau als unternehmerische Aufgabe). *Glückauf* (Essen), 21 June 1979, 115/12 (596-603). Illus. (In German)

"Gas, motor fuel and fuel oil from coal. Technical possibilities and economic aspects of coal conversion" (Gas, Kraftstoff und Heizöl aus Kohle. Technische Möglichkeiten und wirtschaftliche Aussichten der Kohleveredlung). Ziegler, A.; Holighaus, R. *Umschau in Wissenschaft und Technik* (Frankfurt am Main), 1979, 79/12 (367-376). Illus. 7 ref. (In German)

Coalworkers' pneumoconiosis

Coalworkers' pneumoconiosis may be simply defined as the occupational disease caused by prolonged retention in the lung of abnormal amounts of dust derived in the first instance from coalmining operations. Coalworkers' pneumoconiosis exists in two forms, simple and complicated. Complicated pneumoconiosis is often referred to as progressive massive fibrosis or as *masses pseudotumorales*. The difference between simple pneumoconiosis, which is the commoner condition, and complicated pneumoconiosis, which is rarer, will be described later.

Coalworkers' pneumoconiosis is found principally among underground coalminers, particularly those with a history of many years of work on the coalface. It has also been found among workers on coal screens and in coal trimmers but the occurrence of pneumoconiosis among the latter groups is becoming less likely owing to increasing mechanisation.

Aetiology

Simple pneumoconiosis is caused by the inhalation and retention of excessive amounts of airborne dust. In most coal mines in which qualitative measurements of airborne dust have been made, the amount of free silica in the dust produced in coal-getting operations seldom exceeds 10% by weight and usually is less than 5%. Thus silicotic lesions are rare among coalworkers unless, as sometimes happens, they have been employed for long periods on rock work where the strata and the airborne dust contain a significant proportion of free silica.

Generally speaking, the risk of developing pneumoconiosis diminishes as one proceeds from anthracite to hard coal to soft coal and it is not a material hazard in lignite mines. This is probably because the anthracite and hard coal dusts in the respirable size range have a greater mass.

The aetiology of complicated pneumoconiosis is still unknown. Many theories have been propounded to explain its origin—the better known are briefly mentioned below.

(a) The silica theory. This theory became untenable when complicated pneumoconiosis was found in the lungs of workers exposed to pure carbon only.

(b) The tuberculosis theory. This theory arose because it was possible to recover tubercle bacilli in a fair proportion of cases from the lesions after death. But failure to isolate tubercle bacilli from the sputum during life and the lack of response to specific treatment for tuberculosis discredited it. In developed countries as tuberculosis decreased in the general population, the attack rate of complicated pneumoconiosis did not fall correspondingly. It is now presumed that the tuberculous infection of the lesions is a terminal phenomenon.

(c) The general infective theory. This theory gained favour as the tuberculosis theory became less convincing. Its protagonists regard complicated pneumoconiosis as being caused by tubercle bacilli and other organisms. It lacks systematic proof.

(d) The immunological theory. This theory, one of the oldest, was revived when it was appreciated that Caplan's syndrome, which resembles complicated pneumoconiosis in certain respects, had an immunological basis. Moreover, tests of immunological activity tend to be positive more often in complicated pneumoconiosis than in simple pneumoconiosis. But the immunological process, when it occurs, may be a sequel to complicated pneumoconiosis rather than a causal factor.

(e) The dust burden theory. According to this theory, complicated pneumoconiosis originates when the saturation of the lung with dust exceeds a certain level. When allowance is made for individual variation in response to dust, this theory seems tenable as a partial explanation of complicated pneumoconiosis. It is certainly true that in mining populations the risk of acquiring complicated pneumoconiosis increases steadily as simple pneumoconiosis advances.

Pathology

Simple pneumoconiosis of coalworkers is characterised by many small discrete foci or aggregations of dust, which lie like a sleeve along the fine air passages in the lung just before they open into the air-sacs where gas exchange takes place. Strands of reticulin, with an occasional strand of collagen, can usually be seen among the stored coal dust.

Complicated pneumoconiosis develops in lungs already affected by simple pneumoconiosis. One or more masses of fibrous tissue, intermingled with dust, appear in the lung substance, most commonly in the upper parts of the lung. The development of the lesions may be followed in serial radiographs. When first detected they may be quite small, only a few millimetres in diameter. As the synonym "progressive massive fibrosis" implies, they gradually enlarge and eventually may cause gross distortion of the pulmonary architecture (see PNEUMOCONIOSES: INTERNATIONAL CLASSIFICATION). In advanced cases, obliteration of blood vessels in the lung may cause failure of the right side of the heart.

Centriacinar or focal emphysema is frequently mentioned in association with simple pneumoconiosis and is sometimes regarded as an integral part of its pathology. The original descriptions made it clear that focal emphysema is not found in all cases showing dust retention. It has been pointed out that centriacinar emphysema also occurs in the general population. The exact relationship between dust exposure and pulmonary insufficiency due to emphysema or chronic bronchitis has not yet been fully clarified.

Any possible relationship between severe panacinar emphysema with airways obstruction and coalworkers' pneumoconiosis remains uncertain.

Course

Simple pneumoconiosis tends to progress so long as the individual is exposed to excessive amounts of dust. By progression is meant an increase in both the size and number of the dust foci characteristic of the condition. If, however, the individual is removed from a dusty environment, the simple pneumoconiotic process appears to remain relatively static (although complicated pneumoconiosis may still develop) and may even undergo partial regression, unless an element of silicosis is present, in which case progression in the lesion after cessation of exposure to dust is likely to continue.

If simple pneumoconiosis progresses to an advanced stage, lung function tests may reveal some diminution in the ventilatory function of the lungs. This is usually much less than the natural fall in ventilatory function due to advancing age.

When simple pneumoconiosis is moderately advanced, that part of the respiratory function which is connected with gas exchange shows slight abnormality but this appears to have little or no clinical effect. In contrast to simple pneumoconiosis, complicated pneumoconiosis appears to progress, by enlargement of the masses of fibrous tissue and dust, irrespective of further dust exposure, although there are indications that heavy dust exposure accelerates progression.

In advanced complicated pneumoconiosis there is commonly compensatory emphysema (which may be gross), reduction of the pulmonary vascular bed and gradually increasing right heart failure.

Though the progress of complicated pneumoconiosis appears inexorable, there is nevertheless wide variation in the rate of progression from one individual to another and also in a single individual at various periods. Thus several years may elapse without any apparent change and then suddenly for no known reason rapid advance may occur. In advanced complicated pneumoconiosis all lung functions and respiratory haemodynamics are seriously affected.

Simple pneumoconiosis appears to have little or no effect on the expectation of life and it is generally agreed that complicated pneumoconiosis shortens life.

In many coalmining countries, the development of bronchitis has been thought to have some association with pneumoconiosis and there are many disparate reports about the relative prevalence of bronchitis among miners with and without pneumoconiosis and among the general population. One difficulty in establishing the relationship, if any, between pneumoconiosis and bronchitis is the fact that compensation for occupational diseases is often traditional. Thus the miners tend to be particularly aware of respiratory symptoms which determine prevalence in epidemiological surveys. Solution of this problem must await the results of prospective research on the relationship between the attack rate of bronchitis in mining and control populations whose dust exposure has been accurately measured over a sufficient number of years.

Radiographic appearance

When the radiologist is confronted with a radiograph showing opacities suggestive of coalworkers' pneumoconiosis he must make two decisions–first the differential diagnosis, and second, assuming that he has diagnosed pneumoconiosis, the degree or amount of pneumoconiosis present.

A large number of conditions–most of them fortunately are rare–can present an appearance on the radiograph which resembles pneumoconiosis.

A quantitative diagnosis of pneumoconiosis is required for two reasons–first so that adequate advice may be given to the individual, and second, for epidemiological purposes. The first major contribution to the quantitative assessment of coalworkers' pneumoconiosis was the publication of the International Labour Office International Classification of Radiographs of Pneumoconioses in 1958. It is now in general use in most coalmining countries.

In the United Kingdom, it was recognised in recent years that the various categories of the ILO classification of simple pneumoconiosis represent a somewhat coarse gradation along a continuum of abnormality. A modification of the 1958 ILO classification, the United Kingdom National Coal Board elaboration, was evolved which gives a finer gradation of 12 steps in addition to the 4 basic ILO categories. It has already proved of value in epidemiological studies relating dust to the progression of pneumoconiosis. This elaboration has been embodied in the 1968 revised ILO classification as the "extended classification".

A notable aid in the radiological diagnosis of coalworkers' pneumoconiosis is the issue by the ILO of standard films exemplifying the various categories in the classification.

Lastly, radiological technique must be of a high order and it must be consistent.

PREVENTIVE MEASURES

The basic principles of prevention are easier to state than to put into practice. First, the dust in the working environment must be reduced to a level which does not constitute an unacceptable hazard over the working life of the miner. Second, if this proves impossible, the length of time during which miners work in hazardous dust conditions should be restricted. Third, in selected circumstances, e.g. where dust control proves exceptionally difficult, the use of dust respirators, which have been much improved as regards efficiency and comfort, may be justifiable as a temporary measure.

The prevention of coalworkers' pneumoconiosis in an industry requires:

(a) dust control;

(b) dust measurement;

(c) medical supervision.

Dust control. Effective dust control is essentially a matter of good mining engineering. The production of dust can be greatly reduced by well designed coal-producing machines–in this sphere there is great scope for further research and development. Once dust is airborne, it is almost impossible to capture the particles with sprays or aerosols. Water must be applied to the point of coal breakage in sufficient quantity and the machines designed accordingly. Such dust as does become airborne must be diluted by good ventilation.

Dust measurement. It is impossible to gauge the success of dust supression and the hazard to which coalminers are exposed unless the dust in the environment is monitored by a system of regular sampling. Generally speaking, dust sampling is now carried out either by counting the number of particles or measuring their mass. Now that satisfactory instruments for mass measurement are available it should be possible to implement the relevant recommendation of the Johannesburg Pneumoconiosis Conference (1960) which states that, in the case of coal dust, the mass concentration of the respirable dust is the best single descriptive parameter. These instruments also incorporate size selectors which collect only the respirable dust.

Medical supervision. No programme of prevention is complete without medical supervision. The whole mining population at risk should be subject to chest X-ray examination at regular intervals, the period between examinations being chosen according to the degree of hazard likely to be present. Too frequent examinations are likely to show little or no radiological change or to give rise to a situation in which natural variation in radiological interpretation exceeds any real change. On the other hand, if examinations are not frequent enough the opportunity to advise men in good time about their working conditions may be lost and the need for management to intensify dust control may be overlooked. An integral part of any system of medical supervision should be the provision of adequate statistical resources so that epidemiological data illustrating the prevalence, the attack and progression rates of pneumoconiosis can be accurately established.

The success of these technical measures will depend to a considerable extent on the provision of sound administrative arrangements for liaison between the mining engineers, the scientists and the doctors involved. Moreover, the progress of pneumoconiosis prevention should be kept under review by the highest level of management–otherwise it may be regarded in some quarters as an uncongenial interference with the business of coal-getting.

Treatment. There is no specific treatment for coalworkers' pneumoconiosis. But as coalworkers are often anxious about their chests, any concomitant condition should be fully investigated and treated where possible. Careful reassurance where this is justified and in suitable cases transfer to lighter work may well have beneficial results. Nor should the merits of rehabilitation and general supportive therapy be overlooked.

ROGAN, J. McG.
McLINTOCK, J. S.

CIS 77-365 "Respiratory disease in coal miners". Morgan, W. K. C.; Lapp, N. L. *American Review of Respiratory Disease* (New York), Apr. 1976, 113/4 (531-559). Illus. 135 ref.

CIS 79-966 "Complex analysis of factors influencing the development of coal miner's pneumoconiosis" (Kompleksnyj analiz faktorov, vlijajuščih na razvitie pnevmokonioza u šahterov). Suhanov, V. V.; Zinger, F. H.; Velikij, N. I.; Eremenko, E. I. *Ugol' Ukrainy* (Kiev), Oct. 1978, 10 (38-39). 3 ref. (In Russian)

CIS 78-1534 "Coalworkers' pneumoconiosis–Epidemiological data" (Les pneumoconioses du houilleur–Données épidémiologiques). Amoudru, C. *Annales des mines* (Paris), Jan.-Feb. 1978, 184/1, 4 p. 4 ref. (In French)

CIS 78-951 "Lung function and clinical findings in cross sectional and longitudinal studies in coal workers from the

Ruhr area". Smidt, U.; Worth, G.; Bielert, D. *International Archives of Occupational and Environmental Health–Internationales Archiv für Arbeits- und Umweltmedizin* (West Berlin), 17 Oct. 1977, 40/1 (45-70). Illus. 8 ref.

"Objective pathological diagnosis of coal workers' pneumoconiosis". Fisher, E. R.; Watkins, G.; Lam, N. V.; Tsuda, H.; Hermann, C.; Johal, J.; Liu, H. *Journal of the American Medical Association* (Chicago), 8 May 1981, 245/18 (1829-1834). Illus. 28 ref.

CIS 77-657 "Evolution of disability in coalworkers' pneumoconiosis". Lyons, J. P.; Campbell, H. *Thorax* (London), Oct. 1976, 31/5 (527-533). Illus. 17 ref.

CIS 78-62 "Application of epidemiological studies to prevention of dust-induced diseases" (Erkenntnisse epidemiologischer Untersuchungen für den Schutz vor Stauberkrankungen). Reisner, M. T. R. *Glückauf* (Essen), 6 Jan. 1977, 113/1 (21-26). Illus. 18 ref. (In German)

"Non-invasive magnetopneumografic estimation of lung dust loads and distribution in bituminous coal workers". Freedman, A. P.; Robinson, S. E.; Johnston, R. J. *Journal of Occupational Medicine* (Chicago), Sep. 1980, 22/9 (613-618). Illus. 23 ref.

Cobalt, alloys and compounds

Cobalt (Co)

a.w. 58.9
sp.gr. 8.9
m.p. 1 459 °C
b.p. 2 900 °C

a silver-grey, very hard, brittle, magnetic metal.

Cobalt metal fume and dust:

TWA OSHA 0.1 mg/m³
TLV ACGIH 0.05 mg/m³
STEL ACGIH 0.1 mg/m³
IDLH 20 mg/m³

Cobalt and cobalt oxide:

MAC USSR 0.5 mg/m³

Cobalt trivalent compounds are unstable but cobalt is bivalent in most compounds of industrial importance.

Occurrence. Cobalt is a relatively rare metal. The most important mineral sources are the arsenides (smaltite, safflorite, skuterudite and cobaltite), the sulphids (carrolite and linnaeite) and various oxidised forms (asbolite, heterogenite, sphaerocobaltite and erythrite). The main producers are Zaire, Canada, Morocco, Finland, the USSR and Zambia. The principal consumer of cobalt is the United States which uses about half of world production.

Production. Most cobalt is obtained as a by-product during the processing of other metals, mainly copper, nickel and lead. Many different processes are employed, each depending on the special characteristics of the cobalt-containing ore. With one method, cobalt is extracted by dissolving the ore concentrate in hydrofluoric acid, after which lime milk and chloride of lime are added with precipitation of cobalt hydroxide ($Co(OH)_3$) and cobaltic oxide (Co_2O_3) respectively. Another method consists in treating the ore concentrate (which usually also contains nickel, iron, copper and arsenic) with sulphuric acid in autoclaves at a pressure of 8-10 atm. The different metals are then precipitated, copper as copper sulphide with hydrogen disulphide and iron with lime, after being oxidised to trivalent compounds with chlorine gas or chloride of lime. Arsenic is precipitated together with iron; cobalt and nickel can then be disassociated since cobalt hydroxide is more stable than nickel hydroxide.

Uses. Alloyed with nickel and aluminium, cobalt is used in the manufacture of permanent magnets and in combination with chromium, nickel, copper, beryllium and molybdenum, it is used as a high-temperature, high-strength alloy in the electrical, automobile and aircraft industries. It is often added to tool steels to improve their cutting qualities and is of considerable importance in the manufacture of cemented tungsten carbide tools.

Cobalt blue ($CoO.Al_2O_3$)

Cobalt blue is a blue-to-green pigment of variable composition consisting essentially of a mixture of cobalt oxide and alumina. Commercial grades have a variation in cobalt content ranging from approximately 20 to 30%. It is used as pigment in the glass and pottery industry.

Cobaltous acetate ($Co(C_2H_3O_2)4H_2O$)

m.w. 249.1
m.p. 140 °C

red crystals.

It is used as a bleaching agent and as a catalyst.

Cobaltous chloride hexahydrate ($CoCl_2.6H_2O$)

m.w. 237.8
sp.gr. 1.92
m.p. 110 °C
b.p. 1 049 °C (decomposes above 118 °C)

dark red crystals, which are colourless when in very thin layers, or blue powder.

Cobaltous chloride has been used for invisible inks since, when heated, the crystal water is liberated and the almost invisible colour changes to dark blue. Cobaltous chloride is also used in the glass, pottery, photographic and electroplating industries.

Cobaltous nitrate ($Co(NO_3)_2.6H_2O$)

m.w. 291
sp.gr. 1.87
m.p. 55-56 °C
b.p. 55 °C ($-3\ H_2O$)

red crystals, deliquescent in moist air.

Cobaltous nitrate is used as a pigment. When the compound is heated, nitrogen dioxide is liberated. It can react violently with combustible materials.

Cobaltous oxide (CoO)

m.w. 74.9
sp.gr. 6.45
m.p. 1 935 °C

a greyish powder under most conditions, which may, however, form greenish-brown crystals.

MAC USSR 0.5 mg/m³

Cobaltous oxide is used as a pigment in enamels, glazes in the pottery industry, in paints and as a catalyst for the after-burning of engine exhaust gases.

Cobaltous phosphate octahydrate ($Co_3(PO_4)_2.8H_2O$)

m.w. 510.8
sp.gr. 2.77
m.p. 200 °C ($-8\ H_2O$)

a reddish powder.

Cobaltous phosphate octahydrate is used as a blue colouring in the pottery and glass industries and in the manufacture of pigments, enamels and glazes.

Cobalt tetracarbonyl ($Co_2(CO)_8$)

DICOBALT OCTACARBONYL

m.w. 342
sp.gr. 1.87

m.p. 51 °C
b.p. decomposes at 52 °C
v.p. 0.07 mmHg (9.1 Pa) at 15 °C
orange crystals.
MAC USSR 0.01 mg/m³

A catalyst, used mainly in the plastics industry.

Cobalt tetracarbonylhydro (C_4HCoO_4)
COBALT HYDROCARBONYL
m.w. 172
MAC USSR cobalt hydrocarbonyl and decomposition products 0.01 mg/m³

A catalyst.

HAZARDS

Inhalation of cobalt fume and absorption of cobalt salts will produce systemic poisoning with myocardial disorders and irritant effects on the airways, eyes, and the digestive tract; inhalation of cobalt dust has produced an asthma-like disease and fibrotic pulmonary lesions, and allergic dermatitis has also been reported in workers exposed to cobalt.

In the concentration of the cobalt ore, workers are exposed to dust and fumes containing both cobalt and other metals, e.g. arsenic and nickel. Carbon monoxide is formed during melting, and hydrogen sulphide is used for the precipitation of copper. Melting and pouring cobalt before pelletising also produce cobalt fumes. Dust containing cobalt together with tungsten, titanium and tantalum is a potential hazard in the production of cemented tungsten carbides and the grinding and sharpening of cemented carbide tools.

Cobalt carbonyls share the general toxicity of carbonyls because of the direct irritant and systemic action of the compound coupled with the effects of carbon monoxide which is released from their decomposition.

Biological effects and toxicity. Cobalt is an essential trace metal both for man and animals since it takes part in blood formation as a component of vitamin B_{12}. When given to animals, it causes a stimulation of red cell production, a rise in blood proteins, and transient damage to the α-cells of the pancreas with concomitant rise in blood sugar level. Acute oral poisoning with cobalt salts results in diarrhoea, loss of appetite, hypothermia and, ultimately, death. Inhalation of cobalt dust produces pulmonary oedema in laboratory animals. The chronic inhalation of dust containing cobalt together with tungsten, titanium and tungsten carbides resulted, in one animal experiment, in hyperplasia of bronchial cells, focal fibrotic lesions and granulomas of the lungs. Powdered metallic cobalt induces tumours in skeletal muscle of rat; cobalt sulphide has also shown a carcinogenic activity in rats and mice.

Cobalt-induced myocardial disorders. In 1963 and 1964 some breweries in Canada, the United States and Belgium started adding cobalt to their beer to stabilise the froth. A few months later, an "epidemic" of acute forms of cardiomyopathy, the "Quebec beer drinkers' disease", broke out in Canada. Similar disorders were observed a little later in other countries in people drinking beer of certain brands with cobalt additives. The total number of cases of this disease has never been established; some authors speak of 112 cases, 50 of which had a fatal outcome. Autopsy revealed severe degenerative changes in the myocardium without any sign of inflammation. The myocardial tissue contained a high level of cobalt (approximately ten times more than in controls). The synergistic part played by alcohol and lack of proteins in diet was not excluded, but the absence of new cases after the sales of beer with cobalt additives had ceased bears evidence to the involvement of cobalt in the pathogenesis of this type of cardiomyopathy. The clinical picture of the myocardial disorders observed in cobalt ore reduction workers is similar to that of the beer drinkers' cardiomyopathy. One case of fatal cardiotoxic effects of cobalt in an industrial environment has been reported: a smelter worker with four years of exposure to cobalt died of cardiac insufficiency.

The first distinctive myocardial disorders in cobalt production workers were observed by industrial physicians while examining ore processers. Workers employed in the ore roasting and reduction departments are exposed to dust containing insoluble cobalt compounds (mainly oxides, hydroxide and metallic cobalt). Airborne cobalt concentrations sometimes exceed the threshold limit values. In addition, the workers are exposed to intense radiant heat while performing heavy physical work.

The cardiotoxic effects of cobalt have been investigated experimentally. High doses of cobalt lead to the development of necrotic foci in the myocardium. Subacute effects of the metal are characterised by less pronounced damage taking the course of a myocardial protein dystrophy. Chronic administration in doses of 1/20 of the LD_{50} lead to diffuse degenerative and dystrophic changes in the myocardium. Enzyme systems which are most vulnerable to the action of cobalt are those which decarboxylate and dehydrate the keto acids and thus inhibit their oxidative transformation. Disturbances of the catecholamine metabolism, in particular with regard to noradrenaline, play an important part in the cardiotoxic mechanism. A diminution of the noradrenaline level in the myocardium due to the action of cobalt prevents the uptake of the inhibitor monoaminooxidasis. The disturbed permeability of the lysosomal membranes plays a role in this type of myocardial damage.

Clinical picture. The acute phase of the cardiovascular pathology observed is characterised by discrete subjective heart complaints (mainly slight pain in the heart region and dyspnoea) and by pronounced objective changes. Attention is to be paid to tachycardia, heart hypertrophy, signs of serious myocardial involvement and circulatory insufficiency. Other symptoms are: vegetative dysfunction and otorhinolaryngological disorders such as hyposmia, anosmia and chronic subatrophic rhinitis.

Biochemical tests reveal disturbances of the lipid metabolism (increase in total lipids, triglycerides, cholesterol, β-lipoproteins) and of the activity of certain enzymes (lactatedehydrogenase and creatinephosphokinase). A characteristic sign is the disturbed carbohydrate metabolism after application of the glucose tolerance test. There are also changes in the basic haemodynamic indices and in the ECG (mainly in the ST segment and T wave). A hypertrophy of the heart chambers may be observed, more generally of the left ventricle. These myocardial disorders should be classified as cobalt cardiomyopathy.

Industrial exposure to cobalt has also produced an asthma-like disease with cough and shortness of breath. A peculiar form of pneumoconiosis, resembling berylliosis, has been seen in the cobalt/cemented tungsten carbide industry. The lesions consist of nodular conglomerate shadows in the lungs, together with peribronchial infiltration. The disease may be reversible. However, the role played by cobalt here has not been established with certainty since tungsten, titanium and aluminium were also present in the dust.

Transient gastric symptoms, similar to those produced in experimental animals, have occurred after inhalation of excessive amounts of cobalt-containing dust.

An allergic dermatitis has been reported by many authors. It is usually located at the elbow flexures, the ankles and the sides of the neck. Both urticarial eruptions and erythematous papular types have been described. Since minute amounts of cobalt are able to produce this sensitivity, cobalt allergy is seen in many occupations including metalwork, printing, and the pottery, leather and textile industries; cobalt may sometimes be the causative agent in cement eczema.

Diagnosis. The most characteristic signs for early diagnosis of cobalt cardiomyopathy are the repolarisation changes in the final part of the ventricular ECG complex and the deterioriation of the myocardial contractility. In the presence of such disturbances the changes in carbohydrate metabolism revealed by the glucose test are of high diagnostic value. When there are pronounced changes, diagnosis can be based on the results of an objective cardiovascular examination.

SAFETY AND HEALTH MEASURES

Processes which produce cobalt dust or fume, such as grinding, metal spraying, etc., should be provided with an effective local exhaust ventilation. For temporary operations or when ventilation is not practicable, an airline respirator should be worn. If ventilation is not satisfactory, a dust and/or fume respirator can be used.

Careful medical surveillance should be a rule for all cobalt production workers. They should be examined at least once a year with the assistance of an otorhinolaryngologist. Special attention should be given to the cardiovascular system; an ECG should be recorded, and a radiological examination carried out. If cobalt-induced myocardial damage is suspected, the carbohydrate metabolism should be investigated (glucose tolerance test). Cases of cobalt cardiomyopathy must be given hospital treatment. Workers with signs of chronic poisoning by cobalt compounds and discrete disorders of the myocardium should be temporarily (e.g. for two months) transferred to work not involving exposure to cobalt. Cases of relapse and of pronounced poisoning with cobalt cardiomyopathy must be removed from workplaces with cobalt exposure. Contraindications for employment involving exposure to cobalt are all diseases of the cardiovascular system.

Avoiding skin contact may be difficult but protective clothing and barrier creams can be tried. In cases of mild dermatitis, such measures may prove sufficient but severely affected patients must be removed to other occupations. Workers with a history of skin disease should not be employed in jobs where skin contact occurs.

Treatment. No specific antidote exists for cobalt. The most efficient clinical treatment of cobalt-induced cardiomyopathy appears to be a combination of Retabolil (one injection per week during four weeks) with β-blockators in average doses of 60-80 mg Obsidan/24 h. Other symptomatic drugs (potassium salts, diuretics) have proved useful. The use of cardiac glycosides is of little efficiency.

Radioactive cobalt

Radioactive cobalt (^{60}Co) does not exist in nature but is prepared in nuclear reactors and is used as a γ-emitter in industry and medicine. (See RADIATION, IONISING: INDUSTRIAL APPLICATIONS; RADIATION, IONISING: MEDICAL APPLICATIONS.)

SUVOROV, I. M.
ČEKUNOVA, M. P.

General:

CIS 77-1347 *Cobalt.* Hellsten, E.; Henriksson-Enflo, A.; Sundbom, M.; Vokal, H. USIP Report 76-11 (University of Stockholm, Institute of Physics, 113 Stockholm) (Mar. 1976), 44 p. 45 ref.

CIS 77-1360 "Cobalt and mineral compounds" (Cobalt et composés minéraux). Morel, C.; Cavigneaux, A.; Protois, J. C. Fiche toxicologique n° 128, Institut national de recherche et de sécurité. *Cahiers de notes documentaires–Sécurité et hygiène du travail* (Paris), 2nd quarter 1977, 87, Note No. 1064-87-77 (259-262). 19 ref. (In French)

Health impairment:

"Cobalt induced myocardiopathy in occupational pathology" (Kobal'tovye miokardi patii v klinike professional'nyx zabolevanii). Suvorov, I. M.; Uspenskaya, N. V.; Rozina, G. Ju.; Knysh, S. V.; Chekodanova, N. V.; Čekunova, M. P.; Zenkevich, E. S.; Revnova, N. V.; Dobrynine, V. V.; Orlova, T. V.; Ornitsan, E. Ju.; Mishkich, I. A.; Lifyandsky, V. G. *Kliniceskaja medicina* (Moscow), Oct. 1978, 10 (58-63). Illus. 12 ref. (In Russian)

"Cardiotoxic mechanism of certain industrial poisons (metals)" (Mekanizm kardiotoksiceskogo dejstvija nekotoryx promyslennix jadov (metallov)). Čekunova, M. P. *Kardiologija* (Moscow), May 1978, 5 (54-61). Illus. 46 ref. (In Russian)

"Asthma symptoms of chronic bronchitis and ventilatory capacity among cobalt and zinc production workers". Roto, P. *Scandinavian Journal of Work, Environment and Health* (Helsinki), 1980, 6/suppl. 1, 49 p. Illus. 63 ref.

"Evaluation and relevance of isolated test reactions to cobalt". Rystedt, I. *Contact Dermatitis* (Copenhagen), July 1979, 5/4 (233-238). 21 ref.

Preventive measures:

CIS 80-219 *Occupational safety and health rules for the nickel and cobalt industries* (Pravila bezopasnosti v nikel'-kobal'tovoj promyšlennosti). Gosgortehnadzor (Moscow, Izdatel'stvo Metallurgija, 1979), 72 p. (In Russian)

Criteria for controlling occupational exposure to cobalt. NIOSH Occupational Hazard Assessment. DHHS (NIOSH) publication No. 82-107 (Washington, DC, Superintendent of Documents, US Government Printing Office, Oct. 1981), 95 p. Illus. 33 ref.

Cocoa cultivation

The cocoa tree *(Theobroma cacao)* produces a fruit from which cocoa beans are obtained. It grows in the tropics between latitudes 20 °N and 20 °S. It requires a temperature which does not fall below 20 °C, annual rainfall not less than 125 cm and an altitude not exceeding 1 500 m. It originated in Mexico and now thrives in 50 countries situated in Central and South America, Ceylon, the South Pacific Islands, New Guinea and West Africa. Ghana, Nigeria, the Ivory Coast and Cameroon now account [for more than 60% of the world production, which fluctuates around 1.5 million tonnes per year. In Ghana and Nigeria, the main producing countries, there are, for historical reasons, no large plantations and virtually no mechanisation. Large plantations exist in Brazil and Cameroon as well as in former French territories.] Cocoa is used in the manufacture of chocolate and vegetable fat and in the pharmaceutical industry.

Cultivation. To establish plantations, seedlings are raised in nurseries in the dry season, and transplanted to cleared forest lands when the rains become steady. The young plants are pruned to produce umbrella-shaped trees

2.5 m tall. Fruiting, which follows, varies between 3 and 5 years. Harvesting may occur between September and March or throughout the year. [Cultivation operations are mainly manually done even in large plantations, since tractors operate with difficulty on the terrain where cocoa is grown with its abundance of shade trees. Pesticides spraying, against capsid bugs and black pod

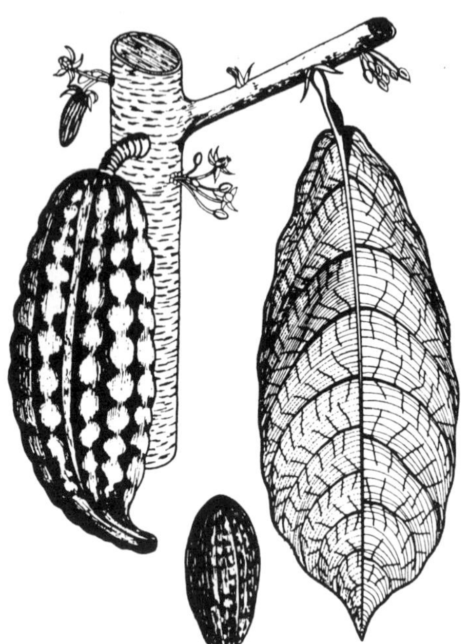

Figure 1. Fruit and leaf of the cocoa tree *(Theobroma cacao)*.

(fungal) disease, is mechanised. Harvesting is necessarily a manual operation, but the removal of the harvested crop by tractors and trailers or by trucks is increasing.] The last stages of production include fermentation, drying, packaging, storing and transportation of beans.

HAZARDS AND THEIR PREVENTION

During forest clearing, fatal injuries may occur from falling trees. Cuts from saws and machetes, and injuries from machinery are frequent. Snake bites are common and fatal injuries have been produced by such dangerous animals as elephants, bush cows, crocodiles and hippopotami. Fire outbreaks occur occasionally.

Chlorinated hydrocarbon and organophosphorus pesticides and copper-based fungicides have caused dermatitis and cases of intoxication. Malaria, dysentery, pneumonia, guinea worm, tetanus, tuberculosis, schistosomiasis, filariasis and sleeping sickness, often related to low socio ecomonic standards, are frequent.

Comprehensive safety and health measures include: controlled forest clearing; provision of safe agricultural implements; tractors with safety cabs and competent, trained drivers; the use of safe spraying equipment and protective clothing and respiratory protective equipment during pesticide application; periodical medical examinations; the provision of medical centres; good housing and sanitary facilities and reasonable social-welfare amenities. (See PLANTATIONS).

SOFOLUWE, G. O.

Industrial and labour problems:

General report. Committee on Work on Plantations, Seventh Session, Geneva 1976 (Geneva, International Labour Office, 1976), 186 p.

Cocoa industry and chocolate production

The stages involved in the processing of the cocoa bean and the manufacture of cocoa powder and chocolate are shown in figure 1.

HAZARDS AND THEIR PREVENTION

The hazards encountered are, to a large extent, the same as those in biscuit making and are related to: the raw materials; lifting and carrying, mechanical handling and storage techniques; greasy and slippery floors; certain dangerous machines common to both industries; high noise levels; atmospheric dust concentrations; heat and hot working conditions.

During the sorting and cleaning of cocoa beans, there is little in the way of hazards, provided moving machinery parts are equipped with adequate machinery guarding. During roasting and drying, there are the hazards commonly associated with the firing equipment used for furnaces and ovens and with the heating of flammable materials. Suitable fire-fighting equipment should be located at strategic points. Roasters and driers should be operated by skilled workers who have received training in the operation and safety of this plant; operating instructions should be displayed in a clearly visible place. Manual handling of cocoa beans may be arduous work.

Cocoa crackers and fanners are enclosed units and present no hazard during normal operation. Manual feeding and discharging may be arduous, and mechanisation by means of skip hoists, belt conveyors and screw conveyors may entail considerable trapping and crushing hazards unless mechanical handling equipment is adequately enclosed.

The only danger points on modern, all-enclosed grinders are the inrunning nips of grinding discs or refining rolls, and these should be fitted with grills or other suitable guards. Pressing and filtering equipment for the extraction of cocoa butter is not dangerous provided loading and filtering operations are interlocked and suitable safety measures are taken during maintenance and cleaning. Cocoa-presscake pulverisers are all-enclosed and present no hazards; however, these machines may produce considerable amounts of dust. Cocoa liquor mills, if not of the modern, fully protected type, may have dangerous unguarded blades and nips which can cause crushing, shearing and tearing injuries.

During intermediate operations and manual lifting, carrying and charging, there is the danger of ill-balanced container lids, various moving parts and the use of stools to gain access to loading points. The various types of conching machines used for refining the mixture of cocoa powder, cocoa butter and fine sugar present a number of hazards in relation to their design; conche loading is often arduous. Chocolate casting machines are dangerous due to the numerous reciprocating mechanisms that they contain, such as transfer chains, pourers and levellers; these machines are also very noisy and the process of mechanical consolidation should be done in a separate soundproof room. The inrunning nips on the steel belt conveyors of chocolate enrobing machines should be suitably protected. Wrapping and packaging machines have their own specific hazards; packaging is often carried out by small teams of workers, and the work is mainly static and many entail fatiguing postures.

VILLARD, R. F.

General:

Labour and social problems arising out of seasonal fluctuations of the food products and drink industries. Second Tripartite

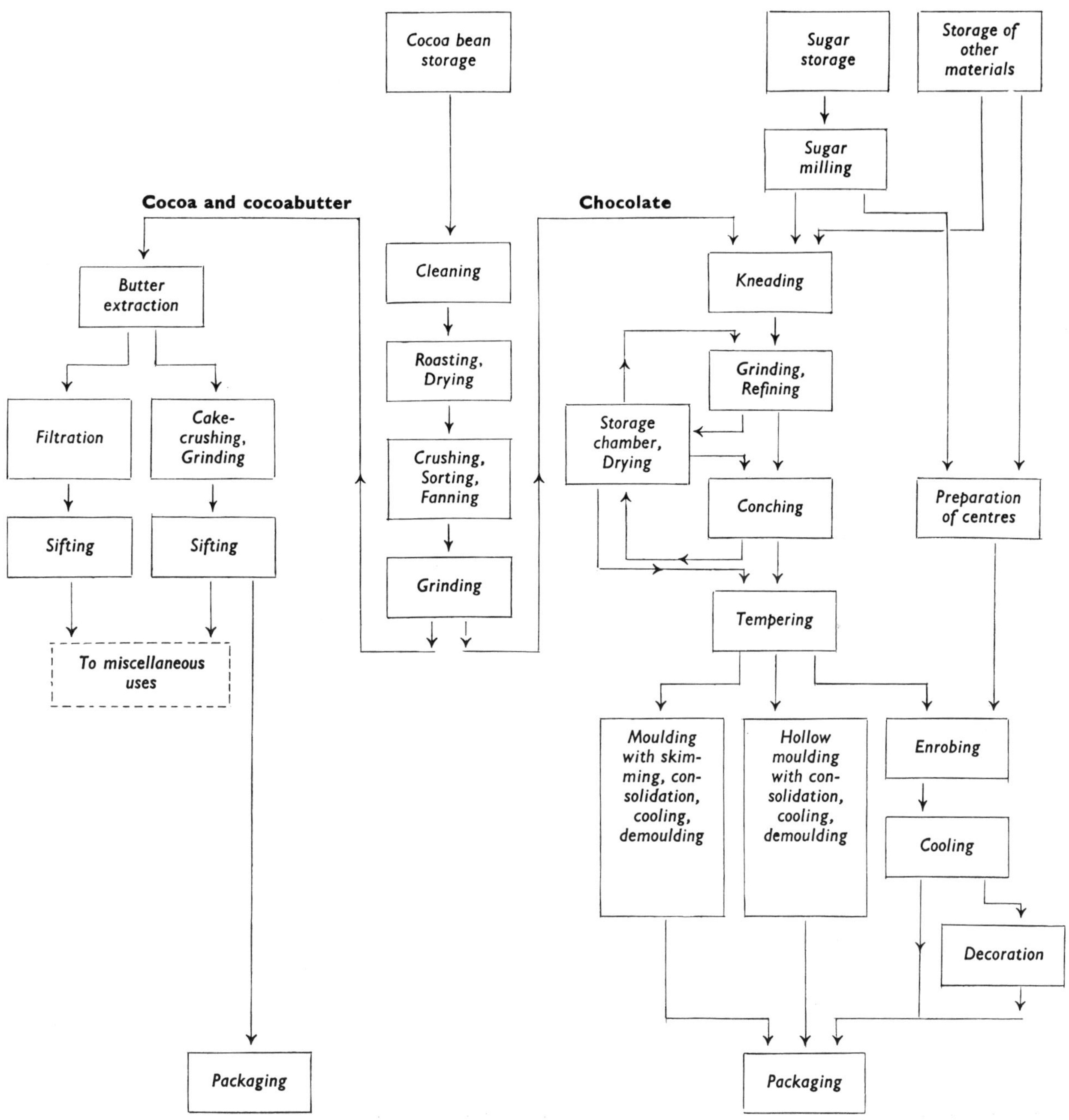

Figure 1. Manufacturing processes in the cocoa and chocolate industries.

Technical Meeting for the Food Products and Drink Industries (Geneva, International Labour Office, 1978), 51 p.

Safety:

Recommendations on the design of machinery and equipment for safe operation: No.22. Confectionery and chocolate making. Commonwealth and State Departments of Labour. 1968 (Department of Labour and National Service, PO Box 2817AA, Melbourne 3001 Victoria) (1968), 10 p.

Health:

"Survey on work in chocolate industries and on relevant diseases. Results of an investigation and documentation data" (Aperçu sur le travail dans les chocolateries et sur la pathologie qui s'y rapporte. Résultats d'une enquête et données de la littérature). Doignon, J.; Lazarini, H. J.; Larche, M.; Perier, F.; Mothe, J. P.; L'Epée, P. *Archives des maladies professionnelles, de médecine du travail et de sécurité sociale* (Paris), Mar.-Apr. 1979, 40/3-4 (569-570). (In French)

Coconut cultivation

The coconut palm *(Cocos nucifera)*, every part of which is put to some use, is very closely linked with the economic well-being of a large proportion of the population of the areas in which it grows. It is found mainly in South-East Asia, the Pacific Islands, East and West Africa, the West Indies and Central and South America. The chief coconut-producing countries are the Philippines, India, Indonesia, Sri Lanka, Malaysia and the Pacific Islands. The fruit is an important foodstuff; it yields an edible oil and the dried meaty endosperm is marketed as copra. See also COIR.

Cultivation. Coconut is generally a smallholder's crop and is grown on a plantation basis for commercial purposes. Inter-planting of other crops within areas of palms with a view to maximum utilisation of land is carried out in some places. Milch cattle and buffaloes used for ploughing and drawing of carts are kept on

Figure 1. Coconut palm *(Cocos nucifera).*

many planatations. The latter are, however, gradually being replaced by tractors.

About 6-10 years from the time of planting the seedling nuts, the tree begins bearing and continues to do so for anything up to 80 years or more with little or no attention. On a plantation basis, however, fertiliser application, soil management and efficient cultivation practices are employed to obtain maximum yields. The trees produce nuts the year round but commercial plucking is most commonly done at 2-monthly intervals. By comparison with most agricultural crops it is far less labour-intensive and resident labour forces are therefore quite small.

HAZARDS AND THEIR PREVENTION

The commonest accidents are those caused by agricultural implements used for cutting and digging. Falls from trees, where plucking is done by climbing, and injuries caused by falling nuts, where knives lashed to long lengths of bamboo are used, occur occasionally. Snake bites are not uncommon. Tapping of the inflorescence for the juice entails the tapper walking from tree to tree on strands of rope, high above the ground. Death from falls are fairly frequent in this, the most perilous of occupations connected with the coconut palm. Development work on an effective safety harness is being carried out.

The chief health problems are malaria, ancylostomiasis, anaemia and the enteric diseases associated with poor environmental sanitation.

Education of labour in careful handling of agricultural implements, use of safety belts by tappers, machinery guarding such as safeguards on tractors and proper training of operators would help cut down accident rates. Better housing, advice on nutrition, provision of safe water supplies and attention to environmental sanitation is needed to reduce illness. (See PLANTATIONS).

FERNANDO, L. V. R.

Codes of practice

National systems for the administration of the legislation necessary for the maintenance of health and safety at work vary and reflect the cultural and legal ethos of the individual country. There are constraints on this system which are imposed by membership of international bodies such as the ILO, whose Conventions and Recommendations are reflected in national legislation, as well as of organisations, such as the European Economic Community, which attempt to harmonise aspects of policy and standard setting in occupational health.

[As discussed in more detail under OCCUPATIONAL HEALTH LEGISLATION, law making aiming at the protection of workers' health in some countries takes the form of detailed regulations while in others it mainly defines objectives. In the latter case statutory obligations have to be supplemented by non-mandatory technical standards providing guidelines for compliance, particularly for the benefit of small and medium-sized enterprises. Such an approach has been followed in Great Britain since 1972.]

In Great Britain occupational health and safety is the responsibility of the Health and Safety Commission, whose operational arm is the Health and Safety Executive. The Commission is made up of representatives of the Confederation of British Industries (employers), the Trades Union Congress (trade unions) and local authorities. The tripartite approach of involving employers, workers and government runs through all levels of legislation and is well expressed in the Report of the Robens Committee (1972): "The primary responsibility for doing something about the present levels of occupational accidents and disease lies with those who create the risks and those who work with them."

Under the Health and Safety at Work, etc., Act, 1974, an attempt has been made to get away from the rigidity of much earlier law making and to provide a framework that can be more responsive to changing technologies and improved expectation of environmental conditions. At the same time it is essential that such flexibility does not introduce any lowering of the level of protection afforded.

The Act provides for three types of control system:

(a) regulations,

(b) approved codes of practice, and

(c) notes of guidance.

Regulations are normally brief and set out the mandatory standards and objectives for control of particular hazards to which all must conform.

Approved codes of practice offer practical guidance on the ways of achieving the mandatory standards and objectives. However, if an employer can achieve these standards and objectives by some other safe and efficient method he can do so and not be in breach of his statutory duty.

Notes of guidance are much more detailed; they provide advice, recommendations and technical information to assist employers and workers in achieving the required levels of health and safety, consistent with any relevant legislation. They are an indication of good practice and while they do not have any specific legal status they can be taken to be illustrative of the types of action required to comply, not only with particular regulations, but also with the general duties imposed by the parent legislation, the Health and Safety at Work, etc., Act, 1974.

Proposals for regulations, approved codes and notes of guidance are developed by the Health and Safety Commission in consultation with technical and professional experts, as well as with employers and workers. The normal practice is for a round of discussions to culminate in the issue by the Health and Safety Commission of a consultative document containing drafts of the regulations, approved codes, and notes of guidance, and for this published document, on which anyone can comment, to be specially drawn to the attention of appropriate employers' and workers' organisations, public authorities and professional societies. After this further round of consultation the drafts are further revised as necessary. Final proposals for regulations are then submitted to the Government for approval and appropriate arrangements made to publish the approved codes and guidance notes. This procedure attempts to ensure the wide measure of agreement among the parties concerned as well as a fullest involvement of those with particular skills and interests.

In addition to the specialised use of the term "approved codes of practice" under the Act, the term "codes of practice" is also used for a very wide range of documents of differing status. For example individual employers may issue such instructions about the conduct of their own operations, and professional and technical associations may similarly use the term to cover advice given to their membership or recommendations for the interest of the public at large. Publications such as these are obviously a source of information which can be drawn on by the Commission in the standard-setting process, but the main characteristic of the United Kingdom approach to approved codes of practice and legislation is the marrying of technical knowledge and skill with the managerial, workforce and public interest to seek the balance which is to be struck in every issue in the health and safety field, in achieving the appropriate levels of protection without inhibiting the proper operation of industries, research laboratories or any other field of human activity.

[At the international level, the ILO has been following a similar pattern for the past 20 years by supplementing systematically Conventions and Recommendations on occupational safety and health subjects by codes of practice, guides and manuals. These documents are not intended to be substitutes for existing national legislation, regulations or safety standards, but rather to provide guidance for governments, employers and workers. They are not compulsory in character and aim at stimulating and guiding those whose duty is to promote at the national level and at the level of the undertaking occupational safety and health in a given area. The codes of practice are examples of national or factory's regulations; the manuals and the guides are of a more descriptive and operational character, are illustrated, and focus on safe methods of work.]

DUNCAN, K. P.

Safety and health at work. Report of the Committee 1970-1972. Chairman Lord Robens. Cmnd 5034 (London, HM Stationery Office, July 1972), 2 vols., 281 and 718 p. 69 ref.

Health and Safety at Work, etc., Act, 1974 (London, HM Stationery Office, 1977), 117 p.

Coffee cultivation

The coffee plant (figure 1), a shrub or small tree of the genus *Coffea*, is a native of Abyssinia but is now widely cultivated throughout the torrid and tropical zones; it is also found south of the Tropic of Capricorn in Brazil and Mozambique. World coffee production was nearly 4 500 000 t in 1977 of which South America produced 46%. The principal South American producers are Brazil and Colombia, which account for 26% and 14% of world production respectively.

Six species of *Coffea* are grown commercially, *C. racemosa*, *C. sterrophyla*, *C. abeotikal*, *C. liberica*, *C. arabica* and *C. canephora*; however, *C. arabica* is grown in over 80% of coffee-producing countries and accounts for around 90% of world production.

Figure 1. The coffee plant (genus *Coffea*).

Cultivation. Coffee is primarily a smallholding crop and there are some 3-4 million coffee-producing units throughout the world. Primitive methods of cultivation are common in these small production units, especially those operating in forest areas where coffee is grown side by side with other crops. However, there is a gradual movement towards the introduction of modern cultivation techniques.

The cultivation cycle comprises: soil preparation (ploughing, hoeing, harrowing, etc.), which may be mechanised or manual; planting of ripe fruits, dry seeds or young plants previously raised in propagation beds; and plant care by the application of fertilisers and pesticides and by irrigation, shading, fire control, etc. In areas where young plants are used as the start of the main cultivation cycle, the growing of young plants from seed may often be a large and important separate industry.

The coffee berries are picked by hand, although picking machines have been developed, and small ladders may be required to reach the ripe fruits.

Preliminary processing. The berries are washed, the pulp and membranes are removed, and the resulting beans are washed again, either in water or in a weak alkaline solution (1% sodium hydroxide) and finally dried, either in the sun or in drying machines at a temperature of 45-70 °C. The dried coffee then goes through a process of benefaction in which the beans are freed of waste material, peeled, screened and classified ready for bagging; these processes are often all carried out in a single machine.

HAZARDS

The incidence of accidents in coffee cultivation is low and those that do occur are mainly snakebites, lightning damage, falls from trees, burns and minor cuts and abrasions. The gradual introduction of machinery, especially tractors for farm transport, is, however, increasing the accident risk. Another current trend is for workers to live in urban areas and travel to and from the plantation by lorry each day; this has increased the hazard of commuting accidents. Drying machines and benefaction machines may cause electrical hazards or crushing and amputation injuries, if moving parts are unguarded.

Coffee plantation workers are exposed to any tropical diseases that are endemic in the region. Pesticides may cause cases of poisoning, and workers may be affected by the local climatic conditions. However, no diseases specific to coffee plantation workers have been reported.

SAFETY AND HEALTH MEASURES

Where snakes prove a menace, active trapping campaigns should be instigated; workers should wear adequate clothing, especially foot and leg protection, and anti-snakebite serum should always be at hand for prompt use. There is little that can be done to protect workers in the field from sudden lightning storms.

Persons handling and applying pesticides should be fully informed of the hazards of the substances; they should be supplied with suitable personal protective equipment and working clothes and adequate sanitary facilities should be provided to allow scrupulous personal hygiene to be practised after exposure.

Drying and benefaction machines should be equipped with suitable machinery guarding and machine operators and tractor drivers should be given thorough instruction and training in the safe use of their equipment. The use of qualified drivers and a high standard of vehicle maintenance will do much to reduce the incidence of commuting accidents where the employer supplies the transport facilities.

Coffee plantation workers should receive regular medical examinations and should be immunised against tetanus; first-aid facilities should be available for the prompt treatment of injuries.

BEDRIKOW, B.

General:

World coffee survey. Krug, C. A.; De Poerck, R. A. Food agricultural studies No. 76 (Rome, Food and Agriculture Organisation, 1968), 476 p.

Housing, medical and welfare facilities and occupational safety and health on plantations. Committee on Work on Plantations, Seventh Session, Geneva 1976 (Geneva, International Labour Office, 1976), 103 p.

Coffee industry

Production. [In 1980, 53 848 000 bags of green coffee, 1 375 000 bags of roasted coffee and 4 160 000 sand bags of soluble coffee were imported from all sources by the countries members of the International Coffee Organisation. In the same year per capita consumption of coffee in the importing member countries was estimated at 4.22 kg in all countries, 4.85 kg in the United States and 5.04 kg in the countries of the European Community.]

The green coffee beans arrive from the coffee plantations in 60 kg hemp bags, which are then stored ready for processing; the beans may be fumigated with bromomethane or other halogenated or organophosphorus pesticides during storage and marked with iron oxide, a process which produces considerable clouds of dust.

Coffee processing. The main stages in coffee bean processing are blending to obtain the desired flavour mix, cleaning to remove light impurities such as lint, hulls, earth, etc., roasting at temperatures of 500 to 1 000 °C for 15-20 min in batch roasters or at around 260 °C for 10-15 min in modern continuous roasters, cleaning by air blast to remove heavy materials such as stones, grinding in a plate or roll mill, and packaging; in many cases, the roasted bean leaves the factory intact and grinding and packaging are carried out at the retail outlet.

Instant coffee production. [It is estimated that in 1982 consumption of instant coffee in industrialised countries will account for about 20% of the total if calculated on the basis of the amount of green coffee used for its manufacture; 30% would appear to be a realistic figure if calculated on the basis of cups of coffee consumed. The figure varies widely between countries; when the use of soluble coffee is calculated through measurement of household purchases, the results are as follows (percentage of instant coffee): Belgium, 6; France, 18; Federal Republic of Germany, 7 to 8; Netherlands, 5; Sweden, 2; Spain, 33; United Kingdom, 95; United States, 30.

Preliminary processing is identical with that for normal coffee. Liquid coffee is made by passing water, at temperatures of up to 150 °C, through ground beans to produce an extract containing about 30% solids. The solution is usually spray dried, although vaporisation and freeze drying (lyophilisation) are now becoming common processes. The instant coffee producer may roast and grind his own coffee or buy ground coffee from a roasting plant.

Decaffeinated coffee production. Decaffeinated coffee, both normal and instant, is prepared from green beans from which the caffeine has been removed by solvent extraction. Trichloroethylene at a temperature of 70 °C is the most widely used solvent and removes up to 95% of the caffeine. The residual solvent is removed by steaming and the beans are then dried and roasted.

HAZARDS

Falls of workers and falls of bags of beans are relatively common during the transportation of bagged coffee in and around roasting or instant coffee plants. The main causes are incorrect mechanical handling procedures and inadequate stacking techniques coupled with a lack of maintenance and good housekeeping. Other causes of falls of workers are incorrect use of ladders, wet and slippery floors and unguarded scaffolds.

In roasting plants, flying particles such as hulls, earth, cloth and other impurities removed from the coffee are a constant hazard. Burns may occur due to contact with hot materials or other equipment in the vicinity of roasters or due to accidents involving hot water or high-pressure steam.

The large quantities of liquid or gaseous fuel required for roasting constitute a considerable fire and explosion hazard. The diphenyl derivative, Dowtherm A, is widely used as a heat transfer medium in this industry and leakage of this substance may lead to an explosion.

The widespread use of machines such as grinding mills, conveyors and packaging equipment increases the danger of mechanical and electrical hazards.

Allergic reactions of the skin, mucous membranes and respiratory system have been encountered in persons handling green coffee. Symptoms observed include nasal discharge with mucous congestion, skin eruptions, dyspnoea, asthma and lacrimation, often with headaches. Severe reactions follow first-time exposure or return-to-work after absence and mild symptoms are encountered in persons who have a long history of exposure. Skin and inhalation tests have been made to determine the causative agent and several groups of workers have been subjected to passive transfer, precipitation reaction and antibody identification studies with different species of coffee; in a few cases, it has been possible to attribute the allergy to the fact that the coffee bags had previously been used for transporting castor bean and that traces of castor bean were still present in the bags. [Microbial foreign components present in contents of bags of green coffee are responsible for a large percentage of allergy cases who do not react to green or roasted coffee. The allergenic potency of such dust decreases on heating.] Allergy testing with fungal and other impurities found in coffee proved negative. Chlorogenic acid has been extracted from raw coffee and identified as a possible allergen. This acid derives from caffeic acid and quinic acid and is found not only in coffee but also in castor bean, oranges and various other plants.

In both normal and instant coffee plants, the high temperatures and the high levels of relative humidity resulting from extensive use of steam may cause fatigue and thermal exhaustion in workers. The noise levels are also high and prolonged exposure may result in occupational deafness.

Cases of pesticide poisoning have been reported in persons fumigating coffee or handling recently fumigated materials, and there is a hazard of solvent poisoning due to the inhalation of the trichloroethylene vapours that occur during solvent extraction of caffeine.

Coffee processing plants may also prove a nuisance to persons living in surrounding residential areas by the emission of odours, and the disposal of solid wastes from instant coffee production into water courses would rapidly cause high levels of water pollution.

SAFETY AND HEALTH MEASURES

Accidents during the transport of sacks of coffee can be prevented by the use of safe mechanical handling and stacking techniques, correct use of well made and well maintained ladders and suitable maintenance and good housekeeping; workers should be provided with hand and arm and foot and leg protection. Oil or gas burners for coffee roasting should be fitted with pilot flames or flame failure protection devices and adequate fire-extinguishing equipment should be available for fire fighting; banks of six or more carbon dioxide cylinders have proved to be the fire extinguishers of choice in this industry.

Effective machinery guarding should be fitted to grinders, conveyors and packaging equipment and this safety equipment should be effectively maintained and regularly inspected. Exhaust and general ventilation should be provided to control high atmospheric concentrations of dust or fumes and to improve the thermal comfort of workers.

Workers should be given pre-employment and periodic medical examinations in order to identify persons who show allergic reactions to green coffee beans or who exhibit noise-induced hearing loss.

BEDRIKOW, B.

Occupational disease:

"Extraction and analysis of coffee bean allergens". Lehrer, S. B.; Karr, R. M.; Salvaggio, J. *Clinical Allergy* (Oxford), 1978, 8 (217-226).

CIS 78-1537 "Coffee roaster's lung" (Le poumon des torréfacteurs de café). Decroix, G.; Fichet, D.; Hirsch-Marie, H. *Revue française des maladies respiratoires* (Paris), 1977, 5 (343-344). 1 ref. (In French)

"Immunological and respiratory changes in coffee workers". Žuškin, E.; Valič, F.; Kanceljak, B. *Thorax* (London), Jan. 1981, 36/1 (9-13). Illus. 15 ref.

Safety:

Occupational safety standards for coffee sack transport (Normas de segurança do trabalho em atividades de transporte de sacas de café). Portaria No. 70 (Ministerio do Trabalho, Brasil, 15 Dec. 1970). (In Portuguese)

Treatment:

"Respiratory function in coffee workers". Žuškin, E.; Valič, F.; Skuric, Z. *British Journal of Industrial Medicine* (London), May 1979, 36/2 (117-122). Illus. 16 ref.

Coir

Coir is the prepared fibre of the husk of the coconut *(Cocos nucifera).* It is an exceedingly strong fibrous material which has many uses in industry for making brushes, ropes, cordage and matting. Production is principally in India, Sri Lanka, the Philippines, Indonesia, Malaysia and the Pacific Islands.

The preparation of the coir fibre is carried out in stages as follows:

Harvesting. The coconuts are collected from the tops of the palm trees which grow to a height of 20-30 m. The husks, consisting of the fibrous covering of the coconuts, are removed by striking and levering them against spikes fixed firmly in the ground.

Retting. The coconut husks (after removal of the nut) are soaked in pits, with 6-10 ft of stagnant water, usually rain collected for 3 to 4 weeks. The removal of the husks from water is done by experienced workers, who often work for 2-3 hours daily in these pits in waist-deep murky water.

Decortication. The husks are then decorticated in machines. The slippery coconut husk is held by the operator with both hands and fed into the in-running nip of a decorticating machine, which is not usually guarded. This has a high risk potential; a momentary loosening of the grip can cause the hand to be pulled into the machine along with the husk, resulting in a crushed hand.

Bleaching process. Coir fibre is sometimes bleached to improve its colour and appearance. The fibres are stacked in a room in which are placed trays of ignited sulphur. After the bleaching is completed the doors are opened and after about 2 h the workers enter to remove the bleached fibre. Reliance is placed on natural ventilation to remove the sulphur fumes.

Fibre processing. The lacerated fibres are then combed in smaller drums to obtain "bristle fibre". The shorter fibres that fall under the drums are sifted, a very dusty operation, to obtain coir or "mattress fibre". The fibre is then pressed mechanically and baled for transport to shippers or agents. There are more than 630 registered fibre mills employing some 15 000 workers.

HAZARDS AND THEIR PREVENTION

The epidemiology of non-occupational diseases of occupational origin such as worm infestations and skin infections among the workers engaged in retting operations are suggested research areas.

The oedema of terminal phalanges, paronychia and similar lesions found in fibre mill workers can be attributed to constant handling of wet coconut husks and pressure on the fingers while feeding the husks into the machines. Subonychial pigmentation could also result from subcuticular petechial haemorrhages. Studies in the epidemiological pathology of these conditions are fruitful in a prevention programme.

Clinical and radiographic studies of workers exposed to fibre mill dust have not revealed the occurrence of an occupational disease. Environmental monitoring combined with a questionnaire survey indicated that workers do not appear to be affected at mean dust concentrations of around 6 mg/m^3. Higher dust levels could be of nuisance value. A tentative threshold limit value of 5 mg/m³ has been proposed for coconut fibre dust.

Accidents are caused by lack of training for workers on the correct use of machines, inadequate guarding and maintenance of machinery, bad housekeeping, overcrowding of machinery and physical exhaustion resulting from continuous piece-rate work. Personal loss of earning capacity and national loss of skilled manpower can be minimised through in-service training, adequate machine guarding and safety surveillance.

Release of sulphur dioxide and acid fumes from fibre bleaching sheds affect the workers in the vicinity. It could also be an environmental problem depending on where the factories are sited. Laboratory experiments in utilising charcoal filter trapping devices to absorb the fumes have been successful. Industrial applications of this technique to prevent pollution should be explored.

PINNAGODA, P. V. C.

CIS 75-652 "A clinical and radiographic study of coir workers". Uragoda, C. G. *British Journal of Industrial Medicine* (London), Feb. 1975, 32/1 (66-71). 15 ref.

"Occupational health surveys in developing countries with examples from Sri Lanka". Pinnagoda, P. V. C. (63-76). *Environmental pollution and human health. Proceedings of the International Symposium on Industrial Toxicology, Nov. 4-7, 1975* (Lucknow, Industrial Toxicology Research Centre, 1977), 909 p. Illus. Ref.

Coke industry

Coke is a coherent, cellular, carbonaceous residue remaining from the dry distillation of coking coal. In the coking process, the volatile components of the natural coals are driven off to form a substance with a substantially higher carbon content.

The world production of coke-oven coke (excluding pitch coke, petroleum coke, semicoke and gas coke) attained 361 958 000 tonnes in 1977. The USSR is the main producer with 86 800 000 tonnes, followed by the United States (52 820 000 tonnes), Japan (42 945 000 tonnes), the People's Republic of China and the Federal Republic of Germany.

Production. The coal is first cleansed and graded. Natural coals with 35% or 15% volatile matter do not coke, but sinter or pulverise. After appropriate crushing, these coals are mixed with coal having volatile matter between 19 and 28%, and oils may also be added. Coking is done in vertical, horizontal or inclined chambers, lined with silica bricks. The retort and chamber are surrounded by combustion flues. Crude or light gas from coke-fed central gas generators, or cracking, town or natural gas is used for heating. The furnaces may be filled by hand, more often semi-automatically; in modern plants, charging is fully automatic. Usually the coal is taken by a coal car from the coal hopper and poured into the various chambers. Depending on the coke quality required and type of coal used, coking is done for 6-20 h at temperatures from 700-1 200 °C. In vertical and

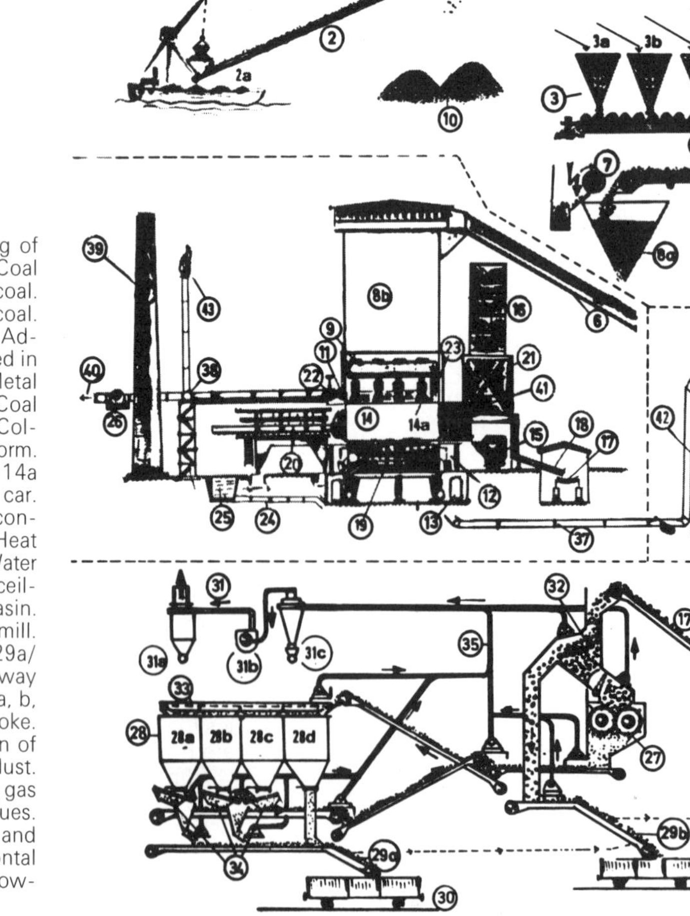

Figure 1.
Coke processing in detail. 1. Dressing of coal. 2. Transportation of coal. 2b. Coal freighter. 3a. Mill for non-coking coal. 3b. Bituminous coal. 3c. Gas coal. 3d. Slack. 4. Coal-mixing screw. 5. Admixture of oil (eventually directly added in the oven). 6. Coal conveyor belt. 7. Metal separation. 8a/8b. Coal bunker. 9. Coal charging car. 10. Coal stores. 11. Collecting main. 12. Exchange platform. 13. Blast pipe channel. 14. Retort, and 14a charging holes. 15. Coke quenching car. 16. Coke quenching tower. 17. Coke conveyor belts. 18. Coke bench. 19. Heat regeneration. 20. Pusher machine. 21. Water spray. 22. De-tarring plant. 23. Oven ceiling. 24. Drains. 25. Quenching basin. 26. Suction blower. 27. Coke mill. 28. Coke bunker (a-d varying sizes). 29a/29b. Conveying of coke. 30. Railway transportation. 31. Sucking off dust (a, b, c). 32. Separation of large-sized coke. 33. Coke separator. 34. Transportation of coke. 35. Sucking off coke dust. 36. Central gas generator. 37. Heating gas flues. 38. Gas pipes. 39. Exhaust gas flues. 40. Flue to the gas compressor station and gas cleaning. 41. Oven doors at horizontal retorts. 42. Cyanogen separator. 43. "Blow-off flue"

inclined ovens, the coke is discharged continuously. Horizontal ovens are discharged at the lateral doors. The coke is usually discharged into a coke-quenching car with a pusher machine and placed under a coke-quenching tower where it is cooled by a water spray. In some plants, "dry" coke quenching is done using an inert gas. The connection current from the large quantities of steam carries coke dust through the quenching tower together with a certain amount of crude gas. Volatile matter and crude gas are pumped out of the coke chamber through a collecting main after a preliminary cooling in the tar separator. The tar may be separated simultaneously by sucking it up with turbo-blowers, or by ultrasonics, chemical absorption or electrostatically. The separation of the fractions of the crude gas is not profitable at the coking plant and hydrogen sulphide, ammonia and cyanogen compounds, for example, are mostly returned to the gas generator for combustion. Otherwise these substances are passed to the drains or the quenching basin.

Uses. The use to which coke is put is determined by its water, ash and sulphur content, its reactivity or crystalline structure, its size and its mechanical strength. Most of it is used for "forging-coke" in the production of metals from their ores, especially iron. It is also being used increasingly in sintering plants. In gas generators, coke is used to produce generator gas, water gas or industrial gas. Other gases are produced in special ovens by means of oxygen-blowers or under special pressure conditions and form the basis of many chemical syntheses, as for example phosgene in the dyestuff industry, Fischer-Tropsch hydrocarbon synthesis, etc. As a raw material, it is used in the direct synthesis of calcium carbide and in the manufacture of graphite and electrodes. Coke has a low volatile matter content and when used as a fuel it is noted for producing little atmospheric pollution.

The raw gas and condensed volatile matter of the coal are the main by-products of coking. In countries with abundant natural gas resources, the crude gas in burnt; elsewhere it is cleaned for use as coking or town gas, and the other by-products are treated for use in tar, pitch and petrochemicals.

HAZARDS

Accidents. Abdominal bruises and lesions of inner organs may occur from crane buckets, from bucket conveyors and from materials falling from conveyors. Bruises of the chest may be caused by sliding coal or coke in the bunkers, in shunting accidents, by the coal-charging car or by the coke-quenching car. Dislocations of the arms and injuries such as fracture of the cervical vertebra may result from being dragged into the conveyor transfer points when crossing over the conveyor belts or when working on running conveyor belts. Fractures and haemorrhage of the kidney and spleen may occur from falling off the batteries of the gas-collecting main. Burns are the most common injury and occur owing to spontaneous combustion or to flames from the charging holes due to a failure of the suction or to unfavourable winds. Burns also occur from coal-dust fires, from explosions and fires caused by leaking gas pipes and drains and during welding work at the heating and gas-collecting mains.

Some accidents are caused by slow reactions, stupor or vertigo, resulting from slight to moderate poisoning by carbon monoxide or hydrogen sulphide inhalation or following consumption of alcohol. Others are caused by insufficient adaptation to hot work or to the effects of ammonia vapours.

Diseases. Carbon monoxide intoxication is the most frequent occupational disease in the coke industry. Acute intoxication occurs if a mistake is made when fitting blank flanges, for example, or when repairing transmission lines and gas-blowers. It may also be caused by inadequate ventilation and incorrect working methods during filling, pricking up or measuring the temperatures of the ovens. In carbon monoxide intoxication, unconsciousness may occur within a few seconds or minutes; in some cases there may even be cessation of respiration. If this occurs when doing repair work on transmission lines and gas-blowers, such cases may be complicated by injuries from falling or burns. The oxygen content of the air may be reduced through displacement of the air by an escape of gas. In small, closed rooms it may fall by oxidation to 10% volume, for example, because of badly regulated gas-heated water heaters.

Injuries to health by the polycyclic aromatic hydrocarbons may occur from the smoke escaping when filling the ovens, from blowing-off and from leaks. These give rise to conjunctivitis and sunburn from ultraviolet rays due to photosensitisation of the skin. Both occur most often in spring with workers not yet adapted to ultraviolet radiation.

Skin cancers are seldom observed in coke plants. It has been shown, however, that the incidence of cancer of the bronchi is statistically significant in workers on the retort or oven ceiling. The main cause appears to be high atmospheric 3,4-benzpyrene concentration (4-200 mg/l) due to badly ventilated oven ceilings. Cancer of the urinary tract (e.g. bladder cancer) is observed frequently amongst these workers, especially with those persons who also work at the tar-collecting main. The excess of illness compared to the rest of the population is statistically significant, but is not as high as with cancer of the bronchi. The responsible carcinogen, 2-naphthylamine, is also a constituent of the tar-aerosol and is found in the air over the collecting main and in workers' clothes.

Silicosis has been found in bricklayers, with many years' experience in rebuilding or repairing the retort or oven chambers with silica bricks. When using fireclay brick with a low free-silica content, silicosis is very rare. Pneumoconiosis from coal or coke dust is not a significant health hazard in coke plants. However, if the bronchial epithelium has been damaged by infectious diseases or smoking, coal and coke dust are no longer tolerated by the respiratory system and, for example, existing bronchitis due to other causes may be aggravated.

Direct damage by radiant and convection heat during the clearing of coking coal is rare. Heat cramps may occur due to perturbations of mineral metabolism in persons not adapted to heat. The strain placed on the heart by hot work may lead to collapse if the coronary vessels have been previously damaged, for example by infectious diseases, influenza, etc. Inflammation of the tendon sheaths and of their insertions are less prevalent in areas where the plant is mechanised. They may result from unilateral strain during heavy work with shovel and pick, when emptying railway wagons of frozen coal or when clearing the chambers.

Hydrogen sulphide intoxications with respiratory paralysis rarely occur in cokeries, although effects of hydrogen sulphide exposure such as irritation of the mucosa, headache, stupor and dizziness do occur. Disfiguring pigmentation of the skin may occur due to coal dust tattooing after a skin lesion.

SAFETY AND HEALTH MEASURES

Medical prevention includes measures to prevent accidents or occupational diseases by inspecting the working place, by medical examination of the workers and instruction in hygiene. First-aid organisation and training will help minimise the consequences of accidents and poisoning. Technical prevention includes instruction in safe working practices and the provision of safety and warning equipment. Regular inspections of the workplace should be carried out by a safety engineer and physician and particular attention should be paid to poisoning hazards.

Technical safety. This includes frequent checks of mechanical handling equipment, including crane grabs, bucket conveyor chains and safety devices on conveyor belts. Where possible, conveyors should be enclosed and should not be cleaned or repaired whilst in motion; under no circumstances should man-riding be allowed on conveyor belts. Conveyors should also be stopped if foreign bodies have to be removed from the conveyor belt directly by the worker. Before conveyor belts are started up, an acoustic and/or visual warning signal should be given.

Where it is necessary for workers to enter silos or bunkers, they should be equipped with safety belts and lifelines and have a colleague in attendance outside.

Coal chutes should be mechanically operated and fitted with chute obstruction clearing devices to reduce the possibility of workers being buried by moving coal.

Fire and explosions. The amount of airborne coal dust should be kept to a minimum by the use of dust-control measures such as exhaust systems fitted with cyclones, scrubbers or electrostatic precipitators. Coal and coke dust bunkers should, in addition, be fitted with explosion venting systems. If gas pipes are to be welded, the gas supply should be cut off and the pipes purged with steam or an inert gas.

Environmental monitoring. Regular analyses should be made of the atmosphere at various points in coke plants to detect the concentrations of explosive and toxic substances, and particular attention should be paid to carbon monoxide concentrations. The use of automatic monitoring equipment is to be recommended; this should be linked to an alarm system which is actuated should the carbon monoxide concentration exceed a pre-set level. Carbon monoxide may be measured by its thermal conductivity or infrared absorption, the 3,4-benzpyrene by a gas interferometer, or fluorometer, the dust by a konimeter. Regular measurement of noise levels in coke plants should also be made and where offensive noise levels are encountered, noise-control measures should be taken or where this is impracticable, workers should be supplied with hearing protection.

Medical examinations. Pre-employment medical examinations should include a medical history and cover the heart, lungs, blood pressure and urine. People suffering from epilepsy, diabetes mellitus, nephritic and cardiac diseases or hypertonia should not work inside the coke plant. The periodic medical examination should take account of the special conditions of the actual working place. Transport workers should be examined for vision, hearing and reflex action. Those working at places where there is a risk of carbon monoxide exposure and those working temporarily with breathing apparatus should be examined for their oxygen tolerance, absolute haemoglobin content, pulse and respiratory rate at rest and under exercise. Electrocardiograms and chest radiograph should be taken.

The simple tests can be repeated at intervals of 1-2 years, whereas for the electrocardiogram a longer interval may be chosen. At the periodic medical examination of workers exposed to tar aerosols, the skin, lung and urinary sediments should be examined. If the cytological examination of bronchial secretion and urine is done every 6 months, there is a chance of recognising precancerosis and cancer in time. Coke-oven bricklayers should have chest radiographs taken at intervals of 6-12 months depending on the type of exposure. People with a damaged bronchial epithelium should not work in dusty conditions. Workers exposed to heat should be examined for disturbances of mineral metabolism and of adrenal and cardiovascular function.

Training and instruction. During the instruction courses in accident prevention and first aid, it should be pointed out that smoking may have an additive effect in intoxications and in initiating cancer, or may cause bronchitis. Furthermore, the combined narcotic effects of alcohol and carbon monoxide should be pointed out. Workers and supervisory staff should be trained in elementary first aid.

Workers should be supplied with respiratory protective equipment, instructed in its correct use and required to wear it when doing repair work where gas leaks are possible. They should also be taught to observe wind direction and heat flow of oven-gases when filling or stirring the oven, in order to avoid intoxications and burns. The need to consume liquids when doing hot work should be explained and an adequate supply of beverages should be provided. Persons exposed to intense radiant heat such as that from retorts should be supplied with asbestos protective clothing and face protection.

ZORN, H.

General:

ECSC round table meeting: Coke oven and coke research. Commission of the European Communities. EUR 7197, Directorate of Health and Safety (Luxembourg, Office for Official Publications of the European Communities, 1981), 361 p. Illus. Ref.

Occupational risks:

CIS 74-1574 "Contribution to the study of the biological effects of coke and graphite" (Zur biologischen Schädigungsmöglichkeit durch die Kohlenstoffmodifikationen Koks und Graphit). Einbrodt, H. J. *Staub* (Düsseldorf), Dec. 1973, 33/12 (474-478). 45 ref. (In German)

CIS 75-1643 "An epidemiological study of exposure to coal tar pitch volatiles among coke oven workers". Mazumdar, S.; Redmond, C.; Sollecito, W.; Sussman, N. *Journal of the Air Pollution Control Association* (Pittsburgh), Apr. 1975, 25/4 (382-389). Illus.

CIS 74-2026 "Long-term mortality study of steelworkers–VI. Mortality from malignant neoplasms among coke-oven workers". Redmond, C. K.; Ciocco, A.; Lloyd, J. W.; Rush, H. W. *Journal of Occupational Medicine* (Chicago), Aug. 1972, 14/8 (621-629). 17 ref.

Coke oven-emissions:

"Pollution by polycyclic aromatic hydrocarbons. 1. Pollution from coke ovens and oil industry" (Pollution par les hydrocarbures aromatiques polycycliques. 1. Pollution par les cokeries et les industries pétrolières). Neuray, M. N.; Stevens, J. M. *Annales des Mines de Belgique* (Brussels), July-Aug. 1980, 7-8 (701-726). 217 ref. (In French)

"Exposure to coke oven emissions". Department of Labour, Occupational Safety and Health Administration. Occupational Safety and Health Standards. *Federal Register* (Washington, DC), 22 Oct. 1976 (46742-46790).

Occupational exposure to coke oven emissions. Criteria for a recommended standard. US DHEW (National Institute for Occupational Safety and Health, 4676 Columbia Parkway, Cincinnati) (1973), 54 p. 63 ref.

Cold and work in the cold

Man's ability to work in the cold is dependent on the functional integrity of brain and limbs. Cooling the brain leads first to confusion and then to incoordination, the classical early signs of exposure. Cooling the limbs, on the other hand, results in numbing and clumsiness, making the performance of intricate tasks difficult.

We are therefore confronted with two different and sometimes contradictory requirements: preservation of body heat and supply of adequate heat to the extremities.

Physiological responses to cold

Acute exposure to cold provokes two primary responses: improved insulation and elevated heat production. Insulation is increased by reduction of blood flow of the skin, occurring particularly rapidly in the fingers, toes, ears and nose. This vasoconstriction allows the extremities to cool; in extreme cooling the vasoconstriction relaxes (cold-induced vasodilation), warming the part, although at the expense of body heat. Cooling the skin provokes shivering, an emergency response elevating heat production by incoordinated contraction of muscles. Fall of brain temperature soon inhibits shivering.

The primary responses to cold are concerned more with survival than with the ability to continue working. Intellectual or behavioural adaptation seem to be greater contributors to keeping man at work.

Adaptation to cold

Adaptation may be thought of as all the changes that occur following periods of cold exposure. These may conveniently be separated into three categories: acclimatisation, habituation and behavioural.

Acclimatisation. If climatic conditions are severe enough, clothing inadequate or activity insufficient, the whole body may be exposed to cold. The extremities, however, because they are more difficult to protect, are more likely to be exposed to cold. The hands in particular present a dilemma. Firstly, their shape–long, thin cylinders–makes it particularly difficult to affect efficient insulation; secondly, thick insulation severely restricts mobility.

As a result, local acclimatisation of the face and hands (and perhaps feet) is more prominent than that of the whole body. Increased blood flow in the fingers in the cold with consequent improved sensitivity and mobility, and an earlier and more profound cold-induced vasodilation, characterise the changes of local acclimatisation to cold. Clearly they are of benefit to an individual working in the cold.

Whole body acclimatisation is recognised by enhanced insulation (probably a change in the circulation) and more efficient extra heat production (non-shivering) on cold exposure; these particular changes seem to be geared to survival rather than to an improved ability to work in the cold.

Habituation. Habituation may be defined as altered sensation in the face of unchanged stimuli. It is well recognised by those who have worked in the cold: they correctly claim that they get used to the conditions, feeling comfortable where once they were too cold. Habituation is dramatically characterised by the disappearance of the pain of immersing the hand in cold water following repeated exposures. The much lower levels of indoor temperatures deemed comfortable by those who have worked in the cold are an example of how habituation can influence general bodily sensations. The stimulus for the acquisition of habituation seems to be repeated exposure; once again the extremities are more likely candidates than the whole body.

Habituation to cold may bring with it a dulling of awareness of being cold, and in consequence carry with it an increased risk of hypothermia or cold injury. Indeed the high incidence of hypothermia in the elderly may in part be caused by the "natural" habituation (dulling of sensitivity) of old age.

Behavioural adaptation. While habituation and acclimatisation play their parts in enabling man to work in the cold, his behavioural adaptations play the star role. Man learns to keep out of the cold, to make effective use of clothing and shelter and to carry out intricate tasks with minimal exposure. He learns only to expose himself to the full rigours of a cold environment while working hard and while producing large amounts of heat to help him keep warm.

The extensive use of technology, ranging from the use of mobile fires by tribes that once inhabited Tierra del Fuego to heated diving suits and air-conditioned living and working quarters, are the acme of man's behavioural responses to the challenge of the cold.

In all these ways he actually avoids cold exposure, and it is interesting to note that, paradoxically, signs of heat acclimatisation may be seen in those working in the cold. Thus there is little doubt that the intellect plays the largest part in man's adaptation to low temperatures. Physiology may save him in a desperate situation, habituation may dull discomfort, but behaviour enables him to perform his tasks in spite of extreme conditions.

Efficiency of work

Energy expenditure is often higher in cold than temperate conditions. Efficiency is only altered when the heat resulting from work is insufficient to maintain body temperature. Under those circumstances additional heat is produced (though shivering may be difficult to detect) without adding to the external work. If adequate protection is provided, this very protection may add to the cost of a task by impeding movement and adding to the external work (5-10%). As a result of either of these, fatigue or exhaustion during physically demanding tasks may occur earlier than expected from predictions using temperate experience.

Performance of skilled jobs requiring manual dexterity pose unusual problems in the cold. Adequate protection of the hands impairs dexterity; exposure leads to cooling and stiffening of joints, loss of sensitivity and muscular dysfunction. Resort to frequent rewarming, itself slowing work, is probably the only practical solution, though thin gloves to prevent direct contact between skin and cold objects can do something to prevent too rapid cooling and injury.

Cold occupations. Those who work in extreme natural or artificial conditions are usually aware of, or trained to consider, the hazards and, as a result, take due precautions. However, the majority at risk are those working in so-called temperate climates where the temperature falls below about 10 °C, especially if it is also wet. It is interesting to note that the 10 °C January isotherm delineates an area of risk which includes nearly all of Europe, most of North America, Asia north of the Indian Sub-Continent and northern China and Japan. Thus about half the world's population are at risk.

HAZARDS

Exposure to extreme cold even for very short periods may result in frostbite. Frostbite is a freezing of tissue most commonly of the periphery. Cheeks, nose and ears are by far the commonest areas to be affected, as the face is rarely covered. More seriously, fingers and toes may be frozen (see FROSTBITE).

Another form of local cold injury is immersion foot. It is a condition caused by chronic cooling, especially long immersion (probably for days) in cold water, and is aggravated by tight footwear. It is fortunately rare in normal work in the cold. The condition is characterised by intense pain and discolouration of the foot; permanent damage may result. Prevention is the real form of treatment: loose, well designed waterproof footwear.

Less dramatic, but with possible very serious consequences, is hypothermia or exposure, which is a loss of body heat. This can occur in relatively mild, particularly cool-wet, climates even during hard physical work. It is also the most frequent cause of death in water immersion (including free diving). Onset is usually insidious, leading to non-cooperative (deliberately obstructive) and bizarre behaviour. This is followed by a general slowing, inability to keep up the work schedule and lethargy. Finally unconsciousness may supervene; death occurs very suddenly at this stage from ventricular fibrillation (heart failure). The slowing and lethargy are the entry into the vicious cycle leading to a loss of body heat which in its turn further reduces activity and heat production.

The most effective treatment is early recognition and removal from the cold, but early signs often go unrecognised, particularly as they tend to make the victim withdrawn and unsocial. In the later stages two principles of treatment are paramount: insulation and heating. Insulation, preventing further loss of body heat, in itself may not be enough to avoid catastrophe. The vicious cycle may have set in and the sufferer may no longer be able to produce enough heat on his own to effect rewarming; thus one has to resort to passive heating using an external source.

In the field, the prevention of convective heat loss may be effected by placing the sufferer in a sleeping bag or under blankets, as far as possible out of the wind. Evaporative heat loss may be an even greater threat to maintenance of body heat, particularly if conditions are wet and cold. An impermeable cover (polyethylene bag) is effective and more efficient than the removal of wet clothing in cutting this threatening avenue of heat loss.

Putting another person into the sleeping bag is an easy and practical way of providing extra heat to the hypothermic patient. Hot-water bottles or hot drinks are the counsel of perfection, while immersion in hot (38-40 °C) water only available to those who have planned ahead, most rapidly restores body temperature.

There is little evidence that conditions such as rheumatism, influenza or even the common cold are more common amongst outdoor workers. Chronic lung affections may be aggravated in the cold and there is an association between acute pneumonia and unusual cooling. On the whole, however, there is little evidence for general ill-effects from working in the cold. Indeed, the proposition that such conditions might actually be beneficial bears examination.

Clumsiness and loss of dexterity, together with a loss of concentration from discomfort might make the occurrence of accidents in the cold more likely. There is some evidence from, for instance, polar regions that minor accidental injuries and indeed accidental deaths are unusually common. In the same way accidents among trawlermen are rather common. Neither group, however, is truly representative of the generality of workers in that the hazards are more associated with their occupations, although they may be exacerbated by the cold.

SAFETY AND HEALTH MEASURES

While the ideal solution to the problems associated with working in the cold is to minimise the cold, in a world increasingly short of energy it may be unrealistic. It is far more economical to heat the individual or keep him warm than to bring his environment to comfortable levels. The emphasis should therefore be based on the triad of protection, training and ergonomics.

The use of lightweight modern materials for clothing, vapour barrier and insulated designs for boots leaves all but the hands adequately protected. Personnel trained in the effective use of their clothing and the proper performance of their tasks can overcome the severest conditions. The design of machinery and tasks avoiding fine manipulation, and low energy expenditure tasks in the cold will enable heavily gloved and adequately clothed workers to operate efficiently.

The design and choice of clothing should take cognisance of three important factors: cold is often accompanied by wind and wet; heat production and work are indissoluably linked; and bulky clothes hamper movement.

Wetting of clothing reduces their insulation and in itself increases heat loss by evaporation. An outer waterproof garment will prevent wetting from the outside and at the same time minimise evaporative cooling, while also keeping out the wind. Contrariwise the same garment will result in wetting of the clothing from the inside. During work body temperature inevitably rises, causing sweating. In "normal" temperate conditions the sweat evaporates, cools the body and tends to restore body temperature to normal levels. A waterproof garment prevents this evaporation, causing the sweat to accumulate in the clothing and avoids the evaporative cooling. Thus a choice has to be made or a compromise reached. The most suitable compromise may be to choose a material for the outer garment that will let out water vapour but will prevent the ingress of liquid water (these materials are also usually windproof). In extremely cold conditions with very low levels of activity, maximum insulation is required and is best provided by an assembly of clothes covered by a vapour-proof garment, on condition that facilities for subsequent drying are adequate. In most practical circumstances, as has already been pointed out, exposure to extreme cold is accompanied by heavy work. Under these conditions overheating is a more urgent problem than cooling. To cover most needs the ideal clothing assembly should be of variable insulative value. Traditionally this is achieved by wearing a number of layers of fairly thin clothes, e.g. vest, shirt, pullover, windproof, which may be removed or donned, one by one, at will. The introduction of quilted or interlined clothes has made this solution less practicable, and variable insulation is achieved by the provision of well designed ways of venting clothes, e.g. zip fasteners, velcro closures, etc.

The provision of comfortable, heated rest facilities is essential for cold workers. While it is not difficult to design clothing for working in the cold, these same clothes would be totally inadequate for resting in the cold, as heat production is related to the level of activity. If there is less heat coming from the inside it must be provided from the outside.

It is probably wise to increase rest allowances for cold workers. While food and hot drinks should be provided there is no evidence that any particular food is more or less beneficial in the cold, though there is some evidence that cold stimulates appetites.

Training in the use of clothing, in the realisation that cold presents hazards, and the recognition of signs and symptoms of early exposure and frostbite, not to say in the performance of the required tasks, are the means by which work may be successfully carried out in the severest climates. Newcomers are always at risk. Experience makes for caution which is transmitted in training.

Machinery and tasks must be carefully designed to make them less hazardous and easier to perform. Attention to the size and spacing of handles and knobs, insulation of metal parts that must be handled, the elimination of sharp protusions are examples of how machinery may be optimised for the cold. Surveillance tasks without activity should be avoided in the cold, while actually increasing the effort required for light tasks could prove beneficial.

There is little evidence that the young are better able to perform in the cold than their seniors. Physical fitness, however, probably makes work in the cold easier and less hazardous. Those suffering from local vascular disease should avoid working in extreme cold. In this regard it is interesting to point out that heavy smoking may mimic some of these diseases. Alcohol, on the other hand, causes peripheral vasodilation. It should be avoided before exposure, but in moderation may have a beneficial effect after exposure.

In summary, hazards to health and safety multiply when warning signals of feeling cold and uncomfortable are ignored. The criterion for adequate protection is comfort.

GOLDSMITH, R.

General:

Man in cold environment. Burton, A. C.; Edholm, O. G. (London, Edward Arnold, 1955), 273 p. Illus. Ref.

CIS 79-1016 "The effects of cold: Frostbite and hypothermia". Coble, D. F. *Professional Safety* (Park Ridge), Feb. 1979, 24/2 (15-18). Illus. 14 ref.

Cold storage rooms:

CIS 77-559 *Guide to refrigerated storage* (International Institute of Refrigeration, 177 boulevard Malesherbes, 75017 Paris) (2nd ed., 1976), 188 p. Illus.

CIS 79-709 "Working in a cold environment". Green, A. *Occupational Health* (London), Aug. 1978, 30/8 (366-371). Illus.

Diving:

CIS 78-1906 "Working in cold environments–Lessons to be learned from diving". Hanson, R. de G. *Annals of Occupational Hygiene* (Oxford), Aug. 1978, 21/2 (193-198). 17 ref.

Mountaineering:

Mountain medicine. A clinical study of cold and high altitude. Ward, M. (London, Crosby Lockwood Staples, 1975), 376 p. Illus. Ref.

Polar regions:

Human adaptability to antarctic conditions. Gunderson, E. K. E. (Washington, DC, American Geophysical Union, 1974).

Exploration medicine: being a practical guide for those going on expeditions. Edholm, O. G.; Bacharach, A. L. (eds.). (London, John Wright, 1965), 410 p. Illus.

Collective agreements (occupational safety and health)

[Since the early 1960s throughout the industrialised world interest in safety and health at the workplace has been on the increase. Different approaches have been followed in various countries by those concerned and different mechanisms have contributed in recent years to improvement of the working environment in many branches of economic activity. In a number of countries new or revised legislation has been enacted to safeguard the health, safety and welfare of persons at work and to control dangerous substances and harmful physical agents by means of precise provisions and by defining the general obligations of employers, employees and manufacturers. In other countries co-operation between employers and workers has led to various forms of industrial democracy, including the setting up of joint industrial safety and health committees often under the supervision of the public authority. In countries where protective legislation or its enforcement has proved inadequate, and industrial relations have been characterised by disputes, collective agreements have played an important part in improving working conditions and environment.

The Italian experience described in this article stands out because of the wide scope of safety and health matters in collective agreements and the consequent importance of these agreements as a supplementary source of standards.

By and large, apart from labour-management relations, trade union rights, grievances and disputes, general conditions of work and wage structure, collective bargaining in occupational safety and health matters has been limited to joint consultation procedures, personal protective equipment, disability benefits in case of occupational injuries, safety training requirements. In Italy, however, collective agreements have often included subjects that in other countries are regulated by mandatory standards promulgated at the national level.]

Collective agreements relating to occupational safety and health were originally concerned with the problems of compensation for occupational injuries and of benefits payable to the victim and his dependants. The idea that efforts should, on the contrary, be concentrated on the prevention of hazards connected with industrial activity, only came to the fore considerably later. Thus the legislative provisions that were adopted during the first half of the present century were predominantly designed to assure the payment of compensation; and, similarly, collective agreements have tended over the years to lay stress on this aspect and on the principle of increased payment for increased hazard (called "danger money"), to the extent of laying down extra payment for work that was particularly fatiguing, hazardous or dangerous to health. It is only recently and in a few cases that legislation has started to move in a different direction. For example, in Italy article 2087 of the Civil Code establishes the duty to provide a safe workplace as one of the fundamental obligations of the employer, and provision is made in article 437 of the Penal Code for sanctions in cases where safety devices or other equipment designed to prevent accidents have not been fitted or have been removed. In 1955 a series of decrees were introduced setting out safety standards to be observed.

Still in Italy, a most important innovation was introduced in article 9 of Law No. 300 of 20 May 1970 whereby the workers were given wide-ranging powers of control over the implementation of safety and health standards for the promotion of research, and for the development and implementation of ideas aimed at achieving higher standards of occupational safety. This was of fundamental importance because it has largely reversed the traditional standard-setting approach, concentrated attention on the preventive approach, and brought to the fore the duties and powers of the workers.

At about the same time the trade unions took a decisive step forward by undertaking commitments excluding any form of "danger money" and by starting to include matters relating to the prevention of accidents and occupational diseases in all collective agreements, both at the national and at the factory levels. Even more important is the fact that collective agreements should not only deal with problems such as the working environment, machinery and equipment, and the substances employed, but should gradually encompass the organisation of work itself.

Recognition has thus been given to the fact that among the basic causes of accidents are to be found the methods and systems that go to make up the work process, in the sense that they are the outcome of the monotony of the work, of fatigue, repetitive movements, or the lack of adequate rest periods. This important step on the part of the trade unions first became apparent in Italy in the framework of the new contracts concluded in 1966, but it was in 1969 after the introduction of the new "workmens' statute" (the above mentioned Law No. 300) that the principle was generally accepted.

A notable addition to the contents of collective agreements was seen when it was agreed to allow the insertion of clauses enabling trade unions to participate in the control of environmental conditions, and, gradually, to insert provisions allowing for the stoppage of work when vapour, dust, or dangerous substances were present in concentrations exceeding the limits agreed to in the contract, usually taken from the list of Threshold Limit Values (TLV) of the American Conference of Governmental Industrial Hygienists (ACGIH). Provisions of this nature were first introduced in the food, printing and paper industries as well as in certain branches of the chemical industry.

In other contracts drawn up later in the light of the special characteristics of different branches of industry, the duties and obligations of the employer concerning safety require him to ensure a healthy working environment and the physical integrity of his employees (e.g. contracts concluded by metalworkers and the publishing industry). In many cases the collective agreements are addressed in particular towards the establishment of safety and health records. These include records of environmental biostatistics and other relevant occupational health data, as well as data on occupational safety and health hazards (e.g. contracts concluded for such industries as food and drink, the state-owned metalworks industry, the hosiery workers, the mining and the chemical industries). These documents go to make up the full history of the working environment and the clinical picture for each worker, thereby providing valuable information concerning the incidence of hazards and data for job allocation.

In other cases—especially the building industry—it was found necessary to introduce suitable provisions to prevent illegal subcontracting and other forms of decentralisation of production which largely contribute towards increasing the risk factor.

Finally, some collective agreements also provide for the employer to pay for all investigations relating to the working environment (motorways, chemical industry, cement, etc.) and also if necessary to set up a system that can be used to provide the trade union representatives with information on safety and health matters.

There is no doubt that this new situation, once it is fully introduced and strengthened, will represent a real improvement, but there are, nevertheless, some noteworthy gaps to be filled and limits to its efficacity. Among the former is the fact that a collective agreement applies only to a specified association or group of people and it cannot legally be extended to cover all those belonging to the given category of workers. Its limits stem from the difficulties that may arise, from the contractual point of view, in regard to problems related to the organisation of work. It is evident that the employer is not going to relinquish his authority to organise work, while the unions plan to reduce or limit this authority by stating that they will not negotiate on the basis of the workers' health, but only on that of work organisation. This may lead to disagreement that is sometimes insurmountable and which will only be resolved by resorting to a trial of strength, whose solution may depend upon the economic situation. It is clear that, at the present time, this is the field in which the trade unions have had the least success. The Italian law introduced in 1978 concerning the health national service seems to provide a fresh impulse to collective agreements, not only because attention is focused largely upon prevention but also because express reference is made to agreements between the two parties. In fact, it is made clear in the last paragraph of article 20 that any changes made in the working environment relating to safety measures that are not provided for in law shall only be introduced on the basis of agreements between the trade union representatives and the employer, following the method laid down in the collective agreement that is applicable. In this way recognition is given to the essential function of collective agreements, enabling at the same time gaps in the existing legislation to be filled wherever special working conditions may warrant it.

In another provision of the same law, by which the government is given authority to issue a new consolidated text relating to safety and health, the employer is required, inter alia, to arrange the process of production in such a way that safety requirements are fulfilled, with particular regard to the layout of plant and to the identification of hazards and the means for reducing them. It would thus seem that the legislator has decided to establish general standards which will be further detailed in the consolidated text. In any case, this valuable legislative procedure certainly cannot be considered as intending to exclude safety and health with respect to standards from collective agreements, in the sense that the relationship between safety and work organisation can only be a matter for the law. On the contrary, collective agreements should relate to the organisation of work with a view to increased safety: the legislator can only dictate standards of a general character and such standards must be detailed at the factory level, to fill in the gap that would otherwise be left open in the framework of an official standard. Thus, there is a wide field available to be covered by collective agreements, and the existence of an official directive or standard will serve to facilitate agreement between the parties.

Finally, the 1978 health reform law amplifies, to a certain extent, the information system foreseen in some collective agreements by requiring certain data that may be useful for preventive purposes to be communicated to the trade union representative (article 20), and joint consultation to take place with the object of establishing appropriate criteria for recording environmental and biostatistical data (article 27). As may be seen, collective agreements drawn up in recent years have come to influence the work of the legislative authority, which in turn has provided new basic instruments from which collective agreements may be worked out.

There is no doubt that the problem will always be a complex one which, if it is to be solved, calls for a decisive role to be played by all concerned using all the means available, including the legislator, the trade unions and various government agencies.

As technology continues to advance, new hazards will make their appearance, and there will thus be a need to reconcile the demands of productivity with those of safety.

It is customary to take into consideration working accidents and occupational diseases; however, the whole range of events involving safety and health and attributable to production have become so complex and cover so wide a field that "work related diseases" can now be distinguished, which are not legally recognised as occupational disease, but which nevertheless have their own causes arising from hazards encountered in work activities. These can also embrace a number of phenomena which are most frequently seen among community diseases, such as premature ageing and early death as a consequence of occupational factors. In fact, it is no longer possible to make a complete distinction between prevention of accidents and occupational diseases at the workplace and community health. There is a need to develop a wider, more global approach embracing all the hazards associated with modern life.

Faced with such a wide and complex field, which will extend beyond the scope of collective agreements, the problem will arise for the trade unions to combine their activities with those of the public health authorities on an over-all basis, and, although most difficult to achieve at the present time, this combination is imperative.

The problems of safety and health are assuming substantially new forms and considerable thought will need to be given to the adequacy of the present structure of the trade unions and to the role to be played in this area by collective agreements. In this connection a co-ordinated, decisive and all-embracing plan of action is needed, which should be founded on the conviction that the maintenance of physical integrity and human life is in the interests of the whole society (see also OCCUPATIONAL HEALTH LEGISLATION).

SMURAGLIA, C.

Occupational safety and its legal protection (La sicurezza del lavoro e la sua tutela penale). Smuraglia, C. (Milan, Giuffre, 1974), 531 p. (In Italian)

"Occupational safety: management policy and co-determination. Example of a works agreement" (Arbeitssicherheit: Unternehmenpolitik und Mitbestimmung im Spiegel einer Betriebsvereinbarung). Heidberg, H. S.; Masanke, H. *Sicherheitsingenieur* (Heidelberg), Jan. 1974, 5/1 (8-13). Illus. (In German)

"Prevention of accidents and occupational diseases in collective agreements" (Note sulla prevenzione degli infortuni e delle tecnopatie da lavoro nei contratti collettivi). Gagliano Candela, F. *Securitas* (Rome), Mar.-Apr. 1976, 61/3-4 (145-157). 32 ref. (In Italian)

"Comparing conditions of work by collective agreement analysis: a case study in the petroleum industry". Evan, H. Z. *International Labour Review* (Geneva, International Labour Office), July 1973, 108/1 (63-81).

Collective bargaining in industrialised countries. Recent trends and problems. Vienna Symposium, 2-9 November 1977. Labour management relations series No. 56 (Geneva, International Labour Office, 1978), 113 p.

Collective bargaining. A workers' education manual (Geneva, International Labour Office, 1978), 142 p.

Colour in industry

Proper colour treatment of the workplace contributes greatly to the efficiency, safety and general welfare of employees. Well chosen finishes on the ceiling, walls and equipment help to produce good seeing conditions and a pleasant, cheerful working environment, which encourages high standards of cleanliness and housekeeping.

Planning colour schemes

Simple schemes in pale colours are usually the best choice for factories and offices. Elaborate schemes with large areas of vivid colour appear attractive at first but may become tiresome when one has to work alongside them all day. Restrained treatments are thus more likely to give lasting satisfaction. However, more elaborate decoration is satisfactory in non-working areas such as entrance halls, lunch rooms and locker rooms.

Warm and cool colours. Colours are sometimes classified as "warm" or "cool" depending on their hue. A scheme based on colours derived from hues in the yellow-red sector of the colour circle shown in figure 1 (such as cream, tan or golden brown) can produce an illusion of warmth. On the other hand, a scheme based on pale blue or pale green can make hot conditions seem a little more tolerable to work in. Under average conditions, warm and cool colours can be used in combination with pleasing results. Apart from this, choice of hue is very much a matter of taste and actually far less important than the colour's light-reflecting property, which should be as indicated in figure 2.

Ceiling and roof surfaces. These should always be made as near white as possible because light diffusely reflected

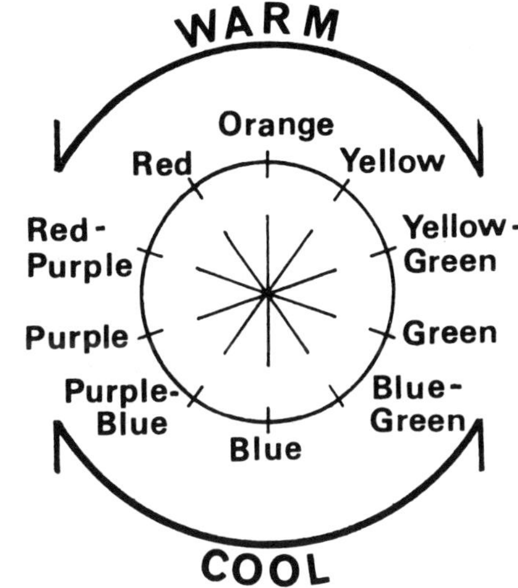

Figure 1. The "warm" and "cool" ranges of hue.

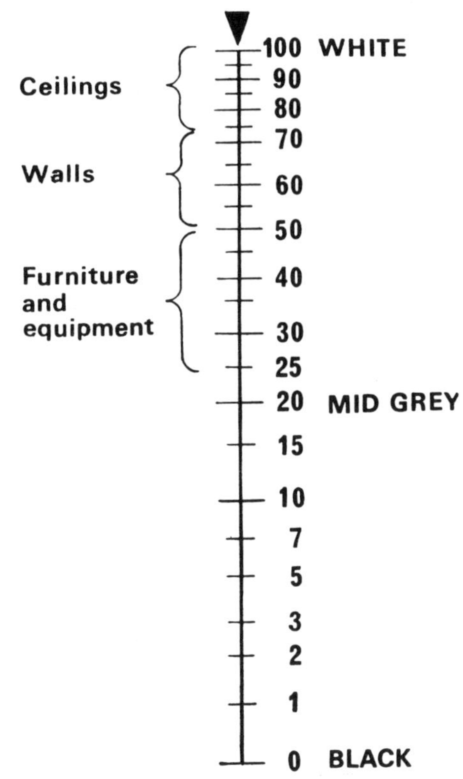

Figure 2. Recommended light-reflectance ranges for various room surfaces. Note that 20% corresponds to a tonal value half-way between white (100%) and black (0%).

from them is spread evenly throughout the interior, dispelling gloom and reducing distracting glitter on shiny surfaces. In buildings in tropical climates it is particularly important to paint the underside of the roof white. The strong daylight and sunlight which is diffusely reflected up from the ground can then penetrate deep inside the building, reducing considerably the need for artificial light.

Walls and floors. White surfaces at, or below, eye level are apt to be glaring; hence pale colours with a reflectance not exceeding 75% are generally suitable for walls, but 50% may be preferable when the wall is very strongly lit. It is advisable that the colour of the floor should normally be slightly darker than that of the ceiling and walls. [It is also advisable to choose "cooler" colours for rooms oriented to the south and "warmer" colours for those oriented to the north. With regard to the tasks to be performed in a room, exciting and diffuse colours are recommended for monotonous work, while soothing colours, i.e. bright, unobtrusive and without contrasts, are preferable for tasks associated with high intellectual requirements and a great deal of mental concentration.]

Plant and equipment. Work benches, machines and desk-tops should normally have reflectances in the 20-40% range. A figure of 20% corresponds to a "middle-grey" half way in tonal value between black and white, as indicated in figure 2. None of the main interior surfaces should be any darker in tone than this, with the possible exception of the floor.

Gloss, semi-gloss or matt finish. Gloss paint generally wears better than matt; however, it gives rise to reflections which show up irregularities in supposedly flat surfaces above eye level. Hence matt, or semi-gloss at least, should preferably be used on walls. Note that the durability of any paint finish depends largely on proper preparation of the underlying surface prior to painting.

Overhead surfaces should be matt whenever possible to ensure maximum diffusion of light. For the same reason, aluminium foil ceilings should have an embossed or finely rippled surface. Aluminium paint (which has a comparatively low reflectance) is not recommended.

Desks and work benches should be finished in a durable, self-coloured, non-glossy material such as linoleum or polyvinyl chloride. Plate glass, sandblasted on the upper surface and painted a suitable colour on the back, can also be used where there is no risk of breakage.

Colour for identification

One advantage of a simple scheme based on pale colours is that it enables objects of special importance to be identified with eye-catching touches of bright colour. Obviously, the fewer there are of these the more conspicuous each will be; hence colour identification must be used sparingly, i.e. only where there is real need and never for mere decoration.

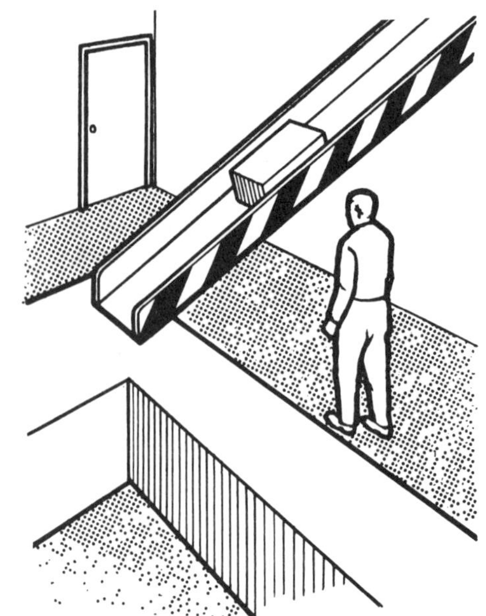

Figure 4. Colour used to pick out a difficult-to-guard hazard.

Fire and safety equipment. Placing a coloured patch on the wall above an appliance ensures its rapid location in an emergency (see SAFETY COLOURS).

Machine controls. The stop button (or equivalent device) is the one part of a machine that should always be clearly marked, because anyone must be able to find it in an emergency. Other controls are usually no more in need of a coloured knob than is the gear lever on an automobile; they should be specially marked only if there is some definite functional reason for it.

Pipes and ducts. When these are of major importance (e.g. in a chemical plant) they may need to be fully coloured. If they are only incidental, colour banding at a few strategic points will be just as effective (see SAFETY COLOURS).

Lubrication points. Small patches of colour indicating the correct lubricant can be given distinctive shapes to show how often each point should be given attention.

Stairways. A single, bold contrasting stripe on each tread assists the safe, speedy descent of stairs. This is more effective than multiple striping, especially in a dim light (figure 3).

Figure 3.
Individual steps are distinguished better by a single bold stripe than by multiple striping.

Hazards. Colour should be used only in those rare cases in which the hazard cannot be eliminated or effectively guarded. Figure 4 shows a situation of this kind. Brightness contrasts catch the eye much more readily than do colour contrasts; for instance, the stripes in figure 4 are conspicuous because a light colour alternates with a dark one. Conversely, the same principle can be used to make unimportant details inconspicuous whilst at the same time avoiding the monotonous effect which results from painting everything the same colour as its background. For example, if a wall column or roof truss is painted a colour which differs in hue from the background but has approximately the same reflectance, it will only be visible when looked at directly and will fade into the background at other times.

LOWSON, J. C.

CIS 76-1288 *Colour schemes for workplaces* (Couleurs d'ambiance pour les lieux de travail). Norme française enregistrée NF X 08-004 (Association française de normalisation, Tour Europe, Paris-la-Défense) (July 1975), 4 p. 3 ref. (In French)

"Design of factory colouring from the viewpoint of occupational health" (Die farbliche Gestaltung des Arbeitsplatzes aus arbeitsmedizinischer Sicht). Hüer, H. H. *Arbeitsmedizin–Sozialmedizin–Präventivmedizin* (Stuttgart), Mar. 1979, 14/3 (59-63). Illus. 19 ref. (In German)

"Additional considerations concerning the effects of 'warm' and 'cool' wall colours on energy conservation". Greene, T. C.; Bell, P. A. *Ergonomics* (London), Oct. 1980, 23/10 (949-954). 14 ref.

Colour at work. Department of Productivity. Occupational Safety and Health Working Environment Series 8 (Canberra, Australian Government Publishing Service, 1980), 32 p. Illus. 7 ref.

Colour vision

Colour vision or chromatic visual perception is the capacity to distinguish the colour of objects. Colour vision comprises at one and the same time a physical and physiological constituent which indicates the response capacity to the light stimulations of different wavelengths, and a subjective psychological constituent which indicates the specific experienced sensation.

Colour and its evaluation

The sensation of colour arises from a psychological experience of the classification of impulses originating from longwave groups of the light reflected or transmitted by objects. In nature a distinction is made between achromatic colours (white, composed of all the visible longwaves, and grey and black) and chromatic colours (which include the seven simple colours of the spectrum, red, orange, yellow, green, blue, indigo and violet, as well as the purples or mixtures of red and violet). The chromatic colours are characterised by their hue or tonality (wavelength) and their saturation (or the proportion of pure colour compared with white). The sensations of colour are, moreover, influenced by the effect of brilliancy or glare of the source of light, by the coefficients of reflection and transmission of objects, by the quality and intensity of the illumination and by simultaneous or successive environmental contrasts. The estimation of colours, either qualitative or quantitative, is made on the basis of the laws of the optic mixtures of the colours of the spectrum in the trichromatic system; an appropriate mixture of three radiations selected at will can be combined in a way to produce any quantitatively clearly defined colour.

Colours are evaluated by colorimeters, and spectrophotometers. A colorimeter is an instrument for determining and specifying colours by reference to other colours, while a spectrophotometer is an instrument for comparing the intensities of the corresponding colours of two spectra. Tables or atlases of colours are also used for colour classification. To describe the appearance of colours by names does not give an accurate classification because they suggest distinctly different shades to different people.

Mechanism of colour vision

Among the theories of colour vision, the most prevalent is that of the three constituents. The theory asserts the existence of three distinct devices of colour perception in the organs of sight, which are excited in a variable degree by the action of the exciters of different wavelengths, an excitation which provokes the perception of all the visible colours. The primary process of the perception of light and of colour is realised mainly by the visual receptors, that is, the cones which perceive daylight and colours and by the rods of the retina adjusted to night and twilight vision. As regards human vision, it is accepted that there are three or more types of cone receptor, the cones of each type having a pigment that selectively absorbs light from one region of the visible spectrum.

The definite formation of the image and of the colour is developed in the visual centres, in the occipital region of the cortical substance of the brain. When the organ of sight receives at the same time the exciters of all the longwaves of the visible spectrum, the eye sees the colour white; if on the other hand the exciters of a particular longwave are predominant in the beam of light, the sensation produced will correspond to the dominant wavelength. The normal visual perception of colours, that is the ability to distinguish normally the three fundamental colours, red, blue and green, is referred to as trichromatopsia, and persons possessing this perception are termed normal trichromats.

Colour blindness

Colour blindness is a condition of faulty colour vision. The commonest form, daltonism, consists in an inability to distinguish between red and green. Even persons with normal sight may be colour-blind to the indigo of the spectrum. Dichromats are partly colour-blind persons whose vision appears to be based on two primary colours rather than the normal three. Dichromasy occurs to about the extent of 8% in men compared with only 0.5% in women. One form of dichromasy is that of protanopia or protanomalopia, a form of partial colour blindness in which the red portion of the spectrum is not perceived so that red lights appear dim and cannot be distinguished from dim yellow or green lights. A second and more frequent form is that of deuteranopia or deuteranomaly in which the green portion of the spectrum is not received. Illnesses or diseases which cause trouble to the eyesight or the central nervous system may result in some form of colour blindness. Changes in colour vision can occur after absorption of drugs or chemicals (table 1).

Colour vision and work

Colour vision is a very important factor in the environmental conditions of all types of factory premises and workplaces. Adequate lighting with an absence of glare and a suitable colour scheme at the workplace and also in restrooms, canteens, etc., can do much to promote the health, safety and well-being of those at work. In addition to the beneficial psychological effect of good

Table 1. Common drugs which produce significant changes in colour vision[1]

Drug	Use	Colour vision change
Barbiturates	Hypnotic and sedative	Transient yellow-green defects
Chloroquine phosphate	Anti-malarial	Blue defect, central scotoma for red
Quinine	Anti-malarial Arthritis	Blue defect, then red-green defect
Cocaine	Anaesthetic and sympathomimetic	Enhances blue sensitivity, reduces red sensitivity
Digitalis		Red and green and blue defects
Oral contraceptives		Blue-green-yellow defects
Adrenaline	Cardiac stimulant sympathomimetic	Enhances green sensitivity, reduces red/orange sensitivity
Atropine	Cyclopegic miotic anti-secretory	Reduced sensitivity to red
Caffeine, coffee, cola		Reduced sensitivity to blue, enhances sensitivity to red
Trimethadione	Anti-epileptic	Red-yellow and blue-yellow defects
Tobacco		Red-green defect particularly red, can be permanent
Thallium		Red-green defect
Sulphonamides	Anti-bacterials	Some reduced discrimination of all colours
Santonin		Blue defects
Snake venoms		Reversible red-green defect
Chlortetracycline		Blue defect
Streptomycin	Antibiotic	Blue defect then green defect
Furaltadone	Antibiotic	Red-green defect
Opium/morphine	Narcotic analgaesic	Red-green defect, blue vision enhanced
Isoniazid, iso-nicotinic acid hydrazide	Anti-tuberculosis	Red-green defect

[1] From "Hazards from colour vision defects" by Voke, J. *(By courtesy of* Occupational Health, *London).*

environmental conditions, there is less risk of eye fatigue, and of faulty work, and usually the efficiency of production is increased.

From the point of view of occupational safety, colour vision is of great importance as many accidents are caused by lack of suitable lighting or by failure on the part of a worker to identify conventional identification colours such as on electric cables, gas cylinders, pipelines, guide marks, control buttons of machines, safety devices and limit signals. It is important that a worker's pre-employment medical examination should include sight testing. Persons suffering from visual defects including colour blindness should not be employed in work in which failure to identify the conventional colours may lead to accidents, as for example in various types of transport–railways, automobiles, aircraft and merchant marine. In certain other occupations, persons with other than normal eyesight should not be employed in paint or colour mixing, in colour printing, art work, dyeing and matching of textiles, geology, chemistry, microscopic work, and in the operation of nuclear reactors.

Training to improve colour vision. As the use of colours and colour signals is increasing in modern life, and as there are many people with congenital colour vision deficiency, action to improve the colour perception functions is of scientific and practical interest. The relatively high sensitivity of the optoneural apparatus and its rapid reaction to external stimuli opens wide perspectives in this field.

One of the main approaches of action on colour vision to bring about an essential improvement of this function is exercise, training. A method (colour tables) has been elaborated for this purpose in the USSR where special training techniques are used to achieve a permanent chromatic adaptation in persons with deficient visual reaction to stimuli of different wavelengths, and of various degrees of saturation and brightness. The length of the training period depends on the individual faculties of the trainee; with two to three weekly sessions of two hours each it generally lasts 10 to 12 months. The first encouraging results obtained with this method show that it is possible to produce a beneficial influence on the natural colour distinction capacity of man and to achieve a partial correction of the congenital and acquired defects of colour vision known as daltonism. See also SAFETY COLOURS; and VISION.

RABKIN, E. B.

CIS 76-104 "Colour vision characteristics". Wright, W. D.; Birch, J.; Palmer, D. A. *Lighting Research and Technology* (London), 1975, 7/3 (155-168). Illus. 44 ref.

CIS 78-1292 "Colour vision defects–Occupational significance and testing requirements". Voke, J. *Journal of the Society of Occupational Medicine* (Bristol), Apr. 1978, 28/2 (51-56). 34 ref.

CIS 79-1183 "Colour bars of an occupational kind–Job problems of visual defectives: How can they be helped?". Voke J. *Occupational Safety and Health* (Birmingham), Feb. 1979 (10-12). Illus. 5 ref.

"Hazards from colour vision defects". Voke, J. *Occupational Health* (London), July 1981, 33/7 (369-376). Illus. 6 ref.

Commerce

The field of commerce covers basically the sale of goods, merchandise, money and securities in wholesale warehouses, retail shops, open-air markets and banks.

HAZARDS AND THEIR PREVENTION

The above establishments employ some workers who, from the point of view of occupational safety and health, have similar working conditions to those of office employees. However, there are other employees who are exposed to very different health and safety hazards. These employees and their working conditions are considered separately by branch of the industry.

Wholesale trade. Here the most frequent accidents involve lifting and carrying and mechanical handling and the most common causes of disease are dangerous substances.

Falls of goods are common during manual or mechanical stacking and shelving at heights and during merchandise retrieval. Many accidents could be avoided by greater care and better stacking techniques. Recent sophisticated methods of stacking goods vertically have introduced new types of danger. Falls of persons are also frequent especially during the use of ladders, steps and stools for access to high storage; the use of such equipment should be avoided by the use of mechanical handling equipment or by provision of elevated storage servicing walkways and, where ladders, etc., must be used, preference should be given to fixed ladders with handrails or ladders which are fixed at the top to a rail along which they may be slid into position. It is bad safety practice to expect a worker to ascend or descend a ladder whilst carrying merchandise in his hands. Above all, the need for care in the use of ladders, etc., should be emphasised to all workers.

In-plant transport equipment such as fork-lift trucks are the cause of collisions and crushing accidents. Truck drivers should receive comprehensive training in the safe use of their vehicles with particular emphasis on loading and unloading techniques. Storage should be spaced sufficiently to allow trucks to pass pedestrians without danger; traffic aisles should be marked and special precautions taken to prevent collisions at crossings; above all, the driver should arrange his load to ensure he has maximum visibility at all times. (See TRANSPORT, IN-PLANT.)

Stationary mechanical handling equipment such as belt conveyors, roller conveyors, overhead trolley conveyors, are now widely used in wholesale warehouses and may cause serious injuries. Appropriate machinery guarding is vital, in particular for drive rollers, transmission belts or gears; where frequent crossing of conveyor belts is necessary, bridges should be provided. Overhead conveyors should be positioned high enough to be above the head height of the tallest workers. Goods lifts are common in multistorey warehouses. (See MECHANICAL HANDLING; LIFTS, ESCALATORS AND HOISTS.)

Goods are often received and despatched by road transport. Reception and despatch bays should be designed with elevated platforms, preferably fitted with guardrails and gates, and at the same height as the vehicle tailgate. Loading bridges should be supplied or, for heavy or bulky goods overhead, mobile or vehicle-mounted hoists should be provided. Manual handling of smaller goods with the resultant sprains and strains, backpain and cuts and abrasions may be largely eliminated by palletisation or containerisation.

All handling techniques and processes in warehouses should be carefully planned and supervised. New employees should be instructed in safety procedures and kinetic lifting and carrying techniques; all employees should be supplied with gloves or other hand and arm protection equipment. (See LIFTING AND CARRYING.)

The fire, explosion and health hazards encountered in wholesale warehouses are as varied as the products handled. Nearly all the products of the manufacturing industries will at some time be stored in a warehouse and many, especially those of the chemical industry, will be highly dangerous. The most valuable factor in the prevention of accidents here is a knowledge of the nature of the substances being stored, and accurate and informative labelling of all dangerous substances is essential. Once the nature of the product has been determined, its hazards can be found from this encyclopaedia or other specialised literature. Of vital importance is knowledge of necessary storage conditions (temperature, humidity, separation from other products) and action to be taken in the event of breakage or spillage; workers should be informed of these factors for all new products they are required to handle. Where dangerous powders or liquids have to be handled in bulk or decanted, workers should be adequately protected from skin contact or inhalation hazards by protective clothing and respiratory protective equipment.

Departmental stores. The most common accidents in large stores are slips and falls, cuts, abrasions and sprains and strains. Many accidents of this type are caused by littered and obstructed aisles and corridors and drawers that are left open, etc.; maintenance and good housekeeping and adequate storage space are essential preventive measures. As in warehouses, great care is required in the use of ladders.

During packaging of goods, cuts may be received from knives, string or the paper itself; parcelling machines should be adequately guarded. Many stores have specialist packers who are specially trained in the safe completion of this work. [Respiratory impairment and eye and throat irritation may result from exposure to the thermal decomposition products of polyvinyl chloride film used for wrapping meat or other perishable foods. The gases are emitted when the film is cut with a hot wire or when adhesives, phthalate based, used in price labelling are heated.]

Noise from traffic, customers, loudspeakers, paging systems and background music may produce sound levels sufficient to produce nervous fatigue. Lighting may be of very high intensity for sales or publicity purposes, with glare and flashing advertisements which may contribute to visual fatigue or even vision disorders.

The majority of shop assistants spend most of their working day in the standing position and varicose veins and postural disorders are not uncommon. Provision should be made for restrooms in which staff can rest seated at periodic intervals (15 min every 2 h for example) or for stools or fold-down seats behind the counter on which staff can sit when not serving. One solution is a seat comprising a plank that slides in and out on tracks fitted between two levels of drawers at the back of the counter and equipped with a spring or counterweight return system. This device has the advantage that, from the customer's position, the assistant appears to be standing. Virtually the same advantage is offered by a small bench fitted in the angle formed by two walls. Discussions between employers' and workers' representatives should help to define precise requirements for seating with regard to local conditions.

[Cash register operators in supermarkets are exposed to nervous fatigue, low back pain and cervicobrachial disorders.]

Small shops. Respiratory tract disorders are relatively common among sales assistants in small shops and kiosks, especially those situated on narrow roads with dense traffic, in which exhaust gases tend to accumulate. The use, in certain countries, of heating stoves in which combustion gases are not fully exhausted, tends to aggravate the situation.

Colds and influenza may be common among staff in such shops, since each time the door is opened in winter they are exposed to a draught of cold air.

Shops with specific hazards. As is the case with the wholesale trade, the fire, explosion and health hazards are as varied as the products dealt in, although of course in retail shops the quantities will be smaller. Even where the products are pre-packaged, leakage and spillage may occur. Particular mention should be made of pharmaceutical chemists' shops, drugstores, gardeners' shops selling pesticides, herbicides and other agricultural chemicals, stores selling pyrotechnic products, etc. Shop assistants may present allergic reactions or hypersensitivity to certain materials they handle, or regular contact with certain products may result in a variety of skin diseases. Such cases require medical supervision to determine the cause so that future exposure to the causative agent may be minimised or job transfer arranged.

In the retail food trade, especially where perishable goods are stored, such as the butchery trade, dairy products, etc., skin reactions and food poisoning may be a problem.

Display and sales counters situated on pavements and market stalls require special precautions. When the temperature falls below 10 °C, heating should be provided (e.g. infrared heaters) and staff should be able to rest in a heated building at regular intervals (e.g. 15 min every hour); during summer, adequate shade from the sun should be available.

Banks. Staff in banks are exposed primarily to the hazards described in the article OFFICES; however, direct contact with the public entails further risks. Bank notes coming from areas of epidemic disease may be infectious. The handling of large sums of money, the need for constant vigilance and attention for cases of fraud or forgery, and the threat or occurrence of hold-ups may create considerable nervous tension; mechanical accounting aids and adequate protection such as bullet-proof glass can help reduce this.

General precautions

Throughout commerce, especially in establishments handling dangerous goods, money and food, personal hygiene is of great importance; adequate washing and sanitary facilities should be provided together with canteen catering or messrooms. Lighting should be of a high standard but staff should be shielded from the glare of publicity lighting; emergency lighting should also be installed.

The ever-increasing use of flammable packaging material has augmented considerably the fire hazard in commercial establishments; adequate waste disposal and refuse collection services should be organised and good housekeeping rigorously enforced to prevent accumulations of litter. Fire protection and prevention measures and equipment should be of a high standard [and specially tailored to meet the situation, e.g. zoned sprinkler system, use of high expansion foam, etc.] and staff should be instructed in fire emergency and evacuation procedures.

Many young persons are employed as shop assistants and the necessary provisions should be made for them. In addition, the hours of work of shop assistants should be subject to careful control especially during periods of "special sales" and festive seasons.

[The International Labour Conference adopted in 1964 a Convention (No. 120) and a Recommendation (No. 120) on Hygiene in Commerce and Offices. The Convention has so far been ratified by 40 member States of the ILO.]

GHERARDI, P.

The salesman in the field. Conditions of work and employment of commercial travellers and representatives. Bell, M. (Geneva, International Labour Office, 1980), 108 p.

Safety in the stacking of materials. Health and safety at work No. 47. Department of Employment (London, HM Stationery Office, 1971), 58 p. Illus.

CIS 911-1970 *Accident prevention from A to Z. A safety handbook for the retail trade* (Unfallschutz von A-Z. Ein Handbuch der Unfallverhütung im Einzelhandel) (Berufsgenossenschaft für den Einzelhandel, Niebuhrstrasse 5, Bonn) (1969), 38 p. Illus. (In German)

CIS 76-550 *Health and safety guide for service stations.* DHEW (NIOSH) publication No. 75-139 (National Institute for Occupational Safety and Health, 4676 Columbia Parkway, Cincinnati) (Feb. 1975), 65 p. Illus.

CIS 76-551 *Health and safety guide for grocery stores.* DHEW (NIOSH) publication No. 75-134 (National Institute for Occupational Safety and Health, 4676 Columbia Parkway, Cincinnati) (Feb. 1975), 47 p. Illus.

CIS 76-549 *Health and safety guide for sporting goods stores.* DHEW (NIOSH) publication No. 75-141 (National Institute for Occupational Safety and Health, 4676 Columbia Parkway, Cincinnati) (Apr. 1975), 82 p. Illus.

CIS 78-2057 "Study of certain categories of supermarket personnel" (Enquête sur certaines catégories de personnel des magasins populaires (supermarchés)). Salord, M. C., et al. *Cahiers de médecine interprofessionnelle* (Paris), 2nd quarter 1978, 18/70 (33-46) Illus. 2 ref. (In French)

CIS 77-1749 "Health hazard among cash register operators and the effects of improved working conditions". Ohara, H.; Aoyama, H.; Itani, T. *Journal of Human Ergology* (Tokyo), Sep. 1976, 5/1 (31-40). Illus. 6 ref.

"Shop cash register operators" (Les caissières de magasins). Godefroy M. *Travail et sécurité* (Paris), May 1982, 5 (254-281). Illus.

Commission of the European Communities (CEC)

On 18 April 1951 six countries, namely Belgium, France, the Federal Republic of Germany, Italy, Luxembourg and the Netherlands, signed the Treaty of Paris setting up the European Coal and Steel Community (ECSC) based in Luxembourg. This treaty envisaged a supranational community responsible for its own budget and exercising control over its own affairs. To carry out this function four separate bodies were created: a High Authority, a Special Council of Ministers, a Court of Justice and a Common Assembly.

1 January 1958 saw the setting-up of the institutions of two new Communities (EEC and EURATOM) following the signing of the Rome Treaties on 25 March 1957. Each of these treaties instituted separate bodies, along the lines of the Coal and Steel Community, but the Court of Justice and the Assembly performed functions for all three communities although the rules of procedure differed. At the inaugural session in March 1958 the new single assembly (common to the ECSC and the two new Communities) resolved to call itself the "European Parliamentary Assembly". Since 1979 the European Parliament had been directly elected. It is worth while noting that the two Communities set up in 1958 were not financially independent as was the ECSC, but depended upon the Council of Ministers for financial support. Both the Commissions were based in Brussels, but under the Euratom Treaty a number of Joint Research Centres were set up in Geel, Karlsruhe, Petten and, the largest, Ispra. The Rome Treaties, but not the Paris Treaty, also created the Economic and Social Committee which has a constitution of members of management, trade unions and other authorities such as consumers.

In 1965 a treaty was signed which resulted in the fusion of the three executive bodies, i.e. the High Authority of the ECSC, the Commission of Euratom and the Commission of the European Ecomonic Community, resulting in the formation of the Commission of the European Communities. This treaty also fused the three Councils into one body. At the same time this treaty fixed the provisional seats of the various bodies either in Brussels, Luxembourg or Strasbourg. It was put into effect on 1 July 1967.

In 1973 Denmark, Ireland and the United Kingdom, and in 1981 Greece, joined the Common Market making a total of ten countries, representing a population of 270 million inhabitants.

Composition and role of the Commission

Today the Commission consists of 14 Commissioners and the permanent staff. There are two Commissioners each from France, Germany, Italy and the United Kingdom, and one from each of the other member States. One of these commissioners is elected President, and each commissioner is responsible for one or more portfolios. The Commission's term of office lasts for four years.

The role of the Commission is to act as guardian of the Treaties with the power of making proposals either based on the treaties or on a decision of the Council of Ministers. Such proposals are submitted to the Council of Ministers, which sends them to the European Parliament and to the Economic and Social Committee. When the two latter bodies have given their opinion the Council of Ministers can then decide on the proposals. When proposals are being drawn up the Commission usually holds several meetings of experts or consultants so that the final document reflects the opinions of the various parties involved.

Although the three Treaties establishing the European Communities have emphasised the economic aspects, a number of dispositions based on these Treaties have enabled the Commission to develop actions on health at Community level that are generally more precise and severe and yet often more extensive than those put forward by other international organisations. The fact that the Commission and the Council of Ministers can adopt directives that are legally binding on the member States distinguishes the activity of the Communities from the other international organisations which mostly act by means of agreements or conventions requiring ratification by States.

European co-operation on health and safety in the coal and steel industries (ECSC)

Article 55 of the Paris Treaty of 1951 permits the Commission to develop research and organise exchanges of knowledge on occupational safety, hygiene, and medicine in three types of industries: coal mines, iron mines, and steel works. Research is particularly directed towards the detection and prevention of occupational disease, the fight against air pollution in steel works, dust in mines, ergonomics, readaptation, and occupational safety. The total amount made available for such research from 1955 to the end of 1979 was over 62 million ECU.

A commission for safety and health in the mines and other extractive industries was established in 1957, consisting of representatives of governments, employers and trade unions. This commission plays an important role in the establishment of recommendations aimed at improving the prevention of occupational risks in the mines; it transmits these recommendations direct to the competent national authorities. More than 500 recommendations have been established since this commission was created, and it represents an important contribution at European level to working conditions in the extractive industries.

A further commission, for safety in steel works, was established in 1964, and although it has not had quite the same impact at the level of the competent national authorities, it nevertheless plays an important role in the analysis of statistics, the establishment of safety rules, and the practical organisation of accident prevention.

European co-operation in the field of nuclear energy (Euratom)

According to article 2 of the relevant Treaty, "the Community shall establish uniform safety standards to protect the health of workers and of the general public and ensure that they are applied". These safety standards were established in 1959, and have given birth to a common policy of protection and a normative action that has no equivalent in any other international body. Based on the 1959 standards, which were completed in 1962 and revised in 1962, 1966, 1972, 1976 and in 1980, there has been constructed a co-ordinated and unified ensemble, at European level, of laws, regulations and administrative practices. This ensemble covers: definitions; scope, reporting and authorisation; limitation of doses for controllable exposures; derived limits; accidental and emergency exposure of workers; fundamental principles governing operational protection of exposed workers; and fundamental principles governing operational protection of the population.

In parallel with this normative and preventive action, the Commission has undertaken programmes of research on radiology and radioprotection, which have provided scientific support to the standard-setting action.

European co-operation in the field of health and safety at work

An important date in the history of the European Communities is that of 29 June 1978, since it marks the adoption by the Council of Ministers of a resolution for a programme of action on safety and health at work, which aims at improving the protection of all workers within the Community, i.e. more than 105 million workers. The improvement of work conditions concerns all sectors of the economy and is based on the participation of the social partners in the initiatives and decisions; and it aims in a global approach to cover the totality of all risks that occur at work. Such a programme should make it possible to achieve the following general objectives:

- *(a)* improvement of the working situation with a view to increased safety and with due regard to health requirements in the organisation of work. There is an urgent need to review and redefine a more effective accident and disease prevention strategy in order to update traditional methods;
- *(b)* improvement of knowledge in order to identify and assess risks and perfect prevention and control methods;
- *(c)* improvement of human attitudes in order to promote and develop safety and health consciousness. Alongside the technical aspects of accident prevention and health protection, a real system of safety instruction and health education must be created. This has yet to be introduced and will be taught in different ways at the various educational levels and at the various levels of responsibility and action within undertakings.

In its resolution the Council agreed that the following actions could be undertaken up to the end of 1982:

Accident and disease aetiology connected with work–research.

(1) Establish, in collaboration with the Statistical Office of the European Communities, a common statistical methodology in order to assess with sufficient accuracy the frequency, the gravity and causes of accidents at work, and also the mortality, sickness and absenteeism rates in the case of diseases connected with work.

(2) Promote the exchange of knowledge, establish the conditions for close co-operation between research institutes and identify the subjects for research to be worked on jointly.

Protection against dangerous substances.

(3) Standardise the terminology and concepts relating to exposure limits for toxic substances. Harmonise the exposure limits for a certain number of substances, taking into account the exposure limits already in existence.

(4) Develop a preventive and protective action for substances recognised as being carcinogenic, by fixing exposure limits, sampling requirements and measuring methods, and satisfactory conditions of hygiene at the workplace, and by specifying prohibitions where necessary.

(5) Establish, for certain specific toxic substances such as asbestos, arsenic, cadmium, lead and chlorinated solvents, exposure limits, limit values for human biological indicators, sampling requirements and measuring methods, and satisfactory conditions of hygiene at the workplace.

(6) Establish a common methodology for the assessment of the health risks connected with the physical, chemical and biological agents present at the workplace, in particular by research into criteria of harmfulness and by determining the reference values from which to obtain exposure limits.

(7) Establish information notices on the risks relating to and handbooks on the handling of a certain number of dangerous substances such as pesticides, herbicides, carcinogenic substances, asbestos, arsenic, lead, mercury, cadmium and chlorinated solvents.

Prevention of the dangers and harmful effects of machines.

(8) Establish the limit levels for noise and vibrations at the workplace and determine practical ways and means of protecting workers and reducing sound levels at places of work. Establish the permissible sound levels of building-site equipment and other machines.

(9) Undertake a joint study of the application of the principles of accident prevention and of ergonomics in the design, construction and utilisation of plant and machinery, and promote this application in certain pilot sectors, including agriculture.

(10) Analyse the provisions and measures governing the monitoring of the effectiveness of safety and protection arrangements and organise an exchange of experience in this field.

Monitoring and inspection–improvement of human attitudes.

(11) Develop a common methodology for monitoring both pollutant concentrations and the measurement of environmental conditions at places of work; carry out intercomparison programmes and establish reference methods for the determination of the most important pollutants. Promote new monitoring and measuring methods for the assessment of individual exposure, in particular through the application of sensitive biological indicators. Special attention will be given to the monitoring of exposure in the case of women, especially of expectant mothers, and adolescents. Undertake a joint study of the principles and methods of application of industrial medicine with a view to promoting better protection of workers' health.

(12) Establish the principles and criteria applicable to the special monitoring relating to assistance or rescue teams in the event of accident or disaster, maintenance and repair teams and the isolated worker.

(13) Exchange experience concerning the principles and methods of organisation of inspection by public authorities in the fields of safety, hygiene at work and occupational medicine.

(14) Draw up outline schemes at a Community level for introducing and providing information on safety and hygiene matters at the workplace to particular categories of workers such as migrant workers, newly recruited workers and workers who have changed jobs.

It is important to point out that an Advisory Committee on safety, hygiene and health protection at work was created in 1974 by the Council of Ministers. The composition is tripartite consisting of 6 members from each member State (2 from Government, 2 from employers, and 2 from workers). In outline the terms of reference are: the Committee shall have the task of assisting the Commission in the preparation and implementation of activities in the fields of safety, hygiene and health protection at work. This task shall cover all sectors of the economy except the mineral extracting industries falling within the responsibility of the Mines Safety and Health Commission and except the protection of the health of workers against the dangers arising from ionising radiations, which is subject to special regulations pursuant to the Treaty establishing the European Atomic Energy Community.

In application of the programme of action a number of proposals for directives are currently being discussed or have been adopted by the Council; these include:

(1) Council Directive of 27 November 1980 on the protection of workers from the risks related to exposure to chemical, physical and biological agents at work (*Official Journal* L327 of 3/12/1980).

(2) Directive on the health of workers exposed to vinyl chloride monomer (*Official Journal* L197 of 22/7/1978).

(3) Proposal for a Council Directive on the protection of workers from harmful exposure to metallic lead and its ionic compounds at work (*Official Journal* C324 of 28/12/1979).

(4) Proposal for a second Council Directive on the protection of workers from the risks related to agents at work: asbestos (*Official Journal* C262 of 9/10/1980).

The first of these proposals is a general Directive which will result in all member States following a similar path in the future. This Directive has two objectives:

- eliminate or limit exposure to chemical, physical and biological agents and prevent risks to workers' health and safety;
- protect workers who are likely to be exposed to these agents.

This Directive which will affect the majority of workers in the Community requires the member States to take short-term and longer-term measures. The short-term measures require that within three years workers and/or their representatives shall have access to appropriate information concerning asbestos, arsenic, cadmium, lead and mercury, and within four years appropriate surveillance of the health of workers exposed to asbestos and lead shall take place. The longer-term measures apply when a member State adopts provisions concerning an agent. In order that the exposure of workers to agents shall be avoided or kept at as low a level as is reasonably practicable, member States should comply with a set of requirements, but in doing so they have to determine whether and to what extent each of these requirements is applicable to the agent concerned.

Table 1. List of agents to which specific requirements apply or will apply

Acrylonitrile	
Asbestos	
Arsenic and compounds	
Benzene	
Cadmium and compounds	
Mercury and compounds	
Nickel and compounds	
Lead and compounds	
Chlorinated hydrocarbon compounds	– chloroform
	– paradichlorobenzene
	– carbon tetrachloride

In addition an initial list of agents (table 1) relates to the implementation of further more specific requirements. This Directive also requires the member States to consult the social partners when the above requirements are being established. With regard to the agents mentioned, the Council will fix in individual Directives that it adopts the limit value(s) as well as other rules. Certain technical aspects concerning specific rules established in the individual Directives can be reviewed by a technical committee in the light of experience acquired and progress made in the technical and scientific fields. The proposal, mentioned above, on metallic lead and its ionic compounds is the first of these individual Directives, and asbestos is the second.

RECHT, P.
HUNTER, W. J.

Council resolution of 29 June 1978 on an action programme of the European Communities on safety and health at work (*Official Journal* C165 of 11 July 1978)

Council Directive of 29 June 1978 on the approximation of the laws, regulations and administrative provisions of the member States on the protection of the health of workers exposed to vinyl chloride monomer (*Official Journal* L197 of 22 July 1978)

Proposal for a Council Directive on the protection of workers from harmful exposure to metallic lead and its ionic compounds at work (*Official Journal* C324 of 28 December 1979)

Council Directive of 15 July 1980 amending the Directives laying down the basic safety standards for the health protection of the general public and workers against the dangers of ionising radiation (*Official Journal* L246 of 17 September 1980)

Proposal for a second Council Directive on the protection of workers from the risks related to exposure to agents at work: asbestos (*Official Journal* C262 of 9 October 1980)

Council Directive of 27 November 1980 on the protection of workers from the risks related to exposure to chemical, physical and biological agents at work (*Official Journal* L327 of 3 December 1980).

Compressed-air work

Pearl and sponge divers who have developed breath-holding techniques which enable them to stay under water for periods of a few minutes at depths up to 30 m illustrate today man's first attempts at working under water. Breathing tubes and diving bells date from quite early times, but creating an environment in which the air pressure is increased to balance hydrostatic pressure so that a man could work under water for long periods was first achieved in the 16th century.

Raising the ambient air pressure results in an increase in the amount of dissolved nitrogen in the body, and if the man returns to normal atmospheric pressure too rapidly, bubbles of gas are formed in the circulation and in the tissues. Serious and fatal results were very common until a rational procedure for entry into and exit from compressed air was adopted about 60 years ago.

The normal atmospheric pressure is 1 kg/cm^2 (0.98 bar or 98.06·10^3 Pa). The usual maximum working pressure in compressed-air work in building and civil engineering adds to this a further 3.5 kg/cm^2 (343·10^3 Pa), but most work in compressed air is carried out at much lower pressures. Men work in increased air pressures in a variety of ways, such as underwater diving, either in conventional suits or with self-contained breathing apparatus; in diving bells; and in caissons or tunnels in water-bearing ground (see also UNDERWATER WORK).

Diving

Deep-sea diving has been associated particularly with naval activities, but it is also required in the construction and maintenance of ports and harbours, cooling-water systems for power stations, bridge piers, immersed-tube tunnels, barrages, effluent outfalls, submarine pipelines and cables, off-shore drilling rigs for oil or gas and salvage work.

Self-contained underwater breathing apparatus is increasingly being employed instead of conventional surface-supported diving suits because of the mobility which freedom from pipelines allows.

The majority of civil engineering construction work requiring divers is at present at depths of 10-30 m. In future the use of compressed air in diving is likely to be limited to a maximum depth of water of 50 m.

Diving bells

A diving bell is a heavily weighted steel chamber, partially open at the bottom, which is lowered through the water by a winch. The bell is supplied with electric lighting and a telephone. Air is pumped in from a compressor at the surface, so that the men can work on the sea bed preparing foundations or carrying out other underwater construction. Access to the bell may be through an airlock, but more commonly the lock does not form part of the bell itself and must be provided on ship or dockside.

Caissons

Construction of foundations for bridges or piers in the sea or rivers, or work in water-bearing ground is often carried out in a caisson (figure 1). This consists of a tubular steel structure open at the lower end where men work, and closed at the top. The working chamber is supplied with air at a suitable pressure and access to the chamber is by an airlock. As material is excavated (figure 2) the caisson sinks into the ground under its own weight. Finally, it can be filled with concrete to make the foundation.

Tunnelling under water

In constructing a tunnel through porous ground under a river or the sea, the open end of the tunnel is sealed by an access lock and air is pumped in to balance the water head. This has the effect of drying and stiffening the working face so that it can be excavated more easily and safely. The power required for compressing large amounts of air, the dangers to the health of the workers which necessitate elaborate medical supervision, and

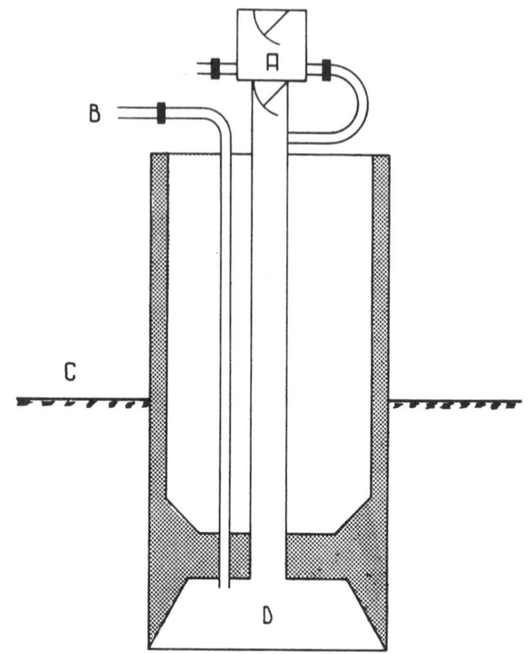

Figure 1. Diagram of a caisson, showing: A. Airlock. B. Compressed air supply. C. Water level. D. Working chamber.

Figure 2. Men excavating material in the working chamber of a caisson. *(By courtesy of Mr. F. V. Rose, and the* Auckland Star.*)*

the long periods of unproductive decompression time make this an expensive method of working. For example, compression in the man-lock at the start of a working shift takes only a few minutes, but if the shift is of 4 h or more current decompression practice may require up to 4 or 5 h decompression time on leaving the working area.

Tunnelling under water or work in caissons is carried out by gangs of men who are engaged in hard physical work in compressed air for shifts which are often of 8 h duration. The problems in this work, although fundamentally the same as in diving, are in practice rather different.

Hyperbaric chambers for medical work

There has been much interest in the use of large therapeutic pressure chambers in which surgical teams can work in air at pressures of 2-3 atm (196-294·10^3 Pa) absolute while the patient breathes oxygen. Hyperbaric oxygen given in this way has been advocated in the treatment of gas gangrene from *Clostridium welchii* infection, carbon monoxide poisoning and air embolism from therapeutic accidents. However, the initial enthusiasm for hyperbaric treatment has waned as alternative methods of treatment can be effective. One-man hyperbaric oxygen chambers are also used in the treatment of cancer by radiotherapy. Medical staff, nurses and technicians who work in these chambers for long periods run some risk of decompression sickness and this is known to have occurred occasionally.

Undersea houses and bases

There has also been great interest in the possibility of using underwater houses in which it is intended that men on underwater research, construction work or prospecting for minerals will live for periods of 1-2 weeks and from which they will emerge with self-contained breathing apparatus to carry out their daily tasks. These are still in an experimental stage.

Normal procedures for compressed-air workers

Entry to the working area of caissons, pressurised tunnels, etc., is through an airlock which is a pressure chamber having two self-sealing doors, one leading to the exterior, the other to the high-pressure area. The internal door can be opened only when air has been pumped into the lock to equal the working pressure. Increasing the atmospheric pressure in an airlock can usually be carried out rapidly and without complications. Normal experienced workers can adjust the rising pressure across their eardrums by opening the eustachian tubes voluntarily or by swallowing. Less experienced men may be unable to carry out this manoeuvre quickly enough, so that slower rates of compression are used for them. There is no bar to normal activity in compressed air, apart from a nasal quality of voice sounds.

At the end of the working period the men again enter the lock which must be at working pressure with the door to the exterior closed (figure 3) and decompression to atmospheric pressure is carried out slowly according to the schedule in use.

In restricted circumstances, such as at the top of a caisson, it may be necessary to carry out a rapid decompression to atmospheric pressure in a small lock. The men then walk to a second and roomier lock in which the pressure is rapidly raised to the working pressure and the normal decompression procedure carried out. This process is sometimes referred to as decanting or surface

Figure 3. Tunnellers in a man-lock during decompression. The entrance to the pressurised working area is through the airtight door in the background.

decompression. The time taken for men to leave the first lock and enter the second should not be more than 5 min.

It is usual to insist that workers should remain on site for at least 1 h after the decompression is completed as this is the period during which decompression sickness is most likely to occur.

HAZARDS

Accidents. Any type of accident may occur to a compressed-air worker, depending on the type of job he carries out. Trauma of various kinds from hand or power tools is, of course, the commonest problem. Falls of rock-bearing material may cause bruising or crushing injuries. Use of large machines, such as excavators, and the manhandling of heavy cast-iron or steel segments of a tunnel lining may lead to injuries. Falls from staging in large tunnels can occur, and wet and slippery ground or catwalks can also be a hazard.

If a tunnel face is unstable it can collapse and men have occasionally been blown through the face into the water. A catwalk at the upper part of the tunnel ensures that if flooding occurs the men can pass freely back to the lock.

Electrical hazards may also exist, as much equipment is electrically powered.

Fire is a serious hazard, particularly if welding or cutting is carried out in compressed air. Clothing, including nylon shirts, may easily ignite from a spark, and it is advisable to wear special non-flammable work-clothes or woollen materials rather than synthetic fibres, and for a man to stand by with a water hose held ready to spray the welder should his clothes catch fire. Fire extinguishers of suitable type and pressure should also be available at strategic points in the tunnel or caisson. A high standard of maintenance and good housekeeping should be ensured and this will help reduce the fire hazard.

Diseases. During compression there may be acute pain in the ears or sinuses if there is an upper respiratory infection, so that men with colds or sinusitis should not be allowed into compressed air. If men are not sufficiently adept at opening the eustachian tubes and if compression is too rapid, they may suffer from dizziness, nausea, disorientation or pain in the ear. A haematoma of the eardrum may result or the drum may rupture. Prophylactic chemotherapy is recommended if this should happen, and further exposure to compressed air avoided until the drum has healed. Sinus pain due to meatal blockage may also occur and decongestants may be used to reduce mucosal swelling if entry to compressed air is essential.

Skin reactions to acid or alkaline materials in the ground may occur when men are digging through industrial wastes in made-up ground.

The most important health hazard which affects compressed-air workers is decompression sickness, which is described in detail in a separate article.

SAFETY AND HEALTH MEASURES

In the conduct of a compressed-air undertaking it is essential for engineers, lock-keepers and medical attendants to have a clear understanding of the risks to which the men are exposed and of the means of reducing the hazards. The safety of men and the successful completion of the work depend on the maintenance of the necessary air pressure continuously until the task is completed. For this reason, where electric compressors are used, it is usual to have standby diesel compressors in a state of readiness in case of a breakdown in electricity supply. Welding or burning may give rise to the production of dangerous oxides of nitrogen or ozone. Air from a compressor is hot and contains oil fumes so that it must be cooled and filtered before being piped to the working area.

The compression chamber or lock in which men are compressed and decompressed must be strong enough for at least the maximum pressures used and the door should be airtight. Men should be able to stand upright without discomfort (minimum headroom 1.8 m) and for each man likely to be in the lock there should be at least 1.2 m^3 of space and 0.6 m of bench seating. Automatic pressure recorders should be fitted to all compression chambers so that a permanent record is made of every pressure change.

At least one separate compression chamber for every 100 men who work in compressed air should be provided for the treatment of decompression sickness if the gauge pressure is over 1 kg/cm^2 ($98 \cdot 10^3$ Pa).

Decompression should be carried out by a properly trained lock-keeper strictly according to the tables. In many cases an automatic valve is used and requires only the selection of a cam suitable for the working pressure. Records of all men entering compressed air and the time of entry are the lock-keeper's responsibility. Telephone communication between the working area and the lock-keeper, and between him and the lock interior is necessary.

Medical prevention. Accepted good practice in the medical aspects of work in compressed air has been published as a Medical Code of Practice and is periodically updated. Selection of suitable men by the physician to the compressed-air work is of great importance, and it should be an inflexible rule that nobody is allowed to enter compressed air who has not recently had a medical examination. Medical attendants, who should be qualified nurses and familiar with the symptoms and clinical signs of decompression sickness, must be on duty day and night while compressed-air work is being undertaken. Compressed-air workers should be required to wear a tag or carry an identity card indicating their profession and the address of the nearest recompression chamber so that prompt treatment can be provided in the event of an attack of decompression sickness.

Legislation

Work in compressed air is controlled by legislation in many but not all countries, depending on the degree of industrial development and the amount of work of this type that is carried out. Regulations can be divided into two main parts: first, the general control of the work from the aspect of health and safety; and second, the schedules governing compression and decompression.

Health and safety. It is usual to specify the safe construction and reliability of airlocks and compression plant, the size and equipment of airlocks, suitable pressure gauges to indicate lock pressure and working pressure and means of accurately controlling pressure inside and outside the lock. Competent lock attendants must be appointed and they must keep an accurate and detailed record by name and works number of each individual who enters or leaves compressed air, the length of his shift, the working pressure and decompression time.

The medical supervision and initial and periodic examination of workers must be specified. Men should be medically examined again if they have an infection of the upper respiratory system or have been off work through illness, to decide whether or not they are fit to enter compressed air.

Changing and locker accommodation, sanitary and washing facilities and provision for men to wait in comfort on the site for about 2 h after decompression are also usual requirements.

One or more medical compression chambers, properly equipped with couch and blankets and a chemical closet, must be provided when the working pressure is about 1 kg/cm² (98·10³ Pa), for the treatment of decompression sickness.

Notification of local police, hospitals, the public health authorities and factory inspectorate that compressed-air work is being undertaken, is required.

Regulations may also describe in detail the type of electrical supply to the working area, telephone and signalling systems, the quantity and quality of the air supply and its temperature and humidity.

Decompression tables. In most cases, decompression will be carried out according to a table which is statutory for the country in which compressed-air work is taking place. Many of these tables have followed the British table of 1958 which is based on Haldane's observation that decompression sickness did not occur after exposure to pressures of less than 1.25 atm (121.5·10³ Pa) above normal atmospheric pressure, no matter how long the exposure or the speed with which the pressure was reduced. Haldane assumed that bubbles are not formed in the body unless the amount of nitrogen supersaturation is more than that of a decompression from a total pressure of 2.25 atm (209.5·10³ Pa), and that the total or absolute pressure, which is the reading on the pressure gauge plus atmospheric pressure, can always be safely halved. This is because the volume of nitrogen released is the same whether the pressure drops from 6 to 3 atm (588 to 294·10³ Pa), from 4 to 2 atm (392 to 196·10³ Pa), or from 2 to 1 atm (196 to 98·10³ Pa). Haldane's investigations provided a scientific basis for decompression, and the hypothesis which he formulated dominated compressed-air procedures all over the world until the last 20 years, when it was questioned and considerably modified.

In many countries the length of shift which is permitted is progressively reduced as the working pressure rises. Another procedure is the split shift in which two working periods, whose length depends on the working pressure, are separated by an interval at atmospheric pressure varying from 30 min to 6 h. This has fallen out of favour because of the increased number of decompressions required and consequent greater risk of decompression sickness.

The decompression time used to bring men to atmospheric pressure varies strikingly in different countries of the same working pressure.

In constructing decompression tables, a compromise has often had to be sought between short enough to be economic and yet still sufficiently long to give reasonable protection to the workers, and this has meant that a proportion of decompression sickness has been tolerated. The realisation during the last few years that bone necrosis is a common sequel of work in compressed air has prompted the use of much longer schedules approaching more those used by the United States and British navies for divers in whom bone necrosis is rare. It is considered now that Haldane's 2 : 1 ratio must be cut back drastically as higher working pressures are used.

In the United Kingdom, the British 1958 table has now been superseded by the Blackpool Decompression Tables published in the Medical Code of Practice for Work in Compressed Air in conventional (lb f/in² ≈ 6.89476·10³ Pa) and S.I. units (bar). These tables, although regarded as an improvement in that decompression sickness rates are reduced, still do not prevent it entirely especially at high pressures, nor do they reduce the risk of bone damage. Recent tables such as that used in Washington State, or the British Blackpool trial tables, decrease the 2 : 1 ratio progressively as the working pressure rises. This results in very long decompression periods at higher working pressures, and consideration must therefore be given to the comfort of the men. A low-pressure chamber to which the men are transferred at some point during a long decompression may allow greater space and comfort, washing facilities or television entertainment.

It is clear that to achieve more complete protection of compressed-air workers against decompression sickness, further experimentation with schedules which both reduce the length of the working shift and increase the decompression time as working pressure goes up, must be undertaken. Inhalation of oxygen during the second phase of decompression when the gauge pressure has dropped to less than 1.8 kg/cm² (176.4·10³ Pa) may reduce decompression time and lessen the rise of decompression sickness. The fire hazards of oxygen at pressure have so far made engineers cautious in using it, but if this problem can be solved satisfactorily, oxygen may be more widely used in future.

McCALLUM, R. I.

Medical code of practice for work in compressed air (Construction Industry Research and Information Association, 6 Storey's Gate, London) (3rd ed. 1980).

CIS 76-1732 *The physiology and medicine of diving and compressed air work.* Bennett, P. B.; Elliot, D. H. (Baillière Tindall, 35 Red Lion Square, London) (2nd ed. 1975), 556 p. Illus. 2 000 ref.

CIS 75-1134 "Work in compressed air" (Le travail en atmosphère comprimée). Susbielle, G.; Jullien, G. *Encyclopédie médico-chirurgicale. Maladies–Agents physiques.* Fascicule 16500 A 10, 1-1975 (18 rue Séguier, 75006 Paris), 12 p. Illus. 45 ref. (In French)

"Some occupational hygiene aspects of man-lock work in relation to the compressed air work legislation" (Enige arbeidshygiënische aspecten van caissonarbeid in relatie tot de Wet op werken onder overdruk (Caissonwert)). Vroege, D.; Jansen, H.; Strijers, R. L. M. *Tijdschrift voor sociale geneskunde* (Leiden), 7 Nov. 1979, 57/22 (737-741). Illus. 6 ref. (In Dutch)

Computers

A computer is any of a large variety of devices capable of performing mathematical calculations on numbers or variables. Computers are broadly classified as either analogue when they represent numbers by continuously varying quantities such as lengths, voltages, etc., or as digital when they represent numbers by devices capable of assuming discrete, countable states. This classification applies to simple devices such as the analogue slide rule and the digital desk-top adding machine; however, the word computer is now usually employed synonymously with "electronic data-processing system".

A computer system (figure 1) unlike previous administrative machinery, constantly processes and converts the data that have been fed into it. The stored data are organised or processed in accordance with a programme ("software") that specifies the processing criteria in logical steps. This, therefore, constitutes a fully automatic process, since the system itself regulates the indications for processing, data output and process control, without external intervention.

The main computer functions are: input, output, storage and processing (sorting; mathematical calculations) (see figure 1). Frequently used input devices are: punched card readers, optical mark readers (for sheets and cards), display keyboards, typewriter keyboards,

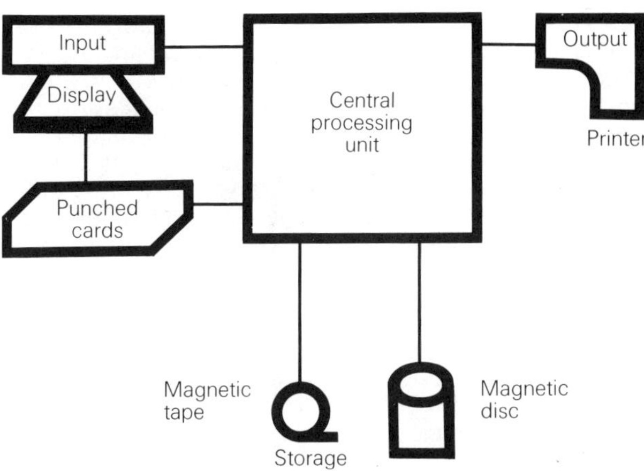

Figure 1. Schematic representation of a computer system.

paper tape readers, cassette readers and floppy disc readers.

Output devices are: displays, printers and plotters. The output can also be punched (on cards or paper tape) or written on a magnetic tape or disc, and in its turn be used for input purposes.

For storage of programmes and information, magnetic discs (random access, very fast) and magnetic tapes (sequential access, slow) are mostly used, containing up to some hundreds of millions of characters.

The size of the organisation and processing requirements determine whether a large computer is needed (in the organisation itself or in a service centre) or whether a small local system will be adequate. Many applications call for local processing which acts as preprocessing for the updating of very large information files in central systems. Data can be communicated by telephone line. Information can be transmitted instantly ("real time") or at certain times ("batch").

The costs of computer systems ("hardware") are decreasing and central processing units and memories are becoming very small as a result of the implementation of miniature circuits (electronic "chips"). In many cases the computer can be built into existing technical equipment (microprocessors). In several applications the computer is constantly connected with the process for which data have to be collected and analysed. In such circumstances interfaces are needed which convert the measurement results into specific input signals.

Uses. Typical business applications of data-processing systems are payrolls, maintenance of accounts and invoicing. In science and engineering in general, computers are used for literature searches, data reduction, mathematical and statistical calculations, recording and maintenance of experimental data, control of automated production lines, etc.

Organising for the computer

Computerisation and automation will have important repercussions on organisational structure and consequently their introduction must be preceded by careful management and systems planning and analysis. It is necessary to determine which information paths are to be automated; it may be necessary to rethink certain empirical attitudes and adopt a more fundamental and analytical approach by reviewing the sequence, relationship and interdependence of data and procedures so that the computer can be adequately programmed.

The design of the system and its operation will involve the collaboration of various people: the user, to formulate and define activities in a manner acceptable to the computer; the computer expert, to choose the equipment and organise the flow and programming of information; the social psychologist, to ensure that all persons encompassed by computerisation are kept fully informed of the situation; and the business economist to ensure optimal utilisation. These specialists should be directed by a project leader who will co-ordinate the team and make decisions.

The computer in medical science and health care

Many general applications of a computer system can be transferred to a medical environment, for example literature searches on medical items, administrative procedures in hospitals, process control in clinical laboratories and medical statistics. More specific applications are: the storage of structured medical records, medical decision-making, processing of biological signals and epidemiology. One of the main difficulties in medical informatics is translating the doctor's partly intuitive and individually oriented way of thinking into the logical and general terms in which computers have to be programmed to allow us to communicate with them. One of the main practical obstacles to the use of computer systems in medicine is the reluctance of doctors to perform their own input and output activities.

Literature searches. The store of knowledge in medical and other sciences has grown exponentially in recent decades. It is humanly impossible to scan the immense store of information in libraries and retain it in the human mind. However, computerised information and documentation systems can be used to classify world literature, search for information on the basis of specific criteria (e.g. keywords) and print out the relevant stored summaries. Examples of this technique are *CIS, Excerpta Medica, Medlars* and *Bio Pascal.*

Administrative procedures. In medical care a great deal of time is spent on administration. Every hospital, family doctor and pharmacy needs a patient file to know the population for which they are responsible or who can ask for their services. The computer can be helpful in scheduling, control of hospital admissions, bed utilisation and billing. Increasing effort is being devoted to improving the administrative procedures in hospitals which form part of a hospital information system, but family doctors' practices too will soon be connected with a computer or have their own small system. One of the specific uses they could make of a computer system is monitoring populations at risk in their patient file. Pharmacists are still using computer systems for monitoring the total amount of prescriptions and the incompatibility of certain drugs.

Process control in clinical laboratories. Certain parts of clinical laboratories can be considered as processes to be controlled. These include such activities as patient identification, the preparation of work lists, quality control, processing of test results, printing of complete patient records, etc. In many clinical laboratories measuring apparatus is directly connected to the computer system, so that human input of test results and the possibilities of making errors can be reduced.

Medical records. Much effort has been devoted to the creation of optimum structures for medical records. These can be considered as a fundamental basis of good medical care. Medical records make their most powerful impact when they apply to several disciplines, so that communication can be improved and speeded up and the duplication of examinations can be avoided.

A central storage system for medical records can be useful in emergency cases, transplantations, etc. Special attention has to be paid to the privacy and protection of patient data. Certain parts of the medical record can be reserved for specific users for both the input and output of data. The computer medical record can contain codes and figures as well as free format text. Stored medical records represent a potential source of medical knowledge, and can be of enormous importance for epidemiological research.

Computerised medical records not only offer a solution to the problem of transverse communication between experts in a multidisciplinary team but also provide a means for the longitudinal evaluation of information in that they can provide a problem/patient history which can be compared and evaluated over a period of time by retrieving data from the past and automatically drawing attention to data to be used in the future.

Signal analysis and medical decision-making. Many biological signals (e.g. recorded by electrocardiography or electroencephalography) are very complicated and their analysis makes great demands on the human mind. Particularly in the field of ECG analysis many computer programmes have been developed to make a diagnosis on the basis of the electrical heart signals, which have to be digitalised for computer input. In this case medical diagnosis is restricted to one kind of signal. Generally speaking, however, different kinds of signals, figures and symptoms have to be combined to form a conclusion: the diagnosis. If we assume that making a diagnosis is mainly a logical process, we can programme the computer to carry out the various steps in thought. It is unreasonable, however, to think that all diseases can be diagnosed by the computer when any given symptom is fed into it. It is only in the case of some specific diseases that we know enough of the relationship between symptoms, signs and diagnosis to put our knowledge in the form of logical and statistical rules. One of the latest contributions to the support of medical diagnosis is computerised tomography. This technique enables the information obtained from a series of X-ray photographs of a part of the body to be combined so that the location of tumours, etc., can be more exactly determined than with the traditional interpretation of X-ray photographs.

Preventive medicine and epidemiology. Preventive medicine is no longer restricted to providing immunisation against communicable diseases. One of the most promising aspects of preventive care is the early detection of a number of chronic diseases. People can suffer from these diseases (e.g. hypertension, diabetes) without being aware of it.

With certain examination methods (screening tests), which are applicable to large groups of people and can be performed by technicians and nurses under medical supervision, a fairly reliable impression of certain aspects of health can be explored. The computer can be of great help in organising these screening examinations, by handling data collection and quality control, and reporting and signalling deviations. The data collected by examinations of subjectively healthy people can be considered as complementary to the data on ill populations registered in the data bases of family doctors and hospitals. In order to find out the "natural history" of the development of diseases, it is extremely important to analyse the figures for these normal populations. In addition, normal values and predictive values of risk factors can be established more accurately with the aid of computers.

Occupational health. Most of the above-mentioned uses of the computer for general medicine are also applicable to the field of occupational health care. In this case the medical record is coloured by the working situation: the physical and organisational environment, the load of responsibilities and specific tasks. Occupational medical records can be correlated with labour absenteeism records. The computer can be used to collect and analyse the characteristics of the working situation which, on the one hand, may be considered as input while, on the other, the biological and behavioural data of patients may be regarded as output. The establishment of relationships, however, is still in its infancy. In occupational health care it is very advantageous to store data on examinations, the medical background to limitations in working capacity, the results of specific periodic examinations and general health testing programmes within the same computer system. This also enables a lifelong record of health in relation to work to be obtained for the employee. Combining the results of many persons considered as groups with certain characteristics will generate valuable epidemiological information.

Since exposure to toxic agents and stressful situations may result in abnormalities only after many years, careful registration over a long period of time is necessary and measuring methods must be kept constant. This can be done by computer quality control. Examinations can be organised more efficiently and be executed on a more standardised basis with the aid of a computer system. Even environmental measuring data gathered by sampling procedures at certain workplaces can be input directly into the computer system.

Ergonomics and computer systems

When the computer is used for the registration of working situations, this will also involve the computer operators themselves. Working conditions in computer rooms frequently leave much to be desired: they may be too noisy, too warm, or unpleasant (temperature, humidity).

Outside the computer room, the display/keyboard workplace, in particular, can cause medical complaints: reflections on the screen and bad contrast can give rise to eye fatigue and headaches. Wrongly designed furniture can cause pain in the neck and back. Aspects of a more psychological nature are: monotonous work in an isolated position, frustration caused by waiting times and the requirement of intense concentration (air traffic control).

Special attention must be paid to the design of registration forms which have to be copied by means of display/keyboard conversation. The high voltages used in cathode ray tubes (display screens) generate some X-radiation outside the apparatus. The intensity, however, is generally too low to harm the user.

DURINCK, J. R.

Safety:

CIS 79-1835 *Safety of data processing equipment.* Amendment No. 1 to Publication 435 (International Electrotechnical Commission, 1 rue de Varembé, 1211 Geneva) (Dec. 1978), 123 p. Illus.

Problems of work:

CIS 79-1468 *Problems of work at computer terminals* (Problèmes posés par les terminaux d'ordinateurs) (Centre de recherche et d'études sur la sécurité, l'ergonomie et la promotion des conditions de travail (CRESEPT), 9 rue de Pascale, 1040 Brussels) (no date), 173 p. Illus. 50 ref.

Uses:

"Data automation". Fischoff, R. L.; Freiberger, F. G. Ch. 4 (99-189). 213 ref. *Patty's Industrial Hygiene and Toxicology,* Vol. III: *Theory and rationale of industrial hygiene practice* (New York, Chichester, Brisbane, Toronto, John Wiley and Sons, 1979), 752 p. Illus. Ref.

"Some comments on the computerised national record: a transitional experience" (En marge du dossier national informatisé: une expérience de transition). Amphoux, M.;

Blaizot, M.; Pavy, F. *Revue de médecine du travail* (Paris), 1978, 6/1 (1-16). Illus. 16 ref. (In French)

"Man-computer interaction". Withfield, D. (472-511). 173 ref. *Proceedings of the Seventh Annual IRA Symposium on Human Engineering and Quality of Work Life* (Ramat Efal, Israel, IRA Memorial Foundation, 1980).

CIS 78-573 "A computerized occupational medical surveillance program". Barrett, C. D.; Belk, H. D. *Journal of Occupational Medicine* (Chicago), Nov. 1977, 19/11 (732-736). 4 ref.

CIS 79-2068 "The computerization of industrial hygiene records". Snyder, P. J.; Bell, Z. G.; Samelson, R. J. *American Industrial Hygiene Association Journal* (Akron, Ohio), Aug. 1979, 40/8 (709-720). Illus. 4 ref.

CIS 76-342 "Reading chest radiographs for pneumoconiosis by computer". Jagoe, J. R.; Paton, K. A. *British Journal of Industrial Medicine* (London), Nov. 1975, 32/4 (267-272). Illus. 7 ref.

CIS 78-874 "Calculation of a lung function index and possibility of using a computer for lung function diagnosis" (Berechnung eines Lungenfunktionsindexes und Möglichkeit des Einsatzes der EDV in der Lungenfunktionsdiagnostik). Reichel, G.; Muhar, F.; Prügger, F.; Raber, A.; Rop, I. *Die Berufsgenossenschaft* (Bielefeld), July 1977, 7 (315-322). Illus. 12 ref. (In German)

Concrete and reinforced concrete work

The aggregates, such as gravel and sand, are mixed with cement and water in motor-driven horizontal or vertical mixers of various capacities, which are generally installed on the construction site, but sometimes it is more economical to have ready-mixed concrete delivered and discharged into a silo on the site. For this purpose concrete mixing stations are installed in the periphery of towns or near gravel pits. Special rotary-drum lorries (figure 1) are used to avoid separation of the mixed constituents of the concrete, which would lower the strength of concrete structures.

Tower cranes or hoists are used to transport the ready-mixed concrete from the mixer or silo to the framework. The size and height of certain structures may also require the use of concrete pumps (figure 2) for conveying and placing the ready-mixed concrete. There are pumps which lift the concrete to heights of up to 100 m. As their capacity is by far greater than that of cranes of hoists, they are used in particular for the construction of high piers, towers and silos with the aid of climbing formwork. Concrete pumps are generally mounted on lorries, and the rotary-drum lorries used for transporting ready-mixed concrete are now frequently equipped with concrete pumps which enable concrete to be delivered from the mixing station and placed without passing through a silo.

Formwork. Formwork has followed the technical development rendered possible by the availability of larger tower cranes with longer arms and increased capacities, and it is no longer necessary to prepare shuttering *in situ*. Prefabricated formwork up to 25 m² in size is used in particular for making vertical structures, such as façades and dividing walls, of large residential and industrial buildings (figure 3). These structural-steel formwork elements, which are prefabricated in the site shop or by the industry, are lined with sheet-metal or wooden panels. They are handled by crane. Before they can be removed after the concrete has set, they must be secured to the crane hook. Depending on the type of building method, prefabricated formwork panels are either lowered to the ground for cleaning or taken to the next wall section ready for pouring.

So-called formwork tables are used to make horizontal structures, i.e. floor slabs for large buildings. These tables are composed of several structural-steel elements and can be assembled to form floors of different surfaces. The upper part of the table, i.e. the actual floor-slab form, is lowered by means of screw jacks or hydraulic jacks after the concrete has set. Special beak-like load-carrying devices have been devised to withdraw the tables, to lift them to the next floor and to insert them there.

Sliding or climbing formwork is used to build towers, silos, bridge piers and similar high structures. A single formwork element is prepared *in situ* for this purpose; its cross-section corresponds to that of the structure to be erected, and its height may vary between 2 and 4 m. The formwork surfaces in contact with the concrete are lined with steel sheets, and the entire element is linked to jacking devices. Vertical steel bars anchored in the concrete which is poured serve as jacking guides. The sliding form is jacked upwards as the concrete sets, and the reinforcement work and concrete placing continue without interruption. This means that work has to go on round the clock, i.e. by day and night shifts.

Climbing forms differ from sliding ones in that they are anchored in the concrete by means of screw sleeves (figure 4). As soon as the concrete poured has set to the required strength, the anchor screws are undone, the form is lifted by the height of the subsequent section to be poured, anchored and prepared for placing the concrete.

So-called form cars are frequently used in civil engineering, in particular for making bridge deck slabs. Especially when long bridges or viaducts are built, a form car replaces the rather complex falsework. The deck forms corresponding to one length of bay are fitted in such a way to a structural-steel frame that the various form elements can be jacked into position and be removed laterally or lowered after the concrete has set. When the bay is finished, the supporting frame is advanced by one bay length, the form elements are jacked again into position, and the subsequent bay is poured (figure 5).

When a bridge is built using the so-called cantilever technique (figure 6) the form-supporting frame is much shorter than the one described above. It does not rest on the next pier but must be anchored to form a cantilever. This technique, which is generally used for very high bridges, often relies on two such frames which are advanced by stages from piers on both sides.

Prestressed concrete is used in particular for bridges, but also in building construction for especially designed structures. Strands of steel wire wrapped in steel-sheet or plastic sheathing are embedded in the concrete at the same time as the reinforcement. The ends of the strands or tendons are provided with head plates in order that the prestressed-concrete elements may be pretensioned with the aid of hydraulic jacks before the elements are loaded.

Prefabricated elements. Construction techniques have been rationalised even further for large residential buildings, bridges and tunnels by prefabricating elements such as floor slabs, walls, bridge beams, etc., in a special concrete factory or near the construction site. The prefabricated elements, which are assembled on the site, do away with the erection, displacement and dismantling of complex formwork and falsework, and a great deal of dangerous work at height can be avoided.

Reinforcement. Reinforcement is generally delivered to the site cut and bent according to bar and bending schedules. Only when prefabricating concrete elements on the site or in the factory are the reinforcement bars tied or welded to each other to form cages or mats which are inserted into the forms before the concrete is poured.

PREVENTION OF ACCIDENTS

Mechanisation and rationalisation have eliminated many traditional hazards on building sites, but have also created new dangers. For instance the fatalities due to falls from height have considerably diminished thanks to the use of form cars, form-supporting frames in bridge building and other techniques. This is due to the fact that the work platforms and walkways with their guard rails are assembled only once and displaced at the same time as the form car, whereas with traditional formwork the guard rails are often neglected. On the other hand, mechanical hazards are increasing and electrical hazards are particularly serious due to the wet environment. Health hazards arise from cement itself, from added substances for curing or water-proofing and from lubricants for formwork.

Some important accident prevention measures to be taken for various operations are dealt with below.

Concrete mixing. As concrete is nearly always mixed by machine, special attention should be paid to the design and layout of switchgear and feed-hopper skips. In particular when concrete mixers are being cleaned it happens that a switch is actuated unintentionally, starting the drum or the skip and causing injury to the attending worker. Therefore, switches should be protected and also arranged in such a manner that no confusion is possible. If necessary, they should be interlocked or provided with a lock. The skips should be free from danger zones for the mixer attendant and workers moving on passageways near it (figure 7). It must also be ensured that workers cleaning the pits beneath feed-hopper skips are not injured by the accidental lowering of the hopper.

Silos for aggregates, especially sand, present a hazard of fatal accidents—workers entering a silo to remove so-called arches without a standby person and without a safety harness and lifeline may be buried in the loose material. Silos should therefore be equipped with vibrators and platforms from which sticking sand can be poked down, and corresponding warning notices should be displayed. No person should be allowed to enter the silo without another standing by.

Concrete handling and placing. The proper layout of concrete transfer points and their equipment with mirrors and bucket receiving cages do away with a standby worker who has to reach out for the crane bucket and to guide it—thus avoiding accidents such as bruised hands.

Transfer silos which are jacked up hydraulically must be secured so that they are not suddenly lowered if a pipeline breaks.

Work platforms fitted with guard rails must be provided for placing the concrete in the forms with the aid of buckets suspended from the crane hook or with a concrete pump. The crane operators must be trained for this type of work and must have a normal vision. If large distances are covered, two-way telephone communication or walky-talkies have to be used.

When concrete pumps with pipelines and placer masts are used, special attention should be paid to the stability of the installation. Agitating lorries with built-in concrete pumps must be equipped with interlocked switches which make it impossible to start the two operations simultaneously. The agitators must be guarded so that the operating personnel cannot come into contact with moving parts. The baskets for collecting the rubber ball, which is pressed through the pipeline to clean it after the concrete has been placed, are now replaced by two elbows arranged in opposite direction. These elbows absorb almost all the pressure needed to push the ball through the placing line and not only eliminate the whip effect at the line end, but also prevent the ball from being shot out of the line end.

When agitating lorries are used in combination with placing plant and lifting equipment, special attention has to be paid to overhead electric lines. These lines must be insulated or guarded by protective scaffolds within the work range to exclude any accidental contact—unless the overhead line can be displaced (figure 8). It is important to contact the power supply station.

Formwork. The assembly of traditional formwork composed of square timber and boards is frequently associated with accidents resulting from falls, because the necessary guard rails and toeboards are often neglected for work platforms which are required for short periods only. Nowadays steel supporting structures are widely used to speed up formwork assembly, but here again the available guard rails and toeboards are frequently not installed on the pretext that they are only needed for so short a time.

Plywood form panels, which are increasingly used, offer the advantage of being easy and quick to assemble. However, having been used several times, they are frequently misappropriated as platforms for rapidly required scaffolds for secondary purposes, and it is generally forgotten that the distances between the supporting transoms must be considerably reduced in comparison with normal scaffold planks. Accidents resulting from breakage of form panels misused as scaffold platforms are still rather frequent.

Two outstanding hazards must be borne in mind when using prefabricated form elements. These elements must be stored in such a manner that they cannot turn over (figure 9). Since it is not always feasible to store form elements horizontally, they must be secured by stays. Form elements permanently equipped with platforms, guard rails and toeboards facilitate slinging the element to the crane hook as well as the assembly and dismantling on the structure under construction. They constitute a safe workplace for the personnel and do away with the provision of work platforms for placing the concrete. Fixed ladders may be added for safer access to platforms. Scaffold and work platforms with guard rails and toeboards permanently attached to the form element should be used in particular with sliding and climbing formwork (figure 10).

Experience has shown that practically no accidents due to falls are recorded where no work platforms have to be improvised and rapidly assembled. Unfortunately, form elements fitted with guard rails cannot be used everywhere, especially not where small residential buildings are being erected.

When the form elements are handled by crane to take them from storage to the structure, lifting tackle of

Figure 1. Concrete agitating lorry. Figure 2. Mobile concrete pump with placer mast. Figure 3. Large-surface form panels made of structural steel with wooden lining. Figure 4. Climbing scaffold on a pair of piers. Figure 5. Form car with advancing frame. Figure 6. Cantilever construction of a viaduct using prefabricated elements. Figure 7. Protective grating round a hopper skip. Figure 8. Protection of trolley wires on subway construction site. Figure 9. Turnover protection for large-surface form panels. Figure 10. Form panel with scaffold platform. Figure 11. Dust control by exhaust ventilation during cleaning of form boards. Figure 12. Removable guard rails to protect workers assembling façade elements. Figure 13. Transport of personnel by crane and cage.

①
②
③
④
⑤
⑥
⑦
⑧
⑨
⑩
⑪
⑫
⑬

appropriate size and strength, such as slings and spreaders, must be used. If the angle between the sling legs is too large, the form elements must be handled with the aid of spreaders.

The workers cleaning the forms are exposed to a health hazard which is generally overlooked: the concrete residues adhering to the form surfaces are more and more frequently removed with portable grinders. Dust measurements have shown that the grinding dust contains a high percentage of respirable fractions and silica. Therefore, dust control measures must be taken, e.g. portable grinders with exhaust devices linked to a filter unit or enclosed form-board cleaning plant with exhaust ventilation (figure 11).

Assembly of prefabricated elements. Special lifting equipment should be used already in the manufacturing plant so that the elements can be displaced and handled safely and without vertebral injury to the workers. Anchor bolts embedded in the concrete facilitate handling not only in the factory but also on the assembly site. To avoid bending of the anchor bolts by oblique loads, large elements must be lifted with the aid of spreaders with short rope slings. If a load is applied to the bolts at an oblique angle, concrete may spill off and the bolts may be torn out. The use of inappropriate lifting tackle has caused serious accidents resulting from falling concrete elements.

Appropriate vehicles have to be used for the road transport of prefabricated elements which must be secured by appropriate supporting structures against overturning or sliding, for example when the driver has to brake the vehicle suddenly. Visibly displayed weight indications on the elements facilitate the task of the crane operator during loading, unloading and assembly on the site.

Lifting equipment on the site should be adequately chosen and operated. Tracks and roads must be kept in good condition in order to avoid overturning of loaded equipment during operation.

Work platforms protecting the personnel against falls from height must be provided for the assembly of the elements. All possible means of collective protection, such as scaffolds, safety nets, overhead travelling cranes erected before completion of the building, should be taken into consideration before recourse is had to personal protective equipment. It is, of course, possible to equip the workers with safety harnesses and lifelines but experience has shown that there are workers who use this equipment only when they are obliged to do so under constant supervision. Lifelines are indeed a hindrance when certain tasks are performed, and certain workers are proud of being capable of working at great heights without using any protection. For details on this subject see FALLS FROM HEIGHTS: PERSONAL PROTECTION AGAINST.

Before starting to design a prefabricated building, the architect, the manufacturer of the prefabricated elements and the building contractor should meet to discuss and study the course and safety of all operations. When it is known beforehand what types of handling and lifting equipment are available on the site, the concrete elements may be provided in the factory with fastening devices for guard rails and toeboards. The façade ends of floor elements, for instance, are then easily fitted with prefabricated guard rails and toeboards before the elements are lifted into place. The wall elements corresponding to the floor slab may thereafter be safely assembled because the workers are protected by guard rails (figure 12).

For the erection of certain high industrial structures, mobile work platforms are lifted into position by crane and hung from suspension bolts embedded in the structure itself. In such cases it may be safer to transport the workers to the platform by crane (which should have high safety characteristics and be manned with a qualified operator) than to use improvised scaffolds or ladders (figure 13).

When post-tensioning concrete elements attention should be paid to the design of the post-tensioning recesses which should enable the tensioning jacks to be applied, operated and removed without any hazard for the personnel. Suspension hooks for tensioning jacks or openings for passing the crane rope must be provided for post-tensioning work beneath bridge decks or in box-type elements. This type of work, too, requires the provision of work platforms with guard rails and toeboards. The platform floor should be sufficiently low to allow for ample work space and safe handling of the jack. •No person must be permitted at the rear of the tensioning jack because of the serious accidents which may result from the high energy released in the event of the breakage of an anchoring element or a steel tendon. The workers should also avoid being in front of the anchor plates as long as the mortar pressed into the tendon sheaths has not set. As the mortar pump is connected with hydraulic pipes to the jack, no person should be permitted in the area between pump and jack during tensioning. Perfect communication among the operators and with the supervisors is also very important.

Training. Thorough training of plant operators in particular and all construction site personnel in general is becoming more and more important in view of increasing mechanisation and the use of many types of machinery, plant and substances. Unskilled labourers or helpers should be employed in exceptional cases only, if the number of construction site accidents is to be reduced.

BACHOFEN, G.

CIS 76-2027 *Safety during construction of concrete buildings–A status report.* Lew, H. S. National Bureau of Standards, Washington, DC. PB-248 683/55WI (National Technical Information Service, Springfield, Virginia 22151) (Jan. 1976), 59 p. Illus. 39 ref.

CIS 80-1744 "Causes and cases of accidents in concrete mixing plants and possibilities of prevention" (Unfallquellen und Unfälle in Transportbetonwerken und Möglichkeiten ihrer Beseitigung und Verhütung). Bott, W. *Industrie der Steine und Erden* (Hannover), Jan.-Feb. 1980, 1 (21-26). 3 ref. (In German)

CIS 80-838 *Concrete grinding and sawing* (Le ponçage et le sciage du béton). Parisot, S. (Caisse régionale d'assurance maladie du Nord de la France, 11 boulevard Vauban, Lille) (1 June 1978), 43 p. Illus. (In French)

CIS 78-1483 "Set of papers on back disorders in concrete reinforcement workers". Wickström, G.; Hänninen, K; Lehtinen, M.; Riihimäki, H.; Nummi, J.; Nurminen, M.; Järvinen, T.; Stambej, U.; Wiikeri, M.; Saari, J. *Scandinavian Journal of Work, Environment and Health* (Helsinki), 1978, 4/Supplement 1 (13-58). Illus. 124 ref.

CIS 80-2043 "Some experiences with epoxy resin grouting compounds". Hosein, H. R. *American Industrial Hygiene Association Journal* (Akron, Ohio), July 1980, 41/7 (523-525). 6 ref.

Conditioned reflexes

The conditioned reflex method elaborated at the beginning of this century by the Russian physiologist I. P. Pavlov is widely used in toxicological research. The field of applied research on nervous functions which deals with the study of the potential harmfulness of chemical compounds differs from pure physiological research in methodical approach, mental representation and interpretation of results. Pavlov perceived the conditioned reflex (CR) as the elementary and universal "unit" of the higher nervous activity (HNA) which enables the organism to adapt itself to the constantly changing environment. The term "conditioning" adopted by American research workers corresponds to the notion of conditioned reflex activity. This term was first used in 1915 by Watson for the study of the HNA and the elucidation of behaviour. The first experiments were carried out by Zavadski in 1908 and Nikiforovski in 1910 to study the effects of alcohol, morphine, cocaine and caffeine on the higher sections of the central nervous system (CNS) of animals.

At present the CR method is employed to determine threshold concentrations of harmful substances after single (rarely repeated) administration, and also to investigate the mechanism of the action of chemical compounds on the CNS.

As regards the application of the CR to small laboratory animals, the motor method relying on food reward has become the most widely used technique because of its high sensitivity. Slightly less sensitive are the method relying on the defence CR and the labyrinth method. In experimental toxicology preference should be given to mixed techniques relying on the association of operations with classical reflexes, and to methods using motor chain reactions (modification by Skinner and Voronin).

The application of the CR method requires the objective representation and automatic recording of the reactions exhibited, the creation of apparatus for the automatic programming and control of the experiment, and machine processing of the results obtained. A typical CR experiment is to study the effect of a harmful factor on animals with an established type of HNA using confirmed conditioned reflexes. It takes several months to prepare such an experiment and to accomplish the programme. According to the specific action of the substance investigated and the aims of the experiment, the functions of the CNS are studied either integrally or in various stages determined by positive or negative conditioned reflexes. The basic mechanisms of the responses of the CNS to the action of toxic agents consist in changes of the response latency, disturbances of the force inter-relations of positive conditioned reflexes, and upset processes of internal restraint.

In order to elucidate the compensated changes and the modified reactivity of the organism during the initial stages of poisoning, various functional stresses are employed: investigation of the reflexes against the background of increased nutritional excitability, administration of an external restraint, prolonged differentiation, changes in the order of succession of the stimuli.

The degree and evolution of the toxic process and its involution in animals depends on particular types of HNA. In animals of the strong balanced type with a good mobility of the nervous functions the intoxication evolves more favourably.

Despite the high informative value of the CR method, its isolated use in toxicological practice is limited, particularly on account of the difficulty of extrapolating experimental data directly to man. It is most useful to employ the CR method in conjunction with integral techniques for determining the threshold of acute action of neurotropic poisons. To elucidate the mechanism of toxic action on the CNS, it is necessary to use certain techniques which interact with the various functional levels of the CNS and reflex systems. The high plasticity of the HNA renders processing of the results of long-term CR experiments difficult. There are fluctuating changes in HNA, alternating periods of perturbation and full restoration in spite of prolonged effects.

The CR technique is more frequently adopted for establishing hygiene standards for harmful substances in environmental water and air than in industrial toxicology. Of 400 MAC values recommended in the USSR for various substances in water, 60% have been established with the aid of the CR technique, and when standards are established on the basis of sanitary and toxicological indices of harmfulness, nearly all MACs are founded on the study of conditioned reflex activity (Krasovskij et al., 1974).

A committee of experts of the WHO has recommended the technique of conditioned reflexes as a highly sensitive and reliable method for the investigation of changes due to toxic substances in the CNS (WHO, 1977).

ULANOVA, I. P.
VESELOVSKAJA, K. A.

Twenty years' experience of the objective study of higher nervous activity (animal behaviour) (Dvadcatiletnij opyt ob"ektivnogo izučenija vysšej nervnoj dejatel'nosti (povedenija ivotnyh)). Pavolv, I. P. (Moscow, Medgiz, 7th ed., 1951). (In Russian)

Methods of establishing permissible levels of action of harmful occupational factors. Report of the committee of experts of the WHO with the participation of the ILO. Technical report series 601 (Geneva, World Health Organisation, 1977), 68 p. Ref.

"Present-day problems of communal toxicology in connection with water pollution by chemical substances" (Aktual'nye problemy kommunal'noj toksikologii v svjazi s himičeskim zagrjazneniem vodoemov). Krasovskij, G. N.; Šigan, S. A.; Egorova, N. A. *Problemy toksikologii*, Vol. 6 (Moscow, VINITI, 1974). (In Russian)

"Methods for the study of the central nervous system in toxicological tests". Pavlenko, S. M. (86-108). 48 ref. *Methods used in the USSR for establishing biologically safe levels of toxic substances* (Geneva, World Health Organisation, 1975), 171 p.

Confectionery industry

Confectionery is, in the strict sense of the word, the industrial or artisanal production of sweetmeats with a sugar base (such as candies, caramels, etc.). Although the productive units involved may be large and independent, they are often closely linked with, and form an important part of, the chocolates, pastry and cake industry due to the large number of intermediate substances common to both (see COCOA INDUSTRY AND CHOCOLATE PRODUCTION).

The raw materials employed are numerous and varied, and the machines and equipment are similar to those found in related industries. These machines include open-flame, steam-jacketed or water-jacketed boilers, grinding machines, dry-fruit and nut processors, kneaders, beaters, and rolling machines, and each has its specific hazards.

HAZARDS AND THEIR PREVENTION

The working conditions are, in many ways, similar to those in biscuit making. The noise levels are lower and the floors less slippery; however, there may be high atmospheric concentrations of sugar or starch dust with the attendant hazards of dust explosions, diseases caused by vegetable dusts and dental diseases due to formation of organic acids when mouth hygiene is neglected.

After the sugar has been boiled, it is poured on to a marble slab and kneaded by machine or hand; the sugar may be worked at temperatures of up to 80 °C and, consequently, adequate hand and arm protection is essential. Sugar-drawing machines are dangerous and require suitable machinery guarding; pastille and candy presses, single- or multi-piston sugar-roasting presses and rotary moulding machines should have safety screens covering the working parts. Pouring machines which are correctly protected (in the same way as biscuit depanners and mogul-type chocolate machines), and series-mounted dragée turbines present no special hazards; sweet-sawing machines require particular attention. As in biscuit making, much of the plant, equipment and special machinery is built by the confectionery firm itself and the safety measures often leave much to be desired.

Wrapping, weighing, folding and packaging machines should be subject to the general regulations on machinery safety, especially as certain versions operate at very high speed (for example, twist-wrappers may have an output of over 600 sweets per minute).

VILLARD, R. F.

CIS 75-1761 "Basic guidelines for preventive check-ups in occupational medicine–Danger of dental disease due to formation of organic acids in the mouth" (Berufsgenossenschaftliche Grundsätze für arbeitsmedizinische Vorsorgeuntersuchungen–Gefährdung der Zähne durch organische Säuren, die sich in der Mundhöhle bilden). *Arbeitsmedizin–Sozialmedizin–Präventivmedizin* (Stuttgart), Feb. 1975, 10/2 (35-36). 1 ref. (In German)

Confined spaces

The term "confined space", as it refers to industrial activity, generally describes a space having a relatively small volume with unfavourable natural ventilation and one into which infrequent or irregular entry is made for purposes of maintenance, repair and/or cleaning. It has been defined as a "space" so enclosed that dangerous air contamination therein cannot be prevented or removed by natural ventilation through opening in the enclosure. Examples of such spaces include silos, tanks, vats, reaction vessels, boilers, sewers, compartments of ships and degreasers. The definition cited can also be taken to include any space not having the inherent safety features of an ordinary workroom. Thus, in addition to the absence of means of natural ventilation, a confined space may be considered as one in which:

(a) the volume is so small that even uniform diffusion of evolving gases or vapours throughout the entire space would not always prevent the formation of a toxic concentration in the worker's breathing zone or flammable concentrations in the space;

(b) there would not be other workers nearby who could observe and rescue a worker suddenly overcome; and

(c) the openings are so remote or small that ready access or egress for removal of an overcome worker is difficult. In some countries legislation prescribes a minimum size of manhole and good practice demands similar arrangements everywhere.

Toxicity in confined spaces

In line with this broad definition of confined space, it should also be appreciated that the terms "toxic" and "non-toxic", as they apply to confined spaces, are both relative. Whereas carbon tetrachloride is generally regarded as a highly toxic agent, methyl chloroform is considered relatively innocuous. However, with adequate controls it is possible to use carbon tetrachloride safely. On the other hand, the use of methyl chloroform without suitable controls may produce harmful and even lethal effects despite the fact that it is a so-called "non-toxic" agent. The following example will demonstrate this paradox.

In a workroom 30 m × 15 m × 6 m (2 700 m³) with no natural or mechanical ventilation, uniform diffusion of a given quantity of methyl chloroform will create a concentration of 350 ppm, the current threshold limit value recommended by the American Conference of Governmental Industrial Hygienists. Under favourable conditions, this concentration can be tolerated safely by a healthy worker for 8 h daily for an indefinite period. The same quantity of methyl chloroform evaporated and uniformly distributed in a confined space measuring 3 m × 3 m × 1.5m (13.5 m³), a volume which may be found inside a tank, will produce a concentration of 70 000 ppm. This concentration will rapidly cause unconsciousness and even death if rescue is not prompt.

Thus the unqualified use of terms such as "toxic", "non-toxic" and "safe", as applied to industrial air contaminants, can create false and dangerous illusions. In an absolute sense there are no non-toxic materials. Rather, there are toxic concentrations or toxic ranges of materials. Similarly, one should never assume that an operation in a confined space is a safe one, insignificant as it may appear. An operation of relatively long duration in a confined space will naturally present the greatest danger, but in many instances exposures to high concentrations even for as short a time as it takes to retrieve a dropped object from inside a tank have caused death.

The substances to be regarded as dangerous in a confined space can cover practically the entire spectrum of gases and vapours found in industry. These agents may be present in the confined space:

(a) as a consequence of the operation which is being performed, as when trichlorethylene vapour is released when cleaning the sludge from the bottom of a degreasing tank;

(b) as an insidious effect incidental to the operation itself, as when oxygen deficiency is caused by fermentation in a grain silo; or

(c) before entry, as in the case of hydrogen sulphide gas in a sewer.

Potentially harmful agents

In a presentation of data on 21 occurrences of illness or death associated with work in confined spaces, a variety of agents was identified as implicated in the morbidity and mortality reported. Lack of oxygen was indicated as the principal single cause of acute illness or death in the majority of the 21 cases. In many instances, oxygen was removed from the air by processes such as fermentation, which simultaneously release carbon dioxide. In some of the cases, excessive carbon dioxide caused a lowering of oxygen content in the confined-space atmosphere and hence contributed to the accident. Frequently, slow oxidation such as that occurring during rusting of the walls in a deep, unused ship compartment will remove oxygen from the air to a point where the atmosphere will not support life. The presence or use of gases such as carbon dioxide or nitrogen may replace the air in a confined space. The presence of carbon dioxide, which is frequently associated with lack of oxygen, aggravates the problem of oxygen diminution. Other causative agents encountered included hydrogen sulphide, which is frequently present in sewers, and solvents such as trichlorethylene, methyl chloroform and dichloromethane, which are used industrially in a variety of cleaning and degreasing processes. Another agent, acetonitrile, while not frequently used, was encountered as an ingredient in a protective coating being applied to the interior of a tank.

Flammable gases and vapours also constitute hazards. Fires or explosions can result if there is a source of ignition in these atmospheres. Where repairs, scaling or inspection work on steam boilers or other steam vessels is being done there is a serious risk of scalding unless the boiler or vessel is effectively disconnected before entry.

In some circumstances work in confined spaces may involve mechanical risks: for example, unless effectively disconnected, moving parts such as paddles of beaters may be started up while repair workers are inside a large mixer.

SAFETY AND HEALTH MEASURES

Despite the fact that the potential danger of a confined space may be obvious, carelessness, lack of good judgement and failure to apply fundamental safety principles may none the less contribute to accidents. To guard against the influence of such factors as the cause of accidents, the dangers inherent in such operations as entering manholes and silos where organic products may be fermenting, and painting or welding inside a tank, should be stressed beforehand to both the workers and supervisors, and acceptable safety precautions should be followed while operations are in progress.

It is of interest that most accidents occurring in connection with work in confined spaces exhibit one or more of several common features. First, the reason for being in the confined space is usually to perform a non-routine function. Repairs, cleaning, painting and similar operations are for the most part done at irregular, widely spaced intervals and in many cases rarely by the same person more than once. Frequently, the person involved is the employee of an outside contractor. Safety indoctrination of such casual personnel is frequently overlooked. Second, few or none of the usual safeguards are provided nor are good practices appropriate for operations in confined spaces followed. Third, personnel having even a rudimentary knowledge of first aid and resuscitation are found to be seldom available when needed.

Recommended safe practices. The safety procedures outlined below, if applied universally, will help to eliminate the deaths and serious illnesses which frequently result from unsafe conditions in confined spaces. These procedures are incorporated in legislation in some countries and are also found in the ILO Model Code of Safety Regulations for Industrial Establishments. They should be binding on all persons concerned, including contractors, and should form part of the safety manual of every large company with a well established safety programme and industrial hygiene or safety engineering staff. Problems often arise in small plants or with small outside contractors who are called in for repair or maintenance work: often such firms and contractors have no proper safety arrangements and supervision is apt to be casual. It is essential that management in such cases should appreciate the necessity of:

(a) providing adequate safety equipment;

(b) setting up and enforcing proper procedures such as those outlined below; and

(c) viewing any entry into a confined space for any purpose and for any length of time as a potentially dangerous operation requiring suitable safeguards.

Pre-entry precautions. Train all personnel working in confined spaces in the potential hazards that exist and in the procedures and precautions to be followed. When necessary, clean the space to remove all residual contaminants such as solvents and organic materials. Check the atmosphere within the space for toxic and flammable airborne contaminants and oxygen concentration. The latter should not be less than 18% for working without breathing apparatus. Numerous direct reading devices are available to monitor for the oxygen concentration and the most commonly encountered contaminants. Special methods should be devised for most of the others. Where work in boilers and furnaces is to be done, allow adequate time for cooling.

Close and lock all valves and switches connected with the operation of the confined space to prevent accidental introduction of contaminants, live steam, hot water, or starting of equipment within the space when it is occupied. Purge the space by ventilating for as long as necessary to reduce any contaminants to safe levels. Remove all possible sources of ignition if flammable or combustible materials may be present or created.

Provide such protective clothing as may be necessary. Provide respiratory protective equipment if the need for it exists or can arise during occupancy. Such equipment must consist of supplied-air or oxygen respirators, either of the hose or self-contained type, to provide protection in oxygen-deficient or grossly contaminated atmospheres.

A "permit-to-work" system may be a useful adjunct to these precautions (see PERMIT-TO-WORK SYSTEMS).

Precautions during occupancy. If possible, provide continuous ventilation during occupancy. Such ventilation is mandatory when a contaminant is being continuously generated. Particular care is needed when oxy-acetylene welding or cutting operations are being carried out. Special purification techniques are also available, such as the use of a chemical which absorbs carbon dioxide and releases oxygen.

Provide the worker with a harness attached to a lifeline constantly held by a second standby worker who should continuously watch the worker in the confined space. This will permit rapid removal of the worker in an emergency. A third worker should be within hailing distance to provide assistance if necessary.

If possible, provide a means of communication between the worker and the outside since the worker may suddenly begin to feel distress and not be able to summon help. Frequently, the bodily positions which are assumed in confined-space work may make it difficult for an outside observer to detect an unconscious worker. One simple method of communication is the use of an alarm which goes off at 5-min intervals and rings until shut off by the man in the confined space. Prolonged ringing will signify trouble.

When flammable vapours may be present or released, all equipment should be spark- and explosion-proof. This includes equipment used for operations as well as for rescue purposes.

Rescue and first aid. The two standby workers should be well trained in rescue techniques, first aid and resuscitation.

Respirators, either of the supplied-air or self-contained type, must be immediately available for use by the rescue workers. If the worker in the tank does not normally use one, one should be available for him as well. These must be maintained in a constant state of readiness.

The worker should be removed from the confined space as soon as any sign of distress becomes evident. Contaminated clothes should be removed and emergency first-aid procedures, including resuscitation if necessary, should immediately be instituted by qualified personnel pending the arrival of a physician or removal to a treatment facility.

KLEINFELD, M.

"Work inside tanks and small rooms" (Travaux à l'intérieur de réservoirs et dans des locaux exigus). Burri, W. *Cahiers suisses de la sécurité du travail* (Lucerne), Nov. 1976-Jan. 1977, 124 (36). Illus.

Entry into confined spaces. Guidance note GS5 from the Health and Safety Executive (London, HM Stationery Office, July 1977), 10 p.

"Lack of oxygen as a cause of accidents in underground works" (Manque d'oxygène comme cause d'accidents dans les travaux souterrains). *Cahiers suisses de la sécurité du travail* (Lucerne), Jan. 1973, 111 (22). Illus. 7 ref.

Working in confined spaces. Criteria for a recommended standard. DHEW (NIOSH) publication No. 80-106 (National Institute for Occupational Safety and Health, 4676 Columbia Parkway, Cincinnati) (Dec. 1979), 68 p. 110 ref.

Contact dermatitis or eczema, occupational

Occupational contact dermatitis, also known as occupational eczema, is a skin disease caused or favoured by exposure to chemical, physical or biological agents present in the work environment (table 1). From the legal standpoint, occupational eczema is any eczema produced or contributed to by noxious agents listed in the regulations concerning occupational diseases.

Occupational dermatoses account for over 50% of occupational diseases receiving compensation, while contact forms of dermatitis account for 80-90% of occupational dermatoses; of the latter, over 50% are attributed to allergy.

Irritant contact dermatitis or irritant eczema

Irritant contact dermatitis (synonyms: irritant eczema, non-allergic contact dermatitis, non-allergic eczema, orthoergic dermatitis) is a skin disease produced by chemical, physical or biological agents which damage the skin by direct action and in the area of contact.

Aetiopathogenesis

Irritant agents capable of producing inflammation may be chemical, biological or physical.

Chemical agents comprise:

(a) oxidisers such as hydrogen peroxide, permanganates, chromic acid and its salts, free iodine, bromine, hypochlorites, persulphates, nitrates, etc.;

(b) dehydrating agents, including strong acids, strong alkali, etc.;

(c) protein precipitants: tannic acid, pycric acid, salts of heavy metals, etc.;

(d) keratolytics: pyrogallol, resorcinol, salicyclic acid, etc.;

(e) degreasing agents: alcohol, ether, chloroform, trichloroethylene, etc.

(f) various organic compounds.

Biological agents include:

(a) agents of vegetable origin: nettles, cactus, latex of figs, primula, etc.;

(b) agents of animal origin: insect bites or contact with certain insects (fleas, Paederus, cantharides, etc.).

Table 1. Potential occupational exposure to contact dermatitis

Occupations	Nature of hazard	
	Irritant	Sensitising
Bricklayers, cement workers, building workers	Cement, lime, moisture	Hexavalent chromium salts, cobalt salts, epoxy resins
Cleaners	Detergents, soap, organic solvents, turpentine, abrasives	Hexavalent chromium salts, turpentine, additives in soap, colophony
Metal industry and engineering workers	Abrasives, detergents, soaps, organic solvents, soluble oils, synthetic coolants	Hexavalent chromium, nickel and cobalt salts, additives in soaps, detergents and oils, triethanolamine, epoxy and other synthetic resins, ethylenediamine
Painters, paint-makers	Lime, organic solvents, turpentine abrasives, soap detergents, nitropaints	Chromates, cobalts salts, azo-dyes, aniline dyes, synthetic resins, turpentine (deltacarene)
Workers in the plastics industry	Organic solvents, acid, detergents, hardeners, dyes	Epoxy, phenolic, acrylic, buthyl-phenolformaldehyde resins, others
Textile workers	Organic solvents, fibre, acids, sodium silicate, bleaches, ammonium sulphide, acetic anhydride	Chromates, dyes, formaldehyde, resins, fungicides
Chemical and pharmaceutical workers	Various chemicals	Chromates, substances of the *para*-nitro group, mercury salts, others
Rubber industry workers	Organic solvents, acids, alkalis, soaps, detergents	Mercaptobenzothiazole, tetramethylthiouramdisulphide, *para*-phenylenediamine derivatives, chromates, diphenylguanidine, phenyl-beta-naphthylamine, carbamates, aniline dyes
Hairdressers, barbers and beauticians	Detergents (synthetic), ammonium thioglycolate, depilatories, soaps, lacquer removers, wave solutions, peroxides	*p*-Phenylenediamine, *p*-toluylendiamine, *p*-aminophenol, resorcinol, Peru balsam, cinnamic aldehyde, hydrocitronella, other fragrances in perfumes
Agricultural workers, gardeners, railroad track and road workers	Pesticides, weed-killers, weeds, flowers, plants, asphalt, pitch	Dithiocarbamates, mercaptobenzothiazole, sodium pentachlorophenate, pentadecylcatechol, alantolactones, sesquiterpenes, other extracts of flowers and plants

Occupations	Nature of hazard	
	Irritant	Sensitising
Health service personnel	Detergents (synthetic), disinfectants, soaps, moisture, various medicaments	Sulphonamides, penicillin, streptomycin, anaesthetics, mercury salts, benzodiazepines, others, rubber gloves, components
Printing workers	Solvents (synthetic), turpentine, detergents, soaps, abrasives, alkalis, gum arabic, inks	Chromates, turpentine, aniline dyes, formaldehyde
Photographers and photoengravers	Acids, alkalis, turpentine	Chromates, *p*-amino phenol, *p*-phenylenediamine, methyl-*p*-aminophenol sulphate, hydroquinone, photographic developers, pyrogallol
Electroplating industry workers	Soaps, acids, alkalis, solvents, moisture, heat, zinc salts, potassium cyanide	Chromium, cobalt and nickel salts, chromium acid
Leather and fur workers	Detergents (synthetic), soaps, insecticides, moisture	Chromates, various glues, buthyl-phenolic resins, aniline dyes, *p*-phenylene-diamine, other resins
Bakers and pastry-cooks	Moisture, dusts, sugar, flour improvers, ammonium, persulphate	Benzoyl-peroxide, azo-dyes, cinnamic aldehyde, Peru balsam
Electricians and electric apparatus makers	Solvents, acids, soaps, pitch, synthetic waxes, chlorinated diphenyls, solder-fluxes	Phenolic resins, epoxy resins, chromates, rubber components
Dentists	Soaps, disinfectants, phosphoric acid	Local anaesthetics, eugenol, oil of clove, mercury, acrylic resins, other resins

Physical agents include:

(a) mechanical injury keratolytic action;

(b) denaturants and coagulants of protein: heat, cold, electricity;

(c) photic agents: all forms of light;

(d) ionising agents: X-rays, radioactive substances;

(e) meteorological agents with general complex effects.

Many chemical irritants, including those capable of precipitating proteins, and some biological factors, e.g. primula, may also act as allergens. The skin reaction is inflammatory in character and commensurate in its severity with the intensity and type of action of the causative agent.

Irritant contact dermatitis usually shows lesions strictly limited to the area of contact with the offending agent (colour plate 8); all the lesions are in the same phase of development in all the areas affected. The most common clinical appearances are those of erythema with vesiculation or peeling, as seen in moderate and subacute eczema, or of erythema with vesiculation and bullae formation with erosion.

The prognosis is good and the lesions regress when the contact ceases.

Treatment

Local treatment consists of applications of weak antiseptic packs, antibiotic ointments (tetracycline or erythromycin) or of inert pastes or ointments. Systemic antibiotics are advisable in the presence of lymphangitis or lymphadenitis.

Allergic contact dermatitis or allergic contact eczema

Allergic contact dermatitis or allergic contact eczema accounts for about 35% of all skin diseases and of this about 20% are of occupational origin.

Bricklayers, cement workers and building workers present the highest incidence of occupational eczema followed by cleaners, metal industry and engineering workers, plastics (colour plate 9) and rubber (colour plate 10) workers, and others.

Aetiopathogenesis

The mechanism of allergic contact dermatitis is that of the delayed type cell mediated hypersensitivity.

Several factors of variable relative importance are involved in the causation of the lesions:

(a) the type and structure of the sensitising agent;

(b) its concentration, amount and vehicle;

(c) the duration of contact;

(d) the site and size of the area of contact;

(e) genetic factors (presence of specific predisposing genes);

(f) individual cutaneous and extracutaneous factors (race, sex, age, type and thickness of the horny layer and of the epidermis, etc.);

(g) factors reducing normal skin defences, which sometimes enhance the penetration of the skin by the sensitising agent;

(h) seasonal factors.

Different substances exhibit different sensitising potential; in all cases the chances of sensitisation increase with the number of exposures.

Enhanced by the predisposing factors, the sensitising agent penetrates the epithelium initially, producing usually a local irritant reaction or primary dermatitis of variable severity lasting 2-4 days (first afferent phase).

The sensitising agent is partly eliminated through the epidermal surface or through the visceral excretor organs and partly binds on to a protein vector to form a complete antigen. This conjugation with protein vectors is necessary for the manifestation of delayed hypersensitivity (second afferent phase). The hapten, or the product of its degradation or enzymatic catalysis, may bind to epidermal or plasma proteins or to blood cells with a covalent bond. The small amount of the complete antigen remaining in the skin provides the information for lymphocytes at the level of the skin through the intermediation of macrophages or of Langerhans cells.

The informed lymphocyte reaches the paracortical area of the regional lymph node where it changes into a lymphoblast. The latter divides and produces a clone of cells known as sensitised lymphocytes.

Should the specific antigen again establish contact with the skin, it comes against the sensitised lymphocyte and transforms it into an activated lymphocyte. The latter releases certain glycoproteins, known as lymphokinins, which are probably responsible for the formation of the inflammatory infiltrate *in vivo*.

Once sensitisation has taken place, the entire skin mass is capable of reacting with the specific antigen. The eczematous contact dermatitis which initially was localised to the site of exposure to the sensitising agent may affect other areas, either as a result of excessive contact with the antigen or of its parenteral introduction.

Substances chemically related to the primary sensitising agent may also induce the specific eczematous reaction by operation of the phenomenon known as *group or cross-sensitisation*. A typical example is the sensitivity to substances of the *para* group, whose formula is based on a benzene ring with an amine group in the *para* position (e.g. *p*-phenylenediamine, benzocaine, sulphonamides, aniline, etc., as used in pharmaceutical, cosmetics and chemical industry). A subject with allergic contact sensitivity to *p*-phenylenediamine may cross-react to other chemically related substances.

Pre-existent allergic contact dermatitis may be complicated by the operation of further new sensitising agents (polysensitisation). For instance, a subject with eczema due to cement may become allergic to topical applications used in the treatment.

Certain chemicals may produce a skin reaction mediated by actinic radiations and thus coming under the heading of phototoxic or even photoallergic reactions.

Symptoms

In general, the symptoms of occupational allergic contact dermatitis do not differ from those of contact dermatitis of non-occupational origin.

In occupational cases, one finds the common eczematous lesions in their various stages: eczemato-oedematous and vesicular lesions, abrasions, lesions oozing serous fluid, sometimes pustulous lesions due to superimposed pyogenic infection, or lichenified or hyperkeratotic and rhagadiform lesions which, at least initially, affect exposed areas of skin that come into contact with the occupational noxa.

Observation of the lesions is in some cases sufficient to suggest the aetiology. For example, cement eczema often starts on the flexor side of the wrist (usually the left wrist) because of the way in which the mortar drips from the hawk and takes on an erythematous, infiltrative, desquamative and rhagadiform character of the fingers, backs of the hands, wrists and forearms, with pronounced nail dystrophy. Basically similar features are often found in housewives' eczematous dermatitis.

In nickel workers' eczema, one finds papulo-vesiculo-follicular or poorly confluent lichenoid lesions whose morphology alone points to the cause. In the case of chrome eczema in electroplating workers there are, firstly, eczematous lesions that are in no way characteristic and secondly, small pitted lesions on the hands that are typical of workers in contact with chromium and its derivatives.

Diagnosis

In the diagnosis of occupational contact dermatitis, reference should be made to the medical history and sites and morphology of the lesions. In determining the occupational origin of the lesions, use should be made of the following critera: knowledge of the manufacturing process, of the substances used and their biological action; knowledge of the fact that, under identical working conditions, the substances frequently induce similar cutaneous changes; the relationship between the time of exposure to the noxa and the onset of the skin complaint; the relationship between the site where the lesions began and are most intense and the sites of greatest exposure to the noxa; the repetition or exacerbation of the manifestations on the patient's return to work; and lastly, the patch tests designed to find out whether the suspect substance is really responsible for contact allergy in the patient under examination.

It is also essential to have available an "allergen library" of the chemical substances commonly used in the principal industries so that solutions and suspensions can be prepared (in suitable non-irritant concentrations) for use in skin testing. The most reliable hypersensitivity tests for the causal diagnosis of contact eczema are patch tests, although in special cases intra-dermoreactions and scratch tests are also useful. In the aetiological diagnosis of occupational eczemas, the best procedure is to conduct different series of tests depending on the patient's occupation.

In view of the dangers of inappropriate hypersensitivity testing, the essential technical precautions and the difficulty of interpreting the results, skin testing should be entrusted to the specialist.

Prognosis and course

The prognosis of contact dermatitis depends upon the course, which may vary from case to case. The eczematous lesions may remit once the causal agent is removed. With subsequent exposure, they may recur or not give rise to other manifestations through spontaneous desensitisation. The recurrence takes the form of a repetition of the eczematous manifestations at the primary site or of an extension to other areas until they generalise and are rarely complicated by an erythrodermal condition that is unlikely to reverse. The chronicity of eczema and its liability to recur depend upon several pathogenic factors in various combinations: the chief ones being persistence of contact with the specific agent, polysensitisation, group sensitisation, microbial agents, drugs and individual disease factors. The highest relapse rates are usually observed in subjects who are hypersensitive to allergens that are more or less ubiquitous.

PREVENTIVE MEASURES

Primary prevention necessitates early identification of newly marketed compounds with high sensitisation potential.

Prevention at the place of work consists in strict observation of industrial health regulations, especially the removal or isolation of the injurious material. Personal prevention is based on observation of personal hygiene, dietetic rules and the use of personal protective equipment that helps maintain the integrity of the skin against all kinds of agents.

Repeated washing with pumice, hard soap, detergent pastes, soda, lye solvents and hard brushes should be discouraged. Neutral soaps and soft soaps rich in fats are less harmful than common soaps which are highly alkaline; pure hypochlorite solutions are relatively harmless. Preference should, however, be given to detergents based on fatty alkalis or alkyl sulphonates to which have been added emollients and buffering compounds. Workers whose hands are exposed to the action of chemical and physical agents should wear hand protection made of cloth, or synthetic polyethylene or polyvinyl plastics. A barrier cream that has been selected to suit the exposure conditions should be employed before work and emollient creams should be applied at the end of the shift.

Treatment

A person found to be suffering from occupational eczema should be immediately removed from exposure to the causative agent. This measure alone is often sufficient to ensure remission without any particular treatment.

Treatment of active lesions involves local medication and systemic treatment. The purpose of the former is to prevent the superimposition of other pathogenic stimuli, especially microbes, to reduce local inflammation and infiltration and to promote recovery of function of the injured skin. Consequently, the affected areas should be swabbed with an antiseptic solution (e.g. an aqueous solution of 3% phenol or 1 per mil silver nitrate).

Local medication will depend upon the clinical stage of the disease. For example, it is advisable not to apply ointments in erythemato-vesicular and exudative conditions and it is preferable to use compresses of weak antiseptic solutions (0.25 per mil potassium permanganate, 1 per mil salicylic acid, etc.), inert powders (zinc oxide, plus borated talc, etc.), water and glycerol pastes, or water emulsified with oil; cortisone creams and ointments (without antibiotics) are also very useful.

In advanced conditions (erythemato-desquamative lesions), use should be made of greasier ointments containing reducing substances (ichthyol, sulphur, dithionol) in a fat vehicle, etc. In refractory, infiltrative conditions X-ray treatment in combination with the administration of reducing ointments has proved effective.

The purpose of systemic therapy is to effect specific desensitisation of the sensitised skin, a complex problem that has as yet found no solution. In practice, a generic systemic desensitisation treatment is based on cortical extracts, synthetic antihistamines and sedatives of the neurovegetative system; corticosteroids should be used only in selected cases.

MENEGHINI, C. L.

"Terminology of contact dermatitis". Wilkinson, D. S.; Fregert, S.; Magnusson, B.; Bandmann, H. J.; Calnan, C. D.; Cronin, E.; Hjorth, N.; Maibach, H. J.; Malten, K. E.; Meneghini, C. L.; Pirilä, V. *Acta Dermatovenereologica* (Stockholm), 1970, 50 (287-292). Illus. 10 ref.

"Legislation on occupational dermatoses". Pirilä, V.; Fregert, S.; Bandmann, H. J.; Calnan, C. D.; Cronin, E.; Hjorth, N.; Magnusson, B.; Maibach, H. I.; Malten, K. E.; Meneghini, C. L.; Wilkinson, D. S. *Acta Dermatovenereologica* (Stockholm), 1971, 51 (141-145).

Manual of contact dermatitis. Fregert, S. (Copenhagen, Munksgaard, 2nd ed., 1981), 140 p.

"Recent trends in the immunology of contact sensitivity. I and II". Polak, L. *Contact Dermatitis* (Copenhagen), 1978, 4 (249-255, 256-263). Illus. 16 ref.

"Evolution of etiological factors and clinical patterns in occupational non-eczematous dermatoses". Meneghini, C. L. (205-207). 27 ref. International congress series No. 289. *Dermatology. Proceedings of the XIV International Congress*. Flarer, F.; Serri, F.; Cotton, D. W. K. (eds.). *Excerpta Medica* (Amsterdam, 1972).

Papers in English on contact dermatitis are currently published in *Contact Dermatitis* (Copenhagen).

Contagious ecthyma

Definition. Contagious ecthyma is a highly infectious viral disease of sheep and goats. The disease has also been called contagious pustular stomatitis. Common names for the same disease are orf and sore mouth. The bovine may be accidentally infected, but no other domestic livestock species is susceptible to the disease. The disease occurs world-wide wherever sheep and goats are raised. It occurs primarily in lambs 3-6 months of age, but also may affect older sheep. Mortality is usually low, but may affect as much as 90% of the flock.

The lesions in sheep are pathologically similar to those of pox and involve mainly the mouth and muzzle. Infected animals may contaminate the environment with material in crust sloughed from lesions, which may remain infectious for other animals for quite long periods of time, even lasting through the winter season.

The aetiologic agent is a pox virus. It was first reported as infectious for humans in 1932, both in Europe and in the United States. Since that time, it has become recognised as a fairly common occupational infection of individuals who have extensive direct or indirect contact with sheep.

Occurrence. Most human cases have histories of direct contact with infected animals. However, contact only with objects in the environment where sheep have been previously kept has also been included in histories of human cases. The virus is apparently unable to penetrate intact skin, as infections are initiated at the site of breaks in the skin.

Occupational exposure may occur during many different activities which involve working with sheep or in the environment where sheep are kept. Handling infected sheep constitutes a high probability for exposure. The prescribed treatment for affected animals is to remove the crust from the lesions and apply amollients or astringents. This requires direct handling of infectious material.

Any activity that may result in breaks in the skin while handling sheep increases the potential for infection. Shearing sheep, retrieving a lamb from briars, and removing burrs from wool are specific examples. Vaccination of sheep is another activity with high exposure potential. The vaccine is autogeneous, a live product prepared by mixing crust from infected animals with glycerol and saline. The animals are inoculated by sclerifying the skin and applying the vaccine to the lesion. The worker may accidentally inoculate an abraided area of his or her own skin in the course of this activity. Cases have also been reported from workers who contacted fencing or other equipment where sheep had been housed. Person to person transmission occurs rarely.

Epidemiology. Few formal studies have been done to precisely define the epidemiology of contagious ecthyma, but several generalisations can be made:

- most cases occur in males, during years of most active labour (18-50 years of age);
- most cases occur in farmers or ranchers who raise sheep;
- most cases have a history of recent direct contact with infected sheep;
- most cases occur in spring or summer, since this is the time when direct contact with animals is most likely to occur.

Symptoms. Lesions usually develop on the hands or forearms, as these parts of the body are most likely to come in contact with infected animals or contaminated objects.

Starting as a papule, the lesion progresses to a vesicle that may ulcerate and form a crust. Tissues around the

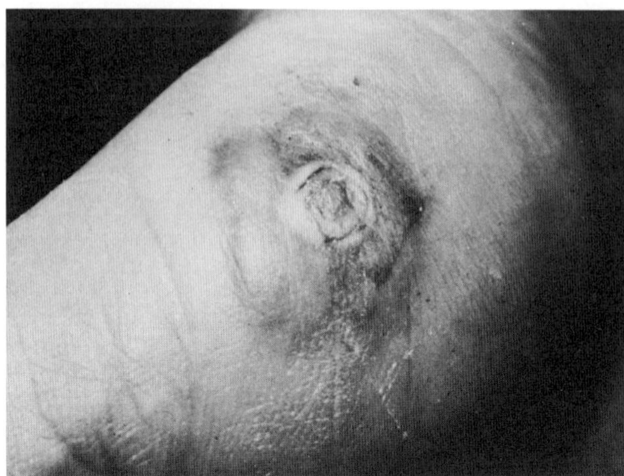

Figure 1. Contagious ecthyma.

lesion are inflamed and oedematous (figure 1). Localised lesions cause little or no pain, but axillary lymph nodes may be swollen or painful. The lesions may be confused with carcinomas or other more serious lesions, and also must be differentiated from those of milker's nodule. Occasionally, generalised infections occur.

Diagnosis. A detailed history accompanied by appropriate symptoms are the two most important things necessary to diagnose contagious ecthyma. Typically the patient is an agricultural worker who has had direct contact or close indirect contact with infected sheep. Diagnostic aids include histologic examination of punch biopsy samples from suspicious lesions. Microscopic findings occasionally include cytoplasmic inclusion bodies in epidermal cells. Electron microscopic examination of samples of crust and fluid from the lesions may exhibit pox viruses, and light microscopic examination of vesicular fluid samples usually demonstrated a large number of neutrophils due to secondary bacterial infection. Isolation may be of value, but serology is not, due to low concentration of circulating antibodies.

PREVENTIVE MEASURES

As with any zoonotic disease in livestock, control of the disease in the animal population is the best way to prevent infection in the human population. Surveillance of the flock, with immediate separation of infected animals, and sound environmental hygiene principles are the keys to control in the animal population. As previously mentioned, the flock may be vaccinated, but this procedure itself may be hazardous to workers unless precautions are taken.

Protective clothing (gloves and long-sleeved shirts) should be worn when working with potentially infected animals. Washing of hands and arms with soap and water following contact with infected or potentially infected animals is beneficial in prevention of infection.

Treatment. There is no specific treatment for infection, but lesions can be treated with antibiotic ointments and dressed for protection and to prevent secondary infection. Spontaneous, uneventful recovery occurs within five to eight weeks, if secondary infections do not occur. Such infections can prolong healing and increase the risk for development of scar tissue.

DONHAM, K. J.

"Contagious ecthyma in humans" (Ecthyma contagiosum beim Menschen). Hübner, G.; Loewe, K. R.; Dittmar, F. K. *Deutsche Medizinische Wochenschrift* (Stuttgart), 1974, 99/47 (2 392-2 394). 27 ref. (In German)

"Orf: report of 19 human cases with clinical and pathological observations". Leavell, U. W.; McNamara, M. F.; Muelling, R.; Talbert, W. M.; Rucker, R. C.; Dalton, A. J. *Journal of the American Medical Association* (Chicago), 1968, 204/8 (657-664). Illus. 21 ref.

Control devices, isolating and switching

Electrical isolating and switching devices comprise installations with which equipment is switched on, switched off or controlled, i.e. devices which can modify the operating conditions of an appliance. From the point of view of safety engineering, we have two main groups of device–those used for switching off appliances and those used to control an appliance.

Switching-off devices

In the case of devices used to switch off an appliance, a distinction should be made between functional switching devices, devices for switching off for mechanical maintenance, emergency switching devices and limit switching devices.

From the safety engineering point of view, functional switching devices are, in general, of no significance. They are intended merely to switch off the appliance, bring it to a standstill or, in general, interrupt its operation.

The second type of switching-off device interrupts the power supply to an appliance so that maintenance work can be carried out on it safely. In this context, maintenance work means any operation not involved in the normal operation of the appliance, i.e. maintenance, repair, cleaning, installation, etc.

An emergency switching device is one that switches off an appliance in order to eliminate as rapidly as possible a dangerous situation that has already arisen. The appliance is already carrying out the task for which it is intended and must be switched off rapidly because a dangerous situation has already occurred.

A limit switching device is used to switch off an appliance when certain threshold values are reached which, if exceeded, would give rise to a dangerous situation, or when safety devices are to be displaced from the safe position, or if the appliance is manipulated in an unsafe situation.

Each of the above isolating and switching devices can take on the function of one or more of the other types of isolating and switching device provided it meets the necessary requirements.

Control of appliances

In the case of devices used to control appliances, a distinction should be made between functional control devices and safety control devices.

Functional control devices are, in general, not of significance from the safety engineering point of view. Their purpose is merely to control the appliance in such a way that it carries out its task in the planned manner.

On the other hand, safety control devices halt working operations which, under certain circumstances, could be dangerous for the personnel operating the appliance. The safety control must therefore meet the relevant safety requirements.

Each of the above types of control device can take on the function of the other devices to the extent that it meets the relevant requirements.

Functional switching devices

With each appliance which is supplied externally with electricity in order to carry out its operation, it must be possible to interrupt the electricity supply. This is achieved by means of a functional switching device. For example, it must be possible to switch off a lamp or a slide projector or a machine tool. However, this switching or isolating capability is not required for safety purposes.

Devices for switching off for mechanical maintenance

In the case of an appliance in which an accident hazard may occur during maintenance, it must be possible to switch off the electricity supply by means of a switch or isolating device. Any electricity accumulated in the appliance must at the same time be dissipated.

Interrupting the supply of electricity. The supply of electricity is cut off by the switch in the narrowest and widest sense. Here a distinction should be made between an isolating device and a switching-off device. The isolating device should cut off the electricity supply from the installation by separating it from every source of electrical energy so as to ensure the safety of the persons carrying out work on or in the vicinity of the installation's electrical equipment. In the "off" position, the isolating distance required by the electrical regulations must be ensured. This isolating distance must be visible; however, it may merely be indicated on the isolating device provided that the "off" indicator can be displayed only if the necessary isolating distance has been established in all conductors. Isolation may be achieved, for example, by means of:

- disconnectors (isolators);
- single-pole or multi-pole switch-disconnectors;
- plug and socket connections;
- fuse-links;
- links;
- special terminals that do not require the removal of a wire.

Suitable means must be provided to ensure that electricity stored in, for example, condensers is discharged.

Devices for switching off for mechanical maintenance are intended to inactivate electrically powered installations. Whereas the isolating device should in most cases be operated only by an electrician, it should be possible for the safety switch-off device to be operated without danger by anybody. The safety switch-off device makes it possible to fully inactivate the electrically powered equipment in general—when it is called the main switch or installation switch—or, in particular, to ensure the inactivation of the installation so that maintenance work can be carried out safely—when it is called the maintenance switching device. As a rule, a maintenance switching device must be capable of disconnecting every circuit from each live supply conductor; however, under certain circumstances, the disconnection of individual conductors may suffice. In the case of a maintenance switching device, too, the open contact must be visible or the "off" position should be displayable only once the contacts have effectively opened. Switching off for mechanical maintenance can be achieved, for example, by means of:

- multi-pole switches;
- circuit breakers;
- control switches operating contactors;
- plug and socket connections.

Here, too, suitable means should be provided, where necessary, to ensure that danger cannot occur owing to accumulated electricity.

Interrupting the supply of pneumatic and hydraulic power. The supply of pneumatic or hydraulic power may be interrupted by closing the supply circuit or switching off the pressure generator. Shut-off of pneumatic or hydraulic systems for mechanical maintenance may be achieved by devices which anybody can operate safely and easily, and in particular, for example:

- rotating valves;
- gate valves;
- switch-off devices for the pressure generator.

In the case of pneumatic or hydraulic systems, it should always be remembered that switching off for mechanical maintenance purposes must not only interrupt the external power supply but must also dissipate or discharge the power already stored in the system.

Interrupting multiple power supplies. For installations supplied by multiple sources of power a single means for switching off for mechanical maintenance should ensure that all energy supplies are interrupted jointly. This is particularly important when the installation is connected to a number of different power sources, for example an electricity and compressed-air supply.

Design and positioning. Every machine or every piece of equipment where mechanical maintenance may present a hazard must be provided with its own switch-off device for mechanical maintenance. Nevertheless, it is permissible, within an installation, to group items into functional units and to provide a safety switch for maintenance purposes for each functional unit. A functional unit may be defined as a group of two or more inter-related machines or items of equipment, provided that all these machines or items of equipment must all be operating continuously together for the operating function to be carried out. Careful selection and judicious design and positioning of maintenance switching devices can make a significant contribution to the safety of maintenance workers and prevent many severe and extremely severe accidents. Electric switch-off devices for maintenance purposes should have red controls. As a rule, they should have only two switch positions—"0" and "1"; where necessary, an additional position marked "Servicing" may be permissible.

Maintenance workers should ensure that the device cannot be unintentionally or inadvertently switched on again, and it must be possible for the switch-off device to be secured in the "off" position by means of a padlock. This requirement may be dispensed with only if, on the one hand, the switched-off installation can be fully supervised from the position of the switching device and if, on the other hand, the means of switching off is constantly under the control of any person performing maintenance. Where a switch-off device is padlockable, provision should be made for the use of at least three padlocks so that, where necessary, a number of maintenance workers can independently lock out the switch (figure 1). When a switch is locked out, it should not be possible to switch it on again by removing covers, switch-panel doors, etc.

The maintenance switching device should be positioned in the vicinity of the machine, equipment or functional unit in such a way that it can readily be reached by maintenance workers. Experience shows, unfortunately, that maintenance switching devices are used only if they are located in the immediate vicinity of the installation, i.e. at the place where the work has to be

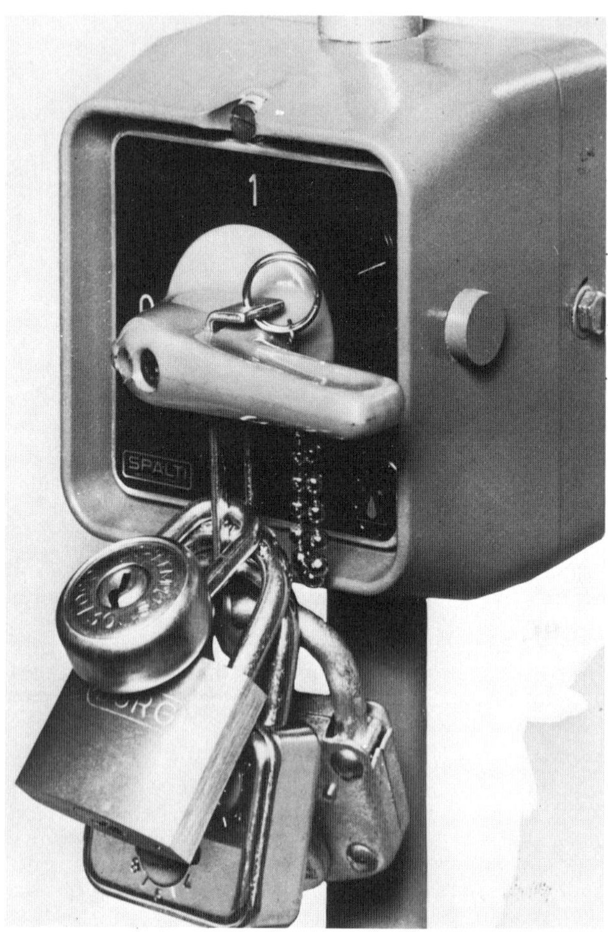

Figure 1. A switch in the "off" position, fitted with padlocks to prevent unauthorised or unintentional actuation.

done. For example, they should never be placed in an adjoining room or on an adjoining floor. In the case of extensive functional units or machines or installations located on several storeys or in several areas, additional switching devices should be provided at all points where work has to be carried out.

Emergency switching devices

Any installation which may present an accident hazard during its operation must be fitted with an emergency switching device which allows it to be returned to a safe condition as rapidly as possible.

Emergency switching devices should be installed where an accident hazard may occur during normal operation of the installations—i.e. not during maintenance. They may not necessarily interrupt the total power supply to the installation. Rather, they interrupt functions which will return the installation as rapidly as possible to a safe condition. For example, in the case of a transmission rotating at an infinitely variable speed, the emergency switching device would return the speed to zero and only then isolate the drive unit from the power supply. However, devices such as clamping units, magnetic holding devices, etc., would not be switched off at all since this would lead to an additional accident hazard. Where emergency switches are actuated by push buttons, these should be equipped with a red mushroom-shaped head.

Emergency switching devices may also be actuated by levers, rotating handles, etc. Actuating handles must be red and the switch should have only two operating positions "0" and "1". When, after actuation, the emergency switch button or lever is returned to the "on" position, this should merely *permit* the installation to be put back into operation; it should not start up the installation.

Emergency switching devices should be designed and positioned in such a way that, in the event of an accident, they can be operated by the endangered persons themselves, for example as in the case of the mushroom-head button, or, wherever possible, in such a way that in the event of an accident they will automatically be actuated by the resultant movement of the victim's body (emergency pull cords, stop barriers). Figure 2 shows the installation of a stop barrier of this type on a stamping press. The stamping press opens and closes continuously. In the open position, the workpiece is removed and the new blank is inserted. If the operator's hand remains for too long between the die plates, it would be crushed if the stop barrier did not halt the operation.

Figure 2. Stop bar on a stamping press.

Limit switching devices

In any installation which produces an accident hazard if a specified limit value is exceeded or when guard devices are removed or displaced from their safe position, or if a part of the operator's body is inserted into a danger zone, it is necessary to instal a limit switching device that will switch the installation off before the dangerous situation develops. Limit values that may require monitoring include, for example, the limit of travel, a maximum pressure, a maximum temperature, etc. Safety devices that may require monitoring include, for example, a safety enclosure, a cover, a barrier, etc. Limit switches which switch off the installation in the event of something being inserted into the danger zone include, for example, no-contact safety guards (light barriers). Limit switches of this type must, without fail, prevent the dangerous situation from occurring. Consequently, they should have a positive disconnection or a valve with a positive blocking system. Where no-contact safety guards are employed, they should be of the type in which the safe operation is automatically tested periodically.

Functional switching devices

In many cases an installation may be required to carry out a series of different functions in the course of its operation. The switching of the various functions is carried out by means of functional switching devices. These may be manually operated or automatically actuated by a computerised system.

As a rule, no safety problem arises here. Under certain circumstances, however, a dangerous situation may arise if a function is actuated at the wrong time or if a dangerous function is started up without all the safety requirements being met. Consequently, in the case of functional switching devices, it is necessary to ensure reliable interlocking of the individual functions or reliable design of the automatic or computerised system.

Control devices with built-in safety

Where it is not possible to eliminate hazards from an installation or provide suitable guards, the operators should be protected by use of a control device with a built-in safety system. Such control devices include:

- 2-handed controls;
- step-by-step controls;
- dead-man's handle controls.

Two-handed controls must be designed in such a way that the operator can actuate them only by using both hands. They must also be constructed in such a way that the hazardous process can be started up each time only when both control organs are actuated once again. Finally, they should be installed only where the dangerous condition can be interrupted as rapidly and reliably as possible.

Step-by-step controls permit the operation of hazardous processes in the installation in well defined short steps. Each following step can be actuated only by releasing and reactuating the step-by-step control. Step-by-step controls are used mainly as setting controls.

Dead-man's handle controls are usually employed as auxilliary safety devices. They ensure that a dangerous movement of the installation can continue only whilst the control lever of the dead-man's handle is being actuated. The control device should be designed and positioned in such a way that the operator has a good view of the danger zone. Where the control device cannot be manually returned to the "off" position and the circuit under control cannot be disconnected, an additional switching device should be located in its immediate vicinity so that the installation can be switched off in the event of a failure in the related control mechanism.

Basic safety principles for isolating and switching devices

Isolating and switching devices should be chosen and installed in such a way that they perform their safety function reliably.

Interruption. The safe function should be actuated by interrupting or halting the supply of electrical, pneumatic or hydraulic power. As a rule, the safety function should not be dependent on the external supply of power.

Reliable operation. The actuation of the control device must reliably ensure the interruption or blocking of the relevant switching device. In the case of mechanical limit switches for electrical circuits, for example, this reliable interruption should be made when the actuating plunger positively displaces the switch unit to the "disconnect" position (figure 3). In the case of manually operated switches for electrical circuits, for example, reliable interruption should be ensured when the actuating unit of the switch operates positively on the switch component (S) (figure 4).

Actuation for safety function. The safety function should be brought into operation by actuation of the switching device. In the case of limit switches that monitor the position of safety enclosures, covers, barriers, etc., the system must be designed in such a way that the switch

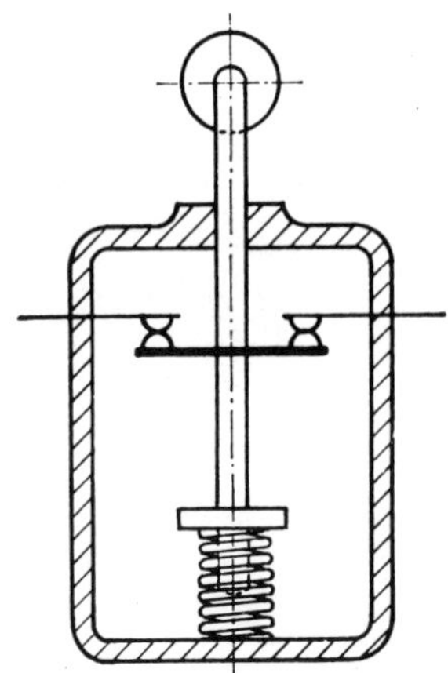

Figure 3. Limit switch in which the switch component is positively opened by the actuating plunger.

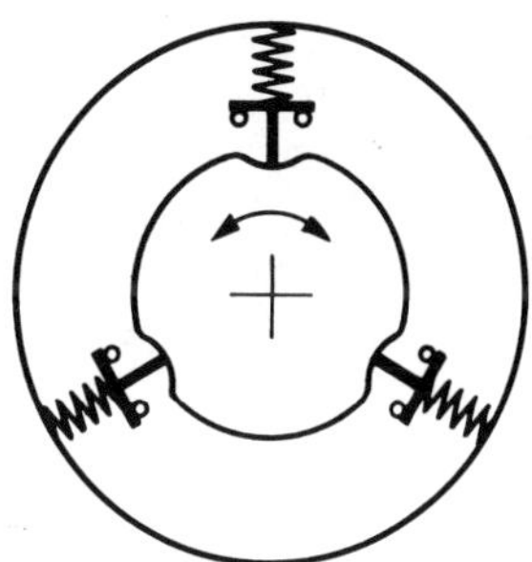

Figure 4. Hand switch in which the switch level operates directly on the rotor and the switching component is positively opened by the rotor.

operating device is actuated when the safety enclosure, cover or barrier is removed or swung out of the safe position (figure 5). In the case of limit switches monitoring a maximum value, the actuating device for the switch must be depressed when the limit value is reached.

Preventing unintentional actuation. It is necessary to prevent the unintentional actuation of a safety device if this might result in an accident hazard.

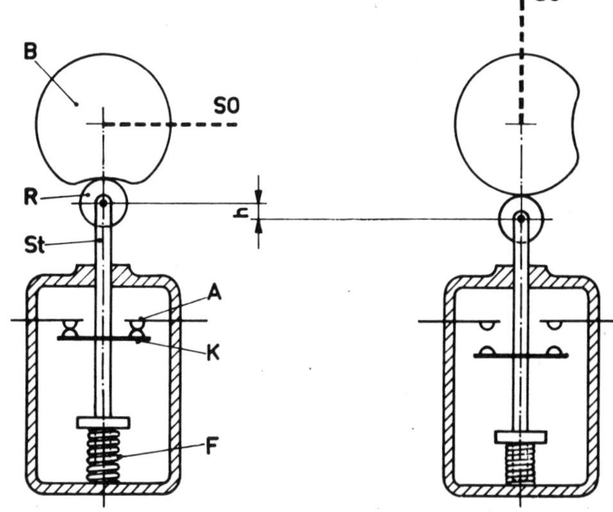

Figure 5. Diagram of a switch with positive mechanical operation and positive disconnection. SO. Protective element. B. Actuating cam. R. Cam follower. St. Switch plunger. F. Return spring. K. Switch contacts. A. Terminal contacts. h. Switch stroke.

Figure 6. Switch lever on a lathe. The switch lever (1) is locked in the "off" position by entry into the notch (3) of the guide plate (4). It is unlocked by swinging the lever out on the linkage (2).

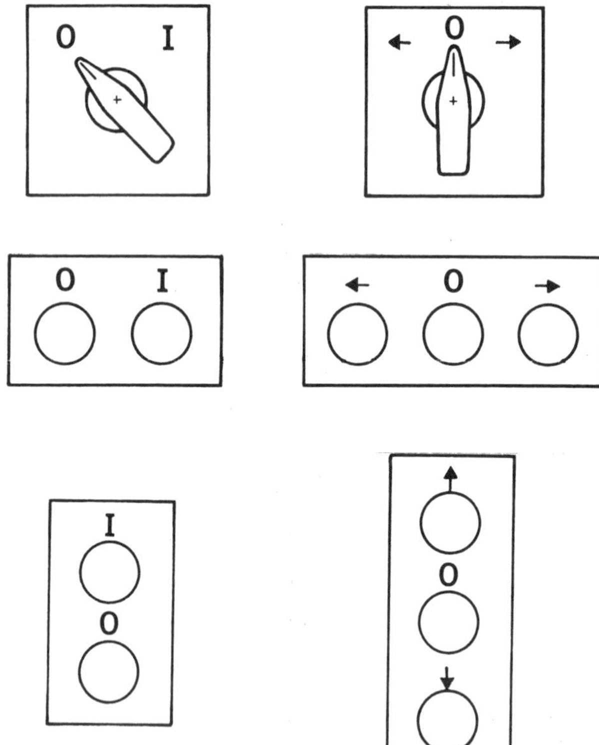

Figure 7. Recommended design of manually operated switching devices.

This applies both to unintentional actuation by a person operating the installation or by a third person or actuation as a result of vibration, etc. (figure 6).

Correct design and layout of actuating devices. The actuating devices of manually operated switching devices must be designed and laid out in such a way that they are logically related to the movements being controlled (figure 7).

Marking. The switching directions must be marked in a clearly legible and comprehensible way so that their relationship to the installation, the function that they control and the position in which they are set can clearly be discerned. In order to avoid misunderstandings and problems of language comprehensibility, the use of symbols is to be recommended.

Layout of switch actuating devices

The correct and uniform selection of switch positions and the arrangement of push buttons can make a significant contribution to their intelligibility.

The following arrangements are recommended: switching on by rotation to the right, i.e. clockwise rotation; "off" position in the centre in an arrangement with right/left movement or upwards/downwards movement; "off" left; "on" right; "off" position in the centre in an arrangement with right/left movement; "off" position below; "on" position above; "off" position in the centre in an arrangement with up/down movement.

TROXLER, R.

Draft publication 364, Chapter46: Isolation and switching. Technical Committee No. 64: Electrical installations of buildings (Central Office of the International Electrotechnical Commission, 1 rue de Varembé, Geneva) (1979), 6 p.

Control technology for occupational safety and health

Control technology in occupational health comprises a set of measures and techniques which aim at the elimination or reduction of exposures to harmful agents in the working environment. The ultimate objective is the prevention of adverse health effects among workers.

The recognition of occupational hazards involves the determination of potentially harmful agents and associated health effects; the evaluation of occupational hazards involves the determination of the degree and conditions of exposure, as well as the comparison of such data with associated health effects and accepted standards. However, these two steps *per se* cannot ensure a healthy working environment, which can only be achieved through adequate control technology. For example, whereas the knowledge that carbon monoxide or benzene or arsine is harmful to health and is present in the workroom air in dangerous concentrations cannot prevent the resulting ill effects, control measures can. Standards for occupational exposure should not only be established but also achieved. The tremendous resources spent in setting up, for example, maximum permissible levels for occupational exposure to chemical agents, would be practically wasted if these values were not put into practice, that is, were not translated into control measures at the workplace.

The correct recognition and careful evaluation of hazards are extremely important and constitute the basis for the design of adequate control measures, which must then be properly implemented, well utilised and maintained. The efficiency of control measures should be periodically checked through routine air monitoring; in certain cases, automatic detectors and alarm systems may be used, particularly when there is danger of a sudden build-up of hazardous concentrations (e.g.

leaks). Besides, techniques for the early detection of health impairment due to occupational exposures and biological monitoring may also be used, when possible, as complementary tools in evaluating the efficiency of control measures.

The ideal is that control technology be incorporated into a workplace at the planning and design stage, and implemented during its construction.

Before planning for a control programme it is essential to study and clearly define all possible sources of health hazards (operations, storage of chemicals, transfer of materials, leaks, etc.). Although many techniques are available and are being successfully used at present for the control of hazards in the work environment, more research and applications are still needed in this field, which requires not only technical knowledge but ingenuity as well.

Although each case can be considered as unique, there are basic principles of control technology which can be applied, either alone or in combination, to a great number of workplace situations. To start with, there are certain questions to be asked, the answers to which are likely to indicate the way to the most suitable and feasible control technology (see OCCUPATIONAL HYGIENE, SYSTEMATIC APPROACH AND STRATEGY OF). These questions include the following:

- Which are the potential hazards, their sources and location?
- Can the hazard sources be removed or completely isolated?
- Can the presence of the hazard, or the possibility of its release, be avoided?
- Is there a less hazardous way to perform a certain operation (different materials, equipment or work practices)?
- Can the hazardous material be in contact with the workroom air, or the workers, less frequently, during less time, or moved over shorter distances?
- Is there a way to prevent or reduce the amount of the hazardous agent coming into contact with the workers (through ventilation, isolation, enclosures) or the workers coming into contact with the hazardous agent (distance, special cabins, personal protection)?
- Can the duration of the exposure be minimised (through adequate work practices or administrative controls)?

The control measures for occupational hazards can be grouped into two main categories: 1 – Environmental control measures; 2 – Personal control measures.

The most important measures for the control of chemical agents and particulates in the work environment are hereby briefly discussed and exemplified. (Control measures for physical agents and for biological agents are discussed under relevant separate articles.)

Environmental control measures

These include changes in the work processes and/or working environment with the objective of controlling the health hazards either by eliminating the responsible agents or reducing them to levels believed not to be harmful to health, as well as by preventing them coming into contact with the workers.

Adequate design and lay-out

The ideal is that health and safety aspects be accounted for at the design stage of equipment, processes and workplaces. After a workplace is established, it is usually difficult to make changes in order to reduce health hazards. Therefore, when selecting processes or equipment, their relative "hazard-producing" characteristics should be weighed together with the other factors affecting the final decision. For example, if a dust-free process can be used instead of a dust-producing one, or a closed system to handle chemicals instead of an open process, the first one should be chosen even if initially it is more expensive. Further need for control measures (e.g. ventilation) may prove to be even more expensive than an eventual difference in cost between the two alternatives. Technical knowledge which can be used for this purpose is available concerning specific branches of activity.

With reference to plant design, in addition to such aspects as illumination and ventilation, lay-out should be carefully considered. An example of the importance of lay-out is the placement of washer and dryer (dry cleaning industry) as close as possible to each other, in order to reduce the time during which the "damp load" can release solvent vapours to the workroom air. The same reasoning can be applied to the transfer of intermediate products such as molten metal (from furnace to moulds), paint and coating mixtures, etc.

Elimination or reduction of the harmful agent at its source

This can be achieved through such measures as the following:

Discontinuation of the process. A process which utilises, produces or leads to the formation of the agent may be discontinued (e.g. discontinuation of benzidine production), or a certain agent banned (e.g. banning of benzene as a solvent; beta-naphthylamine, in the manufacture of dyestuffs and as antioxidant for rubber, and, the removal of beryllium phosphors from fluorescent lamps).

This may prove difficult in practice, owing to considerations of, for example, a socioecomonic nature; however, for certain highly toxic, carcinogenic, mutagenic or teratogenic chemicals it is undoubtedly the safest solution.

Replacement of materials used. Substitution of materials (i.e. solvents, fuels, raw materials, etc.) can be a very effective way of controlling a hazardous exposure; it may involve or not a change in the work process. As classical examples can be mentioned the substitution of phosphorous sesquisulphide for white phosphorus, which eradicated the serious occupational disease known as "phossy jaw", common among match makers in the past; the substitution of mercury-free carroting materials for the mercury compounds previously used in the felt hat industry, which was followed by the disappearance of the "mad hatter", and the use of tritium-activated phosphors instead of radium-based paint for watch and instrument dials, which has greatly reduced the hazards associated with this manufacture.

In order to plan for a substitution, the possible alternatives should be investigated and factors such as technological and economic feasibility, as well as availability of the substitute material in the market, should be considered.

Substitution, although possibly one of the best control measures, can also be misleading. The main pitfall to be avoided is the introduction of a new hazard while removing the old one, as would happen if a dry cleaning solvent of low toxicity but high flammability were utilised in place of another with high toxicity and low flammability. Both toxicological and safety aspects must be taken into account. Even considering only the toxicological properties of a substitute material, care must be taken not to drop a chemical of known toxicity (therefore used under strict control) in favour of another whose toxicological properties are not so well known on the assumption that it is of lower toxicity. This second chemical may prove to be even more dangerous than the first one and may have been used without the required precautions. The substitute material must be of proven much lower toxicity and should not introduce new or higher safety hazards.

As examples of substitution can be mentioned the use of:

- less hazardous solvents instead of toxic ones, e.g. 1,1,1-trichloroethane (methyl chloroform), dichloromethane or a fluorochlorohydrocarbon instead of carbon tetrachloride and toluene, cyclohexane or certain ketones instead of benzene;
- solvents and other chemicals with lower vapour pressures and with higher boiling points instead of equivalent ones (including in degree of toxicity) with higher vapour pressures and with lower boiling points, in order to decrease vaporisation;
- detergent-and-water cleaning solutions instead of organic solvents;
- natural rubber cements with aliphatic hydrocarbon solvents instead of benzene cements;
- water emulsion coatings (e.g. acrylic latex) instead of those containing organic solvents, as well as water-based paints instead of solvent-based paints (this may involve a change in the process);
- argon instead of chlorine in the degassing of aluminium (foundries);
- clean scrap instead of oily and/or painted scrap in foundries;
- freon instead of methyl bromide and chloride as a refrigerant;
- leadless glazes in the ceramics industry;
- leadless paint pigments (e.g. titanium dioxide, zinc oxide, etc.);
- fibreglass instead of asbestos;
- non-silica moulding aggregates instead of quartz sand in foundries;
- steel shot, corundum or silicon carbide instead of quartz sand for abrasive blasting (however, the silica dust hazard would still exist if the parts to be cleaned were, for example, silica-coated castings, regardless of the abrasive used);
- magnesite or aluminium oxide bricks instead of silica bricks for the lining of furnaces and ladles in steelworks;
- synthetic grinding wheels (e.g. aluminium oxide, silicon carbide) instead of sandstone wheels.

Modifications in the process and/or equipment. In this category are included modifications in processes, operations or equipment, leading to appreciable reductions in contaminant generation (e.g. reduced temperatures or speed), the elimination or decrease in the formation of undesirable by-products, the elimination or minimisation of physical contact between workers and hazardous agents (e.g. use of mechanical aids such as tongs, mechanisation, etc.). As in the case of substitution of materials, the new process, operation or equipment, must not introduce a new hazard, and it must be technically feasible and acceptable. Examples include the use of:

- electric motors instead of internal combustion engines;
- mechanical (single or double) pump seals instead of gasket pump seals;
- toxic solids in pellet form rather than as a powder;
- chemicals to suppress or decrease the formation of undesirable agents (e.g. the use of urea as a chemical suppressant of nitrogen dioxide gas formed when nitric acid is utilised in operations such as bright dip and pickling);
- mechanical gauges instead of mercury-containing types;
- "dust-free" cutting equipment in the printing industry;
- airless spray techniques instead of hand-spraying;
- solvents at temperatures as low as technically feasible, in order to decrease vaporisation;
- dip or brush instead of spray painting (it should be kept in mind, however, that this may reduce the hazard during the painting operation while, during drying, the hazard from the evaporated solvent would be the same for both);
- floating plastic balls on open-surface tanks (degreasing, leather finishing, dyeing, etc.), in order to decrease the evaporation surface;
- refractory bricks already purchased in the required dimensions so that sawing in the workplace is not necessary;
- catalysers (e.g. which transform carbon monoxide into carbon dioxide) at the exhaust outlet of internal combustion engines, and air purifiers in compressed-air outlets;
- mechanisation of operations (e.g. machine application of lead oxide to battery grids);
- covered containers to carry materials and products which give off air contaminants.

Maintenance of equipment. This aspect is very important since well maintained, well regulated processes and equipment usually give off less air contaminants. As examples can be mentioned:

- adequate regulation of internal combustion engines, which reduces the amount of carbon monoxide produced;
- best possible combustion in furnaces and ovens, since the more complete the combustion, the less the production of carbon monoxide;
- prevention of leakages in closed systems, valves, pumps, etc.;
- prevention of leakages in drying ovens or dry cleaning units.

Isolation

The harmful agents can be isolated in order not to come into contact with the workers. This can be achieved by interposing, between the agent and the worker, a barrier or a shield (closed systems, enclosures, separating walls), distance or time.

Closed systems. Many toxic chemicals can be used rather safely in closed systems. For example, in the textile industry, the chlorine hazard (dyeing) can be appreciably reduced if the bleaching vats are constructed as closed vessels (with adequate vents which allow a minimum of chlorine escape); *bis*-chloromethyl ether while used in "open kettle" operations was associated with the occurrence of a characteristic carcinoma which, however, has not yet been reported among workers dealing with it in closed systems. In the chemical and petroleum industries, the isolation of processes in closed systems is the usual practice, therefore many hazards are reduced. There are many processes which by nature require a closed system; whenever this is not the case but there is a choice, closed systems should be preferred.

Nevertheless, routine air monitoring is of paramount importance in order to detect any leaks, which should be immediately repaired, otherwise serious hazards may be created. Critical points for leaks are flanges, valves, vents (from process vessels, pneumatic product transfer systems, etc.), relief valves, seals on pumps, compressors, agitator shafts and manways. It may happen that valves, supposedly shut, leak continuously leading to a build-up of air concentrations of chemical agents. In fact, whenever technically feasible, these systems should be in open air in order to reduce this possibility of accumulation of air contaminants.

Vent stacks and relief valves may need to be modified or relocated in order to prevent chemicals re-entering the workplace. In the case of toxic chemicals, care must be taken not to cause air pollution problems by discharging them directly to the atmosphere; systems can be equipped with vent scrubbers or similar installations. All efforts should be made to control any fugitive emissions from a closed system. This matter is well discussed in the specialised literature.

Enclosures. An entire process, part of a process or specific contaminant sources (e.g. pumps that usually leak) can be enclosed to prevent the escape of contaminants into the workroom air. Enclosed spaces should be kept under negative pressure. Enclosures, combined with local exhaust ventilation, constitute one of the best solutions available for the control of very hazardous air contaminants.

Processes or operations which need to be completely enclosed can be mechanised or automated and performed through remote control, or can be handled by means of gloved inlets.

Attention should be paid to the following aspects:

- in a complete enclosure, heat build-up may be a problem depending on the process and this fact should be taken into consideration when the ventilation for the enclosure is designed;
- maintenance and repair work inside enclosed spaces requires special procedures, including the use of adequate personal protective equipment.

Whenever total enclosures are not feasible, partial enclosures may be used; the approach to their design is to imagine a total enclosure and then remove the minimum possible to permit the performance of the operation. These are effective in combination with local exhaust ventilation systems; in fact, the partial enclosure constitutes the hood for the system. Examples of enclosures for different operations and as part of ventilation systems can be found in the specialised literature.

Separating walls (isolated areas and cabins). Whenever there are, in a workplace, operations that are more hazardous than others, they should be localised and separated through adequate isolation.

Hazardous areas can be restricted to a few workers who are then adequately protected (personal protection, limitation of exposure, etc., and subjected to medical supervision). Besides, in a restricted area it may be easier to control the exposures.

Certain highly toxic agents or suspected carcinogens should only be utilised in completely isolated areas which are then marked by signs of "controlled area", indicating the nature of the hazard. Only authorised personnel, adequately equipped, individually protected and under strict medical supervision should be allowed to enter. Special procedures for entry and exit should be carefully followed including adequate locker rooms, shower facilities and used garment disposal. The pressure in such isolated areas should be *negative* so that air can only come in and not out.

The isolation of workers is also possible as is the case with control cabins where a *positive* pressure (by introduction of clean air) ensures that air contaminants do not enter and special walls and windows keep out agents such as radiant heat and noise.

Distance. It may be desirable to perform operations which create health hazards at a distant location. Then the only workers present would be those involved with the operation and they should be individually protected. This is not always technically possible and, even when it is, the problem of environmental pollution should not be overlooked.

Time. It may be desirable to perform certain hazardous operations out of the regular shift hours, when the workers not involved with it do not need to be present. Those performing the hazardous task should be individually protected, and should never be alone. In order to use this measure, the process must permit displacement in time and, besides, there must be administrative provisions for the special work schedule.

Ventilation

Ventilation in workplaces can be used for one of the three following purposes:

- to ensure conditions of thermal comfort;
- to renew the workplace air, therefore diluting eventual air contaminants to acceptable levels;
- to prevent hazardous air contaminants from reaching the workers' breathing zone.

Because of its importance and its specialised nature, industrial ventilation is discussed separately in more detail under VENTILATION, INDUSTRIAL. Only some basic considerations on this extensive subject are presented here.

General or dilution ventilation. From the point of view of the control of air contaminants, general or dilution ventilation aims at the renewal of the air in the work environment so that the possible contaminants are diluted to levels considered to be not harmful to health. However, general ventilation, as a means of controlling exposures to air contaminants, has limitations and can be accepted provided that:

- the contaminants in question are of low degree of toxicity or constitute only a nuisance (control of toxic chemicals and particulates cannot be achieved only through dilution ventilation);
- the quantity of contaminants generated in the workplace is not too great (otherwise the air volumes required for adequate dilution would be impractical);
- workers are far enough from contaminant sources, unless contaminants are given off at very low concentrations (anyway the air concentrations at the breathing zone of workers should be below the maximum permissible levels);
- the evolution of contaminants is reasonably uniform.

Whenever general or dilution ventilation is being planned, besides the calculations such as for required air volumes, fan power, etc., important points to consider are the following:

- lay-out of equipment and operations in relation to air inlets and outlets, or vice-versa (depending on when the system is designed), should be such that the air contaminants are always swept away from the breathing zone of workers;
- the relative location of air inlets and outlets should never permit either short-circuits of air or the formation of strong cross-draughts; the latter may not only cause discomfort but also disturb the performance of local exhaust ventilation systems (if there are);
- the quality of the air introduced in the workplace should be considered, from the point of view of both eventual pollution and temperature (these factors may need to be corrected).

Local exhaust ventilation. Local exhaust ventilation aims at the removal of the air contaminants from the working environment before they can reach the breathing zone of workers in harmful concentrations. Particularly in combination with adequate enclosures, it is the most efficient engineering control measure for airborne chemical agents and particulates in the working environment. Local exhaust ventilation is usually complemented by general ventilation. Since appreciable amounts of air are removed from the workplaces through ventilation, adequate make-up air should be supplied. In fact, supply and exhaust may be combined into "push-pull" systems which can be quite efficient, although expensive.

The basic elements of a local exhaust ventilation system (see EXHAUST SYSTEMS) are: hood, ductwork, fan and collector. Hoods should enclose as much as the process permits (see "partial enclosures"). The most efficient are the enclosing hoods (air contaminants are generated within the confines of the hood while the worker(s) is (are) outside). Whenever exterior hoods are used, capture distances should be as short as possible.

In certain situations, where it may seem difficult to instal a ventilation system because of the mobility of the process and/or equipment, flexible ducts may be the solution. There are also portable exhaust systems (including the air mover) which can be used to ventilate, for example, tanks which contained chemicals before cleaning, underground sewer passageways which need to be inspected through the manholes, welding operations in confined spaces, underground operations that may give off chemicals, etc. (For most of these operations the use of adequate respiratory protection is required in addition.) Whenever this type of portable exhaust ventilation system is utilised, care must be taken with the disposal of the polluted air brought out; there may be need for the use of air cleaning devices. Such portable exhaust units should be used when there is the possibility of throwing the exhausted air outside, and not in situations where the exhausted air would be discharged indoors, since it would be extremely difficult to clean this air to an acceptable standard, particularly when dealing with contaminants in the gaseous state.

Before designing a ventilation system, all hazard sources must be well defined, critical points being the generation of contaminants (e.g. open tanks) and points of dispersion such as transfer of materials (e.g. bag and drum filling and emptying, pouring of molten metal, transfer of solids from conveyor belts to hoppers, etc.) and opening of units such as drying ovens or closed vessels (e.g. reactors).

Types of hoods for specific operations and the design of ventilation systems are well described in the specialised literature. However, the actual design of a ventilation system should be the responsibility of specialised engineers. Errors at this stage will either be noticed after the implementation of the system and, therefore, expensive to repair or, worse, will not be noticed until the manifestation of health effects resulting from exposure to hazardous agents believed to be under control.

In addition to considerations such as hood design, adequate air exhaust volumes, duct velocities, capture velocities, etc., certain aspects should not be overlooked, for example:

- hot processes require special ventilation designs which account for this factor;
- air currents (cross-draughts) may disturb ventilation patterns. When dealing with canopy hoods, for example, even relatively light cross-draughts may require an appreciable increase in duct velocities in order to maintain the required hood suction power. Open windows in summer months may alter the performance of an otherwise efficient exhaust ventilation system;
- there should never be the possibility for the breathing zone of workers to remain between the point of contaminant generation or release and the point of collection. Should this be absolutely necessary, for instance for a repair, either the operation should be stopped or adequate individual protection provided;
- new additions to old systems should be carefully studied and the modified system should be recalculated; even if additions are possible, changes in the fan may be necessary. Also, the need for additional make-up air should not be overlooked;
- it is very unlikely that a system designed to handle air contaminants in the gaseous state could be used for particulates without modifications, since high velocities are necessary for an air stream to hold particles;
- as is the case with any control measure, air monitoring should be carried out periodically in order to ensure the continuous efficiency of the system;
- adequate maintenance is also essential; ducts may become perforated due to rust or corrosive atmospheres, dust may accumulate at elbows, air movers may become clogged, etc.;
- harmful agents removed from the working environment should not be discharged into the general environment, therefore exhaust ventilation systems *must include appropriate collectors* (see AIR POLLUTION).

Wet methods

The control of dust dispersion into the working environment can be successfully achieved, in some instances, through the use of water and wetting agents.

Wet methods are particularly efficient when the water is introduced at the point of dust generation so that the particles become wetted before having a chance to disperse into the ambient air. This is the case of wet drilling which has been widely used to reduce dust exposures in, for example, mines and quarries. Many studies have shown sharp decreases in the occurrence of silicosis in granite quarries and in mines in the years following the introduction of wet drilling. Whenever a choice is possible, wet drilling should be selected over dry drilling. A great variety of wet drills is available in the market, as well as pneumatic jackhammers with continuous-flow water attachments. However, even when drilling is wet, there may be some dust exposure due to the fact that the dust, which is originally dry, is not always completely wetted and retained; besides, for certain positions of the drill (e.g. overhead drilling), the amount of water in the drilling hole may not be sufficient. Therefore, ventilation may be needed as a complement. Also, whenever using wet methods, the evaporation of the dust-laden water may constitute a secondary dust source which must be considered and avoided or controlled.

Another type of wet method is the use of water sprays, which cause the dust to agglomerate in heavy particles and deposit. The water droplets should not be too large in relation to the dust particles (usually not more than 100 times) in order to ensure a good contact. Water sprays are used, for example, in mines after blasting, over rocks and ores which must be transported, in crushers, over transfer points of dusty materials, as a "curtain" to confine dust to certain areas and prevent it from dispersing over large portions of the working environment, and other critical points.

However, the use of water sprays is not always effective, particularly for the control of the very fine "respirable" dust. One problem is that it is difficult to obtain an intimate contact between dust particles and water droplets (unless the dust is coarse); besides, due to the movement of the dusty material (e.g. crushed ores transported on conveyor belts), dry areas may become continuously exposed and dust may be liberated before becoming wet. In such cases continuous application of the water spray as the material moves and dry dust is likely to be released may help the control of, particularly, the less fine dust. Whenever the dust in question may cause a serious health effect by penetrating into the alveoli (e.g. silicosis), the control of the "respirable" fraction is essential and air monitoring should be carried out even if the visual impression is that the water spray does suppress the dust.

Among other techniques which have been successfully used wet abrasive blasting should mentioned. On the other hand, wet grinding is not always efficient to control dust which can escape before becoming adequately wetted, due to the velocity of its generation; in addition, there is the problem that the dust-laden water is thrown off as fine droplets which can evaporate before falling to the floor, thus liberating the dust.

The use of water is very important in the cleaning of dusty workplaces particularly when vacuum cleaning equipment is not available.

Some aspects to be considered when planning the use of wet methods are the following:

- technical feasibility, which also includes the non-interference of water with the process;
- the dust should be "wettable";
- thermal environment, since the increase in the ambient humidity due to the use of wet methods can create or aggravate heat stress problems;
- adequate disposal of the dust-laden water which would eventually evaporate and release the dust thus constituting a secondary dust source.

The efficiency of wet methods depends on how completely the wetting of the particles is achieved. Wetting agents, which improve the spread of water over a surface, can be used. However, monitoring of airborne dust is an essential complement of this method; it should be kept in mind that the most difficult portion of dust to control by wet methods is the "respirable" fraction, which is also invisible to the naked eye and, usually, the most harmful to health. When necessary, the use of wet methods should be complemented by other control measures, mainly ventilation.

Good housekeeping, storage and labelling

Good housekeeping and maintenance. This includes cleanliness of the workplace and machinery as well as adequate waste disposal, and may contribute appreciably to keeping down the exposures to chemical agents and dust (see HOUSEKEEPING AND MAINTENANCE; MAINTENANCE OF MACHINERY AND EQUIPMENT).

As examples of such practices the following can be mentioned:

- cleaning up of spills before they have a chance to evaporate into the workroom air;
- adequate and immediate disposal of solvent-soaked rags as well as of empty containers and bags likely to contain residues of toxic chemicals and dusts;
- periodic cleaning of the workplace (with water or vacuum) in order to avoid accumulation of dust on beams, machinery, window sills, etc.;
- keeping all containers with volatile chemicals tightly closed;
- organisation and general cleanliness, avoidance of obstructed passages and adequate disposal of trash.

Non-compliance with such fundamental and apparently simple good housekeeping rules can lead not only to health but also to safety and fire hazards, besides having a negative effect on the morale of the workers. As an example can be mentioned the case of a plant where a high prevalence of dermatitis was observed among control cabin operators who apparently were not in contact with any dermatitis-producing agents; however, the railings of the stairs used to reach the cabins were covered with grease, due to poor housekeeping.

The importance of good housekeeping cannot be overemphasised and provisions to enable its practice should be made preferably at the building stage and should include:

- smooth surfaces (walls and floors); porous surfaces should be avoided and cracks immediately repaired, particularly where chemicals are used;
- facilities for adequate cleaning, e.g. water supply, steam (if needed), vacuum cleaning (dust should never be blown with compressed air or removed with dry sweeping);
- slightly slanted floors and canalisations to collect washing waters adequately;
- facilities for adequate waste disposal with special provisions for chemical and toxic wastes (e.g. reactive chemicals should not be thrown out together, toxic effluents should be treated, etc.).

Another aspect which must be considered is the maintenance of the workplace, equipment, machinery, etc., as well as of the control measures; it is not enough to build a safe system: it also has to be kept safe.

Storage. Storage of raw materials, chemicals and products in appropriate places and in adequate containers is essential both for health and safety reasons (see DANGEROUS SUBSTANCES, STORAGE OF). Containers should be preferably unbreakable. Another consideration is to avoid any chemical reactions and/or leakages. Particularly for volatile chemicals (e.g. solvents), containers should have well fitted lids.

Whenever storing chemicals special attention should be paid to the possibility of accidental chemical reactions; for example, cyanides and acids should never be stored together.

It is a mistake to assume that storage rooms are hazard-free areas since, depending on the materials or products stored, there may be a build-up of airborne chemicals to dangerous concentrations, particularly if the ventilation is poor (which is often the case). As examples can be mentioned the formaldehyde released during the storage of durable press fabrics which were treated by the "glyoxal-formaldehyde process" or the vinyl chloride (unpolymerised) released during the storage of granulated PVC. Unless the storage area is well ventilated, adequate precautions should be taken when entering, particularly if the chemicals likely to be released are fast acting and have ceiling values for permissible exposure levels. Intermittent exhaust ventilation, to be turned on some time before entering the storage area, can be a solution. Another solution would be the use of personal protection.

Labelling. Adequate labelling of any chemical agent container is of the utmost importance. Labels should indicate, clearly and in the language of the users, the degree of toxicity of the chemical in question, possible routes of entry, main symptoms, safety and fire hazards, possible dangerous reactions, main precautions for use, and first-aid procedures in the case of overexposure or ingestion. Adequate symbols (e.g. fire, corrosive liquids, explosives, etc.) and other visual messages on labels are very important (see DANGEROUS SUBSTANCES, LABELLING AND MARKING OF).

Personal control measures

Work practices

These include specific work procedures designed to minimise the generation of and/or the exposure to hazardous agents in the working environment. Work processes and associated health hazards should be carefully studied with the objective of determining which exposures result from workers' carelessness or mistakes, and which procedures could be changed, and how, in order to decrease the hazards. Experienced workers, if adequately instructed on the potential health hazards, can make valuable contributions to the selection of safe work practices.

Although work practices depend a great deal on the workers' training and collaboration, the responsibility for them lies also with management since there may be need for administrative provisions to allow workers to carry out their tasks in the safest possible way, which is not necessarily the fastest.

Basic principles for good work practices include the following:

- minimisation of the time during which a chemical agent has the possibility to evolve into the workroom air, for example:
  - *(a)* by reducing to a minimum the time during which any container with volatile materials, or reactors (polymerisation), or drying ovens, or dry cleaning washers and dryers, etc., remain open;
  - *(b)* by reducing as much as possible the time during which materials and products which give off air contaminants remain exposed to the workroom air (e.g. in dry cleaning operations, the faster the "damp load" is carried from the washer to the dryer, the better);
- removal of certain products and wastes which liberate air contaminants from the working environment as soon as the process permits, for example:
  - *(a)* in foundries, pre-cleaning of castings by shot blasting (adequately controlled) right after the "shake-out", so that there will be less chance of dust being released into the workroom air during the transportation and handling of castings;
  - *(b)* in certain plastics industries, immediate removal of any crust and other scrap material cleaned from the reactors to an adequate disposal place outside the workroom (concentrations of vinyl chloride up to 100 ppm have been observed in cases when the reactor cleaners just piled up this type of scrap material in the workplace);
- avoidance of possible undesirable chemical reactions and accidental formation of toxic by-products. For example, caustic soda should not be mixed with acid solutions; cleaning agents containing ammonia should not be used together with chlorine bleaches; nitric acid should not come into contact with organic matter, such as wood; arsenic-containing materials should not come into contact with strong acids (unless special exhaust ventilation is provided); in pickling (acid) operations, acid should always be added to water (not the contrary), etc.;
- avoidance of carelessness in closing containers, valves, ovens, etc.; container lids should be tightly closed and kept on while not in use, doors should be completely shut, etc.;
- adequate handling of materials, particularly chemicals, for example, transfers from one container to another as well as transport of such containers, should be carried out with strict precautions in order to avoid losses to the workroom air or accidental skin and/or clothing contacts, thus introducing or aggravating a health hazard. It has occurred that workers who usually carry out a well controlled operation were overexposed to hazardous chemicals during sporadic careless handling (e.g. weighing, loading, transfers, etc.). In such situations, it may happen that the exposures are evaluated during the normal operation with satisfactory results, while serious overexposures occur at particular times;
- suitable speed in performing certain tasks, for example the removal of the basket with cleaned metal parts from the vapour zone in a degreasing tank, should be very slow so that solvent vapour is not entrained into the workroom air;
- leaving adequate time, for instance before opening a drying oven or a dryer (time should be set taking into account the drying time for the conditions in question), or before re-entering a certain area in a mine after blasting (time should be allowed for the dust to settle);
- avoidance of skin contact with chemicals, especially those which can affect the skin (cancer, dermatoses), or which can penetrate through intact skin.

Practices already mentioned under good housekeeping (e.g. immediate cleaning of spills), as well as strict personal hygiene and the adequate use of required personal protection must be part of the work practices adopted.

Training workers in adequate and safe work practices should be a responsibility of management and given during working hours. As already mentioned, administrative controls are also essential.

Personal protective equipment

The worker can be isolated from the hazardous environment by means of personal protective equipment, which may be classified into two categories:

- personal protective equipment required for specific occupations, regardless of the utilisation of environmental control measures; and
- personal protective equipment used to protect workers from hazards which can be efficiently controlled by means of environmental control measures.

The first category would include, for example, hard hats for construction work and heavy industry, safety glasses for work with lathes, grinding wheels and chemical laboratories, protective gloves for handling sharp-edged pieces, face shields and gloves for welding, impervious clothing against chemical splashes and many others, aiming mainly at the control of safety hazards.

The second category would include, for example, respirators to prevent the inhalation of toxic air contaminants which could have been removed by local exhaust ventilation, ear protection for work at a machine which could have been successfully enclosed, etc. The utilisation of this type of personal protective equipment should be acceptable only under the following circumstances:

- while environmental control measures are being designed and implemented, in which case it should be considered as a temporary solution;
- when environmental control measures are technically infeasible, for example painting a bridge or certain types of airport work;
- for operations of short duration, for example entering a generator room to make a check;
- for sporadic operations such as maintenance and repairs, for example replacing the refractory lining in a furnace, cleaning of a tank, welding in a confined space, etc.;
- for operations involving a very small number of workers and which are technically and financially very difficult to control through environmental measures.

In such situations it may be acceptable to isolate the operation and protect the few workers involved through personal protective equipment, limitation of exposure time (e.g. through rotation of workers) and medical supervision.

It should be kept in mind that equipment such as mask-type respirators, ear muffs, impervious clothing, etc., may be extremely uncomfortable to wear, particularly in hot weather. Therefore, there may be need for a reduction in the working hours, at least at the operation requiring the personal protection; administrative provisions should make this possible.

For the different types of personal protective equipment see the relevant separate articles. As regards the philosophy of utilisation of personal protective equipment in the context of a comprehensive control technology programme, it should always be kept in mind that personal protective equipment has to be:

- adequate for the hazards in question. For example, respirators with mechanical filters for dust do not protect against air contaminants in the gaseous state, rubber gloves should not be worn for work with organic solvents, etc. In the case of barrier creams and lotions, special care should be taken in their selection and tests carried out to eliminate the possibility of eventual allergic reactions.
- of proven good quality and efficiency. All personal protective equipment should be adequately tested for efficiency. If the equipment does not meet the minimum performance requirements, the workers wearing it will have a false sense of security and will be more likely to overexpose themselves than if they did not wear any protection.
- resistant to the air contaminants. For example, respirators made of rubber might be attacked by organic solvent vapours, with the resulting formation of cracks and therefore leaks.
- checked for validity. For example, chemical cartridges, activated charcoal filters and similar air-purifying devices utilised in respirators have a limited validity, which should be monitored; mechanical filters for dust become clogged after a certain time, etc. Such devices should be periodically changed.
- fitted to the worker concerned. Mask-type respirators which do not fit the worker's face perfectly permit leakages of contaminated air; loose ear plugs will let the noise pass through, etc.
- well maintained and cleaned, as well as routinely inspected. There must be facilities for cleaning and disinfecting personal protective equipment. If the equipment becomes deteriorated (e.g. cracks, missing pieces, etc.), it should be totally or partially replaced.

In addition, workers utilising personal protective equipment should be adequately trained, educated and motivated.

Limitation of the exposure time

The reduction of the time during which a worker is exposed to a certain hazardous agent may greatly reduce the health hazard involved. This can be achieved through work practices, rotation of workers or administrative controls. One definition of "administrative controls" is "provisions to enable adjustments of the work schedule to reduce exposure". Limitation of the time during which workers wear certain cumbersome types of personal protective equipment (e.g. masks) is also a recommended procedure.

Personal hygiene

Personal hygiene is of particular importance for workers involved with chemical agents and particulate matter. It means cleanliness of both the person and his clothing. Not only should workers be instructed and motivated as to its value but adequate facilities for this purpose should be provided at the workplace. It is useless to require workers to take a shower after work if a sufficient number of showers is not available in the workplace or if, in a cold country, hot water is not provided in winter. And, to take a shower is of limited value if contaminated clothes continue to be worn without adequate laundering.

Adequate locker rooms and, in the case of work with hazardous materials, adequate disposal of and laundry facilities for used garments at the workplace are also essential. Clothing contaminated with toxic materials should never be taken home.

Whenever there is the hazard of a skin effect (e.g. cancer, dermatoses) or when dealing with chemicals which can penetrate through intact skin, workers should wash immediately after any contact and not wait until the end of the shift. Depending on the degree of hazard, contaminated clothing should be immediately changed.

In certain situations there may be need for special soaps and cleaning agents.

Other measures as important for the maintenance and promotion of the workers' health are the pre-employment medical examinations and adequate job placement, periodic medical examinations including biological monitoring and early detection of health impairment, as well as health education–for workers and management–and the application of safety and ergonomic principles. These measures are discussed in detail under the relevant separate articles.

However, control technology and medical control, as well as health education, safety and ergonomic aspects should be integrated into a consolidated programme.

The efficient control of occupational exposure to harmful agents requires a multidisciplinary approach in which environmental and medical sciences complement each other in order to prevent adverse health effects at the workplace.

GOELZER FERRARI, B. I.

Series on evaluation of occupational health hazard control technology published by the National Institute for Occupational Safety and Health, 4676 Columbia Parkway, Cincinnati, Ohio. Among the major industrial sectors and processes covered, see:

Cotton dust control in yarn manufacturing. No. 74-114 (1974), 188 p. *Plastics and resins industry.* No. 78-159 (1978), 242 p. *Control of exposure to metal working fluids.* No. 78-165 (1978), 40 p. *Manufacture and formulation of pesticides.* No. 78-174 (1978), 440 p. *Assessment of selected control technology techniques for welding fumes.* No. 79-125 (1979), 32 p. *Foundry industry.* No. 79-114 (1979), 438 p. *Dry cleaning.* No. 80-136 (1980).

The industrial environment–its evaluation and control. DHEW (NIOSH) publication No. 74-117 (National Institute for Occupational Safety and Health, 4676 Columbia Parkway, Cincinnati) (1974), 719 p.

Workplace control of carcinogens. Proceedings of a topical symposium (Cincinnati, American Conference of Governmental Industrial Hygienists, Oct. 1976), 159 p. Illus. Ref.

Engineering control research recommendations. DHEW (NIOSH) publication No. 76-180 (National Institute for Occupational Safety and Health, 4676 Columbia Parkway, Cincinnati) (1976), 216 p. Illus. 181 ref.

Industrial ventilation–A manual of recommended practice. Committee on Industrial Ventilation (Lansing, American Conference of Governmental Industrial Hygienists, 16th ed., 1980), 328 p. Illus. 108 ref.

Engineering of industrial ventilation (Engenharia de Ventilação industrial). Mesquita, A.; Guimarães, F.; Nefussi, N. (São Paulo, Edgar Blücher, 1st ed., 1977), 442 p. Illus. 39 ref. (In Portuguese)

Copper, alloys and compounds

Copper (Cu)

a.w.	63.5
sp.gr.	8.92
m.p.	1 083 °C
b.p.	2 567 °C

a reddish-brown metal which takes a brilliant polish.

TWA OSHA	dust and mists 1 mg/m³
STEL ACGIH	dust and mists 2 mg/m³
TWA OSHA	fume 0.1 mg/m³
TLV ACGIH	fume 0.2 mg/m³
MAC USSR	dust 1 mg/m³
MAC USSR	copper silicide 4 mg/m³
MAC USSR	Cu-Cr-Ba catalyst 0.01 mg/m³

Copper is malleable and ductile; it conducts heat and electricity exceedingly well and is very little altered in its functional capacity by exposure to dry air. In a moist atmosphere containing carbon dioxide it becomes coated with a green carbonate.

Occurrence. Copper occurs principally as mineral compounds in which ^{63}Cu constitutes 69.1% and ^{65}Cu, 30.9% of the element. Copper is widely distributed in all continents and is present in most living organisms. Although some natural deposits of metallic copper have been found, it is generally mined either as sulphide ores, including covellite (CuS), chalcocite (Cu_2S), chalcopyrite ($CuFeS_2$) and bornite (Cu_3FeS_3); or as oxides, including malachite ($Cu_2CO_3(OH)_2$); chrysocolla ($CuSiO_3.2H_2O$) and chalcanthite ($CuSO_4.5H_2O$).

World mine production of copper exceeded 8 000 000 t in 1977, of which more than two-thirds came from the United States, the USSR, Chile, Zambia, Canada and Zaire. Scrap metal accounts for about 40% of annual world copper consumption.

Extraction. Copper is normally extracted by underground or open-pit mining.

Production. Native copper, which is remarkably pure, is processed by grinding, washing, melting and casting.

The metal is produced from its ores by reduction. Oxides and carbonates can be leached with dilute sulphuric acid and copper may be electrolysed from this solution. Sulphides are crushed and ground; concentrated by flotation, following addition of air and "frothers"; smelted at 1 500 °C with the addition of lime and silica fluxes; and freed of sulphur and iron in a converter. The resulting blister copper, about 98% pure, is fire-refined to a purity of about 99.5% which is suitable for many purposes but not for electrical use. For this, the copper is further refined to at least 99.9% purity by electrolysis.

Uses. Over 75% of copper output is used in the electrical industries. Other applications include water piping, roofing material, kitchenware, chemical and pharmaceutical equipment, and the production of copper alloys. Copper metal is also used as a pigment, and as a precipitant of selenium.

Alloys

The most widely used non-ferrous alloys are those of copper and zinc (brass), tin (bronze), nickel (monel metal), aluminium, gold, lead, cadmium, chromium, beryllium, silicon or phosphorus.

Compounds

Copper sulphate is used to supplement pastures deficient in the metal; as an algicide and molluscicide in water; with lime, as a plant fungicide; as a mordant; in electroplating; and as a component of Fehling's solution to estimate reducing sugars in urine.

Cupric oxide has been used as a component of paint for ship bottoms. Copper sulphate neutralised with hydrated lime, known as Bordeaux mixture, is used for the prevention of mildew in vineyards. Livestock and poultry feeds are frequently supplemented with copper either to promote growth or to provide antibiotic activity.

Copper chromates are pigments, catalysts for liquid phase hydrogenation and potato fungicides. The pigment known as Scheele's green is a complex mixture of cupric oxide and arsenite (see ARSENIC, ALLOYS AND COMPOUNDS), and that called verdigris is cupric oxyacetate. A solution of cupric hydroxide in excess ammonia is a solvent for cellulose used in the manufacture of rayon (viscose).

HAZARDS

Amine complexes of cupric chlorate, cupric dithionate, cupric azide, and cuprous acetylide are explosive but are of no industrial or public health importance. Copper

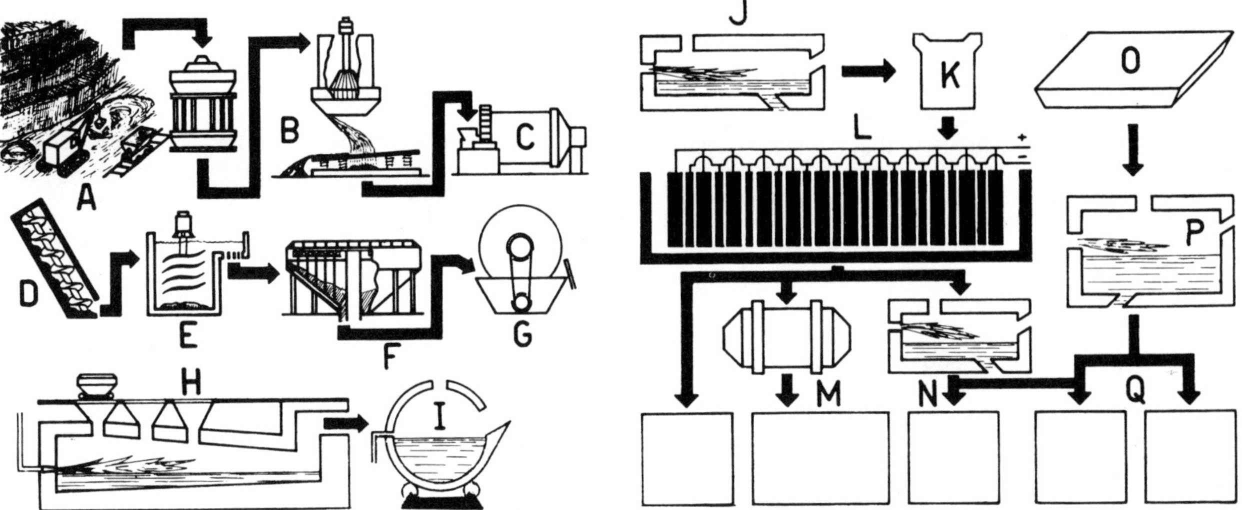

Figure 1. Flow chart of extraction from sulphide minerals. *Mineral processing:* A. Mining, 0.2-5% copper content. B. Grinding. C. Pulverising. D. Classification. E. Flotation. F. Concentration. G. Filtration, roasting. *Smelting:* H. Matte production (reverberatory furnace). I. Converting. *Electrolytic refining:* J. Anode casting. K. Anode. L. Electrolytic tank. M. Cathode refining under protective atmosphere. N. Cathode refining in air. *Thermal refining:* O. Blister copper. P. Oxidation refining. Q. Phosphorus deoxidation.

acetylide was found to be the cause of explosions in acetylene plants and has caused the abandonment of the use of copper in the construction of such plants.

Fragments of metallic copper metal or copper alloys that lodge in the eye, a condition known as chalcosis, may lead to uveitis, abscess and loss of the eye. Workers who spray vineyards with Bordeaux mixture may suffer from pulmonary lesions (known as vineyard sprayer's lung) and copper-laden hepatic granulomas.

Though not proven, the possibility of copper-induced toxicity cannot be dismissed in the following situations:

- the oral administration of copper salts is occasionally employed for therapeutic purposes, particularly in India;
- copper dissolved from the wire used in certain intrauterine contraceptive devices has been shown to be absorbed systemically;
- an appreciable fraction of the copper dissolved from the tubing commonly used in haemodialysis equipment may be retained by the patient and can produce significant increases in hepatic copper;
- copper, not uncommonly added to feed for livestock and poultry, concentrates in the liver of these animals and can greatly increase the intake of the element when these livers are eaten. Copper is also added, in large amounts relative to the normal human dietary intake, to a number of pet animal foods that are occasionally consumed by people. Manure from animals with copper-supplemented diets can result in an excessive amount of copper in vegetables and feed grains grown on soil dressed with this manure.

Acute toxicity. Although works of chemical reference contain statements to the effect that soluble salts of copper are poisonous, this is true in practical terms only if such solutions are used with misguided or suicidal intent, or as topical treatment of extensively burned areas. In the former instances copper sulphate, known as bluestone or blue vitriol, is generally ingested in gramme quantities. Nausea, vomiting, diarrhoea, sweating, and rarely convulsions, coma and death may result. Gastrointestinal irritation, seldom serious, can result following the drinking of carbonated water or citrous fruit juices which have been in contact with copper vessels, pipes, tubing or valves. Such beverages are acidic enough to dissolve irritant quantities of copper. Corneal ulcers, and skin irritation, but little other toxicity, have been noted in a copper-mine worker who fell into an electrolytic bath, but the acidity, rather than the copper, may have been the cause. In some instances where copper salts have been used in the treatment of burns, high concentrations of serum copper and toxic manifestations have ensued. The inhalation of dusts, fumes and mists of copper salts can cause congestion of the nasal and mucous membranes and ulceration with perforation of the nasal septum. Fumes from the heating of metallic copper can cause nausea, gastric pain and diarrhoea.

Chronic toxicity. Chronic human toxicosis due to copper is found only in those rare individuals who have inherited a particular pair of abnormal autosomal recessive genes and in whom, as a consequence, hepatolenticular degeneration (Wilson's disease) develops. Most daily human diets contain 2-5 mg of copper. Almost none is retained and the adult's body content of the metal is quite constant at 100-150 mg. In individuals without Wilson's disease, almost all of the body's copper is present as an integral and functional moiety of one of perhaps a dozen proteins and enzyme systems including, for example, cytochrome oxidase, dopa-oxidase and serum ceruloplasmin. Tenfold, or more, increases in the daily intake of copper can occur in individuals who eat large quantities of oysters (and other shellfish), liver, mushrooms, nuts and chocolate—all rich in copper; or in miners who may work and eat meals, for 20 years or more, in an atmosphere dusty with 1-2% copper ores. Yet evidence of primary chronic copper toxicity (well defined from observations of patients with inherited chronic copper toxicosis—Wilson's disease—as dysfunction of and structural damage to the liver, central nervous system, kidney, bones and eyes) has never been found in any individuals except those with Wilson's disease. However, the excessive copper deposits found in the livers of patients with primary biliary cirrhosis, cholestasis and Indian childhood cirrhosis may contribute to the severity of the hepatic disease characteristic of each condition.

SAFETY AND HEALTH MEASURES

Copper miners generally wear filtering masks when exposed to dust from copper ores. The main purpose here is the retention of free silica in the atmosphere; however, such masks also minimise the inhalation of copper. Particularly in mines where there are water soluble ores as chalcanthite, workers should be particularly careful to wash their hands with water before eating. Food should be kept in covered containers so that it is not exposed to finely divided ore.

Accidental ingestion of soluble copper salts is generally innocuous since the vomiting induced rids the patient of much of the copper.

Medical examination. At least 99.9995% of the world's population is essentially immune to poisoning by copper. But approximately 1 in 200 000 individuals has inherited a pair of abnormal genes, as a result of which copper toxicosis—Wilson's disease—will ultimately develop. This will occur with the ingestion of only a normal diet containing 2-5 mg of copper per day and will probably occur more rapidly and more severely if the individual inhales or ingests more of the metal by working in a copper mine. The disease is progressive and fatal if untreated by a de-coppering regimen.

Screening in pre-employment health examination to exclude the employment of persons suffering from this condition could be accomplished by determining the serum concentration of ceruloplasmin quantitatively since normal individuals have from 20-50 mg/100 cm^3 of this copper protein whereas 97% of patients with Wilson's disease have less than 20 mg/100 cm^3.

Quantitative determinations of ceruloplasmin can be automated using about 1.1 cm^3 of serum but are nevertheless relatively sophisticated procedures, require expensive instrumentation and would be complicated to implement on a wide scale.

SCHEINBERG, H. I.

CIS 77-1063 *Copper.* Scheinberg, H.; Buck, W. B.; Cartwright, G. E.; Davis, G. K.; Dawson, C. R.; Morgan, J. M.; Nelson, K. W.; Price, C. A.; Sternlieb, I.; Boaz, T. D. (National Academy of Sciences, 2101 Constitution Avenue, Washington, DC) (1977), 115 p. 638 ref.

CIS 78-1631 *Copper—Health and hazard.* Bergqvist, U.; Sundbom, M. USIP report 78-05 (Institute of Theoretical Physics, University of Stockholm, Vanadisvägen 9, Stockholm) (Mar. 1978), 224 p. Illus. 346 ref.

"Exposure to copper dust". Gleason R. B. *American Industrial Hygiene Association Journal* (Cincinnati), Sep.-Oct. 1968, 29/5 (461-462). 4 ref.

"Technological progress and occupational hygiene in the metallurgy and pyrorefining of copper" (Tehničeskij progress i voprosy gigieny truda v metallurgii medi). Ljah, G. D. *Gigiena truda i professional'nye zabolevanija* (Moscow), Oct. 1979, 10 (1-5). 2 ref. (In Russian)

CIS 80-831 *At work in copper: occupational health and safety in copper smelting.* Gomez, M.; Duffy, R.; Trivelli, V. (INFORM Inc., 25 Broad Street, New York) (1979). 3 vols., 284+336+484 p. Illus. 1 432 ref.

Copra

Copra is the name given to the dried kernel or "meat" of the coconut *(Cocos nucifera)*, which grows on all coasts and islands in the tropics (see COCONUT CULTIVATION).

The production of commercial copra is estimated at 2.2 million tonnes a year; the Philippines are the major world producer. The coconut oil extracted from copra is used for making soap, candles, cooking oil and margarine, and as an ingredient of many other commercial products including cosmetics, detergents, paints, varnishes, lubricants and plastics. The residue of the copra after the oil has been extracted is known as copra cake or poonac and it is used chiefly as cattle food and sometimes as a fertiliser.

The production of the coconut oil and the copra cake involves the following stages:

Harvesting. The coconuts are picked from the trees [in Malaya, India and Sri Lanka, while in the Southwest Pacific area the ripe coconuts are allowed to fall on the ground, thus suppressing a serious cause of accidents]. The fibrous covering of the nut is stripped off for the manufacture of coir used for rope or matting (see COIR). The stripped coconuts are then split into halves with axes and dried either in the sun or in kilns or in hot-air driers. After drying, the kernel, consisting of the "meat" of the coconut, is easily detached from the hard woody shell. This "meat" contains from 60 to 70% oil.

Processing. The copra is ground into small pieces and the oil extracted in primitive presses, hydraulic expellers or by solvent treatment using solvents such as hexane.

Desiccation. A small quantity of the copra crop is made into desiccated coconut for use in foodstuffs. The preparation consists of stripping the outer shell from the coconuts and washing the kernels. Shredding of the kernels is done by machinery after which the product is placed in a drying oven.

HAZARDS AND THEIR PREVENTION

Falls of workers from trees while harvesting the coconuts are common and may prove fatal. [When the endosperm or "green copra" is removed before drying, the use of the "copra knife" on 500-2 000 coconuts per day is a rather frequent source of severe cuts, mainly of the left hand and wrist.] Machinery accidents often occur, particularly where processing is carried out at small-scale plantations where the machines are often of a primitive nature; efficient machinery guarding is essential.

Finely ground copra dust may form an explosive mixture which is readily ignitible. Consequently dust-control measures should be instituted, use being made, where possible, of natural ventilation; smoking and open flames should be prohibited in danger areas and flameproof electric installations should be provided in all workplaces where this fine copra dust is processed. Numerous fires have occurred in copra bulk-storage warehouses owing to spontaneous combustion, and careful control should be kept of piles of copra to detect any signs of rise in temperature.

Fires and explosions may be caused by solvents used in oil extraction and these solvents may also constitute a health hazard, following inhalation or skin contact, unless correct ventilation is installed.

Workers in desiccated coconut factories, employed full time on washing coconut kernels, may develop skin disorders due to constant immersion of the hands in water. Workers should be employed in rotation on this task and be supplied with protective equipment. Sacks of copra usually weigh between 70 and 75 kg and the health of workers lifting and carrying these sacks may be adversely affected. The ILO's Maximum Weight Recommendation, 1967 (No. 128), specifies a maximum permissible weight of 55 kg to be handled by one person and steps should be taken to ensure this figure is not exceeded.

Copra exported in bulk by ship may become mouldy and infested with mites, especially *Tyroglyphus* and *Rhizolglyphus*, which can cause copra itch to dockers employed in handling such cargo. This itch is produced by a generalised eruption of intensely itchy papules caused by the chitinous cover of the parasite acting as an irritant. Removal from exposure leads to cure.

BULENGO, A. P.

Copy paper, carbonless

A wide variety of symptoms have recently been reported relating to work with carbonless copy papers, such as irritation in the nose and throat, stuffiness, rhinitis, a sense of unpleasant smell and taste, hoarseness, tingling sensations or itching in the face, eyes and on the exposed skin of the hands and arms, dryness and redness of the hands and also eczema. Some people have also reported general symptoms of a non-specific character, such as fatigue, headache, nausea and pains in the joints.

Carbonless copy papers have been used since the 1950s, mainly as form sets. The health effect complaints were first made at the end of the 1960s, when the conclusion was reached that the discomforts might possibly be due to formaldehyde allergy. A few years later, attention was drawn to polychlorinated biphenyls (PCB), because of their presence in carbonless copy papers; since 1972, PCB have not been used in this kind of product.

Today, in some offices using self-copying paper more than 50% of the clerks are complaining, in some others using the same papers and handling procedures there are no complaints. The workshops where the forms are prepared do not seem to have the problem.

The use of carbonless copy papers has been growing and is today two to three times that of a decade ago; the quantity of forms with conventional copy paper is about six to seven times larger and consumption shows the same growth during the past decade.

The donator carbonless copy paper (Coated Back sheet, CB) is an ordinary base paper with a layer of micro capsules made either of aldehyde-denaturated gelatine or of a synthetic polymer. The capsules contain an organic solvent and colour formers and are fixed to the paper with gum arabic and carboxymethyl cellulose. The receptor paper (Coated Front sheet, CF) is ordinary base paper with a layer of clay or polymer acting as a colour developer. The chemicals differ from one brand to another.

Presently available data

Although carbonless copy papers have been used for almost 30 years, few divergent publications and unpublished reports have been presented about health problems connected with this product.

Patch tests were used in these studies, and papers, paper emulsions or paper extracts were tested. Evidence of specific contact allergies could not be demonstrated by any of the investigators. Hannuksels (1974) believed that the troubles were caused by mechanical factors. Magnusson (1974) thought that the troubles were caused by clay dust from the CF layer. Wahlberg (1975) found that the problems were connected with both types of carbonless copy paper dominating the Swedish market at that time. Nilzén (1975) examined symptoms from the mucous membranes and tested his patients not only with patch tests but also with prick tests, eye provocation tests and inhalation tests. Some patients reacted to airborne dust from carbonless copy papers, some others also reacted to ordinary typing paper, and some controls with vasomotor rhinitis developed the same type of irritation of the mucous membranes when exposed. Thus, the symptoms elicited by exposure to carbonless copy papers were non-specific. Nilzén concluded that carbonless copy papers can provoke irritation, and that these reactions primarily occur in persons with a tendency to allergic or vasomotor reactions.

In an inquiry made by the medical services of the County of Stockholm (1977), a connection was noted between the frequency of complaints and the daily handling time of carbonless copy paper. The more time spent using the papers, the more complaints there were.

Calnan (1979) described 18 persons who developed symptoms on exposure to carbonless copy papers. At least 14 of these persons had handled huge amounts of such papers, often in small rooms and when working overtime. Some of the troubles occurred during a heat wave. Calnan concluded that if these symptoms were attributable to carbonless copy papers, the solvents for the colour former would be the most probable reason, and a number of features favoured a toxic or irritant type reaction.

Bekkevold (1980) reported the presence in the paper of traces of Ni and Cr in sufficient quantity to cause skin irritation and allergy.

In a symposium held in February 1981 at the Southern Hospital in Stockholm (under Göthe, C. J. as moderator) it was shown that the occurrence of symptoms at work with carbonless copy paper greatly increased with papers containing monoisopropylbiphenyl or diarylethane + diisopropylnaphthalene as solvent. Papers containing hydrogenated terphenyls gave rise to only few complaints.

In Denmark, Menné et al. (1981) tested 17 paper substances with patch dissolved tests, photo patch tests and open tests in order to reveal a possible phototoxic reaction, and prick tests with the paper substances in petrolatum. No positive reactions were obtained. However, 11 out of 26 persons prick tested with two types of carbonless copy paper reacted positively. They stressed that the majority of the skin symptoms were mild and transient. According to further analytical investigations (Mølhave and Grunnet) the paper sample causing skin problems contained up to 150 times more santasol oil, which consists mainly of hydrogenated terphenyls and solvents. Terphenyls are known for their irritating power. They could be transported as dust-absorbed compounds.

(Adapted from a technical memorandum to the International Occupational Safety and Health Hazard Alert System by The National Board of Occupational Safety and Health of Sweden)

CIS 77-1457 "Can selfcopying paper cause allergic reactions?" (Kan selvkopierende papir gi allergiske reaksjoner?). *Vern og velferd* (Oslo), 1976, 3 (45). Illus. (In Norwegian)

CIS 77-2062 *Selfcopying paper* (Itsejäljentävät paperit). Kiertokirje 1/77 (National Board of Occupational Safety and Health (Työsuojeluhallitus), Tampere) (13 Jan. 1977), 4 p. (In Finnish, Swedish)

"Carbon and carbonless copy paper". Calnan, C. D. *Acta Dermatovenereologica* (Stockholm), 1979, 59/suppl. 85 (27-32). 10 ref.

CIS 81-460 "ASS was mistaken about selfcopying paper: the allergies were due to nickel and chromium" (ASS hade fel om kopiepapperen: Nickel och krom bakom allergierna). Bekkevold, A. *Arbetsmiljö* (Stockholm), 26 Aug. 1980, 10 (55). Illus. (In Swedish)

CIS 81-1393 "Skin and mucous membrane problems from 'no carbon required' paper". Menné, T.; Asnaes, G.; Hjorth, N. *Contact Dermatitis* (Copenhagen), Mar. 1981, 7/2 (72-75). 8 ref.

"Addendum: Headspace analysis of gases and vapors emitted by carbonless paper". Mølhave, L.; Grunnet, K. *Contact Dermatitis* (Copenhagen), Mar. 1981, 7/2 (76). 2 ref.

Unpublished reports and private communications on the subject are available at the National Board of Occupational Safety and Health (S-171 84 Solna, Sweden).

Coral and shell

Coral

This is produced by the calcareous secretion of small marine polyps (of the *Cnidaria* group). It is originally the bone structure and shell of a small larva, which grows during the sexual and organic development of the animal until it assumes the shape of a red (most precious) or pinkish-white, or most seldom black branch, the latter being due to special organic substances. It can be found in warm seas near the coast, in places sheltered from currents, about 40-50 m deep; it lies in banks or on rocks, or in layers or deposits. It is usually either netted from small boats or motorboats or obtained by divers.

Coral has been worked for many centuries in the Mediterranean areas and especially in Italy where there are centres at Livorno, Trapani and Torre del Greco, and also in India and Japan. Much coral jewelry is exported; in India, coral is also used as pharmaceutical powder because of its high content of calcium salts.

Shells

These are used mainly for the production of cameos, which are smooth jewels with figures in relief. They can be worked on precious stones, on coral and, more often, on shells. This type is almost exclusively produced by the craftsmen of Torre del Greco.

Shells used for the manufacture of cameos, jewelry, etc., are usually exotic shells of a porcelaneous type of the *Gasteropoda* group, namely the *Cassis rufa* (which originates from Zanzibar and Madagascar) and the *Strombus gigas* (from Cuba); it is characterised by a spiral, conical shell, with white-opaque upper coat and a lightly dark inner coat. The flake is the useful part.

Mother-of-pearl

This is the inner, iridescent part of the shell of some *Gastropoda* and *Lamellibranchiata* molluscs; it is made of very thin sheets of an organic substance (conchiolina), alternating with lightly wavy sheets of calcium carbonate. The *Meleagrina* species of *Lamellibranchiata* is much in demand. It originates from the East Indies, Eritrea, and Australia, and it is much used in the factories of Torre del Greco; among the *Gasteropoda*, the *Halictis* and *Trochia* species are also in great demand.

Mother-of-pearl has a white-silvery colour, with various hues (yellowish, greenish, blueish, dark, etc.); it is used for ornaments, buttons, cameos, etc.

Nowadays, in many parts of the world, synthetic materials are used to simulate natural coral, shell, or mother-of-pearl at much less cost. Production is on a factory scale but the objects produced have far less intrinsic value.

Processing. Coral-working and cameo-making are usually carried out in small workshops or at home by specialist craftsmen who work on commission or on their own account. Valuable *objets d'art* are produced by fine craftsmanship in engraving but less valuable ornamental products are made on a commercial scale: button making is usually a factory process.

The main stages in processing coral and cameos are the following:

(a) cutting of coral, shaving of the shells by electric circular saws in bowls of soapy water;

(b) wet-grinding by abrasive wheel;

(c) roughing and rounding off by small electric grindstones;

(d) engraving of the pieces with hand or electric burins and scrapers of various types;

(e) polishing with various substances;

(f) finishing.

Additional stages are: marking (of shells) in which the design is traced and then cut on the cameos; the boring of the pieces (coral, cameos), which is done at home by craftsmen using small hand-drills, etc.

HAZARDS AND THEIR PREVENTION

Cutters and, to a lesser extent, grinders are exposed to risk of accident, especially hand injuries; there are also risks of eye injuries from splinters and dust thrown off during cutting and grinding.

Among the occupational diseases, skin affections and changes in the ventilatory system should be mentioned.

Skin diseases consist in callosity, thickening, hyperkeratosis changes of the phalangettes and of the balls of the fingers, and of the base of the right hypotenary points (cutters, grinders): the wounds tend in time to extend to all the skin of the palm (workers on coral, and engravers). Secondary dermatosis is encountered and in some cases angiospastic lesions and maceration are found which are due to frequent immersion of the hands in water. The nails are often torn very short and irregular.

In addition to callosities, rounders and roughers may show rhagadiform lesions to the thumb, the forefinger, and the right middle finger, with irritation all round the nails. In some cases, hyperkeratosis and/or arthrosis between the phalangettes make the complete extension of the fingers difficult or impossible.

Respiratory affections may be produced in craftsmen by the dust emitted during the grinding of shells and mother-of-pearl (cutters, grinders, engravers). The pathogenic action consists in irritation of the mucosae of the upper respiratory tract (rhinitis, pharyngitis, sinusitis) and less frequently, of the bronchial tract. Humidity may sometimes be an additional factor in this condition.

Atrophic rhinitis and sometimes asthmatic syndromes have been found in workers employed in mother-of-pearl processing (button production). The allergic factor is usually represented by the dry residuum of the mollusc body, which adheres to shells which have not been washed well, resulting in positive endoreactions.

Cases of bouts of fever following exposure to shell or mother-of-pearl dust have been reported and the occurrence of ossifying osteoperiostitis of the cubitus, radius, metacarpus and metatarsus has been described in young persons working mother-of-pearl.

The frequency of conjunctival and rhinopharingeal reactions tends to decrease with age, possibly because of progressive adaptation to environment, but bronchial changes increase. These are common among workers on shells, especially mother-of-pearl, in whom chest X-ray examination may reveal radiological changes (reticulation, hilar thickening, discrete basal emphysema).

Solvents used in joining parts together may be toxic if inhaled in sufficient quantity.

Appropriate safety and health measures include:

(a) efficient machinery guarding of the revolving machine parts, especially circular saws and abrasive wheels;

(b) wearing of eye protection;

(c) exhaust ventilation, locally applied, to remove dust or fume generated in processes;

(d) provision of good lighting;

(e) use of barrier creams or lotions to protect hands exposed to frequent immersion in liquids.

SESSA, T.

Cork

Cork is obtained from the bark of the cork oak tree *(Quercus suber)* which is found principally in the western part of the Mediterranean basin. Portugal is the world's largest producer of cork. The exact composition of cork is not yet known. It seems, however, that it comprises a mixture of fatty acids of high molecular weight, some of which are either not saponifiable or insoluble and others of as yet unknown composition. The outer part of the cork contains a certain amount of silica, probably deposited by the wind, since the cork oak grows mainly on poor soils with a high silica content. The ash of cork samples taken from cork processing plants has been found to have a total silica content of 42.2-50.1% and a free silica content of 19.5-26.5%.

Processing and uses. There are two types of cork processing plant. The "natural" cork plant uses cork that has been stripped of its hard outer layer, for the production of bottle corks, Crown closure inserts, cork paper for cigarette tips, balls, rings, buoys, etc. The "reconstituted" cork plant uses the whole cork including the hard outer layer which is ground into particles and bonded together with an adhesive to produce floor coverings and panels and special sections for thermal and acoustic insulation.

HAZARDS AND THEIR PREVENTION

The risks involved in the harvesting of cork are essentially those of agricultural work in general. However, cork processing workers are exposed to a number of accident hazards including mechanical equipment, fires and explosions and to the health hazard of suberosis, a lung condition caused by the inhalation of cork dust.

Accidents. The hazards related to the use of band and circular saws are similar to those encountered in sawmills and the woodworking industry, and the same safety measures apply. Accidents on cork disintegrators are relatively numerous and serious; feed hoppers on disintegrators and mixers should be designed, guarded or positioned in such a way that the worker's hands cannot come in contact with the rotating blades.

Accidents during press moulding are more common but usually less serious; they can be prevented by the use of moulds fitted with handles or by mechanical handling which reduces the danger of crushed hands and feet whilst also rendering the work less arduous. A typical accident during handling and storage is the fall of a bale of cork; safe stacking procedures will reduce this danger but in-plant transport equipment, such as fork-lift trucks, should be fitted with protective canopies as an added precaution.

Fine particulate cork dust, especially when impregnated with binders such as tar, is highly flammable and airborne cork dust may present a serious explosion hazard. During moulding of reconstituted cork there is also a danger of fire when the mass of cork and binder heats up under compression. Dust control and maintenance and good housekeeping are essential (see DUST EXPLOSIONS).

Suberosis. The most serious health hazard encountered in the processing of cork (cutting, rolling, disintegration, classification, pressing, polishing and blending with various chemicals) is the inhalation of the large quantities of dust evolved, since this dust may produce a relatively benign pulmonary fibrosis called suberosis (derived from the Latin name for the cork oak). These changes have been demonstrated experimentally in guinea-pigs and rabbits. [According to recent studies, the disease in man may manifest itself in three different forms: asthma-like (that is obstructive ventilation syndrome), interstitial disease of the type of extrinsic allergic alveolitis and chronic bronchitis with bronchiectasis (the last form appearing only after several years of work). The asthma-like syndrome is characterised by immediate bronchial reaction to inhalational provocation tests; in the blood, eosinophile cells are increased; bronchoscopy shows oedema of the mucosa; there is accumulation of cork dust in histiocytes and within the

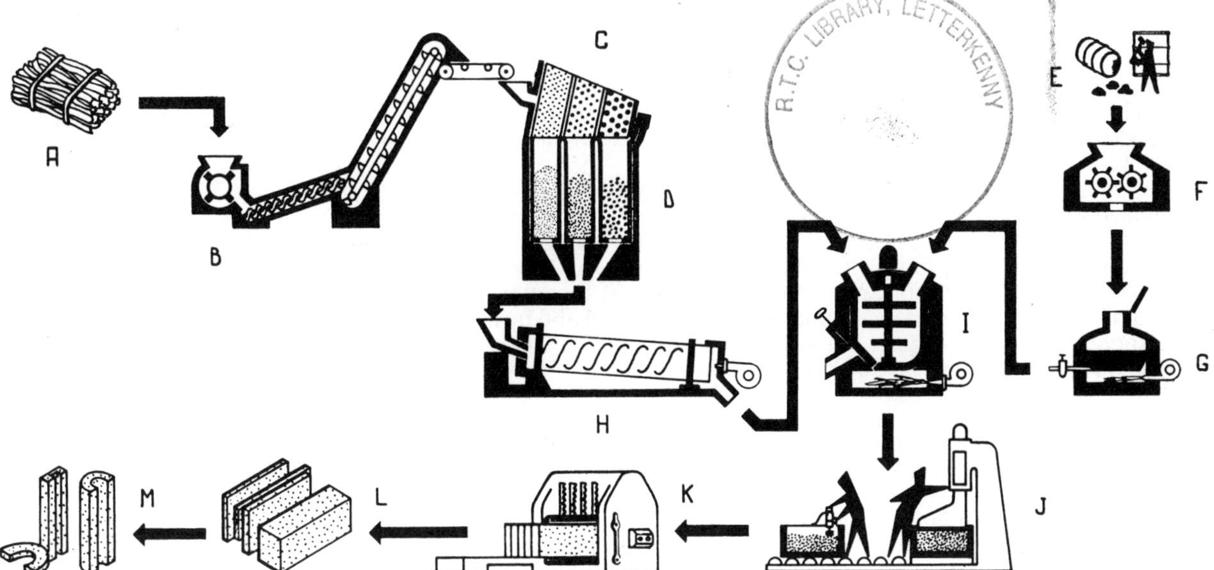

Figure 1. "Reconstituted" cork plant: A. Natural cork. B. Grinding. C. Classifying. D. Silo storage. E. Coal, tar, pitch. F. Grinding. G. Melting. H. Drying. I. Mixing. J. Pressing (block production). K. Slicing, sawing, moulding. L. Blocks, slabs. M. Finished parts.

alveoli. The extrinsic allergic alveolitis may result, as is usual, in acute attacks with fever and pain in the chest, or develop insidiously; gamma-globulins are increased; skin tests show positive reaction to antigens prepared from mouldy cork and *Penicillium frequentans* isolated from it; there is infiltration of alveolar septa sometimes with granulomatous reaction reminiscent of sarcoidosis, and cork dust is found within the lesions. The existence of precipitating antibodies against the fungal species most frequently identified is typical, although the fungus may only be an habitual accompaniment of the inhaled organic material.]

In some workers, on initial exposure, symptoms include irritation of the nasal mucosae, frequent coughing and slight expectoration. These symptoms gradually regress and disappear after 2-3 weeks. More rarely, the symptomatology also includes a sense of oppression in the chest, fever, pain in one side of the chest and flecks of blood in the sputum, but these symptoms regress rapidly.

The chronic form generally appears after 5 years of work in an atmosphere containing large quantities of cork dust, and becomes gradually and progressively worse. There is cough (sometimes spasmodic), expectoration (sometimes copious and mucopurulent), very frequently bloody sputum or even slight haemoptyses, stitch in the side or hemithoracic pain (of no fixed localisation). Working capacity is unaffected at the outset but thereafter gradually diminishes. Approximately one-third of subjects affected by suberosis present severe pulmonary insufficiency whilst in about half the cases insufficiency is accompanied by emphysema.

The first radiological changes usually appear after several years of work in a factory in which the atmosphere is heavily contaminated with cork dust. The chief characteristic is the extraordinary increase in bronchovascular markings which present a generalised arboreal pattern accompanied, in the early stages, by an increase in the hilar shadows. Most cases do not develop beyond this degree of fibrosis, but there are some which show small circular opacities (submiliary, micronodular or nodular) in both lung fields, occupying not more than one-third of each field initially but thereafter extending progressively to cover the whole area of the lung, especially the mid-zone. In rare instances, several nodules coalesce, giving rise to more extensive shadows accompanied by bullous emphysema with compression, distortion or displacement of neighbouring structures. The main information derived from pulmonary angiography has been the marked prolongation of the arterial phase and the delay and prolongation of the venous phase.

Histological examination shows two distinct forms: a picture of early fibrosis and a picture of advanced fibrosis. In the latter case the changes may be limited to the interalveolar septa, to the interalveolar septa and sub-pleural connective tissue or to the supporting connective tissue, or finally extend to all the supporting connective tissue and the alveolar parenchyma.

The complications of suberosis include spontaneous pneumothorax (in about 6% of cases), tuberculosis (frequent but generally with chronic course) and *cor pulmonale* (rare).

The usual development of suberosis is slow, symptoms only appearing after about 5-6 years of work in an atmosphere of cork dust, while radiological signs are not usually seen before 15-18 years in the job. Individual factors have a strong influence, as do the working conditions and the environment. The manifestations are aggravated by continuing to work in a dusty atmosphere but improve if the subject is removed from exposure for a certain length of time. Fibrotic lesions are, of course, irreversible.

There is no specific treatment for suberosis.

Suberosis prevention. Dust-control measures are essential to prevent the evolution of dust (plant design, humidification), or to capture airborne dust as close to the source as possible (local exhaust ventilation). Medical measures include pre-employment and periodical examinations. Workers with respiratory system conditions likely to affect pulmonary ventilation should not be employed on cork-processing operations entailing exposure to cork dust.

Miscellaneous health hazards. There is a definite danger of skin disorders such as eczema and tumours and of systemic poisoning due to exposure to cork binders which often have a tar base. Workers should be supplied with suitable hand and arm protection and working clothes. Strict personal hygiene is essential and adequate washing and sanitary facilities should be provided.

CANCELLA DE ABREU, L. C.

Accidents:

CIS 1695-1966 *The accident picture in the plastics and cork processing industries* (Das Unfallgeschehen bei der Verarbeitung von Kunststoffen und Kork). Veröffentlichung 1756d (Lucerne, Schweizerische Unfallversicherungsanstalt (SUVA)), (May 1966), 42 p. Illus. (In German)

Diseases:

CIS 74-841 "Respiratory disease in cork workers (suberosis)". Pimentel, J. C.; Avila, R. *Thorax* (London), July 1973, 28/4 (409-423). Illus. 22 ref.

Work environment:

CIS 74-2057 "The air spora of a Portuguese cork factory". Lacey, J. *Annals of Occupational Hygiene* (Oxford), Nov. 1973, 16/3 (223-230). 14 ref.

Corrosive substances

The term "corrosive substances" does not denote a specific class of chemical substances with certain common structural, chemical or reactivity characteristics; neither does it denote a group of substances that have a common use. Rather this term should be regarded as a generic designation for those substances which possess the property of severely damaging living (in particular human) tissue and of attacking other materials such as metals and wood.

Initially, the emphasis was placed almost exclusively on the damaging action on living tissue; later on the scope of concern was widened to include damage to inert matter—a development advocated mainly by bodies concerned with the transport of dangerous goods for whom the safety of the carriers (railway trucks, road vehicles, sea and inland waterway vessels, and aircraft) were the prime concern.

The most exhaustive, and currently most widely accepted, definition of "corrosives" is that drawn up in 1956 by the United Nations Committee of Experts on the Transport of Dangerous Goods, which states that:

"These are substances which, by chemical action will cause severe damage when in contact with living tissue or, in the case of leakage, will materially damage or even destroy, other freight or the means of transport; they may also cause other hazards."

This is the definition used by most of the international organisations including the ILO and the Inter-Governmental Maritime Consultative Organisation (IMCO) (with minor changes of wording). The definition employed by the International Air Transport Association (IATA) refers to corrosive liquids only (in general, compressed gases are not accepted for air freight) and contains the statement that these corrosive liquids "are likely to cause a fire when in contact with organic matter or certain chemicals". The Council of Europe, while restricting its definition to substances which destroy living tissue, introduces a further class—"irritants which cause inflammation"—to designate those substances that cannot be considered "corrosive" in the full sense of the word. It is obviously not possible to lay down criteria for distinguishing between damage of only moderate severity and intense irritation, neither can a clear line be drawn between the "severe damage" specified in the United Nations definition and other less severe damage to living tissue. Consequently, the final decision as to whether a chemical substance should be classed as corrosive or not will be an empirical one based on an appreciation of the substance's normal action on living tissue; for this reason, the commonly accepted lists of corrosive substances are subject to constant revision.

Classification is further complicated by the fact that certain corrosive substances have other, more serious, hazardous properties (toxicity, flammability, etc.), and are commonly classed as toxic or flammable, etc., rather than as corrosive; a good example is dimethyl sulphate which, although corrosive, is in most cases classed as toxic only. Certain substances considered non-corrosive in their natural, dry state may become corrosive when in contact with water or the moisture of skin and mucous membranes. These substances are, in general, the readily hydrolysed organic and inorganic halogen compounds such as lithium chloride, chlorosilanes, and halogen fluorides, acetyl chloride, allyl iodide, and benzyl chloride; these substances are usually considered corrosive since, during their hydrolysis, corrosive halogen compounds are released.

Important corrosive substances

Some of the most representative of the corrosive substances of industrial, agricultural or commercial importance are listed below, grouped by general classes, and according to the Council of Europe classification.

Acids and anhydrides. Acetic acid (over 25% concentration), acetic anhydride, chlorosulphonic acid, chromic acid, dichloroacetic acid, fluoboric acid (over 25%), fluosilicic acid (over 25%), hydrobromic acid (over 25%), hydrochloric acid (over 25%), hydrofluoric acid, hydriodic acid (over 25%), nitric acid (over 20%), perchloric acid (over 10%), phosphorus pentoxide, propionic anhydride, sulphuric acid (over 15%), oleum (fuming sulphuric acid), and trichloroacetic acid.

Alkalis. Ammonium hydroxide (over 35% by weight of gas), potassium hydroxide (caustic potash), sodium hydroxide (over 5%, caustic soda).

Halogens and halogen salts. Aluminium chloride, ammonium bifluoride, antimony trichloride and pentachloride, bromine, phosphorus oxychloride, phosphoryl chloride, phosphorus trichloride and pentachloride, potassium bifluoride, sodium bifluoride, sodium hypochlorite (over 10%), stannic chloride, sulphur tetrachloride, sulphuryl chloride, thionyl chloride, titanium tetrachloride, and zinc chloride.

Organic halides, organic acid halides, esters and salts. Acetyl chloride, allyl iodide, benzoyl chloride, benzylamine, benzyl chloroformate and chloroacetyl chloride.

Miscellaneous corrosive substances. The following corrosive substances are widely used but do not fall into any of the above classes: ammonium polysulphide, 2-chlorobenzaldehyde, hydrazine (15-64%), hydrogen peroxide (over 20%), silver nitrate.

Mercuric chloride, commonly called "corrosive sublimate" is not considered to be a corrosive substance.

Special characteristics

Some of these substances meet only one of the criteria of the general definition of corrosive substances given above, whereas others may produce damage to both human tissue and metal or wood. For example dry 1,2-dichloroethane is not corrosive; however, when in contact with moisture at high temperature, this substance will attack iron and certain other metals. Potassium hydroxide or sodium hydroxide will attack aluminium, zinc and tin and will also cause serious damage to human tissue (skin, mucous membranes and eyes); hydrazine, on the other hand, has a corrosive action on human tissue only. Consequently, before a decision is taken on the safety and health measures required in relation to a given substance, full information should be obtained on the relevant properties by reference to the appropriate sections of this encyclopaedia, to standard textbooks or to the manufacturer's specifications.

Uses

It is impossible to describe here the uses of such a vast range of substances. The majority are common basic chemicals used in the chemical industry, textile manufacture, metal smelting and refining, the engineering industry, the plastics and synthetic fibres industries and the pharmaceutical industry; in addition, many of these substances are dealt with in greater detail in separate articles.

SAFETY AND HEALTH MEASURES

When in contact with human tissue, most corrosive substances will produce chemical burns, while certain substances such as chromic acid produce deep ulceration. Many corrosive substances have a defatting action on the skin and may cause dermatitis.

Consequently, preventive measures should be directed primarily at preventing or minimising contact between corrosive substances and skin, mucous membranes and eyes. In addition, corrosive substances should not be allowed into contact with the various materials that they attack.

All the containers, pipes, apparatus, installations and structures used for the manufacture, storage, transport or use of these substances must be of material resistant to corrosive substances or be protected by suitable coatings impervious to and unaffected by corrosives. All containers or receptacles should be clearly labelled to indicate their contents and should bear the danger symbol for corrosives. A high standard of maintenance and good housekeeping is essential. See also DANGEROUS SUBSTANCES, STORAGE OF.

Adequate ventilation and exhaust arrangements, whether general or local, should be provided whenever corrosive gases or dusts are present. Certain strong corrosives may, on contact with organic matter or other chemicals, cause fire and consequently the appropriate measures should be taken.

Before a new process is introduced, it should be ascertained whether any of the raw materials, reagents, intermediate or final products possess corrosive properties or may liberate corrosive compounds under certain conditions and, if so, appropriate measures taken; for instance organic peroxides commonly used as hardeners in the manufacture and utilisation of epoxy and polyester resins are highly corrosive, while polyvinyl chloride, normally considered harmless, liberates hydrogen chloride, an extremely corrosive gas, when involved in a fire; suitable provisions must be made to meet these hazards.

The most satisfactory method of ensuring worker protection is to prevent, from the outset, any planned or accidental contact with corrosive substances by utilising only closed-circuit apparatus. Where such a technique is not possible, it is necessary to rely on suitable personal protective equipment. Such equipment should ideally comprise: corrosion-resistant and impervious suits or overalls, foot protection, hand and arm protection, head protection and eye and face protection; where corrosive gases may be expected, appropriate respiratory protective equipment is required, ranging from simple masks or respirators to air or oxygen lines or self-contained breathing apparatus, coupled with gas-tight goggles and other items of equipment. The scale of the personal protective equipment required depends on the potential dangers present (type and quantity of the corrosive substance handled or liberated) and should be adapted accordingly. No universally suitable material for this personal equipment can be indicated but, on the whole, natural rubbers, synthetic rubbers, polyvinyl chloride, polypropylene or polyethylene either in sheet form or with a fabric backing are suitable. Regular cotton or woollen clothing does not normally afford sufficient protection against most corrosive agents, wool in particular being destroyed by strong alkalis. Where aprons are used instead of protective suits or overalls, these should have bibs; sleeves should be worn outside gauntlets or gloves and, similarly, trouser legs should cover the tops of the shoes in order to prevent splashes from accumulating and penetrating between the different parts of protective equipment or work clothing used. When the exposure to corrosive substances is not severe, suitable barrier creams can often be used instead of gloves.

All workers required to handle corrosive substances or likely to come into contact with them should be fully informed of the hazards involved and trained in the appropriate safe working practices and first-aid measures.

Treatment. The most effective first-aid treatment for cases of contact with corrosive substances is prolonged washing with copious amounts of water; irrigation should start immediately and be applied to all parts affected and contaminated clothing should then be discarded. It is good practice to install emergency showers at all strategic locations; these should preferably be actuated automatically by pressure on a foot plate; bath tubs filled with clean water can also provide valuable service in case of emergency. It is generally advisable to avoid the use of neutralising solutions (weak acid in case of contact with alkalis or weak alkalis in the case of contact with acids); the heat produced by neutralisation may often constitute a greater hazard. The eyes are particularly vulnerable to corrosives and, consequently, eye contact should be treated by prolonged irrigation in a special emergency eyewash fountain operated by a large and readily accessible lever; medical help should always be sought immediately in the case of eye contact.

AMIR, I. D.

Dangerous chemical substances and proposals concerning their labelling (Strasbourg, Council of Europe, 4th ed., 1978), 951 p. + Addendum (Jan. 1979). Illus.

CIS 77-1355 "Corrosive burns–Problems of safe handling" (Les brûlures par produits corrosifs: problème du conditionnement de la manipulation). Ascher, R. *Prévention et sécurité du travail* (Lille), 4th quarter 1976, 110 (30-44). Illus. 35 ref. (In French)

CIS 80-189 *Protective clothing for the hands and arms–Method for determining permeability to acids and alkalies* (Sredstva zaščity ruk–Metod opredelenija kisloto–i ščeločepronicaemosti). Gosudarstvennyj komitet po standartam. GOST 12.4.063-79 (Moscow, Izdatel'stvo standartov, 23 Mar. 1979), 4 p. Illus. (In Russian)

(See also readings suggested under BURNS, CHEMICAL.)

Cosmetics

The word "cosmetics" is defined as any external preparation or application intended to beautify and improve the complexion of skin or hair. The wide range of existing cosmetics may be broadly divided into the following groups:

(a) skin preparations including various creams, lotions, perfumes, colours, powders and soaps;

(b) hair cosmetics comprising shampoos, lotions, oils and creams, fixatives, dyes and bleaches;

(c) nail preparations consisting of various lotions, polishes, colours and hardeners;

(d) miscellaneous cosmetics including deodorants, depilatories, mud packs, etc.

Preparation. A large number of chemical substances may be used in a single cosmetic product. Most of them may be divided into the following five groups: *(a)* emulsions, *(b)* colours, *(c)* perfumes, *(d)* preservatives, and *(e)* special ingredients.

Emulsions

Most cosmetics are either oil-in-water or water-in-oil emulsions. The former are often described as "non-greasy" whereas the latter are termed "greasy". Emulsions are stabilised by using soaps, other synthetic detergents and various chemicals including waxes, alcohols and various solvents.

Lanolin is the commonest, widely used emulsifying agent which has caused adverse reactions. Others like glycerol monostearate and sodium lauryl sulphate have also produced sensitising reactions.

In these and other preparations in cosmetics manufacture, flammable solvents consisting principally of alcohols and esters are used. Many of them are volatile and evolve flammable concentrations of vapour at ordinary temperatures.

Colouring substances

The success of a cosmetic depends to a great extent on the choice and use of colour. Most of the colouring agents are either naturally occurring compounds or belong to the coal-tar group of dyes. The natural colouring matter is of both animal and vegetable origin. The important ones are annatto, cochineal (used in colouring rouge and eye shadow), saffron and alkonet (hair oils). Inorganic colours are mainly iron oxides, ultramarine blue and pink, chrome oxide greens, titanium oxide, zinc oxide and *blanc fixe*. Synthetic colouring agents belong to one of the following groups: *(a)* azo dyes, *(b)* nitro dyes, *(c)* triphenylmethane dyes, *(d)* xanthines, *(e)* quinolines, *(f)* anthraquinones, and *(g)* indigo dyes.

All colouring agents are capable of producing sensitising reactions. In particular "para-type" anthraquinone (used in hair dyes) and azo dyes (face and nail preparations) produce not only sensitisation reactions but also cross sensitivity to each other.

Perfumes

Perfume is an essential component in all cosmetic products. There are three main categories of perfumes: *(a)* plant oils comprising essential oils, flower oils, resins and gums, *(b)* animal secretions like musk, civet, castroeum and ambergris, *(c)* chemical substances, e.g. synthetic materials and plant oil isolates.

Perfumes often produce contact dermatitis and melanosis and the reaction may be acute and severe. Contact dermatitis is usually caused by ionene, balsam of Peru, cloves, oils of bergamot, benzyl alcohol and pine terpenes. Pigmentation may be caused by any perfumed cosmetic and may be localised or diffuse. The diffuse melanosis is to be distinguished from chloasma of pregnancy, Riehl's melanosis and argyria. See PERFUMES AND ESSENCES.

Preservatives

The term is used to refer to a compound which prevents growth of micro-organisms. Cosmetic deterioration may occur due to physical, chemical and enzymatic reactions. All water-containing cosmetics are susceptible to microbial action whereas those containing fats are prone to undergo oxidative deterioration. An ideal preservative should be colourless, odourless, stable and non-toxic. Some of the commonly used preservatives are listed below:

(a) organic acids: benzoic acid, salicylic acid, monochloroacetic acid, propionic acid and citric acid, perhaps in combination with benzoic acid;

(b) alcohols: ethyl alcohol (15% or less), di- and trihydric alcohols and chlorobutanol;

(c) aldehydes: formaldehyde and benzaldehyde;

(d) essential oils;

(e) phenolic compounds: these are the most popular antimicrobials used today and include phenols, chlorinated phenols and quinones, *p*-chloro-*m*-cresol, *p*-chloro-*m*-xylenol, dichloro-*m*-xylenol, hexachlorophene. Phenols are commonly used in soaps, lotions, antiseptic ointments, antiseptic oils and deodorants;

(f) esters of *p*-hydroxybenzoic acid;

(g) miscellaneous preservatives including *o*-phenylphenol, mercury compounds, surface active agents and antioxidants.

Citric acid is a body metabolite and does not present problems but benzoic acid and propionic acid are mild irritants, salicylic acid is a strong irritant and monochloroacetic acid produces severe local reactions of the skin, eye or respiratory tract. Benzaldehyde has an irritant effect on the skin and formaldehyde causes dermatitis, cough, lacrimation and injury to the bronchi. The essential oils have a low order of acute toxicity; they do not accumulate in the body but their vapours in sufficient concentrations will cause irritation of the mucous membrane. Prolonged contact with the skin will cause dermatitis.

Hexachlorophene is present in some toothpastes and cosmetic preparations. Irritant reactions of the skin have been reported. Hexachlorophene is known for its brain toxicity. It rarely produces hypersensitivity skin reactions. The phenols are toxic and can cause serious skin irritation and eye injury. The esters of hydroxybenzoic acid have low acute toxicities and appear to be relatively free from skin-irritant or skin-sensitising properties. Most mercury compounds have dangerously toxic properties. The use of ethyl alcohol is associated with a dangerous fire risk while the less volatile di- and trihydric alcohols are regarded as moderate fire hazards.

SAFETY AND HEALTH MEASURES

The precautions must cover both health and fire risks.

Health precautions. In the cosmetics industry, many of the materials that are capable of causing skin irritation and dermatitis may be handled in small amounts or in low concentration. On the other hand, the materials in question are frequently associated with allergy and sensitisation. To a very great extent, therefore, the precautionary measures must be adapted to particular circumstances. Clearly, if there is a risk of substantial exposure to harmful substances before they have been diluted, appropriate personal protective equipment should be provided and worn.

As in any branch of preventive medicine, recognition and prevention are more effective than treatment. The following procedure should be followed:

(a) Pre-employment examination to detect workers—atopic or otherwise—who may be susceptible to cutaneous or systemic reactions to cosmetics. Preliminary patch tests for primary irritants may be used as screening procedure. If a person proves to be allergic to any of the substances he has to handle, he should be removed from exposure or, if this is not practicable, he should wear suitable protective clothing even when handling diluted ingredients.

(b) Since some of the substances are harmful in the form of dust or vapour, proper cleanliness of the area, hoods and exhaust ventilation should be provided to prevent them from escaping into the air of the workroom.

(c) Cleanliness of the workers and the use of properly fitting masks and protective clothing should be ensured.

(d) Proper covers should be provided for machines, chemicals, glass containers, etc.

(e) Workers should be fully informed of health hazards and examined frequently (at least once a year) by a physician versed in skin test or a dermatologist.

Fire and explosion. Where flammable liquids are used, precautions should be adopted to exclude sources of ignition, and the liquids should be manipulated in such a manner as to avoid dangerous spillage. Suitable carriers and hand trucks should be used when containers of these liquids are moved from the store to the process vessels. If the containers are heavy drums, the trucks should be fitted with hand pumps for the safe transfer of the liquid to the process vessels. Larger quantities of flammable liquid should be fed from storage tanks to process vessels in a safe manner by suitable pumps through pipelines.

Where flammable alcohols and esters are involved, carbon dioxide and dry powder are suitable extinguishers for fire fighting.

SHARMA, O. P.

CIS 76-1147 "Cosmetics, the consumer, the factory worker and the occupational physician". Malten, K. E. *Tijdschrift voor sociale geneeskunde* (Amstelveen, Netherlands), 18 Feb. 1976, 54/4 (121-127). Illus. 41 ref.

CIS 78-475 "Pigmented cosmetic dermatitis and coal tar dyes". Sugai, T.; Takahashi, Y.; Takagi, T. *Contact Dermatitis* (Copenhagen), Oct. 1977, 3/5 (249-256). Illus. 5 ref.

Cotton cultivation

Cotton is a natural fibre obtained from the plant *Gossypium* (figure 1). Its low production cost makes it the most economical natural fibre. Chemically, the fibre

Figure 1. Cotton plant *(Gossypium).*

is about 90% cellulose and 6% moisture; the remainder being impurities. Cotton is also used as a raw material for the production of artificial fibres.

Geographical occurrence. India was the original home of the cotton plant and it was cultivated there as early as 2000-3000 BC. From India it was introduced to China and Korea, and it is now grown widely in Western and Central Asia, the USSR, Turkey, Iran, Arabia, Asia Minor and Southern Europe (Greece and Bulgaria). Egyptian cotton, of New World origin, was introduced to the Nile Delta in lower Egypt about 1820. Cotton has been grown in Brazil and Peru since antiquity, and for a long time in the United States, Mexico and Central America.

In 1978 world production of cotton lint was 12.951 million tonnes, having risen from 10.931 million tonnes in 1961 to a peak of 13.903 in 1974. The major producers in 1978 were the USSR (with 2.64 million tonnes), the United States (2.36), the People's Republic of China (2.1) and India (1.25).

Cultivation cycle. The planting season depends on weather conditions, which vary enormously from country to country and from year to year; in the northern hemisphere planting is done in spring.

Figure 2. Cotton pickers at work.

Cotton is a subtropical plant requiring a long growing season. It can stand more drought than most other commercial crops. It is grown as an annual and as a perennial. Prior to sowing, the ground is thoroughly ploughed and pulverised and then the seed is either broadcast and then covered with the harrow or sown in rows either by hand or by means of a horse planter or tractor. Subsequent operations include tillage, weed control, thinning, topping, defloration and defoliation. When the plants are high, they are weeded and thinned. The plant begins to bloom after 4 months. Creamy-white flowers appear first, change to a reddish-purple in about 2 days and then fall off leaving seed pods that grow to full size.

Numerous species of insects and mites attack the cotton plant, and diseases due to fungi, nematodes, bacteria and viruses are prevalent in cotton in all regions where it is grown. Calcium arsenate, BHC, toxaphene, eldrin, chlordane, dieldrin, heptachlor, DDT and TEPP are some of the pesticides used on cotton plantations.

When the plant has grown to its full height of 0.5-0.7 m, the boll bursts and the fleecy white fibre is exposed; the cotton is now ready for picking (figure 2). However, all the bolls do not burst simultaneously and a cotton field must be gone over several times in a season. Harvested cotton contains seeds, leaf fragments and dirt. In India picking is entirely by hand; however, in the United States and the USSR, mechanical picking, using stripping machines and spindle-type pickers, is practised.

Processing. The lint or fibre is separated from the seed in a ginning machine, either of the roller or saw type. The roller gin separates the lint from the seed by the friction of a leather roller set at a small distance from a knife-edge bar or "doctor blade". The saw gin separates the lint from the seeds by means of saws mounted on a circular shaft which pull the lint whilst the seed is restrained by narrow grids. After ginning, the fibre is blown through ducts into a hydraulic press in which it is compressed into bales.

HAZARDS AND THEIR PREVENTION

The cotton planter, especially in the developing countries, is exposed to inclement weather, fatigue, infection and nutritional deficiencies. He lacks community services like piped water supply, sewage, garbage and waste disposal; water runs the risk of contamination from outdoor latrines, cesspools, septic tanks, animal pans, stables or fodder; infected water, milk or dairy products and food grown on the fields may cause infections of the digestive system. Purulent ophthalmias, trachoma, septic infections of skin, insect-borne diseases, ancylostomiasis, tuberculosis and communicable diseases like cholera, and diphtheria are common. There are dangers due to association with plants and farm animals (dermatitis, glanders, anthrax, trauma from animals, snakebite, rabies, leptospirosis, tetanus, etc.), and hazards from use of insecticides and other pesticides.

Health protection and medical care should be closely linked with the health services under national health programmes for rural populations, especially in developing countries where cotton is grown. Installation of adequate sanitary facilities is essential to reduce the risk of parasitic diseases such as ancylostomiasis, and cotton plantation workers should have adequate housing with an uncontaminated water supply. Workers using pesticides should be informed of the dangers and provided with suitable protective clothing.

The increasing use of agricultural machinery (including tractors) and implements has increased the incidence of accidents. Ginning machines and baling presses present special hazards. In the roller gin, the operator's hand may be caught between the knife and roller. Saw gins are even more hazardous since it is virtually impossible for a worker to pull his hand away once it has come into contact with the saw. Machinery guarding of the fixed interlock type should be fitted to all gins.

Accidents on baling presses may result in limb amputations or even fatal crushing injuries. These can be prevented by an automatic interlock on the baler so that the upper gates cannot operate while the gates are open. The handling of heavy bales is the cause of numerous injuries such as sprains, strains and dislocations, primarily involving the workers' backs. Wherever possible, mechanical handling should be employed.

Fire hazards in the ginning and baling of cotton are very high; adequate fire protection and prevention measures should be applied, especially where gas-fired or oil-fired heaters are used for cotton drying.

GUPTA, M. N.

Farming:

CIS 76-2000 "Occupational safety and health standards for agriculture–Guarding of farm field equipment, farmstead equipment and cotton gins". Code of Federal Regulations, Title 29, Chapter 17, Part 1928: Occupational safety and health standards for agriculture. *Federal Register* (Washington, DC), 9 Mar. 1976, 41/47 (10190-10197).

"Current problems of occupational hygiene for cotton growers" (Aktual'nye voprosy gigieny truda xlopkorobov). Alekperov, I. I. *Gigiena i sanitaria* (Moscow), Oct. 1980, 10 (38-40). 7 ref. (In Russian)

Ginning:

CIS 78-674 "Byssinosis and chronic respiratory disease in U.S. cotton gins". Palmer, A.; Finnegan, W.; Herwitt, P.; Waxweiler, R.; Jones, J. *Journal of Occupational Medicine* (Chicago), Feb. 1978, 20/2 (96-102). 10 ref.

Cottonseed mills:

CIS 77-2058 "Respiratory health and dust levels in cottonseed mills". Jones, R. N.; Carr, J.; Glindmeyer, H.; Diem, J.; Weill, H. *Thorax* (London), June 1977, 32/3 (281-286). 14 ref.

Cotton industry

TWA OSHA	200 µg/m³ in textile yarn manufacturing 750 µg/m³ in textile slashing and weaving 500 µg/m³ in other operations excluding cotton harvesting and ginning, handling or processing of woven or knitted materials or washed cotton
STEL ACGIH	0.6 mg/m³
IDLH	500 mg/m³
MAC USSR	cotton dust (cotton gins) containing more than 10% free silica 2 mg/m³; cotton dust (cotton gins) containing up to 10% free silica 4 mg/m³

The principal cotton textile districts of the world are in Great Britain, the United States, the Federal Republic of Germany, the USSR, France, Italy, Czechoslovakia, Belgium, and the Netherlands. More recently other important centres have developed in Japan, India, the People's Republic of China and Brazil, and the present trend is towards the setting up of textile mills in the countries where the cotton is grown.

Economic importance. Since raw cotton is light and easily compressed into bales, it can be shipped for long distances, and it does not readily spoil. Hence, since the end of the 18th century, it has been an important article of world commerce. The textile mills established during the 19th century in Lancashire and in New England initiated the Industrial Revolution, and their cotton fabrics for many years dominated world markets. More recently this dominance has been increasingly challenged, both by the development of manufacture in other countries and by man-made fibres. Synthetic fibres such as the polyamides and polyesters are often now used in conjunction with cotton for many purposes.

The major producers of cotton yarn (pure and mixed) in 1977 were the USSR (1 597 300 tonnes), the United States (1 114 900), India (1 034 200), Japan (498 300), Pakistan and France. The major producers of woven pure and mixed cotton fabrics in the same year were the USSR (7 461 000 tonnes), India (6 902 000), the United States (3 779 000), Japan and Poland.

Ginning. When picked, the cotton is fed into a machine called a gin to separate the fibres from the seeds. See COTTON CULTIVATION.

Baling. The ginned cotton is compressed and packed into bales, and it is in this form that it reaches the spinning mill. Here the bales are opened, and the cotton is fed into a hopper, where it may be mixed with cotton from other bales. A spiked lattice carries it upwards under a series of rolls; thorough mixing of the fibres occurs and it falls on to a conveyor in a more open condition.

Opening and scutching. These are essentially similar processes, wherein the mixed cotton is further cleaned and opened by the combined action of revolving beaters and air currents against a grid through which the dirt passes. The cotton emerges from the scutcher in a much bulkier state, and passes on to the carding machine.

Carding. This is designed to make the cotton fibres parallel to each other, to break up any remaining hard tufts and remove very short and thin fibres, and to complete the removal of impurities. The cotton passes through a series of toothed and spiked cylindrical rollers, which first attenuate it into a thin film, then comb it, and finally bring the fibres together in a rope-like form known as a sliver. With fine cotton (long staple) there may be a further combing process on a combing machine.

Spinning. The principle of spinning is to elongate and attenuate the slivers by drawing several of them at a time through three or four pairs of suitably spaced rollers, each pair revolving at a higher speed than the preceding pair. A series of these machines is used, with or without refinements depending on the quality and fineness of the yarn it is intended to produce. The end result is a yarn from which short and immature fibres have been removed, together with remaining impurities, and the constituent fibres of which have first been rendered parallel and then had a slight twist imparted to them, sufficient to give the yarn strength for further processing without breaking. At this stage it is the thickness of coarse string.

The final step is further attenuation of the yarn, accompanied by simultaneous twisting to prevent beaking when it is subjected to tension. This may be accomplished either by a machine called a ring frame, or by mules. Mules are now almost entirely obsolete in the cotton industry; the ring frame has a much higher output, and occupies less floor space.

Over the past ten years a new and more productive spinning method has gradually been introduced. This system, known as open-end spinning, is most suitable for spinning coarse yarns, and can replace not only the ring frame but some other initial processes as well.

Weaving. The yarn is now ready for conversion into cloth. All woven fabrics consist of two sets of threads—the warp, which extends throughout the length of the fabric, and the weft, which runs transversely at right angles to it (figure 1).

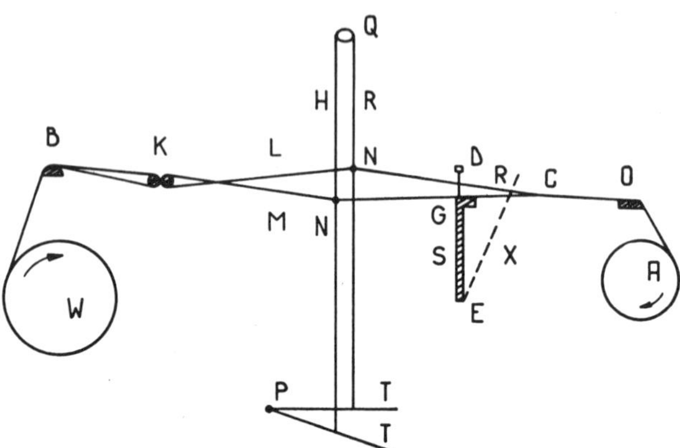

Figure 1. Essential parts of an ordinary loom. W is the warper's beam on which are wound the threads which will form the warp of the fabric. Two of these threads represented by L and M pass over the back bearer B. They diverge from the lease rods K to pass through the two sets of harness H (with a separate heddle eye N for each thread) until all the threads unite again in the fabric at the point C where the last weft thread was inserted. The woven fabric is regularly wound on A. D is the reed or comb through which all the warp threads pass. The two sets of harness are attached to treadles T which are pivoted at P so that one set of warp threads may be raised above the other to form a "shed" at R. At this stage a weft thread is inserted in the shed R by a shuttle impelled along the sley G. The sley sword S swings forward about point E (to position X) and the reed D beats up the weft thread close to the preceding one. The sley sword then returns to its original position whilst the harnesses change their relative positions so that the shuttle can return through a different shed. *(From Hall, A. J., 1965.)*

Warp threads are prepared by winding yarn of the appropriate length on to a large roller (beam) of the same width as the fabric to be made. To prevent breaking during weaving, they are strengthened by coating them with a paste or size, usually made of starch. The weft threads are wound on small tubes called cops (pirns). The beam of warp threads is placed in position on the loom and the ends of the warp drawn through and attached to the roller on which the finished cloth is wound. The weft threads are attached to a shuttle which carries them transversely under and over alternate warp threads as the latter slowly move through the loom. This produces the simplest kind of plain cloth. More elaborate types of woven fabric are made by altering the arrangement of warp and weft and by the use of coloured yarns.

In recent years a number of alternatives to the shuttle for weft insertion have been developed—rapier, water jet, air jet and ripple or wave shedding.

A feature of modern textile technology is the trend towards increasing the number of machines attended by each worker; whereas one worker may at one time have serviced one to four non-automatic looms, it is now likely that he will be responsible for 40-60 automatic looms. This produces two effects: firstly, the worker is subjected to greater disturbance and suffers increased nervous strain through fear of multiple yarn breakages, and secondly he must walk longer distances to locate and mend these faults.

Machine spacing. In most of the textile factories erected more than 30 years ago, the maximum amount of machinery was installed leaving very narrow alleyways and gangways in which the operators have to work. It is necessary to transport cans or skips of bobbins, weaver's beams and bundles of finished cloth either manually or on transporter trucks through these narrow alleyways and gangways; this increases the risk of accidents and makes the work more difficult. In modern textile mills the layout of the machinery is much better, with ample working areas. Many countries have now adopted rules specifying the minimum distances between certain textile machines; for example in the United Kingdom, the minimum distance between cotton weaving looms is 54 cm for the weaver's alley and 2 m for the main alley. In Belgium, France and Japan, the minimum space permitted between machines is 80 cm.

Finishing. When cotton fabrics leave the loom, a further series of operations is undertaken to "finish" the cloth. These vary according to the purpose to which it is to be put, but usually include—

(a) singeing, to remove projecting hairs;

(b) desizing, to remove starch added to the yarn. Steeping in solutions of enzymes, such as malt diastase, is the commonest method;

(c) scouring, which removes fats and waxes. Boiling with hydroxide solution saponifies most of these, and the sodium soaps so formed emulsify the remainder.

The scoured fabric may then be bleached, dyed or printed. A number of bleaching agents are available: hydrogen peroxide is the most widely used, but chlorine derived from sodium chlorite or hypochlorite is also used. The range of dyes is very wide indeed, and is continually growing.

After dyeing, the cloth is dried, smoothed and pressed, and may be submitted to a number of other treatments designed to improve its appearance or wearing qualities or make it waterproof, flame repellent or rotproof. Synthetic resins are among the substances increasingly used for these purposes.

Waste cotton. Waste from raw-cotton factories is often processed separately to provide a lower grade of cotton textile. The preliminary processes are somewhat different: waste spun cotton (hard waste) requires reopening and breaking of the spun fibres before blending with short fibres discarded during preparation (soft waste). Hard and soft waste are blended and thereafter carded on roller and cleaner cards (as distinct from the flat cards used on raw cotton).

HAZARDS AND THEIR PREVENTION

Accidents may occur on all types of cotton textile machinery: the frequency rate is not high but many injuries tend to be serious: for example the rollers of the roller and cleaner card may cause severe mutilation. Efficient machinery guarding of the multiplicity of moving parts presents many problems and needs constant attention; training of operatives in safe practices is also essential, in particular to prevent "picking" or cleaning of machinery in motion, which are common causes of accident.

Byssinosis is a lung disease which may occur in individuals after prolonged exposure to high atmospheric concentrations of cotton dust. Prevention is by dust suppression and extraction. No fully effective method has yet been devised. The hazard is greatest in the handling of raw cotton up to the carding stage (figure 2),

Figure 2. Dust extraction system in a carding machine.

but it has been shown that the disease can occur in workers whose only contact with cotton has been in winding of yarn by high-speed machinery and that waste cotton workers are not immune. See BYSSINOSIS.

Noise is a problem, especially in weaving sheds, where levels over 100 dB are regularly recorded. The long-term solution lies in the design of looms which are quieter than the traditional one. Glass-wool ear plugs afford a measure of hearing protection.

Spinning and weaving processes require high temperatures and artificial humidification of the air; careful attention is always necessary to see that permissible limits (sometimes fixed by law) are not exceeded. Well designed and maintained air-conditioning plants are increasingly used in place of more primitive methods of regulation.

Mule spinner's cancer, an epithelioma of the scrotum due to contamination of clothing by mineral lubricants used on the spindles, is now of only historical interest.

Table 1. Chemical substances commonly used in the processing of cotton

Substance	Application
Sodium hydroxide and sulphuric acid	Scouring and mercerising
Sodium chlorite and hypochlorite, hydrogen peroxide	Bleaching
Synthetic resins, some of which are polymers of ethylene containing vinyl and acrylic groups; others are urea-formaldehyde, melamine formaldehyde and ethylene urea formaldehyde	Finishing (to impart crease-resisting similar qualities)
Detergents and wetting agents	Preparation for dyeing
Dieldrin, sulphonamide, chloro-2,2-chloromethyl diphenyl ether, halogenated diphenyl urea derivatives	Moth-proofing
Pentachlorophenyl laurate, copper naphthenate, salicyl-anilide (Shirlan A), dichlorophen	Rot-proofing
Edetic acid (EDTA)	Dyeing (as a metal-sequestering agent to prevent combination between the dye and metallic impurities in the water)

A number of the substances used in the bleaching, dyeing and finishing of cotton cloth may cause skin diseases and the usual preventive measures are therefore necessary. A number of carcinogenic substances have been used in dyeing.

Table 1 indicates some of the substances more commonly used in the processing of cotton and their application.

TYRER, F. H.

CIS 76-656 *Cotton dust–Proceedings of a topical symposium, November 12 and 13, 1974, Atlanta, Georgia* (American Conference of Governmental Industrial Hygienists, PO Box 1937, Cincinnati, Ohio) (1975), 440 p. Illus. Bibl.

CIS 76-1561 "Chemical properties of cotton dust". Wakelyn, P. J.; Greenblatt, G. A.; Brown, D. F.; Tripp, V. W. *American Industrial Hygiene Association Journal* (Akron, Ohio), Jan. 1976, 37/1 (22-31). 77 ref.

"Bacterial contamination of cotton as an indicator of respiratory effects among card room workers". Rylander, R.; Imbus, H. R.; Suh, M. W. *British Journal of Industrial Medicine* (London), Nov. 1979, 36/4 (299-304). 13 ref.

"Exposure to cotton dust and respiratory disease. Textile workers, brown lung and lung cancer". Heyden, S.; Pratt, P. *Journal of the American Medical Association* (Chicago), 17 Oct. 1980, 244/16 (1797-1798). 11 ref.

CIS 75-1151 *Towards a healthy working environment–First report of the Joint Standing Committee on Health and Welfare in the Cotton and Allied Fibres Industry, 3rd April 1973.* Department of Employment (London, HM Stationery Office, 1974), 35 p. Illus. 11 ref.

Opening processes: cotton and allied fibres. Health and Safety Executive. Health and Safety series booklet HS(G) (London, HM Stationery Office, 1980), 27 p. Illus.

CIS 76-246 *Safety in the cotton and allied fibres industry: Spinning, winding and sizing.* Health and safety at work series, No. 49C, Health and Safety Executive (HM Stationery Office, 1975), 56 p. Illus.

CIS 75-1753 *Safety recommendations–Joint standing committee on safety in the cotton and allied fibres weaving industry.* Department of Employment (London, HM Stationery Office, 1974), 20 p.

Recurrent information:

BTEA statistical reviews (Manchester, British Textile Employers Association).

Coumarins and derivatives of indandione

Coumarin ($C_9H_6O_2$)

1,2-BENZOPYRONE; COUMARINIC LACTONE

m.w. 146.1
sp.gr. 0.93
m.p. 71 °C
b.p. 301.7 °C

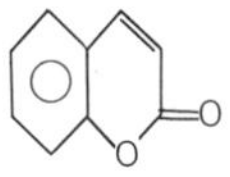

slightly soluble in water;
soluble in ethyl ether, ethyl alcohol, chloroform and volatile oils

colourless crystals, flakes or powder with a fragrant odour similar to that of vanilla, and a burning taste.

Coumarin was first extracted from fermented clover and hay, and fine grades are still isolated from tonka beans. It is made synthetically by heating salicylic aldehyde, sodium acetate and acetic acid anydride. It is used as a deodorising and odour-enhancing agent in perfumes, soaps, tobacco, inks, rubber and other products where an aromatic smell is desirable. It is also used in pharmaceutical preparations.

Of the same family as heparin, coumarin derivatives are synthetic anticoagulants, and, like the indandione derivatives, much more easy to use.

Dicoumarol ($C_{19}H_{12}O_6$)

3,3-METHYLENE-*bis*-4-HYDROXYCOUMARIN; BISHYDROXYCOUMARIN

m.w. 336.3
m.p. 287-292 °C

soluble in alkalis; slightly soluble in chloroform; insoluble in water

creamy-white crystalline powder with a faint, pleasant odour and a bitter taste.

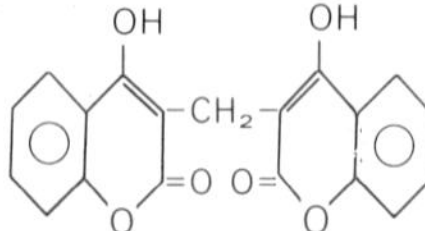

This derivative is now synthesised from methyl acetylsalicylate, sodium and formaldehyde.

Coumafene ($C_{19}H_{16}O_4$)

3-(α-ACETONYLBENZYL)-4-HYDROXYCOUMARIN; WARFARIN

m.w. 308
m.p. 161 °C
b.p. decomposes

soluble in ethyl alcohol and ethyl ether; insoluble in water

colourless, tasteless, odourless crystals.

TWA OSHA 0.1 mg/m³
STEL ACGIH 0.3 mg/m³
IDLH 200 mg/m³

This rodenticide is made by the condensation of benzylideneacetone and 4-hydroxycoumarin.

Coumachlor ($C_{19}H_{15}O_4Cl$)

3-(α-ACETONYL-*p*-CHLOROBENZYL)-4-HYDROXYCOUMARIN

m.w. 342.6
m.p. 164-165 °C

crystalline substance.

Coumachlor is prepared by the reaction of *p*-chlorobenzalacetone and 3-carbethoxy-4-hydroxycoumarin in the presence of triethylamine and water.

Coumafuryl ($C_{17}H_{14}O_5$)

3-(α-ACETONYLFURFURYL)-4-HYDROXYCOUMARIN; FUMARIN

m.w. 298.3
m.p. 124 °C

crystalline substance.

Coumatetralyl ($C_{19}H_{16}O_3$)

4-HYDROXY-3-(1,2,3,4-TETRAHYDRO-1-NAPHTHYL)COUMARIN

m.w. 294.2

Pival ($C_{14}H_{14}O_3$)

2-PIVALYL-1,3-INDANDIONE; PINDONE; BUTYLVALONE

m.p. 109 °C

insoluble in water, soluble in most organic solvents

bright-yellow powder.

TWA OSHA 0.1 mg/m³
STEL ACGIH 0.3 mg/m³
IDLH 200 mg/m³

It is used as a rodenticide, insecticide and pharmaceutical intermediate.

NID ($C_{19}H_{11}O_2$)

1-2-NAPHTHYL-1,3-INDANDIONE; RADIONE; NIDANE-ALPHA

m.w. 271.2

Diphenadione ($C_{23}H_{16}O_3$)

2-DIPHENYLACETYL-1,3-INDANDIONE; DIPHACIN

m.w. 340.4
m.p. 146-147 °C

practically insoluble in water; soluble in acetone, acetic acid and ethyl ether; sodium salt sparingly soluble in water

pale-yellow crystals.

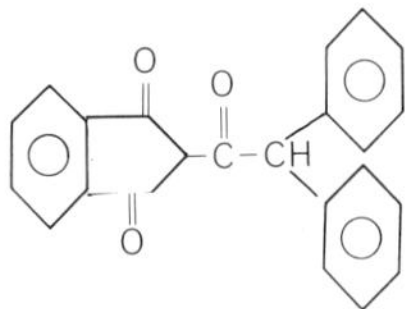

This compound is produced by the condensation of dimethyl phthalate with diphenyl acetone in an inert solvent in the presence of sodium methoxide.

Chlorophacinone ($C_{23}H_{15}O_3Cl$)

2-(α-*p*-CHLOROPHENYL-α-PHENYLACETYL)INDAN-1,3-DIONE

m.w. 374.7

Uses. In agriculture, the above derivatives of coumarin and indan-1,3-dione have superseded most of the classic rodenticides. Effective rat-killers, odourless and flavourless, they cause spontaneous haemorrhages in the rodent which ingests them, causing death from asphyxia. They are effective against brown rat, black rat, mouse, fieldmouse, field vole, and hamster. They are used in the form of ready-made baits (pellets), oily solutions, powder, tablets, or as granules for the preparation of baits or for distribution in the holes and runs frequented by the rodents. They are also used in pulverised form or as fumigants for moles.

HAZARDS

Under the general name of antivitamine K, these derivatives of coumarin and of indan-1,3-dione block the hepatic synthesis of prothrombin at the expense of vitamin K. Some of them also interfere with the synthesis of proconvertin and of factors IX and X without changing the proaccelerin. However, they are not absorbed by the lungs, and they do not penetrate the skin. They present no danger to man under ordinary conditions of use, but they can produce haemorrhagic accidents if they are absorbed in large or repeated quantities, or by persons with a natural or acquired sensitivity. There is a possibility of occupational poisoning in these latter cases through lack of personal hygiene or through food contamination.

In rare cases, kidney, liver, or lymphatic lesions have been observed.

[Many coumarins are mutagenic and high doses of coumarin by mouth produce bile duct carcinomas in rats. No data exist on carcinogenic effects in humans. On the other hand, coumarins antagonise tumour induction by carcinogens and possess potent antimetastatic properties.]

Coumarin is liable to act as a slight allergen in contact with the skin or mucous membrane. The harmful effects from ingestion or inhalation are slight and disappear after the end of exposure. Gross exposure must be avoided by the use, in special circumstances, of protective gloves and respiratory protective equipment.

SAFETY AND HEALTH MEASURES

Care should be taken:

(*a*) not to contaminate foodstuffs intended for animals or men;

(*b*) not to leave the material within reach of children;

(*c*) to use prepared baits rather than scatter the poison;

(*d*) to collect and destroy dead rodents;

(*e*) to bury the baits and the powder when the operation is completed.

In emergencies during manufacture or use when there may be a danger of heavy exposure, suitable respiratory protective equipment should be worn to prevent dust from entering the mouth. Personal hygiene should be practised to ensure that food eaten by people handling the rodenticides does not become contaminated.

Treatment. In case of poisoning by these rodenticides, vitamin K should be administered in high dosage by the intravenous or intramuscular route.

VALLET, G.

"Genetic effects of coumarins". Grigg, G. W. *Mutation Research* (Amsterdam), 1977/78, 47/3-4 (161-181).

Coumarin. Vol. 10 (113-119). 17 ref. *IARC monographs on the evaluation of carcinogenic risk of chemicals to man* (Lyons, International Agency for Research on Cancer, 1976).

Council for Mutual Economic Assistance (CMEA)

The Council for Mutual Economic Assistance—which first met in 1949—is the international organisation for multilateral economic co-operation among socialist countries having similar political and socioeconomic systems and sharing common development aims. Its members at present include ten socialist countries: the People's Republic of Bulgaria, the Hungarian People's Republic, the People's Socialist Republic of Vietnam, the German Democratic Republic, the Republic of Cuba, the Mongolian People's Republic, the Polish People's Republic, the Socialist Republic of Romania, the Union of Soviet Socialist Republics, and the Czechoslovak Socialist Republic.

According to its Statutes, the aim of the CMEA, through the co-ordination of efforts of its member countries, is to assist in:

(*a*) the promotion of mutual co-operation and development of socialist economic integration;

(*b*) the planned development of national economies;

(*c*) the acceleration of economic and technological development;

(*d*) the improvement of the level of industrialisation in the industrially underdeveloped countries;

(*e*) the permanent improvement of the efficiency of labour;

(*f*) the gradual improvement and levelling of economic development; and

(*g*) the continuing growth of wealth in the Council member countries.

The CMEA was established on the basis of sovereign equality of all the member countries, and the co-operation of the member countries is carried into effect in line with the principles of socialist internationalism, respect for national sovereignty, independence, non-interference in domestic affairs, full equality, mutual profit, and friendly assistance.

The Council can invite non-member countries to participate in its activities or co-operate in other ways, on the basis of agreements made with these countries. Such agreements have been signed with the Governments of the Socialist Federal Republic of Yugoslavia, the Republic of Finland, the Republic of Iraq, and the United Mexican States. The representatives of the Lao People's Democratic Republic, the People's Republic of Angola, the Democratic People's Republic of Korea and Ethiopia participate as observers in the activities of various CMEA organs.

The CMEA member countries have equality of representation in the organs of the Council without regard to their population, economic potential, or contribution to the budget of the organisation. Each country defines the scope and extent of its involvement in the activities of the CMEA at its own discretion. The CMEA has the right to make international agreements both with its member countries and with other countries or international organisations.

Structure

The chief organ of the CMEA is the Session of the Council consisting of representatives of all the member countries usually under the leadership of Prime Ministers. The Council Session has full powers to discuss any problems within the range of CMEA activities. The main executive organ of the CMEA is the Executive Committee, in which all the member countries are represented at the level of Deputy Ministers. The Executive Committee controls all activities related to the accomplishment of tasks set forth by the Council Session and, among other things, supervises the fulfilment by the member countries of any obligations resulting from the recommendations of CMEA organs which they have accepted.

There are three committees within the CMEA system for co-operation in the field of planning; for scientific and technological co-operation; and for the provision of material and technology.

These committees are supplemented by:

(*a*) standing committees, established to aid further development of economic relations between the member countries and to organise their multilateral co-operation in specific industrial fields, e.g. building and construction, chemical industries, steel industry, standardisation, etc.;

(*b*) conferences of the representatives (chiefs) of competent bodies from the member countries, e.g. the Conference of Ministers of Labour of CMEA member countries.

The CMEA Secretariat is the administrative and executive body and it acts in accordance with the Statutes of the Council.

Conference of Ministers of Labour

This Conference started its activities in 1969 and became a regular CMEA organ in 1974. It organises multilateral co-operation in specific problem areas such as:

(a) efficiency, organisation, standardisation and working time;

(b) advanced occupational training;

(c) working conditions;

(d) incentives and wages system;

(e) labour legislation and social services.

The Conference arranges and participates in joint consultations and exchanges of opinion and information among the Ministries of Labour of the member countries as well as in the preparation of multilateral agreements (conventions). It has also set up working groups in particular fields, i.e. wages systems and incentives; scientific organisation of work; labour law standards.

A working group on working conditions was set up in April 1972, based on the following terms of reference: "The creation of labour conditions to suit the psychophysical requirements of man and the progress of science and technology, and the continuing improvement of these conditions."

In socialist countries the improvement of occupational safety and health is also covered by the obligatory system of economic planning and forms part of both the long-term and the annual socioeconomic programmes.

The Conference of Ministers of Labour stressed the importance of safe and healthy labour conditions for all the CMEA member countries, including co-operation in the manufacture of personal protective equipment and common safety and health standards pertaining to the import and export of machinery and other equipment.

The task of the working group for working conditions includes the exchange of opinions and information on working conditions and work organisation and technology as regards safety, the evaluation of accidents at work, labour incentives for work in arduous conditions, compensation for occupational injuries, workers' education in occupational safety and health, and advanced training for specialists dealing with problems of occupational safety and health. The working group holds periodical meetings to discuss agreed subjects on the basis of joint reports on the situation in different countries. Draft joint reports are prepared by the co-ordinator of the working group (Poland) on the basis of agreed proposals concerning issues raised in the reports from individual countries. The final version of the report together with conclusions is presented to the Conference of Ministers of Labour for acceptance.

The following subjects have been considered by the working group for working conditions:

1. "Activities aimed at improving labour and social conditions of workers" (1973). The report contained general information on the systems of safety and health in individual countries, on appropriate legislative principles, on measures taken to prevent working accidents and occupational diseases, on both employers' and employees' obligations with a view to ensuring safe and healthy working conditions, and on systems of incentives in connection with specific labour conditions.

2. "Experience of CMEA member countries in the field of women's labour protection and welfare at work" (1974). The report included a detailed status report on the legislative aspects of female labour protection, special rights in the case of maternity, prohibited jobs for women, requirements for sanitary arrangements for women, social help to working women with children.

3. "Provision of personal protective equipment and working clothing" (1975). Details of the manufacture and distribution of protective and working clothing at factory level as well as the rights of the employee and the obligations of the employer in this respect were presented in the report. Proposals were submitted for standardisation and the co-operation of CMEA countries in the manufacture of such clothing and equipment.

4. "Methods and forms of labour safety training, with special stress on training young employees and advanced training of specialists in occupational safety and health in CMEA member countries" (1976). The report discussed current systems of training on occupational safety and health and ergonomics in trade schools, secondary technical schools and high schools. It also included methods of training within industry, which is obligatory for all personnel, including workers, executive staff and management (this obligation ensues from the Labour Code), systems of advanced training for specialists in the field of occupational safety and health, higher management units and design and planning offices, and vocational training for labour inspectors and social labour inspectors.

5. "Planning of projects aimed at the improvement of labour conditions and the reduction of accidents and occupational diseases in CMEA member countries" (1979). This report detailed current methods for the improvement of labour conditions at various levels of the economy, the co-ordination of planning, the competence of planning bodies and Ministers of Labour in this respect, methods of designing equipment to ensure the balance between production and improved working conditions, and the dissemination of information on the improvement of working conditions and on incentive payments in the case of particular working conditions.

The studies of the working group serve as a starting point for a more thorough exchange of opinions and ideas, leading to the joint bilateral implementation of projects on the basis of agreements between the ministries, institutes or other bodies concerned. The findings of the working groups serve as models to enable the best solutions to be used when necessary and possible by individual countries. For instance some countries adopted certain solutions pertaining to regulations forbidding women to perform certain jobs (e.g. unit standards for carried loads, plans for improving working conditions for women, and others).

The working group determined the principles and participation of CMEA member countries in international conferences organised in Poland (1975, 1978, 1982) entitled "Protection of man in his working environment" (SECURA) and consisting of an international symposium and an international exhibition. The representatives of national institutes for labour protection, concerned with multilateral and bilateral co-operation in the field, took part in meetings of the working group on working conditions. The forms of such co-operation cover: joint scientific and research projects, exchange of critical analyses and publications, joint seminars, and the exchange of specialists in various fields in the form of consultations and scholarships. To this end long-term agreements were concluded by the institutes of labour protection.

Scientific co-operation among CMEA countries concerning occupational hazards and worker protection covers the following problems:

(1) Identification and evaluation of occupational hazards, particularly: methods of analysis and evaluation of occupational safety and health status; application of computer systems for processing data concerning safety hazards to man at work; and ergonomic aspects of machinery and equipment.

(2) Methods and instruments used for determining the chemical and physical characteristics of the working environment.

(3) Technological solutions for eliminating or limiting basic hazards at workplaces. These cover the following

problems: engineering solutions to means and methods of transportation, and guiding principles for safety checking of international transport; optimisation of factors such as microclimate, noise level, noxious chemicals, lighting, etc.

(4) Ideas concerning appropriate protective clothing and equipment for various working conditions.

(5) Principles of personnel selection and training for various jobs, namely: training and instruction of safe working methods; training in occupational safety and health; employment of handicapped persons.

Standing Committee for Co-operation in the Field of Health Protection

Individual countries are represented in this committee by permanent delegations led by the Ministers of Health. The committee was established in 1978. Its aim is to encourage new ideas in the field of labour protection, medical sciences, technology and pharmacy. Twelve significant scientific issues have been identified, of which one is labour hygiene and occupational diseases and relates to the following specific problems:

(a) study of the impact of physical environmental factors on workers: control methods and determination of safety standards;

(b) study of the effects on workers of chemical substances in the environment: control methods and determination of safety standards;

(c) standardisation of diagnostics and therapy for occupational diseases;

(d) investigation on work capacity and fatigue: physiological rationalisation of work processes;

(e) labour hygiene in various branches of the national economy.

CMEA member countries co-operate on the standardisation of diagnostic and therapeutic methods for occupational diseases, focusing on the diagnosis of early stages of anthracosilicosis and evaluation of fitness for work, with a view to establishing uniform criteria and methods for quick diagnosis. Diagnostic methods have been prepared for initial stages of bagassosis and lung diseases of workers exposed to sugar cane dust. Similar studies relate to the effects of polycyclic hydrocarbons, organophosphorus pesticides, fluorine and fluorides, toluene, lead and vinyl chloride on living organisms. A number of documents pertaining to toxicology have been issued and standardised; on terms and definitions, classification of toxicity and hazards, toxicology of chemical compounds used in production and social services, methods for testing toxicity under various conditions of work.

Studies have been carried out on problems and strategies for sampling air and biological materials and on methods for isolating organic and inorganic toxic substances from the biological environment. Other studies have covered such subjects as the following: the effects of noise and vibrations on the human organism and methods for the rapid diagnosis of hearing impairment; working ability; job placement; and working efficiency. Recommendations have been prepared for the diagnosis and therapy of diseases resulting from vibrations as well as those caused by the combined effects of noise and vibration and other environmental factors. Studies on occupational pathology have related to different hazards to which workers may be exposed. In the field of labour physiology research has been carried out into fatigue and overwork and its preventive treatment. Draft instructions have been prepared dealing with optimum labour conditions for operators of remote-control machines. Analyses are made of the effect of static work on various physical characteristics, and on the effect of prolonged physical work upon mental states and processes. The analysis of biochemical signs of fatigue and the subjective evaluation of workload are of practical interest. Further research into the mental and emotional effort and motivation for work among workers under increased psychic stress have helped to determine the degree of psychic tension and monotony, with a view to the enrichment of the quality of work.

Ergonomic requirements for a workplace have been defined and investigations made into the impact of automation on the psychophysiological reactions and fatigue of control desk operators, and an optimum arrangement for signal and control devices has been proposed. Recommendations for the implementation of projects to improve working conditions have been made. Certain types of tractors and agricultural implements have been subjected to ergonomic evaluation, and studies made on the working environment and the sick rate and temporary disability rate among hothouse workers. Studies on the impact of manufacturing and other occupational factors on the health and specific functions of women, unborn children and infants enabled procedural recommendations to be made in this field.

Standing Committee for the Chemical Industry

The Scientific and Technological Council operated from 1973 to 1978 in the field of the chemical, cellulose and paper industries. Six meetings were held in various CMEA countries. This council controls and co-ordinates the preparation of technological requirements, standards and regulations, and scientific research on labour protection in this field. It also makes recommendations and disseminates information on the organisation of safe working procedures.

Since 1965 eight international conferences of CMEA member countries have been arranged on "Labour safety and fire protection in the chemical and paper industries".

The subject dealt with by these conferences included: safe methods of carrying out dangerous operations; fire hazards in chemical, paper and cellulose industries; protection against static electricity; safe storage of inflammable liquids; safety techniques for installations working under pressure; limitation of hazards resulting from the use of noxious and poisonous substances; and organisational and practical activities for the prevention of breakdowns in the chemical, cellulose and paper industries.

Standing Committee for the Coal Industry

Occupational health and safety is included among the activities of the CMEA Standing Committee for the Coal Industry, which consists of delegations from CMEA member countries under the leadership of the Ministers for the Mining Industry. Initially, these problems were dealt with by a special organ of the committee–the Council for occupational safety and health–which dealt with about 40 subjects. In 1979 the committee recognised it to be expedient for these problems to be transferred to the scientific and technological councils responsible for the development of underground mines, the enrichment and processing of coal, and the development of quarries. These councils co-operate with the various specialised scientific institutes and research and development centres.

In 1979 the Ministers for the Mining Industry in the member countries signed an agreement on mine rescue. According to this agreement each country is obliged to grant immediate assistance to another country upon request and to take part in consultations aimed at improving the rescue procedures in mines.

Standing Committee for Standardisation

In the field of occupational safety and health the co-operation of CMEA countries concerning standardisation is of special significance. In order to achieve standardisation in labour protection, a working group for occupational safety was established at the 37th Meeting of the CMEA Standing Committee for Standardisation held in Bratislava in July 1975. As a result, the following occupational safety standards were set out:

(*a*) organisational and procedural principles;

(*b*) requirements for hazardous and harmful manufacturing processes;

(*c*) requirements for various manufacturing processes;

(*d*) requirements for personal protective equipment.

These standards relate to all branches of the national economy. Other draft CMEA standards related to:

(*a*) occupational safety and health terminology;

(*b*) classification of standards for hazardous and harmful factors in work processes;

(*c*) methods for the control and enforcement of these standards;

(*d*) general safety requirements for machines and equipment;

(*e*) safety requirements for groups of machines and equipment and methods of meeting these requirements;

(*f*) general safety requirements concerning manufacturing processes;

(*g*) requirements for individual protection of employees and means of enforcement.

The working group prepares programmes for future standardisation in the field of labour protection which are reflected in the forthcoming activities of the CMEA bodies.

For example, within the framework of the CMEA Standing Committee for the Machine Industry, the following obligatory occupational safety and health standards for machine design were prepared: general requirements for machine tool design; standard requirements for electrical systems in machine tools; 11 special standards for specialised groups of machines.

On the basis of these standards, the countries work out their own standards, which are obligatory for the manufacture of equipment intended for export.

Committee for Scientific and Technological Co-operation

The important role of science in health and labour protection is reflected in the scientific and technological co-operation established among CMEA member countries. In this connection, the CMEA Committee for Scientific and Technological Co-operation prepared a programme for scientific and research work on ergonomic standards and requirements. Such co-operation enables CMEA countries to solve scientific and technological problems at lower cost and in less time and to implement the results in various spheres of the national economy. In determining common aims for co-operation, priority was granted to the improvement of living and working conditions.

The inter-branch programme for scientific co-operation covers ten years and embraces theoretical and applied research as well as the preparation of ergonomic standards, requirements, and recommendations. Theoretical problems include various aspects of research on man-machine systems and the ergonomic classification of these systems. Great importance is attached to the ergonomic criteria for creating the best man-machine-environment system, and to planning, production and automatic control systems application. Scientific and technical forecasts of the main trends in applied ergonomics are being prepared. Special research work is done on designing workplaces and proper conditions for handicapped persons. The above research and studies are aimed at establishing a firm scientific basis for all activities related to the preparation of standards and regulations in CMEA member countries. The results of these studies will enable two scientific handbooks on research methods to be prepared, one for designers and manufacturers, the other for ergonomics specialists. Finally, within this committee a Council for Environmental Protection has been established to deal with protection against noise and vibration.

In addition to multilateral co-operation in the field of occupational safety and health among the CMEA member countries, bilateral co-operation also exists. Direct co-operation in the field of occupational health and safety is also carried on among trade unions of the member countries.

NUSBEK, A.

Cowpox and pseudocowpox

Definition. Cowpox and pseudocowpox are two antigenically and structurally distinct pox viral agents that cause skin infections of the teats and udders of cattle. Infections in cattle with these two agents have several similarities: (1) they are usually seen only in milk cows; (2) they are transmitted by hand milking or machine milking and insects; (3) the course of illness and lesions in cattle caused by the two agents is indistinguishable.

Both of these agents cause skin infection in man, but the resulting lesions are different. Cowpox was once common in most countries of the world where cattle were raised. In recent times it is felt that true cowpox infections are rare. There is thought to be some natural relationship between cowpox virus and human smallpox virus. As the incidence of smallpox in the human population decreased by virtue of vaccination, the incidence of cowpox in the bovine population also decreased. Most of the pox-like infections in cattle around the world today are thought to be caused by pseudocowpox (paravaccinia virus); therefore, the remainder of the remarks will be directed at pseudocowpox, unless otherwise indicated.

Occurrence. Pseudocowpox is a relatively common cattle infection in most parts of the world where cattle are raised. Transmission to man is primarily by direct contact with lesions on cattle. Hand milking provides a high risk for exposure to the virus, as does examining and treating infected teats and udders. Machine milking requires less direct contact with teats and udders, and thus the probability for exposure is less.

Epidemiology. Epidemiological data are very superficial and it is impossible to determine prevalence or incidence of infection. However, all reported cases of pseudocowpox in man have been associated with direct contact with cattle exhibiting lesions of pseudocowpox. The reported cases are limited to dairy farmers, cow milkers, veterinarians, and other agricultural workers who may have direct contact with infected animals. There is no seasonal variation with infections in animals or man.

Symptoms. Cowpox and pseudocowpox lesions in man have different clinical appearances. Cowpox infection results in the typical pox lesion, which progresses from a papule to a vesicle to a pustule which becomes covered with a thick crust. This lesion is very similar to that caused by a primary vaccination with vaccinia, which is the laboratory adapated strain of cowpox used for human smallpox vaccination.

Pseudocowpox infections in man are termed milker's nodules, or milker's wart. The lesion progresses from a papule to a vesicle surrounded by erythema and oedema. Pseudocowpox lesions do not progress to a pustule, but instead develop into a firm, elastic nodule consisting of vascular granulation tissue, which starts out blue to red in colour, then progresses to a grey to brown verrucous nodule. These lesions usually heal spontaneously in four to six weeks. In severe cases erythema nodosum and erythema multiforme may accompany pseudocowpox infection.

Diagnosis. Diagnosis of milker's nodule is based on the following findings: (1) the presence of a few localised lesions on hands or arms, which start as reddened papules, progressing to vesicles and nodules; (2) an occupational history of milking or treating infected cows; (3) differentiating lesions from those of contagious ecthyma, a disease contracted from sheep rather than from cattle. Confirmation is based on biopsy and demonstration of the typical pox lesion histologically, with typical pox inclusion bodies. Electron microscopy to demonstrate the specific pseudocowpox morphology is also helpful.

PREVENTIVE MEASURES

Suggestions for prevention of milker's nodule include use of rubber gloves when handling or treating pseudocowpox lesions. Pseudocowpox can be prevented from spreading through herds and presenting an increase threat to humans by isolating infected dairy cows, by exercising sanitary milking procedures, and by controlling flies and other insect vectors.

Treatment. There is no specific treatment for milker's nodule. Topical treatment or systemic antibiotics are sometimes helpful in treating superimposed bacterial infection.

DONHAM, K. J.

"Erythema nodosum and erythema multiforme associated with milker's nodules". Kuokkanen, K.; Launis, J.; Mörttinen, A. *Acta Dermatovenereologica* (Stockholm), 1976, 56/1 (69-72). Illus. 21 ref.

Joint FAO/WHO Expert Committee on Zoonoses. Third report. Technical report series No. 378 (Geneva, World Health Organisation, 1967), 127 p.

Diseases transmitted from animals to man. Hull, T. G. (ed.) (401-405) (Springfield, C. C. Thomas, 5th ed., 1963).

Cramps

There are two types of cramps that may have their origin in working conditions, arising from completely different causes–occupational cramps and heat cramps.

Occupational cramps

These conditions, sometimes known as occupational palsies or professional spasms, are usually found among persons whose work entails uniform, strictly differentiated movements, accompanied by strain in a particular group of muscles. They are commonly in the hand, but may affect the legs and other parts of the body in certain occupations. A wide range of industrial, office and professional workers may be affected, e.g. tailors, ironers, cotton twisters, sorters, writers, switchboard operators, actors, players and musicians.

Symptoms and course. Occupational cramp is a form of neurotic dyskinesia, characterised by limited, selective disturbance of the motor functions involved in the particular activity, all other movements remaining unaffected. Precision in the performance of certain movements is lost owing to sudden weakness in a group of muscles, a forced posture in the extremities or convulsive contraction of the fingers.

The classic form of occupational dyskinesia is cramp in the fingers, or writers' cramp.

The disease generally develops gradually within the age limits of 40-50 years. Persons affected feel awkward when writing and tire rapidly, their writing deteriorates and they lose speed and fluidity of movement. At a later stage, this may be followed by an involuntary positive immobility of the hands and cramp in some fingers, which prevents writing. There is muscular tension in the shoulders, the neck and sometimes in the face. This is the spasmodic form of writers' cramp. The paretic form is characterised by weakening of the wrist muscles at the time of writing; forms of trembling and neuralgia have also been described, in which writing leads to agitation in the fingers or sharp pains. Combinations of the different forms are not uncommon. Disturbance of the motor function is generally accompanied by pain, susceptibility of the neurovascular bundle in the right shoulder and myalgia. All other movements can be carried out correctly without any trouble.

If the primary neurogenic form of occupational cramp is not accompanied by any symptoms of organic lesion of the nervous system, it is considered as an occupational disease; when the syndrome of cramp is the result of organic lesion of the brain, it is considered as a disease of general origin.

Occupational cramp is a largely non-reversible form of dyskinesia and those affected are often obliged to take up different work that does not involve the same movements. For example, sufferers from writers' cramp may be trained to write with the left hand, but this cannot be regarded as fully compensating the defect, as the same symptoms may occur in the left hand unless use is limited.

Pathogenesis. The pathogenesis of dyskinesia is of a complex nature. Many authors regard it as a particular form of neurosis. Manifestations of dyskinesia are caused by loss of synergia in the muscles participating in a complex stereotyped pattern of movement.

The preparation and constant performance of a complex stereotyped pattern providing maximum co-ordination of a strictly differentiated movement entails considerable strain on, and mobility of, the nervous processes.

According to modern concepts of neuroses, dyskinesia results from a break in the dynamic stereotyped pattern of movement. The cause for such a break may be either protracted excessive strain on certain regions of the nervous system permitting the preparation and constant performance of complex patterns of movement or additional endogenous and exogenous adverse effects on the organism whereby the functional state of the nervous system is disturbed (mental disorder, infection, neuroendocrinological and vascular disturbances, occupational fatigue).

Electromyographic investigation of persons thus affected has revealed reduction of mobility, and of the intensity of the process of stimulation and inhibition coupled with inertia of the process of stimulation. Persons with an unstable nervous system, a neuropathic constitution or cerebral sclerosis are more exposed to the risk of dyskinesia.

PREVENTIVE MEASURES

Training in proper methods of work may do much to prevent occupational dyskinesia: indeed prevention of writers' cramp should begin in childhood with instruction in proper ways of writing. Mechanisation has done much to remove some of the causes of cramps. In occupations where it still exists, alternation of work with short rest periods, physical exercises and general strengthening of the parts of the body involved all contribute to prevention.

Treatment. The treatment of occupational dyskinesia is lengthy and involves complete cessation of occupational activities. It is based on inhibiting the focus of congestive excitation in the central regions of the motor analyser and thus breaking the chain of pathological reflexes.

Therapeutic measures include extended sleep, hypnosis and psychotherapy, combined with a range of sedative, psychotropic and restorative pharmaceutical preparations. Various forms of physiotherapy are also recommended. Patients must be trained in correct ways of working, beginning with the separate elements involved, with particular insistence on the need for a correct posture of the hand in the case of writers' cramp, etc.

Heat cramps

Heat cramps are a risk to workers exposed to high temperatures, whether natural, as in sugar-cane cutters, or brought about by process heat, as in furnacemen, stokers, miners, forge and foundry workers, iron and steel workers and glass industry workers. The workers sweat profusely and drink excessive quantities of water to compensate for this. The result is the loss of large quantities of salt from the body and sudden attacks of intensely painful cramps, usually in the calves, upper limbs and abdomen. The provision of saline beverages at work will restore body salt levels and prevent the cramps. It is, of course, essential that all practicable physical means of mitigating high temperatures, e.g. by ventilation and by screening, should be taken (see HEAT DISORDERS).

DROGIČINA, E. A.

Epidemiological studies:

CIS 317-1972 "An Australian study of telegraphists' cramp". Ferguson, D. *British Journal of Industrial Medicine* (London), July 1971, 28/3 (280-285). Illus. 25 ref.

Ergonomics:

CIS 75-1480 "Keyboard operating posture and symptoms in operating". Duncan, J.; Ferguson, D. *Ergonomics* (London), Sep. 1974, 17/5 (651-662). Illus. 8 ref.

Cranes and lifting appliances

Cranes and lifting appliances may be defined as follows:

Crane. A machine for raising and lowering heavy loads and also, in many cases, moving them through a limited lateral distance. Some cranes are power driven and others operated manually. There are hundreds of different types ranging from hand cranes with a capacity of a few kilogrammes to the large floating cranes used in handling exceptionally heavy dock cargoes and capable, in some cases, of lifting up to 250 t. Some cranes are mobile, being either mounted on a motor vehicle or locomotive or running on rails.

Lifting appliances. The expression "lifting appliance" is a generic term which includes cranes, sheer legs, drag lines and also lifting devices such as crabs, winches, pulley blocks, transporters and other appliances.

Construction

All parts and working gear of cranes and lifting appliances, whether fixed or movable, including the anchoring and fixing appliances, should be of good construction, sound material, adequate strength and free from patent defect (figure 1). All parts should be made of metal since wooden members such as crane masts and jibs may undergo considerable deterioration not externally visible, due to rotting. Some countries have now enacted legislation prohibiting the use of wood for cranes or lifting appliances.

Figure 1. Buckled crane tower, probably due to failure of a leg section. *(By courtesy of the Schweizerische Unfallversicherungsanstalt, Luzern.)*

All rails on which a travelling crane moves and every track on which the carriage of a transporter or runway moves should be of proper size and adequate strength and have an even running surface, and such rails or track should be properly laid, adequately supported or suspended, and properly maintained. Safe working load indicators with, e.g. a green light showing for safety and a red light together with an audible warning when the safe working load is exceeded should, whenever practicable, be fitted to all types of cranes. Some countries have legislation making such safe working load indicators compulsory for all cranes above a certain capacity, usually about 1 t.

[Although stress-bearing parts are designed with a comfortable safety factor, they are not intended to withstand abnormally high stresses such as those that may result from excessive overloading (when the load or lifting hook is caught by a substantial obstacle or when the crane ballast is inadequate or improperly secured, or when excessive resistance develops as a result of defective maintenance, for instance). Replacing special assembly bolts by ordinary bolts can cause assembly failures with serious consequences. Manufacturers' instructions and specifications should always be strictly followed, and their advice sought in all cases where major changes are envisaged.]

No crane or lifting machine should be taken into use unless it has been tested by the maker or some other competent person and a certificate of such test specifying the safe working loads has been obtained.

Cranes used in building and works of engineering construction are frequently moved from site to site. It is essential that every time such a crane is re-erected on a new site it should be subjected to an anchorage test as a precaution against overturning or collapsing in use (figure 2).

Figure 2. Tower breakage due to faulty counterweighting. *(By courtesy of the Schweizerische Unfallversicherungsanstalt, Luzern.)*

Very detailed construction specifications for all types of cranes and lifting appliances have been laid down by a number of standards associations or institutes in various countries, with a view to improving safe working.

HAZARDS

There are innumerable hazards associated with cranes and lifting appliances which can result in serious accidents to workmen and also to the destruction of valuable machinery.

[Recent surveys of crane accidents in a number of countries indicate that most fatal and serious injuries occur when workers are crushed between moving parts and fixed objects or struck by falling loads, when persons erecting or dismantling cranes fall from a height or when the driver or another person comes into contact with an overhead line or other live conductor. Overturning of tower or mobile cranes is not uncommon; these accidents are very costly and can also result in serious or fatal injuries. On construction sites erection and dismantling operations entail particularly high hazards and should be carried out with special care and strict adherence to manufacturers' specifications. The correct tensioning and re-tensioning of bolted assemblies assumes particular importance in this connection.]

Overloading. Overloading a crane or lifting appliance may result in a load crashing to the ground with consequent damage to materials and, possibly, injury to workers. Sometimes overloading leads to a complete collapse of a crane. Even if there is no mechanical damage to the crane, there may be serious damage to the crane's electrical equipment and the electrical supply switch gear. The maximum safe working load should be plainly marked on all cranes and lifting appliances. Jib cranes should have capacity markings showing the maximum safe working loads that can be lifted for various jib radii.

Crane drivers, particularly those engaged in dock work, may find that a package or object has to be loaded or unloaded and there is no indication of its weight. Most dock authorities have some responsible person who is able to estimate the weight of the load with a sufficient degree of accuracy, or can arrange for the weight to be measured exactly.

Electrical hazards. Where mobile cranes are in use, particularly on construction sites and in docks, it is important to take special precautions against accidental contact between the crane jib and overhead electric wiring. Electronic devices are available that can be attached to the crane jib or boom and will sound an alarm if the jib or boom comes within a predetermined distance of a live wire. Electrically driven cranes used on building sites should have their electrical equipment efficiently earthed or grounded. An earth or ground connection to a metal spike in a heap of dry earth or sand is not satisfactory. It should be left to an electrician to test the "earth" connection as the crane driver is unlikely to have the necessary competence to do so.

Access to the cabin of an overhead travelling crane is usually by means of a vertical ladder situated at one side of the runway. In climbing this ladder to enter the cabin, there is a danger that the operator may come into contact with the live horizontal trolley wires. Trolley wires should be guarded or, better still, they should be placed on the opposite side to the means of access to the crane cabin. Another electrical hazard from the overhead electrical trolley wires is that maintenance workers or employees of outside firms called in to do repair work or install additional plant may have to approach within easy reach of the live wires without realising the danger involved. There is sometimes a complete lack of communication between the outside contractor's employees and the factory workers. When any such work is undertaken near the trolley wires, the current should be cut off by a responsible person.

Travelling crane track hazards. If any person is employed or working on or near the wheel track of an overhead travelling crane in any place where he would be liable to be struck by the crane, effective measures should be taken by the driver of the crane or otherwise to ensure that the crane does not approach within 6 m of that place. One method of complying with this requirement is to fix a stop on the rail track preventing the crane moving forward to that place. It is important that the crane driver be warned in advance if any person has to work near or on the travelling crane tracks. [Electronic devices are also available to prevent collisions between two cranes travelling on the same runway.]

Fire hazards. A means of emergency escape from the crane cabin (e.g. a rope ladder) should be provided. Serious injuries may occur should fire break out in the crane cabin, leading to inactivation of the electric crane-control switchgear, since the crane driver will be unable to slew his crane into the position in which he can reach his normal means of descent. Every crane cabin should be equipped with a fire extinguisher.

Falls from teagle hoist doorways. The top floor of a multistoried warehouse or factory may have a teagle hoist, which consists of a motor-driven winch supported on a horizontal steel girder projecting about 2 m over the street or factory yard. There is a door opening on each floor directly under this hoist and goods can be loaded or unloaded at each floor landing. There is a danger that a worker may fall through one of the open doorways to the ground level below. When not in use for the passage of goods, the doors should be closed. Secure hand-holds should be provided on both sides of the door openings.

SAFETY MEASURES

Crane drivers (particularly those working on construction sites or in docks) are called upon to operate a variety of lifting appliances having different control layouts. As a result, operational errors may occur which may entail serious consequences. Considerable importance attaches therefore to the standardisation of crane controls in the light of sound ergonomic principles.

[Special attention should be devoted to the design and layout of crane controls and to visibility, access, environmental and seating requirements. No person should be put in charge of a crane unless he has undergone thorough training and is properly certified for the job.]

When a load is lifted, the crane rope or chain should be absolutely vertical and the weight should be taken smoothly and without jerking, otherwise the crane and tackle may be subjected to excessive stress. Mobile cranes on motor vehicles should not operate on sloping ground due to the risk of overturning. All accessible gear wheels and moving parts in a crane cabin should be securely guarded. Operators of hoisting machinery should avoid, as far as possible, carrying loads over the heads of persons. When dangerous loads such as molten metal or objects carried by magnets are carried in workplaces, sufficient warning should be given to workers to permit them to reach safety. Operators should not leave cranes or lifting machinery with suspended loads. Hand-operated winches should be provided with ratchet wheels on the drum shafts with locking pawls, or self-locking worm gears, for holding the loads suspended when the hand cranks are released, and also with braking devices for controlling the descent of the load. The working area of a crane should be provided with adequate lighting during night work. Crane operators should be given instruction in safe working techniques. Signallers and the workers who hook on the loads should also be fully instructed as to the possible dangers. All workers should be warned not to stand or pass under suspended loads.

Inspection and maintenance. Cranes and lifting appliances require regular inspection and maintenance.

Periodic examination of cranes and lifting appliances. A very thorough examination of all parts and working gear should be made at least once a year by a competent person, and in the case of cranes and lifting appliances which are moved from one place to another, as for example to different building sites, there should be an examination every time they are installed in a new place.

Health of crane drivers

[Operating conditions are frequently arduous and crane drivers sometimes complain of cramps or pains in the legs, arms or back, of headaches, stomach pains and general fatigue; stomach ulcers and hypertension are not uncommon among them. Excessive noise levels often present a problem, particularly in the case of internal combustion engines located near the driver's cabin when the latter is inadequately insulated; equivalent sound levels of the order of 85 to 100 dBA in crane cabins are not exceptional.] In iron and steel works, foundries, metal galvanising workshops and processes where offensive and dangerous fumes are evolved, excessive amounts of fumes, smoke, vapours and toxic gases often rise to the height of the crane cabin and constitute a health hazard for the operator. It is not always feasible to provide for the immediate control of a concentration of such contaminants as, for example, when a steel furnace is charged and clouds of smoke and vapours are released. In such cases, individual air conditioning should be provided for the crane cabin. In cold-climate countries, the cabins of cranes used in exposed positions such as on docks and building sites should be provided with adequate shelter against bad weather and some form of heating. Conversely, in hot countries, insulation of the crane cabin roof and provision of an electric fan are necessary to maintain satisfactory environmental conditions. Comfortable seating for the driver is essential and it should be positioned so that he has the maximum view of all lifting operations; crane controls should be within easy reach so that they can be operated without undue effort. A number of studies of the ergonomic problems of crane cabins have already been made by various authorities. Good lighting should be provided in the cabin so that the operator can see the controls.

Medical supervision. Cranes and lifting tackle should be operated only by employees whose physical fitness is good, who have good vision, hearing and muscular co-ordination and who are over 18 years of age. Pre-employment and periodical medical examinations are essential to ensure that persons with psycho-physical impairment are not placed in a job which entails such a high degree of collective responsibility.

Legislation

Many countries have detailed safety regulations governing the construction and use of cranes and lifting appliances. These regulations are contained in Factories Acts and in special statutory rules for various industries and processes such as building and civil engineering work, dock processes, shipbuilding, mining and quarry work. A number of countries have legislation based on the standards of safety stipulated in ILO Conventions such as the Safety Provisions (Building) Convention, 1937 (No. 62). A large number of governments have ratified the Marking of Weight (Packages Transported by Vessels) Convention, 1929 (No. 27), which stipulates that any package or object of 1 t or more consigned for transport by sea or inland waterway shall have its gross weight plainly and durably marked on it on the outside before it is loaded. [ILO Convention 152 concerning occupational safety and health in dock work has been adopted on 25 June 1979 and is now coming into force. It is supplemented by Recommendation 160 concerning occupational safety and health in dock work.]

QUINN, A. E.

Accidents:

CIS 79-902 *Crane accidents: their causes and repair costs.* Butler, A. J. Building Research Establishment current paper CP 75/78 (Garston, Watford, WD2 7JR) (1978), 15 p.

Codes of practice:

Safety and health in dock work. ILO codes of practice (Geneva, International Labour Office, revised ed., 1977), 221 p. Illus.

Safety and health in building and civil engineering work. ILO codes of practice (Geneva, International Labour Office, 1972), 386 p.

Wire rope for lifting appliances–Code of practice for examination and discard. International Standard ISO 4309 (Geneva, International Organisation for Standardisation, 1st ed., 1981), 26 p. Illus.

Crane cabs:

CIS 77-1174 *Human factors in work, design and production.* Maule, H. G.;
Weiner, J. S. (London, Taylor and Francis Ltd., 1977), 138 p. Illus. 51 ref.

Cresols, creosote and derivatives

Cresols ($CH_3C_6H_4OH$)

HYDROXYTOLUENES; METHYL PHENOLS; TRICRESOLS

Pure cresol is a mixture of *ortho-*, *meta-*, and *para-* isomers, while cresylic acid, sometimes used synonymously for a mixture of cresols, is defined as a mixture of cresols, xylenols and phenol in which 50% of the material boils above 204 °C. The relative concentration of the isomers in pure cresol is determined by the source.

	o-Cresol	*m*-Cresol	*p*-Cresol
m.w.	108	108	108
sp.gr.	1.05	1.03	1.03
m.p.	30 °C	11.1 °C	34.8 °C
b.p.	191-192 °C	202.8 °C	201.9 °C
v.d.	3.7	3.7	3.7
v.p. at 20 mmHg ($2.66{\cdot}10^3$ Pa)	0.25	0.15	0.11
f.p.	81.1 °C	94.4 °C	94.4 °C
i.t.	599 °C	559 °C	559 °C
solubility	soluble in water; very soluble in ethyl alcohol, ethyl ether	slightly soluble in water; very soluble in ethyl alcohol	very soluble in ethyl alcohol

a colourless or slightly yellowish liquid which turns brown upon exposure to air.

TWA OSHA (all isomers) 5 ppm 22 mg/m³

IDLH 250 ppm

Production. Cresol is obtained by refining coal tar, by organic synthesis, or from the cracked naphtha of petroleum.

Uses. Cresol has considerable industrial use as an ore flotation agent. In the chemical industry, it is used as an intermediate in many processes, particularly in the manufacture of dyes, plastics, antioxidants, and the plasticiser, tricresyl phosphate. [In the United States over 60% of the total amount produced is used in the production of wire enamel solvents, phosphate esters and phenolic resins. *o*-Cresol is largely used in the production of dinitro-*o*-cresol and *p*-cresol in the production of *tert*-butylcresol.] Grinding and cutting fluid formulations frequently contain small amounts as a bacteriostatic agent. Cresol in soaps and emulsions serves as an antiseptic for minor cuts and wounds and as a disinfectant for instruments and hospital facilities. A 2% solution is issued as a hand wash. Many household disinfecting solutions contain it as a sanitising agent. Its antiseptic potency is about three times that of phenol, with only slight differences among the isomers.

HAZARDS

The toxic effects of cresol are similar to those of phenol. It can be absorbed through the skin, from the respiratory system, and from the digestive system. The rate of penetration through the skin is more dependent upon the surface area than on the concentration.

Like phenol, it is a general protoplasmic poison and is toxic to all cells. Concentrated solutions are locally corrosive to the skin and mucous membranes, while dilute solutions cause redness, vesiculation and ulceration of the skin. [Skin contact has also resulted in facial peripheral neuritis, impairment of renal function, and even necrosis of liver and kidneys.] A sensitivity dermatitis may occur in susceptible people from solutions of less than 0.1%. Systemically, it is a severe depressant of the cardiovascular and central nervous systems, particularly the spinal cord and medulla. Oral administration causes a burning sensation in the mouth and oesophagus and vomiting may result. [Concentrations of vapour that can be produced at relatively high temperatures may cause irritation of the upper airways and nasal mucosa.] Systemic absorption is followed by vascular collapse, shock, low body temperature, unconsciousness, respiratory failure and death. [Pancreatic complications have been described.] The oral toxic dose for small animals averages about 1 mg/kg and specifically 0.6 mg/kg for *p*-cresol, 0.9 mg/kg for *o*-, and 1.4 mg/kg for *m*-cresol. On the basis of its similarity to phenol, the human fatal dose can be estimated to be about 10 g. In the body, some of it is oxidised to hydroquinone and pyrocatechin, and the remainder and largest proportion is excreted unchanged, or conjugated with glycuronic and sulphuric acids. If urine is passed, it contains blood cells, casts and albumin. Cresol is also a moderate fire hazard.

SAFETY AND HEALTH MEASURES

[Cresol should be stored in iron or steel containers, properly labelled.] During the handling of this material, spillage should be avoided [and, should it occur, the spilled cresol should be immediately removed by means of sand or vermiculite, or washed up with large quantities of water]. In order to prevent absorption through the skin, it is advisable to wear rubber hand protection, foot protection and aprons. In case of high vapour concentrations of this material in the air, approved respirators should be worn. Face shields and eye protection should also be provided against droplets or spray. [Emergency showers and eyewashing facilities should be available to workers exposed to cresol.]

[On the assumption that cresol is more toxic than phenol by inhalation, the NIOSH recommended in 1978 that an environmental limit for cresol of 10 mg/m³ as a TWA concentration be established.]

Treatment. The recommended treatment for ingestion is repeated gastric lavage with large quantities of olive oil. Some oil is left in the stomach at the end of lavage to entrain the chemical. Intravenous fluids and other supportive measures are used as indicated.

Creosote

CREOSOTUM; CREOSOTE OIL; BRICK OIL

This is a heavy oily liquid made by the destructive distillation of wood or coal tar at temperatures above 200 °C. It has a heavy smoky smell and is nearly colourless when pure, but the usual industrial product is of a brownish colour. The word creosote is used loosely; in pharmaceutical circles it is considered a wood product, usually from beechwood, while in general commerce it means a coal-tar derivative.

(a) Wood tar creosote. This a mixture of phenolic compounds, being almost completely composed of guaiacol and creosol, which is used in medicine as an expectorant. Its antibacterial potency varies with different samples, but in general is about two to three times that of phenol. It is absorbed from the gastrointestinal tract, and the two main con-

stituents have been identified in the urine as conjugation products of sulphuric and glycuronic acids. In the past, prior to the existence of the new chemotherapeutic agents, wood creosote enjoyed considerable use in the treatment of pulmonary tuberculosis and abscesses of the lung. However, modern therapy has removed this material from most treatment regimes.

(b) Coal tar creosote. This is also known as creosote oil; it is a complex mixture of approximately 160 chemicals, many being of the aromatic series.

Uses. Although its primary use is as a wood preservative, it is also used as pitch for roofing, as an animal dip, as fuel and as a lubricant for die moulds. By controlled combustion it is burned to make lampblack. Wood preservation can be attained by soaking the timber in creosote. Usually, however, in order to attain deep penetration of the preservative, the pieces are placed in a tank from which the air is evacuated by a pump. After this has removed the volatile matter and moisture from the wood, creosote is admitted and allowed to penetrate the wood at temperatures somewhat over 100 °C. In the form of posts and rail ties, sound timber treated in this manner can withstand many years of contact with soil without deteriorating. Timber treated in this fashion is not suitable for use in mines because of the increased fire hazard and possible contamination of the air.

HAZARDS

Ill effects are seen primarily in workers who handle treated wood and timber, use creosote as a material for waterproofing, or as a dip for animals. It causes irritation of the skin, mucous membranes and conjunctiva. The skin can become red, papular, vesicular or ulcerative, depending upon the concentration of the material and length of time of contact. Repeated contact over long periods of time may be responsible for the development of cutaneous neoplasms. Skin cancer is quite easily produced in experimental animals. Photosensitisation can occur. Following ingestion of creosote, irritation of the gastrointestinal tract and cardiovascular collapse may occur. Systemic intoxication is similar to that of phenol, probably because of the large amount of phenolic compounds contained. Some of the material is excreted unchanged in the urine, some is conjugated with sulphuric and other acids, and some is detoxified by oxidation. The fatal dose is about 0.1 g/kg of body weight. Ingestion is followed by nausea and vomiting, salivation, abdominal discomfort, respiratory distress, cyanosis, pupillary changes, convulsive movements, rapid pulse and vascular collapse.

Protective clothing is generally adequate for the prevention of skin contact. Treatment is entirely symptomatic and supportive.

Tricresyl phosphate ($(CH_3C_6H_4)_3PO_4$)

See separate article.

Dinitro-*o*-cresol ($CH_3C_6H_2(NO_2)_2OH$)

See separate article.

HENSON, E. V.

CIS 78-1635 *Criteria for a recommended standard—Occupational exposure to cresol.* DHEW (NIOSH) publication No. 78-133 (National Institute for Occupational Safety and Health, 4676 Columbia Parkway, Cincinnati) (Feb. 1978), 117 p. Illus. 82 ref.

Cryogenic fluids

A cryogenic fluid is one whose vapour must be cooled below room temperature before it can be liquefied by an increase in pressure. (Some authors consider a cryogenic fluid to be one that boils below a specified temperature, for example, −150 °C.) A list of cryogenic fluids considered here is given in table 1. This table includes the following data: molecular weight, M; triple point temperature and pressure, T_{TP} and P_{TP} (the temperature and corresponding pressure at which the solid, liquid, and vapour can coexist); freezing point, T_F; heat of fusion, ΔH_F; boiling point at 1 atm, T_B; heat of vaporisation, ΔH_V; and critical temperature and pressure, T_C and P_C (the maximum temperature at which liquid can exist, and the pressure required for liquefaction at this temperature) – (the λ-point (temperature and pressure at which (normal) liquid helium I, (superfluid) liquid helium II, and vapour can coexist) is given for helium as it does not have a triple point).

By definition each of the fluids considered here except carbon dioxide has a critical temperature below 25 °C (298.16 °K). Carbon dioxide, which is actually a solid at 1 atm, is included here because of its industrial importance (it cannot exist as a liquid below 5.11 atm, the triple point pressure). See also GASES AND AIR, COMPRESSED.

Production. To produce a cryogenic fluid from a gas, it is first necessary to decrease the gas temperature to the boiling point of the fluid and then remove the heat of vaporisation, ΔH_V (table 1). This is achieved in practice by passing the gas through:

(a) a heat exchanger kept at a temperature below that of the gas by a vaporising liquid or a gas (thermal method);

(b) an expansion engine (external work method); or

(c) an expansion valve (internal work method).

The last method can be used only if such an expansion produces a decrease in temperature. This is not the case with neon, hydrogen, and helium at room temperature; these gases must be precooled or passed through an expansion engine until the inversion temperature is attained at the pressure of interest before they can be liquefied (the inversion temperature and pressure define the operating region in which a gas cools on expansion through a valve).

The two basic systems shown in figure 1 illustrate the principles involved in all commercial liquefiers. In the Linde-Hampson system, the gas to be liquefied is first compressed, and then passed through a counterflow heat exchanger and an expansion or Joule-Thomson (J-T) valve where the gas cools. After a number of passes through the compressor, heat exchanger and J-T valve, some of the gas condenses and collects in the liquid storage vessel. After each pass, the uncondensed (cold) gas is returned to the compressor through the low-pressure side of the heat exchanger where it cools the high-pressure gas before it enters the J-T valve. In the Claude system, some of the high-pressure gas is cooled further by performing work in an expansion engine; this cooled gas is returned to the compressor through the low-pressure side of counter-flow heat exchangers 1 and 2 to cool the high-pressure gas still further before it enters heat exchanger 3 and the J-T valve. Again liquefaction is effected by use of an expansion valve inasmuch as most expanders cannot operate satisfactorily on a two-phase (gas and liquid) stream.

Uses. Cryogenic fluids have found widespread use in recent years in commerce and the laboratory. For

Table 1. Physical properties of cryogenic fluids

Fluid	M	T_{TP} °K	P_{TP} atm	T_F °K	ΔH_F cal/g	T_B °K	ΔH_V cal/g	T_C °K	P_C atm
Helium	4.003	2.17[1]	0.05[1]	2 (25 atm)	1.0	4.2	5.5	5.2	2.3
p-Hydrogen	2.016	13.8	0.069	13.8	13.9	20.3	106.5	33.0	12.4
n-Hydrogen	2.016	13.9	0.071	14.0	13.9	20.4	107.1	33.2	13.0
o-Deuterium	4.032	18.7	0.17	18.7	12.0	23.5	68.4	38.3	16.3
Tritium	6.048	20.6	0.21	–	–	25.0	–	43.7	20.8
Neon	20.18	24.5	0.427	24.5	3.9	27.1	20.6	44.4	26.2
Nitrogen	28.013	63.1	0.127	63.1	6.1	77.3	47.7	126.3	33.5
Air	28.96	–	–	–	–	78.8[2]	49.0	132.5	37.2
Carbon monoxide	28.01	68.1	0.151	68.1	7.1[3]	81.6	51.4	132.9	34.5
Fluorine	37.997	53.5	0.002	53.5	9.8[3]	85.2	39.7	144.4	55.0
Argon	39.948	83.8	0.68	83.8	6.6	87.3	39.0	150.9	48.3
Oxygen	31.999	54.4	0.0015	54.4	3.3[3]	90.2	50.9	154.8	50.1
Nitric oxide	30.01	–	–	109.5	18.3	121.4	110.1	179.2	64.3
Methane	16.04	88.7	0.099	90.6	14.0	111.7	121.9	190.7	45.8
Krypton	83.80	116	–	116.5	3.9	119.9	25.8	209.4	54.5
Carbon tetrafluoride	88.01	–	–	89.5	–	145.1	–	227.5	37.0
Ozone	47.998	–	–	80.5	–	162.3	61.7	285.3	54.6
Ethylene	28.05	104.0	–	104.0	28.6[3]	169.3	115	282.8	50.0
Xenon	131.30	133	–	161.3	3.3	165.1	23	290.0	58.0
Carbon dioxide	44.01	216.6	5.11	–	47.5[3]	194.7[4]	36.2[4]	304.2	72.8

[1] λ-point. [2] Bubble point (saturated liquid). [3] At the triple point. [4] Sublimation point (1 atm).

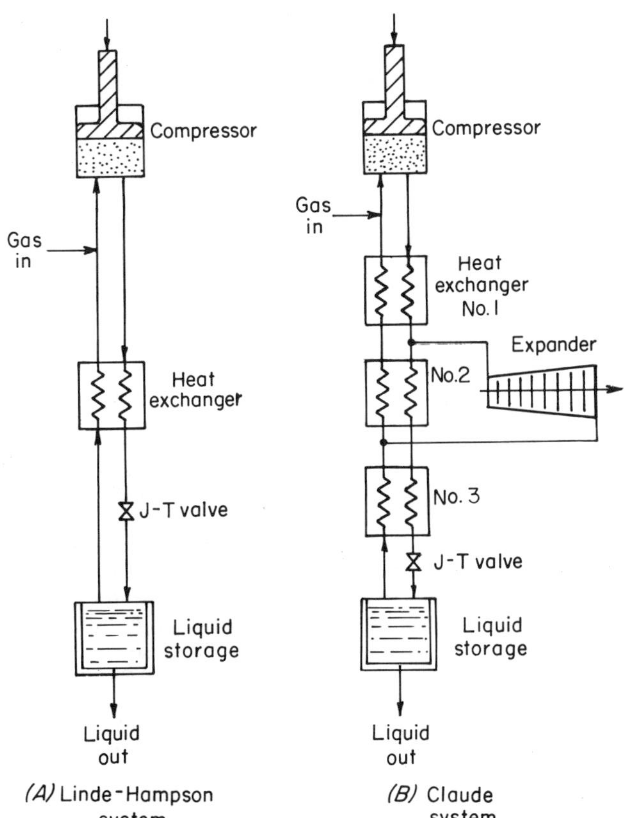

Figure 1. Schematics of the Linde-Hampson and Claude gas liquefaction systems.

example, liquid nitrogen is used in the preparation of frozen foods, in the manufacture of pressure vessels, in space simulation in the aerospace industry (to cryopump or freeze residual gases in a space-simulation chamber), in cryosurgery (selective destruction of tissue, as in the treatment of Parkinson's disease and removal of tumours), and in the preservation of whole blood, tissue, bone marrow, and animal semen for prolonged periods.

Liquid oxygen and fluorine are used as oxidisers in rocket propulsion systems; liquid oxygen also serves as a compact source of gaseous oxygen in breathing apparatus, the chemical industry, and iron and steel mills. Carbon dioxide and liquid nitrogen are common refrigerants in the laboratory and in industry. Methane is liquefied and stored during the summer months for subsequent use as a source of the gas during peak heating demand periods, and liquid hydrogen is used as a particle detector in bubble chambers, as a propellant in chemical and nuclear rocket systems, and as a source of gaseous hydrogen in numerous heat-treating and other applications in which this gas is used. In addition, this liquid and liquid helium are now used routinely to study the behaviour of solids and solid-state devices at low temperatures. (See relevant separate articles.)

HAZARDS AND THEIR PREVENTION

Cryogenic fluid users are faced with certain hazards not ordinarily encountered with other materials. However, these fluids can be used safely if the facilities in which they are to be used are properly designed and maintained, and if operating and support personnel are adequately trained and supervised. Briefly, the user is concerned with three types of hazards: physiological, physical and chemical.

Physiological hazards. These relate to the behaviour of the human body when exposed to the cryogenic fluids and their vapours, and include frostbite, respiratory ailments, and chemical burns. While frostbite usually occurs after prolonged exposure of tissue to temperatures below 0 °C, excessive heat loss may cause the internal organs to malfunction. Further, because a cryogenic liquid can be vaporised rapidly at ambient temperatures, one must consider the consequences of rapidly filling an inhabited space with vapour. All the fluids listed in table 1, except air and oxygen, can rapidly produce hazardous oxygen-deficient atmospheres. In addition, some of the vapours (ethylene, krypton, and xenon) have an anaesthetic effect on the body; carbon monoxide presents a hazard because it combines more readily than oxygen with the haemoglobin of the blood; fluorine and ozone are toxic and can produce severe burns if they contact the skin, eyes, and mucous membranes; and carbon dioxide can affect the body adversely because of its specific role in the respiration process.

Physical hazards. The behaviour of structural and other materials likely to come in contact with cryogenic fluids and their vapours must also be known. Thus, while the tensile strength of most materials increases with decrease in temperature, many (for example, carbon steel) are not usable at low temperatures because they become brittle.

Chemical hazards. Since some of these fluids react violently when combined with each other or with their surroundings, all hazardous combinations must be eliminated. Included here are the obvious fire and explosion hazards that result when gaseous oxidisers (air, oxygen, fluorine, nitric oxide, ozone) are combined with various combustibles (methane, ethylene, construction materials). However, these same oxidisers form explosive combinations with many combustibles in the condensed phase (liquid hydrogen/solid oxygen; liquid oxygen/oil, liquid oxygen/charcoal).

To ensure that these hazards are minimised, buildings and experimental facilities intended to house cryogenic fluids should be constructed or modified only after a study has been made of the hazards presented by the fluids in question. Storage areas should be dry, well ventilated, fire-resistant and free of heat and ignition sources. The safety of the production or storage plant should be considered in detail at the design stage and backed up by an adequate maintenance programme, which should include maintenance of release diaphragms, safety equipment (gas detectors, alarm systems, flame detectors, etc.), ventilation systems, compressors, electrical equipment, etc. All repairs should be double checked and major repairs approved by safety and engineering staff. Checklists and flow-sheets should be drawn up to eliminate all unsafe conditions in start-up, shut-down and normal and emergency operating procedures. Personal protective equipment including respiratory protective equipment, eye and face protection, hand and arm and foot and leg protection should be available for emergency wear but they should not be looked upon as a substitute for adequate protection of plant. In addition, where exposure to cold equipment, vapours and liquid may occur, personnel must be equipped with special clothing designed to prevent the freezing of body tissues.

ZABETAKIS, M. G.

Cryogenics safety manual. A guide to good practice (British Cryogenics Council, 16 Belgrave Square, London) (1970), 122 p. 39 ref.

CIS 78-44 *British standard code of practice for safe operation of small-scale storage facilities for cryogenic liquids.* BS 5429: 1976 (British Standards Institution, 101 Pentonville Road, London) (30 Nov. 1976), 6 p.

CIS 79-628 "Hazard control of liquid oxygen systems". Allison, W. W. *Professional Safety* (Park Ridge), Jan. 1979, 24/1 (21-25). 8 ref.

Cumene

Cumene ($C_6H_5CH(CH_3)_2$)
ISOPROPYL BENZENE

m.w.	120
sp.gr.	0.86
m.p.	−96.9 °C
b.p.	152.4 °C
v.d.	4.1
v.p.	4.5 mmHg (0.58·10³ Pa) at 25 °C
f.p.	43.9 °C
e.l.	0.9-6.5%
i.t.	43.9 °C

$CH_3-CH-CH_3$ (on benzene ring)

soluble in ethyl alcohol, ethyl ether, benzene and its homologues and chlorinated solvents; insoluble in water

a clear, colourless liquid with a sharp, penetrating, aromatic odour detectable at very low levels.

TWA OSHA	50 ppm 245 mg/m³ skin
STEL ACGIH	75 ppm 365 mg/m³
IDLH	8 000
MAC USSR	50 mg/m³

Production. This hydrocarbon is a constitutent of petroleum from which it is obtained by fractionation. Today it is mainly obtained by the alkylation of benzene with propylene catalysed by anhydrous aluminium chloride, sulphuric acid or phosphoric acid. It is a constituent of a number of petroleum distillates.

Uses. Cumene is used as a high octane blending component in aviation fuel, as a thinner for cellulose paints and lacquers and as an important starting material for the synthesis of phenol and acetone and for the production of styrene by cracking. It serves as a constituent of many commercial petroleum solvents in the boiling range of 150-160 °C. It is a good solvent of fats and resins and has therefore been proposed as a replacement for benzene in many of its industrial uses.

HAZARDS

Regard must be paid to certain health and fire hazards when cumene is used in an industrial process.

Health hazards. Cumene is a skin irritant and can be slowly absorbed through the skin. It also has a potent narcotic effect in animals, and the narcosis develops more slowly and lasts longer than with benzene or toluene. In animals it also has a tendency to cause injury to the lungs, liver and kidneys, but no such injuries have been recorded in human beings.

The only reports of the chronic toxicity of cumene have also been related to animal experiments, which have shown hyperaemia of the lungs, liver and kidneys following exposure to 500 ppm daily for 150 days. Even at exposures between 1 300 and 1 400 ppm for 180 days no changes of peripheral blood or bone marrow were found. Absorption of cumene leads to a definite increase in the urinary excretion of glucuronides of compounds formed by the partial oxidation of the alkyl radical. [About 40% of cumene is converted into 2-phenylpropandiol (dimethylphenylcarbinol); phenylpropionic acid and methylbenzylcarbinol are also produced in roughly equal amounts and account for about half of the dose.] The main metabolic pathway of cumene, contrary to benzene, does not involve the production of phenol, which explains the lack of aggressiveness of cumene to the bone marrow.

Fire and explosion. Liquid cumene does not evolve vapours in flammable concentrations until its temperature reaches 43.9 °C. Thus flammable mixtures of vapour and air will be formed only in the course of normal workroom manipulations in the hotter countries. If solutions or coatings containing cumene are heated in the course of a process—in a drying oven for instance—fire and, under certain conditions, explosion readily occur.

SAFETY AND HEALTH MEASURES

Precautionary measures must take account of the harmful and flammable properties of the material.

Health precautions. Depending on the process, enclosure, local exhaust ventilation or general ventilation should be arranged to maintain atmospheric concentrations below safe limits. When there is a possibility of gross skin contact, personal protective equipment should be worn and should include hand and arm protection impermeable to cumene. Contaminated clothing should be removed immediately and laundered before re-use. Where persons are required to enter a vessel which has contained cumene, the measures applicable to work in confined spaces should be implemented.

Workers exposed to cumene should be kept under medical supervision and should be subject to periodic medical examinations in which special attention should be paid to the skin, eyes and mucous membranes [blood, liver and kidney; the urinary excretion of dimethylphenylcarbinol (to be determined by gas chromatography) provides reliable tests of exposure]. If signs of irritation are present, temporary cessation of exposure should be recommended.

Fire precautions. The above precautions to ensure that a harmful concentration of cumene vapour is not present in the working atmosphere are fully adequate to avoid flammable mixtures in the air in normal circumstances. To cover the risk of accidental leakage or overflow of liquid from storage or process vessels, additional precautions are needed such as mounds round storage tanks, sills at doorways or specially designed floors to limit the spread of escaping liquid.

Open flames and other sources of ignition should be excluded where cumene is stored or used.

TRUHAUT, R.

CIS 77-114 "Absorption of cumene through the respiratory tract and excretion of dimethylphenylcarbinol in urine". Seńczuk, W.; Litewka, B. *British Journal of Industrial Medicine* (London), May 1976, 33/2 (100-105). Illus. 12 ref.

CIS 80-1027 "Cumene" (Cumène). *Cahiers de médecine interprofessionnelle*, 4th quarter 1979, 19 (special number, supplement to No. 76) (35). 4 ref. (In French)

Cuts and abrasions

This article deals only with wounds of the skin and subcutaneous cellular tissue and not with wounds affecting muscles, tendons, vessels and internal organs (see TRAUMATIC INJURIES).

These wounds vary considerably depending on—

(a) their site: the hands account for about 30% of these injuries (see HAND INJURIES), and the feet 15%—moreover, a wound close to a joint or a bony protuberance may have more serious consequences;

(b) their shape: cuts may be clean or jagged—cuts from sharp instruments are usually linear and since the edges of the wound are clean and well defined they are brought easily together and knit rapidly, whereas bruised or jagged wounds may be slow in healing;

(c) their extent and the depth of contamination (microbial or viral): puncture wounds require special attention since they drain badly and the damage to underlying tissue may be more extensive than the damage at point of entry—laceration favours infection and may allow foreign bodies to enter; panaris of the right thumb has been found as a sequel to superficial wounds and is due to repeated microtraumatisms such as repeated pressure cycles several thousand times a day in the use of a hand tool;

(d) the traumatising agent: a tool, machine, material or environmental agent.

Course. Two phenomena (cellular destruction and septic contamination) have a dominant influence on the course of a wound, which can be broken down into four distinct phases:

(a) contamination—this may last for several hours but germs do not yet affect interstitial tissue;

(b) penetration of germs into interstitial tissue—this is more intensive and extensive in the case of tissue necrosis (irregular wounds);

(c) inflammation due to infection defence mechanisms—this phase is usually accompanied by pain and occurs about 6 h after injury;

(d) regeneration—lost substance is replaced by temporary tissue until the regenerative process is completed.

Complications. These are due primarily to infection. Tetanus infection is the most serious complication since the infective organism, *Clostridium tetani*, may have a fatal effect even when it enters through the smallest wound. Retention of inflammatory and infectious material may result in an abscess or phlegmon, and infection may be disseminated via the lymph vessels and glands. Formation of scar tissue is a sequel which may have functional or aesthetic repercussions.

Wounds located in the immediate vicinity of joints, where skin and tissue are subject to repeated elastic deformation, are the slowest to heal and the most likely to produce sequelae.

SAFETY AND HEALTH MEASURES

Inadequate and poor general maintenance and housekeeping, especially in building and civil engineering, and the lifting and carrying of glass (especially wired glass), metal sheet and metal sections, such as angle, channel and T-bars are amongst the most frequent causes of cuts and abrasions. Rough edges and burrs on metal should be removed by grinding; wood should be planed to remove splinters; projecting nails or wires should be removed or bent over.

Badly maintained or poor quality hand tools such as cold chisels, punches, drifts and hammers will mushroom or splinter leaving sharp edges; these tools should be deburred regularly and under no circumstances should the shank be hardened. Files should not be used without handles (see HAND TOOLS).

The blades of machine tools and woodworking machines such as saws, planers, millers, etc., should be regularly sharpened; the machines themselves should be fitted with efficient machinery guarding, including a blade guard and riving knife for a circular saw, a cover of the part of an overhead planing machine cutter block not being used, and a cover for the rear side of a spindle moulder cutter block. Push sticks should be employed for feeding small stuff to saws and planers, etc., and machine vices or clamps should be used to hold workpieces on drilling machines, etc.

Personal protection. Wherever material with sharp or jagged edges or rough surfaces is being manipulated, workers should wear suitable hand and arm protection, such as gloves or gauntlets with 3-5 fingers, mittens, finger stalls or finger guards, hand pads and protective sleeves. Hand protection should be well designed and fit correctly in order to ensure that dexterity is not impaired (see HAND AND ARM PROTECTION; FOOT AND LEG PROTECTION). The materials commonly used for hand and arm protection against cuts and abrasions include:

(a) leather–this remains supple and allows the hand to breathe whilst providing protection in the handling of metal components or sheet, and materials with rough surfaces;

(b) natural, artificial or synthetic fibres–these fibres, when woven sufficiently coarse, are suitable for hand protection in manipulating sheet metal or other material with sharp edges;

(c) plastics-coated fabrics–gloves made from these will often require reinforcement at critical points;

(d) metal mesh, woven metal or metal-studded fabric or leather–these materials are used where maximum protection is required such as in the case of butchers or abattoir workers constantly using very sharp knives.

First aid and treatment. In all cases of cuts and abrasions, it is essential to bear in mind the danger of tetanus infection. If the victim has not been immunised against tetanus or if immunisation has not been regularly maintained, tetanus vaccine should be administered. [In all cases of cuts, especially at the hand, it is also essential to ascertain whether there is any deep lesion, in particular a tendon or nerve section or the opening of a joint, or whether foreign bodies are trapped within the wound, because failure to recognise it may involve permanent disability. In any doubtful case the first aider should content himself with applying on the cut a made-up sterile dressing and sending the patient to the doctor. Washing under a tap of drinking water may be the first step required to clean the skin or a small cut; however, unskilled attempts to cleanse a wound must be avoided.]

The wound should be disinfected by topical antibiotic treatment or by the use of antiseptics such as quaternary ammonium or mercurial salt compounds; the presence of a wetting agent facilitates the penetration of antiseptics. The value of antiseptic treatment of puncture wounds may be limited since, after the traumatising agent is removed from the wound, the edges of the broken skin come together and the subsequent penetration of antiseptic is virtually impossible.

The wound should then be covered with a dressing to prevent further contamination, provide protection against impacts and allow the edges of the wound to knit together. In the case of wounds in the vicinity of joints, it is often advisable to immobilise the surrounding sections of the body. An extensive cut may require to be stitched. See also FIRST-AID ORGANISATION.

POLI, J. P.

Cyanogen, hydrocyanic acid and cyanides

Cyanogen (CN)

ETHANE DINITRILE

m.w.	52.04
sp.gr.	0.87
m.p.	−27.9 °C
b.p.	−20.7 °C
v.d.	1.8
e.l.	6.6-32%

very soluble in water, ethyl alcohol, ethyl ether

a colourless gas with an almond-like odour, which is acrid and pungent in lethal concentrations.

TLV ACGIH	10 ppm 20 mg/m³

Cyanogen is produced by adding an aqueous solution of sodium or potassium cyanide to an aqueous solution of copper sulphate or chloride, by heating mercury cyanide, or from hydrocyanic acid with the use of copper oxide. It is used as a fumigant, but it may also be present in blast-furnace gases.

Cyanogen bromide (CNBr)

BROMINE CYANIDE

m.w.	105.93
sp.gr.	2.01
m.p.	52 °C
b.p.	61.4 °C
v.d.	3.6
v.p.	92 mmHg (12.2·10³ Pa) at 20 °C

soluble in water, ethyl alcohol, ethyl ether

colourless crystals.

Cyanogen chloride (CNCl)

CHLORINE CYANIDE

m.w.	61.5
sp.gr.	1.18
m.p.	−6.9 °C
b.p.	12.6 °C
v.d.	2
v.p.	1 000 mmHg (133·10³ Pa) at 20 °C

slightly soluble in water; soluble in alcohol

colourless and very irritant liquid or gas.

TLV ACGIH	0.3 ppm 0.6 mg/m³ ceil

Cyanogen bromide is produced by the action of bromine on potassium cyanide, or by the interaction of sodium bromide, sodium cyanide, sodium chlorate and sulphuric acid. Cyanogen chloride is produced by the action of chlorine on moist sodium cyanide suspended in carbon tetrachloride and kept cooled to −3 °C, followed by distillation.

Cyanogen bromide is used in organic syntheses, as a fumigant and pesticide and in gold extraction processes. It has also been used in connection with cellulose technology. Cyanogen chloride is used in organic syntheses and as a warning agent in fumigant gases.

Hydrocyanic acid (HCN)

HYDROGEN CYANIDE; PRUSSIC ACID

m.w.	27.03
sp.gr.	0.69
m.p.	−14 °C
b.p.	26 °C
v.d.	0.94
v.p.	760 mmHg (101.3·10³ Pa) at 25.8 °C
f.p.	−17.8 °C
e.l.	6-41%
i.t.	538 °C

very soluble in water, ethyl alcohol, ethyl ether

a colourless gas or liquid with a characteristic, faint odour of bitter almonds perceptible to some people.

TWA OSHA	10 ppm 11 mg/m³ skin
NIOSH	5 mg/m³ (10 min) ceil
TLV ACGIH	10 ppm 10 mg/m³ ceil skin
IDLH	50 ppm
MAC USSR	0.3 mg/m³ skin

Hydrogen cyanide, or hydrocyanic acid (HCN) was discovered by Scheele in 1782. He made it by heating sulphuric acid with Prussian blue; hence the old name prussic acid. Hydrogen cyanide occurs in nature as the glucoside amygdalin in some plants, for example almonds.

Methods of production include the reaction of ammonia, methane (and air) in the presence of a platinum catalyst, the reaction of coke-oven gas with sodium carbonate, the reaction of acid with cyanide salts, or the catalytic decomposition of formamide. It is used in the manufacture of synthetic fibres and plastics, mainly via acrylonitrile, for the extermination of rodents and insects, electroplating, and in the production of cyanide salts and nitrites. Variations in the relative hazard of the compounds range from the high toxicity of some of the soluble cyanide compounds to be described, to the only slightly or non-poisonous complex cyanates. The thiocyanates, used at one time as drugs, are in this group of biologically relatively harmless chemicals.

Sodium cyanide (NaCN)
m.w. 49
m.p. 563.7 °C
b.p. 1 496 °C
v.p. 1 mmHg ($0.13 \cdot 10^3$ Pa) at 817 °C
soluble in water; slightly soluble in alcohol
white deliquescent lumps or crystals.

TWA OSHA	(as cyanide) 5 mg/m³ skin
NIOSH	5 mg/m³ (10 min) ceil
IDLH	50 mg/m³
MAC USSR	0.3 mg/m³

Potassium cyanide (KCN)
m.w. 65.1
sp.gr. 1.52
m.p. 634.5 °C
soluble in water, alcohol, glycerol
white deliquescent lumps or crystals.

TWA OSHA	(as cyanide) 5 mg/m³ skin
NIOSH	5 mg/m³ (10 min) ceil
IDLH	50 mg/m³
MAC USSR	0.3 mg/m³

These salts are produced mainly from coke-oven gas. Potassium cyanide, potassium carbonate and carbon are heated in ammonia and the cyanide extracted from the fusion mixture with alcohol. They are used in electroplating, steel hardening, extraction of gold and silver from ores, manufacture of nitriles and fumigation of fruit trees, ships, railway cars, warehouses, etc.

Calcium cyanide ($Ca(CN)_2$)
BLACK CYANIDE
m.w. 92.1
m.p. decomposes > 350 °C
decomposes in water
a white powder.

TWA OSHA	(as cyanide) 5 mg/m³ skin
NIOSH	5 mg/m³ (10 min) ceil
IDLH	50 mg/m³
MAC USSR	0.3 mg/m³

Of several methods of production available, one consists of treating powdered calcium oxide with boiling anhydrous hydrocyanic acid in the presence of an accelerator such as ammonia or water, in order to minimise the loss of hydrocyanic acid by polymerisation. It may also be produced by reacting liquid hydrocyanic acid with calcium carbide. It is used as a fumigant and a rodenticide, in the manufacture of stainless steel and as a stabiliser for cement.

Potassium ferricyanide ($K_3Fe(CN)_6$)
POTASSIUM HEXACYANOFERRATE; RED PRUSSIATE OF POTASH
m.w. 329.24
sp.gr. 1.85
m.p. decomposes
soluble in water; insoluble in ethyl alcohol
bright-red lustrous crystals or powder.

It is produced by the oxidation of ferrocyanide and is used in photography and making of blueprints, metal tempering and electroplating and in the manufacture of pigments.

Potassium ferrocyanide ($K_4Fe(CN)_6 3H_2O$)
POTASSIUM HEXACYANOFERRATE; YELLOW PRUSSIATE OF POTASH
m.w. 422.4
sp.gr. 1.85
m.p. 70 °C ($-3H_2O$)
b.p. decomposes
soluble in water, acetone; insoluble in ethyl alcohol
lemon-yellow crystals or powder which effloresce on exposure to air.

It is produced from gas plant by-products or from alkaline earth cyanides and is used in the tempering of steel, process engraving, the manufacture of pigments, dyeing and as a chemical reagent.

HAZARDS

Acute exposure can cause death by asphyxia [, and this is the result of exposure to adequate concentrations of HCN via either inhalation or percutaneous absorption; in the latter case, however, the dose required is higher]. Chronic exposure to cyanides at levels too low to produce serious clinical complaints are known to cause a variety of problems. Study of workers in the electroplating industry has shown dermatitis to be a problem there. Also reported were itching, scarlet rash, papules, in addition to severe irritation of the nose, leading to obstruction, bleeding, sloughs and in some cases perforations of the septum. Among fumigators, mild cyanide poisoning is recognised as the cause of the symptoms of oxygen starvation, headache, rapid heart rate, nausea, all of which are completely reversed when exposure ceases.

There are a series of cases in the literature indicating that chronic systemic cyanide poisoning may occur, but this is rarely recognised because of the gradual onset of disability and the appearance of symptoms which are consistent with other diagnoses. It has been suggested that excessive thiocyanate in extracellular fluids might explain chronic illness due to cyanide, as the symptoms reported are similar to those found when thiocyanate is used as a drug. The occurrence of symptoms of chronic disease have been reported in electroplaters and silver polishers after several years of exposure. The most prominent effects were motor weakness of arms and legs, headaches and thyroid diseases. All of these findings have also been reported as complications of thiocyanatic therapy.

In spite of the extreme hazard of cyanide exposure, there are far fewer industrial cases of acute poisoning than might be expected. Furthermore, where emergency measures were well understood and contingency plans carefully developed, seriously poisoned workers have been successfully treated.

Toxicity. The cyanide ion of soluble cyanide compounds is rapidly absorbed from all routes of entry, respiratory and gastrointestinal tracts, as well as through unbroken skin. The toxic properties of cyanide depend on its ability to inhibit enzymes required for the respiration of cells. While 42 enzyme reactions have been listed that cyanide can inhibit by forming complexes with heavy metal ions, the most critical action is the inhibition of cytochrome oxidase, which prevents the utilisation of molecular oxygen by all cells through reaction of the ferric ion with the cyanide ion. This action prevents the uptake of oxygen by the tissues causing death by asphyxia. The blood, saturated with oxygen prior to exposure, remains arterial in colour after reaching the venous circulation, producing the well known cherry-red colour of the victim of acute cyanide poisoning. Cyanide combines with the approximately 2% of methaemoglobin normally present–a fact that has helped to develop treatment of cyanide poisoning.

If the initial dose is not fatal, the cyanide ion is gradually released from its link with cytochrome oxidase or methaemoglobin and converted to the relatively harmless thiocyanate ion, which is excreted in the urine. While part of the absorbed cyanide is exhaled unchanged, the greater part is converted to thiocyanate, which remains in extracellular body fluids until excreted. The enzyme rhodanase, which is widely distributed in the body, acts to convert the combination of available thiosulphate ions and cyanide to thiocyanate.

There are variations in biological effects of some of the compounds in this group. At low concentrations, the halogenated cyanide compounds (CNBr, CNCl) behave like irritating gases, causing lacrimation, acute and delayed respiratory effects, including pulmonary oedema, at concentrations in the range of 10 ppm. At very low dosages, symptoms of hydrogen cyanide exposure may be weakness, headaches, confusion, nausea and vomiting. Normal blood pressure with rapid pulse is usual in mild cases. The respiratory rate varies with the intensity of exposure: rapid with mild exposure, or slow and gasping with severe exposure.

SAFETY AND HEALTH MEASURES

Scrupulous attention to proper ventilation is necessary. Where skin can be exposed [either to solution or powder or pellets of cyanide salts or to hydrogen cyanide in gaseous form], protective clothing, including impervious hand protection, should be provided. Because of the low permissible exposure level for hydrogen cyanide, complete enclosure of the process is recommended. [Warning signs should be affixed near entrances to areas containing hydrogen cyanide and where emergencies may occur. All containers (shipping or storage) of HCN or cyanide salts should bear a warning label, including instructions for first aid.]

Those working with cyanide salts should be instructed that contact with acids will release hydrogen cyanide. Where an exposure potential exists, workers should be trained to recognise the odour of hydrogen cyanide and, when this is detected, the work area should be evacuated immediately. Workers entering a contaminated area must wear impervious protective clothing as well as suitable respiratory protective equipment (tables 1 and 2).

All containers of cyanide salts should be kept covered or in an exhausted hood when not in use. Any process that may release hydrogen cyanide should be mechanically exhausted, with provision for a higher rate during emergencies. Direct reading instruments for the determination of hydrocyanic acid are available. Since thiocyanates are contained in certain foods and since tobacco smokers excrete more of these than non-smokers, the amount found in urine bears no dependable relationship to work exposure to cyanides.

[Medical surveillance based on pre-employment and annual medical examinations should be available to all workers occupationally exposed to HCN or cyanide salts.]

Medical control demands teaching artificial resuscitation and the use of the drugs prescribed for emergency treatment of acute poisoning. First-aid kits with drugs and syringes should be placed appropriately and checked frequently. From industrial experience with acute and chronic cyanide poisoning, exclusion of exposure of some individuals may be advisable. Those who are careless and unwilling to accept personal protective devices without question, and perhaps those with emotional problems such as depression and also those who cannot smell concentrations of 10 ppm should probably be excluded. Workers with chronic diseases of the kidneys, respiratory tract, skin or thyroid are at greater risk of developing toxic cyanide effects than are healthy workers.

Table 1. Requirements for respirator usage: hydrogen cyanide[1]

Maximum use concentration (ppm of HCN)	Respirator type for HCN gas
< 90	(1) Type C supplied-air respirator, demand or continuous-flow type (negative or positive pressure), with half or full facepiece (2) Full facepiece gas mask, chin-style canister specific for HCN. The maximum service life of canisters is 1 h
< 200	Full facepiece gas mask, front- or back-mounted type canister specific for hydrogen cyanide. The maximum service life of canisters is 1 h
> 200	(1) Self-contained breathing apparatus in pressure-demand mode (positive pressure) with full facepiece worn under gas-tight suit providing whole body protection (2) Combination supplied-air respirator, pressure-demand type (positive pressure), with auxiliary self-contained air supply and full facepiece; all worn under gas-tight suit providing whole body protection
Emergency (no limit)	(1) Positive pressure self-contained breathing apparatus worn under a gas-tight suit providing whole body protection (2) Combination supplied-air respirator, pressure-demand type, with auxiliary self-contained air supply; all worn under gas-tight suit providing whole body protection
Evacuation or escape (no limit)	(1) Self-contained breathing apparatus in demand or pressure-demand mode (negative or positive pressure) (2) Gas mask, full facepiece or mouthpiece type, with canister specific for HCN

NB: During the use of any respirator with half mask, full facepiece or hood, protective clothing should be worn if there is a chance that liquid HCN may contact any part of the body.

[1] *By courtesy of NIOSH from DHEW (NIOSH) publication No. 77-108.*

Treatment. Since cyanide causes death by asphyxia, the first step in therapy of acute poisoning is artificial respiration if the patient is not breathing, which is likely in other than mild cases. Because methaemoglobin binds cyanide (relatively high concentrations of methaemoglobin in the blood can be tolerated), nitrite therapy is used next. This consists of holding crushed ampoules of amyl nitrite close to the patient's nose, which may suffice to relieve symptoms in milder cases. It is essential to be

Table 2. Requirements for respirator usage: cyanide salts[1]

Maximum use concentration (mg/m^3 expressed as CN)	Respirator type for cyanide salts
< 25	(1) Filter-type respirators, approved for toxic dust, with half-masks (not applicable for $Ca(CN)_2$) (2) Chemical cartridge respirators with replaceable cartridge for toxic dusts and acid gases; with half-mask. Maximum service life 4 h
< 50	(1) Full-face gas mask, chest or back-mounted type, with industrial size canister for toxic dust and hydrocyanic acid gas. Maximum service life 2 h (2) Type C supplied-air respirator, continuous-flow or pressure-demand type (positive pressure) with full facepiece (3) Type A supplied-air respirator (hose mask with blower), with full facepiece
> 50	(1) Self-contained breathing apparatus with positive pressure in full facepiece (2) Combination supplied-air respirator, pressure-demand type with auxiliary self-contained air supply
Emergency (no limit)	(1) Self-contained breathing apparatus with positive pressure in facepiece (2) Combination supplied-air respirator, pressure-demand type with auxiliary self-contained air supply
Evacuation or escape (no limit)	(1) Self-contained breathing apparatus in demand or pressure-demand mode (negative or positive pressure) (2) Full-face gas mask, front- or back-mounted type with industrial size canister for toxic dust and hydrocyanic acid gas

[1] *By courtesy of NIOSH from DHEW (NIOSH) publication No. 77-108.*

certain that all absorption of cyanide has been stopped; for example, contaminated clothes may allow further entry of the poison into the skin.

In more severe cases, intravenous therapy is required. First, 0.3 g of sodium nitrite is given (10 cm^3 of a 3% solution at a rate of 2.5 to 5 cm^3/min) to increase methaemoglobin formation. Following this, thiosulphate is given to hasten and increase the formation of harmless thiocyanate. The dose is 25 g of sodium thiosulphate given by vein (50 cm^3 of a 25% solution at the rate of 2.5 to 5 cm^3/min). The sodium nitrite and thiosulphate therapy by vein should be repeated at half the dose in an hour if the symptoms recur or persist. [Cobalt EDTA has been reported to be very effective in the treatment of HCN poisoning, although with dangerous cardiocirculatory side-effects. It has therefore been suggested for the treatment of comatose patients not recovering with nitrite-thiosulphate therapy.]

Unfortunately some widely distributed handbooks of therapy include the statement that methylene blue is useful in cyanide poisoning. At certain concentrations, methylene blue converts methaemoglobin to haemoglobin, but since methaemoglobin binds cyanide, thus relieving its toxic effect, the use of methylene blue can be dangerous. At other concentrations methylene blue forms methaemoglobin, hence its suggested use in cyanide poisoning. However, the use of methylene blue is not recommended, since the beneficial concentration cannot be determined in emergency conditions (see ANTIDOTES).

HARDY, H. L.
BOYLEN, G. W. Jr.

CIS 79-1123 *Report on an investigation of the use of hydrocyanic acid in greenhouses and related health aspects* (Rapport inzake een onderzoek naar het gebruik van blauwzuur in de glastuinbouw en de veiligheidsaspecten daarbij). Van Nijnatten, P. W. A (Arbeidsinspectie, Directoraat-Generaal van de Arbeid, Balen van Andelplein 2, Voorburg, Netherlands) (Oct. 1978), 55 p. 8 ref. (In Dutch)

CIS 78-1510 "Cyanide exposure in fires". Symington, I. S.; Anderson, R. A.; Thomson, I.; Oliver, J. S.; Harland, W. A.; Kerr, J. W. *Lancet* (London), 8 July 1978, 2/8080 (91-92). 15 ref.

CIS 80-462 "A biological indicator of chronic worker exposure to cyanide derivatives" (Recherche d'un indice biologique d'exposition chronique des travailleurs aux dérivés cyanés). Della Fiorentina, H.; De Wiest, F. *Archives des maladies professionnelles* (Paris), June-July 1979, 40/6-7 (699-704). Illus. 18 ref. (In French)

CIS 77-437 *Criteria for a recommended standard–Occupational exposure to hydrogen cyanide and cyanide salts (NaCN, KCN, and $Ca(CN)_2$).* DHEW (NIOSH) publication No. 77-108 (National Institute for Occupational Safety and Health, 4676 Columbia Parkway, Cincinnati, Ohio) (Oct. 1976), 191 p. 292 ref.

CIS 78-1663 "Hydrogen cyanide poisoning–Treatment with cobalt EDTA". Nagler, J.; Provoost, R. A.; Parizel, G. *Journal of Occupational Medicine* (Chicago), June 1978, 20/6 (414-416). 11 ref.

Cybernetics

Scientists engaged in a variety of different areas of research during the 1940s converged in developing the view that a general theory of communication and control might find application in a wide range of systems. That is to say, it became apparent that there was much in common between the human brain and the electronic computer; between diplomatic cyphers and mechanisms of genetic determination; between automatic pilots and skilled process operators. Norbert Wiener, an American mathematician, and one of the most articulate exponents of this new movement, coined (or, rather, resurrected) the word "cybernetics" to signify the study of "control and communication in the animal and the machine".

There has been a good deal of controversy over the formal definition of the word "cybernetics" and no single trend of development has emerged. It is, however, possible to describe certain characteristics of systems which have been of interest to cyberneticians. Firstly, the emphasis is upon information rather than upon energy. Secondly, it is upon the general principles of system behaviour rather than upon the details of any particular system that cyberneticians have concentrated. Thirdly, a good deal of attention has been devoted to complex probabilistic systems rather than to simple deterministic ones. In what follows here, a few basic notions will be conveyed by way of example.

Cybernetics and mechanical systems

One of the most fundamental concepts in the theory of control is that of feedback, which involves the transmission of information from the output of a system back to the input. One of the earliest examples of a mechanical feedback system was that employing Watt's centrifugal governor to achieve constant speed in the steam engine (see figure 1). The speed at which the locomotive travels

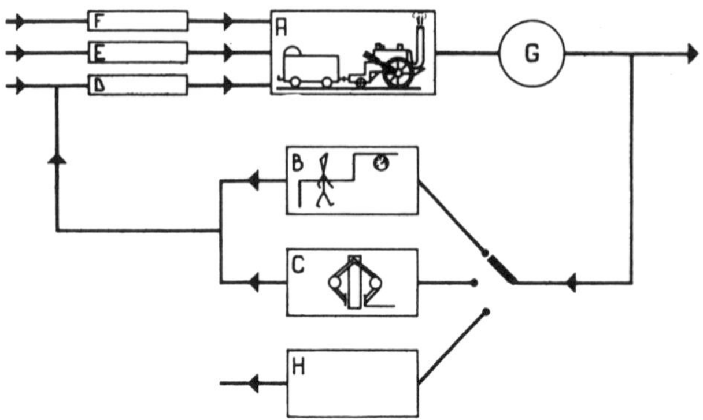

Figure 1. Negative feedback for maintenance of constant speed using either a human controller or a governor. A. "The Rocket" locomotive. B. Engine driver with speed indicator and control levers. C. Watt's governor. D. Regulator setting. E. Gradient. F. Load. G. Speed. H. Open-loop mode.

(G) will be determined by the values of three input variables, viz. the load (F), the gradient (E), and the regulator setting (D). Should it be required to maintain constant speed irrespective of changes in either load or gradient, then two alternative techniques are available. Firstly, a human driver, equipped with a speedometer and a control lever (B), may be employed to introduce compensation for changes in the input values. Secondly, this compensation may be provided by the governor (C), which is constructed in such a way that increases in its own rotational speed lead to a reduction in the steam valve setting, and decreases in speed produce an increment in the valve setting. By either of these methods, the system is controlled by "closing the loop". The feedback is described as negative, since the controlling signal opposes the direction of the error. Negative feedback tends to lead to stability. Positive feedback, conversely, tends to give rise to instability, since deviations from the target value are amplified. If no feedback is provided, in which case the system is said to be in "open-loop mode" (H), the output will vary in ways dictated by changes in load and gradient. This complete dependence upon changes in the environment signifies a lack of control.

In the above example, it has been assumed that the desired speed was predetermined and either conveyed to the human driver or built into the governor. At the next level of sophistication, it might be possible to devise a system which determined its own speed on the basis of, say, minimising the cost of travel. Such a system, comprising numerous interconnected feedback loops, would have outstripped the level of simple self-regulation and have approached the idea of an adaptive system, that is an organisation capable of adjusting its own parameters in order to cope with certain features of its environment.

Cybernetics and human behaviour

Certain analogies between the behaviour of mechanical systems and human behaviour will already be obvious. Biological mechanisms achieve the regulation of such variables as body temperature by means of relatively simple negative feedback control systems. Figure 2 illustrates schematically the neural regulation of muscle contraction. Goal-directed behaviour of a more complex type involves the use of adaptive control systems in which the target value of any single variable (e.g. amount of cash in hand) will vary a good deal from time to time.

The advent of the electronic computer has facilitated progress in the study of large-scale information systems. Computers can not only perform arithmetical operations with speed and precision but may also be programmed to play games, learn, solve problems, recognise patterns and make decisions. The relevance to cybernetics of such abilities is that the computer encourages us to think afresh about the nature of such activities, thus helping to remove the mystery from forms of behaviour long assumed to be the unique province of human thought processes.

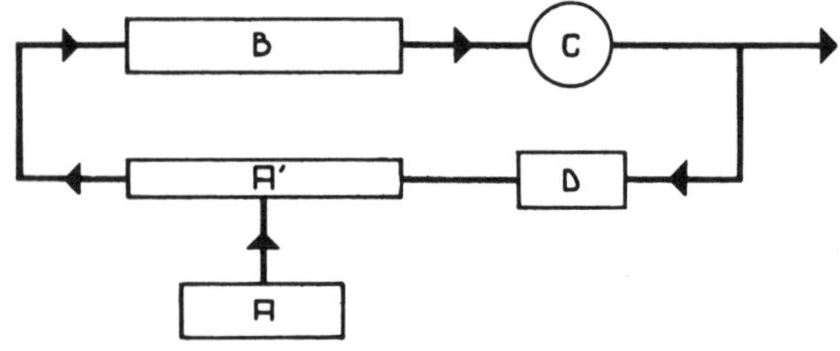

Figure 2. Schematic representation of neural loop for controlling extent of muscle contraction. A. Target value. A'. Brain. B. Muscle. C. Force. D. Force sensor.

In any information system it is necessary for signals to be conveyed from one location to another. Cybernetics has gained considerable impetus from studies in telecommunications where, once again, interesting comparisons may be drawn between hardware and living systems.

Although, as has been stated, the prime concern of the cybernetician is essentially theoretical, the influence of the general theories of control and communication has been substantial in such areas as automation, computer technology, industrial management, medicine and programmed learning.

Bionics. Devices may be produced which simulate the functional mode of operation of biological systems. Bionics is concerned to produce such devices, particularly in relation to the human body. Some progress has been made in initiating the sense organs, limb movements and certain aspects of the central nervous system. Such studies contribute not only to the basic understanding of human anatomy and physiology, but may lead to the production of improved prostheses.

Robotics. Hardware devices capable of carrying out manipulative functions are especially valuable for deployment in hostile environments, such as those containing high levels of radiation. Robotics is concerned with the development of such devices, not necessarily by the direct imitation of human control mechanisms. The availability of powerful data processing equipment makes feasible the production of sophisticated and "intelligent" robots. See ROBOTS AND AUTOMATIC PRODUCTION MACHINERY.

EDWARDS, E.

Design for a brain. Ashby, W. R. (New York, John Wiley, 2nd ed., 1960), 286 p.

The foundations of cybernetics. George, F. H. (London, Gordon and Breach Science Publishers, 1977), 286 p.

Cybernetics and forecasting techniques. Ivakhnenko, A. G.; Lapa, V. G. (Amsterdam, Elsevir North Holland, 1967), 168 p.

Cycloparaffins

CYCLOALKANES

The cycloparaffins are alicyclic hydrocarbons in which three or more of the carbon atoms in each molecule are united in a ring structure and each of these ring carbon atoms is joined to two hydrogen atoms, or alkyl groups. The members of this family include cyclopropane (C_3H_6), cyclobutane (C_4H_8), cyclopentane (C_5H_{10}), cyclohexane (C_6H_{12}) and derivatives of these such as methylcyclohexane ($C_6H_{11}CH_3$) and cyclohexene (C_6H_{10}). From the occupational safety and health point of view, the most important of these are cyclohexane, cyclopropane, methylcyclohexane and cyclohexene.

Cyclohexane (C_6H_{12})

HEXAHYDROBENZENE; HEXAMETHYLENE

m.w.	84
sp.gr.	0.78
m.p.	6.5 °C
b.p.	81 °C
v.d.	2.9
v.p.	40 mmHg (5.42·10³ Pa) at 6.7 °C
f.p.	−20 °C
e.l.	1.3-8%

insoluble in water; soluble in all proportions in ethyl alcohol and ethyl ether

colourless liquid with a sweetish odour similar to that of chloroform; pungent when impure.

TWA OSHA	300 ppm 1 050 mg/m³
STEL ACGIH	375 ppm 1 300 mg/m³
IDLH	10 000 ppm
MAC USSR	80 mg/m³

It is produced by the hydrogenation of benzene in the presence of a nickel-based catalyst at 170-200 °C; in some countries, it is extracted from light ends of cycloparaffinic petroleum oil. It is used as a starting material in organic syntheses, especially for the production of polyamides after oxidation to adipic acid. It is also employed as a solvent; in many cases it is a good substitute for benzene, especially as a cyclohexane/toluene mixture for use as a natural rubber solvent.

Cyclopropane (C_3H_6)

TRIMETHYLENE

sp.gr.	0.72
m.p.	−126.6 °C
b.p.	−33 °C
v.d.	1.88
e.l.	2.4-10.4%
i.t.	500 °C

soluble in water; freely soluble in ethyl alcohol and ethyl ether

a colourless gas with an odour similar to solvent naphtha.

TWA Netherlands 400 ppm 690 mg/m³

MAC Poland 300 mg/m³

Cyclopropane may be prepared by the reduction of 1,2-dibromocyclopropane or from 1,3-dibromopropane or 1,3-dichloropropane. It is used as an anaesthetic.

Methylcyclohexane (C_7H_{14})

HEXAHYDROTOLUENE

m.w.	98.2
sp.gr.	0.77
m.p.	−126.3 °C
b.p.	100.4 °C
v.d.	3.4
v.p.	40 mmHg (5.42·10³ Pa) at 22 °C
f.p.	−4 °C
e.l.	1.2%
i.t.	250 °C

soluble in ethyl alcohol and ethyl ether; insoluble in water

colourless liquid.

TWA OSHA	500 ppm 2 000 mg/m³
TLV ACGIH	400 ppm 1 600 mg/m³
STEL ACGIH	500 ppm 2 000 mg/m³
IDLH	10 000 ppm
MAC USSR	50 ppm 230 mg/m³

Methylcyclohexane is produced by petroleum fractionation. It is used as a solvent for cellulose esters; it is also used in organic syntheses.

Cyclohexene (C_6H_{10})

1,2,3,4-TETRAHYDROBENZENE

m.w.	82
sp.gr.	0.81
m.p.	−103.5 °C
b.p.	83 °C
v.d.	2.8
v.p.	67 mmHg (8.9·10³ Pa) at 20 °C
f.p.	< −7 °C
e.l.	1-5%
i.t.	310 °C

soluble in all proportions in ethyl alcohol, ethyl ether, benzene; insoluble in water

colourless liquid with a sweetish odour.

TWA OSHA	300 ppm 1 015 mg/m³
IDLH	10 000 ppm

It is produced by distillation or dehydration of cyclohexanol and is used in the manufacture of adipic, maleic and cyclohexanecarboxylic acids.

HAZARDS

These cycloparaffins and their derivatives are flammable liquids and their vapours will form explosive concentrations in air at normal room temperature.

They may produce toxic effects by inhalation and ingestion and they have an irritant and defatting action on the skin. In general, the cycloparaffins are anaesthetics and central nervous system depressants, but their acute toxicity is low and, due to their almost complete elimination from the body, the danger of chronic poisoning is relatively slight.

Cyclohexane. Few reports are available concerning the human toxicity of cyclohexane. One study in France indicated that workers exposed to cyclohexane were suffering from blood disorders (two severe cases); however, analysis of the incriminated product showed it had a benzene content ranging from 1.0-3.6 g/100 cm³ and a miscellaneous aromatic hydrocarbon content of 0.9-12.5 g/100 cm³. Studies on workers employing cyclohexane for nickel degreasing in the United Kingdom revealed no harmful effects.

Animal experimentation has shown that cyclohexane is far less harmful than benzene and, in particular, does not attack the haemopoietic system as does benzene. It is thought that the virtual absence of harmful effects in the blood-forming tissues is due, at least partially, to differences in the metabolism of cyclohexane and benzene. Two metabolites of cyclohexane have been determined—cyclohexanone and cyclohexanol—the former being partially oxidised to adipic acid; none of the phenol derivatives that are a feature of the toxicity of benzene have been found as metabolites in animals exposed to cyclohexane, and this has led to cyclohexane being proposed as a substitute solvent for benzene.

[The leakage of hot cyclohexane from a pipe connecting two reactors in June 1974 caused an explosion at a chemical plant at Flixborough (United Kingdom) which killed 28 people, injured 36 and resulted in extensive material damage (see RISK ACCEPTABLE IN AN INDUSTRIAL SOCIETY).]

Methylcyclohexane. This has a toxicity lower than that of cyclohexane. Animal studies show that the majority of this substance entering the bloodstream is conjugated with sulphuric and glucuronic acids and excreted in the urine as sulphates or glucuronides, and in particular the glucuronide of *trans*-4-methylcyclohexanol.

SAFETY AND HEALTH MEASURES

In operations where dangerous concentrations of cycloparaffins may occur in the workplace, exhaust ventilation should be installed and workers should be provided with respiratory protective equipment. Where there is a danger of splashes or prolonged skin contact, eye and face protection and hand and arm protection should be worn. Where workers are required to enter vessels that have contained cycloparaffins, the precautions appropriate to work in confined spaces should be applied.

Strict fire protection and prevention measures are required in the storage, handling and use of these dangerous substances. Storage rooms should be bunded to prevent the spread of escaping liquid and electrical installations should be of flameproof type. Naked flames and smoking should be prohibited at the workplace.

TRUHAUT, R.

Cyclohexane (Cyclohexan). Gesundheitsschädliche Arbeitsstoffe–Toxikologisch-arbeitsmedizinische Begründung von MAK Werten (Verlag Chemie, D-6940 Weinheim) (1977), paper dated 1.2.1971 (1-3). 17 ref.

CIS 80-1363 "Lung uptake and metabolism of cyclohexane in shoe factory workers". Perbellini, L.; Brugnone, F. *International Archives of Occupational Environmental Health* (West Berlin), Mar. 1980, 45/3 (261-269). Illus. 12 ref.

CIS 76-917 *The Flixborough disaster–Report of the Court of Inquiry.* Department of Employment (London, HM Stationery Office, 1975), 56 p. Illus.

Dairy products industry

Dairy produce has formed an important element in human food since the earliest days when animals were first domesticated. Originally the work was done within the home or farm and even now much is produced in small undertakings, although in many countries large-scale undertakings also exist. Co-operatives have been of great importance in the development of the industry and the improvement of its products.

In many countries there are strict regulations governing the preparation of dairy products, for example a requirement that all liquid shall be pasteurised. In most dairies, milk is pasteurised; sometimes it is sterilised or homogenised. There are some variations in butter manufacturing processes and more in the production of the different kinds of cheese. Soft cheeses are produced by souring with lactic acid; hard cheeses by coagulating the protein with rennet, cooking and pressing the curd, ripening with mould or bacteria over a period of months or years. Semi-hard cheeses are produced by a mixture of these processes. Difference in the milk used, in the ripening, cooking and curing determine the particular consistency and flavour of the cheese. The flowcharts shown in figures 1, 2 and 3 indicate the stages involved in the large-scale processing of milk, butter and cheese.

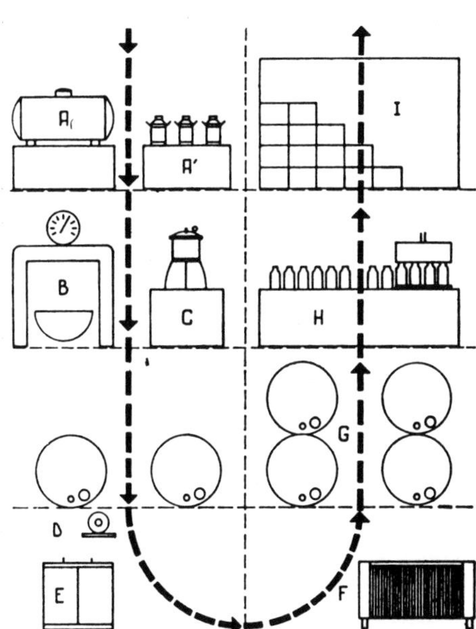

Figure 1. Consumer milk processing. A. Tanks. A'. Cans. B. Weighing machine. C. Gerber centrifuge (control of fat content). D. Pump. E. Pressure filter. F. Plate apparatus (pasteurisation). G. Storage tanks. H. Bottling and sealing machine. I. Cooling room.

Other dairy products include:

(a) Condensed milk which is prepared by vacuum-evaporation of whole or skim milk at 50-60 °C. Sweetened, condensed milk is produced by the addition of cane sugar and needs no further sterilisation. Unsugared, condensed milk (evaporated milk) is sterilised. These products are usually packed in tins.

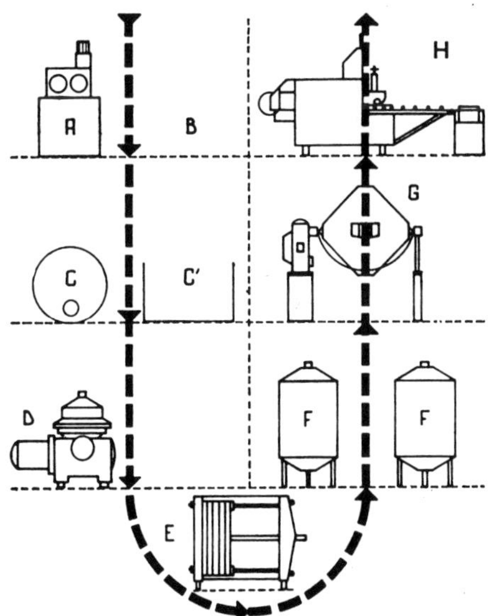

Figure 2. Butter processing. A. Reductase control. B. High-temperature pasteurising (optional). C. Tanks. C'. Vats. D. Separator. E. Cream pasteuriser. F. Souring tanks. G. Churn (kneading, salting). H. Butter moulder.

(b) Powdered milk is prepared either by the cylinder method, when the milk is dried between headed cylinders and the powder scraped off, or by spray-drying, when the milk is atomised in a hot-air drying chamber.

(c) Sour milk products are prepared by self-souring or treatment of the milk in many national varieties: skyr, mazun, yoghurt, leben, kumys, kefir, urda, skuta, whey-champagne.

HAZARDS AND THEIR PREVENTION

Machinery and plant. Modern dairy products are often highly mechanised and contain numerous rotating machine components which require well designed and regularly maintained guards or other devices to ensure safety.

Separators may burst, causing very serious injuries or material damage, and the supplier's instructions for installation, operation and maintenance should be closely followed. Proper assembly is essential and a separator should be halted at the first sign of excessive vibration. Gerber centrifuges should be protected with an interlocked cover and, if steam driven, should be fitted with a speed governor. Rotary churns should be equipped with bar fencing, preferably interlocked with the churn drive unit. The moving parts of ice-crushers, cheese graters, etc., should be provided with adequate machinery guarding to prevent access of hands or fingers to dangerous parts. Serious accidents may occur on bottling machines caused by bursting bottles and flying glass. The machines should, where possible, be fitted with screens and workers provided with eye and face protection, hand and arm protection, aprons, etc. Wide use is made of mechanical handling equipment and particular attention should be paid to protecting the nip between conveyor drums and pulleys.

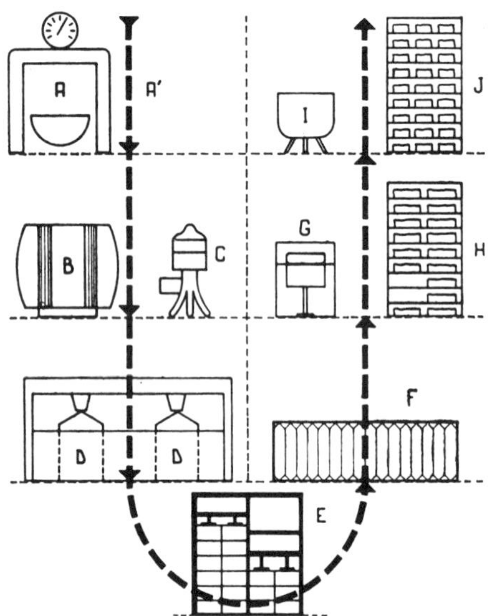

Figure 3. Cheese processing. A. Weighing machine. A'. Control. B. Plate apparatus for pasteurisation at low temperature. C. Cleansing separator. D. Cheese vat with stirrer. E. Cheese press. F. Brine vat. G. Scrapping off. H. Fermentation store. I. Waxing. J. Ripening store.

Steam boilers, steam receivers and pressure vessels of all kinds present hazards which can be kept under control by regular inspection and maintenance. Refrigerating plant may present either explosive or toxic hazards, depending on the refrigerant used. This plant should also be regularly maintained and inspected and respiratory protective equipment should be provided for emergencies.

Falls. Slippery floors and stairways may cause falls, complicated by cuts and abrasions if broken glass is involved. Non-slip floor surfaces and non-slip footwear, cleanliness and good housekeeping will prevent many accidents. Staircases should be provided with railings and treads should be non-slip.

Falls into open milk vats, vessels containing scalding water, etc., can be prevented by guarding: if the rim of a vat is less than 1 m above floor level it should be protected by railings. Fixed ladders should be used for inspection of vats.

Chemical hazards and burns. Strong alkalis and nitric acid used for cleaning may cause injury. The chemical content of branded products should be stated so that workers are aware of the risks. Personal protective equipment (hand protection, eye protection, etc.) should be used and workers should be instructed in safe methods of handling. First aid and medical care should be available for instant treatment of corrosive injuries. Burns and scalds are common. Some accidents can be prevented if batteries for mixing water and steam are so constructed that it is impossible to turn on the steam until after the water has been turned on. Eye protection is advisable during steam cleaning.

Electrical hazards. The prevailing damp and steamy conditions make it necessary to provide special protection for electrical equipment, in particular provision for earthing, double insulation and specially protected portable electric power tools. Where possible, low-voltage electrical installations are desirable.

Fires and explosions. Milk powders are combustible: spray-drying of milk products can cause fires or dust explosions in the drying chambers. The suppliers' rules regarding proper running, temperature and cleanliness should be strictly observed and explosion relief vents should be fitted.

Health hazards. Dairy workers may be exposed to animal infections from untreated milk, e.g. brucellosis and bovine tuberculosis. Many countries have strict regulations designed to protect public health, which also protect dairy workers. Veterinary and medical supervision are important and personal hygiene, coupled with good washing facilities, etc., is essential. In cheese-making, workers may sometimes contract cheese-workers' itch from a cheese maggot.

Much of the work is carried out in cold and damp conditions. Protective clothing, boots, clogs, aprons, gloves and warm clothing for some operations can promote general good health and well-being. Warmed restrooms, canteens, messrooms and the provision of hot drinks are also important.

ØRSTED-MUELLER, A. S.

Milking:

CIS 78-1064 "Rubber as a harmful factor in milker's eczema in socialist agriculture" (Gummi als Schadstoff beim Melkerekzem in der sozialistischen Landwirtschaft). Jung, H. D. *Das deutsche Gesundheitswesen* (Berlin), 1977, 32/39 (1866-1872). Illus. 23 ref. (In German)

Milk processing:

CIS 74-557 "Safety in dairies and dairy-products plants" (Dienstnehmerschutz in Molkereibetrieben und bei der Erzeugung von Milchnebenprodukten). Lonsky, H.; Schneider, K. (80-94). 6 ref. *Amtstätigkeit der Arbeitsinspektorate im Jahre 1970* (Vienna, Verlag des Zentral-Arbeitsinspektorates, July 1971) (In German)

CIS 76-855 *Health and safety guide for fluid milk processors.* DHEW (NIOSH) publication No. 75-152 (National Institute for Occupational Safety and Health, 4676 Columbia Parkway, Cincinnati) (May 1975), 72 p. Illus.

Milk powder:

CIS 75-1217 "Dust and flammability characteristics of settled dust and dust-air mixtures" (Staub- und Zündkennwerte von lagernden Stäuben und Staub-Luft-Gemischen). Keissling, R. *Unser Brandschutz* (Berlin), 5/1974 (71-74). Illus. 5 ref. (In German)

Cheese making:

CIS 78-1611 "Allergic manifestations in cheese workers—Clinical, epidemiological and immunological study" (Manifestations allergiques chez les fromagers—Etude clinique, épidémiologique et immunologique). Molina, C.; Tourreau, A.; Aiache, J. M.; Brun, J.; Jeanneret, A.; Roche, G. *Revue française d'allergologie et d'immunologie clinique* (Paris), 1977, 17/5 (235-245). 20 ref. (In French)

"Dermatological and allergic hazards of cheesemakers". Niinimaäki, A.; Saari, S. *Scandinavian Journal of Work, Environment and Health* (Helsinki), Sep. 1978, 4/3 (262-263). 4 ref.

Dangerous substances

Attempts have been made by individual countries, as well as by regional and international organisations, to classify substances according to their intrinsic hazards. Both definitions and terminology can, however, differ and this may lead to confusion unless there is a specific standard description which can be generally applied.

Within certain limits, the hazard presented by a substance may vary according to the conditions in which it is used or handled. For example, a substance being transported may require precautions to be taken against the risks of fire or of acute poisoning resulting from its loss; in a work situation the use of a substance can give rise to a risk from prolonged exposure, while the same substance being used intermittently for, say, household purposes would not present the same hazard. A typical example is benzene, which for purposes of transport is considered as a fire hazard only, but used industrially, it is also a severe toxic hazard.

It must be recognised that it will perhaps be impossible to achieve a uniform system taking into account all the different terms and criteria against the background of the different uses for which the substance may be required, and the different objectives of the organisations dealing with it, as well as making allowance for difficulties arising from the need for appropriate language to clarify details that are not comprehensible from a simple number. The points that will be considered in this connection include: the hazard of the substance to man, both workers and general population, as well as the environmental risk during transport and handling.

Each classification is thus drawn up in a schematic and simplified form which is not based on strictly scientific principles. Its purpose is to fulfil the need for conveying information to the user in the most direct manner, even if this is somewhat approximate; to stress the need for precautions to be taken against certain hazards; and to outline the necessary protective measures.

EEC classification of hazards

The following provisions are derived from the EEC Council Directive relating to the classification, packaging and labelling of dangerous substances for the protection of the population and of the workers using them. The EEC directive, which has become a law in its nine member States, identifies different categories of hazard associated with the nature of the substance concerned, exception being made in the case of medicinal products, narcotics, radioactive substances and foodstuffs. These categories relate to physical hazards, biological hazards and environmental hazards. They are:

(a) physical hazards:
- explosive
- oxidising
- extremely flammable
- highly flammable
- flammable;

(b) biological hazards:
- very toxic
- toxic
- harmful
- corrosive
- irritant
- carcinogenic
- mutagenic
- teratogenic;

(c) environmental hazards:
- dangerous for the environment.

When one or more of these risk categories are assigned to a substance to be placed on the market, its packaging must be labelled with the danger symbol indicating the danger involved, together with the standard phrases indicating the special risks arising and the safety advice relating to the use of the substance (see DANGEROUS SUBSTANCES, LABELLING AND MARKING OF).

For purposes of comparison, reference may be made to the United States ANSI Standard Z 129.1-1976 which, although not constituting a legal standard, sets out requirements for classification and labelling.

Physical hazards

Explosives. Substances or preparations which may explode under the effect of flame or which are more sensitive to shocks or friction than dinitrobenzene. This definition is taken from the international standard for the transport of dangerous goods and makes use of dinitrobenzene as a standard reference base. This is done in order to overcome difficulties arising from the variables that may influence experimental laboratory determinations. Constant results cannot yet be arrived at with standard instrumentation reproducing shocks and frictions.

The classification of a product as explosive is not always the result of a laboratory test, but results from a global judgement made on the basis of its behaviour. Perhaps it would be more correct to make the distinction between "explosive" in the sense of a product designed expressly for the purpose of creating explosions, and "explosive" in the sense of a product not intended to be used for its explosive property but which because of its liability to explode may constitute a hazard.

Within the framework of the EEC classification, explosives are marked with the symbol of an exploding bomb and the indication "explosives".

Oxidising. These are substances or preparations which give rise to highly exothermic reaction when in contact with other substances, particularly flammable substances. There is no recognised standard method that has been sufficiently proof-tested to enable a substance in this category to be determined as oxidising. The inclusion in this category of a substance is made on the basis of the oxidising properties of its active constituent. By way of example, it includes all organic peroxides not classified as explosives, although it may be that these compounds should, as a matter of principle, be classified as "explosives" in so far as they are unstable and as "oxidising" in so far as they are oxidising.

The EEC directive prescribes a symbol in the form of a flame over a circle with the indication "oxidising". According to the ANSI standard, oxidising substances are defined as "strong oxidisers".

Extremely flammable. These have a flash point lower than 0 °C and a boiling point lower than or equal to 35 °C. Examples of this category include ethyl ether, ethyl formiate, acetaldehyde, and, at the limit of the definition, hydrogen, methane, ethane, acetylene, ethylene, and carbon monoxide, as well as liquefied petroleum gas, ethylene oxide and other products which take the form of gases at ambient temperatures and which form gas-air mixtures whose explosive limits cover a wide range.

The symbol prescribed for this category is a flame with the indication "extremely flammable". It may be accompanied by the name of the liquefied gas.

Highly flammable. This group includes the following:

(a) substances or preparations which may become hot and finally catch fire in contact with air at ambient temperature without any application of energy;

(b) solid substances or preparations which may readily catch fire after brief contact with a source of ignition and which continue to burn or to be consumed after removal of the source of ignition;

(c) liquid substances or preparations having a flash point below 21 °C;

(d) gases which are flammable in air at normal pressure; or

(e) substances or preparations which, in contact with water or damp air, evolve highly flammable gases in dangerous quantities.

It includes, among others, hydrides, aluminium alkyls, phosphorus and most solvents. Its symbol is a flame with the indication "easily flammable".

Flammable. These are liquid substances or preparations having a flash point equal to or greater than 21 °C and less than or equal to 55 °C. There is no symbol, but the indication "flammable". This category includes most of the solvents and many of the petroleum distillates.

The flash point limits prescribed in the EEC directive are derived from the international standards for the transport of dangerous goods which, in a recent amendment, have also laid down that the flame symbol be applied to products having a flash point of up to 55 °C. This is not the only case where differences exist between the standards for the transport and those for the marketing of goods, as a result of which it is possible to find products in circulation with a dual classification and ticketing that is even contradictory. The determination of flammability by means of the flash point makes it possible to assign a classification on the basis of well defined criteria and laboratory tests, and these are described in Annex V of the EEC directive and also in the ANSI standard.

The definitions of the ANSI standard are different, viz.:

(a) extremely flammable liquids: those having a flash point of less than −6.7 °C (20 °F);

(b) flammable liquids: those having a flash point of less than 37.8 °C (100 °F);

(c) flammable gases: those which form a flammable mixture with air at a concentration of 13% or less, or which can be ignited from beyond a fixed distance;

(d) flammable solids: those which catch fire spontaneously or from any other cause and burn vigorously and persistently in such a manner as to create a hazard.

Biological hazards

Toxic and harmful substances. These are defined by the EEC as substances or preparations which, if they are inhaled or ingested or if they penetrate the skin, may involve:

(a) extremely serious, acute or chronic health risks and even death–classified as "very toxic";

(b) serious, acute or chronic health risks and even death–classified as "toxic";

(c) limited health risks–classified as "harmful".

The ANSI standards do not refer to degrees of harmfulness but categorise other substances or preparations such as "strong sensitisers" and "physiologically inert vapour or gas".

The criteria for the assignment of these categories are established by determining the acute toxicity of the substance or preparation in animals, based on the parameters in table 1.

In cases where other short-term or long-term effects are noticed which justify a classification of higher toxicity, the EEC allows for such classification to be made case by case.

Very toxic and toxic products are marked with a skull and crossbones together with an indication of the risk involved, while harmful substances are marked with a St. Andrew's cross and the indication "harmful".

Corrosive and irritant substances. According to the EEC definition, corrosive substances or preparations include those which, on contact with living tissues, can destroy them.

The definition of irritant substances or preparations includes those which, through immediate, prolonged or repeated contact with the skin or mucous membrane, can cause inflammation. These definitions are the same as those adopted in the ANSI standard which, however, makes it clear that they do not apply in the case of inanimate objects or surfaces.

It is not easy to establish a limit between corrosive and irritant, although in the case of attack by a corrosive substance a deep necrosis results, while in the case of an irritant the action should be of a superficial nature. It has been proposed that the classification be correlated with the results obtained from the Draize test, but in fact the values to be found in the literature are generally based on subjective judgements and must inevitably be regarded with a certain amount of suspicion. The EEC have adopted the principle that corrosive substances in a lesser concentration no longer act to the full extent, and can be included among the irritants. Similarly the irritants may not have the same effects when being dealt with in concentrations below those usually encountered. It is understood that the classification of a substance as irritant takes into account the effects over the affected area.

Carcinogenic. The definition adopted by the EEC for carcinogenic substances is the following: substances or preparations which, if they are inhaled or ingested or if they penetrate the skin, may induce cancer in man or increase its incidence (see CARCINOGENIC SUBSTANCES; CANCER, OCCUPATIONAL: LEGISLATION).

In 1978 a paper prepared by the International Agency for Research on Cancer (IARC) notified the existence of 26 substances recognised as carcinogenic, while a further 22 substances were recognised as being

Table 1.

Category	LD_{50} absorbed orally in rat (mg/kg)		LD_{50} by percutaneous absorption in rat or rabbit (mg/kg)		LC_{50} absorbed by inhalation in rat (mg/l/4 h)	
	EEC	ANSI	EEC	ANSI	EEC	ANSI
Very toxic	25	50	50	200[1]	0.5	2[2]
Toxic	25-200	50-500	50-400	200-1 000	0.5-2	2-20
Harmful	200-2 000		400-2 000		2-20	

Note: LD_{50} and LC_{50} represents the dose in mg/kg of body weight and the concentration in mg/l having a lethal effect on 50% of the animals treated.
[1] In 24 hours. [2] in 1 hour.

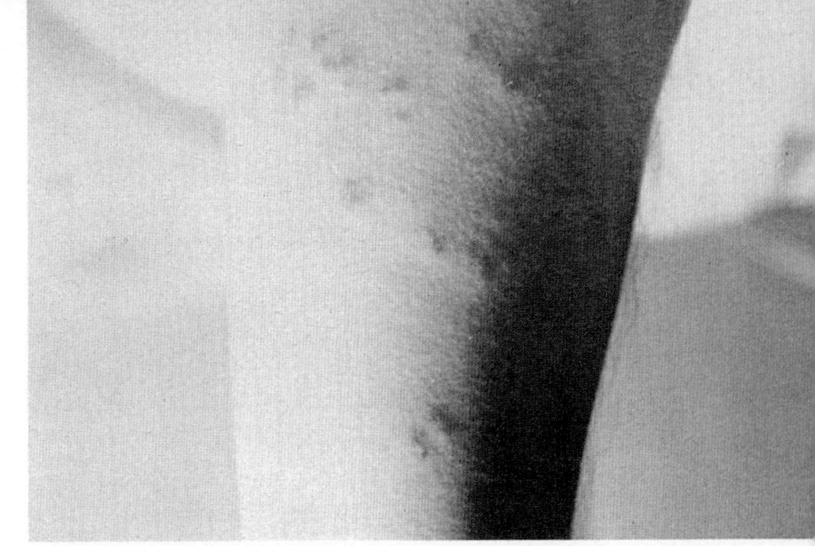

1. Papulo-vesicular dermatitis due to contact with sponges *(Spongia officinalis)*.

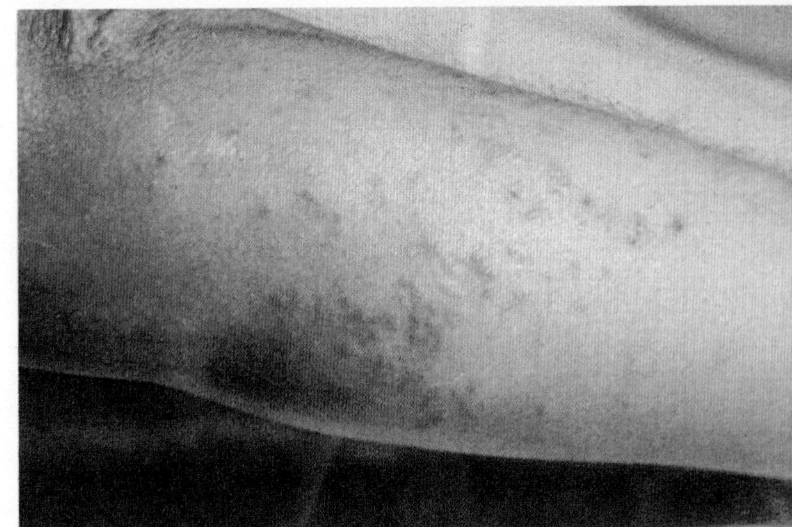

2. Papulo-vesicular dermatitis due to contact with a Sea Anemone.

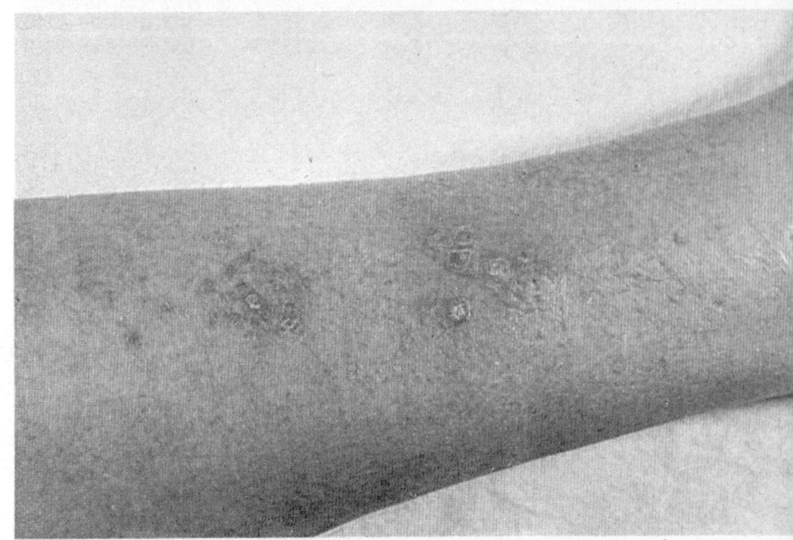

3. Persistent cutaneous granulomata following Sea Urchin puncture.

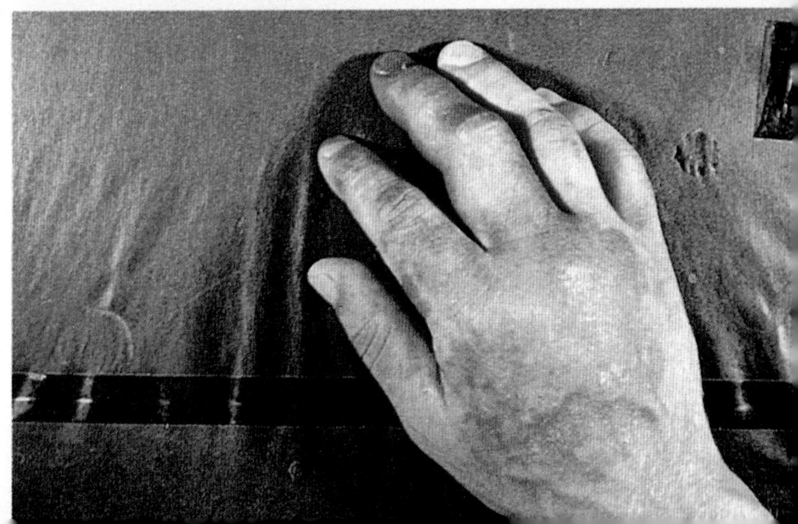

4. Haemorrhagic-oedematous reaction after a wound from a Weeverfish.

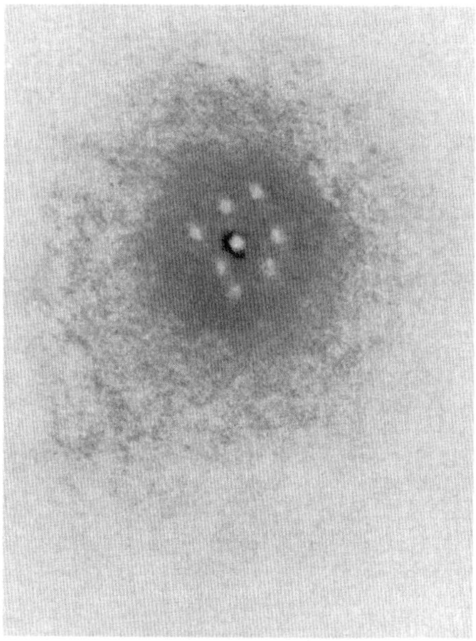

5. Anthrax infection on first or second day. *(By courtesy of the Health and Safety Executive, Employment Medical Advisory Service, London, United Kingdom.)*

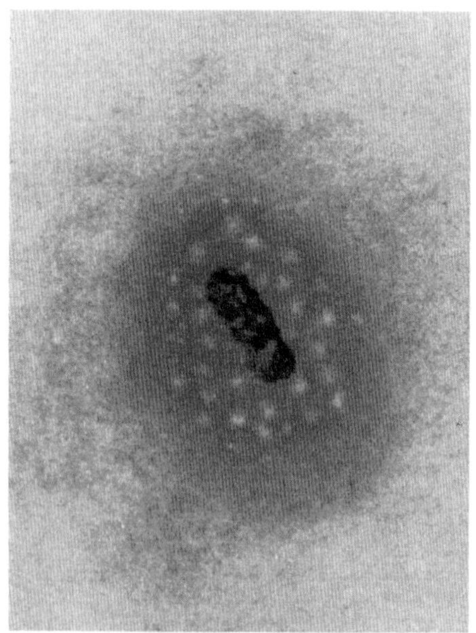

6. Anthrax infection on third or fourth day. *(By courtesy of the Health and Safety Executive, Employment Medical Advisory Service, London, United Kingdom.)*

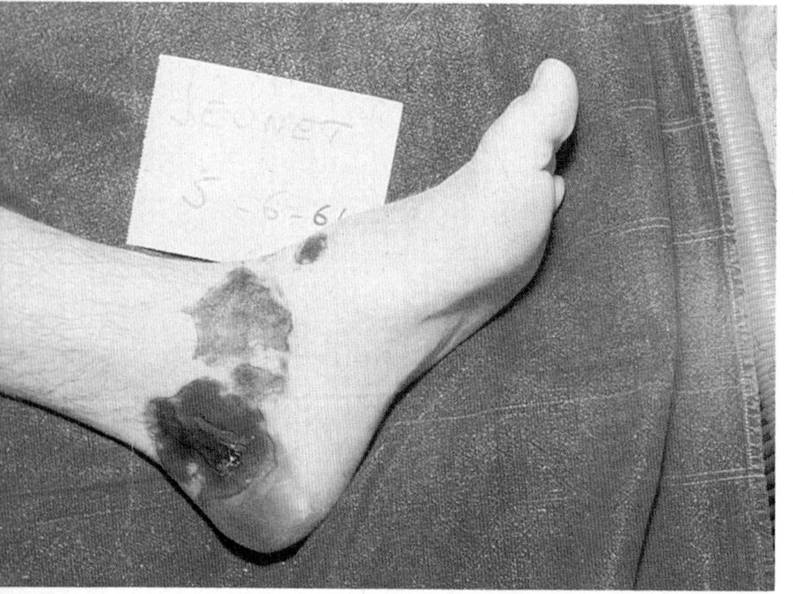

7. A burn caused by a strong acid.

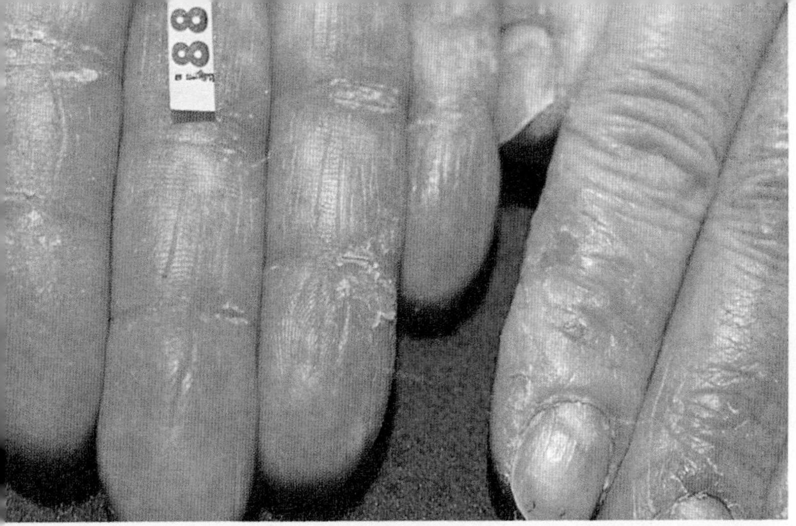

8. Irritant contact eczema in a hairdresser.

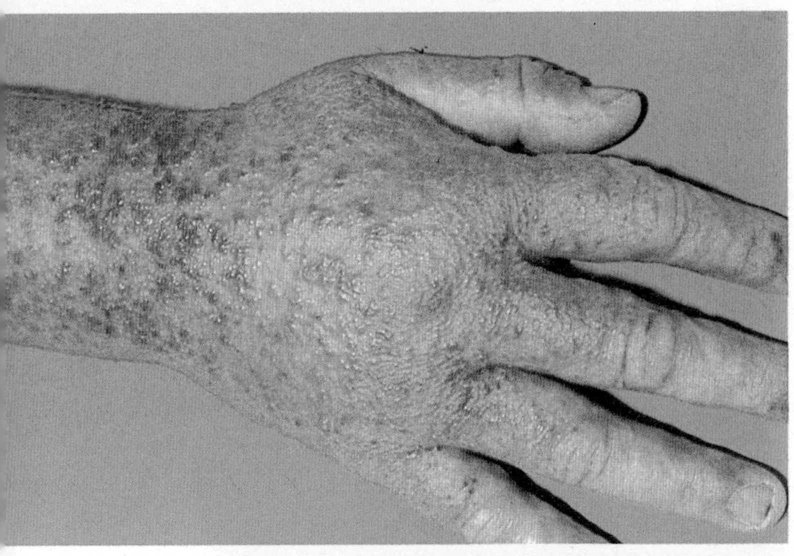

9. Allergic contact eczema in a plastics worker. Patch test positive to epoxy resin.

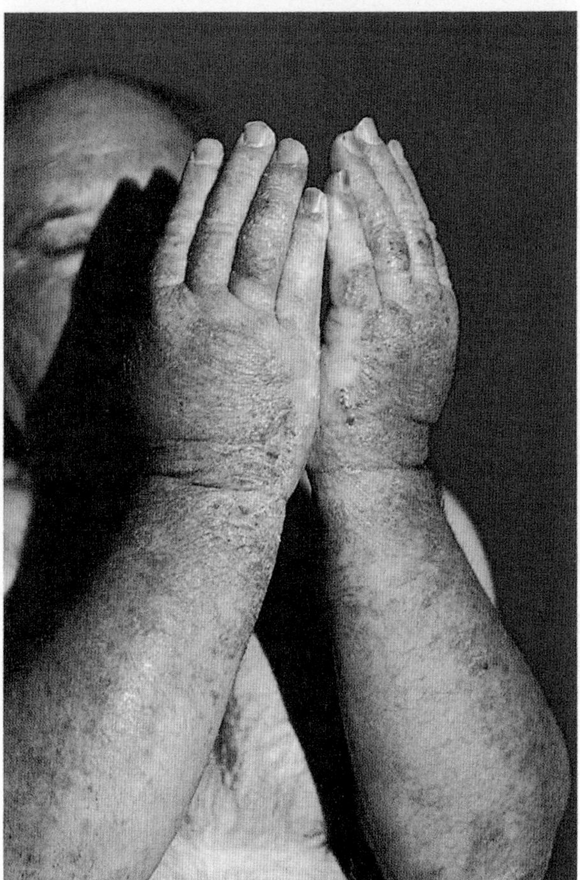

10. Allergic contact eczema in a rubber worker. Extensive pattern. Patch test positive to *p*-phenylenediamine derivatives.

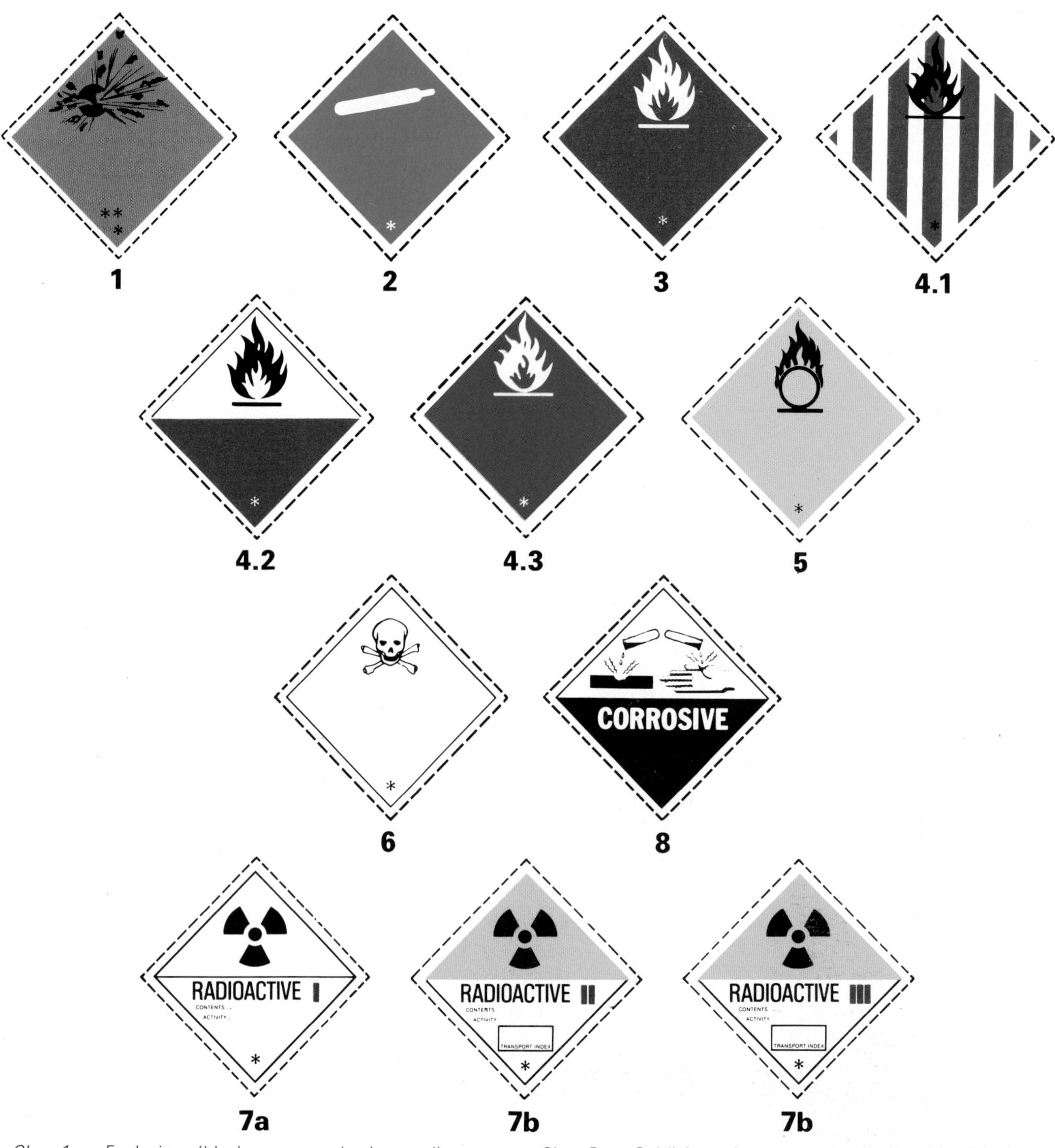

Class 1. Explosives (black on orange background).

Class 2. Compressed non-inflammable gases (black or white on green background).

Class 3. Inflammable liquids (black or white on red background).

Class 4.1. Inflammable solids (black on white background with vertical red stripes).

Class 4.2. Substances liable to spontaneous combustion (black on white background; lower half red).

Class 4.3 Substances which, in contact with water, emit inflammable gas (black or white on blue background).

Class 5. Oxidising substances; organic peroxides (black on yellow background).

Class 6. Division 6.1. Poisonous (toxic) substances (black on white background).

Class 8. Corrosives (black on white background; lower half black with white border).

Class 7. Radioactive substances: *(a)* black on white background with one vertical red stripe in the bottom half; *(b)* black on yellow background—bottom half white with two or three vertical red stripes.

11. United Nations danger labels for the transport of dangerous goods. The labels are all in the form of a square set at an angle of 45° (diamond-shaped) with minimum dimensions of 10×10 cm.

capable of inducing cancer changes in animal experimentation. In addition, 121 substances were described for which insufficient information had been gathered to enable a judgement to be passed.

Discussions held in the United States relating to draft definitions of carcinogens known or suspected as being dangerous to man were based on positive results obtained with various animal species (OSHA Fed. Reg. 4/10/77).

The field of carcinogens is one where it has not yet been possible to bring the administrative framework into conformity with the available scientific evidence. As in the case of general toxicity, the concentration levels below which effects are no longer manifested are determined, and dose/response curves are established also with some carcinogens. This implies that for those substances there must be concentration levels sufficiently low, and not zero, at which the defence mechanism of the organism does not permit the carcinogenic agent to act, or anyway, at least the probability exists that the effect of the agent on the experimental species concerned will disappear. The problem then arises of transposing to man the results obtained in animal experimentation, a problem which has not yet been scientifically resolved. If, on the one hand, the substance does not justify the precise description of "carcinogenic" for a given organism at a stated dose and with a known mode of entry, it would, on the other hand, seem prudent to assume that a substance that has given rise to this effect in any type of experiment is suspect. This may, however, lead to a distorted scale of values in that, in comparison with proved carcinogenic substances, the results obtained may be based on experiments where the exposure does not correspond with that met with in practice.

Mutagenic substances. Here also the EEC has not assigned a definition for this group of substances. At the present time, the term "mutagenic" can only be defined as a substance or agent that can induce biological activity. Whether this activity on cells, bacteria, yeasts or on some other organism of greater complexity can affect man and has undesirable consequences it is not yet possible to say with certainty. It should, however, be treated as being dangerous; meanwhile measures should be introduced to limit occupational exposure and investigations should be extended to find out whether these substances are really dangerous. The problem is more of a scientific nature than one of classification (see MUTAGENIC EFFECTS).

Teratogenic substances. The EEC classification for these substances has been left open, and it has been left to the Scientific Committee to prepare a standard definition after appropriate investigation and in the light of the development of scientific knowledge. The current definition of teratogens refers to substances or preparations having the effect of producing malformation in a fetus and is largely based on the experience of drug use or abuse by pregnant women.

There is also the need to find out more about the extrapolation of animal results to man, as well as to be aware of the importance of adopting preventive measures in good time in particular in the case of occupations employing women of childbearing age. Possibly more than the teratogenic effect, the total consequences for the reproductive process should be considered (see EMBRYOTOXIC, FETOTOXIC AND TERATOGENIC EFFECTS).

Environmental hazards

For the time being, the EEC classifies under this heading all those substances or preparations the use of which presents or may present immediate or delayed risk for the environment. An environmental hazard involves a comprehensive concept which is not always easy to substantiate. It can be a question of causing harm to a given variety of animal or vegetable life in one particular place; there is the risk of eutrophication; there is the risk of biological or physical accumulation which can be balanced through biotic or abiotic degradation; there is above all the negligible risk which becomes significant in case of large amounts. In order to arrive at an assessment of the danger of a product to the environment, a total evaluation must be made which takes into account the intrinsic risk that it represents, the quantity being dealt with, its ultimate destination, the extent to which it will be dispersed into the environment, the nature of the substances which result from its breakdown, and the metabolic functions in which it may be involved, to mention only the basic considerations.

The best defence against environmental hazards, in addition to cleansing and recycling, is to ensure that action is taken to provide full information and education to all those who are responsible for introducing the substance into the environment or for making use of it.

Dangerous preparations

The use of dangerous substances in their pure state is rare in most industrial, trade or domestic applications and they are usually to be found in preparations containing mixtures in varying proportions. It is therefore necessary to introduce a more or less conventional system for the evaluation of the hazard represented by these mixtures, based on the proportion of dangerous substances they contain.

Although the most objective manner of arriving at such a classification would be to base it on experimental data, this is only possible where existing physical parameters are easily measured and can be interpreted against well defined criteria. This is possible in the case of physical hazards such as explosive, flammable or oxidising substances but is not practicable in the case of toxic hazards, for which too many short-term and long-term biological tests would be necessary in order to assess the very large number of preparations on the market or which might be produced. For this reason attempts are being made to introduce a recognised classification in which each substance is given a coefficient or index based on its toxicity. The assignment of the preparation to a given category of toxicity or not is then dependent on a calculation which takes into account its concentration and category of risk. The same procedure has been adopted for the classification of paints, varnishes, inks and adhesives (Directive 77/728, *Official Journal of the European Communities*, 28 Nov. 1977) and has been included as an alternative to the biological testing of acute toxicity of pesticides (*Official Journal of the European Communities*, 29 July 1978). It is expected to be adopted for all preparations (*Official Journal of the European Communities*, 15 Oct. 1979). It is clear that the foregoing is an approximate solution in which the most difficult problem is to fix the coefficient or index of toxicity for each substance, taking into consideration the existence of more than one type of hazard and their possible joint acute and delayed effects.

It is certainly not easy to combine a fixed index for solvents, for which the possible chronic effects of the usual repeated exposure have been considered with the indexes assigned to agricultural pesticides, where the risk is one of acute toxicity, as agreed within many different national standards.

Summing up, the classification based on calculations represents an acceptable compromise which meets the requirements for information and prevention in the light of the limited knowledge at present available.

GARLANDA, T.

"Council Directive of 18 September 1979 amending for the sixth time Directive 67/548/EEC on the approximation of the laws, regulations and administrative provisions relating to the classification, packaging and labelling of dangerous substances". *Official Journal of the European Communities* (Brussels), 15 Oct. 1979, No. L.259/10.

"Dangerous substances and the US worker: current practice and viewpoints". Witt, M. *International Labour Review* (Geneva), Mar.-Apr. 1979, 118/2 (165-177). 23 ref.

Dangerous substances, labelling and marking of

The labelling and marking, by means of warning symbols or texts, of containers or systems that hold dangerous substances are essential safety precautions.

After a dangerous substance has been loaded or packaged and despatched by the manufacturer, and before it is used by the consumer, it may be handled by a number of workers who have no idea of the nature of the product or of the dangers it may present. The workers engaged in the manufacture of the substance are usually well acquainted with the relevant hazards, and they will normally take the appropriate precautions during packing, loading and despatching. However, persons handling the substance during its subsequent transport in the country of origin or the country of destination, or during transit operations cannot be aware of the nature and hazards of the contents unless an identification and warning label or mark is provided. Similarly, persons using the substance may also be unaware of the hazards involved or of the necessary precautions; this may arise, in particular, in the case of artisanal or other small undertakings or in agriculture, where specialist advice on safety and health matters may be difficult to obtain.

Hazard warning is an essential requirement in worker protection but should not be considered to provide complete protection in itself or to make it unnecessary to enforce safety measures commensurate to the risk.

Labelling systems are usually based on some hazardous substance classification scheme and have, as object, to provide workers and the general public with explicit information on the main hazard or hazards of the substances in question. The article DANGEROUS SUBSTANCES indicates the variety that exists in these classification schemes; however, it can be said that most dangerous substances can be adequately classified as either explosive, flammable, toxic, corrosive or radioactive.

Many countries have long-standing regulations or codes governing the labelling of dangerous substances, with particular regard to transport requirements. These provisions were originally drawn up at a national level and, consequently, there was a lack of uniformity between requirements in different countries. As international trade developed, the need for a harmonisation of national requirements and for the establishment of an international system became evident. This need was particularly acute in the case of the symbols used to warn of specific hazards, since such symbols can be used to overcome problems posed by communication in a number of different languages. It has for some time been recognised that, in international trade, graphic symbols are essential for indicating the main hazards of dangerous substances.

Various international recommendations on the classification and labelling of dangerous substances have now been drawn up and, with but few exceptions, all bear considerable similarity to each other. The organisations that have produced recommendations include the ILO, the United Nations (UN), the Inter-Governmental Maritime Consultative Organisation (IMCO), the Council of Europe, the Central Office for International Transport by Rail, the International Organisation for Standardisation (ISO) and the International Air Transport Association (IATA).

The UN labels are shown in colour plate 11 and the ILO labels in figure 1. The ILO system goes no further than to recommend a graphic symbol for each of six categories of dangerous substances; however, the UN system specifies the shape and colour of the labels, and the information they should contain.

Figure 1. Danger symbols proposed by the ILO (1954). 1. Danger of explosion. 2. Danger of poisoning. 3. Danger of ignition. 4. Oxidising agent. 5. Danger of corrosion. 6. Danger of ionising radiation: symbol for packages or consignments containing radioactive sources; packages or consignments containing large sources of radioactive materials, as defined in the IAEA Regulations for the Transport of Radioactive Materials, require a symbol bearing the skull and crossbones.

The labelling of packages for air freight is subject to the IATA system and goods shipped by sea are labelled in accordance with the system devised by IMCO, which are both based on the UN system. Packages for rail freight in Western Europe are subject to the labelling requirements of the Central Office for International Transport by Rail, that form part of the International Convention on the Transport of Dangerous Goods by Rail (CIM); the system used here is also derived from that drawn up by the UN. The same system of labelling is used in the European Agreement concerning the International Carriage of Dangerous Goods by Road (ADR) and the European Provisions concerning the International Carriage of Dangerous Goods by Inland Waterway (ADN).

HELLEN, E.

Problems of safety and hygiene in the chemical industries. a) Classification of dangerous substances b) Labelling of dangerous substances. Chemical Industries Committee, Fourth Session (Geneva, International Labour Office, 1954), 104 p. Illus.

CIS 78-1148 *Transport of dangerous goods.* United Nations Economic and Social Council. Recommendation prepared by the United Nations Committee of Experts on the Transport of Dangerous Goods. Publication ST/SG/AC.10/1/Rev.2 (New York, United Nations, revised ed., 1982), 462 p.

Regulations for the safe transport of radioactive materials. Safety series No. 6 (Vienna, International Atomic Energy Agency, revised ed., 1973), 148 p. Illus.

Dangerous chemical substances and proposals concerning their labelling (Strasbourg, Council of Europe, 4th ed., 1978), 951 p. and Addendum (Jan. 1979). Illus.

European agreement concerning the international carriage of dangerous goods by road (ADR) and protocol of signature. Economic Commission for Europe, Inland Transport Committee (New York, United Nations, 1978), 3 vols. Illus.

"Commission directive of 14 July 1976 adapting to technical progress the Council directive of 27 June 1967 concerning the approximation of the laws, regulations and administrative provisions relating to the classification, packaging and labelling of dangerous substances". *Official Journal of the European Communities* (Luxembourg), 30 Dec. 1976, 19/L360, 424 p.

"Council directive of 18 September 1979 amending for the sixth time directive 67/548/EEC on the approximation of the laws, regulations and administrative provisions relating to the classification, packaging and labelling of dangerous substances". *Official Journal of the European Communities* (Luxembourg), 15 Oct. 1979, 22/L259 (10-29).

IATA restricted articles regulations (Cointrin-Geneva, International Air Transport Association, 22nd ed., 1979), 325 p.

International maritime dangerous goods code (London, Inter-Governmental Maritime Consultative Organisation, 1977), 4 vols.

Dangerous substances, storage of

Explosive, oxidising, flammable, toxic, corrosive or radioactive substances require special storage measures if it is to be ensured that their hazardous properties are not to cause physical injury or material damage. In other cases, it must be ensured that the dangerous substance does not react with other stored substances, or it may be necessary to prevent unauthorised access to the stored materials, for example explosive, narcotic drugs.

Certain substances may present more than one hazardous property; this is the case with benzene which is both toxic and flammable and which, when vaporised, may form explosive mixtures with air. The storage measures necessitated by one hazardous property may, in some cases, be sufficient to ensure that the substance is stored safely from all points of view; however, this is not always so and, consequently, when establishing storage facilities and procedures, all of a substance's hazards should be taken into consideration.

In the paragraphs below the general storage requirements for each of the main classes of dangerous substances will be outlined.

Explosive substances

These include all explosives, pyrotechnics, and matches which are explosives *per se* and also those substances such as sensitive metallic salts which, by themselves or in certain mixtures or when subject to certain conditions of temperature, shock, friction or chemical action, may undergo an explosive reaction. In the case of explosives *per se*, countries have stringent regulations regarding safe storage requirements and precautions to be taken to prevent theft for use in criminal activities. The storage places or magazines should be situated far away from other buildings or structures so as to minimise damage in case of any explosion. Manufacturers of explosives issue instructions as to the most suitable type of storage. The storerooms should be of solid construction and kept securely locked when not in use. No store should be near a building containing oil, grease, gasoline, waste combustible material or flammable material, open fire or flame. In some countries there is a legal requirement that magazines should be situated at least 60 m from any power plant, tunnel, mine shaft, dam, highway, or building. Advantage should be taken of any protection offered by natural features such as hills, hollows, dense woods or forests. Artificial barriers of earth or stone walls are sometimes placed round such storage places. The storage place should be well ventilated and free from dampness. Natural lighting or portable electric lamps should be used, or lighting provided from outside the magazine. Floors should be of wood or other non-sparking material. The area surrounding the storage place should be kept free of dry grass, rubbish or any material likely to burn. Black powder and explosives should be stored in separate magazines and no detonators, tools or other materials should be kept in an explosives store. Non-ferrous tools should be used for opening cases of explosives.

Other substances which are not explosives *per se* should be stored in separate buildings away from the manufacturing plant or, in the case of such goods in transit warehouses, such as in docks, they should be in the part of the warehouse or godown reserved for dangerous goods. Such building should be of fire-resisting construction, well ventilated, with no open lights or flames, and be securely locked when not in use.

Oxidising substances

These are substances which are sources of oxygen, thus supporting combustion and intensifying the violence of any fire. Some of these oxygen-suppliers give off oxygen at storage-room temperature but others require the application of heat. If containers of oxidising materials are damaged, the contents may mix with other combustible materials and start a fire. This risk can be avoided by storing oxidising materials in a separate storage place but this is not always practicable, as for example in dock warehouses for goods in transit.

It is dangerous to store powerful oxidising substances near to liquids of even a low flash point, or even slightly flammable materials. It is safer to keep all flammable materials away from a place where oxidising substances are stored. The storage area should be kept cool, well ventilated and be of fire-resisting construction.

Flammable substances

A gas is deemed to be flammable if it burns in the presence of air or oxygen. Hydrogen, propane, butane, ethylene, acetylene, hydrogen sulphide, coal gas and ethene are among the most common flammable gases. Some gases such as hydrogen cyanide (hydrocyanic acid) and cyanogen are both flammable and poisonous. The fire hazards of flammable liquids are usually classified on the basis of flash points. Flammable materials should be stored in places which are cool enough to prevent accidental ignition if the vapours mix with the air.

The storage area should be situated away from any source of heat or fire hazard. Highly flammable substances should be kept apart from powerful oxidising agents or from materials which are susceptible to spontaneous combustion. When highly volatile liquids are stored, any electrical light fittings or apparatus should be of certified flameproof construction, and no open light or flames should be permitted in or near the storage place.

The storage-room installations should be electrically grounded and periodically inspected, or equipped with automatic smoke- or fire-detection devices. Control valves on storage vessels containing flammable liquids should be clearly labelled and pipelines painted with distinctive safety colours to indicate the type of liquid and the direction of flow. Tanks containing flammable liquids should be situated on ground sloping away from the main buildings and plant installations. If they are on level ground, protection against fire spread can be obtained by adequate spacing and the provision of dikes. The dike capacity should preferably be 1.5 times that of the storage tank, as a flammable liquid may be likely to boil over. Provision should be made for venting facilities and flame arrestors on such storage tanks. Adequate fire-fighting extinguishers, either automatic or manual, should be available. No smoking should be allowed.

Toxic substances

It is almost impossible to obtain a perfect seal on containers and there is always the danger that volatile poisonous substances may leak into the atmosphere; consequently, storage areas should always be well ventilated. If the substances stored are likely to decompose due to contact with heat, moisture, acids or acid fumes, they should be stored in a cool, well ventilated place out of direct rays of the sun, and away from all sources of heat and ignition. Substances which can react chemically with each other should be kept in separate stores.

Corrosive substances

These include strong acids and alkalis and other substances which will cause burns or irritation of the skin, mucous membranes or eyes or which will damage most materials. Typical examples of these substances include hydrofluoric acid, hydrochloric acid, sulphuric acid, nitric acid, formic acid, and perchloric acid. Such materials may cause damage to their containers and leak into the atmosphere of the storage area; some are volatile and others react violently with moisture, organic matter or other chemicals. Acid mists or fumes may corrode structural materials and equipment and have a toxic action on personnel. Such materials should be kept cool but well above their freezing point since a substance such as acetic acid may freeze at a relatively high temperature, rupture its container and then escape when the temperature rises again above its freezing point.

Some corrosive substances also have other dangerous properties; for example, perchloric acid, in addition to being highly corrosive, is also a powerful oxidising agent which can cause fire and explosions. *Aqua regia* has three dangerous properties: firstly, it displays the corrosive properties of its two components (hydrochloric acid and nitric acid); secondly, it is a very powerful oxidising agent and, thirdly, application of only a small amount of heat will result in the formation of nitrosyl chloride, a highly toxic gas.

Storage areas for corrosive substances should be isolated from the rest of the plant or warehouses by impervious walls and floor, with provision for the safe disposal of spillage. The floors should be made of cinders, concrete treated to reduce its solubility, or other resistant material. The storage area should be well ventilated. No store should be used for the simultaneous storage of nitric acid mixtures and sulphuric acid mixtures. Corrosive and poisonous liquids sometimes require to be stored in special types of containers; for example, hydrofluoric acid should be kept in leaden, gutta percha or ceresin bottles. No bottle containing hydrofluoric acid should be stored near to glass or earthenware carboys containing other acids.

Carboys containing corrosive acids should be packed with kieselguhr (infusorial earth), or other effective inorganic insulating material. Any necessary first-aid equipment such as emergency showers and eyewash bottles should be provided in or immediately close to the storage place.

Radioactive substances

See RADIOACTIVE MATERIALS, TRANSPORT AND STORAGE OF.

General storage requirements

The storage of dangerous substances should be supervised by a competent, trained person, and all workers having to enter such storage places should be fully trained in the appropriate safe working practices. Workers with defective vision, hearing or sense of smell and young persons under 18 years of age should not be employed on such work. In the case of explosives, the competent authority may require workers entering explosive magazines to hold a special permit granted only after an examination on the dangers of the work. Persons entering storage areas containing explosive or flammable substances should not carry matches and should be prohibited from smoking. Where necessary, suitable protective clothing should be worn. A periodic inspection of all storage places for dangerous substances should be carried out by a safety officer or some other competent person. Rigorous maintenance and good housekeeping are necessary, and where there is a fire hazard, a fire alarm should be situated in or near to the outside of the storage premises. It is recommended that persons should not work alone in a storage area containing highly toxic substances.

Legislation

Detailed legislation has been drawn up in many countries to regulate the manner in which various dangerous substances may be stored; this legislation specifies the type of building, its location, the maximum amounts of various substances that may be stored in one place, the type of ventilation required, the precautions to be taken against fire, explosion and the escape of any dangerous substances, the type of lighting (flameproof electrical equipment and light fixtures when explosive or flammable materials are stored), the number and location of fire exits, security measures against entry by unauthorised persons and against theft, labelling and marking of storage vessels and pipelines, warning notices to workers as to the precautions to be observed, etc.

In many countries there is no central authority concerned with the supervision of the safety precautions for the storage of all dangerous substances but there are a number of separate authorities such as mine and factory inspectorates, dock authorities, transport authorities, police, fire services, national boards and local authorities who each deal with a limited range of dangerous substances under various legislative powers. Usually in these countries it is necessary for a licence to be obtained from one of the government or local authorities for the storage of certain types of dangerous substances such as petroleum, explosives, cinematograph film, cellulose and cellulose solutions. Such licences are granted only if the storage facilities comply with certain specified safety standards.

QUINN, A. E.

Documents and standards on the storage of dangerous substances usually refer to a specific substance or to a class of substances, and space is not available here for a balanced presentation of data. References to such documents can be obtained through the International Occupational Safety and Health Information Centre (CIS).

Dangerous substances, transportation of

The growth of the chemical industry over the second half of the 19th century led to international interest in the promulgation of safety regulations designed to safeguard workers, the public and property against the hazards presented by dangerous substances during transport. At that time, a large proportion of these substances was carried by rail and, consequently, the first international transport regulations, issued on 14 October 1890, were in the form of an appendix to the International Convention on the Transport of Goods by Rail (CIM). These regulations, which are now known as the International Regulations Concerning the Transport of Dangerous Goods by Rail (RID), are revised and supplemented periodically in line with new developments in railway equipment and dangerous substances. In the socialist countries, the RID has its counterpart in an agreement, concluded after the Second World War by the railway administrations of these countries. Together, the RID and this agreement cover all European and some non-European railway networks.

The Second World War was followed by a considerable expansion in international road transport, and it was found necessary to draw up similar regulations for this transport medium. Accordingly, in 1950, the Inland Transport Committee of the United Nations Economic Commission for Europe (ECE) prepared a Draft European Agreement on the International Transport of Dangerous Goods by Road (ADR).

This Agreement, concluded at Geneva, came into force on 29 January 1968. It provides the legal basis for two technical appendices, one concerning dangerous substances and objects and their packaging, and the other transport equipment and transport conditions.

The Hague Convention of 1 February 1939 on the transport of combustible liquids by inland waterway provided only a partial solution to the problems of the water transport of dangerous substances; consequently, some time after it had prepared the draft ADR, the Transport Committee of the ECE produced a Draft European Agreement on the International Transport of Dangerous Goods by Inland Waterway (ADN). Although this draft has not yet become a formal Agreement, it has been recommended that governments and international river commissions should enforce its provisions.

The provisions of these international agreements have, apart from a few exceptions, all been incorporated into national regulations for domestic transport and consequently both international and national road, rail and inland waterway transport requirements have reached a considerable degree of international co-ordination in Europe.

In the United States, regulations on the transport of dangerous substances have been in existence for many years and have subsequently been adopted in other countries in North America and elsewhere. They differ from European regulations on several points, in particular the classification of goods and the labelling of packages. Packaging specifications are usually very detailed and consist mainly of a compilation of authorisations for new types of packaging that are issued from time to time.

The maritime nations have adopted the International Maritime Dangerous Goods Code that was drawn up by the Inter-Governmental Maritime Consultative Organisation at the request of the International Conference on Safety of Life at Sea (London, 1960) and on the basis of the recommendations of the Committee of Experts on the Transport of Dangerous Goods of the Economic and Social Council of the United Nations; the functions of this Committee are described below. The provisions of the IMCO Code are either incorporated in the national regulations (as is the case with the United Kingdom *Blue Book*) or applied directly.

In the field of air transport, the International Air Transport Association (IATA) has issued world-wide regulations that are based, to a large extent, on requirements in force in the United States; these are amended periodically and submitted for approval to governments of countries whose airlines are members of IATA.

The special problems of transporting radioactive materials are subject, primarily, to the scrutiny of the International Atomic Energy Agency (IAEA) in consultation with other interested international organisations. IAEA recommendations are, subsequently, incorporated into national and international legislation. This subject is dealt with in RADIOACTIVE MATERIALS, TRANSPORT AND STORAGE OF.

In 1952 the United Nations Economic and Social Council appointed a Committee of Experts on the Transport of Dangerous Goods whose recommendations, while not proposed purely and simply as a substitute for national and international regulations, are designed to provide guidance for governments and international organisations that have framed, or are drafting, regulations, so as to achieve at least a minimum of uniformity throughout the world for the various modes of transport.

Classification of dangerous substances

The classification recommended by the UN Committee of Experts is as follows:

Class 1. Explosives

Division 1.1. Explosives with a mass explosion risk:

- Sub-division 1.1.1. Initiating explosives; contrivances which contain both explosives and their own means of ignition;
- Sub-division 1.1.2. Explosive substances other than initiating explosives; contrivances containing explosives but not their own means of ignition;
- Sub-division 1.1.3. Contrivances designed to produce illumination, incendiary, smoke or sound effects; igniters; starter cartridges; small arms ammunition; fireworks liable to explode violently.

Division 1.2. Explosives which do not explode *en masse*, having a projection hazard but minor explosion effect:

- Sub-division 1.2.1. Contrivances containing explosives, with or without their own means of ignition;
- Sub-division 1.2.2. Samples of explosives other than initiating explosives.

Division 1.3. Explosives which do not explode *en masse*, having a fire hazard with minor or no explosion effect.

Division 1.4. Explosives which present no significant hazard.

This division has two sub-divisions. The first (1.4.1) covers items which are so packed or designed as to present only a small hazard in the event of ignition during transport; the second (1.4.2) covers "safety" explosives.

Class 2. Gases—compressed, liquefied or dissolved under pressure

Class 3. Inflammable liquids

Class 4. Inflammable solids; substances liable to spontaneous combustion; substances which, on contact with water, emit inflammable gases

Division 4.1. Inflammable solids.

Division 4.2. Substances liable to spontaneous combustion.

Division 4.3. Substances which, on contact with water, emit inflammable gases.

Class 5. Oxidising substances—organic peroxides

Division 5.1. Oxidising substances, other than organic peroxides.

Division 5.2. Organic peroxides.

Class 6. Poisonous (toxic) and infectious substances

Division 6.1. Poisonous (toxic) substances:

Sub-division 6.1.1. Substances which give off a poisonous (toxic) gas or vapour;

Sub-division 6.1.2. Poisonous (toxic) substances other than those giving off poisonous (toxic) gases or vapours.

Division 6.2. Infectious substances.

Class 7. Radioactive substances

Class 8. Corrosives

Class 9. Miscellaneous dangerous substances

The Committee has also produced packaging recommendations. These draw their inspiration primarily from existing national and international regulations whilst making allowance for the current trend towards replacing detailed packaging specifications, which may vary considerably from country to country, by tests designed to ensure that the packaging around dangerous goods can withstand normal transport conditions and thus offer adequate safety.

As far as labelling is concerned, the Committee has adopted a system of coloured labels designed to avoid dangerous mixed loads; the symbols for the labels are, in the main, taken from the RID. The labelling system recommended by the Committee has been incorporated in the IMCO Code and adopted, in principle, by the RID Committee of Experts.

HALBERTSMA, H. G.

For references concerning UN, IAEA, ADR, IATA, and IMCO regulations see DANGEROUS SUBSTANCES, LABELLING AND MARKING OF.

International regulations concerning the carriage of dangerous goods by rail (RID). Annex I to the International Convention concerning the carriage of goods by rail (CIM) (London, HM Stationery Office, 1978), 446 p. Illus.

Date palms

The date is the fruit of the evergreen date palm *(Phoenix dactylifera)*. It has a high glucose and sucrose content; alcohol can be produced by fermentation. The majority of the large date plantations are in the Northern hemisphere and the main producers are Algeria, the United Arab Republic, the Arabian Peninsula, the United States, Iraq, Iran, Libya, the Sudan and Tunisia.

Cultivation. The date palm requires warmth and abundant irrigation to maintain an adequate and continuous level of moisture in the soil; extensive soil preparation is carried out every 2-3 years. The plant is dioecious (each sex on a separate plant) and the presence of male trees in a proportion of 1-2% is necessary for pollination. This is done during the fertile period (March-April) at an optimum temperature of 30-32 °C in one of the following ways.

(a) by hand, when small branches of male clusters are placed in the centre of each female cluster and the wind disperses the pollen;

(b) mechanically, by a spray sufficiently powerful to project the pollen over the bunches to be fertilised; this method gives the best results.

Ripening of the fruit lasts from the first fortnight of September to December; workers climb the trees twice or three times and gather the ripe clusters. The annual yield of a properly cultivated date palm is of the order of 35-50 kg.

The varieties of dates can be classified into three large categories:

(a) dry dates—called "common" dates;

(b) sweet dates—called "common" dates;

(c) fine dates—these include the "muscat" date cultivated almost exclusively in Algeria.

Processing. After sorting, the dates are subjected to hydration and artificial fruit ripening treatment in which they are exposed to saturated air at up to 60 °C. After cooling in naturally ventilated surroundings, the dates are again graded and then spread out on wattle hurdles and exposed to a stream of hot air (100-110 °C) to glaze the skin and semi-pasteurise the exterior. Finally, they are packaged either in luxury display boxes or in blocks.

HAZARDS AND THEIR PREVENTION

Workers who climb to the top of date palms (often 30 m high) during cultivation and harvesting are often subject to falls resulting in serious fractures. These workers should, therefore, wear safety belts attached by lifelines to a part of the tree strong enough to support their weight in the event of a fall. In the United States extension ladders are used for harvesting. Cuts and abrasions to the

Figure 1. The application of safe working practices to date picking.

hands occur frequently due to contact with the tree's sharp spines. Gloves offer little protection.

Examination of 8 000 medical histories of workers in Southern Algeria has revealed about 40 cases of bronchial asthma, probably due to a sensitivity to the pollen; no preventive measures seem feasible at present. Even more rarely, chronic dry eczema and onychias have been noted: strict attention to personal hygiene of the hand or the wearing of hand protection is sufficient to prevent these. It is desirable that adequate sanitary facilities and first-aid requisites should be provided as near as possible to the place of work.

ABED, D.

Day nurseries and nursery schools

Day nurseries (for children under 3 years of age) and nursery schools (for children over 3 years of age) are public or private establishments intended for the care of healthy children below school age.

In large and medium-sized undertakings and public administrations, the services afforded by these establishments are the result of a threefold need: that of mothers who wish to work full time or part time outside the home; that of employers who wish to keep their women workers after childbirth and at the same time reduce fatigue in working mothers; and that of the children themselves who require special attention and for whom contact with other children is beneficial.

The period of the child's life covered by his attendance at a day nursery or nursery school is of outstanding importance in his future development, and the child's rapid growth during these years imposes a wide range of requirements. Consequently these establishments must be extremely well organised and staff must have specialist training in both medical and teaching disciplines.

In certain countries there are statutory provisions that specify the conditions to be met before an establishment of this type can receive official approval; these provisions usually cover the number and qualification of staff, the type and layout of the premises, safety and health requirements, number of children to be accommodated, and the administrative and sanitary supervision to which they are subject.

The ILO Employment (Women with Family Responsibilities) Recommendation, 1965 (No. 123), proposes that the competent authorities should, in co-operation with the public and private organisations concerned, take appropriate measures to ensure the provision of adequate and appropriate child-care services and establishments. This recommendation also has, as an objective, the establishment of standards concerning the equipment and hygiene requirements of those services and the number and qualification of the staff, together with the supervision of these standards. Finally, it is recommended that the competent authorities should provide or help to ensure the provision of adequate training for the personnel of child-care establishments.

Personnel: choice and medical supervision

Day nurseries and nursery schools require a trained and selected staff. The national education service should be directly or indirectly responsible for education programmes, and the recruitment of teaching staff and the direction and supervision of medical and prophylactic care received by children should be, directly or indirectly, the responsibility of the public health authorities. In countries where day-care services in the undertaking are set up at the employers' initiative, it is desirable that close collaboration should be established between those responsible for the service within the undertaking and the public health and national education authorities.

The supervision of the children's health and hygiene is in the hands of the nursery supervisor assisted by a nurse or nurses, and under the over-all control of a physician. Trained teachers and nursery-school staff are responsible for the children's education, while the general operation of the day nursery and nursery school is the responsibility of the director.

The medical supervision of child-care workers should be similar to that provided for medical services and particular attention should be paid to protection against infectious diseases (both for the staff and the children). The features of this medical supervision will, to a certain extent, be dependent on whether the services are private or public and on their relationship with the undertaking and its medical or social services. Special liaison is necessary in cases where diseases may be transmitted to child-care workers through families and children.

Premises

These should be well lit and possess all the necessary equipment. A well organised outdoor area is necessary with a playground adjacent to the building, surrounded and protected by greenery. Each age group should have its own play area separated by a hedge. It may be useful to have an extra space nearby set aside for breast-feeding. The provision of a swimming pool is recommended but its depth should not exceed 25 cm. The playground should be provided with a covered area and play equipment such as boxes and planks, slides, balls, equipment for playing with sand and water, hurdles, toy cars, tricycles, etc. Arrangements such as these will mean that the children's activities can be safely and rationally organised in the open air.

The internal layout of the premises should be planned to suit the number of groups into which the children are to be separated. Each group should have its own facilities so that it can follow a schedule suited to its own requirements; the basic amenities include a cloakroom, a reception and distribution area, sanitary facilities with toilets, washbasins and showers. A kitchen with pantry and refrigerator, accommodation for administrative and medical staff, linen store and laundry room (with separate treatment facilities for napkins and sterilising equipment for soiled clothing) may serve all the groups. An isolation area is necessary in order to prevent the spread of contagious diseases.

In addition, these premises should meet the safety requirements of a normal workplace and should have suitable fire-protection and prevention equipment, emergency exits, first-aid equipment, arrangements for emergency evacuation to a medical service or hospital, etc. All installations and equipment should, of course, be designed with the safety of the children in mind.

KARDACENKO, V. N.

The care of well children in day-care centres and institutions. Report of a Joint UN/WHO Expert Committee convened with the participation of FAO, ILO and UNICEF. WHO technical report series No. 256 (Geneva, World Health Organisation, 1963), 34 p.

New trends and approaches in the delivery of maternal and child care in health services. Sixth report of the WHO Expert Committee on Maternal and Child Health. WHO technical report series No. 600 (Geneva, World Health Organisation, 1976), 98 p.

DDT

DDT ($C_{14}H_9Cl_5$)

DICHLORODIPHENYLTRICHLOROETHANE; 1,1,1-TRICHLORO-2,2-*bis*(*p*-CHLOROPHENYL) ETHANE

m.w.	355
sp.gr.	1.55
m.p.	108.5-109 °C
b.p.	185 °C decomposes

soluble in most organic solvents; virtually insoluble in water, dilute acids, alkalis; the compound is non-combustible, very stable and degrades only slowly

highly purified DDT consists of large odourless and tasteless glistening needles; the technical product varies in colour from white to deep grey, possesses a fruit-like odour and melts over a considerable range of temperatures.

TWA OSHA	1 mg/m³
STEL ACGIH	3 mg/m³
MAC USSR	0.1 mg/m³ skin

Production. DDT is produced by the condensation of chloral with chlorobenzene.

The technical grade contains about 70% of the *p-p*-isomer with 30% of the *o-p*-isomer and small amounts of several other compounds. Among the latter are a dichloroethane derivative (DDD), a dehydrochlorination product (DDE), *bis*(*p*-chlorophenyl) acetic acid (DDA), and 2,2-*bis*(*p*-methoxyphenyl)-1,1,1-trichloroethane (Methoxychlor). Some of these related compounds have biological activity and have been produced commercially for insecticidal use.

Uses. The compound has been used extensively in the successful control of insects which are parasites or vectors of organisms which cause disease in man. Among such diseases are malaria, yellow fever, dengue, filariasis, louse-borne typhus and louse-borne relapsing fever, which are transmitted by arthropod vectors vulnerable to DDT.

Different DDT formulations are made for specific purposes. For use in agriculture and in the public health control of disease vectors, the compound is available as a dusting powder, a wettable powder for spray use, an emulsifiable concentrate, or as a solution in petroleum distillates. The usual mixture for spray application is 0.75-1.0 of DDT per litre of water. Agricultural dusting compounds usually contain 1-10% DDT, although the most common content is 5%.

Degradation and metabolism. The primary metabolic residues are 1,1-dichloro-2,2-*bis*(*p*-chlorophenyl) ethylene (DDE) and 1,1-dichloro-2,2-*bis*(*p*-chlorophenyl) ethane (DDD) depending upon the animal species into which DDT is absorbed. DDE results from dehydrochlorination and DDD from direct reduction of the DDT. In rats and probably other mammals DDD is dehydrochlorinated to the monochloroethylene analogue, hydrogenated to the ethane derivative and again dehydrochlorinated to *bis*(*p*-chlorophenyl) ethylene (DDNU); this compound is then hydrolysed to the primary alcohol (DDOH), oxidised to *bis*(*p*-chlorophenyl) acetic acid (DDA) and excreted with the urine.

DDE is converted to DDNU also, but by successive oxidation-reduction of the respective ethylene analogues; thereafter it follows the DDOH → DDA pathway. This metabolism proceeds slowly which accounts for the persistence of DDE residues in mammalian adipose tissues.

Adverse effects. Years of world-wide experiences with DDT have exposed undesirable effects which were not foreseen initially. Slow deterioration of the compound and the persistence of its biologically active residues produced increasing environmental contamination. Magnification of residues in the food chain led to toxic levels in some of the higher species of animals, particularly in wet land areas. Practically every species subjected to control by DDT developed a resistance to its toxic effects, a resistance which sometimes carried over to other chemicals.

HAZARDS

There is an extremely wide range of variation in susceptibility to DDT among animals of a given species (and also among species), irrespective of the nature of the solvent or the suspending medium. Absorption of the compound is influenced by the carrier material in which it is administered. In edible oils, the LD_{50} for rats may have a range of 100-350 mg/kg for body weight; however, the LD_{50} for dry powder or aqueous suspension is of the order of 1 400 mg/kg. In oil solutions, the LD_{50} for all solutions in rabbits is 930-1 940 mg/kg.

When absorbed in sufficient quantity, DDT produces a state of hyperexcitability, marked tremors, an increased rate of respiration, weakness and death which is usually preceded by clonic or asphyxial convulsions. Respiration usually fails before the heart ceases to beat but in some animals death may result from cardiac fibrillation during a convulsive episode.

A DDT concentration of 150 ppm in the daily diet of rats over a period of 2 years had no significant effect on the life expectancy of the animals. However, liver cell hypertrophy and hyperplasia were observed in these animals and the concentration of DDT plus related metabolites found in the perirenal fat was 162 ppm. It has also been found that DDT accelerates the action of the drug-metabolising microsomal enzymes in the liver.

SAFETY AND HEALTH MEASURES

Agricultural workers employed on the formulation and application of DDT should be subject to the precautionary measures detailed in the article on PESTICIDES, HALOGENATED and should be informed of the dangers of this product and provided with suitable work clothing and personal protective equipment. Adequate sanitary facilities, including showers, should be provided, and the need for personal hygiene stressed.

Treatment. The treatment of acute DDT poisoning is chiefly symptomatic. In oral poisoning, gastric lavage should be instituted promptly. Amobarbital or pentobarbital has been recommended for the relief of central neurological manifestations and tribromoethanol and paraldehyde are recommended for allaying prolonged convulsions. To lessen damage to the liver, a fat-free diet high in proteins, carbohydrates and calcium has been recommended. Intravenous administration of protein hydrolysates or mixtures of amino acid may prove of value in certain cases.

Legislation

During the 1950s, there was a remarkable growth in the use of DDT for agricultural purposes. Subsequently, realisation that the resistance of DDT to degradation was leading to the accumulation of residues in plant and animal life prompted governments in many countries to establish permissible limits for DDT residues in various agricultural products. More recently, certain countries have drawn up legislation limiting or banning the use of DDT.

For example an official commission set up in the United States reviewed the documented evidence on the danger of DDT residues and recommended, in

November 1969, that within 2 years all uses of DDT and DDD not essential to the preservation of human health and welfare should be prohibited. As a result, the US Department of Agriculture took steps which effectively eliminated the use of DDT for shade trees, tobacco, domestic applications, and for aquatic environments, marshes, etc., except where specifically authorised by public health officers for vector control.

Since 1970, the use of DDT in England, the United States, Japan, the USSR and other countries has been subject to strict governmental control.

WITHERUP, S. O.

General:

CIS 79-1035 *Special occupational hazard review for DDT.* DHEW (NIOSH) publication No. 78-200 (National Institute for Occupational Safety and Health, 4676 Columbia Parkway, Cincinnati) (Sep. 1978), 205 p. Illus. 318 ref.

CIS 77-481 *DDT.* Data sheet on pesticides No. 21 (Geneva, World Health Organisation, and Rome, Food and Agriculture Organisation of the United Nations, Dec. 1976), 11 p. 3 ref.

Chlorinated insecticides. Brooks, G. T. Vol. 11 (Cleveland, CRC Press Inc., 1974).

Metabolism:

CIS 75-173 *Exposure to lindane and DDT and its effects on drug metabolism and serum lipoproteins.* Kolmodin-Hedman, B. (Kungliga Arbetarskyddsstyrelsen, Fack, 100 26 Stockholm, 1974), 48 p. Illus. 125 ref.

"*In vivo* detoxication of *p,p'*-DDT via *p,p'*-DDA in rats". Datta, P. R. *Industrial Medicine and Surgery* (Chicago), Apr. 1970, 39/4 (190-194). Illus. 13 ref.

Exposure limits:

DDT and its derivatives. Environmental health criteria 9 (Geneva, World Health Organisation, 1979), 194 p.

Deafness, occupational

Occupational deafness may be defined as hearing loss caused by specific working conditions. In view of the aetiopathogenic mechanisms involved, a distinction may be made between noise-induced hearing loss; deafness or hypoacusis due to barometric trauma of the middle ear; deafness or hypoacusis due to gaseous emboli; deafness or hypoacusis due to cranial or craniofacial trauma; deafness or hypoacusis due to industrial toxic agents.

Noise-induced deafness or hearing loss

This is by far the most common form of occupational deafness. From the pathological point of view, the main lesions are to be observed in the ciliated cells of the organ of Corti where there is fragmentation and loss of the hairs, breakage of the cellular membrane, leakage of the nucleus, proliferation of the cells of Deiters in substitution of neuroepithelium. These changes are irreversible. The nature and site of the lesions are closely linked to the type of acoustic stimulus. Low-frequency pure tones in the 250-500 Hz range produce damage to the apical spiral of the cochlea whereas high-frequency pure tones (in the 3 000-4 000 Hz range) cause destruction primarily of the neuroepithelium of the basal spiral. The severity of the lesions depends on the level of sound energy. There is, as yet, no explanation of the mechanism by which industrial noise causes selective damage to the cells of the organ of Corti for perception of noise in the 4 000-6 000 Hz range for which, independently of the spectral characteristics of the noise in the working environment, noise-induced hearing loss always has the same characteristics.

Temporary and permanent threshold shift. Exposure of a person with normal hearing to intense noise will cause a hearing loss reflected by an elevated threshold of audibility; quantitatively, this phenomenon is shown by a difference in dB between the auditory threshold when hearing is at rest and that following an auditory stimulus; where this hearing loss is transitory, the term used is temporary threshold shift (TTS). There are two types of TTS:

(1) TTS_2 = physiological hearing fatigue: this is measured 2 min from the termination of noise exposure, it lasts less than 16 h and its level is in linear correlation with the noise intensity and the logarithm of exposure time, whereas the restoration of normal hearing (recovery) is proportional to the logarithm of the time for which the main part of the TTS_2 is recovered in the first 2-3 h;

(2) Prolonged TTS = pathological hearing fatigue = TTS_{16}: this is the condition which persists beyond 16 h following termination of the auditory stimulus, and its recovery is in linear relation to time. TTS_2 and TTS_{16} are probably the expression (at various levels) of a state of functional exhaustion which occurs in the peripheral hearing receptors owing to inadequate energisation in relation to the level of the stimulus. If the functional exhaustion is kept within certain limits, on termination of the exposure complete recovery is possible with a return to the base-line condition; however, if the exhaustion is excessive, with the recovery time being prolonged and the noise exposure being repeated daily, there is no longer the possibility of complete recovery and slowly TTS is transformed into irreversible damage, i.e. permanent threshold shift (PTS), i.e. noise-induced hearing loss.

The distinction between TTS_2 and TTS_{16} is artificial and is derived primarily from considerations relating to the type of work organisation rather than the behaviour of the hearing system. Since occupational noise exposure lasts on an average 8 h and is followed by 16 h of rest, it is certain that if, at the end of this period a TTS persists, it should be considered pathological to the extent that it constitutes the presupposition of permanent damage.

Consequently there seems to be a close correlation between TTS and PTS in that, given a certain type of noise, having measured the TTS produced and knowing the duration of exposure, it should be possible to forecast the PTS. Numerous retrospective and prospective epidemiological studies have been carried out to clarify this relationship but the only conclusion arrived at is that the extent of PTS for a given set of conditions varies significantly from one subject to another to such an extent that the hearing of one individual may be severely damaged whereas that of another is perfectly normal.

In our current state of knowledge it is not yet possible to state that TTS is a personal index to predict future PTS; however, it can be said that if between one noise exposure and the next one there is insufficient time for complete recovery of hearing acuity, it is certain that hearing damage will occur over the long term. Recovery of TTS caused by impulse noise is different from that due to exposure to steady state or oscillating noise. In the case of impulse noise, three phases can be distinguished: initial partial recovery, maximum degradation of hearing 2-6 h after the termination of exposure, slow progressive recovery for a further 100 h. This difference is interpreted by certain authors as proof that impulse noise damages the organ of Corti by a mechanism which is more mechanically than metabolically destructive. Noise-induced hearing loss may be of the chronic type if it develops over a period of years, or of an acute type if it occurs in a relatively limited time, produced by an acoustic stimulus which is intense but of short duration.

Chronic form. The disorder develops insidiously and

unnoticed; four phases can be distinguished in its development:

(1) First phase: this coincides with the first 10-20 days of noise exposure; the subject experiences a ringing in the ears at the end of the work shift, accompanied by a sensation of fullness of the ears, slight headache, and a feeling of tiredness and dizziness.

(2) Second phase: other than for intermittent ringing in the ears, subjective symptoms are completely absent. In this period, which may last from a few months to many years—depending on the noise level, the daily duration of exposure and individual predisposition to hearing damage—the only signs present are those which can be detected by audiometric examination.

(3) Third phase: the subject notices that he no longer has normal hearing, can no longer hear the ticking of a clock, cannot pick out all the components of a conversation especially if there is background noise, and he has to raise the volume of the radio or television, to the complaints of his family.

(4) Fourth phase: the feeling of hearing insufficiency is manifest, any type of communication using acoustic signals is difficult or impossible and, in particular, any conversation is compromised, with severe consequences for the person in question. Any of these four phases may be accompanied by persistent tinnitus indicating impairment to the nerve structure of the cochlea; this not only aggravates the subject's hearing but also severely disturbs his rest, sleep and well-being. Noise-induced hearing loss results not only in "quantitative" reduction in hearing acuity but also in a "qualitative" change, since sounds are perceived in an abnormal manner due to a modification in the relationship between the level of a stimulus and the auditory sensation it provokes.

Noise-induced hearing damage develops in line with a relatively constant model even though the onset and severity are determined by the noise level and individual susceptibility. Initially the high frequencies are affected (3-4-6 kHz) and then the damage extends to the 0.5-1-2 and 8 kHz ranges. The progress of the lesions produces a highly characteristic audiometric curve (see figure 1) and the onset of a hearing deficit which may be described as follows:

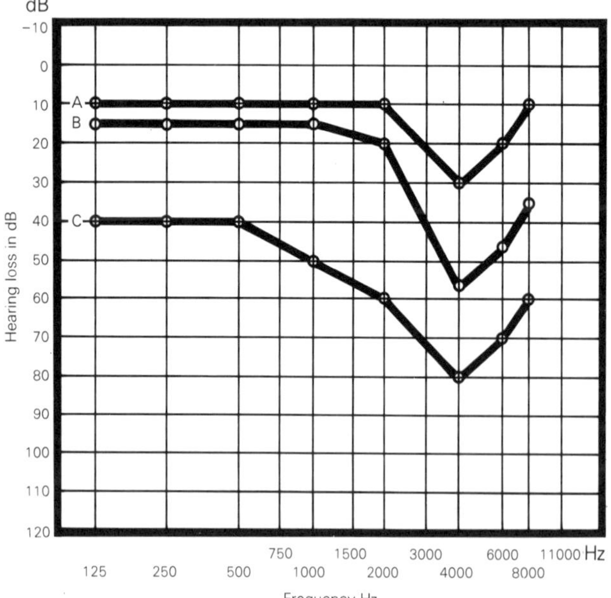

Figure 1. Audiometric curves characteristic of the 2nd (A), 3rd (B) and 4th (C) phases in the course of noise-induced hearing loss.

- deficit with a maximum in the 3-4-6 kHz range;
- bilateral and nearly symmetrical;
- irreversible;
- in the majority of cases the condition does not progress once noise exposure has been terminated;
- recruitment is nearly always present in the initial stages, absent in cases which have existed for some considerable time, in which there is retrocochlear damage in addition to cochlear damage.

Hearing response to acoustical trauma varies considerably between individuals, like all biological parameters having a distribution along a Gaussian curve.

Acute form. The damage affects merely one ear because the head acts as a screen to protect the contra-lateral ear. Immediately after the burst of noise, the subject experiences a piercing pain in the ear, a dazed sensation, hypoacusis or complete deafness with continuing ringing in the ears, sensation of fullness of the ears and frequent bouts of vertigo, probably due to the posterior labyrinth. Otoscopy of the tympanic membrane (TM) may reveal only congestion or lacerations with slight bleeding.

There is a tendency to regression and, in fortunate cases, this is followed by complete recovery.

More frequently, the sequellae persist and are due more to damage to the nerve structures, since there is continuous ringing in the ears and auditory deficit for higher frequencies. The cause of the hearing damage is the rapid change in pressure that occurs to the tympanic membrane and which may even affect the nerve structures of the organ of Corti. Two things may occur:

(a) the pressure change is such that it immediately ruptures the ear drum; the pressure wave, partly dampened by the middle ear, affects only secondarily the nerve structure of the cochlea; or

(b) the pressure is less intense, the ear drum withstands it and the wave is transmitted to the interior of the inner ear with degenerative damage for the acoustic cells of the organ of Corti.

From the audiometric point of view, acute acoustic trauma is characterised by hearing loss either of mixed sensory and conductive type or of a pure sensory type with sudden onset or, much more frequently, monolateral and always accompanied by tinnitus and progressing towards complete or partial cure.

Progressive form. It is widely accepted that noise-induced hearing loss stabilises once noise exposure has terminated, except where there is aggravation attributed to presbycousia. However, it is necessary to partly revise this concept due to a certain number of factors. Clinically, aggravation following termination of exposure to noise is a sufficiently frequent finding even in subjects less than 40 years of age in whom the progress cannot be attributed to presbycousia; experimentally, stabilisation of the lesion following termination of exposure to noise has not been demonstrated.

On the other hand, it may be considered, in line with common concepts of nervous system physiology, that the degenerative lesions affecting the peripheral receptors may have a retrogressive action extending to the centre and reducing the integration potential of the audio message.

Diagnostic criteria. As a rule, there is no problem in making an audiometric diagnosis of perceptive hypoacusia; however, it is more difficult to establish an aetiological diagnosis and, in particular, a diagnosis of

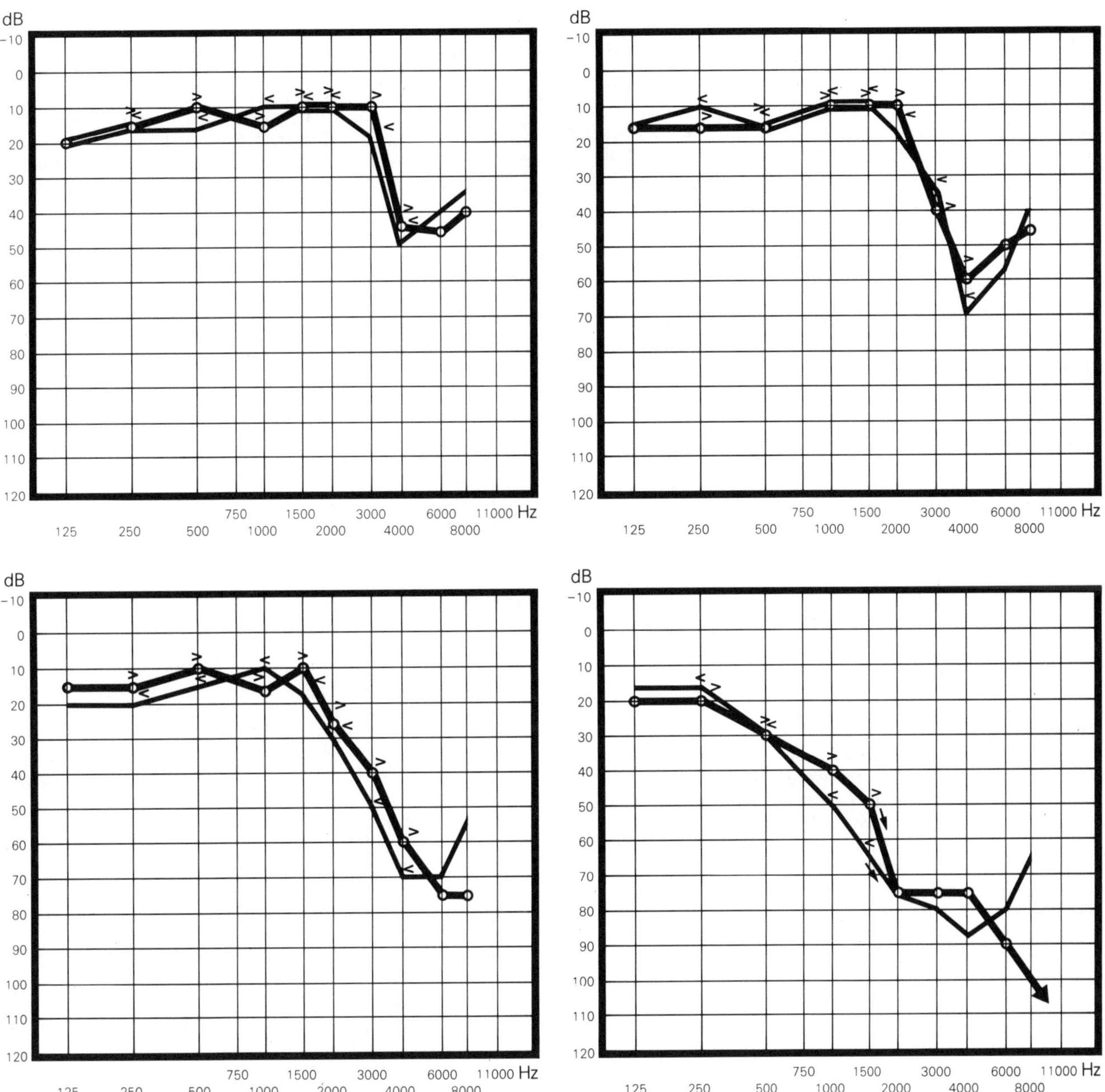

Figure 2. Examples of audiometric curves which can be found in occupational hearing loss. *(By courtesy of Prof. G. Rossi, Director, Institute of Audiology, University of Turin.)*

noise-induced hearing loss. This results from two basic factors: the audiometric characteristics and the patient's history.

In many cases the audiometric characteristics are all present and the comprehensive progress of the auditory threshold is so typical that, if the hypothesis is confirmed by the medical history, there can be no doubt of the diagnosis. In some cases, either due to the part played by pathogens or the extension of the deficit to all frequencies due to the severity of the damage, the audiometric curve is much less typical and the audiometric characteristics less clear. In these cases it is the medical history that will clarify the diagnosis.

SAFETY AND HEALTH MEASURES

Noise-induced hearing loss is an irreversible and incurable disease and it can be corrected only to a small degree by hearing aids; consequently, prevention is of primary importance. Since the disease develops slowly over the years and since the first signs are readily detected by simple audiometric examinations, it can be said that the basic principle for medical prevention is the *periodic audiometric examination*. Clearly, this basic medical practice should be followed up by a series of environmental and organisational preventive measures required to remove the affected worker from the hazardous work environment.

Deafness or hearing loss due to barometric trauma of the middle ear

Sudden variations of barometric pressure associated with changes in the normal function of the Eustachean tube may produce lesions to the tympanic membrane and the middle ear. Barotraumatic lesions may occur in two similar manners; however, the signs differ depending on whether the subject is exposed to a pressure increase or a pressure decrease. This may occur during rapid underwater descent, sudden dives, excessively rapid landings or when a diver ascends to the surface.

When the environmental pressure increases, a relative depression occurs in the tympanic cavity causing the tympanic membrane to bulge inwards; when the environmental pressure decreases, there is a relative pressure increase which makes the tympanic membrane bulge outwards. In both cases the symptoms are similar: a feeling of fullness in the ear, with moderate pain in slight cases and extremely violent pain with bleeding of the ear in cases of tympanic laceration. Generally, the trauma has a benign course with complete functional recovery.

Preventive measures entail correct execution of endotympanic decompression and the treatment of all rhinopharyngeal disorders which may cause stenosis of the Eustachean tube.

Deafness or hearing loss due to gaseous emboli

Breathing air in high atmospheric pressures results in the solution of large quantities of nitrogen in the blood; if the return to normal barometric pressure is not carried out correctly, it is possible that nitrogen emboli will form in the blood and tissues causing vascular obstruction and a wide range of symptoms depending on the site of the lesion. A not infrequent neurosensorial symptom is the development of hearing loss. This is mostly irreversible and sometimes of a very advanced stage; it is of the sensory type with tinnitus and balance disorders due to circulatory changes occurring either in or behind the cochlea.

Deafness or hearing loss due to cranial or craniofacial trauma

Cranial trauma may cause hearing damage not only when it is associated with fractures of the base of the skull and consequently of the *pars petrosa ossis temporalis* with destructive phenomena of the cochlea structures located there; but also when it is characterised by cerebral concussion without lesions to the bone; in this second case, the succussion of the endocochlear liquids is responsible for the hearing damage. The hearing loss that this entails will be of the sensory type, mostly monolateral in relation to the site of the trauma, irreversible and of very varying extent.

A blow to the chin region may be followed by collapse of the mandibular condyle of the external auditory conduit, rupture of the tympanic membrane and fracture of the ear bones; the result is hearing loss of the conductive type.

Deafness or hearing loss due to industrial toxic agents

There are substances with a selective toxic effect on the nerve structures of the ear and substances with a general toxic effect. The first group includes pharmaceuticals (certain antibiotics, antimalarial compounds, salycilates, diuretics, antimitotics, etc.), the action of which has often been observed in patients undergoing therapy; however, there is no evidence in the literature about the incidence of hearing loss in persons involved in the production of these pharmaceutical products, who are potentially exposed to absorption of very small doses but over a very long period.

Among substances with a general toxic effect, there are alcohol and a large number of industrial toxic agents. Desoille makes reference to lead, describing its prevalent vestibular toxicity, and carbon monoxide to which he attributes irreversible deficit in the 4 000 and 8 000 Hz range; mention should also be made of mercury, carbon disulphide, and certain solvents such as benzene and trichloroethylene, which probably produce changes in the acoustic nerve.

Compensation

In a large number of countries noise-induced hearing loss is recognised as an occupational disease and is subject to compensation; however, the approach to recognition and criteria for evaluating auditory deficit are subject to extremely different standards.

In Italy those who contract noise-induced deafness or hearing loss in the course of various types of work scheduled under 22 headings, and subject to insurance legislation, have the right to compensation provided under the law. Acute acoustic trauma is considered to be an occupational accident and is compensated as such. The degree of permanent disability for total bilateral deafness is 60% whereas that for monolateral deafness is 20%.

MERLUZZI, F.

Effects of noise on hearing. Henderson, D.; Hamernik, R. P.; Dosanjk, D. S.; Mills, J. H. (eds.) (New York, Raven Press, 1976), 565 p.

Noise and audiology. Lipscomb, D. M. (ed.) (Baltimore, University Park Press, 1978), 467 p.

Decompression sickness

The symptoms and signs which may follow decompression from exposure to air at a pressure greater than the normal atmospheric pressure (1 kg/cm²) (see COMPRESSED-AIR WORK) are called decompression sickness. In spite of determined attempts to prevent it, acute decompression sickness still occurs after up to 1% of decompressions. The higher the working pressure the greater the frequency of decompression sickness.

Acute decompression sickness can be divided in two main forms, type I and type II (table 1) which, although they are purely descriptive and do not relate to the pathogenesis, are of value in determining prognosis and treatment. They may occasionally occur together. In all forms of decompression sickness, immediate compression in a suitable compression chamber is indicated.

Table 1. Forms of decompression sickness

Form	Symptoms
Acute	
Type I	Mild limb pain
	Severe limb pain
	Skin mottling and irritation
Type II	Vomiting with or without epigastric pain
	Vertigo
	Tingling and numbness of limbs
	Paralysis or weakness of limbs
	Dyspnoea
	Headache
	Epileptiform fit
	Visual defects: flashes of light or scotomata
	Angina, symptoms suggesting coronary dysfunction, irregular pulse
	Collapse, hypotension
	Coma
	Death
Chronic	
	Aseptic necrosis of bone
	Neurological and psychiatric forms, including paralysis

Type I decompression sickness

This is acute pain in a limb or limbs, usually in the region of a major joint. The pain may be mild and trivial or may be severe and incapacitating, requiring urgent treatment for relief. Pain may come on at any time from the end of decompression up to 12 h later. Workers should be encouraged to report even mild discomfort so that they can return to a compression chamber (medical lock) for treatment. Skin mottling or itching, or the presence of gas bubbles in joint spaces may follow decompression but their significance is difficult to evaluate. Although these may occur frequently and apparently with no after-effects, it is safer to treat the men by recompression.

Type II decompression sickness

The serious forms of decompression sickness, which tend to occur soon after decompression, are fortunately relatively uncommon. Treatment is usually urgent and may be lifesaving. The most striking symptoms are those of paralysis of the limbs such as paraplegia or quadriplegia. Much less common is coma, which may be fatal unless the man is quickly recompressed. Post-decompression shock with haemoconcentration is another serious complication. Sensory and eye disturbances or cardiac symptoms can occur and may be misinterpreted unless a history of work in compressed air shortly before their onset is obtained. Alcoholism, head injury or an apparent cerebrovascular accident in a compressed-air worker may be grossly misleading, and it is safer to recompress first and then reconsider the diagnosis if relief of symptoms is not obtained.

Aseptic necrosis of bone (dysbaric osteonecrosis)

Death of areas of bone associated with work in compressed air or diving can occur, even from one apparently satisfactory compression/decompression. Bone necrosis may appear as a late complication in one or more sites such as the head and shaft of the humerus, the head and lower end of the femur, and the head of the tibia. It has a marked tendency to be symmetrical. Aseptic necrosis is likely to be destructive and disabling only if it involves the articular surface of the head of the femur (figure 1) or of the humerus. Fortunately most lesions are symptomless and never cause disability. At present conventional radiography of the limb bones is the only satisfactory way of diagnosing it. However, the lesions can only be detected by radiography from about three months after the bone has died, when deposits of new bone on dead trabeculae give an abnormal shadow in the radiograph. A progressive juxta articular lesion results in rupture of the articular surface and collapse of the joint with subsequent osteo-arthritis (figure 1). In the head of the femur total replacement of the joint with a metal prosthesis is required. Early evidence of bone damage can be difficult to diagnose without considerable experience, and good quality radiographs with clear bone detail and without overlapping bone structures are essential. The prevalence of bone necrosis is related to the maximum pressure experienced, the number of compressions/decompressions and to attacks of the bends, although the latter relationship is not a direct one and men often have frequent attacks of the bends without bone necrosis and vice versa. About 25% of experienced compressed-air workers have areas of bone necrosis but disabling juxta articular lesions only occur in less than 3%. In deep divers between 3 and 6% of men have bone necrosis and less than 0.2% have disabling lesions. In contrast to compressed-air workers divers have few lesions in the femoral heads.

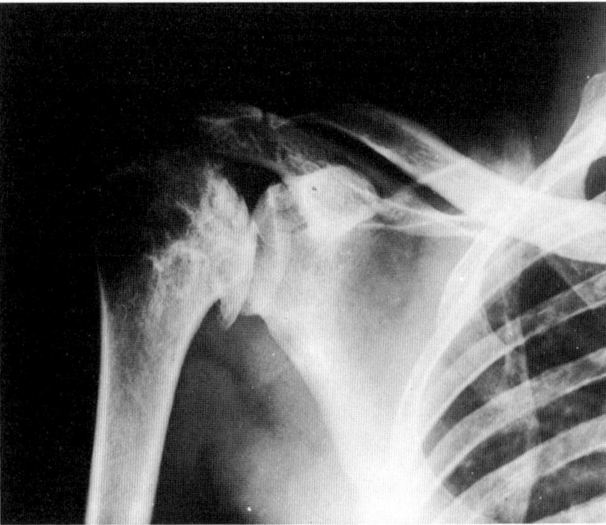

Figure 1. Radiograph of the right shoulder of a compressed-air worker showing a juxta articular lesion which has progressed to destruction of the articular surface and secondary osteo-arthritis.

There is no evidence that particular individuals are prone to develop bone damage, but when a definite juxta articular lesion is diagnosed, even if symptomless at the time, it is prudent to advise cessation of compressed-air work or diving in order to avoid the risk of a second juxta articular lesion.

Aetiology

Decompression sickness is thought to result either from the embolic and pressure effects of gas bubbles formed from supersaturated blood and tissues during compression, or from air embolism from the lungs, or sometimes from both these events. The body is normally fully saturated with nitrogen at atmospheric pressure but during exposure to high gas pressure for a period of hours, as in tunnelling or caisson work, or for short periods in deep diving, the body absorbs more gas in proportion to its partial pressure (Henry's Law). Body fat, for example, can hold more nitrogen than other tissues, and as the blood supply is poor, the nitrogen enters slowly and leaves even more slowly. The oxygen of the air is used up in the normal way and does not present any problem at the pressures usual in civil engineering. Depth (or height of pressure), duration of exposure and rate of decompression are all important factors; age, obesity, activity and ambient temperature also play a part. Decompression tables are designed to reduce the ambient pressure in such a way that either bubble formation is negligible or bubbles do not attain a significant diameter for very long.

The exact mechanism by which type I decompression sickness is produced is still not clear, but the neurological signs of type II decompression sickness may result from the formation of gas bubbles in central nervous system tissue or from obstruction of the epidural vertebral venous system by bubbles, and infarction of the spinal cord. Aseptic necrosis of bone is generally thought to result from blockage of its blood supply by gas bubbles, but so far there is no direct or experimental proof of this. Other theories involve changes in blood constituents secondary to bubble formation such as denaturation of protein, red cell and platelet aggregation and haemoconcentration, osmotic pressure in bone tissue from dissolved gas, or lipid emboli.

Massive air embolism in which the heart, large vessels and the circulatory system are filled with frothy blood containing air is often found in fatal (type II) decompression sickness. The source of the gas is the lungs due to blockage of a bronchus, perhaps with formation of a cyst and its rupture into a small bronchial vessel.

Pathology

No pathological lesion has yet been demonstrated to explain the cause of pain in the limbs but areas of necrosis in the dorsal region of the spinal cord with secondary degeneration have been described and provide a basis for neurological signs.

The pathology of bone necrosis has been studied in detail. In lesions affecting the articular surface of a joint, the necrotic area originally involves nearly the whole head of the bone, but revascularisation reduces the radiologically visible lesion to a much smaller area. Between revascularised and necrotic areas is a dense band of fibrous tissue, and on the revascularised side there is a deposition of new bone on dead trabeculae. Sequestration of part of the articular surface may occur, and secondary osteo-arthritis at a later stage may cause gross changes in the joint.

PREVENTIVE MEASURES

Prevention depends mainly on:

(a) medical selection of workers;

(b) periodic medical examinations during the course of employment in compressed air, or in diving;

(c) adequate safety education and supervision;

(d) decompression according to a pressure/time sequence related to the working pressure and duration of exposure.

Where statutory regulations are in force, all the above points are usually included in them.

Medical selection of men for this work should eliminate the obese, alcoholics and those with chronic ailments, particularly of the respiratory and cardiovascular systems. New workers should not be more than 35 years of age, as older men appear to be more prone to decompression sickness. Men with acute respiratory infections, such as colds, sinusitis or bronchitis, should be temporarily excluded from compressed-air work or diving. Radiographic examination of the lungs and major joints should also be carried out at the initial medical examination. Radiographs of the shoulder, hip and knee joints should be repeated at least once a year up to two years after ceasing to work in compressed air in order to detect aseptic necrosis of bone. Medical examination at regular intervals and after any illness, including head colds, is usually compulsory, and men should not be allowed into compressed air without a medical certificate from the physician to the contract.

Men who have had the more serious form of decompression sickness (type II) should not be permitted to enter compressed air again unless they are key workers, in which case only very short exposures should be allowed.

If ill trained and undisciplined men are employed, serious decompression sickness may occur through ignorance of the risks involved, or negligence. Workers must be helped to understand the hazards of decompression sickness, and particularly of bone necrosis, by written and verbal explanations from the physician and the medical attendants.

Examples of decompression times are given in the article COMPRESSED-AIR WORK; however, decompression according to tables designed on physiological principles to avoid formation of bubbles in the body has not yet succeeded in preventing any form of decompression sickness entirely. There is no evidence that the tables in current use will prevent bone necrosis in a proportion of men as their years of exposure increase. The prevalence of bone necrosis in compressed-air workers may be kept low if the working pressure is low and if shift length is reduced at higher pressures. At present, the decompression tables used in civil engineering can be considered to allow some control of an ill understood situation, and to have kept the complications of decompression to a level tolerable in the past but now no longer acceptable. In diving, the long term effects of saturation work at great depths have yet to be fully evaluated. A follow-up period of several years is required to detect late bone changes.

Treatment. Therapeutic procedures for acute decompression sickness are not usually laid down by statute but are carried out according to the judgement of the physician in charge guided by the response of the patient. Symptoms are relieved by placing the patient in a raised atmospheric pressure as soon as possible. The pressure is then reduced at a much slower rate than that required by the usual decompression table. The pressure reduction may follow, for example, the treatment tables in the United States Navy Diving Manual, or may be empirical. In a severe case of type II decompression sickness, treatment may need to extend over several days. The inhalation of oxygen at low pressure may shorten therapeutic decompression time considerably but oxygen is toxic at pressures greater than 1.8 kg/cm^2 and stringent precautions must be taken against the risk of fire. Other forms of treatment such as intravenous plasma or plasma substitutes, heparin and steroids may also have a place.

Treatment of chronic forms of decompression sickness, such as paralysis or bone necrosis, follows the usual procedures for these conditions arising from causes other than decompression.

McCALLUM, R. I.

Decompression sickness. Vol. 1. *The biophysical basis of prevention and treatment.* Hills, B. A. (Chichester, New York, Wiley Interscience–John Wiley and Sons, 1977), 322 p. Illus. Ref.

CIS 76-1732 *The physiology and medicine of diving and compressed air work.* Bennett, P. B.; Elliot, D. H. (London, Baillière Tindall, 2nd ed., 1975), 566 p. Illus. 2 000 ref.

CIS 75-1701 "Acute decompression sickness". Elliott, D. H.; Hallenbeck, J. M.; Bove, A. A. *Lancet* (London), 16 Nov. 1974, 2/7890 (1193-1199). 92 ref.

CIS 79-544 *Information sheet on the treatment of conditions due to work in pressurised atmosphere (work in compressed air, diving)* (Merkblatt für die Behandlung von Erkrankungen durch Arbeiten in Überdruck (Arbeiten in Druckluft, Taucherarbeiten)). ZH 1/587, Hauptverband der gewerblichen Berufsgenossenschaften (Cologne, Carl Heymanns Verlag, 1978), 14 p. (In German)

Degreasing

Degreasing is a process for the removal of mineral and vegetable oils and greases from the surface of metal, in particular the removal of the film of cutting fluid left on metals after they have been worked on machine tools. The process is an essential preliminary to electroplating, galvanising, tin plating and painting and varnishing.

Degreasing processes. The type of degreasing process will depend on the nature of the impurity to be removed, the metal being degreased and the final surface state required. The most important degreasing techniques are as follows:

(a) degreasing with organic solvents in liquid or vapour form, such as 1,1,1-trichloroethane, trichloroethylene, tetrachloroethylene, etc.;

(b) degreasing with alkaline solutions;

(c) degreasing with emulsifying agents;

[(d) degreasing with safety solvents, i.e. a blend of a flammable solvent and a non-flammable solvent such as perchloroethylene;]

(e) ultrasonic degreasing;

(f) flame degreasing.

HAZARDS AND THEIR PREVENTION

Many of the solvents used for degreasing may be both toxic and highly flammable substances and certain alkaline solutions may have a pronounced corrosive effect when in contact with skin or mucous membranes. The use of ultrasound for degreasing is a relatively recent innovation and experience of possible deleterious effects on workers is still limited.

Solvent degreasing.

[(a) Liquid degreasing. In small shops grease, oil and dirt are usually removed from metal parts by swabbing, brushing, dipping or spraying. The solvents most commonly used are Stoddard solvent, kerosene, mixtures of white spirit and other solvents, and halogenated hydrocarbons. Petroleum products have relatively low toxicity but are highly flammable and their storage, handling and use require special fire-protection and preventive measures (see PETROLEUM PRODUCTS, STORAGE AND TRANSPORT OF). Degreasing should be carried out in premises free from all naked flames, and fitted with flame-proof electrical installations. Rags and waste should be placed in metal disposal cans with self-closing covers. Fire extinguishers should be available.

(b) Vapour degreasing. In this process the degreaser produces a cloud of solvent vapours at its boiling point that condense on the surface of metal. The metal dries when it reaches the temperature of the vapour; then it can be removed from the vapour zone. There are various types of degreasers: the simplest is the straight vapour machine. In the immersion vapour machine the vapour condensation phase is preceded by the immersion of the metal in the boiling solvent and the rinsing with a cool and clean solvent. When soils are only partially soluble or the metal parts are big, the vapour spray machine is used; here the solvent vapour is forcefully sprayed, manually or mechanically, on the metal part.

The proper design of the degreaser can prevent contamination of the atmosphere (figure 1). The control of the solvent vapour is provided by a water jacket plus cooled coils around the walls and/or by thermostatic control. The free-board between the vapour level and the lowest portion of the top of the tank should be at least one-half of the width of the tank. Conveyors should be so designed and operated as to remove only dry parts from the tank. Tanks with an open area of half a square metre or more should also be provided with exhaust ventilation, provided it does not cause turbulence and a consequent dispersion of the vapour outside the vat.

Good housekeeping is essential. When the contamination level reaches 25%, the degreaser should be boiled down and drained. The general cleaning requires the transfer of all solvent, cooling down, draining and thorough ventilation. The man entering the tank should wear harness and lifeline in addition to respiratory protection (see CONFINED SPACES).

The characteristics required of the solvents used are: good solvent powers for all greasy impurities commonly

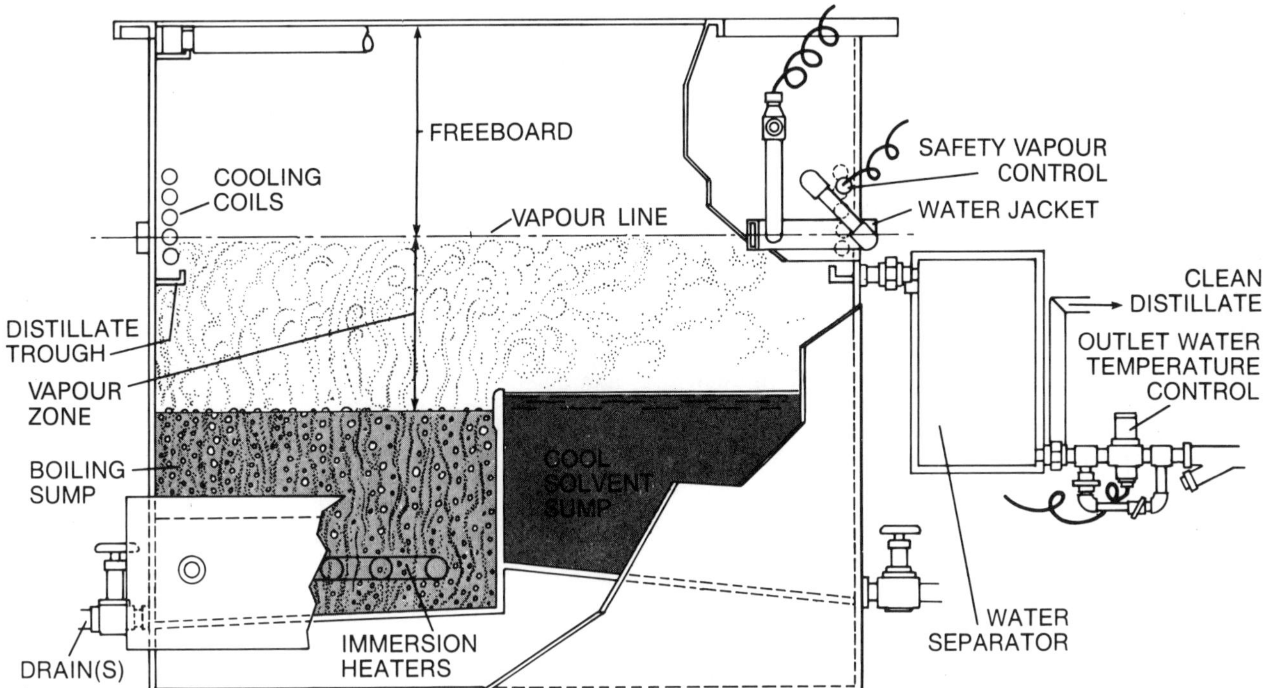

Figure 1. A sketch of an immersion-vapour degreaser that shows the components of a typical vapour degreaser system. *(By courtesy of the National Safety Council, 425 North Michigan Avenue, Chicago.)*

encountered, low boiling point, stability, very low toxicity, non-flammability and ease of regeneration. The solvents commonly used in industry are 1,1,1-trichloroethane (b.p. 74.1 °C), trichloroethylene (87 °C), perchloroethylene (121 °C), dichloromethane (40 °C), trichlorotrifluoroethane (47.7 °C).] Their vapours are non-flammable but are toxic by inhalation and have a pronounced defatting action on the skin. These substances are also relatively unstable when subjected to heat: trichloroethylene may decompose when exposed to sunlight or to hot surfaces at temperatures over 120 °C; tetrachloroethylene decomposes at temperatures above 150 °C. The decomposition products of these substances include hydrochloric acid and phosgene. [When aluminium parts are degreased with trichloroethylene, and especially when aluminium powder is spread out, explosions can occur and phosgene can be produced, probably as a consequence of polycondensation and dechlorination of the solvent. For degreasing aluminium therefore only stabilised solvents should be used.] Large-scale production degreasing with these toxic substances should be carried out in totally enclosed systems fitted with exhaust ventilation at points where the escape of solvent vapour is still possible. Special work or individual items may be dealt with in glove boxes or on grids fitted with exhaust ventilation. Workers should be supplied with hand and arm protection, eye and face protection and impervious aprons where there is a danger of skin contamination; respiratory protective equipment, preferably of the supplied-air type should be provided [to the maintenance personnel before entering the unit and] for use in emergencies. Smoking should be prohibited at the workplace and workers should be instructed in basic first-aid procedures in case of intoxication. Persons working with trichloroethylene, tetrachloroethylene or carbon tetrachloride should receive regular medical examination.

Alkali degreasing. In this technique, the metal is immersed in a bath of alkaline solution such as sodium hydroxide, sodium carbonate or sodium phosphate, which may also contain a wetting agent and emulsifier. The concentration of active ingredients varies between 2 and 10%, and the working temperatures range from 50 to 100 °C; to remain effective, the solution must be maintained above pH 10.

The precautions necessary in alkali degreasing are basically those described in the article ALKALINE MATERIALS and include the supply of hand and arm protection, eye and face protection and impermeable aprons to workers who may be exposed to splashes of caustic liquid.

In electrolytic alkali cleaning, the articles are immersed in an alkaline solution and act as one electrode of an electrolytic cell. The hazards are basically the same as for normal alkali degreasing; however, the gases released during the electrolysis tend to carry minute droplets of the solution with them thus increasing the danger of caustic action on the skin and mucous membranes. Consequently, effective exhaust ventilation above the bath is essential.

Emulsifying agents. The bath is made of soap or detergents usually blended with kerosene and stabilised with phenolic or amino compounds. These agents involve a hazard of dermatitis and skin allergy. In addition such emulsions are flammable.

Safety solvents. The blending of a halogenated hydrocarbon with a flammable solvent still involves a fire hazard when the flammable liquid is highly volatile or when the non-flammable solvent is lost by evaporation.

Ultrasonic degreasing. High-amplitude ultrasound is used widely in the cleaning of small metal parts requiring a high degree of cleanliness such as watch components, immersed in a solvent, since the large acoustic forces actually break off particles and contaminants from metal surfaces. The treatment is applied after pre-cleaning and it is followed by a distillate rinse and final vapour rinse. Ultrasonic cleaners may operate at frequencies of 18-80 kHz, and sound pressure levels of up to 133 dB have been measured at frequencies of 25 kHz. The effects of ultrasound on the body are dealt with in the article ULTRASOUND. In view of the fact that air is a poor conductor of ultrasonic energy, it is considered that work with ultrasonic cleaning equipment can be performed safely provided workers use tongs for inserting and withdrawing components from the cleaning bath. However, where there is a danger of contact with the bath, workers should wear personal protective equipment made of materials which have a good attenuating effect for the frequencies involved, e.g. hand protection made from rubber, leather or bulky textiles.

Flame degreasing. In view of the high temperatures involved in this process and the emission of acrolein and dangerous products of combustion, it is preferable to replace flame degreasing by another cleaning technique. Where the process is used employees should be protected against the danger of burns, and exhaust ventilation should be provided to capture the dangerous gases and vapours given off.

MACHÁČ, D.

Safety:

"Liquid degreasing of small metal parts". Data sheet 537, 1977 revision A—National Safety Council. *National Safety News* (Chicago), Dec. 1977, 116/6 (79-83). Illus. 5 ref.

Vapor degreasers. Data sheet 429, revision A (extensive) (National Safety Council, 425 North Michigan Avenue, Chicago) (1975), 6 p. Illus. 14 ref.

Industrial hygiene:

CIS 1189-1973 "Field studies in solvent plants. First report: Measurement techniques for determining solvent concentrations at various trichloroethylene degreasing installations" (Felduntersuchungen in Lösungsmittelbetrieben. 1 Mitteilung: Messmethoden zur Überprüfung des MAK-Wertes an verschiedenen Trichloräthylen-Waschanlagen). Weichardt, H.; Lindner, J. *Zentralblatt für Arbeitsmedizin und Arbeitsschutz* (Heidelberg) Nov. 1972, 22/11 (323-332). Illus. 19 ref. (In German)

Dentists

Dentistry covers bucco-dental and maxillo-facial treatment and surgery. Odontologists, stomatologists, orthopaedists and pedodontists treat the teeth and the oral cavity and perform surgery on the teeth, the maxillae and the bones of the face. They correct articular deformities and defects, provide specialist care for children's teeth and carry out prophylactic treatment on the denture and dentition. Dentists may work in private surgeries or consulting rooms, public or private hospitals or educational establishments.

HAZARDS AND THEIR PREVENTION

In order to meet all these different requirements, a wide range of equipment and materials are needed. These may be divided into three groups:

(a) surgery installations, which include the rotating instruments (turbines, micro-motors, etc.), lighting, control devices, spitoon, dentist's chair, X-ray apparatus, cupboards and shelving for storage;

(b) small instruments, including surgical instruments, grinding and polishing tools and drills;

(c) pharmaceutical and chemical products.

The use of these instruments in the course of their work sometimes leads to accidents and health problems for dentists. Technical advances over recent decades have led to the introduction of much new equipment for dental surgery which has changed the occupational risks to which the dentist is exposed: radiodermatitis and deformation of the spine in dentists are now less of a problem, while, on the other hand, new hazards have made their appearance such as infections from splashes of septic liquid thrown off by water-cooled drills or auditory problems resulting from the use of ultrasonic turbines.

Radiodermatitis. The use of radiological apparatus may, in certain cases, result in dental practitioners receiving excessive doses of ionising radiation. This radiation may cause haematological effects by its action on the bone marrow. Local effects may include, primarily, finger lesions (hyperkeratosis, atrophy, telangiectasis) and, secondarily, true radiodermatitis with ulceration (which is considered to be the first stage of skin cancer). However, in practice the hazard will be negligible in view of the small doses employed, provided the operator keeps his distance from the radiation source and does not himself hold the radiographic plate, and provided secondary sources are eliminated (prostheses in the mouth). For permissible doses, see RADIATION, IONISING–DETECTORS AND PORTABLE SURVEY INSTRUMENTS.

Septic contamination. The use of water sprays for cooling dental drills, especially of the turbine type, operating at any speed, leads to splashes of liquid being thrown out of the patient's mouth, frequently into the vicinity of the dentist's face. Qualitative and quantitative analyses on the septic content of these splashes have shown that they have three components: droplets, fine particles and aerosols. The droplets and fine particles are carriers of pathogenic and saprophytic bacteria and may reach the dentist's face; however, on such carriers, these bacteria are no more of a hazard than the micro-organisms normally found in the atmosphere.

On the other hand, aerosols present a much greater danger due to their power of penetration which allows them to enter and endanger the bronchi of the dentist, his assistants and the patient. Although immunising processes provide considerable protection against many bacteria, they are ineffective against staphylococci, for example. Moreover, bacteria which are saprophytic in the patient may become more virulent when transferred to the dentist, either because of the transfer itself, or because the new environment is more favourable for development.

The dentist's mouth and nose should, therefore, be adequately protected by the use of a light gauze mask.

Future study of this disease is to be linked directly with the technical development of dental equipment: the total collection of all particles that may be thrown out, particularly through the use of micromotors which work at speeds of 120 000 rpm and allow of greater efficiency, comfort and safety, and which are replacing the turbine-driven instruments.

Noise. While there have been no reports of hearing impairment due to the use of low- and medium-speed drills, cases of aural damage have been attributed to the use of turbine drills.

Audiometric investigations have revealed certain phenomena related to the use of these machines, viz.:

(a) undesirable vibration can be set up by the rotation of these turbines;

(b) drilling with new drills in the smaller sizes (0.8 mm diameter) can give rise to noise within the basic frequency range of 5 000-6 500 Hz;

(c) when worn drills are used the frequency can reach 12 500 Hz, while with large-diameter drills (1.5 mm or greater) or with worn turbines, the frequency can go up to 25 000 Hz with harmonics, which is thus well within the ultrasonic range;

(d) a great deal of the ultrasonic noise is absorbed by the air, however, and its strength is generally insufficient to provoke aural lesions;

(e) as the sound produced is generally of low intensity, around 75 dB, it is unlikely to have a damaging effect.

It can thus be concluded that, in the light of current knowledge, the use of turbines does not harm the hearing provided that small diameter drills are used, that they cannot move within their clamp or casing, and also that the dentist takes up a suitable position (sitting or standing) sufficiently far from the turbine.

Vision. If the operating field is not sufficiently illuminated and the surroundings are badly lit, the dentist may suffer from nodular headaches and loss of visual acuity. To avoid these problems the following lighting standards should be observed:

(a) the ratio between the general lighting and that of the operating field should be 1 : 4, that is about 5 000-8 000 lux in the mouth and about 1 250-2 000 in the surgery;

(b) the lighting ratio between the different working positions and the surgery should be 1 : 10;

(c) all white or dazzling objects or surfaces should be eliminated (chrome, metal supports, towels, etc.).

Mental impairment. The growing number of patients and their increasing demands impose upon the dentist excessive working hours and long periods of concentration, as well as considerable stress from the effort to avoid pain and suffering to the patient. He is also subjected to the strain of having continually to master new techniques. Although not going as far as mental disease, psychic disorders should not be overlooked in relation to the occupational health problems of the dentist.

Cuts and abrasions. Many small, pointed metallic instruments such as probes, reamers, scrapers and nerve extractors are widely used and are a frequent cause of cuts and abrasions to the tips of the dentist's fingers; the handling of pieces of sharp metal such as gold or steel rings, metal laminae, etc., may also cause cuts. Sharp fragments of teeth and metal may be projected by the drill and, should they enter the skin or eyes, they may result in local infection, usually of the streptococcal type. These injuries may range from benign, local inflammation (if they are treated immediately), to phlegmon of the finger or toxic conjunctivitis of the eye. Protection against small cuts is primarily a matter of care in handling sharp instruments; however, the dentist is advised to wear eye protection such as spectacles with either optical or neutral lenses.

Miscellaneous hazards. The dentist may also be exposed, though less commonly, to allergy to medicaments, skin reactions to various medicaments and materials, mercury poisoning, impairment of sight, cardiovascular disorders and varicose veins.

Ergonomic recommendations. A number of general rules should be followed in order to avoid these psychological and physical difficulties:

(a) well designed equipment should be used; instruments should be readily accessible, and turbines replaced by micro-motors; a fully adjustable dentist's chair should be used with the back unobstructed so that the dentist's legs are not blocked when working in the sitting position; correct lighting should be provided;

(b) loud or sudden noises and ultrasonic frequencies (such as the turbine and shrill bells) should be avoided; soundproofing of the surgery is advisable;

(c) the dentist should adjust his rhythm of work to suit his own personality and way of life.

VENDROUX, C.

Hazards:

CIS 79-490 "Myeloneuropathy after prolonged exposure to nitrous oxide". Layzer, R. B. *Lancet* (London), Dec. 1978, 11/8102 (1227-1230). 16 ref.

"Pneumoconiosis hazard to dental technicians" (Le risque de pneumoconiose chez les techniciens-dentistes). Peltier, A.; Moulut, J. C.; Demange, M. *Travail et sécurité* (Paris), Mar. 1979, 3 (166-168). 2 ref. (In French)

CIS 76-920 "Explosion during halothane anaesthesia". May, T. W. *British Medical Journal* (London), 20 Mar. 1976, 1/6011 (692-693).

Prevention:

CIS 78-111 *Control of occupational exposure to N_2O in the dental operatory.* Whitcher, C. E.; Zimmerman, D. C.; Piziali, R. L. DHEW (NIOSH) publication No. 77-171 (National Institute for Occupational Safety and Health, 4676 Columbia Parkway, Cincinnati) (Dec. 1976), 42 p. Illus. 49 ref.

CIS 78-736 "Reduction of mercury vapour in a dental surgery". Wilson, J. *Lancet* (London), 28 Jan. 1978, 1/8057 (200-201). 7 ref.

CIS 77-1609 *Manual on radiation protection in hospitals and general practice.* Vol. 4: *Radiation protection in dentistry.* Koren, K.; Wuehrmann, A. M. (Geneva, World Health Organisation, 1977), 52 p. Illus. 10 ref.

Ergonomics:

CIS 77-569 *Dentistry–Working space of the dentist–Definitions and principles.* International standard ISO 3246-1977(E) (Geneva, International Organisation for Standardisation, 15 Jan. 1977), 3 p. Illus.

Summary of ergonomics applied to odontology (Abrégé d'ergonomie odontologique). Chovet, M. (Paris, New York, Barcelona, Milan, Masson, 1978), 212 p. Illus. (In French)

Detection and analysis of airborne contaminants (chemical laboratory methods)

Chemical analysis is still the basis of the determination of small quantities of contaminants for industrial hygiene purposes. Chemical methods are especially suitable:

(a) to check the calibration of modern techniques such as atomic absorption spectrography, ultraviolet and infrared spectroscopy, gas chromatography, neutron activation, polarography, etc.;

(b) where specialised, automatic instruments are not available;

(c) for intermittent analysis of batches of samples, because only when these are in sufficient number is it economically worth while to use specialised instruments;

(d) when no other method using very specialised instruments is available.

Furthermore the first phase of microchemical analysis is necessary in most cases to prepare the samples to be later subjected to physical analytical methods, or to extraction by using chelating agents, or to enrichment by means of precipitation or chemical conversion.

Choice of method

Many micromethods have been reported for the evaluation of a toxic contaminant in the workroom air or present in a biological sample, some of which have come to be accepted as reference methods. The method to be chosen depends on the following parameters:

(a) the availability of laboratory apparatus and the qualification of the personnel entrusted with making the analysis;

(b) the recommended exposure limit for the substance under examination: the lower the exposure limit, the more difficult the sampling and analytical methods become. For example, if a substance with a limit of 1 000 mg/m³ is being analysed, the method used needs to have a sensitivity of 1-2 mg, whereas for a substance that has a limit of only a few milligrams, a high sensitivity measured in micrograms is required if long duration sampling is to be avoided;

(c) the presence in the sample under examination of one or more substances that might interfere with the result: in some cases the interfering substance reacts similarly to the one under examination, thereby giving a result that is too high; in other cases it may react with the one under examination, giving a result that is too low. There are also substances that combine with a reagent hence hindering the reaction taking place with the substance under examination. In such cases, either a method of analysis that eliminates the interference must be chosen or, as in the case of atmospheric analysis, it must be eliminated at the sampling stage by using an appropriate method. An example would be the case of the oxides of nitrogen, mixed with sulphur trioxide, which are decomposed during the analysis by means of a suitable reagent; the effect of ferric salts, which oxidise quantitatively the sulphur trioxide, is neutralised by the addition of an iron chelating agent to the absorbent liquid; the ketones can easily be identified making use of gas chromatographic analysis.

When two very similar chemical substances coexist, for example, hydrogen chloride and bromic acid, whose exposure limit values are much the same and which it is difficult to separate so as to make a quantitative determination of each of them, a method must be adopted in which the two substances are titrated at the same time and the results expressed in terms of the amount of the substance having the lower exposure value. In such a case a more severe judgement of the environmental conditions would be made than was actually the case. The positive or negative interference in such circumstances, expressed as a percentage, would be

$$\frac{X - Y}{Z} = 100$$

where:

X is the total quantity of the coexistent substances being determined (contaminant and interfering) expressed in ppm,

Y is the quantity of the contaminant substance under examination, expressed in ppm, and

Z is the quantity of the interfering substance, expressed in ppm.

To reduce to a minimum the error due to the interfering substance which may vary from one batch to another and sometimes even within the same batch of a particular production cycle, it is advisable to carry out the evaluation by two or more different methods. Any interference can only be considered as negligible when similar results have been obtained from two different analytical methods.

Finally, in choosing the method of evaluation, the confidence level of the results must be taken into account. The reliability of a method depends upon many different factors such as specificity, accuracy, precision, reproducibility, sensitivity and practical confirmation.

Specificity. This is the degree to which a method is capable of detecting, both qualitatively and quantitatively, a single substance without being affected by other substances present in the same sample. In the absolute sense, truly specific methods are rare in the field of microanalysis. Colorimetric methods are among those which are most sensitive to the presence of other substances, while the physical methods, such as atomic absorption or gas chromatography are capable of eliminating or reducing the effect of interferences to a minimum.

Accuracy. The accuracy of a method refers to the difference between the value which it gives for the concentration of a substance and the true value. It is represented by a positive or negative error which remains as a constant for the particular method. Accuracy can be determined by adding a known quantity of the substance under examination to the material being investigated, or by analysing a series of samples by means of the chosen method in parallel with a reference method and making a statistical treatment of the results thus obtained.

Precision. This is the difference between a number of repeated measurements of the same concentration (which can be the mean value of the results obtained), expressed as the mean deviation of a single result. Precision covers all random variations inherent in the method concerned. The dispersion of analytical values about the mean depends either on factors that are inherent in the method or on factors related to the operator on the one hand or to the quality of the apparatus used on the other.

Reproducibility. The total variation in repeated measurements of the same sample using identical test conditions in the same laboratory.

Sensitivity. The minimum amount of a substance which can repeatedly be measured by a given method of testing.

Practical confirmation. Any method to be used for routine testing must be tried out over a relatively extended period, either in the laboratory or in the field, in order to establish whether it has all the necessary characteristics for routine sampling and whether it will bring to light any possible interference.

Choice and purity of reagents

The purity of the reagents used is of particular importance in chemical analysis, especially when large amounts of acid and/or water are used in preparing the sample. Distilled water often contains impurities which may have a noticeable effect when using colorimetric analysis. In such cases the water must be double or triple distilled or purified by some other technique such as "sub-boiling" or "Milli-Q". The "sub-boiling" technique makes use of a quartz container in which the water is allowed to evaporate without being brought to boiling point; the water vapour condenses on a refrigerated plug and is collected in a suitable container. In the "Milli-Q" process, the water is de-ionised, passed over activated carbon and filtered through a cellulose membrane having a porosity of 0.22 μm. Metallic contamination in water distilled by different methods is shown in table 1.

Table 1. Impurities in distilled or de-ionised water (mg/ml)

Element	Double distilled	De-ionised distillation	Distillation by "sub-boiling"	"Milli-Q" process
Ag	1	0.005	0.002	0.01
Ca	50	1	0.08	0.05
Cr	–	–	0.02	0.1
Cu	50	3	0.01	0.05
Fe	0.1	0.4	0.05	0.02
Mg	8	2	0.09	0.03
Na	1	0.1	0.06	0.07
Ni	1	0.1	0.02	0.1
Pb	50	0.1	0.008	0.005
Sn	5	0.2	0.02	0.1
Zn	10	1	0.04	0.03
Total impurities	176.1	7.9	0.4	0.6

All reagents used should be of the highest possible purity and this should be regularly and carefully checked. In the case of colorimetric analysis there may sometimes be impurities present in the reagents which cause variations in the colour intensity; in the case of sulphur trioxide with *p*-rosaniline, the colour intensity may vary considerably from one batch to another; dithizone also, a basic reagent for several metals, is unstable and may give different colour intensity in the course of time.

When making a number of analyses, it is advisable to use reagents from the same source of supply, since those

Table 2. Concentrations of impurities in perchloric acid of different analytical grades (ng/g)

Element	"Sub-boiling" distillation	Commercial reagent "analytical grade"	Commercial reagent "highest purity"
Pb	0.2	2	16
Tl	0.1	0.1	–
Ba	0.1	1 000	10
Te	0.05	0.05	–
Sn	0.3	0.03	1
Cd	0.05	0.1	4
Ag	0.1	0.1	0.5
Sr	0.02	14	–
Zn	0.1	7	17
Cu	0.1	11	3
Ni	0.5	8	0.5
Fe	2	330	10
Cr	9	10	18
Ca	0.2	760	7
K	0.6	200	9
Mg	0.2	500	4
Na	2	600	–
Total impurities	16 ppb	3 400 ppb	100 ppb

from different suppliers may give rise to results that are not in agreement, as their content in impurities may vary. In a few cases when it is being used for a blank test, the solvent should not only have come from the same supplier, but should be taken from the same bottle as that used for the absorbent.

It may often be found that the purest commercial reagents, known as "analytical grade" contain sufficient impurities to upset the results of an analysis and in such cases it is necessary to resort to further purification of the reagent in the laboratory. Table 2 shows, by way of example, the concentrations of various metals in perchloric acid that has been purified in the laboratory by means of "sub-boiling" distillation, compared with commercial "analytical grade" and "highest purity" reagents.

Table 3 shows the degree of contamination by the metals referred to in table 2 found in acids that are commonly used in industrial hygiene analysis.

Table 3. Total impurities in various acids (ng/g)

Element	"Sub-boiling" distillation	Commercial reagent "analytical grade"	Commercial reagent "highest purity"
$HClO_4$	16	3 400	100
HCl	6.2	820	70
HNO_3	2.3	220	240
H_2SO_4	27	200	–
HF	17	320	–

The possibility of contamination of the water and the reagents used for testing is clear from the above examples and the analyst should thus choose and if necessary purify his reagents so as to reduce errors to a minimum. For the same reasons, one or more blank tests should be performed in order to determine whether impurities are present to an extent that might compromise the precision and accuracy of the tests.

In industrial hygiene practice the values obtained in a blank test should be only a minimum and negligible fraction of the quantities measured in the samples, and consequently determination down to 1 ppm is reliable only if the impurities in the reagent are of the order of 0.005-0.015 ppm.

Approximation of analysis

Requirements for the specificity, accuracy, precision and sensitivity are absolute only for the preparation of calibration curves. In practice there is a loss of reliability in any method due to secondary reactions or to interference in the sample under examination. In many cases the substance to be determined can alter during the sampling or during the period for which it is kept prior to analysis. For example, nitrogen oxide can be further oxidised during sampling, ozone can become altered in certain absorbent liquids, and if sulphur trioxide is absorbed in an alkali, it can become partly oxidised. A sample can also undergo alteration during storage through adsorption by the substance under examination from the walls of the containing vessel. The degree of adsorption taking place under such conditions depends on a number of factors such as the nature and strength of the substance itself, the type of solution and of the walls of the vessel, the acidity of the solution and the length of time it is stored and the temperature at which it is kept. An aqueous solution at pH 6 of a lead salt in a concentration of 0.01 ppm retained in a glass, polyethylene or polypropylene vessel will yield all the lead to the walls within 4-6 days. In an acid solution (pH 2) the adsorption of lead on walls of polypropylene or polystyrene is practically nil, while in the case of glass it would amount to around 60%. Temperature also usually has a negative effect on the stability of substances kept in solution or in the solid state and for this reason the optimum storage temperature for samples is about 4-6 °C, enabling losses of this nature to be minimised. In any case, the least possible time should be allowed to elapse between sampling and subsequent analysis.

Another source of error, especially when very dilute solutions are being analysed, is the contamination that may be introduced in the laboratory itself. Particles of other substances which may be present in the air of the laboratory can find their way into the sample and give rise to a significant and variable amount of contamination. This is especially the case when the laboratory forms part of the factory premises where background contamination can be high. In such cases steps must be taken to eliminate or reduce such contamination by dissolving the sample or evaporating liquid samples within a small container made of glass, quartz or teflon under a stream of filtered air or nitrogen.

It is thus clear that because of the numerous and sometimes unavoidable sources of error, some of the most important of which have been detailed above, test results can only add up to an approximation that is considerably less than that theoretically attainable under ideal conditions, and which amounts to about 3-5%. The analyst thus requires to have a full understanding and control of these matters in order to keep the unavoidable errors within acceptable limits.

In practice, an approximation of about ±10-15% may be considered sufficient in order to evaluate an occupational exposure, taking into consideration the continual and not negligible variations with time of the concentrations of contaminants in the atmosphere of the workplace.

Analysis by titration and gravimetric methods

Determinations made by weighing or by titration are only applicable in very special cases in the field of industrial hygiene. For example for the preparation of a standard mixture of carbon monoxide, this may be done gravimetrically based on the oxidation of the carbon monoxide and the adsorption of carbon dioxide on ascarite, whose increased weight will enable the concentration of carbon monoxide in the mixed gases to be calculated. The sensitivity of this method is of the order of milligrams.

Gravimetric analysis that requires the use of analytical balances of extreme sensitivity may be appropriate in circumstances where the quantity of the substance in question amounts to several milligrams.

Analysis by titration gives greater sensitivity than the gravimetric methods but may be considerably influenced by the subjectivity of the operator in his judgement of the colour change. As an example, sodium hydroxide can be determined by titration with acid; fluorines can be measured by titration with thorium nitrate in the presence of sodium-alizarin-sulphonate as an indicator; sulphuric acid can be determined by titration with barium perchlorate using Thorin as an indicator. Titration of lead using dithizone is a further reference method for the calibration of physical methods.

In most cases, where the amount of the substance available can be measured in micrograms or even in nanograms, gravimetric and volumetric methods have been almost completely replaced by methods which are both more sensitive and more specific, such as spectrophotometric methods.

Spectrophotometry

For micro- and semimicro-determinations, spectrophotometric analyses are now largely used because of their rapidity and ease of execution, not to mention the modest cost of the equipment. Spectrophotometry makes use of the property of different substances to absorb electromagnetic radiation of particular wavelengths, and is based on the laws of Beer, Bourger and Lambert according to which the reduction in intensity (absorption) of a ray of light passing through the solution under examination enables the quantity (concentration) of the light-absorbent solute to be determined.

The Lambert-Beer law may be expressed as:

$$P = P_o 10^{-abc}$$

where:

- P is the radiant energy transmitted
- P_o is the incident radiant energy, measured at the light beam in a blank sample
- a is the specific absorption of the substance, which is characteristic for a given solution and a given frequency of incident light
- b is the length in centimetres of the optical path of the cell
- c is the concentration of the substance expressed in grams per litre.

Transmittance T of light is the ratio of transmitted energy to incident energy, expressed as a percentage, and is given by:

$$T = \frac{P}{P_o} \cdot 100$$

Absorbance A, is given by:

$$A = \log_{10} \frac{P_o}{P}$$

The following instruments are required for spectrophotometric analysis:

- a source of radiant energy (visible, ultraviolet or infrared light);
- an optical system that will provide energy of a fixed wavelength, which can consist of an optical filter that selects a relatively narrow band (as used for photocolorimetry) or a monochromatic light source of a well defined wavelength (as used for spectrophotometry).

A monochromatic light is obtained by a glass prism or a diffraction grating in the case of visible light, a quartz crystal for ultraviolet or crystals of sodium chloride or potassium bromide in the case of infrared light. One or more cells contain the solution under examination and the standard solution; one or more photocells receive the radiant energy transmitted by the solution. A galvanometer takes the measurement.

Photocolorimeters and spectrophotometers may be of two types: with one or with two photocells. The first type enables measurements to be taken of the intensity of the energy which is emitted by the source and which, having passed through the cell containing the reference solution, is caused to pass through the solution being examined by substituting the cell. The degree of precision obtained in the measurement of the intensity of the transmitted light depends on the constancy of the radiant energy supplied during the measurement period.

In the case of instruments with two cells, the radiant energy comes from the same source and passes simultaneously through the reference and the solution under examination and is then picked up by the respective photocell; in this way errors due to fluctuations in the light source are eliminated.

In colorimetric analysis it is frequently found that most of the absorption by the solution takes place over a relatively wide band of the spectrum; in such cases a photoelectric colorimeter can be used. In cases where substances in the solution create interference in a given band of the spectrum, or where the absorption by the substance being examined falls in a very narrow band, it is advisable to measure the absorption in a zone of the spectrum where there is no interference, and in such a case the spectrophotometer should be used.

Photocolorimeters and spectrophotometers generally give values both for transmittance and absorbance. The best conditions for the evaluation of a coloured solution are to be found when the values lie between 0.1 and 1 for absorbance and between 10% and 80% for transmittance based on a blank having a value of zero absorbance and 100% for transmittance.

Visible spectrophotometry

Visible spectrophotometry is used not only for coloured substances, but mainly for those which under a particular radiation produce a colour that is capable of absorbing radiant energy in the range 380 to 760 nm.

In order to establish whether a spectrophotometric method can be used, the following points must be established:

(a) the specificity of any reagents being used;

(b) the validity of Lambert-Beer's law;

(c) the effect of the pH, the temperature, the strength of the reagents, and the sequence in which the reagents should be added;

(d) the time necessary for the colour to develop and to remain stable; and

(e) the range of concentrations for which the Lambert-Beer law is valid.

When measuring absorbance of a solution, a wavelength should be selected that provides the greatest possible difference between that of the sample and that of a blank solution. Generally speaking, the colour selected should be one that is complementary to that of the solution; for example, a green light with a wavelength of around 500 nm or a green filter should be used for a solution having a red colour, and violet light with a wavelength around 400 nm or a violet filter for yellow-coloured solutions. When making an analysis the substance under examination should be of such a concentration that its absorbance will lead to the least percentage error, which will be the case when the value for its absorbance falls in the middle of the calibration curve.

In cases where the absorbance varies significantly with time, the reaction time must be carefully controlled. In fact the colours resulting from the colorimetric reaction can either remain constant indefinitely, or can be stable for a certain time and then change, or can increase or diminish for a certain time and then remain stable, or they may never reach a stable condition. It is thus necessary in all colorimetric analysis to measure the intensity of the transmitted light from the sample being examined after sufficient time has elapsed for the colour to develop and during the period in which the colour remains most stable.

Nephelometric analysis

In a few methods, rather than measuring the light absorbed by a coloured solution, it is the absorption of light diffused by finely divided particles in suspension in a liquid that is measured. But while a solute is always uniformly divided throughout the liquid, thereby always giving a consistent answer, in the case of the suspended particles the amount of light diffused will depend upon their number, size and shape and also the wavelength of the incident light. If there are only a few particles in suspension they react independently and the diffused light will be proportional to their number. If, however, their number is large the light will be diffused from particle to particle and the light transmitted will not be proportional to their total number. This method is therefore more suitable for a homogeneous suspension of particles that are as far as possible of uniform size, but its reproducibility is distinctly inferior to that obtained with the colorimetric technique. The only advantage of nephelometric analysis is its simplicity and rapidity, but in spite of this it is a good rule not to use this method when it can possibly be avoided. Due to its lack of sensitivity and reproducibility it should only be used in industrial hygiene for a few substances.

It is one of the two standard methods proposed by NIOSH for the determination of sulphates following precipitation with barium salts.

Calibration curves

For nearly every photometric method it is necessary to establish a calibration curve, which is prepared with a series of samples. It should be based on known concentrations (standard scalar solutions, five or six at least) and set out on a graph showing the value of the concentrations against the related reading on the instrument. If the transmitted values as measured on the instrument are expressed in terms of absorbance, the graph must be based on a normal scale; if the measurements are expressed as a percentage of the transmittance, the graph must be prepared on a semilogarithmic scale with the transmittance values on a logarithmic scale and concentrations on a normal scale.

It is very important that the calibration curve should be linear, as this shows that there is a constant relationship between the quantity to be measured and the final result of the reaction. It should be kept in mind that the answer given by the photometer is now uniform throughout the scale. For the higher values on the scale, a variation in the concentration is not proportional to the reading given by the instrument, while for very low absorbance values, a small variation in the optical characteristics of the cells or a non-perfect reproducibility of their arrangements will make a difference to the reading obtained.

In both the photocolorimeter and the spectrophotometer the light source and the photocell can suffer a loss of intensity and of sensitivity, for which reason a fresh calibration curve should be established periodically.

In the case of a linear relationship the comparison with the calibration curve can be omitted by making the measurement of only one standard solution; the concentration will be calculated from the following formula:

$$C = \frac{A_c}{A_s} C_s$$

where:

C is the concentration of the substance in the sample under examination

A_c is the absorbance of the sample under examination

A_s is the absorbance of the standard solution

C_s is the concentration of the substance in the standard solution.

For the preparation of a calibration curve, the average of the analytical values for each standard solution must be calculated, since these values will in practice have a certain spread attributable to such factors as the lack of total purity of reagents, glassware not thoroughly clean, etc. Scrupulous attention must be given to all such details by the analyst when preparing a calibration curve.

As has already been pointed out, a blank test must be performed both when preparing a calibration curve and when making a colorimetric analysis. This means that an analysis must be performed using all the absorbant liquids and the reagents with the exception of the substance that is to be examined. This is necessary because in addition to any substances present as impurities in the sample, there may also be various ions, molecules or complexes of elements which have the capacity of light absorption.

As standard solutions can change quite rapidly, it is advisable to store them in a refrigerator and in a recipient of a suitable material in order to reduce to a minimum the phenomenon of adsorption by the walls. It is also necessary to check the stability of these solutions and at the same time to examine one or two samples of standard solution every time an analysis is undertaken.

All these precautions cannot always guarantee the reliability of the analytical results unless the characteristics of the sample to be examined with regard to its stability are taken into account. The analyst should take particular care to ascertain such information. Atmospheric samples collected in an absorbent liquid or on a suitable filter will be in concentrations of the order of a few parts per million and are susceptible to all the possible causes of change that have been described in the case of reagents.

It is thus important to programme the work in such a way that the analysis is made as soon as possible after the samples have been taken, especially when the substance being sought after has been obtained by bubbling through a liquid. Analysis should be carried out as rapidly as possible for the less stable substances such as nitrogen oxide, hydrocyanic acid, hydrogen sulphide, formaldehyde, carbon disulphide, etc.

Basic equipment required for a microanalytical laboratory

The main equipment required for an analytical laboratory is glassware and the usual accessories. In the case of graduated pipettes and measuring flasks, experience has shown that the values indicated on these items seldom correspond with reality, and it is thus necessary to check their accuracy or to use pipettes and measuring flasks that are guaranteed to be exact.

For the analysis of substances such as iron or lead which are often present as impurities in glassware, such articles will require repeated decontamination using heat or mineral acids, followed by checking and should be kept apart from others. For the preparation of chemically pure water and acids the necessary apparatus should be available such as for de-ionisation, for distillation and double distillation of water, for distillation by "sub-boiling" of water and acids and for de-ionisation and filtration of water and acids on activated carbon or membrane filters. In certain cases, as when it is required to determine trace substances or when excessive background contamination is present, small glass, quartz or teflon containers will be used. under filtered air or nitrogen flow. The retention efficiency of the filters used in such chambers must be of the order of 99.9% of particles whose diameter exceeds 0.3 μm so as to ensure that the concentration in the air supplied does not exceed 4 particles per cubic centimetre with a maximum diameter of 0.5 μm.

The refrigerator should be such that it is capable of holding a temperature with a maximum fluctuation of 2 °C, and should also have a deep-freeze compartment with plastic containers for conserving material. Balances are also necessary, including a technical balance and one or two analytical balances having different levels of sensitivity; these delicate instruments should be completely insulated from vibration and protected from the action of corrosive volatile agents and kept away from heat sources.

According to the types of evaluation that it is intended to carry out at the laboratory, it will be necessary to provide a photocolorimeter or a spectrophotometer together with the relevant recording instruments.

Other apparatus which will be required include a sand bath for calcining samples, an electric hot plate, a thermostatically controlled *bain-marie*, a thermostat, a drying oven, a collection of thermometers, and at least one vacuum pump.

METRICO, L.

Methods of air sampling and analysis. Katz, M. (ed.). APHA Intersociety Committee (American Public Health Association, 1015 8th Street, NW, Washington, DC) (2nd ed., 1977), 984 p.

"Accuracy, reproducibility and sensitivity". Prentice, J. *Health and Safety at Work* (Croydon), Sep. 1980, 3/1 (52-53). Illus. 4 ref.

NIOSH Manual of analytical methods (2nd ed.). Vol. 1, publication No. 77.157 A (Apr. 1977), 807 p.; Vol. 2, publication No. 77.157 B (Apr. 1977), 968 p.; Vol. 3, publication No. 77.157 C (Apr. 1977), 879 p. ; Vol. 4, publication No. 78.175 (1978), 528 p.; Vol. 5, publication No. 79.141 (1979), 534 p. (National Institute for Occupational Safety and Health, 4676 Columbia Parkway, Cincinnati).

"Analytical problems in the determination of trace elements" (Problematiche analitiche nel dosaggio di tracce di elementi). Queirazza, G. *Giornale degli Igienisti Industriali* (Milan), Mar.-June 1979, 4/1-2 (8-18). Illus. 22 ref. (In Italian)

Detection and analysis of airborne contaminants (field methods)

Among the most reliable methods for the early assessment of occupational exposure to atmospheric contaminants is the identification and analysis of the contaminants in the workroom air. Valuable information concerning a number of contaminants is also obtained from the analysis of biological samples such as those of blood and urine.

Survey of work environment

The survey of a plant is best performed when carried out step by step.

Collection of basic data. This phase of the survey may consist only of preparing suitable forms to be sent for completion to the management of the plant. All basic data, such as the name of the company, location, number of female and male workers, possible safety and industrial hygiene services, medical services, sickness records, etc., can be collected in this way. The main purpose of this action, however, is to acquire information concerning the materials handled in the plant as well as their by-products and impurities. The number of new materials used in industry is increasing at a high rate; so is the number of substances capable of adversely affecting the health of workers, and it is practically impossible for any industrial hygienist to be familiar with the toxicological and industrial hygiene characteristics of all these materials. The list of materials obtained at this stage, however, enables the industrial hygienist to collect from the literature all pertinent data on potentially dangerous substances, their mode of action and the best methods to be used to evaluate the hazards that may arise from them. He is also able at this stage to acquire a knowledge of the technological processes in the plant, which may well be indispensable in the proper evaluation of a work environment.

Preliminary survey. The preliminary survey is made for the purpose of selecting locations at which to carry out later analytical studies for evaluating possible hazards and accurately assessing whether or not additional control measures are necessary. This step is usually carried out with very limited equipment (of the direct-reading-type only) or with no equipment at all. It consists of following the process from its beginning to its end, including all side-operations. A great many potentially hazardous operations can be detected visually, many vapours and gases by smell. In this phase the surveyor also selects instruments, sampling methods, and methods of analysis or measurement to be used in later analytical studies.

Determination of atmospheric contaminants. Three methods are generally used for the determination of atmospheric contaminants. The contaminant may be determined on the spot by a direct-reading test, it may be separated from a given volume of air and later determined by suitable laboratory analytical methods, or an air sample may be taken to the laboratory and analysed.

Sampling principles

Location of measurement. Depending upon the purpose of the air analysis, there are three locations where air samples should be taken. The first of these locations is the immediate vicinity of the workers, in order to evaluate the exposure level. These samples should be taken at the breathing level of the worker and, if he changes his position during work, the sampling head should move with him, or representative samples should be taken at all his working positions. These samples are the most important in the evaluation of health hazards. The second location is close to the source of emission in order to obtain information on the amount of contaminants emitted into the work environment. These samples are important for the planning of engineering control measures. Finally, it may be necessary to take samples of the general atmosphere of the workroom in order to get the spatial distribution of the pollution concentration.

Types of samples. There are two basic types of air samples, based upon the time element involved:

(a) instantaneous, spot samples collected within short intervals of up to 3-4 min, analyses of which reveal most of the concentration variations occurring during a work shift; and

(b) continuous samples collected over longer time periods giving the average exposure of the worker within the period of sampling.

The equipment needed for continuous samples is more complicated than that for instantaneous samples.

Minimum and optimum volume of sample. In a well planned atmospheric sampling programme it is essential to calculate the "minimum required volume of sample" in order to avoid false results in cases where no contaminant is detected because the air sample is too small. This is dangerous as it leads to an underestimation of hazards.

The minimum required volume of sample is the air volume allowing the determination of the contaminant concentration at the level of its threshold limit value. If lower concentrations are to be determined, proportionally larger air samples must be taken. The minimum required volume is obtained from the following equation:

$$MRV = \frac{S \times 22\,400}{M \times TLV} \times \frac{760}{P} \times \frac{273 + t}{273}$$

where:

MRV – minimum required volume of sample (litres)
S – sensitivity of analytical method (milligrammes)
M – molecular weight of contaminant
TLV – threshold limit value or maximum allowable concentration (ppm)
P – barometric pressure (mmHg)
t – air temperature (° C).

If the air temperature and pressure do not deviate significantly from 25 °C and 760 mmHg respectively, the following simplified expression can be used:

$$MRV = \frac{S \times 24\,450}{M \times TLV}$$

If the threshold limit value is expressed in weight units (mg/m³), the expression for calculating the minimum required volume reads as follows:

$$MRV = \frac{S}{TLV} 1\,000$$

In some countries the minimum volume of sample is modified to that which allows for the determination of lower levels, such as one-half or one-tenth of the TLV as required for action level.

It has been suggested that the optimum duration of sampling can be calculated by a mathematical model. Integration of a proposed differential equation, describing the change of body burden as a function of the resorption and simultaneous elimination rates of a toxic substance in the body, gives the basis for calculating the optimum duration of the collection of a sample for substances of different biological half-times. It has been proposed that this duration should be one-tenth of the biological half-time of the substance. The method, though offering objectivity and some other advantages, is not yet generally accepted; besides, it is not practicable for sampling certain substances.

Duration and time of sampling. There are two possibilities as to the duration of sampling:

(a) the entire period of hazard assessment is covered by air sampling: a *full-period sample* is taken during the whole period, most frequently a work shift, or *full-period consecutive samples*, i.e. two or more samples, are taken consecutively covering the whole period;

(b) the entire period of hazard assessment is not covered by sampling: *partial-period sample* (or samples) are taken over a period shorter than the hazard assessment period or *several instantaneous (grab) samples* are taken, spread over the entire period of hazard assessment.

Regarding the time of sampling in a grab sampling programme, there are four main approaches to this question:

(a) Cyclic sampling. After the job analysis is made the sampling is planned in such a way that each sample or group of samples covers an operation cycle, thus representing the average exposure of the worker during the cycle.

(b) Regular sampling. Samples are taken at regular intervals throughout the work shift. This is justified when concentration levels vary randomly about a mean during the work shift and fluctuations are short in relation to the sampling interval.

(c) Random sampling. Any period of work shift has the same chance of being sampled (random number tables may be used), assuring an unbiased estimate of true exposure. This is the most justifiable way of sampling, particularly for determination of non-compliance with standards expressed as time-weighted averages.

(d) Sampling during maximum expected exposures. Samples for evaluation of non-compliance with standards which are not expressed as time-weighted averages but as ceiling concentrations are best taken during periods of maximum expected exposure levels during work shift.

Number of samples. As the results of air sampling and analysis are subject to random errors of the sampling method, of the analytical method and of the variability of the working environment, particular attention has been given recently to statistics of sampling. It has been shown that, considering the above variances as well as the cost/benefit ratio, two or four consecutive full-period samples per work shift (four or two hours each, respectively) is the best number of samples in a continuous sampling programme, while four to seven short-period samples over the work shift is the optimum in a grab sampling programme.

Samples should be collected during all work shifts in a 24-hour cycle and during various seasons of the year.

Sampling techniques

Three main groups of field instruments are used depending upon:

(a) the type of sample (instantaneous or integrated);

(b) the minimum required volume of sample; and

(c) whether the results are to be obtained immediately.

Air-sample collection. This technique requires the simplest type of equipment, it entails no flow or volume measuring devices and is used for taking instantaneous samples. The main representatives of this group are vacuum tubes, vacuum bottles, gas- or liquid-displacement collectors, metal-type gas collectors, and plastic bags. They are used by being filled with workroom air and transported to the laboratory for analysis. Some of the instruments of this type most often employed are shown in figure 1.

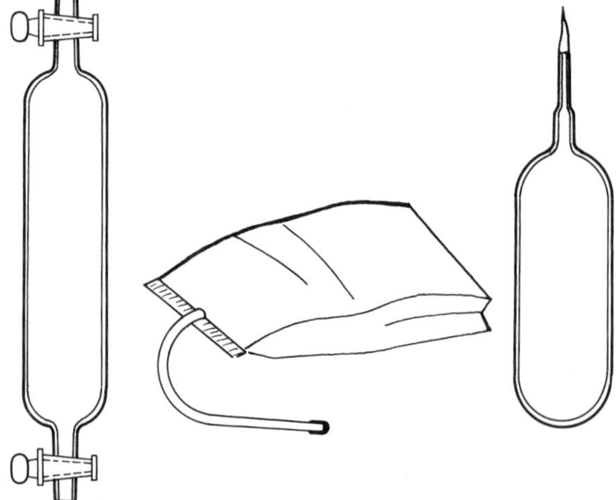

Figure 1. Some instruments for collecting grab samples.

The "wall effect" (losses of contaminants due to absorption on the walls of containers) and chemical reactions (in the case of metal collectors) are the main sources of errors. Conditioning the collector before use, by keeping a concentration of the contaminant to be sampled in the collector before sampling, helps in some cases to reduce the error.

Contaminant-sample collection. Instruments employed for removing a contaminant from a measured volume of air are mainly used for collecting integrated samples. They are more complicated, consisting at least of a suction device, an air-flow or air-volume measuring instrument and a collecting device. The air sample is drawn through the collector at a controlled velocity. The air passes through the collector while the contaminant is retained in the trapping medium by absorption, adsorption or chemical reaction. The main types of collectors are gas washers (simple and multiple contact), adsorbers (usually containing activated charcoal or silica gel), freezing traps and heated traps. A sampling train (a rotameter, a fritted glass gas washer and an electric pump) is shown in figure 2.

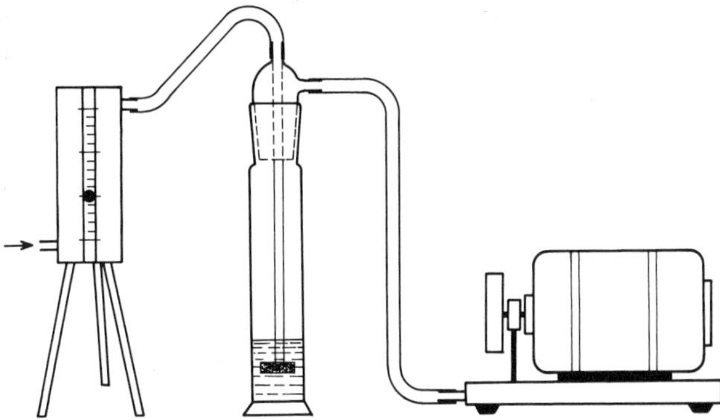

Figure 2. Sampling train for collecting integrated samples consisting of a rotameter, a fritted glass washer and an electric pump.

In this group of instruments, "collection efficiency"–a characteristic disregarded too frequently–should be considered. It is defined as the ratio of the amount of the contaminant in the air to the amount retained in the sampling equipment (usually expressed in percentage). It is not always necessary to use methods of an almost 100% collection efficiency, which is sometimes very difficult to achieve; however, what is necessary is to know the collection efficiency if correct results are to be obtained. It is more practical to use a simple method with a low, but known, collection efficiency than a very complicated method with a collection efficiency over 99%. The best method for determining the collection efficiency is to draw known volumes of air-contaminant mixtures of known concentrations (prepared mixtures in the laboratory) through the sampling equipment at controlled flow rates and controlled temperatures, and to determine the amount of the contaminant trapped in the collector.

The fact should not be disregarded that the solution rate of the contaminant in the collecting medium or the reaction rate with the medium determines the air flow rate through the sampler. Gases and vapours of high vapour pressure will be drawn through the collecting medium and lost if the sampling rate is too high and the sampling time too long.

To know the influence of temperature on the collection efficiency is of particular importance. The chemical reaction rate roughly doubles for each 10 °C temperature rise, which in some cases enhances the collection efficiency for gases and vapours, but at the same time the vapour pressure also increases with temperature, favouring the loss of the contaminant from the collected liquid. In this case, the resultant collection efficiency should be determined as a function of temperature.

Analysis

A sample collected by one of the methods described in the preceding paragraphs may be evaluated in the chemical laboratory by an appropriate analytical method which depends on a chemical reaction between the contaminant and the analytical reagent (see the third section). Chemical tests have been devised and portable apparatus developed to enable the analysis to be performed at the workplace. The airborne contaminant is drawn through the chemical reagent in a combined sampling and analytical operation. These methods have been greatly extended and improved in recent years and are available for a large number of the commonest industrial atmospheric contaminants.

There are three main types of field instruments for this purpose: they are indicating tubes, indicating papers and indicating liquids.

No attempt will be made to describe the very large number of chemical reagents that are used in detector tubes, test papers and bubblers for the assessment of the similarly large number of atmospheric contaminants to which industrial workers are exposed. At one time an even larger number of reagents was used in the laboratory to evaluate samples which had been taken in the factories, but many of the chemical methods in which those reagents were used have now been replaced by the physical methods that are described in the next section. Chemical methods are still essential where quick spot tests have to be made in the field. A very large amount of industrial hygiene investigation is concerned with localised risks. Simple tests to ensure that the concentration of contaminant at the breathing level of the operator is well below the threshold limit value are all that is needed in such cases to determine the adequacy of the precautionary measures. There are many similar situations where refinements of measurement and assessment are not vital if the investigator finds concentrations that are well below the threshold limit values. For such situations chemical methods in the form of detector tubes, test papers and indicating liquids are of great value. These spot tests have other advantages. They give immediate information concerning the concentration of a contaminant, thus enabling remedial measures to be discussed at once.

Indicating tubes

Indicating tubes consist of solid adsorbents, such as silica gel, activated alumina or inert granules, impregnated with detecting chemicals, contained in a small glass tube. The air is aspirated through the tube at a controlled rate (the same as is used during the calibration of the tube). In the case of gases which do not adsorb or react readily, the colour change occurs uniformly throughout the entire length of the impregnated column. The intensity of the colour is taken as an index of concentration. The concentration is obtained by comparing the colour with standard coloured stains (figure 3). In the case of gases which are easily and, consequently, quantitatively adsorbed, the length of stain is taken as the measure of concentration; the higher the concentration, the longer the stain (figure 3). These tubes may be prepared in a laboratory equipped with the necessary facilities or they may be obtained commercially from a number of manufacturers who specialise in their production.

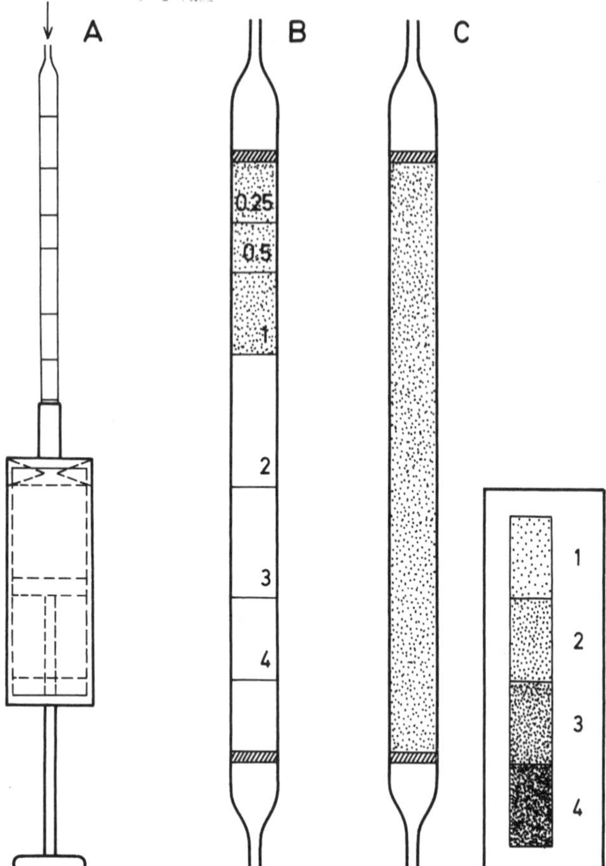

Figure 3. Indicating tubes. A. Indicating tube with a piston pump. B. The length of the stain indicates the concentration of the gas. C. The concentration is obtained by comparing the colour of the indicating substance in the tube with standard colour stains.

There are, however, a number of factors affecting the reliability of this seemingly very simple method, generally believed to require no skilled personnel. Some of these factors will be emphasised.

It is very important to use the same flow rate both during calibration and during measurement. As already mentioned, the collection efficiency varies considerably with the flow rate and a number of indicating tubes are therefore very sensitive to it. Rubber squeeze bulbs, most frequently used for aspirating the air through the tubes, may deteriorate due to the ageing of rubber and valves may get partially clogged thus changing the flow rate through the tube. The flow should, therefore, be checked at intervals, either by timing the period of the bulb expansion or by drawing the air, by means of the squeeze bulb, though a rotameter (or soap-bubble flowmeter) and the tube connected in series. Even shocks during transportation may change the flow resistance of the tubes.

It is also important to check whether the instrument is leakproof in order to avoid drawing through the tube a volume of air different from that drawn during calibration. This is done by inserting an unopened tube into the holder, squeezing the bulb, and measuring the time of expansion or the beginning of the bulb expansion. There should be no evidence of expansion at the end of a 2-min period (or at any other time given in the manufacturer's instructions). In the case of calibrated piston pumps, the handle should be pulled out completely, held in this position for 2 min, and then released. If more than 5% of the piston remains out of its original position, there are leaks. To avoid them, replacing valves, tube connections or the rubber bulb, or greasing the piston is indicated.

Mistakes are also made in the use of the instrument. It is important that the total volume of air should be drawn through the tube, and for this reason enough time must be allowed for each stroke of the pump. The fact must be taken into account that even upon a full expansion of the pump the tube may still be under a vacuum and the air still be slowly drawn through it.

As the collection efficiency depends also on temperature, consideration should be given to the dependence of calibration on temperature. Some tubes should be used within comparatively narrow temperature ranges if mistakes are to be avoided. There are tubes requiring correction by a factor of two for each 10 °C temperature variation. If collecting hot gases (for instance stack gases or engine exhaust gases), previous cooling of gases is essential. This can be done by passing hot gases through a cold metal tube connected to the indicating tubes, but caution should be taken not to introduce errors by the absorption or condensation of some contaminants on the wall of the cooling tube (particularly in the case of solvent vapours).

Interferences should be considered, too. If there are some other contaminants present in the air in addition to the contaminant to be determined, and nothing is mentioned about their interference effects in the manufacturer's instructions, possible interferences should be studied with the mixtures of known concentrations in the laboratory prior to field measurements.

It is useful to check the calibration of each batch of tubes, and particularly after a prolonged storage period.

The storage temperature greatly influences the shelf life of some tubes. As the shelf life roughly decreases exponentially with the temperature increase, it is best to store the tubes in a refrigerator, particularly if a longer period of storage is anticipated. The multiple layer tubes in which the first layers are used for removing interfering gases usually have a shorter shelf life due to the diffusion of different chemicals between layers.

Indicating papers

Indicating papers are filter papers impregnated with specific chemicals yielding a colour reaction with the contaminant to be determined. They may be used in two ways: by suspending them in the air or by drawing a known amount of air through them at a controlled velocity. The first method should be used only for detection purposes and not for quantitative measurement, as the volume of air in contact with the paper (and, consequently, the quantity of the contaminant in contact with the reagent on the paper) depends upon uncontrolled air currents in the room. There should be no colour on the downstream side of the paper in the quantitative analysis. The colour on the downstream side indicates that a part of the contaminant may have passed through the paper without having reacted, due to too high a

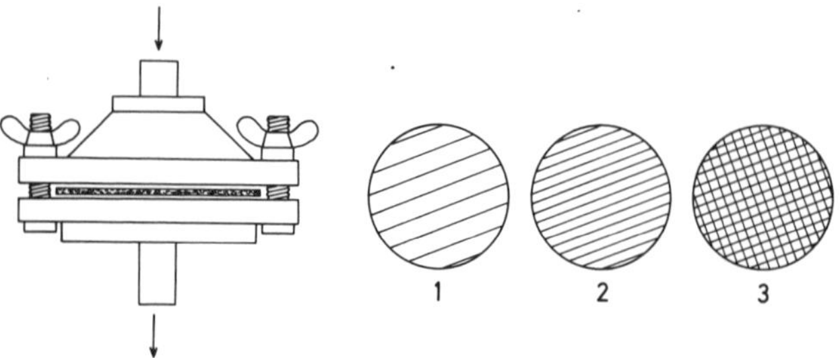

Figure 4. Indicating paper in a holder with a standard stain chart.

velocity of air through the paper. The intensity of colour obtained is compared with separately prepared standard stains (figure 4). There are some stain charts commercially available for the determination of certain gases.

The main advantage of indicating papers is their very low cost and extremely simple handling. This makes it possible to carry out a great many measurements at a low expense and with no highly qualified personnel needed. Indicating papers are particularly recommended for small plants which cannot afford expensive, and in many cases imported, equipment.

Indicating liquids

Indicating liquids are used by aspirating air through the liquid until the colour developed by chemical reaction between the contaminant and the collecting liquid matches a standard colour (air titration), or by aspirating a measured volume of air and comparing the colour obtained with standard coloured solutions prepared separately. There are commercial devices in which the coloured indicating liquid is compared with solid coloured discs in simple optical comparators. The main disadvantage of indicating liquids is the difficulty in handling them in the field.

Interpretation of results

The results obtained are usually compared with "exposure limits" (maximum allowable concentrations, threshold limit values, permissible levels) for airborne substances in the working environment.

It has been calculated that, provided the duration of the individual samples is no more than one-tenth of the biological half-time of the contaminant, the coefficient of variation of a man's body burden is no more than one-fifth of the coefficient of variation of the air concentration. Consequently, by maintaining the average air concentration one standard deviation below the exposure limit, the average body burden is maintained five standard deviations below the threshold body burden. This means that the risk of exceeding the threshold body burden at any time would be less than one in a million. An approximation to the standard deviation for small samples being (range/$\sqrt{n-1}$), it has been proposed that a favourable estimate of the work environment should be given if average concentration + (range/$\sqrt{n-1}$) is lower than exposure limit, where n is the number of the samples taken, and range is the difference between the highest and the lowest result obtained.

Statistics of compliance and non-compliance with occupational hygiene standards have recently been developed by a number of authors. The majority use one-sided statistical tests at the 95% confidence level. Special methods have been developed for full-period single and full-period consecutive, partial-period consecutive, and grab sampling. While almost all authors assume that the results of full-period measurements are distributed along a normal frequency distribution, there is no universal agreement as to the frequency distribution of results obtained by grab sampling. In the latter case, some authors assume a lognormal frequency distribution with consequent more complicated compliance statistics; some, however, prefer to assume, as an approximation, normal frequency distribution even in this case in order to make the corresponding compliance statistics simpler.

Because of the usual variability of environmental conditions at the workplace it is important to use statistical analysis of the results obtained when establishing compliance or non-compliance with standards.

VALIĆ, F. R.

Patty's industrial hygiene and toxicology, Vol. I, Third Revised Edition. Clayton, G. D.; Clayton, F. E. (eds.) (New York, Chichester, Brisbane, Toronto, John Wiley and Sons, 3rd revised ed., 1978), 1 466 p. Illus. Ref.

"Statistical evaluation of the results of measurements of occupational exposure to air contaminants. A suggested method of dealing with short-period samples taken during one whole shift when non-compliance with occupational health standards is being determined". Ulfvarson, U. *Scandinavian Journal of Work, Environment and Health* (Helsinki), Sep. 1977, 3/3 (109-115). 7 ref.

Occupational exposure sampling strategy manual. Leidel, N. A.; Busch, K. A.; Lynch, J. R. DHEW (NIOSH) publication No. 77-173 (National Institute for Occupational Safety and Health, 4676 Columbia Parkway, Cincinnati) (1977), 148 p.

CIS 78-438 *Direct reading colorimetric indicator tubes manual* (American Industrial Hygiene Association, 66 South Miller Road, Akron, Ohio) (1st ed., 1976), 29 p. Illus. 27 ref.

"Decision and estimation procedures for air contaminants". Bar-Shalom, Y.; Segall, A.; Budenaers, D. *American Industrial Hygiene Association Journal* (Akron, Ohio), Aug. 1976, 37/8 (469-473). Illus. 15 ref.

The industrial environment–its evaluation and control. DHEW (NIOSH) publication No. 74-117 (National Institute for Occupational Safety and Health, 4676 Columbia Parkway, Cincinnati) (3rd ed., 1973), 719 p. Illus.

Detection and analysis of airborne contaminants (instrumental methods)

Instrumental methods of analysis have made great advances in recent years. They have, in many instances, superseded traditional wet chemical methods in the area of environmental estimations. The advantages and disadvantages of each of these techniques needs to be carefully considered to ensure that the best method is selected for a particular application.

Advantages and disadvantages

The advantages of physical methods may be summarised as follows:

(a) they offer ease of estimation of trace amounts of contaminants in samples;

(b) they provide the best means of separating complex mixtures of chemically similar substances in airborne contaminants;

(c) with these techniques, it is often possible to automate the analytical system and provide a direct read-out giving multiple evaluation at low cost;

(d) certain methods can be applied to a variety of chemical contaminants having one similar common physical property when the identity of each is known, i.e. one instrument can be used for a number of different substances.

The disadvantages include:

(a) the need for regular calibration (see CALIBRATION);

(b) the need for a high degree of operator skill and experience in the techniques involved;

(c) the possibility of undetected errors due to fault conditions in the system, subsequent to its calibration;

(d) the equipment is usually expensive.

Consequently it is clear that, in general, instrumental methods are to be preferred if large numbers of estimations have to be carried out, since this enables the initial cost of the system to be spread over many determinations.

Instrumental methods also have advantages where chemical methods would prove difficult or laborious because of the need to identify individual substances or to separate substances having similar chemical properties. With increasing awareness that individual exposures are all-important in the field of industrial hygiene, many more samples need to be taken than heretofore. In such circumstances, it is often laborious to process chemically the large numbers of samples involved, and instrumental methods are increasingly being used for the purpose. The most commonly used instrumental methods are considered under separate headings; in each case, a brief description is given of the techniques involved, and the areas in which the method has been found useful are indicated. This will enable a particular type of method to be selected; however, the reader should then consult the relevant textbooks to study the detailed theory underlying the chosen method and should closely scrutinise the manufacturer's specification to see that the instrument has the required sensitivity, specificity and information presentation. It will be necessary to train staff in the use of the instrument and in the detailed techniques to be followed and, in some instances, to set aside a special laboratory or laboratory area in which the instrument can be set up and operated.

Gas chromatography

In this method, a small sample is injected into the top of a chromatographic column. A typical sample is about 1 ml of a gas or vapour mixture or 1 μl of a solvent containing the substances to be determined. A typical column will contain an inert powder, for example kieselguhr, impregnated with a non-volatile liquid, contained in a coiled stainless steel tube 1-3 m in length and a few millimetres in internal diameter.

A stream of gas, usually nitrogen, argon or hydrogen, is passed, at a constant rate, through the column containing the sample, the whole being maintained at a known temperature. The different substances present pass through the column at varying speeds depending on their partition coefficients as between the stationary and mobile phases, i.e. on their volatilities and solubilities. The separated fractions emerge from the column in succession and can be detected by passing them through a device which responds to a suitable physical property of the gas, for example ionisation current, thermal conductivity, or electronegativity. The signals from the detector are amplified and fed into a chart recorder so that a direct read-out is obtained. This read-out takes the general form of a baseline on a moving chart with small variations due to background interference and a series of peaks, each corresponding to a particular substance. The distance between the peaks is a measure of the time taken for the particular substance to travel through the column and is termed the retention time. If, at the same time, a known amount of a standard substance is inserted into the column, it is possible to determine from the position of the peak of this known substance the ratio between the retention time of the substance in the sample and that of the standard substance; this is called the relative retention time and it will be dependent upon the compounds involved and the common stationary phase, i.e. the non-volatile liquid. If this operation is repeated using stationary phases, preferably polar and non-polar, the substances can be positively identified. Alternatively, positive identification can be achieved by separately collecting the various fractions and carrying out a spectroscopic analysis.

Gas chromatography is most useful for estimating organic solvents having boiling points of up to 250 °C. Where complex mixtures of such liquids have to be analysed, it is often the only satisfactory way of separating the various homologues and isomers. This is particularly important when such isomers or homologues have different biological activities. The method has been used, for example, to determine the amount of benzene in toluene and to separate 1-naphthylamine from 2-naphthylamine. Quantitative figures can be obtained by measuring peak areas or heights and then comparing them with the area or height of a known reference substance.

Liquid chromatography

This technique is capable of separating complex organic substances of similar chemical composition, for example the carcinogenic aromatic amines; numerous variations of the technique have been developed. The basic method is similar to gas chromatography except that now the mobile phase is liquid. The most widely used method is to have the solid phase in the form of a sheet, for example a filter paper, or a layer of alumina powder on glass and a mobile liquid phase pass through the solid phase. The contaminant is dissolved in a solvent and a spot of this is placed on the sheet. The moving liquid phase separates the different components across the sheet as they move at different speeds, which are dependent on their partition between the liquid and solid phases. The components may be identified by spraying the sheet with a reagent mixture to produce coloured derivatives. The method can operate with only a very small sample, for example a few microgrammes dissolved in a single drop of solvent, and has a sensitivity of the order of a few nanograms.

A recent development of the method is high performance liquid chromatography (HPLC) in which the solvent is forced at high pressure (about 70 atm) through a short column (about 15 cm) of fine powder. The sample is placed on the front of the column and separation into its components occurs as they pass at different rates down the column. Various types of detectors can be used to monitor the components as they emerge from the column. The most widely used type is UV spectrophotometry, but electrochemical detectors are also being rapidly developed. HPLC is particularly suitable for separating and identifying complex organic molecules at low concentrations.

Spectroscopy

It is possible to use visible, infrared or ultraviolet light or X-rays for spectroscopic analysis. Two general methods may be used:

(a) a spectrographic analysis is made of the emission spectrum of a luminous source; or

(b) a spectrographic analysis is made of the absorption spectrum produced by placing the sample under examination between the source and the spectroscope.

In the latter case, the substance absorbs certain wavelengths from those radiated by the source. The emission or absorption of radiation is due to the transition of the individual atom or molecule from one energy state to another of lower or higher energy.

Emission spectrometry. The emission spectrograph consists of an energy source for exciting the atoms of the substance being examined, a slit forming an optical image through a collimator, a prism or defraction grating to disperse the light into a spectrum, and a photographic plate on which the image is focused. The individual lines of the spectrum are located and their intensity measured by means of a densitometer. Since the intensity is a measure of the quantity of the material under examination, the densitometer reading is an indication of the quantity of the contaminant in the sample. The sample is normally excited by means of an electric arc, and this technique is extremely useful for identifying unknown metals present in the sample. It can also be used quantitatively for the determination of metals, in particular, beryllium. It has the advantage of being very sensitive and specific and is best employed in the analysis of large numbers of similar samples, when automated techniques can be employed. For identification of unknown contaminants considerable operator experience is necessary.

Infrared spectrometry. In the infrared spectrometer radiation from a suitable source, a Nernst filament is passed through a cell containing the vapour or gas to be examined. Selective absorption occurs and characterises the substance. The emergent radiation is analysed by a prism or grating using a suitable detector, for example a solid state sensor. It is usual to provide a double beam of radiation so as to eliminate interference due to atmospheric carbon dioxide, water vapour and instrumental variations. This particular instrument is most useful for the identification of organic substances. A spectrographic index of more than 90 000 of such substances has already been published. However, what is perhaps even more valuable is the general information such an instrument will provide by identifying such features as hydroxy, amine, carbonyl and ethyl groups and aromatic and aliphatic hydrocarbon configurations. In this way, knowledge of the structure of the organic compound can be obtained. This may be particularly important in identifying groups which are biologically active and, therefore, of importance in industrial hygiene estimations, particularly where the individual constituents in the sample under examination are not known with certainty, for example fumes produced in the thermal degradation of organic products. The instrument may also be used for quantitative analysis by employing standard samples for calibration purposes. It will detect contaminants present in quantities as low as 20-100 μg and it is often the best method of positively identifying mixtures of isomers.

Ultraviolet spectrometry. The principle employed here is similar to that for infrared spectrography described above but the light source produces wavelengths in the ultraviolet region. Instruments of this type are commonly used for the automatic determination of mercury and certain hydrocarbons (e.g. benzene) in the air. For the estimation of specific substances, ultraviolet and infrared spectrometers can be simplified by employing a source of radiation which emits the characteristic spectrum of the substance under investigation, for example by exciting a source containing this substance. For example a mercury discharge tube emits the characteristic ultraviolet spectrum of mercury; when this emission is passed through air contaminated with this substance the degree of extinction of the characteristic lines will indicate the concentration present. If other sources of similar radiation are excluded, then a photocell can be used to provide a direct reading.

Atomic absorption spectrometry. This technique is an absorption method in which the sample is atomised in a flame and the absorption of radiation by these atoms is measured. A cold cathode lamp of the element under examination is used as the source of radiation, emitting the characteristic radiation of the particular element chosen (normally in the range 180-850 μm). This characteristic radiation or light is passed through a flame into which a solution of the sample, containing the same element being estimated, is suitably injected. The flame temperature is such that the atoms of the element are in their ground state and will, hence, absorb their characteristic radiation, i.e. that emitted by the source. The method can, therefore, be specific if unwanted radiation is excluded. The light emerging from the flame is passed through a monochromator, which selects one line of the characteristic spectrum, and the atomic absorption is measured by a suitable photosensitive device. The signal from the device can be amplified and used to provide a direct read-out. This technique permits the semi-automatic determination of some 50 elements when they are present in solution. The technique is sufficiently sensitive to determine a few microgrammes of contaminant per cubic centimetre of sample and, with special techniques, quantities in the order of 10^{-8} g/cm^3 can be determined. Various portable atomic absorption spectrography instruments have been developed for the direct determination of certain metals, in particular lead and cadmium, by passing the contaminated atmosphere directly through the flame. Although this technique is sufficiently sensitive to detect fumes at their maximum permissible concentration (i.e. the sensitivity is better than 10^{-10} g for lead and 10^{-11} g for cadmium), it is not suitable for the determination of larger particles; in such cases, the solution technique should be employed. More recent developments have replaced the flame with a high temperature furnace (flameless atomic absorption). This modification greatly increases the sensitivity of the technique.

X-ray fluorescence spectrography. When an atom is bombarded with electrons or an X-ray beam, it is possible to remove an electron. The ionised atom that is formed is unstable and emits energy due to the transition of an electron from one of the lower levels in its depleted shell. The energy so lost is emitted as characteristic X-ray frequencies which can be analysed by techniques similar to those used for the other spectra referred to above. This technique is being increasingly used in hygiene measurements for the estimation of metal ions because it is non-destructive and the sample can be used in further identifications. It is particularly suitable for direct determination of particulate samples on filters since it is most suitable for analysing samples present as thin layers and is a convenient technique for qualitative analysis of samples as well as quantitative determinations.

Polarography

Polarographic analysis depends on the measurement of small currents produced by the reduction of ions at a dropping-mercury cathode, for example Pb^{++} + 2 electrons = Pb, or by oxidation if the electrode is operating anodically. The polarograph consists of an electrolysis cell containing the solution under examination and possessing an anode and cathode of pure mercury. The solution to be analysed is very dilute and, therefore, of high resistance. To support the passage of an electrolysing current, an inert salt such as potassium chloride is added to the solution. It is essential that the electrolyte be free from dissolved oxygen. This may be achieved by bubbling nitrogen through the solution for a few minutes or by the addition of a known amount of sodium sulphide. As the potential across the electrodes is increased, the current through the cell rises and is measured by a suitable chart recorder. Rapid increases over small voltage ranges occur when the voltage across the cell reaches a value where the ions in solution react at the electrode. The voltages at which these increases occur are characteristic of the individual ions and the rise in cell current is a measure of the concentration of that species of ion. The method is a relatively cheap and convenient way of analysing water soluble substances.

X-ray diffraction analysis

This method of analysis depends on the fact that a crystal can act as a three-dimensional diffraction grating for X-rays, producing diffraction patterns depending on the structural form of the crystal and the wavelengths of the X-ray used. The method has become particularly important for the analysis of silica and asbestos in that it can determine the percentage of the various crystalline forms of these minerals (which of course have a different biological activity) in the sample under examination. For quantitative work, it is preferable to use an internal standard, for example calcium fluoride.

The sample is normally mounted in a fine capillary tube or rolled into a fibre of approximately 0.15 mm diameter by the addition of a small amount of adhesive. It is then exposed to the X-ray beam in a powder camera and the diffraction pattern is obtained on a strip of film. Alternatively, a diffractometer may be used. The sensitivity of the method depends on the amount of impurity present, and the sample is normally given a preliminary clean-up with acid to remove gross impurities. If the sample, after clean-up, contains more than 10% of silica or asbestos, quantities as small as 25 μg can be determined by this technique.

This method is also useful for identifying unknown substances which may be present either in air samples or in dusts collected from deposits on the structure or fittings of a building.

Neutron activation analysis

In this technique, the unknown material is exposed to a beam of particles; for example, the material can be placed in an intense thermal neutron flux within a reactor. The interaction of neutrons with various nuclei in the sample produces radioactive isotopes. Following the removal of the sample, these isotopes can be measured by β- or γ-ray detecting devices. If the element to be measured produces radioactive isotopes with a sufficient radiation yield, extremely small (millimicrogramme) quantities can be assayed.

In most cases the measurement is standardised by irradiating known concentrations of the desired material and measuring this standard under conditions identical with those for the unknown. The actual limit of sensitivity in this technique is often set by trace amounts of the chemical to be assayed present in the container or in other chemicals used in preparing the sample.

General methods

In addition to the above physical methods for the determination of particular air contaminants, there are certain general physical methods which can be used to determine a contaminant whose identity is known and when it is the only substance present in the atmosphere which has the particular physical property being used. Such techniques will, of course, determine the total quantity of a number of similar substances present in the atmosphere which behave in a similar way both physically and biologically. Some useful general techniques of this type are given below.

Combustible gas indicators. These instruments operate by passing the atmosphere containing a flammable gas over a small (1 mm diameter) electrically heated element which is coated with a catalyst. The heater of the element is a platinum resistance thermometer which forms one arm of a Wheatstone bridge. The combustible gas is oxidised (burnt) at the surface of the catalyst causing an elevation in its temperature. This, in turn, increases the resistance of the thermometer, causing an out-of-balance current in the bridge circuit. This can be read directly on a meter or fed into a chart recorder. Such a device will measure all the combustible vapours and gases present in the particular atmosphere under test and is, therefore, principally for determining the flammable concentrations of such materials in the air. It is calibrated for this purpose in terms of the percentage concentration relative to the lower flammable limit. Calibration charts are normally supplied by the manufacturers for individual substances. The instrument may also be used for industrial hygiene purposes if a second scale is provided giving a full-scale reading at one-tenth of the lower flammable limit. This scale normally enables readings down to about 100 ppm to be made with a sufficient degree of accuracy for such purposes.

Spectrophotometers. These are employed for measuring the light transmission at a particular wavelength and may be used either directly on a solution of the contaminant or, alternatively, indirectly after the development of a colour produced by a chemical reaction. For quantitative determinations, a calibration curve must be obtained by using known amounts of the substance under investigation with standardised techniques. Particular care must be taken to choose the best wavelength for maximum sensitivity and to avoid interferences. (See DETECTION AND ANALYSIS OF AIRBORNE CONTAMINANTS (CHEMICAL LABORATORY METHODS).)

Flame ionisation detectors. The flame ionisation detector normally used with gas chromatographic apparatus can be employed to indicate the presence of contaminants in the atmosphere directly. The great disadvantage of these detectors is, however, that if high concentrations of contaminants are present, the detector will be poisoned and rendered inaccurate. Such methods are, nevertheless, useful for the detection of atmospheric traces of contaminants which will never be present in excessive concentrations.

Choice and application of method

As has already been indicated, physical methods of estimating air contaminants are undergoing rapid change, and only the more important of the established methods have been described here. In choosing a method, perhaps the most important consideration is one of sensitivity and specificity. In this context environmental standards (including TLVs) may provide an indication as to the sensitivity required which should, where possible, be at least an order of magnitude lower so that the hazard can be properly evaluated.

A knowledge of other likely contaminants derived from a consideration of the materials being processed, including impurities, and the possible formation of intermediates during side reactions which may give rise to other unexpected contaminants will assist in indicating whether or not the method proposed will be sufficiently specific. Further confirmation can be obtained by a qualitative examination of the atmosphere by one or more of the general methods above. It is particularly important to ensure that a less harmful contaminant present at a higher concentration will not invalidate the method which has been selected. If the instrument is sufficiently sensitive and specific, then the next prime consideration is its cost relative to the number of samples to be processed, compared with alternative chemical techniques if these are possible. It is necessary that the instrument be checked regularly against standards. This should be done preferably before and after each run in which the instrument is used; it has been found to be particularly important if physical instruments are used for field work where transportation may disturb the original calibration. It will be seen, therefore, that such instruments are likely to be preferred where there are long runs of routine samples which are analysed at one central point, or where alternative methods are impracticable or non-existent. Accordingly, much work has gone into possible ways of taking samples in the field and transporting them to a central point for analysis. Reference should be made to the section on air sampling, above, for an indication as to how this may best be done.

LUXON, S. G.

(The author acknowledges with thanks the assistance of his colleagues and in particular of Dr. Firth of the Occupational Hygiene Laboratories of the UK Health and Safety Executive.)

General:

The industrial environment–its evaluation and control. NIOSH publication 74-117 (National Institute for Occupational Safety and Health, 4676 Columbia Parkway, Cincinnati) (1974), 719 p. Illus.

Elemental analysis of biological materials. Current problems and techniques with special reference to trace elements. Technical report series No. 197 (Vienna, International Atomic Energy Agency, Feb. 1980), 371 p. Illus. Ref.

Radiation techniques:

Measurement, detection and control of environmental pollutants (Vienna, International Atomic Energy Agency, Sep. 1976), 641 p. Illus. Ref.

Gas chromatography:

"Gas chromatography columns. A syllabus of errors". McCown, S. M.; Tuttle, K. L.; Manos, C. G. *International Laboratory* (Fairfield, Connecticut), Nov.-Dec. 1979 (80-87). 5 ref.

Detergents

The general term "detergents" which originates from the Latin *detergere* (= to wipe off) is applied nowadays to all synthetic washing compounds, although originally it applied only to the soaps made from natural fats and oils. These synthetic compounds have the great advantage over ordinary soap that they do not form a scum with any calcium or magnesium salts which are present in hard water.

Anionic detergents are those which ionise in solution (as soap does), whereas non-ionic detergents do not ionise. Anionic detergents, the active species of which contain a negatively charged polar group such as sulphonate ($-SO_3-$) or sulphate ($-OSO_3-$) attached to a non-polar hydrocarbon chain, are frequently included in products intended for cleaning purposes together with non-ionic detergents, which contain uncharged polar groups such as polyethylenoxy ($-[CH_2CH_2O]_nH$). It is often found that better results are obtained if certain additives known as builders are incorporated. These substances may help the cleaning action for instance by softening the water, buffering the pH of the solution, or preventing redeposition of dirt. Other additives are often included for specific purposes. Some countries now require that all detergents must be biologically decomposable, i.e. that they must be destroyed by bacterial action in watercourses and sewerage purification plants, otherwise foam may cause water pollution in rivers, etc.

Production. The basic raw materials, the non-ionic and anionic detergents, including soap (figure 1), are mixed together with the builders (phosphates, silicates, etc.), and supplementary ingredients such as carboxymethylcellulose (anti-redeposition agent), and optical brighteners which give the fabrics a "shining white" finish. The resultant slurry is then spray-dried, which produces a powder consisting of hollow beads, which flows easily and dissolves rapidly in water. This base powder is then mixed with materials which are too sensitive to be spray-dried such as sodium perborate (bleaching agent containing active oxygen), proteolytic enzymes (valuable for removing certain types of stains) and perfume, to give the finished product.

The preparation of liquid detergents such as dishwashing liquids is relatively simple in that the raw materials are simply mixed together in a tank. These products may contain organic solvents (alcohol), germicides (e.g. hexachlorophene), and dyestuffs. Phosphates are not usually included. Other domestic cleaning agents are similar in basic formulation. In addition to mixtures of detergents (anionics and nonionics) and builders (phosphates), special purpose cleaners may contain abrasives, solvents, germicides, bleaches, etc., and all products may contain perfumes and dyestuffs.

Uses. A further advantage of the synthetic materials is their great versatility, and by blending various types it is possible to produce the emulsifying, wetting, dispersing, or foaming properties which are desired for a given application. Today, synthetic detergents are used not only in domestic and industrial cleaning products but in fields as diverse as leather manufacture (tanning, softening, colouring), textiles (scouring, fibre conditioning, finishing), paper and pulp (preparation and finishing), mining (ore-flotation), chemical industry (production and processing of plastics), dyeing and colouring (disperson of pigments), oil (emulsification, emulsion breaking, lube-oil additives), engineering industry (cutting oil additives, electroplating processes), building and civil engineering (frost protection, breeze concrete production), pharmaceutical industry and cosmetics (non-irritant emulsifiers for creams, pastes, etc.), in agriculture and in the food industry (soil-wetting, application of pesticides, emulsifiers for foods such as mayonnaise), photography (photographic emulsions) and fire fighting (wettability improvement, foam generation).

HAZARDS AND THEIR PREVENTION

In general, injuries that occur during the production or through the use of detergents are few; they are usually confined to people having an allergy either to the product or to one of its components, and to cases of misuse, for example swallowing. [Special risks are, however, involved in the use of sulphuric acid and oleum and in the exposure to sulphuric chlorohydrin, caustic soda and potash during the sulphonation and sulphation processes, in the process of condensation with ethylene oxide, because of its explosivity and that of its decomposition products which, in addition, are toxic, and in the use of catalysts such as anhydrous hydrogen fluoride in the alkylation process.]

Although the production of detergents and cleaners is largely automated, diseases of the respiratory system may occasionally occur if dust control in the working area is not provided for. The presence of certain enzymes, mainly proteases from *Bacillus subtilis*, in the dust given off, especially in packaging and warehouse departments, causes allergic reactions or acute irritation to the respiratory organs. [Outbreaks of rhinitis, laryngitis and even asthma have been described. Some workers suffering from these troubles showed a positive reaction to allergological tests and inhalation of low concentrations of enzyme extracts produced immediate asthmatic reaction. More often, according to certain observations, the airway disturbances were caused by acute irritation. However, the combination of detergents predisposes the personnel to become sensitive to other enzymes. Cases of chronic bronchitis do occur and may present changes

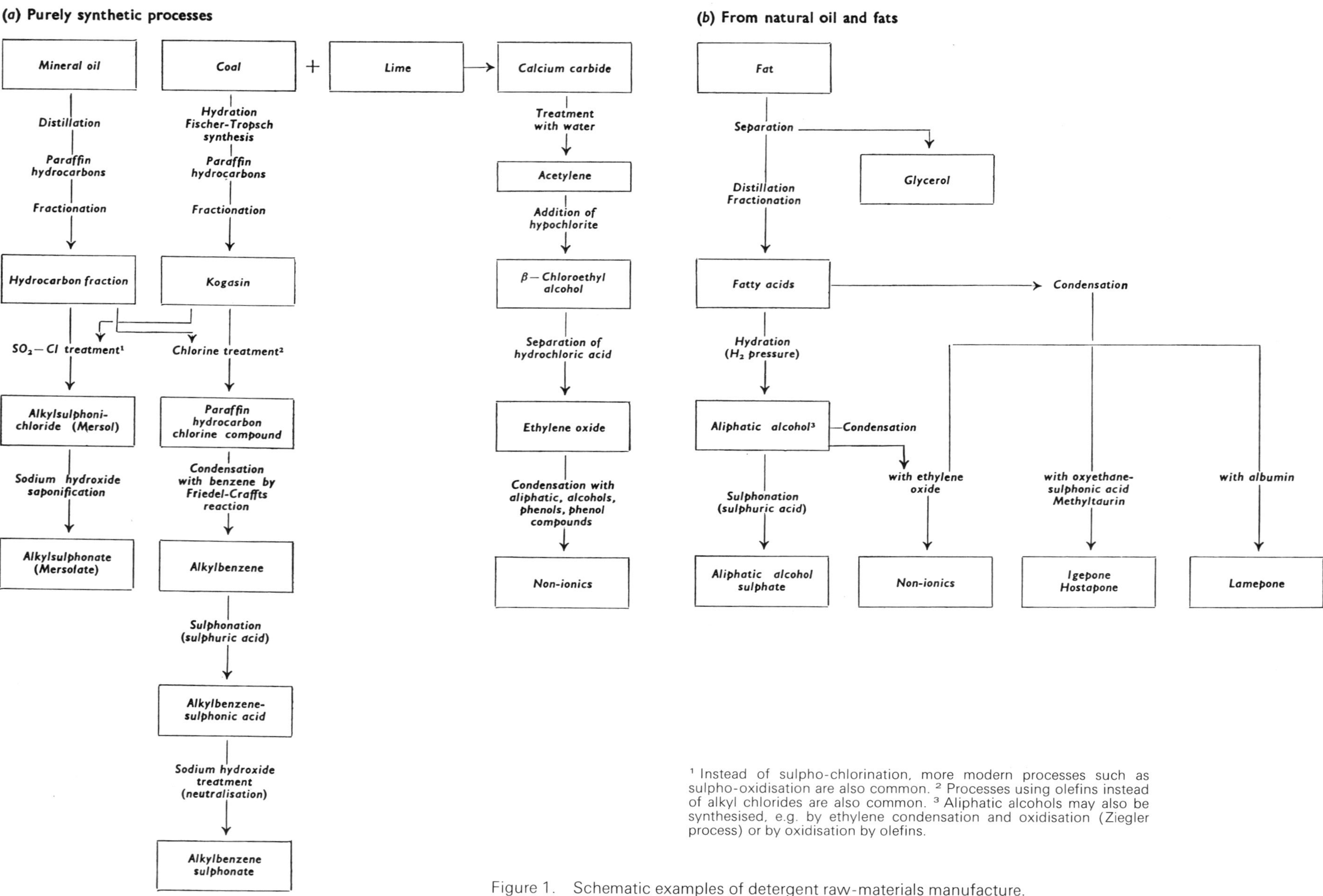

[1] Instead of sulpho-chlorination, more modern processes such as sulpho-oxidisation are also common. [2] Processes using olefins instead of alkyl chlorides are also common. [3] Aliphatic alcohols may also be synthesised, e.g. by ethylene condensation and oxidisation (Ziegler process) or by oxidisation by olefins.

Figure 1. Schematic examples of detergent raw-materials manufacture.

in the serum immunoglobulins and lowering of the antitrypsin level. Exposure to alcalase has been found to be associated with a significant loss of pulmonary elastic recoil. Exhaust ventilation should be provided to remove the airborne dust and a TLV of 0.06 μg/m³ of subtilisin (as 100% pure crystalline proteolytic enzyme) has been recommended as a ceiling by the ACGIH.]

Skin irritations are mainly allergic reactions. They are very seldom caused by the finished products but usually by a single component. At times incompatibilities to perfume additives are observed; the trouble may lie with a single component but synergistic effects do occur. For instance in soaps a certain component may produce an irritation in the presence of a certain perfume, though each substance alone does not cause any irritation.

The cases of sickness caused by swallowing the products should be treated at once by an experienced doctor or with the assistance of a poison information centre. Details of product compositions are registered with the centres, which are able to give precise information and instructions by telephone on a 24-h basis. If such a centre, a hospital or an experienced doctor cannot be reached, first aid must be given as the surface active agents can cause severe damage under certain circumstances. Delay can be most serious. In all cases, medicinal carbon preparations or aluminium hydroxide should be taken. In the case of constipation, a vegetable laxative should be dispensed. If control of foam is necessary, it is recommended to take silicone-oil preparations or liquid paraffin. Before emptying the stomach, a sufficient quantity of foam-depressing agent should be administered in order to ensure that no more foam is present. Extreme care should be taken to eliminate the danger of foam entering the respiratory tract during the pumping operation.

Particularly people who have allergies, but also those having skin irritations, should be treated by a doctor. If a doctor cannot be reached, cooling compresses of weak boric acid or camomile solutions may be applied. Pure drinking water can also be used without hesitation.

Scalds caused by boiling solutions should be treated as burns. If they are minor (without blistering) it is sufficient to hold the affected area, in most cases the hands or forearms, under running water. More serious scalds should be covered with a sterile dressing and medical attention should be sought. If the eyes are affected, they need to be thoroughly washed out and the person should see a doctor immediately.

Pre-employment medical examinations should be carried out to ensure that persons employed are free from diseases of the respiratory organs and from skin allergies; in addition, atopic subjects should not be employed on work involving exposure to enzymes.

SCHAEFER, R.

CIS 77-1648 "*Bacillus subtilis* enzymes: A 7-year clinical, epidemiological and immunological study of an industrial allergen". Juniper, C. P.; How, M. J.; Goodwin, B. F. J.; Kinshott, A. K. *Journal of the Society of Occupational Medicine* (Bristol), Jan. 1977, 27/1 (3-12). 20 ref.

CIS 78-1250 "Methods of measuring biologically active enzyme dust in the environmental air of detergent factories". Bruce, C. F.; Dunn, E.; Brotherton, R.; Davies, D. R.; Hall, F.; Potts, S. C. M. *Annals of Occupational Hygiene* (Oxford), Mar. 1978, 21/1 (1-20). Illus. 22 ref.

CIS 76-1676 "Loss of pulmonary elastic recoil in workers formerly exposed to proteolytic enzyme (alcalase) in the detergent industry". Musk, A. W.; Gandevia, B. *British Journal of Industrial Medicine* (London), Aug. 1976, 33/3 (158-165). Illus. 26 ref.

CIS 79-1344 "Toxicologic properties of fluorescent whitening agents". Gloxhuber, C.; Bloching, H. *Clinical Toxicology* (New York), Sep. 1978, 13/2 (171-203). Illus. 62 ref.

"Toxic contact dermatitis from the ammonium quaternary compound benzalkonium chloride" (Toxisches Kontaktekzem auf die quaternäre Ammoniumverbindung Benzalkoniumchlorid). Gall, H. *Dermatosen in Beruf und Umwelt* (Aulendorf), 1979, 27/5 (139-140). Illus. 6 ref. (In German)

"Investigation of skin damage caused by detergents" (Prüfung von durch Waschmittel bedingten Hautschädigungen). Kiss, Gy. *Dermatosen in Beruf und Umwelt* (Aulendorf), 1981, 29/1 (15-18). 8 ref. (In German)

Diabetics at work

Diabetes mellitus

This article is concerned only with diabetes mellitus, a genetic, hereditary disease characterised by glycosuria caused by an exaggerated rise in blood sugar after ingestion of glucose; this follows an acute lack of endocrine secretion of the β cells of the islets of Langerhans of the pancreas, i.e. of insulin. We are therefore not concerned here with renal diabetes, glycosurias originating in the central nervous sytem which may occur as a result of a skull or cerebral vascular injury, digestive glycosurias especially as a result of liver disease, secondary diabetes arising from the destruction of the pancreas by a neoplastic or inflammatory process and cases of diabetes that are liable to complicate long-term treatment with drugs (corticoids, certain diuretics, oestrogens, progesterones).

The severity of a case of diabetes mellitus at any given moment may vary considerably. It may, for instance, be a case of *prediabetes*, in other words a condition that may remain undetected throughout a person's life or, under various influences, may also develop into a detectable form. Known hereditary factors may provide an indication of its existence. On the other hand, it may be a case of *latent diabetes* which will only come to light in the event of acute or chronic aggressions or under specific physiological circumstances, such as pregnancy.

Systematic tests (urine tests for glucose, post-prandial glycaemia level) may reveal the existence of *asymptomatic diabetes*. Finally, it may be a case of *patent diabetes* accompanied by major clinical symptoms: polyuria, polydipsia, asthenia, weight loss and unmistakable biological criteria such as high-fasting glycaemia, massive glycosuria and, possibly, ketonuria.

It is impossible to foretell the development of the disease from a mild to a serious form. Among young persons, however, the passage from asymptomatic diabetes to a condition presenting a short-term threat to the patient's life may be very rapid. The progress of the disease may be brought about by a variety of aetiological factors, the most prevalent among which include the wrong kind of hygiene and diet and obesity.

In other words, although a person is born a diabetic, the disease can become apparent at any age and can take various forms.

Late-onset diabetes, occurring in a person's fifties, has little socio-occupational relevance and is in fact considered by some to be a different disease from juvenile diabetes.

It is obvious from the foregoing observations that it is difficult to ascertain the *frequency* of diabetes mellitus. This will depend on the life expectancy of the population and the number of calories available in the per capita food intake; it tends to be higher in certain parts of the world where overeating is prevalent. In the United States and European countries, for instance, 1-2% of the population suffer from diabetes. Naturally, statistics for communities under strict medical supervision are likely to show a high rate of asymptomatic diabetes identified as a result of systematic tests.

Progress of the disease and complications

The point has already been made that, as a rule, the progress of any specific case of diabetes is impossible to foretell. Statistically speaking, however, the life expectancy of a diabetic is reduced by one-third. The discovery of insulin and the now standard forms of treatment with drugs and by careful hygiene and diet have made it much easier to make a prognosis of the disease.

Prior to the introduction of insulin therapy, which is required in about 15% of cases, half of the people suffering from the disease died after falling into ketoacidosis and coma. Until the advent of antibiotics and tuberculostatics, the mortality rate for infections, and particularly pulmonary tuberculosis, was extremely high. Nowadays, the vast majority of insulin-dependent diabetics (84% according to V. Conard) die as a result of cardio-renovascular complications; of these, 50% die from cardiac damage (including 37% from coronary lesions), about 10% from cerebral vascular trouble and 13 to 30% from kidney damage, the rate being particularly high in the case of diabetes developing in childhood.

Because of retinopathy, or disease of the retina causing damage to eyesight and even complete blindness, diabetic microangiopathy responsible for kidney damage merits particularly close attention from the industrial physician.

Employment of diabetics

The industrial physician must bear in mind at all times that the diabetic is not an invalid. In most cases he is quite capable of working and possibly of pursuing a career for as long a period as a non-diabetic. Precisely so that the diabetic can continue to work normally, however, the implications of the disease and of its complications and the special kind of treatment that is needed must be borne in mind.

Naturally, the age at which the disease is detected, the need for insulin and the daily dose that is required are all important parameters, but the essential factors that determine a person's ability to engage in working life are bound to be his own awareness of the disease and the kind of discipline he imposes upon himself because he knows how important it is. A special point is often made of *unstable diabetes* which is accompanied by varying degrees of hypoglycaemia; but in fact it would probably be more correct, as a rule, to refer to *unstable diabetics* who have not understood the need for lifelong substitutive therapy or the strict quantitative relationship between the insulin administered and the glucides ingested and who fail to keep strictly to rules which have perhaps not been properly explained to them. Like the general physician, the industrial physician must be a good teacher if he is to be able to convince his patients by using explanations that they can understand and thereby reduce considerably the number of ill balanced diabetics.

There is no psychological profile of the diabetic such as exists (generally to discourage their recruitment or continued employment) for the epileptic, drug addict, alcoholic and even the ulcer-sufferer. Diabetics whose incorporation in the working world is most likely to present a problem, however, are those who combine a behavioural or intellectual handicap with their somatic disease. These can be expected to have a high absence rate, while other diabetics tend to be at least as keen in their job as the average worker, if not more so.

The acute and chronic complications which, more than the disease itself, are responsible for a temporary or permanent unsuitability for work can be retarded by means of good, regular treatment with drugs and a correct diet.

A diabetic's ability to engage in *physical effort* naturally depends on the severity of the disease but it is also very much influenced by the treatment. According to G. Persson, the limit to physical exertion is lower in people who, while not strictly speaking diabetics, suffer from abnormal reactions to hyperglycaemia provocation tests. This does not occur with people being treated with a synthetic hypoglycaemic agent (Tolbutamide). Generally speaking, diabetics should not be expected to engage in heavy work.

Irregular hours, shiftwork and nightwork are not to be recommended, mainly because they interfere with the strict diet rules, by preventing the diabetic from taking his meals and rest at fixed hours, and encourage an excessive intake of glucides during meals taken at the place of work.

Essentially because of the risk of an attack of hypoglycaemia, many diabetics must be looked upon as unsuitable for jobs where *safety* is an important factor and where a sudden loss of consciousness or lack of attention may be the cause of a serious accident to the diabetic or to others. The driving of transport vehicles and lifting appliances, working at heights (scaffolding), operation of valves on pipes carrying dangerous liquids and the use of certain machine tools are the main occupations to be avoided.

Because of the arterial complications to which diabetics are prone, attention is always drawn to the danger of *exposure to cold*. Specifically, habitual access to cold-storage rooms in the food industry should be prohibited, as should exposure to vasculotropic toxic substances (lead, carbon disulphide, etc.).

Traditional agriculture is not a very suitable environment for the diabetic as it implies an irregular workload from one day or season to another, heavy work and exposure to cold, as well as a considerable risk of intoxication. Young diabetics should preferably be steered towards specialised branches of agriculture such as market gardening if they do not wish or are unable to find employment in the secondary or tertiary sectors of the economy.

The unsuitability of work with a high *risk of infection*, such as in hospitals and, above all, bacteriological laboratories and places where animals are kept, is more theoretical than real, since it is generally possible to provide diabetics with preventive vaccination and any failure to do so needs to be fully justified. Moreover, antibiotherapy has modified the prognosis for infectious disease in the case of diabetics just as well as in that of non-diabetics.

In the case of *young diabetics*, who represent the most serious cases, vocational guidance is particularly important. The surest guarantee of their satisfactory incorporation in working life, where they are not obliged to engage in dangerous physical work, is the best possible education and vocational training. Because of the danger of retinopathy, jobs which require 100% vision (watch-making, precision engineering, etc.) should be avoided in favour of the kind of job which it is possible to carry out with reduced vision or even when suffering from blindness.

Medical supervision of diabetics by industrial medicine services

Although routine medical examinations include testing for glycosuria, more is required of the industrial physician than that. As already pointed out, it is important that he enlighten his patient as to the nature of his disease and, specifically, that he explain to him the concept of substitutive therapy when insulin is required. Close co-operation between the industrial physician and the general physician is of course essential in countries where the two functions are quite separate.

It is often easier for the industrial health services to test a person for glycaemia when fasting or after a normal breakfast than to measure the level of glucose in the urine over a period of 24 h.

The following test frequency is suggested:

- well balanced diabetics not requiring insulin: once a year;
- stable diabetics taking a moderate dose of insulin: twice a year;
- all other forms or stages of the disease, particularly among young people: four times a year.

When taking these tests, vascular and renal complications must always be borne in mind. Even in countries where the tuberculosis rate is very low, systematic radiography or radiophotography of the lungs is essential. Beyond the age of 40 or 45, electrocardiograms should be carried out every year.

Because many workers take one or two meals at their place of work, it is important to modify the composition of the traditional mid-morning and mid-afternoon break in order to reduce the intake of glucides. Works' canteens should provide appropriate menus; a self-service system with a fairly wide choice would be a good solution. It is the industrial physician's responsibility to make this clear to those in charge of catering.

Certain rules and regulations include extremely strict provisions which discourage the employment of diabetics and, as such, are altogether unjustified. Only certain tasks need to be prohibited. Although he is often handicapped, *a diabetic worker is not necessarily an invalid*. At the risk of repetition, that is one general observation that cannot be overemphasised.

MEHL, J.

"Exercise test in individuals with borderline glucose tolerance and varying smoking habits" (25-37). 37 ref. *Cardiovascular complications in diabetics and subjects with reduced glucose tolerance*. Persson, G. Acta Medica Scandinavica (Stockholm), 1977, 202/Suppl. 605 (25-37).

CIS 75-882 "Investigation into the fitness for employment of diabetics–Comparative study of diabetic and non-diabetic employees' aptitude to work" (Onderzoek naar de arbeidsgeschiktheid van diabetici–Vergelijking van de arbeidsgeschiktheid van diabetische en niet-diabetische werknemers). Van Kammen-Wijnmalen, E. H. W.; Ten Brink, H. D; Löhnberg-Langelaar, L.; De Kroon, J. P. M. *Tijdschrift voor sociale geneeskunde* (Bussum, Netherlands), 11 Oct. 1974, 52/20 (706-714) and 25 Oct. 1974, 52/21 (754-760). Illus. 7 ref. (In Dutch)

"Diabetes mellitus–also a task for the industrial medical officer" (Diabetes mellitus–auch eine Aufgabe für den Betriebsarzt). Reill, G.; Konietzko, H. *Zentralblatt für Arbeitsmedizin, Arbeitsschutz, Prophylaxe und Ergonomie* (Heidelberg), 30 Oct. 1980, 30/10 (398-401). 1 ref. (In German)

"Work placement and supervision of workers with diabetes mellitus" (Zur Frage des Arbeitseinsatzes und der Überwachung von Werktatigen mit diabetes mellitus). Schunk, W. *Zeitschrift für arztliche Fortbildung* (Jena), 1976, 70/12 (650-651). (In German)

"Physicians' guidelines for employment and placement of the diabetic in industry–An update". Mastbaum, L.; Tetrick, L.; Alexander, W. *Journal of Occupational Medicine* (Chicago), Sep. 1980, 22/9 (601-602). 15 ref.

"Diabetes and flying personnel" (Diabète et personnel navigant). Jorry, F.; Didier, A.; Carre, R.; Bastien, J. *Revue de médecine aéronautique et spatiale* (Paris), 1975, 14/55 (159-163).

Diatomaceous earth

KIESELGUHR; DIATOMITE; TRIPOLI

Diatomaceous earth is a friable, porous, silica material formed by deposits of diatoms, i.e. fossilised algae with siliceous shells. The composition of diatomaceous earth is 65-90% amorphous silica together with small and variable quantities of calcium carbonate, iron oxides, aluminium, magnesium, water and organic matter. In spite of the high silica content of the individual diatoms, diatomaceous earth has a low weight per unit volume due to the very small size of the particles (1 cm^3 of diatomaceous earth contains 2 500 million cells of 5-400 μm in size); the melting point is 1 400 °C and, at 450 °C, conversion to cristobalite (crystalline silica) begins; if calcination takes place in the presence of sodium chloride and calcium carbonate, this conversion is complete at 800 °C.

[The cristobalite content of natural (uncalcined) diatomite is typically less than 1%; that of straight calcined 10 to 20%; and that of flux-calcined products 40 to 60%. Small amounts of quartz may be present, but rarely exceed 4%.]

Sources. Sea and lake deposits of diatomaceous earth have been developed for industrial purposes in all parts of the world including Algeria, Australia, Austria, Canada, Denmark, France, the Federal Republic of Germany, Italy, Japan, Kenya, South Africa, the United Kingdom, the United States and the USSR; the most important deposits are in the United States and Algeria. World production [in 1977 was estimated at 1 800 000 short tons, one-third of which was produced in the United States]. Other major producers are the Federal Republic of Germany and France.

Production. Diatomaceous earth is extracted by opencast or deep mining; it is then dried, calcined at temperatures varying between 450 and 1 000 °C (often with calcium carbonate and sodium chloride), crushed, filtered and bagged. If sodium chloride is employed in the calcination process it leads to the emission of hydrochloric acid vapours.

Uses. Diatomite is an outstanding filtering agent, and filtration accounts for around 50% of diatomite consumption. It is employed for this purpose in sugar refining, breweries, the wines and spirits industry, the beverages and soft drinks industry, water supply and treatment, for mineral and vegetable oils and fats and for antibiotics; it is also used in dyeing, bleaching, and dry cleaning. About 30% of production is used as fillers and body builders in roofing felt, asbestos-cement panel production, fertilisers, pesticides, paints, varnishes, rubbers, asphalt, pitch, and plastics. It is also used in concrete, cement, stucco and plasters, for packing around fragile objects such as bottles of acid during transportation, as a base for catalysts, in explosives, weedkillers and dyestuffs and as thermal insulation around high-temperature objects such as ovens, and boilers, sometimes in combination with asbestos.

HAZARDS

Extraction entails the risks commonly encountered in mines and quarries; however, the risk of silicosis is inexistent at this stage since the silica is amorphous and the material has a high moisture content. However, once diatomaceous earth has been calcined, and the amorphous silica crystallised to cristobalite, it presents a serious silicosis hazard. The use of sodium chloride in the calcination process leads to the emission of hydrochloric acid vapours, which intensify the hazard of pulmonary lesions.

Silicosis in workers exposed to uncalcined diatomite develops only after long-term exposure to heavy dust concentrations. Exposure to calcined diatomite containing crystalline silica (cristobalite) produces silicosis which appears after as little as a few months' exposure and then develops rapidly.

The radiological appearance is characteristic. Sometimes it includes massive opacities with coalescent shadows which develop in spite of the absence of a nodular stage. The opacities are retractile and exert a strong pull on the trachea and ascending branch of the pulmonary artery whose descending branches fall vertically, giving the classical picture of weeping willow. The retraction of the apex is accompanied by a marked emphysema at the bases. The diaphragmatic cupolae are deformed, horizontal or festooned with numerous pleural adhesions. Functional respiratory disorders are accentuated.

SAFETY AND HEALTH MEASURES

Dust control is essential in the production and use of diatomaceous earth, especially the calcined material, which has a high free silica content. Production processes should be mechanised to the maximum, sources of airborne dust should be enclosed and fitted with exhaust ventilation, and exhausted air should be filtered before release into the atmosphere. Where there is a hazard of airborne dust during the use of diatomaceous earth, for example in the refilling of filters, or in pipe or furnace lagging, etc., workers should be supplied with respiratory protective equipment.

The atmosphere in production plants should be monitored to determine the concentration of particles below 5 μm and the free silica content of the dust; the nature of the silica (amorphous or crystalline) may be determined by X-ray diffraction. [A programme of strict dust control, supplemented by the mandatory wearing of respirators by packers, unit loaders and all other workers with a high risk of exposure to cristobalite resulted in the almost complete elimination of new cases of diatomite pneumoconiosis.]

Workers should be subject to pre-employment and periodical medical examinations with particular reference to chest X-ray appearances and lung function. Persons with respiratory system disorders should not be assigned to work in which there may be exposure to diatomite dust, and workers handling this material who show any radiological changes should be removed from exposure.

CHAMPEIX, J.
CATILINA, P.

Occupational exposure:

CIS 78-1535 *Pneumoconiosis in the Cantal (France) diatomite industry, 1949-1975* (Les pneumoconioses dans l'industrie cantalienne des diatomites de 1949 à 1975). Delort, A. (Université de Clermont-Ferrand 1, Faculté de médecine, Clermont-Ferrand) (1976), 118 p. 37 ref. (In French)

"Workers' exposure to free silica dust at the African Diatomite Industries Ltd. in Kariandus. Industrial Hygiene survey". Wambayi, O. N. (55-63). *Finnish-Kenyan Symposium on Occupational Health and Hygiene, 10th Feb. 1981.* International Labour Organisation (Helsinki, Institute of Occupational Health, 1981), 82 p. Illus.

Health effects:

CIS 74-354 "The harmful effects of kieselguhr in man" (Kieselgur und ihre gesundheitsschädliche Wirkung beim Menschen). Einbrodt, H. J.; Grussendorf, J. *Staub* (Düsseldorf), July 1973, 33/7 (273-276). 21 ref. (In German)

"Silicosis in a diatomaceous earth factory in Sweden". Beskow, R. *Scandinavian Journal of Respiratory Diseases* (Copenhagen), 1978, 59/4 (216-221). Illus. 8 ref.

"A 21-year radiographic follow-up of workers in the diatomite industry". Cooper, W. C.; Jacobson, G. *Journal of Occupational Medicine* (Chicago), Aug. 1977, 19/8 (563-566). 5 ref.

Diazomethane

Diazomethane (CH_2N_2)

AZIMETHYLENE

m.w.	42
m.p.	−145 °C
b.p.	−23 °C
v.d.	1.45

decomposes in water; slightly soluble in ethyl alcohol, ethyl ether, benzene

an explosive, yellow gas with a musty odour.

TWA OSHA	0.2 ppm 0.4 mg/m³
IDLH	10 ppm

Production. It is normally prepared from methyl nitrosourethane and sodium methoxide. Alternate methods of preparation include the treatment of nitrosomethylurea with potassium hydroxide and the reaction of potassium hydroxide with a chloroform solution of hydrazine.

Uses. Dissolved in solvents such as benzene or ethyl ether, it has proved to be a valuable agent in the synthesis of organic compounds, since it yields products of methylation quantitatively at room temperature and without the use of other reagents. Owing to the expense of preparation, diazomethane has not found widespread commercial application. However, as a laboratory reagent, the use of this compound has increased to such a large extent that methods for its preparation have been included in elementary manuals of organic chemistry.

HAZARDS

The compound was first described in 1894 by von Pechmann, who indicated that it was extremely poisonous, causing air hunger and chest pains. Following this, other investigators reported symptoms of dizziness and tinnitus. Skin exposure to diazomethane was reported to produce denudation of the skin and mucous membranes, and it was claimed that its action resembles that of dimethyl sulphate. It was also noted that the vapours from the ether solution of the gas were irritating to the skin and rendered the fingers so tender that it was difficult to pick up a pin. In 1930, exposure of two persons resulted in chest pains, fever, and severe asthmatic symptoms about 5 h after exposure to mere traces of the gas. Diazomethane has been described as "an especially insidious poison" and it is possible to work carelessly with it for some time without noticing effects. This may lead, however, to hypersensitivity, so that it is almost impossible to work even carefully with it without being subjected to attacks of asthma and fever.

The first exposure to the gas may not produce any noteworthy initial reactions; however, subsequent exposures may produce extremely severe ones. The pulmonary symptoms may be explained as either the result of true allergic sensitivity after repeated exposure to the gas, particularly in individuals with hereditary allergy, or of a powerful irritant action of the gas on the mucous membranes.

Six cases of acute diazomethane poisoning, including one death, have been reported amongst chemists and laboratory workers. In all cases, symptoms of intoxication included irritating cough, fever and malaise, varying in intensity according to the degree and duration of exposure.

Either in the gaseous or liquid state, diazomethane explodes with flashes and even at −80 °C, the liquid diazomethane may detonate. It has been the general experience, however, that explosions do not occur when diazomethane is prepared and contained in solvents such as ethyl ether or benzene.

Toxicity. The toxicity of diazomethane has been attributed to the intracellular formation of formaldehyde. Diazomethane reacts slowly with water to form methyl alcohol and liberate nitrogen. Formaldehyde, in turn, is formed by the oxidation of methyl alcohol. The possibilities of liberation *in vivo* of methyl alcohol or of the reaction of diazomethane with carboxylic compounds to form toxic methyl esters may be considered; on the other hand, the deleterious effects of diazomethane may be primarily due to the strongly irritant action of the gas on the respiratory system:

$$CH_2N_2 + H_2O \rightarrow CH_3OH + N_2$$
$$RCOOH + CH_2N_2 \rightarrow RCOOCH_3 + N_2$$

Diazomethane has been shown to be a lung carcinogen in mice and rats. Skin application and subcutaneous injection, as well as inhalation of the compound have also been shown to cause tumour development in experimental animals.

Multiple malformations were observed in a stillborn infant whose mother had been in contact with methyl nitrosourethan. The inhalation of methyl nitrosourethan or diazomethane was regarded as a possible cause of the malformations. Experimental studies were undertaken by having pregnant rats inhale diazomethane for 30 minutes. Although there was a significant increase in the number of dead fetuses as well as a decrease in the mean body weight of the rats as a result of the exposure to diazomethane, no correlation between malformations and exposure was observed.

Diazomethane is an effective insecticide for the chemical control of *Triatoma* infestations. It is also useful as an algicide. When the ichthyotoxic component of the green alga *Chaetomorpha minima* is methylated with diazomethane, a solid is obtained which retains its toxicity to killfish.

It is noteworthy that in the metabolism of the carcinogens dimethylnitrosamine and cycasin, one of the intermediary products is diazomethane (figure 1).

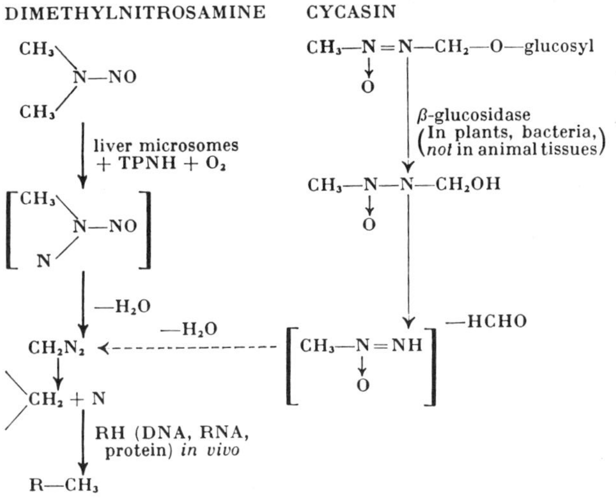

DIMETHYLNITROSAMINE — CYCASIN

$(CH_3)_2N{-}NO$ → liver microsomes + TPNH + O_2 → $[CH_3(N)N{-}NO]$ → $-H_2O$ → CH_2N_2 → $CH_2 + N$ → RH (DNA, RNA, protein) *in vivo* → $R{-}CH_3$

$CH_3{-}N(\rightarrow O){=}N{-}CH_2{-}O{-}$glucosyl → β-glucosidase (In plants, bacteria, *not* in animal tissues) → $CH_3{-}N(\rightarrow O){-}N{-}CH_2OH$ → $-HCHO$ → $[CH_3{-}N(\rightarrow O){=}NH]$ → $-H_2O$ → CH_2N_2

Figure 1. Dimethylnitrosamine and cycasin.

SAFETY AND HEALTH MEASURES

In view of the extreme danger of diazomethane, the lack of warning signs of excessive concentrations and the development of hypersensitivity, all possible precautions should be employed to prevent exposure. Eye protection, i.e. chemical safety goggles, should be worn as well as an approved respirator. It should be used only under a well ventilated exhaust hood. Owing to the explosion hazard, a safety screen should be used and contact with rough glass surfaces must be avoided, as for example ground glass apparatus and glass stirrers with glass sleeve bearings where grinding may occur.

SUNDERMAN, F. W.

"Toxicity of methyl bromide and other gaseous insecticides to *Triatoma infestans*". Castro, J. A.; Zerba, E. M.; Licastro, De, S. A.; Picollo, M. I.; Wood, E. J.; Ruveda, M. A.; De Moutier Aldfio, E. M.; Libertella, R. *Acta Physiologica Latino-Americana* (Buenos Aires), 1976, Vol. 26, 106 p.

"Induction of malformations by N-methyl-N-nitrosourea (MNU)". Warzok, R.; Thust, R.; Schneider, J.; Rupprecht, U. *Experimentelle Pathologie* (Jena), 1977, 13 (11).

"Diazomethane poisoning" (Diazomethan-Intoxikation). Hanusch, A. W.; Schäfer, H.; Hanusch, A. *Zentralblatt für Arbeitsmedizin und Arbeitsschutz* (Darmstadt), Sep. 1966, 16/9 (261-266). Illus. 14 ref. (In German)

"Diazomethane" (223-230). 16 ref. *IARC monographs on the evaluation of carcinogenic risk of chemicals to man.* Vol. 7: *Some anti-thyroid and related substances, nitrofurans and industrial chemicals* (Lyons, International Agency for Research on Cancer, 1974), 326 p.

Dibromochloropropane

Dibromochloropropane ($C_3H_5Br_2Cl$)

1,2-DIBROMO-3-CHLOROPROPANE; FUMAGON; FUMAZONE; NEMABROM; NEMAFUME; NEMAGON; NEMANAX; NEMAPAZ; NEMASET; NEMATOR; NEMAZON; DBCP; BBC 12; OS 1897; SD 1897

m.w.	236.4
sp.gr.	2.05 (20 °C)
m.p.	6.7 °C
b.p.	195.5 °C at 760 mmHg, with decomposition 164.5 °C at 300 mmHg, without decomposition
v.p.	0.8 mmHg (0.10·10³ Pa) at 21 °C
f.p.	76.7 °C (oc)

a colourless to dark-brown liquid slightly soluble in water; miscible with aliphatic and aromatic hydrocarbons, alcohols up to at least isopropanol, acetone, and halogenated hydrocarbons.

OSHA (US Emergency Temporary Standard) 10 ppb TWA concentration for an 8-h workday; 50 ppb for any 15-min period in the workday

Production. Dibromochloropropane is prepared by bromination of allyl chloride at room temperature, with careful control of temperature.

Uses. Dibromochloropropane is applied to soil before planting or around growing perennial plants to control nematodes. DBCP persists in soil for about 40 weeks after injection. Flood irrigation of an area after injection of DBCP into the soil carries the fumigant deeper into the ground. DBCP has been available in the United States as a technical grade chemical, containing not less than 95%

of the pure substance, as emulsifiable concentrates containing 70.7-87.8% of the agent, as a liquid concentrate containing 47.2% of the compound, and as granules containing 5.25-34.0% of the substance. It has been supplied in the USSR as technical grade material comprising 97-98% of the pure substance and 1-3% of other halogenated hydrocarbons. DBCP has also been produced in Benelux, France, Israel, Italy, Japan, Spain, Switzerland and the United Kingdom.

HAZARDS

DBCP has combustible and toxic properties.

Fire. DBCP itself is classified in the United States as a combustible liquid in class IIIA; formulations of DBCP including kerosene or other flammable solvents fall into the flammable range (class IB for formulations made with kerosene).

Health hazards. DBCP has been found to be less lethal to fish than the herbicides CIPC, dimethylamine salt of butyryl analogue of 2,4-D,dimethylamine salt of 4-(2-methyl-4-chlorophenoxy) butyryc acid, and 1,2-dibromoethane. Experiments with mammals have shown that DBCP can cause a decrease in the weight of the testes and a decrease, possibly to zero, in the concentration of sperm in the semen of mice, rats, guinea-pigs and rabbits after intragastric administration, feeding in the diet, or inhalation of vapour. These effects have been associated with increase in the proportion of abnormal sperm in semen, decrease in the duration of motility of the sperm, degeneration of the semeniferous tubules in the testes, and increase in the proportion of Sertoli cells in the testes. The oestrus cycles of female rats were prolonged, and eventually abolished. Two female monkeys exposed 50 and 60 times, respectively, for 7 h/day to 12 ppm of DBCP developed severe leucopoenia and anaemia. Fifty exposures to vaporised DBCP (5 ppm) for 7 h/day, 5 days/week, produced an insignificant decrease in the mean weight of the testes (−18.6%) of the rat. A statistically significant decrease (−49.0%) occurred with the next higher concentration (10 ppm) of DBCP. The latter concentration also resulted in a significant increase in the weight of the kidney (+31.7%).

Exposure to larger amounts of DBCP resulted in degeneration of the proximal convoluted tubules of the kidneys and the centrilobular regions of the liver. The necrotic cells of the testis and the liver have been found to be replaced, at least in part, by fibrous tissue. Other results of exposure of mice and rats to DBCP have been decreases in the phagocytic and digestive activities of leucocytes, the appearance of invasive and metastasising carcinomas in the stomachs of both sexes of mice and rats given repeated doses of DBCP by stomach tube, the development of adenocarcinomas in the breasts of female rats, but not in those of female mice, within 14 weeks after the initiation of gavage of mean daily doses of 29 mg/kg, and failure of induction of tumours of the breast or other organs in female rats after 13 weeks of exposure to an airborne concentration (21.5 ppm) of DBCP that resulted in death of 50% of the rats.

In vitro studies with bacteria have found that DBCP induces mutations of the base-substitution type. Evidence suggesting that this mutagenic activity is due entirely to epichlorohydrin in technical grade DBCP has been presented. This same paper has shown also that incubation of pure, inactive DBCP with a source of microsomal mixed oxidase results in restoration of the ability to produce mutation of bacteria. Recently, male rats exposed to 10 ppm of airborne DBCP for 14 weeks before being mated with unexposed females were reported to have generated an unusually large proportion of resorbed embryos. Another study with the rat has found dose-related increases in the incidences of aberrant metaphases in spermatogonial and bone marrow cells after 5 daily gavage doses of DBCP.

DBCP had only slight irritative actions on intact skin or serous surfaces, but was somewhat more irritative to mucous membranes of the respiratory tract.

During the early part of July 1977 a group of workers at a pesticide-manufacturing plant in California recognised that they had experienced remarkably little success during the period of their employment there in rendering their wives pregnant; epidemiologic study of the situation in this chemical plant, and later in others in various parts of the United States, led to the conclusion that DBCP was the most likely compound to be the cause of the impairment of testicular function. Data collected from three sites of occupational exposure to DBCP revealed that 45.6%, 65.1%, and 50% respectively of workers examined at the three sites had sperm counts less than 40 million per millilitre. Among 9 000 males examined between 1966 and 1977 by J. McCleod, an average of only 26.4% had sperm counts below 40 million per millilitre. In a study in one chemical plant of 126 workers with histories of exposure to DBCP for various periods of time, 35 with no history of exposure included four (11%) with sperm counts below 40 million per millilitre. Groups of workers with exposures of 1-6 months, 6-24 months, 24-42 months, and more than 43 months, had, respectively, 23%, 50%, 67% and 82% with sperm counts less than 40 million per millilitre. A repeat study about three months after the cessation of exposure to DBCP at one site found that whereas about 77% of the workers had sperm counts below 20 million per millilitre at the end of exposure, only about 57% had sperm counts in that range three months after termination of production of DBCP.

SAFETY AND HEALTH MEASURES

Although formulations of DBCP were registered in the United States for use as fumigants in 1964 on evidence that residues of the nematocide would not remain in or on harvested raw agricultural commodities, substantial amounts of DBCP have been found in radishes (up to 0.22 ppm), carrot tops (up to 0.64 ppm), and carrots (up to 1.55 ppm). Tomatoes, grapes, peaches, and water-run treated citrus were free of residues except for trace amounts on peaches. A shank-injected citrus plot, somewhat dry at the time of injection and not irrigated until 10 days later, had 30 ppb of DBCP in the fruit at harvest. The prevention of such residues seems to depend on (1) placing DBCP at least 8-9 in beneath the surface of the soil, (2) immediately sealing the application site with a ring roller, and (3) irrigating for 12 hours. Improved methods of application are being sought.

Of equal importance with protection of the harvester and the consumer is protection of the agricultural user of DBCP. In a trial in California of shank injection in a vineyard, a concentration of 11 ppb was determined at the level of the driver's seat on the tractor moving the injecting rig and of 3 ppb at the middle of the vineyard at 5 ft above the ground. Higher airborne concentrations were measured in citrus groves and a vineyard treated by irrigation. In another trial, in which DBCP was applied by shank injection into a fallow field, the tractor driver and the middle of the field met undetectable concentrations of DBCP but the workers at the loading site met airborne concentrations of 7-131 ppb. For exposure concentrations of the magnitudes mentioned above, a full facepiece mask with either a pesticide cartridge or canister or a source of clean, supplied air (air pump or pressurised tank with reducing valve) is sufficient.

Industrial occupational exposures have the possibility of being much more intensive than those mentioned above. Accordingly, it is important that employees understand the hazards of working with DBCP: anti-spermiogenic and testicular atrophic actions, possible menstrual irregularity, possible malignant tumours of the stomach and breasts after ingestion of the chemical, possible clastogenic, or even mutagenic, effects, and possible degenerative and necrotising lesions of the kidneys and liver. It is important to point out also that during 20 to 25 years of experience in manufacturing and formulating DBCP the only one of the possible toxic actions of the nematocide that has appeared actually is that on the testes. More intensive epidemiologic study of exposed workers might find evidence for other effects. A second important type of information for each employee to understand is that there are no materials for fabrication of gloves, suits or other protective items that are completely impenetrable by DBCP. Protective clothing or gear that becomes contaminated should be washed at once with soap and water or discarded. If the odour of DBCP persists on clothing or protective gear after washing and aeration, the clothing or gear should not be worn. Half-face or full-face respirators with cartridges or canisters for the removal of organic vapours or with supplied air should be worn when working with DBCP in an open system. Transfers of DBCP from one container to another should be made through closed systems, with venting back to the original container or through charcoal or other absorptive or destructive arrangement that will prevent escape of DBCP into the occupational environment. No acceptable chemical decontaminant for DBCP is known; destruction by incineration requires dilution with a flammable solvent and passage of the products of burning through scrubbers to remove the HCl and HBr produced.

During clean-up of small spills or other exposure to liquid DBCP, waterproof boots, gloves and apron and eye protection should be worn in addition to a respirator. Entry into closed spaces in which DBCP has been free should be restricted to individuals with at least respiratory protection until analysis of air has determined that no DBCP is present. Entry into agricultural areas to which DBCP has been applied should be restricted to persons with respirators, protective clothing, waterproof boots and gloves and eye protection for 7 days after the application.

Employers in industrial facilities using DBCP should engage licensed physicians to provide medical advice and supervision to employees required to work with this chemical. Medical supervision should include a pre-exposure physical examination, with an assessment of the employee's ability to wear a respirator for a prolonged period of time, and an assessment of fertility for male workers. The assessment of fertility for all male employees other than those with known azoospermia should be repeated after each period of 30 days of working with DBCP or as requested by the responsible physician. Detailed records of the dates and hours during which each employee worked with DBCP should be included in the employee's medical record along with any records of environmental monitoring performed in the employee's work station.

Engineering controls should be used to keep concentrations of DBCP at work stations well below 1 ppm, and preferably below 0.2 ppm. The latter concentration is the lower limit of the concentrations that had been measured within the chemical plant at which the original cases of human oligo- and azoospermia occurred.

WILLS, J. H.

General:

Dibromochloropropane. A recommended standard for occupational exposure to. DHEW (NIOSH) publication No. 78-115 (National Institute for Occupational Safety and Health, 4676 Columbia Parkway, Cincinnati) (Jan. 1978), 13 p. 15 ref.

Carcinogenic effects:

"1,2-Dibromo-3-chloropropane" (139-148). 28 ref. *IARC monographs on the evaluation of the carcinogenic risk of chemicals to man.* Vol. 15: *Some fumigants, the herbicides 2,4-D and 2,4,5-T, chlorinated dibenzodioxins and miscellaneous industrial chemicals.* (Lyons, International Agency for Research on Cancer, Aug. 1977), 354 p.

Testicular function:

"Testicular function in DBCP exposed pesticide workers". Whorton, D.; Milby, T. H.; Krauss, R. M.; Stubbs, H. A. *Journal of Occupational Medicine* (Chicago), Mar. 1979, 21/3 (161-166). Illus. 10 ref.

"Epidemiological assessment of occupationally related chemically induced sperm count suppression". Milby, T. H.; Whorton, D. *Journal of Occupational Medicine* (Chicago), Feb. 1980, 22/2 (77-82). Illus. 12 ref.

1,2-Dibromoethane

1,2-Dibromoethane ($CH_2Br.CH_2Br$)

ETHYLENE DIBROMIDE; EDB

m.w.	188
sp.gr.	2.18
m.p.	10.0 °C
b.p.	131 °C
v.d.	6.5
v.p.	11 mmHg ($1.43 \cdot 10^3$ Pa) at 20 °C

slightly soluble in water; soluble in ethyl alcohol and ethyl ether

a colourless non-flammable liquid with a chloroform-like odour.

TWA OSHA	20 ppm 155 mg/m³
	30 ppm ceil
	50 ppm/5 min
NIOSH	0.13 ppm 1 mg/m³/15 min ceil
ACGIH	industrial substance suspected of carcinogenic potential for man with no assigned limit–skin
IDLH	400 ppm

Production. It is made from ethylene and bromine and also from acetylene and hydrobromic acid.

Uses. This chemical is used chiefly as a soil fumigant for the control of ground pests (see FUNGICIDES). The second major use of this product is in leaded gasoline. It is less frequently used in fire extinguishers, gauge fluids, as a chemical intermediate and as a special solvent.

HAZARDS

Dibromoethane is a potentially dangerous chemical with an estimated minimum human lethal dose of 50 mg/kg. In fact, the ingestion of 4.5 cm³ of Dow-fume W-85, which contains 83% dibromoethane, proved to be fatal for a 55 kg adult female.

The symptoms induced by this chemical depend on whether there has been direct contact with the skin, inhalation of vapour, or oral ingestion. Since the liquid form is a severe irritant, prolonged contact with the skin leads to redness, oedema, and blistering with eventual sloughing ulceration. Inhalation of its vapours results in respiratory system damage with lung congestion, oedema and pneumonia. Central nervous system depression with drowsiness also occurs. When death supervenes, it is usually due to cardiopulmonary failure. Oral ingestion of this material leads to injury of the liver with lesser damage to the kidneys. This has been found in both experimental animals and the human. Death in these cases is usually attributable to the extensive liver damage. Other symptoms which may be encountered following ingestion or inhalation include excitement, headache, tinnitus, generalised weakness, a weak and thready pulse and severe, protracted vomiting.

[Oral administration of dibromoethane by stomach tube caused squamous cell carcinomas of the fore-stomach in rats and mice, lung cancers in mice, haemoangiosarcomas of the spleen in male rats and liver cancer in female rats. No case reports in man or epidemiological studies on human data are available; however, on the above experimental evidence dibromoethane according to the National Cancer Institute of the United States should be classified as a potential human carcinogen.

Recently a serious toxic interaction has been detected in rats between inhaled dibromoethane and disulphiram taken per os, resulting in very high mortality levels with a high incidence of tumours, including haemoangiosarcomas of liver, spleen, and kidney. Therefore the National Institute for Occupational Safety and Health of the United States recommended that *(a)* workers should not be exposed to dibromoethane during the course of disulphiram therapy (Antabuse, Rosulfiram used as alcohol deterrents), and *(b)* no worker should be exposed to both dibromoethane and disulphiram (the latter being also used in industry as an accelerator in rubber production, a fungicide and an insecticide).]

Fortunately the application of dibromoethane as a soil fumigant is ordinarily under the surface of the ground with an injector, which minimises the hazard of direct contact with the liquid and vapour. Its low vapour pressure also reduces the possibility of inhalation of appreciable amounts.

SAFETY AND HEALTH MEASURES

The greatest potential danger from dibromoethane is in handling the product either during the manufacturing process or during its application by the consumer. Workers should be made aware of the necessity of avoiding direct contact by wearing protective clothing. This includes the use of nylon-neoprene hand protection, which has been found to be the most impermeable to this chemical. Leather foot protection is readily permeated by the liquid form of dibromoethane, so that it should be discarded if it has been exposed to appreciable amounts. The chemical also penetrates most varieties of rubber gloves so that they cannot ordinarily be considered adequate protection; the hands should therefore be washed with soap and water after handling dibromoethane. The eyes should be protected by goggles with side shields. Eye and face protection is imperative for those working in confined spaces where exposure is possible, as for example in a greenhouse. Use of the product in the open air, however, presents no problem as far as vapour inhalation is concerned. The odour of dibromoethane is recognisable at a concentration of 10 ppm, well below the toxic level, and this, in itself, should be adequate warning to the worker. The oral ingestion of this product does not appear to have been a serious problem in the past. Nevertheless, adequate labelling of the containers, with proper recommendations for its handling, are still a strict necessity in view of its high toxicity.

Good general ventilation is necessary in workrooms in which dibromoethane is used. Vessels containing the liquid should be covered and processes involving its use should be provided with local exhaust ventilation.

Treatment. In case of direct contact with the skin surface, treatment consists of removal of covering garments and thorough washing of the skin with soap and water. If this is accomplished within a short time after the exposure, it constitutes adequate protection against development of skin lesions. Involvement of the eyes by either the liquid or vapour can be successfully treated by flushing with copious volumes of water. Treatment of symptoms incidental to inhalation should obviously be directed toward the organs involved, usually the lungs and central nervous system. Evidence of depression of the latter can be treated with stimulants given by mouth or parenterally, as the case dictates. Difficulties in oxygenation due to pulmonary congestion and oedema may require oxygen therapy and, in serious cases, artificial respiration. Antibiotics may be prescribed against the possible development of pneumonia.

Since the ingestion of dibromoethane by mouth leads to serious liver injury, it is imperative that the stomach be promptly emptied and thorough gastric lavage be accomplished. Efforts to protect the liver should include such traditional procedures as a high carbohydrate diet and supplementary vitamins, especially vitamins B, C and K.

OLMSTEAD, E. V.

General:

CIS 78-752 *Criteria for a recommended standard–Occupational exposure to ethylene bromide.* DHEW (NIOSH) publication No. 77-221 (National Institute for Occupational Safety and Health, 4676 Columbia Parkway, Cincinnati) (Aug. 1977), 208 p. Illus. 118 ref.

Carcinogenic risk:

Report on carcinogenesis bioassay of 1,2-Dibromoethane (EDB). Technical background information (Bethesda, US DHEW, National Cancer Institute, 14 Nov. 1978), 11 p.

"Ethylene dibromide" (195-209). 62 ref. *IARC monographs on the evaluation of the carcinogenic risk of chemicals to man.* Vol. 15: *Some fumigants, the herbicides 2,4-D and 2,4,5-T, chlorinated dibenzidioxins and miscellaneous industrial chemicals* (Lyons, International Agency for Research on Cancer, Aug. 1977).

Interaction with disulphiram:

CIS 78-1668 "Ethylene dibromide and disulfiram toxic interaction". Stein, H. P.; Bahlman, L. J.; Leidel, N. A.; Parker, J. C.; Thomas, A. W.; Millar, J. D. *American Industrial Hygiene Association Journal* (Akron, Ohio), July 1978, 39/7 (A-35 to A-37). 3 ref.

Industrial hygiene:

An industrywide industrial hygiene study of ethylene dibromide. Rumsey, D. W.; Tanita, R. K. DHEW (NIOSH) publication No. 79-112 (National Institute for Occupational Safety and Health, 4676 Columbia Parkway, Cincinnati) (Nov. 1978), 57 p. 35 ref.

Dichloromethane

Dichloromethane (CH_2Cl_2)

METHYLENE CHLORIDE; METHYLENE DICHLORIDE

m.w.	85
sp.gr.	1.33
m.p.	−97 °C (freezes)
b.p.	40 °C
v.d.	2.93
v.p.	400 mmHg ($53.2 \cdot 10^3$ Pa) at 24 °C
f.p.	non-flammable, but in presence of heat and moisture may decompose to form hydrochloric acid, carbon dioxide, carbon monoxide and possibly phosgene
e.l.	15.5-66.4 (in oxygen)
i.t.	615 °C

slightly soluble in water; very soluble in ethyl alcohol and ethyl ether

a colourless liquid.

TWA OSHA	500 ppm 1 750 mg/m³
	1 000 ppm ceil
	2 000 ppm/5 min 2 hr peak
TLV ACGIH	100 ppm 360 mg/m³
STEL ACGIH	500 ppm 1 700 mg/m³
IDLH	5 000 ppm
MAC USSR	50 mg/m³

Production. Dichloromethane is prepared by the chlorination of chloromethane or methane at 400 °C in a vertical reactor; the resulting product is then purified by fractional distillation. It is shipped in iron drums and tank cars.

Uses. It is used as a solvent for oils, fats, waxes, bitumen, cellulose acetate and esters, as a paint stripper and degreaser, sometimes mixed with petroleum naphtha and tetrachloroethylene, especially for the cleaning of electric motors. The chemically pure product has been marketed in the Federal Republic of Germany under the name "Solaesthin" as a narcotic, and in the United Kingdom under the name "Solmethin".

HAZARDS

Dichloromethane is highly volatile, and high atmospheric concentrations may develop in poorly ventilated areas producing loss of consciousness in exposed workers. The substance does, however, have a sweetish odour at concentrations above 300 ppm and consequently it may be detected at levels lower than those having acute effects.

Cases of fatal poisoning have been reported in workers entering confined spaces in which high dichloromethane concentrations were present. In one fatal case, an oleoresin was being extracted by a process in which most of the operations were conducted in a closed system; however, the worker was intoxicated by vapour escaping from vents in the indoor supply tank and from the percolators. It was found that the actual loss of dichloromethane from the system amounted to 3 750 l per week.

Symptoms. The principal toxic action of dichloromethane is exerted on the central nervous system—a narcotic, or in high concentrations, an anaesthetic effect; this latter effect has been described as ranging from severe fatigue to light-headedness, drowsiness and even unconsciousness.

The margin of safety between these severe effects and those of a less serious character is narrow. The narcotic effects cause loss of appetite, headache, giddiness, irritability, stupor, numbness and tingling of the limbs. Prolonged exposure to the lower narcotic concentrations may produce, after a latent period of several hours, shortness of breath, a dry non-productive cough with substantial pain and possibly pulmonary oedema. Some authorities have also reported haematological disturbance in the form of reduction of the erythrocyte and haemoglobin levels as well as engorgement of the brain blood vessels and dilation of the heart.

However, mild intoxication does not seem to produce any permanent disability and the potential toxicity of dichloromethane to the liver is much less than that of other halogenated hydrocarbons, in particular carbon tetrachloride, although the results of animal experiments are not consistent in this respect. Nevertheless, it has been pointed out that dichloromethane is seldom used in a pure state but is often mixed with other compounds which do exert a toxic effect on the liver.

[Since 1972 it has been shown that persons exposed to dichloromethane have elevated carboxyhaemoglobin levels (such as 10% one hour after exposure for two hours to 1 000 ppm of dichloromethane, and 3.9% 17 hours later) because of the endogenous conversion of dichloromethane to carbon monoxide. At that time exposure to dichloromethane concentrations not exceeding the American TWA of 500 ppm could result in a carboxyhaemoglobin level in excess of that allowed for carbon monoxide (7.9% COHb is the saturation level corresponding to 50 ppm CO exposure); 100 ppm of dichloromethane would produce the same COHb level or concentration of CO in the alveolar air as 50 ppm of CO.]

Irritation of the skin and eyes may be caused by direct contact, yet the chief industrial health problems resulting from excessive exposure are the symptoms of drunkenness and incoordination that result from dichloromethane intoxication and the unsafe acts and consequent accidents to which these symptoms may lead.

[Dichloromethane is absorbed through the placenta and can be found in the embryonic tissues following exposure of the mother; it is also excreted via milk.]

SAFETY AND HEALTH MEASURES

Workers handling dichloromethane should be informed of its hazards and of the need to handle it with care. [Persons with cardiovascular impairment, pregnant women and lactating mothers should not be exposed to dichloromethane.] Areas in which it is produced or used should be provided with adequate exhaust ventilation, sufficient to keep the atmospheric concentration below the exposure limits.

Special caution should be exercised when entering tanks which may have contained dichloromethane; when such an operation proves necessary, the workers involved should be equipped with oxygen masks or respirators, eye protection, safety belts and lifelines, and be under the constant control of a supervisor. If, in spite of these precautions, one of the workers in the tank should lose consciousness, on no account should he receive epinephrine. Exposure to fresh air and removal of contaminated clothing will normally result in spontaneous recovery; however, should artificial respiration prove necessary, it should be followed by a period of complete rest and reasonable warmth.

Periodic air estimations, using the silica gel absorption method applicable to other halogenated hydrocarbons [or sampling the air in plastic bags and making the analysis by means of infrared spectroscopy or gas chromatography,] should give warning of the presence of dangerous concentrations.

[Biological monitoring of exposure to dichloromethane is now made possible by the high sensitivity of the available analytical methods, which can measure the quantities excreted through the exhaled air, or even by measuring the COHb level or the CO concentration in the alveolar air.]

BROWNING, E.

General:

Dichloromethane (Dichlormethan). Gesundheitsschädliche Arbeitsstoffe. Toxikologisch-arbeitsmedizinische Begründung von MAK-Werten (Weinheim, Verlag Chemie, 1976). 76 ref. (In German)

Effects on CNS:

CIS 968-1968 "Toxic encephalosis in occupations involving contact with methylene chloride" (Toxische Enzephalose beim beruflichen Umgang mit Methylenchlorid). Weiss G. *Zentralblatt für Arbeitsmedizin und Arbeitsschutz* (Darmstadt), Sep. 1967, 17/9 (282-285). 30 ref. (In German)

Conversion to carbon monoxide:

"*In vivo* CO production after dichloromethane exposure. Inhalation experimental study on rabbits: contribution of these data to a possible revision of the TLV of this solvent" (Sur la production in vivo d'oxyde de carbone après exposition au dichlorométhane. Etude expérimentale sur le lapin par inhalation: incidence des résultats obtenus sur la révision éventuelle de la concentration limite tolérable de ce solvant). Boudène, C.; Belegaud, J.; Jouany, J. M.; Truhaut, R. *Archives des maladies professionnelles, de médecine du travail et de sécurité sociale* (Paris), Dec. 1978, 39/12 (657-669). Illus. 26 ref. (In French)

Biological monitoring:

CIS 78-1053 "Modeling the uptake, metabolism and excretion of dichloromethane by man". Peterson, J. E. *American Industrial Hygiene Association Journal* (Akron, Ohio), Jan. 1978 (41-47). Illus. 7 ref.

Mortality study:

"Epidemiologic investigation of employees chronically exposed to methylene chloride. Mortality analysis". Friedlander, B. R.; Hearne, T.; Hall, S. *Journal of Occupational Medicine* (Chicago), Oct. 1978, 20/10 (657-666). Illus. 47 ref.

Fire hazard:

CIS 77-27 "Measurement of flashpoints of chlorinated solvent mixtures and of petroleum products" (Mesure des points d'éclair de mélanges de solvants chlorés et de produits pétroliers). Hervé-Bazin, B. *Cahiers de notes documentaires–Sécurité et hygiène du travail* (Paris), 4th quarter 1976, 85, Note No. 1033-85-76 (537-543). (In French)

Diesel engines, underground use of

The part played by mobile diesel equipment in underground and surface mines has become increasingly important over the past 30 years, not only because of the world-wide expansion in mining operations but also because electrically driven machinery is being replaced by diesel-powered units in many mining applications. Over the same period growing attention has been paid to the working environment, particularly to problems of excessive noise and atmospheric pollution, and within this framework authorities in the major mining countries have been concerned with the question of diesel exhaust emissions in both open-cast and underground mining operations. While by far the greater part of diesel engines are used in surface mining operations, a large and growing number is to be found underground in metalliferous and non-fiery mines and increasing interest is being shown in their use in coal mines. The main hazards represented by diesel equipment underground are the effects upon health, the fire risk, and the explosion danger.

Compression ignition engines

The principal disadvantages of piston-type diesel engines as used in mines arise from the high compression ratio at which they work, which is more than twice that of spark-ignition engines and which gives rise to an increased emission of nitrogen oxides and adds considerably to their operational noise level. On the other hand, the diesel engine has a high thermodynamic efficiency and is thus economical in operation, to which may be added the very low levels of carbon monoxide and hydrocarbons in the exhaust fumes, and these advantages ensure its dominant position in the field of heavy vehicles such as lorries, locomotives, etc. It should be added that a substantial reduction in the NO emissions may be obtained by providing a prechamber or swirl chamber into which the fuel is injected to ensure better mixing before it reaches the cylinder head.

The exhaust gases of a diesel engine may be divided into three main groups: toxic gases, including carbon monoxide and nitrogen oxides; smoke and fumes consisting of unburnt hydrocarbons; and irritant gases comprising partially burnt hydrocarbons, aldehydes and sulphur oxides. A fog type of smoke made up of minute liquid hydrocarbon particles, droplets of fuel, solid carbonaceous particles and some partially oxidised and condensed water vapour may be produced when the engine is running at reduced power. The oxidising properties of this smoke are less important in the case of underground operation since they are the result of a reaction with sunlight. The effect of the irritant gases is also reinforced in the same way. It is thus the toxic products of diesel exhausts that present the main risk to health underground; these are essentially carbon monoxide and nitrogen oxides, although carbon dioxide is also present and may be found in concentrations sufficient to cause a problem.

The design of a diesel engine is such that it should only produce a small amount of pollution. The quantity of air drawn in for each cylinder on each cycle is practically constant, and it is always in excess of the theoretical amount required to ensure complete combustion of the fuel injected, on condition that the injection mechanism is correctly regulated. In spite of this, many miners hold the mistaken belief that diesel engines are one of the principal causes of pollution, because of the unpleasant-smelling, thick smoke which they sometimes give off.

The composition of the exhaust gases discharged from a diesel engine depends upon the operating conditions, on the state of its maintenance, and on the manner in which it is driven. For this reason the responsibility for ensuring that pollution is kept to a minimum falls upon the user, rather than the manufacturer. If there is an over-rich fuel mixture or poor mixing due to defective

Table 1. Future trends in diesel engine design

Characteristics	Present configuration	Mature configuration	Advanced configuration
Type	Prechamber or swirl chamber	Swirl chamber	Swirl chamber
Represented by	Mercedes, Peugeot	(Conjectural)	(Conjectural)
Induction system	Naturally aspirated	Turbocharged, IMPM[1] exhaust tank	Turbocharged IMPM[1]
Timing	Automatic centrifugal	Automatic centrifugal	Automatic centrifugal
Delivery	Throttle-position modulation	Throttle-position, inlet-manifold-pressure and engine-speed modulation	Throttle-position, inlet-manifold-pressure and engine-speed modulation
Compression ratio	Fixed, 18 : 1 to 22 : 1	Fixed, 18 : 1 to 22 : 1	Fixed, approx. 15 : 1
Cylinder head/block material	Ferrous	Ferrous	Ceramic
Piston material	Aluminium	Aluminium	Ceramic
Cooling system	Conventional	Conventional	None
Emission controls	Retarded injection timing	Retarded injection timing, EGR[2]	Retarded injection timing, EGR[2]

[1] IMPM = Intake-manifold-pressure-modulated. [2] EGR = Exhaust gas recuperator.

atomisation resulting from incorrect injection, excessive smoke will be produced. Smoke density also increases as an engine approaches and exceeds full load, due to enrichment of the fuel-air ratio and lengthening of the injection period. The excess fuel can then become vaporised, cracked or even carbonised to a greater or lesser degree, giving off carbon monoxide and unburnt fuel particles which go to make up this disagreeable smoke.

Numerous technical modifications have been introduced by the manufacturers with a view to remedying these defects, and the results of a recent study carried out by the California Institute of Technology on future trends in diesel engine design are shown in table 1. Of particular interest is the possible future use of ceramic material for pistons, cylinder heads and engine block.

Legislative requirements

Regulations governing the use of diesel engines underground have been adopted in the United States and in European and other countries such as Australia, Canada and Japan, and have prescribed exposure limits for carbon monoxide, hydrocarbons, nitrogen oxides and smoke. In most cases until about 1970 it was found possible to comply with these regulations without introducing major changes in the design of the engines by means of improvements in the injection system leading to better combustion of the fuel. Since that time, however, following discussions among the manufacturers, industrial hygienists and other specialists, stricter exposure limits have been introduced. The introduction in a number of countries of these new limits has complicated the problem of the control of exhaust emissions. A further problem, particularly for petroleum-importing countries, is the need for fuel economy resulting from the petroleum crisis of recent years.

Table 2. Limit values for the emission of gaseous pollutants by the engines of vehicles equipped with a positive-ignition engine according to the ECE Agreement of 20 March 1958 as revised in 1978[1]

Reference weight (rw) in kg	Mass of carbon monoxide; grammes per test	Mass of hydrocarbons; grammes per test	Mass of nitrogen oxides in NO_2 equivalent; grammes
rw ⩽ 750	80	6.8	10
750 < rw ⩽ 850	87	7.1	10
850 < rw ⩽ 1 020	94	7.4	10
1 020 < rw ⩽ 1 250	107	8.0	12
1 250 < rw ⩽ 1 470	122	8.6	14
1 470 < rw ⩽ 1 700	135	9.2	14.5
1 700 < rw ⩽ 1 930	149	9.7	15
1 930 < rw ⩽ 2 150	162	10.3	15.5
2 150 <	176	10.9	16

[1] Came into force on 6 March 1978.

With a view to reducing permissible limits below what is at present technically possible, as well as to achieving greater fuel economy, federal legislation in the United States has been directed towards stimulating technological development in the design of vehicles and their engines. A similar policy is being followed in Japan. In Europe, through the United Nations Economic Commission for Europe (ECE), a more cautious policy has been followed, and amendments to the regulations have taken into account local conditions as well as the size and weight of European vehicles compared with those used in the United States, and of the priority given by European countries to stabilising, if not reducing, the consumption of petroleum products.

The main requirements of Regulation 15 of the ECE are shown in table 2. The application of this regulation thus results in an over-all reduction of about 60% in the content of carbon monoxide and hydrocarbons in the exhaust gases of a diesel engine.

A large amount of research into the reduction of harmful exhaust products and into the formulation of recommended standards has been carried out in recent years by the countries of the Council for Mutual Economic Assistance (CMEA). This research, generally based on international experience, has enabled exposure limits for these products to be established and the operating conditions for diesel engines underground and in quarries to be designated.

Figure 1 shows the exposure limits that have been adopted by various countries.

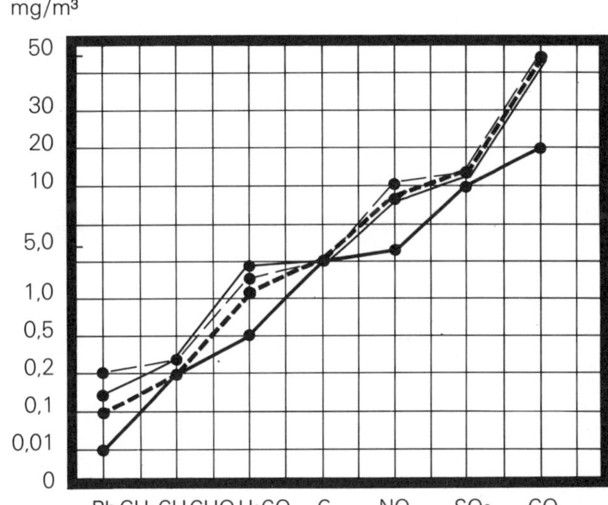

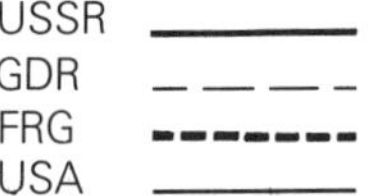

Figure 1. Exposure limits for exhaust products of motor vehicles as applied in various countries.

Safe use of diesel engines

Mining authorities in most countries consider that the primary objective in the safe use of diesel engines in mines is to ensure that their harmful exhaust products are diluted sufficiently to eliminate any health hazard. While various methods may be employed to attain this objective, legislation is based on the establishment of three basic conditions: an exposure limit at the workplace; a minimum ventilation quantity for each type of vehicle and its engine; and a minimum ventilation quantity as a function of rated engine horsepower and the concentrations of noxious gases in the undiluted exhaust.

Concentrations of noxious gas at the workplace are generally limited by various countries to the following values:

CO_2 5 000 ppm 9 000 mg/m³ (0.5%)
CO 50 ppm 55 mg/m³
NO_2 5 ppm 9 mg/m³
O_2 not less than 20%

In recent years, however, extensive research in the United States, the USSR and the countries of the European Coal and Steel Community (ECSC) has resulted in varied and more stringent requirements. In order to meet the second of the above conditions, the minimum ventilation quantity is fixed by the authorities in some countries after examination of the engine concerned, and approval may also require a modification to the engine designed to ensure that a limit is provided to the amount of fuel that can be injected. Under the third condition, approval is preceded by an examination of the type concerned with a view to ensuring that the noxious gases in the undiluted exhaust do not exceed the limit value.

The total exhaust gas produced by a moderate-sized diesel engine is approximately 60 to 70 l/min per rated horsepower at full load. In practice, adequate dilution of these gases will ensure that they do not represent a health hazard, and the minimum ventilation quantities prescribed by the authorities are based on the amount of ventilation required to ensure that the exposure limits for the different gases are not exceeded. The US Bureau of Mines has proposed the following formula for the calculation of the ventilation quantity required:

$$Q = V \cdot \frac{C}{Y}$$

where:

Q = the required volume of air (m^3/min);

V = volume of exhaust gas at full load and rated speed (m^3/min);

C = concentration of noxious substance in the exhaust gas (ppm);

Y = exposure limit for the noxious substance in the air at the workplace (ppm).

Present standards for the amount of air required for the dilution of exhaust gases vary considerably from one country to another: in the United States the figure varies from 0.67 to 2.5 m^3/min/BHP; in the Federal Republic of Germany it is 6.0 m^3/min/BHP; in France from 2.1 to 4.0 m^3/min/BHP; and in Canada from 2.1 to 7.1 m^3/min/BHP. These standards apply in all mines, either underground or surface, wherever a ventilation system either natural or mechanical is in operation. In headings or other localities where auxiliary ventilation is installed, special provisions apply, based on the length of the heading from the point of through ventilation and the time during which the diesel locomotive or other vehicle remains inside. In certain cases, if the vehicle must wait for any length of time, regulations prescribe that the engine must be stopped.

In order to dilute the gases quickly and evacuate them through the used-air vent, special air extractors have been installed in the mines of the European Coal and Steel Community (ECSC). However, for this extra ventilation system to work properly, the increase in pressure in the tunnel caused by the extractor has to be great enough to expel the exhaust gases. This is generally the case only in galleries that are less than 40 m long.

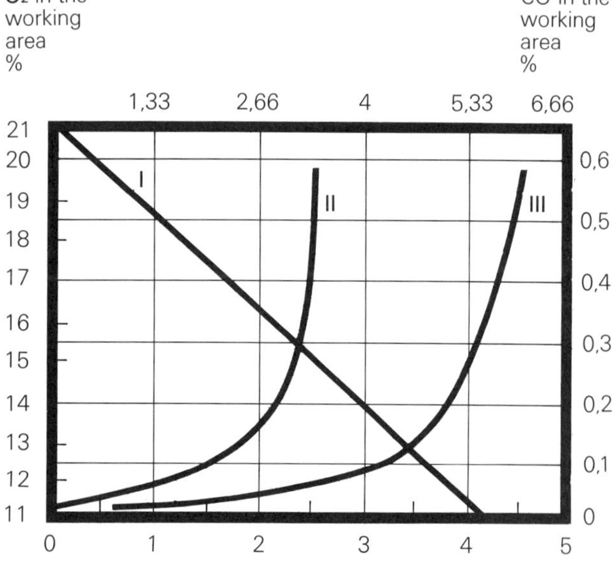

Figure 2. Dangerous concentrations of CO produced by a diesel-powered machine when used in a badly ventilated working area. I. Oxygen. II. CO produced by a machine operating on full power. III. CO produced by a machine operating on low power.

Reduction of harmful or objectionable exhaust constituents

The presence of noxious constituents in a diesel exhaust depends upon the fuel-air mixture, the combustion temperature, and the ignition timing. It can also be affected by the type of fuel supplied. The techniques used to reduce the emission of various pollutants are tabulated in table 3.

Table 3. Techniques for the reduction of pollutants in diesel engine exhausts

Pollutant	Technique
Smokes	Reduced power
Nitrogen oxides	Reduced power Recirculation of exhaust gases Water injection Design of engine
Carbon monoxide Hydrocarbons Odours	Catalytic oxidation

As will be seen, smoke is one of the hazards involved in the running of diesel engines. Apart from engine design, better mixing of the fuel and air and the careful adjustment of the ignition timing, the only practical method for the reduction of smoke lies in the use of additives to the fuel and in placing a limit on the power developed by the engine. The use of additives, however, can lead to the formation of deposits in the engine which may in themselves lead to a further toxic hazard. Tests have shown that good maintenance is an essential prerequisite for keeping smoke to a minimum. This should include careful checking of the air filter, adjustment of the injection pump and injection nozzles and the ignition timing, and regular checking of the compression and back-pressure in the engine.

A reduction of about 40% in the amount of nitrogen oxides liberated into the air of the workplace can be achieved by recirculating about 20% of the exhaust gases in the engine. This is illustrated in figure 3, which compares the NO_x concentrations in an engine without recirculation and a similar engine with recirculation.

In order to maintain the production of carbon monoxide and soot at acceptable levels, the amount of exhaust gas recirculated must be controlled. This also calls for some means of reducing the production of carbon monoxide, for example by incorporating an oxycatalytic afterburner, and it also means that engines with an exhaust recirculation system have to be checked more frequently than those using a standard exhaust system. It has also been shown that recirculation

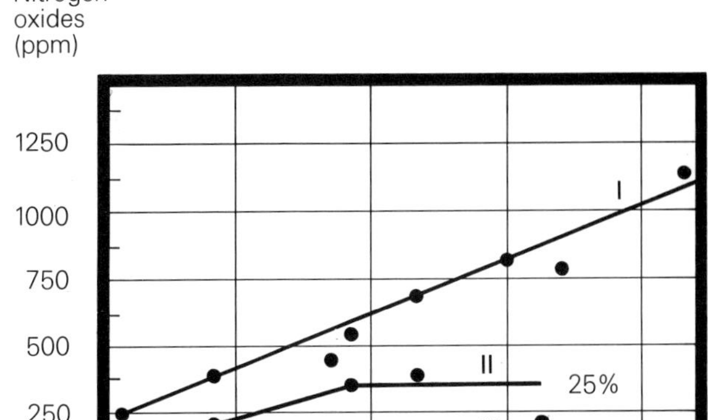

Figure 3. Reduction in the nitrogen oxide concentration by means of exhaust gas recirculation. I. Engine working without recirculation. II. and III. Same engine as above, working with recirculation.

together with advancement of the injection and adjustment of the fuel-air mixture gives rise to a slight reduction in the power developed by the engine. Nevertheless, mining authorities in the ECSC, the United States, the USSR and Canada have adopted these methods, it being considered preferable to ensure a clean atmosphere even at the expense of a slight increase in fuel consumption.

Because of the danger of fires or explosion, most countries require the installation of a system of cooling the exhaust gases to a temperature below 70 °C before they are rejected to the atmosphere. The US Bureau of Mines has extended the same requirement to diesel engines used in mines other than coal mines. The arrangement usually consists of injecting water into the exhaust box which is then collected in a simple scrubber. Some loss of this water takes place, and thus make-up water must be available wherever the engine is being worked so that fresh cold water may be supplied continually during the working shift (figure 4).

The use of afterburning devices for the exhaust gases has attracted considerable interest. They depend on a catalytic reaction using a platinum catalyst and succeed in reducing the carbon monoxide level in the exhaust gases, but are only effective above a certain minimum temperature of operation. Most common is the so-called "Oxycat" which is particularly effective at exhaust gas temperatures exceeding 250 °C; however, in engines fitted with a precombustion chamber using a system of injection with a low fuel-air ratio, the exhaust gas temperature seldom exceeds 150-175 °C. The catalyser converts carbon monoxide to carbon dioxide and partially decomposes nitrogen dioxide (NO_2) into nitrogen and oxygen; this in turn further oxidises the carbon monoxide and also converts some of the sulphur dioxide into sulphurous oxide, which leads to the formation of fog in humid air. For this reason, the use of fuels with a low sulphur content is desirable when a catalyser is fitted. The best results have been obtained when using a catalyst of platinum foam with a honeycomb structure. This system is extensively used underground with a large number of drill carriages and loading and trimming machines.

The air filter is an important part of a diesel engine, since dust-free air is essential for its satisfactory performance. Both dry and oil-moistened filters are in use, and they may be fitted with a pre-separator for large-scale dust particles. All these filters require careful maintenance with regular cleaning or replacement; in the case of paper filters, some sort of vacuum indicators should be furnished to indicate excessive resistance so that they may be replaced in good time.

In order to ensure that the exposure limits for the exhaust gases are not exceeded, regular sampling and measurement is essential. Some doubt exists as to whether it is always possible to obtain exact samples of the exhaust gases under underground conditions, and in terms of the regulations generally applied at the present time, an estimation of the condition of an engine is obtained on the basis of the carbon monoxide concentration in the exhaust gas. This should be completed by the measurement, at regular intervals, of carbon dioxide, nitrogen oxides and hydrocarbons present in the air of the workplace.

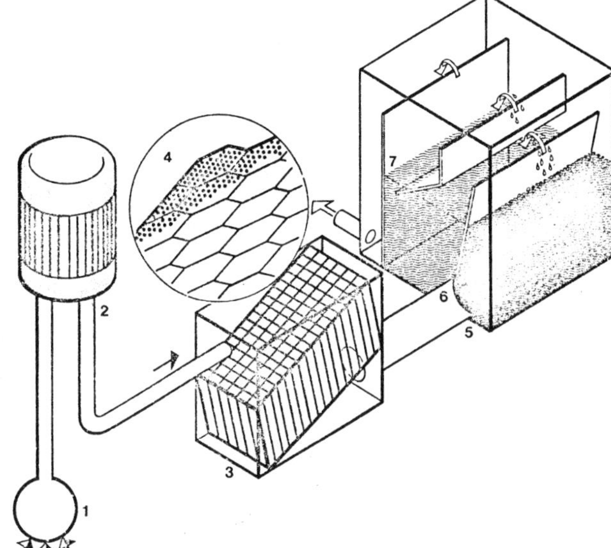

Figure 4. Diagram of diesel engine exhaust-gas cleaning system, the gases being washed in a water bath and/or catalysed. 1. Air filter. 2. Engine. 3. Oxidative catalyser. 4. Honeycomb insert. 5. Gas washer. 6. Split tube. 7. Water bath.

Registration of diesel engine model

As far as safety and health in mines are concerned, the principal requirements for the official registration of a diesel engine model would appear to be as follows:

(1) Composition of exhaust gases, expressed in ppm/hp.

(2) Cooling of exhaust gases by injection of water into the exhaust manifold or gas washer.

(3) Cleaning of exhaust gases by means of gas washers or catalysers.

In the case of mines where there is a risk of firedamp, the following points are also of importance:

(1) Observance of the specified maximum temperature for the surface of the engine and accessories (air-cooling would not seem to be sufficient in this respect).

(2) Flame arresters on the intake side and spark arresters on the exhaust side of the engine.

(3) Assembly of engine to include a sufficient number of suitable gaskets.

(4) Proper resistance of engine to test pressure applied between flame arresters.

The registration documents submitted must be sufficiently complete for any person to be able to determine from them whether an engine of the model concerned, when presented for registration, actually corresponds to that indicated in the file. By way of example, attention can be drawn to the "standard instructions for the registration of vehicles equipped with diesel engines with respect to the emission of polluting agents by the engine" drawn up by a joint ECE/United Nations Working Committee (Addendum 23, Regulation 24 to be annexed to the Agreement).

Operating, maintenance and supervisory personnel

A study of the regulations issued by mining authorities for the use of diesel engines underground shows that, in most cases, they also deal with the maintenance, operation and supervision of the machines. Generally speaking, they take it for granted that the maintenance, operating and supervisory personnel have acquired a certain minimum of skill and received a certain amount of training. In addition, they normally contain maintenance, servicing and handling instructions; persons responsible for supervising the use of the engines are presumed to possess an adequate level of technical skill. Basic technical know-how is provided by the mining authorities themselves through their own trainee schemes and special courses. Arrangements are made from time to time for further training by means of information and technical seminars and round-tables, sometimes accompanied by travelling or permanent exhibitions of personal protective equipment and measuring equipment at the place of work (ECSC, USSR, Poland, Federal Republic of Germany).

The main requirements for the maintenance and supervision of diesel engines which ought to be observed by mining personnel are briefly as follows:

1. The fuel tank must not be allowed to become empty. When a diesel engine runs out of fuel it is necessary to bleed the injection lines before it will start and this can sometimes be a big problem. Fuel must be clean and free of contaminants. Fuel filters must be checked and serviced as required.
2. Frequency of cleaning or change of the intake-air filter will depend on the length of its operation and the amount of dust in the section. This is normally a daily operation. The intake flame arrester must be inspected and serviced periodically. When the intake-air cleaner is serviced properly, once every two to four weeks is ample.
3. The engine lubricating oil must be kept to the indicated level. Oil filters must be changed according to the recommendation of the engine manufacturer.
4. The engine-cooling water supply must be checked and replenished if required, making sure that the radiator and cooling system are functioning properly.
5. The exhaust conditioner must be serviced and the supply tank refilled with clean water.
6. Timing and fuel-injection setting are important on underground diesel-powered equipment and must only be performed by authorised personnel when adjustments are necessary.
7. The emergency air-shutdown control must be checked once a week to assure its proper operation.
8. The mine air must be checked during each shift for toxic gases.
9. Storage and handling of diesel fuel underground will also be under strict regulations.

The example is taken from the regulations applied by the National Mine Service Company of Illinois.

Quality and handling of fuel; fire fighting

Mining authority regulations should devote considerable space to provisions regarding fuel and its handling, from the moment of its delivery above ground to its combustion in the diesel engine. The various regulations concerning transport of fuel, storage underground, filling of diesel engine tanks and the garaging, maintenance and repair of diesel-powered equipment underground appear in many different forms (technical instructions, special instructions, etc.) in mining legislations (Federal Republic of Germany, United States, USSR, France, Belgium). As regards safety, these technical rules, which are generally already recognised and applied in quarries, also apply below ground. Due account must naturally be taken of major safety problems, such as the transport of fuel by pipeline in shafts and galleries, storage below ground and fire hazards (evacuation of the gases produced by fire directly through the ventilation shafts without polluting the parts of the mine occupied by members of the personnel).

Another aspect of occupational safety below ground deserving special attention from mining personnel is the need for diesel engines to be run on an officially approved fuel whose physical and chemical characteristics, especially as regards toxic content, meet safety requirements.

Table 4 gives the fuel specifications established by the Joint EEC/United Nations Committee for official tests and production control.

Table 4. Specifications of reference fuel prescribed for approval tests and to verify conformity of production[1]

	Limits and units	Method
Density 15/4 °C	0.830±0.005	ASTM[2] D 1298-67
Distillation		ASTM D 86-67
50%	min. 245 °C	
90%	330±10 °C	
Final boiling p.	max. 370 °C	
Cetane index	54±3	ASTM D 976-66
Kinematic viscosity at 100 °F	3±0.5 cst	ASTM D 445-65
Sulphur content	0.4±0.1% by weight	ASTM D 129-64
Flash point	min. 55 °C	ASTM D 99-66
Cloud point	max. −7 °C	ASTM D 97-66
Aniline point	69±5 °C	ASTM D 611-64
Carbon residue on 10% bottoms	max. 0.2% by weight	ASTM D 524-64
Ash content	max. 0.01% by weight	ASTM D 482-63
Water content	max. 0.05% by weight	ASTM D 95-62
Copper-corrosion test at 100 °C	max. 1	ASTM D 130-68
Net calorific value	10 250±100 kcal/kg 18 450±180 BTU/lb	ASTM D 2-68 (Ap. VI)
Strong acid number	nil mg KoH/g	ASTM D 974-64

NB: The fuel must be based only on straight-run distillates, hydrodesulphurised or not, and must contain no additives.

[1] Annex 6 to Regulation No. 24 to be annexed to the Agreement concerning the adoption of uniform conditions of approval and reciprocal recognition of approval for motor vehicle equipment and parts, done at Geneva on 20 March 1958–Revision 1 of 11 February 1980. [2] Initials of the American Society for Testing and Materials, 1916 Race St., Philadelphia, Pennsylvania 19103. The figures after the dash denote the year when a standard was adopted or revised. Should any ASTM standards be amended, the standards adopted in the years quoted above will remain applicable unless all parties to the 1958 Agreement which apply this Regulation agree to replace them by later standards.

Unless properly supervised, the handling of fuel underground may constitute a hazard. Due attention must therefore be paid to all the various aspects of safety, organisation of work and fire fighting. The most frequent hazards reported in mining authority reports, from the standpoint of personnel safety, are—

(a) loss of fuel during filling or as a result of a leak;

(b) loss of fuel when tapped by unauthorised persons; and

(c) fires and health hazards caused by uncontrolled and accidental spillage of fuel.

The system of protection of underground personnel against possible fire hazards depends on local mining conditions, the geological characteristics of the deposit, the choice, quality, quantity and regular checking of fire-fighting equipment and its location and use in the event of an incident. Each mine must accordingly have its own alarm system and a plan of evacuation of the working areas endangered by gases produced by fire.

TODRADZE, C.

CIS 79-1918 "Behaviour of exhaust gases from diesel engines when methane is mixed with the air intake" (Das Abgasverhalten von Dieselmotoren bei methanhaltiger Ansaugluft). Mertens, H.; Schlitt, H. G. *Glückauf* (Essen), 7 June 1979, 115/11 (520-525). Illus. (In German)

CIS 80-434 "Diesel exhaust—An occupational carcinogen?". Schenker, M. B. *Journal of Occupational Medicine* (Chicago), Jan. 1980, 22/1 (41-46). 42 ref.

CIS 78-506 "Safe use of diesel equipment in coal mines". Alcock, K. *Mining Congress Journal* (Washington, DC), Feb. 1977 (53-62). Illus. 16 ref.

CIS 77-1134 *Size distribution and mass output of particulates from diesel engine exhausts.* Breslin, J. A.; Strazisar, A. J.; Stein, R. L. Report of Investigations 8141r (Publications Distribution Branch, Bureau of Mines, 4800 Forbes Avenue, Pittsburgh) (July 1976), 10 p.

Digestive system

The digestive system exerts a considerable influence on the efficiency and work capacity of the body, and acute and chronic affections of the digestive system are among the commonest causes of absenteeism and disablement. In this context, the industrial physician may be called upon in either of the following ways to offer suggestions concerning hygiene and nutritional requirements in relation to the particular needs of a given occupation: to assess the influence that factors inherent in the occupation may have either in producing morbid conditions of the digestive system, or in aggravating others that may pre-exist or be otherwise independent of the occupation; or to express an opinion concerning general or specific fitness for the occupation.

Many of the factors that are harmful to the digestive system may be of occupational origin; frequently a number of factors act in concert and their action may be facilitated by individual predisposition. The following are among the most important occupational factors: industrial poisons; physical agents; and occupational stress such as tension, fatigue, abnormal postures, frequent changes in work tempo, shift work, night work and unsuitable eating habits (quantity, quality and timing of meals).

Industrial poisons

The digestive system may act as a portal for the entry of toxic substances into the body, although its role here is normally much less important than that of the respiratory system which has an absorption surface area of 80-100 m² whereas the corresponding figure for the digestive system does not exceed 20 m². In addition, vapours and gases entering the body by inhalation reach the bloodstream and hence the brain without meeting any intermediate defence; however, a poison that is ingested is filtered and, to some degree, metabolised by the liver before reaching the vascular bed. Nevertheless, the organic and functional damage may occur both during entry into and elimination from the body or as a result of accumulation in certain organs. This damage suffered by the body may be the result of the action of the toxic substance itself, its metabolites or the fact that the body is depleted of certain essential substances. Idiosyncrasy and allergic mechanisms may also play a part.

Toxic mechanisms are highly complex and may vary considerably from substance to substance. However, in general, toxic substances impair certain cellular activities. For example, phosphorus acts on the esterases, the lipases and the enzymes involved in glycogenesis and anaerobic glycolysis; mercury, arsenic, selenium, lead and bismuth all act on the oxidation-reduction processes blocking the active sulphydril groups, and mercury also acts on amino, carboxyl and hydroxyl groups; thallium acts on enzymes with flavin and pyridine prosthetic groups and on the succinic-acid dehydrogenases, malic dehydrogenases and zanthine oxidases. The deleterious action of selenium on the liver is a complex trophonotic-type mechanism (blockage of coenzyme I, inhibition of the succinic-acid dehydrogenase activity and of hepatic uricase, displacement of the sulphur in certain amino acids). Some elements and compounds used in industry cause local damage in the digestive system affecting, for example, the mouth and neighbouring area, stomach, intestines, liver or pancreas.

Solvents have particular affinity for lipid-rich tissues and consequently produce acute damage. Chronic poisoning, however, is characterised chiefly by liver involvement. The toxic action is generally complex and different mechanisms are involved. In the case of carbon tetrachloride, liver damage is thought to be due mainly to changes in the permeability of the cellular membrane. In the case of carbon disulphide, gastrointestinal involvement is attributed to the specific neurotropic action of this substance on the intramural plexus whilst liver damage seems to be due more to the solvent's cytotoxic action, which produces changes in lipoprotein metabolism, especially in the event of aneurine deficiency. However, the effect of certain other solvents on the gastrointestinal epithelium is demonstrated by inhibition or retardation of cell renewal. [Methylstyrene, butadiene, and other chemicals used in the production of synthetic rubber have proved to alter fat and protein metabolism in the liver, and to impair the pigment function and the synthesis of hippuric acid.]

Liver damage constitutes an important part of the pathology of exogenic poisons since the liver is the prime organ in metabolising toxic agents and acts with the kidneys in detoxication processes. The bile receives from the liver, either directly or after conjugation, various substances that can be reabsorbed in the enterohepatic cycle (for instance, cadmium, cobalt, manganese). Liver cells participate in oxidation (e.g. alcohols, phenols, toluene), reduction (e.g. nitro-compounds), methylation (e.g. selenic acid), conjugation with sulphuric or glucuronic acid (e.g. benzene), acetylation (e.g. trichloroethylene). Kupffer cells may also intervene by phagocytosing the heavy metals, for example (see LIVER).

Severe gastrointestinal syndromes such as those due to phosphorus, mercury or arsenic are manifested by vomiting, colic and bloody mucus and stools and are usually accompanied by liver damage (hepatomegalia, jaundice). Such conditions are relatively rare nowadays and have been superseded by occupational intoxications which develop slowly and even insidiously; consequently liver damage, in particular, may often be insidious too. A factor of great clinical importance is minor hepatic deficiency—a condition which is common but not easy to diagnose. It may be of occupational origin and may have a definite deleterious effect on general condition and on productivity. The symptoms are non-specific and varied, although in combination they may provide diagnostic indicators; they include asthenia, anorexia, dirty tongue, bitter taste in the mouth especially in the morning, epigastric heaviness after meals, nausea, eructation, meteorism, abdominal irregularities, headache, giddiness, mental torpor, hypotension, skin disorders, etc.

Infectious hepatitis deserves particular mention; it may be related to a number of occupational factors (hepatotoxic agents, heat or hot work, cold or cold work, intense physical activity, etc.), may have an unfavourable course (protracted or persistent chronic hepatitis) and may easily result in cirrhosis. It is frequently anicteric and thus creates diagnostic difficulties; moreover it presents difficulties of prognosis and estimation of the degree of recovery and hence of fitness for resumption of work (see HEPATITIS, INFECTIOUS).

Repeated exposure to irritant gases and vapours (halogens and their derivatives, sulphur dioxide, nitrogen oxides) frequently causes chronic gastritis, and chrome or thallium may produce chronic gastro-enterocolitis leading to atrophic gastritis. Mercury poisoning may display the symptoms of interstitial colitis (elimination colitis) or mucous endocolitis (contact colitis); both are ulcerative, associated with liver disorders and accompanied by moderate-to-intense pain.

Heavy metal or organophosphorus pesticide poisoning may produce pancreatic involvement and the diagnostic problems involved in such cases are considerable.

[Some chemicals may also act on the digestive system through sensitising effects as in the allergic gastrointestinal disturbances caused by dibutylphthalate plasticisers.

An increased prevalence of cancers of the digestive tract has been observed in industrial exposure to asbestos.]

Physical factors

Various physical agents may cause digestive system syndromes; these include direct or indirect disabling traumata, ionising radiations, vibration, rapid acceleration, noise, very high and low temperatures or violent and repeated climatic changes. Burns, especially if extensive, may cause gastric ulceration and liver damage, perhaps with jaundice. Abnormal postures or movements may cause digestive disorders especially if there are predisposing conditions such as para-oesophagic hernia, visceroptosis or *relaxatio diaphragmatica*; in addition, extra-digestive reflexes such as gastrocardiac syndromes may occur where digestive disorders are accompanied by autonomic nervous system or neuro-psychological troubles. Troubles of this type are common in modern work situations and may themselves be the cause of gastrointestinal dysfunction.

Occupational stress

Physical fatigue may also disturb digestive functions, and heavy work may cause secretomotor disorders and dystrophic changes, especially in the stomach. Persons with gastric disorders, especially those who have undergone surgery are limited in the amount of heavy work they can do, if only because heavy work entails higher levels of nutrition.

Nervous gastric dyspepsia (or gastric neurosis) seems to have no gastric or extragastric cause at all, nor does it result from any humoral or metabolic disorder; consequently, it is considered to be due to a primitive disorder of the autonomic nervous system, sometimes associated with excessive mental exertion or emotional or psychological stress. The gastric disorder is often manifested by neurotic hypersecretion or by hyperkinetic or atonic neurosis (the latter frequently associated with gastroptosis). Epigastric pain, regurgitation and aerophagia may also come under the heading of neurogastric dyspepsia. Elimination of the deleterious psychological factors in the work environment may lead to remission of these syndromes.

[Several observations point to an increased frequency of peptic ulcers among people carrying responsibilities such as foremen and executives, workers engaged in very heavy work, newcomers to industry, migrant workers, seamen, and workers subject to serious socioeconomic stress. However, many people suffering the same disorders lead a normal professional life, and statistical evidence is lacking. Drinking, smoking and eating habits and home and social life in addition to working conditions all play a part in the development and prolongation of dyspepsia, and it is difficult to determine what part each one plays in the aetiology of the condition.

Digestive disorders have also been attributed to shift work as a consequence of frequent changes of eating hours and poor eating at workplaces. These factors can aggravate pre-existing digestive troubles and release a neurotic dyspepsia. Therefore workers should be assigned to shift work only after medical examination (see SHIFT WORK).]

Medical supervision

It can be seen that the industrial physician is faced with many difficulties in the diagnosis and estimation of digestive system complaints (due inter alia to the part played by deleterious non-occupational factors) and that his responsibility in prevention of disorders of occupational origin is considerable.

Early diagnosis is extremely important and implies periodical medical examinations and supervision of the working environment, especially when the level of risk is high.

Health education of the general public, and of workers in particular, is a valuable preventive measure and may yield substantial results. Attention should be paid to nutritional requirements, choice and preparation of foodstuffs, the timing and size of meals, proper mastication and moderation in the consumption of rich foods, alcohol and cold drinks, or complete elimination of these substances from the diet.

FRADA, G.

General:

"Digestive system problems as related to fitness for work" (Problemi relativi all'apparato digerente nei confronti della idoneità al lavoro). Frada, G.; Salomone, L. *Securitas* (Rome), 1968, 53/9-10 (63-158). Illus. 158 ref. (In Italian)

Stomach:

"A study of gastric disturbances in industrial workers" (Enquête sur les troubles gastriques observés en milieu industriel). Gaultier, M.; Housset, P.; Martin, E. *Archives des maladies professionnelles* (Paris), Mar. 1961, 22/3 (129-134). (In French)

"Assessment of fitness for work in cases of gastric disease" (Begutachtung und Arbeitsfähigkeit bei Magenkrankheiten).

Koelsch, K. A. *Zeitschrift für ärztiliche Fortbildung* (Jena), 1976, 70/8 (397-400). (In German)

"Social and occupational consequences of partial gastrectomy for gastric or duodenal ulcer" (Incidence sociale et professionnelle de la gastrectomie partielle pour ulcère gastroduodénal). Dupuy, R.; Vallin, G. *Archives françaises des maladies de l'appareil digestif* (Paris), 1975, 64/8 (653-657). (In French)

Dimethylaminoazobenzene

***para*-Dimethylaminoazobenzene** ($C_{14}H_{15}N_3$)

N,N-DIMETHYL-4-PHENYLAZO-BENZENAMINE; DAB; BUTTER YELLOW

m.w. 225.3
m.p. 114-117 °C

yellow leaflets, insoluble in water; soluble in several organic solvents, in inorganic acids and oils. Skin absorption is likely to occur.

Production and uses. It is produced by the action of benzenediazonium chloride on aniline. Early in this century it was largely used as a food colouring (Butter yellow). Many countries in the last half-century have withdrawn DAB from the approved list of food additives and the Joint FAO/WHO Expert Committee on Food Additives declared in 1973 that the use of this substance in food was unsafe. Current production is believed to be small and it is used as a colour and a chemical indicator.

HAZARDS AND THEIR PREVENTION

The metabolism of DAB has been extensively studied and it has been found that it involves reduction and cleavage of the azo group, demethylation, ring hydroxylation, N-hydroxylation, N-acetylation, protein binding and binding of nucleic acids (see AZO AND DIAZO DYES). DAB shows mutagenic properties after activation. It has carcinogenic power by various routes in the rat and mouse (liver carcinoma), and by oral route it causes carcinoma of the bladder in the dog.

The only occupational health observation in man was of contact dermatitis in factory workers handling DAB.

Technical measures should prevent any contact with the skin and mucous membranes. Workers exposed to DAB should wear personal protective equipment and their work should be carried out only in restricted areas. Clothing and equipment after use should be placed in an impervious container for decontamination or disposal. Pre-employment and periodical examinations should focus on liver function. In the United States, DAB has been included by OSHA among the cancer suspect agents for man.

PARMEGGIANI, L.

"Syntheses, toxicities, and carcinogenicities of carcinogenic bifunctional aminoazo dyes". Lin, J. K.; Wu, J. R. *Cancer Research* (Chicago), Sep. 1974, 34/9 (2 274-2 282). 38 ref.

"Changes in polyamine levels and protein synthesis rate during rat liver carcinogenesis induced by 4-Dimethylaminoazobenzene". Perin, A.; Sessa, A. *Cancer Research* (Chicago), Jan. 1978, 38/1 (1-5), Illus. 38 ref.

"*para*-Dimethylaminoazobenzene" (125-146). Illus. 92 ref. *IARC monographs on the evaluation of carcinogenic risk of chemicals to man.* Vol. 8. *Some aromatic azo compounds* (Lyons, International Agency for Research on Cancer, 1975), 357 p.

"4-Dimethylaminoazobenzene" (575-580). *General industry.* OSHA Safety and Health Standards (29 CFR 1910) OSHA 2206 (Revised Jan. 1976) (Washington, DC, US Department of Labor, 1976), 649 p.

Dimethylcarbamoyl chloride

Dimethylcarbamoyl chloride ($(CH_3)_2NCOCl$)

DMCC; DIMETHYL CARBAMYL CHLORIDE

m.w. 107.6
sp.gr. 1.68
m.p. −33 °C
b.p. 166 °C
v.d. 3.73

a liquid which rapidly hydrolyses in water to dimethyl amine, CO_2 and HCl.

ACGIH industrial substance suspected of carcinogenic potential for man

Production and uses. DMCC is obtained by reaction of dimethyl amine with phosgene. It is used as an intermediate in pesticide production (carbamates) and in the pharmaceutical industry.

Toxicity. The acute oral toxicity (LD_{50}) of DMCC is 1 170 mg/kg for rats; the LD_{50} after intraperitoneal administration is 350 mg/kg for mice; the inhalation by rats of 10 ppm during 6 h/day for 15 days caused the death of 51% of the animals. DMCC irritates the mucous membranes of the upper airways and lungs and causes conjunctivitis and keratitis; skin application produces not only irritation but also degenerative changes: of 50 female mice on which 2 mg of DMCC dissolved in acetone was applied between the shoulders 3 times a week over a period of 492 days, 40 developed skin papillomas, 30 of which degenerated into carcinomas. A group of 30 female mice was treated once a week by intraperitoneal injection of 1 mg DMCC over a period of 450 days during which 14 papillary tumours developed, whereas 29 were observed in the control group comprising 100 animals. Subcutaneous injection of 5 mg DMCC in 50 female mice treated for 26 weeks caused the development of 39 neoplasms at the point of inoculation (36 sarcomas, 3 squamous-cell carcinomas).

In a long-term inhalation study, more than 90% of the surviving rats exposed to 1 ppm DMCC presented, after a mean exposure period of 258 days, squamous-cell carcinomas of the nose.

DMCC proved to be mutagenic without metabolic activation in strains of *Salmonella typhimurium*.

Human exposure. Observations made in workers exposed to DMCC are rather scarce. A case of eye irritation and another of liver disorders in DMCC-exposed persons have been reported in the literature. A study covering 65 workers (39 exposed in production and 25 using DMCC) and 42 persons no longer exposed (who had previously worked in contact with DMCC for periods from 6 months to 12 years) did not reveal any death due to cancer nor any radiological evidence of lung cancer. The statistical analysis of chromosomal aberrations in workers exposed for 4 to 17 years yielded no significant differences from those observed in the controls.

In 1980 the US Environmental Protection Agency (EPA) included the substance among those to be labelled as presenting a cancer hazard; in 1981 the ACGIH put it again on the list of suspected carcinogens for man.

SAFETY AND HEALTH MEASURES

Owing to the harmful properties of DMCC it is necessary to take thorough preventive measures with regard to both safety and health. The processes should be carried out in the open air as far as this is feasible; if not, they should take place in well ventilated premises under a slight negative pressure to avoid any diffusion of pollutants. The process plant should be equipped with local exhaust devices, in particular at flanges, valves, etc., where leaks are more likely to arise.

The workers should be specially trained and sufficiently informed about the physical, chemical and toxicological properties of the compound, and they should be supplied with work clothing and underwear kept in lockers separate from ordinary clothing. Taking a shower at the end of the work shift is recommended for adequate protection of the workers' health. Respirators, protective gloves and goggles should be available for daily use. When the substance is handled, not only should the protective equipment suitable for the operations involved be worn, but the equipment used for transferring the substance should also have been previously cleaned and decontaminated. Periodic medical examinations completed by laboratory tests enable the exposed workers' state of health to be kept under surveillance.

ARMELI, G.
DE RUGGIERO, D.

CIS 78-1357 *NIOSH current intelligence bulletin reprints–Bulletins 1 thru 18.* DHEW (NIOSH) publication No. 78-127 (National Institute for Occupational Safety and Health, 4676 Columbia Parkway, Cincinnati) (1 Mar. 1978), 125 p. 104 ref.

CIS 77-702 "Carcinogenic potential of DMCC". *American Industrial Hygiene Association Journal* (Akron, Ohio), June 1976, 37/6 (370-371).

CIS 75-121 "Possible health impairment in the manufacture and processing of dimethylcarbamic acid chloride" (Zur Frage etwaiger Gesundheitsschädigungen bei der Herstellung und Verarbeitung von Dimethylcarbaminsäurechlorid). Hey, W.; Thiess, A. M.; Zeller, H. *Zentralblatt für Arbeitsmedizin und Arbeitsschutz* (Heidelberg), Mar. 1974, 24/3 (71-77). 7 ref. (In German)

"Dimethylcarbamoyl chloride" (77-84). 16 ref. Vol. 12: *Some carbamates, thiocarbamates and carbazides. IARC monographs on the evaluation of carcinogenic risk of chemicals to man* (Lyons, International Agency for Research on Cancer, 1976), 282 p. Ref.

Dimethyl sulphate

Dimethyl sulphate ($(CH_3)_2SO_4$)

METHYL SULPHATE

m.w.	126.1
sp.gr.	1.33
m.p.	−31.7 °C
b.p.	188.5 °C
v.d.	4.4
v.p.	0.5 mmHg ($0.06 \cdot 10^3$ Pa) at 20 °C
f.p.	83.3 °C oc
i.t.	187 °C

soluble in ethyl ether, benzene; slightly soluble in water

a colourless, oily viscous liquid, with a slight smell of onion; the presence of the two methyl groups renders it highly unstable.

TWA OSHA	1 ppm 5 mg/m³ skin
TLV ACGIH	0.1 ppm 0.5 mg/m³: industrial substance suspected of carcinogenic potential for man
IDLH	10 ppm

Production. Dimethyl sulphate is produced industrially by treating methyl alcohol with oleum and distilling *in vacuo*.

Uses. It is used as a methylating agent in the chemical industry, in the manufacture of dyes and dyestuffs, perfumes and pharmaceuticals and as a solvent in the separation of mineral oils. During the First World War, it was used as an asphyxiant gas.

HAZARDS

Dimethyl sulphate is an extremely hazardous poison; it reacts with proteins and attacks the skin and mucous membranes and may cause severe inflammation and secondary necrosis of the affected tissues.

Industrial poisoning is usually acute, occurring more frequently as a result of vapour inhalation; poisoning due to skin contact (contaminated clothing) is less common, and chronic poisoning by vapours is relatively rare and even open to discussion.

Poisoning as the result of a spillage or prolonged direct contact produces generalised symptoms, both ocular and respiratory, intense local irritation and burns with tissue necrosis. Healing is accompanied by the formation of scar tissue.

Toxicity. The pathogenesis of dimethyl sulphate poisoning may be explained by the substance's alkylating properties and the fact that it hydrolyses to form sulphuric acid and methyl alcohol, which enter the bloodstream. These phenomena have been verified by animal experimentation and by the toxicological examination of a case of fatal poisoning in which methyl alcohol was found in the blood and viscera.

Long-term effects on vision are doubtful but are attributed to a metabolic action on the ganglion cells of the retina (methyl alcohol has a pronounced effect on cellular oxidation), and this explains the divergence of symptoms from person to person.

Pulmonary oedema is produced primarily by direct toxic effect on the alveolar structures. Furthermore, the effect of the vapour on the trigeminal and laryngeal nerve endings may result in bradycardia and pulmonary vasodilation.

Dimethyl sulphate has been shown to be carcinogenic in the rat both directly and following prenatal exposure. The inhalation of 1 ppm causes the urinary excretion of methylpurines showing a non-specific alkylation of DNA.

Acute poisoning. Acute poisoning may, in rare cases, have severe onset leading rapidly to asphyxia within a matter of minutes. In general, however, the signs appear suddenly although, in the case of moderate poisoning, there may be a latent period of 6-8 h, during which there is slight throat and eye irritation.

Ocular disorders are serious and include palpebral oedema extending to complete occlusion of the eyes, conjunctivitis and keratitis, photophobia and occasionally amblyopia. Rhinolarynogopharyngeal effects are very frequent and are due to local irritation of the respiratory tract; they include laryngeal dyspnoea, changes in the voice such as hoarseness or aphonia, hydrorrhoea and sensation of increased chest tightness.

If acute poisoning is mild and of short duration, the symptoms may regress without sequelae. However, in cases of massive intoxication, pulmonary oedema occurs, usually 6-24 h after exposure; this alveolar oedema may be acute or subacute and the functional consequences are aggravated by bronchial oedema. In one case, this oedema was of a fulminating type. These manifestations may be accompanied by disorders of the digestive system (dysphagia, vomiting) and urinary disorders (difficulty in micturition).

The ocular manifestations are persistent. Regression may occur within a few days but the functional symptoms (reduction in visual acuity, conjunctivitis) last for some considerable time although, in the case of mild intoxication, objective signs do not persist. Patients continue to complain of rapid visual fatigue for a long time but recovery occurs without sequelae in spite of the signs of mild keratitis observed at the outset. In severe poisoning, keratoconjunctival effects may give rise to corneal opacities and permanent visual disorders.

The upper respiratory tract effects also regress and recovery is complete even when the initial attack was severe. Secondary infection may supervene following acute oedema and the complications may be serious. In persons with pre-existent respiratory disorders, the effects may be more pronounced.

Chronic poisoning. The chronic form is less common than the acute; the effects are similar although usually limited to ocular and respiratory irritation. Vision disorders occur only rarely if at all. Hand and eyelid skin irritation has been observed.

Carcinogenic risk for humans. Three cases of bronchial cancer have been reported from a small enterprise where dimethyl sulphate had been used without the necessary precautions. The best studied case is that of a 47-year-old worker with 11 years' exposure, presenting signs of chronic bronchial irritation; it was an oat cell carcinoma of the upper bronchus.

SAFETY AND HEALTH MEASURES

All industrial operations involving the use of dimethyl sulphate should be carried out in fully enclosed systems. Arrangements should be made for swilling away any spillage. Workers should be supplied with respiratory protective equipment incorporating a filter impregnated with an aqueous sodium or potassium solution.

Workers should be strictly forbidden to attempt to clean up massive spillages such as may occur in the event of container breakage, since nearly all accidents with dimethyl sulphate have been the result of hasty and uninformed cleaning-up action. The inhalation of dimethyl sulphate can be tolerated for several minutes and the pathological effects do not become apparent until some time later.

Treatment. In the event of inhalation, the victim should be made to lie down immediately; even if he is free of symptoms, he should be kept under medical supervision in view of the long latent period (8 h). The eyes should be bathed with a 2% sodium bicarbonate solution and a cortisone-derivative eyewash; where there is internal pain, an analgaesic may be administered. The patient should then be despatched to a resuscitation centre where emergency treatment may prevent acute pulmonary oedema and collapse. This treatment comprises: absolute rest in a humid atmosphere; oxygen therapy; inhalation of aleudrin or long-term bronchodilators; analgaesics which do not have a depressive effect on the nervous system and the respiratory centres; cardiac stimulants, where necessary–intravenous administration of cortisone derivatives; treatment of the ocular mucosae by instillation of non-caustic anaesthetics or antiseptics; in case of pulmonary oedema, morphine and adrenaline should not be administered under any circumstances.

In very serious cases, positive-pressure resuscitation may be valuable; however, in extremely acute intoxications, treatment will be of no avail unless applied with a minimum of delay since, where there is extensive pulmonary oedema and secondary infection, even the most powerful therapy against anoxia, e.g. hyperbaric oxygen, will be ineffective.

Dimethyl sulphate ingestion may be treated with sodium bicarbonate; contaminated skin should be thoroughly irrigated after the victim's clothing has been removed.

FOURNIER, E.

Poisoning:

CIS 78-1657 *Dimethyl sulfate poisoning–13 case studies* (Intoxications au diméthyl sulfate–A propos de 13 observations). Roux, H. (Université Claude Bernard, Lyon II, Faculté de médecine, Lyons) (1977), 94 p. 33 ref. (In French)

CIS 229-1969 "Problems of occupational health associated with dimethyl sulphate poisoning" (Arbeitsmedizinische Fragen in Zusammenhang mit der Dimethylsulfat-Intoxikation). Thiess, A. M.; Goldmann, P. J. *Zentralblatt für Arbeitsmedizin und Arbeitsschutz* (Darmstadt), July 1968, 18/7 (195-204). Illus. 40 ref. (In German)

Cancer:

"Alkylating agents with carcinogenic action. I. The carcinogenic action of dimethyl sulfate in the rat. Dimethyl sulfate as a probable cause of occupational cancer" (Carcinogen alkylierende Substanzen. I. Dimethylsulfat, carcinogene Wirkung an Ratten und Wahrscheinliche Ursache von Berufskrebs). *Zeitschrift für Krebsforschung* (West Berlin), 1966, 68 (103-111). Illus. 17 ref. (In German)

Safety and health:

CIS 77-716 *Dimethyl sulfate* (Dimethylsulfat). Berufsgenossenschaft der chemischen Industrie (Weinheim, Verlag Chemie, 1975), 7 p. (In German)

Dinitro-*o*-cresol

Dinitro-*o*-cresol ($CH_3C_6H_2(NO_2)_2OH$)

DNOC; 3,5-DINITRO-2-HYDROXYTOLUENE; 4-6-DINITRO-2-METHYL-PHENOL

m.p.	86.5 °C
e.l.	0.3 g/l

slightly soluble in water; very soluble in alkaline solution, ethyl ether, acetone

a yellow crystalline solid.

TWA OSHA	0.2 mg/m³ skin
STEL ACGIH	0.6 mg/m³
IDLH	5 mg/m³
MAC USSR	0.05 mg/m³ skin

OH, O_2N, CH_3, NO_2

Dinitro-*o*-cresol has nine isomers of which 4,6-dinitro-*o*-cresol is the most important. It is produced by the sulphonation of *o*-cresol followed by nitration. It has some uses in the dyestuff industry but is employed widely in agriculture both as a pesticide and, much more extensively, as a herbicide.

HAZARDS

DNOC is poisonous to most living things and exerts its effects by disturbing the process of oxidative phosphorylation. The over-all effect of poisoning is a greatly increased requirement by the tissues for oxygen, with the dissipation of much of the unwanted energy thus generated as heat. If such heat production greatly exceeds the heat lost then fatal hyperthermia may result. The effects of DNOC poisoning are acute and, if not fatal, recovery is complete and tissue damage with scarring and other chronic effects are not seen.

Toxic effects on man were first seen when it was introduced for a short period as a drug to stimulate metabolism and promote a rapid loss of weight. Many patients taking DNOC for this purpose died suddenly and others developed cataracts.

Chronic exposure may result in fatigue, restlessness, thirst and weight loss. An increase in chromosome aberrations has been observed in human lymphocytes treated *in vitro* with DNOC and in mouse bone marrow cells treated *in vivo*.

DNOC is reduced in the animal body to aminonitro-*o*-cresol which does not have the toxic action of the parent compound. DNOC is a bright yellow dye which readily stains clothing, skin and hair. It may be detected in the tissues and its concentration accurately determined at low concentrations.

Accidental poisoning. This has occurred during the manufacture of DNOC and during its application as a herbicide. Used in a very dilute solution as an insecticide during the winter season, DNOC has not led to any poisoning. Fatal cases of poisoning among agricultural workers using DNOC as a herbicide have been reported from many countries. In every instance poisoning has occurred in people exposed to DNOC for many hours, usually on several successive days and under hot climatic conditions.

The clinical picture is always the same. Early symptoms are fatigue, excessive sweating, which may be the most striking feature, and unusual thirst and a loss of weight may be noted. These symptoms may be attributed to the effects of the long working hours or hot conditions and not to the effects of DNOC. Further exposure to DNOC may lead to a very rapid deterioration in condition. Weakness and fatigue become severe with an increase in the depth and rate of respiration, tachycardia and rise of body temperature. Death may follow within an hour or two and *rigor mortis* occurs almost immediately. In less severely affected cases, recovery may be virtually complete in 48 h. Exposure to DNOC is usually evident from the nature of the work and the staining of the clothes, skin (particularly of the hands and face) and the hair.

DNOC is absorbed through the skin and readily from the alimentary tract. Exposure during spraying may occur due to spray drift and when cleaning spray nozzles and dispensing the concentrates.

Diagnosis can be made by using a simple method for measuring the DNOC in the blood.

SAFETY AND HEALTH MEASURES

Care in the application of the strong solution of DNOC used as herbicides is essential. Protective clothing with hand protection, foot and leg protection and rubber aprons are needed when dispensing the concentrate. Full protection is difficult when DNOC is applied under hot working conditions, but frequent washing and the limitation of hours of work will help to reduce exposure.

Treatment. There is no specific antidote, and treatment consists of absolute rest, plenty of fluids and cool sponging. Barbiturates should be avoided if sedatives are needed. Atropine sulphate is contraindicated. Thiamine hydrochloride has proved useful in experimental intoxication.

MAGOS, L.
MANSON, M. M.

CIS 79-136 "Studies on exposure to dinitro-*o*-cresol (DNOC) in agricultural chemists" (Untersuchungsergebnisse zur Exposition gegenüber Dinitro-o-kresol (DNOC) bei Agrochemikern). Jastroch, S.; Knoll, W.; Lange, B.; Riemer, F.; Thiele, E. *Zeitschrift für die gesamte Hygiene und ihre Grenzgebiete* (Berlin), May 1978, 24/5 (340-343). Illus. 20 ref. (In German)

"Pesticides". Murphy, S. D. (403-453). *Toxicology—a basic science of poisons.* Casaretti, L. J.; Doull, J. (eds.) (New York, Macmillan, 1975).

"Mutagenic effect of a pesticide containing dinitro-*o*-cresol" (Mutagene Wirkung eines dinitro-o-kresol-haltigen Pflanzenschutzmittels). Nehez, von M.; Selypes, A.; Paldy, A.; Schroeter, C. *Zeitschrift für die gesamte Hygiene und ihre Grenzgebiete* (Berlin), 1 Jan. 1978, 24/1 (20-24). Illus. 30 ref. (In German)

CIS 78-1634 *Criteria for a recommended standard—Occupational exposure to dinitro-ortho-cresol.* DHEW (NIOSH) publication No. 78-131 (National Institute for Occupational Safety and Health, 4676 Columbia Parkway, Cincinnati) (Feb. 1978), 147 p. Illus. 65 ref.

"Animal experiments conducted to determine the protective action of Vitamin B_1 in cases of DNOC intoxication" (Tierexperimentelle Untersuchungen zur Schutzwirkung von Vitamin B_1 bei DNOC-Intoxikation). Kochmann, W.; Bech, R.; Baer, H. P. *Zeitschrift für die gesamte Hygiene und ihre Grenzgebiete* (Berlin, German Democratic Republic), Mar. 1978, 24/3 (163-165). (In German)

Dinitrophenol

Dinitrophenol ($C_6H_3(NO_2)_2OH$)

DNP

m.w.	184.1
sp.gr.	1.68
m.p.	2,3-Dinitrophenol 144 °C
	2,4-Dinitrophenol 112 °C
	2,6-Dinitrophenol 63 °C
v.d.	6.35

OH, NO_2, NO_2

slightly soluble in cold water, freely soluble in hot water, ethyl alcohol, ethyl ether and benzene; steam volatile

yellow crystals.

MAC USSR 0.05 mg/m³ skin

There are six isomers of which 2,4-dinitrophenol is the most important. It is produced by the action of sodium hydroxide on 1-chloro-2,4-dinitrobenzene and used in industry as an intermediate in dyestuff and other chemical manufacture and as a wood preservative.

HAZARDS

Dinitrophenol is an acute poison disrupting cellular metabolism in all tissues by disturbing the essential process of oxidative phosphorylation. If not fatal, the effects are rapidly and completely reversible. Exposure may occur by the inhalation of the vapour, dusts or sprays of solutions of DNP. It penetrates the intact skin but, as it is a brilliant yellow dye, skin contamination is readily recognised. Systemic poisoning has occurred during both production and use.

The DNP solid is explosive and accidents have also occurred during production and use. Care must be exercised when handling it.

Toxicity. Poisoning results first in excessive sweating, a feeling of warmth with weakness and fatigue. In severe cases, there is rapid respiration and tachycardia even at rest and there may be a rise in body temperature. Death, if it occurs, is sudden and *rigor mortis* ensues almost immediately. DNP exerts its toxic effects by a general disturbance of cell metabolism resulting in a need to consume excessive amounts of oxygen in order to synthesise the essential adenine nucleotide required for cell survival in the brain, heart and muscles. If heat production is greater than heat loss, fatal hyperthermia may result. The effects are most severe in hot workplaces.

DNP is readily reduced to the aminophenol which is much less toxic and is excreted in the urine in this form. Since DNP is rapidly metabolised and excreted and since poisoning does not lead to structural changes in tissues, chronic or cumulative effects from small doses absorbed over long periods do not occur.

Diagnosis. It may be difficult to distinguish the effects of mild poisoning–sweating, feeling hot and lassitude–from those of fatigue from hard work, especially in a warm environment. If exposure to DNP is suspected by the nature of the occupation and by staining of the skin, particularly of the hands and round the mouth and nose, individuals should immediately be removed from further exposure, made to lie down and the course of poisoning carefully watched.

Diagnosis of poisoning may be confirmed by finding DNP or aminophenol in the urine by Derrien's test. Methaemoglobinaemia does not develop.

SAFETY AND HEALTH MEASURES

During manufacture of DNP, dust must be kept to a minimum by the use of exhaust ventilation at points where containers are filled or centrifuges emptied. People exposed to the solid or to strong solutions should wear protective clothing, hand protection and, in confined spaces, respiratory protective equipment. Contamination of clothing and the skin is readily detected by the yellow staining, and the need for washing will thereby be indicated. Men regularly exposed to DNP should have their urine regularly tested for DNP or aminonitrophenol by polarography or Derrien's test. In view of the fire and explosion hazard of DNP it should be stored in a cool ventilated place away from areas of acute fire hazard and away from powerful oxidising agents.

Treatment. There is no specific antidote but tepid sponging and oxygen, if there is respiratory distress, may alleviate the condition. Barbiturates, which may exacerbate the biochemical disturbances produced by DNP, are not recommended to relieve the restlessness. If the condition does not continue to deteriorate rapidly after the institution of absolute rest, cooling and oxygen, recovery is likely to be rapid and complete within 12-24 h.

MAGOS, L.
MANSON, M. M.

"2,4-Dinitrophenol–$C_6H_3HO(NO_2)_2$". Morel, C.; Cavigneaux, A. Fiche toxicologique nº 95. *Cahiers de notes documentaires–Sécurité et hygiène du travail* (Paris), 1st quarter 1972, 66, Note No. 784-66-72 (124-129). 22 ref. (In French)

Dioxane

Dioxane ($OCH_2CH_2OCH_2CH_2$)

DIETHYLENE-1,4-DIOXIDE; DIETHYLENE ETHER; DIOXANE

m.w.	88
sp.gr.	1.04
m.p.	11.8 °C
b.p.	101 °C
v.d.	3
v.p.	37 mmHg ($4.92 \cdot 10^3$ Pa) at 25 °C
f.p.	12 °C
e.l.	2.0-22%
i.t.	180 °C

a colourless, volatile liquid, highly soluble in water and most organic solvents.

TWA OSHA	100 ppm 360 mg/m³ skin
TLV ACGIH	(technical grade) 25 ppm 90 mg/m³ skin
STEL ACGIH	100 ppm 360 mg/m³
NIOSH	1 ppm/30 min
IDLH	200 ppm
MAC USSR	10 mg/m³ skin

Production. Dioxane is commercially produced by the dehydrogenation of ethylene glycol, by catalytic dimerisation of ethylene oxide in the vapour phase, or by reaction of *bis*-(2-chloroethyl) ether or 2-chloroethyl-2'-hydroxyethyl ether with strong aqueous sodium hydroxide.

Uses. Dioxane has been in commercial use for almost 50 years, principally as a solvent for fats, waxes, greases, dyes, paints, varnishes, mineral oil, lacquers and similar products where nitrocellulose, cellulose, cellulose acetate, or other cellulose esters or ethers are used. It has been used as a wetting agent and dispersing agent in dye baths, textile processing, and stain and printing compositions. It is also used in adhesives, cosmetics, detergents and as a stabiliser for chlorinated solvents such as 1,1,1-trichlorethane. Other applications include use as a solvent to purify organic compounds, in molecular weight determinations, and in the radioimmunoassay of glucogen after extraction of blood with dioxane.

HAZARDS

Dioxane is a dangerous fire hazard and is a moderate explosion hazard when exposed to heat, flame or oxidising materials. Both acute and chronic health effects may result from inhalation and from penetration of intact skin.

Fire and explosion. At room temperature, liquid dioxane can produce vapour-air mixtures within the explosive range.

Health hazards. Inhalation studies in animals have demonstrated that dioxane vapour can cause narcosis; lung, liver and kidney damage; irritation of the mucous membrane; congestion and oedema of the lungs; behavioural changes; and, increased blood counts. Large doses of dioxane administered in drinking water have led to the development of tumours in rats and guinea-pigs. Chronic inhalation studies with 423 rats exposed at 111 ppm for two years provided no statistically significant evidence ($P < 0.05$) of increased tumour incidence. Animal experiments have also demonstrated that dioxane is rapidly absorbed through the skin producing signs of incoordination, narcosis, erythema as well as liver and kidney injury.

Experimental studies with humans have also shown eye, nose, and throat irritation at concentrations of 200 to 300 ppm. An odour threshold as low as 3 ppm has been reported, although another study resulted in an odour threshold of 170 ppm. Both animal and human studies have demonstrated that dioxane is metabolised to beta-hydroxyethoxyacetic acid. An investigation in 1934 of the deaths of five men working in an artificial silk plant suggested that the signs and symptoms of dioxane poisoning included nausea and vomiting followed by diminished and finally absence of urine output. Necropsy findings included enlarged pale livers, swollen haemorrhagic kidneys and oedematous lungs and brains. In three separate epidemiologic studies, no significant differences were seen between the observed and expected number of cancer deaths. However, in these studies the populations at risk (74, 80 and 165) were small and involved total cancer mortality rather than site or type specific rates. The US National Institute for Occupational Safety and Health has concluded, on the basis of animal studies, that exposure to dioxane may cause cancer.

SAFETY AND HEALTH MEASURES

Dioxane should be stored in tightly closed containers in cool, dry, well ventilated areas. Exposure to heat, sparks, open flames, moisture and strong oxidisers should be prevented. Containers should be bonded and grounded when liquid dioxane is transferred. Employee education as to potential hazards and safe operating procedures should be emphasised.

To prevent acute and chronic health problems, operations involving dioxane should be completely enclosed or local exhaust ventilation used to limit vapour concentrations in workroom air. Entry into confined spaces which have contained dioxane must be carefully supervised with emphasis on minimising vapour exposure and ensuring that an adequate supply of oxygen exists. Where high airborne concentrations of dioxane are encountered the use of positive-pressure respirators is encouraged. Protective clothing to minimise skin absorption should be used where appropriate. Medical surveillance programmes should be directed towards effects on the kidneys and liver. Analytical procedures for the determination of dioxane involve drawing a known volume of air through a charcoal tube to absorb the dioxane vapour, desorption from the charcoal with carbon disulphide and analysis of the desorbed sample by gas chromatography.

Treatment. In acute overexposure situations, appropriate emergency procedures, including removal of workers from the contaminated area, should be implemented. Where appropriate the eyes of exposed workers should be flushed with water, contaminated clothing removed and the skin washed with soap and water. Treatment for shock, with special emphasis on fluids and electrolytes in the presence of oliguria and anuria, may be necessary. Subsequent liver disease should be treated symptomatically.

ROSE, V.

CIS 75-454 "1,4-Dioxane–I. Results of a 2-year ingestion study in rats". Kociba, R. J.; McCollister, S. B.; Park, C.; Torkelson, T. R.; Gehring, P. J. *Toxicology and Applied Pharmacology* (New York), Nov. 1974, 30/2 (275-286). Illus. 10 ref.

CIS 75-455 "1,4-Dioxane–II. Results of a 2-year inhalation study in rats". Torkelson, T. R.; Leong, B. K. J.; Kociba, R. J.; Richter, W. A.; Gehring, P. J. *Toxicology and Applied Pharmacology* (New York), Nov. 1974, 30/2 (287-298). Illus. 12 ref.

CIS 76-1952 "Results of occupational medical examinations of dioxane-exposed workers" (Arbeitsmedizinische Untersuchungsergebnisse von Dioxan-exponierten Mitarbeitern). Thiess, A. M.; Tress, E.; Fleig, I. *Arbeitsmedizin–Sozialmedizin–Präventivmedizin* (Stuttgart), Feb. 1976, 11/2 (36-46). 16 ref. (In German)

CIS 78-429 *Criteria for a recommended standard–Occupational exposure to dioxane.* DHEW (NIOSH) publication No. 77-226 (National Institute for Occupational Safety and Health, 4676 Columbia Parkway, Cincinnati) (Sep. 1977), 193 p. Illus. 122 ref.

Dioxin, tetrachlorodibenzo-*para*

2,3,7,8-tetrachlorodibenzo-*p*-dioxin (TCDD) is not manufactured commercially but is present as impurity in 2,4,5-trichlorophenol (TCP). Minute traces may be present in the herbicide 2,4,5-T and in the antibacterial agent hexachlorophene, which are produced from trichlorophenol.

TCDD is formed as a by-product during the synthesis of 2,4,5-trichlorophenol from 1,2,4,5-tetrachlorobenzene under alkaline conditions by the condensation of two molecules of sodium trichlorophenate (figure 1). When temperature and pressure keeping the reaction in progress are observed carefully, the crude 2,4,5-trichlorophenol contains less than 1 mg/kg up to maximum 5 mg/kg TCDD (1-5 ppm). Greater amounts are formed at higher temperatures (230-260 °C).

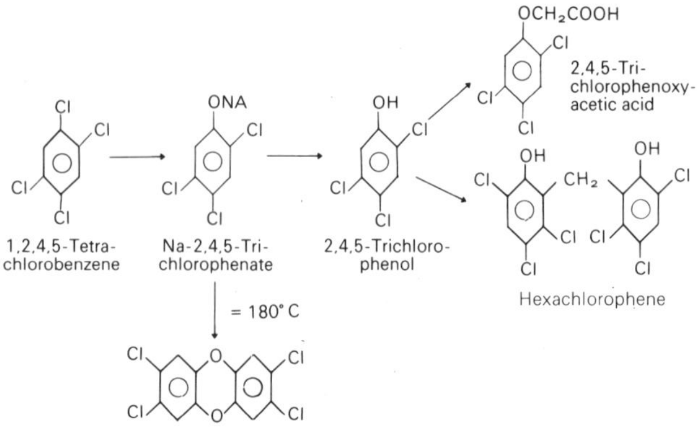

1,2,4,5-Tetrachlorobenzene → Na-2,4,5-Trichlorophenate → 2,4,5-Trichlorophenol → 2,4,5-Trichlorophenoxyacetic acid (OCH_2COOH); Hexachlorophene

Na-2,4,5-Trichlorophenate ↓ = 180° C → TCDD

Figure 1. Formation of 2,3,7,8-tetradibenzo-*p*-dioxin (TCDD) as a by-product in the synthesis of 2,4,5-trichlorophenol.

The chemical structure of TCDD was identified in 1956 by Sandermann et al. who first synthetised it. The laboratory technician working on the synthesis was hospitalised with very severe chloracne.

There are 22 possible isomers of tetrachlorodibenzo-*p*-dioxin. TCDD is commonly used to mean 2,3,7,8-tetrachlorodibenzo-*p*-dioxin, without excluding the existence of the other 21 tetraisomers. TCDD can be prepared for chemical and toxicological standard by catalytic condensation of potassium 2,4,5-trichlorophenate.

TCDD is a solid substance with very low solubility in common solvents and water (0.2 ppb) and very stable to thermal degradation. In the presence of a hydrogen donor it is rapidly degraded by light. When incorporated in the soil and aquatic systems, it is practically immobile.

Occurrence

The only practical source of TCDD formation in the environment is thermal reaction either in the chemical production of 2,4,5-trichlorophenol or in the combustion of chemicals which may contain precursors of the dioxins in general.

Occupational exposure to TCDD may occur during the production of trichlorophenol and its derivatives (2,4,5-T and hexachlorophene): during their incineration and during the use and handling of these chemicals and their wastes and residues.

General exposure of the public may occur in relation to a herbicide spraying programme; bioaccumulation of TCDD in the food chain; inhalation of fly ashes or flue gases from municipal incinerators and industrial heating facilities, during combustion of carbon containing material in the presence of chlorine; unearthing of chemical wastes; and contact with people wearing contaminated clothes.

Toxicity

TCDD is extremely toxic in experimental animals. The mechanism by which death occurs is not yet understood. Sensitivity to the toxic effect varies with the species. The lethal dose ranges from 0.5 μg/kg for the guinea-pig to over 1 000 μg/kg for the hamster by the oral route. The lethal effect is slow and ensues several days or weeks after a single dose.

Chloracne and hyperkeratosis are a distinctive feature of TCDD toxicity which is observed in rabbits, monkeys and hairless mice, as well as in the human being. TCDD has teratogenic and/or embryotoxic effects in the rodent. In the rabbit the major site of the toxic action appears to be the liver. In the monkey the first sign of toxicity is in the skin, whereas the liver remains relatively normal. Several species develop disturbance of the hepatic porphyrin metabolism. Immunosuppression, carcinogenicity, enzyme induction and mutagenicity have also been observed under experimental conditions. The half-life in the rat and guinea-pig is approximately 31 days and the major route of excretion is the faeces.

Table 1. Industrial accidents associated with manufacture of chlorinated phenols

Year	Country	Manufacturer and/or location	Production	Source of exposure	Number of cases
1949	United States	Monsanto/ Nitro, Virginia	TCP	Explosion, occupational	228
1952	Germany (Fed. Rep.)	Boehringer/ Hamburg	TCP	Occupational	31
1953	Germany (Fed. Rep.)	BASF/ Ludwigshafen	TCP	Explosion	75
1954	Germany (Fed. Rep.)	Boehringer/ Ingelheim	TCP 2,4,5-T	Occupational	32
1956	France	Rhône-Poulenc/ Grenoble	TCP	Explosion	17
1956	United States	Diamond Alkali Corp./ N. Jersey	2,4,5-T	Occupational	29
1962	Italy	ICM/ Saronno	TCP	Explosion	16
1963	Netherlands	Philips-Duphar/ Amsterdam	TCP	Explosion	106
1964	USSR	Ufa	2,4,5-T	Occupational	128
1964	United States	Dow Chemical/ Midland, Mich.	2,4,5-T	Occupational	60
1965	Czechoslovakia	Spolana	TCP	Occupational	80
1966	France	Rhône-Poulenc/ Grenoble	TCP	Explosion	21
1969	United Kingdom	Coalite, Chemical Products/ Bolsover, Derbyshire	TCP	Explosion	85-90
1970	Japan		PCP[1] 2,4,5-T	Occupational	25
1973	Austria	Chemical Works/ Linz	2,4,5-T	Occupational	50
1974	Germany	Bayer/ Uerdingen	2,4,5-T	Occupational	5
1976	Italy	Icmesa/ Seveso	TCP	Explosion	187

[1] PCP = pentachlorophenol.

The identification of TCDD as the toxic agent responsible for the lesions and symptoms observed in man after exposure to trichlorophenol or 2,4,5-trichlorophenoxyacetic acid was made in 1957 by K. H. Schulz in Hamburg, who eventually determined in tests with rabbits its chloracnegenic and hepatotoxic properties. In a self-administered skin test (10 μg applied two times), he also demonstrated the effect on human skin. A human experiment was repeated by Klingmann in 1970: in man, application of 70 μg/kg produced definite chloracne.

Toxic effects produced by TCDD in man have been reported as a consequence of repetitive occupational exposure during the industrial production of trichlorophenol and 2,4,5-T and of acute exposure in factories and their environment from accidents during the manufacture of the same products.

Industrial exposure

The annual world production of 2,4,5-trichlorophenol was estimated to be about 7 000 tonnes in 1979, the major part of which was used for the production of the herbicide 2,4,5-T and its salts. The herbicide is applied annually to regulate plant growth of forests, ranges and industrial, urban and aquatic sites. The general use of 2,4,5-T has been partially suspended in the United States. It is prohibited in some countries (Italy, Netherlands, Sweden); in others such as the United Kingdom, the Federal Republic of Germany, Canada, Australia and New Zealand, the herbicide is still in use. The normal application of 2,4,5-T and its salts (2 lb/acre) would disperse no more than 90 μg TCDD on each treated acre at the highest allowed concentration of 0.1 ppm TCDD in technical 2,4,5-T. In the period since the first commercial production of 2,4,5-T (1946-47) there have been several industrial episodes involving exposure to TCDD. This exposure usually occurred during the handling of contaminated intermediate products, i.e. trichlorophenol.

Table 1 lists the 17 episodes of TCDD exposure recorded in the literature, ten of which were associated with occupational exposure.

On eight occasions explosions occurred during the production of sodium trichlorophenate and workers were exposed to TCDD at the time of the accident, during the clean-up or from subsequent contamination from the workshop environment. Four other episodes are mentioned in the literature, but no precise data about the humans involved are available.

Clinical features

About 1 000 people have been involved in these episodes. A wide variety of lesions and symptoms (table 2) has been described in connection with the exposure and a causal association has been assumed for some of them.

Actually only very few cases have been exposed to TCDD on its own. In almost all cases the chemicals utilised for manufacturing TCP and its derivatives, i.e. tetrachlorobenzene, sodium or potassium hydroxide, ethylene glycol or methanol, sodium trichlorophenate, sodium monochloracetate and a few others depending upon the manufacturing procedure, participated in the contamination and might have been the cause of many of these symptoms independently from TCDD. Four clinical signs are probably related to TCDD toxicity because the toxic effects were predicted by animal testing or they have been consistent in several episodes. These symptoms are:

(a) chloracne, which was present in the great majority of recorded cases;

(b) enlarged liver and impairment of liver function, occasionally;

(c) neuromuscular symptoms, occasionally;

(d) deranged porphyrin metabolism in some cases.

Table 2. Toxic effects of TCDD reported in man

Dermatological:	Chloracne
	Porphyria cutanea tarda
	Hyperpigmentation and hirsutism
Internal:	Liver damage (mild fibrosis, fatty changes, haemofuscin deposition and parenchymal-cell degeneration)
	Raised serum hepatic enzyme levels
	Disorders of fat metabolism
	Disorders of carbohydrate metabolism
	Cardiovascular disorders
	Urinary tract disorders
	Respiratory tract disorders
	Pancreatic disorders
Neurological:	*(a) Peripheral:*
	Polyneuropathies
	Sensorial impairments (sight, hearing, smell, taste)
	(b) Central:
	Lassitude, weakness, impotence, loss of libido

Chloracne. Clinically chloracne is an eruption of blackheads, usually accompanied by small, pale-yellow cysts which in all but the worst cases vary from pin-head to lentil size. In severe cases there may be papules (red spots) or even pustules (pus-filled spots). The disease has a predilection for the skin of the face, especially on the malar crescent under the eyes and behind the ears in the very mild cases. With increasing severity the rest of the face and neck soon follow, whilst the outer upper arms, chest, back, abdomen, outer thighs and genitalia may be involved in varying degrees in the worst cases. The disease is otherwise symptomless and is simply a disfigurement. Its duration depends to a great extent upon its severity and the worst cases may still have active lesions 15 and more years after the contact has ceased. In human subjects within 10 days after beginning the application there was redness of the skin and a mild increase in keratin in the sebaceous gland duct, which was followed during the second week by plugging of the infundibula. Subsequently sebaceous cells disappeared and were replaced by a keratin cyst and comedones which persisted for many weeks.

Chloracne is frequently produced by skin contact with the causative chemical, but it appears also after its ingestion or inhalation. In these cases it is almost always severe and may be accompanied by signs of systemic lesions. Chloracne in itself is harmless but is a marker indicating that the affected person has been exposed, however minimally, to a choracnegenic toxin. It is therefore the most sensitive indicator we have in the human subject of overexposure to TCDD.

Enlarged liver and impairment of liver functions. Increased transaminase values in serum over the borderline may be found in cases after exposure. These usually subside within a few weeks or months. However, liver function tests can stay normal even in cases exposed to TCDD concentration in the environment of 1 000 ppm and suffering from severe chloracne. Clinical signs of liver dysfunction such as abdominal disturbances, gastric pressure, loss of appetite, intolerance to certain foods and enlarged liver have also been observed in up to 50% of cases.

Laparoscopy and biopsy of the liver showed slight fibrous changes, haemofucsin deposition, fatty changes and slight parenchymal cell degeneration in some of these cases. Liver damage caused by TCDD is not necessarily characterised by hyperbilirubinaemia.

Follow-up studies in those cases which today still have acneform manifestations after 20 years and more report that enlargement of the liver and pathological liver function tests have disappeared. In almost all experimental animals the liver damage is not sufficient to cause death.

Neuromuscular effects. Severe muscles pains aggravated by exertion especially in the calves and thighs and in the chest area; fatigue and weakness of the lower limbs with sensory changes have been reported to be the most disabling manifestations in some cases. Muscle biopsies revealed no abnormal findings. Biopsy of a peripheral nerve in one case (Suskind) revealed destruction of the myelin sheath and nerve fibres but no evidence of inflammatory cell infiltration. In his cases these symptoms commenced almost concurrently with the eruption of chloracne and kept the cases immobile for as long as two years.

In the animals, central and peripheral nervous systems are not target organs of TCDD toxicity, and there are no animal studies to substantiate the claims of muscular weakness or impaired skeletomuscular function in humans exposed to TCDD. The effect can therefore be related to the concurrent exposure to other chemicals.

Deranged porphyrin metabolism. TCDD exposure has been associated with derangement of the intermediary metabolism of lipids, carbohydrates and porphyrins. In animals TCDD has produced an accumulation of uroporphyrin in the liver with increase of *d*-amino-laevulinic acid (ALA) and of uroporphyrin excretion in the urine. In cases of occupational exposure to TCDD an increased excretion of uroporphyrins has been observed. The abnormality is disclosed by a quantitative increase in the urinary excretion of uroporphyrins and in a change in the proportion with coproporphyrin.

Long-term and other effects of TCDD exposure

Immunocapability. In the Coalite episode three groups of adult males were examined; those exposed to toxic levels of TCDD more than ten years previously and still showing clinical evidence of toxicity in the form of chloracne (38); those exposed to TCDD and showing no more evidence of dermatological toxicity (56) and an unexposed control group (31). The immunological test results suggest that in the exposed groups (chloracne) some changes have occurred. Yet there was no difference in health state between members of the groups.

In Seveso, Italy, the immunocapability was examined in 45 children belonging to the zone of maximum exposure (235 $\mu g/m^2$), 20 of whom had chloracne, and concurrently in a comparable group of 44 children who had not been exposed to TCDD. In Seveso the infective pathology of the people was followed for three years after the accident. It did not exceed the average of the region.

Teratogenic effects. The survey of several thousand women of child-bearing age exposed to a known level of TCDD in Seveso has shown that the frequency of spontaneous abortions per number of calculated pregnancies and per number of women of child-bearing age did not change either immediately after the accident or in the three following years if compared with the previous years. The rate always remained within the world-wide estimated ratio of 15-20 abortions per 100 pregnancies. As for congenital malformations and their pattern, here again their ratio and form did not change after the accident and always remained within the ratio of 2-3% which is that currently occurring world-wide for the major anatomical malformations. An increase in the rate of abortions and malformations has been assumed in these last years in connection with the spraying of 2,4,5-T for weed management and with the use of hexachlorophene in hospital conditions in the United States, New Zealand, Australia and Sweden. The health authorities of the respective governments have examined these cases but could not substantiate the assumption of a causal relationship.

Cancer mortality. Follow-up study of the cancer mortality in workers who have been exposed to TCDD several years ago has been carried out on the workers of Monsanto (1949), Boerhinger (1954), Philips (1963), Dow Chemical (1964) and Coalite (1968). The cancer mortality in these groups remained within the range of frequency expected for the matching population. The 410 workers involved in these accidents have suffered no excess incidence of cancer in a time-span of 14 to 30 years after exposure. For the 71 workers involved in the BASF accident (1953) an increased stomach cancer mortality has been observed.

In Sweden the frequency of soft-tissue sarcoma seems to be higher for railway workers engaged in the spraying of 2,4,5-T and other herbicides. It is not increased in a similar group of people working under the same conditions in Finland. The validity and significance of these observations are the object of further studies.

Chromosome aberrations. Cytogenetic studies have been performed on the plant workers of BASF (1953), Dow Chemical (1964), Coalite (1968) and on samples of the population of Seveso (1976). The measurement of chromosome irregularity in the four groups mentioned, comprising also Seveso children with choracne, as well as plant workers of the former accidents in the Federal Republic of Germany, the United Kingdom and the United States still suffering from chloracne, showed no significant deviation from that observed in the control population.

Diagnosis

The diagnosis of TCDD contamination is actually based on the history of logical opportunity (chronological and geographical correlation) of exposure to substances which are known to contain TCDD as a contaminant and on the demonstration of TCDD contamination of the surroundings by chemical analysis.

The clinical features and symptoms of the toxicity are not sufficiently distinctive to permit clinical recognition. Chloracne, the most sensitive indicator of TCDD exposure, is known to have been produced in the human subject by the following chemicals:

Chlornaphthalenes (CNs)
Polychlorinated biphenyls (PCBs)
Polybrominated biphenyls (PBBs)
Polychlorinated dibenzo-*p*-dioxins (PCDDs)
Polychlorinated dibenzofurans (PCDFs)
3,4,3,4-Tetrachlorazobenzene (TCAB)
3,4,3,4-Tetrachlorazoxybenzene (TCAOB)

Laboratory determination of TCDD in the human organism (blood, organs, systems, tissues and fat) has only just provided evidence of actual deposition of TCDD in the body but the level which is liable to produce toxicity in man is not known.

SAFETY AND HEALTH MEASURES

The experience of occupational exposure to TCDD, either from an accident during the production of trichlorophenol and its derivatives or originating from regular industrial operations, has shown that the injuries sustained may completely incapacitate workers for several weeks or even months. Resolution of the lesions and healing can occur, but in several cases skin and visceral lesions can linger on and reduce working capacity to 20-50% for more than 20 years. Prevention of TCDD toxicity can be obtained if the chemical processes concerned are carefully controlled. By good manufacturing practice it is possible to eliminate the risk of exposure of workers and applicators handling the products or for the population at large. In case of an accident, i.e. if the process of synthesis of 2,4,5-trichlorophenol is running out of control and high levels of TCDD are present, contaminated clothing should immediately be removed, avoiding contamination of the skin or other parts of the body. Exposed parts should be washed immediately and repeatedly until medical attention is obtained. For workers engaged in the decontamination process after an accident, it is recommended that they wear complete throw-away equipment to protect the skin and prevent exposure to dust and vapours from the contaminated materials. A gas mask should be used if any procedure that may produce inhalation of airborne contaminated material cannot be avoided.

All workers should be obliged to take a shower daily following the work shift. Street clothes and shoes should never come in contact with work clothes and shoes. Experience has shown that several wives of workmen affected by chloracne developed chloracne too, although they had never been in a plant producing trichlorophenol. Some of the children had the same experience. The same rules about safety for workmen in case of accident have to be borne in mind for laboratory staff working with TCDD or contaminated chemicals and for medical staff such as nurses and assistants who treat injured workmen or contaminated persons. Animal keepers or other technical personnel coming in contact with contaminated material or with instrument and glassware used for TCDD analysis must be aware of its toxicity and handle the material accordingly. Waste disposal including carcasses of experimental animals requires special incineration procedures. Glassware, benchtops, instruments and tools should be regularly monitored with wipe tests (wipe with filter paper and measure amount of TCDD). TCDD containers as well as all glassware and tools should be segregated and the whole working area should be isolated.

For the protection of the general public and especially of those categories (applicators of herbicides, hospital staff, etc.) more exposed to potential risk, the regulatory agencies throughout the world enforced in 1971 a maximum manufacturing specification of 0.1 ppm TCDD. Under constantly improving manufacturing practice, recent commercial grades of the products contain today (1980) 0.01 ppm of TCDD or less.

This specification is intended to prevent any exposure to and any accumulation in the human food chain of amounts which would pose a substantial risk for the individual. Furthermore, to prevent contamination of the human food chain even of the extremely low concentration of TCDD which might be present on range or pasture grasses immediately following 2,4,5-T application, grazing of dairy animals on treated areas has to be prevented for one to six weeks following application.

Acceptable daily intake and threshold limit value

The "acceptable daily intake" (ADI) is the amount which can be regularly or intermittently absorbed through all possible routes in small amounts or small concentrations for the entire life-span via diet or any other means of environmental contamination. Recently (1980) the Scientific Advisory Panel of the US Environmental Protection Agency has stated that the dose of 0.001 μg/kg/day of TCDD is for all practical purposes a "no observable effect level" including carcinogenic, teratogenic and reproductive risk. At the same time the US Food and Drug Administration has issued a recommendation that fish containing TCDD at a level of more than 100 ppt should not be consumed.

For 2,4,5-T a Joint Committee of the Food and Agriculture Organisation and the World Health Organisation has stated (1980) that the "no-effect level" is 3 mg/kg/day and that the ADI for man is 0.003 mg/kg/day when the TCDD content of 2,4,5-T is 0.05 ppm. No similar calculation has been made for hexachlorophene. The US Environmental Protection Agency has calculated that, taking into account the cumulative exposure (oral, dermal and inhalation exposure), a worker would receive 7 mg/kg of 2,4,5-T and 0.0007 μg/kg of TCDD, which is definitely below the "no observable effect level" for TCDD. Whether these values can be accepted as a permissible limit of exposure at the workplace remains at the moment a matter of controversy.

Disposal of contaminated wastes and residues

Degradation of TCDD can be obtained in the laboratory by physicochemical methods (ultraviolet light or gamma rays in the presence of a H donor), chemical methods (ozonisation, ruthenium salts, etc.) and microbiological methods. None of these has been sufficiently tested in field conditions as yet. The other methods of disposing of wastes and residues containing TCDD are the following.

Incineration on land. Basically such a system includes a large material handling building, tanks for storage of waste materials, a large rotatory kiln with combustion chamber, high-energy scrubber for control of air pollution, waste water treatment facilities and the necessary accessory equipment. A burner at the front of the rotating kiln burns the wastes pumped from the tank to maintain a minimum temperature of 800-1 200 °C. The material has to be exposed to that temperature for times exceeding 30 seconds.

Incineration at sea. This technique is basically the same as above and has been demonstrated on clean chlorinated wastes on the ship *Vulcanus* with a practically 100% combustion efficiency. It cannot handle sludge, slurries and certain types of residues. Impact of fall-out on marine life is said to be negligible.

Vaulting. Vaults are specially constructed underground facilities for preserving containers of dangerous wastes in perpetual care.

Dumping. In case of contaminated soil the procedure consists of excavating a layer of soil to a depth of 20-40 cm, sealing it in proper containers and burying these at a site remote from drainage systems and other potential sources of human exposure and where subsequent excavation does not occur. Prior to burial, the containers should be properly marked with respect to their residual contents and toxicity. The same can be done for reactor, storage tank, pipes and other parts of the equipment and installation that are contaminated.

REGGIANI, G.

Topics in environmental health. Halogenated biphenyls, terphenyls, naphthalenes, dibenzodioxins and related compounds. Kimbrough, R. D. (ed.). (Amsterdam, Elsevier Biomedical Press, 1980).

CIS 80-475 "The mortality experience of workers exposed to tetrachlorodibenzodioxin in a trichlorophenol process accident". Zack, J. A.; Suskind, R. R. *Journal of Occupational Medicine* (Chicago), Jan. 1980, 22/1 (11-14). 15 ref.

Long term hazards of polychlorinated dibenzodioxins and polychlorinated dibenzofurans. Report of a joint NIEHS/IARC working group meeting (Lyons, 10-11 January 1978). IARC internal technial reports No. 78/001 (Lyons, International Agency for Research on Cancer, 1978).

IARC monographs on the evaluation of the carcinogenic risk of chemicals to man. Vol. 15: *Some fumigants, the herbicides 2,4-D and 2,4,5-T, chlorinated dibenzodioxins and miscellaneous industrial chemicals* (Lyons, International Agency for Research on Cancer, Aug. 1977), 354 p.

Diphenyls and terphenyls

Diphenyl ($C_{12}H_{10}$)

PHENYLBENZENE; BIPHENYL; XENENE

m.w.	154
sp.gr.	0.87
m.p.	71 °C
b.p.	256 °C
v.d.	5.3
v.p.	1 mmHg ($0.13 \cdot 10^3$ Pa) at 71 °C
f.p.	112.8 °C
e.l.	0.6-5.8%
i.t.	540 °C

insoluble in water; soluble in ethyl alcohol, ether and benzene; it has a great thermal stability

colourless to white-yellow leaflets, with a pleasant aromatic odour.

TWA OSHA	0.2 ppm 1 mg/m³
STEL ACGIH	0.6 ppm 4 mg/m³
IDLH	300 mg/m³

Production. It is produced by passing benzene vapours over ferroso-ferric or lead oxide at high temperature, in small quantities. It occurs naturally in coal-tar.

Uses. It is used as a heat transfer agent and in organic synthesis for manufacturing chloro-nitro and amino derivatives. It is also used in oil refinery and as an adjuvant for colouring polyester fibres. It acts as a fungicide for the treatment of oranges and is applied to the inside of shipping containers or wrappers for this purpose. It is the starting point for manufacturing PCB. Benzidine is perhaps its most important derivative.

Phenyl ether ($C_{12}H_{10}O$)

DIPHENYL ETHER; DIPHENYL OXIDE

m.w.	170
sp.gr.	1.07
m.p.	28 °C
b.p.	259 °C
v.p.	1 mmHg ($0.13 \cdot 10^3$ Pa) at 66 °C
f.p.	115 °C
e.l.	0.8-1.5%

insoluble in water; soluble in ethyl alcohol and ether

colourless cyrstals with a geranium-like odour.

TWA OSHA	(vapour) 1 ppm 7 mg/m³
STEL ACGIH	2 ppm 14 mg/m³
MAC USSR	5 mg/m³

Production and uses. It is produced by heating sodium phenolate with chlorobenzene, and is used as a heat transfer agent; in organic synthesis; and in perfume manufacture.

Diphenylmethane ($(C_6H_5)_2CH_2$

BENZYLBENZENE; DITAN

m.w.	168.2
sp.gr.	1.0
m.p.	25.3 °C
b.p.	264.3 °C
v.d.	5.8
v.p.	1 mmHg (0.13·10³ Pa) at 76 °C
f.p.	130 °C
i.t.	483 °C

insoluble in water; soluble in ethanol and ethyl ether
orthorhombic needles or a liquid with an odour of oranges.

Production. It is produced by Friedel-Crafts' reaction from benzene and dichloromethane in the presence of aluminium trichloride as a catalyst, or from benzene and formaldehyde in the presence of sulphuric acid.

Uses. It is used as a perfume in the soap industry and as a solvent for cellulose lacquers. It has also some applications as a pesticide. It is the starting point for the manufacture of hexachlorophene. Some of its derivatives (including auramine O) are used as dyes.

1,1-Diphenylethane ($(C_6H_5)_2CHCH_3$)

m.w.	182
sp.gr.	0.99
m.p.	−20 °C
b.p.	272 °C
v.d.	6.28
f.p.	129 °C
i.t.	440 °C

a pale to yellow liquid with an aromatic odour.

1,2-Diphenylethane

sp.gr.	1.82
m.p.	52.2 °C
b.p.	285 °C
v.d.	6.29
v.p.	1 mmHg (0.13·10³ Pa) at 87 °C
f.p.	129 °C
i.t.	480 °C

insoluble in water; very soluble in ether and carbon disulphide
colourless prismatic crystals.

Phenyl ether-biphenyl mixture

DOWTHERM A; DINYL

m.w.	166
sp.gr.	1.03
m.p.	12.2 °C
b.p.	257 °C
v.p.	0.8 mmHg (0.10·10³ Pa) at 25 °C
f.p.	124 °C oc
e.l.	0.5-6.2%

a colourless to straw-coloured liquor with a pungent odour.

TWA OSHA	1 ppm 7 mg/m³	vapour
TLV ACGIH	0.5 ppm 4 mg/m³	vapour
STEL ACGIH	2 ppm 16 mg/m³	vapour
MAC USSR	10 mg/m³ (mixture of 75% phenyl ether and 25% biphenyl)	

Uses. It is used as a heat transfer agent in the liquid or vapour phase at temperatures below 380 °C mainly in the manufacture of synthetic rubber, artificial fibres, plastics and pigments.

Terphenyls ($C_{18}H_{14}$)

DIPHENYL BENZENES; TRIPHENYLS

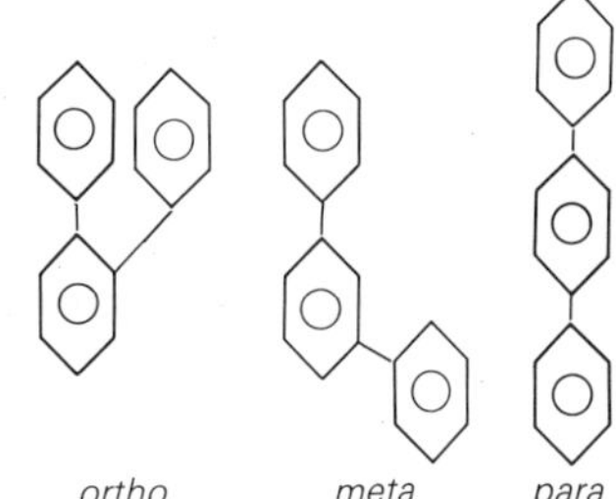

	ortho	meta	para
m.w.	230.3	230.3	230.3
sp.gr.	1.14	1.16	1.23
m.p.	56.2 °C	87.4 °C	
b.p.	332 °C	363 °C	405 °C
v.d.	7.95	7.95	7.95
f.p.	162.8 °C oc	190.5 °C oc	207 °C oc
	prismatic crystals	yellowish needles	yellowish needles

insoluble in water; soluble in ethanol and ethyl ether; commercial preparations contain a mixture of the three isomers

TWA OSHA	1 ppm 9 mg/m³ ceil
TLV ACGIH	0.5 ppm 5 mg ceil
IDLH	3 500 ppm
MAC USSR	5 mg/m³ for a mixture of 63% ortho, 19% meta, 15% diphenyl

Uses. Terphenyls are used as heat transfer fluids and nuclear reactor coolants.

HEALTH HAZARDS

Little information is available about the toxic effects of diphenyl and its derivatives with the exception of polychlorinated diphenyl (PCB). Owing to their low vapour pressure and smell, exposure by inhalation at room temperature does not usually entail a serious risk. However, in one observation, workers engaged in impregnating wrapping paper with a fungicide powder made of diphenyl experienced bouts of coughing, nausea and vomiting. In repeated exposure to a solution of diphenyl in paraffin oil at 90 °C and airborne concentrations well above 1 mg/m³ one man died of acute yellow atrophy of liver and eight workers were found suffering from central and peripheral nervous damage and liver injury. They complained of headache, gastrointestinal disturbances, polyneuritic symptoms and general fatigue; they presented increased serum transaminases, EEG and electroneuromyographic alterations, but no blood or kidney injuries. Molten diphenyl can cause serious burns. Skin absorption is also a moderate hazard. Eye contact produces mild to moderate irritation. Processing and handling of diphenyl ether in ordinary use involves little health hazard. The odour may be very unpleasant, and excessive exposure results in eye and throat irritation. Contact with the substance can produce dermatitis.

The mixture of diphenyl ether and diphenyl at concentrations between 7 and 10 ppm does not seriously affect experimental animals in repeated exposure. However, in humans it can cause eye and airways irritation and nausea. Accidental ingestion of the compound resulted in severe impairment of liver and kidney.

Terphenyl vapours cause conjunctival irritation and some systemic effects. In experimental animals *p*-terphenyl is poorly absorbed by oral route and appears to be only slightly toxic; *meta*- and especially *ortho*-terphenyls are dangerous to the kidney and the latter can also impair liver functions. Morphologic alterations of mitochondria (the small cellular bodies holding respiratory and other enzymatic functions essential to biological synthesis) have been reported in rats exposed to 50 mg/m^3. Heat transfer agents made of hydrogenated terphenyls, terphenyl mixture and isopropyl-*meta*-terphenyl produced in experimental animals functional changes of nervous system, kidney and blood with some organic lesions. A carcinogenic risk has been demonstrated for mice exposed to the irradiated coolant, while the non-irradiated mixture appeared to be safe.

In man terphenyl coolant is a primary skin irritant; airborne concentrations above 10 mg/m^3 produced eye and airways irritation. In 1978 the ACGIH recommended a TLV of 0.5 ppm on account of possible mitochondrial changes.

SAFETY AND HEALTH MEASURES

Diphenyl and its derivatives are moderately flammable and can react with oxidising materials. The danger of fire is negligible at ambient temperature but increases at high temperatures, when the materials are dispersed in the air. These substances must therefore be stored in a cool, well ventilated place, far from open flames and areas where the risk of fire exists. Carbon dioxide and carbon tetrachloride are recommended as fire-fighting materials. Since absorption of diphenyl and its derivatives occurs principally through inhalation, the atmospheric concentration in work areas must be maintained below the exposure limits. The risk of inhalation is low unless these substances are subject to high temperatures, when the process should be enclosed or mechanical exhaust ventilation provided. In the event of spillage or leakage of hot fluids, approved chemical cartridge respirators or masks should be worn for protection against vapours. Skin contact with terphenyls and possibly with diphenyl and derivatives should be avoided by proper handling procedures. Contaminated clothing must be laundered before re-use. Eye protection should be worn if an eye hazard exists.

Treatment. If skin contact with these substances occurs, wash the skin immediately with soap and water. If clothing has been contaminated, remove it promptly. Diphenyl in the eyes should be removed by irrigating with water for at least 15 min. Burns from splashes of liquefied compounds require prompt medical attention. In case of severe exposure, the patient should be taken into the fresh air for rest until the arrival of a physician. Give oxygen if the patient appears to have difficulty in breathing. The majority of persons quickly recover in fresh air, and symptomatic therapy is rarely required.

PARMEGGIANI, L.

"Diphenyl poisoning in fruit paper production—A new health hazard". Häkkinen, I.; Siltanen, E.; Hernberg, S.; Seppäläinen, A. M.; Karli, P.; Vikkula, E. *Archives of Environmental Health* (Chicago), Feb. 1973, 26/2 (70-74). 11 ref.

CIS 74-449 "The LD_{50} and chronic toxicity of reactor terphenyls". Adamson, I. Y. R.; Weeks, J. L. *Archives of Environmental Health* (Chicago), Aug. 1973, 27/2 (69-73). Illus. 14 ref.

CIS 77-1954 "Establishment of the TLVs for hydrogenated terphenyls, a terphenyl mixture and isopropyl-*m*-terphenyl in workplace air" (K obosnovaniju PDK terfenilov GTF, TFS i IPMTF v vozduhe rabočej zony). Hromenko, Z. F. *Gigiena truda i professional'nye zabolevanija* (Moscow), Dec. 1976, 12 (42-44). 6 ref. (In Russian)

CIS 74-1055 "A study of the carcinogenicity for skin of a polyphenyl coolant". Henderson, J. S.; Weeks, J. L. *Industrial Medicine and Surgery* (Fort Lauderdale), Feb. 1973, 42/2 (10-21). Illus. 44 ref.

Dirty occupations

"Dirty" is not a precise term, and any attempt to define "dirty occupations" is bound to be governed by traditions and subjective judgement—even though legislation in certain countries lays down specific measures for this type of work. There are, however, many occupations in which workers' bodies or clothes are contaminated through either direct contact with powders, liquid or viscous materials or through contact with machinery or other objects coated with such substances. Airborne dust may also be a source of contamination.

It is sometimes difficult to draw a clear line between dirt which is merely offensive or obnoxious and dirt which may carry a definite health risk. Even so, it should not be assumed that because no specific risk has yet been proved there is such a thing as "healthy dirt", and any assumption that dirt carries no health risk should be checked against the latest knowledge and experience.

Types of dirty work

Traditionally "dirty" occupations include coal mining, coal processing, coal transport and delivery, chimney-sweeping, boiler- and flue-cleaning, cleaning and maintenance of oil-tankers, oil containers, etc., operations in graphite and rubber works, some work in abattoirs, fish meal and fertiliser manufacture, and loading and unloading of some cargoes at docks.

HAZARDS

Prolonged exposure of the skin to these sources of contamination may lead to various diseases or disorders which, depending on the properties and composition of the dirt, may include pigmentation of the skin and adnexa, photosensitivity, hyperkeratosis and even skin tumours.

Skin pigmentation that resists removal by normal washing techniques (soap and water) is encountered in handling a variety of products. Tetryl may produce green or yellow pigmentation; picric acid turns the hair, nails and skin of the wrists yellow in dark subjects, and the hair of fair subjects green; contact with substances containing tannin turns the skin brown. This pigmentation is limited mainly to the horny layer of the skin and seldom penetrates to the subjacent layers. In some cases it may be accompanied by rashes and fissuring.

The distillation products of petroleum and coal tar may cause folliculitis, photodermatitis and tumours. Cases of folliculitis are encountered in petroleum refinery workers, in engineering industry workers exposed to grinding and cutting fluids and kerosene, in workers employed on the coating of wood with creosote preservative, and in workers using pitch and tar in the manufacture of roofing felts, etc. Photosensitivity can be caused by contact with oil and coal tars, medium and heavy oils and pitch; these substances increase the skin's sensitivity to sunlight which may then produce photodermatitis on exposed parts of the body. The dermatitis is often accompanied by inflammation of the mucous membranes of the mouth and eyes; these manifestations disappear after a few days but a diffuse pigmentation of the exposed parts of the skin will persist.

Wart-like growths and hyperkeratosis, which may become neoplastic, may be caused by petroleum and petroleum products, and by coal tar and a number of its distillates. Studies of workers engaged in the processing of kerosene fractions and exposed to petroleum tars have revealed: brown pigmentation, dryness and burning of the exposed skin, comedones, warts on the face, hands and other areas of the body, and cases of planocellular cancer. The earlier literature reports on cases of scrotal cancer amongst chimney sweeps whose clothes were constantly contaminated with soot.

SAFETY AND HEALTH MEASURES

The first essential when dealing with dirty occupations is to overcome a common attitude amongst both workers and employers that dirt is an unavoidable part of certain trades or industries, and that the need to work in dirty conditions is adequately compensated by the payment of safety incentives or "dirty money".

Technical measures. Dirty processes should be mechanised and enclosed to prevent the formation of dust and, where there is the likelihood of dust escaping from charge or discharge openings, etc., exhaust ventilation should be installed. The application of water sprays to materials such as pitch will prevent the formation of airborne dust. Materials should be conveyed inside the plant by mechanical handling equipment such as pneumatic conveyors, and materials that are to be transported should be packaged in impermeable sacks or containers or carried in closed bulk transporters. Where dirty substances have to be handled by workers, appropriate tools should be supplied such as long-handled shovels or long-handled brushes.

Personal protection. Workers who are employed in dirty occupations should be supplied with working clothes such as overalls or aprons, which should be changed regularly and laundered at the employer's expense. Depending on the type of exposure, it will also be necessary to supply workers with hand and arm protection, foot and leg protection, respiratory protective equipment and head protection. Personal protective equipment should be made of materials which resist the penetration of the dirt, for example of plastics, rubber, oiled-cloth, etc. The use of barrier creams and lotions may prove valuable in many instances.

Personal hygiene amongst workers should be of the highest standard and there should be adequate washing and sanitary facilities including showers and baths. Workers should be encouraged to wash thoroughly at the end of each shift and before meal breaks, and the employer should allow time for this operation. Separate lockers should be provided for work and street clothes. Eating, drinking and smoking at the workplace should be forbidden. A clean and separate messroom or canteen should be provided for meal breaks and rest periods. First-aid equipment should be kept free from contamination and first aid should be administered with scrupulous attention to cleanliness.

The composition of skin cleansing materials will vary depending on the type of contamination. Industrial oils, fuel oils, paints and soot can be removed by a composition comprising 48% ordinary white clay, 8% river sand, 8% kerosene, 32% water, 2.4% sulphuric acid (65-66% concentration), 1.6% oil of vitriol. For badly contaminated hands, use may be made of a paste comprising: 16% alkyl sulphate, 1% sodium hydroxide, 0.5% carboxy methylcellulose, 6% glycerol, 50% powdered pumice, 26.5% water.

Effective synthetic soaps include those made from alkyl phenyl ethers of polyethylene glycol. A suitable cleansing paste for workers in the engineering and painting industry comprises 80 parts of 3% sodium hydroxide, 24 parts glycerol and 16 parts paraffin wax. Pigmentation due to picric acid can be prevented by washing the skin in warm whey. Hands contaminated by tar, paint, etc., can be cleaned with a compound comprising 45% liquid soap, 45% powdered pumice, 5% glycerol and 5% denatured ethyl alcohol.

A compound which has proved effective in the prevention of photodermatitis is a paste comprising 15% resin, 10% phenylsalicylate, 75% denatured ethyl alcohol.

[Ideal cleansers should contain no abrasives or strong alkali and should be neither irritant nor toxic. Because of the potential damaging effect of almost all powerful skin cleansers, their frequent use should be avoided and the utmost care should be taken to prevent skin contamination by means of appropriate protection.]

SADKOVSKAJA, N. I.

Skin cleansing:

CIS 126-1968 "A symposium on skin cleansing". *Transactions of the St. John Hospital Dermatological Society* (London), 1965, 51/2, 126 p. Illus. 93 ref.

CIS 2122-1971 "The use of abrasive products as skin-cleansing agents in industry" (Berufliche Hautreinigung mit abrasiven Präparaten). Tronnier, H.; Martin, U. *Arbeitsmedizin–Sozialmedizin–Arbeitshygiene* (Stuttgart), May 1971, 6/5 (108-110). 6 ref. (In German)

Disability evaluation

(For the definition of disability see under DISABILITY PREVENTION AND REHABILITATION.) Modern social legislation contains provisions for protecting workers and their families from the economic consequences of disability and for compensating workers for the economic loss that results from an occupational accident or disease (see SOCIAL SECURITY; SOCIAL SECURITY AND OCCUPATIONAL RISKS).

Social protection of this type is extremely widespread and is included in even the most primitive forms of legislation. The scope and extent of protection do, however, vary from country to country; in addition, there is constant development in the degree of protection afforded and, in many countries with advanced social legislation, social security in this respect has virtually been achieved. The object of this social protection is not so much to safeguard or repair the health of the worker as to ensure that the worker's income is maintained at an adequate level even though his earning capacity (i.e. his ability to work) is reduced or eliminated either temporarily or permanently. Consequently, to ensure that social assistance is provided at the correct level and for the required period of time, it is necessary to evaluate the degree of disability and the probable duration of this disability accurately.

Duration of disability

Incapacity for work may be defined as temporary or permanent. The term "temporary incapacity" is used in the case of a pathological condition that prevents the worker from following his normal occupation, and thus designates a specific and not a general condition. This provides confirmation of the statement that this form of social protection is concerned less with the morbid condition than with the resultant loss of earning capacity. A worker may, of course, suffer from an injury or illness which does not prevent him following his normal occupation and in such a case there is no right to the benefit payable for temporary disability, which usually takes the form of a daily allowance. In certain countries there is a waiting period (either mandatory or contingent) before benefit becomes payable; the waiting period is contingent if there is provision to make payments retroactively for the waiting period once the contingency deadline has been exceeded. Finally, in most cases of temporary disability, payments are made only where the disability is total; however, in some countries, such as Belgium, there are reduced benefits for partial disability.

Permanent disability or disablement—whether total or partial—means loss or reduction of working capacity, productivity or ability to earn, or, as it has been defined, loss or reduction of biological earning capacity due to the virtually permanent consequences of a pathological event causally related to an occupational accident or disease.

The task of distinguishing between temporary and permanent incapacity may sometimes prove to be very difficult, the more so because the terms temporary and permanent have not been given precise and absolute meanings, but only meanings related to prognosis based on serious consideration. In fact, many legislations provide for the periodical revision of benefit payments, either upwards or downwards, until restoration of the original working capacity following on changes in pathological conditions that were nevertheless originally declared practically permanent. However, it is possible to identify the dividing line between temporary and permanent incapacity, since the main criterion of temporary incapacity is the fact that it is limited in time to a period which, although varying in length, can still be estimated approximately; hence, incapacity may be considered permanent when its duration can no longer be reliably estimated. The following example will render this concept more comprehensible: chronic osteomyelitis may be considered the cause of incapacity of unlimited duration if the condition is found to be unresponsive to medical and surgical therapy; the condition must therefore be considered one of permanent disability although this does not mean that a cure may not one day be found.

Degree of disability

The basic principle that the worker should be compensated for the economic loss he incurs should be achieved in the case of permanent incapacity by paying the worker the real difference between his incomes before and after the pathological event. However, due to practical difficulties, compensation does not deal with the matter in this way.

In a system employing collective bargaining, a decline in fitness is hardly ever accompanied by a decline in remuneration, and employment and level of remuneration depend above all on the state of the labour market. Consequently, it has been found preferable to base compensation not on the effective reduction in earnings, but on the nature and importance of the residual anatomical and functional consequences of the accident or illness, and their effect on working capacity, and to assume that there will be a corresponding reduction in earning capacity. This method has the advantage of basing the evaluation on the objective anatomical and functional criteria constituted by the injury, and hence on a concrete datum that can be verified at any time; in addition, although not perfect, it has nevertheless proved highly practicable.

Disability ratings. This method may be applied in various ways: one can ignore differences between accidents causing equivalent injuries, or one can assign to each injury a value in relation to the importance of the organ affected in the injured person's execution of his work. Opinions diverge as to whether evaluation of loss should be based on work in general or on a specific activity, i.e. on the subject's ability to earn his living doing non-specialised manual work requiring only average working capacity, or on his ability to continue in the occupation he pursued before his accident or illness. The use of the abstract concept of "work in general" will tend to give arbitrary results; this tendency will increase as modern industrial technology demands increasingly specialised workers. On the other hand, the use of a specific activity as a criterion also presents problems; firstly, it makes it impossible to draw up disability ratings for different injuries and, secondly, it produces inconveniences and inconsistencies in cases where, for example, specific working capacity is totally destroyed whilst general capacity remains almost intact.

Perhaps a better system is one that classifies work in large categories on the basis of the main human capabilities required for the work, for example physical strength, manual dexterity or intelligence. Nevertheless, with all systems there remains the problem as to whether evaluation of the degree of disability should be left to the judgement of an assessor, or whether there should be a set schedule of ratings for at least the most common injuries. This latter system is antiscientific, empirical and arbitrary; it takes no account of the victim's personality and is concerned only with the disability and not with the disabled person. Nevertheless, practical requirements such as the need for uniformity and impartiality have usually resulted in a system of standard disability ratings and have led to the incorporation of disability schedules into the legislation of many countries. Criteria for the formulation and application of these schedules vary considerably. In countries such as Italy they are mandatory whereas, for example, in other countries in the European Community they are only indicative. However, in these latter countries the degree of disability, as determined by the schedule, may be raised by the application of a special weighting factor that makes it possible to take into account, in addition to the nature and severity of the injury or the illness, the subject's age, his psychological and intellectual qualities, his aptitudes, and his occupational qualifications. This may be considered a reasonable attempt to make disability evaluation less arbitrary, less rigid and better fitted to viewing the effects of an occupational accident or disease on a subject as a whole.

All disability schedules contain disability ratings for the pathological, anatomical or functional effects of accidents or diseases expressed as percentages of the theoretical capacity of a subject of "normal" capability and efficiency (see TRAUMATIC INJURIES). For example the disability rating for the loss of the right thumb is 25% in Belgium, France and Luxembourg, 20% in the Netherlands and the Federal Republic of Germany and 28% in Italy. Nevertheless, no schedule can cover all the possible injuries and, consequently, there continues to be scope for the assessor to exercise his discretion. In such a case the assessor should endeavour to estimate precisely the extent of anatomical and functional damage so that, in the light of experience and precedent, he can equate it correctly with the appropriate percentage of the theoretical total working capacity of the healthy subject.

Multiple injuries. In the case of multiple injuries, some schedules indicate special degrees of disability (e.g. for loss of all or of various combinations of fingers, eyes, etc.). Where no such information is provided in the schedule, two alternatives for assessment are possible:

(a) the injuries are cumulative, i.e. affect the same limb, organ, organic system or functional system or different systems with mutual effects (two eyes, two lower limbs, sight and hearing, etc.); or

(b) the injuries are merely coexistent, i.e. have no functional correlation (a finger and a toe, a finger and sense of smell, etc.).

In the case of cumulative injuries, it is necessary to make a global evaluation that takes into account the full extent to which the functioning of the organ or system concerned is diminished, allowing 100% for organs or systems whose functioning is in the slightest degree indispensable for the performance of any work, and when it is not, referring to the different values in the scale, or in default of a scale, taking the value indicated by custom and experience. There are those who argue, however, that one should never take the arithmetical sum of the different percentages of incapacity, for a higher total percentage can be arrived at without it. For instance, in different scales, the loss of one eye may be evaluated at 25 or 35%, but the loss of both at 100%; the loss of a thumb may be rated at 25% and a finger at 15%, but the two together, at 42%. Thus, if multiple injuries are to be correctly evaluated, due regard must be paid to their mutual interactions with the functions of members and organs; depending on the circumstances, the result may be either equal to, greater than or less than the arithmetical sum of the individual partial values.

In the case of mere coexistence of injuries, it is widely agreed that the different degrees of incapacity are not additive and the method employed for evaluating total disability usually is that proposed by Balthazard (and also recommended for the official French schedule). The procedure for the application of this method is as follows: when a number of injuries have to be evaluated together, the first injury is assessed at the percentage given in the schedule, and each subsequent injury as a percentage of residual capacity. For example in an evaluation of three injuries which, separately, would represent degrees of disability of 30, 15 and 10%, the first injury will be given its full value of 30%, leaving a residual capacity of 70% (100-30); the second is assessed as a percentage of this 70%, that is 15% of 70 = 10.5%. The residual capacity is now 59.5% (100-30-10.5) and the third injury is assessed at 10% of this, i.e. at 5.95%. The final assessment of disability is thus: 30 + 10.5 + 5.95 = 46.4%, which will be rounded off to 46%. The corresponding arithmetical sum of the three values would be: 30 + 15 + 10 = 55%.

This method has been criticised as being empirical and arbitrary; however, it may be argued in its favour that, although imperfect, it has proved suitable enough for its purpose.

In cases of multiple injury where the arithmetical sum of the partial degrees of incapacity exceeds 100, a method is necessary that would give a reduction of capacity equal to, but never exceeding, 100. It has, of course, been pointed out that, with the Balthazard method, an evaluation may approach, but never quite reach, 100% disability, even when the individual degrees are very high. This objection is unfounded since all fractions are rounded off and, what is most important, the results of the calculations must be considered to have only an indicatory value. It is, nevertheless, true that any method for calculating compensation for disability will be both imperfect and arbitrary, since the quantitative evaluation of a biological fact, which is often subject to change and which results from the most varied interactions of elements, is difficult if not impossible.

For this reason, fine percentage differentiations should be avoided since one is dealing with conventional values and uncertain data that cannot be precisely and objectively determined, if only because of the lack of sufficiently sensitive and accurate instruments for making scientific measurements. Accordingly, in the jurisprudence of several States, there are minima for differences between percentages, amounting in practice to a range of 5 to 10.

The evaluation of incapacity in the presence of a pre-existing incapacity has features in common with the evaluation of multiple injuries. Here too, a distinction must be made between cases in which the injury caused by the accident and the pre-existing injury are cumulative and those in which they are merely coexistent. Where there is cumulation, the direct consequences of the accident will be more serious than if the victim had not been previously incapacitated and, consequently, the evaluation of the incapacity should always be based not on the theoretical capacity of a normal individual (100%), but on a reduced capacity resulting from the previous disability. This criterion has found a place in the jurisprudence of several States, and even in the legislation of some, such as Italy.

In evaluating disability in such a case, it is necessary to determine capacity both before and after the injury; the compensatable disability will be the difference between the two. If C is the capacity before the accident under consideration and C_1 that after the accident and calculated by adding C to the consequences of the accident, the reduction in capacity is given by the formula:

$$\frac{C-C_1}{C}$$

For example, a person who has lost a finger of one hand and consequently has a degree of disability of 15%, subsequently suffers an anatomical or functional injury to the same hand or the other hand, resulting in 20% disability and giving total disability of 35%, then:

$$\frac{C-C_1}{C} = \frac{(100-15)-(100-35)}{100-15} = \frac{(85-65)}{85} = \frac{20}{85}$$

If x is the percentage reduction:

$$20 : 85 = x : 100$$

$$x = \frac{20 \times 100}{85} = 22.52 = 23 \text{ to the nearest integer.}$$

The criticism has been made that this and other formulae give results that are obviously paradoxical and erroneous. Their application may also be arbitrary since, in some cases, it will be difficult to determine the previous degree of incapacity if its characteristics have been obscured by the new accident. Such disadvantages may be overcome if the results of the calculation are considered only as a useful guide, and the evaluation criteria are not applied too strictly. This recommendation has already been made above; it should not be disregarded when dealing with biological data that are so difficult to quantify.

Total disability. If the disability is total and absolute, the assessment will naturally be 100%. However, the terms "total" and "absolute" should not be taken literally, for it is obvious that totally disabled persons also include those who are merely unable to engage in any gainful activity, and thus cannot take advantage of the small portion of capacity that remains to them. For example, a blind person should always be considered totally disabled even if he manages to find some useful occupation.

In certain countries special provision is made for persons who are so seriously handicapped that they require continuous personal assistance to meet the most elementary requirements of inactive life and human relations. The legislation of other countries also includes schedules for evaluating this extreme degree of disability; however, this measure has been criticised on the ground that so serious and important a decision should not be based on preconstituted formulae that may prove inadequate in practice.

Application to compensation

Disability may be compensated by means of a lump-sum payment or by an annuity; these are dealt with in detail under SOCIAL SECURITY AND OCCUPATIONAL RISKS.

The chief merit of the annuity from the point of view of disability evaluation is that, since compensation is a continuing procedure, there exists the possibility of periodic reviewal. Review procedures and intervals vary from country to country but all ensure that the annuity remains commensurate with the degree of disability. Where compensation is on an annuity basis, the disability assessor's task is considerably simplified since he can limit his evaluation to the existing pathological condition and is not required to estimate future developments. On the other hand, where compensation is on a lump-sum basis, evaluation is complicated by the need to assess "future loss" which may be either improbable, possible, probable or certain. Present practice is usually that "future loss" can be taken into account only if its occurrence is rendered certain by current knowledge of the long-term effects or course of a given disease. It is not justifiable to make allowance for every possible reasonable presumption of loss since this would lead to a situation in which aggravating complications (no matter how vague or uncertain) could be invoked for any disease or injury.

In the case of both lump-sum compensation and annuities, the extent of protection varies from country to country. Some countries (e.g. Belgium, France, Luxembourg, the United Kingdom, etc.) compensate all degrees of disability, others provide for compensation only when disability exceeds a specified threshold ranging between 5 and 20%. The justification for this threshold is that very small reductions in work capacity do not result in economic loss.

Occupational diseases

The evaluation of disability due to occupational disease is usually governed by the same principles as those applied to accident injuries. However, evaluation is more difficult in the case of a disease than in the case of injury since quantification entails not only diagnosis but also determination of severity. It is not possible to draw up schedules for diseases, and assessment will depend both on the amount of information available and on the assessor's critical faculty; consequently, accurate evaluation of disability due to industrial disease demands increasingly refined and accurate instrumentation and particularly gifted assessors.

When evaluating reduced working capacity due to occupational disease, it is essential to determine both the extent of the pathological manifestations and their repercussions on the subject's personality. Allowance should be made for pre-existing factors that may affect the significance of the diagnosis, such as the outcome of recent therapy or the effect of specific, contingent climatic conditions (e.g. chronic bronchitis which becomes acute in a person with pneumoconiosis); in cardiovascular function or lung function tests, allowance must be made for the subject's current physical condition, which will depend on whether he is actively employed on moderate or heavy work, or confined to bed. The evaluation of such factors implies reliable methods of investigation and considerable experience on the part of the assessor.

MARANZANA, P.

General:

Social security and disability. Issues in policy research. Studies and research No. 17 (Geneva, International Social Security Association, 1981), 164 p.

Medical problems:

The evaluation of disability in social security within the European Community (L'évaluation de l'invalidité en sécurité sociale dans la Communauté européenne). Geerts, A. (Paris, Leuven, Brussels, Vander, 3rd ed., 1974), 224 p. (In French)

Compensation for bodily harm. A comparative study. Geerts, A.; Kornblith, B. A.; Urmson, W. J. (Brussels, F. Nathan Editions Labor, 1977), 213 p.

Harmonisation of handicap compensation systems and procedures (L'harmonisation des régimes de compensation du handicap): Report of a study group presided by de Forges, J. M. Reporter: Prigent, M. A. (Rapport d'un groupe d'etude présidé par de Forges, J. M. Rapporteur: Prigent, M. A.). (Les publications CTNERHI, 27 Quai de la Tournelle, Paris) (1979), 146 p. (In French)

Comparative study of methods and criteria officially used in France for the assessment of handicap and disability (Etude comparative des méthodes et critères utilisés officiellement en France pour l'évaluation des handicaps et des invalidités). Dambielle, B.; Lecocq, J.; Lesaffre, V. (Les publications du CTNERHI, 27 Quai de la Tournelle, Paris) (1979), 201 p. (In French)

Disability prevention and rehabilitation

The disability process originating from a disease, accident or congenital disorder may be described in the following way:

Disease } → impairment → disability
Accident }
Congenital disorder } → handicap

Impairment is defined as any loss or abnormality of psychological, physiological or anatomical structure or function. In principle it reflects disturbance at the level of the organ. Examples of impairments are: loss of a leg or part of it, stiffness of a joint, hearing loss, visual disturbance due to cataract, etc.

Disability is any restriction or lack (resulting from an impairment) of ability to perform an activity in the manner or within the range considered normal for a human being. Disability reflects disturbances at the level of *the person*, concerning customarily expected activity performance and behaviour. Examples are inability to walk or awkward, inefficient gait; communication problems due to speech disturbances or hearing loss; difficulties in self-care and daily living activities, such as bathing, dressing, feeding; strange behaviour; mobility problems due to blindness or physical impairment; prompt fatigue and diminished endurance; poor memory, poor sensori-motor co-ordination, incapacity to control own work, to plan task and do it, etc.

Handicap is a disadvantage for a given individual, resulting from an impairment or a disability, that prevents or limits the fulfilment of a role that is normal (depending on age, sex, and social and cultural factors) for that individual. Handicap reflects the consequences for the individual–cultural, social, economic and environmental–that stem from the presence of impairment and disability (the interaction and inter-relation between disabled individuals and their social and physical environment).

Examples are:

(a) orientation handicap: the individual's inability to orient himself in relation to his surroundings, including interaction with surroundings;

(b) physical independence handicap: reliance on aids and assistance of others in self-care and daily living activities;

(c) occupational handicap: problems with employment, domestic role, play or recreation;

(*d*) social integration handicap: problems (difficulties) in participation in social life and maintaining of customary social relations due to both individual's impairment disability and the physical environment, including social and physical barriers (attitudes, behaviour of others);

(*e*) economic self-sufficiency handicap: inability or restricted ability to sustain socioeconomic activity and independence; economic reliance to sustain himself and/or inability to sustain others, such as family members.

Disability prevention

Disability prevention is defined as all the measures aimed at preventing the occurrence of impairments (first level prevention), at preventing, limiting or reversing disability caused by impairment (second level prevention) and at preventing transition of disability into handicap (third level prevention). Disability prevention includes all types of health, social, vocational, educational, legislative and other interventions. It should be pointed out that the best results will be achieved only if all these interventions are combined. These interventions consist in:

(*a*) acting upon the *individual* directly;

(*b*) acting upon the *individual's immediate surroundings*: family, community, employer's attitudes and behaviour towards the individual;

(*c*) acting with the broad aim of reducing risks occurring in *society as a whole*.

First level prevention. First level (or primary) prevention includes general education and health education, hygiene and sanitation, vaccinations, legislation to diminish occupational hazards and accidents, work safety measures, improved nutrition and health care, etc.

Second level prevention. Second level (or secondary) prevention comprises early detection and treatment of diseases and disorders, including occupation-related ones: early exercises and ambulation; reduction of continued exposure to hazardous agents or other risk factors; vocational and educational counselling; provision of suitable work, etc.

Third level prevention. Third level (or tertiary) prevention comprises therapeutic measures including surgery, physiotherapy, occupational therapy, speech therapy, etc.; training in self-care; provision of technical aids; social and vocational counselling and guidance; vocational training; provision of education and suitable jobs; education of the public to improve attitudes and behaviour towards the disabled; elimination of physical barriers, etc.

Second level prevention and especially third level prevention measures overlap with those utilised in rehabilitation. In fact, these interventions are used to prevent, reduce or reverse disability when it is preventable or reversible, and to rehabilitate and care for the disabled when disability is established and irreversible.

Rehabilitation

Rehabilitation–as defined by the Third WHO Expert Committee (1981)–is viewed as: all measures aimed at reducing the impact of disabling and handicapping conditions in individuals and at enabling them to achieve social integration. Rehabilitation aims not only at training and adapting individual disabled/handicapped persons but also at intervening in their immediate environment and the society as a whole in order to facilitate their social integration. The disabled/handicapped themselves, their families and their communities should be involved in the planning and implementing of these measures.

Community-based rehabilitation. This is a process involving measures taken at the community level to utilise and build on existing resources in the impaired/disabled/handicapped persons themselves, their families and their environment. Thus it is a *dynamic process* of enabling the individual with disability/handicap to overcome its effects and to achieve social integration. This greatly differs from the traditional concept of rehabilitation as the provision of therapies and services by professionals in institutional settings. In practical terms, it involves a shift from the traditional pattern of services to community-based programmes, building on the considerable rehabilitation resources that exist in the disabled themselves, and in their families and communities. Their active involvement in planning and implementing of rehabilitation programmes is crucial, as is the integration of rehabilitation services into general community services such as health, education and training, employment and social facilities, without which community programmes cannot develop and survive. Community-based rehabilitation aims to enable disabled persons to benefit from essential or basic services, by using the approaches and technologies which are feasible, acceptable, affordable and appropriate to local settings. It relies primarily on members of the local community, who constitute the basic manpower for its implementation.

Family members or other persons in the community have to be trained to undertake the daily tasks related to the rehabilitation of the disabled. Locally recruited supervisors will identify the disabled and motivate the family member to undertake their training, and will instruct, guide and follow up the progress of rehabilitation. They should be members of the community and may have other community functions, such as being primary health care workers, social workers, schoolteachers, etc. Training programmes for local supervisors should be instituted as widely as possible.

The first line of a community-based rehabilitation system should be supported by professionals with sufficient knowledge and skills to be able to train and guide local supervisors and to act as the first level of referral. Existing institutions should be reoriented to support the community-based rehabilitation services in many ways. They should participate in training of first and second level supervisors; they may form mobile rehabilitation teams to act as local consultants for referral cases. And finally, they will constitute part of a two-way referral system for selected patients.

Obviously there is also a need for specialised rehabilitation services; but these should be, in addition to the task mentioned above, reserved for the disabled with special needs, which cannot be met by community services.

Medical rehabilitation. The current view is that medical rehabilitation should be considered very early in the process of medical treatment, and should start as soon as the patient's general condition allows. Techniques such as physiotherapy, occupational therapy and speech therapy are prescribed in order to accelerate the natural regeneration processes and to prevent or reduce sequelae. For disabled persons with morphological or functional destruction, medical rehabilitation must concentrate on the promotion of all physiological processes essential to the development of compensatory mechanisms. Medical rehabilitation is in fact crucial at all levels to reduce the degree of disability and, incidentally, the need to organise special follow-up programmes of

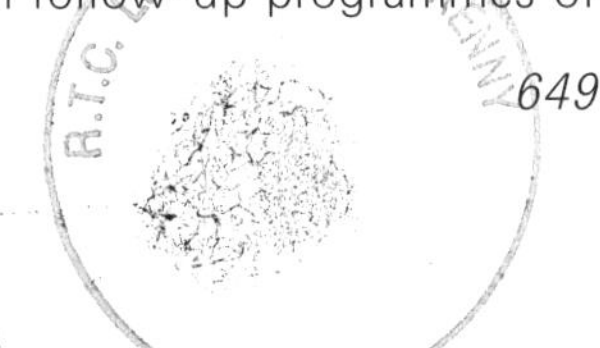

social rehabilitation. It is closely connected with the basic organisation of medical services in every country, and justifies modifications in conventional treatment which should, as stated above, include elements of early rehabilitation–for disability is frequently a result not only of the illness itself but also of static, conventional treatment involving long and often unnecessary immobilisation.

Of course, all efforts must be concentrated primarily on the correct diagnosis and treatment to save the patient's life and keep to the minimum the damage caused by the illness or trauma. Experience has shown, however, that treatment cannot be completely satisfactory unless its social and biological effects are taken into account. Treatment which leaves sequelae of the illness–physical and psychological effects–cannot, in view of the possibilities of medicine today, be considered as acceptable.

For this reason, modern methods of treatment must also cover the over-all rehabilitation aspect. Nowadays, when a physician devises a treatment programme for his patient, he must think not only of the acute stage, when he has to concentrate all his efforts on saving life; he must also introduce in the subsequent stages, when natural recovery following the illness has set in, the appropriate elements of early rehabilitation which prevent the occurrence of sequelae caused by lack of medical control over that important stage, which is often crucial for the patient's future life.

Early prophylactic rehabilitation. Movement and effort are basic to the satisfactory functioning of the human body. When the body is deprived of these two requirements, regressive changes in various tissues occur which are not always connected with the injury. We know today, from experiments on animals and also from experimental clinical examinations, that certain organic disturbances might not have occurred if medical care had also taken into account an appropriate loading of the organism (naturally under the control of various objective medical parameters) and subjected it to specific demands as regards both length and intensity of effort.

Early rehabilitation methods should therefore be introduced in all health service establishments. They are simple, need no extra installations, and can be applied under practically any conditions. They consist of the gradual reintroduction of the active movements of daily life (figure 1), the rate of introduction depending on the condition of the patient and his reactions. Motor training should include a very wide range of breathing and other exercises and gradual general loading. General impairment of the circulatory, respiratory and other systems, which often requires a long regeneration process, will thus be prevented. Individual exercises should involve the diseased organ; they include appropriate movements and a variety of applied muscular isometric training activities in the diseased extremity or in the entire body (if it is a systemic disease case). One of the principal objectives of early rehabilitation is to make the patient leave the bed and thus to reduce the duration of the horizontal position, depending on his capabilities and reactions. Occupational therapy is important as a psychoprophylactic and psychotherapeutic factor, since it activates the attention of every patient at fixed hours of the idle hospital day. Every man has adapted to vocational activities during his life; these take up a large portion of his time, and hospital idleness may cause psychological impairment.

Figure 1. Manipulation of objects in general use.

Whenever possible, occupational therapy is made more attractive; remedial exercises must never be forgotten.

Treatment and care. A fundamental part of the process of medical rehabilitation must be the principle of "progressive care". This not only calls for the stimulation of the entire treatment process by adequate forms of motor and psychological activity on the part of the patient; it also necessitates some fundamental changes in the organisation of in-patient departments. The patient must not stay in the same place at the time when his life is in danger, nor later, when the regenerative processes have begun within his organism. He must enjoy a psychological climate which will dispose him to autotraining; moreover, he must participate in various forms of hospital activities; and he must stay active after leaving hospital in so far as his general condition allows.

This involves a dual conception of prophylaxis in medical care. To a modern physician, good X-ray results or an improvement in clinical tests alone do not mean that a patient is cured. What is important is whether a man can, on the completion of treatment, take up an independent social or vocational life again. In other words, those elements of rehabilitation which concern the sociobiological effects of medical care must be introduced into the fundamental medical care programme.

Comprehensive rehabilitation. Nevertheless, there is a large group of diseases after which there can be no return to active life–those involving serious damage to the morphology of the body, loss of a limb, disturbances in the nervous system, psychological frustration, etc. Persons recovering from such mishaps need an appropriate course of comprehensive rehabilitation which includes all the techniques currently in use and which calls for a large team of specialists of various kinds.

Experience has shown that, in cases involving severe morphological changes in the body, patients should be referred to appropriate comprehensive rehabilitation centres during the days immediately following the occurrence of the injury. The natural compensation process existing in every living organism can be released and controlled in such centres. When a patient does not arrive there until a few months after the injury occurs, the results are never as good, and only training in adaptation is then possible. Some centres therefore provide special wards for acute cases (spinal injuries, peripheral nerve damage, amputation) admitted immediately following the onset of the trauma. As always, early prognosis is of great importance: it saves much time and gives good results.

Comprehensive rehabilitation methods include:

(a) remedial exercises (individual, in water, sports and games, general exercises, training in walking, resistance exercises with weighted loading, etc.);

(b) occupational therapy (psychological and kinetic), with vocational preorientation;

(c) physiotherapy;

(d) psychotherapy (psychotropic drugs, group and individual methods);

(e) reconstructive surgery;

(f) technical aids (in cases of morphological and functional damage).

A comprehensive rehabilitation centre must have vocational preorientation sections with an adequately trained staff, and a social section providing appropriate conditions for medical rehabilitation. Comprehensive rehabilitation must consider all the needs of handicapped persons and prepare their homes adequately.

Compensation or adaptation. After serious morphological body changes or damage to functions of a permanent character, the natural compensation capabilities possessed by every living body must be encouraged, in order to attain the best results in the shortest possible time. The results achieved naturally depend on the individual concerned–on his personality, his social condition and needs, his professional needs–and also on the nature of the morphological damage.

In many instances it is not possible to attain the optimal results; it is then necessary to adapt the patient to his social condition and his morphological situation, which is to be considered as permanent from then on. This task of adaptation is applied, first of all, in the most severe injuries of the body, or in the case of persons whose disability is of a chronic and progressive character. Of course, the optimal result is also possible in the case of invalids who follow adaptation programmes: in most cases, this gives them a modicum of independence and sometimes the ability to resume some restricted vocational activity, for example work at home or some other form of sheltered employment. In all instances, both of early rehabilitation and in the persons in whom we try to release compensation mechanisms and adapt to social conditions, the most crucial factor is undoubtedly the psychological drive, the release in patients of possibilities of acceptance of their respective situations. Acceptance and psychological drive are the most difficult goals to attain in medical rehabilitation; they require a maximum effort on the part of the entire staff, and great determination in their work. Frequently the return to normal social conditions is followed by a collapse of the created psychological drive, with the patient returning to the rehabilitation centre where an attempt is made to change his social conditions. This category of handicapped person probably demonstrates most emphatically the urgent need in the rehabilitation process for a co-ordination of the activities of all the governmental and local authorities and voluntary agencies concerned, so that the desired aim may be reached. Economic studies conducted in certain centres have shown that such co-ordination is possible and that it gives good results, enabling pensions to be reduced, the duration of disability for work to be shortened and a large number of persons who hitherto had to rely on disability benefits only to be activated.

Social rehabilitation

Although the tasks of medical rehabilitation are clear enough, it must be stressed that it is indispensable to combine it, from the earliest stages, with social and vocational rehabilitation. Social rehabilitation has been defined by the World Health Organisation as "that part of the rehabilitation process aimed at the integration or reintegration of a disabled person into society by helping him to adjust to the demands of family, community, and occupation, while reducing any economic and social burdens that may impede the total rehabilitation process". Every member of the community is entitled to benefit from everything the community has to offer as a result of the development of its civilisation. He must not be cut off from normal life because he cannot take advantage of public entertainment or transportation facilities, or move about in a house not planned to take his disability into account, or take part in sports and games, etc. The tasks of social rehabilitation therefore include the adaptation of social conditions so that invalids may join the active members of society. A handicapped person has a right to an active life, not to charity alone. He wants to co-operate in and take advantage of life in the same way as those who are physically and psychologically able to do so. The tasks of social rehabilitation thus include: adaptation of family life and of life in larger social environments; organisation of some technical modifications; and organisation of various social ventures in such a way that handicapped people can participate in them. Even during the period of medical rehabilitation, the social rehabilitation process must be borne in mind, and various services should be organised to take care of the interests of the patient when he is still in hospital and unable to act for himself; this also includes preparing family and professional circles, organising transportation to the place of work, etc.

Vocational rehabilitation

In the ILO's Vocational Rehabilitation (Disabled) Recommendation, 1955 (No. 99), vocational rehabilitation is defined as "that part of the continuous and co-ordinated process of rehabilitation which involves the provision of those vocational services, e.g. vocational guidance, vocational training and selective placement, designed to enable a disabled person to secure and retain suitable employment".

The past 30 years have seen rapid developments in vocational rehabilitation services for disabled persons, often closely associated with medical and social rehabilitation facilities. Such vocational rehabilitation services should aim at obtaining a clear picture of the disabled person's residual capacities for work (vocational assessment); they should include orientation courses to help restore lost confidence, instil good working habits and increase work tolerance (work conditioning); individual help should be given in solving problems of choice of occupation (vocational guidance); vocational training and retraining opportunities and assistance with placement should be readily available if the disabled person is to be resettled in suitable employment at the earliest opportunity. In the case of the severely disabled, special employment arrangements often need to be made through sheltered workshop or homeworker schemes. Moreover, it should be remembered that the provision of technical aids and the adaptation of workplaces and machines can often help to create employment opportunities for disabled persons which otherwise would not have existed. The final phase of the vocational rehabilitation process is the follow-up to ensure that full resettlement has been achieved.

The need for vocational rehabilitation to be combined and co-ordinated with the medical and social aspects of the rehabilitation process cannot be overemphasised. Even during the treatment stage the occupational therapist can often play a vital role in preparing the disabled person for subsequent training and employment. At the same stage, the social worker can help resolve any pressing family or accommodation problems; in addition vocational resettlement or placement officers can discuss possible avenues of training or employment. Prompt action on these lines not only helps to alleviate anxiety about their future which disabled persons invariably experience but ensures that the patient concerned goes forward to active rehabilitation treatment in the best possible frame of mind.

Examples of basic rehabilitation services (tasks)

Simple therapy. Prevention and correction of contractures; training in movements of arms and legs; walking training; orientation and mobility training for the blind; simple speech therapy; training in daily living activities such as eating, dressing, washing, toilet, preparation and cooking of food; etc.

Provision of simple technical aids, made locally. Simple crutches, canes, walkers, pylons, splints and braces, bands, carriages, wheelchairs, etc.

Education/training. Many disabled children who do not go to school can be integrated into normal schools if the teachers are given instruction on how to deal with, for example, a mentally retarded, deaf, blind or physically disabled child.

For some children the problem is to arrange for boarding a child (e.g. with a relative or friend) close to the school, or to arrange for transportation or some assistance to get to school (e.g. for physically handicapped, blind or mentally retarded children).

For vocational training, local craftsmen might be approached or training given at home in farming, horticulture, small husbandry, cottage industries by family members or friends. Nearby training centres for the able-bodied might also be used for the disabled.

Provision of work. Jobs should be given to all disabled who are able to work, thereby allowing them to contribute economically to the family and society. Main areas will probably be farming, handicrafts and cottage industries, although any other appropriate open-market employment should be tried.

Social measures. If a disabled person who is able to support himself through farming has no land, local authorities or the community should consider lending or giving him some land (e.g. for horticulture). If he can be rehabilitated by education and training but has no means, authorities or community organisations should consider paying school fees and other training costs. Tools and equipment may be provided for work at home; loans can be arranged for buying more expensive equipment (e.g. sewing machine) or opening a small business; transportation assistance should be arranged for those who want to participate in training.

The disabled should be given the opportunity to participate in social, cultural, recreational and religious gatherings and events, as well as participate, in conformity with their abilities, in unions, organisations and decision-making bodies of their communities. They should also be given the opportunity to marry and have a family. See also MENTALLY HANDICAPPED, REHABILITATION OF THE.

KROL, J.

Vocational rehabilitation:

Basic principles of vocational rehabilitation of the disabled (Geneva, International Labour Office, 2nd ed., 1976), 53 p.

Vocational rehabilitation of the disabled. An audiovisual sound-slide package (Geneva, International Labour Office).

Co-operatives for the disabled: organisation and development. Proceedings of a seminar (Geneva, International Labour Office, 1978), 238 p.

Medical rehabilitation:

International classification of impairments, disabilities and handicaps (Geneva, World Health Organisation, 1980), 207 p.

Reports on specific technical matters, disability prevention and rehabilitation. A 29/INF.DOC./1 (Geneva, World Health Organisation, 1976).

Training the Disabled in the Community. A manual on Rehabilitation for Developing Countries. Helander, E.; Meidis, P.; Nelson, G. Version 2 (Geneva, World Health Organisation, 1980). Illus.

Expert Committee on Disability Prevention and Rehabilitation, Geneva, 17-23 February 1981. III Report, technical report series (Geneva, World Health Organisation, in press).

"Rehabilitation in the light of technological developments with special reference to the characteristic types and sequelae of injuries caused by employment accidents in an industrialised economy". Nickl, W. *International Social Security Review* (Geneva), 1978, 31/2 (144-172).

Disasters

A disaster is a catastrophic situation in which the day-to-day patterns of life are, in many instances suddenly, disrupted and people are plunged into helplessness and suffering and, as a result, need protection, clothing, shelter, medical and social care and other necessities of life.

Disasters can be divided into two main groups. In the first are disasters resulting from natural phenomena like earthquakes, volcanic eruptions, storm surges, cyclones, tropical storms, floods, avalanches, landslides, forest fires and massive insect infestation. Also in this group, although neither sudden nor violent, must be included drought, for its effects, if not checked, will cause a creeping disaster leading to famine, disease and death. The second group includes disastrous events occasioned by man, or by man's impact upon the environment. Examples are armed conflict, industrial accidents, radiation accidents, factory fires, explosions and escape of toxic gases or chemical substances, river pollution, mining or other structural collapses, air, sea, rail and road transport accidents (e.g. aircraft crashes over built-up areas, collisions of vehicles carrying inflammable liquids, oil spills at sea) and dam failures. Epidemics of infectious diseases may sometimes break out spontaneously, but they can also be the direct result of one of the other kinds of disaster, and can reach catastrophic dimensions in terms of human loss.

The potential impact of disasters of all kinds is constantly tending to increase because of the increasing population of the world. This element is most marked in the developing countries, where a high average rate of population increment (disturbing though it is) masks the much higher rate occurring in the major cities. The industrial development which brings its own hazards in its train also attracts new residential areas to its immediate vicinity, thus exposing even greater numbers of people to the risk. Moreover, these people are often those who can least afford to take any precautionary measures to protect themselves in case a disaster should occur.

There can be no set criteria for assessing the gravity of a disaster in the abstract since this depends to a large extent on the physical, economic and social environment in which it occurs. What would be considered a major disaster in a developing country, ill equipped to cope with the problems involved, may not mean more than a temporary emergency elsewhere. However, all disasters bring in their wake similar consequences that call for immediate action, whether at the local, national or international level, for the rescue and relief of the victims. This includes the search for the dead and injured, medical and social care, removal of the debris, the provision of temporary shelter for the homeless, food, clothing and medical supplies, and the rapid re-establishment of essential services. Once these essential needs for survival have been met in the emergency phase immediately following the disaster, the longer-term problems faced by the stricken community have to be dealt with during the rehabilitation and reconstruction period which may last for many weeks, months or even years. To a greater or lesser degree it will be necessary to take steps for the recovery of land, equipment, personal and household belongings, the reconstitution of livestock herds and flocks, the repair and rebuilding of homes, schools, factories and other public edifices. Many persons may have lost their employment or means of livelihood and will need continued assistance until new occupations can be found for them.

While some regions of the world are more disaster-prone than others, no country, town or village can be considered immune from the possible onslaught of disaster in one form or another. Many of the poorer developing countries lie in the most disaster-prone areas and, in addition to the suffering of the population, their economic development can be impeded or even cancelled out by the effects of repeated disasters.

To help reduce these adverse effects, pre-disaster planning is a necessity, and this also falls under two main headings: preparedness and prevention. Preparedness covers all action taken with a view to organising and facilitating timely and effective rescue, relief and rehabilitation in cases of disaster. Prevention means the formulation and implementation of long-term policies and measures to mitigate the impact of natural phenomena on human settlements, and on economic and social development as a whole.

For the orderly planning of disaster preparedness and relief operations it is essential that the government should set up a national emergency organisation entrusted with complete responsibility for these matters. Organisations of this kind, at varying levels of competence and authority, already exist in a large number of countries, and consideration is being given to their establishment in others. The exact title, composition and location of the organisation within the governmental structure depends on internal factors and administrative practices peculiar to the country, but its functions remain very much the same.

A national plan is drawn up for approval by the government setting forth policies, procedures and guidelines relating to pre-disaster preparedness measures and to emergency relief operations during and after disasters. The plan should provide for co-ordination nationally and define the respective roles and responsibilities of the various ministries, departments and agencies concerned, whether these responsibilities are exercised nationally, regionally or locally. Pre-disaster activities must start with the identification of disaster-prone areas and the types of event, caused naturally or otherwise, most likely to occur. They should continue with provision for public information on these subjects and the establishment of early-warning systems; laying plans for evacuation schemes should these appear to be necessary; training of personnel in rescue and relief operations; setting of procedures for quick deployment of labour for site clearance and identifying projects for relief employment; making special stand-by arrangements with customs authorities and internal transportation companies for the rapid clearance and channelling of relief supplies to disaster sites and distribution to the survivors; stockpiling of relief equipment, foodstuffs, drugs and medicines; contingency planning for transport and communications, earmarking of funds for relief purposes; simulation exercises to test the efficiency of the plan and to make its provisions known to the population; and the collection of data on resources available both nationally and from external assistance.

When a disaster occurs, rescue and relief operations can thus be initiated without delay, and in accordance with procedures laid down in the national plan. The magnitude of the disaster must be quickly ascertained. To do so the authorities must know the exact location and size of the area affected, the number of dead, injured and homeless, the extent of damage to buildings, industry, communications and public services, and in rural areas the animal and crop losses involved. From this information, it is then possible to determine priority needs for relief supplies and assistance. The national Red Cross or Red Crescent Society, the armed forces, civil defence units and other local agencies may all be mobilised during this emergency phase.

After a major disaster the national resources may not suffice to meet relief needs and assistance from the international community will therefore be required. This assistance may be given by or through appropriate specialised agencies of the United Nations, bilaterally by individual governments, or by voluntary agencies operating throughout the world. Since 1972 the Office of the United Nations Co-ordinator for Disaster Relief (UNDRO), which was set up as a focal point in the United Nations system for disaster relief matters, has (following a request for international aid) been responsible for the mobilisation of United Nations assistance and its co-ordination with other external assistance being given to a stricken country. Information on the assessment of remaining relief requirements, which constantly change as the days go by, is thus centralised during the relief operations, and disseminated continuously to all who need to know. In this way UNDRO hopes to ensure the speediest and most adequate response to the needs of the survivors of a disaster.

As the initial emergency period draws to a close and rescue and relief operations come to an end, attention will be increasingly turned towards the repair and reconstruction work to follow. Needs and priorities will of course vary greatly depending on the type of natural phenomenon or man-made accident which caused the disaster, and the different kinds of damage experienced. It will normally be necessary to make safety inspections of damaged buildings, and work must proceed at once on the most urgent repairs and rebuilding needed to permit, as far as may be possible, the resumption of daily life and customary activities. Labour-intensive public works schemes, for site clearance or for urgent projects needed to prevent a second, or secondary disaster, may well have to be organised. Such schemes have a further benefit in that they relieve, in part, problems of unemployment caused by the disaster.

Information derived from the inspection of damaged buildings may well be useful in designing disaster prevention measures to be embodied in reconstruction projects. It is desirable, however, to proceed by way of composite risk vulnerability analyses, for these provide a sound basis for policy decisions on appropriate technical measures in the fields of building, engineering, legislation and physical and economic planning. These analyses should be carried out systematically as an integral part of any new development plan, for no country is altogether immune from every type of disaster.

When a full reconstruction programme has been planned, it may be found necessary to introduce training in new techniques for managers and workers in the construction, transport, civil engineering and public works fields. New agricultural practices, including in particular those concerned with soil erosion and reforestation, may have to be taught. The opportunity may be taken to extend managerial or vocational training into new areas not previously exploited. Finally, attention must be given to the special problems of the newly disabled, for it is almost inevitable that some of the injured survivors will have lost one or more limbs, or sight, or hearing.

Lessons learned from an industrial or similar type of disaster may lead to the enacting of new legislation to improve conditions and safety at work. New environmental measures, especially land-use planning and construction codes, may be used to avoid or reduce the impact of future disasters, and at the same time improve standards of life generally. A disaster can, in fact, provide the impetus not merely to rebuild the old, but to build in such a way as to provide a better life for those who survived, and a worthy memorial to those who did not.

UNDRO

Disaster prevention and mitigation. A compendium of current knowledge. Vol. 1. *Volcanological Aspects;* Vol. 2. *Hydrological Aspects;* Vol. 3. *Seismological Aspects;* Vol. 4. *Meteorological Aspects;* Vol. 5. *Land Use Aspects;* Vol. 7. *Economic Aspects;* Vol. 10. *Information Aspects* (Geneva, United Nations Disaster Relief Office).

Guidelines for disaster prevention. Vol. 1. *Pre-disaster physical planning of human settlements;* Vol. 2. *Building measures for minimizing the impact of disasters;* Vol. 3. *Management of settlements* (Geneva, United Nations Disaster Relief Office).

The great international disaster book. Cornell, J. (New York, Charles Scribner's Sons, 1976), 382 p. Illus.

1001 questions answered about natural land disasters. Tufty, B. (New York, Dodd, Mead and Co., 1969), 350 p. Illus. Bibl.

Red Cross disaster relief handbook (Geneva, League of Red Cross Societies, 1976). A loose leaf publication.

"Communicable diseases and epidemiological surveillance in relation to natural disasters" (Maladies transmissibles et surveillance épidémiologique lors des désastres naturels). De Ville de Goyet, C. *Bulletin of the World Health Organisation* (Geneva), 1979, 57/2 (153-165). (In French)

CIS 77-869 *Major emergencies–Recommended procedures for handling* (Chemical Industries Association Ltd., Alembic House, 93 Albert Embankment, London) (1976), 17 p.

Papers on disasters are currently issued in the following journals:

Disasters. The International Journal of Disaster Studies and Practice (International Disaster Institute, Pergamon Press Ltd., London).

Mass Emergencies: An International Journal of Theory, Planning and Practice (Amsterdam, Elsevier Scientific Publishing Company).

Diving

"Diving" is a collective term used to designate the activity of individuals who operate fully immersed in water and exposed to ambient pressure either as observers, or workers, or in some military capacity.

Diving operations are conveniently separated on the basis of the type of support received by the individuals exposed to the water. Skin divers or, more properly, breath-hold divers, as the name implies, are unsupported by any external supply of respirable gas. All other forms of diving utilise a supplemental source of respirable gas which may be either air or a synthetic mixture made to specifications determined by the diving situation expected and, almost invariably, based predominantly upon the use of helium. Divers using such auxiliary apparatus can be divided into scuba divers, utilising self-contained underwater breathing apparatus and divers supported either from the surface or from a submersible platform by means of some form of an umbilical, which usually provides communications and energy to minimise thermal stress in addition to the gas supply.

Skin diving

Underwater swimmers holding their breath and using primitive devices such as bladders and goggles functioned as combat swimmers throughout antiquity (e.g. the Roman *urinatores*) and, to this day, are employed on collecting food, pearl oysters, coral and shell, etc., in different parts of the world.

Various methods of assisting descent and ascent of the skin diver are used to increase range and maximise useable bottom time, and Japanese and Korean female skin divers *(amas)* commonly descend to depths of 18 m and stay there for around 60 s. Apart from the obvious limitations of oxygen depletion and CO_2 build-up during prolonged apnea, the physiological limitations to breath-hold diving include: poor vision, largely corrected by various types of underwater goggles; the resistance of the chest and lungs to compression; and the accumulation of dissolved inert gases in tissues during repetitive dives leading to the danger of *bends,* which may produce permanent disability and even death if adequate decompression facilities are lacking. In the case of less experienced divers, breath-hold dives entail additional hazards due to deliberate self-training to minimise sensitivity to carbon dioxide and to postpone the break-point at which the diver has to surface. This imposes serious risk of brain damage due to hypoxia, aggravated by the fact that expansion of the lungs during ascent further reduces blood-oxygen partial pressures and may give rise to *shallow water black-out.*

Diving with apparatus

At the minimum, this requires providing a space filled with respirable gas in contact with the respiratory passages. This can be attained either in the form of a spherical helmet or in the form of a mask fitted to the face of the diver. In some cases this is supplemented by an oral-nasal mask, either to minimise dead space or to provide additional safety in case of flooding of the helmet. An alternative is to protect the diver's nose against penetration of water either by a clip or by a mask and to provide a source of respirable gas through a mouthpiece held between the teeth of the diver. The latter form is most commonly used by sports divers, while some variant of the helmet or whole-face mask is almost universally utilised by professional divers. If the respirable gas is supplied from compressed or liquefied gas reservoirs carried by the diver, this form of underwater work is properly designated as scuba diving, which takes its name from the initial letters of "Self-Contained Underwater Breathing Apparatus". If the gas is supplied from the surface or from some variant of the diving bell, one is dealing with umbilical supported diving. The breathing equipment supplied the diver may be supplemented by a diving suit designed to provide protection against heat loss. Diving suits may be subdivided into wet suits, usually manufactured from some such material as neoprene foam, restricting the flow of water past the skin surface of the diver without any attempt to seal tightly, and dry suits, which are designed to completely seal the body of the diver off from the surrounding water. Dry suits are usually provided with some means of introducing gas between the skin of the diver and the suit and thus provide, secondarily, a means of controlling the buoyancy of the diver which can be utilised by the skilled operator to assist in the performance of heavy labour underwater. This type of system, however, entails the danger that the diver may become too light and be forcibly carried to the surface, an accident associated with a high degree of mortality and commonly referred to as *blow-up*.

HAZARDS OF COMPRESSED GAS DIVING

In all forms of apparatus-supported diving, gas pressures in the lungs of the diver must closely approach those prevailing in the environment so that they entail exposure of the diver to high-pressure air or to high-pressure mixed gases. Hazards associated with this situation arise from the behaviour of a body in contact with compressed gas when gas pressures are altered. During descent, these problems are conveniently summed up under the term *squeeze* and may entail serious injuries to the lining of air-filled sinuses and tympanic membrane, as well as injuries to parts of the body surface in contact with wrinkles or unevennesses of dry diving suits. The extreme case of squeeze was encountered during diving operations using primitive helmet gear where it was possible to lose pressure from a hard helmet while the diver was at depth with consequent lethal squeezing of the diver into his own helmet under the influence of the pressure of the surrounding water. During ascent, the hazards of the situation may be divided into two categories. The first of these, once again, is focused upon any gas-filled cavities which do not freely equilibrate with the pressure prevailing in the environment, leading to tissue disruption and collectively known as *barotrauma*. The most frequent accidents in this category are injury to the middle ear and vestibular apparatus. Another related manifestation is *aeroembolism* due to rupture of spaces enclosing gases sequestered in the lungs of individuals who dive in the presence of any kind of inflammatory change in the lung with resultant intrusion of gas into the circulation–often a serious or even fatal accident.

The second category of decompression-related accidents results from the fact that during ascent or decompression following a dive, tissues contain dissolved gases at a total pressure substantially in excess of the actual ambient pressure surrounding the diver, resulting in formation of gas bubbles, which give rise to a complex of pathological changes, referred to collectively as *decompression sickness*. This type of accident can be avoided by timing the reduction of ambient pressure so as to permit progressive desaturation of the tissue with no bubble formation or, at least, without the formation of bubbles of a size that would cause vascular obstruction or tissue disruption. Formulation of theories upon which such decompression can be based has been a major preoccupation of the diving industry for many years, ever since the first attempts in this direction by J. B. Haldane in 1922. In recent times promising attempts are being made to utilise instrumental methods of detecting bubble formation, usually by acoustic means, and to use this information to guide decompression procedures. The time course of pressure reduction compatible with safe recovery of the diver is determined by the gas mixture or mixtures employed, by the duration and profile of the dive, by the thermal and work conditions accompanying it, as well as by the physical characteristics of the individual diver, including body composition and physical conditioning. It is not surprising, therefore, that all standard decompression tables are written with the knowledge that a certain proportion of decompression accidents must be tolerated, and it is the hope that the instrumental methods may eventually permit sufficient individualisation to reduce such accidents to a minimum. Modern decompression practice seeks to eliminate excess inert gas dissolved in the tissue safely, not only by controlling the rate of ascent, but also by carefully timed changes in the make-up of the inert gas components of the respired atmosphere at appropriate stages of decompression, and by maximising oxygen partial pressures within the limits determined by the danger of oxygen toxicity.

Air-breathing divers exposed to pressures in excess of 2 atm can expect to encounter additional hazards due to the fact that the constituents of air exercise progressively severe pharmacologic effects. Prolonged exposure to oxygen at partial pressures of 0.4 or more atm can produce lung injury and at higher pressures changes in brain function which culminate in *oxygen convulsions*. Furthermore, at pressures in excess of those equivalent to 45 m of water, air-breathing divers experience various degrees of *nitrogen narcosis*. These central nervous system manifestations are not unlike the effects produced by alcohol and the degree of impairment at 45 m is roughly comparable to the effects of three or four glasses of whisky. The severity of nitrogen narcosis increases rapidly with depth and most divers are seriously impaired at a depth of 75 m. Training and discipline allow certain individuals to penetrate to a maximum depth of around 100 m. Work beyond this depth is conducted by the use of artificial atmospheres in which nitrogen is replaced by some gas of lower liquid solubility and lower molecular weight. At present helium is by far the most important of such substituents, but both neon and hydrogen are being considered for use in special situations. Using such artificial gas mixtures in which oxygen partial pressure is adjusted to anticipated working depth so as to assure oxygen partial pressures between 0.2 and 0.5 atm (varying according to depth, working conditions and, to some extent, to the as yet somewhat arbitrary judgement of the operator), dives to 300 m have been performed without perceptible changes in diver performance attributable to the gases.

When pressures exceed 25 atm divers encounter a new complex of functional disturbances attributable to the hydrostatic pressure to which they are exposed, the severity of the effects varying, at any given depth, to some extent, with the velocity of compression. The effects observed begin with minor motor disturbances and tremors, build up to rather gross tremors, occasional localised myoclonic events and disturbances of central nervous system function, manifested by changes in vigilance performance and in sleep patterns. The changes are accompanied by electroencephalographic changes, most notably the appearance of theta waves, especially over the parietal regions of the cortex, and in animal experiments have been shown in all vertebrate species including primates to culminate in severe clonic and eventually tonic/clonic generalised convulsions. These effects are collectively known as the *high-pressure neurological syndrome* (HPNS). To some extent, these effects are antagonised by various anaesthetic agents, including inert gas anaesthetics, and various schemes for the addition of nitrogen to helium breathing mixtures have been utilised in attempts to attain greater depths with shorter compression times in divers. At the greatest depths attained in the laboratory to date, 500 m of sea water, both the severity of these HPNS effects and perhaps limitations in respiratory exchange due to the high density of the respired gas atmospheres appear to render the functioning of divers exposed to ambient pressures in excess of 50 atm increasingly problematic.

A considerable step forward from the point of view both of safety and of the amount of work that can be performed by divers is constituted by the development of saturation diving procedures in which divers are provided with some form of a habitat in which they can remain at a pressure in excess of 1 atm for extended periods of time. At the end of such a sojourn, the divers are returned to the surface by means of very slow decompression (saturation table) to take into account the effects of poorly perfused tissues which become saturated during the long sojourn at pressure and which desaturate with half-times that may be in excess of 36 hours. Saturation procedures not only permit optimising the ratio between bottom time and decompression time, but they also appear to provide some relief of those parts of the HPNS which appear at maximal intensity early during, or immediately after, compression. At more extreme depths, there are increasing indications of HPNS components which not only are not relieved by this procedure but may become exacerbated with sojourn at pressure. The usefulness of saturation diving procedures has been greatly increased by the development of excursion tables, that is to say tables which define vertical excursions that a diver saturated to a given depth may undertake with impunity from the point of view of decompression sickness.

All diving procedures involving the use of compressed gases entail a certain risk of long-term injury. Best known in this category is the appearance of *aseptic bone necrosis*, a late effect shown to have been associated with insufficient decompression procedures, especially in early caisson workers. There is reason to suspect the occurrence of this type of injury in association with saturation diving, but development of a satisfactory epidemiology is complicated by the fact that very similar joint and long bone lesions do occur with some frequency in populations not exposed to diving risks.

The possibility that exposure to diving conditions resulting in HPNS symptoms may entail long-term injury cannot be excluded at the present time, although the limited data on this point available as yet from animal experiments fail to provide any conclusive evidence for such late or long-term effects.

In addition to the hazards of diving attributable directly or indirectly to the high pressures of the environment in which the diver must work, a key problem has been the need for *thermal protection*. In general, waters below 100 m can be expected to be at temperatures not much in excess of 5 °C and, in general, a considerable proportion of all diving is undertaken in cold water. As pointed out above, attempts have been made to meet this problem by the use of appropriate garments. At the present time it would appear that divers not provided with protective garments cannot expect to operate safely for any length of time in waters much below 13 °C (and for the majority of divers, probably not much below 18 °C). Wet suits provide adequate protection for short dives in waters down to 5 °C, but are subject to a great decrease in efficiency as divers progress to greater depths because the thickness of the foam insulant decreases rapidly as it becomes exposed to higher pressures. Various forms of dry suits have proved more satisfactory and have permitted exposures up to several hours in waters down to 3 °C or even to 0 °C, though at the risk of increasing discomfort to the diver, whose thermal equilibrium becomes increasingly jeopardised as the time at pressure increases. The efficiency of dry suits is further compromised by percutaneous gas exchange, which converts the atmosphere in the diving suit from air to the much more highly conductant high pressure helium. At the present time most deep sea diving is conducted in suits which provide a constant flow of warm water over the skin of the divers and which, as a result, depend upon a hot water source either in the submersible decompression chamber of a typical industrial diving complex or, in some cases, from the surface. These suits provide for temperature regulation by means of a valve which permits the diver himself to control the flow of hot water to meet his perceived thermal needs. Efforts to automate this system and to reduce the weight of the cumbersome umbilical to which it contributes by providing more nearly autonomous heat sources are under way but effective gear is not yet available. An additional complication is the fact that, as depth is increased, respiratory heat losses rapidly increase in importance, requiring the use of some device for preheating the gas supplied to the diver in cold water at any depth beyond 200 m. The thermal balance of the diver under these conditions evidently deviates considerably from that to which normal central thermoregulatory mechanisms are adapted, and definition of thermal comfort and thermal safety in physiologically meaningful terms applicable to these conditions is, as yet, far from complete. It is known that the comfort zone for divers in high-pressure heliox becomes very narrow indeed, ranging at the most from an ambient temperature of 30 °C to an ambient temperature of 33 °C.

Biological hazards are a problem of considerable significance to the scuba-supplied sports diver and include predatory animals such as sharks or moray eels, venomous animals, including sting rays, some types of jelly fish, hydroids, and sea urchins, as well as, perhaps, certain aquatic plants, of which *Sargasea rosea* appears to be a particular hazard for Mediterranean sponge fishermen (see ANIMALS, AQUATIC).

Selection and supervision. The stresses to which divers are subjected are considerable and the selection procedures must be correspondingly rigorous. The upper age limit for individual divers is determined largely by the condition of their cardiovascular system and by the maximal physical effort of which they are capable. Contingencies of underwater work may at any time impose an unexpected need for very high levels of energy expenditure. Besides the general requirement imposed by this need for a functional reserve, specific attention must be given to the condition of the lungs and bronchi to eliminate subjects with any evidence of local obstruction or cyst formation. Gross obesity is a contraindication since fat tissues act as gas reservoirs. Psychometric testing and psychiatric examination to detect mental disorders may be advisable in certain circumstances since the mental stress to which divers are subject may be considerable, for example persons who suffer from claustrophobia would not be suitable for this occupation.

Attempts are being made to select, for deep diving, persons with presumed high resistance to HPNS effects, but the theoretical basis and the efficacy of such selection procedures, at the time of writing, are still a matter of debate.

BRAUER, R. W.

Underwater medicine. Miles, S.; MacKay, D. E. (London, Adlard Coles Limited, 1976), 330 p. Illus.

The physiology and medicine of diving and compressed air work. Bennett, P. B.; Elliott, D. H. (London, Bailliere Tindall and Cassell, 1969), 532 p. Illus. Ref.

"Human performance in high pressure inert gas environments". Brauer, R. W. *Undersea technology handbook directory.* Ch. 4 (Arlington, Virginia, Compass Publications, 1969).

National plan for the safety and health of divers in their quest for subsea energy. Shilling, C. W.; Beckett, M. (eds.). (Undersea Medical Society, 9650 Rockville Pike, Bethesda, Maryland) (1975).

Underwater physiology. Proceedings of the IV Symposium. Lambertsen, C. J. (ed.). (Baltimore, Williams and Wilkins, 1970).

Underwater physiology. Proceedings of the V Symposium. Lambertsen, C. J. (ed.). (Bethesda, FASEB, 1976).

Underwater physiology. Proceedings of the VI Symposium. Shilling, C. W.; Beckett, M. W. (eds.). (Bethesda, FASEB, 1978).

A guide to the Diving Operations at Work Regulations, 1981. Health and safety series booklet HS(R) 8 (London, HM Stationery Office, 1981), 56 p.

Dock work

Dock work nowadays involves a wide range of operations carried out on board ship, on the quayside and in port warehouses and terminals, and frequently calls for the use of a great variety of equipment, sometimes sophisticated and therefore expensive. Dockworkers are commonly employed in the loading and unloading of ships, the transfer of goods from a ship to a barge or to another means of transport such as a railway car or a lorry and vice versa, the piling and storage of goods in sheds, warehouses and special terminals, the transport of goods and personnel within the confines of the port, the maintenance of cargo-handling appliances and gear, etc.

Ports may be natural or man-made, on the sea or on inland waterways, large or small. Some docks deal with world trade, others may serve a particular factory or installation. In some places the loading and unloading operations may take place in deep water or in midstream, with small vessels providing the link between ship and shore. Ports may handle bulk cargoes or general cargoes; installations intended mainly to receive oil tankers, tankers transporting liquefied gas, vessels carrying coal, ores, cereals, fruit, wine, cattle or motor cars, as well as other ships equipped for carrying special cargoes, can be integrated into the port facilities.

Cargo-handling equipment

As a result of the spectacular increase in maritime shipping since the Second World War, mechanisation of cargo-handling operations has proceeded at a rapid pace in order to speed up the loading and unloading of the larger and more expensive ships and to reduce their turnaround time in ports. This would not have been possible without the aid of advanced equipment and recourse to complex planning and control procedures. A wide variety of equipment is in use for handling the many types of goods carried on board ship: general cargo consisting of crates, boxes, containers, bags, pipes, rolls of paper, bales of cotton, logs, vehicles, machines, animals and so on; bulk cargoes in solid, liquid or gaseous form (coal, ores, sand, grain; mineral oils, latex, heavy chemicals, fertilisers, wines, etc.; propane, butane and natural gas). Heavy and bulky consignments (locomotives, machines, transformers, turbines, etc.) require special handling equipment and methods, whereas smaller consignments are often grouped together on pallets or in containers.

In the larger ports, where ships are lying at the dockside, loading and unloading usually involves the use of electrically driven cranes of various configurations. When there are no cranes on the quay or the ship is moored at some distance from the quay or in midstream, use is made of the ship's derricks or cranes. Bulk cargoes are handled by means of transporters, loading bridges, elevators, and belt or pneumatic conveyors, while liquids and gases are conveyed by pumping. Fork-lift trucks are widely used for handling goods on pallets, rolls of paper or steel, coils of wire, etc., both on shore and in ships' holds. The use of containers conforming to international standards has greatly simplified and accelerated loading and unloading operations; a large part of the packaging and stowing work is now carried out by the export or warehousing and shipping agents and no longer at the docks or on board ship. The impact of containerisation on port operations has been impressive; a modern container berth can handle some 800 000 tonnes of cargo per year, while the same volume of break-bulk cargo would require at least five general cargo berths.

At the same time, ships of entirely new design have appeared: freight container ships, barge-carrying ships (LASH, BACAT, SEABEE, etc.), roll-on/roll-off ships (Ro/Ro), etc. There are also multipurpose carrier ships intended for the transport of two or more types of cargo (e.g. ore and oil, general and bulk cargo, general cargo and containers, lighters and containers, etc.); certain ships may have holds for containers, some deck space for trailers or other cargo on wheels, with roll-on access by side ports, the remainder of the space being reserved for conventionally stowed cargo.

Again, to reduce handling time, goods are increasingly packaged in unit loads. These may be either pre-slung or fitted with baling straps so designed that a number of units can be picked up by specially designed lifting attachments suspended from transporter cranes capable of traversing the length of the deck and having athwartship arms to span the dockside. Freight containers are as a rule stored in terminals specially constructed for the purpose. The four most commonly used container handling systems today are the straddle carrier (figure 1), the trailer, the sideloader and the gantry crane (rubber-tyred or rail-mounted); mixed systems are also in use. Straddle carriers are capable of stacking 12-m containers (weighing up to 30 t) three-high. There have been parallel developments in the design of both port and shipborne lifting gear, particularly as regards mobile cranes, which may have variable-length or telescopic jibs, which in turn may be fitted with fly jibs. The mechanisation and redevelopment of ports is proceeding at a rapid pace, and

Figure 1. A modern Vancarrier for container handling. *(By courtesy of Stovkis Container Handling.)*

more and more ports are capable of receiving the new higher-capacity cargo ships. As regards shipborne equipment, mention should be made of mechanical hatch covers, cargo lifts, and scissor hoists as well as loading doors and ramps and retractable car decks on roll-on/roll-off ships.

When considering all the modern technological developments outlined above, one should not overlook the fact that the proportion of break-bulk cargo handled by conventional equipment and methods is still very substantial and is likely to remain substantial for many years. All over the world, there are still small ports handling only small ships; and all berths and facilities are still maintained in many ports to cater for the conventional type of ships and operations.

Employment of dockworkers. The labour requirement in many ports fluctuates from day to day and, consequently, it is common in many ports to maintain a reserve of workers who receive a fixed wage and can be called upon to work in different docks as required. In other ports reliance is still placed on a pool of casual labour. As a result of increased mechanisation, the dock labour force has been sharply reduced in many of the larger ports (in the developing countries, however, mechanisation is sometimes resisted for fear of unemployment); a similar picture emerges when one considers the number of stevedoring employers.

Figure 2. A mobile crane for cargo and container handling in ports. *(By courtesy of Nellen Kraanbouw, Rotterdam.)*

HAZARDS

Accidents. Injuries may be caused by environmental factors (bad weather, confined spaces, poor lighting, inadequate design or maintenance of the work premises, etc.), by the equipment used (fixed or mobile) and by the goods themselves (fire, explosions, etc.). The hazards are particularly numerous and they are intensified by frequent changes in the place of employment, by the diffusion of responsibility and, in many instances, by the need to work fast.

The commonest accidents as a rule involve slips and falls, falling loads or objects or manual handling and mechanical equipment, although the list may be affected by the type of operation. In some cases poor visibility or traffic congestion may lead to serious and costly accidents (particularly when straddle carriers, side-loaders and fork-lift trucks are used). The majority of injuries are to the hands, the trunk, the lower limbs and the feet; here again, the distribution is influenced by the type of cargo handled (in the handling of containers, for instance, it appears that there are more accidents to the head and to the upper and lower limbs). In the past, workers (and sometimes supervisors as well) were inclined to ignore safety rules and by-pass safety procedures in order to make their work more convenient and remunerative. The situation is changing, however, as both management and workers are becoming increasingly aware of their responsibilities and as the cost of material damage increases as a result of the handling of more expensive goods and the use of more sophisticated equipment. The problem is complicated by the fact that ports and related areas are by no means uniform in layout. Poor surfacing, poor housekeeping of traffic lanes and sharp differences in the level of illumination (when entering a ship's hold, for instance) can create serious hazards. In spite of the progress made, dock work remains a dangerous occupation (as is evidenced by a comparison of accident rates), and the severity rate of accidents occurring in the loading and unloading of ships and related operations is particularly high. Furthermore, in most ports of the developing countries, working hours are long, work is often arduous (particularly when climatic conditions are taken into consideration), overtime is not uncommon and work is still arranged in two or three shifts.

New equipment and new methods of cargo handling require new control measures. As regards ships the problem is aggravated by the fact that remedial measures often involve structural alterations. If many naval architects and engineers in the past have paid too little attention to the requirements of safety at the design stage, it must be recognised that, even today, some container and other ships are built with little regard for the safety of the people who will have to work on them.

Health hazards. Dockworkers may be exposed to dangerous gases, vapours, fumes and dusts that escape from broken or inadequately packaged containers or that are liberated by bulk cargoes in poorly ventilated spaces. Skin or eye contact with corrosive substances may result in chemical burns, and the handling of radioactive materials may entail a radiation hazard. Contact with certain hides or skins may cause anthrax, while some cargoes may harbour venomous animals. Accidents have sometimes been caused by fumigated grain, by oxygen deficiency or by the fermentation of organic matter in holds. Finally, the introduction of the fork-lift truck into ships' holds and port warehouses has created a distinct health hazard as a result of the presence of noxious exhaust gases and fumes in poorly ventilated areas. In the majority of countries, however, the high incidence of low-back disorders still constitutes the major occupational health problem among dockworkers. Fairly recent studies have revealed lumbar complaints or low-back disorders in more than 50% of the population concerned.

Physiological tests conducted to assess the physical workload of dockworkers have shown that even light work can be physically exacting as a result of load peaks and tiring postures. Furthermore, occupational hygiene investigations at various workplaces have revealed high noise levels and inadequate lighting in addition to high concentrations of engine exhaust gases. More recently concern has been expressed by workers' organisations in certain maritime countries regarding the hazards resulting from the inhalation of asbestos dust released during the handling of consignments of asbestos fibre.

SAFETY AND HEALTH MEASURES

The promotion of health and safety at docks is rendered difficult by a number of special factors related to the nature of the work.

In many cases there is no fixed place of employment and it is difficult to locate health and welfare amenities in a suitable place. Work is carried on in all weathers, sometimes in the hours of darkness and at high speed to ensure rapid turnaround of the ship. There is great dispersion of responsibility: stevedores often work with ship's gear for which the shipping company is responsible and the shipping company itself may be registered in a country where safety standards leave much to be desired. Although, in some countries, dock labour is being decasualised, much of the labour force is still hired on a casual basis and is employed only intermittently; consequently, opportunities for training are often inadequate.

Accident prevention. Many maritime countries have legislation stipulating standards of safety and requiring regular inspection of the lifting gear carried by ships visiting their ports. International agreement on standards of examination and maintenance is of great importance. Similarly, international collaboration on the marking of weights of cargoes should do much to ensure the use of the appropriate machinery. Strict legislative control of all lifting and hoisting machinery and equipment used on shore is also required (see CRANES AND LIFTING APPLIANCES). Cranes, hoists, tackle, etc., suffer very heavy wear and tear and a high standard of maintenance is essential. Frequent examination of all gear is necessary and defective equipment should be correctly repaired or discarded at once.

Cranes on the dock should be provided with limit switches and load-limiting devices: attention should be given to remote control for cranes and winches. Loads should not be suspended unnecessarily above areas where men are working, above hatch covers, or vehicle tracks; and loads should always be set down carefully. On board, signals for hoisting should be given by a trained, responsible person posted near the hatches, unless the crane driver has an unrestricted view and the danger area has been fenced off.

All electrical power tools should be carefully maintained and inspected to ensure that earthing arrangements are effective and that defective cables, etc., are replaced immediately. Adequate lighting of all workplaces is a prerequisite of safety on the wharves, on the quays and in the holds. This is particularly important when work is carried on beyond the hours of daylight.

The surface of quays should be well maintained and docksides should be adequately fenced. Safe means of transport from ship to shore should be provided where necessary. Gangways and gangway ladders giving access to ships should be protected at dangerous points by horizontally suspended safety nets. All gangway ladders, etc., should have steps sufficiently wide for the feet. If cargo is stowed on deck, safe passageway for workers should be ensured, for example by walkways with handrails. The use of damaged ladders should be prohibited. Rope ladders should not be used inside the ship.

When work is performed on hatches, all boards should be in good repair. During the loading and unloading of goods, the hatch beams should, if possible, be removed and placed on the dockside: those remaining in the hatches should be secured to prevent moving or falling. Frequent use may impair the beam-securing device and additional precautions such as clamps and stretchers may be necessary. Unless special precautions are taken, vehicles should not be allowed to pass over wooden hatches.

Safety nets should be placed in position when men are working on the upper and lower decks near hatches or when they are working near a partly opened hatch or on a tall stack of goods. Winches should be fitted with a winding device to prevent dangerous loops forming on the winding drums: the steering gear should be operated by a lever and not a wheel.

Handling equipment such as fork-lift trucks driven by petrol engines should not be used in the holds unless an appropriate ventilation plant is installed and continual air analysis made, since there is a risk of poisoning from exhaust gases. If similar machines, driven by other fuels such as diesel fuel, are used the workplace must be adequately ventilated. Modern installations for loading and unloading, many of them highly automated, often require, for safety reasons, that workers remain as far as possible from the danger zones around various lifting gear, especially when mechanisms such as electromagnets and vacuum-lifting devices are used. When container-handling is automated or semi-automated, devices should be provided to indicate whether the containers are firmly secured to the spreaders.

In spite of their apparent simplicity, modern freight container terminals require a high standard of operation. Persons may enter the stacking area unnoticed and may be trapped or run down by moving equipment or cargo; the hazard is particularly great in the case of unauthorised persons (members of ships' crews, visitors to ships, etc.). As complete physical separation is impracticable, it has been found necessary to institute "permit-to-enter" systems similar to the "permit-to-work" procedures adopted in potentially dangerous situations (see PERMIT-TO-WORK SYSTEMS).

Personal protective equipment, including head protection and foot and leg protection, should be provided to prevent injuries to the head and feet. Hand protection of suitable material can prevent many of the common hand accidents; eye and face protection will prevent foreign bodies from entering the eye.

Marking and labelling of dangerous substances to identify goods which require particular care in handling should be supplemented by easily recognised symbols to obviate language and illiteracy difficulties.

Much remains to be done in training dockers in safety. In some countries training schemes have been established which provide the necessary instruction, primarily for workers who operate cranes, winches, hoists and elevators or for signallers and other specialised workers. The appointment of safety officers by combined port enterprises may make a valuable contribution to accident prevention. In addition, safety committees combining representatives of the port inspectorate, port authorities, employers and trade unions, can do valuable work in ensuring the strict application of safety reulations and in spreading information and safety propaganda.

Health protection and welfare. Dock work has to be carried out in all weathers and, in some climates, workers are exposed to extreme conditions where protective clothing is necessary: warm protective clothing may also be necessary for work in refrigerated holds, especially in tropical ports.

The scattered nature of much dock work makes the provision of reasonable health and welfare amenities difficult but this is none the less essential. Sanitary and washing facilities at convenient places on the dock should be provided with efficient arrangements for keeping them in clean and hygienic condition: where dirty cargoes are handled, shower baths are desirable. Catering facilities such as a central canteen serving hot meals are desirable on large docks, and mobile canteens may also be necessary: the minimum requirement is a clean, weather-protected messroom. The availability of non-alcoholic hot and cold beverages will help combat a tendency to excessive consumption of alcohol, detrimental to both health and safety. A central first-aid room, or surgery for treatment of serious injuries should be supplemented by first-aid kits located at the worksite. There should be an adequate number of persons trained in first aid. As a rule, there should be one first-aid attendant for 25 dockworkers and one first-aid room for 100.

Where dockers are exposed to special health risks, basic information on the subject should be provided, for example a card or placard on anthrax for those handling hides or skins from certain countries.

In many countries the large dock companies operate an occupational medical service, and it is desirable that some form of medical supervision should be extended to all dockworkers. Pre-employment medical examinations are important in ensuring that men have the necessary physique for arduous work. Arrangements to deal with emergencies on ship or on shore are essential. Life-saving apparatus should be kept at the water's edge and a notice giving illustrated instructions for resuscitation from drowning or electric shock should be displayed at conspicuous and suitable places; as many people as possible should be trained in methods of artificial resuscitation.

DIMTER, G.
ROBERT, M.

Occupational Safety and Health (Dock Work) Convention, 1979 (No. 152) and Recommendation, 1979 (No. 160) (Geneva, International Labour Office, 1979).

Safety and health in dock work. An ILO code of practice (Geneva, International Labour Office, 1977), 221 p.

Guide to safety and health in dock work (Geneva, International Labour Office, 1976), 287 p. Illus.

"Shipboard safety at a container terminal". Cross, J. P.; Ritchie, J. D. *National Ports Council Bulletin* (London), 1976, 8 (34-57). Illus.

Dockwork (Stockholm, National Board of Occupational Safety and Health, 1978), 96 p. Illus.

"Proposal for a Council directive on the harmonized application of the International Convention for Safe Containers (CSC) in the European Economic Community". *Official Journal of the European Communities* (Brussels), 8 Sep. 1980, No. C 228 (43-63).

Figure 3. Traditional loading and unloading operations. *(By courtesy of the Institut National de Recherche et de Sécurité, Paris.)*

Domestic workers

The expression "domestic workers" is used to denote all the various categories of workers who hire out their services to an employer and his family for the performance of the various tasks involved in the daily running of a house. The definition of "house" recognised by the World Health Organisation (1957) covers the residence and its annexes, garden, yard, garage and other outbuildings, together with the staircases and spaces giving access to the apartments and living rooms, provided that these are reserved for the exclusive use of the tenants. Domestic workers therefore differ from other subordinate workers by the absence of any direct participation on their part in the lucrative, professional activity of their employer. In addition, according to the terminology adopted by certain countries, domestic service need not necessarily mean the performance of manual work only; thus German law uses the expression "domestic worker" to denote both domestic staff in general and those whose services are of a special nature (private teachers and tutors, personal servants, paid companions and the like). In most countries, however, domestic staff are workers who hire out their services for work of an essentially domestic nature.

[Although statistics on domestic employment usually relate only to officially declared or registered workers, who in some countries may be the minority, it can be assumed that today domestic workers do not account for any considerable part of the economically active population. Domestic employment in private households is almost non-existent in socialist countries, it is moderate and decreasing in Western Europe and North America, it is on a fairly wide but indeterminate scale in many developing countries. In most countries women predominate and for them household work is an important field of gainful employment.]

Types of activity

These vary according to the composition and status of the family but include, basically:

(a) kitchen work–preparing meals, waiting on the family at table, care of tableware;

(b) cleaning the residence and its annexes, and care of the furniture and chattels;

(c) washing, ironing and care of clothes and household linen;

(d) outside errands necessary for the performance of domestic duties (shopping for food and other products required for domestic work, etc.).

[The large majority of women in domestic employment are general household workers; some are more specialised in cooking, child care or laundry work; the other main category, made up of part-time domestic workers, is general cleaning women. In the developed countries men are mainly employed in gardening, driving and valet service; in the developing countries men help out with all routine household tasks.]

Living conditions

These vary with the country and time. Originally, being devoted entirely to the service of the master of the house and of the members of his family, the domestic servant lived in the house, was fed and clothed there and was, to some extent, a member of the family, even though his living quarters were quite separate from those of his master's, his food leavings from the family's meals, and his clothes mostly cast-offs from his master's wardrobe. Nowadays, the domestic worker more and more frequently lives away from his place of work, with the members of his own family, and leads a private life independent of his employer's. [Furthermore, in the developed countries there has been a substantial decline in full-time and an increase in part-time domestic employment. In the great majority of countries domestic workers are almost entirely nationals.

Some Western European countries admit foreign young people on an "au pair" basis. Originally the purpose was to learn another language and to share family life, and the "au pair" girls were not regarded as domestic workers. In recent years, however, because of the shortage of domestic workers, there has been a danger of abuse of the "au pair" arrangement and its deterioration into a form of cheap and uncontrolled domestic employment.]

In many countries, this occupation, like any other work, is regulated by codes or legislation. These, although varying somewhat from country to country, generally define:

(a) conditions of employment;

(b) conditions of work during the period of contract [including protection of wages, weekly rest and annual paid vacation, but usually not including working hours, daily rest, sick, maternity or emergency leave, and minimum age for admission to domestic employment in private households];

(c) conditions governing cancellation of the contract to work.

HAZARDS

Domestic workers are only rarely subject to occupational diseases. The diseases that do occur, however, include skin diseases, particularly eczema; chronic paronychia in cleaning women and seamstresses; various rheumatic complaints which are the result of prolonged and repeated immersion of the hands in water while washing clothes or dishes, or working in hot workplaces causing perspiration, particularly in certain kitchens; and psychosomatic disorders. To these must be added risks of contagion from employers. Forms of bursitis and tenosynovitis, such as "housemaid's knee", and lumbago or back pain are relatively common but are gradually becoming less frequent with the introduction of mechanical household equipment.

Occupational accidents are more frequent than occupational diseases and vary with the social organisation and economic development of the country in question. The principal types of domestic accidents in modern societies are accidents caused by the explosion of combustible materials (town gas, for example) and by fire generally; burns caused by corrosive substances or steam; accidental poisoning; falls; electrical accidents and accidents which occur during the cleaning of firearms. Small cuts and abrasions are also common (see ACCIDENTS, OCCUPATIONAL DOMESTIC).

SAFETY AND HEALTH MEASURES

The incidence of paronychia can be reduced by limiting the time that the skin is in contact with polluted water, especially when washing clothes or dishes by hand. The use of hand protection and dish- and clothes-washing machines appears to be contributing to the prevention of these types of affection.

Accident-prevention measures include ergonomic lay-out and equipment of the household, particularly the kitchen; reduction of the amount of work done standing, bending or in other physiologically harmful postures which cause excessive osteomuscular fatigue; care and regard for safe design in the purchase of domestic equipment, particularly electrical appliances, with preference given to those which have seals of approval from safety standards organisations, etc.; measures to reduce physical and mental fatigue with strict observance of regulations on hours of work, rest periods and paid holidays. However, in the majority of countries, 60-80% of domestic workers are not registered with employment boards or labour inspectorates. [They are often young persons especially liable to exploitation, and doubly handicapped by lesser education and training and by the isolated nature of their work.]

KOATE, P.

General conditions of work:

"The employment and conditions of domestic workers in private households: an ILO survey". *International Labour Review* (Geneva), Oct 1970, 102/4 (391-491).

Health impairment:

CIS 77-1391 "Irritation (or housewives') dermatitis" (Les dermites irritatives ou dermites des ménagères). Barrière, H.; Litoux, P.; Geraut, C. *Concours médical* (Paris), 12 Mar. 1977, 99/11 (1603-1607). Illus. (In French)

Medical surveillance:

"Medical surveillance of domestic employees and caretakers. Decree No. 75-882 of 22 September 1975" (Surveillance médicale des employés de maison et gardiens d'immeuble. Décret n° 75-882 du 22 septembre 1975). Marchand, M. *Archives des maladies professionnelles, de médecine du travail et de sécurité sociale* (Paris), Mar. 1977, 38/3 (392-394). (In French)

Doping

A person is said to dope himself when he uses a drug to increase his natural capacity artificially and transiently. This is the definition adopted by French legislation. A more all-embracing definition has been proposed by Rapp: "Any unfair action intended to increase the performance of athletes and which may be harmful to their health".

Since antiquity, man has always sought to improve his natural performance; however, it is the public's craze for sporting records together with the progress of pharmacology that have made doping a social evil during recent years. Secretly and widely used during sporting competitions, reputedly miraculous products are often sold at exorbitant prices by the trainers. It has been called the "cancer of sport", and what is more important, young people are often tempted to follow the example set by the professionals.

The medical profession has drawn attention to the risks, which may be fatal. Following the initiative, in particular, of Dr. Dumas, doctor for the well known "Tour de France" cycling race, a European symposium was held in 1963, leading finally to the adoption of legislation in France (Orders of 1 June 1965 and 10 June 1966) and of various provisions in other countries, as well as to close supervision of sporting events.

Persons may take drugs in any profession (see DRUG DEPENDENCE). In the present context, attention will be devoted to certain sporting professions which have been specifically mentioned in national legislation. Among the many sports in which doping is known to exist, those principally affected are: cycling, swimming, athletics, cross-country cycling, boxing and even shooting.

Drugs involved. A wide variety of drugs have been used, either singly or in combination, including digitalin, because a champion normally has a congenitally slow pulse rate; strychnine, ephedrin and caffeine, to delay the onset of fatigue; anorexic drugs, to aid an athlete to lose weight so that he can get into a lower weight group; and others. In most common use, however, are amphetamines and related drugs, which are taken as stimulants, or tranquillisers, which include barbiturates and meprobromate, to combat anxiety, particularly in shooting competitions. Even strong analgaesic drugs such as morphine or dextromonoramide (palfium) may be used to reduce pain due to excessive effort, as in cycling, or due to injury, as in boxing.

The ease with which chemical tests for doping can be made has tended considerably to restrict the use of these drugs, but in recent times they have been replaced by the corticoids (cortisone affects the cerebral nerve cells and also plays a role in resisting stress, thereby enabling competitors to keep going over long periods of effort), and especially by the anabolic steroids, which by synthesising the tissue substances, particularly proteins, promote the growth of muscle volume when taken during training. The latter substances are sometimes known as "muscle fertiliser".

The question may be asked whether these drugs are really effective. The results of scientific research are contradictory and, taken as a whole, not very convincing; the psychological aspect must not be overlooked, to the extent that an athlete who has taken a drug is conditioned and has great faith in the "miracle drug" that he has received. It has frequently been shown that athletes who have been given a placebo have performed equally well as those that have been given drugs, and the conclusion has been drawn that training is what counts above everything else. These experiments, however, have been made under medical supervision with the administration of doses that are not dangerous, whereas in practice the athlete doses himself in secret, and often takes an excessive dose, leading to the accidents that have been seen.

Symptoms and accidents. Drugs are pain-killers, give rise to euphoria, and suppress the sensation of fatigue which is, physiologically, the danger sign. The athlete surpasses himself—which is the object he is aiming at—but he has made too great an effort and thereby suffers from acute overstrain. In other cases, in a cycle race for example, euphoria makes the athlete oblivious to danger, and reflex changes (or more serious mental disorders) may lead to a serious accident such as a fall into a ravine. Finally, the amphetamines contribute to the development of hyperthermia which can result in heatstroke. In consequence, in the aetiology of the fairly frequent fatal doping accidents in sportsmen, although the direct toxic action of the drug is seldom the single cause, in the absence of the drug the accident would not have occurred.

The use of amphetamines produces the following symptoms: mydriasis, low frequency blinking, tremors, exaggerated reflexes, excitation and, on occasions, speech difficulty. The accidents caused by amphetamines (apart from falls and heatstroke) include the following:

(a) sudden collapse during a race; the athlete totters, falls, gets to his feet again and continues the race, but falls once more, haggard, aggressive and bathed in perspiration;

(b) psychological disorders: a maniacal condition with intense motor excitation, delirium, mental confusion, hallucinations;

(c) epileptiform seizures;

(d) heart failure and coma.

After-effects of hormones. The use of corticoids can lead to duodenal ulcers, oedema, intermittent or premature heartbeats due to loss of potassium, osteoporosis, mental troubles.

Continued heavy dosing with anabolic steroids can lead to aggressive character modifications, hypertension, hypertrophy of the prostate (with a risk of cancer), and alterations in the liver—in some cases even to adenoma or to cancer of the liver.

Testosterone, which is a natural anabolic hormone, may lead to sterility. There is also a risk of virilism in the sportsman or arrest of growth in adolescents.

Prevention of doping. Doping should be suppressed because it is a fraudulent method of obtaining an advantage over competitors and it is the cause of accidents that are often severe and sometimes fatal. It was made a penal offence in Belgium in 1965, followed by France and then Italy. Co-operation was established between the various sporting organisations in order to outlaw doping and this was taken up by the various national and international federations, including the International Olympic Committee, and these have drawn up their own rules.

A list of proscribed medicaments has been published and is brought up to date each year. Checks are made according to the prescribed procedures. These are based mainly on urine analyses, the samples being always divided into two parts in case of contestation. Sanctions include fines or disqualification and are imposed by the sporting organisations without, however, prejudice to later judicial action in certain cases.

Much depends upon the precision of the analytical procedures, which must be performed by qualified experts. Initially they relied largely on gas chromatography, to which mass spectography was subsequently added. Used together, these two techniques offer great promise.

A further check has been included in the regulations which is not a question of doping: certain womens' events have been won by persons of indeterminate sex (hermaphrodite), some of whom have subsequently undergone a sex change to become men. Thus a femininity check consisting of a microscopic examination of the sex chromatin is now carried out before such events.

Other techniques. Apart from doping, other techniques have been tried in order to increase performance. These include:

(a) blood transfusions: hetero- and auto-transfusion have been tried in order to increase the number of red blood corpuscles and improve the oxygenation of the tissues. This practice seems to have been abandoned as the results were disappointing;

(b) electrical stimulation of the muscles: this has been applied during training with the object of developing the muscle volume. It is, however, a risky procedure since it may lead to a ruptured tendon which has not developed in harmony with the muscle–this danger also exists with the use of anabolic steroids; it may also lead to rupture of the aponeurosis or to permanent muscular cramp;

(c) injection of air into the intestine: this has been tried on some swimmers. With increased volume, their weight in water is reduced. There is, however, the danger of rupturing the intestine.

Prevention. Doping should be replaced by physiological and psychological education that will make it possible to improve performance without fraud and without injury to health.

DESOILLE, H.

Doping of sportsmen (Le doping des sportifs). Rapp, J. P. (Paris, Editions médicales et universitaires, 2nd ed., 1978). Illus. 233 ref. (In French)

"Artificial methods for improving performance in championship sports" (Artifizielle Methoden zur Steigerung der Leistungsfähigkeit im Spitzensport). Hollmann, W.; Liesen, H.; Rost, R. *Deutsches Arzteblatt* (Cologne), 1978, 75/20 (1185-1192). (In German)

"Doping" (Doping). Dumas, P. *Gazette médicale de France* (Paris), 26 Jan. 1973, 80/4 (503-514). 8 ref. (In French)

"Position statement on the use and abuse of anabolic-androgenic steroids in sports". *Medicine and Science in Sports* (Madison, Wisconsin), 1977, 9/4 (XI-XII).

Dose-response relationship

Dose

Dose is defined as a precise amount of a substance administered by injection, inhalation or ingestion for experimental purposes. However, in the field of occupational health dose cannot be as precisely measured as in planned experiments. Concentrations of a chemical in the air of a working environment may fluctuate widely during working hours, as well as by day, month, year and decade. Furthermore, workers move about inside and outside the plant, and hence the concentrations of a given chemical to which they are exposed can vary greatly. Thus, "dose" in occupational health is in practice represented in the following two ways:

(1) Occupational exposure may involve the concentration of a substance in air; the rate of inhalation of various factors of deposition, retention and absorption. In addition, skin absorption and exposure via the general environment, for example via ambient air, drinking water, drugs, smoking, consumer products, and other factors may sometimes be a matter of concern.

(2) Alternatively, in occupational health dose may involve the concentrations of a substance in various human biological materials, including blood, urine, hair, secretory fluid, organs, cells, or even subcellular compartments. The concentration in such biological material may give a direct indication of concentration in an organ affected by the chemical in question. Often, however, such measurements indicate no more than the concentration in the indicator medium. Thus it must be kept in mind that the concentrations in indicator media do not necessarily show the level of exposure or the concentration in the organ affected by a given chemical. On the other hand, such measurements can be very useful. For instance lead concentration in blood is a fairly good indicator of lead body burden in soft tissues, though not in the bones. Blood lead concentration has therefore frequently been used for the estimation of body burden or for epidemiological studies of dose-effect or dose-response relationship between lead body burden or exposure and its effects on human beings.

Total occupational exposure to a given chemical can theoretically be expressed by "concentration x time" or $K = C \times t$. This equation has been widely accepted in occupational health, but because the pharmaco- or toxico-kenetics of each chemical is unique it cannot be applied to all chemicals. Usually "time" in occupational health practice is expressed as the length of years of exposure or years of employment in the same job. In some instances the total exposure to a chemical in question may be useful. However, when this equation is used for the estimation of dose or exposure over a long period of years, many other factors must be kept in mind.

Measurement of exposure. The environmental monitoring of a workplace is not necessarily equivalent to the measurement of exposure, but is usually used for the purpose of controlling the work environment. When dose-effect or dose-response relationships are to be studied, the air in the workplace should be sampled by means of personal samplers. Air samplings made at fixed sites in a plant by means of low-volume or high-volume samplers should be used only for reference information concerning exposures. Since the results of sampling by means of low-volume, high-volume, or personal samplers may be significantly different, the methods used should be clearly mentioned in the sampling report. For dusts, it is the respirable dust concentrations that may be very important in a dose-response observation.

Analytical problems. The analysis of trace elements in biological materials may give rise to a wide range of results for the same sample. For instance, lead concentrations in samples of the same blood analysed by different laboratories have shown variations as great as 10 : 1. Thus the measured values of trace elements must be handled with great care.

Effect and response

An effect may reveal various values. For example a certain level of coproporphyrin is excreted in urine by the "standard population", but the urinary excretion of the substance resulting from exposure to lead may indicate levels higher than those observed in the "standard population". Response is defined as the "rate" (percentage) of people (workers) or organisms that exhibit higher or lower ("abnormal") values than the defined "standard" value. Here, "abnormal" needs to be defined. If, for instance, coproporphyrin is taken as a critical effect (see below) in lead exposure, the "normal" value in a non-exposed control group or a "standard population" must first be defined; the "response rate", i.e. the proportion of people in the exposed group who exhibit "abnormal" values, can then be calculated.

Effects may range from those that are extremely adverse, resulting in death, to very minor ones that may merely take the form of the binding of a trace element to ligands of cells or an increase of the body burden of the substance without any functional or morphological evidence of changes. In some instances, a difference of opinion may arise as to whether an effect is or is not adverse for animals or humans. The Committee on Toxicology of Metals of the Permanent Commission of the International Congress of Occupational Health defined "critical concentration" (1973) and "critical effect" (1974) as follows: "The critical concentration for a cell was defined by the Task Group on Metal Accumulation (1973) as the concentration at which undesirable functional changes, reversible or irreversible, occur in the cell. The above-given definition of critical concentration in the critical organ represents a defined point in the relationships between dose and effects in the individual, namely the point at which an adverse effect is present. This is the critical effect." Thus, the critical effect may or may not indicate immediate concern for the health of the entire organism, but it is important for preventing more serious effects on health. As mentioned above, the term "response" is used only in terms of dose-response relationship, where response is the proportion or rate of a population with a critical effect or more severe effects.

Effects may be detected by symptoms and signs. Symptoms are defined as complaints on the part of individuals, whereas signs can be observed by examinations and tests, which may include physiological, biochemical or morphological analyses. Symptoms and signs are caused by morphological, physiological and biochemical changes. The physiological and biochemical effects may be caused by foreign substances binding with proteins and other ligands, or may concern the cell membrane, the energy metabolism system, known enzymic molecules or possible other biochemical changes. In occupational exposure, effects may occur at the site of entry, in a remote organ or organs, or at the site of excretion. Thus, it is important to know the specific effects caused by a chemical to which workers are exposed.

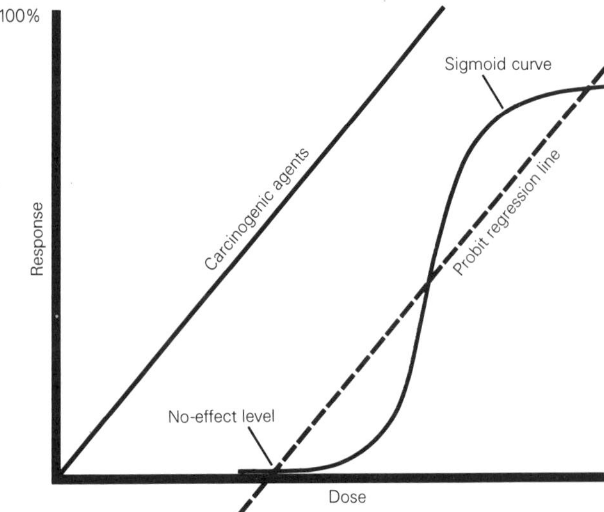

Figure 1. Dose-response relationships.

Dose-response relationship

The dose-response relationship for a chemical may be graphically represented as in figure 1. The dose plotted on the abscissa is indicated by concentrations of exposure in the work environment, or concentrations in the indicator media. Length of exposure (years of employment) may be used as a substitute for dose; although this is not very accurate, it is very useful in some cases, for example in epidemiological studies on occupational cancer. This type of study is performed because of the long incubation period of cancer after exposure, when exact exposure information is not available. Figure 1 shows the curves theoretically considered in the dose-response relationship for toxicological or carcinogenic effects of chemicals. The sigmoid curve of dose-response relationship is observed in animal experiments where the doses are precisely measured and administered. In occupational exposure, a curve exactly fitting the theoretical sigmoid shape is rarely obtained. However, some field studies on working populations may indicate clear dose-response relationships which statistically fit a sigmoid curve fairly well. Figure 2 represents such a case.

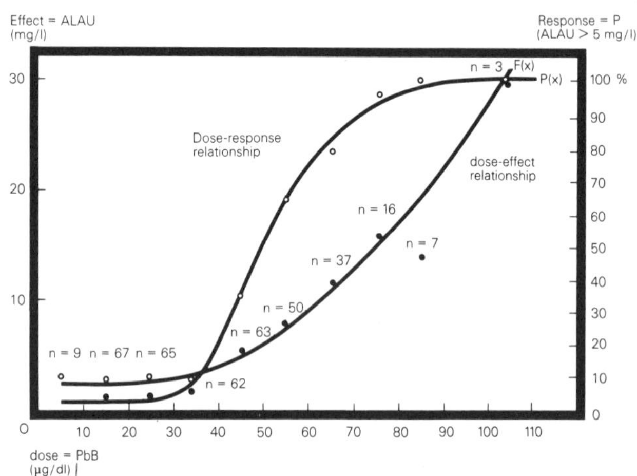

Figure 2. Dose (PbB) response [P(ALAU) > 5 mg/l] relationship and dose-effect (ALAU) relationship in Pb workers. Key: P(x): theoretical dose-response relationship; open circles: observed response; solid circles: observed effect.

In order to calculate the no-effect level from a dose-response relationship, the sigmoid curve, if obtained, can be converted to a "Probit" regression linear curve. The statistical method for this conversion can be found in standard works on pharmacological experiments or statistics. The straight linear curve obtained by the Probit conversion will cross the abscissa at a point which is the no-effect level, as shown in figure 1 (see dotted line). The solid straight linear curve indicates the theoretical dose-response relationship for radiation and carcinogenic chemicals as well. Some toxicologists believe that there is also a no-effect level for a carcinogenic chemical whereas some do not. Even if this is not so in the case of a carcinogenic chemical, a sufficiently low level exists at which cancer would not develop within an ordinary lifetime (e.g. 100 years).

A recent study has shown different no-effect levels derived from data of high and low exposure groups of lead workers. As shown in figure 3, the no-effect level

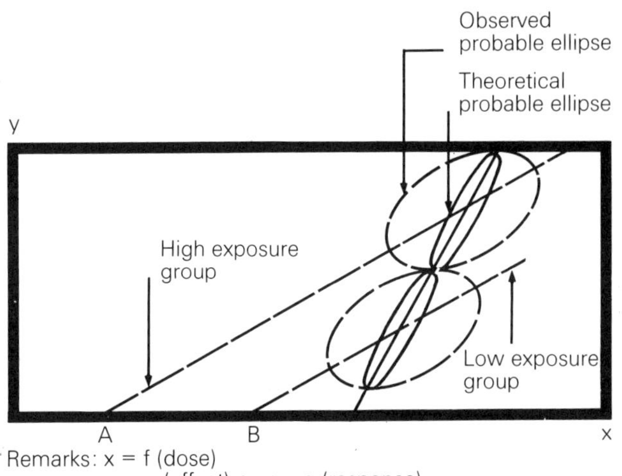

Figure 3. Deviations of no-effect levels from data of high and low exposure groups.

(A) which was calculated for a high exposure group is lower than that calculated for a low exposure group. The difference was caused by measurement errors, particularly of dose and it was shown that according to measurement error both probable ellipses and their major areas differed from those which were expected. It is important to remember that a no-effect level derived for a low exposure group may differ from that of another group (i.e. patients) with higher exposures and higher response rates, and careful evaluation is required in establishing the screening or no-effect level of a given chemical.

TSUCHIYA, K.

Statistical methods in medical research. Armitage, P. (Oxford, Blackwell Scientific Publications, and New York, Halsted Press, 1971), 504 p.

"Intercomparison programme on the analysis of lead, cadmium and mercury in biological fluids". Berlin, A.; Lauwerys, R.; Buchet, J. P.; Roels, H.; Del Castilho, P.; Smeets, J. (2185-2195). *Proceedings. International symposium: Recent advances in the assessment of the health effects of environmental pollution. Paris, 24-28 June 1974* (Luxembourg, Commission of the European Communities, 1975).

Statistical analysis in biology. Mather, K. (London, Chapman and Hall and Science Paperbacks, 1972), 267 p.

Effects and dose-response relationships of toxic metals. Nordberg, G. F. (ed.). (Amsterdam, Oxford, New York, Elsevier Scientific Publishing Company, 1976), 559 p. Ref.

"Dose-response relationships at different exposure levels–Necessity of re-evaluation in establishing no-effect levels". Tsuchiya, K.; Sugita, M.; Sakurai, H. (Vol. I, 123-132). *Chemical hazards. Proceedings of the XIX International Congress on Occupational Health* (Zagreb, Institute for Medical Research and Occupational Health, 1980). Illus. 11 ref.

Double-jobbing

Double-jobbing, or "moonlighting" as it is often called, consists in having an employment over and above one's basic job. This secondary employment may be identical to the first and may entail the skilled worker doing small private jobs on his own account outside the hours he is paid by his normal employer; or the same skilled worker may work a second shift for another employer. On the other hand, the secondary job may have no relation to the first, as in the case of the office clerk who serves behind a bar in a public house or café in the evenings, or the machinist who plays in a dance band on Saturdays. Normally, this secondary employment is remunerated; however, the professional accountant who keeps the books of his sports club may receive not even a nominal payment and yet still be required to devote many hours of intensive labour to the task. The position of the working housewife is closely related to this second group.

Reasons for double-jobbing

These are not confined to material considerations, and may be summarised as follows:

(a) Financial reasons. These may include a feeling of need for security, a desire for a higher standard of living or to maintain social appearances, or pure avidity, [or in some developing countries the need to keep body and soul alive].

(b) Occupational reasons. With the introduction of the shorter working week and the increase in leisure time, some people cannot find a suitable use for their spare time and resort to double-jobbing rather than waste time. [In recent years short-time working has increased opportunities and the need for double-jobbing.]

(c) Social and personal reasons. These may be the desire for the companionship of colleagues, an excuse for escaping from an unpleasant domestic atmosphere, or a method of self-expression. In the latter case the double-jobber is apt to identify himself with the secondary job rather than the primary, because it may bring him greater satisfaction. In addition, there are the voluntary double-jobbers who do social or domestic work and in whom the effect on health and work is often identical.

Effects of double-jobbing

These may be mental or physical and relate to the individual himself, his work, his family or his social environment.

The individual. The over-all effects may vary considerably, the age factor being of paramount importance. In the case of the juvenile, as a general rule, there may be little immediate apparent effect due to the natural resilience of youth. The first major turning point occurs at approximately 30 years of age. From 40 years the changes are dramatic. Varicose veins, hypertension, malignant and coronary diseases are all possibilities apart from menopausal changes, with their consequential effects. From 50 years, there is frequently a slowing down of mental and physical activity, with a sense of awareness by the individual. From 60 there are remarkable variations. Some people remain outstandingly fit, others already showing signs of incipient senility. From 70 some are stronger in spirit and mind than in body, so that great willingness is counteracted by physical incompatibility.

On the credit side the double-jobber may be able to undertake activities denied him in his primary job, for example, a manual worker may be able to participate in mental activities or vice versa. The debit side is influenced by fatigue, and mental factors thus affecting activities. Fatigue leads to insomnia which in turn leads to further fatigue so creating a vicious circle, perhaps rending the individual more accident prone. The main effects are impaired concentration and general lassitude with consequent inefficiency and lowered resistance. Associated irritability frequently renders the individual aggressive, thus affecting social harmony.

It is increasingly recognised that digestive and cardiovascular diseases such as duodenal ulcer and coronary thrombosis are frequently psychosomatic in origin and are not the monopoly of the business tycoon. Pruritus, especially of the genitals and buttocks are also noted. One marked effect is an abnormal liability to minor ailments, with a lowered threshold of resistance, and an above-average reaction or recovery time, the double-jobber frequently projecting his mental state into his physical disability.

Resultant upper respiratory tract infection, which includes coughs, colds and sinusitis is the most common cause of sickness absence among double-jobbers. Orthopaedic disability may arise as a direct result of the second job. Muscles and ligaments geared to standard repetitive movements may as the result of slight unusual movement produce a muscular-skeletal lesion, quite apart from diverse sequelae, such as abdominal herniae. Alimentary factors including irregular meals, unaccustomed diet, and poor facilities for taking meals, may lead to varying gastrointestinal symptoms, frequently associated with disruption of normal toilet habits, with consequent lassitude and headaches.

Skin contact with additional or variable chemical irritants may lead to contact dermatitis, which is frequently attributed to the primary job. This becomes a matter of forensic importance, as almost every job has its own industrial hazards, particularly when unaccustomed precautions are indicated.

The over-all effects of supplementary work on the double-jobber vary from inconsequential to bad. The primary job is usually more likely to suffer but a large firm found that better work was done on an evening shift by double-jobbers whose primary job was different, than by employees of the same firm who returned to work overtime. Sick absenteeism is indisputably higher in double-jobbers. One investigation revealed:

(a) repeated late morning arrival was almost invariably associated with some form of evening activity;

(b) sick absenteeism did not necessarily apply to the secondary job;

(c) lethargy and faulty work frequently brought the double-jobber to notice;

(d) there was some evidence of increased incidence of accidents;

(e) domestic difficulties were often connected with double-jobbing;

(f) it was observed that the double-jobber's rigid trade union principles did not necessarily apply to the secondary job.

Family and general social consequences. With the advent of so many modern domestic utilities, many wives find they have ample free time and so decide to take employment. This may lead to a new social environment which is not always to the advantage of the family unit. Husband and wife may deprive each other of normal companionship, or at home be so tired as to be fit only for sleep, and young people may suffer from lack of parental supervision. On the other hand, the standard of living and the quality of life may be improved by the extra income and interest acquired.

The attitude of fellow workers to double-jobbers varies according to prevailing economic conditions, and only when there is a lack of full employment does one note open resentment. The attitude of management vacillates between complete indifference and absolute dislike, although in some cases, such as night telephonists, double-jobbing is encouraged. Much depends on the nature of the particular work and the state of the labour market. It is often difficult to enforce any legal restrictions on hours of work when separate employers are involved.

As with all medicosocial problems, every case of double-jobbing must be judged on its individual merit. Depending on the individual job and extraneous circumstances the effect may vary from good to bad. Available information can create only an over-all impression, but nevertheless a definite pattern is emerging which may well form the basis of future research. It should always be recognised that many "leisure" activities, such as intensive sports training and participation, hard manual gardening, "do-it-yourself" home maintenance, may have effects very little different from those resulting from paid employment.

SCOTT, M. M.

CIS 1466-1966 "Double jobbing and occupational health—A medico-social survey". Scott, M. M. *Occupational Health* (London), Sep.-Oct. 1965, 17/5 (266-274).

"Clandestine employment: a problem of our times". Grazia, R. de. *International Labour Review* (Geneva), Sep.-Oct. 1980, 119/5 (549-563). 27 ref.

Drilling, oil and water

Drilling consists in boring a hole in the earth's crust for the exploration, discovery and extraction of petroleum, natural gas and water.

Techniques and equipment

Percussion drilling. In this technique, the rock is crushed by the impact of a heavy falling chisel bit; the cuttings are

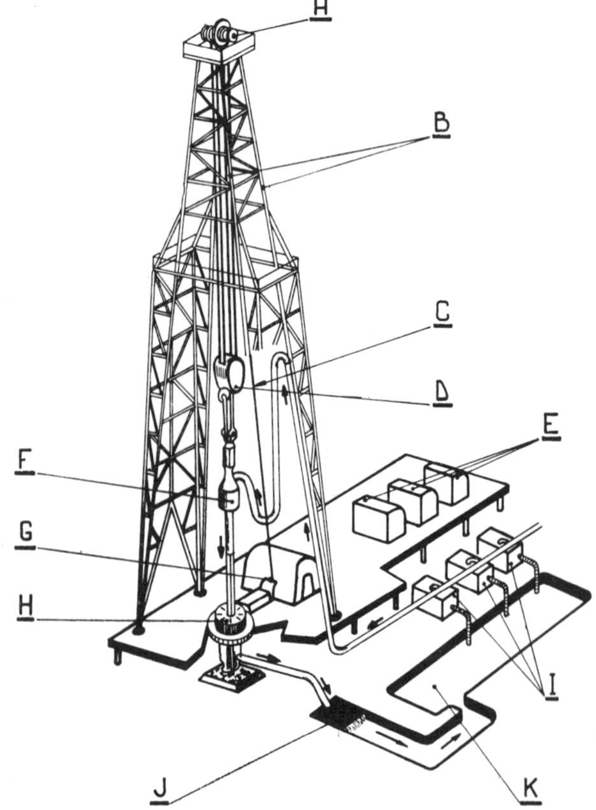

Figure 1. Drilling rig. A. Crown block and sheaves. B. Derrick legs. C. Cable. D. Travelling block. E. Engines. F. Mud injector. G. Winch. H. Rotary table. I. Mud pumps. J. Mud filters. K. Settling tank.

suspended in water and periodically removed by bailing. Progress with this technique is slow, and considerable manoeuvring is necessary to obtain even a shallow depth; it permits an immediate but sudden flow of liquid. Due to the limited depth possible, this technique has virtually been abandoned.

Rotary drilling (figure 1). This is the most common technique. It is used for both exploration and exploitation and can achieve depths of up to 7 000 m. Light drills or borers are used for low-depth exploratory (seismic) work and are usually truck-mounted; however, there is a trend to the use of medium-sized and heavy borers which may be mobile such as the EMSCO 1250, or floating, for offshore work. The equipment comprises a rotary table which turns a square kelly, with a mud swivel at the top and connection to the drilling pipe below. Penetration is by means of drill bits of one of two types: roller bits which have rolling cutters with hard teeth; and drag bits with fixed chisel-type, hard cutting edges. Drills range in diameter from 15 to 60 cm. The drill pipe rotates at a speed of 40 to 250 revolutions per minute. Depth of drilling can be increased by the addition of further drill pipes. The drill is mounted in a derrick 30-40 m in height with a drilling platform on which workers couple and uncouple the drill pipes, a motor, a mud mixer and injector and a wire-line drum-type hoist for providing controlled vertical motion. The pipes are 27 m long and are added to or removed from the drill bit one by one (see PETROLEUM EXTRACTION AND TRANSPORT BY SEA; OFFSHORE OIL OPERATIONS).

Electric drilling. The rotary tables, winches and pumps of heavy borers are often driven by electric motors powered by d.c. diesel-electric generating sets. This system has the advantage of reducing the number of units for transportation, allows the motors to be located outside the danger zone thus reducing noise levels, and enables operation to be more flexible and rapid. Drilling may also be automatically controlled.

Drilling mud or fluid. Mud or other fluids play an important role in drilling by cooling the bit, evacuating cuttings, caking the drill-hole walls and lubricating the drill pipes; a jet of mud or fluid may also be applied to the bottom of the drill hole to accelerate cutting. The fluid comprises water or oil and colloids, lime, sodium hydroxide or baryte, etc.; it is pumped from the mixing tank down the drill pipes to the drill bit and then returns up the space between the pipes and the hole wall to the surface where it is filtered and fed to a storage tank for re-use. Wherever possible, mud is replaced by a lighter fluid such as aerated mud, aerated or non-aerated water, air or natural gas.

Coring. A coring bit is used to obtain a compact, cylindrical sample of rock; the sample or core may be up to 50 m in length.

Casing. This consists in lining the walls of the drill hole to prevent caving and to prevent leakage from the return flow of mud.

Cementation. This improves the leak-tightness of the casing. A cement slurry is pumped down the drill hole and forced up to the surface through the gap between the casing and the surrounding earth; when set, this slurry seals the gap. With the casing in place, drilling is continued with a bit of smaller diameter.

Drilling for gas. The drilling techniques are the same as those for petroleum; however, additional precautions must be taken against blowouts which may occur more rapidly.

Drilling for water, steam and thermal water. The same types of equipment are used but more power is needed for drilling. Drilling and casing are carried out in a similar manner, but boreholes may be over 30 cm in diameter to allow greater flow rates. Drilling depths vary tremendously and range from 10 to 1 500 m. Drilling time is prolonged by the need for special care at the entry of the hole into the reservoir in order to ensure maximal output and operating life.

Work posts

Work on a drilling rig is essentially a team effort. The team comprises six members: the driller, his second, three rotary helpers or roughnecks (assistant drillers) and the cathead man. The latter has the toughest job. When drill rod or pipe is being run or drawn, he is posted on a platform 27 m high. To reach this position, he climbs an almost vertical ladder, sometimes aided by a counterweight and, once in position, attaches himself to the derrick by a safety belt and lifeline. He also assists the drilling crew and operates the mud pump. His task is one of general supervision and is very arduous. The assistant drillers work on the derrick floor (figure 2), attend instruments and perform preparatory work such as shifting equipment and carrying out repairs and maintenance. The driller is responsible for the actual drilling operations during his duty period. The site foreman (tool-pusher) is responsible for the correct progression of the drilling work. Some 20 other persons are employed on the well or field including mechanics, an assistant geologist, drivers and labourers.

Figure 2. The derrick floor.

Working conditions

Work continues in a three-shift system around the clock. The walking and work surfaces are slippery, cluttered and vibrate considerably; the pace of work is often forced, the noise levels are high and the workers are exposed to inclement weather conditions. Levels of energy expenditure are high (2 000 kcal per shift for the cathead man), there is the danger of falls when climbing the rig and of exposure to flying mud and inhalation of engine exhaust gases. Exploration and exploitation sites are isolated workplaces and drilling rigs are frequently remote from the base camp, which entails psychological problems.

HAZARDS

The health hazards in this work are largely the result of exposure to climatic conditions (respiratory tract diseases) and to infections and parasitic diseases in areas where these are endemic. For example the French statutory schedules of occupational diseases contain no mention of diseases specific to drilling workers. [However, periarthritis of the shoulder and shoulder blade, humeral epichondilitis, arthrosis of the cervical spine and polyneuritis of the upper limbs are common among oil rig workers. Their frequency and severity has been found to be proportional to length of service and exposure to adverse working conditions including arduous physical work with wrenches, vibrations and inclement weather.]

The accident hazard is considerable and a typical breakdown of common accident injuries has given the following results: bruising or crushing 40%; sprains, strains 18%; followed by cuts, abrasions, fractures, back pain, amputations. Of these injuries, 52% affect the upper limbs (40% the hands), followed by the trunk, lower limbs, arm and forearm, leg and thigh. The causal agents are mechanical handling 17% (tube and rod handling accounts for 12%); manual lifting and carrying especially of rods and tubes; slippery floors, hand tools; falling objects (spanners, etc.); hoists; falls from heights. Burns caused by fire or steam blowouts are relatively rare but, when they do occur, may be very serious; projections of mud containing sodium hydroxide may cause chemical burns.

SAFETY AND HEALTH MEASURES

Medical prevention. Drillers should receive a pre-employment medical examination. Candidates should be male, right-handed and have the following minimum physical characteristics (due allowance being made for ethnological peculiarities): height 165 cm; weight 65 kg; mean chest circumference 85; pulmonary vital capacity 4 l; vision: binocular 10/10, monocular 6/10 without correction and absence of colour blindness; normal hearing (undamaged eardrums); excellent sense of balance and freedom from vertigo; dynamometric capacity: manual 50, scapular 40, lumbar 125; ability to hold a 5 kg weight at arms length for 45 s.

In evaluating fitness for this work, the arduous nature of the working conditions should be borne in mind and particular attention should be devoted to the examination of organs, systems and functions that might be affected by these conditions: climatic conditions (respiratory system–radiological examination is essential); vibration (kidney stones); hours of work (digestive system, nervous system); general status (absence of organic or mental disorders, normal lumbosacral skeleton–systematic radiography). The medical examination should be backed up by psychotechnical testing. Periodic medical examinations (including radioscopy) should take place every three months during the first year and every 12 months thereafter.

Persons not actually employed on the drilling rig should also be subject to stringent pre-employment and periodic medical examinations if they are to be employed in regions with a harsh climate. All persons employed on the base should be well balanced both physiologically and psychologically.

First-aid organisation at the base and rig should be of a high standard and should include first-aid boxes and kits, emergency radio communication and provision for emergency evacuation by air from isolated workplaces.

Technical prevention. [Weather protection of derricks, mechanisation of work requiring heavy muscular effort, wrenches and uncomfortable posture, and ergonomic design and rearrangement of the equipment are recommended.] Personal protective equipment for drillers should include head protection, hand and arm protection (gauntlets), foot and leg protection (boots with steel toecaps and non-slip soles) and eye protection. When climbing the rig, the cathead man should wear a safety harness and lifeline attached to the counterweight and, when on the gangway, should wear a safety belt with a double lifeline, attached to two anchors.

Workers should receive instruction on the hazards of fire, explosion and asphyxiation; tools and safety equipment should be regularly inspected.

MONTILLIER, J.

CIS 77-1451 "Occupational disease in oil rig workers" (Professional'naja patologija rabočih, zanjatyh bureniem neftjanyh skvăin). Nabieva, G. V. *Gigiena truda i professional'nye zabolevanija* (Moscow), Aug. 1976, 8 (22-24). (In Russian)

CIS 78-576 "Ergonomic assessment of work posture on drilling derricks" (Ocenka ėrgonomičnosti rabočih poz buril'ščika). Panov, G. E.; Artamonov, V. S.; Sorokin, N. A. *Bezopasnost' truda v promyšlennosti* (Moscow), Mar. 1977, 3 (49). (In Russian)

Selected occupational fatalities related to oil/gas well drilling rigs as found in reports of OSHA fatality/catastrophe investigations. Office of Statistical Studies and Analysis (Washington, DC, US Department of Labor, June 1980), 49 p.+Appendices. Illus.

Drilling, rock

The drilling of shotholes and exploratory boreholes in hard rock is an essential part of all mining, tunnelling and quarrying operations.

Until the 1950s mining practice in Western Europe generally favoured the development of galleries and headings that were of smaller cross-section than is the practice today. At that time, jack-hammers of 25 to 30 kg with air-leg mountings were considered most suitable for this purpose and the heavier machines, mounted on "Jumbos" (mobile drilling platforms) were only used for major tunnelling operations. Since that time, mining operations have tended more and more to reverse this practice because of the increasing size of the headings now favoured, which enable heavier and improved types of rock-drill to be used together with more sophisticated varieties of drilling rig.

A number of new techniques have also contributed to this change in practice. These include:

(a) the increasing use of roof-bolts, which require certain changes in drilling technique;

(b) developments in the field of hydraulics, which enable the drilling rigs to be more controllable and are now used as a source of power for the drills themselves making them independent of compressed air;

(c) the perfection of a wide range of drill mountings suitable for different applications;

(d) the appearance on the market of high efficiency "down-the-hole" drills;

(e) the recent breakthrough of electronics into the field of automatic control, which enables an established programme of drilling to be followed based on any desired blasting pattern.

Drilling techniques and equipment

Rock-drills may be divided into three groups: percussive drills, rotary drills and vibrating rotary drills. The selection of the correct equipment should be based on the nature of the ground, the size of the heading, the type of energy source available, the depth, diameter and direction of the holes to be drilled, and environmental problems such as dust production and noise levels.

Certain new techniques are, however, making their appearance, including thermal drilling and drilling with the aid of high-pressure water jets. These methods are at present in the experimental stage and no further reference will be made to them since they are only used in very special cases.

Percussive drilling. Pneumatic percussive drilling. A pneumatic drill, or jack-hammer, is similar to a pneumatic pick in that a blow from the piston is transmitted to a drill rod at the end of which is a bit, while at the same time a rotary motion is imparted to the drill rod which is synchronised with the hammer blow. With the aid of such a machine, short, small diameter holes may be rapidly drilled into the hardest rock. Jack-hammers range in size from the smallest which may weigh about 12 kg up to very heavy machines weighing more than 50 kg.

Hydraulic drills. The use of hydraulic energy to provide the rotary movement and to advance the machine itself first made its appearance in the 1960s, but compressed air has always been used to provide the percussive force. More recently, high performance machines have been developed which are entirely hydraulic in operation: the piston of the machine is driven backwards under pressure from the hydraulic fluid, causing a gas (usually nitrogen) to be compressed within a confined space which in turn forces the piston forwards after the maximum pressure has been reached, thereby delivering the required percussive effect. Compared with the traditional pneumatic drills, this system presents the following advantages: reduced noise level, elimination of fogging caused by the escape of the compressed air/water vapour mixture from the exhaust ports, and greater ease and flexibility in operation, particularly when collaring holes.

Down-the-hole drills. The down-the-hole drill, with its bit, is mounted at the end of a series of connecting rods which, together with the drill and its bit, are caused to rotate by means of a head situated at the top of the hole. The drill itself furnishes a percussive force (figure 1). An increasing number of applications are being found for this type of drill as it enables large diameter holes from 60 mm up to 750 mm to be drilled up to depths of 4 000 m through all types of strata. Holes drilled by this method are less likely to deviate from their true direction because of the reduced thrust force that needs to be applied.

Rotary drilling. In the case of softer, homogeneous rock formations, the percussive force may be replaced by a cutting action in which the bit is continuously rotated as in a milling machine.

Such drills may be driven by compressed air, electricity or hydraulic means. Compressed-air drills are generally light machines either hand-held or mounted on a light stand with a rack-and-pinion and are suitable for small diameter holes in soft rock. They are particularly suitable for drilling water-injection holes in coal. Electrically driven machines, for which asynchronous motors are particularly suitable, enable holes to be drilled in workplaces where no compressed air is available. Hydraulically driven drills are more flexible than electric drills and have a vastly improved performance in comparison with compressed-air drills.

Rotary drills have the advantage, compared with percussive drills, that they produce less dust and noise and have a better coefficient of useful effect and peformance, particularly in softer rock.

Vibrating rotary drilling. In the case of percussive drilling, something like 3 000 blows per minute are struck by the machine, whereas in rotary drilling the bit simply wears away the rock. Vibratory rotary drilling is a combination of these two factors: the bit vibrates continuously but is subjected to vibration having an amplitude of 1 to 2 mm with a frequency of 5 000 to 6 000 blows per minute. This technique which is still subject to further development appears to reduce the wear on the bit but requires greater thrust on the machine thereby making the use of a power-assisted support imperative.

Drill steel. Traditional drill steel of hexagonal section and with an axial channel continues to be most commonly used with light or medium-sized machines. With the introduction of more powerful machines, there is a tendency to use heavier-section steel, while to facilitate the drilling of deeper holes it has become necessary to introduce drill rods with threaded couplings, some of which may be up to 18 m in length and fitted with a mobile support.

In the case of rotary drilling, the drill steel is either of circular section with a rolled-on square-section band or helicoidal with a lozenge-shaped or oval section to facilitate to removal of the cuttings.

Bits. The bit may form an integral part of the drill steel or it may be detachable and joined to the rod by a thread or by a conical joint. In the case of moderately hard ground, the simple chisel bit which can be easily resharpened is most commonly used. Cruciform bits are only used for exceptionally hard ground. The most recent development in this field is the introduction of bits with tungsten carbide insets. These bits provide a faster advance in moderately hard ground and do not need to be reground after use. Different shaped crowns are available for rotary drilling, which may or may not be designed for coring.

Support of the machine. Jack-hammers and rotary drilling machines were formerly hand-held, which represented an arduous task for the operator. It is now customary to make use of mechanical supports and these have undergone remarkable development in recent years.

The first and most simple of these supports was the air-leg, which continues to be generally used in the smaller galleries. It consists of a pneumatic cylinder with a stem supporting the machine at its centre of gravity and whose piston, actuated by an independent compressed-air supply, furnishes the necessary thrust. A special variety, known as a stoper, is adapted for overhead vertical drilling and is particularly suitable for roof-bolting.

"Jumbo" rigs are mobile supports carrying up to seven machines mounted on slides attached to articulated

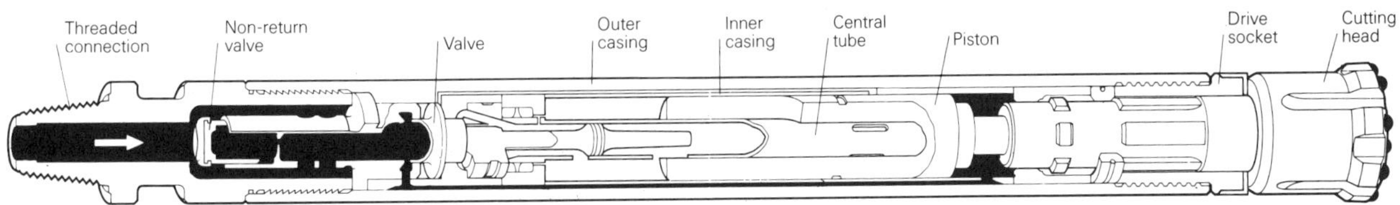

Figure 1. Down-the-hole drill.

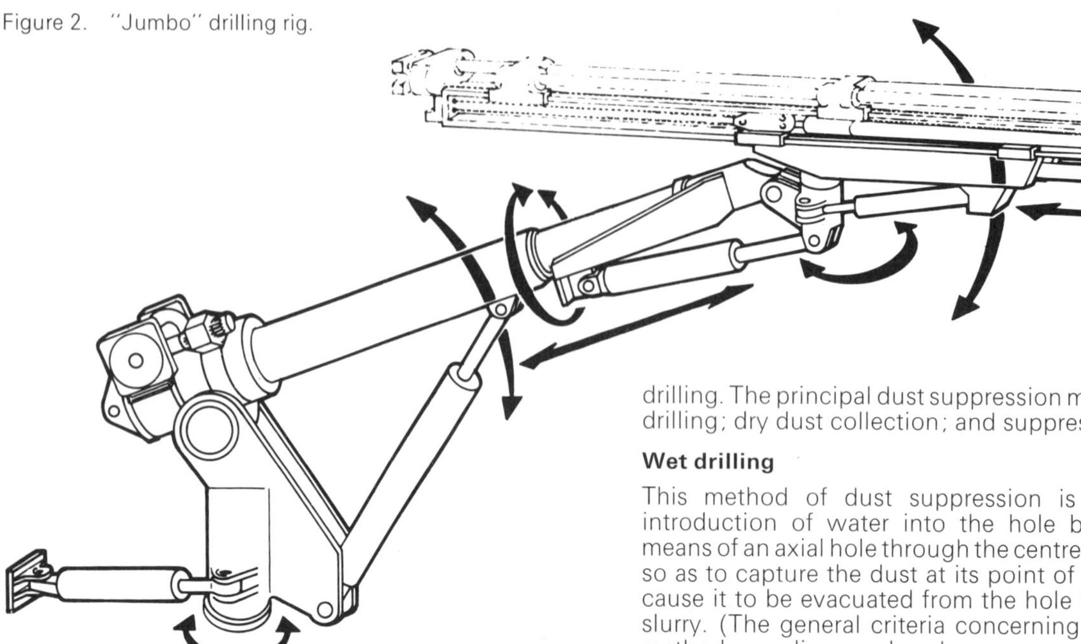
Figure 2. "Jumbo" drilling rig.

arms, which enable the drills to be rapidly brought into position with the slides guiding the advance of the machines. The advancement is effected by a screw thread, a chain or a rack and is either motor- or compressed-air-driven. The articulated arms are provided with hydraulic jacks and can take up any position (figure 2). The whole rig is mounted on a mobile chassis, which may be on rails, on caterpillar tracks, on rubber-tyred wheels or suspended from one or more overhead rails. In some cases the rig may be fitted with a movable nacelle enabling access to be gained to all parts of the face of a large tunnel. Jumbo rigs have contributed largely to both the improvement of drilling performance and a reduction in fatigue for drillers.

Other types of jumbos and mobile drilling rigs exist for use in quarries and underground in mines. They may also be fitted with automatic controls with which the essential features of a required drilling pattern may be reproduced on a matrix that transmits the orders electronically to an electro-hydraulic unit controlling the various jacks and motors.

HAZARDS AND THEIR PREVENTION

Rock-drill operators are exposed to the danger of pneumoconiosis caused by the inhalation of the rock dust produced by the drilling machine. The severity of this problem depends upon the nature of the dust, the size of the particles and their quantity. Noise, vibrations and heat are not specific to drilling; nevertheless, in the case of the larger jumbo rigs mentioned above, the operator can be protected from these risks by closed noise-proof, air-conditioned nacelles.

The widespread use of wet drilling methods, particularly in mines, has resulted in miners being exposed to a high level of humidity, which may be the cause of rheumatic and respiratory complaints.

Methods of dust prevention

Whatever the method adopted for drilling, large quantities of dust will be produced, and various methods of suppression will have to be introduced. In addition, most countries have adopted stringent regulations limiting the amount of free silica dust that may be liberated during drilling. The principal dust suppression methods are: wet drilling; dry dust collection; and suppression by fog.

Wet drilling

This method of dust suppression is based on the introduction of water into the hole being drilled by means of an axial hole through the centre of the drill steel, so as to capture the dust at its point of production and cause it to be evacuated from the hole in the form of a slurry. (The general criteria concerning the use of wet methods are discussed under CONTROL TECHNOLOGY FOR OCCUPATIONAL HEALTH.)

With this method, it has not up to the present been possible to establish clearly the relationship between the dust-collecting capacity and the shape and size of the bit or even the number of outlet holes for the water that should be provided in the bit. It is, however, important that the water should be directed at the bottom of the hole and that there should be no obstruction to the water jet during drilling. The flow of water should be such that the cuttings are satisfactorily eliminated without clogging and for this reason at least two outlet holes should be provided in the bit. It is no longer the practice to use bits with holes at the back as these tend to cause a resistance to the flow to be built up. The water flow must not be interrupted while collaring the hole or throughout the duration of drilling, and for this reason a machine should be so fitted as to prevent its operation unless the water feed is open. Many types of machines, while being provided with a water feed, still do not have a device of this sort; but modern machines are both watertight and efficient and no difficulty should be experienced in convincing operators to use them. This method of drilling is the most effective and least troublesome in operation for the suppression of dust and also has the advantage that wear on the bit is reduced because of the cooling effect of the water.

The addition of a small amount of soluble oil (0.1-3%) creates an emulsion which reduces the surface tension of the water and may also improve the performance and extend the life of the bits. Many studies have been made on the addition of wetting agents to the circulating water with a view to increasing the amount of fine dust collected, but up to the present they have been shown to be expensive with relatively little effect.

In the case of coal mines where gassy seams are encountered, wet drilling must be practised for roof holes where the heating of the bit might otherwise lead to ignition of the firedamp.

Internal water feed. This type of drill was the first developed for wet drilling that gave satisfactory results. It has been and continues to be widely used for drilling in hard rock and since it was first introduced it has undergone many modifications which have improved both its efficiency and its watertightness. The main problems encountered in this type of machine include the leakage of water into its moving parts, inhibiting lubrication; the drop in water pressure resulting from the need to maintain a continuity of flow between the water tube and the axial hole in the drill steel; and the fact that air is sucked into the entrance to the axial hole of the drill steel and released through the hole being drilled, carrying fine dust in suspension.

External water feed. This system overcomes most of the problems encountered with internal water feed machines. Also known as the flush-head machine, it consists of a collar connected to the water supply and fitted over the drill steel (figure 3). An annular groove in the collar coincides with a hole in the drill shank leading to its axial water channel. A water seal made of rubber

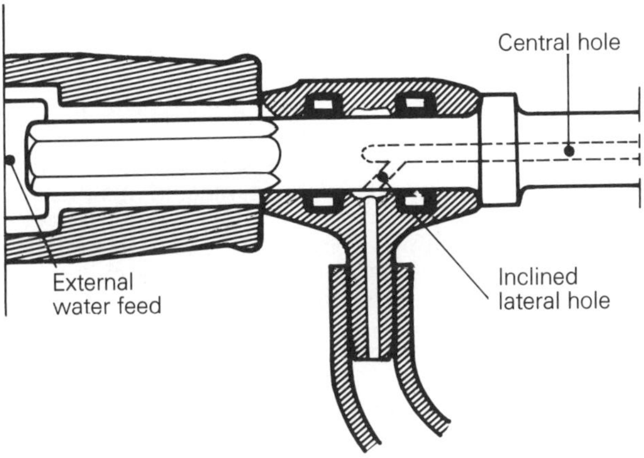

Figure 3. External water feed.

rings or a rubber bobbin prevents the water from escaping while the drill is rotating. This type of water seal has the following advantages:

(a) the exhaust air can be deflected away from the hole being drilled and bubbles carrying fine dust eliminated;

(b) a flush head may be fitted to any type of machine, the only modification required being to the shank of the drill steel;

(c) there is a gain in simplicity of design of the machine with a resultant saving in cost and in maintenance;

(d) it can be used in conjunction with electrically operated drills, since the water can be kept away from the electrical circuit.

One of the main difficulties encountered in perfecting a device of this nature was to bore a hole into the drill steel to allow the admission of water into the axial channel without weakening the steel to such an extent that it breaks during drilling. While this problem may have been overcome for light drills, it continues to represent a difficulty with heavy and extra-heavy machines, which work under greater pressure and where the steel is subjected to a heavier thrust. In spite of its advantages, the external water feed system is thus restricted to the lighter machines and heavy jackhammers are fitted with internal water feed.

Dry drilling

Although the use of water for dust suppression in drilling is both effective and inexpensive, circumstances arise where wet drilling is undesirable either because the machines become clogged, or because there may be objections to wetting the rock, or because the spray and fog released by the machine may affect nearby electrical installations or create undesirable conditions for the operator drilling vertical overhead holes. There may also be workplaces that are not equipped with a water supply or where sub-zero temperatures preclude the use of water. In such cases dry drilling with dust collectors may have to be practised.

The apparatus required for dry drilling must meet certain requirements, the principal among which are the following:

(a) the dust extraction should take place automatically throughout the drilling operation, including during the collaring of the hole;

(b) the collected dust should be disposed of without being liberated into the air of the workplace;

(c) if the extracted air is readmitted to a breathing zone, it should be subjected to adequate filtration, particular attention being paid to particles of less than 5 μm in diameter;

(d) the device should function equally well for holes of all inclinations–upward, downward or horizontal;

(e) the whole apparatus must be capable of being used in the restricted and sometimes rigorous conditions encountered in certain workplaces.

Three main types of dust collector have been devised for use during drilling. The first makes use of a hood or cowl (see figure 4), usually made of rubber, which encloses the drill steel at its point of entry into the hole. The cuttings are collected through a length of flexible hose leading from the hood to the filter with the aid of suction created by means of a compressed-air operated injector or a venturi. The filter consists of a primary stage for the coarse dust, which is collected in an airtight sack; the partially cleaned air then passes through a system of either tubular or flat filters in order to remove the fine dust particles. The tubular filters collect the dust on the inside, whereas the flat ones collect it on the outside, which facilitates cleaning. The collected dust is removed by air pulsations applied in the direction counter to the normal flow and in some models carried out automatically after each drilling operation. The filtered air is generally

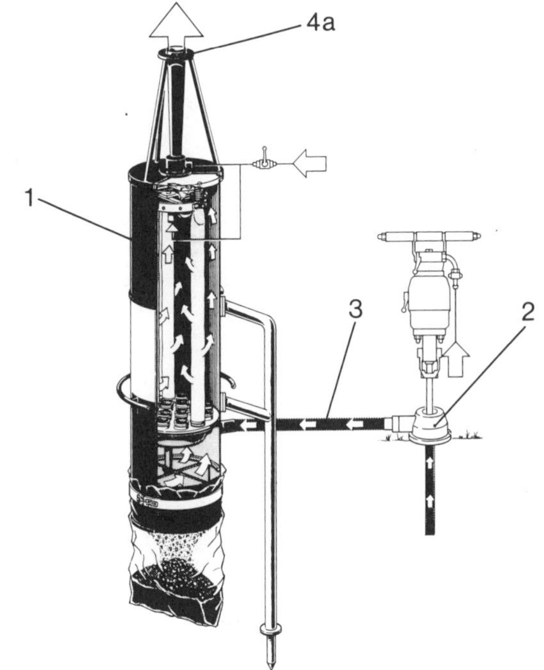

Figure 4. Atlas Copco Deduster. 1. Filtration unit. 2. Collector head. 3. Flexible suction tube. 4a. Ejector.

recirculated into the local ventilation circuit. The whole apparatus may be part of the equipment used for drilling or, in the case of large-capacity systems, it may be installed at distances of up to several hundred metres from the drill itself.

In the second method of dry drilling, suction is applied through the axial hole of the drill steel. This overcomes the disadvantages of the first system, namely that the hood or cowl does not form a dust-tight seal with the rough surface of the rock being drilled and also that the operator is unable to see the point of attack of the drill. There is nevertheless a risk with this system that the axial hole will become blocked, particularly when it passes through more than one type of rock or through damp ground.

The two main disadvantages of these methods, particularly in underground work, are that they entail the use of cumbersome filter apparatus, which may be particularly difficult in a confined space, and that the removal of the collected dust requires strict precautions to be taken to ensure that it is not recirculated into the general atmosphere creating a hazard for persons working in the vicinity. In certain cases such as large-scale tunnels where jumbos are being used, a simple method of removing the drilling dust is to install, in addition to the usual forced ventilation column provided in such cases, an exhaust ventilation column of sufficient capacity to remove all the dust-laden air from the face and deliver it, if necessary, directly to the surface. The area where the drills are working should be closed off with a screen sufficiently airtight to ensure that a negative pressure is maintained inside the screened-off area by means of the exhaust column. This method is only applicable in highly mechanised ends but offers all the advantages and none of the disadvantages of the other dry drilling methods.

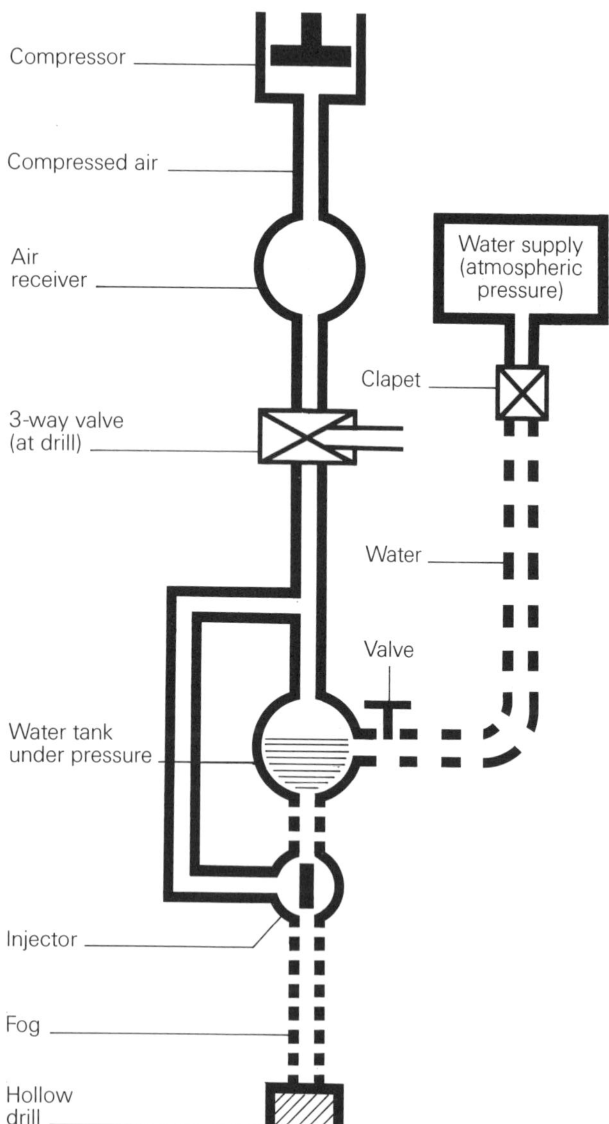

Figure 5. Compressed water fog system.

Fog method

A new method recently introduced in the iron mines of the Lorraine basin makes use of a compressed water fog, which is introduced into the drill steel in place of the usual flushing water. In addition to reducing the quantity of water used in drilling, the fog wets the fine drilling dust sufficiently to carry it out of the hole and to prevent its being released into the air. The system is represented schematically in figure 5. Water is injected at a pressure of 5 to 7 bar, using a 0.75-1.25 mm injector, into the compressed-air supply at a point close to where it enters the drill rod. The water pressure must be maintained slightly above that of the compressed air. An additional injector is fixed at the front of the drill mounting to spray on to the entrance of the hole during drilling. Generally speaking, an amount of under 1 l of water for a hole of 3 m in length is sufficient to maintain a total dust count of below 1 mg/m^3.

In view of the simple and inexpensive nature of the equipment required, the fog method is of interest for workplaces where it is difficult to provide a water supply. As in the case of normal wet drilling, it is not practicable in regions where there is a risk of freezing. The principal problem is to ensure the correct proportion of the air/water mixture and difficulties may also be encountered when drilling through different types of rock strata or through water-bearing strata.

Maintenance

In mines where jack-hammers and other types of drill are in use, the system of dust prevention being applied should be regularly checked by means of sampling and periodical analysis of the airborne dust during drilling.

Simple compressed-air drills should be subjected to regular maintenance, which may be given by the operators themselves. This consists of ensuring that a line lubricator is inserted in the compressed-air feed to the machine and that it is correctly filled with oil which is vaporised by means of the venturi in the lubricator and passes into the machine. The operator should also apply grease to the machine itself and to the moving parts of the rig if in use. He should also ensure that sharp bits are being used, as a worn bit not only reduces the performance but also gives rise to increased amounts of fine dust. In the case of wet drilling, he should also clean the filter on the water inlet regularly. More comprehensive maintenance should be performed in a workshop equipped for the purpose.

The more complicated drills and particularly hydraulically operated drills as well as multiple rigs and jumbos should be maintained by qualified fitters, who should perform a daily check as well as a thorough overhaul at regular intervals.

Special thanks are due to Mr. G. Degueldre, Technical Director of the "Institut d'Hygiène des Mines" at Hasselt and also to INIEX ("Institut National des Industries Extractives" of Belgium) for their valuable assistance.

DELVAUX, A.

"Rock drilling features". Steels, K. *Colliery Guardian* (Redhill, Surrey), Feb. 1979, 227/2 (44-78). Illus.

"Rock drilling in perspective". Edmunds, P. *Civil Engineering* (New York), Sep. 1979 (38-43), Oct. 1979 (29-35). Illus. 3 ref.

"Drilling in quarries" (Forage en carrière). Van Duyse, H. *Annales des Mines de Belgique* (Liège), Sep. 1977, 9 (843-888). Illus. (In French and Dutch)

Drinking water

Drinking water may be defined as water for consumption by humans. It may be ingested directly or indirectly, as when used for making beverages and for other culinary purposes. Drinking water must be free from organisms and from concentrations of chemical substances that may be a hazard to health. In addition, supplies of drinking water should be as pleasant to drink as circumstances permit. Coolness, absence of turbidity and absence of colour and any disagreeable taste or smell are of the utmost importance in public supplies of drinking water.

In 1971 the World Health Organisation published the *International standards for drinking-water*, which specified the minimum requirements as to chemical and bacterial quality that drinking water can reasonably be expected to satisfy. Although these standards do not carry a legal connotation, they have been adopted in whole or in part by a number of countries as a basis for the formulation of national standards, and were cited in the International Sanitary Regulations as applicable in deciding what constitutes a pure and acceptable water supply at ports and airports. The recommended upper limits for various drinking water constituents are given in tables 1 and 2. Limits for toxic substances should, of course, be related to the daily intake of drinking water and values given in table 2 assume an average daily intake of 2.5 l by a man weighing 70 kg.

Table 1. Bacteriological limits

Piped supplies	Limits
(a) Disinfected, entering the distribution system	Coliform organisms: absent in any 100 ml sample
(b) Non-disinfected, entering the distribution system	Coliform organisms: not more than 3 per 100 ml, provided that *E. coli* is absent
(c) Water in the distribution system	*E. coli*: absent in any 100 ml sample
	Coliform organisms: not more than 10 per 100 ml
	Coliform organisms: absent in 95% of yearly samples
	Coliform organisms: absent in any two consecutive samples of 100 ml each
Individual or small community supplies	*E. coli*: absent in any 100 ml sample
	Coliform organisms: preferably less than 10 per 100 ml

Frequent bacteriological examination is required for the hygienic control of drinking water supplies; chemical and radiological examinations are generally required much less frequently. It is important that the required bacteriological, radiological and chemical tests be conducted according to accepted methodologies and that a quality control programme be instituted to validate the results.

Table 2. Chemical, radiological and physical limits

Substance or characteristic	Maximum permissible level[1]
Arsenic	0.05
Cadmium	0.01
Cyanide	0.05
Fluoride	0.6-1.7
Lead	0.1
Mercury (total)	0.001
Nitrate	45
Selenium	0.01
Polynuclear aromatic hydrocarbons	0.0002
Gross alpha activity	3 pCi/l
Gross beta activity	30 pCi/l
Colour	50 units
Odour	Unobjectionable
Taste	Unobjectionable
Suspended matter	25 units
Total solids	1 500
pH	6.5-9.2
Anionic detergents	1.0
Mineral oil	0.30
Phenolic compounds	0.002
Total hardness ($CaCO_3$)	500
Calcium	200
Chloride	600
Copper	1.5
Iron	1.0
Magnesium	150
Manganese	0.5
Sulphate	400
Zinc	15

[1] In mg/l unless otherwise stated.

Daily intake of water may vary with age, sex, ambient temperature and humidity, occupation and physical activity and other ill defined factors related to individual taste, metabolism and the like. The US National Academy of Sciences in its report *Drinking water and health* estimates "that most of those who consume more than 2 l per day still are afforded adequate protection, because the margin of safety estimated for the contaminants is sufficient to offset excess water consumption. Nevertheless, consideration should be given to establishing some standards on a regional or occupational basis, to take extremes of water consumption into account."

WHO DIVISION OF ENVIRONMENTAL HEALTH

International standards for drinking-water (Geneva, World Health Organisation, 3rd ed., 1971), 70 p.

Surveillance of drinking-water quality. Monograph series No. 63 (Geneva, World Health Organisation, 1976), 135 p.

"Study of the relationship between drinking water and health" (Onderzoek naar de relatie drinkwater en gezondheid). Zoeteman, B. C. *Tijdschrift voor sociale geneeskunde* (Amstelveen, Netherlands), 23 Dec. 1981, 59/25 (949-957). Illus. 41 ref.

Drug dependence

A drug is a substance which, when taken into a living organism, may modify one or more of its functions. Since ancient times, man has used drugs in various ways. There is no doubt that the use of certain drugs has made the world a better place in which to live. Notwithstanding the beneficial effects of many drugs utilised in the treatment of sick persons, nearly all effective drugs have some toxic properties. When used in inappropriate ways some may produce adverse, even lethal, effects.

Definition and characteristics of drug dependence

Drug abuse may be defined as persistent or sporadic excessive drug-use inconsistent with or unrelated to acceptable medical practice. Practically all drugs can be abused, but man has usually abused repeatedly only those substances which influenced or altered his psychic activities. Among the drugs long known to man and abused as well as used are alcohol, cannabis, the coca plant and opium. Among the drugs whose abuse has begun in recent years are those of the barbiturate, amphetamine and hallucinogen types.

Drug dependence is a frequent adverse effect of drug abuse. It has been defined by the World Health Organisation Expert Committee on Drug Dependence as "a state, psychic and sometimes also physical, resulting from the interaction between a living organism and a drug, characterised by behavioural and other responses that always include a compulsion to take the drug on a continuous or periodic basis in order to experience its psychic effects, and sometimes to avoid the discomfort of its absence".

People may abuse or become dependent upon a wide variety of substances that produce such central nervous system effects as stimulation, sedation or distortions of perception and/or judgement. When abuse of, or dependence on, alcohol and other drugs is associated with behavioural or other responses which cause adverse physical, mental, social or economic consequences to others as well as to the user, and when such abuse or drug dependence is, or potentially may be, widespread in the population, a public health problem exists (figure 1).

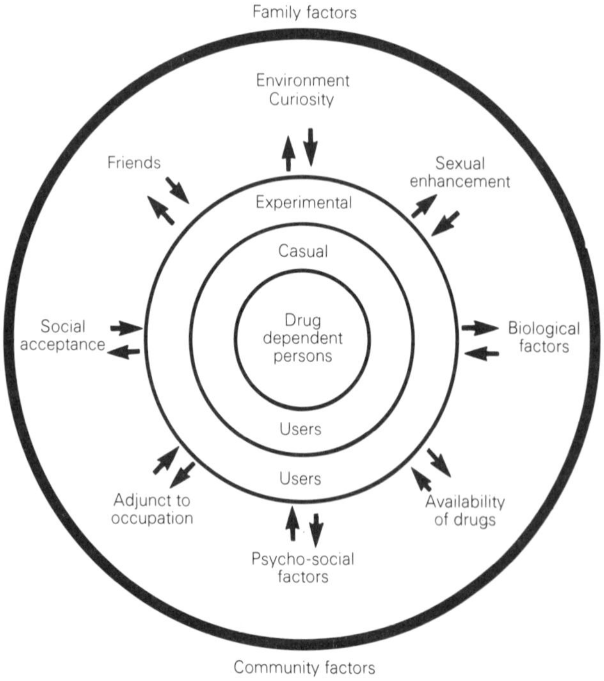

Figure 1. Drug dependence is a multifactorial phenomenon.

Mapping the problem

Historically the use of indigenous drugs within established and stable contexts has often been moderate, characterised by long-standing equilibrium, and more or less constrained by strong but informal controls. During recent decades, the traditional patterns of use have been increasingly modified. The availability of psychoactive ingredients of natural drugs in a purified form, and the arrival of a host of synthetics together with changes in the routes of administration (such as intravenous use), have led to a breakdown in equilibria. Traditional controls exerted by society and culture have been weakened or destroyed by rapid socioeconomic change, and this is often the important background to changes in drug use.

Opium. Opium is used in the Middle East, Far East and South East Asia–Afghanistan, Burma, Bangladesh, Egypt, Hong Kong, Indonesia, Iran, Macao, Pakistan, Singapore, Sri Lanka, Thailand and Viet Nam are countries particularly affected. The bulk of illicit opium production comes from the "Golden Triangle" in Thailand, Burma and Laos, as well as from Afghanistan and Pakistan. In these areas it is mostly older people in rural areas who are users of opium. Opium consumption has a long tradition, and it serves a social as well as therapeutic purpose. It is either smoked or eaten.

Coca and cocaine. The leaf of the coca plant is traditionally chewed, or used in the form of tea. This use is mainly confined to the mountainous area in the Andes in South America, involving parts of Argentina, Bolivia and Peru. Cocaine use has extended to many countries of the world. Cocaine is sniffed or injected intravenously. During recent years coca leaves have been processed into coca paste. Coca paste, taken in the form of a cigarette mixed with tobacco, is used by middle class, and urban populations (particularly young people), in some countries of Latin America.

Khat (Catha edulis Forsk). *Catha edulis Forsk* is a tree of the family Celestraceae, which grows at high altitude in East Africa, Democratic Yemen and Yemen. The leaves and young shoots of this plant have been traditionally used in a variety of ways and for different reasons. Of particular interest is the stimulant effect, which seems to have encouraged its wide abuse in several countries, namely Democratic Yemen, Djibouti, Ethiopia, Somalia, Yemen and, to a lesser extent, within limited areas in Kenya and Tanzania.

Hallucinogens. The extent of hallucinogen use is small in comparison to other drugs. There has been traditional use of plant substances in some cultures, while LSD and phencyclidine are the most widely employed substances today. The peak use of LSD was reported from 1976 to 1978, while more recently its use has declined. Phencyclidine is still reported as being used among young people in the United States, while some countries in Europe have also reported its use among juveniles on a limited scale.

Inhalants and volatile solvents. These substances include glues, thinner, benzine, etc. The users tend to be young. The use of such substances can produce irreversible pathological changes and continuous inhalation can result in paralysis and death. Cheapness and ready availability of these substances create great difficulties in controlling abuse. Fortunately, such abuse is limited to a few countries, particularly Mexico and the Sudan, although it has been reported in the United States and several countries in South and Central America, Europe and Africa.

Heroin. Heroin use is most common in the United States, Europe and Far Eastern countries. Australia, Burma, Canada, the Federal Republic of Germany, Hong Kong, Italy, Macao, Malaysia, Singapore, Sweden, Thailand, the United Kingdom, the United States and Viet Nam are among countries affected to a greater or lesser extent. Since heroin is derived from opium its production partly corresponds to areas of opium production, but it is produced illicitly in some other countries. Most heroin users are young people in urban areas and it is employed most commonly by smoking and intravenous injection. In North America and Europe the route of administration is usually intravenous.

Cannabis. Cannabis is one of the most commonly used drugs throughout the world. It is most frequently smoked, but in India and Pakistan it is commonly taken in infusions or mixed with sweets. It has recently been reported in Europe and the United States that tetrahydrocannabinol (the main active constituent extracted from plant material) is being used. Cannabis has a long history of use among the older age groups of rural populations in some countries of the Middle East and Africa. Other common users today are the young and middle-aged groups in urban populations. There has been much debate on the long-term effects of cannabis and these questions have not yet been resolved.

Amphetamines and other stimulants, barbiturates, other sedatives and minor tranquillisers. Some of these drugs are produced legally but the illicit market is mostly fed by the spillover from drugs produced in developed countries and exported to the developing world. During the past few years the abuse of these drugs has been increasing in many developing countries particularly in Africa and the Middle East. There is growing concern over medical overprescribing of some classes of substances. They are taken orally, but some illicit users inject them.

Unlike opium and heroin users, the users of these drugs cut across social and economic lines in society. There is a tendency for some of these drugs to be used more by females than males and among the middle-aged, although use is common among juveniles in some countries. There have been epidemics of adolescent misuse of amphetamines in the United States and Japan, and of phenmetrazine in Sweden. Barbiturate injection has also become an adolescent problem in some countries.

The excessive use of these drugs can lead to impairment of physical and mental health and sometimes to death. Barbiturates can produce physical and psychological dependence and withdrawal symptoms can be severe. Amphetamine misuse can lead to psychotic reactions.

Alcohol. World patterns of alcohol use are changing in a number of ways (see ALCOHOLISM). The over-all world trend is toward a steep increase in production and consumption, and a recent statistical analysis by the Finnish Foundation for Alcohol Studies and the WHO European Office of data from 97 countries showed that between 1960 and 1972 production of alcoholic beverages rose by rather over 60%. Contributing to this over-all increase is firstly the fact of increased levels of drinking in many Western countries where alcohol is the long-established social drug and production is highly industrialised. Secondly, there is the marked increase in drinking that is taking place in many parts of Africa and Asia where alcohol was previously forbidden for religious reasons, or was relatively inaccessible because of the limitations set by home brewing or distilling, for example. In some societies prohibition is still very effective.

The former traditional national preferences for wine, spirits or beer drinking are tending to lose their distinctiveness as many countries are moving towards a sort of homogenised pattern. Drinking among women and among adolescents is widely reported to be increasing.

Not surprisingly with this increase in alcohol consumption there is widespread evidence of an upsurge in experience of alcohol-related problems, particularly in Third World countries.

Tobacco. There are extremely few parts of the world where tobacco is not used. Over the last decade consumption levels in developed countries have fluctuated around an annual figure of 2 kg per person; but in some countries that have mounted vigorous health education campaigns smoking is becoming rather less frequent. The developing countries show an average annual level which fluctuates around only 0.8 kg per person, indicating the favourable and no doubt tempting markets which will exist for exploitation as levels of wealth rise, unless determined health action is taken to head off this danger.

Patterns of drug abuse

There are remarkable differences as well as similarities in the patterns of drug dependence and related abuse in different geographic areas, countries, and even in different localities within the same country.

The differences have to do with such matters as drug preference, prevalence and degree of use. For example drugs of the morphine and cannabis types appear to be the drugs of choice in certain Middle and Far Eastern countries. Alcohol is preferred in many Eastern and Western countries. Barbiturate and, to a lesser extent, amphetamine- and morphine-type drugs are also used in certain Western countries. This is not to say that multiple drug use is not common. Indeed, the use of a variety of drugs is common in many countries and individual drug-users are tending more and more to multiple-drug use. Data on the incidence and prevalence of drug dependence are difficult to obtain because many of the activities associated with such dependence are illegal. And clearly there are marked differences in the consequences resulting from (1) the experimental use of a dependence-producing drug on a few occasions, (2) the moderate use of such a drug in socially accepted situations, and (3) the use of large amounts for frequent "sprees" or for regular daily consumption.

The similarities in the patterns of drug dependence and related abuse have to do with (1) the communicability of drug use from user to non-user, (2) the use of drugs to achieve some type of intellectual or emotional experience or gratification, and (3) the fact that the consequences of drug-taking behaviour are influenced not only by the nature of the drug taken and the personality of the drug-taker but also by the prevailing sociocultural attitudes regarding the act of taking the drug in question.

Aetiology

The extent and causes of drug dependence and related abuse are not completely known. Certainly, there is no single cause. The abuse of drugs involves complex interactions between the drug-taker, his environment and the drugs taken (figure 1). The nature of the interactions and the mechanisms involved require continuing scientific study.

Dependence-producing drugs are apparently used for a variety of conscious or unconscious reasons. The same or different drugs may be taken by the same persons for differing reasons at various periods of their drug-taking experience. One or more of the following "motives" for drug-taking often seem to be at work:

(a) to escape from something;

(b) to have a new or novel experience;

(c) to achieve improved "understanding" or "insight";

(d) to achieve a sense of belonging; and

(e) to express independence and sometimes hostility.

Among the users of socially unacceptable drugs, there appears to be an over-representation of persons with emotional problems and character disorders. The cause of such personality problems has not been firmly established, but early-life experiences appear to be importantly involved.

Several environmental factors are believed to be associated with the initiation and perpetuation of drug-taking behaviour. Among these are:

(a) the development of rapid communication and transportation systems;

(b) rapid socioeconomic change, especially when accompanied by a modification in the character of family structure and cultural controls; and

(c) the "chemical revolution" which is making available an ever-increasing variety of chemicals.

Psychic dependence, physical dependence, tolerance

Dependence-producing drugs have only one characteristic in common, viz. their capacity to produce psychic dependence in some, but not all, persons. Certain persons find the effects of one or another dependence-producing drug to be so satisfying or gratifying as to prompt them to take it on a periodic or continuous basis in order to experience its psychic effect. When such drug-taking behaviour has become a life-organising force and/or has taken precedence over other means of coping with life situations, psychic dependence has been established.

Certain, but not all, types of dependence-producing drugs have the further capacity to cause physical dependence. This is an adaptive state brought about by the repeated administration of certain drugs and is manifested by the appearance of intense physical disturbances when the administration of the drug is suspended or its action is affected by the administration of a specific antagonist. These disturbances, i.e. the withdrawal or abstinence syndrome, are made up of specific arrays of symptoms and signs that are characteristic for each drug type. The discomfort associated with abstinence phenomena appears to be a powerful factor in reinforcing psychic dependence. Physical dependence clearly occurs in dependence of the morphine, alcohol and barbiturate types. Some authorities believe it occurs in dependence of the amphetamine type. If so, it is clearly of a less intense nature than that associated with withdrawal from drugs of morphine, alcohol and barbiturate types. It does not occur with dependence of the cocaine, cannabis, hallucinogen and khat types.

Many drugs produce tolerance, that is an adaptive state characterised by a declining effect upon the repetition of the same dose and, thus, the need to increase the dose in order to obtain the same degree of pharmacodynamic effect. Tolerance occurs with dependence of the morphine, barbiturate, alcohol, amphetamine and hallucinogen types, but not with those of the cocaine, cannabis and khat types.

Types of drug dependence

The characteristics of drug dependence show wide variations from one generic drug type to another. This makes it particularly important to establish clearly the pattern for each type.

Seven different types of dependence have been described. These are: drug dependence of the morphine, barbiturate-alcohol, cocaine, cannabis, amphetamine, khat and hallucinogen types. The general pattern of response is characteristic for each drug type, even though individual drugs within a group may induce responses differing in some detail.

Dependence of the morphine type is characterised by strong psychic dependence and the early development of physical dependence. Physical dependence increases in intensity as the dosage is increased. Users tend to increase their dose because of the development of tolerance. Once physical dependence is established it is necessary to take a morphine-type drug on a regular basis in order to avoid the discomfort caused by withdrawal. The withdrawal syndrome is self-limited and includes such signs and symptoms as sweating, a runny nose, muscular aching, tremulousness, nausea, vomiting and diarrhoea. This withdrawal syndrome is precipitated not only by withdrawing all morphine-type drugs but also by the administration of an antagonist (e.g. naloxone, cyclazocine, nalorphine).

Dependence of the barbiturate-alcohol type is characterised by the development of psychic dependence of varying degree that may lead to periodic rather than continuous abuse, especially with alcohol. Very characteristic is the definite development of a physical dependence. Barbiturates are able to suppress some alcohol abstinence phenomena, and alcohol will suppress some of the symptoms of barbiturate withdrawal. Thus there is a definite but limited cross-tolerance between these drugs. Alcohol and barbituric acid derivatives are essentially additive and interchangeable in prolonged intoxications. Tolerance to barbiturates and alcohol does develop. However, it is irregular and incomplete and much less marked than the tolerance that develops to morphine-like drugs. The signs and symptoms of barbiturate intoxication are quite similar to the widely known picture of alcohol intoxication. The withdrawal phenomena are also similar and include such signs and symptoms as anxiety, weakness, tremor, confusion and sometimes *delirium tremens* and convulsions. Nausea, vomiting and abdominal pain are also seen.

Dependence of the cocaine and amphetamine types are very similar in that they both produce marked psychic dependence and a highly excited and overactive state during intoxication. Marked tolerance to amphetamine-type drugs can occur rapidly, but tolerance to cocaine is not seen. Cocaine administration does not result in physical dependence. As was earlier noted, a difference of opinion exists as to the ability of amphetamine-type drugs to induce physical dependence. If it does occur, it is clearly much less marked than with drugs of the morphine and barbiturate types, and with alcohol.

In dependence on drugs of the hallucinogen type (e.g. lysergic acid diethylamide, also called LSD-25, or LSD; psilocybin, mescaline, etc.) the degree of psychic dependence varies greatly, but it is usually not intense. Such drugs produce distortions of perception and judgement which sometimes cause those under their influence to indulge in life-threatening behaviour. Panic and psychotic episodes not infrequently result from taking these drugs. A high degree of tolerance to LSD or psilocybin develops and disappears rapidly and there is cross-tolerance.

Dependence of the cannabis type involves only psychic dependence, there being no physical dependence or tolerance. The symptoms of acute intoxication with cannabis are somewhat similar to those of LSD, but generally are milder in intensity when taken in the doses usually self-administered. Some, but certainly not all, persons may become so dependent on cannabis that they devote a major portion or all of their time to its acquisition and use.

Khat effects are somewhat similar to those produced by amphetamine-type drugs but are much milder. There appears to be no tolerance nor physical dependence.

Multidisciplinary approach to services

The nature of any preventive, treatment, rehabilitation and follow-up services established in the field of drug dependence will depend on the character of prevailing conscious and unconscious beliefs and assumptions about (1) the causes of drug dependence, and (2) the immediate and long-term effects of taking drugs. The beliefs and assumptions about drug-taking have to do with the nature of direct and indirect effects on the user himself, on the persons with whom he has contact, and on the social institutions of which he is a part. In the final analysis, however, conscious and unconscious beliefs and assumptions about causes and effects of drug dependence and the characteristics of any programmes developed in this field are profoundly influenced by the much broader system of values prevailing in a particular community. This broad system of values has to do with such matters as:

(a) the relative worth of material possessions and of spiritual, cultural or traditional beliefs and experiences;

(b) the respective importance of individual rights, prerogatives and responsibilities versus those of the society;

(c) the nature of practices which are considered to be "good" or "evil"; and

(d) the meaning and value of life itself.

The health problems associated with the taking of dependence-producing drugs are thus complicated by moral judgements and cultural attitudes.

Since the causes, prevention and control of dependence on alcohol and other drugs and the treatment of drug-dependent persons involve multiple problems that exceed the scope of any one skilled profession or group, and since knowledge of these problems is so imperfect, it is imperative that a multidisciplinary approach be made to their solution.

The professions involved in a treatment programme include: general physicians, psychiatrists, internists (internal medicine), social workers, sociologists, clinical psychologists, nurses and occupational therapists. The practice adopted in some areas of including in the therapeutic team patients who have recovered from dependence on alcohol and other drugs is to be commended.

The prevention of dependence on alcohol and other drugs additionally involves the talents and experience of persons as diverse as sociologists, cultural anthropologists, epidemiologists, economists, educators, industrial and other managers, labour leaders, criminologists, attorneys, legislators, jurists, law-enforcement officers, clergymen and historians.

Occupational health programmes

It is recognised that non-occupational as well as occupational illnesses and injuries greatly affect the health and productivity of workers. Since dependence-producing drugs variously affect neuromuscular coordination, reaction time, state of consciousness and judgement, their excessive use on or off the job, particularly by workers in hazardous occupations, is certainly contraindicated. Physicians carrying out health appraisals of actual and potential workers should be alert to the manifestations of deviant use of drugs, including alcohol. The early detection of such use, especially if followed by prompt counselling and treatment of the persons involved, will help not only to prevent industrial and other accidents but also to improve the health and productivity of the worker.

A study of patterns of absenteeism and accidents may assist in identifying workers who should have a health appraisal because of possible drug abuse. Significant and particularly irregular and fairly rapid changes in mood and patterns of behaviour, if not otherwise readily explained, may likewise call for a prompt health appraisal. It should be recalled that multiple-drug use (simultaneously or in sequence) is not uncommon. Alcohol and barbiturate effects are additive and may lead to a person appearing to be quite intoxicated while having little odour of alcohol about his person and a low blood alcohol level. Other drug combinations such as sedatives and stimulants tend to counteract one another in some respects and may quite consciously be used by drug-takers for that purpose.

At least one drug of dependence—alcohol—is known to aggravate the toxic effects of exposure to carbon tetrachloride.

ARIF, A.

General:

WHO Expert Committee on Drug Dependence 21st Report. Technical report series No. 618 (Geneva, World Health Organisation, 1978), 49 p.

Evaluation of dependence liability and dependence potential of drugs. Report of a WHO scientific group. Technical report series No. 577 (Geneva, World Health Organisation, 1975), 50 p.

International statistics on alcoholic beverages: production, trade and consumption 1950-1972. Finnish Foundation for Alcohol Studies and WHO European Office (Helsinki, Finnish Foundation for Alcohol Studies, 1971).

In industry:

"The working addict". Caplovitz, D. *Journal of Psychedelic Drugs* (Beloit, Wisconsin), 1976, 8/4 (313-316).

CIS 74-0272 "Drug dependence in industry" (Les toxicomanies en milieu industriel). Marchand, M.; Furon, D.; Cabal, C.; Saison, S. *Evolution médicale* (Clermont-Ferrand), 1973, 17/2 (107-113). (In French)

"Effects of certain drugs that could increase the risk of accidental injury". Lewis, S. C. *Medical Bulletin* (New York), 1977, 37/1 (77-84).

Sniffing:

"Gasoline sniffing: a review". Poklis, A.; Burkett, C. D. *Clinical Toxicology* (New York), Jan. 1977, 11/1 (35-41).

"A review of psychological measures relevant to central nervous system toxicity, with specific reference to solvent inhalation". Comstock, B. S. *Clinical Toxicology* (New York), Mar. 1977, 11/3 (317-324).

Dry cleaning

Dry cleaning can be defined as the washing and degreasing by non-aqueous solvents of natural and synthetic fibres (woven or not), furs, leather goods, etc.: the term is most commonly used in reference to the cleaning of garments and soft furnishing. Garment cleaning is often carried on in small establishments, situated in busy town areas, sometimes even in domestic premises ill adapted for the process.

Mechanically, the processes resemble those carried out in laundries. Briefly, dirt is removed in washing machines by means of a solvent which is thereafter extracted from the fabrics by centrifuging in hydro-extractors; the fabrics are then dried in tumbling machines and finished by steam pressing. Stains which may prove resistant to machine cleaning are removed by "hand spotting" and the fabrics are sometimes retextured by silicone treatment. Solvents are recovered for further use by means of filtration and distillation.

At all stages of the cleaning process, solvent vapour may enter the workroom air, especially where there is considerable manual handling of the goods. In recent years, totally enclosed units have done away with much handling, but contamination of the atmosphere may still remain a problem.

Coin-operated automatic dry-cleaning units for use by the public have recently been installed in many towns.

HAZARDS AND THEIR PREVENTION

There are two major hazards—flammability and toxicity: the degree of danger varies with the solvent used. Broadly speaking, progress in reducing hazards depends first on substitution of a less flammable or a less toxic solvent, but this does not eliminate the need for precautionary measures.

Fire and explosions. Formerly, highly flammable solvents such as gasoline were used and very serious fires or explosions occurred. Many countries have now forbidden the use of solvents with a low flash point (e.g. below 90 °C) and even where legislation does not exist, the use of such solvents should not be tolerated.

The substitution of solvents with a higher flash point, such as white spirit, reduces the risk of fire or explosion during washing and hydroextraction, but it is less effective in preventing serious fires or explosions during tumbler drying when temperatures are high. Stringent precautions are therefore still necessary: buildings should be of non-flammable construction and workrooms should never be underground; efficient exhaust ventilation including high-level inlets and low-level discharge outlets should be provided. All sources of ignition should be strictly controlled: electrical installations and equipment should be of the flameproof type and static charges should be removed by earthing. Materials under treatment also need attention and garments made from synthetic fibres should be separated from those made from natural fibres. Solvents should be stored outside the work premises, solvent recovery plants should be kept under strict control and waste solvent never allowed to run into the public drains. Efficient fire extinguishers are necessary and staff should be instructed in procedures to be taken in the event of fire.

Replacement of flammable by non-flammable solvents is the most effective method of fire and explosion prevention and has now been widely adopted: trichloroethylene and tetrachloroethylene are the most commonly used substitutes, but a more recent innovation is trichlorotrifluoroethane. Most of these substitutes are used in totally enclosed plants and, in well ventilated conditions, can be regarded as non-flammable.

Health hazards. Petroleum distillates can cause both acute and chronic poisoning. Efficient exhaust ventilation is required to keep solvent concentrations below permissible limits as much to prevent the toxic as the flammable risks.

Plants should not be installed in underground rooms, and good mechanical exhaust ventilation should be provided in all workplaces, especially at the point of origin of vapour concentrations, for example transfer areas. With vapours that are heavier than air, particular attention is required for tanks and pits. Enclosed plants should be well maintained and regularly checked for leakages. Before garments are pressed, all residual solvent in the fibres should be vaporised. Work at solvent recovery stills should be carried out only by competent persons; sludge removal presents special hazards; solvents should be removed from residues by a passage of live steam, and respiratory protective equipment should be available.

Corrosive or toxic gases, for example phosgene, may be produced if trichloroethylene or perchloroethylene vapours are exposed to high temperatures under certain conditions: all naked flames or red-hot surfaces should be eliminated from the workplace and smoking should be strictly prohibited.

The hygienic conditions in dry-cleaning shops have been greatly improved in the last few years. In the United Kingdom Shipman and Whim (1980) found that all the surveyed dry-cleaning shops use tetrachloroethylene and 92% of them have a concentration below the TLV of 100 ppm. None the less high concentrations may result from one or more of the following causes: old and worn machines, poorly maintained or faulty machines, inadequate or faulty ventilation, lack of concern or supervision resulting in unsafe work practices.

For this reason in the past many workers in dry-cleaning shops may have been exposed to very high levels, such as a mean concentration of trichloroethylene of 631 mg/m^3, ranging from 100 to 2 000 mg/m^3.

This past exposure may account for the results of the epidemiological study by Blair, Decoufle and Grauman (1979), which underlines "the possibility that exposure to dry-cleaning fluids may increase the risk of leukaemia and liver cancer and underscores the need of additional studies of this occupational group". In laboratory animals the carcinogenicity of trichloroethylene and tetrachloroethylene has been demonstrated.

However, apart from the carcinogenicity of trichloroethylene, there was in the past a demonstrated risk of liver disease and toxic neuritis for exposed workers.

Even where the major risks in cleaning and drying have been controlled, there is a temptation to use effective, but highly toxic or flammable solvents in small quantities for the removal of persistent stains by "hand spotting", for example benzene, chloroform, etc.; owing to its high toxicity and flammability, the use of benzene should be totally prohibited; all low flash point solvents should also be eliminated. Even so, strict precautions in the "spotting" department are necessary. Good general ventilation should be provided, coupled with local exhaust ventilation over the "spotting" table if it is in continual use. Solvents should be kept in non-spill containers with clear labelling to indicate the nature of the contents, and only the smallest practicable quantity should be used. Treated materials should be removed from the "spotting room" immediately after treatment to prevent build-up of atmospheric contamination. Prolonged use of powerful solvents may result in skin infections: protective clothing, barrier creams and good washing and other sanitary facilities are necessary.

In general, all workers should be instructed in the possible hazards of their work and the need to report immediately any symptoms of affection that may occur. Medical care at the earliest stages is important.

Coin-operated installations. Use of these installations directly by the public may entail special risks since the degree of control available is lower than that possible in a factory. The solvents used are non-flammable but the toxic risks remain: it is particularly important that time should be allowed for all residual solvent to vaporise before garments, etc., are taken away by the customer.

Mechanical hazards. The dry-cleaning industry uses several machines which consist essentially of a perforated cage revolving, sometimes at very high speed, within a solid case, for example washing machines, hydroextractors, tumbler dryers. Very serious injuries may be caused if a hand is caught between the revolving cage and the casing. Efficient interlocks are necessary to ensure that the machines cannot be set in motion until the covers are closed and that the power supply to the revolving cage is immediately cut off as soon as the cover is opened: braking mechanisms are necessary to deal with residual cage motion. Regular inspection and maintenance are required to prevent the hazards of bursting cages, especially in the case of high-speed extractors (see also CENTRIFUGES).

Steam presses. A variety of steam presses present the hazards of trapping and burns. On some, guarding is possible to prevent access to the danger zone between the press head and the "buck" or table: on all machines of this type, correct positioning of control buttons or levers, careful maintenance and training of operators are of importance in accident prevention. All steam plants or compressed-air plants in use in dry-cleaning establishments demand careful inspection and maintenance.

Waste disposal. Perchloroethylene was found in alveolar air of residents living near dry-cleaning shops. Waste disposal regulations have recently been set by the Environmental Protection Agency in the United States.

RUBINO, G. F.

CIS 80-1632 "Occupational exposure to trichloroethylene in metal cleaning processes and to tetrachloroethylene in the drycleaning industry in the U.K.". Shipman, A. J.; Whim, B. P. *Annals of Occupational Hygiene* (London), 1980, 23/2 (197-204). 6 ref.

CIS 80-867 "Hazards in the dry cleaning industry". *Michigan's Occupational Health*, Aug. 1979, 23/1 (1-6). Illus.

"Launderettes" (14-16). *Health and safety. Manufacturing and service industries, 1976.* Health and Safety Executive (London, HM Stationery Office, 1977), 80 p.

A behavioural and neurological evaluation of dry cleaners exposed to perchloroethylene. Tuttle, T. C.; Wood, G. D.; Grether, C. B.; Johnson, B. L.; Xintaras, C. DHEW (NIOSH) publication No. 77-214 (National Institute for Occupational Safety and Health, 4676 Columbia Parkway, Cincinnati) (1977), 79 p. Illus. 11 ref.

CIS 80-1351 "Causes of death among laundry and dry cleaning workers". Blair, A.; Decoufle, P.; Grauman, D. *American Journal of Public Health*, May 1979, 69/5 (508-511). 24 ref.

"Tetrachloroethylene in exhaled air of residents near dry-cleaning shops". Verberk, M. M.; Scheffers, T. M. L. *Environmental Research* (New York), Apr. 1980, 21/2 (432-437). Illus. 10 ref.

Safety and health standards:

CIS 79-480 *Dry cleaning installations* (Chemischreinigungsanlagen). VGB 66. Hauptverband der gewerblichen Berufsgenossenschaften (Cologne, Carl Heymanns Verlag, 1978), 10+14 p. (In German)

"Standard of performance for new stationary sources; organic solvent cleaners". Environmental Protection Agency. *Federal Register* (Washington, DC), 11 June 1980, 45 (39 766-39 784).

Dupuytren's contracture

This is a disease characterised by a retractile sclerosis of the palmar aponeurosis, which may progress to an irreducible flexion of the fingers.

Aetiology

Although the aetiological conditions are, in general, extremely complex and still ill defined, there is in certain cases a clear correlation between the nature of the subject's work and the appearance of the disease, and statistics show that workers who suffer repeated microtraumata of the hand are often affected. Dupuytren's contracture is six to ten times more frequent in men than in women and the incidence varies with age and length of employment. In a series of French railway workers suffering from Dupuytren's contracture, it was found that:

(a) 73.6% were employed on very heavy manual work (track laying), had 13 years of service and had become aware of the condition at the age of 30;

(b) 22.6% were employed on moderately heavy manual work (maintenance), had 20 years of service and had become aware of the condition at the age of 40;

(c) 3.8% were non-manual workers and had become aware of the condition at the age of 50.

Recent investigations cast doubt on the attribution of an aetiological role to vibration from pneumatic tools or power saws in Dupuytren's disease.

Pathogenesis

The traumatic theory offers an explanation in cases which are thought to be of occupational origin. It is claimed that repeated microtraumata produce chronic inflammation and interstitial microhaemorrhaging. Nevertheless, cases have been reported to have followed on a one-time injury. The appearance of the condition in persons who have evidently never been exposed to hand injuries has been explained as being the result of trophic disorders secondary to nervous-system lesions.

Present thinking is that the condition is the manifestation of a general process that is likely to occur as the result of extremely different factors (traumata, nervous disorders, endocrine disorders, rheumatic conditions, vitamin E deficiency, etc.). It has also been suggested that the disease may be due to a hereditary predisposition.

Clinical picture

The onset is usually insidious. Careful examination of the hand may show a palmar nodule long before any other manifestation. Other nodules may occur progressively along the fourth, fifth and even the third finger.

Pain is exceptional but may be burning, and it is functional difficulties that first draw the patient's attention; gradually, extension of the fingers becomes difficult or impossible. There is thickening of the palmar skin which adheres to subjacent tissue and ceases to perspire.

Localisation of lesions varies and is definitely influenced by occupation. In the 648 cases reported in the literature, the right hand was affected in 227 cases, the left hand in 108 cases and both hands in 313 cases. The most commonly affected fingers are the fourth, fifth and third; the other fingers are seldom affected.

The course may be continuous or intermittent. Spontaneous regression does not occur.

PREVENTIVE MEASURES

Microtraumata of the hand should be prevented by technical safety measures such as the installation of mechanical handling equipment or by the use of hand protection with padded palms. Periodic medical examination is the only way to ensure early detection of the condition so that the worker can be removed from exposure.

Treatment. Medical treatment may be effective if it is applied at an early stage; it should include massive doses of vitamin E, radiotherapy and cortisone injections. Surgery becomes necessary as soon as there is permanent digital flexion and should be followed by functional re-education.

Legislation. In certain countries such as Bulgaria and the USSR, Dupuytren's contracture is recognised as an occupational disease in workers exposed to repeated microtraumata of the palm of the hand, such as tram drivers, locksmiths, miners and post office workers who repeatedly use rubber marking stamps to obliterate mail, etc.

GAVRILESCU, N.

CIS 1579-1970 "The role of traumata in the aetiology of Dupuytren's disease" (Le rôle du traumatisme dans l'étiologie de la maladie de Dupuytren). Larrard, de, J.; Hitier, C. P.; Dervillée, P.; Doignon, J.; Robert, M. *Archives des maladies professionnelles, de médecine du travail et de sécurité sociale* (Paris), Dec. 1969, 30/12 (721-724). (In French)

CIS 77-77 "Incidence of Dupuytren's contracture in workers occupationally exposed to vibration" (Incidence Dupuytrenovy kontraktury u pracujících v riziku vibrací). Landrgot, B.; Hůzl, F.; Koudela, K.; Potměšil, J.; Sýkora, J. *Pracovní lékařství* (Prague), Nov. 1975, 27/10 (331-335). Illus. 18 ref. (In Czech)

"Observations on arsenic intoxication, late damage with porphyria cutanea tarda and Dupuytren's contracture" (Gutachterliche Beobachtungen bei Arsenintoxikationsspätfolgeschaden mit Porphyria cutanea tarda und Dupuytrenscher Kontraktur). Grobe, J. W. *Dermatosen in Beruf und Umwelt* (Aulendorf, Federal Republic of Germany), 1980, 28/4 (116-117). 9 ref. (In German)

"Socio-occupational aspects of Dupuytren's contracture" (Aspects socio-professionnels de la maladie palmaire de Dupuytren). Mehl, J. *Archives des maladies professionnelles, de médecine du travail et de sécurité sociale* (Paris), Nov. 1979, 40/11 (1033-1037). 7 ref. (In French)

"Dupuytren's contracture in manual workers". Bennett, B. *British Journal of Industrial Medicine* (London), Feb. 1982, 39/1 (98-100). 30 ref.

Dust, biological effects of

Dust may be defined as a disperse system (aerosol) of heterogenous solid particles in a gas (air) whose broad size distribution is predominantly that of a colloid. Dust generally originates from the mechanical comminution of a coarser material.

Inhalation, deposition and elimination

Dust particles are carried with the air stream into the lung during inhalation, the majority of them being either exhaled or eliminated by means of the lung clearance mechanism. A small number of these particles may be deposited in the lung, depending on their size, and in accordance with the physical laws governing impaction, sedimentation and Brownian movement. Medical research has shown that particles of 0.1-5 μm can remain in the alveolar passages (respirable dust), while larger particles are retained by the mucous membrane of the nose, the throat, the trachea and the bronchi, and are eliminated by the clearance mechanism. Smaller particles (< 0.1 μm) behave as colloids, of which a typical example is smoke (the affinity of poisonous gases to these particles in suspension must also be taken into consideration). Elimination of the dust particles is largely effected by the mucociliary system of the respiratory tract, which causes the mucous secretion to move towards the mouth aided by the physiological reactions of coughing and sneezing. A smaller fraction of the particles are transported by the lymphatic system after penetrating the interstitial tissue.

In the case of asbestos, inhalation, deposition and elimination depend on the shape and size of the fibres. Very small fibres of 3 μm or less in diameter having lengths of up to 100 μm can also reach the alveoli. The smallest asbestos fibres can reach as far as the pleura and even attain the pleural space.

Biological effects and health hazards

Animal experimentation, together with clinical, pathological, anatomical and epidemiological research have shown that specific illnesses can be linked with certain types of dust. This is the case of quartz dust (silicosis), asbestos dust (asbestosis), allergenic dusts (various allergies), and so on. In the case of most other dusts, no specific clinical picture is presented and such cases are referred to as "chronic non-specific lung disease" (CNSLD), to the extent that animal experimentation and epidemiological observations have shown that it is probable that illness will result from exposure to them. Practical experience has shown that even the so-called "inert" dusts also represent a danger to health. The biological effects of these inert dusts are of a long-term nature and are neither fibrogenic nor carcinogenic, toxic or allergenic. In excessive quantities they will overcharge the protective and scavenging mechanisms, thereby leading to respiratory disease. The extent to which any type of dust represents a health risk thus depends on exposure, which includes the nature of the dust, its concentration and the duration of exposure, as well as upon individual factors such as the general constitution and state of health of the person concerned, including the functional state of the upper respiratory tract, the lung function and its structure, the general immunological status and specific immunological reactivity, and the biochemical reactivity. All these factors will play a part in the onset of disease.

Table 1 shows the main classes of hazardous industrial dust, their occupational occurrence and their biological effects.

Fibrogenic effects

Fibrogenic dusts in the strictest sense comprise those respirable dusts which contain a proportion of quartz (crystalline silica, SiO_2) or asbestos (magnesium silicate fibres) and which can cause a reaction leading to a defined or diffuse fibrosis in the lung tissue and lymph nodes (see PNEUMOCONIOSES). In the wider sense certain metal dusts such as tungsten, titanium carbide, beryllium and aluminium as well as radioactive dusts, talc dust and certain vegetable dusts can produce a fibrotic reaction.

The accumulation in the lung of respirable dust containing quartz gives rise to nodular pulmonary lesions (see SILICOSIS). Irreversible damage to the lung tissue and reduction in lung function leads to a severe reaction in the heart *(cor pulmonale)*. Pathogenetic research is centred on investigation into the cytotoxicity of quartz particles and the study of immunological reaction.

In the case of asbestos the penetration of fine dust into the alveoli causes a diffuse fibrosis with pleural thickening (see ASBESTOSIS). In many asbestos workers hyaline deposits and calcification of the pleura can be seen. Evidence of exposure may be found in the presence of fibres in the sputum or in the lung tissue, but this is not necessarily evidence of illness. The further inhalation of asbestos dust will give rise to malign tumours known as bronchial carcinoma and mesothelioma either with or without symptoms of asbestosis.

In the case of talc dust the fibrogenous effects are different. The rare appearance of a talcosis as a particular form of lung fibrosis may be taken as an indication of the fact that the ill-effects of talc dust are slight and that, where present, they are mainly due to the influence of asbestos or quartz dust. Mineralogical investigations have shown that asbestos may in certain cases be present as an impurity in talc.

Table 1. Main classes of hazardous industrial dusts

Type of dust	Principal occupational occurrences	Reaction type	Lung disease	Observations
I. **Quartz and mixtures containing quartz**				
Coal, mineral ores, fluorspar, rock, sand	Mining, metallurgical engineering, building material and construction, stone cutting, foundrywork (moulders), sand-blasting	Nodular fibrosis	Silicosis, anthraco-silicosis and mixed pneumoconioses	Found also in conjunction with tuberculosis and chronic non-specific lung disease (CNSLD). Crystalline SiO_2 and tridymite act like quartz
Kaolin	Ceramic industries (porcelain, pottery, earthenware, sanitaryware and electrical ceramics)	"	Silicosis	"
Quartzite	Refractories (fireproof clay, silica brick)	"	"	"
Powdered quartz	Plastering (mixing)	"	"	"
Kieselguhr (burnt)	Manufacture of filtering and insulating materials	"	"	"
II. **Asbestos and mixtures containing asbestos**				
Raw asbestos: chrysotile; amphibole	Mining of asbestos, manufacture, treatment and preparation (insulation, textiles, friction materials, packing and jointing materials, fire prevention). Over 3 000 asbestos-containing products are listed	Diffuse fibrosis, carcinoma	Asbestosis, mesothelioma (peritoneum, pleura, pericardium), bronchial carcinoma, carcinoma of upper respiratory tract	Pleural hyalinosis. Chronic non-specific lung diseas (CNSLD). Stomach carcinoma following asbestos dust exposure is open to discussion
Asbestos cement	Building material and construction industry (also used domestically)	"	"	–
Talc	Rubber industry, pharmaceuticals, cosmetics, paint, paper and printing, textiles (also used in insecticides and in health protection)	Diffuse fibrosis, nodular fibrosis in rare cases, carcinoma	Talcosis, mesothelioma, bronchial carcinoma, carcinoma of upper respiratory tract	Frequently due to admixture of asbestos and quartz
III. **Metals and metal compounds**				
Aluminium, aluminium oxide	Pyrotechnics (aluminium powder), manufacture of corundum (aluminium smoke from smelting of bauxite), light metal industries (welding and flame cutting)	Diffuse fibrosis, irritation	Aluminium lung, bauxite smelter's lung, CNSLD	–
Beryllium, beryllium oxide	Metallurgy, manufacture of luminescent tubes	Cellular granuloma, diffuse fibrosis, irritation, immune reaction	Tracheobronchitis, pneumonitis, beryllosis	Often acute development; latent period up to 25 years possible; granuloma may also develop in liver, spleen, skin or muscles
Cadmium, cadmium oxide	Metallurgy, electroplating, paint industry (pigments)	Irritation, systemic poison	Tracheobronchitis, bronchopneumonia, emphysema of the lung	Kidney damage, osteoporosis
Chromium, chromic oxide, chromate	Metallurgy, electroplating, welding and flame cutting, austenitic steel, pigments	Irritation, immune reaction, carcinoma	Ulceration and perforation of nasal septum, bronchial asthma, CNSLD, carcinoma of nasal cavity	Only compounds in the +6 state are carcinogenic (e.g. alkali-chromate, chromic oxide)
Hard metals	Sintering	Part diffuse, part fibrosis	Fibrosis, immune reaction (?)	Pathogenic effects of cobalt not fully ascertained

Type of dust	Principal occupational occurrences	Reaction type	Lung disease	Observations
Iron, iron oxide	Metallurgy, metal working (welding, flame cutting, grinding), paint industry (pigments)	Accumulation	Siderosis	After termination of exposure, X-ray shadows continue to build up
Lead, lead oxide	Metallurgy, accumulator manufacture, lead shot manufacture, paint industry (pigments), glazing, scrap lead working (flame cutting of lead painted material)	Systemic poison effects	–	Lead poisoning (anaemia, colic, polyneurological symptoms, encephalopathy); possible local effects in the lung
Manganese, manganese oxides	Metallurgy, metal working (welding with electrodes containing manganese), preparation and use of manganese ore	Irritation, systemic poison	Manganic pneumonia, CNSLD	–
Nickel, nickel oxides, nickel salts	Metallurgy, electro-plating, chemical industry	Irritation, immune reaction, carcinoma	Bronchial carcinoma, carcinoma of nasal cavity	Systemic poisoning with nickel tetra-carbonyl (liver, dermatitis)
Platinum compounds (salts)	Metallurgy	Immune reaction (type I), irritation	Allergic rhinitis, bronchial asthma	Dermatitis
Vanadium pentoxide	Power stations (cleaning of residue in oilburning furnaces), chemical industry (manufacture of vanadium catalysts)	Irritation	Tracheobronchitis, bronchial asthma, CNSLD	Dermatitis
IV. **Plant and animal dusts (organic)**				
Milled or crushed grain and bran	Grain milling and storage, bakeries	Irritation, immune reaction (type I)	Allergic rhinitis, chronic rhinitis, bronchial asthma, CNSLD	Mould fungi produce similar symptoms (frequently found as impurities)
Wood, particularly exotic types	Manufacture of veneers, furniture industry (polishing), wood turning	Irritation, immune reaction, carcinoma	Allergic rhinitis, bronchial asthma, CNSLD, carcinoma of the nose and nasal cavity	Dermatitis
Animal hides and skins, hair, feathers, and scales	Agriculture, zoo attendants, laboratory animal keepers, furriers and fur dealers	Immune reaction (type I)	Allergic rhinitis, bronchial asthma	Dermatitis
Enzymes (proteases)	Pharmaceutical industry, washing powder manufacture, food and drink industries	Irritation, immune reaction (type I)	Allergic rhinitis, bronchial asthma	Dermatitis
Mouldy hay, straw, cereals and bagasse	Agriculture, grain silos, sugar industry	Immune reaction (type III), allergic alveolitis, diffuse fibrosis	Farmer's lung, bagassosis	Caused by thermophilic actinomycetes *(Micropolyspora faeni, Thermoactinomyces vulgaris)*, rarely by mould fungus *(Aspergillus)*
Excrement from hens, pigeons, parakeets	Poultry-keeping and zoos	Immune reaction (type III), allergic alveolitis, diffuse fibrosis		Caused by protein serum in bird excrement
Cotton, flax, hemp, sisal, jute	Cotton carding, cotton and flax spinning, combing (hackling)	Irritation, release of histamine, immune reaction (?)	Byssinosis, CNSLD	"Monday" symptoms probably due to release of histamines and related substances. Immune reaction uncertain. Chronic symptoms due to long-lasting irritation of mucous membrane

Type of dust	Principal occupational occurrences	Reaction type	Lung disease	Observations
V. Other dusts				
Arsenic, arsenic trioxide, arsenic salts	Mining, metallurgy (lead and zinc smelters), electroplating (metal etching), chemical industry (colouring and finishing materials)	Irritation, carcinoma, systemic toxic effects	Ulceration and perforation of nasal septum, tracheobronchitis, carcinoma of nasal cavity, bronchial carcinoma	Polyneurological symptoms, diseased liver, dermatitis
Maleic anhydride, phthalic anhydride	Chemical industry manufacture of plastics)	Irritation, immune reaction (?)	Conjunctivitis, rhinitis with ulceration, laryngopharyngitis, asthmatic bronchitis	Maleic anhydride and naphthoquinone are often found as impurities in phthalic anhydride, thus immune reaction is a possibility; dermatitis
Carbon dust, soot, graphite	Any work involving soot, rubber industry, manufacture of electrodes	Accumulation, irritation	Graphite pneumoconiosis, CNSLD	Possible fibrogenic effects, possible bronchial carcinoma

Irritant effects

These include irritation of the mucous membrane of the eyes and respiratory tract (reddening, swelling, itching, weeping, sneezing, coughing). The irritation of the sensory receptors of the bronchial mucous membrane leads to a reflex contraction of the smooth muscles (bronchial obstruction). The pathology has not up to the present been fully explained. Some authors have assimilated the irritant effect to local poisoning together with cellular degeneration. A table of the irritation potential of different substances has been published by the World Health Organisation (1977).

The irritant effects of different dusts are not specific to particular substances and thus they cannot be linked by means of a direct causal relationship. Neither clinical nor functional tests nor morphological examination have been able to show a case where these effects have originated from dust. Only in the rarer cases showing acute effects, as, for example, after inhalation of the dust of vanadium pentoxide, can such a close relationship in time provide causal evidence. Outweighing these effects by far, however, are cases where exposure over many years has led to chronic non-specific lung disease (see BRONCHITIS, CHRONIC, AND EMPHYSEMA). Animal experiments and epidemiological findings leave no doubt that exposure to a sufficient concentration over a sufficiently long period of time will result in the appearance of lung disease. Among the factors which tend to increase the frequency of lung disease, tobacco takes precedence over occupational causes as well as those stemming from the general environment.

Allergic effects

Certain dusts, almost exclusively of vegetable or animal origin, have the property of giving rise to allergic reactions (see DUSTS, VEGETABLE). These immune reactions take the form chiefly of a physiological response to an irritation. The pathological implications have to be determined by means of specific tests. Occupational allergies of the respiratory tract generally take the form of bronchial asthma or alveolitis.

Bronchial asthma arising from dust is nearly always of the immediate reaction type, of which the asthma found among bakers and caused by flour is a good example. There is a characteristic build-up of immunoglobulin E following exposure and further exposure after a short time brings on renewed symptoms. The organic dusts responsible include flour, pollen, animal hair, feathers, mould fungus and insects. Salts of the platinum complex can also produce this reaction. The other type of asthma, having a delayed reaction, is seldom seen and is generally caused by various chemicals, of which a good example is the quinone found in wood. There is a characteristic sensitivity of the lymphocytes followed by renewed symptoms after further exposure for several hours. It is frequently associated with other irritant effects such as rhinitis and chronic bronchitis (see ASTHMA, OCCUPATIONAL).

Among the principal allergic forms of alveolitis are "farmer's lung" and "bird breeder's lung". Characteristic of the disease are the precipitating antibodies (immunoglobin G) with interstitial pneumonia, which becomes latent until further exposure takes place giving rise to renewed symptoms of many forms. Of these, dyspnoea, coughing, fever and even fluid in the lung, are typical. A progression to diffuse lung fibrosis is not unusual. The main causes include organic dusts or those containing spores and, less frequently, fungus moulds or animal proteins. It is common to name the disease after the occupation concerned as in the case of farmer's lung, cheese washer's lung, mushroom picker's lung, maple bark stripper's disease, etc. (see FARMER'S LUNG).

Carcinogenic effects

The causes of cancer include both external and internal factors. Cancer resulting from external factors may be considered to be the irreversible effect of a dose of a chemical or physical poison. A carcinogen affects the biochemical regulatory mechanism in such a way as to transform normal cells into malignant cells, or by causing dormant cancerogenous cells to become active. The effect of several carcinogens can lead to additive or more complicated reactions (syncarcinogens). It has also been observed that certain non-carcinogenic substances can cause an increase in the carcinogenic reaction (cocarcinogens). The dose-effect relationship in the causes of cancer is still being investigated.

At the present time it is considered that external effects such as nutrition, living conditions, environmental pollution or occupational influences play a dominant role. Despite many uncertainties, it is clear that in addition to ionising radiation and asbestos dust, further environmental factors including the presence of arsenic, chromium and nickel, can lead to cancerous growths in man. In addition more than 2 000 chemical substances are known to produce cancer in animal experimentation, and these should thus be taken as suspected carcinogens for man. Cancer due to occupational influences can affect either the exposed region or some other organ of the body. Exposure periods of at least ten years with a latent period (induction period) of several decades up to the appearance of the illness has been observed (see CANCER, OCCUPATIONAL).

Systemic toxic effects

The essential indications of systemic poison are that they go beyond the respiratory tracts and manifest themselves either exclusively or additionally in other organs such as the central nervous system, the liver or the kidneys. Thus in the case of manganese and cadmium compounds, a local infection will be observed in the region of the respiratory passages together with simultaneous symptoms of poisoning in other organs. On the other hand, lead only affects organs other than the respiratory tract. Typical toxic agents are the metallic oxides and the metallic salts in the form of dust. Secretions of the alveolar epithelia have a strong affinity for these dusts.

Most harmful substances which cause systemic toxic effects as a result of inhalation may also be ingested with similar effects outside the respiratory tract. Modern technology has ensured that poisoning resulting from inhalation is seen much less frequently (e.g. manganese pneumoconiosis and manganese poisoning). Exposure for many years to low levels, but above the exposure limits, have demonstrated the occurrence of distinct systemic toxic effects.

Skin effects

Dust particles from insulating and reinforcing materials such as glass fibre, stone fibre and certain fire-resistant materials are not infrequently the cause of dermatitis, often of the follicular type and usually with urticarial changes. Dust that is produced in the sawing and particularly polishing of exotic types of wood that are favoured in the furniture industry can lead to irritation or allergic dermatitis. In certain rare cases the inhalation of dust from grain milling can produce an allergic urticaria. Irritative dermatitis can also result from the enzyme of *Bacillus subtilis* used in washing powders, which is a reason why these enzymes should be used in a prilled form.

PREVENTION OF DUST-RELATED DISEASES

Technical measures for the prevention and suppression of dust can provide protection for persons at the workplace (see DUST CONTROL, INDUSTRIAL; DUST SAMPLING; EXPOSURE LIMITS; RESPIRATORY PROTECTIVE EQUIPMENT).

Medical protection is based on regular examination of the exposed persons by the factory doctor. This enables a check to be made on the fitness of the individual for his occupation and makes possible the early detection of health problems and the timely introduction of rehabilitative measures. The factory medical service calls for close co-operation to be maintained between the factory doctors, the responsible medical authority, the trade union representatives, management, safety engineers, technical personnel, local health authorities and regular practising physicians. Effective protection of the exposed worker must be based on accurate information concerning dusty workplaces, since knowledge of this exposure is the most important prerequisite in any investigation to be made by the factory doctor. It enables him to institute a methodical research programme and to form an opinion of the health implications of a particular occupation.

Workers in a dusty occupation should undergo a pre-employment medical examination followed by periodical check-ups. Exposure to dust imposes a heavy load on the different functions of the lung and in particular on the self-clearance mechanism. Hard physical work in dusty surroundings is accompanied by heavy breathing. Workers with functional disturbances or pathological changes in the lung should be excluded from work involving exposure to dust as they would be particularly predisposed to a worsening of the condition or to the onset of pneumoconiosis. Examinations at the time of recruitment and further medical supervision of workers exposed to dust must include a lung-function test including, as a minimum, vital capacity, forced expiratory volume per second, together with an X-ray and clinical examination. In cases where a functional disorder of the lung may be aggravated by further exposure to dust, a change of occupation to a dust-free workplace may be necessary for medical reasons. In the case of exposure to asbestos dust, it is not possible at the present time to fix a medically based time limit, and this is also the case for other carcinogenic substances.

Treatment. The further care and treatment of an affected person becomes the responsibility of the family doctor or a specialist. This care and treatment may be complicated if the victim changes his employer or place of residence. Because of the long latent period before any disease develops, a particularly difficult situation is represented by persons who have left a dusty occupation. For this reason, the follow-up of workers who have been exposed to dust should be concentrated on the most important risk groups, comprising those who were formerly exposed to asbestos and other carcinogenic dusts.

Epidemiological supervision. Since the results obtained from animal experiments can only be applied to humans under certain conditions, the epidemiological aspects and supervision are very important in the case of workers who have been exposed to dust. They provide a valuable starting point for the establishment of scientifically based occupational exposure limits and for verifying their reliability. In many countries epidemiological analysis is limited to the statistical evaluation of occupational diseases and their causes. It should include the periodical evaluation of medical preventive examinations, problem-oriented research based on a complex "work-health" and "work-medical" approach, together with an X-ray screening of the population using an 11 × 11 cm format. For the creation of healthy conditions at workplaces exposed to dust, significant data can be derived from the comparison and simultaneous evaluation of health data concerning dust exposure and medico-clinical data on the state of health and its evolution.

HÄUBLEIN, H. G.
REBOHLE, E.
BECK, B.

"Defensive mechanisms of the respiratory tree" (Mécanismes de défense de l'arbre aérien). Chretien, J.; Huchon, G.; Marsac, J. *Médecine et Hygiène* (Geneva), 23 Apr. 1980, 38/1375 (1458-1473). Illus. 71 ref. (In French)

The mechanics of aerosols. Fuchs, N. A. Transl. from the Russian by Daisley, R. E., and Fuchs, M. Davies, C. N. (ed.). (Oxford, Pergamon Press, 1964), 408 p. Illus. 886 ref.

Toxicology–The basic science of poisons. Casarett, L. J.; Doull, I. (New York, Macmillan, 1975), 768 p. Illus.

Handbook of industrial toxicology. Plunkett, E. R. (London, Heyden, 1976), 552 p.

IARC monographs on the evaluation of carcinogenic risk of chemicals to man. International Agency for Research on Cancer (Lyons, International Agency for Research on Cancer, since 1972).

Methods used in establishing permissible levels in occupational exposure to harmful agents. Report of a WHO expert committee with the participation of ILO. Technical report series No. 601 (Geneva, World Health Organisation, 1977), 69 p.

"Pneumoconiosis" (Pneumokoniosen). Ulmer, W. T.; Reichel, G. (eds.). *Handbuch der inneren Medizin*. Band 4, Teil 1 (West Berlin, Heidelberg, Springer Verlag, 1976), 692 p. Illus. (In German)

Research report: chronic bronchitis and occupational dust exposure–Cross-sectional study of occupational medicine on the significance of chronic inhalative burdens for the bronchopulmonary system. Deutsche Forschungsgemeinschaft (Boppard, Germany (Fed. Rep.), Harald Boldt Verlag, 1978), 503 p. Illus. 299 ref.

Dust control in industry

There are two distinct hazards associated with industrial dust, namely the explosion hazard and the health hazard.

Most combustible dusts can cause explosions if ignited when in cloud form. Extensive damage to buildings and injury to persons has been caused by explosion of dust clouds composed of aluminium, coal, cork, starch, wood, flour or other similar materials. Dust explosions are often in two stages, a primary explosion from a local dust cloud followed by a major secondary explosion due to accumulations of settled dust being dispersed into clouds by the primary explosion. The technicalities of this hazard are dealt with in greater detail in the article DUST EXPLOSIONS.

Serious health risks arise in mining, quarrying, tunnelling, stone crushing, asbestos industries, foundries, steelworks, potteries, chemical and other industries. Risks may also occur in agriculture if fertilisers, pesticides or seed dressings are used in dust form. The effects of dust on the body vary with the nature of the dust. Silica and mixed dusts frequently cause pneumoconiosis. Beryllium and cadmium cause, among other things, severe inflammation of the lung. Vegetable dusts may give rise to asthmatic conditions and byssinosis. Farmer's lung is associated with exposure to dust from mouldy hay or grain. Chromium compounds and asbestos are associated with lung cancer. Lead dust has its toxic effects on internal organs and on the nervous system. Finally, fine particles of toxic substances suspended in air may be deposited on skin and mucous membranes causing irritation, sensitisation, ulceration or even cancer (see DUST (BIOLOGICAL EFFECTS) and separate articles on specific dust sources and associated occupational hazards).

There is strong evidence that cigarette smokers are much more susceptible than non-smokers to the effects of airborne dusts.

Exposure limits

The aim of dust control is to reduce airborne dust to an amount as low as is reasonably practicable and certainly below the exposure limits. These vary according to the toxicity of the dust and range from below 0.1 mg/m^3 total dust to a maximum of 10 mg/m^3 total dust. Much lower limits have to be applied to the respirable fraction of the dust, i.e. the particles under about 5 microns in size. Such fine dust is not easily visible. Visibility of dust to the naked eye depends more on the size of the particles than on their concentration (figure 1).

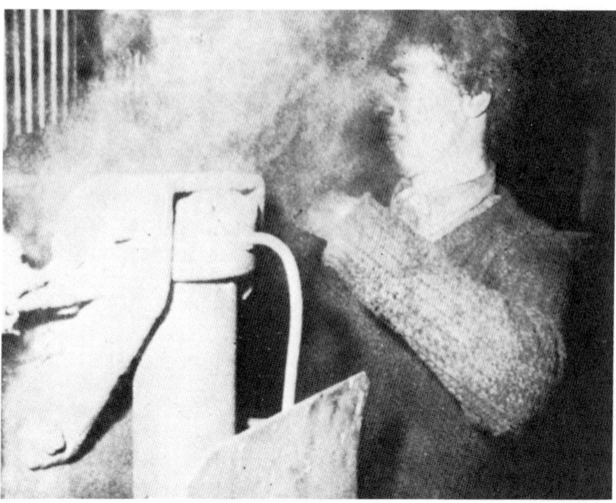

Figure 1. Coal and sand dust from a foundry mixing machine. The dust is visible because of the large particles which it contains.

Dust-control techniques

The basic techniques of dust control are: elimination; the substitution of a less hazardous material; segregation and enclosure of dusty processes; the application of moisture to materials to prevent particles becoming airborne; ventilation; and filtration (see CONTROL TECHNOLOGY).

Elimination. The first question to be asked is whether or not the dust needs to be produced at all. In foundry processes, for example, improved casting techniques may eliminate the need for subsequent dusty fettling processes.

Substitution. If a dust is highly toxic, substitution of a less toxic or non-toxic material must always be considered. Substitution of various materials has been enforced by law in certain countries and, in the United Kingdom for example, the use of sand for sandblasting is prohibited, and metal shot, alumina or some other less toxic material must be used in its place.

In other cases, the cost of implementing rigorous dust-control measures in the use of a dangerous material may be so high that substitution may be attractive on economic grounds as well as desirable from the point of view of worker protection. Thus, glass fibre, slag wool and other insulation materials are replacing asbestos in many applications. The use of metal moulds in place of sand moulds for casting processes is a further example.

Segregation and enclosure. If the evolution of dust cannot be prevented, then segregation or enclosure of the process should be considered. Segregation is important, particularly in dealing with explosive dusts. By placing grinding mills in small, vented enclosures and by placing dust cyclones, etc., outside the building, the risk of damage to property and personnel in the event of an explosion is minimised. Segregation of the dusty processes may also be useful as a health measure. By concentrating the dusty processes in one area it may be much easier to install any necessary control system. Complete enclosure is the best form of segregation. An example of this is the blasting cabinet where the operator controls the blasting operation from outside the enclosure. Enclosure is not normally adequate without some form of exhaust.

Wet methods. Wet methods can be used effectively to prevent the generation of dust. It is obvious that dust is eliminated if powdered material is suspended or dissolved in a liquid. Similarly if such materials are adequately moistened by capillary action or by condensation, the point is reached where they will cease to generate dust. It has been shown that moulding sand is dust-free if it contains about half of its working moisture content. Again, some solid materials such as sandstone can be made to soak up water so thoroughly that dust is avoided in subsequent cutting operations. The two important points in this valuable technique are first that the correct degree of wetting must be achieved before handling or processing and secondly that the moistened material (including waste and spillage) must not be allowed to dry out.

Once a dust cloud has been generated it is not easy to control it by wet methods. High-pressure water jets or sprays are sometimes used and they may achieve some reduction in airborne dust. Nevertheless, difficulty is experienced in wetting fine airborne particles, even though wetting agents may be added to reduce the surface tension of the water.

Ventilation. Local exhaust ventilation is a common method of controlling dust and is most efficient if the source of dust has first been enclosed. The smaller the gaps in the enclosure, the smaller the exhaust rate required for control of the dust. Anyone who has ever used a small vacuum pipe to pick up dust will know that the end of the pipe must be brought very close to the dust in order to collect it. Once a dust cloud has escaped into the general atmosphere it is a difficult matter to gather it up into an exhaust system. Therefore the objective in local exhaust ventilation is to capture the dust as near to its source as possible and to remove it in such a way that it does not enter the breathing zone of any worker (figure 2). In general it has been found that if the source of the dust is enclosed, an inward flow of air at 1 m/s through all openings in the enclosure is sufficient to prevent dust escaping.

Because conventional exhaust systems may have to remove vast quantities of air in order to capture the dust, considerable study has been made of the development of low-volume high-velocity systems where a small volume of air is sucked at a high speed from a point immediately alongside the source of dust. This method is effective for some portable tools and is used for collecting asbestos dust from machining operations, soap dust from wire-drawing operations and for efficient collection of human ashes in crematoria.

General ventilation is seldom the correct solution to the dust problem. It does not guarantee that persons working near the source of dust will breathe clean air. Excessive general ventilation may disturb dust and keep it airborne and can interfere with the correct functioning of local exhaust ventilation by creating unwanted side-draughts. Carefully designed local exhaust systems together with suitable air intakes can usually be so arranged as to afford sufficient air changes within a building.

Filtration and disposal. Once the dust has been captured by an effective local exhaust system, the problem arises as to what is to be done with it. If everybody blew their industrial dust out into the general atmosphere this would simply mean that the risk of disease was being shared with the general population. Persons living near certain types of factory may contract dust disease simply because of the pollution of the surrounding atmosphere.

Various kinds of air-cleaning plant are used industrially, such as fabric filters, electrostatic precipitators and wet collectors of various types. The design and choice of an air-cleaning system is a matter for expert advice. If recirculation of the filtered air is required because of heat loss from the building then the recirculated air should be as pure as practicable. In some circumstances recirculation is permitted provided the contaminant in the recirculated air does not exceed 10% of the exposure limit. Some dusts, for example carcinogenic dusts, are so dangerous that the return of filtered air to the workroom may not be permitted.

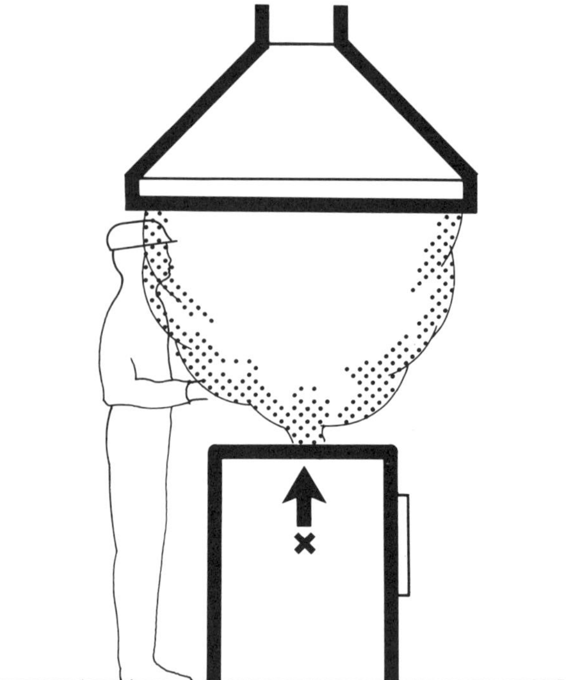

Bad design – dust carried through worker's breathing zone.

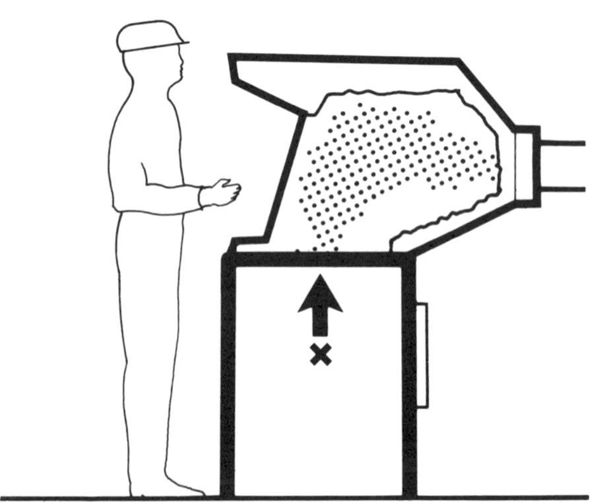

Good design – dust carried away from worker.

Figure 2. The importance of the direction of air flow.

Practical dust-control applications. A good example of the application of dust control to industrial processes can be found in iron foundries; this industry is confronted with a number of important dust problems but, due to extensive research, adequate solutions have been found.

Dusty fettling processes have sometimes been eliminated by improved casting techniques. Substitution has been applied in the foundry by using, for example, non-silicious parting powders in place of silica flour, and steel shot instead of sand in blasting plants. Enclosure has been effectively utilised for sand recovery plants, mixers, etc. Wet methods have also been found effective in the foundry, particularly where sand can be adequately and thoroughly moistened before handling.

Ventilation is applied extensively in foundries, especially in fettling areas, where total dust burdens of over 230 mg/m^3 have been recorded. This problem has been tackled first by precleaning all castings, using an efficient shot-blasting or hydroblasting system. Secondly, the dust sources in the fettling shop itself have been located. These include pedestal grinders, swing-frame grinders and portable grinding machines (figures 3 and 4). If effective local exhaust ventilation is not provided for these machines, then dust will settle on floors, ledges and fixtures only to become redisturbed by the constant movement of persons, vibrations of machines and so on. The pedestal grinders may be fitted with an integral ducted system extracting at exhaust rates up to 1.2 m^3/s depending on wheel speed and size. The swing-frame grinding machines may also have integral extraction systems at exhaust rates of over 0.5 m^3/s or they may be equipped with booth exhaust, which in many ways is easier to maintain and puts less restriction on the movement of the machine. The main problem with booth exhaust is that the air extraction rate may be high, for example 3 m^3/s. For the portable machines, exhausted fettling booths or benches with air velocities of not less than 1 m/s at the open face of the enclosure are likely to be the best remedy for most castings.

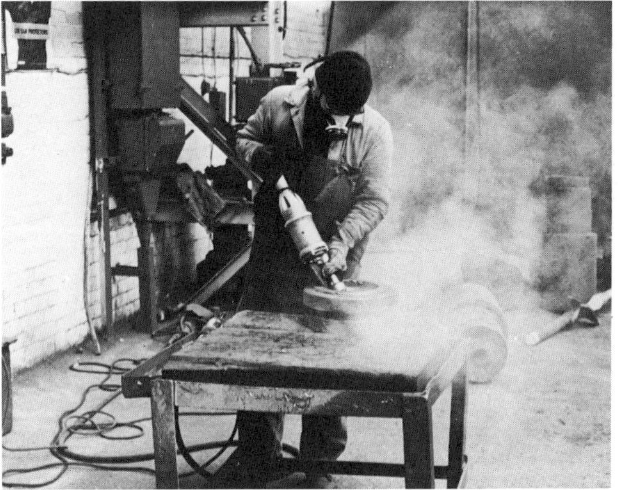

Figure 3. Portable grinding–dust not controlled.

Figure 4. Portable grinding–dust controlled by exhausted booth.

The details of dust-control measures and especially the design data for exhaust ventilation equipment will vary from industry to industry but the basic principles remain unchanged. However, in the choice of an exhaust ventilation system in particular, adequate consideration should be given to the practical problems of the industry in question. It is possible for systems that can be shown to work perfectly under laboratory conditions to be ineffective in practical application. Men whose wages depend on output are often regrettably careless as to their own health. They may be impatient with devices that tend in any way to interfere with the job or to impede the movement of the machine. They will not stop to make adjustments. For this reason booth exhaust systems independent of the worker are sometimes more successful in practice than refined ducted systems (figure 5). This is a pity because the more refined systems would require less power to operate and would remove less air. It is, however, a fact which it is wise not to ignore.

Figure 5. An efficient booth exhaust system (3.5 m^3/s per booth) for steel grinding. *(By courtesy of Swift Levick and Sons, Ltd., Sheffield.)*

For the future we can expect to see more examples of robot machines operating within separate enclosures and controlled by microprocessors, thus enabling workers to be removed completely from unhealthy environments.

Design and construction

Designing and constructing an exhaust ventilation system requires considerable expertise and is dealt with in detail in the article EXHAUST SYSTEMS. Any collection or filtration unit for flammable dust should be provided with explosion relief panels. The relief area may vary between 1 m^2 per 3 m^3 of enclosure to 1 m^2 per 6 m^3 of enclosure depending on the flammability of the dust.

The siting of collection and filtration units needs careful consideration. Advantages of placing the dust collection system outside the workroom include reduction of noise and less trouble in disposing of the collected dust. But these advantages may be outweighed if long distances are involved.

Fabric filters are commonly used for small installations but they can become clogged in use. For this reason, fabric filters are equipped with shaking or cleaning devices designed to release the matted dust from the fabric into the collection chamber. Automatic devices for this purpose are much more satisfactory than manual devices, which tend, in practice, to be neglected. It is important when purchasing a filtration system to know what its eventual performance will be. A filter for example may have a nominal capacity of about 0.5 m^3/s. The user will need to be assured that the fan is sufficiently powerful to maintain this airflow through the restricted exhaust intakes to which the unit may be connected, for example a pedestal grinding machine. He would also need to be assured that the required air flow will be maintained after the filter has been in use for a substantial period (allowing, of course, for the correct functioning of any manual or automatic cleaning arrangements fitted). Filter systems should be so designed that air flow does not fall by more than 10% between shaking or cleaning, and the fans should be on the outlet side of the filtration unit.

Purchasers of exhaust systems should bear in mind that the cheapest plant initially will not normally be the cheapest in the long run. The system should be checked to ensure that the performance is up to the specification before it is accepted.

Verifying dust-control efficiency

Checking the efficiency of dust-control measures begins with elementary visual observation of both atmospheric dust and of dust settled on ledges and fixtures. Such a check is, however, frequently inadequate to identify precise sources of dust, particularly when a number of different dust-producing machines may be in use.

One of the most sensitive methods of checking the effectiveness of dust-control measures is the use of a simple dust lamp. Portable equipment for this purpose is regularly used by the Health and Safety Executive in the United Kingdom. The lamp used is a 100 W 12 V quartz iodine bulb fitted in a parabolic reflector to give a parallel beam of light. It is powered by rechargeable batteries. The dust cloud should be observed by looking up the beam towards the source of illumination (but with the source shielded from the eye), the line of observation being at an angle of approximately 10° to the line of the beam. The usefulness of this method of checking the movement of fine dust clouds not normally visible to the naked eye is greatly enhanced if the work can be undertaken in darkened surroundings.

Local exhaust ventilation should be subject to regular inspection to check that hoods and ducting have not been damaged. Simple dust puffers, or smoke tubes, as well as velometers, manometers and pitot tubes can be used to check that the correct air-flow standards are being maintained. A static pressure gauge permanently mounted on exhaust equipment is useful as a continuous indicator of any change in efficiency. On dry filtration systems, the filter cloths should be inspected and repaired or renewed as necessary. On wet collectors the correct functioning of water-level control, spray jets, etc., should be ensured.

Of equal importance is the need to check by observation that the system is being used as intended and that for example the source of dust has not moved away from its designed position in relation to the exhaust intake.

The instruments used for the sampling and determination of atmospheric dust concentrations are dealt with in the article DUST SAMPLING.

GRANT, S.

CIS 75-1282 *Dust control and air cleaning.* Dorman, R. G. International series of monographs in heating, ventilation and refrigeration, Vol. 9 (Oxford, Pergamon Press, 1974), 634 p. Illus. 569 ref.

CIS 76-647 *Final report on the anti-silicosis project.* Part 3. *Preventive measures* (Silikosprojectets slutrapport–Del 3. Förebyggande åtgärder). Gerhardsson, G.; Isaksson, G.; Andersson, A.; Engman, L.; Magnusson, E.; Sundquist, S. Undersökningsrapport AMT 103/74-3 (Stockholm, Arbetarskyddsstyrelsen, 1974), 266 p. Illus. 18 ref. (In Swedish)

CIS 77-1870 *Dust control measures in crushing plants* (Obespylivanie drobil'nyh cehov). Kalmykov, A. V. (Moscow, Izdatel'stvo "Nedra", 1976), 206 p. Illus. 46 ref. (In Russian)

CIS 78-962 "A new dust control system for foundries". Schumacher, J. S. *American Industrial Hygiene Association Journal* (Akron, Ohio), Jan. 1978, 39/1 (73-78). Illus.

CIS 76-939 *Dust extraction systems in the ceramics industry–Recommendations of the Joint Standing Committee for the Pottery and Allied Industries.* Health and Safety Executive (London, HM Stationery Office, 1975), 25 p. Illus. 11 ref.

CIS 80-963 "Dust control concepts in chemical handling and weighing". Hammond, C. M. *Annals of Occupational Hygiene,* 1980, 23/1 (95-109). Illus. 12 ref.

Dust explosions

Any combustible solid material, in finely divided form, can give rise to a dust explosion hazard. All that is necessary is for the dust to become suspended in air in sufficient concentration and to be subjected to a source of ignition. Flame from the source will spread in all directions through the cloud consuming the combustible content and causing expansion, with resultant pressure effects upon the surroundings. Pressure waves from the initial explosion can whip deposited dust into the air in front of the advancing flame, with the result that the explosion may be extended far beyond the original dust cloud, in the form of a "secondary" explosion. Gas explosions can also be extended by the involvement of combustible dust deposits.

Explosive dusts

The hazard of a dust explosion is encountered with many materials of natural origin (starch, sugar, flour, coal, wood), with plastics and organic chemicals, with light metals (aluminium, magnesium, titanium) and with sulphur. Consequently, potential hazards will exist in agricultural work, in the chemical, metallurgical and process industries, and in coal mining. Mineral dusts are incombustible and therefore not susceptible to explosion. In coal mining, they are in fact used for dust explosion suppression.

Conditions required for explosion

Whether a particular dust is sufficiently combustible and sufficiently finely divided to give rise to an explosion hazard in practice is best determined by test. A number of such tests have been described and some are used for official purposes. Voluminous data have been published on the results of such tests.

The range of concentrations in which combustible dusts are explosible is found above 10 g/m^3. Clouds of dust in such concentrations in a working area would considerably reduce visibility but they commonly occur inside powder-handling plant and equipment. Outside plant, such clouds would not pass unobserved, but they may arise due to accidental spillages or overflows of powder or, in a dusty building, as a secondary effect of a plant explosion.

Explosive clouds are ignited in hot enclosures at temperatures above 400 °C, although there are cases of lower ignition temperatures. If a dust can rest upon a hot surface, however, smouldering is liable to be engendered above 150 °C and such a deposit ignition temperature may be more properly regarded as the highest safe surface temperature in the context of a dust explosion hazard.

Explosive clouds are ignited by electrical discharges in excess of about 10 mJ, but contact with repeated discharges of lower energy could engender smouldering. Dusts should therefore be prevented from entering electrical equipment by the use of dustproof enclosures.

Dust clouds can also be ignited by the effects of mechanical friction or impact, but prolonged friction or repeated impact is generally necessary to be effective. Naked flame will, of course, immediately ignite an explosive cloud.

SAFETY MEASURES

The risk of dust explosion can be avoided by preventing the formation of explosive clouds. Inside powder-handling and powder-storage equipment, this can be achieved in practice only by keeping the equipment filled with incombustible gas so that the oxygen content of the atmosphere is below 5% by volume. Nitrogen, carbon dioxide or washed flue-gas can be used, but for the light-metal dusts, helium or argon are desirable.

Outside the equipment, explosive cloud formation is avoidable:

(a) by constructing the plant in a dust-tight manner;

(b) by maintaining slightly sub-atmospheric pressure in the plant;

(c) by installing exhaust systems to collect all foreseeable escapes of dust (as at feed, or bagging points);

(d) by designing to avoid overflows; and

(e) by maintenance and good housekeeping to ensure that plant structure and any surrounding buildings are kept free from deposited dust.

If the equipment can be accommodated partly or wholly in the open air, hazards external to the plant are reduced.

The second method by which the risk of dust explosion can be avoided is by the elimination of ignition sources. Hot surfaces, high temperatures not necessary for production process, and mechanical equipment that might overheat should be eliminated both inside and outside plant. Lighting and electrical installations and equipment generally should be dustproof and all plant should be electrically bonded and earthed as a safeguard against electrical discharges. Naked flames must be forbidden and no hot work conducted on plant that has not been taken out of commission and thoroughly cleaned.

It is possible to design equipment to contain, without damage, the pressure due to an internal dust explosion. Since the pressures built up inside a closed vessel in the event of a dust explosion may be as high as 8 kgf/cm^2, i.e. 7.84 bar (and occasionally more), the equipment must be robustly constructed. Moreover, travel of the explosion from vessel to vessel can result in still greater pressure and this must be prevented by the insertion of chokes, rotary valves and like devices, which allow passage of the material, but not of the explosion.

The possible explosion pressure can be reduced by fitting explosion reliefs which, in emergency, discharge the explosion flame and pressure to a safe place outside any building that may house the plant. An alternative is to install automatic explosion suppression equipment in which the small pressure rise due to an incipient explosion triggers the release of a suppressant fluid in front of the travelling explosion flame. The actuation of a relief or suppressor should automatically stop further operation of the plant.

BURGOYNE, J. H.

General:

CIS 74-1214 *Dust explosions and fires.* Palmer, K. N. (London, Chapman and Hall, 1973), 396 p. Illus. 301 ref.

CIS 74-1747 "Protection against explosions in the chemical industry" (72-416). *Report of the 2nd International Symposium on the Prevention of Occupational Risks in the Chemical Industry* (Heidelberg, Berufsgenossenschaft der chemischen Industrie, no date), 559 p. Illus. 49 ref.

Relief venting:

CIS 76-1815 "Recent advances in the relief venting of dust explosions". Palmer, K. N. *Journal of Hazardous Materials* (Amsterdam), Jan. 1976, 1/2 (97-111). 33 ref.

Standards:

CIS 77-314 *Dust explosions in industry* (Industrielle støveksplosjoner). Verneregler nr. 25 (Direktoratet for arbeidstilsynet, Postboks 8103, Oslo-Dep) (Mar. 1975), 54 p. Illus. (In Norwegian)

CIS 78-915 *User guide to fire and explosion hazards in the drying of particulate materials* (Institution of Chemical Engineers, 165-171 Railway Terrace, Rugby, Warwickshire) (1977), 96 p. Illus. 20 ref.

Dust sampling

Agencies responsible for the prevention of occupational diseases have to adopt procedures for airborne dust measurement which will enable them to:

(a) detect dust sources and determine their magnitude;

(b) check the efficiency of preventive measures that are adopted; and

(c) in certain cases, monitor the exposure of personnel to airborne dust.

In spite of the interest attached to basic studies on the nature and physical and chemical characteristics of dust, which incidentally require highly specialised equipment and instrumentation, the scope of the present article will be confined to the so-called "routine" measuring instruments. Such instruments need to be suitable for use under all practical conditions: they must be easy to handle, particularly underground in mines; they must be self-contained as far as possible, capable of giving reliable and reproducible results; and they must be safe to use in explosive atmospheres.

Sampling criteria and presentation of results

While not neglecting dust particles larger than 5 µm, which can themselves influence certain respiratory ailments, sampling and measuring instruments, including those for routine controls, must be capable of collecting and analysing particles in the size range below 5 µm, including those of 1-2 µm and, if possible, particles down to 0.5 or even 0.2-0.3 µm in size (see DUST (BIOLOGICAL EFFECTS OF)). These requirements limit the choice of methods available.

Dust samples are usually taken to determine the concentration, size, chemical composition and, possibly, shape of the particles. The different constituents of a given dust are usually expressed in terms of percentage by weight, determined by chemical, physical or mineralogical procedures. Concentrations on the other hand are usually expressed in terms of the number, the surface area, or the volume (mass or weight) per unit volume of air.

In the case of industrial dust, fumes or toxic mists, the results are generally expressed in weight per m^3 of air (mg/m^3). In the case of mineral and coal dusts, i.e. in mining, tunnelling and quarrying operations, the expression of results was often based on the importance attributed to the size of the particles. Determination of dust concentrations by counting may exaggerate the importance of the large number of small particles, while gravimetric determination can be misleading because of the presence of large dust particles; in certain cases the surface area may be considered as a middle course between the two extremes represented by number and mass.

But whatever theory is adopted to explain the origin of silicosis, it is the total surface area or mass of the particles deposited on the pulmonary alveoli and not their number which determines the environmental hazard. It is for this reason that–since the II International Conference on Pneumoconioses held in Johannesburg (9-24 February 1959)–selective sampling instruments have been developed–especially gravimetric–equipped with an elutriator to collect only the particles termed "respirable". That instruments which collect all the airborne particles have been retained in use for so long is explained by the fact that it had not been possible to obtain reproducibility and also by the handling difficulties encountered with instruments that were available at the time.

Different methods of obtaining the particle size distribution are followed depending on the way in which the sampling is effected: either a sample is taken on the spot of all the dust present in the air and a particle size analysis is performed afterwards in the laboratory by whatever means may be appropriate (counting with the aid of a microscope, sedimentation rate, variation in specific weight of the dust in suspension), or a "cross-section" is taken during the sampling itself, making use of selective instruments fitted with a pre-separator or elutriator, which will eliminate the large particles as, for example, those which are normally collected in the upper respiratory tracts, and will not collect the fine, respirable dust.

The instruments that have appeared on the market since 1970 can be considered as being satisfactory in this respect as well as being easily handled. It should be noted, however, that respirable dust is considered in some cases to be that which penetrates the lung, while in other cases, it is only considered as being those particles which are deposited and retained in the pulmonary alveoli.

Selective gravimetric sampling, as employed in pneumoconiosis prevention, seldom takes the shape of the particles into consideration. This is not the case, however, when sampling and examining asbestos particles, where the length and the diameter of some of the fibres has to be taken into account.

To sum up the present situation, therefore, dust concentrations are being increasingly expressed gravimetrically, in mg/m^3, taking into consideration the particles of respirable size, and indicating the proportions by weight of the different constituents such as ash, free silica or quartz, as a percentage.

It is not sufficient, however, to express the results in this way if it is desired to obtain full details concerning the dust conditions in an industrial or mining situation. Various other rules and methods of sampling have to be observed in order to determine the dust exposure of particular workers or the amount of dust produced by any given operation or working method.

If a dust sample in an industrial or mining situation is to be representative, it should not be taken too close to the origin of the dust; in addition, it is important to choose periods of normal activity, to position the sampling head of the apparatus where the air seems to be most homogeneous and to allow sufficient time for the sampling (a minimum of 2 h). Unless it is wished to study the fluctuations in dust concentrations, continuous sampling is, theoretically, better. In total gravimetric sampling, it is advisable to aspirate the dust at a speed comparable to that of the speed of air movement. This is not so imperative for numerical sampling: it is, in practice, impossible in the case of instruments equipped with a pre-separator, and functioning at constant speed.

In addition to these rather general recommendations, different countries have developed their own "sampling strategy" or system of measurement. These are based on the local conditions of work or on a method of mining, and also, to some extent, on established customs. This strategy sets out the duration and frequency of sampling, the sampling position which may be at a fixed point or on the worker himself. Particularly in the case of mines, it is considered that an estimation of the dust conditions requires measurements to be made in the intake and return roadways and even that the working face be subdivided into several sections.

Sampling instruments

These function on the basis of filtration, sedimentation, centrifugation, scrubbing or washing, precipitation by impact, thermal precipitation, electrostatic precipitation or measuring of certain optical properties of dust clouds.

An ILO publication which appeared in 1967 described some 60 routine sampling instruments which were used mainly in mines. Since that time, the number of types of instruments in use has been reduced in favour of those based on filtration with or without sedimentation (using cyclones) and also of those which, in the strictest sense, are not sampling instruments but which measure the amount of light diffused by the dust particles.

A number of countries and dust control agencies have, however, remained faithful to the older techniques, which enable them to maintain comparability with their existing records and which may be justified in situations where the dust concentrations and their size distribution have become stabilised as a result of the intensive implementation of dust control systems.

Filtration devices. These consist of an aspirator (fan, pump, ejector, etc.), an air-flow meter, and a filtering material. The filters are papers in the form of Soxhlet thimbles (figure 1) or simple discs; soluble material;

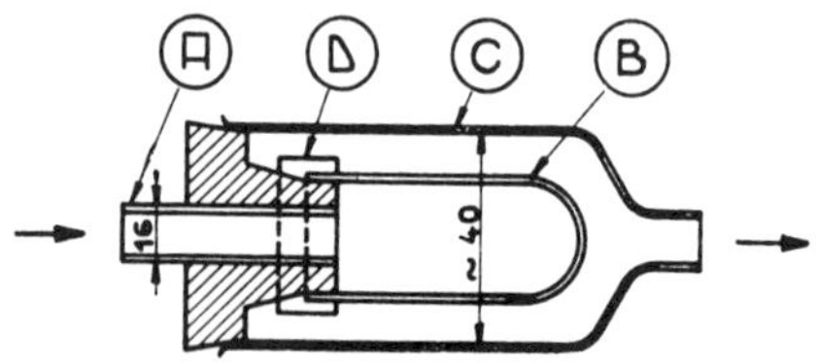

Figure 1. Soxhlet filter container. A. Intake tube. B. Filter. C. Filter container. D. Sleeve for fixing filter.

glass fibre; or cellulose membranes with very fine pores (0.4-0.2 μm).

Some of these instruments are provided with a pre-separator which usually consists of a multi-duct elutriator.

(a) The Soxhlet filter together with the Staser instrument continues to be used for determining the overall gravimetric concentration. Used with a compressed-air ejector, this instrument aspirates 1 m^3/h and the diameter of the entry orifice can be varied to suit the air-flow speed. The other Soxhlet filter instruments such as the Siter (4 m^3/h), the Göthe (12 m^3/h) and the Hexhlet with elutriator (6 m^3/h) have practically disappeared.

(b) Cellulose membranes: these filter materials have a very high retentive capacity and are more and more

used either to sample low concentrations by counting the particles on the filter or to take a sample by weight. Among the devices giving numerical concentration are the Draeger pump (aspiration 100 cm³), the Zurlo apparatus (mercury pump of the elepsyd type: capacity 100 cm³) and the self-contained Morin-Cerchar instrument. The latter aspirates up to 500 cm³/min, according to the adjustment of the electric pump (which must be flameproof for coal mines) and can take up to 20 successive samples for each work shift. The SIMPED personal dust sampler and instruments derived from it, which aspirate about 2 l/min–ranging from 0.8 to 4 l/min–have been used more extensively in research on asbestos and particularly for counting fibres. Instruments using membrane filters for over-all gravimetric samples are still relatively numerous, and these include the Staplex and other types which aspirate from 2 to 10 or 12 and even 20 or more m³/h. They operate by filtration alone, but when fitted with a pre-separator to eliminate non-respirable particles in explosive atmospheres, they work mostly on compressed air if an amount of more than 2 to 3 m³/h is required to be aspirated. One example of these is the SFI - Draeger, and a variation, known as the MPG instrument, which aspirates 3 m³/h and makes a pre-separation which follows the acceptance curve recommended at the Johannesburg Conference. Self-contained instruments which are electrically driven from an accumulator sample a smaller quantity of air, usually about 2.5 to 4 l/min. The flameproof MRE sampler, from the United Kingdom, fitted with a membrane filter is one of the best known of this type. It handles 2.5 l/min and performs a separation which follows the Johannesburg curve (figure 2).

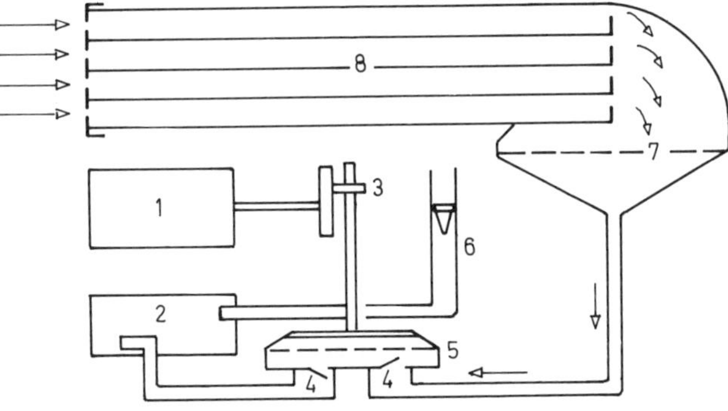

Figure 2. Schematic view of the MRE sampling instrument. 1. Motor. 2. Air flow compensator. 3. Adjusting crank. 4. Valves. 5. Pump. 6. Rotameter. 7. Filter. 8. Horizontal elutriator.

(c) Other filter agents: other filtration materials are used according to particular needs or according to the method by which the collected particles are to be examined. The apparatus produced by the Le Bouchet Laboratories and the Nord-Pas-de-Calais Mines used tetrachloronaphthalene soluble in benzene; the particles collected were counted in the solution or weighed after evaporation of the liquid. Filter papers in disc form, of small dimensions (diameter 10-15 mm), were mounted on clips at the orifice of a hand pump (PRU pump). The dusts were neither weighed nor counted; the examination consisted of making blackening tests, to measure the optical density of the loaded paper or the percentage of light retained by the filter. This method of sampling may continue to be justified for routine measurements when the size distribution and the type of dust remains unchanged. Filter papers made from glass fibre are sometimes used instead of membranes. Such a filter (Whatman) has for several years been used in the MRE self-contained apparatus. The same filter may also be fitted to the MRDE Multishift instrument, which is based on the MRE instrument, and which is designed to take a sample over a duration of up to 5 shifts. Recently, revolving filters of synthetic foam have been used on some instruments. It should be noted, however, that these sampling instruments do not operate by filtration alone.

Sedimentation instruments. These are very rudimentary instruments, consisting essentially of a glass plate which may or may not be covered with vaseline, exposed to the air for a certain period. In Wright's sedimentation apparatus, a sample of air is automatically taken at regular intervals in an enclosed space; the particles settle and are deposited on a glass slide which is then examined under the microscope. These instruments are now used only in exceptional circumstances in laboratory work.

Centrifugal sampling instruments. These may operate with or without filtration or sedimentation and collection is effected by the inertia of particles in suspension in a current of air with a predetermined flow. The classification of dusts is made according to their particle size distribution, their shape and their nature; the instrument retains only those particles with an impact velocity within certain limits, depending on the air flow, the air velocity and the dimensions of the system. The

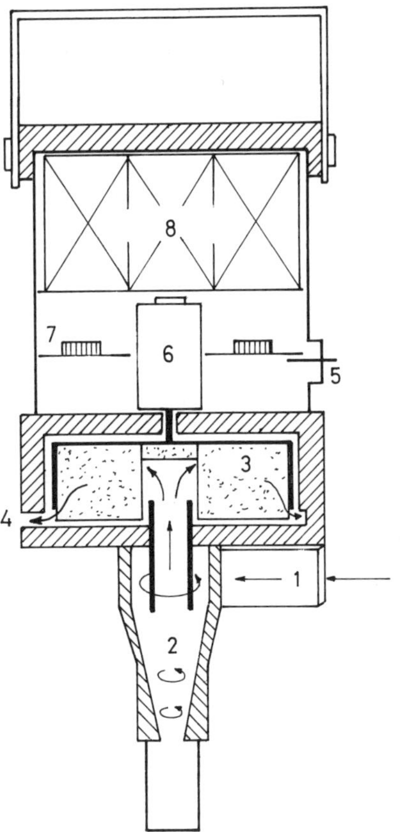

Figure 3. Schematic view of the CPM_3 sampling instrument. 1. Air entry. 2. Cyclone. 3. Rotating filter. 4. Air exit. 5. Switch. 6. Motor. 7. Voltage regulator. 8. Batteries.

Conicycle Selective Sampling System (10 l/min for 8 h) and the Turbo Collector of the Nord-Pas-de-Calais Mines (75 l/h for 40 h) are self-contained devices, equipped with an electric micromotor which turns the sampling head at 8 000 or 6 000 revolutions per min.

The French CPM_3 self-contained instrument from Cerchar, which may be safely used in a firedamp atmosphere, is a new version of the Turbo Collector (figure 3). It consists of a cyclone pre-separator and collects fine dust on a foam filter which revolves at a high speed and thus acts as a fan. Its capacity is 3 m³/h.

By combining the acceptance curve of the cyclone with the efficiency curve of a foam filter, a collection curve for fine dust may be obtained which approaches closely the alveolar retention curve proposed by Hatch and his colleagues.

A comparison between the retention curve and that of various pre-separators may be seen in figure 4, which shows that the fine dust particles, said to be respirable, are not always the same.

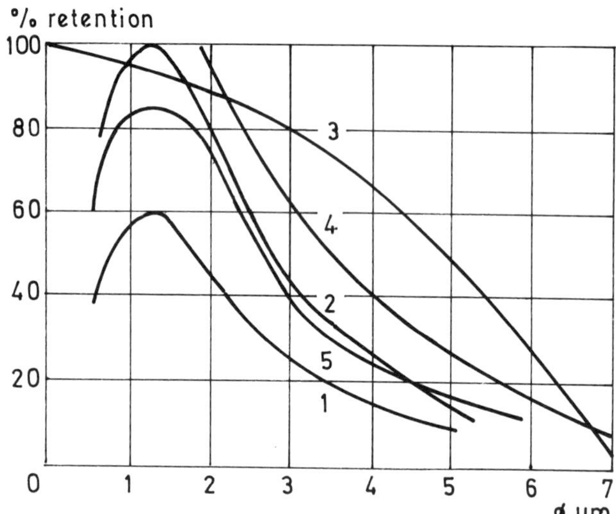

Figure 4. Curves showing alveolar dust retention and typical characteristics for pre-separators. 1. Alveolar retention after Hatch et al. (1950). 2. Curve reduced to unity (i.e. maximum 1 = 100%). 3. Separation curve for MRE instrument (Johannesburg Conference 1959). 4. Separation curve for AEC instrument (Los Alamos Conference 1962). 5. Separation curve for CPM_3 instrument.

The German TBF_{50} instrument from the Mining Research Institute is also safe for firedamp atmospheres and makes use of a cyclone pre-separator which is identical with that of the CPM_3 instrument. It is a self-contained instrument and collects the dust in a second cyclone having a capacity of 3 m³/h (figure 5). A cellulose membrane filter may be added to collect the finest dust particles that pass the second cyclone, in which case it is activated by a compressed-air ejector.

Instruments which sample a large quantity of air, using a cyclone followed by a membrane collector, are still used in Western European mines, although generally for special investigations. Different acceptance curves may be arrived at by varying the quantity of air aspirated, which can run from 50 to 75 or even 100 m³/h. Known as BAT instruments, they run directly off the compressed-air system of the mine.

The Zurlo Pneumo-classifier also collects dust by inertia but does not have recourse to centrifuging; the aspirated air undergoes abrupt changes of direction in a tapering channel, which produces a sharper separation of particles.

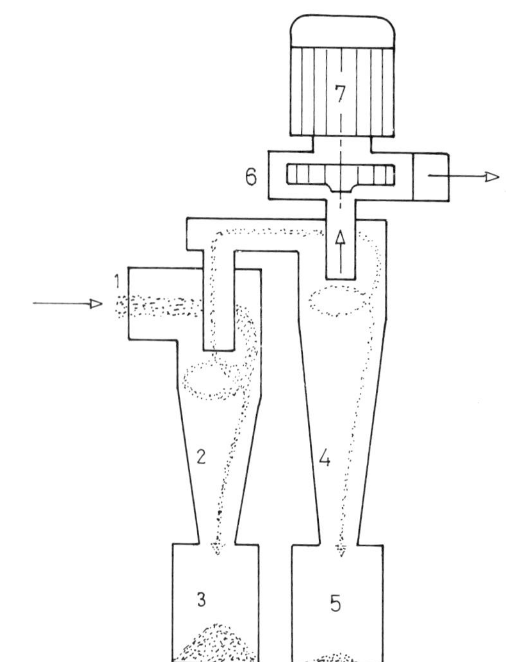

Figure 5. Schematic view of the TBF_{50} sampling instrument. 1. Suction duct. 2. Coarse dust cyclone. 3. Receptacle for coarse dust. 4. Fine dust cyclone. 5. Receptacle for fine dust. 6. Turbine. 7. Motor.

Scrubber devices. The dust-laden air, aspirated by a hand pump, a gas ejector or compressed air, is projected against an obstacle and alters direction abruptly in a liquid (wash bottles containing a solution of low surface tension). The sample is suitable for weighing (after evaporation of the liquid) or for counting in suspension. The best known apparatus is the "midget impinger" (3 l/min). This and other instruments derived from it (e.g. midget scrubber) have poor retentive capacity for particles below 1 µm; they cause disintegration of natural aggregates and even of friable particles.

Jet instruments based on impaction. In this class of devices—commonly known as konimeters—the dust contained in a very small volume of air is aspirated by a hand pump or an automatic pump and projected on to a glass sheet (which may or may not be coated with an adhesive). The collection efficiency varies according to the particle size and concentration and the nature of the dust samples. These disadvantages can be overcome when the apparatus is fitted with a sedimentation chamber which eliminates particles above 5 µm, and an adjustable aspirator is provided, as in the Bergbaukonimeter.

The cascade impactor is an apparatus derived from the konimeter. Continuously operating, it is composed of three or four collectors in series giving a separation of dusts into three of four particle-size fractions.

Electrostatic precipitators. In these instruments, suspended particles are ionised and electrically charged under the influence of an internal electric field; according to this charge, they move towards the anode or cathode where they are collected. Electrostatic samplers collect, by weight, practically all the dust up to 0.1 µm; the MSA apparatus aspirates 85 l/min while the Chamber of Mines (South Africa) apparatus aspirates 600-40 500 l/min. Unless specially constructed (Gast Apparatus), these instruments are not safe in an atmosphere containing firedamp. A further disadvantage of these instruments is that they may give rise to pseudo-

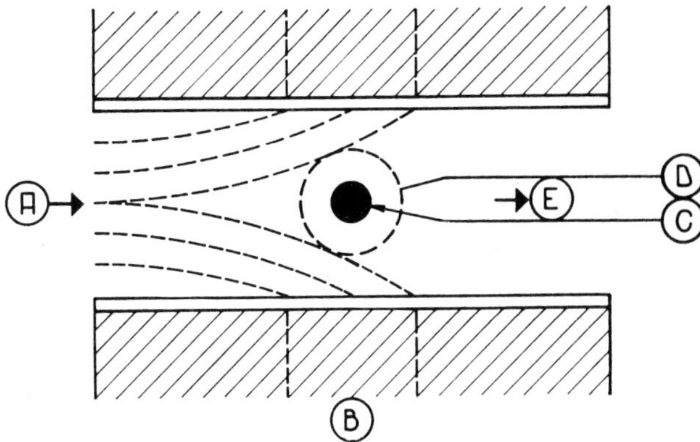

Figure 6. Principle of the thermal precipitator. A. Dust-laden air. B. Zone of precipitation. C. Hot wire. D. Dust-free space. E. Clean air.

aggregates or may break up aggregates that already exist, according to the nature of the dust being sampled.

Thermal precipitators. The operating principles of these instruments are shown in figure 6; the dust is collected by exposing the particles to a thermal field as they pass through the instrument. A sampling head (figure 7) is composed essentially of an electrically heated wire set between two brass blocks separated by a thin bakelite sheet in which has been provided a channel for the passage of air. Two metallic plugs hold in place glass slides (diameter 18 mm) on both sides of the heated wire. After sampling, the image of the wire is reproduced in deposited dust on the slides provided that the air flow does not exceed a given amount and that the temperature of the wire is maintained constant (6 cm^3/min as a maximum and 100-105 °C for the Casella "Mines" type).

As the air flow is constant, the sampling time will be limited by the dust concentration, i.e. the duration of sampling must be adjusted to obtain a deposit of appropriate density. In spite of these drawbacks, every research institute uses the thermal precipitator for the greater part of its laboratory determinations. This instrument is indeed the most dependable, giving, in the range of sizes below 5 μm, an accurate sample from the point of view of concentration, particle size, and the condition of agglomeration of the particles. It is adopted in about ten countries as a routine or semi-routine method. The original concept of the instrument has been modified to make it easier to use by fitting new slide-holders which allow several samples to be taken on the same slide by changing the sampling head so as to collect dust on only one slide and spread out the deposit in such a way as to be able to sample over a longer time, i.e. to avoid the formation of false aggregates. The

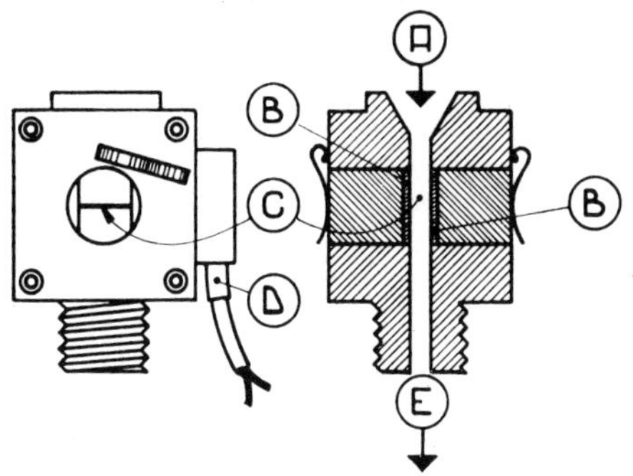

Figure 7. Section through thermal precipitator head. A. Air inlet. B. Glass slide. C. Platinum wire. D. Twin cable. E. Outlet to suction.

thermal precipitator has, however, become less interesting since the introduction of mechanical and automated counting of non-soluble dust particles and the possibility of determining size distribution using very small samples collected on membrane filters.

Optical instruments. These instruments operate by measuring certain optical properties of dust clouds. Under certain conditions, it is possible to determine dust concentration by measuring the intensity of light diffused by the particles when a luminous beam traverses a dust cloud (light absorption and diffusion phenomena, Tyndall effect). A routine instrument, the Leitz Tyndalloscope, based on this principle, comprises a dust chamber traversed by the air currents, a light source and a photometer. If the nature of the dust is known (mineral content of coal dusts) and account is taken of the particle size (by calculating the sedimentation rate before going on to make the measurement), the reading on the Tyndalloscope can be translated into concentration by the use of appropriate nomograms.

This instrument, which came into general use in the mines of the Federal Republic of Germany during the 1950s, requires frequent calibration according to the type of dust being examined and gives false readings if there is a high degree of humidity in the surroundings. It is nevertheless very useful when it is required to study rapid fluctuations in the dust concentration. It was to meet this requirement that the German Mining Research Institution and the Safety in Mines Research Establishment of the United Kingdom recently perfected new instruments based on the Tyndalloscope, known as the TM Digital and the SIMSLIN II, which are capable of following fluctuations in the airborne dust concentrations. These instruments meet the safety requirements for use in coal mines and measure the quantity of light diffused by the respirable dust particles. Results may be recorded on magnetic tapes or may even be transmitted through a digital or analogue system of signals. At the beginning of 1980 it was, however, premature to consider this as a routine instrument.

The weight of the dust may also be measured automatically, making use of low-energy β-rays. These rays are absorbed by collision with electrons whose density is proportional to the mass per unit volume. Penetration by the low-energy β-rays is almost entirely dependent on the mass per unit of surface area of the absorbent and the maximum β-ray energy of the electrons which enter into collision. The absorption is proportional to the mass of the material encountered, independently of its chemical or physical characteristics.

The Massometer model 101 instrument from the GCA Corporation in the United States (1972) aspirates 2 l/min and, with a delay of about 10 s, provides a result expressed in mg/m^3. Some instruments such as the MPSI-IRCHA from France base their operation on a differential measurement by collecting the dust in a given volume of air on a filter band made of glass fibre. It can provide from 2 to 10 readings per hour.

Examination of dust samples

On a routine basis, the examination of dust samples is concerned with determining the concentration.

Weighing. The dust weight is obtained by weighing the filter before and after sampling (with preliminary drying if necessary). To obtain the particle size, the dust is suspended in a liquid, which enables measurement to be made of: the quantity of dust deposited; the variation in the specific weight of the suspended matter; the concentration and relation to time at a given level; and the variation in electrical resistance of an electrolyte in which the particles are diluted.

It should be noted that the size distributions by weight (i.e. the percentage by weight of the different fractions above or below a given diameter) are not necessarily comparable, nor may they be superimposed, because in some cases the volumetric diameter may interfere and in others the equivalent (Stokes) diameter.

Particle counting. For microscopic counting, the particles are counted in several fields and compared with marks of known dimensions placed at the edge of the field or incorporated in an eyepiece graticule. The counting of dust in a suspension is done after homogenisation of the liquid by putting several drops in a counting chamber (height 0.1-1 mm) sealed with a coverslip. Such measurements are not comparable unless a settling time and a specific duration of examination is predetermined.

The counting of dust deposited on glass slides is, at first sight, simple and, in principle, comparable. In fact, the methods used are very different: the slides may or may not be sealed with a coverslip which, itself, may or may not be treated with acid or incinerated (temperature 200-600 °C); the magnification varies from 100 to 3 000 x according to the techniques adopted (light field, dark field and phase contrast).

This situation demands the greatest prudence in comparing "methods of measurement" since the result of microscopic counting is valid only if the method is defined in precise detail. Microscopes of the same over-all enlarging power will not necessarily give identical results even if the slides have undergone the same preliminary treatment; the concentration depends on the aperture of the lens used.

The counting of particles captured on cellulose membranes can be done without removing the holder, after impregnation of the membrane with a suitable liquid. It is essential at all costs to choose a solvent whose refractive index does not mask the presence of certain constituents.

The remarks made concerning the counting of particles on glass slides apply equally to the counting of particles that have been collected on membranes which have been made transparent.

The classical method of counting with the aid of a microscope is now much less frequently used as a result of the increasing use of selective gravimetric sampling. Microscope counting is still used in the case of asbestos fibres, in most instances using a magnification of 400 to 600 x, with phase contrast and fitted with an achromatic objective with a magnification of 40.

Particle size distribution is measured only rarely in routine sampling; this can be obtained by determining the number of particles whose diameters fall within certain limits (0.2-0.5, 1-2.5 or 3-5 μm). Graphical representation gives on the ordinates the percentages of particles smaller or larger than a size indicated on the abscissae or vice versa. For special work, semilogarithmic, logarithmic or log-probability curves are used.

A third method of examination employs photoelectric (extinction measurement, diffused light) or densitometric (blackening tests, transmitted light) procedures. These techniques require numerous gauging samples and very strict attention to technique. They are justifiable when the parameter under investigation is the surface area of the particles and are applicable where the dusts are of the same nature and of a fairly constant particle size.

Dust analysis

In the case of mineral dusts, the measurement of the constituents is confined to the determination of free silica; this is preceded by incineration using a technique which avoids reactions between silica and any silicates present (temperatures of 450-650 °C). With very rare exceptions in the mining industry, the free silica is in the form of quartz.

For routine measurement, chemical analysis can solve all the normal problems. Physical methods, chiefly X-ray diffraction, requires less material and can be automated. The determination of silica content by infrared absorption requires an even smaller weight of the substance concerned, which is why the method is interesting for the analysis of small quantities of respirable dust.

Results measured by weight and obtained either by physical or chemical methods are generally in agreement if the correct standards for the particle size distribution are chosen. Mineralogical analyses according to coloration by immersion phase contrast or limited dark field give numerical results which are often comparable to results by weight when the constituents of the dust have only slightly different specific weights. Except in the case of certain special investigations, this technique is being less and less used in line with the progressive disappearance of numerical sampling on glass slides.

In certain mining regions and for a specific coalfield it is often found that a fixed proportion exists between the ash content and the quartz content which, as far as routine sampling is concerned, enables a simple determination of the ash content to be made. Recently, however, increasing interest is being shown in the composition of the ash itself, since certain substances such as kaolinite, illite and chlorite, either in association or not, appear to play an inhibiting or catalytic role in the onset of pneumoconioses among coal miners.

Comparison of techniques

The comparison of sampling apparatus implies the choice of an instrument and a standard method of analysis and assumes that the reproducibility of the instrument in all possible applications has been checked beforehand.

It would seem that the only results which are directly comparable are over-all gravimetric measurements, or even selective gravimetric measurements, providing that the latter were collected with the aid of pre-selectors or elutriators having the same characteristics.

A recent study made by the coal-producing countries of the European Community compared five routine instruments: the CPM, the MPG, the MRE, the Staser and the TBF. In order to find out whether a correlation existed between the results of each pair of instruments, analytical functions were calculated to enable an estimate to be made of the reading that would be given by one instrument if a reading from one of the other instruments was available, assuming that the first instrument were to be used in accordance with the requirements or recommendations in force in its country of origin regarding the sampling position, duration, analysis, etc. It is interesting to note, by way of example, that, starting with the Staser instrument, for an over-all gravimetric reading of 20 mg/m^3, the following average readings resulted: TBF–3.54 mg/m^3; CPM–3.89 mg/m^3; MPG–4.36 mg/m^3; MRE–5.0 mg/m^3. Each instrument was operated in accordance with the prescribed practice.

Results of counts obtained by sampling on cellulose membranes, or by thermal precipitation are also comparable, provided that the same counting technique is adopted (same optical system and technique for estimating particle aggregates).

In fact, as the methods of sampling and analysis are very closely related, it is preferable to compare not the results of counting but the actual environments, that is to see if the "atmospheres" are categorised (dangerous or not) in the same manner by scrupulously complying with the recognised methods of using each instrument.

Criteria. To determine the harmfulness or the danger of an atmosphere one must determine: the weight of the dust and its mineral content; or the number of particles between two given particle sizes and their free-silica content. Concentration and quartz (or mineral) content are combined to calculate a danger index, establish a threshold limit value, classify atmospheres, etc., or are carried over on to tables which define zones presumed to be dangerous or not. These factors thus go together to make up a criterion of judgement. Most countries have established levels of dust concentrations which may be applied to the measurements made with their own type of instrument in order to define different categories of dust conditions in order of increasing harmfulness.

It should be recognised that in the past any classification of dusty conditions had to be based on the techniques available at the time. In order to be able to assess whether a system of dust prevention was really effective, it was necessary to obtain both medical and engineering data and to await the results of epidemiological research on the subject. Such research has shown that, in addition to the effect of fine dust, other factors play a part in the causation and development of pneumoconiosis, and these are mainly the fibrogenic properties of certain dusts. It was for this reason that the quartz content has always been considered as the second parameter to be taken into consideration when dusty conditions were being classified; similar importance was thus attached to other constituents of the dust. This concept has now been called into question. Intensive research is continuing in this field because, for equal concentrations of dust, it is not necessarily in activities or operations where the highest levels of quartz content occur that the pneumoconiosis risk is the greatest.

With the aid of increasingly advanced epidemiological studies it has been possible to determine the concentrations and the quartz content of dust which should not, as an average over a given period, be exceeded in order to keep a particular radiological image below a stated value. In this connection it is extremely important to note that the dust concentration levels that give rise to the same pneumoconiosis risk vary, often quite considerably, from one factor or from one mine to another. For this reason it can be seen that it is almost impossible to set an exposure limit for dust that would be applicable everywhere, for everyone, and for all circumstances.

Furthermore, in order to fix an exposure limit for dust concentrations, the duration of the intended exposure must be known.

However, in the light of present knowledge, it may be accepted that where the quartz content does not exceed 5-7%, an average concentration of fine, respirable dust consisting of coal or other inert particles that does not exceed 4-5 mg/m^3 may be considered as "acceptable" for normal working conditions for healthy individuals, on the basis of a working life of 30 to 35 years.

DEGUELDRE, G.

International Symposium on the Control of Air Pollution in the Working Environment. Stockholm, 6-8 September 1977. Part I (Stockholm, Work Environment Fund, 1978), 468 p. Illus.

"Method selection and difficulties associated with field sampling of airborne particulates". Corn, M.; Esmen, N. A. *International Laboratory* (Fairfield, Connecticut), Sep.-Oct. 1978 (9-22). Illus. 42 ref.

CIS 74-979 "Coal mine dust". Freedman, R. W.; Sharkey, A. G.; Corn, M.; Stein, F.; Hammad, Y.; Manekshaw, S.; Bell, W.; Penkala, S. J. (7-36). Illus. 37 ref. *Coal workers' penumoconiosis.* Vol. 200. Annals of the New York Academy of Sciences (New York, 1972).

CIS 79-51 "An elutriator with separation curve independent of air flow for determination of the respirable dust fraction" (Ein Vorabscheider mit durchflussunabhängiger Trennkurve zur Ermittlung der lungengängigen Staubfraktion). Mauer, G.; Gast, T. *Staub* (Düsseldorf), May 1978, 38/5 (177-179). Illus. 4 ref. (In German)

CIS 79-380 *SIMSLIN II: An airborne-dust-measuring instrument that uses a light scattering method.* Health and Safety Laboratories, Health and Safety Executive, Sheffield. HSL Technical Information Leaflet: Environmental Contaminants 1 (London, HM Stationery Office, 1978), 2 p. Illus.

CIS 80-948 "Determination of crystalline silica content of dust samples" (Zur Einsatzmöglichkeit verschiedener Untersuchungsmethoden für die Bestimmung der Gehalte an kristallinem SiO_2 in Staubproben). Werner, I. *Zeitschrift für die gesamte Hygiene und ihre Grenzgebiete* (Berlin), Sep. 1979, 25/9 (647-650). 65 ref. (In German)

CIS 79-1562 "Statistical evaluation of dust measurements" (Zur Bewertung von Staubmessergebnissen). Šimeček, J. *Staub* (Düsseldorf), Oct. 1978, 38/10 (409-412). (In German)

Dusts, vegetable

Occupational exposure to vegetable dust is commonly encountered in industry and agriculture, particularly in developing countries. Although there are many reports on various respiratory diseases resulting from exposure to these dusts, the present knowledge on pathogenesis of most of these diseases is still limited.

GENERAL HAZARDS

These are threefold: respiratory effects, skin and eye effects, and fire and explosion.

The main health hazard is respiratory disease. Vegetable dusts may have a direct mechanical and/or chemical irritant action on the respiratory passages causing bronchospasm, cough, tightness in the chest and, following prolonged exposure, chronic obstructive lung disease.

Many of these dusts produce an allergic reaction either through their own content of antigens or through the effects of moulds and fungi that grow in large quantities during storage of agricultural products. Allergic manifestations such as asthma, asthmatic bronchitis and upper respiratory reaction with fever are commonly found in exposed workers. Some dusts are associated with more specific manifestations and symptoms, for example byssinosis resulting from inhalation of cotton, flax and soft hemp dusts, bagassosis from inhalation of bagasse dust and tobacco dust (see separate articles on individual subjects).

The eyes may be affected by irritant dusts from such products as spices and certain woods. Cotton, jute, tea and hemp dust may also irritate the eyes, leading to lacrimation and inflammation of the conjunctiva. Eye irritation may also be due to allergic reaction.

Skin effects are also common. Allergic skin manifestations in the form of urticaria, erythema and itching spots may result from exposure to tobacco dust and certain wood dusts. Other forms of dermatitis may result from dusts which have a chemically irritant action.

Fires and dust explosions are an important hazard in workplaces where there are high atmospheric concentrations of flammable vegetable dusts such as cork, wood, flour or cotton dust; the danger is all the more pronounced when the dust is confined in an enclosed space. This aspect of the hazard of vegetable dusts is dealt with in the article DUST EXPLOSIONS.

SPECIFIC HAZARDS

The pathological effects of dust from different vegetable products vary considerably, and the typical areas of exposure to the most important vegetable dusts and the resultant diseases are described below.

Cereals. Agricultural workers are exposed to these vegetable dusts during harvesting. Although manual winnowing of rice, wheat, barley, etc., is generally performed in the open, there is a danger of exposure to fine particles of husks which may affect the respiratory tract and eyes.

Threshing and milling also produce clouds of dust containing husk, germ and starch particles. Asthma is reported to be common among workers with a long history of dust exposure in threshing and milling. Dockers may also be exposed to high concentrations of vegetable dusts containing husk particles, starch granules and moulds, when unloading grain cargoes. Signs of upper respiratory tract and eye irritation and cases of sensitisation have been encountered, especially among persons exposed for the first time.

Subacute and chronic effects of cereal dusts include extrinsic allergic alveolitis with insidious onset and restrictive respiratory insufficiency (see ALVEOLITIS, ALLERGIC EXTRINSIC). Dust control measures should be introduced in grain handling in order to minimise the level of exposure; in addition to reducing the incidence of disease, such measures would also eliminate the problem of finding alternative work for grain handlers.

Stored cereals, grasses, etc. When these products are stored in damp conditions, the growth of moulds and fungi is considerable. Dust containing moulds and fungi evolved during the handling of stored fodder, etc., may cause a disease known as farmer's lung. This is an allergic alveolitis whose symptoms include an acute response several hours after exposure to mouldy hay, with shortness of breath, persistent cough and fever. The symptoms may subside in a few days but if exposure is repeated prolonged reaction develops with increased breathlessness, chronic cough and permanent lung damage.

Cotton. In the cotton industry, there is exposure to dust containing cellulose fibres, fine plant debris, earth and micro-organisms; the dustiest processes are bale breaking, blowing, carding and waste cleaning. Dust encountered in weaving sheds contains considerable quantities of starch and tamarind seeds and the respiratory disorders encountered in weavers (weavers' cough) are of a special character. A variety of respiratory disorders have been observed in cotton workers but the most common disease is byssinosis; however, prolonged exposure to high concentrations of cotton dust is also associated with a high incidence of chronic bronchitis or asthma which may be superimposed on the symptoms of byssinosis or occur independently.

Flax. Exposure to flax dust occurs when the retted fibres are cleared of binder and prepared for spinning; exposure in flax mills may also be heavy. Typical symptoms of byssinosis are found even more often amongst flax workers than amongst cotton workers. Unretted flax does not cause byssinosis. A recent classification of byssinosis (WHO 1980) includes early stages with dust irritation reaction, typical chest-tightness on return of workers from holidays, eventually leading to chronic respiratory disability with marked reduction in ventilatory capacity.

Jute. High concentrations of jute dust occur during bale breaking, carding and spinning. Studies have shown that workers exposed to jute dust often suffer from chronic bronchitis and asthma. More research is required.

Tea. The final processes of tea sifting and blending are usually accompanied by the emission of particles of fine dust invisible to the naked eye. Exposure to this dust causes irritation of the upper respiratory tract and tightness of the chest.

Tobacco. Bale opening, especially when performed manually, gives rise to large amounts of tobacco dust. This dust may cause dermatitis of the hands and irritation of the eyes; attacks of asthma with fever have been reported in tobacco workers and the term "tobaccosis" has been coined to describe them.

Wood. The sawing, moulding, routing, carving and sanding of wood may also be dusty processes and the dust from certain woods may cause severe irritation of the eyes, skin and respiratory system. The pathological effects of a large number of different types of wood are discussed in the article WOODS.

Soft hemp, sisal. The dusts evolved during the processing of these two plants are not known to cause byssinosis; also, like cotton and flax, more research on sisal is needed.

Sugar cane. Workers handling bagasse, the fibre of sugar-cane stalks remaining after juice extraction, which has been stored outdoors and contains large quantities of moulds or fungi, may suffer from an acute or subacute respiratory condition known as bagassosis, characterised by cough, dyspnoea, fever, and black, scanty sputum. It is believed that the causative agents are moulds or fungi that grow in stored bagasse. This disease appears to be another form of extrinsic allergic alveolitis with reduction in pulmonary function and hypoxaemia.

Cork. Workers exposed to the large quantities of dust produced during the cutting, rolling, disintegration, classification, pressing and polishing of cork may be subject to a relatively benign pulmonary fibrosis called suberosis.

SAFETY AND HEALTH MEASURES

The essential task in reducing or eliminating the hazards of vegetable dusts is effective dust control.

Air sampling and engineering control. The air in dusty plants should be sampled, using counting or gravimetric techniques, to determine the extent of the hazard and the nature of the dust; in many cases it is important to determine the presence of moulds and fungi or of impurities such as free silica. In principle, dangerously dusty processes should be totally enclosed or equipped with local exhaust ventilation. However, where the processes are performed entirely manually, enclosure or exhaust ventilation may not be practicable and, consequently, work should be carried out in the open, or maximum use should be made of natural ventilation; in many instances, workers should be provided with respiratory protective masks. Modern textile mills are designed with built-in continuous downdraught exhaust ventilation. Regular vacuum cleaning and suitable maintenance and good housekeeping are also important. The wetting of dust-producing materials may help in operations performed manually and where a high level of moisture does not affect the industrial process; such a measure has been tried with cotton and flax.

Personal protection. Respiratory protective equipment may be a valuable aid in the prevention of disorders due to the inhalation of vegetable dusts. The simplest equipment consists of a number of layers of fine gauze held by a head-strap across the worker's mouth and nose. Commercial dust masks of simple design are widely available at relatively low cost; it should be possible to adjust them to the contours of the worker's face to prevent leaks of unfiltered air. However, in hot and humid climates, such equipment may be uncomfortable and readily discarded by workers unless close supervision is maintained.

Medical supervision. Workers should undergo a pre-employment medical examination before being exposed to vegetable dusts. Persons who show sensitivity to the dusts in question, who have a history of respiratory disease or who are heavy smokers should not be employed on this type of work. In the pre-employment medical examination, emphasis should be placed on chest examination, clinically, radiologically and, preferably, also by the measurement of ventilatory capacity.

Workers should receive periodic medical examinations for the early detection of effects of dust exposure and those who present skin, membrane or respiratory symptoms should be removed from exposure; the symptoms of incipient disease can usually be reversed in this way.

EL BATAWI, M. A.

Environmental health criteria for vegetable dusts: cotton, flax, soft hemp and sisal (Geneva, World Health Organisation, in press).

"Occupational lung diseases" (219-271). *Diseases of the respiratory system.* WHO/CIOMS (Council of International Organisations of Medical Sciences) (Geneva, World Health Organisation, 1978).

Occupational respiratory diseases from vegetable dusts. Early detection and control. Document OCH (Geneva, World Health Organisation, 1979).

"Intervention studies of cotton steaming to reduce biological effects of cotton dust". Merchant, J. A.; Lumsden, J. C.; Kilborn, K. H.; O'Fallon, W. M.; Copeland, K.; Germino, V. H.; McKenzie, W. N.; Baucom, D.; Currin, P.; Stilman, J. *British Journal of Industrial Medicine* (London), Oct. 1974, 31/4 (261-274). Illus. 31 ref.

"Occupational lung diseases due to natural antigens". Palma-Carlos, A. G. (374-380). Illus. 26 ref. *Allergy and clinical immunology. Proceedings of IX International Congress of Allergology, Buenos Aires, 1976* (Amsterdam, Excerpta Medica International Congress Series No. 414, 1976).

Dyeing industry

Dyeing involves a chemical combination or a powerful physical affinity between the dye and the fibre of the fabric.

Classes of dyes. Acid or basic dyes are used in a weak acid bath for wool, silk or cotton. Some acid dyes are used after mordanting the fibres with metallic oxide, tannic acid or dichromates. Direct dyes which are not fast are used for the dyeing of wool, rayon and cotton. They are dyed at the boil. For dyeing of cotton fabrics with sulphur dyes, the dyebath is prepared by pasting the dye with soda ash and sodium sulphide and hot water. The dyeing is carried out at the boil. For dyeing cotton with azo dyes, naphthol is dissolved in aqueous caustic soda. The cotton is impregnated with the solution of the sodium naphthoxide thus formed and is then treated with a solution of a diazo-compound to develop the dye in the material. Vat dyes are made into leuco-compounds with sodium hydroxide and sodium hydrosulphite. Dyeing is done at 30-60 °C. Disperse dyes are selected for dyeing of all synthetic fibres which are hydrophobic. Swelling agents or carriers which are phenolic in nature must be used to enable the disperse dyes to act. Mineral dyes are inorganic pigments which are salts of iron and chromium. After impregnation, their precipitation is done by addition of hot alkaline solution. Reactive dyes for cotton are used in a hot or a cold bath. Soda ash and common salt are needed for this type of dyeing.

Cotton dyeing. The preparatory processes before dyeing cotton fabrics consist of the following steps. The cloth passes through a shearing machine which cuts the loosely adhering fibres; then, to complete the trimming process, it is rapidly passed over a row of gas flames and the sparks are extinguished by passing the material through a water box. Desizing is carried out by passing the cloth through a diastase solution which removes the size completely. To remove other impurities it is scoured in a kier with dilute sodium hydroxide, sodium carbonate or turkey red oil for 8-12 h at high temperature and pressure.

For coloured woven material, an open kier is used and sodium hydroxide is avoided. The natural colouring in the cloth is removed by hypochlorite solution (figure 1)

Figure 1. Yarn bleaching. *(By courtesy of Sayaji Mills.)*

in the bleaching pits after which the cloth is aired, washed, dechlorinated by means of a sodium bisulphite solution, washed again and scoured with dilute hydrochloric or sulphuric acid. After a final, thorough washing the cloth is ready for the dyeing or printing process.

Dyeing is carried out in a jig or padding machine in which the cloth is moved through a stationary dye solution which is prepared by dissolving the dyestuff powder in a suitable chemical and then diluting with water. After dyeing, the cloth is subjected to a finishing process.

Printing is carried out on a roller printing machine (figures 2 and 3). The dye is thickened with starch or

Figure 2. Fabric dyeing. *(By courtesy of Sayaji Mills.)*

Figure 3. Roller printing of cloth. *(By courtesy of Sayaji Mills.)*

made into emulsion which, in the case of pigment colours, is prepared with an organic solvent. This paste or emulsion is taken up by the engraved rollers, which print the material, and the colour is subsequently fixed in the ager or curing machine; the printed cloth then receives the appropriate finishing treatment.

Nylon dyeing. The preparation of polyamide (nylon) fibres for dyeing involves scouring, some form of setting treatment and, in some cases, bleaching.

The treatment adopted for the scouring of woven polyamide fabrics depends mainly on the composition of the size used. Water soluble sizes based on polyvinyl alcohol or polyacrylic acid can be removed by scouring in a liquor containing soap and ammonia or Lissapol N or similar detergent and soda ash. After scouring, the material is rinsed thoroughly and is then ready for dyeing or printing.

A jigger or winch dyeing machine is used for dyeing polyamide fabric.

Wool dyeing. The raw wool is first scoured by the emulsification process in which soap and soda ash solution are used. The operation is carried out in a washing machine which consists of a long trough, provided with rakes, false bottom and, at the exit, wringers. The wool, after thorough washing, is bleached with hydrogen peroxide or with sulphur dioxide, in which case the damp goods are left exposed to the sulphur dioxide gas overnight. The acid gas is neutralised by passing the fabric through a sodium carbonate bath and then washing thoroughly. After dyeing, the goods are rinsed, hydroextracted and dried.

HAZARDS AND THEIR PREVENTION

Fire and explosion. The fire hazards that characterise a dye works are those caused by the presence of flammable solvents used in the processes and certain flammable dyestuffs. Safe storage facilities should be provided for both. These facilities should comprise properly designed storerooms constructed of fire-resisting materials. Storerooms for flammable liquids should be provided with a raised and ramped sill at the doorway so that escaping liquid is contained within the room and is thus prevented from flowing to a place where it can become ignited. It is preferable that stores of this nature should be outside the main factory building. If large quantities are kept in tanks outside the building, the tank area should be mounded to contain escaping liquid. Similar arrangements should be made when the gaseous fuel used on the singeing machines is obtained from a light petroleum fraction. The gas-making plant and the storage facilities for the volatile petroleum spirit should preferably be outside the building.

Health hazards. Many factories use hypochlorite solution for bleaching; in others the bleaching agent is gaseous chlorine or bleaching powder from which chlorine is evolved when it is charged into the tank. In either case, workers may be exposed to a dangerous atmosphere unless precautions are taken. Chlorine is a skin and eye irritant and a dangerous pulmonary tissue irritant causing delayed lung oedema. To limit the escape of chlorine into the working atmosphere, bleaching vats should be designed as closed vessels provided with vents that allow a minimum escape of chlorine so that the relevant recommended maximum levels are not exceeded, and atmospheric chlorine estimations should be made periodically to check the concentration. The valves and other controls of the tank in which the liquid chlorine is supplied to the dyeworks should be controlled by a competent operator since the possibilities of an uncontrolled leak could well be disastrous.

The use of corrosive alkalis and acids and the treatment of cloth with boiling liquor expose the workers to the risk of burns. Both hydrochloric acid and sulphuric acid are used extensively in dyeing processes. Caustic soda is used in bleaching, mercerising and dyeing. Chips from the solid material fly and create hazards for the workers. Sulphur dioxide, which is used in bleaching and carbon disulphide, which is used as a solvent in the viscose process, can also pollute the workroom. Aromatic hydrocarbons such as benzol, toluol and xylol, solvent naphthas and aromatic amines such as aniline are common dangerous chemicals to which workers are likely to be exposed. Dichlorobenzene is emulsified with water with the help of an emulsifying agent and is used for dyeing of polyester fibres. Many dyestuffs are skin irritants that cause dermatitis and, in addition, workers are tempted to use a harmful mixture of abrasive, alkali and bleaching agent to remove dye stains from their hands. Organic solvents used in the processes and for the cleaning of machines may themselves cause dermatitis or render skin vulnerable to the irritant action of the other harmful substances that are used; furthermore, they may be the cause of peripheral neuropathy as in the case of methyl butyl ketone. Certain dyes have been found to be carcinogenic substances. These include rhodamine B, magenta, 2-naphthylamine and certain bases such as dianisidine. Besides allergies due to the fibre materials and their contaminants, allergy may be caused by the sizing and even by the enzymes used to remove the sizing. Appropriate articles should be referred to regarding the hazards associated with dusts, and physical agents such as heat, humidity and noise.

When a vessel that has contained chlorine or any other dangerous gas or vapour has to be entered, all precautions appropriate to work in confined spaces should be taken.

Suitable personal protective equipment including eye-protective equipment should be provided to prevent contact with these numerous harmful substances. If in certain circumstances barrier creams may have to be used, care should be taken to ensure that they are effective for the purpose and that they can be removed by washing. They can seldom be regarded as providing protection that is equal in reliability to properly designed gloves. Even when all protective measures have been taken, some workers may prove to be particularly sensitive to the effects of these substances and may have to be transferred to other work. Personal hygiene is particularly important for workers in these processes. Protective clothing should be cleaned at regular intervals. Clothing that is splashed or contaminated by dyestuffs should be replaced by clean clothing at the earliest opportunity; sanitary facilities for washing and bathing should be provided and the workers should be encouraged to use them to ensure their personal hygiene.

Accidents. Serious scalding accidents have occurred when hot liquor has been accidentally admitted to a kier in which a worker has been arranging the cloth to be treated. This can occur when a valve is accidentally operated or when the hot liquor is discharged into a common discharge duct from another kier on the range and has entered the occupied kier through an open outlet. When a worker is inside a kier for any purpose the inlet valves should be locked in the closed position, and the kier should be isolated on the discharge end from the other kiers on the range. If the locking device is by means of a key, the person, who would be injured by the accidental admission of hot liquid, should keep the key until he leaves the vessel.

Mechanical hazards such as crushing, nips, cuts and knocks exist mainly at the roller printing machines. Other hazards are connected with handling and in-plant traffic.

It is essential that, in these factories, the factory management make arrangements for the proper observation and recording of industrial illnesses and accidents in order to determine when and how they occur and who is affected. From the statistics thus obtained, the management is able to identify the conditions which are likely to cause illnesses and accidents. Proper placing of workers after pre-employment medical examination is helpful. In addition to this, preventive safety and health education of workers by supervisors is absolutely essential.

NIYOGI, A. K.

General:

Safety and health in the textile industry. Report III. Ninth Session of the Textiles Committee (Geneva, International Labour Office, 1973), 47 p.

Processes:

Technology of textile processing. Shenoi, V. A. (Bombay, Sevak Publications, 1977), 5 vols. Illus.

CIS 76-242 *Study of present-day synthetic materials dyeing procedures* (Etude des procédés actuels de teinture des synthétiques). Gaboreau, M. (Caisse régionale d'assurance maladie du Nord de la France, 11 Boulevard Vauban, Lille) (1974), 6 p. (In French)

Health and safety:

CIS 78-234 "Hazardous substances in the dyeing industry" (Le sostanze pericolose nell'industria tintoria). *Sicurezza nel lavoro* (Rome), May-June 1976, 28/3 (129-144). (In Italian)

CIS 78-1756 "A new product to remove dyes from the hands of workers engaged in textile finishing" (O vnedrenii novogo sposoba udalenija krasitelej s kŏi ruk rabočih otdeločnogo tekstil'nogo proizvodstva). Venediktova, K. P.; Utkina, A. M. *Gigiena truda i professional'nye zabolevanija* (Moscow), Feb. 1978, 2 (8-12). 6 ref. (In Russian)

CIS 75-418 "Peripheral neuropathy in a coated fabrics plant". Billmaier, D.; Yee, H. T.; Allen, N.; Craft, B.; Williams, N.; Epstein, S.; Fontaine, R. *Journal of Occupational Medicine* (Chicago), Oct. 1974, 16/10 (665-671). Illus. 19 ref.

CIS 75-2042 "Roller printing machines for fabrics" (Machines à imprimer à rouleau sur tissus). Comité technique national des industries textiles, Caisse nationale de l'assurance-maladie. *Cahiers de notes documentaries–Sécurité et hygiène du travail* (Paris), 2nd quarter 1975, 79, Note No. 965-79-75 (Note technique No. 117) (281-283). (In French)

Dyes and dyestuffs

The term "dye" or "dyestuff" refers to products which cause reflection or selective transmission of incident light in such a way that they impart colour to the materials to which they have been applied. A distinction is made between dyes and pigments. Dyes are soluble, and the colour tends to attach itself to textile fibres or other materials immersed in the solution. Pigments, on the other hand, are insoluble, and they impart colour to an object by being dispersed throughout it in finely subdivided particles.

In the past, the number of colouring matters available for industrial use was very limited indeed, consisting of only a few organic dyes derived from natural sources (e.g. the tyrian purple of the Roman Emperor's toga, made from molluscs; indigo, made from a plant) and mineral pigments. The majority of dyes and pigments in use today are of synthetic origin. Since the discovery of mauve in 1856, numerous other synthetic dyes have been developed. They have made it possible to colour a wide variety of materials and have been adapted to meet increasingly strict technical requirements both for traditional and new materials (textile fibres, inks, plastics, paints, cosmetics, etc.).

Although many synthetic chemicals are themselves coloured, it does not follow that they have the properties of dyes or pigments. The quality of imparting colour is related to specific chemical structures, of which the most important are triphenylmethane compounds, indigoid and azo structures, azines and thiazines, anthraquinone derivatives and phthalocyanines. These basic chemical structures are shown in figure 1. To meet specific technical requirements, various chemical substituent groups (amino, hydroxy, nitro, sulpho, chloro, carboxy, etc.) are introduced into the base molecules, making it possible to produce differences in shade, fastness, solubility, etc. The final chemical composition of a dye is therefore usually a complex one.

With certain special types of dye the coloured molecule is formed in the mass of the substrate (e.g. by *in situ* oxidation of *p*-phenylenediamine), or is fixed by chemical reaction with the fibre (reactive dyes) thus giving the colour excellent fastness to washing, cleaning, etc.

Manufacture. The synthetic dye industry itself carries out all the manufacturing processes, i.e. it takes the most basic feedstocks (e.g. benzene, naphthalene, anthracene, caustic soda, inorganic acids, etc.) and converts them to a wide range of intermediates of differing degrees of complexity (e.g. aromatic amines, quinones, etc.), which are then used (following one or more intermediate processes) for the manufacture of the dyes themselves. In fact, in a dye factory hazardous exposure arises largely during the synthesis and handling of intermediates and not from contact with the finished dyes themselves.

Inorganic pigments still constitute an important part of the dyestuffs industry. These are mostly salts or oxides of metals such as lead, cadmium, antimony and titanium.

Uses. The purpose for which a dye is used is not primarily related to its molecular structure but to its physical

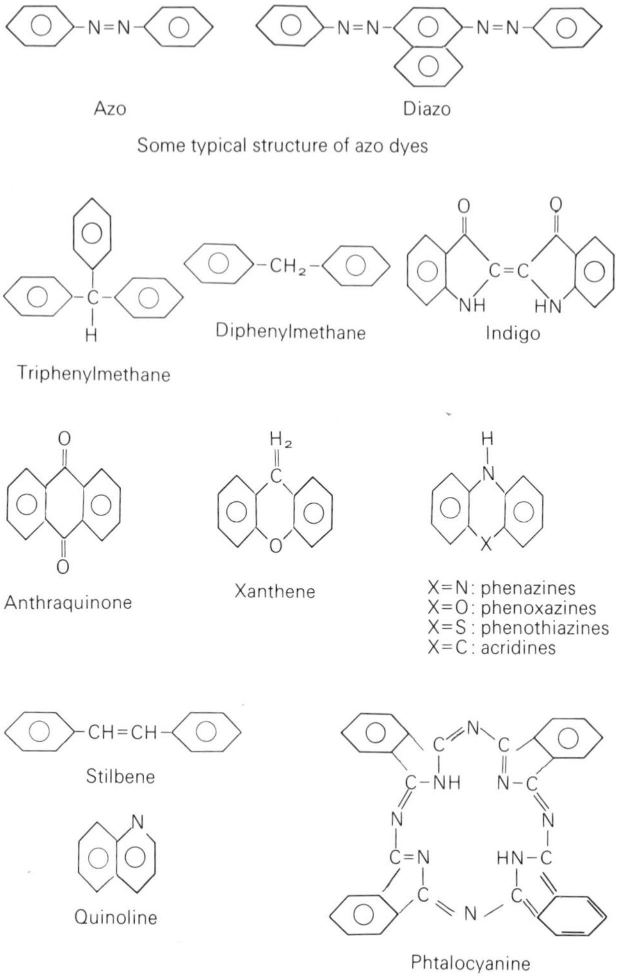

Azo Diazo

Some typical structure of azo dyes

Triphenylmethane Diphenylmethane Indigo

Anthraquinone Xanthene

X=N: phenazines
X=O: phenoxazines
X=S: phenothiazines
X=C: acridines

Stilbene Quinoline Phtalocyanine

Figure 1. Basic structure of some typical dyestuff molecules.

properties. Thus, a particular chemical class of dyestuff (e.g. azo or triphenylmethane) may contain, within the group, colours falling into quite different usage categories, for example acid dyes (used for wool), disperse dyes (used for acetate, rayon and synthetic fibres), solvent dyes (used in spirit inks), pigments (used in paint, rubber, plastics, etc.), and food dyes. Conversely, in any particular usage category, several different types of structure may be represented. Disperse colours may be anthraquinone, azo, or nitroarylamine dyes, and food colours may be azo, triphenylmethane, or indigoid in structure.

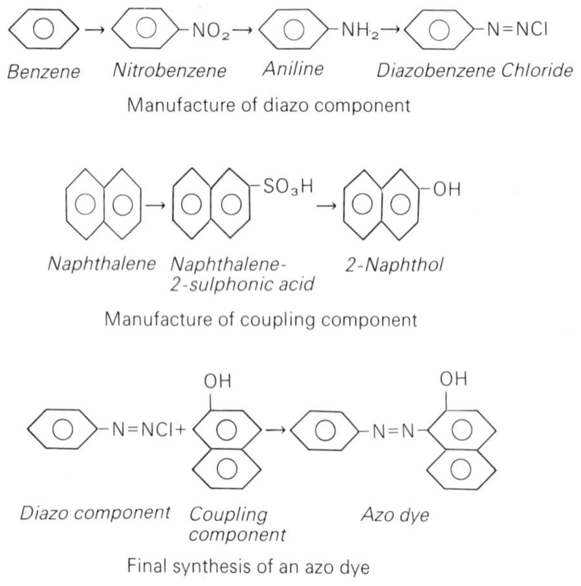

Benzene *Nitrobenzene* *Aniline* *Diazobenzene Chloride*

Manufacture of diazo component

Naphthalene *Naphthalene-2-sulphonic acid* *2-Naphthol*

Manufacture of coupling component

Diazo component *Coupling component* *Azo dye*

Final synthesis of an azo dye

Figure 2. Diagrammatic presentation of the stages in the manufacture of CI solvent yellow 14.

The number of synthetic dyestuffs commercially available runs to several thousand. The need for a comprehensive work of reference is met by the *Colour Index* compiled jointly by the British Society of Dyers and Colourists and the American Association of Textile Chemists and Colourists. Dyes are listed therein by chemical constitution, by arrangement into recognised usage categories, and by an alphabetical list of commercial names, with a simple cross-reference system.

HAZARDS

A distinction should be made between synthetic dyes and inorganic pigments.

Synthetic dyes and pigments. In general, finished products (dyes and pigments) do not present serious toxic hazards to workmen in conditions of industrial exposure. However, the manufacturing processes entail the use of hazardous primary compounds (acids, alkalis, irritant and asphyxiant gases, hydrocarbon derivatives, etc.) or intermediary compounds (e.g. aromatic amines, etc.). (For the hazardous properties of these substances, see the relevant articles: in particular AMINES, AROMATIC; ANILINE; ACIDS AND ANHYDRIDES, INORGANIC; ALKALINE MATERIALS; AMMONIA; AZO AND DIAZO DYES; BLADDER CANCER; CHLORINE AND COMPOUNDS; etc.). There has been a considerable history of occupational diseases in the dye industry. However, modern technology combined with a better understanding of the hazards has led to a significant reduction in individual exposure to hazardous substances and, under normal working conditions, pathological effects are seldom encountered. However, potential hazards should not be forgotten.

Bladder cancer is still the most serious occupational hazard for the synthetic dye worker. There is no doubt that certain specific intermediates are the responsible agents. The production of two dyes—auramine and fuchsine or magenta—seems to have resulted in an increase in the incidence of bladder cancer among the workers involved; however, the epidemiological evidence does not make it clear whether the tumours are caused by intermediates or by the dyestuffs themselves. There is no evidence of cases arising in workmen using these dyes, as opposed to making them, despite the fact that auramine itself has been shown to produce liver tumours in rodents.

Intensified and improved experimental research has brought to light the carcinogenic potential of the various groups, in particular the azo dyes (*p*-dimethylaminobenzene, *o*-aminoazotoluene, citrus red 2, or CI solvent red 80, CI solvent orange 2, trypan blue, 3 R and MX poppy red and, more recently, dyes derived from benzidine) but also derivatives of triphenylmethane such as benzyl violet 4 B and fuchsine (magenta). Difficulties in the interpretation of mutagenicity tests further complicate this matter and there is currently no proof of mutagenicity in man.

Skin damage should be considered from various points of view:

(a) during manufacture, orthoergic dermatitis may result from contact with alkalis, acids or solvents; other substances may produce allergic dermatitis (amino derivatives, nitro and nitroso compounds, etc.);

(b) during the use of dyes, specific dermatitis is rare except in the case of certain substances that cause the formation and fixing of a colour derivative on the substrate, such as for example, *p*-phenylenediamine and various derivatives and soluble salts (diazonium salts stabilised with aromatic amines).

Cases of respiratory intolerance (asthma) to certain of these substances have been reported.

Dyes have virtually no irritant action on the eyes, except for alkaline dyes which, when splashed into the eyes, may produce serious lesions that may lead to the loss of the involved eye.

Inorganic pigments. The most important inorganic pigments are salts and oxides of lead (particularly lead chromate), cadmium sulphide and selenium sulphides, iron oxides, antimony oxide, and titanium dioxide. The potential hazards of these compounds are all related to the biological properties of the parent metal. Thus iron and titanium oxides present very little hazard, but in appropriate conditions the inhalation of dust from lead pigments can readily cause lead poisoning.

Skin cleansing. A major problem among workmen handling dyes is the ease with which the superficial layers of the skin become dyed. The usual type of skin cleaners are virtually ineffective–or even aggressive. Workers often make use of sodium hypochlorite solution, which effectively bleaches the dyed skin. However, sodium hypochlorite is a powerful irritant which can destroy the skin with resultant hyperhidrosis and superimposed infections. When this substance is used for cleaning purposes, it should be in a dilute solution, rinsed off immediately with water and any residues neutralised with sodium bisulphite solution. Workers should be warned of the dermatitis hazard of thoughtless use of this type of cleaning. Certain industrial soap compounds provide relatively satisfactory cleansing. Above all, good process engineering and personal safety measures should limit this type of soiling to a minimum.

Miscellaneous hazards. Dyes used for colouring food now have to conform to increasingly stringent criteria regarding their safety for such a purpose. It is obvious that a dye, or any other chemical, which may be free from hazard in conditions of industrial exposure, is not necessarily safe when it is regularly consumed as part of the diet. Purity specifications also have to be laid down, as most commercial dyes contain significant amounts of metallic and organic impurities. Similar considerations apply to dyes used in cosmetics.

MUNN, A.
SMAGGHE, G.

CIS 79-472 "Prediction, based on chemical classification, of the toxicity of organic dyes used in the textile industry" (Prognozirovanie toksičnosti organičeskih krasitelej, primenjaemyh v tekstil'noj promyšlennosti, na osnovanii himičeskoj klassifikacii). Voronin, A. P. *Gigiena truda i professional'nye zabolevanija* (Moscow), July 1978, 7 (16-23). 27 ref. (In Russian)

CIS 80-1913 *Benzidine-, o-tolidine-, and o-dianisidine-based dyes* (US Department of Labor, Occupational Safety and Health Administration, 200 Constitution Avenue, NW, Washington, DC) (Apr. 1980), 38 p. 35 ref.

CIS 80-1320 *Special occupational hazard review for benzidine-based dyes.* DHEW (NIOSH) publication No. 80-109 (National Institute for Occupational Safety and Health, 4676 Columbia Parkway, Cincinnati) (1980), 60 p. 79 ref.

CIS 78-1959 *Occupational allergic eczema in the textile industry* (Les eczémas allergiques professionnels dans l'industrie textile). Cywie, P. L.; Hervé-Bazin, B.; Fousse-reau, J.; Cavelier, C.; Coirier, A. Notes scientifiques et techniques de l'INRS, No. 11 (Institut national de recherche et de sécurité, 30 rue Olivier-Noyer, Paris) (Mar. 1977), 155 p. Illus. 68 ref. (In French)

CIS 78-475 "Pigmented cosmetic dermatitis and coal tar dyes". Sugai, T.; Takahashi, Y.; Takagi, T. *Contact Dermatitis* (Copenhagen), Oct. 1977, 3/5 (249-256). Illus. 5 ref.

"Chronic cadmium poisoning in a pigment manufacturing plant". De Silva, P. E.; Donnan, M. B. *British Journal of Industrial Medicine* (London), Feb. 1981, 38/1 (76-86). Illus. 12 ref.

Earth-moving equipment

Earth-moving equipment comprises machines for performing earthworks in hydraulic engineering, road-building and industrial, farm, residential, and other segments of construction as well as in mining and mineral resources exploitation. Because of their high efficiency, these machines have found world-wide application, particularly in developing countries. The annual quantities of earthworks handled using earth-moving equipment all over the world are estimated at 100 000 million m³ of soil, and developing countries account for an appreciable portion of that amount. Basic types of earth-moving equipment are cutting and transporting and grading machines, such as bulldozers, scrapers, graders, trenchers, etc., which accomplish cutting and moving or loading of materials with the machine continuously moving forward, and single-bucket excavators which perform the operating cycle with the machine remaining stationary.

Bulldozer. A tracked or wheeled, self-propelled, earth-moving machine with hydraulic or cable control, which essentially is a crawler-mounted or rubber-tyred tractor having a working tool (blade) mounted on a frame or push arms positioned off the undercarriage, most commonly in front of it, whose curved bladed surface faces outwards from the base machine and extends beyond the over-all width of the latter. A bulldozer is designed for stripping in layers, grading, and moving (over haul distances from 60 to 100 m) soil, mineral products, road-building and other materials used when constructing and repairing roads and hydraulic engineering structures, backfilling trenches and holes, etc.

There are different types of bulldozers: equipped with a straight blade mounted perpendicular to the longitudinal axis of the base machine; equipped with an angling blade which can be set at an angle to either side of the longitudinal axis (angle-dozers); multipurpose bulldozers with the blade consisting of two pivoted halves that can be set in a horizontal plane at different angles in relation to the machine's longitudinal axis (track layers); special-purpose bulldozers for doing certain jobs (underground, land reclamation, etc.) or jobs executed under specific conditions of ground (quarry, mountain, desert, underwater, etc.) or climate (low sub-zero temperatures, tropical humidity, etc.). Bulldozers of any type may be fitted with devices for tilting the blade in a lateral plane and used as a unit integral with a ripper attached at the rear. To widen their applicability, the bulldozer blades can be fitted with various attachments (ripper teeth, backslopers, extensions, etc.).

Scraper. An earth-moving and hauling machine with hydraulic or cable control of a working tool (bowl), mounted within the chassis frame of the machine. A scraper is designed for cutting in layers and loading the materials into the bowl, rough grading of the earthwork, hauling the soil, light rocks, or construction materials, and unloading, spreading, and partially compacting them with running wheels.

There are different types of scrapers: self-propelled wheeled, driven by single- or two-axle tractors featuring one or two driving axles and hydromechanical transmission (bowl capacity 6 to 40 m³, speed up to 60-70 km/h, haul distance up to 2-3 km) or diesel-electric with all-wheel drive (25 to 40 m³, 25 to 40 km/h, up to 2 km); self-propelled crawler type, mounted on a special chassis with the bowl built in between the tracks (6 to 16 m³, 10 to 12 km/h, up to 1 km); towed by crawler or rubber-tyred tractors (1.5 to 35 m³, 10 to 12 km/h, up to 0.8-1 km). The performance of scrapers may be improved by using: pushers for loading, particularly under heavy ground and road conditions; dual loading operation, the machines being fitted with a hydraulically operated coupling and cushioned pusher plate; elevator loading, which provides self-loading without pushers; open bowl design, which makes loading by excavator possible; ripping teeth on bowl cutting edges when operating under heavy conditions; positive unloading of the bowl, which makes operation on wet and sticky grounds possible.

Grader, towed, semi-towed, and self-propelled or motor grader. A rubber-tyred machine fitted with a bladed working tool (mouldboard), which is mounted within the wheelbase and may be set at various angles, in both horizontal and vertical planes, and side-shifted or reached when performing such operations as cutting gutters by successive runs, shaping road beds, grading, spreading materials, etc. Two-axle towed and semi-towed graders are towed by crawler or rubber-tyred tractors but their use is steadily declining everywhere. Motor graders are classified as: light (up to 9 t and 120 hp), medium (10 to 13 t and 130-180 hp), heavy (14 to 19 t and 200-280 hp), extra-heavy (over 19 t and 280 hp); fitted with mechanical or hydromechanical drive; fitted with hydraulic or, less frequently, mechanical control. Their travel speed is up to 40-50 km/h. The most commonly used wheel arrangement is 1 x 2 x 3 which provides better blading and traction characteristics. The 1 x 3 x 3 wheel arrangement is used when operating under heavy conditions, and 2 x 2 x 2 to obtain increased manoeuvrability and off-road mobility. These qualities may be obtained by using articulated frame motor graders. To widen their applicability, motor graders are fitted with additional attachments (such as rippers, backslopers, snow ploughs, etc.). In addition to mouldboards, they may also be fitted with scarifiers and dozer blades.

Single-bucket excavator. A self-propelled crawler-mounted or rubber-tyred machine, with a bucket-type attachment rotatable in a horizontal plane by means of which, with the machine remaining stationary, such operations as digging, transporting the drawn out material (by swinging or moving the working tool attachment), dumping, and return to original position (with the possibility of combining individual operations) can be performed. Construction excavators (bucket capacity up to 2.5 m^3, mass up to 60 t) are used for excavating ground, soft rocks, and ripped frozen ground, while quarry excavators (bucket capacity 4 m^3 or more, mass more than 60 t) are used for excavating hard rocks, various ores, and mineral resources. Construction single-bucket excavators are classified by: mass; type of drive (mechanical, hydraulic, electric, and diesel-electric); type of control (cable, hydraulic, or combined); swinging ability (360° or part-circle); design of swing mechanism and front attachments, which come in such configurations as dipper shovels for excavating above the ground level and backhoes, draglines and clamshells mounted on a rigid arm (for excavating below the ground level). Loading buckets, clamshells with an extension, and grapples are used for loading bulk and lumpy materials and loose ground; crane attachments, for handling and erection operations; scraping blades and telescoping boom backhoes, for cleaning-up, sloping, and grading operations; pile-drivers etc. The applicability of excavators may be widened even further if such front attachments as hydraulic breakers, ripper teeth, etc., are mounted on the arm in place of the bucket.

Trenchers and multi-bucket ladder or wheel excavators are special-purpose machines used for digging trenches for pipeline laying, and ditches for land reclamation and irrigation projects; stripping overburden; mining mineral resources; etc.

SAFETY AND HEALTH ASPECTS

In view of the general trends peculiar to earth-moving equipment (increase of unit power, use of pneumatics, hydraulics and automation in control systems, speed-up of operating motions, improvement of performance ratings, etc.), ergonomics and operator's health and safety are the problems of particular importance. To tackle them would contribute to facilitating control and maintenance of the machines, eliminating awkward postures and movements, alleviating fatigue and, as a result, increasing productivity and efficiency of operators (including older ones).

To judge whether earth-moving equipment is adapted to human physical needs the following ergonomic characteristics must be taken into account: hygienic data indicating to what extent the requirements of the operator's normal activity and working capability are complied with; anthropometric data defining the suitability of the operator's control station to his physical dimensions; physiological and psycho-physiological data to check whether machine specifications are adapted to the operator's sensory functions and his muscular and visual-motor abilities.

Hygienic factors

These comprise vibration and noise levels, microclimate, gas and dust concentrations and intensity of illumination.

Vibration level. Vibration levels at seat, cab floor and control levers should be examined separately. When determining the vibration level to which an operator is exposed, the intensity, frequency, direction, and duration of exposure should be taken into consideration. With due regard for all these factors, the following three basic criteria for estimating vibrations should be used: preservation of working efficiency (fatigue-decreased proficiency boundary); preservation of safety and health (exposure limit); preservation of comfort (reduced comfort boundary). As far as the first criterion is concerned, the standard permissible load of vibrations during an 8-h working day within the 4-8 Hz frequency range for vertical vibrations adopted by most nations has been fixed at 0.315 m/s^2. For the second criterion this load may be doubled and for the third it should be decreased by a factor of 3.15. To improve the pattern of vibration, machine springing and cab cushioning, vibration-proof seats, and various vibration control devices are used.

Noise level. Noise levels inside the cab (at the operator's control station) should be distinguished from those within the working area (environment). When estimating noise effect on an operator, noise intensity, frequency, and duration of exposure should be taken into account. There are two groups of noise: steady (sound level changing during 8-h working day by less than 5 dBA) and unsteady noise (fluctuating with time, intermittent, pulse). The sound pressure levels are standardised in octave frequency bands, and sound levels and time-weighted energy average sound pressure levels in dBA. The maximum permissible noise exposure (time-weighted energy average sound pressure level) adopted by many nations has been fixed at 85 dBA. In compliance with international standards, the sound level should be measured within the working area and operator's cab with the machine performing simulated operation cycles (excavator: digging a trench, excavating ground in a cut; crawler bulldozer: moving forward at 4 km/h with the blade in transport position; etc.), the engine running at maximum rpm, on a level and open site with no sound reflecting objects within 20 m from the machine, and with the microphone located at the level of the operator's ear (665 to 818 mm from the seat surface). Sound insulation and sealing of cabs and noise sources (e.g. installation of noise suppressors, i.e. silencers, at exhaust, sound insulation of the space under hood, installation of shields and protection devices on the structural noise propagation pathways, etc.) have been achieved by the use of mastics, liners, and other sound absorbing materials.

Microclimate. This is characterised by temperature, relative humidity, and air flow velocity inside the cab. Under comfortable microclimatic conditions the temperature is 20-22 °C, relative humidity 40 to 60%, and air-flow velocity does not exceed 0.5 m/s. The permissible levels should be defined taking account of the climatic conditions under which the machine is expected to be used. The required microclimate is obtained by sealing and heat-insulation of the cab, and by using tinted glass, air conditioners, heaters, fans, etc.

Gas concentration. This is characterised by the amount of harmful substances (e.g. carbon monoxide and dioxide, gasoline vapours, etc.) in working atmosphere, which must not exceed the exposure limits stipulated in health standards. To reduce the content of harmful substances inside the operator's working area, sealing and ventilation of the cab along with air cleaning may be used, and the exhaust of harmful substances to the atmosphere should be prevented by perfect fuel combustion, the sealing of the fuel and hydraulic systems, etc.

Dust concentration. This is defined as the concentration of airborne dust in the operator's environment with the machine running. Depending on the chemical composition of the dust (primarily its silica content) exposure limits usually ranging from 1 to 4 mg/m^3 are established. The dust concentration can be reduced by sealing and ventilating the cab and by using a dust collection system.

Illumination. Illumination of the machine working tool within its operating area and of the surroundings within the machine working area is obtained by the installation of lighting equipment of suitable rating. Visibility of the working tool position is of major importance and multi-point illumination should therefore be used, with the possibility of controlling both direction and position of the light sources.

Anthropometric data

These data are used to define the internal dimensions of the cab, arrange the operator's control station, and check the visibility of and accessibility to different machine units for maintenance. These data are based on the physical dimensions of small (1 620 mm) and large (1 920 mm) operators in seated and standing positions (statistics suggest that only 5% of all operators are shorter than 1 620 mm or taller than 1 920 mm).

Internal dimensions of the cab. The size of the cab is important for both the efficiency and the comfort of the operator. Reduction of internal over-all dimensions below the permissible limits (width 850 mm, height for a seated operator 1 500 mm and for a standing operator 2 010 mm) results in substantial loss of the operator's working capacity.

Control station. The arrangement of the operator's control station should provide for convenient relative positions of the seat and controls and their most suitable location in relation to the inner walls of the cab. Suitably dimensioned seats should be provided that are adjustable both horizontally and vertically and are fitted with tilting backs and suitably sprung for operators of different weights. The controls should be designed so that those most frequently used are comfortably located and that the others are within the operator's reach, sufficiently separated from one another and not too close to the inner walls of the cab.

Visibility. The operator should be able to see from his control station objects and markers designed to guide operations. Each type of earth-moving equipment has its own particular visibility requirements. For example to operate a motor-grader safely, the front steerable wheels, the side edges of the blade in any operating position, the obstacles beside the machine spaced at a distance not exceeding 0.5 m from the motor-grader sides, road shoulder, etc., should be clearly visible. The required visibility is attained by designing the operator's control station and the components of the entire machine so that the cab is appropriately placed on the machine, by increasing the glazed area, locating the control panel suitably, using adjustable seats, etc.

Access systems. The operator should be able to occupy his control station in the cab, come down to the ground, move about on the machine when inspecting individual assemblies, performing service functions, etc., safely, promptly, and with a minimum of effort. For example the maximum height of the first step from the ground should not exceed 700 mm (preferably 400 mm), step width should be not less than 300 mm to accommodate both feet and not less than 160 mm to accommodate one foot, headroom clearance above all ladders should be not less than 2 010 mm, entrance opening width not less than 680 mm, etc.

Physiological and psycho-physiological factors

These comprise such values as control forces, frequency of lever shifting, etc.

Control forces. The forces required should be measured on both hand and foot controls. Controls may be classified as frequently and infrequently used. The forces for the former should not exceed 60 N on hand levers and 120 N on foot pedals, while for the latter the limits are 120 N and 250 N respectively. The required forces are obtained by installing hydraulic and compressed-air boosters, incorporating automatic control systems, suitably designing the drive of the controls, etc.

Safety features

Safe operation requires the installation of various systems to protect the operator against potential risks: roll-over protective system and falling-object protective system, reserve (emergency) braking system, reserve (emergency) steering system, protective structures and guards, etc.

Roll-over and falling-object protective systems. These are classified, by their design configuration, as protective canopies (combination of structural members mounted on a machine regardless of the availability of the cab) and safety cabs (cabs fitted with a supporting framework), the latter being the more suitable since they require less metal and offer better ergonomic characteristics. The operator must always use a safety belt to ensure that his body remains within the deflection-limiting zones (the space envelope that should remain intact even if the cab has been distorted). Taking into account the probabilistic nature of interaction between an overturning machine and a supporting surface, roll-over and falling-object protective systems are designed to ensure protection of the operator in case of overturn of an earth-moving machine travelling at 16 km/h over firm clay ground on a maximum slope of 30° (the total roll-over angle in relation to the longitudinal axis of the machine being 360° without loss of contact with the supporting surface) and in case of impact by a round-shaped object dropped from a height sufficient to develop an energy of 11 600 J.

Reserve braking system. The reserve braking system of earth-moving equipment (except for excavators) in addition to operating and parking brakes should function properly in the event of failure of the operating braking system. It should be capable of stopping the machine within a determined distance, for example a 23-t scraper travelling at 32 km/h should be brought to a complete halt within a distance not exceeding 55 m.

Steering. An emergency steering system should provide control of an earth-moving machine until it is brought to a complete halt, even with the steering system power source out of action or the engine stalled. This can be attained by incorporating an additional energy source into the system, installing metallic hoses (flexible tubes) capable of withstanding three or four times the permissible hydraulic pressure (which itself is limited by safety devices), and in some cases even by designing for complete redundancy of the system. It is advisable to mount the reserve steering systems independently of other power systems or circuits. If this is not possible they should have priority over all other systems except the reserve braking system. The emergency steering system should be automatically connected in case of break-down of the normal system and an audible or visible alarm signal should be given.

Protective devices and guards. These are designed to protect the operator against the risks of mechanical, chemical, thermal and electrical factors (e.g. fenders above drive wheels or tracks, fan housings, guards for hydraulic system high pressure metallic hoses located in the operator's cab, storage battery casings, etc.). If no cab is provided, fenders protecting the operator against mud and other objects projected by the drive wheels and tracks must be installed.

There are also standards and recommendations stipulating safety requirements for the maintenance, preservation and recommissioning, repair, etc., of earth-moving equipment, as well as safe operation management procedures (operating instructions, manuals on training earth-moving equipment operators and maintenance personnel, procedures to examine their knowledge, etc.).

YARKIN, A. A.
KALMYKOV, V. N.

General:

CIS 79-824 *International Symposium on Hazard Prevention in the Construction Industry–Record of proceedings* (Colloque international de prévention des risques professionnels du bâtiment et des travaux publics–Compte rendu des travaux). Comité international de prévention des risques professionnels du bâtiment et des travaux publics (Association internationale de la sécurité sociale, 2 bis rue Michelet, 92130 Issy-les-Moulineaux) (undated), 448 p. Illus.

Accidents:

CIS 79-588 "Analysis of accidents with driver-controlled self-propelled earth-moving equipment working in the open" (Analyses d'accidents des engins de terrassement automoteurs à conducteurs portés travaillant à l'air libre). Mouret, M. *Cahiers des comités de prévention du bâtiment et des travaux publics* (Issy-les-Moulineaux), July-Aug. 1978, 4 (230-234). Illus. (In French)

CIS 79-630 "Gas pipeline rupture–Holocaust". *ACC Report* (Wellington, New Zealand), Sep. 1978, 3/4 (10-11). Illus.

Noise:

CIS 80-695 *Acoustics–Measurement of airborne noise emitted by construction equipment intended for outdoor use–Method for determining compliance with noise limits.* International Standard ISO 4872-1978(E) (Geneva, International Organisation for Standardisation, 1st ed., 15 July 1978), 11 p. Illus.

Electricity:

CIS 80-318 "Grounding electric shovels, cranes and other mobile equipment". Data Sheet 1-287-79, National Safety Council, revised 1979. *National Safety News* (Chicago), Sep. 1979, 120/3 (79-82). Illus. 1 ref.

Medical supervision:

CIS 80-243 "Eleventh Congress of the National Association of Physicians for Building and Civil Engineering–Driving of heavy vehicles and personnel transport in building and civil engineering" (XIe Journée d'études du groupement national des médecins du bâtiment et des travaux publics–Conduite de véhicules poids lourds et transport du personnel dans le bâtiment et les travaux publics). *Revue de médecine du travail* (Paris), 1979, 7/1, 99 p. Illus. 9 ref. (In French)

Ecotoxicity

Environmental chemicals today

The oceans cover 71% of the Earth's surface and the land only 29%. Thus there is no shortage of water. The mean elevation of the land above the sea level is 200-300 m, the mean depth of the sea 3 800 m, and the Earth is surrounded by the atmosphere, which becomes more tenuous with increasing altitude. With existing technology, water and air, however polluted, can be purified to any desired degree. By contrast, there is no known way of purifying polluted soil. Moreover, for this most limited portion of the Earth, a mere 12% of which is used for agriculture, there are no antipollution laws. We behave, indeed, as if the land was subject to no limitations whatsoever.

Organic chemicals will be the main topic of this article, since here the difference between naturally occurring and synthetic chemicals is much more clear-cut than in the case of inorganic chemicals; we have not the remotest idea what the natural level of inorganic substances is on the Earth's surface. In 1950, 7 million tonnes of organic chemicals were produced, in 1970 the figure was 63 million tonnes, and in 1983 it is expected to be 250 million tonnes; the belief is that this rate of increase will be maintained in the future. A comparison of the tonnage of synthetically produced substances with naturally occurring substances (e.g. those produced by forests) shows that the two are of roughly the same order of magnitude. The production of the biomass lies some orders of magnitude higher. The composition of synthetic production is far less well known. Pesticides are an exception because their application has been controlled by law for some time. A little is known about detergents, but almost nothing about solvents. Little information is available about gaseous basic chemicals and, needless to say, there are not even clues about the many "others". Nothing, for example, is known about lubricants and gasoline additives. The production of plastics is at present evaluated at 40 million tonnes world-wide; they contain something between 1 and 5-6% of additives, that is to say a minimum of 400 000 tonnes annually. These are additives about whose environmental effects no one knows anything at all, and there has been little gain in our knowledge during the last ten years in this area.

Every year, 1 million tonnes of pesticides are produced. Their importation and marketing is subject to licensing in most countries. Quite a lot of information is therefore available about them, and this probably explains why they are so often named in the media. The tonnage of all organic chemicals taken together is more than 150 times greater. They are not subject to controls and little is known about their ecotoxicity and, quite often, even about their plain toxicity. This grossly distorts the whole debate about the environmental effects of chemicals, since a uniform global distribution of all these manufactured substances would give 11.2 kg/ha of general chemicals compared with only 0.07 kg/ha of pesticides.

A few years ago an ad hoc industrial group in the Federal Republic of Germany attempted to compile production figures and application patterns for chemicals with an annual world production of at least 50 000 tonnes. This procedure was continued for some years at great cost and a list of approximately 130 chemicals was drawn up. Unfortunately the pro-

Table 1. World production of chemicals with an annual global production of over 50 000 t[1] ('000 tonnes)

Organics		Inorganics	
Acetaldehyde	2 400	Aluminium oxide	17 000
Acrylonitrile	2 700	Ammonia	40 000
Alkylbenzenes	700	Chlorine	24 600
Benzene	14 400	Iron sulphate	450
Carbon tetrachloride	1 000	Lead oxides	363
Chloroparaffins	270	Nitric acid	23 000
Cyclohexane	2 800	Phosphoric acid	13 000
Dibutyl phthalate	230	Sodium chlorate	600
Phthalic acid anhydride	2 300	Sodium chromate and bichromate	450
Toluene	8 500	Sulphuric acid	108 300
Trichloroethylene	700	Zinc oxide	420
Vinyl chloride	7 730		

[1] Selection of a compilation of 130 chemicals compiled for the OECD by an ad hoc industrial group in the Federal Republic of Germany.

gramme had to be discontinued because no other country and no chemical industry wished to participate in the work. Yet this would be a very worth-while activity, if existing environmental chemicals are ever to be brought under control. As it is the compilation provides some valuable data; for example, that production of chloroparaffin reached 270 000 tonnes. This is an amount which, if it enters the environment, is bound to show up in all biotic and abiotic systems, and it is not surprising that these compounds are found ubiquitously whenever analysis is performed.

Foreseeable development

Is it possible to forecast future trends? Let us start, for example, from the premise that PCBs did not exist on earth 100 years ago and that concentrations measured in ppb (μg/kg) are being found today. Assuming, further, that technology will continue to develop at past rates and that production in the chemical industry will grow at 2% annually–a realistic assumption, with or without an oil crisis–then concentrations of PCBs will increase in the next 100 years from ppb to ppm. In other words, since no noteworthy biological effects caused by ppb concentrations have been reported, but some cases are known in which human beings have suffered direct harm or even died from ppm concentrations, the problem of environmental pollution by PCBs will have to be solved in the next 50 to 100 years. Although there is no immediate world-wide danger, the problem is worth while thinking about because a solution needs to be found in advance: there is no way in which globally dispersed chemicals can be "recaptured", as the example of PCBs clearly shows.

Table 2 portrays a classic picture as revealed in 1966. Much excitement was caused when it was found that pesticides never used in London suddenly appeared there in measurable amounts. At first analytical errors were postulated as the explanation; but then the basic physicochemical principles of our earth and global transport phenomena were remembered; today it is clear that all these organic chemicals disperse uniformly via the air.

Table 2. Insecticide concentrations found in air, rainwater and rivers in the United Kingdom and the United States around 1964-66

Insecticide	Parts per 10^{12} in		
	Air	Rainwater	Rivers
Dieldrin	20 (London, 1966)	9-28 (England, 1964/65)	10 (Scotland, 1966); 0-118 (United States, 1965)
BHC	10 (London, 1966)	12-164 (England, 1964/65)	Detected (Scotland, 1966; United States, 1965)
DDT and analogues	10 (London, 1966)	210 (London, 1965)	5 (Scotland, 1966); 190 (United States, 1965)

Figure 1 presents the comparison made by Appleby of the distribution of radioactive materials after atmospheric nuclear bomb tests with PCBs and similar chemicals. It shows that not only were the concentrations greater in the northern than in the southern hemisphere, as was to be expected, but also that organic chemicals with the chemicophysical behaviour of, for instance, DDT–and that comprises a large portion and even the major proportion apart from plastics–were distributed in the atmosphere in exactly the same way as strontium 90. This underlined, already ten years ago, the importance of ascertaining what it is that we are introducing into the environment, in what amount and where. However, almost nothing is yet being done to that effect.

Harmful effects

One of the first recorded pollution incidents with a fatal outcome was the 1952 fog in London, during which ppm concentrations of sulphur dioxide plus ppm concentrations of dust produced poisoning to the extent that within 4 days 4 000 people more than usually had died. The dust concentration was subsequently reduced and, as is well known, sulphur dioxide emissions have been reduced everywhere with the result that such events have not recurred.

The first two chronic intoxications, the Minamata and Itai-Itai diseases, are also well known. The first was caused by methyl mercury and the second by cadmium. As regards mercury it is well established today that more than 40% and probably as much as 60% of the ADI (acceptable daily intake) is accounted for by natural mercury, i.e. the industrial gap is only 40%. It now seems that the situation with cadmium is entirely similar if available data are correct.

Itai-Itai disease is postulated to arise if the cadmium concentration in the food is around 10 mg/kg. In general we are below this daily absorption figure: typical soil concentrations are 0.1 mg/kg, for example. In Japan a regulation has just been issued banning the use for rice cultivation of soil containing 10 mg/kg cadmium or more. European opinion is that the upper limit should be 5 mg/kg, but if this were to be enforced in Japan, large rice growing areas would have to be abandoned.

The concentration of cadmium in activated sludge exceeds or comes very close to tolerance. Some countries take this situation seriously; Sweden, for example, has excluded all imports of cadmium-containing products since August 1980.

Naturally, the question arises of how these cases of high concentration come about. It seems that a part of the cadmium burden arises from cadmium additives in plastics materials burnt as waste, which leads to the element entering the air, spreading uniformly, and at some stage increasing the soil cadmium concentration. If this is true, it is only logical to restrict the use of cadmium where it is not absolutely essential or, like Sweden, to ban cadmium-containing products absolutely.

This example is an impressive one because, although we have known about Itai-Itai disease since the 1960s and although people died of it, it is still not known in what form cadmium occurs in the organism and in living cells: there are no indications at all, just as there is no evidence whatever about world-wide production figures and patterns of use.

Auto oils and lubricants contain quite a number of additives. A study of their chemistry shows up classes of substances but not single substances. Obviously their content of individual substances first needs to be elucidated. The next step would be to take them apart chemically in order to find out what passes through the engine, and what chemical substances are emitted by it. Samples would have to be investigated in order to verify whether what chemistry teaches really happens. Next, the toxicologists would have to be called in to evaluate the toxicological significance of the data. However, the first step is almost entirely lacking. This is quite startling because our prosperity depends very considerably on the chemical industry.

Such data on consumption as are available apply mainly to the United States, because it is still feasible to get some figures there, whereas in Europe it is almost impossible to do so. For example, 783 000 tonnes of detergent (that is to say alkylphenates, calcium alkylates,

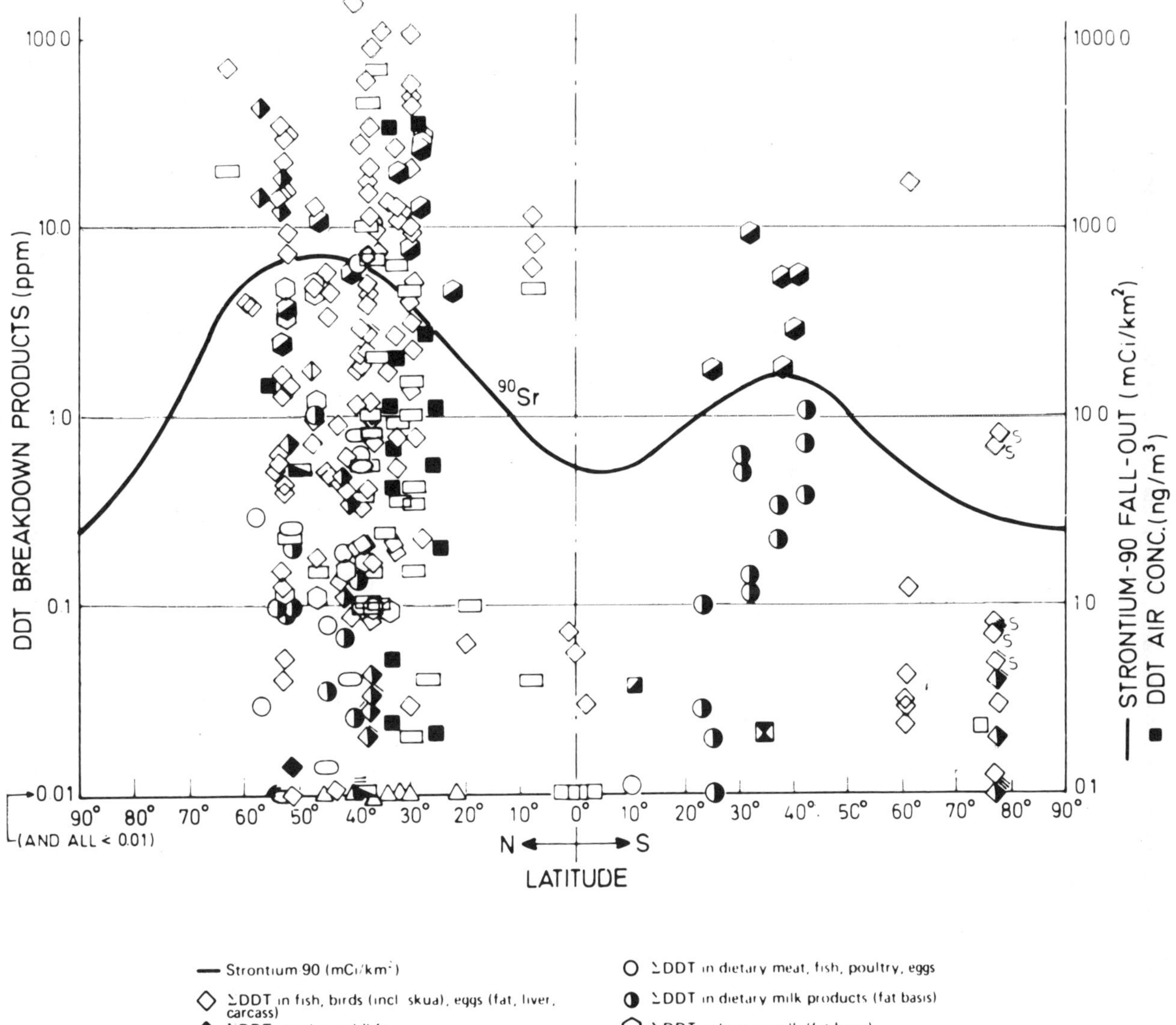

Figure 1. Sr 90 fall-out after an atmospheric nuclear bomb test as compared with DDT air concentration.

etc.) are consumed annually but these are all classes of chemicals and not the individual chemicals required for an evaluation.

Evaluation

In order to make a full evaluation of a given substance, a number of facts have to be known about it: for example, how much is produced; where it is used, i.e. where it enters the environment; its persistence (which can be estimated from its chemical formula); its dispersion (which can be calculated from the strontium 90 model via the transport through the atmosphere); its conversion under biotic conditions (which can be predicted qualitatively but not quantitatively); its behaviour under abiotic conditions (something that has been overlooked so far, though there are already some indications that degradation and structure can be correlated); and—something entirely new—its ecotoxicological behaviour. If we do not obtain these data we shall merely continue in the same irrational vein as before, concentrating our attention on the polychlorinated biphenyls and hexachlorobenzene, the substances which by chance can be readily investigated analytically, instead of proceeding in a rational scientific manner by first examining what we are doing, and what we are introducing into the environment and then reflecting on what the distribution of each individual substance might be and evolving suitable control programmes. Surely our present approach to the problem is scientifically quite untenable.

What remedial concepts are being used today or are being planned for future use? Let us consider environmental monitoring to begin with.

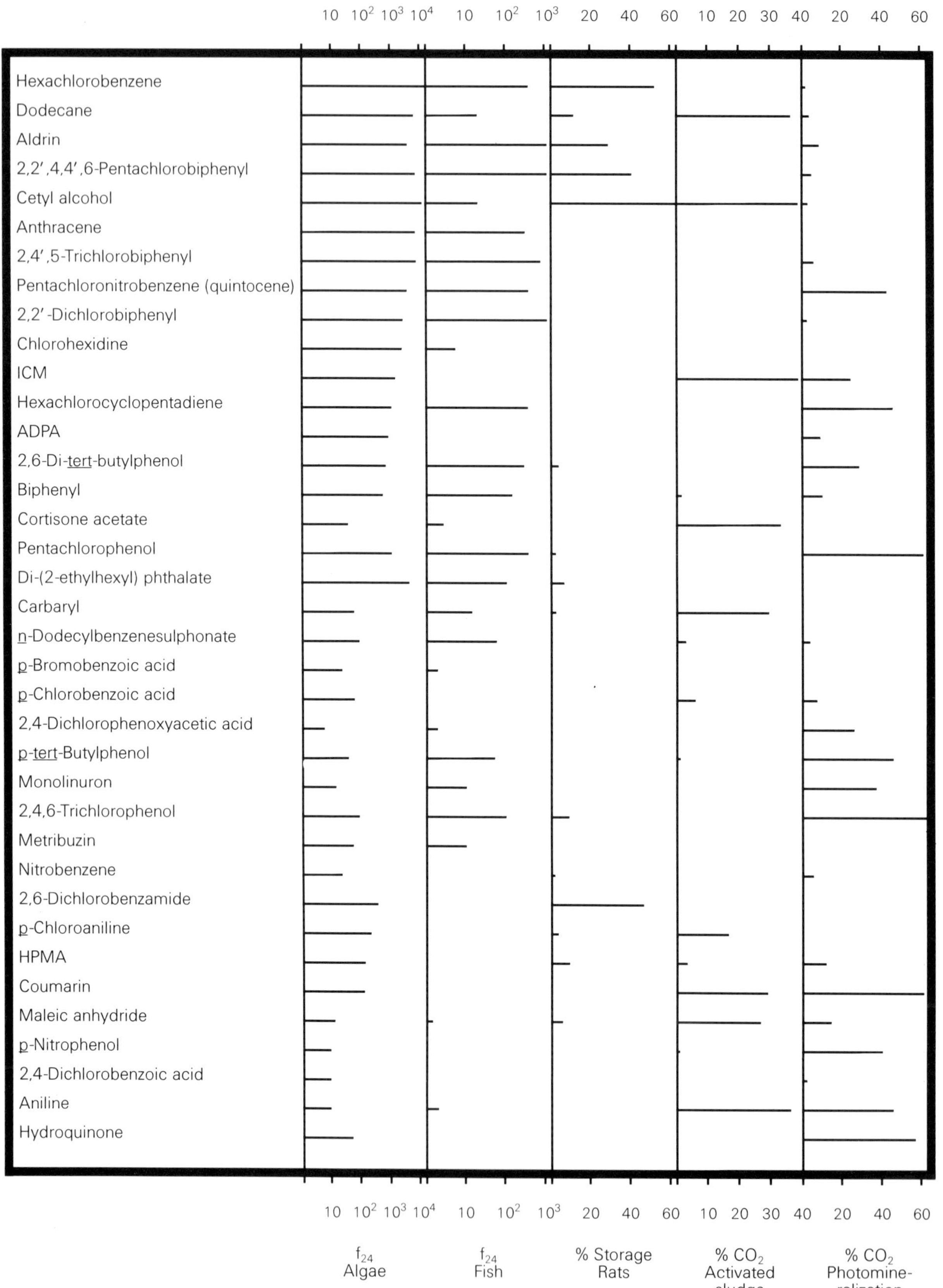

Figure 2. Survey of the results of the ecotoxicology profile analysis. Storage is presented as accumulation factor (f_{24}, calculated for 24 h) or % storage; degradation is presented as terminal breakdown (% CO_2 in relation to the initial amount of chemical). The chemicals are listed in order of increasing water-solubility.

Naturally, there are great advantages in investigating familiar substances, for instance, plant protection agents in food. A start was made in the United States a long time ago. For example, if the diet contains sea food the PCB burden in man may be considerable, whereas while dairy products help to reduce it, a vegetable diet contributes a negligible PCB burden for practical purposes. These data, supplied by monitoring, are very useful because they give an "idea" why substances accumulate in the organism.

A second approach is to build up a predictive evaluation. For the purpose "indicator organisms" (if these exist) must be chosen from human tissue, animals, micro-organisms, plants, and worms and stored in a tissue bank. The problem is that, although tissues can indeed be collected, described and stored, they also have to be analysed reproducibly. Strictly speaking, they should be submitted to a complete chemical analysis, i.e. in respect of all substances. However, suitable techniques and detectors are not yet available to achieve this purpose. First of all an attempt had therefore to be made to see whether organohalogens, the best known compounds so far as analysis is concerned, can be examined together. Tissues were collected, stored and analysed using the fingerprint principle. The aim was to analyse everything and then to follow the height of the individual "peaks" as a function of time. It is only when a peak increases considerably that an attempt is made to elucidate its constitution. The toxicology department is informed and appropriate conclusions are drawn, for example that a careful, specific investigation is indicated. This approach makes it unnecessary to know the identity of the substances on condition that measurements can be reproduced and appreciable peak increases are checked. Relatively large numbers of substances are included.

Not surprisingly this is a very expensive approach. But it is being followed up jointly by the Federal Republic of Germany and the US Environmental Protection Agency, and the analytical aspects do not look quite as hopeless as was at first feared. So it appears that the tissue bank concept is feasible after all and should be pursued with all possible zeal. It has the additional advantage that reference material can be stored over several generations, thus making initial values available which are sadly lacking today.

The third approach involves the simulation of ecosystems. There are, however, doubts and concerns about the conclusions to be drawn from the results, because no one knows what is significant for the evaluation of chemicals in ecosystems. What characteristics should a chemical introduced into the environment definitely not have? The general opinion today is that substances should not be too persistent in mammals, including man; anything that is rapidly eliminated is probably better than anything that accumulates. If this view is even 80% correct it is good enough.

A high degree of accumulation in fish is undesirable because many people and groups depend entirely on fish for food. Accumulation in algae must be controlled. Next, an interesting criterion is whether these substances disappear under photochemical, i.e. atmospheric, effects; it may be that organic chemicals, contrary to earlier expectations, can be degraded to carbon dioxide in the air by solar radiation; and this is indeed the case in an unexpectedly high degree. Experimental evidence should also be obtained on the behaviour of substances in liquid waste; the more the substance is degraded to carbon dioxide the better (degradation always means mineralisation with formation of CO_2, H_2O, etc.). If possible the substance should be degraded to carbon dioxide in the process of refuse composting, a large-scale technical process. Tests have been designed to cover all these aspects so that they closely match real processes in the ecosystem. A basic requisite has been to reduce the experimental concentration down to the ppb level. As a rule the tests are therefore carried out with 50 ppb, thus matching practical conditions as closely as possible. Elimination from mammals (and at the same time the metabolite to non-metabolite ratio) is assessed in a screening test. A test of accumulation in fish takes 3 days, and for algae 24 hours are needed, while photomineralisation requires 17 hours in diffuse daylight. Activated sludge from the treatment plant in Munich has been investigated. Furthermore, we have simulated refuse composting. All these tests are carried out under conditions matching practical conditions, are readily accomplished, and are strictly reproducible.

A prerequisite for such an operation is a uniform method of analysis, and for organic substances carbon-14 analysis is the most uniform. In consequence, all substances we use are labelled with carbon-14. It is the only way to get down to 50 ppb concentrations and yet include the metabolite to non-metabolite ratio.

Figure 2 presents a summary of the ecotoxicological properties of a number of compounds. When one has collected the data, what does one do with them? This is a universal problem. The lowest acceptable value needs to be defined and—even better—the substance needs to be scrutinised as regards its use and site of application, and a possible substitute considered.

I believe that this is the way of the future. However, a prerequisite is a knowledge of what substances are being released into the environment and where they are being introduced. It might after all be worth while attempting to find out whether such a study could once more be undertaken by collaboration between the chemical industry and interested large research establishments. It would seem that this would be the only way of at long last tackling the problem of environmental chemicals rationally in the future.

(The experimental work reported here was carried out in co-operation with the scientists of the Institut für Ökologische Chemie.)

KORTE, F.

Ecological chemistry and ecotoxicology (Ökologische Chemie und Ökotoxikologie). Korte, F.; Klein, W.; Parlar, H.; Scheunert, I. (Stuttgart, Georg Thieme Verlag, 1980). (In German)

CIS 74-806 *Health hazards of the human environment* (Geneva, World Health Organisation, 1972), 387 p. Illus. 1 142 ref.

"Polychlorinated biphenyls (PCB) in foodstuffs" (Polychlorierte Biphenile (PCB) in Lebensmitteln). Sommermann, W.; Rohleder, H.; Korte, F. *Zeitschrift für Lebensmittel-Untersuchung und -Forschung* (Munich), 1978, 166/3 (137-144). Illus. 18 ref. (In German)

"Effects of pesticides and related compounds on the environment". Korte, F. (1 045-1 061). Illus. 7 ref. *Pure and applied chemistry* (Oxford, Pergamon Press), Vol. 50.

Environmental impact assessment. A bibliography with abstracts. Clark, B. D.; Wathern, P. (London, Mansell, and New York, Bowker, 1980), 516 p. Ref.

OECD chemicals testing programme physico-chemical tests. Klein, A. W.; Harmisch, M.; Poremski, H. J.; Schmidt-Bleek, F. *Chemosphere* (Oxford), 1981, 10/2 (153-207). Illus. 10 ref.

Effects, combined

Workers are often exposed simultaneously to different physical or chemical agents which have a "combined effect" that may be different from the simple addition of the effects of various agents applied independently.

A worker may be exposed to industrial dust with known pathological effects but, at the same time, he may become particularly susceptible to the dust because he smokes. Furthermore, the dust exposure may be accompanied by exposure to gases and fumes that complicates the consequences.

The interaction of different external factors is not governed by any known rule. Some combinations are found to be synergistic, some are simply additive, and there are even examples of agents whose effects are "antagonistic". The problem is rendered even more complex by the fact that the effects of simple exposures may vary from one individual to another or from time to time in the same individual, depending on a large number of environmental and human factors, known and unknown.

Combined exposures

It is necessary to distinguish two different types of combined exposure:

(a) combined environmental exposure of physical, chemical, biological or psychosocial factors of persons in "normal" physical and mental health;

(b) the action of environmental exposure on persons whose susceptibility is influenced by certain genetic defects or by such factors as diet and food intake, smoking, alcohol consumption and the use of medicinal drugs.

Exposure in industry. Among the best known exposures in category *(a)* are those involving combinations of chemical and/or physical agents and heat. Exposure to heat results in an increase in the rate of blood circulation, which may augment the uptake of toxic substances in the atmosphere; an increase in metabolism and body temperature, which may influence the metabolism of chemicals; inhibition of organs such as the brain and liver, which may influence their ability to deal with toxic effects.

Furnace operators in the iron and steel industry may be exposed simultaneously to heat stress, irritant gases (e.g. sulphur dioxide), toxic gases (e.g. carbon monoxide) and airborne particles of carbon in smoke; moreover, the intensity of exposure to each of these agents may vary independently or in combination. Workers in the glass industry may be exposed to a combination of heat, furnace fumes and raw material dusts, and combined exposure to heat and dust is relatively common in the cement industry. Other industries where this type of combined exposure is encountered include foundries, the textile industry, viscose industry, pottery industry, laundries, garages, etc.

Exposure in agriculture. Agricultural workers may be exposed to a combination of stresses that include climatic conditions (e.g. heat, humidity, and sunlight), high levels of energy expenditure and agricultural chemicals or vegetable dusts. In many cases in the developing countries additional stresses may be presented by malnutrition and endemic disease.

Air-pollution exposure. Air pollution caused by industrial effluents, automobile exhaust gases and domestic fuel-burning in urban areas may contain toxic, irritant and carcinogenic gases, vapours, fumes and dusts. The individual substances are usually present in such low concentrations that they may independently have no significant harmful effect. However, when acting together, they may cause respiratory disease in the general population (e.g. chronic bronchitis and lung cancer). In cases where special climatic conditions produce a sudden rise in the degree of pollution, there may be increased mortality amongst the aged or amongst persons with a history of cardiac or respiratory disease. In such cases there is complex combination of the effects of the different pollutations on the cardiac or respiratory disability.

Effects of exposure to combined environmental agents

It has been theorised that all agents which act upon the body or any of its parts exert dual effects; the one being specific and the other being non-specific. The specific effect is that which characterises the agent, for example the kind of infection produced by a certain microorganism, or the kind of lesion resulting from the application of a chemical of a defined toxicity. This effect varies with the type of causative agent. The non-specific effect, however, is a common reaction that occurs with all stresses. It involves what is called the "general adaptation syndrome", through which the adrenal cortex is stimulated to produce corticosteroids to sustain the "defence" mechanisms against "damage".

The non-specific reactions to two or more agents may logically be additive; however, the specific effects will probably interact and the net result will depend on whether the effects tend to neutralise or potentiate each other.

Experimental and epidemiological studies

A considerable amount of information on combined effects of chemical substances is available, whereas the combined effect of physical factors are relatively less explored. Animal studies include investigations of the effects of various solvents, groups of pesticides, groups of metals and irritant gases. Observations vary; additive, synergistic and antagonistic effects are found. The initial information based on experimental animal and *in vitro* studies does not necessarily dictate the need for avoiding combined exposures but suggests that appropriate steps should be taken to minimise exposure whenever possible. A review of the subject was made in December 1980 by a WHO Expert Committee on Health Effects of Combined Exposures in the Work Environment.

Epidemiological studies on specific exposure combinations are limited. An example of studies that have been undertaken on effects of exposure to mixtures of toxic substances is that of exposure to a mixture of organic solvents, for example among car painters and varnishers. This was found to cause subjective symptoms of ill-health, behavioural deterioration, neurophysiological and neurological manifestations, but it is difficult to judge whether these findings are less than additive, additive or potentiating (Husman, 1980 and Seppäläinen, 1978).

It is understandable that the mechanisms underlying the combined interactions are complex and require extensive study. Considerable progress, however, has been made recently in establishing the role of the liver in mechanisms of interaction between toxic substances. The induction of microsomal enzymes by some substances capable of changing the metabolism of other substances gives grounds not only for qualitative but also for quantitative assessment of the expected biological effects in combined exposure.

Effects of smoking, alcohol intake, diet and drugs on occupationally related health problems

The possibility of increased health damage resulting from exposure to occupational hazards and from selected personal characteristics, namely smoking, alcohol consumption, dietary intake and intake of drugs has been reported.

Cigarette smoking potentiates the damage resulting from exposure to dusts like cotton, silica and asbestos. There is some information available on combined health effects of smoking and various occupational hazards. Cigarette smoking seems to increase the risk of byssinosis among cotton textile workers exposed to cotton dust, and lung cancer among asbestos workers.

Alcohol intake, although studied in acute exposures, aggravates the impairment of the liver in exposure to chemicals, particularly solvents.

Diet, with its various components has many specific ways of interaction with metabolism of chemicals, including carcinogens. The specific combinations are innumerable but the fact that undernourished human beings are potentially exposed to chemical intoxications in developing countries requires not only research but improvements in nutrition of the working populations at risk.

Drug intake, with its wide and often indiscriminate use, may have various influences on toxicity. Common drugs, like phenobarbital, do at times increase toxicity of some chemicals such as solvents.

The WHO Expert Committee (1980) recognised the extensive gaps in our present knowledge in this field. Our lack of knowledge is also due to the fact that the evaluation of the combined effect of multiple external exposures is more complex and more challenging than the study of single exposures. More research is required to assess the effects of long-term exposure to multiple factors in the workplace, particularly in long-term low level exposures. Experimental studies which focus on exposures simulating those of industrial and agricultural workers, including qualitative studies of the effects of exposure to known airborne concentrations of mixtures, are also important to clarify the mechanisms. In particular, there is a need for information on the exposure levels of mixtures which are not likely to produce adverse health effects in the work environment.

Concerning future evaluation of occupational exposure limits, the Committee proposed the exploration of biological parameters which may give an over-all indication of the degree of response to multiple factors. Such parameters as neuro-behavioural responses, hormonal changes and changes in various physiological and functional phenomena and autoxidation changes, probably arising from the formation of reactive oxygen (singlet oxygen, hydroxyl radical, superoxy anion, etc.) associated with chemical exposure regardless of the specific toxicity of the chemical, should be investigated.

PREVENTIVE MEASURES

The safety and health measures to protect workers from the effects of harmful agents used in industry are usually directed at maintaining exposure below a certain exposure limit that research and experience have shown to produce no adverse health effects. However, where workers are exposed to more than one hazardous agent, care must be taken to ensure that protective measures are taken to deal with all the hazards and not just the most prominent ones. In addition, factors that may increase individual susceptibility to harmful substances should be controlled by careful personnel selection, medical examination, etc.

Nevertheless, should there be any indication that the health of workers is being detrimentally affected, even though exposures are being maintained below recommended limits, investigations will be necessary to determine whether the action of a given agent is being potentiated by that of another.

Finally, in the research that precedes the recommendation of an industrial hygiene standard for any given agent, account should be taken of any interaction with other factors that may influence the outcome of exposure, even though the individual factors have proved normally harmless when acting independently.

EL BATAWI, M. A.

Health effects of combined exposures in the work environment. Report of a WHO Expert Committee (Geneva, World Health Organisation, 1980).

Biochemical Toxicology of Environmental Agents. de Bruin, A. (Amsterdam and New York, Elsevier Scientific Publishing Company, 1976), 1 544 p. Ref.

"Symptoms of car painters with long-term exposure to a mixture of organic solvents". Husman, K. *Scandinavian Journal of Work, Environment and Health* (Helsinki), Mar. 1980, 6/1 (19-32). 27 ref.

CIS 79-780 "Neurophysiological effects of long-term exposure to a mixture of organic solvents". Seppäläinen, A. M.; Husman, K.; Mårtenson, C. *Scandinavian Journal of Work Environment and Health* (Helsinki), Dec. 1978, 4/4 (304-314). Illus. 48 ref.

"Metabolism of trichloroethylene in man. III. Interaction of trichloroethylene and ethanol". Mueller, G.; Spassovski, M.; Henschler, D. *Archiv für Toxikologie* (West Berlin), 1975, 33/3 (173-189).

"A model for toxicity evaluation of combined dangerous substances" (Ein Modell zur Toxicitätsprüfung von Schadstoff-Kombinationen). Szadkowski, D.; Lehnert, G. *Arbeitsmedizin–Sozialmedizin-Präventivmedizin* (Stuttgart), Oct. 1979, 14/10 (217-220). Illus. 6 ref. (In German)

Electrical accidents

Electricity does not cause many accidents, and still fewer fatal accidents, but nevertheless it is a serious source of potential danger. Electrical hazards, unlike many mechanical hazards, are not usually obvious: a live conductor does not differ in appearance from a dead conductor, and the lack of earthing of a metal enclosure may pass unnoticed until too late, when the metalwork is touched and found to be dangerously live. Electric shock can cause death within a few minutes.

Statistics show that, although the number of electrical accidents is usually a small proportion only of the total number of accidents in any particular field of activity (in domestic usage as well as in industrial usage), the percentage of electrical accidents which prove fatal is often much higher than the percentage of fatalities in accidents taken as a whole. This is illustrated by table 1.

The fundamental approach to electrical safety is the adoption, whenever possible, of voltages which are below the level at which lethal current can be passed through the body. Typical applications are for portable lamps and tools, for certain types of test apparatus, and for some forms of welding (e.g. d.c. welding in particularly dangerous situations). The value of "safety extra-low voltage" recommended by the International Electrotechnical Commission is 50 V. In the United Kingdom a very good record of safety has been achieved by the widespread adoption for portable electrical apparatus of 100 V a.c. derived from a double-wound

Table 1. Reported accidents[1] and fatal accidents in the United Kingdom, in all industries covered by the Factories Act (1975-77)

Type of accident	1975	1976	1977
Total accidents (incl. fatal accidents)	243 140	241 685	244 436
Fatal accidents	427	382	357
Percentage of all accidents proving fatal	0.17	0.16	0.15
Electrical accidents[1] (incl. fatal accidents)	761	788	676
Fatal electrical accidents	20	25	14
Percentage of electrical accidents proving fatal	2.62	3.17	2.07

[1] These do not include cases where the only injury was eye-flash from welding.

transformer having the centre point of the secondary (110 V) winding connected to earth. The maximum shock voltage to earth is then 55 V. For much fixed electrical equipment and some portable and transportable apparatus, the use of a non-lethal supply voltage is not possible, for technical or economic reasons. Special attention must therefore be given to other safety methods.

Statistics relating the number of accidents and fatalities to the various ranges of supply voltages in common use are available, and there is ample evidence that most accidents, in a world-wide context, occur on the a.c. electrical systems between 125 and 660 V, which are in most common use in industrial, commercial and domestic premises. This is illustrated by table 2.

Table 2. Electrical accidents in premises in the United Kingdom subject to the Factories Act, occurring within certain ranges of system voltage (1973-77)

System voltage ranges	Accidents	
	Annual average 1973-77	%
All reported electrical accidents	778	100
Accidents in the range 125-250 V a.c.	97	12.5
Accidents in the range 251-650 V a.c.	474	61.0

The important point to note is that the low-voltage ranges are quite sufficiently dangerous to cause fatal shock, despite the fact that at times shocks may be received which cause only minor discomfort and little injury.

Occupational categories of injured persons

Persons exposed to the risk of electrical accident can be divided into two main classes: *(a)* those trained and experienced in electrical work, and engaged in it as their trade or profession; and *(b)* those persons, unskilled in electricity, who use plant, machinery and apparatus energised by electricity and are at risk from fault in the electrical equipment, or by misuse of such equipment (figure 1).

Table 3 shows the distribution of electrical accidents over the various classes of employee in the United Kingdom during a 5-year period.

It might be thought that accidents to skilled persons would be very rare in view of their knowledge of electricity. However, by the very nature of their work, these persons are exposed to electrical danger much more frequently than non-skilled persons. Although it is always preferable to work on equipment which has been made dead, work on low-voltage apparatus with

Figure 1. Damage resulting from taking measurements on live switchgear with a steel rule.

conductors live and exposed is sometimes necessary. Even the most careful workman cannot guarantee that his spanner or screwdriver will never slip in such circumstances. Another common cause of accident is mistakenly switching off the wrong circuit when attempting to isolate, and then failing to test the exposed circuit before starting work on it.

Many of these accidents may cause severe burns. The short-circuit rating of most low-voltage systems is sufficient to produce violent arcing when a fault occurs. The very high temperature of the electric arc constitutes a serious risk, causing instantaneous severe burns to any person in close proximity.

The class of supervisory staff includes engineers authorised to issue and implement permits-to-work, often on high-voltage equipment. They would be responsible for isolating and earthing apparatus to make it safe for others to work on, and for returning the

Table 3. Occupational classification of personnel injured in electrical accidents in factories and industrial processes in the United Kingdom (1973-77)

Classification of personnel	Accidents (annual average 1973-77)
Electrically skilled	
Supervisory staff	32
Switchboard and substation attendants	2
Testing staff	25
Electricians, electrical fitters, wiremen, jointers and mates	254
Electrical apprentices under 18	5
Electrical apprentices over 18	20
Total skilled	338
Non-electrically skilled	
Men	386
Women	52
Total non-skilled	438

apparatus safely to its function when work has been completed. The number of such operations which need to be carried out, especially in the electrical supply industry, is very large, and the exposure of supervisory staff to risk is therefore frequent. In view of this the number of accidents is very small, but none the less disturbing. Such accidents on high-voltage equipment can give rise to severe injuries and extensive damage, as heavy arcing and oil fires are often caused.

Staff concerned with electrical testing are also exposed to risk more frequently than plant operators. Testing work is broadly of two types:

(a) testing of electrical apparatus or plant operated by electricity, in the course of production and prior to sale;

(b) testing on generation, transmission and distribution systems and on consuming apparatus, as a part of routine maintenance and for repair of breakdowns.

Testing in category *(a)* can often be so systematised, with purpose-built testing rigs and apparatus, that it becomes suitable for non-electrically skilled persons. Examples are testing of domestic apparatus and radio, television and electronic apparatus. Provided that the test equipment is properly designed and well maintained (by skilled personnel) and is used according to instructions, the risk of accident can be almost eliminated. These methods, however, do not lend themselves to the testing of higher-powered and especially higher-voltage electrical products, which require the services of fully trained personnel. Testing in category *(b)* will also require to be carried out by fully skilled and authorised persons.

Fires with electrical causation

An analysis of sources of ignition causing fires in occupied buildings in the United Kingdom for the year 1976 indicates that a total of 95 795 fires occurred. Of these, the cause of 12 219 was unknown, and a further 10 811 were due to miscellaneous causes not falling within the 26 main specified sources of ignition. Fires associated with the use of electricity were as shown in table 4.

Table 4. Fires associated with the use of electricity in the United Kingdom (1976)

Electric cooking appliances	12 897
Electric wiring, installation	5 447
Electric space heating	2 382
Radio and television	2 260
Electric blankets	1 544
Electric water heaters and washers	1 539
Electric lighting	1 258
Electric welding and cutting equipment	1 028
Fires (all premises) associated with electricity	28 355

Fires associated with cooking apparatus (irrespective of the source of heat) are usually caused by inattention during the cooking process; this may take the form of allowing oil or fat to overheat and overflow, allowing food to reach burning point, or allowing vessels containing liquids to boil dry. Probably only a small proportion of the cooking fires recorded in table 4 were caused by any failure of the electrical equipment, but such failures would in most cases involve thermostatic devices. Failure of thermostats can also result in explosion of water heaters, although the provision of relief pipes should prevent this. It is significant that, of the fires on electric cooking apparatus, 12 005 occurred in private dwellings; this was by far the largest single cause of fire in private houses.

Of the total number of fires reported, 51 133 occurred in dwelling houses and 11 559 in industrial premises or processes. Industrial fires, however, are usually more destructive of life and property than are fires on domestic premises. Nearly 500 fires in industrial premises and 33 in construction work were caused during electric welding or cutting. As with cooking fires, most fires caused by welding are not due to any electrical fault but merely to the production of great heat. Unless proper care is taken, materials such as oily rags or cotton waste, curtains enclosing welding booths, and spare clothing hung in the vicinity of the workplace, may be ignited.

Investigation of electrical accidents

Most countries now have legislation, administered by a government inspectorate or by an inspectorate of a government-nominated body (such as the electricity supply authority) for the purpose of ensuring that electricity is installed and used safely.

When starting an accident investigation, the inspector should have several aspects in mind:

(a) To decide if statutory regulations appear to have been breached, and if so, by whom. From the evidence available, a decision will have to be made as to whether to take legal action against the alleged offender. This decision will be influenced by the severity of the breach of the regulations; whether this appears to be deliberate, or due to carelessness to a criminal degree; and whether the previous record of the alleged offender in complying with regulations is good or bad. If the decision is to prosecute, the issue will ultimately be decided in a court of law.

(b) If the accident has been caused by a defect in apparatus, the inspector will decide immediately whether to prohibit the use of the apparatus altogether, or to require improvements to be made before further use, or within a given time.

(c) Where no contravention of statute law has occurred, or where the breach is of a minor or marginal nature, statutory orders to prohibit use or require improvement would not normally be appropriate. The inspector may, however, make recommendations designed to improve safety.

(d) During his investigation, the inspector will have particular regard to the level of knowledge and the standard of training and experience possessed by the injured person(s) and by those otherwise involved in the conduct and supervision of the work, in relation to their need for electrical skill to enable them to carry out the work safely. This is of special importance when electrical work is being carried out and also when young persons (in the United Kingdom, persons under 18) are concerned.

(e) Where basic defects in the design or construction of apparatus have been discovered, the inspector will ascertain whether the apparatus is of a special type, of which only a few examples may be in use, or whether it is a standard product made in large quantities. In the former case, it may be possible to trace all the users and advise them to remedy the cause of potential failure. In the latter case, the manufacturer will be asked to advise his agents and distributors of the risk, so that they can interrupt further sales and warn existing users, if necessary by published advertisements. Approaches to manu-

facturers to persuade them to improve defective designs are an important part of the inspector's work.

In order effectively to take action as described above, the inspector must obtain accurate evidence of the circumstances of the accident. In most cases the official inspector will not be the first person to arrive at the scene of the accident. Persons on the spot and members of the electrical maintenance staff, if such are available, will be there first. An official of the electricity supply authority is often the first electrically skilled person to arrive. If the accident is fatal or serious or involves fire, the police and fire service will have been called. The official inspector will probably have legal powers to interview and take sworn statements from all persons whom he considers able to give useful information. Evidence from eye witnesses, including the injured person himself, is sometimes conflicting, as those involved were not prepared for an accident to happen and are often in a state of great confusion afterwards. Occasionally deliberate misrepresentations are made in an attempt to avoid imagined responsibility. The inspector will make his own observations and carry out electrical tests if appropriate, although he will bear in mind the fact that the apparatus may not be as at the time of the accident. He will compare his findings with the evidence taken from the injured person and from witnesses and experts, and so reach his conclusions as to the true cause of the accident.

LEIGHTON, G. L.

Industry and services 1975. Health and safety report. Health and Safety Executive (London, HM Stationery Office, 1975), 92 p. Illus.

Manufacturing and service industries 1976. Health and safety report. Health and Safety Executive (London, HM Stationery Office, 1976), 80 p. Illus.

Manufacturing and service industries 1977. Health and safety report. Health and Safety Executive (London, HM Stationery Office, 1977), 84 p. Illus.

CIS 79-585 "Trends in occupational accidents due to electricity in France" (L'évolution des accidents du travail d'origine électrique en France). Lamrouche, R. *Journal de l'équipement électrique et électronique* (Paris), 25 Oct. 1978, 434 (41-44). Illus. 1 ref. (In French)

CIS 77-1166 "Investigation of electrical accidents" (Zur Problematik der Untersuchung elektrischer Unfälle). Sam, U. *Sicherheitsingenieur* (Heidelberg), Aug. 1976, 7/8 (18-23). Illus. (In German)

CIS 79-1232 "Accidents due to electricity" (Accidents dus à l'électricité). Proteau, J.; Robert, L.; Cabanes, J.; Folliot, D. *Encyclopédie médico-chirurgicale. Maladies causées par les agents physiques.* Fascicule 16515 A10, 2-1979 (Editions techniques, 18 rue Séguier, Paris), 16 p. Illus. 74 ref. (In French)

CIS 81-338 "Electrical accidents" (Erfahrungen mit Elektrounfällen). Hofmann, P. *Zeitschrift für Unfallmedizin und Berufskrankheiten– Revue de médecine des accidents et des maladies professionnelles* (Zurich), 1980, 73/1 (4-37) and 73/2 (50-60). 28 ref. (In German)

Electrical equipment industry

The electrical equipment industry produces a vast number of goods ranging from giant transformers weighing several hundred tonnes and operating at over 300 kV, through heavy industrial motors and domestic appliances such as washing machines and refrigerators, to minute precision devices weighing only a few grammes. The industry has developed rapidly in the last 20 years, and the production of domestic electric appliances, in particular, has been highly automated, with production runs sometimes reaching the hundreds of thousands. On the other hand, the production of certain heavy or specialised equipment may be carried out on a one-off basis by highly skilled technicians, and a single piece of equipment may take many months to complete.

Materials and processes. The range of materials used is extremely wide and includes: metals such as steel aluminium, magnesium, beryllium, lead, cadmium, mercury, selenium and zirconium; insulating materials (paper and plastics tape, dielectric fluids including PCBs); paints, lacquers and varnishes; solvents; acids; alkalis, etc. An increasingly large amount of plastics is being used in the manufacture of electrical equipment.

The manufacturing processes will vary from product to product but, in general, the production flow is as follows: reception of materials; quality control; storage; component manufacture; quality control; electroplating, galvanising or painting of corrodable components; quality control; storage of components; assembly; quality control; testing of finished product; storage of finished product; despatch.

The materials used in the industry must possess certain precisely defined qualities to ensure the soundness of the products and to avoid risks due to material failure. Quality requirements, if not laid down in standards, must be specified when the materials are ordered. Compliance with these requirements must be verified on delivery.

HAZARDS AND THEIR PREVENTION

Fire and explosion. Many of the solvents, paints and insulating oils used in the industry are flammable substances. These materials should be stored in suitable cool, dry premises, preferably in a separate fireproof building. Containers should be clearly labelled and different substances well separated or stored apart as required by their flash points and their class of risk. In the case of insulating materials and plastics, it is important to obtain information on the combustibility or fire characteristics of each new substance used. Powdered zirconium, which is now used in significant quantities in the industry, is also a fire hazard.

The quantities of flammable substances issued from storerooms should be kept to the minimum required for production. When flammable liquids are being decanted, charges of static electricity may form, and consequently all containers should be earthed. Fire-extinguishing appliances must be provided and the personnel of the storeplace instructed in their use. During painting and welding, special fire precautions should be taken.

Accidents. Figure 1 gives a typical breakdown, by site, of the accident injuries that may be expected in the electrical equipment industry; the hands and fingers are the site of about 45% of all injuries. In the Federal Republic of Germany, studies on accident causes and types have shown that only one accident in eight is associated with machinery; figure 2 shows that the largest number of accidents was due to falling objects, flying particles, and striking against objects. The next most frequent causes of injury were falls from ladders, falls from a height and falls on the level due to slippery floors. Electrical accidents accounted for only 1% of all accidents in this industry. The outcome of an analysis of 909 accidents involving machines and resulting in more than three days' absence from work is shown in figure 3.

It has also been found that young persons incur relatively more accidents than older workers.

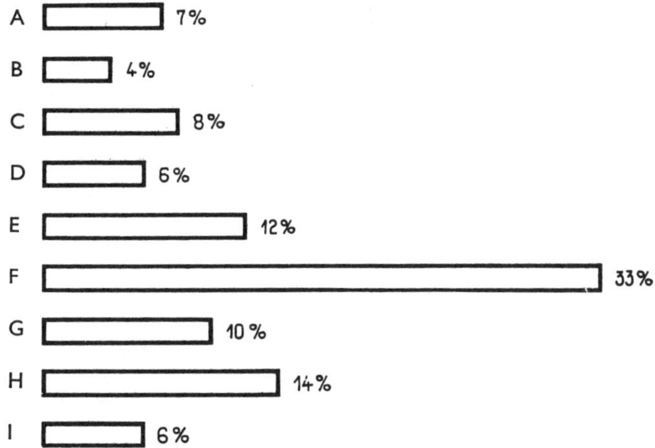

Figure 1. Accident injuries recorded in a study of the electrical equipment industry in the Federal Republic of Germany broken down by site. A. Head. B. Eyes. C. Body. D. Arms. E. Hands. F. Fingers. G. Legs. H. Feet. I. Toes.

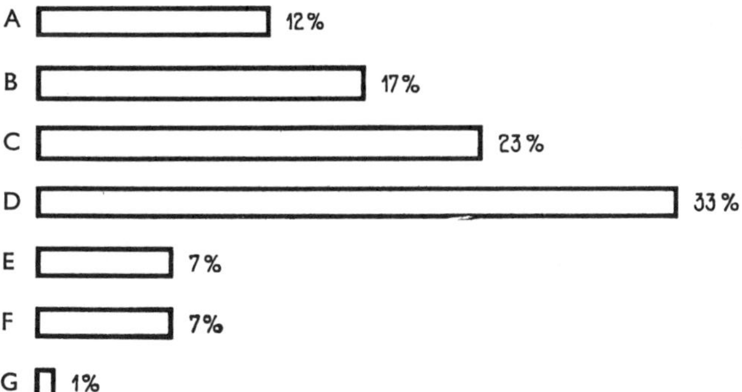

Figure 2. Accident injuries recorded in a study of the electrical equipment industry in the Federal Republic of Germany broken down by type of accident. A. Machines. B. Transportation. C. Falls of persons. D. Falling objects and striking against objects. E. Hand tools. F. Dangerous substances. G. Electricity.

Since numerous different processes are employed in the industry, the accident hazards will vary from shop to shop in the plant. During component production there will be machine hazards in the use of machine tools, power presses, plastics injection-moulding machines, etc., and efficient machinery guarding is essential; during electroplating, precautions must be taken against splashes of corrosive chemicals; accident hazards during component assembly are few; however, the constant movement of components from one process to another means that the danger of accidents due to in-plant transport and mechanical handling equipment is high.

Quality testing does not give rise to any special safety problems. However, performance testing requires special precautions since the tests are often carried out on semi-finished or uninsulated appliances. During electrical testing, all live components, conductors, terminals and measuring instruments should be protected to prevent accidental contact. The workplace should be screened off, entrance of unauthorised persons prohibited, and warning notices posted. In electrical testing areas, the provision of emergency switches is particularly advisable and the switches should be in a prominent position so that, in an emergency, all equipment can be immediately de-energised.

For testing appliances that emit X-rays or contain radioactive substances there are radiation protection regulations. A competent supervisor should be made responsible for observance of the regulations.

There are special risks in the use of compressed gases, welding equipment, lasers, impregnation plant, spray-painting equipment, annealing and tempering ovens, and high-voltage electrical installations.

Health hazards. The incidence of occupational diseases in the electrical equipment industry is low. Of the diseases that do occur, skin conditions due to solvents, cutting oils, epoxy resin hardeners and polychlorinated biphenyls are the most common. [Polychlorinated biphenyls, increasingly used as dielectric fluids mainly in the production of electrical transformers and large capacitors, can be held responsible, in addition to chloracne and other skin diseases, for liver impairment.] Health damage is also due to the inhalation of solvent vapours in painting or degreasing, etc., silica in sandblasting, lead and lead compounds, and noise. However, silicosis in sandblasters is becoming increasingly rare due to improvements in techniques and the use of corundum and steel grit or shot as substitutes for sand.

Wherever possible, highly toxic solvents [and chlorinated compounds] should be replaced by less dangerous substances; under no circumstances should benzene or carbon tetrachloride be employed as solvents. Lead poisoning may be overcome by strict application of safe working procedures, personal hygiene and medical supervision. Where there is a danger of exposure to hazardous concentrations of atmospheric contaminants, the workplace air should be regularly monitored, and appropriate measures such as the installation of an exhaust system taken where necessary. The noise hazard may be reduced by enclosure of noise sources, the use of sound-absorbent materials in workrooms or the use of personal hearing protection.

Technical prevention. Safety engineers and industrial physicians should be called upon at the design and planning stage of new plants or operations, and the hazards of processes or machines should be eliminated before processes are started up. This should be followed up by regular inspection of machines, tools, plant, transport equipment, fire-fighting appliances, workshops and test areas, etc.

Worker participation in the safety effort is essential and supervisors should ensure that personal protective

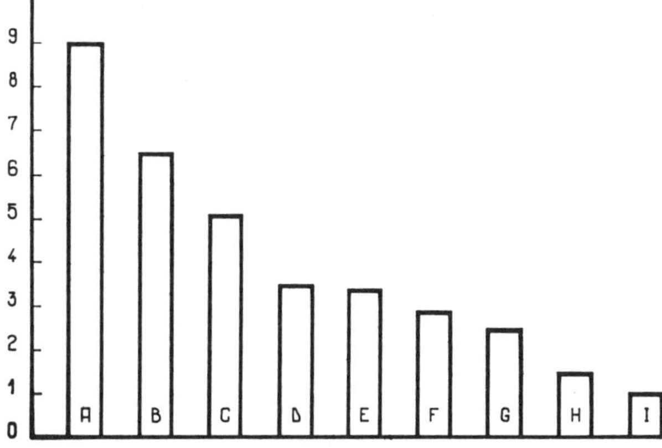

Figure 3. Relative frequency of accidents with different types of machines determined in a study of the electrical equipment industry in the Federal Republic of Germany. A. Woodworking machines. B. Metal-cutting saws. C. Punching presses. D. Drilling machines. E. Lathes. F. Milling machines. G. Grinding machines. H. Miscellaneous machine tools. I. Shears (metal and paper).

equipment is available and worn where necessary. Particular attention should be paid to the safety training of new workers since these account for a relatively high proportion of accidents.

Medical prevention. Workers should receive a pre-employment medical examination and, where there is the possibility of hazardous exposure, periodic examination is necessary. Medical examinations at 3, 6 or 12-month intervals [complemented, as far as possible, by biological monitoring] are desirable where there is exposure to acids, benzene, beryllium, cadmium, carbon monoxide, cemented tungsten carbide dusts, chlorinated hydrocarbons and polychlorinated biphenyls, chromium, cyanogen, fluorine, lead, mercury, methyl alcohol, nitrogen oxides, noise, radioactive materials, sandblasting, selenium, synthetic resins, vibration from pneumatic tools, and X-rays. The frequency of the examination should be related to the nature of the risk. The health department should be informed of the introduction of new substances into production processes so that appropriate measures can be taken. The collaboration of the industrial physician is also of value in the design of new processes, the organisation of welfare facilities and the inspection of operations involving the use of dangerous substances.

Adequate first-aid facilities should be provided and all accident injuries should be recorded and investigated, and the cause of the accident analysed.

KLOST, W.

Safety and health, general:

CIS 74-1766 *Compilation of statutory instruments in force concerning occupational safety and health in the electrotechnical industry* (Sbornik dejstvujuščih pravil tehniki bezopasnosti v ėlektrotehničeskoj promyšlennosti). Saburov, N. Ja. (Moscow, Izdatel'stvo "Ėnergija", 1973), 584 p. Illus. (In Russian)

Special hazards:

CIS 74-1309 "The hazards of chronic lead poisoning in coating electric resistors with a lead-base enamel" (Riziko olova při střikání elektrických odporů olovnatými smalty). Jindřichová, J.; Medek, V.; Boštík, V.; Eminger, S. *Pracovní lékařství* (Prague), June 1973, 25/6 (238-242). Illus. 8 ref. (In Czech)

"Occupational exposure to polychlorinated biphenyls in electrical workers. I. Environmental and blood polychlorinated biphenyls concentrations". Maroni, M.; Colombi, A.; Cantoni, S.; Ferioli, E.; Foà, V. *British Journal of Industrial Medicine* (London), Feb. 1981, 38/1 (49-54). Illus. 12 ref.

"Occupational exposure to polychlorinated biphenyls in electrical workers. II. Health effects". Maroni, M.; Colombi, A.; Arbosti, G.; Cantoni, S.; Ferioli, E.; Foà, V. *British Journal of Industrial Medicine* (London), Feb. 1981, 38/1 (55-60). Illus. 27 ref.

CIS 77-1263 "Synergistic effects of mica dust and resins in the manufacture of micanite products" (K voprosu o kombinirovannom dejstvie pyli sljudy i smol, primenennyh v proizvodstve mikanitovyh izdelij). Dianova, A. V.; Kočetkova, T. A.; Rumjancev, G. I. *Gigiena i sanitarija* (Moscow), July 1976, 7 (33-37). Illus. 4 ref (In Russian)

CIS 79-407 "Two examples of noise suppression" (Deux exemples d'élimination du bruit). *Travail et sécurité* (Paris), Oct. 1978, 10 (516-519). Illus. (In French)

Electrical installations, fixed

Depending on the installed power, an industrial plant will be supplied from a high-voltage transmission line (e.g. 110 kV supply for a refinery, an iron and steel works or a shipyard), from a medium-voltage primary distribution feeder (6-60 kV for small- and medium-sized plants) or from the low-voltage service line (110, 125, 220, 380, 500 or 660 V for small plants or workshops). Increasing electrification favours the use of medium-voltage supplies, except for small artisanal workshops. Alternating current at 50-60 Hz is used predominantly.

Power is distributed inside the plant by means of cables; lead-sheathed cables have so far been preferred for low and medium voltages but aluminium sheathing is now becoming more popular since it is of low weight and cost and, in low-voltage systems in particular, the aluminium sheathing can be used as the return

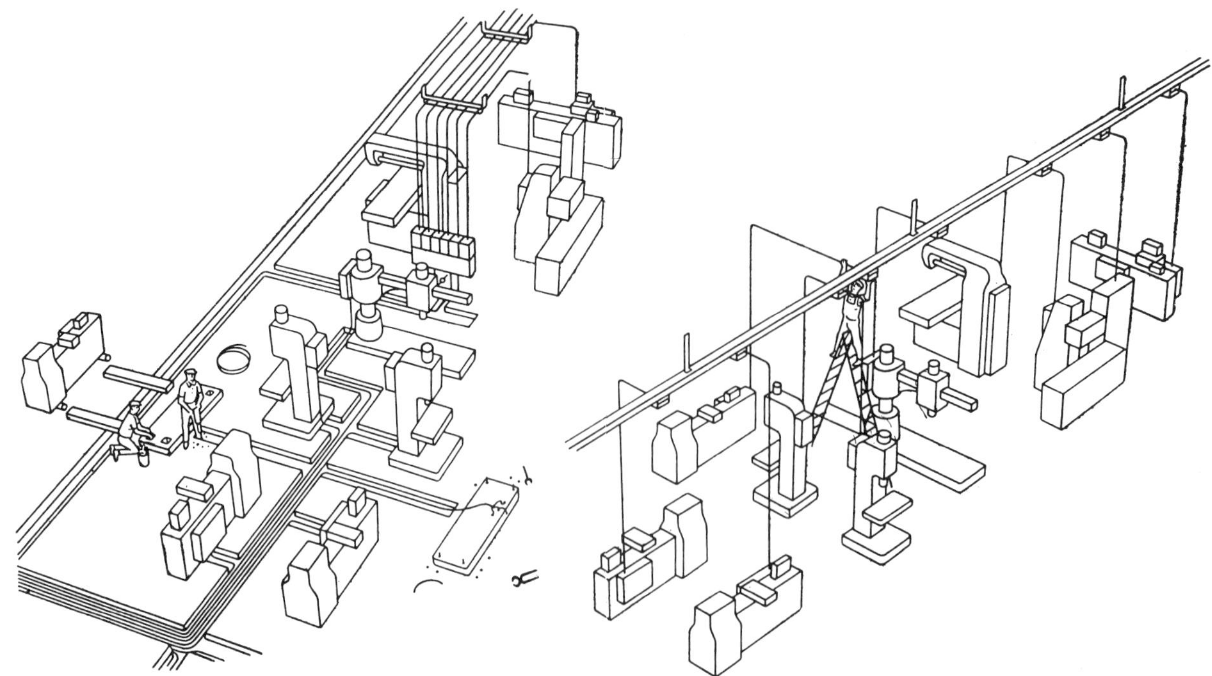

Figure 1. Low-voltage supply by means of an overhead busbar installation *(right)* has outstanding safety, economic and operating advantages over the conventional buried installation *(left)*.

conductor. Plastics-sheathed medium- and low-voltage cables do not usually require cable end seals; this can save space and money, especially where a large number of short cables are needed.

Installation layout should follow traditional well tried principles. It is fundamentally wrong and undesirable from both the economic and safety point of view to adopt unnecessary, novel layouts.

SAFETY REQUIREMENTS

The following requirements must be met if an installation is to offer a satisfactory degree of safety.

(a) All components must be adequately dimensioned to withstand design stresses.

(b) Normally live parts should be insulated in an effective and durable manner, or positioned in such a way (i.e. out of reach, behind screens, etc.) that they cannot be touched. In certain circumstances, this requirement can be dispensed with where live parts carrying voltages below 1 000 V are housed in special, closed premises to which only authorised qualified electricians have access.

(c) Conductive parts of electrical equipment that are accessible but do not form part of the operating circuit (e.g. metal casings), but which may accidentally become live, must be protected so that they do not rise to a dangerous voltage (i.e. 50 V in industrial use and 24 V in agricultural use) (see also LIVE WORK, LOW VOLTAGE).

Overload protection

The rated current capacity of conductors and equipment must be sufficient for the design currents to which they will be subjected. Compliance can be checked by means of load tables.

Over-current protection equipment (e.g. fuses, automatic circuit breakers, thermal overload relays) should be installed to disconnect the system in the event of an overload. This protection equipment must be dimensioned so that it can deal with a short circuit in a whole or part of the downstream network (i.e. break currents of, for example, 1 200, 2 000 or even 10 000 A depending on total plant consumption) and not just the rated current of one or more consumer appliances. If breaking capacity is not adequate, arcing may occur and cause fire and explosion damage or injury to workers. In large plants, protection should be graded and each overload protection device have a rated current one or two stages higher than its downstream neighbour to ensure adequate protection and discrimination in the event of a short circuit (i.e. only the breaker immediately upstream from the fault is tripped; the effects of the fault are immediately localised and other circuits and appliances continue to function). Protection against water should also be selective and matched to the type of hazard, i.e. installations may need no water protection at all or may be drip-proof, splashproof, jetproof or totally waterproof.

Decisions on the type of overload protection that should be built into the system are important factors in installations design. It should also be ensured that under normal operating conditions the voltage drop between the meter or the low-voltage distribution panel and the consumer appliances (e.g. lighting, electric tools) does not exceed 1.5% of the rated voltage; where the consumer appliances are motors, however, a voltage drop of 3% can normally be accepted.

Protective covering

This will normally be achieved by covering or screening. A typical example is the overhead busbar system (figure 1), a highly suitable distribution arrangement for factories, which offers considerable safety and operating advantages provided the danger of moisture is minimal. If production requirements necessitate the movement of a machine to another area and modification of the electrical connections, it is easy to remove the cutout box and fit it to another point on the busbar (figure 2).

Figure 2. In operation the busbar distribution system makes it possible to ensure reliable current supply even with rapid fluctuations in production and frequent shifting of machines, without dangerous connecting and disconnecting in switchgear cubicles and without costly relaying of conductors.

Worker safety is further increased if the cutout-box fuse can be removed and replaced only when the busbar is dead. Motors of machines are fed from the cutout box through flexible tough-rubber or plastics-sheathed multicore cables or, in the case of heavier currents, through individual cables in steel or plastics conduit (figure 3). Where necessary, overhead busbar systems can be used to feed incandescent or fluorescent lamps—and this enables a substantial saving in material and installation time.

Housings for switchgear, cutout boxes, distribution boards, plug sockets and other installation equipment should, wherever possible nowadays, be made from insulating materials. If covers are made from transparent insulating material, it will be possible to keep all appliances and connections under observation during operation and this may speed up fault location; in addition, the danger of a live conductor accidentally energising a metal housing and producing a dangerous contact voltage is thus eliminated.

Contact-voltage protection

The insulation mentioned above offers a high degree of safety in this respect and, what is more, it is not required to meet the high mechanical strength specifications necessary for portable electric power tools. Apparatus enclosed in an insulating casing (i.e. virtually double insulated), such as distribution boards and switchgear, can nowadays be constructed for virtually all amperages encountered in industry. The only limiting factor is the over-all size of the insulating enclosure. A similar trend can be observed in the case of high-voltage switchgear (at the moment up to 30 kV). The potting of small equipment ensures protection against direct contact and

Figure 3. Busbar with adjustable outlets in a press shop.

accidental contact between live conductors and metal casings or control levers, etc.

Extra-low voltage. For small current-consuming equipment such as portable lamps and portable electric-power tools, an extra-low voltage supply (maximum 50 V) offers effective protection against dangerous contact voltages; in agriculture, the maximum is reduced to 24 V (the same as for children's electric toys). Extra-low voltages should be obtained only through one of the following:

(a) an accumulator or a galvanic element—an accumulator battery of higher voltage must not be used;

(b) a transformer with electrically separated windings and additional insulation to prevent a current breakthrough between the two autotransformers (in which part of the primary and secondary winding is common) should not be used for extra-low voltage supply;

(c) a motor-generator set (converter) with electrically separated windings.

Extra-low voltage installations in industry and agriculture should be insulated for voltages of at least 250 V and plugs should be designed to ensure they cannot be accidentally inserted into mains voltage sockets.

Isolating transformers. These are valuable in certain particularly dangerous locations (e.g. for work in boilers, pipelines, etc., where there is limited freedom of movement). Here the consumer appliance is supplied from the mains through an isolating transformer (or a motor-generator set) so that in the event of a fault to earth, a dangerous contact voltage cannot develop. With this system it is necessary to avoid earthing the secondary side and, consequently, only one consumer must be connected to each transformer or generator set; this requirement may be modified in certain circumstances when at least equivalent protection is assured by another means.

Wherever possible, preference should be given to the three measures described above, since these require no protective conductor and have outstanding safety characteristics. When these measures are not practicable, a system with a protective conductor must be employed.

Protective earth (PE). The protective earth conductor is the oldest form of protective conductor and is normally suitable only for small current-consuming plant; only when connected to an extensive water supply network made from conductive material can the protective earth also offer advantages at higher amperages. The effectiveness of this sytem is limited by the fact that when the maximal contact voltage (50 V) occurs, the tripping current must flow through the protective device immediately upstream in order to isolate the consumer appliance in which the hazardous condition exists.

The resistance of the protective earth can be obtained from the formula:

$$R_s \leqslant \frac{50\ \text{V}}{I_A}$$

where I_A is the tripping current of the protective device. It can therefore be seen that, with a tripping current of 75 A, the earth conductor must have a resistance of less than 1 Ω if the tripping time is not to be excessive.

Protective earth neutral (PEN). With this arrangement, the fault current is not returned to earth directly via the earth conductor but along an earthed conductor which forms an integral part of the supply system (in three-phase systems this conductor is usually connected to the neutral point of the network). The resistance of the circuit is, in this way, considerably reduced and, consequently, neutral earthing offers adequate protection against indirect contact with rated current intensities up to several hundred amperes provided the installation is well designed. When the earthed neutral has a conductor cross-section of less than 10 mm² of copper, it is usually divided into a normally current-carrying conductor and a normally dead protective conductor; the latter is connected to the accessible parts of the equipment that are electrically conductive. This arrangement of the neutral conductor, which is imperative in places with a fire risk, has substantial advantages from the safety standpoint: since a break in the neutral conductor cannot occur, disturbances in the operation of other parts of the equipment can often be avoided.

Fault-voltage protection. This also employs a separate conductor. Any contact voltage that develops between the appliance enclosure, or its associated protective conductor, and an auxiliary earthing conductor is monitored by a relay. When a pre-set threshold voltage, which is less than the maximum permissible contact voltage, is reached, the relay is actuated and isolates the appliance in which the dangerous condition has developed. Tripping by extraneous influences, and inadvertent bridging of the trip coil in normal operation, are circumstances that may adversely affect the application of this form of protection.

Protective earth with fault-current protection (differential protection). This system does not have the disadvantages of fault-voltage protection. It employs a normally dead protective conductor and can be used with virtually any system without exerting or being subject to any adverse influences. It operates on the principle that, at the neutral point of any system, the sum of all feed and return currents is zero. If the conductors supplying an appliance pass through a toroidal-core transformer, the magnetic flux in the core will also be zero so long as the connected consuming appliances are free from faults and provided no fault current flows through the appliance enclosure and the protective conductor connected to an auxiliary earthing conductor by another route, namely over the earth. Should such a fault current develop, a magnetic flux is generated in the toroidal core and this induces a voltage in a secondary winding, which, on reaching a pre-set level, actuates a circuit breaker and isolates the faulty appliance.

By its nature, fault-current protection is confined to a.c. installations but, apart from this limitation, which is of little importance in view of the rarity of d.c. installations, this form of protective-conductor leakage protection offers one of the highest degrees of safety against indirect contact. It is used in both fixed and temporary installations and, in many countries, it is virtually compulsory in such places as building sites. The trip current (rated fault current) will be 0.03, 0.3, 0.5 or 1 A according to the circumstances. The 0.03 A switch has the advantage, in certain circumstances, of providing protection in cases of direct contact and of ensuring excellent fire protection. Circuit breakers with even smaller rated fault currents, i.e. down to 5 mA, have little practical importance because they tend to cause unwanted tripping.

Protective-conductor systems (PE). With such a system, applicable only to restricted installations such as factories, all conductive enclosures, accessible metal parts of appliances and all conductive structures that do not form part of the electrical installations (e.g. rails, structural building components) in a given premises or area, are bonded together by a protective conductor; this ensures perfect equalisation of potential. Since the neutral point of the supply transformer is not earthed electrically, no danger can arise for the users if a single fault occurs in the plant, and at the same time the need to switch off the plant or a particular appliance is avoided. A built-in earth-fault indicator signals faults so that a search for the source of the disturbance can be begun in good time.

Installation testing

Before an installation is taken into service it should be tested by the contracting electrician. The tests should include:

(a) mechanical tests of the entire wiring system, and all appliances, switches, distribution switchgear and accessories;

(b) operational tests of the switching, indicating, regulating and control devices and the associated operating appliances, including tests of the condition of the insulation of the installations;

(c) tests of the effectiveness of the protective measures against excessive contact voltages.

This initial testing of a new or extended installation should, in principle, be carried out by a qualified electrician provided by the contractor; in special cases, testing by an independent, possibly officially recognised, expert is also necessary. Installations in fire-risk and explosion-risk areas and in public places require special testing and, in certain cases, acceptance testing by accident and property insurance institutions are required even for ordinary new installations. Installations cannot be put into service until these tests have given satisfactory results. Subsequent tests of electrical installations should be undertaken at intervals suited to the operating conditions; for ordinary industrial undertakings, intervals of 2-4 years may be considered adequate. If necessary, heavy-duty installations and installations exposed to abnormal risks must be tested more frequently. The choice of the testing official should be governed by the same requirements as for the initial tests.

Safety in low-voltage distribution systems

It is not infrequent, for reasons of economy and simplicity of distribution, that in low-voltage distribution systems for public and industrial purposes no special provision is made behind the transformer for circuit breakers or even overload protection. The use of fuses as a tripping mechanism is generally preferred.

This procedure presents certain hazards, particularly when dealing with heavily loaded or direct current circuits. The slow operation of screw-in or clip-type fuses often results in arcing or flashover followed by burning of the contacts, and this can be dangerous for the service personnel concerned. The danger becomes even more acute as the system becomes more heavily loaded or more involved.

At the present time, a large number of low-voltage distribution systems are still to be found which depend on this type of open protection. Certain safety precautions are necessary to protect the service personnel, and the following measures should be observed:

(a) the work must be carried out by experienced electricians who are fully familiar with the technical aspects and the possible dangers that exist, such as arcing. The danger of coming accidentally into contact with live components should be emphasised in all instructions or information given to the persons concerned;

(b) handles or switches for actuating the equipment must be so constructed that the cut-out or fuse cannot be inserted or removed by using makeshift tools. The handles should be provided with hand and lower arm protection of leather or other heat-resisting material. Protection should be afforded at least up to the elbow (figure 4);

(c) the essence of the problem, particularly in the planning stage, is to ensure that as far as possible all existing installations (and certainly all new installations) are laid out in such a way, and technical measures provided, so that arcing will be prevented.

Figure 4. Personal protection of hands and arms to be used against arcing hazards.

This means that overload protection and, whenever possible, protection against accidental contact should be installed (see example in figure 5).

In an installation for low-voltage distribution such as that shown (figure 5) it is not possible to come into accidental contact when the circuit is alive. The fitting of flashproof boxes, which when opened cause the current to be cut, enable this danger to be largely eliminated.

Suitable apertures in the cover permit the introduction of the end of a voltage tester, which, for example, makes it possible to check whether the contacts are alive or not.

Personnel selection

The correct selection and training of electrical fitters employed on the installation, alteration and maintenance of electrical plant are of particular importance. Such work should be done only by a qualified electrician or under his supervision, since specialist experience and familiarity with the relevant regulations are necessary if the work is to be done effectively and all possible dangers recognised. Technical training should be systematic and conclude with an examination. In exceptional cases, several years' practical experience in a particular technical field, such as that of high or low voltage, may suffice to qualify an electrician. A fundamental condition for the safe and reliable operation of electrical equipment is the absolute prohibition of interference with the equipment by persons other than electricians.

Figure 5. Overload protection to prevent arcing.

EGYPTIEN, H.

Modern and statutory electrical installations (Die neuzeitliche und vorschriftsmässige Elektroinstallation). Hösl, A. (Heidelberg, Verlag Dr. A. Hütig, 1977). (In German)

CIS 78-1515 "Protective relays". Hahn, C.; Ungrad, H.; Wildhaber, E.; Vitins, M.; Gantner, J.; Ilar, F.; Ilar, M.; Zidar, J.; Fiorentzis, M.; De Veer, C.; Moser, H.; Hager, W.; Wanner, R.; Magajna, P.; Franzl, M.; Narayan, V.; Ritter, F. *Brown Boveri Review* (Baden, Switzerland), June 1978, 65/6 (345-419). Illus.

CIS 79-18 *Protective measures against excessive contact voltage in low-voltage installations* (Schutzmassnahmen gegen zu hohe Berührungsspannung in Niederspannungsanlagen). Müller, R. (Berlin, Verlag Technik, 6th ed., 1977), 364 p. Illus. 183 ref. (In German)

CIS 79-636 *Safety hints for work on low-voltage electrical equipment and installations* (Conseils de sécurité pour interventions et travaux sur les équipements et installations électriques de la classe basse tension). Edition INRS nº 539 (Institut national de recherche et de sécurité, 30 rue Olivier-Noyer, Paris) (Apr. 1978), 77 p. Illus. (In French)

"Comments on the Decree of 14/11/1962 concerning the protection of workers in establishments where electric current is used" (Recueil des commentaires du décret du 14 novembre 1962 relatif à la protection des travailleurs dans les établissements qui mettent en œuvre des courants électriques). *Bulletin officiel du Ministère du travail et de la participation* (Paris), Fascicule spécial N.79-38 bis, 1979, 76 p. (In French)

Electrical installations, temporary

Numerous machines in agriculture, building, civil engineering, etc., have to be operated at a distance from a permanent electricity supply. Such machines include circular saws, fodder conveyors, concrete mixers, cranes, etc. In certain cases preference will be given to machines powered by an independent internal-combustion engine; however, in others the operator may prefer electrically powered equipment operated from a temporary installation.

This type of electrical equipment is subject to a number of special factors: as the work progresses, the equipment

Figure 1. The rough conditions prevailing at construction and assembly sites require special measures to ensure that all plant is reliably supplied with current.

may be moved to different locations and subjected to different stresses, and there may be exposure to inclement weather and rough treatment. The voltage supply may range from 110 to 250 V either to the earthed neutral conductor or between unearthed active conductors. The current is three-phase 50 Hz in Europe, for example, and three-phase 60 Hz in the United States. For certain large machines, such as heavy cranes, operating voltages may be up to 500 V. Direct current is used less and less.

SAFETY MEASURES

For practical reasons, supply by underground cables is virtually impossible; however, in certain cases, high-voltage underground cables may be employed to supply a substation for a large, long-term project. Where overhead lines are employed, the supporting poles on building sites, for example, should be designed to withstand harsh treatment and may require a steel-reinforced base (figure 1). Overhead conductors (insulated or not) should be positioned so that they cannot be touched by persons working on the site or on scaffolds; where mobile cranes or other tall equipment are moved frequently across a building site, overhead lines should be protected by "goalposts" (figure 2).

Figure 2. Overhead electric lines protected by "goalposts" to ensure safe passage of cranes underneath, and by fences to prevent access from the sides.

Trailing cables

The stipulations for trailing cables vary from country to country, but minimum requirements include multistrand insulated conductors with textile tape covering and an external sheathing of tough, oil-resistant and non-combustible rubber or similar elastomer. All trailing cables should be flexible, i.e. individual conductors should be of the multistrand type, which can withstand frequent and even violent bending stresses. Thermoplastic insulation should not be used for trailing cables since its flexibility is reduced at low temperatures, and brittleness and cracking may occur; in addition, the strength of the insulation is reduced at high temperatures and this may constitute a further hazard. Where a trailing cable would be exposed to excessive stress and wear, i.e. when passing across a traffic lane, it may be run overhead; however, a minimum clearance of 5 m should be ensured between the ground and the lowest part of cable passing over an approach road. If it is necessary to make a joint in an overhead line, the joint should be protected in such a way that it is not subjected to tensile loading; knotting of the joint does not provide adequate protection. In most cases medium-strength, flexible, rubber-sheathed cable will be adequate for the leads of small electric power tools and equipment (drills, portable pumps, etc.).

Equipment design

Installation equipment, including lamps, should be designed for safe operation in humid or wet conditions; quite often, drip-proof construction will prove adequate, but under unfavourable conditions, splash-proof or hose-proof construction may be necessary. This equipment (including plug-and-socket connections) should, where possible, be made from insulating material. Where different voltages, numbers of poles, amperages and frequencies are used on a single site, a different plug design should be used for each combination, in order to avoid confusion and hazardous conditions. Three-phase plug-and-socket connections should always be connected to the supply network in the same phase sequence; this will ensure that unskilled persons are not tempted to rearrange the sequence when a motor is found to be running abnormally.

In order to avoid wrong connections being made, it is recommended that a specially designed plug and socket outlet or connector be used on building and construction sites. Reference should be made in this connection to the relevant recommendations of the International Electrotechnical Commission. The socket outlet is provided with a keyway on its face into which a corresponding lug or key on the plug must fit. The key and the keyway are given a fixed setting according to the nature of the power supply (a.c., d.c., voltage, frequency). In order to ensure that the correct contacts are made, the position of the earth contact-tube is related to a 12-hour clock dial. For example, if the required connection was 380 V, 50 Hz, the keyway would be set at the 6 o'clock position. If the appliance or accessory required 110 V, 50 Hz, the corresponding setting would be at the 4 o'clock position (figure 3).

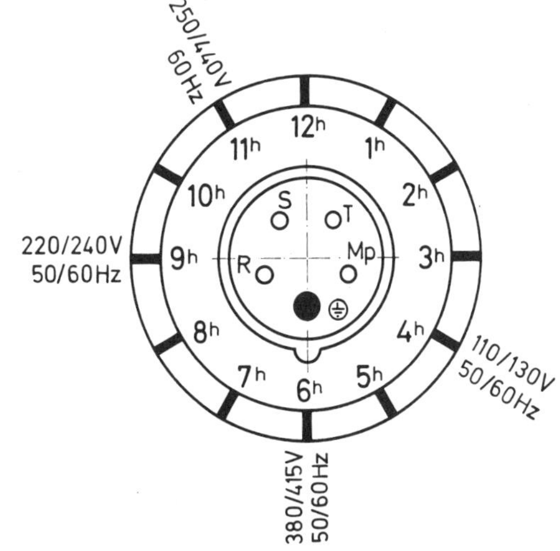

Figure 3. Front view of contact tubes of socket outlet or connector according to DIN 49 462/463.

Appliances driven by an electric motor should be fitted with an omnipolar "on-off" switch; in the case of portable electric power tools, the plug-and-socket device can be considered an omnipolar disconnector with the "on-off" function being performed by a single-pole switch.

EXCESSIVE CONTACT VOLTAGE

Protection against excessive contact voltage requires particular attention in temporary installations. It is considered that, in the event of a fault, 50 V is the highest contact voltage that should be permissible between accessible live parts not forming part of the circuit and other earthed parts. However, in some countries, including the Federal Republic of Germany, the maximum permissible voltage is now 65 V. All electrical equipment on building and construction sites should be supplied from a central distribution point to ensure maximum protection against the development and persistence of excessive contact voltages. Before an installation is taken into service, the installer should make measurements to check the effectiveness of the protection. One of the following types of distribution point should be used:

(a) a building-site distributor, i.e. a transportable switchboard, of at least splash-proof construction, fitted with the necessary terminals, fuses, indicating devices and plug-and-socket connections—this distributor should be used for supplying only the construction site;

(b) a stationary transformer with a low-voltage winding that supplies only the building site—such an installation must meet any special building-site regulations;

(c) a stationary switchboard, in a plant for example which is used wholly or partially for supplying the building site—a distribution arrangement of this type should be suitably marked. If only a limited number of the panel outlets are used to supply the building site, these should be clearly indicated.

The efficacity of the installation protection should be checked before entry into service and at intervals of no less than four weeks thereafter; by preference, the safety of the installation should be verified every week or every day.

In the case of special building-site transformers or outlets from fixed distribution gear, it is also possible to use the following protective measures: protective earthing; connection to neutral; and fault-voltage relay protection. This is permissible when the current-supply network is confined to a limited area and can be kept under surveillance, as for example within the factory grounds, in order to make available to a building site the safety characteristics of a network constantly supervised by specialists. Independently of these considerations, attempts should be made in agriculture, building and construction work to use only fault-current relay protection, perhaps combined with double insulation, extra-low voltage and the use of isolating transformers, since such an arrangement offers considerable safety advantages.

Fault-current relay protection. Experience has shown that certain safety measures adopted for temporary electrical installations are not fully effective. Building-site distributors of the type described above used on sites where the conditions are particularly rough should use only the following types of protection: double insulation; extra-low voltage protection; fault-current circuit breakers; supply through isolating transformer (figure 4). This will also ensure that protective measures will not interfere with one another or even cancel each other out. If all distribution equipment on a building site is fitted with fault-current protection, there can be neither feedback to the supply system nor adverse effects on consuming appliances connected downstream from the distributor and protected by double insulation or extra-

Figure 4. Mains distributor panel with mains fuses and fault-current circuit breakers, showing fuses and socket connections for electricity distribution.

low voltage, etc. Moreover, building-site distributors fitted with fault-current relay protection can be moved easily from site to site and connected with any type of supply network.

Building-site distribution panels with one or more fault-current relays constitute the fundamental measure in all efforts to prevent electrical accidents on temporary electrical installations. Fault-current relays should have a maximum rated fault-current of 0.5 A; this ensures that, at the maximum permissible contact voltage of 50 V, the maximum permissible earth resistance will be 100 Ω (or 48 Ω for the maximum permissible contact voltage of 24 V in agriculture).

It should not be forgotten, however, that if the 0.5 A tripping current resulting from a damaged insulation flows through the human body for more than 0.2 s, cardiac fibrillation may be induced. A fault-protection relay with a lower rating (30 mA) overcomes this drawback. Firstly, the earth resistances required are much lower (a maximum of 1 660 Ω at 50 V and 800 Ω at 24 V contact voltage) and, secondly, dangerous conditions will be prevented both in cases where there is a break in the protective conductor and the occurrence of an excessive contact voltage, and possibly also when direct contact is made with a live conductor with a voltage to earth.

Admittedly, the 30-mA fault-current circuit breaker will be tripped by faults and stray currents that would not be noticed with other protective measures; however, once these potentially dangerous conditions have been eliminated, experience has shown that these circuit breakers work faultlessly.

Confined spaces

Where portable electric equipment such as handlamps and power tools is used in confined spaces like boilers, tanks and so on, special precautions against excessive contact voltage should always be taken. The most important are low or extra-low voltage (maximum 50 V) and circuit separation with every consuming appliance provided with its own isolating transformer. In the case of handlamps, preference should be given to low or extra-low voltage, whereas with handtools, which usually require a high power input, isolating transformers will normally be used.

Qualified personnel

The erection, alteration, maintenance and testing of electrical installations should be entrusted to a qualified electrician and, consequently, unqualified building or agricultural workers should be given strict instructions not to try to carry out running repairs or adjustments to electrical installations or appliances. An electrician should always be on duty on a large building site; on smaller worksites, if a defect occurs in the electrical installation the affected part must be taken out of use until it can be restored to good order by an electrician.

EGYPTIEN, H.

Electrical safety: portable tools and mobile appliances. Occupational safety and health series No. 18 (Geneva, International Labour Office, 1969), 234 p. Illus.

Plugs, socket-outlets and couplers for industrial purposes. Publication 309 (Geneva, International Electrotechnical Commission, 1969), 34 p. Illus.

First Supplement to Publication 309 (1969). Publication 309A (Geneva, International Electrotechnical Commission, 1973), 99 p. Illus.

Electric cable manufacture

Cables carrying eletricity from the distribution point to the consumer can be divided into two basic types: bare and insulated. The former are used in installations erected in the open and the latter inside buildings and underground, etc.

Materials. Both types of cable use copper or aluminium as conducting materials; however, special combinations of these materials such as aluminium and steel and copper-covered steel, etc., are also used in certain cases.

Continuous progress in insulation technology has made available a wide range of insulating materials with different electrical and mechanical properties: cables are protected by various sheaths, some of them providing insulation, others such as lead and steel affording protection against external agents. Many countries require conformity with strict specifications on insulation characteristics.

Some of the insulating materials commonly used are oil for high voltages; paper (impregnated with mixtures of mineral or synthetic oils of varying viscosity) for medium voltages; rubber and plastics for dry cables. Some of the typical insulating materials for dry cables are polyvinyl chloride, polyethylene, and polyamides, natural and synthetic rubber and polymerised chlorobutadiene.

Production. The processes involved in the manufacture of bare copper conductors consist essentially in heating copper bars to red-heat and hot-rolling them on rod rolling mills; after cooling, pickling and descaling, the rod is drawn to obtain the required gauge. The cable may consist of a single strand of wire or of a number of strands twisted or braided together on special cable-making machines.

To produce a sheathed cable, a bare single- or multi-strand conductor is coated or covered with insulating material by wrapping, dipping or brushing.

Electric cable testing. Tests are carried out on the raw materials and on the cables themselves during and following production.

(a) The tests performed on the raw materials cover the conductors, the constituent materials of the insulation and the whole insulation, to ensure that they all have the required characteristics.

(b) The purpose of the tests performed during manufacture is to ensure that the thickness of the insulation is uniform. Other tests measure voltage drop and dielectric strength, etc.

(c) On the finished cables, destructive testing is carried out on samples to determine tensile strength, elongation, flame resistance and chemical and electrical properties, etc.; non-destructive testing includes voltage tests and the determination of dielectric strength and losses, etc., to ensure the absence of manufacturing defects and conformity with specified standards.

HAZARDS AND THEIR PREVENTION

The production of wire for electric cables entails the same hazards as those found in a wireworks. However, the treatment and insulation of the wires present certain specific health hazards directly related to the materials used.

Accidents. Many of the accident risks are the same as in other manufacturing processes (lifting and carrying, impacts and collisions, falling objects and falls of persons). Special machinery hazards are presented by transmission shafting and gearing, nips between wire and drawing blocks, and by wires and cables that may snap and whip. These require secure machinery guarding, preferably by the enclosure of dangerous parts. Metal mesh screens which allow good visibility but prevent access are suitable for many of the machines used (figure 1). Maintenance of machinery and equipment such as cleaning, lubricating, etc., should be carried out only when the machine has been stopped and, wherever possible, interlocking of guards is desirable. Guards should be fitted at die blocks and drawing blocks and automatic stopping devices should be installed on twisting and braiding machines to prevent the cable whipping out in the event of a breakage.

Figure 1. Guarding of a cable-making machine. *(By courtesy of Pirelli S.p.A.)*

One of the principal risks is that of flying metal particles entering the eyes. The basic principle for the prevention of such accidents should be directed towards the points from which the particles are emitted, making use of screens and of extraction or collection systems. Nevertheless, the use of safety spectacles or goggles should be obligatory. These should fulfil two conditions: they must be effective (resistant to impact, to heat and moisture, and of good optical quality), and they must be acceptable to the wearer (comfort, appearance, etc.).

The education of the workers and their active participation in ensuring their use is important.

Maintenance and good housekeeping can do much to reduce the number and severity of accidents. Pre-employment medical examination can ensure that workers are allocated to tasks that match their physical capabilities.

Diseases. Copper and aluminium, as received at the factory, present little hazard although the handling of copper may cause skin disease in sensitive persons. However, this dermatitis rapidly regresses when the affected worker is removed from exposure.

Many of the materials used in sheathing and insulating processes are hazardous, for example lead, plastics, oils and solvents. It has not been found possible to replace lead by a less toxic material in cable production and it is therefore necessary to prevent dangerous concentrations of dust and fume. The metal should be melted in a closed system fitted with efficient exhaust ventilation and the workplace should be kept scrupulously clean. Atmospheric monitoring should be carried out to detect and determine lead concentrations. Strict personal hygiene, good washing and sanitary facilities, provision of suitable working clothes and hand and arm protection for handlers of lead pig, prohibition of eating or smoking in the workroom, are all important; suitable laundry facilities for working clothes should be provided. Pre-employment medical examinations should be carried out to ensure that persons suffering from disorders of the blood, liver, kidneys or nervous systems are not exposed to lead. Periodic medical examinations should include determination of urine and blood coproporphyrins.

Risks arising from the use of rubber, plastics, oils and solvents are dealt with at length in other articles. Benzene should never be used as a solvent and strict attention should be given to ventilation when other solvents and plastics are used. Dermatitis may be incurred in the handling of some of the materials.

Some of the manufacturing processes produce intermittent low-frequency noise which is sufficiently intense to have a harmful effect on hearing. Measurements taken in wire drawing, wire twisting and braiding and in cablemaking shops at an electric cable factory revealed noise levels of 90-102 dB; these are distinctly higher than the limit considered tolerable.

Physical methods of noise control by enclosure or absorption are difficult to apply to the processes, and preventive measures consist mainly in the use of hearing protection equipment and by the provision of rest periods; however, some workers definitely dislike earmuffs and helmets. Audiometry should be included in pre-employment and periodical medical examinations to provide data for the detection and evaluation of occupational noise-induced hearing loss. Audiometry should be carried out more frequently at the beginning of employment in order to ensure early diagnosis of any permanent shift in the auditory threshold.

MALBOYSSON, E.

Work-related diseases:

CIS 78-476 "Working conditions and their influence on gynaecological parameters in women manufacturing enamel-insulated wire" (Uslovija truda i vlijanie ih na nekotorye specifičeskie funkcii ênščin, zanjatyh v proizvodstve êmaljprovodov). Syrovadko, O. N.; Malyševa, Z. V. *Gigiena truda i professional'nye zabolevanija* (Moscow), Apr. 1977, 4 (25-28). 4 ref. (In Russian)

Safety:

CIS 423-1970 *Occupational safety in cable manufacture* (Arbeitsschutz bei der Kabelfertigung). Maibaum, B. (West Berlin, Verlag Tribüne, 1968), 87 p. Illus. (In German)

Occupational health:

"Occupational health at Hidroelectrica Española, S.A." (La medicina de empresa en Hidroelectrica Española, S.A.). Malboysson Correcher; Alvarez Blanco; Bonell Sendra; Gonzalez Llombart; Monzarbeitia Loiti; Ortis Lopez; Rabasa Bayona; Serra Gimenez; Trabuchali Hipola; Vidal Talens. *Medicina y seguridad del trabajo* (Madrid), Apr.-June 1963, 11/42 (48-81). Illus. 27 ref. (In Spanish)

Electric current, physiology and pathology of

The great majority of electrical accidents occur as a result of contact with alternating current, usually at 50 or 60 Hz. This article will therefore be restricted to a discussion of the physiology and pathology of injuries caused by electrical currents in this range. It is not safe to extrapolate the following considerations to other types of current such as direct current or pulsed current and, for these, the specialist literature must be consulted.

Severity of electric shock

The factors that determine the severity of electric shock are the pathway taken by the current through the body and the magnitude of the current.

Current pathway. Because the body behaves as a volume conductor, the current density is greatest along a line joining the two points of contact. It is, therefore, very important to consider the current pathway and the structures along it, because these determine the effects of the shock. The commonest current pathway is between one of the upper limbs and either the opposite upper limb or the lower limbs. This pathway includes the heart and the respiratory muscles of the chest. Only about 3% of fatal shocks pass through the head where the neural centres controlling respiration are located.

Magnitude of current. The effects produced by an electric shock also depend on the current magnitude, but the great majority of electrical accidents occur on installations delivering a fixed voltage (e.g. 115 V, 240 V, 6.6 kV, etc.). Under these circumstances the magnitude of the current depends upon the electrical resistance presented by the human body. Because the human body acts as a volume conductor, currents of 50-60 Hz generally travel through it in a uniform manner, there being little evidence to support the view that the current preferentially flows along blood vessels. At these frequencies, the body behaves as a simple resistance with virtually no capacitive or inductive effect. Most of the body resistance is in the skin, the internal milieu having a fairly constant resistance of about 500 Ω. Skin resistance is variable, being less than 1 000 Ω for large wet contacts and more than 100 000 Ω for thick calloused skin.

Physiology and pathology

The body tissues are very sensitive to electric current; if electrodes are placed directly on it, the tongue can detect 45 μA. A current of 100 μA leaking along a cardiac catheter which was in direct contact with the heart wall has caused ventricular fibrillation.

"Let-go" currents and death from asphyxia. The threshold of sensation for current passing along a conductor held in the hand is about 1 mA. As the current is gradually increased, tingling, heat and pain are felt and, at about 10 mA, the average adult male is unable to let go because the forearm muscles are held in tetanic contraction. Most accidental electric shocks pass across

the chest and, if the current is about 20-40 mA, the chest muscles are held in tetanic contraction and respiration ceases. This may cause death from asphyxia within a very few minutes with hypoxia and a profound mixed acidosis. Because the chest muscles are held in tetanic contraction and the airways remain open, petechial haemorrhages in this type of asphyxial death are very scanty or completely absent. However, if the current is interrupted within 2-3 min, respiration restarts spontaneously, and recovery is usually rapid.

Persistent respiratory arrest. This is caused only by electric shocks passing through the centres controlling respiration which lie in the lower posterior part of the brain. Urquhart and Noble in 1929 demonstrated that passage of an alternating current (a.c.) through the spinal cord produced a condition of temporary block. It was also found that transmission of nerve impulses along a segment of an exposed peripheral nerve could be prevented by previous passage of an alternating current along that segment. This phenomenon was called "a.c. block" and was used to explain those clinical cases in which arrest of respiration persisted after the shock had stopped. It has since been shown that, in the absence of ventricular fibrillation, persistent respiratory arrest occurs only when the current passes through the respiratory centre.

However, as has already been noted, about only 3% of fatal accidental shocks follow this path. Some years ago, Urquhart and Noble's work, together with dramatic descriptions of accident cases, in which current flowed between the head and the limbs, resulting in prolonged respiratory arrest, led to the teaching that prolonged artificial respiration by itself was the correct treatment of electric shock.

Ventricular fibrillation. This is a state of disordered action of the heart and may result from a shock passing through the chest. In ventricular fibrillation, regular heart action ceases, the pulse is absent and circulation is arrested. It is extremely rare for normal heart rhythm to be re-established spontaneously and, consequently, the condition is nearly always fatal unless treated promptly. Several investigators have studied the currents required to cause ventricular fibrillation and two important facts have been established. Firstly, the heart will fibrillate only if the current passes during the period when the ventricles are just relaxing (T wave of the electrocardiogram), a phase which may occupy up to 25% of a single heart cycle. Secondly, the smaller the current, the longer the time for which it must pass to cause fibrillation. A few years ago it was suggested that the relationship between shock durations and threshold of currents to cause fibrillation could be expressed by the formula:

$$\begin{matrix} 5\text{ s} \\ 8\text{ ms} \end{matrix}\left[I = \frac{116}{\sqrt{t}} \right] \quad \text{where } I = \text{current in milliamperes}, \; t = \text{time in seconds}$$

This relationship, which is shown in figure 1, is valid only over the range of shock durations shown, namely from 8 ms to 5 s. It was subsequently shown that if the shock duration was prolonged, even up to 60 s, the threshold of current necessary to cause fibrillation remained about the same as at 5 s and did not continue to fall further.

More recently, it has been claimed that the relationship between the two variables of shock duration and threshold of current to cause fibrillation is Z-shaped, as shown in figure 2. The upper threshold is that of a shock falling on the T wave of the electrocardiogram (as noted above). The discontinuity occurs when the shock duration starts to exceed one cardiac cycle and the lower threshold, for the longer shock durations, represents the reduced threshold of a train of extrasystoles produced by the current. For a fuller description of the physiological mechanism the paper by Biegelmeier and Lee (1980) should be consulted.

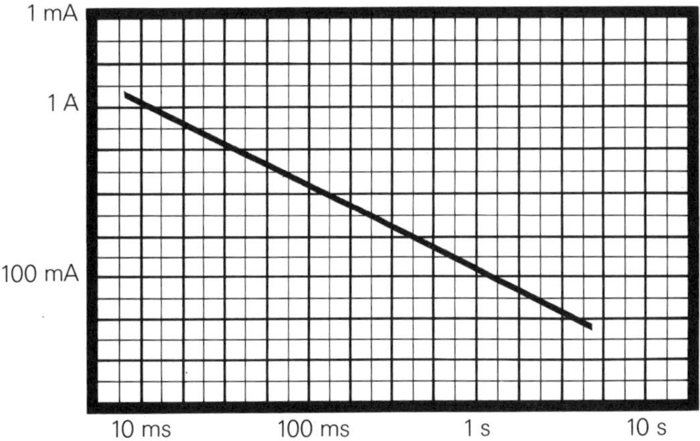

Figure 1. Log-log relationship between shock duration and threshold fibrillating current as proposed by Dalziel and Lee (1968).

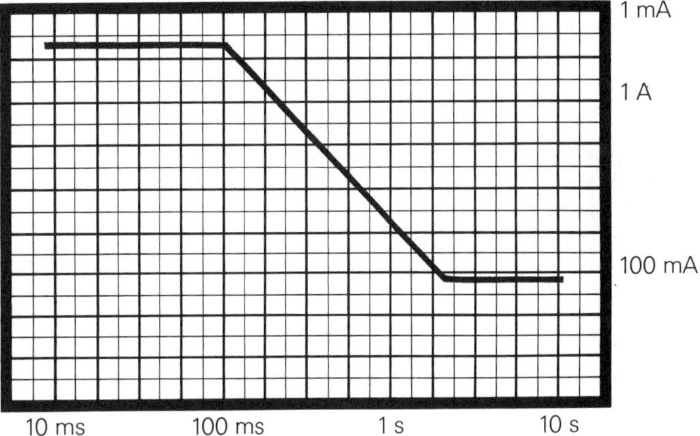

Figure 2. Z-shaped relationship between shock duration and threshold fibrillating current as proposed by Biegelmeier and Lee (1980).

These studies form the theoretical basis for the current-sensitive earth-leakage circuit breaker designed to operate in the event of an electrical accident so as rapidly to disconnect the supply of electricity before ventricular fibrillation is produced.

If a shock current of any given duration is increased above the threshold value, the probability of fibrillation occurring becomes progressively less and, when several amperes are passed, fibrillation is not produced. Thus, ventricular fibrillation occurs only within a certain band of currents. This fact explains why the victims of high voltage accidents often survive the shock (although they may subsequently die from their burns). It also forms the basis of operation of defibrillators used in clinical practice.

Electric burns (Joule burns). The passage of an electric current along any conductor is accompanied by the dissipation of heat. According to Joule's law, the heat dissipated is directly proportional to I^2Rt, where I is the current in amperes, R the resistance in ohms and t the time in seconds. As the skin is the site of highest resistance in the body it is here that burning is most likely to occur when contact is made with a live conductor.

Such burns may be deeper than may first appear on clinical examination. Consequently healing is often slow and may be accompanied by much scarring.

Flash burns. If an earthed conductor is brought close to another conductor at a high voltage, the insulation of the air between them may break down giving rise to a spark. This ionises the air, considerably lowering its resistance which in turn allows the current to increase so that an electric arc is set up. If the earthed conductor is a human being coming too close to a high-voltage line, he will be burned by the arc without actually coming into contact with the conductor. Because of the reduced electrical resistance of the air and the large area of skin burning (which reduces the skin resistance) large currents may flow. Thus the victim is the subject of a double event: a flame burn from the arc and an electric shock from the current which passes. The burns are often made worse as the result of clothing catching alight.

Because flash burn accidents are usually associated with high voltages, the currents which flow are often too great to cause ventricular fibrillation. Also, because the victim does not usually touch the conductor he is practically never "held on" but falls away from the conductor, thus extinguishing the arc. The current, therefore, usually passes through the victim for only a brief time.

Secondary effects. Many after-effects of electric shock have been reported but these reports often describe isolated cases of a disease following a shock and there is no real evidence of a causal relationship. However, several sequelae are well substantiated.

Angina electrica may follow a shock in a relatively young person and is clinically indistinguishable from *angina pectoris*. It nearly always clears up within a few weeks or months leaving no after-effects. Electrical cataract is a permanent condition which may ensue after certain types of electric shock, usually severe and passing through the head.

Chromoproteinuria, sometimes leading to severe disturbance of renal function, may follow a severe electric shock which has caused strong muscular contractions with release of myoglobin. It is analogous in all respects to "crush syndrome" in which muscle proteins may be released into the circulation from damaged muscles.

Various neurological sequelae have been described and it is possible that some of these may be due to injury to the spinal nerves resulting from the violent movement of the vertebrae caused by the strong muscular contractions occurring during a shock. Nevertheless, it should always be remembered that shocks of several hundred milliamperes are passed from one temple to the other in electroconvulsive therapy and that side-effects on the central nervous system from this are very rare.

First aid

It has been shown that the modes of immediate death from electric shock are:

(a) arrest of the circulation from ventricular fibrillation;

(b) less commonly, asphyxia from sustained contraction of the chest muscles; and

(c) persistent arrest of respiration caused by the current passing through the respiratory centre in the lower part of the brain.

First aid should therefore be directed at the deranged system.

Arrest of respiration may complicate arrest of circulation. Therefore, in all cases nothing is lost by prompt and efficient artificial respiration, which should be continued until breathing starts again or the patient is pronounced dead. It must be stressed that those accidents from which electric shock victims have survived after prolonged artificial respiration are generally found to be the unusual ones where the current has passed through the respiratory centre.

The modern first-aid treatment for arrest of circulation is external cardiac massage. It is important to understand that this measure does not restart the heart, but rather that it takes over its function in maintaining the circulation. For this reason, once started it must be continued until medical aid is available. External cardiac massage is potentially a dangerous technique and, if improperly applied, might endanger life, so it should not be used unless the indications for it are clear and it should be given only by those who have been carefully trained. The importance of adequate training, both in the diagnosis of circulatory arrest and in the technique of external cardiac massage, is therefore apparent.

The first-aid treatment for electrical burns varies little from the treatment of thermal burns as described in the article BURNS AND SCALDS. As high-voltage burns may be associated with release of chromoproteins into the circulation and the urine, it has been recommended that immediate alkaline therapy should be given as a preventive measure.

Treatment of electrocution and electric burns

The days when electricity was believed to have mysterious effects have passed. Once the victim of electrocution is in medical care he usually requires treatment only for the conditions and injuries which have resulted. The fact that the cause of the condition was an electric current is usually of not much consequence at this stage. Similarly, victims of electric burns are treated by surgeons along well accepted lines. The deep burning from the "Joule" heating effect of the current requires careful definition and debridement.

LEE, W. R.

Outcome of electric accidents (Folgen elektrischer Unfälle). Posner, G. Medizinischer Bericht 1973-1974 (Cologne, Institut zur Erforschung elektrischer Unfälle, 1974), 87 p. Illus. 81 ref. (In German)

"Importance of current as to the sequelae". Dalziel, C. F. (21-49). Illus. 18 ref. *2nd International Colloquium on the Prevention of Occupational Risks due to Electricity. Technical report, Köln 30.11-1.12.1972* (Geneva, ISSA, no date), 369 p.

"Electrical accidents" (Accidents dus à l'électricité). Proteau, J.; Robert, L.; Cabanes, J.; Folliot, D. *Encyclopédie médico-chirurgicale* (Editions techniques, 18 rue Séguier, Paris), 16515 A10 2-1979, 16 p. Illus. 74 ref.

"New considerations on the threshold of ventricular fibrillation for AC-shocks 50-60 Hz". Biegelmeier, G.; Lee, W. R. *IEEE Proceedings* (New York), Mar. 1980, 127/2, Pt.A.

"Body impedance of living human beings for alternating current 50 c/s". Biegelmeier, G. (11-13). Illus. 11 ref. *ISSA Bulletin* (Geneva), 1980, Vol. 7.

Contribution to first aid and treatment of electrical accidents (Beiträge zur Ersten Hilfe und Behandlung von Unfällen durch elektrischen Strom). Hauf, R. (Freiburg i. Br., Forschungstelle für Elektropathologie, 1981), 121+21 p. Illus. 80 ref.

Electric fields

General. Electric charges, their action on each other and their movement represent an ensemble of phenomena called electricity. Electric charges act upon each other through the electromagnetic field which surrounds them. Fixed electric charges form a constant electric or electromagnetic field.

Strong electric fields are widely used in various technical processes: gas cleaning, electrostatic dust precipitation, separation, electric ore processing, electrostatic spray painting. Electric charges form and accumulate when converting polymers with a high dielectric constant and when producing synthetic fibres. Calendering of paper and certain textiles, sanding of wooden components, pneumatic conveying of bulk materials and transfer of mineral oil products give rise to the formation and accumulation of electrostatic charges and associated static electric fields (see ELECTRICITY, STATIC). Electric lines conducting direct current of high or very high voltage (1 500 kV and more) are sources of strong static electric fields which build up around the conductors.

Biological effects and hazards. Experimental research has shown that electrostatic fields produce biological effects not only on lower forms of living organisms, but also on certain vital functions and basic systems of the human body. These effects are aspecific and, unlike ionising radiation, do not cause disease with distinct symptoms. Strong electric fields may cause induced surface charges to accumulate on the human body. These charges act on the sensory nerve endings in the skin and give rise to reflex modifications. Apart from their direct action on the body, electrostatic fields bring about electric discharges when a man who moves in an electric field comes near to a structure or touches the ground. Spark discharges produce adverse effects on the functions of the central nervous and cardiovascular systems, and cause fright and involuntary reactions which may lead to accidents during pole-top work. Frequently recurring shock currents can cause morphological skin conditions such as hyperaemia, blood effusion, oedema and necrosis. Spark discharges cause painful "puncture" sensations and may lead to arrhythmia, bradycardia and breath holding.

Workers employed on direct-current equipment operating at very high voltage may be exposed not only to electrostatic fields and electric discharges but also to airborne ions produced by corona effect. Airborne ions arising near such d.c. equipment move along the lines of force of the field and may cause a considerable ion flux to pass through the human body. Electrostatic fields and airborne ions are environmental factors acting upon each other. Their action is to be regarded as that of an electric complex comprising ionised air, capacitive charges and electric fields (see AIR IONISATION).

Protective and measuring equipment. Electricians working with highly conductive tools and materials on high-voltage d.c. installations may be exposed for long periods to electrostatic fields of high intensity. The systematic action of the above-mentioned electric complex on the human body may affect health. Measures should therefore be taken to limit the ill-effects of these electric factors. The intensity of the electrostatic field can be lowered by shielding the source or the workplace and by observing a safe distance between the worker and the source of the field.

Means of neutralisation (humidification, ionisation), substitution of highly conductive materials and parts for non-conductive ones, antistatic additives and coverings are widely used to diminish static electricity. Other measures consist in earthing (grounding) and increasing the conductivity of floors and shoe soles.

There is a large number of instruments for measuring electrostatic charges on solid dielectrics. They operate on the principle of measuring the potential (voltage) of the static electric field by electrostatic induction, by deviation of charged particles in a vacuum tube and by electrostatic generators (electrometers).

KALJADA, T. V.

CIS 80-28 *Effect of electric fields on the human body* (Wirkung elektrischer Felder auf den Organismus). Silny, J. Medizinisch-technischer Bericht 1979 (Institut zur Erforschung elektrischer Unfälle, Oberländer Ufer 130, Cologne) (1979), 39 p. Illus. 10 ref. (In German)

CIS 78-721 *Biological effects of electric and magnetic fields of extremely low frequency.* Sheppard, A. R.; Eisenbud, M. (New York, New York University Press, 1977), 270 p. 410 ref.

CIS 78-1297 "Influence on the human body of the electric field caused by high-voltage a.c. current electrical equipment" (Vlijanie ėlektričeskogo polja, sozdavaemogo ėlektroustanovkami vysokogo naprjäenija peremennogo toka, na organizm čeloveka). Krivova, T. I.; Lukovkin, V. V.; Morozov, Ju. A.; Sazonova, T. E.; Asanova, T. P.; Revnova, N. V. *Naučnye raboty institutov ohrany truda VSCPS* (Moscow), 1977, 108 (33-39). 15 ref. (In Russian)

CIS 78-1596 *Exposure to electrical fields–An epidemiological health survey of long-term exposed substation workers* (Exposition för elektriska fält–En epidemiologisk hälsoundersökning av långvarigt exponerade ställverksarbetare). Gamberale, F.; Knave, B.; Bergström, S.; Birke, E.; Iregren, A.; Kolmodin-Hedman, B.; Wennberg, A. Arbete och hälsa–Vetenskaplig skriftserie 1978: 10 (Stockholm, Arbetarskyddsstyrelsen, 1978), 48 p. 35 ref. (In Swedish)

CIS 80-991 *Exposure to electric fields–Mapping of the electrophysical working environment in substations* (Exposition för electriska fält–En kartläggning av den elektrofysikaliska arbetsmiljön i ställverk). Lövstrand, K. G.; Bergström, S. Arbete och hälsa–Vetenskaplig skriftserie 1980: 4 (Stockholm, Arbetarskyddsverket, 1980), 39 p. Illus. 41 ref. (In Swedish)

CIS 78-1000 "Portable screen for protection against electric fields–Protective screen against the effects of electric fields for workers on pylons" (Perenosnoe ėkranirujuščee ustrojstvo dlija zaščity ot vlijanija ėlektričeskogo polja–Ėkran dlja zaščity personala ot vozdejstvija ėlektričeskogo polja pri rabotah na vysote). Stoljarov, M. D.; Moiseev, A. G. *Energetik* (Moscow), July 1977, 7 (15-18). Illus. (In Russian)

Electricity distribution

Electricity is generated in hydroelectric power plants or in thermal power stations which can be fossil fuel burning (coal, lignite, fuel oil, natural gas), nuclear (where the disintegration of uranium or plutonium furnishes the heat source) or solar (presently in the experimental stage).

In the industrialised countries electricity is distributed at high or very high voltages (100 to 765 kV) through interconnected transmission networks or grids which are capable of delivering large quantities of energy over long distances and enable it to be exchanged among neighbouring countries, as well as providing for the delivery of energy direct to the very large industrial consumers.

The actual distribution of electric power to the users is effected by means of a medium-voltage network, which can supply the bigger consumers direct, followed by a low-voltage network, which supplies small consumers and, in particular, domestic customers.

In the newly industrialised countries, and also in the case of large isolated farming developments, it is customary to fall back on local production-distribution installations using self-contained motor-generator sets.

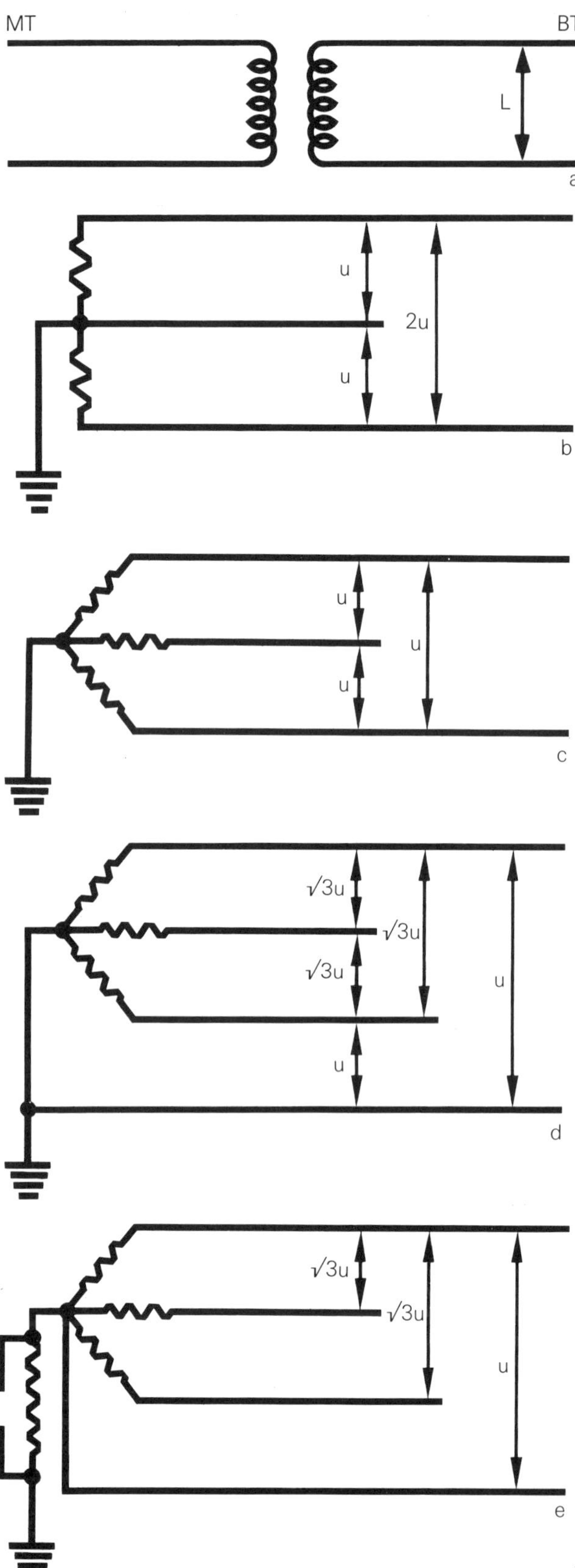

Figure 1. Main a.c. distribution systems. *(a)* single-phase; *(b)* single-phase or two-phase with earthed centre point; *(c)* three-phase star without neutral (or with neutral not distributed) used particularly in factories and workshops; *(d)* three-phase star with neutral earthed directly; *(e)* three-phase star with neutral earthed through an impedance which limits fault currents.

Distribution systems

Network layout and equipment. Electricity may be distributed by a direct feeder, single or double radial feeders, a ring main or a mesh, the choice of network depending on factors such as loads, distances, number of connections, average continuity of the service, etc.

There is a wide variety of equipment. Overhead lines may be mounted on steel pylons or wooden or concrete poles; these may carry single high-voltage or low-voltage conductors, a number of supply or distribution conductors and, in some cases, a telecommunications link. In some countries distribution systems and service lines consist of underground armoured cables.

Distribution of electric power to the consumer (figure 1) is effected by means of:

(a) firstly, a primary, medium-voltage network whose nominal values are usually between 11 and 35 kV for networks having three conductors and between 12 and 50 kV where there are four conductors (as in North America);

(b) secondly, a low-voltage network supplied from the above, for which the International Electrotechnical Commission (IEC) has set the following values:

- three-phase network (3 or 4 conductors): 220/380 V or 240/415 V
- single-phase network (3 conductors): 120/240 V

The standardisation of low-voltage three-phase networks at a single value of 230/400 V is well on the way to general adoption.

Power for domestic consumers is supplied through a low-voltage network (figure 2), either single-phase or three-phase. The choice between these two systems depends upon the customer's installations and his particular needs and also upon the commercial policy of the distributor and the technical possibilities of his network.

Power for industrial consumers is supplied direct from the medium-voltage network and even from the high-voltage and very high-voltage networks in the case of large undertakings.

Layout for low-voltage distribution. Figure 2 shows a simplified distribution system from the transformer to the point of consumption. *G* and *R* indicate the danger points most frequently responsible for accidents.

In fact, distribution systems differ by the type of neutral used for the three-phase system: earthed neutral, isolated neutral or neutral connected to earth via an impedance. Earthing of the neutral point, with or without an impedance connected, is widely used, but all three systems are found in industrial installations. Where the neutral is earthed direct, a fault to earth results in a heavy fault-current flowing (for example the current flowing through a human body would be limited only by the resistance of the body); however, with an isolated neutral, fault currents are weak and the danger is eliminated provided the isolation is permanent. The neutral earthed through an impedance connected to the neutral point is becoming more popular since any fault current to earth has to flow through this impedance and can be limited to such a value that an accidental contact with a metal body connected with earth does not have dangerous effects.

Protection for distribution systems. Certain incidents may occur within a system which may endanger persons in the vicinity. These include: broken or fallen conductors due to bad weather; moving vehicles colliding with the pole or supporting structure; etc. Underground cables run the risk of the insulation being ruptured.

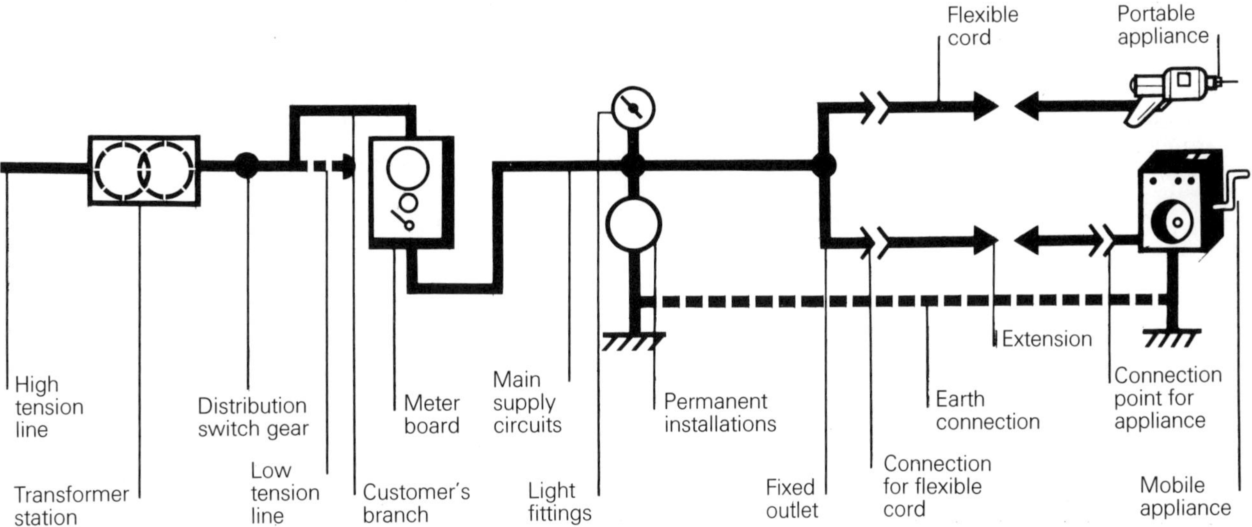

Figure 2. Layout for low-voltage distribution. It shows a simplified distribution system from the transformer to the point of consumption.

Installations such as substations may be badly damaged. Protective arrangements are provided to limit damage and reduce the risk of accidents to persons.

Primary distribution lines are equipped with circuit breakers which ensure automatic selective tripping (disconnection of a defective section from the network), incorporated with arrangements to reconnect the circuit automatically, particularly for the elimination of fleeting faults. Low-voltage secondary networks are provided with protection at their point of departure by means of automatic cut-outs or fuses located at the transformers.

HAZARDS AND THEIR PREVENTION

In the study of the pathology of electrical injuries, a number of different effects on the human body must be considered. A distinction must be made between electrocution as a fatal accident and electric shock as a non-fatal accident. It is on the basis of these studies that a level is fixed at which automatic low-tension protective devices will come into action (based on a function of current intensity and time).

Accidents due to electricity represent no more than 4% of all fatal industrial accidents and only 0.4% of the non-fatal accidents.

The causes of accidents occurring in electrical generation, transmission and distribution undertakings are related to:

(a) workers who are unskilled or insufficiently informed regarding electrical hazards. Most often these are workers in the building and construction industry who are working close to live equipment, generally overhead power lines. They usually come into direct contact with such equipment through cranes or lifting equipment, or when handling lengthy metallic material while at their normal working place;

(b) qualified electricians who may or may not belong to the transmission or distribution undertaking. Accidents occur either while working on a circuit which they know to be live, or while working on equipment which they believe to be dead but is, in fact, live. This may be due to a wrong procedure or to a failure to identify correctly the circuit or equipment concerned. Accidental or premature switching-on of the power is also the cause of many accidents of this type.

It should not be overlooked that further non-electrical hazards may be added which are inherent in the work itself, such as working at a height. In practically every case, accidents result from the fact that safety measures have not been strictly adhered to.

Most accidents in private installations occur as a result of tinkering with live apparatus or while using household electrical appliances; in the latter case, they are usually the result of defective cords or, even more often, of extension leads in which the insulation is ineffective or the plugs badly fitted or where male plugs have been attached at both ends. In factories, accidents occur when workers use electric tools with faulty internal insulation or perform maintenance operations without having previously disconnected the power supply.

In many countries the protection provided to workers who have to work with electrical tools or on electrical circuits and equipment in factories has been considerably improved.

Safety of consumer installations. Safety will depend largely on the design and layout of the equipment. Connections should be clearly subdivided into sections, so that faults can be readily detected and easily repaired. Accessibility is important if falls are to be avoided. Circuits should be separate and each should be equipped with its own switching device. Each individual circuit should be clearly identified. Finally, fuses, earth-leakage circuit breakers, etc., should be provided as a protection against overloading and short circuits.

Monitoring or protection apparatus should be used whenever possible. Simple and easily installed electrical devices can be used to show at a glance whether insulation is effective, by permanent measurement of the insulation resistance of various parts of the equipment in relation to earth. They offer protection, too, against abrupt rises in the voltage to earth of metal casings of electrical apparatus (neutral isolated or earthed through an impedance). Other electronic protection devices set off an alarm signal, or operate a selective cut-off if an insulation fault to earth develops and the residual current exceeds the device's threshold value. A variety of protective devices are used nowadays in factories and on construction sites. They must be carefully selected in the light of the type of distribution system in use and its neutral arrangements, which may be with the isolated neutral point, neutral connected to earth through an impedance, or neutral directly connected to earth.

In installations where the neutral is connected directly to earth use may be made of differential circuit-breakers which are sensitive to the earth-leakage current and will automatically break the circuit at the first sign of a fault. The "high-sensitivity" circuit breakers react to a leakage current of less than 30 mA and have a tripping time of between 0.03 and 0.05 s. This type of device provides effective protection when persons come into contact with a framework or other metallic component which has accidentally become live; they also ensure better protection in certain cases against direct contact with live conductors.

When located on the incoming side of any installation, differential circuit breakers ensure over-all protection. Should they come into operation, they bring about a general disconnection which sometimes makes it difficult to localise the fault. To overcome this inconvenience, while retaining the advantages of a high-sensitivity breaker, manufacturers have perfected simple high-sensitivity circuit-breakers which are not unwieldy and enable protection to be given to individual circuits and even to a single piece of equipment.

The increasing use of thyristors, both for variable speed controls in industrial applications and in domestic machinery, is likely to limit the application of this type of protective system. For low-powered portable equipment, isolating transformers whose output does not exceed 50 V are commonly used.

In addition to these essential safety precautions, the users themselves must be adequately informed. General safety principles should be taught at school, in particular those relating to electrical safety. Modern forms of visual aid, including films, film strips, television, etc., should be used.

Work on electrical installations

In electricity transport and distribution undertakings, safety should be considered under six separate headings: equipment design, work organisation, staff training (general and specific safe working practices), safety equipment, information and publicity, and inspection.

Two types of work on installations in use may be distinguished:

(a) handling of equipment with a view to making changes in the distribution network while maintaining the supply to the customers in order to deal with incidents affecting a circuit, or for maintenance purposes. A distinction may be made between normal working operations, operations to be performed during a breakdown, and work which may be required in the interests of staff safety;

(b) maintenance or repair work and work carried out with a view to extending or increasing the capacity of the supply network.

These operations are subject to strict regulations in every country.

Work on dead equipment. Four basic rules should be strictly applied to ensure safety (figure 3). These are sometimes known as the "Golden rules".

(a) The worker should be able to satisfy himself that there is a visible interruption in the supply, for example, the contacts in a circuit breaker can be seen to have separated. The evidence of a pilot light alone should not be accepted.

(b) Specific steps must be taken to ensure that the supply cannot be switched on again either accidentally or by another worker while the work is in progress. This requires padlocking the switch using a key kept in the possession of the person in charge, together with a prominently placed notice.

(c) The equipment must be tested by the worker (by means of a visual or audible test) to ensure that it is not live.

(d) A good electrical connection must be established between all conductors and earth to deal with any foreseeable short-circuit currents.

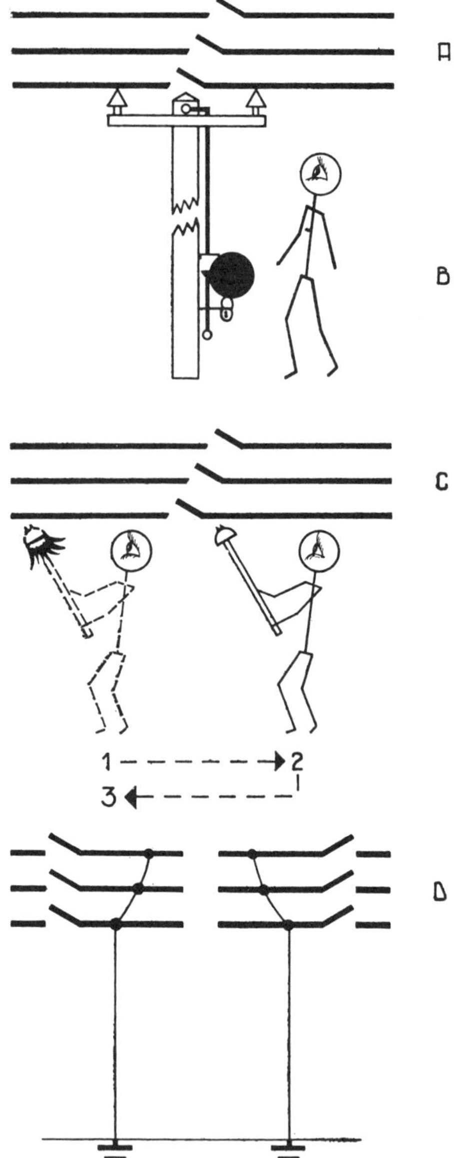

Figure 3. The four golden safety rules for work on dead electrical apparatus. A. See that the electricity supply has been interrupted. B. Ensure that the supply cannot be switched on again accidentally. C. Verify that the circuit is dead. D. Short-circuit and earth.

All installations should be such that these four rules can be followed without difficulty. Nevertheless, certain working positions may be arranged in such a manner that the safety of personnel is ensured by the constructional details, and in such cases the application of these rules is neither possible nor necessary in the interests of safety.

Live work. Two reasons are put forward to explain the increasing interest in live work. In the first place, consumers are using more and more equipment and they rightfully demand that current interruptions be kept to a minimum. Secondly, safety measures for the operating personnel have been improved and there is a great saving in time. Experience has shown that there are dangers attached to hot working because of the possibility of a wrong movement or a failure to carry out instructions to the letter. Hence the slogan "It is better to know you are working on live equipment than to believe you are working on dead equipment."

In all cases work carried out on live installations must be strictly planned throughout: this includes the structural details of the installations, the organisation of the work, special training of the workers in detailed working procedures, individual and collective protective equipment specially designed for the task, and close supervision and careful checking of the personnel concerned. Three basic working techniques exist:

(a) remote working: the operator remains at earth potential and performs the task with the aid of tools attached to the end of insulated rods. This method is used on both high- and low-voltage installations;

(b) working at the same potential as the equipment: the operator is insulated from earth and is in contact with the conductors, as is a bird when perched on a power line. This method is at present used for working on very high voltages, making use of insulated tower trucks or aerial baskets;

(c) working in direct contact with the equipment, but making use of insulated hand protection: the operator remains at earth potential but is insulated from live conductors by means of protective gloves. This method is used with low voltages and, in some countries, up to about 10 kV.

Emergency procedures for electrical accidents. Transmission and distribution workers should be trained in emergency procedures for electrical accidents and in resuscitation techniques for victims of electric shock. It is possible to train a number of workers thoroughly in rescue techniques and in methods of instruction so that, with the necessary equipment, they can pass on their knowledge to other workers.

In the event of an electrical accident the passage of the current through the victim should be interrupted by pulling the victim away from the conductor, by de-energising the installation or by obtaining the assistance of a person capable of doing so. The rescuer must avoid receiving shock himself from a live installation when grasping the victim.

Resuscitation must be started immediately. Death due to electric shock may occur very rapidly and the earlier the resuscitation is started the greater the chance of success. Drinker's curve, for example, shows that where resuscitation is started within a minute of the accident, there is a 95% chance of survival whereas, if resuscitation does not take place before the sixth minute, the chances are reduced to 10%. The victim's clothing should be loosened, and an attempt made to re-establish respiration. Resuscitation should be continued until assistance arrives and should never be terminated until the victim's death has been certified by a physician.

Fire hazards. Fires may not occur often in electrical installations but they can cause extensive damage. The high cost of installations, the probability of damage to plant often of vital importance in supplying electricity to factories and equipment working continuously and in ensuring the uninterrupted operation of public services call for reliable protection and the safe and rapid use of effective extinguishing equipment in case of fire. The safety of workers is itself bound up with the protection of the installations. The causes of fires of electrical origin include: overheating of insulation, arcing between conductors and explosions due to malfunctioning equipment.

The substances used for extinguishing fires are designed to limit temperature rise in combustible materials, and then to cool the material. Fixed and mobile extinguishing apparatus can be used and the commonest techniques are water sprays, sprays of liquids with a very low boiling point, powdered chemicals such as sodium bicarbonate and inert gas such as carbon dioxide which can be introduced in closed circuit; transformer overheating can be reduced by changing the insulating oil. It must be assumed that when the fire breaks out, the equipment is still live, and hence any substance sprayed on the fire must not itself be a conductor. Extinguishing agents such as foams may have a corrosive action on installations and circuiting, and should not be used. In all cases, extinguishing agents should be non-toxic, nor should they release toxic substances when exposed to an open flame.

There are certain standard rules in fire fighting. Action must be immediate to prevent the fire from spreading. Substances used for spraying purposes must not be conductors. The fire-fighting personnel must have proper respirators, since insulating materials, when decomposed by fire, may give off toxic fumes. Automatic devices may be installed to combat fire in fixed plant. If fire fighting is undertaken by manual means, it must begin very quickly. Proper fire-fighting instructions must provide for every contingency, and training exercises for the personnel must be arranged from time to time. For this purpose, the rule must be: cut off the current, put on insulating gloves, and only then attack the fire.

BESSOU, J.
BROULHIET, J.

CIS 77-1826 *National electrical safety code* (Institute of Electrical and Electronics Engineers, 345 East 47th Street, New York, NY) (1977), 348 p. Illus.

Electrical diagrams of industrial machines and safety (Schémas électriques des machines industrielles et sécurité) (Institut national de recherche et de sécurité, 30 rue Olivier-Noyer, Paris) (1979), 46 p. Illus. (In French)

CIS 78-1221 *5th International Colloquy on the prevention of occupational risks due to electricity–Safety for the electrical expert–His basic and advanced training.* ISSA International Section on the Prevention of Occupational Risks due to Electricity (Berufsgenossenschaft der Feinmechanik und Elektrotechnik, 130 Oberländer Ufer, Cologne) (1978), Vol. 1, 251 p. Illus.

"Flexible insulating protective equipment for electrical workers". Data Sheet 1-598-80. *National Safety News* (Chicago), July 1980, 122/1 (59-76). Illus. 8 ref.

Electricity, static

In industry, "static" is the word commonly used to refer to effects produced by electric charges which are trapped by insulators. The charges may be distributed over, or within, an insulating object, or they may be concentrated on a conductor that is insulated from earth. The electric field produced in the surrounding air by the charges is often merely a nuisance, attracting dirt and dust, for example. Most plastics, which are excellent electrical insulators, suffer from such side effects, unless special steps are taken. The presence of a high field can be felt, especially on the back of the hand, as hairs are made to stand on end. If sufficiently high charge levels develop, there will be an electric shock hazard to people and the risk of spark ignition in flammable atmospheres.

The subject has become more important in recent years, partly as a consequence of the greater use of insulating plastics in modern technology. Discovery of the electrostatic ignition mechanism in the disastrous series of explosions aboard petroleum supertankers in 1969 imposed extra urgency. The result is that big advances in our understanding of the problems, in relevant test methods and in safety procedures, have taken place over the past decade.

Sources of charge

Charging may occur in any of the following ways:

Contact charging. There is always some transfer of charge when two dissimilar materials come into contact with each other. After separation, any charges transferred to an insulating material remain trapped on it. In this way plastic films passing over metal rollers (which may be earthed), powders flowing through ducts, or people walking over carpets may become highly charged. Friction generally enhances contact charging.

Liquid flow. When low-conductivity liquids flow through pipes, preferential adsorption of impurity ions of one sign on the walls will leave the liquid charged with the opposite sign. This may readily occur with petroleum products, especially when fine filters are used. The effect can be very large when a second phase, for example a little free water, is present.

Spray electrification. Fragmentation of a liquid surface often produces charged droplets, and for this reason splash-filling of a tank may form a charged mist. In the supertanker explosions the incendive sparks came from charged clouds generated by water sprays used for cleaning the tanks. In chemical plants a wet-steam leak is a notorious source of charge: its spray readily charges any isolated metalwork, for example a section of pipeline that is nearby.

Ironically, a CO_2 fire-extinguisher is also liable to generate hazardous charges as the solid CO_2 particles leave its nozzle, and some disasters have occurred where CO_2 systems were used for the preventive inerting of fuel tanks.

Corona charging. Several modern industrial processes, for example zerographic copying and electrostatic paint or powder spraying, employ deliberate charging by means of air ions.

Induction charging. The potential of an isolated conductor may be raised by the presence of a nearby charge; this may have the effect of concentrating electric fields and initiating sparking.

Charge accumulation

Electrostatic effects are magnified whenever a charge is allowed to collect in one place, and dangerous situations are liable to arise where this proceeds unchecked. Thus winding many layers of charged film on a reel, bulking a large quantity of charged powder in an isolated bin or collecting charged liquid in a tank are operations which frequently give trouble with sparks. It is the same kind of cumulative effect that is exploited in the Van de Graaff generator to obtain very high voltages.

Electrostatic sparks

When charging proceeds so far that the field in the air reaches the breakdown value (approximately 3 MVm^{-1}, depending on the exact gap involved), some sort of air discharge must ensue. The degree of hazard incurred will depend on both the amount of available stored energy and the character of the discharge event. Thus igniting power will mainly depend on the total energy dissipated, but it will be profoundly affected by the distribution of this energy over time and space. A charged isolated conductor, usually a metal object, presents the most potent hazard, because all the stored energy may be released in a single rapid spark from any point on it to earth. Large objects, having a high electrical storage capacity, are liable to produce severe sparks in excess of 100 mJ. It should be noted that a man wearing insulating rubber-soled shoes, and/or standing on an insulating floor, behaves as an isolated conductor, and that he may therefore produce dangerous sparks (up to about 200 mJ in energy) if he becomes charged.

The discharge is modified when a sharp point or edge is present on the conductor or the adjacent earth electrode. The field becomes focused at the tip, where it initiates the ionising process, which remains localised there. The resulting discharge, called corona, is relatively smooth and is much less incendive than a spark.

Sparks from charged insulators are quite different, principally because the main discharge channel can only be fed by means of air ionisation over the surface. In the typical case of a spark to a charged film or sheet, the ionisation can drain charge from no more than about 300 cm^2. Taken together with the inherent limitation that the charge density on the surface cannot exceed about 2 $\mu C.m^{-2}$ before sparking commences, this means that spark energies are no greater than about 4 mJ. (An exception occurs where an insulating sheet carries a high density of opposing charges on its two surfaces, possibly with a conductive backing against one side. If the charge densities are high enough, a progressive discharge may then be triggered, covering a larger area and releasing much more energy.) Again, relatively harmless corona discharges may be produced using a pointed earth electrode.

As mentioned previously, igniting power is chiefly determined by the energy of individual sparks. Most fuel vapours, saturated hydrocarbons and organic solvents only require about 0.2 mJ for ignition, and therefore even sparks to, for example, a plastic bag, may ignite these vapour/air mixtures. (Corona discharges are, however, not sufficiently incendive.) Some chemicals, including unsaturated hydrocarbons, are even more sensitive. If air is replaced by oxygen, all vapours become more sensitive—an oxyacetylene welding torch can be ignited by as little as 0.2 μJ. These figures all refer to the most sensitive vapour/air or oxygen proportions and are called minimum ignition energies (MIE); mixtures that are richer or leaner require larger energies. Table 1 lists representative MIE values for vapour/air mixtures. From

Table 1. Minimum ignition energies (MIE) for vapour/air mixtures

Substance	MIE (mJ)
Hydrogen	0.011
Acetylene	0.017
Coal gas	0.03
Ethylene	0.07
Methanol	0.14
Diethylether	0.19
Benzene	0.2
Ethane	0.24
Petroleum vapour	0.25
Natural gas	0.3
Acetone	1.5

this it is clear that even sparks to insulators are quite capable of igniting common flammable atmospheres, although the risk of this happening will depend on the probability of the insulator becoming fully charged and on the area involved.

Minimum ignition energies for suspensions of particles in air vary widely, from less than 2.5 mJ to more than 5 J, although the majority of powders are much less sensitive than vapours and have values well above 25 mJ. Values decrease rapidly as particle size decreases, however, so that realistic values should relate to the actual form of material that is being used, and fine dusts (particle size less than 75 μm) should be regarded with special caution.

The strength of physiological shock also increases, of course, with discharge energy. Table 2 gives a rough indication of what may be expected.

Table 2. Variation of physiological shock with discharge energy

Discharge energy (J)	Physiological shock
10	burn
1	painful prick
0.1	sharp prick
0.01	prick
0.001	just detectable
0.0001	undetectable

Detection methods

Many commercial field meters are now available for investigating static charges. These instruments are invaluable for locating a source of charge, for telling when charges are approaching sparking levels, and for checking the success of charge control procedures. In making quantitative measurements it is necessary, however, to take account of the geometrical arrangement of meter, charged object, nearby conductors and earth. Caution should be exercised when taking readings in flammable atmospheres, because introduction of the meter itself may trigger sparking.

The occurrence of sparks, for example inside a silo, may be detected by radio receiver. In the dark an image intensifier can also be very useful in revealing just where and how sparking begins in a particular process.

ELECTROSTATIC HAZARD CONTROL

A wide variety of procedures are available to keep electrostatic charges down to a safe level and this section outlines the principal ones. Occasionally it is simpler to avoid having a flammable atmosphere than it is to guarantee no sparking. This is now accepted, for instance, in supertanker cleaning operations, where the tanks are first inerted by filling with engine exhaust gases.

Bonding conductors to earth. In order to avoid any build-up of charge on conductors, they should always be firmly connected to earth. On a chemical plant where flammable atmospheres may occur, all pipework, vessels, sieves, lids, guard rails, etc., must be so treated. This is particularly important where transfer operations are concerned: the transferred material will tend to carry away charge of one sign to the vessel being filled, leaving the vessel being emptied with the complementary charge. Therefore road and rail tankers must be earthed during filling and emptying where insulating materials and flammable atmospheres are involved, and clip leads are normally provided for the purpose.

Earthing within an insulating container. When charged materials are collected in an insulating container, it is very important to provide a route to earth for the charge accumulating on the inside. In powder handling this may be done by means of an earthed metal rod, preferably located at the bulking point, inside the container. In a tank for liquids an earthed metal plate may be located at the bottom on the inside, together with an entry pipe in the form of an earthed metal dip-leg. When insulating linings are used, these should never be separated from the container where a spark could cause an ignition.

Earthing of people. When operators are working where flammable atmospheres might exist, it is important that they remain earthed. This is usually ensured by providing special, electrically conductive footwear, and by maintaining an electrically conductive floor. So-called antistatic shoes are also used: these are not so highly conductive that they present a serious hazard if the wearer accidentally touches a live mains electricity supply. Where either conductive or antistatic footwear is used, it is good practice to have a personnel tester, so that operators can check their resistance to earth before entering a classified area. In the solid-state electronics industry, where electrostatically fragile components are handled, earthing strips are usually provided for each operator, who also wears an outer garment made of antistatic fabric.

Antistatic and conductive materials. Most electrostatic problems are solved if all materials are made sufficiently conductive for charge to be able to leak rapidly away to earth. Now natural materials like cellulosics, for example paper and cotton, and wool, are hydrophylic and absorb water from the atmosphere. This promotes a slight degree of ionic conduction and in most circumstances where the ambient relative humidity is above 40%, this is usually adequate to subdue the effects of static charges. Where the humidity falls to low values, as it may do in heated buildings during a cold winter, or near heated sections of process machinery, the conductivity may become too low. One option is then to increase the humidity artificially to regain the beneficial conductivity. Modern plastics are largely hydrophobic, however, and raising the humidity has little effect on them. One remedy is to apply a chemical antistatic agent, usually a surfactant, to impart surface conductivity. Values in the range 10^{-12} to 10^{-9}S (S is the symbol for the SI unit of conductance called the siemens which is equivalent to the reciprocal ohm) can be obtained, the effect again depending on the presence of moisture. In similar fashion, antistatic additives may be used with insulating liquids to raise the bulk conductivity. It should be emphasised that, where the conductivity of a flammable liquid is less than 100 pSm, special care should be exercised in transfer operations. As a general guideline, all speeds in pipelines should be kept below about 1 ms^{-1} in such cases. Furthermore, where a liquid passes through a fine filter, adequate allowance must be made for dissipation (relaxation) of charge before the filtered liquid reaches a storage vessel where a flammable atmosphere might exist.

A whole range of conductive or antistatic composites for fabrication of bins, pipes, tanks, belts, floor coverings, etc., are also made, usually by incorporation of conductive carbon black into conventional plastics and rubbers.

Static eliminators. When an insulating material becomes charged by passage through a machine and it is impractical or undesirable to render its surface conductive, it is necessary to neutralise the charge by external means. Commercial devices, called static eliminators, do this by confronting the charged material with a supply of

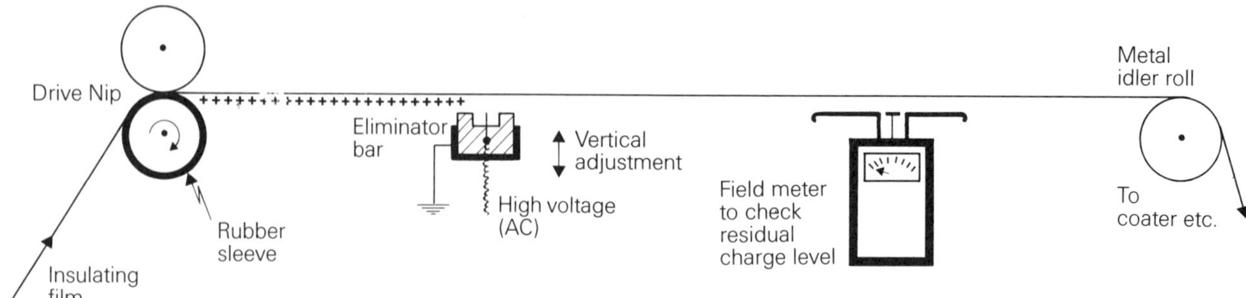

Figure 1. Elimination of static charge on film production line.

ionised air: the self-field from the charge then attracts the required ions for neutralisation. This technique is used extensively on machines which process film and sheet material (figure 1).

The simplest eliminator is a row of earthed metal points arranged in front of, but not touching, the moving charged surface. With a passive device of this kind, the charge itself induces the high field at the points causing the air to ionise there (corona). The system has the advantages of safety–it may be used in flammable atmospheres–and cheapness–a string of conductive tinsel connected to earth serves well. A passive eliminator fails to work, however, in congested areas, where neighbouring earthed metalwork "steals" its field. In that case a separate alternating high-voltage (about 7 kV rms) must be applied to energise the points. These powered eliminators do introduce their own high-voltage hazard, but they are efficient neutralisers. There are "shockless" types which are safe to touch, specially encapsulated ones which may be used in flammable atmospheres, and ionised-air blowers for discharging intricately shaped objects. Radioactivity may also be employed to ionise air and eliminators based on polonium 210 (half-life 138 days) are manufactured. These eliminators may be used safely in flammable atmospheres, but the usual precautions against radiation hazards must be taken.

Whatever the choice of eliminator, its performance will depend critically on its location with respect to the charged surface. It must never be placed opposite a roller, for instance, or other metal part, which will detract from the self-field between the charge being neutralised and the eliminator. It is always advisable to check the performance after installation with a field meter downstream, and to adjust the eliminator accordingly, as indicated in the figure.

BLYTHE, A. R.

Guidelines for avoiding ignition hazards from electrostatic charging (Richtlinien zur Vermeidung von Zundgefahren infolge elektrostatische Aufladung). Richtl. Nr. 4 der Berufsgenossenschaft der Chemische Industrie (Weinheim, Verlag Chemie, 1972), 64 p. (In German)

Electrostatic hazards: their evaluation and control. Haase, H. (Weinheim, Verlag Chemie, 1977), 124 p. Illus. 52 ref. (Translation of German original)

Measurement and elimination of harmful static electricity (Mesure et élimination de l'électricité statique nuisible). Challande, R. (ed.). (Paris, Editions Eyrolles, 1973), 216 p. Illus. 44 ref. (In French)

Recommended practice for protection against ignitions arising out of static, lighting and stray currents. API RP 2003 (Washington, DC, American Petroleum Institute, 1974), 39 p. Illus.

International safety guide for oil tankers and terminals. International Chamber of Shipping Oil Companies International Marine Forum (London, Witherby Marine Publishing, 1978), 162 p. Illus.

Specification for electrical resistance of conducting and antistatic products made from flexible polymeric material. BS 2050 (London, British Standards Institution, 1978).

Specification for electrically conducting and antistatic rubber footwear. BS 5451 (London, British Standards Institution, 1977).

"Long-term exposure to electric fields: a cross-sectional epidemiologic investigation of occupationally exposed workers in high-voltage substations". Knave, B.; Gamberale, F.; Bergström, S.; Birke, E.; Iregren, A.; Kolmodin-Hedman, B.; Wennberg, A. *Scandinavian Journal of Work, Environment and Health* (Helsinki), June 1979, 5/2 (115-125). Illus. 34 ref.

Electric lamp and tube manufacture

There are three basic types of equipment commonly used for electric lighting: the incandescent lamp, the discharge lamp and the tubular fluorescent lamp.

Incandescent lamps

The incandescent lamp consists of a glass envelope which is sealed to the mount, i.e. a small glass plate through which pass two supporting wires carrying the coiled tungsten filament. The wires serve at the same time as conductors of the electric current and are therefore fixed at their other end to the copper cap which encloses the lamp at the bottom. The mount is traversed by a glass tube which serves for evacuating the envelope and for admitting a filling gas, for example argon. After filling, this tube is sealed and pinched off.

HAZARDS AND THEIR PREVENTION

Manufacture of the normal incandescent lamp is largely mechanised but special types still need manual work. The physical stress imposed by the work is generally not great, except where much radiant heat is present. As in any other factory where work with glass is done, there is also here a risk of injury from splinters. Slight injuries to the eyes are fairly frequent. At danger points, the wearing of eye and face protection is necessary. Among people packing the lamps into boxes, a tenosynovitis at the wrist is sometimes seen. Tendinous knots on the back of the hand also occur.

Hydrofluoric acid. The glass envelopes have to be cleaned beforehand with a weak solution of hydrofluoric acid (2-5%). For some lamps, frosted glass is used, which requires treatment with concentrated acid or glass frosting paste containing 20-40% ammonium fluoride. The hands of the workers must be protected against this with strong rubber hand protection. Concentrated solutions of hydrofluoric acid can cause serious etching of the fingers. The eyes should be protected with goggles or a protective shield. When working with concentrated hydrogen fluoride, rubber aprons should be worn. Evaporation of the frosting paste can lead to considerable amounts of hydrogen fluoride in the air. Local exhaust ventilation above the baths and good ventilation of the shop are necessary.

Nitrogen fume. The manufacture of tungsten for the coiled filaments will be discussed elsewhere. In the production of a coil, a tungsten wire is wound around a molybdenum wire: this is subsequently dissolved again in a mixture of concentrated nitric acid and sulphuric acid. Large amounts of the dangerous nitrogen oxides (NO and NO_2) are liberated from the solution. The process should take place in a closed, well ventilated fume cupboard; the vapours should be absorbed after extraction.

Butyl acetate. The envelopes of many lamps are coated with a layer of powder containing pigments and other chemicals. As a rule this powder is first suspended in butyl acetate. When the envelopes are subsequently baked dry, this liquid evaporates and may give rise to objectionable decomposition products. The drying ovens should therefore be provided with an exhaust system. The pigments are rarely toxic. For the production of white envelopes, use is often made of a suspension of aerosil. This material consists of very small particles of amorphous silica and presents no risk of silicosis.

Halogens. The filling gas for projection lamps and some motorcar lamps contains a small percentage of bromine. Provided the tubing and connections of the gas-filling system are regularly inspected, it is possible to prevent the worker from inhaling bromine vapour.

Thorium. The coiled filament of some large lamps is coated with a paste containing thorium oxide. During brushing of the dried filaments, it is necessary to ensure that the radioactive powder is not inhaled. This work should be done in a glove box.

Skin affections. These are rare. The resins of the adhesive with which the copper cap is fixed to the envelope (see section on fluorescent lamps below) sometimes bring about a contact dermatitis. In susceptible individuals, dermatitis of the face and hands may occur as a result of contact with hydrazine hydrobromide or its vapour. This liquid is used for the soldering of the wires to the copper cap. Skin contact should be avoided; the vapour can be directed away from the face by means of a small ventilator.

Discharge lamps

Because of their high luminous efficiency, increasing use is made of gas-discharge lamps. On account of the high temperature which these lamps reach, the envelope is generally made of quartz. The contents may include various toxic materials.

HAZARDS AND THEIR PREVENTION

Blowing and working of quartz envelopes should be done with care. Although the quartz, when dispersed, passes into the amorphous state, local exhaust ventilation is still desirable. The hot flame required for the blowing of quartz is often produced by the burning of a mixture of natural gas and oxygen and this also makes an exhaust system necessary. After the gas-discharge lamp has been evacuated, a few drops of mercury are inserted; sometimes other toxic materials are added also, such as the iodides of indium, thallium or lithium. The glassblower runs the risk of inhaling these materials via his blowpipe. This can be prevented by means of a thin rubber membrane in the mouthpiece or by allowing part of the blowpipe to pass through liquid oxygen, so that the vapours form a sublimate in the pipe. The xenon lamp produces arc light in a quartz envelope containing xenon under high pressure. The short-wave light emitted forms ozone from the atmospheric oxygen. If the lamps are tested in small rooms, good ventilation is necessary. During production, the worker should protect his eyes with safety glasses because of the explosion hazard.

Fluorescent lamps

These light sources consist of a glass tube coated on the inside with a layer of fluorescent powder and provided at either end with a filament. The tube is filled with argon and mercury vapour.

For the manufacture of a fluorescent lamp four processes are necessary:

(a) Coating with powder: the tubes are first washed with hot water and dried. Then a layer of powder is applied to the inside. This powder, which consists of a mixture of chemicals, is ground fine in mills and then mixed with a binder consisting of cellulose nitrate dissolved in butyl acetate. The binder is usually prepared in a separate department, because the handling of cellulose nitrate in the dried state requires special precautions for fire prevention. The mixing of powder and binder should be done under exhaust ventilation. The suspension thus obtained is forced up the vertically standing tubes from the bottom (figure 1). When the suspension drains away, part of it remains adhering to the inside of the tube. Thereafter the tubes are dried in a drying cabinet; the butyl acetate vapourised is removed by exhaust ventilation. Next the binder is removed by heating to high temperatures in a sintering oven. To permit of sealing the tubes at a later stage, the ends are freed from their layer of powder.

(b) The production of the mount: the mount is made on a separate machine. It is a small, round plate of thick

Figure 1. Coating the internal surfaces of tubes with a layer of fluorescent powder.

glass on which a filament is mounted. In its centre is a glass tube (the stem) through which air can be pumped out of the lamp. The mounts close the ends of the lamp and through them the electric current enters and leaves the lamp.

(c) Sealing and pumping: at either end of the coated tube a mount is sealed in. Then the air is pumped out of the tube and replaced by argon and mercury vapour. The mercury is allowed to fall as a drop through the stem and into the tube. Next the tube is evacuated once more and again filled with argon and mercury. Finally the stem is heated, pinched and broken off. All this is done in a machine in which the tubes are set up vertically in a circle and which slowly rotates. Rotary machines in which the tubes are sealed and evacuated horizontally also exist.

(d) Finishing: finally, the lamps are provided at either end with a metal cap in which the ends of the filament are fixed. The cap is cemented to the glass.

HAZARDS AND THEIR PREVENTION

The heat stress to which the worker is subjected in several phases of the production may be appreciable. It consists mainly of radiant heat originating in the machines and the strongly heated glass. Adequate protection against it is difficult, since the various parts of the machine must remain readily accessible. The hot air can be extracted effectively by suction hoods above the sintering oven and the pumps. From the ergonomic aspect, attention should be paid to the working posture. In particular, the insertion of the mounts and the finishing of the tube may involve a tiring posture.

The most important hazards connected with the manufacture of tubular fluorescent lamps are of a toxicological nature. The chemicals used will be discussed individually.

Fluorescent powders. The powders (phosphors) covering the inside of the tube may differ in composition. Almost all phosphors contain calcium halophosphate as the principal ingredient; strontium magnesium phosphate is often added to this. The toxicity of these substances is low. Magnesium arsenate is sometimes used to improve the colour of the light. Phosphors which emit light of a warm colour contain up to 30% arsenate. A 40 W tube coated with this powder contains about 1.8 g of arsenate. It is important to avoid the inhalation of the powder. Since, in the coating of the tubes, some suspension is frequently spilled and then dries on the floor, it is recommended that those engaged in applying the coating should stand on a wooden duckboard above a floor kept wet with water. In a later stage of processing, the ends of the tubes are freed again from the dried powder. For this work it is necessary to have an extraction system.

The grain size of magnesium arsenate is as a rule less than 3 μm. This means that the dust is readily retained in the lungs. A large proportion of the inhaled dust will finally reach the stomach via the throat. In a neutral milieu, magnesium arsenate is only sparingly soluble, but in the gastric juice it dissolves readily. The acute action of magnesium arsenate is less toxic than that of arsenic trioxide but, since in chronic uptake of these substances the deposition in the internal organs shows practically no difference, it is advisable to impose the same maximum allowable concentration in air for magnesium arsenate as for arsenic oxide.

The danger of beryllium and berylliosis, formerly closely related with the manufacture of tubular fluorescent lamps, no longer exists. So far as is known, beryllium has not been added to the phosphors since 1950.

The preparation of the suspension of phosphors with butyl acetate may give rise to irritation of the mucous membranes. Most people quickly get used to the vapour of butyl acetate but those suffering from bronchitis find the irritation troublesome.

Mercury. In the rotary pump, a drop of mercury is inserted twice into the tube. A 40 W lamp contains about 15 mg mercury. Both during batching through the narrow stem and when the batching equipment is filled, mercury is easily spilled. In the hot pump, this quickly evaporates. In spite of exhaust ventilation above the machine, the allowable concentration of mercury is often exceeded. Periodical medical examination of the people at these workstations and control of their urine is necessary. In the urine of staff engaged on repairs of the pumps, mercury has also been detected regularly. Values of from 0.1 to 1 mg/l are not exceptional.

Phenol-formaldehyde resin. The cement used for fixing the metal caps may occasionally give rise to skin complaints. This cement generally contains a phenol-formaldehyde resin and hexamethylene-tetramine; both substances are skin irritants. With modern methods of manufacture, risk of skin contact is slight.

Accidents. In spite of mechanisation of the production process, the tubes have to be transported by hand at several points. Glass cuts and abrasions of hand and wrist therefore remain possible. The risk can be limited if the workers wear leather wrist protectors. Healing of the wounds appears not to be interfered with by contamination with fluorescent powder. To prevent eye injuries, protective goggles should be worn.

GROOT WESSELDIJK, A. T.

CIS 78-1164 *Health and safety guide for manufacturers of lighting fixtures.* DHEW (NIOSH) publication No. 77-228 (National Institute for Occupational Safety and Health, 4676 Columbia Parkway, Cincinnati) (Aug. 1977), 85 p. Illus.

Electric power tools, portable

Almost every type of tool can be worked by means of electric power; the fact that it is portable limits its size, weight and power to that which can be handled comfortably by the average man. The safe operation of a portable electric power tool depends upon its correct selection for the intended work, its proper use and regular maintenance. Adequate inspection and training of the user is essential.

Power tools relieve the workman of most of the physical effort required in performing his work, enabling him to work much faster and with less fatigue and to do work that he could not do without them, owing to the power they provide. The increased productivity that follows the introduction and use of portable electric tools, together with their versatility, will ensure a steady demand for more and more tools. Typical examples of portable electric power tools are: drills, saws, routers, planers, screwdrivers, grinders, etc. Due to the immense variety of these tools, this article will deal, except in a limited number of instances, with merely the electrical safety of the power tool itself.

Electricity supply

This system comprises three elements:

(*a*) the tool;

(*b*) the flexible cable connecting the tool to the electricity supply; and

(*c*) the electricity supply which, for the purpose of this article, is a socket into which a plug is inserted.

Tool. Portable power tools are generally compact, with one or two hand grips. Alongside is fitted a finger-operated non-sustaining trigger switch for starting and stopping; this switch may be sustained in the closed position by the operation of a sustaining button. The operation of the trigger switch must be capable of overriding this sustaining button.

The tool can be single-purpose, such as circular saws, blowers, etc., or multi-purpose, in which case it is usually fitted with a chuck at the drive end which will accommodate twist drills and spindles of various diameters, and the spindles, in turn, may drive different types of tools.

The tool must be of robust construction since, during the course of its life, it will from time to time suffer abuse in the way it is used.

Portable electric tools consist essentially of a universal electric motor with a wire-wound rotor supplied through a commutator and carbon brushes; the rotor shaft drives the chuck through gearing to give a single or multi-speed output.

Tools may be classified into three categories according to the type of protection against electric shock–

Class I: Often referred to as metal clad, have their metal casing insulated from live conductors; all such metal casing should be earthed.

Class II A: All-insulated tools, have an insulated casing that encloses all metal parts; no earth connection is necessary.

Class II B: Double-insulated tools, have double insulation throughout; where a metal casing is fitted, it does not require earthing.

Flexible cable. The flexible cable normally contains three conductors:

(*a*) the live conductor;

(*b*) the neutral conductor; and

(*c*) the earth conductor.

The first two constitute the power supply; the earth conductor connects the exposed metal of a metal-clad tool (i.e. that part that can be touched) to the earth pin of the plug.

Where all-insulated or double-insulated tools are concerned, the earth conductor is not necessary. An additional earth conductor may sometimes be included in a flexible cable in the form of a concentric metal braiding around the three principal conductors; this braiding is covered with insulation and protected.

Colouring of these conductors is subject to national practices, although considerable efforts have been made to standardise the system of colour coding (see also SAFETY COLOURS).

The conductors are secured at one end to the tool and connected to the motor, and at the other end secured to the plug and connected to the pins. A flexible cable must always have a plug at one end and either a tool or a connector at the other–no other arrangement is permitted.

Plug and socket. The electricity supply for the tool must be taken only from a socket outlet into which a plug is inserted; where it is necessary to work at a distance from the supply outlet, a special extension cable fitted at one end with a plug and at the other with a special socket outlet known as a "connector" should be used (figure 1). The supply voltage will vary from country to country (e.g. 110 V, 220 V or 240 V).

The method of connecting a single-phase power tool to the electrical power supply is by means of one conductor connected to the live phase and the other conductor connected to the neutral. In addition there is a third conductor which is connected to the earth conductor of the consumer's installation. In many countries the neutral is also connected to earth at the substation (figure 2). Precise arrangements vary from country to country. Even when the power tool is of the all-insulated or double-insulated type, three-pin plugs may still be used; the third pin is here used as a registering device to ensure that the live and neutral pins are always inserted in the same way.

Figure 1. A typical extension lead consisting of a plug at one end and a socket at the other.

Protection

The electrical energy generated at the power station, distributed over the mains and made available at the socket outlet into which the tool is plugged, via its flexible cable, is considerable. Means must always be instantly available to cut off the supply of electricity to the tool if it develops a fault, otherwise excess energy will be fed into it with serious possibilities of fire and electric shock. Protection against such electrical hazards is particularly important in the case of portable electric power tools which are moved frequently from place to place, often subject to mechanical abuse and used under all types of working and climatic conditions.

Fuse protection. The usual protection is by means of a rewirable or cartridge-type fuse connected in the live conductor of the circuit. The heating effect of an electric current is utilised, in that a fuse of a given rating will melt and so open the circuit, when a current in the circuit exceeds the rating of the fuse. It is recommended that the fuse rating should not exceed twice the normal load current of the tool as indicated on the rating plate.

If, due to a fault in the flexible cable or the tool, excess current is fed into it, this excess current heats up and melts the fuse, thus opening the circuit and cutting off the current, rendering the circuit dead and safe.

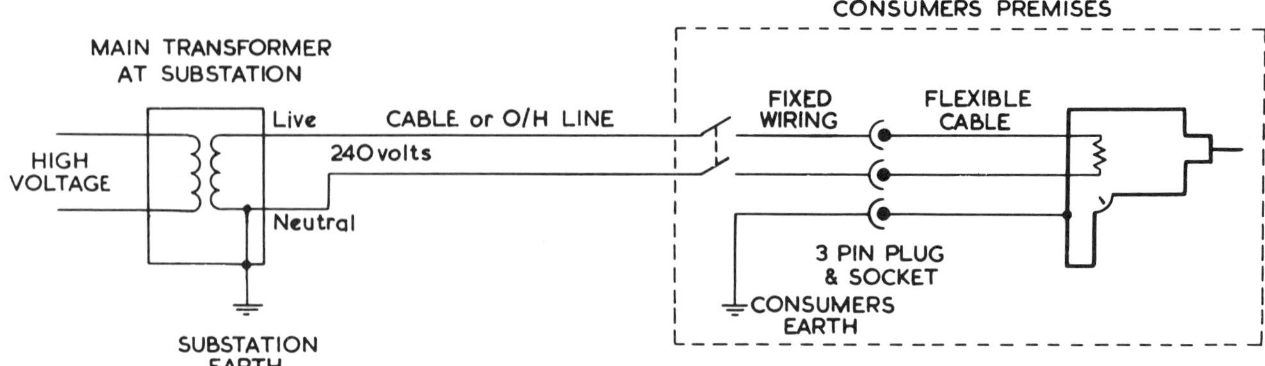

Figure 2. Diagram showing the electric circuit from the high-voltage supply, through a transformer, the medium-voltage distribution, up to the consumer's premises, and from there to a typical hand-held tool.

The types of faults that occur are:

(a) overloading;

(b) contact between live and neutral conductors (short circuit);

(c) contact between the live conductor and the earth conductor usually via the metal casing of the tool (earth fault).

Each of these faults will cause excess current to flow, thus resulting in the fuse melting and opening the circuit. This is dependent upon the correct selection of the fuse rating. In case *(c)* the excess current is dependent upon the resistance of the earth return. High resistance of the earth return will limit the current so that in some cases this will be insufficient to melt the fuse. This is a dangerous situation as there is an uncleared fault on the system which could result in fire or the operator being subjected to an electric shock. In these cases the use of fuses for earth-leakage protection should not be relied upon; instead, earth-leakage circuit breakers should be used. Fuse protection will, however, work correctly for overloads and short circuits.

Earth-leakage circuit-breaker protection. When an electric current flows through a conductor, a magnetic field is created around that conductor. This characteristic is utilised to operate a switch by means of a tripcoil surrounding an electromagnet. When current flows, the electromagnet is energised and in turn operates the switch. This device, known as an earth-leakage circuit breaker, can be designed to operate on low currents, i.e. 30 mA, a level at which the user would be protected from electrocution.

Integrity of earth circuit

The most dangerous type of fault from the operator's point of view is the earth fault, i.e. contact between the live wire and the metal casing of the tool causing this casing to be charged at line voltage: unless the fault is automatically and instantaneously cleared the operator will suffer an electric shock when he touches the case of the tool. The clearance of these faults both in fuse and circuit-breaker protection depends on an unbroken earth circuit. The most likely places for breakages are in the flexible cable, either in the cable itself or in its attachments to the tool or to the plug. The breakage of a live or neutral conductor will make the tool inoperative; however, breakage of the earthwire will pass unnoticed since the tool will continue to function. It is therefore of the utmost importance that frequent tests are made to verify that the current-carrying capacity of the earth conductor between the plug and the tool casing is such that it will carry sufficient current to operate the fuse or circuit breaker.

The integrity of this earth conductor can be monitored continuously and this requires a four-socket outlet and four-pin plug; twin earth conductors are necessary in the flexible cable and usually consist of the normal earth conductor and an over-all braided earth conductor as mentioned above. Both earth conductors are connected separately to the metal casing of the tool and to separate pins on the plug. When the plug is inserted into the socket, a current is circulated at very low voltage through the casing of the tool via the twin earth conductors and through an electromagnet contained in the circuit breaker controlling supply to the tool; the electromagnet is so arranged that when energised its core holds the circuit breaker closed. Should either of the twin earth conductors be broken or detached from the tool, the circulating current ceases, the electromagnet is consequently de-energised and the circuit breaker immediately opens. Thus on the occurrence of damage to the earth conductor, the circuit is automatically made dead and safe, and it cannot be re-energised until the earth conductor has been repaired.

All-insulated tools

As mentioned above, the all-insulated tool dispenses with a metal covering and has, in its place, a covering of strong, insulating material; the chuck, the only exposed metal, has a specially insulated shaft or is driven by nylon gearing. This type of tool is intrinsically safer than a metal-clad tool, in that the earthwire, which if damaged is a serious hazard, is redundant since there is no metal casing to which it requires to be attached; the hazard of the broken earthwire is thus eliminated. Excess current protection is all that is required on this type of tool.

Double-insulated tools

A double-insulated tool is one in which all accessible metal parts are separated from live conductors by two distinct layers of insulation, one being functional, the other protective (figure 3). The provision of a metal case provides protection for tools exposed to arduous working conditions and environments.

Low-voltage tools

In the event of an electrical accident, the severity of the electric shock received depends on the quantity of current (amperes) flowing through the circuit of which the body forms part; this, in turn, depends on the pressure (voltage) forcing the current against the

Figure 3. Cut-away view of a portable electric drill with double insulation. The insulating material between the working elements and casing is clearly visible.

resistance (ohms) of this circuit, which includes the resistance of the body.

In order to reduce the quantity of the current likely to flow through the body due to a fault, it is necessary either to increase the resistance of the circuit by providing operators with insulating hand protection, mats, etc., or to reduce the voltage.

In low-voltage tools, the degree to which the voltage is reduced is a compromise between the safety of the lower voltage and the inconvenience of the increased weight of a low-voltage motor; probably, the most practical arrangement where mains voltage is in the 220-240 V range, is 110 V operation with the centre point of the step-down transformer secondary winding earthed. Under such conditions, the maximum voltage to earth is 55 V which, as far as is known, has not caused a lethal current to flow through a human body. In the United States, standard mains voltage is already only 110 V.

Low-voltage tools are usually meant for use under abnormal conditions, such as are found on building and civil engineering sites, work in humid atmospheres or in close contact with water such as in laundries, or inside metal vessels such as boiler drums. For normal factory work, standard mains voltage tools of 220-240 V have a good history and are quite safe provided they are used correctly and properly maintained.

The low voltage is obtained by means of a small transformer which can be plugged into a mains socket outlet, the low-voltage tool being plugged into a socket outlet on the transformer. It is recommended that the transformer should be of the all-insulated type, i.e. completely encased in robust insulating material with no exposed metal parts or screws. Although all-insulated and double-insulated tools provide protection against electric shock, this protection is limited to the tool itself. In addition, there is the vulnerability of the flexible connection. Thus, for use under any abnormal condition, low-voltage operation is recommended, irrespective of the type of tool.

Three-phase tools

For most situations, the output of a single-phase power tool is adequate. However, three-phase power tools are sometimes necessary to provide the extra power required for certain heavy-duty applications. The normal three-phase operating voltage is 415 V between phases (230 V to earth). For conditions where low-voltage tools are necessary, a three-phase 110 V system (63 V phase to earth) is available to provide increased protection against electric shock.

Operation

Portable electric power tools are simple to operate; however, this simplicity of operation may result in the use of these tools by inexperienced workers and thus be the cause of injury. On no account should an untrained person be allowed to work with one of these tools.

All persons who have to use portable electric tools should be trained in their correct use and maintenance including the need to:

(a) use the tool only for the purpose for which it has been designed;

(b) hold the tool firmly with both hands;

(c) remove the plug from the socket when the tool is not in use;

(d) use the guards provided and check periodically that they are in order;

(e) wear eye protection whenever the work or environment is liable to present hazard to the eye;

(f) report any defects and seek advice before using the tool.

Maintenance

To ensure that power tools operate efficiently and safely, they must be inspected, adjusted and lubricated at regular intervals. The tools should be visually inspected before issue from stores and also by the user before use; such inspections should comprise an examination of the tool, flexible cable and plug for signs of damage. Damaged equipment should be withdrawn from use, labelled as defective and returned for repair.

In addition, power tools should be serviced and tested at regular intervals; servicing should include lubrication and an examination for signs of wear/damage in accordance with the manufacturer's instructions. Worn/damaged parts should be replaced strictly in accordance with manufacturer's instructions. Electrical tests should be carried out to verify the insulation resistance, continuity/resistance of the earth circuit and connection of the flexible cable in the correct polarity. A record should be kept for each tool detailing the nature of all servicing/repair work and the results of all tests.

Guarding

The presence of a cutting or abrasive tool rotating, reciprocating or vibrating at high speed presents a severe accident risk; this risk is even greater when the tool is portable. Consequently, the tool should be fitted, wherever practicable, with an effective guard; this applies particularly to such tools as circular saws and grinders. In the case of saws, the lower part of the guard is pivoted and sprung to allow the teeth of the saw to be exposed when cutting; the guard will spring back into place when the saw is removed from the work; the upper part of the saw should also be fully protected. In the case of portable grinders, the guard usually covers about one-third of the grinding wheel.

BECKINGSALE, A. A.
MASON R.

CIS 76-322 "A preventable death from an electrical hand tool malfunction". Whorton, M. D.; Levine, M. S.; Radford, E. P. *Journal of Occupational Medicine* (Chicago), Sep. 1975, 17/9 (589-591). Illus. 20 ref.

CIS 75-325 "Safe work with portable electric tools" (Mesures de protection lors de l'emploi d'outils électriques portatifs). Abegglen, A. *Cahiers suisses de la sécurité du travail* (Lucerne), July 1974, 116, 16 p. Illus. (In French, German, Italian)

CIS 77-332 "Periodic inspection of portable electric tools" (Vérification périodique de l'outillage électrique portatif). Lamouche R. *Cahiers des comités de prévention du bâtiment et des travaux publics* (Issy-les-Moulineaux, France), May-June 1976, 3 (114-117). Illus. (In French)

Electronics industry

The term "electronics" designates that branch of applied physics that is concerned with the phenomena and technical applications resulting from:

(a) the interaction of electrons among themselves (electronic valves or tubes, photoelectric cells, X-ray tubes); and

(b) the interaction of electrons with the crystal lattice of solids (semiconductors, transistors).

Electronic components. In addition to the actual control and amplifying units (electronic valves and semiconductors), electronics employs three further basic components in its circuitry:

(a) resistors, which consist either of effective resistance wire wound around an insulating form or, for small loads, of a poorly conducting film of carbon or graphite on an insulating cylinder–the unit of resistance measurement is the ohm (Ω);

(b) a capacitor (often called a condenser), which, in general, consists of two metal plates or foils insulated from each other by a dielectric; the capacitance of a capacitor depends on the geometry of the plates and the type and thickness of the dielectric–the unit of capacitance measurement is the farad (F) (in practice smaller units are widely used: microfarad (μF), nanofarad (nF), picofarad (pF));

(c) a coil (also called an inductor or inductive reactor), which consists of turns of wire with or without an iron core–the unit of inductance measurement is the henry (H) (in electronics a smaller unit is used: the millihenry (mH)).

The electronic valve circuits may be considered the first generation of electronic equipment, and the circuits in which semiconductors have replaced the valves as the second generation. This second generation equipment is smaller and more compact and has a lower power loss since, in particular, it is no longer necessary to heat the valves. Progress in semiconductor and ferrite technology, the development of precision etching techniques, etc., have made it possible to miniaturise electronic circuits further and produce complete microcircuits that combine semiconductor circuits, capacitors and inductors in a single unit no more than a few millimetres in size. The microcircuit constitutes third-generation circuitry characterised by the potential of long production runs of standard high-precision electronic circuits and the possibility of manufacturing very compact electronic equipment. These features are of particular value in the manufacture of data-processing equipment and in aviation and astronautical applications.

Uses. In the first half of the 20th century the industrial applications of electronics were confined almost exclusively to those involving electronic valves, photoelectric cells and X-ray tubes; however, about 1950, following development in the field of semiconductor physics, electronics began to break through into the entire domain of technology. This development has not reached its culmination, since new applications for electronics are constantly being discovered, and this new branch of technology is to an increasing extent becoming a decisive factor in economic growth as a whole. Initially, the amplifying and self-regulating characteristics of electronic valves or tubes were used solely in telecommunications; nevertheless, other uses were soon found in the regulation and control of electric machines and subsequently in the automation of complete production processes. The widespread use of electronic devices in telecommunications, data logging and processing, telemetry, safety controls, and measurement techniques accounts for only a small–although widely known–proportion of the total market for electronic components and circuits.

Since 1960, electronic equipment has largely replaced conventional electromechanical equipment in areas such as pressure switches, temperature switches, relays, regulator servocontrols, etc. Initially, the devices were of relatively low power and therefore limited in their applications; however, in this field of power electronics, the semiconductor or electronic valve has already partly replaced contactor circuitry in certain applications.

Nevertheless, it is still in amplification, measurement and control that electronic circuits find their main use. Quantitatively, electronics finds its most important application in entertainment, especially in sound- and picture-reproduction equipment such as radios, television and gramophones, which are mass produced throughout the world. Industrial electronic units are also produced in large quantities but equipment usually has to be adapted to suit special requirements and, consequently, production runs for individual pieces of equipment are relatively short; however, it is considered that reliability is enhanced. Finally, probably the most widely appreciated use of industrial electronic material nowadays is in data processing and its application to automation.

Under this heading are still included the classical uses of electronic units and electronic control circuits. The third generation of electronic equipment has in the meantime been further developed.

Further technological developments have now enabled electronic components to be manufactured successfully so that the use of separate elements for resistance, capacity and inductance can be eliminated and use made of semiconductors and photomechanical processes to produce the required circuits. These have a much greater capacity and also have the advantage that they take up only a fraction of the space required for the earlier systems.

The development of the well known pocket calculator is just one example of this technique. In the meantime, circuits have been miniaturised to such an extent that the control systems for the most complicated processes can now be accommodated in a few cubic decimeters. The peripheral equipment employed for the conversion of impulses from the so-called "microprocessor" into the required output signals still, however, generally needs substantially more space.

These types of microprocessors can be used for the control of various preparatory and finishing operations in machine shops and production lines; they can make it possible for telephone calls to be made throughout the world and can distinguish between the different currencies and furthermore can furnish the correct change after payment. The universally familiar data processing machine provides a further example in that such electronic systems can not only process data but can build up a "memory" that can now even be independently programmed. This means that "behaviour" can be arranged electronically and can thus determine how a regulating device for the selection of components should be varied. In the very near future electronic circuits will greatly simplify expensive programming operations.

With the relatively low manufacturing costs of such microprocessors, it will be possible in the future to add the so-called "microprocessor", for example, in data processing. This means that a central unit with two central processors working in parallel will also have two input and output systems. This will make it possible to increase the programme throughput as well as increasing the operations reliability of the computer. Even in the case of failure of some part of the machine, the system would be able to continue functioning.

In parallel with all these developments has been that of the electronic control of power output, which has become increasingly important in the control of heavy electrical currents. As an example, the use of silicon thyristors led to the introduction of rectifiers, converters and inverters, as well as current, voltage and frequency controllers or regulators. The correction of speed and torque losses, commonly encountered in machine drives, first became possible with the aid of electronic units. The world is just witnessing the start of these new developments which have been made possible through electronics.

HAZARDS AND THEIR PREVENTION

The electronics industry employs both the conventional materials of precision engineering such as steel, copper, aluminium, glass and plastics, and also special materials such as germanium and silicon, which, either in pure or alloy form, are used for semiconductor manufacture.

Circuit components and complete circuits are usually joined together by soldering with a lead-zinc alloy. If soldering is carried out for prolonged periods, it may be necessary to equip the workshop with exhaust ventilation since the fumes may contain lead, zinc and rosin, all of which have a toxic action. The molten-solder tanks of circuit printing machines should also be equipped with exhaust ventilation (figure 1) that extends to the vaporisation and cooling zones.

Polyester and epoxy resin systems are now used widely as insulating materials, and chloronaphthalene is employed as the dielectric and protective sheathing for capacitors. A common procedure is to pot individual components by casting them in an impact-and-water-resistant block of synthetic resin. The polyester and epoxy resin systems used are of the two-component type, the polyesters being cured with a peroxide and the epoxies with a phenol compound; quartz flour is often added to the system to improve strength and appearance. The chloronaphthalenes and particularly the resin hardeners have an irritant effect on the skin, and in the long run can cause severe skin injuries leading to permanent sensitivity. Consequently, where these resins are employed workposts should be equipped with good exhaust ventilation, workers should avoid skin contact by the use of tools, protective clothing and protective gloves, neutral agents for cleansing the skin should be readily available, and waste should be either disposed of immediately or kept in water-filled containers.

Figure 1. A well arranged fully automatic soldering installation for pattern plates.

The organic peroxide hardeners have an extremely damaging effect on the eye and have been known to cause loss of sight; workers should wear eye protection wherever hardeners are being employed undiluted by resin. The use of quartz flour presents a considerable respiration hazard and workers handling loose flour, or grinding and polishing items containing it should be protected by an exhaust system and respiratory protective equipment. Persons should receive a pre-employment examination before being affected to work with resin systems, and a periodic examination should follow at appropriate intervals (e.g. every 6-12 months).

Other occupational hazards are encountered in the preparation of printed circuit cards by the screen printing process and in the manufacture of components, both of which involve the use of numerous dangerous chemicals. An example of this is provided in the case of component manufacture, where silane compounds are used which are both unstable and highly reactive and which require the adoption of very strict safety precautions.

The preparation of printed circuit cards from offset plates is basically an etching process. It starts with a positive or a negative which is used in the printing process to prepare the screen, and which is used directly in the photocopying process.

During the screen printing process, the circuit paths, soldered points and copper surfaces in the circuit outline are given a layer of etchproof paint.

A number of dangerous or toxic substances are used in printing work. They may be explosive, dangerous to health, or corrosive. Care should be taken to:

(a) prevent the build-up of dangerous flammable or explosive mixtures of solvents in the air and to enclose possible sources of a primary explosion risk;

(b) eliminate all possible sources of ignition of such explosive mixtures;

(c) take steps to confine and limit the effects of an explosion should it occur.

In general workrooms only the first two of these precautions can usually be put into effect.

Particular attention should be paid to the health hazards presented by airborne concentrations of toxic substances in cases where repeated measurements have shown that the TLVs have been reached or exceeded and/or where medical research has indicated that such health hazards may be encountered. In such cases protection may be afforded by providing exhaust ventilation for the printing process and by arranging for the washing of the plates to be carried out within an enclosure (figures 2 and 3).

In the photocopying process a light-sensitive coating is applied to the copper surfaces of the printed circuit card on to which the circuit outline is to be copied. The excess copper is removed from the masked printed circuit cards by means of a simple galvanic etching process. The printed circuit outline to be reproduced is protected

Figure 2. Printing machine fitted with exhaust ventilation and guarded against trapping or crushing accidents.

during the exposure and development process and is not attacked by the etching solution.

Workshops where etching takes place should meet the following particular requirements:

(a) acid-resisting, non-slip flooring with proper drainage for handling waste water. Grids and footboards should be acid-resistant and should not create stumbling or tripping hazards;

(b) electrical equipment and material should at least be protected from damp or wet surroundings;

Figure 3. A well guarded washing plant provided with exhaust ventilation for use in the preparation of pattern plates.

(c) exhaust ventilation should be provided around areas where toxic gases or fumes are given off;

(d) eating, drinking and smoking or the storage of foodstuffs should not be permitted in the etching room;

(e) personnel should receive regular instruction at intervals of at least once a year concerning their duties with regard to the hazards and the precautions to be taken;

(f) spilt or used liquids should immediately be washed away using copious amounts of water;

Figure 4. Protective clothing for the face, body, hands and feet should be worn when filling etching baths or vats.

(g) an eye-bath should be kept ready for use in a prominent position among the first-aid equipment together with the requisite neutralising agents, which should be ready-prepared for use;

(h) acids and caustic chemicals should be stored in such a way that they cannot be confused;

(i) process liquids used in etching should be transported in closed containers;

(j) protective clothing including gloves, rubber boots, acid-resistant garments and eye protection must be worn (figure 4);

(k) personnel called upon to handle solid alkalis and dry acids must make use of shovels or tongs;

(l) only cold water should be supplied for filling vats or baths, as experience has shown that a large amount of heat may be produced through exothermic action;

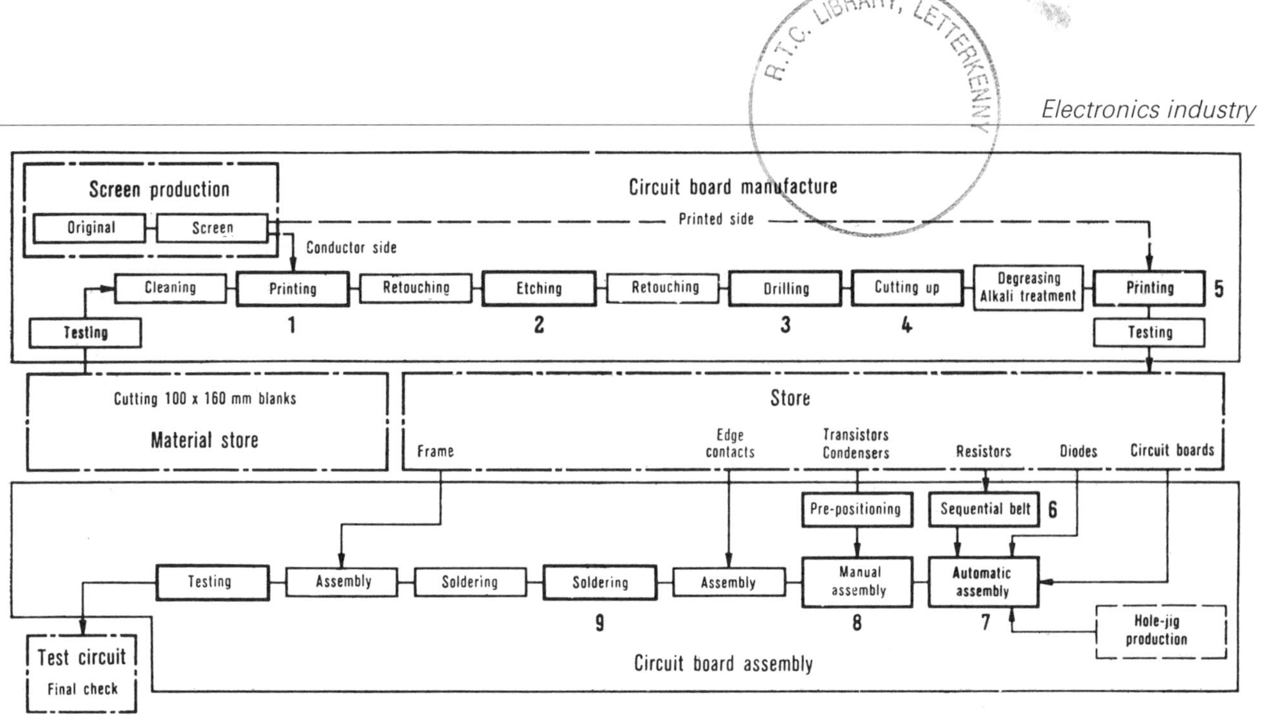

Figure 5. Flowchart of the processes in the production of an electronic printed-circuit module. *(By courtesy of Siemens AG, Berlin and Munich.)* 1. Screen printing machine. 2. Automatic etching machine. 3. Copy drilling machine. 4 Press. 5. Screen printing machine. 6. Component sequence-positioning machine. 7. Assembly machine. 8. Assembly table. 9. Solder vat.

(m) work of this nature must be performed by appropriately qualified and trained personnel only.

The sequence of processes employed in the manufacture of an electronic circuit assembly can be appreciated from figure 5, which contains a flowchart for the production of an electronic module which has edge-board contacts and may be encased in a plastics housing. A printing screen is produced from the circuit diagram and the circuit is then screen-printed on to copper-coated glass-fibre-reinforced epoxy-resin boards. The varnish used for printing is resistant to the etching solution of copper chloride into which the board is subsequently immersed and, consequently, after etching, only that portion of the copper corresponding to the circuit-diagram remains. The board is rinsed and holes and apertures are drilled for the various components and contacts, the varnish on the conductor pathways is washed off and the circuit is tested.

The various components are added to the board in an assembling machine (although certain parts may be assembled by hand) and are soldered mechanically into position. Finally the whole circuit is housed and tested (figure 6).

Figure 6. A typical printed-circuit electronic module.

The danger of accidents during testing is minimised by the automatic nature of the process and the fact that the whole of the circuit is encased in insulating material. The necessary care and cleanliness in the manufacture of products of this kind in the electronics industry are substantial guarantees for a working environment that is quite free from the risks of accident and occupational disease. The heavy expenditure required to ensure the cleanliness, supervision and reliability of all the equipment must be extended to ensure occupational safety and health if the uniform quality of the products is to be maintained for long periods. Experience has shown that in electronics undertakings the quality of the products is high and the loss from shortcomings is small when care is taken to secure the wholehearted and constant co-operation of the entire personnel by maintaining a good standard of industrial hygiene.

EGYPTIEN, H.

Technology:

"Flexible printed circuits". Technical file No. 72. *Engineering* (London), Dec. 1979 (1-8).

The impact of micro-electronics. A tentative appraisal of information technology. Rada, J. (Geneva, International Labour Office, 1980), 109 p. Ref.

Health hazards:

"Irritative dermatitis in female workers of an electromechanical factory" (Epidemia di dermatiti irritative in lavoratrici di una fabbrica elettromeccanica). Arbosti, G.; Marchese, A.; Bragonzi, G.; Nava, C.; Maddalon, G.; Marmondi, E.; Briatico, G.; Brambilla, G. *Medicina del Lavoro* (Milan), Jan.-Feb. 1979, 70/1 (50-66). Illus. 11 ref. (In Italian)

CIS 79-1010 *Microscope work–I. Investigation of visual disorders in workers using microscopes in the electronics industry* (Mikroskoparbete–I. Utredning av ögonbesvär hos mikroskopoperatörer på elektronisk industri). Söderberg, I.; Calissendorff, B.; Elofson, S.; Knave, B.; Nyman, K. G. Arbete och hälsa, Vetenskaplig skriftserie, 1978:16 (Stockholm, Arbetarskyddsverket, 1978), 54 p. Illus. 26 ref. (In Swedish)

CIS 79-1011 *Microscope work–II. An ergonomic study of microscope work in the electronics industry* (Mikroskoparbete–II. En ergonomisk studie av mikroskoparbete pa en elektronisk industri). Söderberg, I. Undersökningsrapport 1978:40 (Stockholm, Arbetarskyddsstyrelsen, 1978), 25 p. Illus. 15 ref. (In Swedish)

CIS 79-796 "Work-related respiratory disease in employees leaving an electronics factory". Perks, W. H.; Burge, P. S.; Rehahn, M.; Green, M. *Thorax* (London), Feb. 1979, 34/1 (19-22). Illus. 6 ref.

Safety in production and use:

Safety in the use of electronic devices and systems. Proceedings of the Fourth International Colloquium on the Prevention of Occupational Risks due to Electricity (Geneva, International Social Security Association, 1976), 529 p. Ref.

Electroplating

Definition. Electroplating is a chemical or electrochemical process of surface treatment. A metallic (more rarely non-metallic) layer is deposited on the base material. In principle two different metals (electrodes), which are immersed in a conductive liquid (electrolyte), are connected externally via a direct-current source. The positively charged cations of the electrolyte (e.g. copper^{2+}) migrate towards the negative cathode where they

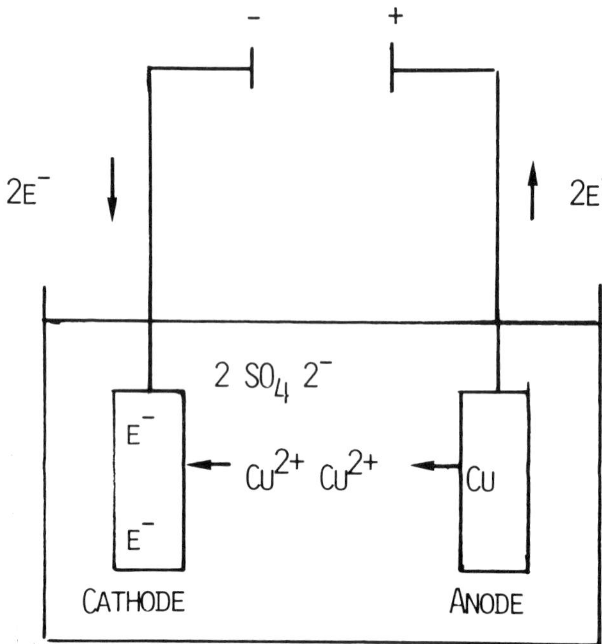

Figure 1. Principle of electroplating.

are reduced by acceptance of 2 electrons to form the neutral metal (Cu^{2+} + 2 electrons → Cu). This metal is deposited on the cathode (the material to be plated) as a thin coating. As compensation, at the positively charged anode, metal is passed by oxidation into the conductive liquid as positively charged cations (e.g. Cu → Cu^{2+} + 2 electrons). See figure 1.

Uses. Electroplating is used for components and equipment in all fields of technology. Three aims are in the foreground:

(a) to protect the material against corrosion (e.g. by a coating of nickel);

(b) to improve the surface properties (e.g. by a coating of hard chromium);

(c) to achieve optimum decorative effects (e.g. by gold or silver plating).

In its narrow sense, galvanisation includes the two main fields of electroplating (electrolytic deposition of layers of metal) and of galvano-plastic art (electrolytic production of metallic objects). In the latter process moulds (e.g. of plaster or plastic) are made conductive by the application of graphite and are connected as cathode so that the metal can be deposited upon them. In recent years anodic treatment of aluminium (eloxation) has become very important. In this process workpieces of aluminium are immersed in diluted sulphuric acid and connected as anode. However, instead of the formation of positive aluminium ions by oxidation, there is deposition of aluminium oxide as the oxygen atoms arising at the anode become bound as an oxide layer. With sulphuric acid as electrolyte the oxide formed is partly dissolved and the surface layer becomes porous. Colouring material or light-sensitive substances can be deposited in these pores (fabrication of nameplates).

Processes. To achieve optimum galvanic deposition the workpieces must be thoroughly cleaned. This can be done by mechanical treatment in the form of grinding, brushing and polishing and a chemical or electrochemical treatment with organic grease solvents or aqueous alkaline solutions or by electrolytic degreasing. In the latter process the grease on the work material is removed in baths containing partly cyanide and concentrated alkali by electrolytically formed hydrogen or oxygen. The final result is metal surfaces free from oxide and grease (so-called blank, metal surfaces). After washing procedures the actual galvanisation is carried out. The electrolyte solutions can be acidic, alkaline or alkaline/cyanidic.

As a rule a galvanisation process runs as follows: mechanical cleaning of the workpiece → rough degreasing with organic hydrocarbons → pickling in concentrated acids or alkalis → washing → electrolytic degreasing (cathodic or anodic) → washing, neutralisation → galvanisation → washing (initial washing and final washing) → drying.

An effective, water-saving washing technique, for example by means of the counterflow principle, is very important since diluted solutions are far more difficult to process than concentrated solutions. The washing effect can be enhanced by blowing in air.

HAZARDS AND THEIR PREVENTION

In the individual steps of the galvanisation process there are different types of danger because of the use of solutions of irritant and toxic chemicals, concentrated acids and alkalis as well as their vapours. In addition, there is exposure to dust from work material and gases and vapours of the organic solvents used in the cleaning procedure. A further hazard is the electric current. As a rule, the equipment for galvanisation should be installed in high rooms with good ventilation. The ventilation system should always take the exhaust air at a low level and supply the fresh air from above.

During the *mechanical cleaning* of the workpiece, for example with polishing and grinding pastes or grinding wheels, metallic or oxidic dust is formed, which must be removed by suction directly at source. Aluminium dust must be collected in a wet trap to avoid possible explosions. Iron and aluminium should be ground in separate units. In the case of persons involved in the mechanical processing, rheumatic complaints frequently arise owing to the static working posture and often as a result of the additional effect of draught. Furthermore, mechanical injury to the hands and eyes as well as toxic-irritative or allergic effects on the skin are to be expected. Protective goggles should be worn and skin protective agents used. The general ergonomic principles must be observed, particularly in the designing of the grinding and polishing units.

In the *degreasing* of the workpieces, in part with heated organic solvents (e.g. chlorinated hydrocarbons), there is the danger of poisoning through inspiration of vapours, skin contact or oral intake. Cumulative effects in the body are particularly dangerous. In the case of acute poisoning, primary narcotic effects occur that in the final stage lead to respiratory paralysis. In chronic poisoning central nervous system effects and injuries to the liver are in the foreground. Protection is achieved by sucking off the solvent vapours–which are heavier than air–from below where they are condensed. The degreasing baths should be installed in a recessed manner so that there is a safety zone of 80 to 100 cm between the breathing region of the worker and the edge of the bath. Bench ventilation must also be installed for the after-treatment of degreased parts. In cases of skin contact the strong degreasing effect of organic solvents should be remembered. Under no circumstances should benzene be used any longer.

In the *pickling process* the chief danger is the corrosive effect of concentrated acids and alkalis on the skin and the mucosa.

After latent periods of several hours nitrous gases from nitric acid can lead to serious poisoning with bronchitis, pneumonia and pulmonary oedema. In such cases hospital treatment is always necessary. Particularly dangerous is the corrosive effect of hydrofluoric acid on the skin and the mucosa, which often does not make itself felt until hours after the contact and can result in very serious injury.

First-aid measures in the event of acid or alkali burns–even to the eyes–always comprise copious washing with water. Special treatment must follow immediately.

As a rule, alkalis lead to more massive skin injury than acids. To prevent injury from acids and alkalis adequate ventilation and/or covering (see below) is also necessary. In addition, acid-proof clothing or aprons, safety shoes and protective goggles must be worn.

When diluting concentrated acids and alkalis the heat of reaction should be borne in mind. The acid should always be poured into the water while stirring continuously, not the other way round. This applies particularly to concentrated sulphuric acid.

In *electrolytic degreasing* baths containing cyanide are frequently used. Cyanides react with acids to form the highly toxic volatile prussic acid. The lethal concentration in air is 300-500 ppm. Fatal poisoning can also be caused by skin resorption or oral intake of cyanides. The first-aid measures consist of immediate transport into the open air, removal of contaminated clothing, washing of the skin with water, if possible oxygen treatment and inhalation of amyl-nitrite from a crushed ampoule.

To prevent accidental cyanide poisoning careful work under optimum hygienic conditions is necessary. Food should not be eaten at the workbench. The hands and clothing must always be carefully cleaned.

There are many dangers at the *galvanic baths* themselves. The effects of chromic acid and its salts as well as nickel compounds must be treated with urgency. In chromium plating a distinction is made between decorative chromium plating (0.1-0.5 μm) with an intermediate layer of copper or nickel, and hard chromium plating (thicker than 10 μm) without intermediate layers. The chromium is deposited from electrolytes containing hexavalent chromium compounds. These compounds cause burns, ulceration and eczematisation of the skin and mucosa; perforation of the nasal septum is typical. Bronchial asthma may also occur. Till now it has not been established clearly whether there is also an increased danger of cancer in the chromium plating industry. There is apparently a greater exposure to chromate in the case of hard chromium plating than in that of decorative chromium plating.

Nickel salts can also cause obstinate allergic or toxic-irritative skin injury and certain nickel compounds are regarded as carcinogenic for humans. Till now, however, no increased rate of malignant tumours has been observed in galvanic plants, although the measurement of nickel concentration in the urine of workers indicates the existence of the exposure. In principle, persons with an allergic case history and lung diseases should not work in an electroplating shop.

When working at galvanic baths, acid-proof clothing and gloves, safety shoes and goggles should be worn. The regular use of skin protective agents is advisable. To improve the air quality in the vicinity of galvanic baths the method of choice is powerful exhaust ventilation in combination with covering of the baths in order to reduce their surface. If possible the exhaust ventilation for cyanide baths, acid baths and alkali baths should be installed separately.

The most effective method is exhaust ventilation at the rim of the bath. The exhaust effect is improved by blowing across the rim. The average value for the quantity of extracted air per dm^2 bath surface is about 1 800-2 700 m^3/h. In many cases the baths can be covered by means of spherical plastic floating bodies or artificially applied foam layers. As a result of decomposition of water during electrolysis hydrogen mixed with air can accumulate on the surface of the bath as an explosive mixture. Here too, exhaust ventilation must be regarded as an urgent protective measure.

Galvanic shops are wet plants; in the planning of them, non-skid floors should therefore be provided–the best solution is plastic duck-boarding. The workpieces should drip above the baths to avoid the formation of a slippery surface on the floor. To prevent electrical accidents the local regulations should be strictly observed. All electric equipment should be properly maintained. In addition, environmental monitoring, and, when possible, biological monitoring are recommended.

ZOBER, A.

"Manufacturing processes: electroplating". Burges, D. C. L. *Journal of the Society of Occupational Medicine* (Bristol), 1977, 27 (114-117).

"Mortality among workers in the metal polishing and plating industry, 1951-1969". Blair, A. *Journal of Occupational Medicine* (Chicago), Mar. 1980, 22/3 (158-162). 24 ref.

"Toxicity of chromic acid in the chromium plating industry". Royle, H. *Environmental Research* (New York), 1975, 10/1 (39-53). 28 ref.

"A study of the difference in chromium exposure in workers in two types of electroplating process". Guillemin, M. P.; Berode, M. *Annals of Occupational Hygiene* (Oxford), 1978, 21/2 (105-112).

"Urinary and plasma concentrations of nickel as indicators of exposure to nickel in an electroplating shop". Tola, S.; Kilpiö, J.; Virtamo, M. *Journal of Occupational Medicine* (Chicago), Mar. 1979, 21/3 (184-188). Illus. 15 ref.

CIS 78-520 *Good work practices for electroplaters.* DHEW (NIOSH) publication No. 77-201 (National Institute for Occupational Safety and Health, 4676 Columbia Parkway, Cincinnati) (June 1977), 32 p. Illus.

Embryotoxic, fetotoxic and teratogenic effects

During the development of the embryo there is a constant interaction between genetic factors and the environment. The genetic information coded in the DNA of the ovum contains the entire programming for the future child, but gene expression depends on the environment which supplies the substances needed for growth and the differentiation of the traits and organs of the embryo. The result is that various external, physical, chemical and infectious factors may perturb morphogenesis, arrest development or produce congenital malformations.

Today, prenatal developmental disorders have become the main cause of perinatal mortality and postnatal morbidity. Before the First World War, around 1910, infant mortality during the first year of life due to prenatal developmental disorders was around 5%, whereas mortality due to infectious diseases exceeded 30%. By 1964 the relative causes of infantile mortality had been virtually inverted; less than 5% of newborn children die from infectious diseases and 20% of deaths are the result of congenital malformations. It is probable that the actual incidence of malformations has not greatly changed; however, as it has proved possible to eliminate the other causes of mortality, the relative importance of malformations becomes predominant.

It has been clearly shown that various exogenous chemical, physical and infectious factors may produce congenital malformations in laboratory animals and in man. The proportion of malformations caused by the environment is difficult to estimate. However, it is calculated that around 20% of malformations are of genetic or chromosomal origin, 10% of exogenous origin and in 60-70% of cases the aetiology of malformation is the result of an interaction between genetic and external factors.

The influence of exogenous agents varies depending on the phase of reproduction during which exposure to the substance takes place (figure 1).

Gametogenesis. It is widely accepted that the action of pharmaceutical substances on the gonads does not produce malformations. Highly toxic pharmaceutical products cause sterility and less toxic substances may reduce fertility without, however, being teratogenic.

Cleavage. This is the stage when embryotoxicity is at its highest.

Embryonic period. Following fertilisation and development of the germ layers, the embryo goes through a period of rapid morphogenesis which terminates in the structuring of the majority of the organs. Significant malformations can then be produced; this is the period of maximum teratogenic sensitivity.

In the human embryo the heart is at its most fragile between the 20th and 40th day; the central nervous system between the 15th and 25th day and the limbs between the 24th and 36th day (figure 2).

However, in addition to this chronological dependency, teratogenic agents may have preferential affinities for a given organ system. For example in the human embryo thalidomide affects primarily the skeleton and limbs; X-rays mainly cause malformation of nerve or eye tissues and rubella causes heart deformities, cataracts and deafness.

Fetal period. During this period, although the fetus is less susceptible than the embryo to pharmaceutical products, its development may still be perturbed; morphological anomalies of the genital tract and histogenic disorders of various organs may be observed, in particular of the nervous system, producing encephalopathies.

Reactions to a pharmaceutical product may vary considerably depending on the subject's genetic constitution.

The physiological or pathological status of the mother also affects the reaction of the fetus to pharmaceutical agents, but to a lesser degree. Various diseases, in particular, chronic disorders and metabolic diseases, diabetes, obesity and hypertension seem to favour the appearance of congenital malformations. In the case of diabetic and prediabetic women, the incidence of

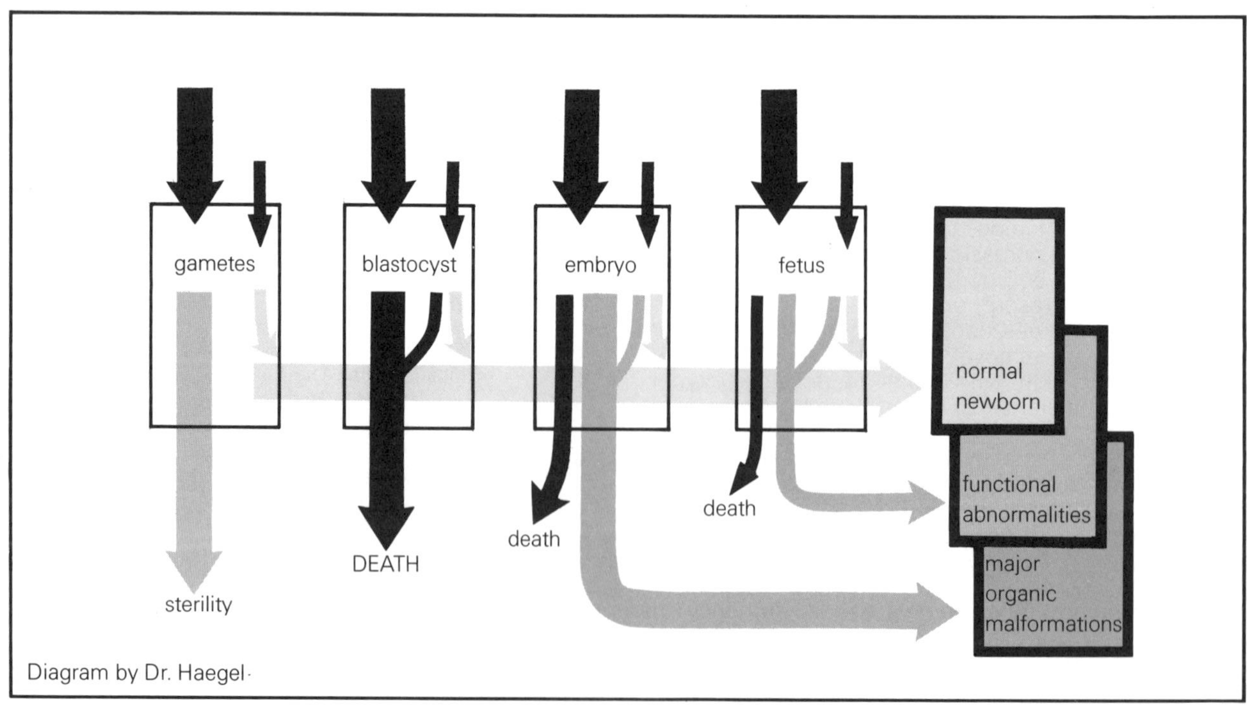

Figure 1. Variety of effects of exposure agents depending on the pace of reproduction during which they act.

abortions, premature births and congenital malformations is reported to be 2-3 times higher than in the general population.

Defects in placental circulation, which are frequent in cases of toxaemia during pregnancy and in hypertensive women, often result in an increased rate of diffusion of various pharmaceutical agents. As a result, their concentration in the fetus may reach much higher levels than normal.

Therapeutic agents

The discovery of embryopathies caused by a sedative previously considered harmless, thalidomide, led to an examination of the effect of a large number of medicines on the fetus.

The main data are summarised in tables 1 and 2.

Occupational teratology

With the increasing number of women employed in industry, environmental pollution due to industrial wastes, the use of insecticides, herbicides and defoliants constitutes a definite potential hazard for the human embryo. It is therefore important to evaluate the hazard that certain occupational activities may constitute for the offspring. These hazards are of a variety of types: retarded development, perinatal mortality, teratogenesis and, possibly, transplacental carcinogenesis such as that observed with diethylstilboestrol.

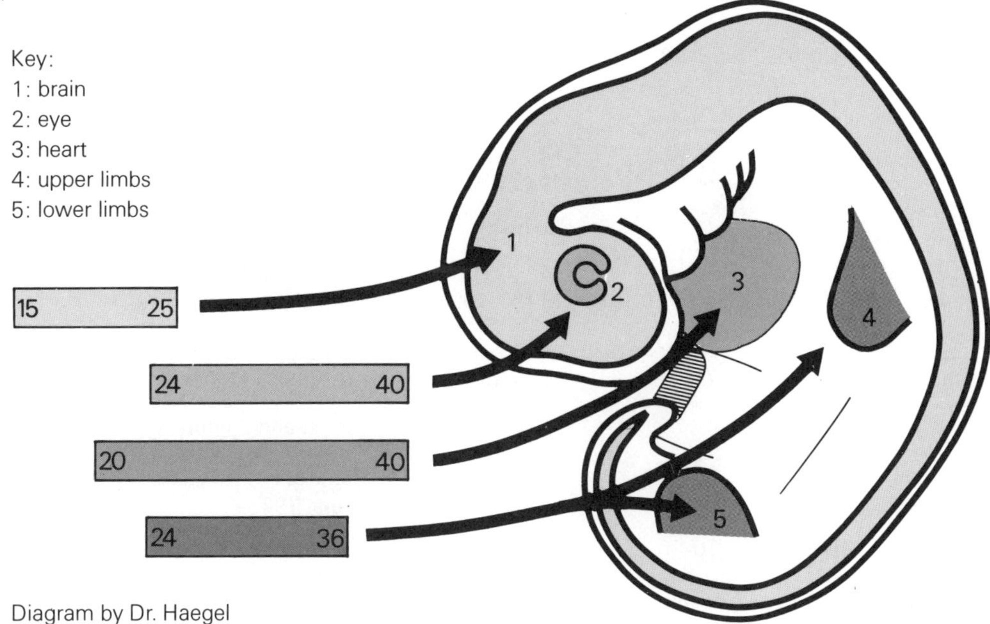

Figure 2. Critical phases of human embryo according to the days of gestation.

Table 1. Medicines which are definitely dangerous for the human fetus

Medicinal agent	Malformations produced
Male hormone and chemical progestogens similar to methyltestosterone (pregnenolone, ethinyltestosterone, ethisterone, norethisterone, norethindrone) used in high doses	Masculinisation of femal fetus
Diethylstilboestrol	Vaginal adenocarcinoma in young girls
Cytotoxic drugs (cyclophosphamide, busulphan, chlorambucil)	Malformations of the central nervous system and skeleton; visceral anomalies
Folic acid antagonists (methotrexate, aminopterine, etc.)	Growth retardation, central nervous system anomalies, visceral and skeletal malformations
Thalidomide	Limb malformations
Mercury salts	Microcephaly, cerebral paralysis

Table 2. Chemical substances which are potentially teratogenic in various species of animal but whose teratogenic action has not been proven in man

Substance	Malformations produced
Anticonvulsants, diphenylhydantoin	Cleft palate, heart malformations
Hypoglycaemic agents (sulphamides, biguanides)	Malformation of the eyes and nervous system
Antiemetics (cyclizine, meclizine)	Malformations of the eyes and skeleton, cleft palate
Corticosteroids–cortisone, hydrocortisone	Malformations of the heart and face
Salicylates–aspirin	Malformations of the face and skeleton
Antitubercular agents (rifamycin)	Malformations of the central nervous system
Antimalarial agents (chloroquine, mepacrine, pyrimethamine)	Malformations of the eyes and ear
Anticoagulants (coumarin)	Haemorrhages
Agents for hypercholesterolaemia and hypocholesterolaemia	Malformations of the central nervous system
Anticancer drugs (actinomycin-D, mitomycin-C, cyctotoxics)	Various malformations
Industrial solvents (formamide, urethane)	Various malformations

Table 3. Teratological testing of industrially important metals

Compound	Species	Dose[1] (mg/kg)	Effects[2] (M = malformations)	Reference	Calculated human dose at TLV (mg/kg/d)[3]
Aluminium-chloride	Rat	40-200	M	Bennett et al. 1974	–
Arsenic-sodium arsenate	Chick	< 6	Growth inhibition	Ridgway and Karnofsky 1952	0.05 (as As)
-sodium arsenate	Hamster	20	M	Ferm and Carpenter 1968a	
-sodium arsenate	Mouse	45	M (63%)	Hood and Pike 1972	
-sodium arsenate	Rat	30	M	Beaudoin 1974	
Cadmium-sulphate	Hamster	2	M (65%)	Ferm and Carpenter 1968b	0.02 (as Cd)
	Mouse	10 (ppm, oral)	M (14%)	Schroeder and Mitchener 1971	
-chloride	Rat	6	M (35%)	Chernoff 1973	
Chromium-sodium dichromate	Chick	< 1.2	M	Ridgway and Karnofsky 1952	0.01 (as CrO_3)
-trioxide	Hamster	15	M (64%)	Gale 1978	
Cobalt-chloride	Chick	< 10	M	Ridgway and Karnofsky 1952	0.01 (as Co)
-chloride	Chick	~ 10	M (2.7%)	Kury and Crosby 1968	
Copper-sulphate	Hamster	7	Embryotoxic	Ferm and Hanlon 1974	0.1 (as Cu)
Lead-nitrate	Chick	~ 2	M (96%)	Ridgway and Karnofsky 1952	0.02 (as Pb)
-nitrate, -chloride, -acetate	Hamster	50	M (90-100%)	Ferm and Carpenter 1967	
-tetraethyl	Mouse	25 (ppm, oral)	M (95%)	Schroeder and Mitchener 1971	
-tetraethyl	Rat	7.5-30 (oral)	Embryotoxic	McClain and Becker 1972	
Mercury, methyl mercury dicyanidamide	Mouse	8	M (50%)	Spyker and Smithberg 1972	0.001 (alkyl Hg)
methylmercury chloride	Hamster	8	M	Harris et al. 1972	
methylmercury chloride	Cat	0.25 (oral)	M (30%)	Khera 1973a	
Nickel-acetate	Hamster	10-30	Embryotoxic	Ferm 1972	0.1 (as Ni)
-carbonyl	Rat	0.3 mg/ l air 15 min	M (28%)	Sunderman et al. 1979	
-chloride	Mouse	4.6	M (51%)	Lu et al. 1979	

[1, 2, 3] See table 6.

Source: K. Hemminki: "Occupational chemicals tested for teratogenicity", *International Archives of Occupational and Environmental Health* (West Berlin), Dec. 1980, 47/3 (191-207) *(By courtesy of Springer Verlag, Heidelberg, and J. F. Bergmann Verlag, Munich.)*

Detection of an "occupational" teratogen is usually extremely difficult, except in the case of a highly dangerous substance which produces in the exposed group a high percentage of malformations of a specific type and during a specific period. This was the case in mercurial contamination in Minamata in Japan, or with polychlorinated diphenyls. The effect of agents with a low teratogenic potential may pass unnoticed for a long time, since it is masked by the normal rate of spontaneous malformations. Chronological diagnosis therefore requires extremely wide surveys of large population groups exposed to the same agents. In this way even slight fluctuations in the number of malformations can be detected.

The problem may also be complicated by other factors: long latent period between exposure to the agent and malformation induction; variations in individual genetic sensitivity; synergic interaction or antagonistic interaction; the quality and strictness of the research workers.

In spite of these difficulties, studies carried out since 1940, at a time when the number of known dangerous agents was no more than four or five (ionising radiation, lead, mercury, arsenic, alcohol) have led to the detection of several substances which are potentially harmful to the human embryo (tables 3, 4, 5 and 6).

A certain number of substances which appear particularly important are analysed below.

Mercury salts. These substances, which contaminated fish in Minamata Bay in Japan, produce severe poisoning in man. A total of 134 cases were observed, 78 adults, 31 children and 25 fetuses. In the fetal cases, Minamata disease produced anomalies of the central nervous system and the skeleton. Very high mercury levels were found in pregnant women and also in female children up to the age of 5 years.

Mercury salts (methyl mercury) used as a fungicide produced a severe accident in Iraq in 1972. Amongst the people who ate bread made with wheat treated with this fungicide, there were 450 cases of fatal poisoning.

Experimentally it has been shown that doses of 2.5-7.5 mg/kg of methyl mercury chloride have an embryolethal and teratogenic effect on mice and rats.

Lead. This may be ingested in food, contaminated fish or drinking water due to the use of lead piping. It is highly teratogenic in the chicken. In the rat, however, it has relatively low teratogenic potential; and a small number of nervous system effects have been observed. Chronic lead acetate poisoning from weaning to copulation and continued throughout gestation has proved to have little or no teratogenic action. Fetal growth is inhibited and sexual maturity of females is delayed by two weeks.

Chelating agents increase the fetal toxicity of the lead. Compounds such as tetraethyl, tetramethyl and trimethyl lead are not teratogenic in the rat. At high doses, there is significant maternal toxicity, however. The fetus is

Table 4. Teratological testing of industrially important organic compounds

Compound	Species	Dose[1] (mg/kg)	Effects[2] (M = malformations)	Reference	Calculated human dose at TLV (mg/kg/d)[3]
Aminoazobenzene	Mouse	200-700	M	Sugiyama et al. 1960	–
Azo dyes, trypanblue, etc.	Rat	~ 50	M (~ 20%)	Beck and Lloyd 1966	–
Detergents					
alkylbenzene sulphonate	Mouse	300	M (minor 34%)	Palmer et al. 1975a	–
alcohol sulphate	Mouse	600	M (minor 30%)	Palmer et al. 1975a	–
olefin sulphonate	Mouse, Rabbit	300-600	M (20-100%)	Palmer et al. 1975b	–
Triton X	Mouse	200	M	Roussel and Tuchmann-Duplessis 1968	–
7,12-Dimethylbenz(a)-anthracene	Rat	0.25	M (~ 100%)	Currie et al. 1970	–
Ethylenethiourea	Rat	40	M (80%)	Khera 1973b	–
Halothane	Rat	0.8%/12 h inhal.	M (skeletal, ~ 50%)	Basford and Fink 1968	–
Hydrazine	Chick	~ 2	M (75%)	Stoll et al. 1967	0.1 (1.3 ppm)
Pentachlorobenzene	Rat	200	M (~ 40%)	Khera and Villeneuve 1975	–
Tetrachlorodibenzo-*p*-dioxine (2,3,7,8-)	Mouse	0.006	M (oral cleft 50%)	Neubert et al. 1973	–
Tetrachlorophenol	Rat	30	Fetotoxicity, delayed ossification	Schwetz et al. 1974b	–
Thiurams					
thiram	Hamster, Rat	250	M (~ 20%)	Robens 1969	0.5 (5 mg/m³)
thiram	Mouse	300-500	M (~ 60%)	Roll 1971	–
disulphiram	Hamster	125	M (~ 60%)	Robens 1969	–

1, 2, 3 See table 6.

Source: K. Hemminki: "Occupational chemicals tested for teratogenicity", *International Archives of Occupational and Environmental Health* (West Berlin), Dec. 1980, 47/3 (191-207). *(By courtesy of Springer Verlag, Heidelberg, and J.F. Bergmann Verlag, Munich.)*

smaller than normal, ossification is delayed but it does not present morphological anomalies.

Clinical data indicate the general toxicity of lead but only seldom indicate a teratogenic potential. In women employed in lead-polluted premises in Japan an increased incidence of abortions was observed. However, other observations in factories producing batteries show no correlation between lead exposure and embryotoxic effects.

Cadmium. This is teratogenic in rodents. It is highly embryolethal and causes malformations of the central nervous system and eye. In the case of mice, about 25% of the fetuses have malformations. Cadmium is also toxic when inhaled. Pronounced reduction of growth was noticed in the litter. Clinically, it has seldom been possible to establish a correlation between cadmium exposure and congenital malformations.

Sodium arsenate. This is teratogenic for birds and produces a high percentage of renal malformations in the rat and hamster. The anomalies are often accompanied by urogenital malformations.

Asbestos. Asbestos is not teratogenic.

Solvents. Since solvents are diffused rapidly throughout the body tissues, one might expect they would be extremely dangerous for the embryo. Experiments on rodents have shown that, with the exception of dimethylsulphoxide and formamide, which have embryotoxic and teratogenic action, other solvents are less dangerous. Chloroform, carbon tetrachloride, trichloroethylene, perchloroethylene, benzene, xylene, cyclohexanone and propylene glycol produce intra-uterine growth retardation and, in certain cases, abortions but not malformations.

Observations on female laboratory workers and printing workers (roto-printing factories) exposed to organic solvents showed an increase in chromosome anomalies. Similar anomalies are reported to have been found in the children born of mothers exposed to these substances during their pregnancies. Erickson et al. (1978) examined several hundred women employed in printing works where they were exposed to various chemicals. They found an increase in the level of malformations amongst the children of these women. However, confirmation of the exactness of this cause of malformation is required.

Detergents. Detergents do not seem very dangerous. One detergent, however, triton W.R. 1339 has been shown to be embryotoxic and teratogenic in the mouse and rat. Nitrilotriacetate has also been recorded as dangerous in certain studies.

Pesticides. Studies on the mouse, rat and rabbit indicate the low hazard of DDT. Other pesticides appear to be more dangerous. Parathion acts electively on the genital tract. Germ cells are severely degraded and sometimes completely destroyed. Ovogenesis is also inhibited.

Table 5. Teratological testing of industrially important solvents

Compound	Species	Dose[1] (mg/kg)	Effects[2] (M = malformations)	Reference	Calculated human dose at TLV (mg/kg/d)[3]
Benzene	Mouse	2 500	M (8%)	Watanabe et al. 1970	3 (10 ppm)
	Mouse, Rabbit	500 ppm/7 h inhal.	M (minor skeletal)	Murray et al. 1979	
Carbon disulphide	Rat	100 mg/m³/8 h inhal.	M (27%)	Tabacova et al. 1978	6 (60 mg/m³)
Chloroform	Rat	100 ppm/7 h inhal.	M	Schwetz et al. 1974a	(50 ppm)
Dimethyl formamide	Mouse	600	M (18%)	Scheufler and Freye 1975	3 (10 ppm)
Dimethyl sulphoxide	Hamster, Mouse	3 300-10 000	M	Staples and Pecharo 1973	–
	Hamster	2 500-8 250	M (33%)	Ferm 1966	
Methylene chloride	Chick	40	M (14%)	Elovaara et al. 1979	174 (500 ppm)
	Rat	1 500 ppm/7 h inhal.	M (minor skeletal)	Schwetz et al. 1975	
	Mouse	1 500 ppm/7 h inhal.	M (minor skeletal)	Schwetz et al. 1975	
Propylene glycol	Chick	850	M (65%)	Gebhardt 1968	–
Trichloroethane 1,1,1-	Chick	13	M (25%)	Elovaara et al. 1979	–
Trichloroethane, 1,1,2-	Chick	13	M (14%)	Elovaara et al. 1979	5 (10 ppm)
Trichloroethylene	Rat	1 800 ppm/6 h inhal.	M (minor)	Dorfmueller et al. 1979	54 (100 ppm)
	Chick	13	M (33%)	Elovaara et al. 1979	
Tetrachloroethane	Mouse	700	M (9%)	Schmidt 1976	3.5 (5 ppm)
Tetrachloroethylene	Chick	17	M (29%)	Elovaara et al. 1979	67 (100 ppm)
Toluene	Rat	1 500 mg/m³/24 h inhal.	M (minor skeletal)	Hudák and Ungváry 1978	75 (750 mg/m³)
Xylene	Rat	1 000 mg/m³/24 h inhal.	M (minor skeletal)	Hudák and Ungváry 1978	44 (435 mg/m³)

1, 2, 3 See table 6.

Source: K. Hemminki: "Occupational chemicals tested for teratogenicity", *International Archives of Occupational and Environmental Health* (West Berlin), Dec. 1980, 47/3 (191-207). *(By courtesy of Springer Verlag, Heidelberg, and J. F. Bergmann Verlag, Munich.)*

Two other insecticides, demeton and penthion, produce abortions and a small number of malformations in mice. In addition, a reduction in cholinesterase activity in the brain is also noted in the fetus.

2,4,5-T (trichlorophenoxyacetic acid) is a pesticide incriminated in the aetiology of two human malformations. This herbicide contains an impurity, dioxine, which is known as one of the most highly toxic substances.

Following the use of Folpet (*n*-trichloromethylthiaphthalamide, a defoliant) between 1962 and 1970, there were reports of an increase in the number of abortions and congenital malformations in the North Vietnamese population. An international scientific commission of 17 members, charged with drawing up a report for the United States Senate, concluded, however, that there was no correlation between Folpet and the appearance of malformations in man. Nevertheless, this commission reported that 3,4,5-T and its main contaminant, 2,3,7,8-tetrachloro-dibenzo-*p*-dioxine (TCDD) was capable of producing malformations in animals. Consequently, the commission recommended that Folpet should be no longer used in the future because of its possible danger for man.

Dioxine. On the other hand, dioxine is a highly toxic substance. Doses of 3 μg/kg given between the sixth and fifteenth day produced a very high percentage of cleft palates and hydronephroses in mice. Doses of 1-10 μg/kg have high embryotoxicity. Experiments with primates using relatively low doses indicate that TCDD may be less toxic for the rhesus monkey than for rodents. The effect of dioxine on human reproduction was studied in detail following the Seveso accident in 1976. Epidemiological studies carried out by the Italian governmental health services and various international specialists indicate that there was no increase in either the rate of spontaneous abortions or of congenital malformations in the dioxine-contaminated zones. Examination of embryos from therapeutic abortions and cytogenetic studies revealed no anomalies or chromosomal aberrations attributable to dioxine. The postnatal development of children of mothers who lived in the Seveso region during their pregnancy was normal.

Other environmental substances that should be mentioned include the phthalic acid esters; high doses (around a third of the LD_{50}) in the mouse and rat inhibit intra-uterine growth and produce embryolethal effects.

Polybrominated biphenyls. These may produce malformations in the mouse.

Table 6. Teratological testing of plastics monomers and additives

Compound	Species	Dose[1] (mg/kg)	Effects[2] (M = malformations)	Reference	Calculated human dose at TLV (mg/kg/d)[3]
Acrylonitrile	Rat	25	M	Murray et al. 1978a	5 (20 ppm)
		80 ppm/6 h inhal.	M	Murray et al. 1978a	
Epoxy resin (Epidian 5)	Guinea-pig	Skin painting	Fetotoxicity	Woyton et al. 1975	–
Methacrylate esters	Rat	5-700	M	Singh et al. 1972a	3.5 (10 ppm)
Phthalate esters	Chick	Unspecified	M	Lee et al. 1974	–
	Rat	300-1 000	M	Singh et al. 1972b	
Styrene	Chick	2.5	M (7%)	Vainio et al. 1977	42 (100 ppm)
	Rat	300-600 ppm/7 h inhal.	M (minor skeletal)	Murray et al. 1978b	
	Rabbit	300-600 ppm/7 h inhal.	M (minor skeletal)	Murray et al. 1978b	
	Mouse	250 ppm/6 h inhal.	Fetotoxicity	Kankaanpää et al. 1980	
	Hamster	1 000 ppm/6 h inhal.	Fetotoxicity	Kankaanpää et al. 1980	
Urethane	Mouse	15	M	Sinclair 1950	
Vinyl chloride	Mouse	50 ppm/7 h inhal.	M	John et al. 1977	0.3 (1 ppm)
	Rat	500 ppm/7 h inhal.	M (minor skeletal)	John et al. 1977	
	Rabbit	500 ppm/7 h inhal.	M (minor skeletal)	John et al. 1977	

[1] Dose referring to the injected amount unless stated otherwise. [2] Figures in parentheses indicate the percentage of malformed animals if stated in the reference. [3] Human dose as calculated by taking the TLV concentration (defined by OSHA 1974) and assuming a daily inhalation of 10 m^3 and absorption of 50% of dose by the lungs of a person weighing 50 kg.

Source: K. Hemminki: "Occupational chemicals tested for teratogenicity", *International Archives of Occupational and Environmental Health* (West Berlin), Dec. 1980, 47/3 (191-207). *(By courtesy of Springer Verlag, Heidelberg, and J. F. Bergmann Verlag, Munich.)*

Polychlorinated biphenyls. These are very toxic for the mother. They inhibit embryo development without, however, causing malformations. Ingestion by pregnant women of rice oil contaminated by polychlorinated biphenyls produces a brownish skin colour amongst the offspring (cola-coloured infants).

Aerosol spray adhesives have been prohibited in the United States on the basis of inaccurate epidemiological observations; these substances are, in fact, not teratogenic either for rodents or the human embryo.

To conclude, the data accumulated over the last 20 years lead to the conclusion that pesticides do not seem to be dangerous for the human embryo when they are used in accordance with regulations. On the other hand, excessive voluntary or accidental use of pesticides may have a dangerous effect on the offspring.

Physical factors

Ultrasound. This does not have embryotoxic or teratogenic effects in man. However, in the mouse, it has been possible to produce malformations.

Hyperthermia. This may produce malformations in various animal species including primates. The malformations are generally polymorphic. In man, it is very difficult to determine the true role of hyperthermia since, in the majority of cases, hyperthermic syndromes are treated by various antibacterial agents. In addition, the exact aetiology of the rare malformation observed in these conditions continues to be indecisive.

Noise. In the mouse it has been possible to produce malformations with audiostimuli. However, it does not seem that the same applies to man.

Ionising radiations. The teratogenic effect of ionising radiation has been recognised since the beginning of this century, and they are dealt with in a separate article.

Hypoxia

Experiments show differences in sensitivity to hypoxia among various animal species. The mouse has very high sensitivity; the rabbit, hamster and, in particular, the rat are more resistant to hypoxia. In severe experimental hypoxia, observations have been made of a large variety of malformations of the central nervous system, the sensory organs, and the skeleton, often associated with functional anomalies. A number of authors have incriminated hypoxia in the aetiology of human malformations. However, epidemiological studies have not confirmed this hypothesis. It has merely been established that perinatal hypoxia is capable of producing (often irreversible) nervous lesions.

Carbon monoxide

This is one of the most dangerous pollutants. In cases of acute poisoning, high fetal mortality or morphological anomalies associated with neuromuscular disorders have been observed.

Other atmospheric pollutants

The influence of other atmospheric pollutants such as vinyl chloride, the fluorides, the arsenicals and the sulphides are well known. As a result of their general toxicity, they may interfere with pregnancy but they do not have a specific teratogenic action. It was reported that the percentage of abortions amongst nurses working in operating theatres was 17% as against 14%

for a control group. The incidence of malformations was also reported as being higher: 9.6% as against 5.9% for unexposed women. Although experimental studies have shown that halothane and nitrous oxide may produce abortions and, more rarely, malformations, the responsibility of anaesthetics in the aetiology of abortions and human malformations does not seem to be sufficiently well established. New prospective studies will be necessary to provide a reply to this preoccupying question.

Tobacco

The harmful effect of smoking has been well established. The most characteristic and most constant result is low birth weight and an increase in perinatal mortality. A clear increase in the rate of abortions amongst female heavy smokers has also been reported: the risk is indicated as being twice as high as in the general population. In addition, tobacco potentiates the danger of alcohol.

Alcohol

The most characteristic symptoms of children born to alcoholic mothers are somatic and psychomotor retardation, head and neck dysmorphias associated with eye anomalies. The clinical picture has been described by the term "fetal alcoholism syndrome". Autopsy showed manifest deficiencies in the central nervous sytem, including anomalies of nervous and glial cell migration. There is retardation of postnatal growth and weight age is below stature age. The malformation hazard is difficult to evaluate. But there is no doubt that this increases with the severity and chronicity of the mother's alcohol impregnation. In general, the rate of malformations is 2-3 per 1 000. It usually affects later siblings in a family up until then unaffected and where the mother has clear signs of alcohol poisoning.

TUCHMANN-DUPLESSIS, H.

"Genetic risks caused by occupational chemicals: Use of experimental methods and occupational risk group monitoring in the detection of environmental chemicals causing mutations, cancer and malformations". Hemminki, K.; Sorsa, M.; Vainio, H. *Scandinavian Journal of Work, Environment and Health* (Helsinki), Dec. 1979, 5/4 (307-327). Illus. 158 ref.

"Occupational chemicals tested for teratogenicity". Hemminki, K. *International Archives of Occupational and Environmental Health* (West Berlin), Dec. 1980, 47/3 (191-207). Illus. 98 ref.

"Experimental methods to detect teratogenic agents" (Méthodes expérimentales pour la détection des agents tératogènes). Tuchmann-Duplessis, H. *Biologie médicale* (Paris).

"Risk to the offspring from parental occupational exposures". Haas, J. F.; Schottenfield, D. *Journal of Occupational Medicine* (Chicago), Sep. 1979, 21/9 (607-613). 49 ref.

CIS 79-1685 "Central-nervous-system defects in children born to mothers exposed to organic solvents during pregnancy". Holmberg, P. C. *Lancet* (London), 28 July 1979, 2/8135 (177-179). 5 ref.

CIS 79-1748 "Major malformations in infants born of women who worked in laboratories while pregnant". Meirik, O.; Källén, B.; Gauffin, U.; Ericson, A. *Lancet* (London), 14 July 1979, 2/8133, 91 p. 5 ref.

CIS 79-1049 "Possible health damage to offspring of operating theatre personnel" (Gesondheidsrisico's voor het nageslacht van operatiekamerpersoneel). Koemeester, A.; Mes, J.; Brimicombe, R. *Tijdschrift voor sociale geneeskunde* (Amstelveen, Netherlands), 21 Mar. 1979, 57/6 (196-200). 25 ref. (In Dutch)

Further bibliographical data on the subject have been provided by the author and are available on request at the ILO.

Emergency exits

Emergency exits offer people in industrial premises, mines, department stores, places of entertainment and hotels the possibility of escaping rapidly to safety in the event of fire, or other dangers. The time required for the evacuation of a large number of people depends on the design of the premises. Rapid escape is possible from single-storey buildings where people may leave through doors and windows. A high level of resistance to fire of the building structure, walls and floor slabs, and unobstructed passages, stairs, doors and corridors are prerequisites for safe evacuation.

The number of emergency exits required for a given building or mine depends on the manner in which people move about: singly or in large groups, in disorder or in streams, freely or in dense crowds, under constraint, for long or short periods. When people move in dense crowds, individual freedom of action is limited. Crowds may have to move under difficult conditions because of heat (more than 60 °C), smoke and toxic combustion products (carbon monoxide may become dangerous in concentrations of more than 0.5% by volume of air).

Badly organised evacuation is often accompanied by panic. Everybody tries to leave the danger zone as quickly as possible and employs the maximum physical force to get as far away as possible from the source of danger. This behaviour blocks movement towards the emergency exit, and causes injuries and loss of life.

The source of danger may be real or imaginary, and its location may be obvious or uncertain. Panic may even arise when there is no real threat to life and health or danger of material damage.

Construction requirements. People may escape safely if enough appropriate emergency exits have been planned for at the design stage. According to the fire protection standards and regulations of many developed countries, emergency exits must afford escape:

(a) from premises on the ground floor either directly or through a corridor, entrance or staircase;

(b) from premises on floors other than the ground floor through a corridor or passage leading to a staircase affording egress either directly or through an entrance separated from the corridors by partitions with doors (including glass doors);

(c) from premises to neighbouring premises on the same level, affording egress of the types mentioned under *(a)* and *(b)*.

Apertures and passages, including those provided with doors, that do not meet the requirements set out above are not regarded as emergency exits and cannot be taken into account.

Premises with a high fire hazard and a large number of occupants (more than 25 people) must have not less than two emergency exits, which should not be more than 40 m from the most remote workplace. The number of emergency exits for escape from very long and spacious premises or mines has to be determined according to the circumstances in each case. The exits are generally arranged at fixed intervals on the sides of buildings. Under normal conditions emergency exits are not used. Therefore, when these exits are designed and laid out, it has to be borne in mind that people will tend to use the usual exits in the event of an emergency. For buildings with large numbers of occupants the normal exits should be used as emergency exits, and the normal exits of industrial buildings with a small number of occupants may exceptionally be used as emergency exits. There should be at least two evacuation routes and one emergency exit per route.

Evacuation process. When people escape from a building or premises with a large number of occupants, three phases of evacuation can be distinguished:

(a) people move from the most remote places of the premises to the exit from these places;

(b) people move from the exits of the workplaces to the exits affording egress from the building, passing through corridors, passages, lobbies, staircases and entrances;

(c) people move through the exits from the building and disperse on the adjacent ground.

The first evacuation phase is of particular importance. In this phase the occupants are close to the source of danger and may be exposed to flames, smoke and toxic combustion products. This phase must be as short as possible so that the occupants may escape from the danger zone before the fire assumes dangerous proportions.

The second phase is less dangerous for human life and may take longer than the first. Safe evacuation in the third phase is ensured by providing free spaces at the exits from buildings.

In the event of forced evacuation single files or elementary streams of people form in the passageways between equipment, structural elements or rows of seats and converge into primary streams in the main passages. Complex streams form in the corridors, lobbies and staircases.

The movement of these streams of people is limited by the width and length of the stream. The width of the stream is determined by that of the escape route. A basic dimension for the design of escape routes is the minimum width of 0.6 m for a single file.

Speed of evacuation depends on the linear density of the stream of people, i.e. the mean length of the escape route for one person. Taking as a basis an average step length of 0.6 m plus a tread of 0.2 m, the theoretical linear density equals 0.8 m per person. The actual linear density depends on the number of people crowding the escape route (occupying 50-100% of its surface). The speed of movement diminishes in proportion to the increase in density of the stream. With a density of 0.8-1 m per person, the speed of movement is 70 m/min; with a density of 0.2-0.85 m per person it is 15-17 m/min. At the point of saturation of the linear density the speed of movement is considered to be around 16 m/min on a horizontal level, 10 m/min downstairs and 6 m/min upstairs.

The maximum time required to evacuate a building is given by the quotient of the length of escape route and the speed of movement. The speed strongly diminishes where the escape route narrows and is controlled by the passage capacity of doors and stairs. Speed of movement through doors and on stairs is determined by the product of the specific passage capacity in persons/m·min and the width of the exit. If a stream of people moves through an exit the width of which is smaller than that of the stream, an arch consisting of several persons is formed. This arch rests on the door frame or on obstacles at the sides of the door opening. Such knots of people form when door openings are around 0.75 m in width. With a width of 0.9 m the movement becomes intermittent, and with a width of 1.2 m and more any arch forming is easily eliminated by the moving stream.

According to the USSR fire protection standards for building construction, the passage capacity of doors and stairs of up to 1.5 m wide is assumed to be 50 persons/m·min, and with widths of more than 1.5 m, 60 persons/m·min. Doors must be at least 0.8 m wide and 2.0 m high and must open in the direction of exit from the premises or building.

The duration of the first escape phase should not exceed 1 to 1.5 min according to fire protection standards, and the total time required for evacuating a building should be 2 minutes for buildings of fire resistance class 3 and up to 6 minutes for buildings of fire resistance classes 1 and 2. The time of escape from a building where a fire has started must be shorter than the critical duration of an uncontrollable source of fire, i.e. the time after which the temperature, the concentration of toxic combustion products and smoke density become dangerous.

Multistorey buildings. The staircases of multistorey buildings with their daylight openings should be constructed of non-combustible materials and be especially fire-resistant so that they can serve as escape routes. The width of the stairways and landings, and also that of the doors, corridors and passages along the escape routes on the various floors should be calculated as not less than 0.6 m per 100 occupants. Spiral staircases and steeply inclined stairs are not admitted as escape ways. Exits to the garrets and roof, and smoke vents, should be provided for at the building design stage.

External fire escapes that can be reached through windows or doors by means of platforms or balconies, and also gangways or ladders between neighbouring balconies, may be used as emergency exits. For industrial and public buildings where cellulose materials are the main combustible hazard the escape should not take more than 2.5-7.5 min according to the size of the premises and the source of fire. This escape time must be reduced to 1.5-2 min for premises where loose combustible fibrous materials and flammable liquids are stored.

Fire escapes should be provided at intervals of 10 m on the façades of industrial buildings. They should be at least 0.7 m wide, must have a railing 1 m in height and steps at least 0.12 m deep and not more than 0.25 m vertically apart. The steps of these stairs and flooring of the platforms must be made of expanded steel sheet or ribbed tread plates.

Escape from trenches, foundation pits, prospecting pits and opencast mine workings should be facilitated by inclines, ramps, gangways and stairways with railings, and underground mine workings should have escape hoists and special exits.

Every building should have approved floor-by-floor escape plans. These must be posted in the workshops, halls and corridors and along the escape route. The workers must be acquainted with the escape rules during fire drills. Their familiarity with the plans and rules should be checked during training sessions. The organisation of rapid escape to safety in the event of fires is a responsibility of the heads of undertakings, institutions, shops, departments and sections, and also of volunteer firemen, fire services and police authorities.

PANOV, G. E.
POLOZKOV, V. T.

Basic fire protection standards in building construction (Osnovy protivopŏannogo normirovanija v stroitel'stve). Rojtman, M. Ja. (Moscow, Stroizdat, 1969).

Building standards and regulations A-A.5-70–Fire protection standards for the design of buildings and plants (Stroitel'nye normy i pravila A-A.5-70–Protivopŏarnye normy proektirovanija zdanij i soorŭenij) (Moscow, Stroizdat, 1978), 16 p.

CIS 79-1247 "Human engineering considerations in emergency exiting". Cohn, B. M. *Professional Safety* (Park Ridge), Apr. 1979, 24/4 (24-28). Illus. 4 ref.

CIS 75-1231 *Life safety code 1973.* NFPA 101-1973 (National Fire Protection Association, 470 Atlantic Avenue, Boston, Massachusetts) (1973), 241 p. 86 ref.

"Evacuation system for high-rise buildings". Data sheet 1-656-81 *National Safety News* (Chicago), June 1981, 123/6 (53-61). Illus. 25 ref.

Employers' and workers' co-operation

The need for co-operation

Safety in the undertaking and the organisation of the prevention of accidents and occupational diseases is a basic responsibility of management. This responsibility covers the layout of factory premises, the machinery and equipment, and the planning and execution of a system of work designed to protect employees to the greatest extent possible against all risks to their life and health. Even though he may delegate all or part of his supervisory duties, a manager remains fundamentally responsible for safety at work.

Although the employer is legally responsible for the safety of work in the undertaking, including the prevention of accidents and occupational diseases, the joint participation of employers and workers and their organisations is indispensable. Starting in the mining industry at the beginning of the present century, the practice of discussing safety questions with the workers spread to the other branches of industry and became legally enforced. There now exists a large body of legislation relating to occupational safety and health and a more or less comprehensive framework for the co-operation of management with the representatives of the workers in the factory.

The participation of the worker is of great importance because he is entitled to the same good health and physical integrity in his working life as he would enjoy throughout a normal lifetime. An accident involves the personal destiny of an individual and thus the claim for increased safety is a basic human requirement and an essential step towards the humanisation of the workplace.

The costs and loss in profits resulting from industrial accidents and occupational diseases do not fall upon the employer alone, but are sufficiently important to become a factor in the over-all political economy of a country. Moreover, the economic situation of a manufacturer and his readiness to invest in, to create and to maintain employment are matters which directly affect the worker.

There is a psychological aspect of co-operation between employers and workers that should not be underestimated. Workers tend to consider that safety measures are a waste of time and a nuisance, despite their recognised necessity, and there is a natural tendency for them to disregard these measures, especially when no clear explanation has been given them of the accident or health hazard involved or if the explanation has not been repeated often enough. Thus, for a safety measure to be accepted and observed, its introduction should be based on a joint decision of management and the workers or their representatives.

Levels and forms of co-operation between employers and workers

Co-operation between employers and workers should take place at a number of mutually complementary levels, namely *(a)* within the undertaking; *(b)* at industry level; *(c)* at the national level; and *(d)* at the international level. In this way, duplication can be avoided and it can be ensured that activities such as research projects, surveys and inquiries can be carried out at the appropriate level, provided that there is constant intercommunication between the different levels.

Co-operation within the undertaking. Co-operation between employers and workers at the factory level enables the everyday practical problems within the plant to be dealt with in a flexible manner. Important basic principles for such co-operation are:

(a) the worker must be kept clearly and fully informed about the accident risks and health hazards liable to be encountered in the course of his work, including the necessary safety measures and first-aid and rescue procedures. Regular explanation and repetition of these matters should be aimed at stimulating safety consciousness and motivating the worker's own individual responsibility;

(b) there should be regular consultation and exchange of information concerning safety and health measures through the safety committee (or other representative body) and the factory medical officer;

(c) over-all medical supervision should be the responsibility of an occupational medical service provided by the undertaking;

(d) certain cases arise where a worker may no longer be able to carry on his former activity because of ill-health. It is important that there should be co-operation between the employer and the workers to ensure that the affected worker does not suffer hardship as a result of his having to obtain treatment or of not being able to continue his previous activity;

(e) the works council or corresponding body elected by the workers has an important role to play in occupational safety and health. The rights of the works council may extend from merely being informed about decisions and measures to be adopted by the employer, through the right to be consulted and express opinions before final decisions are taken, up to, in some cases, the right of veto. In all cases, however, it would be meaningful and advisable if, over and above the legally established rights of participation, the workers' representatives were associated globally with all safety and health measures.

Co-operation at industry level. This may be pursued through employers' and workers' organisations or through independent institutions.

Questions that reach beyond the particular undertaking need to be settled at industry level. The employers' and workers' organisations are of particular importance in this connection. Co-operation is effected through the exchange of practical experience and advice both within their own association and with that of the other party. Discussions, if they are conducted by the competent negotiating bodies, may even lead to the conclusion of collective agreements; for example, wage settlements may be extended to include certain agreed safety and health measures. These agreements may be implemented throughout the industry on the advice of the associations to their members.

In some countries activities in the field of the prevention of accidents and occupational diseases are delegated by the general state and local authority system to legally and organisationally independent institutions jointly managed by employers and workers. These independent institutions are based on the notion that when those most directly concerned work together, there will be increased readiness to accept joint responsibility and to develop their own initiative.

The powers delegated to these institutions can vary greatly. In some cases they are empowered to issue far-reaching instructions relating to safety and health matters and may exercise not only a control and supervisory function but even lay down legally binding standards of their own. Because of their close involvement with the problems, these independent bodies are particularly suited to take charge of the whole field of worker protection. The fact of employers and workers being associated in such bodies enables them to work together both in the administration of the institution and towards the achievement of its aims.

Co-operation at the national level. Co-operation between employers and workers in regard to the protection of workers against occupational hazards can also take place at the national level. Here it takes the form of collaboration in the preparation of standard-setting regulations governing occupational safety and health.

Co-operation at the international level. Co-operation between employers and workers at the international level takes place mostly within the international or supranational organisations that are active in the field of worker protection. The ILO occupies a special place among these international organisations since both employers' and workers' representatives participate in the decision-making processes on an equal footing with government representatives, and act as experts or advisers in the preparation of Conventions and Recommendations.

Numerous organisations sponsored by the employers and workers in the field of the prevention of occupational risks also take part in international activities under the aegis of the International Social Security Association.

Examples of co-operation between employers and workers

Federal Republic of Germany. Regulations governing co-operation between employers and workers in the Federal Republic of Germany in respect of occupational safety and health are contained in the following legal texts:

(a) at the level of the undertaking: the Works Constitution Act of 15 January 1972 and the Industrial Safety Act of 12 December 1973;

(b) at industry level: the Social Code, especially Book I: "General", dated 11 December 1975, and Book IV: "Common social insurance provisions", dated 23 December 1976, and the Federal Insurance Code;

(c) at the national level: the Dangerous Substances Ordinance of 17 September 1971.

Particularly prominent features of the occupational safety and health system in the Federal Republic of Germany are:

(a) the strong position of workers' representatives within the undertaking;

(b) the high degree of independence permitted in the implementation of activities for the prevention of occupational accidents and diseases;

(c) the parallel approach to accident prevention activities adopted by the national factory inspection authority and the independent representatives of the established accident insurance scheme.

The works council, which is the recognised body of the workers, negotiates not only with the employer but also with all other competent bodies and has the right to information regarding all matters relating to occupational safety and health. On the other hand, the works council must support the employer by offering suggestions, advice and information. In this way it is ensured that the employers and the workers' representatives have access to the same information regarding safety and health matters and can therefore co-operate constructively. The works council has a say in the framing of all measures proposed for the prevention of employment accidents and occupational diseases and for the protection of health. In many undertakings the management collaborates with the workers' representatives to prepare and issue works rules in which the statutory provisions relating to accident prevention are spelt out, explained and illustrated. Employers and workers are free to introduce supplementary measures in support of the statutory regulations. An important aspect of co-operation within the undertaking is the establishment of committees comprising the employer, two members of the works council, the works doctors, the occupational safety specialists and the shop floor safety stewards to discuss all matters relating to safety and health.

Statutory accident insurance was introduced in Germany in 1885. Since that time the insurance carriers have devoted considerable efforts to the prevention of occupational accidents and of occupational diseases. Of these insurance carriers, the most important are the industrial and the agricultural employment accident insurance funds (known as industrial injuries mutual insurance institutions: Berufsgenossenschaften). These are genuinely independent institutions in which employers and workers are equally represented. They occupy a special place in the occupational safety and health field thanks to the power granted them to make accident prevention regulations in the form of legal standards binding, both for the members and for the insured persons, within the jurisdiction of the fund issuing them. The technical inspectors employed by the employment accident insurance funds check that the safety regulations are being observed in the factories, and also act in an advisory capacity to employers, workers' representatives, safety specialists and factory doctors.

Co-operation also exists at the national level between employers and workers in regard to the prevention of injuries to health caused by harmful substances. Under the provisions of the Dangerous Substances Ordinance, questions of occupational health and technical matters may be discussed within a committee of the Federal Minister of Labour and Social Affairs. In this way, scientific and technical aspects may be taken into account in the drafting of new legislation.

Other countries. As in the Federal Republic of Germany, the bodies responsible for the statutory accident insurance funds in Austria and Switzerland assume important duties and responsibilities in relation to the prevention of occupational accidents and disease, and they are also independent bodies. Employers and workers have parity on the board of the Austrian General Accident Insurance Institution. The administrative body of the Swiss Accident Insurance Institution is made up of employers' and workers' representatives as well as representatives of government and of the voluntarily insured members.

In addition, among the countries of the European Community, a number enjoy a large measure of independence and have the authority to make recommendations in the field of accident prevention. These include:

(a) Denmark, where the Working Environment Council attached to the Ministry of Labour includes research workers and experts, as well as representatives of government, employers and workers;

(b) the Netherlands, where the National Safety Institute is managed by one executive body composed of representatives of the competent authority together with employers' and workers' representatives in equal numbers;

(c) France, where the managing bodies of the National Institute for Research and Safety (INRS) are composed of employers' and workers' representatives;

(d) the United Kingdom, where representatives of the employers and the workers collaborate with representatives of government in the Safety and Health Commission.

In many other countries the managing bodies of the accident prevention organisations include representatives of the employers and workers. They vary considerably, however, from the points of view of their organisation and their fields of competence. In general, the members of these bodies are appointed on the nomination of the employers' and workers' organisations.

Often the managing boards of the accident prevention organisation legally have no more than an advisory function. Nevertheless, this function often carries great weight politically and socially, and they are entitled *de jure* or *de facto* to put forward proposals, for example for the filling of important posts or during the preparation of new safety and health legislation.

VERSEN, P.

CIS 76-1492 *Role of heads of undertakings and of trade unions in the prevention of occupational accidents* (Mission du chef d'entreprise et rôle des syndicats dans la prévention des accidents du travail). Denison; Renard, P.; Andries, P.; Cordy, A.; Marcorelles, N. E.; Gayetot, J.; Thissen, R. Comité provincial pour la promotion du travail, Namur (Institut international pour les problèmes humains du travail, 117 avenue Gouverneur Bovesse, Jambes, Belgium) (undated), 97 p. Illus. (In French)

CIS 79-890 "Employer-worker co-operation in the prevention of employment injuries". Versen, P. *International Labour Review* (Geneva), Jan.-Feb. 1979, 118/1 (75-88).

Enamels and glazes

Vitreous enamel, sometimes known as porcelain enamel, is used to give a high heat, stain and corrosion resistant covering to metals, usually iron or steel, in a wide range of manufactures, particularly baths, gas and electric cookers, kitchen ware, storage tanks and containers of all kinds, and electrical equipment. In addition, enamels may be used in the decoration of ceramics, glass, jewelry and ornaments. A specialised use of enamel powders in the making of such ornamental ware as cloisonné and Limoges enamels has been known for centuries. Glazes are also applied to pottery ware of all kinds.

Raw materials. The materials used in the manufacture of vitreous enamels and glasses are:

(a) refractories such as quartz, feldspar and clay;

(b) fluxes such as borax (sodium borate decahydrate), soda ash (anhydrous sodium carbonate), sodium nitrate, fluorspar, cryolite, barium carbonate, magnesium carbonate, lead monoxide, lead tetroxide, zinc oxide;

(c) colours such as oxides of antimony, cadmium, cobalt, copper, iron, nickel, manganese, selenium, vanadium, uranium and titanium;

(d) opacifiers such as oxides of antimony, titanium, tin and zirconium, and sodium antimonate;

(e) electrolytes such as borax, soda ash, magnesium carbonate and sulphate, sodium nitrite and sodium aluminate;

(f) floating agents such as clay, gums, ammonium alginate, bentonite and colloidal silica.

Production. The first step in all types of vitreous enamelling or glazing is the making of the frit. Frit making may be divided into three stages: raw material preparation, smelting, and frit handling.

Application. Prior to the application of enamels, metal articles are carefully prepared, for example by pickling, shotblasting and degreasing. The enamel may be applied to the "wet" or "dry" process or by spraying. The ground coats for sheet iron are applied by dipping or slushing. The metal blank is immersed in the enamel slip and then withdrawn and allowed to drain. In slushing, the enamel slip is thicker and must be shaken from the ware. In the dry process the cover-coat enamel is applied by dusting. The ground-coated ware is heated to the enamelling temperature, removed from the furnace, and the dry enamel powder dusted through sieves on to the hot ware. The enamel sinters into place and, when the ware is returned to the furnace, it melts down to a smooth surface.

Spray application is being used to an increasing degree and is usually associated with general mechanisation; in modern works, goods are mechanically conveyed through all stages. Electrical equipment is dipped in an enamel paste before firing. Decorative enamels are applied by hand using brushes or similar tools. Glazes are applied to pottery and porcelain articles, the method of application varying with the type of article. In the domestic pottery industry the glaze may be applied by dipping or spraying. In the dipping process the ware is held in the hand and dipped in a large tub of glaze; the surplus glaze is removed with a flick of the wrist and the ware is then placed in a dryer. Some dipping is now mechanised. Spraying in a cabinet under exhaust ventilation is increasingly common. The ware is then fired in an oven or kiln.

HAZARDS AND THEIR PREVENTION

Accidents. The machines likely to cause serious accidents are those used in the manufacturing process for grinding and mixing the raw materials. Workers may be seriously injured if they are struck by the projections from the surface of ball mills used for grinding or if their hands are trapped between the rotating blades and the container of mixing machines. All driving belts and pulleys should be totally enclosed to a height of at least 2 m above ground level. Gearing also should be totally enclosed. A substantial fence, at least 2 m high, should enclose the ball mills with gates provided for access. The gates should be kept closed when the mills are in motion; interlocking devices are desirable to prevent the opening of the gates before the mill's rotation has ceased.

Mixing machines should be guarded at both the feed and the discharge openings; the risk of hands being trapped at the latter are less obvious but none the less real. Interlocking of mixer covers is essential to prevent workers from putting their hands inside before the mixing arms have stopped rotating.

Fatalities have occurred and many workers have been seriously scalded during the quenching process where the molten enamel is poured into a large tank of water (frit bosch) which reaches boiling point as it fills with frit. If the tank is sunk into the ground, which is the usual arrangement, it should be enclosed by a stout fence, not less than 1 m in height. Suitable personal protective equipment (overalls, eye and face protection and foot and leg protection) should also be provided for any workers liable to be splashed by molten materials.

If furnaces and drying ovens used in the industry are gas fired, special precautions are required to prevent explosions. These may include flame-failure safety devices at burners, low-pressure cut-off valves in the supply lines, and explosion-relief panels in the structure of stoves. The stoves should be adequately ventilated and burners should be protected from clogging caused by dripping enamel.

Diseases. Many workers employed in the preparation and application of enamels and glazes are exposed to two major health hazards: pneumoconiosis and lead poisoning together with a risk of poisoning by the other toxic metal oxides used in the manufacturing processes.

The most important factors affecting the incidence of pneumoconiosis are the size of the dust particles in the air, the chemical composition of the dust inhaled, the concentration of dangerous dust particles in the air breathed and the period of exposure. Among other factors influencing the occurrence of this disease are the conditions of the worker's lungs and a personal constitutional factor.

In the manufacture of enamels and glazes, the processes at which workers are exposed to risk of pneumoconiosis include: (i) the calcining, crushing, and drying of flint, quartz or stone; (ii) the sieving and mixing and weighing out of these materials in a dry state; and (iii) the handling of these materials in the dry state, particularly when furnaces are being charged. Lead poisoning is a risk in the handling, shovelling, weighing and mixing of lead compounds.

In the application of enamels and glazes, workers are exposed to both risks in the following processes: (i) the dusting or "laying-on" and the brushing or "wooling-off" of the enamel in a dry powder form; (ii) the spraying of enamels or glazes; and to a lesser extent (iii) the dipping or slushing of articles in the wet process. In this latter operation the worker's hands and clothing are contaminated, and when the wet enamel or glaze dries, dust is released from the clothing.

Research has shown that the cotton overalls worn by workers are quickly contaminated by the wet glazes and that, when dry, the dust from the clothing is released into the breathing zone of the workers. It was found that the substitution of overalls made from synthetic fibres such as polyester and polyamide would greatly reduce the risk from this source.

The prevention of either pneumoconiosis or lead poisoning is best effected by the elimination of the hazardous material or failing that, for either risk, dust control by local exhaust ventilation or wetting, rigorous cleanliness in the environment and personal hygiene.

In many countries lead has been almost eliminated from the vitreous enamels and from many pottery glazes. The increasing use of low-solubility lead in the manufacture of enamels and glazes and the introduction of leadless glazes has reduced considerably the incidence of lead poisoning amongst workers engaged in their application but, where raw lead is still used, the risk of lead poisoning in the application of enamels and glazes arises in all the processes mentioned above under pneumoconiosis, and the preventive measures mentioned there will also be effective in reducing the risk of lead poisoning.

Either exhaust ventilation or a water spray which effectively suppresses the dust should be provided at the mouth of the calcining kiln to control the dust caused by the fall of the flints and by shovelling, and at the jaw crusher where flint, quartz or stone is being crushed.

The crushed flint or stone usually falls on to a conveyor which carries it to the storage hopper; it is drawn from the hopper as required and is transferred to the grinding mills where it is usually ground in water. The movement of elevators and conveyors, when used for handling crushed material, causes considerable dust to be dispersed. It therefore becomes necessary to enclose the conveyors and elevators and to apply an efficient exhaust draught.

Flint and stone are sometimes dry ground and it is then essential that the grinding plant should be totally enclosed and an efficient exhaust draught applied to the enclosure.

The maximum practicable enclosure is required at the places where dry materials are sieved and measured out before mixing. Mixing machines should also be totally enclosed and mechanically ventilated.

In the rooms in which any enamelling or glazing process is carried on there should be ample air space; at least 150 m^3 should be allowed. Floors and work benches and wall surfaces should be impervious, be maintained in good condition and be washed frequently. Vacuum cleaning is suitable for dry dust removal. Enamelling processes giving rise to dust and spray should be so arranged that the dust or spray is intercepted and prevented from entering the air of the workroom. The processes of dusting or laying-on, and brushing or wooling-off should be done over a grid with a container beneath to collect the dust which falls through.

Firing should preferably be in a room where no other process is carried on, but if this cannot be arranged, other processes should not be done within 6 m of the furnace.

Suitable protective clothing should be provided for all workers employed in processes likely to produce dust, and accommodation should be provided where clothing put off during working hours can be kept in a clean condition. A suitable separate place should be provided for the storage of protective clothing.

Messrooms or canteen accommodation should be provided and workers should not be permitted to remain in workrooms during meal breaks. Where a lead risk exists, food, drink or tobacco should not be taken into the workroom. A high standard of washing facilities is particularly necessary where there is a lead risk.

Some decorative and jewelry work may be done on a craft or home-industry scale, and control of conditions may be difficult but none the less necessary.

Medical supervision. Pre-employment medical examination can prevent the employment of persons likely to be particularly susceptible to either pneumoconiosis or lead poisoning. In particular persons with lung tuberculosis should not be exposed to such work. Periodic medical examinations and chest X-rays can assist in early diagnosis of pneumoconiosis so that persons affected can change their occupation before serious damage to the lungs has occurred. These examinations will also detect tuberculosis in its early stages and so enhance the worker's chance of recovery. Where a lead risk exists, frequent and regular examinations (including possibly blood and urine tests) are necessary to detect lead absorption.

FISH, N.

CIS 913-1964 "Hazards in work with toxic enamelling materials and auxiliary chemicals" (Gefahren bei der Arbeit mit toxischen Emailrohstoffen und Hilfschemicalien). Ankerst, H.; Weimer, G. *Glas-Email-Keramo-Technik* (Hamburg), Apr. 1963, 14/4 (124-126). 17 ref. (In German)

CIS 2542-1972 "Ventilation of work premises and work posts. Examples–6. Handling of dusty lead-based materials" (Ventilation des locaux et postes de travail–Quelques exemples–6. Manipulation de produits pulvérulents à base de plomb). Maisomneuve, J. de; Lardeux, P. *Cahiers de notes documentaires–Sécurité et Hygiène du travail* (Paris), 3rd quarter 1972, 68, Note No. 800-68-72 (279-283). Illus. (In French)

CIS 81-1463 *Prevention of lead poisoning in the pottery industry* (La prevenzione del saturnismo nelle industrie ceramiche). Bertazzini, P.; Candela, S.; Collini, P.; Francesconi, E.; Guidetti, P.; Tonelli, S.; Bolognesi, D.; Busani, R.; Renna, E.; Torreggiani, A.; Vecchi, G. Studi e documentazioni 34 (Dipartimento sicurezza sociale della regione Emilia-Romagna, Italy) (1980), 46 p. 9 ref. (In Italian)

Energy expenditure

Energy expenditure is commonly used to indicate the extent to which the chemical energy available in the human body is consumed, i.e. the number of units of chemical energy converted into mechanical energy and heat per unit of time. A knowledge of the energy expended by the worker in different tasks is valuable in determining the degree of fatigue to which he is exposed, the length of time for which he can be expected to do the work during each shift, the number, frequency and length of the rest periods required, the nutritional requirements involved and so on (see also MUSCULAR WORK). Where all chemical energy is converted into heat–directly or indirectly, i.e. by friction in or near to the body–the level of energy expenditure can be measured in terms of heat production. In principle, it is possible to study the energy expenditure of a human subject in a calorimeter. The intensity of the process can thus be expressed in kilocalories per minute. "Direct calorimetry" such as this is used in some highly specialised laboratories, but has in practice been replaced by other, more practical methods.

The conversion of chemical energy into mechanical energy also requires a supply of oxygen and results in the excretion of carbon dioxide, and efforts have been made to calculate energy expenditure from oxygen uptake or carbon dioxide production. The latter has found little application; however, measurement of oxygen uptake is now the most common method of "indirect calorimetry". By means of the caloric equivalent, it is possible to convert litres of oxygen into kilocalories per minute (1 l of oxygen is the equivalent of 4.7-5.0 kcal/min).

On-the-job measurement

Precise knowledge of energy expenditure in a given physical activity is of use only if the duration of this and other activities on the same day is also known and, consequently, any study of energy expenditure should be preceded by an inventory of the nature and duration of the day's activities. Where expenditure over a 24-h period is to be assessed, some part may be determined by direct measurements and some by the use of tables showing average expenditure for such activities as sleeping, sitting, walking, etc. In many cases it is only the occupational activities that need to be determined by direct measurement since these represent a relatively high percentage of the total daily energy expenditure. Where energy expenditure in a specific activity is to be determined, for example to compare different work methods, it is important to know the significance of changes in energy expenditure on specific activities in the context of all activities.

Direct measurements will be made by determining oxygen uptake, usually on the job, and consequently preference will be given to methods which do not involve heavy laboratory equipment.

Douglas bag. In this technique, the subject wears a nose-clip and a mouthpiece fitted with a respiratory valve, leading through a hose to a bag in which all the expired air is collected over a given period of time. The total volume of expired air is measured by means of a gas meter, and a sample is analysed for oxygen, carbon dioxide and nitrogen content. Measurements are limited to periods of 2-5 min and peak loads are hard to study; nevertheless, this method is still the most reliable available.

Kofranyi-Michaelis gas meter. This device (figure 1) collects and measures expired air continuously and takes samples automatically at regular intervals. Measurements over periods of up to 20-30 min are possible.

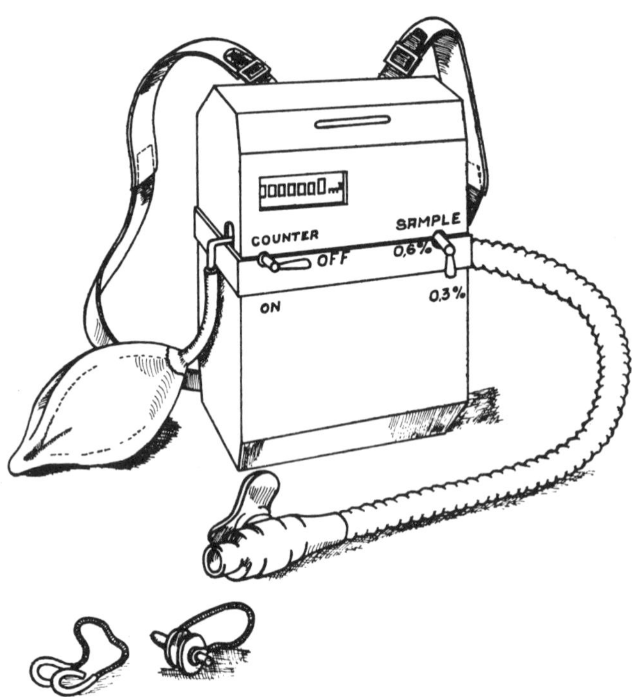

Figure 1. The Kofranyi-Michaelis portable gas meter.

Wolff pneumotachograph. This integrating motor pneumotachograph measures expired air electronically and takes samples by means of an electric pump.

Continuous analysis technique. A recently developed instrument combines continuous measurement of expired air volume with continuous polarographic gas analysis; continuous indication of oxygen uptake can thus be transmitted to the investigator by telemetry. For calibration purposes incidental gas analysis facilities are desirable.

Ventilatory minute volume. To avoid gas analysis, some investigators have taken ventilatory minute volume as an index of oxygen uptake and, consequently, of energy expenditure. However, in addition to energy expenditure, there are both internal and external factors which can influence the ventilatory minute volume and, for this reason, this method has limited applicability.

Heart rate. Similar objections can be made to the use of heart rate as a single index of energy expenditure. Heart rate, however, can be studied throughout a working day by telemetry or using miniature tape-recorders. Therefore, it is attractive to combine a continuous heart-rate recording with intermittent measurements of the energy expenditure. The heart rate reveals all changes in the workload and can be calibrated, to a certain extent, by repeated simultaneous measurements of oxygen uptake. This has less meaning, however, if changes occur in the working posture or the environmental factors and if heart rate is influenced by fatigue, apprehension or cigarette smoking.

Energy expenditure tables

Data have been published on energy expenditure determined in different occupational activities. Using these data and by carefully determining the time spent on different activities in a work situation and adding basal metabolism, it is possible to calculate total energy expenditure. Other data are available showing total energy expenditure for men of 65 kg and women of 55 kg body weight in various occupations or activities.

Motion study

Energy expenditure data for all the activities in question are not always available and, in addition, it may be valuable to relate net energy expenditure to body weight. Data have therefore been collected from the literature and, after comparison, have been grouped in four categories: lying, sitting, standing, walking; within each category it should be stated what extremities are active, what weights should be lifted or carried, what forces should be exerted and so on. The energy expenditure is indicated in small calories per kilogramme of body weight per minute. If the total energy expenditure is to be calculated, the sum of the products of the time spent on each activity and the corresponding metabolic rate should be added to the basal metabolism. The mean value of energy expenditure can be determined by dividing the total by the corresponding time numbered in minutes.

Tables and motion study have a limited accuracy and their use should be based on a thorough time study like the actual measurements of the oxygen uptake.

Acceptable levels

The preceding paragraphs have shown how to determine the mean energy expenditure for certain types of work. The decision whether such a mean value would be acceptable or not has, for many years, been based on fixed criteria. In fact, it was supposed that all workers had the physical working capacity of a healthy 25-year-old European male, and worked 8 h a day, 5-6 days a week. A net caloric expenditure of 4.2 kcal/min (293 W) or a total net energy expenditure of 2 000 or even 2 200 kcal (8 375 or 9 210 kJ) during an 8-h shift was considered an acceptable maximum. About 2 300 kcal (9 630 kJ) were allowed for basal metabolism, 8 h sleep and 8 h spent on personal hygiene, meals, transportation and other activities. There is a tendency in Western European countries, however, to reduce the total net energy expenditure of groups of workers to 1 500 kcal (6 280 kJ).

Energy expenditure and working capacity. Aerobic power, i.e. the degree to which chemical energy can be converted into mechanical energy by virtue of the uptake and transportation of oxygen from the air to the oxygen-demanding tissues has been measured in persons of different race, body build, sex and age; results show that if aerobic power is taken as a criterion, physical working capacity varies considerably as a function of these individual parameters. Consequently, it is not physiologically justifiable to require equal energy expenditure from different workers. One series of experiments indicated that 50% of the maximum working capacity could be used throughout the working day; however, this must have applied to the occupational activities proper and not to the whole work complex comprising such activities as preparation and transport of tools, and material, correction for unforeseen complications and cleaning up. However, a more recent study in the building industry indicates that the over-all average of the energy expenditure in this case amounted to only 38% of the maximum.

In studies involving experimental work which could be done continuously for 8 h, it was found that 35% of the maximum can be sustained for 8 h without signs of undue fatigue. Similar studies carried out elsewhere resulted in a proposal that 33% of maximum energy expenditure should be regarded as the maximum allowable level of energy expenditure for an 8-h day. In practice the aerobic power of each individual worker should be measured to decide whether the required energy expenditure can be considered acceptable. Sometimes it might be preferable to investigate the physical working capacity of groups. Investigations in the Netherlands have revealed that the aerobic power decreases by one-third between the age of 24 and 64 years. An important variable to predict the aerobic power within a group is the lean body mass. This may be assessed from body weight and body composition.

The physical working capacity of the group itself will be influenced by ethnic, environmental, nutritional and social factors. This stresses the importance of group studies prior to any decision being taken on maximum allowable energy expenditure.

Energy expenditure and working time. Maximum energy expenditure under aerobic conditions can be sustained for up to 4 min; the equivalent expenditure level in a healthy, young male is 15 kcal/min (1 050 W). Peak efforts exceeding this level induce anaerobic work and the resulting oxygen debt has to be repaid during the subsequent recovery period. However, what are more important from the occupational physiology point of view are the maximal and maximum allowable energy expenditure levels for aerobic work lasting more than 4 min.

Studies on the endurance time of subjects riding a bicycle ergometer at different oxygen uptake levels showed a rectilinear relationship between oxygen uptake and the logarithm of endurance time. Although these experiments were for periods of 3 h only, they support the hypothesis that a similar rectilinear relationship exists for working times of more than 3 h and that the maximum allowable energy expenditure is governed by the same laws as the 3-h maximal effort experiments.

Knowing that maximal energy expenditure can be sustained for 4 min and that energy expenditure over a period of 480 min should not exceed one-third of the maximal level, it is possible to draw curves showing levels of energy expenditure or oxygen intake considered allowable for different working times in the case of persons with different maximal oxygen uptakes. Figure 2 shows the corresponding curve for a healthy male with a maximal oxygen uptake of 3 l/min. Here, the allowable energy expenditure for 480 min is the equivalent of one-third of 3 l/min, i.e. 1 l/min or approximately 5 kcal/min (349 W).

Rest periods

On the basis of these curves it is possible to determine the

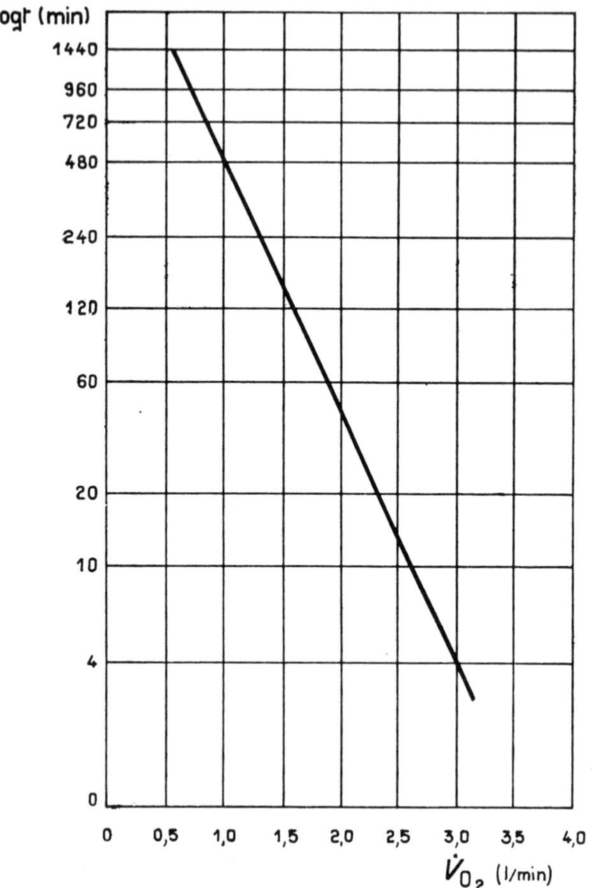

Figure 2. Curve of allowable oxygen intake for a healthy male with a maximal oxygen uptake of 3 l/min.

different types of work requiring known energy expenditure levels. If required energy expenditure exceeds the allowable level for the working time, then the working time should be reduced accordingly.

For example, if the energy expenditure requirement for the job is 6 kcal/min (419 W) and the maximal oxygen uptake is 3 l/min, the curve will show that only 5 kcal/min (349 W) can be maintained for 480 min. A possible solution is to reduce the working time to 300 min of uninterrupted work; such a solution will be attractive if the worksite is not readily accessible or if the work has to be carried out under inclement climatic conditions. However, it is usually more attractive to introduce rest pauses to reduce the mean energy expenditure to the acceptable level. For example, if the 480 min maximum allowable expenditure level is 5 kcal/min (349 W) and 8.5 kcal/min (593 W) is required during work and 1.5 kcal/min (105 W) during rest periods, the work shift could be divided up into 240 min of work (2 040 kcal or 8 540 kJ total expenditure) and 240 min rest pause (360 kcal or 1 505 kJ total expenditure).

In addition, the following formula has been developed to calculate the recovery time required as a function of working time: it is based on an invariable figure of 4.2 kcal/min (293 W) for net energy expenditure:

$$t_r = \left(\frac{\dot{M}}{4.2} - 1\right) \times t_w$$

where:

t_r = recovery time

$\dot{M}$ = net energy expenditure during work

t_w = working time

Theoretically, in the example mentioned above, an energy expenditure of 8.5 kcal/min (593 W) could be made continuously for 80 min before a rest period was necessary. However, this would place an undesirable load on the cardiovascular system and it has been demonstrated that steep increases in heart rate can be avoided by frequent short breaks; a schedule of 20 min work followed by 20 min rest will therefore prove more satisfactory than 80 min work and 80 min rest. This is particularly true in heat or hot work since less heat will accumulate and consequently body temperature fluctuations will be less pronounced.

Energy expenditure and environment

Physical activity is made possible by conversion of chemical energy into mechanical energy. Even in activities characterised by high mechanical efficiency, 80% of the chemical energy involved is converted into heat and only 20% into mechanical energy. In a person riding a bicycle at an oxygen uptake level of 1 l/min, 5 kcal of chemical energy are therefore converted each minute into the equivalent of 1 kcal/min (70 W) mechanical energy and 4 kcal/min (279 W) heat.

Only a fraction of the heat produced can be regarded as useful in maintaining body temperature above ambient temperature. Even moderately high atmospheric temperatures place particular stress on the body when strenuous exercise is being done, because a part of the circulatory capacity has to be used for the internal transport of heat. The higher the ambient temperatures, the more difficult it is for heat to be transferred from the body to the environment.

If balance is not maintained between heat production in the body, heat gain from the environment and heat loss to the environment, heat accumulation will limit further activity within a relatively short time. If, nevertheless, work has to be carried out in high atmospheric temperatures, the balance between heat gain and heat loss must be maintained by limiting energy expenditure. See COLD AND WORK IN THE COLD; HEAT AND HOT WORK; HEAT ACCLIMATISATION.

Energy expenditure and toxic agents

The increased demand for oxygen caused by physical activity entails a rise in ventilatory minute volume which, in the cases of concomitant exposure to airborne toxic substances, will result in a higher uptake of the atmospheric contaminant, through the lungs and a higher "effective exposure".

BONJER, F. H.

General:

"Energy cost of various activities". Astrand, P. O.; Rodahl, K. Ch. 13 (433-452). *Textbook of work physiology* (New York, McGraw-Hill, 1970). 47 ref.

Introduction to work study (Geneva, International Labour Office, 3rd revised ed., 1979), 442 p. Illus. 80 ref.

Methods:

CIS 1329-1972 "Ergonomics guide to assessment of metabolic and cardiac costs of physical work". Ergonomics guides series. American Industrial Hygiene Association Technical Committee on Ergonomics. *American Industrial Hygiene Association Journal* (Akron, Ohio), Aug. 1971, 32/8 (560-564). 11 ref.

CIS 79-281 *Evaluation of work requiring physical effort.* Rohmert, W.; Laurig, W. Commission of the European Communities, Directorate-General "Social Affairs" (Luxembourg, Office for Official Publications of the European Communities, 1975), 168 p. Illus. 78 ref.

Equipment:

Technology for the development of a portable metabolic monitor. James, R. H. DHEW (NIOSH) publication No. 76-

171 (National Institute for Occupational Safety and Health, 4676 Columbia Parkway, Cincinnati) (1976).

"A respirometer for use in the field for the measurement of oxygen consumption. The 'Miser', a miniature indicating and sampling electronic respirometer". Eley, C.; Goldsmith, R.; Layman, D.; Tan, G. L. E.; Walker, E.; Wright, B. M. *Ergonomics* (Cambridge), Apr. 1978, 21/4 (253-264). Illus. 15 ref.

Energy sources, comparative risks of

In the past few years a great many studies have been carried out of the costs and benefits of different sources of energy, with special attention to the risks to health and life of both workers and population. This is such a wide-ranging subject that a satisfactory analysis cannot be provided in this encyclopaedia.

Interested readers may, however, refer to the papers presented at the 1980 IAEA Symposium in Paris and the 1981 WHO-UNEP-IAEA Symposium in Nashville on this subject.

"The respective risks of different energy sources. Selected papers drawn from presentations at the Symposium held in Paris 24-26 Jan. 1980". *IAEA Bulletin* (Vienna), Oct. 1980, 22/5-6 (35-128). Illus. Ref.

Health impacts of different sources of energy. Proceedings of an International Symposium held in Nashville (22-26 June 1981). Proceedings Series (Vienna, International Atomic Energy Agency, 1982), 701 p. Illus. Ref.

Comparative risks of electricity production systems: a critical survey of the literature. Cohen, A. V.; Pritchard, D. K. Health and Safety Executive Research Paper 11 (London, HM Stationery Office, 1980), 31 p. 87 ref.

Engine testing

All internal combustion engines (two- and four-stroke piston engines, rotary piston and turbine and jet engines), after they have been manufactured and assembled and before they are sold and put into service, are tested to check certain basic parameters which indicate whether they are operating correctly or not. In the same way, when a certain degree of engine wear has been reached, the engines are overhauled and another check on their main functions is necessary.

Piston engines are either of the two- or four-stroke type and, for the most part, use gasoline or diesel oil as their fuel. In these engines the tests provide an initial period of running-in, thus reducing the most pronounced frictions at the first stage of operation and allowing operational tuning (distribution, ignition, carburation or fuel injection); at the same time it is possible to check that the variations in oil and water temperature do not exceed set figures. It is also possible to determine certain of the engine's characteristics such as horsepower and specific fuel consumption. The tests carried out on turbine and jet engines, in addition to checking the characteristics mentioned above, are also necessary to record vibration in the engine and thus permit any necessary balancing, and to verify the temperature of the turbine and of the induction and exhaust gases in order to tune all the related thermal, hydraulic and electrical equipment.

Piston engines, which are usually small in size, are mounted for testing on special horizontal benches and connected to a hydraulic brake. The tests may be carried out in any of the factory's workshops (figure 1) or in a special cabin.

Figure 1. Engine testing shop of a large automobile plant.

HAZARDS AND THEIR PREVENTION

Engine testing exposes workers to limited accident hazards, to noise, vibration, heat, solvents and mineral oils and to exhaust gases.

Accidents. Engine testing does not entail any serious accident hazards; however, during certain work processes, workers may be exposed to a variety of injuries. When the engine is being placed on the test bench it is normally transported by means of a hoist or overhead travelling crane: it is therefore necessary to observe the safety measures specified for safe mechanical handling. In the case of large engines, the brake does not form part of the test bench and consequently it also has to be transported by means of a hoist or an overhead travelling crane.

During the tests, and when disconnecting and transporting the engine after the tests, workers are exposed to the hazard of burns due to contact with parts of the engine which are still at very high temperatures. Workers should be provided with hand and arm protection made of heat-resistant material.

Workposts around test benches become impregnated with lubricating and diesel oil and gasoline, which may be a cause of falls. Turbines and jet engines are considerably heavier and more bulky than piston engines and consequently the accident hazards during transportation are greater. Turbines are often tested in pits and

Figure 2. Special installation for the testing of turbines.

workers are required to use gangways and ladders or stairways, from which falls may not be uncommon (figure 2). Here too there is a danger of burns from hot parts of the engines.

Noise. The hazard of noise is more serious during tests with turbine and jet engines owing to the high noise levels produced: the sound pressure level in the direct vicinity of the source is often near to the threshold of pain. When piston engines are tested inside a special cabin, the hazard of hearing damage is virtually eliminated. If the tests are carried out in the factory premises itself, the noise levels attained may be dangerous but, generally, the hazard is a modest one in view of the low pressure levels at high frequencies. The hazard increases with increasing engine speed and preventive measures are based on the acoustic insulation of the noise sources using cabins or by soundproofing the premises with asbestos panels or wall coverings. Workers should be provided with hearing protection equipment such as earmuffs or earplugs.

In the case of turbine engines, the hazard of hearing damage may be reduced by fitting silencers to the turbine air intake; however, these are extremely bulky and can be used only for medium-power turbines. In general, extremely good results can be obtained by use of sound-absorbent coatings on the inside of the intake duct; these consist of cellular structures and vertical diaphragms covered with glass wool. In addition, although the noise levels are extremely high, exposure is intermittent and of short duration. Pre-employment medical examinations should include audiometry to exclude personnel with hearing disorders. This should be followed up by periodical audiometric examinations.

Vibration. Here the hazard is greater in the case of large, low-speed, two-stroke engines (ships' engines). When testing these engines, it is not a single organ but the whole body which is affected and the result is not the localised injury produced by a pneumatic tool for example, but a general disorder which presents a complex symptomatology indicating deleterious effects on different systems (kidneys, osteoarticular system and, in particular, the nervous system). Sources of vibration should be isolated using vibration damping equipment, and pre-employment medical examination should eliminate workers suffering from neuropsychic disorders.

Heat. Engines under test give off considerable heat and may produce a significant rise in the ambient temperature. However, the temperatures recorded are never so high as to present a hazard, and no special measures are required.

Solvents and mineral oils. Diesel oil, gasoline and lubricating oils may cause occupational skin diseases. The most common lesions are of the folliculitis type caused by mineral oils and erythematous squamous lesions which affect mainly the palms of the hand and are caused by diesel oil and gasoline. On rare occasions, lichen eczema, limited to the backs of the hands where there is hair, may be produced by the same substances. The exposure is often more prolonged during engine overhaul where certain materials (diesel oil and gasoline) are commonly used as solvents for greases and other dirt on the engines. However, the most frequently employed solvent is trichloroethylene. Components are usually dipped into open vats of saturated solvent vapour and there is consequently a risk of solvent poisoning. Trichloroethylene degreasing vats should be fitted with cooling coils and local exhaust ventilation. Medical supervision of degreasing workers should consist in periodic medical examinations including urine determinations to detect metabolites of toxic substances (in particular trichloroacetic acid). A less toxic solvent, 1,1,1-trichloroethane, has now come into wide use.

The kerosene used as a fuel for jet engines has no specific pathological effects on testing and overhaul workers.

Exhaust gases. If fuel induction is not properly tuned, dangerous concentrations of exhaust gases may form in test areas which are inadequately ventilated. The main toxic substances found in these gases are carbon monoxide, polycyclic aromatic hydrocarbons and tetraethyl lead.

In addition, these gases contain large quantities of carbon dioxide and, in the case of diesel oil, small quantities of sulphur compounds. The proportion of carbon monoxide found in normal vehicle exhaust gases varies between 10 and 20%. This quantity increases at higher engine speeds and when the engine is worn. The exhaust gases of very worn motors normally have very high concentrations of all these toxic substances. Unless adequate safety measures are taken, the quantities of carbon monoxide given off may produce acute poisoning. When adequate safety measures are taken, it is only under exceptional circumstances that workers are exposed to a poisoning hazard and in most cases the carbon monoxide concentration will be relatively low.

Lead released from the combustion of leaded antiknock agents is seldom the cause of disorders and the carcinogenic polycyclic aromatic hydrocarbons are found only in small quantities in the exhaust gases. Table 1 shows the quantities of the main substances found in the exhaust gases of a 1 200 cm^3 automobile engine.

Table 1. Constituents of automobile exhaust gases (1 200 cm^3 capacity engine)

Distance covered (m)	Exhaust gas volume (l)	Carbon monoxide content (g)	Carbon dioxide content (g)	Hydrocarbon content (g of *n*-hexane)	Fuel consumed (g)
7 990	4 509	275.71	972	8.65	548

As can be seen, the quantity of toxic substances is high; however, with the exception of overhaul work carried out on engines in the workshops of small garages, the exhaust gases are led off through a tube connected to the exhaust pipe (figure 3).

Figure 3. Special extraction arrangements for leading off the exhaust gases during the bench testing of internal combustion engines.

Atmospheric determinations of the main toxic gases should be carried out periodically (carbon monoxide, polycyclic aromatic hydrocarbons) to ensure that the safe limits are not exceeded. One of the best methods of determining the carbon monoxide hazard is to check carboxyhaemoglobinaemia in workers at the end of the shift. Periodic medical examinations should also be aimed at detecting early symptoms of reactions to these substances. The sulphurated substances contained in diesel oil exhaust gases do not constitute an appreciable hazard in engine testing (see also DIESEL ENGINES, UNDERGROUND USE OF; AUTOMOBILES, SAFETY AND HEALTH DESIGN OF).

CROSETTI, L.

CIS 75-970 "Sound-proofing of an engine test bed" (Insonorización de un banco de pruebas de motores). López Cristobal, J. A. *Revista Seguridad* (Avilés, Spain), Apr.-June 1974, 53 (45-48). Illus. (In Spanish)

CIS 77-778 "Development of an air purifier for personnel cubicles". Roehlich, F.; Rodgers, S. J. *American Industrial Hygiene Association Journal* (Akron, Ohio), Oct. 1976, 37/10 (586-589). Illus. 4 ref.

Entertainment industry

In every branch of the entertainment industry, the performance is supported by ancilliary workers such as the backstage men, lighting and projection men, box-office attendants, doorkeepers, usherettes, and refreshment and programme sellers.

Especially in a large theatre, there may be a whole complex of workshops, with teams of painters, carpenters, upholsterers, dressmakers, and shoemakers whose work is solely in connection with the presentation on stage. Scene erectors, property-men, stage hands, and electricians are involved in the performance, the make-up artist and the hairdresser prepare the artistes, the stage manager and his assistants direct activities and the prompter is stationed close to the stage. The same general arrangements with variations and modifications apply to all live entertainment: in circuses, there are also the animal keepers and attendants, and, in the cinema, the key man is the projectionist.

Other workers deal directly with the persons who come to be entertained–in the box offices, the cloakrooms, the bars and restaurants, in selling programmes, ushering, and selling of light refreshments (see ACTORS AND PLAYERS; MUSICIANS; CINEMA INDUSTRY).

HAZARDS AND THEIR PREVENTION

Most of the risks are common to similar occupations outside the entertainment world, but certain special factors may make the work more exacting; in particular, the work has to be carried on at awkward hours, and often at great speed during a performance. The overriding consideration is always that "the show must go on".

Scene shifters and property-men have to deal with a great variety of materials, ranging from iron and wood to plastics and cloth, of different sizes, shapes and weights, with different centres of gravity. The work is more demanding than that of ordinary furniture-removers or porters, and research has shown that the exertion required of stage staff ranges from fairly high to very high. Training in proper methods of lifting and carrying is required. Stage work is a complex of small tasks and cannot be mechanised. The stage-hand has to be always alert and ready to react quickly, he has to adapt himself to rapid changes of light and darkness, to different temperatures and to considerably dusty conditions. In many modern productions, changes are made on the open stage in darkness and each task has to be performed noiselessly.

Much of the success of a show depends upon the lighting workers; they must be competent electricians, able to install lamps and to carry out immediate repairs. They may be exposed to scorching heat from the lamps, in uncomfortably cramped and dusty conditions.

Many of the auxiliary workers in places of entertainment have to work in small kiosks or boxes (ticket sales, prompters, projectionists, etc.) and far too little attention is often given to the adequate ventilation of these workplaces. Studies of projectionists have shown that the carbons used in cinema projector lamps produce varying quantities of gas, vapour and dust which may be harmful to health. Xenon lamps have been recently introduced for projectors; the light from these lamps has an ultraviolet component which ionises oxygen in the air to produce ozone and ozone concentrations of 0.01-0.7 ppm have been measured. Sometimes fatigue is intensified by double-jobbing, as when usherettes are employed in the evenings after they have already completed a full day's work elsewhere. Usherettes may also tend to suffer from varicose veins caused by long periods of standing.

Sanitary and washing facilities often need improvement and there should be arrangements for taking of light meals and hot drinks. Rest pauses may be difficult to arrange but should not be neglected, since fatigue is an important problem in the entertainment industry.

A risk peculiar to animal attendants at circuses is attack from dangerous animals or infections of zoonoses. These are dealt with in separate articles.

The management in the entertainment world is usually without industrial experience and this makes it all the more important that the common industrial risks should not be overlooked and that some competent person should be in charge of safety. Guarding of machinery in the workrooms should be to the highest standards obtaining outside; many of the electrical installations may present hazards and close attention should be paid to all equipment, cables, etc., with special emphasis on earthing arrangements. There should be safe means of access to all places where any person has to work, with special attention to ladders and to guarding of elevated platforms. Above all, every person should have quick access to a safe means of exit in case of fire and adequate training in fire precautions and procedures.

Pre-employment medical examination and medical supervision of all staff are desirable.

GLUCKSMANN, J.

Hazards:

"Places of entertainment". *Health and safety at work* (Croydon), Mar. 1980, 2/7 (33-35).

Safety (equipment):

CIS 76-547 *Stages and studios* (Bühnen und Studios). VBG 70. Hauptverband der gewerblichen Berufsgenossenschaften (Cologne, Carl Heymanns Verlag KG, 1974), 11+16 p. (In German)

Zoos:

"Health and safety in zoos and safari parks". Brooks, P. *Health and safety at work* (Croydon), Jan. 1982, 4/5 (33-35). Illus.

Enzymatic changes

What are enzymes?

All enzymes are proteins composed from a high molecular weight apoenzyme and a low molecular weight constituent (coenzyme, prosthetic group, or activator). The prosthetic group is firmly bound to apoenzyme, while coenzymes are organic fragments that generally do not follow apoenzyme during separation and purification. Likewise, activators (metals, non-specific reducing agents, etc.) are dialysable and thermostable.

Biological role of enzymes

Enzymes enable the organism to exploit the energy of food and its own molecules and macromolecules (fat, glycogen, proteins) in catabolic, exergonic, energy-yielding reactions and to build its parts from molecules thus obtained (anabolic, endergonic, energy-requiring reactions). The cells and the organism as a whole are capable of coupling these two kinds of reactions and thus maintaining life functions, growing and reproducing at the expense of environment. However, some factors of our artificial man-made environment may affect the activity of various enzymes and thus cause modification of cellular functions and even cellular death.

How enzymes function

It takes energy to build large molecules from small ones, complex structures of proteins, nucleic acids and membranes. Most of the reactions would be unable to proceed in the chemically mild environment (temperature, reactant concentrations) of the organism or would proceed too slowly to maintain life functions without enzyme catalysis. There is a barrier even for spontaneous, exergonic reactions known as the *activation energy*. Enzymes are capable of lowering the activation energy of vital chemical reactions, and in so doing to speed up the rate of the reactions without being consumed.

Some enzymes may be reversibly or irreversibly inactivated during the process. Microsomal mono-oxygenases, in particular, catalyse the formation of certain toxic intermediates of xenobiotics that may cause deleterious effects, either immediate (e.g. hepatotoxicity of carbon tetrachloride), or delayed (e.g. the carcinogenicity of aromatic polycyclic hydrocarbons).

Enzymes decrease the activation energy of chemical reactions by forming an enzyme-substrate complex. The part of an enzyme molecule capable of specifically reacting with a substrate is referred to as the *active site*. The observed high degree of specificity of various enzyme-catalysed reactions triggered the "lock and key" hypothesis. This has recently been replaced by the "induced-fit" theory and the "strain" theory, in which the enzyme active site is viewed as a relatively flexible spatial arrangement of enzyme-building blocks (amino acid residue R-groups) that can interact with complementary groups of the substrate rather than as the rigid template of the lock and key hypothesis. This interaction results in a slight modification of the shapes of both substrate and enzyme molecules and decreases the activation energy for the reaction. The strain theory suggests that when a substrate binds, a portion of the intrinsic binding energy is utilised to strain the substrate complex toward the geometry of the transition state—an intermediate state between substrate and product, in which the reactant molecule better fits into the active site than either substrate or product.

Regulation of enzyme activity

Regulation involves co-ordination of many independent chemical reactions and pathways so as to provide not only a given precursor for a following reaction, but also a proper balance of all precursors to ensure their uninterrupted incorporation into the more complex components of the cell's architecture or a complete exploitation of the intrinsic energy. Enzymes that catalyse the rate-controlling steps in a pathway exercise the most profound control over biological function. Regulatory mechanisms may be categorised at three levels: (1) Non-covalent interactions (physicochemical rapid interactions involving direct modulation of an enzyme by substrate or modifier concentration, pH, temperature, ionic strength, or ionic composition). (2) Covalent modification of an enzyme by another enzyme results in the interconversion of forms differing in their catalytic or regulatory properties. (3) Regulatory mechanisms which involve alteration of the total enzyme concentration by influencing the rates of synthesis and degradation of enzymes. According to this last principle there appears to be a reserve of catalytic activity inherent in pathways, since under normal circumstances the actual flux through a pathway is less than the potential maximal flux as determined by the least active enzyme of the pathway.

In the area of xenobiotic-metabolising enzymes, two kinds of modulation of the enzyme activity are most frequently encountered: induction and inhibition. Various drugs (barbiturates, antiepileptics, etc.), environmental pollutants (e.g. carcinogenic polycyclic aromatic hydrocarbons) and industrial chemicals (e.g. benzene and its derivatives) have been proved to increase the activity of these enzymes markedly by inducing their synthesis. On the other hand, when two or more of these substrates are simultaneously present in the organism, a mutual inhibition of their biotransformation may occur, provided their concentrations are high enough. Under conditions of low xenobiotic concentrations in occupational or environmental exposure, although these interactions may hardly influence elimination of the xenobiotics and of their metabolites in urine, significant changes in concentrations of their metabolites in target organs may occur. Either genetically determined or chemically induced higher activity of benzo(a)pyrene metabolising mono-oxygenases producing the diol-epoxides, suspected to be carcinogenic, is considered responsible for individual differences in the carcinogenic risk upon exposure to polycyclic aromatic hydrocarbons.

Inhibitors are an important class of enzyme activity modifiers, usually structurally resembling the substrate. Inhibitors may exert competitive inhibition (e.g. monoamino-oxidase-inhibiting secondary or tertiary amines), non-competitive inhibition (enzymes thus inhibited cannot react with substrate, e.g. RNA-ase inhibition by halogenated acetic acids) or irreversible inhibition (organophosphates inhibit acetylcholinesterase); or they may compete with a cofactor or prosthetic group (pyridoxal-dependent enzymes with hydrazine or carbon disulphide) or act as respiratory poisons (cyanide, azide, hydrogen sulphide, carbon monoxide) or ionic inhibitors (e.g. beryllium competes with Mn and Mg affecting phosphatases and phosphoglucomutase).

Enzyme distribution

Enzymes are organised into multienzyme systems within subcellular particles (e.g. the cytochrome system coupled with oxidative phosphorylation in mitochondria), between the particles and cytoplasm (glycolytic enzymes and pentose shunt outside particles are coupled with the above-mentioned mitochondrial systems). Uncouplers (e.g. dinitrophenol) can disconnect these systems and thus affect the utilisation of energy from glucose. Protein synthesis is localised predominantly in endoplasmic reticulum (ribosomes), while information for the synthesis comes from nuclear DNA via RNA. *w*-Oxidation of fatty acids proceeds in endoplasmic reticulum, while small fragments are metabolised along with glucose fragments in mitochondria and cytoplasm. Biotransformation (xenobiotic-metabolising) enzymes are localised in endoplasmic reticulum (mixed-function oxidases, reductases, glucuronidation), cellular plasma (e.g. glutation-S-transferases), mitochondria and even nuclei. Many enzymes of intermediary metabolism have an ubiquitous occurrence. However, organs vary in both qualitative and quantitative enzyme make-up. Enzymes largely peculiar to the liver are: ornithine carbamyl transferase, sorbitol dehydrogenase, phosphofructo aldolase, arginase, glucoso-6-phosphatase and quinine oxidase. The striated muscle is the primary site of creatine phosphokinase. By virtue of its enzyme equipment liver represents a central organ in the metabolism of glycides, lipids and proteins, and the biotransformation of foreign chemicals.

Serum enzymes

Serum enzymes are of great importance for their easy accessibility. The origin of serum enzymes is apparently in the normal turnover of cells, especially of those in the bloodstream. With the cell damage or disease of various organs it is possible to equate the increased activity to the damage of the cells. The release of soluble, cytoplasmic enzymes usually accompanies the initial stages of toxic action. The escape of structurally bound enzymes (microsomal, mitochondrial, etc.) indicates a more severe attack. Microsomes are the primary source of alkaline phosphatase and gamma-glutamyl transpeptidase; cytoplasm provides aldolase, creatine phosphokinase, glutamic oxalacetic and pyruvate transaminases, lactate dehydrogenase and leucine aminopeptidase; mitochondria are the origin of glutamic oxalacetic transaminase and malate dehydrogenase.

Enzymes in plasma may also be classified into secreted and cellular ones. The secreted enzymes are plasma specific (prothrombin, plasminogen, ceruloplasmin, lipoprotein lipase, pseudocholinesterase) and other (pancreatic and parotid amylases, lipase, prostatic phosphatase, pepsinogen). The cellular enzymes are general metabolic (lactic and malic dehydrogenases, 1,6-diphosphofructoaldolase, glutamic oxalacetic and glutamic pyruvate transaminases, i.e. aspartate and alanine aminotransferases) and organ specific (sorbitol dehydrogenase, glucose-6-phosphatase, 1-phosphofructoaldolase, urea cycle enzymes). Isoenzymes represent a potentially favourable means of identifying organ damage. They possess identical catalytic functions and identical substrate specificity, but physically they are distinct entities. Each organ and even each cell type in the same organ displays its own peculiar isoenzyme pattern for an individual enzyme. When cells are damaged, the isoenzymes are released into blood and the plasma isoenzyme pattern is thus changed, tending to resemble the pattern of the organ of enzyme origin. Enzymes with isoenzymes are: lactate dehydrogenase, isocitric dehydrogenase, GOT, malate dehydrogenase, alkaline phosphatase, sorbitol dehydrogenase, cholinesterase, esterases, etc. Isoenzymes thus serve in localisation of the target organ, the site of damage. For example, LDH possesses isoenzymes 1, 2, 3, 4 and 5. In the serum profile, $LDH_{2,1}$ predominates, while in normal liver $LDH_{5,4}$ and in normal kidney $LDH_{5,4,3}$. For example in CCl_4 poisoning, LDH_5 in serum is elevated due to liver damage.

Enzyme deficiency

Inborn errors of metabolism may be divided into several groups based on the consequences of the deficiency of an enzyme, a receptor, or a transport protein in a membrane. A number of inborn errors of metabolism due to enzyme deficiency lead to the accumulation of the precursor of that particular pathway. The resulting toxic concentrations of such precursor or an end product of a modified pathway may have serious consequences such as mental retardation caused by a deficiency of phenylalanine hydroxylase (phenylketonuria).

Enzymes deficiencies of interest for occupational medicine are apparently similar to those encountered in modified interactions of the organism with xenobiotics (foreign chemicals) in general. These may be classified in two groups: those with deficient or less efficient biotransformation enzymes modifying the fate of the xenobiotic in the body, while receptors mediating its action are normally active; and those, where the xenobiotic fate is not affected, but the abnormal response ensues from different reactivity of receptors.

Examples of the first group include individual differences in the sensitivity to isoniazid, hexobarbital, aminopyrine, antipyrin, phenylbutazone, ethanol, etc., and are caused by individual differences in the rate of their biotransformation to less active metabolites. Typical examples of the second group involve the haemolytic reactions of some individuals after therapeutic doses of aspirin, antipyrin, aminopyrine, antimalarics, sulphonamides, usually connected with an inborn deficiency of glucose-6-phosphate dehydrogenase, allergic reaction to isothiocyanates, etc. Individual differences in the sensitivity of some workers to carbon disulphide appear to be related to individual differences in biotransformation of the xenobiotic, as revealed by different values of the iodine-azide test. Evidence of the involvement of enzyme deficiency in individual differences in the sensitivity to occupational chemicals is still scarce.

The classification and nomenclature of enzymes

According to the "Recommendations of the International Union of Biochemistry (1964)" enzymes are divided into groups on the basis of the type of reaction catalysed and this, together with the name(s) of the substrate(s), provides a basis for naming individual enzymes. A system was devised which provides a classification of enzymes and also a basis for numbering them. Each enzyme number contains four elements, separated by points. The first figure shows to which of the six main divisions of the enzyme list the particular enzyme belongs: 1. Oxidoreductases. 2. Transferases. 3. Hydrolases. 4. Lyases. 5. Isomerases. 6. Ligases (synthetases). The second figure is for the sub-class, the third for the sub-sub-class and the fourth is the serial number of the enzyme in its sub-sub-class. For example lactate dehydrogenase (recommended trivial name) is classified as 1-lactate:NAD oxidoreductase 1.1.1.27. For glutamic-pyruvic transaminase the recommended trivial name is alanine aminotransferase and the systematic name 1-alanine:2-oxoglutarate aminotransferase 2.6.1.2.

Enzymatic changes upon exposure to important industrial and environmental chemicals

Animal experiments have provided a mass of data on enzymatic changes following poisoning by foreign chemicals. The most frequently observed are changes in the activities of serum transaminases (GPT and GOT), lactate dehydrogenase (LDH), sorbitol dehydrogenase (SDH), aldolase (ALD), glutamate dehydrogenase (Gl-

DH), alkaline phosphatase (alkPh) and acetylcholine esterase (AChE).

Examination of workers exposed in industry or agriculture and clinical treatment of cases of poisoning have provided valuable information on human serum enzyme changes. Some of the data concerning important chemicals are listed here:

- carbon tetrachloride: in fatal poisoning an exaggerated rise in ALD, alkPh, GOT, GPT and esterases (in animal experiments soluble supernatant enzymes rise almost instantaneously, structurally bound enzymes rise in the serum after latency usually coinciding with histological changes);
- trichloroethylene: no changes in exposures up to 100 ppm;
- tetrachloroethylene: alkPh, GPT in occupational exposures;
- 1,2-dichloroethane: ALD, human acute poisoning–serum aspartate aminotransferase (GOT), i.e. cytoplastic + mitochondrial enzyme;
- 1,2-dichloroethane + 1,2-dichloropropane: GPT, GOT, LDH, Gl-DH;
- 3-chloroprene-1 (allylchloride): GPT, GOT, LDH, Gl-DH (exposure to 1 ppm);
- chlorinated hydrocarbons hepatotoxicity based on animal experiments decrease in the order: carbon tetrachloride, chloroform, 1,2-dichloropropane, trichloroethylene, tetrachloroethylene;
- polychlorinated biphenyls: increased serum GPT and GOT, alkPh;
- methanol: the formed formic acid couples with folic acid affecting dependent enzymes;
- ethanol: enhances effects of other industrial chemicals;
- organophosphorus insecticides: in addition of AChE inhibition (long-term effect, recovery lasts weeks) occasionally supranormal values of ALD, alkPh, GOT, LDH;
- 2,4-phenoxyacetic acids: acute myopathy accompanied by rise in serum CPK (creatine phosphokinase), ALD, GPT, GOT;
- lead: serum ALD–early sign related to severity of exposure, alkPh, pseudoChE, delta-aminolevulinic acid dehydratase considered too sensitive, examination of coproporphyrin and delta-aminolevulinic acid preferred;
- mercury: GOT, GPT, LDH, alkPh, pseudoChE, LDH decreased, alkPh rise;
- vanadium: increased GPT, GOT;
- thallium: rise in serum CPK due to muscular involvement;
- fluoride: diminished serum alkPh in established fluorosis;
- fluoracetic acid: interferes with the Krebs cycle;
- manganese: rise in GPT, GOT, fall in LDH;
- carbon monoxide: carboxyhaemoglobin, asphyxia follows in elevation of ALD, CPK, GOT, LDH, malate dehydrogenase (MDH), sorbitol-DH;
- benzene: chronic benzolism–LDH increased, ALD, alkPh, GPT decreased, changes in blood cells are more specific;
- trinitrotoluene: excessive exposure–marked increase of alkPh, GOT;
- 1,1-dimethylhydrazine: GPT rise;
- dimethylformamide: elevated GPT, GOT, alkPh, pseudoChE;
- dinitrophenol: uncoupler of phosphorylation–in severe poisoning, fever; in a fatal case, rectal temperature 46 °C;
- disulphiram: interferes with alcohol dehydrogenase;
- carbon disulphide: interferes with vitamin B_6 metabolism and dependent enzymes, affected tryptophan metabolism, increased urine excretion of xanthurenic acid; this test is not specific;
- cadmium: increased activity of delta-aminolevulinic acid dehydratase, muramidase and RNA-ase affected and excreted in urine;
- ionising radiation: elevated GPT, GOT, CPK, decreased pseudoChE, beta-glucuronidase; LDH and MDH fluctuate.

GUT, I.

Biochemical toxicology of environmental chemicals. De Bruin, A. (Amsterdam, Elsevier/North Holland Biomedical Press, 1976), 1 544 p.

Enzymes in industry

Enzymes are complex protein molecules, varieties of which are to be found in all living cells. They are cellular catalysts–that is to say they mediate chemical reactions without themselves becoming involved. They are highly specific in a host of different biological processes and many are capable of being prepared in crystalline form.

The principal enzymes of industrial and pharmaceutical interest owe their importance, in the main, to their properties of acting upon either proteins, carbohydrates or fats, rendering these substances down into simpler forms. Those in most common use are mentioned below, together with the specific industries in which they are employed. Probably the oldest, simplest and best known enzyme is rennin, which is prepared from calves' stomachs and has the property of coagulating milk to form junket. It is foreseen that, in the future, research will make it possible to mount enzymes extracted from living systems on to artificial materials so that they may be exploited in chemical processes in the factory. Also, artificial enzymes are becoming a distinct possibility, i.e. enzymes synthesised and specifically designed to carry out a particular task as predicted.

Industrial production. Currently used enzymes are derived from animal, plant or microbial sources, the latter being easily the largest and most important. Animal and plant supplies are often limited and difficult to obtain, whereas microbial enzymes are not subject to any production and supply limitations. Although individual methods of production may vary in detail, broadly speaking all methods entail extraction, concentration, purification, stabilisation and standardisation.

Extraction entails either the grinding of dry tissues, the grinding or homogenisation of wet tissues or the grinding, lysis, or ultrasonication of microbial cells followed by extraction usually with water, but on occasion with buffer solutions. Finally, debris is removed by filtration or centrifugation.

The aqueous solutions of enzymes are often too dilute for direct recovery, and concentration by vacuum evaporation may be required. Further concentration can be effected where necessary by precipiting the enzymes using an organic solvent, such as ethyl alcohol or acetone, or by inorganic salts such as ammonium sulphate. The precipitated enzymes are then recovered by filtration or centrifugation and dried.

The concentrated products are assayed for potency by the manufacturer before being sold for further use. In the case of liquid products, before despatch they may require stabilisers to prevent microbial growth or loss of activity during storage. Solid products are adjusted to standard potency by the addition of diluents such as starch or lactose. Buffers and other salts may also be required to maintain favourable pH conditions, enzyme activity and stability.

Uses. It will be readily appreciated that enzymes are being utilised in an ever-growing number of industries, and the following list is by no means comprehensive, although it attempts to cover what may be considered to be the most important.

Amylases, proteases and lipoxidases are all used as "intentional" additives in the bakery industry for bread and many other baked goods. Milk contains a large number of natural enzymes and, in the dairy products industry, both lipolytic and proteolytic enzymes enter into cheese production, rennin is used in the formation of milk curd, lactase for frozen milk concentrates and in feed whey, and glucose oxidase in the production of egg albumen and dried egg yolk.

Enzymes are widely employed in the disintegration of fruit pulps and for the clarification of juices and wines. Pectin, glucose oxidase, amylase, invertase and naringinase are all utilised in a variety of different processes in the fruit products industries. Proteolytic enzymes such as papain obtained from the paw-paw fruit are in common use in the meat products industry for meat "tenderising".

The addition of proteolytic enzymes on a large scale to detergent powders in order to digest organic stains and matter from fabrics is a new but rapidly expanding use.

Enzymes are of course involved in the distillation of alcoholic beverages, which are derived mainly from starchy raw materials such as potatoes, rye, wheat, corn and rice in different parts of the world. Barley malt, corn malt, moulds and yeast provide the necessary enzymatic action. Beer is, widely speaking, the alcoholic liquor obtained from fermented cereal mashes of various sorts.

To some extent enzymes are also used industrially in flavour regeneration, the confectionery trade and in textiles and tanning. There is also talk of enzymes being added in the near future to animal feeding stuffs to facilitate and enhance digestion.

In the pharmaceutical field, several digestive enzymes are manufactured for use in the treatment of certain forms of dyspepsia, and a range of gastrointestinal disorders. For these purposes, the main enzymes used are pancreatin, rennin, pepsin, trypsin, catalase and lipase. Chymotrypsin is another proteolytic enzyme used to reduce inflammation and promote healing in chronic ulcers and such conditions as phlebitis, while hyaluronidase, a mucolytic enzyme, facilitates the absorption of injected fluids and the resorption of transudates, for example the products of trauma in the tissues, etc.; lysosyme too is used as an antibacterial and antiviral agent. Streptokinase and streptodornase are used to promote the removal of clotted blood and fibrinous or purulent accumulations. They are mainly of value in facilitating aspiration or drainage. Papain is used as a peeling product in cosmetics.

HAZARDS

There are no particular accident hazards specific to the production of enzymes. However, due to the widespread use of grinding machines, centrifuges and steam-heated plant, particular attention should be paid to machinery guarding and to the lagging or fencing of hot equipment to avoid burns. Accident prevention in this field is dealt with in greater detail in the article FERMENTATION, INDUSTRIAL.

Diseases. As has already been said, enzymes are highly active and specific substances, under appropriate conditions, will exercise their particular action. The proteolytic enzymes are potentially the most dangerous from the point of view of factory operatives as they will, in the presence of moisture, attack skin and the mucous membranes of the eyes and respiratory tract. Soreness, redness, inflammation and even ulceration of the skin, particularly of the finger tips, have been reported as a result of handling enzyme powder. Epistaxis, conjunctivitis and glossitis can also occur.

Additionally, if enzyme dust is liberated and inhaled in the factory, it is capable under certain conditions of sensitising workers by virtue of its protein structure. Trouble of this nature has been encountered and it is possible that, in periods of low humidity and in the added presence of detergent powder, reactions of this type may be potentiated. Respiratory manifestations of two kinds may occur.

Typical allergic asthma may come on rapidly after exposure, and symptoms will include spasm, wheezing and difficulty in breathing. In addition there may be retrosternal discomfort, headache, stomach ache and general malaise. Alternatively, some hours after exposure, general malaise and severe dyspnoea may develop giving a clinical picture similar to that of farmer's lung and allied conditions of extrinsic allergic alveolitis. Workers affected in this way require absence from work for several weeks and their return to normal health will be only gradual. It is also possible that repeated attacks of this nature may lead to permanent impairment of lung function due to fibrotic change.

These two occupational conditions have to be distinguished from other, more common chest complaints such as bronchitis, and in doubtful cases inhalation tests and skin testing may be helpful.

The use of solvents such as acetone in extraction processes may constitute both a health and a fire and explosion hazard.

SAFETY AND HEALTH MEASURES

It follows from what has been said that the aim in factories producing and using enzymes should be the elimination of all dust from the working atmosphere. Well designed plant, adequate enclosure, good ventilation, including exhaust ventilation as necessary, increase of size and weight of airborne particles by means of microagglomerate preparations, and scrupulous attention to plant cleanliness are all essential to avoid trouble. Adequate training of personnel and regular monitoring of the dust in the atmosphere should be carried out. Very low levels of total dust and of enzyme activity in the dust give useful measures of safe working conditions.

During work personnel should wear clean overalls, hand protection such as gloves or gauntlets, eye protection and respiratory protective equipment. Barrier creams should be available and adequate facilities for washing and showering provided.

All workers who are to be assigned to jobs entailing possible exposure to enzymes should be given a pre-employment medical examination and persons found to be atopic, i.e. constitutionally disposed to the common allergies, should be placed in other jobs. Asthma, hay fever and eczema are the main conditions that should be considered contraindications for this type of work, although in doubtful cases, skin testing and assessment of lung function may provide more definite information. In addition, persons with chronic skin disorders, respiratory diseases or with a tendency to sensitisation should not be exposed to work with enzymes.

ROBERTSON, D. S. F.

CIS 77-1648 "*Bacillus subtilis* enzymes: A 7-year clinical, epidemiological and immunological study of an industrial allergen". Juniper, C. P.; How, M. J.; Goodwin, B. F. J.; Kinshott, A. K. *Journal of the Society of Occupational Medicine* (Bristol), Jan. 1977, 27/1 (3-12). 20 ref.

CIS 77-1980 "Diagnostic tests in the skin and serum of workers sensitized to *Bacillus subtilis* enzymes". Belin, L. G. A.; Norman, P. S. *Clinical Allergy* (Oxford), Jan. 1977, 7/1 (55-68). Illus. 19 ref.

CIS 79-138 "The development of a fluorometric method for the assay of subtilisins". Chien, P. T. *American Industrial Hygiene Association Journal* (Akron, Ohio), Oct. 1978, 39/10 (808-816). 27 ref.

Epidemiology

Epidemiology as an approach to health and disease

Epidemiology is the study of the distribution and determinants of health statuses, or, more restrictively, of diseases in human populations. It embraces two bodies of knowledge: one of *epidemiological methods* of investigation and one of *epidemiological contents* generated by the application of such methods. Epidemiological contents are notions on the health profile of communities, on the distribution and aetiology of specific diseases (communicable and non-communicable), on the health effects, at population level, of environmental factors as well as of medical, technological and social interventions, particularly of a preventive nature.

Epidemiology and clinical medicine, which also includes methods of investigation (diagnosis) and substantive notions on diseases, may be seen as parallel but on different levels of observation: the population for epidemiology, the individual for clinical medicine. Correspondence tends to occur between these two plans: for example the clinical observation of a case of asthma associated with a history of exposure to welding fumes may find, sooner or later, its counterpart at the population level in epidemiological studies aimed at testing whether the causal hypotheses (e.g. about the particular fumes and compounds regarded as responsible) prompted by the clinical observation are tenable. This more or less unplanned course of events has happened again and again in the history of medicine in general and of occupational medicine in particular, and has proved essential for the identification of occupational hazards and their subsequent prevention. What has emerged more and more clearly in recent years is the need for a more systematic, rigorous and timely use of the epidemiological approach as a tool to foster health in the workplace. In this sense the epidemiological approach to health and diseases can contribute to at least three main goals:

(*a*) identification of environmental hazards in the workplace, the oldest and still most prominent goal;

(*b*) monitoring of the health effects of changes in the working environment including the enforcement of primary prevention measures (e.g. engineering changes in a plant) and of control limits for toxic exposures;

(*c*) evaluation of the usefulness of secondary prevention measures, for example of cytological screening programmes for the early detection and treatment of occupational diseases.

These goals can be pursued through the use of different types of investigations which are customarily regarded as belonging to one of two distinct, but in fact somewhat overlapping, categories: descriptive and analytical.

Descriptive studies describe the patterns of distribution of a disease or of a symptom (say backache) or of a physiological variable (weight, heart rate) or, more generally, of any health condition (until death) in one or more groups or populations. They include descriptions of variation of disease occurrence in relation to time, to space and to personal characteristics such as age, sex, ethnic group, social class, occupation, etc. They also include correlation studies of disease occurrence with other variables, for example of chronic non-specific lung disease mortality with levels of atmospheric pollution. Finally they include clinical case reports and clinical case series in which the coincidence of a disease (say aplastic anaemia) and an exposure (say to a solvent) has been observed. Descriptive studies are useful to generate hypotheses and to provide suggestive evidence on cause-effect relationships, while they are weak if regarded as a tool for testing aetiological hypotheses. A descriptive study usually represents a form of secondary analysis of existing data primarily collected for purposes other than the study in hand, for example censuses, vital statistics (of deaths, births, infectious diseases, accidents, etc.) or of labour statistics (e.g. of absenteeism).

Analytical studies, on the other hand, provide much more cogent evidence for testing hypotheses about aetiological factors in man. They are usually designed to test in a specific way one or a few hypotheses, and they may take one of three main forms: cross-sectional studies, in which a biological response is measured at the same time as exposure levels, for example forced expiratory flow at one second and air concentrations of a respirable dust are measured for several occupational groups; retrospective or case-control studies, in which previous exposure to suspected pathogens is assessed and compared in cases (say lung cancer cases) and controls (subjects without lung cancer); prospective or follow-up studies, in which subjects differently exposed to suspected pathogens are followed up and their disease (or death) experience is recorded and compared.

Role of the occupational physician in epidemiological investigations

The physician operating within an occupational health service can and should have an important function in epidemiological investigations on the occupational environment. By collaboration with other technical and non-technical parties (such as industrial hygienists, personnel managers and union officers) and by personal initiative he can ensure:

(*a*) that records documenting each worker's job history do exist and are stored in good condition for several decades. There is an obvious need, still very infrequently met, for "exposure oriented" (rather than "administration oriented") records in which the job history is described and codable in a way that would allow for matching with environmental measurements in the workplace. Less ambitiously the simple fact of keeping for a long time whatever administrative records are in current use (personnel books, payroll sheets, etc.) is a most valuable contribution on which the feasibility of many epidemiological studies on long-term pathological effects (e.g. cancer, pneumoconioses) is critically dependent;

(b) that medical documents contain essential information on diseases and occupational exposures prior to enrolment in the present industry as well as on such important personal habits as smoking (recorded in some detail);

(c) that records of different types in current use (administrative, medical, environmental hygiene) are gradually rendered reciprocally compatible so that information pertaining to a given worker and scattered among different records can be easily linked for epidemiological research purposes. Development of compatible records is all the more indicated when computerised systems of records are introduced, as happens in many large modern enterprises. In this development full attention is to be given to guaranteeing the protection of information of a personal and confidential nature;

(d) that biological measurements of exposure, whenever they are deemed relevant, are appropriately performed (in respect to obtaining the biological material and to the analytical procedures) and their results recorded and kept together with any corresponding environmental measurements;

(e) that when medical interviews and examinations are periodically carried out, some monitoring of their quality in time is performed and, if at all possible, a programme of standardisation and quality control of the procedures is adopted.

These points outline what may be considered as a basic instrumental contribution of the occupational physician to epidemiological studies, by ensuring that relevant data of good quality are available. Hopefully he should take a more active role in the actual design, conduct, analysis and interpretation of epidemiological investigations, from simple descriptive to analytical, carried out in the working population under his responsibility in collaboration with epidemiologists. This involvement of the occupational physician may arise on the occasion of an ad hoc study of a problem, or it may take the form of a continuous commitment, if, for example, an on-going programme of epidemiological surveillance of a working population is established in respect to short-term or long-term toxic effects, or both. Such programmes are becoming more numerous. Sometimes other epidemiologically oriented activities are grafted on to these programmes, such as, for example, prophylactic trials to control non-occupational diseases like hypertension in the working population. When a substantial and permanent epidemiological effort is developed, the team in charge ought to include, besides the familiar figures of the occupational physician and the industrial hygienist, that of the medical epidemiologist or statistician epidemiologist.

SARACCI, R.

General:

Foundations of epidemiology. Lilienfield, A. M.; Lilienfield, D. E. (Oxford, Oxford University Press, 1980).

Epidemiology: principles and methods. MacMahon, B.; Pugh, T. (Boston, Little, Brown and Co., 1970).

"Epidemiological strategies and environmental factors". Saracci, R. *International Journal of Epidemiology* (London), June 1978, 7/2 (101-111). 77 ref.

Occupational epidemiology:

Industrial pathology. Epidemiological approach (Pathologie industrielle. Approche épidémiologique). Lazar, P. (ed.) (Paris, Flammarion Médecine-Sciences, 1979), 365 p. (In French)

Occupational epidemiology. Monson, R. (Boca Raton, Florida, CRC Press, 1980), 256 p.

Epilepsy

Epilepsy is one of the world's greatest social problems. It is a serious and widespread disease which may cause disability of varying degrees. Statistics indicate that between 0.5 and 6% of the world's population is affected by the disease—the wideness of the range being the result of the different statistical criteria employed in various countries.

The picture of epilepsy varies considerably depending on the stage and aetiology of the disease. The most dramatic feature is the epileptic fit (the *grand mal*) characterised by loss of consciousness, convulsive seizures and, in some cases, bloody foaming at the mouth; the non-convulsive attack (the *petit mal*) is ostensibly less dramatic. Frequently the patient may suffer alternately from different forms of attack; there may also be mental disorders (excitation with aggressive and destructive tendencies, paranoic episodes). Attacks may occur at any time and are usually preceded by ill defined sensations called aura; it is common for epileptics to suffer from amnesia during the attack and to be confused afterwards. The disease may produce permanent mental changes and even, in severe cases, intellectual degradation; however, most epileptic patients are mentally normal and competent.

Work capacity

There is still wide prejudice concerning the work capacity of epileptics and their increased proneness to accidents. However, recent research has shown that this prejudice is unfounded. Persons in whom epileptic fits are preceded by warning aura and who are not subject to paroxysmal mental disorders readily adapt to work and normal social life, and it has been found that the leading of a normal active life raises the patient's morale, reduces the frequency of epileptic fits and may delay or prevent the onset of intellectual degradation or psychopathological conditions. Moreover, research has also shown that accidents are no more frequent among epileptics than among other population groups. Nevertheless, persons whose form of the disease does not respond to treatment and who suffer from frequent attacks, mental or pseudoneurotic disorders, or severe alcoholism have considerable difficulty in adapting to the work environment.

The possibility of employing an epileptic depends on the form and severity of the disease, the degree of disability, the type of work envisaged and the relevant working conditions. It is generally considered that 75% of all epileptics can be suitably adapted to work. In addition, a large proportion of the remaining 25% can also be employed provided the work is in line with the severity of their disease, and provided suitable social welfare, medicolegal and other conditions are met. The proportion of epileptics who are totally unfit for work is small.

Contraindicated employment. Certain jobs may endanger the epileptic himself or persons working with him. To avoid all danger of an accident in the event of an attack, epileptics should not be employed on work at heights, underwater, or in the vicinity of moving machinery or high-temperature equipment or materials which could cause burns. Moreover, epileptics should not be employed on work involving danger of poisoning due to substances such as tetraethyl lead and carbon monoxide which, as a result of their action on the nervous system, may aggravate or provoke an attack. Work entailing exposure to heat, noise, radiation or glare, etc., work requiring rapid reactions or distribution of attention to several matters at the same time (e.g. despatch, control desk work, belt conveyor work) or work necessitating permanent contact with a large number of people should not be entrusted to epileptics.

Medical supervision

The task of the industrial medical officer in this field is to detect epileptics and to provide for their treatment and rehabilitation. Under normal circumstances diagnosis presents little difficulty and is based on the convulsive nature of attacks and the paroxysmal character of other neuropsychic symptoms; in many cases there may be slight organic symptoms (reflex modifications) and typical personality changes. The hyperventilation test is a valuable diagnostic aid since it will initiate an attack in 50% of cases; electroencephalography can be used to confirm diagnosis.

Working epileptics should be subject to systematic medical supervision and be provided with good working conditions. Changes of job are not advised; however, if such a step is essential, the epileptic should be guided to manual work, preferably where there is contact with only a limited number of people and where the work tempo is regular. Psychological back-up can do much to help raise the epileptic's morale and help him overcome his feeling of inferiority.

LUKANOV, M.

Diagnostic use of EEG:

"Interest and limits of the electroencephalogram in occupational medicine" (Intérêt et limites de l'électroencéphalogramme en médecine du travail). Vercelletto, P. *Archives des maladies professionnelles, de médecine du travail et de sécurité sociale* (Paris), Sep. 1976, 37/9 (702-705). (In French)

Epileptics in employment:

CIS 74-1781 "Epilepsy and work" (Epilepsie et travail). *Réadaptation* (Paris), Mar. 1974, 208 (1-34). Illus. 72 ref. (In French)

CIS 586-1968 *Employing someone with epilepsy.* PL430 (1967) Ministry of Labour and Central Office of Information (London, HM Stationery Office, 1967), 6 p.

Driving:

"Epilepsy and driving licence" (Epilepsie und Führerschein). Bay, E. *Medizinische Welt* (Stuttgart), 1975, 26/13 (596-598). (In German)

Epoxy compounds

Epoxy compounds are a group of cyclic ethers or alkene (alkylene) oxides which have an oxygen atom attached to two adjacent carbon atoms (*oxirane* structure). These ethers will react with amino, hydroxyl, and carboxyl groups as well as inorganic acids to yield relatively stable compounds.

```
      O
     / \
 -C-----C-
  |     |
```

Production. Epoxides are produced through a variety of methods including oxidation of olefins with peroxy acids, oxygen and hydrogen peroxide; cyclodehydrohalogenation (chlorohydrin process); cyclisations involving a leaving group other than a halogen; and many other syntheses. In the chlorohydrin process, simple monoepoxy compounds such as ethylene, butylene and propylene oxide can be prepared from the corresponding unsaturated aliphatic hydrocarbon. Epoxy resins may then be cured by reaction with amino, carboxyl, or hydroxyl groups and inorganic acids. In addition to the resins and final product, many other materials are used in the compounding of plastics and resins. These include accelerators, catalysts, copolymers, dyes, fillers, lubricants, plasticisers, solvents, pigments and stabilisers, all of which may require special consideration regarding potential hazards. Ethylene oxide, the epoxy compound produced in the greatest volume and glycidyl ethers are dealt with in separate articles in the encyclopaedia. (See ETHERS; VINYLCYCLOHEXENE DIOXIDE.)

Uses. Epoxy compounds have found wide industrial use as chemical intermediates in the manufacture of solvents, plasticisers, cements, adhesives, and synthetic resins. The alpha epoxy compounds, with the epoxy group (C–O–C) in the 1,2 position are the most reactive of the epoxy compounds and are primarily used in industrial applications. The epoxy resins, when converted by curing agents, yield highly versatile, thermosetting materials used in a variety of applications including surface coatings, electronics (potting compounds), laminating and bonding together of a wide variety of materials. Epoxy groups are usually cured by reaction with amino, carboxyl or hydroxyl groups and inorganic acids. Several formaldehyde compounds are extensively used as curing agents and may contribute to the occupational health problems associated with epoxy compounds. The more widely used epoxy compounds will be reviewed in this article and include propylene oxide, the butylene oxides, epichlorohydrin, trimellitic anhydride (TMA), diepoxybutane, glycidol and glycidaldehyde.

Epoxy compounds have the potential for adversely affecting the skin, mucous membranes, lungs, central nervous system and liver. The low molecular weight monoepoxides are weakly anaesthetic and are strong irritants. Inhalation can cause pulmonary oedema and secondary pulmonary infection. Skin irritation in varying degrees up to necrosis can result from prolonged contact, usually involving saturated clothing. In addition to pulmonary effects including sensitisation, vapours can also cause eye irritation. Tumours have been produced in experimental animals following both cutaneous and subcutaneous administration of several epoxy compounds. One epoxy compound, epichlorohydrin, has been reported to cause a significant increase of pulmonary cancer in exposed workers.

Propylene oxide (CH_2OCHCH_3)

1,2-EPOXYPROPANE; PROPENE OXIDE

m.w.	58.03
sp.gr.	0.83
fr.p.	−112 °C
b.p.	34 °C
v.d.	2
v.p.	400 mmHg ($53.2 \cdot 10^3$ Pa) at 17.8 °C
f.p.	−37 °C
e.l.	2.1-21.5%

very soluble in water; miscible with acetone, benzene, carbon tetrachloride, ether and methanol

a colourless liquid with ethereal odour.

TWA OSHA	100 ppm 240 mg/m³
TLV ACGIH	20 ppm 50 mg/m³
IDLH	2 000 ppm
MAC USSR	1 mg/m³

Propylene oxide is a highly reactive chemical primarily used as an intermediate in the production of polyether polyols, which in turn are used to make polyurethane foams. The other major use is in the production of propylene glycol and its derivatives. Propylene oxide has been used as a fumigant to sterilise a wide variety of materials including foodstuffs.

HAZARDS

Propylene oxide is, based on animal experiments, approximately one-third as toxic as ethylene oxide when administered by inhalation or ingestion. Primary hazards include irritation of the eyes, respiratory tract and lungs as well as central nervous effects as demonstrated by ataxia, inco-ordination and general depression. Acute overexposure may result in pulmonary infection. Subcutaneous and/or intramuscular administration in rats produced local sarcomas. Other than acute eye (corneal burns) and skin injuries, adverse effects on humans have not been reported. Due to its rapid rate of evaporation, even contact with undiluted propylene oxide is not likely to cause skin irritation. However, when confined to the skin, such as the wearing of contaminated clothing, even dilute concentrations (10%) may cause irritation, blistering and burns. There is some indication that highly dilute solutions (less than 10%) may be more irritating to the skin than undiluted propylene oxide.

Butylene oxides (1,2-epoxybutane and 2,3-epoxybutane) (C_4H_8O)

m.w.	72.1 (1,2-butylene oxide) 72.1 (butylene oxide(s) (mixture))
sp.gr.	0.826 (1,2-butylene oxide) 0.824 (butylene oxide(s) (mixture))
fr.p.	below −60 °C (1,2-butylene oxide) below −50 °C (butylene oxide(s) (mixture))
b.p.	62-64 °C (1,2-butylene oxide) 59-63 °C (butylene oxide(s) (mixture))
v.d.	0.97 (1,2-butylene oxide) 1.36 (butylene oxide(s) (mixture))
f.p.	−26 °C (1,2-butylene oxide) −15 °C (butylene oxide(s) (mixture))
e.l.	1.5-18.3% (1,2-butylene oxide) 1.5-18.3% (butylene oxide(s) (mixture))

The industrially important butylene oxides are the 1,2 isomer and a mixture of the 1,2 and 2,3 isomers. Both compounds are water-white liquids with low boiling points and are highly flammable. The liquids are fairly stable but may react violently with materials having a labile hydrogen. The oxides are used for the production of butylene glycols and their derivatives. They also find use in the manufacture of surface active agents and other industrially important products.

HAZARDS

The discussion of the hazards associated with propylene oxide is essentially applicable to the butylene oxides. The butylene oxides, as compared to propylene oxide, have a lower volatility and based on inhalation studies with rats a lower toxicity. At concentrations likely to cause acute toxic reactions (greater than 400 ppm) the butyl oxides have a disagreeable odour and are irritating to the eyes and nasal passages.

As with propylene oxide, the butylene oxides are primary irritants of the eyes and skin, with skin effects pronounced where saturated clothing remains in contact with the skin. Animal inhalation studies have demonstrated that in addition to pulmonary irritation, secondary effects may include pneumonia.

Epichlorohydrin (CH_2OCHCH_2Cl)

1-CHLORO-2,3-EPOXYPROPANE

m.w.	92.5
sp.gr.	1.18
m.p.	−48 °C
b.p.	116 °C
v.d.	3.21
f.p.	34 °C
e.l.	3.8-21.0%

miscible with water, ethanol, diethyl ether and chlorinated aliphatic hydrocarbon solvents

a colourless liquid at room temperature.

TWA OSHA	5 ppm 19 mg/m³
NIOSH	2 mg/m³
	19 mg/m³ 15 min ceil
TLV ACGIH	2 ppm 10 mg/m³ skin
STEL ACGIH	5 ppm 20 mg/m³
IDLH	100 ppm
MAC USSR	1 mg/m³

Epichlorohydrin is used in the manufacture of synthetic glycerin, epoxy resins, surface active agents, pharmaceuticals, insecticides, solvents and a number of other products.

HAZARDS

As with the other epoxy compounds epichlorohydrin is irritating to the eyes, skin and respiratory tract of exposed individuals. Human and animal evidence has demonstrated that epichlorohydrin may induce severe skin damage and systemic poisoning following extended dermal contact. Exposures to epichlorohydrin at 40 ppm for 1 h have been reported to cause eye and throat irritation lasting 48 h and at 20 ppm caused temporary burning of the eyes and nasal passages. Epichlorohydrin-induced sterility in animals has been reported as have liver and kidney damage.

Subcutaneous injection of epichlorohydrin has produced tumours in mice at the injection site but have not produced tumours in mice by skin painting assay. Inhalation studies with rats have shown a statistically significant increase in nasal cancer. Epichlorohydrin has induced mutations (base-pair substitution) in microbial species. Increases in the chromosomal aberrations found in the white blood cells of workers exposed to epichlorohydrin have been reported. Unpublished results of a long-term epidemiologic study of workers exposed to epichlorohydrin at two US facilities of the Shell Chemical Company have been reported to demonstrate a statistically significant ($p < .05$) increase in deaths due to respiratory cancer. These data were analysed by Dr. Phillip Enterline of the University of Pittsburgh.

Based on the above information, the US National Institute for Occupational Safety and Health has recommended that epichlorohydrin be treated in the workplace as if it were a human carcinogen, and that the permissible exposure limit be set at 2 mg/m³ (0.5 ppm).

Trimellitic anhydride ($HOCOC_6H_3(CO)_2O$)

1,2,4-BENZENETRICARBOXYLIC ACID; 1,2-ANHYDRIDE; TMA

m.w.	192.1
m.p.	165 °C
b.p.	390 °C
v.p.	4×10^{-6} mmHg (0.0005 Pa) at 20 °C

a colourless solid.

TLV ACGIH 0.005 ppm 0.04 mg/m³ (proposed)

Trimellitic anhydride (TMA) is used as a curing agent for epoxy and other resins, in vinyl plasticisers, paints, coatings, dyes, pigments and a wide variety of other manufactured products. These products find applications in high-temperature plastics, wire insulation, gaskets and automobile upholstery.

HAZARDS

Several incidents involving the occupational effects of exposure to TMA have been reported. Multiple inhalation exposures to an epoxy resin containing TMA being sprayed on heated pipes was reported to have caused pulmonary oedema in two workers. Exposure levels were not reported but there was no report of upper respiratory tract irritation while the exposures were being experienced indicating that a hypersensitive reaction might have been involved. In another report, 14 workers involved in the synthesis of TMA were observed to have respiratory symptoms resulting from sensitisation to TMA. In this study three separate responses were noted. The first, rhinitis and/or asthma, developed over an exposure duration of weeks to years. Once sensitised, exposed workers exhibited symptoms immediately after exposure to TMA, which ceased when the exposure was stopped. A second response, also involving sensitisation, produced delayed symptoms (cough, wheezing and laboured breathing) 4-8 h after exposure had ceased. The third syndrome was an irritant effect following initial high exposures.

One study of adverse health effects, which also involved measurements of air concentrations of TMA, was conducted by the US National Institute for Occupational Safety and Health. Thirteen workers involved in the manufacture of an epoxy paint had complaints of eye, skin, nose, and throat irritation; shortness of breath, wheezing, coughing, heartburn, nausea and headache. Occupational airborne exposure levels averaged 1.5 mg/m^3 TMA (range from "none detected" to 4.0 mg/m^3) during processing operations and 2.8 mg/m^3 TMA (range from "none detected" to 7.5 mg/m^3) during decontamination procedures.

Experimental studies with rats have demonstrated intra-alveolar haemorrhage with subacute exposures to TMA at 0.08 mg/m^3. The vapour pressure at 20 C (4 x 10^{-6} mmHg) corresponds to a concentration slightly more than 0.04 mg/m^3.

Diepoxybutane ($C_4H_6O_2$)

BUTADIENE DIOXIDE

sp.gr.	0.96
m.p.	4 °C (*dl*-form), 19 °C (*meso*-form)
b.p.	138 °C; 140-142 °C (*meso*-form)

a colourless liquid.

Diepoxybutane has been used as a polymer curing agent, for cross-linking textile fibres and to prevent spoilage of foodstuffs.

HAZARDS

Short-term (4-h) inhalation studies with rats have caused watering of the eyes, clouding of the cornea, laboured breathing and lung congestion. Experiments in other animal species have demonstrated that diepoxybutane, like many of the other epoxy compounds, can cause eye irritation, burns and blisters of the skin and irritation of the pulmonary system. In man, accidental "minor" exposure caused swelling of the eyelids, upper respiratory tract irritation, and painful eye irritation 6 h after exposure.

Skin application of *dl*- and *meso*-1,2:3,4-diepoxybutane have produced skin tumours, including squamous-cell skin carcinomas, in mice. The *dl* and *l* isomers have produced local sarcomas in mice and rats by subcutaneous and intraperitoneal injection respectively.

Glycidol (CH_2OCHCH_2OH)

2,3-EPOXY-1-PROPANOL; HYDROXYMETHYL ETHYLENE OXIDE; 2-HYDROXYMETHYLOXIRAN

m.w.	74.1
sp.gr.	1.12
b.p.	160 °C
v.d.	2.15
v.p.	0.9 mmHg (0.12·10^3 Pa) at 25 °C, which provides a vapour saturated air concentration of 1 180 ppm

completely soluble in water, ethanol and ether

a colourless liquid.

TWA OSHA	50 ppm 150 mg/m^3
TLV ACGIH	25 ppm 75 mg/m^3
STEL ACGIH	100 ppm 300 mg/m^3
IDLH	500 ppm

Glycidol is used as a stabiliser for natural oils and vinyl polymers, as a dye levelling agent and as a demulsifier.

HAZARDS

Based on experimental studies with mice and rats, glycidol was found to cause eye and lung irritation. The LC_{50} for a 4-h exposure of mice was found to be 450 ppm and for an 8-h exposure of rats it was 580 ppm. However, at concentrations of 400 ppm glycidol, rats exposed for 7 h a day for 50 days showed no evidence of systemic toxicity. After the first few exposures, slight eye irritation and respiratory distress were noted.

Glycidaldehyde ($CH_2OCHCHO$)

2,3-EPOXYPROPIONALDEHYDE

m.w.	72.1
sp.gr.	1.14
fr.p.	−62 °C
b.p.	112 °C
v.d.	2.6
f.p.	31 °C

completely soluble in all common solvents and water

colourless liquid with a pungent odour.

Glycidaldehyde has been used as a cross-linking agent in wool finishing, for oil tanning and fat-liquoring of leather and surgical sutures and for protein insolubilisation.

HAZARDS

Glycidaldehyde has been demonstrated to be moderately toxic following ingestion or inhalation by experimental animals. It is also irritating to the eyes and can cause depression of the central nervous system. Glycidaldehyde can produce severe skin irritation with slow healing, followed by pigmentation of affected areas. It has also produced skin cancers in mice and malignant tumours at the site of subcutaneous injections in both mice and rats.

SAFETY AND HEALTH MEASURES

The primary purposes of control measures for the epoxy compounds should be to reduce the potential for inhalation and skin contact. Wherever feasible, control at the source of contamination should be implemented. This may involve enclosure of the operation and/or the application of local exhaust ventilation. Where such engineering controls are not sufficient to reduce airborne concentrations to acceptable levels, respirators may be necessary to prevent pulmonary irritation and sensitisation in exposed workers. Preferred respirators include gas masks with organic vapour cannisters and high-efficiency particulate filters or supplied-air respirators. All body surfaces should be protected against contact with epoxy compounds through the use of gloves, aprons, face shields, goggles and other protective equipment and clothing as necessary. Contaminated clothing should be removed as soon as possible and the affected areas of the skin washed with soap and water.

Safety showers, eye wash fountains and fire extinguishers should be located in areas where appreciable amounts of epoxy compounds are in use. Handwashing facilities, soap and water should be made available to involved employees.

The potential fire hazards associated with epoxy compounds suggest that no flames or other sources of ignition, such as smoking, be permitted in areas where the compounds are stored or handled.

Treatment. Affected workers should, as necessary, be removed from emergency situations and if the eyes or skin have been contaminated they should be flushed with water. Contaminated clothing should be promptly removed. If exposure is severe, hospitalisation and observation for 72 h for delayed onset of severe pulmonary oedema is advisable.

ROSE, V. E.

"Epoxy compounds". Ch. 32 (2 141-2 257). 236 ref. Hine, C.; Rowe, V. K.; White, E. R.; Darmer, K. I.; Youngblood, G. T. *Patty's Industrial Toxicology.* Vol. 2A Toxicology. Clayton, G. D.; Clayton, F. E. (eds.). (New York, John Wiley and Sons, 1981), 2 878 p.

IARC monographs on the evaluation of the carcinogenic risk of chemicals to humans. Vol. 11: *Cadmium, nickel, some epoxides, miscellaneous industrial chemicals and general considerations on volatile anaesthetics* (Lyons, International Agency for Research on Cancer, 1976), 306 p.

Criteria for a recommended standard–Occupational exposure to epichlorohydrin. DHEW (NIOSH) publication No. 76-206 (National Institute for Occupational Safety and Health, 4676 Columbia Parkway, Cincinnati) (1976), 152 p. 101 ref.

"Chemical pneumonitis secondary to inhalation of epoxy pipe coating". Rice, D. L.; Jenkins, D. E.; Gray, J. M.; Greenburg, S. D. *Archives of Environmental Health* (Chicago), July-Aug. 1977, 32/4 (173-178). Illus. 19 ref.

Current intelligence bulletin reprints 19 thru 30. DHEW (NIOSH) publication No. 79-146 (National Institute for Occupational Safety and Health, 4676 Columbia Parkway, Cincinnati) (1979), 155 p. (Information on trimellitic anhydride and epichlorohydrin)

"Diseases due to epoxy resins" (Erkrankungen durch Epoxidharze). Konietzko, H.; Fischer, G. *Zentralblatt für Arbeitsmedizin, Arbeitsschutz, Prophylaxe und Ergonomie* (Heidelberg), Mar. 1982, 32/3 (86-91). 95 ref. (In German)

Equilibrium

The ability to maintain the body in a desired position in space and to retrieve this position when it is altered by the action of internal or external forces is dependent on the correct operation of the body's complex equilibrium mechanism. This mechanism comprises four essential elements:

(a) the peripheral receptor organs or statocysts found in the inner ear and in the vision, muscle and articulation systems, which are capable of supplying information on the position and movement of the body in space;

(b) the centripetal nervous paths that transmit this information to the cerebral centres for integration and co-ordination;

(c) the cerebral centres in the brain; and

(d) the centrifugal nervous paths that carry the motor impulses from the centres to the muscles.

By interplay of synergic and antagonistic contraction, the muscles move the various parts of the body (head, trunk, joints) in such a way that each assumes a suitable position for maintaining equilibrium whether for rest or for movement. The operation of this complex mechanism is automatic and unconscious, but the will may intervene to regulate and correct if necessary.

Disturbances of equilibrium

The sense of equilibrium may be impaired by pathological changes in the organs on which it depends; disorders of this type are characterised by vertigo. Vertigo, which can be induced experimentally, by suitable stimuli, is a false sensation of movement–usually rotary–of the subject in relation to his surroundings or of the surroundings in relation to the subject. It is accompanied by various symptoms which, in serious cases, will include acute malaise, nausea, blurred vision, pallor and fainting; nystagmus is frequently an accompanying symptom, owing to the connections between the vestibular system and the oculomotor centres.

Three types of vertigo may be distinguished:

(a) true vertigo with objective equilibrium disorders;

(b) malaise and fainting, always related to a disorder of the central nervous system, cardiovascular system, etc.; and

(c) mental vertigo or pseudo-vertigo with no objective clinical manifestations, such as certain "phobic" syndromes (e.g. fear of heights, fear of voids, etc.).

Both occupational and non-occupational factors may produce equilibrium mechanism disorders. In many non-occupational diseases, symptoms may include equilibrium disorders of varying degrees of severity. The diseases in which the symptoms are most significant are certain cardiovascular diseases (hypertension and hypotension, in particular orthostatic; rapid changes in blood pressure; arteriosclerosis) and nervous system diseases (multiple sclerosis, which often commences with vertigo without any other signs, especially in young persons; ataxia; brain stem ischaemia due to vertebrobasal insufficiency; tumours of the posterior cerebral fossa; etc.). However, the sensation of vertigo may also be caused by poisoning due to numerous pharmaceutical agents (salicyclic acid derivatives, barbiturates, streptomycin, antihistamines) or alcohol, tobacco, caffeine, etc. The occupational factors may be physical, chemical or biological and may be specific and direct or general and indirect in their action.

Physical factors. Temperature extremes such as heat and hot work and cold and cold work may produce equilibrium disorders, especially in persons subject to vasomotor disturbances. Vertigo due to changes of atmospheric pressure is a more frequent and often more serious occurrence. It occurs mainly during compression and decompression of compressed-air workers and divers. The worker may be subject to attacks for some considerable time after termination of exposure. In skin or SCUBA (self-contained underwater breathing apparatus) divers, pressure-induced vertigo may cause loss of sense of direction and incoordination of muscular movements. This syndrome, known as nitrogen narcosis, has symptoms closely resembling those of drunkenness.

Compressed oxygen, which is used in some breathing equipment, may produce oxygen poisoning, of which vertigo is also a symptom.

Rapid descent from high altitude may also cause equilibrium disturbances, probably due to a fall in peripheral blood pressure. However, height vertigo, the symptoms of which are dizziness, uncertain gait and fear of falling, implies the psychological factor of anxiety due to the mental image of danger. The vertigo that occurs during high-speed flight is due to a combination of acceleration, centrifugal force, pressure changes, noise and, in particular, loss of orientation resulting from the absence of points of visual reference. Maintenance of posture in the absence of the Earth's gravity (zero G) is one of the most important problems in space flight: weightless, the body tends to float freely in space (see AEROSPACE–SPACE OPERATIONS). Under these conditions, the signals transmitted by the peripheral receptors to the posture-regulating centres undergo fundamental change; the regulating centres are unable to co-ordinate the signals, and disturbances of the equilibrium occur including dizziness, loss of sense of position and vertigo. Usually, adequate adaptation occurs but the disorders reappear when the subject reverts to normal gravity (1 G), and disappear only after two weeks or more. However, the astronauts on the Moon (where the gravitational pull is only one-sixth that on Earth) noted no marked change in their equilibrium in spite of their highly characteristic bounding gait over irregular terrain and their relative weightlessness.

It is widely believed that the equilibrium organs in the ear can be damaged by rapid pressure build-ups due to violent noises, such as explosions and also prolonged exposure to high-intensity noises, probably owing to their high proportion of ultrasonic and infrasonic components and their ability to induce low-frequency vibration. Vertigo has also been experienced in persons undergoing intense long-term exposure to electromagnetic microwave radiation (in telecommuncations and heat generation).

Certain lighting effects may also produce equilibrium disturbances, with vertigo, disorientation and stupor; typical examples are flashing lights, the stroboscopic effect of fluorescent lamps, and the inadequate lighting in mines, which used to cause miners' nystagmus.

Vertigo, usually of a mild form, also occurs in persons doing night work or working night shifts, and is probably due to inversion of circadian biological rhythms.

Chemical factors. Certain toxic substances such as carbon monoxide act almost specifically on the central and peripheral equilibrium organs, whereas substances such as streptomycin have selective allergenic effects. Normally, equilibrium disturbances in such cases form part of the symptomatology of a general intoxication affecting a number of organs.

Various industrial substances, especially anaesthetics or narcotics, such as the chlorinated aliphatic hydrocarbons, bromoethane and chloroethane, vinyl chloride, trichloroethylene, tetrachloroethylene and dichloromethane, act directly on the central nervous system. Acetylene, amyl acetate, ethyl acetates, ketones, such as acetone and butanone, mercaptans, such as ethyl, butyl and methyl mercaptan, methyl alcohol and, to a certain degree, ethyl alcohol, all have similar properties.

Benzene and its homologues, nitrobenzene, phenol, trinitrophenol, phenylhydrazine and boron compounds, such as diborane and pentaborane, may also act on the nervous system to produce equilibrium disturbances, as do also nitroglycerin, ethylene glycol dinitrate, carbon disulphide, gasoline, hydrogen sulphide, phosphine, certain phosphoric acid esters (especially *o*-cresyl-phosphate), nickel carbonyl, cadmium and selenium oxides. Cases of vertigo, some chronic, may also be observed in lead, mercury and manganese poisoning.

Miscellaneous factors. Finally, equilibrium disturbances may be caused by biological agents, as is the case with ancylostomiasis, or by trauma (vertigo as a sequel to head injuries, traumatic neuroses subsequent to electric shock).

SAFETY AND HEALTH MEASURES

These should be based on medical screening, correct job placement and elimination of factors that may cause equilibrium disturbances. It is advisable to include in the battery of routine pre-employment and periodic medical examinations, simple equilibrium functions tests such as standing or walking with the eyes closed, and finger-tip control. These should be applied to persons exposed to hazards that might affect equilibrium mechanism, persons who have contracted a disease which may have repercussions on equilibrium (ear diseases, vision disorders, central nervous system diseases or trauma, diabetes, hypertension, alcoholism, etc.); more comprehensive medical examination and special additional tests will be required (rotational, thermal, galvanic, electronystagmography) in cases of doubt or for workers in jobs for which a normal sense of equilibrium is essential for the protection of both the worker and the community.

There are numerous jobs in which sense of equilibrium is called upon and which entail a responsibility for safety. Apart from theatrical or artistic professions, such as tightrope walkers and acrobats, for whom performing feats of equilibrium under exceptionally difficult conditions is the whole essence of their livelihood, mention should be made of the building and civil engineering industry in view of both the large number of workers involved and the increased hazard presented by the combination of factors having an unfavourable effect on equilibrium. These factors may include: work at heights and on boarding which is often unstable; wind; lifting and carrying of loads such as tools and materials; inclement weather; noise; poor lighting; the handling of dangerous substances; etc. Accident statistics show that the incidence of falls in this industry is twice as high as elsewhere and the majority of these falls are due to the human factor (loss of equilibrium due to a malaise, tripping and stumbling, etc.) rather than to an external event.

Other industries or jobs in which workers require a good sense of equilibrium are: mines and quarries, shipbuilding and repairing, the iron and steel and non-ferrous metals industries, cable and ropeway work, driving or testing vehicles or machines on roads or over rough ground (tractors and other agricultural machinery, road transport, etc.), underwater operations, electricity distribution and generation, telecommunications operation, work with explosives, work where the floor is regularly covered with water, oil, talc or lubricants, work on slippery, sloping or vibrating surfaces, work on cranes or lifting and hoisting appliances in general, chimney building, high-level window cleaning, forestry, signalling, alarm and rescue duties, exposure to toxic or asphyxiant gases, etc.

The placing of persons suffering from equilibrium disorders and the decision whether the worker's condition is compatible with the requirements of a certain job is a delicate task for the industrial medical officer. As a general rule, preference should be given to jobs which are sedentary, do not entail abrupt sidewards movements or frequent stretching of the neck, and that are free from jolting, loud noises, sudden changes of temperature and rhythmic or moving light sources. It is also desirable to ensure discrete supervision by a foreman or workmate since equilibrium disorders may manifest themselves unexpectedly.

Technical prevention should be aimed at eliminating the causes or attenuating the consequences of equilibrium disturbances; improvement of general environmental conditions and safety standards; well designed and surfaced floors and stairways; classification of dangerous jobs and introduction of the appropriate safety measures (such as guard rails, toe boards, railings, etc.); and the installation of safety nets or other fall-arresting devices that will attenuate the consequences of a fall; provision of gangways, etc., that are sufficiently wide and secured against sway; careful operation of transport and lifting appliances that are fast moving or accelerate or decelerate violently; provision of adequate means of access for repair and maintenance of machines and equipment.

Personal protective measures include education and training, limitation of hours of work and arduousness of work for persons predisposed to equilibrium disorders and exclusion of these persons from night work. In work at heights, where collective safety measures are not feasible, use should be made where necessary of personal protective equipment such as safety lifelines with fall arresters; it should not, however, be forgotten that the use of such devices is not lacking in danger and consequently the users themselves must be robust and receive a certain minimum of training (see FALLS FROM HEIGHTS, PERSONAL PROTECTION AGAINST).

DIDONNA, P.

General:

Posture and equilibration (La posture et l'équilibration). Gribenski, A.; Gaston, J. (Paris, Presses Universitaires de France, 1973), 128 p. Illus. (In French)

"Vertigo" (Les vertiges). Valdazo, A.; Cotin, P. *Le concours médical* (Paris), 7 Mar. 1981, 103/10, supplement (1-48). Illus. 29 ref. (In French)

"Vertigo from the point of view of the occupational physician" (Schwindel in werksärztlicher Sicht). *Homburg-Informationen für den Werksarzt* (Frankfurt am Main), 1974, 21/6 (139-147). 20 ref. (In German)

Building industry:

"Problems posed by vertigo sensations to the occupational physician in building and civil engineering" (Problèmes posés par les sensations vertigineuses au médecin du travail du bâtiment et des travaux publics). Charachon, R.; Perret, E. J.; Fecci, R. et al. *Revue de médecine du travail* (Paris), 1977, 5/1-2 (1-101). Illus. 74 ref. (In French)

Aerospace:

"The effects of prolonged exposure to weightlessness on postural equilibrium". Homic, J. L.; Reschke, M. F.; Miller, E. F. Ch.12 (104-112). Illus. 20 ref. *Biomedical results from Skylab.* Johnston, R. S.; Dietlein, L. F. (eds.). (Washington, DC, US Government Printing Office, 1977), 491 p. Illus. 447 ref.

Ergonomics

Ergonomics (meaning literally the study-measurement-organisation of work) is concerned with making purposeful human activities more effective. Most of the activities studied can be called work although there are topics such as "ergonomics of sport", "ergonomics in the home" and "passenger ergonomics". Ergonomics is only just emerging as a science; at present almost all the knowledge and theory entailed can be regarded as derivative from more established disciplines—notably anatomy, physiology and psychology. Nor is it yet a technology, in that the professional practitioners are not commonly accepted as such by the law-courts or the mass-media, although associations aimed at establishing standards of practice and expertise do exist in developed countries. The restrictions are partly to do with the youth of the subject (the term was first used around 1950) and partly to do with the indefiniteness and complexity of the material with which it deals.

The focus of study is the person interacting with the engineered environment. This person has some limitations which the designer should take account of. Complexity arises from the nature of man and the variety of the designed situations to be considered. These latter can vary from relatively simple ones such as chairs, the handles of tools and the lighting of a bench to highly elaborate ones such as aircraft flight-decks, control-rooms in the process industries and the artificial life support systems in space or under the sea. The people studied can vary from fit young men in military systems to middle-aged housewives in the kitchen, from the disabled of many kinds to the very highly selected, such as racing drivers. Sometimes the relevant characteristic is essentially biological and unchanging, except with age, for example the limits of dark adaptation; sometimes it varies with sex and race, for example body dimension; sometimes it varies with degree of economic and social development, for example acceptable working hours. The limits or boundaries of ergonomics itself are not entirely agreed. For example most but not all ergonomists would agree that energy expenditure studies are within ergonomics but the design of diets to provide the energy is not, that physical hazards such as heat stress are part of ergonomics but chemical hazards such as carcinogenic substances are not; personnel selection is not within ergonomics but training techniques might be; design for safety and comfort is ergonomics but design for social satisfaction including quality of working life, industrial democracy and so on is not. The criterion of success for an ergonomics study can vary from a direct increase in knowledge or productivity, for example in the design of a heat-stress index or a work-bench top respectively, to long-term health, for example the design of tractor suspensions and seats to minimise vibration effects. Perhaps most frequently in current practice the objective is the minimisation of human errors, for example for the air-traffic controller. Having made the necessary cautionary statements that ergonomics is elusive and indefinite, that it is not yet fully established as a science or a technology and that the objectives are highly variable we can now proceed by being as definitive as possible about ergonomics as a science and a technology and about its aims.

Ergonomics as a science

Anatomy is a necessary part of ergonomics, providing the conceptual background for anthropometry and biomechanics. *Anthropometry*, the measurement of man, provides the dimensional data needed for the positioning of controls and the size of work spaces. The main dimensions required are shown in figure 1. These dimensions are not easy to obtain or describe, because the body does not have definite end-points. For example what do we mean by the length of a finger? Does it start from the centre of the knuckle, the flat of the back of the hand or the skin between the fingers? Does it end at the tip, the centre of the nail or the centre of the tip? Which finger are we interested in? The first essential in anthropometric data is a clear description of the dimension and the end-points used. The second essential is to be clear about the specific population. Size varies with age, sex and race, and even with social and economic status; for example managers are on average several centimetres taller than unskilled workers. The third essential is to have some expression of variability, usually in the form of percentile data. The fifth percentile is a measure such that 5% of the population considered are smaller than this while 95% are larger in terms of the dimension under consideration; correspondingly, the 95% percentile is such that 95% are smaller while only 5% are larger. The 50% percentile is the same as the average, since such data are normally distributed. For most design purposes the fifth or ninety-fifth percentile are used. For example an escape hatch designed using the fiftieth percentile would trap half the population, and in this case one would use the ninety-fifth percentile; on the other hand one would use the fifth percentile reach distance to determine the maximum height of controls on a panel. It has to be accepted that in practice one can only design for about 90% of the population because of the very long tails on the distribution of dimensions. For example the fact that there are a few men more than 2 m tall does not justify making all doors and ceilings of greater height than this.

Biomechanics is concerned with the application of forces by the human body. This is important for at least two reasons. Firstly, we have lost the capacity to do the right thing by instinct, we have to learn to exert forces effectively and this means that someone has to know in order to be able to teach us. Secondly, again unlike lower animals, the muscular forces we can exert are only of the same order as the body weight. For example if an unskilled person is asked to lift a box off the ground he will invariably use the "derrick" method: that is, he will stand with feet together, bend forward at the hips and grasp the box with straight arms. This illustrates both the points made above. He has instinctively done it wrong and the reason it is wrong is that the main body mass of the trunk must be lifted at the same time as the box. This decreases the lifting potential or increases the muscular force required for a given lift. The first principle of effective application of forces is to try to get the body mass to exert the force rather than the muscles. This is what a boxer is trying to do in "getting his weight behind a punch". The second principle is to use the largest available muscles moving a joint around the central region of its total range. Effective biomechanics requires a knowledge of gross anatomy, in particular the locations of the main muscle groups, their composition (some muscle fibres are suitable for fast action, others for lower speed, higher power) and their modes of action in relation to particular joints.

Physiology also has two main contributions to make to ergonomics; work physiology and environmental physiology. *Work physiology* is concerned with the human process of energy production. It might seem that it is just elementary physics or engineering and that one could calculate the work done by a person by simply measuring forces, distances and times. This is not so. Firstly, there is the problem of static work when a muscle is activated but not moving. According to the engineer, work is done only when a force moves its point of application but it hardly seems adequate to suggest to a person standing in a bus queue carrying two heavy cases that he can't be doing any work because he is not moving. Dynamic work is in fact much easier than static work because a moving muscle encourages blood flow, from which the energy is

Figure 1.
Body size and posture. The main human dimensions needed for the design of workspaces. Taken from W. T. Singleton: *Introduction to ergonomics*, p. 33. *(By courtesy of the World Health Organisation.)*

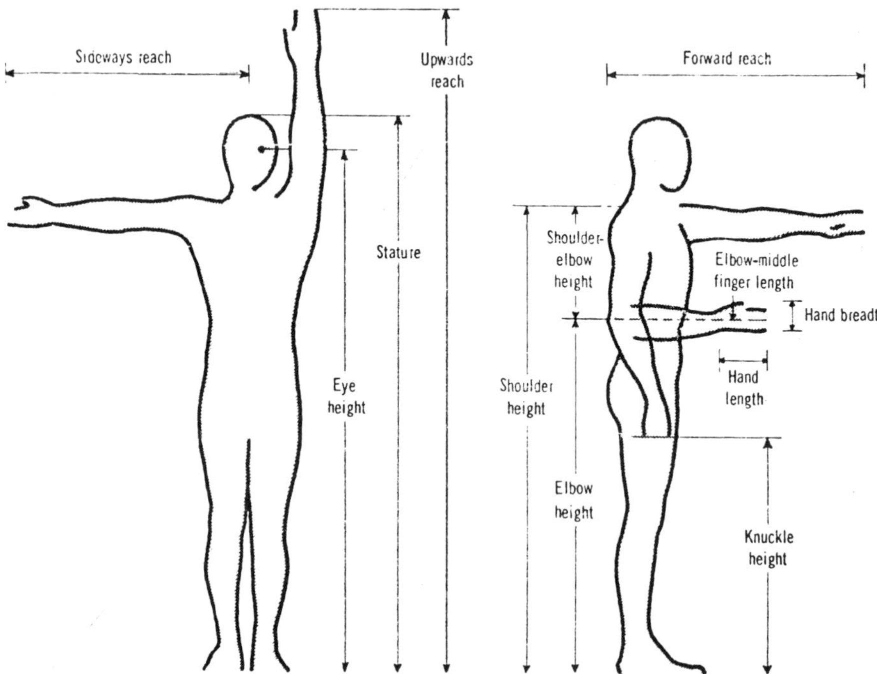

supplied. Secondly, there is an unknown quantity in the efficiency of a particular action. Human efficiency in energy expenditure can vary from 10% to 30%; this is about the same as in the case of other energy conversion systems such as the steam engine, but it is low enough and variable enough to make it essential to have some means of measuring the amount of energy actually generated. In principle this might be done by measuring the energy content of the food input; but this only works on a very long time scale. It might also be done by the change in body temperature, since most of the energy produced is wasted in heat; but this is not very accurate. The method used is to measure oxygen consumption, since there is little storage capacity for this and over a considerable range there is a direct relationship between oxygen consumed and energy produced. Thus the work physiologist specialises in techniques for measuring oxygen uptake—usually by measuring the volume of expired air and its remaining oxygen content (see ENERGY EXPENDITURE). There is a range of techniques for doing this both in the laboratory and under field conditions. Standards of what is reasonable and unreasonable for different conditions and populations are of course necessary. These are obtained by long-term studies where the basic criterion is that the individual should not lose weight because of his energy expenditure.

Environmental physiology depends on the same fundamental method, namely to provide measures of the stress and standards of what is reasonable in terms of these measures. The main environmental parameters are climate, light, noise and vibration although there are others to be found in special environments, for example radiation. In each case the problem is that there is no simple physical measure accurately indicating the effect on the person. In climate, for example, the heat or cold stress depends not only on air temperature but also on radiant temperature, humidity and wind speed. Correspondingly, lighting depends on direction, distribution, colour, contrast, glare, etc., as well as on illumination level. The environmental physiologist's task is to consider all these relevant physical measures and take them into account in determining the quality of the environment. Sometimes he will produce a combined index such as a heat-stress index or he may delineate the effects of the separate parameters and their interactions. He must also consider the kind of person exposed in terms of age, sex, fitness and so on, and the task to be performed in the particular environment.

Psychology has a number of different contributions to make to ergonomics. The central one is human performance theory based on an information model of the human operator. There are others, for example learning theory and skill theory in relation to training, and organisation theory in relation to work design and system design. Almost every psychological theory has something useful to say about human error—why people make mistakes. Much of human work, probably the majority, is now based on the information processing abilities of people. The machine provides the power and the precision but it is still up to the man to select and control the particular operation. This he does on the basis of information received partly from the general environment and partly from artificial displays of information such as dials and charts. Thus the physiologist or anatomist might ask: "What is the man doing?" while the psychologist will ask: "Where is the man getting the information to decide what to do?" Information about the state of the machine or the environment is received mainly through the eyes and ears. A thorough understanding of the visual and auditory mechanisms is necessary in order to design efficient information presentations. Once data has been received through the eyes and ears it must be interpreted and acted upon. These processes are studied by human performance specialists who provide the relevant knowledge on matters such as choice reaction times, information processing capacity, memory, attention and fatigue effects. As in the case of environmental variables there is an optimal range of information flow within which effective performance is maintained: too little flow and we get into the special problems of vigilance and inspection tasks (figure 2), too much and we are into the different problem area of stress and overload. Although ergonomics is primarily concerned with "fitting the job to the worker" the best fit of worker and job essentially demands a two-way approach and some consideration must be given to adapting people to jobs by training. The key to successful training is to provide knowledge of results which, under suitable conditions, will not only orient the operator in the learning process but also has considerable incentive effects. The breakdown of tasks into optimal learning units with the associated provision of knowledge of performance is an important contribution from psychology not only in relation to training but also in relation to the maintenance and improvement of performance by the skilled operator.

Work involves interaction with other people as well as with the physical world and some appreciation is needed of the principles of interpersonal communication, morale, leadership, group behaviour and organisation behaviour.

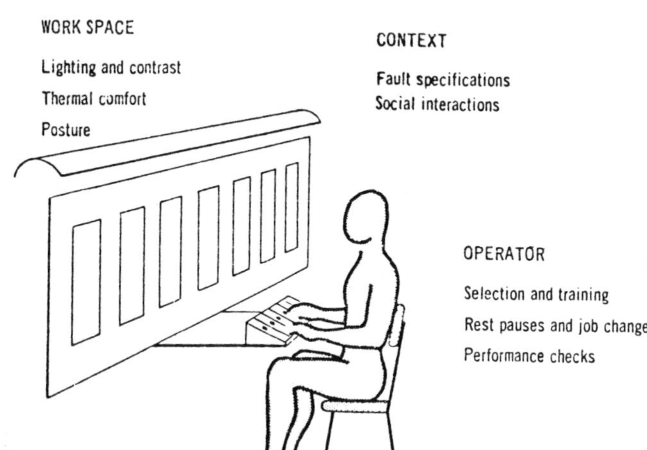

ACCIDENT PREVENTION

Design	Compensate for inherent human and hardware limitations
Programme	Provide adequate instructions and training
Maintenance	Maintain hardware performance and reliability
Operator performance	Ensure adequate motivation, prevent excessive fatigue, use adequate safety margins, maintain basic speed and accuracy requirements

Figure 2. Design factors for effective inspection. Taken from W. T. Singleton: *Introduction to ergonomics*, p. 109. *(By courtesy of the World Health Organisation.)*

Ergonomics as a technology

In ergonomics, as in other applied disciplines, real problems cannot be solved by remaining within the boundaries of a particular science. Invariably an interdisciplinary approach is required and in addition there are new approaches and kinds of knowledge that are not provided by traditional academic disciplines. To encompass these, ergonomics can be reviewed in terms of various aspects of the design of work: the design of systems, workspaces, environments, interfaces and work situations.

Systems design incorporates the principle of man as an integral part of a total man-machine system. The primary design problem is one of allocation of function between man and machine or more generally between men, machines and procedures. The basic reason for the success of the man-machine partnership is that each has characteristics that complement those of the other: machines are powerful, fast and tireless, men are intelligent, versatile and adaptive. The combination is very effective if care is taken to ensure that the main components are allocated functions matching their various advantages and limitations. This has always been done intuitively by good designers, but systems ergonomics aims to ensure that it is done systematically. This involves a set of knowledge and techniques that are new and are not derived from the older human sciences.

Workspace design aims to ensure that the physical surroundings fit the characteristics of the human body so that work can be done without excessive effort within the range of healthy postures either standing, sitting or exerting forces. It incorporates seat design, bench or console design and positioning of displays, controls and materials. Thus it depends largely on the application of anthropometry and biomechanics. It is not sufficient to have these data available in medical formats; there is a great deal of expertise in coding data into a form that will be relatively unambiguous in application. This can involve the use of mannikins (life size or scaled models of the human body, either three-dimensional or two-dimensional), specialised tables and diagrams of data which incorporate the relevant dimensions and indicate importance and tolerance, and computer-based models either of general human operators or of particular problems such as the car-driver workspace design.

Environmental design is concerned with ensuring that the lighting, heating, ventilation, noise, vibration and so on are appropriate to the requirements of the human operator. This is a highly specialised topic. It is possible for a professional practitioner to devote an entire career to any one of these parameters. Lighting design, for example, requires a knowledge of the characteristics of the eye and the current technology of lighting systems as well as the central ergonomics problems of visual protection and visual performance in the context of particular tasks.

Interface design focuses on the exchange of information between the man and the machine/environment. This takes place in both directions: displays present information to the man, controls accept information from the man. *Display design* incorporates such issues as the optimal design of scales, pointers, letters and numbers and the size, positioning and grouping of instruments. The presentation of information is an elaborate topic concerned not just with the design of symbols but also with the rules for their combinations and meanings in relation to different tasks. One aspect of wide current interest is the design of information presentations on video displays. *Control design* has both anatomical aspects concerned with sizes, shapes, positions and forces and psychological aspects concerned with discrimination and identification. As always, the design focuses on the attributes of the operator, some of which are innate and some acquired, such as expected directions of movement—so-called population stereotypes.

Work situation design deals with wider issues such as hours of work, rest pauses, and with special problems such as shift work and interpersonal and organisational aspects of work.

All these areas overlap and although a particular ergonomist may not be an expert on every one of them he is expected to be able to take an over-all view and identify the key design aspects in relation to particular kinds of people engaged in particular tasks.

The aims of ergonomics

The ergonomist as a technologist is concerned to facilitate whatever a person wishes to do and to ensure that he does it efficiently. Efficiency is interpreted widely to mean not only that whatever is done should be effective in the short term but also that in the long term there shall be no detrimental effects on health and that the risk of accidents is minimal. Risks refer not just to the operator but also to others who might be affected by what he does, as for example in the case of any controller of a transport vehicle. Thus, the criterion of success might be as simple as a measure of productivity but more usually, in present society, the objective is to minimise the possibility of human error. Machines can be relied upon to generate adequate productivity in the primary industries, in manufacturing or in the information-based industries. The role of the man is to ensure that this is done with a minimal use of energy and materials and without waste resulting from mistakes. Mistakes can lead to damage to products, distortion of information and most important of all, effects on the safety and health of people. The criterion of success in ergonomics is that these are minimised by minimising human errors. The ergonomist as a scientist is concerned with the development of knowledge and techniques that will further the technology. This might be highly specialised and abstract as, for example, in the measurement of gait, and it is often interdisciplinary as, for example, in the measurement of stress and strain. The combined use of physical, anatomical, physiological and psychological expertise in advancing knowledge is characteristic of ergonomics and so also is the use of the systems approach, which facilitates the consideration of many variables simultaneously in the context of a particular objective.

SINGLETON, W. T.

CIS 77-1172 *Introduction to human engineering.* Kraiss, K. F.; Moraal, J. (Cologne, Verlag TÜV Rheinland GmbH, 1976), 514 p. Illus. Approx. 500 ref.

Human factors in engineering and design. McCormick, E. J. (New York, McGraw Hill, 4th ed., 1976).

Introduction to ergonomics. Singleton, W. T. (Geneva, World Health Organisation, 1972), 145 p. Illus.

Ergonomic principles in the design of work systems. International Standard ISO 6385 (Geneva, International Organisation for Standardisation, 1st ed., 1981), 4 p.

Erysipeloid

SWINE ERYSIPELAS; ERYSIPELOID OF ROSENBACH

Erysipeloid is an infectious condition, predominantly of occupational origin and is officially recognised as an occupational disease in many countries. The disease results from the penetration of broken skin by *Erysipelothrix rhusiopathix*, a thin, non-motile, Gram-positive corynebacterium widely distributed in nature, especially where there is decomposition of nitrogenous material. Although the micro-organism will withstand drying and refrigeration for several months, it is destroyed rapidly by boiling. The organism is pathogenic for a wide variety of animals; in pigs it causes "swine erysipelas"; the infection in man is commonly called "erysipeloid of Rosenbach". The biological activity of the organism may be temperature-related with a seasonal incidence, being most frequent in summer and early autumn and virtually disappearing in winter.

Occupations at risk

The occupations commonly associated with the disease are those entailing contact with fish, meat or poultry, for example abattoir workers, butchers, workers in the fishing industry, road transport and railway workers handling foodstuffs, workers in the canning and food-preserving industry, etc.

Portals of entry

The organism may enter the body through cuts or abrasions, especially small punctures or wounds; it does not appear to be able to penetrate intact skin. The wound may be small enough to pass unnoticed, especially in workers for whom such small abrasions are frequent. Larger wounds demanding immediate treatment are seldom infected with erysipeloid. Pricks with fish and meat bones and knives are common causes of infection in the food industry, but the disease may result from handling soiled containers, etc. The hands are the most common points of infection.

Symptoms

Following an incubation period of 1-7 days (longer if the infection enters through a later but unobserved breach of the skin), the skin lesion appears. This is remarkably constant in appearance and progression. It begins with sensations of heat, itching or burning and later becomes painful with "prickling" tenderness and a feeling of tightness of the affected area. The infected area is swollen, with a pink "blush" which rapidly turns red and finally becomes a reddish-blue colour which is highly characteristic; the edge of the swelling is raised and sharply demarcated. As the lesion develops, the edge advances and may spread to adjacent fingers, and slight pain and stiffness in finger joints near the site of infection is often experienced; lymphangitis and/or lymphadenitis are occasionally detected. After a few days, the centre of infection fades and slight desquamation is observed; no suppuration occurs. The lesion is self-limiting and in most cases disappears after a period of 1-2 weeks or sometimes more. General condition usually remains good; very rarely, *Erysipelothrix rhusiopathiae septicaemia* may occur due to systemic spread of infection with constitutional effects and a generalised purpuric and petechial rash resembling that of meningococcal septicaemia; there may be associated endocarditis. No active immunity to the disease appears to be acquired and subsequent infections are not uncommon.

Diagnosis

This is based on the characteristic clinical picture, and biopsy is not justified. The patient's occupational history may provide a valuable diagnostic lead; occupations in which there is contact with animal products or contaminated material should be immediately suspect. The conditions most likely to be confused with erysipeloid are cellulitis and erysipelas; differential diagnosis should be based on the history, the characteristic appearance of the lesions and the absence of suppuration and systemic symptoms.

PREVENTIVE MEASURES

The most important contribution to the prevention of erysipeloid is the elimination of the decaying nitrogenous material which is the source of infection. In factory premises where infection is liable to occur, walls, floors, tables, containers and the like should be designed to allow rapid and thorough cleaning and maintenance and good housekeeping, and hygiene standards should be applied. Wherever possible, permeable surfaces such as wood should be replaced by impermeable materials such as stainless steel and plastics.

Personnel should wear protective clothing such as aprons and, where practical, hand protection such as chain mail gloves may reduce the incidence of minor cuts as will also the use of correctly maintained and well sharpened knives. Good sanitary facilities should be provided and the workers encouraged to practise good personal hygiene. First-aid facilities should be provided with facilities for the immediate disinfection and dressing of minor cuts and abrasions.

Treatment. Erysipeloid is a self-limiting condition. Irritation of the infected area can be relieved by a dressing of 10% ichtyol in glycerine. Penicillin appears to shorten the duration of the infection and should certainly be given if there is any lymphangitis and/or lymphadenitis. Recourse may also be had to oral administration of tetracyclines. A small number of cases are resistant and tend to relapse; in such cases, the affected areas must be kept dry.

PROCTOR, D. M.

CIS 74-1753 "Factors determining the occurrence and course of erysipeloid" (Faktory obuslovlivajuščie pojavlenie i harakter tečenija ėrizipeloida). Golubev, P. G.; Halitov, G. G. *Gigiena truda i professional'nye zabolevanija* (Moscow), Dec. 1973, 12 (40-42). Illus. (In Russian)

"A report on 235 cases of erysipeloid in Aberdeen". Proctor, D. M.; Richardson, I. M. *British Journal of Industrial Medicine* (London), 1954, 11/2 (175-179). Illus. 7 ref.

"Erysipelothrix infections". Weaver, R. E. *Diagnostic procedures for bacterial mycotic and parasitic infection.* Bodily, H. L.; Updike, E. L.; Mason, J. O. (eds.). (American Public Health Assocation Inc., 1970).

"Erysipelothrix septicaemia without endocarditis". Tawnshind, R. H.; Jephatt, A. E.; Hasaw Yekta, M. *British Medical Journal* (London), 24 Feb. 1973, 1 (464). 7 ref.

CIS 80-1690 "Erysipeloid in slaughterhouse workers" (Erysipeloid bei Schlachthofarbeitern). Marinescu-Dinizvor, G. *Medizinische Klinik* (Munich), 1979, 74/45 (1 686-1 688). 21 ref. (In German)

Esters

Esters are organic compounds which result from the reaction between an organic or inorganic acid and an alcohol with the elimination of water.

$$R\ COOH + HOR' \rightarrow R\ COOR' + H_2O$$

They can be obtained also by other reactions including the following:

(a) from acid halides and alcohols or phenols
$R\ COCl + HOR' \rightarrow R\ COOR' + HCl$;

(b) from acid anhydrides with alcohols or phenols
$(RCO)_2O + HOR' \rightarrow R\ COOR' + R\ COOH$;

(c) from ketenes and alcohols or phenols
$CH_2 : C : O + HOR \rightarrow CH_3COOR$;

(d) from free acids and aliphatic diazo-derivatives, especially diazomethane
$R\ COOH + CH_2N_2 \rightarrow R\ COOCH_3 + N_2$.

The esters of mineral acids are liquids; those of aliphatic acids with saturated alcohols are liquids with a pleasant odour, which are easily soluble in water; aromatic esters are less volatile and have a pleasant odour; benzyl esters are more irritant than the corresponding aliphatic esters. Most aromatic polyesters are liquids with a low vapour pressure and a very high boiling point; their stability makes them suitable for use in the plastics industry.

Methyl silicate ($Si(OCH_3)4$)

TETRAMETOXY SILANE; TETRAMETHYLORTHOSILICATE

m.w.	152.2
sp.gr.	1.03
b.p.	121 °C
v.d.	5.25
v.p.	12 mmHg ($1.6 \cdot 10^3$ Pa) at 25 °C

a clear liquid.

TLV ACGIH	1 ppm 6 mg/m³
STEL ACGIH	5 ppm 30 mg/m³

It has been used in the ceramics industry for closing pores, for coating metal surfaces and as a bonding agent in paints and lacquers. Nowadays its main use is in coating picture screens of television tubes.

HAZARDS

Methyl silicate presents both an inhalation and eye exposure hazard. In concentrations of 200-300 ppm for 15 min it produces minimal corneal lesions; at 1 000 ppm it produces severe corneal lesions, and under certain conditions of humidity it can cause progressive corneal necrosis. Industrial exposure both in the United States and in Europe has been severe, resulting in some cases from eye pain to blindness. Severe exposure can also lead to lung damage, including pulmonary oedema, and kidney impairment with degeneration of convoluted tubules.

Ethyl silicate ($Si(OC_2H)_4$)

TETRAETHYL ORTHOSILICATE

sp.gr.	0.93
b.p.	165.5 °C
v.d.	7.2
v.p.	1 mmHg ($0.13 \cdot 10^3$ Pa) at 20 °C
f.p.	51.7 °C oc

insoluble in water

a colourless liquid with a faint pleasant odour.

TWA OSHA	100 ppm 850 mg/m³
TLV ACGIH	10 ppm 85 mg/m³
STEL ACGIH	30 ppm 255 mg/m³
IDLH	1 000 ppm
MAC USSR	20 mg/m³

This ester is easily hydrolised in water with the formation of ethyl alcohol and silicic acid. It is produced by the reaction between silicon tetrachloride and ethyl alcohol.

By dehydration it forms an adhesive, colloidal silica which, when dried, becomes vitreous, hard, insoluble in water and resistant to high temperature. It is therefore suitable for use in the production of cases and moulds for the casting of metals and as a hardener for water- and weather-resistant concrete. It acts as a heat-resistant adhesive.

HAZARDS

Ethyl silicate is a flammable liquid which produces explosive mixtures of vapour in air, but only at elevated temperatures.

The main acute symptoms of health injury from contact with the liquid or its vapour are irritation to the eyes and respiratory system. Short exposures affect the eyes and nose as follows: 3 000 ppm, extremely irritating and intolerable; 1 200 ppm, lacrimatory and stinging; 700 ppm, mild stinging; 250 ppm, slightly irritant to the eye and nose; and 85 ppm, detectable by odour.

Repeated or prolonged skin contact with the liquid may cause dermatitis owing to its solvent effect. High concentrations can cause severe systemic injury, including tremor, narcosis, liver and kidney damage and anaemia, but at these high concentrations the vapour becomes intolerable.

***n*-Propyl nitrate** ($CH_3(CH_2)_2NO_3$)

m.w.	105
sp.gr.	1.06
b.p.	110.5 °C
v.d.	3.62
v.p.	18 mmHg ($2.4 \cdot 10^3$ Pa) at 20 °C
f.p.	20 °C
e.l.	2-100%
i.t.	175 °C

insoluble in water

a clear water-white to yellow liquid with a sickly odour.

TWA OSHA	25 ppm 110 mg/m³
STEL ACGIH	40 ppm 170 mg/m³
IDLH	2 000 ppm

It is a rocket propellant, fuel ignition promoter and organic intermediate.

HAZARDS

It has an irritant effect on the mucous membranes of the eye and respiratory tract and the skin. Repeated contact produces a yellow discoloration of the skin. There have been no reports of human intoxication. Inhalation of high concentrations in animals produces methaemoglobinaemia, anaemia and hypotension, by a direct action on vascular smooth muscles. Death in animals occurred with a methaemoglobin level of only 4%. In humans severe exposure to produce methaemoglobin is unlikely, since lower concentrations cause sufficient warning in the form of irritation, headache and nausea.

It is a dangerous explosion hazard and may explode on heating.

Methyl formate ($HCOOCH_3$)

METHYL METHANOATE; FORMIC ACID METHYL ESTER

m.w.	60
sp.gr.	0.97
m.p.	−99 °C
b.p.	31.5 °C
v.d.	2.1
v.p.	400 mmHg ($53.2 \cdot 10^3$ Pa) at 16 °C
f.p.	−18.9 °C
e.l.	5.9-20%

soluble in water, ethyl alcohol and ethyl ether

a colourless liquid smelling of ether.

TWA OSHA	100 ppm 250 mg/m³
STEL ACGIH	150 ppm 375 mg/m³
IDLH	5 000 ppm

Methyl formate is obtained by heating methyl alcohol with sodium formate and hydrochloric acid. The methyl formate thus produced is separated by distillation. It is a good solvent for greases, fatty acids, collodion and celluloid. As a general industrial solvent it is often mixed with other esters such as ethyl formate, methyl acetate and ethyl acetate.

HAZARDS

Methyl formate is a flammable liquid the vapour from which forms explosive mixtures with air at normal temperature.

This ester can be absorbed into the body through the respiratory and digestive systems and it may also penetrate the skin. It has an irritant and narcotic effect and, in high concentrations, may cause narcosis, dyspnoea, convulsions, coma and death in experimental animals. In these conditions, the lungs were the most affected organs (congestion, emphysema and oedema), while hyperaemia occurred in the kidneys, adrenals, liver and meninges. Since it is most commonly employed as a component of solvent mixtures, it has been difficult to determine exactly the specific symptomatology of methyl formate. Workers exposed to vapours of a solvent containing 30% methyl formate exhibited irritation of the mucous membrane, a sensation of constriction of the chest, and dyspnoea of varying intensity; some experienced euphoria, others depression. A youth to whose head the ester was applied as a parasiticide became cyanotic in 20 min and died from cardiac insufficiency.

Ethyl formate ($HCOOC_2H_5$)

ETHYL METHANOATE; FORMIC ACID ETHYL ESTER

m.w. 74
sp.gr. 0.91
m.p. −80.5 °C
b.p. 54.3 °C
v.p. 200 mmHg (26.6·10³ Pa) at 20.6 °C
f.p. −20 °C
e.l. 2.7-13.5%

soluble in water; insoluble in ethyl ether

a colourless liquid with an aromatic odour.

TWA OSHA 100 ppm 300 mg/m³
STEL ACGIH 150 ppm 450 mg/m³
IDLH 8 000 ppm

It is produced by heating ethyl alcohol with formic acid in the presence of sulphuric acid, and by distilling the product.

Its principal application is as a solvent for cellulose nitrate, and due to its aromatic properties it is used as a solvent for oils and greases and in the production of lemonade, rum, arrack, and essences. It is also used in chemical synthesis, and as a fumigant.

HAZARDS

Ethyl formate is a flammable liquid and its vapour forms explosive mixtures with air at normal temperatures.

In animal experiments inhalation of vapours caused tremors, progressive depression of the central nervous system and death through circulatory and respiratory insufficiency, the narcotic dose being the same as the lethal dose. No cases have so far been reported of industrial ethyl formate poisoning, although workers at 330 ppm have experienced irritation of the conjunctivae and mucous membranes, which may persist for several hours after contact has ceased.

Butyl formate ($HCOO(CH_2)_3CH_3$)

BUTYL METHANOATE

m.w. 101
sp.gr. 0.91
m.p. −91.9 °C
b.p. 106.8 °C
v.d. 3.5
v.p. 40 mmHg (5.32·10³ Pa) at 31.6 °C
f.p. 17.8 °C
e.l. 1.7-8%
i.t. 322 °C

miscible with water, alcohols, ethers and hydrocarbons

a colourless liquid.

It is a solvent for cellulose compounds and synthetic and natural resins. It is used in the production of lacquers but does not form such an adhesive or resistant coating as those formed with butyl or amyl acetate. It is used in the manufacture of perfumes and flavouring essences and in organic synthesis.

HAZARDS

Butyl formate is flammable and its vapour forms explosive mixtures with air at normal temperatures.

It is irritant to the eyes and mucous membranes. It is absorbed by the lungs and the alimentary canal. Very high concentrations in animal experiments have caused narcosis and pulmonary oedema, but no case of industrial or accidental poisoning in man has been reported.

Allyl formate ($HCOOCH_2CH : CH_2$)

PROPENYL METHANOATE

m.w. 86.1
sp.gr. 0.95
b.p. 83 °C

slightly soluble in water; soluble in ethyl alcohol

a liquid with a mustard-like odour.

Allyl formate is used in the organic chemicals industry and in the pharmaceutical industry for the synthesis of plasmocide. Experiments have established that it is highly toxic to animals, causing liver damage and death. No information is available regarding occupationally induced or accidental poisoning in man.

Miscellaneous formates

The toxicity of some industrial solvents such as amyl formate, cyclohexyl formate, methyl cyclohexyl formate and benzyl formate has not been sufficiently studied for an assessment to be made and no case of industrial poisoning from these esters is known.

Methyl acetate (CH_3COOCH_3)

m.w. 74.1
sp.gr. 0.97
m.p. −98.1 °C
b.p. 57 °C
v.d. 2.8
v.p. 400 mmHg (53.2·10³ Pa) at 40 °C
f.p. −10 °C
e.l. 3.1-16%
i.t. 502 °C

soluble in water; very soluble in organic solvents

a colourless, volatile, flammable liquid with a pleasant odour.

TWA OSHA 200 ppm 610 mg/m³
STEL ACGIH 250 ppm 760 mg/m³
IDLH 10 000 ppm
MAC USSR 100 mg/m³

Methyl acetate is obtained by the esterification of methyl alcohol with acetic acid in the presence of sulphuric acid as a catalyst. The reaction is reversible and must therefore be conducted with heat, eliminating water formed by the reaction.

It is used as a solvent for nitrocellulose, acetyl cellulose, oils, fats, etc. It is generally mixed with acetone and methyl alcohol. It is used in the plastics and artificial leather industries, and in the production of perfumes, colouring agents and lacquers.

HAZARDS

Methyl acetate is flammable and its vapour forms explosive mixtures with air at normal temperatures.

In animal experiments the lethal dose was 31 000 ppm. In man high concentrations of vapour cause irritation to the eyes and mucous membranes. Cases of slight poisoning under industrial conditions have been known. These were manifested by headache, drowsiness, vertigo, eye burns, lacrimation, palpitation, a constricted feeling in the chest and dyspnoea. One case of blindness has been reported but no case of lethal poisoning.

Ethyl acetate ($CH_3COOC_2H_5$)

ETHYL ETHANOATE; ETHYL ESTER; ACETIC ETHER

m.w.	88.1
sp.gr.	0.90
m.p.	−83.6 °C
b.p.	77 °C
v.d.	3.0
v.p.	100 mmHg (13.3·10³ Pa) at 27 °C
f.p.	−4.4 °C
e.l.	2.2-11%
i.t.	426.6 °C

miscible with ethyl alcohol, acetone, chloroform and ethyl ether

a colourless liquid with a pleasant odour characteristic of fruit.

TWA OSHA	400 ppm 1 400 mg/m³
IDLH	10 000 ppm
MAC USSR	200 mg/m³

Ethyl acetate is easily hydrolysed in water, giving a slightly acid reaction.

Ethyl acetate is obtained by the direct esterification of ethyl alcohol with acetic acid, a process which involves mixing acetic acid with an excess of ethyl alcohol and adding a small amount of sulphuric acid. The ester is separated and purified by distillation. In another process the molecules of anhydrous acetaldehyde interact in the presence of aluminium ethoxyde to produce the ester, which is purified by distillation.

It is a good solvent for nitrocellulose, fats and celluloid; it is used in the production of cellulose nitrate lacquers, celluloid, artificial leather, perfumes, in the footwear industry and in the pharmaceutical industry.

HAZARDS

The liquid is flammable and produces a vapour that forms explosive mixtures with air at normal temperatures. Ethyl acetate is an irritant of the conjunctive and mucous membrane of the respiratory tract. Animal experiments have shown that, at very high concentrations, the ester has narcotic and lethal effects; at concentrations of 20 000 to 43 000 ppm, there may be pulmonary oedema with haemorrhages, symptoms of central nervous system depression, secondary anaemia and damage of the liver. In man, concentrations of 400 ppm cause irritation of the nose and pharynx; cases have also been known of irritation of the conjunctiva with temporary opacity of the cornea. In rare cases exposure may cause sensitisation of the mucous membrane and eruptions of the skin. The irritant effect of ethyl acetate is less strong than that of propyl acetate or butyl acetate.

***n*-Propyl acetate** ($CH_3COOC_3H_7$)

m.w.	102.1
sp.gr.	0.89
m.p.	−95 °C
b.p.	101.6 °C
v.d.	3.5
f.p.	14.4 °C
e.l.	2.0-8%

miscible with alcohols, ketones, esters, oils, hydrocarbons and water

clear, colourless liquid with an aromatic odour.

TWA OSHA	200 ppm 840 mg/m³
STEL ACGIH	250 ppm 1 050 mg/m³
IDLH	8 000 ppm
MAC USSR	200 mg/m³

Isopropyl acetate ($CH_3COOCH(CH_3)_2$)

m.w.	102.1
sp.gr.	0.87
m.p.	−73.4 °C
b.p.	93 °C
v.d.	3.5
f.p.	4.4 °C
e.l.	1.8-8%

slightly soluble in water; soluble in most common organic solvents

clear, colourless liquid with an aromatic odour.

TWA OSHA	250 ppm 950 mg/m³
STEL ACGIH	310 ppm 1 185 mg/m³
IDLH	16 000 ppm

These esters are produced by the reaction of acetic acid with the corresponding propyl alcohol in the presence of a catalyst. They are solvents for nitrocellulose in the production of lacquers and are used in the plastics industry, in the manufacture of perfumes and in organic synthesis.

HAZARDS

These two propyl acetate isomers are flammable and their vapours form explosive mixtures with air at normal temperatures.

In man, concentrations of 200 ppm cause irritation of the eyes and greater concentrations give rise to irritation of the nose and larynx. Amongst workers occupationally exposed to these esters, there have been cases of conjunctival irritation and reports of a feeling of constriction of the chest and coughing; however, no cases of permanent or systemic effects have been found in exposed workers. Repeated contact of the liquid with the skin may lead to defatting and cracking.

Butyl acetate ($CH_3COOC_4H_9$)

BUTYL ETHANOATE; ACETIC ACID BUTYL ESTER

m.w.	116
sp.gr.	0.88
m.p.	−77.9 °C
b.p.	126.5 °C
v.d.	4.0
v.p.	10 mmHg (1.33·10³ Pa) at 20 °C
f.p.	22.2 °C
e.l.	1.7-7.6%
i.t.	425 °C

miscible with ethyl alcohol and organic solvents; slightly soluble in water.

TWA OSHA	150 ppm 710 mg/m³
STEL ACGIH	200 ppm 950 mg/m³
MAC USSR	200 mg/m³

This ester forms a constituent of commercial butyl acetate together with isobutyl acetate and *sec*-butyl acetate. It is made by the esterification of *n*-butanol with acetic acid in the presence of sulphuric acid. *n*-Butanol is obtained by the fermentation of starch with *Clostridium acetobutylicum*. Butyl acetate is the most commonly used solvent in the production of nitrocellulose lacquers. It is also used in the manufacture of vinyl resins, artificial leather, photographic film, perfumes and in the preserving of foodstuffs.

HAZARDS

In man, exposures to concentrations of 200-300 ppm cause slight irritation of the eyes and nose and short exposure to a concentration of 3 300 ppm causes extreme irritation of the eyes and nose. Normal occupational exposure, however, does not produce lesions of the eyes or respiratory tract or systemic affections; cases reported appear to have been due to other solvents used with butyl acetate.

***sec*-Butyl acetate** ($CH_3COOC_4H_9$)

ACETIC ACID SECONDARY BUTYL ESTER; 2-BUTANOL ACETATE

m.w. 116.1
sp.gr. 0.86
b.p. 112 °C
v.d. 4
v.p. 10 mmHg ($1.33 \cdot 10^3$ Pa) at 20 °C
f.p. 31 °C oc
e.l. 1.7-9.8%

slightly soluble in water
a colourless liquid with a pleasant odour.

TWA OSHA	200 ppm 950 mg/m³
STEL ACGIH	250 ppm 1 190 mg/m³
IDLH	8 000 ppm

***tert*-Butyl acetate** ($(CH_3)_3COH$)

ACETIC ACID BUTYL ESTER

m.w. 116
sp.gr. 0.86
b.p. 98 °C
f.p. 22 °C
e.l. 1.5%

almost insoluble in water; miscible with ethanol and ether
a colourless liquid with a fruity odour.

TWA OSHA	200 ppm 950 mg/m³
STEL ACGIH	250 ppm 1 190 mg/m³
IDLH	8 000 ppm

***iso*-Butyl acetate** ($C_4H_9OOCCH_3$)

2-METHYLPROPYL ACETATE; METHYLPROPYL ETHANOATE

m.w. 116.2
sp.gr. 0.87
m.p. −98.9 °C
b.p. 118 °C
v.d. 4
v.p. 10 mmHg ($1.33 \cdot 10^3$ Pa) at 12.8 °C
f.p. 17.8 °C
e.l. 2.4-10.5
i.t. 423 °C

slightly soluble in water; miscible with ethanol
a colourless liquid with a fruity odour.

TWA OSHA	150 ppm 700 mg/m³
STEL ACGIH	187 ppm 875 mg/m³
IDLH	7 500 ppm

***n*-Amyl acetate** ($CH_3COOC_5H_{11}$)

AMYL ACETIC ETHER; PEAR OIL

m.w. 130.18
sp.gr. 0.88
m.p. −78.5 °C
b.p. 148 °C
v.d. 4.5
f.p. 25 °C
e.l. 1.1-7.5%
i.t. 373 °C

sparingly soluble in water; soluble in ethanol and ether
a colourless liquid with a banana-like odour.

TWA OSHA	100 ppm 525 mg/m³
STEL ACGIH	150 ppm 800 mg/m³
IDLH	4 000 ppm
MAC USSR	100 mg/m³

Isoamyl acetate ($CH_3COOC_5H_{11}$)

BANANA OIL

m.w. 130.2
sp.gr. 0.87
m.p. −78.5 °C
b.p. 142 °C
v.d. 4.49
v.p. 4 mmHg ($0.52 \cdot 10^3$ Pa) at 20 °C
f.p. 25 °C
e.l. 1-7.5%
i.t. 360 °C

slightly soluble in water; miscible with most organic solvents
a colourless liquid with a banana-like odour.

TWA OSHA	100 ppm 525 mg/m³
STEL ACGIH	125 ppm 655 mg/m³
IDLH	3 000 ppm

In its commercial form amyl acetate is a mixture of isomers notably *n*-amyl acetate and isoamyl acetate. Its composition and characteristics depend on its grade. The flash points of the various grades vary from 17 to 35 °C.

This solvent is made by the esterification of pentyl alcohol with sodium acetate in the presence of sulphuric acid. It may also be obtained by synthesis from pentane.

It is used as a solvent for nitrocellulose in the manufacture of lacquers and, because of its banana-like smell, it is used as an odorant. It is used also for the manufacture of artificial leather, photographic film, artificial glass, celluloid, artificial silk and furniture polish.

HAZARDS

All the isomers and grades of amyl acetate are flammable and evolve flammable mixtures of vapour in air.

Exposure to a concentration of 10 000 ppm for 5 h caused irritation and death in guinea-pigs. In man, exposure to concentrations of 950 ppm for 30 min causes irritation of the nose and eyes, headaches and weakness. The principal symptoms in cases of occupational exposure are headaches and irritation of the mucous membranes of the nose and of the conjunctiva. Other symptoms mentioned include vertigo, palpitations, gastrointestinal disorders, anaemia, cutaneous lesions, dermatitis and affections of the liver. Amyl acetate is also a defatting agent and prolonged exposure may produce dermatitis.

Vinyl acetate

See VINYL COMPOUNDS.

Other acetates

Hexyl acetate ($CH_3COOC_6H_{13}$), benzyl acetate ($CH_3COOCH_2C_6H_5$), cyclohexyl acetate ($CH_3COOC_6H_{11}$), methylcyclohexyl acetate ($CH_3COOC_6H_{11}CH_3$) are all used industrially.

HAZARDS

All these other acetates are flammable but their vapour pressures are low and, unless they are heated, they are unlikely to produce flammable concentrations of vapour.

Animal experiments indicate that the toxic properties of these acetates are greater than those of amyl acetate; however, in practice, due to their low volatility, their effect on workers is limited to local irritation.

Propionates

Methyl propionate ($C_2H_5COOH_3$), ethyl propionate ($C_2H_5COOC_2H_5$), butyl propionate ($C_2H_5COOC_5H_{11}$) are used industrially in the production of lacquers and perfumes, although on a smaller scale than the acetates.

HAZARDS

All these propionates are flammable liquids. The first three are capable of evolving flammable concentrations of vapour at room temperatures and amyl propionate can do so in hot climates.

Limited data are available on their toxicity. No cases of occupational poisoning are known. It appears that their industrial use does not constitute an occupational hazard to health.

Ethyl oxalate $(COOC_2H_5)_2$

DIETHYL OXALATE; DIETHYL ETHANEDIOATE; OXALIC ESTER

m.w.	146
sp.gr.	1.08
m.p.	−40.6 °C
b.p.	185.7 °C
v.d.	5.04
f.p.	75.6 °C oc

miscible in all proportions with ethyl alcohol, ethyl ether, ethyl acetate and other common organic solvents; slightly soluble in water

a colourless, unstable, oily liquid with an aromatic odour.

Ethyl oxalate is produced by a standard esterification procedure, using ethyl alcohol and oxalic acid. It is a solvent for cellulose esters and ethers and for natural and synthetic resins; it is used in perfume preparations as a dye intermediate and in organic synthesis, especially in the manufacture of pharmaceuticals.

HAZARDS

This ester is a combustible liquid, but it does not produce a flammable concentration of vapour at room temperatures.

Ethyl oxalate is hydrolised in the body to form oxalic acid which accounts for its toxic action. No cases of lethal poisoning from ethyl oxalate in industry are known, but its potential toxic action must be taken into consideration. In cases of chronic poisoning in men, a reduction in the number of erythrocytes and leucocytes in the blood has been observed together with slight eosinophilia and neutropenia.

Polyesters

A large group of synthetic resins are produced by reactions between dibasic acids and dihydric alcohols; in some instances, trifunctional monomers such as glycerol or citric acid are used. In general, the first stage of the polymerisation process coincides with the esterification stage. A large amount of synthetic fibre for the textile industry is prepared by the esterification of terephthalic acid with a dihydric alcohol. One such polyester is derived from the condensation of terephthalic acid and ethylene glycol.

Terephthalic acid has no toxic effects. For the hazards associated with ethylene glycol, see GLYCOLS AND DERIVATIVES.

GENERAL SAFETY AND HEALTH MEASURES

Precautions for the use of these esters must take account of the very definite fire danger that is associated with many of them and of the harmful effects that may result from contact with some of them.

Fire and explosion. At ordinary room temperatures, the vapours evolved by many of the esters (notably the formates, acetates and the lower propionates) form explosive mixtures with the air. It is thus important to ensure that open flames and other sources of ignition are not located in the vicinity of processes in which the esters are used. Some esters with higher flash points, like ethyl silicate, amyl propionate and ethyl oxalate, are not so dangerous at room temperature. They are nevertheless readily ignited as flammable liquids, and a source of ignition must not be allowed to come in contact with them. In process vessels, drying ovens, etc., which operate at elevated temperatures, sufficient vapour may be evolved from any of the esters to form an explosive mixture with the air. In these circumstances provision should be made for the removal of the vapours as they are evolved and the means of heating should not be such as to expose the vapour to an open flame (see FLAMMABLE SUBSTANCES).

Health precautions. As has already been indicated, no extensive history of serious injury to health is associated with the industrial use of these esters. Their degree of toxic hazard is closely related to their vapour pressure and consequently the lower members of the group, such as ethyl formate and methyl acetate, which have a relatively high vapour pressure, are more hazardous to health than are the higher members of the group, such as the propionates, which have a relatively low vapour pressure. It is therefore essential that processes employing the lower esters should be equipped with local exhaust ventilation.

Appropriate safety measures are required for work in confined spaces in which an ester has been present. Eye and face protection and respiratory protective equipment should be available for use in case of exposure to these esters in emergencies associated, for example, with a serious escape of liquid.

Separate articles deal with ACRYLIC ACID AND DERIVATIVES; BENZYL CHLORIDE; DIMETHYL SULPHATE; ETHYLENE GLYCOL DINITRATE; GLYCOLS AND DERIVATES; NITROGLYCERIN; PHTHALATES; SILICON AND ORGANOSILICON COMPOUNDS; TRICRESYL PHOSPHATES.

MANU, P.

CIS 75-1755 "Pathology of acetates in the textile industry" (Patologia da acetati nell'industria tessile). Corradini, M. A.; De Rosa, E.; Sarto, F. *Folia medica* (Naples), Sep.-Oct. 1973, 61/9-10 (397-405). 18 ref. (In Italian)

CIS 77-1932 *Amyl acetate.* Data sheet 208, 1977 Revision A (Extensive) (National Safety Council, 444 North Michigan Avenue, Chicago) (1977), 2 p.

CIS 75-1625 "Pulmonary absorption and elimination of ethyl acetate–Experimental study in humans" (Absorption et élimination pulmonaire de l'acétate d'éthyle–Etude expérimentale sur des sujets humains). Fernandez, J.; Droz, P. *Archives des maladies professionnelles* (Paris), Dec. 1974, 35/12 (953-961). Illus. 9 ref. (In French)

CIS 77-203 "Resistance of protective gloves to industrial solvents–Results obtained with ethyl acetate on some hundred commercial gloves" (Résistance des gants de protection aux solvants industriels–Résultats obtenus avec l'acétate d'éthyle sur une centaine de gants du commerce). Chéron, J. *Travail et sécurité* (Paris), Sep. 1976, 9 (421-428). Illus. (In French)

"Ethylene glycol monomethyl ether acetate" (Nitrate de méthylglycol). Morel, C.; Cavigneaux, A.; Protois, J. C. Fiche toxicologique n° 131. Institut national de recherche et de sécurité. *Cahiers de notes documentaires–Sécurité et hygiène du travail* (Paris), 3rd quarter 1977, 88, Note No. 1075-88-77 (383-386). 12 ref. (In French)

CIS 77-183 "Isobutyl acetate" (Acétate d'isobutyle). Morel, C.; Cavigneaux, A.; Protois, J. C. Fiche toxicologique n° 124. Institut national de recherche et de sécurité. *Cahiers de notes documentaires–Sécurité et hygiène du travail* (Paris), 4th quarter 1976, 85, Note No. 1042-85-76 (629-632). 13 ref. (In French)

CIS 74-1947 "n-Propyl acetate–Isopropyl acetate" (Acétate de n-propyle–Acétate d'isopropyle). Morel, C.; Cavigneaux, A. Fiche toxicologique n° 107. Institut national de recherche et de sécurité. *Cahiers de notes documentaires–Sécurité et hygiène du travail* (Paris), 4th quarter 1973, 73, Note No. 874-73-73 (547-550). 13 ref. (in French)

Ethers

Ethers are organic compounds in which oxygen serves as a link between two organic radicals. Most of the ethers of industrial importance are liquids though methyl ether is a gas and a number of ethers, for example the cellulose ethers, are solids.

The following ethers are of industrial importance:

Allyl ether	Eugenol
Anisole	Glycol ethers
bis-(Phenoxyphenyl) ether	Guaiacol
Butyl ether	Hexyl ether
Butyl vinyl ether	Hydroquinone ethers
Cellulose ethers	Isopropyl ether
Chlorinated phenyl ethers	Isopropyl vinyl ether
Chloroethylvinyl ether	Methyl ether
Chloromethyl ether	Methyl ethyl ether
Chlorotrifluoroethyl-methyl ether	Methyl isopropyl ether
Dichloroethyl ether	Methyl propyl ether
Dichloroisopropyl ether	Methyl *tert*-butyl ether
Diphenyl ether	Phenetole
Ethyl butyl ether	Vanillin
Ethyl ether	Vinyl ether
Ethyl vinyl ether	

NB. Glycidyl ethers are treated separately later in this article. The chlorinated methyl and ethyl ethers are discussed in the article ALKYLATING AGENTS.

GENERAL HAZARDS AND THEIR PREVENTION

The lower molecular weight ethers, methyl, ethyl, isopropyl, vinyl, ethyl, and vinyl isopropyl, are highly flammable, with flash points below normal room temperatures. Accordingly, measures should be taken to avoid release of vapours into areas where means of ignition may exist. All sources of ignition should be eliminated in areas where appreciable concentrations of ether vapour may be present in normal operations, as in drying ovens, or where there may be accidental release of the ether either as a vapour or as a liquid. Further control measures should be observed as in the articles on FIRE PROTECTION AND PREVENTION; FLAMMABLE SUBSTANCES.

On prolonged storage in the presence of air or in sunlight ethers are subject to peroxide formation that involves a possible explosion hazard. In laboratories amber glass bottles provide protection, except from ultraviolet radiation or direct sunlight. Inhibitors such as copper mesh or a small amount of reducing agent may not be wholly effective. If a dry ether is not required, 10% of the ether volume of water may be added. Agitation with 5% aqueous ferrous sulphate removes peroxides.

With the notable exception of the chloromethyl ethers and the glycidyl ethers, the ethers as a group have little general toxicological action in industrial use. Their narcotic action causes them to produce loss of consciousness on appreciable exposure and, as good fat solvents, they cause dermatitis on repeated or prolonged skin contact. Enclosure and ventilation are to be employed to avoid excessive exposure. Barrier creams and impervious gloves assist in preventing skin irritation. In the event of loss of consciousness, the person should be removed from the contaminated atmosphere and given artifical respiration and oxygen.

Air analysis can be conducted by a method that has been validated for ethyl ether. A known volume of air is drawn through an activated charcoal tube, 7 cm long and 4 mm in internal diameter, at a rate of 0.2 l/min or less with a maximum sample size of 3 l. The ether is desorbed with ethyl acetate and an aliquot of the desorbed sample is injected into a gas chromatograph equipped with a flame ionisation detector. The area of the resulting peak is determined and compared with areas obtained from the injection of standards. When this method is used for ethers other than ethyl, a prior check should be made against known concentrations of the ether to ensure reliable results.

Methyl ether (CH_3OCH_3)

DIMETHYL ETHER; METHOXY METHANE; METHYL OXIDE

m.w.	46
sp.gr.	0.66
m.p.	−138 °C
b.p.	−24.9 °C
v.d.	1.62
f.p.	41 °C
i.t.	350 °C

a colourless gas with an ethereal odour.

It is produced by the catalytic dehydration of methyl alcohol but is also obtained as a by-product of carbon monoxide and of acetic acid production from carbon monoxide and methyl alcohol. It has limited applications as a refrigerant, an aerosol dispersant, a rocket propellant, an anaesthetic and, in cold weather, a starter for gasoline engines.

HAZARDS AND THEIR PREVENTION

A mixture of 50% methyl ether in air with adequate oxygen is unpleasant to inhale and produces a suffocating response but causes little acute toxicity to humans. No long-term experiments have been conducted. Since no threshold limit values have been proposed for methyl ether, persons repeatedly exposed to appreciable concentrations should undergo periodic medical examinations. The extent of exposures should be determined by air analysis.

Ethyl ether ($C_2H_5OC_2H_5$)

DIETHYL ETHER; ETHOXYETHANE; ETHYL OXIDE; SULPHURIC ETHER

m.w.	74.1
sp.gr.	0.71
m.p.	−116.2 °C
b.p.	34.6 °C
v.d.	2.56
v.p.	442 mmHg ($58.8 \cdot 10^3$ Pa) at 20 °C
f.p.	−27 °C
e.l.	1.8-36%
i.t.	215 °C

a colourless liquid with a characteristic odour.

TWA OSHA	400 ppm 1 200 mg/m³
STEL ACGIH	500 ppm 1 500 mg/m³
IDLH	19 000 ppm
MAC USSR	300 mg/m³

Ethyl ether is obtained from ethyl alcohol by the sulphuric acid process and as a by-product of ethyl alcohol production by the catalytic hydration of ethylene. As one of the first inhalation anaesthetics, ethyl ether continues to be used extensively for this purpose. Its excellent solvent properties cause it to be employed in many industrial operations as a solvent for cellulose acetate, cellulose nitrate, dyes, fats, gums, oils, pharmaceuticals, resins and waxes, and as a reaction medium for certain organic compound syntheses.

HAZARDS AND THEIR PREVENTION

Ethyl ether is highly flammable and forms explosive peroxides in presence of air or in sunlight.

The principal physiological effect of ethyl ether is anaesthesia. Repeated exposures in excess of 400 ppm may cause nasal irritation, loss of appetite, headache, dizziness and excitation, followed by sleepiness. Repeated contact with the skin may cause it to become dry and cracked. Mental disorders following prolonged excessive exposure have been reported, as has kidney damage, the latter being open to question. Few cases of industrial over-exposure have been reported.

For air analysis where gas chromatograph equipment is not available, the air may be passed through a fritted glass bubbler for reaction of the ethyl ether with acidic potassium dichromate and subsequent iodometric determination.

Isopropyl ether ($(CH_3)_2CHOCH(CH_3)_2$)

DIISOPROPYL ETHER; 2-ISOPROPOXYPROPANE

m.w.	102.2
sp.gr.	0.72
m.p.	−85.9 °C
b.p.	69 °C
v.d.	3.5
v.p.	150 mmHg (19.95·10³ Pa) at 25 °C
f.p.	−27.8 °C
e.l.	1.4-7.9%
i.t.	443 °C

miscible with water

a colourless liquid with a sharp ethereal odour.

TWA OSHA	500 ppm 2 100 mg/m³
TLV ACGIH	250 ppm 1 050 mg/m³
STEL ACGIH	310 ppm 1 320 mg/m³
IDLH	10 000 ppm

Isopropyl ether is obtained as a by-product in the preparation of isopropyl alcohol from polypropylene. It is also prepared from isopropyl alcohol by the action of sulphuric acid. In addition to its use in most of the applications listed for ethyl ether, its lower volatility leads to its use also in paint and varnish strippers, rubber adhesives and in aviation gasoline as a blending agent. It is also used as an extraction agent to recover acetic acid from dilute solutions, and to extract nicotine from tobacco.

HAZARDS

Isopropyl ether is highly flammable and also forms explosive peroxides.

The principal physiological response on exposure to excessive concentrations is anaesthesia. A single exposure of rats to 1.6% for 4 h caused some weight loss and blood changes that persisted for several weeks after exposure. A single exposure of 0.3% for 2 h caused no noticeable anaesthesia in experimental animals, nor were any deleterious effects noted after 20 such exposures. Human exposure to 500 ppm over a 15-min period caused no irritation, but irritation of the eyes and nose was noted at 800 ppm for 5 min with some respiratory discomfort. Industrial exposure has caused but few cases of death or serious injury. Repeated skin contact would be expected to cause dermatitis.

Vinyl ether ($CH_2{:}CHOCH{:}CH_2$)

DIVINYL ETHER; ETHENYLOXYETHENE; DIVINYL OXIDE

sp.gr.	0.77
b.p.	39 °C
v.d.	2.4
v.p.	430 mmHg (57.2·10³ Pa) at 20 °C
f.p.	< −30 °C
e.l.	1.7-27%
i.t.	360 °C

Vinyl ether is produced by the action of sodium hydroxide on dichloroethyl ether. As vinyl ether is used principally as an anaesthetic, industrial exposure is limited to its production for this purpose.

HAZARDS AND THEIR PREVENTION

Vinyl ether is highly volatile, highly flammable and also forms peroxides. The anaesthetic grade contains 4% ethyl alcohol and up to 0.025% phenyl 1-naphthylamine, the latter to prevent spontaneous polymerisation. The principal effect of vinyl ether is that of anaesthesia with a greater possibility of respiratory arrest than with ethyl ether. Prolonged exposure may cause liver damage. No cases of industrial injury have been recorded.

Provisions previously noted are applicable to vinyl ether for prevention of both fire and explosion hazards and injury to health. No threshold limit value has been suggested.

Dichloroisopropyl ether ($[ClCH_2C(CH_3)H]_2O$)

bis-(2-CHLOROISOPROPYL) ETHER; 2,2-DICHLOROISOPROPYL ETHER

m.w.	171.1
sp.gr.	1.11
m.p.	−96.8-101.8 °C
b.p.	187.4 °C
v.d.	6.0
v.p.	0.10 mmHg (0.01·10³ Pa) at 20 °C
f.p.	85 °C oc

a colourless liquid.

Dichloroisopropyl ether is obtained as a by-product of the production of propylene glycol from propylene chlorohydrin. Its uses are the same as those of dichloroethyl ether.

HAZARDS AND THEIR PREVENTION

Fire and explosion hazards are comparable to those of dichloroethyl ether.

The toxicity of this ether is somewhat less than that of the dichloroethyl ether but damage occurs in the liver and kidneys rather than in the lungs. This ether causes no primary irritation of the skin but may penetrate the skin sufficiently to cause death. The lowest vapour concentrations to which animals have been exposed is 175 ppm. One of four rats died after an exposure of 3 h. No cases of injury to health of humans have been reported.

Preventive measures for dichloroethyl ether apply to this ether also. No threshold limit value has been suggested.

Guaiacol ($OHC_6H_4OCH_3$)

o-METHOXYPHENOL; 1-HYDROXY-2-METHOXYBENZENE; METHYLCATECHOL

m.w.	124.1
sp.gr.	1.09
m.p.	32 °C
b.p.	205 °C
v.p.	100 mmHg (13.3·10³ Pa) at 20 °C
f.p.	81 °C

soluble in water

a clear pale yellow solid or liquid.

Guaiacol is produced by the pyrolysis of selected hardwoods; it is used in printing inks and surface coatings.

HAZARDS AND THEIR PREVENTION

Guaiacol is combustible but presents little fire hazard due to its high flash point.

Guaiacol is only slightly irritating to the skin but absorption through the skin may result in chills, temperature drop and weakness. Even collapse and death may ensue. Its action is similar to that of phenol though it is only a third as toxic. Excessive absorption may cause muscular weakness, cardiovascular collapse and paralysis of the vasomotor centres. Contact with the eyes may result in damage to the cornea.

Skin contact with guaiacol should be avoided. Should it occur, the affected surfaces should be immediately washed with copious amounts of water. Workers should not be exposed to guaiacol vapour or mist that is irritating to the eyes or respiratory system even though such irritation is not intolerable.

Glycidyl ethers

The term, glycidyl ethers, applies to a group of complex organic compounds with at least one 2,3-epoxypropoxy radical. Most of the glycidyl ethers are liquids but some are solids. The glycidyl ethers are incorporated into epoxy resin systems as reactive diluents to lower the viscosity of the polymer. Occupations in which exposure to the glycidyl ethers may occur include production of glycidyl ethers and epoxy resins, transportation equipment, electrical equipment, fabricated metal goods, rubber and plastic products, communication products and services, and instruments and related products.

Table 1 lists the more important glycidyl ethers together with pertinent data where available.

GENERAL HAZARDS AND THEIR PREVENTION

The glycidyl ethers are flammable or combustible, depending on their flash points, which are generally above normal room temperature (see FLAMMABLE SUBSTANCES).

They are primary skin and eye irritants and may cause sensitisation. Excessive repeated skin contact may cause sufficient absorption to result in systemic effects. Although the low vapour pressure of these compounds tends to keep atmospheric concentrations low, dermal contact being the major source of absorption, there may be sufficient concentrations of vapour in breathing zones to produce significant added absorption through inhalation, especially where heat is produced or applied. No studies of the effects of inhalation of glycidyl ethers by humans have been reported. Many of the glycidyl ethers have produced systemic effects on animal experimentation, such as necrosis, oedema, inflammation, hyperaemia, haemorrhaging and tissue degeneration. No reports have been published of carcinogenic, mutagenic, teratogenic, or reproductive effects of the glycidyl ethers in humans. Such effects have been investigated in animals as are noted under individual glycidyl ether headings.

Measures to prevent injury to health include avoidance of skin contact of liquid glycidyl ethers and enclosure and ventilation of equipment to avoid excessive vapour release into work areas. Pre-employment and periodic medical examinations should be provided for all exposed workers. Informing workers of potential hazards and preventive measures should include an educational programme in addition to labels on containers and warning signs. Periodic monitoring should be conducted both of work areas where glycidyl ether vapour may be present and of the worker exposure. Personal protective clothing should be provided, together with face and eye and respiratory protection.

A method of air sampling and analysis has been validated for *n*-butyl, isopropyl, and phenyl glycidyl ethers. This method is similar to that described for ethyl ether but utilises carbon disulphide as the desorbing agent. Operations involving carbon disulphide should be conducted in a ventilated hood. In order to be applicable to checking the ceiling concentration over a 15-min period, a deviation from the validated method is a sampling rate of 1 l/min over 15 min. The method may be suitable for other glycidyl ethers also with changes in certain conditions such as solvents and gas chromatograph operating conditions, but for other glycidyl ethers, the reliability of the set of conditions selected should be checked against known concentrations before use in the field.

Table 1. Summary data for selected glycidyl ethers

Glycidyl ether	Abbreviated name	No. of exposed US workers[1]	Flash point (° C)	TWA OSHA[3] (mg/m³)	Testicular disorders in animals[4]	Haemopoietic abnormalities in animals[4]
Glycidyl ethers[2]	–	71 000	–	–	–	–
Diglycidyl ether of *bis*-phenol A	–	36 000	–	–	–	–
n-Butyl glycidyl ether	BGE	13 000	54.4	270	+	+
Phenyl glycidyl ether	PGE	8 000	120.0	60	+	+
Resorcinol diglycidyl ether	–	3 000	–	–	–	–
Allyl glycidyl ether	AGE	2 000	57.2 (oc)	22-45	+	+
Octyl-decyl glycidyl ether	–	200	–	–	–	–
Diglycidyl ether	DGE	150	64.0	2.8	+	+
Isopropyl glycidyl ether	IGE	100	33.0	240	–	–
Triglycidyl glycerol ether	–	70	–	–	–	–
Triethylene glycol diglycidyl ether	–	–	79.4	–	+	–
o-Cresyl glycidyl ether	–	–	121.1	–	–	–
Butanediol diglycidyl ether	–	–	–	–	–	+
Diethylene glycol diglycidyl ether	–	–	–	–	–	+

[1] Data collected by the National Occupational Hazards Survey of the US National Institute of Occupational Safety and Health over the period of 1972-74. As a worker may be exposed to more than one glycidyl ether, the exposure estimates are not additive. [2] Where information on the specific glycidyl ether was not available, exposures were listed under the general term of "glycidyl ethers". [3] In the United States the National Institute of Occupational Safety and Health has recommended in its Criteria Document (1978) that the US Federal Standard be changed for BGE from 270 mg/m³ (50 ppm) to 30 mg/m³ (5.6 ppm); for DGE from 2.8 mg/m³ (0.5 ppm) to 1 mg/m³ (0.2 ppm); and for PGE from 60 mg/m³ (10 ppm) to 5 mg/m³ (1 ppm), with all of these being ceiling limits not to be exceeded to avoid irritation effects. [4] The + signs in the table indicate that abnormalities have been found in several different research laboratories in various species of laboratory animals but it is to be noted that none of the laboratory research reports is considered conclusive in respect to the ability of glycidyl ethers to produce permanent changes of the testes or haemopoietic systems in laboratory animals. Although the current animal results are not conclusive, they do constitute a basis for concern. No studies have been reported in available literature of such effects in exposed workers.

In the following section on individual glycidyl ethers, the synonym that carries the asterisk is the name adopted by the International Union of Pure and Applied Chemistry (IUPAC).

Allyl glycidyl ether ($H_2C-\overset{O}{CH-CH_2}-O-CH_2-CH=CH_2$)

OXIRANE, [(2-PROPENYLOXY)METHYL]*; AGE; ALLYL 2,3-EPOXIPROPYL ETHER; 1,2-EPOXY-3-ALLYLOXYPROPANE

m.w. 114
sp.gr. 0.97
b.p. 153.9 °C
v.p. 4.7 mmHg ($0.61 \cdot 10^3$ Pa) at 25 °C
f.p. 57.2 °C oc

soluble in water

a clear colourless liquid with a sweet odour detectable at less than 10 ppm.

TWA OSHA 10 ppm 45 mg/m³ ceil
TLV ACGIH 5 ppm 22 mg/m³ skin
STEL ACGIH 10 ppm 44 mg/m³
IDLH 3 500 ppm

Eye irritation has been reported in one worker and in experimenters exposed to AGE vapour. Exposure of rats to 400 ppm (1 870 mg/m³) 7 h/day, 5 days/week, for 10 weeks caused corneal opacity, emphysema, bronchiectasis and bronchopneumonia. A similar exposure regime at 260 ppm (1 210 mg/m³) caused decreased weight gain, slight eye irritation, and mild respiratory distress of rats for the duration of exposure. AGE has shown mutagenic activity in bacteria but this has not been confirmed by other tests. Intramuscular injection of 400 mg/kg on days 1, 2, 8 and 9 has resulted in haemopoietic abnormalities and focal necrosis of the testis.

Isopropyl glycidyl ether ($H_2\overset{O}{C-CH}-CH_2-O-C(CH_3)_2H$)

OXIRANE, [(1-METHYLETHOXY)METHYL]*; IGE; 3-ISOPROPOXY-1,2-EPOXY PROPANE; (ISOPROPOXYMETHYL) OXIRANE

m.w. 116
sp.gr. 0.92
b.p. 137 °C
v.p. 0.4 mmHg ($0.04 \cdot 10^3$ Pa) at 25 °C
f.p. 33 °C

soluble in water

a colourless liquid.

TWA OSHA 50 ppm 240 mg/m³
STEL ACGIH 75 ppm 360 mg/m³
IDLH 1 500 ppm

No effects have been demonstrated in workers exposed to IGE. In the 400 ppm exposure regime to rats noted under AGE, only slight eye irritation, respiratory distress, and decreased weight occurred.

Phenyl glycidyl ether ($(C_6H_5)-O-CH_2-\overset{O}{CH-CH_2}$)

OXIRANE, (PHENOXYMETHYL)*; PGE; GAMMA-PHENOXYPROPYLENE OXIDE; 2,3-EPOXYPROPYL PHENYL ETHER

m.w. 150
sp.gr. 1.11
b.p. 245 °C
v.d. 4.37
v.p. 0.01 mmHg (1.3 Pa) at 20 °C
f.p. 120 °C

slightly soluble in water

a colourless liquid.

TWA OSHA 10 ppm 60 mg/m³
TLV ACGIH 1 ppm 6 mg/m³

No reports have been published on adverse effects in humans from exposure to airborne PGE. Respiratory tract and skin irritation have been reported in rats exposed repeatedly to 5-12 ppm (30-72 mg/m³), but no effects have been reported at 1 ppm (6 mg/m³). At 10 ppm over the previously noted exposure regime to rats, respiratory tract inflammation and early stages of liver necrosis resulted. PGE has shown mutagenic activity in bacteria, but no dominant lethal or teratogenic effects in mice exposed to 11.5 ppm (71 mg/m³) for 12-19 days. Following exposure of rats to 1.75 ppm, 5.84 ppm, and 11.20 ppm 6 h/day for 19 consecutive days, a marked degree of gonad change in 1 out of 8 rats was found in each exposure group. Application of PGE to shaved backs of rats resulted in a haemopoietic abnormalities. Decreased leucocyte count resulted on exposure of rats to 3 ppm, and on application and intravenous injection of rabbits, and on intravenous injections of dogs.

***n*-Butyl glycidyl ether** ($H_2\overset{O}{C-CH}-CH_2-O-(CH_2)_3CH_3$)

OXIRANE, (BUTOXYMETHYL)*; BGE; 1-BUTOXY-2,3-EPOXYPROPANE; BUTYL 2,3-EPOXY PROPYL ETHER

m.w. 130
sp.gr. 0.91
b.p. 164 °C
v.d. 3.78
v.p. 3.2 mmHg ($0.42 \cdot 10^3$ Pa) at 25 °C
f.p. 54.4 °C

slightly soluble in water

a colourless liquid with a slightly irritative odour.

TWA OSHA 50 ppm 270 mg/m³
TLV ACGIH 25 ppm 135 mg/m³
IDLH 3 500 ppm

No reports have been published on adverse effects in humans from exposure to airborne BGE. BGE has been found to be mutagenic in microbial and mammalian test systems. On application to the skin of male mice, it produced a significant increase in fetal deaths. On inhalation of 300 ppm BGE over the previously mentioned regime, 5 out of 10 rats developed atrophic testes. At 75 ppm, 1 out of 10 rats showed a slight patchy atrophy of the testes. A slight increase in leucocyte count occurred on intramuscular injection of 400 mg/kg for 3 consecutive days.

Diglycidyl ether ($H_2\overset{O}{C-CH}-CH_2-O-CH_2-\overset{O}{CH-CH_2}$)

OXIRANE, 2,2'[OXY-*bis*-(METHYLENE)]*; DGE; DI(2,3-EPOXYPROPYL)ETHER; GLYCIDYL ETHER; DIALLYL ETHER DIOXIDE

m.w. 130
sp.gr. 1.26
b.p. 220-260
v.d. 3.78
v.p. 0.09 mmHg ($0.01 \cdot 10^3$ Pa) at 25 °C
f.p. 64.0 °C

a colourless liquid with an irritant odour detectable above 5 ppm.

TWA OSHA 0.5 ppm 2.8 mg/m³
TLV ACGIH 0.1 ppm 0.5 mg/m³
IDLH 85 ppm

DGE is not widely used in industry and no reports of effects on humans have been published. On animal experimentation, it is the most irritating and the most toxic of the glycidyl ethers. DGE has caused tumorigenic activity in mice and mutations in bacteria. Testicular changes have occurred in rats exposed to 0.3 ppm (1.6 mg/m³). Haemopoietic abnormalities have resulted from skin application of animals and on inhalation of 3 ppm over 19 4-h periods.

Other glycidyl ethers

All glycidyl ethers that have been tested have produced sensitisation and have been mutagenic in bacteria. Triethylene glycol diglycidyl ether has produced lung tumours in mice on receiving intraperitoneal doses in excess of 3.6 g/kg. Only hydroquinone diglicidyl ether has given clearly negative results in a test of its tumorigenicity. Since animal experimentation has shown that glycidyl ethers have the potential to produce harmful effects, precautionary measures suggested for safe handling of certain of the glycidyl ethers should be observed where appropriate for others of these substances for which information on health and safety measures is incomplete or entirely lacking.

COOK, W. A.

CIS 76-1023 "Hazardous chemical reactions–38. Ethers" (Réactions chimiques dangereuses–38. Ethers). Leleu, J. *Cahiers de notes documentaires–Sécurité et hygiène du travail* (Paris), 1st quarter 1976, 82, Note No. 1000-82-76 (127-129). (In French)

Hygienic guide for ethyl ether (American Industrial Hygiene Association, 465 Wolf Ledges Parkway, Akron, Ohio) (Revised 1977-78).

Criteria for a recommended standard–Occupational exposure to glycidyl ethers. DHEW (NIOSH) publication No. 78-166 (National Institute for Occupational Safety and Health, 4676 Columbia Parkway, Cincinnati) (June 1978), 197 p. Illus. 98 ref.

"Effects of *n*-butyl glycidyl ether exposure". Wallace, E. *Journal of the Society of Occupational Medicine* (Edinburgh), Oct. 1979, 29/4 (142-143). 4 ref.

"Contact sensitivity to phenyl glycidyl ether". Rudzki, E.; Krajewska, D. *Dermatosen in Beruf und Umwelt* (Aulendorf), 1979, 27/2 (42-44). 8 ref.

"Phenyl glycidyl ether" (332). 4 ref. *Documentation on threshold limit values 1980* (American Conference of Governmental Industrial Hygienists Inc., Cincinnati) (1980), 486 p.

Ethics

The work of the doctor in industry varies considerably according to the type of industry and the social and administrative structure of the country in which he works. In a situation such as a remote construction site where he is the only doctor available, his role may be predominantly clinical. In an industrialised area where there are other medical resources, his responsibility may be limited to those factors at work which are liable to affect the workers' health. Inevitably, however, there is a combination of the clinical and environmental functions and these impose on the doctor in industry the same ethical obligations as on the medical profession as a whole.

The doctor in industry is concerned, as are his counterparts in hospital and general practice, with the promotion and maintenance of the health of the people in his care. He looks at the subject from two points of view at the same time. On the one side he looks at the work in its physical, chemical, biological and psychosocial aspects as it affects the health of the worker. On the other he looks at the health of the worker as it affects his working capacity.

The doctor is not the only responsible person in an occupational health service. The occupational health nurse, the occupational hygienist, the first aider and other people such as dentists and physiotherapists may also be privy to the secrets of the occupational health service and therefore are equally bound with the doctor in the issue of medical confidence.

Relationship with management

The personnel of the occupational health service are paid by the management and use premises, equipment and material belonging to their employer. Books, journals and even the subscription to professional societies may also be paid by the employer. Various fringe benefits such as pension, car and accommodation may also be provided. The ethical questions which arise are:

(1) To what extent does the management own the services of the doctor and his colleagues and more specifically the records of their work?

(2) To what extent does the doctor owe a duty to his employer in using his professional knowledge for the benefit of the organisation?

The doctor concerned with the health of a community has a somewhat different ethical problem from that of the practitioner dealing with a single patient. For while the latter sees it as his plain duty to promote the interests of his patient without reference to anyone else, the former is obliged to look at the patient against the background of the industrial community in which he works and to take into account the interaction of one with the other.

The nature of the contract, either explicit or implicit, between the doctor and the employer does not involve the buying or selling of anything more than the doctor's professional skill, which necessarily involves an acceptance of his code of ethics. This is different from the implied contract with an insurance company in making a medical examination for insurance purposes. In such a case the patient has accepted the fact that the details of his physical condition are the confidential property of the insurance company which uses this information to determine the nature of its once-for-all contract with the applicant.

The employer, however, is in a different position. The contract, unless it concerns only entry into the firm's pension scheme, in which case it is the same as an insurance contract, is not a once-for-all transaction. The employer needs to know whether the individual is fit to do the job for which he is initially hired, whether he remains fit for the job and whether he is fit to be transferred to another job. Each job has its own set of stresses and requirements and the employee's state of health has to be measured against these to ascertain his fitness.

But this is all that the employer has a right to know, even when there is a condition of employment requiring a medical examination. Naturally, he may be interested in the health of his employee for the best possible reasons and in many circumstances the nature of the disorder may be of prime importance with regard to future employment. He must recognise that, unless the worker agrees to a particular aspect of his health being revealed, these are the limits to his use of the doctor in respect of medical examination.

It has been argued that, as the doctor uses the employer's paper, ink and so on, any report is the property of the employer. No doctor in conscience could accept such an argument. He is required to reveal professional information only in a court of law or other tribunal, unless specifically permitted to do so by the patient. He also has an implicit and binding moral obligation not to reveal the industrial secrets which inevitably in the course of his work he comes to learn.

Relationship with employees and trade unions

The number of personal contacts which the doctor has with workers and their representatives is very much greater than he has with management. Every day he is seeing men at work or in his medical centre and he tends to think of the man in his working environment rather than as a case of a particular disease. This regular contact with people who are not sick gives him a different attitude to the question of their health as compared with the doctor who sees the patient in his consulting room as a sick person. In the former case the relationship is based on mutual trust, while in the latter the patient comes in the role of a suppliant. The doctor in industry must earn respect by his conduct, which is continually being observed by shrewd judges of character.

The conduct of the doctor in industry resolves itself into his general practitioner role, in which he treats the emergencies that arise or gives general medical advice, and his preventive role by which he deals with environmental problems or with the allocation of men to particular jobs. In the first case he must be seen to be competent so as to inspire confidence in his capacity as a doctor. In the second, he must be seen to be concerned for the protection and promotion of the health of workers rather than the protection of the interests of the employer, though the two aspects are not mutually exclusive.

Inevitably in the course of his work the doctor becomes aware of confidential information relating to workers. This may arise out of a statutory medical examination, in the course of dealing with an accident, illness, or during an investigation of a particular group of workers. In the case of the statutory medical examination, the worker has no choice but to accept that any relevant information will be revealed to the employer and the authorities concerned. The operative word is "relevant". The worker must be confident that no other information, not directly related to the circumstances of the case, will be made public.

The same is true in the case of the medical examination under a contract of employment. The worker knows that his fitness for the job is being measured and that if his physical condition disqualifies him, the doctor must say so. But he has a right to expect that the doctor will say no more than that he is fit or not. The details of the consultation are as much a medical secret as between any doctor and his patient. Should there be some condition that prevents the man from doing the job for which he is presenting himself, but does not prevent him from doing some other job in the organisation, he may in consultation with the doctor give him permission to indicate to management the limitations on his working capacity.

In cases of accident or illness, at a counselling session the relationship is entirely one between doctor and patient. The patient is entitled to the same standard of medical ethics as with any other doctor. This presents problems to both. The patient needs the treatment or seeks the advice recognising that the doctor will learn details which could be used to affect his future prospects of employment. This is particularly so in the case of an accident, when either the history given by the patient or the details of the examination may reveal some suggestion of blame. Whatever he may feel, the doctor must not allow himself to reveal, even covertly, his knowledge on this subject unless he is called upon to give evidence in a court.

Trade unions are inherently suspicious of management and the doctor tends to be identified with management because of his manners, mode of dress and the fact that he may usually eat in the management dining room. Starting off with this disadvantage he has a difficult, though not impossible, job to convince the men of his impartiality and his concern for their health. Unless he achieves this confidence he will never be effective and his employer must also appreciate this.

Problems obviously arise, for example when the doctor finds something in the course of an examination which constitutes a contraindication to the work the patient is doing. If it is something which concerns only the worker himself he can be advised to give up the work in his own best interests. If in spite of the advice, he chooses for economic, social or status reasons to continue in work which is liable to damage his health, that is his own affair.

A much greater problem arises when the worker's disorder is likely to affect the safety of others. It is desirable first to obtain a second opinion. If this confirms the occupational physician's judgement the position should be carefully explained to the worker. Only if he remains resolute in spite of the most earnest persuasion should the doctor abandon his primary concern for the welfare of the patient in favour of the greater good of the community for which he is responsible.

Apart from contacts with individual workers as patients, the doctor in industry has frequent contact with workers' representatives. These may be members of the safety committee, shop stewards, or full-time officers of a union. Although it might be assumed that the trade union representative is acting in the workers' best interests, the same standards of conduct apply as in dealing with management. Only at the express wish of the worker involved should any details of his health status be discussed. The extent to which information should be revealed will depend on the circumstances of the case, but the golden rule is, as in other situations, no more than necessary.

Relationships with other practitioners

In cases where the doctor in industry during the course of an examination finds some condition requiring further treatment, it is his duty to refer the patient to his own family doctor except in cases of emergency, when he may take the step of sending the patient to hospital. Conversely, in very many cases the worker is seen by his own family doctor or by a specialist at a hospital and one or other of these may give him some advice about his future work prospects. Unless good relations have been built up between the works doctor and his medical colleagues in the area, he may not be informed about the patient's condition until he returns to work. There are two sources of difficulty. Either the worker comes back too soon because the doctor concerned with his treatment is unaware of the stresses of the job, or he is off work for a longer time than necessary because the treatment doctor does not appreciate that there are resources at the place of employment which will facilitate his recovery.

The doctor in industry should visit his colleagues in the area to acquaint them with the facilities that are available. Even better, he should invite the local practitioners to the factory to see for themselves. Good relations, thus established, are of the greatest help to all concerned. If the worker is aware of confidence between his own doctor and the factory doctor, this increases his trust in both. The doctors learn respect for one another. The general practitioner can refer his patient for ambulatory treatment at the factory and the works doctor can learn something of the family circumstances, which will assist him in handling the patients' problems at the factory.

One of the awkward problems in relation to medical colleagues is sickness absence. The doctor in industry is being unwise if not unethical in attempting to police sickness absence and this is emphasised in the ILO's Occupational Health Services Recommendation, 1959 (No. 112). Cases are bound to arise where the pattern of sickness absence raises doubts about the validity of the certification, but these can usually be resolved by good relations between the doctors concerned.

Conclusions

There are, it might be said, three levels of ethics. In the first place there are the statutory obligations, primarily those concerned with the notification of certain diseases and those of answering the questions in court of the judge or a counsel. In the case of questions from counsel the doctor must always appeal to the judge in matters of conscience for a ruling and, while being as helpful as he can, must provide no gratuitous information that is not of immediate relevance.

Secondly, there is the code of medical ethics which he must respect and guard in his own interests as well as those of his professional colleagues. Certain departures from this code are held to be more heinous than others and these form the substance of the ethical rules governed by the supreme medical body in each country.

Thirdly, there is the need for medical etiquette which is no more than professional good manners. This relates to the doctor's conduct, appearance, demeanour and to the outward evidence of concern, trustworthiness and impartiality by which he earns the respect of others.

MURRAY, R.

Guidance on ethics for occupational physicians (London, Royal College of Physicians, Faculty of Occupational Medicine, 1980), 10 p. 5 ref.

CIS 80-2072 *The occupational physician* (British Medical Association, Tavistock Square, London) (June 1980), 30 p.

"Physicians and patients in the occupational setting: the rules of the game". Schuman, B. J. *Journal of the American Medical Association* (Chicago), 28 Nov. 1980, 244/21 (2417-2418). 1 ref.

"In favour of a 'Code of Ethics' for occupational health" (Em defesa de um 'Código de ética' para a medicina do trabalho). *Fundacentro* (São Paulo), July 1981, 12/139 (5).

Ethyl alcohol

Ethyl alcohol (C_2H_5OH)

ETHANOL

m.w.	46.1
sp.gr.	0.79
m.p.	−177.3 °C
b.p.	78.5 °C
v.d.	1.6
v.p.	43.9 mmHg ($5.84 \cdot 10^3$ Pa) at 20 °C
f.p.	12.8 °C
e.l.	3.3-19%
i.t.	422.8 °C

miscible with water in all proportions

a colourless, volatile liquid with a mild pleasant odour and a burning taste.

TWA OSHA 1 000 ppm 1 900 mg/m³

MAC USSR 1 000 mg/m³

Production. The preparation of beverages containing ethyl alcohol was practised in Egypt and China some 5 000 years ago, and in the 10th century AD the Arabs developed a distillation process for the production of a concentrated alcoholic spirit. Nowadays ethyl alcohol is produced by the fermentation (or by hydrolysis and fermentation) of the sugars in fruit (chiefly grapes), beetroot and sugar cane, changes that are induced by the action of the enzymes in yeast. The starch in wheat, barley, maize, rice, sorghum and potatoes is also converted in stages into a hexose or grape sugar, which is converted to alcohol by fermentation with yeast. Wines and beers with a relatively low alcohol content are thus produced while spirits with a higher alcohol content are obtained from similar products by distillation.

In addition to its wide use in many forms as a beverage, ethyl alcohol is used in the manufacture of many industrial products, and much of the industrial alcohol used for this purpose is produced by chemical synthesis. Ethylene, from natural gas or from the cracking of petroleum products, is treated with concentrated sulphuric acid to form ethyl hydrogen sulphate and diethyl sulphate, which are then hydrolysed to produce ethyl alcohol. In another process ethylene is directly hydrated to the alcohol by treatment with water in the presence of a phosphoric acid catalyst at a pressure of about 70 kgf/cm² and at about 300 °C. The Fischer-Tropsch reaction is an important additional source of industrial ethyl alcohol, which is produced as a by-product of the process (based on the catalytic hydrogenation of carbon monoxide).

Uses. The extensive use of ethyl alcohol in the preparation of other chemicals is exemplified by the production of one company which bases the manufacture of 70 of its products on this one raw material. Acetaldehyde is produced in large quantities by the oxidation of ethyl alcohol in the presence of a silver screen catalyst; ethyl ether is obtained by the catalytic dehydration of the alcohol in the vapour phase; chloroethane is a product from ethyl alcohol which in turn is used in the preparation of tetraethyl-lead, the motor fuel additive used to improve antiknock ratings. The production of butadiene from ethyl alcohol has been of great importance to the plastics and synthetic rubber industry.

Ethyl alcohol is capable of dissolving a wide range of substances and for this reason it is used as a solvent in the manufacture of drugs, plastics, lacquers, polishes, plasticisers, perfumes, cosmetics, rubber accelerators, etc.

HAZARDS

The conventional industrial hazard is exposure to the vapour in the vicinity of a process in which ethyl alcohol is used. Prolonged exposure to concentrations above 5 000 ppm causes irritation of the eye and nose, headache, drowsiness, fatigue and narcosis. Ethyl alcohol is quite rapidly oxidised in the body to carbon dioxide and water and unoxidised alcohol is excreted in the urine and expired in air with the result that the cumulative effect is virtually negligible. Its effect on the skin is similar to that of all fat solvents and, in the absence of precautions, dermatitis may result from contact.

Ethyl alcohol is a flammable liquid and its vapour forms flammable and explosive mixtures with air at normal temperature. An aqueous mixture containing 30% alcohol can produce a flammable mixture of vapour and air at 29 °C and one containing only 5% alcohol can produce a flammable mixture at 62 °C.

While ingestion is not a likely consequence of the use of industrial alcohol, it is a possibility in the case of an addict. The danger of such illicit consumption [depends upon the concentration of ethanol, which above 70% is likely to produce oesophageal and gastric injuries, and upon the presence of denaturants.] These are added to make the spirit unpalatable when it is obtained free of tax for non-potable purposes. Many of these denaturants [(e.g. methyl alcohol, benzene, pyridine bases, methylisobutylketone and kerosene, acetone, gasoline, diethylphthalate and so on)] are more harmful to a drinker than the ethyl alcohol itself. It is important therefore to ensure that there is no illicit drinking of the industrial spirit.

[Recently another potential hazard in human exposure to synthetic ethanol was suspected because the product was found to be carcinogenic in mice treated at high doses. Later on epidemiologic analyses (Lynch et al.) have revealed an excess incidence of laryngeal cancer (on average five times greater than expected) associated with a strong acid ethanol unit. Diethyl sulphate would appear to be the causative agent, although alkyl sultones and other potential carcinogens were also involved.]

SAFETY AND HEALTH MEASURES

Precautions in processes involving the manufacture or use of ethyl alcohol should cover health hazards and the risk of fire and explosion.

Health measures. Good ventilation (preferably local exhaust ventilation) will prevent the formation of harmful concentrations of alcohol vapours.

Personal protective equipment, especially hand protection, should be provided where there is a liability of prolonged skin contact.

[Alcoholics and persons with liver diseases should not be exposed to ethyl alcohol. Eyewashing facilities should be available where projections of the liquid may occur.]

Fire and explosion. The above precautions will prevent the formation of flammable or explosive concentrations in normal circumstances, but dangerous concentrations could form in the event of a massive escape of liquid resulting from a valve failure or a fracture of a pipe or vessel. Arrangements should be made by the provision of sills and curbs and by the design of floors to limit the spread of escaping liquid and to conduct it to a safe place. [The use of compressed air or oxygen to transport alcohol should be prohibited.]

Precautions should be taken, for example by the provision of flameproof electrical installations and equipment, to prevent sources of ignition where large quantities of ethyl alcohol are made or used.

[The most appropriate extinguishers are carbon dioxide and dust; water may be used, provided it is in large amounts.]

MATHESON, D.

Exposure limit:

CIS 1189-1970 "Further investigations on the exposure test of ethyl alcohol". Spasovski, M.; Bencev, I. *Works of the Scientific Research Institute of Labour Protection and Occupational Diseases* (Sofia), Dec. 1969, 17 (59-66). Illus. 10 ref.

Carcinogenic risk:

"An association of upper respiratory cancer with exposure to diethyl sulphate". Lynch, J.; Hanis, N. M.; Bird, M. G.; Murray, K. J.; Walsh, J. P. *Journal of Occupational Medicine* (Chicago), May 1979, 21/5 (333-341). Illus. 40 ref.

Protective gloves:

CIS 77-498 "Resistance of protective gloves to industrial solvents—Results obtained with ethanol on some 100 types of commercial gloves" (Résistance des gants de protection aux solvants industriels—Résultats obtenus avec l'éthanol sur une centaine de gants du commerce). *Travail et sécurité* (Paris), Nov. 1976, 11 (509-516). Illus. (In French)

Ethylene

Ethylene (CH_2:CH_2)

ETHENE

m.w.	28.05
sp.gr.	0.61
m.p.	−169.1 °C (freezes −181 °C)
b.p.	104 °C
v.d.	0.98
e.l.	3.1-32%
i.t.	450 °C

insoluble in water; slightly soluble in ethyl alcohol; soluble in ethyl ether

a colourless gas with a characteristic sweetish odour, which burns with a luminous flame.

ACGIH simple asphyxiant

Production. Practically all commercial ethylene is produced by the thermal cracking of petroleum fractions, the raw materials being refinery gases, liquefied petroleum gas, or light naphthas. The thermal cracking of these fractions is done under pressures that are close to atmospheric, with the temperatures ranging from approximately 700 °C to 800 °C. The ethylene is recovered from the cracked gases by fractionation, generally by low-temperature, high-pressure straight fractionation, although other recovery processes are also available. The cracked gas is compressed, dried over activated bauxite or alumina, and cooled, then piped into the fractionator where hydrogen and methane are removed overhead in the first tower. The bottoms, which are ethylene and heavier components, pass to the second tower where the ethylene is removed overhead.

Uses. The principal industrial use of ethylene is as a "building block" for chemical raw materials which in turn are used to manufacture a large variety of substances and products (figure 1). Ethylene is used also in oxyethylene welding and cutting of metals, as a refrigerant, as an anaesthetic, and also as a plant growth accelerator and fruit ripener. However, the amounts used for these purposes are minor in comparison with the quantities used in the manufacture of other chemicals. Some of the major chemicals and materials derived from ethylene are: polyethylene, which is made by catalytic polymerisation of ethylene and is used for the manufacture of a variety of moulded plastic products (see POLYOLEFINS). Ethylene oxide is produced by catalytic oxidation and in turn is used to make ethylene glycol and ethanolamines. Most of the industrial ethyl alcohol is produced by the hydration of ethylene. Chlorination yields vinyl chloride monomer or 1,2-dichloroethane. When reacted with benzene, styrene monomer is obtained. Acetaldehyde is also made by oxidation of ethylene.

HAZARDS

The major hazard of ethylene is that of fire or explosion. [Ethylene spontaneously explodes in sunlight with chlorine and can react vigorously with carbon tetrachloride, nitrogen dioxide, aluminium chloride and oxidising substances in general.] Ethylene-air mixtures will burn when exposed to any source of ignition such as static, friction or electrical sparks, open flames, or excess heat. When confined, certain mixtures will explode violently from these sources of ignition. Ethylene is often handled and transported in liquefied form under pressure. Skin contact with the liquid can cause a "freezing burn". There is little opportunity of exposure to ethylene during its manufacture because the process takes place in a closed system. Exposures may occur as a result of leaks, spills, or other accidents that lead to release of the gas into the air. Empty tanks and vessels that have contained ethylene are another potential source of exposure.

In air, however, ethylene acts primarily as an asphyxiant. Concentrations of ethylene required to produce any marked physiological effect will reduce the oxygen content to such a low level that life cannot be supported. For example, air containing 50% of ethylene will contain only about 10% of oxygen. Loss of consciousness results when the air contains about 11% of oxygen. Death occurs quickly when the oxygen content falls to 8% or less. There is no evidence to indicate that prolonged exposure to low concentrations of ethylene can result in chronic effects. Prolonged exposure to high concentrations may cause permanent effects because of oxygen deprivation.

Toxicity. Ethylene has a very low order of systemic toxicity. When used as a surgical anaesthetic, it is always administered with oxygen. In such cases, its action is that of a simple anaesthetic having a rapid action and an equally rapid recovery. Prolonged inhalation of about 85% in oxygen is slightly toxic, resulting in a slow fall in the blood pressure; at about 94% in oxygen, ethylene is acutely fatal.

SAFETY AND HEALTH MEASURES

Because of the great hazard of fire and explosion, every precaution should be taken to prevent the accumulation of explosive mixtures. [In principle the concentration of ethylene gas should not be allowed to exceed one-fifth (20% by volume) of the lower explosion limit. This is calculated to be 5 500 ppm.] All sources of ignition should be avoided. All electrical installations and equipment should be explosion-proof. Good ventilation

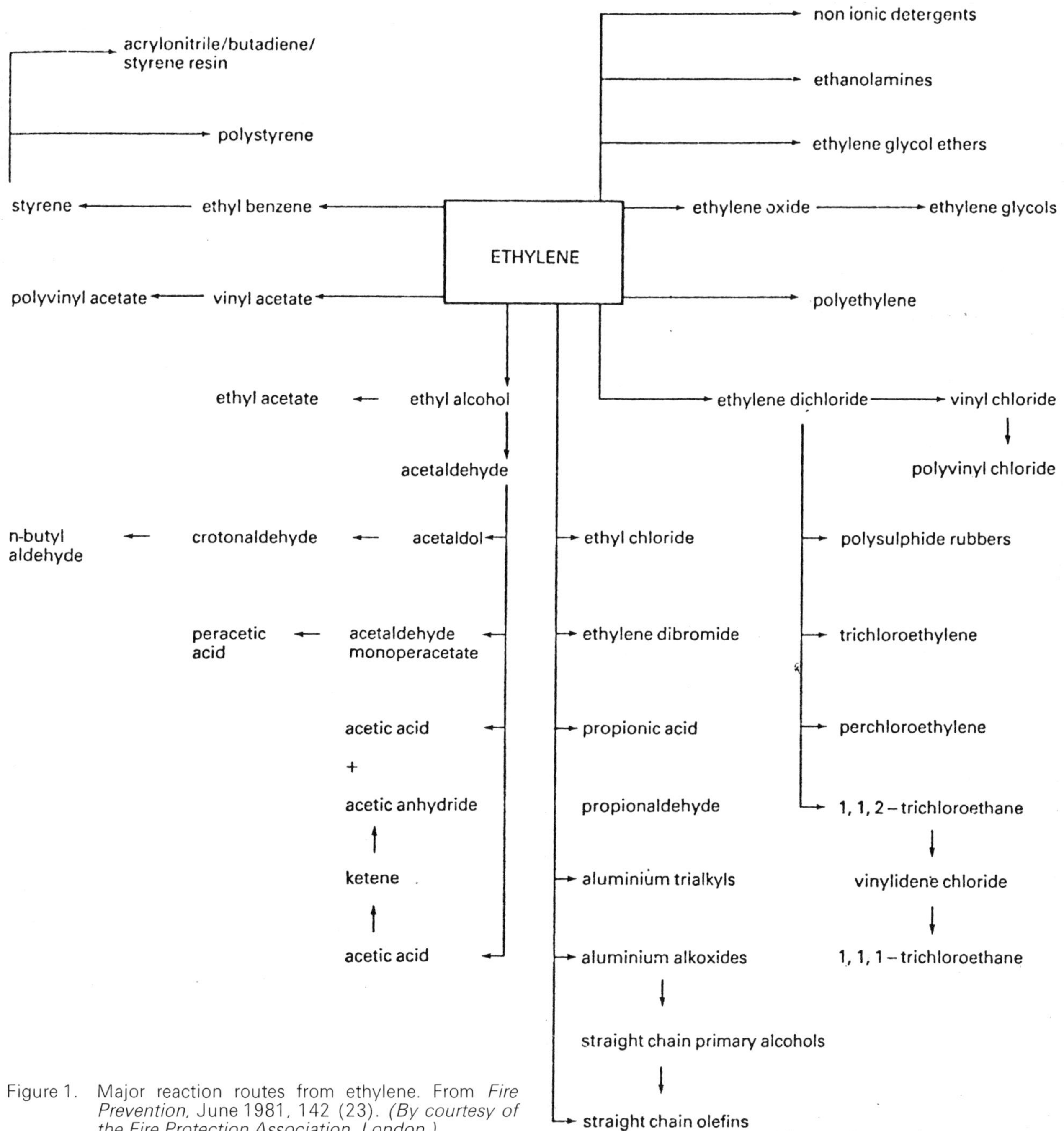

Figure 1. Major reaction routes from ethylene. From *Fire Prevention*, June 1981, 142 (23). *(By courtesy of the Fire Protection Association, London.)*

should be provided in all rooms or areas where ethylene is handled. Entry into confined spaces that have contained ethylene should not be permitted until gas tests indicate that they are safe and entry permits have been signed by an authorised person.

Large ethylene fires are difficult to extinguish. Where possible, the ethylene supply should be shut off, the container cooled by water spray, and the fire allowed to burn itself out.

Impervious hand, eye and face protection should be provided whenever there is danger of contact with liquid ethylene.

Persons who may be exposed to ethylene should be carefully instructed about and trained in its safe and proper handling methods. Emphasis should be given to the fire hazard, the "freezing burns" due to contact with the liquid material, use of protective equipment, and emergency measures.

Treatment. Persons affected by exposure should be removed to fresh air, kept warm and comfortable. Medical care should be obtained as soon as possible. If breathing has stopped, artificial respiration should be instituted immediately on removal from the contaminated atmosphere. Rescuers should never enter a contaminated atmosphere without respiratory protective equipment to rescue anyone who has been affected by exposure to this gas. In the event of eye contact with liquid ethylene, the eyes should be flushed with water for 15 min and medical attention obtained.

DOOLEY, A. E.

CIS 74-1662 *Properties and essential information for safe handling and use of ethylene.* Chemical Safety Data Sheet SD-100 (Manufacturing Chemists Association, 1825 Connecticut Avenue, NW, Washington, DC) (1973), 17 p.

"Safety rules for the use of ethylene in the chemical industry" (Conditions de sécurité pour l'emploi de l'éthylène dans l'industrie chimique). Fiumara, A.; Cardillo, P. (239-257). 18 ref. (In French) *Report of the 3rd International Symposium on the Prevention of Occupational Risks in the Chemical Industry* (Berufsgenossenschaft der chemischen Industrie, Gaisbergstrasse 11, Heidelberg), 598 p. Illus. Ref. (Multilingual)

"Ethylene and polyethylene" (157-177). 105 ref. *IARC monographs on the evaluation of the carcinogenic risk of chemicals to humans.* Vol. 19. *Some monomers, plastics and synthetic elastomers, and acrolein* (Lyons, International Agency for Research on Cancer, Feb. 1979), 512 p.

"Reaction charts of hydrocarbons: ethylene and its derivatives". *Fire Prevention* (London), June 1981, 142 (22-23).

Ethylene dichloride

Ethylene dichloride ($ClCH_2CH_2Cl$)

1,2-DICHLOROETHANE; EDC; BROCIDE; DUTCH LIQUID; GAZ OLEFIANT

m.w.	99
sp.gr.	1.25
m.p.	−35 °C
b.p.	84 °C
v.d.	3.4
v.p.	100 mmHg ($13.3 \cdot 10^3$ Pa) at 29.4 °C
f.p.	13.3 °C
e.l.	6.2-16%
i.t.	412 °C

slightly soluble in water; very soluble in ethanol; miscible with ether in all proportions; when traces of moisture are present, it can corrode steel, iron and other metals; in contact with hot surfaces it decomposes into HCl, CO_2, CO and possibly phosgene

an oily colourless liquid with a sweetish odour.

TWA OSHA	50 ppm 200 mg/m³
OSHA	100 ppm ceil
	200 ppm/5 min peak 3 h
NIOSH	1 ppm/10 h 4 mg/m³/10 h
	2 ppm/15 min ceil 8 mg/m³/15 min ceil
TLV ACGIH	10 ppm 40 mg/m³
STEL ACGIH	15 ppm 60 mg/m³
MAC USSR	10 mg/m³ skin

Production. Ethylene dichloride is manufactured in the greatest tonnage of all chlorinated organic compounds. In 1979 over 5 million tonnes were synthesised in the United States. The US National Institute for Occupational Safety and Health (NIOSH) estimates on the basis of a national survey from 1972 to 1974 that in the United States approximately 2 million workers in 148 165 workplaces were potentially exposed to ethylene dichloride at that time.

Nearly all US manufacturers use a combination of two methods to produce ethylene dichloride. One method is to treat ethylene with chloride gas (chlorination process). The other method is to treat ethylene with oxygen and hydrogen chloride (oxychlorination process).

Uses. Ethylene dichloride is used primarily as a raw material in the production of vinyl chloride, which in turn is polymerised into a number of valuable types of plastics.

In 1978 the major uses for ethylene dichloride were vinyl chloride (86%), methyl chloroform (3%), ethylenamines (3%), trichloroethylene (2%), and perchloroethylene (2%). A further 2% of production is used in the formulation of gasoline lead scavengers. A list of the major uses of ethylene dichloride is given in table 1.

Table 1. Uses of ethylene dichloride

Asphalt processing	Manufacture of–
Bakelite processing	vinyl chloride
Bitumen processing	methyl chloroform
Camphor refining	trichloroethylene
Cosmetics manufacture	perchloroethylene
Cellulose acetate dispersion	vinylidine chloride
Cellulose ester dispersion	ethylene amines
Dry cleaning (patented)	pharmaceuticals
Degreasing operations in–	Ore processing resins:
textiles industry	bakelite
petroleum industry	ehiokiles
electronics industry	rubber
Extraction of–	Textiles:
soybean oil	nylon
fish protein	viscose rayon
caffeine	Paint solvation and stripping
dyes	Pest extermination
camphor	Pesticide processing
Fumigation of grain and seeds	Petroleum refining
Gasoline:	Photography
tetraethyl-lead precursor	Toxicological analysis
lead scavenger	Varnish dilution
anti-knock agent	Water softening
blending	Xerography

Ethylene dichloride has been used as a sealant for the polymethacrylates (Plexiglass). However, because of the carcinogenic hazard associated with it, manufacturers of polymethacrylates no longer recommend it as a solvent-sealant and other solvents have replaced it for this use.

HAZARDS

Ethylene dichloride is flammable and a dangerous fire hazard. It can be absorbed through the airways, the skin and the gastrointestinal tract. It is metabolised into 2-chloroethanol and monochloroacetic acid, both more toxic than the original compound.

Ethylene dichloride has an odour threshold in humans that varies from 2 to 6 ppm as determined under controlled laboratory conditions. However, adaptation appears to occur relatively early and after one or two minutes the odour at 50 ppm is barely detectable.

The worker is exposed primarily through the use of ethylene dichloride. Since its major use is in the manufacture of vinyl chloride, most is contained in a closed process. Leaks from the process can and do occur, however, producing a hazard for the worker so exposed. However, the most likely chance of exposure occurs during the pouring of containers of ethylene dichloride into open vats where it is subsequently used for the fumigation of grain. Exposures also occur through manufacturing losses, application of paints, solvent extractions, and waste disposal operations. Ethylene dichloride rapidly photo-oxidises in air and does not accumulate in the environment. It is not known to bioconcentrate in any food chains or to accumulate in human tissues.

This chlorinated hydrocarbon is appreciably toxic to humans both acutely and chronically; 80-100 ml are enough to produce death within 24 to 48 h. Inhalation of 4 000 ppm will cause serious illness. In high concentrations it is immediately irritating to the eyes, nose, throat and skin.

There are no published studies of carcinogenic, mutagenic or teratogenic effects of ethylene dichloride in humans. The potential for carcinogenicity in humans is based on recent animal studies completed at the National Cancer Institute in the United States, the results of which are summarised in table 2.

Table 2. A comparison of the incidence of various tumours in animals given ethylene dichloride *per os* by intragastric intubation

Animal and tumour	Pooled vehicle control	Matched vehicle control	Dose (mg/kg)	
			50	100
Rats (male):				
Haemangiosarcoma	1/60	0/20	9/50	7/50
Fibroma of sub-cutaneous tissue	0/60	0/20	5/50	6/50
Squamous cell carcinoma	0/60	0/10	3/50	9/50
Rats (female):				
Mammary carcinoma	1/59	0/20	1/50	18/50
Haemangiosarcoma	0/59	0/20	4/50	4/50
Mammary carcinoma or adenoma	6/59	0/20	15/50	24/50
Mice (males):				
Lung adenoma	0/59	0/19	1/47	15/48
Hepatocellular carcinoma	4/59	1/19	6/47	12/48
Mice (females):				
Lung adenoma	2/60	1/20	7/50	15/48
Mammary carcinoma	0/60	0/20	9/50	7/48
Stromal polyp or sarcoma	0/60	0/20	5/49	5/47

Significant increases in tumour production were found in both sexes in mice and rats. Many of the tumours, such as haemangiosarcoma, are uncommon types of tumours, rarely if ever encountered in control animals. The "time-to-tumour" in treated animals was less than in controls. Significant numbers of tumours began to appear by 6 months. Tumours did not appear in any of the controls by this time. Since it caused progressive malignant disease of various organs in two species of animals, ethylene dichloride must be considered potentially carcinogenic in man.

There have been consistent observations of the mutagenicity of ethylene dichloride in bacterial test systems. The experimenters described ethylene dichloride as a "moderate mutagen" without activation and a "potent mutagen" when applied together with liver enzymes. While the relation of mutagenicity to carcinogenicity is not firmly established, the consistent positive mutagenicity findings support the conclusion that ethylene dichloride is a carcinogen.

Vozovaja (1976) reported that in pregnant rats exposed to 15 mg/m^3 ethylene dichloride in air, pre-implantation embryonic deaths were five times higher than in controls. There were haematomas in the region of the head, neck and upper extremities of the fetuses, and total embryonic mortality was increased. Deformities were not reported or discussed.

SAFETY AND HEALTH MEASURES

According to NIOSH (1978) ethylene dichloride should be controlled in the workplace as an occupational carcinogen. In addition to an environmental standard of 1 ppm, NIOSH recommends the following:

1. Medical surveillance procedures.
2. Labelling and posting instructions.
3. Personal protective equipment and clothing requirements.
4. Methodology procedures in informing employees of hazards for ethylene dichloride.
5. Essential work practice requirements including:

(a) engineering procedures;
(b) engineering controls;
(c) regulated areas;
(d) establishment of clean work clothing room;
(e) decontamination procedures;
(f) laundering;
(g) storage;
(h) maintenance;
(i) entry in confined spaces;
(j) disposal.

6. Sanitation procedures.
7. Monitoring the workplace and record-keeping.

The use of ethylene dichloride as a solvent, diluent, or fumigant in open operations should be prohibited. Product substitution should be a paramount consideration, and wherever ethylene dichloride is identified or its presence suspected, it should be replaced by a less harmful substitute.

For sampling workplace air, NIOSH recommends an analytical method that employs absorption of ethylene dichloride on charcoal followed by carbon disulphide desorption, and gas chromatographic measurement.

One part per million represents a feasible level which can be easily detected and which is considered to present no significant hazard to the worker.

GREGORY, A. R.

General:

Ethylene dichloride (1,2-dichloroethane)–Criteria for a recommended standard. DHEW (NIOSH) publication No. 76-139 (National Institute for Occupational Safety and Health, 4676 Columbia Parkway, Cincinnati) (Mar. 1976), 158 p. Illus. 169 ref.

Safety:

CIS 77-773 *Ethylene dichloride–Codes of practice for chemicals with major hazards* (Chemical Industries Association Ltd., 93 Albert Embankment, London) (June 1975), 16 p. Illus.

CIS 77-1065 *Ethylene dichloride.* Data sheet 350, Revision A (extensive) (National Safety Council, 425 North Michigan Avenue, Chicago) (1977), 3 p. 7 ref.

Exposure limit:

Ethylene dichloride (1,2-dichloroethane). Revised Recommended Standard. DHEW (NIOSH) publication No. 78-211 (National Institute for Occupational Safety and Health, 4676 Columbia Parkway, Cincinnati) (Sep. 1978), 33 p. 15 ref.

Carcinogenic risk:

Bioassay of 1,2-dichloroethane for possible carcinogenicity. DHEW (NIH) publication No. 78-1305 (National Institutes of Health, National Cancer Institute, Bethesda, Maryland) (1978), 78 p.

Effects on reproduction:

"The effect of small concentrations of benzene and dichloroethane separately and combined on the reproductive functions of animals" (Vlijanie mal'yx koncentracij benzina, discloretana i ix kombinacii na reproduktivnuju funkciju zivotnyx). Vozovaja, M. A. *Gigiena i sanitaria* (Moscow), June 1976, 6 (100-102). 4 ref.

Ethylene glycol dinitrate

Ethylene glycol dinitrate ($CH_2ONO_2CH_2ONO_2$)

EGDN; NITROGLYCOL (not to be confused with Diethylene glycol dinitrate); 1,2-DINITROETHANEDIOL; ETHYLENE DINITRATE

m.w. 152
sp.gr. 1.49
m.p. −22.3 °C
b.p. 197 ± 3 °C (explodes at 114 °C)
v.d. 5.25
v.p. rises rapidly with temperature; EGDN and nitroglycerin v.p. data are compared below

Temperature (°C)	Vapour pressure			
	EGDN		Nitroglycerin	
	mmHg	Pa	mmHg	Pa
0	0.0044	0.5	–	–
20	0.038	5.0	0.00025	0.03
40	0.026	$0.03 \cdot 10^3$	0.0024	0.3
60	1.3	$0.17 \cdot 10^3$	0.019	2.5
80	5.9	$0.77 \cdot 10^3$	0.10	$0.01 \cdot 10^3$
100	22.0	$2.8 \cdot 10^3$	0.50	$0.05 \cdot 10^3$

f.p. 215 °C

insoluble in water; soluble in ethyl alcohol, ethyl ether and benzene

a colourless, odourless liquid.

TWA OSHA 0.2 ppm 1.2 mg/m³ skin ceil
TLV ACGIH 0.02 ppm 0.2 mg/m³
NIOSH 0.01 ppm 0.1 mg/m³/20 min ceil
STEL ACGIH 0.04 ppm 0.4 mg/m³
IDLH 800 ppm

This is a highly explosive substance which, on explosion, liberates 1 705.3 cal/kg.

Production. Ethylene glycol dinitrate is produced by nitrating a mixture of glycerine and ethylene glycol in the same reactor, in the presence of sulphuric acid.

Use. Ethylene glycol dinitrate is a high explosive but it also has the property of lowering the freezing point of nitroglycerin. At present, in most countries with a temperate-to-cold climate, dynamite is made from a mixture of nitroglycerin and EGDN (the proportion of EGDN varies between 20-90% depending on the climate and the season).

HAZARDS

When EGDN was first introduced into the dynamite industry, the only changes noticed were similar to those affecting workers exposed to nitroglycerin, namely headache, sweating, face redness, arterial hypotension, heart palpitations and dizziness especially at the beginning of work, on Monday mornings and after an absence. Ethylene glycol dinitrate, which is absorbed through the respiratory tract and the skin, has indeed a significant acute hypotensive action. When cases of sudden death started to occur amongst workers in the explosives industry, no one immediately suspected the occupational origin of these accidents until, in 1952, Symansky attributed numerous cases of fatality already observed by the manufacturers of dynamite in the United States, the United Kingdom and the Federal Republic of Germany to chronic EGDN poisoning. Other cases were then observed, or at least suspected, in a number of countries, such as Japan, Italy, Norway and Canada.

Following a period of exposure which often varies between 6 and 10 years, workers exposed to mixtures of nitroglycerin and EGDN may complain of sudden pain in the chest, resembling that of angina pectoris, and/or die suddenly, normally on a Monday or a Tuesday, and most frequently between 4 and 7 o'clock in the morning, i.e. between 30 and 64 hours after termination of exposure, either during the sleep preceding awakening or following the first physical efforts of the day after arriving at work. Death is generally so sudden that it is not usually possible to examine carefully the victims during the attack. Emergency treatment with coronary dilators and, in particular, nitroglycerin, has proved ineffective. In most cases, autopsy proved negative or it did not appear that coronary and miocardial lesions were more prevalent or extensive than in the general population. In general, electrocardiograms have also proved deceptive. From the clinical point of view, observers have noted systolic hypotension, which is more marked during working hours, accompanied by increased diastolic pressure sometimes with modest signs of hyperexcitability of the pyramidal system and mental erethism; less frequently there have been signs of acrocyanosis–together with some changes in vasomotor reaction. Peripheral paraesthesia, particularly at night, has been reported and this may be attributed to arteriolar spasms and/or to peripheral neuropathy. Skin sensitisation has also been reported.

The mechanism of sudden death in chronic exposure to EGDN is controversial. First it was attributed to a coronary spasm occurring when the vasodilating effect has disappeared. Experimental investigations, by showing an increase in the level of catecholamines in the heart and heart hypersensitivity to adrenaline in chronically poisoned rats led to a more complex hypothesis according to which rhythm troubles, or possibly coronary spasm are due either to a sensitisation to cardioinhibitory reflexes or to a sudden release of catecholamines. Others suggested that the repeated absorption of nitro-compounds induces a chronic hypoxaemia by reducing heart oxygen consumption and producing methaemoglobin; when the coronary system ceases to be dilated, hypoxaemia is the cause of death. Direct effects, functional and structural, of nitroglycol on actomyosin have also been observed.

EGDN *in vivo* is metabolised mainly into ethylene glycol mononitrate, which in turn is converted into inorganic nitrates and ethylene glycol and excreted as such in the urine.

SAFETY AND HEALTH MEASURES

The prime consideration in the production and use of EGDN is the prevention of explosions: it is consequently necessary to adopt the same safety measures as those employed in the manufacture of nitroglycerin and in the explosives industry as a whole. Considerable progress in this respect has been achieved by remote control (by optical, mechanical or electronic means) of the most dangerous operations (in particular milling) and by the automation of numerous processes such as nitration, mixing, cartridge filling, etc. Arrangements of this type also have the advantage of reducing to a minimum both the number of workers exposed to direct contact with EGDN and the related exposure times.

In cases where workers are still exposed to EGDN, a variety of safety and health measures are necessary. In particular, the concentration of EGDN in the explosives mixture should be reduced depending on the ambient temperature and–in temperate-climate countries–it should not exceed 20-25% EGDN; during the warm season, it may be appropriate to exclude EGDN completely. However, too frequent changes in the EGDN concentration should be avoided in order to prevent an increased frequence of withdrawals. In order to reduce the inhalation hazard, it is necessary to control the atmospheric concentration at the workplace by means of general ventilation, and, if necessary, air induction; local exhaust ventilation may entail an explosion hazard.

In 1978 the NIOSH recommended an exposure limit of 0.1 mg/m³ of nitroglycerine or nitroglycol or a mixture of the two, measured as a ceiling concentration during any 20-min sampling period, and an action level of 0.05 mg/m³ for any sampling period. In certain Japanese companies work is stopped as soon as the atmospheric concentration exceeds 1.2 mg/m³. Careful control of the temperature of EGDN is extremely important in view of the product's volatility: it is recommended that 22 °C should not be exceeded in areas occupied by workers, and 32 °C in premises where processes are remotely controlled.

Skin absorption may be reduced by the adoption of suitable working methods and the use of protective clothing including polyethylene hand protection; neoprene, rubber and leather are easily penetrated by nitroglycol and cannot provide adequate protection. The equipment should be washed at least twice per week at the employer's expense. Personal hygiene should be encouraged and workers should shower at the end of each shift using a sulphite indicator soap to detect any residual traces of nitroglycerin/EGDN mixture on the skin; work clothing should be completely separated from personal clothing. Respiratory protective equipment may be necessary under certain circumstances (e.g. work in storage areas, etc.).

Medical prevention. This includes a pre-employment examination dealing with the general state of health, the cardiovascular system (electrocardiographic examination at rest and during exercise is essential), neurovegetative reactivity, urine and blood examination. Persons with systolic pressure higher than 150 or lower than 100 mmHg (19.5·10³ Pa), or diastolic pressure higher than 90 (11.7·10³ Pa) or lower than 60 mmHg (7.8·10³ Pa) should not in principle be considered fit for occupational exposure to nitroglycol. Pregnant women and women younger than 20 should also be considered unfit. In addition to periodic examinations, examination of workers returning to work after lengthy absence due to illness is necessary. The electrocardiogram should be repeated at least once a year.

All workers suffering from cardiac diseases, hypertension, hepatic disorders, anaemia or neurovegetative reactivity disorders, especially of the vasomotor system, should not be exposed to nitroglycerin/EGDN mixtures. It is also advisable to move to other jobs all workers who have been employed for more than 5-6 years on dangerous work, and to avoid too frequent a change in the intensity of exposure.

PARMEGGIANI, L.

CIS 79-120 *Criteria for a recommended standard–Occupational exposure to nitroglycerin and ethylene glycol dinitrate.* DHEW (NIOSH) publication No. 78-167 (National Institute for Occupational Safety and Health, 4676 Columbia Parkway, Cincinnati) (June 1978), 215 p. 164 ref.

"Explosive oils (Nitroglycerine, nitroglycol and allied compounds)" (Les huiles explosives (Nitroglycérine, nitroglycol et corps voisins)). Mouret, A. *Archives des maladies professionnelles, de médecine du travail et de sécurité sociale* (Paris), Dec. 1978, 39/12 (671-692). Illus. 53 ref. (In French)

"A cohort study on mortality among dynamite workers". Hogstedt, C.; Andersson, K. *Journal of Occupational Medicine* (Chicago), Aug. 1979, 21/8 (553-556). 19 ref.

"Letters to the Editor". *Journal of Occupational Medicine* (Chicago), Dec. 1978, 20/12 (789-792).

CIS 80-1983 "48-hour ambulatory electrocardiography in dynamite workers and controls". Hogstedt, C.; Söderholm, B.; Bodin, L. *British Journal of Industrial Medicine* (London), Aug. 1980, 37/3 (299-306). 27 ref.

Ethylene oxide

Ethylene oxide (C_2H_4O)

1,2-EPOXYETHANE; OXIRANE

m.w.	44.05
sp.gr.	0.87
m.p.	−112 °C
b.p.	10.4 °C
v.d.	1.5
v.p.	1 095 mmHg (142.3·10³ Pa) at 20 °C
f.p.	−6 °C oc
e.l.	3-100%
i.t.	429 °C

soluble in water, ethyl alcohol, ethyl ether

a colourless vapour at room temperature with an ether-like odour.

TWA OSHA	50 ppm 90 mg/m³
NIOSH	75 ppm/15 min ceil
TLV ACGIH	10 ppm 20 mg/m³
IDLH	800 ppm
MAC USSR	1 mg/m³

Ethylene oxide is manufactured by the catalytic oxidation (silver catalyst) of ethylene with air or oxygen. It is a high-volume chemical used primarily as an intermediate in the production of ethylene glycol, polyethylene, terephthalate polyester film and fibre, di- and triethylene glycol and other organic chemicals. Another major use for ethylene oxide is as a fumigant for foodstuffs and as a sterilising agent for heat-sensitive items in medical facilities.

HAZARDS

Ethylene oxide is highly exothermic and is potentially explosive when heated or in contact with alkali metal hydroxides or highly active catalytic surfaces.

Fire and explosion. As a result of its chemical reactivity and exothermic nature, the handling, storage and use of ethylene oxide presents potentially serious problems. The liquid is relatively stable; however, explosive vapour concentrations of from 3 to 100% are highly flammable. Air mixtures will ignite and explode when exposed to heat or open flames. Polymerisation is possible and is catalysed by anhydrous chlorides of iron, tin and aluminium as well as the oxides of iron and aluminium.

Health hazards. Reports of human exposure to ethylene oxide have primarily been associated with accidental high-level industrial exposures. Acute responses have been reported to include coughing, sustained periodic vomiting and irritation of the eyes, nose and throat. Delayed effects may include headache, nausea, pulmonary oedema, bronchitis, electrocardiographic abnormalities and urinary excretion of bile pigments.

Continuous exposure to low concentrations may result in a numbing of the sense of smell. Slight irritations of the conjunctivae of exposed workers has been reported; however, lacrimatory effects have not been observed. In addition to respiratory responses, skin contact with solutions as low as 1% of ethylene oxide have caused characteristic burns in workers and human volunteers. After a latent period of from 1-5 h, oedema and erythema have been observed followed by vesiculation and desquamation. Complete healing, even without treatment, has been reported, but in some cases residual brown pigmentation has occurred. Skin contact with undiluted ethylene oxide does not cause primary injury to dry skin; however, it can cause frostbite as a result of rapid evaporation. Skin irritation has been observed in workers wearing rubber gloves which had been contaminated with ethylene oxide. It has been suggested that the severity of skin burns is influenced by the length of time of contact and the strength of the ethylene oxide solution. The most hazardous aqueous solutions of ethylene oxide appear to be those in the 50% range. Repeated contact with aqueous solutions have been reported to cause skin sensitivity.

There are few studies available to evaluate the potential effects of long-term exposure to ethylene oxide. One study involving 185 workers engaged in the manufacture of ethylene oxide suggested that exposed workers had higher absolute lymphocyte counts when compared with a non-exposed control group. The difference in the two groups reportedly became smaller as the ventilation in the plant was improved. In addition to increased lymphocyte counts, the exposed group showed lower haemoglobin values and a few cases of slight anaemia. Certain cell abnormalities, three cases of anisocytosis (variation in the size of red blood cells) and one case of leukaemia were also seen in the exposed group but not in the control group. The US National Institute for Occupational Safety and Health has initiated an epidemiologic study of some 2 500 workers involved in manufacture and use of ethylene oxide. The study will include mortality, morbidity and possibly reproductive effects.

Acute and sub-chronic toxicity studies in experimental animals have demonstrated many of the effects seen in man including nausea, vomiting, skin irritation and slight irritations of the eye. No chronic test data or long-term carcinogenicity bioassays have been reported. Experimental attempts to induce mutations have been successful in at least 13 of 14 different species tested, the exception being a bacteriophage of *Escherichia coli*. These studies have demonstrated that several types of genetic damage may be induced by ethylene oxide. In addition to mutations, ethylene oxide has also been demonstrated to cause alterations in the structure of the genetic material in somatic cells of the rat and a covalent chemical reaction with DNA.

An additional hazard associated with exposure to ethylene oxide is the potential for the formation of ethylene chlorohydrin (2-chloroethanol), which may be formed in the presence of moisture and chloride ions. Ethylene chlorohydrin is a severe systemic poison and exposure to the vapour has caused human fatalities.

PREVENTIVE MEASURES

Because ethylene oxide is an extremely volatile, flammable liquid and its vapours form an explosive mixture with air over a wide range (3-100%), stringent safeguards should be taken to prevent fire and explosion. These safeguards should include the control of ignition sources, including static electricity; the availability of foam, carbon dioxide or dry chemical fire extinguishers (if water is used on large fires, the hose should be equipped with a fogging nozzle); the use of steam or hot water to heat ethylene oxide or its mixtures; and storage away from heat and strong oxidisers, strong acids, alkalis, anhydrous chlorides or iron, aluminium, or tin, iron oxide, and aluminium oxide.

Proper emergency procedures and protective equipment should be available to deal with spills or leaks of ethylene oxide. In case of a spill, the first step is to evacuate all personnel except those involved in the clean-up operations. All ignition sources in the area should be removed or shut down and the area well ventilated. Small quantities of spilled liquid can be absorbed on cloth or paper and allowed to evaporate in a safe place such as a chemical fume hood. Ethylene oxide should not be allowed to enter a confined space such as a sewer. Workers should not enter confined spaces where ethylene oxide has been stored without following proper operating procedures designed to ensure that toxic or explosive concentrations are not present. Whenever possible ethylene oxide should be stored and used in closed systems or with adequate local exhaust ventilation.

In areas where ethylene oxide is manufactured or used, proper sanitation practices should be followed. These should include the prompt removal and proper handling of clothing which becomes wet with ethylene oxide, and the availability of showers or work facilities so that contaminated workers can remove ethylene oxide from their skin. Where there is a risk that ethylene oxide may be splashed into the eyes, workers should wear protective goggles, and eye wash fountains should be available.

Continuing medical surveillance of exposed workers might include appropriate histories and examinations with emphasis on the pulmonary, neurologic, hepatic, renal, and opthalmologic systems and the skin. In addition the attending physician might consider the use of complete blood counts including a white cell count, a differential count, haemoglobin, and haematocrit.

Treatment. Affected workers should, as necessary, be removed from emergency situations and if the eye or skin has been splashed with aqueous solutions of ethylene oxide, they should be flushed with water. Contaminated clothing should be promptly removed. If the overexposure is severe, hospitalisation and observation for delayed onset of severe pulmonary oedema is advisable.

ROSE, V. E.

CIS 78-411 *Special occupational hazard review with control recommendations for the use of ethylene oxide as a sterilant in medical facilities.* Glaser, Z. R. DHEW (NIOSH) publication No. 77-200 (National Institute for Occupational Safety and Health, 4676 Columbia Parkway, Cincinnati) (Aug. 1977), 58 p. 139 ref.

CIS 79-1099 "Leukemia in workers exposed to ethylene oxide". Hogstedt, C.; Malmqvist, N.; Wadman, B. *Journal of the American Medical Association* (Chicago), 16 Mar. 1979, 241/11 (1132-1133). 14 ref.

CIS 80-184 "A cohort study of mortality and cancer incidence in ethylene oxide production workers". Hogstedt, C.; Rohlen, O.; Berndtsson, B. S.; Axelson, O.; Ehrenberg, L. *British Journal of Industrial Medicine* (London), Nov. 1979, 36/4 (276-280). 20 ref.

"Ethylene oxide" (157-167). 55 ref. *IARC monographs on the evaluation of carcinogenic risk of chemicals to man.* Vol. 11. *Cadmium, nickel, some epoxides, miscellaneous industrial chemicals and general considerations on volatile anaesthetics* (Lyons, International Agency for Research on Cancer, 1976), 306 p.

"Ministerial memorandum of 7.12.1979 on the use of ethylene oxide for sterilisation" (Circulaire du 7.12.1979 relative à l'utilisation de l'oxyde d'éthylène pour la stérilisation). *Journal officiel* (Paris), 10 Jan. 1980 (307-309). (In French)

Executives

It is difficult to give a definition of the occupation covered by the term "executive". However, for the purposes of this article, the term will be assumed to apply only to individuals in senior positions in line management or similar positions involving decision-making and responsibility for the implementation of these decisions. It is thus characteristic of an executive that his work requires mental rather than physical skills.

Nevertheless, certain executives, for example a works manager, may have a physical role. In these cases, they may also be exposed to the same sort of environmental hazard as the employees on the shop floor. It is easy to overlook such executives and leave them out of the routine precautions that apply to shop-floor workers.

HAZARDS

Due to the exigencies of his job, an executive obtains insufficient physical exercise. In addition, the necessity of business entertaining often leads to a high calorie intake and it is not surprising that many executives are obese. Amongst a sample of businessmen seen at the Medical Centre of the Institute of Directors in the United Kingdom, 30% of the subjects were 10% or more overweight. Moreover, because of the stresses of his work, the executive tends to smoke heavily and overindulge in the consumption of alcoholic beverages.

However, the disadvantages of his way of life are more than counterbalanced by the advantages of his socio-economic status and motivation to work. In the United Kingdom, the average company director has approximately one-third of the sickness absence of the average working man, and similar statistics have been recorded in the United States. In terms of mortality, the advantage is not so good. Studies of the standardised mortality ratio (SMR), which is the number of deaths that occurred in a certain subgroup divided by the number that would have occurred if the subgroup had had the same mortality rates as the whole group have shown that, in the United States, the SMR for managers aged 20-64 years is 89% of the whole population. In the United Kingdom the corresponding figure is nearly 95%.

It has been found that the diseases that are most common among United Kingdom executives are those of overindulgence such as cirrhosis of the liver related to high alcohol intake or coronary thrombosis, stroke and diabetes related to overweight. The manic personality that may lead to great success is also particularly liable to moods of dark despair, which may account for the high suicide rate (see figure 1).

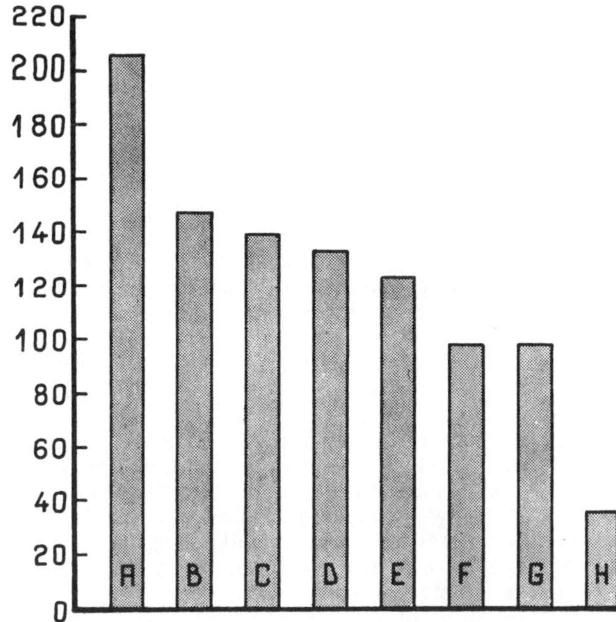

Figure 1. Standardised mortality ratios for different causes amongst higher administrative, professional and managerial workers in the United Kingdom. A. Cirrhosis of the liver. B. Coronary thrombosis. C. Accident and suicide. D. Diabetes. E. Strokes and hypertension. F. All causes. G. Cancer. H. Bronchitis.

Travelling is a major hazard of executive life and may take its toll in two ways. Firstly, constant exposure to the risk of accident inherent in travel accounts for the high mortality found from this cause. Many executives may drive more than 50 000 km a year or fly the equivalent of five to ten times round the world. Secondly there is the problem of travel fatigue which may be the result of repeated fairly short road journeys, especially where the executive drives himself, or may be due to time zone changes in long East-West or West-East flights. It may take the body as long as a week to adjust completely to time changes of 6 h and the performance of any executive, no matter how tough and accustomed he is to such journeys, will fall off markedly for the first couple of days after such a flight (see BIOLOGICAL RHYTHMS).

Apart from these major health hazards, many executives suffer from minor illnesses that are often related to the stressful life they lead. Overwork, excessive travelling, difficulties in interpersonnel relationships, lack of holidays, problems with finance, labour, raw materials, deliveries, production and sales all may cause stress and psychosomatic disease. Frustration may arise from slow promotion, lack of opportunities or inadequate job definition. Personal problems may be important. Wives may resent excessive travel or time spent in business entertaining. Many self-made men, particularly if they marry young, may find that their wives have great difficulty in adjusting to the new social circumstances of their husbands' executive life. All too often, a combination of these factors occurs.

Many symptoms may be related to stress. A study of 343 executives who were considered "stressed" by their examining physician revealed the following prevalence of symptoms: sleep disorders (34% of cases), dyspepsia (23%), fatigue (19%), headache (16%), tenseness or irritability (16%), work disturbance (8%), frank psychiatric disease (17%), miscellaneous disorders such as "allergy" diarrhoea, impotence, etc. (22%).

SAFETY AND HEALTH MEASURES

The preventive measures to be taken will be obvious from what has already been said. Exercise should be encouraged, particularly in any form that the man enjoys. Many companies have sports grounds, gymnasia or other facilities, and executives should be encouraged to use them by good example. Executives should also be encouraged to eat and drink only in moderation, and some companies in the United States provide weight-reducing lunches in their staff restaurants. Careful supervision of expense accounts may detect the heavy drinker. Entertainment of clients should be shared round and not always left to one or two people. It is the responsibility of senior management to establish traditions in a company that lead to health and happiness in their employees.

Much of the psychological and physical load on executives could be reduced by more rational planning of the work they are required to do, and this may often require the services of a management consultant to review procedures for reporting and decision taking. In many cases, it will be found that the executive can delegate much of his administrative work and can receive more of his information in digest form so that more time can be devoted to true executive functions.

Travel obligations should be analysed to determine whether the amount of travelling can be reduced or whether business trips can be better co-ordinated or delegated. Where possible, the use of the self-drive car should be discouraged so that travelling time can be used for work or rest. The amount of overtime worked by the executive should not be allowed to become excessive, and it should be ensured that all executives take at least one substantial holiday per year.

The heavy smoker (20 cigarettes or more a day) has, in terms of life expectancy, cancelled out all the medical advances of the last 30 years. As well as having about twice the chance of dying in any one year, the heavy smoker will, on average, suffer twice as much sickness absence as the non-smoker. All possible encouragement should therefore be given, at all levels, to stop people smoking, particularly cigarettes.

Periodical medical examinations are as important for executives as they are for shop-floor workers. Regular executive health examinations have been popular for some time in the United States, and their use is spreading throughout the world. Experience in the medical examination of over 10 000 executives, many of whom returned several times for repeat visits, has shown that 30% of executives are overweight, 25% are heavy smokers and 10% have an above-average blood cholesterol level (indicating a tendency to coronary thrombosis, stroke and arteriosclerosis). A further 20% had some other previously untreated condition which required treatment. It was found that a considerable number followed the advice of the examining physician and were found to be in better health on subsequent examination. Experience in the United States shows that people taking part in such a programme have a lower-than-expected mortality.

Diagnosis and treatment. Most of the conditions commonly encountered in executives are not specific occupational diseases and may be widely observed in the general population. Consequently they will be readily recognised and diagnosed by the executive's physician, and the necessary preventive or therapeutic measures can be initiated. However, the stress phenomenon, perhaps particularly that due to occupational causes, may be less well understood by the physician. A physician who is well acquainted with his executive patient's personality and work will have his suspicions aroused in the event of changes such as a fall-off in work quality. The executive who complains of a plurality of symptoms or overemphatically asserts that "nothing is wrong" should also arouse suspicion. Diagnosis of executive stress is best made by a psychiatrically oriented physician with an interest in this field and, preferably, direct knowledge of the work situation concerned.

The treatment of executive stress, and diseases related to it, is not solely the province of the physician. The man himself must have some insight into the outside situation and his own reaction to it. The company that employs him may help by improving his working conditions, by providing him with an assistant if necessary, by defining the subject's position and duties more clearly or, in more severe cases, by giving a period of sick leave. A frank discussion between the physician and an interested superior, of course with the patient's consent, may be most helpful.

PINCHERLE, G.
WRIGHT, H. B.

CIS 2767-1972 "Long-term radiotelemetric heart-rate studies and electrocardiograms of industrial managers" (Radiotelemetrische Langzeituntersuchungen von Herzschlagfrequenz und Elektrokardiogramm bei Führungskräften der Industrie). Schäcke, G.; Woitowitz, H. J.; Rietschel, E.; Havla, R. *Internationales Archiv für Arbeitsmedizin–International Archives of Occupational Health* (West Berlin), June 1972, 29/2 (142-158). Illus. 50 ref. (In German)

CIS 482-1968 "Serum uric acid concentration among business executives. With observations on other coronary heart disease risk factors". Montoye, H. J.; Faulkner, J. A.; Dodge, H. J.; Mikkelsen, W. M.; Willis III, P. W.; Block, W. D. *Annals of Internal Medicine* (Philadelphia), May 1967, 66/5 (838-850). Illus. 36 ref.

"Longitudinal evaluation of an exercise prescription intervention program with periodic ergometric testing: a ten-year appraisal". Owen, C. A.; Beard, E. F.; Jackson, A. S.; Prior, B. W. *Journal of Occupational Medicine* (Chicago), Apr. 1980, 22/4 (235-240). 28 ref.

"Occupational stress and managers". Cooper, C. L.; Melhuish, A. *Journal of Occupational Medicine* (Chicago), Sep. 1980, 22/9 (588-592). 13 ref.

Executive health. Goldberg, P. (New York, McGraw Hill, 1978), 272 p. Illus.

Exhaust systems

Local exhaust ventilation is widely used in industry to control hazardous air contaminants. By this method, the noxious agent is carried away with the air drawn from the immediate zone of contaminant production into a closely positioned exhaust hood. Control is thus obtained before the agent contaminates the worker's environment and it is in this basic purpose that local exhaust differs from general ventilation. The latter operates only after the offending substances has escaped into the workroom atmosphere, when a much larger amount of air is required to dilute the noxious agent to a safe breathing concentration (see CONTROL TECHNOLOGY FOR INDUSTRIAL HEALTH; VENTILATION, INDUSTRIAL).

Local exhaust ventilation is restricted in application, however, to processes with sharply defined sources of contamination which can be wholly or partially enclosed or to which closely positioned exhaust hoods can be applied. It is thus well adapted to mechanised processes which, in contrast to commonly scattered manual operations, usually have only one or two fixed points of contaminant release. Generally, too, these processes have high rates of generation of the hazardous substance and, again, local exhaust, with its greater capacity for collecting the offending material, provides a degree of control quite beyond the practical limits of general ventilation.

General principles

An atmospheric contaminant is dispersed from its point of release into the immediately surrounding atmosphere by its own kinetic energy. For gases, this occurs by molecular diffusion; particulates are projected through air in consequence of their velocity of release. This initial dispersion, however, is of very limited magnitude. Molecular diffusion velocities are quite slow and cannot begin to account for the extent to which gaseous contaminants are spread. The fine particles of hygienic interest ($< 10\ \mu m$) have little mass and, even with high initial velocity, the small amount of kinetic energy possessed by a particle is quickly used up in overcoming air resistance. A 1 µm quartz particle with an initial velocity of 5 000 cm/s, for example, will travel through air only 0.04 cm before its kinetic energy is exhausted.

One must therefore look for another dispersal force, apart from the kinetic energy of the offending agent, to account for the wide dissemination of both gaseous and particulate contaminants. The only alternative to movement through air is to be carried with it and this is, indeed, the practical mechanisms of contaminant dispersion. Without air movement through and away from the zone of contaminant production, there would be no significant spread, even into the operator's breathing zone, much less out into the general atmosphere of the workroom. Thus, the function of local exhaust ventilation becomes clear: it is to arrest the outward movement of the air which is carrying the contaminant and to draw it into the exhaust system. The design of an exhaust system, therefore, rests upon an understanding of the causes of air motion around hazardous processes and, in quantitative terms, upon the rate of air flow through the contamination zone and upon the directional pattern and velocity characteristics of the local air movement. Neither the nature of the contaminant nor its rate of release enters directly into the determination of exhaust ventilation requirements.

Design of exhaust systems

A system of local exhaust ventilation usually consists of a number of separate exhaust hoods applied to several different operations and connected by a system of branch pipes and main ducts to a central fan and common air cleaning plant and point of discharge. Each of these four parts–hoods, piping, air cleaning facility, and fan–requires separate consideration in the design of the over-all system. Of the four, the most critical in determining the effectiveness of the system is the design of the exhaust hoods and determination of their respective ventilation needs.

Hood design. The first step in the analysis and design of exhaust hoods is to identify the sources of air movement and to estimate the magnitude of air flow around each operation. Sources of air motion fall into three categories: (1) Inherent in the process. (2) Incidental to the process. (3) External to and independent of the process. The second step is to determine how and to what degree the sources of air movement can be modified, reduced in magnitude or even eliminated in order to reduce the demands for exhaust ventilation. There are many opportunities in industry to make such fundamental improvements by selecting the proper process equipment or by modifying existing equipment to minimise air motion caused by the process itself and by interposing barriers between the process and those incidental and external sources of air motion that are causing trouble. These opportunities should be pursued fully, for the potential benefits are greater than from the most careful design of the exhaust system itself. The third step is to determine the optimum design and placement of the exhaust hood in relation to the source of contamination and the required rate of ventilation to overcome the remaining air movement around the process.

An exhaust hood may take the form of an essentially complete enclosure, a semi-enclosure or booth, or an exterior hood placed near the point of contaminant release. The objective is the same in all three cases: to surround the contamination zone with a barrier (physical and/or aerodynamic) through which the contaminated air cannot penetrate.

Ventilation requirements. There are two requirements to be met in fixing the exhaust ventilation rate. First, it must exceed the rate of entering air flow induced by the process or otherwise created within the contamination zone. This is obvious since, with a lesser exhaust rate, some of the contaminated air is bound to escape. For certain operations, the rate of air displacement can be calculated and in others it may be measured. In most instances, however, the required exhaust rate has to be derived from past experience. The second requirement is that the air velocities established toward the exhaust opening must be high enough to overcome outflowing streams of contaminated air, which may occur with considerable velocity when the pattern of air displacement around the process is not uniform. Under this condition, contaminated air can escape, despite an exhaust rate in excess of the net rate of air flow through the contamination zone. The projected air streams decrease in velocity with distance, however, and finally the directional escape characteristic is lost by turbulent mixing with the surrounding air. Hemeon calls this phenomenon "splash"

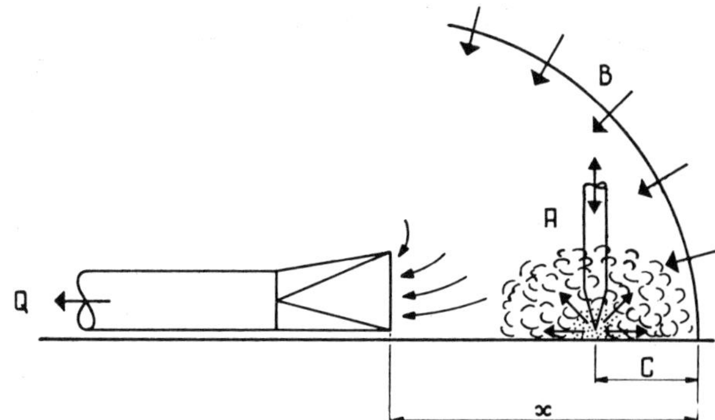

Figure 1. Estimation of exhaust ventilation requirement by Hemeon's method of noting limits of splash. Q must be sufficient to create necessary control velocity at distance x from hood to nullpoint interface. A. Dust-producing operation. B. Control velocity at nullpoint interface > room air currents. C. Nullpoint distance.

and the outward distance at which the directional jet action is lost the "nullpoint distance" (figure 1). Further escape of contaminated air beyond the nullpoint interface is caused by the drag of room air currents. Local exhaust ventilation must create an inward velocity across the interface in excess of external air currents to prevent such escape. When splash occurs within an enclosure, the dimensions of which are small compared with the nullpoint distance, jets of contaminated air will reach and impinge on the walls of the enclosure and some will escape through favourably positioned openings with considerable velocity. To prevent this an equal or greater velocity of inflowing air must be maintained.

Enclosures. A well enclosed process requires the lowest relative ventilation rate since the physical barrier of the enclosing walls provides the major protection. It is clear that the exhaust rate must not be less than the rate of induced air flow but, with carefully restricted openings, little, if any, excess air has to be drawn in to be sure that all the induced flow is captured. Because of this possibility of a nice balance between the induced and exhaust flow rates, however, it is essential that the tightness of the enclosure be maintained at all times, for, with increasing leakage, there will be a rapid loss of effectiveness. Enclosures must therefore be of sturdy construction. Access doors should be designed for easy and positive closure. Equipment bearings and other items requiring regular attention should be outside the enclosure to minimise the need for opening access doors.

Enclosures are particularly adapted to the various pieces of equipment involved in handling and processing solid materials such as stone, sand and ores. Elevators, screw conveyors, transport chutes and storage bins are commonly enclosed and crushers, grinding mills, screens, weighing devices, etc., are easily closed in. Within limits, it is possible to estimate the rate of induced air flow accompanying the movement of material through these handling and processing steps since the major cause of the air induction is the drag upon the surrounding air by the falling material. Hemeon presents equations for calculating rates of air displacement together with graphs and tabulated values for quick reference. In keeping with actual industrial experience, such calculations indicate that very substantial rates of air induction may be encountered, requiring exhaust ventilation up to 100 m^3/min or more on many pieces of equipment.

The splash phenomenon occurs in storage bins and other points where falling material comes suddenly to rest. To offset possible escape of contaminated air an inward air velocity through all wall openings of 100 cm/s is commonly recommended. With a reasonably tight enclosure this addition to the exhaust rate for the enclosure is small, however, compared with that needed to offset the induced flow and can often be ignored. Also, it is possible to minimise the splash effect within the closure and so reduce the need for extra ventilation.

Booths. Processes which require direct manipulation and close attention cannot be completely enclosed. A semi-enclosing exhaust booth with one side open is the next best arrangement. An exhaust booth possesses some of the advantages of a full enclosure and requires less exhaust air flow than an exterior booth. It should always be the aim of hood design to make use of such partial enclosures wherever possible. The booth must be deep enough not only to accommodate the process but also to keep the nullpoint distance well within it so that the splash effect does not extend out to the open face of the booth. A sufficient inward face velocity is required to prevent extraction of contaminated air by outside air currents. For relatively quiet operations in rooms with no unusual air movement, a face velocity of 25-40 cm/s may be sufficient. High velocities up to 100-150 cm/s are required for violent operations in bad locations with respect to room draughts. An exhaust booth is usually large relative to the operation carried on within it and the rate of ventilation needed to produce the required face velocity is correspondingly large compared with the rate of air displacement created by the process. Hence, the latter does not have to be considered in deriving the total exhaust rate, which is simply the product of the face area and the specified velocity through it. The design of exhaust booths for hot processes requires special precautions to prevent leakage in consequence of the thermostatic pressure which is built up within the booth.

Since some booths are relatively shallow, special care must be taken in the design of the exhaust connection at the rear of the structure to ensure a uniform velocity across the open face. Usually there is not space enough to permit a nicely tapered connection. Restrictive baffles are commonly located in front of the exhaust opening to distribute the flow.

Exterior hoods. There are two classes of exterior hoods: the receiving hood and the true exhaust hood. The first applies to processes with well directed movement of contaminated air, such as the rising air stream over a hot process or the induced air flow accompanying the stream of coarse particles projected from a heavy grinding operation. The opening is positioned to receive the air flowing toward it, the flow rate of which determines the minimum rate of exhaust ventilation required. In the presence of disturbing cross draughts in the room, however, extra ventilation may be required.

In contrast to the passive role of the receiving hood, the purpose of the air flow into a true exhaust hood is to envelop the zone of contamination with an aerodynamic barrier of sufficient velocity to prevent extraction of contaminated air by room air currents. Normally, an inward velocity of 25-40 cm/s across the nullpoint interface is sufficient for this purpose; higher velocities are needed with violent processes and unusual room disturbance. It is evident from figure 2 that the required rate of ventilation into an exterior hood will be quite unrelated to the volumetric rate of air flow through the contamination zone. The hood has very low aerodynamic efficiency since much air has to be drawn from wasted areas to ensure sufficient flow through the limited zone of the contamination.

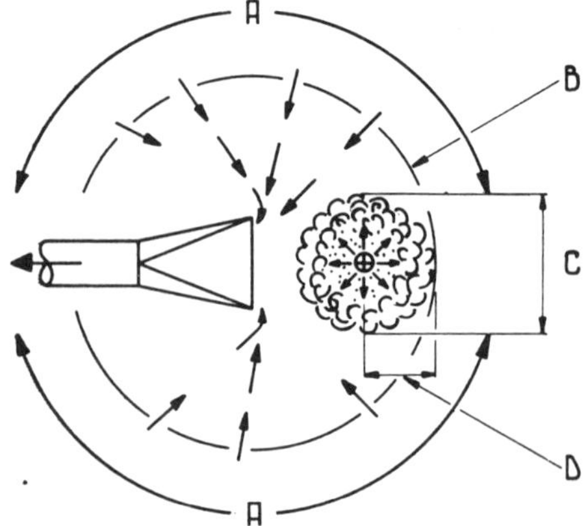

Figure 2. The efficiency of exhaust air flow is low when the dimensions of the contamination zone are small in comparison with the total zone from which air is drawn. A. Wasted air flow. B. Control velocity contour. C. Limits of useful ventilation. D. Nullpoint distance.

Aerodynamic characteristics of exterior hoods. In order to calculate the necessary exhaust ventilation rate into an exterior hood to create the required control velocity at the nullpoint interface, one must know the aerodynamic characteristics of exhaust openings. A hypothetical point source of suction in space will draw air equally from all directions and, because of this spherical distribution of flow, the air velocity toward the point source varies inversely with the square of distance. The ventilation rate required to produce a given velocity, v, at distance x is: $Q = 4\pi x^2 \cdot v$. About a hypothetical line source, the distribution of flow is cylindrical. Velocity varies inversely with distance and $Q = 2\pi Lx.\ v$, where L is the length of the line. Actual exhaust openings have real dimensions and the patterns of flow which they create conform only approximately to the theoretical flow toward point or line sources. Dallavalle's velocity contours for round, square and rectangular openings reveal significant departures from the theoretical flow pattern, especially at close-in distances, as seen in figure 3. The distribution of flow toward the hood and the relative velocity-distance relationship along a centre line out from the open face vary with the geometric shape of the opening but not with the size. Thus, for round and square openings, Dallavalle obtained the following approximate relations between velocity, v_x and distance, x:

$$v_x/v_o = \pi/(10\,m^2 + \pi) = 4/(10n^2 + 4)$$

where $m = x/a$; x = centreline distance out from hood face; a = radius of round opening; $n = 2x/b$; b = side of square opening; v_x = velocity at distance x; v_o = velocity through face of hood = Q/A_f; A_f = area of hood opening.

In the use of these equations it should be noted that the x distance from the hood face to the nullpoint interface is greater than the nullpoint distance itself, since the hood is located some distance beyond the contamination source (figure 2).

The flanged hood shown in figure 4 draws air from a hemisphere and, theoretically, should require only one-half of the flow required by an exhaust opening in free space to produce a given velocity at distance x. Actually, about two-thirds of the free flow is needed. A bench-top exhaust opening (as in figure 5–left) also draws its air from a hemisphere. An exhaust slot at the back edge of an open tank located against a wall draws air from only a quarter cylinder, thus requiring a fourth as much flow as a slot drawing from 360 ° (figure 5). By visualising the zone of influence of an exhaust hood in this manner, one can estimate air-flow requirements in comparison with

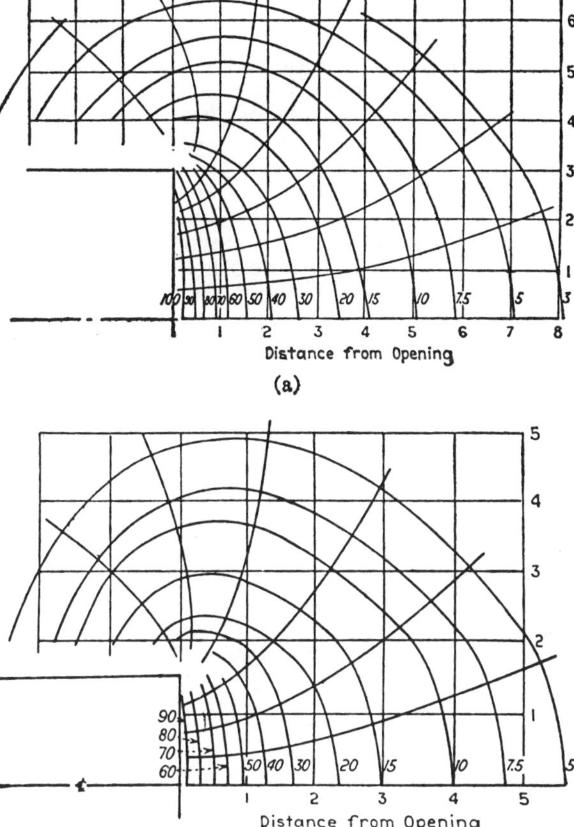

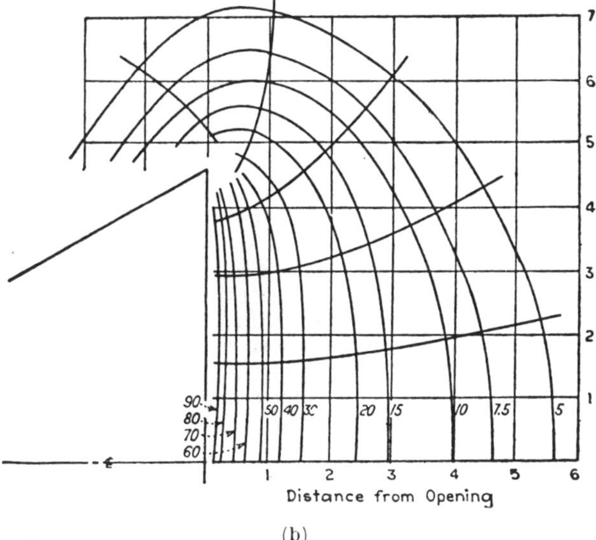

(b)

Contours in centre-plane, perpendicular to short side of opening.

Figure 3. Velocity contours and streamlines for: *(a)* circular hood openings; and *(b)* rectangular hood openings (1 : 3 ratio of sides). *(After Dallavalle.)*

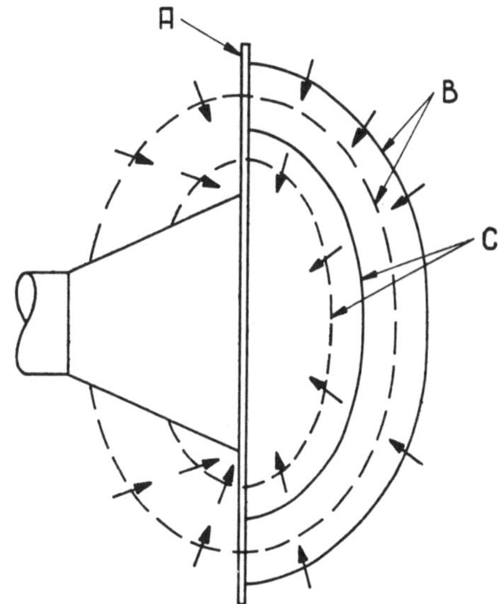

Figure 4. Velocity contours for flanged hood compared with plain opening, showing outward displacement of contours for flanged hood. A. Flange to eliminate wasted flow from rear. B. 10% contours. C. 30% contours; – – – Velocity contours for plain hood; ——— Velocity contours for flanged hood.

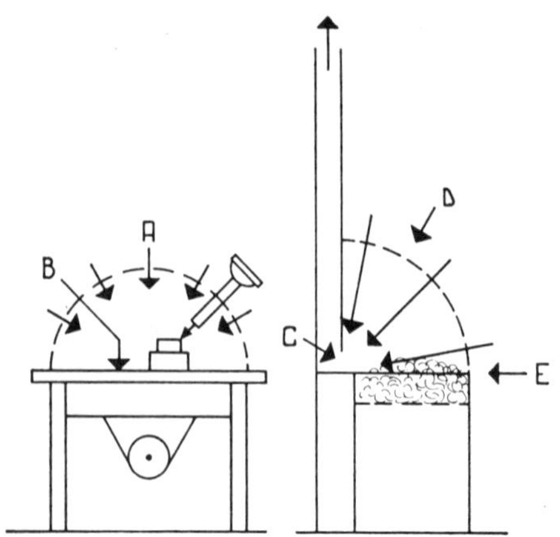

Figure 5. Adjacent plane surfaces exert a confining effect on the pattern of flow into exhaust openings: *on the left*, a down-draught hood on bench, with the work surface acting as a confining plane; *on the right*, an exhaust slot along the back of a tank draws air from only a quarter cylinder because of the confining vertical wall surface and horizontal tank surface. A. Control contour (half-sphere). B. Exhaust grill. C. Vent slot. D. Control contour (quarter-cylinder). E. Open tank.

hoods operating in free space. Hemeon and Dallavalle deal with this matter in considerable detail.

Limited effectiveness of exterior hoods. Since exhaust openings tend to draw air equally from all directions there is very little one can do by shaping an exterior hood to improve its aerodynamic efficiency. Within the limits permitted by the process, full use should be made of flanges and side baffles to minimise air flow from non-contaminated areas. The shape and size of the hood face should also conform reasonably to the dimensions of the contamination zone to take advantage of the relative flatness of the velocity contours close into the exhaust opening (figure 3).

The rapid decrease in velocity with distance outward from an exhaust opening is particularly to be noted. Hood face velocities above 1 500 cm/s are not desirable because of the high entrance pressure loss. A minimum control velocity at the nullpoint is, say, 25 cm/s. Entering these values of v_o and v_x in the equation above we see that the maximum effective control distance for a round opening is only slightly more than two opening diameters (= 4.3). For a slot, the comparable distance is 9.5 times the slot width. These calculations emphasise the need for locating an exterior hood very close to the point of contaminant release.

Standardised ventilation requirements

In the foregoing discussion, exhaust ventilation requirements were presented in terms of the two basic needs which are common to all situations. The practical usefulness of this approach is limited, however, because of the difficulty of identifying the sources of air motion around offending processes, estimating the rate of induced air flow through the contamination zone, visualising the uneven pattern of local air movement and locating the nullpoint interface across which a controlling velocity is to be established. Hemeon describes various procedures and aids for visualising and analysing the patterns and magnitude of air movement around offending processes. It is more common design practice to make use of recommended designs of hoods and standardised exhaust ventilation rates derived from trial and error experiences of the past.

Various official regulations and quasi-official recommendations specify minimum ventilation requirements for standard pieces of equipment, such as stationary grinders, woodworking machines, tumbling mills and other foundry equipment, abrasive cleaning and paint-spraying operations, degreasing and electroplating tanks. Exhaust rates vary with the size of the equipment in recognition of increasing ventilation needs as the magnitude of the operation goes up. These standardised ventilation rates have given satisfactory results, but only in situations where hoods and exhaust connections have also been of standard design. Difficulties arise otherwise because, with significant differences in hood design and in placement relative to the contamination source, ventilation rates, too, must vary in order to produce the required control velocity across the nullpoint interface. With this limitation in mind, such recommended ventilation rates as are found in the manual on industrial ventilation of the American Conference of Governmental Industrial Hygienists and in the various exhaust ventilation standards of the United States of America Standards Institute, as well as in the official health and safety codes of various governmental agencies, may be employed with reasonable assurance of satisfactory performance.

Exhaust piping

It is common practice to connect a number of exhaust hoods through a system of branch and main ducts to a central air cleaning plant and common fan and point of discharge. The sizes of the ducts must be selected to distribute the total air flow among the several hoods according to their respective needs. For dust-control systems it is also necessary to maintain an adequate transport velocity in the pipes to convey the dust without settlement to the central air cleaning plant (around 2 000 cm/s is needed for an average dust).

The simplest design procedure is to calculate the sizes of branch pipes and mains to ensure the proper transport velocity, assuming that the desired air flow will be obtained in each. A system so designed, however, will not usually give the desired distribution of flow. Branches located near the fan will operate above their needed ventilation rates while those at the remote end of the system will get less than their share. To offset this, adjustable gates are installed in the branches and when the system is in operation, they are positioned by trial to throttle down the excessive flow into the near hoods and thus increase the flow into the remote ones. Such gates are subject to rapid wear; they encourage dust settlement and get out of adjustment easily. This method for obtaining a balanced exhaust system is therefore not recommended. A more refined design procedure should be employed.

First, equipment should be located to permit, so far as possible, a symmetrical layout of pipes about the central fan, to minimise inequality in air flow resistance in the several branches. Long branches of low capacity are especially to be avoided because they impose extra-high resistance. However carefully the system is arranged, a certain amount of asymmetry is inescapable. To offset it, without using adjustable gates, a systematic reduction in the diameters of the favourably situated branches must be employed. Selecting the critical branch (usually the most remote) the pipe size is determined from the relation:

$$\text{pipe area} = \frac{\text{desired rate of air flow}}{\text{required transport velocity}}$$

For this flow rate, the total pressure drop through the hood, branch and main duct up to the next branch is then calculated. Similar calculations are made for the next branch up to its junction with the main, also assuming its specified air flow rate. If the pressure loss is less in the second case, then, by inspection and trial, a smaller pipe size is found which gives closest agreement, within the limits of available increments in diameter of fabricated ducts. The actual flow which will be established through the second branch of adjusted size, in joint operation with its upstream neighbour operating at its specified flow rate, is next determined, recognising that there will be equal pressure drops through the two sections up to their common junction. Calculations and adjustments in branch pipe sizes proceed in this manner up to the central fan, always assuming the desired rate of flow in the critical branch. A balanced system should provide rates of flow through the several hoods within ±25% of originally estimated needs. Hemeon presents convenient forms for use in developing the layout and calculating pipe sizes for a multibranch exhaust system, utilising this stepwise design procedure. His illustrative problems make the steps clear.

Air cleaning plant

It was rather common practice in the past to discharge contaminated air directly from exhaust systems to the outside atmosphere, letting wind and natural dilution take care of the atmospheric pollution problem. At most, crude cyclone dust collectors were employed only to remove coarse particles which would otherwise have settled immediately out of the air and created a local nuisance. Today, with widespread concern over community air pollution, industries must go to great lengths to remove both gaseous and particulate contaminants before discharging exhaust air to the outside.

There are so many different air cleaning requirements, depending upon the physical, chemical and toxic properties of the countless number of contaminants, and such a variety of types of air cleaning equipment available that no useful guidelines can be presented in brief form for the selection of the best types, either in general terms or for particular situations. The demands are so great today for control of industrial pollutants that this aspect of exhaust system design should be carried out in close association with an expert in the field of air-pollution control.

Fan selection

An exhaust fan must be selected to produce the rate of air flow required by the final layout and design of the system of exhaust piping. This will be greater than the sum of the originally estimated ventilation rates for the several hoods, depending upon the degree to which the distribution of flow into the hoods is out of agreement with those estimates. This flow must be developed against the total system resistance, including pressure losses through the air cleaning plant and discharge piping as well as those incurred in the hoods, branch and main exhaust ducts and accompanying fittings, such as elbows, branch-main junctions, etc. These two operating values, volume rate of air flow and over-all system resistance, are derived by standard engineering procedures. For the selected fan type, the appropriate size, operating speed and motor size to meet these specifications are easily obtained from fan manufacturers' catalogues.

Fan type. Over-all resistance in a dust-collecting system is usually high (> 10 cm, water gauge) because of the need to maintain a high pipe velocity, and for such heavy duty the so-called paddlewheel or radial-blade type of centrifugal fan is best adapted. The heavy steel or cast iron housing withstands abrasion, and corrosion-resistant coatings must be applied to protect against acid mists and other destructive agents.

Exhaust systems for collection of vapours and gases, on the other hand, can operate with low system resistance and this permits the use of high volume-low pressure fans, such as the squirrel-cage type. Exhaust booths for paint spraying are frequently equipped with simple propeller-type fans. It must be noted in both cases, however, that these fans are permissible only in the absence of air cleaning devices which impose substantial resistance. Modern demands for control of industrial pollutants make it necessary to use high-pressure exhaust fans in these situations as well as in dust-collecting systems.

The exhaust fan should be located downstream of the air cleaning plant to protect it against the abrasive action of the dust or corrosive action of the mist or gas which is being collected.

Exhaust system testing

Proof of satisfactory performance of an exhaust system rests upon the demonstration that an acceptable work environment has been established in which atmospheric concentrations of the hazardous agent are kept within safe tolerance limits. Periodic checking of the environment and even continuous monitoring by appropriate methods of atmospheric sampling and analysis are necessary to make certain that effective protection is being maintained and to detect signs of deterioration early enough so that repairs can be made before injury occurs. Such environmental checks by standard industrial hygiene procedures should be supplemented by appropriate programmes of medical supervision of the exposed workers, however, since the ultimate proof of a safe environment comes only from the demonstrated absence of injury among those exposed to the hazardous agent.

There are certain engineering tests and inspections that should also be applied, both to a newly installed exhaust system, as a check on its initial performance, and as a part of a routine schedule of inspection and maintenance of plant equipment. These are concerned mainly with the measurement of air flow through the several hoods to check on the suitability of air flow distribution and agreement with design expectations. In dust-collecting systems, a check should be made on pipe velocities to make sure that the ducts will remain free of dust settlement. Pipes and fittings should be inspected for signs of excessive abrasion. Performance of the air cleaning plant and of the fan should be measured at regular intervals for comparison with design specifications.

A regular programme of checking on the performance of an exhaust system has considerable indirect benefit. It calls attention to the importance of the protection which it provides and impresses upon management and workers alike the need to maintain the equipment and use it in the fullest degree.

HATCH, T.

See also AIR POLLUTION MONITORING EQUIPMENT in the Appendices.

Principles:

CIS 78-680 "Exhaust ventilation of dust-laden gas–Physical basis and technical execution" (Absaugung staubhaltiger Gase–Physikalische Grundlagen und technische Realisierung). Rausch, W. *Aufbereitungs-Technik* (Wiesbaden), Dec. 1977, 18/12 (659-669). Illus. 14 ref. (In German)

"Local exhaust ventilation principles". Socha, G. E. *American Industrial Hygiene Association Journal* (Akron, Ohio), Jan. 1979, 40/1 (1-10). 19 ref.

Practice:

CIS 77-1000 *Handbook of ventilation for contaminant control (including OSHA requirements).* McDermott, H. J. (Ann Arbor (Michigan), Ann Arbor Science, 1976), 376 p. Illus. 147 ref.

CIS 78-998 *The recirculation of industrial exhaust air–Symposium proceedings Oct. 6-7, 1977.* Research Triangle Institute, Research Triangle Park, North Carolina (National Institute for Occupational Safety and Health, 4676 Columbia Parkway, Cincinnati) (Dec. 1977), 152 p. Illus. 8 ref.

Fire prevention:

CIS 74-417 *Standard for the installation of blower and exhaust systems for dust, stock and vapor removal or conveying.* NFPA No. 91-1973 (National Fire Protection Association, 470 Atlantic Avenue, Boston, Massachusetts) (1973), 29 p. Illus.

Testing:

CIS 77-1302 "Recommended practices for testing open hoods and booths". Cutter, T. J. *Plant Engineering* (Barrington, Rhode Island), 16 Sep. 1976, 30/19 (138-139).

Explosives industry

An explosion may be the result of the release of energy from many different types of reaction, but the explosives industry is concerned with explosive effects that are obtained by chemical reactions. Explosives may be mixtures of combustible and oxidising substances, as in black powder, or they may be single substances, like nitroglycerin, having component groups within each molecule that function as fuel and oxidiser. In actual manufacturing practice, however, industrial high explosives are usually intimate mixtures either of solids or of solid and liquid materials that are susceptible to violent, exothermic gas evolution by detonation following an applied, localised shock initiation of the process (see EXPLOSIVE SUBSTANCES).

The following rather wide range of basic characteristics typifies commercial high explosives:

(a) detonation velocity–1 700-9 200 m/s;

(b) density–0.7-1.9 g/cm^3;

(c) energy of reaction–750-1 500 cal/g (3 135–6 270 x 10^3 J/kg).

The formerly important, deflagrating type of explosive, black blasting powder, develops pressure and energy very slowly by comparison with the detonating type or high explosives and usually is initiated by an incendiary device.

Uses. Explosive energy is as essential as other forms that are used so extensively. Without explosives, it is inconceivable that we could obtain the desired amounts of fossil fuels, metallic ores and rock, or accomplish huge construction projects such as tunnels, dams, and roads.

From the time of the introduction of gunpowder in warfare, explosives of different chemical composition have been produced for many military and civilian applications. Each newly proposed type of application provided reason for developing uniquely effective, economically attractive and safe explosives. Explosive compositions have been varied to produce properties ranging from the slow propellant action of deflagrating explosives to the different shattering or shock effects of detonating explosives.

Explosives are no longer used only for disruptive action but also for metal forming, welding, and cladding and for high-pressure transformations that yield valuable products, including fine diamonds for grinding and polishing.

Blasting explosives

Consideration of safety, cost, and effectiveness has caused black powder to be superseded for most but not all purposes by dynamite, which in turn has been successfully challenged by mixtures based on fuel-sensitised ammonium nitrate, mainly because of safety and cost considerations.

Black powder

The ingredients of this mixture, which is known more colloquially as gunpowder, are charcoal, sulphur and potassium nitrate. It deflagrates at about 450 m/s and produces large amounts of the toxic products, carbon monoxide and hydrogen sulphide. The heat of reaction is about 700 cal/g.

Production. Skill and suitable equipment are needed to produce the different, uniform types of black powder. Both granular and pellet types are marketed.

Before 1900, the production of black blasting powder in the United States greatly exceeded that of high explosives. Production kept up with the increasing demand, but the displacement of black powder by the more effective dynamite was evident and almost continual after 1917 when the consumption of black powder was at a maximum of about 126 million kilograms. By 1968 the continuing decline had brought black powder production down to 194 000 kilograms.

Nitroglycerin

See NITROGLYCERIN.

Ethylene glycol dinitrate

See ETHYLENE GLYCOL DINITRATE.

Cellulose nitrate ($C_6H_7O_2(ONO_2)_2$, empirical)

NITROCELLULOSE; NITROCOTTON; GUNCOTTON

A cotton-like or pulp-like material of variable composition, the products containing approximately 12.5-13.5% nitrogen being used in the manufacture of some dynamites. Cellulose nitrate is obtained by treating cellulose, in the form of linters, cotton waste, etc., with a mixture of nitric and sulphuric acids, the excess of which is removed by washing.

Dynamite

A range of dynamites to serve the requirements of different blasting operations is prepared by mixing nitroglycerin or a mixture of nitroglycerin and ethylene glycol dinitrate with other ingredients (figure 1).

The nitroglycerin is absorbed on a mixture of one or both of the oxidising agents, ammonium nitrate (NH_4NO_3) and sodium nitrate ($NaNO_3$), with carbonaceous combustibles. The oxidiser-combustible mixture provides a source of low-cost energy to supplement that from the nitroglycerin. Explosives high in nitroglycerin content are made by using cellulose nitrate to gelatinise or reduce the fluidity of the nitroglycerin. The stiff, dense mixture of 92/8 NG/NC, called blasting gelatin, is seldom used today. Plastic cohesive, extrudable compositions that contain over 20% nitroglycerin, carbonaceous combustible, minor amounts of cellulose nitrate and large but appropriate amounts of inorganic nitrates are called gelatin dynamites, or gelatins. The so-called straight gelatins contain the largest amounts of nitroglycerin; sodium nitrate is the only inorganic nitrate and oxidiser. The so-called ammonia gelatins may contain both ammonium nitrate and sodium nitrate, the largest amount of ammonium nitrate being present in the highest-strength grades. Excellent water resistance and high density make gelatins uniquely useful. Semi-

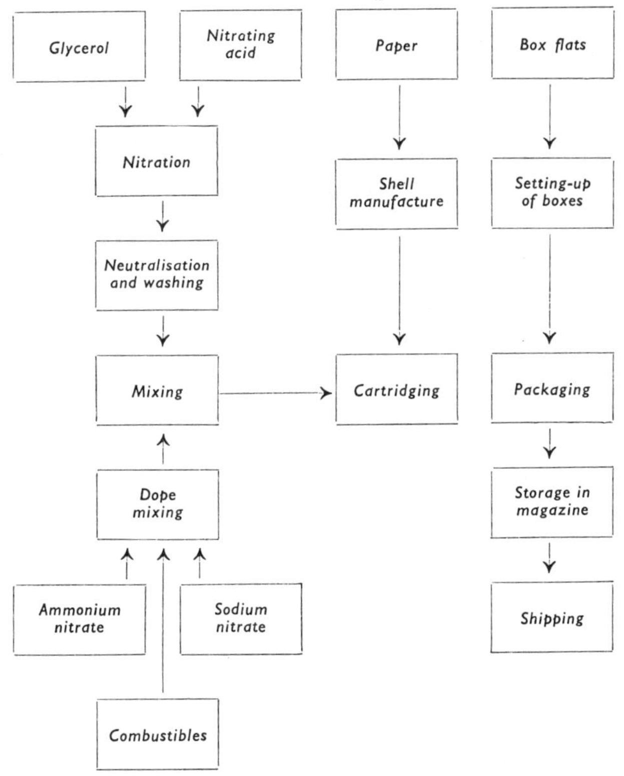

Figure 1. Materials and processes employed in the manufacture of dynamite.

gelatins are a non-extrudable type of moderately cohesive composition that contain less nitroglycerin than gelatins, a large amount of ammonium nitrate and a small amount of cellulose nitrate. Moderate cost and significant water resistance favour their use. Many dynamites, most of which contain no cellulose nitrate, contain appreciable amounts of ammonium nitrate and only 12-15% nitroglycerin. These less expensive products are used wherever possible. The density of these pulverulent dynamites tends to be low, about 0.7-1.3 g/cm³.

Fuel-sensitised, safe explosives

It was known for many years that intimate low-density mixtures of fine ammonium nitrate with hydrocarbons, carbohydrates and other non-explosive materials could be detonated. The cost and safety features of this type

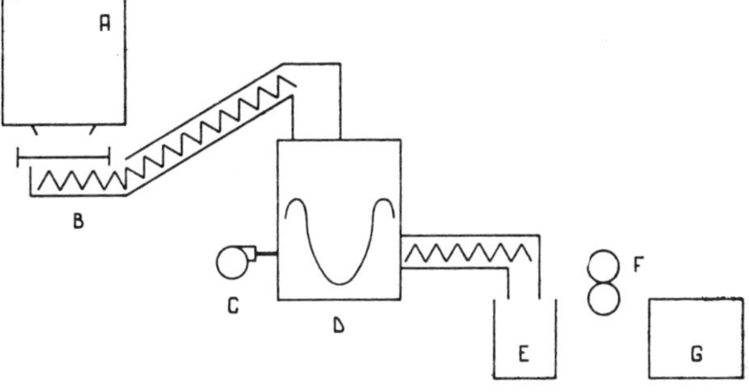

Figure 2. Process flowchart for a plant manufacturing ammonium nitrate/fuel oil (AN/FO) explosive. A. Hopper car for the delivery of ammonium nitrate. B. Unloading conveyor. C. Fuel-oil feed pump. D. Ammonium nitrate/fuel oil mixer. E. Bag filling machine. F. Bag stitching machine. G. Finished multiwall bag containing a mixture of 6% fuel oil and 94% prilled ammonium nitrate.

and slightly modified formulae were attractive, but density was low and water resistance was poor.

Very greatly accelerated use of a combustible-sensitised ammonium nitrate started in 1955 when mixtures were packaged in plastic, waterproof bags for use in large-diameter boreholes, mainly in coalmining stripping operations. Within three years, porous prilled ammonium nitrate, a product of the fertiliser industry, which had a rather large particle size, was tested extensively in both Canada and the United States in approximately a 94/6 ammonium nitrate and fuel oil mixture (figure 2). The mixture, known as AN/FO, was not desensitised by moderately damp conditions, and in wet operations was often protected adequately by enclosure in polyethylene bags. Later on the majority of blasting in the United States was done with AN/FO and similar ammonium nitrate types of explosives.

Military explosives

During the two World Wars, great advances were made in the development and production of new substances and mixtures for military explosives.

Trinitrotoluene

See TRINITROTOLUENE.

Cyclotrimethylenetrinitramine

RDX; CYCLONITE; HEXOGEN

m.w. 222.15
sp.gr. 1.82
m.p. 203.5 °C
a white, crystalline compound.
TWA OSHA 1.5 mg/m³ skin
STEL ACGIH 3 mg/m³
MAC USSR 1 mg/m³

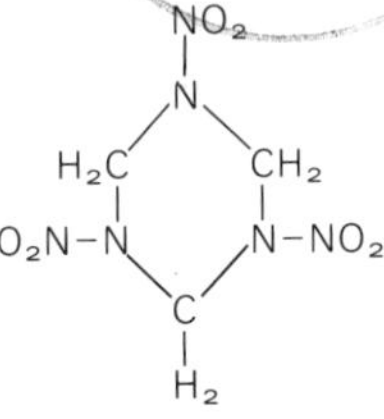

RDX is more powerful as an explosive than TNT with which it is mixed to form a castable mixture often containing 60% RDX and 40% TNT. This explosive is obtained by the nitration (with nitric acid) of hexamethylenetetramine, which is formed by the reaction between ammonia and formaldehyde.

Trinitrophenylmethylnitramine

See TETRYL.

Pentaerythritol tetranitrate

PETN

m.w. 316.2
sp.gr. 1.77
m.p. 140-141 °C
decomposes above 150 °C
explodes at 205-215 °C
a white crystalline substance.

$$C(CH_2-O-NO_2)_4$$

This explosive is made by the esterification of pentaerythritol with nitric acid. It is almost as powerful as RDX and is used mainly in detonating fuses.

Lead azide ($Pb(N_3)_2$)

m.w. 291.3
colourless, crystalline needles.

Lead azide is made by the reaction of sodium azide and a lead salt. It is a very sensitive explosive which must be handled and transported with extreme care, and is used as the primary explosive to detonate the above substances.

Mercury fulminate ($Hg(CNO)_2$)

MERCURIC CYANATE

m.w. 284.7

sp.gr. 4.42
m.p. explodes
a grey, crystalline powder.

It explodes when dry under the slightest friction or shock and must therefore be kept wet until used. It is made by treating mercury with strong nitric acid and ethyl alcohol. Like lead azide it is used as a primary explosive to detonate the less sensitive, secondary explosives.

Propellants

A gun or rocket propellant must not detonate; its function is to produce gas at a desired, moderate rate. Black powder was the first propellant used in military and sporting guns. In most countries the propellant powder now used is smokeless, which consists mainly of nitrocellulose, with or without nitroglycerin. Rocket propellants may be liquid or solid. The liquid propellants for this purpose may consist of an oxidiser like liquid oxygen or nitric acid and a fuel like kerosene or hydrazine. The solid may be a mixture of an oxidiser like ammonium perchlorate and a fuel like thiokol. Many rocket preparations, however, are not representative of the normal products of an explosives factor and are dealt with in the article ROCKET PROPELLANTS.

HAZARDS AND THEIR PREVENTION

Many of the substances manufactured in explosives factories are both toxic and explosive so that regulations must be enforced to minimise both hazards to workers. The explosion risks are paramount, however, and precautions against toxic effects must be effective without in any way jeopardising the provisions for controlling the risk of accidental explosion.

Fire and explosion. Valuable experience in the safe, efficient production of high-quality black powder is, to a degree, applicable to the production of other explosives.

Plants should be sited away from populated areas. To limit the size and the effects of an explosion, it is customary to conduct the operations in a number of small buildings or sheds instead of in one large building. The number of workers and the quantity of explosive material in any one shed is predetermined and firmly limited to that amount. The sheds are of light construction so that an explosion may be readily relieved and be unaccompanied by the flight of heavy missiles comprising structural parts. To prevent the shock, flame and flying fragments associated with an explosion in one shed from starting an explosion in a neighbouring one, the sheds are separated from one another by distance and by mounding which are determined by certain principles based on experience and calculation. Provision should also be made for the directed venting of an accidental explosion. The materials chosen for the structure of the building or for the fittings should be those least likely to be associated with the production of sparks.

Workshops and magazines must be adequately protected against lightning; all safety devices should be inspected periodically and any defect remedied immediately.

Operations should be separated and automated to the maximum possible extent; remote handling equipment and automatic safety devices should be installed. Tools should be made of non-sparking materials and protected from any contact with dust and grit that might produce sparks. Energy sources and methods of applying energy to machines should be designed, used and maintained with due regard to the danger of fire and explosion. It may be that the use of electrical equipment is ruled out by the danger of sparking or flames.

Grit may be a cause of frictional ignition and the floors of sheds and workrooms should be kept scrupulously clean. Special rules about acceptable types of footwear and clothing, and the provision of shoe-cleaning mats prevent the bringing of exposed metal fasteners, loose metal and gritty dirt into explosives buildings. Electrostatic discharge and ignition of the highly sensitive explosives by persons is prevented by the wearing of shoes having conductive soles and by the electrical grounding of both the conductive flooring and the equipment. Smoking and the bringing of incendiary devices into explosives areas is forbidden.

All the usual precautions for a chemical factory should be systematically observed, including those concerning safe entry into confined spaces. The precautions to ensure that a vessel is purged of flammable or explosive contents before heat is applied for welding or cutting are particularly important in an explosives factory.

In large, well run factories of this nature these precautions are systematically codified and enforced without exception on all ranks of personnel.

In many industrialised countries the precautions have now been codified in the form of regulations or of conditions that must be satisfied before a licence to manufacture is granted. These regulations or conditions are numerous and many have a special application to a particular process or substance.

The training of the workers in the purpose of the safety measures is always important, and it is of the very greatest importance in an explosives factory. Personnel selection procedures should also ensure that people suffering from epilepsy or mental disorders, and handicapped and disabled workers, are not employed on work where there is an explosion hazard.

Health hazards. Many of the substances used and produced in an explosives factory can be harmful to health either by inhalation or by skin absorption. Exposure to the vapour of nitroglycerin or to that of the more volatile ethylene glycol dinitrate, or skin contact with these substances can cause violent headaches. Although a certain tolerance develops on continued contact, fatal cardiovascular accidents have been associated with prolonged exposure to these substances. In poor hygienic conditions, serious poisoning fatalities have occurred to workers engaged in the manufacture of trinitrotoluene. Absorption by inhalation and by skin contact may give rise to toxic hepatitis or to aplastic anaemia. Tetryl appears to be less harmful in this respect but it has been responsible for a high incidence of dermatitis, nasal irritation and epistaxis. In the absence of precautions, exposure to dust and fume (inhalation and skin contact) during nitration in the manufacture of cyclotrimethylenetrinitramine has been known to cause irritation, dermatitis and convulsions. In addition to the toxic hazards associated with the products, there are risks associated with the process. The strong acids used in nitration processes may cause burns, and acid fumes are sometimes present in the atmosphere.

Harmful dust and vapour must be prevented, by suitable ventilation arrangements, from accumulating in the working atmosphere. The principles of enclosure and exhaust ventilation are applicable to most of the processes, but the necessity to ensure that explosion precautions are not jeopardised often complicates the ventilation arrangements. It may not be safe to exhaust the vapour or dust of an explosive material through a rapidly revolving ventilation fan and consequently it may often be necessary to have recourse to an ejector or venturi system (to create the necessary draught).

Personal protective equipment including eye protective equipment should be provided for normal process work as protection against eye splashes, acid burns, dermatitis and skin absorption of toxic materials. Personal hygiene is particularly important when harmful nitro-compounds are being produced and manipulated.

In large manufacturing plants where the work is highly organised, all these precautions form a body of internal regulations rigorously imposed on everyone from the management downwards.

WINNING, C. H.
BANERYD, K.

Safety:

CIS 79-339 *Explosives and materials containing explosives (Explosivstoffe und Gegenstände mit Explosivstoff).* VBG 55a + commentary, 67 + 22 p. VBG 55b, 8 p. VBG 55c, 9 p. VBG 55d, 4 p. VBG 55e, 6 p. VBG 55f, 22 p. VBG 55h, 8 p. VBG 55i, 15 p. VBG 55j, 6 p. VBG 55k, 35 p. VBG 55l, 7 p. Hauptverband der gewerblichen Berufsgenossenschaften (Cologne, Carl Heymanns Verlag, 1978). (In German)

CIS 76-1167 *Safety in the Swedish explosives industry—Alternative approaches to safety at work and some results from an attitude study of explosives workers and supervisors.* Baneryd, K.; Kjellén, U. FOA report A 20007-D1 (Försvarets forskningsanstalt, Huvudenhet 2, Stockholm) (Sep. 1975), 42 p. Illus. 22 ref.

Medical surveillance:

CIS 74-0275 "Periodic medical examinations in the explosives industry" (Ärztliche Kontrolluntersuchungen in der Sprengstoff-industrie). Edler, K. *Sichere Arbeit* (Vienna), Dec. 1972, 25/4 (16-19). 9 ref. (In German)

RDX, HMX:

CIS 75-1679 *The toxicology of cyclotrimethylenetrinitramine (RDX) and cyclotetramethylenetetranitramine (HMX) solutions in dimethyl sulfoxide (DMSO), cyclohexanone, and acetone.* McNamara, B. P.; Averille, H. P.; Owens, E. J.; Callahan, J. F.; Fairchild, D. G. AD-780 010/5W1 (National Technical Information Service, Springfield, Virginia) (Apr. 1974), 108 p. Illus. 75 ref.

Explosive substances

Release of energy in a rapid and uncontrolled manner gives rise to explosions. The energy thus released may appear as heat, light, sound and mechanical shock, though in certain cases not all of these forms may be observed. Most frequently the source of energy for an explosion is a chemical reaction, but explosions may also result from the release of mechanical energy or of nuclear energy; examples are respectively the bursting of an overstressed boiler and the fission explosion when a quantity of fissile material is suddenly brought to a critical condition.

Any flammable dust, vapour or gas mixed with air or other supporter of combustion under the proper conditions will explode when ignited. Combustible substances which are capable of exploding may be:

(a) finely divided combustible solids, including certain metals, in the form of dusts or powders;

(b) vapours of flammable liquids; and

(c) flammable gases.

The three prerequisites for an explosion of this nature are:

(a) a combustible material;

(b) air or any other supporter of combustion; and

(c) a source of ignition or temperature above the ignition temperature.

Because of their chemical constitution, certain substances are unstable and may explode, as authorised explosives explode, when they are subjected to percussion, friction or heat. Thus, serious explosions of iodoso- and iodoxy-compounds have occurred in the photographic and pharmaceutical industries, azo- and nitro-compounds have exploded when produced as intermediates in the dyestuffs industry, and there have been accidental explosions of organic peroxides, like benzoyl peroxide, that have been used for bleaching flour and to initiate polymerisation in the manufacture of plastics. In the absence of precautions, the constituents of some mixtures, like thermit mixtures, are capable of reacting with explosive violence.

In general, the products of explosion resulting from chemical reactions are gases or mixtures of gases and solids. Thus, acetylene yields only gaseous products, whereas "black powder" yields both solids and gases. In several cases, such as that of magnesium dust, the products are only solids. The gaseous products of explosion generally have a volume much greater than that of the explosive substance.

Invariably, an explosion results in the liberation of a considerable amount of heat that is sufficient to cause propagation of the explosion from the initial point or area to the entire mass. The liberated heat greatly raises the temperature of the products of explosion, thereby developing a higher pressure than can be applied to doing work. In fact, the work done by an "explosive" is determined primarily by the amount of heat given off during the explosion.

Explosive limits

In the case of most flammable liquids, gases and dusts, there is a minimum concentration of vapour, gas or dust in air or oxygen below which the propagation of flame does not occur on contact with the source of ignition. There is also a maximum proportion of vapour, gas or dust in air above which propagation of flame does not occur. These limit mixtures of vapour, gas or dust with air, which if ignited will just propagate flame, are known as lower and upper explosive or flammable limits. In the case of vapours and gases these limits are usually expressed in terms of percentage volume of vapour or gas in air. In the case of dusts the limits are usually expressed as the weight of dust in a given volume of air.

To determine the explosive limits of a gas or vapour, known concentrations of the gas or vapour in air are prepared by passing, at a flow rate controlled by a rotameter or other metering device, each of the components of the mixture into a combustion chamber of the test apparatus. An electric spark or arc ignites the mixture when the lower flammable limit is reached and flame propagates through the mixture. Flame propagates with increasing speed and increasing rate of pressure rise until the optimum explosive concentration is reached. Thereafter, the speed of propagation decreases until the upper limit is reached after which the flame propagation ceases to take place.

The concentration of a flammable gas or vapour is usually measured by an explosimeter in which the oxidation of the flammable gas or vapour by the catalytic effect of one of the resistance wires of a galvanometer raises the temperature of the wire and thus alters its resistance.

The lower limit of a flammable dust is determined by introducing increasing quantities of dust into the test apparatus, though a constant volume of air is injected at each test. At the lower flammable limit, ignition is effected by an electric spark or arc in one type of apparatus, or by the hot silica wall of the combustion chamber in another apparatus. Because of the difficulty of maintaining a homogeneous dust cloud at high concentrations, upper limits have not been determined with accuracy for flammable dusts.

Limits such as these are valuable for indicating the relative explosion hazard of different materials, so that in certain circumstances it may be possible, for example, to replace a solvent which gives off highly explosive vapours by a substance that is less explosive; they may also indicate the need to institute explosion prevention measures.

Classification of explosive substances

Explosive substances are best classified according to the rates of pressure rise generated by their mixtures with air during explosions, as this is the governing factor to be taken into consideration in devising suitable control measures. Table 1 contains the classification of explosive materials according to National Fire Codes of the United States.

Table 1. Classification of explosive materials according to rates of pressure rise

Class A materials (slow rates of pressure rise)

Metal dusts	*Miscellaneous dusts*
Antimony	Anthracite
Cadmium	Carbon black
Chromium	Coffee
Copper	Coke, low volatility
Iron (impure)	Graphite
Lead	Leather
Tungsten	Tea
Vapours	
1,2-Dichloroethane	

Class B materials (medium rates of pressure rise)

Metal dusts or powders	Phthalic anhydride and its resins
Iron (carbonyl, electrolytic or hydrogen reduced)	Polyethylene
Manganese	Polystyrene
Tin	Urea resins
Zinc	Urea-melamine resin
	Vinyl butyral
Vapours	
Propylene dichloride	*Miscellaneous dusts*
	Bituminous coal
Dusts of grains, spices, etc.	Cork
Alfalfa	Calcium lignosulphonic acid
Cocoa	Coumarone, indene
Grain dust and flour	Dextrin
Mixed grains	Lignin
Rice	Lignite
Soy bean	Peat
Spices	Powdered drugs
Starch, dextrines	Pyrethrum
Yeast	Shellac
Plastics dusts	Silicon
Cellulose acetate	Sulphur
Methyl methacrylate	Tung
Phenolformaldehyde	Woodflour

Class C materials (fast rates of pressure rise)

Metal dusts	*Vapours and gases*
Aluminium	Acetone
Stamped aluminium	Methylethyl ketone
Magnesium	Ethers
Magnesium-aluminium alloys	Alcohols (methyl, ethyl, isopropyl and butyl)
Titanium	Hydrocarbons
Zirconium	Gasoline
Some metal hydrides	Acetylene
	Ethylene
	Carbon disulphide
	Hydrogen

Adapted from: National Fire Protection Association, National Fire Codes, Vol. 9: *Occupancy standards and process hazards; Guide for explosion venting*, NFPA No. 68 (National Fire Protection Association, Boston, Massachusetts).

In general the rates of pressure rise developed by gas-air mixtures are higher than those of dust-air mixtures.

The classification given in table 1 would indicate that the use or production of many types of materials is attended with serious explosion risks. These substances are also widely used in industry. Naturally, the explosion risk is present in many industrial operations. However, some of the important industries which present a serious explosion hazard are mentioned in table 2.

Table 2. Explosion hazards in important industries

Industry	Explosion hazard
Woodworking	Wood dust from sanding machines, etc.
Oils, fats and waxes processing	Unsaturated oils (hydrocarbons)
Solid and liquid fuels processing	Pulverised coal, hydrocarbons and alcohols
Lacquer and varnish	Alcohols, esters, ethers, glycols and petroleum naphtha
Viscose rayon	Carbon disulphide
Rubber	Benzene and other aromatic hydrocarbons
Pharmaceutical	Alcohols, ethers, esters, unstable substances, finely pulverised drugs
Alkalis and heavy chemicals	Hydrogen
Plastics	Formaldehyde, solvents, nitrocellulose, casein and moulding powders
Paint	Aluminium powder, phthalic anhydride and solvents
Metal spraying	Zinc or aluminium in a finely divided state
Paper	Cellulosic material, volatile solvents
Printing	Ink solvents
Lineoleum	Cork and wood flours, unsaturated oils and petroleum naphtha
Textile	Waterproofing with oils such as linseed oil in petroleum naphtha, and coating with flammable rubber and plastics solutions
Dyeing and bleaching	Degreasing solvents such as benzene

HEALTH HAZARDS

Many of the explosive substances which are considered here are also toxic in nature and can present serious hazards to health on exposure. This aspect has been considered under appropriate headings elsewhere.

SAFETY MEASURES

Explosion prevention

This will be directed primarily at preventing the formation of flammable mixtures and/or removing sources of ignition.

Flammable mixtures. It is essential to prevent the formation of explosive mixtures of dusts, vapours or gases with air, particularly in rooms or buildings in which hazardous operations are carried out. This could be accomplished by preventing the escape of dusts, vapours and gases by the provision of good mechanical ventilation, by an effective dust-collection system, by addition of some suitable inert materials such as rock dust or inert gas, depending upon each individual circumstance, and by proper attention to housekeeping. When the risk is that of a dust explosion, experience has indicated that the most disastrous explosions have been secondary explosions in workrooms that have been initiated by a small primary explosion in the plant, which has disturbed and ignited loose dust on ledges, beams, etc. This loose dust may accumulate in spite of the ventilation and such accumulations must be prevented by the scrupulous cleaning of rooms housing dusty plant—preferably by means of suitable vacuum cleaners. (When any material is in a finely divided state, the surface area available for chemical reaction is considerable in magnitude. See DUST EXPLOSIONS.)

Sources of ignition. It is important to eliminate sources of ignition in atmospheres where explosive substances are likely to be encountered. One of the most common sources of ignition is an open flame, for example gas burners, and careless smoking. No open flames or naked lights should be used in the danger zones. Glowing materials also form a common source of ignition. A typical example would be a red-hot iron rivet used in maintenance work on a vessel in which the vapour of a flammable liquid is liable to be present. Danger from this source of ignition must be avoided by the removal of the liquid and the thorough purging of the vessel before such work is started. Many fatal explosions have occurred when heat, such as an oxyacetylene welding or cutting flame, has been applied in the repair of a vessel in which the residue of a flammable substance has been present.

Spontaneous heating of materials often gives rise to serious explosion risks. These risks can be controlled by closely watching the materials in question at all stages of handling to detect any appreciable rise in temperature. It is essential, in this connection, to emphasise that accumulation of waste and oily rags should be avoided by suitable maintenance and good housekeeping.

A spark occurs every time an electric circuit is connected or broken by a conventional switch. Sparking also occurs in electrical equipment due to uneven wear and tear of moving parts, such as the winding of a motor. It is essential, therefore, that explosion-proof electrical equipment and fittings should be used where the atmosphere is liable to become flammable or explosive. Static electricity, generated, for example, when materials flow through pipes or by moving belts and pulleys, can also give rise to explosions in flammable atmospheres. Static sparks can be eliminated by the grounding of moving belts, provision of a common bond between containers during transfer of flammable liquids, the grounding of all electrically driven equipment to maintain common ground potential, and the humidification of the atmosphere so that the thin film of moisture that is formed on many objects will conduct static to ground as fast as it is generated (see ELECTRICITY, STATIC). Heat generated when an exothermic chemical reaction takes place, like that between aluminium and iron oxide, may give rise to explosions if flammable substances are present in the vicinity. It is necessary to take all steps to control exothermic chemical reactions so that they do not become violent (see CHEMICAL REACTIONS, DANGEROUS).

Friction between bodies can produce heat; if this is not quickly dissipated, the temperature rise may be sufficient to ignite the flammable or explosive materials present in the surroundings. This source of hazard can be eliminated by the use of proper bearings and by paying particular attention to the maintenance and lubrication of mechanical parts that are subjected to wear and tear.

Explosion control

In some processes, it is not possible to ensure that flammable concentrations will never occur, and sources of ignition on a large plant can be so numerous and so varied that it is not possible to ensure absolutely that all have been eliminated. Although all the precautions that reasonable foresight may require have been taken, conditions may still develop in which an explosion may be possible. In these circumstances precautions must be applied to limit the spread and effects of an explosion. A small explosion must be prevented from becoming a large one, and an explosion that starts in a part of the plant where it can do no harm must be prevented from spreading to a part where it can cause injury and damage. Explosion-relief devices (in the form of rupture diaphragms or explosion doors, panels or vents) should be provided throughout the plant to relieve the pressure as near as possible to the origin of an explosion. If such devices are provided along with chokes (for dust explosions) and flame arresters (for gas and vapour explosions) an explosion may thereby be prevented from spreading through the plant and from thus increasing in magnitude and violence.

In some instances it is necessary to enclose and isolate plant in which materials that explode with a very rapid rate of pressure rise are processed. This precaution has been adopted for the manufacture of aluminium and magnesium powders. It is also used for plant in which explosively unstable materials are processed. The structure of such an enclosure should be strong and it should be provided with ample explosion relief. It should be segregated from other buildings to minimise damage in the event of the failure of the control measures taken.

HARIHARA IYER, C. R.

CIS 80-633 *Explosible substances—Vol. 1. Properties; Vol. 2. Monographs* (Les explosifs occasionnels—Volume 1. Propriétés; Volume 2. Monographies). Médard, L. Collection Industries, productions, environment (Technique et documentation, 11 rue Lavoisier, Paris) (1979), 2 vols., 872 p. Illus. Bibl. (In French)

CIS 79-620 "Dust clouds and fogs—flammability, explosibility and detonation limits, burning velocities, and minimum ignition energies". Nettleton, M. A. *Fire Prevention* (London), Nov. 1978, 20 (12-18). Illus. 31 ref.

CIS 80-622 *Safety technical data on flammable gases and vapours* (Sicherheitstechnische Kennzahlen brennbarer Gase und Dämpfe). Nebert, K.; Schön, G. (Deutscher Eichverlag, Interbuch, Burgplatz 1, 3300 Braunschweig) (reprint 1978), 176 p. Illus. (In German)

CIS 80-622 "Gas detection and explosion prevention" (Détection de gaz et prévention des explosions). Peissard, W. G. *Ingénieurs et architectes suisses* (Zurich), 6 Mar. 1980, 106/5 (53-58). Illus. 7 ref. (In French)

CIS 80-10 *Explosion venting.* NFPA 68 (National Fire Protection Association, 470 Atlantic Avenue, Boston, Massachusetts) (1978), 108 p. Illus. 91 ref.

"Basic principles of explosion protection" (Grundlagen des Explosionsschutzes) (72-416). Ref. *2nd International Sym-*

posium on Prevention of Occupational Risks in the Chemical Industry, Frankfurt am Main, 21-23June1973: Report. General Topic 1 (Geneva, International Social Security Association, n.d.). (Mainly German, with English summaries)

Exposure limits

This is a new term which was introduced by the International Labour Conference in 1977 in the course of the discussions which led to the adoption of the Working Environment (Air Pollution, Noise and Vibration) Convention (No. 148) and Recommendation (No. 156) of 1978. The term was coined as a general expression that is intended to embrace the various words currently used to refer to air quality limit values in workplaces and, in particular, to replace "Maximum allowable (or permissible) concentrations". The reason for the change is that "allowable" (or "permissible") seems to imply an administrative decision, which is not always the case at the national level, or to suggest a biological harmlessness which in fact is not necessarily a reliable guide.

Another term, which was introduced by the World Health Organisation (WHO) in 1980, is "health-based limit"; but this, as we shall see later, has a different meaning.

For the majority of health effects, there is a close correlation between the amount of a toxic substance absorbed by the body and its effects on health (see DOSE-RESPONSE RELATIONSHIP). This correlation can be expressed as the "uptake-effect relationship" when it relates to the magnitude of a qualitatively specified biological effect in an individual, or as the "uptake-response relationship" when it relates to the proportion of individuals with a quantitatively specified magnitude of a qualitatively specified effect in a group of subjects (WHO, 1977). In the case of occupational exposure, a knowledge of this relationship makes it possible to quantify the aetiology of a large number of occupational health impairments, to evaluate the risk of such impairments and, if necessary, to consider the effectiveness of preventive measures. More specifically, it can be used for the establishment of a safe limit of occupational exposure below which there should be no health hazard for the average worker and no deterioration in the degree of comfort that is required to maintain production and to keep the risk of accidents to a minimum.

The study of exposure limits dates back to the 1930s, when the first Swiss and German contributions on the subject were published. The first lists of maximum allowable concentrations of airborne toxic substances at the workplace were issued between 1933 and 1938 in the USSR (the first country to make them a statutory obligation), the United States and Germany. In the last 15 years, as a result of the improvement of analytical facilities and a growing concern for the risks involved in exposure to chemicals, there has been steadily growing interest in environmental and biological monitoring. Many concepts have evolved, including the basic concepts of health and adverse health effects, and a better understanding has been achieved of why exposure values vary widely from one country to another.

A few definitions concerning the environmental exposure limits will illustrate some of these differences. Biological exposure limits are discussed in a separate article.

Threshold limit value. The term TLV, a registered trade mark of the American Conference of Governmental Industrial Hygienists (ACGIH), refers to concentrations of air contaminants in the working environment to which it is believed nearly all workers may be exposed repeatedly, day after day, without adverse effect. A TLV-TWA is a time-weighted average for a normal 8-h workday or 40-h work-week. Excursions above this limit may occur if they are compensated, during the work-day, by equivalent excursions below the limit.

The above definition has been basically adopted by Western European countries. However, in the Netherlands the National MAC Commission adopted for the 1978-79 list the different definition of maximal accepted concentration set out below.

Maximal accepted concentration (Dutch). The maximal accepted concentration of a gas, vapour or aerosol is that concentration in the air at the workplace which, according to present knowledge, in repeated long-term exposure, even up to a whole working life, does not in general lead to health impairment of either the workers or their offspring. (This definition does not apply to carcinogenic substances.)

USSR maximum allowable concentration. The MAC as defined in the USSR is that concentration of a harmful substance in the air of the working area which, in the case of daily exposure at work for 8 h or other length of working day, throughout the entire working life, will not cause any disease or deviation from a normal state of health of the workers or of their offspring, detectable by current methods of investigation, either during the work itself or in the long term.

USSR tentative safety exposure level or temporary safe reference action level. The TSEL or TSRAL is temporarily established by calculations from physical and chemical properties of the substance or by interpolation and extrapolation in series of compounds with similar structure or from indices of acute hazard. TSELs must not be used for production planning. They are revised 2 years after their approval or are replaced by MACs.

FRG technical reference concentration. The TRC for a dangerous work substance is the concentration of a gas, vapour or airborne particulate to which the requisite protective measures and monitoring of the workplace are to be geared. TRCs are based on both medical data and technical possibilities in view of reducing the hazard of health impairment in exposure to substances for which a MAC cannot be assigned on occupational health grounds. The observance of the TRC should reduce the risk of health impairment but cannot eliminate it entirely. For the time being all TRCs refer to carcinogenic substances and are to be understood as annual average values for a working day usually not longer than 8 h and a working week not longer than 40 h. In plants working four shifts, the relevant figure is 42 h a week, averaged over 4 consecutive weeks. The average concentration over a period of 1 h may not exceed 3 times the TRC.

Technical factors

Time. The concentration of gases, vapours, fumes and dust in the workplace atmosphere may fluctuate widely, according to the phase in the work cycle, the volume of production, variations in ventilation or temperature, possible technical faults, and interruptions in the work. Because these fluctuations may occur at very short intervals, they cannot be overlooked when assessing the concentration of a substance in the working area. The most general trend today is to relate the exposure limit to a *time-weighted average* (TWA), usually the mean concentration over the working day. Every excursion above such a limit must be compensated during the work shift by an equivalent excursion below the limit.

For many substances, however, it would be unsafe to allow unlimited peak excursions above the time-weighted average. Consequently, the American Conference of Governmental Industrial Hygienists (ACGIH) has introduced the concept of a *"permissible excursion"* above the TWA limit which should never be exceeded. The difficulty, however, lies in the assessment of such excursion. The ACGIH used to suggest as a rule of thumb the following formula in an attempt to define the magnitude of the permissible excursion above the limit for substances not given a ceiling designation:

TLV 0-1 excursion factor = 3
TLV 1-10 excursion factor = 2
TLV 10-100 excursion factor = 1.5
TLV 100-1 000 excursion factor = 1.25

This formula has been criticised as an oversimplification, which for analytical reasons permits a higher degree of excursion for the most dangerous substances–those carrying the lowest limit values–and does not take into account the specific biological properties of each individual substance. On the other hand, until individual permitted excursion values have been determined for groups of substances having a similar biological effect, the formula may be applied "as a rule of thumb" and is in fact used as such in several countries.

In the 1980 ACGIH document on TLVs, the Appendix on permissible excursions for time-weighted average (TWA) limits has been deleted and the following explanation added: "The amount by which threshold limits may be exceeded for short periods without injury to health depends upon a number of factors such as the nature of the contaminant, whether very high concentrations–even for short periods–produce acute poisoning, whether the effects are cumulative, the frequency with which high concentrations occur, and the duration of such periods. All factors must be taken into consideration in arriving at a decision as to whether a hazardous condition exists."

There are fast-acting substances that even an instantaneous peak may render dangerous. This is so of substances that are capable of producing severe irritation such as formaldehyde, lung oedema such as cadmium oxide fumes, other often irreversible tissue changes such as hydrogen chloride, or a disabling narcosis such as hydrogen sulphide, or of initiating sensitisation such as organic isocyanates. A *ceiling value*–that is, a concentration that should not be exceeded even instantaneously–is suitable to control such risks. In practice these "instants" may vary according to the time required to take a sufficient sample for analysis, or to determine the substance directly in the air when this is possible. With the range of analytical methods currently in use, sampling implies a duration which is not uniform and, in some cases, may take as long as 2 hours.

Since 1975 the ACGIH has adopted the *short-term exposure limit* (STEL). This is defined as the maximum concentration to which workers can be exposed for a period of up to 15 minutes continuously without suffering from (1) irritation, (2) chronic or irreversible tissue change, or (3) narcosis of a sufficient degree to increase accident proneness, impair capacity for self-rescue or materially reduce work efficiency, provided that no more than four excursions per day are permitted and that at least 60 min elapse between exposure periods and provided that the daily TLV-TWA is also not exceeded. The STEL should be considered a ceiling not to be exceeded at any time during the 15-min excursion period. Ceiling values have been included in a number of national lists of exposure limits, but STELs are for the time being in use only within the ACGIH system. Because the definition of STEL implies that the concentration drops considerably during the intervening period, this limit has been suggested as more appropriate for industrial emissions than for the working environment; it is particularly useful for fast-acting substances such as solvents.

Cumulative substances act over a long period and for these the time unit used for weighting the average may need to be extended beyond the day. Carcinogenic substances are a case in point. In the Federal Republic of Germany, for instance, the technical reference concentrations (TRC) in use for carcinogenic substances refer to an averaging time of as long as one year. A year has also been adopted by the Council of the European Communities in its directive of 29 June 1978 concerning the exposure value of vinyl chloride monomer. Further experience of the practical application of such long-term averages is needed.

To sum up, the following time units for establishing average exposure limits are currently in use:

- one year in the case of the Federal Republic of Germany's technical reference concentrations and in the European Economic Community's directive for vinyl chloride;
- one day for TLVs and for the majority of exposure limits adopted at the national level;
- 15 minutes for the ACGIH's short-term exposure limits (STEL);
- "instants" for the ceiling values of the ACGIH and related lists.

In exceptional working conditions, such as after the occurrence of accidents or disasters involving loss of life and/or severe material damage, an *emergency exposure level* (EEL) has been suggested. This is intended to permit a volunteer to face a highly dangerous toxic concentration for a short time, on the assumption *(a)* that such exposure will never be repeated, *(b)* that any health impairment which may occur as a consequence of such exposure would be reversible, and *(c)* that the possibility of any such health impairment would be fully offset by the anticipated result in terms of human life and fitness and economic benefit. Exposure limits should not be confused with EELs.

Work shift. Exposure limits usually apply to a *reference working time* of 8 h per day and 40 h per week, and to the whole working life. Some flexibility, however, is required for practical reasons, such as when work shifts exceptionally last up to 9 or even 10 h and in countries where the working week varies from 40 to 45 h. Taking into account the usual range of approximation of the exposure limits, such differences do not appear to alter significantly the degree of protection they afford. However, where very long schedules of as much as 10 to 12 h per day (such as have recently become operational in a few industries of certain countries) are worked on a regular basis, the limit values should be modified accordingly.

Environmental conditions. Exposure limits refer to a given *air temperature*. In the United States the limits are for a temperature of 25 °C and the ACGIH TLVs adopted in many other countries refer to the same air temperature. The MAC values adopted in the Federal Republic of Germany are for a temperature of 20 °C; this implies a difference in the conversion formula for figures expressed in terms of weight (mg/m³) and volume (ppm), as follows:

$$\left.\begin{aligned} mg/m^3 &= \frac{\text{molecular weight}}{24.04} \times ppm \\ ppm &= \frac{24.04}{\text{molecular weight}} \times mg/m^3 \end{aligned}\right\} \text{ for } 20\ ^\circ C$$

$$\left.\begin{aligned} mg/m^3 &= \frac{\text{molecular weight}}{24.44} \times ppm \\ ppm &= \frac{24.44}{\text{molecular weight}} \times mg/m^3 \end{aligned}\right\} \text{ for } 25\ ^\circ C$$

Again, these differences may be considered negligible compared with the degree of approximation of the limit values. When the temperature in the working area increases, the increased physiological workload is reflected in an increased breathing rate and, therefore, an increased intake of airborne pollutants not necessarily compensated by increased exhalation. The rate of absorption through the skin may also increase. Although small deviations in the working area temperature are not significant because of the safety factor built into the limit values, corrections do become necessary under severe conditions. Thus, it is recommended to correct the ACGIH TLVs when the air temperature continuously exceeds 32 °C. Atmospheric pressure may also have an influence on health hazards, and exposure limits must therefore be applied most carefully when working under abnormal pressure conditions.

Combined exposure. During the past few years increasing attention has been paid to the fact that exposure limits apply to a given substance in the absence of other contaminants and other occupational and non-occupational agents such as noise, drugs and smoking habits, which may modify the effects of the substance, especially with regard to its long-term effects and the carcinogenic risk involved. There are various approaches to possible *combined effects* in the application of exposure limits.

According to the US and the USSR standards, when two or more hazardous substances are present in the working environment these effects should be considered as cumulative. For each substance the ratio must be calculated between the air concentration and the limit value. If the sum of these fractions exceeds one, the exposure limit for the mixture has been exceeded; if the sum is equal to or less than one, the limit has not been exceeded. When it is not possible to measure the concentration of all the substances, the limit value of one substance should be reduced by a correction factor, whose magnitude will depend on the number, toxicity and relative quantity of the other contaminants ordinarily present.

The above formula is recognised as a useful tool for increasing the protection of workers' health. However, some authorities are reluctant to adopt such a simple mathematical formula to evaluate the risk of combined effects instead of investigating each case separately. Others consider that effects which are usually combined can be included within the range of approximation of the exposure limits.

Sampling. A number of technical factors may interfere with the application of exposure limits. This is especially true of particulates, because the *sampling and analysis* of dust and fibres is one of the most difficult areas of industrial hygiene. The limit values for particulates are so intimately bound up with the techniques being used that they should be considered effective only when applied to data collected by means of the same techniques that were employed for establishing them.

The sampling method in particular must be appropriate to the mode of action of the substances: for those which have a cumulative action, continuous sampling throughout a whole shift is preferable, whereas in the case of fast-acting substances, it is necessary to use a rapid sampling method which will detect peak concentrations, lasting perhaps only a few minutes. The USSR standard divides the harmful substances in four hazard classes and requires that monitoring be carried out continuously for substances of hazard class 1; such a continuous monitoring must be ensured by systems of self-recording automatic instruments which emit a signal when the MAC level is exceeded. For periodical sampling precise requirements are also laid down. In general, the length of the sampling period to be applied for each substance depends on the rate of accumulation and elimination in the human body. The sensitivity and accuracy of the techniques selected should be at least in the same order of magnitude as the measurement that is to be made. The above-mentioned USSR standard specifies that the sensitivity of the monitoring methods and instruments should not be less than 0.5 of the MAC level. Their maximum total should not exceed $\pm 25\%$ of the value to be determined, while the error in measuring the volume of the air sample should not exceed $\pm 10\%$. The technique should always be the same.

Sampling should always be carried out in the atmosphere actually inhaled by the workers because the composition of the atmosphere may vary significantly in the presence of localised sources of contamination even over a distance of a few dozen centimetres. Sampling and analytical methods should be defined in every detail that is liable to reflect on the results, and such data should be standardised at least at the national level. In addition, sampling strategies as regards time, place, duration, amount and number of samplings should also be defined and applied with the utmost care. Progress in this area has been achieved by the use of continuous personal monitors covered by the exposed workers.

Biological factors

1. Exposure limits are hygienic standards and, as such, apply to all exposed persons. The latter, however, are by no means a homogeneous group of average people as regards *susceptibility* to environmental factors. Consequently, in the most widely used definition of exposure limits (the ACGIH TLVs) they are described as relating to the "conditions under which it is believed that nearly all workers may repeatedly be exposed day after day without adverse effects". Theoretically, exposure limits apply only to healthy workers. In practice, therefore, "because of wide variation in individual susceptibility, a small percentage of workers may experience discomfort from some substances at concentrations at or below the threshold limit; a smaller percentage may be affected more seriously by aggravation of a pre-existing condition or by development of an occupational illness" (ACGIH). The increase in susceptibility of healthy workers may derive from genetic changes in the enzymatic system or may come about as a result of an allergy, of organic or functional changes, or of drugs interfering with the substances to which the workers are exposed at the workplace. Observance of the exposure limits does not, therefore, exclude the need for medical surveillance of the workers, especially when the airborne contamination approaches the limit values.

2. The present trend is to ensure that exposure limits prevent health impairment of *both the workers and their offspring*—that is, that they prevent gonadotropic, genetic, embryo- and fetotoxic and teratogenic effects. Knowledge in this area is still scarce, however, and special precautions may need to be taken with certain substances, in addition to the observance of the exposure limits, where pregnant women and even women of childbearing age are involved.

3. Heavy *workloads* may increase absorption of air contaminants as a result of an increase in air intake. The safety factor that is usually built into the value of the

exposure limits is therefore expected to cover adequately the relevant variations of risk.

4. The preventive value of exposure limits would be undermined if dangerous substances could be absorbed in large quantities by routes other than the respiratory tract. Thus, in several lists of exposure limits attention is drawn to the fact that certain substances can be absorbed through the unbroken *skin* including mucous membranes and eyes, particularly by direct contact and even when in suspension in the air. Precautionary measures should be taken to avoid absorption in this way of quantities likely to have toxic effects, in particular by keeping skin, hair, clothing and work surfaces clean; greater precautions are necessary when the skin is broken or abraded. Examples of such substances are aniline and its derivatives, benzene, acrylic compounds, hydrogen cyanide, several alcohols and glycols, pesticides (organophosphorous and halogenated), several organometallic compounds, nicotine, nitroglycerine and thallium soluble compounds.

5. Exposure limits do not necessarily guarantee that the skin or respiratory tract will not be sensitised. The probability of *sensitisation* decreases with decreasing exposure; therefore, for some very active sensitising substances, such as TDI, account is taken of this property in establishing the value of the exposure limit. Examples of sensitisers classified as such in the German MAC list are TDI, chromium, formaldehyde, hydrazine, mercury organic compounds, nickel and paraphenylenediamine.

6. As regards *carcinogenic substances*, trends are rapidly evolving because there is still a great deal of uncertainty in available information and because new data currently being acquired may lead to extensive revisions. Although, in general, carcinogenic substances or substances associated with industrial processes recognised as human carcinogens have no assigned exposure limit, some of them may have an assigned value in certain lists–for example, arsenic, nickel, chromium, asbestos and vinyl chloride in the TLV list of the ACGIH. Substances suspected from animal evidence of having a carcinogenic potential for man usually have an assigned exposure limit, pending evaluation of the degree of their carcinogenic potency (see CARCINOGENIC SUBSTANCES; CANCER, OCCUPATIONAL: LEGISLATION).

Establishment of exposure limits

The legal status of exposure limits varies widely from country to country. Although, generally speaking, they represent nothing more than recommended practice, in the socialist countries, the United States, Belgium and Denmark they are statutory obligations. In some European countries they are in an intermediate position which makes possible their enforcement directly or by indirect means. The trend in developed countries is to make them increasingly compulsory. In certain countries the exposure limits are included in collective agreements by branch of industry.

As regards the establishment of exposure limits at the national level, the values assigned to them in several Western countries have so far been based on the ACGIH list. However, in the last five years the trend has been to refer to values proposed by national experts, on the grounds that the genetic, climatic and industrial conditions under which the ACGIH values are established are different and because the materials used are obtained from different sources or manufactured in different ways. In some countries the procedure for establishing exposure limits is in two stages. The first stage is carried out by experts who study the results of epidemiological investigations, experimental research (whose results are extrapolated with respect to man with the inclusion of an adequate safety factor) (see SAFETY FACTOR IN EXPOSURE LIMITS), case studies and other sources of data. It is only on such medical evidence that proposals are drafted for new exposure limits or for the amendment of limit values already adopted. These proposals are then submitted, as a second stage, to a (usually tripartite) political body, which takes into consideration the technical feasibility and socioeconomic cost of the recommended limit values (including, if necessary, the availability of analytical facilities) and, on this basis, takes its decision.

The *"health-based limits"* introduced by WHO in 1980 in its Technical Report No. 647 on heavy metals are based only on medical scientific data and would therefore correspond to stage one in the two-stage procedure mentioned above. A different procedure is followed in the Federal Republic of Germany for its technical reference concentrations (TRC). These are recommended limit values for carcinogenic substances (to which a maximum acceptable concentration value based only on medical evidence could not apply) and are laid down by a committee composed of representatives of all the interested parties. After checking the analytical methods available and their sensitivity and precision, the committee first decides whether the TRC is to be taken as a short-term or long-term value, a decision that is essential in measurement planning and in the statistical evaluation of the results. Once identified in the light of these analytical and technical considerations, the proposed value is then examined to determine whether there might be any objection to it from the occupational health standpoint.

Exposure limits vary from country to country, depending on the data on which they are based (mainly neurophysiological changes in experimental animals, as in the USSR and socialist countries, or health impairment in exposed workers, as in the United States and Western European countries), the safety factor they incorporate, the account taken of technical and socioeconomic factors, their status as recommendations or statutory provisions and their enforcement.

Differences between values are particularly significant in the case of neurotropic substances; for instance, for ethyl bromide the time-weighted average established by the Occupational Safety and Health Administration of the United States is 890 mg/m^3, whereas the corresponding MAC in the USSR is 5 mg/m^3. Exposure limits are constantly being revised, however, and in the last few years the gap between United States and Soviet limit values has in some cases tended to close.

Application of exposure limits

The effective application of exposure limits depends upon the occupational health background, industrial hygiene facilities (which scarcely exist in the majority of countries), analytical factors, sampling strategy and the practice of monitoring the working environment. Only in a few countries and for a very limited number of substances (such as asbestos, vinyl chloride and chromium) is there currently a statutory obligation to monitor the working environment. However, occasional industrial hygiene surveys are becoming more and more frequent and monitoring is gradually developing into a regular practice.

In large undertakings environmental monitoring is carried out by the company's industrial hygiene laboratory or occupational health service. Industrial hygiene laboratories for field work are being created in most industrialised countries to assist medium-sized and small undertakings to monitor the working environment. Some of these laboratories are attached to universities and other research and teaching institutions, some to professional associations or to semi-private prevention institutes; others are private. In some countries, to ensure that the services offered by these laboratories are of an adequate standard, a formal approval procedure has been adopted by the competent authority.

It is indeed essential that only trained specialists supervise and monitor the working environment and that it be possible for the findings of repeated examinations to be interpreted jointly by the industrial hygienist and the occupational physician. For the protection of workers' health, exposure limits should be applied as widely as possible but should not be interpreted as representing a fine line between safe and dangerous concentrations. As stated in the preface to the ACGIH TLVs for chemical contaminants, "in spite of the fact that serious injury is not believed likely as a result of exposure to the threshold limit concentrations, the best practice is to maintain concentrations of all atmospheric contaminants as low as practical".

PARMEGGIANI, L.

"Safety and health standards". Occupational Safety and Health Administration. *Federal Register* (Washington, DC), 18 Oct. 1972, 37/202 (22139-22142); 27 June 1974, 39 125 (23 502-23 828).

CIS 77-682 *Workplace air–General health and hygiene rules* (Vozduh rabočej zony–Obščie sanitarno-gigieničeskie trebovanija). GOST 12.1.005-76, Gosudarstvennyj komitet standartov (Moscow, Izdatel'stvo standartov, 1976), 32 p. (In Russian)

Harmful substances–Classification and general safety requirements (Vrednye vescestva–Klassifikacija i obscie trebovanija bezopasnosti). GOST 12.1.007-76 (Moscow, State Standards Committee, 10 Mar. 1976), 8 p. (In Russian)

TLVs. Threshold limit values for chemical substances and physical agents in the workroom environment with intended changes for 1980 (American Conference of Governmental Industrial Hygienists, PO Box 1937, Cincinnati, Ohio) (1980), 93 p.

Comparative analysis of the principles and application of control limits in the Member States of the European Community. Parmeggiani, L. 79/33–rev. 1 (Luxembourg, Commission of the European Community, Health and Safety Directorate, 1979), 84+6 p.

Occupational exposure limits for airborne toxic substances. Occupational safety and health series 37. Second (revised) edition (Geneva, International Labour Office, 1980), 290 p.

Methods used in establishing permissible levels in occupational exposure to harmful agents. Report of a WHO expert committee with the participation of the ILO. Technical report series 601 (Geneva, World Health Organisation, 1977), 68 p. Ref.

Exposure limits: biological

Occupational exposure to airborne toxic substances has traditionally been determined by means of chemical analysis of air at the workplace. This method, although it has many advantages for the evaluation of conditions at the workplace, is not ideal since even the most exhaustive air-sampling programme cannot provide a complete assessment of the real level of exposure of workers. This is so, to mention only some of the reasons, when–

(a) exposure is characterised by sudden fluctuations in the concentration of toxic substances;

(b) a substance is absorbed simultaneously via other ways of entry (i.e. through the skin or by ingestion);

(c) workers are also exposed to the same substance outside the workplace (e.g. lead);

(d) combined effects are produced because of multiple exposure (i.e. synergistic or antagonistic effects);

(e) individual differences in working habits, workload and physiological condition exist.

In fact the human organism itself may be regarded as a kind of sampling device. The worker's body represents his own individual collector, register and monitor of his personal exposure. To arrive at an accurate evaluation many additional physiological data are nevertheless required, such as rate of inhalation and quantity of inhaled air, percentage absorption in the lung, possibility of absorption by the skin and (occasionally) the intestinal tract, retention, rate of metabolism and excretion and many other individual characteristics, including genetic differences (see BIOLOGICAL MONITORING).

The biological assessment of exposure (biological monitoring) has been gaining increasing attention recently. Today various "exposure tests" have been developed for numerous substances present in the working or living environment.

Exposure tests

The bioassays used in order to evaluate workers' exposure or detect early signs of health impairment can be divided in two broad groups:

(1) Direct exposure tests involve determination of the toxic substance or its metabolites in various biological specimens: blood, breath, urine, faeces, hair, etc. In the majority of cases, the actual exposure can be estimated, and in some cases the body burden as well.

(2) Indirect exposure tests are based on determination of the products of toxic effects (response) caused by the action of a toxic substance or its metabolites: for example increased porphyrin excretion due to the toxic effects of lead. Procedures applied to evaluate biochemical or physiological response (physiological monitoring) can be included in this category.

Bioassays of this kind provide data that are of value in establishing the exposure/body-burden/excretion relationship, i.e. the relationship between a worker's exposure and the level of toxic substance, metabolites or biochemical or physiological response in body fluids, excreta or breath. These are collective tests applied to a group of workers exposed under similar conditions taking into account individual variability. They may be carried out under actual working conditions or in an exposure chamber.

Analytical methods for exposure tests must be rapid, specific, simple and economical because they are performed in the field on bigger groups of exposed workers. This is especially true of screening test.

Availability of biological specimens

Expired air analysis is the preferred technique for biological monitoring of volatile substances that are not, or are only slightly, metabolised in the organism and are eliminated preferentially by breath, for example volatile solvents, gases, volatile metal derivatives (nickel carbonyl), the radium-derived gas radon, etc. Expired air analysis of the above-mentioned substances can supplement their analysis in blood or urine.

Blood analysis is usually reserved for exposure problems that cannot be solved by expired air or urine monitoring, or for diagnostic purposes. Blood analysis is unavoidable in some cases, for example methyl mercury, determination of blood enzyme activity, etc. Blood is a specimen that may be sampled without danger of external contamination, an essential prerequisite for detecting lead exposure, for instance.

Urine is the most widely used specimen for bioassays because it can readily and repeatedly be obtained from both individuals and groups.

Faecal analysis of a substance with low water solubility or of an ingested substance is preferable.

However, aesthetic considerations may perhaps explain its relatively infrequent use.

Hair analysis has gained great interest recently. During growth the emerging hair accumulates and retains a variety of heavy metals (Pb, Hg, Cd, Zn, As, Se, Cu, Tl), which are firmly bound by the abundant sulphidryl groups in the folicular proteins. Hair grows approximately 1 cm per month, furnishing a permanent record of the heavy metal content of the body and indicating integral exposure and body burden. In the case of multiple exposure to heavy metals it is necessary to take into consideration the possible synergistic or antagonistic effects on accumulation.

Mobilisation from the skeleton can be used for evaluation of the body burden in the case of lead exposure. About 90% of absorbed lead is accumulated in the skeleton by ion exchange with Ca ions. By the use of a suitable chelating agent (CaEDTA, penicillinamine), about 8-10% of lead skeleton burden is mobilised and excreted in the urine over a period of 24 h.

Tentative permissible biological limits

With the development of exposure tests, a tendency has arisen to establish biological exposure limits as a counterpart to atmospheric exposure limits (MAC values, TLVs, etc.). This is valid only for substances that workers inhale in the working environment and that do not contaminate the living environment as well. For economic or social reasons biological exposure limits may be set at a higher level to prevent diagnosable illness, or at a level such that response will be below the level of detectable health impairment. Today occupational health experts are more inclined to the idea that biological limits must mark the integral exposure in both working and living environments that does not produce adverse effects. Non-adverse effects are defined by the US National Academy of Sciences as:

(a) changes that occur with continued exposure and do not result in impairment of functional capacity or ability to compensate for additional stress;

(b) changes that are reversible following cessation of exposure, if such changes occur without detectable decrement in the ability of the organism to maintain homeostasis; and

(c) changes that do not enhance the susceptibility of the organism to the deleterious effect of other environmental influences–whether chemical, physical, microbial or social.

In brief, an adverse or "abnormal" effects is one that is outside the normal range.

The following criteria may be used to evaluate various levels of exposure:

(a) for substances that are transformed in the organism to physiological constituents (e.g. conjugated phenol or sulphate, hippuric acid, etc.), or owing to toxic effects produce an increase in constituents (e.g. porphyrins, delta aminolevulinic acid, etc.) the "normal values" must be established. A similar case is presented by substances that are also present in the living environment (heavy metals). The normal values are influenced by individual differences and therefore have to be established by analysis of a large group of non-exposed persons;

(b) for substances that are not physiologically present in the organism, any positive result indicates that exposure has taken place;

(c) maximum permissible levels for biological specimens must be established marking the non-adverse effect or exposure to the environmental limit in air;

(d) concentrations above the limit indicate excessive or dangerous exposure;

(e) high concentrations indicate health impairment.

The main difficulty in establishing biological exposure limits arises from individual differences among workers. These require a study of large groups of workers over a relatively long period of exposure.

Biological exposure limits could be developed for substances which:

(a) appear as such in body tissue or fluids and breath, excreta, etc.;

(b) appear as metabolites;

(c) cause alterations in kind or amount of some accessible body constituent;

(d) cause alterations in the activity of an enzyme of critical biological importance;

(e) cause alterations in some readily quantifiable physiological function.

However, for the categories of substances discussed below it is not possible to develop exposure tests and hence biological exposure limits:

(a) substances that are constituents of the body, that metabolise to the same (chlorides, phosphates) and that do not produce any measurable alterations in body composition or function;

(b) substances that do not dissolve, are rapidly decomposed or have an aggressive local effect at the site of contact (e.g. corrosives, irritants);

(c) substances producing allergic effects;

(d) substances producing carcinogenic effects. Some scientists suppose that the initial biological effect of carcinogens is at the molecular level, including irreversible changes in the cell, in which case there may not be any threshold. Other scientists suppose that transformation of the cancerous cells or clones of cells has a quantitative threshold.

Today tentative biological exposure limits for the most important toxic substances present in industry have been developed. For the most part they represent a counterpart to atmospheric exposure limits. Some biological exposure limits adopted in Czechoslovakia are presented in table 1.

The most modern approach is to consider the integral exposure resulting from all modes of entry (inhalation, ingestion, skin absorption) including exposure in the living environment. Adopting this approach two WHO study groups recently published "health-based limits" for occupational exposure to some common heavy metals and selected organic solvents. Their conclusions are reported in table 1.

It is advisable to take into consideration some remarks:

- biological exposure limits are mostly expressed as a unique value which sometimes represents average value and sometimes upper limit value. In Czechoslovak literature both values are cited;
- individual exposure should be evaluated, whenever possible, on the basis of all recommended indicators (blood, urine, breath, hair) and not only one;
- biological exposure limits are valid only for standard biological specimens which are sampled and analysed by standard methods.

Finally it must be stressed that environmental monitoring and biological monitoring should not be regarded as

Table 1. Biological exposure limits and corresponding environmental exposure limits in air (Data from Czechoslovak literature, except where WHO health-based limits are cited)

Toxic substance: metabolite or response	Atmospheric exposure limit: a = average c = ceiling (mg/m³)	Normal values			Biological exposure limits		
		Blood	Urine	Other	Blood	Urine	Other
Aniline	a: 5 c: 20	–	–	–	–	–	–
p-amino-phenol		–	–	–	–	100 mg/l	–
Arsenic	a: 0.3 c: 1.0	0.05 mg/l	0.35 mg/l	2 µg/g hair	–	1 mg/l	–
Benzene	a: 50 c: 80	–	–	–	–	–	–
phenol sulphates			20 mg/l 80-95% (inorganic) 5-15% (organic)			100 mg/l 70% (inorganic) 30% (organic)	– – –
Cadmium[1]	a: 10 µg/m³	–	–	–	5 µg/l	5 µg/g creatinine	–
Carbon disulphide[1]	a: 10 (male workers) a: 3 (women of fertile age)	–	–	–	–	–	–
iodine-azide test		–				–[2]	
Carbon monoxide	a: 30 c: 150	–		–	–		0.0025 (breath)[4]
CO-Hb		0.5%			5%[3]		
Fluorine	a: 0.2 c: 1.0	–	1 mg/l	–	–	5 mg/l	–
Chromium	a: 0.05 c: 0.1		10 µg/24 h	–	10 µg/100 ml	25 µg/l	–
Lead[1]	a: 30-60 µg/m³ (according to WHO technical report No. 747)	–	–	400 µg/l (persons over reproductive age) 300 µg/l (women of reproductive age)	–		
protoporphyrin IX (PP)				laboratory "normal" upper limit (mean+2SD)			
δ-aminolevulinic acid (ALA)				laboratory "normal" upper limit (mean+2SD)			
Mercury (metallic)[1]	a: 25 µg/m³	–	–	–	–	50 µg/g creatinine	–
(inorganic)[1]	a: 50 µg/m³	–	–	–	–	–	–
Methanol	a: 100 c: 500	1 mg/100 ml	1 mg/l	–	2 mg/100 ml	3 mg/l	–
Methylene chloride	a: 500 c: 2 500	–	–	–	–	–	–
CO-Hb		0.5%			5%[3]		
Nitro-benzene	a: 5 c: 25	–	–	–	–	–	–
p-nitro-phenol	–		–			5.5 mg/l	

Toxic substance: metabolite or response	Atmospheric exposure limit: a = average c = ceiling (mg/m³)	Normal values			Biological exposure limits		
		Blood	Urine	Other	Blood	Urine	Other
Styrene	a: 200 c: 1 000	–	–	–	–	–	–
mandelic acid			–			1 500 mg/l	
phenyl-glyoxylic acid			–			100 mg/g	
Tetrachloroethylene	a: 250 c: 1 250	– –	– –	– –	– –	– –	– –
trichloro-acetic acid			–			35 mg/l	
trichloro-ethanol			–			20 mg/l	
Toluene	a: 200 c: 1 000	–	–	–	–	–	–
hippuric acid			1 g/l			3 g/l	
benzoic acid			–			2 g/l	
Trichloroethylene[1]	a: 135	–	–	–	–	–	–
trichloro-acetic acid			–			50 mg/l	
Trinitrotoluene	a: 1 c: 3	–	–	–	–	–	–
2,6-dinitro-4-amino-toluene			–			30 mg/l	
Vanadium	a: 0.1 c: 0.3	23 μg/ 100 ml	18 μg/l	–	–	80 μg/l	–

[1] WHO health-based limits. [2] The health-based limit is below the sensitivity level of the iodine-azide test. [3] Smokers: biological exposure limit 10%. [4] Smokers: biological exposure limit 0.005%.

alternatives: they are complementary and should be carried on in parallel.

DJURIĆ, D.

Biological monitoring for industrial chemical exposure control. Lich, A. L. (Cleveland, CRC Press, 1974), 188 p. Illus. 312 ref.

Molecular-cellular aspects of toxicology. Djurić, D. (Belgrade, Institute of Occupational and Radiological Health, 1979), 100 p. Illus. 250 ref.

Exposure tests in occupational toxicology (Exposičny testy v prumyslove toksikologii). Bardodej, Z.; David, A.; Šedivec, V.; Škramovsky, S.; Teisinger, J. (Prague, Avicenum, 1980), 367 p. Illus. 1 311 ref. (In Czech)

Recommended health-based limits in occupational exposure to heavy metals. Report of a WHO Study Group. Technical report series No. 647 (Geneva, World Health Organisation, 1980), 116 p. Illus. 158 ref.

Recommended health-based limits in occupational exposure to selected organic solvents. Report of a WHO Study Group. Technical report series No. 664 (Geneva, World Health Organisation, 1981), 84 p. 186 ref.

Early detection of health impairment in occupational exposure to health hazards. Report of a WHO Study Group. Technical report series No. 571 (Geneva, World Health Organisation, 1975), 80 p. Illus. Ref.

Eye

The human eye resembles a camera which admits light rays through a transparent aperture and focuses them by a lens on to the light-sensitive retina corresponding to a photographic film.

Structure

The eyeball is spherical in shape with a strong white-coloured outer covering–the sclera, which at the front is replaced by the transparent and somewhat raised cornea. The inside of the eyelids and the exposed part of the sclera is lined by a membrane–the conjunctiva. Behind the cornea and separated from it by water fluid–aqueous humour–is the lens, partially covered by the iris, which by contraction or retraction acts as a diaphragm to regulate the amount of light passing through its central aperture (the pupil). The space in front of the lens is the anterior chamber of the eye and the larger space behind, which forms the main globe of the eye, is the posterior chamber. This is filled with a jelly-like substance known as vitreous humour. The posterior chamber is lined by a thin membrane, the retina, which rests on the choroid, a dark layer which prevents internal reflections of light, corresponding to the black colouring of the inside of a camera.

The focusing of objects on to the retina is carried out by alterations in the shape of the lens produced by its elasticity and by the muscular action of the ligaments by which it is suspended. The light rays focused on the retina produce, by chemical reactions in its cells, an image which is transmitted to the brain by the optic nerve from the back of the eyeball. The eye is moved from side to side or up and down by muscles arising from the back of the bony cavity–the orbit–in which the eyeball is placed and attached to the sclera.

Natural protection

The eye is protected from injury by large objects by the bony structures surrounding the orbit, and the nose. The eyebrows form a protection by diverting fluids to their outer side and preventing them from trickling on to the eyelids. Further defence is provided by the eyelashes on the upper and lower lids, which prevent particles entering the eye and which, when touched, cause the lids to close with great speed and form a protective barrier. The exposed surface of the eye is continuously moistened by tears, which are secreted by a gland in the outer corner of the eye and drain from the inner corner into the nose. Any irritation of the eye causes an immediate increase in the secretion and the irritant may be diluted, if it is a fluid, or washed away. This action is aided by the movements of the upper and lower eyelids.

Eye accidents

Occupational injuries to the eye occur frequently. They form about 3-4% of all industrial injuries and about 10% of injuries seen in eye hospitals. They are of all degrees of severity varying from a slight irritation to complete destruction of the eye.

The presence of a foreign body is the most common minor injury. Most small particles will be washed out by tears but some will be retained in the eye, particularly within the margin of the upper eyelid. Some become embedded in the surface of the cornea or sclera, where they cause irritation followed by reddening of the surface and, if not removed, may produce an ulcer and infection. Some foreign bodies are surrounded by a ring or stain on the cornea, which disappears on their removal. Copper and coal particles often produce inflammation. Other minor injuries are abrasions due to rubbing and roughening of the corneal surface, chemical irritation from fumes and bruising of the orbital margins and eyelids.

Major injuries to the eye occur when the eyeball is perforated, burnt by heat or chemicals or disrupted by a blow. The signs of perforation by a small particle are not always obvious and may only be a minute track in the cornea and pupil which is difficult to see. The seriousness of other injuries may be judged by the presence of blood or pus in the anterior chamber, irregularity of the pupil or the appearance in the wound of jelly-like vitreous humour. Severe blows can cause fractures of the bones around the eye, which may produce subsequent distortion of the lids or malposition of the eye leading to double vision. Occasionally, the whole eyeball may be torn out of its socket with injury to the muscles and optic nerve. The possibility of retinal detachment or intra-orbital haemorrhage should also be borne in mind as these cannot be detected without the use of an ophthalmoscope.

There are numerous occupations in which accidents to the eye are a special risk. In agriculture there are many minor injuries from smoke, fertilisers, sawdust, chaff and insects, the more serious injuries occurring from working with hedges and forestry, where perforating wounds are not uncommon. The care of animals can lead to eye injuries of minor degree, as from swinging tails, or to severe injuries, as from horns.

In industry, foundry work, engineering, building construction and work in the chemical industry are the chief causes of eye accidents, of which one-half may be caused by particles or splashes. Of serious injuries in one series, 3 664 were caused by foreign bodies, 1 152 by burns, 968 by surface injuries, and 704 by open wounds. Special dangers arise from the dry grinding of metal, the turning of non-ferrous metals or of cast iron and the use of hand tools or portable electric power tools or pneumatic tools for the cleaning and fettling of metal castings, cutting rivets or bolts, chipping or scaling boilers and breaking stones and concrete. Much risk occurs from the use of chipped tools such as hammers and chisels. Hazards also arise from the manipulation of sodium hydroxide, the handling of calcium oxide pipework or supply lines containing dangerous liquids or gases and the opening of drums or containers of dangerous liquids in which pressure has been caused by dents. Operations at metal hardening plants, where there is a risk of cyanide or nitrite splashing, and work in tinplating and tar and pitch plants are also specially hazardous. The use of compressed air for clearing dust is a major danger.

Ultraviolet light produced by welding and cutting or oxyacetylene burning produces acute inflammation of the eyes (arc eye), particularly in bystanders who may fail to wear personal protective equipment. In this respect the use of contact lenses, whether they are made of plastic or glass, does not present any additional hazard to the cornea and indeed they may well reduce to some extent the penetration of ultraviolet light. It is, of course, possible that if the wearer of contact lenses were to sustain a flash burn the lenses might give rise to additional irritation. A sensible precaution would be for the wearer, if the eyes started to become painful after exposure to flash, to remove the lenses and not to replace them until the irritation had subsided.

Chemical burns are not unusual in industry and of these, burns from alkalis are the most serious, because they penetrate more easily than acids. Ammonia is a particular risk owing to is widespread use. Work in foundries leads to a risk of severe burns and scalds from molten metal.

Some fumes may cause eye irritation in low dilutions and styrene and butanol fumes are not infrequent hazards.

Eye diseases

Occupational diseases of the eye may occur from the direct effect of chemical or physical agents or from an indirect effect due to absorption of the harmful agents in the body.

Chronic irritation of the cornea and conjunctiva may be caused by a great variety of substances, in particular by hydrogen sulphide gas in workers in gas manufacture, paper and pulp mills, excavations, and in the viscose rayon spinning industry, and by ultraviolet light in welding. Severe irritation has the effect of producing coloured haloes, which are particularly pronounced when produced by osmium tetraoxide. A grey-blue effect on vision is caused by exposure to a variety of amines. Fine particles of mepacrine in chemical workers and hydroquinone in photographic workers produce irritation and staining of the cornea which may lead to reduction in vision. Silver deposits can cause a grey discolouration and mercury may be absorbed through the cornea and form a film on the lens.

Epidemic virus infections of the cornea and conjunctiva have occurred in shipbuilding workers and among poultry workers from the virus of Newcastle disease.

The lens of the eye and the related lesions are dealt with in the article CATARACT. Burns of the retina may occur also from the use of lasers. Also gas bubbles in the posterior chamber of the eye may be expanded by continuous exposure to lasers, thus causing the eye to explode.

Grossly inadequate illumination of the coal face in mines produces a chronic oscillation of the eyeballs from side to side—nystagmus—a condition that leads on occasions to complete disability; but with the improvement of lighting in mines this condition is not seen any longer in the United Kingdom. Nystagmus may also be brought about by certain chemicals such as carbon disulphide and dichloroethane, for example.

Disease of the nerve of the eye producing restriction of vision, double vision or blindness has followed heavy exposure to the fumes of chloromethane and bromomethane used in refrigerating plant and fire extinguishers, from methyl alcohol and some organic mercury compounds such as seed dressings. Organophosphorus pesticides used in agriculture produce irregularities of the pupil.

Complaints of eye strain in fine work usually arise because of poor lighting, poor control, the presence of glare or poor accommodation in the individual. The increasing use of visual display units (VDUs) has led to an increase in complaints of eye strain. These are more likely to be ergonomic than organic in origin and can usually be solved by careful design of equipment, such as screen brilliance and character size and style, and of working environment such as light levels, working position and work patterns. In the United Kingdom it was found that the highest level of exposure to ionising radiation from VDUs was 0.05 millirems per hour at certain wavelengths, which amounted to one-tenth of the dose rate permitted by the appropriate British Standard Specification for television equipment.

SAFETY AND HEALTH MEASURES

Many accidents to the eyes may be prevented by simple methods of ensuring that obvious hazards are removed, for example foreign bodies often arise from the use of worn hammers and chisels and supervision over these hand tools would remove a common risk. Poor lighting in factory premises and workplaces is another factor that can be simply remedied, and where a process is found to be an especial risk it may be possible to modify it. Occupational health and safety education of the workpeople is of importance and they should be made aware of risks, particularly where caustic liquids are handled whose dangerous properties may not be obvious to a newcomer.

Special eye and face protection depend upon the circumstances but it should be remembered that these are only the second line of defence and the main defence against eye injuries lies in the careful control of dangerous processes, good housekeeping and a satisfactory general environment. Glass screens may be interposed between the worker and the working point or goggles or face shields worn. In order to protect the eyes properly, safety glasses or goggles must fit. Glasses should also be provided with protective side pieces attached to the frame. Where prescription lenses are normally worn it is advisable, where possible, to incorporate them in the frame, but if this is not possible then the use of full face shields is preferable to the wearing of overspectacles.

Education of workers in the use of all safety equipment is of little use if that equipment is not appropriate to the work; for example, special goggles will be required for protection against radiation from ultraviolet light, as in welding. However, where possible a choice should be offered, as the workman is more likely to wear something he has chosen personally rather than something that has been issued to him. The importance of all grades of persons wearing the appropriate safety equipment when necessary cannot be overstressed. Many countries now lay down standards of eye protection.

Danger from dust or fumes should be averted by the use of good general ventilation or local exhaust ventilation as required. Eye injury or disease will not occur when atmospheric concentrations are maintained below recommended levels.

It is important that statistics of eye injuries be kept, when possible, to indicate where the need for further measures may arise.

Medical prevention. Where work is known to entail a risk of eye injury or disease, medical examinations will prevent those disabled with existing defective function of the natural protective mechanism, such as disease of the eyelids, from undertaking the work. Regular medical examination of workers may also prevent the development of conditions such as heat cataract. Doctors in hospital have an important part to play in prevention by regularly drawing attention to the number of eye injuries and the high percentage that could have been prevented. By arranging for industrial medical officers and occupational health nurses to attend eye clinics for experience, they can ensure that effective treatment can be given at work and serious injuries are immediately recognised.

First aid. The extent of first-aid treatment given must depend upon the skill and experience of the first-aid attendant and the availability of medical assistance, but in no cases should sharp or unclean instruments be used. Antiseptics or local anaesthetics should not be put into the eye without expert medical advice. Any case in which the extent of the injury is in doubt or the patient is shocked should be treated as a serious injury and referred for further advice. Chemical burns, however severe, require immediate treatment by irrigation with water to produce a dilution and flushing of the chemical from the injured tissues. Speed is more important than a search for a supposed antidote and gently flowing tap water from a main supply is quite suitable as the quicker the treatment the less the difficulty in keeping the lids apart for irrigation. In workplaces where the risk of chemical injury to the eye is great, eyewash bottles should be provided and workpeople taught how to use them (figure I).

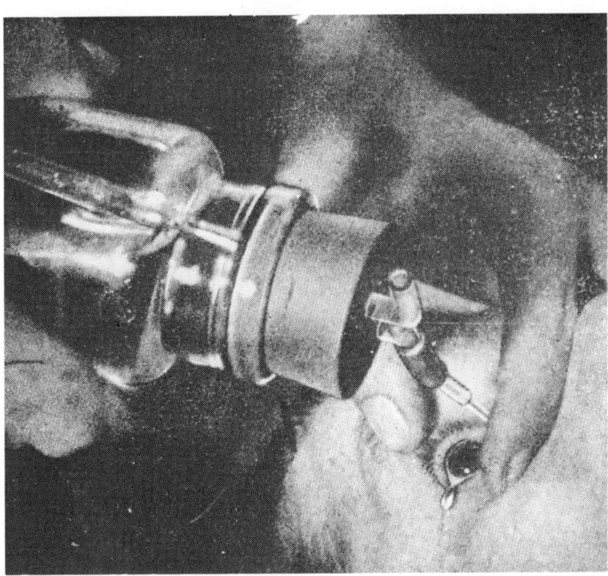

Figure 1. The use of an eyewash bottle. 1. With the patient lying flat on his back, kneel above his head. 2. Holding the bottle in the right hand, tilt it to allow the lotion to flow from the glass tube through the small outlet. 3. Direct the stream of lotion slowly over the patient's cheek to the inner corner of the affected eye and allow it to flow over the eyelids for a few seconds. 4. Without interrupting the flow, gently open the eyelids thus: Place the tip of thumb of the left hand over the centre edge of the upper eyelid and gently pull back the lid, holding it firmly against the bone under the eyebrow. With the forefinger placed at the centre edge of the lower lid, push down and hold against the cheekbone. 5. Tell the patient to move his eye down and up and from side to side. 6. Keep the stream of fluid flowing continuously for 10 min or until the contents of two bottles have been used. 7. A short stream is more gentle than a long stream, so hold the bottle within 2-5 cm of the eye, as shown in the photograph above. *(By courtesy of Imperial Metal Industries (Kynoch) Limited.)*

The great majority of eye accidents are minor in character and will require only first-aid treatment. Foreign bodies can often be seen on the surface of the eye and in such cases an attempt may be made to remove them by gently brushing the foreign body with cotton wool moistened with water once or twice. If this procedure is not successful in removing the foreign body then a pad and bandage should be placed on the eye and the victim referred to a hospital or to a person with expert training. As a general rule, with the exception of the two types of cases mentioned above, the wisest course in the first-aid treatment of eye conditions is to cover the eye and refer the patient to expert help.

SMITH, D. L.

General:

"Note on the physiology of vision" (Rappel de physiologie de la vision). *Revue de médecine du travail* (Paris), 1978, 6/5 (219-276). Illus. (In French)

CIS 74-1477 *Industrial and occupational ophthalmology.* Fox, S. L. (Charles C. Thomas, 301-327 East Lawrence Avenue, Springfield, Illinois) (1973), 203 p. Illus. 87 ref.

CIS 75-175 *Toxicology of the eye.* Grant, W. M. (Charles C. Thomas, 301-327 East Lawrence Avenue, Springfield, Illinois) (1974), 1 201 p. 268 ref.

Accidents:

CIS 78-1079 "A clinical and statistical study of occupational accidents involving foreign intraocular bodies". Schwartzenberg, T. *Medicina del lavoro* (Milan), Nov.-Dec. 1977, 68/6 (450-468). Illus. 19 ref.

"Caustic effects on the eye–Pathogenesis–Treatment–Prophylaxis" (Die Verätzung des Auges–Pathogenese–Therapie–Prophylaxe). Karcher, H. *Zentralblatt für Arbeitsmedizin, Arbeitsschutz, Prophylaxie und Ergonomie* (Heidelberg), 31 Apr. 1981, 31/4 (142-146). 38 ref. (In German)

CIS 77-93 *Blindness from accident.* Douglas, A. A.; MacDonald-MacLean, A.; Cullen, J. F. (W. H. Ross Foundation (Scotland), 20 Lauriston Place, Edinburgh) (1976), 75 p. Illus. 55 ref.

Protection standards:

Factories. The Protection of Eyes Regulations 1974. No. 1681, Statutory Instruments (London, HM Stationery Office, Oct. 1974), 9 p.

Eyewash:

CIS 77-871 *Eyewash and safety shower criteria document.* Weaver, A.; Britt, K.; Pierce, D. (North Carolina Department of Labor, Occupational Safety and Health Administration, Raleigh, North Carolina) (Aug. 1976), 67 p. 61 ref.

Eye and face protection

Eye protection is required in a wide range of occupations, against flying particles and foreign bodies, against chemical fumes and radiation. The whole face may need protection against mechanical or thermal irritation, radiation or chemical irritants. Sometimes the face shield may be adequate also for the protection of the eyes but with some severe risks, specific eye protection is necessary as a complement to the face shield.

The greatest problem is how to provide effective protection which will be acceptable to wear, sometimes over long hours of exertion. Restriction of vision makes for unpopularity of eye protection; the workers' peripheral vision is limited by the side frames, and the nose bridge may disturb binocular vision; misting is a constant problem. Particularly in hot climates or in hot work, additional coverings of the face may become intolerable and may be discarded. Short-term, intermittent use of machines also creates problems as workers may be forgetful and disinclined to use protection. First consideration should always be given to protection of the surroundings–guarding of machines and tools, removal of fumes and dust by exhaust ventilation, screening of sources of heat or radiation, screening of points from which particles may be ejected, for example at abrasive wheels or lathes.

Eye protection

There are four basic types of eye protection:

(a) spectacle type, either with or without side shields (figure 1);

(b) eye-cup (goggles) type (figure 1);

(c) helmet type (figure 2);

(d) hand-shield type (figure 3).

HAZARDS AND SUITABLE PROTECTION

(a) Visible radiation. Transmittance and tolerance in transmittance of various shades of filter lenses and filter plates of eye protectors for high-intensity light sources are shown in table 1 and the standard application of these lenses and plates is shown in table 2. In welding operations, helmet-type protectors and the hand-shield type protectors are generally used, sometimes in conjunction with spectacles or goggles; protection should also be provided for the welder's assistant.

(b) Ionising and microwave radiation. It is essential that ionising and microwave radiation sources be adequately shielded and reliance should not be placed on personal protective equipment for eye safey with this hazard; an exception to this rule may be allowed in the case of persons carrying out medical examinations, when leaded lenses are worn.

(c) Flying particles and other foreign bodies. Spectacles, with or without side shields, eye-cup goggles, plastics eye shields and face shields are used for protection against foreign bodies. Transparent plastic or glass or a wire screen may be used. Where especially violent impacts by foreign bodies are to be expected, a tempered glass lens is recommended.

(d) Chemical hazards. Eye-cup goggles with plastics or glass lenses or plastics eye shields are used for protection against chemicals.

General. Some goggles may be worn over corrective spectacles but it is often better for hardened corrected lenses to be fitted under the guidance of an ophthalmic specialist.

Greater comfort and efficiency is gained if goggles and spectacles are fitted and adjusted by a person who has received some training. Each worker should have the exclusive use of his own protectors but communal provision for cleaning and demisting may well be made in larger works.

Table 1. Transmittance requirements (ISO 4850-1979)

Scale number	Maximum transmittance in the ultra-violet spectrum $\tau(\lambda)$		Luminous transmittance τ_V		Maximum mean transmittance in the infrared spectrum	
					τ_{NIR}	τ_{MIR}
	313 nm	365 nm	maximum	minimum	Near IR 1 300 to 780 nm	Mid. IR 2 000 to 1 300 nm
	%	%	%	%	%	%
1.2	0.0003	50	100	74.4	37	37
1.4	0.0003	35	74.4	58.1	33	33
1.7	0.0003	22	58.1	43.2	26	26
2.0	0.0003	14	43.2	29.1	21	13
2.5	0.0003	6.4	29.1	17.8	15	9.6
3	0.0003	2.8	17.8	8.5	12	8.5
4	0.0003	0.95	8.5	3.2	6.4	5.4
5	0.0003	0.30	3.2	1.2	3.2	3.2
6	0.0003	0.10	1.2	0.44	1.7	1.9
7	0.0003	0.037	0.44	0.16	0.81	1.2
8	0.0003	0.013	0.16	0.061	0.43	0.68
9	0.0003	0.0045	0.061	0.023	0.20	0.39
10	0.0003	0.0016	0.023	0.0085	0.10	0.25
11	Value less than or equal to transmittance permitted for 365 nm	0.00060	0.0085	0.0032	0.050	0.15
12		0.00020	0.0032	0.0012	0.027	0.096
13		0.000076	0.0012	0.00044	0.014	0.060
14		0.000027	0.00044	0.00016	0.007	0.04
15		0.0000094	0.00016	0.000061	0.003	0.02
16		0.0000034	0.000061	0.000029	0.003	0.02

Where the risks of the process allow, some personal choice among different types of protection is psychologically desirable.

Face protection

There are three basic types of face protection:

(a) hood type including diver's helmet type (covering the head completely);

(b) helmet type (shielding the whole front of face);

(c) face-shield type (shielding eye sockets and the central portion of face–figure 4).

HAZARDS AND SUITABLE PROTECTION

(a) Mechanical hazards. The hood type and diver's helmet type as used in sand- and shotblasting fall

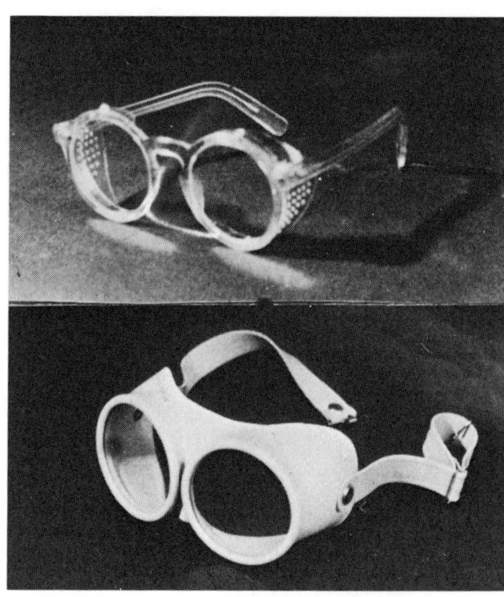

Figure 1. *Above:* spectacle-type eye protection; *below:* eye-cup or goggle-type eye protection.

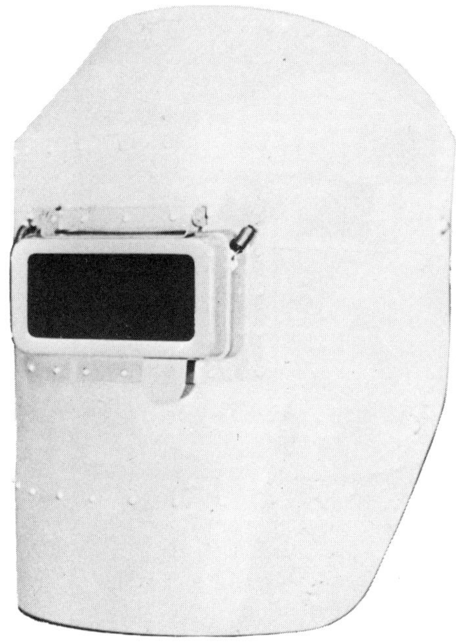

Figure 2. Helmet-type protector.

Table 2. Standard application of filter lenses and filter plates

Scale number	Gas welding, braze welding and plasma arc cutting			Electric arc welding or arc gouging and plasma direct arc welding			
	Gas welding braze welding	Oxygen cutting	Plasma arc cutting	Covered electrodes	Inert gas arc welding on heavy metals	Inert gas arc welding on light alloys	CO_2arc welding
1.2							
1.4							
1.7					Operations receiving scattered light or side light		
2.0							
2.5							
3							
4	I=70						
5	70<I=200	900-2 000					
6	200<I=800	2 000-4 000					
7	I>800	4 000-8 000					
8				20- 40			
9	(I=Flow rate	(Flow rate		40- 80	80-100	80-100	80 40-
10	of acetylene	of oxygen	I⩽150	80-175	100-175	100-175	125 80-
11	l/h)	l/h)	150⩽I<250	175-300	175-300	175-250	175125-
12			250⩽I<400	300-500	300-500	250-350	300175-
13			(I=Current,	500-	500-	350-500	450300-
14			in amperes)			500-	450-
15							
16							

within this category: the helmet type and face-shield type are also used for protection against other mechanical hazards. The materials used are plastics, vulcanised fibres, glass, wire screen and the like.

(b) Thermal hazards. For protection against sparks, flying hot objects, thermal radiation (infrared rays), etc., face protectors of the helmet type and face-shield type are mainly used, the materials being plastics or wire mesh. Aluminised plastics shields give good protection from radiant heat. A face-shield made of wire mesh can reduce thermal radiation by 30-50%.

(c) Ultraviolet radiation. Ultraviolet rays emitted in welding operations constitute the main risk; protectors of the helmet and hand-shield type can protect eyes and face at the same time.

(d) Ionising and microwave radiation. A hose mask or oxygen breathing apparatus equipped with full facepiece is recommended for protection of the face in handling radioactive materials.

(e) Chemical hazards. The diver's helmet or a face-shield made of plastics are used for protecting the face from chemical hazards. A hood made of chemical-resistant material with a glass or plastics window gives good protection.

General. Helmet and hood-type protection may be almost intolerably hot; air lines can be fitted to prevent this. Each worker should have his own hood or helmet. Hoods and helmets should be carefully stored when not in use and should be examined regularly to ensure that they are in good condition.

Figure 3. Hand-shield-type protector.

Figure 4. Face-shield-type protector.

Arc-air gouging	Micro-plasma welding	Furnace operation		Other operations
		Operations of blast furnace, steel heating furnace or ingot casting		Operation receiving reflected light from snow load, roof or sand. Operations using infrared lamp or germicidal lamp
	0.5-1	—	Operations of converter or open hearth furnace	Operation using arc lamp or mercury lamp
	1-2.5			
	2.5-5			
	5-9	Operations of arc furnace	—	—
	9-15			
125-175	15-30			
175-225				
225-275				
275-350				
350-450				
450-				

Testing

Filtering efficiency. Lenses used to filter visible, infrared or ultraviolet radiation should be tested at the following wavelengths:

(a) 400-700 µm for visible radiation;

(b) 700-4 000 µm for infrared radiation; and

(c) 310-400 µm for ultraviolet radiation.

Safety glass or tempered glass lenses should resist, without breakage, the impact by a steel ball 22 mm in diameter and weighing 48 g dropped from a height of 1.3 m on to the specimen. In addition, when the dropping height is increased to cause an impact force sufficient to break the lens, the breakage pattern exhibited by the lens should be such that the lens breaks with radial cracks, not with concentric cracks nor with lines of cleavage parallel to the surface.

MIURA, T.

Personal eye-protectors. Specifications. International Standard 4849 (Geneva, International Organisation for Standardisation, 1981), 8 p.

CIS 78-1094 *Development of criteria and test methods for eye and face protective devices.* LaMarre, D. A. DHEW (NIOSH) publication No. 78-110 (National Institute for Occupational Safety and Health, 4676 Columbia Parkway, Cincinnati) (Aug. 1977), 188 p. Illus. 38 ref.

CIS 78-198 "Use of highly resistant glass for safety spectacles" (Primenenie stekol povyšennoj pročnosti v zaščitnyh očkah). Denisenko, O. N.; Černjakova, T. G.; Sobolev, E. V.; Ščeglova, O. V.; Tihomirova, N. E. *Bezopasnost'truda v promyšlennosti* (Moscow), Jan. 1977, 1 (52-53). (In Russian)

CIS 75-111 "Safety spectacles–Results of tests. Selection criteria for the user" (Lunettes de protection–Bilan des essais. Critères de choix pour l'utilisateur). Ho, M. T.; Mayer, A.; Salsi, S.; Danière, P. *Cahiers de notes documentaires–Sécurité et hygiène du travail* (Paris), 4th quarter 1974, 77, Note No. 924-77-74 (499-520). Illus. 7 ref. (In French)

Personal eye protectors for welding and related techniques–Filters–Utilisation and transmittance requirements. International standard ISO/DIS 4850 (Geneva, International Organisation for Standardisation, 1979), 5 p.

Personal eye protectors–Optical test methods. International standard ISO 4854 (Geneva, International Organisation for Standardisation, 1981), 17 p. Illus.

"The helmet respirator" (173-183). 25 ref. Greenough, G. K. *Recent advances in occupational health.* McDonald, J. C. (ed.) (Edinburgh, Churchill Livingstone, 1981), 290 p. Illus.

Factory premises and workplaces

Types of factory premises vary enormously. At one end of the scale are giant factories employing as many as 100 000 workers in modern buildings constructed to comply with stringent safety and health regulations and provided with good welfare facilities. At the other extreme are small undertakings housed in slum buildings or improvised huts. The contrast is seen most vividly in developing countries where some large factories are of excellent construction with good environmental working conditions and with a wide range of welfare facilities such as a day nursery for the babies of women workers, a health clinic, dental surgery, canteen, swimming pool, basketball and football grounds, library, reading and recreation rooms, gardens, sewing rooms, cinema shows, dormitories for single workers and subsidised housing for married workers. Textile mills and petroleum refineries in developing countries have good reputations for providing such working conditions. In the same country as these modern industrial undertakings one may see many small factories housed in very unsanitary conditions, often in shanty towns with no proper drainage or roads, and no running water laid on. The fire hazard is very high and these premises are a danger to both themselves and the neighbouring houses. The floor is generally of rough earth, the lighting is poor and there is a complete lack of occupational safety and hygiene measures.

In some of these poor factories workers may live on the premises in very primitive conditions. Their children are allowed to play in the factory near running machinery and exposed to the risk of accident. The majority of factories in developing countries are small ones employing up to about 20 workers. Their owners have little capital and cannot afford factory equipment such as air conditioning, expensive dust control and exhaust ventilation systems, and safety and health facilities.

Some governments are taking active steps to build new multistoried tenement factories into which these small factories can move. Only a small rent is charged and good standards of occupational safety and health are maintained because close supervision is exercised by a general manager of the building appointed by the ministry responsible and also by frequent visits of the government factory inspector and the industrial medical officer. A first-aid post is maintained in the building for the use of all the small factory tenants of whom there may be as many as 40, each employing up to about 20 workers.

Siting the industrial undertaking

Many factors are involved in choosing the site of a new industrial undertaking. Among these are the availability of raw materials, a trained labour supply, transport facilities, nearness to the market for the finished goods, and economic factors such as the cost of land and whether the government will make a financial contribution for siting the factory in an area where there is a lack of employment, etc. In addition, it is important to ascertain if there is an adequate supply of water all the year round. Building a foundation for a new building on made ground should only be attempted when the filling is at least 7 years old and then only after engineering advice has been taken. In mining areas or where brine pumping is taking place, the foundations for a building need very careful engineering design.

Siting is particularly important for nuclear establishments since these present a great potential hazard for neighbouring communities. Consequently, before a definite decision on siting is made, an ecological study should be carried out covering factors related to meteorology, winds, rainfall and water courses, geology, microbiology and macrobiology, etc., whilst bearing in mind the magnitude of the maximum credible accident.

Zoning and town planning. A number of countries now have legislation containing very far-reaching provisions prohibiting premises to be used for industrial purposes in certain specified parts of cities and towns and in certain country areas. The object of such legislation is to reserve or zone some parts of both urban and rural districts as residential areas in which no industry or only a few specified light industries may be carried on and allow industrial undertakings to be sited only in "zoned industrial areas". Previous to the coming into force of such legislation it was not unusual for a heavy industry factory to be established in or near a residential area and cause considerable nuisance to the population.

Structural requirements and materials

Practically all countries now have very detailed regulations prescribing standards for the structural requirements, the materials to be used, the site preparation, damp-proofing, air space and ventilation, drainage, waste disposal, fire precautions, the type of sanitary facility fittings, etc. These building regulations are usually enforced by a department of a local authority or by a public health authority. In the case of industrial buildings, there may be additional regulations either administered by the same authority or by the competent authority dealing with occupational safety and health. The plans for the new building have usually to be approved by the appropriate authorities before construction begins. There are similar requirements when an existing building is taken over for industrial use. In countries where there are hurricanes, typhoons, tornadoes, floods, earthquakes or frequent lightning storms, the plans of the building should take into account the necessary protective measures. In the case of high, multistoried buildings the design must take wind pressures into account. Recent experience in some countries has revealed that wind pressures on high buildings are much greater than was estimated a few years ago.

Layout

Layout needs very careful planning to permit the most efficient use of materials and processes and at the same time provide good environmental conditions from the point of view of accident prevention and industrial hygiene. At the design stage of the factory, use should be made of three-dimensional models on a scaled floor plan. The models can be rearranged until the most efficient and safest layout has been devised. Some countries have legislation which specifies the safe distances between certain types of machines; for example that there shall be a specified minimum alley width between looms. The legislation in many countries stipulates a minimum floor area per manual worker–usually not less than 2 m². Excessive crowding in work premises increases the risk of accidents. In general, no agreement has been reached on the minimum space requirements for non-manual workers; however, lack of adequate space may cause claustrophobia.

The ergonomic factors should not be overlooked when layout of the plant and machines is under consideration. Adequate spacing around machinery or pieces of plant is necessary for the provision of access for operation and for the exercise of supervision, to facilitate maintenance work, adjustments and cleaning, and to provide space for the work in progress.

A typical recommendation is that there should be a minimum of 800 mm free space around each machine. The free space should not include portable items such as tool trays, lockers, etc., which should be treated as part of the space occupied by the machinery. Similarly, the free space should not be used for the storage of materials or encroached on by traffic and should be maintained in a non-greasy condition. In addition, a traversing part or materials carried by machinery should not approach within 500 mm of any fixed structure that is not part of the machine if anyone is liable to pass through the space between the moving and fixed parts.

To facilitate cleaning and maintenance work without causing interference to adjacent machinery, suitable platforms, safe means of access and lifting appliance suspension points should be built in where practicable. In undertakings where dirty and dusty processes are carried on, the washing facilities, sanitary accommodation, canteen and rest rooms and first-aid rooms should be sufficiently separated from the process to be kept clean but remain readily accessible to the workpeople.

Foundations and floors should be of sufficient strength to carry the loads which are to be put upon them. If existing premises are taken over for industrial purposes, expert advice should be taken as to the maximum load the floors will carry. Floors should be even and free from holes, splinters, projecting bolts, nails, valves and pipes which might cause stumbling hazards or danger to the trucking of materials. Floors, stair-treads and landings should be made of material which will not become slippery (see FLOORS AND STAIRWAYS). According to international experience, a minimum height of 3 m is necessary. Any height below this leads to difficulties in lighting and ventilation.

Noisy processes. Noisy machines or processes such as hammermills, circular saws, diesel engines, metal stamping and forging, boilermaking and engine testing should be carried out in separate workrooms with adequate standards of noise insulation to prevent the nuisance reaching workers in other parts of the factory.

Approaches and factory yards

All waterways, reservoirs and ditches in the factory compound or near the factory should be fenced off. Workers have been drowned in such places, often during foggy weather or because the surrounding areas were inadequately illuminated in the hours of darkness. Precautions should also be taken to protect workers against traffic accidents from vehicles entering or leaving the premises, or from the movement of railway engines and railway wagons inside or nearby. Maximum speed limits for all vehicles including in-plant transport such as forklift trucks, gantry trucks and tractors should be prescribed.

Where the premises are surrounded by fencing, separate entrance and exit gates should be provided for pedestrian, vehicular and railway traffic. Safe walkways should be constructed in the factory yard along the shortest line between important points.

Underground or basement workrooms

An underground workroom is one which is so situated that at least half of its height, measured from the floor to the ceiling, is below the surface of the footway of the adjoining street or the ground adjoining or nearest to the room. Provided such underground rooms are satisfactory as regards safety and health there is no objection to them being used for industrial processes; but in some buildings such rooms are most unsuitable for use as workplaces. In some countries there is legislation prohibiting the use of underground rooms as workplaces unless the competent health authority has issued a certificate authorising their use. A certificate will not be issued if the underground workroom is of unsuitable construction, of inadequate height (usually not less than 3 m), has insufficient lighting and ventilation or where adequate means of escape in case of fire are not provided. Processes of a hot, wet or dusty nature or which are liable to give off fumes are usually prohibited in underground rooms.

Fire prevention

Fire protection and prevention are a specialised matter and expert advice should be obtained concerning fire hazards at the design stage of a new industrial building or before existing premises are taken over. In many countries, the municipal building authority, the fire services and the government factory inspectorate require plans of industrial buildings to be submitted to them for approval before occupation. In some countries, the fire services or a competent municipal or government authority may issue a certificate of means of escape in case of fire which includes a detailed plan of gangways and fire exits (see EMERGENCY EXITS) with a specification giving the maximum number of persons who may be employed in each specified workroom, and requirements for the safe storage of flammable materials, etc. It is important that an adequate supply of water for fire fighting be available at the premises; this is particularly so in country districts where there is no piped water supply or where there is an acute shortage of water due to periodic droughts.

To stop fires spreading in a large factory, open spaces must be broken down into sections and compartments. For factories with a flow production line this is difficult to achieve. Special attention should be paid to the possibility of isolating stores of flammable substances and processes involving a high fire risk in separate outbuildings.

The importance of fire protection and prevention in industry cannot be overestimated. The cost to industry of fire is very high indeed and it is estimated that 40% of firms losing their premises in a fire go out of business.

Factories with extreme hazards

Combustible materials which, in dust form, are highly explosive substances, such as aluminium, cocoa, cork, magnesium and starch, should be located in outbuildings, situated at least 20 m from all important factory buildings, or if closer, should have blast- or explosion-resisting walls designed for a static load of 7 kgf/cm^2. Where it is not practicable to locate the processes in detached buildings they should be sited adjacent to the exterior walls of the building and surrounded with walls of damage-limiting construction. Such processes should never be carried on in basements where it is impossible to provide adequate explosion venting and which are difficult to enter under fire conditions. One starch factory in Texas was constructed as a multistoried building without walls and sited on the sea coast. In case of a starch dust explosion, there were no walls to be blown out; in addition, the very good ventilation provided by the shore breezes prevented accumulation of dust, thus reducing the risk of an explosion.

Explosives factories

These are subject to stringent government regulations in most countries and are usually required to be situated in non-populated areas. The individual buildings are spaced about 100 m apart, only one process is carried on in a building and there are blast walls or earth embankments surrounding them.

High fire risk factories

Special precautions are required in the design of factory buildings in which there is a high fire risk such as in oil refineries, and in the manufacture of cellulose solutions, foam resins, flammable gases and solvents. Most countries have very stringent regulations regarding the storage, handling and processing of these dangerous substances, and in some cases a licence has to be obtained from the competent government department before such materials can be processed. The most important safety and health factors are the provision of safe storage of these materials, their distribution to the processes by a controlled pumping system and not by manual handling of containers, the absence of open lights, flames or any source of ignition, and flameproof electrical installations and fire-resisting buildings.

Ventilation and heating

Most countries have legislation requiring a minimum cubic capacity of air space per worker, which is generally not less than 10 m^3. To maintain suitable atmospheric conditions in workrooms is often very difficult as there are so many factors involved, such as variations of temperature, harmful draughts, heat or fumes or dust given off by processes, the need to open doors to pass goods into and from the factory, excessive humidity caused by wet processes or excessive dryness caused by hot processes, and the individual idiosyncrasies of the workers. Some workers wear heavier clothing than others in the same workroom and it is not unusual for some workers to complain that it is too cold while others consider it to be either satisfactory or else too warm. A further complication is that, in the same workroom, some workers may have a sedentary occupation while others are very active. In tropical countries, which are often very humid, the problems of ventilation, temperature and humidity are more serious, especially for small and medium-sized factories which cannot afford to install air conditioning. Moreover, many textile factories in hot countries have installed artificial humidification usually resulting in very unsatisfactory conditions. Generally, minimum recommended standards for fresh air requirements are around 10 cubic feet (or 0.28 m^3) per minute per person. It is also possible to relate this requirement to the dimensions of a workroom in which case 0.2 cubic feet per square foot (or 62 l per square metre) of floor area per minute is a useful guide.

Where exhaust ventilation has to be provided for the removal of dusts, fumes, gases, mists or vapours, care should be taken at the planning stage to design the most efficient system. Too often the processes producing obnoxious polluted air are situated at a considerable distance from the outside atmosphere or dust-collecting plant. Long lengths of ventilation ducting are necessary, whereas if the processes are placed near the outside walls it is much easier to remove polluted air. Care should be taken that dust, fumes, gases, etc., exhausted from workrooms do not re-enter any part of the factory. See also AIR CONDITIONING; EXHAUST SYSTEMS.

Lighting

The standards of lighting for industrial processes and even in offices are often very poor, resulting in eye strain, an increased risk of accidents and faults in the work produced. Even in tropical countries where natural daylight presents a dazzling glare in the open air, the lighting inside workrooms is often inadequate because of window blinds to keep out the heat of the sun, the dirty inside walls which reflect very little light and the practice of stacking goods in front of windows. Seldom does a factory management study the illumination problem, know what standards of lighting are required for the various processes or have a light-meter or visibility meter to measure the lighting. Failure to have a programme for the regular cleaning of windows and electric lamps leads to a progressive diminution of light standards. In very large workrooms, workers engaged near the windows often have sufficient light in the daytime but those in the middle of the workroom do not. For the most effective use of natural lighting, there should be a floor/window area relationship, which in the case of a single-storey building should be at least 3:1.

In single-storey shed-type buildings with roof windows, such as are often used for weaving mills, it is usually necessary during the hot season to whitewash the windows for the purpose of mitigating heat or glare. The ratio of window area to floor area is, however, usually as high as 2 : 5 and sufficient natural light is admitted without having to resort to artificial lighting.

In planning the factory premises, attention should be paid to the importance of having the inside walls and ceilings of a light shade of colour to give a high degree of reflection. In some countries there are legally prescribed minimum standards of lighting for certain factory processes. It should be stressed that these are minimum standards and it is advisable in many cases to adopt the higher standards recommended by national safety councils or one of the illuminating engineering societies.

Some modern buildings consist of a steel framework covered with glass windows. These are apt to be overheated during the warm weather unless air conditioning and window blinds are provided.

At the design stage of the building, the problem of window cleaning should be considered as this is one of the most dangerous of occupations. Either provision should be made so that all cleaning can be done from inside the building by having window sashes which swing back into the room to eliminate the washing of windows from the outside, or a properly designed monorail with cage or boatswain's chair should be provided. Where it is necessary for window cleaners to use safety belts and lifelines, fastening anchors or devices should be provided to which the safety belt can be securely attached. It frequently happens that window cleaners employed by outside contractors find, on arrival at the building, that their safety belts cannot be used because there is no anchoring device to which they can be attached. (see FALLS; FALLS FROM HEIGHTS, PERSONAL PROTECTION AGAINST).

Some modern buildings are windowless, it being claimed that it is easier to maintain a constant temperature and relative humidity if no windows are provided. Such buildings must be provided with auxiliary emergency lighting in case of a failure of the electricity supply, and the fire exits need careful planning as there are no windows through which escape can be made in case of fire. Working in a windowless building may have an adverse psychological effect of claustrophobia on some workers.

When considering the lighting of a building a decision must be made about the respective roles of electric light and daylight. In general there are three options. First, to rely on natural light during normal daylight hours and to design the electric lighting system to meet requirements after dark. If heavy clouds cause an exceptional fall in daylight level, electric lights can be used as a temporary supplement to daylight. This kind of system is most applicable to shallow interiors such as small offices where the rapid reduction in daylight level with increasing distance from the windows can be accepted.

Secondly, electric lights can be used as a permanent supplement to daylight. This kind of system is applicable to deep interiors with side windows where, except in perimeter areas, natural light cannot provide the illuminance required for many tasks. However, the system needs to be planned carefully otherwise the illuminance provided by the electric lights at night may be inadequate.

The third and final option is to rely on electric lights only with daylight excluded either because the area is windowless or the windows are very small. This option is usually adopted when, for example, an occupation calls for a fully controlled environment.

Typical modern standards in general suggest that no working space that is to be continuously occupied should have an illuminance less than 200 lx. In interiors without daylight the illuminance should be not less than 300 lx for casual work and 500 lx for other tasks. Areas where detailed and accurate work is carried out may well require illuminance of over 1 000 lx.

Stroboscopic patterns produced in rotating machinery and other moving objects by cyclic variations in light output are annoying if the pattern is seen on a task that is being closely examined and can be dangerous if visible on moving machinery. The effects can be reduced by using lamps with shielded electrodes, by screening the ends of the lamps, by mounting incandescent lamps near the task to supplement the light over the critical area or by using special circuits.

(Further information is provided under LIGHTING.)

Waste disposal

The majority of industries or industrial processes produce some sort of waste materials; provision for their safe disposal should be made at the planning stage for new buildings, or suitable measures should be implemented in existing buildings. Only the basic principles concerning the factory premises are considered here.

Liquid wastes. The problem of waste disposal is becoming more difficult as many governments and public authorities are now taking steps to prohibit the water pollution of rivers, etc., by waste materials of a toxic or obnoxious nature. Where such waste liquids are produced in the manufacturing process they should be diluted, neutralised, filtered, settled or otherwise chemically treated before being discharged into a stream or river, or on to open land. If discharged on open land, as is sometimes the case where no public sewerage system exists, care must be taken that there is no danger of causing pollution to nearby wells used as a source of drinking water. If a public sewerage system is available, permission should be obtained from the authorities to use it for the disposal of liquid wastes. Under no circumstances should toxic, corrosive, flammable, or volatile materials be discharged into a public drainage system.

Radioactive wastes. The disposal of such wastes should conform with the requirements of the competent authority.

Solid wastes. Some wastes can be burned or placed on waste dumps. Combustible materials such as wood and paper can be burned in an incinerator; poisonous and flammable materials require special precautions and procedures.

Smoke, fumes and dust. Public opposition to air pollution is growing and more stringent legislation is being enacted against smoke, fumes and dust from industrial undertakings. Tall factory chimneys are often necessary to diffuse gases into the atmosphere. The efficiency of this method depends upon the nature of the gases or smoke, the location of the plant, atmospheric conditions and the prevailing winds. Rain may absorb some of the harmful gases and cause heavy damage to crops in the neighbourhood. Smoke from factory chimneys in combination with unfavourable atmospheric conditions may cause a nuisance to nearby airports by reducing visibility. To prevent air pollution, waste gases should be purified using scrubbers, bag filters, cyclones or electrostatic precipitators, etc. Often valuable substances may be recovered from the collected material and the waste heat utilised for factory heating.

QUINN, A. E.
WICKAM, E. M.

"Premises of industrial establishments". Ch. II (4-24). *Model code of safety regulations for industrial establishments for the guidance of governments and industry* (Geneva, International Labour Office, 1959), 523 p.

Modern workplaces. Report of a meeting in Dortmund 16-17 May 1978 (Moderne Arbeitstätten. Vorträge der Informationstagung 16-17 Mai 1978 in Dortmund). Schriftenreihe Arbeitsschutz Nr. 19 (Dortmund-Dortsfeld, Bundesanstalt für Arbeitsschutz und Unfallforschung, 1978), 189 p. Illus. Ref. (In German)

Health aspects related to indoor air quality. Report of a WHO Working Group. EURO reports and studies 21 (Copenhagen, World Health Organisation Regional Office for Europe, 1979), 32 p. 54 ref.

Falls

Accidents resulting from falls are of two kinds: those due to a person falling from a height or on the level and those due to a person being struck by a falling object. Such accidents account for a large percentage of total industrial accidents in all countries.

[Table 1 shows the incidence of this type of accident in the United Kingdom.]

Not all countries keep detailed statistical records of occupational injuries classified according to the type of accident but from the available information it appears that falls of objects account for about 20-25% of the total occupational accidents which occur in factory processes, with a higher percentage in building and civil engineering work.

[Table 1. Fatal industrial accidents caused by falls in the United Kingdom, 1972 to 1979

	1972		1973		1974		1975		1976		1977		1978[1]		1979[1]	
	No.	%	No.	%	No.	%	No.	%	No.	%	No.	%	No.	%	No.	%
Falls of persons	68	29.9	54	18.6	52	17.9	44	19.0	39	18.5	42	20.4	30	16.2	28	15.6
Falls and other movement of objects	27	10.3	27	9.3	30	10.3	19	8.2	20	9.5	19	9.2	18	9.7	17	9.5
Total	95	36.4	81	27.9	82	28.3	63	27.3	59	27.9	61	29.6	48	25.9	45	25.1
Incidence rate per 100 000 at risk	1.5		1.3		1.3		1.1		1.0		1.0		0.9		0.8	

[1] Provisional figures not fully comparable with those for earlier years; they are adjusted from a new accident classification first used in 1978.
Note: % = percentage of all fatal industrial accidents.]

In mining and quarrying, falls also account for a high proportion of total accidents [for example in 1978 in the United Kingdom coal mines there was a total of 40 378 underground accidents resulting in disablement for more than 3 days, of which 9 400 (23.3%) were caused by falling objects, including falls of ground, and 9.810 (24.3%) by stumbling, falling or slipping (excluding accidents in connection with haulage operations)]. In some countries the number of accidents in industry caused by falls is greater than that caused by machines.

Classification of falls

The Tenth ILO International Conference of Labour Statisticians in 1962 adopted a classification of industrial injuries according to types of accident, in which all industrial injuries are classified under 10 classes, with subdivisions in each class. The two relevant classes for falls are as follows:

Falls of persons

(a) Falls from heights (trees, buildings, scaffolds, ladders, machines, vehicles) and into depths (wells, ditches, excavations, holes in the ground);

(b) falls of persons on the same level.

Falling objects

(a) Slides and cave-ins (earth, rocks, stones, snow);

(b) collapse (buildings, walls, scaffolds, ladders, piles of goods);

(c) falls of objects during handling.

There is one difficulty regarding the classification of accidents that should be mentioned. An accident is rarely the result of one cause but is usually due to several connected factors and circumstances. Thus, the fall of a person does not itself furnish information on the cause of the fall: it is merely an indication of the manner in which an event occurred. It is more important to know why the person fell: the floor may have been slippery; the victim may have stumbled over an object or been pushed by a moving object; a hole in the ground may not have been properly railed off; the injured person may have violated a safety regulation by neglecting to use guards, etc. However, in numerous national classification systems, such "falls" are considered as "causes".

Types of falls

There are innumerable kinds of falls in all sectors of employment. Among the falls from a height or on the level may be mentioned those caused by falls from buildings or works of engineering construction; falls from trees (coconut and toddy juice collectors); falls through floor openings and unprotected lift shafts, into ship's holds, from ladders, into open vats, pits, or tanks containing scalding, corrosive or poisonous liquid, from teagle openings, or into silos; falls down stairways or from the open side of an unguarded stairway or overhead platform, down stairways, over the side of a fishing trawler, from any overhead or raised workplace, or from vehicles; falls due to slippery or unsafe or uneven floors, or due to wearing unsuitable footwear. Examples of causes of accidents from falling objects are roof falls in mining, falling rock in quarry work, collapse of an overloaded floor in a building, falls of material in the demolition of a building or industrial plant, dismantling scaffolding on a building site, collapse of overloaded lifting machinery, collapse of insecurely stacked goods or waste tips, falls of tools used by overhead workers, falls of unsecured overhead material, falls of coconuts from trees and falling trees in forestry work.

Workers at risk

In some occupations the risk of meeting with an accident due to a fall is very high. The most dangerous industries in this respect are mining and quarrying, building and works of engineering construction. In mining the great danger is from roof falls, while in quarry work large pieces of rock or stone sometimes fall from the quarry face especially when blasting has recently been carried out. Some quarry workers have from time to time to work on a steep inclined quarry face in drilling holes for blasting or in dislodging large pieces of stone. Unless they have safety ropes they may lose their foothold and fall down the quarry face.

The second group of occupations in which there is a high risk of falls includes building and civil engineering (e.g. construction, structural alteration, repair, maintenance, redecoration, external cleaning and demolition of buildings or works of engineering construction).

Accidents from falls are particularly prevalent in building and civil engineering work in many developing countries where makeshift equipment and unsafe working practices are allowed to persist for lack of adequate legislation or law enforcement procedures. In one case, a cinema was being demolished and instead of taking down the first-floor circle piecemeal, the contractor arranged for the periphery to be cut away from the supporting walls. Before the full periphery had been cut away the whole floor collapsed and killed 10 workers below. The safety of equipment on construction sites in some countries is far from adequate. Old bicycle wheels may be used as pulley blocks for lifting material and cranes may have their jibs lengthened to give a wider working radius, thus increasing the risk of the crane overturning because the anchorage is inadequate for the extra load imposed. Night work on building sites is often carried on with very poor standards of lighting, which increases the risk of workers falling into unguarded holes, etc. One very common accident in excavation or trenching work is the caving in or collapse of unshored earth walls and the consequent burial of the workers.

PREVENTION OF FALLS

The prevention of falls from a height necessitates the provision of suitable means of access, adequate fencing and in certain cases personal protective equipment. Falls of material and ground may be prevented by improved working and storage techniques, adequate shoring, and the provision of overhead protection, etc.

Construction sites. Adequate shoring for trenches and excavations is necessary to prevent workers being buried in slides of earth or rock. Many workers and contractors often assume the ground which is being excavated is so firm that it will not collapse. Experience, however, has shown that no matter how solid the earth or rock may appear to be, there is always a risk that some part of it may fall and cause injury to workers. Excavations should be fenced off to prevent workers inadvertently falling into pits or trenches.

During the erection of buildings and works of engineering construction, the provision of guard rails or scaffolding, and the covering over or fencing off of any floor openings are essential. The difficulties of maintaining high standards of safety in this kind of work are due to the fact that conditions change daily and that many different contractors are employed. The workers of one contractor may complete a job and leave it in a dangerous condition so that later an accident is caused to an employee of another contractor. For example a floor opening for a liftshaft is left unfenced in a badly lit part of the building and some worker engaged on other work has to pass nearby and falls down the opening. Falls of tools and the throwing down of used material such as scaffolding also cause many accidents. Ladders should be in good condition and secured at either the top or bottom place of rest. They should rest on firm ground and the top of the ladder should extend at least 1 m above the place of landing.

Workers engaged on work on sloping roofs should be provided with secured cat-ladders or crawling boards or other means of preventing them from falling or slipping, such as safety belts and lifelines, or guard rails (see FALLS FROM HEIGHTS, PERSONAL PROTECTION AGAINST; ROOFING). Safety nets are now widely used in certain classes of construction work, such as bridge building, as a safety provision against falls of workers. Canvas sheets are often used by building, demolition, engineering, painting and other contractors where overhead work is in progress to catch falling tools, materials, debris, etc. These sheets are not intended to safeguard the lives of overhead workers and are also not suitable for use in corrosive atmospheres. Where workers are liable to fall into water during bridge-building or when engaged in dock work, means of support at water level should be provided, by vertical chains at intervals of every few metres or, where the water level is constant, by horizontal chains. Vertical ladders should be provided at reasonable intervals and a supply of life-saving appliances (lifebuoys, etc.) should be at hand.

Docks. All breaks, dangerous corners and edges of a dock, wharf or quay should be fenced to a height of 80 cm. Any waterways, deep streams or canals in or near the approach to an industrial undertaking should be fenced off and well lit during hours of darkness. Accidents usually occur on foggy days when workers lose their direction.

Window cleaning. Safety harnesses consisting of a safety belt and rope are sometimes used by window cleaners and other workers engaged in high-level work. These belts should be kept in very good condition, and periodic examination is necessary. The maximum fall permitted should not exceed about 2 m since the shock of a sudden arrest of a longer fall may cause internal injuries to the worker. These belts require a point of anchorage and many buildings do not have any such anchorage points on the outside, to which the window cleaner may attach the harness. If a worker works above the point of anchorage, his possible fall could be twice the length of the rope and the acceleration attained in such a fall might overstrain the belt and will most likely cause internal injuries. Every year, falls of window cleaners and workers employed on high-building construction work cause many fatal accidents, while the non-fatal accidents often result in severe fractures and paralysis. It is difficult to get workers such as window cleaners and roof coverers to use safety precautions against falls if they are paid on a piece-rate basis. In some countries a window cleaner is not subject to any government safety legislation as regards his own safety but is liable for damages and for a possible prosecution under the police regulations if he causes anything to fall on a passer-by in the street. In other countries there are very detailed government regulations concerning the safety of window cleaners, who are considered to be building operatives within the scope of the building construction safety regulations.

Industrial plants. In factory work the following are the chief safety measures to be adopted. Workers should be trained to observe safety procedures whenever work is to be done in which there is a risk of a fall or a falling object. Safety boots should be provided by the employer where there is a risk of foot accidents. It is difficult to get workers in tropical countries, where the traditional footwear is the open sandal, to adopt heavy reinforced boots. Flimsy shoes and worn-out footwear should not be worn in factory work.

Safe means of access should be provided to every place where work is to be carried out. Overloading of floors, especially in old buildings, should be avoided. Expert advice from an architect or the local building authority should be obtained as to what loading any floor will safely carry.

Floors should be kept in good condition, free from holes, irregularities, uneven surfaces, accumulations of waste, oil or slippery substances. Where wet processes are carried out, adequate drainage or sills should be provided to prevent the water reaching the part of the floor used by workers. Non-skid strips should be placed on floors where a slip or fall might prove exceptionally dangerous, such as at a circular saw. Stairs should be provided with handrails on both sides, and no stairs should have an open side. Non-skid strips on stair treads are useful (see FLOORS AND STAIRWAYS).

Tanks, vats or vessels containing dangerous substances should be covered over or protected by guard rails. The correct stacking of goods to prevent their collapse is important. If any work is carried out near overhead electrical equipment, the current should be cut off. A slight electrical shock, not sufficient in itself to cause physical injury may, however, cause a worker on a ladder or scaffold to lose his balance and be seriously injured by a fall. The manual handling and carrying of excessive loads, very bulky materials or objects difficult to hold should be avoided and mechanical handling be used wherever possible.

In some industries the floors become very slippery from the materials manufactured, for example in soap manufacture. Frequent cleaning is necessary. In some cases it is possible to scatter sand over slippery surfaces such as ships' decks when handling cargo in frosty weather. Safety helmets are a partial protection against falling objects but a worker wearing one can still suffer severe injury to parts of his body not covered by the helmet, and the helmet does not afford 100% protection from very heavy objects falling a great distance. The emphasis in safety should be on the prevention of falling materials. Some tools used by workers at heights can be securely attached by means of a lanyard to the wrist or belt.

Workers in silos or on loose materials such as stacks of wet gravel, sand, small-size coal and grain should be provided with safety harnesses to prevent them being buried should a mass movement of material cause them to lose their foothold.

Other industries. In forestry work, especially in tropical rain forests where large trees are to be felled, the accident risk is very high and special precautions are dealt with in the article FORESTRY INDUSTRY. The articles on MINES, SAFETY IN; MINES, OPENCAST; QUARRIES deal with the hazards of falls in these branches of occupation.

Medical precautions. All workers who have to work at heights from which there is a risk of falling should be given pre-employment and periodic medical examinations to check their suitability. There have been cases of workers subject to epileptic attacks, "blackouts" or similar illnesses who have taken employment involving access to dangerous overhead workplaces and have failed to inform their employer of disability for such work. Workers who suffer from balance disorders, vertigo, etc., should not be employed on this type of work (see EQUILIBRIUM).

QUINN, A. E.

Safe friction coefficient:

CIS 77-606 "Work surface friction coefficients: A survey of relevant factors and measurement methodology". Pfauth, M. J.; Miller, J. M. *Journal of Safety Research* (Chicago), June 1976, 8/2 (77-90). Illus. 27 ref.

CIS 77-1693 "Prevention of falls due to slips–Adherence of safety footwear soles–Results of the second series of tests" (La prévention des chutes par glissade–Adhérence des semelles des chaussures d'atelier–Résultats de la seconde campagne d'essais). Tisserand, M. *Travail et sécurité* (Paris), Apr. 1977, 4 (180-185). Illus. (In French)

Protective screens:

CIS 78-1437 "Protective scaffolding and protective screens" (Fanggerüste und Schutzwände). Quentin, N. *Sicherheitsingenieur* (Heidelberg), Sep. 1977, 8/9 (10-14). Illus. (In German)

Window cleaning:

CIS 75-2051 "Occupational hazards in window cleaning". Ribeiro, B. F. *British Medical Journal* (London), 30 Aug. 1975, 3/5982 (530-532). Illus. 6 ref.

CIS 80-496 "Window cleaning safety". Purchon, D. *Health and Safety at Work* (Croydon), Dec. 1979, 2/4 (44-45). Illus.

Falls from heights, personal protection against

The disturbing nature of falls from a height is clearly shown by national statistics of industrial accidents. Though they are generally less frequent than accidents of other kinds, such as those due to the handling of goods or falling on the level, they often take first or second place on account of their serious consequences. In many countries priority is given to research on building methods or installations designed to eliminate all risk of falling from one level to another. Where no such methods or installations are yet available hand-rails must be used, and if full prevention is impossible resort should be had to collective protection, with measures to limit the height and consequences of falls, such as the use of safety nets or, in exceptional cases, shock-absorbing mattresses. Where such measures are inadequate, however, or while they are being adopted or where they can legitimately be dispensed with, for example where the work at a height is occasional, trifling or of short duration, lasting perhaps a day or less, personal equipment for protection against falls must be provided for workers.

Essential features of protective equipment

Components. Protective equipment ready for use always has a connecting system fastened to a device for attaching the body, the other end of the system being firmly attached to an anchoring point.

Anchoring points. This equipment can be used only if there is a stable and comparatively rigid structure near the raised workplace where accessible and reliable anchoring points can be established.

Automatic working. The purpose of any such equipment is automatically to block or brake the fall of the person who has lost his foothold.

This personal equipment is not to be confused with other devices for individual use that it may be added to in whole or in part for protection against falls, for example the "belts" used in work on the poles or pylons supporting overhead lines or the hoists used in vertical contact work or for rescuing persons (see SAFETY BELTS).

Classification of connecting systems (figure 1 overleaf)

Braking: A person using the equipment properly connected, if he loses his foothold, experiences an initial phase of free fall through a height of Ht_0 during what is known as the time of response t_0 of the equipment, before this begins to stretch and become distorted through elongation (figure 2). He thus acquires a kinetic energy of $M \cdot g \cdot Ht_0$, where M is his mass in kilograms including that of the equipment, possibly increased by any load he may be handling or holding when he loses balance that is not relinquished during the braking process, and g is the acceleration due to gravity, which is about 9.80 ms². This kinetic energy is approximately equal to the potential energy lost during the fall and may be increased by any initial energy of propulsion that may

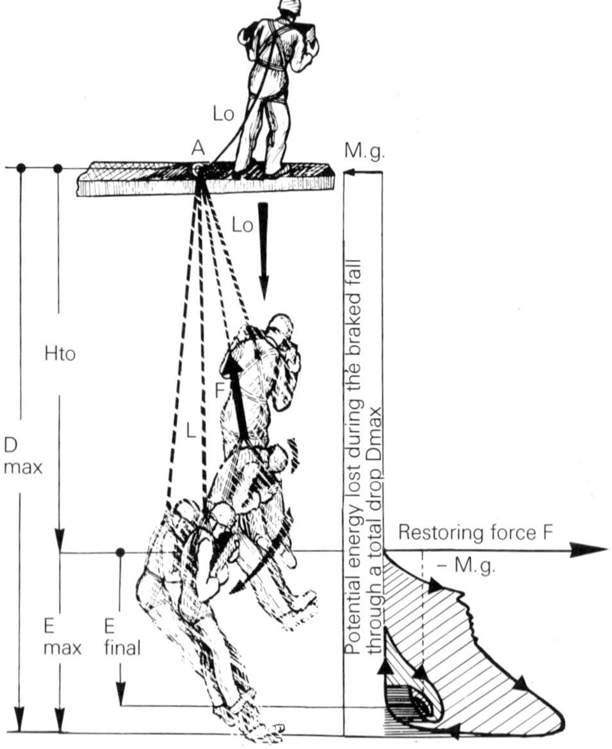

Figure 2. Braked fall of a worker with connected equipment.

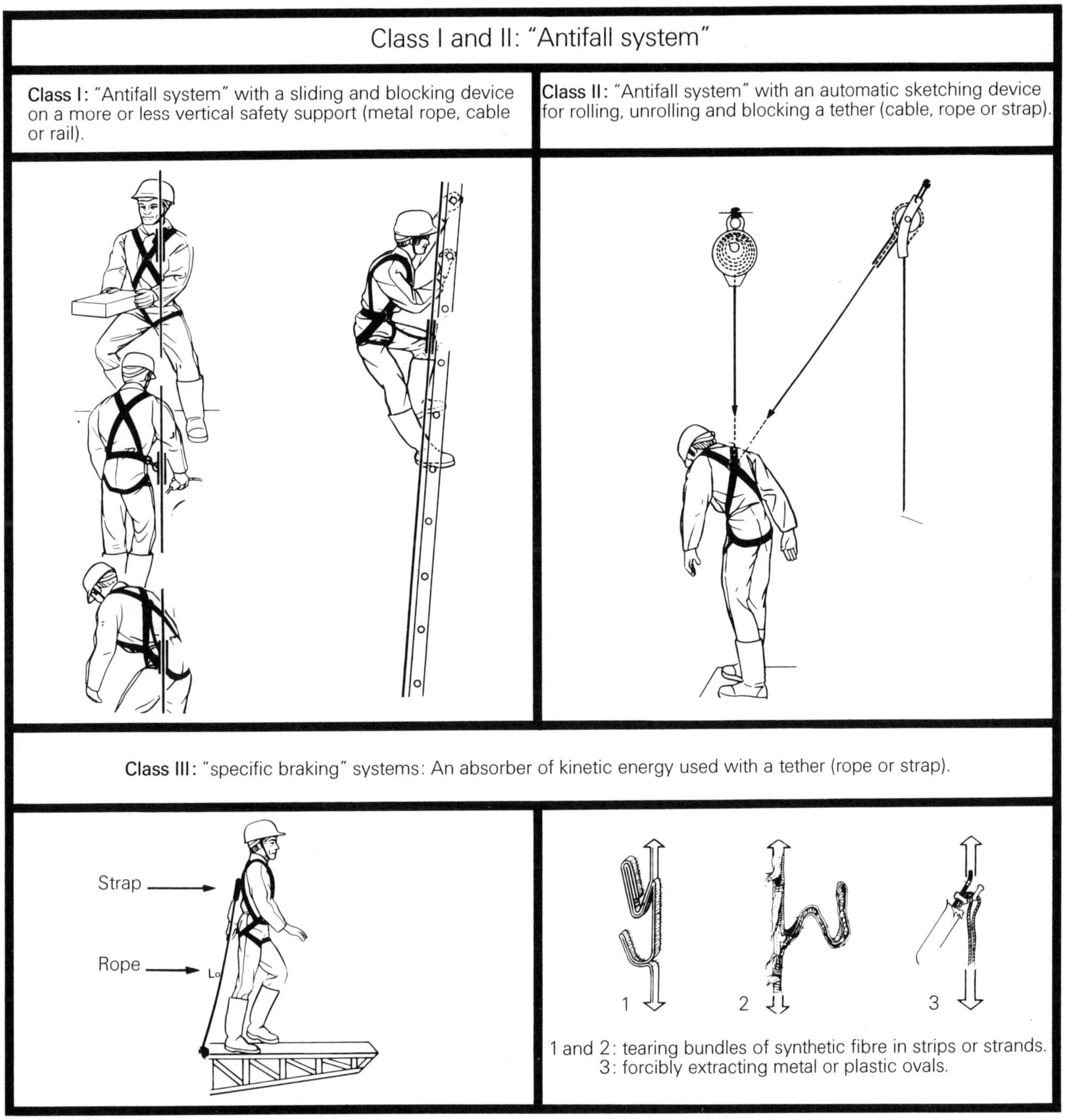

Class I and II: "Antifall system"

Class I: "Antifall system" with a sliding and blocking device on a more or less vertical safety support (metal rope, cable or rail).	**Class II**: "Antifall system" with an automatic sketching device for rolling, unrolling and blocking a tether (cable, rope or strap).

Class III: "specific braking" systems: An absorber of kinetic energy used with a tether (rope or strap).

Strap

Rope

Lo

1 2 3

1 and 2: tearing bundles of synthetic fibre in strips or strands.
3: forcibly extracting metal or plastic ovals.

Figure 1. Classification of commercial connecting systems.

have caused the loss of balance. In other words the user acquires an amount of movement $\overline{M \cdot t_o \cdot g}$ that may be added vectorially to any initial amount of movement due, for example, to loss of balance. The kinetic energy is absorbed and the amount of movement is neutralised by the restoring force (or tension) $\overline{F}$, which acts on the weight $\overline{M \cdot g}$ of the user between the beginning of braking and the moment of maximum elongation. *E* of the equipment. The kinetic energy and the amount of movement form, each respectively, integral parts of the force-elongation diagram and the time range of the restoring force, which can be recorded by means of a suitable installation (figure 3).

In conclusion: (1) The values of the average (or maximum) intensity of the restoring force and/or the duration of its action, the maximum elongation of the equipment and the free space needed below the working

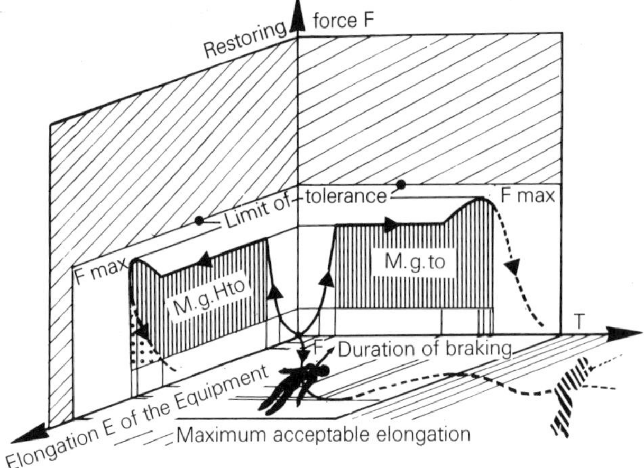

Figure 3. Kinematics and dynamics of braking.

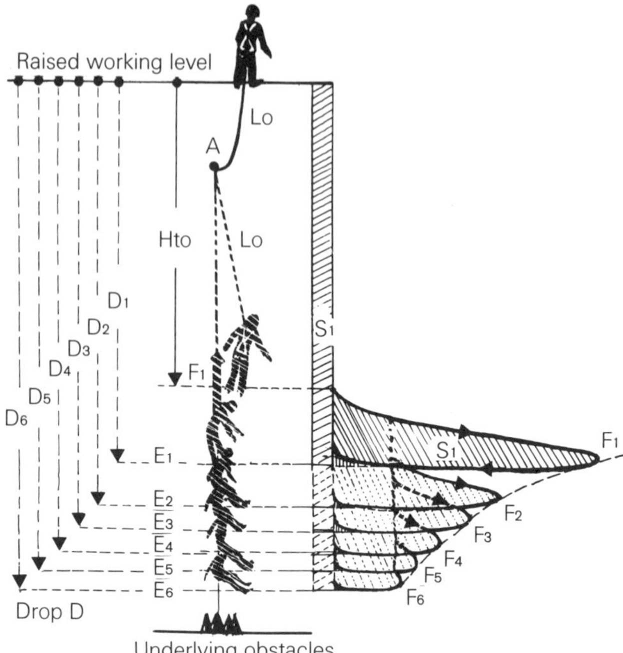

Figure 4. Effect of braking method on the distance the worker falls. The drop is greater as the restoring force exercised by the stretched equipment is weaker.

level may increase as the user (who may be overloaded) is heavier or the response time of the equipment is longer. (2) The user and the response time remaining unchanged, the period of action of the restoring force and the free space needed below the working level are greater for models of equipment with which the restoring force is weaker (figure 4).

Ergonomic criteria

Constraints on the user or his equipment. The user or his equipment may be subjected to various constraints during work and movement from place to place. If the user is hampered to excess he may become overtired or make dangerous movements, and the constraints due to loss of balance may result in bodily injury or damage to the equipment or even its destruction.

Rules for the most suitable compromise. In order to limit the foreseeable effects of such constraints the equipment must be designed with a view to the most suitable compromise between various conflicting factors tending towards the least possible discomfort and the best possible protection of the user.

It is desirable in particular—

(a) to limit the maximum response time of the connecting system without too negative an effect on the user's freedom of movement (radius of action from the anchoring point, for example);

(b) to keep the restoring force within a certain limit of tolerance, at the same time avoiding an excessive braking distance;

(c) to ensure a coefficient of safety (or utilisation) in the maximum acceptable restoring force great enough to leave room for a minimum static braking resistance for the equipment and a minimum static breaking resistance for the anchoring points that is not excessive.

Human tolerance of braking shocks. This depends on the amount of kinetic energy to be absorbed and the method of braking—in other words on the intensity and duration of action of the restoring force—and also on the relative direction of this force and the method of attaching the body. Data concerning human tolerance of shocks occurring when the canopy of a parachute opens are of great interest, despite the obvious differences between deliberate jumps by experienced personnel and the braked fall of a worker who has unexpectedly lost his balance. In France, for example, an analysis of statistics relating to more than a million jumps observed over a ten-year period (1957-66) has shown that injuries caused by the opening of the canopy are extremely rare with what are known as "lines first" parachutes developing in about 2 seconds a maximum restoring force of 6 000 to 7 000 N, for an average value of the mass of the human body of about 75 kg. More recent models develop, other things being equal, a maximum force of 10 000 N for dorsal parachutes and 12 000 N for ventral (reserve) parachutes. These data, which concern only the sub-pelvic method of attaching the body, are all the more useful as the dangers of a simple static suspension of the human body without jolts or any possibility of support, which occurs with abdominal or chest belts, have been clearly shown during tests that have led to the systematic adoption of harness (figure 5).

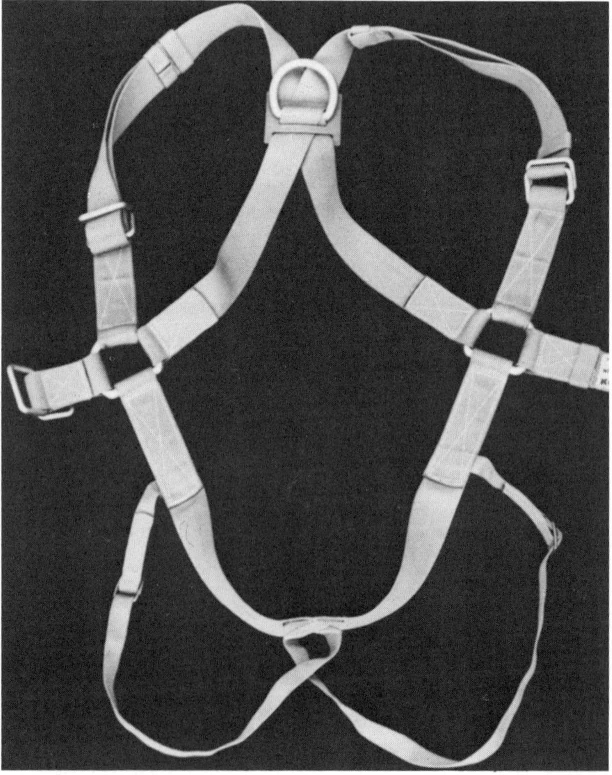

Figure 5. Sub-pelvic harness. *(By courtesy of the INRS, France.)*

Standardisation of equipment

1. Non-injury braking capacity—

(a) Maximum acceptable restoring force (or deceleration).

At present specifications are laid down in the national standards (or advanced draft standards) of very few countries (see table overleaf).

(b) Maximum acceptable drop.

A maximum acceptable value must be specified for the drop of the trial mass (at the start the mass is placed as

Method of attach-ment	United States		United Kingdom	France	
	Max. force	Decele-ration	Maximum decele-ration	Max. force	Decele-ration
Belt	8 000 N	10.g	5.g	–	–
Harness	17 000 N	25.g	10.g	6 000 N	5.g

high above the testing bed as the connecting system permits). In the French standard, for example, the maximum acceptable drop depends on the class of the connecting system (figure 1) and must not exceed in given conditions of temperature (−20°, +20° and +50° C): 0.60 m antifall systems of classes I and II and $2.75 \cdot L_o$ for those of class III of initial length L_o (which must not exceed 2 m).

2. Conversion factor and coefficient of safety (or utilisation)–

(a) Conversion factor.

For given values of the height of the fall Ht_o and the mass to be braked (M = 100 kg, for example), the maximum values of the restoring force *F* and of the elongation of a model of equipment (or its connecting system) depend on the nature and the viscoelastic properties of the mass. When the test for non-injury braking is carried out with a rigid mass *RM*, the maximum force recorded *F (RM)* is equal to or greater than that–*F (DUM)*–obtained with a semi-rigid dummy or the human body. The conversion factor *C* is equal to the value of *F (RM/F(DUM))*, and it may be regarded as a coefficient of safety when it is greater than unity, as it is with connecting systems of classes I and II in particular.

(b) Coefficient of safety (or utilisation).

The choice of too high a limit of tolerance and an adequate coefficient of safety results inevitably in excessive values for the minimum static resistance to breaking required for the equipment and for the anchoring points. The following table shows specifications in force in France and the United States.

Method of attachment	Minimum static resistance to breaking	
	United States	France
Belt	26 600 N	–
Harness	55 600 N	20 000 N

The future

International action to promote standardisation should lead to the unification of the principles in force in the various countries.

Criteria determining choice. The purchaser must specify in detail to the supplier he consults the working conditions that make the use of the equipment necessary and any factors in the environment or any products used that may adversely affect its maintenance in good working order.

The supplier should be required to: (1) deliver models conforming to the specifications of the national standards in force and bearing whenever possible a trademark or a stamp of approval; (2) provide instructions for use and maintenance so that the equipment can be used properly in the prevailing working conditions and kept in good working order.

Recommendations for use at the workplace. For the supervisor or foreman:

(a) The competent person must point out to the user working under him the safe anchoring point to be used and the right model of equipment.

(b) The competent person must verify before and during the work that the available space is clear of any permanent or temporary obstacle that the user or his equipment might come into dangerous contact with following loss of balance.

(c) The fixing (and removal) of the device for attaching the connecting system to the anchoring point and the performance of the work can be carried out only where climatic conditions permit and then under the supervision of the competent person.

(d) The competent person must make sure that the instructions for use and maintenance supplied with the equipment are followed strictly. He must also carry out periodic visual checks and examine the equipment before and after it is used.

For the user:

(a) The user must actually use the equipment and anchoring point provided for him.

(b) In putting on the device for attaching him, he must place, fix and adjust the straps correctly. They are well adjusted when they are neither loose nor too tight for him to make the movements and adopt the positions required by the work.

(c) No modification whatever must be made to the equipment.

(d) It must never be used for any purpose but individual protection against falls.

(e) The user must constantly make sure that his connecting system is never tangled up or rolled round an obstacle; rubbed or jammed for long against an obstacle that is rough or has sharp edges or points; in contact with dirty, hot or corrosive substances; exposed to harmful radiation (such as infrared or ultraviolet rays from a workplace where arc welding or cutting is carried on).

Advice to be followed in case of loss of balance

(1) The person should avoid seizing a neighbouring object that is obviously heavy and may be dragged away and held during the braking process.

(2) The head should immediately be bent forward with the chin pressed firmly against the breastbone and this position should be maintained until the body comes to rest.

(3) Where there is imminent danger of striking against part of the structure, the feet should be used for preference and the legs should be bent to deaden the shock.

(4) At the end of the braking process, where the method of attaching the body permits, the hands should be used to gain support from the connecting system or a neighbouring part of the structure with a view to improving the half seated position on the strap passing under the buttocks or to reducing slightly the tension of the straps. The person who has fallen should try to remain calm while waiting for help and, if it is necessary to attract the attention of the supervisor or some other worker in the vicinity, he should call for help without gesticulating.

ULYSSE, J. F.

"Personal equipment for protection against falls from a height. Test findings, criteria governing choice, recommendations for use (Equipement individuel de protection contre les chutes de hauteur: bilan des essais, critères de choix, recommandations d'emploi). Ulysse, J. F.; Roure, L.; Tisserand, M.; Jayat, R.; Christmann, H.; Schouller, J. F. *Travail et sécurité* (Paris), July-Aug. 1978 (404-431). Illus. (In French)

"Fractures of the spinal column observed after 1 188 155 parachute jumps" (Les fractures du rachis observées au cours de 1 188 155 sauts en parachute). Teyssandrie, M. J.; Dealhaye, R. P.; de France, G. M. *Gazette médicale de France* (Paris), 25 June 1968, 75/18 (3 823-3 828). Illus. 17 ref. (In French)

"Safety equipment in building and public works: apparent lightness and apparent ergonomic soundness in personal equipment do not always mean a life saved after a fall from height" (Le matériel de sécurité dans le bâtiment et les travaux publics: la légèreté et l'ergonomie apparentes ne sont pas toujours synonymes de vie sauvée après une chute dans le vide). Noel, G.; Ardouin, G.; Archer, P.; Amphoux, M.; Sevin, A. *Annales de l'Institut technique du bâtiment et des travaux publics* (Paris), June 1978, 362 (55-68). Illus. (In French)

Personal equipment for protection against falls. Standard NFS 71-020. (Equipement individuel de protection contre les chutes. Norme NFS 71-020). (Paris, AFNOR, July 1978). (In French)

"Selecting fall arresting systems". Sulowski, A. C. *National Safety News* (Chicago), Oct. 1979, 120/4 (55-60). Illus. 12 ref.

CIS 79-1109 "Fall-arresting devices" (Anseilschutz). Amelung, H. U. *Sicherheitsingenieur* (Heidelberg), Sep. 1978, 9 (12-17). Illus. (In German)

Farmer's lung

Farmer's lung is a classical example of allergic disease due to inhaled organic dusts affecting mainly the alveoli. The major causes of the disease are the spores of certain thermophilic actinomycetes, particularly *Micropolyspora faeni* and, to a lesser degree, *Thermoactinomyces vulgaris*; precipitins against these organisms have been found in the serum of affected persons. These organisms, which are found in hay, flourish and produce vast numbers of spores when the temperature of damp hay rises to 40-60 °C as a result of mould and bacterial growth.

Farmer's lung may be described as an "extrinsic allergic alveolitis", a term which may apply to similar diseases in which precipitins have been found against the relevant antigens, such as bagassosis, mushroom picker's lung, maple-bark stripper's disease due to the fungus *Cryptostroma corticale*, cheese-washer's lung due to *Penicillium* spores, maltworker's lung due to spores of *Aspergillus clavatus*, sequoiosis due to spores on mouldy sawdust, woodworker's lung due to sawdusts themselves, bird breeder's lung caused by avian antigens in the dust of bird droppings, and lung disease caused by the wheat weevil *Sitophilus granarius*. Other similar diseases are found when there is exposure to certain vegetable dust, such as paprika-splitter's lung in spice workers and suberosis due to cork dust.

Farmer's lung was registered as an industrial disease in the United Kingdom in 1965 under the following definition: "pulmonary disease due to the inhalation of the dust of mouldy hay or other mouldy vegetable produce and characterised by symptoms and signs attributable to a reaction in the peripheral part of the bronchopulmonary system and giving rise to a defect in gas exchange." It is now recognised that there may also be bronchial involvement.

Incidence

The disease is not rare, occurring most frequently in the winter months, and is commonest in areas of high rainfall. The incidence of the disease was found to be 193.1 per 100 000 amongst the agricultural population in Wales, and 11.5 per 100 000 in the East Anglia region of England. It has been reported in the United States, Iceland, Scandinavia, Switzerland and New Zealand. The introduction of mechanical baling techniques led to an increase in the disease, because it favoured moulding and overheating of pockets of damp hay in the compressed bales.

Symptoms and signs

The classical attack comes on 5-6 h after exposure to the dust, in contrast to the rapid appearance of asthma in atopic subjects sensitive to such dusts. Equally frequently, however, the disease develops insidiously without any observed interval between exposure and the appearance of symptoms. This latter form is perhaps the more dangerous because the relationship to the exposure is not obvious and the patients may present severe irreversible lung damage.

Fever, malaise, rigors or chills with aches and pains occur, and are accompanied by loss of weight. Cough with little sputum is present with occasional haemopthysis. Dyspnoea, frequently very severe, is the more important symptom and is often out of proportion to the crepitant rales. Acute attacks may clear up overnight, but more severe or repeated attacks cause disease lasting weeks or months. Repeated attacks or, on occasions, even a single attack may result in chronic lung disease.

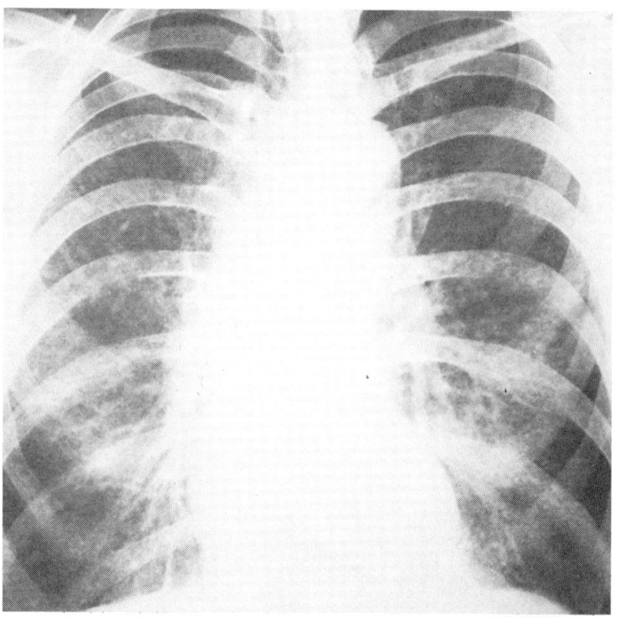

Figure 1. Radiographic appearance in a patient with farmer's lung, showing nodular infiltrations. (See also full-page illustration in Volume II.)

Radiological findings (see figure 1) include widespread nodular shadows, often micronodular giving a ground-glass appearance, most obvious in the lower zones, which are usually seen during the acute stages. Larger opacities may also occur. The hilar lymph nodes are not enlarged. In the chronic disease the appearances are those of fibrosis and "honeycomb" lung, especially in the upper lobe.

Pulmonary function tests show evidence of a restrictive defect, impairment of gas transfer and decrease in compliance, with little or no airways obstruction.

Pathology

In the early acute stages, the alveolar walls are infiltrated with neutrophil, lymphoid and plasma cells, and later by epithelioid cell granulomata. Widespread fibrosis and cystic changes develop, and *bronchiolitis obliterans* may also be present.

Diagnosis

The clear-cut clinical picture of acute farmer's lung may no longer be apparent when the patient seeks medical advice and the subacute and chronic stages may be less easy to diagnose. However, serological, inhalation and skin tests may provide valuable diagnostic indications.

Serological tests. Positive reactions are obtained in agar-gel precipitin tests against extracts of *M. faeni* in more than 90% of cases, and together with the latex-fixation test in about 95%. During active disease, 100% are positive, the test becoming weaker and negative on avoidance of exposure. Positive reactions may be obtained in 18-20% of exposed, but apparently unaffected, subjects. Some investigators claim, however, that almost all subjects with precipitins are affected.

Inhalation tests. The typical systemic and pulmonary manifestations of the disease appear in affected subjects 5-6 h after the inhalation of extracts of *M. faeni* and, in some cases, of *T. vulgaris*, as well as extracts of mouldy hay, but not in response to extracts of good hay or to some of the commoner fungi of mouldy hay, including *Aspergillus fumigatus*.

Skin tests. Some extracts of *M. faeni* have been found to give non-specific reactions to intracutaneous tests and have not yet been found of use.

Differential diagnosis. The acute pyrexial episodes, with or without pulmonary symptoms, may be mistaken for infection, particularly if the relationship to the occupational exposure is not appreciated. Asthma can be distinguished by the reversible wheezing of the bronchial reaction, and by the eosinophilia of the blood and sputum, and, in atopic subjects, by immediate skin test and rapid asthmatic reactions to the relevant allergens. In bronchitis, cough and sputum and rhonchi are present, with other evidence of airways obstruction.

Pulmonary aspergillosis has been confused with farmer's lung. The allergic form of pulmonary aspergillosis occurs in atopic, asthmatic subjects, who show transitory collapse-consolidation shadows of the lungs related to medium-sized bronchi. They cough up plugs of sputum containing eosinophils and fungal mycelium, have a blood eosinophilia, give positive skin test reactions and have precipitins against *A. fumigatus*.

The radiological and histological appearances of pulmonary sarcoidosis may resemble farmer's lung, but hilar node enlargement is absent in farmer's lung. Miliary tuberculosis may have a similar X-ray appearance, but can be distinguished by the culture of the sputum for *M. tuberculosis*. In fibrosing alveolitis (Hamman-Rich syndrome) clubbing of the fingers is commoner, and it is distinguished by the occupational history and serological tests.

An occupational history of work in a silo or exposure to silage during the early stages of fermentation will, however, direct the diagnosis to silo-filler's disease, which is caused by the inhalation of nitrogen oxide from recently filled silos. Cough, dyspnoea and weakness develop and then improve partially until 2-3 weeks later, when severe disease develops, consisting of fever with rapidly progressive dyspnoea, cyanosis and cough due to *bronchiolitis obliterans*.

PREVENTIVE MEASURES

The prime measure is to ensure that the moisture content of the hay is kept low. Hay which has become mouldy and overheated should not be used for any purpose nor handled, since the clouds of dust which come from it contain spores of the thermophilic actinomycetes. Masks may help to reduce the exposure and should be used whenever there is a hazard of exposure; however, since the spores of *M. faeni* are only 1 μm in diameter, they are difficult to exclude effectively by a mask suitable for heavy work.

Treatment

The patient should be removed from exposure to the antigens. Corticosteroid treatment is effective in the acute stages and has been used prophylactically in affected farmers who are obliged to continue working with mouldy hay. Permanent avoidance of mouldy hay is desirable for the prevention of further attacks and the production of increasingly severe irreversible fibrosis.

PEPYS, J.

CIS 77-2022 "Farmer's lung in a group of Scottish dairy farms". Wardrop, V. E.; Blyth, W.; Grant, I. W. B. *British Journal of Industrial Medicine* (London), Aug. 1977, 34/3 (186-195). Illus. 25 ref.

CIS 79-1129 "Farmer's lung disease: Long-term clinical and physiologic outcome". Braun, S. R.; DoPico, G. A.; Tsiatis, A.; Horvath, E.; Dickie, H. A., Rankin, J. *American Review of Respiratory Disease* (New York), Feb. 1979, 119/2 (185-191). 21 ref.

CIS 81-204 "Seasonal variation in the incidence of farmer's lung". Terho, E. O.; Lammi, S.; Heinonen, O. P. *International Journal of Epidemiology* (London), Sep. 1980, 9/3 (219-220). 8 ref.

Fatigue

The two concepts of fatigue and rest are familiar to all from personal experience. The word "fatigue" is used to denote very different conditions, all of which cause a reduction in work capacity and resistance. The very varied use of the concept of fatigue has resulted in an almost chaotic confusion and some clarification of current ideas is necessary. For a long time, physiology has distinguished between muscular fatigue and general fatigue. The former is an acute painful phenomenon localised in the muscles: general fatigue is characterised by a sense of diminishing willingness to work. This article is concerned only with general fatigue, which may also be called "psychic fatigue" or "nervous fatigue" and the rest that it necessitates. (For muscular fatigue see MUSCULAR WORK.)

General fatigue may be due to quite different causes, the most important of which are shown in figure 1. The effect is as if, during the course of the day, all the various stresses experienced accumulate within the organism, gradually producing a feeling of increasing fatigue. This feeling prompts the decision to stop work; its effect is that of a physiological prelude to sleep.

Fatigue is a salutary sensation if one can lie down and rest. However, if one disregards this feeling and forces oneself to continue working, the feeling of fatigue increases until it becomes distressing and finally overwhelming. This daily experience demonstrates clearly the biological significance of fatigue which plays a part in sustaining life, similar to that played by other sensations such as, for example, thirst, hunger, fear, etc.

Rest is represented in figure 1 as the emptying of a barrel. The phenomenon of rest can take place normally if the organism remains undisturbed or if at least one essential part of the body is not subjected to stress. This explains the decisive part played on working days by all

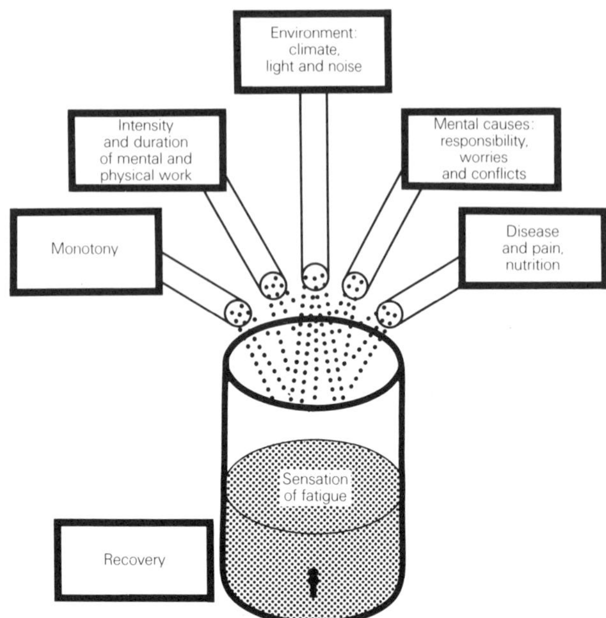

Figure 1. Diagrammatic presentation of the cumulative effect of the everyday causes of fatigue.

workbreaks, from the short pause during work to the nightly sleep. The simile of the barrel illustrates how necessary it is for normal living to reach a certain equilibrium between the total load borne by the organism and the sum of the possibilities for rest.

Neurophysiological interpretation of fatigue

The progress of neurophysiology during the last few decades has greatly contributed to a better understanding of the phenomena triggered off by fatigue in the central nervous system.

The physiologist Hess was the first to observe that electrical stimulation of certain of the diencephalic structures, and more especially of certain of the structures of the medial nucleus of the thalamus, gradually produced an inhibiting effect which showed itself in a deterioration in the capacity for reaction and in a tendency to sleep. If stimulation was continued for a certain time, general relaxation was followed by somnolence and finally by sleep. It was later proved that, starting from these structures, an active inhibition may extend to the cerebral cortex where all conscious phenomena are centred. This is reflected not only in behaviour, but also in the electrical activity of the cerebral cortex. Other experiments have also succeeded in initiating inhibitions from other sub-cortical regions.

The conclusion which can be drawn from all these studies is that there are structures located in the diencephalon and mesencephalon which represent an effective inhibiting system and which trigger off fatigue with all its accompanying phenomena.

Inhibition and activation

Numerous experiments performed on animals and man have shown that the general disposition of them both to reaction depends not only on this system of inhibition, but essentially also on a system functioning in an antagonistic manner, known as the reticular ascending system of activation. We know from experiments that the reticular formation contains structures which control the degree of wakefulness, and consequently the general dispositions to reaction. Nervous links exist between these structures and the cerebral cortex where the activating influences are exerted on the consciousness. Moreover, the activating system receives stimulations from the sensory organs. Other nervous connections transmit impulses from the cerebral cortex—the area of perception and thought—to the activation system. On the basis of these neurophysiological concepts it can be established that external stimuli as well as influences originating in the areas of consciousness may, in passing through the activating system, stimulate a disposition to reaction.

In addition, many other investigations make it possible to conclude that stimulation of the activating system frequently spreads also from the vegetative centres, causing ergotropic conversion of the internal organs, the organism being oriented towards an expenditure of energy, towards work, struggle, flight, etc. Conversely, it appears that stimulation of the inhibiting system within the sphere of the vegetative nervous system brings about a trophotropic conversion (the organism tends towards rest, reconstitution of its reserves of energy, phenomena of assimilation).

By synthesis of all these neurophysiological findings the following conception of fatigue can be established: the state and feeling of fatigue are conditioned by the functional reaction of the consciousness in the cerebral cortex, which is, in turn, governed by two mutually antagonistic systems—the inhibiting system and the activating system. Thus, the disposition of man to work depends at each moment on the degree of activation of the two systems: if the inhibiting system is dominant, the organism will be in a state of fatigue; when the activating system is dominant, it will exhibit an increased disposition to work.

This psychophysiological conception of fatigue makes it possible to understand certain of its symptoms which are sometimes difficult to explain. Thus, for example, a feeling of fatigue may disappear suddenly when some unexpected outside event occurs or when emotional tension develops. It is clear in both these cases that the activating system has been stimulated. Conversely, if the surroundings are monotonous or work seems boring, the functioning of the activating system is diminished and the inhibiting system becomes dominant. This explains why fatigue appears in a monotonous situation without the organism being subjected to any work load (see MONOTONOUS WORK).

Figure 2 depicts diagrammatically the notion of the mutually antagonistic systems of inhibition and activation.

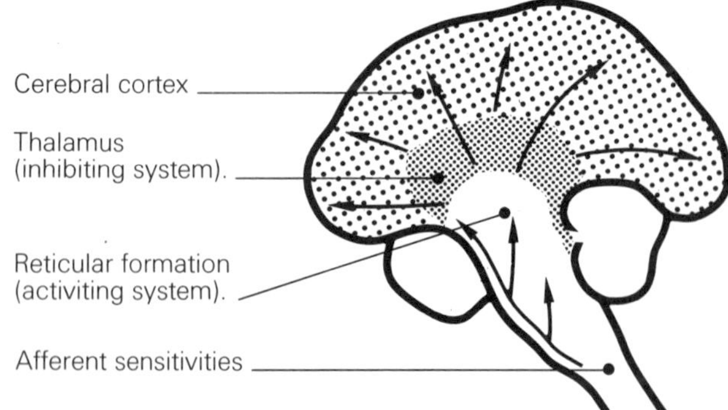

Figure 2. Diagrammatic presentation of the control of disposition to work by means of inhibiting and activating systems.

Clinical fatigue

It is a matter of common experience that pronounced fatigue occurring day after day will gradually produce a state of chronic fatigue. The feeling of fatigue is then intensified and comes on not only in the evening after work but already during the day, sometimes even before the start of work. A feeling of malaise, frequently of an emotive nature, accompanies this state. The following symptoms are often observed in persons suffering from fatigue: heightened psychic emotivity (antisocial behaviour, incompatibility), tendency to depression (unmotivated anxiety), and lack of energy with loss of initiative. These psychic effects are often accompanied by an unspecific malaise and manifest themselves by psychosomatic symptoms: headaches, vertigo, cardiac and respiratory functional disturbances, loss of appetite, digestive disorders, insomnia, etc.

In view of the tendency towards morbid symptoms which accompany chronic fatigue, it may justly be called clinical fatigue. There is a tendency towards increased absenteeism, and particularly to more absences for short periods. This would appear to be caused both by the need for rest and by increased morbidity. The state of chronic fatigue occurs particularly among persons exposed to psychic conflicts or difficulties. It is sometimes very difficult to distinguish the external and internal causes. In fact it is almost impossible to distinguish cause and effect in clinical fatigue: a negative attitude towards work, superiors or workplace may just as well be the cause of clinical fatigue as the result.

Research has shown that the switchboard operators and supervisory personnel employed in telecommunications services exhibited a significant increase in physiological symptoms of fatigue after their work (visual reaction time, flicker fusion frequency, dexterity tests). Medical investigations revealed that in these two groups of workers there was a significant increase in neurotic conditions, irritability, difficulty in sleeping and in the chronic feeling of lassitude, by comparison with a similar group of women employed in the technical branches of the postal, telephone and telegraphic services. The accumulation of symptoms was not always due to a negative attitude on the part of the women affected towards their job or their working conditions.

PREVENTIVE MEASURES

These is no panacea for fatigue but much can be done to alleviate the problem by attention to general working conditions and the physical environment at the workplace. For example much can be achieved by the correct arrangement of hours of work, provision of adequate rest periods and suitable canteens and rest rooms; adequate paid holidays should also be given to workers. The ergonomic study of the workpost can also help in the reduction of fatigue by ensuring that seats, tables and work benches are of suitable dimensions and that the work flow is correctly organised. In addition, noise control, air-conditioning, heating, ventilation and lighting may all have a beneficial effect on delaying the onset of fatigue in workers.

Monotony and tension may also be alleviated by a controlled use of colour and decoration in the surroundings, intervals of music and sometimes breaks for physical exercises for sedentary workers. Selection and training of workers and in particular of supervisory and management staff also play an important part.

GRANDJEAN, E.

Physiology:

"Brain stem reticular formation and activation of the EEG". Moruzzi, G.; Magoun, H. W. *Electroencephalography and Clinical Neurophysiology* (Amsterdam), 1949, 5/1 (1-22).

General:

CIS 76-281 *Stress, fatigue and depression in everyday life* (Stress, fatigue et dépression dans la vie quotidienne). Bugard, P.; Caille, E. J.; Crocq, L.; Ely, B.; Henry, M.; Petrescu, L.; Reinberg, A. (Paris, Editions Doin, 1974) 2 vols., 295 and 302 p. Illus. Approx. 1 100 ref. (In French)

"Fatigue" (La fatigue). Amphoux, L. *Encyclopédie médico-chirurgicale. Instantanés médicaux. Pathologie du travail. Intoxications. Maladies par agents physiques.* N° spécial 1-1976, 16650 A10 (Paris, Les Éditions techniques, 1976), 6 p. 5 ref. (In French)

Assessment:

CIS 78-1484 "A medicobiological approach to workload" (Une approche médico-biologique de la charge de travail). Elias, R. *Cahiers de notes documentaires–Sécurité et hygiène du travail* (Paris), 2nd quarter 1978, 91, Note No. 1118-91-78 (293-306). Illus. 92 ref. (In French)

"Studies on evaluation of mental fatigue". Saito, K.; Kumashiro, M.; Watanabe, H.; Kamita, N.; Takakuwa, E. *Industrial Health* (Kawasaki), June 1978, 16/2 (45-54). Illus. 9 ref.

"Visual flicker fusion and measurement of fatigue: state of the art" (Fréquence de fusion critique visuelle: état de la question). Volle, M. A.; Brisson, G. R.; Dion, M.; Tanaka, M. *Le travail humain* (Paris), 1980, 43/1 (65-86). 106 ref. (In French)

Feathers

There are two main kinds of feather–contour and down. The large contour feathers covering the bird's wings, body and tail have a strong centre shaft; down feathers found underneath the bird are soft and small.

Uses. Feathers have long been used by men for many purposes. The contour feathers from crows, swans and geese were once used as quills for writing. Because of their elasticity, softness and ability to hold up under weight, a mixture of contour and down feathers of geese and ducks are used extensively as stuffing for cushions, pillows and beds. Plumage feathers may also be used to decorate hats and clothes. Subsidiary uses are as holders for sable and camel hair brushes, dusters, artificial flies for fishing, etc. Because of their high protein content, feathers and the residues from feather processing are used in making fertilisers and animal feed.

The feathers and down from geese and other aquatic birds are ideal for bedding. The best feathers are those plucked in spring from the living birds but many feathers are also obtained from the plucking of dead birds in slaughter-houses, often by hand, although modern plant and machinery is also used.

Processing. This depends on the product required. The feathers must always be cleaned, usually by steam, but sometimes by degreasing solvents: bleaching and dyeing may also be necessary. In the making of beds and bedcovers, pillows, cushions, etc., the feathers are rotated in spiked drums and then fall into settling chambers according to size. Stuffing is carried out by mechanical blowers.

The ostrich is the only bird raised exclusively for its feathers: these are cut seasonally from the living animal, cleaned, trimmed and curled. Often, decorative feathers are obtained from the marabou, the osprey, the heron (aigrettes) and birds of paradise (although strict protection laws apply in some countries).

HAZARDS

Fire and accidents. If flammable solvents are used in unsuitable conditions there may be a serious risk of fire. The spiked drums used in bedding factories are dangerous unless secure machinery guarding is provided. Badly maintained hand tools may also be a risk. Steam plant requires regular examination and maintenance.

Diseases. Bird pluckers work in a very dusty atmosphere, containing a mixture of dandruff, earth, dried droppings, feathers, etc. Many people are allergic to these dusts and develop hypersensitive reactions. They may experience nasal obstruction and discharge, wheezy breathing, dyspnoea, fever and even transient increase of eosinophiles in their blood. On removal from the environment, these symptoms disappear quite quickly only to recur on re-exposure. Chest X-rays show fine infiltrations throughout both lung fields at the height of symptoms. These infiltrations disappear when symptoms subside. Lung biopsies often show granulomatous reactions in the lungs. Following repeated exposures for many years, irreversible emphysema and scattered fibrosis of lungs will develop, which may be indistinguishable from effects of smoking and pneumoconiosis due to non-siliceous dusts.

Allergy tests in these susceptible individuals are usually positive. Their intradermal or patch tests to the bird serum and extract of droppings often show an immediate and delayed positive reaction. Inhalation tests with serum and extract of droppings often produce asthma, fever and reaction in the interstitial lung tissue. These tests seem to prove beyond doubt that in patients with "bird breeder's lung", there are precipitins against pigeon or budgerigar serum proteins and against antigens in their excreta. The findings in their lungs are due to hypersensitivity reaction (see EXTRINSIC ALLERGIC ALVEOLITIS).

Feathers frequently contain the spores of histoplasma and men become infected by contact with them. The signs and symptoms of histoplasmosis range from those of a slight self-limited infection to the overwhelming disturbances of fatal disseminated disease. Severe infections are characterised by fever, chest pain, liver and spleen enlargement. The high incidence of positive intradermal reactions to histoplasmosis in healthy persons, however, indicate that most infections are very mild and inapparent. Infected individuals may exhibit mild pneumonitis while others are free from symptoms. The chest X-ray of "chicken-pluckers" often shows numerous, round, well defined calcific densities due to old burned-out histoplasmosis.

SAFETY AND HEALTH MEASURES

Hypersensitive individuals should not be permitted to work with feathers. Any worker who develops asthma, shortness of breath and fever should be immediately suspected of an allergic reaction. Chest X-ray should be taken and the person removed from the environment to see if his symptoms will subside in a few days. Appropriate allergy tests should be carried out if facilities are available. It must be noted that the allergic reactions are probably dose related. An individual who is symptom-free when working with birds for short periods of time may develop severe symptoms if left with birds for long periods of time. Furthermore, allergic reaction may develop in individuals who were not allergic in the past. Medical supervision will ensure that manifestations are recognised and appropriate action taken. The working place must be well ventilated and short shift work is preferable. The best solution is mechanisation, which prevents contact with the feathers.

Workers in bedding factories are also exposed to massive quantities of airborne feathers and dust. Fitting of exhaust ventilation presents difficulties as the feathers themselves may be sucked away: respiratory protection of conventional type may cause great discomfort in hot climates; it is customary in many places for workers to protect the mouth and nose with cloth eye and face protection. Mechanisation and maximum enclosure of processes offer the best solution.

Degreasing and cleaning may involve the use of toxic solvents. Benzene should be eliminated and the least toxic solvent practicable used. Quantities should be kept as small as possible and efficient ventilation maintained.

Workers engaged in cleaning and refilling of used bedding, etc., are subject to risk from infected materials: control lies with the public health authorities in many countries.

At all stages of processing a high standard of personal hygiene, sanitary and washing facilities, and personal protective equipment (overalls, head coverings) is required.

LOWE, W. C.

CIS 78-1868 "Occupational allergic alveolitis due to handling bed-feathers" (Berufsbedingte allergische Alveolitis beim Umgang mit Bettfedern). Schiele, R.; Lutgen, W. *Arbeitsmedizin–Sozialmedizin–Präventivmedizin* (Stuttgart), Feb. 1978, 13/2 (36-39). 17 ref. (In German)

CIS 78-1686 "Coeliac disease–Bird fancier's lung and coeliac disease". Purtilo, D. T.; Bonica, A.; Yang, J. P. S. *Lancet* (London), 29 Apr. 1978, 1/8070 (917-918) and 24 June 1978, 1/8078 (1 357-1 358). Illus. 16 ref.

CIS 75-477 "Budgerigar-fancier's lung: A report of a fatal case". Edwards, C.; Luntz, G. *British Journal of Diseases of the Chest* (London), Jan. 1974, 68 (57-64). Illus. 10 ref.

Felt hat manufacture

Felt hats are usually manufactured from the fur of rodents and other animals (e.g. rabbits, hares, muskrats, coypus, beavers, etc.).

Fur processing. These furs, unlike wool, do not have felting properties, and hence have to undergo a special treatment, called carroting. The best-known carroting agent, and the only one in common use until about the 1950s, is mercury nitrate ($Hg(NO_3)_2$). It has an oxidising action and breaks the disulphide bonds ($-CH_2-S-S-CH_2$) of cystine in the hair keratin. The mercury nitrate solution is obtained by adding 32 kg of mercury to 100 kg of nitric acid. During the operation, which is usually carried on in the open, nitrogen oxides are liberated.

In recent years growing use has been made of non-mercurial carroting agents with either an oxidising action (hydrogen peroxide and perchlorates with organic acids; sodium and barium peroxides with sulphuric acid), or a reducing action (phosphorus salts; tetrahydroxymethylphosphoniochlorate). Mercury-carroted skins are nowadays used only exceptionally for the more expensive or special hats, which account for no more than 5% of total hat production; carroted fur may, in these cases, be used pure or mixed with other hair. Most felt hat manufacturers no longer use mercury carroting but sometimes buy furs that have been mercury-carroted by special carroting firms.

In mercury carroting, the raw pelts are coated with a mercury nitrate solution by hand or mechanically and the treated hair is then removed. However, the process may be carried out by a special carroting firm which supplies treated hair to the hatter. The treated hair is then blown to free it from impurities, carded by drawing it over a metal cup riddled with holes, sprayed with water to promote web adhesion, hardened (first by hand and then by machine) to induce fibre matting, fulled on rotating cylinders over a bath of boiling dilute acid, dyed, moulded to the hat shape, finished and trimmed.

HAZARDS

Apart from the machinery hazards dealt with in the article on the felt industry, the manufacture of felt hats entails exposure to two main hazards: allergy to animal hair and chronic mercury poisoning.

Exposure to rodent hair during storage or, especially, blowing may produce allergic reactions sometimes accompanied by bouts of bronchial asthma. Positive skin tests to hair extracts are frequent but bronchial asthma occurs only rarely.

Chronic mercury poisoning affects primarily workers employed on manual or machine hardening, fullers who inhale steam that contains mercury vapour and, in rare cases, carroters and shapers. During the various wet processes in hat manufacture, mercury-carroted hair loses a certain amount of its mercury content, which is taken up by the process water or steam. This mercury may be in the form of soluble compounds which will be retained in the process water, or in the form of metal which vaporises during storage or in the steam current and accumulates in the workplace atmosphere. Mercury-carroted skins have a mercury content of 1.1-2.4% by weight. During storage, this mercury may be vaporised at a rate of up to 40% per annum.

At present, mercury poisoning has almost completely disappeared from felt hat factories; but there are some older workers still at work presenting symptoms of mercury poisoning, mainly tremor, dating back 20 years or more, when mercury carroting was still widely used.

SAFETY AND HEALTH MEASURES

The only fully effective method of preventing mercury poisoning amongst hatters is the abandonment of mercury carroting; this is a practice that is now observed in most factories. Where mercury is still employed, exhaust systems should be installed to collect vapour given off by boilers and hot felts. In one large Italian hat plant where numerous cases of mercury poisoning were recorded in the period 1942-44, the installation of exhaust ventilation led to the complete disappearance of mercury poisoning—even before the use of mercury carroting processes was abandoned (table 1).

Table 1. Mercury concentrations in the workplace atmosphere of various departments in an Italian hat factory in three different years

Department	Mercury concentration in the air (mg/m^3)[1]		
	1950	1960	1968
Carroting	0.25	0.10	0.05
Cutting hair	0.28	0.14	0.07
Blowing	0.60	0.18	0.06
Carding	0.20	0.08	0.03
Hardening	0.80	0.13	0.09
Fulling	0.60	0.07	0.05
Dyeing	0.40	0.10	0.08
Shaping	0.02	0.02	0.01
Finishing	0.07	0.06	0.02
Trimming	0.04	0.01	0.01

[1] In 1950, exhaust ventilation was insufficient to cope with the mercury emanating from the felt materials; in 1960, the exhaust ventilation was working perfectly; and, in 1968, virtually all the hair was mercury-free. For mercury, the TLV ACGIH is 0.05 mg/m^3 and the MAC USSR, 0.01 mg/m^3.

MOREO, L.
VIGLIANI, E. C.

"Chronic mercury intoxication in felt hat manufacture" (Il mercurialismo cronico nei cappellifici). Baldi, G.; Vigliani, E. C.; Zurlo, N. *La medicina del lavoro* (Milan), Apr. 1953, 44/4 (161-198). Illus. 43 ref. (In Italian)

Felt industry

Felt is usually considered to be a fibrous material obtained through the interlocking of wool, fur or hair fibres under conditions of heat, moisture and friction, to form an unwoven densely matted fabric. There are also needleloom felts where fibres are attached to a loosely woven backing by a needleloom process. The fibres and backing are usually wool or jute.

Raw materials. Wool is the principal raw material and may be unused or reclaimed; wool fibres for felt production should be about 4 cm in length. Fur and hair are used for special felts (e.g. beaver hair is said to be suitable for hats). Jute is used for certain needlefelts and is usually reclaimed from old sacks. Cotton, man-made fibres and silk may be added. Other materials employed in felt manufacture include soap, starch, soda and sulphuric acid.

Manufacturing processes. A typical sequence of processes for non-woven felt production will include the following operations. The materials are sorted and selected, probably by a merchant; if the material is old rags, these will be torn up in a rag-grinding machine—a spiked cylinder which rotates and makes the material fibrous. This is followed by garnetting on a Garnett machine (not to be confused with a Garnett carding machine) which has rollers and cylinders covered with hard saw-toothed wire and opens up the waste wool. The materials are then blended, oiled (where necessary), "willeyed" or teased on a teasing machine which contains spiked or toothed rollers to open up the material further. After teasing, the material is carded on machines which contain a number of rotating cylinders covered with leather or fabric in which fine wire teeth are set. This process further blends and opens up the fibres and arranges them more or less parallel to one another. The material comes off each final cylinder in a fine web which is deposited on a belt travelling the same speed as the web. One web is deposited lengthwise on the belt and the other web is deposited by a slatted conveyor which travels backwards and forwards across the belt. The mechanism is known as a camelback. A different system uses a crossing folder. The two webs build up into a fabric which has roughly equal strength in all directions. The wool is then wound up on a pole and taken to the hardening room. The soft rolls of wool (known as batts) are unrolled on to an apron and sprinkled with water after which the apron is set in motion and passed over a steam-heated table and then pressed between heavy plates. The top plate is lowered on to the batt and vibrates, causing the fibres to curl and cling together. To complete the felting, the material is placed in bowls containing soap or dilute sulphuric acid and is pounded by heavy wooden hammers. The finished material is washed and dewatered by wringers ready for dyeing or final drying. After dyeing, about 90% of the water is extracted by wringers or hydro-extractors (spin dryers); thick felts are dried on heavy frames (tenter frames) by air-heated steam pipes. Drying temperatures vary between 65 °C (soft felts) to 112 °C (hard felts). Chemicals may be added to make the felt rot-proof and some papermakers' felts undergo a tanning process.

The finishing processes consist of shearing, sanding, brushing, pressing and trimming. Soft or thin felts are pressed in cylinder presses and thick felts are pressed between steam-heated plates.

Needlefelt manufacture. The process as far as and including carding is similar to that already described. The raw material may be wool, jute or even synthetic fibres. The needleloom process is essentially a method of attaching a lap or batt of the loose fibrous material to a base. The base may be a loosely woven wool or jute but even paper, rubber or plastics materials have been used. The attaching process is effected by needles having downward facing barbs designed to carry small tufts of the batt through the base. A needleloom has a quickly rising and falling beam on the underside of which is a board containing rows of closely spaced barbed needles. The batt is laid, after it has come from a carding machine, on top of the base which is supported by a slotted bed. It is passed continuously under the rising and falling beam, and the needles drive the fibres through the base. This operation can be repeated to build up a number of batts to give a required density. The material can be reversed and a two-sided product can thus be obtained.

HAZARDS

Accidents. Many of the machines used in felt making have dangerous parts and if accidents are to be prevented, good machinery guarding is essential. Accidents occur at the belts and pulleys driving the machines. Sometimes paste is applied to the moving belts to prevent them slipping. This practice can be dangerous. An accident often occurs at a driving belt when the guard has been removed and either not replaced at all or not replaced correctly. Chain and sprocket drives and gear drives are commonly found at the side of machinery and are capable of producing serious injuries; these elements should be guarded, as should any exposed rotating shaft within reasonable reach. Many of the machine parts are dangerous, including the swift (spiked drum of Garnett machines), the lickers-in of carding machines, heavy wooden rollers on squeezing and drying machines, and the revolving blades of trimming machines.

Fire is a common hazard in the early stages of processing, in the grinding machines, Garnett machines and, sometimes, carding machines. The causes of ignition vary and include metal particles or even boxes of matches caught up in waste fibre, hot running machine bearings or faulty electrical equipment. The electrical equipment should be correctly maintained and appropriate metal parts earthed; this is particularly important in wet dyeing processes.

Health hazards. The industry is a dusty one and not to be recommended for people who suffer from bronchitis or similar ailments but, fortunately, the dust is not associated with any specific disease. Cases of anthrax, however, do occur from time to time and during the period 1900-56, 81 cases of anthrax were reported amongst felt workers in the United Kingdom; in 73 of these cases, the raw material involved was of East Indian origin. The tanning of certain papermakers' felts may involve the use of quinone, which can cause severe local damage to the skin and mucous membranes. The dust or vapour of this substance can cause staining of the conjunctivae and cornea of the eye, which may go on to affect vision if exposure is continued or repeated over a period. This effect can be detected in the early stages by medical examination, which can be a useful check on the degree of exposure in marginal cases. The sulphuric acid used during the fulling process is a very weak solution; however, care should be taken if concentrated acid is bought for subsequent dilution.

SAFETY AND HEALTH MEASURES

Technical prevention. The feed rollers and swift of rag-grinding machines should be provided with a tunnel guard. If the width of the opening at the end of the guard is 10 cm, the distance to the rollers should be at least 56 cm.

As an alternative, an automatic guard can be provided, arranged so that if the guard is displaced the drive for the feed rollers is stopped. Fires occur fairly frequently at these machines and guards should be constructed so that they do not hinder any action necessary to put out the fire. The materials, after processing, are passed into ducts and these usually have removable inspection panels. These panels should be positioned so that when they are removed it is not possible for persons to reach through the opening to dangerous parts, for example revolving spiked drums or fan blades.

The main driving belts of Garnett machines are often guarded by fender guards. Open-rail guarding should not be used and the guards should be fitted with lugs which can fit into sockets on posts fixed in the floor. Special points which need attention include the dangerous intake between stripper rollers and swift. Unless over-all swift covers are provided, the fender guards should extend sufficiently high to guard the swift (spiked drum) and strippers.

The feed rollers and swifts of blending and opening machines should be guarded in a similar way to those of a rag-grinding machine. Inspection panels in any associated ducting should be located at a safe distance from rotating blades, spiked cylinders or other dangerous parts.

The driving belts and pulleys of carding machines will need to be guarded and the same general principles will apply to these parts. It is important to guard the licker-in rollers. Sheet-metal guards can be provided extending across the machine. As an alternative, wing guards of sufficient size may be fitted at the roller ends. Arrangements should be made for allowing parts to be lubricated without it being necessary to remove the guards. Pipe feeds to the bearings and the provision of long-spouted oil cans are often practicable.

In fulling machines the height of the bowl above floor level should be sufficient to prevent people falling in.

Washing and wringing machines have heavy wooden rollers the in-running nips of which should be guarded, for example by fixed wooden guards. In some cases a long glazed pipe (about 15 cm in diameter) is used for a feed opening. A special needle made from stiff rubber, having a slot in one end, is useful for feeding the material to the intake nip of the machine. If hydro-extractors or centrifuges are used, it is desirable that the lid be interlocked so that the machine cannot run until the lid is closed and the lid cannot be opened until the machine is at rest.

The blades of trimming machines usually consist of rotating wheels with sharp edges. As much of the edge of the blade as practicable should be guarded. In modern crosscutting machines, a blade moves across a width of material to cut off a required length. The traversing motion of the blade starts automatically when a predetermined length of material has been produced. Severe accidents to the feet have occurred when people have been standing on the machine in the path of the moving blade and the motion has started unexpectedly. It should be physically impossible for a person to stand in the way of the blade, or a "permit-to-work" system should be introduced to ensure that the driving motor is switched off and cannot start unexpectedly while anyone is in the danger zone.

Mechanical exhaust ventilation is desirable on machinery up to and including the cards. This is a complex problem and much development work needs to be done before a completely satisfactory solution will be found. Simple precautions such as regular cleaning of floors, fixtures and plant are helpful.

Quinone used for tanning is supplied in powder form and should be damped before use to prevent the evolution of dust. Any mixing should be carried out in a fume-cupboard-type of enclosure fitted with mechanical exhaust ventilation. Hand and arm and eye and face protection should be worn by people handling the material in any form.

If supplies of concentrated sulphuric acid are kept for diluting as required, the acid should be poured into a large quantity of water, never the other way round. Eyewash bottles should be kept readily available and protective clothing, including goggles, aprons and gloves, should be provided.

Fire prevention. Access to parts of machines where fires are likely to occur should be easy. There may be a conflict here with the need for secure guarding. It is often possible to arrange for openings covered by pivoted plates to be left in the side of machine covers, through which fire-extinguishing materials can be fed. Although water is an efficient fire-extinguishing agent, its use is not popular as it damages the raw material and causes corrosion of the machine parts. Dry powder extinguishers can be used and, in the case of some modern machinery, automatic carbon dioxide extinguishing systems are installed. A fire-warning system should, generally, be provided. In the special case of the manufacture of felt hats when flammable solvents are used, extra care has to be taken to prevent fires occurring in drying ovens.

Medical prevention. Wool from areas where anthrax is endemic is usually responsible for carrying the bacillus to the worker and the safe condition of imported materials is a national matter usually beyond the control of the user. Early diagnosis is important and the workpeople should know, possibly by illustrated leaflets or posters, what to look for, and be encouraged to report skin conditions when imported wool is being used.

RUSCOE, J. M.

Fermentation, industrial

Industrial fermentation, for the purposes of this article, can be defined as a specialised unit process in the chemical and allied industries covering the manufacture by micro-organisms of a variety of substances, including multiplication of the micro-organisms themselves. It thus includes the production of single cell protein, bakers' yeast, enzymes, alcoholic beverages, certain organic acids including vinegar, antibiotics, certain vitamins and other substances, as well as microbiological steps in the conversion of steroids and other compounds.

The micro-organisms used are species of bacteria, including Actinomycetes (used in the manufacture of many antibiotics), yeasts and filamentous fungi (moulds). Raw materials used as substrates in fermentation processes are usually carbohydrates, such as cane or beet sugar, molasses, glucose or starch in various forms. Fermentations have also been developed based on liquid hydrocarbon fractions, methane, methanol and acetic acid. Nitrogen must be supplied either as a pure chemical, such as urea or ammonium salts, or from animal or vegetable sources; phosphorus, potassium and other elements must also be supplied in some form.

A modern fermentation plant consists of a number of small pressure vessels used as inoculum tanks and much larger ones of similar design for the final production stage. These will be capable of being interconnected either through fixed pipes and manifolds or by flexible hoses. In many cases the vessels are mechanically agitated. In addition the plant may have a section for handling raw materials, vessels for preparing the medium for fermentation, continuous sterilisation equipment or a cooker, high-pressure steam, and for aerobic fermentations provision for large quantities of sterile air. In the manufacture of alcoholic drinks open vats are widely used for the fermentation step. Rarely now other methods of fermentation are used, such as the growth of micro-organisms in solid or liquid media on trays in sterilised rooms, for example the Japanese "koji" process.

The mass of micro-organisms, referred to as mycelium when filamentous moulds or Actinomycetes are the organisms used, is separated at the end of the fermentation from the liquor using rotary filters, centrifuges or other means. The subsequent isolation and recovery of the desired product from the liquor or from the mycelium or cells is carried out by purely chemical means, differing in no way from the recovery of other chemical products.

HAZARDS AND THEIR PREVENTION

Official figures on hazards are not available, as these are hidden in the figures for the industries which use fermentation processes, mostly only as a very small proportion of their total activity. But it can be stated with confidence that the dangers to health and safety are minimal. The use of micro-organisms has never added significantly to the more usual industrial hazards, which in the fermentation industries are not great when properly controlled, when precautions are strictly adhered to and when the plant used has been designed and constructed to eliminate as far as possible any dangerous operations. In general the hazards are similar to those encountered in other parts of the chemical, pharmaceutical and food industries, but in fermentations the materials used, the products made and the operating conditions present fewer potential dangers.

Burns and scalds. To maintain clean sterile conditions live steam is regularly used both for the pre-sterilisation of the fermentation vessels and the pipes and hoses used for transferring liquor, and for keeping sterile sampling cocks and other places on the fermenters where potential contaminating micro-organisms might enter. For this reason the most prevalent accidents are scalds caused by steam and hot water and burns through contact with hot equipment. Prevention is best achieved by proper safety training, by lagging pipes and exposed metal surfaces, and by compulsory wearing of personal protective clothing, including industrial safety gauntlets. Because of danger from high pressure steam any flexible hoses used should be regularly tested against possible bursting.

Slips, falls and cuts. Next in order of frequency is a group of accidents caused by slipping on wet or greasy floors, by tripping over hoses and pipes, by dropping articles on the feet and by cuts sustained in the course of work. The best method of prevention is good housekeeping in mopping up wet floors, proper location of floor drains, elimination of spills and clear marking of pipes and hoses. Safety shoes should be worn at all times.

Gassing and asphyxiation. Carbon dioxide is made in many fermentations. This gas is heavier than air and in anaerobic (unaerated) processes is liable to remain at the bottom of a fermenter when this is emptied. Due to lack of oxygen a person working in such an area can be asphyxiated. Furthermore, carbon dioxide in large quantities is itself toxic, causing dizziness, unconsciousness and death. The safe procedures laid down for entering fermentation vessels, if this should become necessary, must be rigidly followed and should involve the issue of a permit-to-work authority by a qualified person. Before such a vessel is entered it should be isolated from every other source of dangerous gas, vapour or liquid. The area should be tested for oxygen deficiency and for the presence of dangerous gases. A further precaution is to permit entry only when an air line is put inside the vessel to displace any foul air while work is done.

If tests do not indicate that a confined space is free from dangerous gas it should not be entered by anyone unless wearing protective respiratory equipment, which should either be a self-contained type with a compressed air or oxygen supply or should consist of a face piece through which fresh air is fed. Reliance should not be placed on a canister respirator for protection. The person working in such an area should wear a safety belt with a lifeline controlled by another person outside the area; other staff should be on call in case of emergency.

All areas containing open fermenters should be well ventilated and enclosed fermenters should exhaust their fumes outside the building.

Pressure vessels. Enclosed fermenters are pressure vessels and must be equipped with pressure relief valves and vacuum breakers, which must be periodically inspected and kept in proper working condition; subsidiary piping should also be inspected at regular intervals. Inspection may involve hydrostatic testing. On tanks subject to corrosion it may be advisable to test drill and measure the wall thickness at stated intervals to determine that the tanks continue to meet their specification.

Fire and explosion. These hazards are not normally encountered in the fermentation industries. Liquid hydrocarbons used as raw materials usually have a high flash point and will not readily ignite. The use of methane or natural gas might however give explosive mixtures and here special precautions must be observed. Methanol is inflammable. The acetone-butanol fermentation gives off hydrogen and carbon dioxide which is a dangerous mixture. Prevention of fire or explosion is achieved by adequate ventilation and by forbidding smoking, welding and other possible sources of ignition in the area during fermentation. Flameproof switchgear must be installed where any possibility of explosive mixtures exists. With certain raw materials it is necessary to guard against dust explosions in the milling area. Dried fermentation residues with a high fat or oil content have been known to ignite spontaneously. Fire protection such as water hoses and fire extinguishers should be readily available or a sprinkler system installed. Exits should be well marked and personnel well trained in fire and evacuation drill.

Air filters. In aerobic fermentations the air is sterilised by passing under pressure through tightly packed beds of absorbing material. Dust masks should always be worn when repacking the filters and this is particularly important if glass wool or similar fibrous material is used as the packing.

Micro-organisms. Hazards arising from the use of micro-organisms have not in fact proved to be serious. Workers must be prevented from inhaling large numbers of micro-organisms in aerosols. Good ventilation on the plant is therefore essential. In areas where cultures are prepared properly ventilated cabinets should be available for handling them, with no air movement out of the cabinet towards the operator.

Man throughout his life is surrounded by micro-organisms in the air he breathes, the food he eats and the earth under his feet. Relatively few of these are pathogenic and no micro-organisms pathogenic to man are used in industrial fermentation processes. A few, however, are plant pathogens, for example *Ashbya gossypii* for the production of riboflavin (vitamin B_2), *Xanthomonas campestris* for xanthan gum and *Erwinia chrysanthemii* for the enzyme asparaginase. The micro-organisms used in the fermentation industries have all originally been obtained from natural sources, though they may be acquired for use from culture collections. Most strains used will have been developed over many years within the organisations concerned by mutation techniques and by selection to give the required characteristics. The over-all record after many years of experience shows that there is no significant hazard. If there is any doubt about the possible pathogenicity of a newly developed strain of micro-organism to be used in fermentation it should be subjected to thorough toxicity testing before it is used. Micro-organisms pathogenic to man are used on a small scale in the manufacture of vaccines, where very special precautions must be taken, but this is outside the scope of this article.

A few cases of allergic reactions, mostly to mould spores, have occurred to workers in the industry after long exposure, causing asthmatic or bronchial conditions. When this happens the worker should be removed from possible contact with the substance causing the allergy. Allergies can also be caused by certain antibiotics, in particular penicillin. Where they may have cause to handle this, workers should be screened for sensitivity to the antibiotic and, if this is positive, removed from possible contact with it. (See also ANTIBIOTICS; BREWING INDUSTRY; ENZYMES IN INDUSTRY; WINES AND SPIRITS INDUSTRY; CONFINED SPACES; PERMIT-TO-WORK SYSTEM.)

MIALL, L. M.

"Biotechnologies" (Les biotechnologies). *Annales des mines* (Paris), Jan. 1981, 187/1 (1-91). Illus. Ref.

Radiation and radioisotopes for industrial microorganisms (Vienna, International Atomic Energy Agency, 1971), 338 p. Illus. Ref.

International Symposium on continuous cultivation of micro-organisms. Malek, I.; Beranek, K.; Hospodka, J. (eds.). (New York, Academic Press, 1964), 391 p. Illus.

"Poisoning due to prolonged exposure to carbon dioxide" (Die protrahierte CO_2 Vergiftung). Zink, P.; Reinhardt, G. *Beiträge zur Gerichtlichen Medizin* (Vienna), 1975, 33 (211-213). (In German)

Ferroalloys

A ferroalloy is an alloy of iron with an element other than carbon. These metallic mixtures are used as a vehicle for introducing specific elements into the manufacture of steel in order to produce steels with specific properties. The element may alloy with the steel by solution or neutralise impurities which harm the steel by combining with them and by separating from the steel as a flux or slag before the steel solidifies.

Alloys have unique properties dependent on the concentration of the elements in the alloy. These properties will vary directly in relation to the concentration of the individual components and will also depend, in part, on the presence of trace quantities of other elements. The biological effect of each element may be used as a guide to the possible effect of that element in the alloy but there is sufficient evidence for the modification of action by the mixture of elements to warrant extreme caution in making critical decisions based on extrapolation of effect from the single element.

Classification and production. The ferroalloys constitute a wide and diverse list of alloys with many different mixtures within each class of alloy. The trade generally limits the number of types of ferroalloy available in any one class but new metallurgical developments can result in frequent additions or changes. Some of the more common ferroalloys are listed in table 1. Ferrosilicon, ferromanganese, and ferrochromium are the three alloys produced in the largest quantities. Open-hearth steel, for example, requires around 6 kg of manganese as ferromanganese for each tonne of steel.

Table 1. Ferroalloys in common industrial use and their chief constituents

Name of alloy	Typical percentage concentration of non-ferrous elements other than carbon
Ferroboron	16.2% B
Ferrocerium (misch metal)	50% Ce; 45% La (the balance contains a wide range of rare-earth metals)
Ferrochromium	60-70% Cr (may also contain Si, Mn)
Ferromanganese	78-90% Mn; 1.25-7% Si
Ferromolybdenum	55-75% Mo; 1.5% Si
Ferrophosphorus	18-25% P
Ferrosilicon	15-90% Si
Ferrotitanium	14-45% Ti; 4-13% Si
Ferrotungsten	70-80% W
Ferrovanadium	30-40% V; 13% Si; 1.5% Al

HAZARDS

Although certain ferroalloys do have some non-metallurgical uses, the main sources of hazardous exposure are encountered in the manufacture of these alloys and during their use in the production of steel. The wide divergencies in the hazardous nature of the materials employed makes any brief survey impossible but some generalisations can be made. Ferroalloys, in general, have not resulted in major health problems but there are sufficient numbers of individual case reports of illness resulting from excessive exposures to justify a cautious approach. Some ferroalloys are produced and used in fine particulate form; airborne dust may constitute a potential toxicity hazard but the fire and explosion hazard may be of much greater importance. Certain of these alloys may undergo chemical reactions in the presence of moisture, and the gases evolved during these reactions may also constitute a serious hazard.

Ferroboron. Airborne dust produced during the cleaning of this alloy has been found to cause irritation of the nose and throat, which is due, possibly, to the presence of a boron oxide film on the alloy surface. However, dogs exposed to atmospheric ferroborn concentrations of 57 mg/m^3 for 23 weeks suffered no adverse effects.

Ferrochromium. This alloy has been used for many years without adverse effect on the health of men working with it. Materials which appear to be chemically identical may present differences in solubility and biological action and this phenomenon may account for the fact that pulmonary disease has recently been reported amongst workers using ferrochrome in certain plants whereas in other plants, operating under similar conditions, studies have revealed no illness. No dermatitis has been reported amongst workers engaged in the handling and crushing of the finished alloy; however, chrome dust and fume are present in manufacturing operations. [A recent study in Norway on the over-all mortality and the incidence of cancer in workers producing ferrochromium has shown an increased incidence of lung cancer in causal relationship with the exposure to hexavalent chromium around the furnaces. Perforation of the nasal septum was also found in a few workers.]

Ferromanganese. This may be produced by reducing manganese ores in an electric furnace with coke and adding dolomite and limestone as flux. Transportation, storage, sorting and crushing of the ores produce manganese dust in concentrations which can be hazardous. The pathological effects resulting from exposure to dust, from both the ore and the alloy, are virtually indistinguishable from those described in the article MANGANESE. Both acute and chronic intoxications have been observed.

Ferrosilicon. Production of this alloy can result in aerosols of ferrosilicon as well as production of dust. Animal studies indicate that ferrosilicon dust can produce thickening of the alveolar walls with the occasional disappearance of the alveolar structure. The raw materials used in alloy production may also contain free silica, although in relatively low concentrations. There is some disagreement as to whether classical silicosis may be a potential hazard in ferrosilicon production but there is no doubt that chronic pulmonary disease, whatever its classification, can result from excessive exposure to the dust or aerosols encountered in ferrosilicon plants.

Ferrovanadium. Atmospheric contamination with dust and fume is also a hazard in ferrovanadium production. Under normal conditions, the aerosols will not produce acute intoxication but may cause bronchitis and a pulmonary interstitial proliferative process. The vanadium in the ferrovanadium alloy has been reported to be appreciably more toxic than free vanadium as a result of its greater solubility in biological fluids.

[*Leaded steel.* Used for automobile sheet steel in order to increase malleability, this ferroalloy contains approximately 0.35% lead. Whenever the leaded steel is subject to high temperature, as in welding, there is always the danger of lead fume.]

SAFETY AND HEALTH MEASURES

Control of fume, dust and aerosol at each stage of the process of manufacture and use of ferroalloys is essential. Ore piles should be wetted down to reduce dust formation and good dust control is required in the transport and handling of the ores and alloys. In addition to these basic dust-control measures, special precautions are needed in the handling of specific ferroalloys.

Ferrosilicon reacts with moisture to produce phosphine and arsine; consequently this material should not be loaded in damp weather and special precautions should be taken to ensure that it remains dry during storage and transport. Whenever ferrosilicon is being shipped or handled in quantities of any importance, notices should be posted warning workers of the hazard, and detection and analysis procedures should be implemented to check for the presence of these gases in the air at frequent intervals. Good dust and aerosol control is required for respiratory protection. Suitable respiratory protective equipment should be available for emergencies.

Ferromanganese alloys containing very high proportions of manganese will react with moisture to produce manganese carbide, which will, in turn, combine with moisture, releasing hydrogen, and thus constitute a fire and explosion hazard.

Workers engaged in the production and use of ferroalloys should receive careful medical supervision and their working environment should be monitored continuously or periodically depending on the degree of risk. The toxic effect of the various metals in alloy form are sufficiently divergent from those of the pure metals to warrant a more intense level of medical supervision until more data have been obtained on the hazards of ferroalloys. Where ferroalloys give rise to dust, fume and aerosols, workers should receive periodic chest X-ray examinations in order to detect respiratory changes at the earliest possible moment. Lung function testing is also indicated. Studies of the blood enzyme profiles of exposed workers may also be desirable until more detailed information is available on the effects or lack of effects of the particular alloy or alloys being used.

ZAVON, M. R.

"The ferroalloy industry. Hazards of the alloys and semimetallics". Parts I and II. Roberts, W. C. *Journal of Occupational Medicine* (Chicago), Jan. 1965, 7/1 (30-36). 1 ref.; Feb. 1965, 7/2 (71-77). 9 ref.

"A survey of respiratory symptoms and lung function in ferrochromium and ferrosilicon workers". Långard, S. *International Archives of Occupational and Environmental Health* (West Berlin), Apr. 1980, 46/1 (1-9). 18 ref.

"Incidence of cancer among ferrochromium and ferrosilicon workers". Langård, S.; Andersen, Aa.; Gylseth, B. *British Journal of Occupational Medicine* (London), May 1980, 37/2 (114-120). 23 ref.

CIS 77-1963 "Occupational exposure to manganese". Šarić, M.; Markićević, A.; Hrustić, O. *British Journal of Industrial Medicine* (London), May 1977, 34/2 (114-118). 6 ref.

CIS 78-116 "Occupational poisoning by hydrogen cyanide in a ferrosilicon briquetting plant" (Profesionalno trovanje cijanovodonikom u proizvodnji briketa ferosilicijuma). Čremošnik-Pajić, P.; Mikov, M. I. *Arhiv za higijenu rada i toksikologiju* (Zagreb), 1977, 28/2 (187-194). 10 ref. (In Serbocroatian)

CIS 78-834 "Occupational hygiene and new methods of ferrovanadium production" (Gigiena truda pri novyh sposobah proizvodstva ferrovanadija). Kazimov, M. A. *Gigiena truda i professional'nye zabolevanija* (Moscow), June 1977, 6 (8-12). 12 ref. (In Russian)

CIS 78-1031 "Atmospheric lead concentrations during manufacture and processing of lead-alloyed steels" (Concentrazioni di piombo nell'aria degli ambienti di lavoro durante la preparazione e le successive lavorazioni degli acciai al piombo). Venvenuti, F.; Ciccarelli, C. *Securitas* (Rome), Nov.-Dec. 1976, 61/11-12 (500-505). 8 ref. (In Italian)

CIS 77-1035 "Work conditions in lead-alloyed steel production and hygiene measures" (Uslovija truda pri vyplavke svinecsoderaščih stalej i mery po ih ozdorovleniju). Kučerskij, R. A.; Tulin, N. A.; Morozov, N. A.; Pavlov, V. G.; Zaslavskij, A. Ja. *Gigiena truda i professional'nye zabolevanija* (Moscow), Mar. 1976, 3 (1-4). Illus. (In Russian)

The risk to health of microbes in sewage sludge applied to land. Report of a WHO working group. Euro reports and studies 54 (Copenhagen, Regional Office for Europe, World Health Organisation, 1981), 27 p. 30 ref.

Fertilisers

Fertilisers are substances introduced into the soil in order to obtain plentiful and stable harvest crops. According to their origin they are classified as natural fertilisers (manures) and artificial fertilisers.

Manures are mostly organic substances, such as dung, fowl droppings, guano, bone-meal, peat, compost, etc.

Artificial fertilisers are mainly produced in chemical plants by synthesis or by processing of minerals occurring in nature. Therefore, they are also called industrial fertilisers. According to their chemical composition they can be organic or inorganic (mineral) fertilisers, and they may be subdivided into nitrogenous, phosphatic, potash and trace-element fertilisers according to their main constituents (nitrogen, phosphorus, potassium, trace elements). Nitrogenous fertilisers cover saltpetres (ammonium, sodium, potassium and calcium nitrates), ammonium sulphate, calcium cyanamide, ammonia water, urea, etc. Phosphatic fertilisers include superphosphate, ground phosphate rock, basic slag, soot, trisodium phosphate, etc. Potash fertilisers comprise potassium chloride, potassium sulphate, potassium-magnesium sulphate, etc. Apart from simple mineral fertilisers containing one nutrient only, there are compound fertilisers (e.g. ammonium phosphate, nitrogen-phosphorus-potassium fertilisers) containing several nutrients. Trace-element fertilisers include chemical compounds in the form of micronutrients (boron, manganese, copper, zinc, cobalt, molybdenum, iodine, etc.) which can be readily assimilated by plants.

Natural fertilisers

The nature and chemical composition of manures and fertilisers, as well as the technique used in their production, determine the conditions of work and occupational hazards, and the preventive measures to be taken. It is therefore convenient to distinguish the following industrial processes: fertiliser production, storage, and application to the soil.

HAZARDS

Dung collection and removal in farms is associated with a more or less close contact with biological material. The nature of this contact depends on the degree of mechanisation, which varies according to the economic system. However, it should be borne in mind that the manual work involved is in any case considerable, and that agricultural workers are thus exposed to the risk of contracting diseases due to infectious and parasitic agents (common to man and animal) excreted by cattle. Workers cleaning stables also come into contact with the animals themselves, and this contact may give rise to a number of infectious diseases. The following infections are very dangerous for the agricultural worker: Q fever, malignant anthrax, brucellosis, glanders, erysipelas, melioidosis, leptospirosis, bovine tuberculosis, and tularaemia. As regards parasitic infestations, great attention must be paid to ancylostomiasis because hookworm larvae generally penetrate through the unprotected skin of the feet.

Insufficient mechanisation and the use of hand tools are associated with a high injury rate in agricultural work. A particularity of these injuries is the high frequency of infected wounds. Purulent infections are common; tetanus infection is a serious hazard.

As natural putrefaction takes place in dung heaps, workers loading dung onto trailers and taking it out to the fields may be exposed to gaseous decay products such as ammonium, carbon dioxide, carbon disulphide, etc. These gases constitute a serious hazard when manure is kept in enclosed storage structures or in badly aerated premises.

When fowl droppings are collected and removed, there is a danger of contracting psittacosis due to contact with sick fowls.

Guano consists of rests of sea-fowl excrements which are semidecomposed by the action of specific climatic conditions (dry hot climate and no rainfalls). Industrially exploitable guano deposits attaining thicknesses of 30 m and more are encountered on the Pacific coast of South America (Peru and Chile). Guano is extracted by removing the overburden and filling the product in bags. It is used in a number of countries as an organic nitrogenous-phosphatic fertiliser for farming various crops.

Guano contains highly soluble ammonium salts, calcium phosphate and small amounts of potassium salts. It is possible that part of the protein compounds excreted does not decompose but "mummifies" to form a guano constituent. Other types of guano are found on islands of the Caribbean Sea, in Bolivia, in the USSR and elsewhere. These types contain almost no highly soluble ammonium salts because these have been washed out by rain.

As the conditions of work during the extraction, storage and application of guano have practically never been studied, no preventive measures have been elaborated on a scientific basis. The effects of guano on workers are due to its constituents. The inhalation of guano dust may give rise to inflammation of the mucous membranes of the upper airways. If guano comes in contact with the skin, it may cause inflammation, especially when the skin is wet with sweat. As there may be organic inclusions in guano, the possibility of allergic reactions must be borne in mind.

SAFETY AND HEALTH MEASURES

Occupational diseases associated with the extraction, storage and application of natural fertilisers can be prevented by the following measures:

(a) mechanisation of operations with a view to limiting the worker's contact with animals and their excrements;

(b) education of agricultural workers, who should be informed about the health hazards of infectious agents or other unfavourable factors, and should develop habits of personal hygiene;

(c) use of vaccines for active protection of the organism, of sera or drugs for passive protection;

(d) qualified veterinary surveillance;

(e) wearing of protective equipment, i.e. protective clothing, boots, gloves, aprons;

(f) taking showers;

(g) adequate medical surveillance of the workers.

Artificial fertilisers

Plants producing artificial fertilisers differ considerably from each other depending on the raw materials processed and the techniques used, but they also have common features as regards the main production stages, namely: (1) warehousing of the raw materials and their preparation for processing; (2) basic raw material conversion processes and ageing of the intermediate product; (3) finishing processes, weighing and bagging of the finished product. Each of these stages is characterised by typical conditions of work.

HAZARDS

In the warehousing stage large quantities of bulk materials are handled by conveyors, ground and screened, and considerable amounts of phosphate, potassium-salt and other dusts are released into the workplace air.

In chemical processing plants the air is polluted with toxic gases and vapours (fluorine compounds, sulphuric acid fumes, nitrogen oxides, hydrogen chloride, carbon monoxide, ammonia), and with dust from raw materials and finished products. Other harmful factors are high air temperatures in certain production premises, and noise. Some of the toxic agents in the working environment are basic products of the fertiliser industry (ammonia, mineral acid fumes, nitrogen oxides, dust), others are by-products contained in the mineral or other raw materials (fluorine compounds, hydrogen chloride, cyanic acid, cyanuric acid, biuret, etc.).

The degree of pollution depends to a large extent on the more or less careful assembly and operation of the equipment, which is of particular importance if aggressive substances are processed, as these are liable to attack the equipment and cause leakage. The working environment is particularly polluted when inspection covers of processing plant are opened, when plant is cleaned or repaired, and also when ventilation systems do not operate satisfactorily. The air ducts of exhaust systems may become clogged with moist dust or reactor mass.

The storage of finished mineral fertilisers is associated with ageing processes during which gaseous products and above all fluorine compounds are released, a fact that should be borne in mind when planning the layout of storage places, especially in rural localities.

The application of mineral fertilisers gives rise to dust. When organising preventive measures, the seasonal character of this work should be taken into account.

Of the harmful factors encountered in work with mineral fertilisers, fluorine and its compounds are especially dangerous. Fluorine occurs in the form of gases (hydrogen fluoride and silicon tetrafluoride), liquids (hydrofluoric acid and fluosilicic acid) and solids (salts). Gases and vapours of fluorine compounds are released by processing plant (reactors, ammoniation vessels, crystallisers) and during the ageing stage, and dust forms in the preparation and processing departments, and also when the fertiliser is applied to the soil. Workers exposed for longer periods to fluorine and its compounds released into the air develop fluorosis.

The harmful effects of phosphatic and other raw materials are also determined by the fact that they contain 10% or more free silica, which may cause pneumoconiosis in workers inhaling dust-laden air.

The biological effects of fertiliser dust depend on its composition: salts of phosphoric, nitric and other acids. The dust of certain highly soluble fertilisers (ammophos, urea) irritates the mucous membranes of the upper airways.

Workers exposed to fumes or vapours of mineral acids, ammonia, hydrogen chloride, etc., may also develop inflammatory diseases of the upper airways.

As regards trace-element fertilisers, the micronutrients indispensable for plants present no health hazards for agricultural workers. However, industrial workers handling considerable quantities of these fertilisers over longer periods may experience toxic effects.

SAFETY AND HEALTH MEASURES

Basic measures to protect the health of workers employed in the production and application of mineral fertilisers are:

(a) mechanisation and automation of production processes, provision of remote control, careful assembly and operation of equipment, enclosure and heat insulation of plant;

(b) segregation of processes associated with particularly harmful factors, choice of adequate materials for covering walls and floors to limit the absorption of fluorine compounds;

(c) efficient general ventilation of premises, local exhaust ventilation of enclosed plant, cleaning of exhaust air and waste water;

(d) health education of industrial and agricultural workers, personal hygiene;

(e) use of protective equipment, i.e. protective clothing, respirators, rubber gloves;

(f) systematic sanitary supervision of working conditions, observance of safety and hygiene rules;

(g) medical surveillance of workers, pre-employment and periodic examinations, including radiographs of the locomotor system and lungs.

ERŠOV, V. P.

CIS 80-2055 *Hazards in the production of compound mineral fertilisers* (Les risques dans la fabrication des engrais minéraux composés). Rainguez (Caisse régionale d'assurance maladie du Nord de la France, 11 boulevard Vauban, Lille) (10 June 1980), 14 p. Illus. 7 ref. (In French)

"Problems of occupational hygiene in the production of nitrogenous fertilisers" (Voprosy gigieny truda v proizvodstve azotnyh udobrenij) (47-53). Eršov, V. P.; Kotova, N. I. *Gigiena truda v himičeskoj i farmacevtičeskoj promyšlennosti.* Sbornik naučnyh trudov (Moscow, 1976). (In Russian)

CIS 76-2035 *Nitrogen fertilisers* (Les engrais azotés). Grivauval, G. (Caisse régionale d'assurance maladie du Nord de la France, 11 boulevard Vauban, Lille) (3 May 1975), 11 p. (In French)

CIS 79-1726 "Spreading magnesian limestone soil conditioners and magnesian lime in agriculture" (Epandage des amendements calcaires magnésiens et des chaux magnésiennes en agriculture). Comité technique des départements d'outre-mer, Caisse nationale de l'assurance-maladie, Paris, 10 Nov. 1978. *Cahiers de notes documentaires–Sécurité et hygiène du travail* (Paris), 3rd quarter 1979, 96, Note No. 1208-96-79 (Recommendation No. 163) (481-482). (In French)

CIS 80-17 "Explosive behaviour of peat" (Explosionsverhalten von Torf). *Keramik und Glas* (Würzburg, Federal Republic of Germany), 1979, 1 (11-16). Illus. (In German)

CIS 79-1125 *Labour inspectorate instructions concerning liquid manure installations* (Arbejdstilsynets anvisningar for gylleanlaeg). Publikation 76/1978, Directorate of Labour Inspection (Copenhagen), Feb. 1978, 14 p. Illus. (In Danish)

See also the references appended to the article PHOSPHATES AND SUPERPHOSPHATES.

Fettling

The processes of fettling, dressing, chipping, grinding, etc., that are carried out on castings following the knock-out operation in which flasks or mould boxes are removed and the bulk of the sand is cleaned off, are amongst the most dangerous in foundry work. General foundry hazards and sandblasting and shotblasting are dealt with in separate articles.

The various processes involved are variously designated in different places but can be broadly classified as follows:

(a) dressing–which covers stripping, roughing or mucking-off, removal of adherent moulding sand, core sand, runners, risers, flash and other readily disposable matter with hand tools or portable pneumatic tools;

(b) fettling–which covers removal of burnt-on moulding sand, rough edges, surplus metal such as blisters, stumps of gates, scabs or other unwanted blemishes, and the hand cleaning of the casting using hand chisels, pneumatic tools and wire brushes.

Dressing methods in steel, iron and non-ferrous foundries are very similar, but special difficulties exist in the dressing and fettling of steel castings owing to burnt-on fused sand which is not found on iron and non-ferrous castings. Some investigations have shown that the fused sand on large steel castings may contain cristobalite.

A variety of grinding tools are used to smooth the rough casting. Abrasive wheels may be mounted on floor-standing or pedestal machines or in portable or swing-frame grinders. Pedestal grinders are used for the smaller castings which can be easily handled; portable grinders, surface disc wheels, cup wheels and cone wheels are used for a number of purposes, including smoothing of internal surfaces of castings; swing-frame grinders are used primarily on large castings which require a great deal of metal removal.

In addition, a variety of welding processes may be carried out in the fettling shop. The most conventional are oxyacetylene-flame cutting, electric arc, arc-air, powder washing and a newer development, the plasma torch. These methods are principally employed for burning-off headers, for casting repair, and for cutting and washing.

In small foundries, fettling may be carried out in a shop alongside other processes; however, in larger foundries, there is usually a separate fettling shop.

HAZARDS AND THEIR PREVENTION

Accidents. The bursting or breaking of abrasive wheels may cause fatal or very serious injuries: if there is a gap between wheel and rest at pedestal grinders, the hand or forearm may be caught and crushed. Unprotected eyes are at risk at all stages. Slips and falls, especially when carrying heavy loads, may be caused by badly maintained or obstructed floors. Injuries to the feet may be caused by falling objects or dropped loads. In the case of cuts and abrasions, there is always the danger of subsequent infection. Sprains and strains may result from incorrect methods of lifting and carrying. Badly maintained hoisting appliances may fail and cause materials to fall on workers. Electric shock may result from badly maintained or unearthed electrical equipment, especially portable tools.

A high standard of housekeeping must be ensured, well designed and maintained floors and gangways should be provided, and particular attention should be devoted to eliminating obstructions, both permanent and temporary on such floors.

All dangerous parts of machinery, especially of the abrasive wheels, should have adequate machinery guarding. The dangerous gaps between the wheel and the rest at pedestal grinders should be eliminated and close attention should be paid to all precautions in the care and maintenance of abrasive wheels and in regulation of speeds (particular care is required with portable wheels). Strict maintenance of all electrical equipment and earthing arrangements should be enforced. Workers should be instructed in correct lifting and carrying techniques and should know how to attach loads to crane hooks and other hoisting appliances. Suitable personal protective equipment such as eye and face protection and foot and leg protection should also be provided. Provision should be made for prompt first aid, even for minor injuries.

Dust. Numerous studies over many years have shown that dressing and fettling-shop operations expose workers to a serious risk from silica dust. Recent studies have shown that working conditions are still often unsatisfactory. Most of the airborne dust in the cleaning section of the foundry is caused by the disturbance of dry sand and other moulding materials, either on or inside the casting. The dust generated consists of solid matter divided into fine particles and, depending on the temperature of the casting and type of cores, certain gases may be released.

These dusts are rarely of homogeneous chemical composition and generally consist of sand, clay, carbon and the many other additives made to the moulding sand. The free silica content may range from 10-30%; in steel fettling shops, the dust may contain a considerable proportion of cristobalite, the most dangerous form of silica. Working environments contaminated by these dusts will contain particles ranging in size from fractions of a micrometre to 30-50 μm. In most instances, over 90% of the airborne dust particles are of less than 3-5 μm in size.

Silicosis is the predominant health hazard in the steel fettling shop: in iron fettling a mixed pneumoconiosis is more prevalent. Tuberculosis is still often associated with these lung affections although the incidence of this complication is less than in the past.

Many workers suffering from pneumoconiosis also have chronic bronchitis, often associated with emphysema; it has long been thought by many investigators that, in some cases at least, occupational factors have played a part. Cancer of the lung, lobar pneumonia, bronchopneumonia and coronary thrombosis have also been reported as associated with pneumoconiosis in foundry workers.

Prevention of lung affections is essentially a matter of dust control: the usual solution is by providing good general ventilation allied with efficient locally applied exhaust ventilation, low-volume, high-velocity systems being most suitable for some operations.

Blasting or rumbling of castings before dressing disposes of excess sound and dust.

Hand or pneumatic chisels used to remove burnt-in sand produce much finely divided dust and should be used as little as possible. Brushing-off excess materials with revolving wire brushes or hand brushes also produces much dust and exhaust ventilation is required.

Dust-control measures are readily adaptable to floor-standing and swing-frame grinders. Portable grinding on small castings can be carried out on ventilated benches (figure 1) or ventilation may be applied to the tools themselves. Brushing can also be carried out on a ventilated bench.

Dust control on large castings presents a problem, but considerable progress has been made with low-volume, high-velocity systems. Sometimes workers object that vision of the working area is impaired and that the system is cumbersome. Instruction and training are needed to overcome these difficulties.

Dressing and fettling of very large castings, where localised ventilation is impracticable, should be done at a separate or isolated location and at a time when few other workers are present. Suitable personal respiratory protective equipment should be provided for each worker, who should also be trained in its proper use. There should be a regular system of cleaning and repairing this protective equipment. For more detailed information on dust-control measures, see DUST CONTROL IN INDUSTRY.

Welding. Welding in fettling shops exposes workers to metal fume with the consequent hazard of poisoning and metal fever, depending on the composition of the metals involved. The plasma torch produces a considerable amount of fume and generates high levels of noise.

Figure 1. Down-draught ventilated bench for portable grinding and chipping. The operator is wearing earmuffs for noise protection, multiple-layer gloves to reduce the effects of vibration, safety glasses, safety shoes and a protective apron. Note the excellent standards of housekeeping.

An exhaust-ventilated bench can be provided for the small castings. Providing control for the welding or burning operations on large castings can present problems. A successful approach has been to have a central station for these operations. Exhaust ventilation can be provided by means of a flexible duct positioned at the point of welding. This requires the co-operation and education of the worker to move the duct from one location to another. Good general ventilation and personal respiratory protection are an aid in reducing the over-all dust and fume concentrations.

Pre-employment and periodical medical examination, with lung X-rays are most desirable in selection of workers and in detection of any deterioration in the lungs.

Noise and vibrations. Studies of noise and vibrations in the foundry have recently revealed that the highest level of noise is usually associated with knock-out and cleaning operations and is higher in mechanised than in manual foundries: the greatest risk to hearing appears to be in the fettling shop.

Noise levels found in the fettling of steel castings may be in the range of 115-120 dB whereas those actually encountered in the fettling of cast iron are in the 105-115 dB range. The British Steel Casting Research Association established that the sources of noise during fettling include:

(a) the fettling tool exhaust;

(b) the impact of the hammer or wheel on the casting;

(c) resonance of the casting and vibration against its support;

(d) transmission of vibration from the casting support to surrounding structures;

(e) reflection of direct noise by the hood controlling air flow through the ventilation system.

Portable vibrating tools may cause the appearance of Raynaud's phenomenon: this is more prevalent in steel fettlers than in iron fettlers and more frequent in those using rotating tools. The critical vibratory rate for the onset of this phenomenon is between 2 000-3 000 revolutions per minute and in the range of 40-125 Hz.

Noise control is complicated by the size of the casting, the type of metal and the work area available, the use of portable tools and a number of other related factors. Until further research and development solves some of these problems, certain basic measures are available that will aid in reducing noise levels to the individual and his co-workers. In addition, guidelines for safe daily exposure times are available.

Employed singly or in combinations, the following will be of value: isolation in time and space; complete enclosures; partial sound-absorbing partitions; execution of work on sound-absorbing surfaces; baffles, panels and hoods made from sound-absorbing or other acoustical material; personal hearing protection equipment.

A fettling bench has been developed by the British Steel Casting Research Association which reduces the noise of chipping by about 4-5 dB. This improvement is encouraging and, with further development, greater noise reductions will become a reality. Another important feature of this bench is that an exhaust system is incorporated to remove dust.

The amount of vibration transmitted to the hands of the worker can be considerably reduced by: selection of tools designed to reduce the harmful ranges of frequency and amplitude; direction of the exhaust port away from the hand; use of multiple layers of gloves or an insulating glove; shortening of exposure time by changes in work operations, tools and rest periods.

New hazards. Relatively few studies have been carried out to determine what effect, if any, the new core binders have on the health of the de-coring operator in particular. The furanes, furfural alcohol and phosphoric acid, urea and phenol formaldehyde, sodium silicate and carbon dioxide, and the no-bakes, modified linseed oil and methylene diphenyl diisocyanate (MDI), all undergo some type of thermal decomposition when exposed to the temperatures of the molten metals.

No studies have yet been conducted on the effect of the resin-coated silica particle in its relationship to the development of pneumoconiosis. It is not known whether these coatings will have an inhibiting or accelerating effect on lung-tissue lesions. It is feared that the reaction products of phosphoric acid may liberate phosphine. Animal experiments and some selected studies have shown that, when silica has been treated with a mineral acid, the effect of the silica dust on lung tissue is greatly accelerated. Urea and phenol formaldehyde resins can release free phenols, aldehydes, and carbon monoxide. It is believed that the reaction of sodium silicate and carbon dioxide may form cristobalite. The sugars added to increase collapsibility produce significant amounts of carbon monoxide. No-bakes will release isocyanates (MDI) and carbon monoxide.

The processing time and temperature, the size of the core and the amount of resin are most important in the production of the decomposition gases. The release of these gases can occur when hot cores are being de-cored.

The castings should be allowed to cool to about room temperature before de-coring begins. The use of the flexible duct ventilation, as mentioned in the section on welding, will control the dust and gases. If respirators are provided they should be capable of filtering both dust and the specific gases present.

TUBICH, G. E.

Dust:

CIS 76-947 *Foundry dust control–Fettling benches and small adjustable hoods.* Health and Safety Executive (London, HM Stationery Office, 1975), 25 p. Illus.

CIS 74-1117 *Respiratory disease in foundrymen–Report of a survey.* Lloyd Davies, T. A. Department of Employment (London, HM Stationery Office, 1971), 73 p. Illus. 31 ref.

Noise:

CIS 77-1589 *Control of pneumatic chipping and grinding noise.* Visnapuu, A.; Jensen, J. W. Report of Investigations 8223 (Bureau of Mines, 4800 Forbes Avenue, Pittsburgh) (1977), 16 p. Illus. 5 ref.

Vibrations:

CIS 74-1994 "Affections of the tendons in the region of the shoulder and elbow in moulders and fettlers" (Tendopathien im Schulter- und Ellbogenbereich bei Formern und Gussputzern). Gallitz, T. *Arbeitsmedizin–Sozialmedizin–Präventivmedizin* (Stuttgart), Dec. 1973, 8/12 (282-285). Illus. 14 ref. (In German)

CIS 78-696 "Measurement of hand-arm vibration levels caused by chipping hammers of two designs". Redwood, R. A.; Beale, K. P.; Wiseman, A. S. *Annals of Occupational Hygiene* (Oxford), Dec. 1977, 20/4 (369-373). Illus. 4 ref.

Full protection:

CIS 81-1433 "Fettling booths" (Des cabines pour l'ébarbage). *Travail et sécurité* (Paris), Dec. 1980 (652; 661-665). Illus. (In French)

Fibres, man-made

Man-made fibres can be divided into two categories: synthetic fibres (see FIBRES, MAN-MADE, SYNTHETIC) in which the fibre-forming material is derived from petrochemicals or coal chemicals; and artificial fibres in which the fibre-forming material is of natural origin.

Classes of artificial fibre

Artificial fibres may be of one of three types:

(a) viscose rayon fibres which are wholly or mainly regenerated cellulose;

(b) cellulose ester fibres; or

(c) protein fibres.

Glass (see FIBRES, MAN-MADE, GLASS AND MINERAL) and metals are among the inorganic materials used for man-made fibres. The increasing need in industrial fields for fibres which will withstand extremely high temperatures has led to the use of quartz, aluminium oxide and similar refractory materials. Metallic yarns have long been employed to add a decorative metallic glint to fabrics used for clothing and upholstery.

Viscose rayon. This is produced from cellulose sheet which is shredded and converted consecutively into alkali cellulose and cellulose xanthate, and finally extruded into filament fibre. In the early days of regenerated fibre production, this material was known as "artificial silk" due to its resemblance to natural silk; however, in the mid-1920s the term "rayon" was adopted. At that time, the production of cellulose acetate (a cellulose ester) fibres amounted to less than 1% of the total production of man-made fibres. By 1951 the production of cellulose acetate had risen to 34% and, at the request of the rayon industry, the US Federal Trade Commission adopted the rule that the term "rayon" should be used only to describe yarn, thread and cloth made from regenerated cellulose fibres and that the term "acetate" should be used to describe yarn, thread and cloth made from cellulose acetate (i.e. cellulose ester) fibres. Hence rayon in the United States is the description for the two most important types of regenerated cellulose, viscose rayon and cuprammonium rayon, but not cellulose acetate fibres. In other countries there is no hard and fast rule and the term rayon in some countries includes cellulose acetate. The term "artificial silk" is not used now in English-speaking countries but is still in use in some others.

Cuprammonium rayon (cupra rayon). This is made from cellulose dissolved in a solution of copper salts and ammonia which produces cuprammonium liquor. This solution is extruded through spinnerets into a funnel fed with pure soft water which starts a coagulating action. The filaments pass out of a small orifice in the base of the funnel and are subjected to stretching and final coagulation by passing through a bath of acid or metallic salt solution. The filaments are then wound on to skeins or bobbins and washed. Instead of winding on to skeins or bobbins, the filaments from the whole spinning machine, usually about 575 in number are, in some cases, passed through two washing baths and wound on to a beam. This beam can then be used directly on a warp knitting machine, or five or six beams rewound on to a weaver's beam which can be placed in a loom for weaving.

Cuprammonium rayon is sometimes known as Bemberg yarn. In the Federal Republic of Germany it is known as Cupressa when in continuous filament form or as Cuprama when in staple form; in Japan it is called Bremsilkie. In the United States it is sold under about 30 different trade names.

Cellulose acetate. Viscose rayon and cuprammonium rayon are made from regenerated cellulose and their chemical structures resemble cotton or flax. There is, however, another form of cellulosic man-made fibre in which the raw material of cellulose wood pulp or cotton linters is changed into a different substance to make it soluble and spinnable and which after spinning is left in its changed chemical form. This is the fibre known as cellulose acetate. It is prepared by steeping bleached cotton linters or wood pulp in acetic acid to swell the fibres and increase their chemical reactivity. The swollen material is then placed in a mixture of acetic anhydride and a catalyst. The temperature of the mixture is raised until the exothermic reaction commences, after which it is cooled.

Dilute acetic acid is added and the solution left for a long period. Solid cellulose acetate is precipitated by adding water. The precipitate is washed, dried and ground to flakes. The spinning solution is then made by dissolving the cellulose acetate flakes in acetone containing up to 10% water. The solution is pumped through spinnerets into a stream of warm air which evaporates the acetone, leaving the solid filaments, which are wound on to bobbins. This method of spinning is known as dry-spinning whereas in the spinning of viscose rayon and cuprammonium rayon the filaments are coagulated in a bath of acid and the method is known as wet-spinning. Dry-spinning is a simpler operation and gives a greater production but the solvents evaporated in dry-spinning must be exhaust ventilated to avoid the risk of fire and explosion and possible health hazard to the workers because of their toxic nature.

Cellulose triacetate. This is also a cellulose ester fibre. It is made by steeping cotton linters or wood pulp in acetic acid followed by treatment with sulphuric acid and acetic anhydride to form a cellulose triacetate solution. Magnesium acetate and water are added and, after dilution with further water, the cellulose triacetate is precipitated and washed. It is then dissolved in dichloromethane and pumped through spinnerets. This is a dry-spinning process and the dichloromethane evaporates as the threads are formed and has to be removed by exhaust ventilation. Cellulose triacetate yarn is known as Arnel in the United States, Canada and Belgium, as Rhonel in France, and as Tricel in the United Kingdom.

Protein fibres. Textile fibres have been produced from the proteins of milk (casein), seaweed, maize, soya and groundnuts but only on a small scale. The US Federal Trade Commission has adopted the name Azlon for fibres of the regenerated protein type, the official definition being "a manufactured fibre in which the fibre-forming substance is composed of any regenerated naturally occurring protein". The spinning process is similar to that for rayon production. The purified protein is dissolved in alkali, centrifuged and filtered. The spinning solution is extruded into a bath composed of an aqueous acid containing a high concentration of salts such as sodium chloride, sodium sulphate or magnesium sulphate. The salts bring about coagulation of the solution to form the filaments, which are stretched and then treated with formaldehyde to harden the fibre. Washing and cutting into staple fibre follows. The two most common protein fibres are Fibrolane and Merinova, both made from casein.

Filament and staple fibre. Man-made fibres are produced in continuous lengths (filaments) in the same way as natural silk, whereas other natural fibres such as cotton, wool, jute and flax exist only in short staples. Cotton has a staple of about 2.6 cm, wool of 8-10 cm and flax of 30-50 cm. Some man-made fibres produced in continuous lengths are cut up after spinning into short lengths or staples of 3-30 cm on special cutting or stapling machines. This short staple fibre can then be processed and re-spun on ordinary cotton or woollen spinning machinery, producing a better-looking yarn, free from the glassy appearance common to some artificial yarns. In some re-spinning processes, a mixture of artificial fibres and natural fibres or of different kinds of artificial fibres such as viscose rayon and cuprammonium rayon may also be made. [According to the UN *Statistical Yearbook*, in 1977 world production of rayon and acetate continuous filaments was in the order of 1 150 000 metric tons and that of rayon and acetate discontinuous fibres in the order of 2 100 000 metric tons.]

HAZARDS AND THEIR PREVENTION

Fire and explosion. The fire and explosion risks are very high in most artificial-fibre factories due to the use of large quantities of flammable substances such as solvents. All flammable raw materials should be stored in out-buildings separate from the main production workrooms, and automated pipeline feed systems should be installed between storage facilities and processing plant. Where flammable solvents are stored or used, there should be no open lights, flames, or hot surfaces, the electrical equipment should be of certified flameproof construction and smoking should be prohibited. Static electricity may develop in the spinning of some types of fibre and particularly in the carding of staple acrylic fibres. The dangers of electrostatic charges can be prevented by either providing controlled humidity or by installing static eliminators. Very good exhaust ventilation is required to keep the concentration of flammable solvents in the working atmosphere down to a minimum. High standards of fire prevention are essential.

Accidents. Falls on wet and slippery floors are common in artificial-fibre production; steps should be taken to prevent spillage of water or coagulating solution from the spinning machines on to the floor. In addition to the ordinary machinery guarding requirements for belt drives, pulleys and shafting of machines used in this industry, there are special parts of machinery which require guarding, such as feed rollers and fluted rollers of staple fibre-cutting machines, and hydro-extractors or centrifuges. Where staple spinning is carried out, the standards of safety for the carding machines, the winding, warping and preparation machinery and looms should be similar to those in the cotton and woollen industry.

Health hazards. (For detailed information on health hazards see specific articles and in particular CARBON DISULPHIDE; ESTERS.) The artificial-fibre industry uses many toxic chemicals including carbon disulphide, benzene, acetic acid, hydrogen sulphide and various inorganic acids. Exhaust ventilation is essential to ensure that the toxic fumes in the working atmosphere are kept within safe exposure limits. In some acetylation processes, the ripening pans require to be emptied and cleaned every few hours. Workers engaged on this task should be provided with suitable respiratory protective equipment against the acid fumes given off. Similar precautions should be taken when cellulose xanthate churns are cleaned out. The respirators provided should also protect the eyes from the irritating effects of the fumes.

Workers should be fully instructed as to the dangers from the toxic and flammable materials used and as to the precautions to be observed. Protective clothing should be provided for workers engaged in wet processes or in handling toxic and corrosive materials. As far as possible, plant and machines should be in airtight enclosures and fitted with automatic feeding and discharging devices.

Pre-employment and periodic medical examinations of workers is highly desirable.

QUINN, A. E.

CIS 77-1739 *Occupational safety and health techniques in chemical fibre production* (Tehnika bezopasnosti i promyšlennaja sanitarija v proizvodstve himičeskih volokon). Zak, S. L.; Kuznecov, V. A. (Moscow, Izdatel'stvo "Himija", Stromynka 13, 1976), 215 p. Illus. 29 ref. (In Russian)

CIS 79-1242 "Machines used in the man-made yarn and fibre industry" (Machines de l'industrie des fils et fibres artificiels). Comité technique national des industries textiles, Caisse nationale de l'assurance maladie. *Cahiers de notes documentaires–Sécurité et hygiène du travail* (Paris), 2nd quarter 1979, 95, Note No. 1187-95-79 (Recommendation No. 153) (311-314). (In French)

CIS 79-1695 "Literature survey of toxicology problems in the viscose industry" (Bibliografische studie van de toxicologische problemen in de viscose industrie). Vanhoorne, M. *Cahiers de médecine du travail–Cahiers voor arbeidsgeneeskunde* (Gerpinnes), Mar. 1979, 16/1 (83-99). 136 ref. (In Dutch)

5th International Symposium on Occupational Health in the Production of Artificial Fibres. Abstracts. Belgirate, Italy, 16-20 Sept. 1980. Cavalleri, A. (ed.). (Permanent Commission and International Association on Occupational Health, Scientific Committee for Occupational Health in the Production of Artificial Fibres, 1980), 84 p.

Fibres, man-made glass and mineral

Production. These fibres are made from glass, natural rock or any readily fusible slag and are all amorphous silicates, in contrast to naturally occurring mineral fibres such as asbestos, which is crystalline in structure and differs chemically. Glass fibres are made from either borosilicate or calcio-alumina silicate glass.

Wool for insulation products is made by allowing molten material to run through a series of small openings which are situated in a revolving spinner in the modern rotary process (figure 1), and then attenuating the primary filament by air or steam blowing (figure 2). Continuous filament for textiles or reinforcement in materials such as resin or cement is manufactured by mechanical drawing at very high speed. A third process, flame attenuation (figure 3), is used mainly for specialised productions of very fine diameter fibres which are

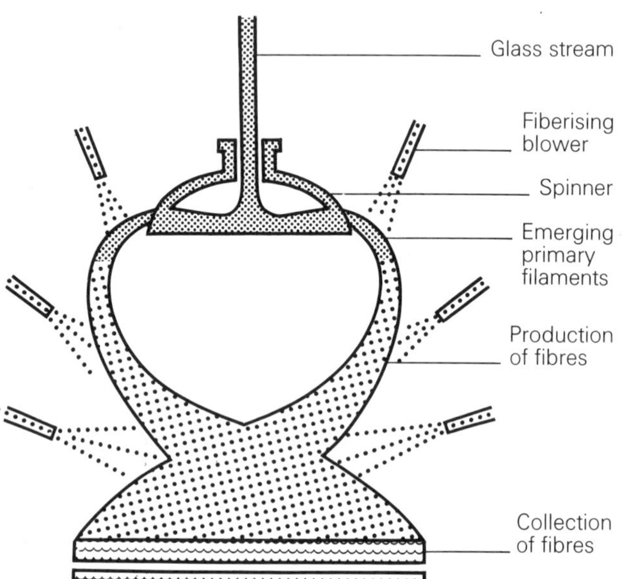

Figure 1. Rotary process. In recent years, rotary processes have largely replaced direct blowing processes for the commercial production of glass fibre insulation products. These rotary processes all employ a hollow drum, or spinner, mounted with its axis vertical and rotating about that axis. The vertical wall of the spinner is perforated with several thousand holes uniformly distributed around the circumference. Molten glass is allowed to fall at a controlled rate into the centre of the spinner from where some suitable distributor forces it to the inside of the vertical perforated wall. From that position, centrifugal force drives the glass radially outwards in the form of discrete glass filaments issuing from every perforation. Further attenuation of these primary filaments is achieved by a suitable blowing fluid emerging from a nozzle or nozzles arranged around, and concentric with, the spinner. The net result is the production of fibres with a mean fibre diameter of 6-7 μm. The blowing fluid acts in a downwards direction and so, as well as providing the final attenuation, it also deflects the fibres towards a collecting surface situated below the spinner. On the way to this collecting surface the fibres are sprayed with a suitable binder before being uniformly distributed across the collecting surface.

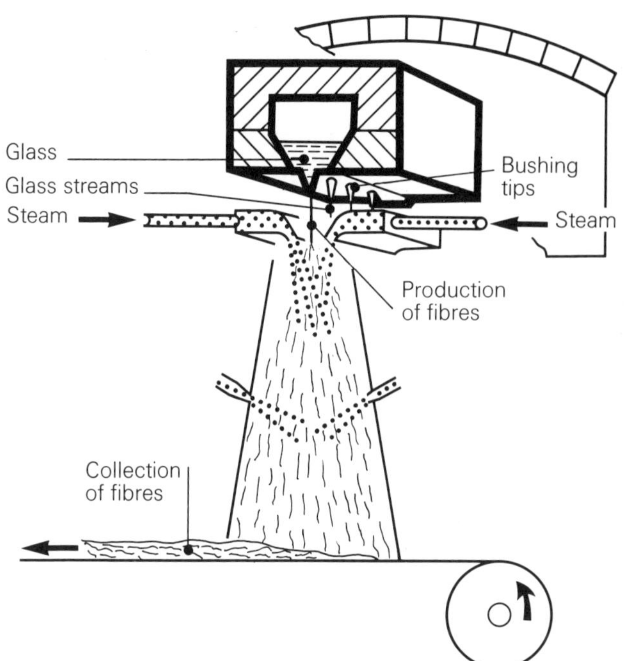

Figure 2. Steam blowing process. Developed during the early 1930s, the steam blowing process was the first process for fiberising glass to be commercially exploited on a world-wide scale. The process employs a special alloy crucible, or bushing, which is electrically heated and into which glass flows from a furnace. In the base of the bushing is a series of small holes, or tips, from which streams of glass emerge, under the influence of gravity. These emerging streams of glass are attenuated into fibres by the combined action of two steam jets, blowing essentially downwards, but also converging slightly to meet the glass streams just below the tips. Friction between the steam jets and the glass, plus the subsequent turbulence, result in the production of fibres of a mean diameter of about 15 µm. A small percentage of the glass, possibly up to 10%, remains unfiberised and emerges as tiny glass beads. The resulting fibres are drawn by suction to a collecting conveyor, receiving a spray of bonding agent just prior to collection.

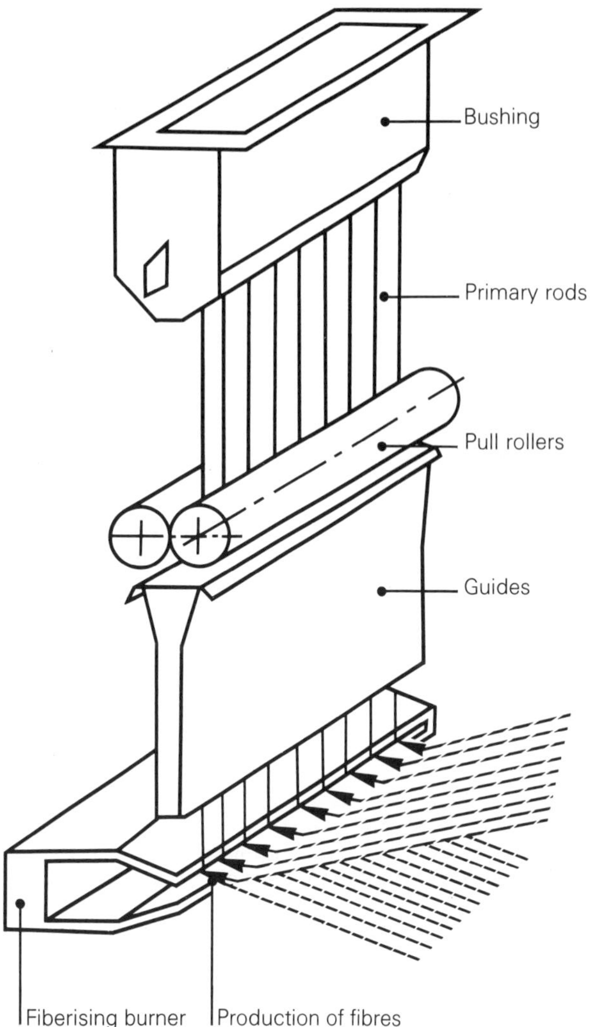

Figure 3. Flame blowing process (flame attenuation). An alternative approach to fiberising, developed mainly for specialised applications, is the flame blowing process. The first stage of this process, the production of fine glass rods, is achieved by drawing molten, viscous glass, emerging from holes in the base of a special alloy bushing, between pull rollers. These fine rods are then guided at right angles into a high-velocity, high-temperature gas stream emerging from the exhaust nozzle of an internal combustion burner. This gas stream progressively remelts the fine rods and attenuates them into very fine fibres which may have a mean fibre diameter as low as 1 µm. The orientation of the fiberising components is usually such that the final fibres are produced in a horizontal direction. Subsequent to fiberising, a spray of binder is applied from above and below the fibre stream and the resulting fibre and binder collected on an inclined conveyor.

produced for a very specialised market for purposes such as scientific filter papers. It consists of mechanical drawing of molten glass rods followed by further attenuation in a stream of hot gas. The fibres then receive a thin, often molecular, coating to bind them together. The composition of this binder differs considerably both between manufacturers and according to the requirements of the end-product. With rock and slag wools (figure 4) the binder may be a simple oil, but in the case of glass fibres it is usually a complex formulation containing a lubricant, a resin (usually thermosetting) for bonding, and one or more cation active agents for adhesion. The coated fibres are then usually heat cured to polymerise the resin, after which it is considered chemically inert for practical purposes.

Physical characteristics and uses. The most important physical characteristic of these fibres is that, in contrast to asbestos, they cannot fracture longitudinally into finer fibrils and are subject to transverse fracture because they are glassy in nature, the result of which is to reduce them to fibres of shorter length but the original diameter of manufacture remains unchanged.

Man-made mineral fibres are manufactured to controlled nominal diameters, appropriate to the end use and fall into three groups. Continuous filament is manufactured to a large nominal diameter, approximately 9-25 µm and the method of production results in a very narrow spread of diameter around the nominal size. Levels of airborne respirable fibres are of the order of 0.002 µm. When the material is incorporated as a reinforcement and subjected to grinding or sawing, the dust evolved would be predominantly that of the reinforcement but, because of their diameter, respirable fibres are not evolved, except in negligible quantities, although some glassy fibres are fragmented to a particulate dust of respirable diameter.

Insulation wool is produced to a nominal diameter of approximately 6 µm but the method of production gives

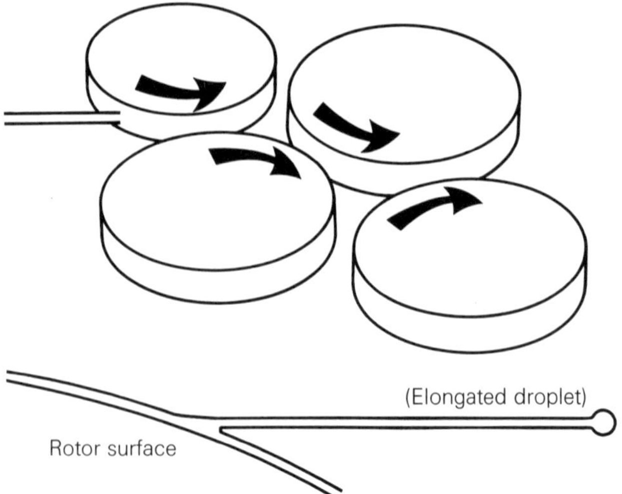

Figure 4. Span wool process. A set of rotors are mounted in a cascade formation and rotate very rapidly. A stream of molten stone is continuously transferred to one of the upper rotors and from this rotor distributed on the second and so on. The melt is uniformly spread on the outside surface of all the rotors. From the rotors, droplets are thrown out by centrifugal force. The droplets are attached to the rotor surface by elongated necks which under further elongation and simultaneous cooling develop into fibres. The elongation is of course followed by a decrease of diameter which in turn causes an accelerated cooling. Thus there is a lower limit for the diameter among fibres produced in this process. A normal distribution of fibre diameters around the mean value is therefore not expected.

rise to a much wider distribution curve and the product contains numbers of respirable fibres. Typical levels in the manufacturing industry are 0.5-0.2 respirable fibres/cm^3. Some insulation wools are produced with nominal diameters of 1-2 μm for special application, such as acoustic insulation. The very fine diameter fibres usually at or below 1 μm in diameter are usually made by flame attenuation and represent about 1% of world production and are used for purposes such as scientific filter papers. These fibres may not be coated with a binder and may give rise to a far greater number of fibres/cm^3, and these will of course all be in the respirable range.

It is important to note that whilst the very fine fibres in the latter group have only come into commercial production since the 1960s, the glass wool industry started in the 1930s and some of the mineral fibres have been made since the late 1800s. Whilst, over the years, there has been a reduction from about 15 to 6-9 μm in the nominal diameter of wool products, because of the very wide distribution curve the effect of this change in the mean can be expected to have a minimal effect on the number of fine respirable fibres contained in the products, and examination of old products shows them to contain roughly the same number of respirable fibres as present-day products.

Limited information is available on exposure levels in the user industries. The application of wool products for insulation in house lofts shows levels of about 1 fibre/cm^3, but since professional installers do this job intermittently the application of a time-weighted average reduces this figure to about 0.5 fibres/cm^3. Gravimetric estimation of total dust is very variable indeed and it is characteristic that where there are large quantities of extraneous dust, gravimetric levels may be very high whilst the fibre count remains of the levels quoted. Because of the inability of these fibres to split longitudinally, thereby producing large numbers of finer fibres, electron microscopy levels show a ratio of approximately 1 : 2 with optical counts, in contrast to the situation with asbestos, where the electron microscopy count can be very much higher.

HAZARDS AND THEIR PREVENTION

Noise. Fans and steam jets constitute a serious acoustic hazard and noise control measures and hearing protection may be necessary, together with a hearing conservation programme.

Radiant heat. Where workers are exposed continually or for lengthy periods in their working day to infrared emissions from molten glass, a heat absorbing glass screen or other suitable means of protection should be employed.

Injury. Needle-shaped fragments of unattenuated fibre (slugs) may cause skin penetration. In manufacture they are also an eye injury hazard and eye protection is necessary.

In continuous filament manufacture, multiple skin penetration may occur from the back lash of a broken strand. These fibres are difficult to see and to remove. If the area is covered by a piece of adhesive strapping most fibres will adhere and be removed immediately with the strapping.

Skin irritation. Skin conditions are the major health problem of glass fibre handling. Contact with uncured resins, hardeners and accelerators, usually in user applications, may give rise to sensitivities. Glass fibre itself, however, is not a sensitising agent. Most people handling glass fibre for the first time suffer from a transient irritation of the exposed parts of the skin. This is usually accompanied by a punctate itching erythema which passes off with continued exposure in the course of a few days or a few weeks. The skin is then said to be hardened, but if the exposures are interrupted by sickness absence or are not continuous enough to maintain hardening, the rash will recur and the process will start once again. Topical treatment is by *lotio calamine cum penol* during work combined with topical steroids at home and if the rash has not settled within 2-3 weeks, medical opinion should be sought, both to establish the diagnosis and because some predisposed patients may proceed to a neurodermatitis. If this occurs, the patient should be regarded as unsuitable for further work with man-made mineral fibres. This new-starter fibreglass rash represents a histamine reaction and is associated with fibres greater than 5 μm in diameter.

Workers should be made aware of these problems and preventive measures should be based on good hygiene practice. Sanitary facilities should be of the highest standards and hands and arms should be rinsed in cold or lukewarm water before soap is applied. Protective overalls and changing rooms should be provided, but overalls must be laundered separately from other clothing to prevent contamination. Clothing should be loose at the waist, neck, wrist and elbows to prevent friction effects. Some workers prefer short-sleeved loose shirts while others prefer full-length sweaters, etc. Workers should be encouraged to report all signs of skin irritation at an early stage when treatment is usually easy and effective.

The classical fume-sensitive rash is sometimes seen in workers exposed to formaldehyde or resin fumes and consists of peri-orbital oedema with erythema and itching of the face and commonly the hands. Whilst eczematide reactions should be excluded, this condition dictates immediate removal from the source followed by suitable patch-testing 6 weeks after the rash has entirely abated. Since only 50% or so of patch tests provide conclusive evidence a firm clinical diagnosis should be sufficient to remove the worker from further exposure to the sensitising agents. Particularly in continuous filament production, the hands may be exposed to combinations of moisture, heat and chemicals in relatively low dilutions and these agents may react in combination to produce a contact eczema. This is best managed by treating the patient whilst retaining him at work and needs to be accompanied by advice directed to the avoidance of skin contact. Barrier creams have a limited part to play in this situation. Close supervision of this type of skin condition is nevertheless required to ensure that it is adequately healed and that work practices are sound.

From what has been said, it will be clear that patients with any of the itching dermatoses or with bad personal hygiene are unsuitable for work in fibreglass. Other conditions like psoriasis are not adversely affected.

Dust and fumes. Glass and man-made mineral fibres contain no free silica. Since glass fibre manufacture began between 1920 and 1930, no evidence of pneumoconiosis has been found. Good local exhaust and general ventilation should, however, be provided where substantial quantities of dust are emitted. Local exhaust ventilation is also necessary to remove phenol and ammonia fumes generated in the mixing of resin emulsions and from curing ovens. The increasing complexity of sizes which is occurring in continuous filament manufacture may give rise to aerosols and air-knives, or other means of protection may be required according to the specific chemicals being used.

Nose and throat irritation of a transitory nature may occur occasionally and is most often seen when dust concentrations have become unusually high and have been due to an alteration in the binder content of the fibres.

Chest disease. As with any other non-specific dust, pre-existing chest conditions like asthma, bronchiectasis and bronchitis may be exacerbated. However, extensive cross-sectional prevalence studies have shown no changes in chest X-rays or pulmonary function and no excess reports of symptoms of chronic bronchitis in workers exposed to man-made mineral fibres. A limited post mortem survey has shown no significant difference between the lungs of glass fibre workers and those of city dwellers. Fibrosis is not regarded as a human hazard of these fibres. The work of Pott and Stanton showed that if fibres were injected intrapleurally or intraperitoneally into rats, mesotheliomas were induced and that the reaction occurred irrespective of the chemical nature of the fibre but was significantly related to diameter and length, the highest tumour potential being associated with fibres narrower than 0.5 μm and longer than 10 μm. Sometimes the fibres used are not known to be associated with human pathology and inhalation experiments in rats showed only a macrophage reaction with man-made mineral fibres, i.e. a typical non-specific dust reaction. No tumours are induced by inhalation of glass fibres although the inhalation of asbestos produces all the conditions, including tumours, which that fibre produces in man. An extensive literature has developed in which there is agreement between all observers that there is no increase in over-all mortality and no increase in mortality due to cancer of the respiratory tract. Although suspicions have been raised in the category "non-malignant respiratory disease, excluding pneumonia and influenza", critical evaluation raises doubts about this finding, which is not in agreement with the animal or cross-sectional human prevalence studies and has failed to be confirmed by a subsequent and much larger study by Enterline. For a fuller discussion of this topic see Hill, 1979.

These negative findings are not surprising in view of the extremely low levels of exposure experience.

HILL, G. W.

"Fibrous glass dermatitis". Possick, P. A.; Gillin, G. A.; Key, M. M. *American Industrial Hygiene Association Journal* (Akron, Ohio), 1970, 31/1 (12-15). Illus. 19 ref.

"Glass fibres: absence of pulmonary hazard in production workers". Hill, J. W.; Whitehead, W. S.; Cameron, J. D.; Hedgecock, G. A. *British Journal of Industrial Medicine* (London), Apr. 1973, 30/2 (174-179). Illus. 15 ref.

"New knowledge on glass and mineral fiber hazards" (Neuere Erkentnisse über Gefährdungen durch Glas- und Mineralfasern). Mayer, P. *Zentralblatt für Arbeitsmedizin* (Dortmund), 30 Aug. 1980, 30/8 (280-285). Illus. 13 ref. (In German)

"The lungs of fiberglass workers: comparison with the lungs of a control population" (249-263). *Occupational exposure to fibrous glass.* DHEW (NIOSH) publication No. 76-151 (National Institute for Occupational Safety and Health, 4676 Columbia Parkway, Cincinnati) (1976), 420 p.

"Review of the epidemiology of man-made mineral fibres". Hill, J. W. (979-983). 5 ref. *Biological effects of mineral fibres.* Vol. 2. Wagner, J. C. (ed.). IARC scientific publication No. 30, INSERM symposia series, Vol. 92 (Lyons, International Agency for Research on Cancer, 1980), 1 007 p. Illus. Ref.

"Mortality of workers in the man-made mineral fibre industry". Enterline, P. E.; Marsh, G. M. (965-972). 7 ref. *Biological effects of mineral fibres.* Vol. 2. Wagner, J. C. (ed.). IARC scientific publication No. 30, INSERM symposia series, Vol. 92 (Lyons, International Agency for Research on Cancer, 1980), 1 007 p. Illus. Ref.

"Mortality patterns of rock and slag mineral wool production workers: an epidemiological and environmental study". Robinson, C. F.; Dement, J. M.; Ness, G. O.; Waxweiler, R. J. *British Journal of Industrial Medicine* (London), Feb. 1982, 39/1 (45-53). Illus. 32 ref.

Fibres, man-made synthetic

Man-made fibres can be divided into two categories: artificial fibres in which the fibre-forming material is of natural origin, and synthetic fibres in which the fibre-forming material is derived from petrochemicals or coal chemicals.

Classes of synthetic fibre

Synthetic fibres are made from polymers that have been synthetically produced from chemical elements or compounds obtained from the petrochemical industry. The following are the main groups of substances from which the fibres are produced: polyamides, polyesters, polyvinyl derivatives, polyolefins, polypropylenes, polyurethanes and polytetrafluoroethylenes.

Polyamides. The most important in this class are the nylons. "Nylon" is the generic term for any long-chain polymeric amide. The names of the many different kinds of nylon fibre are distinguished by a number which indicates the number of carbon atoms in their chemical constituents, the diamine being considered first. Thus the original nylon produced from hexamethylene diamine and adipic acid is known in the United States and United Kingdom as nylon 66 or 6.6, since the diamine and the dibasic acid both contain six carbon atoms. However, in the Federal Republic of Germany this polyamide is marketed as Perlon T, in Italy as Nailon, in Switzerland as Nylsuisse, in Spain as Anid and in the Argentine as Ducilo.

Nylon 6 is made from caprolactam which contains six carbon atoms. In Japan, nylon 6 is marketed as Amilan, in Spain as Dayan, in Italy as Lilion, in the USSR as Kapron, in the Federal Republic of Germany as Perlon. In the United States the majority of producers employ the term nylon 6 (see POLYAMIDES).

Polyesters. The first polyester was produced in 1941. The process of manufacture is similar to that for nylon. Ethylene glycol and terephthalic acid are reacted together so that their short molecules link up into long chains forming a plastic material which is pumped in molten form through spinnerets to make the filaments which harden in cold air; a drawing or stretching process follows. Polyester fibre is available under trade names such as Terylene in the United Kingdom, Dacron in the United States, Tergal in France, Terital and Wistel in Italy, Lavsan in the USSR, and Tetoran in Japan (see POLYESTER FIBRES).

Polyvinyl derivatives. The most important member of this group is polyacrylonitrile or acrylic fibre. This was produced for the first time in 1948 and various types are now available under a variety of trade names in different countries, e.g. Acrilan and Orlon, etc., in the United States, Crylor in France, Leacril and Velicren in Italy, Amlana in Poland, Courtelle in the United Kingdom, and many others (see POLYACRYLONITRILE FIBRES).

Carbon fibres are made from a special grade of polyacrylonitrile filament and are eight times as strong and stiff as high-tensile steel. Carbon fibres were used for the first time in 1964 and are still undergoing development; however, they are already in use in the manufacture of aircraft engine parts, aerospace components and pressure vessels. Combined with glass fibres they are used for making yacht masts, skis, golf club shafts, compressed-gas cylinders and the bodywork of racing cars. Carbon fibres are available under such trade names as Morganite and Thornel.

Polyolefins. The most common fibre in this group is known as Courlene in the United Kingdom. The filaments are made by an extrusion process similar to that used for nylon. Molten polythene at 300 °C is forced through spinnerets and cooled to form the fibre in either air or water. A drawing or stretching process follows (see POLYOLEFINS).

Polypropylenes. Propylene, a common by-product of petroleum cracking can be polymerised to fibre-forming polypropylene. The polymer is melt-spun, stretched or drawn and then annealed. The fibre is known under such trade names as Hostalen in the Federal Republic of Germany, Meraklon in Italy and Ulstron in the United Kingdom (see POLYPROPYLENE FIBRES).

Polyurethane. The first polyurethane fibre produced in 1943 was known as Perlon U. It was spun from the polyurethane produced by the reaction of 1,4-butanediol with hexamethylene diisocyanate. In recent years polyurethanes have become the basis of a new type of synthetic fibre known as spandex. Spandex fibres are highly elastic and are sometimes called snap-back or elastomeric fibres on account of their rubber-like elasticity. The fibre is manufactured from a linear polyurethane gum, which is cured by heating at very high temperatures and pressures to produce a "vulcanised" cross-linked polyurethane which is extruded as a monofil. This thread can be covered by rayon or nylon fibre to improve the appearance while the thread itself serves as an elastic core. It is used extensively in garments requiring a high degree of elasticity. Trade names for spandex yarns include Lycra, Vyrene and Glospan in the United States, Spandell in the United Kingdom and several others (see ISOCYANATES).

Polytetrafluoroethylenes. These polymers were first used for the production of fibres in 1954; they are available under such trade names as Teflon. They are not used for garment fabrics but they have properties which make them very suitable for certain industrial uses. The fibres are spun from an aqueous colloidal dispersion, being extruded into a 5% solution of hydrochloric acid in water. The filament is sintered on a metal roller at 385 °C for a few seconds, then quenched in water. In the subsequent drawing process it is stretched to four times its original length (see POLYFLUORINES).

Synthetic fibres for work clothing

Some of the earlier synthetic fibres were unsuitable for use in work clothing, since they had a low melting point and inadequate resistance to abrasion and corrosive substances, or were flammable or subject to electrostatic charging. Subsequent research has led to the development of fibres with properties that make them highly suitable for working clothes or protective clothing. The made-up fabrics have a unit weight considerably lower than that of rubberised or plastic-coated materials but offer a similar degree of protection in many cases; the finished garments are often more comfortable to wear, especially in hot atmospheres.

The chemical resistance of polyester cloth makes this material suitable for protective clothing in acid-handling plants. Polyolefin fabrics are suitable for clothing used to give protection where there is prolonged exposure to both acids and alkalis. High-temperature-resistant nylon is particularly suitable for the manufacture of work clothes in fire-hazard areas (see HEAT PROTECTIVE CLOTHING); it also has good resistance at room temperature to solvents such as acetone, benzene, trichloroethylene and carbon tetrachloride. Certain polypropylene fibres are resistant to a wide range of corrosive substances and are suitable for work and laboratory clothing and protective clothing for certain hazards.

In choosing protective clothing made from synthetic fibre, care should be taken to determine the generic name of the fibre and to obtain details on properties such as shrinkage, sensitivity to light, dry-cleaning agents and detergents; resistance to oil, corrosive chemicals, common solvents and heat; and susceptibility to electrostatic charging.

Special processes

Stapling. Silk is the only natural fibre that is available as a continuous filament; other natural fibres exist only in short lengths or staples. Cotton has a staple of about 2.6 cm, wool of 8-10 cm and flax 30-50 cm. Man-made fibres are produced in continuous lengths, but some are passed through a cutting or stapling machine for cutting into short staples similar to the natural fibres. They can then be re-spun on cotton or wool-spinning machinery with the object of producing a better finish free from the glassy appearance associated with some synthetic fibres. In some re-spinning processes, a mixture is made of synthetic and natural fibres or of different kinds of synthetic fibres.

Crimping. In order to give a synthetic fibre the appearance and feel of wool, the cut or stapled fibres (in a twisted or tangled state) can be passed through a crimping machine in which they are given a permanent crimp by means of hot, fluted rollers. Crimping can also be done chemically by controlling the coagulation of the filament in such a way as to create a fibre of asymmetrical cross-section, one side being thick-skinned and the other thin-skinned. When the fibre is wet, the thick side has a tendency to curl and produce a crimp. Another form of crimping is that used for making crinkled yarn. The synthetic yarn is knitted into a fabric, heat-set and then wound from the fabric by backwinding. In the United States crimped yarn is known as non-torque yarn. The latest crimping method consists of passing two nylon threads through a heater to raise them to a temperature of 180 °C. They then pass through a high-speed revolving spindle to insert twist. The first machine for this process had spindles running at 60 000 revolutions per minute but the latest machines have spindles revolving at 1.5 million times a minute.

HAZARDS AND THEIR PREVENTION

Accidents. Machinery guarding should be applied to traditional hazardous machine parts such as drive belts, pulleys and shafting and also to the feed and fluted rolls of fibre-cutting machines; hydroextractors or centrifuges also need special guards and interlocking devices. Where staple spinning is carried out, the carding, spinning, preparation, winding and warping machines should have the same high standard of fencing as similar machines used in the cotton industry or wool industry. In wet spinning there is danger of falls due to wet floors; steps should be taken to reduce spillage from vats and to ensure good drainage.

Fire and explosion. The synthetic fibres industry uses large quantities of toxic and flammable substances. Storage facilities for flammable substances should be located, for preference, either in the open or in a separate building of fire-resistant construction; bunds or dikes should be provided to contain spillage in the event of a leak or fire. Automation of the delivery of toxic and flammable substances by the use of pumps and piping systems will reduce the danger of exposure and fire; the storage of drums, etc., of dangerous substances near to production plant is particularly hazardous. High standards of fire protection and prevention are essential.

In dry spinning the spinning solution is forced through spinnerets and the extrusion solidifies as the solvent vaporises. The large quantities of solvent vapour that are given off during spinning constitute a considerable explosion and poisoning hazard and should be removed by exhaust ventilation; the exhausted vapours may be distilled and recovered for further use but should not be allowed to escape to the environmental atmosphere. The rate of drying for the extruded fibre should be such that, if air is used as the drying agent, the vapour concentration in the spinning cell at no time falls within the solvent's explosive limits. Where flammable solvents are used, smoking should be prohibited, open lights, flames and sparks should be eliminated and electrical equipment should be of certified flameproof construction. Machines should be earthed to prevent the build-up of static electricity which may cause a dangerous spark when a worker opens a spinning cell to remove a bundle of filament. Workers should wear footwear with rubber soles and heels. Electrostatic charges may build up in carding machines for nylon staples; this problem may be overcome by the installation of static eliminators or by atmospheric humidification.

Health hazards. The steps taken to prevent dangerous concentrations of flammable solvents will normally be sufficient to eliminate any poisoning hazard due to these solvents. Where a toxic solvent presents no explosion risk, exhaust ventilation should be designed to keep concentrations of vapour below the relevant maximum permissible concentration. Respiratory protective equipment should be available for use in emergencies due to leaks or spillage and where maintenance and repair workers may encounter hazardous concentrations. Where solvents or corrosive substances may come into contact with the worker's body, adequate personal protective equipment should be provided.

QUINN, A. E.
MATTIUSSI, R.

CIS 78-1552 "Clinical, industrial hygiene, and experimental study on the biological effects of synthetic polymer dust" (Kliniko-gigieničeskie i èksperimental'nye issledovaanija biologičeskogo dejstvija pyli sintetičeskih polimerov). Martynova, A. P.; Mashulija, E. Š.; Alekseeva, O. G.; Voroncov, R. S.; Kudinova, O. V.; Lopuhova, K. A.; Popova, N. G. *Gigiena i sanitarija* (Moscow), Mar. 1978, 3 (37-41). (In Russian)

CIS 81-555 "Occupational hygiene in textile spinning processes" (Esperienze igienico-preventive nelle lavorazioni di filatura nell'industria tessile). Perrelli, G.; Giachino, G. M.; Botta, G. C.; Scansetti, G.; Pettinati, L.; Rubino, G. F. *Medicina del lavoro* (Milan), July-Aug. 1980, 71/4 (334-342). Illus. 9 ref. (In Italian)

CIS 75-10 "Resistance of fire of Trevira-cotton work clothes" (Das Brennverhalten von Arbeitskleidung aus Trevira/Baumwolle). Rieber, M. *Die Berufsgenossenschaft* (Bielefeld), Mar. 1974, 3 (103-106). Illus. (In German)

Fibres, natural

A natural fibre is any flexible substance, having a thread-like shape, with a diameter of some hundredths of a millimetre and a length a few thousand times greater. There are several hundred natural fibres of animal, vegetable or mineral origin, but only a very small number of them have physical and chemical properties suitable for commercial use. For industrial purposes a natural fibre must be cheap and available in large quantities, have adequate strength, have a fibre sufficiently long to be capable of being spun, woven or felted and bleached, dyed or printed, the minimum length ranging between 0.5 and 1 cm. Also if intended for making clothing, the fibre must not cause bodily discomfort. The most important natural fibres are:

(a) animal–silk, wool and its related fibres such as alpaca, mohair, goat and camel hair;

(b) vegetable–cotton, flax, hemp, jute, ramie, sisal;

(c) mineral–asbestos and other mineral fibres.

The vegetable fibres may be divided into three classes:

(a) bast fibres–these are obtained from the stem of the plant and have to be extracted by retting or soaking in water followed by hand or mechanical separation of the fibres from the outer covering, the most important being flax, hemp, jute and ramie;

(b) leaf fibres–these are obtained from the leaves or leaf structures of the plant by crushing and then scraping the leaf tissue from the strands of fibre;

(c) fibres from seeds or fruit–the most important are cotton, kapok and coir. These are obtained by picking from the plant and then separating the seeds from the fibre. In the case of coir fibre obtained from the coconut, the coconut husks require prolonged soaking in water before the outer fibre covering can be stripped off mechanically.

(See also ASBESTOS; BAGASSE; COIR; COTTON CULTIVATION; COTTON INDUSTRY; FELT INDUSTRY; FIBRES, NATURAL MINERAL; FLAX AND LINEN INDUSTRY; HEMP; JUTE; KAPOK; RAFFIA; SILK INDUSTRY; SISAL; TEXTILE INDUSTRY; WOOL INDUSTRY.)

QUINN, A. E.

Fibres, natural mineral

Definition. A natural mineral fibre may be defined in simple terms as "a particle with a length to diameter ratio (aspect ratio) ⩾ 3 : 1". This definition of a fibre eliminates the difficult task of testing its flexibility as well as its ability to split into fibrils, which, from the mineralogical point of view, are essential supplementary criteria. Hence, acicular fragments of a series of minerals, for instance, are not distinguishable from asbestos. Acicular fragments are produced by perfect prismatic cleavage upon crushing and are neither flexible nor composed of fibrils. Alteration of the aspect ratio to 5 : 1 or 10 : 1 to avoid misinterpretations is under scientific discussion. For occupational health and safety purposes, however, the aspect ratio ⩾ 3 : 1 should be maintained in order to be on the safe side when assessing risk at workplaces, particularly since present knowledge of the mechanism of the biological activity of fibrous minerals is still unsatisfactory. It has still to be clarified whether, besides the shape of fibrous particles, only their surface properties or, in addition, their flexibility and their ability to split into fibrils are important determinants.

Main natural mineral fibres

Natural mineral fibres are formed either as a direct result of the growth process or from the fragmentation of larger, more massive, crystals. The following will be devoted to those natural mineral fibres which have commercial significance, are important from a geological standpoint or have been shown to be biologically active in man or experimental animals.

Minerals which fit the above description are asbestos, wollastonite, attapulgite, sepiolite and zeolites.

Not mentioned are minerals with rare fibrous occurrences or fibrous minerals that are found only in small amounts. Included are those which are relatively common throughout the world, as well as those which are important rock-forming minerals.

Asbestos is encountered in the form of–

(*a*) amphiboles: actinolite, amosite, anthophyllite, crocidolite, tremolite; and

(*b*) serpentines: chrysotile.

Among the zeolites the natrolite group, with tetrahedal linkages formed in chain-like structures, is of particular importance. These minerals, which tend to have an elongate habit and be subject to cleavage include natrolite, mesolite, scolicite, thompsonite, gonnardite and eddingtonite.

Moreover, among the group of framework zeolites, erionite is of importance, since it occurs in fibrous habit and is a very common constituent of the local soil in Turkey. This mineral has received a great deal of attention in recent months because of its reported association with a cluster of mesothelioma cases.

Biological effectiveness

A health risk exists only upon exposure (dependent on concentration and time) to respirable fibres. By respirable fibres is meant fibres of a diameter smaller than 3 μm, since these may reach the alveoli of the lung and from there may penetrate the lung tissue. Of these fibres, those shorter than 5 μm may be eliminated by the clearance mechanism within the respiratory system, whereas it is mainly fibres longer than 5 μm that are retained, those with lengths of about 20 μm being considered to be the most dangerous. Figure 1 shows the influence of fibre dimensions on biological activity.

By health risk is meant the development of fibrosis, lung cancer or mesothelioma induced by the inhalation of fibres. There is ample evidence that fibre-related diseases follow a dose-response curve, either linear or quadratic. Still open to debate is the question whether there is a limit value for fibre concentrations below which no diseases will occur and, if so, at what level it may be established. A basic requirement for biological effectiveness is certainly also the durability of the fibres in the biological system.

Sampling, identification and measurement

For assessment of fibre concentrations in the air at workplaces filter instruments are used. A defined volume of air is sucked through a filter (membrane or nuclepore with a pore size of about 1 μm) over a certain period with a flow rate of 1 l/min. The precipitated fibrous dust will be analysed in a light or electron microscope and the fibres ($d < 3$ μm, $l \geqslant 5$ μm, $l : d \geqslant 3$) are counted numerically. The result is the fibre concentration in fibres/ml.

Sampling strategy differentiates between: static sampling and personal sampling. For epidemiological investigations it is the latter that should be used.

Figure 1. Graphical representation of the influence of fibre dimensions on biological activity (from Pott, 1978).

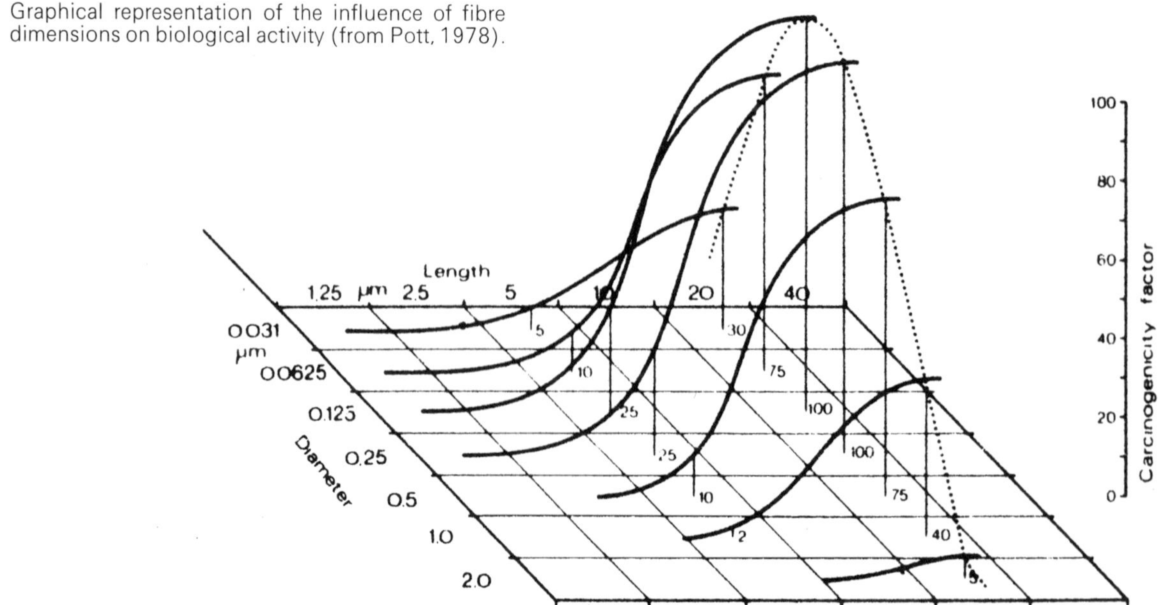

There are three ways of identifying the various types of mineral fibres:

(a) infrared-spectrographic or X-ray-diffractometric analysis of the dust precipitated on a filter regarding its mineral composition;

(b) micro-elementary analysis of a single fibre by application of the scanning or transmission electron microscope with microprobe (energy-dispersive X-ray analysis); or

(c) electron-diffraction analysis in the scanning or transmission electron microscope.

It is impossible to identify the mineral type of single fibres down to a diameter of 0.25 µm by using the dispersion staining light microscopic method; here the lower detection limit is a diameter of 1 µm. However, the first value has to be reached as this is the detection limit of the phase-contrast light microscopical method routine used for counting fibres sampled at workplaces.

Application of gravimetric measuring instruments makes it possible to determine the entire fine dust (fibrous and non-fibrous) in mass units per m^3. The dust can be analysed as described under *(a)*. This method does not consider the biologically relevant geometric fibre dimension and can therefore be used only as a relative method.

ROBOCK, K.
LEINEWEBER, J. P.

Asbestos. Vol. 1: *Properties, applications and hazards.* Michaels, L.; Chissick, S. S. (eds.). (Chichester, John Wiley and Sons, 1979), 553 p. 937 ref.

CIS 79-1275 "Some aspects on the dosimetry of the carcinogenic potency of asbestos and other fibrous dusts". Pott, F. *Staub* (Dusseldorf), Dec. 1978, 38/12 (486-490). Illus. 27 ref.

Asbestos. Measurement and monitoring of asbestos in air. Health and Safety Commission, Second Report by The Advisory Committee on Asbestos (London, HM Stationery Office, 1978), 29 p.

Short course in mineralogical techniques of asbestos determination. Ledoux, R. L. (ed.). (Quebec, Mineralogical Association of Canada, 1979).

Reference method for the determination of airborne asbestos fibre concentration at workplaces by light microscopy (membrane filter method). Recommended international technical method No. 1, Health and safety publication (Asbestos International Association, 68 Gloucester Place, London) (1979), 15 p. Illus. 18 ref.

Second International Colloquium on Dust Measuring Technique and Strategy. Washington, DC, Oct. 11-13, 1978 (Asbestos International Association, Asbestos Information Association, 1745 Jefferson Davis Highway, Arlington, Virginia) (Oct. 1978), 223 p. Illus. Ref.

Third International Colloquium on Dust Measuring Technique and Strategy. Cannes, France, June 10-12, 1980 (Asbestos International Association, 68 Gloucester Place, London) (Nov. 1980), 323 p. Illus. Ref.

Fire

Combustion is the violent combination of oxygen or another combustion-supporter with a combustible substance. This chemical reaction is exothermic. In the case of certain combustible substances, the reaction may occur at normal ambient temperatures due either to the direct action of atmospheric oxygen or to the promoting action of certain agents; however, except under special circumstances, such reactions do not usually result in a fire because the reaction is too slow and, consequently, even if the substance has poor thermal conductivity, the heat that is generated can still be dissipated without a significant rise in the temperature of the substance.

Initially, the amount of heat produced by the reaction remains limited and is readily dissipated, and the temperature of the substance rises only slightly. If the substance is heated, the reaction accelerates as the temperature rises; at a certain temperature, which depends on the nature of the substance and various other factors, combustion becomes more active and the heat of combustion is added to the heat from the external source. The equilibrium temperature rises sharply and, once the ignition temperature has been reached, the substance ignites. For ignition to occur, i.e. combustion with the production of flames, the substance must emit flammable gases or vapours by vaporisation, distillation or chemical reaction (e.g. by the formation of carbon monoxide); if this does not occur, combustion will take place without flame.

Flammable liquids

See FLAMMABLE SUBSTANCES.

Complete and incomplete combustion

For combustion to be complete, sufficient oxygen must be available for the complete transformation of the fuel into its saturated oxides. Where the air supply is inadequate, only a part of the substance is oxidised; the remainder decomposes and gives off large amounts of smoke, and there is the formation of carbon monoxide. The smoke is composed of solid or liquid particles which remain in suspension in the combustion gases and move with them; the quantity and density of the smoke depends on the nature of the combustible substance; substances with decomposition products that contain a large proportion of heavy residues, such as tars, give off very dense smoke.

In a fire there is always incomplete combustion and, consequently, there will be evolvement of smoke and carbon monoxide that will hinder fire fighting by reducing visibility or creating a toxic atmosphere.

An excess of air, on the other hand, will cool the combustion gases; where the amount of combustible material is small, cooling will be sufficient to bring about extinction of the fire by reducing the temperature to below that necessary for ignition. This is the case when a candle is blown out; however, violent wind blowing on a forest fire will have an intensifying effect since the mass of combustible material and the volume of combustion gases are too great for sufficient cooling to take place.

Phenomena producing fire

Flames and incandescent materials. When a solid is exposed to a flame, its temperature rises, the process described above is initiated and a fire may ensue. The likelihood of a fire occurring depends on:

(a) the nature of the solid which may be highly, moderately or only slightly combustible;

(b) the mass of the solid–it is obvious that a small amount of material will not give off sufficient heat of combustion to propagate the fire;

(c) the state of the solid–it is easy to ignite wood shavings or loose sheets of paper with a match because of the large surface area that these materials expose to the air and, consequently, to their greater speed of oxidation, whereas a larger flame will be required to ignite a log of wood or a tightly packed stack of paper;

(d) the manner in which the flame is applied to the combustible solid–if the solid is placed vertically above the flame, it will ignite more rapidly than if it were placed horizontal to the flame.

An incandescent material whether combustible (e.g. coke) or incombustible (e.g. red-hot metal) may cause a fire when in contact with a combustible solid, provided that its mass is great enough to ensure that cooling is not too rapid and that the combustible solid is in a state which permits rapid oxidation (sawdust, wood shavings, uncompacted paper, etc.); in the latter case, a splash of incandescent molten metal produced by a flame-cutting torch may suffice to start a fire. A solid which is heated above the temperature at which it incandesces (e.g. when an electric conductor is overloaded), may ignite material with which it is in contact if the heat is not dissipated sufficiently fast.

Radiation. It is not necessary for physical contact to occur between the combustible substance and the flame or incandescent material. All sources of heat emit visible and infrared radiation, i.e. electromagnetic waves. When these waves encounter an obstacle, they transfer to it their energy which is converted to heat. The body receiving the radiation is therefore heated; if cooling is insufficient, the body will eventually reach its combustion temperature and ignite. Wood piled near a red-hot stove (sustained incandescence) may therefore ignite and cause a fire.

Explosions of vapours or gases. Any mixture of combustible gas or vapour with air will flash when in contact with an incandescent body and the resultant flame will spread if the concentration of the gas or vapour lies within the flammable or explosive limits; these limits vary, depending on the nature of the substance. The rate of flame propagation depends on the nature of the combustible material, ambient temperature and pressure, and may vary between 1 and 2 000 m/s; it is the factor that determines rate of expansion of the heated gases and, as a result, the damage that an explosion can produce.

Explosions of dust or atomised liquids. Dusts of combustible substances or minute droplets of liquids suspended in air behave in much the same way as gas/air or vapour/air mixtures, and may also explode. See DUST EXPLOSIONS.

Sparks. A spark of sufficiently high temperature may ignite a flammable mixture of gas, vapour or dust and air; such a spark would not ignite a flammable solid since it has insufficient energy and the heat released would dissipate in the solid. Sparks may be produced by electric current in a number of ways: when an energised circuit is broken by means of a switch or circuit breaker, the withdrawal of a plug from a socket, or the breakage of a conductor; by sliding contacts such as trolley wires or motor collector rings; by electrical discharge between electrodes with a considerable difference in potential such as gasoline-engine spark plugs, oil burner igniters, etc. Sparks may be produced by static electricity resulting from friction between two moving parts (transmission belts on pulleys), between moving objects and air (transmission belts) and between a non-conducting liquid or gas and the pipe through which it is passing (a typical example being the filling of a fuel tank during which an electrostatic discharge may occur between the filler hose and the tank–the danger here is greater with heavy fuels since light fuels vaporise so quickly that the vapour mixture may be too rich to ignite). Sparks produced by the impact between two objects may also ignite a flammable gas/air or vapour/air mixture. In order to cause ignition, a spark of electric or mechanical origin must have an energy of at least 0.1 mJ.

Finally, sparks produced by friction between two hard surfaces, in which particles are projected (e.g. in grinding non-ferrous metals) may also be a dangerous source of ignition; the lighter flint is a common example.

Spontaneous combustion. Combustion may take place in a heap of solid mineral fuel or organic material if the circulation of air is sufficient to allow oxidation but insufficient to dissipate the heat generated. This phenomenon is promoted by moisture; in the case of minerals, the presence of certain substances (e.g. iron) may catalyse the process, and in the case of organic substances, bacterial activity is an important factor.

The majority of oils oxidise easily, especially vegetable oils; the amount of heat given off is determined by the surface exposed to the air. It is small in the case of a pool of oil, for example, but if this same oil is absorbed by rags or sawdust, the exposed surface is considerably increased and the heat given off is greater since the oil-impregnated materials are poor heat conductors. Heat is accumulated and spontaneous combustion may often occur within a relatively short time.

Chemical reaction. Certain chemical reactions generate sufficient heat to cause fire: yellow phosphorus oxidises very rapidly and ignites on contact with air; fine particulate iron (pyrophoric iron) incandesces in the presence of air and may ignite combustible materials; calcium carbide in contact with water decomposes exothermically liberating acetylene which may be ignited by the heat evolved; sodium and potassium react violently with water, liberating hydrogen which may ignite if the reaction heats the water to a temperature higher than 40 °C; nitric acid coming into contact with organic materials may ignite them; celluloid decomposes at about 100 °C, may ignite at about 150 °C and, due to its oxygen content, may burn even in a sealed container. Oxidising agents such as hydrogen peroxide, chlorates, perchlorates, borates, perborates, etc., which liberate oxygen when heated, actively promote oxidation phenomena and cause the ignition of oxidisable products. Even where there is no external source of heat, an oxidising agent may provoke the ignition of an organic material, especially if this material is in particulate form or in close contact with the oxidiser. Pure oxygen, especially when compressed, may cause a fire or explosion when in contact with a combustible substance; consequently, oil and grease should never be used on oxygen cylinders and valves.

Miscellaneous phenomena. Friction between two materials generates heat, the rate of generation rising as the coefficient of friction increases. Where heat is generated faster than it can be dissipated, there is the danger that combustible materials will be ignited (e.g. fires caused by the overheating of poorly lubricated bearings). Finally, adiabatic compression of a gas produces a rise in the gas temperature; consequently, inadequately cooled compressors or compressed-gas cylinders may explode as a result of the spontaneous ignition of lubricating oil.

LEPAGE, L.

CIS 75-1524 "Model for evaluating fire hazard". Smith, E. E. *Journal of Fire and Flammability* (Westport, Connecticut), 5 July 1974, 5/3 (179-189). Illus. 1 ref.

CIS 79-9 *Flash point index of trade name liquids.* NFPA No. SPP-51 (National Fire Protection Association, 470 Atlantic Avenue, Boston, Massachusetts) (9th ed., 1978), 308 p.

CIS 74-1221 *Oxygen index of materials.* Hilado, C. J. Fire and flammability series, Vol. 4 (Technomic Publishing Co. Inc., 265 W State Street, Westport, Connecticut) (1973), 219 p. Illus. 224 ref.

CIS 74-1219 *Surface flame spread.* Hilado, C. J. Fire and flammability series, Vol. 5 (Technomic Publishing Co. Inc., 265 W State Street, Westport, Connecticut) (1973), 149 p. Illus. 99 ref.

CIS 76-1816 "Some fires and explosions in liquids of high flash point". Kletz, T. A. *Journal of Hazardous Materials* (Amsterdam), Jan. 1976, 1/2 (165-170). 8 ref.

CIS 79-8 *Occupations at high burn risk—A summary report* (National Institute for Occupational Safety and Health, 4676 Columbia Parkway, Cincinnati) (1978), 152 p. Illus. 10 ref.

Fire fighting

No matter how high the standard of fire protection and prevention measures in a plant, the possibility of a fire breaking out is still present, and consequently arrangements must be made to obtain immediate assistance from the local or regional fire brigade in the event of an outbreak of fire, and also to deal with outbreaks still at the initial stage, evacuate workers and contain larger fires until external assistance arrives.

Local or regional fire brigades

Personnel. Fire-brigade personnel may be engaged on a full-time, part-time, paid-on-call or unpaid, volunteer basis—or on a combination of these systems. The type or organisation employed will, in most cases, depend on the size of the community, the value of the property to be protected, the types of fire risk and the number of calls answered. Cities of any appreciable size require regular fire brigades with full crews on duty equipped with the appropriate apparatus.

Smaller communities, residential districts, and rural areas having few fire calls usually depend upon volunteer or paid-on-call firemen for a full manning of their fire-fighting apparatus or to assist a skeleton force of full-paid regulars.

Although there are a great many efficient, well equipped volunteer fire departments, full-time, paid fire departments are almost essential in the larger communities, for the following reasons. A call or volunteer organisation does not lend itself as readily to the continuous fire-prevention inspection work that is an essential activity of modern fire departments. Using volunteer and call systems, frequent alarms may call out workmen, causing a loss of time with seldom any direct benefit to employers, thus making it difficult to obtain good call-men for the fire service. Where full-time firemen are not employed, the apparatus must wait for drivers before response can be made to a call.

In addition, audible public fire signals used to summon call-firemen or volunteers may disturb the community and attract numerous spectators, causing traffic difficulties which may impede the response of fire apparatus.

A city unable to operate a brigade on a full-time basis, with all fire companies adequately manned, can best be served by providing six to eight regular men on duty at a centrally located station housing at least one pumper and one ladder truck. In smaller localities, where a ladder truck is not required, two to four permanent men should be provided at all times.

It is essential, however, that where there are only a few regulars, a supplementary group of well trained call or volunteer firemen should be provided. There should be a reserve apparatus that can be made available for the response of neighbouring departments on a mutual-aid basis.

Capability. The minimum force is one which is capable of extinguishing fires in dwellings and other relatively small buildings and also of preventing extension of the fire to other buildings not initially involved.

The fire department force available at any fire should be sufficient to make possible at least a strong two-position attack on a fire, as well as the force necessary to perform rescue operations. An effective fire force must be able to make prompt use of sufficient hose lines and perform the related operations to make these extinguishing facilities effective, in addition to rescue operations. A more precise basis for judging the required capability would be the size of the structure involved or the extent of the fire. The capability defined can be applied to other arrangements of buildings which present fire-fighting situations of approximately equivalent difficulty.

As dwellings become larger or have less space between them, or mercantile and manufacturing premises replace private dwellings, the capability of the fire-fighting force must also increase. Some of the required increase can be offset by limitations incorporated into the building structures, usually by building code provisions; these would include area and height limitations, use of fire walls and other cut-offs, and installation of automatic sprinkler protection, particularly where areas are large or relatively inaccessible, like sub-basements or interiors of buildings some distance from the street.

Efficient and safe fire-fighting operation requires that adequate apparatus should be available and that personnel should be equipped with the proper appliances, masks and protective clothing. Fire-brigade areas, regardless of population, which have unusual situations of building density, building size or occupancy hazard, require a larger fire department in order to ensure effective fire-fighting forces at each individual fire.

The minimum necessary for a company is at least one piece of fully equipped mobile apparatus. The type of apparatus will vary with the normal responsibilities assigned to the company and may include more than one type of apparatus. Figure 1 shows a typical mobile fire-fighting apparatus.

Figure 1. Mobile fire-fighting apparatus. A foam gun capable of delivering 1 900 l of foam per minute. *(By courtesy of the Tokyo Metropolitan Fire Board.)*

Structure. Generally, each fire-fighting vehicle together with its on-duty or available personnel is designated as a fire company; this is the basic operating unit of a fire brigade. A company, regardless of size or nature, should be organised with a company officer and the necessary personnel on duty or available to utilise the apparatus and equipment effectively at a fire.

The operations of a company should be under the direction of the company officer, and those of the fire-fighting force at the fire, regardless of the number of companies involved, should be under the direction of a chief officer.

In-plant fire-fighting organisation

In an emergency, particularly where there is danger to or loss of life, the management of an industrial plant has to take the following action:

(a) evacuate the danger area;

(b) account for every person on the payroll, and rescue anyone who may be trapped in the plant;

(c) call the public fire department and get the plant fire brigade into action;

(d) issue releases to the press, radio and television news services, giving the facts so far as they are known and of public interest;

(e) report the extent of damage and the possible time the plant will be shut down, to top management, particularly where there is a central management, so that necessary arrangements for transferring business to other plants and other adjustments in operations can be promptly made.

All these measures should be incorporated in an established emergency plan.

Every processing plant must have facilities to help prevent explosions, combat fires and contribute to safe and orderly operation. They are not necessarily the same for each plant, but they all serve the same purpose–to prevent injury to personnel and damage to plant facilities.

To extinguish any fire, all that is necessary is to remove any one of the three ingredients–fuel, heat or oxygen. Lowering the concentration of any of these ingredients below the minimum requirements will also extinguish and prevent a fire.

Most plants use four means of protection: fire-fighting systems; snuffing and smothering systems; by-pass systems; shut-down systems.

Fire-fighting systems. In most plants adequate fire hydrants are strategically placed throughout the plant. The main water lines are also looped. Block valves are placed so that sections of the main line can be isolated in case of a break. Thus, in most cases, water is available at all times even when repairs to the line are being carried out. A 4-h supply of water for the pump should be made available through a water-storage tank or a reservoir of some type. In addition to the water supply, liquid foam may be made available in the plant.

For plants located in the vicinity of a town with fire equipment available, the hose threads on the fire trucks should adapt to the hydrants in the plant. In all new plants advance contact should be made with the local fire department to ensure that adapters are available for this purpose.

Metal hose boxes should be placed in and around the process area and the housing area. A fog nozzle and foam nozzle may also be placed in each box. The hose should be permanently hooked up to the fire hydrant. With this arrangement, the hydrant can be opened, the pump started, and the fire fighter can proceed to the fire with his hose.

Portable fire extinguishers are also necessary. An individual study should be made of each new plant and any existing plant that may be purchased to determine the quantity, type and proper spacing of portable fire extinguishers in the process area.

Snuffing and smothering systems. This type of system can control and extinguish a fire by reducing the temperature of the burning material to below its ignition temperature. When water is used, it is discharged through fog nozzles to produce a very large number of tiny drops of water. These drops of water are directed towards the hottest part of the fire where the water is turned into steam, thereby taking away the heat from the fire and extinguishing it. When steam, nitrogen or carbon dioxide are used, the extinguishing method is one of blanketing, since the air is kept away from the fire until the amount of oxygen present falls below combustion needs.

By-pass systems. Where a fire or emergency in a plant necessitates a shut-down, some means has to be provided for safely venting flammable gases or for ensuring that they by-pass the plant. This is generally done by installing automatic valves in the gas lines into and out of the plant. These valves are connected to a separate system of electric or pneumatic switches located at key spots within a plant. In the event of a major fire, one of these switches should be closed. This will automatically close off the gas supply to and from the plant, and open the by-pass line around the plants. In some cases this operation is not automatic and certain valves have to be closed and opened manually.

Shut-down systems. These are used to eliminate, minimise, or control damage to plant, personnel or property only after all other types of preventive action have failed, since, when a unit or process is to be shut down, it means that all or most of the plant is so severely affected that it, too, has to be shut down.

Transport fire fighting

All transport facilities have special fire problems which cannot be dealt with using standard procedures for fire safety in fixed structures. Weight and space are the chief restricting factors and motion introduces special dangers for the life of the crew and passengers. Sudden impacts with fixed structures on the ground often produce distinct fire dangers since hazards which would be segregated in fixed structures are here placed in close proximity. The degree of public fire protection available for transportation facilities is often slight. All these considerations must be offset as far as possible by superior construction and maintenance.

Marine fire safety. Marine fire prevention and protection may logically be divided into the following aspects:

(a) vessel design, construction and installed fire-protection systems;

(b) vessels in course of construction, undergoing repair and during lay-up;

(c) stowage of hazardous cargo, and cargo handling;

(d) operation of marine terminals;

(e) fire fighting on shipboard, at sea and in ports;

(f) special problems of tanker vessels.

Fixed carbon dioxide systems are preferable to the older method of steam smothering for controlling fires in cargo holds, cargo tanks, bunkers, machinery spaces, stores spaces, paint rooms, etc. High-pressure water-spray systems are also approved for certain locations.

All seagoing ships should be provided with pumps available for fire use and to supply water to standpipe connections and automatic sprinkler systems which may be provided in enclosed passenger and crew sections (not cargo holds, machinery, spaces, toilets, bathrooms and similar spaces of fire-resistant construction).

Since the majority of fires aboard ships at sea and in ports originate in cargo holds or machinery spaces, the

Figure 2. Fire boat of the Tokyo Metropolitan Fire Board.

fire-fighting requirements are highly specialised. The type of fuel or cargo involved and exposed, the access provided to fire fighters, the ventilation of the area burning, the character of the restraining bulkheads, the installed protection facilities, the buoyancy and stability problems and the influence of the employment of large capacity water lines, and many similar problems must be considered (see figure 2).

Aviation fire safety. Built-in fire-extinguishing systems in aircraft are designed to provide protection for most fire hazards. Uncontrolled fire in flight can very rapidly cause disastrous structural failure of aircraft supporting surfaces and control facilities. Immediate fire detection, and the prompt application of effective extinction procedures thus assume life-saving importance.

Extensive research is still in progress on "continuous" and "unit" type fire- and heat-detector systems for aircraft power-plant use. Types recognised as suitable include those that operate on a rate-of-temperature-rise principle, in response to flame contact, or when a calibrated fixed temperature is reached. The detector system must also be designed to give a visible and audible alarm to the flight crew, indicating the zone in which the fire condition exists. Smoke detectors are used principally in inaccessible fuselage areas other than engine nacelles. One other type of detector being experimented with is the combustible vapour analyser adapted for aircraft use, which would advise the crew of the existence of hazardous vapour concentrations.

Present-day built-in protection systems are so installed that the extinguishing agent is distributed by specially designed piping to the prescribed "fire zones" from a fixed supply source. Carbon dioxide and bromide compounds such as bromomethane or chlorobromomethane are the extinguishing agents most commonly used in built-in aircraft extinguishing systems. Both are effective on flammable liquid and electrical fires of the type liable to be encountered; however, carbon dioxide has to be available in large quantities and is therefore heavy whereas bromomethane is both toxic and corrosive.

Effective extinction of engine fires in aircraft in flight depends heavily on the proper implementation of emergency procedures by the pilot as soon as the fire is detected. Considerable refinement has been achieved in airborne extinguishing systems which facilitate the pilot's task and contribute to maximum extinguishing efficiency.

The special fire hazards of aircraft raise acute problems in the event of ground accidents. There are two distinct types of accidents to consider. One type involves high-impact forces, when the death of all occupants is almost certain to be instantaneous and there is major failure of the aircraft structure. The other type involves relatively low-impact forces, when it can be expected that survival rates will be high and occupants' injuries non-fatal.

When fire occurs and fire-fighting measures are inadequate, rescue is often a tragically delayed formality. Facilities must therefore be available to attack aircraft fires quickly and to hold the fire in check, at least until all the occupants can be rescued.

There is no single unit of rescue and fire-control equipment available which has sufficient fire-extinguishing efficiency to handle successfully every type of crash-fire emergency involving large multi-engine aircraft under all conditions of terrain and weather. A combination of units, staffed with well trained specialist fire fighters, is necessary and will continue to be required as long as aircraft are powered with low flash-point fuels, lack impact-actuated fire-extinguishing equipment and possess severe fire hazards in lubricants, hydraulic fluids and electrical systems. The systems commonly employed include combinations of two or more of the following extinguishing agents: carbon dioxide, water fog or spray and foam or fog-foam. Figure 3 shows a blazing aircraft being attacked with a foam extinguisher.

Figure 3. A blazing aircraft being attacked with a vehicle-mounted foam gun. *(By courtesy of the Tokyo Metropolitan Fire Board.)*

Motor-vehicle fire safety

In the event of a fire breaking out in a motor vehicle, the ignition should be switched off (and the battery disconnected if possible), all the vehicle occupants should be evacuated and the fire brigade summoned. The fire may then be attacked with a portable extinguisher from outside the vehicle—the fire always being approached from the windward side. If the fire is in the engine it is generally necessary to raise the bonnet (hood) in order to apply the extinguishing agent to the base of the fire. If the bonnet cannot be raised, the stream of extinguishant should be introduced through any apertures that may exist. If the fire originated from a fuel-tank leak, the part of the fire closest to the fuel tank should be attacked first. When the fire occurs in upholstery, the affected area should be saturated and then inspected to ensure that no hidden sparks remain. Tyres which have caught fire should be flushed with large quantities of water.

After the fire has been brought under control, the vehicle should be thoroughly aired; if the fire occurred in the engine, the vehicle should not be started up again until the cause of the fire has been eliminated. It is recommended that every motor vehicle should be equipped with a suitable fire extinguisher.

Forest fire safety

Forest fires are a very important part of the national fire problem in countries with large timber reserves, and involve a public interest far beyond the immediate financial loss, owing to the effect on future timber supplies, water-flow or rivers, soil erosion, fishing, hunting and other recreational uses of forest areas. Forest fires are also of concern to the fire-protection engineer because forest or brush fires frequently spread to buildings, and because in some localities the same fire-fighting forces are called upon to handle both building fires and forest fires.

Organisation and training are essential to the suppression of forest fires. Under some circumstances, large numbers of untrained labourers must be employed to fight fires which have grown to large dimensions. At such times it is important that trained men be available to organise and supervise this force. Lookout men and other fire guards also require a high standard of training.

Adequate equipment suited to the locality is also essential. Hand tools used include shovels, axes, hoes and rakes of several different designs. Water cans equipped with hand pumps can be carried on men's backs provided a water supply is within reasonable distance.

Prompt detection of forest fires is essential to effective control. In areas where there is a comprehensive organisation for the detection and reporting of fires, the great majority of fires are extinguished in the incipient stage with a minimum of damage. Methods used include observation towers, aircraft and ground patrols.

NAITO, M.

CIS 79-1830 *Maintenance and inspection of fire protection equipment.* Safety guide SG-13 (Chemical Manufacturers Association, 1825 Connecticut Avenue, NW, Washington, DC) (1978), 8 p.

CIS 79-1514 *Technical fire-fighting equipment–Safety rules* (Tehnika pŏarnaja–Trebovanija bezopasnosti). GOST 12.2.037-78. Gosudarstvennyj komitet po standartam (Moscow, Izdatel'stvo standartov, 1979), 10 p. (In Russian)

CIS 79-931 "Fire fighting" (Lutte contre le feu). Fiche de sécurité A.601. Organisme professionnel de prévention du bâtiment et des travaux publics. *Cahiers des comités de prévention du bâtiment et des travaux publics* (Issy-les-Moulineaux), Nov.-Dec. 1978, 6, 4 p. (detachable insert). Illus. 8 ref. (In French)

CIS 79-1826 "Taking the mystery out of the fire hose". Murray, W. S. *National Safety News* (Chicago), June 1979, 119/6 (61-65). Illus.

Firemen

The profession of fire fighting has hazards unmatched by any other occupation. It exerts great demands on its personnel and the only reward is the satisfaction of knowing that one has served one's fellow man at a time of great crisis.

Qualifications

Not everyone can or should be a fire fighter. The person who chooses this profession must possess certain physical and mental qualifications. Among the more important requisites are mental alertness, mechanical aptitude, physical health, strength and agility.

Mental alertness. It is essential that the recruit should be educated to a certain standard and able to pass a written examination designed to assess his suitability, general intelligence, powers of observation and ability to understand orders.

Physical health. Because of the demands of the profession, the recruit must meet definite physical requirements of height and weight and pass a medical examination. His age should be between 20 and 30 years at the start of his career and the maximum age in service should be 55 years, although this requirement may vary from country to country. His visual acuity should be at least 20/30 Snellen without glasses, and his hearing normal. Upon examination it should be determined that he is free from lung, heart and skeletal system defects. Particular care must be taken to ascertain that he is free from hypertension and cardiovascular and respiratory defects.

Strength and agility. Physical performance tests of strength, agility and endurance form an important part of the examination. They should include weight-lifting tests in which arm, back, chest and abdominal muscles are utilised, long jumping and an obstacle course.

Training

Since fire fighters are not chosen on the basis of experience, the recruit must undergo a comprehensive training programme covering the physical, practical, and theoretical aspects of fire fighting. A probationary period should be established to give him time to prove his competence in fire techniques and related duties and for reflection and decision regarding his choice of career. Probationary training should be divided into theoretical, physical and practical phases. Emphasis should be placed throughout on the axiom that the correct method of operating is the safest and most efficient.

Theoretical training. This should familiarise the recruit with the theory of combustion, heat transmission, fire hazard, fire prevention, building construction, basic hydraulics, fire department signals and communication, use of fire extinguishers, sprinkler systems, and uses and limitations of personal protective equipment.

Physical training. Throughout the training period, the recruit must develop general physical fitness and body tone with particular emphasis on his arms, legs, thighs and endurance. A programme of general callisthenics and track work is recommended.

Practical training. The recruit must be introduced to fire-fighting methods and related activities. The following subjects should be taught:

- *(a)* Knots–methods of making the various types used in fire-fighting operations.
- *(b)* Fire department tools and appliances and the proper methods of using them.
- *(c)* Fire hose–proper care, storage and prevention of damage.
- *(d)* Methods of extinguishing fire and the proper selection of extinguishing agent.
- *(e)* Fire streams–the essentials of good streams and how to obtain such streams.
- *(f)* Forcible entry–familiarisation with all types of forcible entry tools, with emphasis on their effective use and limitations.
- *(g)* Standardised routine fire procedures–hose stretching operations, stretching hose lines to roofs or upper floors via the outside of buildings, replacing

burst lengths of hose, stretching hose lines via fire escapes, operating fog nozzles, drafting water with pumps, using hose lines from standpipes, foam operations, manoeuvring portable ladders to roofs, operating aerial ladders for rescue operations and lowering persons by means of a roof rope. Each of these routines should provide for specific tools, manoeuvres and quota of personnel.

(h) Smoke discipline–it is highly desirable for a recruit to be subjected to heat and smoke at mock fire operations. This exposure familiarises him with the conditions in which he will work, and gives him the confidence needed to overcome his natural fear and anxiety in smoke and flame.

(i) Ventilation and rescue operations–these are critical in any fire operation. The recruit must be made aware of the importance of ventilation in effecting extinguishing operations, control of fire spread and prevention of smoke explosion. Search and rescue operation techniques should be taught so that the recruit will be able to locate and remove trapped and/or unconscious persons.

(j) Salvage and overhauling–material other than that involved in the fire must be protected from unnecessary damage. A recruit must be taught proper methods of removing run-off water, use of tarpaulins and the reduction of unnecessary damage. He must also be taught correct overhauling practices to prevent rekindles.

(k) First aid–a recruit must be competent in first aid in order to cope with the many emergency situations which will confront him as a fire fighter; he must be familiar with the method of rescue breathing, external cardiac massage, and the handling and transportation of injured persons.

(l) Respiratory protective equipment–a systematic training programme should be established which will enable a recruit to gain familiarity with the breathing equipment used by his department, learn its limitations and understand his responsibility for its maintenance: the fire fighter's life will depend on the proper use and condition of his breathing equipment.

HAZARDS

Cardiovascular and respiratory system diseases are the principal occupational diseases of fire fighters. [More than 30% of injuries, fatal and non-fatal, sustained by fire fighters in the United States in 1977 resulted from exposure to combustion products. Smoke contains carbon monoxide and irritant and/or toxic combustion and pyrolysis products (see PYROLYSIS) likely to include, when they are emitted from plastics, hydrogen cyanide and hydrogen chloride; sometimes it may even contain radioactive substances. In 72 structural fires the carbon monoxide concentration exceeded 500 ppm for 29% of the time and in 10% of the incidents its peak concentration exceeded 5 500 ppm. Exposure to smoke can produce a drop of FEV and a reduction of myocardial oxygen supply while the exercise of fire fighting increases the demand. Repeated exposures are followed by an increased prevalence of chronic non-specific respiratory diseases and ischaemic heart disease. Hearing loss is also a hazard to firemen because of the combined noise of the motor, siren, air horn and radio speaker, which may exceed 115 dBA.]

The fire fighter is exposed to and expected to operate in all types of hazardous conditions. Because of this and the very nature of the man who puts service to his fellow man above personal safety, the injury rates for fire fighters are among the highest of all occupations.

Road transport accidents, a prime injury producer, occur mainly when fire fighters are responding to or returning from fires and emergencies. Serious injury, death and costly repairs are the result. Safe driving practices should be constantly stressed by each fire department.

While a fire fighter's injuries include all types, the main ones are strains, sprains, exhaustion and smoke inhalation, lacerations, cuts and abrasions and burns.

Examples of common causes of accidents to fire fighters are: improper use of tools and equipment, operating directly below an area where hose, tools and equipment are being used, operating close to the edge of roofs, using improper or badly worn tools, operating hose lines that are overpressured, improper climbing of ladders, wearing protective equipment improperly, improper method of pulling ceilings, tripping over hose lines, failure to hold on to apparatus, jumping off moving apparatus and improper methods of stopping traffic.

SAFETY AND HEALTH MEASURES

In order to reduce the incidence of injury and disease to the fire fighter it is the fire authority's responsibility to establish a formal training programme, and a safety programme, and to provide its fire fighters with the safest and most efficient tools, equipment, clothing and apparatus.

The prevention of accidents (including occupational diseases) requires the employment of the three basic principles of accident prevention: education, engineering, enforcement. Accident investigation and analysis will reveal which of these basic principles should be stressed in solving the problem.

Research and development projects will uncover what personal protective equipment, tools, apparatus, fire station design and training will be needed to ensure the safety of the fire fighter. It is recommended that a fire department set up a safety unit within its organisation to cope with the safety problems and needs of its firemen. The people within this unit should be consulted on the purchase of fire apparatus, equipment or building in order to prevent undue hazards.

Persuasion and appeal are important elements of a safety programme. Beginning with recruit training, safety should be included in all segments of the training process. It is well to begin indoctrinating the recruit with the principles of safety before the practical training has begun; he should be trained to be safety conscious and instruction should be given in the safety aspects of protective clothing and equipment, as well as the safe methods of accomplishing the many aspects of fire department manoeuvres.

During the entire practical training process the recruit should be closely supervised and made aware that proper performance is the safest performance. Proper wearing of clothing must be insisted upon so that good safety habits are formed.

Periodic attention can be brought to specific problems by means of posters, bulletins and tailor-made safety presentations. The principles of defensive driving should be taught in driver training. A continuing programme of instruction should be instituted in the fire stations either through direct approach of a visiting safety conference leader or by audiovisual aids. Incentives in the form of rewards may also be used to advantage.

Personal clothing and equipment

A fire fighter must be properly equipped with adequate personal protective equipment (figure 1). Head protec-

Figure 1. Fireman at work. *(By courtesy of the New York Fire Department.)*

tion should be chosen for its shielding and impact resistance. Eye protection is mandatory. A fire coat should provide adequate insulative qualities and be lightweight and fire-resistant. Foot protection should have a steel insole and steel toe protection. Hand protection should be worn to reduce hand injuries. Breathing apparatus should be of the self-contained demand type. It should be considered in terms of its weight and bulkiness and its air supply should be of sufficient duration to permit the fire fighter to perform his assigned tasks.

Recent safety trends

The basic tools, equipment, clothing and extinguishing methods of the fire services have remained unchanged for many decades. At present industry does not show enough interest in the needs of fire fighters, because there is general ignorance about what those needs, in fact, are.

In the United States, the National Safety Council has developed a Fire Division dedicated to the safety of fire fighters. This organisation will combine the ideas of all fire authorities throughout the United States as well as abroad with the ultimate aim of providing the fire fighter with the safest and most efficient means to fight fires.

This organisation has promoted a uniform method of measuring, recording, and reporting accidents and injuries in the fire service. It is hoped that, with full co-operation, it may become a centre for the safety of firemen everywhere.

TRAVELL, J. R.

"Physical work capacity of firemen with special reference to demands during fire fighting". Kilbom, A. *Scandinavian Journal of Work, Environment and Health* (Helsinki), Mar. 1980, 6/1 (48-57). Illus. 15 ref.

CIS 79-916 "Pulmonary function in firefighters: Acute changes in ventilatory capacity and their correlates". Musk, A. W.; Smith, T. J.; Peters, J. M.; McLaughlin, E. *British Journal of Industrial Medicine* (London), Feb. 1979, 36/1 (29-34). Illus. 23 ref.

CIS 77-1674 " 'Ischemic' heart disease in fire fighters with normal coronary arteries". Barnard, R. J.; Gardner, G. W.; Diaco, N. V. *Journal of Occupational Medicine* (Chicago), Dec. 1976, 18/12 (818-820). 20 ref.

CIS 78-558 "The role of exertion as a determinant of carboxyhemoglobin accumulation in firefighters". Griggs, T. R. *Journal of Occupational Medicine* (Chicago), Nov. 1977, 19/11 (759-761). 12 ref.

CIS 77-1411 "Minimum protection factors for respiratory protective devices for firefighters". Burgess, W. A.; Sidor, R.; Lynch, J. J.; Buchanan, P.; Clougherty, E. *American Industrial Hygiene Association Journal* (Akron, Ohio), Jan. 1977, 38/1 (18-23). 4 ref.

CIS 76-1699 *Protective clothing for structural fire fighting 1975.* NFPA No. 1971 (National Fire Protection Association, 470 Atlantic Avenue, Boston, Massachusetts) (1975), 26 p. Illus. 14 ref.

"Compulsory antitetanic vaccination of firemen" (A propos de la vaccination antitétanique obligatoire chez les sapeurs-pompiers). Prim, L. C. *Sécurité civile et industrielle* (Paris), Dec. 1979, 292 (5-12). 20 ref. (In French)

Fire prevention and protection

Fire prevention and protection include everything relating to the prevention, detection and extinguishing of fire and cover both the safeguarding of human life and the preservation of property. Faced with a steady increase in the amount of combustible property, intensification of fire prevention and protection measures is required to reduce losses to a minimum. Fire prevention should not be considered as being synonymous with fire protection but, instead, a term to indicate a means directed toward avoiding the origin of fire.

The prevention of fire and the reduction of fire and casualty losses depend upon five fundamental principles:

(a) prevention of personal injuries resulting from fire or panic;

(b) fire protection engineering;

(c) regular, periodic inspections;

(d) early detection and extinction; and

(e) damage control to limit the damage resulting from fire and fire extinction.

Fire safety in plant layout and design

Since fires may and do occur, one of the basic fundamentals of all design is isolation to minimise damage and to permit effective control and extinction. A good approach to this is to carefully evaluate the occupancies involved and the layout required for production facilities. For example such occupancies might include offices, cafeterias, shops, stores, dispensaries, yard storage and other services in administrative buildings. These should be grouped within a certain area of the plant, with suitable spacing between the individual occupancies. Provision for proper fire extinguishing media should be made in accordance with standard practice.

The design of plant structures and buildings should provide for fire resistance in accordance with the severity of the hazard involved. For example a building should be so constructed that it will resist exposure to fire for a reasonable length of time, for example 2 h, without collapse.

Fire-resistant construction should be used in the following cases:

(a) in hazardous occupancies to minimise the chance of fires spreading throughout the building and to adjoining buildings;

(b) in vital occupancies such as storage rooms, transformer rooms, electrical substations, laboratories, etc.;

(c) in high-value occupancies where any serious fire would cause excessive property damage and production loss; and

(d) in multistorey buildings.

Recommended practice and experience indicate that fire-resistance construction should be provided throughout a processing plant. Thus, employees have a safer place in which to work, morale is higher, and management-employee relations are better. Further, the best protection is obtained at the lowest cost because greater security is guaranteed against major fire. In addition, maintenance costs are minimised and lower insurance rates are possible.

Geographical situations have considerable bearing on the architectural features of an industrial plant. Where processes are contained within buildings for operational comfort or for product cleanliness, added fire protection and control measures become necessary. Where a danger of explosion is present, the building design should provide for explosion venting by means of a light roof construction or explosion relief panels of an area of approximately 1 $m^2/10\ m^3$ building volume, to prevent an excessive build-up of pressure and the resulting structural damage. Forced-air ventilation systems will help to prevent the formation of flammable vapour-air mixtures; however, for the more hazardous operations where flammable vapours are continuously released, direct vapour removal or inerting systems are necessary.

Water sprays or sprinkler systems are often required to prevent damage or to extinguish fire when it occurs.

Building construction and materials

Many features of building construction are closely related to fire safety. When new buildings or alterations are planned, construction features affecting fire safety should always be considered during the initial stages of design. This procedure eliminates the need for making expensive changes later, and ensures that the completed buildings can withstand fire, explosion and wind with minimum damage and disruption to operations.

The relationship between construction and fire safety is discussed in a general way. There are five standard types of construction that are considered fire-safe.

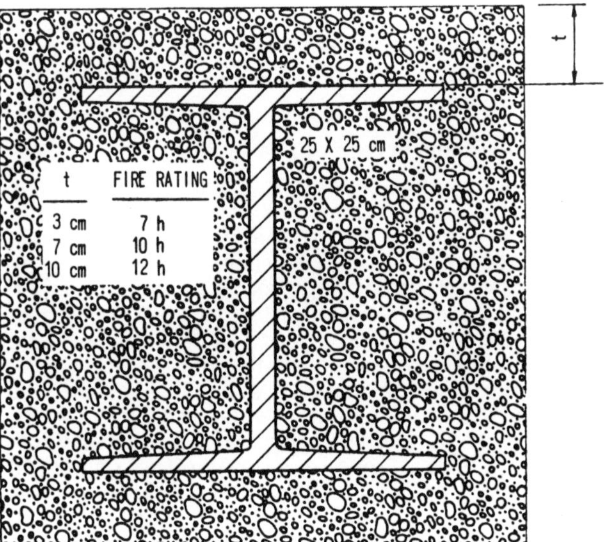

Figure 1. A steel column completely encased in concrete. Fire ratings are for Class I concrete (i.e. calcareous aggregate with less than 10% of siliceous component). Lower-grade aggregates will give lower fire ratings.

Figure 1 shows an example of the fire resistance of protected steel columns and figure 2 shows the heavy timber construction type.

Although the modern types of construction generally contain much less combustible material than the old plank-on-heavy-timber or joisted brick-walled buildings, they will collapse much sooner when exposed to fire temperature; unprotected steelwork will fail at relatively low temperatures. It is therefore essential that automatic sprinkler protection be provided wherever there are combustibles in the occupancy or the construction since automatic sprinklers keep small fires from becoming large ones and they also cool the structural steelwork.

Smoke and hot gases confined within a building during a fire increase the hazard to life and interfere with access and visibility for fire fighting and rescue. The principle of effective smoke and heat ventilation depends upon the installation of curtain boards (figure 3) to produce heat-banking areas under a roof and the installation of roof vents (figure 4) for the release of smoke and heat through the roof.

Combustibility of building materials. A widely used method for evaluating combustibility of materials is the "tunnel test" in which a sample 45 cm wide by 8 m long is placed on the underside of a removable cover that, when in place, forms the top of the 8-m-long test tunnel. One end of the sample is subjected to a gas-flame controlled under regulated constant fuel and draught conditions. Results are reported for flame spread, fuel consumed and smoke developed. For each of these factors, numerical ratings are used which rate the product tested on a scale where cement-asbestos board is 0 and red oak is 100. Flame-spread is observed through windows in the side of the tunnel. Fuel contributed by the sample is related to the temperature rise at the downstream end of the test sample. Smoke

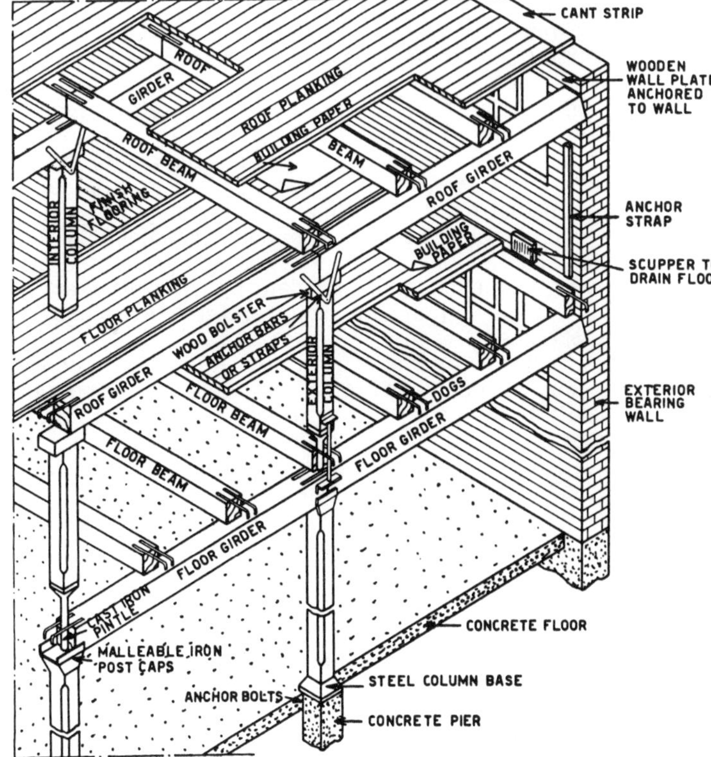

Figure 2. Section of a heavy timber building showing floor framing; this type of structure is known as a semi-mill.

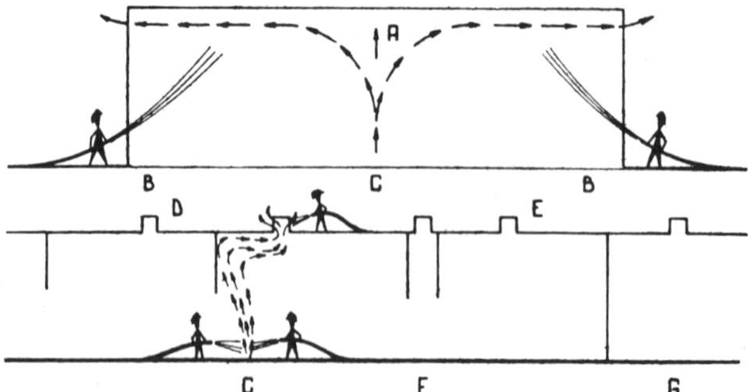

Figure 3. *Above:* The behaviour of the gases under a flat roof; *below:* the behaviour of hot gases under a monitored and curtain roof. The drawings are not to scale but they do show how a monitored and curtained roof permits hot gases to be vented locally, allowing firemen to attack the fire from the roof. Monitors in the roof allow hot gases and smoke to escape. Curtain boards (extending down from the roof) prevent the lateral spread of smoke and heat. A. Hot gases. B. Maximum hose-stream reach of 15-20 m. C. Origin of fire. D. Curtain boards extending 2.5 m down from the roof. E. Fire barrier or fire wall. F. In a small, special hazard area, the curtain boards may extend to within 1.5 m of the floor. G. Large special hazard area.

density is measured by photoelectric means. Many agencies treat any product having a flame spread of 25 or less as a non-combustible material.

Fire inspections

The best time to stop a fire is before it starts. Even though buildings are properly designed and provided with protective devices and construction elements intended to provide fire-safety characteristics, only regular periodic inspections can assure the full value of fire protection. In addition to inspections made by insurance companies and by the fire protection services of public fire departments, every industrial plant should include periodic self-inspections in its fire-safety programme.

In many plants the responsibility for finding and reporting fire hazards is entrusted to the safety committee or to one of its special subcommittees. The function of these committees is to inspect for common fire causes, such as poor housekeeping, improper storage of flammable materials, smoking violations, excessive accumulations of dust or flammable material, defective electrical equipment, etc. Fire inspections should also cover the fire detection and alarm systems, fire-warning notices, fire-fighting equipment and emergency lighting arrangements.

The inspector, fire chief, or other individual in charge of fire prevention and protection should have a complete list of all the items that should be inspected at regular intervals. If fire-brigade members are assigned to this duty, they should receive special instructions. Where possible, a roster of inspectors should be drawn up so that the inspection is not always done by the same person. In some high-hazard occupancies, daily inspections are required; otherwise, thorough weekly or monthly inspections are satisfactory, especially if the plant safety committee inspects for simple fire hazards.

Industrial fire-alarm systems

Fire-alarm systems are of two types:

(a) manual systems which enable a person discovering a fire to call help promptly by operating a manual fire-alarm switch; and

(b) automatic systems which detect a fire and sound the alarm without human intervention.

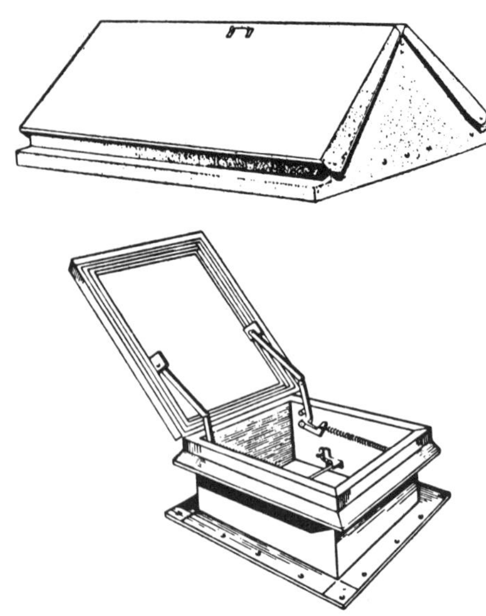

Figure 4. Typical unit-type roof vents for releasing smoke and hot gases to the atmosphere in the event of a fire.

Either system is a valuable part of a plant's fire-protection system, but neither should be considered a substitute for automatic sprinklers.

Manual fire-alarm systems are widely used at industrial plants; they provide a ready means for warning of a fire and its location, and some municipal authorities stipulate the provision of manual fire-alarm systems in certain types of buildings. Manual systems are specifically recommended for large wood pulp and lumber-storage yards.

Automatic fire-alarm systems are used for occupancies highly susceptible to smoke or water damage where an alarm is desirable even before sprinklers operate, and in unsprinklered areas where a complete watch service is not maintained. Similar equipment is also used for a wide variety of occupancies to perform such operations as actuating special extinguishing systems, shutting down fans and conveyors, and closing fire dampers.

Fire-alarm circuits may be opened and unsupervised or closed and electrically supervised. In the latter the failure of the power supply or a break in the circuit causes a fire alarm and consequently this system is not recommended, or it causes a fault signal but not a fire alarm.

Means of escape in case of fire

Ideally all buildings should provide at least two escape routes in opposite directions from a fire occurring anywhere in the building, so that no one should need to go towards the fire in order to escape. It is imperative that all routes of escape be kept free from obstruction, and that they be clearly visible or the routes to reach them conspicuously marked in such a manner that their location can easily be found so that the way to a place of safety outside is unmistakable.

Multistorey buildings will require escape corridors, lobbies and stairways enclosed by structures with a minimum fire resistance of 30 min and equipped with fire-resisting or smoke-stop doors and emergency lighting.

The generally accepted value for the maximum travel distance is about 40 m although in many low fire risk occupancies in buildings of fire-resisting construction it is considered that this can safely be increased to as much as 50 m. On the other hand, where the risk of spread of fire and smoke is high, it may be necessary to reduce this distance to 30 m or even less. Travel distance should take into account the actual distance a person must travel and often cannot be measured in a straight line because of the presence of the contents of buildings.

One of the factors limiting the speed of evacuation from a building is the number of exits available and their width. It has been established experimentally that when people are moving in groups, about 40 persons can pass through an opening just over 0.5 m wide—approximately the width across the shoulders of an average person—in 1 min. This measurement is commonly known as the unit of exit width for means of escape purposes. Two units' width per exit is a minimum requirement except for stairways serving less than 50 persons.

The number of units of exit width required will depend on what is regarded as a reasonable amount of time for this evacuation. The usually accepted evacuation time is 2.5 min although it is regarded as reasonable to clear the less hazardous areas in most factories within 4 min.

In-plant fire protection equipment

This is dealt with in a separate article.

Fire-prevention regulations

Fire-prevention legislation is usually drafted by States and cities in one of two forms: detailed regulations; or regulations outlining fire-prevention principles. The more common practice of including detailed regulations to control fire hazards has been responsible for laws which, although offering satisfactory protection when enacted, are frequently rendered inadequate by hazards introduced as a result of subsequent inventions and new industrial processes. A more effective method, and one which is used increasingly, outlines the general principles of fire prevention in the law and provides for application of nationally accepted standards for guidance of law-enforcement officials. Legislation of this type can be kept up to date with a minimum of revision. Adoption of a specific code in this way has been generally accepted as legally proper although provision for automatic adoption of any future revisions of the code has been criticised as constituting a delegation of legislative power. It is customary for courts to uphold fire-prevention legislation that is reasonable and in the public interest.

The fire marshal's office is the primary fire legislation enforcement agency in most States. It has responsibility for enforcement of laws relating to fire prevention, arson, installation and maintenance of fire-prevention equipment and exits from public buildings. Authority for inspections is usually delegated by the fire marshal to local fire departments.

Municipal fire-safety laws are enforced by various city agencies. Fire, police, building departments, electrical inspectors and others usually have authority in their respective fields. Effective enforcement depends upon competent personnel and co-operation between agencies. Education of the public is the most important fire-prevention function of city departments and is accomplished principally by building inspectors.

Fire protection services

Public fire department personnel may be on a full-time, part-time or volunteer basis, or on a combination of these—the size of the community and number of fire alarms that must be answered determining the organisation adopted. Small communities having few fires usually depend upon volunteers or part-time firemen. Many localities retain a skeleton force of permanent firemen augmented by call-men at major fires. The number and distribution of fire stations and apparatus depends on several factors. Among them are the geographical size and topography of the community, its population, and the amount of manufacturing, commercial and residential development. Stations are distributed to assure quick response of at least one unit to any location and to provide prompt response of enough units to handle fires where the life hazard is severe or property values are high.

Private fire-protection services are a necessary supplement to public protection services. Private fire brigades hold an important place in many commercial and industrial establishments. They are organised and equipped to extinguish fires quickly if possible, or to hold them in check until the regular fire department arrives, when full authority is immediately assumed by the public fire officer. Members of the brigade may be employed as full-time firemen or may be volunteers regularly employed at other jobs in the plant. To be effective, brigades must command the confidence and respect of management and other employees, maintain rigid discipline, and have necessary equipment and proper training. The discipline and training are based on that of a public fire department. The type and amount of equipment is determined by the size of the property to be protected and a knowledge of the hazards present.

NAITO, M.

CIS 77-916 *National fire codes* (National Fire Protection Association, 470 Atlantic Avenue, Boston, Massachusetts) (1977), 16 vols., approx. 11 800 p. Illus.

CIS 78-316 *National fire code of Canada 1977*. NRCC No. 16034 (Ottawa, National Research Council of Canada, 1977), 171 p.

CIS 79-1224 *Handbook of industrial fire protection and security* (Morden, Surrey, Trade and Technical Press, 1st ed., 1979), 423 p. Illus.

CIS 79-341 *Fire prevention and fire fighting* (Prévention et lutte contre le feu). Edition INRS No. 310 (Institut national de recherche et de sécurité, 30 rue Olivier-Noyer, Paris) (1977), 88 p. Illus. 26 ref. (In French)

CIS 79-1518 *Basic principles of fire prevention* (Prevención básica de incendios). López, J. E.; Torres Díaz, M. (Madrid, Editions Minerva SA, 1977), 143 p. Illus. (In Spanish)

Fire protection equipment, in-plant

Fixed fire protection equipment

Fixed in-plant fire protection equipment includes water equipment such as automatic sprinklers, hydrants, standpipes, hoses, water-spray fixed systems, and special pipe systems for dry chemicals, carbon dioxide, halogenated extinguishing agents or foam, etc. Special pipe systems are applicable to areas of high fire potential where water may not be effective, such as those containing tanks for storage of flammable liquids, or electrical equipment. Fixed systems, however, must be supplemented by portable fire extinguishers. These often can eliminate the action of sprinkler systems because they can, if correctly handled, prevent a small fire from spreading, as well as provide a means of rapid extinction in the early stages of the fire.

Sprinkler systems. Automatic sprinkler systems of one type or another have been designed to extinguish or control practically every known type of fire in practically all materials in use today. It is essential, though, to use

Figure 1. A typical sprinkler installation showing all common water supplies, outdoor hydrants, and underground piping.

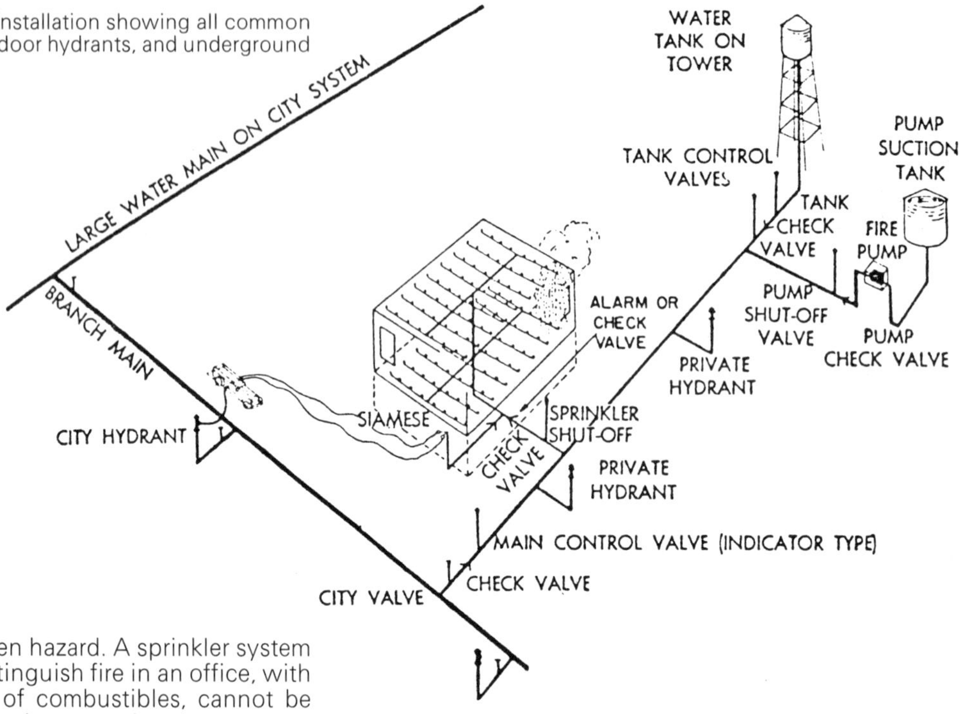

the proper system for a given hazard. A sprinkler system designed to control and extinguish fire in an office, with a relatively small amount of combustibles, cannot be expected to be as effective in protecting a hazardous manufacturing process involving considerable combustible materials, or a storage area where the fire risk is severe.

The terms "sprinkler protection", "sprinkler installations", and "sprinkler systems" usually signify a combination of water discharge devices (sprinklers); one or more sources of water under pressure; water-flow controlling devices (valves); distribution piping to supply the water to the discharge devices; and auxiliary equipment, such as alarms and supervisory devices. Outdoor hydrants, indoor hose standpipes, and hand hose connections are also frequently a part of the system that provides protection. Figure 1 shows these major components except the indoor standpipes and hose connections.

Automatic sprinklers are the most extensively used installations of fixed fire-extinguishing systems. These systems are simple and have proved so effective that most fire-protection engineers consider them the most important type of fire-fighting equipment. There are six basic types of automatic sprinkler systems: wet pipe, dry pipe, preaction, deluge, combination dry pipe and preaction, and limited water supply.

Water spray. This is effective on all types of fires where there is no hazardous chemical reaction between the water and the material that is burning. These systems are independent of, and supplementary to, other forms of protection—not a replacement for automatic sprinklers. Fixed water-spray systems are similar to the standard deluge system except that the open sprinklers are replaced with spray-type nozzles and the water supply to the system can be controlled automatically or manually. They are generally used to protect flammable liquid and gas tankage, piping and equipment, cooling towers, and electrical equipment, such as transformers, oil switches and motors. The type of water spray required depends upon the nature of the hazard and the purpose for which the protection is provided. Water-spray systems can be designed effectively for any one or any combination of the following: extinction of fire, control of fire where extinction is not desirable (e.g. gas leaks), exposure protection (i.e. to remove or reduce heat transfer) and prevention of fire.

In larger plants where parts of the plant are at a considerable distance from public fire hydrants, or where no public hydrants are available, hydrants should be installed at convenient locations within the plant yard. The number needed depends on the fire risk and the hose-laying distance to the built-up plant areas. Exterior fire department connections serving sprinkler or hose systems should be kept accessible and unobstructed. The discharge points should be at least 45 cm above the ground or floor level.

The general patterns of water discharge from the standard sprinklers are shown diagrammatically in figure 2.

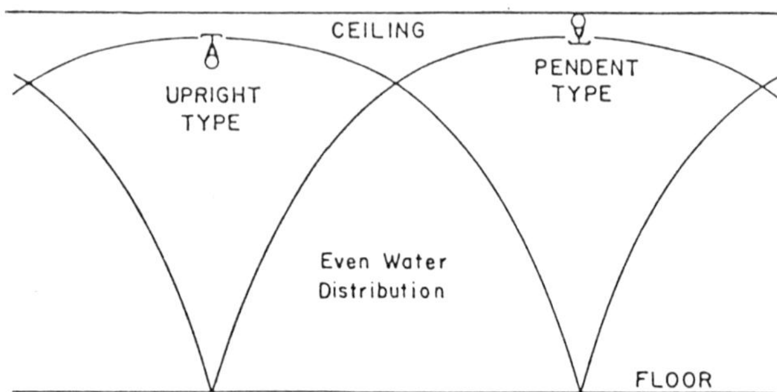

Figure 2. Principal distribution pattern of water from standard sprinklers.

Dry chemicals. These have been found to be an effective extinguishing agent for fires in flammable liquids, and in certain types of ordinary combustibles and electrical equipment, depending upon the type of chemical used.

Dry chemical extinguishing systems can be used in situations where quick extinction is desired and where re-ignition sources are not present. They are used primarily for flammable liquid fire hazards such as dip tanks, flammable liquid storage rooms, and areas where flammable liquid spills may occur. Where it is necessary to extinguish a flammable liquid or gas fire being fed by fuel under pressure, dry chemical hand hose line systems can be used.

Since dry chemicals are electrically non-conductive, extinguishing systems using them can be used on electrical equipment that is subject to flammable liquid fires, such as oil-filled transformers and oil-filled circuit breakers. Dry chemical systems are not recommended, however, for delicate electrical equipment, such as telephone switchboards and electronic computers: they are subject to damage by dry chemicals and, because of the latters' insulating properties, may require excessive cleaning to restore operation.

Hand hose line systems containing regular dry chemicals (bicarbonate base) have been used to a limited extent for quick-spreading surface fires on ordinary combustible material. Fixed systems containing multipurpose dry chemicals (ammonium phosphate base) are now available and have the added ability of being suitable for the protection of ordinary combustibles provided the dry chemical can reach all burning surfaces.

Carbon dioxide. This can be used to extinguish fires in practically all combustible materials, except a few active metals and metal hydrides, and materials that contain available oxygen, such as cellulose nitrate. However, its most widespread use is in extinguishing flammable liquid fires because it can rapidly form a temporary inert atmosphere above the surface of the liquid. As carbon dioxide is electrically non-conductive, it is also used extensively for the protection of electrical equipment. The non-damaging characteristics of carbon dioxide make it attractive as protection for rooms with high-value contents, such as fur and record vaults, computer rooms, and computer tape storage rooms. Where prompt resumption of operations after a fire is important, carbon dioxide is often used to protect production equipment and processes. Carbon dioxide does no damage to equipment or materials in process, and, as there are no liquid or solid residues to clean up, damage and downtime after a fire is held to a minimum. Carbon dioxide has the advantage of providing its own pressure for discharge through pipes and nozzles, and being a gas it can penetrate and spread to all parts of the hazard.

There are two general methods of applying carbon dioxide to extinguish fire. One is to create an inert atmosphere in the enclosure or room in which the hazard is located. In some cases it is necessary to maintain this inert atmosphere in the enclosure for some time until extinction is complete. This method is known as total flooding.

The other method is to discharge carbon dioxide at the surface of liquids or non-combustible surfaces coated with flammable liquids or light deposits of combustible residues. This method is known as local application. No enclosure is needed about the hazard but it is essential in local application that all fire be completely extinguished without possibility of re-ignition during the period of carbon dioxide discharge.

Discharging large quantities of carbon dioxide to extinguish fires may create hazards to personnel. Carbon dioxide "snow" in the discharge may seriously interfere with visibility during and immediately after the discharge period. In addition, the noise of discharge may frighten people who are not used to it. Oxygen deficient atmospheres can be produced where extinction depends on creating such an atmosphere in an enclosed space or room. They also may be produced by any large volume of carbon dioxide drifting and settling in adjacent low spaces, such as cellars, tunnels or pits.

Hazards of oxygen deficient atmospheres can be controlled by the installation of warning systems and devices, by establishing specific emergency procedures, by delayed discharge of the carbon dioxide, and by other similar steps.

The principal property limiting the use of carbon dioxide is its low cooling capacity compared with water. The problem of providing enough carbon dioxide to extinguish some types of fire completely without danger of re-ignition may rule out using a carbon dioxide system despite other factors that may make its use desirable. Flammable liquid storage rooms and fur storage vaults are examples of enclosures where surface and deep-seated fire hazards, respectively, can be protected by total flooding types of carbon dioxide systems. Figure 3 shows how a total flooding system can be used.

Foam. The quantity of foam required to extinguish a fire varies widely, depending upon conditions. In the case of

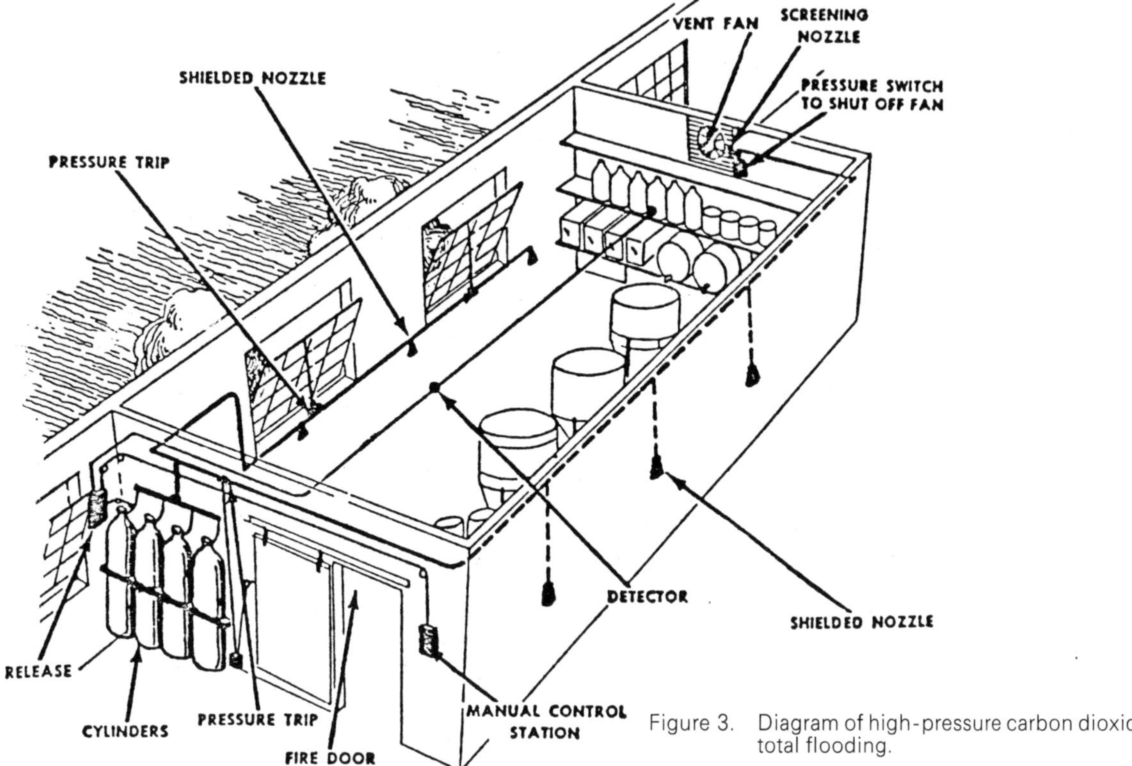

Figure 3. Diagram of high-pressure carbon dioxide system for total flooding.

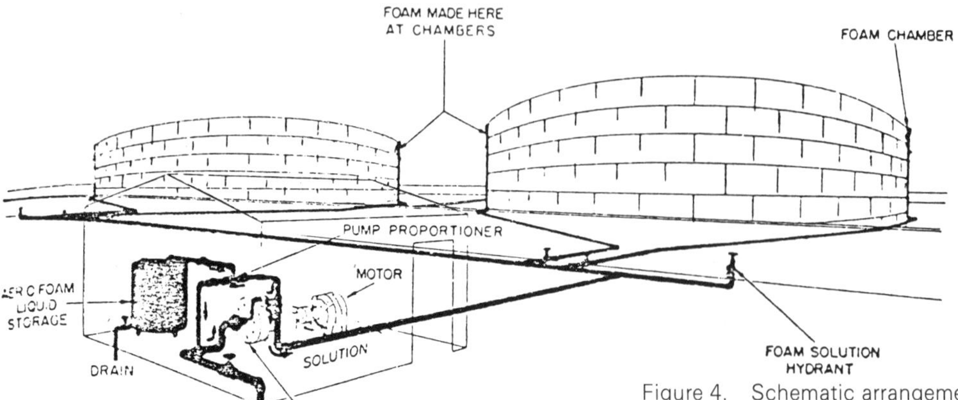

Figure 4. Schematic arrangement of a mechanical (air) foam fixed system for protection of large flammable liquid storage tanks.

fires in small indoor tanks of flammable liquids, foam to cover the liquid surface to a depth of a few inches may be sufficient, while for fires in large outdoor tanks, foam to a depth of several feet may be necessary. The amount of foam required to blanket a spill of flammable liquids on the ground will depend on the area of the spill and the technique used to apply it.

In general, four variables determine the amount of foam-producing materials required to protect a specific hazard: (1) the required rate of application, (2) the area of the surface to be protected, (3) the presence of any obstructions to the distribution of the foam over the surface of the burning material, (4) the length of time the protection of the foam must be maintained to extinguish the fire. Two other allied influences are: (1) the proper placement of fixed foam discharge devices or the skill of the operator handling portable playpipes and (2) the quality of the foam. For many hazards protected by fixed foam discharge outlets, a supplementary supply of foam-producing materials is desirable for foam hose stream application.

There are two methods of producing foam: mechanical and chemical.

Mechanical (air) foam is produced by first mixing a foam liquid concentrate and water in proper proportions by means of a proportioner, introducing air into the water-concentrate solution, and then mixing the air and solution.

There are four basic methods of producing mechanical (air) foam: (1) nozzle aspirating systems, (2) in-line foam pump systems, (3) in-line aspirating systems, and (4) in-line compressed air systems.

Figure 4 shows the schematic arrangement of a mechanical (air) foam fixed system.

Chemical foam systems are gradually being replaced by mechanical (air) foam systems. However, there are still a number of chemical foam systems in use. The four principal types of equipment for producing chemical foam are self-contained units, closed-type generators, hopper-type generators, and stored solution systems.

Foam system installation should be individually engineered for the particular hazard protected. To ensure satisfactory results, such work should be done only by experienced installers, including cases where an existing installation is being extended or converted to a different type.

Foam equipment, like all other fire protection equipment, requires maintenance to keep it in reliable operating condition.

Portable fire extinguishers

A portable fire extinguisher is designed to permit the discharge of a contained amount of fire extinguishing agent at the will and direction of a human operator. Such a device can either be easily carried to the scene of a fire or wheeled on a mobile carrier. Successful utilisation depends upon: (1) the proper choice of extinguisher and extinguishing agent for the type of fire concerned; (2) knowledge of the correct techniques for application of the extinguishing agent to the fire; (3) adequacy of the amount of extinguishing agent supplied for the size of the fire; and (4) the proper functioning of the device as influenced by its design and maintenance.

Relation of extinguishers to classes of fires. From the viewpoint of the proper use of extinguishers on different types of fires, fires can be classified into four types as follows:

(a) fires involving ordinary combustible materials (such as wood, cloth, paper, rubber and many plastics) requiring the heat-absorbing (cooling) effects of water, water solutions, or the coating effects of certain dry chemicals which retard combustion;

(b) fires involving flammable or combustible liquids, flammable gases, greases, and similar materials where extinction is most readily secured by excluding air (oxygen), inhibiting the release of combustible vapours, or interrupting the combustion chain reaction;

(c) fires involving energised electrical equipment, where safety to the operator requires the use of electrically non-conductive extinguishing agents. When electrical equipment is de-energised, the use of extinguishers of types *(a)* or *(b)* may be indicated;

(d) fires involving certain combustible metals, such as magnesium, titanium, zirconium, sodium, potassium, etc., requiring a heat-absorbing extinguishing medium not reactive with the burning metals.

Some portable fire extinguishers are of primary value on only one type of fire; some are suitable for two or three types; none is suitable for all four types of fire.

Selection of extinguishers. Selection of the best portable fire extinguisher for a given situation depends on: (1) the nature of the combustibles which might be ignited; (2) the potential severity (size, intensity, and speed of travel) of any resulting fire; (3) matching the effectiveness of the extinguisher to the hazard; (4) the ease of use of the extinguisher; (5) the personnel available to operate the extinguisher and their physical abilities and emotional reactions as influenced by their training; (6) the ambient temperature conditions and other special atmospheric considerations (wind, draft, presence of corrosive fumes); (7) suitability of the extinguisher for its physical environment; (8) any anticipated adverse chemical reactions between the extinguishing agent and the burning materials; (9) any health and operational safety concerns (exposure of operators during fire-control efforts); and (10) the upkeep and maintenance requirements for the extinguisher.

Distribution of extinguishers. Fire extinguishers are the first defence against fires, and they have maximum effectiveness when used during the early stages of a fire. These truisms indicate the great importance of ensuring the immediate availability of extinguishers in any property or area. "Availability" means proper distribution and access. In any fire emergency where an extinguisher is relied upon, some person must travel to obtain the device and return to the fire in order to begin extinguishing operations. This connotes "time" with the number of seconds or minutes governed mainly by the "travel distance" involved in securing the extinguisher and placing it in operation. Sometimes extinguishers are purposely kept at hand, as during welding operations, but mostly they are strategically positioned in recognition of the fact that the location of an outbreak of fire usually cannot be prejudged. Travel distance cannot be measured in a straight line; it is the actual distance traversed by the potential user of the extinguisher. Consequently, it will be affected by partitions, location of doorways, aisles, piles of stored materials, machinery, and the like.

Operation and use of extinguishers. The method of operation of most fire extinguishers is apparent from their design features. In addition, the label on each extinguisher gives simple operating instructions to ensure that any user can put it in service with minimum delay. Some extinguishers have similar exterior appearances but are operated or used differently.

Most hand portable fire extinguishers can easily be carried to a fire in an upright position. Wheeled extinguishers are balanced so that they are easily pushed or pulled. Most extinguishers are operable from an upright position, although some must be inverted to become functional. The latter type will commence to operate within seconds of inversion and thus should not be inverted until the user is in a proper position to fight the fire; these devices, and some operable from an upright position, discharge their entire contents once placed in service with no shut-off capability. The duration of full discharge from hand fire extinguishers may be as short as 8 s, while that of wheeled types is a maximum of 3 min. To gain maximum effectiveness the available quantity of fire fighting agent must not be wasted; for this reason training in the use of portable fire extinguishers is of importance.

Extinguisher use and training. Knowledge of the types of extinguishers does not assure maximum effective usage. The same extinguishers in the hands of different operators can produce widely different results on the same fire depending on the skills used in applying the available quantity of extinguishing agent. Each extinguisher is designed to facilitate effective use but the following basic factors are involved in practice:

(a) recognition of the device as an extinguisher;

(b) selection and suitability of an extinguisher for the existing fire condition;

(c) transport of the extinguisher to the fire;

(d) actuation of the extinguisher; and

(e) application of the extinguishing agent on the fire.

Inspection and maintenance of extinguishers. Once a purchaser of a fire extinguisher has installed it in his premises, it is his responsibility, or that of his agent, to maintain the device so that it remains fully operational. To fulfil this responsibility there must be a programme to provide for: (1) periodic inspection of each extinguisher, (2) effective maintenance of fire extinguishers, (3) recharging each extinguisher following discharge and as may be specified on the extinguisher label, and (4) hydrostatic testing of each extinguisher which requires such checks.

NAITO, M.

CIS 79-1828 *Water spray fixed systems for fire protection 1977.* NFPA No. 15-1977 (National Fire Protection Association, 470 Atlantic Avenue, Boston, Massachusetts) (1977), 79 p. Illus.

CIS 80-11 *Care and maintenance of sprinkler systems 1978.* NFPA No. 13A (National Fire Protection Association, 470 Atlantic Avenue, Boston, Massachusetts) (1978), 46 p. Illus. 5 ref.

CIS 78-1840 "Sprinkler systems". *National Safety News* (Chicago), June 1978, 117/6 (49-56). Illus. 15 ref.

CIS 79-342 *Portable fire extinguishers* (Les extincteurs mobiles). Edition INRS No. 380 (Institut national de recherche et de sécurité, 30 rue Olivier-Noyer, Paris) (1977), 32 p. Illus. (In French)

"Selecting, maintaining and using portable fire extinguishers". Katzel, J. A. *National Safety News* (Chicago), June 1979, 119/6 (52-58). Illus.

"Fire detectors: is there anything new?" Shernediak, S. *National Safety News* (Chicago), June 1981, 123/6 (38-42). Illus. 1 ref.

First-aid organisation

Where safe working practices are strictly enforced and closely supervised, the number of accidents will be maintained at a low level; nevertheless, no matter how satisfactory the working conditions, there is always the possibility that an injury may occur. The most common injuries are cuts and abrasions, sprains and strains, bruises, eye injuries, and small burns (caused by either heat or chemicals). The number of injuries will depend on the standards of safety in the undertaking and the type of work, and nature of injury will be influenced by the working processes and operations.

Consequently, first-aid practice in industry will differ considerably from the conventional pattern of first aid and the majority of patients that have to be treated will be suffering from minor cuts and sprains, and minor illnesses. Many of these patients can be adequately treated by a trained first aider without any further help from a nurse or a doctor, and when a nurse or doctor is not easily available the first aider must, and can, be taught to undertake an even wider range of treatment. The first aider in industry is therefore, of necessity, often required to undertake more responsibility than the standard first-aid teaching course equips him for. Frequently he must make the decision when to refer for medical or nursing help, and he may have to be prepared to undertake what is really the full treatment of a wide range of minor injuries and illnesses. He must also be able to give immediate treatment to serious injuries and illnesses and to deal with such emergencies as severe bleeding, concussion, fractures, asphyxia, extensive burns, unconsciousness, etc.

The responsibilities he has to accept will of course depend on the availability of nursing and medical care and, where full-time nursing staff is employed, his duties may be limited. In countries where there is an adequate occupational health service, the need for good first aiders is less, though it is possible to save much skilled nursing time by using first aiders within a comprehensive health service, to give initial treatments. In most countries, even where there is an adequate supply of nurses and doctors, no special industrial casualty service is provided. The trained first aider helps to fill this gap, and the cost of training him is low.

However, the first aider should never be regarded as a substitute for a nurse or doctor and both he, his employer, and the employees he treats must be aware of his limitations. If he has to assume additional responsibilities because of a lack of medical facilities, then it should be clearly recognised that this is not a desirable state of affairs and should not continue for any longer than is necessary. A good first aider should be part of a factory health team and not a single-handed worker. He may, in a smaller factory, be working on his own, but preferably with visiting medical help and advice, whether provided by the State or by owners of small factories combining together to provide a group medical service.

First-aid personnel

First aiders must be interested in the work, and must therefore be interested in people and their problems. Certain individuals are temperamentally quite incapable of coping either with an emergency or with a serious injury. Others are neurotically obsessed with illness and are equally unsuitable.

The best first aider is a genuine, reasonably intelligent volunteer who is willing to spend a little off-duty time studying, who has a job in the factory that can be left at a minute's notice, and who is likely to remain calm in a crisis. He should be paid while he is training, but the question of special payment for his extra responsibilities must be a matter for the firm and unions concerned. He should never be financially worse off through losing individual production bonuses, etc. It is important to have at least one woman first aider if there are any female employees.

The number of first aiders required in individual plants is often specified in national legislation; however, if no such stipulations exist, the following practice is to be recommended. There should be two or more trained first aiders in every location, and on every shift of any size, to cover illness and holidays. There should be an additional first aider if there are more than 150 employees, and thereafter one for every 200 employees. In a high-risk industry employing more than 400 workers, it may be desirable and more economical in production time (if no nurse is available) to employ a full-time first aider.

First-aid equipment

All workplaces should be equipped with materials required for administering first aid in the event of injury or illness. In nearly all countries the provision of these materials is a statutory requirement that applies to all work premises and sites, and may also apply to public places such as railway stations and even to vehicles used for the transport of persons or goods.

The type and minimum quantity of materials required are usually laid down in official regulations or may be recommended by semi-official or private bodies such as insurance associations, safety and health organisations, or the local Red Cross or Red Crescent.

This first-aid equipment must be kept in dustproof containers, ideally a box of wood or metal which can be fixed to a wall, or sited on a suitable bench or table surface. These first-aid boxes ought to be available in every reasonable size shop and every high-risk department, and they should contain a minimum of dressings and other equipment needed to treat the particular type of injury that might be expected to occur within the area. Minimum standards laid down by regulations obviously must be followed, and these will vary according to the number of workers employed. Provided the legal requirements as to the number of fully stocked first-aid boxes are met, it is preferable to have additional smaller ones so that immediate and possibly urgent treatment can be given quickly and easily on the spot. For instance, if chemicals or solvents are used there must be immediate means of washing out the eye and rinsing the skin.

In addition to first-aid boxes it is desirable for smaller kits to be provided contained in a suitable carrying bag such as a haversack, and which would be available for the first aider to carry to the scene of a serious accident when the patient cannot be moved. This is particularly necessary on scattered sites or where numbers of people might be working at a height, down a shaft, in mines, on construction sites and in other areas where it is impossible to have a fixed first-aid box. Haversacks should either be dustproof or the contents should be enclosed in a plastic bag.

Contents of first-aid boxes and kits

The majority of wounds will be minor ones and therefore equipment must provide for the complete treatment of all types of minor injuries, i.e. cuts, grazes, puncture wounds, burns, blisters, bruises and sprains. It is also necessary to have a sufficient number of larger dressings to deal with severe injuries where there may be haemorrhage that requires immediate control, or large open wounds, compound fractures or extensive burns. There must, therefore, be enough sterile pressure dressings and sterile slings, wrapped and available for immediate use. In addition to this, simple medicaments should be available for minor ailments, and elasticated bandages for the support of strains and sprains.

There must also be equipment for the immediate washing out of eyes, and the removal of simple dust particles not adherent to the cornea. First-aid boxes should therefore contain–

(a) sufficient pressure bandages, large, medium and small, ready sterilised in packets;

(b) sterile slings in packets;

(c) small individual dressings, wrapped, or contained in a dustproof plastic bag. Such dressings should either be provided in a variety of sizes or an alternative strip dressing should be included, one that can be cut up to suit the size of the individual wound. In some areas waterproof or oil-resistant dressings will be required;

(d) a sufficient supply of bandages of varying width;

(e) strip plaster for fixing bandages and dressings;

(f) additional larger sterilised gauze dressings if the workplace has no proper first-aid room and is some way from a central clinic;

(g) safety pins and scissors;

(h) a suitable antiseptic, preferably ready diluted for immediate use;

(i) cotton wool for cleaning wounds;

(j) a suitable plastic receptacle for containing disinfectant, instruments, etc.;

(k) suitable dressings for small burns, for example Tullegras, ready sterilised;

(l) ointment for insect bites such as calamine or proprietary creams;

(*m*) suitable simple medicines, for example analgaesics (the safest available which can be purchased by the general public), possibly also antacids for indigestion, and throat tablets for throat infections (nothing should be provided that is not available for purchase by the general public and its issue should be controlled through the first aider);

(*n*) tweezers for removal of splinters;

(*o*) for the eyes, cotton wool buds for removing harmless dust particles (only trained first aiders should attempt this); eye bath for washing out harmless dust particles; eye wash bottle, which should invariably be used for washing out chemical splashes or dust that might be contaminated with chemicals.

First-aid kits

These are intended for emergency use only and need to be light as they will have to be carried by the first aider. They need only contain pressure dressings and slings for dealing with severe haemorrhage or covering extensive wounds, and labels and pencil for sending information on with the casualty, so that the local clinic or hospital can be informed as to the cause of the injury and the initial treatment and response of the patient.

Individual first aiders may wish to equip themselves with more first-aid dressings, but kits to be used only in emergencies should be kept as simple as possible.

It is very important that it should be clearly laid down whose responsibility it is to look after the first-aid box and replenish it when required. The employer has the ultimate responsibility, but it may well be delegated to the first aider in charge of the box or kit.

A first-aid instruction leaflet or simple first-aid book should be included in every box in case treatment has to be given by unqualified people.

Where special antidotes are required, for example when cyanide is being used, or hydrofluoric acid, then these antidotes are better provided in a separate box clearly labelled with exact instructions as to their use.

All boxes and kits must be labelled with a red or green cross and the words "First Aid".

Accommodation requirements

Any factory with a high accident risk and any moderately sized factory (e.g. employing over 100 persons) should have a first-aid room, however small, in which patients can be seen and treated. In smaller factories the corner of a shop can be screened off; on the other hand, provided separate facilities can be arranged for men and women, part of a washroom could be used, in which case it would be necessary to screen or partition off a corner.

Wherever possible, hot and cold running water should be provided in the first-aid room or by the first-aid box; where this is not possible, a jug of clean water, a basin and soap should be kept at the ready. A supply of industrial skin cleanser is useful for removing oil and dirt. If water supplies are not clean, then arrangements must be made for boiling the water before use.

First-aid room. This must be sited so as to be easily accessible from all parts of the workplace. Access routes should be signalled and the door must be clearly marked with the conventional sign appropriate to the country in question, for example red, blue or green cross. The door to the first-aid room must be large enough to allow the safe passage of a stretcher. The floor and walls should be covered with a smooth washable material which is easy to keep clean. Adequate lighting, heating and ventilation must be provided. When the room is unmanned the name and means of calling the nearest first aider or his relief must be clearly marked on the door. A loud speaker system or an alarm bell for contacting the first aider is valuable.

The first-aid room should have an examination couch or bed. A couch is a much more suitable height for giving treatments; if a bed is used, the legs should be as high as possible. There should be pillows with pillow cases (and spares), two or three blankets or sheets, depending on the environmental temperature. There should also be one or more cupboards with an easily cleaned working surface on top, or a smooth surfaced table, together with a chair, preferably with a headrest to facilitate examination of the eyes and face, plus a footrest or another chair. If the first-aid room is to be used, even only occasionally, by a nurse or doctor, then obviously more cupboard space will be required for additional equipment, and it will be necessary to have a table or desk, an extra chair, and a cabinet for filing records. A portable screen should be provided, which can be put round the couch or bed, should privacy be desirable for a patient needing to lie down. The best arrangement is to have a supplementary, small examination or rest room opening off the first-aid room. This need contain only a couch or bed, a chair and a small bedside table or locker. In a smaller factory, if space is short, a screen will suffice. Where there are many women employees, it is far more satisfactory to partition off a separate room.

There should also be a footbath with running water and waste pipe, or a large bowl for washing and soaking feet, and a sink, wash basin, floor drain, or easy access to a means of disposing of waste water; if this is not possible, a covered waste-water bucket must be provided. Clean towels are essential, and should be either disposable or washable. Extra towels should be provided in case they are required for special patients who may be sick, or who may need protection for their clothes during treatment.

A first-aid room must contain all the dressings and equipment that have been recommended for general first-aid use. The quantities required will vary according to the needs of the factory; where the service is provided or supervised by a nurse or doctor, the equipment may considerably exceed the basic minimum requirements. See also TRAUMATIC INJURIES.

The "treatment corner". Where there is no regular first-aid room, a treatment corner should be installed. A bench or table is required with a washable surface, and there should be two chairs. The equipment should be kept in a first-aid box, so designed that all the dressings are easily available directly the lid is opened. An upright box is best, with a lid opening downwards, to provide a ready-made working surface. If the box is kept locked, then the eyewash bottle should be left standing beside it, available for emergency use.

Transport of patients

It is essential to have a stretcher available or a suitable carrying chair for patients who are too ill or too seriously injured to come into the first-aid room.

There are different types of approved stretchers. The simplest and most common is made from a wooden frame over which is stretched a piece of strong canvas. Some stretchers are fitted with shoulder straps for the bearers and safety belts for the wounded person. Special types of stretchers have been designed for rescuing linesmen who have been electrocuted whilst working on electricity distribution pylons or poles, for the transport of injured persons along mine roadways or shafts, for evacuating the injured from confined spaces or through manholes, for transport up and down staircases, sea rescue, air transport, etc. However, in cases of necessity an improvised stretcher can always be made from a plank, a ladder or a door. It is also possible to use a sack by cutting holes at each of the four corners through which are passed two stout sticks; a stretcher may also be made from coats by doing up the buttons, turning the sleeves inside out and passing two sticks through them. When using an improvised stretcher it is essential to test its strength before lifting the injured person.

Carrying chairs. Light chairs are available on which a patient can be easily carried by two people for quite considerable distances, including up and down stairs without the difficulties encountered in getting a stretcher round corners and in and out of doorways. Such a chair is not suitable for an unconscious person, but any large organisation will find it a useful addition.

Blankets will be required, at least two in number, for each stretcher or carrying chair.

Ambulance. An ambulance is normally used for transporting injured persons from the scene of the accident (or the first-aid room) to the hospital, clinic, etc. Except in special cases (large undertakings, isolated workplaces, etc.), the firm will not have its own ambulance and, consequently, the address and telephone number of the doctor, the local hospital and the ambulance service should be clearly posted in the first-aid room. Isolated workplaces should have their own vehicle suitably equipped for transporting an injured worker.

Helicopters are now being used in such sites as mines, civil engineering or forestry camps, large plantations, or on islands, etc. These helicopters are specially equipped and operated by civil or military rescue centres or regional services.

Whenever an injured worker is transported from the workplace to a hospital, clinic, etc., a note stating briefly the nature of the accident and the first aid given should always accompany him; this is particularly important in cases of acute poisoning, so that appropriate treatment may be given immediately.

Other equipment

Splints. These can be purchased in a variety of patterns. They are, in general, all suitable provided the first aider knows how to use them. Splints can also be improvised from a light piece of wood, a stick, a rigid magazine, etc., provided padding material is also available to stop them rubbing. If patients are going to have to be transported any distance, then first-aid equipment should include one large splint for fractures of the thigh, two medium-sized splints for immobilising the arm or leg, and two small splints for the hand or foot. They are not necessary when distances involved are small and roads are good.

Resuscitation equipment. In many instances regulations require certain industries to have resuscitation equipment on the spot, for example mines, chemical works, etc. There are many resuscitators on the market. The simplest provide only a limited amount of oxygen, usually up to about a 20-min supply, but are better for general use as they are more or less foolproof, having escape valves in case they are inadvertently used on a collapsed patient who is already breathing. Should, however, the workplace be some distance from a clinic or hospital, then the resuscitator will need to be capable of supplying oxygen for a longer period of time and one of the more complicated respirators should be obtained. In either case it is important for first aiders to familiarise themselves with this equipment, and in the case of the more complicated respirators, regular practice sessions should be held so that first aiders retain the necessary expertise in handling the equipment.

Training for occupational first aid

Training must be designed to suit local requirements. Some countries have standards of qualification prescribed by an authorised body such as a voluntary first-aid institution. Ideally, the industrial first aider should be taught by both a doctor and a nurse who are fully aware of the risks in his particular factory. He should always be taught by someone with a knowledge of industrial conditions so that proper emphasis can be placed on problems likely to be met with in practice. The experience and knowledge of the teachers are of the utmost importance, and all those teaching first aid who are not familiar with industry should be prepared to acquaint themselves with the special risks involved.

The minimum practicable length of a course of instruction is about 10 h, preferably spread over a period of a few weeks and including at least 3 h supervised practice in bandaging, splinting and artificial respiration. In remote areas, it may be necessary to provide a short, full-time course, though such concentrated teaching is more difficult to assimilate. Instruction must be given in:

- *(a)* the complete treatment of minor cuts, burns and strains;
- *(b)* the complete treatment of minor illnesses such as mild gastrointestinal upsets, minor upper respiratory infections, i.e. the kind of illness normally dealt with at home;
- *(c)* the initial first-aid treatment of serious injuries and serious illness, including those resulting from exposure to toxic or irritant chemicals;
- *(d)* the initial treatment of such emergencies as asphyxia, gassing, electric shock, unconsciousness, severe bleeding, severe burns;
- *(e)* the treatment of eye injuries;
- *(f)* special problems relating to the particular factory in question, for example chemical risks, antidotes that may be required, heat and humidity problems, etc.;
- *(g)* the lifting, carrying and transport of patients.

Where the first aider has little opportunity of obtaining advice from a doctor or nurse, some general instruction should be given on the prevention of accidents and occupational diseases. In countries where there is a shortage of medical aid and nursing personnel, it may be desirable to include some instruction in personal hygiene and health, such as diet, sanitation, clean water supplies, and protection against endemic diseases. During instruction, the limits of the first aider's responsibilities must be clearly defined.

Refresher courses. These should be arranged about every three years and should include revision of emergency first aid and discussions of new methods of treatment and prevention.

Teaching manual. In the majority of countries, the national Red Cross organisation has published its own training manuals which cover the general teaching of first aid.

These are good general reference books, but do not always contain information specific to industrial conditions. Copies can be obtained from the headquarters of the local Red Cross or Red Crescent organisation, etc. Advice on current Red Cross publications can always be obtained from the International Red Cross, Geneva.

Advice on publications about industrial first-aid practice will usually be obtainable from the national Red Cross organisation, the national factory inspectorate, the regional occupational health service, etc. Suggested publications are also listed in the bibliography to this article.

It must be emphasised, however, that training should not be divorced from the factory or workplace. The standard Red Cross teaching manual must be supplemented by special instruction in local needs. No single course can cover all aspects of occupational first aid without the risk of muddling the trainee. It is better to teach the basic emergency first aid and treatments of simple injuries, and then to add teaching designed to cater for local conditions and local work. General principles of industrial work will cover most eventualities; however, in many industries, special knowledge of certain risks is needed, for example chemicals, gases, spinal injuries, transport underground, etc.

Teaching aids. In addition to the use of simple equipment such as bandages, dressings, splints, etc., which are likely to be available within the factory, a variety of audio-visual aids can be employed to assist the instructor. A dummy is extremely useful in the teaching of resuscitation techniques and a number of different models are available commercially. Film, film-strips, and display cards on mouth-to-mouth resuscitation, in particular, are now available and are mentioned in the bibliography.

First-aid legislation

Labour legislation in most countries contains reference to the treatment of industrial accidents and makes it compulsory for the employer to supply dressings and equipment for this purpose. The majority of laws are not too specific, merely requiring a sufficient supply of material under the control of a suitable person.

Obviously, the legislation must be considered in relation to other factors—the existence of an occupational health service, the efficiency and availability of the local medical services, and the effectiveness of the inspectors who enforce the legislation. Detailed specifications are of little use if not enforced. General provisions are satisfactory only if recognised standards are high.

In the USSR and certain Eastern European countries, the State health services have given priority to an efficient occupational health service. A large number of health clinics or polyclinics are available where workers can obtain a full range of treatment. Except in the smallest factories, first-aid equipment and first aiders fulfil only the traditional role—literally the "first" and "immediate" treatment. The first-aid treatment may be in many instances carried out by a nurse, and not by a first aider. In France the statutory occupational health service is based on the factory or workplace rather than on a polyclinic. Again, the first-aid facilities have been influenced by the medical services available, and are correspondingly of less importance. The Federal Republic of Germany has a similar, purely voluntary set-up. In both these countries, workers and employers have a joint responsibility for the appointment of a medical officer, and therefore indirectly for the standard of medical and first-aid services.

In Norway, Sweden and Finland, legislation as to equipment and facilities is of a general nature; however, the standards enforced are high, and medical and nursing services are provided in the majority of high-risk and medium- and large-sized factories. In Sweden, for example, the law requires all such factories to employ someone trained in nursing duties. There is no requirement as to the training of a first aider in the smaller factories, but he must be a suitable person and this is generally interpreted as meaning that he must have some knowledge of first aid. Large factories must provide a first-aid room.

In the United Kingdom legislation is very detailed and all workplaces are required to have both first-aid equipment and trained first aiders. The standards of equipment are laid down in great detail for shops and offices, factories, mines and building sites, and all trained first aiders must attend refresher courses every 3 years. The standards recommended by the law are the minimum acceptable ones, and are concerned with dressings rather than equipment. The training course given must be recognised as adequate by the Health and Safety Commission.

ELLIOTT, P. M.

Codes of practice and manuals:

First aid at work. Health and Safety series booklet HS(R) 11. (London, HM Stationery Office, 1981), 46 p. Illus.

First aid and rescue. Commission of the European Communities. Industrial health and safety. EUR 5928 (Brussels-Luxembourg, ECSC-EEC-EAEC, 1978), 66 p.

New advanced first aid. Ward Gardner, A.; Roylance, P. J. (Bristol, J. Wright and Sons Ltd., 1977).

Manual of first aid. Industrial chemical non pharmaceutical products (Manuel de premiers soins d'urgence. Produits chimiques industriels non pharmaceutiques). Lefèvre, M. J. (Gembloux, Belgium, Editions J. Duculot, 1974), 219 p. (In French)

Medical first aid guide for use in accidents involving dangerous goods. IMCO-WHO-ILO (London, Inter-Governmental Maritime Consultative Organisation, 1973), 147 p. Illus.

Teaching:

CIS 76-2062 "First aid: training, plant program, cardiopulmonary resuscitation, fire blanket, emergency medical technicians, rescue gear". *National Safety News* (Chicago), Aug. 1976, 114/2 (55-67). Illus.

Rooms:

CIS 78-867 "First-aid rooms". Swanson I. *Occupational Health* (Basingstoke), July 1977, 29/7 (292-301). Illus.

Kits:

Minimum requirements for industrial unit-type first-aid kits. ANSI Z308.L-1978 (New York, American National Standard Institute, 1978), 12 p.

Fishing

The fishing industry employs about 2 million workers and is of importance to almost all countries. [The world catch of fish in 1976 from both sea and inland fisheries, but excluding whaling, was 74.7 million tonnes. Japan was the largest catcher with 10.662 million tonnes, followed by the USSR (10.134), the People's Republic of China (6.88), Peru (4.343), Norway (3.435) and the United States (3.101).] Japan and the USSR are the leaders in the mechanisation and industrialisation of fishing.

The products of the fishing industry (fish, molluscs, crustaceans) are intended primarily for human consumption; the by-products (meal, oil, liver, skin, shell) are used as cattle fodder, fertiliser, in medicine and for the production of soap and glue.

Parallel with industrialised fishing, artisanal and family fishing continues in many parts of the world—especially in Africa, Asia and South America. In sea fishing, a distinction is usually made between coastal fishing, high-sea fishing, and ocean fishing.

Coastal fishery. This type of fishing, which is practised offshore within about 60 miles of land, varies considerably depending on the climate and season. It is geared primarily to supplying fresh fish (sardine, mackerel, spring herring), shrimp and shell fish to coastal villages in which fish is the principal food.

In the developing countries, use is still made of light wooden rowing or sailing boats. In the industrialised countries, the boats used are generally motorised. This type of fishing is practised mainly by young persons or men over 50 years of age; the boat has a crew of two or three who have no fixed hours of work and who may double-job—often in agriculture.

High-sea fishery. For European countries, this covers the area between the Equator and latitude 60 °N, longitude 10 °E and 60 °W, and provides 40% and more of catch of fish while employing 25-30% of fishermen. The main species of fish caught are herring (in autumn), tuna, plaice, hake, cod, mackerel, sardine, sole and dab. High-sea fishery is primarily a daytime operation carried out with motorboats of up to 400 hp and 120-150 t and having a crew of 15-20; large sail boats with auxiliary engines and refrigeration plant are also employed. Although trawling is by far the most common technique employed nowadays, other techniques such as fixed nets, single or multiple hooks and lines and live bait, etc., are also used. In the Mediterranean, tunny are fished in special fixed, compartmental nets which may be left in place, offshore for 8-10 days. This tunny fishing technique is extremely arduous for the fishermen since the caught tunny have to be hoisted aboard manually using iron hooks. [The development of high-sea fishery may be affected by the widening of territorial waters up to 200 miles.]

Ocean fishery. This is practised outside the high-sea fishing grounds (Newfoundland, Iceland and Arctic waters), and the main catch is whale and cod; the boats used have steam or diesel engines, are in the 200-1 500 t range and may have refrigeration equipment on board (see WHALING).

Traditionally cod was fished from small, flat-bottomed and not very stable dories using deep lines carrying hundreds of hooks, which were cast in the evening and reeled in early in the morning. Nowadays the cod fishers use trawlers of 200-500 t with a crew of 40 or so hands who sail and fish day and night. When double-rig trawling, each boat will have a crew of about 15.

Factory ships range from 10 000 to 40 000 t and may have as many as 200 hands. Using this type of vessel, it is possible to fish, take on board fish from other boats and prepare the catch for sale. Factory ships are also equipped for the production of fishmeal and, in the case of whalers, for the preparation of by-products, the machines for which may be extremely large and noisy. They carry their own engineers and may have one or more male nurses and perhaps even carry a ship's surgeon for the whole fishing fleet. The latest in factory ships are equipped with radar and reconnaissance seaplanes.

HAZARDS

For general problems of working and living aboard see MERCHANT MARINE.

Accidents. There is a lack of adequate statistical data on fishing accidents; however, it has been pointed out that over the period 1959-64 the mortality rate among trawl fishermen in Great Britain was twice that for miners and 20 times as high as that for the manufacturing industries. [Later data showed an accident incidence rate eight times higher than in the manufacturing industries.]

The study of accidents among fishermen brings to light a difference between coastal fishing and high-sea fishing. In coastal fishing, the crew hands are usually mature and well trained; serious accidents are much less numerous than slight accidents (cuts and abrasions, puncture wounds, sprains and strains, bruises, etc.). In high-sea fishing, especially on factory ships, work is increasingly mechanised; the men are often signed on from outside the traditional fishing regions and may lack training; their work is intense and they suffer bruising and other severe injuries similar to those encountered in industry. They may spend several months at sea in bad weather conditions facing violent storms with the risk of shipwreck and falls overboard. The crew members at greatest risk are the deck hands, engine room personnel and galley crew. The main accident hazard on a trawler is that of tripping over ropes or of being struck by the whiplash of broken warps and being knocked overboard. Other very dangerous operations include the removal of waste adhering to the nets, holding the one end of a net when fishing double-rigged, pulling in the nets and securing the after trawl door in bad weather. The most frequent injuries are skull and chest wounds, limb fractures, finger amputations and acute intoxications. Slight accidents and injuries include: splinter wounds, falls on to the deck, against handrails and levers, frostbite of the feet due to standing on frozen fish or ice; prolonged pressure of the knees against the bulwarks in men hauling in nets may cause bursitis, and prolonged use of gutting knives may provoke tenosynovitis.

Diseases. The most common diseases are those of the digestive system, which are usually functional and caused by irregular hours of work, excessive spicing of foods, nervous tension, consumption of alcohol and cold food, etc. Next in line are respiratory diseases, in particular chronic bronchitis and bronchiectasis in heavy drinkers and smokers whose condition is aggravated by the cold. Sinusitis and dental caries are frequent, and older workers may suffer from back pain. Nervous disorders may be provoked by the very special working conditions (isolation, lack of leisure, constant presence of danger); fits of violence may be seen in epileptics due to a lack of medical attention, excessive alcohol intake, etc. Tapeworm infestation may occur in fishermen who consume raw fish that are vectors of *Diphyllobothrium*; tapeworm is relatively common among Finns and Japanese who are prolific raw fish eaters. Poisonous fish, found in particular in the coastal waters of Chile and Argentina, may cause envenomed bite or sting wounds when captured.

Contact with diesel oil may be the cause of dermatitis. In cold climates, skin diseases, fissuring, chapping and skin wounds may take a particularly long time to heal. Disorders of the peripheral circulation, in particular varicose veins, are relatively frequent.

Figure 1. Unloading a fishing vessel.

Occupational hearing loss may be encountered amongst workers on highly mechanised vessels where the noise levels produced by the engines and the refrigeration, ventilation and air-conditioning compressors are particularly high.

[The risk of cancer of the lip (squamous cell carcinoma) has been shown in Newfoundland (Canada) to be 4.5 times higher among fishermen than among controls. However, no specific work activity or a carcinogen has been identified so far.

Among the workers unloading industrial fishing vessels (figure 1) episodes of loss of consciousness, including a few cases of death or permanent disability due to brain damage, keratoconjunctivitis and skin eruptions have been reported from several countries. High concentrations of hydrogen sulphide, carbon dioxide, ammonia, diethylamine and N-butylamine and lack of oxygen were detected, without a clear relation to the type of fish in cargo or the type of preservation (ice/formaldehyde).]

SAFETY AND HEALTH MEASURES

Technical prevention. A Code of Safety for Fishermen and Fishing Vessels has been published jointly by the Food and Agriculture Organisation of the United Nations (FAO), the Inter-Governmental Maritime Consultative Organisation (IMCO) and the ILO. Part A of this Code deals with safety and health practice for skippers and crews. It covers such matters as navigation, vessel safety, deck safety, safety during fishing operations and safety in machinery spaces and of mechanical equipment. Part B concerns safety and health requirements for the construction and equipment of fishing vessels. It contains numerous provisions on such subjects as hull and equipment, stability, machinery and equipment, fire protection, protection of the crew, crew accommodation and navigational equipment. In addition, the three organisations have published *Voluntary Guidelines for the Design, Construction and Equipment of Small Fishing Vessels.* This publication, which is intended for vessels of less than 24 m in length, covers subjects similar to those found in Part B.

Safety measures should be aimed primarily at the safety of the ship in the face of storms, fog and collision hazards. Certain jobs, such as those on deck, are always hazardous and should never be undertaken during storms; breastrails, handrails and bulwarks should be high or made more effective by ropes, and the number of men on deck should be kept to a minimum. Requirements are specified for the protection of steam pipes, electric conduits and trawl warps. Refrigeration spaces must be designed so that workers cannot be accidentally locked inside. Measures must be taken to prevent the escape of refrigerants, and motors and compressors must be acoustically insulated to minimise the transmission of noise and vibration; floors of refrigerated spaces should be non-slip.

Minimum space requirements per crew member, sanitary facilities, nutrition and leisure activities are dealt with in the article MERCHANT MARINE.

Skippers must ensure the careful maintenance of fishing gear, engines, refrigeration equipment, electricity installations and winches, etc. Inspections should be carried out periodically in accordance with standard practices.

In many countries legislation specifies the number of life-boats, life-rafts and life-jackets that must be provided on board fishing vessels; this number is usually dependent on the size of the vessel and the size of the crew; requirements are also laid down concerning fire extinguishers, bilge pumps and signalling equipment. Fishing vessels, especially those used for high-sea fishing, are nowadays widely fitted with radiotelegraphy equipment.

Crew members must wear appropriate personal protective equipment including: gloves or mittens lined with cotton or nylon, which should be kept clean to prevent abrasions and eczema; headwear that protects the nape of the neck; boots with reinforced toecaps to protect the feet from falls of frozen fish; and waterproof leggings if the worker is required to stand on ice, frozen fish or salt. Workers exposed to high levels of machine noise should be provided with hearing protection.

Medical prevention. The ILO Medical Examination (Fishermen) Convention, 1959 (No. 113), specifies that no person shall be engaged for employment in a fishing vessel unless he produces a certificate attesting to his fitness for the work for which he is to be employed at sea, signed by a medical practitioner. The Convention deals solely with the pre-employment medical examination; however, fishermen should receive periodic examinations to ensure that they are still fit for work.

First-aid organisation requires special attention, and regulations in most countries stipulate that each vessel must possess a first-aid kit or ship's pharmacy, the contents of which will depend on the number of days to be spent at sea. International standards for first aid at sea are laid down in the *International Medical Guide for Ships* published jointly by the Inter-Governmental Maritime Consultative Organisation, the World Health Organisation and the ILO.

Various countries such as Canada, the Federal Republic of Germany, the United Kingdom and the United States have well developed rescue and coastguard services and a network of medical radio stations to receive distress signals and calls for assistance from vessels, direct medical aid to the required location and, where necessary, despatch a fast boat, helicopter or other aircraft to evacuate persons requiring intense medical care.

NOGALES PUERTAS, B.

"Fatal poisoning and other health hazards connected with industrial fishing". Dalgaard, J. B.; Dencker, F.; Fallentin, B.; Hansen, P.; Kaempe, B.; Steensberg, J.; Wilhrdt, P. *British Journal of Industrial Medicine* (London), July 1972, 29/3 (307-316). Illus. 13 ref.

CIS 76-476 "The occupation of fishing as a risk factor in cancer of the lip". Spitzer, W. O.; Hill, G. B.; Chambers, L. W.; Helliwell, B. E.; Murphy, H. B. *New England Journal of Medicine* (Boston), 28 Aug. 1975, 293/9 (419-424). 46 ref.

"Deaths from asphyxia among fishermen". Glass, R. I.; Ford, R.; Allegre, D. T.; Markel, H. L. *Journal of the American Medical Association* (Chicago), 1980, 244/19 (2 193-2 194).

"Analysis of deep-sea fishermen's satisfaction with work" (Analyse der Arbeitszufriedenheit von Hochseefischern). Ulrich, H. *Verkehrs-Medizin und ihre Grenzgebiete* (Berlin), 1981, 28/2 (47-54). 22 ref. (In German)

Code of safety for fishermen and fishing vessels. Part A. *Safety and health practice for skippers and crews.* Part B. *Safety and health requirements for the construction and equipment of fishing vessels* (FAO/ILO/IMCO, 1975), 108 and 158 p.

Voluntary guidelines for the design, construction and equipment of small fishing vessels (FAO/ILO/IMCO, 1980), 52 p.

Report of the Working Group on the Occupational Safety of Fishermen. Department of Trade (London, HM Stationery Office, 1979), 22 p.

Recommended code of safety for fishermen. Department of Trade (London, HM Stationery Office, 1978), 64 p.

Code of safety of fishermen (London, City Publications).

Fitness for employment

Testing of fitness for employment is an essential feature of any form of preventive action designed to protect the physical and mental health of workers, which is the very purpose of occupational medicine as defined by an ILO/ WHO Joint Committee in 1950. Whether it is directed towards vocational selection or towards appropriate guidance, the aim of fitness testing is to seek a balance between the man and his work in order to provide better protection with less strain on the individual.

Fitness in this context cannot be defined in absolute terms because it reflects the relationship between two factors, namely the demands of the job and the abilities of the person who is to do the job. The nature of the idea is essentially variable for the same individual. It can correspond only to one phase in the relationship between the work and the worker's state of health, both of which are highly subject to change. Consequently, any assessment of fitness for employment is constantly open to review and is always a compromise relating to one point in time.

Basic considerations in testing

For any relationship to exist, the two factors to be linked must be comparable: the demands of the job must be seen from the physiological and psychological points of view; and the individual's physical and mental capacity must be evaluated within the same functional approach and through identical criteria. Obviously, there can be no such thing as fitness for employment in general; it can be defined only in terms of a particular job or type of work.

Knowledge of the working conditions. The first stage in this process of determining a person's fitness or unfitness consists of precisely defining conditions of work. Knowledge of conditions of work is based largely on a study of individual jobs (see WORK STUDY). It involves precise analysis of the level of the work to be done and the physical effort involved, in both kinetic and static terms; job study may consist merely of estimating weights to be moved or the effort required for a specific task, or it may entail the application of complex ergometric techniques, and the interpretation of polygraphic patterns or measurement of energy cost (oxygen uptake, calorimetry, etc.).

In addition to these general physiological factors, there are others no less important and even more delicate, relating to the environment (temperature, humidity, lighting, presence of dust, toxicity of the atmosphere), as well as requirements relating to neurosensory functions (visual, auditory, olfactory and behavioural); all of these values are affected to a greater or lesser extent by hours of work and the speed and rate of working.

In order to classify these conditions and make them comparable, the results obtained are expressed in figures. The ergonomist can do his job only if the features of the situation can be quantified. Job sheets have therefore to be compiled. These investigations, which demand the collaboration of a wide range of specialists (technicians, workers, physicians, psychologist, laboratory specialists, physiologists, biochemists or even sociologists) cannot be restricted to theory: they must be directly compared with the actual situation at workplaces.

Knowledge of personal factors. The factors relating to the individual are even more complex, and they are certainly not fully understood.

It might appear quite easy to determine a person's physical working capacity. Biometric study gives a preliminary idea of his constitution, but mechanical strength cannot be reduced to terms of structure. Physical working capacity can be determined by a number of physical tests (cardiotachometry, lung function, spirometry, blood-gas analysis, electrocardiography); however, these tests do not always give a clear-cut indication of capacity and sometimes circumstances preclude their use. Moreover, a subject's performance in this kind of test does not reveal how he will stand up to regular and normal work 8 h a day for months on end.

Of greater somatic significance is the analysis of individual tolerance to certain occupational hazards, i.e. the resistance or fragility of various organs and systems to various occupational stressors. The use of these somatic means is just as important as their existence, and this is the whole problem of ability and psychomotor behaviour. This is a sphere where tests predominate. Psychomotor tests are particularly valuable in determining fitness for any manual occupation (lathe operators' tests, for example) where dexterity and speed of reaction are of special importance alongside various other factors. Certain more precise tests may put the subject into the actual occupational situation he will have to cope with. In any event, these tests are well standardised and codified.

Intelligence tests are more subtle, and character or "projection" tests are difficult to interpret. Provided its limits are recognised, a well designed battery of tests can be most valuable. However, although it is fairly easy to obtain valid and informative results for manual trades which are simple, or at least not specially complex, there is greater difficulty in assessing fitness for occupations demanding a high level of intelligence or for supervisory posts, or jobs involving considerable responsibility. These cases call for psychologists fully familiar with a wide range of techniques.

In addition, it is essential to fill in the picture with the help of data on the social and environmental background. The individual's proneness to fatigue and his vigilance capacities may be influenced by school record (in case of young workers), family situation and responsibilities, housing conditions, commuting time and difficulties, double jobbing and leisure activities, including sports.

Application of findings

The use made of such information and the value of the conclusions have a considerable effect on the life of the undertaking. Such important operations cannot be left to chance, particularly where they concern the engagement or dismissal of a key employee. But the way in which things are done will vary greatly according to the circumstances. Easy as it is to correlate human characteristics and job requirements in manual trades where psychomotor demands can be fairly easily measured, it is extremely hard to produce any firm conclusion where there is no glaring incompatibility between a person's mental and moral characteristics and the qualities needed for a job involving great responsibility.

When should fitness for employment be tested? Vocational guidance for young persons presents a special problem. Only a few years ago, testing attempted to determine the young person's fitness for a particular occupation, but this approach is now regarded as somewhat questionable, and the tendency is to consider fitness for a particular occupational level and the individual's ability to adapt to modifications in employment necessitated by economic and industrial change.

The concept of fitness for employment becomes clearer when it concerns the hiring of production workers. Whether intentionally or not, some sort of selection takes place, but the harshness of its effects can be diminished by guiding rejected applicants to another type of work.

However, fitness for employment should also be assessed following a disease or injury that may result in some degree of permanent disability.

Who should determine fitness? The importance of the problems raised by the physical or mental health of the individual justifies the pre-eminent position enjoyed by the industrial medical officer, who is fully qualified to appreciate the somatic and psychopathological issues involved. It may often be necessary to consult a psychologist or a specialist in psychotechnics; however, psychotechnical testing entails close collaboration between the specialist and the industrial medical officer. Provided strict medical secrecy is observed, the assistance of the personnel manager may also prove of value.

Methods. In order to make comparative analysis more rapid and impressive, findings can be represented in graph form. If job-requirements graphs (job profiles) and individual aptitude or inaptitude graphs (personal profiles) are drawn up using the same coding system, juxtaposition or superimposition of the two profiles and comparison of the areas of agreement or disagreement will enable a definite conclusion to be reached. The apparent simplicity of this technique offers certain attractions; however, considerable precision is required in drawing up and applying the numerical criteria for establishing the individual's profile, and slight deviations should not be regarded as significant. A comparison of graphs or profiles should be viewed as having only indicative and, in particular, temporary value; it does not dispense with the need for individual critical analysis.

Aptitude profile. Information that might prove of value to the undertaking's labour or personnel department should be extracted from the results of these studies and compiled on a personal profile form. This form should provide all relevant details on personal aptitudes or contraindications for specific types of work, but should not infringe professional secrecy.

The contents of personal profiles will obviously vary considerably depending on the undertaking and the degree of precision required. There is often some reluctance to give a categorical ruling on a particularly delicate point in borderline cases since our means of investigation may often be faulty and result in errors of judgement: moreover, individual adaptability is often unpredictable. In the final analysis, the only true solution is on-the-job testing and, consequently, a system of trial periods has been adopted by a number of undertakings. Once it has been possible to observe the inter-reaction of the worker and his working environment, a genuinely valid decision on fitness for employment can be reached.

In determining the fitness for employment of a diseased or physically handicapped person, two major risks should be avoided: the first is to overestimate functional disability by failing to allow for any adaptation of the job to the worker, while the second is to underestimate an intelligent and determined person's ability to overcome his disability and produce satisfactory results in a job that might appear beyond his powers. Here, fitness for work is closely linked with ergonomics and functional and vocational rehabilitation.

Great care is needed in assessing the fitness for employment of the mentally handicapped. The equilibrium achieved by intense medical therapy and the beneficial environment of a psychiatric clinic may be illusory; the patient may be declared fit for work in the hope that a regular job will exercise a beneficial effect; however, the fitness of the mentally handicapped is even more than normally dependent on the type of work (activity, tempo of work, responsibility), the reactions of his fellow workers (colleagues, supervisors, etc.) and family attitudes. Success depends on combining so many favourable conditions that such attempts all too frequently run into trouble sooner or later.

Consequences

No matter how it is reached, certification of fitness for employment expresses a commitment, a declaration that a person can do a given job without endangering himself or others (contagion, aggressivity, lack of care, etc.), and that the employer will profit from the use of his services.

In the case of jobs on which the safety of others may depend, relevant, specialised and often complex examinations may be necessary. However, the initial examination may not always reveal even serious defects; an epileptic's lack of capacity for sustained vigilance, or another person's lack of willpower or efficiency may be either deliberately or unconsciously concealed in the desire to obtain or keep a job. Even if there is wilful concealment by the person concerned, an employer or medical officer who fails to carry out the examinations required by basic safety considerations could be liable to prosecution in the event of a disabling accident.

Similarly, the personnel manager or the medical officer may be held responsible should the health of a worker who has not received an adequate pre-employment examination suffer as a result of exposure to a specific occupational hazard (e.g. to toxic substances) or as the result of physical exertion (an undiagnosed fracture) and should it be proved that lack of adequate examination had aggravated the worker's condition.

These two examples illustrate the consequences of decisions of fitness for employment and the way that personal profiles may be misused by administrative staff; they also emphasise the significance of the conclusions that may be reached and the practical consequences. In ruling that a person is fit for employment, it is essential to maintain constant vigilance and to keep the original decision permanently under review. This cannot be done single-handed: it demands a team of people who will keep a mind open to the various problems involved in the reciprocal adaptation of man and his work.

BOURRET, J.

CIS 74-572 "Present situation and trends of working capacity and job fitness assessment in occupational health" (Stand und Entwicklung arbeitsmedizinischer Tauglichkeits- und Eignungsdiagnostik). Thiele, H.; Seeber, A. *DDR-Medizin-Report* (Berlin), 1973, 2/6 (507-527). Illus. 142 ref. (In German)

Flammable substances

Flammable substances are combustible substances or materials which, after having been ignited by a source of ignition, continue burning after its removal. The relationship between the amount of heat set free by a burning sample and the amount of heat produced by the source of ignition is used as the index of flammability. This index enables highly flammable substances to be distinguished from poorly flammable ones. Highly flammable substances may readily ignite outdoors or indoors without previous heating upon brief contact with low-energy ignition sources, whereas the ignition of poorly flammable substances requires a prolonged action of a high-energy ignition source. The flammability index of combustible substances ranges from 0.5 to 2.1, and that of highly flammable substances exceeds 2.1.

The class of flammability of a substance is determined by a number of indexes relating to chemical composition, state of matter and other parameters, the values of which are specified in standards and technical specifications used to classify flammability of chemical substances and compounds (pure and commercial grades), mixtures of chemical substances (formulations), natural or artificial materials, industrial intermediates and by-products, and industrial wastes.

For the purpose of evaluating the flammability of substances, those having a saturated vapour pressure of 0.3 Pa or more at 50 °C are considered as gases, and those with a melting (or dew) point above 50 °C are considered as liquids.

The definition of the principal properties of flammable substances are given below.

Flash point. This is the lowest temperature at which, under special test conditions, vapours or gases form on the surface of the flammable substance and flash upon application of an external source of ignition (electric spark or gas-burner flame). In this case the substance does not continue burning as the evaporation intensity is too small to release a sufficient quantity of combustible gas. The flash point determines the conditions under which a combustible substance becomes flammable.

According to international recommendations, highly flammable liquids comprise all combustible liquids having flash points below 61 °C as determined by the closed-cup method or 66 °C with the open-cup method. The flash point is taken into account when industrial plants, premises and installations are classified by fire and explosion hazard categories.

Flammable substances are divided into three classes by flash point as follows:

Class	Designation of substance	Flash point in °C	
		Closed cup	Open cup
I	Very hazardous	⩽ −18	⩽ −13
II	Moderately hazardous	−18 to +23	−13 to +27
III	Hazardous at elevated temperatures	23 to 61	27 to 66

Examples of flash points are as follows: A-72 motor fuel (petrol, gasoline) −32 °C; acetone −17.8 °C; benzene −11 °C; methanol +11 °C; gas oil +40 °C.

Fire point. This is the temperature at which a combustible liquid gives off vapours or gases at such a rate that they continue burning after ignition. This temperature characterises the ability of a substance to burn independently.

Auto-ignition temperature. Also known as spontaneous ignition temperature, this is the lowest temperature of a flammable substance at which the exothermic reaction is abruptly accelerated and changes to combustion with flame. This temperature is not constant for a given substance and depends on the test conditions (amount of heat applied, heat transfer, volume of mixture, presence of catalysts). The lowest values are obtained when the test is carried out in a spherical glass flask. The standard auto-ignition temperature is determined by uniformly heating mixtures of combustible gases or vapours with air, without applying an external source of ignition.

Examples of standard auto-ignition temperatures are as follows: methane +537 °C; acetone +465 °C; A-74 motor fuel +300 °C; diesel oil +250 °C.

This temperature is taken into account when explosion-proof electrical equipment is chosen and when the conditions of safe heating to high temperatures in process engineering are determined.

Flammable substances are divided into the following classes according to auto-ignition temperature:

T1 > 450 °C (e.g. methane, ammonia, benzene, ethane, propane);
T2 300 to 450 °C (e.g. butane, motor fuel, acetylene);
T3 200 to 300 °C (e.g. hexane, heptane, petroleum, gas oil, hydrogen sulphide);
T4 135 to 200 °C (e.g. dioxane);
T5 100 to 135 °C (e.g. carbon disulphide);
T6 85 to 100 °C.

Some flammable substances tend to auto-ignite upon heating to relatively low temperatures when they are in contact with other substances or when heat is generated by micro-organisms. A distinction is therefore made between thermal, chemical and microbiological auto-ignition. The tendency to thermal auto-ignition is characterised by the auto-ignition temperature, i.e. the lowest temperature at which practically discernible exothermic processes go on in the flammable substance which may cause auto-ignition. This temperature is taken into account when the safe conditions for prolonged heating of flammable substances (to not more than 90% of the auto-ignition temperature) are determined.

Flammable (or explosive) limits. These define the range of concentrations of a substance in a mixture with an oxidant, i.e. from the minimum (lower limit) to the maximum concentration (upper limit). In this range a flammable substance may ignite on application of a source of ignition, and the combustion of the mixture may spread to any distance from this source. The lower concentration limit of flammability is taken into account when industrial plants are classified by degree of fire hazard and when admissible non-explosive concentrations are calculated for process plant, premises and workplaces where open flames and sparking tools are used. The admissible non-explosive concentration is determined, with a degree of reliability ranging from 0.999 to 0.999,999 for the non-flammability of the mixture of the combustible substance with air, by dividing the lower flammable limit by a safety factor varying from 1.2 to 4 (for different substances).

Flammable vapours and gases with a low concentration limit of up to 10% by volume of air, and airborne particulates with a lower limit up to 15 g/m^3 are considered to present particularly great explosion hazards.

The temperature range of the flammability of vapours is defined by a minimum (lower limit) and maximum temperature (upper limit). Within this temperature range saturated vapours of flammable substances form in a given oxidising atmosphere concentrations ranging between the lower and upper concentration limits of flammability. The temperature range of flammability is taken into account when safe operating temperatures of process plant are calculated. A temperature considered to be safe with regard to the possible formation of explosive gas/air mixtures is one that is 10 °C below the lower or 15 °C above the upper temperature limit of flammability of a given substance.

Examples of flammable limits are as follows (in percentage of air by volume): acetone 2.6 and 12.8; acetylene 2.5 and 81.0; hydrogen 4.1 and 74.2; butane 1.9 and 8.5; benzene 1.3 and 7.1; motor fuel 0.96 and 4.96; methane 5.3 and 14; methanol 6 and 34.7; and (in temperature range): diesel oil +27 and +69 °C; light petroleum −21 and −8 °C.

If a process plant is to operate at hazardous temperatures even for a short time, inert gases are used. The addition of inert gas to a flammable mixture reduces the range of its flammability or completely eliminates the possibility of ignition. Flammable substances do not burn if the oxygen concentration is below the minimum limit presenting an explosion hazard in a mixture of oxygen and inert gas.

Examples of minimum oxygen concentrations presenting an explosion hazard in mixtures of flammable substances with inert gases are as follows: acetylene with carbon dioxide 14.9%; acetylene with nitrogen 11.9%; methane with carbon dioxide 15.6%; methane with nitrogen 12.8%; benzene with carbon dioxide 14.4%; benzene with nitrogen 11.5%.

Maximum explosion pressure. This is the highest pressure building up when a flammable substance is ignited in a closed vessel at normal atmospheric pressure. The explosion pressure is taken into account when the explosion resistance of process plant, of enclosures for flameproof electrical equipment, as well as of safety valves and bursting discs is calculated.

Examples of maximum explosion pressure (in kgf/cm^2) are as follows: acetylene 10.3; acetone 8.93; butane 8.6; hydrogen 7.39; methane 7.2.

Maximum safe gap. Narrow gaps, clearances and small-diameter tubes are obstacles to the transmission of fire, explosion or heat. The maximum clearance between the flanges of enclosures through which an explosion (taking place in the enclosure) is not transmitted to the surrounding atmosphere (at any concentration of flammable substances in the air) is called the maximum safe gap. Flammable substances are divided into the following categories according to the width of this gap:

I. gap > 1 mm: methane (mine gas);
IIA. > 0.9 mm: industrial gases and vapours (ammonia, acetone, benzene, carbon monoxide);
IIB. 0.5 to 0.9 mm: hydrogen sulphide;
IIC. < 0.5 mm: hydrogen, carbon disulphide.

The maximum safe gap is taken into account for the selection of explosion-proof electrical equipment.

Minimum ignition energy. This is the smallest amount of energy of an electric discharge spark which is sufficient to ignite a flammable gas, vapour or dust in mixture with air. Flammable substances are divided into the following categories according to the magnitude of this energy:

1. 0.3 mJ and more (methane);
2. 0.18 to 0.8 mJ (ethane, butane, motor fuel);
3. 0.06 to 0.18 mJ (ethylene, methanol);
4. < 0.06 mJ (hydrogen, acetylene).

This energy is taken into account for the design of intrinsically safe explosion-proof electrical equipment, and also for the elaboration of safety measures against electrostatic hazards.

Minimum extinguishing concentration. This is the smallest concentration of an inert gas in air which extinguishes practically instantaneously (under test conditions) the diffusion flame of the substance. This concentration is taken into account for calculating the volume of gaseous extinguishing agents required, for example 57% by volume of carbon dioxide or 70% of nitrogen for extinguishing acetylene; 26% of CO_2 or 39% of N_2 for extinguishing methane; 62% of CO_2 or 76% of N_2 for extinguishing hydrogen.

Rate of combustion. The normal rate of combustion of gas/air mixtures is the velocity with which the flame front moves in relation to the non-combustible gas. This rate is taken into account when the spread of a fire is determined and when gas burners are designed and set to operate with a continuous, regular flame.

Examples of combustion rates in m/s are: acetylene 1.57; hydrogen 2.67; methane 0.338; methanol 0.572.

Rate of total combustion. This is determined by the mass of combustible substance which burns in a unit of time on a defined area. It is taken into account for the calculation of the duration, heat release and temperature of fires. The rate at which the level of a combustible liquid lowers during combustion is 20 cm/h for acetone, 24 for gas oil, 30 for motor fuel, 12-15 for petroleum and 15 for ethanol.

Rate of heating. The rate of heating of a combustible liquid is determined by measuring the velocity with which the thickness of the homothermal layer (at a temperature equal to the boiling point of the liquid) increases. This rate is taken into account when the time required for extinguishing a fire is calculated.

Examples of rates of heating (in cm/h) are as follows: acetone 60; motor fuel 70; petroleum 25 to 40.

Type of reaction. The type of reaction of a burning substance with water-base extinguishing agents (water jets, water sprays, foam, steam) determines the choice of fire-fighting strategies, the efficacy of fire extinction and the possibility of complications during fire fighting (effervescence, projections, violent chemical reactions).

Evaluation of flammability

The flammability of gases is evaluated by determining the flammable limits in air, the maximum explosion pressure, the auto-ignition temperature, the category of the explosive mixture, the type of reaction of the flammable substance with water-base extinguishing agents, the minimum ignition energy and oxygen content representing an explosion hazard, the normal combustion rate, the maximum safe (quenching) gap or diameter.

The flammability of combustible liquids is evaluated by determining the flash point of the vapours, the fire point, the minimum quenching concentration of bulk extinguishing agents, the combustion rate and the rate of temperature rise during combustion.

The flammability of combustible solids is evaluated by determining the category of ignitability, the ignition and auto-ignition temperatures, and the type of reaction of the combustible substances with water-base extinguishing agents. For porous, fibrous and bulk solids, the temperatures of spontaneous heating, smouldering and auto-ignition are also determined. If the solids are powdery and may form clouds of dust, additional parameters to be determined are the lower limit of flammability, the maximum explosion pressure, the minimum energy required to ignite the airborne dust and the minimum content of oxygen presenting an explosion hazard.

To evaluate the flammability of a substance it is necessary to study its properties and to take into account the possibility of their change with time and when used under certain conditions. This is particularly important in cases where flammable substances come in contact with other active substances or are exposed for longer periods to heat, radiation or other external effects which may bring about changes in their physical and chemical properties.

The evaluation of the flammability of combustible substances is generally carried out in research laboratories. Certain parameters may exceptionally be determined in the pilot production stage. Industrial and pilot plants, storage and transport facilities can be designed only after data on the fire hazard of combustible substances to be used as construction materials or to be processed have been obtained.

PANOV, G. E.
POLOZKOV, V. T.

Fire hazards of substances and materials used in the chemical industry (Pŏarnaja opasnost' veščestv i materialov, primenjaemyh v himičeskoj promyšlennosti). Rjabov, I. V. (ed.). (Moscow, Izdatel'stvo, "Himija", 1970), 336 p. (In Russian)

CIS 80-1522 *Prevention of damage in chemical plants* (Predupredenie avarij v himičeskih proizvodstvah). Besčastnov, M. V.; Sokolov, V. M. (Moscow, Izdatel'stvo "Himija", 1979), 390 p. Illus. 93 ref. (In Russian)

CIS 80-24 *Research methods on ignition hazards of substances* (Metody issledovanija pŏarnoj opasnosti veščestv). Monahov, V. T. (Moscow, Izdatel'stvo "Himija", 2nd ed., 1979), 423 p. Illus. 234 ref. (In Russian)

CIS 79-620 "Dust clouds and fogs–flammability, explosibility and detonation limits, burning velocities, and minimum ignition energies". Nettleton, M. A. *Fire Prevention* (London), Nov. 1978, 20 (12-18). Illus. 31 ref.

CIS 80-622 *Safety technical data on flammable gases and vapours* (Sicherheitstechnische Kennzahlen brennbarer Gase und Dämpfe). Nebert, K.; Schön, G. (Braunschweig, Deutscher Eichverlag GmbH, 1978), 176 p. Illus. (In German)

CIS 79-9 *Flash point index of trade name liquids*. NFPA No. SPP-51 (National Fire Protection Association, 470 Atlantic Avenue, Boston, Massachusetts) (9th ed., 1978), 308 p.

"Destructive capacity of flammable liquids" (La puissance destructrice des liquides inflammables). Tremblay, S. *Prévention* (Montreal), Dec. 1980, 15/9 (2-15). Illus. 9 ref. (In French)

Flax and linen industry

Linum usitatissimum, a member of the *Linaceae* or flax family, has probably been cultivated for over 4 000 years, and other members of the species may have been cultivated before this. The fibres of the flax plant stalk are used to make linen cloth and towelling and flax twines, nets and ropes. The fibre itself is light, strong and very absorbent, and its strength increases on wetting.

Cultivation. The flax plant requires a damp soil and humid atmosphere for optimum growth. Thus it is mainly grown in temperate countries with moderate rainfall. About four-fifths of the total world crop is grown in the USSR followed by Poland, France, Czechoslovakia, Turkey, Romania, the German Democratic Republic, Japan and the Netherlands.

Flaxseed is sown in late March or early April and, with a growing period of around 100 days, is ready for harvesting in mid-July. The plants, usually 60-120 cm in height, are harvested by mechanical pullers, although in some countries hand pulling is still common. After harvesting the stalks are tied into bundles, called beets, in preparation for fibre extraction.

Figure 1. The flax plant *(Linum usitatissimum)*.

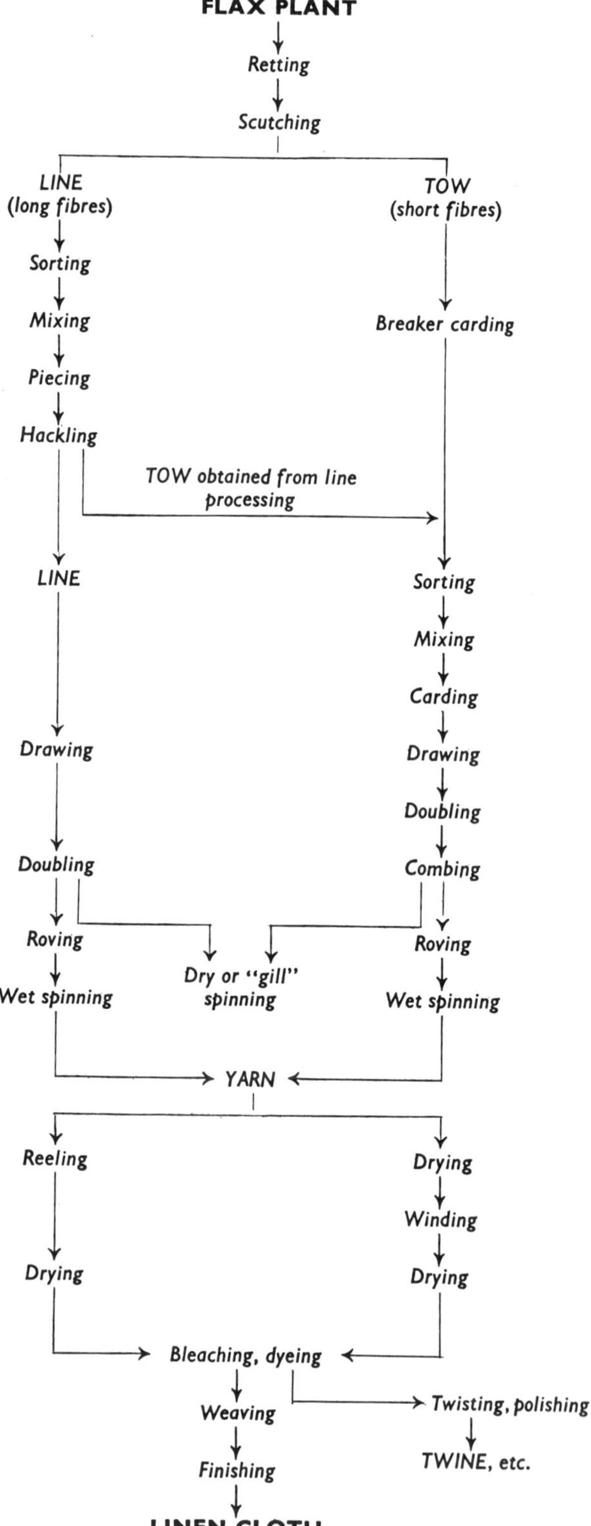

Figure 2. The processing of flax.

Processing (figure 2). Flax fibres are more difficult to prepare than wool or cotton, and machinery was developed and introduced into the industry later than was the case with the wool and cotton industry. Indeed, in Egypt and some other countries, flax processing is still a handicraft or homework industry.

High atmospheric humidity also facilitates the processing of flax, particularly the drawing out and spinning of the fibres. Ireland has a climate which proved ideal in this respect and during the second half of the 19th century and the first two decades of the 20th, probably at least one-third of the world's flax was processed in Northern Ireland. But the industry there has contracted markedly over the last 20 years. Nowadays, since humidity in factory premises can be readily controlled, flax is processed in over 40 countries including a number in Africa, Asia and North and South America.

Retting. Flax fibres, which are up to 1 m long, occur in bundles around the periphery of the stem of the plant. The bundles are bound by pectins to the main fibrous and woody parts of the stem, and they can only be separated from the rest of the stem after the whole plant has been subjected to a rotting or "retting" process. Four types of process exist: water retting in ponds, rivers or tanks; dew retting in fields; chemical retting with alkaline solutions; and by mechanical means ("green" flax), with later chemical treatment of the binding materials in the preparing and spinning. Dew retting is the commonest practice in Europe often supplemented by a chemical treatment of the fibre before spinning. Following retting the stems are crushed and beaten ("scutched") to remove the woody and other parts of the stem ("chive") from the fibres. This process separates most of the short fibres ("tow") from the longer fibres ("line"). Retting and scutching usually take place in the area of cultivation and the flax is then transported.

In some mills the flax is chemically treated to soften the natural gums prior to processing, enabling the fibres to be processed on systems other than those designed for handling flax; the fibres are very much cleaner than those treated in the usual way. This system could lead to much improved conditions in processing rooms throughout the industry.

Hackling. In the flax mills, fibres of different qualities are sorted, mixed if necessary, and separated out into small bundles ("piecing"). These bundles are fed into a machine which combs the fibres with a series of pins which get progressively finer ("hackling").

The short fibre, or tow, is produced during both scutching and hackling. Tow, particularly that which is produced during scutching, contains much dirt and chive, and this is removed by a "carding engine". Tow, after "mixing" if necessary, is fed in a loose sheet into this machine where it is opened up, agitated and combed ("carding") and it emerges as a loose band or "sliver" of fairly clean parallel fibres. Recent technical advances will probably lead to the replacement of conventional hackling by sliver formation and a breaking process which gives a fibre with a consistent staple length capable of fast machine processing.

Spreading and drawing. Following hackling, bundles of line are spread end to end, overlapping slightly, on moving belts to give a continuous sliver. The spreading process is illustrated in figure 3.

The fibres in this sliver of line, and those in the sliver of tow following carding, require repeated separating and straightening. This is done on "drawing frames", in which slivers pass from between two rollers over a series of slowly moving pins, to a second pair of rollers, the speed of which is greater than the sliver, and which therefore draw it out ("drawing"). Several slivers are then superimposed to give one of more uniform thickness ("doubling").

Figure 3. The spreading of line fibre in the production of linen.

Roving and spinning. The sliver of line or tow which has been drafted and doubled three or more times is then put through a machine which, in addition to drawing it out, gives a preliminary twist to form "rove", and this is wound on to large bobbins ("roving"). In the spinning of coarse yarns, a sliver may be much more finely drawn out than in roving, and fully spun at this stage ("dry spinning"). This process is usually completed with the sliver and the yarn in a dry state, but occasionally the yarn, after spinning, is passed over a damp roller to smooth down protruding fibres ("gill", "damp" or "half-dry" spinning). Very fine yarn can be produced by passing the rove through a trough of hot water ("wet spinning"). However, as the range of dry spun yarns is acceptable for most industrial users, wet spinning is becoming less common. Both "wet" and "dry" spinning are carried out on modern ring spinning frames.

After spinning, yarn is either wound into hanks ("reeling") or, more usually now, wound on hollow tubes into cylindrical packages known as "cheeses" or "cones" ("winding"). Cheeses are sometimes wound from yarn which has been dried, and this sequence is usual when high-speed winding machines are used. These machines, because of the speed at which the yarn is handled, may, if dry yarn is being wound, produce considerable quantities of fine dust.

Finishing. Yarn finishing processes take place in only a few flax-preparing mills. In some, yarns are bleached and dyed before despatch to weaving factories, and in others, thread is produced by twisting together several yarns ("twisting") and occasionally the yarn or thread is polished.

Linen is normally woven from the yarn on automatic looms, although in some countries it is still a hand loom industry. Resin finishes may be applied to make the linen crease-resistant, flame-resistant, stain-repellent or water-repellent. Blending with viscose and polyester is becoming a common practice.

HAZARDS AND THEIR PREVENTION

Accidents may occur on many types of linen industry machines, especially crushing rollers and hackling machines, which may cause severe mutilation. Efficient guarding of nip points between rollers and of pulleys, drive belts and other moving parts is essential. Training of operatives in safe working practices is also necessary, in particular to prevent "picking" or cleaning of machinery in motion which are common causes of accidents. In addition, operatives should not wear loose clothing, especially neckties, and long hair should be retained beneath headwear.

The chief health hazard in the flax and linen industry is caused by the dust which is produced at almost every stage of fibre processing, but particularly during the early preparing stages. Mill fever and weaver's cough have long been associated with flax handling. Flax dust also appears to be the cause of byssinosis and chronic bronchitis. Reports from many countries, including Northern Ireland, the Netherlands, France and Egypt, indicate that byssinosis is a problem wherever flax is handled.

During the early cleaning or "pre-preparing" processes, heavy debris falls through the machines into pits, which are cleaned out regularly, except in hackling in which the debris and tow are combed out together and are separated later by carding. During all these processes, and in particular carding, very large quantities of fine dust are also produced, much of which is dispersed into the atmosphere, though some fine plant and other debris remains in the fibres and is gradually loosened and dispersed during the later processing stages. The drawing out of sliver of dry fibres, which takes place in the "preparing" processes and in dry or gillspinning, also loosens much fine dust which until then adhered to the fibre. However, during the drawing and twisting stages of wet spinning, no dust is produced and the fibres are wet.

Yarn finishing processes involve the handling of dry yarn and some remaining dust may be loosened from the fibres and scattered into the air, though, apart from bundling, which is sometimes a very dusty process, the amount of dust produced is probably negligible.

Mean dust levels with the associated byssinosis prevalence found in a survey in Northern Ireland mills are shown in table 1. Dust concentrations varied widely within each category of room, although in general the pre-preparing rooms had the highest levels, followed by other preparing rooms, other finishing rooms and wet finishing rooms, in that order. In some parts of the process, for example feed points in hackling and carding, levels can rise above 20 mg/m³ unless controlled. In manual flax processing in small workshops and homes in Egypt high mean levels of total dust (hackling 17.4 mg/m³; spinning 8.54 mg/m³) have been reported, although, for a number of possible reasons (e.g. lower general air pollution and differences in exposed populations), the prevalence of byssinosis recorded was lower than that reported in Northern Ireland. It is of interest that in Egypt the prevalence of syndromes associated with flax exposure have been reported to be higher in seasonal workers than in those permanently exposed to flax dust. No TLV has been promulgated by the ACGIH for flax dust but a TLV of 0.2 mg/m³ for cotton dust has been published. Although the cases are not directly comparable, since cotton dust is probably more byssinogenic than flax, it is obvious that levels of flax dust should be markedly reduced in the industry if the prevalence of byssinosis is to drop to a few per cent. Biologically retted flax dust appears to present a greater respiratory hazard than that from chemically or "green" retted flax, and future standards will probably recognise this. Dust measurements should be made at least every 6 months, representative of the workrooms and workshifts, by a trained industrial hygienist. Total dust samples appear suitable and have the advantages of simplicity and reliability.

Table 1. Flax dust levels and byssinosis prevalence[1]

Occupation	Mean total dust concentrations (mg/m³)	Byssinosis prevalence (% all grades)
Pre-preparers	6.7	44.0
Preparers	2.7	30.0
Wet-finishers	0.6	3.6
Other finishers	1.4	0.7

[1] From Carey et al., 1965.

Dust control is of the greatest importance throughout the industry, and while some measure of control is economic and enforcement of regulations is possible in factories, in areas where the industry is still a handicraft or home industry, the whole problem of dust control is much more formidable. Most of the systems of dust removal used in factories depend simply on the filtration of air removed from the points on the machines where the greatest quantities are liberated, usually points where the fibres are combed, or drawn out. The air removed is usually filtered through a canvas bag or a series of coarse "brush" fibres, and then recirculated into the workroom. Other systems depend on extractors of the cyclone type but, although cyclones are efficient at removing coarse dust, fine dust is removed less efficiently than with a good filter.

Most of the airborne dust is liberated from the hackling machines and carding engines. This may be prevented by constructing enclosing hoods around the machines and applying exhaust ventilation to the enclosures. The extracted air should be filtered and passed to the atmosphere through a stack above roof height. Bag filters are commonly used. If work has to be carried out where dust concentrations are very high (e.g. maintenance and repair), effective respiratory protection should be provided.

No successful treatment is known for byssinosis. However, in the early stages of the disease, considerable improvement takes place if the individual changes his job to a less dusty one. Periodic medical examinations should be made, if possible at intervals of one year, and should include a standardised questionnaire relating to the characteristic symptoms of byssinosis, and a test of ventilatory capacity. Smokers appear to be at excess risk of developing byssinosis and should be advised of this risk and preferably not be employed in the more dusty processes.

Noise is a problem, especially during weaving, and this hazard may be reduced by applying sound absorption treatment to walls, ceilings and floors and by fixing vibration isolating mounts to machines. Processes should be segregated as far as possible, with heavy partitions to insulate adjoining rooms. Muff-type hearing protection is a temporary measure but provides adequate protection if the workers can be persuaded to wear it. Many new plants are being installed in developing countries. Properly designed plant which eliminates noise and dust hazards should be specified in all orders and be correctly installed and maintained.

ELWOOD, P. C.

"A basis for hygiene standards for flax dust". British Occupational Hygiene Society Committee on Hygiene Standards, Sub-Committee on Vegetable Textile Dusts. *Annals of Occupational Hygiene* (Oxford), 1980, 23/1 (1-26). 23 ref.

"HLA antigen frequencies in flax byssinosis patients". Middleton, D.; Logan, J. S.; Magennis, B. P.; Nelson, S. D. *British Journal of Industrial Medicine* (London), May 1979, 36/2 (123-126). 14 ref.

"Byssinosis, chronic bronchitis and pathology of ventilatory function in a population of workers exposed to flax dust" (Bissinosi, bronchite cronica e patologia della funzione ventilatoria in una popolazione di operai esposti a polvere di lino). Rossi, A.; Fabbri, L.; Mapp, C.; Moro, G.; Brighenti, F.; De Rosa, E. *Medicina del lavoro* (Milan), 1978, 69/6 (698-707). Illus. 45 ref. (In Italian)

CIS 75-1857 "Dust exposure in manual flax processing in Egypt". Noweir, M. H.; El-Sadik, Y. M.; El-Dakhakhny, A. A.; Osman, H. A. *British Journal of Industrial Medicine* (London), May 1975, 32/2 (147-154). Illus. 30 ref.

CIS 77-1156 *Flax in Normandy and conditions of work in scutching sheds in the Eure Department* (Le lin en Normandie et les conditions de travail dans les teillages de l'Eure). Revel, P. (Laboratoires Miles, rue des Longs-Réages, 28230 Epernon, France) (1976), 56 p. 22 ref. (In French)

Byssinosis in flax workers in Northern Ireland. Carey, G. C. R.; Elwood, P. C.; McAulay, I. R.; Merrett, J. D.; Pemberton, J. (London, HM Stationery Office, 1965).

Floors and stairways

It is impossible to imagine a factory or commercial building without traffic aisles or stairways, and the accidents caused by the condition or littering of these surfaces or by shortcomings related to them (lighting, ventilation, maintenance, good housekeeping, warning signs) may account for between 20 and 25% of all occupational accidents.

HAZARDS

Floors

Falls, especially those caused by slipping, are often due to the type of floor construction and its coefficient of friction, and to the materials and waste products that are allowed to litter traffic aisles, such as dust, grease, solid and liquid waste, etc. The risk of falls may be significantly increased by the gutters or kerbs found around production equipment such as vats, hoppers, hoists, etc. These devices are often situated in the immediate vicinity of traffic aisles in constant use and are frequently the cause of loss of balance; the same applies to openings in the floor (chutes, inspection plates, etc.) which, having been opened, are either not surrounded by a fence or guard rail to protect unwary persons or prevent material falling through to the floor below, or are not closed again after use.

Another hazard of floors is that of non-conductivity, especially in premises where explosive atmospheres may occur; in such cases, a discharge of static electricity may result in a spark which is sufficient to ignite the explosive mixture. Particular attention in this respect should be paid to heavier-than-air flammable or explosive gases or vapours, which may collect at floor level where there is minimum air movement. Floors made of stone, concrete or steel may also present a spark hazard if workers wear hobnailed or steel-tipped boots or drop steel tools on the floor.

Stairways

Falls are the main hazard on stairways; such factors as the dimensions of tread and riser, stair-rails, stairway gradient, cleanliness, and type, straight or turning flight and the nature of the stair covering may all play a part in breaking the rhythm of the user's movement, which leads to loss of balance and finally a fall. In addition to the nature and type of the stairway, obstructions, signs, lighting, changes in direction may all constitute hazard factors in falls due to hitting against or tripping over objects.

PREVENTION OF ACCIDENTS

Floor design and construction

Floors and floor coverings should be designed to offer maximum safety in relation to the type of plant in which they are installed. Various criteria related to service conditions have to be met such as: resistance to wear and loads (pedestrians or wheeled traffic, weight and incidence of loads); resistance to environmental factors (humidity, water, condensation, heat, vibration and the meterological conditions, e.g. sunlight, frost, snow or rain); resistance to materials used in the production process (chemicals such as acids and alkalis, solvents, corrosive liquids and gases, oils and greases, etc.). Finally, consideration must be given to appearance, surface degradation, dust formation and ease of maintenance.

The following factors should be considered when selecting the type of flooring for industrial or commercial premises:

Strength. A floor should be sufficiently strong to resist crushing due to any static load such as that imposed by machine tools or plant and also that due to any dynamic load such as may result from traffic or handling of materials. In either case its resistance to penetration must be given particular attention when high density loads are to be expected. In the case of materials handling, care must be taken to ensure that the wheels on all mobile materials-handling equipment are of adequate width and diameter having regard to the loads being handled and the nature of the floor concerned. Floors should not be submitted to excessive elastic deformation.

Resistance to wear and abrasion. In places where materials are frequently handled and where there is considerable pedestrian or wheeled traffic, the floor must have sufficient resistance to abrasion to withstand normal use over a period of several years without deterioration and without excessive signs of local wear.

Resistance to chemicals. It is important that the floor should be resistant to chemicals wherever there is a risk of oils, solvents, acid or other chemicals being spilt. This applies particularly in the chemical and petrochemical industries. In premises where corrosive substances are to be used, preliminary tests should be carried out to ensure that all floors to be constructed or floor coverings to be used are sufficiently resistant to the substances in question.

Resistance to fire. In general, floors which fulfil the foregoing requirements possess a considerable resistance to fire. In case of fire in a multi-storey building, the floors need to be sufficiently strong to enable the occupants to be evacuated and equipment to be protected.

Special factors. In areas where an electrostatic discharge may cause an explosion the floor or floor covering should be slightly conductive; however, care must be taken under these conditions to ensure that a further electrical hazard is not created.

Comfort. The floor should have low thermal conductivity and absorb noise and vibration well, since these phenomena have a direct effect on the occurrence of fatigue.

Cost. This factor should be considered not only in relation to initial installation but also with regard to maintenance and cleaning.

Types of floor surface

Cement-based floors are either cast *in situ* or are composed of prefabricated slabs; special hardeners or non-slip agents are incorporated in the mass or in the surface layer, for example certain synthetic resins such as the epoxy compounds, aluminium oxide, etc. Tiled floors are easy to wash down and their non-slip characteristics may be improved by using tiles with a ribbed or corrugated surface. Steel sheets are often used for floors that are subjected to heavy wear or abrasion in places such as rolling mills or for large areas where there is little traffic, for example in power stations; the non-slip characteristic of solid sheet can be improved by use of a hard abrasive surface coating; perforated sheet steel is often used for traffic aisles, stairs and gangways; in heavy engineering works, this perforated sheet can be bonded into the concrete floor with the concrete appearing through the perforations to improve the non-slip characteristics. Wooden floors are used only where mechanical strength is of little importance, or as intermediary surfaces (e.g. for duckboards in electroplating shops, etc.).

The main floor coverings are asphalt or bitumen panels (for areas where wear is minimal), epoxy resin-based coatings, non-slip canvas, synthetic fibre carpeting, plastics tiles or sheet, cork, linoleum, etc.

Floor maintenance and cleaning

However strong a floor may be, it will suffer deterioration and deformation under conditions of heavy use. The resulting unevenness will interfere with the movement of persons and materials, increase the accident hazard, and reduce productivity. Several types of floor coverings are now available for smoothing out irregularities; these include synthetic resin-based, mineral fillers and hydraulic binders, as well as other thermo-setting resins made of specially treated oil- and dust-resistant phenolic resins, or even self-smoothing coatings made from polyester products.

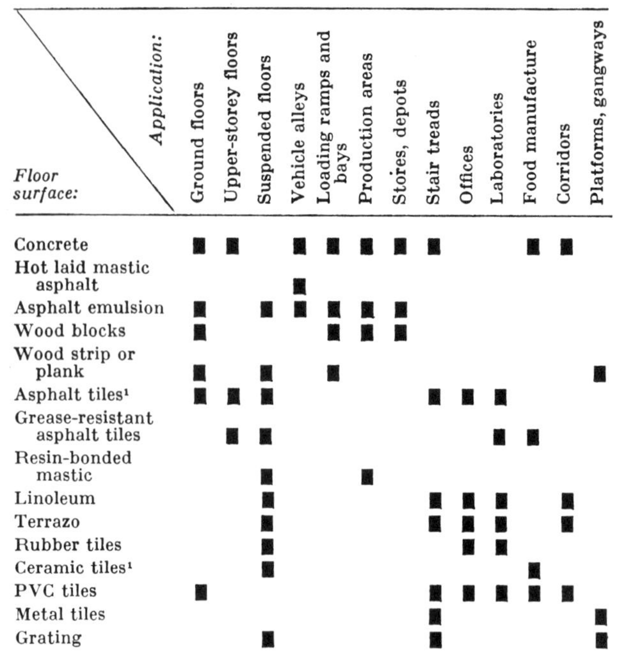

Floor surface: / Application:	Ground floors	Upper-storey floors	Suspended floors	Vehicle alleys	Loading ramps and bays	Production areas	Stores, depots	Stair treads	Offices	Laboratories	Food manufacture	Corridors	Platforms, gangways
Concrete	■	■		■	■	■	■	■			■	■	
Hot laid mastic asphalt				■									
Asphalt emulsion	■		■	■	■	■	■						
Wood blocks	■				■	■	■						
Wood strip or plank	■		■		■								■
Asphalt tiles[1]	■	■	■					■	■	■			
Grease-resistant asphalt tiles		■	■							■	■		
Resin-bonded mastic			■			■							
Linoleum			■					■	■	■		■	
Terrazo			■					■	■	■		■	
Rubber tiles			■						■	■			
Ceramic tiles[1]			■								■		
PVC tiles	■							■	■	■	■	■	
Metal tiles								■					■
Grating			■					■					■

[1] Also available in slightly conductive or anti-static versions for explosion-hazard zones.

Figure 1. Suitable floor surfaces for different applications.

Table 1. Recommended and prohibited products for use on different floors

Type of floor	Dust control or cleaners recommended	Top dressing recommended	Sealers recommended	Avoid
Cork	Neutral detergent, impregnated mop sweeper, wax sweeping powder, solvent cleaner	Solvent wax polish, emulsion dressing over a sealed surface	All except water-based	Water, alkali, acids
Concrete and granolithic	Any type	Solvent wax polish, emulsion dressings over a sealed surface	Any type	Acids
Linoleum	All types except strong alkalis and abrasive sweeping powder	Solvent wax polish, emulsion dressings	Any type	Water, alkali, acids
Magnesite	All types except solvents or abrasive sweeping powder	Solvent wax polish, emulsion dressings	Any type	Acids, alkalis
Quarry tile and ceramic	Neutral detergent, impregnated mop sweeper	None	None	Acids, alkalis
Rubber	Neutral detergent, impregnated mop sweeper, wax sweeping powder	Emulsion dressings	None	Solvents, oils
Thermoplastic	Neutral detergent, impregnated mop sweeper, wax sweeping powder, alkali cleaner	Emulsion dressings	None	Solvents, oils
Vinyl and vinyl asbestos	Neutral detergent, impregnated mop sweeper, wax sweeping powder, alkali cleaner	Emulsion dressings	None	Solvents, oils
Wood	All types exept alkaline and abrasive cleaners	Solvent wax polish, emulsion dressings over a sealed surface	All except water-based	Alkali, water, acids
Terrazzo	Neutral detergent, impregnated mop sweeper	Emulsion dressings	Water-based resin, emulsion	Alkali, acids
Mastic	All except white spirit solvent	Emulsion dressings	Water-based sealers	Solvents, oils

The cleaning process should not damage the floor and should not affect adversely its strength or its anti-slip properties. In general, the following products may be used: soaps and neutral detergents, alkaline cleaning agents and abrasives, damp floorcloths, abrasive powders, waxes, solvents and white spirit.

Table 1 gives examples of recommended and prohibited products for use on different types of floors and floor-coverings.

Floor slope

Floors which are frequently washed down with water should have a slight, even gradient of 1-2% towards a drain to ensure that the water flows away from the traffic areas. The slope of a workshop floor should never exceed 4-5% since, above this figure, a traffic aisle becomes a ramp and should be marked as such.

Design of stairways

In addition to the question of the surface material for the stair tread, there is the tricky problem of the biomechanics related to the stair gradient. Between the horizontal and the vertical, it is possible to distinguish different gradient zones; flat floor, inclined surface, stairway and ladder (figure 2). If a stairway is to be safe and easy to climb it should be designed according to the following formula: $a + s = 46$ cm, where a is the depth of the step and s the height of the riser. The recommended height of the riser is 16-20 cm; the width of the staircase should not be less than 75-80 cm. Stairways with only one step should be avoided and groups of a minimum of five steps are to be preferred; the optimum gradient for a staircase on which the traffic is dense is between 30 and 35%. When designing a stairway, care should be taken to ensure that the steps and landings are uniform in dimension so that there is no need for a break in the natural rhythm of movement when climbing or descending them. Wherever possible and, in particular in large premises, turning staircases should be avoided; the landings between flights of stairs should also be straight and their length may be determined from the formula: $p = a + n(a + 2s)$, where a is the depth of the step, n the number of steps and s the height of the riser. Steps in outdoor staircases should have a slope of 1-2% towards the nose to promote the drainage of rainwater. All stairways having four or more risers should be equipped, on any open side, with stair-railings, the height of which, from the upper surface of the top rail to the surface of the tread in line with the face of the riser at the forward edge of the tread, should be not less than 76 cm; if the railing is used as a handrail the height should be not more than 86 cm.

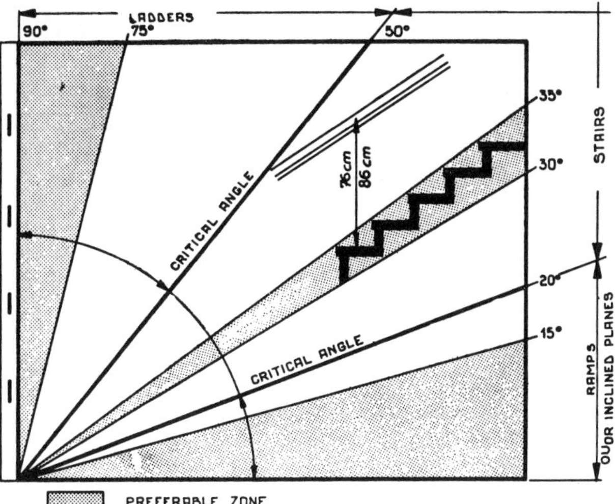

Figure 2. Preferred and critical angles of inclination for ramps and inclined planes, stairs and ladders.

Stairways and fire precautions

The problems of fire precautions are closely related to the provision of suitable stairways, which must be such as to provide an evacuation route for personnel in case of fire.

In principle all stairways should be constructed of fireproof material. Even better, they should be continuous throughout, should not be open, and should have a high resistance to heat.

They should consist of straight flights joined by landings and their size should be based upon the number of persons that may have to use them in case of evacuation being necessary. To calculate the width of a stairway, a figure of 1.25 cm should be allowed for each person occupying a given floor at one time in cases where persons may be called upon to descend, and of 2 cm per person in cases where they have to ascend the stairway.

Since a stairwell always acts as a chimney, particular attention must be given to keeping them clear of smoke. This object may be achieved by providing ventilation openings at the top and at the bottom of the stairwell and by installing fire-doors on each landing. These doors should be self-closing and in certain cases double doors should be provided to form an airlock. In tall buildings or in premises where there is a high fire risk, arrangements may be made to place the stairways under positive ventilation pressure in case of fire, thereby keeping them clear of smoke.

Signs and guards

Floor hazards, especially in traffic aisles, should be clearly marked, with particular mention of changes in gradient and the presence of obstacles. Sloping surfaces and stairways should be provided with guard, hand and intermediate rails; this structure must be able to withstand a lateral force equal to a man's weight. The top and bottom of stairways and the existence of landings should also be clearly marked.

VAN ROOSBROECK, A.

CIS 75-338 "Defining a safe walking way surface". Doering, R. D. *National Safety News* (Chicago), Aug. 1974, 110/2 (53-58). Illus. 3 ref.

CIS 78-906 "Directives for workplaces" (Arbeitsstätten-Richtlinien). Bundesministerium für Arbeit und Sozialordnung. *Arbeitsschutz* (Stuttgart), May 1977, 5 (98-103). Illus. (In German)

CIS 77-606 "Work surface friction coefficients: A survey of relevant factors and measurement methodology". Pfauth, M. J.; Miller, J. M. *Journal of Safety Research* (Chicago), June 1976, 8/2 (77-90). Illus. 27 ref.

CIS 79-1513 *Data sheet on tiled flooring in workplaces and working areas where there is a high risk of slipping* (Merkblatt keramische Bodenbeläge für Arbeitsräume und Arbeitsbereiche mit erhöchter Rutschgefahr). ZH 1/571, Hauptverband der gewerblichen Berufsgesnossenschaften (Cologne, Carl Heymanns Verlag, 1979), 21 p. Illus.

Flour milling

Milling constitutes the series of operations in the processing and grinding of cereals to produce flour for human and animal consumption.

Raw materials. The products most commonly milled include cereals such as wheat, maize, rye, oats, barley and millet; however, flour can also be made from such leguminous plants as beans, peas and chick peas and from fruit and nuts.

Flours obtained from different products vary in unit weight, compressibility, moisture content, etc. The typical composition of an 85% extraction wheat flour used for breadmaking is 14.37% protein, 1.9% oil, 1.3% cellulose, 1% minerals and 81.3% non-nitrogenous substances.

Processes. Although once an artisan activity, milling has become increasingly industrialised; in a modern mill, the whole process may be automated and controlled from a central station, the product being moved from process to process by pneumatic conveyors. The individual stages in processing of flour for human consumption and animal feedstuffs are:

- *(a)* discharge from ships' holds, railway wagons or road vehicles by pneumatic, mechanical and manhandling methods;
- *(b)* storage in grain silos;
- *(c)* hoppering, weighing and primary cleaning;
- *(d)* milling to different particle sizes and separation of by-products for animal feeding stuffs;
- *(e)* storage of flour in flour silos and addition of additives, for example mineral salts;
- *(f)* storage in silos for delivery in bulk containers or for delivery in hessian and paper bags where smaller quantities are required;
- *(g)* conveying the flour to loading vehicles.

HAZARDS AND THEIR PREVENTION

Accidents. The incidence of accidents in modern flour mills has been reduced to a minimum by the advent of automatic control. Those accidents that do occur take place during the unloading of grain and the loading of the finished products. During maintenance and repair work, precautions should be taken to ensure that men entering silos are equipped with safety belts and lifelines and that machinery cannot be operated until the work has been terminated.

Fire and explosion. Dry flour presents a constant hazard of fire and dust explosion, especially during grinding, and in conveyor cyclones and storage silos. All sources of ignition should be eliminated from hazardous areas; electrical equipment should, where possible, be located outside the dusty zone or be of an approved explosion-proof design. Sources of friction should be eliminated and aspirators and separators should be employed to remove stones, sand or metal particles from the grain before processing. Welding, cutting and smoking in dusty areas should be prohibited. In certain cases, a full explosion relief system may be required. To prevent the risk of secondary explosions, strict cleanliness is essential and all deposited flour dust should be removed at regular intervals.

The mill should have adequate fire exits; fire-fighting equipment should be strategically located and workers should be trained in fire drill.

Health hazards. Exposure to flour may cause respiratory system disorders and skin diseases, but these are only rarely severe.

The respiratory effects include: flour allergies; chronic rhinitis; chronic bronchial asthma; miller's chronic spastic bronchitis; and eosinophilic pulmonary infiltration. These manifestations are due partly to the allergic action of airborne dusts and individual sensitivity and partly to the mechanical action of dust particles that are deposited on the mucous membranes or enter the respiratory tract. Relatively severe bronchical affections have been observed.

The X-ray picture of chronic bronchitis in millers is of bronchial markings which are widespread and symmetrically distributed. The flour dust has no fibrogenic action and the disease is characterised by a slow development, a benign early stage and by progressive changes in respiratory function in the later stages. The subject suffers from frequent asthmatic attacks that may be accompanied by coughing fits, copious expectoration and dyspnoea on exertion.

Eosinophilic pulmonary infiltration is found primarily in grinding workers and is caused by a reaction of the pulmonary tissue (in particular the eosinophilic cells) to dust that has entered the lung.

Beans and certain cereals may cause pruritus and generalised papular lesions. The most common conditions are those caused by cereal parasites, especially *Pediculoides ventricosus*; however, conditions caused by *Rhizipus nigricans*, *Sphoerogyna cerealella* and *Tyroglyphus farinae* have also been observed. These parasites feed on the larvae found on growing cereal but once their normal food supply is terminated, they will infest man and animals. Vanilla flour and coconut flour may be infested with *Tyroglyphus longior*, which causes similar lesions and, together with wheat, vanilla and coconut flour, are the main sources of "grain itch", as the disease is often called.

In mill workers the parts of the body most frequently affected are the arms and the neck (the neck being particularly prone in persons carrying sacks on the back or shoulder). The pruritus is severe and is most evident at night. There may also be erythemato-vesicular eruptions with pinhead elements having a clear central zone; coalescence is rare and pemphigoid or purpuriform varieties are seen only seldom. There is no pronounced effect on the general state of health. The condition usually subsides after about 10 days, although there may occasionally be temporary, slight, residual pigmentation. *Pediculoides ventricosus* remains in the superficial skin layers and never penetrates deeply into the body. Diagnosis is easy. Grain itch can be distinguished from scabies by the fact that the parasite does not penetrate deeply into the skin, and from filariasis from the fact that the *Onchocerca volvulus* embryo does not penetrate the dermis.

Mill workers may also suffer from allergic skin conditions caused by moulds such as *Aspergillus glaucus* and *Penicillium glaucum* that develop in stored flour.

The prime measure in the prevention of respiratory and skin disease in this industry is process automation, which completely isolates the majority of workers from exposure. However, where sacks of grain or flour are manually handled, potentially dangerous exposure will take place; mechanical handling is the preferred solution but much can be done by the use of protective clothing, well designed sacks, disinfestation of grain on arrival and strict personal hygiene. Flour should be kept at 60% relative humidity and at a temperature of 15 °C. Humidity control also prevents mould formation and it has been recommended that the relative humidity of the wheat should be reduced by drying immediately after harvesting; good ventilation of storage areas also helps reduce mould formation.

Aflatoxins may also develop in stored groundnuts, cottonseed and certain varieties of wheat, soya, sorghum and barley. They enter the body, accumulate in the liver and are excreted in the urine and faeces. They have been found to be carcinogenic in animals but their action on man is still under study.

Workers should be given a pre-employment medical examination with particular attention to the lungs and skin. Persons suffering from bronchial conditions, chronic rhinitis, allergic asthma and eczematous or erythematous dermatitis should not be exposed to grain or flour. Periodic medical examinations should be carried out to detect any onset of lung or skin conditions.

Treatment of grain itch. This consists of washing with a soap solution and application of emollient creams; clothing should be disinfected.

NUNZIANTE CESARO, A.
GRANATA, A.

Occupational pulmonary disease: focus on grain dust and health. Dosman, J. A.; Cotton, D. J. (eds.). (New York, Academic Press, 1980), 615 p. Ref.

CIS 77-947 "Simultaneous exposure to airborne flour particles and thermal load as cause of respiratory impairment". Beritič-Stahuljak, D.; Valić, F.; Cigula, M.; Butković, D. *International Archives of Occupational and Environmental Health–Internationales Archiv für Arbeits- und Umweltmedizin* (West Berlin), 5 July 1976, 37/3 (193-203). 39 ref.

"Respiratory abnormalities among grain handlers: a clinical, physiologic and immunologic study". doPico, G. A.; Reddan, W.; Flaherty, D.; Tsiatis, A.; Peters, M. E.; Rao, P.; Rankin, J. *American Review of Respiratory Disease* (New York), June 1977, 115/6 (915-927). Illus. 36 ref.

"Nonsmoking grain handlers in Saskatchewan: airways reactivity and allergic status". Gerrard, J. W.; Mink, J.; Cheung, S. S. C.; Tan, L. K. T.; Dosman, J. A. *Journal of Occupational Medicine* (Chicago), May 1979, 21/5 (342-346). 14 ref.

"Workroom inhalation test (AIT), acetylcholine test (ACH-T) and radioallergosorbenttest (RAST) as diagnostic criteria of an obstructive airways disease caused by flour powder" (Arbeitsplatzbezogener Inhalationtest (AIT), Acetylcholintest (ACH-T) und RAST als Beurteilungskriterion einer berufsbedingten obstruktiven Atemwegserkrankung durch Mehlstaub). Triebig, G.; Thürauf, J.; Zober, A.; Weltle, D. *Arbeitsmedizin–Sozialmedizin–Präventivmedizin* (Stuttgart), Oct. 1979, 14/10 (232-235). Illus. 20 ref. (In German)

CIS 78-1529 "Acute bronchopneumopathy with precipitins following occupational inhalation of mouldy cereals" (Broncho-pneumopathies aiguës à précipitines après inhalation professionnelle de céréales moisies). Stévenet, A.; Stévenet, P.; Esnault, D.; D'Arlhac, M.; Maurisset, O.; Durvy, P. *Revue française d'allergologie et d'immunologie clinique* (Paris), Jan.-Mar. 1978, 18/1 (5-10). 18 ref. (In French)

CIS 76-542 *Health and safety guide for grain mills.* DHEW (NIOSH) publication No. 75-144 (National Institute for Occupational Safety and Health, 4676 Columbia Parkway, Cincinnati) (Apr. 1975), 87 p. Illus.

"Allergic broncho-alveolitis following inhalation of maize dust" (Bronchoalvéolites allergiques liées à l'inhalation de poussière de maïs). Vergnon, J. M.; Pacheco, Y.; Perrin-Fayolle, M. *Médecine et hygiène* (Geneva), 13 Jan. 1982, 40/143 (158-162). 13 ref. (In French)

Fluorine and compounds

Fluorine (F)

m.w.	38
sp.gr.	1.11 (liquid)
m.p.	−219.6 °C
b.p.	−188.1 °C
v.d.	1.69
v.p.	> 1 bar

a highly irritant, yellow gas.

TWA OSHA	0.1 ppm 0.2 mg/m³
TLV ACGIH	1 ppm 2 mg/m³
STEL ACGIH	2 ppm 4 mg/m³
IDLH	25 ppm

Fluorine is the most electronegative of all the elements and, in pure concentrated form, will react violently with a wide range of substances. Fluorine is a halogen but, unlike chlorine, bromine and iodine, will form only fluorides and not fluorates and perfluorates; except for small amounts of manufactured elemental fluorine, it is as fluorides that this element is generally encountered. Fluorine ignites bromine, iodine, sulphur, crystallised silicon, alkaline metals and a number of organic substances (benzene and ethyl alcohol). It explodes on contact with hydrogen and attacks chlorides, bromides and iodides. It decomposes water with production of ozone and hydrofluoric acid. The inorganic compounds are used in larger quantities in industry than the organic compounds; however, production of the latter is rising rapidly.

Sources. The majority of fluorine and its compounds is obtained directly or indirectly from calcium fluoride (fluorspar) and phosphate rock (fluorapatite) or chemicals derived from them. In 1965 fluorspar production was over 3 million t and phosphate rock production over 64 million t. The fluoride in phosphate rock limits the usefulness of this ore and, therefore, the fluoride must be removed almost completely in the preparation of elemental phosphorus or food grade calcium phosphate and partially in the conversion of fluorapatite to fertiliser. These fluorides are recovered in some cases as aqueous acid or as calcium or sodium salts of the liberated fluoride (probably a mixture of hydrogen fluoride and silicon tetrafluoride), or released to the atmosphere.

Production. Elemental fluorine is most commonly produced by electrolysis of a hydrogen fluoride/potassium fluoride solution.

Uses. The major use of fluorine is in the conversion of uranium tetrafluoride to uranium hexafluoride. However, its use as an oxidiser in rocket fuel systems, along with chlorine trifluoride, nitrogen trifluoride and oxygen difluoride, may be destined for a sharp increase.

Calcium fluoride (CaF_2)

FLUORITE; FLUORSPAR

m.w.	78.1
sp.gr.	3.18
m.p.	1 360 °C
b.p.	2 500 °C

soluble in ammonium salts; slightly soluble in acids
a white powder or cubic crystals.

TWA OSHA	2.5 mg/m³ (fluorides)
IDLH	500 mg/m³
MAC USSR	1 mg/m³ (fluorides as HF)

Production. This is the most important industrial fluorine compound. It is obtained from the mineral fluorspar, which must be processed to remove such impurities as calcite and clay. The degree of purification varies depending on the use made of the finished product and, of the three grades most commonly used in industry, the metallurgical grade contains 60% or more CaF_2, the ceramic grade 85-97% and the acid grade 97% or more. Certain amounts of impurity may be removed by washing but flotation is normally required to obtain the purer grades.

Uses. The major uses of fluorspar are: as the raw material for the production of hydrofluoric acid (i.e. the base for nearly all the fluorides of commercial importance), steelmaking, glass manufacture, iron founding and special fluxes, including coatings for welding rods.

Hydrofluoric acid

See separate article.

Cryolite (Na_3AlF_6)

SODIUM ALUMINIUM FLUORIDE; SODIUM FLUOROALUMINATE

m.w. 210
sp.gr. 2.90
m.p. 1 000 °C

slightly soluble in water

snow-white, semi-opaque masses.

TWA OSHA 2.5 mg/m³ (fluorides)
IDLH 500 mg/m³
MAC USSR 1 mg/m³ (fluorides as HF)

Production. This is the third most important fluoride used in industry. The original source was the large natural cryolite deposits found in the Urals and in Greenland. Cryolite mining is now of limited importance and the majority of the cryolite used today is synthetic. It is commonly produced by mixing sodium aluminate solution with liquid or gaseous hydrogen fluoride. Cryolite production is frequently located near a Bayer process alumina plant to ensure a convenient supply of sodium aluminate solution. In one plant in the United States, fluosilicic acid recovered from by-products of phosphate rock processing has been used to produce cryolite. Since the phosphate and silica content of cryolite must be kept at very low levels, the processes may be complex and, consequently, relatively little of the potentially available by-product fluosilicic acid is used for this purpose.

Uses. The major portion is used as the electrolyte in the manufacture of aluminium. The exact cryolite production figures are unknown since much of that used in aluminium manufacture is captured and re-used. It has also been used as a pesticide and in the manufacture of glass and ceramics.

Aluminium fluoride (AlF_3)

ALUMINIUM TRIFLUORIDE

m.w. 84
sp.gr. 2.88
m.p. 1 291 °C
b.p. 1 537 °C

hexagonal crystals.

TWA OSHA 2.5 mg/m³ (fluorides)
IDLH 500 mg/m³
MAC USSR 1 mg/m³ (fluorides)

Production. Aluminium fluoride is frequently produced by combining hydrogen fluoride with aluminium trihydrate. In one process, gaseous hydrogen fluoride may be passed through a bed of aluminium hydrate lumps or pellets at a temperature above 540 °C to yield aluminium fluoride directly. In another, wet process, aluminium hydrate may be added to 15% aqueous hydrogen fluoride; after filtration, the aluminium fluoride trihydrate crystals are calcined to yield anhydrous aluminium fluoride. These processes are usually located near a plant calcining aluminium trihydrate in connection with alumina manufacture.

Uses. It is used in the pottery industry, as a flux in metallurgy, as an inhibitor of fermentation and as a catalyst in organic reactions.

Fluosilicic acid ($H_2SiF_6.nH_2O$)

HYDROGEN HEXAFLUOROSILICATE; HYDROFLUOSILICIC ACID; SILICOFLUORIC ACID

m.w. 144.1
sp.gr. 1.46 (61% solution)

soluble in water and alkalis

a liquid with a strong, pungent odour when anhydrated, dissociates almost instantly into silicon tetrafluoride and hydrogen fluoride.

The water used to capture fluorides released in some phosphate rock processing plants yields fluosilicic acid and small quantities of hydrofluoric acid. The common commercial grades of fluosilicic acid contain 30-35% H_2SiF_6. A 1-2% solution is widely used for sterilising equipment in brewing and bottling plants. It is also used in the electrolytic refining of lead, in the treatment of hides and skins and as a timber preservative.

Sodium hexafluorosilicate (Na_2SiF_6)

SODIUM FLUOSILICATE; SILICON SODIUM FLUORIDE

m.w. 188.1
sp.gr. 2.68
m.p. decomposes

slightly soluble in water

a white, granular powder.

This salt is made by neutralising fluosilicic acid with sodium hydroxide or sodium carbonate. The production of sodium fluosilicate exceeds that of all the other fluosilicates together. It is used in enamels and glazes in the pottery industry, as a rodenticide and as a fluoridation agent for drinking water (there is good evidence that the incidence of dental caries can be reduced by the ingestion from birth of water containing up to 1 ppm of fluoride).

Miscellaneous inorganic fluorides

Several dozen other fluorides are manufactured in small quantities. These include boron trifluoride, the fluoborates, chlorine trifluoride, nitrogen trifluoride, oxygen difluoride, the phosphorus fluorides and sodium fluoride.

Organic fluorine compounds

The fluorinated carbon compounds comprise the fluorine-containing hydrocarbons and hydrocarbon derivatives and the fluorocarbons and their derivatives. The former include a number of dyes but, on the whole, these compounds are difficult to produce and are of little commercial importance. The latter substances are dealt with in the articles HYDROCARBONS, HALOGENATED ALIPHATIC; FLUOROCARBONS; POLYFLUORINES; TETRAFLUOROETHYLENE; FLUOROACETIC ACIDS AND COMPOUNDS.

HAZARDS

Fire and explosion. Many of the fluorine compounds present a fire and explosion hazard. Fluorine reacts with nearly all materials, including metal containers and piping if the passivating film is broken. Absolute cleanliness is required in conveying systems to prevent localised reactions and subsequent fire hazards. Special lubricant-free valves are used to prevent reactions with lubricants. Oxygen difluoride is explosive in gaseous mixtures with water, hydrogen sulphide or hydrocarbons.

Acute toxicity. Elemental fluorine, chlorine trifluoride and oxygen difluoride are strong oxidisers and may be highly destructive. At very high concentrations, they may have an extremely corrosive effect on animal tissue. However, nitrogen trifluoride is strikingly less irritating. Hydrogen fluoride is highly corrosive and produces pronounced irritation of mucous membranes.

Ingestion of quantities of soluble fluorides in the range of 1-4 g has been reported to have been fatal to man. Human fatalities have been reported in connection with the ingestion of hydrogen fluoride, sodium fluoride and fluosilicates. Non-fatal illnesses have been reported due to ingesting these and other fluorides, including the sparingly soluble salt, cryolite.

Industrial exposures to fluorides have rarely caused acute illness but one fatal inhalation of an aqueous sodium fluoride aerosol has been reported. There have been reports on subacute illnesses among cryolite workers that were probably due to the swallowing of significant quantities of cryolite derived from airborne dust.

Chronic toxicity. The sparse data available on industrial exposure of workmen to fluorine suggest that no hazards will be encountered if the irritancy level of 0.1 ppm is not exceeded. There is even less experience of long-term exposure to chlorine trifluoride and oxygen difluoride; however, there is little reason to expect chronic changes in tissues of workmen if a level of 0.1 ppm is not exceeded.

Inhalation of fluorine and oxygen difluoride has been found to produce irritation and damage to the respiratory tract, pulmonary oedema, haemorrhage and kidney damage. Inhalation of pure calcium fluoride and fluorspar has been known to produce pulmonary fibrosis.

Chronic exposures to the other airborne inorganic fluorides have generally involved a mixture of fluorides. Examples are a mixture of hydrogen fluoride and a particulate fluoride, a mixture of different gaseous fluorides, or a mixture of different fluoride dusts. The exact role of the individual fluoride involved and the extent to which it has been absorbed by either the respiratory or gastric route has not been fully clear in many cases.

However, enough information has been collected from residents in areas where elevated fluoride levels in drinking water were ingested over a number of years, from measured amounts of fluorides administered to human experimental subjects and from observations on persons exposed to fluorides at their work, to clarify certain chronic aspects of fluoride metabolism. Absorbed fluoride is generally rapidly eliminated from the circulating blood by a combination of renal excretion and skeletal deposition.

Sporadic allegations of hazards to soft tissue from fluorides in drinking water have continued. Although some such allegations may appear frivolous, the Public Health Service of the United States investigates the possible tissue hazards that have been claimed. Recently one outstanding commission of leading scientists, having investigated the basis of such allegations, concluded that fluoridation of water was not allergenic, mutagenic or carcinogenic.

The question of allergic types of reactions among workmen exposed to fluorides was extensively reviewed at a conference on health protection in primary aluminium reduction. In these operations workers are exposed to airborne fluorides but usually along with exposure to hydrocarbons. These may include polynuclear hydrocarbons. Allergic reactions were reported among some of the aluminium reduction workers in Norway, Australia and the Netherlands but similar reactions were not found during studies of workmen in 18 reduction plants in the United States.

Fluorosis or chronic fluorine intoxication has been widely reported to produce fluoride deposition in skeletal tissues of both animals and man. Fluorosis of the bones and ligaments was first reported in 1933. The symptoms found included increased radiographic bone opacity, formation of blunt excrescences on the ribs, and calcification of intervertebral ligaments. Dental mottling is also found in cases of fluorosis. The exact relationship between fluoride levels in urine and the concurrent rates of osseous fluoride deposition is not fully understood. Enough is known, however, for urine levels of fluoride to serve as a guide in safeguarding workmen potentially exposed to fluorides. It can be stated that: provided urinary fluoride levels in workmen are consistently no higher than 4 ppm, there is little need for concern; at a urinary fluoride level of 6 ppm more elaborate monitoring and/or controls should be considered; at a level of 8 ppm and above, it is to be expected that skeletal deposition of fluoride will, if exposure is allowed to continue for many years, lead to increased osseous radio-opacity.

As from 1901, the residents of Bartlett, Texas, used a drinking water source containing 8 ppm of fluoride. This level was lowered to 1 ppm in 1952. A medical survey in 1943 found that, among all residents having been 19 years or longer in Bartlett, 10-15% showed an increase in osseous radio-opacity. Re-examination 10 years later showed that the frequency of these changes had not altered and, of great importance, there was still no evidence of damage, injury, disability or discomfort associated with the increases in osseous radio-opacity.

In addition, no alleged disability related to fluoride-induced osseous changes has been reported among workmen since the case of the Danish cryolite workers and recent clinical trials on the use of massive oral doses of fluoride to combat osteoporosis raise doubts as to the validity of the concern expressed about the slight increases in osseous radio-opacity due to occupational fluoride absorption.

In industry fluoride-bearing dusts play a part in a considerable proportion of cases of actual or potential fluoride exposure, and dust ingestion may be a significant factor. Occupational fluoride exposure may be largely due to gaseous fluorides but, even in these cases, ingestion can rarely be ruled out completely, either because of contamination of food or beverages consumed in the workplace or because of fluorides coughed up and swallowed. In exposure to a mixture of gaseous and particulate fluorides, both inhalation and ingestion may be significant factors in fluoride absorption.

The fluoborates are unique in that absorbed fluoborate ion is excreted almost completely in the urine. This implies that there is little or no dissociation of fluoride from the fluoborate ion and, hence, virtually no skeletal deposition of that fluoride would be expected.

Burns. The majority of the fluorides are primary irritants. At very high concentrations, elemental fluorine, chlorine trifluoride and oxygen difluoride may cause tissue damage that resembles thermal burns. When in contact with the skin, hydrogen fluoride may not produce immediate pain but severe delayed damage may occur. Gaseous fluorine in contact with water forms hydrofluoric acid, which will produce severe skin burns and ulceration.

Mild burns of the skin by hydrofluoric acid may be treated by applying iced hyamine chloride or Sephiran solution over periods of 1 to 4 h or as long as required to eliminate blanching of the burned area. For severe burns, additional treatment may require local infiltration of the skin around the burn with 10% calcium gluconate solution and, if necessary, block anaesthesia of the affected part.

Environmental hazards. Industrial plants using quantities of fluorine compounds, such as iron and steelworks, aluminium smelters, superphosphate factories, etc., may emit fluorine-containing gases, smokes or dusts into the atmosphere. Cases of environmental damage have been reported in animals grazing on contaminated grass, including fluorosis with dental mottling, bone deposition and wasting; etching of window glass in neighbouring houses has also occurred.

SAFETY AND HEALTH MEASURES

Fire prevention. This is mainly a question of equipment design. All materials that come into contact with dangerous concentrations of fluorine compounds for any length of time should be made of, or coated with, a non-reactive material. Both water and powder have been recommended for the extinguishing of fires involving fluorine and fluorine compounds.

Health protection. To prevent lesions of the respiratory system, concentrations of fluorine and its compounds should not exceed the recommended maximum permissible levels. Processes in which there is a potential exposure hazard should be equipped with local exhaust ventilation and should, where possible, be mechanised. Workers handling dangerous substances should be supplied with eye and face protection, respiratory protective equipment, protective clothing and foot and leg protection. Additional protection may be provided by the use of lanolin as a barrier cream. Workers should not consume food or beverages in the workplace and rigorous personal hygiene should be observed before meals are taken.

Medical supervision. The relationship between recommended maximum fluoride levels in air and very early fluoride-induced changes in osseous radio-opacity have not been precisely established. Consequently, biological monitoring of workmen should be considered along with the prevention of concentrations of airborne fluorides exceeding recommended levels. Fluoride levels in urine should be checked periodically and all workers should be subjected to periodical skeletal X-ray examinations, particularly of the pelvis.

LARGENT, E. J.

Occupational exposure:

"Occupational fluoride exposure". Hodge, H. C.; Smith, F. A. *Journal of Occupational Medicine* (Chicago), Jan. 1977, 19/1 (12-39). 164 ref. Illus.

Health effects:

Fluorosis. The health aspects of fluorine compounds. Largent, E. J. (Columbus, Ohio State University Press, 1961), 140 p. 204 ref.

"Hydrofluoric acid burns of the hand". Dibbell, D. G.; Iverson, R. E.; Jones, W.; Laub, D. R.; Madison, M. S. *Journal of Bone and Joint Surgery* (Boston), July 1970, 52-A/5 (931-936). Illus. 12 ref.

CIS 76-783 "Some aspects of industrial fluorosis in Switzerland–I. The industrial hygienist's viewpoint; II. Radiology and bone fluoride (Preliminary study)" (Quelques aspects de la fluorose industrielle en Suisse–I. Le point de vue du médecin du travail; II. Radiologie et fluor osseux (Etude préliminaire). Maillard, J. M.; May, P.; Boillat, M. A.; Dettwiler, W.; Rouget, A.; Curati, W.; Demeurisse, C. *Archives des maladies professionnelles* (Paris), July-Aug. 1975, 36/7-8 (409-420). Illus. 14 ref. (In French)

Risk of accidents:

CIS 75-724 "Hazardous chemical reactions–18. Fluorine" (Réactions chimiques dangereuses–18. Fluor). Leleu, J. *Cahiers de notes documentaires–Sécurité et hygiène du travail* (Paris), 3rd quarter 1974, 76, Note No. 917-76-74 (421-427). 1 ref. (In French)

Prevention:

Health protection in primary aluminium production. Proceedings of a seminar. Copenhagen. Hughes, J. P. (ed.). (London, International Primary Aluminium Institute, 1977), 158 p. 146 ref.

Occupational exposure to inorganic fluorides. Criteria for a recommended standard. DHEW (NIOSH) publication No. 76-103 (National Institute for Occupational Safety and Health, 1975), 190 p. Illus. 263 ref.

Fluoroacetic acids and compounds

Sodium fluoroacetate (FCH_2COONa)
COMPOUND 1080; SFA
m.w. 100
b.p. decomposes at 200 °C
soluble in water
a fine, white, odourless powder.
TWA OSHA 0.05 mg/m^3 skin
STEL ACGIH 0.15 mg/m^3
IDLH 5 mg/m^3

Fluoroacetic acid (FCH_2COOH)
MONOFLUOROACETIC ACID
m.w. 78.9
m.p. 35.2 °C
b.p. 165 °C
soluble in water
a fine, white, odourless powder.

[**Fluoroacetamide** (FCH_2CONH_2)
AFL; 1081
m.w. 77
m.p. 109 °C
soluble in cold water and in ethyl and methyl alcohols
a white crystalline powder, volatile at room temperature.]

[**Difluoroacetic acid** (CHF_2–$COOH$)
m.w. 96
b.p. 134 °C
soluble in water and many organic solvents
a strong acid and a colourless liquid which emits fumes in the presence of air.]

[**Trifluoroacetic acid** (CF_3–$COOH$)
m.w. 114
m.p. −15.4 °C
b.p. 72.4 °C
a colourless liquid which emits fumes in the presence of air.]

Production. Fluoroacetic acid is prepared commercially by reacting purified ethyl chloroacetate with dry, powdered potassium fluoride in a stirred autoclave at 200 °C for about 11 hours. The resultant ethyl fluoroacetate is converted to sodium fluoroacetate with methanolic sodium hydroxide. Sodium fluoroacetate thus becomes a valuable, commercially available intermediate for the synthesis of the free acid and many derivatives. Other compounds which can give rise to this acid through biochemical processes, such as hydrolysis and β-oxidation, include simple derivatives (esters, amides, etc.) and ω-fluorocarboxylic acids containing an even number of carbon atoms.

[Fluoroacetamide is produced from sodium fluoroacetate by alkaline hydrolysis. Difluoroacetic acid is prepared by oxidating halogenated olefins with potassium permanganate. Trifluoroacetic acid is obtained by fluorinating acetyl fluoride.]

Uses. Sodium fluoroacetate is used as a rodenticide and a general mammalian pesticide. Its favourable characteristics for control purposes include: high toxicity, excellent acceptance, absence of objectionable taste and odour, chemical stability, non-volatility and little or no development of tolerance on ingestion of sub-lethal quantities. Less well established applications of fluoroacetates include their proposed use as chemical warfare agents and as insecticides.

[Fluoroacetamide is used as a systemic aphicide and as a rodenticide. The vinyl ester of difluoroacetic acid ($CHF_2-COOCH=CH_2$) is the starting point for the preparation of many fluorinated polymers. Trifluoroacetic acid is a good solvent of several aromatic and aliphatic compounds and is used in chemical synthesis. Sodium trifluoroacetate is used in the preparation of pesticides.]

HAZARDS

[Di- and trifluoroacetic acids have a lower level of toxicity than monofluoroacetic acid. The last named and its compounds are stable, highly toxic and insidious.] At least four plants in South Africa and Australia owe their toxicity to this acid *(Dichapetalum cymosum, Acacia georginae, Palicourea marcgravii)* or to ω-fluoro-oleic acid *(Dichapetalum toxicarium)*, [and recently more than 30 species of *Gastrolobium* and *Oxylobrium* have been found in Western Australia to contain various amounts of fluoroacetate]. Since sodium fluoroacetate became available commercially, a few dozen cases of accidental or deliberate poisoning have been recorded, of which at least 16 have terminated fatally; undoubtedly many undetected cases have also occurred. It is thus of importance that medical and industrial health workers are aware of these very dangerous compounds, referred to, with justification, as being "in the category of the most poisonous substances known". In addition, there are other types of potentially dangerous organic fluorine compounds, but the biological mechanisms of these differ from those of the fluoroacetates.

The biological mechanism responsible for the symptoms of fluoroacetate poisoning involves the "lethal synthesis" of fluorocitric acid, which in turn blocks the tricarboxylic acid cycle by inhibiting the enzyme aconitase. The resultant deprivation of energy by stopping of the Krebs cycle is followed by cellular dysfunction and death. It is impossible to be specific about the toxic dose of fluoroacetic acid for man; a likely range lies between 2 and 10 mg/kg; but several related fluoroacetates are even more toxic than this. In general, a drop or two of the poison by almost any route of administration [that is by inhalation, ingestion and absorption through skin cuts and abrasion or even undamaged skin] is apt to be fatal.

Symptoms. From a study of hospital case histories, it is apparent that the major toxic effects of fluoroacetates in man involve the central nervous system and cardiovascular system. Severe epileptiform convulsions alternate with coma and depression; death may result from asphyxia during a convulsion or from respiratory failure. The most prominent features, however, are cardiac irregularities, notably ventricular fibrillation and sudden cardiac arrest. These symptoms (which are indistinguishable from those frequently encountered clinically) are usually preceded by an initial latent period of up to 6 h characterised by nausea, vomiting, excessive salivation, numbness, tingling sensations, epigastric pain and mental apprehension; other signs and symptoms which may develop subsequently include muscular twitching, low blood pressure and blurred vision.

SAFETY AND HEALTH MEASURES

[According to the precautionary measures recommended by the UK Ministry of Agriculture, Fisheries and Food, fluoroacetamide and sodium fluoroacetate and baits made from them should be kept under lock and key and each issue recorded. They should be handled only in closed or draught-proof places and operators should wear dust-masks and rubber or PVC gloves, apron and boots. All contacts by mouth should be avoided, therefore hands and exposed skin should be carefully washed before eating, drinking and smoking after work. Bait containers should be marked "Poisons" and all rodent bodies should be burned. As a rodenticide the chemical should be used only on ships or in sewers where access by the general public or animals can be completely prevented during treatment, and only experienced operators should use it. In agriculture the use is acceptable on a non-edible crop and when proper precautions are applied, including a four-week interval between the last application and harvesting, as well as on broad beans, brassica, sugar beet and strawberries.]

Diagnosis. Histopathological studies contribute little of value in detecting the cause of death. Certain biochemical changes may develop as corroborative evidence of fluoroacetate poisoning. Of these the most characteristic and reliable is the massive increase of citrate in certain tissues, particularly the kidneys. Since citrate may be estimated reliably and without much difficulty, such an increase could be considered as *a priori* evidence of fluoroacetate poisoning.

The unequivocal demonstration of organically bound fluorine in the body would surely be the most reliable proof of poisoning. The two procedures most likely to be successful are the following:

(a) nuclear magnetic resonance spectroscopy;

(b) chemical analysis.

It should be mentioned that a general medical practitioner would probably not have the necessary facilities and specialised knowledge to make a correct diagnosis. Hence, in the interests of safety, recourse must still be had, by all those handling these poisons, to such time-honoured methods as detailed inventories and signed records of sales.

Treatment. First-aid measures [consist of removal of the poison from the stomach and the intravenous injection of 20-50 g of acetamide in 500 ml of 5% glucose solution. Definitive hospital treatment includes intravenous injection of acetamide as above, 2-3 times a day, followed by intravenous injections of 5-10 g of acetamide in 20-40 ml of 40-50% glucose solution, and symptomatic therapy as required.]

PATTISON, F. L. M.

CIS 1604-1971 "Acute ethyl fluoroacetate poisoning" (Akutní otrava fluoroctanem etylnatým). Pazderova, J. *Pracovni lékarstvi* (Prague), Sep. 1970, 22/7 (251-254). Illus. 11 ref. (In Czech)

"Subacute fluoroacetate poisoning". Peters, R. A.; Spencer, H.; Bidstrup, P. L. *Journal of Occupational Medicine* (Chicago), Feb. 1981, 23/2 (112-113). 4 ref.

"Preclinical experiments on the antidote against monofluoroacetic acid derivatives poisoning in mammals". Hashimoto, Y.; Makita, T.; Mori, T.; Nishibe, T.; Noguchi, T.; Watanabe, S. (59-62). *Whither rural medicine* (Tokyo, Japanese Association of Rural Medicine, 1970).

Biochemical lesion and lethal synthesis. Peters, R. A. (Oxford, Pergamon Press, 1963).

Fluorocarbons

The fluorocarbons are derived from hydrocarbons by the substitution of fluorine for some or all of the hydrogen atoms. Hydrocarbons in which some of the hydrogen atoms are replaced by chlorine or bromine in addition to those replaced by fluorine (i.e. chlorofluorohydrocarbons, bromofluorohydrocarbons) are generally included in the classification of fluorocarbons, for example bromochlorodifluoromethane ($CClBrF_2$).

The first economically important fluorocarbon was dichlorodifluoromethane (CCl_2F_2) which was introduced in 1931 as a refrigerant of much lower toxicity than sulphur dioxide, ammonia or chloromethane, which were the currently popular refrigerants.

The fluorocarbons of greater economic importance are listed below.

Uses. The fluorocarbons are used as refrigerants, aerosol propellants, solvents, foam blowing agents, fire extinguishants, anaesthetics and polymer intermediates.

Production. The most widely employed fluorocarbons, trichlorofluoromethane and dichlorodifluoromethane, are produced by the reaction of hydrofluoric acid with carbon tetrachloride in the presence of a chlorofluoroantimony catalyst.

HAZARDS

The fluorocarbons are, in general, lower in toxicity than the corresponding chlorinated or brominated hydrocarbons. This lower toxicity may be associated with the greater stability of the C–F bond; and perhaps also with the lower lipoid solubility of the more highly fluorinated materials. Because of their lower level of toxicity, it has been possible to select fluorocarbons which are safe for their intended uses. And because of the history of safe use in these applications, there has grown up a popular belief that the fluorocarbons are completely safe under all conditions of exposure. This is not so.

To a certain extent, the volatile fluorocarbons possess narcotic properties similar to, but weaker than, those shown by the chlorinated hydrocarbons. Dichlorodifluoromethane (CCl_2F_2) if inhaled at 5% by volume concentration induces dizziness in man. If inhaled at 15% concentration, loss of consciousness results. At the ACGIH TLV of 1 000 ppm, narcotic effects are not experienced by man.

Toxic effects from repeated exposure, such as liver or kidney damage, have not been produced by the fluoromethanes and fluoroethanes. The fluoroalkenes, such as tetrafluoroethylene, hexafluoropropylene or chlorotrifluoroethylene can produce liver and kidney damage in experimental animals after prolonged and repeated exposure to appropriate concentrations.

Even the acute toxicity of the fluoroalkenes is surprising in some cases. Perfluoroisobutylene

Fluorocarbon	Formula	m.w.	b.p.[1] (°C)	Uses	TLV ACGIH		STEL ACGIH	
					ppm	mg/m³	ppm	mg/m³
Trichlorofluoromethane/ FLUOROCARBON 11	CCl_3F	137.4	23.8	refrigerant, propellant, solvent, blowing agent	1 000	5 600	1 250	7 000
Dichlorodifluoromethane/ FLUOROCARBON 12	CCl_2F_2	120.9	−29.8	refrigerant, propellant, blowing agent	1 000	4 950	1 250	6 200
Chlorotrifluoromethane/ FLUOROCARBON 13	$CClF_3$	104.5	−81.4	refrigerant	–	–	–	–
Dichlorofluoromethane/ FLUOROCARBON 21	$CHCl_2F$	103	8.9	refrigerant	10	40	–	–
Chlorodifluoromethane/ FLUOROCARBON 22	$CHClF_2$	86.5	−40.8	refrigerant	1 000	3 500	1 250	4 375
Trifluoromethane/ FLUOROCARBON 23	CHF_3	70	−82	fire extinguishing agent, propellant	–	–	–	–
Tetrachlorodifluoromethane/ FLUOROCARBON 112	$(CCl_2F)_2$	203.8	92.8	solvent	–	–	–	–
Bromotrifluoromethane/ FLUOROCARBON 13B1	$CBrF_3$	149	−57.8	fire extinguishing agent, refrigerant	1 000	6 100	1 200	7 300
Dibromodifluoromethane/ FLUOROCARBON 12B2	CF_2Br_2	210	24.5	polymer intermediate, fire extinguishing agent	100	860	150	1 290
Trichlorotrifluoroethane/ FLUOROCARBON 113	CCl_2FCClF_2	187	47.6	refrigerant, solvent, intermediate	1 000	7 600	1 250	9 500
Dichlorotetrafluoroethane/ FLUOROCARBON 114	$(CClF_2)_2$	171	3.6	refrigerant, propellant	1 000	7 000	1 250	8 750
Chloropentafluoroethane/ FLUOROCARBON 115	$CClF_2CF_3$	155	−38.7	food aerosol propellant, refrigerant	–	–	–	–
Chlorodifluoroethane/ FLUOROCARBON 142b	CH_3ClF_2	100.5	−9.5	refrigerant	–	–	–	–
Octafluorocyclobutane/ FLUOROCARBON C-318	F_2C-CF_2 / F_2C-CF_2 (ring)	200	−5.8	food aerosol propellant, dielectric gas	–	–	–	–
Chlorotrifluoroethylene/ FLUOROCARBON 1113	$CClF{:}CF_2$	116.5	−27.9	polymer intermediate	–	–	–	–
Tetrafluoroethylene/ FLUOROCARBON 1114	$CF_2{:}CF_2$	100	−76.3	polymer intermediate	–	–	–	–
Fluoroethylene/ FLUOROCARBON 1141	$CH_2{:}CHF$	46	−72.2	polymer intermediate	–	–	–	–
Hexafluoropropylene/ FLUOROCARBON 1216	$CF_2{:}CFCF_3$	150	−30.5	polymer intermediate	–	–	–	–
Bromochlorotrifluoroethane/ FLUOROCARBON 123B1	$CF_3CHBrCl$	197.4	50.2	anaesthetic	–	–	–	–
Difluoroethylene/ FLUOROCARBON 1131	$CH_2{:}CF_2$	64		polymer intermediate	–	–	–	–

($(CF_3)_2C:CF_2$) is an outstanding example. With an LC_{50} of 0.76 ppm for 4-h exposures for rats, it is more toxic than phosgene. Like phosgene, it produces an acute pulmonary oedema. On the other hand, vinyl fluoride and vinylidene fluoride are fluoroalkanes of very low toxicity.

Like many other solvent vapours and surgical anaesthetics, the volatile fluorocarbons may also produce cardiac arrhythmia or arrest under circumstances where an abnormally large amount of adrenaline is secreted endogenously (e.g. anger, fear, excitement, severe exertion). The concentrations required to produce this effect are well above those normally encountered during the industrial use of these materials.

All fluorocarbons will undergo thermal decomposition when exposed to flame or red-hot metal. Decomposition products of the chlorofluorocarbons will include hydrofluoric and hydrochloric acid along with smaller amounts of phosgene and carbonyl fluoride. The last compound is very unstable to hydrolysis and quickly changes to hydrofluoric acid and carbon dioxide in the presence of moisture.

The three commercially most important fluorocarbons (trichlorofluoromethane, dichlorodifluoromethane and trichlorotrifluoroethane) have been tested for mutagenicity and teratogenicity with negative results. Chlorodifluoromethane, $CHClF_2$, which received some consideration as a possible propellant, was tested by Du Pont's Haskell Laboratory and Imperial Chemical Industries and was found by both to be mutagenic and teratogenic in a battery of tests. The results have not been published in scientific journals, but point to caution when dealing with fluorocarbons that are not fully halogenated. No further consideration was given to the use of chlorodifluoromethane as a propellant.

Recently attention has been directed to a potential toxic effect attributable to the destruction of certain stable halofluorocarbons in the stratosphere. In 1974 F. S. Rowland and M. J. Molina postulated that stable fluorocarbons, chiefly dichlorodifluoromethane and trichlorofluoromethane, which are widely used as propellants and refrigerants, would when eventually released into the atmosphere slowly diffuse upward into the stratosphere, where intense ultraviolet radiation could cause the molecules to release free chlorine atoms. It was further postulated that the chlorine atoms would react with ozone atoms as follows: $Cl + O_3 = ClO + O_2$ and, $ClO + O = Cl + O_2$ with the net result $O + O_3 = 2O_2$. Since the chlorine atoms are regenerated in the reaction, they would be free to repeat the cycle; the net result would be a significant depletion of stratospheric ozone, which shields the Earth from harmful solar ultraviolet radiation.

The National Academy of Sciences of the United States has issued three interim reports which state that no significant fault has been found with the proposed chemistry. On the other hand, no significant decrease in stratospheric ozone has been detected to date. In April 1979, however, the Consumer Products Safety Commission in the United States banned nearly all aerosol products containing chlorofluorocarbons from interstate commerce. European countries are following the United States in banning chlorofluorocarbon propellants. Eventually the validity of the Rowland-Molina hypothesis will be tested by measurements in the stratosphere which are under way.

SAFETY AND HEALTH MEASURES

Due to their low order of toxicity, prevention of injury resulting from overexposure is relatively easy.

Sufficient exhaust and general ventilation should be provided to keep vapour concentrations below recommended levels. Excessive skin contact with the liquid fluorocarbons should be minimised to prevent defatting of the skin and possible skin absorption. Until further information on teratogenic and mutagenic properties of CF 22 is obtained, exposure to it of women of child-bearing age should be limited.

Treatment. Victims of fluorocarbon exposure should be removed from the contaminated environment and treated symptomatically. Adrenaline should not be administered, because of the possibility of inducing cardiac arrhythmias or arrest.

ZAPP, J. A. Jr.

"Detection of chemical mutagens by the dominant lethal assay in the mouse". Epstein, S. S.; Arnold, E.; Andrea, J.; Bass, W.; Bishop, Y. *Toxicology and Applied Pharmacology* (New York), Oct. 1972, 23/2 (288-325). Illus. 20 ref.

"Absence of teratogenic effect of fluorocarbons in rat and rabbit" (Absence d'effet tératogène des fluorocarbones chez le rat et le lapin). Paulet, G.; Desbrousses, S.; Vidal, E. *Archives des maladies professionnelles, de médecine du travail et de sécurité sociale* (Paris), June 1974, 35/6 (658-662). (In French)

"Mutagenicity and chromosomal aberrations as an analytical tool for *in vitro* detection of mammalian enzyme-mediated formation of reactive metabolites". Greim, H.; Bimboes, D.; Egert, G.; Göggelmann, W.; Krämer, M. *Archives of Toxicology* (West Berlin), 30 Dec. 1977, 39/1-2 (159-169). 60 ref.

CIS 76-432 "Concentrations of fluoroalkanes associated with cardiac conduction system toxicity". Flowers, N. C.; Hand, R. C.; Horan, L. G. *Archives of Environmental Health* (Chicago), July 1975, 30/7 (353-360). Illus. 14 ref.

CIS 79-1622 *Data sheet on handling of fluorocarbons* (Merkblatt für den Umgang mit Fluorkohlenwasserstoffen). ZH 1/409. Hauptverband der gewerblichen Berufsgenossenschaften (Cologne, Carl Heymanns Verlag, 1978), 12 p. (In German)

"Protection against depletion of stratospheric ozone by chlorofluorocarbons" (Washington, DC, National Research Council, National Academy of Sciences, 1979), 392 p. Illus. Ref.

"Is freon 113 neurotoxic? A case report". Raffi, G. B.; Biolante, F. S. *International Archives of Occupational and Environmental Health* (Heidelberg and West Berlin), Nov. 1981, 49/2 (125-127). 8ref.

Foam resins

Foam resins are plastic materials rendered light and porous by the inclusion of minute gas bubbles. They may be rigid or flexible in structure. Commercially important foam resins are produced from polystyrene, polyurethane or polyvinyl chloride and, less commonly, from polyolefins and epoxy resins. Natural or synthetic rubber can also be foamed.

HAZARDS AND THEIR PREVENTION

During the production of foam resins there are certain toxic hazards, together with the danger of explosion due to excessive build-up of pressure. The finished foam resins constitute a considerable fire hazard.

In the manufacture of polyurethane, polyethers or polyglycols are combined with aromatic isocyanates, which constitute a toxic hazard and may enter the body through the skin or the respiratory system. Strict precautions must be taken against either form of contact. Metal soaps are used in relatively small quantities as catalysts in resin manufacture, and the foam structure is obtained by the flash vaporisation of superheated solvents, by the incorporation of chemical blowing agents or by mechanically mixing a gas into the resin during compounding. The use of closed moulds may result in the development of high pressures. Solvents may offer a fire hazard and blowing agents, which release large volumes of gas when heated, entail the risk of bursting containers if accidentally exposed to high temperatures. [A catalyst (dimethylaminopropionitrile: DMAPN) has been recently shown to have neurotoxic effects mainly on the bladder, with symptoms of urinary retention and decreased libido in both men and women soon after exposure. The same effects have been observed in the use of Niax catalyst ESN, which contains 15% DMAPN.]

Foam polystyrene, polyvinyl chloride and polyolefins are manufactured by dissolving a blowing agent in the polymer melt and depressurising. Foam rubber can be produced from rubber latex or the uncured elastomer, and flash vaporisation of water or blowing agents is employed to effect expansion. Thiazoles, guanidines and aldehyde amines used as accelerators and phenolic compounds used as antioxidants may also constitute health hazards (see RUBBER INDUSTRY, NATURAL; POLYSTYRENE).

The finished foam resins and rubbers are, themselves, free from health hazards, unless comminuted or strongly heated, because they burn readily and give off [very rapidly large volumes of hot, dense smoke, toxic gases such as CO and HCN, and fractions which have been reported to possess mutagenic properties (from rigid polyurethane foam). The smoke] from flexible polyurethane is less black and dense than that from other resins, and cold-cured polyether foam burns less readily. Most foamed resins melt when they burn and fire spread is accelerated by the running and dripping of the flaming melt.

All foam resins are excellent thermal insulators and a risk of spontaneous ignition may arise if freshly manufactured material is not allowed to cool thoroughly before stacking in bulk. Foam polyurethane is particularly hazardous in this respect since it takes some time before curing is complete. Foam resins are not readily ignited by a small moderate-temperature source of heat such as a burning cigarette since the foam melts and recedes from the source; however, they are readily ignited by a small flame. [In addition the smoke and vapours from smouldering are flammable and can produce an explosive atmosphere; the rate of release of flammable products increases with time; in a closed space flaming can then produce an explosion.] It is important to ensure that storage areas for foam resins are kept free of paper or textile which could be ignited by a smouldering object and then act as a fuse to carry fire to the foam product.

Once a fire becomes established in a stock of foam resin or rubber it will develop rapidly and the application of large quantities of water at an early stage is necessary to effect extinguishment. For this reason, the installation of a sprinkler system in a warehouse used for storing foam resins is an advisable precautionary measure but an above-average density of sprinkler heads is necessary. Adequate fire exits and escapes should be provided and arrangements should be made for venting the large quantities of smoke produced, preferably by the use of automatic or remotely controlled equipment.

Measures should be taken to limit the dripping and flow of the burning melt.

The fire hazard of foam resins can be reduced to a certain degree by incorporating fire-retardant additives such as halogen and/or phosphorus compounds in the polymer. Chemical incorporation affords permanent protection but is likely to affect the flexibility of the finished product and is normally applied only to rigid foams; physical incorporation has a less permanent effect. However, the main effect of the retardant is merely to render the product self-extinguishing after the limited application of a small igniting flame. In the presence of the flame or in an established fire the burning will continue and the halogen and/or phosphorus compounds released in the fire smoke have the disadvantage of being irritant and corrosive.

Certain special chlorobutadiene-based formulations are now employed to produce synthetic flame-resistant foam rubber; however, natural foam rubbers with fire-retardant properties are still at the development stage.

[The application of foam resins by spraying or moulding is not without risks; certain polyurethane foams used for insulation may give off isocyanates, tertiary amines, and chloral, which readily combines with water vapours into the highly toxic chloral hydrate. In all spraying operations, indoor or out, personal protective equipment should consist of a supplied air respirator of the continuous flow, positive pressure type and impervious hood, footwear and protective clothing.]

BURGOYNE, J. H.
RUTLEDGE, P. V.

Health impairment:

"An epidemic of urinary retention caused by dimethylaminopropionitrile". Keogh, J. P.; Pestronk, A.; Wertheimer, D.; Moreland, R. *Journal of the American Medical Association* (Chicago), 1980, 243/8 (746-749).

Experimental alert No. 4 concerning the occupational hazard due to Niax catalyst ESN. International Occupational Safety and Health Hazard Alert System (Geneva, International Labour Office, 1980), 5 p.

"Mutagenicity of aerosols from the oxidative thermal decomposition of rigid polyurethane foam". Zitting, A.; Falck, K.; Skyttä, E. *International Archives of Occupational and Environmental Health* (West Berlin), Oct. 1980 47/1 (47-52). 26 ref.

Fire risk:

CIS 77-607 *Fire risk in the storage and industrial use of cellular plastics.* Health and Safety Executive, Guidance note, General series/3 (London, HM Stationery Office, Dec. 1976), 7 p. Illus. 56 ref.

Safe use and storage of flexible polyurethane foam in industry. Health and Safety Executive, HS(G) series (London, HM Stationery Office, 1978). Illus. 56 p.

Workers education:

CIS 77-548 *Urethane foams—Good practices for employees' health and safety.* DHEW (NIOSH) publication No. 76-154 (National Institute for Occupational Safety and Health, 4676 Columbia Parkway, Cincinnati) (Apr. 1976), 26 p. Illus.

Food-borne infections and intoxications

Food contaminated by micro-organisms may cause toxic infections or food poisoning.

Food-borne intoxications of bacterial origin

The toxin is formed in the food before ingestion as a result of the development of a sufficiently large number of micro-organisms.

Botulism. The thermolabile toxin is produced by *Clostridium botulinum*, an anaerobic organism that has heat-resistant spores. They can multiply in various foodstuffs with a pH of 4.5 or more, but not in acid foods (fruits, marinades, etc.). The toxin is produced at temperatures of between 10 and 40 °C; however, *Cl.*

botulinum type E will grow at temperatures as low as 3 to 4 °C.

Foods most readily affected are raw ham and raw fish preparations. The spores have high resistance to heat and, consequently, home-made preserves, which may not be sufficiently cooked, should be suspect; factory-made preserves are rarely contaminated. Tins of preserves contaminated with *Cl. botulinum* usually have a bulge.

Two clinical forms of botulism have been described:

(a) The acute form–there is an incubation period of 12-36 h and the symptoms include hoarse voice, headache and sometimes nausea and vomiting. The victim's condition then deteriorates with eye disorders (mydriases and diplopia), insufficiency of saliva and nasal section, and motor and respiratory paralyses in serious cases. Death occurs in 50% of cases.

(b) The mild form–the incubation period is longer (sometimes a few days) and the symptoms are more or less mild ocular or secretory disorders (dry mouth and nose).

Treatment is based on the use of antitoxic serums, initially polyvalent and subsequently monovalent when the toxin type has been determined by laboratory testing. Success in the treatment of severe cases depends on the rapidity of diagnosis.

Staphylococcal enterotoxin food poisoning. Certain staphylococci produce a thermostable enterotoxin, which is encountered in dangerous amounts, in particular, in foodstuffs that are prepared and then consumed only 12-24 h later (especially if left unrefrigerated); the foodstuffs most commonly incriminated are, by descending order of frequency: processed meat products (meat pies, hams, etc.); ice creams; pastry goods; sweetened condensed or powdered milk; tinned sardines. Any food containing large numbers of staphylococci (100 000 or more per gramme) should be considered suspect. The food may be contaminated at source, such as in the case of cows with staphylococcal mastitis; however, in most cases, contamination originates in intermediate handling by persons with staphylococcal skin lesions–or even by healthy persons.

The incubation period may vary between 1 and 6 h but is normally 2-4 h. Symptoms include nausea followed by vomiting and profuse diarrhoea; temperature remains normal and the symptoms disappear after 1-2 days.

There is no specific medication; in fact, it is preferable to abstain from medication although rehydration should be practised in cases of dehydration.

Alimentary mycotoxicoses

Literally hundreds of toxic compounds produced by actinomycetes and other fungi are known. Several of these species have been clearly implicated in incidents of toxicity in man, for instance *Claviceps purpurea* in ergotism, and several species, for example *Fusarium poae, F. sporotrichoides* and *Cladosporium epiphyllum*, in alimentary toxic panmyelophthisis.

Numerous other moulds which can develop in a wide variety of foodstuffs have been shown to be capable of producing substances toxic to various animal species. The most outstanding are *Penicillium islandicum*, a species primarily involved in toxic yellow rice, which produces the toxins islanditoxin and lukeoskyrin, and strains of *Aspergillus flavus*, which produce a group of closely related chemical substances known as the aflatoxins.

The degree to which various mycotoxins–with the exception of the above-mentioned toxins causing ergotism and alimentary toxic panmyelophthisis–represent a health hazard to man has not yet been fully explored. The results of animal experiments, however, have been so alarming (e.g. aflatoxin B1 has been shown to be one of the most potent hepatocarcinogens yet discovered) that several countries have accepted the evidence from animal experiments as sufficient justification for legislation prohibiting the sale of foods contaminated with aflatoxin.

Increasing attention is being paid to the possibility that aflatoxins or similar food contaminants may be involved in the aetiology of liver disease, including primary liver carcinoma. An accurate assessment of the importance of mycotoxins in this regard has not yet been made and will be possible only by systematic surveys for their presence in diets of affected populations, characterisation of their biological effects, etc.

Food-borne infections

These are caused by pathogenic bacteria which usually produce toxic symptoms.

Salmonella and Arizona infections. Salmonelloses are the most common types of food-borne infections, and are due primarily to *Salmonellae*, over a thousand serotypes of which have been distinguished according to their antigens.

Some, like *S. typhi* and *S. paratyphi* A, B and C, cause typhoid and paratyphoid fever, whilst others, such as *S. typhi murium* and *S. enteritidis* cause food poisoning. Foodstuffs most likely to be dangerously contaminated are meat, especially minced meat, processed pork meats, frozen or powdered egg (ducks' eggs are particularly susceptible), ice cream and pastry cream, fruits such as strawberries and, occasionally, butter and margarine. It has been found that contamination in shellfish is most often due to serotypes that produce the typhoid or paratyphoid forms of salmonelloses.

The food may be contaminated either at source, as is the case with diseased animals or infected products (meat, eggs), by contact with a contaminated environment (fresh or sea water contaminated by waste water, tools, workbenches or production lines), or by contact with other contaminated foodstuffs in a refrigerator or pantry or with crop manures and dressings. Contamination may also be traced to insects (flies), rodents (rats, mice) or domestic animals that have been in contact with contaminated food.

The incubation period is generally 12-24 h or more rarely 8-72 h. Symptoms appear violently and include colic, vomiting, diarrhoea and persistent fever (39-40 °C). The severity and duration of the infection vary from person to person. Fatalities are rare due to antibiotic treatment; however, antibiotic-resistant strains may be encountered and laboratories should be asked to determine effective antibiotics for particular cases. The most commonly used antibiotics are tetracycline, oxytetracycline, chlorotetracycline, chloramphenicol, neomycin and kanamycin.

Arizona causes disease similar to that caused by *Salmonella*; however, cases are rarer.

Clostridium perfringens infection. Clostridium perfringens is an anaerobic, sporogenous bacterium that has been claimed to account for 8-12% of all cases of food poisoning. Infection is nearly always contracted from meat since, when a large piece of meat is roasted or boiled, the temperature at the centre may not be sufficiently high to destroy the spores, and this may give rise to large numbers of vegetative forms during cooling; meat which is left to be eaten either cold or after moderate reheating may contain millions of *Cl. perfrin-*

gens per gramme at the moment of consumption. Contamination is due primarily to human carriers, such as butchers and cooks, from whose faeces the pathogens can be extracted. The strains of *Cl. perfringens* are so widely distributed that they are potential contaminants of nearly all foods.

The incubation period is usually 10-12 h, sometimes 8-22 h. The symptoms are violent and include colic followed by diarrhoea, but vomiting is exceptional; there is no fever and the disorder is over in 1-2 days. As with staphylococcic enterotoxicosis, it is advisable to abstain from treatment.

Vibrio parahaemolyticus infection. Vibrio parahaemolyticus is a halophile (i.e. it is capable of flourishing in a salt culture medium) and has been responsible for producing, in Japan, outbreaks of food poisoning resembling salmonellosis. The infection results from eating raw sea fish or shellfish during the hot season; salted cucumbers may also be a source of infection.

Bacillus cereus infection. B. cereus is aerobic, sporogenous and may have spores that are highly resistant to heat. The most common sources of infection are mashed potatoes and pastry creams. Ingestion is followed after 10-12 h by colic and watery diarrhoea; vomiting is rare and fever is absent. Symptoms disappear within 12 h and, in view of the benign nature of the infection, it is preferable to abstain from therapy.

Infections due to miscellaneous micro-organisms

Sometimes, the only noteworthy fact in a case of food poisoning is the abundance of bacteria in the food ingested; there may be over one thousand million per gramme. One of the species may predominate or there may be a mixture of innocuous species with no known pathogenic or toxic action. It is not everyone who can safely eat food contaminated to such an extent, and some people may suffer from vomiting and diarrhoea. The symptoms are usually mild and regress rapidly without treatment, although some fatal cases have been recorded.

Group food poisoning

Food-borne infections are becoming increasingly common and affect increasingly large numbers of the population, largely as a result of the industrialisation of food production, processing and distribution.

The most common types of group food-borne infections and intoxications are those caused by the *Salmonellae*, staphylococcal enterotoxin and *Cl. perfringens*; botulism occurs only rarely. Two factors are of particular importance in food contamination by *Salmonella*, *Staphylococci* and *Clostridium*: the origin of the contamination and the rate of multiplication.

The food may be contaminated at source; for example, meat may have come from an animal which was diseased or a carrier on arrival at the abattoir, or the milk may have been taken from a cow with staphylococcal mastitis, etc. Food may also be contaminated when it is handled by a person who is diseased or is merely a carrier. Other vectors of pathogenic or toxic bacteria, such as insects, rodents, cats and dogs may cause contamination when in contact with food for human consumption. Microbial foci in uncleaned corners and crevices of food processing machines such as mincers and grinders, and in dirty handtowels, are also important sources of infection.

Due to the way in which the production and distribution of meat and, in particular, eggs is organised, contamination by *Salmonella* is unusually intense and difficult to combat. In the case of certain *Salmonellae*, it is not until the bacterial concentration exceeds an order of magnitude of 1 000 per gramme that there is a risk of food poisoning. However, certain *Salmonella* serotypes (*S. typhi murium*, in particular) are more intensely pathogenic than others and consequently will produce harmful effects in man at lower concentrations. The same applies to the *Staphylococci*, which have to be present in large numbers before a dangerous quantity of endotoxin can be produced. For example, an ice cream which is contaminated with a small number of *Staphylococci* during production will remain harmless provided it is kept refrigerated; however, if it is allowed to remain for several hours at room temperature, the micro-organisms multiply rapidly and cause disorders in the consumer.

The food prepared in catering facilities such as restaurants and canteens should be analysed periodically to determine the degree of bacterial contamination. Food handlers should be screened to detect persons who are diseased or germ-carriers; healthy carriers should be examined by stool analysis *(Salmonella, Cl. perfringens)* or throat swab analysis *(Staphylococci)*.

The cleanliness and hygiene of premises, equipment and personnel should be of the highest standard. Kitchen utensils should be washed in clean water and detergent immediately after use and then rinsed in clean, slightly chlorinated water. Dish and floor cloths are a constant source of contamination; they should be replaced by disposable paper wipers. Tableware should be washed in very hot water or be disposable. Kitchen and serving staff should be taught to practise scrupulous personal hygiene, washing after having used the toilets, avoiding coughing or sneezing over foodstuffs and wearing working clothes that are changed and laundered at frequent intervals.

Certain elementary precautions should be taken to prevent the multiplication of micro-organisms:

(a) Cook immediately before eating. Cooking destroys vegetative forms of bacteria but sometimes leaves spores intact. Staphylococcal enterotoxin is thermostable, and has been shown to withstand boiling and even higher temperatures; however, even well cooked food can undergo secondary contamination and must thus be handled hygienically.

(b) Refrigerate. At a temperature of around 0 °C, multiplication of pathogenic and toxic micro-organisms is adequately inhibited; however, multiplication recommences freely at above 6 °C and even at 3-4 °C in the case of *Cl. botulinium* type E. Refrigerator temperatures should be checked regularly. The ideal conditions in a refrigerated compartment are a temperature of +2 °C and a relative humidity of not more than 85%.

(c) Deep-freeze. This is more effective than refrigeration and should be used for all foods that are not adversely affected by the temperatures involved. Deep-frozen food should be thawed rapidly and immediately cooked or consumed cold since micro-organism multiplication recommences once the inhibiting effect of the low temperature is terminated.

"Left-overs" should be consumed within 2 h or disposed of. Raw food of unknown origin and "semi-preserves" should be viewed with suspicion, except in the case of acidulated foods (pH less than 4.5) in which pathogenic and toxic micro-organisms cannot multiply.

In the event of a case of group food-borne infection or intoxication, the public health authority should be informed immediately. The effectiveness of diagnosis and treatment depends, to a large extent, on the speed of the epidemiological investigation.

BUTTIAUX, R.

Food hygiene. Fourth Report of the Expert Committee on Environmental Sanitation. WHO technical report series No. 104 (Geneva, World Health Organisation, 1956), 28 p.

European Technical Conference on Food-borne Infections and Intoxications. WHO technical report series No. 184 (Geneva, World Health Organisation, 1959), 18 p.

Food-Borne Disease. Methods of sampling and examination in surveillance programmes. Report of a WHO study group. WHO technical report series No. 543 (Geneva, World Health Organisation, 1974), 50 p.

Microbiological aspects of food hygiene. Report of a WHO expert committee with the participation of FAO. WHO technical report series No. 598 (Geneva, World Health Organisation, 1976), 103 p.

Environmental carcinogens. Selected methods of analysis. Vol. 1: *Analysis of volatile nitrosamines in food.* IARC scientific publications No. 18 (Lyons, International Agency for Research on Cancer, 1978), 212 p. Illus.

"Natural Products" (145-177). *IARC monographs on the evaluation of carcinogenic risk of chemicals to man.* Vol. 1 (Lyons, International Agency for Research on Cancer, 1972), 184 p. Ref.

Food industries

The term "food industries" covers a whole series of industrial activities directed at the treatment, preparation, conversion, preservation and packaging of foodstuffs. The raw materials used are generally of vegetable or animal origin and produced by arable agriculture or gardening, stock farming and breeding, and fishing. This article deals in a general way with the whole complex of food industries: many other articles deal with particular branches and particular hazards.

Demographic pressure, the uneven distribution of agricultural resources and the need to ensure preservation of food products to facilitate their better distribution, explain the rapid technical evolution in these industries, which are moving from the artisan to the industrial stage.

In practice, to satisfy population requirements, there is a need not only for a sufficient quantity of foodstuffs, which presupposes an increase of production, but also strict control and hygiene to obtain the quality essential to maintain the health of the community. Only modernisation of techniques and rational organisation can realise these conditions.

Statistical data on the manpower employed in the food industries in various countries are very variable, and valid comparison is not possible. It appears, however, that a large section of the working population is employed in these industries, on average between 10% and 15% of the effective force of the combined manufacturing industries of the various countries. The proportion of women employed is noteworthy, ranging between 20% and 55%.

The food industries are extremely diverse and the techniques used are as varied as the types of food produced (table 1).

Many of the food industries depend almost entirely on local agriculture or fishing. This means seasonal production, and difficulties arise from periodic fluctuations in the employment of seasonal workers. Wherever possible, and especially in the food-preserving trades, it is desirable to group different products in the same factory so that production can be spread over the seasons.

General processes

In spite of the extreme diversity of the food industries, the preparation processes can be divided thus: handling and storage of raw materials; extraction; processing; conservation.

Handling and storage. Manipulation of the raw materials, the ingredients during processing and the finished products, is varied and diverse. It is the current trend to minimise manual handling by rational organisation, by mechanisation, by "continuous processing" and by automation.

Mechanical handling may involve: self-propelled in-plant transport with or without palletisation; conveyor belts (beet, grain, fruit, etc.); bucket elevators (grain, fish, etc.); spiral conveyors (confectionery, flour, etc.); air fluming (unloading grain, transport of flours, etc.).

Storage of raw materials is most important in a seasonal industry (sugar refining, brewing, flour milling, canning, etc.). It is usually done in silos, tanks, cellars, or cold stores. Storage of the finished products varies according to their nature–liquid or solid–the method of preserving and the method of packaging (loose, in sack, in bundles, in bottles), and the respective premises must be planned to suit the conditions of handling and preserving (traffic aisles, ease of access, temperature and humidity suited to product, cold-storage installations).

Extraction. To extract a specific food product from fruit, cereals or liquids, any of the following methods may be used: crushing, pounding or grinding, extraction by heat (direct or indirect), extraction by solvents, drying and filtration.

Crushing, pounding and grinding are usually preparatory operations, for example the crushing of cocoa beans, the slicing of sugar beet. In other cases it may be the actual extraction process, as in flour milling.

Heat can be used directly as a means of preparation by extraction, as in roasting (cocoa, coffee, chicory); however, in manufacturing it is usually used directly or indirectly in the form of steam (extraction of edible oils or extraction of sweet juice from thin slices of beet in the sugar industry).

Oils can be extracted equally well by combining and mixing the crushed fruit with solvents that are later eliminated by filtering and reheating.

The separation of liquid products is carried out by centrifuging (turbines in sugar refinery) or by filtering through filter presses in breweries, and oil and fat production.

Production processes. Operations in processing food products are extremely varied and can only be described after individual study of each industry, but the following general procedures are used: fermentation, cooking, dehydration and distillation.

Fermentation, obtained usually by addition of a microorganism to the previously prepared product, is practised in bakeries, breweries, the wine and spirits industry and cheese products industry.

Cooking occurs in many manufacturing operations: canning and preserving of meat, fish, vegetables and fruits; in bakeries, biscuit making, breweries, etc. In other cases, cooking is done in a vacuum-sealed container and produces a concentration of the product (sugar refining, tomato-paste production).

Besides the drying of products by the sun, as with many tropical fruits, dehydration can be carried out, in hot air (fixed dryers or drying tunnels), by contact (on a drying drum heated by steam, such as in the instant coffee industry and tea industry), vacuum drying (often combined with filtering) and lyophilisation (freeze drying), the product being first frozen solid, and then dried by vacuum in a heated chamber.

Distillation is used in the making of spirits: the fermented liquid, treated to separate grain or fruit, is vapourised in a still; the condensed vapour is then collected in the form of ethyl alcohol.

Table 1. The food industries, their raw materials and processes

Industry	Materials processed	Storage requirements	Processing techniques	Preserving techniques	Packaging of finished products
Meat processing and preserving	Beef, lamb, pork, poultry	Cold stores	Slaughtering, cutting up, boning, comminuting, cooking	Salting, smoking, refrigeration, deep-freezing, sterilisation	Loose or in cans, cardboard
Fish processing	All types of fish	Cold stores, or salted loose or in barrels	Heading, gutting, filleting, cooking	Deep-freezing, drying, smoking, sterilisation	Loose in refrigerated containers, or in cans
Fruit and vegetable preserving	Fresh fruit and vegetables	Processed immediately; fruits may be stabilised with sulphur dioxide	Blanching or cooking, grinding, vacuum-concentration of juices	Sterilisation, pasteurisation, drying, dehydration, lyophilisation (freeze drying)	Bags, cans, or glass or plastics bottles
Milling	Cereals	Silos	Grinding, sifting	–	Silos (conveyed pneumatically), sacks or bags
Baking	Flour	Silos, sacks	Kneading, fermentation, baking	–	–
Biscuit making	Flour, cream, butter, sugar, fruit	Varies from case to case	Mixing, kneading, laminating, moulding, baking	–	Bags, boxes, cans
Pasta manufacture	Flour, eggs	Silos	Kneading, grinding, cutting, extrusion or moulding	Drying	Bags, packets
Sugar processing and refining	Sugar beet, sugar cane	Silos	Crushing, maceration, vacuum-concentration, centrifuging, drying	Vacuum-cooking	Bags, packets
Chocolate making and confectionery	Cocoa bean, sugar, fats	Silos, sacks, conditioned chambers	Roasting, grinding, mixing, conching, moulding	–	Packets
Brewing	Barley, hops	Silos, tanks, conditioned cellars	Grain milling, malting, brewing, filter-pressing, fermentation	Pasteurisation	Bottles, cans, barrels
Distilling and manufacture of other beverages	Fruit, grain, carbonated water	Silos, tanks, vats	Distillation, blending, aeration	Pasteurisation	Barrels, bottles, cans
Milk and milk products processing	Milk, sugar, other constituents	Immediate processing; subsequently in ripening vats, conditioned vats, cold store	Skimming, churning (butter), coagulation (cheese), ripening	Pasteurisation, sterilisation or concentration, desiccation	Bottles, plastics wrapping, boxes (cheese) or unpackaged
Processing of oils and fats	Groundnuts, olives, dates, other fruit and grain, animal or vegetable fats	Silos, tanks, cold stores	Milling, solvent or steam extraction, filter-pressing	Pasteurisation where necessary	Bottles, packets, cans

Preservation processes. It is important to prevent any deterioration of food products, as much for the quality of the products as for the more serious risk of contamination or threat to the consumers' health.

The micro-organisms which cause so much deterioration are destroyed by heat in certain conditions; freezing retards their growth; isolation of the finished product from the surrounding atmosphere prevents their propagation. Current methods of preservation are based on high-temperature and low-temperature processes, although experiments with ionising radiation are relatively advanced.

(a) High-temperature processes:
cooking;
sterilisation (mainly used in canneries) – submitting the already canned product to the action of steam, generally in a closed container such as an autoclave;
pasteurisation – particularly reserved for liquids

such as fruit juice, beer, milk or cream, carried out at a lower temperature and for a short time;
smoking–carried out mainly on fish, ham and bacon, ensuring dehydration and giving a distinctive flavour.

(b) Low-temperature processes:
storage in a cold store–the temperature determined by the nature of the products (meat, fish, vegetables);
freezing and deep-freezing–which allow foodstuffs to be preserved in their naturally fresh state, by various methods of slow or rapid freezing.

HAZARDS AND THEIR PREVENTION

Accident hazards. An examination of statistics reveals a certain consistency in the distribution of accidents amongst various food industries. The following is a typical breakdown of accidents in France (1977): 40-45%: handling, falling objects; 25-30%: falls on ground; 15-20%: hand tools; 8-10%: machinery and transmission; 3-5%: vehicles. The rate of accidents is especially high in cold stores, meat processing and drink industries.

It should be noted that although accidents involving transmission machinery are relatively infrequent, they are likely to be serious.

Risks due to machines and handling must be studied individually in each industry; handling problems can be met by rational organisation of work, or by mechanisation, and also by provision of appropriate personal protection, such as foot and leg protection, hand and arm protection and eye and face protection.

Falling accidents are most often caused:

(a) by the state of the floor, uneven, wet or made slippery by the type of surface, by products, or by fatty or dusty waste; (anti-slip floors, easy to clean and well maintained, proper arrangements for maintenance and good housekeeping will prevent many of these);

(b) by uncovered pits or drainage channels–maintenance of covers or fencing is necessary;

(c) by high-level work, access to or examination of equipment, silos, use of ladders, etc. (provision of safe means of access, sound ladders, safety belts and lifelines can prevent many of these);

(d) by steam or dust, which may not only make the floor slippery but prevent good visibility;

(e) by the absence of, or insufficient, lighting.

Good ventilation, adequate illumination and reduction of the noise level are prerequisites of safety.

The other categories of accidents are related either to specific risks of the type of food handled, for example the use of knives (serious cuts) in abattoirs and the butchery trade, and in the canning of meat and fish, or to general risks such as burns and scalds from steam, and even more serious accidents due to explosion of boilers or autoclaves from lack of regular examination or poor maintenance.

Electrical installations, especially in wet or damp places, require efficient earthing and good maintenance. Airborne concentrations of dusts produced during the processing of certain foodstuffs may constitute a considerable explosion hazard; explosions may also occur in gas or oil-fired ovens if they are not installed correctly and provided with the essential safety devices (especially in baking and biscuit making).

Health and safety committees, appointment of safety officers, and safety education all have their effect in reducing risks.

Health hazards. Apart from ailments caused by using chemicals which may be irritant, caustic or even toxic (notably refrigerants), which are just as often met in other industries, the majority of occupational ailments met with in the food industry (some of which are not considered as compensatable occupational diseases in certain countries) are of two kinds:

(a) infections and infectious or parasitic diseases spread by animals or the waste products of animals used in manufacture; these zoonoses include anthrax, brucellosis, the leptospiroses, tularaemia, bovine tuberculosis, glanders, erysipeloid, Q fever, foot and mouth disease, rabies, etc.;

(b) the contact dermatoses and allergies of the skin or respiratory system caused by organic products, animal or vegetable.

Cold-store workers may suffer impairment of health through exposure to cold if adequate protective clothing is not supplied.

Strict industrial hygiene control is vital at all stages of food processing, including slaughterhouses and abattoirs. Personal hygiene is most important in guarding against infection or contamination of the products. The premises and equipment should be designed to encourage personal hygiene, through good, conveniently situated sanitary and washing facilities, showerbaths, where practicable, provision and laundering of suitable protective clothing and provision of barrier creams and lotions.

Medical supervision of workers is desirable; many food factories are small and membership of a group medical service may be the most effective way of securing this.

MALAGIE, M.

General:

CIS 78-1449 *General report.* Report I, Second Tripartite Technical Meeting for the Food Products and Drink Industries, Geneva, 1978 (Geneva, International Labour Office, 1978), 121 p.

Accident prevention:

"Occupational safety in food industry, trade and services" (La sécurité du travail dans les industries, commerces et services de l'alimentation). Godefroy, M. *Travail et sécurité* (Paris), May 1981, 5 (230-251). Illus.

CIS 76-854 *Food, drug, and beverage equipment.* ANSI-ASME F2.1-1975 (American Society of Mechanical Engineers, 345 East 47th Street, New York), 16 p.

CIS 76-912 *Accident hazards in merchandise delivery in the food trade* (Risques d'accidents lors de la livraison de denrées dans les commerces de l'alimentation). Edition INRS No. R 110 (Institut national de recherche et de sécurité, 30 rue Olivier-Noyer, Paris) (no date), 19 p. Illus. (In French)

"Food machinery guarding". Daniels, G. G.; Taylor, J. *Health and safety at work* (Croydon), Aug. 1980, 2/12 (27-30). Illus.

Industrial hygiene:

CIS 78-353 *Airborne dust suppression in establishments in the food industry* (Očistka vozduha ot pyli na predprijatijah piščevoj promyšlennosti). Štokman, E. A.; Kadaševskij (Moscow, Izdatel'stvo "Piščevaja promyšlennost", 1977), 304 p. Illus. 105 ref. (In Russian)

Technology:

Combination processes in food irradiation. Proceedings of a Symposium, Colombo, 24-28 November 1980, jointly organized by IAEA and FAO (Vienna, International Atomic Energy Agency), Sep. 1981, 467 p. Illus. Ref.

Ergonomics:

CIS 76-1745 *Colloquium on "Man–machine–environment"* (Kolloquium "Mensch–Maschine–Umwelt") (Berufsgenossenschaft Narhungsmittel und Gaststätten, Steubenstrasse 46, 6800 Mannheim) (no date), 85 p. Illus. (In German)

Foot and leg protection

Injuries to the feet and legs are common to many industries. Falls of heavy objects may damage the foot, particularly the toes, in any working place, more especially in the heavier industries, mines, metal manufacture, engineering, building and construction work. Burns of the lower limbs from molten metal or sparks, or corrosive chemicals occur frequently in foundries, iron and steelworks, chemical works, etc.; dermatitis or eczema may be caused by a variety of substances, acids and alkalis. The feet may also be damaged by striking against objects or by stepping on sharp protrusions: in the construction industry, upturned nails penetrating the sole are a constant hazard. More general foot affections may be caused by working on damp or hot floors; risks of falls on slippery or wet floors may be aggravated by wearing unsuitable shoes.

Types of protection

The type of protection should be related to the risk. In some light industries, it may be sufficient to persuade workers to wear ordinary well made shoes: a tendency, particularly among women workers, to wear out unsuitable footwear such as old slippers, etc., should be combated: very high or worn-down heels, flapping soles, etc., may be the initial cause of accidents.

Sometimes a safety shoe or clog is adequate, sometimes a boot or leggings will be required (figures 1, 2 and 3). The height of the boot–ankle, knee or thigh length–depends on the hazard, although comfort and mobility will also have to be considered. Thus a shoe and gaiters may in some circumstances be preferable to a high boot.

Safety shoes and boots may be made from leather, rubber, synthetic rubber or plastic and may be made by sewing, vulcanising or moulding. Since the toes are most vulnerable to impact injuries a steel toe-cap is the essential feature. For comfort the toe-cap must be reasonably thin and light and carbon tool steel is therefore used. These safety toe-caps may be incorporated in many types of boots and shoes. In some trades, where falling

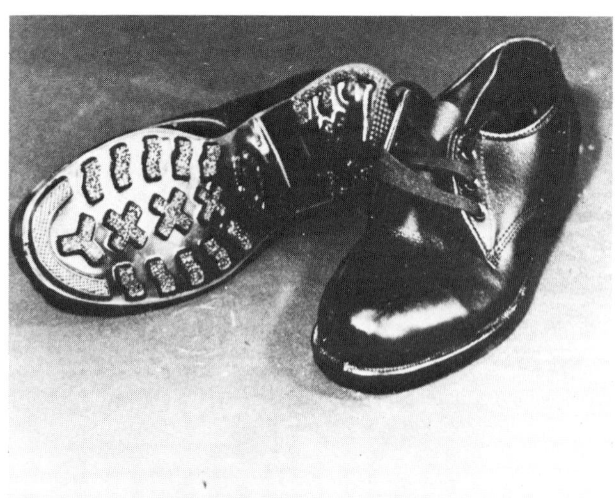

Figure 1. Safety shoes.

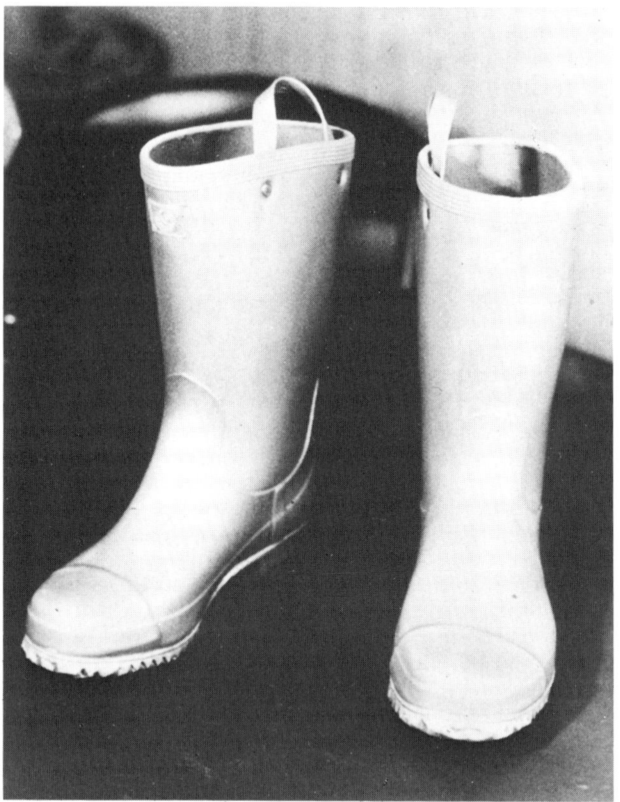

Figure 2. Heat protective boots.

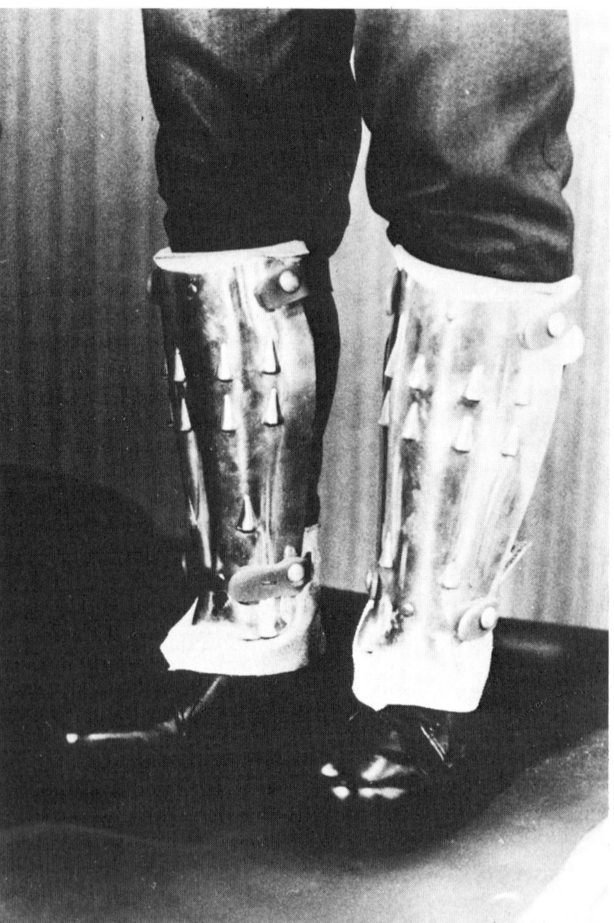

Figure 3. Protective leggings.

objects present a particular risk, metal instep guards may be fitted over safety shoes.

Rubber or synthetic outsoles with various tread patterns are used to prevent risk of slipping: this is especially important where floors are likely to be wet or slippery. The material of the sole appears to be of more importance than the tread pattern and should have a high coefficient of friction.

Reinforced, puncture-proof soles are necessary in such places as construction sites: metallic insoles can also be inserted into other types of footwear.

Where an electrical hazard exists, shoes should be entirely stitched or cemented to obviate the use of nails. Where static electricity may be present, safety shoes should have electrically conductive rubber outsoles.

Wooden clogs or shoes with wooden soles are useful in damp conditions and also give a degree of protection against falling objects or protrusions. Where the heat is not immoderate, wooden soles also protect against discomfort from working on hot floors.

Synthetic rubber boots are useful protection from chemical injuries: the material should show not more than 10% reduction in tensile strength or elongation after immersion in a 20% solution of hydrochloric acid for 48 h at room temperature.

Especially where molten metals or chemical burns are the hazard, it is important that shoes or boots should be without tongues and that the fastenings should be easily undone. For similar reasons, trousers should be pulled over the top of the boot and not tucked inside.

Some countries have standards of testing to which all safety toe-caps should conform; an example is given below:

(a) Compression test. Using the compression-testing apparatus the toe-cap shall be inserted between the two parallel horizontal surfaces of the apparatus, and a static load of 1 100 kg shall be applied. Immediately after the test, the clearance between the base surface of the toe-cap and the most dented inside of the rear edge of the arch of the toe-cap shall be not less than 22 mm and, in addition, there shall be no cracks.

(b) Impact test. Using the impact-testing apparatus, a weight weighing 23 kg and having a striking surface of hemispherical shape of 25 mm in axial length and 25 mm in radius shall be freely dropped along appropriate guides from a height of 460 mm on to the toe-cap. The instantaneous clearance caused by the impact between the base surface of the toe-cap and the most deformed inside of the rear edge of the arch of the toe-cap shall be not less than 22 mm and, in addition, there shall be no cracks.

Rubber, asbestos or metallic spats, gaiters or leggings may be used to protect the leg above the shoe line, especially from burning risks. Protective knee pads may be necessary, especially where work involves kneeling, for example in some foundry moulding.

Aluminised heat-protective shoes, boots or leggings will be necessary near sources of intense heat.

Use and maintenance

All safety footwear should be kept clean and dry when not in use and should be replaced as soon as necessary. In places where the same rubber boots are used by several people, there should be regular arrangements for disinfection between use to prevent spread of foot infections. There is the risk of foot mycosis resulting from the unnecessary use of too tight and heavy types of boots or shoes.

The success of any protective footwear depends upon its acceptibility and this is now widely recognised in far greater attention to styling: this is particularly important to young workers and women. Comfort is a prerequisite and the shoes should be as light as is consistent with their purpose: shoes weighing more than 1 kg should be avoided.

Sometimes safety foot and leg protection is required by law to be provided by the employers. Where the use is good practice rather than legal obligation, it is often found very effective to have some arrangements for easy purchase at the place of work. If protective wear can be obtained at cost price, sometimes on extended payment terms, it is much more likely to be used than if it is left to individual purchase; not only that, but the type of protection worn can be better controlled.

MIURA, T.

"Foot protection. More effective safety shoes" (Protection des pieds. Des chaussures de sécurité plus efficaces). Chesnay, C. *Revue de la sécurité* (Paris) Apr. 1980, 16/167 (10-17). (In French)

CIS 2446-1973 *Personal protective equipment. Instructions for manufacturers, dealers and users. VI. Foot and leg protection* (Personlige beskyttelsesmidler. Asvininger for fabrikanter, forhandlere og brugere–VI Fod- og benvaern). Publication No. 16 (Copenhagen, Direktoratet for arbejdstilsynet, 3rd ed., 1973), 9 p. (In Danish)

CIS 2359-1971 *Specification for gaiters and footwear for protection against burns and impact risks in foundries.* BS 4676 (British Standards Institution, 2 Park Street, London W1A 2BS) (15 Mar. 1971), 10 p.

CIS 78-1404 *Development of test methods and procedures for foot protection.* Scalone, A. A.; Davidson, R. D.; Brown, D. T. (National Institute for Occupational Safety and Health, 4676 Columbia Parkway, Cincinnati) (Dec. 1977), 177 p. Illus.

Shoes and foot hygiene (Hakimono to Ashi no Eisei). Miura, T. (Tokyo, Bunka Publishing Bureau, 1978), 182 p. 118 illus. (In Japanese)

CIS 81-286 *Work injury report–An administrative report on accidents involving foot injuries* (Washington, DC, Bureau of Labor Statistics, Department of Labor, Feb. 1980), 35 p.

"Testing safety shoes" (Les chaussures de sécurité à l'examen). *Cahiers suisses de la sécurité du travail* (Lucerne), Mar. 1981, CSST No. 136, 11 p. Illus. (In French)

Foot and mouth disease

STOMATITIS APHTHOSA EPIZOOTICA; APHTHAE EPIZOOTICAE

A contagious virus disease affecting mainly cattle but also, although to a lesser degree, pigs, goats and sheep, it is characterised by fever and the formation of vesicles in the mouth and on the skin. The effects, particularly on cattle, are disastrous, with high economic losses. The disease is encountered in Europe, Africa, Asia and South America.

Human infection

This is rare and serious only when contracted by young children (due to infected milk). Occupational infection is exceptional but has been known to occur in cowherds, butchers, veterinarians and laboratory workers.

The disease is always transmitted through direct contamination by animals and never from man to man; it is characterised by headache and listlessness following an incubation period of 2-6 days. These symptoms are often associated with pharyngitis and are followed by the appearance of blisters on the fingers, the palms of the hands and soles of the feet. There is sometimes inflammation of the nailbeds and loss of the nails. Aphthae of the mucous membrane of the mouth desiccate and slough off leaving a reddish surface, which exudes serous fluid. The disease is always benign in its course.

Diagnosis

It is insufficient to base the diagnosis on the clinical picture and current epizootic position alone, since widespread human enteroviruses occasionally cause herpangina with similar clinical signs. Foot and mouth disease can be diagnosed by laboratories experienced in techniques of virus isolation and identification, and serological tests for the demonstration of complement fixing or neutralising antibody. Blister material for virus isolation and a first sample of serum should be collected as early as possible; a second and third serum sample should be collected about 4 and 8 weeks after the onset of the disease.

PREVENTIVE MEASURES

Persons in contact with infected animals should practise scrupulous personal hygiene, which is necessary for protection from other zoonoses as well, and follow the rules of disinfection in foot and mouth disease control. Suspected cases should be subjected to restriction of movement in an agricultural area and be treated in isolation to prevent possible spread of infection to animals and children. Prophylaxis of the disease in animals comprises vaccination, observation and restriction of movement or slaughtering of diseased, exposed and directly threatened animals.

LAMBERT, G.

CIS 965-1964 "Foot-and-mouth as an occupational disease in cattle farms" (Berufsbedingte Maul- und Klauenseuche bei Angestellten in Viehzüchtereien). Schwann, T. *Berufsdermatosen* (Aulendorf), Dec. 1963, 11/6 (309-319). Illus. 27 ref. (In German)

Serologic evidence of vesicular stomatitis in slaughterhouse workers in Antioquia, Colombia (Evidencia serológica de estomatitis vesicular en empleados de matadero, Antioquia, Colombia). Hanssen, H.; Zualuaga, F.; Hanssen, G. *Boletín de la Oficina Sanitaria Panamericana* (Washington, DC), Feb. 1979, 58/2 (141-147). Illus. 13 ref. (In Spanish)

Footwear industry

The term footwear covers a vast range of products made from many different materials: boots, shoes, sandals, slippers, and clogs are made wholly or partly of leather, rubber, synthetic and plastics materials, canvas, rope, and wood. This article deals with the footwear industry as generally understood, i.e. based on traditional manufacturing methods. The manufacture of rubber boots (or their synthetic equivalents) is essentially a section of the rubber industry which is covered in the article RUBBER INDUSTRY, NATURAL; RUBBER, SYNTHETIC.

Egyptian sandals made of papyrus or palm leaves are the earliest known footwear, although a type of esparto grass sandal has been found among prehistoric remains in Spain. Shoes, boots and sandals made from leather, felts and other materials have been made by hand over the centuries. Fine shoes are still made wholly or partly by hand by craftsmen but in all the industrialised countries there are now large mass-production plants. Even so, some work may still be given out to be done as home work. The village cobbler is a traditional figure and most towns and cities have a profusion of small shoe-repairing shops.

Boot and shoe plants grow up close to leather-producing areas (i.e. near cattle-raising country); some slipper and light-shoe making developed where there was a plentiful supply of felts from the textile trade and in most countries the industry tends to be localised in its original centres. Leathers of different type and quality, and some reptile skins, formed the original materials, with a tougher quality for the soles. In recent years leather has been increasingly displaced by other materials, in particular rubber and plastics. Linings may be made of wool or polyamide (nylon) fabric or sheepskin; laces are made of horsehair or synthetic fibres; paper, cardboard and thermoplastics are used for stiffenings. Natural and coloured wax, aniline dyes and colouring agents are used in finishing.

Processes. There may be over a hundred operations in the making of a shoe and only a brief summary is possible here. Mechanisation has been applied at all stages but the pattern of the hand process has been closely followed: introduction of new materials has modified the process without changing the broad outline.

In the making of the uppers, the leather or other materials are sorted and prepared and the uppers are then cut out on stitching (or dinting) presses by shaped, loose-knife tools. The parts, including the linings, are then "closed", i.e. sewn or stuck together. Perforating, eyeletting and button-holing may also be carried out.

For making the bottom stock, soles, insoles, heels and welts are cut out in revolution presses using loose-knife cutters, or in sole-moulding presses; heels are made by compression of leather or wood strips, etc. The stock is trimmed, shaped, scoured and stamped.

The uppers and bottom stock are assembled on the last and then stitched, glued, nailed or screwed together; these operations are followed by shaping and levelling between rollers. The final finishing of the shoe includes waxing, colouring, spraying, polishing and packaging.

Boot- and shoe-repair shops will often be equipped with stitching and finishing machines for buffing and waxing, and repairers may use a variety of solvents in adhesives and for cleaning purposes.

Among the raw materials used in the manufacturing process, the most important from the point of view of occupational hazards are the adhesives. These include natural solid and liquid adhesives (see ADHESIVES) and adhesive solutions based on organic solvents.

Glues and adhesives falling into the first two groups do not appear to give rise to any toxic effects. A considerable amount of toxicological experience has been recently gained on those included in the group based on organic solvents. These can be divided into the natural rubber elastomers (table 1), and the polychloroprene elastomers (table 2).

HAZARDS AND THEIR PREVENTION

The use of flammable liquids constitutes a considerable fire hazard, and the wide use of presses and assembling machines has introduced a new risk of accidents into this industry. The main health hazards are those of toxic solvents and high atmospheric dust concentrations.

Fire. The solvents and sprays used may be highly flammable. In the first place, solvents with the highest practicable flash point should be used but, even so, strict attention is necessary to:

(a) reducing the flammable vapour in the atmosphere by good general ventilation and by local exhaust ventilation in spraying cabinets and at other points

Table 1. Constituents of natural rubber elastomer adhesives

Group	Parts by weight %	Components
Solvents	90.3	Benzene, petrol (gasoline), hexane-toluene, petrol (benzine)
Natural rubber	8	Crepe, smoked sheets
Resins	1.5	Colophony, colophony esters, modified phenols
Antioxidants	0.2	–

Table 2. Constituents of polychloroprene elastomer adhesives

Group	Parts by weight %	Components
Solvents	500	Benzene-ketone-esters, petrol-benzene-esters, hexane-toluene-esters, hexane-toluene-ketones, benzene-toluene
Chloroprene	100	Various
Resins	40-45	Phenol reagents, terpene reagents, modified phenols, cumenes, colophony esters
Use may also be made of:		
Magnesium oxide	4	–
Zinc oxide	5	–
Antioxidants	2	–
Bindings		Epoxide resins

where a high concentration may be expected—racks of treated footwear should not be left to dry off in the vicinity of workplaces;

(b) maintaining unobstructed exits and gangways (here again the disposal of racks is important as they are often left blocking essential gangways);

(c) keeping the amount of flammable liquids in the workroom as low as possible and in closed containers—a fireproof store outside the building should be maintained;

(d) removing residues from cabinets and workbenches, providing closed receptacles for waste and rags—celluloid is now rarely used for toe puffs but many other materials, offcuts, waste and dust from polishing and finishing may increase the fire risk and good housekeeping and cleanliness are essential.

Flameproof electrical installations may be necessary in some areas and polishing machines may have to be earthed to prevent the build-up of static electricity. Fire extinguishers and fire alarms should be provided, and all staff should be instructed in procedures in case of fire. Serious fires have been reported where solvents were used by homeworkers in unsafe surroundings, for example in the kitchen; conditions of use in small repairing shops require strict control.

Accidents. Many of the operating parts of the machines present serious hazards, in particular presses, stampers, rollers and knives. The loose-knife cutters at stitching and revolution presses can cause serious injury: the appropriate precautions include two-hand controls (perhaps in addition a photo-electric cell device) (figures 1 and 2), the reduction of stroke to a safe level

Figure 1. Two-handed electrical control on a textile dinting press. Since both hands are required to operate the controls, the press ram cannot descend whilst the operator's hands are in the danger zone.

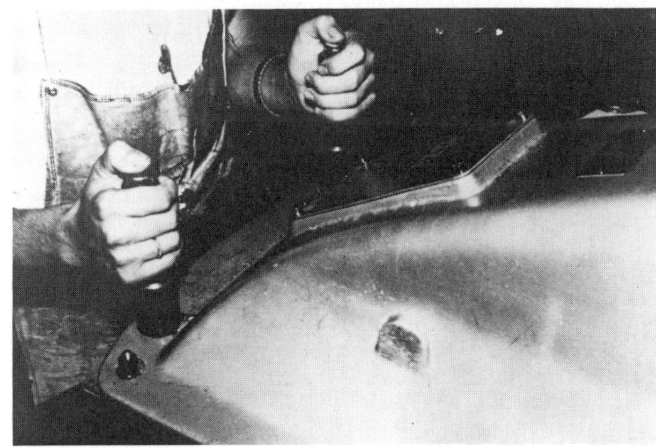

Figure 2. Two-handed electronic control on a leather dinting press.

in relation to the size of the cutter and the use of well designed, stable cutters of adequate height, with flanges fitted perhaps with handles. Sole-moulding and heel presses should be guarded to prevent access of the hand: stamping machines can cause burns as well as crushing injuries unless access of the hand is prevented by guarding. Nips of rollers and knives of milling and shaping machines should be fitted with suitable machinery guarding. The shading and polishing wheels of finishing machines and the spindles on which they are mounted should also be guarded. Among the electrical hazards, mention should be made of static electricity, particularly in relation to polishing machines that have not been adequately earthed.

Health hazards. Outbreaks of a disease known popularly as "shoemakers' paralysis" have appeared in a number of factories, presenting a clinical picture of a more or less severe form of paralysis. This paralysis is of the flaccid type, it is localised in the limbs (pelvic or thoracic) and gives rise to osteo-tendinous atrophy with areflexia and no alteration in superficial or deep sensitivity. Clinically, it is a syndrome resulting from functional inhibition or injury of the lower motor neurons of the voluntary motor system (pyramidal tract). The common outcome is the neurological regression with extensive proximo-distal functional recuperation. In spite of the seriousness of some cases, no fatalities have occurred (see *n*-HEXANE).

Other types of poisoning by aromatic solvents have resulted in anaemia, and, occasionally, leukaemia (see BENZENE). Good general ventilation and exhaust ventilation at the point of origin of the vapours should be provided to maintain concentrations well below maximum permissible levels. If these levels are observed, the fire risk will also be diminished.

Finishing machines produce dust, which should be removed from the atmosphere by exhaust ventilation. Some of the polishes, stains, colours and polichloroprene glues may carry a dermatitis risk. Good washing and sanitary facilities should be maintained and personal hygiene encouraged. Many of the machines produce considerable noise which should be eliminated at the source or damped by noise-control measures.

Prolonged work on nailing machines which produce high levels of vibration may produce "dead hand". It is advisable that spells of work at these machines should not exceed 4 h per day.

Pre-employment medical examinations are desirable to determine aptitude, evaluate work capacity and detect possible contagious diseases or affections which could be aggravated by the type of work to be performed. Periodical medical examinations are also desirable, especially for workers using toxic substances.

Appropriate personal protective equipment, overalls, head coverings, etc., should be provided.

CONRADI, F. L.

General:

The effects of technological progress on working conditions and working environment in the leather and footwear industry. Second Tripartite Technical Meeting for the Leather and Footwear Industry (Geneva, International Labour Office, 1979), 69 p.

Occupational diseases:

CIS 79-1971 "Polyneuropathy due to glues and solvents" (La polineuropatia da collanti e solventi). Mazzella di Bosco, M. *Rivista degli infortuni e delle malattie professionali* (Rome), Nov.-Dec. 1978, 117/6 (1 163-1 169). 33 ref. (In Italian)

CIS 79-1408 "Methyl ethyl ketone polyneuropathy in shoe factory workers". Dyro, F. M. *Clinical Toxicology* (New York), Oct.1978, 13/3 (371-376). 8 ref.

CIS 76-1091 "Benzene-induced acute aleukaemic leucosis, with diffuse leukaemic infiltration of the parenchyma" (Leucosi acuta aleucemica benzolica con diffusa infiltrazione parenchimale leucemica). Pollini,G.; Biscaldi,G. P.; Carosi, L. *Lavoro umano* (Naples), Jan.1976, 28/1 (10-16). Illus. 5 ref. (In Italian)

CIS 79-1646 "Allergic properties of p-tert-butylphenol-formaldehyde resins" (Zur Allergennatur der para-tert. Butylphenolformaldehydharze). Schubert, H.; Agatha, G. *Dermatosen in Beruf und Umwelt* (Aulendorf), 1979, 27/2 (49-52). 22 ref. (In German)

CIS 80-1979 "Adenocarcinoma of the nose and paranasal sinuses in shoemakers and woodworkers in the province of Florence, Italy (1963-77)". Cecchi, F.; Buiatti, E.; Kriebel, D.; Nastasi, L.; Santucci, M. *British Journal of Industrial Medicine* (London), Aug. 1980, 37/3 (222-225). 21 ref.

Safety and health:

CIS 76-2042 "Footwear manufacture" (Fabrication de la chaussure). Oeconomos J. *Cahiers de médecine interprofessionnelle* (Paris), 4th quarter 1975, 60 (33-38). (In French)

CIS 77-243 "Risk of electrostatic build-up during glue production" (Riziko elektrostatických nábojů při výrobě lepidel). Procházka, J.; Kvasnička, B.; Zeman, M. *Kožařství* (Prague), 1975, 25/12 (343-344). Illus. 10 ref. (In Czech)

Forestry industry

Forestry work comprises the establishment, maintenance, harvesting and transport of forest products. The most important of these operations is usually the felling and transport of wood, which is also called "logging".

Before the Second World War, large-scale forest activities were concentrated almost entirely in the temperate countries of the northern hemisphere, where coniferous forests are abundant, supplying large quantities of raw materials for industrial utilisation as lumber, pulp and paper. During the past two decades, however, the exploitation of tropical hardwood forests has also become increasingly important owing to the world's rapidly growing timber demand and because of the improved accessibility to natural forests.

Traditionally forest work requires heavy physical effort, especially during manual felling, cross-cutting and transportation. In the industrially advanced countries the forest worker's workload has now been lightened to some extent by mechanisation. However, in developing countries heavy manual work is still widespread and, for socioeconomical reasons, is bound to continue for some time.

Forestry in both temperate and tropical countries differs considerably from other industries. Its principal features are usually the isolated and frequently changing working places, the dispersion of relatively few men over wide areas, work in the open air subject to bad weather conditions and seasonal changes, and the need to work on difficult terrains.

HAZARDS AND THEIR PREVENTION

Statistics collected for the Tripartite Meeting of the Timber Industries held by the ILO in Geneva in 1974 indicate that accident frequency and severity rates in forestry are near those in coal mining.

In Canada the fatality rate in 1969 was 13.0 for mining, 10.5 for forestry, 4.5 for construction, 3.0 for transportation, 1.1 for manufacturing and 0.6 for agriculture. For logging workers alone the fatality rate was 14.1. According to statistics on accidents occurring in public forests in Switzerland during 1973-74, forestry ranks highest in accidents in comparison with all other activities.

Among workers employed for their whole working life in Swiss forestry 1 in 3 can expect partial permanent disability and 1 in 17 death, as a consequence of a forestry accident; the average forestry worker will experience 18 accidents. In the developing countries the accident frequency and severity rates are often several times higher. By far the largest number of forestry accidents occur during logging operations. In the State Forest Services of the Federal Republic of Germany during 1975 to 1977 their share ranged between 70 and 82%, for Sweden in 1975 it amounted to 81%, for Hungary in 1978 a figure of 85% was quoted. The remaining 15-30% occurred in silvicultural and miscellaneous activities.

In undertakings which have only limited silviculture interests, the predominance of logging accidents will be even greater, whereas where work is concentrated purely on forest establishment and protection the proportion of logging accidents will be negligible. Logging accidents if related to working time are by far more frequent and more severe than accidents in silvicultural operations.

Until recently the health hazards to which forestry workers were exposed were limited to the arduous nature of the work, the need to operate in all types of inclement weather conditions and the danger of contracting endemic diseases in tropical regions. However, the widespread use of machines such as chain saws and tractors has introduced the danger of disorders due to noise and vibrations.

Figure 1. Traditional tree felling by hand tools.

Logging accidents: struck by timber

Almost all severe logging accidents occur due to sliding, rolling or falling timber during felling, skidding and loading; the times at which accidents are most likely to occur are when a tree is toppled and when hanging trees are pulled down. Fatal accidents during logging work are usually caused by falling trees or branches or by moving logs. Stringent safe work practices must be applied if such accidents are to be prevented. Logging should always be supervised by a competent person fully instructed in safe working procedures. Logging workers should wear hard hats. While a tree is being felled, persons not involved in the work should remain at a minimum distance of twice the length of the tree being felled.

The felling direction should be controlled by a proper under-cut and back-cut, leaving a hinge of sufficient strength to guide the tree during its fall. Hanging trees should be released by tools such as cant hooks and picaroons, by a tractor winch or by block and tackle. They should not be released by throwing other trees against them, felling the tree in which they are caught or cutting off branches. Tree felling should be prohibited in stormy weather and during low visibility. On steep terrain workers should not work one above another. Logs should be adequately secured against sliding downhill especially during branching, barking and cross-cutting. During skidding and loading, workers should always be kept safe from swinging logs and should use a clear set of signals for the principal working phases. Tractors should be fitted with an adequate safety cab to protect the driver if the tractor turns over. When steel cables are used during skidding, they should be handled by workers wearing leather gloves with double lining.

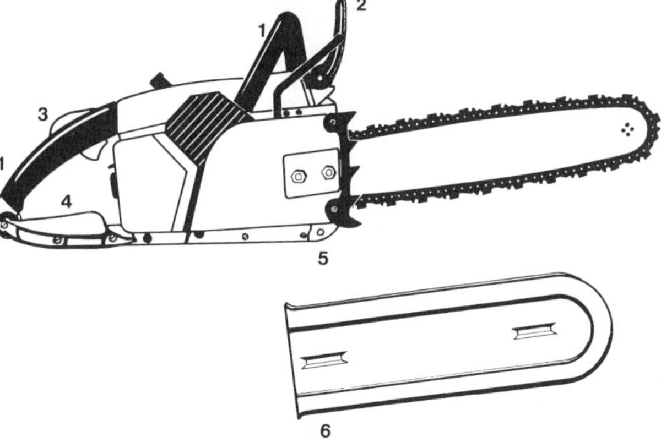

Figure 2. Chain saws for forestry work should be fitted with the following safety devices: 1. Anti-vibration handles. 2. Front handle guard combined with chain brake. 3. Throttle control latch. 4. Rear handle guard. 5. Chain catcher to restrict whip backwards if chain breaks. 6. Protective cover for blade during transport.

Logging accidents: cutting tools

In manual logging (figure 1), the axe is usually the most dangerous tool and may cause 20% or more of all logging accidents, especially during branching; wounds are usually of the open type and affect predominantly the lower limbs.

Axes should be properly maintained and fitted with adequate handles. The axeman should adopt a safe working posture and obtain a firm foothold; the axe should have a clear path free of branches, and should not be swung towards the legs (e.g., during branching, the tree should be kept as protection between the legs and the branches being cut). Workers should be well separated and wear safety boots with steel toe-caps and instep guards, and knee pads. During transport, axes should be fitted with safety guards. In mechanised logging the chain saw has taken the place of the axe as the most dangerous tool. In North-West Germany in 1966, 123 forest accidents were caused by axes and 38 by chain saws. In 1977 the axe accounted for 28 accidents and the chain saw for 68 accidents. In Finland accidents caused by chain saws amounted to 20% of all forest accidents in 1968 and to 39% in 1975. According to a study carried out in France in 1974, of all the chain saw accidents 42% occurred during branching, 28% during cross-cutting and 20% during felling. In the coniferous forests of the northern hemisphere the proportion of chain saw accidents during branching is even higher, whereas in tropical hardwood logging they are insignificant.

The most dangerous part of the saw is the moving chain. However, a large percentage of saw accidents are due to kickbacks when obstacles are encountered. To reduce the danger of kickbacks, safer chains and guide bars have been developed. Furthermore, chain saws should be fitted with front handle guards and a chain brake which stops the chain within 0.1 seconds in case of kickback (figure 2).

The professional chain saw operator should wear close-fitting clothing, hard helmet with face shield and ear protection, gloves, trousers with ballistic leg protection and safety boots (figure 3).

During sawing the saw should be held firmly at a point close to the body, and cutting with the return side or tip of the saw blade should be avoided as far as possible. During branching a clear working space should always be kept, tension being released if necessary when advancing and the saw being supported as far as possible on the tree. Workers operating saws should keep at least 2 m away from other persons. Saw chains should be maintained in good condition, and be discarded when worn.

During power saw work wedges made of wood, plastic or soft metal should be used. The motor should be shut off during transport, except for short distances from cut to cut, and a protective cover used for the saw blade if the chain saw is carried over longer distances.

Figure 3. Chain saw operator branching a coniferous tree, wearing tight-fitting clothing with leg protection, a safety helmet combined with eye and hearing protection and leather gloves.

Logging accidents: falls

Falls of persons may account for 20-30% of all forestry accidents, depending on terrain and weather conditions. Work on steep terrain or on slippery surfaces, for example after rain, snow or frost, is particularly dangerous. Statistics from Czechoslovakia for 1966 showed that 18.7% of all forestry accidents were due primarily to unfavourable terrain and 8% to unfavourable weather conditions.

Workers should be provided with footwear having non-slip soles, with calked boots, detachable calked soles or spikes when working on steep terrain during winter. If the ground becomes too slippery, operations should be suspended and work preferably so planned that difficult terrain is avoided during unfavourable climatic conditions. Where possible, operations such as branching, cross-cutting and barking should be partly or wholly transferred to the roadside, to lower landings or to a timber yard; work on difficult terrain can thus be restricted to the minimum.

Logging accidents: highly mechanised operations

In recent years, especially in North America, Scandinavia and the USSR, chain saws are being increasingly replaced under easier terrain conditions in the large-scale harvesting of rather uniform forest stands by complex harvesting machines for felling, branching and cross-cutting (e.g. feller-bunchers, processors). Such machines reduce the accident risk and the physical workload in logging. However, they tend to increase the mental workload, to create monotony and social isolation and to lead to the introduction of shift work which altogether may result in reduced job satisfaction. Job rotation and job enrichment are measures which help to overcome such effects.

Silviculture accidents

Falls are the most common cause of accidents in silviculture work. According to a Swiss investigation in 1973-74, 81.3% of all accidents in planting were caused by slipping and falling. Steep terrain and unfavourable weather played a major part in this respect. To reduce accidents through falls the same precautions apply as in logging. Cutting tools such as sickles, scythes and bill-hooks used for clearing are as dangerous as axes used in logging. Safe working techniques similar to those for axe work are essential. Powered brush cutters have to some extent replaced manual clearing tools. They can be most dangerous if safety regulations are not scrupulously respected. Work with brush cutters is a one-man operation. Cutting must stop immediately if any person comes nearer than 10 m.

Commuting accidents

Commuting between living accommodation and the worksite is an increasingly important hazard in all types of forestry work. This risk is becoming more pronounced with the tendency for fewer workers to live in forestry camps; the density of motor traffic and the distances covered make it increasingly important that forestry roads and vehicles should satisfy minimum safety standards for passenger transport. It is especially important that there should be sufficient visibility of approaching traffic and that speed limits should be established which correspond to the width, curves, gradients and surface of forest roads.

Noise and vibration

Portable forestry machines such as power chain saws and brush cutters and mobile machines such as skidders, forwarders and complex harvesting machines may be the source of dangerous levels of noise and vibration.

A large number of chain saw operators suffer from "white finger" or "white hand", a condition resulting from spastic constriction of the blood vessels in the hands, that becomes most pronounced during cold weather (see RAYNAUD'S PHENOMENON); the condition may be accompanied by pain or numbness in the wrist, elbow and shoulder.

Notably in Czechoslovakia, Hungary, Poland and Japan alarming numbers of forest workers showing advanced stages of such symptoms had to give up chain saw operation and received workmen's compensation. In some countries this resulted in regulations limiting chain saw work (e.g. during a single shift to a maximum of 2 h in Czechoslovakia and Japan).

Thanks to collaboration between researchers and manufacturers chain saws are now provided with efficient anti-vibration dampers. Their widespread application has led to a considerable reduction of the symptoms of "white finger". Medical examination of several thousand forest workers carried out in North Sweden showed that 38% of all chain saw operators had "white finger" in 1967 as against less than 10% in 1977. A more difficult problem is to keep low-frequency whole-body vibration of operators of cross-country vehicles such as skidders and forwarders within tolerable limits, in spite of the use of vibration-damping seats ergonomically adjusted to the operator's body measurements. An increased incidence of back trouble has been noted, besides pains in the neck and the shoulders. The best remedy available is regular rotation between jobs.

The noise level of most chain saws used in regular forestry work exceeds 100 dB(A). Chain saw operators are exposed to this noise level daily for about 2-5 h. To avoid hearing damage the use of ear protectors, preferably fitted to the hard helmet, is essential (figure 4). In many countries hearing losses of chain saw operators have been reported. An inquiry carried out in 1978 revealed that in nine European countries the percentage of chain saw operators annually claiming compensation for loss of hearing ranged from 0 (France, Czechoslovakia, Sweden) to 0.04 (Poland), 0.05 (Finland), 0.1 (Federal Republic of Germany), 0.3 (Netherlands) and 2.7 (Hungary). These differences can largely be explained by the use of different criteria in order to establish hearing losses as an occupational disease.

Figure 4. Brush saws for clearing work require eye and hearing protection. Other persons must keep a minimum distance of 10 m when the brush saw is running.

As a general rule the transfer of operations such as barking or cross-cutting from the field to the roadside, to lower landings, or to central conversion yards is a good way of reducing exposure to noise and vibration while at the same time diminishing accident risks. Care must be taken, however, not to reduce the job of tree fellers in the field to unrelieved chain-saw work nor to subject workers transferred to semi-stationary or stationary worksites to an excessive mental workload coupled with monotony and lack of communication. Here again job rotation offers the solution.

Climatic conditions

For work in the open air, especially under unfavourable climatic conditions, adequate working clothes are essential. In cold weather special clothing should be provided to keep the worker's body warm and dry, to allow evaporation of perspiration and to protect him against thorns, whipping branches and irritating plants without interfering with his freedom of movement in handling tools or operating machines.

In warmer climates only light clothing is required. Of particular importance is the use of adequate footwear by forestry workers in tropical countries, as can be seen from the high rate of foot injuries amongst workers working with bare feet.

Furthermore, shelters should be provided for use during rest periods and rain. In many countries mobile shelter huts are now widely used for forest workers if the working places are accessible by road; this has significantly reduced the high incidence of respiratory disease during rainy periods.

Physical workload

Forestry work may be extremely arduous. Manual logging will entail, on the average, energy expenditure levels of 2 000 kcal/day (8 370 kJ/day), i.e. 4 kcal/min (16.740 kJ/min), although levels up to 4 000 kcal/day (16 740 kJ/day) have been measured; tree planting is less arduous but will still entail expenditures of around 1 600 kcal/day (6 690 kJ/day) or 3.2 kcal/min (13.380 kJ/min).

The operation of portable machines such as power saws may require even greater energy expenditure than manual logging because of their considerable weight. In addition, the operation of portable machines tends to include a high proportion of static work, which is more tiring than dynamic work with manual tools, where muscles are alternately contracted and relaxed.

The energy cost of driving a self-propelled machine is only moderately high, although the driver may be subjected to a considerable mental stress and also to static work stress resulting in a relatively high pulse rate, for instance during grapple loading.

Because of the considerable physical stress of forest work, efforts should be made to organise operations ergonomically and, by increasing efficiency, reduce the physical workload. Distances to be covered on foot to the workplace or during work should be minimised. Heavy lifting and carrying should be mechanised wher-

Figure 5. Log loading on trucks in developing countries is still, to a large extent, done by primitive and dangerous methods. *(By courtesy of Mr. A. Berry, Dehra Dun, India.)*

ever possible (figure 5). Workers should be trained in kinetic lifting techniques and physiologically efficient working techniques, i.e. when hand sawing, to use long saw strokes propelled by a swinging movement of the whole body; and when power sawing, to rest the weight of the saw as far as possible on the tree.

Full working capacity can be achieved only if workers receive adequate nutrition. This is a special problem in many developing countries, where the population is traditionally used to low levels of energy expenditure and where food intake is insufficient for 8 h of heavy work per day. For moderately heavy forest work, the daily food intake should correspond to approximately 3 000 kcal (12 555 kJ) and, for heavy forest work, to 4 500 kcal (18 830 kJ) (aggregate net calorie requirements for work plus basic metabolism and activities outside work). In some tropical countries where the average weight of the male worker is lower, these values may be reduced to 2 700 kcal (11 300 kJ) and 4 000 kcal (16 740 kJ). Workers should be provided with adequate rest and meal breaks.

Medical supervision

Workers should receive a pre-employment medical examination during which it should be ensured that their physical work capacity is adequate for the job they will be required to do. Male workers under 18 years and over 50-55 years of age, and all women workers, should preferably be engaged in nursery and plantation work and not in logging operations. Periodical medical examinations are also desirable.

In recent years several countries have introduced special occupational health services for forestry workers. Their efforts have resulted in encouraging improvements of the ergonomic conditions of forestry work, notably in Sweden.

Especially in the developing countries, it is advisable for the larger logging companies to employ a full-time medical staff and provide general medical care, particularly where diseases such as ancylostomiasis, cholera, malaria, schistosomiasis and typhoid fever are endemic. Proper sanitation in logging camps is essential, and the provision of safe drinking water is especially important in order to control gastrointestinal infections, which are particularly frequent during rainy seasons when drinking water is derived from contaminated surface water. There should be an adequate first-aid organisation adapted to the difficulties of the worksite and to the injuries that must be expected. First-aid equipment and portable kits should be provided, and workers trained in first-aid procedures.

STREHLKE, B.

Guide to safety and health in forestry work (Geneva, International Labour Office, 1968), 223 p. 51 illus.

Safety and health in forestry work. ILO codes of practice (Geneva, International Labour Office, 1969), 165 p.

Conditions of work and life in the timber industry. Report II. Second Tripartite Technical Meeting for the Timber Industry (Geneva, International Labour Office, 1973), 115 p.

Safe design and use of chain saws. ILO codes of practice (Geneva, International Labour Office, 1978), 71 p.

Ergonomics in tropical agriculture and forestry. Proceedings of International Symposium, Netherlands, 1979 (Wageningen, Centre for Agricultural Publishing and Documentation, 1979), 135 p.

Chain saws in tropical forests. FAO/ILO training manual (Rome, Food and Agriculture Organisation, 1980), 96 p. 47 illus.

"Forestry fires" (Les incendies de forêts). *Protection civile internationale* (Geneva), Oct. 1981, 316 (2-7). (In French)

Forges

Forging is a technique used for the plastic deformation of metals and alloys, either hot or cold, by the application of multiple compressive forces (hammer drop or impact forging) (figure 1), or a single compressive force (press or roll forging) (figure 2). Hammer and drop forging are carried out on hot metal only, whereas press forging can also be done cold. Forging may be done manually or mechanically.

Figure 1. Hammer forging.

Figure 2. Press forging.

The processes involved include cutting material to size, heating, forging, heat treatment (figure 3), cleaning and inspection. In small, manual forges, all these operations are carried out in a confined space by a limited number of forgers and hammermen, exposed to the same harmful influences and occupational hazards; in large forging shops, the hazards vary with the work post.

Figure 3. Metal heating in a forging shop.

Working conditions

These vary from forge to forge but have a certain number of common features: physical effort of moderate intensity, hot and dry microclimate, noise and vibrations, air polluted by fumes and gases.

Energy expenditure. This varies between 3 and 6 kcal/min (12.56-25.12 kJ/min) with the type of forge and the level of mechanisation. In mechanised forges it has been found that 70% of the working day is devoted to forging with a mean energy expenditure of 4.5 kcal/min (18.84 kJ/min), 10% to subsidiary tasks with a mean expenditure of 2 kcal/min (8.37 kJ/min) and 20% to organisation activities and rest periods with a mean expenditure of 1 kcal/min (4.19 kJ/min). This gives a total expenditure of 1 750 kcal/day (7 324 kJ/day), which places forging at the upper limit of moderately heavy work. The neuropsychic effort involved is also at the limit between moderately and very heavy work.

Microclimate. A prominent feature is the high air temperature due to the hot furnace walls (100-180 °C) and charging doors (220-260 °C), hot metal (800-900 °C) and tools, and to the radiant heat from the floors on which hot metal has been placed (over 35 °C). In a poorly ventilated forge, the air temperature may rise to 40-43 °C on hot days. In large forges, radiant heat is of high intensity (0.8-5 kcal/cm^2·min or 3.35-20.93 kJ/cm^2·min and above) and exposure may last for 55% of the working day; in small forges, the heat intensity is lower (0.8-3 kcal/cm^2·min or 3.35-12.56 kJ/cm^2·min) but exposure is more prolonged (85% of the working day). Relative humidity is usually 15-50%, air speed varies between 0.4 and 1.0 m/s in summer, whereas in winter it may be as high as 6 m/s near the doors.

Workers are exposed simultaneously to high air temperatures and radiant heat; this results in a heat build-up in the body, which, when added to metabolic heat, may cause thermolysis disorders and pathological changes. Sweat excretion over 8 h of work may range between 1.5 and 5 l or even higher depending on the microclimate, physical exertion and degree of heat adaptation. The Belding and Hatch thermal stress index is usually 55-95 in smaller forges or at points distant from heat sources but may be as high as 150-190 for workposts close to furnaces or drop-hammers in large forges, and cause salt depletion and heat cramp. Exposure to changes in microclimate during the cold season may favour adaptation somewhat—but rapid and excessive changes may constitute a health hazard.

Atmospheric contamination. The workplace air may contain smoke, carbon monoxide, carbon dioxide, sulphur dioxide and perhaps acrolein—the concentration depending on the type of furnace fuel and its impurities and the efficiency of combustion, draught and ventilation.

Noise and vibration. Large forging hammers produce predominantly low-frequency noise and vibration—although there may be a definite high-frequency component. Sound pressure levels range between 95 and 115 dB. Exposure to vibration in forges may cause organic and functional disorders which may impair work capacity and affect safety.

HAZARDS

Accidents. These may result from shortcomings in tooling or environmental conditions or from unsafe work practices, poor work organisation, or the absence or inadequacy of safety devices and personal protective equipment.

Serious injuries may be caused at drop-forging hammers by unexpected falls of the tup, either during production or during die-changing. Many can be prevented by efficient props and catches to support the tup. Shrouded pedals prevent accidental operation of the machines.

Cold metal used to dislodge hot forgings may be projected violently from between tools, and strike operatives. Flying hot scale may damage eyes if eye protection is not worn. Protective, heat-resistant armlets, gaiters and aprons can prevent many burns to arms and legs, and safety footwear will protect the feet against crushing injuries.

Diseases. Although harmful factors in the work environment may produce pathological conditions that may appear of an occupationally induced nature, their normal effect is on general morbidity by favouring the appearance of certain diseases or by reducing the body's overall resistance. The most important nosological groups in the morbidity of forge workers are chronic rheumatism, burns, digestive disorders (enteritis), respiratory system disorders and inflammatory skin diseases. Disorders of occupational origin include hearing loss due to exposure to intense noise and vibration, and local vibration-induced disorders.

SAFETY AND HEALTH MEASURES

Good plant layout can do much to improve working conditions: furnaces and forging machines should be correctly positioned, congestion avoided, the flow of work rationalised, finished forgings removed from the shop, processes mechanised where possible, and good housekeeping practised. The furnace should have a good draught, and furnace gases, fumes and hot air should be led off from the shop. Sources of radiant heat and hot air should be thermally insulated by means of water curtains, reflective or insulating screens, etc. There should be efficient general ventilation of the forging shop (well designed natural ventilation will often suffice), local exhaust systems at the furnaces, and cold air showers at hot workplaces together with air curtains around the doors. Rest rooms protected against radiant heat should be provided, and these should be equipped with air and water showers, etc. Investigations have shown that in working conditions thermal comfort at forges is obtained between 19 and 24 °C air temperature, 30-50% relative humidity and an air velocity about 0.5 m/s. To avoid excessive stress an effective temperature of 27 °C should not be exceeded.

Sources of dangerous noise should be enclosed or surrounded with noise-absorbent panels, and workshops should be kept at a distance from residential areas. In order to suppress vibrations, installations should be placed on deep and massive foundations laid below the level of the building foundations and separated from all structural components.

Workers should receive a pre-employment and periodical medical examinations: they should be provided with personal protective equipment (in particular hearing protection). The work tempo should be reasonable and beverages should be distributed during work to ensure replacement of water, salts and vitamins lost due to sweating. Workshops should be equipped with adequate sanitary facilities and workers should receive a sound training in safety.

CADARIŬ, G.

CIS 77-835 "Results of an accident analysis at a drop forging shop" (Ergebnisse aus der Untersuchung des Unfallgeschehens in einer Gesenkschmiede). Skiba, R.; Kröger, U. *Stahl und Eisen* (Dusseldorf), July 1976, 96/15 (717-723). Illus. 8 ref. (In German and Spanish)

CIS 75-385 "Noise surveys and exposure tests in the metallurgic industry" (Lärmuntersuchungen und Expositionstests in der Metallindustrie). Schreiner, L.; Eder, H. *Zentralblatt für Arbeitsmedizin und Arbeitsschutz* (Heidelberg), May 1974, 24/5 (148-153). Illus. 8 ref. (In German)

CIS 76-831 *Occupational safety and health in forging shops and metal working with forging presses* (Tehnika bezopasnosti i promyšlennaja sanitarija v kuznečno-pressovyh cehah). Zlotnikov, S. L.; Kazakevič, P. I.; Mihajlova, V. L. (Moscow, Izdatel'stvo "Mašinostroenie", 1974), 296 p. Illus. 9 ref. (In Russian)

Formaldehyde and derivatives

Formaldehyde (HCHO)

FORMIC ALDEHYDE; METHANAL

m.w.	30
sp.gr.	1.0
m.p.	−92 °C
b.p.	−21 °C
f.p.	50 °C (15% methanol-free)
e.l.	7.0-73%
i.t.	430 °C

very soluble in water, alcohol and ether

a colourless gas at ordinary temperatures, with a pungent odour which is perceptible at concentrations even lower than 1 ppm.

TWA OSHA	3 ppm, 5 ppm ceil, 10 ppm peak
NIOSH	0.8 ppm 1.2 mg/m³ 30 min ceil
TLV ACGIH	2 ppm 3 mg/m³ ceil
IDLH	100 ppm
MAC USSR	0.5 mg/m³

Trioxymethylene ($(CH_2O)_3$)

METAFORMALDEHYDE; TRIOXANE

m.w.	90.08
sp.gr.	1.17
m.p.	61 °C
b.p.	114.5 °C
v.d.	3.1
v.p.	13 mmHg (1.73·10³ Pa) at 25 °C
f.p.	44 °C
e.l.	3.6-28.7%
i.t.	410 °C

soluble in water, alcohol and ether

colourless, needle-shaped crystals with a faint odour of ethyl alcohol.

A formaldehyde polymer with a cyclic structure.

Tetraoxymethylene ($(CH_2O)_4$)

A formaldehyde polymer which has physical and chemical properties similar to trioxymethylene.

Paraformaldehyde ($(CH_2O)n.H_2O$) (where n = 8 to 100)

PARAFORM; TRIFORMOL; POLYOXYMETHYLENE

This is the most common commercial polymer obtained from formaldehyde and consists of a mixture of products with different degrees of polymerisation. It is a white, amorphous powder soluble in water and insoluble in most organic solvents. It readily depolymerises, liberating formaldehyde, especially when heated or when in the presence of acids or alkalis.

Production. Formaldehyde is obtained industrially by the oxidation of methyl alcohol in the presence of a catalyst:

$$CH_2OH + \frac{Cu}{[O]} = HCHO + H_2O$$

The catalyst exercises a dehydrogenating action on the methyl alcohol vapours, and the liberated hydrogen burns with steam. The process is entirely exothermic. Copper or silver in dense, spirally wound netting is used as a catalyst. Catalysts with a molybdate or vanadate base can also be used.

Trioxymethylene is obtained by distillation of a 60% formaldehyde solution containing 2% sulphuric acid:

$3CH_2O \rightarrow$ (cyclic trimer: H_2C, CH_2, CH_2 linked by O atoms)

Tetraoxymethylene is obtained by heating polyoxymethylenes under the action of acetic acid.

Paraformaldehyde is obtained by concentrating aqueous solutions of formaldehyde.

Uses. Formaldehyde polymerises readily in both liquid and solid state. This polymerisation process is delayed by the presence of water and, consequently, commercial formaldehyde preparations (known as formalin or formol) are aqueous solutions containing 37-50% formaldehyde by weight; 10-15% methyl alcohol is also added to these aqueous solutions as a polymerisation inhibitor.

Formaldehyde is widely and increasingly used in plastics production (e.g. urea-formaldehyde, phenol-formaldehyde, melamine-formaldehyde resins); it is also used in the photographic industry, in dyeing, in the rubber, artificial silk and explosives industries, tanning, etc. Formaldehyde is also a powerful antiseptic, germicide, fungicide and preservative.

The polymethylenes are used in the form of pastilles for atmospheric disinfection, as a non-aqueous source of formaldehyde. Industrially they are used in the manufacture of resins, adhesives, fungicides and bactericides.

HAZARDS

Formaldehyde and its solutions are flammable and the vapours may form explosive concentrations in air; however, the fire hazard of aqueous formaldehyde solutions is reduced by the presence of water.

Formaldehyde is toxic by ingestion and inhalation and it may also cause skin lesions. [It is metabolised into formic acid.]

[Trioxane and paraformaldehyde are flammable substances and on heating may give off flammable gases and explode. They react violently with strong oxidisers and decompose. Finely dispersed particles of formaldehyde polymers are explosive and easily carry electrostatic charges.]

Health hazards. Relatively large quantities of ingested formaldehyde can be tolerated by man: consumption of 22-200 mg per day for 13 consecutive weeks did not cause any significant toxic effects; higher doses cause irritation of the digestive system with vomiting and dizziness; ingestion of very large quantities causes convulsions and possibly death.

Exposure to low atmospheric concentrations of formaldehyde causes irritation, especially of the eyes and respiratory tract. Due to the solubility of formaldehyde in water, the irritant effect is limited to the initial section of the respiratory tract. A concentration of 2-3 ppm causes slight formication of the eyes, nose and pharynx; at 4-5 ppm, discomfort rapidly increases; 10 ppm is tolerated with difficulty even briefly; between 10 and 20 ppm, there is severe difficulty in breathing, burning of the eyes, nose and trachea, intense lacrimation and severe cough. Exposure to 50-100 ppm produces a feeling of restricted chest, headache, palpitations and, in extreme cases, death due to oedema or spasm of the glottis. Eye burns can also be produced. Asthmatic symptoms may occur due to allergic sensitivity to formaldehyde, even at low concentrations as well as urticaria. [Kidney injury may occur in excessive and repeated exposure.]

[Formaldehyde reacts with hydrogen chloride and it was reported that such reaction in humid air could yield a non-negligible amount of bischloromethyl ether, a dangerous carcinogen. Further investigations have shown that at ambient temperature and humidity, even at very high concentrations, formaldehyde and hydrogen chloride do not form bischloromethyl ether at the detection limit of 0.1 ppb. However NIOSH has recommended that formaldehyde be treated as a potential occupational carcinogen because it has shown mutagenic activity in several test systems and has induced nasal cancer in rats and mice, particularly in the presence of hydrochloric acid vapours.]

There have been reports of both inflammatory and allergic dermatitis including nail dystrophy due to direct contact with solutions, solids or resins containing free formaldehyde. The inflammatory form follows even short-term contact with large quantities of formaldehyde; the allergic form is usually the result of allergic sensitisation, and may follow contact with only very small quantities.

The toxicity of polymerised formaldehyde is similar to that of the monomer since heating produces depolymerisation.

SAFETY AND HEALTH MEASURES

Formaldehyde should be stored in a cool, well ventilated place, away from open flames or areas of fire hazard. Containers must be securely closed and clearly labelled. In the event of spillage, cleaning-up operations should be carried out by workers equipped with suitable respiratory protective equipment. In the event of very high atmospheric concentrations, air-line or self-contained air-supply-type respirators should be employed. Fires should be fought with carbon dioxide or carbon tetrachloride fire extinguishers.

Personnel in contact with solid material containing free formaldehyde or with concentrated solutions of formaldehyde, or exposed to formaldehyde vapours, should be protected by suitable exhaust or general ventilation and be supplied with hand and arm protection and respiratory protective equipment; barrier creams may also provide valuable skin protection.

During periodical medical examination of formaldehyde workers, a careful watch should be kept for signs of sensitivity which might give rise to asthmatic attacks.

[The determination of formic acid in blood and urine has been suggested for the biological monitoring of workers regularly exposed to formaldehyde.]

Treatment. Persons who have ingested formaldehyde should be made to swallow milk or water containing ammonium acetate, and vomiting should be induced. This should be followed by gastric lavage with a weak (0.1%) ammonia solution which will convert the formaldehyde to relatively inert pentamethylenetetramine (10 l of 0.1% ammonia solution converts 4 g of formaldehyde). Gastric lavage is warranted only in the first 15 min following ingestion.

Contaminated skin should be washed with soap and water; splashes in the eye should be treated by irrigation for 15 min.

ZURLO, N.

General:

Formaldehyde—An assessment of its health effects. Committee on Toxicology, National Research Council (Washington, DC, National Academy of Sciences, 1980), 38 p. 131 ref.

CIS 78-1630 *Formaldehyde and its polymers (medical problems)—Chief uses of formaldehyde in industry, hazards and prevention* (L'aldéhyde formique et ses polymères (problèmes médicaux)—Principaux emplois de l'aldéhyde formique dans l'industrie, ses risques, sa prévention). Frimat, P.; Furon, D.; Haguenoer, J. M.; Ascher, R. (Caisse régionale d'assurance maladie du Nord de la France, 11 boulevard Vauban, Lille) (11 Oct. 1977), 7 and 34 p. 14 ref. Illus. (In French)

Contact allergy:

"A note on formic aldehyde-contact allergy" (Aktuelles zur Formaldehyd-Kontaktallergie). Kleinhaus, D.; Dayss, U. *Dermatosen in Beruf und Umwelt* (Aulendorf), 1980, 28/4 (101-103). 5 ref. (In German)

Carcinogenic risk:

CIS 74-750 "Investigations of the formation of bis-chloromethyl ether in simulated hydrogen chloride-formaldehyde atmospheric environments". Kallos, G. J.; Solomon, R. A. *American Industrial Hygiene Association Journal* (Akron, Ohio), Nov. 1973, 34/11 (469-473). Illus. 5 ref.

CIS 81-1027 *Formaldehyde: Evidence of carcinogenicity.* DHHS (NIOSH) publication No. 81-111 (National Institute for Occupational Safety and Health, 4676 Columbia Parkway, Cincinnati) (23 Dec. 1980), 15 p. 27 ref.

Biological monitoring:

CIS 77-725 "Blood and urine formaldehyde and formic acid levels in man after formaldehyde exposure" (Der Formaldehyd- und Ameisensäurespiegel im Blut und Urin beim Menschen nach Formaldehydexposition). Einbrodt, H. J.; Prajsnar, D.; Erpenbeck, J. *Zentralblatt für Arbeitsmedizin, Arbeitsschutz und Prophylaxe* (Heidelberg), Aug. 1976, 26/8 (154-158). 10 ref. (In German)

Exposure limit:

CIS 77-1076 *Criteria for a recommended standard–Occupational exposure to formaldehyde.* DHEW (NIOSH) publication No. 77-126 (National Institute for Occupational Safety and Health, 4676 Columbia Parkway, Cincinnati) (Dec. 1976), 165 p. 222 ref.

Formaldehyde (Formaldehyd). Gesundheitsschädliche Arbeitsstoffe. Toxikologisch-arbeitsmedizinische Begründungen von MAK Werten (Weinheim, Verlag Chemie, 1971), 6 p. 35 ref. (In German)

Foundries

Founding consists in pouring molten metal into a mould which is made to the outside shape of a pattern of the article required and contains, in some cases, a core which will determine the dimensions of any internal cavity. Foundry work comprises: making the pattern; making and assembling the mould; melting and refining the metal; pouring the metal into the mould; and, finally, removing all adherent sand and superfluous metal from the finished casting. The basic principles of foundry technology have changed little in thousands of years. However, in recent times the founding process has been subject to a considerable amount of development and modernisation. Processes have become more mechanised and automatic, wooden patterns have been increasingly replaced by metal and plastic ones, new substances have been developed for the many new processes of producing cores and moulds and a very wide range of alloys are used to supplement the output of base metal castings.

Foundry metals and materials. The traditional cast metals are iron, steel, brass and bronze. Developments in recent years have widened the scope considerably, so that cast metals and alloys may now contain aluminium, titanium, chromium, nickel and magnesium and even toxic metals such as beryllium, cadmium and thorium. Moulds used to be made from silica sand bound with clay and cores were traditionally produced by baking silica sand bound with vegetable oils or natural sugars to achieve the necessary setting. Modern advances in founding technology have led to the development of a wide range of new techniques to produce moulds and cores.

Foundry processes (figure 1). On the basis of the designer's drawings, a pattern is constructed which conforms to the external shape of the finished casting. In the same way, a corebox is made that will produce suitable cores to dictate the internal configuration of the

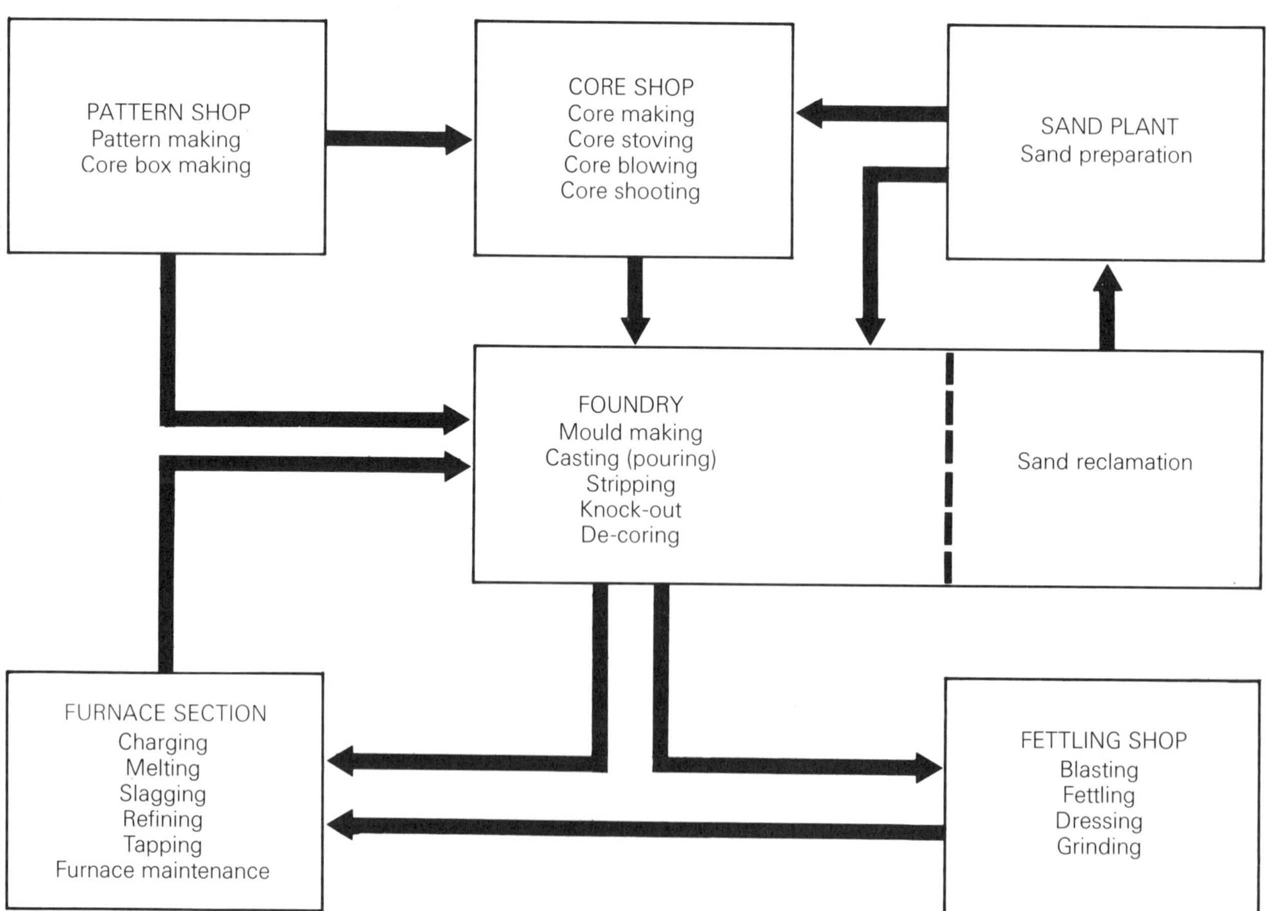

Figure 1. The flow of operations in a typical iron foundry.

final article. Sand casting is still the most widely used method, but a great many other techniques are available to the founder. These include: permanent mould casting using moulds of iron or steel; die casting, in which the molten metal, often a light alloy, is forced into a metal mould under pressures of 70-7 000 kgf/cm^2; investment casting, where a wax pattern is made of each casting to be produced and is covered with refractory which will form the mould into which the metal is poured.

The metals or alloys are melted and prepared in a furnace which may be of the cupola, rotary, reverberatory, crucible, electric arc, channel, or coreless induction type. After relevant metallurgical or chemical analyses have been performed to test the quality of the molten metal, it is poured, either via a ladle or directly, into the assembled mould. When the metal has cooled the mould and, if present, core material are removed (stripping or knockout) and the casting is cleaned and dressed (despruing, shot-blasting or hydro-blasting and other abrasive techniques). Certain castings may require welding, heat treatment or painting before the finished article will meet the specifications of the buyer.

HAZARDS AND THEIR PREVENTION

There are certain hazards common to the majority of foundries, irrespective of the particular casting process employed. The danger arising from the presence of hot metal is one such example. In addition there are hazards specific to a particular foundry process, for example the use of magnesium presents certain flare risks not encountered in other metal founding industries. The main emphasis of this article will be on iron foundries as this industry can be said to represent most of the typical foundry hazards. For simplicity, an iron foundry can be presumed to comprise the following six sections: *(a)* metal melting and pouring; *(b)* moulding; *(c)* coremaking; *(d)* knock-out; *(e)* fettling; *(f)* miscellaneous. In many foundries, almost any of these processes may be carried out simultaneously or consecutively in the same workshop area. Figure 1 illustrates the general sequence of operations in an iron foundry.

Melting and pouring

The iron founding industry still relies heavily on the cupola furnace to satisfy its metal melting and refining requirements. The cupola is essentially a tall, vertical furnace, into the top of which is charged coke, pig iron, limestone and scrap iron or steel. All these materials are continuously fed into the cupola during its operation and must be stored close at hand, usually in compounds or bins in the yard adjacent to the charging machinery. Tidiness and efficient supervision of the stacks of raw materials is essential to minimise the risk of injury from slippages of heavy objects. To reduce the scrap metal to manageable sizes for charging into the cupola, and also to fill the charging hoppers themselves, cranes with large electromagnets or heavy weights are often used. Strict rules are necessary to reduce the risk of injury from flying fragments. The crane cab should be well protected and the operators properly instructed.

Men handling raw materials should wear hand leathers and protective boots. Careless charging can overfill the hopper and can cause a dangerous spillage. If the charging process is found to be too noisy, the noise of metal-on-metal impact can be reduced by fitting rubber noise-dampening liners to storage skips and bins. The charging platform is necessarily above ground level and can present a hazard unless level, with a non-slip surface and strong rails around it and any floor openings.

By its very nature, the cupola generates large quantities of carbon monoxide, which may leak from the charging doors and be blown back by local eddy currents. Carbon monoxide is an insidious poison because it is invisible, odourless and can quickly produce toxic effects if ambient levels are high enough. For these reasons, men working on the charging platform or surrounding catwalks should be well trained to recognise the dangers of carbon monoxide poisoning. Breathing apparatus and resuscitation equipment should be kept and maintained in readiness and operators should be instructed in its use. When emergency work is carried out a system of working in pairs or under observation should be developed and enforced.

Cupolas are usually sited in pairs and operated alternately, the one being in use while the other is being repaired. These repairs necessarily involve the presence of employees inside the cupola shell itself, as refractory linings have frequently to be mended or renewed. Positive precautions should be taken to prevent the discharge of material through the charging doors at such times. To protect the men from falling objects, they should wear safety helmets and, if working at a height, safety belts as well.

Workers employed in tapping cupolas, i.e. in transferring molten metal from the cupola well to a ladle, must observe rigorous personal protection measures. Goggles and protective clothing are essential. The eye protectors should be of such a standard that they can resist both high velocity impact and molten metal. The most extreme caution should be exercised in order to prevent remaining molten slag (the unwanted debris removed from the melt with the aid of the limestone additives) and metal from coming into contact with water for this can result in an explosive reaction. It is the duty of tappers and foremen to ensure that any person not involved in the operation of the cupola remains outside the danger area, which is delineated by a radius of about 4 m from the cupola spout. Under the British Iron and Steel Foundries Regulations of 1953, delineation of a non-authorised no-entry zone is a statutory requirement.

When the cupola run is at an end the cupola bottom is dropped to remove the unwanted slag and other material still inside the shell before employees can carry out the routine refractory maintenance. Dropping the cupola bottom is a skilled and dangerous operation requiring trained supervision. A refractory floor or layer of dry sand on which to drop the debris is essential. If a problem occurs, such as jammed cupola bottom doors, great caution must be exercised to avoid risks of burns to workers from the hot metal and slag.

Where white hot metal is visible, there can be a danger to workers' eyes due to the emission of infrared and ultraviolet radiation, extensive exposure to which can cause cataracts.

The metal and pouring sections of the foundry are those in which personal protection is of the greatest importance and too much care cannot be taken to ensure that eye protection, aprons, gaiters or spats and boots are provided and that adequate instruction and enforcement of use are maintained. In all areas where molten metal is being manipulated, the encouragement of high standards of good housekeeping may be the most fruitful of safety measures.

When metal is poured into the mould, quantities of visible fume are evolved as the coal dust, water and chemical binders from the mould and cores are evaporated or decomposed. Poor sealing and weighting of boxes can cause a run out of molten metal during pouring. Where large ladles are slung from cranes or overhead conveyors, positive ladle control devices should be employed to ensure that spillage of metal cannot occur if the operator releases his hold.

Moulding

The process of moulding in the iron founding industry invariably involves sand and other additives. Sand has been part of the iron founding process for centuries and its hazards are well documented, but the number and nature of the additives recently developed are many and varied and the risks involved in their use have been less well investigated. These additives are generally used to increase the strength of the sand, by acting as binders. The traditional green sand mould is made from natural or silica sand and coaldust. The size and shape of the sand grains and other additives used impart different strengths and mould properties and thus deciding which types of sand to use for a particular casting is of great importance. Apart from the use of clay as a binder, there are three general methods of mould production: thermosetting, cold self-setting and gas-hardened.

The thermosetting process used in iron foundries is that of shell moulding and coremaking. Here, a metal pattern heated, usually by gas, to around 450 °C is clamped to a hopper, or dump box, containing the sand coated with a phenolic, thermosetting resin. The unit is inverted, so that the resin sand falls onto the upturned pattern, and held for about 30 s to allow bonding to take place. After a further short curing time the shell has hardened considerably and can be pushed clear of the pattern plate by means of inbuilt ejector pins. Shell sand may also be blown into the coreboxes. Clearly there is a risk of exposure to the thermal decomposition products when the phenolic resin is in contact with the heated pattern plate and there may also be volatile organic compounds present that can evaporate during the setting and curing processes. It is essential to avoid the contact of phenol or phenolic resins with the skin or eyes because they are irritants or sensitisers that can lead to dermatitis. Copious washing with hot water will help to alleviate the problem. If any of the resin is ingested, medical attention should be sought immediately.

In the cold-setting category of moulding processes several types of hardening system are presently in use. These include: acid-catalysed furanes and phenolics; alkyd and phenolic isocyanates; Fascold; self-set silicates; Inoset; cement sand and fluid or castable sand. All these systems are similar in that they do not require external heating to effect the setting reaction. The isocyanates employed in binders are normally based upon methylene bisphenyl isocyanate (MBI) which, if inhaled, can act as a respiratory irritant or sensitiser, producing asthmatic symptoms [for which reason it is the subject of strict legislation in most countries]. It is advisable to use barrier creams and protective goggles or spectacles when handling or using these compounds. Local exhaust ventilation is recommended when dealing with sand mixtures containing resins, especially where the sand is hot. The isocyanates themselves should be carefully stored in sealed containers in dry conditions at a temperature between 10 and 30 °C. Any empty storage vessels should be filled and soaked for 24 h with a 5% sodium carbonate solution in order to neutralise any residual chemical left in the drum. Most general housekeeping principles should be strictly applied to resin moulding processes using the phenolic resins and alkyd oils but the greatest caution of all should be exercised when handling the catalysts used as setting agents. The catalysis for the phenol and oil isocyanate resins are usually aromatic amines based on pyridine compounds, which are liquids with a pungent smell. They can cause severe skin irritation and renal and hepatic damage, and can also affect the central nervous system. These compounds are supplied either as separate additives (3-part binder) or ready mixed with the oil materials, and local exhaust ventilation should be well maintained at the mixing, moulding, casting and knock-out stages. For certain other no-bake processes the catalysis used are phosphoric or various sulphonic acids which are also toxic, and accidents during transport or usage should be adequately guarded against.

The last category of moulding processes is the gas-hardened type. Essentially these comprise the CO_2-silicate and the Isocure (or "Ashland") processes. The CO_2-silicate process has been in use since the 1950s, and many variations of it have been developed. This process has generally been used for the production of medium to large moulds and cores (see next section). The type of sodium silicate binder used in this method is a range of chemicals formed by the reactions between sodium oxide and silica dioxide. They are usually modified by adding such substances as molasses as breakdown agents. Sodium silicate is an alkaline substance, and can be harmful if it comes into contact with the skin or eyes or is ingested. It is advisable to provide an emergency shower close to areas where large quantities of sodium silicate are handled, and barrier cream should always be worn. A readily available eye-wash bottle in any foundry area where silicates are used can be very effective in remedying what might otherwise be a very serious accident. The carbon dioxide can be supplied as a solid, liquid or gas. Where it is supplied in cylinders or pressure tanks a great many housekeeping precautions should be taken, such as cylinder storage, valve maintenance, handling, etc.; but there is also the risk from the gas itself to be guarded against.

Whatever the type of moulding process employed, where sand is used there may be a risk from inhaling dust. At the mould-making stage, however, the sand is usually either damp or mixed with liquid resin, and is therefore less likely to be a significant source of respirable dust. To promote the ready removal of the pattern from the mould, a parting agent is sometimes added. The substance most commonly used is talc, the respirable fraction of which can give rise to a risk of talcosis, a type of pneumoconiosis. The use of parting agents is more widespread where hand moulding is employed; in the larger, more automatic processes it is rarely seen.

In order to achieve a casting within a finer surface finish, chemicals are sometimes sprayed on to the mould surface, suspended or dissolved in isopropyl alcohol which is then burned off to leave the compound, usually a type of graphite, coating the mould. This involves an immediate fire risk and all employees involved in applying these coatings should be provided with fire-retardant protective clothing and hand protection, as organic solvents can also cause dermatitis. They should be applied in a ventilated booth to prevent the organic vapours from escaping into the workplace. Strict precautions should also be observed to ensure that the isopropyl alcohol is stored and used with safety. It should be transferred to a small vessel for immediate use and the larger storage vessels should be kept well away from the burning-off process. Smoking should be prohibited in the immediate vicinity of this area of the iron foundry.

Mould making can involve the manipulation of large and cumbersome objects. The moulds themselves are heavy, as are the moulding boxes. They are often lifted, moved and stacked by hand. Employees should be aware that lifting heavy objects can exert a great strain on the back and care should be taken to ensure that they do not attempt to lift objects too heavy to be carried safely.

Coremaking

Cores are made, and inserted into the mould, in order to determine the internal configuration of a hollow casting. The core must be strong enough to withstand the casting process but at the same time must not be so strong as to resist removal from the casting during the knocking-out stage. Traditional core mixtures comprise sand and binders, to give the necessary strength, such as linseed oil, molasses or dextrin; these are dried in an oven and produce a core which, although initially firm and dry, becomes fragile when exposed to molten metal and then breaks down, facilitating core removal during knockout. Where the cores are baked in an oven there will be a risk from the possible evolution of harmful fumes and a suitable, well maintained chimney system above the core oven should be provided. Normally convection currents within the oven will be sufficient to ensure satisfactory removal of fumes. Placing the finished cores on racks to cool, after removal from the oven, can give rise to a small amount of fume but the hazard is minor, although in some cases small amounts of acrolein in the fume may be a considerable nuisance. Cores may be treated with a "flare-off coating" to improve the surface finish of the casting which calls for the same precautions as in the case of moulds.

In finishing cores, it may be necessary to use a hand file. The dust produced in this way is too large to constitute a risk of pneumoconiosis. Core filers should wear hand protection to prevent the possibility of abrasion.

Many of the processes used in mould production are also applied to the manufacture of cores. Again, these can be divided into the three categories: thermosetting, self-setting and gas-hardened. The materials used as, and with, binders in these processes can include phenol formaldehyde, urea formaldehyde, furfuryl alcohol, phenol formaldehyde furfuryl alcohol, phenol formaldehyde/urea formaldehyde, polyurethane, triethylamine, organic sulphonic acids and, in the recently developed SO_2-process, methyl ethyl ketone peroxide and a host of other organic compounds. The cores produced by the thermosetting processes (air-set, shell moulding, hot-box and traditional core binding) can generate a multitude of thermal degradation products and care should be taken to avoid too much exposure to these largely uninvestigated compounds. The same precautions should be taken in dealing with the cold self-setting and gas-hardened systems of coremaking as in the case of mould making, as many of the hazards are identical. One example of a process used to manufacture both cores and moulds with equal facility is the Isocure, or Ashland, process. This is a gas-setting system in which a resin, frequently polyurethane, is mixed with a diisocyanate (MDI) and then gassed with an amine, usually either triethylamine or dimethyl ethylamine to effect the crosslinking, setting reaction. The amines, often sold in drums, are colourless liquids with a strong smell of ammonia. The vapours are heavier than air and will thus tend to accumulate near the ground. There is a very real risk of fire or explosion and extreme care should be taken, especially where the material is stored in bulk. The physiological action of these amines is primarily directed towards the central nervous system, where they can cause convulsions, paralysis and, occasionally, death. Should some of the amines come into contact with the eyes or skin, first-aid measures should include washing with copious quantities of water for at least 15 min and immediate medical attention. In the Isocure process the amine is applied as a vapour. This renders it more dangerous, as the risks of leakage are higher. Great care should be taken at all times when handling this material and suitable extraction equipment should be installed to remove any fumes from the working areas.

Knock-out

After the molten metal has been poured into the mould, previously assembled with cores if necessary, and allowed to cool, the rough casting must be removed. This is a noisy process, often giving rise to an equivalent continuous sound level well in excess of the recommended 90 dBA L_{eq} over an 8-h working day. Hearing protectors should be provided if it is not practicable to reduce the noise output. The main bulk of the mould is separated from the casting usually by jarring impact. Frequently the moulding box, mould and casting are dropped onto a vibrating grid. The sudden impact and continued shaking dislodges much of the sand, which then drops through the grid into a hopper or onto a conveyor where it can be subject to magnetic separators and recycled for milling, treatment and re-use, or merely dumped. The casting is then removed and transferred to the next stage of the knock-out operation. The sand has been in contact with molten metal at temperatures around 1 500 °C. It is therefore very dry and has a much greater tendency to give rise to dust. If resins or oils have been used, either in the moulds or in the cores, then the thermal breakdown products can still be present at the knock-out stage. The metal itself is still very hot, as is the sand. Eye protection is provided and used whenever there is a risk of injury to eyes from hot sand. The usual housekeeping practices should be strictly adhered to in this part of the foundry. The presence of large pieces of hot metal can constitute a great hazard to the unwary employee. It is well known that, of the accidents that occur in the iron foundry, the commonest consist of falling onto, off, or against objects commonly seen lying around a foundry. The problem is exacerbated if the object happens to be metal, jagged and hot, or if there is an abundance of dry, slippery sand lying around.

Fettling

This is the process of stripping away unwanted metal to leave the finished cast product and can be defined to include the processes of grinding, shotblasting, water or water/sand blasting, tumbling, despruing and chipping.

The first of these dressing operations is despruing. As much as half of the metal cast in the mould is superfluous to the final casting, but vital to the casting process itself. The mould must obviously be filled with metal to ensure a complete cast object and, to make sure that the right regions of the mould solidify in the right order, a feeder is included in the mould. This essentially is a reservoir or cavity, which can be relatively large, to ensure that a sufficient head of pressure will be present to produce the correct metal distribution. The mould cavity and any associated feeders are filled via the sprue, which is the vertical channel of metal between the pouring basin and the rest of the casting. Usually the sprue can be removed during the knock-out stage, but sometimes the nature of the casting dictates that this must be carried out as a separate stage of the fettling or dressing operation. When this is the case, the despruing is done by hand, usually by knocking the casting with a hammer. To reduce noise the metal hammers can be replaced by rubber covered ones and the conveyors lined with the same, noise-damping rubber. There may be a danger from fragments of hot metal being thrown off, in which case eye protection must again be issued, maintained and used. Care should be taken when discarding the detached sprues, for they can pose a risk as they may well have jagged edges. They should normally be returned to the charging region of the melting plant and should not be permitted to accumulate at the despruing section of the foundry. After despruing (but sometimes before) most castings go through a cleaning process to remove unwanted mould materials and perhaps to improve the surface finish. An old but still practised cleaning process uses rumbling barrels. Castings are put into barrels sometimes with small pieces of cast iron to improve cleaning. The cylindrical barrel is rotated so that castings are cleaned by attrition. An excellent burnished surface can be achieved but obviously the method can only be used for small castings not liable to damage. The barrel may be rotated for 30 minutes or so and generates high noise levels. Enclosures may be necessary, which can also provide local exhaust ventilation. Again there is the obvious risk from moving many pieces of metal. If the tumbler is charged by hand, then protection should be worn, as even small pieces of metal can do a surprising amount of damage to the body.

Water or water/sand or pressure shot blasting may be used to remove adherent sand by subjecting the casting to a high pressure stream of either water or iron or steel shot. Sand blasting used to be used for this purpose, but has now been banned in the United Kingdom because of the pneumoconiosis risk as, during the abrasion process, the sand particles become finer and finer and thus the respirable fraction continually increases. The water or shot is discharged through a gun and can clearly present a risk to personnel if not handled correctly. All blasting enclosures—and it is vital that blasting is always carried out in an isolated, enclosed space—should be inspected at regular intervals to ensure that the dust extraction system is functioning as it should, and that there are no leaks through which shot or water could escape into the foundry. Blasters' helmets, which must be of a type approved by the Chief Inspector of Factories, should be carefully maintained. It is advisable to post a notice on the door to the booth, warning employees that blasting is under way and that unauthorised entry is prohibited. In certain circumstances delay bolts linked to the blast drive motor can be fitted to the doors, making it impossible to open the doors until blasting has ceased.

Most castings are cleaned by airless shotblasting. The shot is centrifugally propelled at the casting and no operator is required inside the unit. Extraction is standard practice in these units and only when there is a breakdown or deterioration of the shot-blast cabinet and/or the fan and collector is there a dust problem. (See also FETTLING.)

Miscellaneous

Pattern making. This is the process of forming, traditionally in wood but nowadays often in metal, plastic or foam, a likeness to the final casting. It will not be identical because allowances have to be made for the shrinkage of the metal in the mould. Pattern making is a highly skilled trade. The pattern maker is a craftsman who has to translate the two-dimensional design plans to a three-dimensional object. The traditional wooden patterns are made in standard workshops containing hand tools and electric cutting and planing equipment. The pattern shop is the only part of the foundry to be subject to statutory requirements concerning the exposure of workers to occupational noise. Here all reasonably practicable measures should be taken to reduce the noise to the greatest extent possible, and suitable ear protectors must be provided. It is important that the employees are aware of the advantages of using such protection.

Power driven wood cutting and finishing machines are obvious sources of danger and often suitable guards cannot be fitted without preventing the machine from functioning at all. Employees must be well versed in normal operating procedure and should also be instructed in the hazards inherent in the work.

Where wood is sawn there can be a nuisance from dust. An efficient ventilation system should be fitted to eliminate wood dust from the pattern shop atmosphere. In the past, in certain industries using hard woods, the heat from the cutting blades has caused thermal degradation of natural resins within the wood, associated with specific types of cancer. This has not been a problem in the founding industry.

Cranes and slingers. Cranes are used during several stages of the iron founding operation, either for cupola charging or for moving heavy objects within the iron foundry. Careful selection and training of crane operators and slingers should be undertaken, for a mistake by one of these employees can have serious consequences not only for himself, but also for other workers in the iron foundry. Crane drivers and slingers must be fully aware of the correct signals to use during a moving operation.

Electrical work. This is not a specific foundry hazard, but much equipment in a foundry is electrically powered. It is important that all employees are familiar with all markings on such equipment and are instructed in methods of ensuring that the current is turned off before any maintenance is attempted. It is also desirable that employees are able to deal with victims of electric shock. All electricians and supervisors should be able to administer respiratory resuscitation, as immediate treatment is essential if breathing has stopped as a result of a shock.

Painting. Some castings must be painted before despatch to the buyer. Both dipping and spraying are practised. If the paint is spirit based, then precautions should be taken to ensure that any fumes are removed from the painting area. Certain paints also present a fire risk; employees should be made aware of this, and smoking should be prohibited. It is advisable to use this type of paint in small quantities, collecting it from larger storage vessels kept at a safe distance from the area of application.

Permanent-mould process

An important development in the foundry has been the practice of casting in permanent metal moulds as in die-casting. In this case, pattern making is largely replaced by engineering methods and is really a die-sinking operation. Most of the pattern making hazards are thereby removed and the risks from sand are also eliminated but are replaced by a degree of risk inherent in the use of some sort of refractory material to coat the die or mould. In modern die-foundry work, increasing use is made of sand cores, in which case the dust hazards of the sand foundry are still present.

Mechanised founding

The mechanised foundry employs the same basic methods as the conventional iron foundry. When moulding is done by machine and castings cleaned by shotblasting or hydroblasting, etc., the machine usually has built-in dust control devices, and the dust hazard is reduced. However, sand is frequently moved from place to place on an open belt conveyor, and transfer points and sand spillage may be sources of considerable quantities of airborne dust; in view of the high production rates, the airborne dust burden may be even higher than in the conventional foundry. The installation of exhaust hoods over transfer points on belt conveyors combined with scrupulous housekeeping should be normal practice. Conveying by pneumatic systems is sometimes economically possible and results in a virtually dust-free conveying system.

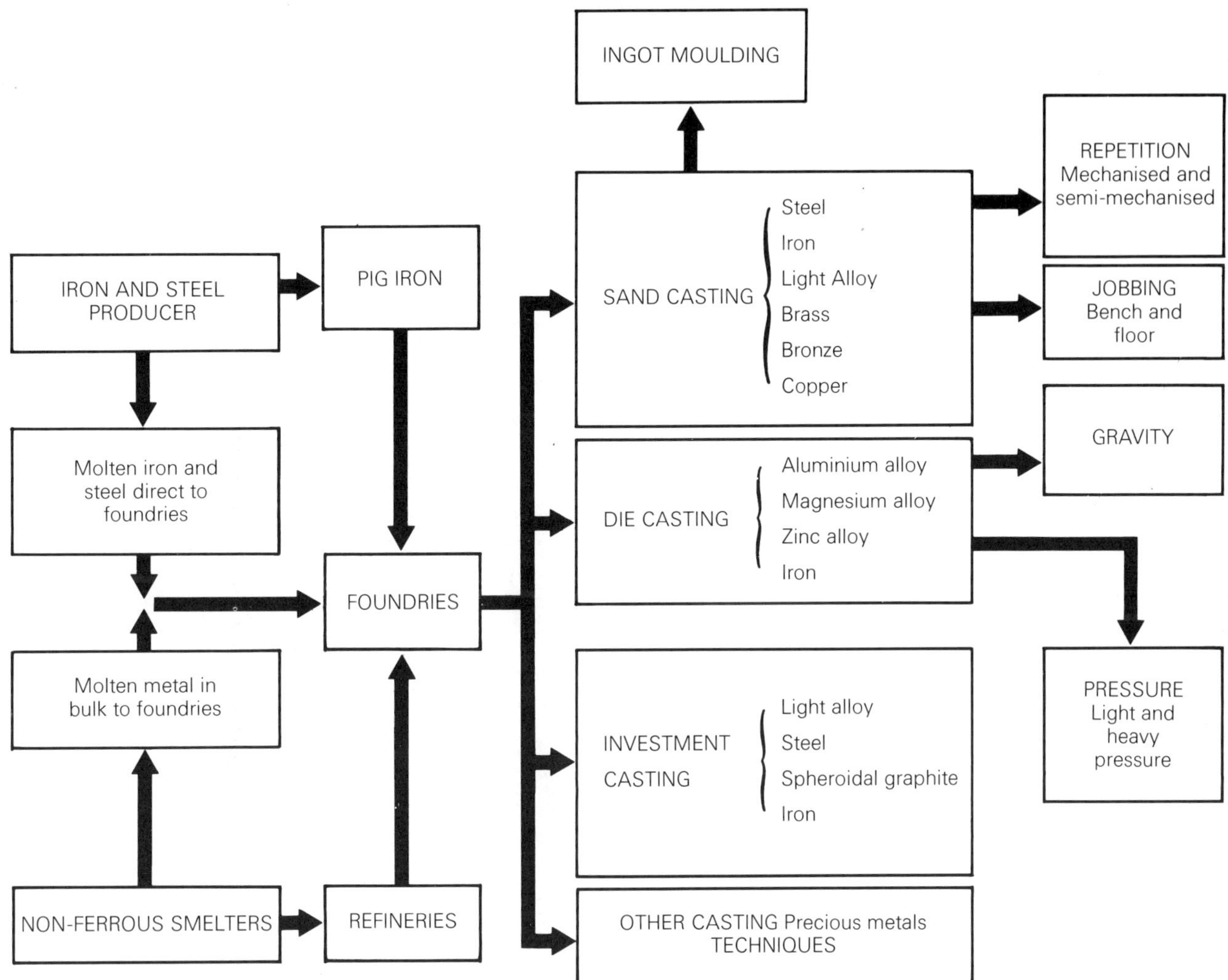

Figure 2. Relationship of the foundry industry to the iron and steel industry and the non-ferrous metal refinery industry. Although the metal founding industry may be assumed to start by remelting solid material in the form of metal ingots or pigs, the iron and steel industry in the large units may be so integrated that the division is less obvious. For instance, the merchant blast furnace may turn all its output into pig iron but in an integrated plant some iron may be used to produce castings, thus taking part in the foundry process, and the blast furnace iron may be taken molten to be turned into steel, where the same thing can occur. There is in fact a separate section of the steel trade known for this reason as ingot moulding. In the normal iron foundry, the remelting of pig iron is also a refining process. In the non-ferrous foundries the process of melting may require the addition of metals and other substances and thus constitutes an alloying process.

Steel founding

The pattern of production in the steel foundry is similar to that in the iron foundry; however, the metal temperatures involved are much higher. This means that eye protection with coloured lenses is essential and that the silica in the mould is converted by heat to tridymite or crystobalite, two forms of crystalline silicon dioxide which are particularly dangerous to the lungs. Sand often becomes "burnt on" to the casting and has to be removed by mechanical means, which give rise to dangerous dust and, consequently, effective dust exhaust systems and respiratory protection are essential (see figure 2).

Light-alloy founding

The light-alloy foundry uses mainly aluminium and magnesium alloys. These often contain small amounts of metals which, under certain circumstances, may give off toxic fume. Where the alloy contains such components, the fume should be analysed to determine its constituents.

In aluminium and magnesium foundries melting is commonly done in crucible furnaces. These are often fired by gas or oil, although electric furnaces are occasionally encountered. It is advisable to have exhaust vents around the top of the pot for removing fumes. In oil-fired furnaces, incomplete combustion due to faulty burners may result in the products of incomplete combustion being released into the air. Where fuels with a high sulphur content are used, there is the danger of high atmospheric concentrations of sulphur dioxide and, consequently, low sulphur fuel should be used or good ventilation ensured. Furnace fume may contain complex hydrocarbons, some of which may have carcinogenic properties. During furnace and flue cleaning there is the hazard of exposure to vanadium pentoxide concentrated in furnace soot from oil deposits.

Fluorspar is commonly used as a flux in aluminium melting and significant quantities of fluoride dust may be released to the environment. In certain cases barium chloride has been used as a flux for magnesium alloys; this is a significantly toxic substance and, consequently, considerable care is required in its use. Light alloys may occasionally be degassed by passing sulphur dioxide or chlorine (or proprietary compounds that decompose to produce chlorine) through the molten metal; exhaust ventilation and respiratory protective equipment are required for this operation. In order to reduce the cooling rate of the hot metal in the mould, a mixture of substances (usually aluminium and iron oxide) which react highly exothermically is placed on the mould riser. This "thermit" mixture gives off dense fume which has been found to be innocuous in practice. When the fume is brown in colour, alarm may be caused due to suspicion of the presence of nitrogen oxides; however, this suspicion is unfounded. The finely divided aluminium produced during the dressing of aluminium and magnesium castings constitutes a severe fire hazard, and wet methods should be used for dust collection, etc. Although fine particulate aluminium is recognised as a respiratory hazard (aluminosis), no cases of this disease in foundry work have been reported.

The manufacture of magnesium castings entails a considerable potential fire and explosion hazard. Molten magnesium will ignite unless a protective barrier is maintained between it and the atmosphere; molten sulphur is widely employed for this purpose. Foundrymen applying the sulphur powder to the melting pot by hand have been found to suffer from dermatitis and they should be provided with, and wear, gloves made of fireproof fabric. Since the sulphur in contact with the metal is constantly burning, considerable quantities of sulphur dioxide are given off. Where this is likely to exceed recommended permissible levels, exhaust ventilation should be installed, even though workers are said to "harden" to the exposure after a few weeks. Workers should be informed of the danger of a pot or ladle of molten magnesium taking fire; they should be instructed in the outstandingly dangerous consequences of sudden combustion of a quantity of metal which may give rise to a cloud of finely divided magnesium oxide thus producing an impenetrable fog and causing a panic and impeding fire fighting. Protective clothing of fireproof materials should be worn by all magnesium foundry workers and clothing coated with magnesium dust should not be stored in lockers without humidity control, otherwise spontaneous combustion may occur. French chalk is used extensively in mould dressing in magnesium foundries; the dust should be controlled to prevent talcosis. Penetrating oils and dusting powders are employed in the inspection of light-alloy castings for the detection of cracks. Dyes have been introduced to improve the effectiveness of these techniques; certain red dyes have been found to be absorbed and excreted in sweat, thus causing soiling of personal clothing; although this condition is a nuisance, no effects on health have been observed.

Founding copper-based alloys

In the founding of brass, bronze, etc., attention should be paid to the possibility of emissions of toxic fumes from metals used in the alloys. Some high-duty alloys contain cadmium and the melting of lead-bronze may lead to the emission of dangerous quantities of lead fume. The lead hazard in furnace cleaning and dross disposal is particularly acute.

The fumes of zinc and copper (the constituents of bronze) are the most common causes of METAL-FUME FEVER (dealt with in a separate article), although the condition has also been observed in foundrymen working with magnesium, aluminium, antimony, etc.

Die-casting

Pressure die-casting machines present all the hazards common to power presses. In addition, the worker may be exposed to the mist of oils used as die lubricants and must be protected against the inhalation of these mists and the danger of oil-saturated clothing. The fire-resistant hydraulic fluids used in the presses may contain toxic organophosphorous compounds and particular care should be taken during maintenance work on hydraulic systems.

Zinc is a common die-casting metal and the hazard of metal-fume fever should be constantly controlled.

Precision founding

An important process in precision foundries is the investment or lost-wax casting process in which patterns are made by injecting moulding wax into a die; these patterns are coated with a fine refractory powder which serves as a mould-facing material, and the wax is then melted out prior to casting or by the introduction of the casting metal itself.

The removal of wax presents a definite fire hazard and wax-removal stoves must be adequately ventilated. Tri-

Table 1. Injuries encountered in the foundry

Type of injury	Part of the body affected	Agent of injury	Comments
Burns	Hands and arms; legs and feet	Molten metal or hot sand	Burns do not usually come high on the list of injury frequency
Foreign bodies	Eyes; arms	Sparks, blown sand; chips and dust from power tools; fragments from worn percussion tools	A common cause of lost-time injuries. Eye protection must be positively enforced
Bruises and contusions	Legs and feet; arms and hands	Striking and falling against objects; hand tools	Often the commonest cause of injuries encountered in a foundry
Fractures	Fingers and toes; hands and feet	Trapping in machines and dropping objects	This is more common in mechanised foundries
Severe trauma and multiple injury	Head, neck and trunk, etc.	Crane accidents; metal explosions; gas or steam explosions; burst grinding wheels	These accidents are rare, but when they do occur they can cause a number of casualties
Radiation effects (heat and ultraviolet)	Eyes; skin (sunburn effect)	Welder's arc; molten metal glare; argon arc welding	These risks are usually well appreciated and the need for protection understood
Frostbite	Fingers	Leaking carbon dioxide cylinders or valves	A fairly rare occurrence, with symptoms often mistaken for a burn

chloroethylene has been used to remove the last traces of wax; there is the danger that this solvent may collect in pockets in the mould or be absorbed by the refractory material and vaporised or decomposed during pouring. There has been a trend towards the inclusion of asbestos investment casting refractory materials. The introduction of effective safeguards for this material is so difficult in foundries that its inclusion in the mixtures should be abandoned.

MacBAIN, G.
STRANGE, R. C.

Health hazards:

CIS 81-532 *Health aspects of the foundry industry–Bibliography* (Industrial Health Foundation, Inc., 5231 Centre Avenue, Pittsburgh) (1980), 25 p. 260 ref.

Proceedings of Conference "The working environment in iron-foundries" (Alvechurch, Birmingham, British Cast Iron Research Association (BCIRA), 1977), 10 vols.

CIS 80-347 "Lung contamination among foundry workers". Kalliomäki, P. L.; Korhonen, O.; Mattsson, T.; Sortti, V.; Vaaranen, V.; Kalliomäki, K.; Koponen, M. *International Archives of Occupational and Environmental Health* (West Berlin, Heidelberg, 1979), 43/2 (85-91). 16 ref.

CIS 79-215 *Respiratory disease among workers in iron and steel foundries* (TNO Research Institute for Environmental Hygiene, PO Box 214, Delft, Netherlands) (1978), 103 p. Illus. 42 ref.

CIS 80-161 "Lung cancer mortality among iron foundry workers". Tola, S.; Koskela, R. S.; Hernberg, S.; Järvinen, E. *Journal of Occupational Medicine* (Chicago), Nov. 1979, 21/11 (753-760). Illus. 23 ref.

Safety:

Guide to safe working practices in ironfoundries. Council of Ironfoundry Associations (London, Ditchling Press, 1978), 55 p.

Reports of the Joint Standing Committee on Health, Safety and Welfare in Foundries. Health and Safety Executive (London, HM Stationery Office, various dates).

CIS 79-2019 *Foundries* (Giessereien). VBG 32, Hauptverband der gewerblichen Berufsgenossenschaften (Cologne, Carl Heymanns Verlag, 1979), 26+20 p. (In German)

Frostbite

Frostbite is a local morbid condition caused by a fall in the temperature of cutaneous tissue, which leads to a perturbation of the peripheral circulation. Frostbite may occur even if temperatures to which the subject is exposed are not extremely low; it may occur even at temperatures above 0 °C if the subject is wearing wet shoes, or clothing which is damp or too light for the season, or if he is required to remain for some considerable time in unheated premises. By and large, any factor limiting local blood circulation may contribute to the occurrence of frostbite.

Frostbite may be considered an occupational disease in any worker who is exposed to cold or damp in a region with a harsh or even temperate climate: for example, forestry workers, agricultural workers, seamen, fishermen, marsh-drainage workers, cold-store workers.

Symptoms

In most cases frostbite is localised to the fingers, toes, nose and earlobes; only in rare cases does the condition extend to the whole of the hands and feet, the face, ankles and wrists. Frostbite affecting other parts of the body such as the buttocks, abdomen and genitals occurs only under special circumstances, for example where persons have been buried in snow for many hours or have been mounted on horseback for long periods at sub-zero temperatures.

A special feature in the clinical course of frostbite, which is not encountered in other traumata, is the two-phase development of the condition. In the first phase there are signs of tissue circulation disorders (pallor, local hypothermia, loss of feeling). If exposure is terminated within 12-24 h, the process is reversible and circulation and tissue will revert to their normal state. If exposure is prolonged, the second phase occurs, in which, in spite of the subsequent application of heat, there is tissue necrosis with more or less severe inflammation at the periphery.

Depending on the depth of tissue necrosis, four degrees of severity can be distinguished. In the first two degrees, necrosis is insignificant and does not extend beyond the basal layer of the skin. There may be simple superficial hyperaemia (first degree) or blistering (second degree). The third degree is characterised by complete cutaneous necrosis and the fourth degree by deep-tissue necrosis (tendons, bones, joints).

The clinical picture includes skin cyanosis of the distal aspects of the affected area, which then take on the blackish colour of dead tissue whereas the adjacent zones are oedematous and hyperaemic. The patient's general condition deteriorates in line with the severity of the frostbite–and there may be chill and fever of varying intensity.

In the great majority of cases, frostbite is not severe and recovery is complete. In only third- and fourth-degree frostbite is there the danger of skin loss or amputation of the extremities. In very serious cases or in sick or debilitated persons, there may be complications such as acute lymphangeitis with lymphadenitis, superimposed infection of the damaged areas, and, in rare cases, anaerobic infection or tetanus.

Persons who have suffered from severe frostbite remain sensitive to the cold and are particularly vulnerable to further lesions. Many complain of irritation and pain in the affected areas.

PREVENTIVE MEASURES

Frostbite prevention is based on the principles of modern hygiene and on the wearing of waterproof clothing and footwear which are not too tightly fitting and which provide protection against the cold whilst still allowing circulation of air around the body.

In isolated workplaces in country or mountain regions, workers exposed to low temperatures should be provided with heated messrooms where they can dry their clothing. A diet rich in fats, proteins and vitamins should be provided. The application of grease, oil, creams or similar substances to the skin as a preventive measure is valueless.

Where there is a risk of frostbite, workers should not consume alcoholic beverages.

First aid

The victim's extremities and body should be warmed as quickly as possible and the subject should be placed in a warm room or immersed in a warm bath. This treatment should be accompanied by light massage of the affected areas until the victim's body has warmed up and the skin of the affected parts, which was white or blueish, becomes pink and warm. Under no circumstances should the subject be immersed in water for more than 20-25 min. Subsequently, he should be put to bed with the frostbitten extremities elevated. Warm drinks, light food and general tonifying treatment should be administered. Contrary to earlier theories, it is dangerous to delay application of heat, to rub the victim with snow or to remove his clothes in an unheated room. It is merely necessary to compare the temperature of the victim's body (which will not fall below 30-32 °C) with that of the frostbitten extremities (10-15 °C) to realise that such action would only lead to a further fall in the victim's temperature.

Treatment

This will depend on the severity of the injury; it should be conservative for first- and second-degree frostbite but plastic surgery or amputation may be necessary for third- and fourth-degree damage. Physiotherapy and medical gymnastics may prove of value in rehabilitation.

ARYEV, T. Y.

CIS 79-1016 "The effects of cold: frostbite and hypothermia". Coble, D. F. *Professional Safety* (Park Ridge, Illinois), Feb. 1979, 24/2 (15-18). Illus. 14 ref.

Frozen food industry

The frozen food industry comprises all methods of deep freezing fresh food at temperatures below their freezing point, thus forming ice crystals in the watery tissues. The food may be frozen raw or partially cooked, for example animal carcasses or made-up meat dishes, fish or fish products, vegetables, fruits, poultry, eggs, ready-made meals, bread and cakes. Otherwise perishable products can be transported over long distances and stored for processing and/or sale when demand arises, and seasonal products can be available at all times.

Food for freezing must be in prime condition and prepared under strict hygienic control. Packaging materials should be vapour- and aroma-proof and resistant to low temperatures. The quality of the product depends on the rate of freezing: if too slow, the structure of the food may be damaged by large ice crystals and enzymatic and microbiological properties destroyed. Small items, such as shrimps and peas, can be frozen quickly which makes for an improvement in quality.

Processes. The various methods of freezing include: air freezing, blast freezing, fluid-bed freezing, fluid freezing, contact freezing, liqui-freezing and dehydro-freezing.

Air freezing in its simplest form involves placing food in trays on shelves in a cold store at approximately −30 °C for a time varying according to size from a few hours to 3 days. Blast freezing, a more complicated technique, uses a rapidly circulating stream of cold air, sometimes combined with cold spirals, which remove heat by means of radiation. Temperatures range between −40 °C and −50 °C and the maximum air speed is 5 m/s. Blast freezing may be carried out in tunnel freezers, often equipped with conveyors to carry the food through to cold-storage rooms. When the freezer is adjacent to the cold store the tunnel is often closed with an air curtain instead of doors.

Fluid-bed freezing is used for chopped or sliced vegetables, peas, etc., which are placed on a perforated belt through which a stream of air is blown. Each item is coated with ice and thus retains its shape and separateness. The frozen vegetables may be stored in large containers and repacked when needed in small units. In fluid freezing (one of the oldest known methods) the food, usually fish, is immersed in a strong solution of brine. Salt may penetrate unwrapped goods and even wrappings, affect the flavour and hasten rancidity. This method had declined in use but is now gaining ground again as more effective plastic wrapping materials are developed. Poultry is frozen by a combination of the fluid and air-freezing methods. Each bird, packed in polyethylene or similar material, is first sprayed or immersed in a fluid to freeze its outer layer; the inside is afterwards frozen in a blast freezer.

Contact freezing is the common method for foodstuffs packed in cartons, which are placed between hollow shelves, through which a cooling fluid is circulated: the shelves are pressed flat against the cartons, usually by hydraulic pressure.

In liqui-freezing the product is placed on a conveyor belt which is passed through a tank of liquid nitrogen (or occasionally liquid carbon dioxide) or through a tunnel where liquid nitrogen is sprayed. Freezing occurs at a temperature as low as −196 °C and not every type of product or wrapping can withstand this. Dehydro-freezing, which removes some of the water before freezing, is used for certain vegetables and fruits. A considerable reduction of weight is achieved, involving lower transport, storage and wrapping costs.

During cold storage, the product must be maintained at a temperature of between −25 and −30 °C and good air circulation must be maintained. Transport of frozen goods has to be in refrigerated wagons, lorries, ships, etc., and during loading and unloading, the goods must be exposed to as little heat as possible. Usually, firms producing frozen food also prepare the raw material but sometimes this treatment is carried out in separate establishments.

HAZARDS AND THEIR PREVENTION

Preparation of materials for freezing corresponds with work carried on in other food factories, slaughterhouses, canneries, bakeries, etc., and the hazards are described in the articles dealing with these activities. Injuries caused by knives in meat and fish preparation can be minimised by design and maintenance, selection of the right knife for the job, provision of tough protective aprons and correct training of workers. Risks from machinery can be prevented by secure machinery guarding. Mechanical handling equipment, especially conveyors, is widely employed and particular attention should be paid to inrunning nips on such equipment.

Falls may occur on wet and slippery floors, and suitable floor surfaces, maintenance, good housekeeping, and non-slip footwear will help in prevention. Curbs round machines will prevent water flowing on to the floor but good drainage should be provided to remove rapidly any spillage that does occur. Scalds from steam or hot water can be prevented by careful maintenance of all steam plant, protective lagging and railing off of open vessels. Lifting of heavy weights, for example carcasses, may cause back injuries, and training in correct methods of handling is important.

Electrical hazards are high and should be countered by efficient earthing and careful maintenance of portable electric power tools and inspection lamps; portable electric equipment should be of the low-voltage type.

Animal infections (anthrax, brucellosis, etc.) may occasionally be contracted from slaughterhouses; dermatitis may be caused by vegetable dust, flour, sugar, etc., and erysipeloid infection may occur due to contact with rotting nitrogenous material. Good washing facilities, hand and arm protection, barrier creams and protective clothing will considerably reduce the danger of infection. However, regular medical supervision of the health of workers is essential.

Refrigerants such as ammonia, methyl chloride and halogenated aliphatic hydrocarbons used in freezing and cold storage bring risks of poisoning and chemical burns. Respiratory protective equipment, protective clothing, arrangements for warning signals and quick exits are essential. Ammonia, and even more so other refrigerants less frequently used, such as propane, butane, ethane and ethylene, are flammable and explosive. Adequate measures should be taken for explosion prevention and fire fighting (see REFRIGERATING PLANTS).

The industry involves general ill effects from exposure to cold unless appropriate warm, protective clothing is provided and well maintained. In large tunnel freezing plants, it may be fatal for workers to stay in the rapidly moving stream of air, even if dressed in polar clothing. It is particularly important to prohibit entry into a tunnel freezer in operation and to make effective interlocking arrangements to ensure that freezers cannot be started up whilst workers are still inside them.

Warm messrooms and provision of hot drinks will mitigate the effects of cold work (see COLD AND WORK IN THE COLD).

JENSEN, G.

CIS 80-1291 *Freezing rooms, cold-storage rooms and drying rooms* (Fryserom, kjølerom og tørkerom). Bestillingsnr. 374 (Direktoratet for arbeidstilsynet, Postboks 8103 Dep., Oslo 1) (May 1979), 5 p. (In Norwegian)

CIS 79-1019 *Occupational safety and health in cold and deep-frozen storage installations* (Arbeitsschutz bei der Kühl- und Gefrierlagerung). Dömland, D.; Hoffmann, R.; Sambleben, H. (Berlin, Verlag Tribüne, 1978), 48 p. (In German)

CIS 79-1906 "Preventive aspects of the cold-storage industry" (Aspetti preventivi nell'industria del freddo). Melino, C.; Messineo, A.; Carlesi, G.; De Luca, L. *Rivista degli infortuni e delle malattie professionali* (Rome), Nov.-Dec. 1978, 117/6 (1 081-1 093). 32 ref. (In Italian)

Fruit ripening

In former times, fruit was eaten or preserved immediately it was ripe but there is now a large consumer demand for fresh fruit at all seasons. Temperate countries provide an ever-increasing market for tropical fruits which, in turn, form a very valuable export crop in many developing countries.

Operations. Temperate fruits are stored in their area of origin for gradual release or are transported from one temperate zone to another, for example Australia to the United Kingdom.

Many types of tropical fruit are transported great distances by sea for consumption in temperate countries. The fruit is gathered fully grown but not yet ripe, packed in refrigerated chambers in the ships and brought artificially to full ripeness at its place of destination. By far the most important example is the banana: when the ships reach the importing country, the fruit is transferred to banana ripening rooms. It is kept there at a temperature of 20 °C and a relative humidity of 90-95%. Ethylene is generated naturally in the fruit during ripening but sometimes it is introduced artificially to accelerate the ripening process; other gases, such as coal gas or acetylene, have also been used for ripening.

A different process is necessary with temperate fruits, such as apples and pears; these are kept in chilled storage to preserve them for later market demands. When they are required for sale they are removed from the storage chambers and allowed to mature in warmer conditions. A very delicate regulation of atmosphere in the storage chambers is necessary, including removal of the ethylene generated by the fruit, reduction of the quantity of oxygen in the air, and scrubbing of carbon dioxide from the air. Refrigeration may be necessary for prolonged preservation. Similar methods are used to preserve fruits transported by sea or land over long distances.

HAZARDS AND THEIR PREVENTION

Use of gases, such as coal gas and acetylene, in ripening rooms is prohibited by law in some countries and the risks of these explosive substances are high. Ethylene gas may cause explosions in ripening chambers unless appropriate precautions are taken, especially to prevent excessive build-up of the gas.

Refrigerant gases may be toxic or explosive. If explosive, all electrical installations and equipment should be flameproof. Respiratory protective equipment against fumes from refrigerants and also against fumes generated during storage should be kept for emergencies. Warning light signals should be maintained to show when anyone is inside a refrigerated chamber and locks should be designed to allow exit from the inside. Warm protective clothing should be provided.

Handling of some types of fruit, especially the citrus varieties, may cause allergies or other skin disorders: good sanitary and washing facilities are essential, barrier creams and hand protection may give additional protection. Medical supervision is desirable.

DE PUTTER, P.

Fuel and oil additives

Additives are substances introduced into a basic medium to enhance, modify, or suppress some property of that medium. The safety and health hazards of fuel and oil additives are determined by the individual agents and their combinations. Literally hundreds of chemicals are used today as petroleum additives and examples of the most important of these are given below, grouped by the functions they perform.

Antiknock agents and scavengers. The organic lead compounds are the principal antiknock agents used in gasoline and are dealt with in the article LEAD ALKYL COMPOUNDS. The commercial antiknock preparations also contain halogenated hydrocarbons as scavengers, kerosene and/or toluene as solvents, and minor amounts of organic dyes and antioxidants.

Pour-point depressants, viscosity-index and flow improvers. These additives improve the viscosity-temperature relationship of lubricating oils and their flow characteristics at low temperatures. Substances commonly used for these purposes are isobutylene and methacrylate polymers and wax-naphthylene and wax-phenol condensation products. These agents are physiologically inert.

Antioxidants and corrosion inhibitors. Zinc alkyl or aryl dithiophosphates, methyl ditertiary butylphenol, terpene-phosphorus pentasulphide addition products and *p*-phenylenediamine compounds are typical oxidation inhibitors. The zinc alkyl dithiophosphates are slightly toxic orally and may be mildly irritating to skin and highly irritating to the eyes. Methyl ditertiary butylphenol is moderately toxic on ingestion and irritating on skin contact. The terpene-phosphorus pentasulphide addition products are not hazardous. *p*-Phenylenediamine derivatives are potent irritants and sensitisers. Other agents used as antioxidants and corrosion inhibitors are sulphurised esters, terpenes and olefins; they are non-hazardous.

Detergents and dispersants. The most commonly encountered detergents are metallic alkyl sulphonates and alkyl phenolates, in which the metals are usually barium, calcium or magnesium. The concentrated compounds are only slightly toxic orally; they are slightly irritant to the skin but more severely irritant to the eyes.

A barium-olefin-phosphorus pentasulphide reaction product with detergent properties has low toxicity by oral and skin absorption and a mildly irritant action on the skin and conjunctivae. An ashless dispersant, alkyl polyamide, similarly has low systemic toxicity and slightly irritating properties.

Extreme-pressure agents. The commonly used extreme-pressure additives include tricresyl phosphate, sulphurised animal oils, lead soaps (naphthenates), zinc dithiophosphates and chlorinated paraffins, naphthalenes and diphenyls. The *ortho* isomer of tricresyl phosphate presents a definite toxicity hazard but the other two isomers are considerably less toxic. Lead naphthenate and other lead soaps may cause systemic intoxication on ingestion or prolonged skin contact. Chlorinated diphenyls and naphthalenes are well recognised causes of chloracne, while the chlorinated paraffins are relatively innocuous.

Metal deactivators. These additives counteract the catalytic effects of metals (especially copper) in the oxidation of hydrocarbons in fuels. The most common deactivator in use is N,N'-disalicylidene-1,2,-diaminopropane. This additive exhibits moderate systemic toxicity on ingestion and inhalation, and is mildly irritating to skin and eyes.

Miscellaneous additives. These include products such as bactericides, anti-fouling and anti-icing agents and cetane, combustion and odour improvers and alcohol fuel extenders.

Additive preparation and blending. Additive manufacturing involves hundreds of processing techniques. Ingredients range from simple to complex, innocuous to lethal, and include metals, other elements and a tremendous variety of inorganic and organic compounds. Catalysts, solvents, intermediates, and by-products may be present along with the desired product.

Additives in commerce are frequently in solution or dispersion and are further diluted when added to the fuel or petroleum product. In certain cases the blending of an additive into the base stock must be carried out with great precision.

The amount of additive blended is determined by weight or volume. In batch operations, predetermined units (bags, drums, kilogrammes, litres, etc.) are released into the medium, which may be agitated to achieve mixing. In a continuous process, in-line blending is regulated by metered weights or volumes. Health and safety factors may determine the addition technique; for example highly toxic lead alkyls are blended into gasoline by eductors which keep the additive under negative pressure.

Additive applications. Common uses of additives in petroleum products are shown in table 1.

HAZARDS

To ensure the safety of workers engaged in the manufacture, storage, transport and use of additives, it is necessary to evaluate the physical, chemical and toxicological properties of each substance used.

Accidents. Certain additives and their diluents may cause fires or explosions. This risk is increased if the flammable agent or mixture has a low flash point or high vapour pressure, or is subject to oxidation reactions.

Diseases. Additives range in toxicity from innocuous to extremely poisonous. While serious and fatal systemic intoxications do occur, occupational diseases are relatively uncommon. Dermatitis is the disorder most frequently seen, since a number of additives are primary irritants and sensitisers.

SAFETY AND HEALTH MEASURES

While concentrated additives may require stringent technical control measures, the dilute concentrations in final commercial use generally present no particular problems in safe handling. Additive mixtures which are explosive or flammable substances must be protected from open flames, sparks, heat sources and oxidising agents. Vessels, containers and transport systems must be designed to withstand the pressure, temperature, solvent and chemical actions of the contents. Highly toxic materials are best controlled in a totally enclosed system. Appropriate precautionary labels are needed for additive containers.

Good hygienic practices call for storage and handling of additives in clean, uncluttered, well ventilated areas. Workmen should be educated as to the hazard involved. Personal protective equipment (working clothes, eye and face protection, respiratory protective equipment) must be available and used when needed. Medical examinations (preplacement, periodic and toxicity screening) are an important aspect of the occupational health programme of workers exposed to oil and fuel additives.

WEAVER, N. K.

"Fuel and lubricant additives". Cummings, W. M. *Lubrication* (New York), 1977, 63/1 (1-12). 15 ref.

"The toxicity of some typical lubricating oil additives". Clark, D. G. *Erdoel, Kohle, Erdgas, Petrochemie vereinigt mit Brennstoff-Chemie* (Leinfelden), Dec. 1978, 31/12, 584 p.

Table 1. Common applications of fuel and oil additives

Application	Corrosion inhibitors	Anti-oxidants	Detergents	Dispersants	Metal deactivators	Bactericides	Flow and viscosity improvers	Other agents
Well drilling	x		x			x		
Refining	x	x		x		x		Anti-fouling
Gasoline and aviation fuel	x	x	x	x	x	x		Anti-icing; anti-knock
Diesel fuel	x	x		x	x	x	x	Cetane improver; odour improving; combustion improving
Furnace oil	x	x		x		x	x	Combustion improving
Residual fuel oil	x			x			x	Combustion improving
Lubricating oils	x	x	x	x			x	Extreme pressure

Fungicides

Definition

Some fungi, such as rusts, mildews, moulds, smuts, storage rots and seedling blights, are able to infect and cause diseases in plants, animals and man. Others can attack and destroy non-living materials such as wood and fibre products. Chemicals called fungicides are applied to prevent these diseases.

Fungi causing plant diseases can be arranged into four sub-groups which differ by the microscopic characters of the mycelium, the spores and the organs on which the spores were developed. The first group of fungi, known as Phycomycetes are soil-borne organisms causing club rot of brassicae, wart diseases of potatoes, etc. The second group consists of Ascomycetes, perithecia-forming powdery mildews and fungi causing apple scab, black currant leaf spot and rose black spot. The third group are the Basidiomycetes, including loose smut of wheat and barley, and several rusts species. To the fourth group, known as *Fungi imperfecti*, belong the genera Aspergillus, Fusarium, Penicillium, etc., that are of great economic importance, as they cause significant losses because of mould invasion during plant growth, at harvest, and after harvest. (Fusarium species infect barley, oats and wheat; Penicillium species cause brown rot of pomaceous fruits).

Fungicides have been used for centuries. Copper and sulphur compounds were the first to be used, and Bordeaux mixture was applied in 1885 to vineyards. At present a great number of organic compounds and complex metal salts have been synthesised and are being used in practice.

Uses. Fungicides are used by spraying, dusting, seed dressing, seedling and soil sterilisation and fumigation of warehouses and greenhouses.

According to their mode of action, fungicides can be classified into two groups:

(a) protective fungicides–applied at a time prior to the arrival of the fungal spores (such as sulphur and copper compounds);

(b) eradicant fungicides–applied after the plant has become infected (such as mercury compounds and nitroderivatives of the phenols).

The fungicides either act on the surface of the leaves and seeds, or penetrate into the plant and exert their toxic action directly on the fungi (systemic fungicides). They can also alter the physiological and biochemical processes in the plant and thus produce artificial chemical immunisation. Examples of this group are the antibiotics and the rodananilides.

Fungicides applied to seed act primarily against surface-borne spores. However, in some cases they are required to persist on the seed coat long enough to be effective against the dormant mycelium contained within the seed. When applied to the seed before sewing, the fungicide is called seed disinfectant or seed dressing, though the latter term may include treatment not intended to counter seed-borne fungi or soil pests.

To protect wood, paper, leather and other materials, fungicides are used by impregnation or staining.

Special drugs with fungicidal action are also used to control fungal diseases in humans and animals.

Classification

A great number of widely differing chemical compounds with fungicidal action are used in many countries. A number of them, classified by composition, are given in table 1.

Table 1. List of the most widely used fungicides, with common names and chemical names

Copper compounds
Copper chloride
Copper oxychloride
Cupric oxide
Bordeaux mixture (copper sulphate and calcium hydroxide)
Copper-8-quinolinolate
Copper carbonate, basic
Copper naphthenate
Copper sulphate
Copper chromate
Copper oleate

Mercury compounds
Calomel (mercurous chloride)
Mercuric oxide
Mercury lactate
Mercuram (methoxyethyl mercury acetate)
MEMC (methoxyethyl mercury chloride)
PMA (phenyl mercury acetate)

Tin compounds
Fentin acetate (triphenyl tin acetate)
Fentin chloride (triphenyl tin chloride)
Fentin hydroxide (triphenyl tin hydroxide)
Butyl tin oxides
Plictran (tricyclohexyl tin hydroxide)

Zinc compounds
Zinc chloride
Zinc chromate
Zinc naphthenate
Zinc oleate

Other metallic compounds
Potassium permanganate
Cadmium chloride
Ferrous sulphate
Neo asozin (ferric monomethyl arsonate)
Rhizoctol (methylarsin sulphite)
Urbacid (*bis*(dimethylthiocarbamoylthio)methylarsine)
Chromium naphthalenate

Sulphur
Sofril (flowers of sulphur)
Lime sulphur (calcium polysulphides)

Organophosphorous compounds
Phyrazophos, O,O-diethyl-O-(5-methyl-6-ethoxycarbonyl-pyrazolo(1-5,a)-pyrimid-2-y)thionophosphate
IBP, kitazin, O,O-diisopropyl-S-benzylthiophosphate
Edifenphos, O-ethyl-S,S-diphenyl phosphorodithionate
Ditalimfos (O,O-diethyl phthalamido)phosphonothioate

Dithiocarbamates
Zineb, zinc ethylene *bis*(dithiocarbamate)
Maneb, manganese ethylene *bis*(dithiocarbamate)
Mancozeb, zinc co-ordination product of maneb
Nabam, disodium ethylene *bis*(dithiocarbamate)
Thiram, tetramethylthiuram disulphide
Ferbam, ferric dimethyl dithiocarbamate
Ziram, zinc dimethyl dithiocarbamate
Bunema, potassium N-hydroxymethyl-N-methyl dithiocarbamate
Vapam, sodium methyl dithiocarbamate
Metiram, complex of zineb and polyethylenethiuram disulphide
Methylmetiram, complex of zinc (propylene) *bis*-dithiocarbamate and polyethylenethiuram disulphide

Carbamates

Thiophanate, 1,2-*bis*(3-ethoxycarbamoyl-2-thiureido)-benzene

Thiophanate methyl, 1,2-*bis*(3-methoxycarbamoyl-2-thiureido)benzene

Halogenated hydrocarbons

1,1-Dichloromethane

Dibromomethane

Bromomethane

Chloropicrin

Carbon tetrachloride

p-Dichlorobenzene

Hexachlorobenzene

Chloroneb, 1,4-dichloro-2,5-dimethoxybenzene

Dodecylammonium bromide

Hexachlorophene, 2,2'-methylene *bis*(3,4,6-trichlorophenol)

Pentachlorophenol

Isobac, monosodium salt of hexachlorophene

Aromatic nitro compounds

Dinitrophenol

Nitrobiphenyl

4,6-Dinitro-*o*-cresol (DNOC)

Dinobuton, 2-(1-methyl-2-propyl)-4,6-dinitrophenyl isopropyl-carbonate

Tecnazene, 2,3,5,6-tetrachloro-3-nitrobenzene

Binapacryl, 2-*sec*-butyl-4,6-dinitrophenyl-3-methyl-2-butenoate

Dinocap, 2,4-dinitro-6-(2-octyl)phenyl crotonate

Nirit, 2,4-dinitrophenyl thiocyanate

Brassicol, pentachloronitrobenzene

Quinones

Chloranyl, 2,3,5,6-tetrachloro-1,4-benzoquinone

Dichlone, 2,3-dichloro-1,4-naphthoquinone

Benzoquinone

Dithianone, 2,3-dicyano-1,4-dihydro-1,4-dithioanthraquinone

Anilides

Benodanil, 2-iodobenzanilide

Pyrocarbolid, 2-methyl-5,6-dihydro-4-H-pyran-3-carboxylic acid anilide

Carboxin, 5,6-dihydro-2-methyl-1,4-oxathiin-3-carboxanilide

Oxycarboxin, 5,6-dihydro-2-methyl-1,4-oxathiin-3-carboxanilide-4,4-dioxide

Salicylanilide

Guanidine compounds

Dodine, dodecylguanidine acetate

Guzatine, 9'-axa-1,17-diguanidinoheptadecane sulphate or acetate

Phthalimides

Folpet, N-(trichloromethylthio)phthalimide

Captan, N-(trichloromethylthio)-4-cyclohexene-1,2-dicarboximide

Captafol, *cis*-N-(1,1,2,2-tetrachloroethyl(thio)-4-cyclohexene-1,2-dicarboximide

Chlorothalonile, tetrachloroisophthalonitrile

Dimethachion, N-(3,5-dichlorophenyl)succinimide

Pyrimidine

Dimethirimol, 5-butyl-2-dimethylamino-4-hydroxy-6-methylpyrimidine

Ethirimol, 5-butyl-2-ethylamino-4-hydroxy-6-methylpyrimidine

Bupirimate, 5-butyl-2-ethylamino-6-methylpyrimidin-4-yldimethylsulphamate

Pyridines

Piperalin, 3-(2-methylpiperidino)propyl-3,4-dichlorobenzoate

Thiodaiazoles

Dazomet, tetrahydro-3,5-dimethyl-2H-1,3,5-thiadiazine-2-thione

Terrazole, 5-ethoxy-3-trichlorodimethyl-1,2,4-thiadiazole

Milneb, 3,3-ethylene-*bis*(tetrahydro-4,6-dimethyl)-2H-1,3,5-thiadiazine-2-thione

Triazines

Anilazine, dichloro-N-(2-chlorophenyl)triazine amino

Triadimefon, 1-(4-chlorophenoxy)-3,3-dimethyl-1-(1,2,3-triazol-1-yl)butan-2-one

Isoxazolones

Hymexazol, 3-hydroxy-5-methylisoxazole

Drazoxolon, 4-(2-chlorophenylhydrazono)-3-methyl-5-isoxazolone

Imidazoles

Glyodine, 2-heptadecyl-imidazoline acetate

Bencmyl, methyl-1-(butylcarbamoyl)-2-benzimidazole carbamate

Thiabendazole, 2-(4-thiazolyl)benzimidazole

Triforine, N,N-(1,4-piperazinediyl-*bis*(2,2,2-trichloro-ethylidene)bisformamide)

Carbendazim, 2-(methoxycarbonylamino)benzimidazole

Other heterocyclic compounds

Tridemorph, 2,6-dimethyl-4-tridecylmorpholine

Chinomethionate, 6-methyl-2-oxo-1,3-dithiolo-(4,5-b)-quinoxaline

Antibiotics

Blasticidin

Gliotoxin

Griseofulvin

Polyoxin

Phytobacteriomycin

Narimycin

Kasugamycin

Validamycin

Oils

Anthracene

Ammonium naphthenate

Aldehydes, ketones, oxides

Formaldehyde

Paraformaldehyde

Allyl alcohol

Ethylene oxide

Propylene oxide

o-Phenylphenol

Miscellaneous

Rodamine, 2-thioxo-4-thizolidinene

Trapex, methyl isothiocyanate

Dichlofluanid, N'-dichlorofluoromethylthio-N,N-dimethyl-N'-phenylsulphamide

Fenaminosulf, sodium (4-(dimethylamino)phenyl)diazene-sulphonato

Ridomyl, 2,4,6-dimethylphenyl-(N-2-methylacetyl)-alaninmethylether

---

Otherwise, as regards the field of application, fungicides can be classified as follows.

Seed dressing. This is a simple and economically efficient method for the control of plant diseases. The pests are destroyed on the seeds and in the soil during the development of the seed. Despite the availability of efficient alternative compounds, the mercury fungicides are still used to a considerable extent for this purpose. Of the remaining compounds the dithiocarbamates and particularly thiuram are the most important. Chloranil and dichlone of the quinone group, hexachlorobenzene, formaldehyde and some antibiotics are also used for seed dressing. The seeds can be treated by either the dry or the wet method.

Soil disinfection. This is carried out with fungicides which have a more general action and are incorporated into the soil as solid or liquid formulations liberating volatile or easily soluble components. Such compounds are chloropicrin, methyl bromide, dibromomethane, formaldehyde, vapam, dazomet, allyl alcohol, pentachloronitrobenzene, chloroneb and others. These fungicides are used most intensively on greenhouse soil.

Application on plants. In order to control airborne diseases fungicides are applied on annual field crops, fruit trees and berry crops. Almost all fungicide groups are used for this purpose. Copper compounds, dithiocarbamates, aromatic nitro derivatives, quinones, phthalimides, guanidines and chlorinated hydrocarbons are the most frequently used, as well as some heterocyclics, nickel compounds and some antibiotics.

A large number of fungicides are used in industry. Metallic salts and mineral oils are mostly used for the protection of wooden materials, textiles and paper.

HAZARDS

The fungicides cover a great variety of chemical compounds differing widely in their toxicity. Highly toxic compounds are used as fumigants of foods and warehouses, for seed dressing and soil disinfection, and cases of poisoning have been described with organomercurials, hexachlorobenzene and pentachlorobenzene, as well as with the slightly toxic dithiocarbamates. See separate articles on ANTHRACENE AND DERIVATES; ARSENIC AND COMPOUNDS; BROMOMETHANE; CADMIUM AND COMPOUNDS; CARBAMATES AND THIOCARBAMATES; CARBON TETRACHLORIDE; CHLOROPICRIN; COPPER, ALLOYS AND COMPOUNDS; DINITRO-*o*-CRESOL; ETHYLENE OXIDE; FORMALDEHYDE AND DERIVATIVES; LIMESTONE AND LIME; PHENOLS AND PHENOLIC COMPOUNDS; SULPHUR; TIN, ALLOYS AND COMPOUNDS; ZINC, ALLOYS AND COMPOUNDS.

Hexachlorobenzene (C_6Cl_6)

m.w. 284.8
sp.gr. 1.57
m.p. 230 °C
b.p. 326 °C
v.d. 9.8
v.p. 1 mmHg (0.13·10³ Pa) at 114.4 °C
f.p. 242 °C 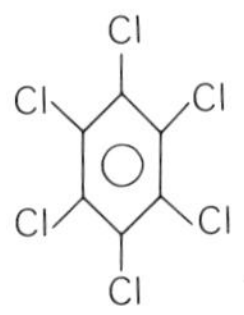
insoluble in water; soluble in organic solvents
a stable, colourless-white powder or needles.
MAC USSR 0.9 mg/m³
Oral LD_{50} 3 500-10 000 mg/kg

Hexachlorobenzene is stored in the body fat. It interferes with porphyrin metabolism, increasing the urinary excretion of coproporphyrins and uroporphyrins; it increases also the levels of transaminases and dehydrogenases in the blood. It can produce liver injury (hepatomegaly and cirrhosis), photosensitisation of the skin, a porphyria similar to *porphyria cutanea tarda*, arthritis and hirsutism (monkey disease). It is a skin irritant. In hexachlorobenzene poisoning, EDTA has been used with some success. Chronic poisoning needs long-term treatment, mainly symptomatic, and it is not always reversible on cessation of exposure.

2-(1-Methyl-2-propyl)-4,6-dinitrophenyl isopropyl-carbonate ($C_{14}H_{18}N_2O_7$)

DINOBUTON; ACREX

m.w. 326
m.p. 60 °C
scarcely soluble in water
a yellowish solid.
MAC USSR 0.2 mg/m³
Oral LD_{50} 140 mg/kg (for rat)
Dermal LD_{50} 2 500 mg/kg (for rabbit)

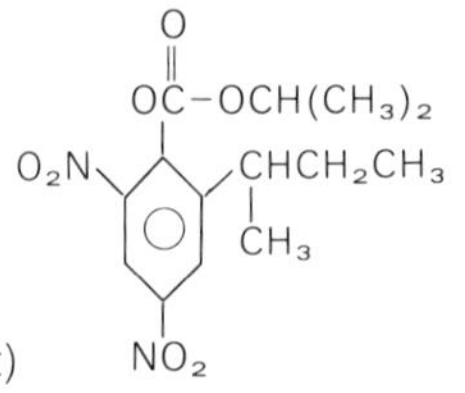

Like dinitro-*o*-cresol, dinobuton disturbs cell metabolism by inhibiting the oxidative phosphorilation, with the loss of energy-rich compounds such as adenosintriphosphoric acid. It can cause severe liver distrophy and necrosis of the convoluted tubules of the kidneys. The clinical manifestations of the intoxication are high temperature, methaemoglobinaemia and haemolysis, nervous disturbances, irritation of the skin and mucous membranes.

Dinitrocapryl crotonate ($C_{12}H_{20}N_2O_6$)

DINOCAP; KARATHANE

m.w. 288.3
v.d. 9.95
practically insoluble in water; soluble in most organic solvents
a dark-brown liquid
formulation: wettable powder.
MAC USSR 0.2 mg/m³
Oral LD_{50} 800 mg/m³

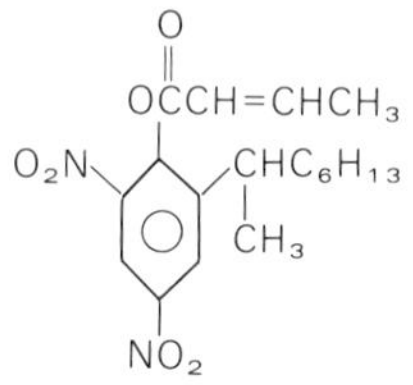

Exposure to dinocap can increase the blood level of alkaline phosphatase. Dinocap is a moderate irritant of the skin and mucous membranes; it produces distrophic changes in the liver and kidney and hypertrophy of myocardium. In acute poisoning, disturbances in thermoregulation, clonic cramps and breathing difficulties have been observed.

2,4-Dinitrophenylthiocyanate ($C_6H_3(NO_2)_2NCS$)

NIRIT

m.w. 230
m.p. 138 °C
insoluble in water; readily soluble in aromatic hydrocarbons
a crystalline yellow powder.

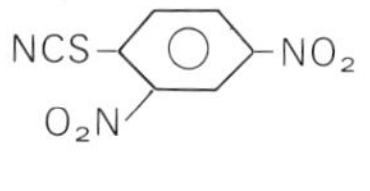

This compound has haemotoxic properties, causing anaemia and leucocytosis with toxic granulation of the leucocytes, in addition to degenerative changes in the liver, spleen and kidneys.

***n*-Dodecyl guanidine acetate** ($C_{12}H_{25}NHC(:NH)NH_2CH_3COOH$)

DODINE; CYPREX

m.w. 287
m.p. 136 °C
slightly soluble in water; soluble in acids and alcohols
white crystals
formulation: wettable powder, liquid.
MAC USSR 0.1 mg/m³
Oral LD_{50} 660 mg/kg (female rat); 1 000 mg/kg (male rat)

Dodine has irritant properties and may easily produce, in occupational exposure, a readily reversible keratoconjunctivitis. Dermatitis and irritation of the respiratory and gastrointestinal tract have also been observed.

cis-N-(Trichloromethyl)thio-4-cyclo-hexene-1,2-dicarboximide ($C_9H_8O_2NSCl_3$)

CAPTAN

m.w. 300.6
m.p. 170 °C
insoluble in water; soluble in organic solvents; stable except under alkaline conditions
white crystalline solid
formulations: wettable powder, 40% aqueous suspension, dust for field use and seed treatment.
TLV ACGIH 5 mg/m³
STEL ACGIH 15 mg/m³
MAC USSR 0.5 mg/m³
Oral LD_{50} 9 000 mg/kg (rat)

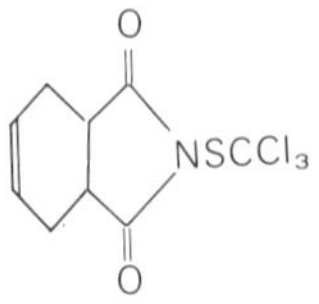

Captan can induce chromosome alterations in cell culture and has proved to be teratogenic in hamsters and rabbits. In chronic exposure it has gonadotropic effects. It is irritant to the eye, nose, mouth and skin. It interferes with oxidative phosphorilation, inhibits nucleic acid synthesis, protein synthesis and glucolysis processes, and reacts with thiol groups. It has systemic effects, mainly in the central nervous system of experimental animals. Other observations were decrease of body temperature, loss of weight, hypochromic anaemia, leukopoenia, alterations of proximal and distal tubular cells of the kidney.

Similar effects are shown by Captafol (MAC USSR 0.1 mg/m³) and Folpet (MAC USSR 1 mg/m³).

Tetrachloroisophthalonitrile ($C_8N_2Cl_4$)

CHLOROTHALONILE

m.w. 265.8
m.p. 250 °C
insoluble in water; soluble in organic solvents, stable
white crystalline solid
formulation: wettable powder.
Oral LD_{50} 10 000 mg/kg (rat)

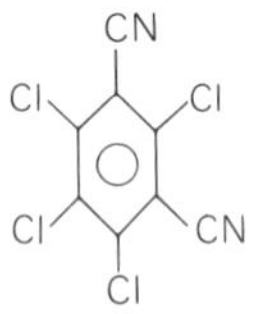

This compound has irritant properties to the gastrointestinal tract and the skin; the most significant effects are obtained in the kidney.

Tetrahydro-3,5-dimethyl-2H-thiadiazine-2-thione ($C_5H_{10}N_2S_2$)

DAZOMET; MYLONE; CRAG FUNGICIDE 974

m.w. 162.3
m.p. 107 °C
a crystalline dust.
Oral LD_{50} 608 mg/kg (rat); 120 mg/kg (rabbit); 79 mg/kg (cat)

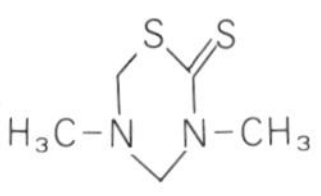

It is used also as a nematocide and a slimicide. This compound and its decomposition products are sensitisers and mild irritants of the eye, nose, mouth and skin. Poisoning is characterised by a variety of symptoms, including anxiety, tachycardia and quick breathing, hypersalivation, clonic cramps, impaired movement co-ordination, sometimes hyperglycaemia and cholinesterase inhibition. The main pathomorphological findings are enlargement of the liver and degenerative changes of the kidney and other internal organs.

3,3-Ethylene-*bis*(tetrahydro-4,6-dimethyl)-2H-1,3,5-thiadiazine-2-thione ($C_{12}H_{16}S_4N_4$)

MILNEB

m.w. 344
m.p. 140 °C
insoluble in water; slightly soluble in alcohol
white crystals
formulation: wettable powder.
MAC USSR 4 mg/m³
Oral LD_{50} 5 600 mg/kg (rat)

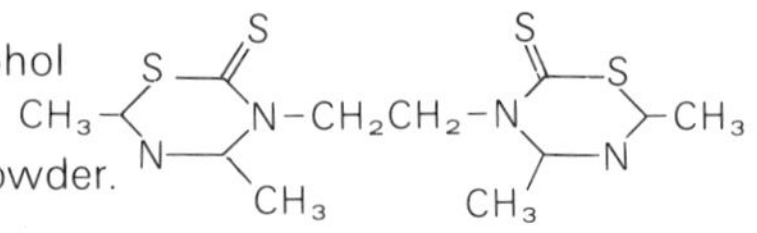

This compound can cause gastrointestinal disturbances, weakness, decrease of the body temperature and leukopoenia.

Quinones

Quinones in general cause blood disturbances (methaemoglobinaemia, anaemia), affect the liver, disturb vitamin metabolism, particularly that of ascorbic acid, and are irritant to the respiratory ways and the eye.

Treatment: ascorbic acid, cysteine or vitamin B_6; nicotinic acid. (See PHENOLS AND PHENOLIC COMPOUNDS.)

Chloranil and dichlone are the quinone derivatives most largely used as fungicides.

2,3,5,6-Tetrachloro-1,4-benzoquinone ($C_6O_2Cl_4$)

TETRACHLOROQUINONE; CHLORANIL

m.w. 245.9
m.p. 290 °C
almost insoluble in water
yellow crystals.
MAC USSR 1 mg/m³
Oral LD_{50} 4 000 mg/kg (rat)

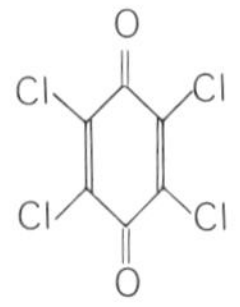

It is irritant to the skin and the upper respiratory tract; it can also cause depression of the central nervous system and dystrophic changes in the liver and kidney. The biological monitoring of exposed persons has shown an increased level of the urinary phenols, both free and bound.

2,3-Dichloro-1,4-naphthoquinone ($C_{10}H_4O_2Cl_2$)

DICHLONE; PHYGON

m.w. 227.1
m.p. 193 °C
v.d. 7.8
insoluble in water; moderately soluble in organic solvents; stable in non-alkaline media
yellow crystals.
MAC USSR 0.5 mg/m³
Oral LD_{50} 1 300 mg/m³ (rat)

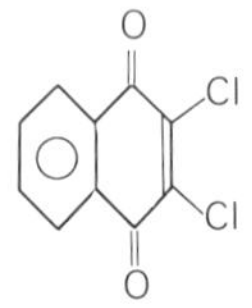

In addition to sharing the irritant and blood depressant properties common to quinones, 2,3-dichloro-1,4-naphthoquinone is under consideration as an experimental animal carcinogen.

2-(4-Thyazolyl)-benzimidazole ($C_{10}H_7N_3S$)

THIABENDAZOLE

m.w. 201
m.p. 304 °C
almost insoluble in water; slightly soluble in organic solvents
white crystalline powder.
Oral LD_{50} 3 000 mg/kg (rat)

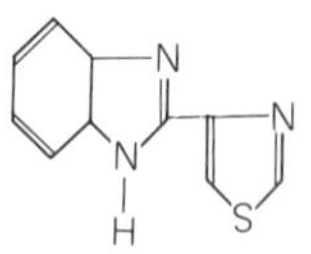

It is used also as anthelmintic in cattle. In experimental animals thiabendazole has caused thymus involution, colloid depletion in the thyroid and increase in liver and kidney size.

6-Methyl-1,3-dithiolo(4,5b)quinoxalino-2-one ($C_{10}H_6N_2OS_2$)

CHINOMETHIONATE

m.w. 234
m.p. 172 °C 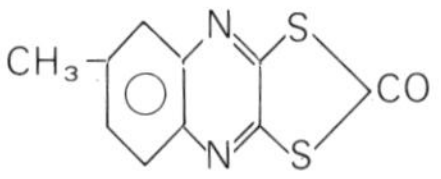
a crystalline solid
formulations: wettable powder, dust, smoke generator.
Oral LD_{50} 500 mg/kg (rat)

Chinomethionate has a high cumulative toxicity and inhibits thiol groups and some enzymes containing them; it lowers phagocytic activity and has antispermatogenic effects. It is irritant to the skin and the respiratory system. It can damage the central nervous system, the liver and the gastrointestinal tract. Glutathione and cysteine provide protection against the acute effects of chinomethionate.

Dichlorofluoromethylthio-N,N-dimethyl-N-phenylsulphamid ($C_9H_{12}N_2O_2S_2Cl_2F$)

DICHLOFUANID

m.w. 333.8
m.p. 105 °C 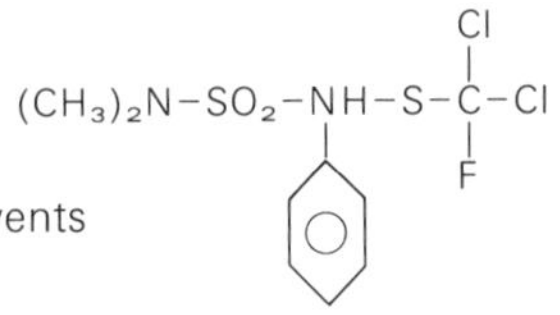
insoluble in water; slightly soluble in organic solvents
a white powder.
MAC USSR 0.5 mg/m³
Oral LD_{50} 1 000 mg/m³ (rat)

This compound inhibits thiol groups. In experimental animals it caused histological changes in liver, proximal tubules of the kidney and adrenal cortex, with the reduction of the lymphatic tissue in the spleen. It is a moderate irritant of the skin and mucous membranes.

SAFETY AND HEALTH MEASURES

The adverse effects of fungicides on health can be prevented by adequate safety measures. These measures vary according to the toxicity and hazards of a particular fungicide and the manner of its application. Regulations concerning any particular fungicide should include instructions about specific safety measures. Some general rules, however, can be formulated that are applicable to all chemical fungicides (see PESTICIDES):

(1) People handling fungicides should receive thorough instruction. Fumigation with highly toxic chemicals should be carried out only by trained people. Thorough preliminary training and continuous briefing during work are indispensable, particular attention being paid to health risks and safety measures.

(2) Adequate facilities for personal hygiene should be provided, including protective equipment, mask and gloves. Suitable equipment should be used in production plants and in agriculture, as well as during fumigation, care being taken to prevent contact with fungicides.

(3) Preplacement and periodic medical examinations should be organised in order to prevent the exposure of workers suffering from diseases that might be affected by fungicides, such as liver, kidney, nervous system and other disorders. Health monitoring would help discover people with hypersensitivity and early signs of intoxication, provided the medical check-up includes clinical and laboratory examination. Environmental and biological monitoring should be carried out to evaluate the extent of exposure of workers.

Treatment. Fungicide poisonings are usually treated symptomatically. Care should be taken to discontinue exposure by taking off the victim's contaminated clothing, washing him and removing him to fresh air, thus preventing further absorption of the chemical. In severe cases of acute poisoning hospitalisation may be necessary.

KALOYANOVA-SIMEONOVA, F.

Evaluation of pesticides residues in food. FAO/WHO Joint Meetings on Pesticide Residues. Technical reports and monographs (Geneva, World Health Organisation, 1965-1977).

Pesticides: toxic action and prophylaxis (Pesticidi: tokichno deistvie y profilaktika). Kaloyanova-Simeonova, F. (Sofia, Bulgarian Academy of Sciences, 1977), 307 p. (In Bulgarian)

Insecticides and fungicides handbook. Martin, H. (Oxford and Edinburgh, Blackwell Scientific Publication, 3rd ed., 1969), 387 p.

Pesticide handbook (Spravochnik po pestizidam). Medved, L. I. (ed.). (Kiev, Urozhai, 2nd ed., 1977), 375 p. (In Russian)

Pesticide dictionary. Farm Chemicals (Willoughby, Ohio, Meister Publishing Company, 1978).

Toxic and hazardous substances in industry (Toksicheskie veshtestva v promishlenosti). Lazarev, N. V. (ed.). (Leningrad, Khimia, 7th ed., 1976 and 1977), 3 vols., 590+624+608 p. (In Russian)

Furfural and derivatives

Furfural (C_4H_3OCHO)

2-FURALDEHYDE; FURFURALDEHYDE

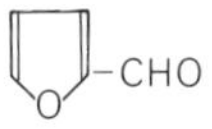

m.w. 96.1
sp.gr. 1.16
m.p. −38.7 °C
b.p. 161.7 °C
v.d. 3.31
f.p. 60 °C
e.l. 2.1-19.3%
i.t. 315.6 °C
very soluble in alcohol, ether, benzene; slightly soluble in water
an amber-coloured liquid with an aromatic odour.
TWA OSHA 5 ppm 20 mg/m³
TLV ACGIH 2 ppm 8 mg/m³ skin
IDLH 250 ppm
MAC USSR 10 mg/m³

Furfuryl alcohol ($C_4H_3OCH_2OH$)

2-FURYLCARBINOL; FURFURAL ALCOHOL

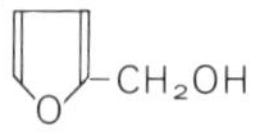

m.w. 98.1
sp.gr. 1.13
m.p. 14.8 °C
b.p. 171 °C
v.d. 3.4
v.p. 1 mmHg (0.13·10³ Pa) at 31.8 °C
f.p. 75 °C oc
e.l. 1.8-16.3%
i.t. 490.5 °C
very soluble in water and most organic solvents
a colourless liquid with an ether-like odour and bitter taste.
TWA OSHA 50 ppm 200 mg/m³
TLV ACGIH 5 ppm 20 mg/m³ skin
STEL ACGIH 10 ppm 20 mg/m³
IDLH 250 ppm

Tetrahydrofurfuryl alcohol ($C_4H_7OCH_2OH$)

TETRAHYDRO-2-FURANCARBINOL; THFA

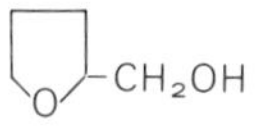

m.w. 102.1
sp.gr. 1.05
b.p. 177.5 °C
v.d. 3.5
v.p. 3 mmHg (0.39·10³ Pa) at 50 °C
f.p. 75 °C oc
e.l. 1.5-9.7%
i.t. 282.2 °C
soluble in water and most organic solvents
a colourless hygroscopic liquid with a mild odour.

Tetrahydrofuran

See separate article.

Production. Furfural is obtained from raw materials containing pentosans (e.g. oats, straw, bran and other cereals) by hydrolysis and dehydration with sulphuric acid. Furfuryl alcohol, which naturally occurs in coffee oil and in yellow tobacco leaves, is obtained by catalytic hydrogenation of furfural under pressure. Tetrahydrofurfuryl alcohol is produced by catalytic hydrogenation of furfuryl alcohol.

Uses. Furfural and its derivatives are used widely as solvents for oils, synthetic and natural resins, cellulose and derivatives, dyes, polymers and other organic chemicals. They are also used as intermediates in the production of plastics, herbicides, pesticides and fungicides and as vulcanisation accelerators in the rubber industry.

[The consumption of furfuryl alcohol is increasing steadily. Almost 90% of the production is used in the manufacture of furan resin and furfuryl alcohol-formaldehyde resin for binding foundry core sand in the so-called "hot box" and "no bake" methods (see FOUNDRIES). It is also used as a solvent and a plasticiser.]

HAZARDS

Furfural and its derivatives are moderate fire hazards and their vapours may form explosive mixtures in air.

Toxicity. The hazard of poisoning by furfural and its derivatives is limited in view of the low volatility of these products at normal temperatures. Experiments have shown that administration of a single lethal dose produced a pronounced inhibitory effect on the bulbar vegetative centres and the encephalic nuclei with signs of congestion in the liver, kidney, lungs and brain and with degenerative lesions in the liver and kidney. Rabbits exposed to furfural vapours several hours daily manifested hepatic and renal lesions and modifications in blood picture.

Furfural vapours are a strong irritant to the skin and mucous membranes of the eyes and respiratory tract. Human exposure to high concentrations of furfural has produced eye and respiratory tract irritation including lung oedema. Irritant dermatitis, which may become eczematous, has occurred and cases of skin sensitisation have been reported.

Persons having suffered prolonged exposure to furfural vapours complain of headache, fatigue, loss of the sense of taste and numbness of the tongue, and may present irritation of mucous membranes and nervous disorders such as tremors and dizziness.

[Animal exposure to high concentrations of furfuryl alcohol caused eye irritation. Death occurred as a consequence of respiratory paralysis; a depressive action on the CNS and the heart was also observed. In workers exposed to both furfuryl alcohol and formaldehyde in foundries lacrimation was reported at 16 ppm of furfuryl alcohol. No medical findings were detected in workers after repeated daily exposure to furfuryl alcohol and furfural.

In contact with acids furfuryl alcohol can produce violent explosions.]

SAFETY AND HEALTH MEASURES

Furfural and its derivatives should be stored in well ventilated areas, sheltered from direct sunlight and away from heat and other sources of ignition.

During the handling or processing of furfural, exhaust ventilation and general ventilation should be provided to keep the atmospheric concentration below recommended levels. Persons required to handle furfural and its derivatives should wear personal protective equipment to avoid skin contact, including eye and face protection and hand protection. Where brief exposure to high atmospheric concentrations is possible, workers should wear respiratory protective equipment.

Measures to prevent inhalation hazards will normally be sufficient to prevent the formation of explosive mixtures; nevertheless, the products should be confined as much as practicable, especially when heated, to prevent the escape of vapour, and sources of ignition should be eliminated.

Medical prevention. It is inadvisable to employ workers with a history of contact dermatitis on work with furfural. [The determination of the total furoic acid in the urine of the whole shift has been suggested as a reliable test for the biological monitoring of workers exposed to furfural.]

Treatment. Eye splashes should be copiously flushed with water. Contaminated clothing should be removed and skin washed with soap and water. Oxygen therapy may be given following acute inhalation. Ingestion should be treated with gastric lavage.

CASTELLINO, N.

CIS 79-170 "The absorption, metabolism and excretion of furfural in man". Flek, J.; Šedivec, V. *International Archives of Occupational and Environmental Health* (West Berlin), 1978, 41/3 (159-168). Illus. 8 ref.

CIS 79-180 "Biologic monitoring of persons exposed to furfural vapors". Šedivec, V.; Flek, J. *International Archives of Occupational and Environmental Health* (West Berlin), 15 Sep. 1978, 42/1 (41-49). 8 ref.

CIS 75-1092 "Experimental data on the toxicity of the monomer FA" (Ėksperimental'nye dannye po toksikologii monomera FA). Tŭilina, L. V. *Gigiena truda i professional'nye zabolevanija* (Moscow), May 1974, 4 (53-55). 6 ref. (In Russian)

Criteria for a recommended standard–Occupational exposure to furfuryl alcohol. DHEW (NIOSH) publication No. 79-133 (National Institute for Occupational Safety and Health, 4676 Columbia Parkway, Cincinnati) (Mar. 1979), 60 p. Illus. 72 ref.

Fur industry

The origin of the use of animal furs to provide articles of clothing for man is lost in antiquity. Nowadays the fur industry provides a variety of outer garments such as coats, jackets, hats, gloves and boots and also provides trim for other types of garments. There is a great variety of fur-producing animals. They include aquatic species such as beaver, otter, muskrat and seal; northern land species such as fox, wolf, mink, weasel, squirrel, bear, badger, marten and raccoon; and tropical species such as leopard, ocelot and cheetah. Certain animals such as cattle, horse, pig, caribou, goat, etc., may be denoted hide-bearing animals since the skin is usually stripped of hair and tanned to produce leather. However, most of these, especially when young, may alternatively be processed to produce furs.

Fur farming has increased enormously over the past two decades. Fox was the original farmed species. Now fur farming is almost totally devoted to mink production, though a few other species such as chinchilla are farmed to a minor extent.

The production of fur of one kind or another is nearly world-wide. However, the number of farmed mink pelts greatly exceeds any other type of fur, and in 1975 the combined mink pelt output of the Scandinavian countries was nearly 10 000 000, followed by Russia with 8 000 000 and the United States with about 3 000 000.

Treatment of furs

Primitive means of preserving furs have been used since very early times and are still practised in many parts of the world. Typically, after the pelt is scraped and cleaned by washing, the skin is impregnated with animal oil, which serves to preserve it and make it more pliable. The pelt may be beaten or chewed after the oil treatment in order to effect better impregnation by the oil.

In the modern fur industry pelts are obtained from fur farmers, trappers or hunters. At this stage they have been stripped from the carcass, flesh and fatty deposits have been removed by scraping, and the pelts have been stretched and air dried. The fur industry then grades them according to factors such as the general condition of the pelt, fur length, curl and patterning. The pelts are soaked in salt water (figure 1) for several hours to re-soften them, after which excess water is removed in revolving drums. Next the under side is drawn across razor-sharp knives by workers known as fleshers. This operation, known as fleshing (figure 2), removes the areolar tissue from the under side of the dermis. The object is to remove, as far as possible, any tissue which is not involved in the attachment of the fur, thus producing the maximum degree of lightness and flexibility of the pelt.

The pelts are now ready for tanning and are soaked in alum solution acidified somewhat with hydrochloric or sulphuric acid. The alum treatment may be carried out in either an aqueous or an oil solution. Excess liquid is extracted and the pelts are dried to set the skin collagen. They are then treated with an oil solution in a kicking machine or similar type of machine to force the oil into the skin. They are then cleaned in rotating drums charged with sawdust which absorbs moisture and excess oil.

Although dyeing of furs was at one time not looked upon favourably, it is now an accepted part of fur preparation and is practised extensively. The usual procedure involves treatment of the pelts with a weak alkaline (e.g. sodium carbonate) solution to remove dirt and oil residues. The pelts are then soaked in a mordant (e.g. ferric sulphate) solution, after which they are steeped in the dye (e.g. an aniline dye) solution until the desired colour is obtained. They are then repeatedly rinsed and drum-dried with the aid of sawdust.

Pelts contain guard hairs as well as the softer fur fibres. The guard hairs are stiffer and longer than the fur fibres and, depending on the type of fur and the final product desired, these hairs may be either partially or totally removed by machine or by hand plucking. Some pelts also require shearing (figure 3).

Before being made into garments, pelts may be cut and "let out". This involves making a series of closely spaced diagonal or V-shaped slits in the skin, after which the pelt is pulled in order either to lengthen or to broaden it according to requirements. The pelt is then re-sewn (figure 4). This type of operation requires great skill and

Figure 1. The soaking department in a fur processing works. *(By courtesy of* Office du film du Québec.*)*

Figure 2. The machine fleshing of lamb skins. *(By courtesy of* Office du film du Québec.*)*

Figure 3. Shearing operation on Canadian beaver. *(By courtesy of* Office du film du Québec.*)*

Figure 4. Operators engaged on the machine sewing of skins. *(By courtesy of* Office du film du Québec.*)*

experience. The pelts are next thoroughly moistened and then laid out and tacked on a board according to a chalked-on pattern, left to dry, and sewn together. Finally, lining and other finishing steps complete the garment.

HAZARDS AND THEIR PREVENTION

Accidents. Some of the machines used in fur processing present serious hazards unless sufficient guarding is maintained: in particular, all drums should be protected with an interlocking gate and the centrifuges used for extraction of moisture should be fitted with interlocking lids; fur clipping and fur cutting machines should be totally enclosed except for the feed and discharge openings.

Vats should be covered or effectively railed to prevent accidental immersion. Falls on wet and slippery floors can be largely prevented by maintenance of sound, impervious surfaces, well drained and frequently cleaned. Dyeing vats should be surrounded by drainage channels. Accidents caused by hand tools can be reduced if the handles are well designed and the tools well maintained. In the manufacturing sector, sewing machines require similar protection to those used in the garment trade, e.g. guarding of transmission machinery and of needles.

Health hazards. The use by the fur industry of such a large proportion of pelts from animals bred in captivity has considerably reduced the likelihood of transmission of animal diseases to fur workers. Nevertheless, anthrax may occur in men handling carcasses, skins, hides or hairs from infected animals: a vaccine may be administered to all likely to have contact. All should be aware of the risk and trained to report any suspicious symptoms immediately.

Various chemicals used in the fur industry are potential skin irritants. These include alkalis, acids, alum, chromates, bleaching agents, oils, salt, and the compounds involved in the dyeing process which comprise various types of dyes as well as mordants.

Unpacking of bales which have been treated with dusting powder in their countries of origin (Europe and Asia), drumming, plucking, unhairing and shearing can all produce irritant dust. In dye houses and dye kitchens, where salts of lead, copper and chromium are weighed and cooked, there is also a risk of ingestion of toxic dusts. Injurious fumes may arise from degreasing solvents and fumigating chemicals. There is also the possibility of development of contact sensitisation or allergy to some of these chemicals or to one or more of the types of fur being handled.

The main protection against the hazards of dust and fume is the provision of locally applied exhaust ventilation: good general ventilation is also necessary throughout the process. Personal respiratory protective equipment may be necessary for short-term jobs or in addition to local exhaust at particularly dusty operations.

Protective clothing appropriate to the process is necessary at most stages of fur processing. Rubber hand protection, foot and leg protection and aprons are required for wet processes and as a protection against acids, alkalis and corrosive chemicals, for example at the dye and mordant vats. Good sanitary and washing facilities including showers should be provided. Bleaches and strong alkali soaps should not be used for hand cleansing.

Pre-employment medical examination can assist in the prevention of dermatitis by selection of employees without a history of sensitivity. Medical supervision is desirable; well maintained first-aid provisions in the charge of trained personnel are essential. Strict attention to hygiene, ventilation and temperature are necessary in the many small workrooms in which much of the making up of fur garments is done.

BRAID, P. E.

A candid view of the fur industry. Prentice, A. C. (Bewdley, Ontario, Clay Publishing Co., 1976), 319 p. Illus.

Joint FAO/WHO Expert Committee on Zoonoses. Third report (Geneva, 1966). Technical report series No. 378 (Geneva, World Health Organisation, 1967), 127 p.

"Multiple disease of mink handler following exposure to Aleutian disease". Henry, L. W. *Cancer* (Philadelphia), July 1979, 44/1 (273-275). 20 ref.

CIS 79-783 "Severe epidemic of skin disease due to Ursol in fur industry workers" (Hromadný výskyt kožních onĕmocnení z ursolu u pracujících v kŏešnickém průmyslu). Štĕpànek, O.; Hassman, P.; Hassmanovà, V.; Chytilová, M.; Kŭelova, M.; Skutilová, I. *Pracovní lékařství* (Prague), Aug. 1978, 30/7 (268-270). Illus. 10 ref. (In Czech)

Furnaces, kilns and ovens

Practically every industrial product involves melting, refining, drying, cooking, baking, the promotion of chemical reactions or some other operation requiring the use of heat at some stage in its manufacture. Furnaces, kilns and ovens are industrial heating devices used in such operations. The terms "furnace", "kiln" and "oven" have no precise definitions, having been adopted by different industries to define what is very similar equipment used for very different purposes. For example a fuel-fired or electrically heated batch-type furnace suitable for heat treating steel could be used for firing or glazing ceramic products such as porcelain and would be called a kiln by the ceramics manufacturer. On the other hand, some types of kilns have been designed solely for firing brick and have no exact counterpart in other industries. A rotary kiln is a long, inclined rotating cylinder fired from its lower end and charged at its upper end; it has uses in several industries and is generally called a rotary kiln by all of them.

All three classes of industrial heating equipment have in common the characteristic that they provide an enclosure to which heat can be applied in a controlled manner to raise the temperature in the enclosure to the desired degree. Some are equipped to provide a controlled atmosphere in the enclosure.

In general, it may be said that furnaces and kilns operate at much higher temperatures than ovens (coke ovens are an exception) and, for this reason, require the lining of their interiors with heat-resistant ceramic or other materials called refractories. Ovens operating in the higher temperature ranges may employ insulating refractories to improve heat retention, while other insulating materials may be used when oven temperatures are not too high.

Refractories are products designed to withstand heat, erosion, abrasion and chemical attack. They include brick, special shapes, granular products, mouldable plastic forms that can be rammed into place, types that when mixed with water can be poured into place like concrete, and other forms. Special mortars and cements are made for installing refractories.

The materials from which a refractory is made are selected to give the necessary physical and chemical characteristics. For example modern steelmaking processes require the use of a chemically basic slag for absorbing the impurities removed from the molten metal in the refining process. Such furnaces must be lined with chemically basic refractories to prevent the basic slag from reacting with and rapidly destroying the lining.

Chemically basic refractories include those called chrome, chrome-magnesite, magnesite-chrome, magnesite, dolomite and alumina refractories, and owe their properties to the oxides of chromium, magnesium, calcium and aluminium that they contain. The principal chemically acid refractory material is silica. Numerous refractory products are made from fireclays (hydrated aluminium silicates). Carbon and graphite are sometimes used as refractories where special conditions warrant their use.

The heat required by a furnace, kiln or oven may be generated by the combustion of a fuel (solid, liquid or gaseous) as listed under "common heat sources" below. Heat also may be generated by the use of electricity in arcs or resistance elements or, less commonly, by electromagnetic induction, microwave radiation and electric-powered sources of infrared radiation. Some of the ways in which electricity can be used to generate heat for furnaces operating at high temperature are summarised in the following table with special reference to their adaptability in the steel industry.

An interesting exception to the use of fuel or electricity to generate heat in steelmaking is the basic oxygen process, in which oxygen is blown on to the surface of a bath of molten pig iron and steel scrap in a refractory-lined barrel-shaped furnace (figure 1). Heat for refining the bath to produce steel is derived from the oxidation of carbon and other elements and no external source of heat is required.

Table 1. Classification of electric furnaces according to principles of heating employed

A. Resistance furnaces

1. *Indirect heating:* Current is passed through resistors to generate heat which heats the furnace charge by radiation and convection. Such furnaces are used for heat treating and maintaining temperature of molten metal but not for melting steel.
2. *Direct heating:* Current from low-voltage transformers is passed through the steel to be heated. Used to heat steel for hot working, but not successful for melting steel.
3. *Induction heating:* Current is induced in the steel by an oscillating magnetic field.
   (a) Low-frequency induction furnaces: Employ the principle of a transformer with the metal charge forming the secondary circuit and a coil with an iron core forming the primary circuit.
   (b) High- and medium-frequency induction furnaces: Current of high or medium frequency is passed through a coil surrounding a crucible containing the charge.

B. Arc furnaces

1. *Indirect- or independent-arc furnaces:* The metal charge is heated by an alternating-current arc passing from one electrode to another above the metal. Furnaces may be stationary, oscillating, or rolling.
   (a) Single-phase furnaces:
      (i) Rolling furnaces with horizontal electrodes.
      (ii) Furnaces for special purposes.
   (b) Two-phase furnaces:
      (i) Straight arcs. Not used for making steel.
      (ii) Deflected arc. Not used for making steel.

C. Three-phase furnaces

   (i) Straight arcs. Not used for making steel.
   (ii) Repel-arc. Not used for making steel.

2. *Direct-arc furnaces:* The current passes through arcs from the electrode or electrodes to the metal charge.
   (a) Direct-current arc furnaces: The direct-current arc principle is used in consumable-electrode furnaces.
   (b) Alternating-current series-arc furnaces: Current passes from one electrode through an arc to the metal charge, through the charge, then from the charge through an arc to another electrode. Although single-, two-, and three-phase current can be used, furnaces employing three-phase current are used almost exclusively for steelmaking.
   (c) Alternating-current single-arc furnaces: Current passes from one electrode through an arc and the charge to an electrode in the bottom of the furnace. Single-, two-, or three-phase current can be used.
3. *Combination direct-arc and resistance furnaces:* Current passes from an electrode through the arc to the charge, through the charge to the bottom, and leaves the furnace through the bottom. The charge is heated by the arc and by heat generated in the refractory bottom material by the resistance of the bottom to passage of the electric current. Such furnaces employ two- or three-phase alternating current.

Figure 1. Basic oxygen process steelmaking furnace in operating position, with oxygen being blown into the furnace at the beginning of a heat. Gases generated by the oxidation reactions are captured by the hood, cooled and cleaned before discharge into the atmosphere. *(By courtesy of the United States Steel Corporation.)*

Types. Industrial heating devices may be of the batch type or continuous type. In batch-type ovens and furnaces, for example, a batch or lot of the material is placed in the furnace and heated for a suitable time until it attains the desired degree and uniformity of temperature, after which it may be withdrawn for further processing; by discontinuing the supply of heat, the product may be allowed to cool in the furnace. In continuous ovens and furnaces, on the other hand, the work is moved into, through and out of the heating chamber by some sort of conveying system.

Some of the principal operations performed in furnaces, ovens and kilns are given below. Stills and other types of heated processing equipment in the chemical and petroleum industry are excluded, although much of the later discussion applies to them also.

(a) Furnaces: smelting and reduction, melting and/or refining, heat treating (not involving melting) (figure 2), brazing and soldering, heating for hot working, boiler furnaces, and incinerators.

(b) Kilns: cement kilns, lime kilns, ceramic kilns (brick, tile, refractories, etc.), and drying kilns.

(c) Ovens: drying (moisture elimination) and curing, baking, decorating, and solvent-evaporation (paint drying).

Figure 2. Rotary-hearth heat-treating furnace. *(By courtesy of Aluminium Company of America, Pittsburgh, Pa.)*

Common heat sources. Many types of heat sources have been used, including:

(a) gaseous fuels—natural gas and manufactured gases such as producer gas, water gas, carburetted water gas, coal gas, oil gas, reformed natural gas, butane and propane;

(b) by-product gases—blast-furnace gas, coke-oven gas and oil-refinery gas;

(c) liquid fuels—fuel oil, tar and pitch;

(d) solid fuels—coal (lump or pulverised), semi-coke, coke, charcoal, lignite and peat;

(e) electricity—arc, resistance, induction, microwave and infrared.

Maintenance. Furnace, oven and kiln structures must have adequate strength to resist operating loads. In the case of high-temperature furnaces, the refractory brickwork is supported by a structural-steel framework or, in some cases, the entire furnace is encased in steel plate. Many furnaces, kilns and ovens are thermally insulated for reasons of heat economy, the maintenance of comfortable temperatures in surrounding working areas, and safety. All furnaces, kilns and ovens should be examined at suitable intervals during operation to make certain that no progressive damage is occurring that could result in unexpected failure.

Materials-handling equipment (cranes, conveyors, pushers, hearth rollers, etc., and their drives) should be designed to handle the required loads (and occasional overloads due to emergency conditions) safely and with a minimum of attention except during regular inspection and maintenance programmes.

The automatic controls that regulate electric-power input or fuel and air supply and ensure the correct temperature should be maintained in good condition; these controls should be calibrated at frequent intervals.

HAZARDS AND THEIR PREVENTION

Contact with hot parts of equipment and around any piece of industrial heating equipment can cause burns which may prove serious. The product removed from a furnace or oven may, after a short period, seem to have cooled to ambient temperature but may still be dangerously hot. In the case of furnaces containing molten metal, bubbling or splashing may present a hazard (figure 3). The presence of water is dangerous where there is molten metal or slag, since an explosive generation of steam can occur if the metal or slag contacts the water. During fuel handling careful attention should be given to storage facilities, piping, metering equipment and burners to prevent leaks, spillage or other dangerous conditions that could result in fire or explosion. Some gaseous fuels and products of combustion contain sufficient carbon monoxide to be dangerous. Where controlled atmospheres are used in furnaces, etc., these may contain toxic ingredients or be explosive; hydrogen in particular presents a serious explosion hazard.

Figure 3. A furnace operator skims dross from a 27 000 kg holding furnace. *(By courtesy of Aluminium Company of America, Pittsburgh, Pa.)*

Proper precautions must be observed in lighting a fuel-fired furnace to prevent flash-back or explosion. Automatic controls should shut off the fuel supply quickly in the event of flame failure.

Due safety precautions must be observed when material is fed into or withdrawn from a furnace, kiln or oven. Workers should not be allowed to enter furnaces, etc., until the temperature has fallen below a specific level. Dust, fumes, spills and splashes that arise during loading, operation or unloading must be guarded against by respiratory protective equipment, eye and face protection, heat protective clothing, or other adequate equipment. Exhaust hoods or other devices may be employed to carry dangerous dust and fume emissions from the working area to recovery equipment, where disposal can be effected safely. Particular mention should be made of hazardous substances that may be present in scrap charged into the furnaces, the alloys used to produce specific products, and certain fluxes which are potentially dangerous unless proper ventilation is maintained.

Adequate training programmes will make personnel aware of dangers inherent in the operation of high-temperature equipment. Safety clothing (including proper head, foot and hand protection and heat protective clothing) is essential in many types of operations. Respiratory protective equipment and eye protection such as goggles, tinted eyeglasses, and other safety equipment are required to protect against fumes, dust, toxic gases and glare. Permanent eye damage, including cataract, can result from excessive exposure of the unprotected eye to the glare of high-temperature sources.

In some operations, heat and hot work in the vicinity of furnaces, kilns or ovens may lead to heat disorders.

With paint-drying, lacquer-drying and plastics-curing ovens, which operate at relatively low temperatures, the evaporated solvents may present an explosion hazard in the ovens used for such operations, or may cause discomfort or actual injury to personnel (skin diseases, disorders of the respiratory system, eye irritation).

The hazards of and safety measures to be applied in the maintenance and repair of furnaces, kilns and ovens are dealt with in BOILERS AND FURNACES, CLEANING AND MAINTENANCE OF. It should be noted, however, that in the "tear-down" and rebuilding of furnaces, kilns and ovens the material is often a ceramic material. Depending upon the free silica content, the risk of development of silicosis over years of exposure must be dealt with and precautions of ventilation and personal protection (respirators) must be taken. It is also likely that various types of asbestos may be used as a shield, insulator or heat resisting material. Should this be found, then precautions in eliminating the exposure (substitution of material) or using ventilation and personal protection (respirators) must be taken. Medical surveillance and environmental monitoring of these potential exposures are required. More detailed descriptions of the materials are dealt with in REFRACTORIES; ASBESTOS. See also IRON AND STEEL INDUSTRY.

BUNDY, M.

CIS 80-14 *Standard for ovens and furnaces–Design, location and equipment.* NFPA No. 86A-1977 (National Fire Protection Association, 470 Atlantic Avenue, Boston, Massachusetts) (1977), 123 p. Illus.

CIS 74-1211 *Recommendations on the use of flammable gases in kilns, ovens and dryers* (Recommandations pour l'emploi des gaz combustibles dans les étuves et séchoirs). Association technique de l'industrie du gaz en France (Société du Journal des usines à gaz, 62 rue de Courcelles, Paris) (3rd ed., Nov. 1972), 22 p. Illus. (In French)

CIS 80-1412 *Steelworks environment–Stripping and relining of ladles* (Stålverkens arbetsmiljö–Rivning och infodring av skänkar). Broms, G. TRITA-AML-SA 15 (Stockholm, Arbetsmiljölaboratoriet, KTH, Sep. 1979), 106p. Illus. (In Swedish)

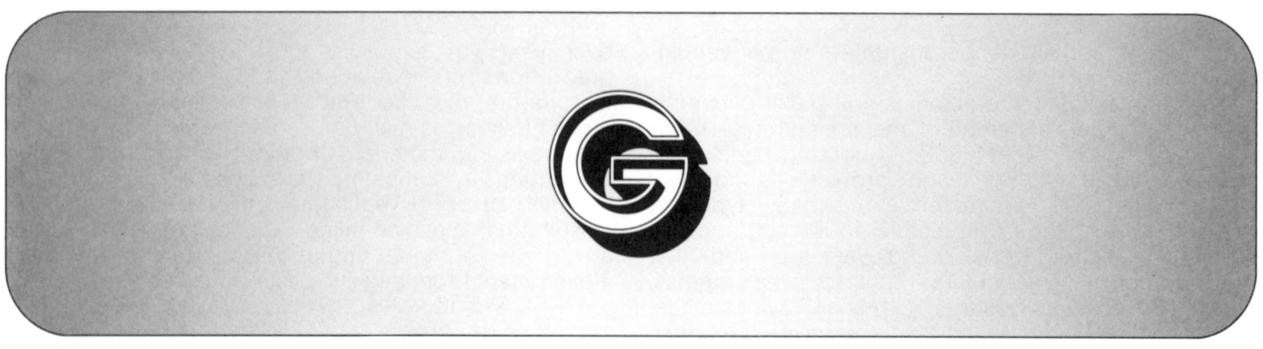

Gallium and compounds

Gallium (Ga)
a.w. 69.7
sp.gr. solid 5.90, liquid 6.09
m.p. 29.8 °C
b.p. 2 403 °C
a light, brittle, greyish-blue metal.
MAC USSR GaAs 2 mg/m³

Chemically gallium is similar to aluminium. It is not attacked by air and does not react with water. Cold gallium reacts with chlorine and bromine and, when heated, with iodine, oxygen, and sulphur. There are 12 known artificial radioactive isotopes, with atomic weights between 64 and 74 and half-lives between 2.6 min and 77.9 h.

Occurrence. The richest source is the mineral germanite, a copper sulphide ore which may contain 0.5-0.7% gallium and is found in south-west Africa. It is also widely distributed in small amounts together with zinc blendes, in aluminium clays, feldspars, coal and in the ores of iron, manganese and chromium.

Production. The metal can be obtained from by-products in the production of alumina and zinc, or from the residual anodic alloy produced in the refining of aluminium. In the Bayer process, bauxite is dissolved in sodium hydroxide and aluminium oxide trihydrate is precipitated. After this process is repeated often enough, gallium builds up in the recycled liquor. Following removal of the alumina, gallium is precipitated by carbon dioxide treatment. The metal is obtained from the precipitate by electrolysis. The USSR, France and the Federal Republic of Germany are the main producers. ^{67}Ga is obtained by irradiation of natural zinc with protons in a cyclotron.

Uses. On a relatively small scale, the metal, alloys, oxides, and salts are used in industries such as machine construction (coatings, lubricants), instrument making (solders, washers, fillers), electronics and electrical equipment production (diodes, transistors, lasers, conductor coverings), and in vacuum technology.

In the chemical industries they are used as catalysts. Today gallium arsenicum is widely used for semiconductor production. Other applications include the use of ^{72}Ga for the study of gallium interactions in the organism and ^{67}Ga [as a tumour scanning agent; because of the high affinity of macrophages of the lymphoreticular tissues for ^{67}Ga, it can be used in the diagnosis of Hodgkin's disease, Boeck's sarcoid and lymphatic tuberculosis].

Gallium compounds

When gallium is dissolved in inorganic acids, salts are formed, which change into insoluble hydroxide $Ga(OH)_3$ with amphoteric properties, if the pH is higher than 3. The three oxides of gallium are GaO, Ga_2O, and Ga_2O_3.

HAZARDS

Since gallium is not widely used in industry, relatively little information is available about its hazards. Inhalation exposures of dusts are possible during the production of the oxides and powdered salts ($Ga_2(SO_4)_3$, Ga_3Cl), and in the production and processing of monocrystals of semi-conductor compounds. The splashing or spilling of the solutions of the metal and its salts may act on the skin or mucous membranes of workers. [Grinding of gallium phosphide in water gives rise to considerable quantities of phosphine requiring preventive measures.] Gallium compounds may be ingested via soiled hands and by eating, drinking and smoking in workplaces.

Occupational diseases from gallium have not been described [but for a case report of a petechial rash followed by a radial neuritis after a short exposure to a small amount of fume containing gallium fluoride]. The biological action of the metal and its compounds has been studied experimentally. The toxicity of gallium and compounds depends upon the mode of entry into the body. When administered orally in rabbits over a long period of time (4-5 months), its action was insignificant and included disturbances in protein reactions and reduced enzyme activity. The low toxicity in this case is explained by the relatively inactive absorption of gallium in the digestive tract. In the stomach and intestines, compounds are formed which are either insoluble or difficult to absorb, such as metal gallates and hydroxides. The dust of the oxide, nitride, and arsenide of gallium was generally toxic when introduced into the respiratory system (intratracheal injections in white rats), causing dystrophy of the liver and kidneys. In the lungs it caused inflammatory and sclerotic changes. Gallium nitrate has a powerful caustic effect on the conjunctivae, cornea, and skin. The high toxicity of the acetate, citrate and chloride of gallium was demonstrated by intraperitoneal injection, leading to death of animals from paralysis of the respiratory centre.

SAFETY AND HEALTH MEASURES

In order to avoid contamination of the atmosphere of workplaces by the dusts of gallium dioxide, nitride, and semi-conductor compounds, precautionary measures should include enclosure of dust-producing equipment and effective local exhaust ventilation. Personal protective measures during the production of gallium should prevent contact of gallium compounds with the skin, and their ingestion. Consequently, good personal hygiene and the use of personal protective equipment are important. In view of the toxicity of gallium and its compounds, as shown by experiments, all persons involved in work with these substances should undergo periodic medical examinations, during which special attention should be paid to the condition of the liver, kidneys, respiratory organs and skin.

ROSCINA, T. A.

CIS 1445-1973 "Gallium fluoride poisoning. A probable case with skin effects and neurological sequelae". Meigs, J. W. *Journal of Occupational Medicine* (Chicago), Dec. 1972, 14/12 (925-926). 7 ref.

"Problems of occupational hygiene related to the industrial application of gallium and indium compounds" (Problemy gigieny truda v svjazi s promeneniem v narodnom xozjajstve coedinenenij gallia i indija). Taracenko, N. Ju.; Fadeev, A. I. *Gigiena i sanitarija* (Moscow), Oct. 1980, 10 (13-16). 4 ref. (In Russian)

"Tumour scintigraphy with Gallium-67". Langhammer, H.; Hör, G.; Kempken, K.; Pabst, H. W.; Heidenreich, P.; Kriegel, H. (69-81). Illus. 62 ref. *Tumour localization with radioactive agents* (Vienna, International Atomic Energy Agency, 1976), 142 p.

CIS 76-1663 "Hazards of phosphine formation during polishing of gallium phosphide plates" (Fosforovodík jako riziko při mechanickém opracování destiček fosfidu galia). Čakrtova, E.; Knîzek, M.; Lepší, P. *Pracovní lékařství* (Prague), Sep. 1975, 27/7-8 (250-254). Illus. 14 ref. (In Czech)

Galvanising

Galvanising is a process by which a zinc coating is applied to a wide variety of steel products to provide protection against corrosion. The base metal to be coated must be clean and oxide-free for adherence of the coating and, therefore, all steel products undergo several, generally similar, cleaning, rinsing, drying and/or annealing processes before entering the galvanising bath.

Two basic methods of galvanising are in use:

(a) hot-dip galvanising, i.e. dipping or passing the steel product through a bath of molten zinc; and

(b) cold electrogalvanising, i.e. zinc electroplating.

In hot-dip galvanising, the batch method is used for fabricated production and the continuous strip method for steel strip, sheet or wire. Fluxes have been largely discontinued in continuous strip galvanising but are still used for wire.

Hot-dip galvanising. The processing steps involved in typical batch galvanising are shown in the flow diagram in figure 1. The product to be galvanised must be thoroughly cleaned. This is done by alkaline cleaning, usually with sodium hydroxide, followed by a water rinse and then pickling in acid, usually sulphuric or hydrochloric. As cleaning solutions are more effective when hot, cleaning tanks are maintained at about 65-77 °C. The concentration of acid also varies in the range between about 7-12% by weight.

When a flux is employed, its function is to maintain satisfactory cleaning of both the product to be galvanised and the zinc bath, and to dry the product. This may be done by a prefluxing step followed by a flux cover on the surface of the zinc bath or by the flux cover alone. In batch galvanising, ammonium chloride is used as the flux cover on the molten zinc bath. In pipe galvanising, zinc ammonium chloride is the flux and the pipe is immersed in a hot solution of this compound following acid cleaning and prior to entering the molten zinc bath.

Continuous strip hot-dip galvanising may be classified into four types, the major difference being the method of cleaning the product and the ability to clean in-line. In the first type, cleaning of the strip is done by flame oxidation of the surface oils with subsequent reduction in the furnace. Annealing is done in-line. The second type employs electrolytic alkali cleaning prior to in-line

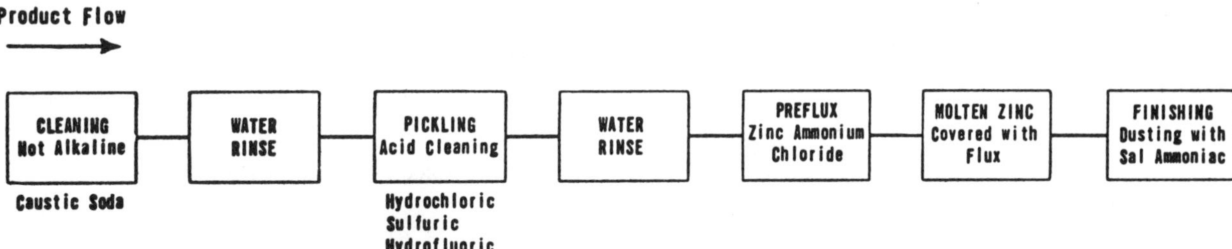

Figure 1. A typical non-continuous or batch-type galvanising process. *(From* Air Pollution Engineering Manual, *US Department of Health, Education and Welfare.)*

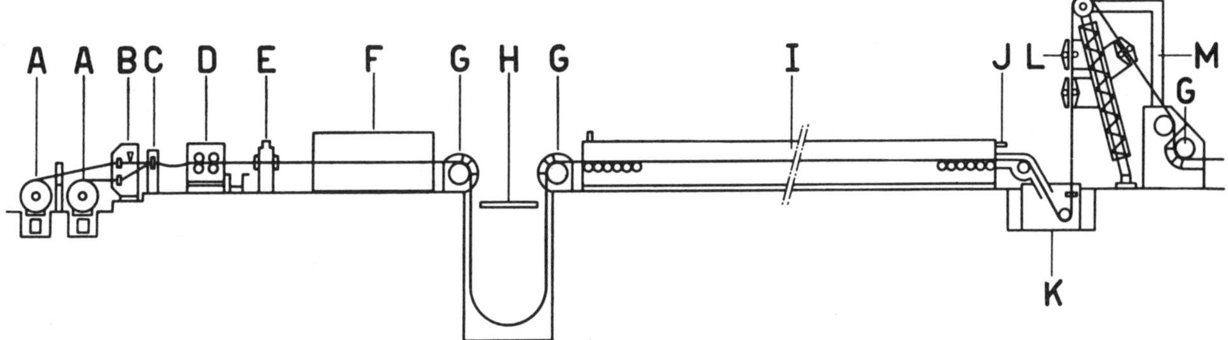

Figure 2. The process section of a continuous galvanising line for light-gauge steel strip. A. Uncoiler. B. Pinch rolls and shear. C. Pinch rolls. D. Side trimmer. E. Pinch rolls and lap welder. F. Cleaning unit. G. Bridle roll. H. Looping pit. I. Heat-treating chamber with controlled atmosphere–approximately 60 m in length. J. Furnace-atmosphere inlet. K. Galvanising and pig-melting pots. L. Forced-cooling hoods. M. Cooling tower.

annealing. The third type cleans the metal by acid pickling and alkali cleaning. A flux is employed prior to the preheat furnace. Most coils are annealed before entering the line, in a furnace shaped like a box. The fourth type of hot-dip galvanising employs both acid pickle and alkali immersion for cleaning the strip. No flux is used and the steel is preheated in a reducing gas before galvanising. Some coils are also box annealed as in the previous method.

The process section of a continuous galvanising line for light-gauge steel strip is shown in figure 2. Of note is the general discontinuance of fluxes in favour of the heat treating and annealing chambers. Further, this line omits acid pickling and utilises an alkaline cleaning process. Maintenance of clean surface is accomplished by keeping the heated strip in the chamber or furnace under a reducing atmosphere of hydrogen until it passes below the surface of the molten zinc bath.

In the galvanising of wire about 30 parallel strands of wire first pass through certain preparatory steps, then through a zinc bath to take-up frames on which each strand of wire is separately coiled. Generally, wire to be galvanised must be annealed to remove the effects of cold working. One of the continuous annealing processes is employed. Usually this is done by a molten lead pan in front of the cleaning and galvanising tanks. The molten lead also serves to remove the wire-drawing lubricant. Following this annealing step, the wires are cooled in air or in a water tank and thereafter cleaned in hot dilute hydrochloric acid. Next the wire is rinsed in hot water and a flux is applied by passing it through a hot solution of zinc chloride or zinc ammonium chloride, following which it passes over a dryer, then into the molten zinc bath. The flux coats the wire with a film of zinc chloride which protects it from oxidation.

A dross, which is an alloy of iron and zinc, settles to the bottom of the molten zinc bath and must be periodically removed. The zinc bath is generally covered with various types of materials to prevent rapid oxidation of the molten zinc on the surface of the bath, except at the point of entry and exit of the wires. At these uncovered points, frequent skimming is necessary.

Electrogalvanising. Electrogalvanising is the process of providing any metal with a zinc coating by means of an electric current, and is actually zinc electroplating. The equipment employed is similar to that used in other types of electroplating, and the process requires the same cleaning of the metal prior to galvanising as employed in hot-dip galvanising.

HAZARDS AND THEIR PREVENTION

Modern continuous galvanising lines present minimum health problems. No true poisoning occurs from handling zinc compounds: metal fume fever caused from inhalations of zinc oxide fume may occasionally occur but the molten zinc bath is kept at a lower temperature, 450-480 °C, than molten zinc in brass foundries and the surface of the bath is often covered with various materials which minimise the emission of zinc fumes. Similarly, the temperature of the bath and the covering of the surface prevent appreciable emission of lead fumes. Impurities, such as lead, antimony and cadmium, can be present in the zinc used in galvanising; however, none of these exceeds 0.75%.

The main source of air contaminants in galvanising is the use of fluxes, principally in batch galvanising, and the cleaning operations utilising hot acid solutions. Zinc ammonium chloride and ammonium chloride fluxes are thermally decomposed to form hydrogen chloride and ammonia gas both of which are readily detected by the upper respiratory tract at concentrations below levels considered hazardous. Both the acids employed for pickling are likewise upper respiratory irritants in quite low concentrations.

So-called fumeless fluxes have been in use for a number of years; however, ventilation still should be applied to galvanising lines. Control of fume by exhaust ventilation in the roof may be adequate since thermal updrafts from the hot processes will direct the contaminants towards the roof. Exhaust ventilation applied by exhaust slots extending along both the long sides of the tank is the necessary and most effective control measure for pickling and cleaning tanks or vats. Exhaust hoods are required for skimming and drossing operations over the pots or containers into which the dross or skims are dropped. The latter type of exhaust ventilation is especially necessary for lead containing drosses. Careful handling of the dross or skimmings is also required: the materials should be carefully placed in the containers to prevent heavy emission of dust, which may arise if they are thrown in from the top rim.

Protective railings should be provided for both galvanising and pickling lines, particularly where workers are on walkways along the top of the tank.

Workers in contact with molten lead or lead dressings should be subject to medical supervision with periodic blood and urine examinations.

Throughout the process suitable protective clothing should be provided including foot and leg protection, aprons, hand protection and eye and face protection such as face shields to guard against the risk of burns, and splashes from molten metal and acid. Good sanitary facilities, including showers, should be provided.

MORSE, K. M.

"Galvanic processes of metal surface" (Galwaniczna obrobka powierzchni metali). Pawelczak, M. *Ochrona Pracy* (Warsaw), Aug. 1979, 79/8 (16-18). 15 ref. (In Polish)

CIS 76-1422 *Hot-dip galvanising shops–concentration of harmful substances–safety engineering* (Feuerverzinkereien–Schadstoffsituation–Technische Prophylaxe). Engels, L. H. STF Report Nr. 1-76 (Staubforschungsinstitut des Hauptverbandes der gewerblichen Berufsgenossenschaften, Langwartweg 103, 5300 Bonn 1) (1976), 27 p. Illus. 15 ref. (In German)

CIS 74-0214 "Occupational physiology and safety in galvanising" (Arbeitsphysiologische und sicherheitstechnische Analyse der Arbeit an Zinkbädern). Nemecek, J.; Grandjean, E. *Zeitschrift für Präventivmedizin–Revue de médecine préventive* (Zurich), May-June 1973, 18/3 (131-133). Illus. (In German)

Garages

Garage-keeping is one of the most widespread and diverse of employments: any one or all of the following activities may be carried on in a garage:

(a) housing (or parking) of vehicles for a fee or in connection with an undertaking (e.g. bus garages);

(b) supply of motor fuel and maintenance services (washing, greasing, tyre maintenance);

(c) electrical or mechanical repairs (major or minor repairs to the electrical system, transmission, steering, engine, gear box, differential, rear suspension and brakes);

(d) bodywork repairs including painting and surface treatment of metal.

At one end of the scale, a garage may be a large, well equipped establishment, employing a hundred or more employees and often belonging to a chain, and at the other, a very small undertaking, concerned mainly with selling fuel and effecting minor running repairs with the proprietor working alongside one or two employees. Many small premises consist of a single building with one side continually open to the air.

Even a small garage will carry a variety of appliances and equipment, including possibly: jacks, hydraulic platforms, portable electric drills and lamps, a few small machine tools, a collection of spanners and other hand tools, a compressed-air system, welding equipment, inspection pits, perhaps a paint spray booth, and a battery charging section.

HAZARDS AND THEIR PREVENTION

Such a range of activities and equipment carries inevitably a wide range of risks and it is only possible to indicate the most common of them. Broadly speaking, risks are easier to control in the larger establishments where work areas and working personnel can be separated than in small places where use of any part of the plant is intermittent and the staff has no specialised skill. The keynotes for prevention are inspection and maintenance of equipment, good housekeeping, good storage, good ventilation, adequate sanitary and washing facilities, protective clothing, training and supervision.

Fire and explosion. Fires and explosions are ever-present hazards, not only the conflagration but more often the small incident involving grave personal injuries: many such injuries are caused by the accidental ignition of petrol-soaked clothing or of gasoline vapour. In general, the aim must always be to keep down the amount of petrol vapour in the air at every point, to prevent or control spillage, to eliminate sources of ignition, to keep a high standard of ventilation and cleanliness, to maintain efficient fire-fighting appliances and rapid means of escape, to provide a regular supply of clean overalls, and to train workers in safe methods of work and procedures in case of fire.

In particular: most countries have national or local regulations governing storage of gasoline and siting of pumps: these should be strictly observed; during filling, the engine should be stopped and the mouthpiece of the hose should be pressed firmly into the tank pipe. Gasoline tanks in cars should be drained and purged before repairs are carried out on them: ignition should be kept switched off, as short circuits may produce an arc capable of igniting gasoline vapour. Great care is necessary if welding or cutting is carried out: no work should be done on tanks or other receptacles until they have been thoroughly purged of all inflammable vapour. Where there is any risk of concentration of vapour, electrical installations and equipment should be of flameproof type. Safe storage of all cans of flammable solutions, whether full or empty, is necessary. The amount of any flammable fuel or solvent in a workroom should be kept as low as possible and containers should be kept covered. Workers should not be allowed to use gasoline for cleaning parts, floors, clothes or hands. Smoking should be prohibited in areas where gasoline vapour may arise.

Inspection pits present special dangers: gasoline may spill into the enclosed space and fires or explosions result; they should not be used for draining gasoline from tanks, which should have been previously purged; good ventilation and protected electrical equipment will cut down the risks. Flammable cellulose solutions may bring an additional risk if bodywork is spray-painted unless the work is carried out in a special area or booth of fireproof construction with exhaust ventilation. Acetylene gas cylinders should be carefully inspected, maintained and stored.

Lifting machines and tackle. Injuries are caused by defects in cranes, lifting machinery or tackle, failure of chain slings, falls of vehicles and overturning or slipping of platform-type hoists or jacks. Thorough regular inspection and maintenance of all plant is a prerequisite and, particularly when use is intermittent, the equipment should also be carefully inspected before use. In the case of hydraulic platforms, stops should be provided to control the rise of the ram, and end-stops on channels will prevent movement of the vehicle.

Electrical equipment. Apart from the need for flameproof electrical equipment where flammable vapours are present, robust equipment is necessary everywhere. Much use is made of portable equipment, especially portable electric power tools, and electric shock is always a risk unless efficient earthing, tough conductors, and specially designed hand lamps are used. Makeshift and improvised portable equipment is to be condemned.

Compressed-air plant. Regular inspection and maintenance will guard against risks of bursting of the air receiver. Workers should be instructed in careful use of all appliances and particularly against horseplay with compressed air.

Tyre maintenance. Bursting tyres may project fastenings violently into the air and cause serious injuries during repair and maintenance, especially during inflation. Tyres should be inflated within a cage or behind a screen and only competent persons should carry out the work.

Hand tools. These should be well maintained: spanners should be used for their proper purpose.

Falls. Floors should be kept clean and good housekeeping will prevent many falls and stumbles.

Inspection pits should be fenced or covered when not in use and should be protected during use: one method is shown in figure 1. Non-slip surfaces should be provided for pit stairs.

Traffic control. Where many vehicles are in use, strict rules for traffic control should be observed.

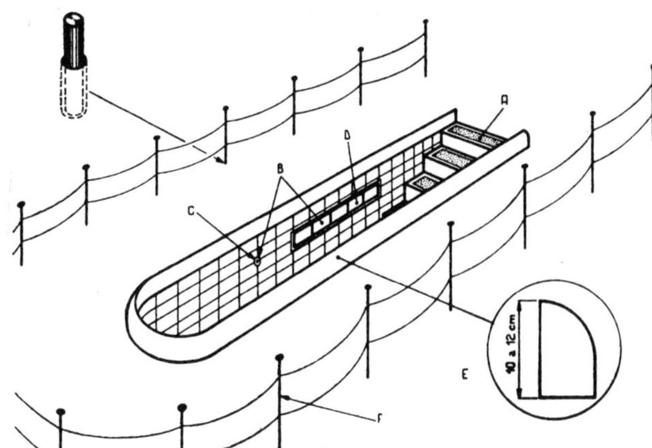

Figure 1. Safe design for a vehicle-inspection and repair pit. A. Non-slip stair treads. B. Protected electrical installation. C. Electrical power point. D. Lighting covered with wire mesh. E. Cross-section of raised pit edge. F. Removable barrier.

Health hazards. The exhaust fumes of internal combustion engines may cause carbon monoxide poisoning. Good ventilation is always essential: many garages are open to the air but in enclosed premises, especially where vehicles are used, exhaust ventilation may be necessary (see tables 1 and 2).

Table 1. Volume of air to be extracted from parking place in relation to the area and number of vehicles that can be parked

Number of vehicles per parking level[1]	Amount of air per m² of floor (m³/min)
Up to 60 vehicles	$\frac{195 - \text{number of cars}}{200}$
Over 60 vehicles	0.55

[1] From 20 to 25 m² is allowed per vehicle.

Table 2. Volume of air to be extracted from exhaust pipe of a vehicle in relation to its type and power

Type and power of vehicle	Minimum amount of air to be extracted from exhaust pipe (m³/min)
Motor car and truck with engine up to 200 bhp	3
Motor car and truck with engine over 200 bhp	6
Diesel-engined vehicle	12 or more

Where any dust is generated in quantity, local exhaust ventilation should be provided. Special care should be paid to asbestos fibres, where brake and clutch parts are repaired or linings and discs are dressed. These operations should be carried out in a booth with good exhaust ventilation. For the application of undersealing compounds containing asbestos it is required to use appropriate respiratory protection.

Epoxy resins as used in lacquers and with reactive curing agents in glues, are irritants to the eyes and skin, and are known to produce dermatitis and sensitisation. Therefore special care should be taken in handling these compounds and the workers involved should always wear personal protection. Some epoxy resins are suspected of producing cancer.

Rubber adhesives and solvents used in vulcanisation may constitute a toxic risk. Use of benzene should be prohibited and all work should be done in well ventilated conditions. Gasoline and oil may cause skin affections, dermatitis or oil acne: leaded gasoline brings a hazard of lead poisoning. Good washing facilities, barrier creams and instruction of workers are essential. Medical supervision is desirable.

Manual handling. Training in methods of lifting and carrying heavy objects will reduce risk of hernia and strains.

Protective clothing. Protective clothing is important to prevent both accident and health hazards. Overalls, rubber foot and leg protection, hand and arm protection as well as eye and face protection may all be necessary.

Temperature. In the colder countries, it may be difficult to maintain a reasonable working temperature in open-sided garages: sufficient means of heating of a safe type should always be provided: improvised heating such as braziers brings grave risk of fire and also of carbon monoxide poisoning. A heated messroom, even if small, is always desirable.

Lighting. The lighting of all areas of the garage should be adequate.

First aid. Many injuries are slight in the first place: immediate first aid from a well stocked first-aid kit by a person trained in first-aid principles reduces the risk of more serious developments. Medical supervision is desirable for preplacement and regular examinations of workers.

Hours of work. To meet public needs, garages tend to be open for very long hours, during rest days and at holiday periods. Careful scheduling of hours of work to avoid over-work is necessary, particularly where young persons are employed.

SCHMIDT, E.

Accidents:

CIS 74-1764 "Poisoning and explosion hazards in motor car repair workshops, garages and car-wash tunnels" (Risques d'intoxication et d'explosion dans les ateliers de réparation d'automobiles, dans les garages et dans les tunnels de lavage). Burri, W. R. *Schweizerische Blätter für Arbeitssicherheit–Cahiers suisses de la sécurité du travail* (Lucerne), Mar. 1974, 114, 32 p. Illus. (In French, German, Italian)

Work environment:

CIS 77-1950 "Heavy metal pollution among autoworkers. I. Lead–II. Cadmium, chromium, copper, manganese, and nickel". Clausen, J.; Rastogi, S. C. *British Journal of Industrial Medicine* (London), Aug. 1977, 34/3 (208-220). Illus. 63 ref.

CIS 77-1269 "Dust concentrations in the manufacture of brake linings containing asbestos, and their use in garages" (Staubkonzentrationen beim Herstellen asbesthaltiger Reibbeläge und deren Bearbeitung in Kfz-Werkstätten). *Staub* (Dusseldorf), Dec. 1975, 35/12 (433-436). Illus. 17 ref. (In German)

Hygiene and safety:

CIS 2712-1971 *Report of industrial hygiene surveys in garages and auto body repair shops.* Rivera, R. A. Report TR-68 (Bureau of Occupational Safety and Health, Public Health Service, 1014 Broadway, Cincinnati, Ohio) (June 1970), 11 p. 3 ref.

Motor garages and service stations. Safe operations in industry No. 5 (Adelaide, South Australia Department of Labour and Industry, 1975), 40 p. Illus.

CIS 81-565 *Repair and maintenance of motor vehicles–Prevention of occupational accidents* (Réparation et entretien des véhicules automobiles–Prévention des accidents du travail). INRS 607 (Institut national de recherche et de sécurité, 30 rue Olivier-Noyer, Paris) (July 1980), 92 p. Illus. 27 ref. (In French)

"Safe motorcar maintenance" (Sichere Fahrzeuginstandhaltung). *Sicher Arbeiten* (Dusseldorf), Feb. 1980, 1 (9-16). Illus. (In German)

Gardening and market gardening

The hazards encountered in gardening are very similar to those met with in agriculture (see AGRICULTURAL WORK). However, gardeners and market gardeners usually work small plots and use intensive methods. They have more direct contact with the soil than does the average agricultural worker.

HAZARDS AND THEIR PREVENTION

Insects. The stings and bites of insects, such as wasps, hornets, bees, mosquitoes, horseflies, flies, fleas and ants, to which gardeners and market gardeners are exposed, can cause local chemical or allergic reactions (see ANIMALS, DANGEROUS). Horseflies, frequently encountered on dunghills, have a long proboscis with which they can pierce clothes; their sting is painful and leads to a swelling surrounded by an oedematous zone, followed by adenitis, and their larvae cause a red linear eruption which grows by 2-3 cm per day. Horseflies can transmit tularaemia and filariasis. Certain species of fly, such as the simuliidae (blackflies), which live in small water-filled furrows, have a sting even more painful than a mosquito bite; it may raise a hard swelling which will last for several weeks. Fleas are found in areas inhabited by contaminated persons or animals. One type of flea, *Tunga penetrans*, or jigger flea, attacks the feet and causes irritation, swellings under the toe nails, and pseudo warts; it may also carry plague. Ants cause local or multiple urticarial lesions, accompanied at first by a burning sensation due to the formic acid in the sting, then by itching. These lesions may ultimately become eczematous or indurated.

One method of ensuring protection against insects is the use of repellents; these are odourless substances which are applied to the skin or clothes and remain active for several hours (*n*-butyl, dimethyl and ethyl phthalate or N,N-diethyl-*m*-toluamide against fleas; universal repellents such as M-1960, composed of benzylbutylacetanilide and 2-ethyl-1,3-hexanediol). Insecticides can be used to exterminate insects, or the sexual reproduction of insects can be stopped by using radioactive substances to sterilise the males (such methods are used in treating large areas, under technical supervision). Personal protective equipment can be worn (long clothes, gloves, boots and hoods of woven textile mesh against bees and of transparent plastics against caterpillars). The ground, together with plants and animals, may be sprayed or otherwise disinfected.

Hairs, cocoons and nests of insects. Caterpillar hairs, which are impregnated with liquids such as formic and uric acid, can cause urticarial dermatitis with intense itching, oedema with or without dermatitis, conjunctivitis and keratitis if there is eye contact, rhinitis, pharyngitis, and bronchitis as a result of penetration of the respiratory tract. The most dangerous are the pine-tree processionary caterpillar *(Thaumetopoea pityocampa)* which lives on and contaminates the ground and the *Euproctis chysorrhoea* which awakes in spring and falls on gardeners. *E. chysorrhoea* cocoons occur in such large numbers that they have to be pulled off, and contact produces erythematous or urticarial rash, accompanied by itching. The same is true of caterpillar nests, in which dejecta accumulate. Certain butterflies, such as *Limantria dispar*, lay their eggs in a sort of nest formed of a bag or pouch, which may cause conjunctivitis, itching, rashes, rhinitis and pharyngitis when in direct contact with the skin or following wind dissemination.

Plants. Numerous plants cause dermatitis in exposed parts of the body or in parts of the body touched by contaminated hands. Nettles and poison ivy produce small wounds through which formic acid and other substances penetrate the skin, giving rise to a painful burning sensation, redness and nettle-rash, which may spread and cause fever. The bulbs of tulips, hyacinths, narcissi, asphodels, etc., contain calcium oxalate which may cause the skin on the fingers to peel (usually under the finger nails), a process which is sometimes painful. The *Primulaceae* can cause eczematous dermatitis. [Daffodils and other bulbs of the Amaryllidaceae family can also cause contact dermatitis.] Certain thorny plants can cause inflammation of the cell tissue, which is painful and may be complicated by secondary infection followed by the formation of whitlows or boils, or the penetration of vegetable poisons.

The resin of *Rhus toxicodendron* causes eczematous dermatitis of an erysipeloid type, and the turpentine contained in the resins of conifers may cause reddening and blisters, while the creosote in beech tar is irritating to the skin.

Celery and parsley may cause urticaria or erythematous dermatitis. Thyme and rosemary can cause dermatitis in the palms of the hands and on the fingers. The dust of cereals produces papulous urticaria. Vanilla and myroxylon, which contain Peruvian balsam, can cause erythema and blisters. Virginia creeper, arnica, *Arundo donax*, asparagus, geraniums, oranges and lemons may also cause dermatitis.

Some plants may cause poisoning. Crocus saffron, if inhaled, can cause vomiting and even painful diarrhoea. The berries and leaves of the deadly nightshade are poisonous; however, it is not poisonous to snails or rabbits, and cases of poisoning have been reported resulting from eating the flesh of animals fed on this plant. Hemlock contains alkaloids. Certain solanaceae *(Solanum nigra)* may also be poisonous. There are a variety of other poisonous plants: *Artemisa cina*, which causes digestive, urinary, nervous and visual disorders; the bark and the root of the pomegranate tree contain alkaloids; *Chenopodium* can lead to abortion and deafness; rue *(Ruta graveolens)* and savin *(Juniperus sabina)* contain abortifacient oils irritating to the skin and to the nervous and digestive systems; *Bryona dioica* causes very serious digestive troubles; aconite, sometimes confused with celery or horseradish, causes poisoning, accompanied by sensory disorders.

Certain people are allergic to the pollen of plants, and contract hay-fever, which may evolve into asthmatic disease (pollinosis).

It is important to know these plants and the dangers they present, and to keep away from the harmful ones. Disturbances due to pollen may make it necessary for the subject to find some other occupation. Protective clothing and cream can be used against dermatitis. Sesame oil or yellow wax are a protection against calcium oxalate, while dilute ammonia is a protection against formic acid.

Miscellaneous hazards. [Garden soil is a substrate of saprophitic vegetation, and skin microtrauma in gardeners can facilitate the occurrence of dermatophytoses due to geophilic microsporia.] Fertilisers sometimes exert a chemical action but, more important, when of animal origin they may carry tetanus, typhoid fever, cholera, dysentery and tuberculosis. Soot, used to destroy slugs, and arsenical products can cause cancer of the skin. Ethyl ether, used to produce early flowering, may cause fires or explosions, or produce toxicomania by narcosis. The handling of pesticides and insecticides, too, can be dangerous (see PESTICIDES; PESTICIDES, HALOGENATED; PESTICIDES, ORGANOPHOSPHORUS).

[Lawn mowers can injure inexperienced operators, or bystanders, especially children, who fall under a walking mower or carelessly insert hands or feet under the rotating blades. Bystanders should keep a safe distance and children should be forbidden to operate such machines.]

Tools may cause many accidents, and should therefore be maintained in good condition and disposed of when no longer serviceable; cutting tools should be sheathed when not in use. All implements should be carefully stored away in a suitable place. See AGRICULTURAL IMPLEMENTS.

Incorrect working postures may lead to hernias or back pain (caused by an effort), as well as to rheumatism.

Work should not be carried on in an uncomfortable posture, and a single position should not be maintained too long. Too much digging may lead to dorsal or lumbar disorders. Long use of a bill-hook may lead to teno-synovitis. But provided such activities are interspersed with others, there is little likelihood of trouble.

Water used to water plants is frequently contaminated by large numbers of parasites (larvae, leptospirae) and germs. Water should not be allowed to stagnate, and it should be disinfected if necessary. The soil may contain threadworms, tapeworms, hookworms and tetanus bacilli.

Mushrooms may produce mycosis of the skin or intestines. The stalks of plants which may be contaminated should never be chewed, and efforts should be made not to inhale the dust given off by straw or dry grass.

Gardeners should practise rigorous personal hygiene, carefully wash their hands before eating or smoking, and wear suitable working clothes, waterproof boots and gloves.

CAPELLA, F. E. P.

Safety of lawn mowers:

CIS 77-1239 "Power lawn mowers". Data Sheet 464, Revision B (Extensive). *National Safety News* (Chicago), Apr. 1977, 115/4 (75-78). Illus. 2 ref.

Gas cylinders

For convenience in handling, transportation and storage, gases are commonly compressed in metal gas cylinders at pressures that range from a few atmospheres overpressure to 200 bar or even more. Alloy steel is the material most commonly used for the cylinders, but aluminium is also widely used for many purposes, for example for fire extinguishers.

The hazards met with in handling and using compressed gases are:

- normal hazards entailed in handling heavy objects;
- hazards connected with pressure, i.e. the amount of stored energy in the gases;
- hazards from the special properties of the gas content, which may be flammable, poisonous, oxidising, etc.

Cylinder manufacture. Steel cylinders may be seamless or welded. The seamless cylinders are made from high-quality alloy steels and carefully heat treated in order to obtain the desired combination of strength and toughness for high-pressure service. They may be forged and hot drawn from steel billets or hot formed from seamless tubes. Welded cylinders are made from sheet material. The pressed top and bottom parts are welded to a cylindrical seamless or welded tube section and heat treated to relieve material stresses. Welded cylinders are extensively used in low-pressure service for liquefiable gases and for dissolved gases such as acetylene.

Aluminium cylinders are extruded in large presses from special alloys that are heat treated to give the desired strength.

Gas cylinders must be designed, produced and tested according to strict norms or standards. Every batch of cylinders should be checked for material quality and heat treatment, and a certain number of cylinders tested for mechanical strength. Inspection is often aided by sophisticated instruments, but in all cases the cylinders should be inspected and hydraulically tested to a given test pressure by an approved inspector. Identification data and the inspector's mark should be permanently stamped in the cylinder neck or in another suitable place.

Most standards or codes are national, but much work is being done in order to reach international agreement on gas cylinders and their use.

Periodic inspection. Gas cylinders in use may be affected by rough treatment, corrosion from inside and outside, fire, etc. National or international codes therefore require that they shall not be filled unless they are inspected and tested at certain intervals, which mostly range between 2 and 10 years, depending on the service. Internal and external visual inspection together with a hydraulic pressure test is the basis for the approval of the cylinder for a new period in a given service. The test date (month and year) is stamped into the cylinder.

Disposal. A large number of cylinders are scrapped every year for various reasons. It is equally important that these cylinders are disposed of in such a way that they will not find their way back into use through uncontrolled channels. The cylinders should therefore be made completely unserviceable by cutting, crushing or a similar safe procedure.

Valves. The valve and any safety attachment must be regarded as a part of the cylinder, which must be kept in good working condition. Neck and outlet threads should be intact, and the valve should close tight without the use of undue force.

Shut-off valves are often equipped with a pressure relief device. This may be in the form of a resetting safety valve, bursting disc, fuse plug (melt plug) or a combination of bursting disc and fuse plug. The practice varies from country to country, but cylinders for low-pressure liquefied gases are always equipped with safety valves connected to the gas phase.

HAZARDS AND THEIR PREVENTION

Different transport codes classify gases as compressed, liquefied or dissolved under pressure. For the purpose of this article, it is more useful to use the type of hazard as a classification.

High pressure. If cylinders or equipment burst, damage and injuries may be caused by flying debris or by the gas pressure. The more a gas is compressed, the higher is the stored energy. This hazard is always present with compressed gases and will increase with temperature if the cylinders are heated. Hence it is necessary to—

(a) avoid mechanical damage to the cylinder (dents, cuts, etc.);

(b) store cylinders away from heat and not in direct sun;

(c) remove cylinders from fires;

(d) connect cylinders only to equipment conditioned for the intended use;

(e) protect the cylinder valve with the cap during transport;

(f) secure cylinders in use against falling, which may knock off the valve;

(g) avoid tampering with safety devices;

(h) handle cylinders with care to avoid mechanical shocks in very cold climates, since steel may become brittle at low temperature;

(i) avoid corrosion, which reduces the strength of the shell.

Low temperature. Most liquefied gases will evaporate rapidly under atmospheric pressure and may reach very low temperatures. A person who has got such liquid on himself may sustain injuries in the form of "cold burns". (Liquid CO_2 will form snow particles when expanded.) Proper protective equipment (gloves, goggles) should therefore be used.

Oxidation. The hazard of oxidation is most evident with oxygen, which is one of the most important compressed gases. Oxygen will not burn on its own, but is necessary for combustion. Normal air contains 21% oxygen by volume.

> *All combustible materials will ignite more easily and burn more vigorously when the oxygen concentration is increased.*

This is noticeable already with a slight increase in oxygen concentration, and utmost care must be taken to avoid oxygen enrichment in the working atmosphere. In confined spaces small oxygen leaks may lead to dangerous enrichment.

The danger with oxygen increases with increasing pressure to the point where many metals will burn vigorously. Finely divided materials may burn in oxygen with explosive force. Clothing that is saturated with oxygen will burn very rapidly and be difficult to extinguish.

Oil and grease have always been regarded as dangerous in combination with oxygen. The reason is that they react readily with oxygen, their existence is common, the ignition temperature is low and the developed heat may start a fire in the underlying metal. In high-pressure oxygen equipment the necessary ignition temperature may easily be reached by the compression shock that may result from rapid valve opening (adiabatic compression).

It is therefore necessary to—

(a) operate valves slowly;

(b) keep all oxygen equipment clean and free from oil and dirt;

(c) use only materials that are proven to be safe with oxygen;

(d) refrain from lubricating oxygen equipment;

(e) avoid entering confined spaces where oxygen may exist in higher concentration; check the atmosphere;

(f) avoid strictly the use of oxygen instead of compressed air or some other gas.

Flammability. The flammable gases have flash points below room temperature and will form explosive mixtures with air (or oxygen) within certain limits known as the lower and upper explosion limits.

Escaping gas (also from safety valves) may ignite and burn with a shorter or longer flame depending on the pressure and amount of gas. The flames may again heat nearby equipment, which may burn, melt or explode. Hydrogen burns with an almost invisible flame.

Even small leaks may cause explosive mixtures in confined spaces. Some gases, such as liquefied petroleum gases (LPG), mostly propane and butane, are heavier than air and are difficult to vent away, as they will concentrate in the lower parts of buildings and "float" through channels from one room to another. Sooner or later, the gas may reach an ignition source and explode.

Ignition may be caused by hot sources, but also by electrical sparks, even very small ones.

Acetylene takes a special place among the combustible gases because of its properties and wide use. If heated, the gas may start to decompose with the development of heat even without the presence of air. If allowed to proceed, this may lead to cylinder explosion.

Acetylene cylinders are, for safety reasons, filled with a highly porous mass which also contains a solvent for the gas. Outside heating from a fire or welding torch, or in certain cases internal ignition by strong backfires from welding equipment, may start a decomposition within the cylinder. In such cases:

(a) close the valve (using protective gloves if necessary) and remove the cylinder from the fire;

(b) if part of the cylinder becomes hotter, put it into a river, canal or the like to cool down, or cool it with water sprays;

(c) if the cylinder is too hot to be handled, spray it with water from a safe distance;

(d) continue cooling until the cylinder stays cool by itself;

(e) keep the valve closed, because a gas flow will accelerate decomposition.

Acetylene cylinders in several countries are equipped with fuse (melting) plugs. These will release the gas pressure when they melt (usually at about 100 °C) and prevent cylinder explosion. At the same time there is a risk that the released gas may ignite and explode.

Common precautions to observe in respect of combustible gases are as follows:

(a) store cylinders separately from other gases in a well ventilated area above ground level;

(b) do not use leaking cylinders or equipment;

(c) store and use liquid gas cylinders in an upright position. Larger quantities of gas will come out if liquid is expelled through the safety valves instead of gas. The pressure will be reduced more slowly. Very long flame will result if the gas ignites;

(d) in case of leaks, avoid any possible ignition source and ventilate;

(e) do not allow smoking where flammable gases are stored or used;

(f) the safest way of extinguishing a fire is usually to stop the supply of gas. Merely extinguishing the flame may cause the formation of an explosive cloud, which may re-ignite in contact with a hot object.

Toxicity. Certain gases, if not the most common, may be toxic. At the same time, they may be irritating or corrosive to the skin or eyes.

Persons who handle these gases should be well trained and aware of the danger involved and the necessary precautions. The cylinders should be stored in a well ventilated area. No leaks should be tolerated. Suitable protective equipment (gas masks or breathing equipment) should be used.

Inert gases. Gases such as argon, carbon dioxide, helium and nitrogen are widely used as protective atmospheres to prevent unwanted reactions in welding, chemical plants, steel works, etc. These gases are not labelled as being hazardous and serious accidents may happen because only oxygen can support life.

When *any* gas or gas mixture displaces the air so that the breathing atmosphere becomes deficient in oxygen, there is a danger of asphyxiation. Unconsciousness or death may result very rapidly when there is little or no oxygen, and there is no warning effect.

Confined spaces where the breathing atmosphere is deficient in oxygen must be vented before entering.

When breathing equipment is used, the person entering must be supervised. Breathing equipment must be used even in rescuing operations. Normal gas masks give *no protection* against oxygen deficiency.

The same precaution must be observed with large, permanent fire-fighting installations, which are often automatic, and those who may be present in such areas should be warned of the danger.

Cylinder filling. Cylinder filling involves the operation of high-pressure compressors or liquid pumps. The pumps may operate with cryogenic (very low-temperature) liquids. The filling stations may also incorporate large storage tanks of liquid gases in a pressurised and/or deeply refrigerated state.

The gas filler should check that the cylinders are in acceptable condition for filling and fill the correct gas in not more than the approved amount or pressure. The filling equipment should be designed and tested for the given pressure and type of gas and protected by safety valves. Cleanliness and material requirements for oxygen service must be observed strictly. When filling flammable or toxic gases, special attention should be given to the safety of the operators. The primary requirement is good ventilation combined with proper equipment and technique.

Cylinders which are contaminated with other gases or liquids by the customers constitute a special hazard. Cylinders with no residual pressure may be purged or evacuated before filling. Special care should be taken to ensure that medical gas cylinders are free from any harmful matter.

Transport. Local transport tends to become more mechanised through the use of fork-lift trucks, etc. Cylinders should be transported only with the caps on and secured against falling from the vehicles. Cylinders must not be dropped from trucks directly onto the ground. For hoisting with cranes, suitable lifting cradles should be used. Magnetic lifting devices or caps with uncertain threads should not be used for lifting cylinders.

When cylinders are manifolded into larger packages, great care should be taken to avoid strain on the connections. Any hazard will be increased because of the greater amount of gas involved. It is good practice to divide larger units into sections and to place shut-off valves where they can be operated in any expected emergency.

The most frequently occurring accidents in cylinder handling and transport are injuries caused by the hard, heavy and difficult-to-handle cylinders. Safety shoes should be worn. Trolleys should be provided for longer transport of single cylinders.

In international transport codes compressed gases are classified as dangerous goods. These codes give details about which gases may be transported, cylinder requirements, allowed pressure, marking, etc.

Identification of content. The most important requirement for safe handling of compressed gases is the correct identification of the gas content. Stamping, labelling, stencilling and colour marking are the means that are used for this purpose. Certain requirements for marking are covered in ISO standards. The colour marking of medical gas cylinders follows the ISO standards in most countries. Standardised colours are also used in many countries for other gases, but this is not a sufficient identification. In the end only the written word can be regarded as a proof of the cylinder content.

Standardised valve outlets. The use of a standardised valve outlet for a certain gas or group of gases strongly reduces the chance of connecting cylinders and equipment made for different gases. Adapters should therefore not be used, as this sets aside the safety measures. Only normal tools and no excessive force should be used when making connections.

Safe practice for users. The safe use of compressed gases entails applying the safety principles outlined in the preceding sections. This is not possible unless the user has some basic knowledge of the gas and the equipment that he is handling. In addition he should take the following precautions:

- Use gas cylinders only for the purpose for which they are intended and not as rollers or work supports.
- Store and handle the cylinders in such a way that their mechanical strength is not reduced, for instance by severe corrosion, sharp dents, cuts, etc.
- Remove cylinders from fires or excessive heat.
- Keep no more than the necessary number of gas cylinders in working areas or occupied buildings. Keep them preferably near doors and not in emergency escape routes or difficult-to-reach places.
- Clearly mark and return to the filler (owner) any cylinders that have been exposed to fires, since they may have become brittle or lost their strength.
- Store cylinders in a well ventilated place, away from rain or snow and any combustible storage.
- Secure cylinders in use against falling.
- Identify the gas content positively before use.
- Read labels or instructions carefully.
- Connect only to equipment meant for the particular service.
- Keep the connections clean and in good order. Check their conditions at regular intervals.
- Use good tools (normal length, fixed wrench).
- Leave loose valve keys in place when the cylinder is in use.
- Close the valve when the cylinder is not in use.
- Remove cylinders or connected equipment from confined spaces when not in use (even during shorter breaks).
- Check the atmosphere for oxygen content and if possible for flammable gases before entering confined areas and also during prolonged working periods.
- Observe that heavy gases may concentrate in lower areas and that they may be difficult to remove by ventilation.
- Protect cylinders against contamination from pressurised equipment, since backflow of other gases or liquid may lead to serious accidents. Use proper non-return valves, block-and-bleed arrangements or the like.
- Return empty cylinders to the filler with the valves closed and the caps on. Always leave a little residual pressure in the cylinder to prevent contamination with air and moisture.
- Notify the filler of any faulty cylinders.
- Use acetylene only at a properly reduced pressure.
- Use flame arrestors in acetylene lines where acetylene is used with compressed air or oxygen.
- Have fire extinguishers and heat-protecting gloves available with gas welding equipment.

- Store and use liquid gas cylinders in an upright position.
- Allow poisonous and irritating gases, such as chlorine, to be handled only by well informed operators with personal safety equipment.
- Do not keep unidentified cylinders in stock.

Fixed installations, with the gas cylinders connected in separate gas centrals, are safest where gases are used regularly.

MATHISEN, K. R.

Gas cylinders for industrial use. Marking for identification of content. ISO 448-1977 (Geneva, International Organisation for Standardisation, 1977), 2 p.

Dissolved acetylene cylinders. Basic requirements. ISO 3807-1977 (Geneva, International Organisation for Standardisation, 1977), 7 p.

Valve outlets for gas cylinders–List of standard provisions or those in use. (Formerly ISO/DATA 4.) ISO/TR 740-1978 (Geneva, International Organisation for Standardisation, 1978), 9 p.

"Compressed Gases. Part I–Cylinders". *Safety Surveyor* (London), May 1980, 8/1 (4-12). 18 ref.

See also references appended to: DANGEROUS SUBSTANCES, LABELLING AND MARKING OF.

Gases and air, compressed

Gases in their compressed state, and particularly compressed air, are almost indispensable to modern industry, and are also used widely for medical purposes, for the manufacture of mineral waters, for underwater diving, and in connection with motor vehicles.

For purposes of the present article, compressed gases and air are defined as being those with a gauge pressure exceeding 1.47 bar (1.5 kgf/cm²) or as liquids having a vapour pressure exceeding 2.94 bar (3 kgf/cm²). Thus, consideration is not given to such cases as town gas distribution where the working pressures are only slightly above atmospheric and which is dealt with in the article GAS MANUFACTURE.

Among the gases commonly encountered, the following may be considered as being dangerous because of their flammable or toxic properties:

Acetylene*	Helium
Ammonia*	Hydrogen*
Butane*	Hydrogen chloride
Carbon dioxide	Hydrogen cyanide*
Carbon monoxide*	Methane*
Chlorine	Methylamine*
Chlorodifluormethane	Neon
Chloroethane*	Nitrogen
Chloromethane*	Nitrogen dioxide
Chlorotetrafluoroethane	Nitrous oxide
Cyclopropane*	Oxygen
Dichlorodifluoromethane	Phosgene
Ethane*	Propane*
Ethylene*	Propylene*
	Sulphur dioxide

* These gases are flammable.

All the above gases present either an irritant, asphyxiant, or highly toxic respiratory hazard and may also be flammable and an explosive when compressed. Most countries provide for a standard colour coding system whereby different coloured bands or labels are applied to the gas cylinders to indicate the type of hazard to be expected. Particularly toxic gases, such as hydrogen cyanide, are also given special markings.

All compressed gas containers are so constructed that they are safe for the purposes for which they are intended when first put into service. However, serious accidents may result from their misuse, abuse or mishandling, and the greatest care should be exercised in the handling, transport, storage and even in the disposal of such cylinders or containers (see GAS CYLINDERS).

Characteristics and production

Depending on the characteristics of the gas, it may be introduced into the container or cylinder in liquid form or simply as a gas under high pressure. In order to liquefy a gas, it is necessary to cool it to below its critical temperature and to subject it to an appropriate pressure. The lower the temperature is reduced below the critical temperature, the less the pressure that is required.

Certain of the gases listed above have properties against which precautions must be taken. For example, acetylene can react dangerously with copper and should not be in contact with alloys containing more than 66% of this metal. It is usually delivered in steel containers at about 14.7-16.8 bar (15-17 kgf/cm²). Another gas that has a highly corrosive action on copper is ammonia, which must also be kept out of contact with this metal, use being made of steel cylinders and authorised alloys. In the case of chlorine, no reaction takes place with copper or steel except in the presence of water, and for this reason all storage vessels or other containers must be kept free from contact with moisture at all times. Fluorine gas, on the other hand, although reacting readily with most metals, will tend to form a protective coating as, for example, in the case of copper where a layer of copper fluoride over the metal protects it from further attack by the gas.

Among the gases listed, carbon dioxide is one of the most readily liquefied, this taking place at a temperature of 15 °C and a pressure of about 14.7 bar (15 kgf/cm²). It has many commercial applications and may be kept in steel cylinders.

The hydrocarbon gases, of which liquefied petroleum gas, known as LPG, is a mixture formed mainly of butane (about 62%) and propane (about 36%), are not corrosive and are generally delivered in steel cylinders or other containers at pressures of up to 14.7-19.6 bar (15-20 kgf/cm²). Methane is another highly flammable gas that is also generally delivered in steel cylinders at a pressure of 14.7-19.6 bar (15-20 kgf/cm²).

HAZARDS AND THEIR PREVENTION

Storage and transport. When a filling, storage and despatch depot is being selected, consideration must be given to the safety of both the site and the environment. Pump rooms, filling machinery, etc., must be located in fire-resistant buildings with roofs of light construction. Doors and other closures should open outwards from the building. The premises should be adequately ventilated and a system of lighting with flame proof electrical switches should be installed. Measures should be taken to ensure free movement in the premises for filling, checking and despatch purposes, and safety exits should be provided.

Compressed gases may be stored in the open only if they are adequately protected from the weather and direct sunlight. Storage areas should be located at a safe distance from occupied premises and neighbouring dwellings.

During the transport and distribution of containers, care must be taken to ensure that valves and connections are not damaged. Adequate precautions should be taken to prevent cylinders from falling off the vehicle and from being subjected to rough usage, excessive shocks or local stress, and to prevent excessive movement of liquids in large tanks. Every vehicle should be equipped with a fire extinguisher and an electrically conductive strip for earthing static electricity, and should be clearly marked "Flammable liquids". Exhaust pipes should have a flame control device and engines should be halted during loading and unloading. The maximum speed of these vehicles should be rigorously limited.

Use. The main dangers in the use of compressed gases arise from their pressure and from their toxic and/or flammable properties. The principal precautions are to ensure that equipment is used only with those gases for which it was designed, and that no compressed gases are used for any purpose other than that for which their use has been authorised.

All hoses and other equipment should be of good quality and should be examined frequently. The use of non-return valves should be enforced wherever necessary. All hose connections should be in good condition and no joints should be made by forcing together threads that do not exactly correspond. In the case of acetylene and combustible gases, a red hose should be used; for oxygen the hose should be black. It is recommended that for all flammable gases, the connection-screw thread shall be left-handed, and for all other gases, it shall be right-handed. Hoses should never be interchanged.

Oxygen and some anaesthetic gases are often transported in large cylinders. The transfer of these compressed gases to small cylinders is a hazardous operation which should be done under competent supervision and making use of the correct equipment in a proper installation.

Compressed air is widely used in many branches of industry and care should be taken in the installation of pipelines and their protection from damage. Hoses and fittings should be maintained in good condition and subjected to regular examinations. The application of a compressed air hose or jet to an open cut or wound through which air can enter the tissues or the bloodstream is particularly dangerous; precautions should also be taken against all forms of irresponsible behaviour which could result in a compressed air jet coming in contact with any openings in the body–the result of which can be fatal. A further hazard exists when compressed air jets are used to clean machined components or workplaces: flying particles have been known to cause injury or blindness, and precautions against such dangers should be enforced.

Labelling, marking. This should be in accordance with standard practice in the country or region in question. The use of one gas in mistake for another or the filling of a container with a gas different to that which it previously contained, without the necessary cleaning and decontamination procedures, may cause serious accidents. Colour marking is the best method of avoiding such errors and consists of painting specific areas of containers or piping systems in accordance with the colour code stipulated in national standards or recommended by the national safety organisation.

TURKDOGAN, A.

Compressed air:

CIS 79-359 "Air compressors and air receivers" (Compresseurs d'air et récipients d'air comprimé). Busenhart, H. *Cahiers suisses de la sécurité du travail* (Lucerne), Nov. 1978, 129, 20 p. Illus. (In French, German, Italian)

CIS 80-701 "Air equipment and noise". Moffatt, R. A. *National Safety News* (Chicago), Dec. 1979, 120/6 (55-56).

Compressed gases:

CIS 75-1544 *Safe work with oxygen, acetylene, butane and propane* (Sécurité dans l'utilisation de l'oxygène, l'acétylène, le butane et le propane). Edition OPPBTP No. 158A (Organisme professionnel de prévention du bâtiment et des travaux publics, 2 bis rue Michelet, 92130 Issy-les-Moulineaux) (1st ed., 1973), 64 p. Illus. (In French)

Aerosol sprays:

CIS 79-358 "Technical rules for compressed gases" (Technische Regeln Druckgase (TRG)). Bekanntmachung, Bundesministerium für Arbeit und Sozialordnung *Arbeitsschutz* (Stuttgart), July-Aug. 1978, 7-8 (281-284). (In German)

Gases and vapours, biological effects of

A gas may be defined as a substance in a state which is governed by the gas laws at atmospheric temperature and pressure. A vapour is the gaseous phase of a substance which is liquid at ordinary temperature and pressure. Gases and vapours mix with the normal components of air, and many foreign gases may be present in the atmosphere at the same time. The concentration of a gas or vapour in air is expressed in parts per million by volume, or in milligrammes per cubic metre, usually at a barometric pressure of 760 mmHg or 1 013 mbar and a temperature of 25 °C (as in the United States) or of 20 °C (as in the Federal Republic of Germany).

Foreign gases and vapours are inhaled with the air, and many are hazardous or injurious to health. These effects, however, vary with the substance.

The physical characteristics of a gas or vapour of greatest significance in determining its biological effect are its concentration in air and its solubility in blood and tissues. The higher these are, the greater will be the absorption of the gas or vapour in the body. In addition, its toxicity, sites and modes of entry also influence its effect on health.

Some gases and vapours are highly flammable, including the majority of the industrial solvent vapours, and the utmost precautions and measures are necessary to prevent accidental fires or explosions from these.

Classification of gases and vapours

Almost all known gases and vapours are now produced or used in industry. The common ones may be classified as follows:

(a) industrial organic solvent vapours, such as benzene, xylene, alcohols, acetone, trichloroethylene, carbon tetrachloride and carbon disulphide;

(b) upper respiratory irritant gases, such as ammonia, acetic acid, sulphur dioxide and formaldehyde;

(c) pulmonary irritant gases, such as chlorine, phosgene and nitrogen oxide;

(d) chemical asphyxiant gases, such as carbon monoxide and hydrocyanic acid;

(e) simple asphyxiant gases, such as nitrogen, carbon dioxide, methane, its homologues and acetylene;

(f) other gases, inorganic and organic, such as hydrogen sulphide, arsine and agricultural pesticides.

Absorption, metabolism and elimination

There are three principal sites of entry into the body: the respiratory system, the digestive system and the skin.

The most important site of entry is the respiratory tract. Gases and vapours reach the lungs by inhalation, are dissolved in the alveoli, and diffuse into the blood vessels in the lungs. Highly soluble substances are rapidly absorbed by the superficial tissues of the nose and the trachea, and may be eliminated with the nasal mucous and sputa by sneezing or coughing. Less soluble gases and vapours penetrate to the lungs, where they may be absorbed and distributed by the blood. The absorption and subsequent distribution of gases and vapours increases as the rates and depths of respiration and the pulmonary blood flow are increased through muscular exercise.

The general law governing absorption of inert gases through the lungs was described by Henderson and Haggard, who theoretically deduced the following equation for the time (t_o) required to reach a certain fraction (x) of blood saturation:

$$t_x = 2.303 \log_{10} \frac{1}{1-x} \cdot \frac{(BK + L)W}{LB}$$

where B = amount of blood circulation through the lungs per minute; K = coefficient of distribution of gas between equal volumes of air and blood in the lungs; L = efficient pulmonary ventilation volume per minute; and W = body weight.

The equation holds true only with inert gases. If other gases and vapours which are not inert are reactive with various constituents of blood and tissues having tendencies to metabolise in either of them, the equation should be modified to a certain extent depending upon the nature of the gases and vapours. Nevertheless, the equation gives a basic concept for considering absorption of many gases and vapours in the body of industrial workers.

Whereas the Henderson-Haggard law was primarily for the case of absorption, it is also applicable in an exact reverse sense for the case of elimination of gases and vapours from the lungs. Thus measurement of the concentration of a gas in the expired air makes it possible to evaluate the concentration of the same gas in the blood (e.g. carbon monoxide).

The digestive tract constitutes the second main site of entry. Gases and vapours may be swallowed directly or may be absorbed in food and saliva. This is one reason why eating and smoking may be prohibited in working places.

The third main site of entry is the skin. The skin is at all times either directly or, through the clothing, indirectly in contact with the contaminated air. Fat-soluble gases and vapours are absorbed by the lipoid constituents of the skin, and thus penetrate the skin and are transferred to the blood. Gases or vapours which do not penetrate the skin may remain in hair follicles or sebaceous glands of the skin and eventually cause dermatitis. Cleaning and protecting the skin is one of the important measures for the protection of workers' health. The surface areas considered effective for absorption by the respiratory and digestive tracts and the skin are respectively 100 m^2, 10 m^2 and 1.5 m^2 approximately.

The metabolism of gases and vapours in the body and the identification of their metabolites are established by toxicological studies. This knowledge is important in the prevention and treatment of occupational diseases. Elimination of gases and vapours takes place through the lungs, the large intestine, the biliary duct, the hair and the sebum, corresponding to the sites of entry. In addition to these, the kidneys form a further channel of elimination, that is through the urine.

Mode of action

When gases, vapours and their metabolites are absorbed and distributed into an organ and, as a consequence, its functions and/or morphological features indicate abnormalities, the individual is said to exhibit signs or symptoms of poisoning.

It may be difficult to detect any immediate signs, at the stage of absorption. However, at the stage of poisoning, there should be some definite symptoms and where possible, the gas, vapour or its metabolites should be demonstrable in biological samples taken from the patient. The occupational history is valuable for differential diagnosis since the same signs or symptoms may appear in conditions other than those of occupational origin.

Poisonings may be categorised into two types: acute and chronic.

Acute poisoning is sometimes termed accidental poisoning, as it may result from a fire, explosion or other accident such as a massive leakage of gas or vapour when tanks or containers are being cleansed. The number of injured is usually limited. However, their symptoms may be grave and severe, and may be complicated by unconsciousness and extensive burns that could result in death if the treatment is delayed or inadequate.

Chronic poisoning occurs among workers who are continuously exposed to noxious gases and vapours for an extended period of time. The amount inhaled at one time is too small to cause immediate damage. However, when the exposure is repeated, the gas or vapour accumulates to a sufficient amount to cause the symptoms of poisoning. Chronic poisoning is also called repeated, intermittent or accumulative poisoning, depending upon the exact mode of exposure. The duration of the chronic poisoning is measured in terms of weeks and months, and sometimes in terms of years, in the case of cancer of occupational origin, for example.

The number of workers involved in chronic poisoning is usually larger than in acute poisoning. Special signs or symptoms are not readily detected by the workers, so that early detection and prompt treatment of the poisoning is made more difficult.

Local and systemic effects

Adverse effects from gases or vapours are observed as symptoms. The symptoms are caused by functional and morphological damage to tissue cells, including cell membrane, protoplasm, minute particles formed therefrom, enzymes and other constituents. When gases and vapours are absorbed by only a small part of the body and the actions are accordingly limited to a small area, it is called a local effect. Examples are the upper respiratory irritant gases or pulmonary irritant gases. Many gases, vapours and their metabolites are, however, distributed to several organs, depending upon their affinities, causing extensive morphological and functional damage with symptoms. These are called systemic effects. They occur, for example, from exposure to many industrial organic solvent vapours.

PREVENTIVE MEASURES

Less toxic gases or vapour should, if possible, be substituted for the toxic substances in use. Failing this, the process may be totally enclosed. Where this is not practicable, segregation of certain processes from other parts of the working place is usually helpful in limiting the number of people exposed to the toxic substance.

Dilution ventilation and exhaust ventilation systems should be applied to reduce the concentration to levels below the maxima recommended by authoritative hygiene standards. Personal protective equipment, such as eye and face protection, protective clothing, etc., may be necessary.

Even with the enforcement of the above-mentioned preventive measures, some workers may develop the symptoms of poisoning. Through periodical medical examination properly conducted, early detection of the symptoms, prompt treatment of the disease and further arrangements for the worker's job are feasible.

HARASHIMA, S.

Noxious gases. Henderson, Y.; Haggard, H. W. (New York, Reinhold, 1943).

"Routes of entry and modes of action". Stokinger, H. E. (11-42). Ref. *Occupational diseases. A guide to their recognition* (National Institute for Occupational Safety and Health, 4676 Columbia Parkway, Cincinnati) (Revised ed., 1977), 608 p.

Gases and vapours, irritant

The generic term "irritant gases and vapours" covers a whole series of chemicals whose common characteristic is the action they exert on the respiratory system and conjunctivae.

General characteristics

This category is made up of gases and volatile liquids which, apart from their irritant action, have very disparate characteristics. The intensity and severity of this irritant action depends on the chemical structure of the substance, the concentration in the respired air and the length of exposure; the solubility of the substance may also be a factor since it will determine the region of the respiratory tract that is exposed to the irritation. Water-soluble irritant gases and vapours, such as ammonia, affect primarily the conjunctivae and the upper sections of the respiratory tract; the irritation is incisive and there is a burning of the eyes, profuse lacrimation and coughing. This acts as a warning to the subject who can withdraw from exposure before the irritant substance penetrates further along the respiratory tract.

The most dangerous gases and vapours are those which have very low solubility (nitrogen oxides, phosgene) and those which have no warning odour. Some, like chlorine and hydrogen sulphide, have an irritant action on the whole of the respiratory tract (upper and lower).

Nevertheless, the severity of the lesions depends fundamentally on the concentration and length of exposure. At high concentrations, ammonia may also cause acute pulmonary oedema and nitrogen oxides may irritate the upper respiratory tract. However, high concentrations of irritant gases and vapours do serious damage even when exposure is of only very short duration.

Many irritant gases and vapours may also have a pronounced general toxic action; in some substances, the irritant action is the fundamental toxicological characteristic whereas in others it is of only subordinate significance.

The most important gases whose action is primarily irritant include: chlorine and derivatives (hydrochloric acid, phosgene, chlorine dioxide, etc.); fluorine and derivatives (hydrofluoric acid, silicon tetrafluoride, fluosilicic acid, etc.); bromine and iodine; sulphur dioxide and sulphuric acid; nitrogen oxides (nitrous oxide, nitric oxide, dinitrogen tetroxide, nitrogen dioxide), nitrogen dioxide being the most dangerous; ozone; ammonia; acrolein; acetic acid; and formaldehyde.

Other toxic substances which also have an irritant action include: benzene and its homologues; bromoethane; bromomethane; butadienes; cobalt, molybdenum, nickel; diolefins; ketones; methyl-chloroacrylate (methyl ester of 2-chloroacrylic acid); polyvinyl chloride; phosphine; hydrogen selenide; acrylonitrile; dimethyl sulphate; tetrachloroethylene; carbon tetrachloride; trichloroethylene; metal mercury vapour.

A number of fumes and aerosols, also acting on inhalation, have strong irritant properties on airways; these include those of antimony compounds, beryllium and compounds, cadmium (lung oedema), hexavalent chromium compounds, molybdenum trioxide, nickel carbonyl, osmium tetroxide, vanadium pentoxide and zinc chloride. This explains in particular the health hazards involved in the large inhalation of welding and metal cutting fumes.

HAZARDS

Irritant gases are of widely varying chemical structures and are employed in nearly all branches of industry, handicrafts and agriculture.

They can be found in chemical and plastics plants and laboratories, the petroleum, metalworking, mining and refrigeration industries, painting and varnishing shops, dry cleaners and laundries, dye works and tanneries, and are used for disinfection and rat extermination, etc.

During the preparation and use of irritant gases, high concentrations may be found in the workplace atmosphere. Gas escaping from leaky piping or plant or as the result of the explosion of a pressure container is particularly dangerous. Simultaneous inhalation of a number of irritant substances seems to intensify their action. Poor micro- and macro-climatic conditions may have the same effect.

Following prolonged exposure to significant concentrations of these substances, certain workers may present less pronounced irritation or show no irritation at all; this may be interpreted as inurement or habituation to the harmful agent. In fact, there has been a weakening of the defensive reflexes due to changes in sensitivity following damage to the nerve endings in the mucous membranes of the conjunctivae and upper respiratory tract. This may prevent workers from becoming aware of a dangerous increase in the atmospheric concentration of the toxic substance.

Mode of action. The irritating effects are attributed essentially to excitation of the neural receptors in the conjunctivae and the mucous membranes of the respiratory apparatus which initiate painful sensations (burns, stings) and a series of reflexes (motory, secretory and vascular). An inhibitory reflex may slow down respiration to such an extent that it stops temporarily, to the accompaniment of bradycardia and arterial hypertension. Irritation of the glottis induces a reflex polypnoea which is irregular, spasmodic and accompanied by a sensation of suffocation and anxiety. Irritation of the trachea and bronchi is associated with bronchoconstriction reflex and cough. Injury to the lung parenchyma may evolve into pneumonitis with or without pulmonary oedema.

Acute pulmonary oedema results from a change in the permeability of the pulmonary vessels followed by release of histamine, which initiates bronchoconstriction and a rise in pressure in the pulmonary capillaries with transudation of considerable quantities of serous fluid into the interstices and pulmonary alveoli.

Main syndromes of acute poisoning. Congestion of the nasal mucosa and the paranasal sinuses leads to a very violent frontal headache accompanied by nasal obstruction, rhinorrhoea, and sometimes epistaxis. When the inflammation reaches the larynx there is also hoarseness which may develop into loss of voice. Oedema of the glottis and laryngeal spasms are reflected in intense dyspnoea, cyanosis and anxiety. The clinical signs of irritant bronchitis are: aggravation of coughing; a sensation of tightness in the chest; accentuated dyspnoea; and cyanosis. Some irritants (dichloroethylene ether, toluene diisocyanate, etc.) may induce bouts of asthmatic bronchitis.

The most serious condition, however, is acute pulmonary oedema. Depending on the nature of the irritant, the atmospheric concentration and the duration of exposure, pulmonary oedema may take one of three forms.

(a) The paralysing form, very rare, is characterised by the sudden loss of consciousness preceded by a very brief period of irregular breathing.

(b) The severe form whose symptoms are a very intense irritation of the upper respiratory tract (in the case of substances such as chlorine, sulphur dioxide, ammonia), or very mild irritation (in the case of dimethyl sulphate, phosgene, nitrogen oxides). This is followed by a period of remission, for a few hours in the case of phosgene and for 24-48 h in the case of nitrogen oxides. The pathological condition proper begins with an intensification of coughing, dyspnoea and cyanosis. Expectoration becomes abundant, frothy, aerated, white, yellow or pink. Acceleration and enfeeblement of the pulse indicate progressive weakening of the heart.

(c) The mild form develops more discreetly, with symptoms of irritation of the upper respiratory tract predominating.

Effects of chronic exposure. Chronic irritant bronchitis is the principal clinical sign of prolonged and repeated exposure to low concentrations of irritant gases and vapours. Chronic irritation of the eyes leads to conjunctivitis, blepharo-conjunctivitis and sometimes pterygium. Some irritants may cause chronic keratitis.

SAFETY AND HEALTH MEASURES

These consist primarily in the enclosure of industrial process equipment, widespread process mechanisation, capturing of gases at the point of emission, local exhaust ventilation of processes and general ventilation of premises, automatic gas-leakage or hazardous-concentration monitoring devices which actuate an alarm when a danger level is reached, etc.

It may also be necessary to provide workers with respiratory protective equipment, the various types of which are described in a separate article.

Pre-employment medical examination should be practised to detect disorders which may potentiate the action of irritant substances, such as: atrophic rhinitis; chronic bronchitis; pulmonary emphysema; pulmonary sclerosis; and allergic diseases of the respiratory tract, etc. Persons suffering from pulmonary tuberculosis (even stabilised) or from heart conditions should not be employed on work where there is a danger of exposure to irritant substances. Workers should also receive periodical medical examinations.

First aid and treatment. The victim of the exposure should be immediately removed to the fresh air; all muscular effort on the part of the victim should be avoided, and he should be carried flat on a stretcher. Artificial respiration should not be applied since this may aggravate the risk of acute pulmonary oedema. Low-pressure oxygen therapy should be practised as soon as possible. The victim should be conveyed to the hospital promptly, or a doctor should be called immediately and informed of the substance responsible for the poisoning. The victim should be covered with a blanket or the like and given hot drinks to prevent circulatory failure. He should rest and remain under medical supervision for at least 48 h after acute poisoning, even in apparently benign cases, so as to avoid acute pulmonary oedema.

GAVRILESCU, N.

Health impairment:

"Primary pulmonary responses to toxic agents". Witschi, H.; Coté, M. G. *CRC Critical reviews in toxicology* (Cleveland), 1977, 5/1 (23-66). Illus. 423 ref.

CIS 2408-1971 "Diseases of the upper airways caused by occupational exposure to harmful chemicals" (Erkrankungen der oberen Luftwege durch chemische Schadstoffe im Arbeitsprozess). Zenk, H. *Arbeitsmedizin–Sozialmedizin–Arbeitshygiene* (Stuttgart), June 1971, 6/6 (144-147). 34 ref. (In German)

CIS 75-784 "Inventory and interpretation of pulmonary oedema due to lesions observed in clinical toxicology" (Inventaire et interprétation des œdèmes pulmonaires lésionnels observés en toxicologie clinique). Fournier, E.; Gervais, P.; Diamant-Berger, O.; Efthymiou, M. L. *Revue française des maladies respiratoires* (Paris), 1974, 2/1 (31-44). Illus. 19 ref. (In French)

Irritation data:

CIS 74-1920 "Range-finding toxicity data: List VIII". Carpenter, C. P.; Weil, C. S.; Smyth, H. F. *Toxicology and Applied Pharmacology* (New York), May 1974, 28/2 (313-319). 7 ref.

Testing of the irritative power:

CIS 74-743 "Sensory irritation of the upper airways by airborne chemicals". Alarie, Y. *Toxicology and Applied Pharmacology* (New York), Feb. 1973, 24/2 (279-297). Illus. 32 ref.

CIS 78-1958 *Irritative power of common acids and bases on the rabbit skin and eye mucosa* (Pouvoir irritant des acides et bases usuels sur la peau et les muqueuses oculaires du lapin). Duprat, P.; Delsaut, L.; Gradisky, D. Notes scientifiques et techniques de l'INRS, No. 10 (Institut national de recherche et de sécurité, 30 rue Olivier-Noyer, Paris) (Aug. 1976), 63 p. 27 ref. (In French)

Gas manufacture

With the growth of industry in the late 17th and early 18th centuries, supplies of wood for manufacturing iron and steel became short and were replaced by coke made from coal. During the manufacture of coke it was observed that the volatile material driven off the coal would burn with a luminous flame. The gas industry began as a way of using these gases for the lighting of towns, theatres and large houses.

For many years the industry remained based on coal. By the mid-1950s, hydrocarbons derived from the oil industry were rapidly replacing coal in gas manufacture. At the same time the use of natural gas, which had hitherto been confined to such places as the United States and Canada, where abundant supplies were available close to populated areas, began to be extended. In many cases this natural gas was reformed to make a gas of similar burning quality to that previously manufactured from coal, but now natural gas is increasingly used in a purified but otherwise unaltered condition. Long distance high-pressure pipelines are being built to transport natural gas on a continental scale, and special ships have been constructed to carry low-temperature liquid natural gas across the oceans.

In the 1970s the increasing cost of oil and the diminution in the amount of natural gas available started a trend back to coal and allied fossil fuels as a source of synthetic natural gas, and the development of total gasification processes able to use types of coal previously unsuitable.

Coal gas. Coal is a complex substance containing some hundreds of hydrocarbons: to make gas, it is heated, usually in the presence of steam, at atmospheric pressure at a temperature of 750-1 000 °C. Gas, tar and coke–a solid residue of carbon and mineral ash–result.

The composition of crude gas can be varied considerably, principally by altering the amount of steam present, but also by the addition of light hydrocarbon oils and by varying the time for which the coal is treated. Two main types of process are involved–continuous and intermittent. The continuous process allows a very exact control of the quality of the gas and coke. Coal from a hopper is fed into the top of a vertical retort heated by producer gas and travels down by gravity, the coke at the bottom being discharged into another hopper. Steam is fed into the bottom of the retort and light oil can also be introduced. The rate of travel of the charge and the amount of steam and oil can be varied as required. This process produces a coke which is easily ignitable and suitable for domestic use.

Intermittent retorts may be either horizontal or vertical. The coal charge is put into the retort and heated for a predetermined period (usually about 10-12 h); at the end of this period the retort is opened, the coke discharged and the retort recharged. Steaming of the charge can be carried out in vertical retorts. The coke produced is hard and low in volatile substances, and is more suitable for industrial and commercial use. During the charging and discharging of the retorts considerable quantities of dust and fumes escape into the atmosphere.

The approximate composition of crude gas is shown in table 1. However, considerable variations are possible. Apart from nitrogen, other substances include saturated and unsaturated hydrocarbons and a tarry vapour containing cyclic hydrocarbons, for example benzene, toluene, polycyclic hydrocarbons (mainly naphthalene, but also anthracenes), ammonia, hydrogen sulphide and various cyanides. These impurities must be removed before the gas is distributed. The tar is precipitated, usually electrostatically; the gas is then washed to remove the ammonia and water soluble substances. Finally the hydrogen sulphide is removed, usually by passing over iron oxide.

Producer (lean) gas. This is used for heating the retorts and occasionally for diluting coal gas. It is made by burning coke in a restricted air supply with the addition of some steam or water vapour. It has a low calorific value and is highly toxic.

Water gas. Water gas (or blue water gas as it is sometimes called) is used to provide additional gas at peak load periods, to dilute gas to required calorific value or for certain industrial processes. It is made by reacting coke and steam:

$$C + H_2O = CO + H_2 \quad \xrightarrow[\text{Endothermic}]{}$$

The process is cyclic. A large vessel is filled with coke, which is lit and air blown through until it is white hot. Steam is then blown in and the resultant gas collected until the coke has fallen to a dull red heat, when the steam is shut off and air blown through again to raise the temperature. Ash is removed as slag from the bottom of the vessel, and coke added as required.

Carburetted water gas. A refinement of the above process is to produce a gas of higher calorific value and to make use of the heat and gas of the air-blow cycle by partially reforming oil or other petroleum feedstocks.

The gases of the air-blow phase are passed into a checker brickwork chamber. Oil is sprayed into the hot gas stream and decomposes on the hot bricks, reacting with the CO_2:

$$C_nH_m + nCO_2 = 2nCO + \frac{m}{2} H_2$$

During the steam-blow phase excess steam reacts with the oil:

$$C_nH_m + nH_2O = nCO + \left(\frac{m}{2} + n\right) H_2$$

and this gas is added to the gas produced by the reaction above. The practicalities are somewhat more complex than this description, but beyond the scope of this article.

Both water gas and carburetted water gas are extremely toxic, having a high carbon monoxide content as can be seen from table 1.

Hydrocarbon reforming. Many processes are available and they fall into two main groups:

(a) low-pressure cyclic processes similar to the carburetted water-gas process described above; and

(b) high-pressure continuous processes.

In both processes the hydrocarbon feedstock is reacted by heat with steam to give $CO-CO_2-H_2-CH_4$, and in both cases catalysts are commonly used to stabilise the reactions, inhibit the formation of carbon and reduce the amount of process steam needed. The active catalytic agent is almost invariably nickel and the amount present varies from less than 5% up to 80% according to the process (figures 1 and 2). The original processes produced gas consisting mainly of hydrogen and some methane. Now a rich gas process has been developed to produce synthetic natural gas by the reaction $CO + 3H_2 --- CH_4 + H_2O$.

Table 1. Approximate composition of various types of gas

Type of gas	Hydrogen H_2	Methane CH_4	Carbon monoxide CO	Carbon dioxide CO_2	Oxygen O_2	Nitrogen N_2	Hydrocarbons C_nH_m
Crude coal gas (intermittent)	50-53%	25-30%	5-10%	2-3%	0.2-0.8%	15-20%	3-4%
Crude coal gas (continuous)	52-57%	20-25%	12-16%	2-3%	0.2-0.8%	15-20%	2-3%
Producer gas	< 10%		20-35%	10%		50-60%	
Water gas	50%	–	41%	5%	–	3.5%	0.5%
Carburetted water gas	43%	–	33%	7%		7%	11%

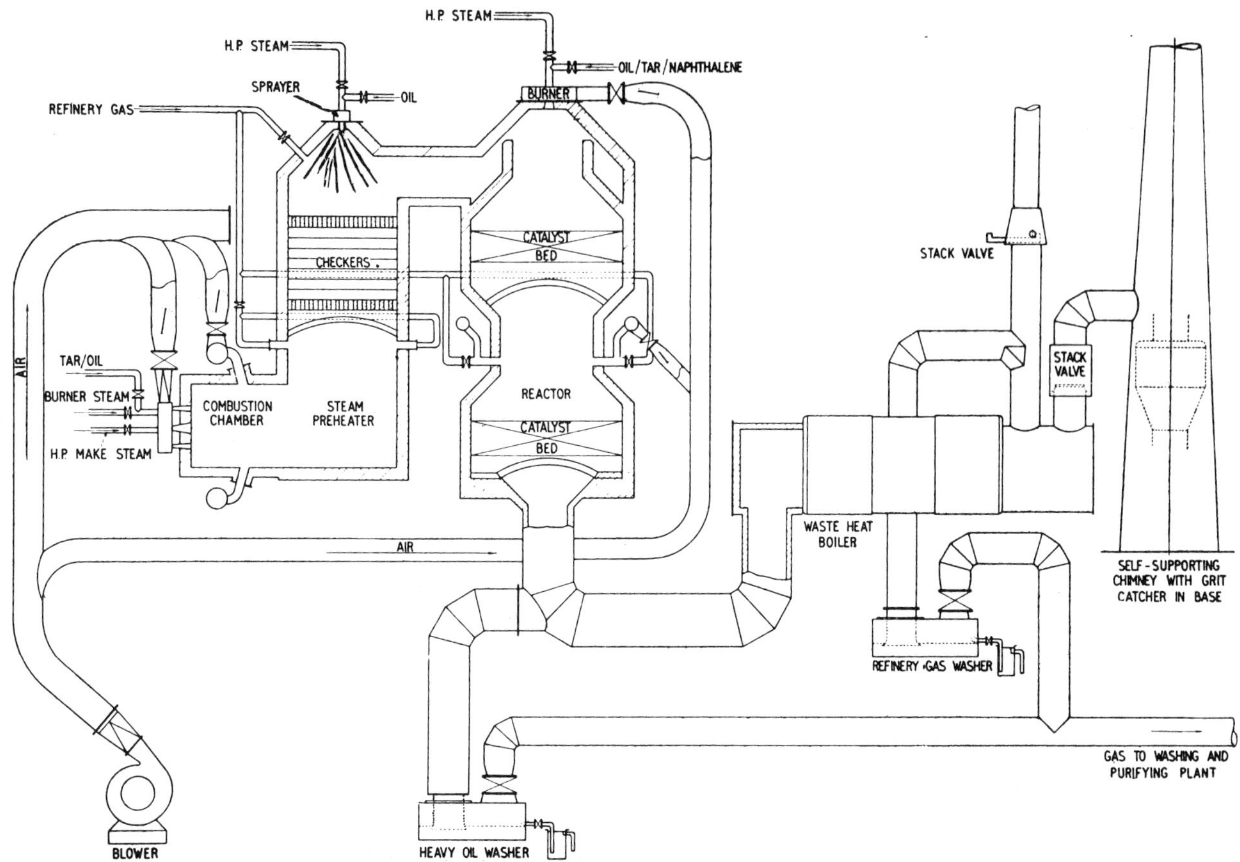

Figure 1. Plant for the catalytic reforming of oil and refinery gas–INIA-GEGI process. *(By courtesy of Humphreys and Glasgow Ltd.)*

Total gasification of coal. The Lurgi and its development, the British Gas slagging gasifier, are examples of the new processes which are practicable today.

Coal, oxygen and steam are reacted in a pressure chamber at about 25 atmospheres. The slagging gasifier operates at slightly higher temperatures and pressure than the Lurgi and has the advantage of using less steam and being able to consume its own by-products, oils and tars, in the process.

The crude gas leaving the chamber contains tars and oils and some impurities, mainly sulphur compounds depending on the feedstock. These have to be removed as in conventional plants. The composition of the resulting gas (H_2 28%, CO 60%, CH_4 7%, CO_2 3%, other gases approximately 2%) is suitable for methanation as above.

Uses. Today, the gas industry is engaged in producing a high-quality fuel of great flexibility. The use of gas for lighting is insignificant. Half the gas distributed is for heating and cooking in the home.

Industrially, gas is mainly used as a fuel, with a few specialised exceptions such as providing inert or

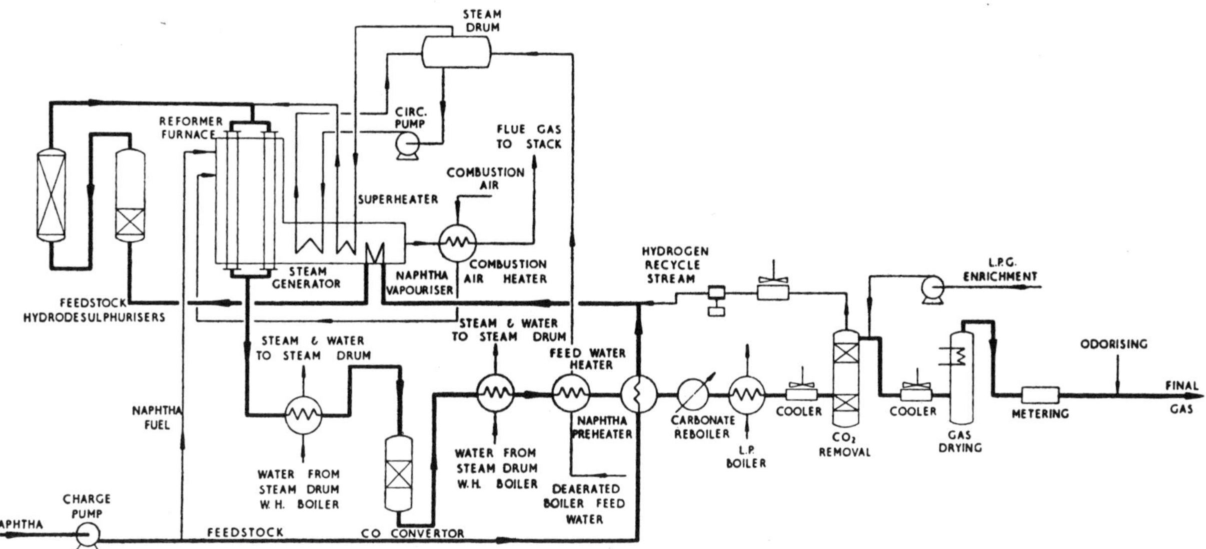

Figure 2. Simplified flow diagram of a continuous naphtha reforming plant. *(By courtesy of Humphreys and Glasgow Ltd.)*

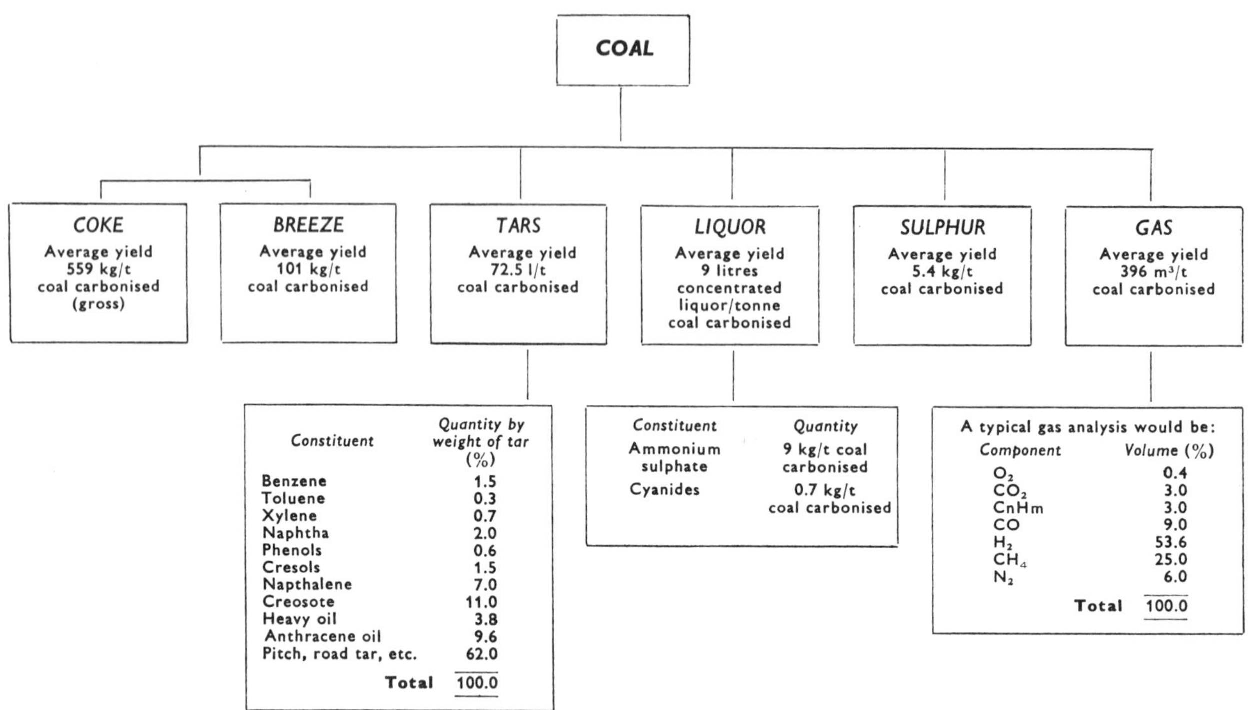

Figure 3. Coal and its derivatives.

reducing atmospheres for heat treatment. The increased availability of natural gas opens up the possibility of its use as a chemical feedstock, for example, in the manufacture of synthetic fibre or artificial fertilisers.

By-products. The by-products obtained during coal carbonisation are shown in figure 3.

HAZARDS

Natural or manufactured gas is flammable and may form explosive mixtures with air. The same applies to liquefied petroleum gas (propane or butane) which may be used to enrich or augment gas supplies. The storage and handling of liquid methane, liquid petroleum gases and petroleum feedstocks present special problems with regard to fire hazards, and vapours from them may accumulate and cause asphyxia. "Cold burns" are also a possibility with refrigerated liquid gases, particularly liquid methane. Inhalation of very cold gases may cause pulmonary oedema.

Coal may ignite spontaneously in stockyards or bunkers and the fumes from such fires are particularly dangerous as they contain a high content of carbon monoxide. Spent oxide (iron sulphide) can combine exothermically with atmospheric oxygen and start a fire, as can reduced nickel if a catalyst is "dropped" in an unoxidised state.

Work in hot workplaces is often necessary, so heat exhaustion may be a problem and there is a risk of burns from hot parts of the plant. High pressures are common in reformers, compressor houses and transmission mains. Heavy work may be necessary in restricted spaces. Town gas and its by-products are toxic and catalysts and other substances may be dangerous to health. It is not always possible to shut down the plant, and emergency work often has to be done with the plant in operation or the main transmitting gas. Accidents are rare and this is due to codes of safe working practices drawn up over the years and kept under constant revision. Written procedures are advisable.

Total gasification processes are considerably less hazardous. There is little heavy physical work involved. The process is totally enclosed and the operators are not exposed to fume nor are they likely to come into physical contact with any hot part of the plant. Measurements of polycyclic aromatic hydrocarbons, particularly benz(a)pyrene, on a plant show that the operator has less exposure than may be experienced in the centre of an industrial city.

The pitch and by-products are virtually identical with those of a conventional plant and must be handled with the same care.

The International Gas Union, which represents the gas industry throughout the world, has committees at which delegates from all countries meet to ensure that recommendations and standards are of universal application.

Carbon monoxide. Acute carbon monoxide poisoning is surprisingly rare in manufacturing stations. It most often occurs on producer gas plants, water gas plants and oxide purifying boxes.

Ammonia. This gas may be liberated in large quantities from tanks containing the liquid after washing the gas.

Hydrogen sulphide. Crude gas contains 1-2% of hydrogen sulphide (H_2S) and significant quantities of H_2S may escape if there are leaks. However, since the CO content is 5-10 times higher than the H_2S content, it is difficult to say whether symptoms produced by exposure to a gas leak are due to H_2S or CO poisoning.

Tar. Tar is an important by-product in gas manufacture and skin contact with tar in gas workers can cause numerous skin conditions.

(a) Acute erythematous dermatitis–tar acts as a photosensitiser and people exposed to vapour or pitch dust, particularly if they are not accustomed to the work, develop painful erythema on skin exposed to sunlight. Pigmentation of the skin does

not protect–severe cases have been seen in negroes.

(b) Subacute erythematous dermatitis–after 10-30 years' exposure, the skin becomes reddened with multiple telangectases, particularly on the face; this condition is harmless and painless.

(c) Chronic tar dermatitis–"shagreen skin"–the skin is mottled with reddish patches, white atrophic patches and blackish brown hyperkeratotic patches. This skin condition is likely to progress.

(d) Tar keratosis–rough patches of hard keratin build up on the skin of the face, hand and arms, often deeply pigmented. They grow slowly for a while, then remain stationary; they usually drop off, leaving a roughened patch from which the processes recommence; sometimes the lesion does not recur, and the skin returns to normal.

(e) Tar mollusca–these lesions are identical with mollusca sebacea. Round shiny papules with a small cap or plug of keratin at the apex–sometimes just a dimple. They grow rapidly up to 2 cm in diameter (usually about 1 cm). At any stage they may stop growing and remain stationary for months or years, eventually regressing and dropping off, leaving at most a tiny scar. Mollusca may occur in skin which does not show any marked changes of tar dermatitis.

(f) Tar epitheliomata–both mollusca and tar keratoses may become malignant. Tar keratoses, which appear to be in rather than on the skin, or in which the skin is raised round the lesion, are suspect. Any keratosis in which the centre is weeping or occupied by a scab is also suspect, even if a history of trauma is obtained. The clinical distinction between mollusca and epitheliomata is difficult. Biopsy is not to be relied upon; the section taken may not include the precancerous part of the lesion. If histological confirmation of the diagnosis is necessary, it is advisable to excise the lesion entirely or to follow the biopsy with radiotherapy.

(g) The skin of the scrotum is peculiarly susceptible to epitheliomatous change. Any tar keratosis or molluscum on the scrotum is likely to progress to an epithelioma; one authority states that he has never known of a scrotal molluscum regressing spontaneously. Active treatment at an early state is advised.

(h) Other skin lesions. Tar acne is very similar to oil acne, mainly affecting the thighs. Acute folliculitis may affect the thighs and flexures of the elbows and is usually mild. Purulent generalised folliculitis is rare, but may follow gross contamination with pitch dust or tar vapour in an individual with poor standards of hygiene.

Dust. The dust from refractory bricks may cause silicosis in bricklayers and retort sealers. Although maintenance on a plant is seasonal and exposure is not great, many of these men, when younger, have been engaged on constructional work for contractors, where exposure to silicaceous dust may be continuous and heavy. Asbestos dust from lagging is also a hazard.

Vanadium. In oil reforming, sludge oils and heavy oils, particularly from Venezuela, may contain vanadium; this will be deposited on the checker brickwork and the catalyst in cyclic reformers as a black glassy ash. Respiratory protection and covering of the skin is necessary for workers removing the catalyst or working on the brickwork. Acute poisoning is characterised by metallic taste, greenish-black discoloration of tongue and teeth, tracheitis and bronchitis. Serious poisoning has not been reported in the gas industry. Vanadium compounds are also used for water treatment.

Nickel. Nearly all the catalysts used in oil reforming are based on nickel. Adequate respiratory protection is necessary. Nickel dermatitis is a risk and skin covering is advised. Nickel carbonyl can be formed if the operating conditions are incorrect; leaks from the reforming tubes may contain this substance, and tubes should be thoroughly purged with inert gas before opening.

Uranium. This has also been used in catalysts. The radioactive risks from the catalysts are small. Radiation from uranium has low penetration and elaborate shielding is unnecessary. The uranium used is insoluble and the risk of chemical poisoning is minute. Precautions should be directed against the retention of insoluble uranium in the lung and protection of the skin against surface radiation.

Miscellaneous chemicals. Any gas manufacturing station is a chemical works and a wide range of chemicals is used. For the most part these chemicals are used for water treatment and cleaning, and consequently cause skin troubles rather than systemic poisoning. Medical officers must be continually alert.

SAFETY AND HEALTH MEASURES

Fire, explosion and asphyxia are the major risks for workers in the gas industry. Protection from these hazards is achieved by adherence to strict codes of safe working practices. It is impossible to detail these, but an outline of the principles on which they are based is given below.

Repairs and maintenance. Where possible the plant should be shut down or isolated from the main gas supply. It is then "purged" with a non-flammable gas, which may be CO_2 and air or, better, pure nitrogen. Purging is continued until tests show that the gas has been eliminated and the purge gas is then replaced with air. When work is completed, the plant is recommissioned by replacing the air with purge gas before allowing gas to return. The removal of gas does not eliminate the risk of fire entirely as crude gas mains, tar scrubbers, gas holders, etc., are often coated with flammable oils and tar, and discretion must be exercised in the use of hot cutting and welding methods.

It is frequently necessary to work with "live gas": this particularly applies to gas holders and transmission mains. The gas must be kept at a pressure slightly above atmospheric to avoid the aspiration of air and the formation of an explosive mixture inside the main or holder. Hot cutting or gas welding is forbidden but electric welding may be used under certain conditions.

Transmission and distribution of gas

The last 25 years has seen a transformation of this aspect of the industry. High-pressure pipelines transmit gas for hundreds of miles and the control of supply is done by computer telemetry from one or two central control rooms. The high pressure not only reduces energy losses entailed in transmitting enormous volumes of gas; it can also reduce the need for low-pressure storage in the conventional types of gas holder.

For distribution these pressures have to be reduced but even here the tendency has been to increase the pressures. The increased pressures and lack of moisture in the gas have led to problems over leaks and new methods of sealing joints and the replacement of cast iron, ductile iron and steel pipes with plastic tubing have evolved to deal with these problems.

Transmission. High-pressure pipes must be constructed to the highest possible standards. Radiographic inspection of welds is standard practice and must be carried out by skilled operators, and all unauthorised persons excluded from the site (see RADIATION, IONISING, INDUSTRIAL USES OF). Radioactive sources may also be used in instrumentation and inspection. Pipes have to be painted or wrapped to resist corrosion; bitumen or tar are common substances used for these purposes and "tar smarts" have occurred. Resin coatings have hazards, (see below).

Obstructions, solid or liquid, in the pipeline are usually dealt with by pigging. (A pig is a free piston driven down a pipe by the pressure differential behind and in front of it.) "Intelligent pigs" carrying instruments, often with radioactive sources, can be used for internal inspection. Sometimes it is necessary for someone to go into the pipe to inspect (see below).

Hazards arise in commissioning and inspecting pipelines.

In commissioning, pipelines have to be tested under pressure. This has to be done with water because of the enormous potential energy involved; otherwise should the pipe fail the results would be equivalent to an explosion. The water must subsequently be removed by pumping and pigging. The pipe is then dried with methanol; if any traces of moisture are left methane will form solid deposits of methane hydrate. All air must then be displaced with an inert gas, usually nitrogen; finally the nitrogen is displaced by gas and the pipe brought up to operating pressures.

Other hazards include noise during venting, at compressor stations or during pressure reduction; asphyxia if the inert gas can accumulate in a pit or excavation; poisoning from methanol; fire from methanol or escaping gas; and physical injury from the force of the escaping gas or the pig, which is equivalent to a projectile in a gun. Correct working procedures will ensure safety, in addition protective clothing should be provided: safety helmets, eye protection, breathing apparatus under certain conditions, impervious clothing when methanol is being used, etc.

Inspection of the pipeline is often done from the air and by the use of pigs (see above), but it is still necessary to "walk the line". Pipeline inspectors may be exposed to severe weather conditions and must be given proper instruction as to how to behave, on the necessity of adequate clothing, emergency rations, etc. Pipelines under the sea raise all the dangers of marine and diving operations.

Distribution. Distribution work still involves a lot of heavy manual labour and strains of the back are a significant cause of lost time.

Pipe joints used to be sealed by fibre packing and lead oxide and finished with molten lead. The workman melting the lead was at risk from the fumes, but this process is now obsolete.

Much use is now being made of polyethylene tubing in appropriate sizes. This can be joined by heat and this is the method most commonly used. Correct temperatures are essential to get a good weld and at these temperatures no dangerous fumes are evolved.

Even if polyethylene catches fire the fumes are relatively harmless. The only significant components being small quantities of acrolein, apart from the usual products of combustion.

Solvents for cement jointing are potentially toxic; tetrahydrofuran is a common component. Cleaners are often organic halogen compounds. In the confined space of a deep trench or manhole dangerous concentrations of vapours can be reached. The pipe is safe, the associated operations may not be.

Maintenance and repair of old mains. Modern gas is usually dry and where yarn joints were used the yarn shrinks and the main leaks; to avoid this, swelling agents are introduced as a fog into the gas stream. Most of these are skin irritants and care should be taken to avoid skin and eye contact.

Leaks in the mains are now often encapsulated with polyurethane, the hardener being a diisocyanate. Toluenediisocyanate must not be used; methyldiisocyanate or other non-volatile isocyanates are suitable.

Epoxy resins may be used for sealing and joint encapsulation and also for protection against corrosion. In an uncured state the components are skin irritants and sensitisers.

Dust in old mains, particularly those that may have carried unpurified coal gas, may contain significant amounts of benz(a)pyrene or naphthalene. Samples should be taken and analysed before removing and disposing of the deposits, so that proper precautions can be taken.

An excess of urinary tract cancer has been reported in distribution workers in Hamburg.

Entry into large mains for internal sealing is a new technique, made possible by the new resins. It is not a particularly hazardous operation but it is essential that the length of pipe or main is cleared of all traces of gas and dust deposits removed as far as possible. Large volumes of air must be supplied down the pipe, not only for the respiratory needs of the worker but also to carry away solvent vapours, etc., from the sealing resins. Medical examination is necessary to exclude individuals likely to suffer from sudden collapse, epilepsy, diabetes, hypertension, cardiovascular disease, etc. It is also probably wise to exclude anyone with a history of psychiatric illness. Individuals may be working 500 m from the nearest entry or exit point in a tube 0.4-0.5 m in diameter. Despite the fact that they may be only a few metres from the surface, rescue may take an hour or more. If a man has to enter he must wear suitable breathing apparatus, a life line must be attached to him, a second man must be outside also equipped with breathing apparatus, and suitable resuscitation apparatus should be at hand.

Natural gas is free of carbon monoxide, as is most of the gas distributed nowadays; nevertheless the use of breathing apparatus whilst working with live gas is still necessary since asphyxia can occur from oxygen lack; a breathing apparatus also offers protection to the face and respiratory tract should there be ignition of escaping gas. Before any closed space, pit or any place where there is not free natural ventilation is entered, the atmosphere must be tested for three things:

(a) that there is no risk of explosion;

(b) that there is sufficient oxygen; and

(c) that there are no toxic gases.

HEALTH AND SAFETY

Any large undertaking in the gas industry should employ the services of an occupational health physician, occupational health nurses and qualified safety advisers. Experience in many countries has shown that where such advice is available the incidence of accident and industrial sickness is reduced to a very low level.

There is still much heavy physical work in the gas industry. Safety helmets, eye protection, industrial gloves and armoured boots are usually necessary. The health and safety staff should give advice and training to all employees on safe handling and lifting.

Asphyxia. Carbon monoxide remains a hazard when gas is manufactured. In modern high-pressure plants it should not escape, but the atmosphere should be continuously monitored at all the operators' stations and an alarm set off if a level of 50-100 ppm is reached.

Similar systems are not always practical on older plants. Particular care must be taken with producer gas and water gas, and when opening box purifiers even when purged.

Hydrogen sulphide is present in crude coal gas but always in association with carbon monoxide.

Other gases act as simple asphyxiants and are a potential hazard where free ventilation is prevented.

Breathing apparatus should be worn if there is risk of asphyxia. The simplest type consists of a mask with valves and a tube down which air is drawn by the wearer; the limiting factor is the length of the tube: 9 m. This can be doubled by the use of a high volume pump, manually or mechanically operated. The simplicity and cheapness of the apparatus makes it suitable for most routine work on mains and services. Compressed-air breathing apparatus, either self-contained or long-line, is suitable for other tasks. Oxygen breathing apparatus is not recommended for general use, its disadvantages in the circumstances of the gas industry outweighing its merits.

Dust. Many operations, from changing catalysts to grit blasting mains before encapsulation, produce much dust. Most of this is nuisance dust but nickel, in catalysts, silica asbestos lead, etc., may all be encountered. Dust masks must be effective and fitted with proper filters. Pad masks are cheap and easy to wear and hence popular with management and workers. However, they are ineffective and dangerous, as the user is not usually able to discriminate between different types of dust. For certain operations, for example grit blasting and catalyst changing, a hood supplied with air at high volume and low pressure provides a more comfortable alternative.

Skin contact. Many substances—solvents, cleaners and yarn swelling agents for example—are powerful irritants, and impervious aprons and gloves must be provided. Care must be taken as some of the chemicals are capable of penetrating many kinds of protective gloves. For some operations, for example joint encapsulation, throw-away gloves may be suitable. For some dust, for example nickel from catalysts, the risk of sensitising is such that complete body protection by a closely woven overall fitting over gloves and boots is necessary.

Tar photosensitivity needs a screen against the sun. Titanium dioxide in an ointment base will make a very effective screen. The dead white appearance is unsightly but can be mitigated by suitable coloration.

Barrier creams should not be relied on for protection but may be useful as a second line of defence.

Cancer. Skin cancer was the first recognised occupational cancer and the first chemical carcinogens were isolated from coal tar.

In the gas industry skin cancer is almost entirely preventable and readily cured if diagnosed sufficiently soon.

Baths and showers should be provided and workers should shower after each shift and change clothes completely. A medical service to which there is easy and informal access can identify lesions in the early stages and arrange treatment. Workers should be told how to recognise skin lesions and the importance of reporting them at once, and should be educated in personal hygiene.

Lung cancer. There is evidence that retort-house workers have an increased incidence of lung cancer compared with the general population. This is probably associated with a type of process which is obsolescent. Prevention is difficult, though good ventilation of retort houses to remove fumes and an active campaign to discourage smoking amongst workers can be advocated. Health screening, including routine chest X-rays, is unlikely to have any significant effect on the incidence or prognosis of the disease.

Bladder cancer. Alpha and beta naphthylamines have been found in tars and the atmosphere of retort houses, and carcinogens have been found in deposits of some mains dusts. The incidence of renal tract cancer in the industry is very low but is significantly greater than in the normal population. Prevention is as for lung cancer; in addition, when dust in mains has been analysed and shown to contain carcinogens, masks and overalls should be provided to protect the worker. The case for routine urinary cytology is somewhat doubtful. On a cost-effective basis the incidence of disease is so low as to render the examination uneconomic and it has not yet been established that diagnosis of symptomless patients is giving significantly better results on treatment than in the case of those who have been picked up when symptoms have developed and been reported without delay.

If a chemical test is discovered this may resolve the dilemma. At the moment the expense and diversion of resources from areas where cytology is of proven value do not seem to justify this particular investigation as a routine measure in gas workers.

Mortality of gas workers. An investigation into the mortality of gas workers was made by members of the British Gas Medical Committee and the regius professor of medicine of Oxford University. The over-all mortality per 1 000 from all causes was found to be 14.97, the expected rate being 16.51 for the general population. For retort-house workers the mortality from all causes was 17.09: cancer of the lung 2.72 (expected 2.03); cancer of the bladder 0.23 (expected 0.15) were the two classes for which the excess mortality was significant. It is worth noting that there were no deaths from skin cancer in the group and the deaths from accidental injuries were not different from the general population. This is an indication of what can be achieved by attention to health and safety.

BUCKLEY, A. R.

General:

Conditions of work and employment in water, gas and electricity supply services (Geneva, International Labour Office, 1982), 62 p.

Technology:

Treatment of liquid effluents from coal gasification plants. A report by Economic Assessment Service established under the auspices of the International Energy Agency. EAS report No. B2/79 (London, IEA, Mar. 1979), 64 p. Illus. 45 ref.

Health and safety:

"Mortality of gasworkers. Final report of a prospective study". Doll, R.; Vessey, M. P.; Beasley, R. W. R.; Buckley, A. R.; Fear, E. C.; Fisher, R. E. W.; Gammon, E. J.; Gunn, W.; Hughes, C. O.; Lee, K.; Norman-Smith, B. *British Journal of Industrial Medicine* (London), Oct. 1972, 29/4 (394-406). Illus. 15 ref.

CIS 78-169 *Cancer as a factor of mortality in town gas workers* (Krebs als Todesursache bei Beschäftigten der Gasindustrie). Manz, A. Forschungsbericht Nr. 151 (Bundesanstalt für Arbeitsschutz und Unfallforschung, Martener Strasse 435, 4600 Dortmund-Marten) (1976), 119 p. Illus. 84 ref. (In German)

CIS 79-838 "Coal gasification and occupational health". Young, R. J.; McKay, W. J.; Evans, J. M. *American Industrial Hygiene Association Journal* (Akron, Ohio), Dec. 1978, 39/12 (985-997). 102 ref.

CIS 78-1734 *Recommended health and safety guidelines for coal gasification pilot plants.* DHEW (NIOSH) publication No. 78-120 (National Institute for Occupational Safety and Health, 4676 Columbia Parkway, Cincinnati) (Jan. 1978), 239 p. Illus. 173 ref.

CIS 76-43 *Gas transmission and distribution piping systems.* ANSI B31.8-1975 (American Society of Mechanical Engineers, 345 East 47th Street, New York) (Standard approved on 6 Mar. 1975), 118 p. Illus. 99 ref.

Genetic manipulation

Genetic manipulation is the term used to describe the process whereby sequences of DNA carrying genetic codes may be excised from genes and transferred to the genome of other organisms, thus altering their genetic structure. According to the Health and Safety (Genetic Manipulation) Regulations, 1978, in the United Kingdom it means "the formation of new combinations of heritable material by the insertion of nucleic acid molecules, produced by whatever means outside the cell, into any virus, bacterial plasmid, or other vector system so as to allow their incorporation into a host organism in which they do not naturally occur but in which they are capable of continued propagation."

The most favoured application, and the one easiest to perform, is the transfer through these recombinant techniques of material from higher eukaryotes to lower forms and to prokaryotes in which the rapid replication of the host organism leads to an increase in the quantity of the genetically active material in a form in which it may be studied or used. Contrary to eukaryotes, prokaryotes, which include *Bacteriaceae* and *Cyanophyceae*, do not possess a nuclear membrane, can reproduce very quickly and do not present cellular differentiation.

In this way research into the genetic process may be forwarded, and as the science progresses it may be possible to produce, by the expression of alien genetic material produced in micro-organisms, biological products of higher organisms, including man, *in vitro*. Human insulin is often mentioned in this connection. Throughout the world this type of research activity is in progress, mainly as an academic exercise, but increasingly with a view to commercial development. Inevitably there have been reactions to this type of activity, not just on the part of uninformed groups, but also amongst scientists themselves working in recombinant DNA chemistry.

The superstitious fear that this work must inevitably lead to evil consequences for man's own genetic structure need not be considered in detail. The handling of genetically active purified sequences of DNA might in itself pose some dangers, supposing such sequences could become active in a biological or environmental milieu, but there is at present no evidence of this occurring. Many of the host organisms used to replicate transferred genetic material are capable of existence and growth in man and animals, or in plants. There is thus a clear danger that such organisms if allowed to establish themselves outside the strict confines of the laboratory might produce hazards to man or his environment. This hazard is particularly important in view of the fact that most workers on recombinant DNA are not specially trained in the handling of living micro-organisms.

Another possible source of danger is that genetic manipulation might result in the emergence of new strains of micro-organisms (due to unforeseen recombinations) and that these might produce new or widespread diseases in man, animals or plants. None of these hazards has so far been reported as having occurred, and most scientific thought at the moment is veering away from the earlier alarmist attitudes, towards a more rational type of risk assessment.

Most nations have considered these matters at a national level with varying results in administrative or legislative measures. There is an international consensus that degrees of occupational and environmental risk can be assessed and that the work can be carried out under certain safeguards. The basic principles of these safeguards are physical and biological containment. Biological containment implies the use of host organisms, and where possible vectors, which have little or no infective potential and preferably cannot survive outside an artificial environment.

The successful development of biological containment confines the hazard to the laboratory, where physical containment, that is the use of closed laboratories, safe systems of working and monitoring of the health of personnel, can be absolute. International differences in the application of these principles reflect the differing organisational arrangements for biomedical research and relationships between government, universities and industry. Most countries where this work is carried out have instituted some form of control body, notification system, licensing arrangements or regulations based either on new or existing legislation. In 1979 there was a general tendency to recognise that many of the laboratory activities in recombinant DNA research are frequent natural occurrences, and as such impossible to regulate.

There is still a need for a cautious approach to genetic manipulation research and particularly when developed genetic systems are exploited on a large scale by industry. The effects of "scaling-up" biological processes are well understood, using the organisms nature has permitted to be developed by conventional genetics. There may be no reason to anticipate any greater danger from organisms whose genetics have been modified by the newer technology.

HAINES, D. O.

"Evaluation of the risks associated with recombinant DNA research". Working Group on Revision of the Guidelines. *Recombinant DNA Technical Bulletin* (Washington, DC), Dec. 1981, 4/4 (166-176). 47 ref.

"NAS ban on plasmid engineering". *Nature* (London), 19 July 1974, 250 (175).

First report of the genetic manipulation advisory group. CMND 7215 (London, HM Stationery Office, 1978), 77 p.

The Health and Safety (Genetic Manipulation) Regulations, 1978. Statutory Instrument No. 752/1978 (London, HM Stationery Office, 1978), 3 p.

"Guidelines for research involving recombinant DNA molecules". *Federal Register* (Washington, DC), 22 Dec. 1978, 43 (60 108).

Administrative practices supplement to the NIH guidelines for research involving recombinant DNA molecules (Bethesda, Maryland, National Institutes of Health, Apr. 1980), 22 p.

"Recombinant DNA research; physical containment recommendations for large-scale uses of organisms containing recombinant DNA molecules". *Federal Register* (Washington, DC), 11 Apr. 1980, 45/72 (24 968-25 370).

"Recombinant DNA research; actions under guidelines". *Federal Register* (Washington, DC), 14 Apr. 1980, 45/73 (25 366-25 370).

Genital system and sex-linked occupational characteristics of women

Working conditions may have an influence on the female genital system; they are not, however, the only cause of disorders. Pre-disposing circumstances include pregnancies, deliveries, condition of connective tissue, obesity and nutrition.

Direct mechanical trauma on the interior part of the genital system is rare, because of its protected position in the small pelvis. Injuries to the outer genitals as a result of occupational hazards occur occasionally.

Emotional reactions may also influence the genital system of female workers: dysmenorrhoea, polymenorrhoea and metrorrhagia can be caused by psychic influences on the endocrine system.

Menstrual disorders. As is well known from sports medicine, the female muscular efficiency is not particularly affected during menstruation. Working capacity increases by approximately 6% during the post-menstrual follicle-phase and is reduced by approximately 10% during the pre-menstrual phase. Women suffering from pre-menstrual tension also show reduced working capacity. In cases of dysmenorrhoea and intermenstrual dysmenorrhoea, mental capacity during menstruation decreases. Cases of severe dysmenorrhoea leading to disability for work are rare and must be medically treated. Disabling dysmenorrhoea in women unwilling to submit to routine medical treatment usually also has a psychic component. Reduced concentration makes women more accident-prone during the period of dysmenorrhoea and their productivity may be reduced.

Sanitary installations in some plants are not geared to the specific needs of female employees during menstruation. Lack of menstrual hygiene may result in infection of the small pelvis.

Prolapse of the genital organs. Prolapse is more frequent in multiparous women. Weak connective tissue at the base of the pelvis and obesity are pre-disposing factors. The descent of the genital organs can be aggravated by physical exertion. Heavy postnatal physical work done by farmers and peasant women has been observed to be particularly damaging. Lifting of heavy loads can lead to increased symptoms of prolapse, and a higher incidence is to be expected after menopause because of atrophy of the genitals. The subjective complaints do not always correlate with the degree of prolapse. In mild cases relative urinary stress incontinence as well as drawing pain in the lower abdomen and low back pain may occur. More severe cases may lead to prolapse of the interior genital organs, ulcerations as well as dysuria. Severe genital prolapse impairs working capacity considerably. In these cases surgical correction becomes necessary in order to restore the anatomy as well as the function of these organs.

Inflammatory disorders of the small pelvis. These disorders are usually not of direct occupational origin; however, they can be aggravated by occupational activities. Inflammatory processes and irritations in the outer genital area can be of mechanical, chemical or bacterial origin. Tight fitting clothes may be the cause of mechanical irritation and chemical irritation may be caused by environmental pollution. Specific inflammations occur in poorly controlled diabetics and atrophic inflammations are found in older women. These inflammations may lead to reduced working capacity and require medical attention and treatment. Colpitis and vulvitis cases caused by bacteria and/or parasites can impair occupational efficiency by causing leucorrhoea and irritation of the genital organs. The inflammatory processes may ascend into the salpinx, ovary and the small pelvis. Parametritis and peritoneal irritation in the small pelvis may be due to continuous motion or vibration, which may worsen the condition and delay the healing process. Temporary complete rest should be recommended for these cases. Localised endometritis is not too painful; however, it causes a bloody discharge and should be vigorously treated in order to prevent further spread of the inflammation. Chronic inflammatory processes may lead to genital tumours causing peritoneal irritation and displacement of adjacent organs. In order to maintain full working capacity, removal of these tumours may be indicated. In addition to bacterial inflammation the genital mucous membranes can also show inflammatory changes due to viruses which may produce temporary changes in the menstrual cycle.

Gynaecological tumours. These tumours are classified as benign or malignant. Benign tumours are either neoplastic or inflammatory. Neoplasms most frequently found on the ovaries are cysts and solid tumours. On the uterus they are myomas, fibromas and myofribromas. Polyps occur in the female genital area rather frequently. Neither these tumours nor the malignant tumours of the uterus can be considered work-connected. Irritations in the vulvar area may be caused by occupational carcinogenic substances; however, occurrence of malignant neoplasms is primarily due to advanced age, which makes the occupational origin rather difficult to prove. Papers have been published claiming that breast tumours may be favoured by chronic trauma. Cases found in the textile industry were described in which osseous giant cell tumours had possibly been caused in this way. In order to detect gynaecological tumours as early as possible, industrial medical services should urge women to submit voluntarily to periodic examinations and provide the necessary services for these examinations.

Deformities of the pelvis and lumbar spinal column. Severe deformities have become comparatively rare in industrialised countries and are in most cases the result of accidents. In developing countries, however, pelvic deformities and rachitic hip joint luxations occur frequently owing to lack of prophylaxis as well as child labour. The anatomical structure of the female pelvis and lumbar region has certain unique characteristics. Because of the wider pelvic canal the hip joint is located more laterally in females than in males. Therefore, in walking leverage is more pronounced and should be taken into consideration. From this anatomical difference a tendency towards a valgus-position in the knee-joint area results and limits the load-carrying capacity somewhat. The length of the lumbar spine in females is greater than in males. This not only favours swaying of the hips in walking, but also leads to a steadily increasing intervertebral motility. In addition, females show an increased pelvic incline resulting in a more pronounced lumbar lordosis. Discomfort while standing may be the result of weakness of the connective tissue, joint fixation or intervertebral joints, caused by endocrine disorders. Back aches in females indicate changes or dysfunctions in the lumbar spinal column and pelvis rather than changes in the genital organs.

During pregnancy a loosening of the pelvic joints may result from horizontal changes. Extreme physical stress at this time may therefore have permanent harmful consequences. Long-lasting, severe mechanical strain of the lumbar spine causes wear and tear in the intervertebral disc area and the small vertebral joints. The subjective complaints are usually more severe during the early stages of the changes than after a pronounced ankylosis. Women doing heavy physical work in industry, in agriculture or in nursing (especially in geriatric departments) reported sore spots in the sacrum and lumbar spine area. Congenital hip-luxation and hip-dysplasias are more frequent in females than in males, especially in advanced age. Dysplasia of the hip joint may lead to decreased capacity or total inability to work because of the resultant severe pain.

The female skeletal structure should be taken into consideration, especially in terms of ergonomics. The average body measurements vary between males and females. The average female body is approximately 10 cm shorter. Particularly pronounced differences exist in the length of the tibia and the upper arm. These variations should be taken into consideration when working areas are designed. Chairs that are adjustable according to requirements are particularly favourable. The back of the chair should be constructed in such a way that it does not support body-weight passively, but permits sitting while making full use of the back-musculature. In a sedentary position while the back muscles are relaxed the strain on the spinal column is increased with a resultant greater susceptibility to minor trauma. Working in an upright position placing maximal strain on one leg for a long period of time is notably harmful for women, since they show signs of fatigue more rapidly than men. A short rest period every 45 minutes is considered optimal when working produces great strain on the female support apparatus. Variable work alternating between sedentary, upright or walking positions is preferable to a stationary occupation. Sitting for long periods of time and doing work which requires full concentration (working with cash registers, typewriters, punch presses, as adjusters, etc.) may cause back aches, tiredness, loss of sleep as well as paraesthesias and pain in the shoulders, arms, hands or fingers. An optimally designed working area and diversified work can be of help in relieving these conditions.

Neuro-endocrine disorders. They occur frequently in women but are often difficult to diagnose because of underlying emotional problems. Women are emotionally more closely bound to other people than men and suffer far more from an unfavourable working climate. In addition, working women have to carry a double or triple load which makes excessive psychic and somatic demands on them: they have to attend to household duties and possibly have to take care of children. Pregnancy may increase their difficulties. The early months of pregnancy usually mean a period of increased emotional instability, whereas in the second half (usually starting with the experience of fetal movements) emotional stability develops and, with it, preparation for future maternal duties.

Industrial absenteeism because of sickness is influenced more by the age of the employee than by family status or the number of children. Age seems to be an especially important factor in respiratory, circulatory, uro-genital and gynaecological disturbances. The number of children, however, does influence various periods of absenteeism, especially in employees aged between 25-40 years.

Ambivalent reactions, emotional and physical stress and vegetative exhaustion vary from case to case. They occur mostly in the form of depressions, neuroses and neurasthenic or psychosomatic reactions. Genital manifestations developing frequently by way of the endocrine system are: amenorrhoea, hypo-menorrhoea, metrorrhagia, increased dysmenorrhoea, spastic pelvic disease and other painful pelvic disorders. Menopausal women present certain characteristic conditions: one-third have rather severe climacteric complaints, whereas another third have lesser difficulties. In addition, productivity is slightly reduced at this time of life and the ability to concentrate is also reduced because of psychic involvements. This may lead to increased stress at work, since this group of women is particularly concerned about the loss of their jobs. A departmental change or a reassignment of work taking achievement into consideration (without giving the employee the feeling of being demoted) produce favourable results. For the single working woman the menopause presents a series of special psychic problems. It marks the end of fertility and hence of all possibility of establishing a family. Lack of satisfaction in private and family life may lead to overcompensation in working life. Under these circumstances a stress situation may develop, since our society is not always willing or able to recognise these manifestations. The duration as well as the intensity of climacteric complaints vary. Approximately 25% of all women are in need of medical treatment in order to maintain the balance between work output and emotional equilibrium.

Notwithstanding the provision made by law for prenatal care, freely chosen pregnancy or the prevention of conception are of great importance for the working woman. Uncertainty or an undesired pregnancy may contribute to temporary dysphoria. It is therefore important for industrial medical services to provide efficient family planning consultations.

OCCUPATIONAL HAZARDS

By and large occupational hazards have similar consequences for men and women. Chemicals occupy a special place among factors likely to affect the genital system, and in the early stages of pregnancy teratogenesis is of special importance for the female organism (see EMBRYOTOXIC, FETOTOXIC AND TERATOGENIC EFFECTS; ANAESTHETISTS).

To recognise occupational hazards and differentiate between them and non-occupational hazards is somewhat difficult. A careful anamnesis and examination of the worker is a prerequisite. In addition, the workplace should be thoroughly investigated and evaluated in the light of ergonomic criteria. Substances which could possibly be harmful should be analysed. Identification of potentially harmful substances as well as biological testing is required. In order to detect minute amounts of these substances more detailed laboratory procedures such as radio-immunoassays, etc., may be necessary. For differential diagnostic purposes the domestic surroundings and living conditions of the patient should also be investigated.

PREVENTIVE MEASURES

Prevention of overstrain and damage to the health of working women entails three main aspects:

(a) sociopolitical considerations;

(b) working conditions; and

(c) domestic circumstances.

In the sociopolitical area, protection during pregnancy is very important. Labour protection laws should include special provisions for juveniles as well as older workers. In younger workers as well as in older ones, psychic instability has to be taken into consideration and, in addition, the fact that increasing age reduces the work capacity has to be noted.

In industrialised countries the extended family is a rarity and outside help should therefore be provided for working women to take care of their children. It is most important for management to provide adequate working conditions. Ergonomic and industrial pre-employment medical examinations should establish the applicant's suitability for a particular job in order to avoid unnecessary stress later on. Wherever possible, the work rhythm should be adjusted to the productive capacity of the employees. Occasional voluntary activities, not work-connected, may improve working conditions.

The excessive burden of carrying on household duties and raising of children in addition to employment should be overcome by alleviating the domestic situation. An intact family structure and understanding of the work situation by all family members is necessary.

No general suggestions can be made concerning suitable working conditions for women with gynaecological problems. Job applicants should be assigned to work individually on the basis of industrial physiological investigations. Women with severe dysmenorrhoea should not, for example, be employed as public transportation drivers or at jobs having high accident rates. Women with pronounced prolapse of the genital organs should not lift heavy loads. After surgical prolapse correction, increased endurance at light physical work can be expected.

Women suffering from chronic inflammatory processes should not be exposed to cold air or draughts during working hours. In order to prevent recurrent inflammations, hormone therapy can be recommended, possibly combined with planned parenthood measures. Women with pelvic and spinal column deformities require individual special consideration regarding their place of work. Sedentary employment is generally preferable in these cases. In neuro-endocrine problems a stress-free job is to be recommended as far as possible, taking personal preferences of the workers and their productive capacities into consideration.

Since, however, private and family conflicts are beyond the influence of management, psychiatric or psychotherapeutic care is to be suggested. Overtaxed and overburdened women are frequently found on the job. In a field survey of 50% of men as well as women reported that they had satisfactory working conditions; ergonomic measurement showed, however, that this assessment was justified in the case of only 66% of male and 35% of female employees.

STOEGER, H.

CIS 77-798 "Menstrual and reproductive functions and gynaecological morbidity in women occupationally exposed to petrol" (Menstrualna, generativna funkcija i ginekologična zabolevaemost eni v professionalen kontakt s benzin). Panova, Z. *Letopsi na higienno-epidemiologičnata služba* (Sofia), 1976, 20/1 (53-56). 6 ref. (In Bulgarian)

CIS 78-484 "Occupational health problems of women working in the mercury production industry" (Voprosy gigieny truda enščin pri proizvodstve rtuti). Gončaruk, G. A. *Gigiena truda i professional'nye zabolevanija* (Moscow), May 1977, 5 (17-20). 4 ref. (In Russian)

CIS 76-1284 "Particular effects of vibration on blood circulation in female lower abdominal organs at different phases of the menstrual cycle" (Osobennosti vozdejstvija vibracii na krovosnabenie organov malogo taza enščin v raznye periody mentrual'nogo cikla). Frolova, T. P. *Gigiena truda i professional'nye zabolevanija* (Moscow), Dec. 1975, 12 (14-18). Illus. 3 ref. (In Russian)

General:

"Occupational health hazards of women: an overview". Stellman, J. M. *Preventive Medicine* (New York), Sep. 1978 (281-293). 18 ref.

CIS 78-1495 *Proceedings–Conference on women and the workplace* (Society for Occupational and Environmental Health, 1714 Massachusetts Avenue, NW, Washington, DC) (1977), 364 p. Illus. 301 ref.

Genital system, male

The male genital system comprises the gonads (testes), a duct apparatus and the annexe glands. The genital system is regulated by intrinsic factors (genetic and neurohormonal) and influenced by extrinsic factors (e.g. nutrition, infection, intoxication). In adult life the principal function of the genital system is to assure reproduction. The testes play the prime role here by their germinal and hormonal production. The main male hormone is testosterone, which induces pubertal changes and the development of secondary sexual characteristics and the male psyche. Continuous testosterone secretion during the post-pubertal period is a prerequisite of normal spermatogenesis, external genitalia aspect, sexual hair and sexual activity. The functions of the testicle are controlled by the anterior pituitary gonadotrophins and the hypothalamus. Sexual function is controlled by the pituitary-testicular system and by the central and peripheral nervous system.

HAZARDS

A variety of hazardous agents that man encounters in his work may have deleterious effects on the genital system and sexual function; such effects include sexual insufficiency, gonadic insufficiency, scrotal cancer, and infectious diseases of occupational origin.

Sexual insufficiency. This is evidenced by decreased libido and functional disorders and may be a manifestation of neurasthenic syndrome in severe chronic poisoning due to lead, mercury, manganese, cadmium, organic solvents or carbon monoxide, or the result of chronic exposure to ionising and microwave radiation. Sexual insufficiency may also occur in advanced silicosis with respiratory insufficiency.

Gonadic insufficiency. A large number of occupational hazards may induce gonadic insufficiency (hormonal and/or germinal); however, only recently has research thrown light on the nature of such disorders.

Testicular atrophy has been observed following prolonged, severe chronic lead poisoning, and studies indicate that the incidence of spermatogenic disorders (asthenospermia, teratospermia) is high in cases of lead poisoning. In chronic manganese and carbon monoxide poisonings the decrease in 17-ketosteroid secretion suggests the existence of hormonal insufficiency, and experimental studies have shown that cadmium salts cause severe disorders of the seminiferous tubules and, in some cases, total testicular necrosis. See also DIBROMOCHLOROPROPANE; BENZANTHRONE.

Carbon disulphide has been reported to cause histopathological lesions, sexual insufficiency and germinal and hormonal insufficiency due to its combined effects on the hypothalamus and pituitary systems and on the gonads. Intoxication due to many other organic solvents may result in gonad insufficiency; for example benzene, trichloroethylene and methyl chloride have been shown to cause testicular hormonopoietic insufficiency and severe spermatogenic disorders. Workers employed on oestrogen production have also been reported to suffer from sexual and gonadic insufficiency (see HORMONES, SEX).

The physical hazards known to cause gonadic insufficiency include ionising radiations, microwaves and high temperatures. Experimental studies have demonstrated that spermatogenic cells are extremely sensitive to ionising radiation and it has been found that doses of 10-30 rad may reduce fertility, doses of 200-400 rad may induce temporary sterility and 600-800 rad, permanent sterility. Internal radioisotope contamination may also induce sterility.

Experimental exposure to microwaves has induced sterility in rats and lowered enzymatic activity in the germinal epithelium, and studies on workers exposed to microwaves demonstrated a high incidence of asthenospermia and teratospermia without testicular hormone secretion.

Heat and hot work may result in reduced fertility, and men exposed to high temperatures may present a high incidence of oligospermia, asthenospermia or teratospermia.

Scrotal cancer. Cancers of the scrotal epithelium have been described in workers exposed to arsenic, tar and creosote (e.g. chimney sweeps, gas industry workers) and shale oil (mule spinners), etc.

Infection. Persons who are occupationally exposed to persons suffering from mumps or to animals with brucellosis have sometimes been found to develop orchitis parotidea and brucellar orchiepididymitis respectively.

Medical supervision. Workers presenting gonadic insufficiency should not be exposed to hazards that have an effect on the genital system. Persons exposed to such hazards should receive a periodic medical examination that makes special reference to this aspect.

LANCRANJAN, I.

Chemicals:

CIS 78-1995 "Endocrinological functions and sexual behaviour, neurological and neurophysiological impairment, and behavioural alterations in carbon disulphide workers". Cirla, A. M.; Bertazzi, P. A.; Tomasini, M.; Villa, A.; Graziano, C.; Invernizzi, R.; Gilioli, R.; Bulgheroni, C.; Cassitto, M. G.; Jacovone, M. T.; Camerinio, D. *Medicina del lavoro* (Milan), Mar.-Apr. 1978, 69/2 (118-150). 56 ref.

CIS 76-1065 "Oestroprogestogenic syndrome in a worker involved in production of a contraceptive pharmaceutical" (Sindrome da estroprogestinici in operaio addetto alla produzione di un farmaco anticoncezionale). Gambini, G.; Farina, G.; Arbosti, G. *Medicina del lavoro* (Milan), Mar.-Apr. 1976, 67/2 (152-157). Illus. 12 ref. (In Italian)

CIS 76-1664 "Gonadal and reproductive changes in white rats after oral intoxication with maneb" (Promeni v gonadite i v sposobnostta za reprodukcija na beli plăhove sled oralna intoksikacija s maneb). Ivanova-Čemišanska, L. Vălčeva, V.; Čakărov, E.; Načev, Č.; Takeva, C. *Problemi na higienata* (Sofia), 1975, 1/1 (25-30). Illus. 11 ref. (In Bulgarian)

CIS 77-782 *Effects of administration of cadmium sulfate on the hypothalamo-pituitary-testicular axis in the rat* (Effets de l'administration de sulfate de cadmium sur l'axe hypothalamo-hypophyso-testiculaire du rat). Granet, P. Université René-Descartes, Faculté de médecine Necker–Enfants-Malades, Paris) (1976), 50 p. Illus. 91 ref. (In French)

Radiations:

CIS 76-407 "Gonadic function in workmen with long-term exposure to microwaves". Lancranjan, I.; Măicănescu, M.; Rafailă, E.; Klepsch, I.; Popescu, H. I. *Health Physics* (Oxford), Sep. 1975, 29/3 (381-383). 6 ref.

CIS 76-408 "Elimination of pituitary gonadotropic hormones in men with protracted irradiation during occupational exposure". Popescu, H. I.; Klepsch, I.; Lancranjan, I. *Health Physics* (Oxford), Sep. 1975, 29/3 (385-388). Illus. 28 ref.

Geology and safety

A common feature of all civil engineering structures is that they have more or less deep foundations in the ground. Their construction is therefore subject to geological conditions which determine to a great extent the type of foundation and techniques of construction to be adopted. From the point of view of construction site safety, the following geological factors must be taken into account:

(a) the topography of the ground resulting from the evolution of the contours;

(b) the composition and stability of the ground, which depend on the structure of the underlying rock and the nature of the loose superincumbent soil;

(c) underground water.

Geological conditions liable to cause accidents. There are many geological phenomena presenting hazards for construction site workers. The following aspects of the problem have to be considered:

(a) the types of phenomena likely to occur, such as landslips, falls of rock, gushes of water under pressure, outbursts or explosions of natural gas;

(b) the possibility of predicting these events; in view of the present state of the art and apart from purely statistical predictions, only very elaborate studies enable such phenomena to be forecast;

(c) the detrimental effects of human activities, i.e. excavation and tunnelling operations, which are particularly dangerous since they disturb the natural balance of rock grounds and soils.

Natural instability of slopes. The equilibrium of a slope is never permanent. The behaviour of a slope depends mainly on the constitution of the subsoil, which may consist of stable rock or loose ground.

Rocky slopes at the surface of which stratified formations crop out offer a more durable stability if the strata are horizontal or inclined opposite to the direction of the slope; if the strata are inclined parallel to the slope, there is a great risk of a sudden slip.

However, even in favourably inclined strata the creepage of argillaceous rocks may induce shearing and cause a landslip. Furthermore, all types of rock, even the hardest ones, often have fissures cleaving rock masses which may completely break loose under the action of frost and give rise to falling stones, a phenomenon which is normal and almost periodical in mountains, where it contributes to the evolution of the contours. It is impossible to predict when and where this phenomenon will take place, but the thaw period is most dangerous.

Slopes consisting of loose ground generally have characteristic declivities and profiles. As regards cohesionless soil, the natural declivity is determined by the mean granulometry: the finer the particles, the smaller will be the angle of slope. The latter may attain 45 ° with gravel, but will never be more than 35 ° with sand. In cohesive soils the natural slope depends on the fineness of the particles and their cohesion. In general the following angles of repose may be considered to yield stable slopes:

- up to 60 ° for ancient moraines consisting of coarse elements;
- 30 ° to 40 ° for moraines with medium-sized to fine elements and for rubble;
- up to 75 ° for loess;
- up to 62 ° for ancient coarse alluvial deposits;
- 30 ° to 40 ° for recent alluvial deposits.

However, an increase in water content can considerably reduce these angles and give rise to landslips, which may either evolve slowly and gradually or occur suddenly by disruption. Such slips are normal events in the geomorphological evolution of mountainous regions. They are generally foreseeable but difficult to prevent.

Opencast work. Modifications of equilibrium are very often caused directly by civil engineering work, and preventive measures must be taken.

In the case of rocky ground, slopes without vegetation having a mean declivity of less than 45 ° are suspect, especially if rocky ridges or needles emerge from them. The same is true of recent rubble, which is always a sign of periodic falls of rock. Cuttings made in mountain slopes for the construction of roads and canals or for quarries may remain stable over long periods if their angle of slope is judiciously chosen. Artificial slopes in rocky ground are generally made at an angle of less than 60 °. Isolated falls of small blocks can be arrested by wire mesh laid on the rock walls.

The stability of loose grounds (soils) is much smaller than that of rock; any artificial increase in slope angle may cause a slip of ground. large backfills which have insufficiently settled also present poor stability. All work in excavated cuttings must be performed under the protection of timbering or sheeting adapted to locally prevailing conditions.

As regards bearing resistance, a few examples for foundations in alluvial deposits are given hereunder:

Ancient, well settled deposits consisting of pebbles and gravel	6-8 kg/cm²
Recent cohesionless deposits	4-6 kg/cm²
Gravel and sand deposits	3-5 kg/cm²
Coarse to medium-sized sand deposits	2-4 kg/cm²
Fine sand and silt deposits	2-2.5 kg/cm²
Argillaceous deposits (according to water content)	0.2-2 kg/cm²

The principal physical and mechanical characteristics of soils are determined in geotechnical laboratories on intact samples taken from trial trenches or boreholes. Measurements made *in situ* enable the actual mechanical behaviour of the ground to be checked. The nature of the subsoil can also be investigated beforehand to avoid excessive loads being applied to very plastic and compressible soils such as peat or clay.

Accidental overloading, the passage of heavy plant and its vibration, and the flow of water are also dangerous factors to be avoided by all means near the edges of excavated cuttings. Excavations in the proximity of groundwater tables must be preceded by careful investigations.

Underground work. The conditions of underground work, such as excavation of tunnels, sinking of wells or shafts, are particularly dangerous for the workers and therefore require the use of special techniques and equipment. As for opencast work, a distinction can be made between excavation in rock and that in loose ground. Particular precautions must be taken when work is performed in loose ground, and the more so when water is present. Whatever solution is adopted for the final lining of the excavation, its perfect stability during construction is of the utmost importance. The workers must be reliably protected—even against falls of isolated stones—by efficient supports. To prevent the phenomenon of disruption of argillaceous rocks due to atmospheric humidity and the falling of small blocks, the walls and ceiling of the excavation should be "gunited" (shotcreted). Both temporary supports and linings and the definitive lining will be the more effective the sooner they are carried out after excavation of the rock. As long as the conditions of safety are not absolute, the workers must be protected by mobile shelters (shields) (see TRENCHING).

Hazards from surface water. Percolating and flowing water diminishes the mechanical resistance and stability of soils and rock. The mere presence of surface water (trickling, river, lake and sea water) is even under normal conditions a hindrance to the accomplishment of certain tasks (e.g. bridge-pillar foundations, piers, embankments, coast protection structures).

The slowly but continuously proceeding erosion phenomena due to the mechanical action of flowing water are at times intensified by heavy rainfalls, which deteriorate slopes, or by storms, which cause abrasion to coasts. Exceptionally high river waters give rise to inundations and accelerated alluvial deposits covering tens or even thousands of square kilometres and causing the collapse of embankments, bridges and other structures, sometimes accompanied by loss of human lives. Dam construction sites, especially when located at the bottom of deep gorges, may be easily inundated by sudden floods. They should be protected by erecting a cofferdam upstream and by diverting the course of the river in a diversion canal.

Hazards from volcanoes and earthquakes. Regions with recent or ongoing volcanic activity are well known. However, eruptions occur at an unforeseeable rhythm. The devastations they cause in populated areas are nearly always catastrophic, and man has only little means of protection and defence.

There are regions where the probability of major earthquakes can be predicted according to the frequency with which they occurred in the past. Furthermore, certain local premonitory signs are sometimes observed. However, the first shock, which often comes completely unexpectedly, may be the most intense and devastating. The best protection against earthquakes is to adapt the design of buildings and other structures to this phenomenon. Special regulations have been laid down to this end in the countries most threatened by earthquakes.

The first salvage operations in devastated areas are generally very dangerous on account of the risk of collapse of damaged buildings, and also because there may be more seismic shocks.

Harbours liable to be destroyed by tidal waves (tsunamis) must be protected by seawalls.

PERETTI, L.

CIS 77-1442 *Stability of unretained slopes and embankments* (Stabilité des pentes et talus non soutenus). Edition INRS No. 514 (Institut national de recherche et de sécurité, 30 rue Olivier-Noyer, Paris) (July 1976), 98 p. Illus. 15 ref. (In French)

CIS 79-1740 "Earthmoving, excavation and working opencast quarries" (Travaux de terrassement et d'exploitation de carrières à ciel ouvert). Comités techniques nationaux des industries des pierres et terres à feu, et du bâtiment et des travaux publics, Caisse nationale de l'assurance-maladie. *Cahiers de notes documentaires–Sécurité et hygiène du travail* (Paris), 3rd quarter 1979, 96, Note No. 1204-96-79 (Recommendation No. 159) (463-464). (In French)

CIS 75-1720 "Health and safety aspects of mine dump recovery" (L'hygiène et la sécurité dans l'exploitation des terrils). Gribauval, G. *Prévention et sécurité du travail* (Lille), 1st quarter 1975, 103 (17-22). Illus. (In French)

Germanium, alloys and compounds

Germanium (Ge)

a.w. 72.6
sp.gr. 5.32
m.p. 937 °C
b.p. 2 830 °C

a greyish-white, lustrous, brittle metalloid.

MAC USSR (Ge and oxides) 2 mg/m³

Germanium dioxide (GeO_2)

m.w. 104.6
sp.gr. 4.23
m.p. 1 115 °C

a white powder.

MAC USSR (Ge and oxides) 2 mg/m³

Germanium tetrachloride ($GeCl_4$)

m.w. 214.4
sp.gr. 1.84
m.p. −49.5 °C
b.p. 84 °C

a colourless liquid which decomposes in water.

MAC USSR 1 mg/m³ (as Ge)

Germanium tetrahydride (GeH_4)

MONOGERMANE

m.w. 76.6
sp.gr. 1.52 (liquid)
d. 3.43 (gas)
m.p. −165 °C
b.p. −88.5 °C

almost insoluble in water

a colourless gas which splits into Ge and H_2 at 200 °C.

TLV ACGIH 0.2 ppm 0.6 mg/m³
STEL ACGIH 0.6 ppm 1.8 mg/m³
MAC USSR 5 mg/m³

Occurrence. Germanium is always found in combination with other elements, never in the free state. Among the most common germanium-bearing minerals are argyrodite (Ag_8GeS_6), containing 5.7% of germanium, and germanite ($CuS.FeS.GeS_2$), containing up to 10%. Extensive deposits of germanium minerals are rare, but the element is widely distributed within the structure of other minerals, especially in sulphides, most commonly in zinc sulphide, and in silicates. Small quantities are also found in different types of coal.

Production. It may be obtained as a by-product in zinc production, in the processing of coal and, to a lesser extent, from the minerals containing it. If recovered from zinc metallurgy Ge is extracted from cadmium dust, retort residues and pitch after leaching out the cinders. The Ge concentrate obtained from these waste products contains 4-7% Ge. Treatment of the concentrate with strong hydrochloric acid leads to the formation of germanium tetrachloride ($GeCl_4$), which distills off at 86 °C. To separate admixtures of arsenium chloride, $GeCl_4$ is redistilled in a stream of chlorine. The highest degree of purity is obtained by distilling $GeCl_4$ in fractionating columns. The purified germanium tetrachloride is hydrolysed in water to obtain a precipitate of $GeO_2 + H_2O$. After dehydration, germanium dioxide (GeO_2) is left. If extracted from the by-products of coal (ash, soot), the fusion of soda, lime, copper oxide and coal dust is followed by chlorination, acidification and distillation to give a technical tetrachloride of germanium. Metallic germanium is obtained by reducing germanium oxide with hydrogen at 600-700 °C to a powder which is melted in the presence of nitrogen at 1 000 °C, or it can be reduced with carbon and sodium cyanide at 1 200 °C. Highly purified metal is obtained by multiple-zone melting and fractional distillation in order to concentrate impurities at one end of the ingot, which can then be cut off.

Uses. Germanium is of particular importance in the electronics industry, where its semi-conductive properties make it a valuable component in rectifiers and amplifiers required for radio engineering, radar stations, remote controls and automated systems. Ge also serves as a catalyst. It is also used in the production of alloys, of which germanium-bronze is characterised by high corrosion resistance. Germanium dioxide is used in the manufacture of optical glass and in cathodes for electronic valves. Some compounds are used for medical purposes, such as in dermatology.

HAZARDS

Occupational health problems may arise from the dispersion of dust during the loading of germanium concentrate, breaking up and loading of the dioxide for reduction to metallic germanium, and loading powdered germanium for melting into ingots. In the process of producing metal, during chlorination of the concentrate, distillation, rectification and hydrolysis of germanium tetrachloride, the fumes of germanium tetrachloride, chlorine and germanium chloride pyrolisis products may also present a health hazard. Other sources of health hazards are the production of radiant heat from tube furnaces for GeO_2 reduction and during melting of germanium powder into ingots, and the formation of carbon monoxide during GeO_2 reduction with carbon.

The production of single crystals of Ge for the manufacture of semiconductors brings about high air temperatures (up to 45 °C), electromagnetic radiation with field strengths of more than 100 V/m and magnetic radiation of more than 25 A/m, and pollution of the workplace air with metal hydrides. When alloying germanium with arsenic, arsine may form in the air (1-3 mg/m³), and when alloying it with antimony, stibine or antimonous hydride may be present (1.5-3.5 mg/m³). Germanium hydride, which is used for the production of high-purity germanium, may also be a pollutant of the workplace air. The frequently required cleaning of the vertical furnaces causes the formation of dust, which contains, apart from germanium, silicon dioxide, antimony and other substances.

Machining and grinding of germanium crystals also give rise to dust. Concentrations of up to 5 mg/m³ have been measured during dry machining.

Animal experiments on the effects of germanium and its compounds have shown that dust of metallic germanium and germanium dioxide causes general health impairment (inhibition of body weight increase) when inhaled in high concentrations. The lungs of the animals presented morphological changes of the type of proliferative reactions, such as thickening of the alveolar partitions and hyperplasia of the lymphatic vessels round the bronchi and blood vessels. Germanium dioxide does not irritate the skin, but if it comes into contact with the moist conjunctiva it forms germanic acid, which acts as an eye irritant. Prolonged intra-abdominal administration in doses of 10 mg/kg leads to peripheral blood changes. The effects of germanium concentrate dust are not due to Ge, but to a number of other dust constituents, in particular SiO_2. The concentrate dust exerts a pronounced fibrogenic effect resulting in the development of connective tissue and formation of nodules in the lungs similar to those observed in silicosis.

The most harmful germanium compounds are germanium hydride (GeH_4) and germanium chloride. According to Gus'kova, E. I. and Roščin, A. V., GeH_4 is a dangerous toxic substance which may provoke acute poisoning. Morphological examinations of organs of animals which died during the acute phase revealed circulatory disorders and degenerative cell changes in the parenchymatous organs. GeH_4 is a polytropic poison which also affects the nervous functions and peripheral blood.

$GeCl_4$ is an industrial irritant with a specific action. Its threshold of irritation is 13 mg/m^3. In this concentration it depresses the pulmonary cell reaction in experimental animals. In stronger concentrations it leads to irritation of the upper airways and conjunctivitis, and to changes in respiratory rate and rhythm. Animals which survive acute poisoning develop catarrhal-desquamative bronchitis and interstitial pneumonia a few days later. $GeCl_4$ also exerts general toxic effects, as may be induced from morphological changes observed in the liver, kidneys and other organs of the animals.

SAFETY AND HEALTH MEASURES

Basic measures during the manufacture and use of germanium should be aimed at preventing the contamination of the air by dust or fumes. In the production of metal, continuity of the process and enclosure of the apparatus is advisable. Adequate exhaust ventilation should be provided in areas where the dust of metallic germanium, the dioxide, or the concentrate is dispersed. Local exhaust ventilation should be provided near the melting furnaces during the manufacture of semiconductors, for example on zone refining furnaces, and during the cleaning of the furnaces. The process of manufacturing and alloying monocrystals of germanium should be carried out in a vacuum, followed by the evacuation of the formed compounds under reduced pressure. Local exhaust ventilation is essential in operations such as dry cutting and grinding of germanium crystals. Exhaust ventilation is also important in premises for the chlorination, rectification and hydrolysis of germanium tetrachloride. Appliances, connections and fittings in these premises should be made of corrosion-proof material. The workers should wear acid-proof clothing and footwear. Respirators should be worn during the cleaning of appliances. Workers exposed to dust, concentrated HCl, germanium hydride and $GeCl_4$ and its hydrolysis products should undergo periodic medical examinations.

MOGILEVSKAJA, O. Ja.

Germanium:

"Germanium and its compounds as harmful occupational factors" (Germanij i ego soedinenija kak faktory professional'-noj vrednosti). Roščin, A. V.; Gus'kova, E. I. *Kazanskij medicinskij žurnal* (Kazan), 1976, 6 (583-587). (In Russian)

Germanium dioxide and organic compounds:

CIS 525-1963 *Changes in haematopoiesis due to germanium dioxide and organic germanium compounds* (Veränderungen der Hämatopoese unter Germaniumoxyd und organischen Germaniumverbindungen) (Zurich, Pharmakologisches Institut der Universität Zürich, 1962), 35 p. Illus. 89 ref. (In German)

Germanium hydride and germanium tetrachloride:

CIS 77-1058 "Comparison of reactions of the organism to repeated exposure to germanium hydride and tetrachloride" (Sravnenie reakcij organizma pri povtornom vozdejstvii gidrida i tetrahlorida germanija). Ivanov, N. G.; Germanova, A. L.; Gus'kova, E. I. *Gigiena truda i professional'nye zabolevanija* (Moscow), Jan. 1976, 1 (34-38). 17 ref. (In Russian)

CIS 74-1953 "Occupational safety and germanium hydride" (O bezopasnosti raboty s gidridom germanija). Gus'kova, E. I.; Babina, M. D. *Bezopasnost'truda v promyšlennosti* (Moscow), Jan. 1974, 1 (40-41). (In Russian)

"Toxicology of germanium hydride" (K toksikologii gidrida germanija). Gus'kova, E. I. *Gigiena truda i professional'nye zabolevanija* (Moscow), Feb. 1974, 2 (56-57). (In Russian)

Glanders

MALLEUS

An infectious disease found in animals and man with an acute or chronic course.

Aetiology

The causal micro-organism, *Actinobacillus mallei* of the *Brucellaceae* family, is a slender, slightly curved rod with rounded ends, 2-3 μm long and 0.5 μm broad. The bacteria stains easily with all aniline dyes and is Gram-negative. It is sensitive to light, heat and desiccation, but will survive up to one month in animal excreta. It is relatively stable in water and may survive up to 30 days. *A. mallei* is very sensitive to disinfectants; at 100 °C, it is destroyed within a few minutes, at 55-60 °C, within 1-2 h.

Epidemiology

Glanders is a typical zoonosis and man is affected only sporadically. *A. mallei* is pathogenic to equines (horse, mule, ass), camels and felines. In animals, infection is spread by feeding on contaminated food or water and may occur in various forms: acute, chronic, hidden and symptomatic.

In man, the main danger of infection is contact with animals suffering from the acute form, especially when treating sick animals, in disposing of their corpses or excreta, or in the laboratory when preparing cultures or by contact with experimental animals. The persons most frequently affected are veterinarians, agricultural workers, and laboratory staff.

A. mallei can penetrate unbroken human skin, the nasal mucosa, respiratory system, conjunctivae and less frequently the digestive tract. Contamination is usually by contact, although in laboratories, infection by inhalation of airborne bacterium is possible.

Symptoms

Glanders in man may be acute, which is usually fatal, or chronic which lasts several months, sometimes several years, and with the victim only rarely recovering his health.

In the acute form, the incubation period lasts 2-3 days, seldom any longer. Normally the onset of the disease is brutal with shivering, a temperature of 38-39 °C, headaches and muscular pains. Temperature then fluctuates appreciably and pains occur in the joints which tend to swell. A reddish purple papule, surrounded by an erysipelous reddening, forms at the point of inoculation and later develops into a bloody pustule and finally an ulcer. Pneumonia accompanied by cough and expectoration of bloody sputum may also develop. Lesions of the nasal septum are accompanied by purulent greenish secretions sometimes flecked with blood; similar lesions may develop in the mucosa of the mouth. The victim's condition becomes extremely serious and is further aggravated by purulent lesions of the joints, diarrhoea and signs of heart disorders. The outcome is often fatal.

The chronic forms of glanders in man are cutaneous, pulmonary or nasal. The inflammatory process is slow, improvements being followed by aggravations and vice versa. The most common cutaneous form is characterised by the appearance of pustules which develop into abundantly suppurant purple ulcers and heal slowly forming large scars. Other symptoms include lymphangitis, lymphadenitis and muscular abscesses. Relapse and aggravation are common and, in 50% of cases, the outcome is fatal.

Diagnosis

Three important steps in diagnosis are clinical examination, epidemiological history and, in particular, laboratory analysis of sputum, blood, pus, the mucosae of open, external ulcers and the contents of closed nodules or of abscesses. The basic diagnostic reaction for glanders is the complement fixation reaction, which is usually positive throughout the course of the disease whilst the micro-organism is in the blood. A change from a positive to negative reaction normally indicates that the disease has been cured.

The diagnosis of glanders in man by the allergic method is usually made by intradermal injection of 0.1 cm^3 of mallei filtrate consisting of 4-8 month old, dead culture of *A. mallei*, to which 4% glycerine solution has been added. The positive reaction takes the form of an inflammation of varying intensity and appears during the second or third week of the disease, depending on the course of the infection. The reaction is specific but does not allow differentiation between glanders and melioidosis.

PREVENTIVE MEASURES

In the event of an outbreak of glanders, all the horses in the area in question should be given a mallein test. Every animal with clinical evidence of glanders and all those giving a positive mallein and complement fixation tests should be slaughtered; the carcasses should be buried in a pit and covered with chloride of lime (bleaching powder) and the infected premises and equipment must be thoroughly disinfected. Animals suffering from an asymptomatic form of glanders, which react positively to the mallein test, should be placed under isolation.

The animal can be considered healthy if the mallein test applied three times at 6-day intervals proves negative. To prevent infection of persons working with animals reacting positively to the mallein test, equipment should be carefully disinfected and personal protective equipment should be worn. All animal excreta used as fertiliser should be disinfected regularly, as should stalls, stables, harnesses, etc. Stable personnel should be provided with adequate sanitary facilities and liquid disinfectants for the hands, etc., and scrupulous personal hygiene should be encouraged.

Laboratory work with *A. mallei* should be strictly controlled. Laboratories and related premises, especially those containing infected animals, should be fitted with exhaust and general ventilation; equipment which may produce atmospheric contamination (e.g. dryers, centrifuges, etc.) should be effectively enclosed. Laboratory workers should change their clothes on entering and leaving areas in which there is a danger of infection. Workers involved in the manipulation of *A. mallei* should wear eye and face protection and respiratory protective equipment (face masks made of six layers of gauze or gas masks with filters).

Persons who have been exposed to infection should be kept under close medical supervision for 21 days and should not be considered definitely free from *A. mallei* until after several negative mallein and complement fixation reactions have been obtained during subsequent months.

Persons suffering from glanders should be hospitalised in special infectious disease establishments. For chronic cases, isolation is necessary at every aggravation; persons having been in contact with glanderous patients should be subject to medical supervision. Hospital wards and equipment should be carefully disinfected.

No effective vaccines are available against glanders, and therapeutic measures are not recommended for the treatment of glanders in animals.

Treatment

Persons suffering from glanders require a rich diet, close medical supervision and symptomatic treatment. Recent clinical experience with human glanders indicates that the disease can now be treated effectively with sulphonamides and antibiotics on a prolonged regimen.

BLUMEL, N. F.

"Glanders". Steele, J. H. (339-362). *CRC Handbook Series in Zoonoses*, Vol. I (Boca Raton, Florida, CRC Press, 1979), 643 p. Illus. 22 ref.

Glass industry

Glass has physical properties intermediate between liquid and solid: when cooled from the hot molten state, it gradually increases in viscosity without crystallisation over a wide temperature range until it assumes its characteristic hard, brittle form: cooling is controlled to prevent crystallisation, or high strain.

Any compound having these physical properties is theoretically a glass. There are a very large number of chemical compositions which fall into three main types:

1. Soda-lime-silica glasses. These are the most important glasses in terms of quantity produced and variety of use, including almost all flat glass, containers, low-cost mass produced domestic glassware and electric light bulbs.

2. Lead-potash-silica glasses. These contain a varying but often high proportion of lead oxide. Optical glass manufacture makes use of its high refractive index, hand-blown domestic and decorative glassware, its ease of cutting and polishing, electrical and electronic applications, its high electrical resistivity and radiation protection; the lead content may be as high as 90%.

3. Borosilicate glasses. Borosilicate glasses having a low thermal expansion are resistant to thermal shock. This makes them ideal for domestic oven and laboratory glassware.

Raw materials. These include: silica in the form of sand or crushed rock quartz; soda ash (anhydrous sodium carbonate) or in some cases salt cake (sodium sulphate); potassium carbonate or nitrate; crushed limestone or dolomite; red lead or litharge; boric acid or anhydrous borax; cullet which consists of broken or crushed glass.

Not all these constituents are contained in every type of glass and other chemicals may be added in small amounts for special purposes. Some of these substances present a potential toxicity hazard, for example:

(a) salts of chromium, cobalt, cadmium, manganese, nickel and selenium used as colouring agents;

(b) arsenic and antimony salts employed to remove bubbles from molten glass;

(c) fluorine, calcium fluoride or sodium silicofluoride, which may be added to accelerate melting;

(d) barium salts to increase electrical resistivity and refractive index;

(e) thorium and rare earth metals such as lanthanum, which may be used to improve optical properties.

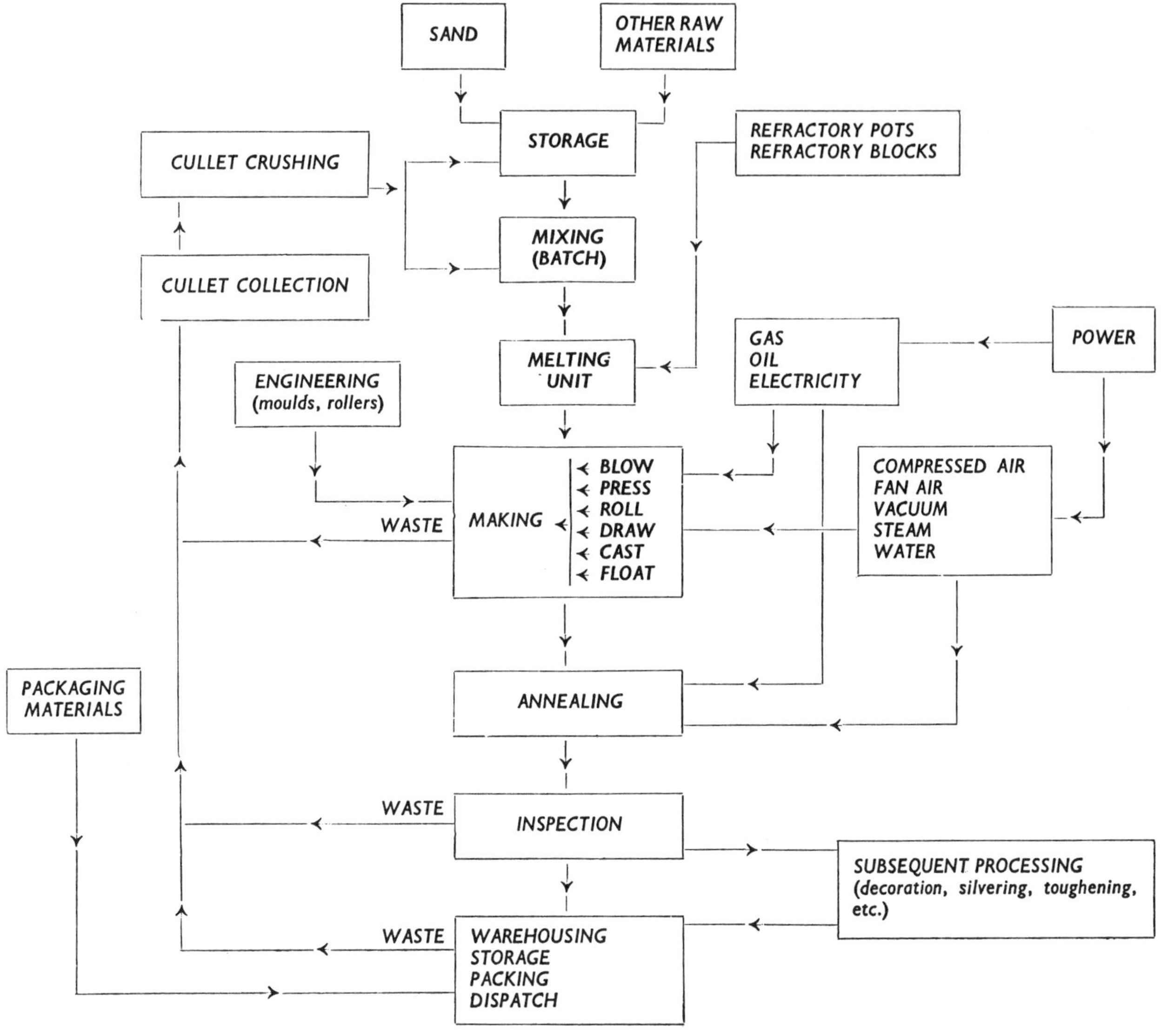

Figure 1. The processes and materials involved in the manufacture of glass.

Processing. Figure 1 illustrates the basic principles of glass manufacture. The raw materials are weighed, mixed and after the addition of broken glass (cullet), are taken to the furnace for melting. Small pots of up to 2 t capacity are still used for the melting of glass for hand-blown crystalware and special glasses required in small quantity. Several pots are heated together in a combustion chamber.

In most modern manufacture, melting takes place in large regenerative or recuperative furnaces built of refractory material and heated by oil or natural or producer gas. The hottest part of the furnace may be at 1 500 °C, from where controlled cooling reduces this to about 1 000 °C at the point at which the glass leaves the furnace. All types of glass are subjected to further controlled cooling (annealing) in a special oven or lehr. Subsequent processing depends upon the type of manufacture.

In addition to the traditional hand-blown glass, automatic blowing is used on machines for bottle and lamp bulb production. Simple shapes, such as insulators, glass bricks, lens blanks, etc., are pressed. Some manufacture uses a combination of mechanical blowing and pressing. Wired and figured glass is rolled. Sheet glass is drawn from the furnace by a vertical process which gives it a fire-finished surface. Owing to the combined effects of drawing and gravity, some minor distortion is inevitable.

Plate glass passes through water-cooled rollers into an annealing lehr. It is free from distortion but surface damage requires removal by grinding and polishing. This process has largely been replaced by the float glass process.

Float glass. In recent years a new process—the float process (figure 2)—has made possible the manufacture of a glass combining the advantages of both sheet and plate. Float glass has a fire-finished surface and is free from distortion.

In the float process a continuous ribbon of glass moves out of a melting furnace and floats along the surface of a bath of molten tin. The glass conforms to the perfect surface of the molten tin. On its passage over the tin, the temperature is reduced until the glass is sufficiently hard to be fed on to the rollers of the annealing lehr without marking its under surface. An inert atmosphere in the

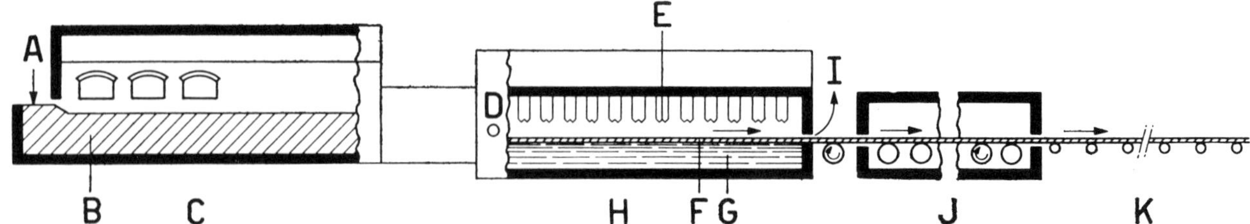

Figure 2. The float process for the manufacture of float glass. A. Raw material feed. B. Molten glass. C. Glass-melting furnace. D. Controlled inert-atmosphere inlet. E. Heaters. F. Glass ribbon. G. Molten metal. H. Float bath. I. Atmosphere outlet. J. Annealing lehr. K. Cutting section.

bath prevents oxidation of the tin. The glass, after annealing, requires no further treatment and lends itself to disposal by automatic cutting and packing.

Surface modification. This is a new technique, which uses an electrochemical system to drive metallic ions into the float glass ribbon whilst it is passing through the bath. In this way the glass can be coloured and given controlled physical properties in respect of heat and light transmission. The potential exists for the application of a wide range of metals by this method.

Other surface modification processes use organic based sprays of, for example, silane compounds on the hot glass surface. Amorphous silica may be evolved. Lung fibrosis has been reported from amorphous silica exposure but seems likely to be due to the presence of alpha quartz crystals within the amorphous spheres. X-ray diffraction is required to detect the presence of crystalline silica.

Some optical glasses are chemically strengthened by processes which involve immersing the glass for several hours in high-temperature baths containing molten salts of, typically, lithium nitrate and potassium nitrate.

Safety glass. Safety glass is of two types:

1. Toughened glass is made by prestressing, as a result of heating, and then rapidly cooling pieces of flat glass of desired shape and size in special ovens.
2. Laminated glass is formed by bonding a sheet of plastic (usually polyvinyl butyral) between two thin sheets of flat glass.

HAZARDS

In an industry so widely spread and so various, hazards also differ in type and intensity: fully mechanised modern plants have eliminated some of the risks which still remain in older factories using traditional methods.

Accidents. Accidents during handling of glass are a major hazard in the flat-glass industry and may cause lacerated wounds often involving tendons and peripheral nerves. They occur to a lesser extent in other glass manufacture, where burns are the most frequent injury.

Flying glass may cause penetrating wounds and serious eye injury: a special risk exists when toughened glass "explodes" during manufacture.

Mechanisation and automation is playing a major part in the prevention of severe injuries in producing and handling flat glass. It is important to realise that with automation the risk area is often transferred from the glass handling personnel to the maintenance team. Due to the inherent risk in handling glass it is essential to both glass handling and maintenance personnel that full consultation is encouraged to establish correct methods of work in risk areas and personnel adequately trained in these agreed methods.

In some areas, it is still essential to provide personal protective clothing, which must provide adequate protection but must also be comfortable in order to be acceptable to wearers. Knitted wire sandwiched between two layers of material has been found to be a satisfactory compromise, giving a high level of protection whilst remaining acceptable to the workers.

Eye protection is required when working on glass or in handling cullet. Again, it is essential to provide comfort and this is best attained by personal fitting and prescription toughened lenses where required.

At all times in the glass industry housekeeping must be kept at a high level.

Silica. This is the largest component in commercial glass manufacture; it is used in the form of natural sand from which the fine particles have been removed during washing, and is not a cause of silicosis. There is a tendency, because of ease of melting, to use finer raw materials. Fine sand which has been well agitated in sand dryers can produce a hazardous airborne siliceous dust.

When a very high order or purity is required, crushed quartz may be substituted for natural sand. This dust is particularly dangerous.

It is not uncommon for glass manufacturers to make their own refractory clay-work for use in the tanks. Silica mortar is also widely used. The making of refractories and the cutting, sawing and chipping to size can give rise to dust containing a dangerous amount of alpha quartz, or the more dangerous cristobalite. Many cases of silicosis have occurred in the preparation of special glasses for processing in small refractory furnaces.

Lead. Lead poisoning may occur from the mixing of raw materials containing red lead or litharge and subsequent filling into the pot or furnace. Certain lead-containing glasses are water soluble to some extent and contamination from the fine spray produced during cutting may require control. Lead glazes (usually borates or borosilicates) are used in the decoration and colouring of glass. The glaze is fused to the glass by heat.

Alkaline dusts. Alkaline materials such as soda ash and potash can cause nasal ulceration and perforation of the nasal septum. They are skin irritants and a small lump falling on sweating skin may cause a whole-thickness caustic burn.

Other raw materials. These are too numerous to mention individually: some with potential toxicity are mentioned above in the list of minor ingredients, some such as arsenic are poisons and are potentially skin and lung carcinogens, whilst being skin irritants. It has not yet proved possible to replace them with safer materials. Danger arises almost entirely during handling in concentrated form and during mixing.

Fuels and products of combustion. The use of fuel oils for firing provides a potential risk of scrotal carcinoma during skin contact, and good hygiene practice is necessary together with regular medical examinations. The oil-fired process may give rise to irritant concentrations of sulphur dioxide under the roof of the tank house and under the stacks of the regenerators. Other situations are less important as they do not represent places where men work. Vanadium deposits in the flues may be a hazard during cleaning and repair; suitable ventilation, wet methods and protective clothing, including masks, may be required. Producer gas may create a risk of carbon monoxide.

Miscellaneous hazards. A common method of etching or obscuring glass is by the application of hydrofluoric acid. The problem is twofold–burns due to skin contact and a fume hazard when the hydrofluoric acid is spread in quantity over the glass. There are often several uses of asbestos in the industry. A particular problem in sheet-glass manufacture is the making, and after-use stripping of asbestos rollers. Fumes from lubricating oils at presses may be dense and unpleasant.

Heat and radiant energy. The working environment around a glass furnace is hot but the significant heat problems arise during maintenance and emergency repair work. Most of the heat is in the form of radiant energy. The black-bulb temperature in areas where men are doing routine maintenance is frequently in the range 120-160 °C. Under emergency repair conditions it may reach 200 °C These conditions may give rise to physiological strain, which is normally part of the adaptive process. In hot countries, in particular, various forms of heat illness may result from excessive exposures. Mechanisation of processes, together with shielding in the form of cold-water reflectors and reflective screening for very hot areas are effective and water-cooled platforms and cooling rooms are desirable.

Glass blowing. Glass blowing by mouth may cause deformity of the cheeks and damage to the mouth and teeth. Emphysema, and chancre of the lips caused by multiple use of the same apparatus, have been largely eliminated.

Heat cataract. Glassmakers are one of the groups exposed to the risk of this posterior polar cataract. A modern concept is that the lens changes are secondary to heat effects on the iris and ciliary body. The decline in the handworking of glass, the enclosure and increased mechanisation of modern glass manufacture together with the use of scientific temperature measurement in place of the traditional visual colorimetric estimations have so reduced the exposure that this condition has been virtually abolished from the modern glass industry. Where these methods are not possible, screening and the use of face and eye protection remain the most effective prevention.

Noise. Harmful noise levels of over 100 dB, with a significant high-frequency component, are met with at some glass-pressing machines such as are used in making glass bottles, and are mainly produced by high-pressure compressed-air cooling jets. Ear protection is a necessary part of a hearing conservation programme.

SAFETY AND HEALTH MEASURES

Accident prevention. Good housekeeping is of prime importance in the prevention of injuries from broken glass projecting from tables, racks, or bins or lying on the floor. An efficient safety organisation is most important. Increasing mechanisation and automation greatly diminishes risks from manual handling. The resulting increase in size and packaging, particularly where racks and stillages are involved, may result in a greater hazard and both design and safe systems of work require close examination.

Figure 3. Workers wearing protective clothing for glass handling.

Personal protective clothing designed for the purpose can prevent many injuries in glass handling or from flying glass: knitted-steel wire mesh between two layers of cloth can be used in making jackets, sleeves, leggings, etc.; increased flexibility can be obtained if polypropylene strands are substituted for wire in one axis. Efficient eye protection is essential where risk of flying glass exists and against glare (figure 3).

Dust and fume control. Enclosure or efficient locally applied exhaust ventilation are essential wherever toxic dusts, fumes or vapours are evolved: silica flour can be safely handled only in an enclosed system. Regular monitoring of the working environment is desirable wherever toxic materials are in use.

Risks are significantly reduced in modern glassmaking. Bulk delivery of raw materials in tankers or wagons, bottom-opened into a hopper or pneumatically emptied, together with automatic weighing, mixing and transfer by enclosed methods have removed the hazards of dust in manual handling. Substitution of granular lead monosilicate for other finely powdered lead preparations brings about a marked reduction in the evolution of dust.

For emergency short-term work, for example trimming refractories on-site during tank repair, the use of efficient respiratory protection is necessary. Tank furnaces are so hot that, after filling, the upward convection in a well ventilated tank house usually takes dust and fumes away from the working atmosphere. Maintenance men cleaning or changing light fittings or doing other short-term jobs in the upper parts of tank houses should wear suitable cartridge respirators against sulphur dioxide fumes.

Hydrogen fluoride. Efficient exhaust is required to remove the fume, and safe handling is obtained by the use of polyethylene containers and syphon methods of transfer. Full personal protection against the risk of splashing is required, including goggles, hand protection and apron. In burns, there is often a latent period between skin contact and the onset of pain. Washing facilities and magnesium oxide paste should be immediately available as first-aid measures. The subcutaneous injection of 10% calcium gluconate under the burn gives dramatic relief from symptoms.

Heat. The risk of heat illness is greatly reduced by acclimatisation. An adequate supply of saline beverages at work should be available to all workers around glass furnaces to enable replacement of salt and water loss by sweating. Considerable protection from radiant energy can be provided by screening. The best is highly polished, heat-reflecting metal. An efficient portable screen can be made of two layers of anodised aluminium–corrugated for additional strength–thermally insulated from each other and separated by a 10-15 cm space to allow a chimney effect. Regular cleaning to maintain the reflecting surface is essential. Where vision is necessary, heat-reflecting or heat-absorbing glass can be used.

The traditional blowing of cool air on workers at glass-processing machines is still used. Modern protective clothing is made of suitable material coated with a vacuum-deposited reflective aluminium surface. To this, when necessary, is added air cooling or water cooling. The latter is most efficient. If a water-cooled undersuit is worn beneath reflective asbestos clothing, physiological equilibrium can be maintained for long periods (figure 4).

Figure 4. A worker wearing a water-cooled heat-protective suit carrying out a hot repair on a glass furnace.

Medical supervision. Manual dexterity and upper limb joint movements are important in glass handlers. Lead workers should be under regular medical supervision at such intervals as are mandatory or as the nature of the exposure dictates. Routine chest radiography of those exposed to silica dust should be carried out annually or at such longer periods as the degree of exposure requires.

First-aid treatment. An efficient first-aid organisation, with the members specially trained in the control of haemorrhage and the emergency dressing of wounds, can minimise the effects of injury. Large flat-glass manufacturing units can with advantage provide treatment and rehabilitation facilities. Where there is a possibility of a retained glass foreign body, an X-ray examination of the part should be arranged: almost all glass is sufficiently radio-opaque to be identified on conventional X-ray photographs.

Washing facilities of high standard, including showers, should be maintained.

Legislation. Continuous production requires shift working in many processes: national legislation may control the length and arrangement of hours that may be worked, especially by women and young persons. The ILO Sheet-Glass Works Convention, 1934 (No. 43) and Reduction of Hours of Work (Glass-Bottle Works) Convention, 1935 (No. 49) both relate to this aspect. National legislation may also exclude women or young persons from certain dangerous, heavy or unpleasant processes associated with the glass industry.

CAMERON, J. D.
HILL, J. W.

CIS 74-0254 *Safety and health in the glass industry* (Arbeitsschutz in der Glasindustrie). Polleschner, B.; Koch, K.; Weise, A. (Berlin, Verlag Tribüne, 1972), 187 p. Illus. 21 ref. (In German)

CIS 77-248 "Pulmonary emphysema in glass blowers" (Das Lungenemphysem bei Mundglasbläsern). Wagner, F. *Arbeitsmedizin–Information* (Berlin), 1975, 2/576 (28-32). (In German)

CIS 77-960 *Dust abatement in glass works air* (Snĭenie zapylennosti vozduha na stekol'nyh zavodah). Suhareva, A. I. (Leningrad, Strojizdat, 1976), 137 p. Illus. 45 ref. (In Russian)

CIS 81-378 "Noise evaluation and noise control in the glass industry" (Lärmerfassung und -bekämpfung in der Glasindustrie). Kilian, A.; Zaunick, U. *Die Technik* (Berlin), Feb. 1979, 34/2 (101-104). Illus. 6 ref. (In German)

CIS 77-1746 "Using a cool spot to improve the thermal comfort of glassmakers". Sims, M.; Gillies, G.; Drury, R. *Applied Ergonomics* (Guildford), Mar. 1977, 8/1 (2-6). Illus. 3 ref.

CIS 76-1100 "Work in glass works–Protective clothing against infrared radiation" (Travail dans les verreries–Un vêtement de protection contre le rayonnement infrarouge). Aubertin, G.; Granjon, M. *Travail et sécurité* (Paris), Aug.-Sep. 1975, 8-9 (402-409). Illus. (In French)

Glove manufacture

Gloves may be used for adornment, warmth or protection from injury, and may be made from a wide range of materials. Those for adornment may be from fabric of natural or artifical fibre, knitted or stitched, leather, or plastics; those for warmth are usually made of leather, wool and even plastics; and those for protection, from leather, cotton, rubber, asbestos, fabric-backed polyvinyl chloride material, or chain mail, according to the type of protection required.

Manufacturing processes. The essential processes in the manufacture of gloves differ little from the processes employed in the manufacture of similar articles from the same materials in the textile industry, except in the manufacture of leather dress gloves. For this purpose, sheep and goat skins of fine quality are preferred. Whereas, at one time, the whole process of preparation of the skin was carried out by the glove manufacturer, it is now usual for skins to be received fleshed, tanned and dyed from tanneries. After receipt the skin is cut into rough shape with its transverse measurement in excess of the measurement required by about 25%. Each piece is stretched by working it across an edge, commonly the edge of a table, until the length is increased and the width decreased to the desired distance. The shape of the glove is then cut out. Other pieces of the skin are used for the gussets at the side of the fingers and experience is needed for exact matching in colour and quality. The leather is hand or machine clicked in the same way as in the manufacture of other leather articles. The gloves are assembled by hand or machine sewing.

HAZARDS AND THEIR PREVENTION

Accident hazards. Clicking machines can cause injury to the hands: two-handed control and safety cutters are useful in prevention.

As in other trades, the driving mechanisms of power-driven sewing machines present a hazard unless the belts and (where it is used) shafting have secure machinery guarding. Occasional needle injuries may be caused especially as fine stitching is often involved. The process of working the skins requires some manual force but this is not rotational and tenosynovitis is uncommon. Prevention of injury depends largely on adequate training.

Health hazards. In the earlier years of the century medical publications described defects, most of which were a result of the environmental conditions under which the process was carried out rather than a result of the process itself. Attention is still required to environmental conditions, especially to maintenance and good housekeeping and good lighting.

The making-up of gloves lends itself to homework, which is usually either outside, or only partially within, statutory control. Bad environment, employment of women and young persons for long hours and low rates of pay may still have a detrimental effect in some places.

Sometimes toxic or flammable solvents may be used in small quantities for cleaning of fabrics or skins. Non-flammable and non-toxic solvents should be substituted whenever possible: in any event the smallest possible quantity should be used in well ventilated conditions.

Good visual acuity and colour vision for correct matching are required in the manufacture of leather dress gloves. These should normally be checked in any pre-employment medical examination.

COPPLESTONE, J. F.

Glycerol and derivatives

Glycerol ($CH_2OHCHOHCH_2OH$)

GLYCERINE; GLYCERIN; 1,2,3-PROPANETRIOL

sp.gr. 1.26
m.p. 18.6 °C
b.p. 290 °C
v.d. 3.17

$$\begin{array}{l} CH_2OH \\ | \\ CHOH \\ | \\ CH_2OH \end{array}$$

v.p. 0.0025 mmHg (0.3 Pa) at 50 °C
f.p. 177 °C oc
i.t. 370 °C

soluble in all proportions in water and ethyl alcohol; virtually insoluble in ethyl ether, chloroform and benzene

a colourless, odourless, hygroscopic, viscous liquid with a sweet taste. The usual article of commerce contains up to 1% of water, and as such has a freezing point of 15 °C. However, the liquid is very prone to supercooling, and may remain liquid at lower temperatures until sudden crystallisation occurs. The resulting solid is difficult to liquefy and may require heating to 20-25 °C before melting occurs.

TLV ACGIH (glycerin mist) 10 mg/m³ of total particulate or 5 mg/m³ of respirable particulate

It is an alcohol and, consequently, the term glycerol is preferred for the pure chemical; however, the commercial product is usually called glycerin.

Production. Until the late 1940s all glycerol was produced by the hydrolysis of animal and vegetable oils and fats consisting primarily of glycerides—i.e. the esters of glycerol and fatty acids; but increasing quantities are now being produced synthetically from petroleum products. Additionally, some glycerol is produced by hydrogenolysis of refined sugar and by fermentation of glucose, but these processes are of lesser commercial importance.

Preparation from oils and fats. The chief constituents of virtually all animal and vegetable oils and fats are triglycerides of fatty acids, for example 1-stearo-2-palmito-3-laurin:

$$\begin{array}{l} CH_2OCOC_{17}H_{35} \\ | \\ CHOCOC_{15}H_{31} \\ | \\ CH_2OCOC_{11}H_{23} \end{array}$$

Hydrolysis of such a glyceride yields free fatty acid and glycerol. Two hydrolysis techniques are used, namely alkaline hydrolysis (saponification) and neutral hydrolysis (splitting).

Saponification. Fat is boiled with sodium hydroxide and sodium chloride. This releases the fatty acid constituents as their sodium salts (soaps) which are salted out of solution by the sodium chloride and separated into an upper layer. The lower aqueous layer, known as "spent lye" contains glycerol, water, sodium chloride and unchanged sodium hydroxide. After separation from the soap, the spent lye is subjected to a preliminary treatment to remove impurities. The liquor remaining is then subjected to a two-stage evaporation process to remove water and to yield "soap-lye" crude glycerin which contains about 80% of glycerol. This residual glycerin is filtered while it is still hot to remove salts including sodium chloride, most of which is precipitated in the course of the evaporation process. The remainder of the sodium chloride and a number of other impurities such as glycols, acids, esters, aldehydes, nitrogenous and coloured matter are removed by fractional distillation under reduced pressure and treatment with activated carbon to ensure that the purity of the product is adequate for its use in foods, pharmaceuticals and cosmetics.

As an alternative to distillation, ion exchange or ion exclusion techniques are sometimes used to refine glycerin.

Neutral hydrolysis. The old technique of splitting, the "Twitchell process", involved heating fat with water in the presence of an acid catalyst, but nowadays most fat splitting is carried out at elevated pressure without a catalyst. The fats are hydrolysed by a batch or semi-continuous process in a high-pressure autoclave, or by a continuous countercurrent technique in a high-pressure column.

The products of hydrolysis are insoluble fatty acids and an aqueous solution of glycerol known as "sweet water". The sweet water is drawn off and cleaned of residual fatty acids which are precipitated as heavy metal salts. The glycerin in the clean sweet water is concentrated and refined in a manner similar to that already described except that there is no sodium chloride to remove.

Synthesis. There are two main processes for the synthesis of glycerol, both starting with propylene ($CH_3CH{:}CH_2$). In one process, propylene is treated with chlorine to give allyl chloride; this reacts with sodium hypochlorite solution to give glycerol dichlorohydrin from which glycerol is obtained by alkaline hydrolysis.

$$CH_3CH{:}CH_2 \xrightarrow{Cl_2} CH_2ClCH{:}CH_2 \xrightarrow{HOCl}$$

$$CH_2ClCHClCH_2OH \xrightarrow{H_2O} CH_2OHCHOHCH_2OH$$

In the other process, propylene is oxidised to acrolein ($CH_2{:}CHCHO$), which is reduced to allyl alcohol ($CH_2{:}CHCH_2OH$). This compound may be hydroxylated with aqueous hydrogen peroxide to give glycerol directly, or treated with sodium hypochlorite to give glycerol monochlorohydrin which, upon alkaline hydrolysis, yields glycerol.

$CH_2{:}CHCH_3OH \xrightarrow{H_2O_2} CH_2OHCHOHCH_2OH$, or alternatively

$$CH_2{:}CHCH_2OH \xrightarrow{HOCl} CH_2OHCHClCH_2OH \xrightarrow{H_2O} CH_2OHCHOHCH_2OH$$

Uses. Next to water glycerol is probably the most versatile compound known. This versatility is due to a number of factors. The three hydroxyl groups confer a hydrophilic character on the molecule, thus enabling glycerol to be used in aqueous systems. Association of glycerol molecules conveys low volatility and hence a high flash point. The fact that glycerol is safe to handle and virtually non-toxic makes it suitable for use in food, pharmaceuticals, toiletries and cosmetics. The hygroscopic nature of glycerol renders it ideal as a humectant in such diverse products as tobacco, confectionery icing, skin creams and toothpaste, which would otherwise deteriorate on storage by drying out. Some measure of the versatility of glycerol can be gained from the fact that some 1 700 uses for the compound and its derivatives have been claimed.

In medicine, glycerol is currently used mainly for its emollient and laxative action and in cosmetics. Probably the largest single use of glycerol is in the production of alkyd resins for surface coatings. These are prepared by condensing glycerol with a dicarboxylic acid or anhydride (usually phthalic anhydride) and fatty acids. A further major use of glycerol is the production of explosives (nitroglycerine, dynamite, etc.).

HAZARDS AND THEIR PREVENTION

Glycerol has a very low toxicity (oral LD_{50} (mouse) 31.5 g/kg) and is generally considered harmless under all normal conditions of use. However, when present as a mist, it is classified by the American Conference of Governmental Industrial Hygienists (ACGIH) as a "particulate nuisance", and as such a TLV of 10 mg/m^3 has been assigned. In addition, the reactivity of glycerol makes it dangerous, and liable to explode in contact with strong oxidising agents such as potassium permanganate, potassium chlorate, etc. Consequently it should not be stored near such materials. There are also a number of hazards associated with its production.

In saponification, the process of boiling fat with sodium hydroxide solution may produce dangerous splashes. If an accident does occur, speed in removing the sodium hydroxide from the skin is essential; the affected area should be immediately and copiously irrigated with water and a physician should be consulted for further treatment. At other stages in the saponification process and throughout the process of neutral hydrolysis, hot liquids are manipulated. The risk of splashing is reduced very considerably by the use of modern plant in which arrangements are made to ensure the correct sequence of valve, heating and condenser controls. The standard of maintenance should be high, particularly in high-pressure hydrolysis, where the leakage of steam or hot fat could cause a serious accident. As in all processing, good housekeeping is essential. Spillage of fat onto floors and stairways should be avoided, but if any does occur the area should be cleaned immediately to prevent injury from slipping and/or falling on greasy surfaces.

In the synthetic processes, the starting material, propylene, is a flammable gas which forms explosive mixtures with air in concentrations between 2.0 and 11.1%. It is important therefore to prevent even quite small leakages from the plant. The intermediate products, allyl chloride and allyl alcohol, are both highly toxic and flammable liquids. Both materials give rise to irritant vapours and are rapidly absorbed through the skin. The irritation which they cause should be treated as a warning of exposure which, if ignored, could lead to serious injury. Oxidising substances such as hydrogen peroxide and sodium hypochlorite are harmful to the skin and eyes and their presence tends to intensify a fire once it becomes general.

Persons engaged in the synthesis processes should wear suitable personal protective equipment and eye protective equipment. As with plant used for saponification and hydrolysis a high standard of equipment maintenance should be observed. General maintenance and good housekeeping are also important.

Glycerol derivatives

Glycerol is a trihydric alcohol and undergoes reactions characteristic of alcohols. The hydroxyl groups have varying degrees of reactivity, and those in the 1- and 3-positions are more reactive than that in the 2- position. By using these differences in reactivity and by varying the proportions of reactants, it is possible to make mono-, di- or tri- derivatives.

Glyceryl monostearate ($C_{17}H_{35}COOCH_2CHOHCH_2OH$)

GLYCEROL MONOSTEARATE; MONOSTEARIN

m.w. 358
sp.gr. 0.98
m.p. 57-65 °C depending on concentration of monoester

slightly soluble in ethyl alcohol and hydrocarbons; insoluble in water

a pure-white or cream-coloured, wax-like solid with a faint odour and a fatty, agreeable taste.

Glyceryl monostearate is produced by the alcoholysis of the fat with glycerol, yielding a mixture of the mono-, di- and triglycerides in which the monoglyceride is predominant. Monoglycerides can also be made by esterification of fatty acids with glycerol. It is used as an emulsifying agent for margarine, oils, waxes and solvents and as a protective coating for hygroscopic powders. It acts as an opacifier, detackifier and a resin lubricant. It is extensively used in cosmetics.

Industrial exposure to glyceryl monostearate does not present any hazards.

Glyceryl diacetate ($HOCH(CH_2O_2CCH_3)_2$)

DIACETIN

m.w. 176.2
sp.gr. 1.18
m.p. 40 °C
b.p. 280 °C

soluble in water

a colourless, mobile liquid, used extensively as a food flavour solvent and also in the foundry industry.

It is produced by the reaction of glycerol and acetic acid. It is of low toxicity (oral LD_{50} (mouse) 8.5 g/kg) but may cause slight irritation in sensitive individuals.

Glyceryl triacetate ($CH_3CO_2CH(CH_2O_2CCH_3)_2$)

TRIACETIN

m.w. 218.2
sp.gr. 1.16
m.p. −78 °C
b.p. 258-260 °C
f.p. 149 °C

(data provided by the author)

very soluble in ethyl alcohol and other organic solvents; slightly soluble in water

a colourless liquid with a slight, fatty odour and a bitter taste.

Glyceryl triacetate is produced by the action of acetic acid on glycerol. It is used as a camphor substitute in the pyroxylin industries, as a plasticiser, a fixative in perfumes, in the manufacture of cosmetics and to remove carbon dioxide from natural gas.

Glyceryl triacetate appears to be innocuous when swallowed, inhaled or in contact with the skin, but may cause slight irritation to sensitive individuals (oral LD_{50} (rat) 3 g/kg).

Glycerophosphoric acid ($CH_2OHCHOHCH_2OPO(OH)_2$)

sp.gr. 1.60
m.p. −20 °C

soluble in water and ethyl alcohol

a colourless, odourless liquid.

Glycerophosphoric acid is produced by the interaction of glycerol and phosphoric acid. Salts of glycerophosphoric acid are used as tonics in debilitated conditions and in convalescence.

Certain precautions are recommended for the manipulation of phosphoric acid. See PHOSPHORUS AND COMPOUNDS.

HAZARDS AND THEIR PREVENTION

This material in contact with the skin causes redness and swelling. It is intensely irritating to the eyes and mucous membranes. When heated to decomposition, or in contact with acid, it emits highly toxic fumes of sulphur compounds. The material should be prepared in enclosed plant provided with exhaust ventilation to prevent the escape of harmful fumes through the openings provided for the purposes of manipulation. Suitable personal protective equipment, including eye and face protection, should be worn; respiratory protective equipment should be available.

FENLON, A. T.

CIS 75-1052 *Dangerous properties of industrial materials.* Sax, N. I. (New York, Van Nostrand Reinhold Company, 5th ed., 1979), 1 118 p. Illus. Ref.

Further information is available in the documentation published by the United Kingdom Glycerine Producers' Association.

Glycols and derivatives

The commercially important glycols are aliphatic compounds possessing two hydroxyl groups per molecule and are colourless, viscous liquids that are essentially odourless. Ethylene glycol and diethylene glycol are of major importance among the glycols and their derivatives.

The toxicity and hazard of certain important compounds and groups are discussed in the final section of this article. None of the glycols or their derivatives that have been studied have been found to be mutagenic, carcinogenic or teratogenic.

The glycols and their derivatives are flammable liquids. Their flash points are above normal room temperature, however, and the vapours are liable to be present in concentrations within the flammable or explosive range only in plant such as ovens, etc., where the temperature is above that of the flash point. For this reason they present no more than a moderate fire risk.

Production. The parent glycols are commonly made by the hydrolysis of the appropriate oxide. Ethylene glycol is also made by direct oxidation of ethylene in acetic acid to form the diacetate, which is then hydrolysed to the glycol. Polyglycols may be obtained by the catalytic reaction of the appropriate glycol with an appropriate alkylene oxide. The molecular weight of the product is controlled by varying the proportions of the reactants, including water. The numerous glycol ethers are produced by reacting the parent glycol with a suitable alcohol or by direct alkylation with dialkyl sulphate. Dialkyl ethers are derivable from the mono-ether's alkyl halides. Glycol esters are prepared by esterifying the selected polyol with the appropriate organic acid.

Uses. Many of the applications of glycols are dependent upon their property of being completely water-soluble organic solvents. In addition, the two chemically reactive hydroxyl groups make the glycols important chemical intermediates. Among the many uses of glycols and polyglycols, the important ones include freezing point depression, lubrication, solubilisation, explosive and alkyd resin formulations, as constituents of cosmetics and indirect and direct additives to foods. The ethylene glycol ethers are used widely as solvents for resins, lacquers, paints, varnishes, dyes and inks, as well as components of painting pastes, cleaning compounds, liquid soaps, cosmetic and hydraulic fluids. Propylene and butylene glycol ethers are valuable as dispersing agents and as solvents for lacquers, paints, resins, dyes, oils and greases. The acetic acid esters, diesters and ether-esters are very useful solvents for oils, greases, and inks. In addition, they are used in the formulation of lacquers, enamels, and adhesives to dissolve resins and plastics.

TOXICITY AND HAZARDS OF SPECIFIC GLYCOLS AND GROUPS

Common glycols

Ethylene glycol ($HOCH_2CH_2OH$)

1,2-ETHANEDIOL; GLYCOL ALCOHOL

m.w. 62.1
sp.gr. 1.11
m.p. −13.2 °C
b.p. 198 °C
v.d. 2.14
v.p. 0.05 mmHg (6.5 Pa) at 20 °C
f.p. 111.1 °C
e.l. 3.2%
i.t. 405 °C

very soluble in water, ethanol and ether

a colourless sweetish liquid.

TLV ACGIH (particulate) 10 mg/m³
(vapour) 50 ppm 125 mg/m³ ceil
STEL ACGIH (particulate) 20 mg/m³

The oral toxicity of ethylene glycol in animals is quite low. However, it has been estimated that the lethal dose for man is about 100 cm³ per person or about 0.6 g/kg, thus indicating a greater toxic potency for man than for laboratory animals. Typical effects of excessive oral intake of ethylene glycol are narcosis, depression of the respiratory centre, and progressive kidney damage. Monkeys have been maintained for 3 years on diets containing 0.2-0.5% of ethylene glycol without apparent adverse effects; no tumours were found in the bladder, but there were oxalate crystals and stones. Primary eye and skin irritation is generally mild in response to ethylene glycol, but the material can be absorbed through the skin in toxic amounts. Repeated exposure of rats and mice to levels comparable to a saturated atmosphere at room temperature (360 to 400 mg/m^3) failed to induce organic injury. Consequently, repeated exposures of man to vapours at room temperature should not present a significant hazard. Ethylene glycol does not seem to present a significant hazard from the inhalation of vapours at room temperatures or from skin or oral contact under reasonable industrial conditions. However, an industrial inhalation hazard could be generated if ethylene glycol was heated or vigorously agitated, or if appreciable skin contact or ingestion occurred over an extended period of time. The primary health hazard of ethylene glycol is related to the ingestion of large quantities.

Diethylene glycol ($(HOCH_2CH_2)_2O$)

2,2'-OXYDIETHANOL; DIGLYCOL

m.w. 106.1
sp.gr. 1.12
m.p. −10.5 °C
v.d. 3.66
v.p. 1 mmHg (0.13·10^3 Pa) at 91.8 °C
f.p. 123.9 °C
i.t. 227 °C

soluble in water, ethanol, ether
a colourless syrupy liquid.

Diethylene glycol is quite similar to ethylene glycol in toxicity and mode of action. When excessive doses are ingested, the typical effects to be expected are diuresis, thirst, loss of appetite, narcosis, hypothermia, kidney failure and death, depending on the severity of exposure. While not of practical concern, when fed at high doses to animals, diethylene glycol has produced bladder stones and tumours, probably secondary to the stones. These may have been due to monoethylene glycol present in the sample. Diethylene glycol does not seem to present a significant hazard from the inhalation of vapours at room temperatures or from skin or oral contact under reasonable industrial conditions. However, an industrial inhalation hazard could be generated with the material, if heated or vigorously agitated, or if appreciable skin contact or ingestion occurred over an extended period of time. The primary health hazard of diethylene glycol is related to the ingestion of large quantities.

Propylene glycol ($CH_2OH\text{-}CHOH\text{-}CH_3$)

1,2-PROPANEDIOL; 1-2-DIHYDROXYPROPANE

m.w. 76.1
sp.gr. 1.04
m.p. supercools
b.p. 187.2 °C (760 Torr)
v.d. 2.62
v.p. 0.08 mmHg (10.4 Pa) at 20 °C
f.p. 100 °C
e.l. 2.6-12.6%
i.t. 371 °C

soluble in water, ethyl alcohol, ethyl ether
a colourless, odourless hygroscopic liquid.

Propylene glycol presents essentially no toxicity hazard. Long-term exposures of animals to atmospheres saturated with propylene glycol are without measurable effect. As a result of its low toxicity, propylene glycol is used widely in pharmaceutical formulations, cosmetics and, with certain limitations, in food products.

Dipropylene glycol $HO(C_3H_6O)_2H$

2,2'-DIHYDROXYISOPROPYL ETHER

m.w. 134.2
sp.gr. 1.02
m.p. supercools
b.p. 232 °C (760 Torr)
v.d. 4.63
v.p. 1 mmHg (0.13·10^3 Pa) at 73.8 °C
f.p. 122-139 °C

Dipropylene glycol is of very low toxicity. It is essentially non-irritating to the skin and eyes and, because of its low vapour pressure and toxicity, is not an inhalation problem unless large quantities are heated in a confined space.

Butanediol ($C_4H_{10}O_2$)

DIHYDROXYBUTANE; BUTYLENE GLYCOL

Four isomers exist; all are soluble in water, ethyl alcohol and ether (see table 1).

The butanediols (butylene glycols) are used in the production of polyester resins. With the exception of the 1,4-isomer, they create no significant industrial hazard. In rats, massive oral exposures of 1,2-butanediol induced deep narcosis and irritation of the digestive system. Congestive necrosis of the kidney may also occur. Delayed deaths are believed to be the result of progressive renal failure, while acute fatalities are probably attributable to narcosis. Eye contact with 1,2-butanediol may result in corneal injury, but even prolonged skin contact is usually innocuous with respect to primary irritation and absorption toxicity. No adverse effects of vapour inhalation have been reported.

1,3-butanediol is essentially non-toxic except in overwhelming oral doses, in which case narcosis may occur.

1,4-butanediol is about eight times as toxic as the 1,2-isomer. Acute ingestion results in severe narcosis and possibly renal injury. Death probably results from collapse of the sympathetic and parasympathetic nervous systems. It is not a primary irritant, nor is it easily absorbed percutaneously.

Ethylene glycol ethers

Most of the ethylene glycol ethers are more volatile than the parent compound, and, consequently, less easily

Table 1

Isomer	*m.w.*	*sp.gr.*	*m.p.* °C	*b.p.* °C	*v.d.*	*v.p.* (at 20 °C) mmHg	Pa	*f.p.* °C	*i.t.* °C
1,2-	90.1	1.002	–	193-195	3.1	–	–	90	–
1,3-	90.1	1.006	<−50	207	3.2	0.06	7.8	121	394
1,4-	90.1	1.020	16	230	3.1	–	–	121 oc	–
2,3-	90.1	1.009	19	182	3.1	0.17	18.1	85	402

controlled with respect to vapour exposure. All of the ethers are more toxic than ethylene glycol and exhibit a similar symptomatological complex.

Ethylene glycol monomethyl ether ($CH_3OCH_2CH_2OH$)

METHYL CELLOSOLVE; DOWANOL EM; 2-METHOXYETHANOL

m.w.	76.1
sp.gr.	0.96
m.p.	−85.1 °C
b.p.	125 °C
v.d.	2.6
v.p.	9.7 mmHg ($1.3 \cdot 10^3$ Pa) at 25 °C
f.p.	46.1 °C
e.l.	2.5-14%
i.t.	289 °C

a colourless liquid with a faint odour.

TWA OSHA	25 ppm 80 mg/m³ skin
STEL ACGIH	35 ppm 120 mg/m³
IDLH	2 000 ppm

The oral LD_{50} for ethylene glycol monomethyl ether in rats is associated with delayed deaths involving lung oedema, slight liver injury, and extensive kidney damage. Renal failure is the probable cause of death in response to repeated oral exposures. This glycol ether is moderately irritating to the eye, producing acute pain, inflammation of the membranes, and corneal clouding which persists for several hours. Although ethylene glycol monomethyl ether is not appreciably irritating to skin, it can be absorbed in toxic amounts. Experience with human exposure to ethylene glycol monomethyl ether has indicated that it can result in the appearance of immature leucocytes, monocytic anaemia and neurological and behavioural changes. Thus, it is evident that the monomethyl ether of ethylene glycol is a moderate toxic compound and that repeated skin contact or inhalation of vapour must be prevented.

Ethylene glycol monoethyl ether ($C_2H_5OCH_2CH_2OH$)

CELLOSOLVE SOLVENT; DOWANOL EE; 2-ETHOXYETHANOL

m.w.	90.1
sp.gr.	0.93
b.p.	135 °C
v.d.	3.1
v.p.	5.3 mmHg ($0.68 \cdot 10^3$ Pa) at 25 °C
f.p.	94.4 °C
e.l.	1.8-14%
i.t.	235 °C

a colourless, odourless liquid.

TWA OSHA	200 ppm 740 mg/m³
TLV ACGIH	50 ppm 185 mg/m³ skin
STEL ACGIH	100 ppm 370 mg/m³
IDLH	6 000 ppm

Ethylene glycol monoethyl ether is much less toxic than the methyl ether. The most significant toxic action is on the blood, and neurological symptoms are not expected. In other respects it is similar in toxic action to ethylene glycol monomethyl ether, which is described above. Excessive exposure can result in moderate irritation to the respiratory system, lung oedema, central nervous system depression and marked glomerulitis.

Other ethylene glycol ethers

Mention of ethylene glycol monobutyl ether is also in order because of its extensive use in industry. In rats, deaths in response to single oral exposures are attributable to narcosis, whereas delayed deaths result from lung congestion and renal failure. Direct contact of the eye with this ether produces intense pain, marked conjunctival irritation and corneal clouding, which may persist for several days. As with monomethyl ether, skin contact does not cause much skin irritation, but toxic amounts can be absorbed. Inhalation studies have shown that rats can tolerate 30 7-h exposures to 54 ppm, but some injury occurs at a concentration of 100 ppm. At higher concentrations, rats exhibited haemorrhaging in the lungs, congestion of the viscera, liver damage, haemoglobinuria, and marked erythrocyte fragility. Enhanced erythrocyte fragility was evident at all exposure concentrations above 50 ppm of ethylene glycol monobutyl ether vapours. Humans appear to be somewhat less susceptible than laboratory animals because of apparent resistance to its haemolytic action.

Both the isopropyl and *n*-propyl ethers of ethylene glycol present particular hazards. These glycol ethers have low single-dose oral LD_{50} values and they cause severe kidney and liver damage. Bloody urine is an early sign of severe kidney damage. Death usually occurs within a few days. Eye contact results in rapid conjunctival irritation and partial corneal opacity in the rabbit, with recovery requiring about one week. Like most other ethylene glycol ethers, the propyl derivatives are only mildly irritating to the skin but can be absorbed in toxic amounts. Furthermore, they are highly toxic via inhalation. Fortunately, ethylene glycol monoisopropyl ether is not a prominent commercial compound.

Diethylene glycol ethers. The ethers of diethylene glycol are lower in toxicity than the ethers of ethylene glycol, but they have similar characteristics.

Polyethylene glycols. Triethylene, tetraethylene, and the higher polyethylene glycols are particularly innocuous compounds of low vapour pressure. Essentially no hazard exists with respect to skin and eye contact, or even prolonged inhalation.

Propylene glycol ethers. Propylene glycol monomethyl ether is relatively low in toxicity. In rats, the single oral dose LD_{50} caused death by generalised central nervous system depression, probably respiratory arrest. Repeated oral doses (3 g/kg) over a 35-day period induced in rats only mild histopathological changes in the liver and kidneys. Eye contact resulted in only a mild transitory irritation. It is not appreciably irritating to the skin, but confinement of large amounts of the ether to rabbit skin causes central nervous system depression. The vapour does not present a substantial health hazard if inhaled. Deep narcosis appears to be the cause of death in animals subjected to severe inhalation exposures. This ether is irritating to the eyes and upper respiratory tract of humans at concentrations that are not hazardous to health, hence it does have some warning properties.

Di- and tripropylene glycol ethers exhibit toxicological properties similar to the monopropylene derivatives, but present essentially no hazard with respect to vapour inhalation or skin contact.

Polybutylene glycols. Those that have been examined can cause kidney damage in excessive doses, but they are not injurious to the eyes or skin and are not absorbed in toxic amounts.

Acetic esters, diesters, ether esters. These derivatives of the common glycols are of particular importance since they are employed as solvents for plastics and resins in diverse products. Many explosives contain ester of ethylene glycol as a freezing point depressant. With respect to toxicity, the glycol ether fatty acid esters are considerably more irritating to mucous membranes than the parent compounds discussed previously. However, the fatty acid esters have toxicity properties essentially identical to the parent materials once the former are absorbed, because the esters are saponified in biological environments to yield fatty acid and the corresponding glycol or glycol ether.

[The most important acetic esters of glycols are the following:

Ethylene glycol monomethyl ether acetate
($CH_3CO_2CH_2CH_2OCH_3$)
METHYL CELLOSOLVE ACETATE

m.w.	118.2
sp.gr.	1
b.p.	143 °C
v.d.	4.07
f.p.	44 °C
e.l.	1.7-8.2%

miscible with water and most of the organic solvents
a colourless liquid.

TWA OSHA	25 ppm 120 mg/m³ skin
STEL ACGIH	35 ppm 170 mg/m³
IDLH	4 500 ppm

Ethylene glycol monoethyl ether acetate
($CH_3CO_2CH_2CH_2OC_2H_5$)
CELLOSOLVE ACETATE; ETHOXYETHANOL ACETATE

m.w.	132.2
sp.gr.	0.97
b.p.	156.4 °C
v.d.	4.72
v.p.	1.2 mmHg ($0.16 \cdot 10^3$ Pa) at 20 °C
e.l.	1.7%
i.t.	379.4 °C

soluble in water
a colourless liquid with a mild ester-like odour.

TWA OSHA	100 ppm 270 mg/m³
TLV ACGIH	50 ppm 270 mg/m³ skin
STEL ACGIH	100 ppm 540 mg/m³
IDLH	2 500 ppm]

SAFETY AND HEALTH MEASURES

Since ethylene glycol and diethylene glycol present no significant hazards under normal industrial conditions, no need exists for special industrial hygiene measures. If, however, the materials are agitated or heated, or if skin contact is extensive and prolonged, it is advisable to protect the personnel by enclosing the process or providing local exhaust ventilation. Skin and eye protection should also be worn if the possibility of such contact exists. The main hazard associated with these substances, that of ingestion, can usually be prevented by keeping them in properly labelled containers. When using ethylene glycol monomethyl ether, the workers should wear chemical safety goggles, and adequate ventilation is necessary. Eye protection is also recommended whenever the possibility of such contact exists with ethylene glycol monobutyl ether. Inhalation of its vapours and skin contact should be avoided.

Treatment. In case of ingestion of ethylene glycol, the emergency treatment consists of induction of vomiting followed by gastric lavage with 1 : 5 000 potassium permanganate and medical attention. Evidence is also accumulating to indicate that acute intoxications from ethylene glycol can be treated with ethyl alcohol. Rats, monkeys, and humans have been successfully treated. The principle involves the greater affinity of alcohol dehydrogenase for ethyl alcohol than for ethylene glycol, thus permitting the excretion of ethylene glycol unchanged in the urine. Unfortunately, ethyl alcohol appears to have no such beneficial effect in the treatment of diethylene glycol intoxication. If diethylene glycol has been ingested, vomiting should be induced and the patient should receive immediate medical attention.

ROWE, V. K.
TORKELSON, T. R.

"Research into the toxicity of glycol derivatives" (Contributii la cercetarea toxicității derivatilor glicolici). Goldstein, L.; Dumitru, E.; David, V.; Melinte, L. *Igiena* (Bucarest), Apr. 1970, 19/4 (209-218). 40 ref. (In Romanian)

"Ethylene and diethylene glycol toxicity". Winek, C. L.; Shingleton, D. P.; Shanor, S. P. *Clinical Toxicology* (New York), Sep. 1978, 13/2 (297-324). Illus. 25 ref.

CIS 77-454 "Mechanism of the toxic action of vinyl glycol ethers" (K mehanizmu toksičeskogo dejstvija vinilovyh ėfirov glikolei). Gadaskina, I. D.; Rudi, F. A. *Gigiena truda i professional'nye zabolevanija* (Moscow), Feb. 1976, 2 (31-35). 5 ref. (In Russian)

"Transcutaneous ethylene glycol monomethyl ether poisoning in the work setting". Ohi, G.; Wegman, D. H. *Journal of Occupational Medicine* (Chicago), Oct. 1978, 20/10 (675-676). 9 ref.

Gold, alloys and compounds

a.w.	197.2
sp.gr.	19.3
m.p.	1 063 °C
b.p.	2 966 °C

soluble in aqua regia, hot sulphuric acid, alkali cyanides; insoluble in alkalis and in strong inorganic acids
a soft, ductile, malleable, yellowish-red metal.

Gold has a low degree of chemical activity and under normal conditions it does not combine with oxygen, hydrogen, nitrogen or carbon, even at high temperatures; when heated above its melting point, it volatilises forming a yellowish-green vapour. It forms alloys with a number of metals including silver, copper, platinum, palladium, manganese, zinc, magnesium, tellurium, antimony and mercury. Gold easily forms complex compounds in which it is a monovalent or trivalent metal. The complex salt formed through reaction with a potassium cyanide solution is the basis for industrial processes for the extraction of gold from ores:

$$4Au + 8KCN + 2H_2O + O_2 = 4KAu(CN)_2 + 4KOH.$$

Gold has one stable isotope and 14 radioactive isotopes.

Occurrence. Gold occurs widely throughout the world but usually very sparsely. Native, or metallic gold and various telluride minerals are the only forms of gold found on land. Native gold may occur in veins among rocks and ores of various other minerals or it may be scattered in sands. Depending on the elements accompanying the gold, a distinction is made between gold-wolfram, gold-iron, gold-arsenic, gold-copper-lead-zinc, gold-silver and gold-tellurium ores. Gold-bearing deposits occur in the vicinity of gold lodes, either on the actual site of destroyed rock or along river valleys. Since rivers tend to change their courses over a long period of time, deposits are liable to occur some distance from the present river beds.

There are major gold lodes in South Africa (Transvaal), while the large gold deposits of Canada and Australia have already been worked out. A large gold-bearing belt stretches along the Pacific shore of North America from California to Alaska, comprising both gold lodes and rich deposits. An extensive zone of minor deposits runs through Japan, the Philippines, Indo-China and Burma. India has one of the largest gold fields at Kolar. There are no major deposits in Europe with the exception of Transsylvania and the South of France. The USSR has a large number of gold lodes and deposits in the Urals, Kazakhstan, eastern and south-eastern Siberia and elsewhere.

Extraction. There are two main methods for the extraction of gold from ore. These are the processes of amalgamation and cyanidation. The process of amalgamation is based on the ability of gold to alloy with metallic mercury to form amalgams of varying consistencies, from solid to liquid. The gold can be fairly easily removed from the amalgam by distilling off the mercury. In internal amalgamation, the gold is separated inside the crushing apparatus at the same time as the ore is crushed. The amalgam removed from the apparatus is washed free of any admixtures with water in special bowls. Then the remaining mercury is pressed out of the amalgam. In external amalgamation, the gold is separated outside the crushing apparatus, in amalgamators or sluices (an inclined table covered with copper sheets). Before the amalgam is removed, fresh mercury is added for disintegration. The purified and washed amalgam is then pressed. In both processes the mercury is removed from the amalgam by distillation.

Extraction of gold by means of cyanidation is based on the ability of gold to form a stable water-soluble double salt $KAu(CN)_2$ when it is combined with potassium cyanide or sodium cyanide in association with oxygen. Gold is readily precipitated from this solution by the addition of metallic zinc. The pulp resulting from the crushing of gold consists of larger crystalline particles, known as sands, and smaller amorphous particles, known as silt. The sand, being heavier, is deposited at the bottom of the apparatus, and easily allows solutions to pass through. The extraction of gold is based on this characteristic. The process consists of feeding the ore, ground into a fine sand, into a leaching tub and filtering a solution of potassium or sodium cyanide through it.

The silt is not precipitated from the solutions nor does it filter them. For gold to be extracted from silt it must be mixed with cyanide solutions and then separated by means of different processes. The solutions obtained in this manner are fed into thickeners, dehydrated in vacuum filters and then passed on to an extraction section where the gold is precipitated on a stream of zinc. Under the influence of carbonic acid, water and air, as well as the acids present in the ore, the solutions decompose and give off hydrogen cyanide which contaminates the air of the working premises. In order to prevent hydrolisation of the cyanide solutions, alkali is added (lime or caustic soda).

Both amalgamation and cyanidation produce metal that contains a considerable quantity of admixtures, the pure gold content rarely exceeding 900 per mil fineness, unless it is further electrolytically refined in order to produce a degree of fineness of up to 999.8 per mil and more.

Uses. The hard solutions formed with some metals (e.g. with copper) increase the hardness and firmness of gold and are therefore used in the production of coins, jewelry and false teeth.

A large percentage of the gold produced is employed for monetary purposes. The industrial use is limited except for the manufacture of jewelry. It is used in an alloy with platinum for the production of chemically resistant apparatus, and in alloys with platinum and silver in electrical engineering. It is used in photography in the form of chemical preparations including a complex salt: sodium tetrachloroaurate dihydrate ($NaAuCl_4 2H_2O$). Gold is also used in dentistry and some inorganic and organic compounds of gold have therapeutic application because they stimulate the reticulo-endothelial system, promote the formation of antibodies, and possess chemotherapeutical properties with regard to *Streptococcus haemolyticus*. Radioactive gold ^{198}Au is used in curing carcinoma of the prostate and other malignant tumours.

HAZARDS AND THEIR PREVENTION

Gold ore occurring in great depths is extracted by underground mining. This necessitates measures to prevent the formation and spread of dust in mine workings. The separation of gold from arsenical ores gives rise to arsenic exposure of mine workers and to pollution of air and soil with arsenic-containing dust (see ARSENIC AND COMPOUNDS).

In the mercury extraction of gold, workers may be exposed to airborne mercury concentrations when mercury is placed in or removed from the sluices, when the amalgam is purified or pressed, and when the mercury is distilled off; mercury poisoning has been reported amongst amalgamation and distilling workers.

In amalgamation processes the mercury must be placed on the sluices and the amalgam removed in such a manner as to ensure that the mercury does not come in contact with the skin of the hands (by using shovels with long handles, protective clothing impervious to mercury, etc.). The processing of the amalgam and the removal or pressing of mercury must also be as fully mechanised as possible, with no possibility of the hands being touched by mercury; the processing of amalgam and the distilling off of mercury must be carried out in separate isolated premises in which the walls, ceilings, floors, apparatus and work surfaces are covered with material which will not absorb mercury or its vapours; all surfaces must be regularly cleaned so as to remove all mercury deposits, and all premises intended for operations involving the use of mercury must be equipped with general and local exhaust ventilation. These ventilation systems must be particularly efficient in premises where mercury is distilled off. Stocks of mercury must be kept in hermetically sealed metal containers under a special exhaust hood, workers must be provided with the personal protective equipment necessary for work with mercury; and the air must be monitored systematically in premises used for amalgamation and distilling.

Contamination of the air by hydrogen cyanide in cyanidation plants is dependent on air temperature, ventilation, the volume of material being processed, the concentration of the cyanide solutions in use, the quality of the reagents and the number of open installations. Medical examination of workers in gold-extracting factories has revealed symptoms of chronic hydrogen cyanide poisoning (high frequency of allergic dermatitis, eczema and pyoderma).

Proper organisation of the preparation of cyanide solutions is particularly important. If the opening of drums containing cyanide salts and the feeding of these salts into dissolving tubs is not mechanised, there is substantial contamination by hydrogen cyanide dust and gases. Cyanide solutions should be fed in through closed systems by automatic proportioning pumps, etc. In gold cyanidation plants, the correct degree of alkalinity must be maintained in all cyanidation apparatus; in addition, cyanidation apparatus must be hermetically sealed and equipped with local exhaust ventilation backed up by adequate general ventilation. All cyanidation apparatus, and the walls, floors, open areas and stairs of the premises must be covered with non-porous materials and regularly cleaned with weak alkaline solutions.

Smoking should be prohibited and workers should not be allowed to eat and drink at the workplace. First-aid equipment should be available and should contain material for immediately removing any cyanide solution that comes in contact with workers' bodies. Workers must be supplied with personal protective clothing impervious to cyanide compounds. Sulphuric acid and hydrochloric acid in concentrations from 10 to 30% are used in order to break down zinc chips in the processing of gold slime and may give off hydrogen cyanide and arsine. These operations must therefore be performed in specially equipped and separated premises, with the use of exhaust hoods.

GADASKINA, I. D.
RYŽIK, L. A.

"Arsenic accumulation in people working with and living near a gold smelter" (627-642). 17 ref. Jervis, R. E.; Tiefenbach, B. *Nuclear activation techniques in the life sciences 1978. Proceedings of a symposium, Vienna 22-26 May 1978* (Vienna, International Atomic Energy Agency).

CIS 77-1538 "Massive fibrosis in gold miners—A radiological evaluation". Solomon, A. *Environmental Research* (New York), Feb. 1977, 13/1 (47-55). 10 ref.

"Occupational contact eczema due to potassium auric cyanide" (Berufliche Kontaktekzeme durch Kaliumgoldzyanid). Schmollack, E. *Dermatologische Monatsschrift* (Leipzig), 1971, 157/11 (821-824). (In German)

CIS 80-481 "Role of cyanides in the pathogenesis of allergic dermatoses in gold ore processors" (Značenie cianidov v vozniknovenii allergičeskih dermatozov u rabočih zolotoizvlekatel'nyh fabrik). Kolpakov, F. I.; Prohorenkov, V. I. *Vestnik dermatologii i venerologii* (Moscow), July 1978, 7 (76-79). 15 ref. (In Russian)

CIS 80-1045 "Hazardous chemical reactions—67. Copper, silver, gold" (Réactions chimiques dangereuses—67. Cuivre, argent, or). Leleu, J. *Cahiers de notes documentaires—Sécurité et hygiène du travail* (Paris), 1st quarter 1980, 98, Note No. 1234-98-80 (123-125). 16 ref. (In French)

Graphite

PLUMBAGO; BLACK LEAD; MINERAL CARBON

a.w.	12
sp.gr.	2.25
m.p.	3 652-3 697 °C
b.p.	4 200 °C

compact, crystalline masses of a black or grey colour with a metallic lustre.

TWA OSHA	(graphite natural) 15 mppcf
TLV ACGIH	(graphite synthetic) 30 mppcf 10 mg/m³ of total dust < 1% quartz or 5 mg/m³ respirable dust
MAC USSR	(graphite fireclay) 2 mg/m³

Graphite was named by the German geologist Abraham Werner in 1789 after the Greek word "to write". The three types of natural graphite, lump, amorphous and flake, are so designated because of geological occurrence and not mineral configuration.

Occurrence. Graphite is found in almost every country of the world, but the majority of production of the natural ore is limited to Austria, the USSR, Mexico, the Malagasy Republic, the Federal Republic of Germany, Sri Lanka, Norway and Korea. World production is less than 1 million t with all but 6-10% in the above-named countries.

Lump graphite. This is found in veins which cross different types of igneous and metamorphic rock containing mineral impurities of feldspar, quartz, mica, pyroxine, zircon, rutile, apatite and iron sulphides. The impurities are often in isolated pockets in the veins of ore. Mining is commonly underground with hand drills for selective mining of narrow veins. Air drills are used in development headings. Sorting, grading and classification are done on the surface.

Amorphous graphite. Deposits of amorphous graphite are also underground, but usually in much thicker beds than the veins of lumps. It is commonly associated with sandstone, slate, shale, limestone and adjunct minerals of quartz and iron sulphides. The ore is drilled, blasted and hand-loaded into wagons and brought to the surface for grinding and impurity separation.

Flake graphite. This is usually associated with metamorphosed sedimentary rock such as gneiss, schists and marbles. The deposits are often on or near the surface. Consequently, normal excavating equipment such as shovels, bulldozers and scarifiers are used in opencast mining and a minimum of drilling and blasting is necessary.

Production. This is accomplished by grinding, flotation, screening and kiln drying. In the Federal Republic of Germany low-grade graphite is upgraded to a purity exceeding 99% by removing silica with hydrofluoric acid and sodium hydroxide.

Uses. The principal uses for natural graphite include foundry facings, steelmaking, lubricants and refractories. Manufactured graphite is used for electrodes, anodes, bricks, blocks, cylinders, engineering and chemical applications, jet engine throat liners and moderators in nuclear reactors. The "lead" in pencils is also graphite.

HAZARDS

Mining and processing natural graphite can result in employee exposures to dusts, explosive chemicals, noise, carbon monoxide, physical trauma, acids and alkalis.

Before 1940 it was generally accepted that graphite caused a benign pneumoconiosis. Subsequently, a number of studies have indicated that graphitosis is a progressive and disabling disease and that the presence of crystalline silica and some silicates have a synergistic effect. Graphite pneumoconiosis does not appear to cause a predisposition to tuberculosis nor does graphite provide a protective mechanism against tuberculosis. Graphite workers have reported symptoms of headaches, coughing, depression, low appetite, dyspnoea and black sputum. Others have remained asymptomatic until they began to spit up graphite and were then suddenly disabled. Workers suffering from graphite pneumoconiosis have generally worked in the industry for long periods, 10 years or more. There are, however, reported cases of typical graphitosis in men who have worked as little as 4 years. There is strong evidence that individual susceptibility exists. Employment in non-dusty trades for long periods after graphite exposure has not prevented onset of symptoms.

Roentgen examinations of graphite workers have shown varying classifications of pneumoconioses. Microscopic histopathology has revealed pigment aggregates, focal emphysema, collagenous fibrosis, small fibrous nodules, cysts and cavities. The cavities have been found to contain an inky fluid in which graphite crystals were identified.

There are few data available as to the concentration of graphite dust which causes pneumoconiosis. It was reported that dust concentrations in a Japanese plant averaged 57.6 mg/m³ and 967 particles/cm³. Dust studies in the United States reported concentrations ranging from 1 165-7 837 particles/cm³. Dust studies have not been included with most medical reports of graphite pneumoconiosis.

SAFETY AND HEALTH MEASURES

Present data strongly indicate that both natural and manufactured graphite dust concentrations should be controlled to the threshold limit value. Most, if not all, natural graphite ores contain crystalline silica and silicates. The amount and type of impurities determine, in part, the potential hazard associated with inhalation of the dust. It is therefore essential that periodic analyses are made of the raw ore and airborne dust for crystalline silica and silicates, with special attention to feldspar, talc and mica. Acceptable dust levels must be adjusted to accommodate the effect these disease-potentiating dusts may have on the health of the workers.

Graphite pneumoconiosis is progressive even after the worker has been removed from the contaminated environment. Workers may remain asymptomatic during many years of exposure and disability often comes suddenly. Prevention of the disease is, therefore, essential. Engineering controls, wet drilling, exhaust ventilation, remote handling, etc., are recommended. Where effective engineering controls are not possible, workers should wear respiratory protective equipment approved for use against pneumoconiosis-producing dusts.

As in all industries, an employee health maintenance programme must consider all environmental conditions associated with the job. In the mining of natural graphite, this may include noise generated by drills, excavating equipment, ore crushing and processing. Explosive chemicals and their decomposition products, carbon monoxide, nitrogen oxides, etc., may, if not controlled, result in excessive worker exposure. Graphite purification utilises hydrofluoric acid and sodium hydroxide. In addition to the corrosive action of both chemicals, overexposure to fluorides can result in serious disease. Worker protection from physical hazards is an essential part of any health programme.

PENDERGRASS, J. A.

Pneumoconiosis:

CIS 76-953 "Pneumoconiosis in graphite curers". Uragoda, C. G.; Rajendra, M. *Indian Journal of Industrial Medicine* (Calcutta), July 1975, 21/3 (105-113). 14 ref.

"Pneumoconiosis in persons engaged in the production of carbon electrodes" (Pylica pluc u zatrudnionych przy produkcji elektrod weglowych). Zahorski, W.; Potoska-Skowronek, Z.; Pierzchala, W. *Medycyna Pracy* (Lodz), 1975, 26/1 (1-8). 12 ref. (In Polish)

Experimental pathology:

CIS 74-1574 "Contribution to the study of the biological effects of coke and graphite" (Zur biologischen Schädigungsmöglichkeit durch die Kohlenstoffmodifikationen Koks und Graphit). Einbrodt, H. J. *Staub* (Dusseldorf), Dec. 1973, 33/12 (474-478). 45 ref. (In German)

Industrial hygiene:

CIS 81-669 *Dust collection in plants manufacturing electrodes and artificial graphite* (Obespylivanie na êlektrodnyh i êlektrougol'nyh zadovah). Koptev, D. V. (Moscow, Izdatel'stvo "Metallurgija", 1980), 127 p. Illus. 63 ref. (In Russian)

Grinding and cutting fluids

In metal-cutting machine tools, a hardened tool edge removes chips by different mechanical means, such as turning, planing, shaping, drilling, milling, boring and broaching. Most of the work performed in metal-cutting is translated into heat, which, if not dissipated, can cause welding of metal to tool point, loss of hardness of the tool point, and distortion of the workpiece. Grinding is a kind of cutting in which metal is removed by a grinding wheel, containing abrasive grains that act as miniature cutters.

Cutting and grinding fluids, also known as lubricoolants, have two primary functions:

(a) to cool the workpiece and tool, thus preventing heat damage; and

(b) to lubricate, thus reducing frictional heat and the chip-tool interface and between the tool and the freshly cut surface.

Secondary functions are to flush away chips and swarf, reduce strain hardening of the machined metal, and protect the workpiece against rusting. On high-speed machines, such as automatic lathes, considerable quantities of these fluids may be needed.

Types of fluid. The three basic types of cutting and grinding fluids are: straight cutting oils; oil-in-water emulsions and aqueous solutions. Their ingredients are listed in tables 1, 2 and 3.

Table 1. Composition of straight cutting oils (insoluble oils)

1. Paraffinic or naphthenic mineral oils 60-100%. Sometimes blended with vegetable or animal oils.
2. Extreme pressure (EP) additives
   (a) Sulphur-free or combined with base oil or fats
   (b) Chlorinated fats and paraffin oils, 1-3%
   (c) Sulphochlorinated fats, 1-3%
3. Biocides, sometimes added to prevent rancidity.

Table 2. Composition of soluble oils (water-miscible oils or emulsions)

Depending on intended use, concentrate is diluted with water at a rate of between 1 : 5 and 1 : 50; pH 8.0-9.5.

1. Mineral oil 60-85% of concentrate.
2. Emulsifiers, for example, petroleum sulphonates, soaps, non-ionic surfactants.
*3. Biocides–thiocarbamates, phenols, adamantane salts, isothiazolinones, nitromorpholine, *tris*-(hydroxymethyl) nitromethane, triazines, pyridinethiol-l-oxide.
4. Corrosion inhibitors, for example alkanolamines.
*5. Extreme pressure (EP) additives.
6. Antifoaming agents, for example silicones.
7. Dyes, for example fluorescein, petroleum azo dyes.
8. Water conditioners, for example polyphosphates, trisodium phosphate, borax, sodium carbonate (soda ash). Sometimes added to water before formation of emulsion.

* Present in premium soluble oils (emulsifiable concentrates).

HAZARDS

Slippery floors caused by oil thrown off from machine tools can considerably increase the likelihood of falls, which, if they occur in the proximity of fast-running machinery, may have very serious results. However, the main hazards are those which affect the machine operator's health. The hands and arms of machine operators, tool setters and swarf removers are constantly exposed to cutting oils: splashes and mist contaminate skin and clothing and saturated wiping rags are kept in pockets. Where large quantities of mist are formed there may be an inhalation hazard.

Table 3. Composition of synthetic and semi-synthetic cutting fluids

A cutting fluid concentrate which contains little or no oils. The concentrate is diluted at a rate of between 1 : 10 and 1 : 200, depending on intended use; pH 8-10.

1. Oils: 5 to 30% in semi-synthetic fluids; none in synthetic fluids.
2. Corrosion inhibitors, for example alkanolamines, carboxylic acids, mercaptobenzothiazoles, borate esters.
3. Anionic and non-ionic surfactants, for example soaps, sulphonates, ethoxylated alcohols or phenols.

*4. Blending agents, for example glycols.

*5. Biocides.

*6. Water conditioners, for example EDTA, borates and phosphates.

*7. Antifoaming agent, for example silicones.

*8. Dyes.

* Additives not present in all formulations.

During the first half of the 20th century, the most common cutaneous problem of machinists was oil acne. However, with the increasing use of water-base cutting and grinding fluids, which account for about two-thirds of the market, eczematous contact dermatitis has become the most common occupational skin disease of machinists. Chloracne, keratoses, squamous cell cancer and melanosis can also be caused by cutting oil exposure.

Oil acne. In oil acne a variety of lesions may be seen–comedones, folliculitis and furuncles. The lesions usually occur on the extensor surfaces of the forearms and thighs, but are sometimes seen on the temples, nape, and backs of the hands, i.e. areas contaminated by oil or oil-soaked clothing and infrequently washed. Individuals with hairy or oily skin have a greater tendency to develop oil acne. Occlusion of the follicular orifice, the antecedent of the lesions of oil acne, is caused by overgrowth of keratin, which is stimulated by retention of oil. Inflammatory and pustular reactions which occur around the follicles may or may not contain bacteria. When staphylococci contribute to the folliculitis, they should be regarded as secondary invaders from the skin and not as pathogens from the cutting oil.

It is generally supposed that bacteria present in cutting fluids cause folliculitis and furuncles, but this is contrary to bacteriologic evidence. Fresh insoluble oil is sterile, but the presence of water, such as condensation from the machines, will allow survival and limited growth of bacteria, usually Gram-negative. Emulsifiable oil and synthetic coolants are also sterile when manufactured; but after they are diluted with water and come in contact with air and human refuse, a variety of bacteria and fungi can be isolated. The predominant bacteria, pseudomonads, are frequently found in association with *Achromobacter* and *Vibrio* species, *Bacillus subtilis*, and *Proteus vulgaris*. Coliform bacteria and enterococci, indications of faecal pollution, are occasionally found. Pathogenic *Staphylococcus*, *Salmonella* and *Shigella* species are found only rarely. Biocides are added to soluble oils and semi-synthetic and synthetic coolants primarily to prevent bacterial decomposition and the formation of odours, and not for prevention of skin infections.

Paradoxically, the water-base lubricoolants, which frequently contain bacteria, are not usually associated with bacterial infections of the skin, whereas the insoluble oils, which are relatively free of bacteria, are associated with folliculitis.

Chloracne. Chloracne is a severe form of occupational acne caused by exposure to certain highly toxic halogenated aromatic chemicals (naphthalenes, biphenyls, dibenzofurans, dibenzo-*p*-dioxins and chlorobenzenes). Chloracne from cutting oils is seldom seen because of the elimination of chloronaphthalenes and biphenyls as extreme pressure additives. However, chloracnegens may, in rare cases, be formed during thermal decomposition of chlorinated cutting oils. Chloracnegenic chemicals cause metaplasia of sebaceous glands with the production of keratinous cysts on exposed or contaminated skin. Comedones, papules and pustules may also occur. Although chloracne represents one of the most sensitive indicators of biological exposure to these chemicals, systemic toxicity may also occur, especially liver damage, hyperlipidaemia and nervous system disturbances.

Eczematous contact dermatitis. Subacute and chronic eczematous contact dermatitis, manifested by varying degrees of erythema, scaling, fissuring and thickening of the skin, is usually attributed to repeated and prolonged exposures to emulsions with soap-like action and/or to alkaline aqueous solutions, both of which can act as low-grade irritants. With some of the milder products it would probably be more accurate to presuppose previous damage to the skin such as chapping due to low humidity in winter or use of harsh cleaners or solvents. Often this type of dermatitis will clear after the onset of warm weather or the use of an emollient.

Less common causes of eczematous contact dermatitis are irritation from and sensitisation to bactericides, corrosion inhibitors, extreme pressure oil additives, and sometimes the oil itself. When sensitisers and strong irritants are involved, the dermatitis is usually acute, with erythema, oedema, papules and vesicles. Low-grade irritants, such as low-boiling distillates similar to light machine oil and kerosene, tend to produce subacute or chronic dermatitis. As the boiling point increases skin irritance may lessen. A common mistake in the maintenance of an emulsion-type cutting fluid is the addition of biocide in excess of that necessary to prevent bacterial growth. Excess biocide can be irritating, and the likelihood of sensitisation is increased.

Although said to be rare, cases of allergic contact dermatitis to cutting fluid additives, breakdown products or metal salts (nickel, chromium or cobalt) are often overlooked.

Allergic contact dermatitis from the following additives has recently been reported: ethylenediamine, butylphenol (antioxidant), chlorocresol, *tris*-nitrol and triazine (antimicrobials), hydrazine (stabiliser), balsam of Peru (fragrance), and Proxel CRL (benzisothiazolin-3-one).

The diagnosis of allergic contact dermatitis requires accurate performance and interpretation of patch tests to the fluid itself and to individual additives. This procedure should only be performed by a physician experienced with this test. The confidential nature of most fluid formulations makes such testing extremely difficult to perform accurately. Dilution of the sensitising chemical additive in the final cutting fluid formulation may obscure identification of the specific allergen.

Photosensitivity may occur from certain wavelengths of artificial light which react with the skin and/or chemicals in contact with the skin producing eczema on exposed body areas. Photoallergic contact dermatitis may rarely occur from halogenated salicylanilides, phenothiazines, or certain fragrances added to cutting fluids.

Keratoses and squamous cell skin cancer. Cutaneous neoplasia has occurred from contact with European shale oil, but not from North American shale oil. Few cutaneous cancers from cutting oil exposures have been reported in Canada and the United States, but studies in the United Kingdom have shown a significant incidence, especially of cancer of the scrotum. The incidence appears to be greatest in automatic bar lathe workers and especially among setters and setter operators exposed to neat cutting oils, but cases are unevenly distributed among the persons at risk.

Oil melanosis. This rare pigmentary disorder is seen primarily in machinists of Mediterranean origin. The perifollicular hyperpigmentation may result from direct irritation caused by the insoluble cutting oil, or it may be secondary to folliculitis. Limitation of the pigmentation to exposed areas indicates a relationship to sun exposure.

Granulomas. Granulomas are chronic, indolent, focal inflammations which tend to heal with scarring. Foreign body granulomas may develop after penetration of metallic swarf or splinters. Granulomas may develop from beryllium alloys, and oil granulomas may develop from grease and oil gun injuries.

Pulmonary hazards. Although occupational exposures to oil mists generated during cutting operations have been numerous and frequently heavy, reported cases of respiratory system disorders (lipid pneumonia) have been rare. Nuisance and subjective complaints do occur, however, especially when the oil mist is visible (i.e. at concentrations greater than 5 mg/m^3). No relationship has been established between inhalation of oil mist and lung cancer although warnings have been issued against cutting fluids containing nitrites and secondary or tertiary amine additives because of the formation of carcinogenic nitrosamines. Also it would seem prudent to limit exposure to aromatic base oils. Some European countries, such as the United Kingdom, allow only low (i.e. less than 10%) aromatic oils to be used in straight cutting oils as well as water-miscible cutting fluids. It should be noted that high concentrations of oil mist are also dangerous from the safety point of view. Flash fires have occurred from general contamination of the work area.

SAFETY AND HEALTH MEASURES

In the prevention of cancer due to cutting oils, the most effective measure is the use of non-carcinogenic oil, for example cutting oil from which the carcinogens have been removed by solvent extraction.

Environmental control measures should include:

(a) careful identification, by generic name, of all ingredients in the cutting fluid used;

(b) programmes to keep the coolants free of tramp oil, foreign particles, and dirt through the use of effective filters and redesigning the coolant flow system to eliminate "eddies" and "backwaters" of coolant;

(c) daily programmes to monitor coolant characteristics, such as pH, bacteria count, etc.;

(d) daily programmes, such as hosing down, to keep machinery clean;

(e) redesigning spray application to minimise coolant splash and spray;

(f) using splash goggles and curtains;

(g) use of local exhaust systems and oil collectors to reduce airborne oil mist.

Further reduction in contact exposure can be expected from increased automation of metal-cutting operations.

When contact cannot be avoided, protective clothing, including impervious aprons to protect the groin, and arm protection such as sleeves should be worn if not a safety hazard (gloves are generally considered a safety hazard and should not be worn); barrier creams of the oil- or water-repellent type are helpful in keeping the skin clean (amine-containing creams may react to form nitrosamines in the presence of nitrites). Cleansing soiled skin at work breaks followed by thorough drying and the application of an emollient cream is a hygienic procedure applicable to all types of lubricoolant exposures. Suitable and readily accessible washing and sanitary facilities are essential: shower baths are desirable. For removing oil, powdered cleansers containing a non-irritating scrubber, sulphonated oil soaps and waterless skin cleansers are effective. The use of petroleum solvents and abrasive cleaners for cleaning the skin should be avoided. Wiping rags and waste should be clean and free of metal chips. Work clothing and underwear should be changed daily. The practice of keeping wipers in trouser pockets should be discouraged in every way. There should be routine arrangements for laundering and good changing room accommodation. The workers should be thoroughly instructed in the need for scrupulous personal hygiene.

Pre-employment medical examination will serve to exclude from exposure those persons most likely to be affected, for example acne sufferers (especially young persons), persons with follicular hyperkeratosis, those showing any skin intolerance to oil and those with eczematous dermatitis of the hands.

Periodical medical examination will serve to control the development of skin affections and enable sufferers to be transferred to other work when the condition fails to respond to treatment.

In dealing with the risk of scrotal cancer, it is important to watch for early symptoms and it should be ensured that workers are aware of the dangers and of the necessity of seeking early advice and treatment.

KEY, M. M.
TAYLOR, J. S.
YANG, C.

"Manufacturing exposure to coolant-lubricants". Welter, E. S. *Journal of Occupational Medicine* (Chicago), Aug. 1978, 20/8 (535-538). 8 ref.

"Dermatoses associated with metal-working fluids". Taylor, J. S. *Proceedings. Second International Conference; Lubrication Challenges in Metal-working and Processing* (Chicago, Illinois Institute of Technology Research, 1979), 239 p.

"Bacteria and soluble oil dermatitis". Rycroft, R. J. G. *Contact Dermatitis* (Copenhagen), Jan. 1980, 6/1 (7-9). 4 ref.

"Allergic contact sensitization to Epoxide 7 in grinding oil". Rycroft, R. J. G. *Contact Dermatitis* (Copenhagen), Aug. 1980, 6/5 (316-320). 2 ref.

CIS 80-111 "Allergic skin reactions in metal machining shops" (Allergische Hautschäden bei der Metallbearbeitung). Ippen, H. *Dermatosen in Beruf und Umwelt* (Aulendorf), 1979, 27/3 (71-74). Illus. 28 ref. (In German)

Nitrosamines in cutting fluids. Current intelligence bulletin No. 15. (National Institute for Occupational Safety and Health, 4676 Columbia Parkway, Cincinnati) (6 Oct. 1978), 4 p.

CIS 79-532 *Control of exposure to metalworking fluids.* O'Brien, D.; Frede, J. C. DHEW (NIOSH) publication No. 78-165 (National Institute for Occupational Safety and Health, 4676 Columbia Parkway, Cincinnati) (Feb. 1978), 36 p. 17 ref.

Grinding and polishing

Grinding in the widest sense of the word is a form of processing by means of a tool, the active part of which is composed of hard, sharp grains which cut loose small particles from the workpiece. The object is to give the work a certain shape, correct its dimensions, increase the smoothness of a surface or improve the sharpness of cutting edges.

Grinding is the most comprehensive and diversified of all machining methods, and is employed for many varying materials, predominantly iron and steel but including other metals and wood, plastics, stone, glass, pottery, etc. The term covers other methods of producing very smooth and glossy surfaces, such as polishing, honing, whetting and lapping.

The tools used are wheels of varying dimensions, grinding segments, grinding points, sharpening stones, files, polishing wheels, belts, discs, etc. In grinding wheels and the like, the abrasive material is held together by bonding agents to form a rigid, generally porous body. In the case of abrasive belts the bonding agent holds the abrasive secured to a flexible base material. Polishing and finishing are usually carried out by fabric mops or buffs to which an abrading or polishing material has been applied.

The natural abrasives, natural corundum, emery, diamond, sandstone, flint and garnet have been largely superseded by artificial abrasives including aluminium oxide (fused alumina) silicon carbide and synthetic diamonds. A number of fine-grained materials such as chalk, pumice, tripoli, tin putty, and iron oxide are also used, especially for polishing.

Aluminium oxide is most widely used in grinding wheels, followed by silicon carbide. Natural and artificial diamonds are used for important special applications. Aluminium oxide, silicon carbide, emery, garnet and flint are used in grinding and polishing belts.

Both organic and inorganic bonding agents are used in grinding wheels. The main type of inorganic bonds are vitrified silicate and magnesite. Notable among organic bonding agents are phenolic resin, rubber and shellac. The vitrified bonding agents and phenolic resin are completely dominating within their respective groups. Diamond grinding wheels can also be metal bonded. The various bonding agents give the wheels different grinding properties as well as different properties with regard to safety.

Abrasive and polishing belts and discs are composed of a flexible base of paper or fabric to which the abrasive is bonded by means of a natural or synthetic adhesive.

Different machines are used for different types of operation, such as surface grinding, cylindrical including centreless grinding, internal grinding, rough grinding, and cutting. The two main types are:

(a) those where either the grinder or the work is moved by hand; and

(b) machines with mechanical feeds and chucks.

HAZARDS AND THEIR PREVENTION

The major accident risk in the use of grinding wheels is that the wheel may burst during grinding. Normally grinding wheels operate at high speeds. There is a trend towards ever-increasing speeds. Most of the industrialised nations have regulations differing in detail regarding the maximum speeds at which the various grinding wheels may be run.

The fundamental protective measure is to make the grinding wheel as strong as possible; the nature of the bonding agent is most important. Wheels with organic bonds, in particular phenolic resin, are tougher than those with inorganic bonds and more resistant to impacts. High peripheral speeds may be permissible for wheels with organic bonds.

Very high speed wheels, in particular, often incorporate various types of reinforcement, for example certain cup wheels are fitted with steel hubs to increase their bursting strength. During rotation the major stress develops around the centre hole. To strengthen the wheel, the section around the centre hole, which takes no part in the grinding, can thus be made of an especially strong material which is not suitable for grinding. Large wheels with a centre section reinforced in this way are used particularly by the steel works for grinding slabs, billets and the like at speeds up to 80 m/s (figure 1).

Figure 1. Rough-grinding steel slab using resin-bonded grinding wheels with reinforced centres operating at a peripheral speed of 80 m/s.

The most usual method of reinforcing grinding wheels, however, is to include glass fibre fabric in their construction. Thin wheels such as used for cutting may incorporate glass fibre fabric at the centre or at each side while thicker wheels have a number of fabric layers rising with the thickness of the wheel.

With the exception of some grinding wheels of small dimensions, either all wheels or a statistical sampling of them must be given speed tests by the manufacturer. The wheels are run over a certain period at a speed exceeding that permitted in grinding. Test regulations vary from country to country, but usually the wheel has to be tested at a speed 50% above the working speed.

In some countries regulations require special testing of wheels that are to operate at higher speeds than normal at a central testing institute. The institute may also cut out specimens from the wheel and investigate their physical properties. Cutting wheels are subjected to certain impact tests, bending tests, etc.

The manufacturer is also obliged to ensure that the grinding wheel is well balanced prior to delivery.

Bursting accidents. The bursting of a grinding wheel may cause fatal or very serious injuries to anyone in the vicinity and heavy damage to plant or premises. In spite of all precautions taken by the manufacturers, occasional wheel bursts or breaks may still occur unless proper care is exercised in their use. Precautionary measures relate to the following aspects:

(a) Handling and storing. A wheel may become damaged or cracked during transit or handling: moisture may attack the bonding agent in phenolic resin wheels, ultimately reducing their strength. Vitrified wheels may be sensitive to repeated temperature

variations: irregularly absorbed moisture may throw the wheel out of balance. Consequently it is most important that wheels are carefully handled at all stages and kept in an orderly manner in a dry and protected place.

(b) Checking for cracks. A new wheel should be checked to ensure that it is undamaged and dry, most simply by tapping with a wooden mallet. A faultless vitrified wheel will give a clear ring, an organic bonded wheel a less ringing tone but either can be differentiated from the cracked sound of a defective wheel. In case of doubt the supplier should be consulted.

(c) Testing. Before the new wheel is put into service it should be tested at full speed with due precautions being observed. After wet grinding, the wheel should be run idle to eject the water, otherwise the water may collect at the bottom of the wheel and cause imbalance, which may result in bursting when the wheel is next used.

(d) Mounting. Accidents and breakages occur when wheels are mounted on unsuitable apparatus, for example on spindle ends of buffing machines. The spindle should be of adequate diameter but not so large as to expand the centre hole of the wheel; flanges should be not less than one-third the diameter of the wheel and made of mild steel or of similar material.

(e) Speed. In no circumstances should the maximum permissible operating speed, specified by the makers, be exceeded. A notice indicating the spindle speed should be fitted to all grinding machines and the wheel should be marked with the maximum permissible peripheral speed and the corresponding number of revolutions for a new wheel. Special precautions are necessary with variable speed grinding machines and to ensure the fitting of wheels of appropriate permissible speeds in portable grinders.

(f) Work rest. Wherever practicable, work rests of adequate dimensions and rigidly mounted should be provided. They should be adjustable and kept as close as possible to the wheel to prevent a trap in which the work might be forced against the wheel and break it or, more probable, the operator's hand could be caught and injured.

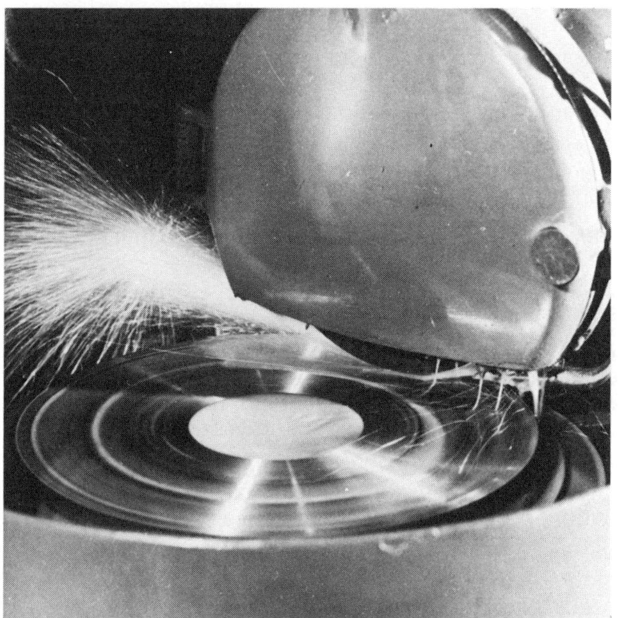

Figure 2. A well guarded, vitrified abrasive wheel mounted in a surface grinder and operating at a peripheral speed of 33 m/s.

(g) Guarding. Abrasive wheels should be provided with guards strong enough to contain the parts of a bursting wheel (figure 2). In some countries detailed regulations exist regarding the design of the guards and the materials to be used. In general, cast iron and cast aluminium are to be avoided. The grinding opening should be as small as possible and an adjustable nose piece may be necessary.

Exceptionally, where the nature of the work precludes the use of a guard, special protective flanges or safety chucks may be used.

Eye injuries. Dust, abrasive, grains and splinters are a common hazard to the eyes in all dry-grinding operations. Effective eye protection by goggles or spectacles and fixed screens at the machine is essential: fixed screens are particularly useful when wheels are in intermittent use, for example for tool grinding.

Polishing machines. The spindles and tapered ends of double ended polishing machines can cause entanglement accidents unless they are effectively guarded.

Fire. Grinding of magnesium alloys carries a high fire risk unless strict precautions are taken against accidental ignition and in the removal and drenching of dust. High standards of cleanliness and maintenance are required in all ducting from any exhaust plant to prevent risk of fire and also to keep ventilation working efficiently.

Health hazards. Although modern grinding wheels do not themselves create the serious silicosis hazard associated in the past with sandstone wheels, highly dangerous siliceous dust may still be given off from the materials being ground, for example sand castings. Certain resin-bonded wheels may contain fillers which create a dangerous dust; lead compounds need particular attention. In any event the volume of dust produced by grinding makes efficient local exhaust ventilation essential. It is more difficult to provide local exhaust for portable wheels, although some success in this direction has been achieved; prolonged work should be avoided and respiratory protective equipment provided. Exhaust ventilation is also required for most belt sanding, finishing and polishing and similar operations: dust of varying degrees of danger may be thrown off from the material.

Protective clothing, good sanitary and washing facilities with showers should be provided and medical supervision is desirable especially for metal grinders.

WELINDER, K.

CIS 79-653 *Safety requirements for the use, care, and protection of abrasive wheels.* ANSI B7 1-1978 (American National Standards Institute, 1430 Broadway, New York) (Standard approved on 5 Jan. 1978), 106 p. Illus.

CIS 78-711 *Ventilation requirements for grinding, buffing, and polishing operations.* Bastress, E. K.; Niedzwecki, J. M.; Nugent, A. E. DHEW (NIOSH) publication No. 75-107 (National Institute for Occupational Safety and Health, 4676 Columbia Parkway, Cincinnati) (Sep. 1974), 188 p. Illus. 21 ref.

Abrasive wheels: a reading list. Crawshaw, M. (London, Health and Safety Executive Library and Information Service, May 1981), 9 p.

Training advice on the mounting of abrasive wheels. Health and Safety Executive, Guidance note PM 22 (London, HM Stationery Office, 1981), 4 p. Illus.

Code of practice. Safeguarding grinding and honing machines (Machine Tool Trades Association, 62 Bayswater Road, London), 40 p. Illus. 27 ref.

Group medical services

Small undertakings in the manufacturing and service industries

In both developing and industrialised countries, and in market as well as some socialist economies, small undertakings are a vital force in production.

Taking, for example, an industrialised country such as France, it will be found that:

(a) the average number of employees per firm is 15-16;

(b) around half the total working population are in firms employing fewer than 50 workers;

(c) small and medium-sized firms (up to 500 employees) account for 98% of all industrial firms;

(d) these firms (excluding building, civil engineering, agriculture and food industries) account for 41.7% of gross turn-over.

Nearly the same situation is encountered in the United States, where the average number of employees per firm is 17-18 persons.

Small firms are even more important in the developing countries. (See HANDICRAFTS AND CRAFTSMEN).

The extensive health requirements of small firms

Income per worker is often lower than in large firms and this has both individual and group repercussions: nutritional deficiencies, promiscuity and contagion; poorly mastered technology, low investment, ignorance of hazards, absence or inadequacy of personal and group safety equipment, non-existent ergonomic organisation of work posts.

Workers in small firms have a clear need for medical supervision—both clinical and biological supervision and protection against occupational hazards. In no way is the small size of the firm a favourable health factor. Occupational medicine should be available to all workers no matter what the size of the firm. Any numerical limit below which there is an exemption for the provision of occupational medical services is a nonsense from the health point of view. However, this is nevertheless the case in many countries.

Obviously the dispersion of labour in a population of small firms does not facilitate health supervision. The objective of group medical services is to overcome this difficulty and provide small firms with an occupational medical service that will meet doctrinal and statutory requirements.

Meeting the occupational health needs of small firms

Clearly, small firms cannot set up their own independent medical services, and one of two possible situations is encountered depending on the country:

(a) either no occupational medical service is provided for small firms (in many countries there is still a numerical limit below which firms are not covered); or

(b) medical needs are met by a group medical service, which may be either a state service, a municipal service, an employer-financed service or a mixed-finance service.

The procedures differ greatly depending on the country. The range of services offered often includes organisation of first aid, minor health care, medical supervision of workers, and technical assistance in the fields of occupational hygiene, safety and ergonomics.

The French example

France is the Western country in which group occupational medical services have achieved maximum development: the Act of 11 October 1946 made it compulsory for all firms in the private sector, even if they have only one employee, to provide an occupational medical service; this had resulted, by January 1975, in the establishment of 585 group medical services covering a total of 8 242 590 workers.

The outlay per worker per year for these services was around 100 French francs in 1978.

The statutory and technical requirements are laid down basically in Decree No. 79/231 of 20 March 1979 (*Journal Officiel* of 22 March 1979).

The services to be provided are specified as follows:

Clinical and complementary supervision. This includes, if necessary, specialist medical examination. The main objective is to monitor and maintain working capacity for the purpose of improving health, not to carry out selection.

Supervision and optimisation of working conditions. This primary preventive task is developing to such a degree as to account for one-third of the occupational medical officers' time.

Distribution of functions and responsibilities. The firm is required to assume the organisational and financial cost of occupational medicine by joining a group medical service as a full member of a "non-profit association" as stipulated under French law. The occupational medical officer has the technical responsibility. The Minister of Labour and the workers have the right of supervisory control.

The majority of French group medical services have joined together in an association called the Information Centre on Company and Group Medical Services (Centre d'Information des Services Médicaux d'Entreprises et Interentreprises—CISME).

The aim of the group medical services is to provide a dynamic and personalised occupational medical service as close as possible to a model one, but nevertheless within a bearable budgetary framework.

Other examples of group medical services

Australia. A feature here is the importance of the work of specially trained nurses in the day-by-day implementation of occupational medicine.

Belgium. The regulations in Belgium are similar to those in France. However, systematic medical examinations are required only where workers are exposed to a specified occupational hazard.

Federal Republic of Germany. The scheme covers firms employing more than 50 persons and the service is extensive with numerous benefits; the cost is borne by the firm. Scientific studies are undertaken.

Finland. This country has a comprehensive service which includes the provision of nurses to undertakings. The national social security system covers 60% of the cost.

Japan. Japan has group co-operative associations and the university institutes have an eminent role in occupational medicine.

Netherlands. Here, the cost of up to three annual medical examinations is borne exclusively by the undertaking. Nurses play an important role.

Portugal. Annual medical examinations are required. Firms employing more than 100 workers are affiliated and there are specialised consultations, training and counselling on occupational safety.

Sudan. The service here is under development and stems from the Occupational Medical Department of the Ministry of Health. This country still has numerous small traditional firms.

Sweden. Experience here is highly advanced and the organisation is geographical or by branch of industry. A consultancy service on safety and ergonomics provided with a technician is being developed. The scheme illustrates the value of the team approach incorporating the physician, nurse and safety engineer.

United Kingdom. The services are financed by the firms themselves and the services provided are in line with international recommendations; affiliation is not compulsory. Emphasis is placed on the training of first-aid workers.

United States. Group medical services are being developed, particularly since the promulgation of the Occupational Safety and Health Act (1970). (See also SMALL UNDERTAKINGS.)

Practical organisation

In all countries group medical services are faced with a major problem: the geographical scatter of the workplaces they are required to cover, and this has led to three approaches:

(a) groupings by occupation–possible in urban zones;

(b) groupings on a geographical basis–essential in rural zones;

(c) mobile units as a back-up for stationary units, to ensure that occupational health services are as close to the users as possible, to ensure better service and the reduction of indirect costs (travel and lost time).

Small Industries Scientific Committee of the Permanent Commission and International Association on Occupational Health. The foundation of this Committee in 1971 and its subsequent development (25 members in 1979) shows clearly the topicality of this problem and the difficulties encountered in dealing with it. The Committee holds meetings at international congresses and organises symposia periodically.

Approaches in group medical services. The objective is to provide orthodox, coherent, effective occupational medical services as close as possible to those provided by a good independent service.

A group service should aim at achieving the output of a firm of equivalent size. However, it has an additional stimulating factor: its product is occupational medicine, which imposes the basic obligation of product quality.

This quality demands well co-ordinated service taking into account: general health, individual and group occupational health, safety, the reduction of excessive workload and, finally, work comfort.

PARDON, N.

CIS 75-1776 "International symposium on occupational medicine in small industries" (Un symposium international sur la médecine du travail dans les petites industries). El Batawi, M. A.; Cho, K. S.; Lee, T. J.; Hakkinen, I.; Rothan, A.; Pardon, N.; Sumamur; Jongh, J.; Osman, Y.; Forssman, S.; Jones, W.; Elliot, P. M. *Cahiers de médecine interprofessionnelle* (Paris), 1st quarter 1975, 15/57 (5-51). (In French)

CIS 79-1770 "Occupational medicine in small and medium-sized undertakings" (Arbeitsmedizinische Tätigkeit in mittleren und kleinen Betrieben). *Arbeitsmedizin–Sozialmedizin–Präventivmedizin* (Stuttgart), Apr. 1979, 14/4 (77-100). 8 ref. (In German)

CIS 76-2064 "New aspects of occupational health services in small workplaces in Finland". Tolonen, M. *Scandinavian Journal of Work, Environment and Health* (Helsinki), June 1976, 2/2 (45-56).

Gum arabic

It is not possible to give an exact chemical definition of a gum. The term "gum" in commerce is often applied wrongly to resins, but the so-called "varnish gums" are resins with no chemical relation with gums. The various gums are closely related to the carbohydrates.

The principal constituents of the various types of gum arabic, including gums from the Blue Nile district, from Kordofan, Sennar, Somalia and Ethiopia, are calcium (with traces of magnesium and potassium), salts of arabin and arabic acid $(C_6H_{10}O_5)_2$ and water. Gum arabic also contains some tannin, sugar and enzymes, and its composition is similar to that of the pectins.

Sources and properties. Gum arabic is also known as acacia gum, mimosa gum or African gum. The main source is the Sudan where it is obtained from the bark of various kinds of acacia trees or shrubs, the gum being exuded from the stems and branches. The gummy exudation loses about 15% of water by evaporation and is exported in the form of round or oval tears having a matt surface and which breaks with a vitreous fracture similar to glass. The substance is odourless and is either colourless or has a yellowish or brownish hue. Mixed with twice its weight of water it dissolves slowly forming a viscous liquid which is very sticky. The addition of ethyl alcohol causes it to coagulate into a gelatinous form.

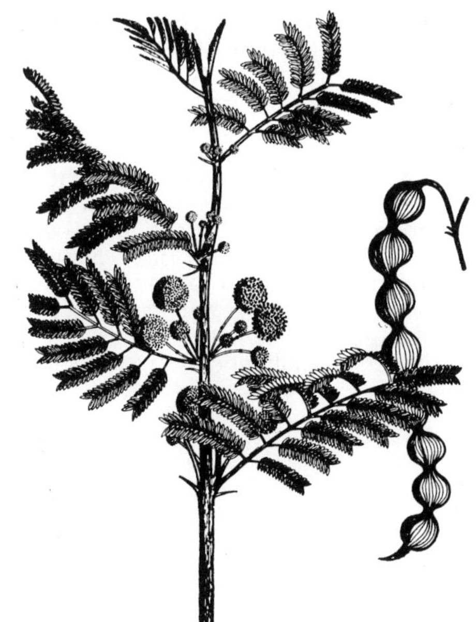

Figure 1. The acacia species are the prime source of gum arabic.

Uses. Gum arabic is used as office adhesive and as a binding medium for the flammable materials of matches and for the manufacture of water colours. It is used in the finishing processes for cotton and silk cloth, in the preparation of lozenges, and as a demulcent in cough mixtures and emulsions. It has also been widely employed in colour printing as an anti-offset agent: an aqueous solution of gum arabic was sprayed on the wet sheets of paper delivered from the printing machine, the object being to prevent colour from one sheet of paper passing on to the next sheet added to the pile.

HAZARDS

Gum arabic may produce respiratory symptoms such as asthma. Printing workers exposed to this mist have been found to suffer from allergic reactions often known as "printer's asthma", the frequency of the allergic symptoms depending mainly on the atmospheric gum arabic concentration. A study carried out in printing works in Stuttgart revealed symptoms such as incipient or clearly defined asthma, catarrh and conjunctivitis of the nasal mucus, sinus, throat, the respiratory tract and bronchus. Cases of fainting fits among women exposed to gum arabic have been recorded.

Examination of 37 printing workers revealed 13 cases of marked dyspnoea, soon after exertion. A distinguishing feature was that the trouble appeared shortly after arrival at work and did not occur on non-working days. In 20 of the 37 printers, there were well defined radiological findings in the lungs, with occasional chronic bronchitis, and pulmonary congestion. In 9 cases, vital capacity was appreciably reduced, even in workers who were otherwise apparently without subjective disturbances. Intracutaneous injection of 1% gum arabic solution produced positive reaction in 16 of the 37 workers.

An attempt at passive transmission using the serum from a patient suffering from printer's asthma has proved positive, and 11 employees of a printworks who had never had any ill effects also showed positive reactions. Further studies have shown that sensitisation occurred in about 50% of workers. The course of the allergy is in two periods; there is first an antigen-antibody reaction, sometimes without symptoms (silent sensitisation). This is followed, in many cases, by clinical disorders due to another mechanism, possibly the release of histamine.

SAFETY AND HEALTH MEASURES

The provision of better ventilation in printing shops using gum arabic has not given satisfactory results. Therefore the use of gum arabic anti-offset solutions has been abandoned in most countries in favour of powdered chalk. Replacement of gum arabic by less hazardous substances has led to the total elimination of new cases of printer's asthma, and in all cases of asthma, cessation of exposure has led to the disappearance of symptoms in persons suffering from the disease.

HOSCHEK, R.

Gypsum

HYDRATED CALCIUM SULPHATE

A mineral with the chemical composition $CaSO_4.2H_2O$. However, it is rarely found pure and gypsum deposits may contain quartz, pyrites, carbonates and clayey and bituminous materials. It occurs in nature in five varieties: gypsum rock; gypsite (an impure earthy form); alabaster (a massive, fine-grained translucent variety); satin spar (a fibrous silky form); and selenite (transparent crystals). It occurs throughout the world.

Processing. Gypsum rock may be crushed and ground for use in the dihydrate form, calcined at 190-200 °C (thus removing part of the water of crystallisation) to produce calcium sulphate hemihydrate or plaster of Paris, or completely dehydrated by calcining at over 600 °C to produce anhydrous or dead-burned gypsum.

Uses. Ground dihydrate gypsum is used in the manufacture of Portland cement and artificial marble products, as a soil conditioner in agriculture, as a white pigment, filler or glaze in paints, enamels, pharmaceuticals, paper, etc., and as a filtration agent.

Calcium sulphate hemihydrate of various types is used for builders' plaster and the manufacture of plaster building materials (mouldings, panels, etc.); in medicine, plaster of Paris is used for surgical casts or supports or the taking of impressions (e.g. dentures).

The anhydrous form is used in cement formulations and as a paper filler; due to its hygroscopic nature, it is also used as a drying agent.

HAZARDS AND THEIR PREVENTION

Workers employed in the processing of gypsum rock may be exposed to high atmospheric concentrations of gypsum dust (especially during crushing, grinding, etc.), furnace gases and smoke. In gypsum calcining, workers are exposed to high environmental temperatures, and there is also the hazard of burns. Crushing, grinding, conveying and packaging equipment presents a danger of machinery accidents.

Gypsum dust has an irritant action on the mucous membranes of the respiratory tract and eyes, and there have been reports of conjunctivitis, chronic rhinitis, laryngitis, pharyngitis, impaired sense of smell and taste, bleeding from the nose and reactions of the tracheal and bronchial membranes in exposed workers. [Tobacco smoking has been found to be an important co-factor in the causation of chronic respiratory diseases in workers exposed to gypsum dust.] Experimental animals exposed to gypsum dust developed pneumonia and interstitial pneumosclerosis, and blood and lymph circulation disorders occurred in the lungs.

[In their fibrous form, calcium sulphate particles can be confused with amphibole asbestos unless they are examined by sophisticated techniques (scanning electron microscope and X-ray energy dispersive analysis); whether it is the physical character (durable fibre with a diameter < 0.5 μm and a length > 10 μm) rather than the chemical composition of fibre which controls the ability to cause disease, especially mesothelioma, is still to be ascertained.] Dust formation in gypsum processing should be controlled by mechanisation of dusty operations (crushing, loading, conveying, etc.), addition of up to 2% by volume of water to gypsum prior to crushing, use of pneumatic conveyors with covers and dust traps, enclosure of dust sources and provision of exhaust systems for kiln openings and for conveyor transfer points. Good housekeeping and maintenance are essential in preventing the accumulation of deposited dust, and it is advisable, in the workshops containing the calcining kilns, to face the walls and floors with smooth materials to facilitate cleaning. Hot piping, kiln walls, drier enclosures, etc., should be lagged to reduce the danger of burns and to limit heat radiation to the work environment.

Workers should wear close-fitting working clothes of dust-tight material and, where necessary, should be supplied with respiratory protective equipment and goggles. It is advisable for gypsum and gypsum-products workers to receive a pre-employment examination followed by periodic examinations each year [possibly completed by appropriate ventilatory tests].

ROSCINA, T. A.

CIS 78-1252 "On the occurrence of fibres of calcium sulphate resembling amphibole asbestos in samples taken for the evaluation of airborne asbestos". Middleton, A. P. *Annals of Occupational Hygiene* (Oxford), Mar. 1978, 21/1 (91-93). Illus. 7 ref.

"Skin damage in construction workers caused by cement, lime and plaster" (Branchenspezifische Einwirkungsprobleme von Zement, Kalk und Gips auf die Haut). Weiler, K. J.; Rüssel, H. A. *Dermatosen in Beruf und Umwelt* (Aulendorf, Federal Republic of Germany), 1980, 28/6 (182-185). 35 ref. (In German)

"The traditional plasterer. An ergonomic study" (Le plâtrier traditionnel. Essai d'étude ergonomique). Dalmais, A.; Deramond, J.; Durand, J. P. *Revue de médecine du travail* (Paris), 1980, 8/4 (191-218). Illus. (In French)

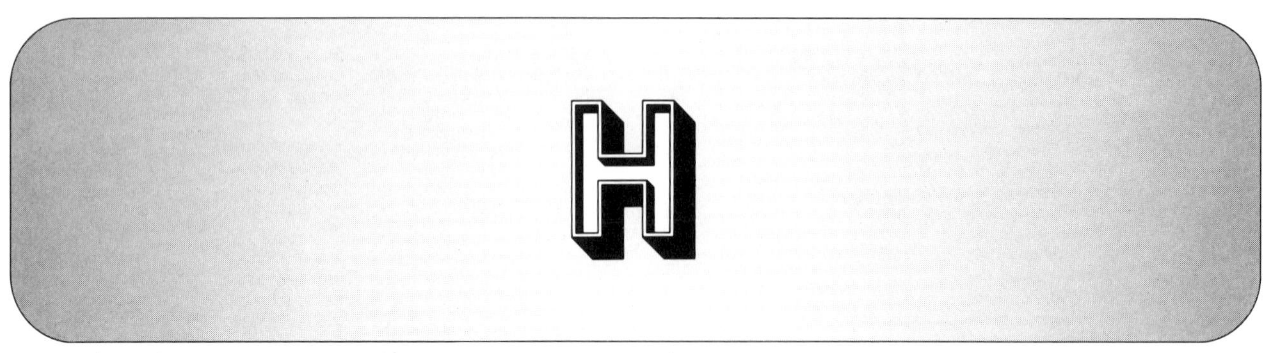

Hair and bristle

Hair is found as a covering and protection on the bodies of nearly all mammals. Bristle is the name given to the thick, long and hard hairs found on the tails of animals such as the boar. Horsehair, as its name implies, is obtained from the horse–usually from the mane; however, the name vegetable horsehair is applied to the fibres of various plants (hemp, jute, flax, esparto grass, etc.), and rayon and synthetic fibres may be used to produce synthetic horsehair. Synthetic horsehair is rapidly replacing vegetable horsehair since it is often easier and less expensive to produce.

Processing. After disinfection, bristle is sorted by colour, weight and length and then combed and carded. The material is degreased, dyed by boiling in dye (usually an aniline black) for over 3 h, dried, lightly combed and cut to length. Hair and bristle which is to be bleached is treated with sulphur fumes, hydrogen peroxide or hypochlorite. The processing of hair is basically the same as that of bristle.

The main processes in the treatment of vegetable horsehair are: drying, retting, batting and carding (see HEMP).

Uses. Hair and bristle are made up into finished products together with materials such as wood, horn, ivory, metal, celluloid, cotton and wool; other materials used in the manufacturing processes include dyes, pigments and adhesives. The commonest products made from hair and bristle include: brushes, threads, bags, cords, fishing nets, sieves, fabrics (camel hair), rugs, blankets, transmission belts, wigs, bowstrings for musical instruments, paint brushes, canvas shoes, cleaning materials, upholstered furniture, etc.

HAZARDS

The main hazard is the dust given off during processing–especially by vegetable horsehair–which may cause chronic respiratory disorders including the so called "Monday fever". Pathogenic micro-organisms found on hair and bristle may cause infections or skin eruptions, erythema and furuncles in man; material contaminated with anthrax spores is particularly dangerous. A characteristic dermatitis has been described amongst bristle cleaners. Various other disorders have been reported, including finger callosities due to the use of trimming shears, finger cramps, and cracks in the skin of the left thumb.

There is also a fire and explosion hazard due, in particular, to the presence of airborne concentrations of flammable dust, and a poisoning hazard in the use of toxic dyes and bleaching agents.

SAFETY AND HEALTH MEASURES

Animal products should be disinfected immediately on delivery, and superheated steam is usually employed for this purpose. Storage areas must be disinfected periodically and then sterilised, and scrupulous good housekeeping must be enforced.

Sulphur dioxide and hypochlorite vapours should be prevented from contaminating the workplace atmosphere by the use of exhaust ventilation, and bleaching processes should be mechanised to prevent direct contact with bleaching agents. Suitable dust control techniques should be implemented, in particular during horsehair processing. All open flames and other sources of ignition should be eliminated in fire and explosion hazard areas, and electrical equipment should be of the flameproof type.

Workers should receive a pre-employment medical examination during which particular attention should be given to the detection of diseases that might be aggravated by the work; periodic examinations are also necessary.

Workers should be provided with suitable protective clothing; storage and bleaching workers should be equipped with hand, eye and respiratory protective equipment.

CONRADI, F. L.

Sensitisation:

CIS 515-1969 "Eczema and asthma caused by cattle and horse hair" (Ekzem und Asthma durch Rinder- und Pferdehaare). Roth, W. G. *Berufsdermatosen* (Aulendorf, Federal Republic of Germany), Oct. 1968, 16/5 (278-282). Illus. 21 ref. (In German)

CIS 75-1991 "Occupational sensitisation of the respiratory tract by animal hair and dust" (Berufsbedingte Sensibilisation der Atemwege durch Tierhaare und Tierstäube). Czecholinski, K.; Veltman, G. *Berufsdermatosen* (Aulendorf, Federal Republic of Germany), June 1975, 23/3 (87-96). 19 ref. (In German)

Hair-cutting and shearing

Hair-cutting is the removal of portions of hair from a live animal in order to maintain the cleanliness or comfort of the animal or to prepare it for exhibition; sheep, goats, cattle and horses are the animals most commonly dealt with in this way. Shearing is the removal of the body hair in the form of a fleece for subsequent processing, usually by the textile industry. In commercial practice hair is removed from the hide of a dead animal by the use of chemical depilatories. Other articles dealing with related aspects are TANNING AND LEATHER FINISHING; FUR INDUSTRY; FELT INDUSTRY; WOOL INDUSTRY.

Processes. Hair-cutting is carried out by using a hand-operated type of scissors, or hand or motorised reciprocating serrated combs and cutters. Shearing is carried out similarly, although the hand-piece with reciprocating cutters is by far the most common instrument. It can be integrally powered with an electric motor or, more usually, remotely powered by an electric or internal combustion motor, connected to the hand-piece by articulated revolving shafts. Before shearing, it is usual to remove hair encrusted with faeces, a process known as dagging or crutching. After shearing, the fleeces are trimmed by hand according to the quality and staple of the hair, and are then compressed by hand-operated screw or a hydraulic ram into packs suitable for transportation.

HAZARDS

These may be physical, mechanical, electrical, toxicological or postural; there is also the danger of infection and parasitic disease.

Physical hazards are chiefly those of cuts and abrasions. Immediately prior to shearing, the animal is caught in a pen, dragged to the stand, laid on its back and held in this position between the legs of the shearer. For the inexperienced, there is some hazard of injury from the hooves and horns of a struggling animal. Thorns and other vegetable matter in the fleece can cause painful conditions of the hands and arms. Shearing sheds are often organised on several levels to facilitate the handling of animals and fleeces. The flow of animals may be regulated by counter-weighted gates. There is a risk of injury due to falls on faulty steps and uneven floors, and the danger of being struck by a counter-weight due to the failure of a frayed supporting rope.

Mechanical hazards are common where mechanical transmission systems are employed. Belts are often used to carry power from the motor to an overhead transmission line, and these must be fitted with adequate machinery guarding. The articulated transmission line to each shearer has a tight leather guard over the elbow. If this becomes loose or worn, the exposed cogs can catch in the hair, jamming the machine and causing the down-tube to fly in any direction. This can also happen if the cutter tension is too tight, causing it to lock in the hair. Mechanical hazards may also arise elsewhere in the shed in the use of unguarded grinders for sharpening cutters, and when using hydraulic presses for compressing fleeces.

Electrical hazards may arise when electricity is used to power the hand-pieces. These portable electric power tools can be particularly dangerous if the earth wire in the flexible lead is damaged or disconnected or if the insulation of the hand-piece itself becomes faulty.

Toxicological hazards may occur due to the inhalation of exhaust fumes from internal combustion motors. There is also a danger of pesticide poisoning when shearing animals very recently treated for lice with an organophosphorous pesticide.

Postural hazards are usually confined to the shearer's back and are due to the particular posture adopted when catching and tipping the animal; figure 1 shows this posture, in which the back is in a highly flexed position. The holding of the animal between the legs tends to accentuate the strain on the back, and, in the process of shearing, torsional movements are produced. This happens particularly with inexperienced men who tend to sweep the cutter with a twisting motion of the back, rather than use the full stretch of their arms. In manual shearing, tenosynovitis occasionally occurs.

The diseases that may be contracted during this work include tetanus, salmonellosis from dagging and crutching, leptospirosis, anthrax and parasitic diseases.

SAFETY AND HEALTH MEASURES

The precautionary measures for the majority of hazards are dealt with in the corresponding articles. However, fundamental requirements include regular inspection of the workplace, adequate training of workers and proper supervision. It is also essential that good lighting should be provided, preferably above and 2 m in front of the shearer's work point so that he is not working in shadow. Good sanitary and washing facilities, which should be used before eating and drinking, are also essential.

Figure 1. Shearing a sheep. Note the posture adopted by the worker and the leather guard over the elbow of the transmission line.

Medical supervision. Hair-cutting and shearing are tasks which require considerable and unpredictable levels of energy expenditure due to the handling of heavy animals and to the postures adopted. Since it is a seasonal activity which is usually carried out under pressure of time, workers require good physique and also need to keep in training during the off-season. Any man with a history of back disorders should be excluded. The work also calls for good vision and a high degree of muscular co-ordination. It is also wise to exclude men with skin disease of the hands, arms and legs, as such lesions can be mechanically irritated or exacerbated by grease from the hair.

First-aid boxes should be provided in shearing sheds and, in first-aid organisation, provision should be made for the transport of persons injured in isolated workplaces.

COPPLESTONE, J. F.

Hairdressers

The work of a man's hairdresser or barber includes the cutting, washing, treatment and conditioning of hair and the shaving of beards. The work of the woman's hairdresser is more complicated, requires a wider technical knowledge due to modern developments in hair care, and covers additional tasks such as bleaching, tinting, permanent waving and setting, and face and hand care. The tools of the trade include combs, brushes, scissors, razors, clippers, driers, electric waving irons, curlers, setting clips.

Cleaning, waving and dyeing of wigs or hairpieces are also done by the hairdresser.

Working conditions. Hairdressers are required to work in a standing position in salons which vary considerably in their size, layout, microclimate, lighting, equipment and conditions of hygiene. The hours of work are often irregular and overtime may be necessary at peak periods; shift work is not common.

Hair-care and beauty products. Over recent years, there has been considerable development in this field and hair-care products now include bleaches, dyes, permanent-wave lotions, shampoos, depilatory creams and lotions, and setting lacquers. See also COSMETICS.

Bleaches usually consist of poultices containing 20-30% hydrogen peroxide by volume in a mineral or vegetable base. Vegetable dyes are rarely employed nowadays and metallic salts are widely prohibited due to their hazardous properties. The "permanent" organic dyes derived from *p*-phenylenediamine have been prohibited in certain countries (e.g. Switzerland in 1936 and France in 1951) and have usually been replaced by *p*-toluenediamine. In the organic dyes with a tar base, the colouring agent (a liquid, cream, gel or powder) is normally mixed, at the time of use, with an oxidising agent (e.g. hydrogen peroxide or other peroxides), which acts as a developer; stabilisers such as hydroquinone and resorcinol, alkalis such as ammonia, and wetting agents, which improve penetration, may also be added.

Permanent wave treatment makes it possible to produce hair waves which are virtually indistinguishable from natural ones. Cold-wave treatment, which has practically replaced the hot process, comprises two operations:

(a) rolling the hair on curlers and impregnating it with a thioglycolate solution (1-10% sodium or ammonium thioglycolate, alkaline sulphite and wetting agent) for 10-45 min;

(b) rinsing and neutralisation with a solution containing oxidising agents such as hydrogen peroxide, perborates or persulphates, to which have been added citric, tartaric or acetic acid. In this process the thioglycolate breaks down the cystine bonds in the hair keratin, which are then reconstituted by the oxidising action of the neutralising solution to hold the hair in the position imposed by the rollers. A more recent innovation, the "neutral" permanent wave, uses no thioglycolates and has a neutral pH. The technique is, however, complicated and requires the use of a highly acidic oxidising agent which may damage hair keratin.

Shampoos are used to clean the hair and scalp. They are now mainly liquids based on surface active agents and are either: anionic (ammonium lauryl sulphate or triethanolamine) often with added thickeners, acidifiers and oiling agents (sodium sulphoricinate, lecithin); cationic with a quaternary ammonium base (high cationic with a quaternary ammonium base (high affinity for hair keratin, and an anti-dandruff action); or non-ionic, which do not degrease or denature the hair. Several cheap shampoos may contain formaldehyde, used to prevent growing of various Gram-negative bacteria.

Depilatory agents usually contain calcium thioglycolate in an alkali medium although various sulphides are also employed. Hair lacquers are normally sprayed from pressurised aerosol containers; they contain polyvinylpyrrolidone (PVP), PVP and polyvinyl acetate copolymer, shellac, dimethylhydantoin formaldehyde resin modified with polyacrylic acid, and lanolin. The various perfumes employed are solutions of natural perfumes and essences (either simple or complex aromatic aldehydes) in alcohol. A distinction is made between: grease lipsticks which contain no eosine and in which the colouring consist of lacquers, various pigments and, in some cases, titanium dioxide; emulsified lipsticks in which the colouring consists of soluble, alkaline eosine; and colloidal lipsticks in which the colouring is alkaline eosine.

Make-up creams are complex emulsions which usually contain the following substances: distilled water and perfumes (aqueous phase), glycerol, glycolsorbitol (humidifying agents), various vegetable oils, surfactants, alcohols or esters (stabilisers), hydroquinone or tetrahydrodimethyldiphenyl (antioxidants), salicylic or benzoic acid (antiseptics).

The composition of nail varnishes is as follows: a base of nitrocellulose or acetylcellulose, solvent (acetone, ethyl alcohol, pentyl alcohol, benzyl alcohol, various acetates, etc.); thinner (toluene, xylene, sometimes even benzene); plasticisers (various phthalates, phenyl phosphates); resins (copal, urea-formaldehyde, vinylic, acrylic, methacrylic, etc.) which give a gloss to the varnish and ensure adhesion; colouring agents; and perfumes. Nail varnish solvents usually have an acetone or acetate base.

HAZARDS

These may stem from the working posture and movements, the hours of work, the physical and psychological environmental conditions and the substances and equipment employed.

The work itself is considered light but does, nevertheless, impose a significant static load since the worker must remain in a more or less fixed position for long periods; this may also favour the development of varicose veins. Nervous fatigue may result from working hours which are often irregular and prolonged by overtime, from the workplace environment which is often badly ventilated, hot and humid, and from the constant need to be pleasant to customers. Digestive system disorders may be caused by meals eaten at irregular hours and under bad conditions.

Cuts and abrasions, with the possibility of infection, may be sustained from scissors, razors, etc., and faulty electrical appliances may produce electrical accidents. Particles of hair may lead to irritation of the interdigital spaces and nail root, and irritation of the respiratory tract may result from the inhalation of dry hair, hair sprays or other products. The harmful effects of hair sprays on the respiratory tract have been studied and the name "thesaurosis" has been adopted to describe them; however, these effects have not so far been confirmed. A number of cases of asthma due to persulphate bleaching agents has also been reported.

Skin diseases are the main occupational hazard in hairdressers and include contact eczema, toxic dermatitis and alkali dermatitis. Recent statistics from the Federal Republic of Germany indicate an incidence of 2.3 reported cases of skin disease for 1 000 hairdressers; 0.18 cases per 1 000 hairdressers are recognised.

The most frequent allergy is to *p*-phenylenediamine and to nail varnishes. A few cases of dermatitis caused by shampoos containing formaldehyde have been reported. Other products have only a low sensitisation potential. The primary irritants are mainly depilatory agents and permanent-wave lotions. Certain perfumes may produce photosensitivity and various substances employed may produce nail and hair fragility.

Several studies have been carried out on the carcinogenic and mutagenic risk of certain dyes containing, in particular, aromatic amines, aminophenols and their nitro- and hydroxy-derivatives. The results of these studies vary according to the methods that have been used. However, it would appear that several dyes are mutagenic to bacteria and can cause chromosome changes in human cells. Evidence has been provided that a dye, the 2,4-diaminotoluene, can produce liver carcinomas in rats. It is premature to extrapolate these observations to man; however, certain epidemiological studies suggest the existence of a correlation between hairdressing and an increased incidence of cancer of the urinary tract.

SAFETY AND HEALTH MEASURES

A well designed and equipped salon is essential if the safety and health of hairdressers is to be ensured. The floor area should be adequate and in proportion to the number of seats, and the premises should have adequate general ventilation and lighting. Seats of adjustable height mounted on castors, etc., should be provided. To facilitate maintenance, good housekeeping and personal hygiene, floors should be of easily washable material, there should be a constant supply of freshly laundered working clothes for staff and towels and napkins for clients; there should also be closable waste bins and adequate washing and sanitary facilities with hot and cold water. All equipment should be regularly disinfected and electrical equipment should be checked for safety periodically; for hand-held equipment weighing over 500 g, the possibility of spring-loaded overhead suspension should be investigated.

The use of the following substances in hair care and other cosmetic preparations should be abandoned: arsenic, antimony, barium, lead, cadmium, mercury, selenium, thallium, ionisable chromium compounds, aniline and its aromatic derivatives containing up to six carbon atoms, *m*-, *o*- and *p*-nitroaniline, 2-naphthylamine and its homologues, *m*-, *o*- and *p*-nitrophenol, 3,4-benzpyrene and other carcinogenic substances, and carbon tetrachloride. In addition, the use in hair-care and beauty products of flammable substances such as ethyl ether, acetone, benzene, gasoline, nitrobenzene, chlorinated hydrocarbons and methyl alcohol should be avoided; the use of ethyl alcohol and isopropyl alcohol should, however, be permitted. Acetone, acetic acid esters, and ethylene glycol and butyrolactone derivatives should be the solvents of choice for nail varnish preparations and removers. It should be ensured that hair dyes do not contain harmful organic compounds such as, in particular, *p*-phenylenediamine, polymerised formaldehyde (trioxymethylene, etc.) or pilocarpine. The thioglycolic or thiolactic acid used in cold-wave preparations should be very dilute with a pH as near to neutral as possible; for example in Switzerland it is stipulated that for thioglycolic acid the solution should be 7.5% with a maximum pH 9.5, or 9% with pH 8; in the case of thiolactic acid, the solution should be 8.5% with maximum pH 9.5 or 11.5% with maximum pH 8.

Medical prevention. In the personnel guidance and selection of hairdressers, it should be remembered that the following requirements and aptitudes are necessary for this profession: full complement of limbs; absence of varicose veins and deformation of the skeleton; good vision and hearing; freedom from skin disorders, allergy, hyperhidrosis and seborrohea.

Personal hygiene is of importance in this profession. Persons using products containing thioglycolic acid should wear hand protection. Where persons are found to have allergic reactions to certain hair-care products it may often be necessary to advise a change of job. Efforts should be made to ensure that hairdressers' hours of work are kept regular and that overtime does not become excessive.

Finally, epidemiological studies on the causes of death and diseases in hairdressing are urgently needed in order to ascertain whether the carcinogenic risk is a real hazard to hairdressers.

LOB, M.

Asthma:

CIS 77-1981 "Asthma due to inhaled chemical agents–Persulphate salts and henna in hairdressers". Pepys, J.; Hutchcroft, B. J.; Breslin, A. B. X. *Clinical Allergy* (Oxford), July 1976, 6/4 (399-404). Illus. 11 ref.

Dermatitis:

"Hairdresser's hand". Hannuskela, M.; Hassi, J. *Dermatosen in Beruf und Unwelt* (Aulendorf, Federal Republic of Germany), 1980, 28/5 (149-151). Illus. 11 ref.

CIS 77-1992 "Clinical aspects and pathogenesis of hairdresser's eczema" (Zur Klinik und Pathogenese des Friseurekzems). Czarnecki, N.; *Zeitschrift für Hautkrankheiten* (West Berlin), 1 Jan. 1977, 52/1 (1-10). Illus. 16 ref. (In German)

Thesaurosis:

CIS 77-570 "Occupational disease due to the use of hair sprays" (La pathologie professionnelle due à l'utilisation de laque pour cheveux). Silbert, R. (Université de Paris VI, Faculté de médecine Saint-Antoine, Paris) (1975), 100 p. 80 ref. (In French)

Mutagenic and carcinogenic risk:

"Examination of the potential mutagenicity of hair dye constituents using the micronucleus test". Hossack, D. J. N.; Richardson, J. C. *Experientia* (Basel), 1977, 33 (377-378). 20 ref.

"Cancer mortality in male hairdressers". Alderson, M. *Journal of Epidemiology and Community Health* (London), Sep. 1980, 34/3 (182-185). 30 ref.

2,3-Diaminotoluene:

"2,4-Diaminotoluene" (83-95). *IARC monographs on the evaluation of the carcinogenic risk of chemicals to man. Vol. 16: Some aromatic amines and related nitro compounds–Hair dyes, colouring agents and miscellaneous industrial chemicals* (Lyons, International Agency for Research on Cancer, Jan. 1978), 43 ref.

Halogens and compounds

Fluorine, chlorine, bromine, iodine and the more recently discovered radioactive element, astatine, make up the family of elements known as the halogens. Except for astatine the physical and chemical properties of these elements have been exhaustively studied. They occupy group VII in the periodic table and they display an almost perfect gradation in physical properties.

The family relationship of the halogens is illustrated also by the similarity in the chemical properties of the elements, a similarity which is associated with the arrangement of seven electrons in the outer shell of the atomic structure of each of the elements in the group. All the members form compounds with hydrogen, and the readiness with which union occurs decreases as the atomic weight increases. In like manner, the heats of formation of the various salts decrease with the increasing atomic weights of the halogens. The properties of the halogen acids and their salts show as striking a relationship; and the similarity is apparent in organic halogen compounds, but, as the compound becomes chemically more complex, the characteristics and influences of other components of the molecule may mask or modify the gradation of properties.

HAZARDS

The similarity which these elements exhibit in chemical properties is apparent in the physiological effects associated with the group. The gases, fluorine and

	Atomic weight	Physical state	sp.gr.	m.p.	b.p.	v.d.	TWA ppm	OSHA mg/m³	MAC USSR mg/m³
Fluorine (F)	19	gas	1.10 (−188 °C)	−219.6	−188.1	1.7	0.1	0.2	–
Chlorine (Cl)	35.4	gas	1.56 (−34 °C)	−101	−34.6	2.5	1 (ceil)	3	1
Bromine (Br)	79.9	liquid	3.12	−7.2	58.8	3.5	0.1	0.7	–
Iodine (I)	126.9	solid	4.98	113.5	184.3	–	0.1 (ceil)	1	1
Astatine (At)	210	solid	–	302	337	–	–	–	–

chlorine and the vapours of bromine and iodine are irritants of the respiratory system; inhalation of relatively low concentrations of these gases and vapours gives an unpleasant, pungent sensation, which is followed by a feeling of suffocation, coughing, and a sensation of constriction in the chest. The damage to the lung tissue which is associated with these conditions may cause the lungs to become overloaded with fluid, resulting in a condition of pulmonary oedema which may well prove fatal.

The halogenated hydrocarbons tend to have narcotic properties and some have been used as anaesthetics in surgery. A number of these organic halogen compounds are notorious for the injury they have caused to the liver and kidneys.

Because of their great industrial and toxic significance, halogens and their compounds are dealt with in separate articles (see BROMINE AND COMPOUNDS; CHLORINE AND INORGANIC COMPOUNDS; FLUORINE AND COMPOUNDS; IODINE).

Astatine (At)

ATOMIC NUMBER 85

This element was first idientified as a product (^{211}At) arising from the bombardment of bismuth by high-energy α-particles. About 20 isotopes have since been prepared by nuclear reactions, the longest lived being ^{210}At with a half-life of 8.3 h. The element occurs naturally in uranium minerals but the total amount of astatine existing in the Earth's crust is assessed to be less than 30 g. It is a radioactive material, the only member of the halogen family without stable isotopes. It resembles most the member next above it in the group, namely iodine. Like iodine, it is readily taken up by the thyroid gland and can cause the selective destruction of the thyroid tissue by ionising, short-range α-particles.

MATHESON, D.

Hand and arm protection

The human hand is probably the most valuable and versatile tool used in industry. Considering also that hand and arm injuries account for approximately 25% of all industrial injuries, the need for protection becomes obvious, if for no other reason than one of economics.

Hand and arm hazards

The causes and types of hand and arm injuries sustained are as numerous as the tasks they perform, but most can be assigned to one or more of the following categories: fire and heat, cold, electromagnetic and ionising radiations, electricity, chemical substances, impacts, cuts and abrasions and infections.

Mechanical hazards are amongst the most common and can usually be foreseen. Electricity and radiation hazards may have very serious consequences; their dangers are usually fully recognised by the exposed workers and safety measures and protective equipment form an integral part of the work processes. However, the dangers arising from contact with a large number of toxic, irritant or allergenic chemical substances are not well recognised and the effects of exposure to such substances have to be assessed both in relation to the risk of skin lesions and their severity and by making allowance for the fact that the skin may act as the portal of entry for a toxic agent (see also HAND INJURIES).

Types of hand and arm protection

Any hazard is best controlled by modifying the environment, process or procedure, either to eliminate the hazard completely or to reduce it to a level where alternative preventive measures are more easily implemented: for example guarding of dangerous parts of machinery, screening of sources of heat and radiation, provision of well designed handles for hand tools, effective exhaust ventilation to remove dust and fumes. Where a hazard is known or suspected but cannot be effectively controlled by these methods, the use of protective devices must be considered.

A large variety of hand and arm protection devices is available, many of them designed for a specific purpose. The terminology applied to the several kinds of protectors may also vary with the user, so a description of each follows in the listing below:

Gloves. The thumb and each of the fingers are in separate divisions for complete finger movement and control,

Figure 1. Heavy duty cotton gloves with additional hand and arm sleeve for protection against cuts and abrasions in steel mill. *(By courtesy of Jomac Inc.)*

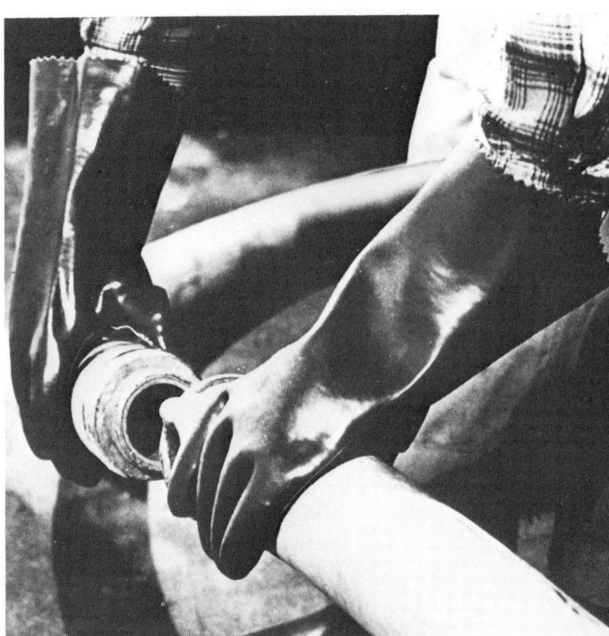

Figure 2. Polyvinyl chloride coated cotton gloves provide protection against acids—used in assembly and materials handling. *(By courtesy of Jomac Inc.)*

generally used where finger dexterity is necessary, such as in picking up small sharp objects or in handling certain materials. They also provide protection to the fingers, hands, and sometimes the wrists and forearm, from cuts, abrasions, chemicals and other hazards.

Mitts (mittens). A covering for the whole hand with the thumb in a separate division—for work not requiring precise, individual finger movement. Mittens are also made with separate coverings for the thumb, index finger, or other fingers. They are used in the same situation as gloves and for the same type of protection.

Finger guards or cots. These provide protection for a single finger or finger tip. Sometimes two or more fingers are protected by a single cot. They are designed for protection against skin friction and pressure, moisture, chemicals and other liquids, cuts, and hot materials.

Thumb- and finger-guard thimbles. These are designed to afford protection to the thumb and index finger or the thumb and first two fingers.

Hand pads. These are used to protect the palm of the hand against cuts and abrasions or against burns caused by direct contact with hot objects. They are generally heavier and less flexible than gloves or mittens.

Sleeves. These are for protection of the wrist and arm. Specific designs are used to protect all or any portion of the wrist and arm up to the shoulder. They are generally worn along with gloves and are made to give protection against flame and heat, impact or cuts, splashing liquids, energised conductors, and general skin abrasion, friction and pressures.

Materials. In addition to selecting the shape and size of the protector to give adequate cover for the area in danger, the fabricating material chosen must be able to withstand the specific hazard involved. A summary of common arm and hand hazards and suitable protective materials follows:

Heat—asbestos, asbestos reinforced with leather, leather, aluminised fabric, glass fibre, aluminised asbestos. (See HEAT PROTECTIVE CLOTHING.)

Flame—asbestos, leather, fire-resistant duck, aluminised fabric, glass fibre.

Sparks—asbestos, asbestos reinforced with leather, fire-resistant duck, leather, glass fibre.

Hot metal splashes—leather, fire-resistant duck, glass fibre, aluminised asbestos.

Moisture and water—natural and synthetic rubber, coated fabric, coated glass fibre, plastic.

[Mild acids and alkalis—polyvinyl chloride, butyl rubber, neoprene, natural rubber, nitrile rubber, coated glass fibre.

Strong acids and alkalis—polyvinyl chloride, butyl rubber, nitrile rubber (only for alkalis).

Hydrocarbons and chlorinated hydrocarbons—polyvinyl alcohol.

Ketones, esters—butyl rubber, neoprene, natural rubber.

Alcohols—natural and synthetic rubber.

Oxidants—polyvinyl chloride.]

Chips, abrasives—fabric, leather, coated fabric.

Cuts and severe blows—leather reinforced with steel, metal mesh, mesh and moulded plastic.

X-rays—leaded rubber, leaded leather, leaded plastics.

Electricity—leather gloves over rubber gloves for flash burns; lineman's gloves and rubber sleeves for electric shock.

Apparel for chemical resistance. When selecting apparel for chemical resistance, information on the following factors should be established:

(a) ability of the apparel to resist penetration of the chemical;

(b) chemical composition of the solution to be handled;

(c) degree of concentration of the solution at various processing stages;

(d) abrasive effects of the apparel material on the skin;

(e) temperature conditions;

(f) time cycle of usage.

Protection against dermatitis. The choice is not always a simple one when protection against dermatitis-producing agents is required. If the irritant substance is in dry form, plain cotton gloves and other apparel may serve as well as any and have the added advantage of easy laundering. Disposable paper and plastic garments can also be used.

When the irritating substance is a liquid, the material chosen for the apparel must be impervious to the specific agent, otherwise the substance may penetrate through the material and then be maintained in a high concentration in prolonged contact with the skin. Since there is always the chance of some of the substance getting inside the garment, frequent cleaning is important.

If the apparel chosen is not absolutely impervious to the irritating agent, it will eventually penetrate to the inside and be trapped next to the skin, causing repeated exposure every time the garment is worn.

Protective creams. Specially formulated barrier creams and lotions (see separate article) may sometimes offer the only practical protection, for example:

(a) when personal protective equipment is too cumbersome or uncomfortable;

(b) where manual dexterity is all-important;

(c) where protective apparel might constitute a hazard near moving machinery.

Skin cleansers. Germicidal and antiseptic soaps, detergents and cleansing creams are used to remove dirt, grease, oil and chemicals from the skin after exposure.

Machine operations. In general, gloves should not be worn when operating machinery: there is an added risk of entanglement and the hand may be severely damaged if the gloves are caught by the moving parts. This particularly applies to revolving shafts and spindles, drills, in-running rolls and gears. Protection should be by secure guarding of the dangerous parts.

Testing

To ensure that the apparel selected is providing the necessary protection, testing procedures have been developed and should be used where data from the manufacturer is insufficient, or where there is reason to doubt the usefulness of the protector after a period of exposure.

Procedures similar to the following may be used or modified to provide information on the performance of impervious rubber or plastic gloves.

First, the gloves are tested for leaks by filling them with air and immersing them in water. They are then washed, dried, turned inside out, weighed and measured (polyvinyl alcohol gloves are wiped with dry paper towels and stored in a dessicator to await testing). The re-usable gloves are filled with a measured amount of a given solvent and after 24 h are measured for length and their gross weights are determined. They are then emptied, wiped as dry as possible, reweighed, washed and examined for physical condition. With this procedure, information is obtained concerning the permeability of the glove, solubility of the glove material, resistance to change of shape, hardening characteristics and resistance to stretching and tearing.

Other types of protectors could be tested for other reasons using modified versions of this technique. Manufacturers and government agencies often provide guidance information for testing materials.

Maintenance

Continued good performance of protective apparel depends largely on the care it receives. In the first place, proper fitting is important, especially with gloves. Not only does this ensure greater ease in working but the gloves can be put on and taken off without unnecessary damage. The gloves and sleeves should not be left folded in an unnatural way as this places the folded part under stress and weakens it.

Periodic cleaning and maintenance to remove the build-up of solvents, degreasing agents, etc., lengthens the life of the protector. Gloves with rough finishes require thorough cleaning because the irregular surface forming the finish traps solutions which may cause deterioration. If gloves swell, they should be taken out of use to permit solvents to evaporate and the original shape to be restored.

Some gloves made of rubber, cotton or leather may be reconditioned to prolong their useful life. Reconditioning may include cleaning, sterilisation and reshaping as well as repair.

Defective protectors should be scrapped and replaced at once.

Ensuring regular use of protection

It is essential to secure the co-operation of the workers in the constant use of the protection provided. Consultation and discussion in accident-prevention committees, safety campaigns, suggestion schemes and posters can all play their part; the user can often make a valuable contribution in discussing the merits or demerits of a particular type of protection. Supervision is necessary to ensure that protectors are worn and receive proper care.

PROWSE, W. A.
WILLIAMS, D.

CIS 78-2001 *Protective gloves against aggressive chemicals* (Guantes de protección frente a agresivos químicos). Dirección general de trabajo. Norma MT-11, Colección textos legales n° 020.00.14 (Madrid, Servicio social de higiene y seguridad del trabajo, 1978), 33 p. Illus. 8 ref. (In Spanish)

CIS 77-1107 "Resistance of protective gloves to industrial solvents–Recapitulation" (Résistance des gants de protection aux solvants industriels–Tableaux récapitulatifs). Chéron, J.; Guenier, J. P.; Moncelon, B.; *Travail et sécurité* (Paris), Dec. 1976, 12 (554-561). (In French)

CIS 77-1114 *The development of firefighters' gloves–Vol. I. Glove requirements; Vol. II. Glove criteria and test methods.* Coletta, G. C.; Arons, I. J. Ashley, L. E.; Drennan, A. P. DHEW (NIOSH) publication No. 77-134-A and 77-134-B (National Institute for Occupational Safety and Health, 4676 Columbia Parkway, Cincinnati) (Feb. 1976), 97 p. (Vol. 1), 179 p. (Vol. 2). Illus. 65 ref.

"Hand protection". Bennett, D. C. *Health and Safety at Work* (Croydon), May 1981, 3/9 (32-35). Illus.

"How to select proper hand protection". Dionne, E. D. *National Safety News* (Chicago), May 1979, 119/5 (44-53). Illus. 8 ref.

Handicapped and disabled persons

Various types of illness and accident produce disability. In many cases, however, the sufferer recovers completely after appropriate treatment and returns to his or her normal occupation; the disability has been purely temporary, has had no effect on the person's employability and there is little or no need for vocational rehabilitation. It is long-term or permanent disability that produces an occupational handicap, however, which is of direct concern to the worker, his family and his employer and which calls for the prompt application of rehabilitation if an early return to work is to be achieved. Occupational handicap will usually be due to the disability, but the handicap will differ in each individual case–even though the disability may be the same–depending on such factors as age, intelligence, education, social circumstances, aptitudes and interests of the individual person, and his determination and capacity for self-adjustment. For example the amputation of a leg is a physical disability but the occupational handicap imposed by an amputation is very different in the case of an office worker with a completely sedentary job and, say, a professional footballer.

The ILO Vocational Rehabilitation (Disabled) Recommendation, 1955 (No. 99), defines the term "disabled person" as "an individual whose prospects of securing and retaining suitable employment are substantially reduced as a result of physical or mental impairment".

In so far as the scope of vocational rehabilitation services is concerned, the Recommendation states that "such services should be made available to all disabled persons, whatever the origin and nature of their disability and whatever their age, provided they can be prepared for, and have reasonable prospects of securing and retaining, suitable employment".

The identification of disabled persons requiring rehabilitation services often presents some difficulty in the early development stages of the rehabilitation programme. It should, however, be remembered that there is no clear line of demarcation between the able-bodied and the disabled and that any definition or selection procedure adopted must be flexible and emphasise residual abilities, not limitations (see also REHABILITATION in the Index).

Size of the disablement problem

It is estimated that some 40 million people in the world are blind, 80 million are deaf or hard of hearing, 15 million are suffering from leprosy, and some 40 million are mentally retarded to such an extent as to require some assistance with routine activities of daily living. Moreover, the World Health Organisation estimates that one person in ten is likely to suffer from some form of mental illness at some time during his life. Poliomyelitis is still a major orthopaedic problem in most developing countries of the world; hundreds of thousands of new cases of paralysis occur each year and the number of untreated patients requiring orthopaedic care totals several millions. Leaving aside the enduring problem of malnutrition, which itself may put a large proportion of the world's population at risk of disabling diseases, various estimates indicate that at least 400 million people representing 10% of the world's population are disabled as a result of congenital defect, injury, chronic disease (especially cardiovascular diseases, diabetes and diseases of the musculoskeletal system), acute disease, or drug or alcohol abuse.

Employment of handicapped persons

Studies and experience have shown that disabled persons can be satisfactorily employed in a wide variety of occupations provided they are properly rehabilitated and trained. Successful placement is dependent on the recognition that each disabled person has individual likes and dislikes, individual qualifications, experience and aptitudes, and that most have more abilities than disabilities. Placement should be carried out by matching a disabled person's residual abilities with the requirements of the job so that both worker and employer are satisfied.

Seeking employment opportunities for the disabled should be based on the principle of equal opportunity with the non-disabled to perform work for which they are qualified. Establishing lists of jobs suitable for the disabled in general or for specific groups is now considered a restrictive and outdated method.

The first avenue of approach in placing a disabled worker is employment in normal competitive industry provided he has the physical and mental requirements for the job, is not a hazard to himself and does not jeopardise the safety of others. Such placement is often carried out by a special (disablement) placement officer in close collaboration with rehabilitation agencies, voluntary organisations concerned with the handicapped, employer and trade union associations, industrial medical officers, and personnel managers in industrial firms.

Some handicapped persons prefer to work on their own in a small business. Success in such an enterprise depends mainly on such factors as business ability, willingness and capacity for hard work, knowledge of the demand for the goods to be made or sold and sufficient capital to make a start.

In many countries, sheltered employment schemes have been established for those severely disabled who are unable to work under normal industrial conditions. Such schemes provide productive activities in sheltered or production workshops and home-work schemes in which the severely disabled can make a living by working in their own homes.

Another employment scheme which has been successfully developed in Poland and some other countries is organisation of co-operatives for disabled persons. Large numbers of disabled people are engaged in such schemes making a variety of products, sometimes in their own homes, which are marketed through a sales organisation, with governmental guidance and assistance.

Work and workplaces for handicapped persons

One method of increasing employment opportunities for disabled workers is that of job adaptation to reduce the strain imposed by certain jobs and workplaces or assist in increasing productivity.

In designing sheltered workshops, architectural barriers such as narrow doorways, entrance steps, etc., should be avoided so as to enable all kinds of handicapped workers, including wheelchair cases, to enter and work at the premises without difficulty. The use of braille lettering on walls and doors helps the blind to locate their position in a building.

The basic principle governing all job adaptations is to remove completely or reduce significantly the restrictions imposed by a disabling condition. In many cases only slight adaptations are necessary and can be carried out inexpensively. For example typewriters for one-armed typists may be fitted with a knee-activated key-shift mechanism. A specially designed orthotic device will enable a left-arm-amputee welder to hold the flux in his artificial left arm while operating the welding equipment with the right hand.

Particular attention should be paid to the safety of disabled workers. While there is no real evidence that they are more accident-prone than other workers, they may encounter more difficulty in the early stages of new employment. However, with proper introduction to the new job by a placement officer experienced in dealing with the disabled, these difficulties will be minimised. Where signals related to the work or to safety are normally conveyed audibly, arrangements will have to be made to convey them visually to deaf workers. Similarly, in the case of blind workers, visual signals must be replaced by audible ones.

Successful rehabilitation and the successful promotion of job-adaptation schemes are a matter of teamwork. Vocational rehabilitation workers, employers and design engineers cannot be expected to know the precise physical limitations caused by a particular disability in a particular person; this can be ascertained only by obtaining authoritative medical advice.

Legislation and governmental programmes

In various countries, legislation has been passed to promote the absorption of handicapped workers into active and remunerative employment. In the United Kingdom, the Disabled Persons (Employment) Act, 1944, is the foundation for employment services for disabled people. Under this Act, the Department of Employment maintains a register of disabled persons, provides a specialist employment placement service, enforces the obligations placed upon employers with 20 or more workers to employ a quota of registered disabled persons, offers courses of industrial rehabilitation and vocational training and provides sheltered employment for the severely disabled. Throughout the country, there are 1 100 disablement-resettlement officers who can be contacted at every employment exchange. Any disabled person, whether registered or not, can call upon these officers for advice and assistance, and all medical authorities are encouraged to refer to them any patient who has an employment problem.

Other countries in which quota systems of employing disabled persons are in operation include Austria, Belgium, Brazil, France, the Federal Republic of Germany, Greece, Italy, Japan, Luxembourg and Portugal.

Sweden possesses an integrated employment programme for all categories of persons handicapped by physical, mental, emotional or social conditions and who are on the fringes of the labour market. The programme is closely associated with an active manpower policy and the Swedish Employment Service seeks organised co-operation with hospitals, social services, alcoholic-care, penal and corrective institutions, welfare agencies, organisations of and for the handicapped, and the social and sickness funds. To meet the temporary or permanent employment needs of severely handicapped persons, Sweden has resorted to sheltered employment schemes and also to financial measures in the form of wage subsidies or tax incentives to encourage employers to hire these workers in spite of their handicaps.

In Portugal a scheme for promoting the employment and re-employment of the disabled centres on three government-sponsored agencies: the Rehabilitation Service; the National Employment Service; and the Accelerated Vocational Training Institute (figure 1).

International activities

Responsibility for promoting services for the handicapped at the international governmental level is shared by the United Nations, ILO, WHO, UNESCO and UNICEF, each covering their respective areas of interest.

In the past 20 years the ILO, in close association with the United Nations and its specialised agencies and with governments and non-governmental organisations, has developed its technical co-operation activities to such an extent that some 80 developing countries have received direct ILO assistance in the form of experts, fellowships and equipment. With ILO Recommendation No. 99 as a firm basis, the ILO has helped to develop national programmes of vocational rehabilitation for the disabled in many countries of Africa, Asia, the Middle East and Latin America, with a view to creating a new understanding of the dignity of work and the right of the handicapped to be considered an integral part of the productive labour force of their country.

MONIZ, A. E.

Vocational rehabilitation of the mentally retarded (Geneva, International Labour Office, 1978), 200 p.

Basic principles of vocational rehabilitation of the disabled (Geneva, International Labour Office, 2nd. ed., 1976), 53 p.

Adaptation of jobs for the disabled (Geneva, International Labour Office, 1978), 65 p. Illus.

Co-operatives for the disabled. Organisation and development (Geneva, International Labour Office, 1978), 230 p.

"Problems related to handicapped people in occupational health" (Problèmes posés par les handicapés en médecine du travail). Gillon, J. J. *Encyclopédie médico-chirurgicale. Instantanés médicaux* (Paris), 1981, 5/76, 16506 A 10, 10 p. 31 ref.

CIS 75-1189 "XIIth National Seminar on Occupational Diseases, Marseilles, 10-14 May 1972" (XIIes Journées nationales des maladies professionnelles, Marseille, 10-

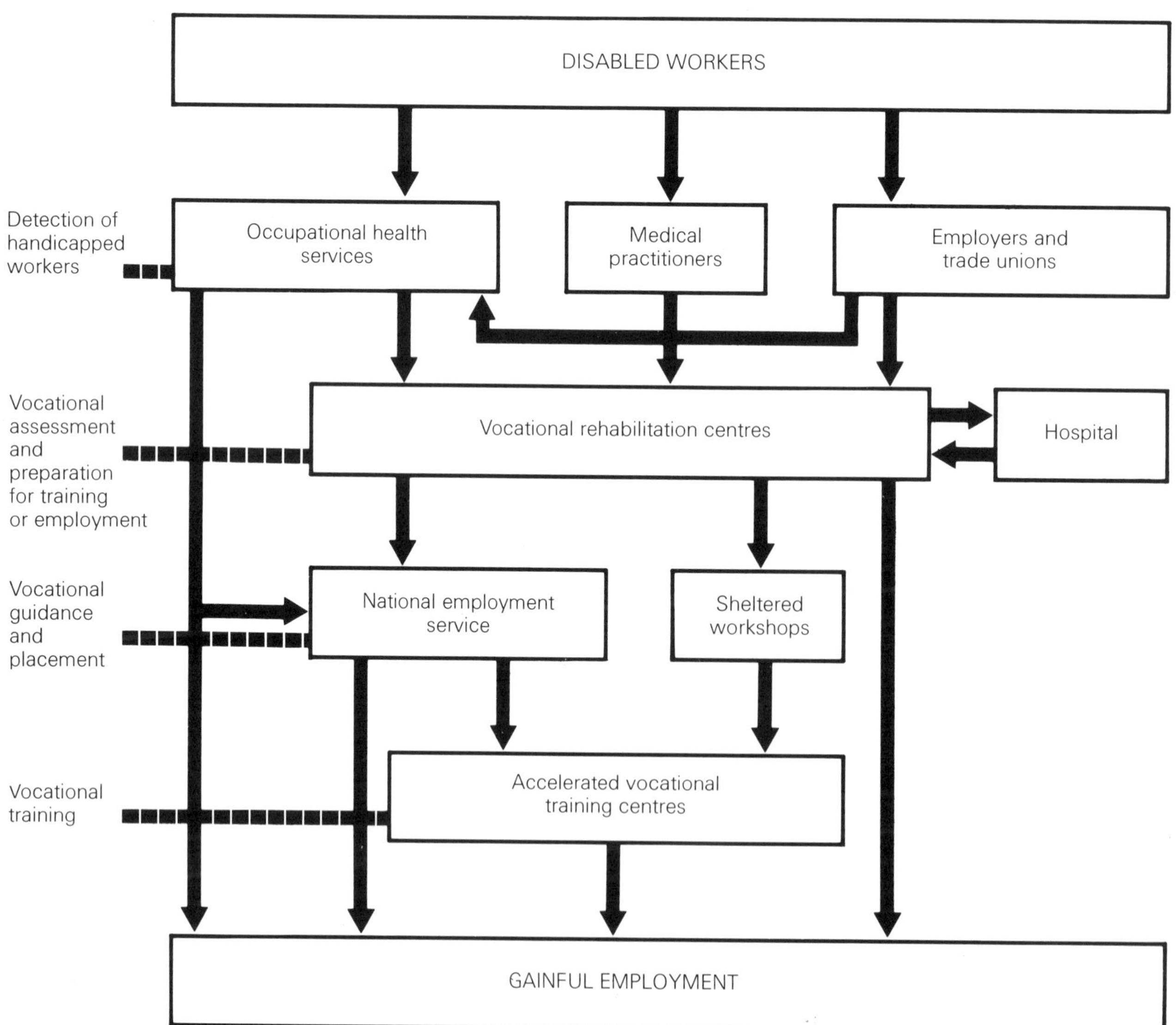

Figure 1. The outline of a typical service for promoting the employment and re-employment of disabled workers.

14 mai 1972). *Archives des maladies professionnelles* (Paris), Jan.-Feb. 1974, 35/1-2 (1-296). Illus. (In French)

"Job analysis applied to the disabled" (Ermittlung der Arbeitsanforderungen für Leistungsgeminderte). North, K.; Rohmert, W. *International Archives of Occupational and Environmental Health* (West Berlin), Nov. 1980, 47/2 (143-154). Illus. 11 ref. (In German)

Specifications for making buildings and facilities accessible to and usable by physically handicapped people. ANSI A117.1.1980 (American National Standards Institute Inc., 1430 Broadway, New York) (3 Mar. 1980), 68 p. Illus.

Handicrafts and craftsmen

Craftsmen or artisans may be defined as workers who:

(a) exercise a craft aimed at the production of goods or the supply of services in their own locality, or a craft of an artistic nature;

(b) themselves manage their own work;

(c) work alone or with small establishments of less than ten persons;

(d) have partners often belonging to the family or with tribal links in developing countries;

(e) employ a modest amount of capital locally invested, and work with a little equipment or manually;

(f) transmit skills from master to partners and in family business from father to sons or between brothers; and/or

(g) must demonstrate, in some countries, the mastership of their craft by an apprenticeship or long practice of their profession.

Handicrafts flourished in Antiquity and the Middle Ages and probably reached their climax during the Renaissance in the schools of artists. It was in studying the diseases of craftsmen that, in the 18th century, Ramazzini established the basis of occupational health. In spite of the introduction of machines during the Industrial Revolution, the merging of firms, the expansion of markets and the opposition of employees and employer, handicrafts are still a living force, especially in the developing countries. In 1960 it was estimated that a tenth of the population of Morocco earned its living primarily from work of a craft nature. In the industrialised countries the craftsmen pursue artistic crafts and contribute to the preservation of folk-art; they play an important role in the economy of small communities where they ensure the flexibility of production and maintenance necessitated by the distance from the large commercial axes. According to data gathered in 1977 and 1978 by missions of the ILO's International Programme for the Improvement of Conditions of Work and Environment (PIACT), in Peru 65% of the manufacturing labour force is engaged in handicrafts, and in Morocco, where there are 71 000 undertakings with fewer than ten employees, some 10% of the population earn their living by the practice of handicrafts; in Italy somewhat more than 2% of the population do the same.

Normally, the work is carried out in the home or, in temperate climate countries, may even be done in the open air; craftsmen may also be itinerant.

Range of handicrafts. There are over a thousand trades that may be practised as handicrafts. In western countries the work is usually of an artistic nature (goldsmithing, painting, sculpture, glass or pottery production, photography, etc.) or of a traditional utilitarian nature (flour milling, baking, laundering, joinery and cabinet-making, tailoring, dressmaking, knitting, shoemaking, hairdressing, bookbinding, glazing, etc.).

In the developing countries handicrafts cover a much wider field and include crafts related to folk-art (weaving, carpet-weaving, leather-working, pottery, wood, ivory and stone carving); utilitarian crafts (smithing, horseshoeing, masonry, carpentry, butchery, tanning, bootmaking, coopering, spinning, ropemaking, hammock and broom-making); crafts related to rural activities such as tobacco curing, oil extraction and nut-shelling; modern activities resulting from the country's economic and industrial development (automobile servicing and repair, panel beating, garage operation, vulcanising, radio fitting and repair, etc.).

HAZARDS AND THEIR PREVENTION

The hazards of specific types of handicraft work are dealt with in the separate articles covering the relevant occupation or trade. The prime feature of handicraft work is that the craftsmen have less social security than workers in large undertakings, if they have any at all, especially as far as occupational safety and health is concerned. Handicrafts workshops may be considered as small undertakings; however, they have certain intrinsic features that distinguish them.

In the organisation of a family-type business, the demands of production are usually placed above the need to arrange working hours, work breaks and holidays to prevent fatigue, and to lay out the working place to meet health and welfare requirements, and to ensure the safety of tools and materials. Consequently, a craftsman who has practised his trade for a number of years may present occupational stigmata and postural deformations not found in wage earners doing the same type of work. On the other hand, the craftsman's environment is particularly propitious to the safety training of workers. The relationships between the craftsman and his apprentices are those of teacher and pupil in vocational training or of a master passing on his artistic tradition to his disciples. The interest of a worker in a task which is often artistic, free of all monotony, and for which he is responsible in its entirety—from raw material to finished product—and the independence which he enjoys are all factors which motivate the worker and make him conscious of the intrinsic value of his task, and contribute appreciably to the development of safety consciousness.

Public authority activity. In some countries handicrafts workers enjoy a certain degree of legal protection. In France, for example, the Labour Code covers family undertakings, and a number of general provisions such as the prohibition of the use of certain harmful substances also apply to handicrafts workers. In Italy occupational accident and disease insurance also applies to handicrafts workers.

Handicrafts, especially in the developing countries, make a valuable contribution to the economy and to the improvement of conditions of life. The public authorities pay constant attention to the workers employed in this field and the ILO has on many occasions been called upon to provide technical assistance in the vocational training of craftsmen or in the organisation of craftsmen's co-operatives at a national or regional level. It is, however, essential that these public authorities ensure that, in the efforts made to increase productivity, sight is not lost of the need to protect the worker from accidents and diseases and to make safety and health instruction an integral part of vocational training. Craft unions and co-operatives can contribute to instructing craftsmen in the safety and health requirements of their trade.

Although efforts are made to bring small firms into the scope of national safety and health legislation, handicrafts businesses of the family type are still beyond reach. For example the Swiss Federal Labour Act of 13 March 1964 applies to persons employing even only a single worker, even on a temporary basis and not using special equipment or premises; however, it does not apply to family businesses, except in the case of certain provisions which can be made applicable by ordinance to the young persons belonging to the family of the head of the firm, in cases where it is necessary to protect their life or health or preserve their morality.

In several countries legislation applies to and labour inspection is competent for handicrafts; however, for a number of reasons handicrafts usually escape surveillance and labour legislation is only partially applied.

A step forward in the safety and health of craftsmen is the institution of sickness insurance and occupational accident and disease insurance. Occupational disability evaluation is particularly difficult in the case of handicrafts workers; whereas wage earners can be put on to other work after an accident or a disease, it is very difficult for the craftsman to change his trade and, consequently, it should be advisable here to evaluate specific disability rather than general disability.

PARMEGGIANI, L.

"Federal law on labour in industry, handicrafts and commerce (Labour Act 13.3.1964)" (Loi fédérale sur le travail dans l'industrie, l'artisanat et le commerce (loi sur le travail du 13.3.1964)). *Recueil de lois fédérales* (Berne), 24 Jan. 1966, 4 (85). (In French)

CIS 74-846 *Occupational safety and health in small industries and handicrafts* (Tehnika bezopasnosti i promsanitarija v mestnoj promyšlennosti). Moiseev, A. E.; Laptij, V. A. (Moscow, Izdatel'stvo "Legkaja industrija", 1972), 176 p. 5 ref. (In Russian)

See also SMALL UNDERTAKINGS.

Hand injuries

In relation to the wrist joint, the hand is the terminal anatomical segment, whereas, from the functional and surgical points of view, it must be considered as forming a whole with the forearm since its vascularisation and innervation, as well as its position and movements, are closely linked with the anatomy and physiology of the forearm; any injury to the latter will also have a more less serious effect on the functioning of the hand.

The hand comprises some 104 anatomical formations, all of which have a clearly defined role and contribute to the harmony of its movement. Its functions are prehension (involving strength and delicacy) and sensation (touch, temperature, pain, depth), in particular, stereognostic sense. For the hand to perform its functions properly, all its osteo-articular, tendinous, vascular and nervous structures must be intact, since an impairment of any one of these may have a deleterious effect on the hand's function.

Hand accidents

Because of its important role in everyday life and work, the hand is the part of the body most frequently injured in accidents.

Injuries to the hand may be closed or open. Closed injuries may affect the bones, joints and tendons and, less frequently, the nerves. Open injuries are, in principle, more serious because they are more frequent, may affect all parts of the hand and consequently may require more complex treatment.

The nature of hand injuries varies considerably depending on the type of traumatic agent and mechanism. In the home the most frequent causes of hand traumata are cuts and puncture wounds; cuts may involve only cutaneous tissue, or both cutaneous tissue and other formations, or may result in complete or partial amputation, and puncture wounds may be complicated by the inclusion of foreign bodies in the wound. These are followed by contused wounds (with or without injury to deeper parts of the hand), bites and tear wounds (in which rings, bracelets or, less frequently, the hand itself may be torn off by pointed objects, machinery, etc.).

Due to their frequency and their severity, hand burns, sometimes accompanied by burns to other parts of the body, form a special category of lesion.

In agriculture, accidents involving injury to the hand are relatively common, especially where primitive working methods are still employed. Cuts and puncture wounds are the most frequent lesions but bites from farm animals can prove very serious. The use of agricultural machines by persons lacking in experience of their operation has led to serious injuries to both the hand and other parts of the upper extremities. Traumatic exarticulations and stripping of skin and tissues, accompanied by the destruction of deeper tissues, are a frequent result of accidents on husking machines, etc.

In industry, hand traumata are the most common results of accidents and entail the largest number of days lost due to disability. The types of lesion, by common order of gravity, are as follows: cuts, puncture wounds, explosion injuries, exarticulations, crushing and grinding injuries (see also INJECTION, ACCIDENTAL).

The great variety of hand injuries makes it practically impossible to make an aetiological classification. Cuts and crushing wounds depend on the nature of the causal agent, the force with which it is applied and the position of the hand at the moment of injury. The same is true of injuries caused by such agents as pneumatic hammers, presses, etc. Extremely serious injuries may be caused when hands are trapped in gears, rollers, mixers, transmission belts, etc. Animal bites produce irregular lesions or may even amputate part or all of the hand.

The gravity of burns depends on the causative agent (heat, chemicals, electricity), the duration of exposure, the surface area affected and the depth of the wound.

Rings and bracelets may be torn from the hands by machinery or other agents, stripping skin and tissue from the finger or from the entire hand, and baring the deeper parts of the organ.

In this type of injury the degree of hand soiling at the moment of the accident is of prime importance. In general, wounds that occur in slaughterhouses, meat-processing plants and tanneries readily become infected, whereas in the metallurgical industry infection is less likely. Infection is always a possible complication.

For repeated microtraumata, see DUPUYTREN'S CONTRACTURE.

SAFETY AND HEALTH MEASURES

Safety training for workers and adequate machinery guarding are fundamental measures for the prevention of hand injuries. Personal hygiene is essential to ensure that hand injuries are not complicated by sepsis. In some cases hand and arm protection may provide a certain degree of safety. Careful disinfection is necessary for all hand injuries in order to prevent tetanus infection; systematic prophylactic tetanus immunisation is desirable.

Treatment. Due to their extreme variety, hand injuries pose complex problems of therapy.

Treatment should be directed towards maximal restoration of anatomical function. Satisfactory repair of bones, tendons or nerves can be achieved only if the tissues are in a good condition. Consequently, the first task of hand surgery is to restore damaged or lost tissue and modern surgical techniques make it possible to reconstruct amputated fingers and in some cases to achieve partial restoration of taction or even to replace the totally severed member.

It is clear that such problems are of a highly specialised nature, and as a result such work can only be undertaken by specialists who are particularly qualified in this field. At the same time the closest attention must be paid to emergency treatment, even in the case of what may appear to be uncomplicated injuries. A careful and exact diagnosis is essential so that all injuries to tendons or nerves can be identified and, similarly, sterilisation must be of the highest standard in order to eliminate all possible sources of infection.

Because of the functional value of the hand and the fact that surgery is to be aimed at preserving this value to the maximum extent possible, a special type of emergency treatment is applied, which consists, firstly, of cleansing and surgical clean-up of the wound, identification of serious injuries and, if possible, clearing them up. Secondly, if there is no infection or necrosis (which is usually the case if the first part of the treatment has been properly carried out), the wound should be closed up by whatever methods may be necessary (straightforward suture or stitches).

Advances in modern technology have led to the development of a wide range of plain and powered prostheses and to the design of a whole series of special tools for use by handicapped workers in different professions.

IONESCO, A.

General:

"The hand" (La main). *Travail et sécurité* (Paris), Feb.-Mar. 1980, special issue, 184 p. Illus. (In French)

Injuries:

"Aetiological and clinical remarks on some mutilating injuries of the hand and their prevention" (Considérations étiologiques et cliniques sur quelques lésions traumatiques invalidantes de la main et leur prévention). Narakas, A. *Sozial- und Praeventiv-Medizin* (Zurich), Feb. 1982, 27/1 (43-46). (In French)

CIS 647-1973 "Industrial finger injuries". Goulston, E. *Medical Journal of Australia* (Glebe, N.S.W.), 2 Sep. 1972, 59/2 (530-532). 2 ref.

CIS 78-2007 "Hand and body injuries: Prime targets in the war on accidents". Nemec, M. M. *Occupational Hazards* (Cleveland), Mar. 1978, 40/3 (37-41). Illus.

Micro-surgery, a socio-economic need (La microchirurgie, une nécessité socio-économique) Merle-Deroide. Cahiers des Comités de Prévention du Bâtiment et des Travaux Publics (Boulogne-Billancourt, France), Mar.-Apr. 1980, 2 (89-99). Illus.

"Upper limbs reimplantation" (Réimplantation de membres supérieurs). Le Viet, D.; Mitz, V.; Vilain, R. *Concours médical* (Paris), 6 Feb. 1982, 104/6 (720-732). Illus. 23 ref. (In French)

Hand tools, ergonomic design of

From an ergonomic viewpoint certain basic requirements for the design of an efficient hand tool can be defined, namely:

(a) it should effectively perform its intended function;

(b) it should be properly proportioned to the dimensions of the user;

(c) it should be appropriate to the strength and endurance of the user;

(d) it should minimise user fatigue;

(e) it should provide sensory feedback.

Role of anthropometry and biomechanics

Anthropometry is the science of human measurement; biomechanics, in this context, is concerned with the analysis of human motion. Each has application to tool design, not only with respect of compatibility with human dimensions but also with respect to use. Thus for example the hand cannot be effectively operated in extreme positions; also, a limb will develop fatigue when held away from the body, a situation that is aggravated by the additional requirement to support the weight of a heavy tool.

It must be recognised also that human dimensions vary from person to person and from time to time; thus a measure on one person at one time cannot be used to characterise all measures. Fortunately the average of a large number can generally serve the purpose in establishing dimensions for tool design. It is better practice, however, to ensure that specifications for a design for human use should accommodate 95% of the population, that is, excluding the top and bottom 2.5%.

Certain specific factors bear special examination:

Grasp. The prehensile capacity of the human hand can be largely defined in terms of a power grasp and a precision grasp (figure 1). Occasionally a hook grasp may be used.

In a power grasp the object is held in a clamp with counter pressure applied by the thumb, as in holding a heavy hammer. In a precision grasp the object is pinched between the flexor aspects of the fingers and the opposing thumb, as in tapping with a light hammer. The two varieties of grip are not mutually exclusive. One may yield

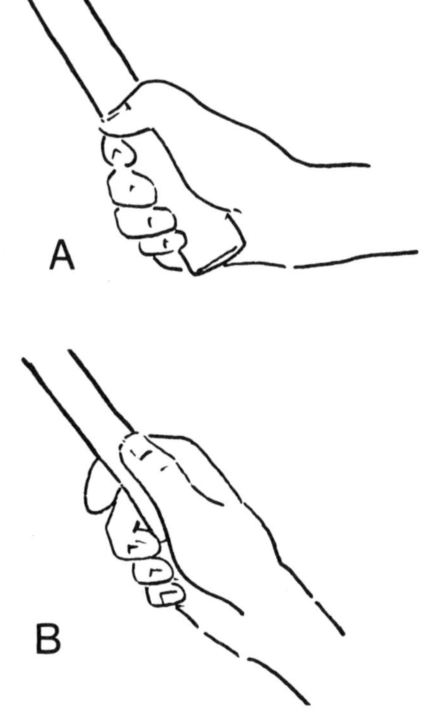

Figure 1. A. Power grasp. B. Precision grasp.

to the other in the course of an activity. The design of a tool may have to meet the needs of both. The hook grasp is used where something heavy has to be carried suspended from the flexed fingers, with or without the support of the thumb.

Handedness. For single-handed activity about 90% of persons prefer the right hand. A few are ambidextrous and all can learn to function with either hand. Thus, wherever reasonable, tools should be designed for use equally by left- or right-handed persons.

Hand strength. Hand strength is significant in the operation of manual tools. In the US population, for which data are available, hand strength ranges from 41.9-59.8 kg (male) and 24.5-35.0 kg (female).

Sex. The foregoing numbers indicate a sex difference, significant in tool design. Upper extremity strength measurements for women average 60% of those for men. In addition many common tools, such as soldering irons, metal shears, hammers, and planes have been described by women as being unsuitably designed for their use.

Clothing. Tools are not always operated with bare hands. Woollen or leather gloves add 5 mm to hand thickness and 8 mm to hand breadth at the thumb. Heavy mittens add more.

Design criteria

A tool comprises a head and a handle with sometimes a shaft or a body. The handle, however, is the focus of human interaction.

Shape of handle. The shape should conform to the natural holding position of the hand. Broadly, it should correspond to the form of a cylinder, or a truncated, inverted cone, or occasionally a section of a sphere.

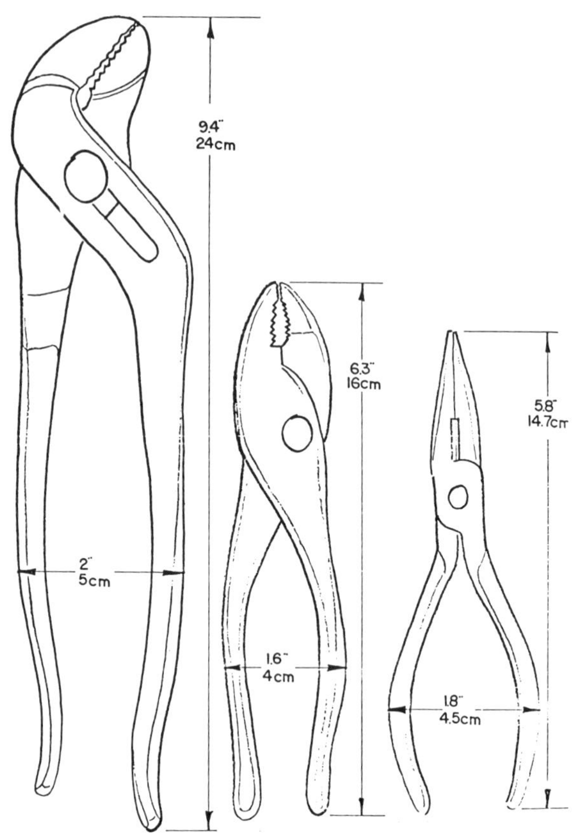

Figure 3. Pliers, shears, etc.; representative dimensions.

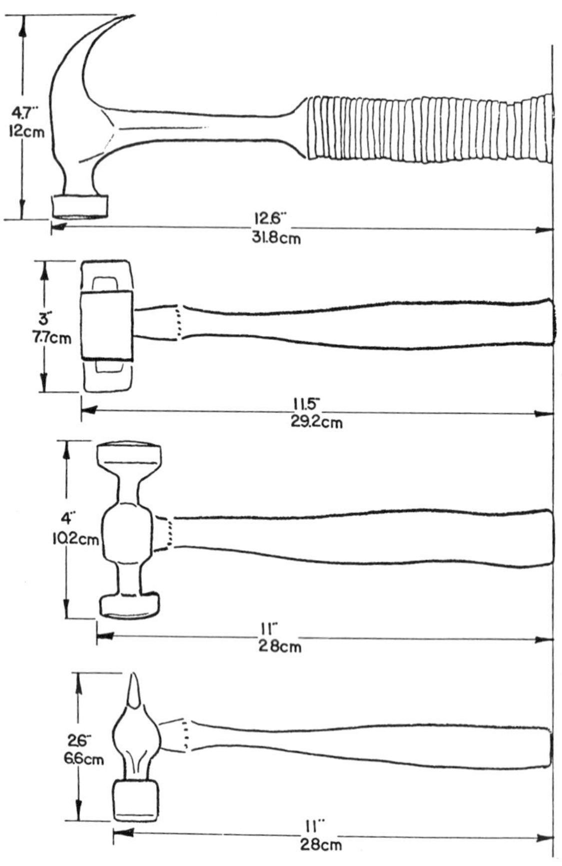

Figure 2. Hammers, representative dimensions.

Because of its attachment to the body of a tool it may also take the form of a stirrup, T-shape, or L-shape.

The space enclosed by the grasp is, of course, not cylindrical but complex. Exact matching, however, is undesirable since this would grossly limit its usefulness. In fact any form of specific shaping, such as ridges and valleys for fingers, is undesirable. Shapes should be basic, sectors of spheres, flattened cylinders, long contoured curves, and flat planes. A hexagon section can also be of value, paticularly for small calibre handles, where it improves the stability of the grip. Contact surface between handle and hand should be kept large enough to minimise high-compression stresses.

Diameter of handle. The handle diameter for a power grip should range between 25 and 40 mm. Above 50 mm the capacity to exert grip force diminishes. For a precision grip diameters of less than 6 mm tend to cut into the hand and do not give sufficient control. For a hook grip a diameter of 20 mm is recommended.

Length of handle. In some tools, for example a hammer, the handle merges into the body. In others the length should accommodate the dimensions of the hand. The handle of a power tool, or a saw, should accommodate the maximum width of the closed grasp at the 97.5th percentile, approximately 100 mm, while a heavy screwdriver must accommodate the length from the palm to the flexed knuckle of the forefinger, also approximately 100 mm.

Angulation of handles. Angulation of handles, for example in power tools or single-handed shears, should not only reflect the axis of grasp (i.e. about 78° from the horizontal) but also be so oriented that the eventual axis of use is in line with the extended index finger.

Texture. Texture is not merely aesthetic, it is also functional. A tool handle requires a readily identifiable texture to provide an input to the sensory nervous system to assist in maintaining the grip. Dull roughening is superior to flutings and ridges. All exterior edges of a tool which are not part of the functional operation, and which meet at an angle of 135° or less, should be rounded with a radius of at least 1 mm.

Specific criteria

Hammer. The hammer is normally used in a power grip but may require a precision grip for light work. A straight cylindrical handle of 25-40 mm calibre is generally effective, with mean maximum length of 60 cm and mean maximum head weight of 6.5-7.5 kg. Representative dimensions are shown in figure 2. Wrapped metal handles are common, but wood (ash or hickory) is superior. The handle should be lightly contoured with enlargement at the head of the shaft. The same principle applies to axes, adzes, mallets, and mauls.

Screwdrivers. A common effective shape for handles of screwdrivers, files, scrapers and hand chisels is that of a modified cylinder, dome-shaped at the end to receive the palm. A recent successful design has a triangular section. Effective calibres range between 1 and 5 cm.

Pliers. The general specifications for pliers, wire strippers, pincers, nippers, single-handed cutters, shears, and scissors are shown in figure 3. The shape of the handles at rest should approximate the position of grasp, with a working hand width of 90 mm for men and 80 mm for women. The length of handle without angulation should be 110 mm for men and 100 mm for women. Angulation of the handle (figure 4), with thickening of the legs, provides a more comfortable tool.

Hand saws and power tools. Curiously, heavy hand saws and power tools require essentially the same form of "pistol grip" handle. For the saw the handle is frequently closed (figure 5), whereas the power tool commonly uses an open handle (figure 6). In either case requirements are met by providing a handle of standard calibre, lightly contoured to the volume enclosed by the grasp, and angled at approximately 78° with the horizontal.

With respect to power tools, placement of the handle is significant. The handle and the head should be so placed with respect to the body that the line of action is the imaginary line of the extended index finger. The head may thus be eccentric with respect to the central axis of the body.

For fret saws, light hacksaws, and light hand drills a handle comparable to that of a screwdriver meets the requirements.

FRASER, T. M.

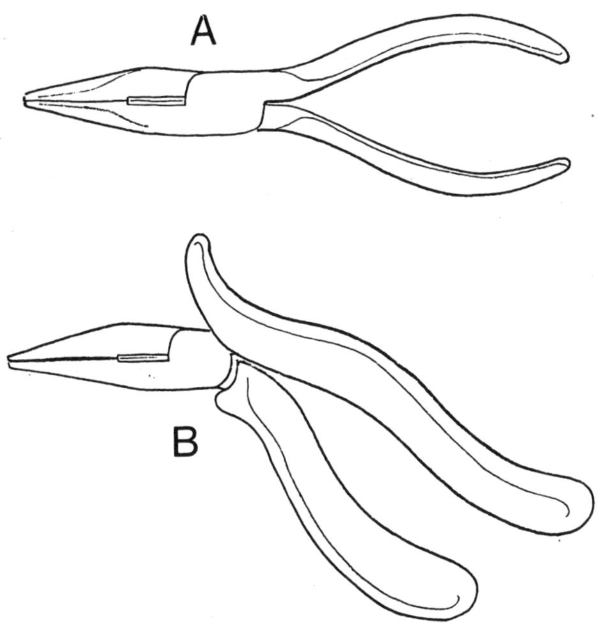

Figure 4. A. Standard pliers. B. Ergonomically designed pliers.

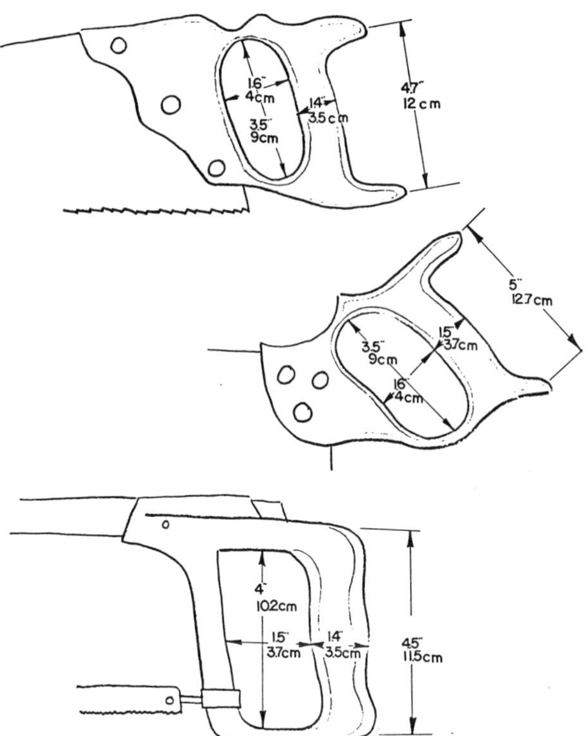

Figure 5. Saw, representative dimensions.

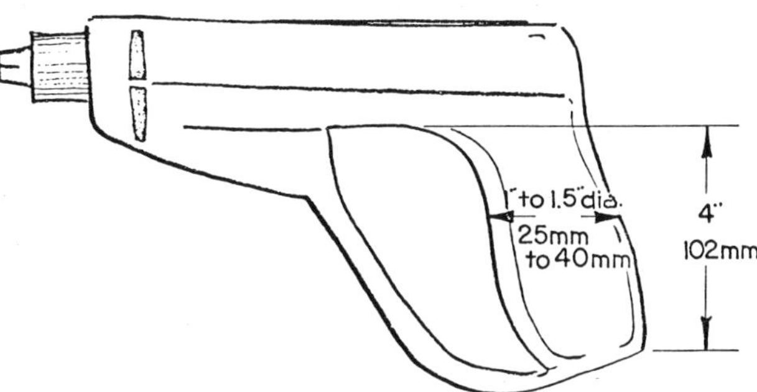

Figure 6. Pistol grip handle, representative dimensions.

CIS 81-1226 *Ergonomic principles in the design of hand tools.* Fraser, T. M. Occupational safety and health series No. 44 (Geneva, International Labour Office, 1980), 93 p. Illus. 84 ref.

CIS 79-272 *Workers and their tools—A guide to the ergonomic design of hand tools and small presses.* Greenberg, L.; Chaffin, D. B. (Midland, Michigan, Pendell Publishing Co., 1977), 155 p. Illus. 127 ref.

CIS 79-861 *Ergonomic principles basic to hand tool design.* Tichauer, E. R.; Gage, H. Ergonomics guides (American Industrial Hygiene Association, 66 South Miller Road, Akron, Ohio) (Nov. 1977), 13 p. Illus. 13 ref.

CIS 80-269 "Ergonomic design of tools and controls" (Auch Arbeitsmittel ergonomisch gestalten). Warnecke, H. J.; Kern, P.; Solf, J. *Die Tiefbau-Berufsgenossenschaft* (Munich), June 1979, 91/6 (448-451). Illus. (In German)

Hand tools, safety of

In many trades work is done with tools that rely on the hands for motive power to move, transform or adapt materials. Tools of this type can truly be called "hand tools". Other kinds of tools held in the hands, but driven by electricity, by mechanical means, pneumatically or by explosive charge, would perhaps be better designated as "portable power tools" and are dealt with in the articles BOLT GUNS; ELECTRIC POWER TOOLS, PORTABLE; PNEUMATIC TOOLS.

The designation "hand tools" covers a wide range of appliances, including such items as picks, shovels, axes, knives, wrenches, crowbars, saws, chisels, hammers, spanners and many others that are generally known and used; there are also tools designed for special purposes.

It might be thought that powered tools only were serious accident producers, but hand tools are responsible for a very high proportion of accidents among their users. Among the factors that lead to the accidents are:

(a) failure to keep the tools in a completely serviceable condition;

(b) use of an incorrect tool for the work being done;

(c) carelessness and wrong usage; and

(d) bad storage of the tools.

General design and construction

Economically it is unsound, and it contributes to accident causes, if cheap tools, made from unsuitable materials and of bad workmanship are used. Hammers made from poor quality steel may chip or shatter when a blow is being struck. Cheap chisels, punches or drifts will mushroom and quickly become unserviceable; poor-quality knives will soon lose their edge. Wrenches and spanners of unsuitable material and poor manufacture will open out under normal working pressure, probably causing hand injuries at least, as well as damage to the work through slipping.

It does not follow, however, that hand tools should be of the finest materials that it is possible to obtain; this also may not be economically sound practice. Tools must be made of materials best suited to the use for which they will be required. Hand hammers, for example, require special steel with a controlled grain size. When the steel is ordered the suppliers should be informed that it is for the manufacture of hand hammers.

Irrespective of the fact that the material selected is correct for the manufacture of a certain type of tool, the final product will not be satisfactory unless it has been correctly processed. Impact tools require to be hard enough to withstand repeated blows, but not so hard that they will easily splinter or chip in use. They must not, however, be such that they mushroom quickly in use. Much of the suitable processing is accomplished by specific hardening and tempering that leaves the tool body soft enough, so that it will not fracture, the cutting or working edges being hard enough to stand up to the work without quickly losing their usefulness and without chipping or splitting.

Handles fitted to tools must be able to stand up to the type of work for which they are designed; this applies especially to wooden handles, which should be straight-grained and usually of ash or hickory. The handles must be of smooth finish, and properly seasoned. The fitting of wooden handles is the job of a skilled workman, so that the handles will be properly and securely fixed; this will reduce the possibility of accidents and also increase the efficiency of those using the tools.

Special operations will possibly call for special tools, certainly where work is being done in explosive atmospheres or near flammable liquids, where a spark generated by ordinary steel tools could possibly result in an explosion or fire. Tools for these conditions would probably be of non-ferrous metal, possibly of beryllium-copper alloy.

There are available special hammers for work on hard steel, case-hardened tools or where the item or surface being struck must be protected from chipping, splintering or damage. For this work a hammer with a non-ferrous metal head, such as copper or lead, could be used. There are hammers with striking faces of specially prepared rawhide, plastic or rubber, the type of work on which they are to be used dictating the most suitable striking face; "soft hammers", as these tools are called, are designed to act as shock absorbers while the force of the blow is transmitted, in order that damage will not be sustained by the tool or material that is being struck. In some cases, drifts of non-ferrous metal are used between a normal hammer head blow and the tool or working surface, instead of a "soft hammer" being used.

Screwdrivers

Screwdrivers are possibly the most frequently used type of hand tool and very often the most misused. Many accidents are caused through the misuse of screwdrivers as levers, chisels, reamers, and on other work for which they are not designed. In addition, screwdrivers are very often used by unskilled or semi-skilled workmen who do not keep the tools in a satisfactory, safe working condition. Through lack of training or carelessness, these workmen use the wrong size or type of screwdriver for the work being done. A frequent cause of accidents is holding the workpiece in the palm of the hand while tightening up screws; a slip can result in a serious injury through the screwdriver penetrating the hand or wrist. The piece being worked upon should not be held in one hand and the screwdriver used in the other: work should be secured in a vice or held on other firm support.

A screwdriver tip must be properly ground to fit the slot in the screw head, and it must be the correct size for the head. Handles must be sound and smooth; they must not be hammered as this will mushroom or split the handle. Pliers or grips should not be used on the shanks of screwdrivers unless these are of a type designed to withstand the strain.

Screwdrivers must not be used for work other than that for which they are designed. For electrical work, screwdrivers with insulated handles should be used, but these must not be taken as affording absolute protection against electric shock. Screws should not be overtightened, as slipping can occur. Where screws are of a special type with serrated or cruciform heads, screwdrivers must be of a kind to mate with the particular design and size of slot. Screwdrivers should not be carried in pockets.

Metalworking chisels

The mushroom heads of chisels can often produce flying chips or splinters which can be very dangerous, especially to the eyes. Chisel heads should be ground when mushrooming starts to develop. A slight taper ground round the periphery of the head helps to reduce the tendency toward mushrooming.

Eye protection must be worn at all times when chisels are being used; other men working nearby should also wear protection against any flying particles or suitable screens should be provided for their protection.

Cutting edges of chisels must always be kept sharp so that the original shape and angle of the cutting edges is maintained. The resharpened chisels should be suitably hardened and tempered.

The right size chisel must be selected big enough for the job in hand. The hammer used should be heavy enough for the chisel being employed. The chisel should be held in the fingers, with a steady but relaxed grip. Vision should be focused on the cutting edge of the chisel rather than on the chisel head, the depth of cut being controlled by the angle at which the chisel is held.

Small workpieces to be chipped should always be firmly clamped in a vice before work is commenced on them with a chisel; chipping being done towards the solid or stationary end of the vice. Work with a chisel should always be done in a direction away from the worker.

If sledgehammers are being used to strike chisels, tongs must be used to hold the chisels. Workers holding the tongs should be in a position at right angles to the striker with the sledgehammer. Both workers should wear eye protection and screens should be provided to shield other workers or passers-by from flying particles.

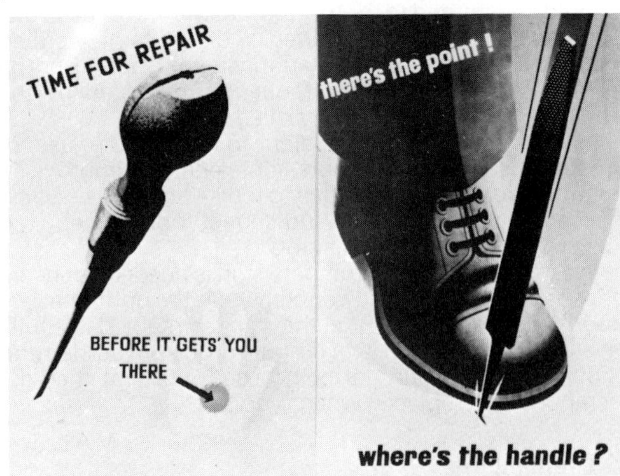

Figure 1. Safety posters for tool handling.

Files and rasps

When files or rasps are being used they must be of a suitable size and shape for the work; the teeth size and cut must also be satisfactory. Files should never be used without a handle, for the sharp pointed tang can be driven into the hand. The handle selected must have a ferrule and should be of a proper size with the hole correct for the file tang. The tang should be inserted into the hole in the handle, which is then tapped on a hard surface to drive the tang securely into the handle. The file must not be driven into the handle with a hammer or other hard object as this may break the file or split the handle. If the tang of a file is particularly pointed it should be rounded off slightly, in order to prevent hand wounds when handling the file.

Files should not be used as punches or drifts as the highly tempered metal is too brittle; for that reason also, files should not be struck with a hammer or other hard object, as the file will undoubtedly shatter and pieces of the metal will fly. Files must not be used as prying bars or levers; shattering of the metal will result.

When the teeth of a file become clogged they should be cleared by the use of a file card. Hard objects wear the teeth smooth and soft objects clog the teeth, in both cases the smoothness of the file may cause it to slip, resulting in damaged hands or perhaps more serious injuries. Small objects being filed should be held in a hand vice or bench vice. Files must not be twisted in slotted work as the files may break.

Files should not be carried in the pocket, most especially if a handle is not on the tang. When files are stored, each one should be wrapped in a piece of cloth or paper which must be kept dry to avoid rusting. Files must be kept clean and free from oil or grease.

Edged tools

Some hand tools require a sharp cutting edge to ensure efficiency and the safety of the user; many of these are wood-cutting tools, such as axes, draw-knives, adzes, wood-chisels and planes; knives are used in many trades and processes. If tools of this description are not maintained with a good cutting edge, there is a greater tendency towards accidents owing to tools slipping, possibly because greater effort is required to make the cutting action. The handles of knives require particular attention; they should be firmly attached and, where likely to become greasy as in the meat trades, shields between the blade and the handle or specially designed handles may be necessary.

Many accidents are occasioned by the misuse of sharp edged tools and by failure in guarding the cutting edge adequately when the tools are stowed away or when the tools are being carried about. When they are not in use these tools should always be put away in special racks or the edges guarded by proper protective sheaths or cases. Where an unguarded sharp edge tool has to be carried about in the performance of a job, the greatest care must be exercised.

Even in the sharpening of hand tools, great care must be taken to perform the task correctly, in order to avoid accidents. If grinding wheels are being used they must be correctly guarded and eye protection must be worn. When hand steels or stones are being used they should have a suitable guard to prevent the tool slipping and cutting the hand. Protection can also be obtained by the use of hand guards made of suitable material, possibly fibre, leather or metal. Oil-stones should be placed on a flat stable surface and they should not be used dry. An oil-stone should be kept clean, for in use a tool can slip on a dirty stone, causing cut hands or fingers.

Spanners and wrenches

The use of spanners or wrenches of the wrong size or with strained jaws can be the cause of accidents. Spanners and wrenches should be of the right size for the job and well maintained: if a spanner or wrench is too large or is worn it may slip and be the first cause of a series of events leading to a serious injury. Tube extensions should never be fitted to spanners or wrenches: their use may mean excessive leverage which may result in cracks to the jaws of the tool.

Hand tools used at machines

The use of hand tools in conjunction with power-driven machinery should be strictly controlled. In particular, hand-held files and scrapers are often wrongly used in lathe work when the operation could more safety be performed by a tool mounted in the tool post. Serious injuries have been caused in this way especially when the hand tools used were without handles.

Training and supervision

The potential hazards that exist in the use of hand tools have been mentioned only briefly and the examples quoted for certain implements are by no means the sum total of dangers that exist.

With all types of hand tools there is possible danger unless training is given in their use or unless they are used under skilled supervision. A skilled man will usually have his own tool kit and take pride in the maintenance of his tools. Where hand tools are issued to unskilled men, they should be returned to the tool store at the end of each day so that they can be examined and any fault rectified before reissue. Failing this arrangement, tools used by unskilled men should be frequently inspected by the supervisors, who should be skilled men.

Supervision should be alert to detect the use of unauthorised and unsuitable implements or improvised modifications and extensions to tools. It may be revealed that additional authorised and correct tools are required to carry out some particular job.

Adequate training and supervision is needed to ensure that hand tools are used correctly, that the right ones are used for specific work, that they are maintained in a fully serviceable condition, that they are properly guarded and stowed safely when not being used and that they are scrapped and replaced when worn.

MAY, J.

Hand tools (Handwerkzeuge). Merkblatt 1012/1967 Sicher Arbeiten (Lucerne, Schweizerische Unfallversicherungsanstalt, 1967), 12 p. Illus. (In German; published also in French and Italian)

"Regulations respecting the design, construction, operation, use and maintenance of hand tools and portable power tools in federal works, undertakings and businesses (Canada Hand Tools Regulations)". Department of Labour. *Canada Gazette*, Part II (Ottawa), 8 Dec. 1971, 105/23 (1 981-1 984).

"Portable tools. Part I: Manual tools". Lahey, J. *National Safety News* (Chicago), July 1981, 124/1 (31-36). Illus.

Hardeners

A hardener or curing agent can be a pure chemical, a mixture, or complex polymer which, when added to an appropriate macromolecular substance in greater than catalytic amounts, will react with it and thereby convert it irreversibly into a polymer possessing well defined characteristics not otherwise achievable. [The typical change is the transformation of a linear prepolymer into a tridimensional macromolecule following the setting up of bridges between the chains.]

The four systems of industrial importance that use hardeners and that will be considered below are: epoxy resin systems; unsaturated polyester systems; isocyanates; and vulcanised rubber.

Epoxy resin systems

By far the largest group of epoxy resins used in industry is that made by condensation of diphenylol propane

$$HO-C_6H_4-C(CH_3)_2-C_6H_4-CH \quad \text{and epichlorhydrin} \quad CH_2(O)CH-CH_2Cl$$

By varying the proportions of the reactants, a range of resins is obtained, varying in molecular weight, which can be either liquids or solids. The idealised formula of such resins would be:

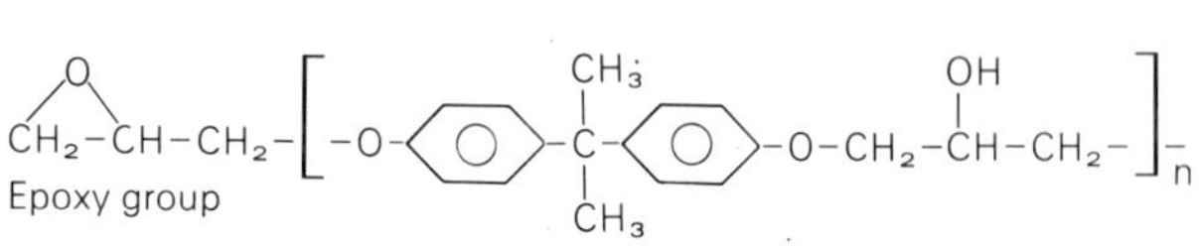

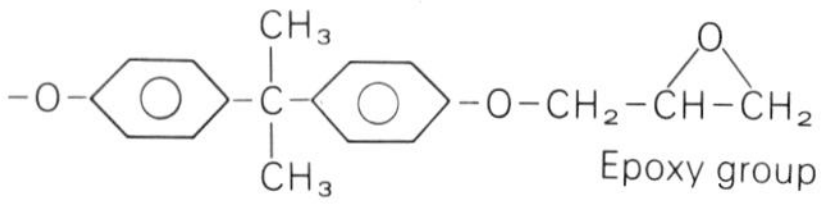

$$\underbrace{CH_2(O)CH}_{\text{Epoxy group}}-CH_2-\left[-O-C_6H_4-C(CH_3)_2-C_6H_4-O-CH_2-CH(OH)-CH_2-\right]_n$$

$$-O-C_6H_4-C(CH_3)_2-C_6H_4-O-CH_2-\underbrace{CH(O)CH_2}_{\text{Epoxy group}}$$

where the repeating unit "n" ranges from 0 to 12. These resins by themselves do not exhibit any worthwhile technological properties until they are effectively hardened by chemical cross-linking reactions. The principal mechanisms of hardening reactions are summarised below:

(a) polymerisation through the epoxy groups; this reaction is facilitated by the catalytic action of tertiary amines, i.e. compounds which do not possess reactive hydrogen;

(b) addition reaction with primary amines (RNH_2)–in which one epoxy group reacts with one reactive aminohydrogen atom at a time;

(c) esterification with fatty acids (mono-carboxylic acids)–which is principally an addition and condensation reaction;

(d) reaction with acid anhydrides (e.g. phthalic anhydride)–whereby the hydroxyl group of the resin reacts with the CO group of the anhydride;

(e) reaction with phenol formaldehyde (resole) resins–in which the phenolic hydroxyl and methylol groups of the phenolic resin react with epoxy groups;

(f) reaction with amino resins (urea formaldehyde, melamine formaldehyde)–in which methylol ($-CH_2OH$) or butylated methylol groups react with epoxy groups and with resin–OH (hydroxyl) groups as in *(e)*, together with a primary and secondary amine (RNH_2 and $-NH-$) reaction as under *(b)*;

(g) reaction of the hydroxyl groups with isocyanates–in which the –OH group of the resin reacts with the N:C:O isocyanate group.

All reactions with epoxy groups are exothermic; *(a)*, *(b)* and *(g)* can take place at ambient temperatures whereas the others require the application of heat to effect chemical cross-linking. The products of reaction *(c)*–epoxy esters–are useful resins in their own right and require curing (by oxidative air drying) or by chemical cross-linking (condensation) with amino resins (urea formaldehyde, melamine formaldehyde) at elevated temperatures. The exothermic nature of the reactions increase the hazard from the volatile components.

Common epoxy resin hardeners

These hardeners may be grouped as follows: amines; amine adducts; polyamides; aromatic and cycloaliphatic anhydrides; miscellaneous; and formaldehyde resins.

Amines. These may be: primary, secondary or tertiary aliphatic amines; aromatic polyamines; or cycloaliphatic amines.

In general, the liquid amines react at room temperature and are all irritant to skin and mucous membranes. Primary and secondary amines are more irritant than tertiary amines, and aliphatic more than the aromatic. The aliphatic polyamines are the most commonly used room-temperature hardeners and the following are of importance: ethylene diamine (EDA); diethylene triamine (DTA); triethylene tetramine (TETA).

All these are volatile liquids and the liquids will irritate exposed body surfaces and may even cause a corrosive-type skin burn with prolonged contact, or corneal damage from eye splashes. The vapours, similarly, will irritate the skin (particularly the face and eyelids) and mucous membranes (conjunctivae and upper respiratory passages) if exposure is prolonged. They are all skin sensitisers and occasionally, though rarely, cause asthma.

The following polyamines are examples of solids or powders at room temperature: diamino diphenyl methane (DDM); metaphenylene diamine (MPD); dicyandiamide (DICY). They all require applied heat to react. The members of this group are less irritating than the liquid amines and have low vapour pressures but, as will all amines, they may be absorbed through the skin and are toxic on ingestion. Some of the aromatic amines may cause methaemoglobinaemia or carcinoma of the renal tract on absorption into the body. DDM contamination of flour caused an outbreak of hepatitis in England ("Epping Jaundice") in 1965. MPD is a skin sensitiser and may cause allergic bronchial asthma.

Amine adducts. These are partially reacted mixtures of resins with excess of amine; they have the advantage of diluting the irritancy potential of the mixture and are of minimal volatility.

Polyamides. These compounds act similarly to aliphatic polyamines but have a low vapour pressure, react at room temperature, and are less irritating to the skin and mucous membranes.

Aromatic and cycloaliphatic anhydrides. These require elevated temperature for reaction. The commonest agent is phthalic anhydride–a solid which sublimes into an irritant fume on heating. It is thought to have caused asthma and digestive disturbances, but these may have been due to impurities. Irritancy of the anhydrides in the dry state is low but is increased by moisture or heat, and acid burns to the skin and eyes may result.

Miscellaneous. Boron trifluoride complexes are liquids or solids, require heat for successful reaction and, at elevated temperatures, give off the irritant boron trifluoride fume. The amine salts have less irritating and less sensitising properties than the amines. Others in this group include ketimines, modified amines and imidazoles.

Formaldehyde resins. These include the amino resins (urea and melamine formaldehyde), phenolic resins (phenol formaldehyde), and their reactions have been discussed above. Amino resins are always in solution (butyl alcohol, aromatic hydrocarbon mixtures). Butyl alcohol has a pungent irritating vapour (see FOUNDRIES).

Applications

Ambient temperature hardeners include the aliphatic primary polyamines, amine adducts, modified amines, and polyamide resins, and are used in two-component, high-performance, chemical-resistant anti-corrosion paints, adhesives, stoppers, surfacing compositions, repair compounds, and in laminating by wet lay-up.

The elevated-temperature hardeners include DDM, 4,4'-sulphonyl dianiline, DICY, boron trifluoride complexes, cycloaliphatic and aromatic amines and acid anhydrides, and are used in electrical and mechanical laminates, special adhesives, powder coatings, electrical casting and impregnating compounds. Amino resins and phenolic resins are used in industrial, high-performance stoving primers, lacquers and finishing enamels.

HAZARDS

In general, the hardeners themselves are not primarily flammable or explosive substances. However, those that are handled in solvent mixture may be flammable depending on the composition of such solvents.

As indicated above, all the hardeners should be regarded as irritant to any body surface with which they may come in contact, either as liquids or, when in a volatile state, as vapours. Many are sensitisers, that is after a period of use or, more commonly, after one or more episodes of physical irritation (e.g. dermatitis), certain individuals become hypersensitive. In this event, with minimal further contact with liquid hardener or the vapour, an acute skin reaction may result extending further than the area of contamination and often including the face and eyelids. Some persons may develop asthma from spray droplets, or vapour. There is no definite evidence of any other toxic action from industrial-type exposures.

When investigating the cause of a contact dermatitis or eczema in a person working with a resin system, it must be remembered that the system used often contains other irritant chemical components. Thus, apart from the uncured resin and the hardener, there may be viscosity reducers (such as phenylglycidyl ether), solvents, fillers, modifying resins (such as formaldehyde resins) and accelerators; these substances may themselves cause similar conditions. In addition, cleansers (e.g. methyl ethyl ketone), excessive sweating from wearing impervious gloves, or a sensitivity to a chemical in the rubber of a glove may also be the cause of dermatitis of the hands. When only the uncured resin and the hardener are involved, it is more likely to be the hardener which is the cause of the trouble.

Uncured resins are of low or negligible volatility, and irritancy diminishes as viscosity increases. Nevertheless, some dermatologists feel that the resin is usually the irritant or sensitiser, and one series of studies showed that out of 96 cases, 78 showed positive patch tests to the epoxy resins used.

SAFETY AND HEALTH MEASURES

Since all the various components used in the mixing and application of epoxy resin systems are potential irritants, safety and health measures should be directed at the process as a whole and not purely at the hardener. Consequently, it will be necessary to instal local exhaust systems and general ventilation, provide workers with hand and arm protection and eye and face protection; respiratory protective equipment may be necessary where resin systems are applied in confined spaces, etc.

In addition to these general measures, certain specific precautions are necessary. Regular supervision, precise training on safe practices, shop maintenance and good housekeeping and "no-touch" techniques are of paramount importance throughout all stages of the process.

Plant design. Lay-out should be such as to allow systematised procedures, and the maintenance of cleanliness in the work area. Liquid hardeners should be dispensed by gravity flow direct into mixing pots when batch systems are used. The plant areas should, if possible, be kept physically separate from other work areas, so as to limit the numbers of workpeople exposed to an essential minimum. In the case of volatile substances, exhaust ventilation should be provided during weighing, mixing and reaction processes, and particular regard to this must be held when applied heat is required at the reaction and cure stages. Ovens used for curing should be exhausted to the outside of the building.

Fitness for employment. Persons with a history of allergic conditions (with the possible exception of hay fever) or of recurrent skin conditions of any type should not be employed on work involving the use of hardeners. The evidence that fair-skinned, fair-haired individuals are more susceptible to skin irritation than their darker colleagues is slight. On the positive side, it is advisable to choose neat, hygienically careful, methodical persons for such work.

Personal hygiene. Smoking, eating and drinking must be forbidden in work areas, and hands must be washed before these activities and before use of toilets. Washing and sanitary facilities (hot and cold water, soap and paper towels) should be provided nearby and used immediately following skin contamination by a hardener or any other chemical or reaction product up to the stage of cure.

Treatment. Any person who contracts any type of skin condition, or bronchitis (including asthmatic symptoms) should temporarily be removed from work with resin systems. Even if the cause is not occupational, he is "at risk" during the illness and should not resume work until recovered. When it is established or suspected that a contact (or primary) dermatitis is due to a hardener, treatment should be by a medical practitioner. The rash usually responds quickly to corticosteroid creams. At least 2 weeks, ideally 6 weeks, should elapse after recovery before the patient is allowed to resume work, but with the provison that any "breakthrough" in the precautions in his particular case are first remedied. If a further attack occurs, then it is advisable that the man be transferred to other work. If a man becomes sensitised (i.e. has an allergic type eczema or asthma), he must be permanently removed from hardeners and resin systems.

Unsaturated polyester systems

Unsaturated polyesters are usually hardened (or cured) by the addition of the co-reactant solvent styrene in the presence of a catalyst, resulting in a hard insoluble thermoset solid. The commonly used catalysts which may be looked upon as the hardeners in the reaction complex are organic peroxides of general type R–O–O–R'.

Common hardeners for unsaturated polyesters

Organic peroxides. These comprise a large group which is dealt with in a separate article (see PEROXIDES, ORGANIC); consequently, only brief mention will be made here. Benzoyl peroxide is a representative of one subgroup. It requires either the application of heat to initiate the thermosetting reaction in the system, or the addition of an accelerator (metallic compounds, tertiary amines) to cure at room temperature. Although non-volatile it is explosive with accelerators and is irritating to the skin and eyes. Skin sensitisation may occur. Mixed peroxides (or hydroperoxides), are another important subgroup, and two common ones, of which the commercial forms are mixtures, are cyclohexanone peroxide (HCH) and methyl ethyl ketone peroxide (MEKP). They react at room temperature. HCH is normally supplied as a paste in dimethyl phthalate, and MEKP as a solution in dimethyl phthalate.

Allyl compounds. The commonest used is triallyl cyanurate which, though of low volatility, may irritate the skin and eyes. It is not known whether it can be absorbed through the skin.

Applications

Hardening may be carried out by automatic processes such as machine casting, by semi-automatic methods as in extrusion or spraying, or by non-automatic means such as brush painting of lay-up moulding, and the possibility of skin contamination increases in this order.

SAFETY AND HEALTH MEASURES

The organic peroxides constitute about only 2% of the resin system so their irritancy potential is much reduced during the reactions. Nevertheless, as they are very reactive, face shields should be worn during open mixing and pouring. Non-pressure or gravity eye-wash bottles or eye-wash fountains should be near at hand and all operators should be instructed in their use.

These hardeners, in general, are not as irritant or as sensitising as the epoxy hardeners. However, as styrene–a volatile and toxic compound–is an integral component of the thermosetting reaction, the same safety and health measures as for the epoxy hardeners are applicable.

Isocyanates

These compounds are dealt with in separate articles (see ISOCYANATES; FOAM RESINS) and only their use as hardeners will be discussed here. The nitrogen atom of the isocyanate group (N:C:O) combines with a reactive hydrogen from a resin to effect cross-linking at the carbon atom position of the isocyanate. The resins used industrially are the polyols and polyesters, and the products so formed find application as paints, polyurethane joining compounds and foams. The isocyanates may also be used in certain epoxy resin systems.

Vulcanisation of rubber

Vulcanisation is a cross-linking of rubber unsaturated polymer molecules by sulphur or by sulphur-containing compounds. By this method, plastic raw rubber is hardened (or cured) into a resilient, tough, vulcanised product. The sulphur-containing agents used include aryldithiols, alkylene dithiols, amine disulphides, tetra-alkyl thiuram disulphides and aryl dinitroso compounds. The newer rubbers, for instance ethylene propylene rubber, are hardened by means of peroxides (e.g. dicumyl peroxide).

The hazards and safety measures relevant to vulcanisation as a whole are dealt with in the article RUBBER INDUSTRY, NATURAL.

BROUGHTON, W. E.

"Epoxy resins and hardeners" (Epoxidharze und Härter). Ebeling, U. *Sicherheit* (Leipzig), Feb. 1971, 17/2 (31-35). Illus. (In German)

CIS 75-1388 "Allergic pathology of epoxy resins" (La patologia allergica da resine epossidiche). Nava, C.; Marchissio, M.; Briatico-Vangosa, G.; Arbosti, G. *Securitas* (Rome), 1974, 59/7 (469-486). Illus. 18 ref. (In Italian)

CIS 78-1654 "Occupational allergic contact dermatitis to isophorone diamine (IPD) used as an epoxy resin hardener". Lachapelle, J. M.; Tennstedt, D.; Dumont-Fruytier, M. *Contact Dermatitis* (Copenhagen), 1978, 4/2 (109-112). 6 ref.

"Recently reported causes of contact dermatitis due to synthetic resins and hardeners". Malten, K. E. *Contact Dermatitis* (Copenhagen), Jan. 1979, 5/1 (11-23). Illus. 47 ref.

CIS 77-1942 *Hazards of thermal decomposition of polyurethane paints and their polyfunctional hardeners* (Gefahren bei der Hitzezersetzung von Polyurethanlacken und deren polyfunktionellen Härtern). Seeman, J.; Wölcke, U. Forschungsbericht Nr. 152 (Dortmund-Marten, Bundesanstalt für Arbeitsschutz und Unfallforschung, 1975), 59 p. Illus. 24 ref. (In German)

CIS 78-13 "A severe accident in polyester concrete production" (Schwerer Unfall bei der Herstellung von Polyesterbeton). Gehrke, H. *Zentralblatt für Arbeitsmedizin, Arbeitsschutz und Prophylaxe* (Heidelberg), Jan. 1977, 27/1 (7-10). Illus. (In German)

Hazardous areas, electrical apparatus and wiring in

Many gases, vapours, liquids and dusts produced or used in industry are flammable. Fuels, solvents, coal dusts and dusts of organic materials such as flour and sugar can be ignited when mixed with air in suitable concentrations, and the mixtures will burn readily, often with considerable explosive force. Since the possible sources of ignition include sparks and hot surfaces created by electrical apparatus, strict control must be exercised on the use of such apparatus where flammable materials may be present. Many standards and codes of practice have therefore been produced on this subject. Certain specific activities present very special problems, and are usually covered by separate, and often statutory, requirements. Coal mining is one example, and although the principles of apparatus design are similar to those for above-ground use, the standard area classification techniques are not used and special installation requirements apply, which vary from country to country. The handling of explosives is also a special application and, again, direct reference has to be made to national regulations.

Explosions of flammable dusts can be disastrous, but although some national standards exist for control of electrical installations where these hazards exist, there are as yet no international standards and codes. The international area classification principles applied to gas and vapour hazards cannot be used directly as these take into account dispersal of the flammable materials after release, by natural or artificial ventilation, whereas dusts settle out. The only distinction is usually as to whether an ignitable dust cloud can be expected in normal operation or if the normal hazard is the slow accumulation of layers of dust.

As with all locations where flammable atmospheres can arise, electrical apparatus should be installed where dust is a hazard only when it is impractical to do otherwise. The principal precautions when it is so installed are *(a)* to use apparatus whose surface temperatures, when blanketed with dust, are below the values required to ignite the dust and *(b)* to prevent dust entry by the use of gaskets, seals and similar techniques. As far as sealing is concerned, apparatus is described as dust-proof or dust-tight according to the extent to which dust entry is prevented. Appropriate tests are given in the International Electrotechnical Commission (IEC) Publication 529, which describes both the tests and a system of markings to indicate the degree of protection against dust entry.

In the United States apparatus which has been satisfactorily tested for both temperature rise and dust entry is described as dust-ignition-proof, but there is no internationally agreed designation.

Flammable vapours and gases

Contrary to the situation for dust hazards, there are international electrical standards dealing with the precautions to be taken against the risk of ignition of flammable gases and the vapours derived from flammable liquids. There are also national standards which generally follow the same principles but which are not completely aligned.

A "hazardous area" (apart from its more general meaning) is given a specific definition for electrical engineering purposes by the IEC in its Publication 79-10 (1972): "An area in which explosive gas-mixtures are, or may be expected to be, present in quantities such as to require special precautions for the construction and use of electrical apparatus."

Not all such areas are equally hazardous. The vapour space in a fixed-roof storage tank may contain an explosive mixture continuously, depending on the flash point of the stored material in relation to the storage conditions. On the other hand, in some open-air installations an explosive mixture may only occur for a brief time in the event of occasional accidental leakage. In order to reduce the necessity for using the most highly protected electrical apparatus in the less dangerous area, the IEC has sub-divided hazardous areas into three zones:

Zone 0: in which an explosive gas-air mixture is continuously present or is present for long periods.

Zone 1: in which an explosive gas-air mixture is likely to occur in normal operation.

Zone 2: in which an explosive gas-air mixture is not likely to occur, and if it occurs it will only exist for a short time.

Thus the storage tank and the open-air installation quoted above are examples of Zone 0 and Zone 2 respectively.

As the first objective in any hazardous installation must be to minimise the likelihood of creating an explosive atmosphere, not only because of the risk of explosion but also because of the toxicity of many flammable substances, Zone 0 areas are usually a few well defined enclosed spaces, and Zone 1 areas are of limited extent compared with the much larger Zone 2 areas. Factors which influence classification as Zone 1 or Zone 2 areas, and their extents, include whether the source of release is a normal (primary grade) or an abnormal (secondary grade) source, the degree of natural or artificial ventilation present, the quantities of flammable materials that might be released and their special characteristics such as flash point and vapour density, the last being particularly important if there are below-ground trenches or enclosed roof spaces.

It has not yet been found possible to produce detailed guidance for the classification of hazardous areas and the determination of their extents because of the wide range of types and sizes of occupancies. There are codes of practice for some occupancies in which installations do not vary greatly, such as oil refineries, and for very specific industrial processes. References to some of these may be found in IEC Publication 79-10, but such codes must be used very carefully and with consideration of all the different circumstances if they are used for situations other than those for which they were originally intended. National codes may also differ as, for example, in the United States, where Zones 0 and 1 are combined as Division 1, and Zone 2 becomes Division 2.

Electrical apparatus

Appropriate precautions must be taken in designing and selecting the electrical apparatus which is installed in the above hazardous areas. The principles adopted are largely the same, irrespective of the premises or occupancy concerned, whether this be a large oil refinery or a roadside filling station for cars, a chemical plant or a small paint-spraying area in a factory.

First, electrical apparatus should be placed in a non-hazardous area whenever practicable, and otherwise in the least hazardous area possible. If placed in a hazardous area it must be specially protected, installed in a safe manner and maintained in a safe condition.

International standards for the design of various types of explosion-protected electrical apparatus have been produced by IEC, as listed at the end of this article. These standards are based on the requirements that surface temperatures must be below the values which can ignite the flammable atmospheres and that electrical sparks are very unlikely to be produced or, if produced, will not ignite the atmosphere concerned.

In Zone 0 it is rarely necessary to instal electrical apparatus except for instrumentation such as tank level gauges. Only "intrinsically safe" apparatus is normally permitted in Zone 0. This apparatus is designed so that any sparking which may occur normally, or with up to two simultaneous faults present, is incapable of igniting the explosive atmosphere, as demonstrated in ignition tests in which a specially designed test apparatus is used to produce sparks in the circuits. Intrinsically safe apparatus is necessarily of low power and the type of protection is used principally for instrumentation.

A second category of intrinsically safe apparatus, in which only one fault is taken into account instead of two, is permitted in Zone 1, together with several other types of protection. The electrical apparatus, whether normally sparking or not, can be mounted in a "flameproof enclosure", known in the United States and some other countries as an explosion-proof enclosure. Such an enclosure is designed to withstand the pressure of an explosion of a flammable atmosphere inside it without transmitting the explosion, through any joints in its walls, to a surrounding flammable atmosphere. Thus, the enclosure has to be strong enough to withstand pressures up to 8-10 bar, and the dimensions of joints must comply with specified limits of width and gap. Alternatively, if apparatus does not produce sparks in normal operation, it can be made adequately safe for use in Zone 1 if special measures are taken to reduce the likelihood of sparks due to electrical or mechanical faults. This is achieved by specifying a high standard of insulation, a lower than normal operating temperature to increase insulation life, specific requirements for mechanical construction and, for motors, careful attention to overcurrent protection. Such apparatus is described as "increased safety" or "type of protection e" apparatus.

Individual items of electrical apparatus in Zone 1, or complete rooms, may be "pressurised". An overpressure of air or, in apparatus, an inert gas is used to prevent any external atmosphere from reaching sparking or potentially sparking parts. The enclosed apparatus can be standard industrial apparatus in some situations but precautions in the form of interlocks or alarms are necessary in case of failure of the pressurisation system. Other less commonly used types of protection are "oil-immersion" and "sand (or powder) filling", in which those components liable to spark are covered by an adequate depth of oil or sand, respectively, to prevent ignition of a flammable atmosphere.

Any of the above types of protection may, of course, be used in Zone 2 but lesser precautions are permitted in many countries, even though international standards for apparatus of the so-called "type of protection n" are still under discussion. Usually, standard good quality non-sparking industrial apparatus is used but limited special requirements are sometimes applied. Unprotected sparking components are allowed only if they are incapable of causing ignition in normal operation.

Temperature classification

In selecting suitable electrical apparatus the nature of the potentially hazardous atmospheres must be considered. The temperature at which these can be ignited by hot surfaces varies widely and it would be highly restrictive to limit all equipment to a surface temperature which is safe for all possible atmospheres. Therefore all explosion-protected apparatus is marked with a Temperature Classification to indicate the maximum temperature of any surface to which the gas has access, but only where ignition could be a hazard, for example surfaces inside a flameproof enclosure are not considered. The IEC marking for these Temperature Classes consists of the symbols T1, T2 ... T6, for maximum surface temperatures from 450 °C down to 85 °C, but other systems are in use.

Apparatus groups

The minimum ignition energies and currents and the gaps through which explosions may be transmitted also vary with the nature of the flammable atmosphere. It would again be impractical to design intrinsically safe and flameproof apparatus, whose safety depends on these factors, to be suitable for all atmospheres so a system of apparatus groups is adopted, with IEC designations of Group I (for firedamp in mines), Group IIA (for many hydrocarbons and solvents), IIB (for ethylene and a few other compounds) and IIC (originally for hydrogen but now extended to include acetylene and carbon disulphide).

Apparatus identification

For ease of selection and identification in use, the IEC has developed a standard system of marking in which all explosion-protected apparatus carries the mark "Ex". It also has a letter symbol indicating the type of protection as follows:

Intrinsic safety ia (for use in Zone 0)
Intrinsic safety ib (for use in Zone 1)
Flameproof d
Increased safety e
Pressurised p
Oil-immersed o
Sand-filled q

Thus, a typical complete marking, including apparatus group and temperature class, would be Ex ia IIB T3.

Installation and maintenance

The marking of apparatus also assists in ensuring its maintenance in a safe condition after installation. In addition to normal maintenance, it is essential that the special safety features of explosion-protected apparatus are retained in service, and the marking "d", for example, identifies equipment whose safe joints have to be correctly maintained. Particular care is needed for identification of intrinsically safe apparatus, as its use for instrumentation systems often results in some part of an intrinsically safe system being located in a non-hazardous area where it might not be realised that maintenance operations could be affecting circuits in a hazardous area.

Special precautions also have to be taken in installing apparatus, in addition to compliance with normal installation practice: earthing and bonding of metalwork is specially important, to avoid the occurrence of sparking due to unequal potentials arising from stray or fault currents. Cables have to be mechanically protected, by conduit, armour or similarly effective means and those for intrinsically safe systems have to be segregated from power cables to avoid the induction of unsafe voltages and currents. IEC is currently preparing international guidance for these and other installation rules, but pending their publication national rules have to be carefully followed.

IEC publications

The IEC publications relating to electrical apparatus for hazardous areas are listed below. However, many other standards exist, published nationally or by international organisations outside IEC, many of them for specific occupancies or applications. A few examples of the marked differences which exist between these and the IEC documents have been mentioned, and until international rules have been fully adopted, national regulations will specify the standards and codes of practice with which compliance is required.

RIDDLESTONE, H. G.

IEC Publication 79. *Electrical apparatus for explosive gas atmospheres.*

79-0. Part 0: *General introduction* (1971), 9 p.

79-1. Part 1: *Construction and test of flameproof enclosures of electrical apparatus* (1971), 52 p. Amendment (1979), 12 p.

79-1A. *First Supplement: Appendix D: Method of test for ascertainment of maximum experimental safe gap* (1975), 14 p.

79-2. Part 2: *Pressurized enclosures* (1975), 18 p.

79-3. Part 3: *Spark test apparatus for intrinsically safe circuits* (1972), 21 p.

79-4. Part 4: *Method of test for ignition temperature* (1975), 19 p.

79-4A. *First supplement* (1970), 5 p.

79-5. Part 5: *Sand-filled apparatus* (1967), 36 p.

79-6. Part 6: *Oil-immersed apparatus* (1968), 13 p.

79-7. Part 7: *Construction and test of electrical apparatus, type of protection "e"* (1969), 39 p.

79-8. Part 8: *Classification of maximum surface temperatures* (1969), 9 p.

79-9. Part 9: *Marking* (1970), 15 p.

79-10. Part 10: *Classification of hazardous areas* (1972), 13 p.

79-11. Part 11: *Construction and test of intrinsically safe and associated apparatus* (1976), 39 p.

79-12. Part 12: *Classification of mixtures of gases or vapours with air according to their maximum experimental safe gaps and minimum igniting currents* (1978), 18 p.

Head protection

Head injuries

Head injuries are fairly common in industry and account for some 10% of all industrial injuries. They are often severe and result in an average loss of work of about 3 weeks. The injuries sustained are generally the result of blows caused by angular objects such as tools, bolts, etc., falling from a height of several metres; in other cases they may result from falls of persons hitting the ground or some fixed object with their heads.

A number of different types of injury may result:

(a) perforation of the skull, which results from excessive force being applied over a very localised area, as for example in the case of direct contact with a pointed or sharp-edged object;

(b) fracture of the skull or of the cervical vertebrae occurs when excessive force is applied over a larger area, stressing the cranium beyond the limits of elasticity or compressing the cervical vertebrae;

(c) brain lesions without fracture of the skull are the result of the brain being displaced suddenly within the skull, which may lead to contusion, concussion, haemorrhage of the brain or circulatory problems.

There is general agreement that such lesions depend on the rate of acceleration imparted to the head.

An understanding of the physical parameters that are responsible for these various types of injury is difficult, although of fundamental importance, and there is considerable disagreement among the extensive literature which exists on this subject. Some specialists consider that the force involved is the principal factor to be considered, while others claim that it is the energy, or the amount of movement that takes place; further opinions relate the injury to acceleration, to rate of acceleration, or to the degree of shock sustained. In practice, it is probable that in most cases each one of these factors is involved to a greater or a lesser degree. It may be concluded that knowledge of the shock to which the head may be subjected is only partial and is controversial. The degree of shock at which damage to the head is sustained is determined by means of experimentation on cadavers or on animals, and it is not easy to extrapolate these values to living man.

On the basis of the results of analyses of accidents sustained by building workers wearing safety helmets, however, it would seem that these injuries occur when the amount of energy involved in the shock is in excess of about 100 J.

Other types of injury are less frequent but should not be overlooked. They include burns resulting from splashes of hot or corrosive liquids, molten material, or accidental contact by the head with bare live electrical conductors.

Safety helmets

The purpose of a safety helmet is to protect the head of the wearer against shocks. It may also provide additional protection against certain other mechanical and electrical hazards.

A safety helmet should fulfil the following requirements in order to reduce the destructive effect of shocks to the head:

(1) It should limit the pressure imposed on the skull by spreading the load over the greatest possible surface. This is achieved by providing a sufficiently large lining that matches closely various shapes of skull, together with a hard shell strong enough to prevent the head from coming into direct contact with objects falling accidentally and to provide protection in case of collision with a hard surface. The shell must therefore resist deformation and perforation and the material used should therefore be chosen with these objects in mind.

(2) It should deflect falling objects by having a suitably smooth and rounded shape. Helmets with protruding ridges tend to arrest falling objects rather than to deflect them and thus retain slightly more kinetic energy than helmets that are perfectly smooth.

(3) It should dissipate and disperse energy that may be transmitted to it in such a way that the energy is not passed totally to the head and neck. This is achieved by means of the harness lining which must be securely fixed to the hard shell so that it can absorb a shock without becoming detached. It must also be flexible enough to undergo deformation without touching the inside surface of the shell. It is this deformation which absorbs most of the energy of a shock and it is limited by the amount of free space between the hard shell and the skull and by the stretch imparted to the harness webbing before it breaks. Thus the rigidity or stiffness of the harness should be a compromise between the maximum amount of energy that it is hoped to absorb and the progressive rate at which the shock is to be allowed to be transmitted to the head.

There may be other requirements of a helmet in the case of certain special jobs. These include protection against splashing of molten metal in the iron and steel industry and protection against electric shock by direct contact for helmets worn by electricians and maintenance staff.

Materials used in the manufacture of helmets and harnesses should retain their protective qualities over a long period of time and under all foreseeable climatic conditions including sun, rain, heat, freezing, etc. Helmets should also be fairly fire-resistant and should not break when allowed to drop on to a hard surface from a height of a few metres.

Performance tests

ISO International Standard No. 3873-1977 was published in 1977 as a result of the work of the subcommittee dealing with "industrial safety helmets". This standard was approved by practically all the member States, and sets out the essential features required of a safety helmet together with the related testing methods. These tests may be divided into two groups (see table 1); namely–

(a) obligatory tests, to be applied to all types of helmets for whatever use they may be intended: shock-absorbing capacity, resistance to perforation, resistance to flame; and

(b) additional optional tests, intended to be applied to safety helmets designed for special groups of users: resistance to splashing with molten metal, dielectric strength, resistance to lateral deformation.

Additional requirements under the first group include the resistance to ageing of the plastics materials used in the manufacture of helmets. A simple test consists of exposing the helmets to a high-pressure, quartz-envelope 450-W xenon lamp over a period of 400 h at a distance of 15 cm, followed by a check to ensure that the helmet can still withstand the perforation test.

Among the optional tests, it is recommended that helmets intended for use in the iron and steel industry be subjected to a test for resistance to molten metal splash. A quick way of carrying out this test is to allow 300 g of molten metal at 1 400 °C to drop on to the top of a helmet and to check that none has passed through to the interior.

Choice of a safety helmet

The ideal helmet providing protection and perfect comfort for every situation has yet to be designed. Thus, in the choice of a helmet, an element of compromise cannot be avoided, and this must take into account the different characteristics of helmets, the hazards against which protection is required, and the conditions under which the helmet will be used.

General considerations. It is advisable to choose helmets that conform to the recommendations of ISO Standard No. 3873 (or its equivalent) and that meet the following requirements:

(1) A good safety helmet for general use should have a strong outer shell resistant to deformation or puncture (in the case of plastics, not less than 2 mm thickness), a webbing harness fixed in such a way as to ensure that there is always a free space of 40 to 50 mm between its upper side and the shell, and an adjustable headband fitted to the inner lining to ensure a close fit and stability of the helmet. Helmets with the harness only attached to the outer shell by a lace are to be avoided as they offer a false protection against falling objects (figure 1).

(2) The best protection against perforation is provided by helmets made from thermoplastic materials (polycarbonates, ABS, polyethylene, polycarbonate-glass fibre) and fitted with a good harness. Helmets made from light metal alloys do not stand up well to puncture by sharp or angular objects.

(3) Helmets with projections inside the shell should not be used, as these may cause serious injuries in the

Table 1. Safety helmets: testing requirements of ISO Standard 3873-1977

Characteristic	Description	Criteria
Obligatory tests		
Absorption of shocks	A hemispherical mass of 5 kg is allowed to fall from a height of 1 m and the force transmitted by the helmet to fixed false (dummy) head is measured The test is repeated on a helmet at temperatures of −10 °C, + 50 °C and under wet conditions	The maximum force measured should not exceed 500 daN
Resistance to penetration	The helmet is struck within a zone of 100 mm in diameter on its uppermost point using a conical punch weighing 3 kg and a tip angle of 60° Test to be performed under the conditions which gave the worst results in the shock test	The tip of the punch must not come into contact with the false (dummy) head
Resistance to flame	The helmet is exposed for 10 s to a Bunsen burner flame of 10 mm in diameter using propane	The outer shell should not continue to burn more than 5 s after it has been withdrawn from the flame
Optional tests		
Dielectric strength	The helmet is filled with a solution of NaCl and is itself immersed in a bath of the same solution. The electric leakage under an applied voltage of 1 200 V, 50 Hz is measured	The leakage current should not be greater than 1.2 mA
Lateral rigidity	The helmet is placed sideways between two parallel plates and subjected to a compressive pressure of 430 N	The deformation under load should not exceed 40 mm, and the permanent deformation should not be more than 15 mm
Low-temperature test	The helmet is subjected to the shock and penetration tests at a temperature of −20 °C	The helmet must fulfil the foregoing requirements for these two tests

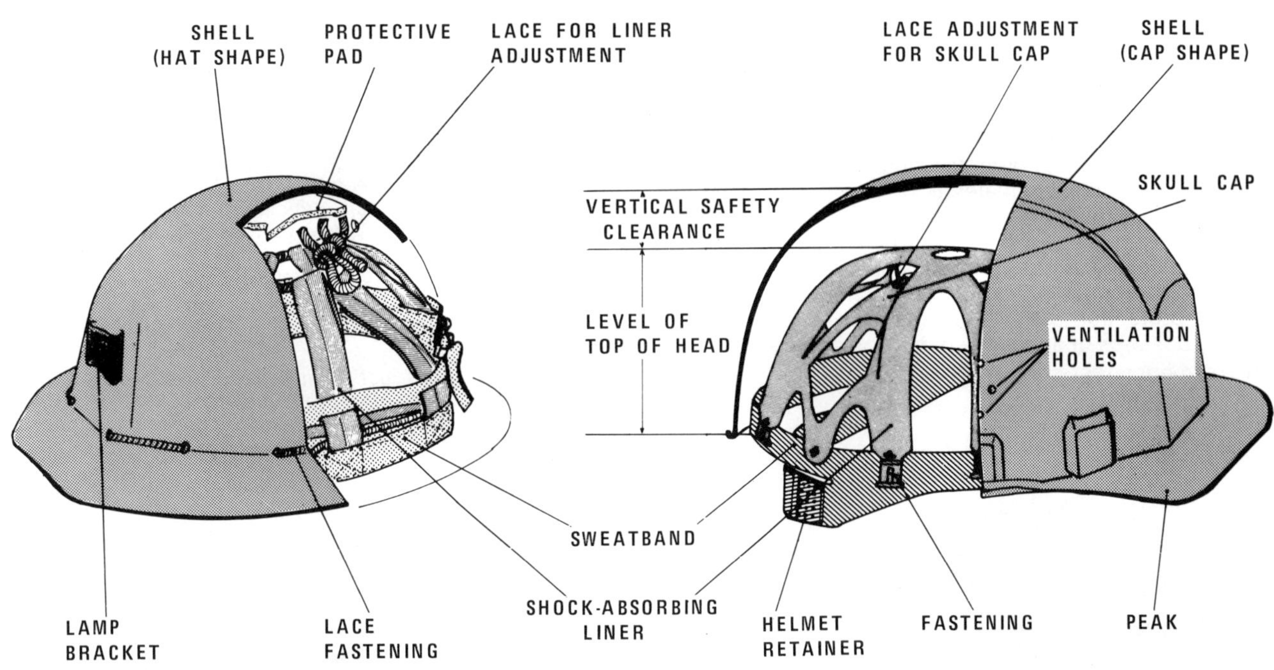

Figure 1. General details for safety helmets.

case of a sideways blow; they should be fitted with a lateral shock-absorbing band; this must not be flammable nor should it melt under the effects of heat. A padding made of fairly rigid foam, 10 to 15 mm thick and at least 4 cm wide will serve this purpose.

(4) Helmets made of polyethylene, polypropylene or ABS have a tendency to lose their mechanical strength under the action of weather (heat, cold and particularly strong sunlight). If such helmets are regularly used in the open air they should be systematically replaced at least after 3 years' use. In countries where strong sunlight is encountered it is recommended that polycarbonate, polyester or polycarbonate-glass fibre helmets be used, as these have a greater resistance to ageing. In every case any evidence of discoloration, cracks, shredding of fibres, or of creaking when subjected to twisting, should cause the helmet to be discarded.

(5) Any helmet that has been subjected to a heavy blow, even if there are no evident signs of damage, should *be discarded*.

Particular considerations. Helmets made of light alloys or having a brim along the sides should not be used in any workplace where there is a danger of splashing from molten metal. In such cases helmets of polyester-glass fibre, phenol textile, polycarbonate-glass fibre or polycarbonate are to be recommended.

In cases where there is a danger of contact with low tension electricity, only helmets made of thermoplastic material should be used. They should not have ventilation holes and no metal parts such as rivets should be apparent on the outside of the shell.

Helmets for persons working overhead, and particularly in the case of steel frame work erectors, should be provided with chinstraps. The straps should be about 20 mm in width and should be such that the helmet is held firmly in place at all times.

Helmets made largely of polyethylene are not recommended for use in high temperatures. In such cases helmets of polycarbonate, polycarbonate-glass fibre, phenol textile, or polyester-glass fibre are preferable. The harness should be of a woven fabric. Where there is no risk of electrical contact, ventilation holes in the helmet shell may be provided.

Situations where there is a danger of crushing call for helmets made of glass fibre reinforced polyester or polycarbonate and having a rim with a width of not less than 15 mm.

Comfort considerations. In addition to the safety aspect, consideration should also be given to the physiological aspects of comfort for the wearer.

The helmet should be as light as possible and certainly not more than 500 g in weight. Its harness should be flexible and permeable to liquid and should not irritate or injure the wearer; for this reason harnesses of woven fabric are to be preferred to those made of polyethylene. A full or half sweatband of leather should be incorporated in order to provide absorbency and reduce skin irritation and this should be replaced for hygienic reasons several times during the life of the helmet. Careful adjustment of the helmet to fit the wearer is important in order to ensure its stability and to prevent its slipping and interfering with his vision. Various shapes of helmet exist, the most common being the "cap" shape with a peak and a brim around the sides; for work in quarries and on demolition, the "hat" type of helmet with a wider brim provides greater protection. A "skull-cap" shaped helmet without a peak or a brim is particularly suitable for persons working on overhead erection as these avoid a possible loss of balance caused by the peak or brim coming into contact with joists or girders among which the workers may have to manoeuvre.

Accessories and other protective headgear

Helmets may be fitted with eye or face shields made of plastic material, metallic mesh or filter glass. Hearing protectors and attachments to keep the helmet firmly in position may also be fitted, together with woollen neck protectors or hoods against wind or cold (figure 2). For use in mines and underground quarries, an attachment for a headlamp and a cable holder are fitted.

Other types of protective headgear include those designed for protection against dirt, dust, scratches and bumps. Sometimes known as "bump caps", these are made of light plastic material or linen. For persons working near machine tools such as drills, lathes, spooling machines and suchlike, where there is a danger of the hair being caught up, linen caps with a net, peaked hair-nets or even scarves or turbans may be used. To provide protection against splashing of liquids, hoods, waterproof caps or sou'westers may be provided.

Figure 2. Main accessories for attachment to safety helmets.

Hygiene and maintenance

All protective headgear should be cleaned and checked regularly. If splits or cracks are apparent, or if a helmet shows signs of ageing or deterioration of the harness, it should be discarded. Cleaning and disinfection is particularly important if the wearers sweat abundantly or if more than one person is obliged to wear the same headgear.

Substances adhering to a helmet such as chalk, cement, glues or resins may be removed mechanically or with the aid of an appropriate solvent that does not attack the material from which the outer shell is made. Warm water with a detergent may be used with a hard brush.

To disinfect headgear, the article should be dipped in a suitable disinfecting solution such as 5% formalin or a solution of sodium hypochlorite.

MAYER, A.

Industrial safety helmets. International Standard ISO 3873 (Geneva, International Organisation for Standardisation, 1st ed., 1977), 8 p. Illus.

CIS 79-1716 *Head protection.* Data Sheet H-4 (Canada Safety Council, 1765 St. Laurent Blvd., Ottawa, Ontario) (1979), 30 p. Illus. 17 ref.

Industrial safety helmets. Result of tests, principal manufacturing details, criteria for selection (Casques de protection pour l'industrie: bilan des essais, principales données constructives pour le fabricant–Critères de choix pour l'utilisateur). Mayer, A.; Salsi, S.; Grosdemange, J. P. Les notes scientifiques et techniques, No. 14 (Paris, INRS, 1974), 102 p.

"Head protection is on top of worker safety". Dionne, E. D. *National Safety News* (Chicago), July 1979, 120/1 (41-47). Illus. 1 ref.

Safety helmets on construction sites. Recommendations of the Construction Industry Advisory Committee on their provision and use. Discussion Document (London, Health and Safety Commission, 1979), 12 p.

Health physics

Health physics, which is known in different countries under a variety of labels, covers a wide variety of activities that are concerned with improving methods of protection against exposure to ionising radiation. The people who work in this field are qualified experts who, in many countries nowadays, are recognised as such by the competent authorities and are directly involved in setting up and operating an effective system for protecting man and his environment from every kind of undesirable radiation. Certain nuclear plants have their own health physics departments whose size varies according to the degree of risk to which the workers may be exposed.

The main principles of health physics were defined in some detail in 1977 by the International Commission on Radiological Protection, whose recommendations have been widely followed in international or national rules and regulations.

Radiation protection is based on three general principles, justification, optimisation, and limitation of man's exposure to radiation.

First, any exposure to ionising radiation has to be justified by the potential benefits to man or society. Medical radiation and the use of nuclear power for the production of electricity are outstanding examples.

Second, the principle of optimisation means that any exposure must be kept as low as is readily achievable.

Finally, the limitation of exposure concerns the creation of a system of protection of the individual and the population such that the doses experienced or the possibility of radiation remain within the limits of the basic standards of radiation protection which represent the maximum admissible exposure. In addition to these standards other limits, often applicable to specific cases, have been devised that represent the practical limits to be observed in the effective organisation of radiation protection.

Health physicists. Health physicists work hand in hand with industrial physicians, hygiene experts, safety engineers and various administrative and health authorities in ensuring compliance with these three general principles of radiation protection: justification, optimisation, and limitation of exposure. Particularly close collaboration is required at the place of work with the doctors responsible for the medical supervision of the workers.

Qualifications. Radiation protection activities are conducted by qualified experts recognised by the competent authorities who have acquired the necessary know-how and training to carry out physical or technical tests or to act as consultants so as to ensure the effective protection of individuals and/or the proper operation of protective devices. Where the need arises, these experts must be capable of interpreting the rules and regulations and taking whatever measures are required for compliance with radiation standards.

They must be able to assess and interpret the doses to workers—particularly committed and effective individual doses—so as to make sure that they are kept within admissible limits and that the collective doses that apply under the optimisation procedures are respected.

Because of the difficulty of foreseeing and assessing the biological risk represented by the presence of radioactive substances in the ecosystem and in the food cycle, radiation protection experts are expected to be familiar with numerous branches of science. In the absence of absolute criteria for safety and protection, professional judgement based on experience and training is often essential.

Activities. Radiation experts are frequently involved in operations entailing a risk of radiation even *before* they begin. Their work is, in fact, essentially of a preventive nature. Projects for nuclear activities or nuclear plants responsible for ionising radiation must first come in for critical examination from the point of view of radiation protection. Before they can be authorised, new plants need to be controlled and the protective devices checked. Periodically, experts are called upon to verify that the measuring apparatus is in proper working order and is being correctly employed.

One of the more delicate problems that experts are generally asked to deal with is the assessment of the health risk involved in the various foreseeable types of accident. The kind of preventive measures that are adopted and the arrangements that are made for taking action in the event of an accident depend on the probability of their occurrence and on the technical repercussions and health hazards that may ensue.

In operational terms, worker protection is based on the classification of clearly defined working areas in accordance with the corresponding risk of exposure, the division of the workers into categories according to the likelihood or otherwise of their being exposed to a certain fraction of the maximum admissible dose, and the adoption of appropriate measures and control procedures for each of these areas and categories. The job of the expert is to determine what hazard is entailed in the work to be carried out and to mark off the working areas, with particular attention to the controlled areas. Depending on the nature and seriousness of the risk, he is responsible for keeping a constant check on radiation hazards in the atmosphere of the working areas, drawing up a programme for measuring activities, doses and sources of radiation, recording his findings, planning working procedures in terms of the radiation hazard, making sure that the sources of radiation in the place of work are properly marked, drawing attention to the hazards which the radiation sources inside the controlled areas represent and controlling access to these areas. In addition, the expert is responsible for examining and testing the protective apparatus and measuring instruments.

In any assessment of internal contamination, measurements of the working environment are particularly important, and one of the major difficulties of radiation protection continues to be the interpretation of the findings in terms of body burden and committed dose. This is a problem for the industrial physician who is responsible for determining the radiation status of the worker and for keeping a close eye on any developments. In this, the active collaboration of the radiation expert is essential.

A record is maintained of contamination or exposure measurements and the documents containing the assessment of the individual doses absorbed are kept, along with the worker's medical file, for at least 30 years after completion of the job on which he was exposed to ionising radiation.

Even when outlined in broad terms as they have been in this article, the list of activities involved in health physics leaves no doubt as to their importance. In every instance where there is a significant danger of radiation, a proper system of radiation protection must be set up by the management of the plant or by the competent authorities.

The radiation expert's job is not an easy or simple task. The need to see that safety regulations are applied and that they are adapted to new methods of work, and the preventive action that is required to avoid accidents and technical mishaps mean that he has to keep every place of work under constant observation and maintain a virtually uninterrupted flow of scientific information.

Health physics has evolved out of a considerable fund of accumulated theoretical and practical know-how. Because of the new uses that are being found for ionising radiation and the introduction of radioactive substances whose properties have not yet been fully explored, however, it would be a mistake to think that the risk of radiation has been overcome once and for all. Radiation protection must continue to be developed in conjunction with the progress being made in nuclear science, and the trend is in fact for its scope to be extended to the entire spectrum of ionising radiation.

RECHT, P.

Implications of Commission Recommendations that doses be kept as low as readily achievable. A report by Committee 4 of the International Commission on Radiological Protection. ICRP publication 22 (Oxford, New York, Toronto, Pergamon Press, 1973), 18 p.

Recommendations of the International Commission on Radiological Protection. ICRP publication 26. *Annals of the ICRP* (Oxford), 1977, 1/3, 53 p.

A large amount of relevant information on the subject is available in the publications of the ICRP, International Atomic Energy Agency (IAEA), and other international and national agencies.

Hearing protection

Industrial noise levels are increasing. Effective protection of workers against the harmful results of noise is therefore a problem of great actuality. Ear protectors prevent excessive sound energy from entering the external ear canal. This simple and inexpensive protection is used particularly in cases where technical control measures are unsuccessful in reducing noise to acceptable levels. The systematic and correct use of ear protectors can prevent hearing loss and other noise-induced health impairment. Individual ear protectors include earplugs, earmuffs and helmets.

Earplugs. These are inserted in the external ear canal and remain in position without any special fixing device. There are numerous types of plugs made of fibrous materials impregnated with oils or consisting of wax-like mastics, rubber and other polymers, or of solid cores covered with soft materials. The different shapes and dimensions of external ear canals render the use of earplugs difficult and make it necessary to manufacture them in various standard sizes. An exception are disposable one-time earplugs made of dense organic wadding which is frequently impregnated with paraffin, wax or other binders, and of recent date also plugs of organic materials which become plastic at skin temperature.

Earmuffs. These are almost hemispheric cups made of light alloys or plastics, and filled with fibrous or porous sound absorbents; to ensure a comfortable and tight fit around the ear they are provided with sealing rims of thin synthetic film often filled with air or liquids with high internal friction (glycerine, mineral oil). The sealing rim damps at the same time the vibration of the muff shell itself, which is important when the noise spectrum comprises low-frequency sound vibration. Earmuffs give better attenuation at high frequencies than at low ones, and the average attenuation they afford at frequencies below 1 000 Hz is generally poorer than with earplugs. Earmuffs may prove heavy and cumbersome, and they are more expensive than earplugs. They are held by headbands which press the muff shells against the ears and allow for adjustment to different head shapes and dimensions.

Hearing protective helmets. They are the most bulky and expensive form of personal hearing protection and are used, often in combination with earmuffs or earplugs, when noise levels are very high.

Ear protectors of whatever type should ensure efficient noise attenuation, and be comfortable and safe in use as well as aesthetic in style.

The efficiency of ear protectors is expressed by the degree of attenuation of the noise penetrating into the external ear canal. There are certain methods enabling this degree to be evaluated; they are based on threshold audiometry. The methods most frequently used are those of the absolute hearing-threshold shift, of threshold masking and of loudness balance. These psycho-physiological methods of human experimentation correspond more precisely to the actual attenuation afforded by an ear protector than purely physical methods which make use of acoustic probes, an artificial ear or an artificial head. The method of the absolute hearing-threshold shift, which is the most simple technique, is used more frequently than the other methods. Its efficiency is expressed by the difference between the threshold levels determined for unprotected and protected ears exposed to pure tones or noise bands.

Noise attenuation by ear protectors is limited by bone conduction for high frequencies and by skin resistance to low ones.

A distinction must be made between the requirements for individual ear protectors against powerful noise occurring for short periods during the work shift, against noise with clearly defined low- or high-frequency spectra and against broad-band industrial noise to which workers are exposed during the entire shift.

Protection against powerful noise can only be achieved with highly efficient ear protectors, i.e. helmets (which are now made of rigid materials) with built-in earmuffs tightly fitting to the head. The attenuation must attain 15-25 dB in the low and medium frequency bands and 40-45 dB in the high ones.

The requirements to be met by ear protectors for pronounced low- or high-frequency noise spectra differ from each other. While earmuffs for low-frequency noise must have large-volume and relatively heavy muff shells, those for high-frequency noise must be light with small shells filled with sound-absorbent material, well fitting around the ear with their sealing rims. These requirements show that it is difficult to design ear protectors against strong noise which can be comfortably worn for longer periods.

In the USSR noise attenuation specifications have been elaborated for the most widespread types of industrial broad-band noise to which workers are exposed during the entire shift, and maximum values have been established for the weight of earmuffs and the pressure they may exert on the head.

The attenuation specifications for ear protectors fall into the groups A, B and C for both earmuffs and earplugs (see table 1).

These specifications allow for existing types of ear protectors and should permit for wearing them in comfort for 4 h without interruption. Earplugs for repeated use should be available in several standard sizes to ensure good individual fit; their weight is not specified but should not exceed 10 g.

Ear protectors of any type should be selected according to the formula:

$$L_n - (L_{en} - \Delta L_n) \leqslant N_n$$

where:

- L_n is the sound pressure level in dB in the nth octave band measured at the workplace;
- L_{en} is the attenuation in dB afforded by the ear protector in the nth octave band according to the type of protector;
- ΔL_n is the correction factor of the protector reliability in dB (in the USSR this factor equals 8 dB for the octave bands up to 500 Hz and 10 dB for the bands of 1 000 Hz and higher);
- N_n is the permissible sound pressure level in dB in the nth octave band.

Ear protectors have been correctly chosen if the difference calculated exceeds the permissible level in none of the octave bands.

Table 1. Attenuation and comfort specifications for ear protectors in the USSR

Type of ear protector	Attenuation group	Attenuations (dB) corresponding to frequencies (Hz) of							Weight of muffs and head band (kg)	Earmuff pressure (kg) [1]
		12	250	500	1 000	2 000	4 000	8 000		
Earmuffs	A	12	15	20	25	30	35	35	< 0.5	< 0.8
	B	5	7	15	20	25	30	30	< 0.3	< 0.5
	C	–	–	5	15	20	25	25	< 0.2	< 0.4
Earplugs	A	10	12	15	17	25	30	30	–	–
	B	5	7	10	12	20	25	25	–	–
	C	5	5	5	7	15	20	20	–	–

[1] These pressure values if expressed in Pascal would correspond to < 800, < 500, < 400 Pa x 10^2.

Well designed, correctly selected ear protectors which are systematically worn prevent hearing impairment and health damage due to extra-auditory effects of noise. It is important to use ear protectors from the very first day of work in a noisy environment. A great deal of information work must be undertaken among workers to have them accepted.

ŠKARINOV, L. N.

"Effects of intense persistent noise on humans and animals" (Vlijanie na organizm čeloveka i ivotnyh intensivnyh i dlitel'nyh šumovyh vozdejvij). Suvorov, G. A.; Škarinov, L. N. Ch. 6 (240-291). *Adaptacija čeloveka k ěkstremal'nym uslovijam sredy* (Moscow, Nauka, 1979).

Individual hearing protectors. General technical rules (Sredstva individual'noj zaščity organa sluha. Obščie tekničeskie uslovija), GOST SSBT 12.4.051-78 (Moscow, Gosstandart, 1979).

CIS 80-698 "Guide for the design of hearing protectors for general industrial use". BOHS Technology Committee, Working Party on Hearing Protectors. *Annals of Occupational Hygiene* (Oxford), 1979, 22/3 (203-211). 11 ref.

CIS 79-1300 *Personal hearing protection* (La protection individuelle de l'ouïe). Schmuckli, F. CCST nº 130, Cahiers suisses de la sécurité du travail (Lucerne, Schweizerische Unfallversicherungsanstalt, 1979), 28 p. Illus. 24 ref. (In French, German, Italian)

CIS 79-1006 *Method for determining a suitable band of noise levels for the choice of personal hearing protection for use in noisy areas* (Methode zur Bestimmung eines geeigneten Pegelbereiches zur Auswahl von Gehörschützern für Lärmbereiche). IFL Report No. 1-79 (Institut für Lärmbekämpfung des Hauptverbandes der gewerblichen Berufsgenossenschaften, Postfach 2430, 6500 Mainz 1) (no date), 100 p. Illus. 12 ref. (In German)

CIS 80-972 *The danger of evaluating hearing protectors on their attenuation alone* (Bilsom AB, 260 50 Billesholm, Sweden) (1980), 15 p. Illus. 4 ref.

Heat acclimatisation

Acclimatisation to high temperatures is the result of processes by which the subject adapts himself to living and working in a climate which is hot and perhaps humid. It is manifested as a reduction in the heart rate and internal body temperature at the expense of increased sweating.

Acclimatisation is always relative and specific. The individual acclimatises to a specific dry or humid atmosphere and to a specific workload. Any increase in this load or in the thermal burden may result in health damage. An absence from work of one week may result in the worker losing between one-quarter and two-thirds of his acclimatisation and a 3-week absence from exposure, whether in summer or winter, will mean virtually total loss of acclimatisation unless he is very athletic and in good physical condition. In such a case the loss of acclimatisation is slight and the worker very quickly regains his initial condition when he goes back to the heat. A person may be acclimatised to both cold and heat at one and the same time.

Thermal regulation and acclimatisation physiology

Three factors should be considered here: the process of thermal regulation itself; cardiac output and heart rate; and sweating.

Thermal regulation. The maintenance of a body temperature of approximately 37 °C is achieved by constant adjustment of the processes of thermogenesis and thermolysis. In accordance with clinical custom, the rectal temperature is taken to represent the central temperature of the body. Oral temperatures are less reliable as they are influenced by pulmonary hyperventilation, ambient temperature and by the fact of having taken cold drinks. Thermoreceptors in the skin, the spinal marrow, the viscera and the walls of the vessels draining the muscles emit nervous reflex signals which stimulate the supra- and pre-optic anterior hypothalamic centres. In hot climates these reduce heat generation by inhibition of the centres located in the posterior part of the hypothalamus and increase thermolysis by means of the sympathetic nervous system by adjusting blood flow, muscular tonus, breathing and sweating. Thermolysis is essentially physical and may take place by conduction, convection, radiation or sweat evaporation.

Cardiac output and heart rate before acclimatisation. When work is being performed at a submaximal rate, the heart rate is much higher in hot conditions than at normal ambient temperatures. Cardiac output does not change. This is surprising at first sight, as it is known that exposure to heat leads to an increased peripheral blood flow, which may reach 2.6 l/min·m², as this promotes the loss of calories by means of convection and radiation. This general peripheral vasodilation is mainly compensated for by vasoconstriction in the splanchnic region and in the renal arteries, where the blood flow decreases by from 40 to 80%. This reduction in the splanchnic and renal blood flow enables 600 to 800 ml of blood per minute to be redistributed to the skin. The muscular blood flow is itself reduced, since arterial lactacidaemia rises much quicker than under normal temperature conditions. Among other things these circulatory modifications account for the risk of hepatic or renal lesions and the reduced work capacity of a person exposed to heat.

During maximum short duration effort, of say 5 to 20 min, oxygen consumption does not significantly lessen but internal temperature greatly increases. There is a large exchange between the peripheral and the muscular circulation, and work capacity is temporarily maintained at the expense of the process of thermal regulation. However, if the maximum oxygen consumption is measured during a longer period of effort or if the subject is overheated before measuring maximum $\dot{V}O_2$, there is a reduction in this maximum $\dot{V}O_2$ of 3 to 27% according to different authors. According to Rowell, a reduction of 3% in the maximum $\dot{V}O_2$ enables 600 ml of blood to be redistributed from the muscles to the skin.

Cardiac output and heart rate after acclimatisation. The heart rate slows down but remains higher than it would be at normal temperatures; the systolic output increases and cardiac output remains unchanged. These circulatory modifications after acclimatisation are assisted by an increase of at least 10% in extracellular plasma volume as a result of the increased secretion of anti-diuretic hormone, and also because of an increase in venous vascular tonicity and a reduction by close to half of the peripheral circulation as a result of increased sweating.

Sweating. Liquid evaporation is the predominant mechanism in preventing hyperthermia—it is, in fact, the only mechanism when the ambient temperature exceeds 35 °C. This liquid evaporation is achieved mainly by sweating, which starts as soon as skin temperature rises above 33 °C—although the threshold temperature may be higher in the case of dehydrated subjects. However, it may be incomplete if the climate is humid, as in many mines where the relative humidity may exceed 80%. Drops of sweat then form on the skin, which are of no help from the point of view of thermal regulation while at the same time the ambient humidity causes additional wetting of the skin, and this inhibits the ability of the sweat pores to allow the further flow of sweat. This reduced ability for sweating is partially reversed on passing from humid heat to dry heat.

Acclimatisation increases and prolongs the activity of the sweat glands, estimated at 2 500 000 for a man of 70 kg, by reducing the number of inactive glands. Acclimatisation affects more the sweat glands of the back than those of the chest. Where the maximum sweat rate for a non-acclimatised subject in an environment with low relative humidity is only 1.5 l/h, this figure rises to 3 l/h after 10 days of acclimatisation and 3.5 l/h after 6 weeks. There is also a change in the electrolytic content of the sweat; whereas the sodium chloride loss in the first days is 15-25 g per day, after 6 weeks acclimatisation, this loss has fallen to 3-5 g per day. Due to the effect of aldosterone, the sodium content of sweat decreases after acclimatisation and falls from 4 g/l to 1 g/l, although individual variations occur.

In the subject who has been very well acclimatised for a number of years, the sweating mechanism produces exactly the right quantity of liquid needed to maintain correct body temperature and consequently there is no longer formation of beads of sweat which do not contribute to temperature regulation but merely aggravate liquid and electrolyte loss. Sweating will also begin as soon as skin temperature exceeds 33 °C and without waiting for a rise in rectal temperature.

Methods of acclimatisation

Heat acclimatisation must be carried out when the dry-bulb and wet-bulb temperatures are between 33-35 °C and 25-28 °C respectively. There has been some discussion as to the relative merits of acclimatisation in a natural environment (such as a tropical region) and in a climatic chamber. The differences that have been observed between the two methods are based purely on psychological factors. While acclimatisation procedures may seem to be expensive, the results obtained are beneficial: in particular the physical condition of the workers is improved, the incidence of fatal heat stroke is reduced and non-adaptable workers can be eliminated in good time.

To achieve good acclimatisation for heavy work under hot conditions, it is better to subject the individual to very heavy work under moderately warm conditions than to subject him to light work under very severe climatic conditions. However, in order to obtain good acclimatisation within a reasonable period of time, the duration during which exertion is called for must be sufficiently long. Working for one hour a day at a high temperature will only result in partial acclimatisation at the end of two weeks and both the heart rate and the rectal temperature will continue to be too high. On the other hand, according to South African writers, moderate work requiring an oxygen consumption of 1.0 to 1.4 l/min for 4 h per day and at a wet-bulb temperature of 32 °C, over a period of 8 consecutive days, will ensure complete acclimatisation for the performance of the hardest tasks in a mine and will reduce the risk of fatal heat stroke to 0.02%. The work performed in the climatic chamber consists of stepping on to and off a stool whose height is calculated as a function of the weight of the subject. A more general method for rapid acclimatisation consists of subjecting the individual to half his theoretical workload and to half of the thermal burden for the first day. These are then increased by 10% each day and the acclimatisation is finished in 6 days. For the reacclimatisation of a worker who has been absent from work for a period of more than one week, for reasons of holiday, sickness or other, the procedure would be to subject him to 50% of both his usual workload and the thermal burden for the first day and to increase these by 20% each succeeding day so that reacclimatisation is thus completed in 4 days.

At least 2% of the subjects will be found not to tolerate high temperatures because of circulatory problems. The increase in their plasma volume does not take place properly and has been found to be 60% less than that of a heat-tolerant subject for an effort equal to 50% of maximum oxygen consumption. The reason for this is unknown.

Factors influencing acclimatisation

There are individual differences in acclimatisation to high temperatures. These are mainly dependent on age and sex. Subjects of more than 60 years of age are more likely to suffer from heat stroke than those who are younger. After the age of 40 the onset of sweating is distinctly retarded, the volume of sweat is less and the cardiovascular system has a reduced capability for adaptation. Women are less able to stand heat than men as they commence to sweat later, while both their skin and internal body temperatures are higher; they also sweat less even though they have an increased number of sweat glands both in absolute terms and per cm² of skin than men; after acclimatisation, the amount of sweat produced is half that of a man.

Although Black subjects do not become acclimatised better than Whites, it would seem that morphology and the colour of the skin may play a part. In this respect it is of interest to note the tall, thin stature of the people of the Sahelian region where the temperatures are extremely high. With such a build, the ratio of body surface to volume is improved. In the same way, the most effective cutaneous cover would be that which contained sufficient melanin to impede the ultraviolet rays, and not so dark as to have a heat absorption coefficient that would be too high: once again, this corresponds with the brownish skin of some of the Sahelian populations.

Stout and obese persons are less adaptable to heat. Under such conditions they are less capable of working and more likely to succumb to heat stroke, which in their case would be three to four times more likely to be fatal than in the case of a person of normal weight. They are also at a disadvantage because of the low values of their maximum $\mathring{V}O_2$ per kg of body weight and of the ratio of skin surface area to body weight. The level of fitness of a thin or undernourished subject is not, however, superior to that of the stout person.

Dehydration reduces the flow of sweat, and adds to the rise in central temperature and to the increase in heart rate by reducing the systolic cardiac output as a result of a reduction in the volume of blood circulating. It also shortens the duration of maximum workload of a subject in good physical training, and even upsets the submaximal effort, lengthens the recuperation time after effort and delays the onset of acclimatisation. It is not unusual for a subject who does not receive a drink to lose 3 to 4 l of water per working shift. Conversely, hyperhydration prior to work in high temperatures and administration of d-aldosterone at a dose-rate of 1 mg per day for 3 days before exposure and during the first day of exposure will favour acclimatisation.

Although it can in no way replace exposure to thermal stress, intensive training in normal environmental conditions favours acclimatisation by improving the tonus of the cardiovascular system. It is also possible to predict individual physiological reaction to hot work on the basis of the oxygen consumption per min per kg body weight at a heart rate of 170 beats per min ($\mathring{V}O_{2,170}$ kg body weight) at normal temperature and pressure. A $\mathring{V}O_{2,170}$ of 35 cm³/min · kg body weight may be taken as the line dividing subjects more suitable for hot work from those less suitable for hot work. It has been shown that the same is true for a $\mathring{V}O_{2\,160}$ of 32 cm³/min·kg and $\mathring{V}O_{2\,150}$ of 30 cm³/min·kg. Using these criteria, it is possible to avoid submitting to an expensive acclimatisation programme those subjects whose reactions are predictably mediocre. Training in a normal thermal environment also has an effect in that the dissipation of heat from the body is improved as a result of stimulation of the sweat rate. A subject in good physical training sweats faster and more copiously following any increase in central body temperature, while the latter rises less quickly than in the case of a sedentary subject.

Finally, certain medicaments such as salicylates, meprobamate, pilocarpine, hyoscine, atropine, spironolactones and amphetamines inhibit acclimatisation and consequently promote the risk of accidents due to heat exposure. Beta-blocking agents have no effect on thermal regulation.

Acclimatisation evaluation criteria

Evaluation may be based on sweat rate and on central body temperature. An acclimatised subject should not lose more than 1 l per hour or more than 5 l per working shift from sweating; the rectal temperature should not exceed 38 °C. As far as the heart rate during work is concerned, it is to be noted that this is not only a reflection of the intensity of the work being performed. Further, the standards for heart rate that are applicable for normal work are not valid for work at high temperatures, even in the case of acclimatised subjects. Standards for maximum permissible heart rate at high temperatures have not yet been established.

Sweating is normally checked by weighing the subject or by measuring sweat secreted over a 14 cm² area of the back; there is good correlation between these two measurements. However, it is often difficult to measure sweat secretion under actual working conditions. According to certain authorities, rectal temperature is the most reliable criterion for evaluating the degree of acclimatisation; as soon as rectal temperature reaches 39 °C, the subject is liable to collapse. Rise in body temperature is also accompanied by a fall in vigilance and a consequent increase in the likelihood of accidents. In hot work, there is a difference in the relationship between the rise in rectal temperature and sweat rate depending on whether the subject is acclimatised or not. In the acclimatised subject, a smaller rise in rectal temperature will provoke sweating and for a given rise in rectal temperature, maximum sweating will be most profuse.

Practical conclusions

The following measures should either facilitate acclimatisation to high temperatures or improve the behaviour under such conditions of subjects who are not so well acclimatised or, being well acclimatised, are faced with exceptionally onerous working conditions.

Selection. Persons of more than 45 years of age, those suffering from obesity and those whose body weight is less than 50 kg or whose maximum oxygen consumption is less than 2.5 l/min should definitely be excluded from acclimatisation programmes. Those who are suffering from tropical infections, such as malaria or bilharzia, and those who are suffering from any acute infection, particularly respiratory, should be temporarily eliminated until a cure has been effected.

Drinks. Liquids should be taken in small quantities and often from the start of exposure to high temperatures: 100 to 150 ml of water every 15 to 20 minutes. The quantity to be drunk should be calculated on the basis of the fluid loss, since the thirst mechanism does not at all furnish an appropriate basis for compensating the important factor, which is fluid loss. The degree of thirst felt is always less than the actual loss that occurs. Recommended drinks are: plain (non-carbonated) cool water (9-12 °C); cold lemon tea; well diluted fruit juice; etc. Carbonated drinks, undiluted fruit juice, milk and especially any alcoholic drinks should be forbidden. Intolerance to high temperatures due to dehydration disappears completely when, for example, miners exposed to a wet-bulb temperature of more than 29 °C drink at least 3 l of water per shift; their capacity for work is even increased by the addition of 100 g of sugar to the drinking water distributed during the working day.

Food. The intake of fatty foodstuffs should be reduced. The administration of additional salt is only justified in the case of unacclimatised workers who are newly assigned to a hot workplace; in such cases the additional NaCl should be given in the form of a salty liquid such as meat bouillon or tomato juice with 20 g/l of added salt. In the case of acclimatised subjects their salt requirements are largely catered for in most industrial regions by food, which frequently contains too much salt. A temporary deficiency may, however, exist in cases where there has been an excessive consumption of alcohol, and then it would be advisable on the following day to take 3 g of NaCl per litre of sweat lost. In conclusion it is to be noted that desert populations consume very little salt and content themselves with water.

Vitamins B and C. A supplementary dose of 100 mg of thiamin, 8 mg of riboflavin, 5 mg of pyridoxin, 25 mg of cobalamin, 100 mg of niacin and 30 mg of pantothenic acid will delay the appearance of fatigue during work in high ambient temperatures. In the same way a daily supplementary dose of 250 mg of vitamin C during 10 days or so will enable a man to withstand heat better and will hasten the process of acclimatisation.

Heat-insulated clothing. The use of special heat-reflecting clothing or refrigerated jackets using air, or better still, water is to be recommended. In the latter case cooling is effected by means of circulating water or through contact with ice inserted into pockets in the work jacket. Much of the clothing recommended in the literature, however, has little practical interest, because it is not strong enough, too expensive or uncomfortable. The jacket proposed by Wyndham, which contains 4.5 l of water distributed among 28 ice pockets, is, however, a practical working garment which has passed the stage of being an experimental prototype.

Work breaks. Many countries, including Belgium, have adopted in recent years the WBGT index (Wet-Bulb Globe Temperature) recommended by the American industrial hygienists to establish an exposure limit value for work in a hot environment. This index sets the duration of work and rest periods so as to ensure that the central body temperature does not rise above 38 °C, based on the workload and the resultant temperature of the index calculated according to the following formula: 0.7 *thn* + 0.3 *tg*, or, 0.7 *thn* + 0.2 *tg* + 0.1 *ts*, according to whether solar radiation is present or not. *Thn*, *tg* and *ts* represent respectively the natural wet-bulb temperature, the black-globe temperature, and the dry-bulb temperature.

Values for WBGT in °C are as follows:

Work/rest	Workload		
	Light	Medium	Heavy
Continuous work	30.0	26.7	25.0
75% work–25% rest	30.6	28.0	25.9
50% work–50% rest	31.4	29.4	27.9
25% work–75% rest	32.2	31.1	30.0

In the view of the American writers, the limit values in this table are such that the rise in the central body temperature linked to the work period would be balanced during the following rest period, the duration of the total work/rest cycle being 1 h. The above values are recommended when the temperature conditions at the resting place differ little from those at the workplace. If the resting place temperature is equal to or less than 24 °C WBGT, the rest periods may be shortened by 25%.

It must not, however, be forgotten that this index has been proposed for young male subjects in good physical condition and acclimatised. Care should therefore be taken when applying it in industrial situations where both men and women of all ages and in different states of physical condition will be found. It has in fact already been established that the proposed values are too high and the thermal load too great in still air conditions with a high degree of humidity, since the WBGT index does not take sufficient account either of the ventilation effect or of the degree of humidity of the air.

LAVENNE, F.
BROUWERS, J.

"Research in the human sciences in the gold mining industry". Yant memorial lecture. Wyndham, C. H. *American Industrial Hygiene Association Journal* (Akron), Mar. 1974, 35/3 (113-136). Illus. 86 ref.

"Human cardiovascular adjustments to exercise and thermal stress". Rowell, L. B. *Physiological Reviews* (Washington, DC), Jan. 1974, 54/1 (75-159). Illus. 422 ref.

Heat and hot work

Man must keep the temperature of his vital organs within narrow limits if he is to survive exposure to intemperate environments. As heat impinges upon man, his first response is a sensation of discomfort. This discomfort increases as thermoregulatory adjustments are made to counteract thermal stresses on the body. Inefficiency in the performance of non-physical tasks, an increased propensity to minor accidents and changes in the emotional tone of workers are found in association with these changes in sensation and body temperature.

Maximum permissible limits must be set for the thermal severity of workplaces if men performing hard physical work are to maintain their thermal balance either throughout a working day or over the duration required for completion of a specified task. If the combination of workload and environmental heat is so great that thermal balance cannot be maintained, workers will become susceptible to heat collapse. Variation between men, between workloads, and between environmental thermal characteristics must be taken into consideration when recommendations are made of the durations of exposure over which men will be protected from heat collapse. In workplaces with extremely high environmental temperatures, the exposed skin surfaces and respiratory organs of workers may be subjected to extreme discomfort, pain or tissue damage. Limits must be placed on the duration of exposure or on the environments to be entered by unprotected men.

Thermal equilibrium

The interactions of five sources of heat gains and heat losses determine whether the body can maintain thermal equilibrium or is subjected to increasing storage of heat. An equation expressing this relationship is:

heat storage = metabolism ± radiation ± conduction ± convection − evaporation.

Metabolic heat is created during digestive processes and during all muscular work from that needed to sit upright (about 90 J/s) to that involved in extremely hard physical activity (about 695 J/s). In hot workplaces, radiation to the man from all surfaces at temperatures higher than those on the surface of the body produces heat gains. Physical contact with hot surfaces produces heat gains by conduction. Heat may be gained also by convection when warmer air displaces that in immediate contact with the body.

Total heat gains during exposure to a hot workplace can be reduced by lowering the total amount of work done and by the provision of conditioned or unconditioned protective clothing which creates a cooler micro-environment around the man. Heat shields may be used to reduce radiant heat transfer to the man.

Changes in peripheral circulation of blood and the onset of sweating characterise physiological responses to thermal imbalance. With cooler surrounding surfaces and air, heat may be lost to the environment by radiation, conduction and convection. The evaporation of sweat from the entire body surface area is man's main source of heat loss. With maximal sweating care must be taken to ensure that it does not lead to serious water and salt deficiency. Water loss through sweat may be more than a litre an hour and up to 20 g of salt can be lost during a working day through sweat and urine.

The ease or difficulty with which heat balance is achieved in hot workplaces is related to the acclimatisation of exposees, their physical fitness and general health and the job demands made upon them. In general, men who are acclimatised to heat, physically fit, young, in good health and skilled at their jobs tolerate exposure to hot workplaces better than others. The environmental conditions which produce feelings of lassitude, irritability, reduced mental and physical efficiency or acute heat illness and collapse will vary with the individual characteristics of the exposees. However, the interaction of man and his thermal environment can be classified roughly into six categories which reflect increasing levels of thermal stress:

(a) discomfort;

(b) job inefficiency;

(c) continuous work without physiological risk;

(d) continuous work for specified duration;

(e) heat collapse;

(f) painful exposure.

Discomfort

The mildest forms of heat stress are those which cause exposees to feel uncomfortably warm. Several studies have attempted to establish the upper limits of warmth below which a majority of exposees will be free from the experience of discomfort. These limits vary with the population sampled, the method used to assess comfort or discomfort, activity levels during exposure, clothing worn, season of the year and proportion of exposees to be protected. A range of upper limits has been suggested. These limits have been expressed in terms of values on the scales of Corrected Effective Temperature (CET). A CET describes all those environments which produce an equivalent sensation of discomfort or warmth as an environment which has an air temperature of the same value, 100% relative humidity and still air. An upper limit of 17 °C CET for workers in light industry in winter in Britain may be contrasted with an upper limit of 24 °C CET for an inactive sedentary population in summer in the United States. Within this range of recommendations, lower limits are required for the protection of more active exposees in winter clothing and higher limits are permissible for less active exposees in light summer clothing.

Failure to keep environmental temperatures below these limits results in a greater proportion of exposees who will experience and complain about discomfort. Attempts to avoid discomfort in warm workplaces have been based upon the assumption that an uncomfortably warm worker is less efficient at his job particularly when its performance involves higher mental processes.

Job inefficiency

Systematic studies of the effects of exposure to uncomfortably warm environments have suggested that inefficiency in the performance of tasks which require little physical effort is due to increased numbers of errors, slower work rates and the lack of continuity at work which results from increased minor accidents, sickness and labour turnover. Field studies in industry have produced findings which suggest that sudden exposure to environments at a level of 23-25 °C CET is associated with a pronounced increase in frequency of minor accidents and absenteeism and a loss of workshop productivity. It appears that an important factor which influences the responses of workers exposed to environments at this level is their previous thermal experience over the preceding days or weeks. The experience of environments warmer than usual may be a more important determinant of industrial inefficiency than the temperature of the environmental conditions, as such, at these levels of warmth.

Laboratory studies of volunteer subjects employed on a variety of tasks have produced findings which suggest that marked deterioration in performance is unlikely at temperatures lower than 27 °C CET. Beyond this limit decrements have been noted in tasks which require the skills of fine motor manipulations, vigilance and decision making. The performance of less-skilled operatives is more at risk than the performance of more-skilled operatives. The stress imposed by the task itself on an unskilled but highly motivated performer may serve as a primer to the adverse effects of environmental stress.

Continuous work without physiological risk

The tasks described in the preceding section require little physical effort and providing environments are kept within the limits described they induce little physiological strain in the operative. In many hot workplaces, however, the combination of environmental stress and physical work requires limits to be set according to the conditions which enable workers to maintain a deep body temperature at or below 38 °C throughout an 8-h shift. For operatives wearing light summer clothing, who are physically fit and fully acclimatised to work in heat, permissible heat exposure Threshold Limit Values (TLV) have been suggested. These are based upon the Wet-Bulb Globe Thermometer (WBGT) Index of environmental heat stress, which combines humidity, air and radiant temperature values. Operatives will rarely work continuously throughout an 8-h period, and for those who are permitted to rest for 25% of each hour, the TLV limits for the environmental stress are 30.6 °C (WBGT) for light work (up to 235 J/s), 28.0 °C (WBGT) for moderate work (235-405 J/s). At these limits, the proportion of each hour spent at rest may be reduced if the resting environment is cooler than the working environment. With longer rest periods in each hour, or with especially heat-tolerant operatives, the WBGT values for permissible heat exposure may be increased.

Higher work rates or environmental temperatures impose additional strain on man's thermoregulatory systems. Unless measures are taken to reduce these sources of strain by the introduction of rest pauses in cooler surroundings, there will be an increasing storage of heat within the body, which will result in the eventual heat collapse of the exposee.

Continuous work for the specified duration

The limits of work rate and environmental warmth suggested in the preceding sections should permit daily exposure to hot workplaces without incurring undue physiological hazard. Other limits have been expressed which would permit men to work in a warm environment at a given work rate for a specified length of time. These limits define those working conditions in which thermal equilibrium may be lost but heat storage rates are sufficiently slow for a task to be completed in the time allowed. A physiological index of thermal strain called the Predicted 4-h Sweat Rate (P4SR) has been developed which reflects the stress of working in hot environments for specified periods of time.

For fit, young, acclimatised men a P4SR value of 4 l has been suggested as the upper limit of thermal stress which would permit operatives to complete a task lasting about 4 h without the onset of acute heat illness. For unacclimatised men, the limit of a P4SR value of 4 l should be reduced to an upper limit of 3 l. Below this value, physical work tasks can be completed by the majority of operatives. Above the two limits (for acclimatised and unacclimatised men), an increasing number of men will find the conditions intolerable and present symptoms of physiological distress.

Heat-collapse limits

When the combination of environmental and metabolic heat gains produces a thermal strain which cannot be counteracted adequately by heat losses, there is an elevation of heat storage in the body and an increase in the temperature of vital organs to a point where circulatory failure induces heat collapse in exposees. The duration of time after which this heat collapse occurs varies with the work and environmental heat load on the exposee. The duration is determined also by individual characteristics of exposees. Several studies have examined this variation between men in their "tolerance times" and have suggested limits of continuous exposure, expressed in minutes, which would protect specified proportions of exposees from heat collapse.

On average, unacclimatised men can endure climates for 60 min up to about 39.5 °C saturated if their metabolic heat production is only 90 J/s; up to about 36.5 °C saturated if their metabolic heat production is only 210 J/s; up to about 35.5 °C saturated if their metabolic heat production is 315 J/s; and up to about only 31.5 °C saturated if their metabolic heat production is 420 J/s. Fit men acclimatised to work in hot conditions can usually improve upon the endurance times of unacclimatised or less fit men working in similar conditions. In order to avoid heat collapse in half the populations exposed for the average durations suggested by research findings, recommendations have been made that exposures be limited to proportions (66-75%) of average endurance times.

In industrial practice it is probably more common to find either that conditioned garments have been provided as protective clothing, which permits continuous work of longer duration to be performed, or that work is performed in short exposures of only a few minutes alternated with periods of rest in cooler conditions. Little evidence is available from research studies about the optimal work/rest cycles for particular levels of work and environmental thermal stress.

Painful exposures

At very high temperatures, workplaces can be tolerated for only a brief time, if at all, because exposure produces severe discomfort or unbearable pain in exposed skin surfaces or in respiration. When skin temperatures reach 44 or 45 °C, pain is experienced. If skin temperature is permitted to rise higher still, then skin becomes burned, blistered and finally destroyed.

Inspiration of hot air can lead to pain if the air is humid rather than dry. Air at temperatures above about 50 °C produces severe discomfort in the oral, nasal and bronchial passages if it is close to saturation with water vapour. Air can be tolerated at temperatures considerably in excess of 50°C provided it is very dry.

SAFETY AND HEALTH MEASURES

Industrial workers may be protected from heat stress in hot workplaces in accordance with two principles. One relates to the management of exposees and the other relates to manipulation of the environment. Because heat stress in workers results from the interactions of environmental and human factors, protective measures based upon the two principles are complementary to each other.

The first of these two principles states that heat stress will be less when exposees are specially selected and prepared for work in heat. Thus men who have been acclimatised to the particular working conditions to which they will be subjected, men who are healthy and physically fit and men who have been well trained in the job they have to do will be less at risk in heat than other exposees. Even among these workers the risk of heat stress can be further reduced by selecting those individuals who show, on exposure to the conditions, the most favourable reactions (see HEAT ACCLIMATISATION).

During and between bouts of work in hot environments the provision of cool environments in which men may rest and take cool beverages (and salt replacement for heavily sweating men) (see BEVERAGES AT WORK) may minimise or prevent the stresses of heat storage, dehydration and salt depletion if the exposed workers are encouraged to make use of these facilities.

The second of these two principles of heat protection states that heat stress will be less when exposees are provided with a cooler micro-environment within the hot workplace. Reduction of thermal severity requires minimisation of heat gains and maximisation of heat losses for the exposee.

The effects of radiant heat may be reduced by placing barriers between its source and the workers. These barriers may take the form of:

(a) layers of insulation over the heat source;

(b) reflectant shields, which, to be effective, must be highly polished and maintained in a state of cleanliness;

(c) absorbent shields, which can be prevented from becoming sources of heat by being specially cooled with air or water; or

(d) personal protective garments (heat protective clothing), which provide a layer of insulation over the wearer and, if coated with aluminium, a surface which will reflect infrared radiations.

Convective heat gains may be reduced and evaporative heat losses facilitated if an adequate system of ventilation introduces cool dehumidified air into the hot workplace. Where this is not practicable, the plugging of steam leaks, the guiding of hot air from machinery spaces into the outer air and the provision of localised air or water cooling at the exposees' actual places of work may produce a sufficient reduction in the thermal strain placed on the workers. Conditioned protective garments which provide the wearer with a micro-environment cooled by an air or water distribution system may be necessary in some hot workplaces.

Whether protective measures such as these will be found in any particular hot workplace will depend upon a number of factors—including economic ones—which determine the nature of the situation at hand.

BELL, C. R.

"Temperature". Ch. 5 (85-104). *Men at work.* Bell, C. R. (London, Allen and Unwin, 1974), 119 p.

"Development of permissible heat exposure limits for occupational work". Dukes-Dobos, F. N.; Henschell, A. *American Society of Heating, Refrigeration and Air-Conditioning Engineers Journal* (New York), Sep. 1973, 15/9 (57-62).

"Isodecrement curves for task performance in hot environments". Ramsey, J. D.; Morissey, S. J. *Applied Ergonomics* (Guildford, Surrey, England), 2 June 1978, 9/2 (66-72). Illus. 23 ref.

CIS 77-1301 "Principles for occupational preventive medical examinations—Heat hazards" (Berufsgenossenschaftliche Grundsätze für arbeitsmedizinische Vorsorgeuntersuchungen—Gefährdung durch Hitze). Hauptverband der gewerblichen Berufsgenossenschaften. *Arbeitsmedizin—Sozialmedizin—Präventivmedizin* (Stuttgart), Oct. 1976, 11/10 (258-260). 11 ref. (In German)

CIS 79-90 *Evaluation of the work load in hot environments.* Vogt, J. J.; Metz, B. (Luxembourg, Office for Publications of the European Communities, 1977), 110 p. Illus. 36 ref. (In English, French, German)

Heat disorders

A classification of disorders caused by exposures to high levels of environmental heat is as follows:

(a) systemic disorders—heat-stroke (hyperpyrexia), heat exhaustion (from circulatory deficiency: heat syncope), water deficiency, salt deficiency, heat cramps or sweating deficiency;

(b) skin disorders—prickly heat (miliaria rubra), cancer of the skin (rodent ulcer);

(c) psychoneurotic disorders—mild chronic (tropical) heat fatigue, acute loss of emotional control.

Only a brief account of these disorders and their treatment is given here. It cannot be overemphasised that the diagnosis and treatment of these disorders must be the responsibility of a medical team trained in special skills in dealing with heat disorders.

Systemic disorders

Heat-stroke (hyperpyrexia). Risk of heat-stroke occurs whenever the combination of workload (metabolic heat production) and thermal environmental stress is sufficient to produce a degree of thermal strain with which the body cannot cope. Heat storage increases and high levels of deep body temperature are rapidly attained. The risk is significantly greater when workers are unacclimatised, have a poor capacity for work, are obese, are unsuitably dressed adding heat gains and obstructing heat losses, are dehydrated or short of water, have consumed alcohol, or have a history of cardiovascular disease or recent prickly heat. In over two-thirds of all cases of heat-stroke, there is a rapid rise of body temperature to between 40 and 43 °C with, frequently though not invariably, a hot, dry skin and cessation of body sweating. Collapse is preceded by disorientation, delirium, struggling and convulsions. The majority of fatalities occur within 24 h and the remaining fatalities within 12 days of the onset of heat-stroke.

The essential treatment is immediate reduction of deep body temperature (usually rectal temperature) to about 39 °C but not lower. This can be accomplished by sponging with cold water, wrapping with cold wet sheets or towels or blowing cool, dry air over the patient. These procedures may produce vasoconstriction, which should be counteracted by massaging the body surface and limbs during cooling to promote blood circulation. Body temperature must be monitored constantly so that it does not fall below 39 °C. Rapidly induced reduction of body temperature to below this level may produce a state of shock. Once 39 °C is reached, body temperature should continue to fall, without active intervention, until it reaches about 37.5 °C.

Heat exhaustion. Circulatory deficiency *(heat syncope)* can occur at levels of thermal stress less severe than those giving rise to hyperpyrexia. Body temperature need not be abnormally high. Fainting may occur with a slow, weak pulse; a cold, moist skin; and a fall in blood pressure. Unfit, unacclimatised individuals are particularly at risk. Collapse from circulatory deficiency is sometimes preceded by general tiredness, giddiness, nausea or paradoxical cold feelings; sighing, yawning, and shallow or irregular breathing; stumbling and a loss of motor control; facial pallor with sometimes a blueish tinge on the face (cyanosis).

Treatment requires taking the patient into cooler conditions and allowing rest there in a recumbent posture with the knees raised or in a seated position with the head down. Symptoms are usually short-lived and recovery is rapid.

Water deficiency *(dehydration)* occurs when intake of water has been insufficient to replenish water losses through the kidney (urine), lungs (expired air) and, particularly after long working exposures to thermally stressful conditions, skin (sweat). A mild thirst reflects uncorrected water loss of less than 5% of body weight. There may be increases in pulse rate and body temperature, a decrease in output of urine (oliguria), a loss of working efficiency, complaints of restlessness, irritability, lassitude or drowsiness and of thirst when uncorrected water loss amounts to 5-8% of body weight. No work can be performed by patients with a water loss deficit of about 10% of body weight. Death occurs when depletion of blood volume (oligaemic shock) results from water loss in excess of 15% of body weight.

Treatment comprises confining the patient within a cool environment and replacing water loss by drinking. The patient's replenishment of water should always be monitored carefully as during recovery of a positive water balance the patient's subjective impressions are often a poor and inadequate guide to the correction of the dehydration.

Salt deficiency or salt depletion occurs particularly in unacclimatised individuals and results from long periods of continuous sweating with insufficient replacement of salt. There may be no abnormal levels of body temperature or complaints of thirst. Salt depletion may produce headaches, tiredness, irritability and muscular weakness and may be accompanied by nausea and vomiting. A standing posture may be difficult to maintain and dizziness experienced. An exceptionally high pulse rate may be evident. Pathological signs of salt deficiency are found in low levels of salt in urine and blood, reduced plasma and extracellular fluid volumes, and increased levels of blood urea.

Treatment requires reduction in sweating by removal from hot environments and cessation of physical work, and the replacement of salt loss by the consumption of saline drinks. In some emergency conditions it may be necessary to use intravenous injections. Prevention or retardation of salt depletion in workers in hot environments requires the provision of moderately saline drinks. These will be not only acceptable but in fact attractive to those at risk from salt deficiency.

Heat cramps often occur in conjunction with conditions of salt depletion when levels of sodium chloride circulating in the blood fall below a critical level. The attacks of severe painful spasms in limb and abdominal muscles may last for several hours, days or even weeks.

Treatment, and prevention, consists in raising and maintaining an adequate salt intake in food, tablet form or in saline drinks.

Sweating deficiency (anhydrosis) or suppression is a condition in which the efficiency of man's principal channel of heat loss—the evaporation of sweat—is reduced. The impairment occurs over a large area of the body surface and the worker feels hot and exhausted. If the worker attempts any physical exertion the condition may become exacerbated. Over-breathing, a rapid pulse rate and heat collapse may occur. The distressed state of the patient may be reflected in facial sweating, abnormally frequent urination and the appearance of non-itchy rashes anywhere on the body surface. Cases of anhydrosis often have a previous history of prickly heat episodes.

Treatment consists in reduction of the thermal stress imposed on the patient by removing him to a cool environment.

Skin disorders

Prickly heat (miliaria rubra). This is a condition in which dysfunction of the sweat glands prevents sweat from reaching the skin surface and evaporating. The patient suffers from distressing sensations of prickling, tingling and burning over his skin surface as the body tries to lose heat by sweating. Red, itchy rashes appear on parts of the body usually covered by clothing.

Treatment requires removal to a cool environment. Cool showers, thorough gentle drying and the application of calamine lotion may attenuate the patient's distress. Even with treatment, however, the itchy, red rashes may persist for several days or even weeks.

Skin cancer (rodent ulcer). Prolonged exposure to the ultraviolet radiations in sunlight may induce carcinoma of the skin cells. The lighter the skin and the less amenable to the protective pigmentation of melaninisation, the greater is the risk. Fair-skinned, red-haired, blue-eyed, freckled visitors to, and workers in, tropical sunlight appear to be particularly at risk. Incidence is also known to increase with age.

Preventive treatment comprises protection of areas of exposed skin in those at risk from prolonged exposure to ultraviolet radiation from the sun.

Psychoneurotic disorders

Tropical fatigue. Chronic effects upon Europeans of working for long periods in the tropics have been reported. Loss of motivation, lassitude, irritability, sleeplessness appear to constitute the symptoms of a condition for which little physiological evidence has been found. Though the phenomena of tropical fatigue are quite real to sufferers, it appears that their occurrence is more closely related to psychological factors of the individual's intolerance of boredom, monotony, thermal discomfort and heat illnesses than to a specific physical basis.

Acute distress. There have been instances of sudden, dramatic loss of emotional control in workers subjected to acute thermal stress. Uncontrollable weeping or outbursts of violent anger appear to characterise the phenomenon. It is not known whether there are particular predisposing or precipitating factors.

Treatment comprises removal of the patient from the conditions of thermal stress as rapidly as possible without exacerbating the condition of a violently angry patient.

BELL, C. R.

"Terrestrial animals in dry heat: man in the desert". Lee, D. H. K. Section 35 (551-582). 88 ref. *Handbook of physiology.* Dill, D. B. (ed.). (Washington, DC, American Physiological Society, 1964), 1 056 p.

"Disorders due to heat". Leithead, C. S. (127-236). *Heat stress and heat disorders.* Leithead, C. S.; Lind, A. R. (London, Cassell, 1964), 304 p.

"Heat disorders". Weiner, J. S. Section 31 (505-521). 24 ref. *Medicine in the tropics.* Woodruff, A. W. (ed.). (London, Churchill Livingstone, 1974), 623 p. Illus.

Man–Hot and cold. Edholm, O. G. Institute of Biology's Studies. Biology No. 97 (London, Edward Arnold, 1978).

"Heat-stroke: report on 18 cases". Khogali, M.; Weiner, J. S. *The Lancet* (London), 9 Aug. 1980, 2/8189 (276-278). Illus. 9 ref.

Heating of workplaces

Fundamentals

A person's need for heat at his workplace depends on environmental and human factors. The first includes the latitude, altitude, climate and exposure of the workplace, and the protection provided by the building structure against the outside environment. The second includes the energy expenditure of the worker and the pattern of the work, the clothing worn and personal preference.

In all the normally inhabited parts of the world the need for heating is either non-existent or seasonal. Although heating is therefore often given low priority, cold environmental conditions are known to affect health, efficiency, accuracy and productivity of workers, as well as increasing the risk of accidents.

The need to maintain a controlled thermal environment may be set by process or equipment requirements (for example, in the manufacture of textiles or in computer rooms), but this contribution only considers control to maintain human heat balance.

The fundamental purpose of any industrial heating system is to keep workers warm so as to optimise production and health. In the past it has often been inefficient and ineffective through imprecise design and control; buildings have been heated rather than workers warmed. Until recently, the cost of heating has represented only a small fraction of total manufacturing costs, but recent changes in the pricing of petroleum-based fuels is causing extensive reassessment of heating needs and systems.

Human physiology permits quite wide variation of thermal conditions to be tolerated and workers tend largely to be self-regulating in thermal control.

Thermal control is often achieved in stringent conditions, such as outdoor work in cold climates or in cold stores, by a combination of warm clothing and restriction of working time in the cold conditions. It is then dependent on the body's thermal capacity, heat recovery being provided in heated areas or buildings. This method is increasingly used in many large-scale process plants, which are constructed in the open air but operated from heated control rooms. In certain specialised occupations heat may even be supplied to the clothing itself by electrical power, providing what is probably the most efficient and least convenient method of warming the worker.

The most precise control of heating is required where workers are sitting at continuous repetitive work, requiring little energy expenditure, and low circulation of blood to warm the extremities. At manual tasks or where variation of workplace or job is possible, less strict control is needed as workers have more opportunity to adjust the human factors.

Thermal standards

Preferred thermal conditions for workers have been established in a number of countries, but, as these are affected by acclimatisation and clothing habits, they cannot be universally applied. From the work of Bedford the following criteria are suggested for workers engaged in light manual work in the United Kingdom:

Environmental factor	Suggested standard
Air temperature	21 °C
Mean radiant temperature	⩾ 21 °C
Relative humidity	30-70%
Air movement	0.05-0.1 m/s
Temperature gradient (foot to head)	⩽ 2.5 °C

Some of these tabulated factors are interactive. Thus, where radiant heat is significant (for example, near a furnace), then the air temperature may be lower. Where air movement is high (for example, near extract ventilation) the air temperature should be higher.

The criteria may be applied in most countries if adjustment of air temperature is made. It should be raised (1) where air movement is high, (2) for non-manual work, (3) where light clothing is worn, or (4) where people are acclimatised to higher indoor temperatures in winter, or outdoor temperatures in summer. It should be lowered for heavier manual work, or where heavy clothing is worn.

For a sensation of comfort, it is desirable that the temperature of the surroundings should be rather greater (and certainly not less than) the air temperature, and the flux of radiant heat should be reasonably uniform in all directions and not excessive on the head. The increase of air temperature with height (temperature gradient), should be minimal if feet are to remain warm without the head becoming "stuffy". Both these factors are only essential for long-term sedentary work, but are often difficult to achieve with a heating system designed for low capital cost and high flexibility.

Minimum occupational thermal standards are legislated in a number of countries, but are not necessarily optimal, and in recent years maxima have been established in some countries to conserve energy. Typically, a maximum air temperature of 20 °C has been set, and has induced a return to heavier clothing worn in the past by workers in many industrial countries. A typical comfort zone for light work is shown in figure 1.

Recent developments of comfort assessment have become either more complex, or greater simplicity has been attempted. In the latter case measurements may be

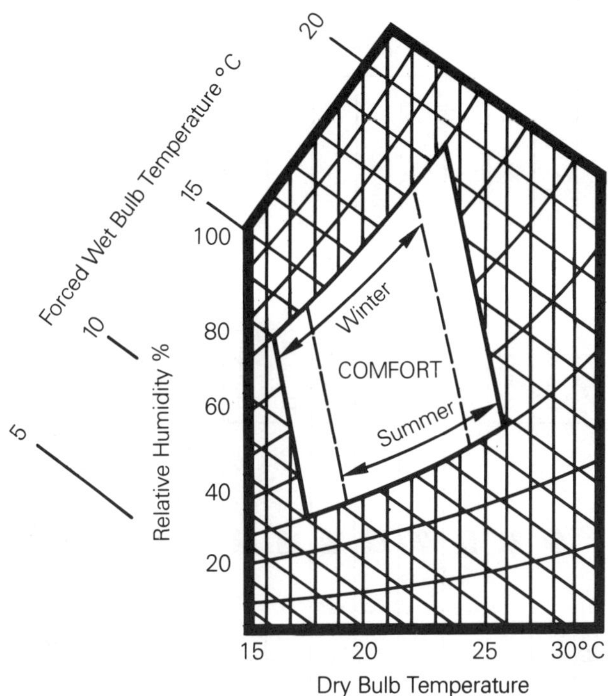

Figure 1. Comfort zone for light work.

restricted to globe thermometer and air movement, and the scales suggested in figure 2 should be applicable wherever heating is needed.

The above values should be valid where the monthly mean outdoor temperature is about 0 °C. Where mean outdoor temperature is about −10 °C or −15 °C, indoor temperature should be raised about 1 °C, and by about 2 °C where outdoor temperature is −20 °C.

The choice of clothing habitually worn indoors can affect the preferred temperature by as much as 5 °C.

Heating systems

Any system for keeping people warm at work must be closely related to the work being undertaken, and to the design of the enclosing building.

At the design stage of an industrial building, requirements for warming workers are seldom stipulated as processes and working positions have often not been defined. It is therefore usual to provide a system giving maximum freedom for future process changes and to stipulate only the total required heat capacity to maintain the building interior at a specified temperature. This usually takes no account of heat distribution within the building, the location of working positions, or the production of cold zones near the floor and close to openings in the fabric of the building. In a number of countries until recent years this has led to minimum cost design of building structures with inadequate insulation, high infiltration of cold air, and installation of a minimum number of unit heaters (fan and heater batteries) discharging a minimum quantity of air at the maximum practicable temperature. Comfortable thermal conditions have seldom been achieved. Aspects of the building design to be considered for successful heating include:

(a) insulation for energy economy and to minimise temperature gradient within the building;

(b) reduction of infiltration of cold air to minimise cold draughts and to restrict temperature variations over the working area;

(c) control of air contamination, principally by local extract ventilation, supplemented by displacement or dilution ventilation;

(d) control of heat emission by processes and its distribution to occupied areas.

Sources of heat for warming workers are generally coal, oil, gas, or electricity, though increasing attention is now being given to solar energy as a supplementary source. Heat may be generated by combustion of fuel at a central point in the district (typically a thermoelectric power station), and distributed to factories in the form of low-pressure steam, or it may be generated at a central boiler plant on the factory site and distributed to individual buildings and work areas, as low-pressure steam or hot water. Alternatively, fuel may be burned locally, and heat delivered to the working zone, commonly in the form of hot air, but sometimes also as radiation. For all except the smallest units, liquid or gaseous fuel is supplied by pipe from a central source. If heating is provided by flueless burners, then the hazard of inhaling the products of combustion must first be assessed. Limits are generally set for sulphur dioxide produced from oil, and nitrogen oxides produced from gas. Carbon monoxide may be a significant hazard from both, and from coke, and is a particular risk in confined areas as the concentration builds up rapidly if combustion efficiency is reduced due to the presence of carbon dioxide or reduction of oxygen. Selection of a particular system for heating is often governed by national regulations or commercial practices. For example, a supply of cheap off-peak electricity may make electric storage heaters economic, particularly if a chemical heat storage system (e.g. sodium sulphate) can be developed as an

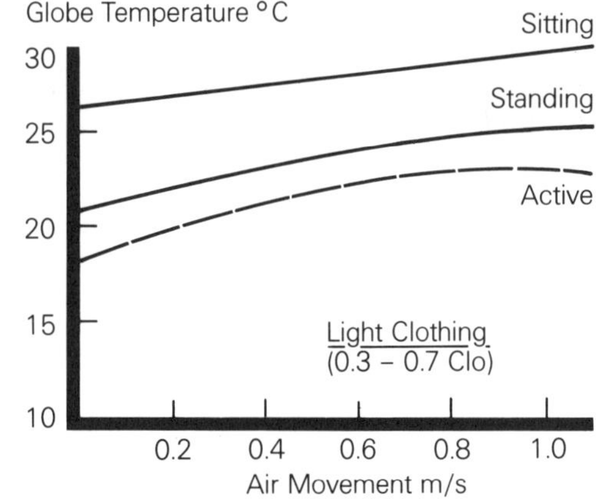

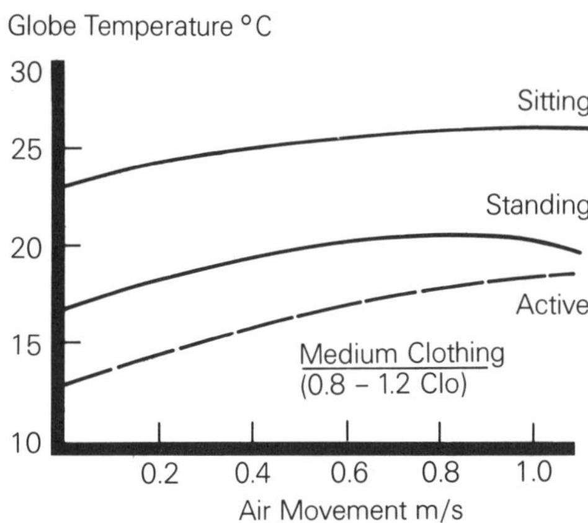

Figure 2. Comfort zones based on measurements restricted to globe thermometer temperature and air movements.

The heating systems shown below are those typically used in industrial areas. Heated floors, conventional "radiators" and air conditioning systems may be more appropriate for sedentary work in offices or in some light assembly areas.

KEY	
Positive features (advantage)	■
Negative features (disadvantage)	●
Neutral features	—
Variable features	?

	SYSTEM	Low capital cost	Low operating cost	Free floor area	Quick response	Steam/water system needed	Little maintenance	Draught free	Low temperature gradient	Uniform heating over area	Summer air movement	Can warm fresh air	Amenable to zoning or local control
Convective (Unit) Heaters	Radiant strip Truss mounted 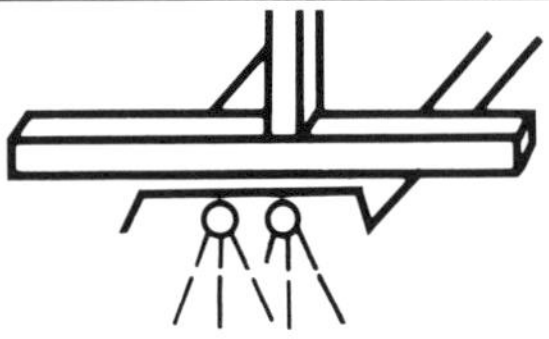	●	■	■	●	■	■	■	■	■	●	●	●
	Vertical panel Stanchion mounted 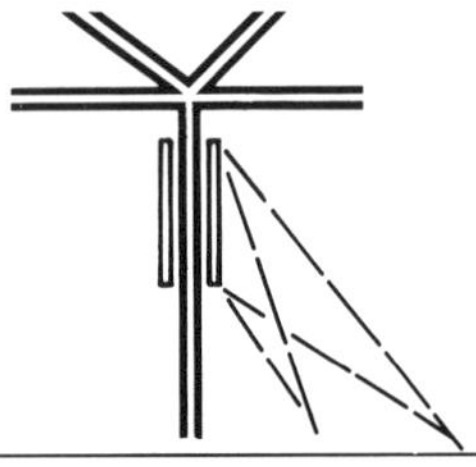	●	■	■	●	■	■	■	■	■	●	●	—
	Incandescent heater, gas or electric 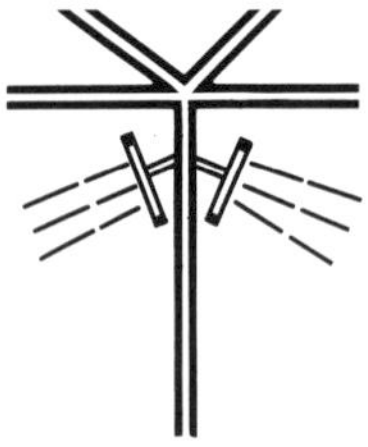	■	—	■	■	●	—	■	?	●	●	●	■
Radiant Heaters	Vertical discharge Mounted in apex 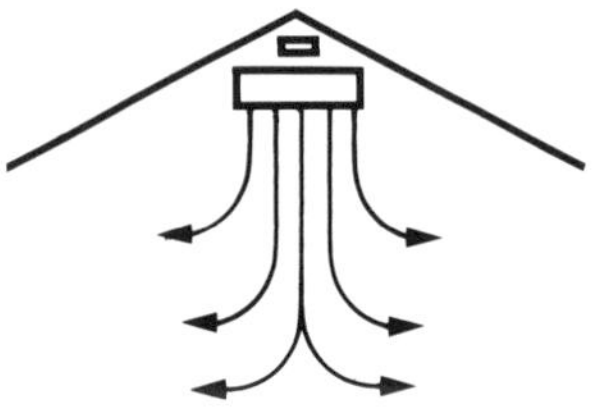	—	●	■	■	■	—	—	■	■	■	■	—
	Horizontal discharge Truss mounted 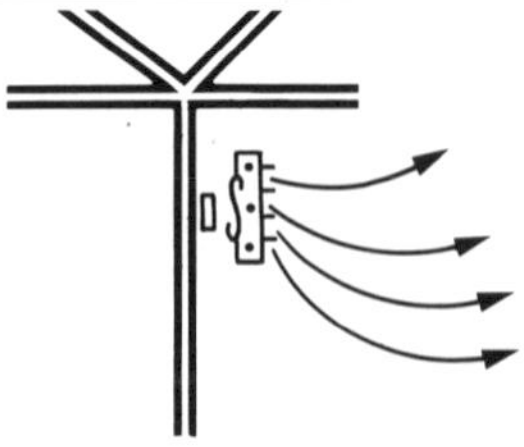	—	●	■	■	■	—	●	●	●	■	■	■
	Horizontal discharge Floor mounted 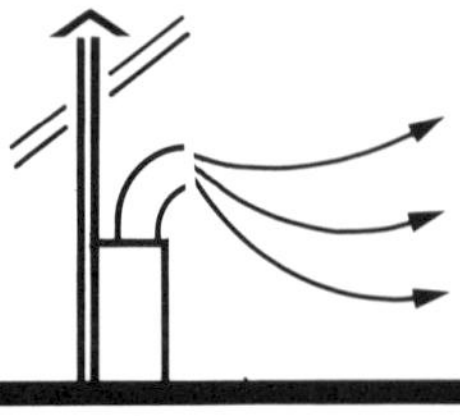	■	●	●	■	●	●	●	●	●	■	?	■

Figure 3. Key characteristics of the main heating systems used at workplaces.

economic system. A surcharge on natural gas at peak periods can lead to the introduction of dual-fired heaters (gas and oil) as a method of reducing operating costs. To use electricity directly for heating is a waste of high grade energy, but may be economic where flexibility of heating is essential. However, heat released from lighting systems may be economically used in some office buildings. Electrically powered heat pumps can offer a more economic solution to the heating of working areas as they produce about four times the amount of useful heat as electricity used direct, but few applications have been made in the past as the capital and maintenance costs have been high.

The design of a heating system is always a compromise between low capital cost, flexibility, fuel economy and reliability. In figure 3 some typical systems are illustrated and their key characteristics indicated.

Prevention of cold draughts. One of the most common problems is the creation of cold areas or localised draughts due to the entry of cold air, either when personnel or vehicle doors are opened, or to replace air removed by flues or extract systems. The first can be controlled by provision of "heat-locks" at all regularly used openings. These comprise an isolated space between two sets of doors so designed that cold air cannot enter working areas directly. Alternatively a warm air-curtain can be provided by fan heaters actuated by the door opening. Incoming make-up air should preferably be warmed at inlet, but can sometimes be introduced successfully at high level so long as downdraughts are prevented.

Portable heaters. For temporary work, or to warm a solitary worker in a large area (such as a clerk in a warehouse), portable man-heaters can be used, burning oil or gas as a fuel. These may be incandescent radiant heaters or hot air blowers. Neither type is provided with a flue and they should not be used in confined spaces.

Energy conservation. For existing installations the greatest contribution to energy conservation can usually be made by installation of a programmable controller using a microprocessor to provide optimum start control. Heat can be provided in the amounts and at times needed to give comfort with operating savings on existing installations of up to 30%.

In some processes greater use can be made of waste heat to provide thermal comfort, but high capital costs of specialised installations often make such systems uneconomic.

SHERWOOD, R. J.

Basic principles of ventilation and heating. Bedford, T.; Chrenko, F. A. (eds.). (London, H. K. Lewis, 3rd ed., 1975). Illus.

Heating and cooling for man in industry (Akron, American Industrial Hygiene Association, 2nd ed., 1980).

Thermal, visual and acoustic requirements of buildings. Building Research Establishment Digest 226 (London, HM Stationery Office, 1979).

Handbook and Product Directory: Fundamentals (1977); *Applications* (1978) (New York, American Society of Heating, Refrigerating and Air Conditioning Engineers (ASHRAE)).

Environmental Criteria for Design. Guide Part A1; Heating, Guide Part B1 (London, Chartered Institution of Building Services (CIBS), 1978).

CIS 74-705 "Assessment of man's thermal comfort in practice". Fanger, P. O. *British Journal of Industrial Medicine* (London), Oct. 1973, 30/4 (313-324). Illus. 48 ref.

Heat protective clothing

The most suitable type of heat protective clothing for a given hot environment is determined, to a large extent, by the work to be performed and the characteristics of the heat load. Because of this, it is convenient to classify hot environments according to the method by which the heat load is imposed:

(a) high radiant temperature from a localised source (unidirectional) or from all around (omnidirectional);

(b) high air temperatures–either low wet bulb (dry heat) or high wet bulb (wet heat);

(c) a combination of *(a)* and *(b)*–an extreme example of which is flame lick;

(d) conduction from hot solid objects;

(e) condensation (a rare situation arising from contact with saturated air at temperatures higher than body temperature).

Clothing materials

Many of the materials used in everyday clothing can be used for heat protection, the main restriction being that of their flammability. However, wool, flameproofed cotton, and asbestos dominate, although synthetics are also used in some situations. Wool will not readily burn, it has good compressional resilience and wicking properties. Cotton can be tightly woven and flameproofed whilst it is, at the same time, one of the cheaper materials. Wool and cotton are somewhat hygroscopic in that they take up water from the air at high relative humidities and

Table 1. Suitability of various fibres for use in heat protective clothing

Wool	Flammability medium to low. It can be flameproofed. It can be made up into high density felts and will stand somewhat higher temperatures than cotton.
Cotton	Burns unless flameproofed; singes at about 230 °C.
Asbestos	Garments made from it are heavy; they are used for high temperatures as it does not start to degrade until 500 °C. Temperatures in excess of this can be tolerated for a limited time.
Glass	Used for high temperatures but starts to degrade at about 450 °C.
Nylon (polyamide)	Softens at about 235 °C and melts at about 260 °C (nylon 11 at 180 °C and nylon 6.6 at 250 °C).
Flame-resistant and fire-retardant aromatic polyamide (e.g. Nomex)	A member of the nylon family of fibres. It does not melt but degrades at 371 °C. After 1 000 h in dry air at 260 °C the breaking strength at room temperature is about 65% of that before exposure.
Polyacrylonitrile	Slightly higher melting point than nylon, about 280 °C.
Polyester	Reached about the same as wool for flammability, melts at about 260 °C, softens at about 235 °C.
Polyurethanes	They can be used up to 100 °C-120 °C and will take peaks of 200 °C for very short periods. They are available in a self-extinguishing form.

release it at low humidities. The latent heat changes involved dampen the rate of temperature change in the fabric.

Asbestos and glass fabrics are required when temperatures are reached which degrade wool and cotton. Leather, polyurethane foams and polyester film or fibre are also used. Some of the synthetics such as the polyesters are readily aluminised and lend themselves to the construction of permeable, heat-reflecting fabrics. Care has to be exercised in the use of synthetics as some such as the polyamides soften at about 200 °C and eventually melt. However, the deterioration of a fabric depends not only on the temperature to which it is exposed, but also on the duration of exposure. Fabrics of high initial strength, e.g. polyamides, are therefore not necessarily the best for garments which are exposed to high temperatures for long periods. Table 1 summarises the properties of a number of materials.

Thermal insulation characteristics. In general, the thermal insulation of a fabric is proportional to its thickness, rather than to the nature of the fibre used. High-density structures have higher conductivities than low density structures since some of the heat flow through a textile material is *via* the fibres themselves (figure 1).

It is the air trapped in the material that is mainly responsible for its insulating value but the conductivity of air increases by 0.28% per degree rise above 0 °C. Since the temperature of the outer layers of a protective-clothing assembly can easily rise to 100 °C in hot environments, the insulation provided can decrease appreciably. Fabrics are diathermous in that thermal radiation can penetrate the fabric to be converted to heat on absorption well below the outer surface. It is, therefore, advisable to have a closely woven outer layer, or an aluminised fabric to reflect radiant heat, to ensure that radiation is absorbed as close to the surface as possible. Over the face and eyes, masks of gold-coated transparent materials such as polyesters or glass are often used.

Unconditioned garments

This type of garment, which is aimed at extending the wearer's exposure time, can be used to protect the whole

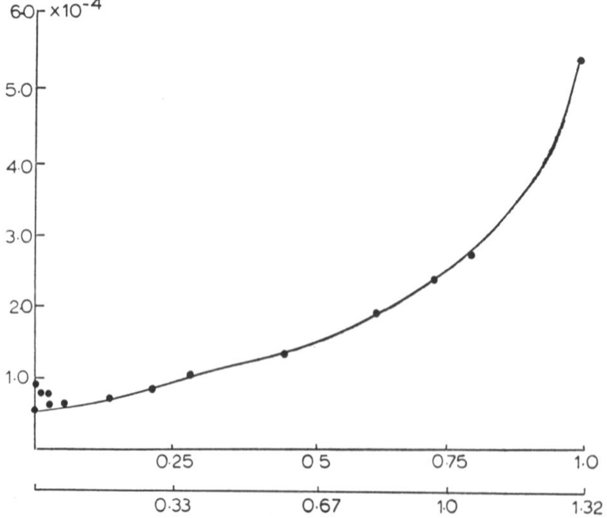

Figure 1. The thermal conductivity of wool/air mixtures increases as the percentage of the fabric volume occupied by fibres increases. Most wool fabrics would have densities below 0.4. Ordinate: thermal conductivity (CGS units); abscissa: upper scale, volume fraction of fibre; lower scale, density. *(Adapted from Baxter, S., 1964.)*

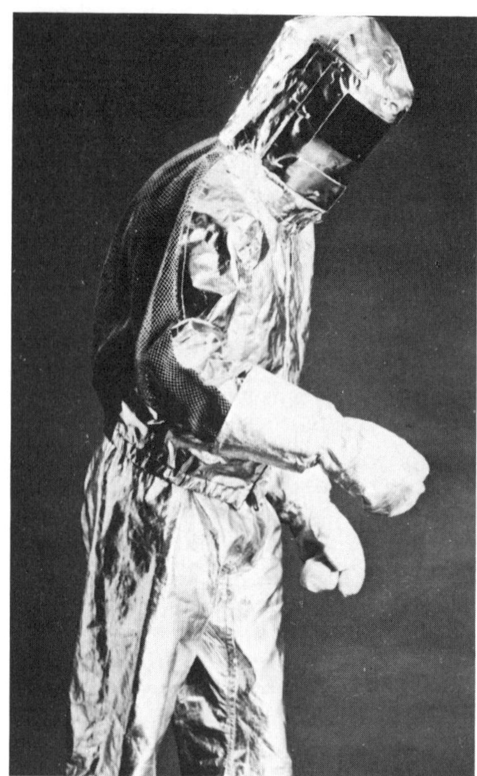

Figure 2. The garment illustrates the use of a heat-reflecting surface to protect the wearer from a unidirectional source of radiant heat. The back of the jacket is made from a net-like fabric to facilitate evaporation of sweat. *(By courtesy of C. H. Greville Ltd.)*

body. Protection for just that part of the body exposed to a localised heat source is to be preferred as it permits evaporative and convective and radiant cooling from the rest of the body's surface (figure 2). Unconditioned garments are designed to make use of:

(a) aluminised fabrics to reflect radiant heat;

(b) a thick, though not dense, layer of material for its insulation properties;

(c) the material's thermal capacity to absorb heat;

(d) compression-resistant layers of insulation to prevent hot surfaces touching the wearer;

(e) permeable materials which permit evaporation of water and wick sweat from the wearer's body;

(f) vapour barriers which prevent scalding of the wearer by water vaporised on contact with the hot outer layers of the garment; and

(g) materials, such as wool, which do not readily burn but degenerate into a hard layer affording some protection from hot-metal splashes or sparks.

Based on these principles, garments can be designed for particular circumstances. Aprons, sleeves, gloves, helmets with wire mesh or heat-reflecting vizors can protect the wearer from unidirectional radiant heat. The garments should be aluminised and constructed from material like asbestos or wool if contact with hot objects is likely, otherwise lighter materials may be used. The apron should, if possible, be supported in such a way that air circulates freely between it and the body.

Complete body protection from omnidirectional radiant heat can be achieved by loose-fitting aluminised coats and leggings. The choice of material is governed by the same considerations as above. With low air temperatures, evaporative heat loss should be facilitated by suitable design and, in high air temperatures, thick clothing is required to provide a layer of insulation. In dry heat conditions, the use of permeable fabrics facilitates the evaporation of sweat. With a combination of radiation and high air temperatures, complete bodily protection is recommended by the use of garments which are aluminised and have a thick layer of insulation (figure 3). Garments should be designed for quick removal. Where contact with hot objects is likely, protective clothing materials should be heavy and resistant to complete collapse under pressure as the insulation they provide is proportional to the thickness.

Ordinary clothing, having the advantages of cost, comfort and permeability to air and water vapour, is used in many hot working situations. It provides a layer of insulation and at the same time blocks a considerable proportion of the radiant heat which it absorbs in the surface layers. The specialised clothing is used where conditions are too extreme for conventional garments.

Conditioned garments

Protective clothing can be considered to be conditioned if it incorporates a system, however simple, which regulates changes in the microclimate. Such garments are more complex and expensive than unconditioned ones and are used where work cannot be done safely or satisfactorily without their protection, such as in the inspection of furnaces after shut-down, in environments containing harmful dusts and gases and in extremely cold environments.

Figure 3. An aluminium-coated fire-fighter's garment designed to provide flame protection.

A conditioned garment is an assembly of components or "systems" and for this reason is generally referred to as a "protective clothing assembly". Each system has a particular function and is designed to meet a certain specification. For example, fabrics used in a hot-entry assembly are selected to meet a temperature specification which, if exceeded, may result in their destruction. Conditioned garments are used primarily for heat and hot work, but there are garments available for cold work.

The main components of the assembly are an outer garment, a system for distributing the conditioning medium, a circulating pump, a cooling and conditioning system for the medium and a link between the medium source and the garment. The outer garment may be in one or more pieces, partially or completely covering the body and is generally designed for a specific situation which determines the type of materials used and the design details. For example, protection against high radiant temperatures indicates the use of radiant heat reflecting surfaces. The presence of toxic chemicals and dust necessitates designs which enclose the head and enable a positive pressure to be maintained within the garment.

Where the cooling medium is water, the distribution system is worn as a separate unit next to the skin; air-cooling systems are normally a part of the outer garment and lie between the protective garment and the underclothing. At its simplest, the system for cooling and conditioning the air or water may be ice or a supply of cold water from the water mains, and an appropriate heat exchanger. The cooling power of the system should be matched to the anticipated heat load on the microclimate. The type of cooling unit used determines to a large extent the capital and running costs of a conditioned assembly. For example, vortex tubes are inexpensive but their running costs are high.

When the conditioned garment is tethered and not self-contained, there is a connecting link between the garment and the external source of cooling medium. The main advantage of the tethered assembly is that it can be designed to tolerate very high thermal loads indefinitely. The limited distances that these garments can penetrate into hot environments arise from the increase in temperature of the air or water as it flows along the hose.

Conditioned garments can be classified according to whether they are self-contained or tethered and whether they are air, water or electrically conditioned.

Self-contained assemblies. These achieve their independence by using a heat sink or reservoir such as ice batteries, or any convenient power source or substance in which phase changes are associated with large latent heat changes.

Water or air is circulated in a closed system through the garment microclimate where it absorbs heat and then through a heat exchanger which is in contact with the heat sink. Water has the advantage that the power requirement of the circulation pump is small compared to that required by an air blower; in addition, the tubing diameter of the water system is much smaller. The heat sink normally worn as a back-pack may weigh 10 kg or more depending on design. Its weight and size impose some restrictions upon the wearer, such as the size of hole or door which can be entered. Backing-up facilities are required for the storage and production of the heat sinks.

The working limits of any assembly are determined by the thermal insulation value of the garment, fabric temperature limits, the capacity of the heat transfer system and the thermal capacity of the heat sink. If the assembly is water-cooled, the evaporation of sweat by the wearer will be limited and hence also its use as a reserve mechanism for heat loss. If air is being used in a closed system, although sweat can evaporate, this does not increase the thermal capacity of the system. In an open or partially open system where ambient air of low water-vapour pressure is drawn in and cooled, sweat evaporation would increase the system's thermal capacity.

Tethered assemblies. In a heat protective assembly where water is used straight from the mains at a fixed temperature, the heat removal capacity of the system is determined by the flow rate. If refrigeration is used, then temperature can be employed to control the rate of heat removal. The equation for heat removal by a water conditioned system is: Heat removed = Mass flow × Difference between inlet and outlet temperatures. When using water-conditioned systems, the body must be allowed to warm up before low inlet temperatures are used to avoid vasoconstriction, which impedes heat loss from the body.

Air is a suitable medium for a wide range of uses including garments for protection against toxic substances. Air has the advantage that it enables sweating to be effective as a method of heat loss. A cooling system is invariably required for air-ventilated heat protective assemblies. The most commonly used cooling system is the vortex tube. Conventional expansion units and turbine coolers are also used. These latter are expensive but efficient.

Tethered air-conditioned garments can be divided into three main groups. Firstly, there are garments that do not include thermal insulation and are used in environments at room temperature. The quantity of air required is that necessary to remove metabolic heat, preferably without the wearer sweating, and to prevent the ingress of contaminants. Between 0.5 and 1.4 m^3/min of air with a water vapour pressure of 10 mmHg ($1.33 \cdot 10^3$ Pa) or less may be required. Ventilation is normally of the axial type where air flows parallel to the body surfaces and exhausts at valves. Secondly, there are garments that incorporate thermal insulation and in which the air flow is parallel to the body surface (axial ventilation); the air exhausts at garment openings. Thirdly, there are garments designed to use a system of insulation called "dynamic insulation" when the air is made to flow out through the permeable insulating material of the garment (radial ventilation) and a heat counter-flow system is set up in the material. Cool air picks up much of the heat before it gets through the material and so reduces substantially the quantity of heat penetrating the garment.

Protection of the extremities. Adequate protection of the hands, feet and face is necessary in conditioned clothing to prevent pain and to prevent these extremities short-circuiting heat into the microclimate where it has to be dissipated by the ventilating air. Because of the difficulty of providing the extremities with an adequate thickness of insulation, heat reflecting surfaces have to be used. This reflectivity can be maintained by using self-adhesive aluminium foil or reflective polyester tape which is easily removed and replaced as required.

Special protection requirements

Any hot work for which clothing is used should be assessed for special requirements; for example, confined spaces may hamper the recovery of collapsed personnel and a built-in harness would facilitate this. Hot, underfoot surfaces not only require suitably insulated boots but, because the inside of the sole can continue to increase in temperature after leaving the hot area, footwear should be designed for rapid removal. Fastenings of metal become very hot and should be avoided if possible.

CROCKFORD, G. W.

CIS 80-1113 "Protective clothing for industrial workers subjected to high temperatures". Hartwig, P. *Protection* (London), Aug. 1979, 16/8 (27-29; 38). Illus.

CIS 80-490 *Fabrics and materials for protective clothing—Methods for determining protective qualities and resistance against infrared radiation* (Tkani i materialy dlja specodědy—Metody opredelenija zaščitnoj sposobnosti i stojkosti pri vozdejstvii IK-izlučenija). Gosudarstvennyj komitet SSSR po standartam. GOST 12.4.074-79 (Moscow, Izdatel'stvo standartov, Novopresnenskij per.3, 8 Aug. 1979), 10 p. Illus. (In Russian)

CIS 81-514 *Study of the absorptance coefficient of aluminised fabrics* (Contribution à l'étude du coefficient d'absorption des tissus aluminisés). Cornu, J. C. Notes scientifiques et techniques 21 (Institut national de recherche et de sécurité, 30 rue Olivier-Noyer, Paris) (Aug. 1979), 98 p. Illus. 13 ref. (In French)

CIS 80-1256 "Health hazards of asbestos protective clothing" (Pathologie de l'amiante et vêtements de protection). Sors, C.; Heintz, C.; Cabasson, B. *Archives des maladies professionnelles* (Paris), Nov. 1979, 40/11 (987-995). 24 ref. (In French)

"Cooling and respiratory protective equipment—an astronaut suit just for mine rescuers?" *South African Mining and Engineering Journal* (Johannesburg), Oct. 1980, 91/4167 (60-75). Illus. 4 ref.

Clothing for protection against heat and fire. Method of evaluation of thermal behaviour of materials and material assemblies when exposed to a source of radiant heat. International Standard 6942 (Geneva, International Organisation for Standardisation, 1st ed., 1981), 15 p. Illus.

"Heat protective clothing. Measurement of total hemispheric absorption factor of aluminous fabrics" (Vêtements antithermiques. Mesure du coefficient d'absorption hémisphérique total des tissus aluminisés). Cornu, J. C.; Aubertin, G. *Cahiers de notes documentaires* (Paris), 3rd quarter 1979, 96 (357-371). Illus. 13 ref. (In French)

Helium-group gases

NOBLE GASES

The helium-group gases consist of helium, neon, argon, krypton, xenon and radon (see RADON AND THORON). Until 1962 it was generally assumed that the then-considered "noble gases" were incapable of forming simple inorganic compounds. Only the existence of hydrates (e.g. $Kr \cdot 6H_2O$) and clathrates (e.g. $4Xe \cdot 12C_6H_5OH$) was recognised. In that year, however, it was shown that xenon combined with fluorine to form xenon fluorides like XeF_2, XeF_4 and XeF_6 and by hydrolysis yielded a xenon oxyfluoride, $XeOF_4$ and a xenon oxide, XeO_3. It was shown that Xe^{8+} exists in the stable, colourless solid sodium perxenate, a highly oxidising substance. Only one binary compound of krypton exists, $-KrF_2$. New compounds are still being made (XeO_2F_2, $XeCl_2$, 1967).

Occurrence. On earth, argon and helium are the most abundant, occurring in the earth's crust at 3.5 and 8×10^{-3} ppm by weight, and in the earth's atmosphere at 9 340 and 5.24 ppm by volume respectively. The atmosphere serves as a commercial source of neon, argon, krypton and xenon, whereas natural gas wells in the United States are a source of helium.

Production. Production methods are by cryogenic separation, radon production being by separation from the gases evolved from radium.

Uses. Uses of the helium-group gases are confined mainly to helium, neon and argon. In addition to rather

	Atomic number	Atomic weight	m.p.	b.p.	sp.gr.
Helium (He)	2	4.0	−272.2 °C	−268.9 °C	0.18
Neon (Ne)	10	20.2	−248.6 °C	−245.9 °C	0.90
Argon (A)	18	39.9	−189.2 °C	−185.7 °C	1.78
Krypton (Kr)	36	83.8	−156.6 °C	−153.3 °C	3.74
Xenon (Xe)	54	131.3	−111.9 °C	−107.1 ± 3 °C	5.89
Radon (Rn)	86	222	−71 °C	−61.8 °C	9.73

extensive applications in the field of cryogenics, where helium produces the lowest temperatures known, and to use for military aircraft and rocketry (helium) and for signs and fluorescent lamps (neon), one of the major uses is in metal smelting, refining and welding, where helium and argon have presently a dominant role in providing an inert gas "shield". Hydrogen-argon mixtures allow temperatures in excess of 10 000 °K in plasma-jet torches; and large quantities of argon are used for the purging of carbon monoxide during the vacuum degassing of certain high-grade steels. [There is increasing use of helium-neon lasers to replace optical alignment appliances, mainly in the building industry.] Xenon is an effective anaesthetic agent.

HAZARDS AND THEIR PREVENTION

In all of these numerous applications, however, exposure to the helium-group gases is not of a type to cause serious health impairment. The only present important application in which biological effects are manifested is as a breathing gas for divers and in underwater (deep sea) activities. Helium, by virtue of its greater diffusibility (low atomic weight), not only permits divers to descend to greater depths and enhances their performance at those depths, but also reduces the time for decompression. Associated with these desirable features, however, is the production of characteristic high-pitched voice, and, more important, unintelligible speech. This latter becomes more serious in the excitement of an emergency. Because adaptation to this effect is slight even on prolonged exposure, the remedy for this serious drawback to the use of helium-oxygen mixtures is being sought in the development of an electronic speech unscrambler.

The only other use of these inert gases, at present very minor, in which biological effects are manifested, is as anaesthetic agents. In this role, xenon excels all others in the group and indeed even nitrous oxide. The narcotic property of xenon is approximately equivalent to that of ethylene. Inert-gas narcosis is directly related to the pressure of the inert gas: with increasing gas density, helium < argon < xenon, increasing brain carbon dioxide retention occurs, potentiating the narcosis. No significant clinical or biological effects in man result, however, from breathing continuously helium-oxygen mixtures at 258 mmHg total pressure for 58 days.

On the other hand, metabolic increases have been noted in animals breathing helium-oxygen mixtures to be followed by depression below normal (7-11%) on return to normal atmosphere, with no evidence of metabolic acclimatisation. Second generation animals showed reduced body weight gains or even weight losses. A striking relationship between the biological activity of the helium-group gases and their capacity to take part in weak intermolecular reactions was found in the gases' ability to prevent the growth of a mould; inert gas partial pressures for 50% growth inhibition were: helium, ca. 300 atm; neon, 35 atm; argon, 3.8 atm; krypton, 16 atm; and xenon, 0.8 atm. Presumably this interaction, leading to narcosis, occurs through the formation of the inert-gas hydrates or clathrates mentioned above—the larger the atom (e.g. Xe) the larger geometric disturbance in the normal crystal hydrate lattice surrounding nerve endings. Physiologically, helium has been found to reverse the effects of air-way resistance (as in asthma obstruction) proximally, but not distally, in the lungs.

Although exposure to these gases does not have serious toxic effects, their use, like that of nitrogen, may in certain circumstances be associated with asphyxiating atmospheres, and death has been caused by the presence of leaking argon from welding equipment in a confined space. An ample supply of fresh air should be provided when a welding operation is being conducted in a confined space.

[Helium-neon lasers do not present a burning hazard; they are, however, dangerous for the eye and require adequate eye protection.]

STOKINGER, H. E.

Helium-neon:

CIS 77-696 "Use of lasers on construction worksites" (Lasergeräte auf Baustellen). *Mitteilungsblatt der Bau-Berufsgenossenschaft Wuppertal* (Wuppertal-Elberfeld), June 1976, 2 (52-59). Illus. 12 ref. (In German)

Argon:

CIS 76-241 *Safety manual for personnel of plants producing acetylene and gases extracted from the atmosphere—Manuel de sécurité pour le personnel ouvrier travaillant dans les usines de production d'acétylène et des gaz extraits de l'air.* Permanent International Commission for Acetylene, Gas Welding and Allied Industries (Publications de la soudure autogène, 32 boulevard de la Chapelle, Paris) (2nd ed., 1973), 22 p. Illus. (In English, French)

Krypton, xenon:

Separation, storage and disposal of Krypton-85. IAEA technical report series No. 199 (Vienna, International Atomic Energy Agency, 1980), 66 p. Illus. 48 ref.

CIS 74-1300 *Krypton 85—A review of the literature and an analysis of radiation hazards.* Kirk, W. P. (Eastern Environmental Radiation Laboratory, Box 61, Montgomery, Alabama) (Jan. 1972), 68 p. Illus. 280 ref. (Available from National Technical Information Service, 5285 Port Royal Road, Springfield, Virginia) (Accession No. PB 207 079).

CIS 77-1014 *Safety and health of the environment in use of ^{133}Xe in hospitals* (Veiligheids- en milieu-aspecten bij het toepassen van Xe-133 in ziekenhuisen). Sturm, J. (Kernfysiche Dienst, Directoraat-Generaal van de Arbeid, Voorburg, Netherlands) (July 1976), 20 p. Illus. 12 ref. (In Dutch)

Helminthiasis

At one time helminthic infections were considered to be "exotic" diseases limited to the tropical belt. Now many professions including businessmen, diplomats, consultants, journalists, hunters, soldiers, anthropologists, volunteers, entertainers and transient workers who travel can acquire virtually any helminthic infection prevalent in a visiting area depending on their lifestyle and physical environment. The pathogenic effects may not be apparent for a long time. An awareness of the problem, accurate history taking and appropriate diagnostic techniques are essential for correct diagnosis. Although individual treatment of cases is essential, the best way to deal with the problem is by preventive measures.

Helminths are divided into two large groups—the platyhelminths and the nemahelminths.

Platyhelminths include the *Trematoda* (flukes) and the *Cestoda* (tapeworms). The trematodes have a flattened leaf-like appearance and most of them are hermaphrodites with the exception of the *Schistosoma*. All trematodes use the snail as an intermediary host. Prodigious asexual reproduction takes place within the snail. Cercaria escape from snails, and in the case of schistosomes, they are infective to human beings. In all other parasitic trematodes the cercaria undergo another stage of development, metacercaria, on a plant or in another host, which may be fish or crab. Man becomes infected by ingesting the metacercaria.

The *Cestoda* or tapeworms are parasitic animals elongated and chain-like. They lack a digestive tract and absorb nutrients through their integument. The scolex has suckers and hooks to attach the tapeworm to the intestinal mucosa. The posterior segments contain the reproductive organs and the terminal segments are essentially sacs full of eggs which detach so that the eggs are passed in the stool of the host. With the exception of *Hymenolepis nana*, all cestodes that are human parasites have complicated life cycles involving intermediate hosts. For some species man can also serve as an intermediate host and the presence of the larval stage may produce serious disease.

Nemahelminths *(Nematoda)* are cylindrically shaped worms, have an alimentary canal, a body cavity and separate sexes. Nematodes are the most ubiquitous parasites of man. Some nematodes, such as insect transmitted filarial worms, utilise an intermediate host as part of the life cycle. Others have direct life cycle. The larvae that are released from the eggs have to moult and develop into the filariform larvae that are infectious or develop an infective larva in the egg. The adult form of the worm develops within the vertebrate host.

(For schistosomes (blood flukes) see separate article on SCHISTOSOMIASIS.)

Clonorchis sinensis and other flukes that have two intermediate hosts—a snail and a fish—are contracted by consumption of raw, frozen or pickled fish. This is not very important as an occupational infection.

Among the tapeworms the most common are *Taenia solium* (the pork tapeworm) and *T. saginata* (the beef tapeworm). Both these are significant as occupational hazards among meat industry workers and butchers. *Echinococcus granulosus* or the hydatid worm can cause serious illness for certain occupational groups. The adult worm occurs in dogs. Eggs in dogs' faeces are the source of infection for man and herbivorous animals. The larval stage develops in man in the form of hydatid cysts in the liver, lungs or any other tissue. The hydatid cysts grow to considerable size, causing severe symptoms and even death. Cases have been reported from Australia, New Zealand, Argentina, Uruguay, Chile, Southern Brazil and the Mediterranean countries. The infection occurs in dogs when they consume the uncooked viscera of herbivorous animals. Rigid control of slaughtering, extermination of wild canines and treatment of infected dogs are some of the preventive measures. The other hydatid worm, *Echinococcus multilocularis*, which causes liver echinococcosis, frequently fatal, may be a serious occupational infection among fur industry and fur animal farm workers as well as in hunters. The infection is transmitted mainly by eggs in fox faeces.

The most widely occurring nematodes are *Ascaris lumbricoides*, *Ancylostoma duodenale*, *Necator americanus* and *Trichuris trichiura*. ANCYLOSTOMIASIS is discussed in another section of the encyclopaedia. *Ascaris* and *Trichuris* are spread by direct or indirect contamination of food with embryonated eggs from contaminated soil. They can be controlled by the sanitary disposal of human faeces and by improved personal and food hygiene.

Strongyloides stercoralis is a nematode that has a life cycle very similar to the hookworm. Workers who come into contact with infective larvae in soil get infected. As the larvae migrate through the lungs they may produce a very severe and chronic pulmonary disease. There is an increase in eosinophil count and the patient suffers from severe bronchial asthma or broncho-pneumonia. Larvae can be detected in the sputum, pleural fluid and lung tissue. The distribution of the parasite is similar to the hookworm and prevention is by the improvement of environmental sanitation.

The nematodes that are transmitted through an anthropod vector are an important group of parasites. Best known amongst this group are *Wuchereria bancrofti* and *Brugia malayi* that can cause filariasis in humans. The adult worm is in man and the larvae circulating in the bloodstream are transmitted from man to man by the mosquito vector. Repeated infection is necessary for the disease to develop. It is widely distributed in Asia, Africa, Latin America and the Pacific islands. In endemic areas whole communities are victims of the infection. Control is by elimination of the mosquito vector, which is difficult. Mass treatment of communities can reduce the illness to a very low level.

Onchocerca volvulus is another nematode which has a very significant impact as an occupational hazard especially in young adults working in agriculture. The disease caused by this parasite is known as river blindness. Some of the most fertile lands along the rivers in Africa cannot be cultivated because of this illness. The disease is spread by the black fly, *Simulium*, and its subspecies. The major occurrence of this disease is in West Africa. It is also present in East Africa, Yemen and in Central America and Brazil. The infected person develops skin changes and eye lesions which cause permanent blindness in a great percentage of the exposed population. Eradication of the flies is very difficult and most of these communities do not have the political or economic resources to control this disease.

MANOHARAN, A.

General parasitology:

Parasitic zoonoses. Report of a WHO expert committee with the participation of FAO. Technical report series No. 637 (Geneva, World Health Organisation, 1979), 107 p. 34 ref.

Parasitic zoonoses. Clinical and experimental studies. Soulsby, J. C. (ed.). (New York, Academic Press, 1974), 402 p. Illus.

Parasitology. Noble, E. R.; Noble, G. A. (Philadelphia, Lea and Febiger, 4th ed., 1976), 566 p. Illus. Ref.

Medical parasitology. Marbell, E. K.; Voge, M. (Philadelphia, W. B. Saunders, 4th ed., 1976).

Elements of medical parasitology (Eléments de parasitologie médicale). Golvan, Y. L. (Paris, Flammarion, 3rd ed., 1978), 616 p. Illus. (In French)

Diagnosis:

Ova and parasites. Desowitz, R. S. (Hagerstown, Maryland, Harper and Row, 1980).

Laboratory procedures for diagnosis of intestinal parasites. Melvin, D. M.; Brooke, M. M. (Atlanta, Georgia, DHEW, Center for Disease Control, 1974).

Helminthiasis:

Intestinal protozoan and helminthic infections. Report of a WHO scientific group. Technical report series No. 666 (Geneva, World Health Organisation, 1981), 152 p. Ref.

Hemp

A herbaceous plant of the *Cannabinaceae* family, the principal species of which is *Cannabis sativa*; there are various sub-species, the most important being *C. indica*. *C. sativa* is grown widely throughout the world. Its cultivation is favoured by a somewhat warm and humid climate and in Europe, for example, it flourishes in the Mediterranean region but is found as far north as latitude 66 °N on the Baltic coast. It is also grown in Africa, Asia, North and South America and Australia.

Cultivation. Hemp is an annual; it is planted at the end of the spring frosts since about 4 months of frost-free weather are required for fibre growth. Hemp is a plant which requires a fertile, well drained soil.

Processing. At harvesting time it is cut, spread out to dry and then retted (partially rotted) in pits of static or running water for several days or even weeks to destroy the gums which bond the fibres. The stems are then crushed to separate out the non-fibrous material, either by a threshing machine or manual pounding, and "batted" by holding the sheaf of fibre in one hand over a board of hardwood and hitting it with a stick held in the other hand, in order to release the remaining impurities. The resulting fibres are carded to separate the short from the long fibres, packed and despatched to factories where they are made up into the finished product.

Uses. The main use of hemp is for textiles, which are made from the fibres in the stalk; it is also used for string and rope manufacture. Various types of alkaloid are extracted from *Cannabis indica.* These are in the top of the stem and in the resin of the glandular hair, especially the leaves and bracts. The more important kinds–cannabina and tetanocannabina–are narcotics and used in pharmaceutical and inks manufacture. They can also be used for the production of alcoholic beverages and may be smoked pure or mixed with other materials under such names as "hashish" or "marijuana". The seed is used as a foodstuff or processed to produce oil. Finally, due to its density and fertility, the plant may be grown as a soil steriliser.

Figure 1. Hemp plant.

HAZARDS AND THEIR PREVENTION

The principal health hazard in hemp processing is byssinosis or cannabosis–a respiratory disease due to the inhalation of hemp dust. The processes of breaking and spinning and especially those of batting and carding liberate fine dust which is inhaled by the worker. This dust contains substances which affect the respiratory tract. A sharp reaction in the first few weeks of exposure is rare; however, after prolonged exposure a gradually increasing bronchial reaction takes place showing some allergic characteristics.

It is usually not until after at least 10 years' exposure that the worker reaches his maximum bronchial hyperactivity with obstructive characteristics; these are most evident on the first day back at work after a break and are, consequently called "Monday syndrome". After 2-3 h exposure, the hemp worker develops an itching in the throat, feels hot and flushed and complains of tightness of the chest, difficulty in breathing and a sensation of "bloatedness"; his work capacity gradually falls off and eventually he is obliged to stop work altogether. In some workers there is an appearance of inebriation. Lung function tests demonstrate a large reduction in vital capacity in the course of a day's work and the forced expiratory volume ($FEV_{1.0}$) may fall to less than 1 l.

No improvement occurs until 4-5 h after termination of exposure; and by the following day, although recovery of the $FEV_{1.0}$ base line is not complete, the worker is able to resume work. The disease is progressive with continued exposure over many years. It takes about 20 years for the development of advanced byssinosis or, more specifically, cannabosis, bronchial catarrh and pulmonary fibrosis. It may reduce working capacity by anything from 25-40% and eventually develop into a general condition rendering the victim incapacitated and emaciated with cor pulmonale; death results from a weakening of the right ventricle. Statistics indicate that the average hemp worker has a life expectancy of no more than 39 years whereas the age at death of the normal agricultural worker is 67 years.

The consumption of hashish and marijuana results in toxicomania and produces behavioural disorders which will in turn affect social well-being. These substances also give rise to several kinds of permanent disorders including obstructive arteritis (Buerger's disease).

The prevention of cannabosis necessitates dust control measures at all stages of production and handling or the use of suitable dust masks, although the latter is a less desirable method. High concentrations of dust may be prevented by the application of good general ventilation techniques and by the mechanisation of processes in combination with local exhaust ventilation.

Periodical medical examinations should be made, where possible, at yearly intervals and should include a standardised questionnaire relating to the characteristic symptoms of byssinosis and a test of ventilatory capacity.

No successful treatment is known for cannabosis. [However, the inhalation of sodium cromoglycate or administration of ascorbic acid before exposure can inhibit the bronchoconstriction caused by the inhalation of hemp dust.] In the early stages of the disease, considerable improvement takes place if the individual changes his job for a less dusty one. Education and social guidance may help in persuading men to abandon the use of hashish and marijuana.

In plants processing the hemp fibre, there are a large number of dangerous heavy textile machines such as bale openers, hackling, carding and teasing machines, etc., which should be fitted with adequate machinery guarding. Where artificial humidification is employed in hemp mills, temperature and humidity should be controlled to ensure that they do not exceed an acceptable or statutory limit.

The hemp fibre and dust is highly flammable and maintenance and good housekeeping as well as fire-prevention measures should be enforced to minimise accumulation of waste materials and fire hazards.

BARBERO-CARNICERO, A.

CIS 76-352 "Byssinosis: Airway responses in textile dust exposure". Zuskin, E.; Bouhuys, A. *Journal of Occupational Medicine* (Chicago), June 1975, 17/6 (357-359). 19 ref.

CIS 77-640 "Allergological and functional tests in workers exposed to hemp dust" (Alergologični i kliniko-funkcionalni proučvanija na rabotnici v kontakt s konopen prah). Demirova, M. *Letopisi na higienno-epidemiologičnata služba* (Sofia), 1976, 20/1 (63-65). 15 ref. (In Bulgarian)

CIS 74-633 "Lung static recoil and airway obstruction in hemp workers with byssinosis". Guyatt, A. R.; Douglas, J. S.; Zuskin, E.; Bouhuys, A. *American Review of Respiratory Disease* (New York), Nov. 1973, 108/5 (1 111-1 115). Illus. 13 ref.

Hepatitis, infectious

Virus hepatitis; VH

Virus hepatitis is one of the most frequently encountered of the infectious diseases. It is found throughout the world. The distribution of the disease among the population shows that it has a distinct tendency to become rarer towards the northern latitudes, being most prevalent in the tropical and sub-tropical regions of the world and least in northern Europe and North America. Because of its generally protracted course, and because a number of the cases are marked by later complications, the illness should always be treated as a serious one. Typical symptoms are not always apparent and a high percentage of undetected cases must be reckoned with. In recent years VH has been regarded as an increasingly important occupational disease, especially as it affects medical personnel. In many countries where there is a proven occupational risk of infection it is recognised as an occupational disease for which compensation is paid.

Definition of the different types of VH

On the basis of intensified virological and immunological research in recent years, new information has been brought to light on the aetiology of VH, so that today it is possible to distinguish between virus hepatitis A (VHA), virus hepatitis B (VHB) and virus hepatitis non-A-non-B.

It has been shown that at the onset of the illness, particles of 27 μm in size exist in the stools of patients. This virus was found to be passed in the stool in large numbers before the outbreak of jaundice; throughout the further course of the illness and at its climax, excretion of the virus is no longer evident. Up to the present time the virus has not been found in the bloodstream of patients. During the acute phase of the illness an antibody known as anti-HA is built up and is manifest for several years after the illness. The anti-HA antibody limits any remaining illness for a number of years, perhaps even maintaining an immunity against VHA throughout the remaining lifetime. Chronic carriers of the HA virus have not up to the present time been reported. The incubation period for VHA varies from 15 to 45 days. During the phase when the virus is excreted in profusion, and thus before the outbreak of jaundice, the patient is particularly infectious. Transmission of VHA follows the faecal-oral route. Of particular significance in this respect is direct contact with the patient through the common use of toilets, towels and eating utensils. Transmission may also take place by way of contaminated food and drinking water following excretion of the virus.

The causative agent of VHB is probably the establishment in the serum of the patient of identical Dane particles. These are spherical particles of about 45 μm in size, surrounded by a double membrane and having a characteristic spherical inner body whose size is about 27 μm. The so-called HB_s antigen (HB surface antigen, previously also known as Australia antigen) is found on the upper surface of the Dane particle, and the HB_c antigen (HB core antigen) in the centre of the Dane particle. A further antigen is known as the HB_e antigen, which is probably situated between the outer double membrane and the inner body of the Dane particle. The HB_s antigen has been much more frequently identified in the serum of a VHB patient than the complete Dane particle. After infection by the HB virus there is a very early appearance of the HB_s antigen (up to 3 weeks before the clinical manifestation of the illness). On the other hand, the HB_s antigen can persist for many years after the VHB illness, and this status is known as an HB_s antigen carrier. The simultaneous appearance of HB_e antigen and DNS polymerase may be taken as a guide to the degree of infection in such persons. Antibodies are built up against the three antigens observed in VHB (anti-HB_s, anti-HB_c, anti-HB_e); anti-HB_c is clearly demonstrable at the early stages of the illness, a little later the anti-HB_e appear, while HB_s are mostly found 2-3 months after the VHB illness has died down. The incubation period for VHB is stated to vary between 25 and 180 days. The infection is essentially due to inoculation by infected matter (an inoculation of 0.0004 ml of infected blood is sufficient to transfer VHB). The inoculation mechanism involves a blood or plasma injection, through the use of unsterilised or inadequately sterilised syringes or instruments, and possibly also through dentists' equipment, as well as the contamination by infectious material of the skin and mucous membrane. On the other hand, contrary to earlier interpretations, inoculation is not the only manner of contracting a VHB infection, it is also possible to transmit it by way of the faecal-oral route. The HB_s antigen is not only demonstrable in the blood, but also in the stool, urine, tears, saliva, sweat, bile and seminal fluid (e.g. transmission can occur during sexual intercourse). The danger can be present not only with acute VHB patients, but also with clinically healthy HB_s antigen carriers, where the coexistence of HB_e antigens and DNS polymerase has been demonstrated. But where anti-HB_s or anti-HB_e have been shown to be present in HB_s antigen carriers, it is probable that they are not infectious. While a terminated VHB illness will result in a build-up of defensive antibodies, recurrences of the illness have been seen. Cross-immunity between VHA and VHB does not exist.

It has been observed that many patients who have received blood transfusions have fallen victim to a form of VH, which serologically can be associated neither with VHA nor with VHB. This type of hepatitis is typified as non-A-non-B. The causative agent has not yet been identified nor has the immunological reaction been sufficiently investigated. The illness can be transmitted experimentally to chimpanzees. In North America more than 80% and in Germany more than 40% of hepatitis cases following blood transfusion are probably of the non-A-non-B type.

Symptoms and course of the illness

It is not possible to differentiate the three types of hepatitis on the basis of clinical symptoms. The illness starts with an uncharacteristic early phase of debility, anorexia and nausea. Frequently joint and limb pains as well as sub-febrile conditions are displayed, and also a high fever, particularly in the case of children and young people. During the further course of the illness the urine turns brown, the stool becomes pale and at the same time jaundice becomes apparent (yellowing of the conjunctiva of the eyes and later of the skin). The liver is distinctly enlarged and painful on pressure. Laboratory results show the following changes: bilirubin serum concentration is high in varying degrees, values of up to 30 mg/dl of bilirubin i.S. being found. However, an appreciable number of VH cases can run their course without an increase in bilirubin (this non-jaundice type of VH is particularly frequent among children and young people). Serum glutamic pyruvic transaminase (SGPT) and serum glutamic oxalacetic transaminase (SGOT) can reach 10-200 times their normal value. Alkaline phosphatase (AP) and gamma glutamyl peptidase (gamma-GT) values are also high. Gamma-GT is the last of the blood enzymes to return to its normal range, and is thus an important indicator in clinical checking for the cure of VH. The concentration of iron in the serum is generally raised to a value of between 150 and 300 μg/dl. The blood picture shows a normal or slightly high leucocyte count and relative lymphocytosis with atypical lymphocytes. Serum electrophoresis during the further progress of the illness is marked by an increase in gamma globulin. Laparoscopy and biopsy play an important part in the diagnosis. This method is of particular value for the recognition and differentiation of chronic forms of VH. The jaundice phase of uncomplicated VH lasts from 2 to 6 weeks, normalisation of the serum enzyme follows the disappearance of jaundice. Return to full health after VH takes 6 to 12 months. After an acute attack debility, digestive disturbances and slight enlargement of the liver may give cause for complaint for a longer period. The most feared complication is an acute liver dystrophy, with continuing destruction of the cell tissue leading generally to death. The death rate with VHA varies from 0.1 to 0.4% with children and young people and from 1.2 to 1.5% in adults. In the case of VHB it amounts to 2%. In 70-90% of the cases a cure is effected without a sequel. The most serious sequel is a transition to chronic VH, in which a distinction must be made between chronic persistent and chronic aggressive hepatitis based on histological evidence obtained from microscopical examination of liver cell tissue. Chronic hepatitis is to be seen particularly in patients having had VHB and hepatitis non-A-non-B. The frequency of transition to chronic hepatitis is given as from 1 to 15%. Chronic persistent hepatitis is generally curable, however, only after a number of years, without permanent damage. The prognosis for a chronic aggressive hepatitis is doubtful. In a high percentage of cases the progress of this illness cannot be prevented and can terminate in a transition to cirrhosis of the liver. In the case of VHA, progress towards chronic hepatitis is improbable. Recently discussions have been taking place as to the existence of a causative relationship between VHB and certain forms of liver carcinoma.

Occupational risk

The risk of infection by hepatitis is particularly severe for all medical and paramedical personnel. Epidemiological investigations have shown that the incidence of the illness among this occupational group is significantly higher than that found among the general population. The risk of infection is highest for personnel employed in dialysis and infection wards. An increased risk must also, however, be assumed for all the other branches of the medical profession, particularly those exposed to contact with blood or excreta from these patients. The danger exists not only with patients who are suffering from an acute positively diagnosed VH, but also with patients where the progress of the illness is not typical and often therefore not diagnosed as hepatitis, and with the numerous hepatitis B carriers (it is estimated that there are 120 million carriers in the world). There is also an enhanced risk of infection for dental surgeons and dentists. Personnel working in medical laboratories are at risk from exposure due to unavoidable contact with blood and excretory matter; in this connection, it is important to note that even the control sera used in statistical quality control may be infectious. Occupational hepatitis has been seen among kindergarten staff, mostly related to infectious outbreaks of VHA, particularly among the children. Because of the high incidence of VH in tropical countries there is an enhanced risk of infection even among persons who only have to spend a short time in such countries in the course of their work (e.g. personnel in UN development agencies, military personnel, and technical personnel in industrial projects). There is naturally a particularly acute risk to medical personnel in tropical areas, but as a result of the poor hygiene and bad conditions, members of other groups are also exposed to risk. Mention should also be made of the possibility of the transfer of VH by insect bites. Yet another occupational group that is at risk are those who are involved with a plasma fractioning process, work which involves the use of blood plasma in the manufacture of various serum preparations. In order to reduce the danger of infection by an occupational hepatitis, it has been suggested that work situations where there is a special VHB risk should be manned by persons in whose blood anti-HB_s have been shown to exist.

In addition to the risk of occupational infection of persons employed in certain activities, a further problem concerning the communication of hepatitis has recently come under discussion. It concerns the question of the communication of infection by HB_s antigen-positive medical personnel. Doctors, dentists and hospital nurses, who are VHB carriers, can under certain circumstances infect the patients they are attending. According to the WHO definition, a VHB carrier is a person in whom HB_s antigens have been shown to reside for longer than 6 months. Such VHB carriers who are employed as medical personnel and are in close contact with the population should not present any particular risk of infection provided the necessary preventive measures are observed; nevertheless reports continue to be heard of patients being infected by HB_s antigen-positive personnel. In some instances therefore, the competent health authorities have already imposed a ban, usually temporary, on doctors and dentists who are HB_s antigen-positive. In order to estimate the degree of infectiousness of a carrier, as has been mentioned, the presence of HB_e antigens and DNS polymerase must be proved and possibly that of the more significant Dane particles as well.

Medical prevention and hygiene

In the absence of aetiological therapy, particular importance must be attached to preventive medicine. In the case of VHA, the administration of gamma-globulin before infection or during its early incubation period (up to about 10 days after the infection) can prevent or mitigate the illness. Gammaglobulin therapy is ineffective. An intramuscular injection should be from 0.02 to 0.12 ml per kg of body weight of a 16% solution of normal human immunoglobulin. In the case of a continuing risk of infection a repeat injection is advisable after 6-12 weeks (e.g. in the case of a prolonged stay in an endemic zone under poor hygienic conditions). Immunoglobulin should not be given to persons in whom the presence of VHA antibodies (anti-HA) has been shown to exist. In the case of VHB normal immunoglobulin offers no protection. In such cases recourse should be made to immunoglobulin preparations that contain increased quantities of HB_s antibodies (anti-HB_s). Their use is to be recommended in cases of inadvertent inoculation or exposure to mucous tissue (e.g. a scratch, splashes in the eye, swallowing) while using substances containing HB_s antigen. The preparation should be given as an intramuscular injection 6-12 h after the incident. A repeat injection could be given after 4-6 weeks.

Over and above these passive immunisation precautions, work has been going on over a long period to prepare an effective vaccine for active immunisation, but this, unfortunately, is not yet available.

In addition, hygiene measures to reduce the risk of infection continue to be indispensable. The patient should be isolated and all excretory matter as well as laundry, eating vessels and utensils should be disinfected. Medical personnel should wear protective clothing and disinfect their hands thoroughly. Disposable gloves should be worn during the examination of patients, while taking blood specimens and when handling objects that may have been infected. The use of sterile disposable syringes and cannulae is strongly recommended. Hands should be disinfected by rubbing them for 5 min using a swab soaked in a solution of 80% ethyl alcohol or 70% isopropyl alcohol or 60% propyl alcohol. For disinfecting articles contaminated with blood or serum a strong solution of hypochlorite that has 10 000 ppm available chlorine (e.g. a 0.5% solution of sodium hypochlorite), and for disinfecting other articles, a weak hypochlorite solution, having 1 000 ppm available chlorine (e.g. a 0.05% sodium hypochlorite solution). The solutions must be freshly prepared every day and their efficacity tested using iodine papers (deep blue colouring). For metal objects a 2% solution of glutaric aldehyde is more suitable as it is non-corrosive. Bed linen and underclothing should be soaked for 12 hours in a 1.5% solution of formalin; boiling them for 30 min is also effective. Dishes and eating utensils should be disinfected by boiling thoroughly in a soda solution. Sterilisation can be effected by placing articles in a current of hot air at 180 °C for at least 30 min or in steam at 120 °C under a pressure of 1 bar for 15 min. Articles that do not stand heat should be placed in a 3% formalin solution for at least 6 h.

Treatment

Up to the present time, no aetiological therapy is available. The patient should be confined to bed until the serum enzymes have returned to normal (getting up too soon will cause a worsening of the clinical picture). A diet that provides an adequate caloric intake should be rich in carbohydrates and vitamins with a relatively low fat content. During the phase of loss of appetite at the onset of jaundice the protein intake should also be reduced. Neither the various medicaments that are said to strengthen the liver, nor intravenous administration of levulose and glucose preparations, result in any reduction in the length of the illness. Alcohol, analgaesics, sleeping tablets, laxatives and sulphonamide are to be avoided. In serious cases or where coma is imminent gluco-corticoids (suprarenal hormone preparations) are indicated; they should not, however, be given in cases of mild or only moderately severe illness.

WEISS, G.

"Viral hepatitis". Krugman, S. Ch. 4 (61-78). *Human diseases caused by viruses. Recent developments* (New York, Oxford University Press, 1978), 361 p. Ref.

"Viral hepatitis". Ch. 14 (175-192). *Diagnostic Methods in Clinical Virology.* Grist, N. R.; Bell, E. J.; Follett, E. A. C.; Urquhart, G. E. J. (Oxford, London, Edinburgh, Melbourne, Blackwell Scientific Publications, 3rd ed., 1979), 229 p. Ref.

"Viral hepatitis: an occupational disease of hospital personnel" (Hépatites virales, maladies professionnelles du personnel hospitalier). VIII Journées nationales de Médecine préventive du personnel hospitalier. *Archives des maladies professionnelles, de médecine du travail et de sécurité sociale* (Paris), Sep. 1978, 39/9 (543-576). Illus. Ref. (In French)

"The question of infectiousness of HB antigen-positive medical personnel" (Zur Frage der Infektiosität von HB_s-Antigen-positivem medizinischem Personal). Scheiermann, N.; Kuwert, E. K.; Dermietzel, R. *Deutsche medizinische Wochenschrift* (Hannover), 16 Mar. 1979, 104/11 (381-384). 24 ref. (In German)

"Prohibition of practice for dentists with unfavorable hepatitis B serology?" (Berufsausübungsverbot für Zahnarzt bei ungünstiger Hepatitis-B-Serologie?). Kuntz, E.; Kuntz, A. *Muenchener Medizinische Wochenschrift* (Munich), 27 Oct. 1978, 120 (1 407-1 410). 22 ref. (In German)

"The chronicle of viral hepatitis". Zuckerman, A. J. *Abstracts on Hygiene* (London), Nov. 1979, 54/11 (1 113-1 135). 119 ref.

Herbicides

Grassy and broad-leaved weeds compete with crop plants for light, space, water and nutrients. They are hosts to bacteria, fungi, and viruses and hamper mechanical harvesting operations. Losses in crop yields as a result of weed infestation can be very heavy, commonly reaching 20-40%. Weed control measures such as hand weeding and hoeing are ineffective in intensive farming. Chemical weedkillers or herbicides have successfully replaced mechanical methods of weed control.

In addition to their use in agriculture in cereals, meadows, open fields, pastures, fruit growing, greenhouses and forestry, herbicides are applied on industrial sites, railway tracks and power lines to remove vegetation. They are used for destroying weeds in canals, drainage channels, and natural or artificial pools.

Herbicides are sprayed or dusted on weeds or on the soil they infest. They remain on the leaves (contact herbicides) or penetrate into the plant and so disturb its physiology (systemic herbicides). They are classified as non-selective (total)–used to kill all vegetation–and selective–used to suppress the growth or kill weeds without damaging the crop. Both non-selective and selective can be contact or systemic.

Selectivity is true when the herbicide applied in the correct dose and at the right time is active against certain species of weed only. An example of true selective herbicides are the chlorophenoxy compounds, which affect broad-leaved, but not grassy plants. Selectivity can also be achieved by placement, i.e. by using the herbicide in such a way that it comes into contact with the weeds only. For example paraquat is applied to orchard crops where it is easy to avoid the foliage. Three types of selectivity are distinguished:

(a) physiological selectivity, which relies upon the plant's ability to degrade the herbicide into non-phytotoxic components;

(b) physical selectivity, which exploits the particular habit of the cultivated plant (e.g. the upright in cereals), and/or a specially fashioned surface (wax-coating, resistant cuticule) protecting the plant against herbicide penetration;

(c) positional selectivity, in which the herbicide remains fixed in the upper soil layers adsorbed on colloidal soil particles and does not reach the root zone of the cultivated plant, or at least not in harmful quantities. Positional selectivity depends on the soil, precipitation and temperature as well as the water solubility and soil adsorption of the herbicide.

EFFECTS ON ENVIRONMENT AND HEALTH

Herbicides are decomposed chemically or biologically by bacteria and fungi in the soil, from which they disappear usually within several months but not later than 1-2 years. Soils of high organic or clay content can delay breakdown. Herbicides applied to foliage are rapidly destroyed photochemically or chemically within 2 to 6 weeks.

The toxicity of most herbicides to mammals, birds and bees is low, so no important direct effects on wildlife have been observed. No adverse actions on soil organisms have been reported.

Herbicides are generally of low acute and cumulative toxicity. When properly used, they seldom present a hazard. A few cases of acute or chronic occupational poisonings have been reported. Irritation of the skin, eyes and upper respiratory ways may result if protective measures are not taken. In certain samples of 2,4,5-T contamination by 2,3,7,8-tetrachlorodibenzo-*p*-dioxin (TCDD) was found to exceed by a factor of 50 the limit allowed in commercial formulation (see DIOXIN).

Nowadays strict chemical and biological control of TCDD in chlorophenoxyacetic herbicides is required. Cases of severe poisoning due to paraquat ingestion from non-standard containers (e.g. soft drink or wine bottles) have been reported. The new high-viscosity formulation should prevent such accidental ingestion in the future.

Aminotriazole

See separate article.

Atrazine ($C_8H_{14}N_5Cl$)

2-CHLORO-4-ETHYLAMINE-6-ISOPROPYLAMINO-S-TRIAZINE

m.w. 215.6
m.p. 174 °C

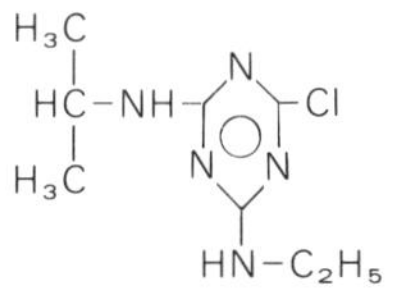

H_3C–CH(CH_3)–NH–[triazine ring: Cl; NH–C_2H_5]

slightly soluble in water and organic solvents; slightly volatile; decomposes in soil after 1-2 years

a white powder.

TLV ACGIH 10 mg/m³

MAC USSR 2 mg/m³ in working areas

Oral LD_{50} for rats 1 400-3 330 mg/kg, mice 1 570 mg/kg. Skin irritation and dermal toxicity after repeated application of 1 000 mg/kg in rabbits. Toxic concentration 1 200 mg/m³ in the air. Moderate cumulative toxicity.

Gives rise to decreased body weight, anaemia, disturbed protein and glucose metabolism in rats. Causes occupational contact dermatitis due to skin sensitisation.

Barban ($C_{11}H_9Cl_2NO_2$)

4-CHLORO-2-BUTYLYNYL-N-(3-CHLOROPHENYL)CARBAMATE; *m*-CHLOROCARBANILIC ACID 4-CHLORO-2-BUTYNYL ESTER; CARBYNE

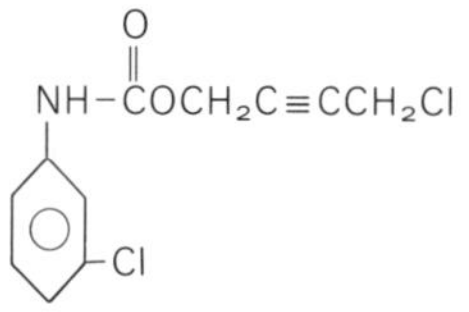

NH–C(=O)OCH₂C≡CCH₂Cl on benzene ring with Cl

m.w. 258.1
m.p. 78 °C

slightly soluble in water; soluble in organic solvents; slightly volatile; decomposes within 28-38 days of the last plant spraying

white crystals.

MAC USSR 0.5 mg/m³

Oral LD_{50} for rats, mice, rabbits 600 to 820 mg/kg. Dermal LD_{50} for rabbits more than 1 000 mg/kg. Threshold concentration after single inhalation in rats 80 mg/m³. Repeated contact with 5% water emulsion results in severe skin irritation in rabbits. Provokes skin sensitisation in both experimental animals and agricultural workers.

Causes anaemia, methaemoglobinaemia, changes in lipid and protein metabolism. Ataxia, tremor, cramps, bradycardia and ECG deviations are found in experimental animals.

Chlorpropharm ($C_{10}H_{12}ClNO_2$)

ISOPROPYL-N-(3-CHLOROPHENYL)CARBAMATE; ISOPROPYL-*m*-CHLOROCARBANILATE

m.w. 213.7
m.p. 41 °C
b.p. 247 °C

slightly soluble in water (80 mg/l); soluble in organic solvents; slightly volatile; decomposes in soil 4 months after use

white crystals.

MAC USSR 2 mg/m³

Oral LD_{50} for rats 1 500-3 000 mg/kg. Slight dermal irritation and penetration. Low cumulative toxicity. Threshold concentration after single aerosol inhalation 45 mg/m³.

Causes anaemia, methaemoglobinaemia, reticulocytosis. Chronic application causes skin carcinoma in rats.

Cycloate ($C_{11}H_{21}NOS$)

S-ETHYL-ETHYLCYCLOHEXYL-THIOCARBAMATE; RO-NEET; HEXYLTHIOCARBAM

m.w. 215
b.p. 140 °C

slightly soluble in water; soluble in organic solvents; slightly volatile; decomposes 2 days after the application

clear oily liquid.

MAC USSR 1 mg/m³

Oral LD_{50} for rats and mice 2 300 mg/kg. Dermal LD_{50} for rabbits more than 4 640 mg/kg. Slightly irritant for skin, more so for eyes. Toxic concentration after single inhalation 500 mg/m³ in cats. Low cumulative toxicity.

Causes polyneuropathia, liver damage in experimental animals. No clinical symptoms are described after occupational exposure of workers for 3 consecutive days.

2,4-D ($C_8H_{16}Cl_2O_3$)

2,4-DICHLOROPHENOXYACETIC ACID

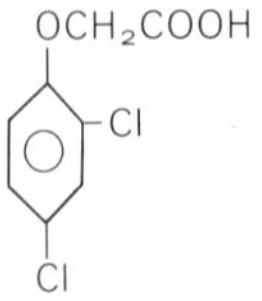

m.w. 221
m.p. 141 °C
b.p. 160 °C
v.d. 7.6

almost insoluble in water; highly volatile; decomposes in the soil 1-2 months after application

white crystals.

TWA OSHA	10 mg/m³
IDLH	500 mg/m³
MAC USSR	1 mg/m³

Oral LD_{50} for rats 730 mg/kg, mice 360 mg/kg. Moderate dermal toxicity and skin irritancy. Highly irritant for eyes. Moderate cumulative toxicity.

Acute poisoning in workers provokes headache, dizziness, nausea, vomiting, raised temperature, low blood pressure, leucocytoxic, heart and liver injury. Chronic occupational exposure without protection may cause nausea, liver functional changes, contact toxic dermatitis, irritation of airways and eyes, as well as neurological changes. Four sprayers had no clinical symptoms during a week of work with mean 24-h excretion of 2,4-D in the urine of 9 mg.

Analogous effects are shown by other 2,4-D salts (ammonium salt with MAC USSR 0.5 mg/m³) or esters (butyl ester with MAC USSR 1 mg/m³ and octyl and chlorooctyl esters with MAC USSR 1 mg/m³). Some of the derivatives are embryotoxic and teratogenic for experimental animals in high doses only.

2,4,5-T ($Cl_3C_6H_2OCH_2COOH$)

2,4,5-TRICHLOROPHENYXOYACETIC ACID

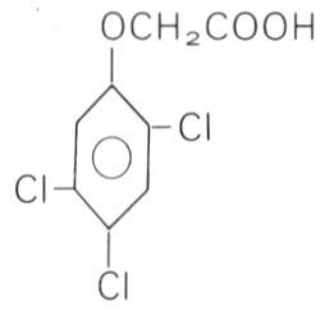

m.w. 255.5
m.p. 158 °C

insoluble in water

white crystals.

TWA OSHA	10 mg/m³
STEL ACGIH	20 mg/m³
IDLH	5 000 mg/m³

2,4,5-TCPPA ($Cl_3C_6H_2OCH_2CH_2COOH$)

2,4,5-TRICHLOROPHENOXYPROPIONIC ACID

m.w. 269.5
m.p. 182 °C

These two are of similar toxicity and require the same hygiene standards. Because of their pronounced irritation, embryotoxic, teratogenic and carcinogenic effects in animals, as well as data on gonadotoxic action in women, their use is prohibited in many countries. The limit of TCDD as contaminant in the technical products is less than 0.5 ppm. [According to recent investigations occupational exposure to phenoxy acids increases the risk of soft-tissue sarcomas.]

Dalapon-Na (CH_3HCl_2COONa)

SODIUM-2,2-DICHLOROPROPIONATE

m.w. 165
m.p. 166 °C

slightly soluble in water; slightly soluble in organic solvents; slightly volatile

a white powder.

Oral LD_{50} for mice 7 100 mg/kg, rats 9 250 mg/kg. Slight skin and eye irritancy. Low cumulative toxicity.

Causes depression, unbalanced gait, decreased body weight, kidney and liver changes, thyroid and pituitary dysfunctions, and contact dermatitis in workers.

Diallate ($[(CH_3)_2CH]_2NCOSCH_2C(Cl){:}CHCl$)

S-2,3-DICHLOROALLYL-N,N-DIISOPROPYL-THIOCARBAMATE; AVADEX

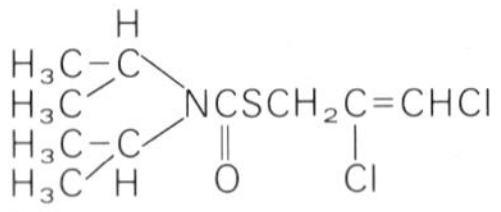

m.w. 270.2
m.p. 30 °C

slightly soluble in water; soluble in organic solvents; slightly volatile; decomposes in soil 2-3 months after application

a liquid with an amber colour.

Oral LD_{50} for rats 395 mg/kg, dogs 510 mg/kg. Dermal LD_{50} for rabbits 2 000 mg/kg. The concentrate has dermal toxicity and causes irritation to the skin, eyes and mucous membranes.

Diquat ($(C_5H_4NCH_2)_2Br_2$)

1,1'-ETHYLENE-2,2-BIPYRIDYLIUM DIBROMIDE

m.w. 344
m.p. 335 °C

very soluble in water; slightly volatile

white crystals.

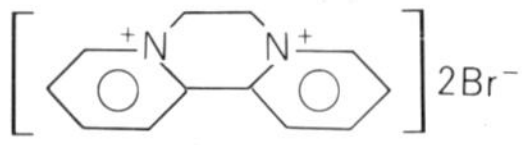

TLV ACGIH	0.5 mg/m³
STEL ACGIH	1 mg/m³
MAC Bulgaria	0.1 mg/m³

Oral LD_{50} for rats and rabbits 230 mg/kg, mice 80 mg/kg. Dermal LD_{50} for rabbits and rats 730 mg/kg. Low cumulative toxicity. Irritant for the skin, eyes and upper respiratory ways. Causes a delay in the healing of cuts and wounds, gastrointestinal and respiratory disturbances, bilateral cataract and functional liver and kidney changes.

Dinoseb ($C_{10}H_{12}N_2O_5$)

DNBP; 2-(1-METHYL-N-PROPYL)-4,6-DINITROPHENOL

m.w. 240.2
m.p. 40 °C

slightly soluble in water; soluble in organic solvents

orange-brown liquid.

MAC USSR 0.1 mg/m³

Oral LD_{50} for rats 58 mg/kg. Dermal LD_{50} for rabbits 80-200 mg/kg. Dangerous because of its dermal toxicity. Moderate skin and pronounced eye irritation. Moderate cumulative toxicity. The fatal dose for man is about 1-3 g.

Causes CNS disturbances, vomiting, erythema of the skin, sweating, high temperature after acute poisoning in man. Chronic exposure without protection results in decreased weight, contact (toxic or allergic), dermatitis and gastrointestinal, liver and kidney disturbances.

Because of its adverse effects not used in many countries.

Fluometuron ($C_{10}H_{11}F_3N_2O$)

1,1-DIMETHYL-3-3-TRIFLUOROMETHYLPHENYL UREA; COTORAN; LANEX

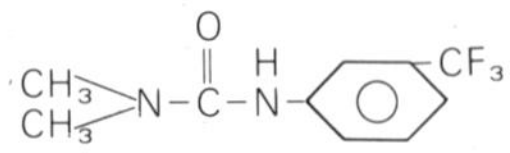

m.w. 232
m.p. 163 °C

slightly soluble in water; soluble in organic solvents; stable in the soil

white crystals.

MAC USSR 0.5 mg/m³

Oral LD_{50} for rats 8 900 mg/kg. The concentrate causes irritation to the eyes and mucous membranes. Low cumulative toxicity. Moderate skin sensitiser in guinea-pigs and men.

Caused decreased body weight, anaemia, liver, spleen and thyroid gland disturbances. The biological action of diuron is similar.

Linuron ($C_9H_1 Cl_2N_2O_3$)

3-3,4-DICHLOROPHENYL-1-METHOXY-1-METHYLUREA; HOROX

m.w. 214
m.p. 94 °C

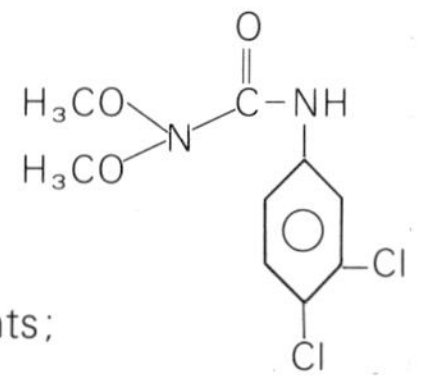

slightly soluble in water;
soluble in organic solvents;
decomposes in the soil 4 months after application

MAC USSR 0.5 mg/m³

Oral LD_{50} for rats and mice 2 400 mg/kg. Dermal LD_{50} for rabbits more than 2 000 mg/kg. Causes mild irritation to the skin and eyes. Low cumulative toxicity. Threshold value after single inhalation 29 mg/m³.

Causes CNS, liver, lung and kidney changes in experimental animals, as well as thyroid dysfunction.

MCPA ($C_9H_9O_3Cl$)

4-CHLORO-2-METHYLPHENOXYACETIC ACID; METHAXONE

m.w. 200.6
m.p. 120 °C

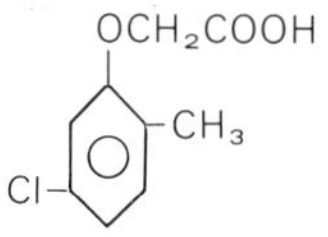

slightly soluble in water;
stable for 1 year in the soil after application, but decomposes in the plants

white solid.

Oral LD_{50} for rats 700 mg/kg, mice 550 mg/kg. Highly irritant to skin and mucous membranes. Low cumulative toxicity. Embryotoxic and teratogenic in high doses in rabbits and rats.

Acute poisoning of man (about 300 mg/kg) resulted in vomiting, diarrhoea, cyanosis, mucus burns, clonic spasms, myocardium and liver injury. Provokes severe contact toxic dermatitis in workers. Chronic exposure without protection results in dizziness, nausea, vomiting, stomach aches, hypotonia, enlarged liver, myocardium dysfunction and contact dermatitis.

Molinate ($C_9H_{17}NOS$)

S-ETHYL-N,N-HEXAMETHYLENE-THIOCARBAMATE; ORDRAM

m.w. 187
b.p. 137 °C

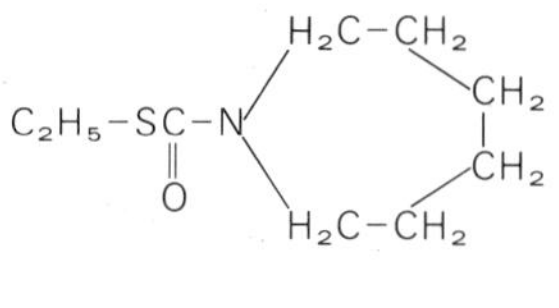

slightly soluble in water;
slightly volatile;
decomposes in the soil 45 days after application

light-brown liquid.

MAC USSR 0.5 mg/m³

Oral LD_{50} for rats 530-657 mg/kg. Dermal LD_{50} for rabbits 1 167 mg/kg. Low cumulative toxicity. Toxic concentration after single inhalation 200 mg/m³ in rats.

Causes liver, kidney and thyroid disturbances. Gonadotoxic and teratogenic in rats. No clinical symptoms or laboratory tests deviations were observed 3 days after daily occupational exposure to 0.2-5 mg/m³. Moderate skin sensitiser in man.

Monuron ($C_9H_{11}ClN_2O$)

CMU; 3-(*p*-CHLOROPHENYL)-1,1-DIMETHYLUREA

m.w. 198.6
m.p. 171 °C

slightly soluble in water; stable in the soil 2 years after application

white crystals.

MAC USSR 2 mg/m³

Oral LD_{50} for guinea-pigs 1 500-3 700 mg/kg. Low cumulative toxicity. High doses result in liver, myocardium and kidney disturbances. Causes skin irritation sensitisation.

Similar effects are shown by monolinuron, chloroxuron, chlortoluron, dodine (for which MAC USSR is 0.1 mg/m³).

Nitrofen ($C_{12}H_7O_3NCl_3$)

2,4-DICHOROPHENYL-4-NITROPHENYL ETHER

m.w. 284.1
m.p. 70 °C

slightly soluble in water and slightly volatile

a white solid.

MAC USSR 1 mg/m³

Oral LD_{50} for rats 700 mg/kg, mice 450 mg/kg. Low dermal and cumulative toxicity. Strong skin and eye irritant.

Chronic occupational exposure without protection results in CNS disturbances, anaemia, raised temperature, decreased body weight, fatigue, contact dermatitis.

Paraquat ($C_{14}H_{20}N_2O_8S_2$)

1,1-DIMETHYL-4,4-DIPYRIDIUM DIMETHYLSULPHATE or DICHLORIDE; METHYL VIOLOGEN

m.w. 408

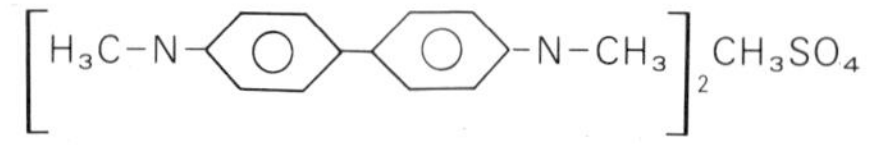

soluble in water; slightly volatile

white crystals.

TWA OSHA 0.5 mg/m³
TLV ACGIH 0.1 mg/m³, respirable sizes
IDLH 1.5 mg/m³
MAC Bulgaria 0.01 mg/m³

Oral LD_{50} for rats 140 mg/kg, mice 104 mg/kg. Dermal LD_{50} for rabbits 236 mg/kg. Has dermal toxicity and irritant effects on skin or mucous membranes. Causes nail damage and nose bleeding in occupational conditions without protection.

Accidental oral poisoning with paraquat is the result of leaving it within reach of children or transferring it from the original container into a bottle used for a beverage. Early manifestations of the intoxication are corrosive gastrointestinal effects, renal tubular damage and liver dysfunction. Death is due to circulatory collapse and progressive pulmonary damage (pulmonary oedema and haemorrhage, intra-alveolar and interstitial fibrosis with alveolitis and hyaline membranes), clinically revealed by dyspnoea, hypoxaemia, basal rales, roentgenographic evidence of infiltration and atelectasis. The renal failure is followed by lung damage, and accompanied in some cases by liver or myocardium disturbances. Mortality is higher with poisoning from liquid concentrate formulations (87.8%), and lower from granular forms (18.5%). The fatal dose is 6 g paraquat ion (equivalent to 30 ml "Gramoxone" or 4 sachets of "Weedol") and no survivors are reported at greater doses, irrespective of the time or vigour of treatment. Most survivors had taken less than 1 g paraquat ion.

Repeated gastric lavages with administration of Fuller's earth suspensions in saline, coupled with vigorous purgation, is the initial therapy. Additional measures are: forced diuresis (not in oliguric cases with renal damage), haemodialysis and haemoperfusion, together with corticosteroids and immunosuppressive agents. No oxygen is allowed in case of respiratory failure. Patients should be treated within 6 h of ingestion: the critical factor of doses lower than 6 g paraquat is the time at which treatment started. Results are poorer with a delay beyond 5 h whereas the chances are better with 2-h interval only.

Potassium cyanate (KOCN)

m.w. 81.1
sp.gr. 2.05
m.p. 700-900 °C (decomposes)
soluble in water
colourless crystals.

Oral LD_{50} for mice 841 mg/kg. High inhalation and dermal toxicity in experimental animals and man, related to a metabolic conversion to cyanide.

Prometryen ($C_{13}H_{19}N_5S$)

2,4-*bis*-ISOPROPYLAMINO-6-METHYLTHIO-1,3,5-TRIAZINE

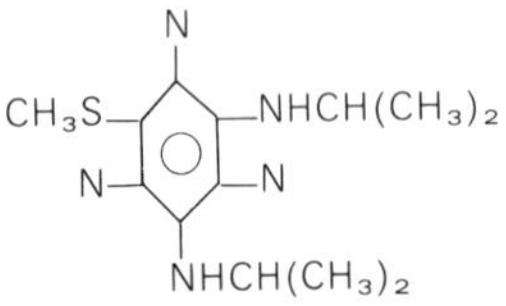

m.w. 241.1
m.p. 120 °C
slightly soluble in water and slightly volatile
white solid.
MAC USSR 5 mg/m³

N
CH_3S — NHCH$(CH_3)_2$
N — N
NHCH$(CH_3)_2$

Oral LD_{50} for rats and mice 1 800-3 150 mg/kg. Moderate dermal toxicity and skin and eye irritation. Low cumulative toxicity.

Provokes decreased clotting and enzyme deviations in animals. Embryotoxic in rats. Exposed workers may complain of nausea and sore throat.

Analogous effects are shown by propazine (MAC USSR 2 mg/m³), and desmetryne (MAC USSR 0.4 mg/m³).

Propachlor ($C_{11}H_{14}ONCl$)

2-CHLORO-N-ISOPROPYLACETANILIDE; RAMROD

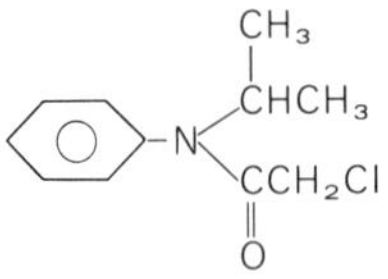

m.w. 211.5
m.p. 79 °C
slightly soluble in water; soluble in organic solvents
white crystals.
MAC USSR 0.2 mg/m³

CH_3
|
CHCH$_3$
—N
CCH$_2$Cl
‖
O

Oral LD_{50} for rats 1 100 mg/kg, mice and rabbits 500 mg/kg. Dermal LD_{50} for rabbits 580 mg/kg. Toxicity is doubled at high environmental temperatures. Skin and mucous membrane irritation and mild skin allergy. Toxic concentration after single inhalation 18 mg/m³ in rats. Moderate cumulative toxicity.

Causes polyneuropathia; liver, myocardium and kidney disturbances, anaemia and damage to testes in rats. During spraying from the air the concentration in the cabin is about 0.2-0.6 mg/m³.

Similar toxicity is shown by propanil (MAC USSR 0.2 mg/m³).

Propham ($C_6H_5NHCOOCH(CH_3)_2$)

IPC; ISOPROPYL-N-PHENYLCARBAMATE

m.w. 179
m.p. 87 °C
slightly soluble in water; soluble in organic solvents; slightly volatile
white crystals.
MAC USSR 2 mg/m³

Oral LD_{50} for rats 2 700-7 400 mg/kg. Dermal LD_{50} for rabbits more than 2 000 mg/kg. Threshold concentration after single inhalation 54.4 mg/m³ in rats. Moderate cumulative toxicity.

Haemodynamic disturbances and liver, lung and kidney changes are found in experimental animals.

Simazine ($C_7H_{12}N_5Cl$)

2-CHLORO-4,6-*bis*-ETHYLAMINO-1,3,5-TRIAZINE

m.w. 201.7
m.p. 227 °C
slightly soluble in water and organic solvents; slightly volatile; stable in the soil for 2 years
white crystals.
MAC USSR 2 mg/m³

Oral LD_{50} for rats 1 390 mg/kg, mice 4 100 mg/kg. Dermal LD_{50} for rabbits more than 10 200 mg/kg. Slight irritation of the skin and mucous membranes. Low cumulative toxicity. Moderate skin sensitisation in guinea-pigs.

Causes CNS, liver and kidney disturbances and has mutagenic effect in experimental animals. Workers may complain of weariness, dizziness, nausea and olfactory deviations after application without protective equipment.

Trifluralin ($CF_3(NO_2)_2C_6H_2N(C_3H_7)_2$)

TRIFLUORO-2,6-DINITRO-*n*,*n*-DIPROPYL-*p*-TOLUIDINE

m.w. 335
m.p. 48.5 °C
b.p. 139 °C
slightly soluble in water; slightly volatile
yellow-orange solid.
MAC USSR 3 mg/m³

Oral LD_{50} for rats and mice 500-5 000 mg/kg. Slight irritation of skin and mucous membranes. Low cumulative toxicity. An increased incidence of liver carcinoma has been found in hybrid female mice probably due to contamination with N-nitroso compounds.

Causes anaemia and liver, myocardium and kidney changes in experimental animals. Extensively exposed workers have developed contact dermatitis and photodermatitis.

GENERAL SAFETY AND HEALTH MEASURES

Adverse effects may be avoided by following the instructions given on the labels of these products. Their general toxicity, irritation and sensitisation potentials and the mode of use determine the protection necessary in each case. Skin contamination is a hazard especially in countries where heat and humidity preclude the use of protective clothing. Irritation of the skin and mucous membranes may result from unwashed spillings of the commercial concentrates. Extended contacts are hazardous when the highest recommended application rates of working dilutions are used in summer or in greenhouses. The following general rules are important:

(a) People handling herbicides should undergo strict training. Preliminary medical examination to detect

chronic diseases of the CNS, liver, heart, kidneys, lungs and skin, as well as endocrinological or immunological disturbances, should eliminate susceptible individuals. The manner of use should be chosen only by a specialist. Leaking knapsack sprayers are dangerous. Herbicides should be applied only as indicated on the label and should be stored in original labelled containers out of the reach of children.

(*b*) The use of personal protective equipment such as glasses, synthetic gloves and face masks is important. Only clean clothing and underwear should be worn and clothing should be changed daily for freshly washed clothes. Contaminated equipment should not be taken home. Adequate sanitary facilities and washing water should be provided for workers to wash before meals. Smoking and the consumption of alcoholic drinks before or during the handling of herbicides should be forbidden. Contaminated clothing should be removed immediately and a hot bath taken if possible. Personal hygiene should be encouraged.

(*c*) Herbicides should not be removed from their original containers. Injurious concentrates should be used in small containers that are easy to handle and to destroy. Empty packages should be rinsed, paper bags and fibre drums burnt; and workers should keep clear of the resulting smoke. Non-combustible containers should be crushed and buried under more than 40 cm of soil. They should not be taken home or used at home.

(*d*) Periodical medical examination of internal organs, skin and eyes is important to avoid chronic occupational intoxications. It should include laboratory tests and patch tests if necessary.

The rules mentioned above apply equally to factories producing herbicides. Fully enclosed processes, remote control and exhaust ventilation should be used.

In case of overexposure to herbicides the victim should be immediately removed from exposure to the fresh air, thoroughly washed, and his clothing changed. Complete rest and qualified medical care should be provided. Treatment is symptomatic and should be carried out in hospital.

BAINOVA, A.

Phytosanitary index (Index Phytosanitaire). Bailly R.; Duboid G. (eds.). (Association de coordination technique agricole, 149 rue de Bercy, Paris) (5th ed., 1979), 407 p. (In French)

Pesticide Handbook (Spravochnik no Pesticidam). Medved, L. I. (ed.). (Kiev, Urozai, 2nd ed., 1977), 375 p. (In Russian)

Pesticide manual. Martin, H. (ed.). (British Crop Protection Council, Clacks Farm, Boreley, Worcester) (3rd ed., 1972), 535 p.

Pesticide Dictionary. Shepard, H. H. (ed.). (Willoughby, Ohio, Meister Publishing Co., 1978), 291 p.

Report on the potential dangers of the use of chemical herbicides. Brugnon, M. Parliamentary Assembly of the Council of Europe. 18 Oct. 1977, Doc. 4067, 20 p.

"Soft-tissue sarcomas and exposure to chemical substances: a case-referent study". Eriksson, M.; Hardell, L.; Berg, N. O.; Moller, T.; Axelson, O. *British Journal of Industrial Medicine* (London), Feb. 1981, 38/1 (27-33). 30 ref.

Chemical classification of most widely used herbicides (Common names and chemical names)

Anilides

Alachlor, 2-chloro-2′,6′-diethyl-N-(methoxymethyl)-acetanilide

Carbetamide, D-1-(ethylcarbamoyl) ethyl phenylcarbamate

Difenamid, N,N-dimethyl-diphenyl-2,2-acetamide

Monalide, N-(4-chlorophenyl)-2,2-dimethylvaleriamide

Napromid, 2-(alpha-naphtoxy)-N,N-diethyl-propionamide

Pentachlor, N-(3-chloro-4-methyl phenyl)-2-methyl-pentamide

Propachlor, 2-chloro-N-isopropyl acetanilide

Propanyl, N-(3,4-dichlorophenyl)-propionamide

Benzonitriles

Bromoxynil, 3,5-dibromo-4-hydroxy-benzonitrile

Chlorthiamid, 2,6-dichloro thiobenzamide

Dichlobenil, 2,6-dichloro-benzonitrile

Ioxynil, 4-hydroxy-3,5-diiodo-benzonitrile

Propyzamide, 3,5-dichloro-N-(1,1-dimethyl-2-propyl) benzamide

Bipyridyliums

Diquat, 1,1′-ethylene-2,2′-bipyridylium dibromide

Paraquat, 1,1′-dimethyl-4,4′-bipyridylium dichloride or dimethylsulphate

Carbamates

Asulam, methyl(4-aminobenzen-sulphonyl) carbamate

Barban, 4-chlorobut-2-ynyl-N-(3-chlorophenyl) carbamate

Chlorbufam, 1-methylprop-2-ynyl-N-(3-chlorophenyl) carbamate

Chlorpropham, isopropyl-N-(3-chlorophenyl) carbamate

Methiocarb, 4-methylthio-3,4-xylyl-N-methyl carbamate

Pebulate, S-propyl-buthylethiol carbamate

Phenmedipham, methyl-3-(*m*-tolyl-carbamoyloxy)-phenyl carbamate

Terbutol, 2,6-di-*t*-butyl-4-methyl phenyl-N-methyl carbamate

Propham, isopropyl-N-phenyl carbamate

Thiocarbamates, dithiocarbamates

Butylate, S-ethyl-N,N-diisobutyl-thiocarbamate

Cycloate, S-ethyl-N-cyclohexyl-N-ethyl-thiocarbamate

Diallate, S-2,3-dichloroallyl-N,N-diisopropyl-thiocarbamate

Metham–sodium, methyl dithiocarbamic acid

Molinate, S-ethyl-N,N-hexa-methylene-thiocarbamate

Nabam, disodium ethylene-*bis*-dithiocarbamate

Triallate, S-(2,3,3-trichloro-allyl-N,N-di-(isopropyl)-thiocarbamate

Sulfallate, 2-chloroallyl-N,N-diethyl-dithiocarbamate

Diazines

Bentazone, 3-isopropyl-(1H)-benzo-2,1,3-thiadiazin-4-one-2,2-dioxide

Bromacil, 5-bromo-6-methyl-3-*s*-butyluracil

Brompyrazon, 5-amino-4-bromo-2-phenyl-pyridasin-3-one

Dazomet, tetrahydro-3,5-dimethyl-2H-1,3,5-thiadiazine-2-thione

Lenacil, 3-cyclohexyl-6,7-dihydro-1H-cyclopentapyrimidine-2,4-(3H, 5H)-dione

Maleic hydrazide, 1,2-dihydro-3,6-pyridazinedione

Terbacil, 3-t-butyl-5-chloro-6-methyl-uracil

Nitro compounds

Dinoseb, 2-(1-methyl-*n*-propyl)-4,6-dinitrophenol

Dinoterb, 2-*t*-butyl-4,6-dinitrophenol

DNOC, 2-methyl-4,6-dinitrophenol

Nitrofen, 2,4-dichlorophenyl-4-nitrophenyl ether

Phenoxy compounds

2,4-D, 2,4-dichlorophenoxyacetic acid

2,4-DB, 4-(2,4-dichlorophenoxy) butyric acid

Dichlorprop, 2-(2,4-dichlorophenoxy) propionic acid

Chemical classification of most widely used herbicides (Common names and chemical names)

Fenoprop, 2,4,5-trichlorophenoxy propionic acid

MCPB, 4-(4-chloro-2-methyl-phenoxy) butyric acid

Mecoprop, 2-(4-chloro-2-methyl-phenoxy) propionic acid

2,4,5-T, 2,4,5-trichlorophenoxy acetic acid

Toluidines

Benfluralin, N-butyl-N-ethyl-2,6-dinitro-4-fluoromethylaniline

Butralin, dinitro-2,6-*tert*-butyl-4-N-*sec*-butylaniline

Dimetachlor, 2,6-dimethyl-N-(2-methoxyethyl)-chloro-acetanilide

Ethalfluralin, 2,6-dinitro-N-(2-methyl-2-propyl)-4-trifluoro-methylaniline

Fluorodifen, 2,4-dinitro-4-tri-fluoromethyl diphenyl ether

Nitralin, 4-(methylsulphonyl)-2,6-dinitro-N,N-dipropylaniline

Penoxaline, N-(ethylpropyl-1)-dinitro-2,6-xylidine-3,4

Trifluralin, 2,6-dinitro-N,N-dipropyl-4-trifluoromethylaniline

Triazines

Ametryne, 2-ethylamino-4-isopropyl amino-6-methylthio-1,3,5-triazine

Atrazine, 2-chloro-4-ethylamino-6-isopropylamino-1,3,5-triazine

Cynatrine, ethylamino-methylthio-*s*-triazine-2-yl-amino-2-methyl-propionitrile

Cynazine, 2-(4-chloro-6-ethylamino-1,3,5-triazon-2-ylamino)-2-methyl-propionitrile

Desmetryne, 2-isopropylamino-4-methylamino-6-methylthio-1,3,5-triazine

Diprometryn, 6-ethylamino-2,4-*bis*-(isopropylamino)-*s*-triazine

Metribuzin, 4-amino-6-*tert*-butyl-4,5-hydro-3 methylthio-1,2,4-triazin-5-one

Prometon, 2,4-*bis*-isopropylamino-6-methoxy-1,3,5-triazine

Propazine, 2-chloro-4,6-*bis*-iso-propylamino-1,3,5-triazine

Simasine, 2-chloro-4,6-*bis*-(ethyl-amino)-1,3,5-triazine

Terbutilazine, 2-*tert*-butylamino-4-chloro-6-ethylamino-*s*-triazine

Terbutryne, 4-ethylamino-2-methyl-thio-6-*tert*-butylamino-1,3,5-triazine

Urea compounds

Chloroxuron, N-4-(4-chlorophenoxy) phenyl-N,N-dimethylurea

Chlorotoluron, 3-(4-4-chlorophenoxy phenyl)-1,1-dimethylurea

Cycluron, N-cyclo-octyl-N,N-dimethylurea

Diuron, 3-(3,4-dichlorophenyl)-1,1-dimethylurea

Isonoruron, N-/1-or 2-(3a,4,5,6, 7,7a-hexahydro-4,7-methanoidanyl)/-N,N-dimethylurea

Isoproturon, isopropylphenyl-N,N-dimethylurea

Fenuron, 1,1-dimethyl-3-phenylurea

Fluometuron, 1,1-dimethyl-3-(3-fluoromethylphenyl) urea

Linuron, 3-(3,4-dichlorophenyl)-1-methoxy-1-methylurea

Methazole, 2-(3,4-dichlorophenyl)-4-methyl-1,2,4-oxadiazolidine-3,5-dione

Metobromuron, 3-(4-bromophenyl)-1-methoxy-1-methylurea

Metoxuron, 3-(3-chloro-4-methoxy-phenyl)-1,1-dimethylurea

Monolinuron, 3-(4-chlorophenyl)-1-methoxy-1-methylurea

Monuron, 3-(4-chlorophenyl)-1,1-dimethylurea

Neburon, 1-*n*-butyl-3-(3,4-di-chlorophenyl)-1-methylurea

Noruron, 3-(hexahydro-4,7-methano-indan-5-yl)-1,1-dimethylurea

Phenobenzuron, benzoyl-dichloro-phenyl-1,1-dimethylurea

Siduron, 1-(2-methylcyclohexyl)-3-phenylurea

Miscellaneous

Aminotriazole, 3-amino-1,2,4-triazole

Benazolin, 4-chloro-2-oxobenzo-thiazolin-3-yl-acetic acid

3,6–DCP, 3,6-dichloro-picolinic acid

Dalapon–Na, sodium 2,2-di-chloropropionate

Dicamba, 3,6-dichloro-2-methoxy-benzoic acid

Dimexan, di-(methoxythiocarbonyl)-disulphide

Ethofumesate, 2-ethoxy-2,3-dihydro-3,3-dimethylbenzofuran 5-yl-methanesulphonate

Flurecol-butyl, 9-hydroxyfluorene-9-carboxylic acid

Glyphosate, N-phosphonomethyl-glycine

Hexaflurate, potassium hexafluoro-arsenate

Naptalam, N-1-naphthylphthalamic acid

Oils, petrolatum products

Oxadiazon, 5-*tert*-butyl-3-(2,4-dichloro-5-isopropoxyphenyl)-1,3,4-oxadiazol-2-one

Picloram, 4-amino-3,5,6-trichloro-picolinic acid

Propazon, 5-amino-4-chloro-2-phenyl-3-pyridazone

Sodium chlorate

Sodium metaborate

Sulphuric acid, brown oil or vitriol

2,3,6–TBA, 2,3,6-trichloro-benzoic acid

Source: "Toxicology of herbicides with special reference to the bipyridiliums". Rose, M. S.; Smith, L. L.; Wyatt, I. *Annals of Occupational Hygiene* (Oxford), 1980, 23/1 (91-94).

Hernia

Hernia is the abnormal protrusion of viscera outside the abdominal wall. There are many anatomical variations of hernia that come within the province of the surgeon and many different factors may play a part in hernia causation; consequently, the occurrence of hernia may give rise to numerous medico-legal problems.

The abdominal walls have orifices (inguinal canals) which may become enlarged or progressively distended and allow the passage of a hernia, the onset of which may be slow or imperceptible. The process is often the result of age and independent of the subject's work. This is the "hernia of weakness" or "disease hernia", a chronic complaint, without acute symptomatology and often without an immediate precipitating cause or following a slight effort such as an attack of coughing.

Conversely, the much rarer "stress hernia" or "accidental hernia" is the immediate result of a violent effort made whilst the body is badly positioned; it is a surgical emergency with dramatic symptoms.

Occupational causes

Generally, hernia is of mixed origin; effort increases or reveals a distension but is not the sole cause, for over and above the effort habitually expended by the patient, other, sometimes contradictory factors, are involved; individual anatomical predisposition, heavy work (stress hernia) but also sedentary work (weakness hernia) where lack of exercise may cause the abdominal walls to be weak. Good examples are the risks to bulldozer drivers when fatigued abdominal walls are submitted to the pressure of continual shocks from vibrations and the risks involved in changes in occupation as when a sedentary worker undertakes occasional heavy work.

Hernias are attributed, more or less correctly, to a wide variety of jobs. These most frequently incriminated include heavy manual work, including lifting and carrying and moving heavy objects, especially when these jobs are incidental to the main occupation. Hernias are encountered in persons with such differing heavy professions as metal industry worker, miner, porter, all-terrain vehicle driver, etc. However, even a slight effort may suffice to produce a hernia in a sedentary worker with a weak abdominal wall.

PREVENTIVE MEASURES

Prevention should comprise regular medical examinations and studies of work conditions and practices. Careful and unhurried pre-employment and periodical examinations will ensure the early detection of an incipient hernia or of a major predisposition. These examinations, supplemented by job studies, will ensure optimum placement of workers. Heavy tasks, especially those involving effort in abnormal positions (mining or quarrying, certain types of manual handling, etc.), or repeated jarring are contraindicated for known or potential sufferers. This contraindication should be clearly pointed out. The wearing of a truss, or a change of employment (often difficult to arrange and not easily accepted) is appropriate for the elderly. Surgery will restore working capacity in the young.

Much can be done to remove the conditions likely to produce hernia. Many jobs can be mechanised, many methods of work can be improved by applying ergonomic principles; if manual handling cannot be eliminated, workers can be trained in the best lifting and carrying procedures. The doctor should collaborate with the engineers and other responsible persons in devising improvements. In many cases, long and exacting efforts are required to find a solution. It is most important to secure the confidence of the workers in dealing with a subject that may have both medical and legal consequences.

Legal problems

Although hernia is generally a clinically benign condition, it may involve delicate legal questions since it is often difficult to determine with precision the part that occupational factors have had to play in the development. The circumstances surrounding the accident that led to the injury are of decisive importance in determining the true causes—and these circumstances are frequently disputed.

The legal concept of "occupational hernia" differs from country to country, and extensive case law on the subject may have been constituted. In some countries, occupational attribution is admitted when the person concerned (who has the onus of proof) can verify the existence of an effort made during his usual occupation and protrusion of a hernia as a direct result of this effort, even if he showed a predisposition beforehand. In others, the accident is not recognised unless the plaintiff personally received a direct order to perform some exceptional task.

CHAUDERON, J.

Economic implications:

"Sickness absence after inguinal herniorrhaphy". Griffiths, M.; Waters, W. E.; Acheson, E. D. *Journal of Epidemiology and Community Health* (London), June 1979, 33/2 (121-126). Illus. 11 ref.

"Time off work after herniorrhaphy". Semmence, A. *Journal of the Society of Occupational Medicine* (Bristol), 1973, 23/2 (36-48).

CIS 1057-1965 "Some economic aspects of inguino-femoral hernias". McCaffrey, J. F. *Medical Journal of Australia* (Sydney), Apr. 1965, 1/14 (491-494). 6 ref.

Hexachlorobutadiene

Hexachlorobutadiene (CCl_2:CClCCl:CCl_2)

HLBD

m.w.	260.8
sp.gr.	1.68
m.p.	−21 °C
b.p.	215 °C
v.d.	8.99
v.p.	376 mmHg (50·10^3 Pa) at 20 °C
i.t.	610 °C

insoluble in water; soluble in alcohol and ether.

TLV ACGIH 0.02 ppm 0.24 mg/m^3, industrial substance suspect of carcinogenic potential for man

Production. This compound is obtained as a subproduct of processes involving the chlorination of aliphatic hydrocarbons, especially in the preparation of tetrachloroethylene.

Uses. It is used as a solvent, as an intermediate in lubricant and rubber production, and as a pesticide for fumigation.

Toxicity. The most acute toxicity parameters are: oral LD_{50} for rats−90 mg/kg; intraperitoneal LD_{50} for rats−175 mg/kg; oral LD_{50} for mice−110 mg/kg; intraperitoneal LD_{50} for mice−76 mg/kg; dermal LD_{50} for rabbits−1 211 mg/kg. A single dose of 100 mg/kg administered to rats caused tubular necrosis in the kidneys, increase in urinary secretion, hyperproteinuria and selective damage to the organic cation transport; 100% of the rats inhaling 135-150 ppm for 6-7 h died; after inhaling 161 ppm for 88 min or 34 ppm for 3.5 h, 100% of the animals survived.

Subchronic toxicity experiments in rats inhaling hexachlorobutadiene in various concentrations disclosed no effect at 5 ppm for 6 h, delayed growth at 10 ppm for 6 h, loss of weight and dyspnoea at 25 ppm for 6 h, irritation of the nose and eyes, dyspnoea and slight anaemia at 100 ppm for 6 h, irritation of the nose and eyes, dyspnoea and kidney and cortex degeneration at 250 ppm for 4 h.

In long-term experiments (2 years) the administration of HCBD with the diet in doses of 0.2, 2 and 20 mg/kg caused adenomas and carcinomas of the uriniferous tubules at the higher doses. The IARC attributes limited evidence of carcinogenicity to this compound, and the EPA put it on its 1980 list of substances to be labelled as presenting a cancer risk.

HCBD proved to be mutagenic in *Salmonella typhimurium* TA 100. Its toxicity for new-born rats has been demonstrated after the mother had been administered a single dose of 20 mg/kg before copulation. The same dose given orally over a period of 90 days before copulation brought about changes in the fertility of rats.

Effects on humans. Observations on occupationally induced disorders are scarce. Agricultural workers fumigating vineyards and simultaneously exposed to 0.8-30 mg/m^3 HCBD and 0.12-6.7 mg/m^3 polychlorobutane in the atmosphere exhibited hypotension, heart disorders, chronic bronchitis, chronic liver disease and nervous function disorders. Skin conditions likely to be due to HCBD were observed in other exposed workers.

SAFETY AND HEALTH MEASURES

Any process in which HCBD is used should be considered potentially hazardous and should therefore be carried out in plant providing the necessary guarantee of health protection and safety. Apart from health engineering measures, the workers' health should be protected by an adequate theoretical and practical training, by laying down rigorous rules of behaviour, by supplying work clothing and protective equipment (respirators, goggles and gloves), by respecting the rules of personal hygiene (taking showers at end of shift, washing hands before eating, no smoking in work premises, separation of plain clothes from working clothes, etc.).

Periodic medical examinations, complemented by laboratory tests should complete the engineering and hygiene measures of prevention.

ARMELI, G.
DE RUGGIERO, D.

"Short-term toxicity and reproduction studies in rats with hexachloro-(1-3)-butadiene". Harleman, J. H.; Seinen, W. *Toxicology and Applied Pharmacology* (New York), 1979, 47/1 (1-14). Illus. 24 ref.

"The acute toxic effects of hexachloro-1:3-butadiene on the rat kidney". Lock, E.; Ishmail, J. *Archives of Toxicology* (West Berlin), Oct. 1979, 43/1 (47-57). Illus. 26 ref.

"Percutaneous toxicity of hexachlorobutadiene". Duprat, P.; Gradiski, D. *Acta pharmacologica et toxicologica* (Copenhagen), 1978, 43/5 (346-353). Illus. 15 ref.

"Chronic toxicity and reproduction studies of hexachlorobutadiene in rats". Kociba, R. J.; Schwetz, B. A.; Keyes, D. G.; Jersey, G. G.; Ballard, J. J.; Dittenber, D. A.; Quast, D. F.; Wade, C. E.; Humiston, C. G. *Environmental Health Perspectives* (Research Triangle Park), Dec. 1977, 21 (49-53). 14 ref.

"Carcinogenicity of halogenated olefinic and aliphatic hydrocarbons in mice". Van Duren, B. L.; Goldschmidt, B. M.; Loewengart, G.; Smith, A. C.; Melchionne, S.; Seidman, I.; Roth, D. *Journal of the National Cancer Institute* (Washington, DC), 1979, 63/6 (1 433-1 439). 31 ref.

n-Hexane

***n*-Hexane** ($CH_3(CH_2)_4CH_3$)

n-H

m.w.	86.18
sp.gr.	0.66
b.p.	68.9 °C
v.d.	2.97
v.p.	100 mmHg (13.3·10³ Pa) at 15.8 °C
f.p.	−21.7 °C
e.l.	1.2-7.5%
i.t.	227 °C

insoluble in water; very soluble in ethanol

a colourless liquid with a faint odour.

TWA OSHA	500 ppm 1 800 mg/m³
NIOSH	100 ppm
	510 ppm/15 min ceil
TLV ACGIH	50 ppm 180 mg/m³
	other isomers 500 ppm 1 800 mg/m³
STEL ACGIH	other isomers 1 000 ppm 1 800 mg/m³
IDLH	5 000 ppm

n-Hexane (*n*-H) is a saturated aliphatic hydrocarbon (or alkane) with the general formula C_nH_{2n+2} and one of a series of hydrocarbons with low boiling points (between 40 and 90 °C) obtainable from petroleum by various processes (cracking, reforming). These hydrocarbons are a mixture of alkanes and cycloalkanes with five to seven carbon atoms (*n*-pentane, *n*-hexane, *n*-heptane, isopentane, cyclopentane, 2-methylpentane, 3-methylpentane, cyclohexane, methylcyclopentane). Their fractional distillation produces single hydrocarbons that may be of varying degrees of purity.

n-H is sold commercially as a mixture of isomers with six carbon atoms boiling at 60 to 70 °C. The isomers most commonly accompanying it are 2-methylpentane, 3-methylpentane, 2,3-dimethylbutane and 2,2-dimethylbutane. The term "technical hexane" in commercial use denotes a mixture in which are to be found not only *n*-hexane and its isomers but also other aliphatic hydrocarbons with five to seven carbon atoms (pentane, heptane and their isomers).

Hydrocarbons with six carbon atoms, including *n*-H are contained in the following petroleum derivatives: petroleum ether, petrol (or gasoline), naphtha and ligroin, and fuels for jet aircraft.

Uses. The principal use is as a solvent in glues, cements and adhesives for the production of footwear, whether from hide or from plastics. *n*-H has also been used, in combination with other hydrocarbons, as a solvent for colours in the graphic arts. It has been used as a solvent for glue in the assembling of furniture, in adhesives for wallpaper, as a solvent for glue in the production of handbags and suitcases in hide and artificial hide, in the manufacture of raincoats, in the retreading of car tyres and in the extraction of vegetable oils.

It is not possible to list all the occasions when *n*-H may be present in the working environment. It may be advanced as a general rule that its presence is to be suspected in solvents and grease removers based on hydrocarbons derived from petroleum.

Exposure to *n*-H may result from occupational or non-occupational causes. In the occupational field it may occur through the use of solvents for glues, cements, adhesives or grease-removing fluids. The *n*-H content of these solvents varies. In glues for footwear it may be as high as 40 to 50% of the solvent by weight. The uses referred to here are those that have caused occupational disease in the past, and it is possible that they have to some extent been discontinued. Occupational exposure to *n*-H may occur also through the inhalation of petrol (gasoline) fumes in fuel depots or workshops for the repair of motor vehicles. The danger of this form of occupational exposure, however, is very slight, because the concentration of *n*-H in petrol for motor vehicles is maintained below 10% owing to the need for a high octane number.

Non-occupational exposure is found mainly among children or drug addicts who practise the sniffing of glue or petrol. Here the *n*-H content varies from the occupational value in glue to 10% or less in petrol.

HAZARDS

Health impairment. *n*-H may penetrate the body in either of two ways: by inhalation or through the skin. Absorption is slow by either way. In fact measurements of the concentration of *n*-H in the breath exhaled in conditions of equilibrium have shown the passage from the lungs to the blood of a fraction of the *n*-H inhaled of from 5.6 to 15%. Absorption through the skin is extremely slow.

The effects of *n*-H on higher organisms fall into two main classes: irritation of the skin and the mucous membranes, and post-absorption effects. The latter can themselves be divided into two classes: acute effects and chronic effects.

Contact of *n*-H with the skin and the mucous membranes produces the types of irritation common to all fat solvents: erythema, oedema, vesicles, sensations of burning. Very high concentrations in the atmosphere produce the same symptoms in the mucous membranes.

Acute or chronic post-absorption effects are almost always due to inhalation. Acute effects occur during exposure to high concentrations of *n*-H vapours and range from dizziness or vertigo after brief exposure to concentrations of about 5 000 ppm, to convulsions and narcosis, observed in animals at concentrations of about 30 000 ppm.

Chronic effects occur after prolonged exposure to doses that do not produce obvious acute symptoms and tend to disappear slowly when the exposure ends, sometimes leaving traces after death. These chronic toxic effects are manifest in the nervous system, typically through paralysis of the muscles of the lower limbs and, less often, the upper limbs as well. A symptomatology of the preparalytic phase exists, with slight gastroenteric symptoms, reduced appetite, general asthenia and loss of weight.

The usual initial nervous symptoms are hyposthenia of the legs and most often a dorsal extension of the foot. Hyposthenia of the hands and arms is much less common. Troubles of sensitivity are comparatively uncommon and are manifest in distal paraesthesia or cramp-like pains. When the symptoms are marked there is a motor deficiency of varying importance that may be serious, and some badly affected patients have to keep to their bed or a wheel-chair and be helped to eat. Motor troubles may be confined to certain of the nerves, particularly the anterior tibial. The muscles of the thigh may also be paralysed, especially the quadriceps. The symptoms are generally bilateral and symmetrical, though one side sometimes predominates. The tendon reflexes of the paralytic regions are feeble or non-existent and there are generally no signs of impairment of the pyramidal system. The motor troubles are accompanied, particularly in serious and protracted cases, by a certain atrophy of the paralytic muscles.

Troubles of the objective sensitivity of the upper limbs are very rare, whereas there are sometimes more or less extensive areas of tactile, thermal and dolorific hyposthesia of the legs.

Symptoms in the central nervous system (defects of the visual function or the memory) have been observed in serious cases of intoxication by *n*-H and have been related to degeneration of the visual nuclei and the tracts of hypothalamic structures.

The course of the disease is generally very slow. After the appearance of the first symptoms a deterioration of the clinical picture is often observed through an aggravation of the motor deficiency of the regions originally affected and their extension to those which have hitherto been sound. The extension generally takes place from the lower to the upper limbs. In very serious cases ascending motor paralysis appears with a functional deficiency of the respiratory muscles. Counting from the start of the disease, the paralytic pattern usually begins to disappear after from 1 month to 8 months and full clinical cure takes place after from 3 to 21 months, depending on the seriousness of the case, the average period being 9 or 10 months. Recovery is generally complete, but a diminution of the tendon reflexes, particularly that of the Achilles tendon, may persist in conditions of apparent full well-being.

With regard to laboratory tests, the most usual haematological and haemato-chemical tests do not show characteristic changes. This is also true of urine tests, which show increased creatinuria only in serious cases of paralysis with muscular hypotrophy.

The examination of the spinal fluid does not lead to characteristic findings, either manometric or qualitative, except for rare cases of increased protein content. Lastly, only the nervous organs show characteristic modifications. The EEG is usually normal. In serious cases of disease, however, it is possible to detect dysrythmias, widespread or subcortical discomfort and irritation. The most useful test is the EMG. The findings indicate myelinic and axonal lesions. The motor conduction velocity (MCV) and the sensitive conduction velocity (SCV) are reduced, the distal latency (LD) is modified and the sensory potential (SPA) is diminished.

Differential diagnosis respecting the other peripheric polyneuropathies is based on the symmetry of the paralysis, on the extreme rareness of sensory loss, on the absence of changes in the cerebrospinal fluid, and, above all, on the knowledge that there has been exposure to solvents containing *n*-H and the detection of more than one case in the same establishment.

The anatomical modifications of the nerves underlying the clinical manifestations described above have been observed, whether in laboratory animals or in sick human beings, through muscular biopsy. The first convincing *n*-H polyneuritis reproduced experimentally is due to Schaumberg and Spencer (1976). The anatomical

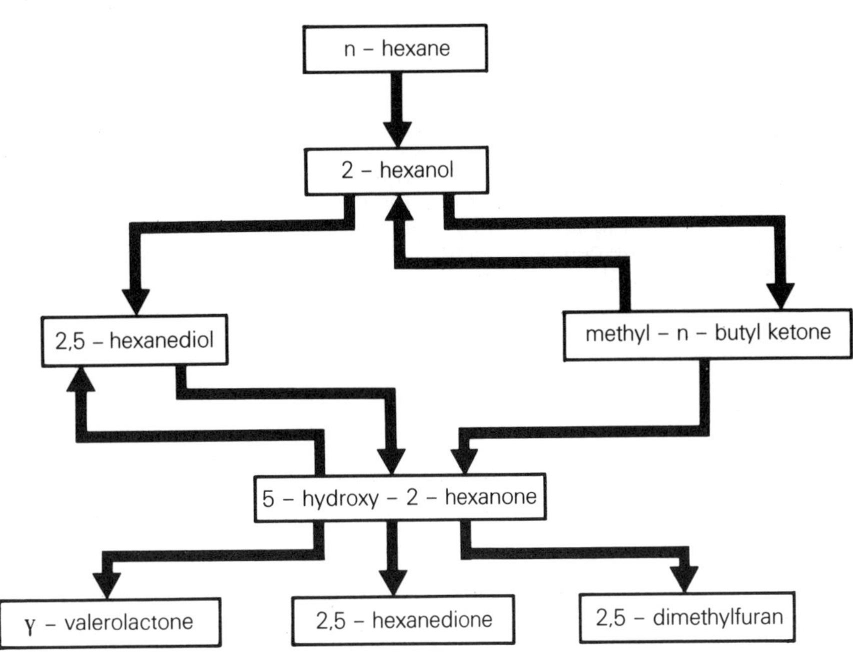

Figure 1. *n*-Hexane metabolic pathways.

modifications of the nerves are represented by axonal degeneration. This axonal degeneration and the resulting demyelination of the fibre start at the periphery, particularly in the longer fibres, and tend to develop towards the centre, though the neurone does not show signs of degeneration. The anatomical picture is not specific to the pathology of *n*-H, for it is common to a series of nervous diseases due to poisons in both industrial and non-industrial use.

A very interesting aspect of *n*-H toxicology lies in the identification of the active metabolites of the substance and its relations with the toxicology of other hydrocarbons. In the first place it seems to be established that the nervous pathology is caused only by *n*-H and not by its isomers referred to above or by pure *n*-pentane or *n*-heptane.

Figure 1 shows the metabolic pathway of *n*-H and methyl *n*-butyl ketone (MBK) in human beings. It can be seen that the two compounds have a common metabolic pathway and that MBK can be formed from *n*-hexane. The nervous pathology has been reproduced with 2-hexanol, 2,5-hexanediol and 2,5-hexanedione. It is obvious, as has been shown, moreover, by clinical experience and animal experiment, that MBK is also neurotoxic. The most toxic of the *n*-H metabolites in question is 2,5-hexanedione. Another important aspect of the connection between *n*-H metabolism and toxicity is the synergistic effect that methyl ethyl ketone (MEK) has been shown to have in the neurotoxicity of *n*-hexane and MBK. MEK is not by itself neurotoxic either for animals or for man, but it has led to lesions of the peripheral nervous systems in animals treated with *n*-H or MBK that arise more quickly than similar lesions caused by those substances alone. The explanation is most likely to be found in a metabolic interference activity of MEK, which leads from *n*-H and MBK to the neurotoxic metabolites referred to above.

Accidents. *n*-H is a highly flammable substance and its mixtures with air within given limits are explosive. Being heavier than air, it may travel along the ground and provoke distant ignition. Furthermore, because of its low conductivity, it can easily generate electrostatic charges.

SAFETY AND HEALTH MEASURES

It is clear from what has been observed above that the association of *n*-H with MBK or MEK in solvents for industrial use is to be avoided.

With regard to TLVs in force for *n*-H, modifications of the EMG pattern have been observed in workers exposed to concentrations of 144 mg/mc (40 ppm) that have been absent from workers not exposed to *n*-H. The medical monitoring of exposed workers is based both on acquaintance with the data concerning the concentration of *n*-H in the atmosphere and on clinical observation, particularly in the neurological field. It is premature to give the search for 2,5-hexanedione in the urine as a means of carrying out the biological monitoring of exposure, but this is nevertheless almost certainly the best indicator of exposure.

It does not seem advisable to adopt mass EMG unchecked by the findings of neurological clinical examination as a means of monitoring those exposed to *n*-H, for there are often slight anomalies in EMG findings that are not of toxic origin.

As is evident from the physical and chemical characteristics set forth at the beginning of this article, the flammable and explosive nature of *n*-hexane and its mixtures with air calls for care. The prevention of exposure hazardous to health is thus not the only reason why containers of glues, cements and other materials in industrial use must not be placed near sources of heat or left open. These substances should be put in suitable distributors that reduce to a minimum the contact of the surface containing *n*-H with the atmosphere.

Products that have been treated with materials containing *n*-H may also be dangerous during the evaporation stage. They must be left to dry in suitable places under a hood and not in the ordinary workplace.

GAFFURI, E.

CIS 79-112 *Occupational exposure to leptophos and other chemicals.* Xintaras, C.; Burg, J. R.; Tanaka, S.; Lee, S. T.; Johnson, B. L.; Cottrill, C. A.; Bender, J. DHEW (NIOSH) publication No. 78-136 (National Institute for Occupational Safety and Health, 4676 Columbia Parkway, Cincinnati) (Mar. 1978), 148 p. Illus. 62 ref.

CIS 81-505 *Synthesis of recent facts about polyneuritis due to hexane* (Synthèse des faits récents concernant les polynévrites à l'hexane). Fichelson, F. (Université de Paris VII, Faculté de médecine Lariboisière-Saint-Louis, France) (1980), 71 p. 83 ref. (In French)

CIS 81-784 "Peripheral nervous system functions of workers exposed to *n*-hexane at a low level". Sanagi, S.; Seki, Y.; Sugimoto, K.; Hirata, M. *International Archives of Occupational and Environmental Health* (West Berlin), Oct. 1980, 47/1 (69-79). 46 ref.

CIS 79-1710 "Urinary excretion of *n*-hexane metabolites in rats and humans". Perbellini, L.; Brugnone, F.; Pastorello, G.; Grigolini, L. *International Archives of Occupational and Environmental Health* (West Berlin), 1979, 42/3-4 (349-354). Illus. 12 ref.

CIS 81-1352 "Urinary excretion of the metabolites of n-hexane and its isomers during occupational exposure". Perbellini, L.; Brugnone, F.; Faggionato, G. *British Journal of Industrial Medicine* (London), Feb. 1981, 38/1 (20-26). Illus. 10 ref.

CIS 80-630 "*n*-Hexane". Fire Protection Association. H83, Information sheets on hazardous materials. *Fire Prevention* (London), Sep. 1979, 132 (45-46).

High risk groups

The increased susceptibility of certain subgroups within the population to the adverse health effects of toxic substances or work stresses is well known. Why some people become adversely affected while others apparently exposed to the same substances and/or performing the same task do not is a major issue today. A number of general factors may affect one's susceptibility to environmental agents. These include developmental/ageing effects, nutritional status, genetic factors, pre-existing diseases and lifestyle/personal habits. With respect to occupational health practices the three last-named factors have received the most attention.

About 30 genetic conditions that may enhance the susceptibility of individuals to the adverse effects of toxic substances have been identified. However, in practical terms it is considered that many of these conditions (e.g. sulphite oxidase deficiency) exist in such a low frequency in the population that they do not constitute a true public health problem or that the genetic disease itself is not inherently dangerous. The prerequisites for hypersusceptibility testing, according to Stockinger and Scheel, include *(a)* identification of the specific genetic effect, *(b)* availability of an inexpensive and simple test which indicates the presence of the genetic defect, *(c)* a relatively high occurrence of the defect in the worker population, *(d)* a connection with pollutants usually encountered in industry, and *(e)* compatibility of the defect with an apparently normal life until industrial exposure occurs.

There are five disorders that satisfy the prerequisites for the above stated hypersusceptibility tests. They include (1) glucose-6-phosphate dehydrogenase (G-6-PD) deficiency, (2) carbon disulphide (CS_2) sensitivity, (3) serum alpha antitrypsin (SAT) deficiency, (4) reaginic antibodies to allergic chemicals, and (5) haemoglobin S in sickel cell anaemia. Other genetic conditions such as atypical cholinesterase, pyruvate kinase deficiency, Wilson's disease, and abnormal porphyrin metabolism do not satisfy all the criteria given above. For example, detection of subnormal amounts of cholinesterase activity has not been associated with adverse health in pesticide workers, while the remaining three diseases are considered too rare for genetic screenings. However, a closer look at the five genetic conditions which supposedly meet the criteria for screening reveals that *(a)* a total SAT deficiency is very uncommon (< 1/1 000); certainly this does not meet even the most liberal definition of a "relatively" high occurrence in the worker population, *(b)* sickle cell anaemia is also not frequent in the Black population (0.5%), and an even smaller percentage is able to seek employment as adults because of the manifestation of overt illness before adulthood. However, the heterozygous condition (i.e. sickle cell trait) is estimated to be around 7 to 13% of American Blacks of African origin. Yet, despite this relatively high percentage, there are no data that indicate that those with the trait would be at increased risk to toxic substances. It has been suggested that persons with sickle cell abnormalities may be at increased risk to anaemia producers (e.g. benzene, Cd, Pb), methaemoglobin producers (i.e. aromatic amino and nitro compounds) and cyanide and carbon monoxide, which reduce the blood oxygen tension. However, without evidence to indicate such enhanced susceptibility, it appears that the inclusion of sickle cell trait in screening programmes would not be appropriate.

The genetic condition which most closely fulfils the above criteria is G-6-PD deficiency, because of its high frequency in the American Black male population (13%) and of industrial documentation of haemolytic crises in persons exposed to oxidant stressor agents such as the aromatic nitro compounds. However, despite the concern that G-6-PD deficients may be at greater risk to over 30 haemolytic agents used in industry, the documentation of the American Conference of Governmental Industrial Hygienists (ACGIH) Threshold Limit Values (TLVs) states that for only two substances (carbon monoxide and naphthalene) the TLV may not be sufficient to protect those with hereditary blood disorders. No evidence was presented as to why the ACGIH felt that those with blood disorders may be at potentially increased risk at or below the TLV value nor was any mention ever made in the documentation of TLVs as to whether persons with a G-6-PD deficiency were ever evaluated, either *in vitro* or *in vivo*, with respect to enhanced susceptibility to any of the haemolytic compounds commonly used in industry. At this point it is appropriate to quote the reasonable assessment of Cooper:

"What is the current status of tests of hypersusceptibility? There is insufficient epidemiologic evidence to support the use of any of them as a criterion for employability, without many qualifications. On the other hand, there is ample scientific evidence to support wider testing. Premature assumptions as to the necessity for such tests, or over-optimistic claims for their benefits, can actually impede testing. On the basis of what we now know, no employer should be regarded as liable or derelict for not choosing to screen his employees. If he screens all employees, he would have to consider whether he would be regarded as liable to criticism for using a positive test to deny employment, or conversely, for jeopardising the health of an individual permitted to work with a positive test. If it is clearly understood that the appropriate application of tests of hypersusceptibility is still moot, still on trial, then progress can be made in studying them. To clarify this point, it would be well if NIOSH developed positions, including consideration of their possible value when appropriate in criteria documents developed for control of occupational exposure."

Beyond genetic considerations

It should be emphasised that the concept of increased susceptibility is considerably broader than genetic predisposition and has been employed within industry for many years. For example persons with certain respiratory and cardiovascular disorders have often been restricted from certain jobs, as well as persons with X-ray diagnosed lower back disorders such as spondylolisthesis. A new approach in dealing with a specific high risk group has been initiated by the Johns Manville Company in the United States, which does not hire persons who smoke cigarettes because such behaviour is known to enhance the carcinogenic potential of asbestos markedly. Furthermore, a recent comprehensive assessment of how nutritional status affects susceptibility to pollutant induced toxicity (and/or carcinogenicity) suggests that this area will gain considerable importance in explaining differential susceptibility to occupational disease. In addition, a new area which remains to be more fully explored is how taking certain medications and drugs may enhance the adverse effects of toxic substances. For example it has recently been proposed that consumption of certain oral contraceptive agents may enhance CS_2 toxicity. It is obvious that the concept of increased susceptibility should never lead to the total exclusion of any worker from employment, but should help in providing for the employment of each worker, in spite of his deficiencies, in work which he is capable of performing, taking into account the employment opportunities in the establishment. (The same principle was adopted by an ILO Tripartite Technical Conference in 1948 as guidance for the medical examination of workers.)

Standard setting

Knowledge of high risk groups has played a significant role in the setting of permissible exposure limits for community air and water pollutants as well as for industrial toxicants. Information on differential susceptibility to toxic pollutants is essential for decision-makers since it enables them to establish dose-response relationships for the various subsegments in the exposed population.

CALABRESE, E. J.

"Indicators of susceptibility to industrial chemicals". Cooper, W. C. *Journal of Occupational Medicine* (Chicago), Apr. 1973, 15/4 (355-359). 34 ref.

"Hypersusceptibility and genetic problems in occupational medicine. A consensus report". Stokinger, H. E.; Scheel, L. D. *Journal of Occupational Medicine* (Chicago), July 1973, 15/7 (564-573). 43 ref.

Methodological approaches to deriving environmental and occupational health standards. Calabrese, E. J. (New York, John Wiley and Sons, 1978), 402 p. Illus. Ref.

Nutrition and environmental health. The influence of nutritional status on pollutant toxicity and carcinogenicity. Vol. 1. *The vitamins* (New York, John Wiley and Sons, 1980), 585 p. Illus. Ref. Vol. 2. *The minerals and macronutrients* (New York, John Wiley and Sons, 1981), 468 p. Illus. Ref.

"Pollutants and high risk groups. Conference proceedings". Calabrese, E. J. (ed.). *Environmental Health Perspectives* (Research Triangle Park, N. Carolina), Apr. 1979, 29 (1-176). Illus. Ref.

Histoplasmosis

Definition. Histoplasmosis is a systemic fungal infection caused by inhalation of spores of the organism *Histoplasma capsulatum.* Serological studies indicate that infection is common in endemic areas, as antibodies to the organism may be demonstrated in 30-80% of the population. Clinical illness occurs in the minority of infections, and symptoms vary from a mild influenza-like illness to a severe chronic disseminated illness depending on degree of exposure, particular organ or tissues involved, and degree of host immune competence.

The organism is a soil-borne organism, with fairly exacting growth requirements, but may be found in many parts of the world. Generally, it may be expected to be found in most river basins in temperate and tropical zones between latitudes 45 °N and 45 °S. More exacting growth requirements (all of which are not fully known) within these regions account for high endemic areas. However, certain soils enriched with faeces from birds or bats have been well documented as local areas of high concentrations of *H. capsulatum.* Many areas in South America, Africa, India, Burma, Thailand, Cambodia, Indonesia, Europe, Australia and East Central United States and known to be endemic areas.

Occurrence. The spores of *H. capsulatum* can be found in soil in and around farm structures where faecal material from avian species has accumulated. Soils around silos, chicken houses, and old abandoned farm buildings are particularly likely places for the organism. A farm worker working in and around these places may become exposed by inhalation of spores when the soil is disturbed resulting in an aerosol. Cleaning out old abandoned poultry houses or silos are particular activities that may result in exposure (figure 1). The trend in agriculture towards larger, fewer, and more specialised farms in certain Western countries has left many abandoned farm structures. Renovation or destruction of these structures would provide a likely means of exposure. Any activity which disrupts soil in or around an area where large numbers of birds have been roosting creates a likely exposure source. Soil exposure seems to be a common factor in many agriculturally associated cases. The essential factor for exposure is the production of an aerosol of soil or dust in an area likely to contain large numbers of the organism.

Figure 1. Cleaning out old abandoned poultry houses or silos are particular activities that may result in exposure.

Epidemiology. The epidemiology of histoplasmosis is poorly recorded on a world-wide basis. Most of this information comes from a few studies in the United States. A review of 15 outbreaks occurring between 1957 and 1969 included three exposures to chicken coops and one silo as sources of infection. Eight of the cases were associated with starling or blackbird roosts. A 1957 review of outbreaks included 18 cases where chicken manure was indicated as the source, and only one case which indicated starling roosts as a source. It was suggested that a possible trend is present in which starlings and other birds may be becoming more important in the disease picture as compared to chickens. However, epidemiological investigations indicate that starling and blackbird roosts are primarily responsible for urban outbreaks rather than agricultural cases.

A greater number of acute cases are usually involved in urban outbreaks, as compared to rural outbreaks, because of the greater concentration of people. However, most outbreaks are rural in origin. The 1957 review indicated that the disease was an occupational hazard of outdoor workers, and most cases were in young males. Case histories reveal that one or more of the following circumstances are commonly associated with histoplasmosis: (1) the patient is a rural resident or visitor to a farm; (2) the patient had exposure to soil imported from a farm; (3) the patient had exposure to farm structures where there were bird roosts; (4) the patient had exposure to an open urban area.

Seroepidemiological studies indicate that agricultural workers have the highest seroconversion rates (28%) as compared to general rural residents (19%) and urban residents (9.5%). In high endemic areas the seroconversion rate may be as high as 70% but, in comparison, rates in urban areas in the same state are much lower.

Symptoms. Human infections present a variety of clinical courses depending on an individual's immune status and degree of exposure to fungal spores. Most symptomatic cases are acute respiratory infections. These occasionally progress to chronic respiratory infections or to disseminated infections.

Acute respiratory infections range from a mild influenza-like illness to a moderately severe illness with fever, malaise, prostration, chest pain, cough, and possibly erythema multiforme. Recovery is spontaneous but slow.

Previous infection with *H. capsulatum* may be noted on a chest X-ray film as small, scattered, radio-dense nodules in the lungs, mediastinal lymph nodes, and spleen.

Acute disseminated infections resemble miliary tuberculosis. They present as a febrile response with hepatosplenomegaly. The course of the disease is rapid.

Chronic respiratory infections resemble chronic pulmonary tuberculosis. The disease progresses over a period of months to years, possibly with periods of remission. This form is most common in males over 40. Death may result. A chronic disease is accompanied by variable symptoms, depending on the organs involved. Unexplained fever, anaemia, pneumonia and mucosal ulcerations of the mouth, bowel or stomach may be seen.

Diagnosis. Laboratory confirmation is obtained, first, by direct examination of body tissues or fluids, often sputum or scrapings of lesions; second, by histopathologic examination of several tissues (tissues of choice being bone marrow, liver, spleen and lung, stained with special fungal stains); third, by isolation through culture of the fungus from sputum, blood, bone marrow, biopsy tissues, lesion scrapings or other body tissues and fluids; fourth, serologic tests can be used; fifth, a histoplasmin skin test can be used. The histoplasmin skin test is primarily an epidemiologic tool to define endemic areas. Its diagnostic value is limited because it does not distinguish between past and present infection, and non-specific reactions can result in false positives.

Preventive measures. Prevention is based on education and measures to avoid inhalation of dust from a likely site where the organism may be growing in high numbers. Wetting down soil or dust before cleaning out an old chicken coop or other area where birds have been roosting is useful. Wearing a well fitted dust mask may be helpful when disturbing potentially contaminated soil. Formalin has been found to be an effective chemical to decontaminate soil containing *H. capsulatum.*

Treatment. Specific antifungal treatment is usually not indicated in acute respiratory infections. Supportive therapy is usually adequate, with recovery expected in 10-30 days. Amphotericin B is the antibiotic of choice to use in chronic pulmonary or disseminated histoplasmosis.

DONHAM, K. J.

"An outbreak of histoplasmosis at an Arkansas courthouse, with five cases of probable reinfection". Dean, A. G.; Bates, J. H.; Sorrels, C.; Sorrels, T.; Germany, W.; Ajello, L.; Kaufman, L.; McGrew, C. H.; Fitts, A. *American Journal of Epidemiology* (Baltimore), July 1978, 108/1 (36-46). Illus. 14 ref.

"Ecologic studies of histoplasmosis". Menges, R. W.; Furcolow, M. L.; Selby, L. A.; Habermann, R. R.; Smith, C. D. *American Journal of Epidemiology* (Baltimore), Jan. 1967, 85/1 (108-119). Illus. 10 ref.

"Chemical decontamination of soil containing Histoplasma capsulatum". Tosh, F. E.; Weeks, R. J.; Pfeiffer, F. R.; Hendricks, S. L.; Chin, T. D. Y. *American Journal of Epidemiology* (Baltimore), Mar. 1966, 83/2 (262-270). Illus. 7 ref.

"Acute histoplasmosis. Description of an outbreak in Northern Louisiana". Storch, G.; Burford, J. G.; George, R. B.; Kaufman, L.; Ajello, L. *Chest* (Chicago), Jan. 1980, 77/1 (38-42). Illus. 29 ref.

Home work

In practically all countries, a small percentage of the workers employed by an industrial or agriculture establishment carry on the work in their own home. These employees are termed homeworkers or outworkers; an alternative name for home work or home industries is cottage industries.

From the health and social point of view the word "homeworker", means a worker who is employed for an industrial or commercial undertaking, and implies the exclusion of the homeworker whose activity is undertaken on his own account, such as the small tradesman (see HANDICRAFTS AND CRAFTSMEN). It must be admitted that the distinction between the two groups is not very clear and that they are exposed to similar risks. For legislative purposes "home work" may be considered as all work executed with a view to gain carried out in a room having, at the same time, another use (bedroom, dining-room, kitchen) without distinction being made as to whether it is the employer who possesses the premises or not, or work executed by a wage earner in a workroom not comprising a part of his house, but for which he is obliged to pay rent, lighting charges, etc.

Types of home work

Common examples of such employment are clothing manufacture, embroidery work, lace making, tailoring, glove making, footwear manufacture, laundry work, making of artificial flowers, sack making and repairing, polishing of precious stones, coral and shell work, making lampshades, toy making, brush and basket making, manufacture of fireworks (pyrotechnics) and matchboxes, hand-loom cloth or carpet weaving, umbrella making and repairing, furniture making, file making, clerical work, handicrafts, etc. Such employment may be skilled or unskilled, full-time or part-time, and the workers may be able-bodied, handicapped, disabled or war invalids or even young persons or children. Sometimes factory workers are given extra work to take home, which is then done by either themselves or with the aid of other members of the household including old people and children. In one country, several textile factories producing a high-grade cloth on power-driven looms decided a few years ago that there would be a better market for this cloth if it could be advertised as "handwoven". They closed down their weaving mill and the looms were distributed among the workers, each of whom installed one loom at his house; the loom was, however, no longer power-driven but converted to a handloom. Even factories engaged in the manufacture of highly technical products or precision instruments may employ some outworkers.

Some governments in developing countries have encouraged the growth of home industries, for example handloom weaving in India; while the sale to tourists of hand-made goods produced by home industries in many non-industrialised countries provides an appreciable source of national income. In one South-East Asian country, the government has provided, at low cost, handlooms to small peasant rice farmers to enable them to obtain a supplementary income. They work on the handlooms during the period when no work has to be carried out in the rice paddy fields. The yarn is supplied by merchants who buy the woven cloth. Despite the fact that outwork saves the employer considerable overhead costs such as rent, rates, heating and lighting and interest on capital such as buildings and machinery, it becomes less and less able to compete against a well equipped and efficiently organised factory and where there is legislation for minimum wage rates it can no longer flourish on low wages. The system has, however, some advantages, since, for example, it enables a widow with young children under school age to earn a living and at the same time attend to her family. Some handicapped or disabled persons are unfit to travel to workplaces or are unable to leave their home so that there is no other work they can do.

Social aspects

In some cases the homeworker is as well paid and his conditions of employment are as good as those of the factory worker engaged in the same type of work, but in others such work is the source of a series of evils designated by the term "sweated labour"; this implies unduly low wages, long hours of work, night work, no paid holidays, overworking of women and children, bad environmental conditions, a quality of work inferior to that produced in factories, health risks for the workers, members of their families and even for the consumers of the products. The outworker is, in many countries, not included in the workmen's compensation schemes so that in the event of an accident or an occupational disease contracted by reason of his work, he and his family will be in a very unfortunate position. This exclusion seems to have been based on the fact that the employer does not have control or management of the premises in which outwork is done, and so has no means of guarding against accidents; it has consequently been considered unfair to place upon him any onus for accidents or diseases that may occur. These considerations do not apply to national insurance industrial injuries benefit schemes which have in some countries replaced workmen's compensation, because industrial injury benefits are now paid for occupational accidents and diseases irrespective of whether or not the employer is to blame.

Generally speaking, legislative measures relating to the employment of women in factories do not apply in the case of outworkers, so that they are not entitled to a period of rest either before or after childbirth. It is difficult to organise outworkers into trade unions and consequently they may be unable to obtain certain of their rights and press for justified wage increases.

SAFETY AND HEALTH HAZARDS

The worker in some cases provides his own tools and machinery but, generally, these are provided by the employer. If he does provide his own, however, it may be that they are not of the best type for safe working or not kept in a good state of repair. If they are provided by the employer, there is most likely no periodic check to see if safe standards are being observed. In the majority of cases, neither the employer nor his managerial staff are ever requested to visit the premises of the outworker since the latter will himself collect the raw materials from the factory or shop and return the finished product. The employer is most unlikely to be concerned with the conditions in the home workshop.

Environmental conditions may be unsatisfactory due to inadequate ventilation, unsanitary premises, poor lighting and overcrowding. Excessive hours of work including late night work and Sunday work, with no holidays are common in some countries. In some cases, children of 6 or 7 years of age are employed outside school hours, or even full-time in countries where there are no schools. Pre-employment and periodic medical examinations of workers are not usually carried out even where dangerous toxic substances are used.

The Federal Republic of Germany State Labour Inspectors have found outworkers using dangerous liquids and solvents such as trichloroethylene under conditions in which the relevant maximum allowable atmospheric concentrations had been substantially exceeded; in addition, these dangerous substances were kept in bottles originally intended for beverages, and this constitutes a risk of accidental ingestion.

A most serious health hazard is the use of radioactive materials in home industries without any safety precautions and without the worker being aware of the risks involved. Cases have occurred of workers being supplied with a radioactive luminous compound for painting the figures and hands of watches and clocks, with no checks on the safe use and storage of this dangerous material, thus creating a health risk to both the worker and members of his family. There have also been cases of lead poisoning among outworkers making pottery and among polishers of precious stones who have used a lead grinding wheel.

One serious hazard in home work is that of explosion and fire, particularly where organic solvents and flammable materials are used. In one Asian country, the manufacture of fireworks is carried out mainly as a home industry and almost every year one or more houses are completely destroyed by explosion, usually killing the family and some passers-by and neighbours. When the home of an outworker is situated in an apartment block there may be a nuisance to neighbouring tenants caused by noise, dust or fumes. This has become a serious problem in one Asian country where many new multistoried buildings 12 or more storeys high and consisting of 400-500 separate apartments may house several outworkers. While it is illegal in this country to carry on factory processes in domestic buildings of this type, it has been found very difficult to take action which would mean throwing these workers out of work as there is an acute shortage of alternative accommodation.

Public health risks

Health risks to the consumers or users of the goods produced by outworkers may be very serious. Infectious diseases may be spread, especially in the case of work connected with food industry products, tobacco, toys, etc.

SAFETY AND HEALTH MEASURES

In a number of developing countries, the staff of medical inspectors of factories, health officers, factory and labour inspectors is so small that they can only devote their efforts towards securing safe and healthy working conditions to the large and medium factories. The small factories and outworkers are consequently left unvisited without any check as to the conditions of work. In some countries, the labour legislation applies only to industrial establishments employing more than a certain number, usually 10, of workers. In the more industrialised countries with larger staffs of officials concerned with occupational safety and health and public health, the degree of control over working conditions of outworkers varies considerably, some countries exercising a greater degree of control than others.

The most common type of control is aimed at the prevention of the spread of infectious diseases from the outworkers to the consumers of the products. This control is usually exercised by the public health authorities in agreement with the factory inspectorate. The employer is required by law to send a list of all his outworkers to the local authority health department twice a year. If the health officer receives notification of certain infectious diseases such as smallpox or scarlet fever, he searches the lists of outworkers, and if he finds that the case is at the dwelling house where an outworker is employed, he can at once make an order forbidding any work of a certain class to be given out to any person living or working in that house either for a specified time or subject to disinfection of the house or to the adoption of other appropriate measures.

In some countries, additional powers give the local authority or the factory inspectorate the right of inspection; if the premises in which outwork is carried on are likely to be injurious or dangerous to health, notice is served on the employer prohibiting him from giving out any further work to be done on these premises unless they are made safe and healthy.

[In Italy employers intending to give out home work must have their names entered in a register of home work employers kept by the competent labour authority. Any employer who arranges for work to be performed outside his undertaking must keep a register containing the names and addresses of persons working outside the production unit and the type and quantity of the work to be performed.]

Instruction of workers

Safety and health instruction and training is very important, especially where dangerous materials are used but, in practice, very little has been done in this respect, due to the shortage of factory and health inspectors and the absence of legislation in many cases.

Enforcement of safety and health measures

A number of countries are gradually extending the application of their industrial safety and health legislation to outwork. Switzerland now prohibits the use of radioactive material by outworkers, the Federal Republic of Germany has authorised the Ministry of Labour to prohibit entirely, or to provide strict measures of protection for, home work involving health risks, and a number of American States have detailed regulations concerning this class of work. So far, no international labour standards have been established on safety and health of outworkers but there are three ILO Conventions which include outworkers in social security schemes; these are the Invalidity Insurance (Industry, etc.) Convention, 1933 (No. 37), the Old-Age Insurance (Industry, etc.) Convention, 1933 (No. 35), and the Survivors Insurance (Industry, etc.) Convention, 1933 (No. 39). Generally speaking the present legislation in most countries is inadequate or non-existent and there is a lack of uniform universal standards.

QUINN, A. E.

Home work. The case of Umbria (Il lavoro a domicilio. Il caso dell'Umbria). Crespi, F.; Segatori, R.; Bottacchiari, V. (Bari, De Donato, 1975), 158 p. (In Italian)

"Theoretical approach and methods for the study of industrial home work in Mexico" (Notas teóricas y metodológicas para el estudio del trabajo industrial a domicilio en México). De Avelar, S. M. *Revista Mexicana de Sociologia* (Mexico City), Oct.-Dec. 1977, 39/4 (1 227-1 249). 26 ref. (In Spanish)

Homeworking (London, Trades Union Congress, 1977), 38 p.

"Act No. 877 to issue new rules governing home work. Dated 18 Dec. 1973". *ILO Legislative Series*, Jan.-Feb. 1975–It. 2.

Hormones, sex

The sex hormones occurring in the animal body (natural sex hormones) are steroids: they contain a cyclopentaphenanthrene nucleus.

The natural sex hormones–oestrogens, gestagens and androgens–are secreted by the gonads (ovaries and testes), the adrenal cortex and the placenta.

Gestagens (C_{21} steroids) belong to the pregnane series in which a side chain, $C_{20}-C_{21}$, is attached at C_{17}. Androgens (C_{19} steroids) are derived from the androstane nucleus in which the $C_{20}-C_{21}$ side chain of pregnane is absent. Oestrogens (C_{18} steroids), derived from the estrane nucleus, lack the $C_{20}-C_{21}$ side chain but also a methyl group, at C_{19} (see figure 1).

The synthesis of sex hormones starts with cholesterol produced by the endocrine glands themselves or available from other sources. Six carbon atoms are removed from the cholesterol side chain to produce pregnenolone (a C_{21} steroid with a two carbon, $C_{20}-C_{21}$, side chain), the last common step of natural sex hormone synthesis. Afterwards, two pathways may be followed leading to androstenedione, depending upon the type of cell involved.

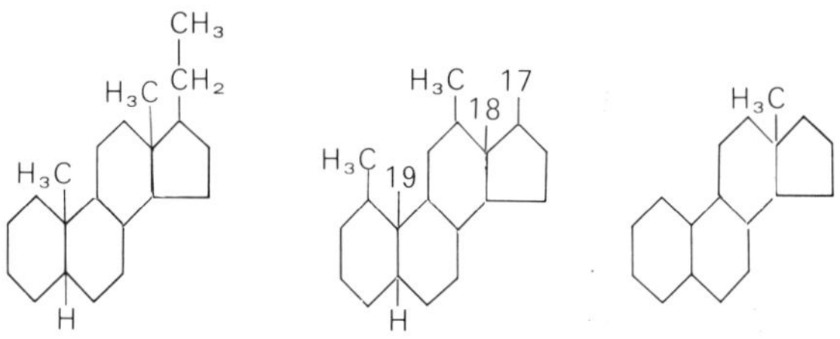

Pregnane Androstane Estrane

Figure 1. Basic chemical structure of sex hormones.

Androstenedione (Δ_4-adrostene-3,17-dione) may be further converted into testosterone or estrone, which may be converted into estradiol. Androstenedione and testosterone, and estrone and estradiol, are interconvertible. Androgens may also be converted into oestrogens in tissues other than the endocrine glands, e.g. in the adipose tissue and the brain (in both sexes). In the brain, oestrogens may be converted into catechol-oestrogens that seem to have a role in the neuroendocrine activity. Some progestagens may also be converted into oestrogens. These facts may explain the occurrence of feminisation symptoms after exposure of men to any sex hormone.

The plasma level of sex hormones is the result of the activity of the endocrine nuclei of the hypothalamus that secrete the gonadotropin releasing hormone. This hormone activates the secretion of gonadotropins (the follicle stimulating hormone–FSH–and the luteinising hormone–LH) by the anterior hypophysis. FSH, in its turn, stimulates the growth of the ovarian follicles in females and spermatogenesis in males. The luteinising hormone stimulates ovulation, the formation of corpus luteum and its hormonal activity in females and the secretion of testosterone by the Leydig cells in men. The gonadal hormones, in a certain amount, suppress the hormonal activity of the hypothalamo-pituitary axis which is depressed when the gonadal hormones level decreases (feedback mechanisms). The liver and other tissues (adrenals, adipose tissue, uterus, placenta, brain, etc.) involved in the metabolism and the homeostasis of sex hormones play a role in the determination of the blood level of the sex hormones. The interplay of the hypothalamo-hypophyso-peripheral glands is carried out by activation and suppression of hormonal syntheses that depend in the last analysis on specific enzymes induced in appropriate cells by appropriate stimulation (hormones, biogenic amines, prostaglandines, etc.).

The hormones of the hypothalamo-pituitary axis act at the level of cell membranes of peripheral endocrine glands by activating adenylate cyclase, and thus initiate the synthesis of specific hormones. Steroid hormones diffuse across the cell membranes of target tissues and, after binding to a cytoplasmic receptor, reach the nucleus, where they initiate the synthesis of specific proteins.

Oestrogens are compounds able to induce estrus (period of sexual receptivity in non-primate, female mammals). The most important natural oestrogens (17β-estradiol, estrone and estriol) are produced by the Graffian follicle, the testes, the adrenal cortex and the placenta. Oestrogens are responsible for the development and maintenance of the female accessory sex organs and of the secondary sexual characteristics (specific fat deposition, hair distribution, etc., of the female type). Oestrogens also stimulate the secretion of adrenal androgens and augment the amount of some blood proteins.

The principal *gestagens* (natural progestins) are: progesterone, 20α-hydroxyprogesterone, 20β-hydroxyprogesterone and 17α-hydroxyprogesterone. The latter is a precursor of other steroidal hormones. Gestagens are secreted by the luteinised granulosa cells of corpus luteum, by the adrenal cortex and the placenta. They prepare the uterus for the implantation of the conceptus, maintain the pregnancy, adapt the immunological system for the acceptance of the conceptual "graft", and participate in the development of the lactating breast. An initial estrogenic activity is needed for the progestational activity to take place by inducing progesterone receptors and gestagens antagonise the activity of oestrogens by decreasing the number of oestrogen receptors. Progestins may thus prevent carcinoma of the endometrium brought about by increased oestrogen levels. Progestins may interact with androgen receptors in target tissues. This may explain their androgenic or antiandrogenic effects.

Androgens (testosterone), androstenedione, dehydroepiandrosterone, and other related steroids are secreted by the Leydig cells, situated in the interstitial tissue of the testis (especially testosterone, the most potent androgen), by the theca interna of the Graffian follicle of the ovary, as precursors of oestrogens, and by the adrenal gland, a source of androgen and oestrogen in both sexes. Testosterone may act as such in some target cells (internal genitalia, muscle and bone) or has to be converted in the cytoplasm of other target cells (external genitalia, prostate, hair follicles), into its active metabolite, dihydrotestosterone, by an enzyme, 5α-reductase, in order to become active. Testosterone and dihydrotestosterone are coupled in their target cells to a cytoplasmic receptor and transferred to the nucleus where they bind to chromatin and initiate specific protein synthesis. The most important metabolites of testosterone are the 17-ketosteroids. They are conjugated with glucuronic or sulphuric acid and excreted in urine. Androgens develop and maintain the male reproductive tract and secondary sexual characteristics (muscle development, fat and hair distribution, voice characteristics). Androgens also stimulate skeletal growth before puberty. Androgens and FSH promote spermatogenesis. An important anabolic effect of androgens, the synthesis of proteins, led to the manufacture of anabolic androgens in which the androgenic activity is smaller.

Uses. Oestrogens are used in replacement therapy, the chemotherapy of prostatic carcinoma, the formulation of oral contraceptives, the fattening of fowl and cattle. Progestagens are used in replacement therapy, oral contraception, endometrial cancer. Androgens are used in replacement therapy. Anabolic androgens are used to stimulate anabolism in bedridden patients, to stimulate muscular development in athletes, erythropoiesis in aplastic anaemia, etc.

Production. The principal oestrogens produced in industry are:

(a) natural steroidal oestrogens of animal origin: extracted from the urine of pregnant mammals (women, mares);

(b) synthetic steroidal oestrogens;

(c) semi-synthetic steroidal oestrogens. Ethynyl estradiol and 3-methyl-ether-ethynyl estradiol constitute, together with a progestagen, the ingredients of oral contraceptives. Premarin and similar products contain sodium estrone sulphate, piperazine estrone sulphate, etc.;

(d) non-steroidal oestrogens: synthetic such as the stilbene (diphenylethylene) derivatives.

The progestins produced in industry are derivatives of: progesterone, 17α-hydroxyprogesterone and derivatives, retroprogesterone, 19-nortestosterone and derivatives and the more recent, highly potent norgestrel.

The principal androgens produced in industry are: testosterone and derivatives, methyl testosterone, anabolic steroids, antineoplastic androgens.

HAZARDS

Adverse effects of oestrogen in women. Prolonged oestrogen therapy has caused malignant endometrial changes in predisposed persons. Endometrial hyperplasia and endometrial carcinoma occurred also after exposure to diethylstilbestrol. Clear cell adenocarcinoma of the vagina and cervix uteri occurred in young women exposed prenatally to diethylstilbestrol or other non-steroidal oestrogens. Women treated with oestrogen may complain of menstrual disorders, nausea, headaches, etc. Similar effects were observed during occupational exposure to natural or synthetic oestrogens and oral contraceptives. Menstrual disorders were frequent. Metrorrhagia was observed in women who started to work. Menorrhagia occured in women workers with endometrial hyperplasia. In menopausal women diethylstilbestrol may produce abnormal uterine bleeding which often leads to suspicion of cancer of the uterine body. Manifestations like metrorrhagia, nausea and headache usually disappear after discontinuance of exposure.

Excess of progesterone may be responsible for weight gain, acne, mastalgia and breast enlargement and recurrent monilial vaginitis. Toxicity may also include headache, nausea, chloasma, breakthrough bleeding, weight gain, loss of libido, cholestatic liver damage; sterility may also occur.

Adverse effects of androgen in women. Women occupationally exposed to androgens may present nausea, oedema, gain in weight, a degree of masculinisation, acne, baldness, hirsutism (growth of facial hair and excessive body hair), deepening of the voice, enlargement of the clitoris, increase in libido, and when the exposure is high, menstrual irregularities, oligomenorrhoea or amenorrhoea, and decrease in size of the breasts. A differential diagnosis to exclude other causes of virilisation in women may be necessary (androgen secreting tumours or hyperplasia of the ovary or of the adrenal gland). The oral androgens (alkylated androgens) are also hepatotoxic.

The large female population taking oral contraceptives for up to nearly 20 years has made it possible to carry out epidemiological studies concerning the adverse effects of these drugs. An increased incidence of benign liver tumours occurred in these women. Some hepatocellular carcinomas were also reported. An increased incidence of cervical carcinoma, skin cancer and malignant melanoma was found in long-term oral contraceptives users. Oral contraception, like pregnancy, may disclose latent metabolic defects, especially diminished carbohydrate tolerance of the diabetic type, or increased levels of plasma cholesterol and triglycerides, and increased blood pressure. Minor side effects complained of include irritability or depression, changes of libido, nausea, headache, migraine, hyperpigmentation of the skin, fluid retention, menstrual irregularities, increased varicosities, and gastric disturbances. Infertility may also occur. In the absence of specific risk factors (hyperlipidaemia, diabetes, hypertension, obesity, cigarette smoking) the danger to young, non-pregnant women is relatively small. The above-mentioned data point to the adverse effects that may occur in women occupationally exposed to the ingredients of "the pill", which should be monitored by careful periodical examination.

Adverse effects of oestrogens in men. Anorrhexia, nausea, vomiting, oedema, a feminisation syndrome characterised by gynaecomastia (uni- or bi-lateral), increased pigmentation of the areollae, tenderness of the nipples, with or without secretion, slight loss of libido with difficulty in erection, with or without involution of the secondary sex organs and sterility (by inhibition of spermatogenesis), may occur. Urinary oestrogens are increased. A differential diagnosis of breast tumours in men occupationally exposed to oestrogen is needed. Gynaecomastia may occur in cases of functional insufficiency of organs involved in the metabolism of oestrogens (chronic liver or kidney diseases) or of increased endogenous oestrogen synthesis by neoplastic growths such as pituitary, adrenal and testicular tumours or ectopic secretion of gonadotropins or prolactin by lung tumours. Endocrinopathies may be accompanied by gynaecomastia. The conversion of androgen into oestrogen may explain the occurrence of gynaecomastia in the case of gonadotropin or androgen excess in men. Some drugs (spironolactone, digitalis, etc.) may induce gynaecomastia. Physiological gynaecomastia at birth, puberty and old age, as well as the possibility of the occurrence of breast cancer in men, should be also considered.

Progestagens in excess, in men, besides the general symptoms described in women, may inhibit spermatogenesis.

Adverse effects of androgen in men. Androgen may cause nausea, oedema, inhibition of testicular activity (by inhibition of gonadotropin secretion) with sterility, painful gynaecomastia, prostatic hypertrophy, polycythemia. The alkylation of androgens at C_{17} increases their oral efficacy but also their toxicity. Serum levels of bilirubin, alkaline phosphatase and transaminases are increased. Prolonged treatment with anabolic steroids in some pathologic conditions, and in athletes treated for considerable periods of time to stimulate muscular development, caused liver hyperplasia, liver adenomas and hepatocellular carcinomas. Liver function tests return to normal after cessation of exposure. Spontaneous regression of the tumours followed, generally, but not always, the withdrawal of these steroids.

Adverse effects of sex hormones in children. Hyperestrogenism has been noted in workers' children. Gynaecomastia may be present at birth in boys whose mothers were occupationally exposed to oestrogens during pregnancy. Pseudo *pubertas praecox* has been noted in girls. A feminising syndrome has been seen in boys whose parents worked in an atmosphere contaminated by oestrogens and were not strict in observing occupational hygiene. Skeletal maturation may be accelerated in these children. Children affected may also show strong pigmentation of the sexual organs. Exposure of children to androgens caused a precocious puberty in boys and female pseudohermaphroditism in girls and inhibition of longitudinal growth in both sexes.

Adverse effects of sex hormones in prenatal life. A small number of human female fetuses exposed prenatally to oral contraceptives with androgenic metabolites, or to androgens had masculinised genitalia. The degree of masculinisation is higher when exposure takes place during the embryonal stage of the intrauterine life. Exposure of the conceptus to oestrogen alone or to a progesteron-oestrogen product may also result in specific birth defects or an increase in the incidence of malformations in genetically predisposed subjects.

PREVENTIVE MEASURES

The working environment may be contaminated during sex hormone manufacture, especially during the extraction and purification of natural steroid hormones, grinding of raw materials, handling of powdered products and recrystallisation. Airborne particles of sex hormones may be absorbed through the skin, ingested or inhaled. Enteric absorption results in quick inactivation of sex hormones in the liver. The rate of inactivation is decreased for the oral, alkylated steroid hormones (methyl testosterone, anabolic steroids, etc.). Sex hormones may accumulate and reach relatively high levels even if their absorption is intermittent. Consequently, repeated absorption of small amounts may be detrimental to health.

Intoxication by sex hormones may occur in almost all the exposed workers if preventive measures are not taken. To this effect the industrial sector is more successful than the agricultural one (chemical caponising of cockerels by stilbestrol implants and incorporation of oestrogens in feed for body weight gain promotion in beef cattle), where measures taken are summary and the number of cases of intoxication is consequently bigger.

Adequate preventive measures, taken in the pharmaceutical industry, have succeeded in eliminating the occurrence of intoxications almost completely. The air-polluting processes are isolated in areas having adequate exhaust facilities and hermetically sealed machines. The isolated areas are entered only by workers wearing special clothing including underwear, socks, long-sleeved overalls (with no pockets) buttoned to the neck and with ties at the bottom of the trousers for tying over the boots, rubber gloves, head cover and dust respirators. Air-supplied vinyl suits may be used by the groups at highest risk. When workers leave the polluted area they should undress, take a shower, wash their hair, clean their nails and put on their own clothes if the working day is over. Workers indirectly exposed to hormonal dust (i.e. mechanics who change the filters in the ventilation system) must also be provided with adequate protective equipment. The contaminated work clothing should be thoroughly cleaned. Disposable paper garments can be burned. Gloves should be rinsed with acetone or methanol, then washed and dried. Respiratory protection equipment should also be cleaned before re-use. Workrooms must be kept very clean. An alkaline detergent has been found to be most suitable for washing clothes and wiping surfaces. Mixers, stirring rods, spatulas, glassware, dishes, etc., should be rinsed after use with acetone or methanol in a hood or near a vacuum, and then washed in the conventional manner. Methanol should be used on plastic items.

It is important to inform workers of the risk represented by the workplace and to win their co-operation in lowering of occupational exposure; for example, not putting on rubber gloves after having worked barehanded, not rubbing the face or nose with contaminated gloves, not contaminating the inside of respirators by leaving them exposed to processing dust, not using outside the work area items which have been inside it (cigarettes, pipes, handkerchiefs) and being alert to detect deficiencies in preventive measures. In some pharmaceutical companies the workers alternate one week in the polluted area with two weeks in another place of work. In other companies rotation is carried out only when signs of intoxication occur. If women are employed the alternating work environment should respect their cyclic hormonal pattern. Workers who develop symptoms of chronic intoxication in the presence of a low concentration of hormones in the air must be forbidden to return to the polluted area.

The effectiveness of preventive measures should be checked by analyses of the amount of hormonal compounds in the air of the working environment and in the plasma of the employees, and by clinical examinations.

An industrial hygiene guideline of 0.05 μg airborne oestrogen dust per cubic metre of air, for an 8-h workday (time weighted average) was adopted at one manufacturing company.

Persons directly involved in the hazardous sectors of manufacture should be examined every two weeks. Persons engaged in the pharmaceutical stage of manufacture should be examined monthly. The frequency of the physical examination of the employees should increase with the age of the worker.

Feminisation in men, masculinisation or menstrual disorders in women, changes in certain metabolic parameters and other symptoms detrimental to health are indications for changing the workplace.

Before admission of a new employee a very careful health examination will serve to exclude persons at risk, namely women of childbearing age, epileptics (steroid hormones may increase the frequency of seizures by changes in fluid retention) and persons with hepatic insufficiency, for example.

Intoxication by sex hormones is generally reversible when exposure ceases; however, several months or even years may elapse before clinical phenomena disappear entirely. Treatment is indicated when natural involution of the clinical signs does not occur during an observation period after leaving the polluted area. Treatment with androgens was needed in some persons with signs of feminisation after exposure to oestrogens. Surgical intervention may be necessary for gynaecomastia resistant to hormonal treatment.

The fact that cases of chronic intoxication may occur in spite of considerable efforts to minimise occupational exposure to sex hormones suggests the need for a closed system of the entire production process.

WASSERMANN, D.
WASSERMANN, M.

Sex hormones:

"Sex hormones". Vol. 6. *IARC monographs on the evaluation of carcinogenic risk of chemicals to man* (Lyons, International Agency for Research on Cancer, 1974), 243 p. Ref.

CIS 79-1050 "Essential hormones as carcinogenic hazards". Hickey, R. J.; Clelland, R. C.; Bowers, E. J. *Journal of Occupational Medicine* (Chicago), Apr. 1979, 21/4 (265-268). 22 ref.

CIS 76-1065 "Oestroprogestogenic syndrome in a worker involved in production of a contraceptive pharmaceutical" (Sindrome da estroprogestinici in operaio addetto alla produzione di un farmaco anticoncezionale). Gambini, G.; Farina, G.; Arbosti, G. *Medicina del lavoro* (Milan), Mar.-Apr. 1976, 67/2 (152-157). Illus. 12 ref. (In Italian)

CIS 81-1085 "Urinary monitoring for diethylstilbestrol in male chemical workers". Shmunes, E.; Burton, D. J. *Journal of Occupational Medicine* (Chicago), Mar. 1981, 23/3 (179-182). 14 ref.

Other hormones:

CIS 79-1346 "Steroids can backfire". Symington, I. S. *Occupational Health* (London), Apr. 1979, 31/4 (207-208). 1 ref.

CIS 77-753 *Occupational diseases in workers manufacturing hormone products* (Pathologie professionnelle du personnel fabriquant des produits hormonaux). Jamet, P. (Université de Bordeaux II, Unité d'enseignement et de recherche des sciences médicales et pharmaceutiques, Bordeaux) (1976), 61 p. 21 ref. (In French)

"Occupational exposure to drugs: hormones" (L'esposizione professionale a farmaci: Ormoni). Farina, G.; Alessio, L.; Banderali, S. *Medicina del Lavoro* (Milan), Mar.-Apr. 1977, 68/2 (105-117). 29 ref.

Horn

Horn is a natural product obtained from the defensive horns and antlers of animals. Horns of domestic animals (cattle, sheep and goats) come from slaughter-houses; antlers of deer, elk, etc., are obtained from hunters and trappers. In Latin America, Portugal and Spain, the horns of fighting bulls are much prized for their size and appearance.

Uses. Horns and antlers have been adopted since the earliest times as utensils and tools and are still used for some of those purposes. The finest quality horns and antlers in size, shape and appearance are selected for hanging as trophies or for making into furniture, utensils and ornaments, usually for country houses; chairs, tables, lampstands, screens and coat hooks, are assembled, polished and sometimes varnished.

Horn of lower quality, off-cuts, etc., are shaped and cut, often by handicrafts workers, to form a variety of small objects, such as tool handles and shafts, walking sticks, umbrella handles, reeds for musical instruments, buttons and flexible rods. The bony core is removed by steam before manufacture: this has also a certain degreasing and cleaning effect.

Because of its high nitrogen content (9-12%), manufacturing waste and misshapen or imperfect horn is degreased, crushed and calcined to make fertilisers: it may be mixed with the bones and entrails of the animals. Horn is also used to produce bone-black and in the making of methylene blue.

HAZARDS

Accidents may be caused directly by the horn itself during handling (pricks, splinters, cuts and abrasions, foreign bodies in the eyes, etc.), or by machine and tools during processing. Burns may result from use of alkalis and acids.

Diseases may be caused by pathogenic germs from animals (staphylococcus, streptococcus, brucella, tuberculosis, tetanus, anthrax, clostridium, etc.).

The direct action of powder from the horn on the skin and mucous membranes may produce phenomena associated with irritation (rhinitis, conjunctivitis), together with allergic manifestations in the respiratory system. Solvents used for degreasing may have toxic effects.

SAFETY AND HEALTH MEASURES

Exhaust venitlation should be provided at dust-producing processes, as near as possible to the source. The least toxic solvents should be used for degreasing and the process carried on under enclosure and mechanical exhaust ventilation. Secure guarding should be maintained for all dangerous parts of machinery. Appropriate personal protective equipment should be provided where necessary, for example hand protection and eye protection; good sanitary and washing facilities should be maintained.

Vaccination of workers may be effective against certain diseases transmissible from animals to man. Workers should be carefully selected. Pre-employment medical examinations may determine possible allergic responses to powder emanating from the horn. Periodic medical examinations are desirable to check adaptation to the work and the effectiveness of preventive measures adopted.

VACAS ZAMORA, M.

Hospitals

The progress made in public health over the last few decades has led to a corresponding growth in hospital services. In France, for example, there are at present 5.2 beds per 1 000 inhabitants available in public hospitals alone, and the numbers of persons employed in this field has increased considerably.

Characteristic of the non-nursing personnel is the wide variety of trades and occupations represented, as may be seen in table 1, which relates to a regional hospital centre of about 4 000 beds.

Clearly, table 1, being based on a particular case, is not generally applicable: transport requirements would vary according to the layout of the hospital and the related buildings; some hospitals are not self-supporting in various technical fields such as heating, building maintenance and cleaning, laundry, etc.

Finally it should be stressed that the management problems of a hospital become increasingly complex in relation to the size of the establishment and grow faster proportionally than the number of beds. It is not unusual to find a staff of 6 000, excluding doctors, in a regional hospital centre of 4 000 beds. Thus the management problems that have to be faced are in many ways similar to those encountered in a large industrial undertaking.

Table 1. Occupations encountered in a large hospital

Heating services: pipefitters, welders, boiler-makers, heating engineers, mechanics, plumbers, maintenance mechanics, fitters, electricians, electronics engineers, boiler attendants	46
Breakdown services: electrical fitters and mechanics, general mechanics	14
Mechanical maintenance section: fitters, electricians, incinerator attendants	17
Transport section (internal combustion and electrically propelled vehicles): mechanics and drivers	23
Electrical and electronic maintenance section; telephone services:	41
Building services: masons, plasterers, tilers, painters, plumbers, welders, carpenters, carpet-layers, window-cleaners, gardeners, cleaners, handymen	86
Housekeeping services: porters and doorkeepers, baggage orderlies, housekeepers, launderers and ironing-women, seamstresses, mattress-repairers, printers	254
Catering services: butchers and pork-butchers, cooks and assistant cooks, bakers and pastry-cooks, storemen, serving staff	218
Ambulance drivers–first-aid personnel	33
Total	732

WORKING CONDITIONS AND OCCUPATIONAL HAZARDS

Medical and paramedical staff. The hospital patient of today demands not only the benefit of the most elaborate medical techniques in furtherance of his right to good health, but also the greatest personal comfort, which is why the public authorities are at pains to make hospitals more humane. The combination of these two requirements considerably increases the physical and mental workload of the staff in spite of their increasing numbers. An example of this is the provision of private wards or wards with two beds only, which represent a considerable advance from the patient's point of view since they eliminate the indiscriminate mixing of patients and also enable visitors to be received outside normal visiting hours. Nevertheless, they add to the complications of supervision by the nursing staff, of providing medical care and of serving meals, and increase the distances which the staff have to cover.

The layout of a hospital and its interior arrangements as well as the organisation of work should be based on ergonomic studies similar to those which have been used to great advantage by industry. Starting from the design stage, it is clear that building layout must be based on functional requirements, and not only on capacity in terms of the number of beds, initial cost or aesthetic aspects. When the architects have, with greater or less success, incorporated ergonomic principles, the best use must be made of the existing arrangements and for this reason each job should be studied, including the daily and weekly workload and the general movements of patients, staff and material. In addition to the shortage of staff, about which so much is heard, it will be found that staff is seldom organised according to actual needs, which are continually varying.

Occupational accidents. The most frequent accidents are those resulting from lifting too heavy loads or using faulty lifting techniques (traumatic lumbago), slips and falls particularly on slippery floors, cuts from sharp and pointed tools, and those caused by various physical agents including heat, electricity, X-rays and explosive mixtures of gases (see table 2).

Occupational diseases. These may be grouped under three main headings: infectious or parasitic diseases; contact dermatites, particularly allergic eczemas caused by pharmaceutical products and disinfecting agents; and diseases due to ionising radiations (see table 2). (See also MEDICAL PRACTITIONERS; NURSES; LABORATORY WORK.)

It should be noted that, from the aetiological point of view, infectious occupational diseases may be compared with occupational accidents; however, the exact date of infection and the circumstances surrounding the accident being usually unknown, medico-legal criteria lead to their being considered as occupational diseases.

Non-nursing staff. The hospital administrative, engineering and general staff is, to a certain degree, exposed to the same hazards as the medical and paramedical staff: for example, reception office workers come into direct contact with patients, laundry and disinfection staff handle infected material and the necessary disinfectants, electricians and mechanics may be exposed to ionising radiations during maintenance and inspection work.

In addition to the hazards encountered in hospital work, there are also the risks specific to each individual profession contained in the far from exhaustive list shown in table 2 and dealt with in separate articles.

SAFETY AND HEALTH MEASURES

A hospital, in the same way as any industrial undertaking, requires an efficient safety and health service to review working practices, pick out hazards and arrange for their elimination, to provide pre-employment and periodic medical examinations, to ensure that work is ergonomically designed and that workers are assigned to jobs for which they are suited, to carry out accident investigations and implement the safety conclusions. To be effective this service must fit into the normal hospital routine, its task must be clearly recognisable and it must be sure of the necessary co-operation, particularly from

Table 2. Hazards encountered in the various professions practised by non-medical hospital staff

Service	Occupational accidents	Occupational diseases
Catering services:		
Butchers	cuts from sharp and pointed tools; septic wounds from bone splinters; handling accidents	
Bakers, confectioners		cutaneous staphylococcal infections; bakers' asthma
Cooks, kitchen assistants, dishwashers	burns	eczemas caused by dishwashing products
Storekeepers, food porters	handling accidents	
Building and garden-maintenance services:		
Masons, builders labourers	falls; bolt-gun accidents	cement dermatitis
Gardeners	tetanus	tetanus; pesticide poisoning
Tinsmiths, plumbers, fitters	falls; hand-tool injuries; arsine poisoning	lead poisoning
Painters	falls	benzene poisoning; contact eczema
Joiners	injuries from wood-working machines (electric saws, planers and spindle moulders)	dermatitis due to paints, varnishes, stains, woods and synthetic adhesives
Engineering services:		
Maintenance engineers	injuries from hand tools, machine tools and grinders (foreign bodies in the eye); "arc eye" in arc welding	folliculitis; eczema due to trichloroethylene and other solvents
Heating engineers	burns; acute carbon monoxide poisoning	carbon monoxide poisoning
Electricians	falls; electrical accidents	
General services:		
Laundry workers	infections and contaminations; burns	eczemas due to washing products and disinfectants
Mattress cleaners		infection; dust-induced asthma
Needlewomen		dorsal, cervical and postural lumbar pains
Porters (including stretcher bearers)	traffic accidents; traumatic lumbago; handling accidents	infection
Administrative services:		
Doormen, keepers		infection
Typists, business machine operators		dorsal, cervical and postural lumbar pains; wrist tenosynovitis
Miscellaneous:		
Mortuary and pathology laboratory workers	cuts and septic puncture wounds	

the hospital hygiene laboratory, the personnel department and the safety and health committee.

The success attendant on the efforts of the occupational health physician depends upon the influence he can bring to bear and can only be achieved through an exhaustive knowledge of all the jobs of the many different groups of workers.

The systematic medical supervision of staff. Although this may often be thought to be the principal task of the occupational health physician, he is also responsible for checking new arrivals to ensure that they are assigned to jobs for which they are suited, and, by means of subsequent examinations, for detecting signs of unsuitability or illness at the earliest possible stage. Certain groups should be subjected to increased medical supervision.

In view of the omnipresent danger of infection and contamination, one of the service's prime tasks will be worker vaccination. The vaccinations required will depend on the type and incidence of the infectious diseases treated in the hospital. In Western Europe, it is considered essential to vaccinate against smallpox, poliomyelitis, diphtheria and tetanus and provide protection against tuberculosis by means of BCG. The value of vaccination against typhoid and paratyphoid, although required by law in countries such as France, is open to question nowadays. Since a large number of women of childbearing age are employed in hospitals, the problem of rubella infection deserves special attention; there is still no effective and safe vaccination for this disease and it is, therefore, advisable to check for rubella antibodies during the pre-employment medical examination; if the serological examination proves negative the person in question should not be employed where there is a danger of exposure, especially in children's wards. Some countries have done away either temporarily or permanently with the compulsory smallpox vaccination of their nationals; it would seem that an exception to this rule should be made in the case of nursing personnel in hospitals, because they are in the front line as regards exposure if there should be sporadic outbreaks of this disease; additionally, they may act as a vector for infection.

Instruction and training. Safety and health efforts will prove ineffective unless the full support of the staff is obtained. The safety and health service must therefore provide instruction and education concerning the insidious hazards such as radiations and contamination resulting in infectious hepatitis, and covering the lower echelons such as ward orderlies. Posters and circulars are of only limited effectiveness and direct personal action is essential. Collaboration between the physician in charge of safety and health, heads of services, the hospital hygiene service, the welfare offices, and the joint safety and health committee is essential; here the physician should act as a stimulator and co-ordinator.

Staffing. The work to be done by the occupational health physician, such as vaccinations and supervision of workers exposed to ionising radiation, entails a considerable administrative workload: an adequate secretarial and nursing staff should consequently be provided.

MEHL, J.

See also readings suggested on ANAESTHETISTS; MEDICAL PRACTITIONERS; NURSES.

Main hazards:

CIS 78-1454 *Pilot study–Working conditions in the medical service* (Health and Safety Executive, Baynards House, 1 Chepstow Place, London W2 4TF) (1978), 115 p.

Occupational diseases:

CIS 78-469 "Occupational pathology of medical and paramedical personnel" (Pathologie professionnelle des professions médicales et paramédicales). Mehl, J. *Encyclopédie médico-chirurgicale. Intoxications–Pathologie professionnelle* (Paris), Fascicule 16545 A 10, 9-1977. 7 p. (In French)

Pregnancy hazards:

"Delivery outcome in women employed in medical occupations in Sweden". Baltzar, B.; Ericson, A.; Källén, B. *Journal of Occupational Medicine* (Chicago), Aug. 1979, 21/8 (543-548). Illus. 16 ref.

Psychological problems and criteria of fitness:

"IX National Congress of Occupational Medicine of Hospital Staff" (IXes Journées nationales de médecine du travail du personnel hospitalier). *Archives des maladies professionnelles, de médecine du travail et de sécurité sociale* (Paris), 1980, 41/5 (235-312). (In French)

"The preplacement medical evaluation of hospital personnel". Schneider, W. J.; Dykan, M. *Journal of Occupational Medicine* (Chicago), Nov. 1978, 20/11 (741-744). 15 ref.

Ergonomics, work schedules, viral hepatitis:

"VIII National Congress of Preventive Medicine of Hospital Staff" (VIIIes Journées nationales de médecine préventive du personnel hospitalier). *Archives des maladies professionnelles, de médecine du travail et de sécurité sociale* (Paris), Sep. 1978, 39/9 (511-585). Illus. Ref. (In French)

Operating theatres:

CIS 78-473 "Occupational hazards for operating room-based physicians–Analysis of data from the United States and the United Kingdom". Spence, A. A.; Cohen, E. N.; Brown, B. W.; Knill-Jones, R. P.; Himmelberger, D. U. *Journal of the American Medical Association* (Chicago), 29 Aug. 1977, 238/9 (955-959). 10 ref.

Conditions of work of nurses:

CIS 78-1174 "Convention and Recommendation concerning employment and conditions of work and life of nursing personnel". Convention 149 and Recommendation 157, International Labour Conference, Geneva, 1977. *Official Bulletin*, International Labour Office (Geneva), 60/A/3 (142-146 and 151-166). (In English, French, Spanish)

Hotels and restaurants

The keeping of inns for travellers and the selling of prepared food and drink are among the oldest and most universal of commercial services, but in modern times development has been very rapid. Quick transport and paid holidays have enormously increased the number of travellers; tourism has become of major importance in the economic life of many countries: more and more people are able to afford the pleasures of eating and drinking in restaurants, cafés and bars. In some countries a very large labour force is employed in one way or another in the hotel and catering industry.

Developments in recent years are changing the pattern of hotel services: in many places the long-stay guest is replaced by the motorist who seeks only simple accommodation for one or two nights. The popularity of winter sports and beach life make it profitable to construct large hotels in suitable places but may also create a problem of strictly seasonal employment. Many countries have difficulties in recruiting labour: migrant workers, seasonal and part-time workers are often employed.

The industry covers a wide range of premises from the great international hotels in large cities to small country inns, from world-famous restaurants to the roadside cafés or snack bars. One general distinction can be made between those places which provide only food and drink (restaurants) and those where living and sleeping accommodation is provided (hotels), but apart from these the widest possible differences exist between the types of services offered, the numbers employed, and the occupational hazards to which they are exposed.

A large hotel is almost a self-contained community with its own repair and maintenance workshops, its own laundry and sewing rooms, its own painting and decorating staff. Even in medium-sized hotels there may be as many "back" workers, i.e. those not serving the customers directly as "front workers", those engaged in direct services. The number and diversity of people employed shade off in the smaller hotels until in some inns the work may be shared among members of the proprietor's family with little outside help. Similarly a large restaurant will have a large kitchen with much mechanised equipment, a cellar and storeroom staff, waiters and waitresses, but a small café or bar will have the minimum of equipment and few staff.

Categories of occupation, gradings and titles differ from country to country but, broadly speaking, divide by function. The housekeeper and housekeeping department is responsible for the orderliness and cleanliness of the premises with particular attention to public rooms and guest rooms, making of beds, care of linen, upkeep and arrangement of furniture. Porters, lift-men, doormen, page-boys, etc., assist with luggage, transport and other services. Office staff includes receptionists and switchboard operators. In the kitchens there may be a hierarchy of chefs and cooks responsible to the chief chef and supported by vegetable and salad preparers. In the sculleries, others will clean vegetables and wash up. In large establishments much of the work will be done mechanically. Other staff will be employed in storerooms and wine and liquor cellars.

HAZARDS AND THEIR PREVENTION

Accident hazards. In the workshops of the large hotels the accident hazards are the same as in the same trades carried out elsewhere: dangerous parts of the machinery must be guarded, electrical plant must be efficiently earthed and protected, good housekeeping must be observed and new workers must be trained in safe practices. In building and maintenance work, attention must be paid to safety of ladders, scaffolding and working platforms. Laundry and sewing machinery must be guarded to the same standard as in outside industries.

Two items of plant need particular attention–the boilers and the lifts or hoists. The boilers should be under regular examination and well maintained and all provisions regarding safe working pressure observed. All hoists and lifts, whether for passengers or goods, should be provided with effective interlocking gates or doors and regularly examined and maintained. Lift attendants should be trained to ensure that safe working loads are never exceeded.

Adequate fire exits, fire alarms, fire-fighting equipment and training procedures in case of fire are as essential for the safety of employed persons as for the clients.

The machinery used in the kitchens of hotels and restaurants is the same as that used in other food trades (bakery, confectionery, butchery and vegetable preparation). Mixers, beaters, mincers, slicers, etc., require the same protection. There is the same danger from the use of knives in preparing food: sharp blades and suitable protective handles are required. Refrigerating plant requires the same high standard of maintenance and the same provision for emergencies. A recently encountered hazard has been the introduction of microwave ovens, which may expose kitchen staff to dangerous levels of non-ionising radiation if not properly protected.

Slips and falls are a major hazard: kitchen and scullery floors become greasy, damp and slippery. Floors must be kept scrupulously clean and tidy; the floor surface itself is important: floor tiles impregnated with an abrasive and adhesive-backed abrasive strips or blocks have been used with success in many service kitchens in the United States. In other parts of a hotel, precautions against slips are necessary, especially as persons may be carrying trays, bottles, etc.; for example the fastenings of stair carpets, the maintenance of all service stairways, in particular the stairways to wine and liquor cellars, where badly lit and badly maintained stairways may lead to serious injuries from falling.

Burns from hot stoves; ignition, spattering and spilling of hot fat; scalds from steam and boiling liquids are a common hazard in kitchens. Prevention is by proper planning of kitchens, with ample space around danger areas, use of safe equipment, training in safe methods of work and procedure in case of ignition. Pilot lights and flame-failure devices installed in gas appliances will prevent explosion risks.

The domestic and service workers are particularly liable to sprains and strains from lifting and carrying heavy luggage and furniture, from turning mattresses, etc. Hand trucks and trolleys should be used where possible and staff trained in safe methods of lifting and handling.

Cleaners of bathrooms and bedrooms may suffer severe cuts from discarded razor blades: receptacles should be provided and staff should be trained not to put their hands into waste-baskets.

Health hazards. In many countries public health regulations require defined standards of hygiene in premises where food is prepared. They may also prohibit the employment of persons suffering from certain communicable diseases. In any event, medical selection and supervision is most desirable on both counts. They can also control the other hazards to which hotel and restaurant workers are particularly susceptible.

Barmen and publicans are particularly susceptible to alcoholism: circulatory diseases and arthritis may afflict waiters, doormen, counter staff and others subject to long hours of standing; skin diseases may be caused by cleansers and detergents used in room cleaning or in sculleries; some workers may be allergic to certain foods and vegetables.

Kitchen workers are often exposed to high temperatures, radiant heat, and a variety of strong smells: efficient ventilation without draughts is essential, usually with local exhaust hoods over stoves. It is unfortunate that in towns many kitchens and other workrooms and laundries are situated in basements and here it is particularly important to overcome defects of ventilation and lighting. Kitchens are often liable to be overcrowded at busy times and this is to be avoided by careful planning.

Good provision for separate sanitary and washing facilities with showers and for taking meals are essential, especially for kitchen staff. Where workers are housed on the premises, their general health will depend a great deal on the standard of accommodation they are given.

Appropriate protective clothing is necessary for most hotel and restaurant workers. Good housekeeping and cleanliness are the basis of all other health requirements.

In some countries hours of work are controlled by law but most hotel workers are subject to serious pressure at certain times, irregular meal hours and rest periods, and overtime; shift work is sometimes necessary. It is essential that adequate rest breaks should be given with proper facilities for taking meals.

ELLIS, R. C. Jr.

General:

Conditions of work and life of migrant and seasonal workers employed in hotels, restaurants and similar establishments. Report II. Second Tripartite Technical Meeting for Hotels, Restaurants and Similar Establishments (Geneva, International Labour Office, 1974), 54 p.

Hotels:

CIS 78-1161 *Health and safety guide for hotels and motels.* DHEW (NIOSH) publication No. 76-112 (National Institute for Occupational Safety and Health, 4676 Columbia Parkway, Cincinnati) (Aug. 1975), 84 p. Illus.

Restaurants:

CIS 77-564 *Health and safety guide for eating and drinking places.* DHEW (NIOSH) publication No. 76-163 (National Institute for Occupational Safety and Health, 4676 Columbia Parkway, Cincinnati) (May 1976), 63 p. Illus.

CIS 75-1219 *A study of restaurant fires.* NFPA No. FR74-1 1974 (National Fire Protection Association, 470 Atlantic Avenue, Boston, Massachusetts), 19 p. Illus. 52 ref.

CIS 74-1896 "Ventilation systems for restaurant kitchens" (Lüftungstechnische Anlagen in gewerblichen Küchen). Fischer, H. *HLH* (Dusseldorf), Feb. 1974, 2 (55-59). Illus. 6 ref. (In German)

Hours of work

In early days (and still today in some countries), the tradition in agriculture and handicrafts was to work from dawn till dusk: the only established rest days resulted from the respect of weekly religious observances and other religious festivals. However, the intensity of work often varied with the seasons, leaving most workers underemployed some of the time. Handicraft workers could, to some extent, control the use of their time, and in all occupations the speed of work was determined by human, or sometimes animal, capacity.

The Industrial Revolution and the rapid development of the factory system gave rise at first to grave abuses: for a time, the predominant consideration was to use the newly installed machinery to its fullest capacity and even young children were kept at work for 14 hours a day, 6 days a week.

Limitation of hours of work

As the deplorable results became apparent, the efforts of reformers to obtain State regulation of hours of work gathered momentum and by the middle of the 19th century most of the industrially developed countries had begun rather tentatively to limit hours by statute. Although often legislation limited only the working hours of women and children, the effect extended also to the hours worked by adult males. At the same time, pressure from the workers themselves and the growth of trade unionism often led to agreements between employers and employed about maximum hours of work. Even so, a 10-hour day and a 60-hour week were widely permitted until the First World War, although in practice the 48-hour week was already gaining ground, especially in advanced industrial countries where collective bargaining was a strong force. In the emergencies of war, restraint was loosened and very long hours were again worked. Studies made on the effect on output during those years showed clearly that excessive hours of employment had an adverse effect on production.

Various attempts had been made to bring the discussion of hours on to an international level, for instance the campaign for the 8-hour day at the First Congress of the International Association of Working Men in 1866, but it was with the constitution of the International Labour Office in 1919 that the question for the first time achieved a firm international basis. There had been great pressure from workers' organisations that the peace settlement after the First World War should incorporate provisions ensuring better conditions of work and it was fitting that the first Convention of the ILO should deal with the question of hours.

Certain effects of hours of work

Little statistical evidence is available to show a precise connection between long hours of work and ill health but there is strong evidence of the deleterious effect of long hours of work on production and their connection with absenteeism. This gives an indication of the impact of fatigue. Similarly, it is difficult to prove that long hours directly affect the incidence of accidents although studies made in the United Kingdom in wartime showed a drop in accidents when excessive hours were reduced. It is indeed reasonable to assume that a tired man or woman is more likely to incur injuries, especially those caused by momentary inattention. It is generally recognised that long hours and fatigue are to be avoided, particularly where the person concerned has responsibility for the safety of others, as in all forms of transport. The General Conference of the International Labour Organisation has recognised such a need by the adoption in 1979 of the Hours of Work and Rest Periods (Road Transport) Convention (No. 153) and Recommendation (No. 161).

Of the effect of long hours on general well-being and the quality of life, there can be no argument: leisure is needed for the full development of the personality, of the arts of living and of deeper social relationships.

All problems of fatigue do not disappear with shorter hours: the increasing speed of some processes may make the stress of shorter hours no less than that of longer hours at a lower speed; with many modern techniques there may be less muscular but more nervous strain. Another factor to be reckoned with is the practice sometimes known as "double-jobbing". Already a substantial number of people whose hours have been reduced choose to spend their free time in a secondary occupation, probably on their own account. To some this may be as satisfactory as a hobby, and not always more strenuous, and it is undoubtedly financially attractive. In many countries married women form an increasing part of the labour force: most of them have family and domestic responsibilities and are inevitably "double-jobbers" (see DOUBLE JOBBING). Flexible patterns of working, sometimes part-time, are often necessary to enable them to make their contribution to the economic life of the country without undue strain or fatigue. Indeed it is likely that greater flexibility in hours of employment will develop generally with a greater choice for the individual between more leisure and more pay. The question now exercising many authorities and voluntary associations is how best to provide recreational facilities of every kind for the leisure hours. The reduction of hours of work is also seen by some as a means of preventing unemployment and increasing employment.

ILO activity

Between 1919 and 1962, the International Labour Conference adopted a total of 14 Conventions and 10 Recommendations on the question of hours of work (apart from instruments concerning related aspects, such as weekly rest periods, night work, and provisions concerning hours of work in instruments relating to general aspects of conditions of work in specialised industries and types of work).

Of these the most important are:

(a) The Hours of Work (Industry) Convention, 1919 (No. 1) (ratified by 44 States). This lays down the standard of a 48-hour week with an 8- or 9-hour day in mining, manufacturing, construction and transport.

(b) The Hours of Work (Commerce and Offices) Convention, 1930 (No. 30) (ratified by 29 States). This lays down a 48-hour week and an 8-hour day but with the possibility of a 10-hour day, for commercial and trade establishments and administrative services.

(c) The Forty-Hour Week Convention, 1935 (No. 47) (came into force in 1957 and was ratified by seven States). This embodies the principle of the 40-hour week to be applied in such a manner as not to reduce the standard of living.

(d) The Reduction of Hours of Work Recommendation, 1962 (No. 116). This recommends an immediate reduction to 48 hours where that standard is not already enforced and the progressive attainment of 40 hours.

These basic provisions of the ILO reflect the progress of thought on hours of work over the period they cover: whereas in 1919, 48 hours was the goal, within 20 years it had become 40. In the 1930s, the 40-hour week was seen partly as a measure against the massive unemployment then prevalent, but by 1962 it had become desirable in itself. The methods of implementation envisaged had also become wider, the earlier Conventions looking to legislation but the 1962 Recommendation including collective bargaining and arbitration.

Ratification is not the only criterion of the effect of a Convention: many ratifications are withheld for minor reasons and the impact of international standards extends to all member States and even beyond.

Attainments at national level

It is impossible to give here a comprehensive review of the enormous variety of laws and agreements covering hours of employment throughout the world. However, the available information indicates that—although in some countries 48 h is the general standard—there is a progressive tendency towards the adoption of the 40-hour week, especially in heavy industries and those with high risks, in night work, or as regards women and young persons. On the whole, the more industrially advanced the country or the industry, the shorter the hours of work. Many developing countries plan or have started a reduction to 40 hours by stages over different sectors of the economy. It has to be remembered that in many countries small undertakings, and especially activities carried out on a home-work basis, fall outside regulations and that many other exceptions vary the general tendency at different times and in different places.

In assessing the actual number of hours, the following factors have to be taken into account:

(a) collective agreements often modify legislative provisions so that hours in practice may be considerably less than the maximum required by law;

(b) overtime may be worked so frequently that several hours are consistently added to the weekly total; hours may fluctuate seasonally or with changes in the economic situation and sometimes they are averaged over a period;

(c) meal hours are usually not included in working time but there are exceptions.

Distribution of working hours

However, it is not only the length of the working week but the distribution of hours within that framework that have to be reckoned with and this entails consideration of the length of the working day; meal breaks and rest pauses; daily and weekly rest; and special arrangements for shift work.

The length of the working day. It is of doubtful advantage to reduce the number of days worked per week if this means undue lengthening of the hours worked each day. This applies especially in large conurbations where travelling time may extend absence from home substantially. Similarly, overtime beyond the standard day may threaten any beneficial effect; in general, the longer the working day the more rapidly will output fall towards the end of the afternoon spell and a tendency to accidents in this period may be expected. Overtime may be financially advantageous to the worker but if unduly extended may be expensive for the employer. One reckoning is that over 8 h a day and 48 a week (on non-automated work), it requires 3 h overtime to achieve 1 h normal output in light work and 2 h overtime in heavy work.

Rest pauses and meal breaks. ILO Conventions prohibit the employment of women and young persons in industry at night with a specification that the night interval should include 11 h for women and 12 for young persons but for most workers the length of the working day itself governs the period of night rest. Special arrangements are necessary when continuous shifts are worked to ensure a proper break at the change-over. In employment where fatigue might endanger the safety of others, e.g. in aviation or road transport, special arrangements for adequate rest between spells of work are necessary.

The midday or midshift break is usually required by law or agreement but its length may be governed by local customs or climate. Its value as a restorative is bound up with the existence of suitable arrangements for taking meals and the provision of catering or canteen facilities. It is always desirable that this break should include a change of environment even where there are no special health risks at the workplace.

Apart from the main meal break, agreements, law or good practice often require short breaks in the spell of work. Their frequency may be determined by the nature of the work: sometimes work is broken by periods of waiting as a result of technical factors, sometimes workers can pause in their work and will naturally do so. Where, however, a machine has to be constantly attended or a workplace always filled, experience shows the importance of set pauses. Where work requires great concentration as in air traffic control or instrument scanning, there may be need for a long pause or diversion to other work to prevent perceptual fatigue. In any short breaks, it is desirable that light beverages should be speedily available.

Shift work. See the article on SHIFT WORK.

Rest days. The necessity of a weekly rest is almost universally accepted though not everywhere rigorously enforced. The ILO Weekly Rest (Industry) Convention, 1921 (No. 14) has been ratified by 94 countries. When, in emergencies, the weekly rest day has been abandoned in the interests of production, the measure has rapidly proved to defeat its own object. The standard 5-day week has become widespread in industry and the provision of two rest days in the week has become an accepted principle. These days are particularly welcomed by women employees with domestic responsibilities and shopping needs.

Holidays. At all times and in all places there have been some holidays, religious or national festivals which have meant cessation of all but the most essential tasks, but the right to prescribed annual holidays with pay is a recent development which has spread widely especially among organised workers. The length of the paid holiday is usually between two and three weeks in any year but may be increased with seniority to four or five or even more weeks. Proposals for an international standard of at least three weeks came before the International Labour Conference in 1970.

More recently (1974) the ILO General Conference adopted the Paid Educational Leave Convention (No. 140) according to which ratifying member States will formulate and apply a policy designed to promote the granting to workers of leave for a specified period during working hours, with adequate financial entitlements, for the purpose of: training at any level, general social and civic education and trade union education.

Enforcement

No arrangements of working hours can be maintained without effective enforcement. In most places this is the task of a labour inspectorate and may involve not only the checking of forms and records but also unannounced inspection of enterprises, perhaps at unusual times. Trade unions may also exercise a vigilant check on the application of agreements.

EVANS, A. A.
DE GRAZIA, R.

Adapting working hours to modern needs. The time factor in the new approach to working conditions. Maric, D. (Geneva, International Labour Office, 1977), 50 p.

Hours of work in industrialised countries. Evans, A. A. (Geneva, International Labour Office, 1978), 164 p.

Housekeeping and maintenance

In any undertaking, good housekeeping and maintenance are the essential routine supports of industrial safety and health. They are complementary and, on some points, it may be difficult to draw a line between them, but a general distinction may be made.

Maintenance covers the work done to keep building, plant, equipment and machinery in safe and efficient working order and in good repair, the upkeep of all sanitary and welfare facilities and the regular painting and cleaning of walls, ceilings and fixtures. Good housekeeping includes day-to-day cleanliness, tidiness and good order in all parts of the undertaking. Good housekeeping is almost impossible without good maintenance of machinery and equipment; for example it is difficult to keep a badly worn floor clean or to keep it dry if there is leakage from a broken roof or some ill maintained piece of plant. On the other hand, good day-by-day housekeeping will considerably cut down the amount of maintenance work required.

Many accidents can be attributed, in the first place, to defective maintenance; falls on broken floors or worn-down steps or stairs, falls of or from defective ladders, stools or chairs; access to dangerous parts of machinery through broken or badly fixed guards; scalds from leaking steam pipes or burns from unlagged hot pipes. Workers may be trapped by fire if exit doors do not open quickly or outside staircases are damaged or obstructed; unrepaired damage to electrical equipment, earthing arrangements, plugs, flexes, etc., may bring a risk of electric shock. Lack of maintenance may be the prime cause of failure in lifting plant or explosion of pressure vessels. Badly maintained hand tools cause many injuries. Dirty windows or light fittings may so much reduce the level of illumination that accidents occur through failure to see danger.

More insidious dangers to health can be caused by badly maintained exhaust ventilation systems which allow dangerous fumes or dust to escape into the air; a badly worn workbench will be difficult to keep clean (see CONTROL TECHNOLOGY). Health and well-being can be affected by defective plumbing, failure of hot water or drinking water supply (see MAINTENANCE OF MACHINERY AND EQUIPMENT).

Good housekeeping

A very large number and a very wide range of accidents may be caused, in part at least, by bad housekeeping, e.g. falls on floors left slippery, greasy, or damp; striking against or falling over machine parts, material or other obstructions left lying in passageways; cuts from objects left protruding from benches; and, especially on construction sites, punctures by nails protruding from wood.

Risks from internal transport are intensified if passageways are not kept clear or if vision is impeded by stacking of materials. Badly stacked materials may fall and cause serious injury. Fires may be started if combustible waste is not regularly removed or if excessive quantities of flammable materials are kept in workshops. Accidents occur in offices when filing cabinet drawers are left open. Health risks from dangerous dusts or chemicals are much increased unless all working surfaces and surroundings are kept rigorously clean.

Good housekeeping cannot be left to the unplanned activities of persons employed but is, in its broad outlines, a responsibility of management. The undertaking should be laid out in such a way that it is easy for order and cleanliness to be observed. Aisles, walkways, traffic areas and exits should be properly marked and defined. Special areas should be set aside for storage of raw materials, finished work, tools and accessories. Racks for hand tools or implements above workbenches, an underbench slot or other simple provision for storage of small personal possessions will keep working areas clear. Adequate receptacles for waste and debris should be conveniently placed. Floors and workbenches should be constructed of materials suitable for the work done and also easy to clean; non-slip surfaces allied with non-slip polishing methods will obviate the risk of slipping which sometimes occurs in public offices.

Many machines are liable to eject quantities of oil, swarf or water but much can be done by screening and simple physical devices to prevent deposit on the surrounding floors. Wet processes or plant should be provided with drainage channels and sometimes isolated by curbs.

In an establishment of any size, day-to-day cleanliness cannot be left to the last few minutes of the working day of the production workers. It may be reasonable to expect a man to leave his immediate workplace in a clean and tidy condition but the general cleaning of workshops, sanitary facilities and yards can be effectively carried out only by special cleaning gangs, employed or hired for that express purpose.

In many trades this cleaning will take place after the end of the working day with a thorough weekly clean during the weekend pause but special arrangements will be necessary where continuous shifts are worked or where cleaning during the working day is essential. Where dangerous dusts, such as those containing free silica, may be evolved into the air, continuous cleaning by vacuum methods may be necessary.

Tidiness and good order throughout working time are especially difficult to maintain in trades where there is rapid production of finished goods and/or waste. Finished goods should be removed to a proper storage area and regular arrangements made for removal of waste and emptying of waste receptacles. Particular care is needed to prevent accumulation of debris and clutter under benches.

Each type of industry has its particular housekeeping problem, from large steel works to a small dressmaker's workroom. Construction sites present serious difficulties: only the most rigorous supervision and the co-operation of all employees can keep the site, work platforms, etc., free from tools, bolts, planks (including upturned nails) and other objects likely to cause serious accidents.

Supervision and cleaning arrangements for sanitary facilities, bathrooms, cloakrooms and messrooms are as essential as for the working areas. Co-operation of all concerned may be ensured to some extent by admonition but good relationships are at least as important.

The worst risk to persons engaged in housekeeping activities is that excessive zeal may lead to cleaning-up in proximity to machinery in motion, normally safe by position. Cleaning should never be carried out when any risk of entanglement is present. Removal of waste and debris may involve serious risk of cuts from scrap metal, swarf or broken glass. Detergents and cleaning materials may cause dermatitis and skin affections. Personal protective clothing, overalls, hand protection and foot protection are often necessary; respiratory protective equipment may be required where dangerous dusts are present. Washing facilities and first-aid treatment should be available to cleaners employed outside normal working time.

MAY, J.

CIS 79-2055 "The annual spring-cleaning" (Le grand nettoyage annuel). Sladden, J. *Promosafe* (Brussels), Aug.-Sep. 1979, 7/32 (10-15). (In French)

Housing of workers

Since early history, when caves and similar refuges provided man with shelter from the elements and protection from his enemies, almost all fundamental human activities have been centred in the home. It is there traditionally that man has installed his wife, that children have been born and raised, and parents cared for in their old age. The seeds of instruction and religion have been nurtured there and the traditions of society begun. Social progress throughout history has merely modified these functions. The dwelling is still the place of rest and relaxation, of family life and recreation while its value as a social unit still persists. Sound dwelling life, therefore, is important not only to the individual and his family, but also to the physical, social and cultural health of society.

Standards of housing

The importance of housing was recognised by the 45th Session of the International Labour Conference which adopted the Workers' Housing Recommendation, 1961 (No. 115) concerning the "housing of manual and non-manual workers, including those who are self-employed and aged, retired or physically handicapped persons". This Recommendation, which deals with national housing policies, the responsibility of public authorities and employers, the financing of housing programmes, and more rational building construction and planning, suggests that, as a general principle, competent housing authorities should take the following factors into account:

(a) protection against the elements and natural disasters, such as earthquake, typhoon and fire, heat, cold, damp, noise, disease-carrying animals and insects;

(b) minimum space per person or per family expressed in terms of floor area, cubic volume or size and number of rooms;

(c) the supply of safe water in quantities sufficient to provide for all personal and household uses;

(d) adequate sewage and household waste disposal systems;

(e) adequate sanitary and washing facilities, ventilation, cooking and storage facilities and natural and artificial lighting;

(f) a minimum degree of privacy for members of the household as a whole and between individuals; and

(g) suitable separation between rooms occupied for living purposes and accommodation assigned to animals.

The "Köln Standard", established in 1957 and hitherto regarded as the international standard for dwellings, indicates similar requirements. It stipulates a living room, dining room and kitchen, one bedroom for husband and wife, and one private bedroom for every two children. In addition, it stipulates that there should be a bathroom, lavatory and storeroom. This standard specified the following minimum areas: 50 m^2 for a family of three; 55-60 m^2 for a family of four; and about 70 m^2 for one of five.

Each country has its minimum standards, regarding sanitary and constructional requirements for housing, but to ensure a healthy physical environment for its citizens, it is essential to go beyond these, while failure to observe even these minimum requirements would result in a disastrous decline in the morale and physical health of the society.

Workers' housing is part and parcel of their working and living conditions and, as such, has to be seen in close correlation with other factors. In other words, the concept of "adequate and suitable housing" is inseparable from a person's wage, the location of his accommodation and the kind of work on which he is employed. Workers at the lower end of the scale need special attention as their jobs are often arduous and their housing conditions among the poorest. In addition, more and more jobs nowadays involve a certain amount of nervous strain, which makes it all the more important—and all the more difficult—for their hours of sleep to be properly restful. Overcrowding may make housing that is otherwise quite suitable unhealthy or inadequate; and overcrowding, of course, is a problem that arises in every big city.

Living environment

Housing standards cannot be maintained and improved by the provision of adequate buildings and equipment alone, however, and few families can achieve and maintain high living standards by their own efforts. Public authorities must therefore devote considerable efforts to creating the necessary conditions in which living standards can rise. Health and sanitation are today matters of public concern.

The role of public authorities is particularly evident in modern urban communities, in which the benefits of urbanisation are to a large extent offset by its obvious disadvantages. On the one hand, modern technology and economic development make housing construction safer and raise domestic standards of sanitation; on the other, external urban environment, in many cases, appears to be deteriorating with increasing industrialisation and urbanisation.

Obviously undesirable and harmful consequences of urban growth are air and water pollution. Springs and rivers are poisoned by industrial effluent and detergents. Smoke and fumes from factory and domestic chimneys are reinforced by the stale air from ventilation systems serving buildings housing populations as large as those of former towns. The construction of ever taller buildings robs neighbourhoods of sunshine, causes stagnation of the air and brings new dangers. Unsuspected risks exist where the continual extraction of water from the subsoil for industrial use brings the danger of subsidence to adjacent residential buildings or where a gas leak in one appartment may cause an explosion which endangers the whole block of flats.

Housing development

Unless systematic measures are taken at the community level to halt this deterioration in urban conditions, it is difficult to see how the standards of workers' housing can be improved. In fact, both national and municipal authorities are making efforts to protect workers' housing standards by legislation and through planning programmes aimed at separating residental and industrial, and even educational areas. Building permission may even be withheld in the case of new plant which may constitute a public nuisance and smokeless zones are declared. In this area, occupational health and public health interests run parallel.

As far as is practicable, and taking into account available public and private transport facilities, workers' housing should be within easy reach of places of employment and close to community amenities, such as schools, shopping centres, recreation areas, religious facilities, medical facilities and nurseries, and should be designed to form attractive and well laid-out neighbourhoods, including open spaces. From the standpoint of workers' health, however, where the nature of industrial plants constitutes a hazard to workers' safety and health, it is preferable to locate dwellings at an adequate distance from their workplace. Careful thought has to be given to all these factors (transport, community facilities, etc.) at every stage of urbanisation.

Special housing requirements

In the case of certain categories of workers, special conditions must be taken into account. Migrant and immigrant workers are frequently separated from their families and find themselves in a strange and sometimes, regrettably, hostile environment. Housing for workers such as these is particularly important because on it depends the possibility of the family being united and because, along with integration in the working environment, it is a key factor in workers' ability to adapt to life in the host country. Agricultural, plantation, lumber and construction workers must sometimes be housed at considerable distances from large centres of population and from their own homes. Because there are sometimes no community services at all, it may be necessary to provide medical and educational facilities and outlets for basic products, especially food, as the accommodation arrangements are made (see WELFARE IN INDUSTRY). In all these cases, housing must frequently be provided by the employer or by a responsible authority and, since these workers are usually without their families, this housing is generally collective. Minimum standards for accommodation must clearly be established and every effort made to prevent communal living adversely affecting the health and morale of workers. Basic requirements would probably include:

(a) a separate bed for each worker;

(b) separate accommodation for different sexes;

(c) an adequate supply of safe water for drinking and ablutions;

(d) adequate drainage and sanitary conveniences;

(e) adequate ventilation and, where necessary, heating; and

(f) common dining rooms, canteens, rest and recreation rooms and health facilities, where these would not otherwise be available.

Where, despite the remoteness of the locality, employment is not of a temporary nature, workers' housing and communal amenities should be of durable construction and should take account of local climatic and other conditions, e.g. the likelihood of earthquakes or hurricanes.

In all these cases and in all other circumstances where housing is provided by the employer, certain safeguards for workers should be ensured either by law or by collective or other binding agreement. For instance, a worker or his family should be entitled to a reasonable period of continued occupancy to enable satisfactory alternative accommodation to be obtained when he leaves his employment through sickness, incapacity, occupational injury or retirement. A similar safeguard should be provided for his family in the event of his death. Provisions should also be made for compensation to workers for improvements made to housing or amenities which they vacate on leaving their employment.

Further categories of workers for whom special provision must be made are shift workers and night workers. These include factory workers, public transport workers, hospital staff and workers generally in the public sector of industry. The need to commute between their homes and places of work at very early or very late hours, and to sleep during the day amidst the noise of normal daytime activity and in daylight poses distinct problems. One solution is the provision of collective housing in the form of boarding houses where workers can rest and sleep before returning to their own homes for proper rest. Nurses are frequently housed in residential quarters attached to the hospitals in which they work. This eliminates the need to commute between home and workplaces, which is an especial advantage in the case of night work and eliminates tiredness caused by travelling. Special requirements affecting such forms of collective housing relate to the need to be able to ensure darkness and quiet for sleeping during daylight hours. Ventilation and heating are other important requirements.

As regards ships' crews, two Conventions (the Accommodation of Crews Convention (Revised), 1949 (No. 92), and the Accommodation of Crews (Supplementary Provisions) Convention, 1970 (No. 133)) adopted by the ILO contain a number of relevant provisions. These have been supplemented by two ILO Recommendations (Nos. 140 and 141) concerning the air conditioning of crew accommodation and certain other spaces on board ship and control of harmful noise in crew accommodation and working spaces on board ship.

Planning and financing

Public authorities have a dual responsibility with regard to workers' housing. Having due regard to the constitutional structure of the country, the competent national authorities should create a central body to plan and co-ordinate activities in this sector. This central body should evaluate the needs for workers' housing and related community facilities, and formulate workers' housing programmes. In addition, it should direct slum-clearance schemes and the rehousing of occupiers of slum dwellings. All possible measures should be taken to provide the skilled labour force, materials, equipment and finance required for the construction of housing.

As far as possible, any plans for a new factory should provide for the simultaneous construction of housing for workers in a healthy environment within easy reach of the plant.

Public and private facilities should be made available for loans at moderate rates of interest and should be supplemented by other methods of direct and indirect financial assistance such as subsidies, tax concessions and grants. The creation of co-operative and similar non-profit-making housing societies should be encouraged. Where a sound credit market exists, national mortgage insurance systems or public guarantees of private mortgages should be established as a means of promoting the construction of workers' housing. Measures should be taken in accordance with national custom to stimulate saving and encourage investment. Above all, workers' housing built with assistance from public funds should not be allowed to become an object of speculation.

Only by meeting all these divergent requirements—structural, topographic, sanitary and financial—will it be possible to provide housing of a standard to ensure the physical and moral health of workers.

UMEZAWA, T.
CLERC, J. M.

The physiological basis of health standards for dwellings. Goromosov, M. S. Public health papers No. 33 (Geneva, World Health Organisation, 1968), 163 p.

Use of epidemiology in housing programmes and in planning human settlements. Report of a WHO committee on housing and health. Technical report series 544 (Geneva, World Health Organisation, 1974), 64 p. 106 ref.

Housing, the housing environment and health. An annotated bibliography. Martin, E.; Kaloyanova, F.; Maziarka, S. WHO offset publication No. 27 (Geneva, World Health Organisation, 1976), 113 p.

Convention (133) concerning crew accommodation on board ship (supplementary provisions). International Labour Conference (Geneva, International Labour Office, 1970), 9 p.

Convention (143) concerning migrations in abusive conditions and the promotion of equality of opportunity and treatment of migrant workers. International Labour Conference (Geneva, International Labour Office, 1975), 7 p.

Recommendation (140) concerning air conditioning of crew accommodation and certain other spaces on board ship. International Labour Conference (Geneva, International Labour Office, 1970), 1 p.

Recommendation (141) concerning control of harmful noise in crew accommodation and working spaces on board ship. International Labour Conference (Geneva, International Labour Office, 1970), 2 p.

Recommendation (151) concerning migrant workers. International Labour Conference (Geneva, International Labour Office, 1975), 7 p.

Human engineering

Human engineering, or its equivalent, human-factors engineering, means, most simply, designing for human use. More precisely, it is the application of information about human characteristics, capacities and limitations to the design of machines, machine systems and environments so that people can live and work safely, comfortably and effectively. The term also designates the profession that deals with such problems.

The use of the terms human engineering and human-factors engineering is confined almost exclusively to the North American continent. Throughout the rest of the world, except possibly for the USSR where the term engineering psychology seems to be preferred, this area of specialisation is called ergonomics. The ergonomist tends to be somewhat more biologically or physiologically oriented than his American counterpart. This difference in orientation appears to be in part the result of historical accident and in part a result of the somewhat greater emphasis that ergonomists place on the well-being of the worker or the operator and on his workload. Although human-factors engineers are also concerned with the comfort and safety of operators, their work tends to be directed more towards the design of machines and machine systems and the successful integration of man into systems. In short, ergonomists are often more concerned with production, and human-factors engineers with products. None the less, ergonomics and human-factors engineering are more alike than they are different and, for most practical purposes, the two fields may be considered equivalent. See also ERGONOMICS.

As is true of many modern technical fields, small subgroups with specialised interests within human engineering have devised special names to refer to their own specific fields of activity. Some of these are: applied experimental psychology, biomechanics, biotechnology, engineering psychology, environmental engineering, life-sciences engineering, man-machine systems research, psychotechnology, and work physiology. Clear-cut distinctions between these various subfields are difficult to make and it is correct to say that they may all be subsumed under the more general and inclusive heading of human engineering.

Goals of human engineering

Although increased safety and, consequently, reduction in the number of accidents are prime goals of human engineering, they are not the sole important objectives of this field. Human engineers also include among their major aims: increasing the efficiency with which machines can be operated; increasing productivity in industrial and system operations; decreasing the amount of human effort required to operate machines; and increasing human comfort in man-machine systems.

To attain these aims, the human engineer may have recourse to methods, data and principles from such scientific and professional disciplines as: biology, education, engineering, industrial design, medicine, physical anthropology, physiology, psychology, sociology, and toxicology.

The main activities of the field include basic and applied research on human performance to generate new knowledge, as well as the application of this knowledge to the design and evaluation of systems.

Research methods

In their research human engineers make extensive use of the classical laboratory methods used widely in the human and biological sciences. In addition they have adapted some research methods borrowed from other disciplines and have devised new methods of their own for tackling the complex problems of man at work. Some of these special techniques are described below.

Critical incident methods. Critical incident studies of accidents and near accidents have been especially productive in identifying error-provocative situations and in locating problems that should be explored with more sophisticated research methods.

Observational methods. Human engineers have adopted a variety of observational methods, principally from industrial engineering, for studying man at work. Some of these are activity sampling, function analysis, and flow analysis.

Simulation and modelling. Because of the complex nature of the machine systems with which they frequently deal, human engineers often resort to simulations in their work. As used here, the term simulation includes a number of techniques and devices: simulators, mock-ups, training devices, and models of one sort or another.

Simulators serve a number of functions in human engineering research. By modelling closely some operational system a simulator helps to achieve the greater realism that is required for extrapolation from laboratory studies. Since simulators are usually much cheaper than the systems they mimic, realistic research can be carried out at a fraction of the cost that would be required if it were done on an operating system. Simulators, or mock-ups, can also be built before the system itself has actually been constructed. In this way human engineers can carry out research in sufficient time for their research findings to be applied to the system before it has reached its final form. Although simulation still remains as much art as science, some principles of successful simulation are slowly being accumulated.

Subjective reports. Finally, human engineers use a variety of techniques for sampling operator and user opinions. These range from simple unstructured interviews, through elaborate questionnaires, to highly refined rating scales, psychophysical judgements and similar measurement devices.

Fields of application

In the United States human engineering specialists are employed in a number of government departments and agencies; in several dozen university engineering and psychology departments; in many independent research and consulting organisations; in a variety of industries, principally those engaged in the design and production of aerospace systems, automotive vehicles, communications equipment, computers, electronic equipment and farm machinery; and in every branch of the military service.

One of the most impressive and dramatic uses of human engineering has been in the design of vehicles, suits and tools for the exploration of space. Human engineering has also, however, been applied to many more commonplace and ordinary devices, such as highway signs, telephones, typewriters, computer consoles and data-processing systems, machine tools, automobiles and kitchen stoves. In these applications, human engineering does not, however, address itself directly to problems of "consumer motivation".

Most recently human engineers have turned their attention to a variety of urban problems, such as street and highway design, rapid-transit facilities, hospitals and urban health facilities, architecture and housing, pollution control, education, law enforcement, postal services and airports. Underlying these recent developments is a new awareness that, in the final analysis, man-made systems are effective only in so far as they serve man safely, comfortably and conveniently. Technology must be designed for human use and against possible misuse.

Subject-matter

Human engineers draw upon data from the full spectrum of human and biological sciences and acquire additional data, as required, through experimental and survey research. The amount of information needed for any particular design problem depends on the problem itself. In the design of a deep-sea habitat, almost everything that is known about man is important and relevant: body dimensions; physiological reactions to environmental stresses such as those produced by unusual mixtures of gases used for breathing, pervasive cold and dampness, and greatly increased atmospheric pressures; sensory capacities; decision-making abilities; psychomotor performance and the ability to make precise, rapid, and correct control movements; learning and the ability to modify behaviour through training; purely biological considerations associated with eating, drinking, and waste disposal; and psychological factors associated with fatigue, emotion, isolation, and the constant threat of danger. By contrast, the design of a household appliance or of a computer console requires information about a more restricted set of human characteristics.

The man-machine model

Human engineers look upon man and the machine he operates as interacting components in an over-all system. A simplified model illustrating the basic interactions between man and machine is illustrated in figure 1. In effect this model considers the human operator as an

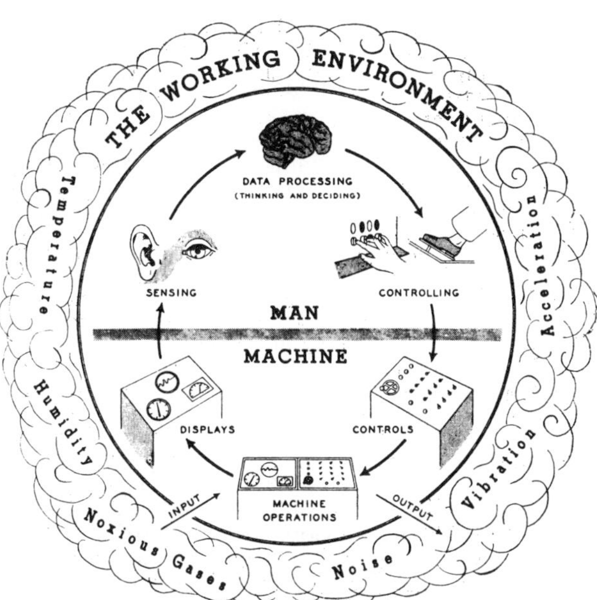

Figure 1. A simplified model of a man-machine system.

organic sensor, data-processor and controller located between the displays and controls of a machine. An input of some kind is transformed by the machine into a signal, which appears on a display. The information so displayed is sensed by the operator, processed mentally, and translated into control responses. The control actions in turn alter the behaviour of the machine to produce an output and further changes in the displays, thus causing the whole cycle to be repeated.

A man driving an automobile is a good example of such a system. The driver reacts to inputs from the speedometer and other displays on his dashboard, inputs from the road and outside environment, noise from the engine, feedback to his muscles from the steering wheel, and other stimuli. From these inputs he makes decisions to take certain control actions. These control actions affect the movements of the automobile, which in turn furnish new and different inputs to the driver.

This model of a simple man-machine system provides a framework for some of the things with which human engineers are concerned. Some important areas of work in human engineering are: the allocation of functions between men and machines; task analysis; the design of information displays, controls, workplaces and passenger stations; design for maintainability; and control of the work environment.

Functional allocation. The human engineer is called upon to assist in allocating functions between men and machines, i.e. to "divide up a job" or decide who should do what in a system. When we ask about the allocation of functions between men and machines we ask two essential questions: What functions of the system should be assigned to men and what to machines? What kinds of things can and should human operators be doing in man-machine systems? Decisions of this kind are ideally made after a careful consideration of the kinds of things that man can do better than machines (for example, perceiving, responding to low probability emergency situations) and those things that machines can do better than people (for example, computing, handling large amounts of information). These decisions are also ideally made early in the design process because they affect all the later design in the system.

Task analysis. This entails drawing up a very detailed and explicit list of the functions that men will be doing in the completed system. The task analysis provides data essential for making decisions about selection standards, training requirements, workloads, the numbers of people who will be needed in the system, and the design of equipment to support the operator in his task.

Display design. Displays are those devices by which machines communicate information to man. The first problem is the choice of the sensory channel to use for displaying information. Information may be displayed visually (by a dial, gauge, video-screen, or print-out), aurally (by a buzzer, bell, gong, or voice), or through some other sense channel (for example by the "feel" of

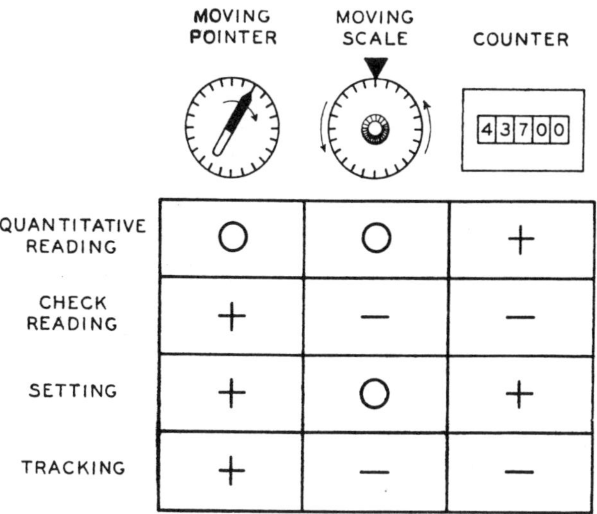

	MOVING POINTER	MOVING SCALE	COUNTER
QUANTITATIVE READING	O	O	+
CHECK READING	+	–	–
SETTING	+	O	+
TRACKING	+	–	–

Figure 2. Recommended usages of three types of mechanical indicators. Usages marked with a (+) are recommended, those marked with a (0) have questionable value, those marked with a (–) lead to frequent errors.

a control). Some kinds of information are better displayed to one particular sense channel, other kinds of information to another.

Once a decision has been made about the sense channel that is to be used, the human engineer is then concerned with the design of the display itself. This includes a great variety of design factors such as the size, shape, colour, brightness, markings and illumination of visual displays; and the frequency, intensity, duration and signal-to-noise ratio of auditory signals. A great deal of precise information is available about the design of displays and figure 2 shows typical recommendations of this type dealing with one aspect of display design.

Control design. Controls are devices that enable man to communicate and transmit information to machines. Controls have to be selected so as to match human actions and the functions required of the machine. Some important factors that influence the design of controls are the control-display ratio, the compatibility of direction-of-movement relationships, safeguards against accidental activation, and control coding. A great deal of information is available about these problems and figure 3 gives some design recommendations concerning direction-of-movement relationship.

CONTROL MOVEMENT	SYSTEM RESPONSE				
	UP	RIGHT	FORWARD	CLOCKWISE	"ON", "GO" OR INCREASE
UP	✓	X	?	X	✓
RIGHT	X	✓	X	?	✓
FORWARD	X	X	✓	X	✓
CLOCKWISE	?	?	?	✓	✓

Figure 3. Recommended direction-of-movement relationships or various system responses. The check marks show recommended usages, question marks show usages of dubious value, and Xs show usages that lead frequently to errors and accidents. *(After Morgan et al., 1963.)*

Workplace and passenger-station design. The topics that have to be considered here are workplace dimensions, the location of controls and displays, seat and panel design, the design of doors and accesses for easy entry and exit, and protective devices, such as seat belts, restraining harnesses, and visors, that may be required for certain situations. Figure 4 illustrates some of the many recommendations drawn up by human engineers working in this field.

Design for maintainability. As machines become more complex, problems of repair and maintenance become more critical. Human engineering design for maintainability includes such things as the design of efficient fault-finding strategies, location of units for easy access, the design of auxiliary tools and test equipment, and the production of simple, easy to use maintenance manuals.

Work environment. Some of the important factors influencing the work environment that are studied by the human engineer include: lighting; noise; vibration; acceleration and zero-gravity effects; variations in temperature, humidity, and barometric pressure; noxious gases, fumes and other contaminants; and radiation.

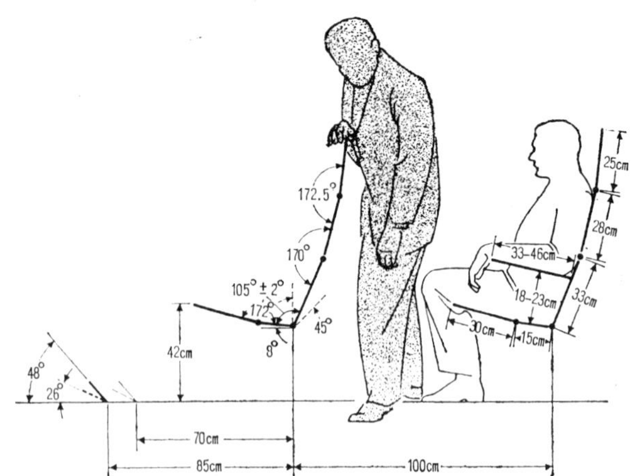

Figure 4. Some recommendations for the design of aircraft passenger seats. *(From Woodson and Conover, 1966.)*

The personnel subsystem

However important it may be to match a single operator to his machine, some of the most challenging and intricate problems of human engineering arise in the design of large man-machine systems and the integration of man into these systems. Examples of such large systems are a nationwide telephone network, a nuclear submarine, an air-traffic control system, a computerised banking system, an automated post office or an automated factory.

In dealing with such large systems the human engineer has to combine certain elements of man-machine design with some conventional aspects of industrial psychology. These problems are sometimes collectively referred to as the "personnel subsystem" as distinguished from the "hardware subsystem". The design of personnel subsystems includes not only all those factors that have been discussed above in connection with the design of machines, but also personnel requirements, selection and training, and operating procedures.

Personnel. A man-machine system is obviously not complete unless there are suitably qualified people available to operate and maintain it. In staffing a system, important questions that have to be faced and answered are: What kind of people will be needed to operate the system? How will these people be selected?

Traditionally, personnel or occupational psychology accepted the job as a constant factor. The job was a fixed quantity with certain definite requirements and the personnel psychologist merely had to find people who could do that particular job. The systems idea has changed that. In systems design the job is considered to be a variable, something to be changed and manipulated until a final design has been realised. This means that there is a genuine interaction between personnel selection and job design. If a job is designed in such a way that it is complex, the standards applied in staff selection must be high. By redesigning the job to make it simpler, selection standards may be reduced.

Training. Personnel training cannot be separated from personnel selection and job design. In general, the more rigorous the criteria applied during selection, the less training the operators will need to do a job. Conversely, where selection is less rigorous, more training will be needed. Both selection and training, in turn, interact with the complexity of the system. The more complex the system, the more stringent must be the selection and training requirements. By redesigning a system so as to make it simpler, selection and training requirements may be relaxed. Good systems design includes the development of training techniques, the design of an appropriate training programme, and sometimes the design of training devices and aids to provide training for operators on the tasks they are to perform.

Operating procedures. Instructions, operating rules and procedures set forth the duties of each operator in a system. Operating rules that have been well thought out and carefully drafted will contribute significantly to safe and orderly operations; such rules have to be prepared specifically for the people who will operate the system and written in such a way that those people can easily understand what they are supposed to do and how they are to do it.

CHAPANIS, A.

CIS 76-876 *Ethnic variables in human factors engineering.* Chapanis, A. (Baltimore and London, Johns Hopkins University Press, 1975), 290 p. Illus. Bibl.

CIS 77-1172 *Introduction to human engineering.* Kraiss, K. F.; Moraal, J. (Cologne, Verlag TÜV Rheinland, 1976), 514 p. Illus. Approx. 500 ref.

See also readings suggested under ERGONOMICS.

Human relations in industry

Human relations have been studied since the dawn of history; however, it is only in the last half-century that more systematic and scientific research has become available.

In industry, social scientists and doctors can help; first, because they have had some systematic training in individual and group psychology—whereas the knowledge of managers and trade unionists is mostly drawn from unsystematised, though often immense personal experience; and second, because they have studied in detail some of the disasters of human relations and discovered what errors produced them (in particular, dissatisfactions at work, and tensions with superiors, and their consequences).

Progressive managers have eagerly accepted their help in getting information and acquiring skill, and have co-opted them in formal and informal training. But others have been very reluctant to seek such help, and have justified their reluctance with reasonable request for validated results or sometimes admitted flatly they they were prejudiced from the start. Trade unionists have, in general, been even more reluctant, and have even regarded social scientists as tools of management brought in to exploit the worker or to obscure the basic issues and conflicts of interest by a smokescreen of jargon.

Yet productivity, satisfaction at work and health are all intimately interwoven; managers, trade unionists and doctors whose primary interests lie in these three fields, should learn from each other how satisfaction at work can be increased, and vicious circles of aggression, ill-will and ill health avoided.

Satisfaction at work

Much has been written on wages and their determination, which often seems to lead to ever-increasing vicious circles of inflation, disgruntlement and self-seeking splinter groups. Far less has been thought about the emotional satisfactions of the job itself, and of the working group. Yet the lack of these satisfactions lead to recognisable and foreseeable symptoms of personal illness and unhappiness, and of social disharmony. If the needs of different individuals can be estimated and compared with the capacity of a post to provide these satisfactions, it should be a step towards diminishing neurotic illness, absenteeism, labour turnover and social ill-will.

Some of the emotional satisfactions from the "working group" (the other people regularly concerned in the job, above, below and on the same level) are obvious and well known; some are less obvious, and their importance is not evident until they are lacking. They fall into several groups.

The first of these, in time, is the need to belong. Each human being needs the acceptance and support of his fellows. Without this, other needs can scarcely find fulfilment. There are many people who do not "belong". They may have started as unwanted children, or later become "odd man out", a difficult, unhappy, uncooperative person, permanently or temporarily.

Secondly, once he belongs, an individual can afford to express his own personality and to seek fulfilment of his talents; he generally needs the appreciation of his fellows and superiors, not only as a man doing a worthwhile job well, but also as a likeable person. Unfortunately, in industry verbal appreciation and frank praise are rarely given. Hence there is preoccupation with money (as the only form of appreciation) and with rises in wages. The resulting pattern of differentials, overstated claims, threats, strikes, loss to national productivity and inflation is only too well known. Conceivably more praise might diminish preoccupation with pay rises; this preoccupation has been less in the services and professions, probably because of their extra status, though recently a diminution of this status has led to arguments about pay and even to strikes. There are reports—not in the complicated pattern of Western industrial relations—of strikes that have been settled by giving not more pay, but more appreciation. Clearly, this appreciation must not be false but must be coupled with an opportunity for genuine achievement.

Thirdly, if an individual belongs and feels appreciated, it is natural for him to desire to take more responsibility—within his capacity—and to pull his weight better. If he is denied this, he becomes irresponsible. Possibly, the irresponsibility of "the average worker today"—so often blamed—is the result of this denial. Many managements today have moved to far greater sharing of responsibility with workers; the extreme case of course being the control of management by elected workers. But this does not necessarily solve all psychological problems, for individuals can still feel unwanted and unappreciated and behave irresponsibly if they have no emotional contact and support from their fellows, even though politically and legally they have every opportunity for a share of control. Psychological barriers are even stronger than "iron curtains".

If his needs are not satisfied, the worker (or manager) is left seeking an outlet for his energies and aggression. The results vary according to his personality, his past history, his present opportunities and the help and direction others give him.

Aggression and illness

Aggression may be focused on other people, and they in their turn will react. Sometimes this merely results in an explosion which may "clear the air", but more often it leads to resignation or discharge, or to a chain of criticism downwards, which each justifies by rationalisation; or the blame may be put on some other group, especially an irritating minority, with an increase of race prejudice or class war. Recent studies have demonstrated the correlation between poor mental health, low job satisfaction and race prejudice.

Another sequel is possible: if outlet is barred either by fear of the consequences, or by shame of expressing aggression, a state of unpleasant tension remains, which may be repressed and promote chronic anxiety or depression and physical changes which provoke various psychosomatic illnesses. These are on the increase in industrialised countries–especially those rapidly industrialising–and especially in frustrated people. Similar conditions may be found among managers and trade unionists, caught as they are between the pressures of their superiors and their subordinates and between their image of what is expected of them (by their own conscience or by others) and their incapacity to carry it out.

Some neuroses are trifling, others are extremely incapacitating: depression may lead to suicide, and some psychosomatic diseases are fatal. Tension can predispose to accidents in industry and on the roads, which can also serve as escapes from intolerable situations; and ill advised attempts to relieve tension may lead to worse consequences, e.g. chronic alcoholism and drugs. Finally, much of the toll of domestic unhappiness, divorce and broken homes stems from people frustrated at work. Yet industrial "progress" may increase the number of frustrating jobs for some and increase pressures on others.

Need for education

From all this, it is not just industry but society which suffers, and the need is evident for more understanding and more control. This means more education, and some of it is industry's responsibility–if satisfaction at work has decreased–and industry's opportunity, especially in the induction of new and young workers. But the bulk of it ought to be part of any country's educational system, for children and adolescents. It seldom is. Technological training–physical facts and skills in handling tools and machines–is increasing everywhere; sociological training–psychological facts and skills in handling emotions and people–lags far behind.

Communication

In any education–for children, for adolescents or for adults–communication of ideas is vital. This needs psychological research, for the general awareness of the skills of communication is low.

Communication today occurs by many media, of which face-to-face contact is less important than previously. Modern methods are wider but the chances of misunderstanding or, of course, of deliberate deception are greater. The more the man in the street–and the man in the factory–understands the forces that really determine communication, the better. These are not simply logic; they include the previous relation of speaker to listener, his prestige, his skill in presentation, the way in which the idea impinges on the listener's existing interests, the mood and suggestibility of the latter, and perhaps most important of all, the opportunity given to both speaker and listener to develop a new idea together.

Educators must give a basic account of individual and group psychology; in particular the emotional sources of behaviour must be recognised, as must the human mind's capacity to defend itself against pain and regain its "peace of mind" by various built-in defence mechanisms such as the displacement of aggression; repression; attempts to deal with guilt by blaming others; regression to childish attitudes of appealing helplessness; and preventing opposing ideas from open conflict by dissociating them. All may be seen in healthy people from day to day, but carried to extremes, and persisted in, they may cause illness and irreversible behaviour in an individual or group. It is therefore vital to recognise them early and to help one's colleague (or oneself) deal with them constructively or at least to release the tension harmlessly and avoid reactions which are destructive or may set up vicious circles.

Mental first aid

Experience in physical first aid has shown that much damage after a serious accident, and sometimes death, can be prevented by the combination of simple knowledge, practical skill and common sense, but that to "leave it to nature" is often tragically useless. In mental or emotional first aid, there is less general acceptance of the policy of using these three qualities: yet here, to stand aside is even more tragic, constituting, as it must, another blow to the vulnerable patient in rejecting his need. These three qualities, therefore, need to be practised.

Skills in handling tension are of many kinds; good listening certainly ranks high, and most people know how much they can be relieved by getting something "off their chest". Much can be done by a listener who adopts a neutral but sympathetic attitude intent only on letting the speaker ventilate and clarify his views. From this stems the whole system of "professional counselling". It may by itself allow tension to be dissipated and sound judgement regained.

Some people, however, need more–some need hope and reassurance, and others a physical outlet; some need help to regain their self-esteem, others can do so by taking pride in their ability to accept injustice.

Some have, in fact, found very constructive or at least harmless outlets for tension; sport, music, art, and even the space-race may act as safety valves. However, it is important, if ironic, here to consider that the prevalent management-worker antagonism, which is undoubtedly deplorable in its effect on productivity and co-operation, may yet serve as a useful outlet for tension if it provides a legitimate–and highly reputable–channel to express anger against the other side–"tyrannically selfish exploiters" on the one hand or "obstinately selfish idlers" on the other. So too with race or class prejudice if this antagonism is to be diminished, some other outlets must be found instead or the situation may "blow up" before any long-term improvement can be achieved.

Skills in handling individual and group tensions take time to learn, and longer if one is older. Some insight is needed into one's own conflicts, and this too may be hard to achieve. Moderately sophisticated techniques are now being used to this end, such as sensitivity training by T groups or Blake's Grid; these can be of value to managers and others in understanding and resolving complicated behaviour patterns. Long-term education, begun earlier, could do more.

Collaboration

In all this there is room for collaboration between all involved–managers, trade unionists, scientists, workers, teachers and doctors, with their various skills and experience. But collaboration needs understanding of the roles of each other, of their training, special skills, strengths and weaknesses; and to learn this needs discussion as between equals, and not dictation nor manipulation. To obtain mutual confidence, patience and persistence are required and, what is harder, the sacrifice of points of view, interests and even status.

TREDGOLD, R. F.

"Human relations effects on safety". Hikson, R. R. *National Safety News* (Chicago), July 1980, 122/1 (52-54). 5 ref.

CIS 81-1785 *Occupational health as human ecology*. Wolf, S.; Bruhn, J. G.; Goodell, H. (Springfield, Illinois, Charles Thomas, 1978), 115 p. Illus. 213 ref.

Hunting and trapping

Even though great numbers of hide-bearing animals are raised for food and leather production and fur-bearing animals for fur production, hunting and trapping of wild animals are of practically world-wide occurrence. The terms "fur-bearing" and "hide-bearing" are not precise, so that there is a certain amount of overlapping.

Wild hide-bearing animals are frequently rather large and include species such as deer, moose, elk, antelope, caribou, bison, goat and numerous others. In general they are taken by shooting since the wound produced by the weapon does not interfere with the use to which the prey is put—food, head trophies and articles of leather. Wild fur-bearing animals include many different species such as leopard, lynx, ocelot, lion, wallaby, beaver, muskrat, nutria, seal, rabbit, raccoon, opossum, skunk, fitch, fisher, wolverine, marten, ermine, mink, fox, bear and wolf. Although a number of valuable fur types come from warmer parts of the world, the majority of prime furs originate in the colder parts of the northern hemisphere, including North America, Asia and Europe. With few exceptions such species are valued entirely for their pelts, which must be intact and undamaged to be of value to the fur trade. Therefore, they are obtained by trapping: the essential function of the animal trap is to ensure that pelts will be obtained intact.

Hunting techniques. In order to ensure the maintenance of adequate animal populations, governments commonly establish "open season" and "closed season" periods. In addition hunters and trappers are usually required to be licensed. The number of sport hunters who take advantage of the open season is increasing steadily in some parts of the world. Among them there are growing numbers of inexperienced individuals who are inclined to shoot without clearly identifying the quarry. As a result many injuries and fatalities occur in the forests each year. An experienced hunter not only identifies the quarry, but shoots only if he is satisfied that the shot will kill it. Thus he avoids the necessity of tracking an injured animal through the forest or of encountering a live injured animal at close quarters.

Trapping techniques. In northern regions trapping is a winter activity, since that is the time when pelts are at their best. Typically, the professional northern trapper obtains his supplies for the season at a trading post and lives in a cabin in an isolated area. He sets up his trap line in a circuit out from the cabin and back. This line may cover a long distance and the work may sometimes involve nights spent in the woods away from the cabin. The trapper must therefore understand the basics not only of trapping animals and preparing pelts but also of wilderness living. Since trapped animals may fall prey to wolves or other predators, regular patrolling of the trap line is essential.

Originally trapping was done by means of snares, deadfalls and other devices made from materials at hand. To some extent such devices continue to be used, particularly among native peoples. For many years the professional trapper has depended on the leg-hold trap which is made in a variety of sizes. Recently, there has been a trend toward more humane traps such as the Conibear trap, which kills the animal when sprung. Some fur-bearing species are not amenable to trapping techniques. Seals, for example, are usually either shot or clubbed, depending on age.

Pelt treatment techniques. Most pelts are case-skinned from the carcass, a slit being made up the inside of one hind leg, across and down the other. The pelt is then pulled toward the head and is removed inside out as a tube. All flesh and fatty deposits must be scraped from the pelt before it is stretched on some type of stretching board and left in a cool place to dry. A few pelts such as beaver are stretched flat by means of lacing from the edges of the pelt to a hoop. The pelts are later stacked and kept in the cool until marketed.

The large proportion of farmed mink for the fur industry and the vagaries of demand and prices for various types of wild fur has a marked effect on the professional trapper. At times the financial returns do not compensate the considerable effort and inconvenience involved in trapping. Nevertheless, there remain those who are attracted to this way of life.

HAZARDS

The hazards of hunting and trapping are numerous—falls, drownings, frostbite, animal trap injuries, animal bites, woodcutting wounds, sun glare, nutritional diseases, and others. However, it is usually the less experienced who suffer such mishaps.

The most important factor contributing to the severity of occupational hazards is isolation and distance. Hunters and trappers frequently work alone in areas remote from centres in which medical treatment is possible, and their exact whereabouts may often be unknown to anyone. A wound or animal bite or other accident that would otherwise be a minor matter can have more serious consequences under such circumstances.

Accidents. Since professional trappers work mainly in the winter season in northern climates, sun glare from snow can produce eye injuries and cold can produce frostbite. Crossing frozen lakes and rivers requires caution, because breaking through a thin area of ice can result in drowning or freezing. Exhaustion in conditions of severe cold can lead to euphoria and lethargy with fatal consequences if not recognised in time.

Other accidents include gunshot wounds, wounds from skinning and wood-chopping, the accidental tripping of traps and bites from trapped animals. In addition to the possibility of infection from wounds, there is that of contracting certain animal diseases.

A poor diet may give rise to nutrition-related disease, although this is much more uncommon nowadays than in the past. The symptoms usually disappear with the end of the trapping season.

Diseases. Rabies is the most serious disease which can be contracted from wild animals. The wild animal population is a reservoir of the rabies virus, the disease being endemic in certain areas and occurring in many animal species including wolves, foxes, dogs, cats, squirrels, bats, raccoons, skunks, bears, rabbits and beavers as well as larger animals such as caribou and moose. The rabies virus affects the brain; therefore, any wild animal which appears to lose its fear of man or to show any other unusual behaviour should be suspected. Any hunter or trapper who is bitten by a suspected animal should seek medical assistance without delay and should try to obtain the head of the animal for testing.

Tularaemia, also known as deer fly fever and rabbit fever, is a bacterial disease that can be transmitted by the deer fly and other biting flies, by bites of infected animals or by handling carcasses, furs and hides of infected animals. It can also infect water supplies. Its symptoms are similar to those of undulant fever and plague. In areas in which the disease is suspected, water supplies should be disinfected, and wild game should be thoroughly cooked before eating. Arms and hands should be kept clean and disinfected. Rubber gloves should be worn if there are any cuts or abrasions. The area in which carcasses, hides and pelts are handled should be kept clean and disinfected.

Anthrax may cause disease in trappers and hunters, since it is enzootic in both wild and domesticated animals in most parts of the world. A cutaneous infection from contact with contaminated skins and hides is the most frequent form of anthrax. The inhalation form rarely occurs. Treatment should be sought at once.

Animals may occasionally be a source of direct tuberculosis infection for hunters. Monkeys are highly susceptible to both human *(Mycobacterium tuberculosis)* and bovine *(Mycobacterium bovis)* tuberculosis and can transmit the disease. *M. tuberculosis* can cause disease in parrots, dogs, goats, elephants, tapirs and other animals. *M. bovis* has been isolated from cats, goats, dogs, camels and elephants, and is believed to have been transmitted to man on a number of occasions.

Fungal diseases caused by *Trichophyton verrucosum* and *T. mentagrophytes* are the main ringworm agents affecting man. In wooded areas, small mammals have been shown to be carriers of several types of dermatophytes that may infect the hunter and trapper. Also the dog serves as a reservoir for *Microsporum canis*, the principal cause of animal ringworm in man.

In addition hunters and trappers may be exposed to other viral, bacterial and parasitic infections that are at times found in wild animals. Standard reference works may be consulted for details. Wood ticks are considered an important agent, often transmitting infections from animals to man.

There have been references in the literature to a condition among seal hunters called "seal finger" (North Atlantic), "Spaek finger" (Norway), etc. It usually occurs in the spring of the year while hunters are out on the drifting ice. Incubation time varies as well as the duration. Pain and swelling are characteristic and sometimes a deformity persists at the distal interphalangeal joint. The disease, although the cause is unknown, has been reported in Northern Europe and Canada.

BRAID, P.

Trapper's Guide (Guide du Trappeur). Provencher, P. (Brussels, Les Editions de l'Homme, 1977). (In French)

"An ecological study of bear hunting activities of the Matagi: Japanese traditional hunters". Takeda, J. *Journal of human ergology* (Tokyo), 1972, 1/2 (167-187).

"Changes in fitness level of humans attributable to hunting activities". Docherty, D.; Eckerson, J.; Collis, M. L.; Hayward, J. *Journal of sports medicine and physical fitness* (Turin), 1977, 17/3 (315-320).

"Aetiology of sealer's finger". Thjotta,T.; Kvittingen, J. *Acta Pathologica et Microbiologica Scandinavica* (Copenhagen), 1949, 26/3 (407-411). 2 ref.

Hydrazine and derivatives

Hydrazine (NH_2NH_2)

m.w. 32
sp.gr. 1.01
m.p. 1.4 °C
b.p. 113.5 °C
v.d. 1.1
v.p. 14.4 mmHg (1.91·10³ Pa) at 25 °C
f.p. 37.8 °C oc
e.l. 4.7-100%

very soluble in water; soluble in ethyl alcohol; insoluble in hydrocarbons

a colourless, oily liquid with an ammoniacal, fishy, or amine-like odour, which fumes in air.

TWA OSHA	1 ppm 1.3 mg/m³
TLV ACGIH	0.1 ppm 0.1 mg/m³ skin; industrial substance suspected of carcinogenic potential for man
IDLH	80 ppm
MAC USSR	0.1 mg/m³ skin

Production. Hydrazine is usually prepared by oxidising ammonia or urea with sodium hypochlorite and treating it with sulphuric acid to give hydrazine sulphate, which, in turn, is treated with sodium hydroxide and distilled to form hydrazine hydrate and then refluxed with a dehydrating agent to yield anhydrous hydrazine.

Monomethylhydrazine (NH_2NHCH_3)

METHYLHYDRAZINE; MMH

m.w. 46
sp.gr. 0.87
m.p. < −80 °C
b.p. 87 °C
v.d. 1.6
v.p. 36 mmHg (4.79·10³ Pa) at 20 °C
f.p. 26.7 °C

soluble in water and ethyl ether; very soluble in ethyl alcohol

a colourless liquid with an ammoniacal odour.

TWA OSHA	0.2 ppm 0.35 mg/m³
TLV ACGIH	0.2 ppm 0.35 mg/m³ ceil skin; industrial substance suspected of carcinogenic potential for man
IDLH	5 ppm
MAC USSR	(hydrazine and derivatives) 0.1 mg/m³

Production. Monomethylhydrazine is prepared from primary amines through substituted ureas by treatment with nitrous acid, reduction and hydrolysis with concentrated HCl.

1,1-Dimethylhydrazine ($NH_2N(CH_3)_2$)

ASYMMETRICAL-DIMETHYLHYDRAZINE; UDMH; DIMAZINE

m.w. 60.1
sp.gr. 0.79
b.p. 63 °C
v.d. 2.0
v.p. 156.8 mmHg (20.85·10³ Pa) at 25 °C
f.p. −15 °C
e.l. 2-95%
i.t. 248.9 °C

very soluble in water, ethyl alcohol and ethyl ether

a hygroscopic, mobile liquid that fumes in air and gradually turns yellow.

TWA OSHA	0.5 ppm 1 mg/m³ skin
STEL ACGIH	1 ppm 2 mg/m³
IDLH	50 ppm
MAC USSR	(hydrazine and derivatives) 0.1 mg/m³

It is prepared from dimethylnitrosoamine by reduction.

Phenylhydrazine ($C_6H_5NHNH_2$)

See separate article.

Uses. The industrial uses of hydrazine and its derivatives are manifold. They are strong reducing agents and therefore useful as intermediates in organic synthesis. Hydrazine has been used in photography, metallurgy, and a variety of chemical preparations such as anticorrosives, textile agents and pesticides, and as a starting material for many drugs. Since they are highly reactive volatile compounds with a high energy content, the methylated derivatives as well as hydrazine, either singly or in various mixture ratios, are in large scale use as liquid rocket propellants. The most often used methylated derivates are monomethylhydrazine and 1,1-dimethylhydrazine; the arylhydrazines, especially phenylhydrazines, are frequently used as intermediates in chemical synthesis and in the preparation of drugs (see ROCKET PROPELLANTS).

HAZARDS

The major hazards in using hydrazines are flammability, explosion and toxicity. For example when hydrazine is mixed with nitromethane, a high explosive is formed which is more dangerous than TNT. All hydrazines discussed here have sufficiently high vapour pressures to present serious health hazards by the inhalatory absorption route. They have a fishy, ammoniacal odour which is repulsive enough to indicate the presence of dangerous concentrations for brief accidental exposure conditions. At lower concentrations, which may occur during manufacturing or transfer processes, the warning properties of odour may not be enough to preclude low level chronic occupational exposures in fuel handlers.

Moderate to high concentrations of hydrazine vapours are highly irritating to the eyes, nose and the respiratory system. Generally, concentrations of approximately 2% in air represent the lower explosive limit, but irritation at this level would be unbearable. Skin irritation is pronounced with the propellant hydrazines; direct liquid contact results in burns and even sensitisation type of dermatitis, especially in the case of phenylhydrazine. Eye splashes have a strongly irritating effect and hydrazine can cause permanent corneal lesions.

In addition to their irritating properties, hydrazines also exert pronounced systemic effects by any route of absorption. After inhalation, skin is the second most important route of intoxication. All hydrazines are moderate to strong central nervous system poisons resulting in tremors, increased central nervous system excitability and, at sufficiently high doses, convulsions. This can progress to depression, respiratory arrest and death. Other systemic effects are in the haematopoietic system, the liver and the kidney. The individual hydrazines vary widely in degree of systemic toxicity as far as target organs are concerned. As a quick reference, these variations are summarised in table 1. A rating of 0 to +4 is used to depict the absence or severity of a specific effect.

The haematological effects are self-explanatory on the basis of haemolytic activity. These are dose dependent and, with the exception of monomethylhydrazine, they are most prominent in chronic intoxication. Bone marrow changes are hyperplastic with phenylhydrazine, and extramedullary haemotopoiesis has also been observed. Monomethylhydrazine is a strong methaemoglobin former and blood pigments are excreted in the urine. The liver changes are primarily of the fatty degeneration type, seldom progressing to necrosis, and are usually reversible with the propellant hydrazines. Monomomethylhydrazine and phenylhydrazine in high doses can cause extensive kidney damage. Changes in the heart muscle are primarily of fatty character. The nausea observed with all of these hydrazines is of central origin and refractory to medication. The most potent convulsants in this series are monomethylhydrazine and 1,1-dimethylhydrazine. Hydrazine causes primarily depression, and convulsions occur much less frequently.

All hydrazines appear to have some kind of oncogenic activities in some laboratory animal species by some route of entry (feeding in drinking water, gastric intubation or inhalation). Currently there is no published evidence of carcinogenic activity in humans. However, even in laboratory animals, with the exception of one derivative not discussed here, 1,2-dimethylhydrazine (or symmetrical dimethylhydrazine), there is a definite dose response and, in that sense, they are considered to be weak carcinogens. Whether hydrazine and its derivatives are truly carcinogens or promoters, or are metabolised to the causative agents is unknown at present. Therefore, any exposure of humans should be minimised by proper protective equipment and decontamination of accidental spills.

SAFETY AND HEALTH MEASURES

For personal protection, vinyl-coated hand protection, natural- or reclaimed-rubber foot protection, rubber aprons, and plastic eye and face protection can be used when working with small quantities. Where the possibility of gross splashing exists, full protective clothing made of rubber, Neoprene or vinyl-coated materials should be worn. For respiratory protection in situations where the recommended tolerance limits are exceeded respiratory protective equipment such as approved canisters or a gas mask or self-contained breathing apparatus must be used. Appropriate shipping and storage instructions including labelling of containers should be strictly observed, as determined by the physical/chemical characteristics of the hydrazines. Since they are strong reducing agents, they are hypergolic when in contact with strong oxidisers; therefore, proper quantity-distance criteria must be observed. Under proper storage conditions they are fairly stable.

In the laboratory hamster and the rat there is evidence of nasal polyp formation with a relatively small percentage of malignant tendency after prolonged exposure to neat hydrazine. In addition, contaminants remaining from starting materials in the chemical synthesis process may also pose some oncogenic hazard. Much of the 1,1-dimethylhydrazine manufactured in the past contained varying small amounts of 0.1-0.5% of N-nitrosodimethylamine, which is considered a very potent carcino-

Table 1. Differences in systemic toxicity of hydrazines

	Haemolysis	Anaemia	Methaemoglobinaemia	Heinz body formation	Bone marrow damage	Liver damage	Kidney damage	Heart damage	Nausea	Convulsions	Depression
Hydrazine	+1	+2	0	0	0	+4	+2	+1	+3	+1	+4
Monomethylhydrazine	+3	+2	+4	+4	0	+1	+4	0	+4	+4	±
1,1-Dimethylhydrazine	±	+1	0	0	0	±	0	±	+4	+4	+1

gen. Consequently, consideration should be given in periodic physical examinations to these potential health hazards.

Treatment. Skin contamination must be immediately removed with large quantities of water. If eye contamination occurs, flushing of the eyes for 15 min should begin immediately and take precedence over other self- and first-aid measures. Severe exposure cases where vomiting or central nervous system irritation is likely to occur should be treated by a physician. Treatment of hydrazine intoxication is primarily symptomatic and supportive, with the exception of the methylated derivatives, where pyridoxine hydrochloride in large doses may be effective. Prognosis for acute intoxication is usually good and most changes are reversible. Approximately 50% of the absorbed dose is excreted in 24 h. The industrial safety record is excellent due to the strict enforcement of handling and precautionary measures.

THOMAS, A. A.

"Occupational hazards of missile operations with special regard to the hydrazine propellants". Back, K. C.; Carter, V. L. Jr.; Thomas, A. A. *Aviation, Space, and Environmental Medicine* (Washington, DC), Apr. 1978, 49/4 (591-598). 41 ref.

CIS 79-1064 "Problems in the use of propergols in aeronautics" (Problèmes posés par l'utilisation des propergols en aéronautique). Pepersack, J. P. *Cahiers de médecine du travail–Cahiers voor arbeidsgeneeskunde* (Gerpinnes, Belgium), Dec. 1978, 15/4 (281-284). (In French, Dutch)

CIS 79-440 *Criteria for a recommended standard–Occupational exposure to hydrazines.* DHEW (NIOSH) publication No. 78-172 (National Institute for Occupational Safety and Health, 4676 Columbia Parkway, Cincinnati) (June 1978), 269 p. 229 ref.

"Pyridoxine and phenobarbital as treatment for Aerozine-50 toxicity". Azar, A.; Thomas, A. A.; Shillito, F. H. *Aerospace Medicine* (Washington, DC), Jan. 1970, 41/1 (1-4). Illus. 18 ref.

Hydrazoic acid and azides

Hydrazoic acid (HN_3)

AZOIMIDE; HYDROGEN AZIDE

m.w.	43
sp.gr.	1.09
m.p.	−80 °C
b.p.	37 °C

a colourless, explosive, mobile liquid with a strong odour, which forms stable solutions in water and in many organic solvents.

MAC (Yugoslavia) 1 ppm 1.8 mg/m^3

Production. Hydrazoic acid is commonly produced from sodium azide, which is prepared by the reaction between sodamide and nitrous oxide. The sodium azide is treated with sulphuric acid, and an aqueous solution of hydrazoic acid is obtained by distillation. An anhydrous solution of the acid is obtained by extracting the aqueous solution with ether and drying the ether solution over calcium chloride. The acid may also be obtained by the hydrolysis of acyl azides and by treating hydrazine with nitrous acid or its ethyl ester. Salts of the acid are prepared by neutralising the aqueous acid with appropriate hydroxides. Lead and silver azides, however, are usually made by the controlled addition of a solution of sodium azide to solutions of the corresponding nitrates or acetates.

Uses. Hydrazoic acid is encountered mainly as a by-product in the photographic-chemicals and explosives industry. Lead azide is employed as a primer.

HAZARDS AND THEIR PREVENTION

Hydrazoic acid has highly explosive and toxic properties.

Explosion hazard. In its pure form or in highly concentrated solution the acid is explosively unstable and sensitive to heat or shock. The production of the concentrated solution containing 91% of hydrazoic acid should be conducted with small batches behind protective shields. More dilute solutions, however, are stable, and the majority of reactions can be carried out with the less dangerous solutions.

The salts vary in their explosive properties from the inert alkali metal azides to the highly shock-sensitive silver, copper, mercury and lead azides. It is thus advisable to avoid the use of lead and copper in the design of plant intended for reactions in which hydrazoic acid or its salts are used or may be produced. It is especially important that maintenance workers should be informed of the hazards associated with such plant.

Health hazard. In animal experiments, the major effects of intraperitoneal injections of hydrazoic acid were shown to be excessive stimulation of respiration, fall in blood pressure, and generalised convulsions followed by depression. Inhalation of vapour caused acute bronchiolar inflammation and pulmonary oedema, while continued exposure to higher concentrations gave rise to convulsions and death.

In men who were exposed to the vapour in the course of their work the acute effects were eye and nose irritation, bronchitis, headache, fall in blood pressure, dizziness, faintness and weakness or collapse. Hypotension has also been observed in exposed workers.

These conditions have been known to develop as a result of exposure to concentrations of vapour in the air of little more than 1 ppm, and it has been suggested that any concentration above this figure should be prevented from developing in the working atmosphere. Any operation, therefore, from which the vapour may be evolved should be conducted in well ventilated conditions. If local exhaust ventilation is provided for this purpose, it may well be desirable to effect this by means of an induced draught or venturi arrangement rather than risk contact of unstable azides with the moving parts of a fan mechanism (see EXPLOSIVE SUBSTANCES).

MATHESON, D.

Dangerous reactions:

CIS 78-1059 "Hazardous chemical reactions–49. Hydrazoic acid and azides" (Réactions chimiques dangereuses–49. Acide azothydrique et azotures). Leleu, J. *Cahiers de notes documentaires–Sécurité et hygiène du travail* (Paris), 1st quarter 1978, 90, Note No. 1103-90-78 (137-148). (In French)

Sampling and detection:

CIS 1859-1972 "Detection of hydrogen azide in air" (Azoimid meghatarozasa levegöben). Arato-Sugar, E.; Bittera, E. *Munkavéedelem* (Budapest), 1972, 18/1-3 (33). (In Hungarian)

Hydrocarbons, aliphatic

Aliphatic hydrocarbons are compounds of carbon and hydrogen. They may be saturated or unsaturated open chain, branched or unbranched molecules, the nomenclature being as follows:

(a) paraffins (or alkanes)–saturated hydrocarbons;

(b) olefins (or alkenes)–unsaturated hydrocarbons with one or more double bond linkages;

Paraffin

	Chemical formula	m.w.	d.	b.p. (°C)	v.d.	v.p. (mmHg (10^3 Pa) at °C)	f.p. (°C)	e.l. (%)	i.t. (°C)
Methane	CH_4	16.05	0.41	−161.5	0.6	–	gas	5.3-14	537
Ethane	C_2H_6	30.07	0.45	−88.6	1.0	–	gas	3.0-12.5	515
Propane	C_3H_8	44.09	0.58	−44.5	1.6	–	gas	2.2-9.5	466
n-Butane	C_4H_{10}	58.1	0.60	−0.5	2.0	–	gas	1.9-8.5	405
n-Pentane	C_5H_{12}	72.15	0.63	36.1	2.5	400 (53.2) at 18.5	−49.3	1.5-7.8	309
n-Hexane	C_6H_{14}	86.2	0.66	68.7	3.0	100 (13.3) at 15.8	−21.7	1.1-7.5	261
n-Heptane	C_7H_{16}	100.2	0.68	98.4	3.5	40 (5.32) at 20	−4	1.2-6.7	223
n-Octane	C_8H_{18}	114.2	0.70	125-126	3.9	10 (1.33) at 19.2	13.3	1.0	220

	TWA OSHA		TWA NIOSH ceilings		TLV ACGIH		STEL ACGIH		IDLH
	ppm	mg/m³	ppm	ppm	ppm	mg/m³	ppm	mg/m³	ppm
Methane					simple asphyxiant				
Ethane					simple asphyxiant				
Propane	1 000	1 800			simple asphyxiant				20 000
n-Butane					800	1 900			
n-Pentane	1 000	2 950	120	610	600	1 800	750	2 250	5 000
n-Hexane	500	1 800	100	510	25	90	100	350	5 000
n-Heptane	500	2 000	85	440	400	1 600	500	2 000	4 250
n-Octane	500	2 350	75	385	300	1 450	375	1 800	3 750

(c) acetylenes (or alkynes)—unsaturated hydrocarbons with one or more triple bond linkages.

The general formulae are C_nH_{2n+2} for paraffins, C_nH_{2n} for olefins, and C_nH_{2n-2} for acetylenes.

The smaller molecules are gases at room temperature (C_1-C_4). As the molecule increases in size and structuralcomplexity it becomes a liquid with increasing viscosity (C_5-C_{16}) and finally the higher molecular weight hydrocarbons are solids at room temperature (above C_{16}).

The olefins and acetylenes are treated in separate articles, HYDROCARBONS, ALIPHATIC: OLEFINS; ACETYLENE.

Production. The aliphatic hydrocarbons of industrial importance are derived mainly from petroleum, which is a complex mixture of hydrocarbons. They are produced by the cracking, distillation and fractionation of crude oil. Methane, the lowest member of the series, comprises 85% of natural gas, which may be tapped directly from pockets or reservoirs in the vicinity of petroleum deposits. Large amounts of pentane are produced by fractional condensation of natural gas.

Uses. The saturated aliphatic hydrocarbons are used in industry as fuels, lubricants, solvents and after undergoing processes of alkylation, isomerisation and dehydrogenation, as starting materials for the synthesis of paints, protective coatings, plastics, synthetic rubber, resins, pesticides, synthetic detergents and a wide variety of petrochemicals.

The fuels, lubricants and solvents are mixtures which may contain many different hydrocarbons. Natural gas has long been distributed in the gaseous form for use as a town gas. It is now liquefied in large quantities, shipped under refrigeration and stored as a refrigerated liquid until it is introduced unchanged or reformed into a town gas distribution system. Liquefied petroleum gases, consisting mainly of propane and butane, are transported and stored under pressure or as refrigerated liquids, and are also used to augment town gas supply. They are used directly as fuels, often in high-grade metallurgical work in which a sulphur-free fuel is essential, in oxypropane welding and cutting, and in circumstances where a heavy industrial demand for gaseous fuels would strain public supply. Storage installations for these purposes vary in size from about 2 t to several thousands of tons. Liquefied petroleum gases are also used as propellants for many types of aerosols, and the higher members of the series, from heptane upwards, are used as motor fuels and solvents.

The hydrocarbons used as starting materials of intermediates for synthesis may be individual compounds of high purity or relatively simple mixtures.

HAZARDS

The hazards associated with the use of these hydrocarbons in industry are those of fire and explosion and those associated with injury to health.

Fire and explosion. The development of large storage installations first for gaseous methane and later for liquefied petroleum gases has been associated with explosions of great magnitude and catastrophic effect, which have emphasised the danger when a massive leakage of these substances occurs. The flammable mixture of gas and air may extend far beyond the distances that are regarded as adequate for normal safety purposes, with the result that the flammable mixture may become ignited by a household fire or automobile engine well outside the specified danger zone. Vapour may thus be set alight over a very large area and flame propagation through the mixture may reach explosive violence. Many smaller—but still serious—fires and explosions have occurred during the use of these gaseous hydrocarbons.

The danger associated with the liquids when they are used as motor fuels and solvents is that of vapour explosions and liquid fires. Disastrous fires and explosions have occurred after massive escapes of liquid or liquefied gas, caused by the malfunctioning of a valve or by the failure of a pipe or of the vessel itself. Many deaths have been caused by explosions in vessels to which heat has been applied for the purpose of cutting, welding, brazing or soldering when the vessels have contained residues of these light petroleum fractions.

Health hazards. The first two members of the series, methane and ethane, are pharmacologically "inert", belonging to a group of gases called "simple asphyxiants". These gases can be tolerated in high concentrations in inspired air without producing systemic effects. If the concentration is high enough to dilute or exclude the oxygen normally present in the air, the effects produced will be due to oxygen deprivation or asphyxia. Methane has no warning odour. Because of its low density, methane may accumulate in poorly ventilated areas to produce an asphyxiating atmosphere.

Ethane in concentrations below 50 000 ppm (5%) in the atmosphere produces no systemic effects on the person breathing it.

Pharmacologically, the hydrocarbons above ethane can be grouped with the general anaesthetics in the large class known as the central nervous system depressants. This includes such well known chemicals as ethyl alcohol, diethyl ether, and acetone. The vapours of these hydrocarbons are mildly irritating to mucous membranes. The irritation potency increases from pentane to octane.

The liquid paraffin hydrocarbons are fat solvents and primary skin irritants. Repeated or prolonged skin contact will dry and defat the skin, resulting in irritation and dermatitis. Direct contact of liquid hydrocarbons with lung tissue (aspiration) will result in chemical pneumonitis, pulmonary oedema, and haemorrhage. Chronic intoxication by these compounds may involve polyneuropathy.

Propane causes no symptoms in man during brief exposures to concentrations of 10 000 ppm (1%). A concentration of 100 000 ppm (10%) is not noticeably irritating to the eyes, nose or respiratory tract, but it will produce slight dizziness in a few minutes.

Butane gas causes drowsiness, but no systemic effects during a 10-min exposure to 10 000 ppm (1%).

Pentane is the lowest member of the series that is liquid at room temperature and pressure. In human studies a 10-min exposure to 5 000 ppm (0.5%) did not cause mucous membrane irritation or other symptoms.

Hexane is three times more acutely toxic than pentane. In man 2 000 ppm (0.2%) hexane produced no symptoms during a 10-min exposure, whereas 5 000 ppm (0.5%) caused dizziness and a sensation of giddiness. [In the late 1960s and early 1970s, attention was drawn to outbreaks of sensorimotor polyneuropathy among workers exposed to mixtures of solvents containing *n*-hexane in concentrations mainly ranging between 500 and 1 000 ppm with higher peaks. Various working processes were involved such as glueing in sandal, shoe, belt and furniture manufacturing, polyethylene laminating, and cleaning tablets in pharmaceutical plant. Paraesthesia, numbness and weakness of distral extremities were complained of, mainly in the legs. Achilles tendon reflexes disappeared, touch and heat sensation was diminished. Conduction time was decreased in the motor and sensory nerves of the arms and legs. Muscular atrophy and signs of denervation-type injury could be present. Experimentally, technical grade *n*-hexane produced peripheral nerve disturbances in mice at 250 ppm and higher concentrations after 1 year's exposure. Metabolic investigations indicated that in guinea-pigs *n*-hexane and methyl butyl ketone are metabolised to the same neurotoxic compounds (2-hexanediol and 2,5-hexanedione) (see HEXANE).]

Heptane caused slight vertigo in men exposed for 6 min to 1 000 ppm (0.1%) and for 4 min to 2 000 ppm (0.2%). A 4-min exposure to 5 000 ppm (0.5%) heptane caused marked vertigo, inability to walk a straight line, hilarity, and incoordination. These systemic effects were produced in the absence of complaints of mucous membrane irritation. A 15-min exposure to heptane at this concentration produced a state of intoxication characterised by uncontrolled hilarity in some individuals and in others it produced a stupor lasting for 30 min after the exposure. These symptoms were frequently intensified or first noticed at the moment of entry into an uncontaminated atmosphere. These individuals also complained of loss of appetite, slight nausea, and a taste resembling gasoline for several hours after exposure to heptane. [No peripheral nerve disturbances have been attributed directly to *n*-heptane. However, this compound was present in several mixtures of alkanes producing polyneuropathy and it has been shown that alkane toxicity tends to increase as the carbon number of alkanes increases. In addition straight chain alkanes are more toxic than the branched isomers. For these reasons NIOSH has recommended the same exposure limits, on a volume/volume basis, for *n*-heptane as for *n*-hexane.]

Octane in concentrations of 6 600-13 700 ppm (0.66-1.37%) caused narcosis in mice within 30-90 min. No deaths or convulsions resulted from these exposures to concentrations below 13 700 ppm (1.37%).

SAFETY AND HEALTH MEASURES

Where these substances are stored and used, precautions must be provided to prevent fire and explosion and injury to health.

Fire prevention. The worst disasters have occurred when large quantities of liquid have escaped and flowed towards a part of the factory where ignition could take place, or have spread over a large surface and evaporated quickly. To reduce the likelihood of accidental leakage from a storage or process vessel, reserve valves and automatic closure devices should be provided to cover the risk of pipe fracture or of the failure of an operating valve. If liquid escapes in spite of these precautions, the spread and flow of the escaping liquid or liquefied gas should be controlled in such a manner that the surface from which vapour may be evolved is as small as possible, and the liquid is led from the vicinity of other storage and process vessels to a safe place where its accidental ignition would not endanger the flammable liquid in neighbouring tanks. Sources of ignition should be prevented and fire-fighting equipment should be provided.

Before heat is applied for the purpose of cutting or welding a vessel that has contained any of these substances, all remaining residues must be drained from the vessel, which must then be thoroughly purged with steam. Only when this precaution has ensured that an explosive concentration of vapour and air will not be formed, may the cutting or welding flame be applied.

Health precautions. Good ventilation will prevent the formation of harmful concentrations of these vapours in normal workplaces. In confined spaces like a process or storage vessel or a garage inspection pit, high concentrations capable of causing unconsciousness or death have been known to develop (see CONFINED SPACES).

[Because it is likely that in an alkane mixture the components have additive toxic effects, NIOSH has recommended keeping a threshold limit value for total alkanes (C_5-C_8) of 350 mg/m^3 as a time weighted average, with a 15-min ceiling value of 1 800 mg/m^3.]

GERARDE, H. W.

CIS 77-1372 *Criteria for a recommended standard—Occupational exposure to alkanes (C_5-C_8).* DHEW (NIOSH) publication No. 77-151 (National Institute for Occupational

Safety and Health, 4676 Columbia Parkway, Cincinnati) (Aug. 1977), 129 p. Illus. 119 ref.

CIS 76-1893 *Vapor-phase organic pollutants–Volatile hydrocarbons and oxidation products* (National Academy of Sciences, 2101 Constitution Avenue, NW, Washington, DC) (1976), 411 p. Illus. 1 364 ref.

CIS 79-1710 "Urinary excretion of *n*-hexane metabolites in rats and humans". Perbellini, L.; Brugnone, F.; Pastorello, G.; Grigolini, L. *International Archives of Occupational and Environmental Health* (West Berlin, 1979), 42/3-4 (349-354). Illus. 12 ref.

Evaluation of portable direct-reading hydrocarbon meters. Willey, M. A.; McCammon, C. S. Jr. DHEW (NIOSH) publication No. 76-166 (National Institute for Occupational Safety and Health, 4676 Columbia Parkway, Cincinnati) (1976), 83 p. Illus. 6 ref.

Hydrocarbons, aliphatic: olefins

Separate articles deal with: ETHYLENE; BUTADIENE.

Production. The light mono-olefins are manufactured from petroleum naphthas by a thermal or catalytic cracking process. Higher olefins may be recovered from the cracked liquid phase from the process. Specific mono-olefins can be produced by cracking paraffin wax. The important diolefins, butadiene and isoprene, are made respectively by catalytic dehydrogenation of *n*-butylenes and by steam cracking higher hydrocarbons from refinery streams.

Uses. The olefins are a series of chemically reactive unsaturated hydrocarbons commercially important as starting materials for the manufacture of numerous chemicals and polymers (resins, plastics, rubbers, etc.) (see POLYOLEFINS). The vast production of the petrochemicals industry is based on the reactivity of these substances.

HAZARDS

Consideration has to be given to the risk of fire and explosion and to that of injury to health.

Fire and explosion. The lower members of the series, ethylene, propylene, and butylene are gases at room temperature and highly flammable or explosive when mixed with air or oxygen. The other members are volatile, flammable liquids capable of giving rise to explosive concentrations of vapour in air at normal working temperatures. When exposed to air, the diolefins can form organic peroxides which, upon concentration or heating, can detonate violently. Most commercially produced diolefins are generally inhibited against peroxide formation.

Health hazards. Olefins generally have a low order of toxicity. Like their saturated counterparts, the lower olefins are simple asphyxiants, but as the molecular weight increases the narcotic and irritant properties become more pronounced than those of their saturated analogues. Ethylene, propylene and amylene have been used as surgical anaesthetics but require large concentrations (60% or more) and for that reason are administered with oxygen. The diolefins are more narcotic than the mono-olefins and are also more irritating to the mucous membranes and the eyes. The available toxicological data show them to be practically non-toxic to relatively harmless.

Substance		m.w.	d.	m.p. (°C)	b.p. (°C)	v.d.	v.p. (mmHg (10^3 Pa))	f.p. (°C)	e.l. (%)	i.t. (°C)
Ethylene ETHENE	$CH_2{:}CH_2$ a colourless gas with a sweet odour	28	0.61	−169.4	−104	0.98			3.1-36	450
Propylene PROPENE	$CH_3CH{:}CH_2$ a colourless gas	42.1	0.58	−185	−47.8	1.5	7 600 (1011) at 19.8 °C	108	2.0-11.1	497
α-*Butylene* 1-BUTENE	$CH_3CH_2CH{:}CH_2$ a colourless gas with a faint odour	56.1	0.66	−185.3	−6.3	1.9	3 480 (462.8) at 21 °C	108	1.6-9.3	384
α-*Amylene* 1-PENTENE PROPYL-ETHYLENE 2-METHYL-BUTENE	$CH_3CH_2CH_2CH{:}CH_2$ a colourless liquid with an unpleasant odour	70.4	0.64	−124	29.2	2.4		−18 (oc)	1.5-8.7	273
β-*Hexylene* 2-HEXENE	$CH_3(CH_2)_2CH{:}CHCH_3$ a colourless liquid	84.2	0.67	−139.9	68	3.0	310 (41.2) at 38 °C	−7		
β-*Heptylene* 2-HEPTENE	$CH_3(CH_2)_3CH{:}CHCH_3$ a colourless liquid	98.2	0.70	−10	98	3.4		−1		375
*1,3-Butadiene** (uninhibited) ERYTHRENE	$CH_2{:}CHCH{:}CH_2$ a colourless liquid with an aromatic odour	54.1	0.62	−113	−4.4	1.9	1 840 (246.7) at 21 °C	−76	2.0-11.5	429
*Isoprene*** 2-METHYL-1,3-BUTADIENE	$CH_2{:}C(CH_3)CH{:}CH_2$ a colourless liquid	34	0.68	−146.7	34	2.4	400 (53.2) at 21 °C	−54		220

* TWA OSHA 1 000 ppm 2 200 mg/m³; STEL ACGIH 1 250 ppm 2 750 mg/m³; IDLH 20 000 ppm; MAC USSR 100 mg/m³
** MAC USSR 40 mg/m³

It must be emphasised that while the olefins are quite innocuous toxicologically, the possibility of asphyxia by their lowering the oxygen content in the working atmosphere must be guarded against. To prevent an explosion hazard, the maximum permissible olefin concentration in the air should never exceed one-fifth of the lower explosive limit.

SAFETY AND HEALTH MEASURES

The precautions that should be observed are similar to those described in the article on HYDROCARBONS, ALIPHATIC.

PABST, A. C.

CIS 75-451 *Properties and essential information for safe handling and use of propylene.* Chemical safety data sheet SD-59 (Manufacturing Chemists' Association, 1825 Connecticut Avenue, NW, Washington, DC), 1974, 17 p.

CIS 79-336 *Propylene (cylinders).* H75, Information sheets on hazardous materials. Fire Protection Association, London. *Fire Prevention* (London), Dec. 1978, 128 (47-48).

CIS 79-320 "Explosive limits of propylene-oxygen mixtures at high temperatures and pressures" (Predely vzryvaemosti smesej propilena s kislorodom pri povyšennyh temperaturah i davlenijah). Zakaznov, V. F.; Kurševa, L. A.; Fedina, Z. I. *Himičeskaja promyšlennost'* (Moscow), June 1978, 6 (427-429). 11 ref. (In Russian)

"Propylene and Polypropylene" (213-230). 60 ref. *IARC monographs on the evaluation of the carcinogenic risk of chemicals to humans.* Vol. 19: *Some monomers, plastics and synthetic elastomers, and acrolein* (Lyons, International Agency for Research on Cancer, 1979), 513 p. Ref.

Hydrocarbons, aromatic

Aromatic hydrocarbons are those hydrocarbons that possess the special properties associated with the benzene nucleus or ring in which six carbon-hydrogen groups are arranged at the corners of a hexagon (see BENZENE AND DERIVATIVES). The bonds joining the six groups in the ring exhibit characteristics intermediate in behaviour between single and double bonds. Thus, although benzene can react to form addition products such as cyclohexane, the characteristic reaction of benzene is not an addition reaction but a substitution reaction in which a hydrogen is replaced by a substituent, univalent element or group. Aromatic hydrocarbons and their derivatives are compounds whose molecules are composed of one or more stable ring structures of the type described and can be considered as derivatives of benzene according to three basic processes:

(a) by replacement of hydrogen atoms with aliphatic hydrocarbon radicals (see TOLUENE AND DERIVATIVES; ETHYLBENZENE AND STYRENE; CUMENE; XYLENE):

CH_3 CH_2CH_3 $CH{:}CH_2$ $H_3C{-}CH{-}CH_3$

Benzene Toluene Ethylbenzene Styrene Cumene

Monoalkylderivatives

CH_3 $-CH_3$

Xylene or Dimethylbenzene

(b) by linking of two or more benzene rings, either directly:

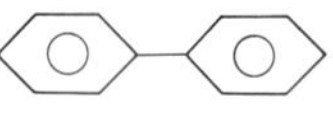

Diphenylmethane

or by intermediate aliphatic chains or other radicals:

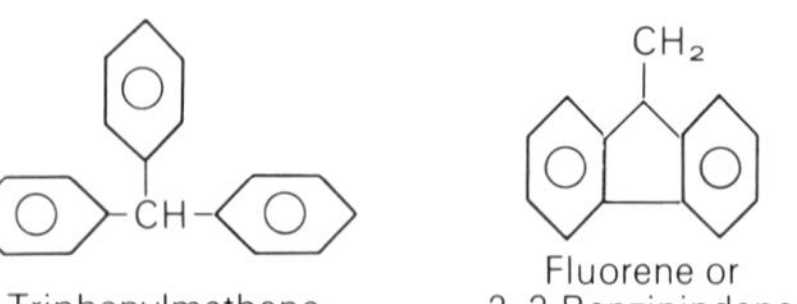

Triphenylmethane   Fluorene or 2,3-Benzinindene

(c) by condensation of benzene rings:

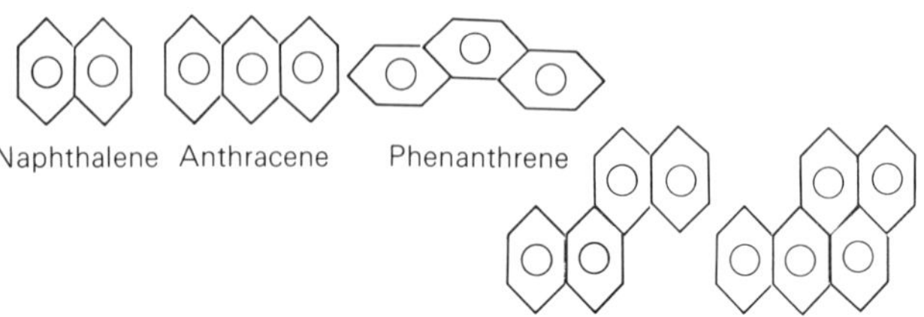

Naphthalene Anthracene Phenanthrene Chrysene Benz(a)pyrene

Each of the ring structures can form the basis of homologous series of hydrocarbons in which a succession of alkyl groups, saturated or non-saturated, replaces one or more of the hydrogen atoms of the carbon-hydrogen groups.

Because of their importance, a large number of aromatic hydrocarbons and their derivatives are given individual treatment in separate articles in addition to those referred to above (see ANILINE; AMINES, AROMATIC; BENZANTHRONE; CHLOROBENZENE AND DERIVATIVES; CHLORONAPHTHALENES; CRESOLS, CREOSOTE AND DERIVATIVES; DINITRO-*o*-CRESOL; DINITROPHENOLS; DIPHENYLS AND TERPHENYLS; NITROBENZENE; NITRO-COMPOUNDS, AROMATIC; PHENOLS AND PHENOLIC COMPOUNDS; PHENOTHIAZINE; PHENYLHYDRAZINE; PHTHALATES; PHTHALIC ANHYDRIDE AND SOME DERIVATIVES).

The main sources of the aromatic hydrocarbons are the distillation of coal and a number of petrochemical operations, in particular catalytic reforming, distillation of crude oil, alkylation of lower aromatic hydrocarbons (see PETROLEUM AND PETROLEUM PRODUCTS). Essential oils, containing terpenes and *p*-cymene, can also be obtained from pines, eucalyptus and aromatic plants and are a by-product in the papermaking industry using the pulp of pines. Polycyclic hydrocarbons occur in the smoke of urban atmospheres.

The economic importance of the aromatic hydrocarbons has been steadily increasing since coal tar naphtha was used as a rubber solvent early in the 19th century. Now the main uses of the aromatic compounds as pure products are the chemical synthesis of plastics, synthetic rubber, paints, dyes, explosives, pesticides, detergents, perfumes and drugs; they are used mainly as mixtures in solvents and constitute a variable fraction of gasoline.

Absorption takes place by inhalation and in small quantities through the intact skin. In general the monoalkyl derivatives of benzene are more toxic than the dialkyl derivatives and the derivatives with branched chains are more toxic than those with straight chains. Aromatic hydrocarbons are metabolised through the bio-oxidation of the ring; if there are side chains, preferably of the methyl group, these are oxidised and the ring is left unchanged. They are, in large part, converted into water-soluble compounds, then conjugated with glycine, glucuronic or sulphuric acid and eliminated in the urine.

Monoalkyl derivatives of benzene

(See also CUMENE; ETHYLBENZENE AND STYRENE; TOLUENE.)

Propylbenzene ($C_3H_7C_6H_5$)

m.w.	120.2
sp.gr.	0.86
m.p.	−99.5 °C
b.p.	159.2 °C
v.d.	4.1
v.p.	10 mmHg (1.33·10³ Pa) at 43.4 °C
f.p.	30 °C
e.l.	0.8-6%
i.t.	450 °C

$CH_2-CH_2-CH_3$

insoluble in water; miscible with alcohol, ether and benzene
a light liquid with a mild pleasant odour.

It is produced by distillation of coal tar and it is used in textile printing and as a solvent for cellulose acetate. It is a depressant of the central nervous system with slow but prolonged effects.

In the body, following oxidation of its side chain, it is converted into benzoic acid, conjugated with glycine and excreted in the urine as hippuric acid.

Sodium dodecylbenzene sulphonate
($C_{12}H_{25}C_6H_4SO_3Na$)

m.w. 348

a colourless liquid with an aromatic odour used as an anionic detergent.

It is produced by catalytic reaction of tetrapropylene with benzene, acidification with sulphuric acid, and treatment with caustic soda. Repeated contact with the skin may cause dermatitis; in prolonged exposure it might act as a bland irritant of mucous membranes.

Dialkyl derivatives of benzene

(See also ETHYLBENZENE AND STYRENE; XYLENES.)

***p*-Cymene** ($C_{10}H_{14}$)

ISO-PROPYLTOLUENE; 4-ISOPROPYL-1-METHYLBENZENE

sp.gr.	0.86
m.p.	−68 °C
b.p.	177 °C
v.d.	4.6
v.p.	1 mmHg (0.13·10³ Pa) at 17.3 °C
f.p.	47 °C
e.l.	0.7-5.6%
i.t.	436 °C

CH_3

CH_3-C-CH_3
H

a colourless transparent liquid with an aromatic odour.

p-Cymene occurs in several essential oils and can be made from monocyclic terpenes by hydrogenation. It is a by-product in the manufacture of sulphite paper pulp.

It is used chiefly with other solvents and aromatic hydrocarbons as a thinner for lacquers and varnishes.

HAZARDS

Health and fire risks are associated with the use of this liquid.

p-Cymene is a primary skin irritant; contact with the liquid can cause dryness, defatting and erythema. There is no conclusive evidence that it can affect the bone marrow.

SAFETY AND HEALTH MEASURES

Drying ovens and other plant which are heated to a temperature above 47 °C should be ventilated to prevent vapours from accumulating to flammable concentrations. The permissible atmospheric concentration, though not established, is suggested as 50 ppm (Gerarde).

***p-tert*-Butyltoluene** ($C_{11}H_{16}$)

TBT; 1-METHYL-4-*tert*-BUTYLBENZENE

m.w.	148
sp.gr.	0.86
m.p.	−52 °C
b.p.	193 °C
v.d.	4.6
v.p.	< 1 mmHg (< 0.13·10³ Pa) at 20 °C
f.p.	68.5 °C

CH_3

$H_3C-C-CH_3$
CH_3

insoluble in water; very soluble in ether
a colourless liquid, with an aromatic odour.

TWA OSHA	10 ppm 60 mg/m³
STEL ACGIH	20 ppm 120 mg/m³
IDLH	1 000 ppm

It is produced by alkylation of toluene with isobutylene, and used as a solvent for resins and in chemical synthesis.

HAZARDS

Acute exposure to concentrations of 20 ppm and above may cause nausea, metallic taste, eye irritation and giddiness. Repeated exposure was found responsible for decreased blood pressure, increased pulse rate, anxiety and tremor, slight anaemia with leukopenia and eosinophilia. In repeated exposure it is also a mild skin irritant because of fat removal. Animal toxicity studies show effects on CNS, with lesions in *corpus callosum* and spinal cord.

SAFETY AND HEALTH MEASURES

Those suggested for industrial solvents (see SOLVENTS, INDUSTRIAL).

Trialkyl derivatives of benzene

Trimethylbenzenes (C_9H_{12})

In the trimethylbenzenes three hydrogen atoms in the benzene nucleus have been replaced by three methyl groups to form a further group of aromatic hydrocarbons.

Hemimellitene CH_3 CH_3 CH_3
1,2,3-trimethylbenzene

Mesitylene CH_3 H_3C CH_3
1,3,5-trimethylbenzene

Pseudocumene CH_3 CH_3 CH_3
1,2,4-trimethylbenzene

	Hemimellitene	Mesitylene	Pseudocumene
sp.gr.	0.89	0.86	0.88
m.p.	−25.4 °C	−44.7 °C	−43.8 °C
b.p.	176.1 °C	164.7 °C	169.4 °C
v.d.	4.1	4.1	4.1
f.p.	−	50 °C	45.5 °C
TWA OSHA	25 ppm 120 mg/m³	25 ppm 120 mg/m³	25 ppm 120 mg/m³
STEL ACGIH	35 ppm 170 mg/m³	35 ppm 170 mg/m³	35 ppm 170 mg/m³

insoluble in water; miscible with alcohol, ether, and benzene
colourless liquids with an aromatic odour.

These substances are present in certain petroleums and in distillates of coal tar.

Pseudocumene is used in the manufacture of perfumes and mesitylene and pseudocumene are used also as dyestuffs intermediates, but the chief industrial use of these substances is as solvents and paint thinners.

HAZARDS

The risk of injury to health and a fire risk are associated with the use of these liquids.

Fire and explosion. All three isomers are flammable. The flash point of pseudocumene is 45.5 °C but the liquids are commonly used industrially as constituents of coal tar solvent naphtha which may have a flash point anywhere in a range from 32 °C to below 23 °C. In the absence of precautions, a flammable concentration of vapour may be present where the liquids are used in solvent and thinner operations.

Health hazards. The main information as to the toxic effects of mesitylene and pseudocumene, both on animals and also on human beings, has been derived from studies of a solvent and paint thinner which contains 80% of these substances as constituents. They act as depressants of the central nervous system and can affect the blood coagulation. Bronchitis of an asthmatic type, headache, fatigue and drowsiness were also complained of by 70% of the workers exposed to high concentrations. A large proportion of mesitylene is oxidised in the body into mesitylenic acid, conjugated with glycine and excreted in the urine. Pseudocumene is oxidised into *p*-xylic acid, then excreted as well in the urine.

SAFETY AND HEALTH MEASURES

The danger arising from the toxic effects of these substances and the risk of fire and explosion require that the concentration of vapour in the atmosphere be kept below 25 ppm by means of effective ventilation or, preferably, locally applied exhaust ventilation (see SOLVENTS, INDUSTRIAL).

Dicyclic aromatic hydrocarbons

(See also NAPHTHALENE.)

Indene (C_9H_8)

INDONAPHTHENE

m.w.	116.2
sp.gr.	1
m.p.	−1.5-1.8 °C
b.p.	182.6 °C
v.d.	4.01
f.p.	78.3 °C
e.l.	0.9%

insoluble in water; miscible with alcohol, ether and benzene

a colourless liquid with a pungent odour.

TLV ACGIH 10 ppm 45 mg/m³

STEL ACGIH 15 ppm 70 mg/m³

Indene occurs in the tars from coal, lignite and crude petroleum. It can be produced by catalytic thermal treatment of tetrahydronaphthalene. It is used in the production of coumarone-indene resins, in chemical synthesis and as an insecticide.

HAZARDS

Indene is readily flammable and as a vapour will easily form explosive mixtures with air. From its chemical composition and animal toxicity studies it is expected to act as an irritant of mucous membranes and a skin defatting agent, and to affect the liver. In the animal body it undergoes hydroxylation of the 5-membered ring to indane-1-2-diols, which are then conjugated with glucuronic and sulphuric acids.

HEALTH AND SAFETY MEASURES

See SOLVENTS, INDUSTRIAL.

PARMEGGIANI, L.

General:

Toxicology and biochemistry of aromatic hydrocarbons. Gerarde, H. W. Elsevier monographs on toxic agents No. 8 (Amsterdam, London, New York, Princeton, Elsevier Publishing Company, 1960), 329 p. Illus. 201 ref.

CIS 74-1046 "Aromatic hydrocarbons" (Hydrocarbures aromatiques). Girard, R.; Prost, G. H.; Barthelémy, G.; Bourret, J.; Tolot, F. *Encyclopédie médico-chirurgicale. Intoxications* (Paris), Fascicule 16046 A1°, 6-1973. 10 p. Illus. (In French)

CIS 80-114 "Use of the relation chemical structure/biological activity to predict the toxicity of benzene derivatives" (Primenenie zakonomernostej himičeskaja struktura-biologičeskaja aktivnost' dlja prognozirovanija parametrov toksičnosti proizvodnyh benzola). Krasovskij, G. N.; Egorova, N. A.; Žoldakova, Z. I. *Gigiena i sanitarija* (Moscow), June 1979, 6 (7-11). 10 ref. (In Russian)

Benzene alkyl derivatives:

CIS 79-1659 "Reciprocal metabolic effect of benzene and its methyl derivatives in rats.—I. Study in vivo.—II. Study in vitro". Mikulski, P.; Wiglusz, R.; Galuszko, E.; Delag, G. *Bulletin of the Institute of Maritime and Tropical Medicine in Gdynia* (Gdynia), 1979, 30/1 (77-95). Illus. 25 ref.

C_9-C_{12} fractions:

CIS 1406-1966 "C_9-C_{12} fractions obtained from petroleum distillates. An evaluation of their potential toxicity". Nau, C. A.; Neal, J.; Thornton, M. *Archives of Environmental Health* (Chicago), Mar. 1966, 12/3 (382-393). Illus. 10 ref.

Analysis:

CIS 78-121 "Chromatographic determination of aromatic hydrocarbons in ambient air". Esposito, G. G.; Jacobs, B. W. *American Industrial Hygiene Association Journal* (Akron), Aug. 1977, 38/8 (401-407). Illus. 12 ref.

Hydrocarbons, halogenated aliphatic

The large and very important group of halogenated aliphatic hydrocarbons can be divided on the one hand according to the constitution of the original hydrocarbons (saturated and unsaturated aliphatic hydrocarbons; halogenated paraffins, olefins, diolefins, etc.) and on the other hand according to the hydrogen-replacing halogen (fluorinated, chlorinated, brominated or iodised hydrocarbons) and the number of replaced hydrogens (mono-, di-, etc., up to polyhalogenated aliphatic hydrocarbons). Since the boiling point of substances with a higher molecular mass is generally higher and is then further raised by halogenation by the heavy halogens, only the not very highly fluorinated aliphatic hydrocarbons (up to and including decafluorobutane), chloromethane, dichloromethane, chloroethane, chloroethylene and bromomethane are gaseous at normal temperatures. The derivatives with a higher molecular weight are liquids, and the heavy chlorinated compounds as well as tetrabromomethane and triiodomethane are solids. The odour of hydrocarbons is often strongly enhanced by halogenation and several volatile members of the group have a more than merely unpleasant odour. Some halogenated aliphatic hydrocarbons, e.g. chloroform and heavily halogenated derivatives of ethane and propane, have a pronounced sweet taste.

Production. The chlorinated aliphatic hydrocarbons are produced by chlorination of hydrocarbons, by the addition of chlorine or hydrogen chloride to unsaturated compounds, by the reaction between hydrogen chloride or chlorinated lime and alcohols, aldehydes or ketones, and exceptionally by chlorination of carbon disulphide or in some other way. In some cases more steps are necessary (e.g. chlorination with subsequent elimination of hydrogen chloride) to obtain the derivative needed, and usually a mixture arises from which the desired substance has to be separated. Brominated aliphatic hydrocarbons are prepared in a similar manner, while for iodised and particularly for fluorinated hydrocarbons other methods (electrolytic production of iodoform and substitution of chlorine for fluorine in the production of fluorocarbons) are preferred.

Uses. The uses of aliphatic halogenated hydrocarbons are manifold, of paramount importance being the application of some chlorinated compounds as solvents (e.g. trichloroethylene, tetrachloroethylene, carbon tetrachloride in dry cleaning and metal degreasing, dichloromethane in the production of cellulose derivatives, dichloroethane in the raffination of mineral oils) and of several chlorinated and fluorinated hydrocarbons as plastics intermediates (e.g. vinyl chloride for polyvinyl chloride, tetrafluoroethylene for polytetrafluoroethylene, and chloroprene for synthetic rubber). The low-boiling-point members of the group (e.g. chloromethane, and particularly fluorocarbons) are very widely employed as refrigerants, and the less toxic ones as aerosol propellants as well. Numerous halogenated aliphatic hydrocarbons are used as chemical intermediates (e.g. alkylating agents), for the substitution of halogen for other groups is mostly a rather easy reaction. Several halogenated aliphatic hydrocarbons have been used as anaesthetics (chloroform, chloroethane, bromochlorotrifluoroethane) or have found some other (nowadays mostly obsolete) medical application (e.g. triiodomethane and tetraiodoethylene as antiseptics, tetrachloromethane and tetrachloroethylene as antihelminthics, tribromomethane as an antitussic). Other members of the group have importance as fumigants (e.g. bromomethane, dibromoethane and propylene dichloride), as fuel and oil additives (e.g. dibromoethane and chlorinated paraffin), as fire extinguishing agents (tetrachloromethane and bromochloromethane), as ingredients in chemical smoke mixes (hexachloroethane), as gauge fluids and for ore separation (tetrabromoethane).

HAZARDS

Fire and explosion. Only the higher members of the series of halogenated aliphatic hydrocarbons are not flammable and not explosive. Some of them do not support combustion and are used as fire extinguishers. In contrast the lower members of the series are flammable, in some instances even highly flammable (e.g. 2-chloropropane) and form explosive mixtures with air. Besides, in the presence of oxygen, violently explosive peroxide compounds may arise from some unsaturated members (e.g. dichloroethylene) even at very low temperatures. Toxicologically dangerous compounds may be formed by thermal decomposition of halogenated hydrocarbons.

Health hazards. The production and use of halogenated aliphatic hydrocarbons involves serious health problems. They possess local as well as many systemic toxic effects, the most serious of the latter being carcinogenicity and mutagenicity, effects on the nervous system, and injury of vital organs particularly the liver. Despite the relative chemical simplicity of the group, the toxic effects vary greatly and the relation between structure and effect is not easy to establish.

The local irritant properties of these substances are particularly pronounced in the case of some of the unsaturated members; surprising differences exist, however, even between very similar compounds (e.g. octafluoroisobutylene is enormously more irritating than the isomeric octafluoro-2-butene). Lung irritation may be the major danger in acute inhalation exposure to some compounds belonging to this group (e.g. allyl chloride), and a few of them are lacrimators (e.g. carbon tetrabromide). High concentrations of vapours or liquid splashes may be dangerous for the eyes in some instances; the injury caused by the most used members, however, recovers spontaneously and only prolonged exposure of the cornea gives rise to persistent injury. Several of these substances are definitely irritating and injurious to the skin, causing reddening, blistering, and necrosis even on brief contact (1,2-dibromoethane, 1,3-dichloropropane). Being good solvents, all of them damage the skin by degreasing it and making it dry, vulnerable, cracked and chapped, particularly on repeated contact.

For several halogenated aliphatic hydrocarbons (chloroform, carbon tetrachloride) experimental evidence of carcinogenicity was found rather a long time ago. More recently vinyl chloride has been associated with the occurrence of cancer in humans, and at present the carcinogenicity of many members of the group is attracting increasing attention. So far some evidence of carcinogenicity for methyl iodide, trichloromethane, tribromomethane, carbon tetrachloride, 1,2-dichloro- and 1,2-dibromoethane, 1,1,2,2-tetrachloroethane, 1- and 2-iodopropane, 1,2-dibromo-3-chloropropane, 2-chloro- and 2-bromobutane, 1-bromo- and 2-bromo-2-methylpropane, vinyl chloride, vinyl bromide, 1,1-dichloroethylene, trichloroethylene, and *trans*-1,4-dichloro-2-butene has been found in animals. With the exception of vinyl chloride, however, the results of epidemiological studies or case reports are either not available or do not provide sufficient evidence on which to base a solid evaluation of the risk to humans at the present time. Similarly, the mutagenic and teratogenic effects of some halogenated aliphatic hydrocarbons are under investigation, 2-chloro-1,3-butadiene (chloroprene) and 2-bromo-2-chloro-1,1,1-trifluoroethane (halothane) being the most suspect members of the group.

Depression of the central nervous system is the most outstanding effect of many of the halogenated aliphatic hydrocarbons. Inebriation and excitation passing into narcosis is the typical reaction and, for that reason, many of them have been used as anaesthetics or abused, unfortunately, especially by young people. The narcotic effect of one compound may be very pronounced and of another only weak; in severe acute exposure, however, there is always the danger of death from respiratory failure or cardiac arrest, for the halogenated aliphatic hydrocarbons make the heart more susceptible to catecholamines. In several instances (e.g. methyl chloride and bromide and other brominated or iodised compounds), the effect of exposure, particularly of a repeated or chronic exposure, on the central nervous system cannot be described as a simple depression. Headache, nausea, ataxia, tremors, difficulty in speech, visual disturbances, convulsions, paralysis, delirium, mania or apathy give evidence of a more complicated affection, leading sometimes to a very slow recovery or even to permanent brain changes. Rather vague denominations have been used for these conditions (e.g. methyl chloride encephalopathy, chloroprene encephalomyelitis). The peripheral nerves may also be more or less affected (e.g. tetrachloroethane and dichloroacetylene polyneuritis).

Harmful effects on the liver, the kidney, and other organs are common to all the halogenated aliphatic hydrocarbons, though the extent of damage varies substantially from one member of the group to another. Since the signs of injury do not appear immediately, these effects have sometimes been referred to as delayed effects. The course of acute intoxication has often been described as biphasic, the signs of a reversible effect right at the beginning of the intoxication (narcosis) being seen as the first phase and the signs of injury to the organs (including sometimes the nervous system), which become apparent in a few days (e.g. jaundice), as the second phase. It is not always possible, however, to make a sharp distinction between the toxic effects of chronic or repeated exposure and the delayed effects of acute intoxication. There is no simple relation between the intensity of the immediate and the delayed effects of particular halogenated aliphatic hydrocarbons, for the delayed effects depend in the main upon metabolic products and have not much in common with the rather physico-chemical causes of narcosis. It is possible to find substances in the group with a rather strong narcotic potency and weak delayed effects (e.g. trichloroethylene), usually described as only slightly toxic, and substances that are very dangerous because they may cause irreversible organ injuries without showing very strong immediate effects (e.g. tetrachloroethane). Almost never is only a single organ or system involved; in particular, injury is rarely caused to the liver or kidneys alone, even by compounds which used to be regarded as typically hepatotoxic (e.g. carbon tetrachloride) or nephrotoxic (e.g. methyl bromide). Although harmful effects on the pancreas (e.g. carbon tetrachloride), heart (e.g. pentachloroethane), blood vessels (e.g. vinyl chloride), spleen (e.g. 1,1-dichloroethylene), adrenal glands (e.g. perchlorobutadiene), and bone marrow (e.g. methyl chloride) are much less frequently mentioned, such damage, even if limited in extent, cannot be completely excluded in most halogenated aliphatic hydrocarbons. The harmful effects of halogenated aliphatic hydrocarbons on the liver have been investigated very thoroughly; several model hepatotoxic substances belong to this group. Of the commonest compounds carbon tetrachloride is the most harmful, 1,1,2,2-tetrachloroethane being only slightly less harmful, followed by trichloroethylene and tetrachloroethylene, the least toxic in experimental animals being 1,1,1-Trichloroethane.

Fluorinated aliphatic hydrocarbons

See FLUOROCARBONS.

Chlorinated aliphatic hydrocarbons

Saturated

Chloromethane

METHYL CHLORIDE

See separate article.

Dichloromethane

METHYLENE CHLORIDE

See separate article.

Trichloromethane

See CHLOROFORM.

Tetrachloromethane

See CARBON TETRACHLORIDE.

Chloroethane (CH_3-CH_2Cl)

ETHYL CHLORIDE

m.w.	64.5
sp.gr.	0.92
m.p.	−139 °C
b.p.	12.3 °C
v.d.	2.22
v.p.	1 000 mmHg ($133 \cdot 10^3$ Pa) at 20 °C
f.p.	−50 °C
e.l.	3.8-15.4%
i.t.	510 °C

a gas with a characteristic ethereal odour.

TWA OSHA	1 000 ppm	2 600 mg/m³
STEL ACGIH	1 250 ppm	3 250 mg/m³
IDLH	20 000 ppm	
MAC USSR	50 mg/m³	

Because of long and extensive use in medicine (local anaesthetic by freezing, inhalation anaesthetic), the effects of ethyl chloride are comparatively well known. Its narcotic effect is not too strong; inhalation of 0.1% by volume is believed to be without any signs of narcosis; at concentrations of 2 to 4% incoordination and a weak analgaesia rather quickly occurs with subsequent loss of consciousness. The irritating effect is very slight; at a concentration of 4% only slight irritation of the eyes is caused. However, depression of the circulation and respiration is considerable and, particularly for this reason, the use of ethyl chloride as an inhalation anaesthetic cannot be recommended. The probability of liver and kidney injury is very low in acute as well as in chronic exposure. The substance is absorbed through mucous membranes and the skin and is quickly eliminated, mainly through the lungs. Liquid ethyl chloride may damage the skin by freezing and may injure the eye if splashed into it. The epithelial damage to the cornea is not attributed to the temperature-lowering effect of ethyl chloride but to its solvent or chemical action. Ethyl chloride is much less toxic than methyl chloride and many other lower chlorinated aliphatic hydrocarbons and it does not constitute a serious health hazard in the industry. The major problems are fire and explosion.

1,1-Dichloroethane

ETHYLIDENE CHLORIDE

1,2-Dichloroethane

See ETHYLENE DICHLORIDE.

1,1,1-Trichloroethane

METHYLCHLOROFORM

1,1,2-Trichloroethane

VINYL TRICHLORIDE

See TRICHLOROETHANES.

1,1,1,2-Tetrachloroethane

1,1,2,2-Tetrachloroethane

ACETYLENE TETRACHLORIDE

See TETRACHLOROETHANES.

Pentachloroethane ($CHCl_2-CCl_3$)

PENTALIN

m.w.	202.3
sp.gr.	1.67
m.p.	−29 °C
b.p.	161 °C

a colourless liquid with a sweetish chloroform-like odour at appreciably higher concentrations than those causing systemic poisoning.

Pentachloroethane acts as a strong central nervous system depressant, its narcotic effect being greater than that of chloroform. It has a local irritating effect on the eyes and the upper respiratory ways, too. In animals, liver, kidney and lung injury was observed following repeated inhalation exposures.

Hexachloroethane (CCl_3-CCl_3)

PERCHLOROETHANE; CARBON TRICHLORIDE; CARBON HEXACHLORIDE

m.w. 236.8
sp.gr. 2.1
m.p. 186.6 °C (sublimes)
v.d. 8.2
v.p. 1 mmHg (0.13·10³ Pa) at 32.7 °C

crystalline colourless solid with camphor-like odour.

TLV ACGIH 1 ppm 10 mg/m³ skin
STEL ACGIH 3 ppm 30 mg/m³

Hexachloroethane possesses a narcotic effect; however, since it is a solid and has a rather low vapour pressure under normal conditions, the hazard of a central nervous system depression by inhalation is low. It is irritating to skin and mucous membranes. Irritation has been observed from dust, and exposure of operators to fumes from hot hexachloroethane has been reported to cause blepharospasm, photophobia, lacrimation and reddening of the conjunctivae but not corneal injury or permanent damage. May cause dystrophic changes in the liver and in other organs as demonstrated in animals.

1-Chloropropane ($CH_3-CH_2-CH_2Cl$)

PROPYL CHLORIDE

m.w. 78.5
sp.gr. 0.89
m.p. −122.8 °C
b.p. 47.2 °C
v.d. 2.7
e.l. 2.6-11.1
i.t. 520 °C

a colourless liquid.

Toxicological information on propyl chloride is scarce, and is based on its insignificant use as a parasiticide and anaesthetic. It has narcotic effects and may injure the liver and kidneys.

2-Chloropropane ($CH_3-CHCl-CH_3$)

ISOPROPYL CHLORIDE

m.w. 78.6
sp.gr. 0.85
b.p. 35.3 °C
v.d. 2.71
f.p. −32 °C
e.l. 2.8-10.7%
i.t. 593 °C

a colourless liquid.

Isopropyl chloride is a potent anaesthetic; it has not been used more widely, however, because vomiting and cardiac arrhythmia have been reported in humans, and injury to liver and kidneys has been found in animal experiments. Its irritant effects are not great and no serious injury would be expected to result from splashes on the skin or into the eyes. It is a severe fire hazard.

1,2-Dichloropropane ($CH_3-CHCl-CH_2Cl$)

PROPYLENE DICHLORIDE

m.w. 113
sp.gr. 1.16
b.p. 96.8 °C
v.d. 3.9
v.p. 40 mmHg (5.32·10³ Pa) at 20 °C
f.p. 15.5 °C
e.l. 3.4-14.5%
i.t. 556 °C

slightly soluble in water

a colourless mobile liquid with a chloroform-like odour.

TWA OSHA 75 ppm 350 mg/m³
STEL ACGIH 110 ppm 350 mg/m³
IDLH 2 000 ppm
MAC USSR 10 mg/m³

In animals central nervous system depression and, following chronic exposure to vapour, injury to the liver, kidneys and adrenal glands have been reported. Its irritating properties are moderate; no serious skin or eye injury is likely to result from contact.

1,2,3-Trichloropropane ($CH_2Cl-CHCl-CH_2Cl$)

ALLYL TRICHLORIDE; GLYCEROL TRICHLOROHYDRIN; TRICHLOROHYDRIN

m.w. 147.44
sp.gr. 1.39
m.p. −14.7 °C
b.p. 156.2 °C
v.d. 5.0
v.p. 3.4 mmHg (0.49·10³ Pa) at 20 °C
f.p. 82.2 °C (oc)
e.l. 3.2-12.6%
i.t. 309 °C

a colourless liquid with a strong acid odour.

TWA OSHA 50 ppm 300 mg/m³
STEL ACGIH 75 ppm 450 mg/m³
IDLH 1 000 ppm

At a concentration of 0.01%, the unpleasant odour of 1,2,3-trichloropropane is already objectionable and some eye and throat irritation occurs. In animal experiments liver and kidney injury was noted.

Scarce information (mostly only LD and/or LC values in laboratory animals) exists on the effects of 1,1-dichloropropane, 1,1,1- and 1,1,2-trichloropropane, 1,1,1,3-tetrachloropropane, 1,1,2,3,3-pentachloropropane, 1-chlorobutane (butyl chloride), 2-chlorobutane (carcinogenic in animals), 1-chloro-2-methylpropane, 2-chloro-2-methylpropane, 1,4-dichlorobutane, 1,2-dichloro-2-methylpropane (isobutylene dichloride), 1,2,3-trichlorobutane, pentachlorobutane, hexachlorobutane, 1,1,1,5-tetrachloropentane, 1,1,1,7-tetrachloroheptane, 3-chloromethylheptane (1-chloro-2-ethylhexane, 2-ethylhexyl chloride), 1,1,1,9-tetrachlorononane, chlorodecane and 1,1,1,11-tetrachloroundecane.

Unsaturated

Chloroethylene (VCM)

See VINYL CHLORIDE.

1,1-Dichloroethylene

See VINYL COMPOUNDS.

1,2-Dichloroethylene ($CHCl=CHCl$)

ACETYLENE DICHLORIDE

m.w. 97
sp.gr. 1.27
m.p. −80.5 °C *cis* −50 °C *trans*
b.p. 59 °C *cis* 49 °C *trans*
v.d. 3.34
v.p. 400 mmHg (53.2·10³ Pa) at 41 °C *cis* 30.8 °C *trans*
f.p. 4 °C
e.l. 9.7-12.8%
i.t. 460 °C *trans*

a colourless liquid with an ether-like odour.

TWA OSHA 200 ppm 790 mg/m³
STEL ACGIH 250 ppm 1 000 mg/m³
IDLH 4 000 ppm

The major response to both of the stereoisomers of 1,2-dichloroethylene is one of central nervous system depression. The irritating properties are minor, but corneal opacity may be caused by 1,2-dichloroethylene vapours, particularly perhaps by the *trans*-isomer. In animals 1,2-dichloroethylene seems to be a little less toxic than 1,1-dichloroethylene peroral, the acute inhalation toxicity of the *cis*-isomer being somewhat greater than the toxicity of the *trans*-isomer. The odour of *trans*-1,2-trichloroethylene is noticeable at a concentration of 1 100 mg/m³ (278 ppm), a concentration of 3 300 mg/m³ (834 ppm) is tolerable for half an hour without unpleasant effects; from 6 800 to 8 800 mg/m³ (1 718 to 2 224 ppm) causes vertigo and sleeplessness in 5 min. Liver and kidney injury does not appear to be of great significance in exposure to 1,2-dichloroethylene, although study of this has been limited during recent years.

Trichloroethylene

See separate article.

Tetrachloroethylene
PERCHLOROETHYLENE

See separate article.

Dichloroacetylene ($CCl{\equiv}CCl$)
DICHLOROETHYNE
m.w. 94.9
a colourless gas.
TLV ACGIH 0.1 ppm 0.4 mg/m³ ceil

Dichloroacetylene, which may arise by thermal decomposition of trichloroethylene, is reported to have caused a number of serious and even fatal intoxications. After an incubation period ranging from 1 to 3 days, extreme nausea occurred among persons exposed for some time to concentrations as low as 2 to 4 mg/m³ (0.5 to 1 ppm). Cases of violent headache, pain in the jaw *(polyneuritis cranialis)*, cranial nerve palsey and labial herpes have been reported. In fatal cases oedema of the brain was found by autopsy.

1-Chloro-2-propene
ALLYL CHLORIDE

See ALLYL COMPOUNDS.

1,3-Dichloropropene ($CH_2Cl{-}CH{=}CHCl$)
1,3-DICHLOROPROPYLENE
m.w. 111
sp.gr. 1.22
b.p. 107 °C
v.d. 3.8
f.p. 35 °C
a colourless liquid.
MAC USSR 5 mg/m³

Some narcotic effect and a strong irritant effect has been observed. On contact of the liquid with the skin blisters may arise.

1,2-Dichloro-2-propene ($CH_2{=}CCl{-}CH_2Cl$)
1,2-DICHLOROPROPYLENE
m.w. 111
b.p. 94 °C
v.d. 3.8
a colourless liquid.

Depression of the central nervous system and severe irritation of the skin have been reported as the main effects of this compound.

Hexachloropropene ($CCl_3{-}CCl{=}CCl_2$)
m.w. 248.8
b.p. 209 °C
a colourless liquid.

Narcotic effects are minimal; the irritant effect and injury to the liver may be the main hazard.

1-Chloro-2-butene ($CH_3{-}CH{=}CH{-}CH_2Cl$)
CROTYL CHLORIDE
m.w. 90.5
sp.gr. 0.93
b.p. 80 °C
v.d. 3.13
e.l. 4.2-19%
a colourless liquid.

Irritation of eyes and respiratory passages caused by both *cis*- and *trans*- stereoisomers is very significant.

1-Chloro-2-methyl-1-propene ($(CH_3)_2C{=}CHCl$)
1-CHLOROISOBUTYLENE
sp.gr. 0.92
b.p. 68 °C
a colourless liquid.

Narcotic and local irritating effects of 1-chloroisobutylene have been reported. It is readily absorbed through the skin.

1-Chloro-2-methyl-2-propene ($CH_2{=}C(CH_3){-}CH_2Cl$)
METHALLYL CHLORIDE
m.w. 90.55
sp.gr. 0.93
b.p. 71 °C
v.d. 3.12
v.p. 101.7 mmHg (13.53·10³ Pa) at 20 °C
f.p. −12 °C
e.l. 3.2-8.1%
a colourless liquid.
MAC USSR 0.3 mg/m³

In animals inhalation exposure caused signs of depression of the central nervous system and irritation of mucous membranes. The liquid does not irritate the skin or damage the eyes seriously.

1,3-Dichloro-2-butene ($CH_2{=}ClCHCClCH_3$)
m.w. 125
b.p. 123 °C
v.d. 4.3
f.p. 26.7 °C
a colourless liquid.
MAC USSR 1 mg/m³

The irritant effect of 1,3-dichloro-2-butene is markedly stronger than that of chloroprene. Its narcotic potency is weak and no organ effects have been found in animal experiments.

1,4-Dichloro-2-butene ($CH_2Cl{-}CH{=}CH{-}CH_2Cl$)
m.w. 125
sp.gr. 1.18
m.p. 1-3 °C
b.p. 152 °C *cis* 155 °C *trans*
both the *cis*- and *trans*- stereoisomers are colourless liquids.

Irritating effects predominate in inhalation exposure in animals. Blisters may result from skin contact. There is evidence of carcinogenicity of the *trans*-isomer in animals.

1,3-Dichloro-2-methylpropene ($CH_2Cl-C(CH_3)=CHCl$)
1,3-DICHLOROISOBUTYLENE
m.w. 125
b.p. 130 °C
a colourless liquid.
MAC USSR 0.5 mg/m³

An irritant that may injure the liver.

2-Chloro-1,3-butadiene
CHLOROPRENE

See separate article.

2,3-Dichloro-1,3-butadiene ($CH_2=CCl-CCl=CH_2$)
m.w. 125
b.p. 98 °C
a colourless liquid.
MAC USSR 0.1 mg/m³ skin

In animal experiments no significant injury to the liver has been demonstrated. It is not absorbed through the skin in appreciable amounts and is less irritating to the skin than chloroprene.

Hexachloro-1,3-butadiene ($CCl_2=CCl-CCl=CCl_2$)
PERCHLOROBUTADIENE
m.w. 260.7
b.p. 215 °C
v.d. 8.99
i.t. 610 °C
thick colourless liquid.

In animal experiments narcotic and irritating effects have been observed and some changes in the liver, kidneys and adrenal glands have been found. There is also some evidence of embryotoxicity and carcinogenicity in animals.

The results of rare studies of the toxicity and effects of 1-chloro-1-propene, 1,2-dichloropropene, 1,2,3-trichloropropene, 1,1,2,3-tetrachloropropene, 1,3-dichloro-2-butene and tetrachlorohexatriene also exist.

Brominated aliphatic hydrocarbons

Saturated

Bromomethane
METHYL BROMIDE

See separate article.

Dibromomethane (CH_2Br_2)
METHYLENE BROMIDE
m.w. 174
sp.gr. 2.48
b.p. 97 °C
v.d. 6.05
a colourless-to-yellowish liquid.
MAC USSR 10 mg/m³

Methylene bromide has a narcotic, nephrotoxic and hepatotoxic effect in animals. Exposure to methylene bromide in industry has been reported to give rise to complaints of headache, dizziness, muscle pains, and loss of appetite. Hypotension, loss of weight, enlarged and painful liver and glomerular filtration changes have also been reported.

Tribromomethane ($CHBr_3$)
BROMOFORM
m.w. 252.77
sp.gr. 2.89
m.p. 6-7 °C
b.p. 149 °C
v.p. 5 mmHg (0.65·10³ Pa) at 20 °C
colourless heavy liquid with chloroform-like odour and a sweetish taste; decomposes gradually, acquiring a yellow colour; light and air accelerate the decomposition, and commercial articles are generally preserved by the addition of alcohol.
TWA OSHA 0.5 ppm 5 mg/m³
TLV ACGIH 0.5 ppm 5 mg/m³ skin
MAC USSR 5 mg/m³

Much of the experience in poisoning cases in humans has been from the oral administration and it is difficult to determine the significance of the toxicity of bromoform in industrial use. Bromoform has been used as a sedative and particularly as an antitussive for years, ingestion of quantities above the therapeutic dose (0.1 to 0.5 g) having caused stupor, hypotension, and coma. In addition to the narcotic effect, a rather strong irritant and lacrimatory effect occurs. Exposure to bromoform vapours causes a marked irritation of the respiratory passages, lacrimation and salivation. Bromoform may injure the liver and the kidney. In mice, tumours have been elicited by intraperitoneal application. It is absorbed through the skin. On exposure to concentrations of up to 100 mg/m³ (10 ppm), complaints of headache, dizziness and pain in the liver region have been made and alterations in the liver function have been reported.

Carbon tetrabromide (CBr_4)
TETRABROMOMETHANE
m.w. 331.7
sp.gr. 3.42
m.p. 94 °C
b.p. 189.5 °C
v.d. 11.6
v.p. 40 mmHg (5.32·10³ Pa) at 96.3 °C
insoluble in water
colourless monoclinic crystals, that turn yellow on decomposition.
TLV ACGIH 0.1 ppm 1.4 mg/m³
STEL ACGIH 0.3 ppm 4 mg/m³

In inhalation exposure the irritant effect of carbon tetrabromide is predominant. The vapours are highly irritant to the eyes, and heavy exposure may cause oedema of the lungs. Severe injury to the eyes may result from topical application. Reports on the effect of carbon tetrabromide on the skin and on the possibility of liver and kidney injury are conflicting.

Bromoethane (CH_3-CH_2Br)
ETHYL BROMIDE
m.w. 109
sp.gr. 1.45
m.p. −119 °C
b.p. 38 °C
v.d. 3.76
v.p. 400 mmHg (532·10³ Pa) at 21 °C
f.p. −16 °C
e.l. 6.7-11.3%
i.t. 511 °C
a volatile liquid with an ether-like odour; colourless when fresh, it turns yellowish on exposure to air and light.
TWA OSHA 200 ppm 890 mg/m³
STEL ACGIH 250 ppm 1 110 mg/m³
IDLH 3 500 ppm

In the concentration of 3 to 10% the vapours of ethyl bromide lead to narcosis and, despite its side effects, it has been used as a general anaesthetic. The irritant effect on the respiratory system is noticeable; injury of central nervous system (particularly extrapyramidal), heart and intestine, fatty degeneration of liver and kidney have been reported. It is regarded as distinctly less toxic than methyl bromide but as noticeably more toxic than ethyl chloride. In chronic exposure sleepiness, dizziness and weakness, followed by gait disturbances, hyper-reflexia, spastic parapareses, and, in chronic exposure, speech disturbances, nystagmus and pathologic reflexes have been reported, giving evidence for the marked neurotoxicity of ethyl bromide and indicating the health hazard of the substance in occupational exposure.

1,2-Dibromoethane

See separate article.

1,1,2,2-Tetrabromoethane ($CHBr_2-CHBr_2$)

ACETYLENE TETRABROMIDE; TBA

m.w. 345.7
sp.gr. 2.96
b.p. 151 °C
f.p. −18 °C
i.t. 346 °C

a colourless liquid.

TWA OSHA 1 ppm 14 mg/m³
STEL ACGIH 1.5 ppm 20 mg/m³
IDLH 10 ppm
MAC USSR 1 mg/m³

Its toxicity and effects are not well known. In one near fatal case of intoxication with liver injury the predominant finding was reported to have been caused by one day's work in concentrations averaging 2 ppm with a peak exposure for about 10 min at 16 ppm. Other workers in the area contaminated complained of slight irritation of the eyes and nose, with headache and lassitude later on. It is, however, possible that the concentrations involved were higher and that absorption through the skin was an important factor with the operator most seriously afflicted. At all events, acetylene tetrabromide has to be regarded as a dangerous hepatotoxic substance.

1-Bromopropane ($CH_3-CH_2-CH_2Br$)

PROPYL BROMIDE

m.w. 123
sp.gr. 1.35
m.p. −110 °C
b.p. 70.9 °C
i.t. 490 °C

a colourless liquid when fresh.

A depressant of the central nervous system and an irritant. In animals exposure to anaesthetic concentrations resulted in injury to the lungs and liver. It is absorbed through the skin.

1-Bromobutane ($CH_3-CH_2-CH_2-CH_2Br$)

BUTYL BROMIDE

m.w. 137
sp.gr. 1.27
m.p. −112.9
b.p. 101.4
v.d. 4.72
f.p. 18.3 °C (oc)
e.l. 2.6-6.6%
i.t. 269 °C

a colourless liquid when fresh.

In rats irritation of the respiratory ways and injury to the liver and kidneys have been reported to result from repeated exposures to butyl bromide vapours.

Limited information exists on the toxic properties of 1,1-dibromoethane (ethylidene bromide), 1,2- and 1,3-dibromopropane, about 2-bromobutane (carcinogenic in animals), 1- and 2-bromo-2-methylpropane (both carcinogenic in animals), 2-bromopentane, 1-bromohexane and 1-bromooctane.

Unsaturated

Bromoethylene ($CH_2=CHBr$)

VINYL BROMIDE

See VINYL COMPOUNDS.

Tribromoethylene ($CHBr=CBr_2$)

m.w. 264.7
b.p. 162 °C

a colourless liquid when fresh.

The narcotic effect of tribromoethylene is comparatively weak; it is, however, more toxic than trichloroethylene and injury to the heart, kidneys, liver and respiratory ways has been observed in animals subsequent to repeated inhalation exposure. It is a skin irritant.

3-Bromo-1-propene

ALLYL BROMIDE

See ALLYL COMPOUNDS.

3-Bromo-1-propyne ($CH\equiv C-CH_2Br$)

PROPARGYL BROMIDE

m.w. 118.9
b.p. 89 °C

a colourless liquid with a sharp odour.

Propargyl bromide has a pronounced irritating effect.

1,4-Dibromo-2-butene ($CH_2Br-CH=CH-CH_2Br$)

m.w. 213.8
m.p. *cis*- 53 °C, *trans*- 54 °C

both *cis*- and *trans*-2,4-dibromo-2-butene are solids.

This compound is very irritating to the eyes. Contact with the skin causes blistering. The *trans*-isomer seems to be the slightly more toxic one.

Iodised aliphatic hydrocarbons

Saturated

Iodomethane (CH_3I)

METHYL IODIDE; MI

m.w. 142
sp.gr. 2.3
m.p. −66.1 °C
b.p. 42.5 °C
v.d. 4.9
v.p. 400 mmHg (53.2·10³ Pa) at 25 °C

a colourless liquid with a characteristic odour, reacts with water.

TLV ACGIH 2 ppm 10 mg/m³ skin
STEL ACGIH 5 ppm 30 mg/m³

In animals signs of narcosis and pulmonary oedema were reported after exposure to high concentrations of methyl iodide vapours. Some injury to the liver and kidneys was also found in animal experiments. In rats and mice tumours were caused by the application of methyl iodide and the carcinogenic determination is considered to be positive. The irritation effects are moderate. Methyl iodide is absorbed through the skin. Years ago two cases of professional intoxication due to vapour exposure were reported. The first showed symptoms of vertigo, ataxia and diplopia, developed delirium and serious mental disturbances. The second was found to be drowsy, unable to walk and with slurred incoherent speech, and died a week after exposure. See also ALKYLATING AGENTS.

Triiodomethane (CHI_3)

IODOFORM

m.w. 393.8
sp.gr. 4.1
m.p. 119 °C
a yellow solid with a disagreeable odour.
TLV ACGIH 0.6 ppm 10 mg/m³

Absorption of significant amounts of iodoform results in central nervous system depression, visual disturbances, and injury to the heart, liver and kidneys. The possibility of some health hazard in industry could be inferred from these findings, although the problems associated with iodoform are related to its topical application and oral administration in medical use.

Iodoethane (CH_3-CH_2I)

ETHYL IODIDE

m.w. 156
sp.gr. 1.90
m.p. −108 °C
b.p. 72 °C
v.d. 5.38
v.p. 100 mmHg ($13.3 \cdot 10^3$ Pa) at 18 °C
liquid.

Ethyl iodide may cause central nervous system depression and affect the kidneys, liver, thyroid gland and lungs.

1-Iodopropane (propyl iodide) and 2-iodopropane (isopropyl iodide) have some narcotic and a significant irritant effect. In both, experimental evidence of carcinogenicity has been noted. 1-Iodobutane (butyl iodide) as well as 1-iodo-2-methylpropane (isobutyl iodide) are also narcotic and irritating, but only butyl iodide is included in the list of experimental carcinogens. Scarce information is available referring to the toxicity of 1-iodo-3-methylbutane, 1-iodopentane and 1-iodooctane.

Unsaturated

1-Iodo-2-propene

ALLYL IODIDE

See ALLYL COMPOUNDS.

According to some scanty, long-standing information, local irritation is the predominant effect of 1,2-diiodoethylene and diiodoacetylene.

Halogenated aliphatic hydrocarbons with two or more different halogens

Fluorine containing substances of this type

See FLUOROCARBONS.

Saturated

Bromochloromethane (CH_2BrCl)

METHYLENE CHLOROBROMIDE; CHLOROBROMOMETHANE

m.w. 129.4
sp.gr. 1.93
b.p. 69 °C
v.d. 4.46
a colourless liquid with a sweet odour.
TLV ACGIH 200 ppm 1 050 mg/m³
IDLH 5 000 ppm

Though somewhat more toxic than methylene chloride, methylene chlorobromide is one of the less toxic halogenated derivatives of methane. In animals rather prolonged anaesthesia was observed in inhalation trials at high concentrations. After repeated inhalation exposures fatty degeneration of the liver, kidneys, and occasionally heart, lipoid depletion of the adrenal cortex, interstitial pneumonitis, and, in animals that died, opacity of the eyes was noted. Injury to the liver of the type seen after carbon tetrachloride exposure was, however, not observed even after many inhalation exposures and bromochloromethane has probably only a slight capacity to produce a reversible liver injury. It has a distinctive odour at 400 ppm, this fact being considered as a good warning at a concentration well below the acutely hazardous level, though at this concentration it is not disagreeable enough to drive anyone from the contaminated area. Several cases of acute poisoning caused by brief exposure at probably very high concentrations were reported in fire fighters using the substance as a fire-extinguishing agent. They were characterised by severe headache, loss of consciousness after exposure, gastric upset and loss in weight in the ensuing course and a rather slow recovery. It is believed that the injury to the heart and kidneys which may be caused by very severe exposure is not progressive.

1-Bromo-2-chloroethane (CH_2Cl-CH_2Br)

ETHYLENE CHLOROBROMIDE

m.w. 143.4
sp.gr. 1.73
b.p. 107 °C
v.d. 4.94
v.p. 40 mmHg ($5.32 \cdot 10^3$ Pa) at 29.7 °C
a colourless liquid with a chloroform-like odour.

Headache, gastric discomfort, pallor, mucous membranes irritation, lowering of the blood pressure and body temperature, and ultimately coma have been described as signs of acute ethylene chlorobromide intoxication. In chronic exposure injury to the liver, kidneys and eyes may occur.

1,2-Dibromo-3-chloropropane ($CH_2Cl-CHBr-CH_2Br$)

DPCP; FUMAZONE; NEMAGON; NEMAFUME

See separate article.

Little information is available on the toxicity and effects of several other members of this series. Bromodichloromethane and dibromochloromethane are not hepatotoxic while the harmful effect on the liver of bromotrichloromethane and dibromodichloromethane is considerable.

Unsaturated

1-Bromo-3-chloro-2-propene ($CHCl=CH-CH_2Br$)

CBP; CHLOROALLYL BROMIDE

m.w. 155.4
b.p. 130 °C
a liquid.

Local irritation is the predominant response to chloroallyl bromide, its irritating capacity being comparable with that of allyl alcohol. At concentration of 0.1 ppm, the irritation is slight; at 2 ppm, it is very severely irritant and unbearable for more than a few minutes.

SAFETY AND HEALTH MEASURES

From an examination of the flammability as well as the health effects of different halogenated aliphatic hydrocarbons, it is obvious that they cannot be lumped together from the point of view of the hazards connected with their industrial use, either quantitatively or qualitatively. For this reason, each member of the group and each particular operation using it has to be considered separately. Different safety and health measures at different levels have to be adopted and, particularly as regards the more dangerous compounds, very strictly observed.

The use of the most dangerous compounds of the group should be avoided entirely, in particular if it does not stem from their chemical constitution (e.g. intermediates) but from their technical properties or economic advantage (e.g. solvents), when they can be replaced by less harmful substances.

In most instances the fire and explosion risk associated with the use of flammable members of the group will be covered by precautions designed to ensure that no health damage will occur in persons exposed.

Harmful concentrations of the vapours of halogenated aliphatic hydrocarbons should not be allowed to build up in the atmosphere in which operators are working and if this is unavoidable (e.g. work in confined spaces) adequate personal protective equipment must be provided and appropriate safety measures adopted. To avoid harmful concentrations in the workroom air, total enclosure of the process should be used if practicable. Otherwise an appropriate ventilation system must be designed taking account of operating temperature, exposed liquid surface, evaporation rate, air flow and work practices of the operators. Full awareness of the significance of the possible consequences of exceeding the permissible exposure limits for each individual halogenated aliphatic hydrocarbon is necessary for the intelligent application and monitoring of the standards established.

In handling the liquid substances, every precaution must be taken to prevent spillage, and contact with the skin and eyes must be avoided since most halogenated aliphatic hydrocarbons have irritant properties and/or are absorbed through the skin and mucous membranes.

The possibility of misuse and dependence must not be overlooked in substances with a narcotic and not too strong irritative effect.

When any of the compounds having harmful effects on the organs is used, persons with pre-existing damage or suspected of being susceptible to damage of the respective organ or system should not be employed. Alcoholism and liver diseases are the most frequent contraindications of this sort. In periodic medical examinations particular attention should be devoted to the organs and systems at risk.

MARHOLD, J. V.

General:

Registry of toxic effects of chemical substances. US Department of Health, Education, and Welfare, Public Health Service, Center for Disease Control, NIOSH (Microfiche on a quarterly base. Subscription from Superintendent of Documents, Government Printing Office, Washington, DC).

Health hazards:

CIS 80-730 *Some halogenated hydrocarbons.* Vol. 20. *IARC monographs on the evaluation of the carcinogenic risk of chemicals to humans* (Lyons, International Agency for Research on Cancer, Oct. 1979), 609 p. Illus. 2 166 ref.

"Toxicological evaluation of halogenated ethylenes" (Die toxicologische Beurteilung halogenierter Aethylene). Bolt, H. M. *Arbeitsmedizin–Sozialmedizin–Präventivmedizin* (Stuttgart), Mar. 1980, 15/3 (49-53). Illus. 11 ref. (In German)

Safety:

"New safety rules for the use of chlorinated aliphatic hydrocarbons and their mixtures" (Neue Sicherheitsregeln für den Umgang mit Chlorkohlwasserstoffen). *Die Tiefbau Berufsgenossenschaft* (Stuttgart), Jan. 1979, 91/1 (24-29). (In German)

CIS 80-134 "Hazardous chemical reactions–64. Halogenated hydrocarbons (Part 2)" (Réactions chimiques dangereuses–64. Hydrocarbures halogénés (2e partie)). Leleu, J. *Cahiers de notes documentaires–Sécurité et hygiène du travail* (Paris), 4th quarter 1979, 97, Note No. 1219-97-79 (603-612). (In French)

Hydrochloric acid

Hydrochloric acid (HCl)

MURIATIC ACID; AQUEOUS HYDROGEN CHLORIDE

m.w.	37
sp.gr.	1.19 (commercial "concentrated" or fuming acid with 38% HCl)
m.p.	−114.8 °C (pure HCl)
b.p.	−84.9 °C (pure HCl)
v.p.	> 1 atm (> 10^5 Pa)

soluble in water, ethyl alcohol and ethyl ether; insoluble in hydrocarbons

a clear, colourless or slightly yellow, fuming, pungent liquid.

TWA OSHA	5 ppm 7 mg/m³ ceil
IDLH	100 ppm
MAC USSR	5 mg/m³

Anhydrous hydrogen chloride is not corrosive; however, aqueous solutions attack nearly all metals (mercury, silver, gold, platinum, tantalum and certain alloys are exceptions) with release of hydrogen. Hydrochloric acid reacts with sulphides to form chlorides and hydrogen sulphide. It is a very stable compound but at high temperatures it decomposes into hydrogen and chlorine.

Production. Hydrochloric acid is prepared industrially by several different processes, e.g.:

(a) by the reaction of sodium chloride or, in some cases, potassium chloride with sulphuric acid in a muffle or mechanical furnace at temperatures up to 600 °C;

(b) by the Meyer process in which sodium bisulphite is reacted with sodium chloride at 400-800 °C;

(c) by the Hargreaves process using sulphur dioxide, salt and steam in an exothermic reaction;

(d) by synthesis, in which hydrogen is burnt in chlorine;

(e) as a by-product of the chlorination of organic compounds.

These processes are followed by the elimination of suspended solids, cooling, condensation and purification.

Uses. Hydrochloric acid is used in the manufacture of fertilisers, dyes and dyestuffs, artificial silk and pigments for paints and in the refining of edible oils and fats, electroplating, leather tanning, ore refining, in the photographic industry, soap refining, textile industry, rubber industry, petroleum extraction, pickling of metals, etc.

Anhydrous hydrogen chloride is used in the production of vinyl chloride from acetylene and alkyl chlorides from olefins; it is also used in polymerisation, isomerisation, alkylation and nitration reactions.

HAZARDS

The special hazards of hydrochloric acid are its corrosive action on skin and mucous membranes, the formation of hydrogen when it contacts certain metals and metallic hydrides, and its toxicity.

Hydrochloric acid will produce burns of the skin and mucous membranes, the severity being determined by the concentration of the solution; this may lead to ulcerations followed by keloid and retractile scarring. Contact with the eyes may produce reduced vision or blindness. Burns on the face may produce serious and disfiguring scars. Frequent contact with aqueous solutions may lead to dermatitis.

The vapours have an irritant effect on the respiratory tract causing laryngitis, glottal oedema, bronchitis, pulmonary oedema and death. Digestive diseases are frequent and are characterised by dental molecular necrosis in which the teeth lose their shine, turn yellow, become soft, pointed and then break off.

SAFETY AND HEALTH MEASURES

Hydrochloric acid should be manufactured in closed-circuit processes; particular attention should be paid to the detection of leaks in view of the product's highly corrosive nature. Storage areas should be well ventilated and have a cement floor, and shelter from direct sunlight and heat should be provided. The acid should not be stored in the vicinity of flammable or oxidising substances, e.g. nitric acid or chlorates, or near metals and metal hydrides which may be attacked by the acid with the formation of hydrogen, explosive limits of which are 4-75% by volume in air. Electrical equipment should be flameproof and protected against the corrosive action of the vapours.

When handling hydrochloric acid (loading, unloading and decanting), measures should be taken to avoid splashes or the inhalation of vapours, e.g. by use of decanting pumps, pouring frames, etc. Difficult operations should be carried out in fume cupboards or under exhaust ventilation and an abundant supply of water should always be available in the near vicinity. Workers should never enter tanks or other vessels that have contained hydrochloric acid until they have been cleaned. Workshops in which hydrochloric acid is frequently handled should be equipped with emergency showers and eyewash equipment, etc. The acid should never be disposed of into the sewage system until the pH has been brought to 5.5-8.5.

Workers should wear acid-resistant protective clothing, including hoods, eye and face protection, acid-resistant hand and arm protection, and foot and leg protection; where there is a danger of vapour inhalation, workers should wear respiratory protective equipment of the self-contained or canister type depending on the concentration. Adequate sanitary facilities should be supplied and workers encouraged to wash thoroughly before meals.

Medical prevention. Persons suffering from skin, respiratory or digestive diseases should not be employed on work entailing exposure to hydrochloric acid.

Treatment. Splashes in the eye should be treated with copious irrigation with water. In the event of skin contact the affected area should be washed with a cloth and large amounts of water. Contaminated clothing should be removed. Finally, the skin should be treated with a 5% solution of triethanolamine.

Persons who have inhaled HCl should be removed immediately from the contaminated zone and prevented from making any effort. They should be put in the care of a physician immediately. In the event of accidental ingestion, the victim should be given a neutralising substance (magnesium oxide with large amounts of water) and gastric lavage should be carried out. Vomiting should not be induced since this may make the injury more widespread.

FERNANDEZ-CONRADI, L.

Environmental pollution:

CIS 77-469 *Chlorine and hydrogen chloride* (National Academy of Sciences, 2101 Constitution Avenue, Washington, DC) (1976), 282 p. Illus. 459 ref.

Safety:

CIS 77-774 *Hydrogen chloride (anhydrous) – Codes of practice for chemicals with major hazards* (Chemical Industries Association Ltd., Alembic House, 93 Albert Embankment, London) (Feb. 1975), 20 p. Illus.

Properties and essential information for safe handling and use of hydrochloric acid, aqueous, and hydrogen chloride, anhydrous. Chemical safety data sheet SD-39 (Manufacturing Chemists' Association, 1825 Connecticut Avenue, NW, Washington, DC) (revised May 1970), 27 p. Illus.

CIS 81-1297 "TLV for hydrochloric acid in pickling plants" (MAK-Wert für Chlorwasserstoff in Beizereien). Mappes, R. *Zentralblatt für Arbeitsmedizin, Arbeitsschutz, Prophylaxe und Ergonomie* (Heidelberg), May 1980, 30/5 (172-173). (In German)

Hydrofluoric acid

Hydrofluoric acid (HF)

HYDROGEN FLUORIDE

m.w.	20
sp.gr.	0.99 (liquid)
m.p.	−83.1 °C
b.p.	19.5 °C
v.d.	1.27
v.p.	400 mmHg (53.2·10³ Pa) at 2.5 °C

very soluble in water

a colourless, fuming liquid.

TWA OSHA	3 ppm 2 mg/m³
TWA NIOSH	2.5 mg/m³
NIOSH	5 mg/m³ 15 min ceil
IDLH	20 ppm
MAC USSR	0.5 mg/m³

Production. Hydrofluoric acid is made by the action of concentrated sulphuric acid on fluorspar (calcium fluoride).

Uses. Hydrofluoric acid is used in the production of fluorocarbons and inorganic fluorides, in the refining of certain metals, as a catalyst in organic chemical reactions, and in etching glass and pottery products.

HAZARDS

Hydrofluoric acid is strongly corrosive, highly irritating and poisonous. Both liquid and vapour forms cause severe and painful burns on contact with the skin, eyes, or mucous membranes. Burns from dilute solutions of hydrofluoric acid or low concentrations of the vapour may not be immediately painful or visible. Inhalation of the vapour may cause lung oedema which may not become apparent until 12-24 h following exposure. Nosebleeds and sinus troubles have reportedly occurred among metal workers exposed to very low concentrations of a fluoride or fluorine in air.

SAFETY AND HEALTH MEASURES

The safe handling of hydrofluoric acid requires that well designed equipment be properly operated and maintained by well trained, adequately protected, responsible personnel. Specifications for the proper design and construction of equipment used in the manufacture, storage and transportation of hydrofluoric acid depend on whether anhydrous or aqueous solutions are involved. Anhydrous hydrofluoric acid must be stored and shipped in steel pressure vessels, while the aqueous acid (70% or less in concentration) may be stored or shipped in rubber-lined or plastic vessels. Tanks and other containers should be protected from heat and direct rays of the sun. Although hydrofluoric acid is non-flammable, its corrosive action on metals can result in the formation of hydrogen in containers and piping to create a fire and explosion hazard; potential sources of ignition should therefore be excluded from areas around equipment containing this acid. Shipping containers, including drums, cylinders, tank trucks and tank cars, should bear proper warning labels indicating the hazardous nature of the compound, the proper action to be taken in case of accidental spillage and emergency first-aid instructions.

Important factors in the prevention of injury by hydrofluoric acid are:

(a) to provide adequate ventilation in operational areas at all times and ensure that the atmospheric concentration of hydrogen fluoride does not exceed permissible levels;

(b) to prevent all contact of vapour or liquid with eyes, skin, digestive system or respiratory system;

(c) to ensure the use of suitable personal protective equipment made of materials that are resistant to hydrofluoric acid such as polyvinyl chloride or chlorobutadiene rubber (Neoprene). This equipment should include overalls with sleeves to the wrists, eye and face protection such as face shields or chemical safety goggles, hat, hand protection and foot protection. For particularly hazardous duties such as opening equipment which may contain hydrofluoric acid, the use of lightweight gloves, boots, jumper and airline hood are also recommended;

(d) to ensure that employees who work with hydrofluoric acid are in good health, free from chronic disease of the respiratory system and not severely handicapped;

(e) to ascertain that employees who work with the acid are fully cognisant of the hazardous nature of the product and are thoroughly trained in routine operations and for emergencies;

(f) to provide adequate, well marked safety showers and eye-wash fountains.

Treatment

The type of treatment will vary depending on the manner of exposure.

Skin contact. Flush off acid immediately, using large quantities of cool water. Remove contaminated clothing as quickly as possible while flushing. Immerse the burned part in an iced 70% solution of ethyl alcohol for 1-4 h or make a compress by inserting ice between layers of gauze; apply to burned area and keep it continually soaked with the alcohol solution for 1-4 h. If blisters form, complete debridement is necessary. Burns around the fingernail may require splitting the nail from the distal end in order to relieve pain and to facilitate drainage, prior to soaking in the solution of iced alcohol. Following soaks, dry the part gently and apply magnesium oxide-glycerine paste or vitamin A and vitamin D ointment mixture and cover with a compression dressing. Reapply ointment daily for several days, as indicated by the appearance of the burn. The treatment preferred by some practioners involves the use of an iced, saturated solution of magnesium sulphate for soaks and the intracutaneous and subcutaneous injection of a 10% solution of calcium gluconate into and around the burn. [Recently inunction of calcium gluconate gel has also proved to be effective in first aid of liquid splashes and vapour burns and to present some practical advantages over the injection treatment. Replacement of serum electrolytes in burns involving skin areas greater than 160 cm^2 is recommended. Injection of corticosteroids under and around the lesion is useful.]

Eye contact. In the case of contact with the acid or high concentrations of its vapour, flush the eyes immediately with large quantities of clean water while holding the eyelids apart. Continue flushing for 15 min. Pain can be relieved with 2 or 3 drops of a topical anaesthetic opthalmic solution. A specialist should be consulted immediately in other than minor cases of injury.

Vapour inhalation. Immediate removal of the patient to an uncontaminated atmosphere and prompt medical attention are required. Generally, patients respond satisfactorily to early unpressurised inhalation of oxygen with a respiratory-type mask. The oxygen inhalation may be discontinued after 2-4 h if there are no signs of pulmonary oedema or other respiratory distress. These patients should be kept under observation for 24-48 h. They should be kept warm and at complete rest throughout the treatment. Bronchodilators and systemic steroids may be used as required.

Ingestion. The patient should drink a large quantity of water as quickly as possible, to be followed by milk or milk of magnesia. Gastric lavage with lime water or milk may be performed by a physician only. Do not induce vomiting.

REINHARDT, C. F.

CIS 79-501 "Hydrofluoric acid skin corrosion—Review of 68 cases" (Flussäureverätzungen der Haut—Übersicht über 68 Fälle). Schmidt, C. W.; Metze, H. *Das deutsche Gesundheitswesen* (Berlin), 1978, 33/16 (761-766). Illus. 104 ref. (In German)

"Fatality due to acute systemic fluoride poisoning following a hydrofluoric acid skin burn". Tepperman, P. B. *Journal of Occupational Medicine* (Chicago), Oct. 1980, 22/10 (691-692). 7 ref.

CIS 76-1963 *Methods for the detection of toxic substances in air—Hydrogen fluoride and other inorganic fluorides.* Booklet No. 19, Health and Safety Executive, HM Factory Inspectorate (London, HM Stationery Office, 1975), 15 p. Illus. 5 ref.

CIS 78-1023 "Renal fluoride excretion as a useful parameter for monitoring hydrofluoric acid-exposed persons". Zober, A.; Geldmacher von Mallinckrodt, M.; Schaller, K. H. *International Archives of Occupational and Environmental Health—Internationales Archiv für Arbeits- und Umweltmedizin* (West Berlin), 17 Oct. 1977, 40/1 (13-24). 24 ref.

CIS 75-565 "The treatment of hydrofluoric acid burns". Browne, T. D. *Journal of the Society of Occupational Medicine* (Bristol), July 1974, 24/3 (80-89). Illus. 44 ref.

Hydrogen

Hydrogen (H)

sp.gr.	0.09
m.p.	−259.1 °C
b.p.	−252.5 °C
v.d.	0.07
e.l.	4.1-74.2%
i.t.	400 °C

slightly soluble in water
a colourless, odourless, highly flammable gas.
ACGIH simple asphyxiant

Tritium ^3_1H is a β-emitter with a lifetime of 12.2 years. DAC IAEA, 2×10^{10} Bq/m³. Under normal conditions hydrogen has a low degree of chemical activity, but at high temperatures it reacts with many elements acting as a reducing agent. When heated with alkalis, alkaline earth and certain other elements, it forms hydrides. In the presence of catalysts it may combine with fatty, unsaturated, and aromatic compounds. Hydrogen does not promote combustion, but it burns with a pale blue flame in the presence of oxygen. It ignites easily and in combination with oxygen (2 : 1) it forms a detonating gas. A mixture of hydrogen and oxygen or air explodes when an electric spark is passed through it or when in contact with an open flame. A mixture of hydrogen and chlorine explodes in light.

Occurrence. Hydrogen is one of the most widespread elements in nature. In the earth's crust (water, organic compounds, etc.) the hydrogen content is about 1% by weight. The amount of free hydrogen in air is about 1 ppm. The gas is a vital component in living organisms. Most organic compounds contain it and the active hydrogen ions (pH) are extremely important in chemical and biochemical processes. It is also involved in most metabolic reactions in the organism.

Production. Water electrolysis is the simplest process for producing hydrogen (and oxygen) and it is used when a high degree of purity is desired. However, owing to the high energy consumption, this process is now seldom used for producing large quantities of the gas. A common method involves the separation of gaseous mixtures by means of deep freezing or conversion. The most economical method is the conversion of hydrocarbon gases. The interaction of methane (or other lower paraffin hydrocarbons) with water vapour yields hydrogen according to the following reaction:

$$CH_4 + H_2O \rightleftarrows CO + 3H_2 - 48.9 \text{ kcal}$$

In order to increase the quantity of hydrogen produced, the carbon monoxide formed must then be converted:

$$CO + H_2O \rightleftarrows CO_2 + H_2 + 9.7 \text{ kcal}$$

Hydrogen is stored and transported in steel gas cylinders under a pressure of 150 atm.

Uses. Enormous quantities of hydrogen are used in the production of synthetic ammonia, in the hydrogenation of liquid fuels, in extracting liquid fuels from coal, and in the hydrogenation of plant oils in order to obtain solid fats. It is further used in many organic syntheses for reduction reactions. Oxyhydrogen flame is used for welding lead and platinum and in the processing of quartz. Hydrogen is used in welding and cutting steel, in reducing metal oxides at high temperatures and in the production of several metals which resist fusion, such as molybdenum, bismuth, etc. A considerable quantity of hydrogen goes into the production of methyl alcohol and other alcohols. In view of its very low boiling point, liquid hydrogen is widely used in producing very low temperatures.

HAZARDS

It is a physiologically inert, non-toxic gas. Gaseous hydrogen in large enough quantities can cause asphyxiation by replacing the oxygen in air. It can also exert a narcotic effect at very high pressures. The relation between the narcotic effect of nitrogen and that of hydrogen is 1 : 0.26.

The greatest danger involved in the production, use, and storage of hydrogen is the ease with which it ignites and explodes. It is extremely dangerous for even a small quantity of oxygen or air to come into contact with liquid hydrogen. Since the melting point of oxygen is higher than the temperature at which liquid hydrogen is stored, oxygen and air freeze according to different patterns as compared with liquid hydrogen and they may gradually accumulate to a quantity at which the mixture of liquid hydrogen and the solid oxygen or air explodes violently.

SAFETY AND HEALTH MEASURES

Vessels containing liquid hydrogen should be kept away from heat since the low evaporation point requires little heat in order to emit a considerable quantity of gas. The empty portions of vessels should be filled with an inert gas before the hydrogen is introduced, since any remaining oxygen or air may cause an explosion. Hydrogen cylinders should never be stored together with flammable materials. Large hydrogen tanks must be separated by a safe distance from each other and from dwellings, as is determined by the size of the tanks. The measures for preventing fires and explosion can be summarised as follows:

(a) strict observance of technical standards;

(b) ensuring that apparatus is airtight and that leaks are repaired;

(c) installation of water seals, safety valves, regulators, etc., in hydrogen plants;

(d) provision of ventilation in premises where hydrogen is produced or used;

(e) use of inert gases in hydrogen plants;

(f) lightning protection of apparatus used in the production or storage of hydrogen;

(g) disconnecting hydrogen containers from the gas supply network prior to maintenance work;

(h) filling hydrogen vessels to a specified limit;

(i) observance of temperature regulations, heat insulation of cylinders, and protection from the sun;

(j) protection of cylinders from jolting;

(k) maintaining a pressure of not more than 150 atm in cylinders;

(l) labelling of containers.

GADASKINA, I. D.

CIS 76-1026 *Review of hydrogen accidents and incidents in NASA operations.* Ordin, P. M. National Aeronautics and Space Administration, Lewis Research Center, Cleveland, Ohio. N74-28457/1/1WI (National Technical Information Service, Springfield, Virginia 22151) (1974), 41 p.

CIS 77-922 *The explosion at Laporte Industries Ltd., Ilford, 5 April 1975.* Health and Safety Executive (London, HM Stationery Office, 1976), 24 p. Illus.

Hydrogen in steelworks and safety. The use of hydrogen in the steel industry. Production, storage and distribution of hydrogen. Safety problems. Decker, A.; Arragon, Ph. Report EUR 6200 (Luxembourg, Commission of the European Communities, 1979), 128 p.

CIS 75-314 "Explosion hazards of hydrogen-chlorine mixtures" (Vzryvoopasnost' smesej vodoroda s hlorom). Antonov, V. N.; Frolov, Ju. E.; Rozlovskij, A. I.; Mal'ceva,

A. S. *Himičeskaja promyšlennost'* (Moscow), Mar. 1974, 3 (45-48). Illus. 15 ref. (In Russian)

Behaviour of tritium in the environment. Proceedings of a Symposium, San Francisco 16-20 Oct. 1978 (Vienna, International Atomic Energy Agency, 1979), 711 p. Illus. Ref.

Hydrogen peroxide

Hydrogen peroxide (H_2O_2)
HYDROGEN DIOXIDE

Hydrogen peroxide has numerous uses, most of which derive from its properties as a strong oxidising or bleaching agent. It is also used as a reagent in the synthesis of various chemical compounds.

Hydrogen peroxide is commercially available in aqueous solutions, usually 35%, 50% (industrial strength), 70% and 90% (high strength), by weight, but also available in 3%, 6%, 27.5% and 30% solutions. It is also sold by "volume strength" (meaning the amount of oxygen gas which will be liberated per millilitre of solution). The following table correlates the two standards:

% by weight	Volume strength
3.0	10
6.0	20
27.5	100
30.0	110
35.0	132
50.0	208
70.0	264
90.0	415

m.w. 34.01

	Concentration H_2O_2 (% by weight)			
	35%	50%	70%	90%
sp.gr.	1.13	1.20	1.29	1.39
b.p. (oc)	108	114	125	141
fr.p. (oc)	−33	−52	−39	−11
v.p. (mmHg at 30 °C)	0.36	0.74	1.5	2.50

not flammable but a powerful oxidiser

a clear, colourless liquid with a pungent odour, miscible with water in all proportions.

TWA OSHA 1 ppm 1.5 mg/m^3
STEL ACGIH 2 ppm 3 mg/m^3
IDLH 75 ppm

Hydrogen peroxide is stabilised during manufacture to prevent contamination by metals and other impurities, although, if excessive contamination occurs, the additive cannot inhibit decomposition.

Decomposition occurs at a slow rate even when the compound is inhibited and thus it must be stored properly and in vented containers. High-strength hydrogen peroxide is a very high-energy material. When it decomposes to oxygen and water, large amounts of heat are liberated leading to an increased rate of decomposition, since decomposition is accelerated by increases in temperature. This rate increases about 2.2 times per 10 °C temperature increase between 20 to 100 °C. Although pure hydrogen peroxide solutions are not usually explosive at atmospheric pressure, equilibrium vapour concentrations of hydrogen peroxide above 26 mol% (40 weight %) become explosive in a temperature range below the boiling point of the liquid.

Since the compound is such a strong oxidiser, when spilled on combustible materials it can set fire to them. Detonation can occur if the peroxide is mixed with incompatible (most) organic compounds.

Solutions less than 45% concentration expand during freezing; those greater than 65% contract.

Production. The most important process for producing hydrogen peroxide involves the reduction (hydrogenation) of alkyl anthraquinones to corresponding anthrahydroquinone, then oxidation to produce hydrogen peroxide and the original alkyl anthraquinone, which is recycled. Another method of manufacture used by small plants throughout the world is production by electrolysis of acidic sulphate solutions. This yields persulphates, which are then hydrolysed, producing hydrogen peroxide and the starting sulphate. The use of acidic ammonium sulphate solutions is preferred. One company in the United States uses isopropyl alcohol and oxygen to yield acetone and hydrogen peroxide. This process was discontinued in mid-1980. Other processes not used in the United States are: reaction of mineral acids with barium peroxide, hydrolysis of organic peracids, autoxidation of various hydrocarbons, and direct combination of hydrogen and oxygen.

Uses. Various grades of hydrogen peroxides have different uses: 3% and 6% solutions are used for medicinal and cosmetic purposes; the 27.5% solution was once used for commercial purposes; the 30% solution is used for laboratory reagent purposes, the 35% and 50% solutions for most industrial applications, the 70% solution for some organic oxidation uses and the 90% solution for some industrial uses and as a propellant for military and space programmes. Solutions of over 90% are used for specialised military purposes.

The largest market for hydrogen peroxide is in the production of chemicals: inorganic peroxygen compounds, glycerin, plasticisers, organic peroxygen compounds for bleaching agents, pharmaceuticals and cosmetics and drying agents for fats, oils, and waxes, and amine oxides for home dish-washing detergents. It is used in the textile industry for bleaching textiles, particularly cotton, and in the pulp and paper industry for bleaching of mechanical wood pulps. A new and fast growing use for hydrogen peroxide is waste-water treatment. This has particular value, as it can reduce the use of chlorine with its potential for formation of toxic residues. It degrades some compounds which are resistant to other water treatment compounds. Its major asset is the ability to convert hydrogen sulphide to elemental sulphur and water. This property also enables hydrogen peroxide to be used to remove hydrogen sulphide from steam at geothermal power plants. This is a new application of hydrogen peroxide which may expand as the use of geothermal power increases. In mining, hydrogen peroxide is used to make uranium soluble in the leaching solution. It is used for metal etching/oxidising in the electronics industry and for treating metal surfaces in other applications. Other uses are related primarily to its oxidising and bleaching abilities. These include uses in cosmetics, and as an antiseptic and disinfectant; processing of food and wine; rocket fuel, gas, oil, and wood bleaching; and production of some chemicals not already mentioned.

SAFETY AND HEALTH HAZARDS

Hazards include the possibility of explosion when contaminated or mixed with organic compounds, rapid and violent decomposition if contaminated by metallic ions or salts, fire if rapid decomposition takes place near combustible materials, and severe irritation of skin, eyes, and mucous membranes. Hydrogen peroxide solutions in concentrations greater than 8% are classified as corrosive liquids.

Fire and explosion. Hydrogen peroxide is not itself flammable but can cause spontaneous combustion of flammable materials and continued support of the combustion because it liberates oxygen as it decomposes.

It is not considered to be an explosive; however, when mixed with organic chemicals, hazardous impact-sensitive compounds may result. Materials with metal catalysts can cause explosive decomposition.

Decomposition. Contamination of hydrogen peroxide by such metals as copper, cobalt, manganese, chromium, nickel, iron, lead, and their salts, by dust, dirt, oils, various enzymes, rust and undistilled water results in increased rate of decomposition. Decomposition results in the liberation of oxygen and heat. If the solution is dilute, the heat is readily absorbed by the water present. In more concentrated solutions the heat increases the temperature of the solution and its decomposition rate. This may lead to an explosion. Contamination with materials containing metal catalysts can result in immediate decomposition and explosive rupture of the container if it is not properly vented.

Health hazards. Human exposure by inhalation may result in extreme irritation and inflammation of nose, throat and respiratory tract; pulmonary oedema, headache, dizziness, nausea, vomiting, diarrhoea, irritability, insomnia, hyper-reflexia; tremors and numbness of extremities, convulsions, unconsciousness and shock. The latter symptoms are a result of severe systemic poisoning.

Exposure to mist or spray may cause stinging and tearing of the eyes. If splashed into the eye, severe damage such as ulceration of the cornea may result; sometimes, though rarely, this may appear as long as a week after exposure.

Skin contact with hydrogen peroxide liquid will result in temporary whitening of the skin; if the contamination is not removed, erythema and vesicle formation may occur.

Although ingestion is unlikely to occur, if it does the hydrogen peroxide will cause irritation of the upper gastrointestinal tract. Decomposition results in rapid liberation of O_2, leading to distension of the oesophagus or stomach, and possibly severe damage and internal bleeding.

The ACGIH TLV of 1 ppm (1.4 mg/m^3) is based on studies by Oberst et al. in which dogs were exposed 6 h per day, 5 days per week to a vapour concentration of 7 ppm of 90% hydrogen peroxide. The dogs developed sneezing, tearing of the eyes, bleaching of hair and irritation of their skin. There was thickening of the skin, but no hair follicle destruction at autopsy. Lungs were also irritated, but no other abnormal effects were noted.

The production of hydrogen peroxide via ammonium peroxidisulphate involves a risk of bronchial and skin sensitisation.

SAFETY AND HEALTH PRECAUTIONS

Equipment for handling hydrogen peroxide solutions must be thoroughly cleaned and passivated before use to prevent contamination. Storage tanks should be constructed of high-purity aluminium alloy. Undesirable contaminants must not become imbedded in the aluminium surface, and extreme care should be taken to ensure proper welding techniques. Hydrogen peroxide should be stored only in original containers or in containers of compatible materials which have been properly designed and thoroughly passivated. Materials which are satisfactory for use with hydrogen peroxide include: glass, porcelain, vitreous stoneware, Teflon, polystyrene, polyethylene, Tygon 3400, Koroseal 700, properly cleaned pure block tin, and properly passivated 99.6% pure aluminium and aluminium alloys and stainless steel. Once removed from the original container, the hydrogen peroxide must not be returned to it. Storage containers should be properly vented and kept away from direct heat and sun and combustible materials. Adequate ventilation is necessary as is an ample water supply for washing away spills. Fires caused by the compound are best controlled by large amounts of water. Chemical extinguishers should not be used as they hasten decomposition of the peroxide. Fire fighters should wear goggles and self-contained breathing apparatus.

Personnel using hydrogen peroxide should be thoroughly instructed in its hazards, carefully instructed in the proper safety measures for handling it, and the understanding that copious quantities of water should be used for spills or personal contact. Special emphasis should be placed on prevention of contamination of the peroxide by foreign materials. Storage areas should have posters describing hazards and precautions.

Proper protective equipment, safety goggles, gloves of polychloroprene rubber, polyvinyl chloride, polyethylene, etc., should be used. Aprons and foot and leg covering of similar materials should be used when working with concentrated solutions.

Medical control and treatment. Preplacement screening for personnel who are going to work with hydrogen peroxide should emphasise detection of any history of chronic respiratory, skin, or eye disease, as such persons may be at increased risk from exposure.

Appropriate treatment for exposure to hydrogen peroxide is removal from the source to fresh air and immediate flushing of eyes and skin with water. If inhalation has been prolonged, a physician should be summoned. If clothes are contaminated they should be removed and washed thoroughly with water. If the compound is allowed to dry in the fabric, there may be a fire hazard, particularly if the clothing is also soiled. If the hydrogen peroxide is accidently ingested, vomiting and belching should be induced. Give lukewarm water freely and call a physician.

WOODBURY, C. M.

Industrial waste treatment. FMC reports on hydrogen peroxide as an effective effluent-treatment agent (FMC Chemicals, FMC Corporation, Industrial Chemical Division, 633 Third Avenue, New York).

Chemical Economics Handbook. SRI International, CEH Product Review, Hydrogen Peroxide. Davenport, R. E.; Kamatari, O.; Wilson, E., 1978.

Chemical hazards of the workplace. Procter, N. H.; Hughes, J. P. (Philadelphia and Toronto, J. B. Lippincott Co., 1978), 533 p.

"Inhalation toxicity of ninety percent hydrogen peroxide vapor". Oberst, F. W.; Comstock, C. C.; Hackley, E. B. *AMA Archives of Industrial Hygiene and Occupational Medicine* (Chicago), Oct. 1954, 10/4.

Toxic and hazardous industrial chemicals safety manual for handling and disposal, with toxicity and hazard data (Tokyo, International Technical Information Institute, 1979).

CIS 78-1362 "Comparison of inhalation toxicity and dermal toxicity of hydrogen peroxide" (O sravnitel'noj toksičnosti parov perekisi vodoroda pri ingalacionnom i kŏnom putjah vozdejstvija). Kondrašov, V. A. *Gigiena truda i profession-al'nye zabolevanija* (Moscow), Oct. 1977, 10 (22-25). 16 ref. (In Russian)

Hydrogen sulphide

Hydrogen sulphide (H_2S)

SULPHURETTED HYDROGEN

m.w.	34
d.	1.54
m.p.	−85.5 °C
b.p.	−60.7 °C
v.d.	1.19
v.p.	20 atm (2026·10³ Pa) at 20 °C
e.l.	4.3-46%
i.t.	260 °C

soluble in water, ethyl alcohol, gasoline, kerosene and crude oil

a colourless gas with a characteristic rotten-egg odour.

TWA OSHA	20 ppm ceil
	50 ppm/10 min peak
NIOSH	10 ppm/10 min peak
TLV ACGIH	10 ppm 15 mg/m³
STEL ACGIH	15 ppm 27 mg/m³
IDLH	300 ppm
MAC USSR	10 mg/m³

Production. It is produced by reacting iron sulphide with dilute sulphuric or hydrochloric acid or by reacting hydrogen with vaporised sulphur. It is also released during the decay of sulphur-containing organic materials and, as such, may be encountered in mines (where it is often called "stink damp"), gas wells, sewers, etc. It occurs as the by-product of many chemical processes, particularly those involving viscose rayon, synthetic rubbers, petroleum products, dyes and leather, and in the processing of sugar.

Uses. Hydrogen sulphide is used for the production of various inorganic sulphides, sulphuric acid and organic sulphur compounds such as thiophene and mercaptans. It is widely used as an analytical reagent. In agriculture it is used as a disinfectant.

HAZARDS

Fire and explosion. Hydrogen sulphide is a flammable gas which burns with a blue flame giving rise to sulphur dioxide, a highly irritating gas with a characteristic odour. Mixtures of hydrogen sulphide and air in the explosive range may explode violently; since the vapours are heavier than air, they may accumulate in depressions or spread over the ground to a source of ignition and flash back.

When exposed to heat, it decomposes to hydrogen and sulphur and, when in contact with oxidising agents such as nitric acid, chlorine trifluoride, etc., it may react violently and ignite spontaneously.

Extinguishing agents recommended for the fighting of hydrogen sulphide fires include carbon dioxide, chemical dry powder and water sprays.

Health hazards. Even at low concentrations, hydrogen sulphide has an irritant action on the eyes and respiratory tract. Intoxication may be hyperacute, acute, subacute or chronic.

Low concentrations are readily detected by the characteristic rotten-egg odour; however, prolonged exposure dulls the sense of smell and makes the odour a very unreliable means of warning. High concentrations can rapidly deaden the sense of smell. Hydrogen sulphide enters the body through the respiratory system and is rapidly oxidised to form compounds of low toxicity; there are no accumulation phenomena, and elimination occurs through the intestine, urine and the expired air.

In cases of slight poisoning, following exposure to from 10 to 500 ppm, a headache may last several hours, pains in the legs may be felt and rarely there may be loss of consciousness. In moderate poisoning (from 500 to 700 ppm) there will be loss of consciousness lasting a few minutes, but no respiratory difficulty. In cases of severe poisoning the subject drops into a profound coma with dyspnoea, polypnoea and a slate-blue cyanosis until breathing restarts, tachycardia and tonic-clonic spasms.

Inhalation of massive quantities of hydrogen sulphide will rapidly produce anoxia resulting in death by asphyxia; epileptiform convulsions may occur and the individual falls apparently unconscious, and may die without moving again. This is a syndrome characteristic of hydrogen sulphide poisoning in sewermen; however, in such cases, exposure is often due to a mixture of gases including methane, nitrogen, carbon dioxide and ammonia.

In subacute poisoning, the signs may be nausea, stomach distress, foetid eructations, characteristic "rotten-egg" breath, and diarrhoea. These digestive-system disorders may be accompanied by balance disorders, vertigo, dryness and irritation of the nose and throat with viscous and mucopurulent expectoration and diffuse rales and ronchi.

There have been reports of retrosternal pain similar to that found in *angina pectoris* and the electrocardiogram may show the characteristic trace of myocardial infarction, which, however, disappears quite rapidly. The eyes are affected by palpebral oedema, bulbar conjunctivitis, and mucopurulent secretion with, perhaps, a reduction in visual acuity—all of these lesions usually being bilateral. This syndrome is known to sugar and sewer workers as "gas eye". It has been found in the viscose rayon industry that the severity of this keratoconjunctivitis is directly related to the atmospheric concentration of hydrogen sulphide and not to that of carbon disulphide which lowers the danger threshold level of the former. Experimental studies have shown that the atmospheric concentration of hydrogen sulphide must be kept below 10 mg/m³ if keratoconjunctivitis is to be prevented.

The existence of chronic hydrogen sulphide intoxication is denied by certain authorities, whereas others claim that this disease is characterised by headaches, asthenia, eye disorders, chronic bronchitis and a grey-green line on the gums; as in acute poisoning, the ocular lesions are said to predominate. Reports of nervous system disorders including paralysis, meningitis, polyneuritis and even psychic troubles have also been made.

In rats, exposure to hydrogen sulphide has given rise to teratogenic effects.

Metabolism and pathology. Hydrogen sulphide has a general toxic action. It inhibits Warburg's respiratory enzyme (cytochrome oxidase) by binding iron, and the oxydo-reduction processes are also blocked. This inhibition of enzymes essential for cellular respiration may be fatal. The substance has a local irritant action on the mucous membranes since, on contact with moisture, it forms caustic sulphides; this may also occur in the lung parenchyma as a result of combination with tissue alkalis. Experimental research has shown that these sulphides may enter into the circulation, producing respiratory effects such as polypnoea, bradycardia and hypertension, by their action on the vasosensitive, reflexogenic zones of the carotid nerves and Hering's nerve.

Post mortem examination in a number of cases of hyperacute poisoning has revealed pulmonary oedema and congestion of various organs. A characteristic autopsy feature is the odour of hydrogen sulphide that emanates from the dissected corpse. Other features of note are the haemorrhages of the gastric mucosae, the greenish colour of the upper regions of the intestine and even of the brain.

Detection and analysis. A prime indicator of hydrogen sulphide presence in the air is the characteristic odour, which can be detected at even minute concentrations.

A variety of methods exist for detecting hydrogen sulphide in the atmosphere, including the use of detector papers impregnated with lead sodium acetate and detector tubes, and various other reagents such as cadmium sulphide, lead sulphide and *p*-dimethylphenylenediamine. A very sensitive method (1 ppb) with sampling on silica gel and spectrophotometric analysis has recently been described.

SAFETY AND HEALTH MEASURES

Technical prevention. Persons working on processes in which hydrogen sulphide is either employed or given off as a by-product should be informed of the dangers of this substance. Processes should be enclosed and exhaust ventilation applied to possible escape areas. The atmospheric concentration of hydrogen sulphide around these processes should be monitored.

Where it is necessary to enter a confined space that may contain hydrogen sulphide (such as a process plant or sewer), the space should be purged and the hydrogen sulphide concentration determined before entry and at frequent intervals during the course of the work; under no circumstances should reliance be placed on the sense of smell to detect the presence of the gas.

Where the presence of hydrogen sulphide has been detected, the worker entering the confined space should wear suitable respiratory protective equipment of a self-contained or airline type, a safety belt and lifeline, and should be observed from the outside by a responsible worker. Hydrogen sulphide may be dissolved or trapped in sludge in sewers or process vessels and will be released into the atmosphere during sludge agitation. Workers exposed to hydrogen sulphide should also wear chemical safety goggles.

It has been suggested that calcium chloride or a mixture of ferrous sulphate and lime should be added to process washing water as a neutralising agent each time the development of hydrogen sulphide occurs.

Hydrogen sulphide cylinders should be stored in a well ventilated, fire-resistant structure, protected from the weather. Smoking and naked flames should be prohibited in areas where hydrogen sulphide is stored or used and electrical equipment should be of the flameproof type. During transport the cylinders should be suitably restrained and should bear an appropriate warning label.

Medical prevention. Persons required to work in areas where hydrogen sulphide may be encountered should receive a pre-employment medical examination. Persons with eye and nervous disorders, in particular, should not be assigned to work entailing exposure to hydrogen sulphide; the pre-employment examination should be backed up by periodic examinations (preferably at intervals of 6 months). In Italy these examinations are compulsory for oil refinery workers, viscose spinners and sewermen.

Diseases caused by hydrogen sulphide are recognised as occupational diseases in Czechoslovakia, Finland, the Federal Republic of Germany, Italy, Japan, Mexico, Spain, Switzerland, etc.

Treatment. In the event of acute poisoning, the victim should be removed from exposure and transported to the nearest resuscitation centre for hyperbaric oxygen treatment. When no such facilities are available it is helpful to carry out artificial respiration with inhalation of oxygen. The respiratory centre may be stimulated by injections of lobelin and nikethamide (1 cm^3 and 5 cm^3 respectively). Vitamin C may be injected intravenously.

Eye exposure should be treated with boric acid solution or isotonic physiological solutions; instillation of a drop of olive oil has also been recommended as an immediate measure. For the more serious cases, recourse may be had to 1% adrenalin solution drops and the application of hot or cold compresses.

CACCURI, S.

Occupational health:

CIS 77-1371 *Criteria for a recommended standard–Occupational exposure to hydrogen sulfide.* DHEW (NIOSH) publication No. 77-158 (National Institute for Occupational Safety and Health, 4676 Columbia Parkway, Cincinnati) (May 1977), 149 p. 171 ref.

"Hydrogen sulphide poisoning. Medical emergency at the enterprise" (Intoxications par hydrogène sulfuré. Urgence médicale dans l'entreprise). Demaret, D.; Fialaire, J. *Archives des maladies professionnelles, de médecine du travail et de sécurité sociale* (Paris), Dec. 1978, 39/12 (761-767). (In French)

Safety:

CIS 78-753 *Hydrogen sulfide.* Data Sheet 284, Revision A (National Safety Council, 425 North Michigan Avenue, Chicago 60611) (1977), 4 p. Illus. 7 ref.

Detection and analysis:

CIS 77-1350 *An evaluation of portable, direct-reading H_2S meters.* Thompkins, F. C.; Becker, J. H. DHEW (NIOSH) publication No. 77-137 (National Institute for Occupational Safety and Health, 4676 Columbia Parkway, Cincinnati) (July 1976), 167 p. Illus.

CIS 79-1319 "Hazards in the work environment–Hydrogen sulfide". Waernbaum, G.; Wallin, I. *Scandinavian Journal of Work, Environment and Health* (Helsinki), Mar. 1979, 5/1 (31-34). 3 ref.

CIS 80-1008 *H_2S safety handbook* (Safety Oilfield Services, PO Box 52722, Lafayette, Louisiana 70505) (1978), 32 p. Illus.

Hydroxylamine

Hydroxylamine (NH_2OH)

OXAMMONIUM

m.w.	33
sp.gr.	1.20
m.p.	33.0 °C
b.p.	56.5 °C
v.p.	10 mmHg (1.33·10³ Pa) at 47.2 °C
f.p.	129.4 °C (explodes)

soluble in water, ethyl alcohol and acids

very hygroscopic white crystals or colourless liquid with alkaline reaction; in alkaline medium, hydroxylamine is a powerful reducing agent; in an acid medium, it acts as an oxidising agent.

Production. Mainly by electrolytic reduction of ammonium chloride and alkaline decomposition of the resultant hydroxylamine hydrochloride, or by reducing sodium nitrite with sodium bisulphate and sulphur dioxide in solution. Hydroxylamine salts are obtained by direct reduction of NO with hydrogen in the presence of a platinum catalyst.

Uses. Hydroxylamine and its salts (sulphate and hydrochloride) are used in the manufacture of caprolactam and in many of the organic syntheses of interest to the pharmaceutical and perfume industry, rubber, chemicals, dyes and dyestuffs, as strong reducing agents, photographic developers, analytical reagents and in the preparation of oximes by reaction with aldehydes and ketones. Hydroxylamine sulphate is also used to purify aldehydes and ketones. Dilute hydroxylamine solutions are used for the treatment of skin diseases (lupus, sycosis).

HAZARDS

Hydroxylamine and its derivatives are toxic. They inhibit the nervous system of animals, and high concentrations cause spasms, paralysis and death by respiratory standstill. Hydroxylamine forms methaemoglobin. A single application of 0.02 ml of a 10% hydroxylamine solution to the skin of guinea-pigs increases the permeability of the blood vessels, and repeated application of a 50% solution for 5-8 days gives rise to considerable hyperaemia; later, sensitisation to hydroxylamine develops. In workers, increases of the methaemoglobin content of the blood by up to 25% have been reported at the end of the working day. Its level depends on concentration and duration of exposure and is particularly high after operations involving the release of considerable hydroxylamine concentrations such as the loading and unloading of crystallisers and centrifuges and the packaging of the finished product. When hydroxylamine chloride comes in contact with the skin, it causes irritation accompanied by itching, which intensifies at night. Allergic dermatitis of the erythematous-squamous type and eczema may develop after prolonged exposure to hydroxylamine chloride. The main body regions affected are the trunk and the lower third of the forearms. These disorders appear 1-2 weeks after starting work in the plant, but sometimes also after longer periods (up to 3 to 5 years). Hydroxylamine is a powerful skin sensitiser. In the body it is decomposed into ammonia and nitrogen.

The major hazards of hydroxylamine and its salts are, however, fire and explosion.

PREVENTIVE MEASURES

The material should be handled with caution to avoid contact, and complete elimination of manual processes by mechanisation is recommended. It should be stored in a cool well ventilated place away from fire hazards and heat sources. In the event of a fire, workers should immediately evacuate the premises in view of the explosive nature of the substance, and fire fighting should be undertaken from a safe distance.

Sources of gas and dust must be provided with local exhaust ventilation. All measures of personal hygiene must be scrupulously observed, i.e. showers after work, change of protective clothing, etc. Wearing protective equipment such as respirators and skin protection is advisable.

During the pre-employment medical examination, workers should be subjected to skin sensitivity tests and, during periodic medical examinations, persons found to be suffering from persistent skin disease should be removed from exposure. The use of desensitising drugs, for 2 weeks twice a year, has been suggested to prevent sensitisation.

SYROVADKO, O. N.

CIS 78-617 "Hydroxylamine salts (Hydrochloride and sulphate)". H66, Information sheets on hazardous materials. Fire Prevention Association. *Fire Prevention* (London), Feb. 1978, 123 (49-50).

CIS 77-141 "Hazardous chemical reactions–42. Hydrazine–Hydroxylamine" (Réactions chimiques dangereuses–42. Hydrazine–Hydroxylamine). Leleu, J. *Cahiers de notes documentaires–Sécurité et hygiène du travail* (Paris), 3rd quarter 1976, 84, Note No. 1025-84-76 (429-430). (In French)

Hygiene, personal

Health and well-being at work depend on an interaction between the physical environment and the personal habits of the employee. The most scrupulous personal hygiene will be of little avail against an environment contaminated with dangerous dust or fumes; but the benefits of a good environment can be frustrated by lack of personal responsibility on the part of either management or workers. It is difficult to practise personal hygiene unless adequate sanitary facilities are provided but the best fittings may be neglected or misused. Indeed the cultivation of good standards of personal hygiene often involves human relations, personnel management and joint consultation. (The importance of personal hygiene in a comprehensive programme of occupational health and safety is also emphasised under CONTROL TECHNOLOGY.)

Personal hygiene is always necessary; it is essential for workers exposed to toxic compounds, especially those which may be absorbed through the skin, to substances which are dangerous, allergenic or radioactive, to high temperatures, and to dirty work or the risk of infection.

Eating and drinking

In general, drinking water must comply with public health standards; fountains with an upward jet avoid the need for cups, but in any event communal cups should be avoided; disposable cups or individual cups should be provided. Drinking water should always be clearly indicated. Serious injuries have occurred through drinking contaminated water or other colourless fluids by mistake.

Where dangerous materials are used or where the conditions of manufacture demand, eating or drinking should be prohibited in workrooms. The corollary of prohibition is that messrooms and canteens or other catering facilities should be provided. A canteen serving wholesome food can also be a valuable means of supplementing the workers' diet and can play its part in health education. Especially where toxic materials are handled, there should be arrangements for washing and for deposit of working overalls, etc., outside the messroom or canteen. Hot and cold beverages may be provided outside the main meal breaks but again away from the workplace where toxic risks exist (see CATERING, INDUSTRIAL; BEVERAGES AT WORK).

Smoking

Control of smoking may involve difficult questions of personnel management and supervision; it is usually as well to recognise the habit and provide opportunities for occasional smoking outside the working area; lack of such opportunities may lead to illicit smoking and abuse of ablution blocks and sanitary conveniences. Workers must be informed of the hazards of cigarette smoking, in particular if consumption exceeds 15-20 cigarettes per day, both in everyday life and in occupational exposure to carcinogens such as asbestos, aromatic amines, and alkylating agents, and to irritant substances.

Skin hygiene

The skin, being the most exposed part of the body, is most vulnerable. While the injured skin may be the focus of entry for various micro-organisms, intact skin itself could give access to various industrial systemic poisons into the body. Chemical agents are the most common causes of industrial dermatitis. Biological agents may cause bacterial, fungal or parasitic infections of the skin. Great emphasis has to be given to the care of the skin. Cleanliness of the skin is of prime importance. This must be ensured by daily baths, and regular washing of the exposed parts of the body (see SANITARY FACILITIES).

Oral hygiene

Dusts, fumes and gases of lead, mercury, chromium, etc., are liable to enter the mouth and nose of workers exposed to them, resulting in irritation, inflammation or even cancer. Dental erosions occur from acids or by mechanical means among glass blowers, etc. Dental caries is a common occurrence. Daily cleaning of the teeth and mouth is necessary to prevent teeth from decaying. Factory workers exposed to dust, fumes, gases and chemicals should wear protective masks to prevent their entry into the mouth. After work, rinsing the mouth and gargling will remove chemicals that are likely to enter the mouth. Periodic check-up of the teeth by a dentist will help in the early detection of tooth decay.

Working clothes

Proper clothing plays an important role in protecting the health of the industrial worker. In selecting the type of clothing required, special consideration should be given to the conditions under which the wearer will work, and the effect it will have on the worker's efficiency.

Arrangements should be made to launder working clothes, towels, etc. Where a large number of clothes, particularly contaminated clothes, require cleaning, the factory should have a laundry of its own, on the premises, to ensure that clothes contaminated with toxic substances are detoxified before laundering. Changing rooms with separate accommodation for working clothes and outdoor clothing are necessary. Where dangerous dusts are present, e.g. silica in pottery manufacture, shaking and brushing of overalls should be prohibited.

Hygiene education

Where housing is provided for workers, as on plantations, there is a further responsibility and opportunity to provide conditions conducive to personal hygiene. Even in advanced countries, good habits acquired in the workplace may be jeopardised by lack of facilities in the home; similarly workers coming from bad or primitive home surroundings may find it difficult at first to acquire good habits.

In addition to the provision of facilities, it is part of the duty of the employer to see that workers make good use of them; the workplace has often an essential educational role to play. It is to be hoped that most young entrants will have learned the elements of personal hygiene at home or at school but they will need to have their attention directed to special needs in their occupations. Much can be done by an industrial medical officer or works nurse to instruct on risks and advise on personal behaviour; however, tact and sympathy are necessary. Simple instruction in the particular risks of any occupation and the appropriate preventive measures are needed for new workers of any age; inculcation of cleanliness, including care of the hair and teeth, cleaning in folds in skin and under nails, etc., and discouragement of offensive habits, such as spitting, are particularly important for young workers. Placards and posters may be useful but it is all important that confidence should be gained so that advice is sought immediately, and irrespective of any periodical medical examination, if any suspicious symptoms develop.

Beyond all preventive work, much can be done by positive health instruction to encourage a healthy use of leisure and rest. Sedentary workers, in particular, should participate in compensatory physical activities in line with their age and physical condition; physical exercise is valuable in the prevention of cardiovascular and metabolic disease and in the stimulation of mental health (see PHYSICAL TRAINING).

DE SOUZA, L. S.

Itinerant workers:

CIS 76-566 *Notification No. 11/1975 concerning hygiene and welfare arrangements for itinerant workers* (Meddelelse nr. 11/1975 om sundheds- og velfaerdsforanstaltninger for arbejdere uden fast arbejdssted). M 11/1975 (Direcktoratet for Arbejdstilsynet, Rosenvaengets Allé 16-18, 2100 København) (27 Aug. 1975), 4 p. (In Danish)

Hypoxia and anoxia

Where the oxygen supply is unable to meet tissue demand, or where the cells are unable to utilise the available oxygen, a state of hypoxia or anoxia will develop. The older term "anoxia" (a state of deficiency of oxygen in the tissues) and the newer term "hypoxia" (a state of decreased oxygen supply to the tissues) can normally be considered as synonyms; however, "anoxia" should be applied only to a serious deficit of tissue oxygen, and hypoxia to conditions which are better tolerated. In the following discussion the term "hypoxia" will be given preference. The term "hypoxaemia" (anoxaemia), which is used to describe a condition in which there is an oxygen deficiency in the blood, is not a synonym of hypoxia, rather it refers to the type of hypoxia of prime importance from the occupational safety and health point of view.

Numerous factors, either separately or in various combinations, may lead to hypoxia. Low oxygen content of the inspired air or gas mixture, diseases of the respiratory organ or failure of the respiratory functions may result in disturbances of oxygen uptake. Oxygen transport may be impeded by central and peripheral circulatory disturbances or by limited performance of the circulatory system, as well as by decreased carrier capacity due to changes in the composition and/or oxygen-binding ability of the blood. Factors influencing cellular metabolism may bring about failures of oxygen utilisation.

Factors such as physical exertion, or sex and age may also contribute to the development of hypoxia. Whatever the mechanism by which hypoxia develops, it is the decrease or cessation of tissue oxidation alone that produces all the acute and chronic consequences. Nevertheless, from the point of view of aetiology, diagnosis, therapy and prophylaxis, it has proved useful to classify hypoxia according to the mode of development. The description of the following categories in this classification deals purely with the occupational health aspects (see also ASPHYXIA).

Hypoxaemia

This is a type of hypoxia in which the oxygen carried by the arterial blood is below the normal range.

Hypotonic hypoxaemia. This results from a reduction of both the arterial oxygen concentration and the arterial oxygen tension and may occur due to any of the following conditions.

(a) Inspiration of air or a gas mixture with a subnormal oxygen partial pressure such as when the oxygen content has been reduced below 17% by contamination with a simple asphyxiant.

(b) Bronchial obstruction or the occurrence of pulmonary processes that impede oxygen uptake. Insufficient oxygen may reach the alveoli as a consequence of inadequate ventilation resulting from respiratory paralysis (e.g. in poisoning), insufficient respiratory movements (thoracic trauma), or obstruction in the respiratory tract (glottal spasm or oedema, or bronchial spasm). Alveolo-capillary oxygen diffusion may be impeded by a decrease in the respiratory surface, or by thickening of the aveolar capillary membrane. This first mechanism may play a role in atelectasis, emphysema, silicosis, and talcosis, rarely in other pneumoconioses; the latter mechanism may intervene in pulmonary oedema (e.g. phosgene, nitrogen oxide intoxication, etc.), asbestosis, beryllosis, pulmonary fibrosis and diffuse pulmonary disease.

(c) Impaired pulmonary circulation in which alveolar perfusion decreases, and the arterialisation of the blood is impeded by dynamic factors. This form may occur due to cardiac decompensation in the pulmonary circulation.

(d) The admixture of venous blood due to congenital cardiac defects of various forms.

Isotonic hypoxaemia. This is characterised by a decrease in the amount of oxygen carried by the arterial blood as the result of anaemia (in which the blood haemoglobin concentration is subnormal) or of impairment of the oxygen-combining capacity of haemoglobin due to intoxication by substances which combine with haemoglobin to form carboxyhaemoglobin (carbon monoxide) or methaemoglobin (aniline, etc.).

Hypokinetic hypoxia

This form is brought about by local or generalised insufficiency of the systemic circulation.

Figure 1. A diagrammatic survey of the factors contributing to the development of hypoxia.

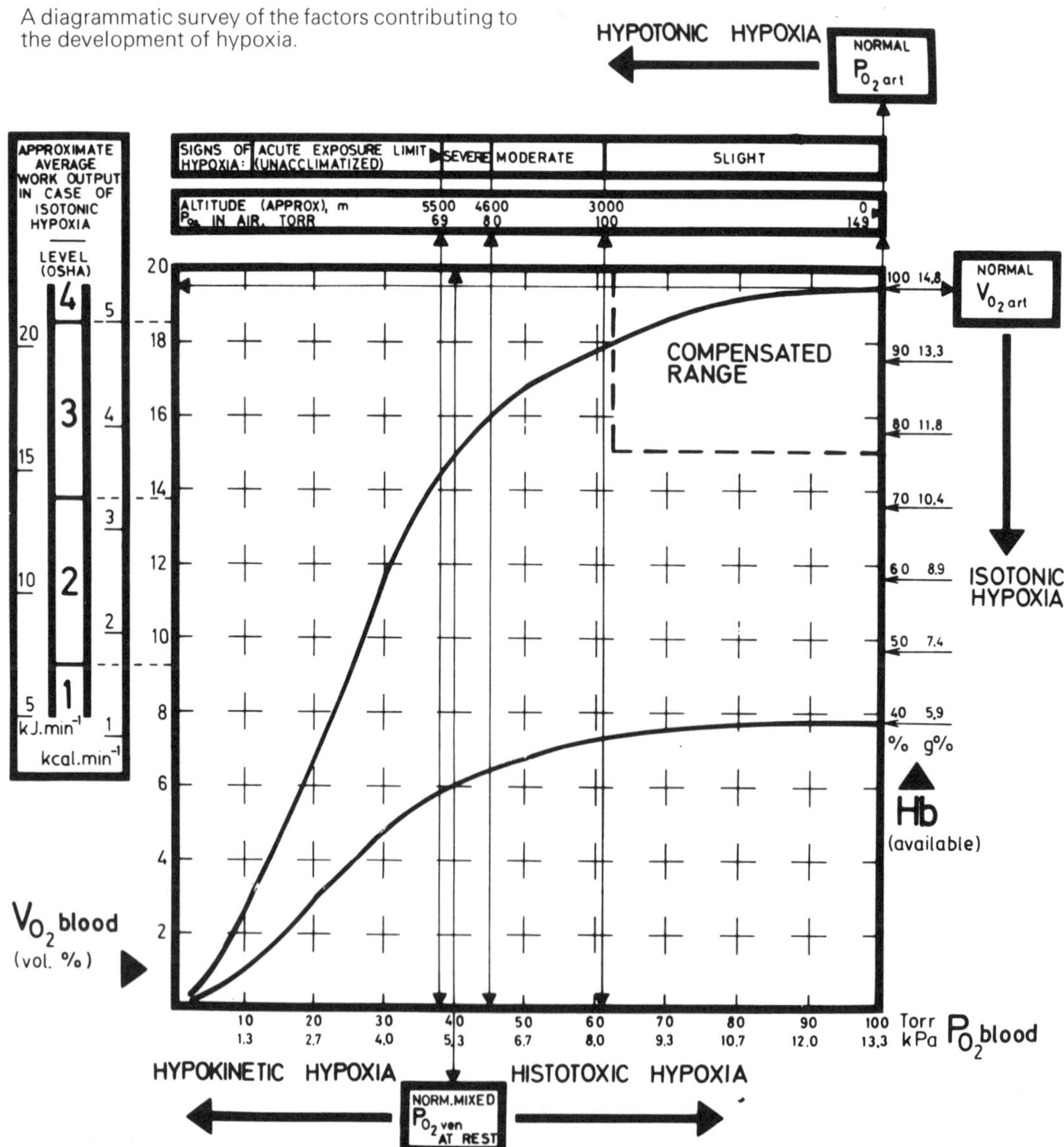

Ischaemic hypoxia. This is due to an absolute or relative insufficiency of the arterial circulation. Local ischaemic hypoxia develops if the arterial inflow to a given peripheral vascular bed is impeded or blocked, and the two forms that are of special importance in occupational hygiene are the local ischaemia due to sustained muscular contractions, and the peripheral vascular injury brought about by chronic local vibration. General ischaemic hypoxia may occur if intense physical activity or combined factors, such as high energy expenditure in a hot environment, over-stress the cardiovascular system. This type of hypoxia may develop even in healthy subjects if the cardiovascular stress is extremely high. Additional factors, however, such as decreased oxygen content or pressure of inspired air, disturbances of oxygen uptake, anaemia, cardiac diseases, etc., may provoke this condition even during moderate energy expenditure.

Histotoxic hypoxia

In this form of hypoxia, the oxygen uptake and transport and the oxygen content and pressure of the blood are all normal. The oxygen consumption of the cells, however, is severely decreased or entirely blocked, due to damage to the oxidative-enzyme system. Cyanide and sulphide poisoning are typical causes of histotoxic hypoxia.

Effects and symptoms of hypoxia

Since the symptoms of high-altitude sickness do not become manifest until levels of 4 000-4 500 m are reached, it was at one time considered that smaller equivalent falls in oxygen tension were of negligible effect. However, it has now been found that even comparatively small reductions in oxygen partial pressure may produce physiological changes—the symptoms of which are, however, much milder than those of high-altitude sickness.

The cardiovascular system at rest can adapt itself asymptomatically to changes in the atmospheric oxygen content over the range of 16 to 21% by volume. However, in an atmosphere containing e.g. 19% by volume of oxygen, moderate physical exercise will raise the respiratory minute volume to a level higher than that occasioned by the same effort in an atmosphere containing 21% by volume of oxygen. If the atmospheric oxygen content is 14-15% by volume, then the arterial oxygen tension is about 60 mmHg ($8 \cdot 10^3$ Pa) and the haemoglobin saturation is still almost 90%. There are no clinical signs but this is at the expense of an increased load upon the cardiorespiratory system. In this case, the load at rest is equal to a physical performance of about 3 kcal/min and the range of physical working capacity is narrowed.

If the atmospheric oxygen content is between 14 and 16% by volume and PO_2 in the tracheal air 100-107 mmHg ($13.3 \cdot 10^3$-$14.2 \cdot 10^3$ Pa), healthy but non-acclimatised persons are subject, even at rest, to a moderate increase in heart and respiratory rate, may have a feeling of fatigue, limitation of certain higher psychic functions (mainly of judgement and decision, which may increase the accident hazard) and a fall-off in physical working capacity. Experiments have shown that inhalation of air containing 12.5-14.5% by volume of oxygen significantly increases the relative frequency of errors in task performance; the more difficult the task, the greater the percentage of errors. The outstanding features of hypoxia at this level indicate damage to the central nervous system—with the cortical functions being most severely affected. The post-hypoxic symptoms are dominated by neural injury of a permanent nature.

When the atmospheric oxygen content falls to 10.5-14% by volume and PO_2 in the tracheal air to 76-100 mmHg ($10.1 \cdot 10^3$-$13.3 \cdot 10^3$ Pa) clinical signs of severe hypoxia will be seen, and when the oxygen content falls below 7-8% by volume and PO_2 to 50-57 mmHg ($6.65 \cdot 10^3$-$7.58 \cdot 10^3$ Pa) hypoxia becomes intolerable (figure 1).

The clinical symptoms of hypoxia vary depending on whether the condition is acute or chronic, and on whether the acute condition is of sudden or gradual onset.

Acute hypoxia with sudden onset. A sudden decrease of the oxygenation of the blood leads to loss of consciousness within 1 min without any warning symptoms. Such a situation may occur in mines after a methane outburst, in a fire, following failure of an oxygen respirator, or in carbon monoxide or cyanide intoxication, etc. Emergency measures are successful only if started not more than 2-3 min after the accident.

Acute hypoxia with gradual onset. If the oxygen content of the air decreases gradually to a level of 14% or less by volume, the following symptoms develop: tachycardia, tachypnoea, dizziness, sensation of warmth in the face and limbs. These preliminary symptoms are followed by a gradual decrease of physical and psychic capacities commencing with changes in the sensory and psychic functions (vision, hearing, cutaneous sensation, labyrinthine function, perception, attention, memory, thinking, judgement, self-criticism, voluntary activity). Although cyanosis is a common symptom of the more severe forms of hypoxia it does not occur in all types of hypoxia and, for example, there is no cyanosis in anaemia, carbon monoxide intoxication and histotoxic hypoxia. Loss of consciousness then follows, anoxic cardiac decompensation develops and the subject dies.

The duration of the diffrent phases depends on the magnitude of the hypoxia and on the speed of its onset, and might be influenced by the fitness of the subject and by the intensity of physical work previously performed. Where hypoxia develops relatively gradually, the clinical signs may be negligible at first. The greatest danger here is that, due to the disturbances of mental function, especially with regard to criticism and judgement, the subject does not realise that his condition is deteriorating and cannot take the necessary emergency measures. As a consequence, hypoxic decompensation may set in suddenly.

Chronic hypoxia. In cases in which oxygen tension in the arterial blood falls repeatedly and for prolonged periods to around 60 mmHg ($8 \cdot 10^3$ Pa) or less, and in cases where, although there is a smaller decrease in the arterial oxygen tension, the subject performs intense physical work, a chronic form of hypoxia may develop. This condition may also occur with severe pulmonary or cardiac diseases, or with their milder manifestations if the subject is engaged in strenous physical work. Chronic hypoxia, too, is characterised primarily by psychic and sensory symptoms but there is also a decrease in physical working capacity. Chronic hypoxia brings about a certain degree of acclimatisation but even the most perfect acclimatisation is only relative. If the hypoxia is of greater magnitude, the compensation over-runs the physiological limits and may be realised only at the cost of damage to certain structures or functions. In extreme cases, these lesions may be irreversible.

When the oxygen tension of the inspired air is 100-107 mmHg ($13.3 \cdot 10^3$-$14.2 \cdot 10^3$ Pa), maximal working capacity is considerably less than normal; at lower oxygen partial pressure, maximal working capacity may be only a fraction of the normal level. However, well acclimatised persons are capable of performing fairly intense muscular work even if the PO_2 of the tracheal air is around 71 mmHg ($9.34 \cdot 10^3$ Pa), e.g. in certain mines at high altitude (5 000-5 500 m). This condition can be tolerated only temporarily, and a deterioration of both physical and psychic capacities develops after a while.

PREVENTIVE MEASURES

For the technical preventive measures, see ASPHYXIA.

In all conditions or diseases which themselves may lead to hypoxia, or may impede the uptake or transport of oxygen (e.g. anaemia, cardiorespiratory diseases, etc.), the maximal work output level corresponding to the condition of the subject must be very carefully controlled.

Treatment. Victims of hypoxia must be immediately rescued from the hazardous atmosphere. If the atmosphere is explosive, rescuers must on no account use open flames or spark-producing tools, or operate electric switches, etc. Before rescue operations start, the danger area should be ventilated if possible. Rescuers should wear self-contained oxygen breathing equipment–filter masks offer no protection against oxygen-deficient atmospheres. If no breathing equipment is available, the rescuer should fully ventilate his lungs before entering the danger area, holding his breath until returning to safety. Under no circumstances should the rescuer operate alone or without supervision by someone outside the danger area. If rescue operations are undertaken in premises of large dimensions, the rescue team entering the danger zone must comprise at least two persons. The members of the team must keep continuous contact with each other and with the members of the outside group (with the aid of ropes, by holding hands, etc.), and the way back must be secured. The victim, having been removed from the dangerous area, needs complete rest in fresh air. Physical activity of any type must be avoided. Cooling of the body must be prevented. In cases of acute anoxia and absence of spontaneous respiratory movements, artificial respiration must be started immediately. Breathing of pure oxygen is the most effective measure. If there is no oxygen respirator, mouth-to-mouth or mouth-to-nose ventilation must be applied, and maintained, until the return of spontaneous respiration (see RESUSCITATION).

TIMAR, M.

"Anoxia, altitude, and acclimatisation". Lambertsen, C. J. (1 358-1 562). *Medical physiology.* Mountcastle, V. B. (ed.). (Mosby, St. Louis, 13th ed., 1977).

"Aviation physiology–the effects of altitude". Luft, U. C. (1 099-1 145). *Handbook of physiology.* Sect. 3 Respiration, Vol. II (Washington, DC, American Physiology Society, 1965).

Immunisation and vaccination

Immunity is the state of resistance of the body to agents foreign to it, among which environmental and, especially, infectious agents are the best known. Immunity is not necessarily an absolute nor a permanent state. It may be natural, dependent on the species, or the individual; in the latter case it may be genetic (e.g. resistant strains of animals), actively acquired (e.g. subclinical infections) or passively acquired (e.g. by means of maternal antibodies passively transferred through placenta or colostrum). Immunity may also be provided artificially by active or passive immunisation. Active immunisation by means of vaccines provides usually effective and long-lasting immunity, while passive immunity acquired by injection of ready-made antibodies is short-lasting and less effective.

The mechanisms of immunity are very complex. The first line of defence by mammalian hosts against monocellular microbes generally consists of natural barriers (such as intact integument and motile cilia), phagocytosis of the offending organisms by polymorphonuclear leucocytes and mononuclear macrophages, and the complement system.

When the first line of defence cannot cope with an invader, the second line of defence is mobilised. This involves a specific immune response to antigenic components of the invading organism. There are two kinds of immune responses: cell-mediated immunity and humoral antibody production.

Cell-mediated immunity plays a major role in destroying bacterial, viral or fungal agents that replicate intracellularly and is also of a great importance in tumour immunity, delayed-type hypersensitivity, graft rejection and many instances of autoimmune diseases. This kind of immunity is mediated by special lymphoid cells, named T cells because of their intimate association with the thymus gland, which is their site of differentiation.

Antibodies are particularly effective in combating infections by micro-organisms that replicate outside cells and are thus free in body tissues and fluids, and in neutralising bacterial toxins. They are proteins (immunoglobulins) produced by another type of lymphoid cell, designated B cells because of their relations to the bursa of Fabricius in chicken and the bone marrow in man. Antibodies belong to the different classes of immunoglobulins and function in a variety of different ways to diminish the pathogenic effects of the invading organisms. They appear in the blood some time (1-4 weeks) after the first contact with the foreign agents. Later their level in the blood will decline and they may even disappear entirely, unless renewed contact with the antigens once again stimulates their production.

Active immunisation

The objective of immunisation is to produce specific immunity without the inconveniences of the first attack of the disease. This objective is achieved by injecting the antigen into the body with a view to stimulating the production of antibodies without endangering the organism. Some immunisation is carried out by oral route, e.g. immunisation against poliomyelitis with live polio-virus vaccine. Since a certain time is required before immunity is achieved, immunisation is essentially a prophylactic technique.

Nowadays vaccines are available against a wide range of diseases and include:

(a) live, attenuated micro-organisms (weakened by laboratory procedures), e.g. oral poliomyelitis (Sabin), measles, rubella or mumps vaccines;

(b) whole, inactivated (killed) micro-organisms, e.g. bacterial vaccines against cholera, pertussis or typhoid, viral vaccines against poliomyelitis (Salk) or influenza;

(c) inactivated bacterial toxins (toxoids), e.g. diphtheria or tetanus vaccines;

(d) fractions of whole killed organisms, e.g. vaccines against pneumococcal pneumonia or meningococcal meningitis; and

(e) cross-reactive organisms, i.e. live microbes related immunologically to the disease organism which, when given to man, produces a milder infection that protects against a serious one, e.g. smallpox vaccine, derived from cowpox, and BCG for tuberculosis, derived from the bovine tuberculosis organism.

Vaccination may be used both to prevent the disease or to modify its course favourably. Some vaccines provide lifelong immunity, whereas others provide immunity for only limited periods and must be reinforced by periodic booster injections. The duration of effectiveness varies greatly with the individual, age and the specific disease. However, the personal protection offered by vaccination should not be considered adequate in itself but should always form part of a total health protection programme which, for example in the case of zoonoses, will be directed primarily at eradicating animal reservoirs of infection.

Because of their importance to the working population a number of vaccinations are of direct concern to the industrial physician.

Anthrax. See separate article.

Brucella. See BRUCELLOSIS.

Cholera. Staff of infectious diseases hospitals and medical laboratories, and workers engaged in waste disposal in newly infected areas *may* be vaccinated, though education on personal and food hygiene is more important. Primary immunisation consists of two doses of vaccine administered subcutaneously.

Diphtheria. Hospital staff, young nurses in particular, are usually exposed to this infection, especially those working in infectious diseases wards. It may be noted that adults in non-endemic areas are usually susceptible. Primary immunisation consists of two subcutaneous doses of (combined tetanus and) diphtheria toxoid, with re-immunisation in cases of serious danger of exposure.

Influenza. See separate article.

Insects and centipedes. See ANIMALS, VENOMOUS.

Leptospira. See LEPTOSPIROSIS.

Plague. Health personnel and hospital staff may be exposed to this infection, as well as persons, hunters in particular, working in areas where infected rodents are to be found. Initial immunisation consists of 2 or 3 doses of killed, or one dose of live vaccine administered subcutaneously or cutaneously, with a booster dose 6-12 months later, and regular immunisation every 1-2 years depending on the degree of exposure.

Q fever. See separate article.

Rabies. See separate article.

Scorpions. See ANIMALS, VENOMOUS.

Smallpox. In December 1979 the Global Commission for the Certification of Smallpox Eradication was convened by the World Health Organisation in Geneva and concluded that smallpox had been eradicated throughout the world. The last case of naturally occurring smallpox was recorded in Somalia in October 1977. Hence, smallpox vaccination is no longer necessary. Accordingly, the Commission recommended that, considering the risk of complications, nobody should be vaccinated against smallpox, except for investigators at special risk. These investigators include laboratory workers who might have direct contact with orthopox viruses pathogenic to man.

Snakes. See ANIMALS, VENOMOUS.

Tetanus. See separate article.

Tuberculosis. See BCG.

Tularaemia. See separate article.

Typhoid. Waste-disposal workers, military personnel and hospital and laboratory staff are particularly exposed to this infection. Immunisation should be carried out with acetone-dried or phenol-inactivated potent vaccine. Initial immunisation should consist of two subcutaneous doses, with re-vaccination (single dose) every 3-5 years.

Typhus (epidemic, louse-borne). Labour forces moving into typhus endemic areas, health personnel dealing with people coming from infected areas, and workers in laboratories potentially at risk may be vaccinated with two doses of inactivated vaccine given at an interval of not less than 7 days, and a reinforcing dose annually while the risk continues. In epidemics in general populations, reliance for control should be placed on application of suitable insecticides to deal with louse infestation.

Yellow fever. Yellow fever vaccine seems highly effective: one dose will provide immunity for at least 10 years, possibly much longer. All laboratory workers who may come into contact with the virus must be vaccinated. In the endemic countries of the Americas and Africa, persons in rural areas should be vaccinated if their occupation brings them into the forests (where the animal hosts and the insect vectors are found). Persons who are to visit such areas should also be vaccinated. The vaccination of labour forces coming into an endemic area to work on development projects is particularly important.

Passive immunisation

When the infectious disease has started, it is usually too late for vaccination and it is necessary to raise the body's resistance immediately by passive immunisation, i.e. by injecting the patient with specific ready-made antibodies. Passive immunisation can also be applied to patients who, although exposed to infection, still show no signs of disease.

Specific passive immunisation is obtained by injecting into the patients the serum of animals specially treated with bacteria or toxins, or the serum of humans recovering from the same infectious disease.

Of particular interest to the industrial physician is the use of passive immunisation for the prevention of tetanus in injured workers who have not been sufficiently protected by active immunisation (see TETANUS).

When injecting horse serum even years after the previous administration, there is a risk of allergic reactions, which may be so severe as to produce fatal shock. These reactions may be prevented by testing the patient's sensitivity before injecting the serum; persons found to be sensitive should receive human gammaglobulin, purified serum or sheep serum, or should be treated by the Besredka anti-anaphylactic technique. Late anaphylactic reactions do not usually fall within the scope of the industrial physician.

Passive immunisation is still essential in the treatment of rabies, snake bite, and mushroom poisoning. In a number of infectious diseases, antibiotics are complementary to passive immunisation.

WHO DIVISION OF COMMUNICABLE DISEASES

Vaccination certificate requirements for international travel. Published periodically by the World Health Organisation, 1211 Geneva 27, Switzerland.

"Immunocompetent cells in resistance to bacterial infections". Campbell, P. A. *Bacteriological Reviews* (Baltimore), June 1975, 40/2 (284-313). 372 ref.

"Immunity, allergy and related diseases". Hong, R. (582-598). *Nelson textbook of pediatrics.* Vaughan, V.; McKay, R.; Behrman, R. (eds.). (Philadelphia, London, Toronto, Saunders, 1979).

"Problems with immunization against infectious diseases". Mortimer, E. A. *Pediatric Research* (Baltimore), May 1979, 13/5 (684-687).

Final Report of the Global Commission for the Certification of Smallpox Eradication (Geneva, World Health Organisation, 1980), 122 p.

Incentives and productivity

Productivity incentives or incentive wage rates form the basis of systems of wage payment under which the worker's earnings are related directly to some measurement of the work done by himself or by the group or working unit to which he belongs. The chief characteristic of these systems of "payment by results", under which the worker's reward varies in the same proportion as his output, is that any gains or losses resulting directly from changes in his output accrue to him (leaving to the employer any gains or losses in overhead costs per unit of output). In contrast, when a worker is paid by the hour or by the day, all gains or losses resulting from changes in his output accrue to the employer. The primary objective of systems of payment by results is to increase output, reduce overheads, increase profits both to the employer and worker, and consequently improve standards of living. However, the negative aspects of these incentives, namely the increased risk to life and limb that may be caused by the worker speeding up his output using inherently dangerous machinery and/or operations, may negate the benefits that might otherwise result from these productivity systems.

Other somewhat similar incentives or bonuses are paid for work under special or unusual conditions; for example the payment made to a worker as compensation for the need to work in dirty conditions ("dirty money"); payments for long service; acquisition of specialised qualifications.

When systems of payment by results were first introduced, the whole of the worker's income depended on his output. However, modern incentive plans provide for a guaranteed minimum wage often in the form of a guaranteed hourly minimum rate. At present, in certain industries payment by results accounts for as much as 50-60% of wages; however, the general trend is towards wage stability, and the guaranteed minimum wage tends to account for an ever larger proportion of total income with the variable incentive premiums accounting for about 20%. Figure 1 shows graphically the breakdown of a worker's income, in the organised sector (where trade union action for minimum wages has been effective), under a wage system employing payment by results and other incentives. In the unorganised sector it is not unusual that the terms of employment and payment of wages are restricted to "piece rates" depending on the units of production per day.

Figure 1. A typical breakdown of the wages earned by an industrial worker.

Payment-by-results systems

Typical methods include the straight piece-work system and the standard-time system in which salaries are proportional to output, and systems in which a fixed wage is supplemented by a bonus depending on output or factors influencing production (machine utilisation time, down-time, etc.). In general, it is possible to distinguish between three types of system, depending on whether the incentives induce a worker or the group to attain a given output, whether they are proportional to work done, or whether they encourage as high a level of output as possible.

The systems of payment by results that have been devised are numerous and varied; some of the most common include: straight piece-work system (more often with a guaranteed basic wage calculated on the hours of work); the standard-time system (based on a standard time calculated for each individual job); the Halsey system (based on set standard times, with the benefit of the time saved shared between the worker and employer); the Rowan system (on the basis of set standard times, with a bonus proportional to the ratio between standard time and time saved); the Taylor, Merrick and Gantt systems (on the basis of set standard times with the hourly rates varying in proportions which differ at different levels of output); and the Bedaux system (based on a codification of the different parts of the task with a bonus for improvement on the set standard).

Standard performance

Where workers are paid by results, it is necessary to be able to measure these results; consequently the work processes and all the individual tasks must be studied in detail. This work study or job evaluation serves as a basis for calculating the rates of payment and for determining the standard performance with which actual output is compared, for the purpose of bonus payments. The times set will be those applicable to the "average worker" at "normal workspeed and effort" and under "normal working conditions". They must not be set so high as to make the worker go too fast and impair quality nor so low that costs per unit of production rise excessively.

The work study undertaken for establishing performance standards may also be employed to improve work processes, the operation of the man-machine system, the ergonomic layout of the work station and the suitability of tools, thus providing not only a higher output but making a general contribution to better working conditions.

In the transfer of technology and sophisticated machinery to industrially developing countries, the terms "average worker", "normal workspeed" and "effort" are not defined and interpreted in the context of existing local conditions. Due to anthropometrical variations, poor nutritional levels, inadequate community health services and low medico-social background of the workers in those countries, the norms that are specified by the technology-supplying countries to define the terms stated above are not valid in the technology-receiving countries. The risks to health of employed persons, especially young and inadequately trained workers, are enhanced by attempting to achieve the targets specified for those in industrially advanced countries. There is therefore a vital need to establish achieveable performance standards in developing countries. It should really be a prerequisite to introduction of any incentive systems and the standards should be established with due regard to ergonomics, anthropometry and existing working conditions.

HAZARDS

Payment by results has a number of disadvantages. Product quality may fall (unless suitable quality control is introduced), employee/employer conflicts and rivalries between workers may arise leading to a degeneration of human relations in the firm. If the performance standards or wage rates are not established with due regard to all the relevant circumstances, this friction may be disastrous to the good functioning of the undertaking and the personal happiness of the workers.

Payment by results may have serious safety and health repercussions: workers may over-exert themselves in an attempt to increase their output and income, leading to stress and fatigue and to even more dangerous consequences in elderly workers or cardiacs, etc.; the resultant diminished vigilance, reflexes and acuity may increase the liability to human error and accidents (human factors); individual workers wishing to raise output may inactivate safety equipment, remove machinery guarding or disregard safe working practices; group incentives, in particular, may create pressures which lead to dangerous behaviour. Finally, a combination of work study and payment by results may lead to the implementation of piece-work systems, conveyor-belt work or inflexible and excessively high work rates with consequences such as nervous tension, distaste for the work and effects on mental health and emotional equilibrium.

In the developing countries, and particularly in small-scale industries where minimum safety standards are often not enforced, physical exhaustion resulting from continuous piece-rate work contributes to raising the rate of accident injuries at work. Apart from the loss of personal earning capacity and the lower national productivity that result, the partial or total incapacitation of the sole breadwinner of a family is a serious social and human tragedy.

Special bonuses

There are a number of these paid for work under unusual conditions: hot money (for work in high temperatures), dark money (for work in poorly lit workplaces), dirty money, night-work bonuses (especially for night shifts). Special bonuses for safety achievement are dealt with in the article INCENTIVES, SAFETY.

The principle underlying these special bonuses, which constitute compensation for work under substandard conditions, is incompatible with good safety, health and welfare practices. It not only allows undesirable conditions to be maintained through the payment of compensation, but it may also tend to encourage workers to prefer exposure to certain obnoxious or dangerous conditions in order to swell their wage packet; in addition, should conditions for which bonuses have been paid be rectified, there may be discontentment among the men, who see their wages reduced by the suppression of the bonus.

Payment of incentive bonuses for working in arduous situations (e.g. in industries with a high radiant heat load) and for arduous tasks (e.g. lifting excessive weights to complete a given task within specified periods) are contrary to the pratice of occupational health, as these conditions of work are known to cause occupational lesions or aggravate existing non-occupational diseases among the exposed workers.

Incentives versus safety and health

The unequivocal right of every wage earner to safe and healthy working conditions is guaranteed by several ILO Conventions, in particular the Minimum Wage Convention, 1970 (No. 131), the Guarding of Machinery Convention, 1963 (No. 119), the Maximum Weight Convention, 1967 (No. 127), and the Working Environment (Air Pollution, Noise and Vibration) Convention, 1977 (No. 148). These Conventions along with the United Nations Covenant on Economic, Social and Cultural Rights (Article 7*(b)*) should ensure that wage earners are not enticed by the offer of incentives to work in conditions that may adversely affect their health. Productivity assessment that takes into consideration the concept of "total loss control" (which includes loss to industry through employment injury and occupational diseases) will eventually ensure increased production through reassurance, greater stability and uniformity in the whole production process.

DUCREY, L.
PINNAGODA, P. V. C.

CIS 78-1107 "The effect of piecework on accident rates in the logging industry (incorporating a different approach to the exposure problem). Mason, K. *Journal of Occupational Accidents* (Amsterdam), July 1977, 1/3 (281-294). 1 ref.

CIS 74-1800 "Performance efficiency and injury avoidance as a function of positive and negative incentives". McKelvey, R. K.; Engen, T.; Peck, M. B. *Journal of Safety Research* (Chicago), June 1973, 5/2 (90-96). Illus. 17 ref.

Incentives, safety

How to secure the active co-operation of the workers in accident prevention is a burning problem. Many accidents involve human errors, departures from safe practice or inattention, and modern safety concepts emphasise the need to develop safe attitudes at all levels.

The first responsibility of management is to ensure that, as far as is reasonably practicable, the workplaces, equipment and processes under its control are safe and the workers properly trained and supervised. Providing this has been achieved, it may be of advantage to offer financial and other incentives to encourage employees to work more safely. From the purely economic viewpoint, it is logical to offer rewards to reduce the financial losses resulting from occupational accidents, in the same way as schemes that insurance companies grant reductions in premiums to safe drivers or companies with accident rates below the branch average. It must be stressed, however, that safety incentives are no substitute for a properly planned and implemented safety programme and should not be considered unless management has already fulfilled all its obligations in respect of safety and health. The knowledge that management from the top down is keenly and actively concerned with safety constitutes in itself an incentive to employees. If, in addition, a good safety record is taken into consideration when making promotions, this constitutes a powerful incentive, particularly to foremen and supervisors.

On the other hand, if safety always runs a poor second to production, incentive schemes will have very little effect on accident frequency rates or at best will be regarded as fanciful additions to the real business of the enterprise.

Granted these preconditions, it is still possible to debate what form an incentive scheme should take. Opinions and experience differ in this respect from country to country and from plant to plant. Broadly, schemes may be based on:

(a) rewards for achievement of low or reduced accident frequency rates; or

(b) rewards for positive contributions to safety by suggestions, design of new or improved safety devices, posters, etc.

Types of award

Various methods of making awards for a low accident rate have been used, e.g.:

(a) group reward for an accident-free period of predetermined duration;

(b) group reward for a reduction in accident frequency rate over, say, a year;

(c) individual reward for accident-free performance over a given period;

(d) competitive reward for lowest accident frequency rate among groups within an enterprise, or among different undertakings in the same branch of economic activity.

Incentives for groups have advantages over individual incentives as they create a corporate interest; charge hands or supervisors should be included in the groups as they then have an extra incentive to ensure that the men under their control perform their work in the safest possible manner. The nature of the reward itself can vary from a badge or small trophy to a substantial money prize or bonus. Rewards in kind are also used: sometimes a successful group may be given a dinner or an outing in which members of the family can join.

Initiation and conduct of incentive schemes

Whatever the type of scheme, the rules governing awards should be fair, simple and clearly stated. A scheme is better instituted after discussion in a works' safety committee or with workers' representatives; these bodies should preferably also be associated in judging the results.

Persons who have been injured as a result of an occupational accident during the period of operation of the scheme should be able to participate, except in exceptional circumstances (cases of gross negligence, etc.). In the case of group rewards the size of the groups should not be too small; it should not, on the other hand, exceed about a hundred persons.

Any incentive scheme must obviously be properly organised and carried out. In some undertakings, especially smaller ones, an ad hoc reward for some outstanding performance may have a more stimulating effect than a more formalised scheme.

Safety incentive schemes are open to serious objection if they lead to concealment of accidents. If a worker does not report for first-aid treatment or continues to work when unfit, for fear of damaging his own or his group's safety record, the final result may be detrimental to health and safety; this is particularly true of rewards for "no accident" records.

Other questions to be decided when organising an incentive scheme are:

(a) how can the differences in occupational hazards between various workplaces (in the case of individual rewards) or between groups be taken into account when judging the results? Some adjustments are required in specific instances (maintenance as opposed to clerical workers, for example);

(b) how far is the accident severity rate, as distinct from the frequency rate, to be taken into consideration?

(c) how to determine the optimum time a competition should run—too short a time will not give a true picture, too long may generate boredom?

With any incentive scheme, one of the greatest problems is to prevent its becoming stale. It is all too easy for both management and employees to become bored with a scheme that runs without variation for too long. It is necessary to be alert to the need to initiate new ideas and new forms of incentive; seasonal and local events can often form a nucleus for variations.

Results achieved

The literature on safety incentive schemes consists mainly of short articles and notes containing in general little information on the conditions under which the schemes concerned have been planned and conducted and on their results. From the experience quoted, however, it appears that when incentives are part of a well conceived safety programme, they contribute to a reduction in accident frequency rates. There is general agreement also that firms which have used incentive schemes will continue to do so.

An interesting study on the effectiveness of safety bonuses was carried out by Bartels in the Federal Republic of Germany, following a survey of 13 plants selected from a group of 140 that had applied an incentive scheme in the recent past or were applying one at the time of the survey; these plants included establishments of different sizes (600-26 000 employees) and represented seven branches of economic activity. In the great majority of the plants surveyed, accident frequency rates had decreased after the introduction of the scheme, but severity rates had tended to increase. Safety bonuses in most cases achieved a significant decrease in the number of accidents ascribable to unsafe behaviour, but at the same time they resulted in the concealment of minor injuries (a result which may be corrected by taking into consideration not only the accident frequency rate but the severity rate as well). It was further found that unsafe acts performed in an attempt to gain time were less affected by safety bonuses or other forms of incentive than acts which are performed rather unconsciously or which do not result in time saving. The effect of bonuses on workers' safe behaviour is all the more marked as the amount of the bonus is higher and as workers are better informed of the hazards inherent to their work during the period of operation of the scheme. Incentive schemes should be adapted to the peculiarities of each plant and their duration should be limited so as to avoid boredom. It was found that group rewards were more effective than individual rewards, probably as a result of the pressure which the members of a group exert on each other.

(As regards means of stimulating employers' interest see SOCIAL SECURITY AND OCCUPATIONAL RISKS; OCCUPATIONAL SAFETY IN UNDERTAKINGS.)

DICICCO, A.

CIS 78-243 *The effectiveness of safety bonuses* (Über die Wirksamkeit von Arbeitssicherheitsprämien). Bartels, K. Forschungsbericht Nr. 163 (Dortmund-Marten, Bundesanstalt für Arbeitsschutz und Unfallforschung, 1976), 217 p. Illus. 64 ref. (In German)

"The Western approach to worker motivation". Martin, G. *Job Safety and Health* (Washington, DC), Nov. 1977, 5/11 (19-29). Illus.

"The use of incentives to motivate safety". *National Safety News* (Chicago), Jan. 1979, 119/1 (50-52).

"Motivating workers with incentives". Dionne, E. D. *National Safety News* (Chicago), Jan. 1980, 121/1 (75-79).

CIS 78-1460 "Psychological and material incentives for the improvement of workplace conditions" (Moral'noe i material'noe stimulirovanie ulučšenija uslovij truda na proizvodstve). Karasina, N. I. (Moscow, Vsesojuznyj central'nyj naučno-issledovatel'skij institut ohrany truda VCSPS, 1977), 44 p. 74 ref. (In Russian)

Indicators and control panels

The complexity of the instrumentation needed in modern technological equipment makes it essential that instruments and control panels should be designed with the abilities and limitations of the human operator constantly in mind. Failure to do so can cause expensive inefficiencies and, sometimes, disastrous results.

The following gives some guidelines for the efficient layout of indicators and control panels; but, throughout, it should be remembered that the most important factor is detailed consideration of the function of the display or control which is to be designed.

Choice of indicator

Where information is to be conveyed to an operator, it is best to convey it by the simplest type of indicator compatible with the information required. All irrelevant notations on indicators should be removed. The designer should consider whether quantitative or qualitative information is to be conveyed and design accordingly, i.e. often an operator wants to know only if a certain value has been exceeded but is not interested in absolute values.

A warning indicator is often best revealed by a flashing light, which can be colour coded. A flash rate of 3 flashes per second with a dark : light ratio of about 1 : 1 is optimal. Specialised literature exists on all important indicator categories and should be referred to for specific details.

Control panel design

Whenever possible, control panels should be designed so that the operator can sit while operating. The correct use of anthropometric data of the relevant population should assure that all controls are within reach, that displays and controls are unobstructed and can be viewed without gross parallax errors. Generally, displays on a control panel should be at or slightly below eye level. Where the panel is, of necessity, other than a simple vertical surface i.e. V-shape or U-shape, the display should still be in direct line of sight for the operator. A mock-up of the control panel will help reduce any inadequacies.

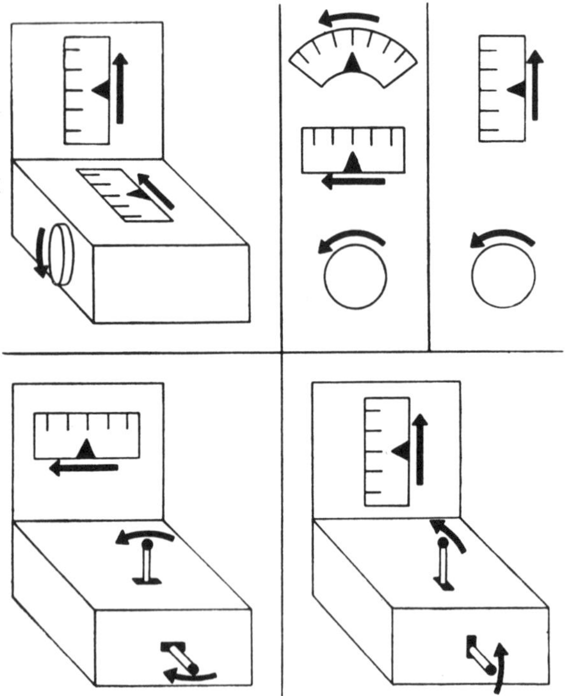

Figure 1. Expected relationships of movement between controls and displays. *(By courtesy of the Controller of HM Stationery Office, London.)*

When an operator is seated, the forward line of sight is normally 10° below the horizontal. Cathode ray tubes should be scanned from 36-44 cm. Dials can be best read at 44-75 cm. Normally, controls are best situated slightly below the optimal display area and should be placed at elbow height.

The main principles to bear in mind in the design and construction of indicator and control panels are:

(a) the functional principle, in which displays and controls with similar functions are grouped together;

(b) the importance principle, in which the most important displays or controls are placed in the "best" positions ("best" and importance can be decided only by reference to all relevant factors);

(c) the sequence of use principle, in which displays or controls are grouped in the sequence in which they are most typically used;

(d) the frequency of use principle, in which the most frequently used displays or controls are positioned in the preferred locations.

Emergency controls should be clearly identified and within easy reach; they should generally be placed away from the controls governing normal functions. The emergency controls can often be recessed to avoid accidental activation. An auditory signal is often effective in drawing attention to an emergency situation when the visual sense is already heavily loaded. Care should be taken to ensure that any display is not hidden from the operator's sight during the operation of another control. This applies particularly when controls are mounted near associated displays in a vertical line, i.e. above one another.

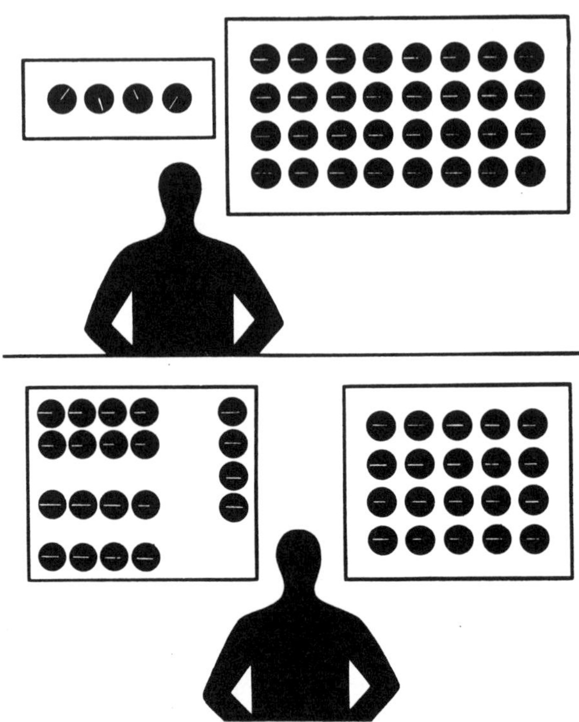

Figure 2. *Top:* because the 32 dials of the right hand bank are aligned to point in the same direction when indicating "Normal", the operator can check them just as quickly as he can the four dials on the left, which are not aligned. *Bottom:* logical groupings within a bank of dials–left–enable the operator to identify a particular dial easily. *(By courtesy of the Controller of HM Stationery Office, London.)*

The relationship between controls and displays becomes vitally important where the operator is subjected to serious stress. Certain relationships exist between some controls and their displays and these relationships are "expected" by the operator. Where the controls and displays are not as expected, the operator reverts to his normal or expected mode of operation in an emergency.

The use of labels on controls and displays is sometimes needed. When this is so the labels should be placed above the control or display. Most controls can be operated equally well with either hand. The preferred hand is better for fine accurate adjustment and if possible such controls should be placed so as to suit both left- and right-handed people.

DAVIES, B. T.

CIS 77-1773 "CAPABLE–A computer program to layout controls and panels". Bonney, M. C.; Williams, R. W. *Ergonomics* (London), May 1977, 20/3 (297-316). Illus. 18 ref.

CIS 76-1161 *Ergonomic characteristics of display systems and control devices* (Ergonomiczne właściwości przyrządów sygnalizacyjnych i sterowniczych). Paluszkiewicz, L. (Warsaw, Instytut Wydawniczy CRZZ, 1975), 176 p. Illus. 86 ref. (In Polish)

CIS 74-1173 "Dials v counters: Effects of precision on quantitative reading". Nason, W. E.; Bennett, C. A. *Ergonomics* (London), Nov. 1973, 16/6 (749-758). Illus. 16 ref.

"Ergonomic trends in the design and use of machine tools for the motor industry" (L'ergonomia nella progettazione e nell'impiego delle macchine utensili per la fabbricazione di autoveicoli). Parmeggiani, L.; Rovetto, G. P. *La Medicina del Lavoro* (Milan), Nov.-Dec. 1976, 67/6 (568-580). Illus. 11 ref. (In Italian)

Indium, alloys and compounds

Indium (In)

a.w.	114.8
sp.gr.	7.3
m.p.	156.6 °C
b.p.	2 080 °C

a malleable metal, softer than lead, with a brilliant silvery-white lustre.

TLV ACGIH	0.1 mg/m^3
STEL ACGIH	0.3 mg/m^3

Occurrence. In nature, it is widely distributed and occurs most frequently together with zinc minerals (sphalerite, marmatite, christophite), its chief commercial source. It is also found in the ores of tin, manganese, tungsten, copper, iron, lead, cobalt and bismuth, but generally in amounts of less than 0.1%.

Production. Its close association with zinc enables indium to be recovered commercially from zinc residues and smelter slags. Of the many methods of separation available, the most common involves the dissolution of indium in acid, from which a crude preparation of the metal is made either by sponging with zinc, or by forming insoluble hydroxides, sulphites, or orthophosphates. The crude sponge or cake is next repurified in solution and the metal is usually electroplated on rods, which can result in a purity of more than 99.999%.

Uses. Indium is generally used in industry for surface protection or in alloys. A thin coat of indium will increase considerably the resistance of metals to corrosion and wear. In bearings it will prolong the life of moving parts and, consequently, finds wide use in the aircraft and automobile industries. As an alloy of many metals it increases their hardness and, as a result, it is used in dental alloys and bearings. Its "wettability" makes it ideal for plating glass. Owing to its resistance to corrosion, it is utilised extensively in making motion picture screens, cathode ray oscilloscopes and mirrors. When joined with antimony and germanium in an extremely pure combination, it has wide use in the manufacture of transistors and other sensitive electronic components.

Compounds

In addition to the metal, the most common industrial compounds of indium are the trichloride ($InCl_3$), used in electroplating; the sesquioxide (In_2O_3), used in glass manufacture; the sulphate ($In_2(SO_4)_3 \cdot 9H_2O$); [and the antimonide (InSb) used as a semiconductor].

HAZARDS

Probably the greatest current potential hazard comes from the use of indium together with antimony and germanium in the electronics industry. This is due primarily to the fumes given off during welding and soldering processes in the manufacture of electronic components. Any hazard arising from the purification of indium is probably attributable to the presence of other metals, such as lead, or chemicals, such as cyanide, used in the electroplating process. Exposure of the skin to indium does not seem to present a serious hazard. The tissue distribution of indium in various chemical forms has been studied by administration to laboratory animals.

The sites of highest concentration were kidney, spleen, liver and salivary glands. [After inhalation widespread lung changes were observed such as interstitial and desquamative pneumonitis with consequent respiratory insufficiency.]

The results of animal studies showed that the more soluble salts of indium were very toxic, with lethality occurring after administration of less than 5 mg/kg by way of parenteral routes of injection. However, after gavage, indium was poorly absorbed and essentially non-toxic. Histopathological studies indicated that death was due primarily to degenerative lesions in the liver and kidney. Minor changes in the blood have also been noted. [In chronic poisoning by indium chloride the main change is a chronic interstitial nephritis with proteinuria.] The toxicity from the more insoluble form, indium sesquioxide, was only moderate to slight, requiring up to several hundred mg/kg for lethal effect.

SAFETY AND HEALTH MEASURES

Preventing the inhalation of indium fumes by the use of proper ventilation appears to be the most practical safety measure.

Treatment. Intoxication in rats from indium-induced hepatic necrosis has been reduced considerably by administration of ferric dextran, the action of which is apparently very specific. The use of ferric dextran as a prophylactic treatment in humans has not been possible owing to a lack of serious cases of industrial exposures to indium.

THOMAS, R. G.

CIS 1518-1966 "The toxicological characteristics of indium antimonide and gallium arsenide–two new semiconductor materials" (Toksikologiceskaja harakteristika antimonida indija i arsenida gallija-novyh poluprovodnikovyh materialov). Roscina, T. A. *Gigena truda i professional'nye zabolevanija* (Moscow), May 1966, 10/5 (30-33). 5 ref. (In Russian)

CIS 74-1692 *Experimental poisoning of guinea-pigs by indium chloride–Preliminary study* (Intoxication expérimentale par le chlorure d'indium chez le cobaye–Etude préliminaire). Le Gallo, J. (Université de Paris V, Faculté de médecine Necker–Enfants-Malades, Paris) (1974), 44 p. Illus. 31 ref. (In French)

"Problems of occupational hygiene related to the industrial application of gallium and indium compounds" (Problemy gigieny truda v svjazi s primeneniem v narodnom xozjajstve coedinenij gallia i india). Taracenko, N. Ju.; Fadeev, A. I. *Gigiena i sanitaria* (Moscow), Oct. 1980, 10 (13-16). 4 ref. (In Russian)

CIS 80-1047 "Hazardous chemical reactions–69. Boron, gallium, indium, thallium" (Réactions chimiques dangereuses–69. Bore, gallium, indium, thallium). Leleu, J. *Cahiers de notes documentaires–Sécurité et hygiène du travail* (Paris), 1st quarter 1980, 98, Note No. 1236-98-80 (131-133). 16 ref. (In French)

Industrialisation, impact on health and safety of

The impact of industrialisation on the health and safety of workers and on the environment is one of the subjects currently being studied throughout the United Nations system, and specifically in the International Labour Organisation (ILO) and the World Health Organisation (WHO). Although this impact varies greatly from one country to another, the present findings suggest that there are certain requirements that have to be met if industrialisation is not to exact the same toll of accidents and diseases on the 20th century as it did on the 19th.

Industrialisation has been defined as "the process of change in the mode of production to utilise more capital per unit of output, higher levels of technology and management, widening markets with cost economies of scale and specialised location of plant, type of plant and labour skills". An industrialised country has been defined as one that derives more than 30% of its gross domestic product from manufacturing, while a semi-industrialised country derives from 20 to 30%, an industrialising country between 10 and 19%, and a non-industrial country less than 10% from the same source. The industrialisation trend between 1960 and 1975 is shown by the evaluation of 38 developing countries made by the United Nations Industrial Development Organisation (UNIDO) set out in table 1.

Table 1. Industrialisation status of 38 countries in 1960 and 1975

Industrialisation status	1960		1975	
	No.	%	No.	%
Industrialised	2	5.3	10	26.3
Semi-industrialised	12	31.6	15	39.5
Industrialising	20	52.6	12	31.6
Non-industrial	4	10.5	1	2.6
Totals	38	100	38	100

When investigating the effect of rapid industrialisation on workers' health, one is immediately struck by the limited availability of statistical data at the national level and, quite often, by the notable disproportion between the volume of information forthcoming from the central statistical services on economic subjects and the sparseness of data on the state of health of the workers. This is particularly true of countries where the reporting of occupational injuries is either not compulsory or not enforced, where insurance against occupational injuries is not mandatory, or where, because of the level of underemployment, it is in the interest neither of the worker nor his employer to make an official declaration of disability. It is not always easy to have the occupational nature of an injury and particularly a disease recognised. Furthermore, such data, when they exist, concern only the organised sector which, in a developing country, usually accounts for only a small proportion of the active population: in Senegal for example, out of a population of over 5 million and an active population of almost 2 million, the organised sector in 1975 consisted of a mere 141 500 workers (7%), while in Ethiopia in the same year, out of an active population of 9.2 million, the organised sector comprised only 54 818 persons (0.6%).

Impact of industrialisation on occupational accidents

There is more information available on occupational accidents than on occupational diseases, and a series of national surveys have drawn attention to the high level of accidents that goes with industrialisation. The rate of occupational accidents that were reported to social security institutions in Bolivia and Chile in 1972 was 34 per 100 exposed workers; in Columbia the corresponding figure for 1971 was 42 per 100 (involving permanent invalidity in 47% of the cases reported); between 1966 and 1973 the accident rate per million man-hours in Iran ranged from 25 to 49, compared to between 10 and 11 in the United States; the comparable accident rate in Brazil in 1975 was 86.4 in the mining sector and 105.1 in construction, as against 25.5 and 29.5, respectively, in the United States in 1974. Furthermore, a great number of accidents do not find their way into the statistics in developing countries.

The breakdown of accidents by undertakings reveals, as regards machinery, two principal causes: *(a)* lack of protective devices on the dangerous parts of machines and transmission systems, and *(b)* poor maintenance and control of machinery prior to use or total absence of maintenance and control. In the various industrial sectors, certain causes of accidents are particularly common in developing countries. In the extractive industry, the most frequent cause of accidents is cave-ins: piece-work has often been found to induce miners to be more careless about timbering; certain tunnelling techniques, such as the old method of chambers and pillars, are particularly dangerous; it is also quite common for miners not to have adequate personal protection and even to work barefoot in the tunnels. The construction and civil engineering sector,which comes immediately after the mining industry as a source of serious accidents, generally has its highest accident rate in demolition and excavation work, which is carried out by untrained workers. Instances of suffocation in confined spaces occur year after year in farms, shipyards and the entire manufacturing industry. In some countries, especially in the Near and Middle East, road transport accidents are regularly among the three principal causes of lost days of work.

Although agriculture tends to be less dangerous than industrial work, so many more people are exposed to its hazards in industrialising countries that the absolute number of victims of agricultural accidents is often higher. Here, the main cause of accidents is not mechanisation, as machinery is usually operated by skilled workers, but the more insidious risk posed by the widespread use of chemicals in a rural environment that has received no background training. However, the work in the open air and the continuous changes of hazard reduce the danger of acute poisoning. Moreover, after the initial stages, in the long run the conditions of work and life in the rural environment can be expected to improve as incomes rise and the general level of know-how increases.

In industrialisation, while the severity of accidents seems to be mainly influenced by technical factors, their frequency is more generally attributable to the human element. Preventive measures must therefore concentrate *(a)* as far as undertakings with sufficient capital are concerned, on the purchase or importation of fully protected—in other words, properly designed—machines, *(b)* as far as small enterprises are concerned, on the creation of technical prevention advisory services, which could be backed by social security institutions or occupational injuries insurance schemes (because in the long term, it is in their interest to promote prevention and they do, in fact, have the means to do so), and *(c)* on facilities for carrying out a periodic control of lifts and elevators, boilers and pressure vessels. At the same time, workers involved in the industrialisation process must be made as safety conscious as possible. Although the organised sector may represent only a fraction of the active population, it is an excellent target for such measures as it is easily accessible, being concentrated in undertakings, it is more open to new techniques bringing with them a new sociocultural outlook, and it is ideally situated to set an example for the rest of the population as to the proper approach to adopt towards the hazards that industrialisation is introducing into the country. For various reasons, however, an efficient safety and health inspectorate does not always exist in these countries.

Impact of industrialisation on occupational diseases

Because of their severity, pneumoconioses keep the first place among occupational diseases. Systematic data on the subject are compiled by the ILO in its international reports on the prevention and suppression of dust in mining, tunnelling and quarrying. Its findings point to high levels of pneumoconiosis in several industrialising countries despite the very high annual labour turnover in some mines (up to 100%) which shortens the duration of exposure but at the same time cuts out medical supervision and chances of being registered. Even so, the recorded frequency of silicosis and other forms of pneumoconiosis in miners at work can be as high as 10-12%. In some of the smaller mines, especially, pneumatic drilling is still carried out in dry conditions, sometimes because wet methods cannot be used owing to the shortage of water or to the excessive humidity of deep mines in tropical areas; ventilation does not always keep pace with production increases; often, the parameters of the silicosis hazard and, sometimes, the actual presence of free silica in the ore have not been identified and, as a result, not even the most elementary precautions or reliable preventive measures are taken. Medical prevention is compulsory in a fair number of countries but compensation, social security and rehabilitation facilities are so hopelessly inadequate that in many cases workers suffering from silicosis have no option but to continue working as long as possible. Rapid labour turnover increases the risk of contracting tuberculosis or silicotuberculosis, and of course the mere fact of dismissing workers who are found to be suffering from the disease does nothing to solve the medical problem.

The use of heavy metals and various chemicals in manufacturing has a direct bearing on the appearance of numerous cases of poisoning which are not generally covered by national statistics but come to light in the course of surveys carried out at the level of the undertaking. The development of a mining, smelting and manufacturing complex was responsible for many cases of lead poisoning and of an excessive absorption of lead even in the workers' families. The concentration of lead in the atmosphere in lead retrieval and battery factories has been found to be up to 200 times the exposure limit. Although the number of cases of lead poisoning is not known, the melting down of old pieces of lead in Asian and African villages without any kind of protection is current practice. Several countries have introduced medical supervision regulations for workers exposed to lead but the laboratory tests they provide for are sometimes so complicated and costly as to be impracticable.

The use of toxic solvents without any protective measures has also been found to be dangerous. Turkey's footwear industry started using benzene instead of benzine, which was more expensive and a less efficient solvent for rubber, between 1955 and 1960. The first instances of aplastic anaemia were diagnosed in 1961 and the first cases of leukaemia from 1967 onwards (see table 2). Most cases were discovered in small, badly ventilated workshops operating under poor conditions of health where the concentration of benzene did not drop to the maximum exposure limit even outside working hours and, in the course of the day, could be as high as 200 ppm or more. The adoption of a law banning the use of benzene as a solvent in 1969 resulted in the disappearance of these diseases. In Morocco numerous cases of motor and sensory polyneuropathy were discovered in the footwear industry following the transfer of a technology which had already been widely responsible for the disease in European countries.

Table 2. Annual cases of leukaemia among footwear manufacturers in Istanbul, 1967-75

Year	Cases of leukaemia	Year	Cases of leukaemia
1967	1	1972	5
1968	1	1973	7
1969	3	1974	4
1970	4	1975	3
1971	6		

For the same volume of work, risks are often greater in industrialising than in industrialised countries. In India, for example, where the consumption of organochlorinated insecticides per hectare of cultivated land is roughly 30 times lower than in the United States or the United Kingdom, the concentration of organochlorines in the fatty tissue of the population is 2 to 6 times higher. This may be caused by genetic factors or by the type of food taken or, more simply, by poor hygiene or the use of improper working methods, such as fertiliser spreading by hand.

Mechanisation is often a cause of excessive noise levels, and health surveys in industrialising countries have registered levels of between 112 to 118 dBA in shipyards, 106 to 112 dBA in boilerworks, 94 to 108 dBA in a cement factory, 94 to 102 dBA in sawmills, 98 to 107 dBA at a thermal power station, 88 to 102 dBA in a spice mill and 95 to 97 dBA in the textile industry. In most countries no tests are yet carried out for hypacousia, which usually is not even registered as an occupational disease. Comparative investigations, however, suggest that the prevalence of hypacousia is very high compared with that of other occupational diseases (see table 3).

Table 3. Cases of various occupational diseases registered in the Republic of Korea in 1974

Total	2 862
Pneumoconioses	817
Poisoning:	
metals	144
gases and solvents	127
Hypacousia caused by noise	1 332
Dermatoses	406
Others	36

Several hazards are made particularly severe by a combination of working conditions and local factors, such as tropical heat that precludes the use of a good deal of personal protective equipment including safety glasses. Thus, in 1968, foreign bodies in the eyes affected 17% of a population of 23 000 persons working in open-cast mines in New Caledonia, where the sun, wind and dust account for the increased frequency of pterygiums in damaged eyes. Schistosomiasis is a frequent result of the construction of dams, artificial irrigation and the migration of infected workers being a major contributing factor where no elementary measures of hygiene are taken. A similar aetiopathogenic explanation can be found for an epidemic of sleeping sickness which occurred among workers on the Kariba dam in Zambia.

On other other hand there are diseases connected with the working and living environment that are very much more common in industrialised than in developing countries, such as cancer of the lung, rectum, prostate, breast and uterus (cf. Higginson and Tomatis). Whether this is a direct effect of exposure to the pollutants used in industrial processes or an indirect result of a higher standard of living involving unknown pathogenic mechanisms is not certain and research so far has not been able to provide an answer. It is, however, possible that a high intake of meat, fat and refined carbohydrates has some bearing on the aetiology of recto-colic cancers.

Another set of occupational diseases to which workers in developing countries may be less susceptible than those of industrialised countries is occupational dermatosis, since a more pigmented and thicker skin is more resistant to attack; however, this advantage may be cancelled by less hygienic working conditions.

Summing up the meagre data available at the level of the undertakings, occupational diseases, of which national statistics almost invariably contain no trace, are more common in relation to the number of persons exposed and, above all, more serious in the industrialising countries.

Of the two fundamental methods of health protection at the place of work–industrial hygiene and medical supervision–the former tends to be particularly neglected during periods of industrialisation. Medical supervision, of course, does not exist everywhere and is often neither as efficient nor as specific as it should be, but it is sometimes the only means of protection against occupational diseases and, in some areas, the only form of medical assistance that is available to workers. It is particularly important for the more vulnerable groups, such as those suffering from endemic diseases or undernourishment, pregnant women, adolescents and migrants.

A specific long-term International Programme for the Improvement of Working Conditions and Environment (PIACT) was recently launched by the International Labour Organisation. Even before this programme, many industrialising countries were receiving international co-operation in the establishment of properly equipped and staffed industrial hygiene laboratories. Since the prerequisite for the prevention of diseases caused by the use of chemicals is the generalisation of medical analyses rather than their absolute accuracy, preference should here be given to simple methods which are nevertheless adequate considering the magnitude of possible errors. At the national level, moreover, these laboratories should be designed as the essential machinery for long-term routine work rather than as more or less sophisticated research instruments with limited objectives.

Further upstream, there are the problems inherent in the transfer of technology and technical know-how, material resources and socioeconomic options of the country. The sphere of action of industrial physicians and safety technicians is therefore not broad enough on its own, and constant attention must be paid to the education of the workers and the general awareness of employers, as well as to their appreciation of the close link that exists between working conditions, working environment and productivity (see PRODUCTIVITY AND SAFETY AND HEALTH).

Working environment and conditions

Generally speaking, major undertakings in industrialised countries are more aware than small industrial enterprises of occupational safety and health issues and are more inclined to conform to such laws and regulations as exist concerning occupational health services and enterprise safety and health committees. Some of them have their own industrial hygiene laboratories and, in the more backward regions, their medical services are available to the surrounding population. On the other hand, much of the arduous and particularly dangerous work is often contracted out to undertakings whose personnel are covered neither by a permanent contract nor by social security, have no specific training in the risks involved in their work and have no personal protective equipment. Furthermore, because there is no medical supervision and workers disappear from the labour market as soon as they become disabled, nothing is known about the type of illnesses they may have contracted.

Most new undertakings are in the middle-size bracket and usually employ modern safety techniques; however, as far as noise levels are concerned they are often unsatisfactory. Factories that belong to multinational enterprises are in a separate category as they usually tend to introduce the same techniques and methods of organisation as in the country of origin. As a result there is a striking contrast in the least developed countries between the working environment in these enterprises and the sometimes deplorable safety and health conditions prevailing in the older, national enterprises.

Small undertakings employing up to 50 workers are much more common in industrialising than in industrialised countries. As industry begins to take off and to create employment opportunities for more and more workers, the economic benefits, which at the beginning are restricted to a small minority of industrial workers who are the only ones with a guaranteed wage, tend to spread first to the services sector, which develops more quickly, and subsequently to the informal sector. Gradually the standard of living around the factory rises and education improves. The sacrifices endured by the first generation are to the advantage of the following generation of workers, who have greater security of employment, are less anxious for their families, are better educated and are more aware of their rights, of the value of health and of the importance of looking after it.

The working conditions of women vary greatly from one country to another. Usually, the surplus supply of male manpower restricts women to jobs to which they are thought to be particularly well suited–in textiles, clothing, canning, electronics assembly, office work. Elsewhere women tend to be looked upon as marginal manpower. The worst conditions of all are usually to be found in the home-work sector (see HOME WORK).

Overtime is a particularly widespread practice in industrialising countries. While it enables workers to increase their income and employers to cope with temporary increases in production without having to take on extra staff, however, the system is also open to abuse. One finds, for example, alternating shifts of 12 h, without any break or meal, being kept up for several months in extremely dangerous construction work and 14-h uninterrupted shifts in agricultural enterprises at the height of the season. Situations such as these are responsible for a higher rate of accidents and chronic exhaustion (as well as a higher labour turnover and absenteeism), especially when the workers involved are undernourished and employed in harsh climates, sometimes without food or drink for the entire day because of ritual fasts. Even leaving aside such flagrant cases of abuse, workers in small enterprises and family undertakings often work more than 48 h a week on a regular basis.

Ergonomics

Industrial equipment, which is built on a larger scale in industrialised countries, is often imported directly into industrialising countries and used without any modifications. The difficulties this poses are numerous. To begin with, equipment may not be suited to the physical build of the worker, as is all too clear in the case of the driving cabins of heavy vehicles and the controls of agricultural machinery. Problems also often arise in terms of the interface of man and machine, as regards signals and controls, danger markings, warnings written in the language of the country of origin and safety posters, which may be quite incomprehensible to the workers. The frequent absence of even the slightest familiarity with machinery, too, may be the cause of errors that would be unthinkable in industrialised countries.

In other instances, when the machines are installed no allowance is made for local weather conditions or for the need for ventilation because of cramped premises. Technical errors regarding the ventilation of places of work are frequent and sometimes so serious that a system which is supposed to protect the worker in fact has the opposite effect of further raising the temperature. Because of a failure even to understand the problem, laziness may be blamed for a situation that is caused by conditions that make work quite impossible.

The lack of the most basic study of working conditions sometimes accounts for unnecessary and arduous physical effort: a technical co-operation mission visiting an agricultural undertaking in Africa recently observed a woman bending down to a container of water on the floor to rinse the cloth that she was using to wipe fruit passing on a conveyor belt at a height of 1.20 m, whereas raising the container by means of any kind of stool would have saved her the need to repeat the operation 600 to 700 times a day.

If it is to be effective, the ergonomic approach in these countries must at least be broad enough to encompass the principles of social psychology and occupational health. As far as industry employing advanced technology is concerned, the training must be directed first and foremost at the people who hold the key posts in labour programming and in the design of industrial plant. For the rest, the training of middle management personnel and trade union representatives must be pursued so that working methods can be improved, often by very simple measures that can be implemented immediately: the introduction of elementary checklists could be particularly useful at this level.

Living conditions

Urbanisation is generally associated with rapid industrial development. The hope of finding better living conditions through the diversified activities that a town can offer draws vast numbers of people to the large urban areas in search of work. Over the past 10 years, the urban population of industrialising regions has increased by 16 million units on average every year, half of which is attributable to the rural exodus; in the city of São Paulo, the population trebled between 1950 and 1970, two-thirds of the increase being the result of immigration.

Municipal services are not always able to keep pace with such rapid expansion, and this usually leads to a shortage of drinking water. Water shortage, the population density and the absence of sanitary measures are responsible for the high rate of gastrointestinal infections; poor housing contributes to the spreading of disease. In 1974, according to United Nations data, the proportion of the population living in virtually inhuman conditions was 90% in Addis Ababa, 67% in Calcutta, 61% in Accra, 60% in Bogotá and Ankara. In 1970, 52% of the population of Kenya was living more than three people to a room.

When worksites are isolated, workers' housing is generally provided by the employer on plantations, in the extractive industry (mining and petroleum), in the building sector and in public works (dams, roads, railways), but the facilities do not always conform to minimum standards of hygiene and comfort or may be overcrowded. Further problems arise when workers have to leave their homes as soon as their employment comes to an end, even though this may be because of an accident or disease.

On the other hand, there are a number of major undertakings that have not only built decent housing units for the workers and their families but have also established all the basic social services. The arrival of an undertaking in a backward area is often marked by the laying on of drinking water, the construction of a drainage system, the organisation of refuse disposal, the opening of supermarkets, hospitals and schools and the disappearance of malaria and trachoma.

For workers in industrialising countries the journey from their home to their place of work is often a source of physical and nervous strain that adds to the fatigue of a day's work: in many African countries workers sometimes have several hours' walk to their jobs, which means that there is virtually no opening for women.

Prevention

Legislation. In the history of the industrialised countries, occupational safety and health legislation has played a major role in the past and continues to do so today, and international progress in labour protection has been marked by the adoption of certain conventions and recommendations. In industrialising countries laws and regulations have often been drafted very much along the lines of those of the industrialised countries. In many cases the legislation introduced by the colonial powers has been kept virtually unchanged since independence. Elsewhere, new legislation has been drafted with the assistance of European experts who were not necessarily familiar with the special problems of developing countries. Consequently, the legislation does not always correspond to the specific needs of the country but, rather, to a particular institutional concept borrowed from distant models–though perhaps applicable at some time in the future. A purely preventive occupational health system is no use so long as there is a shortage of doctors; safety and health legislation should not be allowed to provide over-protection for workers in the organised sector by comparison with that of workers in the informal sector or with general health conditions in the country.

Safety and health legislation in some countries is virtually non-existent, and almost nowhere is it kept up to date with developments in the industrial field.

Invariably, the most serious difficulties arise with the enforcement of the legislation, especially if it contains detailed technical provisions. Thus, medical provisions are at least formally applied whereas often technical regulations governing occupational safety and health, if they exist at all, are not.

Moreover, it is a common weakness of official inspection services that they suffer from a shortage, or even complete lack, of personnel with proper technical training and have no sampling and analytical equipment for measuring and controlling hazards at the workplace. Even so, providing a budget for inspection services and for the training of inspectors is a major problem. Despite these difficulties the adoption of legislation in industrialising countries is still necessary and fundamentally useful, the basic issue being not so much whether it exists but what it says and how it is applied. This is an area in which international co-operation can be of great assistance to the developing countries.

Insurance. The preventive role of insurance against occupational injuries has been amply demonstrated. Information of employers and the application of differential rates for insurance premiums according to the degree of risk have frequently had good results. More recently, the substitution of social security schemes for the former system of employer responsibility has put an end to multiple premiums, and insurance institutions have found other ways to show their interest in prevention. They may, for example, finance prevention institutes at the national level, such as the Fundacentro in Brazil. Institutes such as these, which provide employers with technical advice and have no powers of enforcement, do not serve the same purpose as official inspection services and, if they are given the necessary resources, can ideally be made responsible for the more technical aspects of prevention. The frequently tripartite nature of the social security administration is a promising basis for the tripartite management of these activities, which is usually a positive factor in public education and awareness and guarantees that concrete measures are taken in the undertaking. Recent improvements have been made in several countries in their insurance schemes, with the more socially advanced countries gradually moving towards a more liberal interpretation of the concept of industrial accidents and occupational diseases (see SOCIAL SECURITY). In some countries, however, insurance institutions do not engage in preventive activities and do not even compile statistics for preventive purposes. In such cases, disability compensation often falls short of the basic needs of the victims of occupational injuries; because of the surplus of manpower, rehabilitation is non-existent; medical supervision of workers exposed to hazards becomes meaningless and works physicians are led to operate on a purely selective basis.

Education and training. Many employers and workers in small undertakings in industrialising countries are unfamiliar with even the most rudimentary standards and principles of prevention. This explains not just the high rate and severity of accidents and poisonings but, above all, the fact that these accidents occur again and again and that the more serious they are the more fatalistically they tend to be accepted. One can come across small sawmills, for instance, where dozens of workers have lost fingers and yet the machines are still not protected. Finding ways of reaching these people, however, is no easy matter.

A possible line of action that seems to hold out some hope, judging from the findings of PIACT, involves two main stages: first, a better social policy which recognises that a proper health and safety infrastructure must accompany the process of industrialisation and not follow it and, second, safety and health training for workers. Obviously this is bound to be a long process. The immensity of the task, to which international co-operation can contribute in all kinds of ways, means that a simultaneous effort must be made to train instructors, introduce the teaching of basic principles at school and use the mass media to inform the public. The effectiveness of the latter in developing countries should not be underestimated, if one is to judge from the encouraging results of television programmes on the prevention of cancer in women shown in Singapore and a series on food sponsored by a Middle East petroleum company.

PARMEGGIANI, L.

The impact of industrialization on environment and health. Paper prepared by the UNEP Secretariat with the co-operation of and input of ILO, WHO and UNIDO. III General Conference of UNIDO. New Delhi, India, 21 Jan.-8 Feb. 1980. ID/CONF/4.13, 5 Dec. 1979, 16 p.

Influenza

Influenza is an acute respiratory disease of epidemic occurrence caused by one of the influenza viruses. The onset, after an incubation period of 24-72 h, is abrupt with fever, severe muscle aches, scratchy, sore throat, and cough. The constitutional symptoms are often more prominent than the respiratory ones. The course is usually self-limited with fever lasting only 3-4 days. Bacterial pneumonia is the most important complication. The importance of the disease lies in its rapid spread in epidemics involving a large part of the population. It may have a severe or even fatal outcome in the elderly or those debilitated by chronic diseases of the heart, respiratory tract, etc.

Epidemiology

This disease of man is the principal one that currently retains the capacity to cause world-wide outbreaks called pandemics. At various intervals of time pandemics have occurred which have been characterised by rapid spread, high attack rates and an increase in the death rate from respiratory and cardiac complications. In other years epidemics confined to one or more countries have occurred. Epidemics of influenza have been recorded and described since before the 13th century. Between 1510 and 1979 about 34 pandemics can be recognised. The last occurred in 1957-58. The pandemic of 1918-19 was the largest and most severe ever experienced. Twenty million of the world's population perished during this visitation.

The disease is prevalent at epidemic levels for only 6-8 weeks in any given locality. In years of high attack rates (25 to 40%), the economic loss from the disease is great owing to the disruption of industrial production and other business activity because of the large number of personnel who are ill.

The presence of an epidemic in a community can be strongly suspected on epidemiological grounds alone. Confirmation is by virus isolation and by demonstration of antibody rise in a sample of typical cases. The epidemic begins abruptly, reaches its peak incidence in 3 weeks and is over in another 3-4 weeks. The highest incidence is usually in school-age children and after the age of 20 the incidence declines progressively. At times the disease may linger at very low frequency and reappear in epidemic proportions in the community after a few months interval, as a second wave. The behaviour was observed in some countries in 1957-58 and was a prominent feature of the 1918-19 pandemic.

The usual season for influenza epidemics in temperate climate regions is from late autumn to early spring but outbreaks have been observed in summer. During pandemics and epidemics, the disease spreads from the initial focus along routes of travel and trade. In 1957 when the pandemic first appeared in the Far East, it rapidly spread and seeding of the other countries of the world occurred during the summer and early autumn. In many countries the seeding was indicated only by sporadic cases and focal outbreaks, whereas the full-blown epidemic did not begin until late autumn or early winter.

Aetiology

There are three types of influenza viruses; influenza virus types A, B and C. These have similar physical, chemical, and biological properties, but are antigenically unrelated. Types A and B are the aetiologic agents of epidemic influenza; influenza A is also the cause of pandemics. Influenza virus type C has been implicated only in sporadic outbreaks and single cases.

One of the hallmarks of influenza A viruses is their ability to vary their antigenic make-up at frequent intervals. This has resulted in numerous variant strains which tend to group themselves serologically into families or subtypes with closer resemblances among family members than to strains of the other families. Five subtypes have been recognised in the human population. The period of prevalence of each family has been 10 or more years. The "swine" family was prevalent from 1918-28 and was the virus of the pandemic of 1918-19. Hong Kong influenza virus H_3N_2 was the only influenza A subtype prevalent in man from 1968 to 1977. In 1977, strains of the HIN family which had not been seen since 1957 reappeared first in China and then in Russia and, in early 1978, in many countries. To date (summer 1979) it has been isolated from many outbreaks of disease, almost solely confined to children and young adults with no previous experience with the subtype. However, strains of the Hong Kong H_3N_2 family continue to be isolated. This is unique, since prior experience has been that when a new subtype emerges, the old vanishes.

Changes in the antigenic composition of strains occur from year to year during the prevalence of a family but are of lesser degree than the change giving rise to a new subtype.

Antigenic variation also occurs among strains of influenza B viruses, but the degree of variation observed has been much less than among influenza A viruses.

Influenza in industry

During an epidemic the incidence of influenza illness among workers will, in most industries, be very similar to the incidence found for those of the same age in the general population. Influenza viruses are transmitted from individual to individual by large droplets of respiratory secretions. This type of transmission requires a close association between the infected and susceptible persons and thus the family unit is the principal source of infection for adults. However, in industries in which close contact between workers is a requirement or in overcrowded and underventilated plants, the incidence rate may become much higher than that observed in the population at large.

Prevention

It has been shown repeatedly that influenza virus vaccine affords a high degree of protection against influenza A and influenza B. When potent vaccines containing strains of influenza virus that are closely related to the epidemic strain have been tested in controlled field trials, the incidence of influenza has been reduced by 80% or more in the vaccinated group. In many countries a killed influenza virus vaccine containing both A and B viruses is used while a live, attenuated one is employed on a large scale in the USSR. Both are equally effective. Two types of killed vaccine are licensed in some countries, one a purified, whole virus vaccine, and the other a sub-unit vaccine where the immunising antigens have been extracted with various lipid solvents. The great advantage of the sub-unit vaccine is that it is much less reactive, and systemic reactions are rare. In experienced individuals, a single dose of vaccine gives an adequate antibody response; however, in inexperienced individuals, satisfactory antibody levels may only be achieved after two doses. The amount of antigen in a vaccine is given in terms of the international unit (IU) which is based on a standard dried reference preparation established by the WHO. Immunisation of a group should be completed in the autumn before the expected epidemic appears. Yearly vaccination is recommended for essential industries, as prediction of epidemics is very uncertain. If repeated annually, vaccination may considerably reduce sickness absenteeism due to the disease. In many countries the use of vaccines is limited to special occupational groups such as transport workers, hospital staff and other public service employees; however, the old and sick in whom the disease tends to be more severe possibly constitute the most important group for vaccination at present.

It should be stressed that unsatisfactory conditions of work, such as overcrowding and poor ventilation, should be dealt with rigorously, not only because of the likelihood of infection but because of the effect on the general health and well-being of the workers.

There is no specific therapy for influenza.

HENNESSY, A. V.

"Influenza". Davenport, F. M. Ch. 12 (273-296). *Viral infections of humans.* Evans, A. S. (ed.). (New York, Plenum Publishing Corporation, 1979).

Influenza: the virus and the disease. Stuart-Harris, C.; Schild, G. C. (London, Arnold, 1976), 242 p.

"The search for the ideal influenza vaccine". Davenport, F. M. *Postgraduate Medical Journal* (Oxford), 1979, 55 (78-86). 83 ref.

"Revised requirements for influenza vaccine (inactivated). Requirements for influenza vaccine (live)" (148-194). 7 ref. *WHO Expert Committee on Biological Standardization. 30th Report.* Technical report series 638 (Geneva, World Health Organisation, 1979), 199 p.

Information and documentation

The increasing complexity and diversity of technological processes, the growing number of chemical compounds that find their way into industry, agriculture and other branches of economic activity and, last but not least, the development of new tools and new techniques to identify, evaluate and control occupational hazards compel the occupational safety and health specialist to make arrangements for the systematic gathering of a substantial amount of information. The knowledge acquired by an industrial physician, a safety engineer or an industrial hygienist during the course of his basic and subsequent training periods is no longer adequate to enable him to provide satisfactory answers to all the problems with which he is likely to be confronted in his day-to-day activity. He will therefore be required to supplement this knowledge by drawing on various sources of information.

This need is less obvious but none the less present in the case of undertakings, institutions and persons who assume occasional or peripheral duties in the field of occupational safety and health. It is perhaps even greater in countries where literature and experience in this field are scant or non-existent and there is no possibility of obtaining competent advice on the spot.

Problems sometimes arise not because information is lacking or difficult to obtain, but because its mass is overwhelming. It is thus understandable that quality and selection of information are more important than quantity. If one excepts the larger corporations or institutions that are well endowed with library or similar facilities, difficulties arise in many cases where reliable, specific and reasonably adequate information has to be located and obtained at fairly short notice. The elimination or minimisation of occupational hazards calls for prompt and efficient action, and the time element therefore assumes vital importance.

The information cycle

The findings of research workers and the observations of specialists in the field can make no practical contribution to progress in occupational safety and health unless they are brought to the attention of those responsible for the implementation of safety measures and the supervision of safe working conditions. For this goal to be attained, three fundamental conditions must be fulfilled:

(1) The information must have been published in a "document", the expression "document" being taken in its widest sense.

(2) This document must be brought to the knowledge of persons whom it is likely to interest.

(3) These persons must make proper use of it.

The complete information cycle, therefore, comprises three main phases: the emission, transmission and use of information; these phases correspond to sectors (1), (2) and (3) of figure 1.

This diagram can be completed by indicating the secondary loops (2'), (4) and (5). Because of the proliferation of scientific and technical publications of all kinds, specialised documentation centres are coming to play an increasingly important part in the information field, sometimes themselves producing so-called secondary publications which periodically draw attention to significant work (loop 2'); this will be dealt with below. In addition, certain periodicals publish communications sent in by readers, and thus introduce an element of feedback (loop 4). Finally, the user may himself produce information (loop 5) which will be introduced in turn into the cycle.

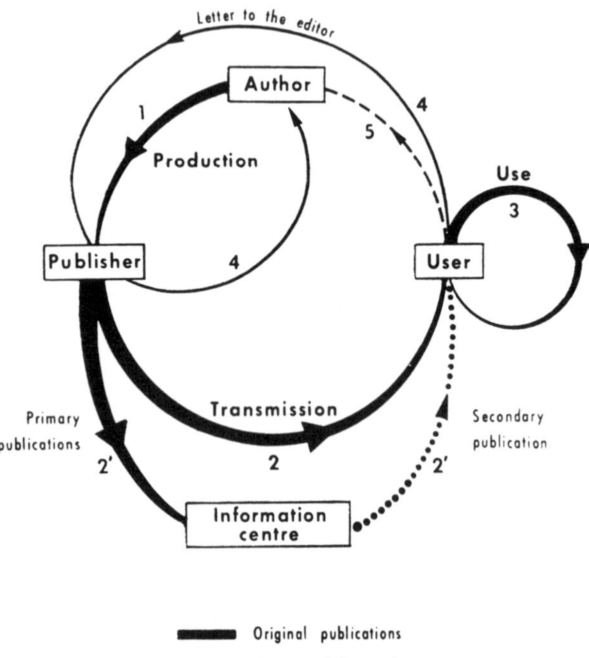

Figure 1. The information cycle.

For the information produced to fulfil its purpose completely, it is clearly not sufficient for the cycle outlined above simply to be closed: the information must also conform to certain criteria of quality, be presented in an appropriate form, be used systematically and with discrimination and, once classified and stored, must be easily retrievable. Practical suggestions relating to improved liaison between authors, editors and users of periodicals, the presentation and intrinsic quality of articles, and the possibilities of improving readers' personal information work were put forward and discussed as long ago as 1969 at a meeting of editors and users of occupational safety and health periodicals from 21 countries (International Symposium on Occupational Safety and Health Periodicals, ILO, Geneva, 7-8 July 1969).

The nature and sources of information

The time is long past when personal communications between individual research workers formed the main and sometimes only channel for the exchange of information. Periodicals (technical and medical), research reports, specialised works, proceedings of symposia and other meetings are now the main vehicles for the exchange of occupational safety and health information, without, however, containing all the useful information which is available. It is sufficient to mention laws and regulations, standards and directives, information sheets, toxicological data sheets, films, slides, theses, patents, manufacturers' catalogues, etc., to give some idea of the vast, heterogeneous mass of documentation which exists. Moreover, this mass, which already amounts to more than 40 000 documents a year, continues to grow rapidly.

The fact that these documents may be published in different languages—the main ones being English, French, German and Russian—creates another formidable obstacle to their utilisation. In any case the vast volume of documentation relating to the detection, analysis, evaluation and control of occupational hazards is such that exhaustive searches based on the original documents themselves have become practically impossible.

Reports of pure or applied research carried out at universities or specialised centres, or by learned societies, generally provide the most up-to-date information. Books are much less up to date at publication, while in this respect, periodicals occupy an intermediate position. These differences very largely explain the increasing favour enjoyed by meetings of specialists, which offer the possibility of spontaneous reciprocal exchanges. Their growing number and the fact that they are becoming increasingly scattered geographically, however, make it impossible for specialists to participate in as many of them as they would like.

For the great majority of specialists, therefore, periodicals continue to offer the most convenient means of keeping abreast of the latest developments and trends. Approximately 70% of useful documentation relating to occupational safety and health consists of articles appearing in periodicals and 10% of the periodicals in question—which can be considered as "basic" periodicals—contain almost two-thirds of the useful information, the remaining one-third being very widely scattered in more than 1 300 titles.

It would be wrong, however, to conclude from this that all periodicals hold the same interest for the occupational safety and health specialist. The quality of specialised periodicals is in fact very uneven, even allowing for the varied interests of their readership; some of them publish a considerable proportion of repetitive and trivial articles and information of secondary interest which frustrate the readers.

Generally speaking the reader will be wise not to assume that mere publication of information is a guarantee of its accuracy or value, or to accept an article uncritically simply because the author is well known through having published many articles. He would be well advised to place more reliance on work published by institutions with a solid reputation (labour inspectorates, public health inspectorates, specialised occupational safety and health organisations, accident insurance institutions, chemical manufacturers' associations, standards institutions, etc.) or under their auspices.

The role of documentation services

The increasing volume and complexity of the documentation relating directly or indirectly to one or another aspect of occupational safety and health, together with the difficulty of locating, obtaining, scrutinising, classifying (indexing), filing and retrieving the documents themselves, confers more importance to the documentation or information officer as well as to the so-called "secondary" publications.

Whereas the task of the librarian formerly was to store documents which provided replies to inquiries addressed to him (passive information), the modern documentalist must be capable of anticipating the needs of customers, with whose basic interests and temporary preoccupations he should therefore be familiar (active information). Since his task is to provide information discriminately and promptly, he must master information-processing techniques and have sufficient scientific knowledge to assess the value of the documents which he scans and carry on a meaningful dialogue with users. It is equally necessary for him to know from which sources he can fill the gaps in his own service by contacting other documentation centres.

Secondary publications

The factors leading to the setting up of documentation and information services in establishments with sufficient resources to afford these has also led to the establishment of centralised systems to cope with the mass of primary documentation (that is to say, the original documents themselves) and pass on its substance to a large circle of potential users. In this way the latter are freed from the onerous task of scanning which would normally be theirs and they can concentrate their attention on those documents which seem to them to be the most useful.

This has resulted in the appearance of a number of "secondary" publications, the purpose of which is to draw attention to the publication of important primary documents; generally, a summary or abstract of the document is provided, accompanied, where appropriate, by some indexing symbol to facilitate classification and subsequent retrieval of the information provided.

These secondary publications—whether abstract bulletins or cards, or simple lists of bibliographical references—are generally very selective. Selection is determined not only by the intrinsic value of the documents themselves, but also, and to a very large extent, by the availability of the resources required for scanning in a wide range of disciplines and languages, the cost of which is relatively high. Certain sectors (e.g. ionising radiations and nuclear energy, mines) and certain languages (English and Russian) enjoy a privileged position. In the field of occupational safety and health, the most comprehensive computerised abstracting service at present is that provided by the International Occupational Safety and Health Information Centre (CIS), established in 1959 by the International Labour Office (see separate article). Another widely used service is that provided by the Occupational Health and Industrial Medicine series of *Excerpta Medica*.

Making use of information

However great the effort made by publishers and editors in compiling, preparing and presenting information, a substantial effort is still required of the user, despite the wealth and refinements of the modern data-processing techniques at his disposal.

The practical value of documentation increases proportionately with the degree of relevance and rapidity with which individual documents can be retrieved; the method by which the information is abstracted and classified obviously plays a determining role.

There are *a priori* no good or bad classifications; the systems chosen are valid for a given group of users only and by virtue of their ability to provide the services required. In the field of occupational safety and health, where the volume of information to be processed, despite everything, remains within definite limits, experience allows a number of general rules to be established.

The terms retained for the purpose of indexing a document can be determined subjectively or statistically. At the present time the process of abstracting and classifying documents remains a clearly intellectual task. Moreover, indexing based on consideration of the title alone is a crude and deceptive process which should be abandoned.

Any form of indexing, whatever its depth, results in practice in an appreciable loss of information. Certain information systems have adopted two levels of indexing—one (using primary descriptors only) for the compilation of annual subject indexes, and another (including secondary descriptors allowing for a greater indexing depth) for the performance of more specific or more comprehensive searches.

The better the classification system and the method of searching, the easier and more efficient will information retrieval be. It is essential that each inquiry be formulated as clearly and precisely as possible, so as to restrict the search to relevant aspects alone.

For example an inquirer interested in cadmium proteinuria among workers in battery factories should direct his search in the first place towards documents which combine the three facets "cadmium", "proteinuria" (or "kidney disorders") and "battery manufacture" (or any equivalent descriptor), rather than considering all documents relating to other hazards associated with battery manufacture (1, 2, 3, 4), to other substances likely to cause kidney disorders, but which are not encountered in battery manufacture (5), to other industrial applications of cadmium (6) or to other pathological disorders which could be caused by cadmium (7).

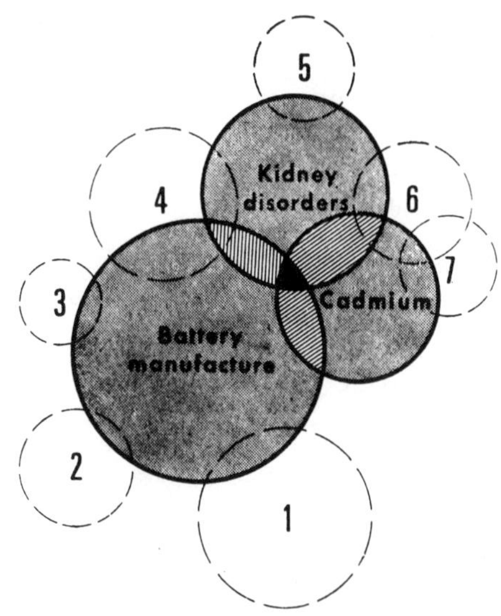

Figure 2. Search strategy.

If, in the light of the information found, he now wishes to extend the scope of his search, he can include documents in which the three facets "cadmium", "kidney disorders" and "battery manufacture" are combined in pairs (shaded areas in figure 2).

Finally, if he wishes to make an exhaustive search, he can next complete his investigations by including documents which include one of the facets "cadmium", "kidney disorders" and "battery manufacture" (grey areas in figure 2). The value of this search strategy obviously increases as the mass of documentation to be searched also increases and the user's interest becomes more specialised.

In countries which do not yet have sufficient resources to produce or, at least, to assemble adequate documentation, persons responsible for occupational safety and health may experience considerable difficulty in locating, selecting and processing the information which they need. In this case they will find it to their advantage to approach either the regional offices of the competent international organisations (ILO and WHO, in the first place) or a specialised institution of a country whose language is accessible to them. They may wish to make use also of the various information networks providing access to computerised data bases in the field of occupational health and safety or related fields. Examples of such data bases are:

- TOXLINE, operated in the United States by the National Library of Medicine and containing more than 500 000 references to human and animal toxicity studies, effects of environmental chemicals and pollutants, adverse drug reactions, etc.;
- RTECS (Registry of Toxic Effects of Chemical Substances), compiled by the National Institute for Occupational Safety and Health (NIOSH) in the United States and providing toxicity and related data (TLVs, recommended standards in air, etc.) for more than 30 000 substances;
- CHEMLINE, operated jointly by the National Library of Medicine and Chemical Abstracts Service;
- ECDIN (Environmental Chemicals Data and Information Network) developed by the European Communities in Ispra, Italy, to provide information on chemicals potentially dangerous to the environment and human health;
- CANCERLINE, operated by the National Cancer Institute in the United States;
- MEDLINE, operated by the US National Library of Medicine and containing more than 600 000 references to biomedical articles from more than 3 000 journals; etc.

Computers in documentation work

Whenever any task appears to be beyond the capacity of humans, it is fashionable nowadays to assume that it must be entrusted to a machine. The development of electronic data processing and the availability of increasingly powerful computers have led to their use in documentation work. According to circumstances, this may be fully warranted, desirable, premature or totally unjustified on economic grounds. Moreover, there is a general tendency to underestimate the amount of preparatory work input required for storing data in a computer's memory.

If computers are remarkable for numerical calculation, they have not been designed for automatic information processing. The only logical operators available being "and", "or", "except", the result is that even if the search equation is correctly put, an appreciable percentage of documents are missed or are irrelevant. This situation, which is tacitly accepted, imposes the need for a complementary "human" search and check.

It must further be recalled that in the great majority of cases, electronic data-processing techniques provide only references to documents; they do not eliminate the need to consult these documents to locate the information required. A thorough and objective evaluation should therefore be undertaken, covering all the operations involved, so as to determine in each case the economic feasibility of using a computer. It will be found in many instances, where the number of documents to be handled does not exceed say 30 000 to 40 000, that conventional methods of information indexing and retrieval yield results which, from an over-all point of view, are more satisfactory to the user and at any rate more economical in their operation.

ROBERT, M.

List of periodicals abstracted. International Occupational Safety and Health Information Centre (CIS) (Geneva, International Labour Office, 1978), 95 p.

CIS Thesaurus. Occupational Safety and Health. International Occupational Safety and Health Information Centre (CIS) (Geneva, International Labour Office, 1976), 100 p. (plus 4 supplements, 1978, 1979, 1980 and 1981).

International Symposium on Occupational Safety and Health Periodicals. Reports and conclusion (Geneva, International Labour Office, 1969), 139 p.

"Recent advances in information storage and retrieval relevant to occupational hygiene". Wright, R. B. *Annals of Occupational Hygiene* (Oxford), 1981, 24/3 (313-322). 6 ref.

Injection, accidental

Injection injuries of the hand are apparently increasing with the multiple uses of high-pressure spray and injection equipment. It seems that such injuries occur more frequently than is commonly recognised.

The pressures applied to the supply containers are often deceptively small, 70-200 N/cm², but final delivery through a nozzle or spray head with a fine aperture creates pressures from 340-7 000 N/cm². These conditions are met by several common types of devices:

(a) the fuel injection systems of internal combustion engines, particularly diesel engines;

(b) airless spray paint guns;

(c) pressure lubrication devices (grease guns);

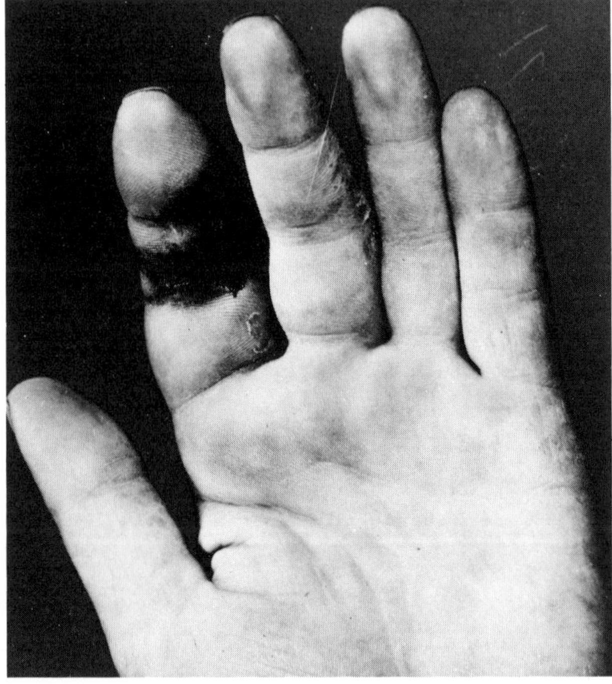

Figure 1. Grease gun injury 72 hours post injury. Note wound of entrance near flexion crease. Tendon sheath filled with grease.

(*d*) semi-solid propelling devices for asphaltic sound deadening and rustproofing materials for automobiles;

(*e*) injection moulding machines in the plastics processing industry.

In this context one is dealing with pressures that are most often considerably in excess of the 520 N/cm² used to create injuries experimentally, or 450 N/cm² known to have caused injury clinically.

At these pressures the jet is capable of penetrating skin and the soft tissue of the hand from some distance.

Accidental injections are probably most common in garages and service stations. A grease gun may slip from the nipple on a car chassis especially during the greasing of flexible brake cables; or it may be accidentally directed on to the hand, especially during testing or cleaning.

Injection injuries

Except in certain areas, the average doctor or surgeon is unlikely to have had experience in dealing with the injuries produced.

The initial injection injury often appears innocuous, usually a simple puncture; however, if the wounding device strikes the skin tangentially it can cause severe laceration. The injection factor is not usually appreciated for several hours when increasing pain caused by ischaemia is manifest. If the quantity of injected material is not sufficient to cause ischaemia then several days may elapse before the thermal or chemical irritation of the injected material is evident. Systemic symptoms may appear due to the solvents in wax or paint. Even at this juncture the physician may be unaware of the true nature of the problem.

Safety measures

Most of the devices involved can be modified to minimise the possibility of injury. As an example, the airless paint sprayer which delivers a solid stream of paint atomised by passage through a spray head with a fine aperture can be rendered relatively safe by the addition of a diffuser which minimises the possibility of injection if discharge occurs when the spray head is removed (figure 2). Semi-solid sealant guns can be made safer by the use of angled baffles which create turbulence, decreasing the efficiency of the device. Plastics injection moulders can be made so that the injection nozzle pressure is cut off if the dies are not properly in place. Pooling of information about safety modifications among manufacturers is much to be desired.

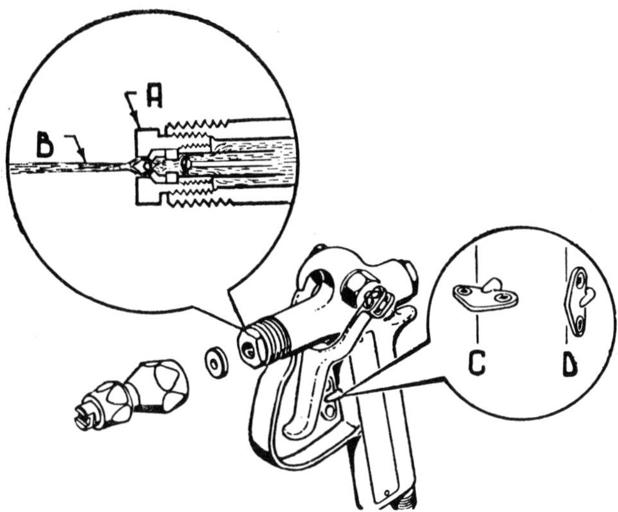

Figure 2. An airless paint gun modified so injury factor is reduced if spray head is removed. *(By courtesy of Gray Company, Inc., Minneapolis.)*

The problem of lubricating guns remains, however, because of the necessity of delivering the greases under high pressure through a small aperture. The aperture can be enlarged for safety, but if blocked partially or completely, the pressure can build up, creating a serious hazard. Even a simple hand-powered gun can deliver 7 000 N/cm². In addition to modification of the equipment, training and supervision of the workers is essential. Many operators have no conception of the inherent dangers of the appliances and suppliers do not usually stress the need for safety measures. Large industrial concerns with safety engineers or officers can foresee and prevent many injuries by demanding machine modifications from the manufacturers and by training of operators, but the problem is difficult in small undertakings, especially in such places as service stations and garages. Workers may be careless in using the equipment and may wound a fellow worker in an attempt to be humorous. Safety sleeves can be fitted but strict supervision is needed to ensure that they are used. Good lighting under cars lessens the risk.

Treatment

As first-aid, an Eschmarch bandage or rubber tubing should be wrapped tightly around the lower forearm and continued distally in an effort to force the injected material out of the site of entry. After removal of the reverse tourniquet, the hand should then be cooled to decrease the metabolic demand of the tissues during the period in which definitive care is being arranged.

The surgeon should obtain roentgenograms for soft tissue detail. The digit or involved area of the hand should be explored promptly and completely. The threat of unnecessary surgery is small compared to the damage inflicted by even relatively small amounts of injected grease or other materials. It is imperative that the tendon sheaths in the area be opened to rule out damage, which may extend far beyond the point of entry.

The injected material should be removed as completely as feasible. Expendable damaged tissue should be excised in so far as possible. Important functional structures may have to be selectively sacrificed to obtain reasonable salvage of remaining undamaged portions of the hand.

After the injected material has been excised, the wound should be closed. Electing to leave the hand wounds open should be a last resort and secondary debridement and delayed closure by suture of skin graft attempted in 4-5 days.

Unfortunately, many patients do not reach the surgeon until some days have elapsed. The hand at this stage is often inflamed and swollen and the need for decompression plainly apparent. Decompression will often arrest impending gangrene.

In some late injuries, digital or ray amputation may be the only logical course; in others, conservative excision may be productive of satisfactory salvage.

KELLY, A. P. Jr.

CIS 78-338 "High pressure paint gun injuries". Booth, C. M. *British Medical Journal* (London), 19 Nov. 1977, 2/6 098 (1 333-1 335). Illus. 18 ref.

CIS 76-1692 "Periocular injury due to projection of liquid under pressure–Prognosis–Course–Pathology" (Lésions périoculaires par projection de liquide sous pression–Pronostic–Evolution–Pathologie). Vérin, P.; Michelet, F. X.; Vildy, A.; Gilbert, D. *Archives des maladies professionnelles* (Paris), Oct.-Nov. 1975, 36/10-11 (593-600). 7 ref. (In French)

CIS 76-570 "High-pressure injection injury–Potential hazard of 'enhanced recovery". O'Reilly, R. J.; Blatt, G. *Journal of the American Medical Association* (Chicago), 11 Aug. 1975, 233/6 (533-534). Illus. 7 ref.

CIS 81-644 "Airless paint spraying techniques" (La pulvérisation sans air/airless–Techniques d'utilisation). *Journal de la construction de la Suisse romande* (Lausanne), 15 July 1980, 54/14 (36-37) and Aug. 1980, 54/16 (29-30). Illus. (In French)

Inks

Ink making is identified with the graphic arts but is classified as a part of the chemical industry. It is paradoxical that a substance so easily defined as the fluid or viscous material used for writing and printing has become so complex and specialised that nearly 1 million new formulae are compounded each year.

Manufacture. The ingredients used in the manufacture of printing inks fall into three main groups:

(a) the pigment or solid ingredients;

(b) the vehicle or fluid ingredients; and

(c) the miscellaneous other ingredients, principally driers and compounds which are used to impart special characteristics to the ink.

A relatively new class of inks which are "ultraviolet (UV)-cured" also contain the more conventional pigments. However, the components of the vehicle represent yet another spectrum of chemical compounds (table 1).

Table 1. Components of UV-cured ink vehicle

	Common additives
Pigment	All conventional pigments
Polyfunctional acrylic monomer	Pentaerythritol triacrylate (PETA)[1, 2] Trimethylolpropane triacrylate (TMPTA)[1, 2] Hexanediol diacrylate (HDODA)[1, 2]
Monofunctional acrylic monomer	Methyl methacrylate (MMA)[1]
UV reactive unsaturated polymer	Acrylated urethane polyester oligomer Acrylated epoxy resin oligomer[1]
Photoinitiator	Benzophenone[3] Amyl dimethylaminobenzoate (isomers)[3, 4]
Diluent	Primary and aliphatic alcohols[1] Phthalates[1]
Hydrogen transfer agents	Triethanolamine[1]
Miscellaneous compounds	Stabilisers; surfactants; fillers; flattening agents; polymerisation inhibitors

[1] Associated with allergic and irritant contact reactions. [2] Reported allergen or irritant in UV-cured ink handlers. [3] Associated with phototoxicity reactions. [4] Reported phototoxin in UV-cured ink handlers.

The first step in the manufacture of printing inks is the mixing of the pigments, vehicles, compounds and driers. This is usually accomplished in a change-can or dough-type mixer. The varnish is first placed in the mixer and then the pigment is added gradually and mixed in until it has been "wet down".

The manufacturing of marking and writing ink involves the mixing of a vehicle, a dye, and a resin, although a pigment may occasionally be used with a dye or by itself. Ink for ball-point pens is essentially of the same nature as printing ink and comprises a solution of pigments or resins.

The types of pigment vary with the colours desired. Common inorganic pigments include lead chromates, cadmium compounds, titanium dioxide, mercuric sulphide and earth pigments such as ochre, sienna, umber and complex iron compounds. Organic pigments are insoluble colours manufactured from natural or synthetic dyestuffs. They include Hansa and benzidine yellows, methyl violet and peacock blue, eosin, phloxine, toluidines, lithol reds, and phthalocyanine blues. These organic pigments are made by long and complicated chemical procedures from raw materials obtained from coal tar. Silver and gold metallic effects are achieved by incorporation of flakes of metals such as aluminium, brass and copper.

Driers act as catalysts and speed up the oxidation and drying of the varnish. They consist principally of lead, manganese and cobalt soaps. To a lesser extent, compounds of calcium iron, copper, zinc, and zirconium are also used.

Waxes are also added, either cooked directly into the varnish or prepared as a compound and added to the ink. Waxes serve primarily to prevent the smearing of ink from freshly printed matter to another surface with which the undried ink comes in contact. The most commonly used waxes are paraffin wax, beeswax, carnauba wax and ozokerite.

Greases and lubricants are added to help reduce the tack of an ink or its stickiness and pull. They also help lubricate the ink so that it will disperse and transfer properly. Commonly used greases and lubricants are cup grease, lanolin, petrolatum and tallow. Thin-bodied reducing oils are used in much the same way as the greases. High-boiling solvents and thinners may be used in some writing and marking inks and in letterpress and lithographic inks to reduce the tack. Body gum and varnish are used to add viscosity to an ink. They "pull" the ink together and help it to print sharply.

Antioxidants are sometimes used to reduce excessive drying and the formation of a film on the surface of a quantity of ink after a period of standing. These are very active chemically and should be used with caution. Corn starch and other dry powders such as dry magnesium oxide are used to "body up" an ink. In addition, surface active agents effect better dispersion of pigments.

After manufacture, inks are filled into bottles and packaged for despatch.

In the ultraviolet-cured ink process, a photoinitiator is exposed to an intense, continuous emission of UV radiation between the wavelengths 300 and 400 nm. The absorption of UV radiation leads to photoexcited states which promote the polymerisation of the vehicle in which the pigments are dispersed. Under optimal conditions this results in an almost instantaneous cure of the ink film.

HAZARDS AND THEIR PREVENTION

Many of the solvents used for ink manufacture are flammable or explosive substances and the fire hazard is a constant problem.

A wide range of mixing machines is used, and these machines may cause accidents if not adequately fenced. Finally, many of the ingredients are highly volatile or used in fine particulate form; there is a definite danger of intoxication if high atmospheric concentrations of certain of these substances are inhaled.

For the UV-cured inks, an ultraviolet light encasement is housed within the printing presses and coating machines. Exposure of the printed or coated surface to ultraviolet light results in immediate drying which permits increased press speed, more complicated multi-colour operations, elimination of smudging of printed surfaces and a significant reduction of storage space required for printed materials. Perhaps the greatest advantage, however, is the complete elimination of gas ovens needed for metal printing, thus conserving energy and doing away with costly engineering required to clear oven exhausts of solvent vapours that evolve with many conventional ink processes.

Fire and explosion. There is a serious risk of fire and explosion when flammable vehicles or solvents are used. Many flexography and rotogravure inks and most aromatic hydrocarbon writing inks fall into this category. Safety must be a prime consideration from the storage stage to waste disposal.

The vehicles should be stored externally, preferably in underground tanks, and piped into a special room where the mixing of inks with combustible vehicles is performed. All electrical installations and equipment should be flameproof, and ventilation should be adequate to prevent concentrations of flammable vapour. It is often advisable to have an exhaust system built directly over mixing vats so that the flammable vapours are led to the open. The mixing rooms themselves should be well ventilated. Any significant waste should be stored in underground tanks since disposal into conventional sewerage systems will almost certainly cause harmful water pollution. Exhaust ventilation is also required in filling and packaging.

Adequate, unobstructed fire exits and gangways, fire alarms, fire extinguishers and training of staff in fire procedures are essential.

Accidents. Ink manufacture has three mechanical processes where special attention to secure machinery guarding is required:

(a) the mixing arms of the mixing machines should be protected by an interlocking cover;

(b) the nips of the roller mills should be efficiently protected; and

(c) the revolving cages of the centrifuges should be provided with interlocking lids.

Eye accidents due to splashes may be an important hazard especially during filling. Eyewash devices and solutions should be kept close at hand. Personal protective equipment, including hand protection and eye protection, should be provided where necessary.

Health hazards. In the ink industry, health hazards may be due to either hazardous materials, hazardous conditions or a combination of the two. The hazards may be systemic, acting especially on the respiratory, cutaneous and nervous systems or related to a special organ such as the eye. Hazardous materials used include the chlorinated hydrocarbons, acids, antioxidants, acrylates, photoinitiators, azo dyes (some of which are mutagenic) and many caustic additives. Inhalation of vapours or direct contact with these materials are irritating and can produce conjunctivitis, rhinitis, dermatitis, dizziness, and headache. In high concentrations the aromatic hydrocarbons will produce acute narcotic effects. In the case of very large doses death may occur within a few minutes. Chronic exposure to the chlorinated hydrocarbons can produce toxic reactions in the liver, kidneys and other organs, depending on the compound in question. The use of benzene is highly dangerous; it is prohibited by law in some countries and should, in any case, be eliminated. The least toxic solvents practicable should be used in conjunction with efficient exhaust ventilation.

It should be noted that risks from solvents also occur in lithographic printing works where ink is supplied as a stiff mixture and adjusted with solvents by the users.

The fine particulate pigments used in the mixing process may be the source of high concentrations of atmospheric dust. Exhaust ventilation should be provided where necessary and respiratory protective equipment and eye protection should be made available for specific operations.

Many of the dyes, pigments, varnishes, waxes, driers, and minor additives may be both primary irritants and potential skin sensitisers. Compounds used in cleaning the skin of workers can also produce irritant reactions and sensitisation. Good sanitary and washing facilities, barrier creams and emollient creams for use after washing should be provided and thorough instruction in their use given.

As the applications for UV-cured inks continue to expand, it has become apparent that workers exposed to the multifunctional acrylic monomers may develop cutaneous sensitisation as well as irritation of the skin and conjunctiva (table 1). Allergic contact dermatitis to PETA has been reported in two independent studies, in addition to which TMPTA and HDODA have been recognised as sensitisers. Phototoxic reactions to compounds used in ink formulations as photoinitiators also have been reported among workers employed in the manufacture of UV-cured inks.

Certain particular materials used also have a primary irritant action on the respiratory system and most pigments, which are relatively inert dusts, produce low-grade lung tissue reaction but do not affect pulmonary function.

The prevention of many of the major hazards in the ink industry is linked inseparably to factory planning and requires design of special handling arrangements for volatile solvents and vehicles, flameproofing of rooms and provision of high standards of ventilation to ensure that atmospheric concentrations of toxic substances are kept below maximum permissible levels.

While UV-cured ink processes circumvent problems associated with the disposition of volatiles, the use of adequate personal protective gear for workers engaged in the newer ink technology is still essential to prevent potentially serious allergic and irritant reactions.

COHEN, S. R.
SAMITZ, S. H.
SHMUNES, E.

Technology:

"Inks". Dunn, H.; Burachinsky, B.; Ely, J.; Grubel, P. *Encyclopaedia of Chemical Technology.* Kirk, R. E.; Othmer, D. F. (eds.). Vol. 611 (New York, John Wiley and Sons, 1966).

Printing ink handbook. Technical and Educational Committee of the National Association of Printing Ink Makers (New York, National Association of Printing Ink Makers Incorporated, 1958).

Skin diseases:

"Phototoxicity occurring during the manufacture of ultraviolet cured ink". Emmett, E.; Taphorn, B.; Kominsky, J. *Archives of Dermatology* (Chicago), June 1977, 113/6 (770-775). 9 ref.

"Skin problems associated with multifunctional acrylic monomers in ultraviolet curing inks". Nethercott, J. R. *British Journal of Dermatology* (Oxford), May 1978, 98/5 (541-552). Illus. 42 ref.

Inland navigation

Although inland navigation was practised in antiquity through the use of rafts and other shallow river craft, it was only in the 15th century that rivers in Europe were linked by man-made canals to permit more continuous navigation. In the 16th century the first locks were built, enabling vessels to negotiate differences in water levels, and since that time rivers and canals have continually been widened and deepened in order to allow the passage of longer, faster and more economic vessels. Inland navigation developed very considerably in the 19th century, with the coming into use of steam as a means of propulsion. Whereas 100 years ago the largest inland waterway craft were only about 38 m in length with a tonnage of approximately 300 t deadweight, most rivers and canals in Europe can now accommodate the so-called "European barge" of 1 500 t and 80 m in length. In the United States, which is at present the most advanced country in this field, barges of 3 000 t or more are common. But there are still small sailing vessels on the rivers of Egypt and certain Asian countries.

Despite the severe competition it has suffered from railways for the last 100 years and from road transport during the past 50 years, and although it tended sometimes to be considered as slow and old-fashioned, inland navigation is still the cheapest method of inland transport available in terms of fuel consumption per ton of cargo transported. Further arguments in favour of the inland water transport industry are the ever rising cost of fuel and the need to save the world's organic energy resources. In industrialised countries inland waterways are far from being saturated and will continue to play an important role in the process of economic decentralisation undertaken by many countries, while developing countries are beginning to realise that the development and extension of existing inland waterways would greatly contribute to a better flow and distribution of goods and require less capital investment than the creation of extensive rail and road networks.

Types of inland waterway vessels

Self-propelled barges. These are motorised barges varying in length from 38 to 90 m and in tonnage from 300 to 2 000 t. They are found in Europe and certain developing countries and carry either general cargo or dry goods in bulk. Some are specially equipped for transporting petroleum and chemical products and a variety of other cargoes. There are also roll-on, roll-off barges with an access ramp (based on the landing barges used during the Second World War) enabling cargo to be rolled on or off such a barge without using quayside lifting equipment, and barges specially shaped for carrying containers.

Dumb barges. These are non-propelled barges of between 300 and 2 000 t (up to 3 000 t or more in the United States) for the transport of bulk or specialised cargoes; they are either pulled or pushed in groups by tow boats.

Shipborne barges. In recent years increased use has been made by some highly industrialised countries of the technique of carrying small barges (lighters) aboard large ocean-going vessels by a system, known as LASH (lighter aboard ship) for intercontinental transport, thereby avoiding any intermediate handling of the goods so transported.

Tow boats. Until 1960 the traditional method of transporting barges in Europe was by tugs towing a string of one to eight barges, depending on the characteristics of the particular waterway. Tows of up to eight barges, carrying a total of approximately 10 000 t and stretching over a distance of more than a kilometre are still a standard practice on the Rhine.

Push-towing (pusher tugs). In the 1930s a new method was developed in the United States on the Mississippi River; it consisted of pushing a train of barges rigidly connected together and to a powerful and specially designed pusher unit equipped with several steering and flanking rudders to ensure better manoeuvrability. This method, introduced in Western Europe in the late 1950s, is gradually gaining more importance, and tends to replace the conventional method of towing in Europe and certain other regions.

Coasters. Another method of avoiding intermediate handling of goods is the use of ocean-going coastal vessels of up to 1 500 t with collapsible funnels and masts (and sometimes telescopic wheelhouses), which ply in coastal waters, rivers and canals.

Living and working conditions of inland boatmen

The majority of the small self-propelled vessels in Europe (of the "péniche" type) are run by artisan bargee owners who live on board with their family. Sometimes there is also a hired deckhand on board but very often the bargee's wife has to act in that capacity. Although these artisan bargees enjoy a certain amount of independence and freedom in their work, there are also many disadvantages such as: restricted living quarters which do not permit a harmonious family life; accidents, illnesses, medical or surgical treatment of the bargee or a member of his family which cause delays and loss of earnings; the very itinerant nature of the work and the absence of fixed social amenities; the permanent risk of accidents for small children living on board and, later on, the financial burden of sending them to boarding schools; and finally the lack of social contacts with people living ashore. The crews of the large self-propelled barges and push-tugs live on board, without their families, for periods of about two weeks followed by leave and usually enjoy more comfortable living quarters.

But in any case the bargee's hours of work are long, 8 to 12 hours a day and often 7 days a week. The crew members are responsible for mooring and stemming the boat at quays and in locks, for lashing or unlashing the barges and checking that they are firmly attached and for keeping watch along the waterway during the voyage. The introduction of day and night navigation, often with the assistance of radar, imposes further strains. Fog, rain and snow create additional dangers, particularly at night. Floods and droughts, by modifying the depth and width of navigable rivers, also render navigation more dangerous and increase the risk of collisions. Crew members spend most of their time in the open exposed to the weather, and their constant activity on deck, quays, locks and riverbanks expose them constantly to the risk of falling into the water and drowning.

HAZARDS AND THEIR PREVENTION

As indicated above, falls contribute the major hazard to inland boatmen as a result of the constant use of ladders, narrow stairways combined with narrow or restricted passageways and work areas. Other hazards are created by the handling of ropes and cables (cuts and abrasions) winches and other ships' gear and hatches (crushing of fingers, contusions). Crew members are also exposed in a general way to noise, vibration, fumes, gases, etc.

As an example, table 1 shows the distribution of accidents by origin of the 636 accidents in inland navigation reported in France for the year 1978.

Table 1. Distribution by origin of accidents in inland navigation in France, 1978

Origin of accidents	Number of accidents	Percentage
Ships' gear	202	31.76
Movements on board	167	26.26
Embarkation and disembarkation	65	10.22
On land	75	11.79
Equipment external to the ship	60	9.43
Maintenance or repairs	58	9.12
Others (not specified)	9	1.42
Total	636	100.00

A variety of measures are applied or proposed to reduce the number of these accidents and to increase safety on board. They include:

(a) the provision of non-slip surfaces on floors and decks;

(b) the use of non-slip footwear;

(c) the use of protective clothing (including helmets, goggles and gloves) during certain loading and unloading operations;

(d) the provision of respiratory masks when handling dangerous goods;

(e) the wearing of life-jackets;

(f) the use of hearing protection in the engine-room;

(g) the proper guarding of machinery;

(h) the installation of railings for passages and ladders;

(i) better training in the handling of dangerous goods;

(j) encouragement for the crew members to learn to swim;

(k) training in safe practices.

As for safety measures, laws and regulations on occupational health of inland boatmen vary greatly from one country to another and leave room for much improvement.

TISSOT, H.

"European inland waterways". *Lloyd's List* (London), 28 May 1979 (5-12). Illus.

Statistics on occupational accidents occurred in 1978 (Statistiques sur les accidents du travail survenus en 1978) (Association de prévention des accidents du travail de la batellerie, 9 place des Vosges, Paris) (1979), 16 p. (In French)

Règles du C.I.P.A. (Comité international de prévention des accidents du travail de la navigation intérieure, 1 place de Lattre, 67000 Strasbourg). (In French)

Inspection, safety and health

The history of labour legislation and of the machinery for its enforcement is inseparable from the development of technologies which, since the end of the 18th century, have completely transformed the means of production and, with them conditions of employment. Perhaps surprisingly at a time of economic laissez faire, the labour legislation that first made its appearance in Europe in the mid-19th century was designed to offset some of the human and social consequences of the Industrial Revolution, as people came to realise that the rapid growth of mining and manufacturing, because it involved a large volume of manpower and an increasing number of women and children, posed a serious threat to the physical and mental health of workers.

The first attempts to legislate the hours of work of women and children between 1830 and 1850–essentially in Great Britain, France and Prussia–were not, however, very successful, despite the appointment of supervisory bodies in 1833 in Great Britain and, shortly afterwards, in other industrialised European countries. These bodies did not normally possess the necessary powers to enforce the earliest protective measures, yet they were able to amass valuable information on new working methods and on their effect on health. They were thus the forerunners of what was later to be termed labour inspection.

It was in 1890 that an international approach to the enforcement of social laws was first developed in the course of a conference held in Berlin, at which the representatives of 15 European States met to draw up international conventions on labour regulations. A protocol that was adopted at the end of the conference recommended that the implementation of the measures taken in each State should be supervised by a sufficient number of specially qualified public servants, appointed by the government and independent of both the employers and the workers. Part XIII of the Treaty of Peace of Versailles (1919), setting up the International Labour Organisation, laid down the principle that "each State should make provision for a system of inspection in which women should take part, in order to ensure the enforcement of the laws and regulations for the protection of the employed."

In 1947 the International Labour Conference of the ILO adopted the first international standards concerning the organisation of labour inspection–the Labour Inspection (Industry) Convention (No. 81) and Recommendation (No. 81) and the Labour Inspection (Mines and Transport) Recommendation (No. 82). The adoption of the Labour Inspection (Agriculture) Convention (No. 129) and Recommendation (No. 133) followed in 1969.

In their efforts to organise and improve their labour inspection system, whose sole or principal object is to enforce laws and regulations concerning occupational safety and health, industrialised and developing States are able to refer to the principles of international law formulated by the International Labour Organisation. Generally speaking, such efforts are directed first and foremost at defining new ways of encouraging co-operation between employers and workers, public services and public or private specialised institutions involved in the prevention of employment injuries and occupational diseases. At the same time the traditional approach to safety and health has evolved and the protection of men and women at work nowadays comprises the relevant aspects of pathology, psychology and sociology and influences not just working methods but the working and social environment as well.

Status of inspection services

Inspection services have responded directly to the broadening scope and objectives of occupational safety and health legislation. Initially designed to enforce a small number of regulations, these small teams of inspectors–only later looked upon as public servants–were compelled to begin recruiting experts when, towards the end of the last century, several industrialised European countries introduced the first general health and accident prevention regulations of a technical nature, as well as to co-ordinate their activities on a rational basis. These were the first supervisory services attached to a central authority, initially the Ministry of

Industry or Trade and then, at the beginning of the century, the Ministry of Labour–a new specialised ministerial department born out of a political will to promote a programme of social action on behalf of the employed. The supervisors, known as labour or works inspectors, were given the authority to visit places of work and to draw employers' attention to their obligation to conform to provisions aimed at protecting the health and safety of their staff. In several countries they were empowered to notify the judicial authority of any infringements.

In 1947 the ILO's Labour Inspection (Industry) Convention (No. 81) and Labour Inspection (Mines and Transport) Recommendation (No. 82) helped to confirm the public service status of labour inspection services and to reinforce their prerogatives.

The labour inspection system was also influenced by the political structure of certain countries. In a number of Eastern European socialist States, for instance, occupational safety and health is partly or entirely the responsibility of the Central Committee of Workers' Trade Unions, sometimes backed by a parallel government inspection system.

In federal States, there are three types of inspection service. In a few countries, such as Austria and Cameroon, labour inspection is the exclusive domain of the central government authority. Much more often, however, it comes under the federal authority and the government of each federated unit. Finally, the United States, where for many years the government of each State has had sole responsibility in this area, has now (1970) set up a federal inspection department which works in conjunction with the services of each State.

Structure and terms of reference

Except in very small countries, it is rare for just one inspection service to cover all economic sectors. As a rule, there are several specialised services whose terms of reference are based on one or both of the following criteria.

The first criterion relates to the branch of the economy. Countries nearly always have a mines and quarries inspection service under the authority of the Ministry of Industry, very often an inspection service for agriculture and, occasionally, for transport as well. The terms of reference of the Labour Inspectorate, normally attached to the Ministry of Labour, vary accordingly. There is some hesitation as to whether the nuclear industry should have its own specialised service, or whether the regular industrial inspection service should recruit nuclear safety experts, although the second solution seems more likely to prevail if radioactive sources are employed in an increasing number of industrial enterprises. Further specialisation is possible, too, a separate inspection unit being assigned to each branch of industry–chemicals, metals, textiles, etc.

The second criterion involves distinguishing between the various objectives of labour inspection, so that the activities of several inspectors–specialising for instance in social, technical, chemical or medical questions–can be combined. A different trend in some countries has been to create a "general" inspectorate in a given sector or branch which is responsible for all aspects of safety and health, and sometimes for such other areas as hours of work, wages and labour-management relations as well. These two methods of organisation have long been looked upon as tending in opposite directions, the first on an Anglo-Saxon model (Federal Republic of Germany, Scandinavia, United Kingdom) and the second on a French model (France, Portugal, Spain and French-speaking Africa). In point of fact, recent attempts to improve efficiency have drawn attention to the increasing similarity of the two systems. Neither can hope to monitor all potential hazards and develop effective preventive action without reference to the numerous, complementary disciplines that are involved (labour law, technology, toxicology, mechanics, etc.). Consequently, the standard approach is more and more one of co-ordinated action. Where several specialised inspectors work side by side, one is now generally appointed to co-ordinate the work of the team and to act as spokesman in discussions with the highest decision-making level of the enterprise, while in a general inspectorate technical advisers or experts (doctors, chemists, physicians, legal experts, etc.) are usually attached to an inspector who is responsible for over-all accident prevention.

Finally, inspection services are no longer the only bodies involved in the working environment and various public or private institutions now provide enterprises with methodological or technical and advisory support or conduct supervisory work with the occasional possibility of enforcement. Quite often, the activities of institutions such as these have to be co-ordinated with those of the public inspection service.

Objectives

Labour inspection has always been defined by two types of laws and regulations, those stipulating the obligations of employers as regards occupational safety and health and those laying down the terms of reference and powers of the inspection services. In most States inspectors are required by law, and according to procedures laid down by their central authority, to supervise the application of laws and regulations concerning the safety and health of workers. This juridical framework has gradually come to acquire considerable significance, both because of the steadily growing body of legislation and the increasingly broad objectives assigned therein to the prevention of occupational hazards and because of the progress being made in the scientific and statistical study of their distribution, origin and contributory factors.

In industrialised and developing countries alike there are three current trends in safety and health legislation. Technical regulations continue to be drawn up specifying the material, organisational or medical means of preventing the principal mechanical or chemical hazards inherent in each of the major types of technology. Meanwhile, the more innovative legislations are extending the concept of worker protection to cover not just the workstation itself but the whole complex issue of "working conditions" and "working environment". One industrialised European country has recently (1977) decreed that "the working environment must be such as to ensure that workers enjoy total safety in respect of physical and mental hazards and that technical standards, occupational health and general welfare conform at all times to the technological and social progress of society". Provisions such as these do not specify the means to be used, as in the past, but lay down far-reaching objectives with an eye to future research and experimentation. The third trend in current legislation has been to involve the workers or their representatives and the works' managers in the design and implementation of preventive measures and of improvements in working conditions.

Obviously, then, safety and health inspection is no longer merely a matter of ensuring that working methods and environment conform to legislative provisions expressed in terms of means. New, more complex and more difficult inspection techniques are needed to encourage the various partners in an enterprise to combine their efforts to create a working environment which conforms to legal requirements that are often defined in the broadest terms. Inspection services are already or will very shortly be controlling the rationality of the objectives of co-operation among the partners—employers, safety experts, production managers, workers' representatives—and the adaptation of these objectives to the particular characteristics of the enterprise. If necessary, they will have to try to reconcile the opposite viewpoints of these partners.

Powers

In keeping with the principles of international law and, more specifically, with the international Conventions and Recommendations of the ILO on labour inspection, national legislations have by and large instituted the right to enter and visit establishments, laid down conditions and limits for such visits (access to documents, sampling of products for analysis, measurement of the working atmosphere, and so on) and stipulated the circumstances under which penal or administrative sanctions can be imposed or recommended by inspectors. In several States inspection services are empowered to interrupt particularly dangerous activities, either directly or by recommendation to a higher administrative authority or the judiciary.

Instructions to inspectors on how to exercise their powers invariably hark back to the principle embodied in paragraph 2 of Article 17 of International Labour Convention No. 81, which leaves it "to the discretion of labour inspectors to give warning and advice instead of instituting or recommending proceedings". In fact most national inspection systems afford a wider choice than this, based either on official procedures or on established practice and accumulated experience. For example, some legislations make a distinction between measures that are immediately applicable and measures which require due notice to be given to the employer. In the first instance, the inspector may use the discretion provided for in the aforementioned Convention to communicate to the employer such information or advice as might incite him to comply with regulations or to order their immediate implementation and, if necessary, initiate proceedings leading to penal or administrative sanctions. In the second case the inspector may use his discretion simply to draw the employer's attention to his obligations or to extend or shorten the period of notice. In both cases the inspector's decision is based on a rational appreciation of the consequences for the workers of any failure to comply with regulations and on his degree of confidence in the employer's ability to act upon a simple warning.

The exercise of an inspector's legal powers must also be mentioned in connection with Article 6 of Convention No. 81, which provides that the status and conditions of service of the inspection staff shall be "such that they are assured of stability of employment and are independent of ... improper external influences". The concept of improper external influences has led many countries to insist that safety and health inspectors should have specific powers and to define the conditions under which they may exercise those powers. Often, an inspector's independence is looked upon as investing him with the legal authority to take administrative steps—in the juridical sense—without any prior censorship, subject to appeal. On the other hand, this independence is not normally thought of as entitling an inspector to interpret laws and regulations or to choose his own scientific criteria for assessing working conditions and issuing orders to the manager of an enterprise. In most inspection services, moreover, a higher authority exists to define the inspection programmes and supervise the way they are carried out.

Finally, recent legislation in several States has further widened the terms of reference of inspection services by requiring them to supervise industrial, commercial or agricultural construction projects so as to ensure that they conform to occupational safety and health objectives at the design stage. Various procedures exist but, in many cases, they seem so far to be experimental.

Organisation of inspection visits

Article 16 of the Labour Inspection Convention, 1947 (No. 81), provides that "workplaces shall be inspected as often and as thoroughly as is necessary to ensure the effective application of the relevant legal provisions". The extent to which this principle is applied varies considerably, as do the administrative and technical facilities at the disposal of inspection services in the various countries. Moreover, the terms of reference of each inspection service may have a direct bearing on the frequency and quality of the inspection visits. Whatever the structure of the inspection services, they carry out their functions essentially on the occasion of visits during which they examine the data relating to working methods and environment, evaluate the relevance of these data to occupational safety and health and notify the manager of any preventive steps to be taken. In practice, these visits are never identical. Some are part of systematic, usually annual, programmes which have been drawn up by the inspector and/or a higher authority. Other visits can be traced back to a particular incident, such as the presentation of a complaint by the workers or their representatives, a request by the manager of an enterprise for advice or the occurrence of an employment injury or occupational disease justifying an immediate investigation.

In most countries inspectors have a certain amount of very tiring desk work, much of which is spent with workers or union representatives and employers who come to them for information or advice or to complain of infringements that call for an on-the-spot inquiry or a letter to the manager. The preliminary inspection of construction projects referred to above also promises to entail much desk work, though this may be extremely valuable for all the workers employed there.

In most countries the frequency of occasional inspection visits and the growing volume of desk work seems likely to jeopardise seriously the possibility of carrying out systematic inspection programmes. As a result, more and more research is being carried out into ways of maintaining a reasonable balance between systematic programmes, occasional visits and desk work, and specifically into techniques for selecting priority objectives. These new techniques are expected, on the one hand, to reinforce activities in areas where potential or observed hazards are known to be most frequent and, on the other hand, to promote greater equality before the law by replacing systematic but blind "door-to-door" visits by a more selective approach. The most sophisticated methods of selecting priorities are based on the use of statistics and inspection reports and on methods of analysis or assessment that can indicate where supervision, advice and guidance are required.

Inspection methods and facilities

The reform of inspection methods in many countries has brought improved technical and juridical documentation facilities, a methodological definition of the various stages of inspection and, in particular, an analysis of the causes of hazards, a set of procedures for notifying employers of their obligations and an operational approach to the filing of records. The need for these improvements derives from the considerable increase in the objectives and volume of safety and health legislation, in the scientific findings and body of precedents and in the objectives assigned to the inspectorate in laws and administrative instructions. In the specific case of developing countries, the desire to improve methods is generally dictated by a need to give inspection services a better standing among managers, who are concerned first and foremost with the development of production capacity and therefore tend to be more interested in the creation of jobs than in better working conditions.

The documentation which inspectors are frequently provided with by their central authority consists of a set of laws and regulations brought up to date periodically, instructions for interpreting and enforcing these texts, and the necessary technological, scientific and juridical publications dealing specifically with the prevention of occupational hazards. A number of current projects involve feeding all this information into computers and providing inspectors or groups of inspectors, at an as yet unspecified date, with their own terminals. In the more immediate future computers could probably be used to keep on file the most important data concerning the largest enterprises, such as size of personnel, main potential hazards, distribution of observed hazards and their causes and the purpose and findings of inspection activities. They could in fact be used as regular operational files, but inspectors will first have to develop a very great capability to carry out quantitative and qualitative assessments of the relevant aspects of an enterprise and of its performance in the field of prevention.

Inspection visits are often facilitated by the use of guides or plans prepared by the central authority. Because there are so many types of enterprise, however, documents such as these cannot necessarily provide a comprehensive framework for such visits and tend rather to be used as a general guideline. On the other hand, the central inspection authority apparently quite often lays down a fairly precise procedure for inquiries into the causes of employment injuries. The most striking innovations in this respect are designed to enable inspectors to analyse not just the ultimate cause of the accident but the series of causes that led up to it. Beyond the material cause, such as a faulty protective device on a machine, an inspector can look into the underlying reasons for the accident, such as poor organisation of prevention and the respective role of the central design and decision-making echelon (the employer) and the intermediate echelons (methods bureau, production service, occupational safety and health service, supervisory staff). In any investigation of this nature the inspector must consider how his approach (information, advice and guidance) might need to be adapted to the category he is dealing with.

Inspectors have all kinds of technical facilities at the place of work. Here there seem to be two approaches. The first is to provide the inspectors with detection and measuring apparatus (for air pollution, noise level, etc.) or with samples for subsequent analysis by laboratories attached either to the enterprise or to the inspection service. The other approach is based on the principle that sampling, analysing and measuring must be the responsibility of the employer rather than of the inspectorate. As a result, legislation has been passed to set up bodies approved by the Ministry of Labour to carry out such tests or measurements as an inspector may require an employer to undertake. In such cases the employer officially informs the inspector of the findings of the approved body. Some countries combine both approaches and provide the inspectors with simple devices which they can use to determine whether more precise tests need be carried out by the approved bodies.

In the now numerous countries where enterprises are required by law to set up consultation machinery such as safety and health committees or functional units such as medical or safety and health services, labour inspectors are generally called upon to supervise their smooth running and, where necessary, to act as adviser and conciliation expert to make them more efficient.

To help them in the complex tasks assigned to them inspectors often receive technical assistance. Doctors working for the inspectorate or as consultants, consultant engineers and technicians may, for instance, either be assigned specific inspection or inquiry missions with an independent capacity to inform the enterprise of the measures that need to be taken or be called upon to co-operate with the inspectors and to provide them with any advice or information they may need.

Recruitment and training of inspectors

Inspectors are recruited in many different ways, generally on the basis of higher university diplomas—with or without the requirement that they should be in scientific or technological subjects—and experience in the industrial sector—preferably at the middle-management level, sometimes as trade union officials. Selection differs from country to country and may depend either on a national or regional competition or on individual assessment. Where no previous experience in industry is required, new recruits often begin work as trainees in inspection services and in industrial, mining, commercial or agricultural enterprises.

The growing complexity of safety and health activities has led a number of States to introduce a specific initial training scheme for inspectors, followed by regular further training periods to acquaint them with new technologies and the most recent developments in production methods. Background knowledge and the ability to analyse a working environment are further developed by the regular circulation of juridical, economic and technical publications and sets of technical instructions defining the methods of inspection in each particular field.

Collaboration with other bodies

When labour inspectors first crossed the threshold of industrial enterprises, they were very much pioneers endeavouring to make the best possible use of regulations that were still at an embryonic stage and struggling to overcome the distrust and sometimes outright hostility of the employers and the scepticism of the workers. The administrative records that have been preserved by some European inspection services are a tribute to the competence and ingenuity of those early inspectors who, perhaps equipped with regulations that were not suited to the prevention of the kind of hazards encountered in the course of painstaking analyses, were hard put to it to convince employers of the desirability of taking measures to protect the safety and health of their workers. Initial successes were achieved at the actual workplace, and these gradually gave rise to the adoption of broader measures affecting whole occupations. At the instigation of the employers and their occupational organisations, associations were created to provide enterprises with appropriate technical and documentation facilities, among which one might mention the associations of owners of steam-driven equipment that were responsible for testing boilers and high-pressure apparatus in general. Later, these same associations set up special services for testing electrical installations. Private or public insurance agencies nowadays consider that financial coverage of the consequences of an accident must go hand in hand with the adoption of appropriate preventive measures. These agencies have their own machinery for advising enterprises and may possess laboratories for carrying out analyses, measurements and research as well. Workers' trade unions have often provided their members with documentation services and organised safety training. Universities, too, devote part of their research to the study of toxicology and, more recently, ergonomics.

Quite apart from the activities of public inspection services, therefore, enterprises receive methodological and technical assistance from all kinds of other bodies, which is particularly useful because the tremendous development of industrial techniques has greatly extended the range of potential employment injuries and occupational diseases and because the laws and regulations are more and more severe.

The central services that are responsible for safety and health inspection now have to co-ordinate these various public and private initiatives. Without taking away the latter's autonomy, the aim is to prevent general or specialised activities in this field from becoming so disseminated as to jeopardise their efficiency. Countries which conduct systematic research into the principles and practice of co-ordination and co-operation generally emphasise the complementary nature of the different forms of action, and it is quite common for the various bodies involved to consult one another on exchanges of information and their choice of priority objectives.

These central services are also increasingly anxious to convince higher education and vocational training institutes of the need to include in their syllabuses the study of human sciences and occupational accident prevention techniques. In turn these establishments very frequently elicit the co-operation of labour inspectors to lecture on the concrete aspects of work and outline the objectives of existing legislation.

International co-operation

For over 60 years the International Labour Organisation has been devoting a substantial proportion of its activities to improving safety and health at the workplace and in the working environment. The result has been the adoption of a large number of international Conventions and Recommendations, the publication of numerous technical studies, the convocation of committees open to accident prevention experts and the storage on computer and dissemination of technical data on the subject. The ILO's standard-setting activities, which are intended to help in the drafting of national legislation, have been supplemented by the adoption of several Conventions and Recommendations on labour inspection which have already been discussed. Reference might also be made to the Organisation's extensive technical co-operation programme in the field of accident prevention in developing countries. A major step forward was taken in 1974 when the ILO launched the International Programme for the Improvement of Working Conditions and Environment (PIACT). The ILO's over-all action is unquestionably useful as a means of helpful inspection services to adapt their objectives, structure and methods to the requirements of the world of work. Its most recent initiative will provide further technical assistance in respect of the most serious hazards. This is the creation of an "international occupational safety and health hazard alert system", so far involving over 60 countries, which will serve as a means of exchanging information on serious hazards, particularly chemical or toxicological hazards but also those of a physical nature. The corresponding national centre of each country will be able to obtain information from this international system at any moment.

Finally, mention should also be made of a number of international bodies that are engaged in the study of accident prevention methods and facilities, such as the International Social Security Association (ISSA) and the International Association of Labour Inspection (IALI) and the European Federation of Associations of Industrial Safety and Medical Officers (FAS) (see INTERNATIONAL SOCIAL SECURITY ASSOCIATION; INTERNATIONAL ASSOCIATION OF LABOUR INSPECTION).

BOIS, P.

CIS 79-2094 *Field operations manual.* Vol. 5. US Department of Labor, Occupational Safety and Health Administration (Washington, DC, US Government Printing Office, 1978), 275 p. Illus.

CIS 80-1794 "A computerized safety management information system for state OSHA inspection and enforcement programs". Ayoub, M. A.; Kushner, K. J. *Journal of Safety Research* (Spring 1980), 12/1 (21-35). Illus. 12 ref.

CIS 79-1793 *Occupational safety and health manual intended for the Labour Inspectorate* (Manual de seguridad e higiene ocupacionales para la Inspección del trabajo). Cabrera, I. T. REG. 35/SH-23, Interamerican Center for Labor Administration (Lima, Centro interamericano de administración del trabajo, CIAT, 2nd ed., 1978), 3 vols., 573 p. Bibl. (In Spanish)

CIS 80-1196 *Safety and health: An inspectorate, what for?* (Sécurité et hygiène du travail: un service d'inspection, pour quoi faire?). D/1980/1205/4 (Commissariat général à la promotion du travail, 53 rue Belliard, 1040 Brussels) (1980), 135 p. 3 ref. (In French, Dutch)

CIS 79-892 *Current situation and trends in worker protection* (Der Arbeitsschutz und seine Entwicklung). Mertens, A. (Dortmund, Bundesanstalt für Arbeitsschutz und Unfallforschung, 1978), 227 p. Illus. (In German)

CIS 79-900 "Act No. 833 of 23 December 1978 – Establishment of a National Health Service" (Legge 23 dicembre 1978, n.833. Istituzione del servizio sanitario nazionale). *Gazzetta ufficiale della Repubblica italiana* (Rome), supplemento ordinario n.360 (3-48). (In Italian)

International Agency for Research on Cancer (IARC)

The International Agency for Research on Cancer came into being by resolution of the World Health Assembly in 1965. The statutes of the Agency set out its mission as "planning, promoting and developing research in all phases of the causation, treatment and prevention of cancer", but limitations imposed by its budget and manpower resources necessitated a more selective approach to cancer research, compatible with the size and with the international role of the Agency.

Since the start of its research activity, the Agency has devoted itself to studying the causes of cancer present in the human environment, in the belief that identification of a carcinogenic agent was the first and necessary step towards reducing or removing the causal agent from the environment, with the aim of preventing the cancer that it might have caused.

In order to achieve this aim, the Agency has been established as an independent research institute, but within the framework of the World Health Organisation. It has its own Governing Council, consisting of representatives of all the contributing countries and the Director-General of the WHO. In January 1982, the world 12 participating States were: Australia, Belgium, Canada, France, the Federal Republic of Germany, Italy, Japan, the Netherlands, Sweden, the United Kingdom, the USSR and the United States. The budget of the Agency is financed by the annual contributions of the member governments, with additional support coming from certain national contracts.

The Director, Dr. Lorenzo Tomatis, and the staff of the Agency are guided in the selection of their research programmes by the Scientific Council, whose 12 members are elected by the Governing Council from among leading scientists of international repute in the field of cancer research. The Council meets annually to advise on programmes that should be terminated, those that should continue, and new programmes to be undertaken.

The headquarters of the Agency is in Lyons, France, in a building housing laboratories, epidemiological offices and computer facilities, as well as the administrative machinery of an international organisation. The staff totals 150, of whom 45 are scientists, 50 technicians and the remainder administrative staff. The 14-storey building was erected for the Agency by the French authorities, in response to the invitation and initiative of the municipality of Lyons.

The Agency's research activities fall into two main groups–epidemiological and laboratory-based experimental–but there is considerable interaction between these groups in the actual research projects undertaken.

The scientists of the Agency's staff are responsible for the design of their research projects but in almost every case the execution depends very largely on collaboration with scientists working in national institutions with whom research agreements have been concluded. In this way the Agency not only carries out its own research programme but at the same time develops international collaborative research.

Epidemiological activities

The epidemiologists in the Agency have developed programmes in both descriptive and analytical epidemiology.

The most important aspect of descriptive epidemiology carried out by the Agency is the co-ordination of cancer registry data in collaboration with the International Association of Cancer Registries for which the Agency acts as Secretariat. Cancer incidence data collected from 79 cancer registries covering 104 populations are published in standardised form every 5 years in a volume entitled *Cancer incidence in five continents*. The groups are also preparing atlases showing the geographical distribution of cancers and studying time trends–that is changes in the pattern of cancer distribution in a given population over a passage of several decades.

Analytical epidemiology studies are aimed at identifying risk factors for cancer. The term "risk factor" is a very general one and includes not only specific carcinogens such as are encountered in occupational exposure, but also dietary habits–which might, for example, be associated with cancers of the gastrointestinal tract–infection with hepatitis B virus–which might be associated with liver cancer–high consumption of alcoholic beverages–which might also be associated with liver cancer.

The Agency has research projects in all these topics but in the field of occupational health, they are of special interest.

One study is on the long-term effects of pesticides on human health in Colombia, where the potential health hazards including carcinogens, are being studied in a group of 9 000 workers in the flower-growing industry who have been exposed to a wide variety of different pesticides. The study,which will run for several years, has been started with a survey of congenital malformations among children born to women working in floriculture or to wives of men working in this industry.

A second investigation, now in its final phase, stresses the special role that the Agency can play in co-ordinating large-scale international studies in occupational epidemiology, which until now have been remarkably rare. The investigation includes 13 factories producing man-made mineral (vitreous) fibres in seven European countries. A total of about 25 000 workers ever employed in these factories have been followed up from entry into the plants to present date and their mortality and cancer incidence experience is being compared with that of the general population of the same age and sex. Environmental measurements of airborne fibres carried out within each plant allow further comparisons of mortality and cancer incidence of workers grouped according to level of exposure to fibres.

A further development in the area of international collaboration in occupational epidemiology is the project, now at the initial exploratory stage, of a cohort study of workers exposed to phenoxyacid herbicides, trichlorophenols and pentachlorophenols, particularly in the production industries.

Evaluation of the carcinogenic risk of chemicals to humans

This programme, which started in 1971, is of direct relevance to occupational health, as a substantial number of compounds present in the working environment are evaluated with respect to a possible carcinogenic risk for humans. Chemicals are selected for evaluation on two grounds:

(a) human exposure to the compound is known to occur; and

(b) there is some suspicion (derived from human data, animal data, or other laboratory data) that the chemical may be carcinogenic.

The evaluation process for each compound begins with an extensive bibliographic search, followed by the preparation of a draft monograph which constitutes the raw material on which an international working group of scientists from a variety of relevant fields (chemistry, experimental and human pathology, mutagenesis, toxicology, epidemiology, etc.) works for 6-7 days, with the twin purpose of achieving a final version of the monograph and expressing an evaluation on what the experimental evidence and the human evidence described in the monograph means in terms of carcinogenicity. On average, three working groups are convened per year, each in charge of examining the data of some 20 compounds.

The corresponding monographs are published in volumes (IARC monographs on the evaluation of the carcinogenic risk of chemicals to humans) which, at May 1982, number 29. After volume number 20 a supplement (No. 1, 1979) to the collection was published, summarising the evaluation for all compounds considered; a similar supplement is in preparation covering all compounds included in volumes 1 to 29. A recent and important expansion of the programme is the inclusion not only of individual chemicals but also of complex mixtures, often poorly defined, as they occur for example in occupational settings: thus the leather, wood and rubber industries have formed the object of monographs and have been evaluated for carcinogenic risk.

SARACCI, R.

International and regional organisations

During the past 35 years a considerable number of international and regional organisations have been established to deal with various social, economic, cultural, trade development, health, scientific and other problems. In addition many non-governmental international organisations are concerned with particular activities such as air and sea transport, standardisation, social security, learned professions, trade unionism, employers' interests, industrial development, etc.

While the International Labour Office is the only international organisation taking constitutional obligations for the whole protection of workers' health in all branches of economic activities, including occupational safety and health and ergonomics, many organisations and bodies listed below are concerned with specific aspects of these matters.

International organisations

United Nations Development Programme (UNDP) (One United Nations Plaza, New York, NY, 10017, USA). The UNDP is engaged in a large number of projects designed to assist developing countries to build up their nascent economies and raise their living standards. Several thousand internationally recruited experts are kept steadily at work in the field. In 1979, 8 445 expert missions were carried out. Several amongst these projects are devoted to the improvement of occupational safety and health standards in industry and other walks of economic life, the implementation of which is entrusted to the ILO and WHO. Such field projects may range from the provision of short-term consultancy to more massive assistance over a period of several years for the establishment of fully fledged occupational safety and health institutes designed to provide training, applied field research and direct service to places of employment. Projects supporting occupational health and safety programmes or laboratories have been undertaken with UNDP financing in Bolivia, Ecuador, Egypt, Greece, Guinea, India, Indonesia, the Republic of Korea, Singapore, Syria, Turkey and Viet Nam, and more are under consideration in other developing countries.

International Labour Organisation (ILO). See separate article.

World Health Organisation (WHO). See separate article.

Food and Agriculture Organisation (FAO) (Via delle Terme di Caracalla, I-00100 Rome, Italy). The aims of the Organisation are: to raise the levels of nutrition and standards of living of the peoples under the respective jurisdiction of the member governments; to ensure improvements in the efficiency of production and distribution of all food and agricultural products; to better the conditions of rural populations, thus contributing towards an expanding world economy and freeing humanity from hunger.

The structure of the Organisation is based on: the Conference; the Council; the statutory subsidiary bodies. The Conference appoints the Director-General who is assisted by an international staff.

A number of activities of FAO are directly or indirectly concerned with occupational health and safety and ergonomics in agricultural, forestry and fishery work.

In fishery activities FAO collaborates at the secretariat level with ILO and the Inter-Governmental Maritime Consultative Organisation (IMCO) on the IMCO Sub-Committee on Safety of Fishing Vessels and participates actively in the work of the IMCO Sub-Committee on Standards of Training and Watchkeeping, with particular reference to training and certification of crews on board fishing vessels. FAO collaborates with ILO in regard to conditions of work in the fishing industry.

In forestry activities the FAO/ECE/ILO Committee on Forest Working Techniques and Training of Forest Workers deals at the interagency level with health and safety matters. Field projects and publications in this area cover such aspects as safety in logging and industry and heat stress in forest work.

In the agricultural field some of the diseases of economic importance in livestock also present hazards to persons handling livestock and animal products (e.g. brucellosis, tuberculosis, leptospirosis, anthrax, rabies, Rift Valley fever). For these disease-related activities close liaison is maintained with WHO through joint committees. FAO is also concerned with the harmonisation of registration requirements for pesticides and the assessment of pesticide residues in food and in the environment. As regards atomic energy in food and agriculture, programmes are co-ordinated with the IAEA in order to assist scientists of developing countries to make safe and effective use of relevant isotope techniques, e.g. the use of radio-labelled enzyme substrates for detecting occupational exposure to insecticides.

International Atomic Energy Agency (IAEA). See separate article.

Inter-Governmental Maritime Consultative Organisation (IMCO) (101-104 Piccadilly, London W1V OAE, United Kingdom). IMCO deals with technical and related aspects of international shipping, particularly from a safety angle (e.g. carriage of dangerous goods and bulk cargoes, life-saving appliances, safety of navigation—including traffic separation schemes, ship design and equipment and stability). It is also concerned with preventing marine pollution caused by ships in the marine environment (including legal aspects). It has produced an International Maritime Dangerous Goods Code, and various other Codes and Recommendations. It is responsible for convening maritime conferences and drafting international maritime conventions, including the International Conventions for Safety of Life at Sea, for Prevention of Pollution of the Sea, on Facilitation of International Maritime Traffic, on Load Lines and on Tonnage Measurement of Ships. As an executing agency of the United Nations Development Plan in the maritime field, it provides experts, sponsors training projects and has a fellowship programme.

International Civil Aviation Organisation (ICAO) (International Aviation Square, 1 000 Sherbrooke St. West, Montreal, P.Q. H34 2R2, Canada). Much of the work of ICAO is in the interest of safety of flight. It has established and, in the light of technical development, reviews: international standards for the operation of aircraft; competence and health of air crew and persons in other aeronautical trades and professions; publication of aeronautical information; investigation of accidents, information on which is published; design of aerodromes; aids to navigation; air traffic control; aeronautical telecommuncations; aeronautical meteorology; and search and rescue. It also, in the light of the development of international air traffic, reviews and revises regional air navigation plans which set out in detail the air navigation services required.

Regional organisations

UN Economic Commission for Europe (ECE) (Palais des Nations, CH-1211 Geneva 10, Switzerland). The Inland Transport Committee of this Commission has been working in the field of road traffic safety and safety in vehicle construction in co-operation with various international and regional organisations including the ILO.

Two Conventions on Road Traffic and on Road Signs and Signals (1968) were prepared at the United Nations Conference on Road Traffic which was held in Vienna in 1968. These Conventions entered into force respectively in 1977 and 1978. They have been supplemented by the 1971 European Agreements on Road Traffic and on Road Signs and Signals, both of which entered into force in 1979. Furthermore, a Protocol on Road Markings, additional to the latter Agreement, was drawn up in 1973; this is not yet in force.

Road traffic safety is the competence of two subsidiary bodies of the ITC Working Party on Road Transport, namely the Groups of Experts on Road Traffic Safety (GE20) and on the Construction of Vehicles (WP29). Their membership includes generally all European countries as well as Canada and the United States.

The Group of Experts on the Construction of Vehicles (WP29), in which representatives of the Governments of Australia and Japan also participate, has a number of groups of rapporteurs dealing with subjects such as noise (GRB), pollution and energy (GRPE), lighting and light signalling (GRE), protective devices (GRDP), brakes and running gear (GRRF), safety provisions on motor coaches and buses (GRSA), crash-worthiness (GRCS) and general safety provisions (GRSG). It elaborates draft regulations and makes recommendations in the field of vehicle construction, aiming at improving their safety and protecting the environment. So far 40 regulations are in force as annexes to the 1958 Agreement and are kept regularly up to date and about 20 others are in the course of preparation.

The Group of Experts on Road Traffic Safety deals with different aspects of road traffic rules, road signs and signals, and road and vehicle use. Its current work programme covers, among other subjects, the promotion of a uniform application of the existing international instruments in the field of road traffic and road signs and signals, the standardisation of speed limits, the education and instruction of road users, and the marking of certain categories of vehicle, through the adoption of recommendations. Member governments are expected to include such recommendations in their national legislation.

All the recommendations approved so far in the field of road traffic, signs and signals as well as the construction of vehicles are being grouped systematically in two consolidated resolutions.

The activities of these subsidiary bodies of the Working Party on Road Transport include the elaboration or modification of other international instruments and recommendations, issued under the auspices of the ECE and having an impact on road safety, such as the European Agreement concerning the Work of Crews of Vehicles engaged in International Road Transport (AETR, 1970), the European Agreement on Main International Traffic Arteries (AGR, 1975), and the European Agreement on Minimum Requirements for the Issue and the Validity of Driving Permits (APC, 1975).

Council of Europe. See separate article on PARTIAL AGREEMENT.

Commission of the European Communities (CEC). See separate article.

Organisation for Economic Co-operation and Development (OECD). See separate article.

Non-governmental organisations

International Organisation for Standardisation (ISO). See separate article.

International Electrotechnical Commission (IEC) (1 rue de Varembé, CH-1211 Geneva 20, Switzerland). The IEC is the body which deals with world-wide standardisation in the electrical and electronic fields. It is composed of national committees from 43 countries representing 80% of the world population producing and consuming 95% of the world energy.

The IEC issues publications, including recommendations in the form of international standards. The preparation of these standards is carried out by 190 technical committees and sub-committees and some 500 specialised working groups.

All aspects of occupational health and safety are covered in areas such as electrical equipment of industrial machines, the effects of current through the human body, classification of electrical and electronic equipment with regard to protection against electric shock, degrees of protection by enclosures, electrical safety of laser equipment, tools and equipment for live working at all voltages, electrical installations and equipment for use in:

(*a*) buildings including construction and demolition sites;

(*b*) hospitals;

(*c*) explosive atmospheres; and

(*d*) open-cast mines and quarries.

Special mention must also be made of the standardisation of safety standards for household and similar electrical appliances, microwave ovens, refrigeration, air conditioning, commercial catering equipment, hand-held motor-operated tools, disposal, packaging and labelling of askarels for transformers and capacitors including benefits and risks arising from their use, colours for push-buttons, colour identification for flexible electric cables, safety isolating transformers, fire hazard testing, reactor and radiation protection instrumentation, detection, alarm and monitoring systems for the protection of persons and property, arc welding equipment, industrial electro-heating equipment, data-processing equipment and office machines, and mains operated electronic and related apparatus for household and similar general use.

International Social Security Association (ISSA). See separate article.

Permanent Commission and International Association on Occupational Health (PCIAOH). See separate article.

International Ergonomic Association (IEA). See separate article.

International Commission on Radiological Protection (ICRP) (Clifton Avenue, Sutton SM2 5PU, Surrey, United Kingdom). The International Commission on Radiological Protection (ICRP) has been functioning since 1928, when it was established under the name of the International X-ray and Radium Protection Committee by the Second International Congress of Radiology. It assumed the present name and organisational form in 1950 in order to cover more effectively the rapidly expanding field of radiation protection. As one of the commissions established by the International Congress of Radiology, the ICRP has continued its close relationship with succeeding Congresses of Radiology, and it has also been looked to as the appropriate body to give general guidance on the more widespread use of radiation sources caused by the rapid developments in the field of nuclear energy. The Commission continues to maintain its traditional contact with medical radiology and the medical profession generally, and it also recognises its responsibility to other professional groups and its obligation to provide guidance within the field of radiation protection as a whole. The policy adopted by the Commission in preparing its recommendations is to consider the fundamental principles upon which appropriate radiation protection measures can be based, while leaving to the various national protection bodies the responsibility of formulating the specific advice, codes of practice, or regulations that are best suited to the needs of their individual countries. The Commission consists of 13 members who are elected on the basis of their recognised authority in the field of radiation protection and related matters without regard to nationality. It works by means of committees and ad hoc task groups and it meets every year. The radiation protection criteria adopted by the ILO and many inter-governmental organisations are largely based on recommendations of the ICRP.

The International Radiation Protection Association (IRPA) (Secretariat: c/o B.P. 35, F 92260 Fontenay-aux-Roses, France). The International Radiation Protection Association was founded in 1966 to provide a medium whereby international contracts and co-operation may be established among those engaged in radiation protection work. The aim of the IRPA is to provide for the protection of man and his environment from the hazards caused by ionising radiation, and thereby to facilitate the medical and industrial exploitation of radiation and atomic energy for the benefit of mankind. Protection against non-ionising as well as ionising radiation is pursued through contact and collaboration with international agencies.

The Association encourages the establishment of radiation protection societies throughout the world, provides for and gives support to international meetings, encourages international scientific publications, scientific research and educational opportunities in the science of radiation protection and the development of international standards. Information is disseminated through the journal *Health Physics* and a quarterly *Bulletin*. There are 9 000 members in 26 Associate Societies in more than 30 countries. An international congress is held every 3 or 4 years to review the whole field of radiation protection. Regional congresses also provide a forum for specific radiation protection topics and are usually organised by a Regional Group of Associate Societies.

International Association of Agricultural Medicine and Rural Health (IAAMRH) (Secretariat: c/o Saku Central Hospital, 197 Usuda-machi, Minamisaku-Gun, Nagano Pref. 384-03, Japan). This association started its activities in 1961 with the organisation of the First International Congress of Rural Medicine at Tours. Since then it has organised a congress every three years (at Bad Kreuznach, Bratislava, Usuda, Varna, Cambridge, and in 1978, Salt Lake City). The Assocation aims at concentrating efforts on problems of agricultural medicine and rural health in all countries around the world. In view of the global nature of this aim, the Association works in close collaboration with the United Nations and its specialised agencies, such as UNESCO, WHO, FAO and ILO. The present President is Professor Pavel Macuch (Czechoslovakia) and the Secretary-General is Dr. Toshikazu Wakatsuki (Japan). The Association has set up a number of commissions dealing with such topics as anthropozoonoses, ergonomics and safety, living and working conditions, medical care, nutrition and toxicology. It has regional chapters in Asia, Europe and Latin America.

International Association of Labour Inspection (IALI). See separate article.

International Air Transport Association (IATA) (PO Box 160, CH-1216 Cointrin-Geneva, Switzerland). The IATA, which is the world organisation of scheduled airlines, was founded in 1945. Its members carry the bulk of the world scheduled international air traffic under the flag of some 85 nations.

In the field of occupational safety and health, the IATA has undertaken to seek agreement amongst its members on a list of dangerous goods and on regulations (including provisions for labelling) concerning the transport of such goods by air.

The IATA annually updates and issues its regulations in English, French, Spanish and German. They contain a list of dangerous goods and provide detailed provisions (including labelling) for handling, controlling and expediting the movement of hazardous merchandise by air. The main publications of the IATA in this field are: IATA restricted articles regulations (see DANGEROUS SUBSTANCES, LABELLING AND MARKING OF), IATA slide presentation of restricted articles (which includes 80 colour slides, a sound cartridge with English version programme track and a programme manual instructor's guide) and IATA programmed instructions on restricted articles, a training manual.

International Committee for Lifts Regulations (CIRA) (c/o President Hans Egli, Kalchbülstrasse 33, CH-8060 Zurich, Switzerland). This Committee was formed in 1957 with the support of the ILO to promote the development of safety regulations for passenger and goods lifts. It works in close co-operation with the ILO.

International Association of Labour Inspection (IALI)

The International Association of Labour Inspection (IALI) was founded in 1972 to serve as an international instrument of documentation and comparative studies, which was designed ultimately to be able to provide labour administrations with information on the methods and practices employed by labour inspectors in enterprises. The need for this kind of exchange of information stemmed from the increasing complexity of labour inspection activities as a result of developments in labour law, on the one hand, and the working environment, on the other. There has been such a constant extension of national laws and regulations relating to general conditions of employment, occupational safety and health, labour-management relations and wages, in both industrialised and developing countries, and production technologies and processes have become so extraordinarily diversified that labour inspection services nowadays are permanently having to find answers to complex juridical, technical, economic and human problems so that they can carry out their work of inspection, information and advice as efficiently as possible. As a result, labour administrations are having to review the methods employed by their inspection services, a procedure which quite often entails adapting their structure and the administrative and technical facilities at their disposal.

The programmes of the IALI, whose statutes preclude any political, trade union or religious activity or the passing of any judgement on the labour law and inspection systems of States, are devoted entirely to the dissemination of information deriving from replies to international inquiries conducted by means of questionnaires, international or regional symposia and an international Congress held in Geneva every three years during the sessions of the International Labour Conference. The Congress and most of the symposia are organised in collaboration with the International Labour Office, which accorded the IALI the status of a non-governmental international consultative organisation in 1978.

Since 1974 all the IALI's three-year programmes have concentrated essentially on the study of labour inspection practices in the field of occupational safety and health. The items placed on the agenda of the first three Congresses (1974, 1977 and 1980) and of some of the regional symposia include the methods of selecting priority objectives for the prevention of occupational hazards, methods of examining the construction projects of industrial establishments, an analysis of the circumstances and causes of labour accidents, methods of co-operation between labour inspectors and industrial physicians, the role of labour inspectors in the promotion of co-operation among the various partners in an enterprise (management, experts, workers' representatives) with a view to the prevention of hazards and the improvement of working conditions. The 1981-83 three-year programme will be very much geared to the study of accident prevention methods but will also cover a number of other aspects, such as enforcement of the principle of equality of wages of men and women engaged in comparable work, and methods of detection of clandestine labour. Several international inquiries conducted by means of questionnaires will also be carried out to compile information on the organisation of labour inspection services, particularly as regards the organisation of the central authority of inspection services, the use of computer technology for documentation purposes and for evaluating inspection service activities, and the organisation of training and further training programmes for labour inspectors. Several regional symposia will be held to study means of co-operation between labour inspectorates and public or private bodies concerned with the prevention of occupational hazards.

The IALI has two categories of members: regular members, which consist of national groups of labour inspectors (associations, trade unions, study committees, etc.), and correspondent members, which are the labour administrations themselves (Labour Inspectorate, Labour Directorate, Occupational Safety and Health Directorate, etc.). On 31 July 1980 the membership of the IALI was as follows: regular members: Belgium, Chile, France, the Federal Republic of Germany, Greece, Ireland, Italy, Luxembourg, Morocco, Senegal, Spain, Switzerland and Togo; correspondent members: Israel, Italy, Mexico, Netherlands, Portugal, Spain and the United Kingdom.

Though not members of the IALI, the labour inspection services of other countries have co-operated in its programmes by sending in documentation or participating in the Congresses. These include Austria, Canada, Gabon, Hungary, Ivory Coast, Japan, Libya, Norway, Poland, Sri Lanka, Turkey, Upper Volta, Venezuela, Yugoslavia and Zaire.

The statutory bodies of the IALI are the General Assembly, Executive Committee, Permanent Secretariat and regional delegates. Within the General Assembly, the delegates of regular members and correspondent members have the same voting power; however, decisions must be taken by a simultaneous majority of both groups of members. The Executive Committee appointed for the period 1981-83 consists of the president (France) and three vice-presidents of the IALI (Federal Republic of Germany, Switzerland, United Kingdom). The Permanent Secretariat is the executive, administrative and technical body of the IALI. The regional delegates, more of whom are being appointed in Africa, Latin America, Asia, etc., maintain liaison between the labour administrations of each group of States and the IALI Executive Committee; one of their main tasks will be the organisation of regional symposia.

The IALI's publications are circulated in three languages by the Permanent Secretariat to over 60 countries. They consist of an *Information Bulletin* and the reports on the international inquiries, international and regional symposia and international Congresses.

BOIS, P.

International Atomic Energy Agency (IAEA)

Origin of the IAEA and statutory objectives

The impulse to create the IAEA came from a speech of President Dwight D. Eisenhower addressed to the eighth regular session of the General Assembly of the United Nations in December 1953, followed by detailed consultations and debate in the United Nations which lasted just short of three years and culminated in the formulation of the statute. The statute entered into force on 29 July 1957 and thereby the International Atomic Energy Agency itself automatically came into existence on that date.

Article II of the statute entitled "Objectives" indicates the two inter-related purposes for which the Agency was established: to further the peaceful uses of atomic energy throughout the world, and to do so in such a manner that the Agency's assistance is not misused for any military purpose. The methods by which these activities are to be carried out are set forth in articles VIII to XII, dealing respectively with "Exchange of information", "Supplying of materials" (i.e. the receipt of nuclear and other materials), the furnishing to the Agency of "Services, equipment and facilities", "Agency projects" and "Agency safeguards".

The IAEA fosters and encourages, guides and advises the development of the peaceful uses of atomic energy throughout the world. It organises meetings, publishes books, establishes safety standards for all types of nuclear activity, prepares feasibility and market studies, operates three laboratories and applies safeguards to nuclear facilities and materials to ensure that they are used only for their intended peaceful purposes. Upon request, it advises governments on atomic energy programmes, awards fellowships for advanced study, arranges the loan of equipment, finances research and acts as an intermediary in arranging the supply of nuclear materials. It also advises member States on the physical protection of nuclear materials. The work of the IAEA is carried out in close co-operation with dozens of other organisations, both national and international.

Structure of the IAEA

Although autonomous, the IAEA is a member of the United Nations system and sends reports on its work to the General Assembly of the United Nations and, when appropriate, to the Security Council, the Economic and Social Council (ECOSOC), and other United Nations organs on matters within the competence of these organs. It is directed by a Board of Governors, which is composed of representatives from 34 member States, and a General Conference of the entire membership of 110 States. The IAEA has its own programme, approved by the Board of Governors and the General Conference, and its own budget, currently about $81 million a year, financed by contributions from member States. Its staff are international civil servants, and administratively the IAEA is part of the common United Nations system.

The Board is one of the policy-making organs of the Agency. It considers all major questions, including applications for membership and the Agency's programme of work. The annual budget and report require its approval as well as that of the General Conference. Under its own authority, the Board approves all safeguards agreements, important projects and safety standards. It normally meets four times a year. Twelve of its members are designated by the Board itself and 22 are elected by the General Conference. The designation criteria (level of advancement in nuclear technology and equitable geographic representation) ensure continuity of membership.

The General Conference is convened once a year for a general debate on the Agency's policy and programme, to review and approve the budget and the annual report, to approve applications for membership and to elect new and replacement members to the Board of Governors.

The Secretariat, consisting of about 1 500 staff members recruited from more than 60 member States, is headed by the Director General, who is responsible for the administration and implementation of the Agency's programme. The Director-General is assisted by five Deputy Directors General, each head of a Department (Administration, Research and Isotopes, Safeguards, Technical Assistance, Publications and Technical Operations). The Director General is advised on scientific and technical matters by a Scientific Advisory Committee of, at present, 15 distinguished scientists, whose appointments are for three-year terms and who represent all fields of nuclear science. Another important committee, the Standing Advisory Group on Safeguards Implementation (SAGSI), provides advice on technical aspects of safeguards. The organisational structure of the IAEA is shown in figure 1.

Radiation protection of the worker

The objectives of the Agency in the field of nuclear safety and environmental protection is to ensure the safe utilisation of nuclear energy and the protection of man and his environment from the harmful effects of nuclear radiation as well as from radioactive and non-radioactive releases from nuclear facilities. These objectives stem from the Agency's statute whereby under article III A.6 the Agency is authorised to establish or adopt, in consultation and, where appropriate, in collaboration

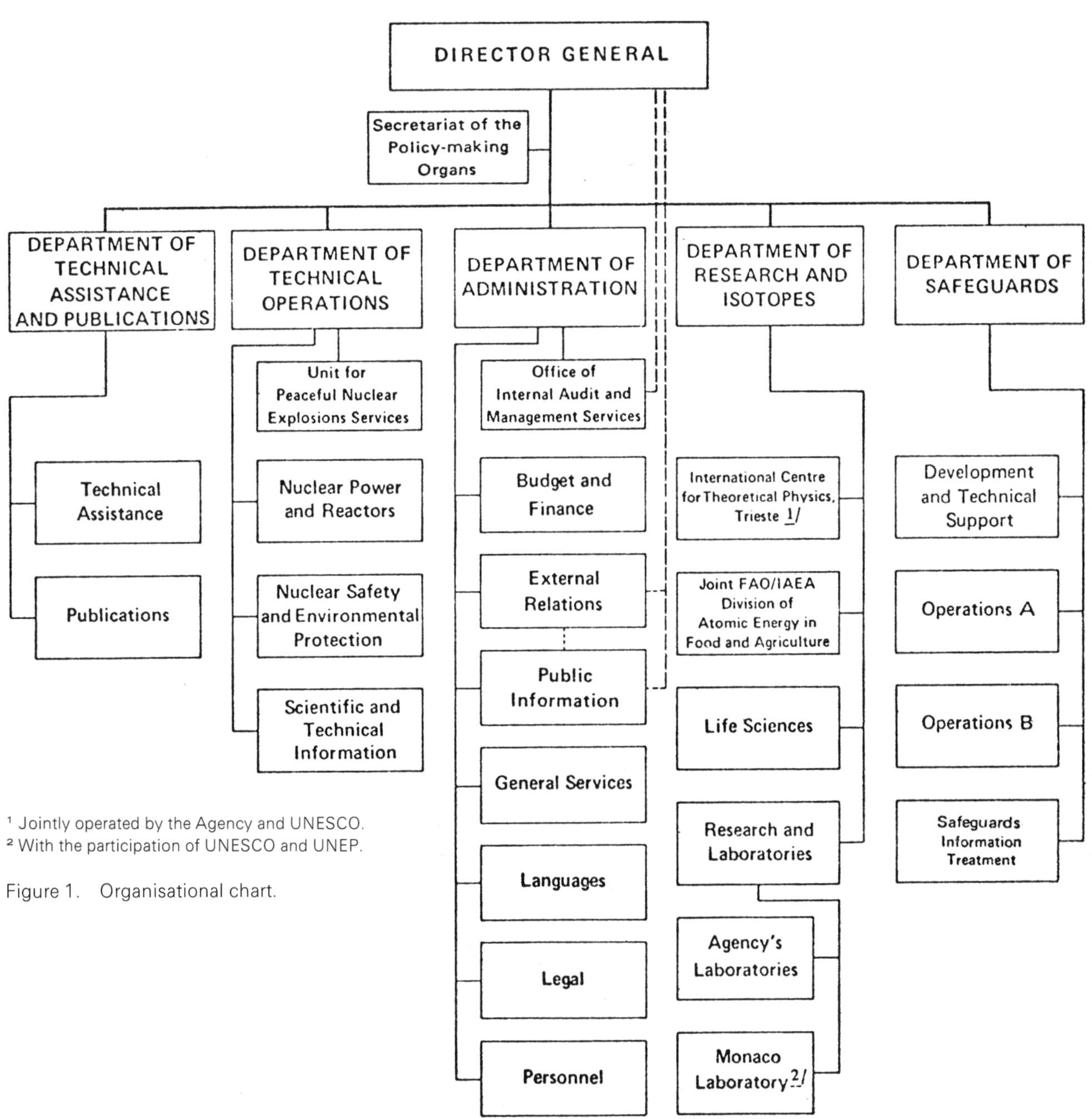

[1] Jointly operated by the Agency and UNESCO.
[2] With the participation of UNESCO and UNEP.

Figure 1. Organisational chart.

with the competent organs of the United Nations and with the specialised agencies concerned, standards of safety for protection of health and minimisation of danger to life and property (including such standards for labour conditions), and to provide for the application of these standards to its own operations as well as to the operations making use of materials, services equipment, facilities and information made available by the Agency or at its request or under its control or supervision, and to provide for the application of these standards, at the request of the parties, to operations under bilateral or multilateral arrangement, or, at the request of a State, to any of that State's activities in the field of atomic energy. As can be seen, these broad guidelines in the statute include provision for protection of the worker.

The Agency from its inception in 1957 initiated a programme on radiation protection by establishing in its Secretariat a Division of Health, Safety and Waste Management. Later, in 1971, it renamed this division the Division of Nuclear Safety and Environmental Protection to elaborate and strengthen its role in the environmental field.

The activities involve assistance to member States, through the provision of standards, recommendations, guidance and practical assistance, in the further development and harmonisation, within member States, of practices for the protection of workers and the general public against the harmful effects of ionising radiation arising in the peaceful utilisation of atomic energy and to provide effective radiological protection services in connection with the Agency's activities or Agency-assisted activities.

The detailed activities of the Agency in the area of radiological protection of workers involves the development of safety standards, preparation of safety guides, codes of practice, manuals, holding of scientific meetings for exchange of information or preparation of manuals or technical guidebooks, organising training courses, visiting seminars, and study tours, development of technical expertise in developing member States through the awards of research contracts and fellowships, and helping the developing member States in the organisation of radiation protection programmes through the provision of technical assistance, experts' services, advisory missions, and advisory services on nuclear law regulatory matters. These activities are geared to provide full guidance on radiation protection such that their application or follow-up will ensure the provision of safe working conditions for radiation workers. Many of these activities involve the co-operation of other international organisations. Their principal objective is to ensure optimisation of protection of the workers and the assessment of individual and collective doses.

Safety standards

The objective of this programme is to prepare and bring up to date, as necessary, safety standards for the protection of health and minimisation of risk of life and property from the effects of ionising radiation and to promote the harmonisation of the existing standards. This activity was initiated in 1958. The Agency has to date issued a number of safety standards. These have been prepared on the basis of the recommendations of the International Commission on Radiological Protection (ICRP) and have been periodically revised to keep them in line with revisions and the needs of the time, most important of which is the Basic Safety Standards. Since 1976 the major emphasis has been on the revision of the 1967 edition of the Agency's Basic Safety Standards for Radiation Protection (SS No. 9) as a co-sponsored activity with the WHO, ILO and the Nuclear Energy Agency (NEA) of the Organisation for Economic Co-operation and Development to take into account the recommendations in Publication 26 (1977) of the ICRP.

Of the safety standards, the Regulations for the Safe Transport of Radioactive Materials and the Management of Radioactive Wastes Produced by Radioisotopes Users, mainly concern the public environment. However, a significant part involves the working environment: the packaging of radioactive materials and handling of transport packages and, in the case of radioactive wastes, the treatment and handling of radioactive wastes before storage or disposal.

Safety guides and manuals

The objective of the preparation of safety guides and manuals is to provide guidance on measures for the radiological protection of occupationally exposed persons, on the provision of adequate physical and medical surveillance and on the assessment of occupational exposure. Many safety guides, codes of practice and manuals have been published by the Agency covering various aspects of the radiological protection of workers. The subject areas are the organisation of radiation protection programmes, safe operation of nuclear facilities, safe handling of radioactive materials, physical and medical surveillance of workers, personnel and area monitoring, diagnosis and treatment of radiation injury to workers, design and construction of nuclear reactors and power plants, handling of radiation accidents, treatment of radioactive wastes, radiation protection procedures, calibration of instruments, etc.

Scientific meetings

The main objective of holding scientific meetings is to provide means for the exchange of information between scientists of the member States, both developed and developing countries. This helps in better understanding of the problems, and improvement of the techniques and methods of radiation protection, as well as in better organisation of radiation protection programmes. The Agency has held many symposia and seminars on subject-matters related to occupational radiation protection. New meetings are planned when new problems on radiation protection are encountered or foreseen or when certain developments on a subject area require attention. Meetings are also planned when an existing publication requires revision or when exchange of information is considered useful in the light of the new developments.

Development of technical expertise in developing countries

The availability of trained manpower is one of the most essential parts of an effective radiation protection programme. The Agency organised training courses in radiation protection in various regions of the world. Visiting seminars and study tours with participants from the developing member States are organised. The purpose of these training programmes is to provide training and education in radiation protection which can be applied by the participants in professional work in their own countries. The Agency has awarded research contracts to member States for research and development in radiation protection instrumentation, techniques and methods of measurements, assessment of organ and body radiation doses, etc. It has also awarded fellowships to developing member States for training of their radiation protection personnel. These efforts have contributed towards the development of technical manpower in radiation protection mainly in the developing member States.

Technical assistance

Technical assistance has been provided to member States in the form of radiation protection equipment, and experts, as well as fellowships for the training of manpower as mentioned in the preceding section. Such technical assistance has been very effective in helping many member States in the organisation of their radiation protection programmes.

Advisory services

The Agency has sent advisory missions to many member States to advise on various matters of nuclear safety and radiological protection. These missions have carried out on-the-spot examinations of the problems in question and advised the requesting member States accordingly. They have been very useful not only in providing advice to the member States but also in receiving a feed-back of information on various problems which has contributed towards the planning of subsequent programmes of the Agency in radiation protection that have served the interests of many other member States.

Advisory services on legal and regulatory matters

Licensing and regulatory control of the uses of radiation sources and radioisotopes and of the construction and operation of nuclear facilities are effective means of ensuring safe working conditions for radiation workers. Over the past ten years the Agency has been providing advisory services to its member States, in particular developing countries, in the framing of requisite legislation and in tackling various regulatory and organisational matters involved in radiation protection and nuclear safety. In addition, it has held training courses and meetings on nuclear law and regulatory matters and will continue such activities as and when necessary or as requested by member States.

Radiological protection services for the Agency's laboratories and staff

The Agency provides radiological protection services for its laboratories and for staff who may be exposed to radiation in the course of their work. Radiological protection rules and procedures, based on the Agency's safety standards, for all work by Agency staff members and technical assistance experts which may involve exposure to radiation, have been established and are kept up to date. This is now the task of an interdepartmental Radiation Protection Committee. Personnel monitoring services for external radiation, bioassay and whole-body monitoring services are provided for the Agency's staff at laboratories and for the safeguards inspectors as well as for other staff.

Health and safety inspection missions

These missions are sent to certain member States to advise and assist them in regard to the safe operation of their nuclear reactors and other nuclear facilities and on nuclear health and safety matters in general. They are periodically sent to those facilities in member States to which the Agency's safety standards and measures apply under the relevant Project Agreements between the Agency and the member States concerned.

Emergency assistance with regard to nuclear accidents

The programme of the Agency has an over-all objective to help member States in assessing and improving their emergency preparedness programme as well as their accident handling capabilities and to enable the Agency to assist in co-ordinating the provisions of additional emergency assistance which member States may require. This activity was initiated in 1961. A document (Document WP/35 prepared in collaboration with the WHO, ILO and FAO and distributed to all member States) outlines the nature of the emergency assistance which member States might be willing to make available on request and indicates the proposed channels of communication; this additional document is brought up to date periodically (most recently in late 1980). This programme activity includes assistance to requesting member States for the diagnosis, treatment or follow-up of workers who may have been overexposed to radiation in accident situations.

The Agency will provide special missions on the adequacy of emergency plans and participate in emergency training exercises including aspects related to radiation workers.

INIS and information

According to its statute, the IAEA is to "foster the exchange of scientific and technical information on peaceful uses of atomic energy". In doing this, it serves as both a collector and a disseminator of information.

In 1970 the IAEA established the International Nuclear Information System (INIS). INIS is the first operational information system to employ a technique of decentralised input preparation combined with centralised processing of information. Information is collected by the IAEA member States and submitted to the IAEA headquarters in Vienna. Here it is checked and merged, using modern computer methods. It is then redistributed to national information centres for use by organisations and individuals. At present more than 50 countries and 13 international organisations participate in INIS by providing input. It is estimated that they are responsible for at least 95% of world publications on nuclear science and technology. The output of the system consists of magnetic tapes, which give bibliographic descriptions, key words and abstracts of all documents reported to the system, and a printed abstracts journal, *INIS Atomindex*. Both are distributed twice per month.

Thus INIS gives all member States a current awareness of the latest publications in nuclear science, including information concerning the radiological protection of the worker as well as information on improving the working environment from a radiological protection point of view. INIS benefits member States by making more accessible the results of current scientific research, whilst avoiding expensive duplication of information processing activities. Developing countries, in particular, receive through INIS information on new developments in nuclear science to which they might not otherwise have access. Through the INIS training programme, the staff of information centres in developing countries are trained in the most up-to-date techniques of information handling. Through INIS therefore authorities concerned with radiation protection of the worker receive up-to-date information on the state of the art.

The Nuclear Safety Standards Programme (NUSS)

The Agency has established a wide-ranging programme to provide member States with guidance on the many aspects of safety associated with thermal neutron nuclear power reactors. The programme, at present involving the preparation and publication of about 50 books in the form of Codes of Practice and Safety Guides, has become known as the NUSS programme. The publications are being produced in the Agency's Safety Series and each one will be made available in separate English, French, Russian and Spanish versions. They will be revised as necessary in the light of experience to keep their contents up to date.

The Codes of Practice and Safety Guides are recommendations issued by the Agency for use by member States in the context of their own nuclear safety requirements. A member State wishing to enter into an agreement with the Agency for the Agency's assistance in connection with the siting, construction, commissioning, operation and decommissioning of nuclear power plants will be required to follow those parts of the Codes of Practice and Safety Guides that pertain to the activities covered by the agreement. Part of the NUSS programme deals with the requirements for the operational radiation protection of the worker. In particular, two safety guides are now in preparation, one dealing with "Radiological protection during operation of nuclear power plants", and the other with "Design aspects of radiological protection for operational states of nuclear power plants".

The radiological protection of the worker is, in addition, naturally implicit in the safety design and operational criteria required by the various Codes of Practice and Safety Guides of the NUSS programme.

Future trends

Existing radiological safety standards will be brought up to date as required and new standards will be prepared as the need arises. This up-dating is required in the light of the new ICRP dose limitation system published in 1977. Advice will continue to be given through regional seminars and advisory services on the harmonisation of national legislation and standards for radiological protection.

The Agency's safety standards will continue to be applied to Agency-assisted projects, and health and safety advisory and inspection missions will be sent as required to facilities in member States that have concluded agreements with the Agency.

Attention will continue to be given to the special radiation protection problem arising in the handling of such radioactive substances as the transuranium elements and tritium. Efforts will also be made to encourage radiation protection measures in order to reduce the contribution of occupational exposure to the collective dose to the population. Emphasis will be placed on the organisation of regional seminars and training courses and on the provision of advisory services. Attention will also be given to the radiological safety aspects of the design and operation of facilities for handling transuranium elements and of facilities in the nuclear fuel cycle.

DAW, H. T.
AHMED, J. U.

The law and practices of the International Atomic Energy Agency. Szasz, P. C. IAEA legal series No. 7. STI/PUB/250 (Vienna, International Atomic Energy Agency, 1970), 1 180 p.

International Ergonomics Association (IEA)

The aim of the International Ergonomics Association is to bring together organisations and people throughout the world concerned with the study and application of ergonomics and human factors, i.e. the relation between man and his occupation, equipment and environment in the widest sense including work, play, leisure, home and travel.

To further these aims it sets up meetings, establishes liaison with other bodies, issues publications and generally tries to promote the knowledge and practice of ergonomics internationally, especially in areas of the world where there are, as yet, no federated societies.

The Association has a membership of–

(a) federated ergonomics societies: these are societies which have as their main aim the promotion of ergonomics. They are bodies which elect their governing body from within their own membership and encourage the free publication of research and other material. In each geographical area there is only one federated society;

(b) affiliated societies: these are societies which have a major interest in a topic but whose main aims are in an associated area or which are ineligible for federated society status because there is already an existing one for their area;

(c) associated society: these are international organisations which have an indirect interest in ergonomics such as telecommunications, and design organisations which are concerned with the application of the subject;

(d) sustaining members: these are organisations which have an interest in the Association and its aims and wish to support it by the payment of an annual subscription.

Origins of the Association

Two organisations have had a major role in the setting up of the Association: the Ergonomics Research Society (now the Ergonomics Society) of the United Kingdom and the European Productivity Agency (EPA), which was an agency of the Organisation for European Economic Co-operation (later changed to the Organisation for European Co-operation and Development).

The Ergonomics Research Society, which was founded in 1949, was the first national ergonomics society to be formed and, as such, had attracted a large international membership. In 1951, 14 of the total of 80 members of the Society were from outside the United Kingdom. This balance of membership was also represented in the attendance at the early meetings of the Society. It was clear, however, that a purely national and only English-speaking Society could not meet the needs of a subject that was, even then, becoming international in scope. This view was naturally not held unanimously amongst the members of the Ergonomics Research Society, some of whom felt that it should be possible to develop the existing society into an international body.

In Western Europe the EPA, which had a strong emphasis on technology and applied science, became interested in ergonomics and in the early 1950s set up a co-operative project on "Fitting the job to the worker". This project involved the sending in 1950 of a team of nine scientists from the seven European countries of the OEEC to the United States, the organisation of an international conference of experts with representatives from Western Europe and the United States held at Leiden in the Netherlands in 1957, and the running of a meeting of experts and industrialists in Zurich in 1956. It was at the Leiden meeting in 1957 that the need was expressed for some form of organisation which could continue and extend the international co-operation which was effectively started then by the EPA. As a result a committee was set up under the chairmanship of the late Professor G. C. E. Burger of Philips and the Netherlands Institute for Preventive Medicine to explore ways of renewing international contact.

The International Ergonomics Association can be said to have officially come into being during the Annual Conference of the Ergonomics Research Society in Oxford on 6 April 1959.

History of the Association

When the Association was first set up, very few national ergonomics societies had been formed and it was agreed that the Association would initially be an association of individuals. This was a decision which was reluctantly accepted by many of those concerned and caused some friction over the succeeding years. From 1964 onwards it was accepted that national societies should be federated members of the IEA, but they could not be said to be in control of it until 1976.

Between 1961 and 1976 the main decision-making body of the Association was the General Assembly, which is held during the Triennial Congress of the Association. The 1961 and 1964 General Assemblies were of individual members only but from 1967 to 1976 they were composed of representatives of federated societies in proportion to their size together with representatives of the individual members. It was also agreed in 1967 that individuals who were members of federated societies could not also be individual members of the IEA. The General Assembly continued to be the main decision-making body until 1976 but the Council, which met more frequently, was responsible for the main discussions surrounding the decision. The Council, however, at that time was not representative of the federated societies because they did not all have delegates on it. Since 1976 the Council has been the governing body, its membership being entirely made up of representatives of all the federated societies, each being represented by from one to three people according to its size. The General Assembly has now become an open meeting for the transmission of information and the collection of views and has no power to make decisions.

The Association, which started up as a Western European grouping, now has a truly international spread with members from the United States, Japan and the socialist countries. Although for short periods groups were accepted which did not meet the criterion of self-governing societies, in that they were state-run organisations, all the federated societies now in membership do fulfil that requirement. There are now no individual members of the Association. As most of the federated societies accept members from any country it was felt that the most effective way to deal with applications for individual IEA membership was to suggest that they join any one of the federated societies which allows them to get the benefits of the IEA as well as those of the society they join at minimum cost.

Present membership

The following ergonomics societies are federated members (the approximate numbers of individual members of each Society is given in parentheses): the Ergonomics Society, UK (700), the Ergonomics Society of Australia and New Zealand (300), the Gesellschaft für Arbeitswissenschaft (450), the Human Factors Society, United States (1 600), the Human Factors Association of Canada (150), the Japan Ergonomics Research Society (200), the Nederlandse Verenigung voor Ergonomie (370), the Nordic Ergonomics Society (620), the Polish Ergonomics Society (220), the Société d'Ergonomie de Langue française (300), the Società Italiana di Ergonomia (100), the Hungarian Society for Organisation and Management Science (80), and the Yugoslav Ergonomics Society (60). At present there is one affiliated society, the Sociedad Española de Psicología with 100 members, and one Sustaining Member, the American Telegraph and Telephone of the United States. There are as yet no associated societies.

Subject area

Ergonomics is a subject area with imprecise boundaries and there are different ideas as to where those boundaries lie in the different federated societies. Each society also puts the emphasis on different aspects. In societies in countries where there is a concern for developing technology and complex systems, the emphasis tends to be on the more psychological aspects of the subject, such as the kinds of stresses which can affect people working in such systems and the ways in which the equipment can be designed to reduce those stresses and the risks of human errors. In areas where work is more basic, the concern has been more with physiological problems such as safe working loads, maximum energy expenditure levels, etc., with the aim of reducing physical strain on the worker. There is also an increasing interest in the effects of mass production with its paced repetitive work on the health and well-being of the individual worker in many countries with industrialised manufacturing systems.

The social environment and the degree to which work should take account of the individual's psychological needs are on the boundaries of the subject. These are becoming of increasing interest to many workers in the field, but because they are less amenable to objective and experimental study they have tended to be considered outside the area by many more traditional ergonomists. This situation is changing as society tries to find solutions to the problems that modern industrial work causes in these areas.

Meetings

The main activity of the IEA has been the organisation of Triennial International Congresses. Three meetings have also been held, each with a small number of specially invited participants, with the aim of producing publications. These have been on (1) sitting posture, (2) physiological and psychological criteria for the study, design and validation of man-machine systems, and (3) ergonomics for underdeveloped countries (this last actually sponsored by NATO).

In April 1973 a seminar was held on "Ergonomics and standards" to discuss how the IEA could contribute to the development of international standards, as it was then starting to discuss with the International Organisation for Standardisation the setting up of an ISO technical committee on ergonomics.

In 1977 an open meeting was held in the United Kingdom jointly with the Ergonomics Society on "Ergonomics and job satisfaction", with the dual aims of making ergonomists more knowledgeable about work in the area of job satisfaction and the need to consider individual psychological needs and of bringing the ergonomic approach to the notice of people working in allied disciplines on these problems. As a result of this meeting a book was published containing the invited papers and some specially commissioned articles by some of the participants based on their work and the discussion at the meeting.

A meeting was also held in Oslo in August 1980 organised by the Nordic Ergonomics Society on behalf of the IEA on "Ergonomics in action from theory to practice" to look at the factors that make for success in the application of ergonomics.

Discussions are in hand for an open meeting on the relevance of ergonomics to the developing countries, which it is hoped will be organised by Société d'Ergonomie de Langue française on behalf of the IEA and in co-operation with the ILO.

The Association is hoping that federated societies will take increasing initiatives in the future to organise meetings on its behalf which will establish links with other groups and help bring the importance of ergonomics to their notice. Such meetings could be based on the application of ergonomics to a particular activity or industry, organised in co-operation with the appropriate body for that activity or industry.

Publications

When the IEA was formed in 1961 the journal *Ergonomics* was already being published. Although this was mainly an English-language publication it had a large international circulation and it was therefore agreed that it should act as the official journal of the Association and it has continued in that role ever since.

Since the 1967 Congress the invited papers have been published in special Congress issues, which have been available to delegates at or before the meeting. In addition to the publications resulting from the special meetings covered in the previous section, the Human Factors Society published in full the papers given at the 1976 Congress in Maryland.

A major problem for any international organisation is that of communicating both with its members and to the outside world, and the Association is trying to remedy this by the publication of a regular newsletter. Two issues have been provided to date and it is hoped that as this becomes more regular it will make the Association more visible to those concerned with the study and application of ergonomics. It has also produced a brochure in English, French and German.

To help those who wish to pursue a course of study in ergonomics, an international directory of educational programmes in the subject has been produced listing the different courses available throughout the world giving information on their level, coverage, etc.

Contacts with other bodies

As one of the Association's main aims is to develop the international understanding and application of ergonomics it has to look outside its own boundaries and develop contact with other international bodies.

It has been a co-operating organisation with the ILO for many years and has sent representatives to many ILO meetings which are of common interest. Members of federated societies have acted as advisers to the ILO on a number of topics.

The international organisation with which the IEA has had most recent contact is the International Organisation for Standardisation which since 1975 has had a Technical Committee (ISO/TC/159) on Ergonomics with which the IEA has a category A liaison. The Association considers that this is an important activity and has now appointed someone to have special responsibility for it. The Technical Committee has been very active and the IEA is able to comment on proposals as they arise and to have representatives at all meetings of the Technical Committee and its subcommittees to ensure that the standards established take account of the ergonomics information available.

The IEA is also a non-governmental organisation in liaison with the World Health Organisation and sends representatives to the relevant meetings of that body.

The Association is concerned that there should not be two different international groupings in ergonomics, one representing the West and one the East and, as can be seen from the membership list above, it has members from all parts of Europe and the world. The countries that are members of the Council for Mutual Economic Assistance, for their part, carry on some joint co-operative activities in the ergonomics field, and the IEA has been officially represented at meetings where this work has been discussed.

The IEA is also trying to encourage the formation of new ergonomics societies in parts of the world where there are, as yet, none in existence which meet the criteria for membership of the Association. Recent discussions have been held with individuals in Brazil, India, Israel and the USSR regarding developments in their countries which it is hoped will lead in time to societies there becoming Federal Members of the Association.

Discussions have been held with the International Council for Societies in Industrial Design regarding possible co-operative ventures.

Future plans

The IEA has many problems. It represents around 4 500 people spread throughout the world and has to do its work through the voluntary efforts of its officers and others who take on special responsibility. Finance is a problem in that international travel is expensive and it is difficult to raise funds from such a small base.

The main aims for the future are to increase the visibility of the IEA to both ergonomists and their clients and to demonstrate the importance of the subject to those who should be involved in its application by issuing more publicity and arranging more meetings together with those outside the Association but concerned with its applications, such as designers, physicians and managers.

It will do this by issuing more publications and by co-operating with international organisations representing people who have such an interest in the application of ergonomics in the organisation of meetings and other joint work.

(The views expressed in this article are those of the author and not necessarily the official views of the Association.)

SELL, R. G.

Sitting posture. Grandjean, E. (ed.). (London, Taylor and Francis, Ltd., 1969), 262 p.

Measurement of man at work. Singleton, W. T.; Fox, J. G.; Whithfield, D. (eds.). (London, Taylor and Francis, Ltd., 1970), 268 p.

Ethnic variables in human factors in engineering. Chapanis, A. (ed.). (Baltimore, Johns Hopkins University Press, 1970), 290 p.

Satisfaction in work design. Sell, R. G.; Shipley, P. (eds.). (London, Taylor and Francis, Ltd., 1979), 202 p.

International directory of educational programmes in ergonomics/human factors. Pearson, R. G. (ed.). (International Ergonomics Association, c/o Human Factors Society, Box 1369, Santa Monica, California) (1979), 16 p.

International Labour Organisation

"Universal and lasting peace can be established only if it is based on social justice."

These, the opening words of its Constitution, are the *raison d'être* of the International Labour Organisation, a specialised agency of the United Nations which bands together in a common endeavour the governments, employers and workers of 146 countries.

This call for social justice was first heard in the history of international relations in 1919, in Part XIII of the Versailles Peace Treaty, the instrument which brought the ILO into being.

The call is still as meaningful and relevant as ever. More than ever before, world peace is still jeopardised by the persistence of inequalities. Despite economic growth and rapid technological progress, the problem of the poor has not yet been solved. Hundreds of millions of men and women are still living on paltry incomes in complete and utter poverty, while some of the better endowed peoples are looking for ways to cut down waste. Even in the developed countries there are still serious problems of social injustice and infringement of workers' rights.

Nowadays, efforts to achieve a New International Economic–and social–Order give concrete expression in a contemporary context to the idea that has inspired the patient achievements of the ILO for more than 60 years: "Poverty anywhere constitutes a danger to prosperity everywhere."

The progress of an idea

This idea had been gaining ground in many minds long before the creation of the ILO.

From the end of the 19th century onwards, under the influence of social theorists, efforts had been made by the trade unions and also by some industrialists wishing to improve the lot of the proletariat without distorting competition between countries, with a view to promoting social progress by means of international agreements.

An International Association for the Legal Protection of Workers, the forerunner of the ILO, established in 1890, continued until the First World War and enabled two international labour Conventions to be passed by diplomatic conferences.

At the end of the war, at the urging of the trade unions, the Versailles Peace Conference was to bring the International Labour Organisation officially into being; indeed, Part XIII of the Peace Treaty, which sets out the basic principles, aims and main operating procedures of the Organisation, still remains the essential part of the ILO Constitution.

An autonomous organ of the League of Nations until the Second World War, the ILO became in 1946 the first specialised agency to be associated with the United Nations. The number of States Members has gradually risen from 42 in 1919 to 58 in 1948 and 148 in 1982.

The International Labour Office, the secretariat of the Organisation, has its headquarters in Geneva.

The mandate of the ILO

When the first session of the International Labour Conference to be held after the end of the Second World War met in Philadelphia in 1944, it set out to redefine and specify the aim and purposes of the ILO while reaffirming the fundamental principles proclaimed in 1919.

Every sovereign State which joins the Organisation subscribes to the Constitutional affirmation that "all human beings, irrespective of race, creed or sex, have the right to pursue both their material well-being and their spiritual development in conditions of freedom and dignity, of economic security and equal opportunity".

It is this over-all objective that the Organisation endeavours to translate into reality by seeing to it that the fundamental rights of workers are respected throughout the world and by supporting the efforts of the international community to achieve full employment, to raise living standards, to distribute the rewards of progress fairly, to protect the life and health of the workers, and to foster co-operation between employers and workers in all fields of common interest.

Immense field of action

The activities of the Organisation cover an immense field, what with the diversity of the occupations and individuals concerned and the diversity of the problems to be solved.

There are, by definition, no trades or occupations that fall outside the mandate of the ILO. It is true that the word "workers" in the sense in which the pioneers of the Organisation understood it–and in which it is still generally understood–meant primarily the impoverished proletariat that was born of the Industrial Revolution, a population of wage earners engaged in mostly manual labour and confronted with the problems of the oncoming Machine Age: quarrymen, miners, smelters, employees in factories of all kinds–glassworks, cement plants, paper and textile mills, etc.–building or transport workers, dockers, seafarers, shop hands, office clerks, employees in electricity and gasworks, farm hands, and so on.

Protection of these categories of workers is still a priority concern of the ILO, even though their employment and working conditions have changed out of all recognition, and though new trades have sprung up in industries ranging from chemicals to electronics and cybernetics, from the oil driller in the wastes of the Arctic to the girl assembling chips in micro-processors. A whole range of manual jobs unheard of 60 years ago has come into being and continues to expand.

But help and protection are also due to office workers; to civil servants whose numbers have increased considerably throughout the world; to postal and telecommunications workers and those employed in modern transport, from railwaymen to air traffic controllers, from airline pilots to truck-drivers; to health personnel whose role is constantly growing; to teaching personnel at all levels and research workers; to authors, artists and performers worried about the technological revolution, and so forth.

Similarly, the dividing line between handicrafts and industry has sometimes become very hazy in such sectors as clothing, leather and footwear, furniture and food products. Indeed, even self-employed workers themselves, farmers, craftsmen or small tradesmen, need to be safeguarded against the hazards of their callings and to be entitled to the various social security benefits, including retirement pensions.

Certain classes of workers are especially vulnerable and require particular attention. There is the immense mass of the rural poor, wage earners and otherwise, in the Third World countries, there are the workers in the informal sector in cities and on the land, whose humble callings are nevertheless essential to the working of the economy; the travelling and migrant workers forced by necessity to seek a living in foreign lands, where often their rights are trampled underfoot.

Regard must also be had to the special difficulties that may be encountered in the working world by women, who now represent more than one-third of the total labour force and should be able to reconcile their outside work with their traditional role and enjoy equality of opportunity and treatment; by young people whose health, vocational training and guidance constitute constant subjects of concern; by other workers who require adjustments in their working routine to prepare them for retirement; and by the physically and mentally handicapped, who might be enabled by appropriate rehabilitation services to play their full part in the economy and in society.

Different fields

The diversity of the problems to be tackled is no less great than the diversity of occupations and individuals concerned.

One major goal is to ensure that everyone gets a fair wage or decent livelihood through the implementation of dynamic manpower policies and programmes to combat unemployment and foster productive employment. For wage and salary earners this goal may involve such matters as career policies, job classification and evaluation of work performed, collective bargaining, job security, and so on. For self-employed workers an assured livelihood may be keyed to the organisation of production structures, the establishment of co-operatives, an increase in productivity, the availability of loans, or problems of marketing.

Another major goal is to achieve occupational safety and health and to bring about better general working conditions. In this framework lies the campaign for the reduction and adjustment of hours of work, the control of night work and home work, and the fight against "moonlighting" and clandestine working generally. Attention must also be paid to improving the working environment, abolishing or alleviating the more arduous tasks, controlling dust and all harmful airborne substances, combating excessive noise and vibration and imposing stringent checks and inspections of machinery and the countless substances used so as to reduce or eliminate the risks of diseases and accidents.

Yet another goal is to secure the respect of the workers' fundamental rights: trade union freedom, the right of association and of collective bargaining, the abolition of forced labour, and the elimination of any discrimination against vulnerable groups.

Mention must also be made of the interest that the ILO has to take in neighbouring or "borderline" fields, since certain options have inevitable repercussions on the labour scene, e.g. national planning, agrarian policy, technological choices, international trade, regulation of the activities of multinational corporations, medical and scientific research, and civil and political liberties. In all matters relating to these fields the ILO co-operates closely with the specialised agencies in the United Nations system and with the relevant governmental and non-governmental organisations.

Means of action

To carry out the mission entrusted to it in these various fields, the ILO has basically three means of action at its disposal, ranging from research and the framing of needed action to its formulation and implementation, and supervision of its application.

Studies, research and meetings. Among its 1 500 officials and 660 experts on mission throughout the world, the International Labour Office has specialists in all the fields concerned. Its library possesses a collection of more than 150 000 books, which are accessible to research workers and the public at large by means of a computerised data scanning and dissemination process hooked up to a world-wide library network. The many studies published by the ILO on a broad range of topics are primarily of a practical character, being designed to throw light on labour problems, to point to solutions and work out the action needed to implement them. The same applies to the reports issued in preparation for various meetings and for the International Labour Conference, and to the articles published in the *International Labour Review*, a bi-monthly journal with a wide circulation. The ILO also publishes the *Year Book of Labour Statistics*, a basic reference work brought up to date quarterly by a *Bulletin of Labour Statistics*, and various other periodical publications such as *Women at Work*, the *Labour Information Record* and a bulletin entitled *Labour Education*.

The necessary confrontation of ideas takes place in a whole series of conferences and meetings: the International Labour Conference, which is the "annual general meeting" of the Organisation, but also the regional conferences which make it possible to hammer out policies common to a whole group of countries on labour issues; the Industrial Committees and tripartite technical meetings; the Joint Commissions, and meetings of experts.

International labour standards and supervision of their application. Standard-setting is the ILO's oldest activity and it remains its fundamental task. It consists in the adoption of international labour Conventions–which are subject to ratification–and Recommendations defining standards that are acceptable to all countries in the various fields covered by the Organisation, from protection of the fundamental rights of workers to the regulation of conditions of work or the promotion of social security. Between 1919 and 1980 the ILO adopted 153 Conventions and 162 Recommendations. It is, however, important to see to the application of these instruments and to redress the sometimes serious breaches of their principles that come to light. To this end the ILO has its own unique supervisory machinery, which includes a committee of independent experts and review and appeals procedures available to all its members, including the employers' and workers' organisations. In recent years the ILO has increasingly sent direct contact missions with a view to assisting member States to overcome the difficulties they may have encountered in applying ILO standards.

Technical co-operation activities. These cover all the operational assistance which the ILO makes available to its member States to further their progress in the social and labour field, principally in the form of missions of experts, the award of fellowships and gifts of equipment. These activities developed mainly since the 1950s in connection with the various programmes launched by the United Nations and in particular the United Nations Development Programme (UNDP) which still puts up the greater part of the financing for ILO projects for the development of human resources, including such matters as vocational training and management development, employment promotion and planning, the development of social institutions and living and working conditions. Technical co-operation expenditure at the present time is running at a rate of more than $100 million a year, and increasingly the distinction between "donor" and "receiving" countries is fading away to make room for a system of mutual co-operation on a global scale, especially at the level of the major geographical regions, in which every country has something to learn and something to "receive". The International Institute for Labour Studies established in 1960 in Geneva, and the International Centre for Advanced Technical and Vocational Training, opened in 1965 in Turin, are two more of the instrumentalities available to the ILO in the fields of research and action.

The rules of the game

However, the deployment of these various capabilities and their efficacy or otherwise depend very much upon the conditions in which the Organisation operates. Several factors come into the picture.

The first and fundamental point is the *sovereignty* of member States. The ILO naturally endeavours to get the sovereign States of which it is composed to join forces and co-operate within the Organisation, and indeed admission to the Organisation presupposes an undertaking to abide by the principles enshrined in its Constitution. But the authority of the ILO is essentially moral authority, and it cannot in practice encroach upon the authority of national governments. It has no right to order them about, nor has it the power to take action of its own accord in the territory of this or that government. All it can do is to estimate and back up national action.

The *universality* of the Organisation is and has always been a "must", but it is also a tower of strength. From its inception, at a period when the number of member States was scarcely one-third of what it is today, the ILO has striven to extend its influence to all peoples. Since the 1950s a large number of new States have attained independence and the ILO's composition has become practically universal. This universality lends unquestionable relevance and authority to the decisions taken, particularly by the General Conference, but it is not without raising some difficult problems–

(a) in connection with the framing, adoption and policing of standards, which must have regard to differing levels of development in the various regions and countries and yet set goals that are valid for all. There cannot be "sub-standards" for "sub-human beings";

(b) as regards the composition of the executive organs of the ILO, especially the Governing Body of the International Labour Office, which directs the work of the Office and appoints the Director-General. It is important that the Governing Body should reflect as faithfully as possible the political and geographical diversity of member States, and for this reason a process of enlarging and re-balancing the Governing Body is now in hand. The outcome should enable the countries of the Third World to have a large say in the deliberations and the decisions, with due regard both to diversity of economic, political and social systems, and to the economic importance of each region of the world.

The *tripartite system* is the distinguishing feature of the Organisation and accounts very largely for the practical influence it has been able to wield over social policies in its member States. Governments, employers and workers are represented on a footing of equality in practically all ILO bodies and no decision can be taken without discussion, negotiation and if necessary voting in a tripartite assembly. In this way the ILO has been able to keep abreast of the concrete realities of the labour world and with the main economic and social forces in each country.

Tripartism, however, is an exacting principle that is by no means easy to apply. It involves efforts to overcome the conflicts of interests between employers and workers, and in addition, if it is to function properly, it must be alive to the social and political realities of the present-day world. For instance, some employers in countries where the free-enterprise system prevails challenge the representativity of the delegates of heads of undertakings in the socialist countries, who for their part consider that they should be more heavily represented on the Governing Body. Similarly, there are several different conceptions of trade unionism, corresponding to particular political and social options. But these difficulties of interpretation and operation do not in any way detract from tripartism itself, which is generally regarded as an objective to be reached rather than something that is achieved once and for all time. The problems thrown up by the practical application of this rule are being considered at the same time as the question of enlarging the composition of the Governing Body in the discussions on the adaptation of the structure of the ILO.

Concrete results

In more than 60 years of existence, the ILO has shown itself to be an efficient tool in the service of social justice throughout the world.

This is because its work is not confined to theoretical research, reports, meetings, speechmaking and recommendations that are not followed up. It leads to concrete commitments on the part of member States and to specific instruments, the application of which is strictly and impartially supervised.

This supervision is the task of a Committee of Experts which can point year after year to scores of changes for the better in the legislation and practice of many countries so as to secure more complete implementation of Conventions and Recommendations. For instance, 1 300 cases of changes were registered between 1963 and 1980. For millions and millions of workers this means a step forward–or several steps forward–towards a more dignified and freer life in keeping with their aspirations.

Another example of effectiveness: over a period of three years more than 250 imprisoned trade union leaders have been freed in African, Latin American, Asian, European and Middle-Eastern countries as the result of direct or indirect action by the ILO. In one country a general amnesty made it possible for a large number of trade unionists to be released or to return to their country.

In key areas of the labour world, at the request of its member States, the Organisation has launched extensive programmes of combined action which employ all the means at its command–studies and research, standards and technical co-operation. This is the case of the World Employment Programme launched in 1969, which has enabled many countries to incorporate a real strategy to combat unemployment into their national plans and at the same time has given the ILO a unique testing ground for a better understanding of the links between employment, growth and the meeting of basic needs. The same is true of the International Programme for the Improvement of Working Conditions and the Working Environment, which was launched in 1976 on the strength of several years of ILO research, information supplied by numerous member States and the suggestions of employers' and workers' organisations. Nearly 50 countries have requested assistance under this programme, and 20-odd have been visited by a multidisciplinary team to examine the whole problem of working conditions.

Whether in these major fields or in other more specific aspects of social and labour policy throughout the world, the concrete achievements to the credit of the ILO are practically countless: here, poor rural workers were able to band together in trade unions and establish production or sales co-operatives; there, increased resources have been allotted to labour inspection and several governments have joined forces to set up a Regional Labour Administration Centre; elsewhere, ILO experts have helped in the establishment and, later, the development and regional harmonisation of social security systems; yet elsewhere, all over the Third World, vocational training centres are opening up and instructors are being given further training so that they can play their part to the full. In a given African or Asian country, a really comprehensive policy for the vocational rehabilitation of the handicapped has been framed and put into effect; throughout the poorer rural areas of the world, labour-intensive public works projects are being put in hand to provide a minimum of earnings for the populations concerned; a world-wide information and co-operation network is being set up to eliminate dangerous products used in industry, and so on in an endless list.

Generally speaking, the ILO's work for development in all its aspects–industrialisation, agricultural development, employment promotion, increases in earnings and purchasing power, assistance in reconversion in the industrialised countries—contributes to the restructuring of international economic relations.

To carry out this long-lasting mission successfully, the ILO can count on the assets of its universality and its tripartite structure, but also on the close ties it has with the other organisations in the United Nations system. By thus uniting its efforts with those of the whole international community in favour of peace, it can widen the scope of the battle it is waging against poverty and injustice in the world.

For its work as a whole the ILO received the Nobel Peace Prize in 1969, the year of its fiftieth anniversary.

FROMONT, M.

The ILO and the world of work (Geneva, International Labour Office, 1979), 64 p. Illus.

International Labour Organisation (application of Conventions on occupational safety and health)

States Members of the ILO that have ratified international labour Conventions concerning occupational safety and health have an obligation to apply them under article 19, paragraph 5*(d)* of the ILO Constitution, which provides that when a Convention has been ratified the Member "will take such action as may be necessary to make effective the provisions of such Convention".

International supervision

The constitutional instruments of the Organisation provide for international supervision of compliance with this obligation as well as with the detailed requirements of each particular Convention. This supervision is designed to promote wider ratification of ILO Conventions and their full implementation. It facilitates the development of appropriate national legislation and practice by means of the follow-up of the measures requested from or announced by the governments concerned to ameliorate national laws, regulations and practice and also by technical assistance from the ILO to the interested countries. The supervisory procedures facilitate the participation of workers' and employers' organisations in the formulation, application and enforcement of national laws and practical measures designed to give effect to the ratified instrument.

The ILO system of supervision comprises mainly the Committee of Experts on the Application of Conventions and Recommendations appointed by the Governing Body of the ILO and the Committee on the Application of Conventions and Recommendations appointed by each session of the International Labour Conference. The principal task of these supervisory bodies, particularly the Committee of Experts, is to establish, on the basis of the national legislation, and the reports of the governments of ratifying States, and taking into account any observations made by workers' and employers' organisations, whether the requirements of the ratified Conventions have been complied with. The findings of the Committee take the form either of requests addressed directly to the governments concerned, or of observations published in its report, which can then be discussed in the Conference Committee on the Application of Conventions and Recommendations.

Main problems of application

The analysis carried out by the supervisory bodies may lead to the conclusion either that a ratified Convention is fully applied in law as well as in practice or that discrepancies exist between the national legislation and actual situation on the one hand, and the requirements of the Convention on the other.

The variety of problems that arise in the application of Conventions may be reduced to two main types—those involving the existence and the adequacy of the means of application and those relating to the effect given to an instrument in practice.

In some instances, especially shortly after ratification, national provisions for the application of a Convention may be more or less inadequate, either as regards the scope of their coverage or as regards the level of protection they afford to workers.

The national legislation may not adequately give effect to the Convention in terms of *branches* of industry or of the economy covered. The main reason why gaps appear in the coverage of certain branches to which a Convention applies is probably the fact that labour codes may contain no provisions for, or do not apply to, the types of activity (e.g. mining, shipping, railways, air and road transport) for which special regulations exist outside the labour code; the decision to ratify is sometimes based on the code, insufficient attention being given to regulations for particular branches, which may not fully conform to the standards of the Convention.

Adequate *territorial coverage* is not ensured when laws and regulations exist only for one part of the country; in one case the safety and health legislation for the construction industry applied only to the areas of the national capital, and more than half of the nation's construction workers in the rest of the country did not enjoy the legal protection required.

The legislation may also fail to ensure the *personal scope* of coverage required by a Convention, as was the case in several instances with regard to the prohibition of employment of persons under 16 years of age in work involving ionising radiations. Some members did not initially have such a prohibition, but at the request of the Committee of Experts, their laws have been modified to comply with the requirement of the Radiation Protection Convention, 1960 (No. 115).

National legislation may also be insufficient in respect of the *level of protection* afforded to workers. This aspect is illustrated by cases in which not all the dangerous parts named in the Guarding of Machinery Convention, 1963 (No. 119), are required by law to be supplied with appropriate guards, or the content of benzene in the air of workplaces is fixed nationally at a level above that prescribed by the Benzene Convention, 1971 (No. 136).

Insufficient coverage or inadequate protection may also result from recourse to exceptions which go beyond those allowed by a Convention.

Discrepancies between national law and practice and the terms of a Convention sometimes take the form of *insufficient powers* attributed to the bodies or persons responsible for inspection in the field of safety and health. Thus, the laws and regulations of several ratifying countries did not empower labour inspectors to initiate measures with immediate executory force in order to remedy defects in plant or working methods that constitute a threat to the health or safety of the workers, as required by the Labour Inspection Convention, 1947 (No. 81).

Occasionally the supervisory bodies have to deal with cases where there is a more or less *total absence* of measures needed to apply the provisions of a ratified Convention. The reasons for a ratification not adequately supported by national legislation may be many. For instance the national texts may be considered nationally as applying the Convention when in fact they do not relate to its subject-matter. Such was the case of the ratification of the Maximum Weight Convention, 1967 (No. 127), on the basis of national rules that laid down the remuneration of workers in proportion to the weight of loads carried but did not in any way limit their maximum weight. Consequently, the supervisory bodies asked for appropriate measures to be taken to give effect to the Convention.

Sometimes ratifying States indicate in their reports that the non-existence of the legislation or other measures required by the Convention is due to the fact that the activity with which the instrument is concerned is not carried out in their territory and consequently there is no need for regulation. The Committee of Experts may note such a statement and ask for information on any changes in the situation. This approach is adopted when the actual position is bound to remain unchanged for some time, as in the case of absence of legislation to apply the Medical Examination of Young Persons (Underground Work) Convention, 1965 (No. 124), in a country where there is no underground work of this kind. When, on the contrary, the actual situation is such that it may change at any time, the Committee would ask for laws or regulations to be adopted to deal with such an eventuality. This was the case, for example, when the reports from States that had ratified the White Lead (Painting) Convention, 1921 (No. 13), said that white lead was not used in their territory in painting work of an industrial character. The Committee's comments resulted in the adoption in a number of countries of legislation formally prohibiting the use of white lead in conformity with the Convention. Finally, the Committee may point out the activities that may have been judged by the national authorities not to come within the scope of a Convention, whereas in fact the instrument covers such activities. Thus when government reports stated that ionising radiations, which are the subject of the Radiation Protection Convention, 1960 (No. 115), were not used in their countries, the Committee pointed out that the instrument applies also to radiological work by doctors and in criminal investigations. Appropriate national regulations were subsequently adopted in a number of countries.

Along with cases where the necessary legislation was initially lacking there have been situations in which laws and regulations that had ensured the application of a given Convention were subsequently repealed as a result of an overhaul of national legislation. In one such instance comments by the Committee led to a revision of the newly adopted law to reinstate the chapter on the guarding of machinery.

The absence of any national official texts to deal with the matters coming within the scope of the Convention is a clear-cut case of a "premature" ratification, when viewed in the light of the constitutional obligations to apply the ratified instrument. A more elaborate analysis of the situation is required when official texts do exist; their binding force must then be verified.

The Conventions adopted before 1960 require that their provisions be given effect through legislation to the exclusion of other means of application. Texts not endowed with a sufficient measure of legal authority, such as the recommendations of safety and health institutions, codes of practice, etc., do not therefore qualify as adequate means of implementing the provisions of these Conventions. Where this situation prevails, the ratifying State is invited either to recast existing provisions in the form of laws or regulations or at least to supplement the existing rules with an appropriate system of inspection and with effective sanctions.

The more recent Conventions, along with laws or regulations, admit other methods of application consistent with national practice and conditions. However, because within these more flexible instruments there are important provisions that rely solely on laws and regulations for their implementation, most ratifying countries have chosen to give effect to the whole of such Conventions through laws or regulations.

Particular problems arise in the application of safety and health Conventions by countries where, because of their constitutional position, the ratification of an international instrument gives it the force of national law. This method of giving effect to a safety and health Convention, as to ILO Conventions generally, has been accepted with an important proviso: the Articles of the Convention thus applied must be capable of automatic incorporation in the national legal sphere. Such provisions, which are called self-executing, do not require additional action on the part of national authorities and can be directly enforced by the courts. In some safety and health Conventions there are more self-executing provisions than in others, but no such Convention is wholly self-executing. Therefore in addition to the act of ratification other measures must be taken to ensure application, notably by laying down appropriate sanctions for non-compliance with the provisions of the Convention and instituting an appropriate system of inspection. Probably as a result of the experience with the application of previous Conventions, some countries have adopted specific legislative measures of application, even though by virtue of their constitutional systems ratified Conventions are considered as incorporated in national law. The drift away from reliance on the self-execution of safety and health Conventions is further reinforced by the changes in the ILO standard-setting techniques as a result of which the proportion of self-executing provisions in the newer safety and health Conventions has in general decreased.

Ensuring effective application

Besides ensuring legislative conformity with the terms of a given Convention, the supervisory bodies evaluate the effect given to provisions designed to ensure application of the Convention's requirements in practice. For instance, they request information on the measures taken to give effect to Articles of the instruments which require administrative, organisational or financial action by the national authorities.

Extracts from the reports of the inspection services and information on the number and nature of the contraventions reported are necessary to give a more precise picture of the working of this or that instrument. The same is true with regard to the number and nature of accidents reported. A high incidence of occupational accidents or their gravity among the workers covered by the particular Convention may suggest that additional measures must be taken in the country concerned to ensure more adequate protection.

Role of workers' and employers' organisations

A special effort has been made by the ILO to involve organisations of workers and employers more closely in the implementation and supervision of safety and health Conventions. Nearly all of them provide for the consultation of workers' and employers' organisations before enactment of legislation and in the course of its application or envisage the creation of safety and health bodies with the participation of workers' and employers' representatives. The communication to the workers' and employers' organisations of the governments' reports concerning the application of ratified Conventions in conformity with article 23, paragraph 2, of the ILO

Constitution allows these organisations to formulate comments on the manner in which the relevant Conventions are applied. They may also make a representation under article 24 of the ILO Constitution if a member State has failed to secure the effective observance of a safety and health Convention to which it is a party.

The application of Conventions concerning occupational safety and health thus involves complex problems of a legal, administrative and even a financial nature. They are but a reflection of the complexities of the real world of work, and should be viewed as such. Neither the national law nor the international labour Conventions are "art for the art's sake"; the Conventions and the system of international supervision over the compliance by States with their provisions are there to further the protection of workers' life and health. It is with this aim that hundreds of reports are sent annually by the ratifying States to the ILO and are analysed by the supervisory bodies, who each year address over a hundred requests to the countries concerned to take measures for the effective application of Conventions on safety and health matters. Some 80 cases of improvement in the national safety and health legislation and practice of various countries during the last 20 years are the concrete result of the working of the system of supervision. Another, even more important, achievement is the ever-increasing awareness of governments, workers' and employers' organisations of the capital importance of the legal regulation of safety and health for the well-being of the working people.

KHOKHLOV, I.

Report of the Committee of Experts on the Application of Conventions and Recommendations. Volume A. General Report and observations concerning particular countries. International Labour Conference, 64th Session (Geneva, International Labour Office, 1978), 259 p.

International labour law. Valticos, N. (Deventer, Netherlands, Kluwer, 1979), 267 p.

"The changing pattern of ILO supervision". Samson, T. K. *International Labour Review* (Geneva), Sep.-Oct. 1979, 118/5 (569-587). 37 ref.

"International labour Conventions and the USSR". Ivanov, S. A. *International Labour Review* (Geneva), Apr. 1966, 93/4 (401-413).

"The influence of international labour Conventions on Indian labour legislation". Menon, V. K. R. *International Labour Review* (Geneva), June 1956, 77/6 (551-571).

"International labour Conventions and Bulgarian legislation". Mrachkov, V. *International Labour Review* (Geneva), Mar.-Apr. 1979, 118/2 (205-221). 33ref.

International Labour Organisation (occupational safety and health activities)

The protection of the worker against "sickness, disease and injury arising out of his employment" is one of the essential tasks assigned to the International Labour Organisation under its Constitution. From its earliest days, the ILO was therefore concerned with the prevention of occupational hazards, but primarily in attempting to eliminate the most dangerous conditions and the most serious hazards. After the Second World War, the economic changes and social progress to which it had itself made a considerable contribution in the labour field led the ILO to adopt a more positive attitude to workers' health promotion by adapting work to the worker. At its First Session in 1950, the Joint ILO/WHO Committee on Occupational Health adopted the following definition:

"Occupational health should aim at: the promotion and maintenance of the highest degree of physical, mental, and social well-being of workers in all occupations; the prevention among workers of departures from health caused by their working conditions; the protection of workers in their employment from risks resulting from factors adverse to health; the placing and maintenance of the worker in an occupational environment adapted to his physiological and psychological equipment and, to summarise: the adaptation of work to man and of each man to his job."

The entry of numerous newly independent African and Asian countries lent new stimulus to the ILO's action and directed it, in particular, towards the needs of developing countries.

Unlike other inter-governmental agencies such as the World Health Organisation or the International Atomic Energy Agency, which are purely governmental, the ILO brings together representatives of governments, employers and workers, that is to say the great majority of those directly concerned in the prevention of occupational accidents and diseases. This tripartite structure has remarkable advantages for the ILO, enabling it to direct its activities towards practical forms of action and establish realistic priorities at an international level, and also facilitating application of the Organisation's decisions at the workplace.

The ILO has adopted the principle that occupational safety and health are two aspects of one and the same problem and is the only international organisation which is concerned, under the specific terms of reference laid down in its Constitution, with all the biological and technical aspects involved in the protection and promotion of workers' health.

The level of employment accidents and occupational diseases throughout the world has reached alarming proportions: in 1970 these injuries resulted world-wide in over 100 000 deaths and more than 1.5 million cases of permanent disability; it may be that at the present time the number of occupational fatalities is smaller; however it is hard to believe that the over-all level of occupational injuries has decreased since then, because several national statistics show significant increases. In other words, many countries are hit far harder by occupational hazards than they have been by the consequences of war. Traditional occupational diseases are far less numerous than occupational accidents and, as a result of industrial hygiene, are gradually declining both in severity and in prevalence. Nevertheless, tens of thousands of workers still suffer disablement from occupational diseases, which, in the light of current knowledge about their aetiology and with the available means of technical prevention and biological monitoring, could be eradicated. Such an aim is today realistic in the new factories at least. However, beside improvements with regard to traditional occupational hazards, the new technology is introducing new forms of fatigue and psychosocial stressors whose occupational nature is not always easy to identify and which require new multidisciplinary approaches.

During the first half-century of its existence, the ILO diversified its action in order to assist countries in their own safety and health programmes and the range of this action is shown in table 1.

At the present time, the ILO safety and health activities present two principal facets: *(a)* contribution to in-plant prevention of occupational accidents and diseases by the means mentioned in table 1 in order to face traditional occupational hazards, and *(b)* contribution to the improvement of conditions of work and life in order

Table 1. How the activities of the ILO have been designed to back up national action in occupational safety and health

National action	ILO action
Legislation	Conventions, Recommendations and advice on the drafting of legislation
Regulations	Model codes, codes of practice, technical advice
Technical and medical inspection	Manuals, guides, technical publications, the CIS[1] services
Activities by safety and health institutes, training and information of specialists	Fellowships, courses, symposia, congresses, technical advice and co-operation, the CIS[1] services
Information for employers	Seminars, publications, the CIS[1] services
Workers' education	Seminars, publications, audio-visual aids, the CIS[1] services

[1] The International Occupational Safety and Health Information Centre.

to adapt ILO action to new industrial realities through a more flexible, open, and global programme. The latter is discussed in detail under PIACT (INTERNATIONAL PROGRAMME FOR THE IMPROVEMENT OF WORKING CONDITIONS AND ENVIRONMENT). Needless to say, the means of action of the ILO in the pursuance of such objectives are only indirect.

The ILO is guided by two priority criteria in this area: firstly, the number of workers at risk, and secondly, the level of risk present in a given process or industry; the second criterion may be used to weight the first. Application of the first criterion brings to the fore the largest classes of workers of both sexes, in particular those employed in agriculture and small undertakings, which, in some countries, may account for over half the labour force. To these should be added young workers and aged workers, each group reacting specifically to various occupational risks and harmful factors in the workplace environment. Application of the second criterion brings to the fore workers exposed to carcinogenic or highly toxic substances, ionising radiation, industrial dusts, etc., and workers in particularly dangerous industries such as mining and civil engineering.

The types of action shown in table 1 fall under three major, traditional headings: standards; research and information dissemination; and technical co-operation. These are obviously closely interlinked: research prepares the ground for standards and is generally developed as a result of technical co-operation; standards prepare the way for technical co-operation and stimulate research; technical co-operation promotes the preparation of new standards. In recent years, particular stress has been laid on:

(a) the activities most suited to the needs of developing countries, namely education, information and training;

(b) collaboration with national agencies, official services and safety and health institutes as well as with other international or regional organisations concerned;

(c) the integration of safety, health and ergonomics in all branches of activity.

Safety and health standards

The preparation of international standards for the protection of workers' health was the principal aim of the ILO's safety and health activities before the Second World War, and it remains one of its most important forms of action in this field. Of the 156 Conventions and 165 Recommendations adopted between 1919 and 1981, 58 Conventions and 60 Recommendations relate directly or indirectly to occupational safety and health and 22 Conventions and 21 Recommendations concern safety and health and the working environment (see appendices in Volume 2). Without detailing these various standards, there are some which call for special attention.

Accident prevention. Recommendation No. 31, although adopted 50 years ago, points to most of the essential principles followed in modern employment accident prevention, such as the need for close collaboration between industry and society, the part played in accident prevention by workers' skills and motivations, and the importance of collaboration between employers and workers in safety matters, especially through safety and health committees.

Convention No. 119 (1963), which is supplemented by Recommendation No. 118, concerns the guarding of new or second-hand power-driven machinery; it provides that the sale and hire of such machinery should be prohibited if certain specified dangerous parts are not equipped with appropriate safety devices, and that the use of such machinery should be prohibited if any dangerous part is without an appropriate guard. Responsibility for the guarding of machinery is thus transferred from the user to the manufacturer, which means that guards must be an integral part of the machine. This requirement has already had considerable repercussions on international trade in machinery and protects countries importing even second-hand machinery, which include a large number of developing countries. Convention No. 119 as at 1 January 1981 had been ratified by 35 member States.

To help governments prepare safety regulations, the ILO published in 1949 a Model Code of Safety Regulations for Industrial Establishments for the Guidance of Governments and Industry, a fundamental work covering safety as well as the essential aspects of occupational health; some of its sections have since been brought up to date. The Model Code is at present under revision.

Labour inspection. Accident prevention and the effective application of safety and health regulations depend largely, however, on the existence and activities of labour inspection services. The Labour Inspection Convention (No. 81) and Recommendation (No. 81), both of which were adopted in 1947, are among the most important instruments for the protection of workers' health. The 98 member States which had ratified the Convention as at 1 January 1981 undertook to establish a system of labour inspection, at least in industrial workplaces; commerce, mining and transport may be exempted from these obligations. The Convention defines the duties and powers of labour inspectors and stipulates that governments should secure the collaboration of duly qualified experts and technicians including specialists in medicine, engineering, electricity and chemistry. In 1969, the ILO adopted the Convention (No. 129), ratified by 21 Members as at 1 January 1981, and Recommendation (No. 133) concerning labour inspection in agriculture.

To facilitate the application of these standards, the ILO, after studying law and practice in certain countries, has published a series of documents and manuals on the role, the functions and the responsibilities of labour inspectors, and more particularly of medical inspectors of health.

Occupational health. In the 1960s, a number of international standards were adopted in specialised areas of occupational health such as radiation protection (Convention No. 115, ratified by 35 member States as at 1 January 1981, and Recommendation No. 114), hygiene in commerce and offices (Convention and Recommendation No. 120, the former ratified by 40 member States as at 1 January 1981), medical examination of young persons employed underground (Convention No. 124, with 33 ratifications as at 1 January 1981), and the maximum permissible weight to be carried by one worker (Convention No. 127, ratified by 20 member States as at 1 January 1981, and Recommendation No. 128); the general problems relating to protection of workers' health at the workplace and the organisation of occupational health services had already been dealt with in 1953 by Recommendation No. 97 and in 1959 by Recommendation No. 112.

Recommendation No. 97 is the first international instrument to deal with occupational safety and health measures in the workplace with a true professional approach instead of administrative restrictions. It covers the two basic methods for the protection of workers' health: technical measures for hazard control, concerning premises, working environment and equipment, including personal protective equipment on the one hand, and medical surveillance of the individual worker on the other. The instrument also makes provision for compulsory notification of occupational diseases and first-aid facilities.

In 1980 the International Labour Conference adopted the amended list of occupational diseases appended to the Employment Injury Benefits Convention, 1964 (No. 121), on the basis of a report prepared by a meeting of experts held in Geneva in 1980 in co-operation with the WHO.

In the 1970s the subjects of international instruments adopted by the International Labour Conference in the field of occupational health were of a more technical nature, requiring an integrated approach of technical, environmental and medical prevention. These were: benzene, dealt with by Convention No. 136 (1971), ratified as at 1 January 1981 by 23 member States, and by Recommendation No. 144; occupational cancer, dealt with by Convention No. 139 (1974), ratified as at 1 January 1981 by 17 member States, and by Recommendation No. 147; air pollution, noise and vibration, dealt with by Convention No. 148 (1977), ratified by 8 member States as at 1 January 1981, and by Recommendation No. 156; occupational safety and health in dock work, dealt with by Convention No. 152 (1979), ratified by 2 member States as at 1 January 1981 and by Recommendation No. 160.

Occupational health services. Recommendation No. 112 defines the role of occupational health services in places of employment, providing that such services may be established, according to the circumstances, by laws and regulations, or by collective agreement between the employers and workers concerned and that they may be organised either for a single undertaking or for a number of undertakings. Their principal functions consist of surveillance of all factors which may affect the health of the workers, pre-employment and periodic medical examinations, emergency treatment, job study and surveillance of the adaptation of jobs to workers, statistical surveys, and health education. Recommendation No. 112 establishes the fundamental principle of the works physician's professional and moral independence; it excludes from the physician's functions all verification of justification of absence on grounds of sickness and it also stresses the importance of preventive action as a proper part of occupational medicine. This instrument, which inspired the recommendation adopted by the European Economic Community in 1962 concerning occupational health in workplaces, as well as the work of the Social Committee of the Council of Europe, has contributed to the development of national law and practice. By the beginning of the 1980s the establishment of occupational health services had been laid down by law in more than 30 countries. In others, there is legislative provision for the performance of the essential functions of these services, whilst in still others, employers have established such services, at least in large undertakings; unfortunately, smaller undertakings and rural communities do not always possess the resources needed for establishing an occupational health service. In developing countries, only limited funds are available and specialised personnel are in short supply; under such conditions, the occupational physician will be obliged to delegate some of his functions to medical auxiliaries and himself be prepared to administer treatment.

The global approach. In 1981 the International Labour Conference adopted the Convention (No. 155) and Recommendation (No. 164) on safety and health in the working environment. These instruments reflect the new national awareness of the need to protect workers' health, laying the foundations of a national policy and of the action which is necessary at the level of the undertaking to achieve a coherent and global occupational health and safety system.

Information dissemination and research

The two main tools available to the ILO for the collection and dissemination of information are the International Occupational Safety and Health Information Centre (CIS), which is dealt with in a separate article, and the present encyclopaedia (see the Prefaces for the historical background). These are complemented by meetings and symposia, which have been particularly numerous during the last 20 years.

Meetings and symposia. The ILO's work in information dissemination is not limited to the distribution of documents; it is increasingly involved in the organisation of scientific and technical meetings such as the International Occupational Safety and Health Congress organised in Geneva on the occasion of the Organisation's fiftieth anniversary in 1969; the World Congresses on the Prevention of Occupational Risks (in collaboration with the International Social Security Association), which are held every three years; and the International Conferences on Pneumoconioses (1930, 1939, 1950, 1971 and 1978).

International symposia are organised by the ILO in collaboration with various national agencies. Among the most recent are those which have discussed occupational safety policies (Turin, 1976), the safety and health of migrant workers (Cavtat-Dubrovnik, 1977), the control of air pollution in the working environment (Stockholm, 1977), workers' protection against noise (Dresden, 1979), occupational cancer (Helsinki, 1981), education and training policies in occupational safety, health and ergonomics (Sandefijord, 1981). Special mention must be made of symposia dealing with ergonomics: Prague (1967), Rome (1968), Hamburg (1969), Geneva (1971), Bucharest (1974), and Istanbul (1978).

Activities in the field of ergonomics. The ILO is greatly interested in promoting the application of ergonomics since this can provide both national economic planners and individual employers with an incentive for progressively improving safety and health conditions; this is a consideration that applies to both the industrialised countries and those in the process of industrialisation. The practical value of ergonomics is that it enables physiological, psychological and medical knowledge to be presented in quantitative terms that can be used by designers and employers. It is a stage beyond conventional industrial hygiene, which aims at providing protection by creating environmental conditions in which the levels of any hazardous factor are maintained below certain limits that the average man is considered able to tolerate for a given period of time. Ergonomics, on the other hand, aims at creating conditions in which the worker can achieve optimum performance within the limits of his anthropological, physiological and psychological characteristics.

Dust prevention. The ILO acts as a centre for the collection and dissemination of information in the field of prevention and suppression of dust in mining, tunnelling and quarrying. It compiled international reports from the information collected by member countries during the years 1952-54, 1955-57, 1958-62, 1963-67 and 1968-72. The report covering the period 1973-77 is under preparation. These reports deal with legislation and its application; pneumoconiosis statistics; technical, administrative and educational measures against dust; the provision of water; ventilation; roof control; working precautions; removal of airborne dust; maintenance of equipment; education of personnel; and sampling and analysis of dusts.

Pneumoconioses. The area of ILO research most directly affecting occupational health relates to the pneumoconioses and efforts to combat the effects of dust. An International Classification of Radiographs of Pneumoconioses was adopted in 1958 and revised in 1968 and 1978; it is illustrated by a set of standard films. It comprises a classification of the radiological appearances of regular rounded opacities applicable to silicosis, to coal miners' pneumoconiosis and to pneumoconioses caused by mixed mineral dusts, as well as of irregular opacities such as those due to asbestosis and berylliosis.

Exposure limits. In recent years the ILO has devoted considerable attention to the question of permissible limits of toxic substances in industry and has been responsible for research, publications and meetings on this subject. The question was discussed in particular at the Sixth Session of the Joint ILO/WHO Committee on Occupational Health. At present, several member States have introduced statutory limits, while in other countries appropriate values have been recommended to industry by scientific bodies. All of the studies in this connection result in steady extension of knowledge, which definitely contributes to the scientific and technical progress of occupational health and, provided that these limits are respected, to a progressive improvement in working and living environment.

Expert committees and panels of consultants. The ILO organises ad hoc meetings of experts on an occasional basis to discuss technical subjects in preparation for the adoption of international agreements or to draw up guidelines for ILO action; meetings of this type were organised to lay down guidelines for the preparation of the previous edition of this encyclopaedia. In the field of occupational health, the ILO and the WHO set up a Joint Committee in 1950; this generally meets at 3- to 5-year intervals and its sessions up to now have related to the following main areas:

1st Session (1950) – ILO and WHO activities in the field of industrial hygiene; training in occupational medicine of doctors and auxiliary medical personnel.
2nd Session (1953) – Measures of general protection of workers in places of employment.
3rd Session (1957) – Training of physicians in the field of occupational health. Scope and organisation of occupational health institutes.
4th Session (1962) – Occupational health problems in agriculture.
5th Session (1966) – The organisation of occupational health services in developing countries.
6th Session (1968) – Permissible levels of toxic substances in the working environment.
7th Session (1975) – Safety and health of migrant workers.
8th Session (1981) – Education and training in occupational safety, health and ergonomics.

The reports of the Committee and the preparatory documents are published in the ILO's Occupational Safety and Health Series.

For subjects requiring repeated exchanges of views with specialists from the different regions, the ILO has recourse to panels of technical consultants reflecting the Organisation's tripartite structure; these panels replace the former Correspondence Committee on Occupational Safety and Health, but with more restricted terms of reference.

Technical standards, codes and guides. In order to help authorities and employers to draft technical standards of safety and health in conformity with Conventions and Recommendations, the ILO prepares codes of practice which are added to the Model Code previously mentioned. These codes of practice relate to particularly hazardous branches of activity such as agriculture, forestry, construction, the operation of tractors, the design and use of chain saws, mining and underground work, the prevention of explosions in coal mines, fire and electricity accident prevention, building and civil engineering, protection against noise and vibration and against airborne substances harmful to health, work involving the use of ionising radiations, electric passenger, goods and service lifts, shipbuilding, accident prevention on board ship, safety for fishermen and fishing vessels, dock work, and offshore installations. They are followed up by guides and manuals providing practical advice on surmounting the various problems faced by technical managers, safety officers, labour inspectors and workers' representatives.

Employment injury statistics. A further essential instrument in accident and disease prevention consists of statistics on employment accidents and occupational diseases, which are the only means of identifying hazards, evaluating their severity and forming an idea of the effectiveness of safety and health measures. Since its first meeting in 1923, the International Conference of Labour Statisticians organised by the ILO has on several occasions dealt with this problem. The most important landmarks in this connection were the Sixth and Tenth Sessions which met in 1954 and 1962 respectively; in 1962 the Conference adopted international definitions of accidents, permanent disability and temporary disability, as well as a model classification of employment accidents. A further refinement in reporting was achieved by the ILO Meeting of Experts on Statistics of Occupational Injuries held in Geneva in 1980.

Fatal accident rates for mining and quarrying, coal mining, manufacturing, construction and railways in a number of countries are published by the ILO in its *Year Book of Labour Statistics* (since 1940 for the majority of data).

Vocational rehabilitation. The ILO has never neglected the vocational rehabilitation of disabled persons. Recommendation No. 99, adopted in 1955, defines vocational rehabilitation as "that part of the continuous and co-ordinated process of rehabilitation which involves the provision of those vocational services, e.g. vocational guidance, vocational training and selective placement, designed to enable a disabled person to secure and retain suitable employment". The Recommendation further states that such services "should be made available to all disabled persons, whatever the origin and nature of their disability and whatever their age, provided they can be prepared for, and have reasonable prospects of securing and retaining, suitable employment". Special provisions cater for disabled children and adolescents. The ILO and WHO collaborate in the development of medical rehabilitation and have helped several countries to organise such service.

Technical co-operation

The operational programmes undertaken by the ILO with a view to supporting individual countries' efforts to overcome underdevelopment were launched with the institution of the United Nations Expanded Programme of Technical Assistance in 1949. They are part of the general action consisting of standard setting, technical publications, meetings, information and advice which is vital in preparing a national structure and providing the sort of conditions without which direct action in a field that is both technical and social, as is the case of occupational safety and health, would have no prospect of lasting success.

Between 1950 and 1980, over 60 countries received ILO technical assistance in matters of safety and health, either directly or through the United Nations Development Programme (UNDP).

In the early stages, it was a matter of sending experts to a country or providing fellowships or technical equipment, but more recently the policy has been to place ILO decentralised staff or a whole team of experts at the disposal of several countries in a region both in order to respond rapidly to calls for assistance and in order to help these countries to create a nucleus of specialists who can set up a system of technical and medical inspection, an industrial hygiene laboratory and an advisory centre serving industry. The developing countries suffer from a great shortage of technical and medical personnel, but as soon as they have their own specialists they can use the resources available under technical co-operation programmes, including allocations under the United Nations Special Fund, the UNDP or multi-bilateral programmes, and their own contributions, in order to establish occupational safety and health institutes or centres serving a single country or several countries within the same region.

Several major projects of this kind have already been put into effect and administered by the ILO in India, the United Arab Republic, Turkey, Iran, and the Republic of Korea. In other countries such as Brazil, Algeria, Sri Lanka and Kenya, ILO co-operation has been instrumental in the implementation of the programmes of such institutes. The most interesting example from the point of view of integrating safety and health within the economic process is certainly provided by the Regional Labour Institutes created in India. The purpose of these institutes is to train specialists, to provide regular technical services to the Government and industry, and to carry out research into safety, health and ergonomics. They comprise a safety, health and welfare centre, an industrial hygiene laboratory, an information centre with its own library, and a section for the organisation of practical training courses and seminars. For the ILO they represented a pilot project on the basis of which other projects have been launched. Progress in occupational safety and health is closely linked to industrial development and to economic and social progress, and every effort must be made to ensure that promotion of the workers' health finds its proper place even in the earliest stages of industrialisation.

For the development and operation of the PIACT, which has given a special impetus to technical co-operation with both developing and industrialised countries, and for the most recent activities of the INTERNATIONAL OCCUPATIONAL SAFETY AND HEALTH HAZARD ALERT SYSTEM see the separate articles on these subjects.

Future prospects

The ILO's occupational safety and health activities during its first 62 years of existence reflect the development of worker protection activities throughout the world during a period of rapid technological and social progress. One of the aspects of this progress is the international character which these activities now have; apart from exchange of knowledge and some degree of co-ordination of research programmes, this is also true of operational activities with particular reference to technical co-operation harmonisation of legislation, the gradual levelling-up of conditions of work, and even incipient standardisation in the organisation of services. Occupational physicians and safety officers in coming generations should not neglect this combination of effort, which is one effective means of improving protection of the workers' health in so many countries.

PARMEGGIANI, L.

Activities of the ILO. Reports of the Director-General to the International Labour Conference (Geneva, International Labour Office).

Publications on occupational safety and health. D.15/1979 (Geneva, International Labour Office, 1981), 27 p.

"Fifty years of international collaboration in occupational safety and health". Robert, M.; Parmeggiani, L. *International Labour Review* (Geneva), Jan. 1969, 99/1 (3-54). Ref.

"International Labour Organisation (ILO)" (Organisation internationale du Travail (OIT)). Parmeggiani, L. *Feuillets de médecine du travail* (Paris), 1974, 6 (1-12). Illus. 4 ref. (In French)

International Occupational Safety and Health Hazard Alert System

The International Occupational Safety and Health Hazard Alert System has the capability to disseminate rapidly, through a world-wide network of designated bodies, scientific and technical information on newly discovered or suspected occupational hazards, and possibly on newly developed methods of prevention or protection. It enables a country to issue an alert or request information on safety and health hazards that are found to be increasing. It is part of the ILO International Programme for the Improvement of Working Conditions and Environment (PIACT).

The establishment of the Hazard Alert System was suggested by the Government of the United States and the proposal was endorsed by the International Labour Conference in 1976, which adopted a resolution on working conditions and environment requesting that the highest priority be given to the setting up of the system.

During the process of consultation which defined the structure and operation of the system, all comments stressed its usefulness and its particular interest in relation to the wider issues of the co-ordination of research, the assessment of chemicals according to internationally accepted standards, appropriate transfer of technology and industrialisation. It is to be viewed within the framework of the co-ordinated efforts of the UN system towards the improvement of working conditions and the environment. It is linked with the relevant programmes of other international organisations, particularly with the International Programme on Chemical Safety (IPCS), a joint WHO/ILO/UNEP venture, the International Register of Potentially Toxic Chemicals (IRPTC) of the United Nations Environment Programme, the activities of the Commission of the European Communities and the OECD Complementary Information Exchange Procedure.

The Hazard Alert System is intended to convey, in a coherent and co-ordinated manner, original scientific or technical information concerning the safety or health of workers which warrants attention and is considered sufficiently important to require action at the national level. It deals with all aspects of safety and health in the working environment. Thus it covers not only chemical risks but also physical and biological ones; it also covers safety problems such as those relating to machinery and working procedures, as well as the occupational hazards due to new technologies or to an insufficient implementation of ergonomic principles. It is designed to assist countries in the exchange of information on occupational safety and health hazards and their prevention. It is not, however, intended to satisfy requests for information on published material; these can be addressed to the ILO's International Occupational Safety and Health Information Centre (CIS) under its established procedures.

The Hazard Alert System operates in co-operation with government designated bodies. In 1982, 99 countries had designated competent bodies to issue, transmit or receive communications. The members of the network are expected to maintain liaison with all interested parties, such as the appropriate government institutions, the industries involved, workers' and employers' organisations, professional associations and learned societies, teaching and research centres, etc., in their countries. Through the designated bodies, the information circulated within the system is intended to reach all those having responsibilities in the field of occupational safety and health.

Three types of communications may be circulated in the system:

(a) alerts relating to hazards which are confirmed, described in detail, and well documented. Alerts warrant immediate corrective action;

(b) information concerning evidence of the existence of an occupational hazard that is not yet fully documented;

(c) requests for information regarding a process or the use of a chemical substance suspected of presenting an occupational hazard on which more information is required.

Four requests for the launching of hazard alerts were received by the ILO from the Government of the United States and processed on an experimental basis in co-operation with a limited number of countries, taking into account the views expressed by the experts at the Meeting of Experts on the Limits of Exposure to Harmful Airborne Substances (Geneva, November 1977). The first of these related to dibromochloropropane (DBCP), a pesticide which, according to the information then available, could diminish male fertility. The other three alerts processed on an experimental basis related respectively to the explosion hazard in grain elevators, the risk of occupational cancer due to acrylonitrile and the occupational hazard due to a chemical catalyst known under the name of "NIAX catalyst ESN" used in the manufacture of expanded polyurethane. A further health hazard alert, on the occupational risks due to 2-nitropropane, was circulated to all the designated bodies participating in the Hazard Alert System. Laboratory findings had confirmed that exposure to 2-nitropropane, which is widely used as solvent in the application of coatings and in the production of printing inks and adhesives, causes cancer in animals and is suspected to be a human carcinogen. A request for information on occupational hazards related to carbonless copy papers was then circulated at the initiative of the National Board on Occupational Safety and Health, a designated body part of the system in Sweden.

One of the main features of the system agreed upon during the consultations with designated bodies is that communications–both incoming and outgoing–should be assessed at the national level and that this assessment should involve tripartite consultations. As regards alerts, the ILO Governing Body approved, at its 220th Session in May-June 1982, the structure and procedures of a tripartite mechanism for this assessment at the international level. The Hazard Alert System is a dynamic system designed to promote preventive action. Such action may consist in the setting up of an inquiry, a research project, a safety compaign, an alert at the national level, or a training programme, and/or the preparation of guidelines, laws or regulations. The Occupational Safety and Health Recommendation, 1981 (No. 164), adopted by the International Labour Conference at its 67th Session, indicated that the competent authority should secure good liaison with the International Occupational Safety and Health Hazard Alert System in relation to the implementation of national policy on occupational safety, occupational health and the working environment.

COPPEE, G. H.

Proceedings of the International Workshop on Occupational Safety and Health Hazard Alert System, Luxembourg, 18-19 December 1979. EUR 6981 EN (Brussels, Commission of the European Communities, 1980), 154 p.

Report of the Meeting of Designated Bodies Part of the International Occupational Safety and Health Hazard Alert System in the American Region, Lima, 26-27 May 1980. AS/LIMA/REP (Geneva, International Labour Office, 1980), 9 p.

Report of the Meeting of Designated Bodies Part of the International Occupational Safety and Health Hazard Alert System in the Asian and Pacific Region, Bangkok 18-19 December 1980. AS/BANGKOK/REP (Geneva, International Labour Office, 1981), 9 p.

Report of the Nordic Workshop on the International Occupational Safety and Health Hazard Alert System, Helsinki, 24-25 March 1981 (Helsinki, National Board of Labour Protection, in preparation).

International Occupational Safety and Health Information Centre (CIS)

Although there are scores of organisations specialising in one or other branch of information on the prevention of industrial accidents and occupational diseases, together with innumerable clearing-houses for the collection and analysis of such information, the accident prevention officer at any level is far from being able, even with the assistance of computerisation, to find out everything that is known on a given subject of concern to him whenever he needs to do so.

In an attempt to fill this gap, the International Labour Office in 1958 took upon itself to consult some of the most active European accident prevention institutions as to the best ways of collecting and disseminating information on occupational safety and health. Three years before, at the First World Congress on the Prevention of Occupational Risks which was held in Rome, it was proposed that the ILO should be responsible for co-ordinating on a world-wide level the work of gathering and distributing such information. Once a rough inventory had been made of the organisations and facilities existing in the various countries of the world, representatives of nine countries met in Geneva at the joint invitation of the ILO and the International Social Security Association to lay the foundations for what might be an international service especially designed for that purpose, and in May 1959 the Governing Body of the International Labour Office decided to set up the International Occupational Safety and Health Information Centre, which it placed under the authority of the Director-General of the Office and which came to be known by its French initials as the CIS.

Realising that it would be presumptious to attempt such an extensive and complex task unaided, the CIS set about securing the active co-operation of national accident-prevention institutions, beginning with those of the nine European countries that had given it the "go-ahead". This was the first nucleus of the so-called National Centres. The CIS quickly grew and attracted other members, eventually grouping some 40-odd national institutions in every continent at the present time. It also had the support of other international agencies: the World Health Organisation took a hand in launching it and the European Coal and Steel Community gave it financial support from the beginning. This support has continued ever since and is currently provided by the Commission of the European Communities, the successor organisation.

The above-mentioned National Centres in each country collect the various publications dealing with the prevention of work accidents and occupational diseases—making a preliminary selection between original information and repeated items, and between material of strictly local interest and material of universal interest—and forward them to the CIS in Geneva. The CIS effects a second screening and arranges for the final writing of the bibliographical analyses that form the substance of the *CIS Abstracts*, their translation into the languages of the Geneva-produced editions, their indexation according to an original classification system devised by the Centre itself and their publication in the *CIS Abstracts* in English and French. For many years the National Centres in Spain, Italy and the Soviet Union have been responsible for the translation of CIS publications into Spanish, Italian and Russian respectively, and for their appropriate distribution.

The following shows in summary form the various stages in the development of the Centre over the years.

1959 Representatives of the ILO, ISSA and the national accident prevention institutions of nine countries draw up the project

1960 The first instalment of the international edition of *CIS Abstract Cards* is printed in English, French and German

The Italian National Centre publishes the Italian edition of *CIS Abstract Cards*

1961 Publication of the "facetted" *CIS Classification*

1962 Publication in English, French and German of the first issue of the *CIS Information Sheet*

In collaboration with the French National Centre and a co-operating agency in that country, the CIS organises an International Symposium on Electrical Accidents in Paris

Publication in English, French and German of the first number in the *CIS Bibliographies* series

1963 CIS abstracts are also published in brochure form in the three languages of the international edition

1965 CIS has extensive consultations with its users by means of questionnaires and individual talks to determine the directions in which it should develop

1967 The Romanian National Centre puts out a Romanian-language edition of the bulletin

1969 The Spanish and USSR National Centres launch editions of the bulletin in Spanish and Russian

Inception of the CIS Microfiches Programme, the complete collection of which now amounts to some 30 000 microfiches

Organisation of the First International Symposium on Periodical Information on Occupational Safety and Health

1974 Computerisation of CIS operations. The combination of abstract cards and bulletin is superseded by a single publication, the *CIS Abstracts* bulletin, each issue of which contains a subject index, while the last issue in each year is a cumulative author and subject index. A five-yearly consolidation of the annual indexes is published at intervals of two years

Introduction of a bibliographical research service on keyboard terminals to meet individual requests

1976 The CIS becomes a Registered Service in the INFOTERRA international reference system set up in Nairobi by the United Nations Environment Programme and collaborates actively in the constitution under UNEP auspices of the International Register of Potentially Toxic Chemicals (IRPTC)

1977 Publication of the *CIS Thesaurus* in French and English, the second generation of the completely remodelled facetted *CIS Classification*, the 3 000 entries of which are replaced by more than 10 000 terms

1979 In co-operation with the Bulgarian National Centre, progressive computer storage of the CIS documentation base prior to 1974 in reverse order

Under special arrangements with SPIDEL, a French provider of computerised data bases, and also with the Swedish National Labour Protection Administration, the Swedish National Centre, on-line access on display terminals to the CIS data base is made available to users in France and (in English) to users in the Scandinavian countries

The CIS assists in the preparation of hazard alert files as part of the experimental phase of the International Occupational Safety and Health Hazard Alert System set up by the International Labour Office.

What the service consists of

Continually diversifying its production to give better service to the user, the CIS has endeavoured at the same time to facilitate the use of the information processed by availing itself of the various facilities at its disposal as the result of advances in documentation technology. The main constituent of the service is still the bibliographical abstracts–published on index cards between 1960 and 1973, and computer stored since 1974 for subsequent publication in a single bulletin–which cover all notable publications on occupational safety and health throughout the world.

As the originals of all documents reported on are systematically filed at Headquarters, the Centre is at all times able to place its reproduction service at the disposal of users who cannot obtain publications at source. This service is the active counterpart to the *List of Periodicals Abstracted*, which gives the postal addresses of all the publications mentioned, is regularly updated and is published at intervals for distribution with the regular service. The latest edition of this list contained some 1 400 titles.

The foregoing summary of the various stages in the development of the Centre also illustrates the very active part which the National Centres take in the work of the International Centre, not only by sending in regularly the publications–books, brochures, research reports, annual reports of specialised agencies, dissertations, reports of conferences, seminars, round tables, etc., handbooks, statutes, regulations, standards, operating manuals, codes of practice and guides, data sheets, statistical reports, articles in periodicals, films and other audio-visual devices (including any translations they know of) which come out in their respective countries, but also by giving assistance in some particular operation such as the publishing of national editions or the indexing of the documentary stock prior to computer storage.

The new computerised service

Starting in 1974 the whole of the service was computerised, the first consequence being a change in the vehicle for the information published by the Centre. Instead of the old international library-format index cards published since 1960 (some 30 000 in all), a single bulletin with a detailed index was introduced, the aim being to save the user the trouble of filing the cards each time a new batch arrived and, at the stage of subsequent consultation, the labour of searching the card index with all the risks of error and omission that this implies.

Under the new system, the entering of information on a keyboard terminal makes it possible, in a single operation, to memorise the information–which constitutes the storage function–and to produce the magnetic tape that is used periodically for the photocomposition of successive issues of the *CIS Abstracts* bulletin–which constitutes the publishing process. At the same time, the computer uses this single entry to prepare the index of each number of the bulletin and, at the end of the year, even the number containing the cumulative annual subject-matter and author indexes. It also runs the cumulative tapes used by the external networks–the SPIDEL data base facility in France and the Swedish National Labour Protection Administration for the Scandinavian countries, at the present time–to feed their own computers and thus allow users in those countries to have on-line access to the CIS data base. Thus, at a time when telematics, the happy marriage of data processing and telecommunications which upsets habits acquired since the beginnings of information, is taking root in the industrialised countries and is about to spread to the others, the CIS has seized the opportunity to improve the transmission of the information it puts out and at the same time to afford the user better access to it. This means that the health and safety specialist is in a position to call up at a moment's notice and with an unprecedented degree of reliability all the information which the CIS has published for the last eight years or so on the subject he is working on. He can do so directly, by himself, provided that he has the use of a display terminal, or indirectly by applying to his National Centre or to CIS headquarters at Geneva if he is not able to have on-line access to the data base. The time thus saved in card-index searching necessarily makes for greater efficiency of action.

The *CIS Thesaurus*

Another consequence of computerisation was the discarding of the facetted *CIS Classification*, under which all the documents reported on by the CIS between 1960 and 1973 were indexed. That classification, however, with its 3 000 logically ordered terms, provided the outline plan for the *CIS Thesaurus*, which now contains more than 7 000 "descriptors" and a further 3 000 "related terms". From the *Thesaurus* the CIS takes the univocal terms or descriptors which it assigns for indexing purposes to each abstract with a view to subsequent computer retrieval of the information it contains. These descriptors are arranged in a two-tier order, consisting of "main" descriptors, which reflect the major themes of the material abstracted and which the user will find listed in alphabetical order in the subject index at the end of each issue of the *CIS Abstracts* bulletin, and "secondary" descriptors, which do not appear in the printed indexes but exist only in the computer files for in-depth indexing and hence the eventual retrieval of the information.

The facetted *CIS Classification* could not be carried over as it was into the new system, because it had been designed for succinct code indexing with a view to the filing of information on index cards by means of letter codes, and did not lend itself to precise indexing. It was likewise not possible to adopt or adapt one or other of the thesauruses already published on occupational safety and health, either because the existing ones were too limited or because their structure was incompatible with the CIS system. The choice was between two forms of thesaurus: either an alphabetical list of descriptors which the computer would compile from the free indexing of the literature abstracted, or else a structured and hierarchical classification of deliberately chosen, single-concept descriptors derived from the initial *CIS Classification*, i.e. a "facetted thesaurus" supplemented by an alphabetical index containing the descriptors themselves and their most widely used synonyms ("related terms"). The second solution was finally chosen, since it shared with the first the advantage of lending itself to the compilation of an open and flexible thesaurus, but excluded the risk of cluttering up the system with descriptors more or less synonymous with those already used, thereby impairing the reliability of later information retrieval. It also had the advantage of being less elaborate in structure and accordingly less expensive. Features taken over from the former classification–which incidentally is still in use in specialised documentation centres in several countries–include the general articulation of the subjects and also the use of letter codes. Although it was necessary to remodel the codes in the original classification, there were a number of substantial advantages in keeping alphabetical codes, particularly those of linking the present editions of the *Thesaurus* with any future editions in different languages and of simplifying references in the event of telegraphic requests for a search of the data base. Chosen so as to reflect the correct hierarchical ranking within the "facets" and "subfacets", these letter codes (or numerical codes to designate countries, territories and international organisations) render superfluous the indications "narrower term" (NT) and "broader term" (BT) as often found in the cross-references in other thesauruses, because first the length of the code and secondly the extent to which the descriptor is indented in the typographical layout enable the user to see at a glance the NT or BT relationship applying to the case in point. This being so, cross-references to "related terms" (RT) are all that is needed.

As computer programming is not compatible with letter codes where groups of four letters precede groups of three letters or less, it was necessary to discard the principle of the former classification which went from the particular to the general. The structure of the *CIS Thesaurus* thus embodies a progression from the general to the particular: the former facet 2 becomes facet A, and the general concepts which were placed at the end of the former facet subdivisions are now at the head of the new "subfacets".

Incorporating chemicals–chiefly organic compounds–was a special problem, because the systematic inclusion of all compounds would have greatly increased the size of the *Thesaurus*. In facets C and D (inorganic and organic chemicals) only the major classes of compounds and only the particular chemicals of industrial toxicological interest or of relevance to occupational health are shown. By courtesy of the US Chemical Abstracts Service (CAS) which allots a registry number to each chemical compound, thereby compiling the world's most comprehensive catalogue on the subject, it has been possible to establish an internal chemical thesaurus of all the chemicals mentioned in the literature reported on in the *CIS Abstracts* bulletin with their CAS classification numbers, which are memorised as secondary descriptors.

Direct access to CIS data via display terminals

Another consequence of computerisation is the fact that since 1974 the information published by the CIS has constituted an automated data base which may be made directly accessible to users over conventional telecommunication networks such as the telephone, telex, etc. This combination of data processing and telecommunications has taken on the name of "telematics" or "teleinformatics".

Under a 1979 agreement with the SPIDEL department of the Société pour l'informatique, a French server of data bases, the CIS offers its users in France this kind of direct access. In the same year the Swedish National Labour Protection Administration, the CIS National Centre for Sweden, launched a pilot project to disseminate CIS information over the Scandinavian SCANNET network in conversational mode to users in four Scandinavian countries (Denmark, Finland, Norway and Sweden), who can thus interrogate the CIS data base in English via a cathode-ray terminal. Other hook-ups of the same kind are under consideration in various countries to develop this new form of service. The printed edition of the *CIS Abstracts* bulletin will still have its utility, however: a user who has searched the data base on his terminal can note the relevant reference numbers and then turn to his collection of *CIS Abstracts* for a more leisurely perusal of the abstracts which his interrogation of the computer has brought to light. In addition, the retrospective memorisation of the pre-1974 stock of documentation–some 30 000 abstracts–began in 1979, working backwards over the years, 1973, 1972 and so on.

Future prospects

For the occupational safety and health professional, as for any other specialist, information is of increasing importance. The International Programme for the Improvement of Working Conditions (PIACT), launched by the International Labour Organisation in 1976, sets much store by the exchange of experience and information. As occupational safety and health is a component of this Programme, the CIS will fit neatly into this sector of its activity. Similarly, when the International Occupational Safety and Health Hazard Alert System, which is also being set up by the ILO, has passed the experimental stages in which the CIS has co-operated, the Centre will be particularly well equipped to assist it in processing and compiling the hazard alert files that are drawn up. The Centre already contributes to schemes initiated by other international organisations such as the INFOTERRA international reference service and the International Register of Potentially Toxic Chemicals (IRPTC) set up under the United Nations Environment Programme (UNEP).

What with its subscribers to the two international editions, to whom must be added the subscribers to the three national editions and the regular or casual users in government departments, undertakings, libraries and documentation centres, the CIS at present serves between 20 000 and 30 000 direct users. About one-half of the subscribers to the service are recruited in industry.

It is hardly surprising that a secondary information service like the CIS should still be most widely used in the industrialised countries, where one finds the heaviest concentration of scientific and technical information facilities, both for disseminating and communicating the information and for turning it to account. It is also in these countries that one finds the heaviest concentration of men and women working full time on the prevention of industrial accidents and occupational diseases, i.e. the potential users of such a system.

The CIS is, however, always seeking ways and means of giving its information a wider audience and more efficient application in the developing countries. The developed world is redoubling its efforts–in transfers of technology or advocating appropriate technologies–to spare the industrialising countries the costly lessons it learnt by trial and error, and to help them to move swiftly on to harmonious development. That development will only be harmonious if the inputs of plant, equipment, technology and know-how go hand in hand with ways of saving the developing countries the tribute in human life, disablement and disease which their forerunners paid to technological progress. Clearly, information on occupational safety and health has no small role to play, and the CIS has a responsibility to see to it that the developing countries may turn to the best advantage the information of all kinds that it has amassed over the years and is continuing to collect day by day.

JUVET, G.

Occupational safety and health thesaurus (CIS, International Labour Office, Geneva, 1976), 100 p. (+ 4 supplements 1978, 1979, 1980, 1981).

Information on CIS publications is provided in the ANNEX.

International Organisation for Standardisation (ISO)

The International Organisation for Standardisation (ISO) is the specialised international agency for standardisation, at present comprising the national standards bodies of 87 countries. The object of the ISO is to promote the development of standards in the world with a view to facilitating international exchange of goods and services, and to developing mutual co-operation in the sphere of intellectual, scientific, technological and economic activity. The results of ISO technical work are published as International Standards.

The scope of the ISO is not limited to any particular branch; it covers all standardisation fields except standards for electrical and electronic engineering, which is the responsibility of the International Electrotechnical Commission (IEC).

The ISO brings together the interest of producers, users (including consumers), governments and the scientific community in the preparation of International Standards.

ISO work is carried out through some 2 000 technical bodies. More than 100 000 experts from all parts of the world are engaged in this work which, to date, has resulted in the publication of nearly 4 000 International Standards, representing some 30 000 pages of concise reference data in each of the ISO's official languages.

Origin and membership

International standardisation started in the electrotechnical field more than 70 years ago. While some attempts were made in the 1930s to develop International Standards in other technical fields, it was not until the ISO was created that an international organisation devoted to standardisation as a whole came into existence.

Following a meeting in London in 1946, delegates from 25 countries decided to create a new international organisation "whose object shall be to facilitate the international co-ordination and unification of industrial standards". The new organisation, the ISO, began to function officially on 23 February 1947.

A member body of the ISO is the national body "most representative of standardisation in its country". It follows that only one such body for each country is accepted for membership of the ISO. Member bodies are entitled to participate and exercise full voting rights on any technical committee of the ISO, are eligible for membership of the Council and have a seat in the General Assembly. By June 1980 the number of member bodies was 71. More than 70% of the ISO member bodies are governmental institutions or organisations incorporated by public law. The remainder have close links with the public administration in their respective countries.

A correspondent member is normally an organisation in a developing country which does not yet have its own national standards body. Correspondent members do not take an active part in the technical work, but are kept fully informed of it. Normally, a correspondent member becomes a member body after a few years. Nearly all the present correspondent members are governmental institutions. By June 1980 the number of correspondent members was 16.

Basic data on each ISO member body are given in the publication *ISO member bodies.*

Technical work

The technical work of the ISO is carried out through technical committees (TC). The decision to set up a technical committee is taken by the ISO Council, which also determines the scope of the committee. Within this scope, the committee determines its own programme of work.

The technical committees may, in turn, create subcommittees (SC) and working groups (WG) to cover different aspects of the work. Each technical committee or sub-committee has a secretariat, assigned to an ISO member body. At the end of 1979 there were in existence 160 technical committees, 589 sub-committees and 1 303 working groups.

A proposal to introduce a new field of technical activity into the ISO working programme normally comes from a member body, but it may also originate from some other international organisation. Since the resources are limited, priorities must be established. Therefore, all new proposals are submitted for consideration by the ISO member bodies. If accepted, either the new work will be referred to the appropriate existing technical committee or a new committee will be created.

Each member body interested in a subject for which a technical committee has been authorised has the right to be represented on that committee. Detailed rules of procedure are given in the *Directives for the technical work of ISO.*

International Standards

An International Standard is the result of an agreement between the member bodies of the ISO. It may be used as such, or implemented through incorporation in national standards of different countries.

A first important step towards an International Standard takes the form of a *draft proposal* (DP)–a document circulated for study within the technical committee. This document must pass through a number of stages before it can be accepted as an International Standard. This procedure is designed to ensure that the final result is acceptable to as many countries as possible. When agreement is finally reached within the technical committee, the draft proposal is sent to the Central Secretariat for registration as a *draft International Standard* (DIS); the DIS is then circulated to all member bodies for voting. If 75% of the votes cast are in favour of the DIS, it is sent to the ISO Council for acceptance as an International Standard. Normally the fundamental technical issues are resolved at technical committee level; however, the voting by member bodies and the Council provides an important assurance that no important objections have been overlooked.

The greater part of the work is done by correspondence, and meetings are convened only when thoroughly justified. Each year some 10 000 working documents are circulated. Most standards require periodic revision. Several factors combine to render a standard out of date: technological evolution, new methods and materials, new quality and safety requirements. To take account of these factors, the ISO has established the general rule that all ISO standards should be reviewed every five years. On occasions it is necessary to revise a standard earlier.

A full list of all published ISO standards is given in the *ISO Catalogue.*

ISO work in the field of occupational safety

Every ISO International Standard is prepared with concern for safety: the safety factor is an integral part of the work of the ISO.

The 4 000 International Standards already published by the ISO cover a wide spectrum, from aerospace, aircraft and agriculture to building, fire tests, containers, medical equipment, mining equipment, computer languages, the environment, personal safety, ergonomics, pesticides, nuclear energy, etc.

Many International Standards are easily recognised as important in preventing occupational risks: examples are the basic symbol for signifying ionising radiation or radioactive materials (ISO 361), safety symbols and signs (ISO/R 557), and the industrial safety helmet (ISO 3873) specified for medium protection in mining, quarrying, shipbuilding, structural engineering and forestry. Other International Standards are not so easily identified as being directly relevant, but have an equal impact on the prevention of occupational accidents and

Table 1. ISO technical committees most concerned with prevention of occupational accidents and diseases

No.	Title	Typical example of ISO standard	
21	Equipment for fire protection and fire fighting	ISO 3941	Classification of fires
23	Tractors and machinery for agriculture and forestry	ISO 3776	Agricultural tractors. Anchorage for seat belts
35	Paints and varnishes	ISO 3679	Paints and varnishes. Rapid test for determination of danger classification by flash point
43	Acoustics	ISO 4872	Measurement of airborne noise emitted by construction equipment intended for outdoor use. Method for checking compliance with noise limits
44	Welding and allied processes	ISO 2405	Recommended practice for radiographic inspection of fusion welded butt joints for steel plates 50 to 200 mm thick
59	Building	ISO 3571/1	Passenger lift installations– Part 1: Residential buildings. Definitions, functional dimensions and modular co-ordination dimensions
80	Safety colours and signs	ISO/R 557	Symbols, dimensions and layout for safety signs
82	Mining	ISO 3155	Stranded wire ropes for mine hoisting. Fibre components. Characteristics and tests
85	Nuclear energy	ISO 1709	Nuclear energy. Fissile materials. Principles of criticality safety in handling and processing
86	Refrigeration	ISO/R 1662	Refrigerating plants. Safety requirements
92	Fire tests on building materials, components and structures	ISO 1716	Building materials. Determination of calorific potential
94	Personal safety–Protective clothing and equipment	ISO 2801	Clothing for protection against heat and fire. General recommendations for users and for those in charge of such users
96	Cranes, lifting appliances and related excavator equipment	DP 4306/5	Crane terms–Part 5: Instruments and safety devices
101	Continuous mechanical handling equipment	ISO 1819	Continuous mechanical equipment. Safety code. General rules
108	Mechanical vibration and shock	ISO 2631	Guide for the evaluation of human exposure to whole-body vibration
144	Air distribution and air diffusion	ISO 3258	Air distribution and air diffusion. Vocabulary
146	Air quality	DIS 4220	Air quality. Determination of a gaseous acid pollution index. Titrimetric (or potentiometric) method

diseases; one example is ISO 2631–"Guide for the evaluation of human exposure to whole body vibration", which grades the "reduced comfort boundary", the "fatigue–decreased proficiency boundary" and the "exposure limit" according to varying levels of vibration frequency, acceleration magnitude and exposure time, and according to the direction of vibration relative to recognised axes of the human body. This standard, like all others, is continuously updated in the light of research and experience and relates to such forms of transport as dumpers, tractors, excavators and many other vehicles and worksites.

The ISO technical committees listed in table 1 are among the most prominent in the work for safety and accidents and disease prevention.

These technical committees, and others, have prepared or are preparing International Standards concerned with occupational risks in such areas as building construction sites, factories, docks, agriculture and forestry, nuclear installations, handling of materials and personal protective clothing and equipment.

The field of building provides a very clear example of the intensive concern for accident and disease prevention in the work of the ISO. Of the 37 ISO technical committees in this field, 8 deal with the problems of the working environment. The physical factors in the building field cover aspects such as personal safety, vibration and shock, noise, plant and equipment, earth-moving machinery, cranes and lifting devices and ergonomics. The chemical factors cover air quality, paints and varnishes, protection of welding workers and protective clothing and equipment.

ISO TC 127, Earth-moving machinery, has set up a sub-committee to deal specifically with safety requirements and human factors in respect of all the current basic types of earth-moving machinery such as tractors, loaders, dumpers, tractor scrapers, excavators and graders. Standards are already in existence for safe access to driving cabs via steps, ladders, walkways and platforms and the dimensions of cabs have been established for both large and small operators, sitting or standing and in arctic clothing or not, as appropriate.

Sitting positions and the sizes and shapes of seats for different operators are also the subject of International Standards. Sitting positions are now being related to areas of comfort and to reach for both hand and foot controls, and drafts are being prepared to determine the field of view available to operators of earth-moving machines, based upon determination of the shape, size and position of areas of invisibility caused by obstructing parts of the machines.

To prevent machines from crushing their operators in the event of accidental overturning, roll-over protective structures (ROPS) have been developed and standardised. Falling rocks, trees and parts of buildings in the process of demolition can prove hazardous, so falling-object protective structures (FOPS) have been standardised so as to minimise the possibility of injury to the operative.

In preparation is a standard containing approximately 120 symbols concerning the operation of machine components, the operation of controls other than components, and for general information. When completed, the international adoption and use of such a standard should not only assist operators in their work, but should help training and be of value to banksmen (the men who direct the operator on the ground), engineers and management alike.

The ISO work in the building field is both intensive and extensive, just as it is in other ISO fields. (The scope of ISO includes most industrial, agricultural and maritime activities except the electrotechnical field, handled by the International Electrotechnical Commission, and pharmaceutical products, handled by the World Health Organisation.)

On the factory floor International Standards take on a special meaning as persons seeking work migrate from one country to another, and often to jobs where they cannot speak or read the local language. Easily recognised graphic symbols for controls on machinery that conform to International Standards are vital here as in the building industry; so are standardised locations for foot and hand controls, and International Standards for guards to moving parts.

An ISO safety code for compressors, now approaching completion, covers a wide range of safety and environmental factors, such as the prevention of oil inhalation and the control of toxic oil inhibitors, the prevention of oil coke ignition and of crankcase explosion, and the use of relief and safety valves.

The safety of continuous mechanical handling equipment is the subject of nearly 40 International Standards. They cover such aspects as safety and safety codes for the different kinds of equipment, such as belt conveyors, vibrating feeders, overhead chain conveyors, hydraulic conveyors, pneumatic handling equipment, and roller and screw conveyors.

In the field of agriculture and forestry, the ISO has developed important International Standards that protect the worker. Anchorages for seat belts for farm tractors are the subject of a well known standard that is making the import-export trade easier for manufacturers as it is implemented, replacing a plethora of national standards and regulations on the subject. ISO standards even provide rules for presenting operators' manuals and technical publications for agricultural tractors and machines, making them easy to read and understand.

On the docks the worker will soon be protected by International Standards that determine the stability of cranes and mobile cranes in action, and determine the effect of wind loads on crane structures. Standards for indicators and for safety devices that will operate in the event of an operator's misjudgement are being prepared and have reached an advanced stage. Also under preparation are standards for indicators such as wind gauges, overvoltage annunciators, and mass, slope and slew indicators. Standards for "automatic cut-off", such as derricking limiters, load-lifting capacity limiters and slack rope stops are being developed. The standards produced and in preparation should not only assist operators in their work, but enhance the working environment by inspiring confidence in all works personnel moving under and around cranage. A related International Standard under preparation will provide discard criteria in relation to wear, corrosion, deformation and wire strand breaks and is intended to guide competent persons involved in the maintenance and examination of cranes and lifting appliances.

Safety for the worker and others at or near nuclear installations is covered by a dozen International Standards, and work is being accelerated in this area. Subjects covered are methods for testing exposure meters and dosimeters, a test for contents leakage and radiation leakage, and the general principles for sampling airborne radioactive materials.

International Standards for protective clothing and equipment are the responsibility of TC 94. In addition to the standard for industrial safety helmets, it has developed a standardised vocabulary for personal eye-protectors, established utilisation and transmittance requirements for infrared filters for eye-protectors, and general recommendations for users and those in charge of users of clothing for protection against heat and fire.

The production and use of ISO International Standards such as these, produced through world-wide co-operation, have unquestionably improved the quality of the workplace.

NORBRINK, B.

The International Programme on Chemical Safety (IPCS)

In 1977 the 30th World Health Assembly, concerned about the growing use of chemicals—in public health, industry and agriculture, in food processing and in the home—and the increasing number of incidents entailing the unintentional, and sometimes catastrophic, release of toxic chemicals into the environment, requested the Director-General of the World Health Organisation (WHO) to devise long-term strategies to control and limit the impact of such chemicals on human health and the environment, bearing in mind existing activities and the possibilities for international collaboration.

This concern was reiterated by the 1978 Assembly, and in January 1979 the 63rd Session of the WHO Executive Board (Resolution EB63.19) endorsed a plan of action that had been prepared in the meantime for implementing an International Programme on Chemical Safety (IPCS). Hence the IPCS, based on existing programmes and activities, operates with the active participation of national institutions under the guidance and co-ordination of a Programme Advisory Committee and the WHO Central Unit. Resolution EB63.19 also called for negotiations with other UN agencies in order to secure their collaboration and to co-ordinate all activities on chemical safety.

The Executive Heads of the United Nations Environment Programme (UNEP), the International Labour Organisation (ILO) and WHO signed a memorandum of understanding in April 1980, whereby the IPCS became a co-operative venture of the three organisations, to evaluate the effects of chemicals on health and the environment, on human and non-human targets.

The IPCS became operational in June 1980 when the Central Unit (CU) was formed at WHO Headquarters.

Objectives

The aims and activity of the IPCS are not new. What is perhaps new is the attempt to group and co-ordinate under one heading the many existing activities concerned with chemical safety being conducted by various organisations and agencies with a view to—

(a) evaluating the effects of chemicals on human health and the environment and disseminating the results;

(b) assembling background data for the establishment of exposure limits (e.g. acceptable daily intakes or maximum permissible or desirable levels in air, water, food and the working environment) for chemicals such as household products, cosmetics, food additives, industrial chemicals, toxic substances of natural origin, plastics, packaging materials, and pesticides;

(c) developing methods that will produce internationally comparable results in epidemiological and experimental laboratory studies, in evaluating exposure to multiple chemicals, and in extrapolating experimental data to effects on human subjects;

(d) co-ordinating laboratory testing and epidemiological studies, when an international approach is appropriate, and promoting research on dose-response relationships and on mechanisms of the biological action of chemicals;

(e) developing know-how for coping with chemical accidents, and promoting effective international co-operation in this field;

(f) promoting technical co-operation in dealing with specific problems in member States caused by toxic substances;

(g) promoting training and manpower development in the field of toxicology.

Organisation

Structurally, the IPCS comprises–

(a) the Inter-Secretariat Co-ordinating Committee (ICS), composed of representatives of the Executive Heads of the co-operating organisations, which reviews and decides on the work proposals of the IPCS, and advises and keeps under review the staff of the Central Unit;

(b) the Programme Advisory Committee (PAC), composed of some 20 members appointed by the Director-General of WHO in consultation with the Executive Heads of the other two co-operating organisations, which advises them on policy questions and in setting the over-all goals and priorities of the programme;

(c) the Technical Committee (TC), consisting of the directors of the Lead Institutions (LI–see below), which prepares the annual workplans and sets operational priorities bearing in mind the policy goals of the PAC.

The operation of the programme rests on the Central Unit, the Inter-Regional Research Units (IRRUs), and the network of Lead and Participating Institutions (LI and PI). More specifically, the functions of the CU are *(a)* to develop workplans, *(b)* to co-ordinate the programme components delegated to the national institutions (LI and PI) and to the IRRUs, and ensure liaison with other international organisations, and *(c)* to provide technical and scientific support for the programme.

The IRRUs are intended to support the Central Unit and operate under the general direction of its Manager. They can be located at LIs (e.g. there is an IRRU at the National Institute of Environmental Health Sciences, Research Triangle Park, in the United States). The network of LIs and PIs has been designed to facilitate the distribution of work among member States actively participating in the IPCS. LIs are designated by the Executive Heads of the co-operating organisations after negotiation with the respective governments to ensure that IPCS commitments are met and that adequate support (including sufficient national staff) is provided, internationally recognised competence in a specific field being the criterion of choice.

Sub-networks of PIs assist the LIs in specific activities, co-ordinated by the LIs in question.

In addition, there are 2 international LIs: the International Agency for Research on Cancer (IARC), for chemical carcinogenesis, and the International Register of Potentially Toxic Chemicals (IRPTC) of UNEP, for the collection, retrieval and dissemination of information.

Present and expected outputs

The main outputs anticipated from the IPCS can be summarised as follows.

Evaluation of the effects of chemicals on human health and the environment. In principle, evaluations will be compiled in the first instance by LIs, in accordance with their experience and resources, but will be submitted, for review, to other LIs of the IPCS, other agencies and selected individual specialists, and finally to a task group before being passed by the Central Unit for publication and dissemination. Under this heading come pre-existing activities, namely the *Environmental Health Criteria* (EHC), the Joint FAO/WHO Expert Committee on Food Additives (JECFA), and the Joint FAO/WHO Meeting on Pesticide Residues in Food (JMPR). In addition to the already published criteria dealing with 13 classes of chemicals or individual chemicals, some 25 further EHC studies are at various stages of preparation. Evaluations of food additives and pesticide residues in food resulting from JECFA and JMPR constitute another output under this heading.

As a new activity, the IPCS has embarked on the preparation of short evaluation documents to respond as quickly as possible to the demands of member States.

Guidelines on appropriate methods for exposure measurement and assessment, on toxicity testing, epidemiological studies, and risk assessment and hazard evaluation. Consistency in testing will facilitate comparability and acceptance of data obtained in different countries and may promote the standardisation of control measures. The IPCS is therefore working closely with other organisations and groups engaged in this field–the Organisation for Economic Co-operation and Development (OECD), the Council for Mutual Economic Assistance (CMEA), the Commission of the European Communities (CEC), the International Commission for Protection against Environmental Mutagens and Carcinogens (ICPEMC) and the Scientific Group on Methodologies for the Safety Evaluation of Chemicals (SGOMSEC), the European Chemical Industry Ecology and Toxicology Centre (ECETOC) and others–to ensure co-ordination and avoid duplication of effort. The IPCS has launched three far-reaching international studies under this heading with the task of reviewing, harmonising and validating existing methods for assessing the mutagenic effects of chemicals, for assessing neurobehavioural toxicity, and for the comprehensive evaluation of risk associated with exposure to chemicals during pregnancy, while the SGOMSEC, co-sponsored by IPCS and the Scientific Committee on Problems of the Environment (SCOPE), has set up two working groups for evaluating the effects of chemicals on reproductive function and for studying the quantitative aspects of risk assessment.

Manpower development. The control of chemicals implies the availability of toxicologists and allied personnel to carry out routine measures and to be ready to deal with any emergencies arising from chemical accidents. The WHO Regional office for Europe (EURO) has assumed global responsibility for this aspect of the programme.

The participating countries

To date, some 23 countries have expressed interest in participating—most of them already are participating—in the IPCS. Of these countries, some have designated LIs and pledged contributions, in cash or in kind, over and above their statutory budgetary dues to the three organisations in question.

MERCIER, M.

International Social Security Association (ISSA)

The aim of the International Social Security Association (ISSA) is to co-operate at the international level in the protection, promotion and development of social security throughout the world, basically through its technical and administrative improvement. Its members are government departments, centralised institutions and national federations of institutions administering one or more branches of social security or of mutual benefit societies. At the beginning of 1980 it numbered 315 organisations among its membership, from 115 countries, representing nearly 1 000 million insured persons.

The principal activities of the ISSA are:

(a) the organisation of periodical meetings of its members;

(b) the exchange of information as well as the ideas and experience acquired by its members in the course of their activities, and the provision of technical assistance to each other;

(c) the performance and promotion of research work and inquiries in the field of social security;

(d) the publication and dissemination of documentation concerning social security.

All the activities of the ISSA fall under the direction of the General Assembly, which is composed of the member institutions and meets every three years. The Council is composed of titular members, each delegate representing one country, and gives effect to the decisions taken by the General Assembly. It also elects the President, Treasurer and the Secretary-General, as well as the 24 members of the Bureau, which generally meets once a year.

The ISSA has set up Permanent Technical Committees and Study Groups in relation to its technical activities each of which works in a specific field and prepares reports that are submitted for examination by the General Assembly.

Up to 1980, 11 such Permanent Committees had been established for the different branches or technical aspects of social security.

The prevention of social risks is today considered to form an inseparable part of social security. In this connection, it is of interest to note that the concept of prevention already existed in the minds of the early pioneers working in the field of social insurance, when they caused it to be included among the basic principles in the General Declaration adopted more than 50 years ago by the International Conference of Sickness Insurance Funds and Mutual Benefit Societies, which was the predecessor of the ISSA. In the words of one of these principles which deals with the need for obligatory social insurance, preventive action constitutes the most effective part of any social activity, whose aim must be to work systematically towards "prevention and reparation of any loss of worker's productive ability". Over the past 30 years, however, the part played by prevention in the different branches of social security has increased considerably in importance, particularly in the case of sickness insurance and, more recently, in that of unemployment insurance, as may be seen by reference to the activities of the ISSA Permanent Committees. Nevertheless, as is to be expected, the initial impact of prevention programmes is most clearly felt in that sector of activity where its economic benefits are most apparent, i.e. in the field of occupational safety and health. Questions concerning safety at work are a matter of continuing concern for all those who have a responsibility for the safety and welfare of workers, and the ISSA, with a view to furthering the implementation of prevention at the workplace, has established a Permanent Committee for the Prevention of Occupational Risks.

Three-quarters of all the members affiliated to the ISSA are responsible for the administration of insurance schemes covering occupational risks, and one-third of these are directly involved in preventive activities; the remainder participate in such activities in other ways, principally by providing financial grants to specialised institutions; several others, mainly outside Europe, are engaged in providing education and the material means required to promote preventive activities in line with the industrial and economic development of the countries concerned. Of the associate members, some two-thirds are engaged in preventive activities.

The ISSA has thus extended its work in the field of occupational risks under two different and complementary aspects which fall within the scope of its competence—promotional activities related to prevention, and technical activities. From the point of view of structure, these activities take place within the framework of two types of organisation: the Permanent Committee and the International Sections.

The Permanent Committee

The Permanent Committee for the Prevention of Occupational Risks dates back to 1954 and is governed by a special rule under which it is charged with the task of undertaking, at the international level, promotional activities for the prevention of occupational risks. These include, in particular, the exchange of information among institutions working in the field and the publication of such information; the organisation of meetings of working groups at the international level, as well as of symposia and congresses; the establishment of a programme relating to education and propaganda; the conduct of inquiries; the encouragement of research and all other activities that the Permanent Committee may deem appropriate. In addition, the Permanent Committee co-ordinates the activities of the International Sections.

The Permanent Committee meets regularly every two years as well as on the occasion of the World Congresses on prevention of occupational accidents and diseases which it organises every three years in collaboration with the International Labour Office and the national organising committee of the country in which the congress is held. Between 1955 and 1980, congresses were held in: Italy, Belgium, France, the United Kingdom, Yugoslavia, Austria, Ireland, Romania and the Netherlands.

It is not easy to quantify the extent to which the World Congresses have kept pace with the different stages of development in the prevention of occupational risks that have coincided with the social, economic and industrial progress of the past 25 years or the extent to which they have given a lead to or encouraged this development. There is, however, no doubt that the exchange of ideas and information relating to recent research and to its practical application in different countries, both at the national level and within industry, have enabled a large number of the participants at these congresses to take note of the many changes that are being introduced. This, in turn, has enabled them to make a greater contribution within their particular field of activity. The published proceedings of the congresses, with the contributions of the ILO and ISSA which deal with recent advances and thinking in the matter of prevention, the reports of experts on technical subjects of the day, and the numerous contributions from participants coming from all parts of the world covering a wide range of different aspects of occupational safety and health and harmful conditions encountered in many different industrial occupations and their prevention, provide a lasting source of valuable information.

Apart from the World Congresses, the Permanent Committee organises, upon demand and in collaboration with member institutions of ISSA, international symposia dealing with specific subjects. Under this heading, mention may be made of the following:

- Warsaw, 1963: the prevention of occupational risks in agriculture; the training of supervisory personnel; the work of research institutes; very high frequency currents and the protection of workers; and methods of personal protection;
- Vienna, 1965: tractor accidents; the economic repercussions of employment accidents; the prevention of mechanical vibrations dangerous to drivers of heavy engines; and the creation and development of a safety spirit on building and public worksites;
- Hamburg, 1966: accidents in the home; and commuting accidents;
- Helsinki, 1968: occupational risks in forestry; and the organisation of occupational safety in the industrial undertaking.

Inquiries, research and technical studies fall directly within the scope of the Permanent Committee. Among the subjects dealt with between 1959 and 1980, mention may be made of: the installation of machinery, the economic repercussions of employment accidents, the psychophysiology of accidents, safety training, very high frequency electric currents, personal protective equipment, agriculture, mechanisation and automation, training of safety personnel, standardisation in the prevention of occupational risks, the prevention of occupational risks and the protection of the environment, and the use of the mass media in the prevention of occupational risks.

In certain cases some of these studies have been prepared in collaboration with other permanent committees of the ISSA. Two typical examples of these are a study carried out on the use of statistics in the prevention of occupational risks by a joint working group from the Permanent Committee on the Prevention of Occupational Risks and the Permanent Committee of Actuaries and Statisticians. The second example concerned a study of the possibilities of calling upon the resources of accident and occupational sickness insurance schemes to promote industrial safety carried out jointly by the Permanent Committee on the Prevention of Occupational Risks and the Permanent Committee on Insurance against Employment Accidents and Occupational Diseases.

The International Sections

In addition to the general safety problems that are common to all branches of industry, there are certain particular industries that are faced with specific problems determined by the nature of their activities. For this reason the Permanent Committee has, over a period of time, created specialised groups dealing with the specific problems of these industries. There are International Sections on the prevention of occupational risks in agriculture, in the construction industry, in the chemical industry, in iron and steel manufacturing, and in mining.

Following the favourable results obtained with the International Sections in these specific industries, the same formula has been adapted to deal with certain broadly based problems such as occupational risks due to electricity, machinery guarding, research on the prevention of occupational risks, and information and propaganda in the field of occupational safety and health. Personal protective equipment is also dealt with by a special working group.

These International Sections are characterised by the fact that their secretariats are furnished by various member institutions of ISSA that are specialised in the given fields. They function under a standard set of regulations which enables them to maintain liaison with the Permanent Committee and which sets out their objectives and means of action. As in the case of the Permanent Committee, they are expected to further the prevention of occupational risks by exchanging and publishing information, organising international meetings, carrying out inquiries and studies, and instituting programmes of research and publicity.

In order to carry out this work, the Sections may set up technical sub-committees and working groups, and these may admit scientific and technical institutions or industrial establishments that are working in the relevant field of activity. Designated experts may also be admitted as correspondents.

Under the auspices of the Permanent Committee these Sections have organised the following symposia or colloquia in their particular fields of activity:

- the International Section on the Prevention of Occupational Risks in Agriculture has organised symposia, round table meetings and other international gatherings in Vienna (1971), Hanover (1972), Paris (1973), Jönköping (1975), Vienna (1976), Perugia (1978), and Paris (1979). It also collaborated in organising specialised meetings on occupational risks in agriculture during the World Congresses in Dublin (1974) and in Amsterdam (1980);
- the International Section on the Prevention of Occupational Risks in the Construction Industry has organised symposia in Wiesbaden (1970), Vienna (1971), Stockholm (1973), Athens (1975) and Bucharest (1977), and took part in the organisation of specialised meetings on occupational risks in the construction industry during the World Congresses in Dublin (1974), and in Amsterdam (1980);
- the International Section on the Prevention of Occupational Risks in the Chemical Industry organised symposia at Frankfurt on the occasion of the ACHEMA exhibition in the years 1970, 1973, 1976, 1979, as well as during the World Congress at Bucharest in 1977;
- the International Section for the Prevention of Occupational Accidents and Diseases in the Iron and Manufacturing Industry organised symposia in Vienna in 1976 and in Dusseldorf in 1978;

- the International Section for Safety in the Mining Industry organised symposia in Karlovy Vary in 1972, Bratislava in 1976, and in Ostrava in 1977 and 1980;
- the International Section on the Prevention of Occupational Risks due to Electricity organised symposia in Cologne (1972), Paris (1974), Marbella (1975), Rome (1976), Lucerne (1978), and Vienna (1980);
- the International Section on Machinery Guarding organised a symposium at Aix-la-Chapelle in 1978;
- the International Section for Research on Prevention of Occupational Risks maintains an inventory of institutions carrying out research on occupational risks and their prevention, as well as of the subjects under study at each institution. This information is put together and published periodically in an International Directory, its publication being timed to coincide with the World Congresses;
- the International Section for Information and Propaganda has produced over the years several series of film transparencies reproducing safety posters and other informative or publicity material from many countries. Interested persons may obtain these as a source of new ideas or for comparison with their own work. The Section also follows the technical safety journals with a view to maintaining closer contact and more frequent exchanges of information between their editors;
- the Sections dealing with research and information and propaganda also made an active contribution towards the organisation of specialised meetings both for the staff of research institutions and for the editors of safety and health journals during the Congress at Amsterdam in 1980.

The foregoing list provides some indication of the great quantity of work accomplished by specialists from all over the world relating to the occupational risks affecting large numbers of workers. The outcome of this work is to be found in the published proceedings or in technical papers that have been prepared by the International Sections. Additional information has appeared in bulletins and information sheets distributed by certain Sections such as those for the construction and chemical industries, electricity and guarding of machinery.

Finally, over and above the subjects dealt with by the symposia, a number of matters of current interest have been studied on an ad hoc basis by a growing number of Sections, including those concerned with construction, iron and manufacturing, mines and machinery guarding.

Regional activities

Another aspect of the ISSA's technical work concerns its activities in the different regions of the world. In collaboration with the Inter-American Committee on Social Security, it has established a Regional Committee for the Prevention of Occupational Risks under whose auspices five inter-American congresses have been held in Mexico (1963 and 1979), Venezuela (1966), Colombia (1969), and Porto Rico (1975). In addition to these congresses, ISSA has furnished reporters or speakers for the Permanent Committee and for the International Sections.

Problems of occupational risks which inevitably follow the development of industry and increasing mechanisation of agriculture are among those which affect large numbers of workers in Africa. The social security institutions, particularly those administering accident insurance schemes, are contributing towards the needs of the population in adapting to these problems by assisting the authorities to establish safety and health organisations. ISSA made an active contribution to the first inter-African Congress held in Algiers in 1971 and continues to lay stress on the need for prevention of occupational risks in the course of the African regional conferences and on other suitable occasions.

Publications

In addition to the proceedings of the World Congresses and those of other international meetings held under the auspices of ISSA, which are published by the host countries, further publications include:

- articles which appear from time to time in the *International Social Security Review*;
- reports published in the proceedings of the General Assembly;
- proceedings of symposia organised by the Permanent Committee.

Reference should also be made to the publications issued by the International Sections, namely:

- reports of symposia organised by these Sections;
- technical monographs and studies on subjects dealt with by the Sections or their working groups;
- information bulletins issued by certain Sections.

Some idea of the variety and the quantity of literature stemming from numerous sources and made available to all those concerned with safety and health matters may be gained from the foregoing list.

ANDREONI, D.

ISSA and the prevention of occupational risks: 1957-1977 (Geneva, International Social Security Association, 1977), 56 p.

Forty years in the service of social security. Stack, M. (Geneva, International Social Security Association, 1967), 61 p. Illus.

Iodine

Iodine (I)
m.w. 253.8
sp.gr. 4.93
m.p. 113.5 °C
b.p. 184.3 °C
v.p. 1 mmHg ($0.13 \cdot 10^3$ Pa) at 38.7 °C
slightly soluble in water
blueish-black plates or scales with a metallic lustre, characteristic odour, and a sharp acrid taste.
TWA OSHA 0.1 ppm 1 mg/m³ ceil
IDLH 10 ppm

Occurrence. Iodine does not occur free in nature, but iodides and/or iodates are found as "trace" impurities in deposits of other salts. Chilean saltpetre deposits contain enough iodate (about 0.2% $NaIO_3$) to make its commercial exploitation feasible. Similarly, some naturally occurring brines, especially in the United States, contain recoverable quantities of iodide. Iodide in ocean water is concentrated by some seaweeds (kelp), the ash of which was formerly a commercially important source in France, the United Kingdom and Japan.

Production. The main production processes are:

(a) Chilean saltpetre is dissolved in water and concentrated by evaporation, causing sodium nitrate to

precipitate. Sodium sulphite and sodium hydrogen sulphite are then reacted with sodium iodate in the mother liquor to precipitate iodine. The iodine crystals are collected, dried, and purified by sublimation.

(b) Seaweed is either destructively distilled to obtain iodine directly, or its ash is heated with water, which is evaporated to crystallise other salts. The mother liquor is heated with sulphuric acid and gradual addition of manganese dioxide causes the evolution of iodine vapour, which is condensed to the crystals.

(c) Brine salts are oxidised in any of several ways to iodine, which is usually absorbed on activated charcoal and sublimed, or sublimed directly.

Potassium iodide is prepared from ferrous or cuprous iodides by reaction with potassium carbonate. The ferrous and cuprous iodides are prepared by direct reaction of the elements.

Uses. Because it is relatively rare and expensive, as compared to other halogens such as bromine and chlorine, iodine is used industrially only when less expensive substitutes cannot be found. Therefore, organic compounds containing iodine usually have only temporary industrial importance; continual (but minor) importance is restricted almost entirely to the pure element and to potassium and sodium iodides.

In analytical chemistry iodine acts as an oxidising agent to estimate the quantity of a reducing agent. In preparative organic chemistry iodine-containing organic compounds may be used because of their high reactivity (methyl iodide, for instance, is an excellent laboratory-scale methylating agent). The function of modern photographic film is based upon the sensitivity to light of silver bromide, to which a small quantity of silver iodide has been added. Tincture of iodine (a solution of iodine and potassium iodide in ethyl alcohol), iodoform and other iodine-containing materials have long been used medicinally, but they have generally been replaced by antibiotics. Iodine, an essential element in human nutrition, is commonly provided in iodised salt (containing 0.023% potassium iodide) to prevent goitre. In the final stages of their manufacture, zirconium and hafnium are purified either as the tetrachlorides or the tetraiodides.

HAZARDS

Iodine vapour, even in low concentrations, is extremely irritating to the respiratory tract, eyes, and to a lesser extent, the skin. Concentrations as low as 0.1 ppm in the air (the current TWA OSHA) may cause some eye irritation upon prolonged exposure. Concentrations higher than 0.1 ppm cause increasingly severe eye irritation along with irritation of the respiratory tract and, ultimately, pulmonary oedema. Other systemic injury from the inhalation of iodine vapour is unlikely unless the exposed person already has a thyroid disorder. Iodine is absorbed from the lungs, converted to iodide in the body, and then excreted, mainly in urine. Iodine in crystalline form or in strong solutions is a severe skin irritant; it is not easily removed from the skin and, after contact, tends to penetrate and cause continuing injury. Skin lesions caused by iodine resemble thermal burns except that iodine stains the burned areas brown. Ulcers, which are slow to heal, may develop because of iodine remaining fixed to the tissue. Iodine is a powerful oxidising agent; an explosion may result if it contacts materials such as acetylene or ammonia.

Toxicity. Accidental ingestion of 2-3 g of iodine may be fatal to man. In general, iodine-containing materials (both organic and inorganic) appear to be more toxic than analogous bromine or chlorine-containing materials. In addition to "halogen-like" toxicity, iodine is concentrated in the thyroid gland (the basis for treating thyroid cancer with ^{131}I) and therefore metabolic disturbances are likely to result from overexposure. Chronic absorption of iodine causes "iodism", a disease characterised by tachycardia, tremor, weight loss, insomnia, diarrhoea, conjunctivitis, rhinitis, and bronchitis. In addition, hypersensitivity to iodine may develop, characterised by skin rashes and possibly rhinitis and/or asthma.

Radioactivity. Iodine has an atomic number of 53 and an atomic weight ranging from 117 to 139. Its only stable isotope has a mass of 127 (126.9004); its radioactive isotopes have half-lives from a few seconds (atomic weights of 136 and higher) to millions of years (^{129}I). In the reactions that characterise the fission process in a nuclear reactor, ^{131}I is formed in abundance. This isotope has a half-life of 8.070 days; it emits negatron and gamma radiation with principal energies of 0.606 MeV (max) and 0.36449 MeV, respectively.

Upon entering the body by any route, inorganic iodine (iodide) is concentrated in the thyroid gland. This, coupled with the abundant formation of ^{131}I in nuclear fission, makes it one of the most hazardous materials that can be released from a nuclear reactor either deliberately or by accident. The International Committee on Radiation Protection (ICRP) in 1978 adopted a new model for predicting doses of ^{131}I (and ^{137}Cs) to a population from such a release. Information contained in that publication will be very useful in analysing the consequences of any release of or exposure to this isotope.

SAFETY AND HEALTH MEASURES

Technical. The warning (irritating) properties of iodine vapour cannot be relied upon to prevent excessive inhalation. Iodine should be handled in closed systems. Local exhaust ventilation is necessary when its vapours may escape into the breathing zone of workers. If skin contact is at all likely, rubber (or plastic) hand protection, foot protection and aprons should be worn. Eye protection should consist of a full-face gas mask (with an acid vapour canister), air supplied hood, etc., which will also provide respiratory protection. If a half-face gas mask is used, the eyes should be protected by chemical safety goggles. Air sampling consists of bubbling the air containing iodine vapour through a complexing solution, such as 3% potassium iodide in water, toluene, or *o*-toluidine and measuring the absorption of ultraviolet light by the formed complexes.

Medical. Persons with bronchitis, emphysema, or other lung or thyroid disorders should not work with iodine without having undergone a pre-employment medical examination. Persons with thyroid disorders should not work with compounds containing iodine until the approval of a physician has been secured.

Treatment. An eye exposure to solid iodine should be followed immediately by irrigation with copious amounts of water for at least 15 min and medical assistance should be obtained. If swallowed, the irritant action of iodine usually causes prompt vomiting, which is one of the reasons for the rarity of such acute intoxications. Treatment consists of administering large amounts of starch (a thin paste of flour or laundry starch), followed by gastric lavage with 1-5% sodium sulphate or the induction of vomiting. Finally, 15 g of sodium thiosulphate and milk or eggs should be administered.

Supportive treatment (relief of pain, maintenance of electrolyte balance) may be necessary.

PETERSON, J. E.

Iodine:

CIS 75-1056 "Hazardous chemical reactions–21. Iodine" (Réactions chimiques dangereuses–21. Iode). Leleu, J. *Cahiers de notes documentaires–Sécurité et hygiène du travail* (Paris), 4th quarter 1974, 77, Note No. 932-77-74 (603-604). (In French)

CIS 78-1999 "Occupational eye disorders caused by iodine and their prevention" (Professional'nye zabolevanija organa zrenija, vyzyvaemye jodom, i ih profilaktika). Alieva, Z. A. *Gigiena truda i professional'nye zabolevanija* (Moscow), Apr. 1978, 4 (16-20). (In Russian)

Radioiodine:

Radionuclide release into the environment: assessment of doses to man. Annals of the International Commission on Radiological Protection, ICRP publication No. 29 (Oxford, New York, Frankfurt, Pergamon Press, 1979), 76 p. Illus. 62 ref.

Compounds:

CIS 75-135 "Hazardous chemical reactions–15. Hydriodic acid. Iodides. Iodic acid. Iodates" (Réactions chimiques dangereuses–15. Acide iodhydrique. Iodures. Acide iodique. Iodates). Leleu, J. *Cahiers de notes documentaires–Sécurité et hygiène du travail* (Paris), 2nd quarter 1974, 75, Note No. 902-75-74 (271-274). (In French)

For methyl iodide (iodomethane) see ALKYLATING AGENTS.

Iridium and compounds

Iridium (Ir)

a.w.	192.2
sp.gr.	22.4
m.p.	2 410 °C
b.p.	4 130 °C
hardness	Mohs 6.5

Iridium, the name of which (iris = rainbow) relates to the colouring of its salts, belongs to the platinum family. It is very hard and the most corrosion-resistant metal known; however, it is attacked by some salts. It occurs in nature in the metallic state, usually alloyed with osmium (osmiridium), platinum or gold and it is produced from these minerals. $^{192}_{77}Ir$ is used as a gamma emitter (0.31 MeV at 82.7%) and beta emitter (0.67 MeV at 47.2%) in industrial radiology, the metal is used to manufacture crucibles for chemical laboratories and to harden platinum.

Toxicity. Very little is known about the toxicity of iridium and compounds. According to Hardy, its lack of harmfulness to man may be due to the fact that it is used only in small amounts. Soluble iridium compounds such as iridium tribromide and tetrabromide and iridium trichloride were suspected of being toxic; however, no reports have thus far confirmed the toxicity of the metal component. Inhaled aerosol of metallic iridium is deposited in the upper respiratory ways of rats; the metal is then quickly removed via the gastrointestinal tract and approximately 95% can be found in the faeces. In man the only reports are those concerning radiation injuries due to accidental exposure to $^{192}_{77}Ir$.

PARMEGGIANI, L.

"Retention and fate of iridium-192 in rats following inhalation". Casarett, L. J.; Bless, S.; Katz, R.; Scott, J. K. *American Industrial Hygiene Association Journal* (Akron), Oct. 1960, 21/5 (414-418). Illus. Ref.

CIS 77-1605 "^{192}Ir over-exposure in industrial radiography". Jacobson, A.; Wilson, B. M.; Banks, T. E.; Scott, R. M. *Health Physics* (Oxford), Apr. 1977, 32/4 (291-293). Illus. 2 ref.

CIS 79-103 "Radiation injury from acute exposure to an iridium-192 source: Case history". Annamalai, M.; Iyer, P. S.; Panicker, T. M. R. *Health Physics* (Oxford), Aug. 1978, 35/2 (387-389). Illus. 7 ref.

Iron and compounds

Iron (Fe)

a.w.	55.85
sp.gr.	7.86
m.p.	1 535 °C
b.p.	2 750 °C

soluble in acids; insoluble in water

a malleable, tough, silvery-grey magnetic metal.

TWA OSHA	(iron oxide fumes) 10 mg/m³
TLV ACGIH	(iron oxide fumes) 5 mg/m³
MAC USSR	6 mg/m³

Exposure limit for ferric salts soluble (as Fe) 1 mg/m³ (Australia, Finland, Switzerland).

Occurrence. Iron is second in abundance amongst the metals and is fourth amongst the elements, being surpassed only by oxygen, silicon and aluminium. The most common iron ores are: haematite or red iron ore (Fe_2O_3) containing 70% of iron, limonite or brown iron ore ($FeO(OH).nH_2O$) containing 42% of iron, magnetite or magnetic iron ore (Fe_3O_4) which has a high iron content, siderite or spathic iron ore ($FeCO_3$), pyrite (FeS_2), the most common sulphide mineral, and pyrrhotite or magnetic pyrite (FeS). [World production of iron ore in millions of metric tons in 1978 was 838 (after a peak of 906 in 1976). The production of the main iron ore producing countries in 1978 was: the USSR 244, Australia 82, Brazil 82, the United States 81, China 70.]

Extraction. Because iron ore deposits are usually near the surface, most of the mining is done in opencast systems. In general the procedure is economical when the covering layer is less than 100 m thick. Underground mining is adopted when the ore is at a deeper level.

Production. See IRON AND STEEL INDUSTRY.

Alloys. Iron itself is not particularly strong but the great increase in strength which results when it is alloyed with carbon and rapidly cooled to produce steel accounts for its outstanding industrial position amongst the metals. Certain characteristics of steel–soft, mild, medium and hard–are largely determined by the carbon content which may vary from 0.10% to 1.15%. About 20 other elements are used in varied combinations and proportions in the production of steel alloys with many different qualities–hardness, ductility, corrosion resistance, etc. The most important of these are manganese (ferromanganese and spiegeleisen), silicon (ferrosilicon) and chromium (see FERROALLOYS).

Compounds. Industrially the most important compounds of iron are the oxides and the carbonate, which constitute the principal ores from which the metal is obtained. Of lesser industrial importance are cyanides, nitrides, nitrates, phosphides, phosphates and the carbonyl.

HAZARDS

Industrial dangers are present during the mining, transportation and preparation of the ores, during the production of the metal and alloys in iron and steel works and their use in foundries and during the manufacture and use of certain compounds. [Inhalation of iron dust or fumes occurs in iron ore mining, arc welding, metal grinding, polishing and working, and in boiler scaling.]

Ores. Mechanical accidents are liable to occur during the mining, transportation and preparation of the ores because of the heavy cutting, conveying, crushing and sieving machinery that is used for this purpose. Injuries may also arise from the handling of explosives used in the mining operations.

Pneumoconiosis may result from the inhalation of dust containing silica or iron oxide (see SIDEROSIS).

[Mortality studies of haematite miners have shown an increased risk of lung cancer, generally among smokers, in several mining areas such as Cumberland, Lorraine, Kiruna and Krivoi Rog. An increased incidence of lung cancer has also been reported, but less significantly, among iron and steel foundry workers and metal grinders; among welders results are controversial. In experimental studies ferric oxide has not been found to be carcinogenic; however, the experiments were not carried out with haematite. The presence of radon in the atmosphere of haematite mines has been suggested to be an important carcinogenic factor.]

Metal and alloys. Serious mechanical accidents are liable in iron processing, and burn accidents occur in the course of work with molten metal (see IRON AND STEEL INDUSTRY). Finely divided freshly reduced iron powder is pyrophoric and ignites on exposure to the air at normal temperatures. Because of this property, fires and dust explosions have occurred in ducts and separators of dust-extraction plant associated with grinding and polishing wheels and finishing belts used for grinding and polishing steel piano wire, surgical instruments, etc., when sparks from the grinding operation have ignited the fine steel dust in the extraction plant.

Compounds. Iron compounds as a class are not associated with any particular industrial risk. The inhalation of iron oxide fume or dust, however, is the cause of a pneumoconiosis (see SIDEROSIS). For the rest, the dangerous properties of an iron compound are usually due to the radical with which the iron is associated. Thus ferric arsenate ($FeAsO_4$) and ferric arsenite ($FeAsO_3.Fe_2O_3$) possess the poisonous properties of arsenical compounds. Iron carbonyl ($FeCO_5$) is one of the more dangerous of the metal carbonyls, having toxic and flammable properties.

Ferrous sulphide (FeS). In addition to its natural occurrence as pyrite, this compound is occasionally formed unintentionally when materials containing sulphur are treated in iron and steel vessels–e.g. in petroleum refineries. If the plant is opened and the deposit of ferrous sulphide is exposed to the air, its exothermic oxidation may raise the temperature of the deposit to the ignition temperature of gases and vapours in the vicinity. A fine water spray should be directed on such deposits until flammable vapours have been removed by purging. Similar problems may occur in pyrite mines where the air temperature is increased by a continuous slow oxidation of the ore.

SAFETY AND HEALTH MEASURES

The precautions for the prevention of mechanical accidents include the fencing and remote control of machinery, the design of plant which, in modern steelmaking, includes computerised control, and the safety training of workers.

The danger arising from toxic and flammable gases, vapours and dusts is countered by local exhaust and general ventilation coupled with the various forms of remote control. Protective clothing and eye protection should be provided to safeguard the worker from the effects of hot and corrosive substances, heat and other forms of radiation.

It is especially important that the ducting at grinding and polishing machines and at finishing belts should be maintained at regular intervals to maintain the efficiency of the exhaust ventilation as well as to reduce the risk of explosion.

SCHULER, P.

"Haematite and iron oxide" (29-39). 41 ref. *IARC monographs on the evaluation of carcinogenic risk of chemicals to man.* Vol. I (Lyons, International Agency for Research on Cancer, 1972), 184 p.

CIS 74-818 "A study of mortality from lung cancer among miners in Kiruna 1950-1970". Jörgensen, H. S. *Work–Environment–Health* (Helsinki), 1973, 10/3 (126-133). Illus. 18 ref.

"Bronchial cancer in iron ore miners. Review of selected literature relating to epidemiological surveys and experimental studies" (Le cancer bronchique chez les mineurs de fer. Sélection bibliographique d'enquêtes épidémiologiques et d'études expérimentales). Cavelier, C.; Mur, J. M.; Cericola, C. *Cahiers de notes documentaires* (Paris), 3rd quarter 1980, 100 (363-371). 17 ref.

"Magnetite pneumoconiosis". Morgan, W. K. C. *Journal of Occupational Medicine* (Chicago), Nov. 1978, 20/11 (762-763). Illus. 7 ref.

"Iron pentacarbonyl" (229-230). 4 ref. *Transactions of the 32nd Annual Meeting of the American Conference of Governmental Industrial Hygienists, Detroit, May 12, 1970* (Cincinnati, American Conference of Governmental Industrial Hygienists Inc., 1970).

Iron and steel industry

Iron is most widely found in the crust of the earth, in the form of various minerals (oxides, hydrated ores, carbonates, sulphides, silicates, etc.). Starting from prehistoric times, man has learned to prepare and process these minerals by various washing, crushing and screening operations, by separating the gangue, by calcining, sintering and pelletising, in order to render the ores smeltable and to obtain iron and steel. In historic times, a prosperous iron industry developed in many countries, based on local supplies of ore and the proximity of forests to supply the charcoal for fuel. Early in the 18th century, the discovery that coke could be used in place of charcoal revolutionised the industry, making possible its rapid development as the base on which all other developments of the Industrial Revolution rested: great advantages accrued to those countries where natural deposits of coal and iron ore lay close together.

Steelmaking was largely a development of the 19th century with the invention of melting processes, the Bessemer (1855), the open hearth, usually fired by producer gas (1864), and the electric furnace (1900). Since the middle of this century, oxygen conversion, pre-eminently the LD (Linz-Donowitz) process by oxygen lance, has made it possible to manufacture high quality steel with relatively low production costs.

Today, steel production is an index of national prosperity and the basis of mass production in many other industries such as shipbuilding, automobiles, construction, machinery, tools, industrial and domestic equipment. The development of transport, in particular by sea, has made the international exchange of the raw materials required (iron ores, coal, fuel oil, scrap and additives) economically profitable. Therefore, the countries possessing iron ore deposits near coal fields are no longer privileged, and large smelting plants and steelworks have been built in the coastal regions of major industrialised countries to be supplied with raw materials from exporting countries which are able to meet the present-day requirements for high-grade materials.

During the past decades, so-called direct reduction processes have been developed and have stood their test. The iron ores, in particular high-grade or upgraded ores, are reduced to sponge iron by extracting the oxygen they contain, thus obtaining a ferrous material replacing scrap.

Iron and steel production. The world's pig iron production was 526 million tonnes in 1979, of which 80 million tonnes were produced in the United States, 109 million tonnes in the USSR, 98 million tonnes in the European Community (35 million tonnes in the Federal Republic of Germany) and 83 million tonnes in Japan. The world's ingot steel production was 796 million tonnes in 1979, of which 126 million tonnes were produced in the United States, 149 million tonnes in the USSR, 140 million tonnes in the European Community (46 million tonnes in the Federal Republic of Germany and 24 million tonnes in Italy), 112 million tonnes in Japan and 34 million tonnes in China.

The tendency today is towards large integrated steel work where everything is done from the receipt of ores and coals to the production of the finished article. Integrated steel works with production capacities in the order of 10 million tonnes a year are common in the main steel producing countries of the world. There remain many small enterprises where the production may be iron, steel or one of the specialised steels.

Ironmaking. The essential feature is the blast furnace where iron ore is melted (reduced) to produce pig iron. The furnace is charged from the top with iron ore, coke and limestone: hot air frequently enriched with oxygen is blown in from the bottom, and the carbon monoxide produced from the coke transforms the iron ore into pig iron containing carbon: the limestone acts as a flux. At a temperature of 1 600 °C the pig iron melts and collects at the bottom of the furnace and the limestone combines with the earth to form slag. The furnace is tapped regularly and the pig iron may then be poured into pigs for later use, e.g. in foundries, or into ladles where it is transferred, still molten, to the steelmaking plant.

Some large plants will have their own coke ovens on the same site. As already mentioned above, the iron ores are generally subjected to special preparatory processes before being charged into the blast furnace (washing, reduction to ideal lump size by crushing and screening, separation of fine ore for sintering and pelletising, mechanised sorting to separate the gangue, calcining, sintering and pelletising). The slag removed from the furnace may be converted on the premises for other uses, in particular for making cement.

Steelmaking. Pig iron contains large amounts of carbon as well as other impurities (mainly sulphur and phosphorus). It must, therefore, be refined. The carbon content must be reduced, the impurities oxidised and removed, and the iron converted into a highly elastic metal which can be forged and fabricated. This is the purpose of the steelmaking operations. There are three types of steelmaking furnaces: the open-hearth furnace, the converter, and the electric furnace. In gas- or oil-fired open-hearth furnaces (figures 1 and 2), which are gradually being replaced by converters, steel is made from pig iron and scrap iron. In converters, steel is made by blowing air or oxygen into molten pig iron. In electric furnaces, scrap iron of high quality and sponge-iron pellets are the chief raw material.

Figure 1. Open-hearth furnace–charging floor.

Figure 2. Charging an open-hearth furnace.

Special steels are alloys in which other metallic elements are incorporated to produce steels with special qualities and for special purposes, e.g. chromium to prevent rusting, tungsten to give hardness and toughness at high temperatures, nickel to increase strength, ductility and corrosion resistance. These alloying constituents may be added either to the blast-furnace charge or to the molten steel (in the furnace or ladle).

Steel from the furnace is poured into moulds to form ingots which are stored in soaking pits, i.e. underground ovens with doors, where ingots can be reheated before passing to the rolling mills or other subsequent processing (figure 3). Rolling mills, foundries, forging and pressing are dealt with in other articles.

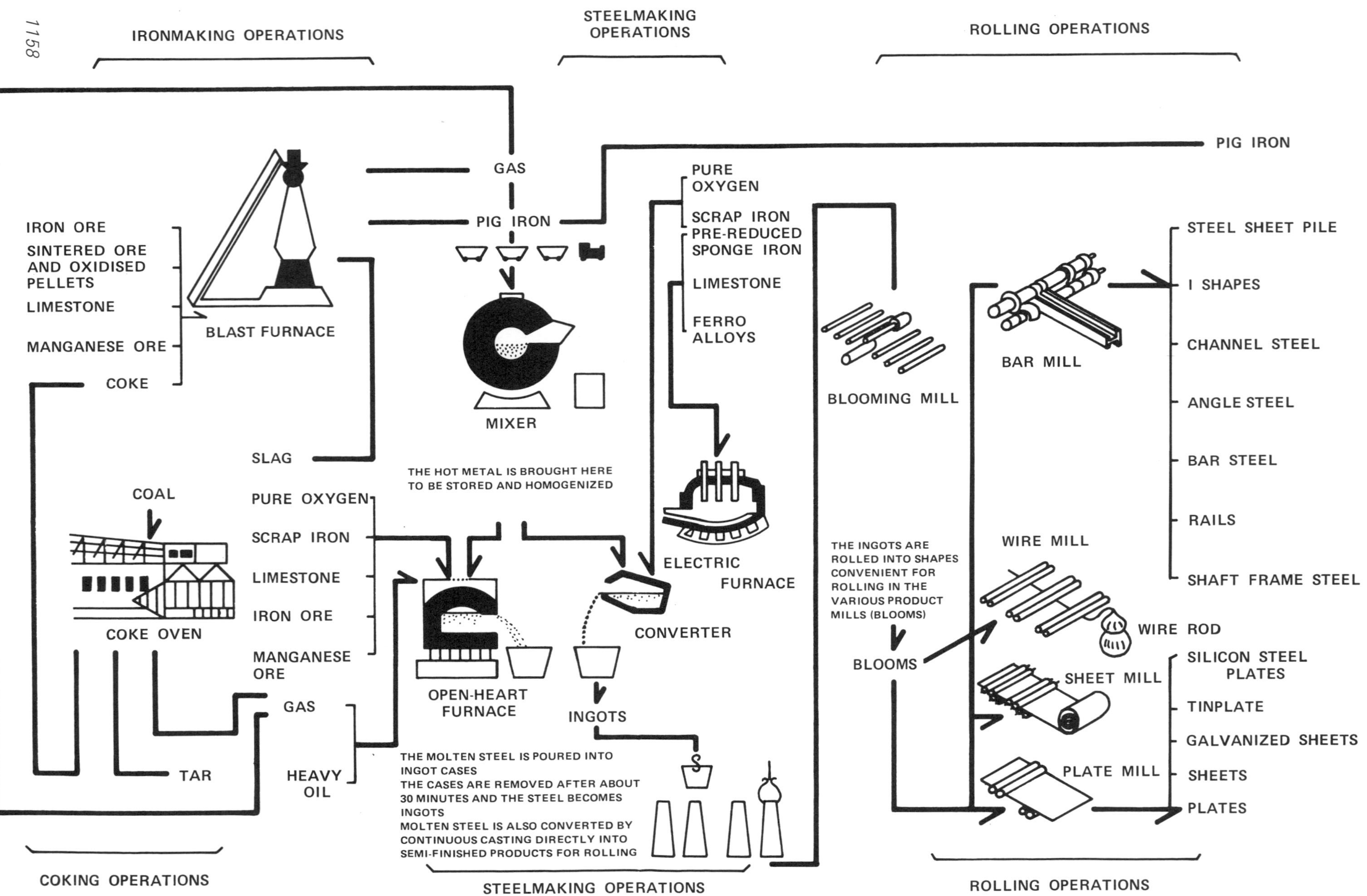

Figure 3. Processes in an integrated steelworks.

HAZARDS

Accidents. More than in most industries, the risks in iron and steelmaking are inherent in the processes–splattering, gas explosions, projections and break-outs of molten metal or slag, movement of locomotives, wagons, bogies, of furnace chargers, of cranes and of ladles and other heavy loads suspended from them, falls of heavy objects, obstruction of floors and passageways. Often the risks are multiple, e.g. a ladle of molten metal slung from an overhead travelling crane, moving above a badly obstructed floor.

Burns may occur at many points: at the front of furnaces, during tapping, from streaming molten metal or slag; from the tipping or falling of the ladle, in the pouring of ingots and in the soaking pits; a stumble may throw a man into or on to hot or molten metal; eyes and other parts of the body may be damaged by spatters or sparks.

Explosions in metal or slag ladles, sometimes caused by insertion of a damp implement, may spatter hot metal or material over a wide area. The increased use of oxygen in modern steelmaking has brought a new explosion risk in its transport, storage, distribution and use.

Mechanical transport is essential in iron and steel manufacture, and in large works, locomotives and lines of rails abound. Running down through warning and signalling failures, especially during shunting, traps between wagons, failure of couplings, and overturning of wagons or bogies may cause fatal or very serious injuries. Men may be run down or trapped by rail-mounted furnace chargers.

Breakage or failure of crane parts, lifting tackle, slings or hooks, may cause tipping or falling of ladles, ingots, etc.; faulty slinging and lack of communication between crane drivers and slingers may have similar results. Accidents may also occur on the traverse-way of an overhead crane, or through entanglement with its driving mechanism; crane drivers may be at risk because of unsafe means of access to their cabs.

Floors and passageways can quickly become obstructed with materials and implements. Tools are subject to very heavy wear and tear and soon become defective and perhaps dangerous in use. Although mechanisation has greatly lessened the amount of manual handling, there are still many occasions on which strains may occur.

Programmed maintenance is of particular importance for accident prevention. Its purpose is to keep up the efficiency of the equipment, the failure of which may cause accidents, and to maintain guards fully operative.

Carbon monoxide poisoning. Large quantities of gases are produced in the blast furnaces, converters and coke ovens in the process of iron and steel manufacture. After the dust has been removed, these gases are used as fuel sources in the various plants, and some are supplied to chemical plants for use as raw materials. They contain large amounts of carbon monoxide (blast furnace gas: 22-30%; coke oven gas: 5-10%; converter gas: 68-70%).

Carbon monoxide sometimes emanates or leaks out from the tops or bodies of blast furnaces or from the many gas pipelines inside the plants, accidentally causing acute monoxide poisoning, and most cases of acute carbon monoxide poisoning occur during work around blast furnaces, and especially during repair work. Others occur during work around hot stoves, during tours of inspection around the furnace bodies, or during work near the furnace tops, or are caused by gas released by explosions during work near the cinder notches or the tapping notches. Poisoning may also result from gas released from water-seal valves or seal pots in the steelmaking plants or rolling mills, from sudden shutdown of blowing engines, boiler rooms, or ventilation fans, from leakage, from insufficient gas removal during cleaning of electrostatic precipitators, and during closing of pipe valves.

Exposure to heat. In ironmaking (operations in front of blast furnace), steelmaking (furnace-front, ingot-making and continuous casting operations), and coking (furnace-front and furnace-top operations), there are many cases where strenuous work must be done in a hot environment. Heat-stroke is a constant risk, especially during hot seasons of the year. Heat cramps are caused by lack of salt due to excessive perspiration.

Dust. Much dust is generated at many points in the manufacture of iron and steel in the preparation processes, especially sintering, in front of the blast furnaces and steel furnaces, and in ingot-making. Dusts from iron ore or ferrous metals do not readily cause pulmonary fibrosis, and pneumoconiosis is infrequent; some lung cancers are thought to be connected with carcinogens in some of the fumes. Dense fumes emitted during the use of oxygen lances and from the use of oxygen in open-hearth furnaces may particularly affect crane drivers.

There is a severe risk of silicosis to men engaged in lining, relining, and repairing blast furnaces and steel furnaces with refractory bricks which may contain as much as 80% silica; ladles are lined with fire-brick or bonded crushed silica and this lining requires frequent repair. It should, however, be borne in mind that the silica contained in refractory bricks is partly in the form of silicates, which do not cause silicosis, but pneumoconiosis. Workers may be exposed to heavy clouds of dust.

Miscellaneous hazards. Glare from furnaces and so on may result in damage to eyes unless suitable eye protection is provided and worn. Blower plants, oxygen plants and gas discharge blowers produce high levels of noise.

High-power electric furnaces are sources of considerable noise levels which may cause hearing damage. The furnace operators should therefore be protected by enclosing the source of noise with sound-deadening material or by providing for sound-proofed shelters or also by reducing the exposure time. Hearing protectors (earmuffs or earplugs) should only be supplied if the other measures prove to be insufficient.

SAFETY AND HEALTH MEASURES

Safety organisation. Safety organisation is of prime importance in the iron and steel industry, where safety depends so much on the reaction of the workers to the potential hazards of their environment. The first responsibility is on management to provide the safest possible physical conditions but in this industry it is more than usually necessary to obtain the co-operation of all persons employed. Accident-prevention committees, workers' safety delegates, safety incentives, competitions, suggestion schemes, slogans, warning notices and posters can all play an important part. Accident statistics will reveal danger areas and the need for further physical protection and for greater stress on housekeeping and maintenance. The value of different types of protective clothing can be assessed and the advantages brought home to the workers concerned.

Training. Training should include safe methods of work, avoidance of risks, and the wearing of personal protective equipment. When new methods are introduced such as the LD process, it may be necessary to retrain even men with long experience on older types of furnace. Training courses for all levels of personnel are particularly valuable. They should aim to familiarise the personnel with safe working methods, unsafe acts to be proscribed, safety rules and the chief legal provisions concerning the prevention of accidents, or should serve as refresher courses. They should be conducted by qualified experts and should make use of effective audiovisual aids.

Engineering measures. All dangerous parts of the machinery including lifts, conveyors, long travel shafts and gearing on overhead cranes should be securely guarded. A regular system of inspection, examination and maintenance is necessary for all parts of the plant and in particular for cranes, lifting tackle, chains, hooks, etc. Defective tackle should be scrapped. Safe working loads should be clearly marked and tackle not in use should be stored in an orderly fashion. Means of access to overhead cranes should, where possible, be by stairway; if a vertical ladder has to be used, it should be hooped at intervals. Effective arrangements should be made to limit the travel of an overhead crane when persons are at work in the vicinity. It may be necessary, as required by law in certain countries, to instal appropriate switchgear on overhead cranes to prevent collisions if two or more cranes travel on the same runway.

Locomotives, rails, wagons, bogies and couplings should be of good design and maintained in good repair; and an effective system of signalling and warning should be in operation. Riding on couplings or passing between wagons should be prohibited. No operation should be carried on in the track of a furnace charger.

Great care is needed in oxygen storage arrangements. Supplies to different parts of the works should preferably be piped, and the pipes clearly identified. All lances should be kept clean.

There is a never-ending need for good housekeeping. Falls and stumbles on obstructed floors or caused by implements and tools left lying carelessly can not only cause injury in themselves but throw the person against hot or molten material. All materials should be carefully stacked, and storage racks, conveniently placed, should be provided for tools. Lighting of all parts of the shops and guards should be of a high standard.

Industrial hygiene. Good general ventilation throughout and local exhaust ventilation wherever substantial quantities of dust and fume are generated or gas may escape are necessary in association with the highest possible standards of cleanliness and housekeeping. Gas equipment must be regularly inspected and well maintained so as to prevent any gas leakage. Whenever any work is to be done in an environment likely to contain gas, carbon monoxide gas detectors should be used to ensure safety. When work in a dangerous area is unavoidable, oxygen respirators or supplied-air masks should be worn and an additional worker should be at hand to watch for and, if necessary, deal with any danger arising. Oxygen cylinders should always be kept in readiness, and the operatives should be thoroughly trained in methods of operating them.

With a view to improving the work environment, induced ventilation should be installed to supply cool air. Local blowers may be sited to give individual relief, especially in hot working places. Heat protection can be provided by installing water screens or air curtains in front of furnaces or by installing heatproof wire screens. Heat-protective clothing and masks and air-cooled suits can be worn. A suit and hood of heat-resistant material with air-line breathing apparatus gives the best protection to furnacemen. As work in the furnaces is extremely hot, cool-air lines may also be led into the suit. Fixed arrangements to allow cooling time before entry into the furnaces are also essential. Repairs to furnaces making special steels sometimes bring additional risks, e.g. from vanadium.

Acclimatisation leads to natural adjustment in the salt content of body sweat. The incidence of heat affections may be much lessened by adjustments of the workload and by well spaced rest periods, especially if these are spent in a cool room, if necessary air-conditioned. As a palliative, a plentiful supply of saline drinks, water, tea and other suitable beverages should be provided and there should be facilities for taking light meals. The temperature of cool drinks should not be too low and men should be trained not to swallow too much cool liquid at a time; light meals are to be preferred during working hours. In cold climates care is required to prevent the ill-effects of sudden and violent changes of temperature such as are caused by passing from the working area into the open air to reach canteen or sanitary facilities; canteen, washing and sanitary conveniences should preferably be close at hand. Wherever possible, sources of noise should be isolated and ceilings and walls made sound-absorbent. Remote central panels remove some operatives from the noisy areas; hearing protection should be available to those in the worst areas. Washing facilities should include shower baths; changing rooms and lockers should be provided. As excessive noise levels at the workplaces are an ever-increasing problem, noisy machinery should be enclosed with sound-absorbing material or the workers should be protected by sound-proofed shelters or screens. In extreme cases earplugs or earmuffs should be supplied for hearing protection.

Personal protective equipment. All parts of the body are at risk in most operations but the type of protective wear required will vary according to the location. Those working at furnaces need clothing that protects against burns—overalls of fire-resisting material, spats, boots, gloves, helmets with face shields or goggles against flying sparks and also against glare. Safety boots and hard hats are imperative in almost all occupations and gloves are widely necessary. The protective clothing needs to take account of the risks to health and comfort from excessive heat, for example a fire-resisting hood with wire mesh visor gives good protection against sparks and is resistant to heat; various synthetic fibres have also proved efficient in heat resistance. Strict supervision and continuous propaganda are necessary to ensure that personal protective equipment is worn and correctly maintained.

Ergonomics. The ergonomic approach, i.e. investigation of the man-machine-environment relationship, is of particular importance in the iron and steel industry, where a large number of working conditions need to be thoroughly examined to adapt them to the worker. An appropriate ergonomic study is necessary not only to investigate conditions while a worker is carrying out various operations, but also to explore the impact of the environment on the worker and the functional design of the machinery used.

Medical supervision. Pre-employment medical examinations are of great importance in selecting persons suitable for the arduous work in iron and steelmaking. For most work, a good physique is required: hypertension, heart diseases, obesity and chronic gastroenteritis disqualify from work in hot surroundings. Special care is needed in the selection of crane drivers, both for physical and mental capacities. Tuberculosis disqualifies from work with refractory materials.

Medical supervision should pay particular attention to those exposed to heat stress, to periodic chest examinations for those exposed to dust and audiometric examinations for those exposed to noise; crane drivers should also receive periodic medical examinations to ensure their continued fitness for the job.

Constant supervision of all resuscitory appliances is necessary and also training of workers in first-aid revival procedure.

A central first-aid station with the requisite medical equipment for emergency assistance should also be provided. If possible, there should be an ambulance for the transport of severely injured persons to the nearest hospital under the care of a qualified ambulance attendant. In larger plants first-aid stations or boxes should be located at several central points.

DABALA, G.

CIS 78-1876 *Technical control of pollution in the iron and steel industry–Research progress report, 30 June 1977.* Maurer, H.; Will, G. Commission of the European Communities, Directorate-General for Employment and Social Affairs (Luxembourg, Office for Official Publications of the European Communities, 1978), 89 p.

"The iron and steel industry and the environment". *Industry and environment* (Paris), Apr.-June 1980, 3/2, 16 p. Illus. Ref.

Environmental control in the iron and steel industry (Brussels, International Iron and Steel Institute, 1978), 113 p. Illus.

CIS 79-215 *Respiratory disease among workers in iron and steel foundries* (TNO Research Institute for Environmental Hygiene, PO Box 214, Delft, Netherlands) (1978), 103 p. Illus. 42 ref.

CIS 79-897 *Comprehensive accident control for preventing accidents causing injury.* Jankovsky, F.; Cavé, J. M. (Luxembourg, Office for Official Publications of the European Communities, 1977), 65 p. Illus. 6 ref.

Isocyanates

Definition. Isocyanates, which are more commonly called polyurethanes, enter into the composition of most industrial products known by that name. They form a group of neutral derivatives of primary amines with the general formula $R - N = C = O$.

The isocyanates most used at present are tolylene 2,4-diisocyanate, tolylene 2,6-diisocyanate, also known as Desmodur T, and diphenylmethane 4,4'-diisocyanate. Hexamethylene diisocyanate and 1,5-naphthylene diisocyanate are less often used.

Isocyanates react spontaneously with compounds containing active hydrogen atoms, which migrate to the nitrogen. Compounds containing hydroxyl groups spontaneously form esters of substituted carbon dioxide or urethanes.

Production. Polyurethane foam is obtained by mixing an isocyanate (tolylene 2,4- (or 2,6-) diisocyanate) with a polyol, which leads to the immediate release of carbon dioxide. Amines are added to the volatile liquid mixture and act as agents cross-linking the hydrocarbon chains and producing a felting comparable to that of microfibres. Other surface-active substances or hardeners give the mixture the exact consistency desired. The same principle is used in the other applications of isocyanates.

Uses. Polyurethane foam is used as a filling material to replace certain textiles, for example in furniture, coachwork, packing and insulation. Isocyanates are used in the manufacture of elastomers of polyurethane, which form the basis of a widely varied range of products, including plastics, synthetic rubber, adhesives and glues, anticorrosives and the plastics insulation of cables and wires. They are also entering increasingly into the composition of certain very widely employed paints, varnishes and lacquers.

HAZARDS

Isocyanates are irritating to the skin and the mucous membranes, the skin conditions ranging from localised itching to more or less widespread eczema. Eye affections are less common and, although lacrimation is often found, conjunctivitis is rare. The commonest and most serious troubles, however, are those affecting the respiratory system. The great majority of writers mention forms of rhinitis or rhinopharyngitis, and various lung conditions have also been described, the first place being taken by asthmatic manifestations, which range from minor difficulty in breathing to acute attacks, sometimes accompanied by sudden loss of consciousness.

These manifestations come within the field of pathological allergies, though the antibodies have not yet been found. The substances are often volatile and the vapour can then be detected by smell at a concentration of 0.1 ppm, but this is already dangerous for some persons.

The hazards associated with the various isocyanates differ and are better described separately.

2,4-Tolylene diisocyanate ($CH_3C_6H_3(NCO)_2$)

TDI; 2,4-TOLUENE DIISOCYANATE

m.w.	174.16
sp.gr.	1.22
m.p.	21.7 °C
b.p.	238 °C
v.d.	6.0
v.p.	0.05 mmHg (0.007·10³ Pa) at 25 °C
f.p.	135 °C (oc)
e.l.	0.9-9.5%

CH₃ — NCO — NCO

a water-white liquid which turns straw-coloured on standing, with a fruity pungent odour.

OSHA	0.02 ppm 0.14 mg/m³ ceil
TWA NIOSH	0.005 ppm 0.035 mg/m³
	0.02 ppm/10 min ceil 0.14 mg/m³/10 min ceil
STEL ACGIH	0.02 ppm 0.15 mg/m³
IDLH	10 ppm
MAC USSR	(all isomers) 0.05 mg/m³

This is the substance that is most widely used in industry and that leads to the greatest number of pathological manifestations, for it is highly volatile and is often used at considerable concentrations. The symptomatology of the troubles due to inhaling it are stereotypic. At the end of a period ranging from a few days to two months the symptoms include irritation of the conjunctiva, lacrimation and irritation of the pharynx, and later there are respiratory troubles, with an unpleasant dry cough in the evening, chest pains, chiefly behind the sternum, difficulty in breathing and distress. The symptoms become worse during the night and disappear in the morning with a slight expectoration of mucus. After a few days' rest they diminish, but *a return to work is generally accompanied by the reappearance of the troubles*: cough, chest pains, moist wheezing dyspnoea and distress. Radiological and humoral tests are usually negative, and the patient nearly always recovers without sequelae when kept out of the polluted atmosphere.

In other cases there may be recurrent common cold or a particularly pruriginous eczema that may occur on many different parts of the skin. Some victims may suffer from cutaneous and respiratory troubles at the same time.

In addition to these characteristic consequences of the intoxication, there are rather different effects resulting from exposure to very low concentrations over a long period running into years; these combine typical asthma with expiratory bradypnoea and eosinophilia in the sputum.

The physiopathology of the intoxication is still far from being fully understood. Some believe that there is a primary irritation, others think of an immunity mechanism, and it is true that the presence of antibodies has been shown in some cases. Provocative tests reproducing experimentally the symptoms produced by various vapours and other chemicals are in any case of great interest. Under the method described by J. Pepys, which seems the most suitable and the least dangerous, the substances are tested in the form they are used in at the workplace, which avoids the danger of violent reactions. In the field of functional tests, the FEV_1/FVC ratio seems to be the most convenient way of expressing defective respiration. The usual functional examinations carried out away from a place of exposure to the hazard are normal, however, and allergological tests (with acetylcholine or the standard allergens, for example) are generally negative.

Diphenyl methane 4,4'-diisocyanate ($CH_2(C_6H_4NCO)_2$)

MDI; METHYLENE *bis*(4-PHENYL) ISOCYANATE

m.w.	250.25
sp.gr.	1.19
m.p.	37.2 °C
b.p.	172 °C
v.p.	0.001 mmHg (0.13 Pa) at 40 °C
f.p.	202 °C (oc)

light yellow to white crystals or fused solid, slightly soluble in water.

OSHA	0.02 ppm 0.2 mg/m³ ceil
TWA NIOSH	0.005 ppm 0.05 mg/m³
	0.02 ppm/10 min ceil 0.2 mg/m³/10 min ceil
IDLH	10 ppm

This substance is less volatile and its fumes become harmful only when the temperature approaches 75 °C, but similar cases of poisoning have nevertheless been described. They occur mainly with aerosols, for MDI is often used in liquid form for atomising.

Hexamethylene diisocyanate ($(CH_2)_6(NCO)_2$)

HDI

m.w.	168.2
TWA NIOSH	0.005 ppm 0.035 mg/m³
	0.02 ppm/10 min ceil 0.14 mg/m³/10 min ceil
MAC USSR	0.05 mg/m³ skin

This substance, which is less widely used, is highly irritating to the skin and eyes. The commonest troubles due to it are forms of blepharoconjunctivitis.

1,5-Naphthylene diisocyanate ($C_{12}H_6N_2O_2$)

NDI; 1,5-NAPHTHALENE DIISOCYANATE

m.w.	210.2
TWA NIOSH	0.005 ppm 0.04 mg/m³
	0.02 ppm/10 min ceil 0.17 mg/m³/10 min ceil

This isocyanate is little used in industry. Poisoning after exposure to the vapour heated to over 100 °C has been reported.

SAFETY AND HEALTH MEASURES

It is essential to make provision for the effective exhaustion of isocyanate vapour. The mechanism must be placed as close as possible to the source of the vapour, which must be evacuated towards a place outside where there is no danger of its causing the secondary forms of intoxication that have sometimes been described.

If the work is carried out in the open, the wind must be taken into account, for it may bring down the toxic substances elsewhere.

In certain cases it is better to provide the workers with protective clothing, gloves and goggles. Respiratory protection may be provided by masks incorporating a prefilter, but the best form of protection in work carried on in a high concentration of isocyanates is air-supplied breathing apparatus.

Industrial waste must not be destroyed by burning. Where there is a fire involving isocyanates, carbon dioxide or powder extinguishers must be employed. Firemen must be equipped with self-contained breathing apparatus.

Concentrations of tolylene diisocyanate may be measured by the colorimetric method of Marcali or that of Ravita.

Medical prevention. The pre-employment medical examination must include a questionnaire and a thorough clinical examination in order to prevent the placing of persons with allergic cutaneous or respiratory antecedents in jobs presenting a hazard. Exposed workers must be kept under regular observation (see SUSCEPTIBILITY AND HYPERSENSITIVITY). Working clothes and protective material must be cleaned at intervals varying with the nature of the work. The sanitary facilities at the disposal of the workers must include showers.

Treatment. In the event of poisoning, once the victim has been removed from the contaminated area the treatment is symptomatic. A bronchodilator such as Ventolin is generally used, but resort may be had to oxygen in acute attacks. The person must be kept away from the working environment long enough for the respiratory symptoms to disappear. The trouble is cured quickly and normally leaves no sequelae, but there is sometimes a special sensitisation in workers who have already suffered from the intoxication.

Where the substance attacks the skin or the mucous membranes, the affected areas must be washed copiously, and if the irritation persists a symptomatic treatment must be undertaken.

Compensation. French legislation provides for the compensation of certain of the troubles in question in Schedule of Occupational Diseases No. 62 of the general scheme (No. 43 of the agricultural scheme) headed: "Occupational diseases caused by organic isocyanates".

The Commission of the European Communities in its European Schedule of Industrial Diseases includes "Diseases liable to be caused in industrial environment by organic isocyanates" (Appendix II–Particular No. A 9b).

HADENGUE, P.
PHILBERT, M.

CIS 79-1413 "Health damage due to diisocyanates" (Erkrankungen durch Diisocyanate). Lubach, D. *Dermatosen in Beruf und Umwelt* (Aulendorf, Federal Republic of Germany), 1978, 26/6 (184-187) and 1979, 27/1 (5-10). 73 ref. (In German)

CIS 79-1072 "Respiratory allergy to polyurethanes–Usefulness of provocation tests–A case study" (Allergie respiratoire aux polyuréthannes–Intérêt des tests de provocation–A propos d'une observation). Liot, F.; Philbert, M.; Dessanges, J. F.; Hadengue, P.; Briotet, A.; Seitz, B.; Lemaigre, D. *Archives des maladies professionnelles, de médecine du travail et de sécurité sociale* (Paris), Dec. 1978, 39/12 (713-719). Illus. 14 ref. (In French)

"Field investigations with specific RAST in case of exposure to TDI" (Feldstudie mit spezifischen Radio-Allergo-Sorbens-Testen (RAST) bei Isocyanat-Exponierten (TDI)). Diller, W.; Alt, E.; Baur, X.; Fruhmann, G. *Zentralblatt für Arbeitsschutz, Prophylaxe und Ergonomie* (Heidelberg), Apr. 1980, 4 (100-103). 18 ref. (In German)

CIS 79-1631 *Isocyanates: toxic hazards and precautions.* Health and Safety Executive, Guidance Note EH 16 (London, HM Stationery Office, Feb. 1979), 7 p.

CIS 80-1005 *Nordic Group of Experts for TLV Documentation–9. Diisocyanates* (Nordiska expertgruppen för gränsvärdesdokumentation–9. Diisocyanater). Arbete och hälsa–Ventenskaplig skriftserie 1979:34 (Stockholm, Arbetarskyddsverket, Aug. 1979), 54 p. 106 ref. (In Swedish)

CIS 79-1337 *Criteria for a recommended standard–Occupational exposure to diisocyanates.* DHEW (NIOSH) publication No. 78-215 (National Institute for Occupational Safety and Health, 4676 Columbia Parkway, Cincinnati) (Sep. 1978), 138 p. 156 ref.

CIS 80-735 "Use of organic isocyanates in industry–Some industrial hygiene aspects". Hardy, H. L.; Devine, J. M. *Annals of Occupational Hygiene* (Oxford), 1979, 22/4 (421-427). 14 ref.

Isolated work

Work in isolation is generally understood as the "performance of individual tasks which are outside other people's field of vision or range of hearing for more than a few minutes". However, there are other situations where individuals working together with others perform their tasks in conditions of isolation (e.g. in telephone exchanges or crane cabins). Furthermore, it often happens that persons are required to work alone for longer periods in isolated premises or locations, and sometimes also at night. Under these conditions one of the major problems of work organisation is safeguarding the worker's mental health and physical integrity. In the event of an accident or a fit of faintness, it may happen that a person working in isolation is left to himself for some time before help arrives.

It is therefore necessary to equip isolated workers with adequate signalling devices to call for help. For these reasons regulations have been issued which lay down that certain work should be performed by two or more persons, even if it could practically be done by one worker, e.g.:

- work to be performed under dangerous conditions which may necessitate the immediate aid of other persons (work in underground mines, tunnels or excavations where there is a danger of cave-in, work in confined spaces with a hazard of harmful emissions, work on live electrical equipment, work in the mountains);
- work involving responsibility and on which depends the safety of other people and public utilities (aircraft pilot, air or railway traffic controllers).

Types of isolated work. Isolated work may be classified according to where it takes place and to the tasks to be performed.

Location:

- indoors (maintenance tasks in large factories, work in sewers, in cold-storage rooms, in basements, in laboratories, etc.);
- outdoors (monitoring of dams, logging, agricultural operations, forestry, hunting and fishing, work on stacks, façade cleaning on skyscrapers, guard duties, etc.).

Tasks:

- work performed by single persons in isolation (assembly lines, telephone exchanges, control cabins, etc.);
- preponderantly autonomous work performed by single persons, independent of organisational aspects (analysts, plant operators, plant supervisors). Although these isolated workers co-operate with others, they rarely come into contact with the other workers.

HAZARDS AND THEIR PREVENTION

Hazards of work in isolation. The accident hazards associated with work in isolation are theoretically the same as those associated with any other work; however, the consequences of accidents in isolated locations (injuries, dislocations, sprains and fractures due to falls, wounds caused by tools or machines, loss of consciousness, heart attacks, attacks of appendicitis, cerebral haemorrhage, etc., due to physical factors) are likely to be much more severe.

Accidents which are considered as minor under normal circumstances may therefore become dangerous for isolated workers, as for instance a twisted ankle, which may put a worker into a critical situation if he is not adequately equipped.

Safety measures. It is particularly important that workers who are currently employed on isolated tasks are physically fit and mentally healthy. They should therefore undergo, after their pre-employment examination, periodic medical examinations at fixed intervals (at least once a year) during which the physician should make sure that:

- vision and hearing, as well as the other sense organs, are in a good state;
- there is no cardiac or respiratory impairment;
- the worker has no fits of giddiness and no fear of heights;
- the worker is free from mental disorders;
- the worker does not suffer from any psychosomatic disorder;
- the worker is not liable to have cramps;
- there is no epilepsy;
- the limbs are sound and free from amputations;
- the worker is free from any dependence such as alcoholism or toxicomania.

These examinations and any other tests the physician may consider necessary are an essential prerequisite for safety at work in isolation. All appropriate measures should be taken to reduce or eliminate the consequences of potential accidents. The isolated worker should not only be equipped with everything he needs to perform his task safely, but it is also necessary to reduce as far as possible any danger inherent in the working environment and equipment by adequate design, construction and maintenance of plant and equipment, and by adopting appropriate working methods.

In addition to the normal safety measures, the isolated worker should be equipped with special signalling devices enabling him to call for assistance in the event of an accident or emergency. Devices generally used for this purpose are:

- service telephones;

- radio communication systems;
- walkie-talkies;
- intermittent signalling devices which have to be activated at regular intervals.

These signalling and alarm devices may be stationary, as would be the case for personnel working in isolation in premises connected to a control centre, or portable for personnel working in isolated outdoor locations.

Organisation of communications and emergency assistance. Unless exceptional or improvised isolated work is to be carried out, the communication systems are generally linked to a centre from which assistance can rapidly be despatched. In such cases the isolated worker uses his signalling equipment to communicate his position at regular intervals to the centre. If there is no communication after a fixed interval has elapsed, the alarm is given and help is sent.

In Italy, for instance, the dam monitors of the state electricity board have to communicate to a centre, by service telephone at fixed hours of the day and night, the data they have read on the displays in the dam. If no data are transmitted, it is first ascertained whether or not the communication system is out of order to avoid false alarms; thereafter, the emergency plan is automatically applied, using helicopters for locations of difficult access.

Another very practical system, which is, however, limited to areas with a radius of about 1 000 m, consists in attaching to the belt a small transmitter which emits radio signals. These signals are received in a control centre where there are several visual displays, each covering an area of 600-800 m^2 and lighting up as soon as a signal is received from its area. The area from which the emergency signal is emitted being small, no precious time is lost in searching.

This system is based on the following principle. If the transmitter, which is attached to the belt and is normally vertical, is inclined by more than 50° it emits a radio signal. The signal is delayed by approximately 5 s to allow for sudden movements of the wearer. If the inclination remains unchanged, a weak signal is received by the control centre, lasting from the sixth to thirtieth second after the inclination started. This weak signal is not regarded as an actual alarm, and the worker is informed about it by an acoustic signal tripped after the thirtieth second. If the worker does not stop the "pre-alarm" signal, the actual alarm signal is given which enables the accident area to be located on the visual display, and a rescue team is immediately despatched.

The transmitter is, of course, attached to the worker's body in the position which corresponds to his work posture (standing, lying, on hands and knees, etc.). Moreover, the transmitter is equipped with a position reversing switch which can be actuated by the worker in accordance with the posture he takes.

Psychological and social effects. Work in isolation involves psychological effects which differ considerably according to whether the individual works for long periods in an isolated location, or whether the task is performed in isolation but in co-operation with other workers (telephone exchange operators, crane drivers in cabins, etc.). It is true that the latter category of workers can to some extent attenuate their isolation, for example by the use of hand and other signals to communicate with others.

In any case the conditions under which work in isolation is carried out, and the corollary difficulties arising from social and cultural factors of the working environment tend to aggravate certain forms of stress which may, in particular cases or for predisposed subjects, easily give rise to forms of neurosis, to psychosomatic disorders, to relational troubles that risk leading to a loss of contact with reality.

In isolated premises the psychological effects depend on the possibility of having social contacts (even indirect ones, e.g. by phone, television) and on the length of time passed in isolation.

The social isolation in which the subject lives in such a case has an influence on his general state, on his output and on the development of his personality (in particular with very young workers). Prolonged periods of isolation may also lead to a denial of social community, e.g. to loss of the need to be with other people because a habit of being alone has developed that may become a fondness for loneliness.

Finally, particular attention should be paid to the problem of isolated work in confined spaces, as this type of environment may have negative effects on subjects suffering from claustrophobia.

It should be borne in mind that all the aforementioned effects arise more easily in subjects who have been sorely tried or who have a labile personality. To avoid these problems it is necessary to exercise great care in selecting workers for employment on tasks to be performed in isolation.

GIOVANARDI, V.

"Working alone in R and D laboratories". Data Sheet 1-670-78. *National Safety News* (Chicago), Aug. 1978, 118/2 (73-76). Illus. 7 ref.

CIS 81-1478 "The safety of persons working in isolation–New alarm and paging equipment" (La sécurité des travailleurs isolés–Nouveaux appareils d'alarme et de recherche de personnes). *Travail et sécurité* (Paris), Dec. 1980 (647-651). Illus. (In French)

CIS 78-199 "Automatic alarm system to detect isolated workers in distress" (Système d'alarme automatique pour le repérage des travailleurs isolés victimes d'accidents ou de malaises). Radiquet, B. *Travail et sécurité* (Paris), June 1977, 6 (324-325). Illus. (In French)

"The one-man workplace" (Der Ein-Mann-Arbeitsplatz). Merkl, K. *Sichere Arbeit* (Vienna), 1974, 27/3 (9-11). (In German)

"Safety at work" (La sicurezza lavorativa). Spaltro, E. *Psicologia e lavoro* (Milan), 10 Nov. 1970, 3/special issue (119-138). (In Italian)

Ivory

Ivory is obtained from the tusk of the elephant and certain other animals. The male and female African elephants and the male Asiatic (Indian) elephants produce good-sized tusks; the female Indian elephants produce much smaller tusks or none at all. In the USSR, tusks from prehistoric mammoths, preserved in parts of Siberia, furnish a supply.

Uses. Ivory carving has been an art since the earliest civilisations and it is still most widely used in the production of a variety of articles ranging from statues to tiny figures, jewelry, and ornaments. Ivory is also used for toilet articles (especially hairbrush backs and combs), knife handles, billiard balls and piano keys but cheaper artificial materials have largely taken its place.

Treatment. Ivory is divided into two types, hard and soft, depending on the region of origin but can be hardened or softened as required by simple procedures. It is an elastic substance and very susceptible to changes of temperature. Its pliability can be controlled to some extent by immersion in hot or cold water.

Ivory can be bleached by exposure to sun and moisture, by hydrogen peroxide or by washing alternately with potassium permanganate and oxalic acid. It takes dyes well.

Artificial ivory. Many artificial substitutes have been developed: chips and shavings from carvings bonded together, celluloid, a casein called galalith and various plastics. Ivory can be distinguished from its synthetic imitations by exposure to filtered ultraviolet light, when it shows a characteristic white fluorescence that is absent in the synthetic substitute.

HAZARDS AND THEIR PREVENTION

Ivory dust contains virtually no silica. However, bone is sometimes shaped on sandstone grinding wheels, the dust from which could in time produce silicosis. Persons exposed to organic dusts also seem to have a susceptibility to pneumonia. Ivory dust may attack the air passages and thus give rise to an increased incidence of diseases of the respiratory system among workers exposed to it. During processing, ivory is usually sawn in water which prevents the dust becoming airborne; if this is not done, exhaust ventilation is required as also at other dust-producing points.

Occasionally animal dusts may carry infection, especially anthrax. Although elephants in the wild state may die of anthrax and their tusks come on the market, it is almost unknown for anthrax to occur in ivory workers. However, one case has been recorded in which a man died after infection from an elephant tusk used in making piano keys.

Where chemical solutions are used for cleaning, bleaching or dyeing, local exhaust ventilation is required to remove any injurious or offensive fumes.

KHIN MAUNG NYUNT

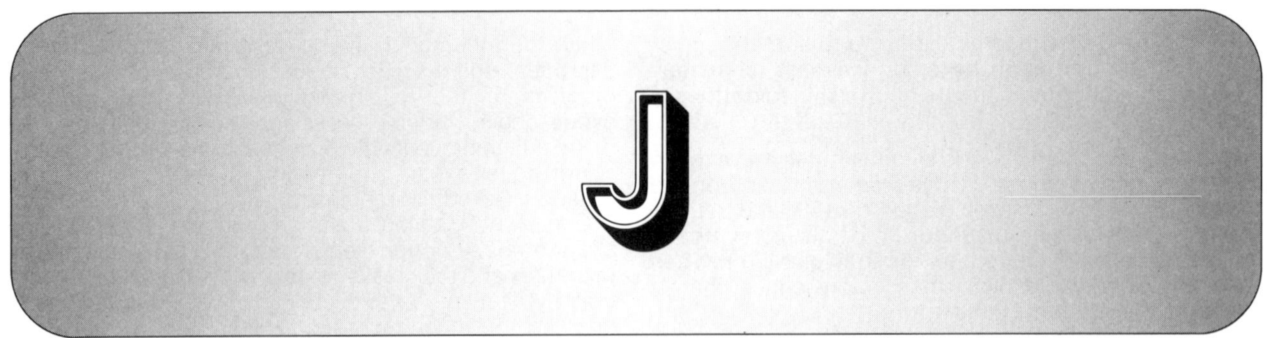

Jewelry manufacture

Work in the jewelry industry consists essentially of taking precious, semi-precious or synthetic stones, cutting and shaping them, and setting them in a frame which may be of precious metal or of an alloy of precious metal with base metal or of base metal to which a precious metal has been applied by electroplating. Adhesives may be used in the setting of stones. Materials commonly used in the jewelry industry include: gold, silver and platinum and their alloys and compounds, anodised aluminium, precious stones, both natural and synthetic, ivory, plastics, coral and shell pearls, enamels, etc. (see PRECIOUS STONES; CORAL AND SHELL; PEARLS).

The cutting of stones is carried out by small saws, and is followed by polishing. This work is visually demanding and is usually carried out under a magnifying lens. The metal settings are sometimes made by injection moulding but, in better class jewelry, are frequently cast using the "lost wax" process. Heated wax is poured into a rubber mould and allowed to harden. The mould is removed and the wax model is encased in plaster of Paris with holes left in the top and bottom of the plaster block. The block is then fired in an oven and the melted wax from the model drains from the bottom of the block and is recovered for later use. The drain hole is then plugged and the block is filled with molten metal. When the block has cooled, the plaster is knocked away and the metal piece is trimmed and polished. In a variation on this process the wax is replaced with an amalgam of mercury and cadmium.

The stones are set into the metal and may be held in place by spurs which form part of the metal frame. However, it is more usual in modern practice to bond the stone to the metal with an adhesive; epoxy resins are often used to ensure a good bond between the metal and the smooth surface of the stone. These and related adhesives are usually of the quick-drying type, and this is achieved by the evaporation of low flash point solvents.

Another method of decorating a casting or pressing is by enamelling. This is done either by printing on the glaze and fixing it, or by painting on quick-drying paints. Enamels may contain several metals including nickel and lead.

HAZARDS

The accident rate in this industry is usually low. The small saws and abrasive wheels used in polishing can cause skin abrasions, and burns can be caused by spillages of hot metals and contact with firing ovens.

Silicosis and silicotuberculosis, due to exposure to mixed dusts, have been described among jewelry foundry workers. Apart from these conditions, most of the occupational diseases which may arise are related to exposure to metal fumes. The mercury-cadmium amalgam process carries with it a considerable mercury poisoning hazard due to mercury exposure during the pouring of the amalgam into the moulds, in the room where the amalgam is melted out of the blocks, and in the cleaning of exhaust ventilation ducts.

Epoxy resin adhesives have caused cases of contact dermatitis of the sensitisation type. The solvents used in these adhesives are usually of the less toxic type but may give rise to skin sensitisation. Under poorly ventilated conditions, high concentrations may be encountered and if there is a large volume of work, explosive concentrations may arise.

The hazards of enamelling are slight provided that brushes are kept away from the mouth and that the skin is not soiled by paint. A solvent hazard may arise from quick-drying paints. The condition of "enameller's cramp" has been described and is attributed to repeated small movements of the hand.

SAFETY AND HEALTH MEASURES

In the design of any workroom for the manufacture of jewelry, attention has to be paid to the installation of exhaust ventilation for the removal of dusts from cutting and polishing processes and of metal and solvent fumes. With some metals and alloys condensation occurs on the inside of the ducting and this has to be cleansed in a safe manner from time to time, with the use of respirators of the cartridge type. Ducting carrying solvent fumes should be earthed to avoid any static electricity explosion hazard. Techniques for the application of adhesive should avoid skin contact.

Much of the fine work in preparing and finishing jewelry is visually demanding and good lighting is essential. The level at the workpoint should be at least 270 lx, and care must be taken to provide a suitably contrasting background and to site lights to prevent glare from highlights. Such conditions are best achieved by local lighting that can be adjusted by the worker.

Routine medical examinations are advisable to ensure that no absorption of metals is taking place, and are essential where mercury is being used. Routine chest X-rays for those cutting and polishing stones are advisable. In pre-employment examinations the demands made on vision by many jobs in this industry should be considered. Workers who are to be exposed to metallic fumes should be free of any type of kidney disorder and those who will be using adhesives should be free from any allergic disorder.

COPPLESTONE, J. F.

"Cause of death among jewelry workers". Sparks, P. J.; Wegman, D. H. *Journal of Occupational Medicine* (Chicago), Nov. 1980, 22/11 (733-736). 19 ref.

"Subacute cadmium intoxication in jewelry workers: an evaluation of diagnostic procedures". Baker, E. L. Jr.; Peterson, W. A.; Holtz, J. L.; Coleman, C.; Landrigan, P. J. *Archives of Environmental Health* (Chicago), 1979, 34/3 (173-177).

CIS 677-1973 "Silicosis in a jewelry factory" (Silikose in einer Schmuckwarenfabrik). Finzel, L. *Arbeitsmedizin–Sozialmedizin–Arbeitshygiene* (Stuttgart), Nov. 1972, 7/11 (331). Illus. (In German)

CIS 322-1968 "Lead poisoning hazard in a precious metals recovery plant" (Il rischio di saturnismo tra gli operai di una industria per il recupero di metalli preziosi). Capellini, A.; Cavagna, G. *Lavoro e medicina* (Genoa), 1967, 21/1 (1-6). 6 ref. (In Italian)

CIS 1554-1967 "Mercury exposure in a jewelry molding process". Jones, A. T.; Longley, E. O. *Archives of Environmental Health* (Chicago), Dec. 1966, 13/6 (769-775). Illus. 6 ref.

Jute

Jute fibre is obtained from the bark of the two cultivated species of the genus *Corchorus* i.e. *C. capsularis* and *C. olitorius* of the family *Tiliaceae*. It is usually grown in soils of alluvial origin around the Bay of Bengal, in India, Pakistan and Burma. A total rainfall of 150 cm per year is required with humidity varying from 65-95%. Harvesting is usually done when the plants are in small pods. The jute is then retted in running or stagnant water and afterwards dried and tied into bundles.

Processing. This may take place in the country of origin, but there is also a large export trade to other industrial countries. In the factory raw jute first undergoes batching by passing through the softening machine. Oil and water emulsion is sprayed on to the jute: in Burma sodium alkyl phosphate (Teepol) is used. The manufacturing processes are similar to those in other textile trades, although the terminology may be different. After preparation, the fibres are carded (or sometimes combed), drawn and spun; cop-and-spool winding, weaving, finishing, cropping, cutting and lapping complete the processing.

Uses. The most important use of jute is for making fabrics (hessians and sackings) for containers in which a large variety of commodities are transported or stored. Additional uses can be found in horticulture, agriculture, industry and transport, particularly as coverings. It is also used in the making of carpets and carpet underlays, in upholstery, soft furnishings, clothing and footwear.

HAZARDS AND THEIR PREVENTION

Accidents. The accident rate in jute manufacture is relatively high and the proportion of machinery accidents is also high. Provision and maintenance of secure machinery guarding for dangerous parts of the machinery are essential, in particular for gearing, in-running nips of rollers or bowls, spindles and shafts, knives and cutters. Guards for flying shuttles at looms are also necessary as are efficient stopping and starting arrangements for all machines.

Fire. There is a high fire risk. Friction and heat may ignite dust; particularly at carding engines and looms and, in the dusty conditions found in the industry, fire will spread rapidly. Maintenance of adequate exits with clear access, fire alarms, fire drills and suitable fire extinguishers are all essential. Cleanliness and dust control, so important for other reasons also, will contribute to prevention of outbreaks. Spontaneous combustion may occur in storage go-downs where the jute is kept wet; ventilation will do much to prevent this.

Health hazards. Pesticides used in the cultivation of jute may cause poisoning from ingestion, inhalation or skin contamination. Substitution of less toxic insecticides, training of personnel in safe techniques and personal hygiene and provision of personal protective equipment are all useful methods of prevention.

Jute dust is a non-toxic vegetable dust. Most particles are large enough to be trapped in the nose and upper respiratory system. Only the small dust particles reach the lungs and contribute to the causes of certain non-specific lung diseases like bronchitis, emphysema and pneumonia.

Workers in the jute industry (as in other textile trades) are liable, especially on first employment, to "mill fever", which is characterised by febrile attacks, violent coughing, and prostration. The condition lasts two to three days and frequently clears up without treatment. Byssinosis has not been identified in jute workers.

Dust is given off in bale opening and yarn preparation sections. Efficient local exhaust ventilation should be provided as near as possible to its point of origin. Good ventilation and temperature control are required throughout. Artificial humidification is necessary in some countries like the USSR, England and France, and is restricted by regulations.

Many cases of skin troubles, e.g. dermatitis, arise from jute dust, batching oil and the dyes with which jute is impregnated. Dermatitis also occurs among jute boilers using solutions of soap and sodium hydroxide and potassium hydroxide and also among the dyers. Workers who handle the threads impregnated with the solutions of yellow soap, impure fish oil, and mineral oil may develop skin troubles consisting of swelling, eczemas and often suppuration. Dermatitis can also result from colouring materials and oils. Jute can also convey tetanus germs owing to the contamination of the soil and dirt with which jute is always mixed. Good sanitary and washing facilities, including showers where practicable, are needed.

Excessive noise in weaving operations is associated with gradual irreversible hearing loss. Hoarseness and throat troubles like pharyngitis, noticeable in workers employed in preparing and spinning rooms, are ascribed partly to the dust and partly to fatigue of the voice caused by the necessity of speaking loudly because of noise. Noise should be controlled at source by padding, equipment suspension, mufflers, sound absorbents and so on. As an additional measure, hearing-protection equipment may be applied to the ears.

Lifting and carrying of heavy loads (bales of jute and yarn beams) may cause strains, sprains and back pain. This is especially likely in developing countries, where mechanical handling has not been installed in many places. Mechanisation is the best prevention but, in its absence, lightening of the loads and training in methods of lifting and carrying are doubly important.

Skeletal deformities of the legs have also been described among jute workers who started work when very young. This has been thought to be due mainly to constant standing on continually vibrating floors.

Pre-employment medical examination, particularly of young workers, is desirable and also regular medical supervision.

KHIN MAUNG NYUNT

Accidents:

CIS 76-248 *Accidents in jute textile factories.* Krishnan, G.; Purushothama, S. (Lake Town, Calcutta, Regional Labour Institute, 1975), 39 p.

Diseases:

CIS 76-699 "Hearing loss in female jute weavers". Kell, R. L. *Annals of Occupational Hygiene* (Oxford), Sep. 1975, 18/2 (97-109). Illus. 19 ref.

CIS 78-1241 "Respiratory function and symptoms in workers exposed simultaneously to jute and hemp". El Ghawabi, S. H. *British Journal of Industrial Medicine* (London), Feb. 1978, 35/1 (16-20). 19 ref.

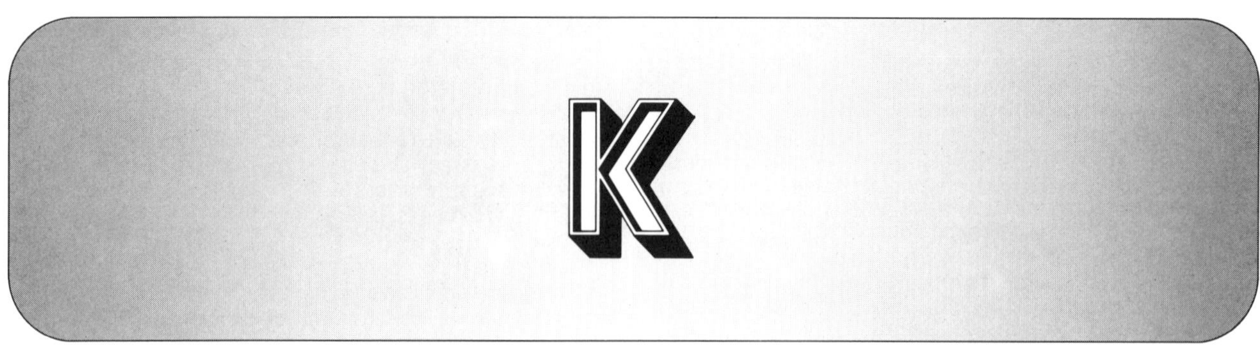

Kapok

The kapok tree *(Ceiba pentandra)* is a perennial plant about 15 m high, the fruit of which yields a soft cotton-like fibre called "silk cotton" or "silk floss". It is grown in India, Sri Lanka, the Philippines, Indonesia, Thailand and other countries in the tropics. The fibre has been known and used in tropical countries for stuffing cushions, pillows, mattresses, etc., for many centuries, but was not exported to European countries until 1850.

Cultivation. Seeds or stems are raised in greenhouses for 6-12 months and then planted out. When the tree is 3-5 years old, it bears fruit which when mature is elliptical in shape and has a hard and fragile shell containing light, soft and lustrous fibres.

Processing. After picking, the fruit is field dried for 2-3 days to render the shell more brittle (figure 1), the fruit is then broken open manually using simple hand tools such as hammers or a length of steel pipe. Still in the field, the fibres are ginned to remove the seeds, simply by shaking and stirring; the fibres are then packed by hand in jute sacks (figure 2), which are then batted to make them more soft and consistent (figure 4), after which they are baled (figure 5).

Figure 1. Pods of the kapok tree *(Ceiba pentandra)*.

Uses. The most important product of the kapok tree is the fibre, which is soft and light and suitable for stuffing mattresses, cushions, pillows and lifebelts. Kapok seeds may be used for animal feed, the manufacture of edible oil and fertiliser.

HAZARDS

The branches of the kapok tree are fragile and, when workers climb to the tree tops during harvesting, falls from broken branches are not uncommon. Minor hand

Figure 2. Kapok sack in the store house.

Figure 3. Once unpacked, the kapok fibre is hand fed into the hole of the separator. Fine nylon veils are used to protect the eyes and nose from irritation.

Figure 4. After blowing up, the big lump of the fibre is carried to the packing machine by the workers.

Figure 5. The fibres are packed tightly by a packing machine and then wrapped by jute sack and strong belts.

injuries may occur during shell cracking, and the stored fibre is a highly flammable material and constitutes a considerable fire hazard.

Kapok beaters and stuffers may be exposed to high atmospheric concentrations of vegetable dust, which may produce allergic reactions such as urticaria, hay fever, rhinitis, asthma or conjunctivitis. The shorter, softer fibres are small enough to reach the alveoli and may produce symptoms of acute and chronic bronchitis and chronic emphysema; bronchopneumonia, fibrosis and bronchiectasis may develop following prolonged exposure.

SAFETY AND HEALTH MEASURES

Dusty processes such as batting should be mechanised or equipped with an exhaust ventilation system. Stuffing and ginning workers should be provided with respiratory protective equipment. Strict fire protection and prevention procedures should be applied in fire hazard areas.

Pre-employment and periodic medical examinations should pay particular attention to respiratory system disorders and allergic reactions.

EGTASAENG, U.

Kapok ginnery:

CIS 77-1868 "An investigation into the health of kapok workers". Uragoda, C. G. *British Journal of Industrial Medicine* (London), Aug. 1977, 34/3 (181-185). 13 ref.

"Kapok allergy". Farrerons-Co', J. (53-54). *Occupational allergy* (Leiden, H. E. Stenfert-Kroese N.V., 1958), 329 p. Illus.

Kepone

Kepone $(O-Cl_{10}-C_{10})$

CHLORDECONE; 1,1a,3,3a,5,5a,5b,6 DECA CHLOROOCTAHYDRO-1,3,4-METHENO-2H-CYCLOBUTA(cd) PENTALEN-2-ONE

m.w. 490.7

m.p. decomposes at 350 °C

slightly soluble in water (2-4 ppm at pH 4-9); soluble in alcohols and ketones

greyish white crystals

non-flammable, non-combustible.

TWA OSHA (recommended) 100 µg/cm³ 8 h

TWA NIOSH (recommended) 1 µg/cm³ 10 h (suspected carcinogenic agent)

Production. Hexachlorocyclopentadiene (HCP) and sulphur trioxide are combined in a steam reactor with antimony pentachloride as a catalyst. The resulting slurry is water-quenched, followed by caustic soda neutralisation. Filtration and drying produces a final technical grade which is approximately 90% kepone and 10% water.

Use. This substance has been used as an ant and roach insecticide, usually in the form of a 0.125% mixture and for the control of banana borer weevils in Central and South America and to produce a kepone conjugate (Kelevan) for control of potato beetles in Europe. At present there are no known producers or users of kepone. The last known production of kepone was discontinued in 1975 after an investigation of the sole world producer had revealed significant employee health problems.

HAZARDS

Health effects attributable to kepone poisoning and identified during the above-mentioned investigation included: nervousness, hand tremors, ataxia, visual difficulties, opsoclonus, pleuritic pain, weight loss, joint pain, sterility, abnormal liver function and hepatomegaly. Animal studies have shown similar findings, while mice and rats fed kepone in dietary concentrations ranging from 8 ppm to 40 ppm for 80 weeks also developed a statistically significant increase in the incidence of hepatocellular carcinoma.

Laboratory analysis of exposed worker blood specimens for kepone showed concentrations ranging from less than 0.1 to 11.8 ppm for "worker cases" and from non-detectable to 4.1 ppm for "worker non-cases". The mean level for cases was 2.5 ppm with a median of 1.8 ppm. The mean level for non-cases was 0.60 with a median of 0.2.

While there are no known antidotes capable of reversing the effects of kepone the use of an ion exchange resin, cholestyramine, has been shown to increase its faecal excretion.

SAFETY AND HEALTH MEASURES

Since kepone represents a significant hazard through all routes of entry (inhalation, ingestion, and skin absorption) precautions must be taken to protect exposed employees against each.

Emissions of airborne particulates (dust, mist, spray, etc.) of kepone should be controlled at the sources of dispersion by means of effective and properly maintained methods, such as fully enclosed operations and local exhaust ventilation. Employees working in areas where contact of skin or eyes with kepone (wet or dry) may occur should wear impervious full-body protective clothing, protecting neck, head, hands and feet. At the end of the work-day protective clothing should be removed and a shower taken before putting on street clothing. Employees should be prohibited from wearing or carrying work clothing home to avoid the spread of contamination.

Precautions should be taken to ensure that employees do not eat or smoke in areas where kepone is handled, processed or stored so as to prevent ingestion of kepone. Where personal respiratory protective equipment is needed to reduce employee exposure, equipment suitable for the use and exposure to be encountered should be provided.

Spills and disposal. Kepone has been shown to be very persistent in the environment and difficult to destroy. Because of this, it should be disposed of only in approved disposal sites. While efforts to incinerate kepone waste have been successful, the cost has been very high. The discharge of kepone into sewage disposal systems may destroy the bacteriological system and should be prohibited. Spills should first be removed dry by vaccuum cleaners, then washed. All waste should be gathered into suitable containers for disposal. Workers engaged in clean up or maintenance work should be especially cautious to follow all previously mentioned measures regarding personal protective equipment.

STRAUB, W. E.

Citations and inspections regarding exposures to kepone. OSHA Program Directive No. 200-39 (Washington, DC, US Department of Labor, Occupational Safety and Health Administration, 13 January 1976).

"Epidemic kepone poisoning in chemical workers". Cannon, S. B.; Veazey, J. M. Jr.; Jackson, R. S.; Burse, V. W.; Hayes, C.; Straub, W. E.; Landrigan, P. J.; Liddle, J. A. *American Journal of Epidemiology* (Baltimore), June 1978, 107/6 (529-536). 14 ref.

"Liver structure and function in patients poisoned with chlordecone (Kepone)". Guzelian, P. S.; Vranian, G.; Boylan, J. J. et al. *Gastroenterology* (Baltimore), 1980, 78/2 (206-213).

CIS 76-1947 "Carcinogenesis bioassay of technical grade chlordecone (Kepone)". National Institutes of Health, Department of Health, Education and Welfare. *Federal Register* (Washington, DC), 8 Apr. 1976, 41/69 (14914-14915).

CIS 78-801 "Treatment of chlordecone (Kepone) toxicity with cholestyramine–Results of a controlled clinical trial". Cohn, W. J.; Boylan, J. J.; Blanke, R. V.; Fariss, M. W.; Howell, J. R.; Guzelian, P. S. *New England Journal of Medicine* (Boston), 2 Feb. 1978, 298/5 (243-248). 24 ref.

Ketones

Organic compounds that contain a carbonyl group (=CO) linked to two carbon atoms are known as ketones. They are represented by the general formula RCOR′ where R and R′ are usually alkyl or aryl groups. Considerable similarity exists between the methods used for their production and also between their properties–biological as well as chemical.

Production. Ketones are produced in industry by catalytic dehydrogenation or oxidation of secondary alcohols; these are usually obtained by hydration of olefins in the petrochemical industry.

Uses. Ketones are produced mainly for use as solvents in the production of plastics, artificial silk, explosives, cosmetics, perfumes and pharmaceuticals. They are intermediates in some chemical syntheses. They are widely used as solvents for dyes, resins, gums, tars, waxes and fats, and in the extraction of lubricating oils.

HAZARDS

Ketones are flammable substances and the more volatile members of the series are capable of evolving vapours in sufficient quantity at normal room temperatures to form explosive mixtures with air. Although in industrial exposure airways are the main route of absorption, a number of ketones are readily absorbed through the

Table 1. Irritation produced by some ketones in humans[1]

Ketone	Concentration/ duration	Effects
Acetone	800-1 000 ppm/8 h	Slight to moderate eye irritation
"	500 ppm/6 h	Irritation to eyes, nose, throat and trachea
"	500 ppm/2-4 h	No symptoms
Methyl ethyl ketone	33 000 ppm/ momentary	Intolerable eye and nose irritation
"	3 300 ppm/ momentary	Moderate eye and throat irritation
Diisobutyl ketone	100 ppm/3 h	Slight eye and throat irritation, slight headache
"	50 ppm/3 h	Slight transitory eye and nose irritation

[1] From *Criteria for a recommended standard–Occupational exposure to ketones.* DHEW (NIOSH) publication No. 78-173 (National Institute for Occupational Safety and Health, 4676 Columbia Parkway, Cincinnati), p. 136.

intact skin. Usually they are rapidly excreted, for the most part in the expired air. Their metabolism generally involves an oxidative hydroxylation, followed by reduction to the secondary alcohol. Ketones possess narcotic properties when inhaled in high concentrations. At lower concentrations they can provoke nausea and vomiting and are irritating to the eyes and respiratory system (see table 1). Sensory thresholds correspond to even lower concentrations (table 2). These physiological properties tend to be enhanced in the unsaturated ketones and in the higher members of the series.

Table 2. Sensory thresholds in humans for ketones[1]

Ketone	Highest satisfactory concentration[2] (ppm)	Irritating concentration[3] (ppm)		
		Eyes	Nose	Throat
Acetone	200	500	500	500
Methyl ethyl ketone	200	350	350	350
Methyl isobutyl ketone	100	200	>200	>200
Diisobutyl ketone	25	50	>50	>50
Cyclohexanone	25	75	75	75
Mesityl oxide	25	25	50	>50
Diacetone alcohol	50	100	>100	100
Isophorone	10	25	25	25

[1] From *Criteria for a recommended standard–Occupational exposure to ketones*, op. cit., p. 30. [2] Concentration judged by majority of exposed volunteers to be satisfactory for an 8-h exposure. [3] Concentration that caused irritation in the majority of subjects.

In addition to CNS depression, effects on the peripheral nervous system, both sensory and motor, can result from excessive exposure to ketones. They are also moderately irritant to the skin, the most irritant being probably methyl *n*-amylketone.

Acetone (CH_3COCH_3)

See ACETONE.

Methyl ethyl ketone ($CH_3CH_2COCH_3$)

2-BUTANONE; MEK

m.w. 72
sp.gr. 0.81
m.p. −87 °C
b.p. 79.6 °C
v.d. 2.5
v.p. 100 mmHg (13.3·10³ Pa) at 25 °C
f.p. −6.1 °C
e.l. 1.8-10%

a clear volatile liquid with a minty, somewhat pungent odour.

TWA OSHA 200 ppm 590 mg/m³
STEL ACGIH 300 ppm 885 mg/m³
IDLH 3 000 ppm
MAC USSR 200 mg/m³

It is produced by the dehydrogenation or selective oxidation of *sec*-butyl alcohol. It is used as a solvent for cellulose compounds, in particular nitro-cellulose, in the manufacture of acrylic and vinyl surface coatings, and in pharmaceutical and cosmetic manufacture, and is a common constituent of dewaxing compositions.

Short exposure of workers to 500 ppm of MEK in air provoked nausea and vomiting; throat irritation and headaches were experienced at somewhat lower concentrations. Because the metabolic pathways of this ketone include methanol, it is not surprising that several cases of peripheral neuropathy have been observed as a result of acute prolonged or repeated exposure. Neuropathy is usually symmetrical, painless with sensory lesions predominating; it may involve upper or lower limbs; in some cases the fingers have been affected following immersion of bare hands in the liquid. Dermatitis has been reported both after immersion in the liquid and after exposure to concentrated vapours.

Methyl *n*-propyl ketone ($CH_3(CH_2)_2COCH_3$)

2-PENTANONE; ETHYLACETONE; MPK

m.w. 86
sp.gr. 0.81
m.p. −77.8 °C
b.p. 102 °C
v.p. 16 mmHg (2.13·10³ Pa) at 25 °C
f.p. 7 °C
e.l. 1.5-8.2%

a water-clear liquid with a smell similar to that of acetone.

TWA OSHA 200 ppm 700 mg/m³
STEL ACGIH 250 ppm 875 mg/m³
IDLH 5 000 ppm
MAC USSR 200 mg/m³

This ketone is produced by the oxidation of 2-pentanol.

Methyl *n*-butyl ketone ($CH_3(CH_2)_3COCH_3$)

2-HEXANONE; PROPYLACETONE; MBK

m.w. 100
sp.gr. 0.81
m.p. −57 °C
b.p. 126 °C
v.d. 3.5
v.p. 3.8 mmHg (0.49·10³ Pa) at 25 °C
f.p. 35 °C (oc)
e.l. 1.2-8%

a colourless liquid with a smell similar to that of acetone.

TWA OSHA 100 ppm 410 mg/m³
TLV ACGIH 25 ppm 100 mg/m³ skin
STEL ACGIH 40 ppm 165 mg/m³
IDLH 5 000 ppm

On an industrial scale, this ketone is manufactured from acetic acid and ethylene reacting under the influence of a catalyst and pressure. It is mainly used in the lacquer industry.

Cases of peripheral neuropathy have been attributed to the exposure to this solvent in a coated fabric plant where methyl *n*-butyl ketone had been substituted for methyl isobutyl ketone at printing machines before any neurological cases were detected. This ketone has two metabolites (5-hydroxy-2-hexanone and 2,5-hexanedione) in common with *n*-hexane, which has also been regarded as a causative agent of peripheral neuropathies.

Methyl isobutyl ketone ($(CH_3)_2CHCH_2COCH_3$)

HEXONE; ISOPROPYLACETONE

m.w. 100.2
sp.gr. 0.80
m.p. −84.7 °C
b.p. 116.8 °C
v.d. 3.5
v.p. 7.5 mmHg (0.97·10³ Pa) at 25 °C
f.p. 17 °C
e.l. 1.4-7.5%

a colourless liquid with a smell characteristic of ketones.

TWA OSHA 100 ppm 410 mg/m³
TLV ACGIH 50 ppm 205 mg/m³
STEL ACGIH 75 ppm 300 mg/m³

It is produced by the selective catalytic hydrogenation of mesityl oxide.

Exposure to 50-105 ppm for 15-30 min provoked gastrointestinal disturbances and CNS impairment in a few workers.

Methyl *n*-amyl ketone ($CH_3(CH_2)_4COCH_3$)

2-HEPTANONE; *n*-AMYL METHYL KETONE

m.w. 114
sp.gr. 0.82
b.p. 150 °C
f.p. 44.4 °C (oc)

a colourless liquid with a mild odour of bananas.

TWA OSHA 100 ppm 465 mg/m^3
TLV ACGIH 50 ppm 235 mg/m^3
STEL ACGIH 100 ppm 465 mg/m^3
IDLH 4 000 ppm

It is produced by catalytic dehydrogenation of 2-heptanol. It is used mainly for metal roll coating. It is irritant to the skin (above the threshold limits) and produces narcosis at high concentrations, but it is not neurotoxic.

Ethyl butyl ketone ($CH_3(CH_2)_3COCH_2CH_3$)

3-HEPTANONE

m.w. 114
sp.gr. 0.80
m.p. −39 °C
b.p. 147 °C
v.d. 3.93
v.p. 4 mmHg (0.52·10^3 Pa) at 20 °C
f.p. 46 °C (oc)
e.l. 1.4-8%

a colourless liquid with a mild fruity odour.

TWA OSHA 50 ppm 230 mg/m^3
STEL ACGIH 75 ppm 345 mg/m^3
IDLH 3 000 ppm

5-Methyl-3-heptanone ($CH_3CH_2CHCH_3CH_2COCH_2CH_3$)

ETHYL *sec*-AMYL KETONE; AMYL ETHYL KETONE; EAK

m.w. 128
sp.gr. 0.82
b.p. 157 °C
v.p. 2 mmHg (0.26·10^3 Pa) at 20 °C
f.p. 59 °C

a colourless liquid with a mild fruity odour.

TWA OSHA 25 ppm 130 mg/m^3

It is used as a solvent for nitrocellulose-alkyd, nitrocellulose-maleic and vinyl resins.

Diisobutyl ketone ($((CH_3)_2CHCH_2)_2CO$)

2,6-DIMETHYL-4-HEPTANONE; ISOVALERONE; VALERONE

m.w. 142
sp.gr. 0.81
m.p. 43.9 °C
b.p. 166 °C
v.d. 4.9
v.p. 1.7 mmHg (0.22·10^3 Pa) at 20 °C
f.p. 60 °C
e.l. 0.8-6.2%

a colourless liquid with a mild fruity odour.

TWA OSHA 50 ppm 290 mg/m^3
TLV ACGIH 25 ppm 150 mg/m^3
IDLH 2 000 ppm

Diacetone alcohol ($(CH_3)_2COHCH_2COCH_3$)

DIACETONE; 4-HYDROXY-4-METHYL-2-PENTANONE

m.w. 116
sp.gr. 0.94
m.p. −42.8 °C
b.p. 169.2 °C
v.d. 4.0
v.p. 1.2 mmHg (0.16·10^3 Pa) at 25 °C
f.p. 64.4 °C
e.l. 1.8-6.9%

a colourless liquid with a minty odour.

TWA OSHA 50 ppm 240 mg/m^3
STEL ACGIH 75 ppm 360 mg/m^3
IDLH 2 100 ppm

This ketone is produced by condensation of acetone with an alkaline catalyst. It is a solvent for cellulose compounds and vinyl and epoxy resins. It is a constituent of hydraulic brake fluids.

It has irritant properties to eyes and upper airways; at higher concentrations it causes excitement and sleepiness. Prolonged exposure may result in liver and kidney damage and in blood changes.

Mesityl oxide ($CH_3COCH{=}C(CH_3)_2$)

4-METHYL-3-PENTENE-2-ONE; METHYL ISOBUTENYL KETONE; ISOPROPYLIDENE ACETONE

m.w. 98
sp.gr. 0.85
m.p. −59 °C
b.p. 130 °C
v.d. 3.38
v.p. 10 mmHg (1.33·10^3 Pa) at 26 °C
f.p. 30.6 °C
e.l. 1.4%
i.t. 344.4 °C

a colourless to yellow liquid with a strong peppermint odour.

TWA OSHA 25 ppm 100 mg/m^3
TLV ACGIH 15 ppm 60 mg/m^3
STEL ACGIH 25 ppm 100 mg/m^3
IDLH 5 000 ppm
MAC USSR 1 mg/m^3 skin irritant

Mesityl oxide is produced by distilling diacetone alcohol in the presence of iodine. It is used as a good solvent for nitrocellulose-vinyl resins and gums, and in the preparation of methyl isobutyl ketone.

Mesityl oxide is a strong irritant both on contact with the liquid and in the vapour phase and can cause necrosis of the cornea. Short exposure has narcotic effects; prolonged or repeated exposures can damage liver, kidneys and lungs. It is readily absorbed through the intact skin.

α-Chloroacetophenone ($C_6H_5COCH_2Cl$)

2-CHLOROACETOPHENONE; PHENACYL CHLORIDE; TEAR GAS

m.w. 154.6
sp.gr. 1.19
m.p. 56 °C
b.p. 244 °C
v.d. 5.2
v.p. 0.012 mmHg (1.6 Pa) at 0 °C
f.p. 117.7 °C

white crystals with a fragrant odour.

TWA OSHA 0.05 ppm 0.3 mg/m^3
IDLH 100 mg/m^3

α-Chloroacetophenone is prepared by the chlorination of acetophenone. It is used as a lacrimatory warfare gas. This ketone is a strong irritant of the eyes, inducing lacrimation. On heating it decomposes in toxic fumes.

Cyclohexanone ($C_6H_{10}O$)

PIMELIC KETONE; KETOHEXAMETHYLENE; ANONE

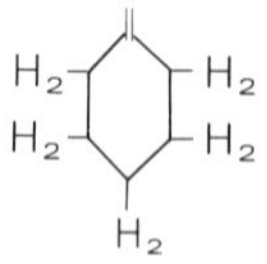

m.w. 98.1
sp.gr. 0.95
m.p. −16.4 °C
b.p. 156 °C
f.p. 43.9 °C

a colourless liquid with a smell similar to that of acetone and peppermint.

TWA OSHA 50 ppm 200 mg/m^3
MAC USSR 10 mg/m^3

It is produced by the catalytic dehydrogenation of cyclohexanol or by the oxidation of cyclohexane.

It is an intermediate in the manufacture of the adipic acid for nylon; it is used as a solvent for a variety of compounds including resins, nitrocellulose, rubber and waxes.

High doses produced in experimental animals degenerative changes in liver, kidney and heart muscle; repeated administration on the skin produced cataracts; cyclohexanone also proved to be embryotoxic to chick eggs.

2-Methylcyclohexanone ($C_7H_{12}O$)

o-METHYLCYCLOHEXANONE

m.w. 112
sp.gr. 0.93
b.p. 167 °C
v.d. 3.86
v.p. 1 mmHg ($0.13 \cdot 10^3$ Pa) at 20 °C
f.p. 47.7 °C

a colourless liquid with a mild peppermint odour.

TWA OSHA	100 ppm 460 mg/m³
TLV ACGIH	50 ppm 230 mg/m³ skin
STEL ACGIH	75 ppm 350 mg/m³
IDLH	2 500 ppm

On contact it is a strong irritant to eyes and skin; by inhalation it is irritant to the upper airways. Repeated exposure can damage kidneys, liver and lungs. Methylcyclohexanone reacts violently with nitric acid.

Isophorone ($C_9H_{14}O$)

3,5,5-TRIMETHYL-2-CYCLOHEXENE-1-ONE; ISOACETOPHORONE

m.w. 138
sp.gr. 0.92
b.p. 214 °C
f.p. 84.4 °C
e.l. 0.8-3.8%

O
H_2 H
$(CH_3)_2$ CH_3
H_2

a colourless liquid with a peppermint odour.

TWA OSHA	25 ppm 140 mg/m³
TLV ACGIH	5 ppm 25 mg/m³ ceil

It is obtained by condensation of acetone either by heating with alkaline catalyst or under high temperature and pressure.

Repeated exposure in experimental animals caused toxic effects on lungs and kidneys; single exposure to high doses can produce narcosis and paralysis of the respiratory centre.

SAFETY AND HEALTH MEASURES

Fire and explosion. Measures recommended for flammable substances should be applied (see FLAMMABLE SUBSTANCES).

Health precautions. Work practices and industrial hygiene techniques should minimise the volatilisation of ketones in the workroom air in order to ensure that the exposure limits are not exceeded.

In addition, as far as possible, ketones with neurotoxic properties such as methylethylketone and methyl *n*-butylketone should be replaced by products which have not been found to be neurotoxic.

Preplacement and periodic medical examinations are recommended with particular attention to the CNS and peripheral nervous system, respiratory system, the eyes, kidney and liver function. An electrodiagnostic examination with electromyography and nerve conduction velocity is appropriate particularly for workers exposed to methyl *n*-butylketone.

Skin contacts should be avoided by the use of protective gloves; goggles may also be necessary to prevent eye contact.

PARMEGGIANI, L.

CIS 81-167 "Aspiration toxicity of ketones". Panson, R. D.; Winek, C. L. *Clinical Toxicology* (New York), 1980, 17/2 (271-317). Illus. 49 ref.

Criteria for a recommended standard–Occupational exposure to ketones. DHEW (NIOSH) publication No. 78-173 (National Institute for Occupational Safety and Health, 4676 Columbia Parkway, Cincinnati) (1978), 244 p. Illus. 186 ref.

2-Butanone (2-Butanon). Gesundheitschädliche Arbeitsstoffe. Toxikologisch-arbeitsmedizinische Begründung von MAK-Werten (Weinheim, Verlag Chemie, 22 Nov. 1976), 6 p. 24 ref. (In German)

"Sniffing addiction: chronic solvent abuse with neurotoxic effects in children and juveniles" (Schnüffelsucht. Chronischer Lösungsmittelmissbrauch bei Kindern und Jugendlichen mit neurotoxischen Folgen). Altenkirch, H. *Deutsche medizinische Wochenschrift* (Stuttgart), 1979, 104/26 (935-938). (In German)

2-Hexanone (2-Hexanon). Gesundheitschädliche Arbeitsstoffe. Toxikologisch-arbeitsmedizinische Begründung von MAK-Werten (Weinheim, Verlag Chemie, 30 Apr. 1975), 5 p. 13 ref. (In German)

"Data for basing the maximal permissible concentration of methyl isobutyl ketone in the atmosphere of the producing area" (Materialny k obosnovaniju predel'no dopustimoj koncentravii metilizobutilketona v vozduxe rebocej zony). *Gigiena truda i professional'nye zabolevanija* (Moscow), 1973, 17/11 (52-53). 3 ref. (In Russian)

Isophorone (Isophoron). Gesundheitschädliche Arbeitsstoffe. Toxikologisch-arbeitsmedizinische Begründung von MAK-Werten (Weinheim, Verlag Chemie, 31 May 1976), 4 p. 8 ref. (In German)

Kidney

The functioning unit of the kidney is the nephron, which is composed of the glomerulus and the tubule. Each kidney contains more than a million glomeruli, representing a total filtering surface of about 1 m², permeable to water and solutes, but impermeable to colloids; filtration depends on the blood pressure on the glomerular capillaries, which is of the order of 55-60 mmHg. The glomerular filtrate is therefore the ultrafiltrate of the plasma. The urinary tubule takes part in urine formation by reabsorbing essential substances (water, electrolytes, glucose, amino acids, etc.) from the glomerular filtrate and allowing certain waste substances (catabolites) and those which are not normal constituents of the plasma to pass through. The tubule comprises three sections (proximal, intermediate, or Henle's loop, and distal) each of which have different functions. The proximal section absorbs 80% of the filtered water, 80% of the sodium, chlorine and bicarbonate and 100% of the urea. The kidney has a dense vascular network and receives 1 200 cm³ of blood per minute (this represents one-fifth of the cardiac output or 650 cm³ of plasma); 20% of this is filtered producing around 110 cm³ of glomerular filtrate per minute.

The basic functions of the kidney are as follows: excretion (elimination of almost all the terminal products of nitrogen metabolism); regulation of the volume and composition of body fluids (preservation of the dynamic equilibrium between the absorption and elimination of water and electrolytes); regulation of acid-base balance.

Renal detoxication

Being the main excretory organ, the kidney plays an important part in the excretion of industrial poisons, and the majority of poisons which are water soluble or can be converted into water-soluble compounds are eliminated in the urine.

The precise nature of renal excretion is still obscure (glomerular filtration, tubular reabsorption or excretion). The heavy metals are probably bound to plasma proteins; they are, however, filtered in only small quantities by the glomeruli, though there is every reason to suspect that they are actively secreted or concentrated by the cells of the tubules. Several organic mercurial compounds, for example, are secreted by the tubules in combination with cystine. Mercury, bismuth, cadmium, uranium and other metals are concentrated in the cells of the tubules, where they combine with certain enzymes (especially those containing sulphydryl groups). Other toxic materials are excreted by glomerular filtration, e.g. the thiocyanates. Even the metabolic products of trichloroethylene (especially trichloroacetic acid) probably undergo a definite and regular renal clearance.

The qualitative and quantitative investigation of toxic substances in the urine is of great importance in confirming the aetiology of acute poisoning in a worker exposed to a toxic hazard of undefined or ill defined nature, and in measuring the level of absorption in prolonged, continuous exposure.

Lead provides a typical example of tissue storage with slow and delayed elimination. Lead accumulates in the tissues (in bone in particular) but its elimination, though it begins early, does not equal the absorption, at least not until a certain saturation is reached. As a result of its accumulation in the body, lead continues to be excreted even after exposure has ceased, and the existence of the accumulated metal can be determined by measuring lead in urine after administration of a chelating agent such as calcium disodium edetate (EDTA).

Study of the urinary concentrations of different toxic substances has made it possible to establish maximum biological levels for a number of these substances. These maximum biological levels are the concentrations of the substance (or of its metabolites) in biological fluids, below which any immediate or future risk of poisoning can be discounted in the case of workers exposed for a normal work schedule (see EXPOSURE LIMITS, BIOLOGICAL).

In order to avoid erroneous interpretations of analytical data, the urine analysis should be carried out on samples collected at the end of the work shift. The values obtained should be corrected per gramme of creatinine, or at least by specific gravity, as has been the practice for a number of years and recommended again in 1965 by Elkins and Pagnotto. (The method consists in multiplying the measured value of the substance by the ratio between 24 and the last two digits of the specific gravity of the specimen; if this is, for instance, 1 012, the correction factor will be 24/12 = 2.)

Diseases of occupational origin

Acute renal disease. Table 1 shows the main types of acute anuric renal insufficiency of occupational origin, divided into three groups according to the dominant pathogenic mechanism.

For example renal ischaemia is a primary factor in acute renal insufficiency arising from post-traumatic shock; tubular necrosis or degeneration is characteristic of acute insufficiency arising from massive doses of toxic substances; and tubular blockage by casts (of haemoglobin or myoglobin origin) is the usual result of the crush syndrome, severe burns and special forms of poisoning. Another example is that of tetrachloroethane, one of the most toxic of the chlorinated aliphatic hydrocarbons, which has a TLV ACGIH of 5 ppm compared with 10 ppm for carbon tetrachloride, 50 ppm for chloroform and 100 ppm for trichloroethylene; this substance causes severe renal lesions often unappreciated because of the dominance of hepatic disorders. Carbon tetrachloride itself is a well known nephrotoxic agent in both industry and agriculture; it causes acute renal insufficiency in almost all severe cases. With both tetrachloroethane and carbon tetrachloride, oliguria or even anuria occur within a few hours or days, preceded by gastrointestinal symptoms and especially by signs of liver involvement; 10% of these cases are fatal but, in more favourable circumstances, the oliguria ceases after 4-10 days and diuresis subsequently returns to normal. These lesions are due not only to the direct toxic effect but also to other factors such as the circulatory imbalance (shock) and electrolytic disorder resulting from the disturbances of the primary toxic phase. Each parameter finally returns to normal within 3-6 months.

Table 1. Principal causes of acute renal insufficiency of occupational origin

Renal ischaemia	Tubular necrosis	Haemoglobinuria, myoglobinuria
Traumatic shock	Mercury	Arsine
Anaphylactic shock	Chromium	Crush syndrome
Acute carbon monoxide poisoning	Arsenic	Struck by lightning
	Oxalic acid	
Heat stroke	Tartrates	
	Ethylene glycol	
	Carbon tetrachloride	
	Tetrachloroethane	

Subacute renal disease. Changes in renal function occur when lead poisoning reaches the colic stage. The renal blood flow is reduced by 40-50% due to vasoconstriction of the afferent arterioles. The result is reduction of the glomerular filtrate and this accounts for the oliguria in the colic stage. Oliguria is very often the only symptom; occasionally the more severe forms present slight haematuria or proteinuria and slight but transient hyperazotaemia; these symptoms led to an erroneous diagnosis of "lead nephritis" in certain cases. Glycols are highly nephrotoxic; however, since they have very low vapour pressure they are rarely inhaled in significant quantities. For ethylene glycol (a classic acute nephrotoxic poison), the danger level is so high (100 mg and above) that it is virtually impossible that such an intoxication could occur by inhalation under working conditions.

Chronic renal disease. Prolonged lead poisoning or repeated acute episodes may cause a chronic kidney condition similar in nature to arteriosclerotic renal disease. Persistence of the intrarenal circulatory phenomena characteristic of acute poisoning gradually results in a permanent reduction in renal vascularisation. This progressive reduction of renal function is accompanied by a corresponding rise in arterial pressure, the clinical picture resembling that of arterial hypertension accompanied by renal insufficiency of vascular origin.

Another example of chronic kidney disease can be found in chronic carbon disulphide poisoning, which has been found to cause diffuse arterial changes similar to those encountered in arteriosclerosis. The lesions can affect various areas of the cardiovascular system but have a predilection for the cerebral, coronary and renal circulations. In one series of observations it was found that renal arteriosclerosis occurred in 50% of cases of carbon disulphide poisoning. In cases of severe renal involvement diffuse glomerular sclerosis has been indicated by clinical data and confirmed by histological examination.

Chronic cadmium poisoning has special clinical and anatomical features. Studies of alkaline battery plant workers suffering from chronic cadmium poisoning revealed cases of proteinuria involving proteins of low molecular weight (20 000-30 000), i.e. albumin and globulins (alpha, beta and gamma). It has been suggested that this is the result of a chronic, cadmium-induced tubular lesion which prevents or inhibits tubular reabsorption of low molecular weight proteins, whereas the principal function of the tubule cells is to concentrate the glomerular filtrate. Histological examination has, in fact, shown tubular lesions of a degenerative type.

Chronic mercury poisoning, unlike the acute condition, seldom results in renal involvement; nevertheless, it has been reported that intracutaneous therapeutic administration of inorganic mercury or occupational inhalation of metallic mercury over a period of months or years may lead to glomerular lesions of the membranous glomerulonephritis type, with the corresponding clinical signs which are not accompanied by symptoms of nephrosis and which occur in around 5-10% of the exposed persons. In such cases, urine mercury levels may exceed 1 mg/l, i.e. over ten times higher than the accepted maximum biological concentration.

Effect of work on renal function

This problem should be considered from two aspects: the effects of increased energy expenditure on renal circulation; and the effects of unfavourable workplace microclimate on renal function.

Muscular work entails an increase in oxygen consumption directly proportionate to energy cost. To cope with this, the heart increases the flow of blood in direct proportion to the expected energy expenditure. Provided the work is not too demanding, the increase in cardiac output is sufficient to provide the necessary blood supply to the muscles without sacrificing the flow to the viscera. But if the physical effort is considerable the blood is withdrawn from the splanchnic zone to the muscles, in particular at the expense of the kidneys. This reduction in renal blood flow may be as much as 80% and can cause nutritional disorders and consequently functional disturbances. The most outstanding example of this is effort proteinuria. Naturally this diminution in blood flow has more serious effects where the kidney is damaged. Therefore, workers with kidney disease should not be employed on work requiring moderate or high energy expenditure; the pre-employment medical examination should include a careful urine examination and, if necessary, comprehensive kidney function tests to ensure that workers with a kidney function deficiency are not employed on work that is physically too demanding.

The main factor in the workplace microclimate that may have a deleterious effect on kidney function is environmental heat. When the body is exposed to high temperatures, secondary cutaneous vasodilatation leads to a reduction in the splanchnic and, consequently, the renal circulation; excessive sweating, resulting in dehydration, further reduces the renal blood flow.

[It has recently been suggested that the first stage in the pathogenesis of autoimmune glomerulonephritis may be a chemical injury to lung or kidney. The antibodies to glomerular basement membrane found in this syndrome would be considered a secondary response to chemical injury to lung or kidney basement membranes.]

Detection of renal disease

The techniques for the detection of renal disease used by the industrial physician are no different from those in current use in clinical medicine: they are based on the systematic examination of the urine, backed up by more detailed analyses (e.g. renal clearance) where findings indicate the existence of renal lesions.

Urinalysis should form part of the pre-employment examination to ensure that persons with renal disease are not employed on types of work contraindicated by their condition, e.g. heavy physical effort, exposure to high temperatures or toxic substances. Urinalysis should also form part of periodic medical examinations in order to detect incipient kidney lesions. In mercury or cadmium workers, the detection of even slight haematuria should be considered a warning signal; in workers exposed to aromatic amines, particular attention should be paid to the detection of haematuria.

Rehabilitation of persons with renal disease

The problems of rehabilitation differ according to whether the patient suffers from an acute disease, or one of the chronic forms, either uraemic or compensated.

In the case of chronic uraemia, the condition is so serious that it was considered hopeless; nowadays, periodic haemodialysis offers hope of satisfactory rehabilitation and there have been cases of patients who have been able to return to work provided this entails only moderate physical effort, e.g. housekeeping, shopkeeping, etc.

In well compensated chronic kidney diseases (except in extreme cases nearing uraemia), the patient can resume light work; the more benign the clinical picture and the better the therapeutic prospects, the less stringent will need to be the limitation of physical effort.

In acute kidney diseases the prospects for rehabilitation depend on the extent of recovery. Resumption of normal work is more difficult for patients who, after an acute phase, still show signs of residual haematuria or proteinuria; however, when the only evidence of the renal disorder is that obtained in urinalysis, a gradual return to work can be permitted; where possible, only light work should be done and the patient should receive periodic examinations for at least two years.

CREPET, M.

"Occupational exposure and chronic nephritis" (Berufliche Exposition und chronische Nephritis). Schollmeyer, P.; Zimmermann, W. *Arbeitsmedizin–Sozialmedizin–Präventivmedizin* (Stuttgart), Jan. 1979, 14/1 (6-8). 16 ref. (In German)

Renal disorders of occupational origin (Les manifestations rénales d'origine professionnelle). Lejeune, A. (Lille, Université de Lille, Faculté de Médecine, 1972), 116 p. 135 ref. (In French)

"Glomerulonephritis associated with hydrocarbon solvents". Beirne, G. J.; Brenman, J. T. *Archives of Environmental Health* (Chicago), Nov. 1972, 25 (365-369). Illus. 9 ref.

CIS 81-1407 *Organic solvents and kidney function–A methodologic and epidemiologic study.* Askergren, A. Arbete och hälsa 1981:5 (Solna, Arbetarskyddsverket, 1981), 83 p. 306 ref.

"Occupational and post-traumatic kidney diseases. Clinical and medico-legal aspects" (Nefropatie professionali e post-traumatiche. Aspetti clinici e medicolegali). *Rivista degli infortuni e delle malattie professionali* (Rome), May-June 1981, 68/3 (239-331). Illus. Ref. (In Italian)

CIS 534-1976 "Is the 24-hour urine sample a fallacy?" Elkins, H. B.; Pagnotto, L. D. *American Industrial Hygiene Association Journal* (Detroit), Sep.-Oct. 1965, 26/5 (456-460). Illus. 6 ref.

Kienböck's disease

This is a relatively rare condition of aseptic semi-lunar osteonecrosis, normally unilateral, which occurs mainly in young males doing work which exposes them to microtrauma of the carpus or to repeated forced extensions of the wrist. [Sometimes it is attributed to low-frequency vibrations of pneumatic hammers, riveters and other engines. However, the role of vibrating tools is now considered to be that of an aggravating rather than a determining factor. Most likely the disease results from an occasional trauma followed by an unnoticed fracture of the semi-lunar or other carpal bones and is due to localised vascularisation disorders.]

Symptoms

The first clinical symptoms appear only after prolonged exposure and include carpal pain, oedema and erythema, accompanied by functional incapacity. When the hand is at rest, the symptoms recede; however, following a latent period ranging in length between a few months and several years, the pain recurs. Carpal osteoarthrosis may complicate the disease.

Diagnosis

Radiological diagnosis presents no problems (figure 1); the semi-lunar is flattened and irregular with varying degrees of opacification. The other carpal bones, in particular the large bone and the scaphoid, may show clearly defined vacuolar lesions related to osteoporosis.

[Semi-lunar fractures are early diagnosed with the aid of profile tomography. Clinically, lesions of the semi-lunar bone should be suspected in case of local pain on palpation of its posterior horn, immediately below the posterior rib of the articular face of the radius, in line with the third metacarpal bone, when the wrist is half flexed.]

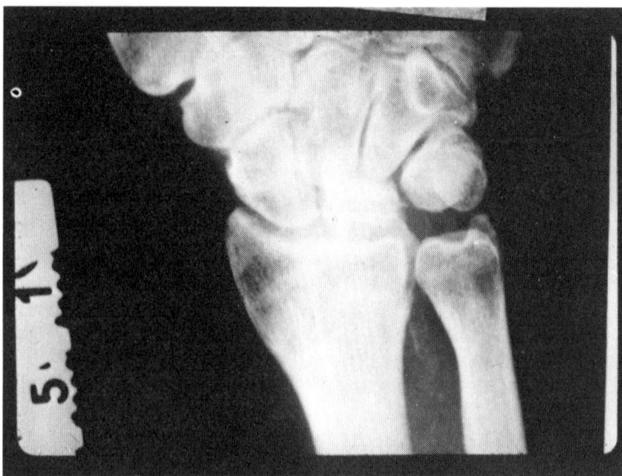

Figure 1. Kienböck's disease.

PREVENTIVE MEASURES

The measures applicable to the prevention of Raynaud's phenomenon should be implemented.

Treatment. Physiotherapy is recommended although semi-lunar ablation may be necessary if improvement is limited.

LAMBERT, G.

CIS 74-0185 "Kienböck's disease—An occupational accident?" (La maladie de Kienböck: un accident de travail?). Amphoux, M.; Gentax, R.; Poli, J. P.; Sevin, A. *Archives des maladies professionnelles, de médecine du travail et de sécurité sociale* (Paris), June 1973, 34/6 (309-320). Illus. 5 ref. (In French)